Fourth Edition

CHEMISTRY

Matter and Its Changes

BRADY • SENESE

Prepared for
Chemistry Department
The College of New Jersey

BICENTENNIAL
1807
WILEY
2007
BICENTENNIAL

WILEY
CUSTOM SERVICES

Bicentennial Logo Design: Richard J. Pacifico

To order books or for customer service, please call 1(800)-CALL-WILEY (225-5945).

Printed in the United States of America.

ISBN 978-0-470-89549-8

10 9 8 7 6 5 4 3 2 1

I BRIEF CONTENTS I

Periodic Table of the Elements

Atomic number → 1 H 1.00794 ← Atomic mass

Group designation

Periods	IA (1)	IIA (2)	IIIB (3)	IVB (4)	VB (5)	VIB (6)	VIIB (7)	VIIIB (8)	VIIIB (9)	(10)	IB (11)	IIB (12)	IIIA (13)	IVA (14)	VA (15)	VIA (16)	VIIA (17)	VIIIA (18)
1	1 H 1.00794																	2 He 4.002602
2	3 Li 6.941	4 Be 9.012182											5 B 10.811	6 C 12.0107	7 N 14.0067	8 O 15.9994	9 F 18.99840	10 Ne 20.1797
3	11 Na 22.98977	12 Mg 24.3050											13 Al 26.98154	14 Si 28.0855	15 P 30.97376	16 S 32.065	17 Cl 35.453	18 Ar 39.948
4	19 K 39.0983	20 Ca 40.078	21 Sc 44.955910	22 Ti 47.867	23 V 50.9415	24 Cr 51.9961	25 Mn 54.938049	26 Fe 55.845	27 Co 58.93320	28 Ni 58.6934	29 Cu 63.546	30 Zn 65.409	31 Ga 69.723	32 Ge 72.64	33 As 74.92160	34 Se 78.96	35 Br 79.904	36 Kr 83.798
5	37 Rb 85.4678	38 Sr 87.62	39 Y 88.90585	40 Zr 91.224	41 Nb 92.90638	42 Mo 95.94	43 Tc 98.9072	44 Ru 101.07	45 Rh 102.90550	46 Pd 106.42	47 Ag 107.8682	48 Cd 112.411	49 In 114.818	50 Sn 118.710	51 Sb 121.760	52 Te 127.60	53 I 126.90447	54 Xe 131.293
6	55 Cs 132.90545	56 Ba 137.327	57 *La 138.9055	72 Hf 178.49	73 Ta 180.9479	74 W 183.84	75 Re 186.207	76 Os 190.23	77 Ir 192.217	78 Pt 195.078	79 Au 196.96654	80 Hg 200.59	81 Tl 204.3833	82 Pb 207.2	83 Bi 208.98037	84 Po 208.9824	85 At 209.9871	86 Rn 222.0176
7	87 Fr 223.0197	88 Ra 226.0254	89 †Ac 227.0277	104 Rf 261.1089	105 Db 262.1144	106 Sg 263.118	107 Bh 262.12	108 Hs 265.1306	109 Mt (268)	110 Ds (271)	111 Uuu (272)	112 Uub (285)		114 Uuq (289)				

*

58 Ce 140.116	59 Pr 140.90765	60 Nd 144.24	61 Pm 144.9127	62 Sm 150.36	63 Eu 151.964	64 Gd 157.25	65 Tb 158.92534	66 Dy 162.50	67 Ho 164.93032	68 Er 167.26	69 Tm 168.93421	70 Yb 173.04	71 Lu 174.967

†

90 Th 232.0381	91 Pa 231.0359	92 U 238.0289	93 Np 237.0482	94 Pu 244.0642	95 Am 243.0614	96 Cm 247.07003	97 Bk 247.0703	98 Cf 251.0796	99 Es 252.083	100 Fm 257.0951	101 Md 258.0984	102 No 259.1011	103 Lr 262.110

TABLE OF ATOMIC MASSES AND ATOMIC NUMBERS

Data were obtained from the National Institute for Standards and Technology. Values are for the elements as they exist naturally on earth or for the most stable isotope, with carbon-12 (the reference standard) having a mass of exactly 12 u. The estimated uncertainties in values, between ±1 and ±9 units in the last digit of an atomic mass, are in parentheses after the atomic mass.
(Source: http://physics.nist.gov/PhysRefData/Compositions/index.html)

Element	Symbol	Atomic Number	Atomic Mass (u)		Element	Symbol	Atomic Number	Atomic Mass (u)	
Actinium	Ac	89	227.0277	(L)	Mercury	Hg	80	200.59(2)	
Aluminum	Al	13	26.981538(2)		Molybdenum	Mo	42	95.94(1)	(g)
Americium	Am	95	243.0614	(L)	Neodymium	Nd	60	144.24(3)	(g)
Antimony	Sb	51	121.760(1)		Neon	Ne	10	20.1797(6)	(g, m)
Argon	Ar	18	39.948(1)	(g, r)	Neptunium	Np	93	237.0482	(L)
Arsenic	As	33	74.92160(2)		Nickel	Ni	28	58.6934(2)	
Astatine	At	85	209.9871	(L)	Niobium	Nb	41	92.90638(2)	
Barium	Ba	56	137.327(7)		Nitrogen	N	7	14.0067(2)	(g, r)
Berkelium	Bk	97	247.0703	(L)	Nobelium	No	102	259.1011	(L)
Beryllium	Be	4	9.012182(3)		Osmium	Os	76	190.23(3)	(g)
Bismuth	Bi	83	208.98038(2)		Oxygen	O	8	15.9994(3)	(g, r)
Bohrium	Bh	107	264.12	(L)	Palladium	Pd	46	106.42(1)	(g)
Boron	B	5	10.811(7)	(g, m, r)	Phosphorus	P	15	30.973761(2)	
Bromine	Br	35	79.904(1)		Platinum	Pt	78	195.078(2)	
Cadmium	Cd	48	112.411(8)	(g)	Plutonium	Pu	94	244.0642	(L)
Calcium	Ca	20	40.078(4)	(g)	Polonium	Po	84	208.9824	(L)
Californium	Cf	98	251.0796	(L)	Potassium	K	19	39.0983(1)	(g)
Carbon	C	6	12.0107(8)	(g, r)	Praseodymium	Pr	59	140.90765(2)	
Cerium	Ce	58	140.116(1)	(g)	Promethium	Pm	61	144.9127	(L)
Cesium	Cs	55	132.90545(2)		Protactinium	Pa	91	231.03588(2)	
Chlorine	Cl	17	35.453(9)	(m)	Radium	Ra	88	226.0254	(L)
Chromium	Cr	24	51.9961(6)		Radon	Rn	86	222.0176	(L)
Cobalt	Co	27	58.933200(9)		Rhenium	Re	75	186.207(1)	
Copper	Cu	29	63.546(3)	(r)	Rhodium	Rh	45	102.90550(2)	
Curium	Cm	96	247.0703	(L)	Rubidium	Rb	37	85.4678(3)	(g)
Darmstadtium	Ds	110	271	(L)	Ruthenium	Ru	44	101.07(2)	(g)
Dubnium	Db	105	262.1144	(L)	Rutherfordium	Rf	104	261.1089	(L)
Dysprosium	Dy	66	162.500(1)	(g)	Samarium	Sm	62	150.36(3)	(g)
Einsteinium	Es	99	252.083	(L)	Scandium	Sc	21	44.955910(8)	
Erbium	Er	68	167.259(3)	(g)	Seaborgium	Sg	106	263.1186	(L)
Europium	Eu	63	151.964(1)	(g)	Selenium	Se	34	78.96(3)	
Fermium	Fm	100	257.0951	(L)	Silicon	Si	14	28.0855(3)	(r)
Fluorine	F	9	18.9984032(5)		Silver	Ag	47	107.8682(2)	(g)
Francium	Fr	87	223.0197	(L)	Sodium	Na	11	22.989770(2)	
Gadolinium	Gd	64	157.25(3)	(g)	Strontium	Sr	38	87.62(1)	(g, r)
Gallium	Ga	31	69.723(1)		Sulfur	S	16	32.065(6)	(g, r)
Germanium	Ge	32	72.64(1)		Tantalum	Ta	73	180.9479(1)	
Gold	Au	79	196.96655(2)		Technetium	Tc	43	98.9072	(L)
Hafnium	Hf	72	178.49(2)		Tellurium	Te	52	127.60(3)	(g)
Hassium	Hs	108	265.1306	(L)	Terbium	Tb	65	158.92534(2)	
Helium	He	2	4.002602(2)	(g, r)	Thallium	Tl	81	204.3833(2)	
Holmium	Ho	67	164.93032(2)		Thorium	Th	90	232.0381(1)	(g)
Hydrogen	H	1	1.00794(7)	(g, m, r)	Thulium	Tm	69	168.93421(2)	
Indium	In	49	114.818(3)		Tin	Sn	50	118.710(7)	(g)
Iodine	I	53	126.90447(3)		Titanium	Ti	22	47.867(1)	
Iridium	Ir	77	192.217(3)		Tungsten	W	74	183.84(1)	
Iron	Fe	26	55.845(2)		Ununbium	Uub	112	277	(L)
Krypton	Kr	36	83.798(2)	(g, m)	Unununium	Uuu	111	272	(L)
Lanthanum	La	57	138.9055(2)	(g)	Ununquadium	Uuq	114	289	(L)
Lawrencium	Lr	103	262.110	(L)	Uranium	U	92	238.0289(1)	(g, m)
Lead	Pb	82	207.2(1)	(g, r)	Vanadium	V	23	50.9415(1)	
Lithium	Li	3	6.941(2)	(g, m, r)	Xenon	Xe	54	131.293(2)	(g, m)
Lutetium	Lu	71	174.967(1)	(g)	Ytterbium	Yb	70	173.04(3)	(g)
Magnesium	Mg	12	24.3050(6)		Yttrium	Y	39	88.90585(2)	
Manganese	Mn	25	54.938049(9)		Zinc	Zn	30	65.409(4)	
Meitnerium	Mt	109	268	(L)	Zirconium	Zr	40	91.224(2)	(g)
Mendelevium	Md	101	258.0984	(L)					

(g) Geologically exceptional specimens of this element are known that have different isotopic compositions. For such samples, the atomic mass given here may not apply as precisely as indicated.
(L) This atomic mass is the relative mass of the isotope of longest half-life. The element has no stable isotopes.
(m) Modified isotopic compositions can occur in commercially available materials that have been processed in undisclosed ways, and the atomic mass given here might be quite different for such samples.
(r) Ranges in isotopic compositions of normal samples obtained on earth do not permit a more precise atomic mass for this element, but the tabulated value should apply to any normal sample of the element.

eGrade Plus

with EduGen

www.wiley.com/college/brady

Based on the Activities You Do Every Day

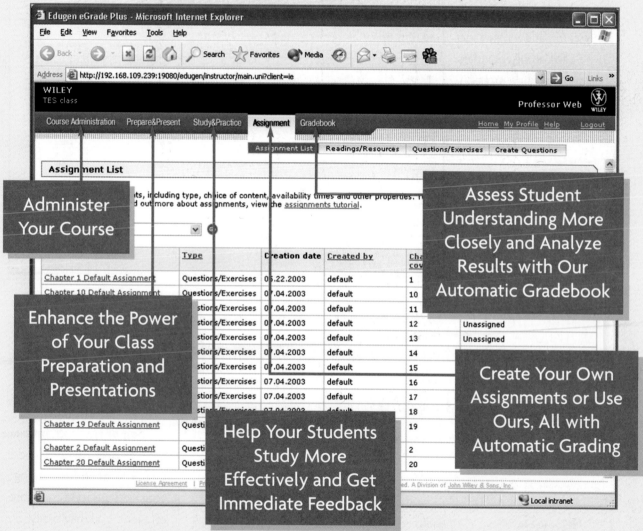

Administer Your Course

Enhance the Power of Your Class Preparation and Presentations

Help Your Students Study More Effectively and Get Immediate Feedback

Assess Student Understanding More Closely and Analyze Results with Our Automatic Gradebook

Create Your Own Assignments or Use Ours, All with Automatic Grading

All the content and tools you need, all in one location, in an easy-to-use browser format. Choose the resources you need, or rely on the arrangement supplied by us.

Now, many of Wiley's Book Companion Sites are available with EduGen, allowing you to create your own teaching and learning environment. Upon adoption of EduGen, you can begin to customize your course with the resources shown here. eGrade Plus with EduGen integrates text and media and keeps all of a book's online resources in one easily accessible location. eGrade Plus integrates two resources: homework problems for students and a multimedia version of this Wiley text. With eGrade Plus, each problem is linked to the relevant section of the multimedia book.

Administer Your Course

Course Administration tools allow you to manage your class and integrate your Wiley website resources with most Course Management Systems, allowing you to keep all of your class materials in one location.

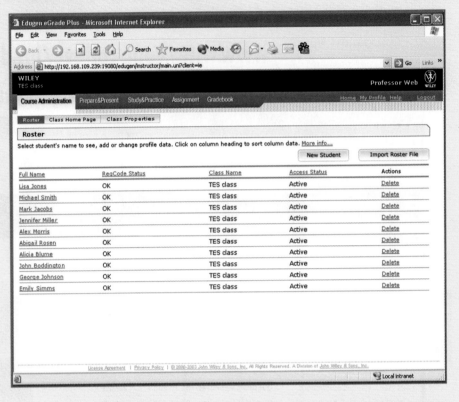

Enhance the Power of Your Class Preparation and Presentations

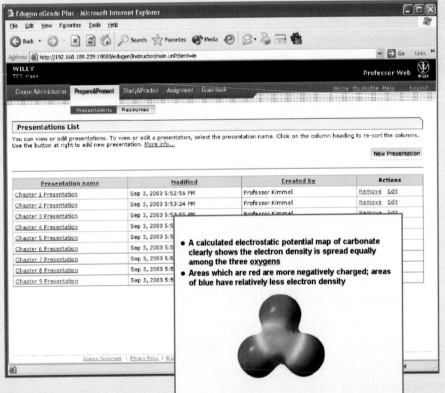

- A calculated electrostatic potential map of carbonate clearly shows the electron density is spread equally among the three oxygens
- Areas which are red are more negatively charged; areas of blue have relatively less electron density

A **"Prepare and Present" tool** contains all of the Wiley-provided resources, such as **a multimedia version of the text, interactive chapter reviews,** and **PowerPoint slides,** making your preparation time more efficient. You may easily adapt, customize, and add to Wiley content to meet the needs of your course.

Create Your Own Assignments or Use Ours, All with Automatic Grading

An **"Assignment"** area allows you to create **student home-work** and **quizzes** that utilize **Wiley-provided question banks,** and an **electronic version of the text.** One of the most powerful features of Wiley's premium websites is that student assignments will be automatically graded and recorded in your grade-book. This will not only save you time but will provide your students with immediate feedback on their work.

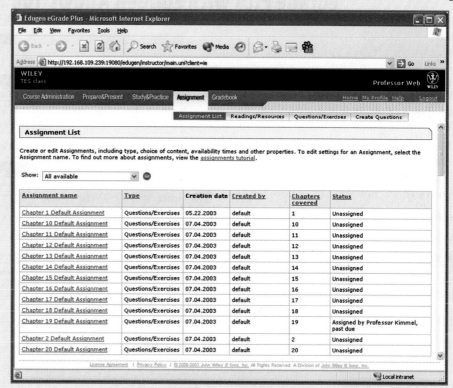

Assess Student Understanding More Closely

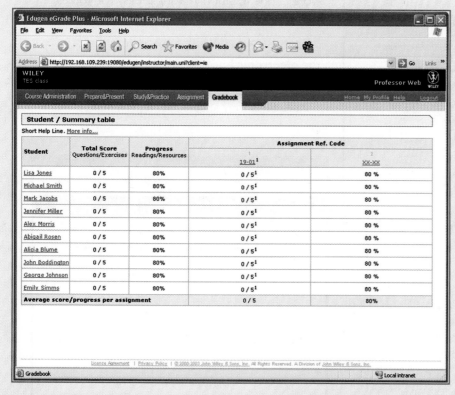

An **Instructor's Gradebook** will keep track of your students' progress and allow you to analyze individual and overall class results to determine their progress and level of understanding

Students,
eGrade Plus with EduGen Allows You to:

Study More Effectively

Get Immediate Feedback When You Practice on Your Own

Our website links directly to **electronic book content,** so that you can review the text while you study and complete homework online. Additional resources include **self-assessment quizzing** with detailed feedback, **Interactive Learningware** with step by step problem solving tutorials, and **interactive dialogs** to help you review key topics.

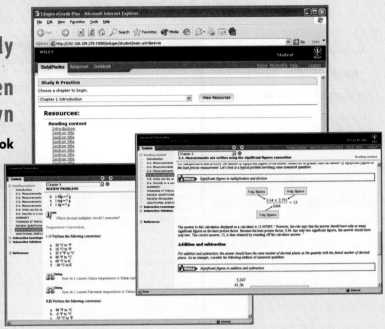

Complete Assignments / Get Help with Problem Solving

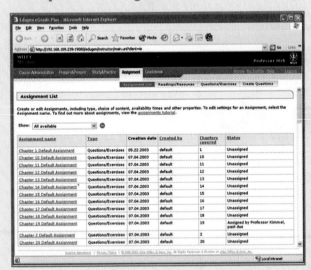

An **"Assignment"** area keeps all your assigned work in one location, making it easy for you to stay on task. In addition, many homework problems contain a **link** to the relevant section of the **electronic book,** providing you with a text explanation to help you conquer problem-solving obstacles as they arise.

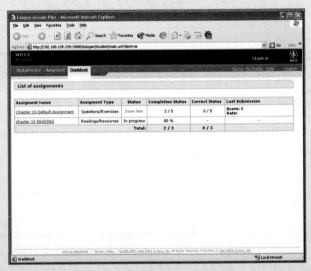

Keep Track of How You're Doing

A **Personal Gradebook** allows you to view your results from past assignments at any time.

Chemistry
MATTER AND ITS CHANGES

4th EDITION

James E. Brady
St. John's University, New York

Fred Senese
Frostburg State University, Maryland

WILEY

John Wiley & Sons, Inc.

EXECUTIVE EDITOR	Deborah Brennan
SENIOR DEVELOPMENT EDITOR	Ellen Ford
SENIOR MARKETING MANAGER	Robert Smith
SENIOR PRODUCTION EDITOR	Elizabeth Swain
SENIOR DESIGNER	Kevin Murphy
TEXT DESIGN	Circa 86
COVER DESIGN	Norm Christiansen
ILLUSTRATION EDITOR	Anna Melhorn
PHOTO EDITOR	Jennifer MacMillan
PHOTO RESEARCHERS	Elyse Rieder, Teri Stratford

COVER PHOTO: Boron Nitride image courtesy of Zetti Research Group, University of California at Berkeley, and Lawrence Berkeley National Laboratory.

This book was set in 10/12 Times Ten by Progressive Information Technologies and printed and bound by Von Hoffmann Press. The cover was printed by Von Hoffmann Press.

This book is printed on acid free paper.   ∞

To order books or for customer service please, call 1(800)-CALL-WILEY (225-5945).

Library of Congress Cataloging-in-Publication Data:
Brady, James E.
    Chemistry: matter and its changes / James E. Brady, Fred Senese.
    4th ed.
        p. cm.
    Includes index.
    ISBN 978-0-471-21517-2 (cloth)
    ISBN 0-471-44891-5 (Wiley International Edition)
    1. Chemistry. I. Senese, Fred. II. Title.

QD33.2.B73 2003
540--dc21

                                                                              2003049673

Printed in the United States of America

10 9 8 7 6 5 4

|PREFACE|

It is with pride that we welcome Fred Senese as co-author of the text. Fred is a respected educator whose web site, *General Chemistry Online,* has received millions of visitors since it was launched in 1997 and is used daily by chemistry students and teachers at hundreds of institutions from around the world. He brings a new perspective and fresh ideas to the development of this edition and is the architect of the innovative instructional media described later in this preface. His work greatly enhances this new edition.

Goals of the fourth edition

Although not immediately apparent from a quick perusal of the table of contents, this edition represents a significant change in direction for the text. While we have retained our commitment to providing accurate, easy to read, and understandable discussions of difficult topics, our approach to problem solving has broadened in both intent and method of delivery. The web resources have been specifically designed to challenge strong students *and* provide appropriate support for those who need it. The result is high quality, individualized instruction for *every* student who visits the site.

Reaching ALL students

Many teachers have confirmed a common difficulty. Students taking chemistry often fall into either of two groups, one consisting of those who are well prepared and able to grasp problem-solving concepts relatively quickly, and the other of those who struggle valiantly but are never quite able to succeed. Traditionally, textbooks have been aimed at one audience or the other. Either the text is "high level" and aimed at the stronger student, leaving the struggling student behind, or the text is written for the struggling student at the risk of failing to challenge the stronger student.

The goal of our teaching package, the centerpiece of which is this textbook, is to address this issue through an integrated system of support designed to provide scaffolding for students when they need it without burdening them when they do not. We accomplish this by building on the past strengths of our approach to teaching problem solving and critical thinking skills, which are now complemented by media support that is tied closely to the text, giving students tutorial help and instruction precisely when they need it.

What's new in this edition

General changes

- **The approach to problem solving has been bolstered** in this new edition (see page vi for more details).

- **A new two-page cumulative test section,** the *Test of Facts and Concepts,* has been added after every three to four chapters. Here students have the opportunity to synthesize and use information from several chapters. This section encourages and reinforces the cumulative development of knowledge so critical to success in the General Chemistry course.

- We use **descriptive phrases as section headings** to provide students with a better preview of section content and to give them a valuable study tool.

- **Many new "Facets of Chemistry" and "Chemistry in Action" boxes** have been added to help students see the relevance of what they are learning.

- The **experimental foundation for theoretical models of nature have been emphasized.** In Chapter 1, for example, we have added discussion about experimental evidence for atoms and included a brief discussion of the scanning tenneling microscope, which provides images that suggest atoms on the surface of metals. In Chapter 7, we examine tabulated experimental data to discover the relationship between heat, mass, and temperature change.

- **Nomenclature is now presented on a "when needed" basis** so that students feel less overwhelmed by the subject.

Chapter-by-chapter changes

- **Chapters 1 through 3 have been extensively reorganized.** The first two chapters now include an introduction to foundational topics such as elements and compounds, the basic structure of the atom, the periodic table, etc.

- A **brand new Chapter 3 deals with measurements,** experimental uncertainty, significant figure, units of measurement and unit conversions. It is presented in a way that allows the instructor to cover the material at any convenient time, permitting better coordination with the lab. Some schools may wish to cover this chapter in their recitation sections in order to free more time to discuss chemical concepts in lecture.

- We have **expanded discussion of the mole concept in Chapter 4.** Worked examples now emphasize finding "critical links" between given and desired quantities.

- The **Energy and Thermochemistry chapter (Chapter 7) has been completely reorganized.** Calorimetry is introduced immediately after the concepts of heat and temperature, building a strong foundation for discussion of the first law of thermodynamics discussed later in the chapter. We emphasize the process of science by using tabulated experimental data to discover relationships between heat, mass, and temperature change.

- **Chapter 8, Atomic and Electronic Structure, has also been revamped.** We explain wave/particle duality by building a simple model of an electron confined to a small space. The model is used to naturally explain topics that are presented abstractly and out of context in most other texts (such as the meaning of quantum numbers, the true significance of the de Broglie equation, the existence of zero point energy, energy quantization, and the resolution of the collapsing atom paradox). More importantly, our presentation provides another example of how model-building is used to explain the behavior of matter at the atomic level.

- In **Chapter 9 we have expanded discussion on the role that the lattice energy** plays in determining the nature of ions formed by metals and nonmetals. The discussion of dipole moments has also been expanded.

- **Chapter 10 now opens with an engaging discussion of the importance of molecular shap**e and charge distribution on biological activity. The chapter presents valence shell electron pair repulsion theory (VSEPR) in terms of Gillespie's revised electron domain model.

- **Chapter 11 now begins with a qualitative analysis of how observable properties of gases fit with the atomic model** and the kinetic molecular theory. Discussions of the gas laws has been extensively rewritten. The discussion of real gas behavior has been expanded so it now includes an explanation for the correction terms in the van der Waals equation of state.

- We have added some additional explanation in **Chapter 12, showing how estimating the relative strengths of intermolecular forces can help predict physical properties.**

- **Chapter 13 is a brand new chapter on solids.** It includes some material from our old Chapter 11 (Intermolecular Attractions and the Properties of Liquids and Solids) as well as some new topics. We have expanded the treatment of lattices and illustrated the use of crystal structure data in calculating atomic sizes. New sections on polymers, liquid crystals, advanced ceramics, and nanotechnology round out our discussion of the topic.

- In **Chapter 14 we discuss spontaneous mixing as a tendency toward more probable states rather than a tendency toward disorder,** in preparation for an improved treatment of entropy in Chapter 20.

- The kinetics chapter **(Chapter 15) now stresses the distinction between average and instantaneous rates.** The discussion of the experimental determination of reaction orders has also been simplified—it now emphasizes a general technique for finding patterns in experimental data.

- **Chapter 16 on Chemical Equilibrium includes many new worked examples.**

- A new **Facets of Chemistry,** *Swimming Pools, Aquariums, and Flowers,* **has been added to Chapter 17** (Acids and Bases).

- **Chapter 18, Equilibria in Solutions of Weak Acids and Bases, includes some sentence-level rewrites** to further clarify the topic for students.

- We have added a new **Facets of Chemistry,** titled *No More Soap Scum— Complex Ions and Solubility,* to our **Solubility and Simultaneous Equilibria chapter (Chapter 19).**

- **We have changed our approach to presenting Thermodynamics (Chapter 20).** This chapter now explains entropy as a measure of the number of equivalent ways to spread energy through a system. Entropy is introduced using a simple but quantitative molecular model of heat transfer. The model demonstrates the role of probability in determining the direction of a spontaneous process, showing that spontaneous processes proceed toward states of high probability.

- **Chapter 21 (Electrochemistry) has been completely reorganized.** We now treat galvanic cells first, followed by electrolysis for a better flow of topics. We have modernized the treatment of practical galvanic cells (batteries) by describing the chemistry of the nickel-metal hydride battery and lithium cells.

- **A discussion of the law of radioactive decay has been added to Chapter 22.**

- We have **added a section on metallurgy to Chapter 23** (Properties of Metals and Metal Complexes).

- Chapters 24 and 25 remain essentially intact from the last edition with some rewrites and reorganization.

Audience

In preparing this revision, we continued in our commitment to providing a book appropriate for the mainstream university general chemistry course for science majors (e.g., chemistry, biology, pre-med). Our intent was to fashion a teaching tool that is capable of challenging the better student while providing a learning environment that will enable *all* students to succeed. To accomplish this, we have built on what has proven successful in previous editions, such as the writing style, student-friendly attitude, and the clarity and thoroughness of our explanations of difficult concepts. We studied comments and suggestions by users of the previous edition as well as comments by teachers who have been using other texts. With these as a basis, we modified the topic sequence somewhat, expanded the treatment of some topics and trimmed the treatment of others. As in the previous edition, students are not assumed to have had a previous course in chemistry and mastery of only basic algebra is expected.

Overall philosophy and goals

The philosophy of the text continues to be based on our conviction that a general chemistry course serves a variety of goals in the education of a student. First, of course, it must provide a sound foundation in the basic facts and concepts of chemistry upon which the theoretical models can be constructed. The general chemistry course should also give the student an appreciation of the central role that chemistry plays among the sciences as well as the importance of chemistry in society and day-to-day living. In addition, it should enable the student to develop skills in analytical thinking and problem solving.

Learning features

Aware of student difficulties in problem solving and analytical thinking, we have adopted a unique approach to developing thinking skills. We distinguish three types of learning aids: those that enhance problem-solving skills, those that further comprehension and learning, and those that extend the breadth and knowledge of the student.

Features that enhance problem-solving and critical thinking skills

Many students entering college today lack experience in analytical thinking. A course in chemistry should provide an ideal opportunity to help students sharpen their reasoning skills because problem solving in chemistry operates on two levels. Because of the nature of the subject, in addition to mathematics many problems also involve the application of theoretical concepts. Students have difficulty at both levels, and one of the goals of this text has been to develop a unified approach that addresses each level.

Chemical tools approach to problem analysis Students are taught a variety of basic skills, such as finding the number of grams in a mole of a substance or writing the Lewis structure of a molecule. Problem solving often involves bringing together a sequence of such simple tasks. Therefore, if we are to teach problem solving, we must teach students how to seek out the necessary relationships required to obtain solutions to problems.

We use an innovative approach to problem solving that makes an analogy between the abstract tools of chemistry and the concrete tools of a mechanic. Students are encouraged to think of simple skills as tools that can be used to solve more complex problems. When faced with a new problem, the student is urged to examine the tools that have been taught and to select those that bear on the problem at hand.

To foster this approach to thinking through problems, we present a comprehensive program of reinforcement and review:

The *Tools Icon* in the margin calls attention to each chemical tool when it is first introduced and is accompanied by a brief statement that identifies the tool. Following the Summary at the end of the chapter, the tools are reviewed under the heading **Tools You Have Learned,** preparing students for the exercises that follow.

TOOLS **YOU HAVE LEARNED**

The following table lists the concepts that you've learned in this chapter which can be applied as tools in solving problems. Study each one carefully so that you know what each is used for. When faced with solving a problem, recall what each tool does and consider whether it will be helpful in finding a solution. This will aid you in selecting the tools you need. If necessary, refer to this table when working on the Thinking-It-Through problems and the Review Exercises that follow.

TOOL	HOW IT WORKS
Chemical formula (page 43)	Subscripts in a formula specify the number of atoms of each element in one formula unit of the substance. This gives us *atom ratios* that we will find useful when we deal with the compositions of compounds in Chapter 4.
Rules for naming molecular compounds (page 54)	They allow us to write the name of a molecular compound given its chemical formula and to write the formula given the name.
Rules for writing formulas of ionic compounds (page 60)	The rules permit us to write correct chemical formulas for ionic compounds. You will need to learn to use the periodic table (see below) to remember the charges on the cations and anions of the representative metals and nonmetals. You also should learn the ions formed by the transition and post-transition metals in Table 2.4, and it is essential that you learn the names and formulas (including charges) of the polyatomic ions in Table 2.5.
Rules for naming ionic compounds (page 63)	These rules allow us to write the name of an ionic compound given its chemical formula and to write the formula given the name.
Periodic Table (pages 51 and 58)	From a nonmetal's position in the periodic table we can write the formula of its simple hydride. For the nonmetals and the metals in Groups IA and IIA, we can use the elements' positions in the periodic table to obtain the charges on their ions.

Worked Examples follow a three step process. Each begins with an *Analysis* that describes the thought processes and critical links involved in selecting the tools needed to solve the problem, as well as how information will be assembled to achieve the solution. In many of the example problems in this edition, the analysis step has been expanded to include greater detail. Next comes the *Solution* step in which the problem is solved according to the plan developed in the *Analysis*. Examples conclude with a section titled *Is the Answer Reasonable?*, in which the answer is studied to see whether it "makes sense." Here, students are taught to perform approximate arithmetic to get ballpark estimates of the answers to numerical problems. They are also warned of common errors and other ways to check their work.

Practice Exercises follow most worked examples to enable the student to apply what has just been studied to a similar problem.

EXAMPLE 4.4
Converting Grams of Element to Moles of Element

NASA's Cassini–Huygens space probe used plutonium dioxide pellets as fuel. There was some concern that if the craft burned up in earth's atmosphere, it would disperse small inhalable particles of plutonium dioxide that could cause lung cancer. To estimate the effects of radiation on the structural integrity of the pellets, we must first calculate the number of moles of Pu-238 in a single fuel pellet. If each pellet contained 0.880 g of Pu-238, how many moles of Pu-238 were in the fuel pellet?

ANALYSIS: The problem can be stated as follows:

$$0.880 \text{ g Pu-238} \Leftrightarrow ? \text{ mol Pu-238}$$

In other words, 0.880 g of Pu-238 is equivalent to how many moles of Pu-238? *The critical link between moles and mass is the molar mass.* The atomic mass gives the molar mass of an element. Note that this is an isotope of plutonium, so we'll use the isotope mass rather than the average atomic mass given on the periodic table. For Pu-238, then, the molar mass is approximately 238 g. We can write

$$238 \text{ g Pu-238} \Leftrightarrow 1 \text{ mol Pu-238}$$

Now two possible conversion factors become clear; one is the tool we need for the grams-to-moles calculation.

$$\frac{238 \text{ g Pu-238}}{1 \text{ mol Pu-238}} \quad \text{and} \quad \frac{1 \text{ mol Pu-238}}{238 \text{ g Pu-238}}$$

If we multiply the given, 0.880 g of Pu-238, by the second conversion factor, grams will cancel to yield moles of Pu-238.

SOLUTION: Draw in the cancel lines yourself.

$$0.880 \text{ g Pu-238} \times \frac{1 \text{ mol Pu-238}}{238 \text{ g Pu-238}} = 3.70 \times 10^{-3} \text{ mol Pu-238}$$

In other words,

$$0.880 \text{ g Pu-238} \Leftrightarrow 3.70 \times 10^{-3} \text{ mol Pu-238}$$

If you are writing in your own cancel lines, *be sure to show them canceling the full unit.* For example, cancel "g Pu-238," not just "g."

Is the Answer Reasonable?

- One mole of Pu-238 would weigh 238 g.
- One-thousandth of a mole would weigh one-thousandth of 238 g, or 0.238 g.
- Our answer was about four-thousandths of a mole, which should be 4×0.24 g or about 0.92 g. This isn't far from the mass of Pu-238 we started with (0.88 g).

PRACTICE EXERCISE 5: How many moles of sulfur are present in 35.6 g of sulfur?

Thinking it Through is a selection of end-of-chapter problems that shifts the emphasis of problem solving away from the answer itself. Instead, these problems ask students to focus on the information and methods needed to solve them. The intent of the Thinking-It-Through problems is to develop the student's ability to solve conceptual problems and to further accustom the student to analyzing a problem before trying to carry out any computations. The problems are divided into two categories. Students who have developed reasonably good thinking skills can skip the Level 1 problems and proceed directly to the more challenging problems in Level 2.

End-of-Chapter Exercises are divided into three sections. **Review Questions,** classified according to topic, enable students to gauge their progress in learning concepts presented in a chapter. **Review Problems,** also classified according to topic, are presented in pairs of similar problems, with the first member of each pair having its answer in Appendix B. These problems provide routine practice in the use of basic tools as well as opportunities to incorporate these tools in the solution of more complex problems. **Additional Exercises** at the ends of the problem sets are unclassified. Many problems are cumulative, requiring two or more concepts, and several in later chapters require skills learned in earlier chapters. In this edition we have added a number of more difficult problems to both the **Review Problem**s and **Additional Exercises.** These serve to increase the range of problem difficulty available to the instructor when assigning homework.

Test of Facts and Concepts are problem sets placed between chapters at strategic intervals. These enable the student to review and gauge their cumulative progress. Many of the problems in these sets incorporate concepts developed over two or more chapters.

CHAPTERS 1–4 | **TEST OF FACTS AND CONCEPTS**

Many of the fundamental concepts and problem-solving skills that were developed in the preceding chapters will carry forward into the rest of this book. Therefore, we recommend that you pause here to see how well you have grasped the concepts, how familiar you are with important terms, and how able you are at working chemical problems. Don't be discouraged if some of the problems seem to be difficult at first. Most students of chemistry have experienced the frustration of initially not remembering what had only one or two weeks earlier seemed clear and straightforward. But if such frustrations take you back to a review, you will also experience what so many others have—what was once learned well (but since forgotten) comes back very quickly.

Some of the problems here require data or other information found in tables throughout this book, including those inside the covers. Freely use these tables as needed. For problems that require mathematical solutions, we recommend that you first assemble the necessary information in the form of equivalencies and then use them to set up appropriate conversion factors needed to obtain the answers.

1 A rectangular box was found to be 24.6 cm wide, 0.35140 m high, and 7,424 mm deep.
(a) How many significant figures are in each measurement?
(b) Calculate the volume of the box in units of cm^3. Be sure to express your answer to the correct number of significant figures.
(c) Use the answer in (b) to calculate the volume of the box in cubic feet.
(d) Suppose the box was solid and composed entirely of zinc, which has a specific gravity of 7.140. What would be the mass of the box in kilograms?

2 What is the difference between an atom and a molecule? What is the difference between a molecule and a mole?

3 If a 10 g sample of element X contains twice as many atoms as a 10 g sample of element Y, how does the atomic mass of X compare with the atomic mass of Y?

4 How did Dalton's atomic theory account for the law of conservation of mass? How did it explain the law of definite proportions?

5 If atom A has the same number of neutrons as atom B, must A and B be atoms of the same element? Explain.

6 Construct a conversion factor that would enable you to convert a volume of 3.14 ft^3 into cubic centimeters (cm^3).

7 The atoms of an isotope of plutonium, Pu, each contain 94 protons, 110 neutrons, and 94 electrons. Write a symbol for this element that incorporates its mass number and atomic number. Write the symbol for a different isotope of plutonium.

8 An atom of an isotope of nickel has a mass number of 60. How many protons, neutrons, and electrons are in this atom?

9 A solution was found to contain particles consisting of 12 neutrons, 10 electrons, and 11 protons. Write the chemical symbol for this particle, consulting the periodic table as needed.

10 Classify each of the following either as a substance or as a particle of which a substance can consist.
(a) Ion   (e) Compound
(b) Mixture   (f) Molecule
(c) Isotope   (g) Element
(d) Atom   (h) Nucleus

11 For the following, is it possible to have a visible sample of each? If not, explain.
(a) An isotope of iron
(b) An atom of iron
(c) A molecule of water
(d) A mole of water
(e) An ion of sodium
(f) A formula unit of sodium chloride

12 Make a sketch of the general shape of the modern periodic table and mark off those areas where we find the metals, metalloids, and nonmetals.

13 Which of the following elements would most likely be found together in nature: Ca, Hf, Sn, Cu, Zr?

14 Match an element on the left with a description on the right.

Calcium — Halogen
Iron — Noble gas
Helium — Alkali metal
Gadolinium — Alkaline earth metal
Iodine — Transition metal
Sodium — Inner transition metal

15 Define *ductile* and *malleable.*

16 Which metal is a liquid at room temperature? Which metal has the highest melting point?

17 What is the most important property that distinguishes a metalloid from a metal or a nonmetal?

18 Give the symbols of the post-transition metals.

19 Give chemical formulas for the following.
(a) potassium nitrate   (h) copper(II) perchlorate
(b) calcium carbonate   (i) bromine pentafluoride
(c) cobalt(II) phosphate (j) dinitrogen pentaoxide
(d) magnesium sulfite   (k) strontium acetate
(e) iron(III) bromide   (l) ammonium dichromate
(f) magnesium nitride   (m) copper(I) sulfide
(g) aluminum selenide

20 Give chemical names for the following.
(a) $NaClO_3$ (e) ICl_3 (i) $MnCl_2$
(b) $Ca_3(PO_4)_2$ (f) PCl_3 (j) $NaNO_2$
(c) $NaMnO_4$ (g) K_2CrO_4 (k) $Fe(NO_3)_2$
(d) AlP (h) $Ca(CN)_2$

Features that further comprehension and learning

Macro-to-micro illustrations To help students make the connection between the macroscopic world we see and events that take place at the molecular level, we have increased the number of illustrations that combine both views. We have also increased the number of illustrations that visualize reactions at the molecular level. The goal is to show how models of nature enable chemists to better understand their observations and to get students to visualize events at the molecular level.

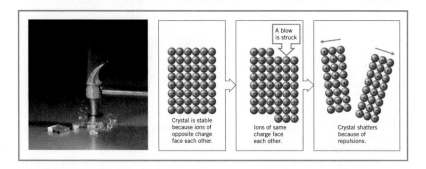

Margin comments These comments make it easy to enrich a discussion, without carrying the aura of being essential. Some margin comments jog the student's memory concerning a definition of a term.

Periodic table correlations One of our goals was to call particular attention to the usefulness of the periodic table in correlating chemical and physical properties of the elements.

Boldface terms These terms alert the student to "must-learn" items. Especially important equations are highlighted with a yellow background.

Problem analysis at a glance Where appropriate, figures contain flow-charts that summarize the relationships involved in solving problems, the approach used to analyze the method of attack on problems, or the approach to applying rules of chemical nomenclature.

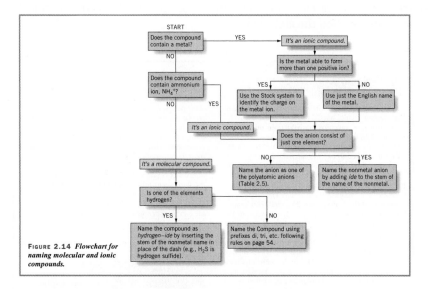

FIGURE 2.14 *Flowchart for naming molecular and ionic compounds.*

Chapter summaries Summaries use the boldface terms to show how the terms fit into statements that summarize concepts.

Features that extend the breadth of knowledge of the student

Facets of Chemistry These short essays serve several purposes. Most provide descriptions of real-world, practical applications of chemistry in industry, medicine, and the environment. Others serve to supplement topics discussed in the text or provide historical background.

FACETS OF CHEMISTRY 17.1

Swimming Pools, Aquariums, and Flowers

Something these have in common is the need for a proper pH. The water in swimming pools is more than just water; its a dilute mix of chemicals that prevent the growth of bacteria and stabilize the pool lining. For the pool chemistry to be properly balanced, the optimum pH range is between 7.2 to 7.6. Aquariums have to have their pH controlled because fish are very sensitive to how acidic or basic the water is. Depending on the species of fish in the tank, the pH should be between 6.0 and 7.6, and if it drifts much outside this range, the fish will die.

The pH can also affect the availability of substances that plants need to grow. A pH of from 6 to 7 is best for most plants because most nutrients are more soluble when the soil is slightly acidic than when it is neutral or slightly basic. If the soil pH is too high, metal ions such as iron, manganese, and boron that plants need will precipitate and not be available in the groundwater. A very low pH is not good either. If the pH drops to between 4 and 5, metal ions that are toxic to many plants become released as their compounds become more soluble. A low pH also inhibits the growth of certain beneficial bacteria that are needed to decompose organic matter in the soil and release nutrients, especially nitrogen.

There are chemicals that can be added to swimming pools, aquariums, and soil that will raise or lower the pH, but to know how to use them, it's necessary to measure the pH. Commercial test kits are available to enable you to do this, and they involve the use of acid–base indicators. Kits for testing the pH of pool water (Figure 1) use *phenol red* indicator, which changes color over the pH range 6.4–8.2. This places the intermediate color of the indicator right in the middle of the desired range of 7.2–7.6. To use the kit, a sample of pool water is placed into the plastic cylinder and five drops of the indicator solution are added. After shaking the mixture to ensure uniform mixing, the color is compared to the standards to gauge the pH. It can then be decided whether chemicals have to be added to either raise or lower the pH.

Similar test kits are available to test the water in an aquarium, except that the indicator used is bromothymol

FIGURE 1 *Swimming pool test kit.* This apparatus is used to test for both pH and chlorine concentration. The color of the indicator on the right tell us the pH of the pool water is approximately 7.6, which is slightly basic.

blue, which changes color over a pH range of 6.0–7.6. This generally matches the desired pH range for the water in the aquarium. Soil test kits, such as that shown in Figure 2, use a "universal indicator," which is a mixture of indicators that enable estimation of the pH over a wider range. A sample of the soil to be tested is placed in the plastic apparatus that comes with the kit. Water is added along with a tablet of the indicator. The mixture is shaken and then the color of the water is compared with the color chart. The color that most closely matches the color of the solution gives the estimated pH.

FIGURE 2 *Testing soil pH.* The indicator reveals that the soil sample being tested has a pH of approximately 6.5, which is just slightly acidic.

Chemistry in Practice These are very brief descriptions of practical applications of chemistry interspersed throughout the text.

CHEMISTRY IN PRACTICE

At one time, most of the street lighting in towns and cities was provided by incandescent lamps, in which a tungsten filament is heated white-hot by an electric current. Unfortunately, much of the light from this kind of lamp is infrared radiation, which we cannot see. As a result, only a relatively small fraction of the electrical energy used to operate the lamp actually results in visible light. Modern streetlights consist of high-intensity sodium or mercury vapor lamps in which the light is produced by passing an electric discharge through the vapors of these metals. The electric current excites the atoms, which then emit their characteristic atomic spectra. In these lamps, most of the electrical energy is converted to light in the visible region of the spectrum, so they are much more energy-efficient (and cost-efficient) than incandescent lamps. Sodium emits intense light at a wavelength of 589 nm, which is yellow. The golden glow of streetlights in scenes like that shown in the photograph at the right is from this emission line of sodium, which is produced in high-pressure sodium vapor lamps. There is another, much more yellow light emitted by low-pressure sodium lamps that you may also have

seen. Both kinds of sodium lamps are very efficient, and almost all communities now use this kind of lighting to save money on the costs of electricity. In some places, the somewhat less efficient mercury vapor lamps are still used. These lamps give a bluish-white light.

Fluorescent lamps also depend on an atomic emission spectrum as the primary source of light. In these lamps an electric discharge is passed through mercury vapor, which emits some of its light in the ultraviolet region of the spectrum. The inner wall of the fluorescent tube is coated with a phosphor that glows white when struck by the UV light. These lamps are also highly efficient and convert most of the electrical energy they consume into visible light.

Park Avenue in New York City is brightly lit by sodium vapor lamps in this photo taken during the Christmas season.

Illustrations The illustrations, which are distributed liberally throughout the text, have been drawn using modern computer techniques to provide accurate, eye-appealing complements to discussions. Color is used constructively rather than for its own sake. For example, a consistent set of colors is used to identify atoms of the elements in drawings that illustrate molecular structure. These are shown in the margin.

Photographs The many striking photographs in the book serve two purposes. One is to provide a sense of reality and color to the chemical and physical phenomena described in the text. The photographs also serve to illustrate how chemistry relates to the world outside the laboratory. The chapter-opening photos, for example, call the students' attention to the relationship between the chapter's content and common (and often not-so-common) things.

Supplements

A comprehensive package of supplements has been created to assist both the teacher and the student and includes the following:

Study Guide by James E. Brady. This guide has been written to further enhance understanding of concepts. It is an invaluable tool for students and contains chapter overviews, additional worked-out problems giving detailed steps involved in solving them, alternate problem-solving approaches, as well as extensive review exercises.

Solutions Manual by Nicholas Drapela of Oregon State University. The manual contains worked-out solutions for text problems whose answers appear in Appendix B.

Laboratory Manual for Principles of General Chemistry, Seventh Edition by Jo Beran of Texas A&M University, Kingsville. This comprehensive laboratory manual is for use in the general chemistry course. This manual is known for its broad selection of topics and experiments, and for its clear layout and design. Containing enough material for two or three terms, this lab manual emphasizes techniques, helping students learn the time and situation for their correct use. The accompanying Instructor's Manual presents the details of each experiment, including overviews, an instructor's lecture outline, and teaching hints. The IM also contains answers to the pre-lab assignment and laboratory questions.

Instructor's Manual by Mark A. Benvenuto of University of Detroit–Mercy. In addition to lecture outlines, alternate syllabi, and chapter overviews, this manual contains suggestions for small group active-learning projects, class discussions, and short writing projects.

Test Bank by Raymond X. Williams of Howard University. The Test Bank contains over 1,800 questions including; multiple choice, true-false, short answer questions, and critical thinking problems.

Computerized Test Bank IBM and Macintosh versions of the entire Test Bank are available with full editing features to help the instructor customize tests.

Four-Color Overhead Transparencies Over 125 four-color illustrations from the text are provided in a form suitable for projections in the classroom.

Instructor's Resource CD-ROM This CD-ROM contains PowerPoint lecture slides covering key topics in the text, incorporating text art and images.

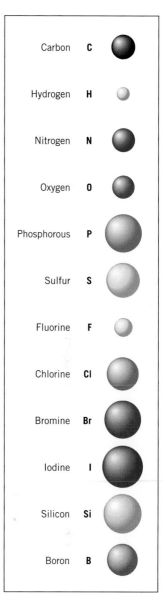

Carbon	C
Hydrogen	H
Nitrogen	N
Oxygen	O
Phosphorous	P
Sulfur	S
Fluorine	F
Chlorine	Cl
Bromine	Br
Iodine	I
Silicon	Si
Boron	B

Instructor's Solutions Manual by Alison Hyslop of St. John's University. Contains worked-out solutions to all end of chapter problems.

Digital Image Archive Text web site includes downloadable files of text images in JPG format.

Integrated technology solutions: Helping teachers teach and students learn (www.wiley.com/college/brady)

The Brady/Senese web site was developed by co-author Fred Senese. Fred is also the developer of General Chemistry Online (www.generalchemistryonline.com), one of the most popular general chemistry resources in the world.

This new site is specific to the Brady/Senese text and contains a wealth of instructor and student resources, such as interactive dialogs (alternate presentations of topics from the book), self-assessment quizzes, web-based activities, quiz and test banks, and other rich media content. The site provides a seamless integration of all of the media and helps keep all of the book's online resources in one easily accessible location.

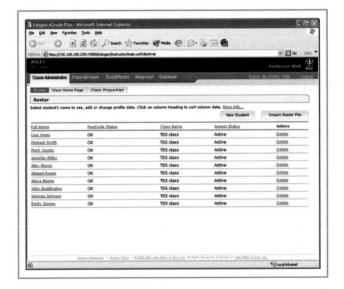

In addition, a premium version of the site is available which allows you to create your own teaching and learning environment. This premium web site, powered by Wiley's EduGen technology platform, provides many additional tools and resources for instructors and students. It allows for the complete integration of the text with all of the various media resources. Upon adoption of the premium web site, you can begin to customize your course with the following resources.

For instructors:

- **Course Administration** tools allow you to manage your class and integrate your Wiley web site resources with most Course Management Systems, thereby allowing you to keep all of your class materials in one location.
- A **"Prepare and Present"** tool contains PowerPoint presentations and a full digital archive of all the images from the book. You may easily adapt, customize, and add to Wiley content to meet the needs of your course.

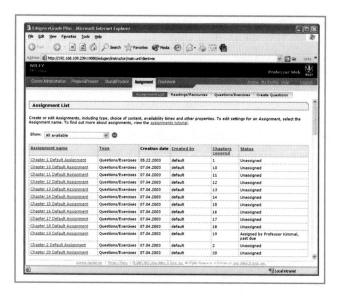

- **An "Assignment" area** allows you to create and manage student homework and quizzes online by using the Wiley-provided question banks, or by writing your own. You may also assign readings, activities, and other work you want your students to complete. One of the most powerful features of Wiley's premium web sites is that student assignments will be automatically graded and recorded in your gradebook. This will not only save you time but will provide your students with immediate feedback on their work, allowing them to determine right away how well they understand the course material.

- **An Instructors Gradebook** will keep track of your students' progress and allow you to analyze individual and overall class results to determine their progress and level of understanding.

For students

Wiley's premium web sites provide immediate feedback on student assignments and a wide variety of rich support materials. This powerful study tool will help your students develop their conceptual understanding of the class material and increase their problem solving skills.

- **A "Study and Practice"** area links directly to text content, allowing students to review the text while they study and complete homework assignments. Resources include:

  E-text. A complete version of the text online with links to a glossary, homework, quizzes, tutorials, animations, virtual experiments, and other interactive media.

  Dialogs. The dialogs are alternate presentation for topics in the book using interactive examples, virtual experiments, animations, learning games, etc. These dialogs guide students to construct knowledge themselves through a series of questions. They were written by Fred Senese, Mark Benvenuto of University of Detroit–Mercy, Nancy Mullins of Florida Community College at Jacksonville, Jim Giles of Mesa Community College, and Steve Lower of Simon Fraser University.

  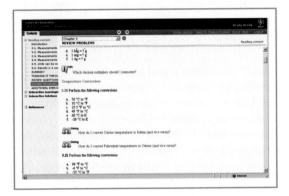

  Interactive LearningWare. Created by David Anderson of the University of Colorado at Colorado Springs, these step-by-step interactive tutorials help develop problem-solving skills.

  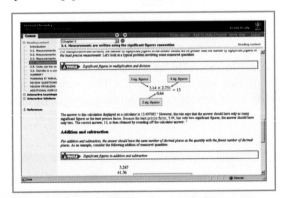

  Self-assessment quizzes. These quizzes, written by Dan Huchital of Florida Atlantic University, have detailed feedback based on the common errors the students make.

- **An "Assignment"** area keeps all the work you want your students to complete in one location, making it easy for them to stay "on task." Students will have access to a variety of interactive problem-solving tools, as well as other resources for building their confidence and understanding. In addition, many homework problems contain a link to the relevant section of the e-text, providing students with context-sensitive help that allows them to conquer problem-solving obstacles as they arise.

- **A Personal Gradebook** for each student will allow students to view their results from past assignments at any time.

|ACKNOWLEDGMENTS|

We begin by taking this opportunity to acknowledge with gratitude John Holum's contributions to this text in its previous editions and to wish him well in his retirement. John has always been a gifted writer, a superb teacher, and most of all, a good friend.

We express our fond thanks to our wives, June Brady and Lori Senese, whose words of encouragement kept us going through difficult periods and tight deadlines. We are also grateful for the constant support, understanding, and patience of our children, Mark and Karen Brady and Kai and Keiran Senese. They have been, and continue to be a steady source of inspiration for us both.

It is with particular pleasure that we thank the staff at Wiley for their careful work, encouragement, and sense of humor, particularly our editor, Debbie Brennan; developmental editor, Ellen Ford; media editor, Martin Batey; media project manager, Bridget O'Lavin; marketing manager Bob Smith; photo editor, Jennifer MacMillan; designer, Kevin Murphy; illustration editor, Anna Melhorn; editorial program assistants, Cathy Donovan and Justin Bow; and the entire production team, especially Elizabeth Swain. Our thanks also go to Kim Martin, Colleen Franciscus, and others at Progressive Publishing Alternatives for their tireless efforts toward changing a manuscript into a book.

We express gratitude to the colleagues whose careful reviews, helpful suggestions and thoughtful criticism of previous editions as well as the current edition manuscript have been so important in the development of this book. Our thanks go out to the following reviewers of the current edition:

Shawn Abernathy
Howard University

Patricia Amateis
Virginia Polytechnic Institute

David S. Ballantine
Northern Illinois University

Susan Bates
Ohio Northern University

Mark Benvenuto
University of Detroit, Mercy

Charles Carraher
Florida Atlantic University

Walter K. Dean
Lawrence Technological University

William Donovan
University of Akron

Nicholas Drapela
Oregon State University

Nancy Gardner
California State University, Long Beach

Marie G. Hankins
University of Southern Indiana

David Harris
University of California, Santa Barbara

Carl Hoeger
University of California, San Diego

Alison Hyslop
St. John's University

Colleen Kelley
Pima Community College

Peter Krieger
Palm Beach Community College

Bette Kreuz
University of Michigan, Dearborn

Gerald Lesley
Southern Connecticut State University

Nancy Mullins
Florida Community College, Jacksonville

Renee Muro
Oakland Community College

Robert Paine
Rochester Institute of Technology

Cynthia Nero Peck
Delta College

Susan Rutkowsky
Drexel University

Richard Schwenz
University of Northern Colorado

Erich Steinle
Southwest Missouri State University

Keith Stine
University of Missouri, St. Louis

Russell Tice
California Polytechnic State University, San Luis Obispo

Edmund Tisko
University of Nebraska, Omaha

Anthony Toste
Southwest Missouri State University

Daniel Williams
Kennesaw State University

Kurt Winkelmann
Florida Institute of Technology

This edition was built upon the feedback of many reviewers of the previous three editions and their contributions are still clearly a part of this book. These include:

Roma Advani
Prairie State University

Hugh Akers
Lamar University

Robert D. Allendoerfer
State University of New York, Buffalo

Mark Amman
Alfred State College

Dale Arlington
South Dakota School of Mines and Technology

George C. Bandik
University of Pittsburgh

Wesley Bentz
Alfred University

Mark A. Benvenuto
University of Detroit–Mercy

Keith O. Berry
Oklahoma State University

William Bitner
Alvin Community College

Simon Bott
University of North Texas

Donald Brandvold
New Mexico Institute of Mining and Technology

Robert F. Bryan
University of Virginia

Steven W. Buckner
Columbus State University

C. Eugene Burchill
University of Manitoba

Barbara A. Burke
California State Polytechnic University-Pomona

Jerry Burns
Pellissippi State Technical College

Jidhyt Burstyn
University of Wisconsin

Deborah Carey
Stark Learning Center

Jefferson D. Cavalieri
Dutchess Community College

Michael Chetcuti
University of Notre Dame

Ronald J. Clark
Florida State University

Wendy Clevenger
University of Tennessee at Chattanooga

Paul S. Cohen
The College of New Jersey

Kathleen Crago
Loyola University

Henry Daley
Bridgewater State College

Diana Daniel
General Motors Institute

John E. Davidson
Eastern Kentucky University

William Davies
Emporia State University

William Deese
Louisiana Tech University

David F. Dever
Macon College

David Dobberpuhl
Creighton University

Joseph Dreisbach
University of Scranton

Barbara Drescher
Middlesex County College

Wendy Elcesser
Indiana University of Pennsylvania

William B. Euler
University of Rhode Island

James Farrar
University of Rochester

John H. Forsberg
St. Louis University

David Frank
Ferris State University

DonnaJean A. Fredeen
Southern Connecticut State University

Donna Friedman
Florrisant Valley Community College

Ronald A. Garber
California State University, Long Beach

Paul Gaus
College of Wooster

John I. Gelder
Oklahoma State University

David B. Green
Pepperdine University

Thomas Greenbowe
Iowa State University

Michael Guttman
Miami-Dade Community College

Peter Hambright
Howard University

Henry Harris
Armstrong State College

Daniel T. Haworth
Marquette University

Carl A. Hoeger
University of California, San Diego

Paul A. Horton
Indian River Community College

Thomas Huang
East Tennessee State University

Peter Iyere
Tennessee State University

Denley Jacobson
Purdue University

Andrew Jorgensen
University of Toledo

Wendy L. Keeney-Kennicutt
Texas A & M University

Janice Kelland
Memorial University of Newfoundland

Henry C. Kelly
Texas Christian University

Reynold Kero
Saddleback College

Ernest Kho
University of Hawaii-Hilo

Louis J. Kirschenbaum
University of Rhode Island

Nina Klein
Montana Tech

Larry Krannich
University of Alabama-Birmingham

Robert M. Kren
University of Michigan-Flint

Russell D. Larsen
Texas Technical University

Ken Loach
SUNY-Plattsburgh

Glen Loppnow
University of Alberta

David Marten
Westmont College

Barbara McGoldrick
Union County College

Jeanette Medina
SUNY-Geneseo

William A. Meena
Rock Valley College

Patricia Moyer
Phoenix College

Robert Nakon
West Virginia University

James Niewahner
Northern Kentucky University

Brian Nordstrom
Embry-Riddle Aeronautical University

Sabrina Godfrey Novick
Hofstra University

Robert H. Paine
Rochester Institute of Technology

Naresh Pandya
Kapiolani Community College

Les Pesterfield
Western Kentucky University

Jerry L. Sarquis
Miami University

Paula Secondo
Western Connecticut State University

Ronald See
St. Louis University

Karl Seff
University of Hawaii

Edward Senkbeil
Salisbury State University

Venkatesh Shanbhag
Mississippi State University

Ralph W. Sheets
Southwest Missouri State University

Anton Shurpik
U.S. Merchant Marine Academy

Reuben Simoyi
West Virginia University

Mary Sohn
Florida Institute of Technology

S. Paul Steed
Sacramento City College

Darel Straub
University of Pittsburgh

Agnes Tenney
University of Portland

Roselin Wagner
Hofstra University

David White
State University of New York, New Paltz

S.D. Worley
Auburn University

Warren Yeakel
Henry Ford Community College

David Young
Ohio University

E. Peter Zurbach
St. Joseph's University

Of critical importance to this edition is the input and hard work contributed by a superb team of media reviewers, testers, contributors and consultants. Their dedication and expertise has made the media component the rich resource that it is. The media team includes:

Zeki Al-Saigh
State University of New York, Buffalo

Red Chasteen
Sam Houston State University

Ana Ciereszko
Miami Dade Community College

Jeff Coffer
Texas Christian University

Lin Coker
Campbell University

William Donovan
University of Akron

Karen Eichstadt
Ohio University

Gregory Ferrence
Illinois State University

Carl Hoeger
University of California, San Diego

Laura Kibler-Herzog
Georgia State University

Melvin Lesley
Southern Connecticut University

Sunil Malapati
Ferris State University

Garrett McGowan
Alfred University

Jalal U. Mondal
University of Texas, Pan American

Alex Nazarenko
Buffalo State University

Cynthia Peck
Delta College

Lee Pedersen
University of North Carolina, Chapel Hill

James Penner Hahn
University of Michigan

Mary Sohn
Florida Institute of Technology

Thomas Sorensen
University of Wisconsin, Milwaukee

Additional thanks go to the following who participated in special ways to the media development by creating content such as storyboards for dialogs, writing questions and problems, and/or extensive reviewing and consulting: David Anderson, University of Colorado, Colorado Springs (Interactive Learning Ware problems); Mark Benvenuto, University of Detroit, Mercy (dialog author and consultant); Jim Giles, Mesa Community College (dialog author); Dan Huchital, formerly at Seton Hall University presently at Florida Atlantic University (self-assessment quizzes); Steve Lower-Simon, Fraser University (dialog author); Barbara Mowery, Thomas Nelson Community College (on-line homework problems); Nancy Mullins, Community College of Florida, Jacksonville (dialog author).

Additional assistance was provided by dedicated attendees of focus groups who discussed their courses, their students, and their needs and concerns. Their honest and forthright comments allowed for important face-to-face feedback and discussion of how to best reach the general chemistry student. These focus group participants included: Chandrika Kulatilleke, Baruch College; Charles Carraher, Florida Atlantic University; Conrad H. Bergo, East Stroudsburg University; Sheila Cancella, Raritan Valley Community College; John Sheridan, Rutgers University—Newark.

TO THE STUDENT

You are about to begin what could be one of the most exciting courses that you will undertake in college. It offers you the opportunity to learn what makes our world "tick" and to gain insight into the roles that natural and synthetic chemicals play in nature and in our society. This knowledge will not come without some effort, however, and this book has been carefully designed and written with an awareness of the kinds of difficulties you may face. Therefore, before you begin, let us outline some of the key features of the book that will aid you in your studies.

Chemical vocabulary

Every specialty has its own unique vocabulary, and chemistry is certainly no exception. In this text you will find new terms set in bold type when you first encounter them. *We have not assumed that you've had a previous course in chemistry,* so these terms will not be used until they have been clearly defined. New terms are also set in bold type in the Summary that accompanies each chapter, and they are included in a Glossary at the end of the book, so you can look up their meanings at a later time if necessary.

Problem solving

Students often equate problem solving with doing mathematical calculations. It is really much more than that. Solving a problem involves *thinking* and *analyzing information.* This is true whether the problem relates to chemistry, physics, psychology, or to some aspect of your private life. Learning to solve problems is learning *how* to think and *how* to incorporate information into the solution.

When first faced with a genuine problem, many people have no idea where to begin. One of the goals of this text is to provide you with a way to think about problem solving, so you can be successful. You may find it difficult and slow going at first, but you *can* succeed if you're willing to put in the effort. And it is an effort that will pay off, not only in your chemistry course, but in other endeavors as well.

The *Chemical Toolkit* approach to problem solving

Imagine for a moment what it would be like to be an auto mechanic. Among other things, you would need to be familiar with an assortment of tools. Then, when faced with a repair job, you would study the problem and select the proper tools for the work.

Solving chemistry problems is not much different from repairing a car; we simply use different tools. When faced with a problem, we study it to see which of the tools we have learned fit the job. After we've selected the necessary tools, we use them to obtain the solution.

This is such a useful approach to solving problems that we have organized much of the book around it. In each chapter, you will be introduced to certain important concepts that will form the basis of problem-solving tools. Sometimes they

◀**TOOLS**

will be mathematical in nature and other times they will involve theoretical concepts. The icon that appears in the margin will be used to highlight the tools as we go along, and at the end of each chapter there is a summary of the tools you have been taught.

Within the text you will encounter worked-out examples that illustrate how the "chemical tools" are applied. Each of the examples begins with an *Analysis* section that describes the thought processes involved in solving the problem. We then work through the *Solution*. At the end of each example we perform a *Check* to see if the answer is reasonable. When you work problems on your own, try to follow these same steps: analysis, solution, check the answer. Following worked-out examples are Practice Exercises that give you an immediate opportunity to apply what you've just learned.

At the end of a chapter you will find two sets of exercises. The first set, titled *Thinking It Through*, asks you to describe *how* you would solve various problems. The goal is not the answers themselves, but rather descriptions of the tools and methods needed to obtain the answers. Their purpose is to emphasize the importance of the analysis step in problem solving. Some of these problems are simple, while others require a lot of thinking. Be sure to work on them all, however, because once you've learned *how* to solve a problem, obtaining the answer is simple.

Following the Thinking It Through exercises is a comprehensive set of *Review Questions* and *Review Problems* that test your knowledge of all the topics discussed in the chapter. The Review Problems are given in pairs, with the answer to the first problem in each pair given in Appendix B. Finally, after groups of chapters there are problem sets called *Test of Facts and Concepts* that give you an opportunity to review what you've learned.

Companion web site

Extensive support is available online to help you study, prepare for tests and quizzes, and deepen your understanding of critical concepts. We urge you to visit the site to check it out (www.wiley.com/college/brady).

Here is some of what you'll find:

- Online homework drawn from your end of chapter problems.
- Over 600 self-assessment quizzes with detailed feedback based on the common errors students make.
- A complete e-text with content from the book available instantly as a resource.
- Interactive Learningware step-by-step tutorials to help develop your problem-solving skills.
- Dialogs which help you explore topics from the book in a different way via animations, virtual experiments, learning games, interactive examples, etc.

|FACETS OF CHEMISTRY|

|BRIEF CONTENTS|

CONTENTS

Atoms and Elements: The Building Blocks of Chemistry

The slow, almost imperceptible, wearing away of stone steps, similar to the ones seen here at ruins near the Parthenon in Greece, led ancient philosophers to ponder whether matter was continuous and could be forever divided into smaller and smaller pieces, or whether matter was ultimately composed of indivisible particles. From these reflections, the concept of atoms was formed, a concept we now take for granted in our explanations of chemical phenomena. The atomic theory and our modern view of atoms are discussed in this chapter.

THIS CHAPTER IN CONTEXT This chapter has three principal goals. The first is to provide you with an appreciation of the central role that chemistry plays among the sciences. The second is to have you understand the way scientists approach the study of nature and how they construct mental pictures of the microscopic world to explain the results of experimental observations. And third, we will begin to discuss how chemistry views the world around us and how chemists organize the enormous amounts of information they've acquired.

If you've had a prior course in chemistry, perhaps in high school, you're likely to be familiar with many of the topics that we cover in this chapter. Nevertheless, it is important to be sure you have a mastery of these subjects, because if you don't start this course with a firm understanding of the basics, you may find yourself in trouble later on.

In our discussions, we do not assume that you have had a prior course in chemistry. However, we do urge you to study this chapter thoroughly, because the concepts developed here will be used in later chapters.

1.1 ▶ Chemistry is important for anyone studying the sciences

Much of the quality of life that we and others in the world enjoy today can be traced to the successes of science, and in particular to chemistry. Almost everything we touch or eat or drink or take into our bodies for nutrition or to cure illness bears the marks of chemical research. Synthetic fibers make up much of the fabric of the clothes we wear or form the threads that hold more "natural" fabrics together. Synthetic adhesives are used to bind together natural materials such as wood chips and fibers to yield construction products that are stronger, more stable, and more resistant to decay than traditional wood products. Many foods contain preservatives that retard spoilage, artificial flavors that enhance taste, or substances that improve texture and color. And, of course, drugs developed by pharmaceutical companies cure illnesses today that would have been fatal only a short time ago.

Chemistry[1] is a science that studies the composition and properties of **matter,** which we define *as anything that takes up space and has mass.*[2] All of the chemicals that make up tangible things, from rocks to people to pizza, are examples of matter. Chemists seek to answer fundamental questions about how the properties of substances are affected by their compositions. As part of this quest, they seek to learn the way substances change, often dramatically, when they interact with each other in *chemical reactions.* And permeating all of this is a search for knowledge about the basic underlying structure of matter and the

---

[1]Important terms will be set in bold type to call them to your attention. Be sure you learn their meanings.

[2]Mass is a measure of the amount of matter in an object. It is similar to weight, except that weight refers to the force with which an object of a given mass is attracted by gravity. We discuss mass and the units used to express it in Chapter 3.

TABLE 1.1	NAMES OF SOME OF THE DIVISIONS OF THE AMERICAN CHEMICAL SOCIETY	
Agricultural & Food Chemistry	Environmental Chemistry
Agrochemicals	Fertilizer & Soil Chemistry
Biochemical Technology	Fluorine Chemistry
Biological Chemistry	Fuel Chemistry
Business Development & Management	Geochemistry
Carbohydrate Chemistry	Industrial & Engineering Chemistry
Cellulose, Paper & Textile	Medicinal Chemistry
Chemical Health & Safety	Nuclear Chemistry & Technology
Chemical Toxicology	Petroleum Chemistry
Chemistry & the Law	Polymer Chemistry
Colloid & Surface Chemistry	Polymeric Materials: Science & Engineering
Computers in Chemistry	Rubber

forces that determine the properties we are able to observe through our senses. From such knowledge has come the ability to create materials never before found on earth, materials with especially desirable properties that fulfill specific needs of society.

Although you may not plan to be a chemist, some knowledge of chemistry will surely be valuable to you in whatever branch of science you study. In fact, the involvement of chemistry among the various branches of science is evidenced by the names of the various divisions of the American Chemical Society, the largest scientific organization in the world (see Table 1.1).

The reason chemistry holds such a unique place among the sciences is because all things are composed of chemicals. A cricket, for example, is made up of a complex set of chemicals that possess the unique quality we call life. Chemists are interested in these chemicals for their complex structures and the way they behave toward each other. On the other hand, a biologist might wish to study how the cricket metabolizes nutrients and how it derives energy from the process, while an engineer or a physicist might be interested in the mechanics of motion of the limbs of the cricket that allow it to jump. In their activities, the biologist, engineer, and physicist are likely to draw on knowledge gained by chemists. In this way, chemists, biologists, engineers, and physicists may all have interests in the cricket and study the creature from slightly different points of view. Today, in fact, the lines between traditional disciplines such as chemistry and biology have become blurred, so there is little difference, for example, between a molecular biologist and a biochemist.

1.2 ▶ The scientific method helps us build models of nature

Chemistry is a dynamic subject, constantly changing as new discoveries are made by researchers who work in university, industrial, and government laboratories. The general approach that these people bring to their work is called the **scientific method.** It is, quite simply, a commonsense approach to developing an understanding of natural phenomena.

Observations and conclusions are not the same thing

A scientific study normally begins with some questions about the behavior of nature. To search for answers, we begin by examining the work of others who have published in scientific journals. As our knowledge grows, we begin to plan our own experiments, and an essential part of these is the recording of observations. An

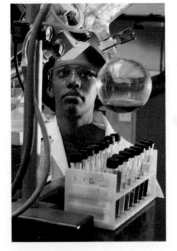

A scientist working in a chemical research laboratory.

observation *is a statement that accurately describes something we see, hear, taste, feel, or smell.*

For a scientific observation to be useful, it must be the same no matter who the observer is. Because of this, experiments are performed using well-defined procedures, under carefully controlled conditions, so they are *reproducible*. In this way, others can repeat and confirm the observations. In fact, the ability to obtain the same results when experiments are repeated is what separates a true science from a pseudoscience such as astrology.

Observations gathered during an experiment often lead us to make conclusions. A **conclusion** *is a statement that is based on what we think about a series of observations.* To understand the difference between observations and conclusions, consider the following statements about the fermentation of grape juice to make wine:

1. Before fermentation, grape juice is very sweet and contains no alcohol.

2. After fermentation, the grape juice is no longer as sweet and it contains a great deal of alcohol.

3. In fermentation, sugar is converted into alcohol.

Statements 1 and 2 are observations because they describe properties of the grape juice that can be tasted and smelled. Statement 3 is a conclusion because it *interprets* the observations that are available. Now, consider what happens when we add a fourth statement:

4. During fermentation, bubbles of a colorless, odorless gas form in the grape juice.

This is an observation because the bubbles can be seen directly. Making this additional statement doesn't affect observations 1 and 2, but it may very well cause the conclusion (statement 3) to be revised:

3. In fermentation, sugar is converted into alcohol *and a colorless gas.*

Thus, we see that conclusions interpret observations, and when new observations are made, the interpretations may need to be revised.

Empirical facts lead to scientific laws

The observations we make in the course of performing experiments provide us with **empirical facts**—so named because we learn them by *observing* some physical, chemical, or biological system. These facts are referred to as **data.** For example, if we study the behavior of gases, such as the air we breathe, we soon discover that the volume of a gas depends on a number of factors, including the mass of the gas, its temperature, and its pressure. The observations we record relating these factors are our data.

One of the goals of science is to organize facts so that relationships or generalizations among the data can be established. For instance, one generalization we would make from our observations is when the temperature of a gas is held constant, squeezing the gas into half its original volume causes the pressure of the gas to double. If we were to repeat our experiments many times with numerous different gases, we would find that this generalization is uniformly applicable to all of them. Such a broad generalization, based on the results of many experiments, is called a **law** or **scientific law.**

Laws are often expressed in the form of mathematical equations. For example, if we represent the pressure of a gas by the symbol P and its volume by V, the inverse relationship between pressure and volume can be written as

$$P = \frac{C}{V}$$

where C is a proportionality constant. (We will discuss gases and the laws relating to them in greater detail in Chapter 11.)

When reporting scientific results, it is important not to confuse observations with conclusions.

Observations are statements that directly describe what we measure or perceive and don't need revision when additional data become available.

Webster's defines *empirical* as "pertaining to, or founded upon, experiment or experience."

We would say that the pressure of the gas is inversely proportional to its volume—the smaller the volume, the larger the pressure.

Hypotheses and theories are models of nature

As useful as they may be in summarizing the results of experiments, laws can only state what happens. They do not explain *why* substances behave the way they do. *Why,* for example, are gases so easily compressed to a smaller volume? More specifically, *what must gases be like at the most basic, elementary level for them to behave as they do*? Answering such questions when they first arise is no simple task and requires much speculation. But gradually, scientists build mental pictures, called **theoretical models,** that enable them to explain observed laws.

In the development of a theoretical model, tentative explanations, called **hypotheses,** are formed. These explanations are then tested by performing experiments that test predictions derived from the model. Sometimes the results show the model is wrong. When this happens, the model must be abandoned or, as often happens, modified to account for the new data. Eventually, if the model survives repeated testing, it gradually achieves the status of a theory. *A* **theory** *is a tested explanation of the behavior of nature.* Most useful theories are broad, with many far-reaching and subtle implications. It is impossible, however, to perform every test that might show a theory to be wrong, so we can never be *absolutely* sure a theory is correct.

The sequence of steps just described—observation, explanation through the creation of a theoretical model, and the testing of the model by additional experiments—constitutes the **scientific method.** Despite its name, this method is not used only by those who call themselves scientists. An auto mechanic follows essentially the same steps when fixing your car. First, tests are performed (*observations* are made) that enable the mechanic to suggest the probable cause of the problem (*a hypothesis*). Then parts are replaced and the car is checked to see whether the problem has been solved (*testing the hypothesis by experiment*). In short, we all use the scientific method as much by instinct as by design.

From the preceding discussion you may get the impression that scientific progress always proceeds in a dull, orderly, and stepwise fashion. This isn't true; science is exciting and provides a rewarding outlet for cleverness and creativity. Luck, too, sometimes plays an important role. For example, in 1828 Frederick Wöhler, a German chemist, was heating a substance called ammonium cyanate in an attempt to add support to one of his hypotheses. His experiment, however, produced an unexpected substance, which out of curiosity he analyzed and found to be urea (a component of urine). This was an exciting discovery because it was the first time anyone had knowingly made a substance produced only by living creatures from a chemical not having a life origin. The fact that this could be done led to the beginning of a whole branch of chemistry called *organic chemistry.* Yet, had it not been for Wöhler's curiosity and his application of the scientific method to his unexpected results, the significance of his experiment might have gone unnoticed.

As a final note, it is significant that the most spectacular and dramatic changes in science occur when major theories are proved to be wrong. Although this happens only rarely, when it does occur, scientists are sent scrambling to develop new theories, and exciting new frontiers are opened.

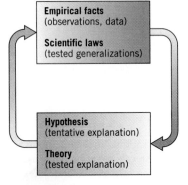

The scientific method is cyclical. Observations suggest explanations, which suggest new experiments, which suggest new explanations, and so on.

Many breakthrough discoveries in science have come about by accident.

The atomic theory is a model of nature

Virtually every scientist would agree that the most significant theoretical model of nature ever formulated is the atomic theory. According to this theory, which we will discuss further in Section 1.5, all chemical substances are composed of tiny particles that we call **atoms.** Individual atoms combine in diverse ways to form more complex particles called **molecules.** Consider, for example, the substance water. Experimental evidence suggests that water molecules are each

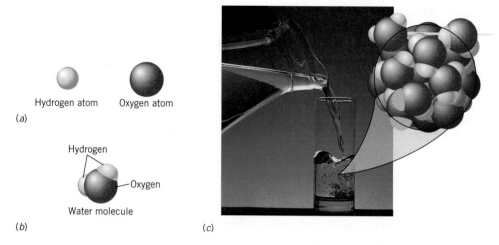

FIGURE 1.1 *Atoms combine to form molecules.* Illustrated here are molecules of water, each of which consists of one atom of oxygen and two atoms of hydrogen. (*a*) Colored spheres are used to represent individual atoms, gray for hydrogen and red for oxygen. (*b*) A drawing that illustrates the shape of a water molecule. (*c*) A glass of liquid water contains an enormous number of tiny water molecules jiggling about.

composed of two atoms of hydrogen and one of oxygen. To aid in our understanding and to help visualize how atoms combine, we often use drawings such as Figure 1.1, which is an attempt to depict a collection of water molecules. Notice that we show each molecule as composed of the same atoms (hydrogen and oxygen) in the same proportions. In a similar fashion, each of the different chemicals we find around us is composed of atoms of various kinds in specific combinations.

The concept of atoms and molecules enables us to visualize what takes place when atoms combine in various ways. As you will see, it also enables us to understand, explain, and sometimes predict a wide range of observations concerning the behavior of nature. To help us, we will frequently use drawings to illustrate molecules; Figure 1.2 shows some of the ways molecules can be represented.

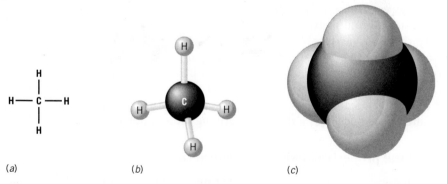

FIGURE 1.2 *Some of the different ways that structures of molecules are represented.* (*a*) A structure using chemical symbols to stand for atoms and dashes to indicate how the atoms are connected to each other. The molecule is methane, the substance present in natural gas that fuels stoves and Bunsen burners. A methane molecule is composed of one atom of carbon (C) and four atoms of hydrogen (H). (*b*) A *ball-and-stick model* of methane. The dark gray ball is the carbon atom and the light gray balls are hydrogen atoms. (*c*) A *space-filling model* of methane that shows the relative sizes of the C and H atoms. Ball-and-stick and space-filling models are used to illustrate the three-dimensional shapes of molecules.

We use the term *macroscopic* to mean the world we observe with our senses, whether it be in the laboratory or the world we encounter in our day-to-day living.

We will expand on the atomic model as we proceed in our discussions, and we will frequently make the connection between what we physically observe in our large, *macroscopic* world and what we believe takes place in the tiny, submicroscopic world of atoms and molecules. Learning to recognize these connections should be one of your major goals in studying this course.

1.3 ▶ Properties of materials can be classified in different ways

If you had lost your chemistry book and were asked to describe it, you would probably list its size, color, and the printing on its cover. These are the characteristics that books have that help you identify them and distinguish among them. Similarly, in chemistry we use the characteristics, or *properties,* of materials to identify them and to distinguish one kind from another. To help organize our thinking, we classify properties into different types.

Properties can be classified as physical or chemical

One way to classify properties is based on whether or not the chemical composition of an object is changed by the act of observing the property.

Physical properties are observed without changing the chemical makeup of a substance

A **physical property** *is one that can be observed without changing the chemical makeup of a substance.* For example, one of the physical properties of gold is that it is yellow. The act of observing this property (color) doesn't change the chemical makeup of the gold. Neither does observing that gold conducts electricity, so electrical conductivity is another physical property of gold.

Sometimes, observing a physical property does lead to a change. Melting point is an example. To measure this property, we observe the temperature at which melting begins. This does not lead to a change in chemical composition, however. For instance, both ice and liquid water are composed of water molecules; melting just permits molecules that are locked in place in the solid to become mobile and move about in the liquid.

The change we observe during melting is a **physical change** because *it is not accompanied by a change in chemical makeup.* Another physical change is observed when a substance dissolves in a liquid (for instance, sugar dissolving in water). This ability of sugar to dissolve in water is a physical property because the act of dissolving the sugar doesn't alter its chemical composition. Figure 1.3 illustrates how the molecules of sugar and water simply intermingle when the sugar dissolves. By evaporating the water we can show that the chemical makeup of the sugar has not been affected; the sugar is recovered chemically unchanged. Because the sugar isn't altered chemically as it dissolves, forming the solution is a physical change.

Liquid water and ice are both composed of water molecules. Melting the ice cube doesn't change the chemical composition of the molecules.

A chemical property describes a chemical change

Chemistry is especially concerned with learning and understanding changes that alter the chemical makeup of substances, changes that we call chemical reactions. In a **chemical reaction,** chemicals interact with each other to form entirely *different* substances with different properties. An example is the rusting of iron, which involves a chemical reaction between iron, oxygen, and water. Iron is a metallic solid that is attracted by a magnet. Oxygen is one of the components of air and is a gas. Water, of course, is a liquid. When these react to form rust, the product no longer looks like iron, oxygen, or water. It is a brown solid that doesn't look at all like a metal and it is not attracted by a magnet. (See Figure 1.4.)

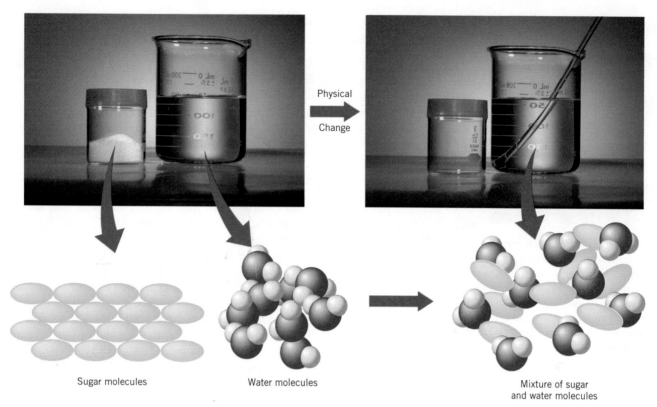

Sugar molecules   Water molecules   Mixture of sugar
                                                                      and water molecules

FIGURE 1.3 *Formation of a solution of sugar in water.* As the sugar mixes with the water, sugar molecules (shown here in a very simplified way) mingle with water molecules. The chemical makeup of the individual sugar and water molecules does not change, however, so no chemical reaction has occurred.

A **chemical property** *describes a* **chemical change** *(chemical reaction) that a substance undergoes.* Thus, one chemical property of iron is that it forms rust in the presence of oxygen and moisture. When we observe this property, the reaction changes the iron, oxygen, and water into rust, so after we've made the observation we no longer have the same substances as before. Another example is the change that occurs when we melt sugar in a pan and heat it to a high temperature. We observe that the color of the sugar darkens as it begins to decompose into carbon and water. The decomposition of sugar at high temperatures is a chemical property, because the only way we can observe it is by having the chemical reaction occur. It is a reaction that leads to a permanent change in chemical composition, which is not reversed by cooling the sugar to room temperature.

When we observe a chemical property of a substance, the substance undergoes a reaction and changes into something else. When we observe a physical property, the substance is not changed chemically.

Properties can also be classified as intensive or extensive

Another way of classifying a property is according to whether or not the property depends on the size of the sample under study. For example, the volumes of two different pieces of gold can be different, but both have the same characteristic shiny yellow color, both conduct electricity, and both will begin to melt if heated to the same temperature. Volume is said to be an **extensive property**—*a property that depends on sample size.* Color, electrical conductivity, melting point (and boiling point) are examples of **intensive properties**—*properties that are independent of sample size.*[3]

Volume is an extensive physical property. Color is an intensive physical property.

[3]Color can often be a useful property for identification, but there are instances where you can be fooled, particularly when particle size is very small. For example, silver is a white metal with a high luster; however, in a very finely divided state, as in the image on black-and-white photographic film or paper, metallic silver appears black.

FIGURE 1.4 *Chemical reactions cause changes in composition.* Here we see a coating of rust that has formed on iron nuts and bolts. The properties and chemical composition of the rust are entirely different from those of the iron.

Some kinds of properties are better than others for identifying substances

A job chemists are often called upon to perform is chemical analysis. They're asked, "What is a particular sample composed of?" To answer such a question, the chemist relies on the properties of the chemicals that make up the sample.

For identification purposes, intensive properties are more useful than extensive ones because every sample of a given substance exhibits the same set of intensive properties. For instance, if you were asked to decide whether a particular liquid sample was water, measuring its volume wouldn't help in your identification because volume is not an unvarying property of a substance. By taking appropriate amounts, it is possible for *any* liquid to have a given volume. However, if you observe that the liquid is clear and colorless and then by measurement find that it freezes at 0 °C and boils at 100 °C, you might be fairly confident that the liquid is water.[4] This is because these are properties that all samples of water have in common. On the other hand, if the liquid *did not* freeze and boil at 0 °C and 100 °C, respectively, then you would be very confident the liquid was *not* water.

Color, freezing point, and boiling point are examples of physical properties that can help us identify substances. Chemical properties are also intensive properties and also can be used for identification. For example, gold miners were able to distinguish between real gold and fool's gold (a mineral also called pyrite, Figure 1.5) by heating the material in a flame. Nothing happens to the gold, but the pyrite sputters, smokes, and releases bad-smelling fumes because of its ability, when heated, to react chemically with oxygen in the air.

1.4 ▶ Materials are described by their properties

It's natural for us to look for threads of organization among apparently unrelated facts. When we examine the myriad of different materials that surround us, we therefore seek to classify them, and one way to do this is according to their properties.

[4]We usually use the Celsius temperature scale in the sciences. We discuss this in more detail in Chapter 3.

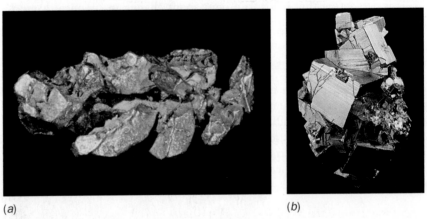

(a)   (b)

FIGURE 1.5 *Identification of substances by their properties.* Gold and the mineral pyrite (also called iron pyrite) have similar colors, which is why some miners mistook pyrite for gold. (*a*) A nugget of pure gold. (*b*) A sample of iron pyrite. The color of the mineral accounts for its nickname, "fool's gold."

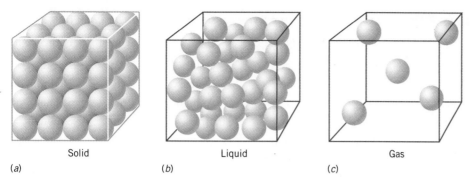

Solid (a)   Liquid (b)   Gas (c)

FIGURE 1.6 *Solid, liquid, and gaseous states of matter as viewed by the atomic model of matter.* (*a*) In a solid, the particles are tightly packed and cannot move easily. (*b*) In a liquid, the particles are still close together but can move past one another. (*c*) In a gas, the particles are far apart with much empty space between them.

Solids, liquids, and gases are states of matter

As you probably know, ice, liquid water, and steam are *not* different substances. Although they have quite different appearances and physical properties, they are just different forms of the same substance, water. **Solid, liquid,** and **gas** are the most common **states of matter.** As with water, most substances are able to exist in all three of these states, and the state we observe generally depends on the temperature.

The obvious properties of solids, liquids, and gases can be interpreted at a sub-microscopic level according to the different ways the individual atomic-size particles are organized (Figure 1.6). In a solid, they are packed tightly together and are unable to move about, so solids are rigid. In a liquid, the particles are still close together, but they are capable of moving past each other and a liquid is therefore able to flow. In gases, the particles are far apart with much empty space between them. This makes it easy to see through gases and also explains why they are so easily compressed; pushing the particles together just reduces the empty space between them.

Elements cannot be decomposed by chemical means

When a chemical reaction changes one substance into two or more others, a **decomposition** has occurred. For example, if we pass electricity through molten (melted) sodium chloride (salt), the silvery metal, sodium, and the pale green gas, chlorine, are formed. In this example, we have decomposed sodium chloride into two simpler substances. No matter how we try, however, sodium and chlorine cannot be decomposed further by chemical reactions into still simpler substances that can be stored and studied.

In chemistry, *substances that cannot be decomposed into simpler materials by chemical reactions are called* **elements.** Sodium and chlorine are two examples. Others you may be familiar with include iron, chromium, lead, copper, aluminum, sulfur, and carbon (as in charcoal). Some elements are gases at room temperature, including chlorine, oxygen, hydrogen, nitrogen, and helium. Elements are the simplest forms of matter that chemists work with directly. All more complex substances are composed of elements in various combinations.

Chemical symbols are used to identify elements

So far, scientists have discovered 90 existing elements in nature and have made 23 more, for a total of 113. Each element is assigned a unique **chemical symbol,** which can be used as an abbreviation for the name of the element. As we will dis-

TABLE 1.2		ELEMENTS THAT HAVE SYMBOLS DERIVED FROM THEIR LATIN NAMES			
Element	Symbol	Latin Name	Element	Symbol	Latin Name
Sodium	Na	Natrium	Gold	Au	Aurum
Potassium	K	Kalium	Mercury	Hg	Hydrargyrum
Iron	Fe	Ferrum	Antimony	Sb	Stibium
Copper	Cu	Cuprum	Tin	Sn	Stannum
Silver	Ag	Argentum	Lead	Pb	Plumbum

Latin was the universal language of science in the early days of chemistry.

cuss in the next chapter, chemical symbols are also used to stand for atoms of elements when we write chemical formulas. In most cases, the symbol is formed from one or two letters of the English name for the element. For instance, the symbol for carbon is C, for bromine it is Br, and for silicon it is Si. For some elements, the symbols are derived from the Latin names given to those elements long ago. Table 1.2 contains a list of elements whose symbols come to us in this way.[5]

Regardless of the origin of the symbol, the first letter is always capitalized and the second letter, if there is one, is always written lowercase. Thus, the symbol for copper is Cu, not CU. Be careful to follow this rule so you can avoid confusion between such symbols as Co (cobalt) and CO (carbon monoxide). The names and chemical symbols of the elements are given on the inside front cover of the book. Although the list may seem long, many elements are rare, and we will be most interested in only a relatively small number of them.

PRACTICE EXERCISE 1: As you will learn in Chapter 2, chemical symbols are used to write formulas for compounds. For each of the following, identify the elements present on the basis of their chemical symbols (the meaning of subscripts will be made clear in Chapter 2). (a) Fe_2O_3 (rust), (b) Na_3PO_4 (TSP, a cleaning agent), (c) Al_2O_3 (the major component of the gem, ruby), (d) $CaCO_3$ (found in seashells)

Compounds are composed of two or more elements in fixed proportions

By means of chemical reactions, elements combine in various *specific proportions* to give all the more complex substances in nature. Thus, hydrogen and oxygen combine to form water, and sodium and chlorine combine to form sodium chloride (common table salt). Water and sodium chloride are examples of compounds. A **compound** *is a substance formed from two or more **different elements** in which the elements are always combined in the same fixed (i.e., constant) proportions by mass.* For example, if any sample of pure water is decomposed, the mass of oxygen obtained is *always* eight times the mass of hydrogen. Similarly, when hydrogen and oxygen react to form water, the mass of oxygen consumed is always eight times the mass of hydrogen, never more and never less.

Water and salt must be more complex than elements because they can be decomposed to elements.

In our description of the makeup of a compound, there is a subtle but significant point to be understood. For example, we say that the compound carbon dioxide, a substance formed in the combustion of many fuels, is *composed of* carbon and oxygen. However, in this compound these two elements are not present in the same form as we find them in their pure states. We can see this by comparing the properties of the compound with those of the elements from which it's formed. Ordinarily, carbon is found as a black solid, as in coal and charcoal. But at room temperature, carbon dioxide isn't a solid, it's a colorless gas. Oxygen is also a gas,

[5]The symbol for tungsten is W, from the German name *Wolfram*. This is the only element whose symbol is neither related to its English name nor derived from its Latin name.

Orange juice, Coca-Cola, and pancake syrup are mixtures that contain sugar. The amount of sugar varies from one to another because mixtures can have variable compositions.

but it supports combustion, whereas carbon dioxide is used to extinguish fires. Thus, the properties of carbon dioxide are much different from those of both carbon and oxygen. When carbon and oxygen combine to form carbon dioxide, the properties of carbon and oxygen disappear and in their place we find the unique properties of the compound carbon dioxide.

Mixtures can have variable compositions

Elements and compounds are examples of **pure substances.**[6] The composition of a pure substance is always the same, regardless of its source. All samples of table salt, for example, contain sodium and chlorine combined in the same proportions by mass. Similarly, all samples of water contain the same proportions by mass of hydrogen and oxygen. Pure substances are rare, however. Usually we encounter mixtures of compounds or elements. Unlike elements and compounds, **mixtures** *can have variable compositions.* Carbon dioxide and water are examples of compounds because each consists of elements chemically combined in definite proportions. But, when we dissolve carbon dioxide in water to give a carbonated beverage such as seltzer, we can vary the amounts of the two compounds. As a result, we can vary the amount of "fizz" in the beverage.

Mixtures can be either homogeneous or heterogeneous. A **homogeneous mixture** *has the same properties throughout the sample.* An example is a thoroughly stirred mixture of sugar in water. We call such a homogeneous mixture a **solution.** Solutions need not be liquids, just homogeneous. Brass, for example, is a solid solution of copper and zinc, and clean air is a gaseous solution of oxygen, nitrogen, and a number of other gases.

A **heterogeneous mixture** *consists of two or more regions, called* **phases,** *that differ in properties.* A mixture of olive oil and vinegar in a salad dressing, for example, is a two-phase mixture in which the oil floats on the vinegar as a separate layer (Figure 1.7). The phases in a mixture don't have to be chemically different substances like oil and vinegar, however. A mixture of ice and liquid water is a two-phase heterogeneous mixture in which the phases have the same chemical composition but occur in different physical states.

An important way that mixtures differ from compounds is in the changes that occur when they form. Consider, for example, the elements iron and sulfur, which are pictured in powdered form in Figure 1.8*a*. We can make a mixture simply by dumping them together and stirring them. In the mixture (Figure 1.8*b*), both elements retain their original properties. The process we use to create this mixture

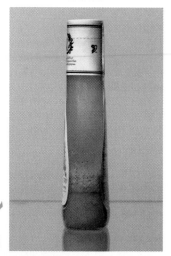

FIGURE 1.7 *A heterogeneous mixture.* The salad dressing shown here contains vinegar and vegetable oil (plus assorted other flavorings). Vinegar and oil do not dissolve in each other but form two layers. The mixture is heterogeneous because each of the separate phases (oil, vinegar, and other solids) has its own set of properties that differ from the properties of the other phases.

---

[6]We have used the term *substance* rather loosely until now. Strictly speaking, **substance** really means *pure substance.* Each unique chemical element and compound is a *substance;* a mixture consists of two or more substances.

(a)   (b)

FIGURE **1.8** *Formation of a mixture of iron and sulfur.* (*a*) Samples of powdered sulfur and powdered iron. (*b*) A mixture of sulfur and iron is made by stirring the two powders together.

FIGURE **1.9** *Formation of a mixture is a physical change.* Here we see that forming the mixture has not changed the iron and sulfur into a compound of these two elements. The mixture can be separated by pulling the iron out with a magnet.

involves a *physical change,* rather than a chemical change, because no new chemical substances form. To separate the mixture, we could similarly use just physical changes. For example, we could remove the iron by stirring the mixture with a magnet—a physical operation. The iron powder sticks to the magnet as we pull it out, leaving the sulfur behind (Figure 1.9). The mixture also could be separated by treating it with a liquid called carbon disulfide, which is able to dissolve the sulfur but not the iron. Filtering the sulfur solution from the solid iron, followed by evaporation of the liquid carbon disulfide from the sulfur solution, gives the original components, iron and sulfur, separated from each other.

As we noted earlier, formation of a compound from its elements involves a chemical reaction. Iron and sulfur, for example, combine to form a compound often called "fool's gold" because of its appearance (Figure 1.5, page 8). In this compound the elements no longer have the same properties they had before they were combined, and they cannot be separated by physical means. Just as the formation of a compound involves a chemical reaction, so also does its decomposition. The decomposition of fool's gold into iron and sulfur is therefore a chemical change.

The relationships among elements, compounds, and mixtures are shown in Figure 1.10.

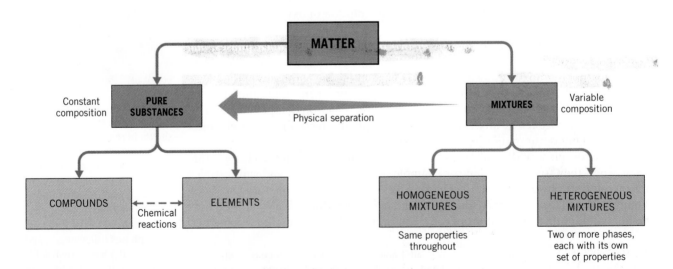

FIGURE **1.10** *Classification of matter.*

1.5 ▶ Atoms of an element have properties in common

In our discussion of elements in the preceding section, no reference was made to the atomic nature of matter. In fact, the distinction between elements and compounds had been made even before the atomic theory of matter was formulated. In this section we will examine how the atomic theory began and take a closer look at elements in terms of our modern view of atomic structure.

Dalton's atomic theory explained chemical laws

In modern science, we have come to take for granted the existence of atoms and molecules. In fact, we've already used the atomic theory to explain some of the properties of materials. However, scientific evidence for the existence of atoms is relatively recent, and chemistry did not progress very far until that evidence was found. Therefore, let's take a brief look at how the atomic theory evolved.

The concept of atoms began nearly 2500 years ago when certain Greek philosophers expressed the belief that matter is ultimately composed of tiny indivisible particles, and it is from the Greek word *atomos,* meaning "not cut," that the word *atom* is derived. The philosophers' conclusions, however, were not supported by any evidence; they were derived simply from philosophical reasoning.

Laws of chemical combination evolved from experimental observations

The concept of atoms remained a philosophical belief, having limited scientific usefulness, until the discovery of two quantitative laws of chemical combination—the *law of conservation of mass* and the *law of definite proportions.* The evidence that led to the discovery of these laws came from the experimental observations of many scientists in the eighteenth and early nineteenth centuries.

Law of Conservation of Mass
No detectable gain or loss of mass occurs in chemical reactions. Mass is *conserved.*

Law of Definite Proportions
In a given chemical compound, the elements are always combined in the same proportions by mass.

◀ **TOOLS**

Laws of definite proportions and conservation of mass

Notice that both of these laws refer to the masses of substances. Earlier, in our definition of matter, we noted that mass is a measure of the amount of matter in an object. In the sciences, we measure mass in units of **grams** (symbol, **g**). In Chapter 3 we will discuss units of measurement, including those of mass, in greater detail, but for now grams are sufficient.

In the United States, we commonly measure mass in units of ounces or pounds. One pound equals 453.6 g. A penny has a mass of about 3 g.

The **law of conservation of mass** means that if a chemical reaction takes place in a sealed vessel that permits no matter to enter or escape, the mass of the vessel and its contents after the reaction will be identical to its mass before. Although this may seem quite obvious to us now, it wasn't quite so clear in the early history of modern chemistry. Not until scientists made sure that *all* substances, including any that were gases, were included when masses were measured could the law of conservation of mass be truly tested.

We actually used the **law of definite proportions** on page 10 when we defined a compound as a substance in which two or more elements are chemically combined in a *definite fixed proportion by mass.* Thus, if we decompose samples of water (a compound) into the elements oxygen and hydrogen, we always find that the ratio of oxygen to hydrogen, *by mass,* is 8 to 1. In other words, the mass of oxygen obtained is always eight times the mass of hydrogen.

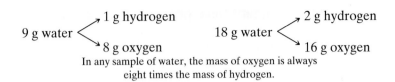

9 g water $\Big\langle$ 1 g hydrogen
8 g oxygen

18 g water $\Big\langle$ 2 g hydrogen
16 g oxygen

In any sample of water, the mass of oxygen is always
eight times the mass of hydrogen.

Similarly, if we form water from oxygen and hydrogen, the mass of oxygen consumed will always be eight times the mass of hydrogen that reacts. This is true even if there's a large excess of one of them. For instance, if 100 g of oxygen are mixed with 1 g of hydrogen and the reaction to form water is initiated, all the hydrogen would react but only 8 g of oxygen would be consumed; there would be 92 g of oxygen left over. No matter how we try, we can't alter the chemical composition of the water formed in the reaction.

100 g oxygen
1 g hydrogen

$\xrightarrow[\substack{\text{occurs consuming} \\ \text{8 g oxygen} \\ \text{1 g hydrogen}}]{\text{chemical reaction}}$

9 g water
92 g oxygen

mixture before
reaction

mixture after
reaction

Let's look at a sample calculation that shows how we might use the law of definite proportions.

EXAMPLE 1.1

Applying the Law of Definite Proportions

The element molybdenum (Mo) combines with sulfur (S) to form a compound commonly called molybdenum disulfide that is useful as a dry lubricant, similar to graphite. It is also used in specialized lithium batteries. A sample of this compound contains 1.50 g of Mo for each 1.00 g of S. If a different sample of the compound contains 2.50 g of S, how many grams of Mo does it contain?

A Word about Problem Solving

This is the first of many encounters you will have with solving problems in chemistry. Helping you learn how to approach and solve problems is one of the major goals of this textbook. We view problem solving as a three-step process. The first is figuring out what has to be done to solve the problem, which is the function of the *Analysis* step described below. The second is actually performing whatever is required to obtain the answer (the *Solution* step). And finally, we examine the answer to determine whether it seems to be *reasonable*. For more information on the aids that are available to assist you in problem solving, we recommend that you read the "To The Student" section at the beginning of the book.

ANALYSIS: Much of the effort in solving a chemistry problem is devoted to determining which concepts have to be applied. We view these concepts as *tools*, each with its specific uses when applied to problem solving. Our goal in this Analysis step is to identify which tools we need and how we will apply them.

We begin by examining the problem and asking a question: What have we learned that relates the masses of elements in two samples of the same compound? The law of definite proportions states that the proportions of the elements by mass must be the same in both samples, so the law of definite proportions is the tool we need to apply. The law tells us that the ratio of grams of S to grams of Mo must be the same in both samples. To solve the problem, then, we will set up the mass ratios for the two samples. In the ratio for the second sample, the mass of molybdenum will be an unknown quantity. We'll equate the two ratios and solve for the unknown quantity.

SOLUTION: Now that we've determined what we need to do to solve the problem, the rest is pretty easy. The first sample has a Mo to S mass ratio of

$$\frac{1.50 \text{ g Mo}}{1.00 \text{ g S}}$$

In the second sample, we know the mass of S (2.50 g) and we want to find the mass of Mo, so let's call this quantity x. The mass ratio of Mo to S in the second sample is therefore

$$\frac{x}{2.50 \text{ g S}}$$

Now we equate them, because the two ratios must be equal.

$$\frac{x}{2.50 \text{ g S}} = \frac{1.50 \text{ g Mo}}{1.00 \text{ g S}}$$

Solving for x gives

$$x = 2.50 \text{ g S} \times \frac{1.50 \text{ g Mo}}{1.00 \text{ g S}} = 3.75 \text{ g Mo}$$

Is the Answer Reasonable?
To avoid errors, it's always wise to do a rough check of the answer. Usually, some simple reasoning is all we need to see if the answer is "in the right ball park." This is how we might do such a check here: Notice that the amount of sulfur in the second sample is more than twice the amount in the first sample. Therefore, we should expect the amount of Mo in the second sample to be somewhat more than twice what it is in the first. The answer we obtained, 3.75 g Mo, is more than twice 1.50 g Mo, so our answer seems to be reasonable.

PRACTICE EXERCISE 2: Cadmium sulfide is a yellow compound that is used as a pigment in artist's oil colors. A sample of this compound is composed of 1.25 g of cadmium and 0.357 g of sulfur. If a second sample of the same compound contains 3.50 g of sulfur, how many grams of cadmium does it contain?

The atomic theory was proposed by John Dalton

The laws of conservation of mass and definite proportions served as the *experimental foundation* for the atomic theory. They prompted the question, "What must be true about the nature of matter, given the truth of these laws?" In other words, what is matter made of?

At the beginning of the nineteenth century, John Dalton (1766–1844), an English scientist, used the Greek concept of atoms to make sense out of the laws of conservation of mass and definite proportions. Dalton reasoned that if atoms really exist, they must have certain properties to account for these laws. He described such properties, and the list constitutes what we now call **Dalton's atomic theory.**

Dalton's Atomic Theory

1. *Matter consists of tiny particles called atoms.*
2. *Atoms are indestructible. In chemical reactions, the atoms rearrange but they do not themselves break apart.*
3. *In any sample of a pure element, all the atoms are identical in mass and other properties.*
4. *The atoms of different elements differ in mass and other properties.*
5. *When atoms of different elements combine to form compounds, new and more complex particles form. However, in a given compound the constituent atoms are always present in the same fixed **numerical** ratio.*

Dalton's theory easily explained the law of conservation of mass. According to the theory, a chemical reaction is simply a reordering of atoms from one combination to another. If no atoms are gained or lost and if the masses of the atoms can't change, then the mass after the reaction must be the same as the mass before. This explanation of the law of conservation of mass works so well that it serves as the reason for balancing chemical equations, which we will discuss in the next chapter.

The law of definite proportions is also easy to explain. According to the theory, a given compound always has atoms of the same elements in the same numerical ratio. Suppose, for example, that two elements, *A* and *B*, combine to form a compound in which the number of atoms of *A* equals the number of atoms of *B* (i.e., the *atom ratio* is 1 to 1). If the mass of a *B* atom is twice that of an *A* atom, then every time we encounter a sample of this compound, the mass ratio (*A* to *B*) would be 1 to 2. This same mass ratio would exist regardless of the size of the sample, so in samples of this compound the elements *A* and *B* are always present in the same proportion by mass.

The atomic theory led to the discovery of the law of multiple proportions

Strong support for Dalton's theory came when Dalton and other scientists studied elements that are able to combine to give two (or more) compounds. For example, sulfur and oxygen form two different compounds, which we call sulfur dioxide and sulfur trioxide. If we decompose a 2.00 g sample of sulfur dioxide, we find it contains 1.00 g of S and 1.00 g of O. If we decompose a 2.50 g sample of sulfur trioxide, we find it also contains 1.00 g of S, but this time the mass of O is 1.50 g. This is summarized in the following table.

Compound	Sample Size	Mass of Sulfur	Mass of Oxygen
Sulfur dioxide	2.00 g	1.00 g	1.00 g
Sulfur trioxide	2.50 g	1.00 g	1.50 g

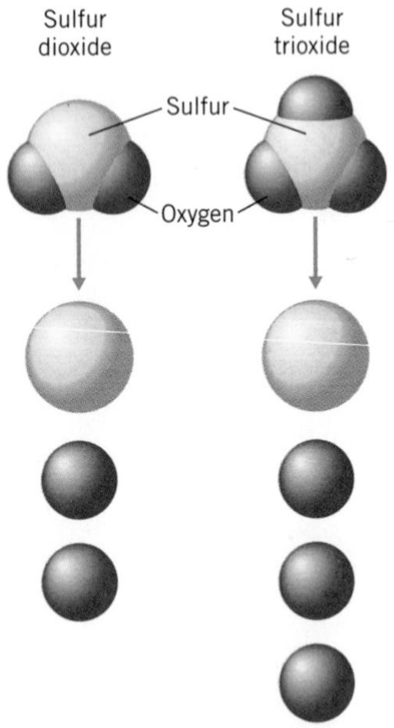

First, notice that sample sizes aren't the same; they were chosen so that each has the *same mass of sulfur*. Second, the ratio of the masses of oxygen in the two samples is one of small whole numbers.

$$\frac{\text{mass of oxygen in sulfur trioxide}}{\text{mass of oxygen in sulfur dioxide}} = \frac{1.50 \text{ g}}{1.00 \text{ g}} = \frac{3}{2}$$

Similar observations are made when we study other elements that form more than one compound with each other, and these observations form the basis of the **law of multiple proportions.**

> **Law of Multiple Proportions**
> Whenever two elements form more than one compound, the different masses of one element that combine with the same mass of the other element are in the ratio of small whole numbers.

Dalton's theory explains the law of multiple proportions in a very simple way. Suppose a molecule of sulfur trioxide contains one sulfur and three oxygen atoms, and a molecule of sulfur dioxide contains one sulfur and two oxygen atoms (Figure 1.11). If we had just one molecule of each, then our samples each would have one sulfur atom and therefore the same mass of sulfur. Then, comparing the oxygen atoms, we find they are in a numerical ratio of 3 to 2. But because oxygen atoms all have the same mass, the mass ratio must also be 3 to 2.

The law of multiple proportions was not known before Dalton presented his theory, and its discovery demonstrates the scientific method in action. Experimental data suggested to Dalton the existence of atoms, and the atomic theory

Sulfur dioxide

Sulfur trioxide

— Sulfur —

— Oxygen —

FIGURE 1.11 *Oxygen compounds of sulfur demonstrate the law of multiple proportions.* Illustrated here are molecules of sulfur trioxide and sulfur dioxide. Each has one sulfur atom, and therefore the same mass of sulfur. The oxygen ratio is 3 to 2, both by atoms and by mass.

suggested the relationships that we now call the law of multiple proportions. When found by experiment, the existence of the law of multiple proportions added great support to the atomic theory. In fact, for many years it was one of the strongest arguments in favor of the existence of atoms.

Modern experimental evidence exists for atoms

Atoms are so incredibly tiny that even the most powerful optical microscopes are unable to detect them. In recent times, though, scientists have developed very sensitive instruments that are able to map the surfaces of solids with remarkable resolution. One such instrument is called a **scanning tunneling microscope.** It was invented in the early 1980s by Gerd Binnig and Heinrich Rohrer and earned them the 1986 Nobel prize in physics. With this instrument, the tip of a sharp metal probe is brought very close to an electrically conducting surface and an electric current bridging the gap is begun. The flow of current is extremely sensitive to the distance between the tip of the probe and the sample. As the tip is moved across the surface, the height of the tip is continually adjusted to keep the current flow constant. By accurately recording the height fluctuations of the tip, a map of the hills and valleys on the surface is obtained. The data are processed using a computer to reveal images such as that shown in Figure 1.12.

Relative atomic masses of elements can be found

One of the most useful concepts to come from Dalton's atomic theory is that atoms of an element have a constant, characteristic **atomic mass** (or **atomic weight**). This concept opened the door to the determination of chemical formulas and ultimately to one of the most useful devices chemists have for organizing chemical information, the periodic table of the elements. But how can the masses of atoms be measured?

Individual atoms are much too small to weigh in the traditional manner. However, the *relative masses* of the atoms of elements can be determined *provided we know the ratio in which the atoms occur in a compound.* Let's look at an example to see how this could work.

Hydrogen (H) combines with the element fluorine (F) to form the compound hydrogen fluoride. Each molecule of this compound contains one atom of hydrogen and one atom of fluorine, which means that in *any* sample of this substance the fluorine-to-hydrogen *atom ratio* is always 1 to 1. It is also found that when a sample of hydrogen fluoride is decomposed, the mass of fluorine obtained is always 19.0

FIGURE 1.12 *Individual atoms can be imaged using a scanning tunneling microscope.* This STM micrograph reveals the pattern of individual atoms of palladium deposited on a graphite surface. Palladium is a silver-white metal used in alloys such as white gold and dental crowns.

times larger than the mass of hydrogen, so the fluorine-to-hydrogen *mass ratio* is always 19.0 to 1.00.

F-to-H atom ratio: 1 to 1

F-to-H mass ratio: 19.0 to 1.00

How could a 1-to-1 atom ratio give a 19.0-to-1.00 mass ratio? *Only if each fluorine atom is 19.0 times heavier than each H atom.*

Notice that even though we haven't found the actual masses of F and H atoms, we do now know how their masses compare (i.e., we know their *relative masses*). Similar procedures with other elements in other compounds are able to establish relative mass relationships among the other elements as well. What we need next is a way to place all these masses on the same mass scale, but before we study this, let's take a closer look at one of Dalton's hypotheses.

Isotopes are atoms of the same element with different masses

The cornerstone of Dalton's theory was the idea that all of the atoms of an element have identical masses. Actually, most elements occur in nature as uniform mixtures of two or more kinds of atoms that have slightly different masses, which we call **isotopes.** An iron nail, for example, is made up of a mixture of four iron isotopes, and the chlorine in salt is a mixture of two chlorine isotopes.

Even the smallest laboratory sample of an element has so many atoms that the relative proportions of the isotopes is constant.

The existence of isotopes did not affect the development of Dalton's theory for two reasons. First, all the isotopes of a given element have virtually identical *chemical* properties—all give the same kinds of chemical reactions. Second, the relative proportions of its different isotopes are essentially constant, regardless of where on earth or in the atmosphere the element is found. As a result, every sample of a particular element has the same isotopic composition, and the *average* mass per atom is the same from sample to sample. Therefore, in the laboratory elements *behave* as though their atoms have masses equal to the average.

Carbon-12 is the standard on the atomic mass scale

To establish a uniform mass scale for atoms it is necessary to select a standard against which the relative masses can be compared. Currently, the agreed-upon reference is the most abundant isotope of carbon, called carbon-12 and symbolized ^{12}C. One atom of this isotope is assigned *exactly* 12 units of mass, which are called **atomic mass units.** Some prefer to use the symbol **amu** for the atomic mass unit. The internationally accepted symbol is **u,** which is the symbol we will use throughout the rest of the book. By assigning 12 u to the mass of one atom of ^{12}C, the size of the atomic mass unit is established to be $\frac{1}{12}$ of the mass of a single carbon-12 atom:

The atomic mass unit is sometimes called a dalton.

$$1 \text{ u} = 1 \text{ dalton}$$

1 atom of ^{12}C has a mass of 12 u (exactly)

1 u equals $\frac{1}{12}$ the mass of 1 atom of ^{12}C (exactly)

In modern terms, the atomic mass of an element is the average mass of the element's atoms (as they occur in nature) relative to an atom of carbon-12, which is assigned a mass of 12 units. Thus, if an average atom of an element has a mass twice that of a ^{12}C atom, its atomic mass would be 24 u.

The definition of the size of the atomic mass unit is really quite arbitrary. It could just as easily have been selected to be $\frac{1}{24}$ of the mass of a carbon atom, or $\frac{1}{10}$ of the mass of an iron atom, or any other value. Why $\frac{1}{12}$ of the mass of a ^{12}C atom? First, carbon is a very common element, available to any scientist. Second, and most important, by choosing the atomic mass unit of this size, the atomic masses of nearly all the other elements are almost whole numbers, with the lightest atom (hydrogen) having a mass of approximately 1 u.

Chemists generally work with whatever *mixture* of isotopes comes with a given element as it occurs naturally. Because the composition of this isotopic mixture is very nearly constant regardless of the source of the element, we can speak of an *average atom* of the element—average in terms of mass. For example, naturally occurring hydrogen is a mixture of two isotopes in the relative proportions given in the margin. The "average atom" of the element hydrogen, as it occurs in nature, has a mass that is 0.083992 times that of a ^{12}C atom. Since 0.083992 × 12.000 u = 1.0079 u, the average atomic mass of hydrogen is 1.0079 u. Notice that this average value is just a little larger than the atomic mass of ^1H because naturally occurring hydrogen also contains a little ^2H.

Hydrogen Isotope	Mass	Percentage Abundance
^1H	1.007825 u	99.985
^2H	2.0140 u	0.015

Average atomic masses can be calculated from isotopic abundances

Originally, the relative atomic masses of the elements were determined in a way similar to that described for hydrogen and fluorine in our earlier discussion. A sample of a compound was analyzed and from the formula of the substance the relative atomic masses were calculated. These were then adjusted to place them on the unified atomic mass scale. In modern times, methods have been developed to measure very precisely both the relative abundances of the isotopes of the elements and their atomic masses. This kind of information has permitted the calculation of more precise values of the average atomic masses, which are found in the table on the inside front cover of the book. Example 1.2 illustrates how this calculation is done.

Naturally occurring chlorine is a mixture of two isotopes. In every sample of this element, 75.77% of the atoms are ^{35}Cl and 24.23% are atoms of ^{37}Cl. The accurately measured atomic mass of ^{35}Cl is 34.9689 u and that of ^{37}Cl is 36.9659 u. From these data, calculate the average atomic mass of chlorine.

EXAMPLE 1.2

Calculating Average Atomic Masses from Isotopic Abundances

ANALYSIS: This problem is a little like calculating the average height of a group of people. Let's see how we might do that and then apply the same principles to the problem at hand.

Suppose 10% of a group of people are 4 ft tall and the other 90% are 6 ft tall. The average height of the group is *not* 5 ft (an average of 4 ft and 6 ft). To calculate the average, we have to weight it according to the proportions of 4- and 6-footers. If there were 100 people in the group, 10 of them (which is 10% of 100) would be 4 ft tall and the other 90 would be 6 ft tall. To compute the average, we add all the heights of all the people and then divide by the total number. Let's see how this works out when we set up the math.

$$\frac{(10 \times 4 \text{ ft}) + (90 \times 6 \text{ ft})}{100} = \frac{(10 \times 4 \text{ ft})}{100} + \frac{(90 \times 6 \text{ ft})}{100}$$

$$= (0.1 \times 4 \text{ ft}) + (0.9 \times 6 \text{ ft})$$

$$= 0.4 \text{ ft} + 5.4 \text{ ft} = 5.8 \text{ ft}$$

Notice that in this calculation, multiplying 4 ft by 0.1 is the same as taking 10% of 4 ft. Similarly, multiplying 6 ft by 0.9 is the same as taking 90% of 6 ft. Thus, we could say that the "average person" in this group (who really doesn't exist) has a height that's 10% of 4 ft plus 90% of 6 ft. With this in mind, let's see how we can apply it to calculating the average atomic mass.

In a sample containing many atoms of chlorine, 75.77% of the mass is contributed by atoms of ^{35}Cl and 24.23% is contributed by atoms of ^{37}Cl. This means that when we calculate the mass of the "average atom," we have to weight it according to both the masses of the isotopes and their relative abundances. As in the discussion above, to do this it is convenient to imagine an "average atom" to be composed of 75.77% of ^{35}Cl and 24.23% of ^{37}Cl. (Keep in mind, of course, that such an atom doesn't really exist. This is just a simple way to see how we can calculate the average atomic mass of this element.)

SOLUTION: We will calculate 75.77% of the mass of an atom of ^{35}Cl, which is the contribution of this isotope to the "average atom." Then we will calculate 24.23% of the mass of an atom of ^{37}Cl, which is the contribution of this isotope. Adding these contributions gives the total mass of the "average atom."

decimal form of 75.77% — $0.7577 \times 34.9689 \text{ u} = 26.50 \text{ u}$ (for ^{35}Cl)

decimal form of 24.23% — $0.2423 \times 36.9659 \text{ u} = 8.957 \text{ u}$ (for ^{37}Cl)

total mass of average atom = 35.46 u (rounded)

The average atomic mass of chlorine is therefore 35.46 u.

Is the Answer Reasonable?

Once again, the final step is a check to see if the answer makes sense. Here is how we might do such a check: First, from the masses of the isotopes, we know the average atomic mass is somewhere between approximately 35 and 37. If the abundances of the two isotopes were equal, the average would be nearly 36. But there is more ^{35}Cl than ^{37}Cl, so a value somewhere between 35 and 36 seems reasonable; therefore, we can feel pretty confident our answer is correct.

PRACTICE EXERCISE 3: Aluminum atoms have a mass that is 2.24845 times that of an atom of ^{12}C. What is the atomic mass of aluminum?

PRACTICE EXERCISE 4: How much heavier than an atom of ^{12}C is the average atom of naturally occurring copper? Refer to the table inside the front cover of the book for the necessary data.

PRACTICE EXERCISE 5: Naturally occurring boron is composed of 19.8% of ^{10}B and 80.2% of ^{11}B. Atoms of ^{10}B have a mass of 10.0129 u and those of ^{11}B have a mass of 11.0093 u. Calculate the average atomic mass of boron.

1.6 ▶ Atoms are composed of subatomic particles

The earliest theories about atoms imagined them to be indestructible and totally unable to be broken into smaller pieces. However, as you probably know, atoms are not quite as indestructible as Dalton had thought. During the late 1800s and early 1900s, experiments were performed that demonstrated that atoms are composed instead of simpler **subatomic particles.** (For some of the details about these experiments, see Facets of Chemistry 1.1 and 1.2.) From this work the current theoretical model of atomic structure evolved. We will examine it in general terms in this chapter. A more detailed discussion of atomic structure will follow in Chapter 8.

Protons, neutrons, and electrons are subatomic particles

Experiments have shown that atoms are composed of three principal kinds of subatomic particles: **protons, neutrons,** and **electrons.** Experiments also revealed that at the center of an atom there exists a very tiny, extremely dense core called the **nucleus,** which is where an atom's protons and neutrons are found. Because they are found in nuclei, protons and neutrons are sometimes called **nucleons.** The elec-

Experiments Leading to the Discovery of Subatomic Particles

Our current knowledge of atomic structure was pieced together from facts obtained from experiments by scientists that began in the nineteenth century. In 1834, Michael Faraday discovered that the passage of electricity through aqueous solutions could cause chemical changes, which was the first hint that matter was electrical in nature. Later in that century, scientists began to experiment with *gas discharge tubes* in which a high-voltage electric current was passed through a gas at low pressure in a glass tube (Figure 1). Such a tube is fitted with a pair of metal *electrodes,* and when the electricity begins to flow between them, the gas in the tube glows. This flow of electricity is called an *electric discharge,* which is how the tubes got their names. (Modern neon signs work this way.)

The physicists who first studied this phenomenon did not know what caused the tube to glow, but tests soon revealed that negatively charged particles were moving from the negative electrode (the *cathode*) to the positive electrode (the *anode*). The physicists called these emissions *rays,* and because the rays came from the cathode, they were called *cathode rays.*

In 1897, the British physicist J. J. Thomson constructed a special gas discharge tube to make quantitative measurements of the properties of cathode rays. In some ways, the *cathode ray tube* he used was similar to a TV picture tube, as Figure 2 shows. In Thomson's tube, a beam of cathode rays was focused on a glass surface coated with a phosphor that glows when the cathode rays strike it (point 1). The cathode ray beam passed between the poles of a magnet and between a pair of metal electrodes that could be given electrical charges. The magnetic field tends to bend the beam in one direction (to point 2) while the charged electrodes bend the beam in the opposite direction (to point 3). By adjusting the charge on the electrodes, the two effects can be made to cancel, and from the amount of charge on the electrodes required to balance the effect of the magnetic field, Thomson was able to calculate the first bit of quantitative information about a cathode ray particle—the ratio of its charge to its mass (often expressed as e/m, where e stands

for charge and m stands for mass). The charge-to-mass ratio has a value of -1.76×10^8 coulombs/gram, where the coulomb (C) is a standard unit of electrical charge and the negative sign reflects the negative charge on the particle.

Many experiments were performed using the cathode ray tube and they demonstrated that cathode ray particles are in all matter. They are, in fact, *electrons.*

Measuring the Charge and Mass of the Electron

In 1909, a researcher at the University of Chicago, Robert Millikan, designed a clever experiment that enabled him to measure the electron's charge (Figure 3). During an experiment he would spray a fine mist of oil droplets above a pair of parallel metal plates, the top one of which had a small hole in it. As the oil drops settled, some would pass through this hole into the space between the plates, where he would irradiate them briefly with X rays. The X rays knocked electrons off molecules in the air, and the electrons became attached to the oil drops, which thereby were given an electrical charge. By observing the rate of fall of the charged drops both when the metal plates were electrically charged and when they were not, Millikan was able to calculate the amount of charge carried by each drop. When he examined his results, he found that all the values he obtained were whole-number multiples of -1.60×10^{-19} C. He reasoned that since a drop could only pick up whole numbers of electrons, this value must be the charge carried by each individual electron.

Once Millikan had measured the electron's charge, its mass could then be calculated from Thomson's charge-to-mass ratio. This mass was calculated to be 9.09×10^{-28} g.

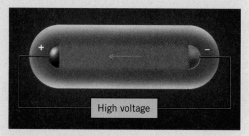

FIGURE 1 *A gas discharge tube.* Cathode rays flow from the negatively charged cathode to the positively charged anode.

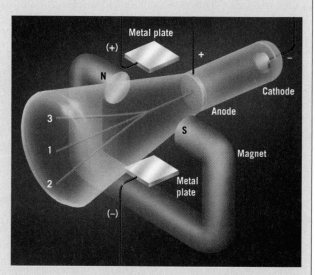

FIGURE 2 *Thomson's cathode ray tube,* which was used to measure the charge-to-mass ratio for the electron.

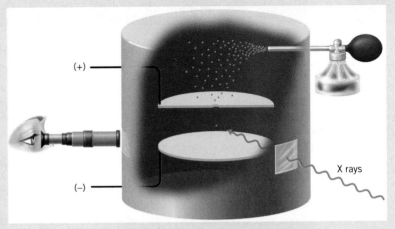

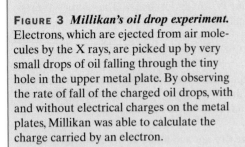

Figure 3 Millikan's oil drop experiment. Electrons, which are ejected from air molecules by the X rays, are picked up by very small drops of oil falling through the tiny hole in the upper metal plate. By observing the rate of fall of the charged oil drops, with and without electrical charges on the metal plates, Millikan was able to calculate the charge carried by an electron.

(+)

(−)

X rays

More precise measurements have since been made, and the mass of the electron is currently reported to be $9.1093897 \times 10^{-28}$ g.

Discovery of the Proton

The removal of electrons from an atom gives a positively charged particle (called an *ion*). To study these, a modification was made in the construction of the cathode ray tube to produce a new device called a *mass spectrometer*. This apparatus is described in Facets of Chemistry 1.2 and was used to measure the charge-to-mass ratios of positive ions. These ratios were found to vary, depending on the chemical nature of the gas in the discharge tube, showing that their masses also varied. The lightest positive particle observed was produced when hydrogen was in the tube, and its mass was about 1800 times as heavy as an electron. When other gases were used, their masses always seemed to be whole-number multiples of the mass observed for hydrogen atoms. This suggested the possibility that clusters of the positively charged particles made from hydrogen atoms made up the positively charged particles of other gases. The hydrogen atom, minus an electron, thus seemed to be a fundamental particle in all matter and was

named the *proton*, after the Greek word *proteios*, meaning "of first importance."

Discovery of the Atomic Nucleus

Early in the twentieth century, Hans Geiger and Ernest Marsden, working under Ernest Rutherford at Great Britain's Manchester University, studied what happened when *alpha rays* hit thin metal foils. Alpha rays are composed of particles having masses four times those of the proton and bearing two positive charges; they are emitted by certain unstable atoms in a phenomenon called *radioactivity*. Most of the alpha particles sailed right on through as if the foils were virtually empty space (Figure 4). A significant number of alpha particles, however, were deflected at very large angles. Some were even deflected backward, as if they had hit stone walls. Rutherford was so astounded that he compared the effect to that of firing a 15 in. artillery shell at a piece of tissue paper and having it come back and hit the gunner! From studying the angles of deflection of the particles, Rutherford reasoned that only something extraordinarily massive and positively charged could cause such an occurrence. Since most of the alpha particles went straight through, he further reasoned

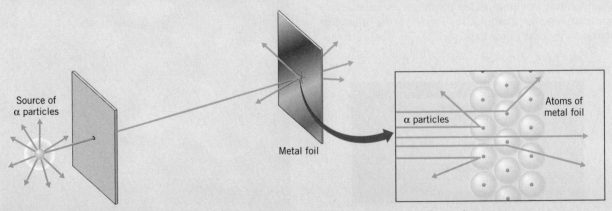

Source of α particles

Metal foil

α particles

Atoms of metal foil

Figure 4 Alpha particles are scattered in all directions by a thin metal foil. Some hit something very massive head-on and are deflected backward. Many sail through. Some, making near misses with the massive "cores" (nuclei), are still deflected, because alpha particles have the same kind of charge (+) as these cores.

that the metal atoms in the foils must be mostly empty space. Rutherford's ultimate conclusion was that virtually all of the mass of an atom must be concentrated in a particle having a very small volume located in the center of the atom. He called this massive particle the atom's *nucleus.*

Discovery of the Neutron

From the way alpha particles were scattered by a metal foil, Rutherford and his students were able to estimate the number of positive charges on the nucleus of an atom of the metal. This had to be equal to the number of protons in the nucleus, of course. But when they computed the nuclear mass based on this number of protons, the value always fell short of the actual mass. In fact, Rutherford found that only about half of the nuclear mass could be accounted for by protons. This led him to suggest that there were other particles in the nucleus that had a mass close to or equal to that of a proton, but with no electrical charge. This suggestion initiated a search that finally ended in 1932 with the discovery of the *neutron* by Sir James Chadwick, a British physicist.

trons in an atom surround the nucleus and fill the remaining volume of the atom. (*How* the electrons are distributed around the nucleus is the subject of Chapter 8.) The properties of the subatomic particles are summarized in Table 1.3, and the general structure of the atom is illustrated in Figure 1.13.

Notice that two of the subatomic particles carry electrical charges. Protons carry a single unit of **positive charge,** which is one type of electrical charge. Electrons carry one unit of the opposite charge, a **negative charge.** These electrical charges have the property that like charges repel each other and opposite charges attract, so negatively charged electrons are attracted to positively charged protons. In fact, it is this attraction that holds the electrons around the nucleus. Neutrons have no charge and are said to be electrically neutral (hence the name *neutron*).

Because of their identical charges, electrons repel each other. The repulsions between the electrons keep them spread out throughout the volume of the atom, and it is the *balance* between the attractions the electrons feel toward the nucleus and the repulsions they feel toward each other that controls the sizes of atoms.

Protons also repel each other, but they are able to stay together in the small volume of the nucleus because their repulsions are apparently offset by powerful nuclear forces that involve other subatomic particles we will not study.

Matter as we generally find it in nature appears to be electrically neutral, which means that it contains equal numbers of positive and negative charges. Therefore, *in a neutral atom, the number of electrons must equal the number of protons.*

The proton and neutron are much more massive than the electron, so in any atom almost all of the atomic mass is contributed by the particles that are found in the nucleus. (The mass of an electron is only about 1/1800 of that of a proton or neutron.) It is also interesting to note, however, that the diameter of the atom is approximately 10,000 times the diameter of its nucleus, so almost all the *volume* of an atom is occupied by its electrons, which fill the space around the nucleus. (To place this on a more meaningful scale, if the nucleus was 1 ft in diameter, it would lie at the center of an atom with a diameter of approximately 1.9 miles!)

Physicists have discovered a large number of subatomic particles, but protons, neutrons, and electrons are the only ones that will concern us at this time.

Protons are in all nuclei. Except for ordinary hydrogen, all nuclei also contain neutrons.

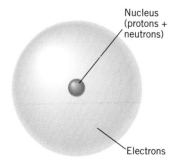

TABLE 1.3	PROPERTIES OF SUBATOMIC PARTICLES			
Particle	Mass (g)	Mass (u)	Electrical Charge	Symbol
Electron	9.109383×10^{-28}	$5.485799095 \times 10^{-4}$	$1-$	$_{-1}^{0}e$
Proton	$1.6726217 \times 10^{-24}$	1.0072764669	$1+$	$_{1}^{1}H^{+}, _{1}^{1}p$
Neutron	$1.6749273 \times 10^{-24}$	1.0086649156	0	$_{0}^{1}n$

FIGURE 1.13 *The internal structure of an atom.* An atom is composed of a tiny nucleus that holds all the protons and neutrons, plus electrons that fill the space outside the nucleus.

Nucleus (protons + neutrons)

Electrons

FACETS OF CHEMISTRY 1.2

The Mass Spectrometer and the Experimental Measurement of Atomic Masses

When a spark is passed through a gas, electrons are knocked off the gas molecules. Because electrons are negatively charged, the particles left behind carry positive charges; they are called *positive ions*. These positive ions have different masses, depending on the masses of the molecules from which they are formed. Thus, some molecules have large masses and give heavy ions, while others have small masses and give light ions.

The device that is used to study the positive ions produced from gas molecules is called a *mass spectrometer* (illustrated in the figure at the right). In a mass spectrometer, positive ions are created by passing an electrical spark (called an *electric discharge*) through a sample of the particular gas being studied. As the positive ions are formed, they are attracted to a negatively charged metal plate that has a small hole in its center. Some of the positive ions pass through this hole and travel onward through a tube that passes between the poles of a powerful magnet.

One of the properties of charged particles, both positive and negative, is that their paths become curved as they pass through a magnetic field. This is exactly what happens to the positive ions in the mass spectrometer as they pass between the poles of the magnet. However, the extent to which their paths are bent depends on the masses of the ions. This is because the path of a heavy ion, like that of a speeding cement truck, is difficult to change, but the path of a light ion, like that of a motorcycle, is influenced more easily. As a result, heavy ions emerge from between the magnet's poles along different lines than the lighter ions. In effect, an entering beam containing ions of different masses is sorted by the magnet into a number of

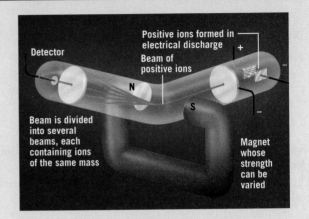

beams, each containing ions of the same mass. This spreading out of the ion beam thus produces an array of different beams called a *mass spectrum*.

In practice, the strength of the magnetic field is gradually changed, which sweeps the beams of ions across a detector located at the end of the tube. As a beam of ions strikes the detector, its intensity is measured and the masses of the particles in the beam are computed based on the strength of the magnetic field and the geometry of the apparatus.

Among the benefits derived from measurements using the mass spectrometer are very accurate isotopic masses and relative isotopic abundances. These serve as the basis for the very precise values of the atomic masses that you find in the periodic table.

Atomic numbers and mass numbers describe isotopes

What distinguishes one element from another is the number of protons in the nuclei of its atoms, because *all the atoms of a particular element have an identical number of protons.* In fact, this allows us to redefine an **element** as *a substance whose atoms all contain the identical number of protons.* Thus, each element has associated with it a unique number, which we call its **atomic number (Z),** that equals the number of protons in the nuclei of any of its atoms.

$$\text{atomic number, } Z = \text{number of protons}$$

What makes isotopes of the same element different are the numbers of neutrons in their nuclei. *The* **isotopes** *of a given element have atoms with the same number of protons but different numbers of neutrons.* The numerical sum of the protons and neutrons in the atoms of a particular isotope is called the **mass number (A)** of the isotope.

$$\text{mass number, } A = (\text{number of protons}) + (\text{number of neutrons})$$

Therefore, every isotope is fully defined by two numbers, its atomic number and its mass number. Sometimes these numbers are added to the left of the chemical symbol as a subscript and a superscript, respectively. Thus, if X stands for the chemical symbol for the element, an isotope of X is represented as

$$_{Z}^{A}X$$

The isotope of uranium used in nuclear reactors, for example, can be symbolized as follows:

mass number (protons + neutrons) $\longrightarrow$ $_{92}^{235}U$

atomic number (number of protons) $\longrightarrow$

uranium-235

As indicated, the name of this isotope is uranium-235. Each neutral atom contains 92 protons and $(235 - 92) = 143$ neutrons as well as 92 electrons. In writing the symbol for the isotope, the atomic number is often omitted because it is really redundant. Every atom of uranium has 92 protons, and every atom that has 92 protons is an atom of uranium. Therefore, this uranium isotope can be represented simply as ^{235}U.

In naturally occurring uranium, a more abundant isotope is ^{238}U. Atoms of this isotope also have 92 protons, but the number of neutrons is 146. Thus, atoms of ^{235}U and ^{238}U have the identical number of protons but differ in the numbers of neutrons.

In general, the mass number of an isotope differs slightly from the atomic mass of the isotope. For instance, the isotope ^{35}Cl has an atomic mass of 34.968852 u. In fact, the *only* isotope that has an atomic mass equal to its mass number is ^{12}C; *by definition* the mass of this atom is exactly 12 u.

> It is useful to remember that for a neutral atom, the atomic number equals both the number of protons and the number of electrons.

PRACTICE EXERCISE 6: Write the symbol for the isotope of plutonium (Pu) that contains 146 neutrons.

PRACTICE EXERCISE 7: How many protons, neutrons, and electrons are in each atom of $_{17}^{35}Cl$?

1.7 ▶ The periodic table is used to organize and correlate facts

When we study different kinds of substances, we find that some are elements and others are compounds. Among compounds, some are composed of discrete molecules. Others, as you will learn, are made up of atoms that have acquired electrical charges. For elements such as sodium and chlorine, we mentioned metallic and nonmetallic properties. If we were to continue on this way, without attempting to build our subject around some central organizing structure, it would not be long before we became buried beneath a mountain of information in the form of seemingly unconnected facts.

The need for organization was recognized by many early chemists, and there were numerous attempts to discover relationships among the chemical and physical properties of the elements. A number of different sequences of elements were tried in the search for some sort of order or pattern. A few of these arrangements came quite close, at least in some respects, to our current periodic table, but either they were flawed in some way or they were presented to the scientific community in a manner that did not lead to their acceptance.

Mendeleev created the first periodic table

The periodic table we use today is based primarily on the efforts of a Russian chemist, Dmitri Ivanovich Mendeleev (1834–1907), and a German physicist, Julius Lothar Meyer (1830–1895). Working independently, these scientists developed

Periodic refers to the recurrence of properties at regular intervals.

similar periodic tables only a few months apart in 1869. Mendeleev is usually given the credit, however, because he had the good fortune to publish first.

Mendeleev was preparing a chemistry textbook for his students at the University of St. Petersburg. Looking for some pattern among the properties of the elements, he found that when he arranged them in order of increasing atomic mass, similar chemical properties were repeated over and over again at regular intervals. For instance, the elements lithium (Li), sodium (Na), potassium (K), rubidium (Rb), and cesium (Cs) are soft metals that are very reactive toward water. They form compounds with chlorine that have a 1-to-1 ratio of metal to chlorine. Similarly, the elements that immediately follow each of these also constitute a set with similar chemical properties. Thus, beryllium (Be) follows lithium, magnesium (Mg) follows sodium, calcium (Ca) follows potassium, strontium (Sr) follows rubidium, and barium (Ba) follows cesium. All of these elements form a water-soluble chlorine compound with a 1-to-2 metal to chlorine atom ratio. Mendeleev used such observations to construct his **periodic table,** which is illustrated in Figure 1.14.

The elements in Mendeleev's table are arranged in order of increasing atomic mass. When the sequence is broken at the right places and stacked, the elements fall naturally into columns, called **groups,** in which the elements of a given group have similar chemical properties. The rows themselves are called **periods.**

Mendeleev's genius rested on his placing elements with similar properties in the same group even when this left occasional gaps in the table. For example, he placed arsenic (As) in Group V under phosphorus because its chemical properties were similar to those of phosphorus, even though this left gaps in Groups III and IV. Mendeleev reasoned, correctly, that the elements that belonged in these gaps had simply not yet been discovered. In fact, on the basis of the location of these

Periods

	Group I	Group II	Group III	Group IV	Group V	Group VI	Group VII	Group VIII
1	H 1							
2	Li 7	Be 9.4	B 11	C 12	N 14	O 16	F 19	
3	Na 23	Mg 24	Al 27.3	Si 28	P 31	S 32	Cl 35.5	
4	K 39	Ca 40	—44	Ti 48	V 51	Cr 52	Mn 55	Fe 56, Co 59 Ni 59, Cu 63
5	(Cu 63)	Zn 65	—68	—72	As 75	Se 78	Br 80	
6	Rb 85	Sr 87	?Yt 88	Zr 90	Nb 94	Mo 96	—100	Ru 104, Rh 104 Pd 105, Ag 108
7	(Ag 108)	Cd 112	In 113	Sn 118	Sb 122	Te 128	I 127	
8	Cs 133	Ba 137	?Di 138	?Ce 140	—	—	—	———
9	—	—	—	—	—	—	—	
10	—	—	?Er 178	?La 180	Ta 182	W 184	—	Os 195, Ir 197 Pt 198, Au 199
11	(Au 199)	Hg 200	Tl 204	Pb 207	Bi 208	—		
12	—	—	—	Th 231	—	U 240	—	——— ———

FIGURE 1.14 *The first periodic table.* Mendeleev's periodic table roughly as it appeared in 1871. The numbers next to the symbols are atomic masses.

gaps Mendeleev was able to predict with remarkable accuracy the properties of these yet-to-be-found substances. His predictions helped serve as a guide in the search for the missing elements.

The elements tellurium (Te) and iodine (I) caused Mendeleev some problems. According to the best estimates at that time, the atomic mass of tellurium was greater than that of iodine. Yet if these elements were placed in the table according to their atomic masses, they would not fall into the proper groups required by their properties. Therefore, Mendeleev switched their order and in so doing violated his ordering sequence. (Actually, he believed that the atomic mass of tellurium had been incorrectly measured, but this wasn't so.)

The table that Mendeleev developed is in many ways similar to the one we use today. One of the main differences, though, is that Mendeleev's table lacks the column containing the elements helium (He) through radon (Rn). In Mendeleev's time, none of these elements had yet been found because they are relatively rare and because they have virtually no tendency to undergo chemical reactions. When these elements were finally discovered, beginning in 1894, another problem arose. Two more elements, argon (Ar) and potassium (K), did not fall into the groups required by their properties if they were placed in the table in the order required by their atomic masses. Another switch was necessary and another exception had been found. It became apparent that atomic mass was not the true basis for the periodic repetition of the properties of the elements. To determine what the true basis was, however, scientists had to await the discoveries of the atomic nucleus, the proton, and atomic numbers.

The modern periodic table arranges elements by atomic number

Once atomic numbers had been discovered, it was soon realized that the elements in Mendeleev's table are arranged in precisely the order of increasing atomic number. In other words, if we take atomic numbers as the basis for arranging the elements in sequence, no annoying switches are required and the elements Te and I or Ar and K are no longer a problem. The fact that it is the atomic number—the number of protons in the nucleus of an atom—that determines the order of elements in the table is very significant. We will see later that this has important implications with regard to the relationship between the number of electrons in an atom and the atom's chemical properties.

The modern periodic table is shown in Figure 1.15 and also appears on the inside front cover of the book. We will refer to the table frequently, so it is important for you to become familiar with it and with some of the terminology applied to it.

Special terminology is associated with the periodic table

As in Mendeleev's table, the elements are arranged in rows that we call **periods,** but here they are arranged in order of increasing atomic number. For identification purposes, the periods are numbered. We will find these numbers useful later on. Below the main body of the table are two long rows of 14 elements each. These actually belong in the main body of the table following La ($Z = 57$) and Ac ($Z = 89$), as shown in Figure 1.16. They are almost always placed below the table simply to conserve space. If the fully spread-out table is printed on one page, the type is so small that it's difficult to read. Notice that in the fully extended form of the table, with all the elements arranged in their proper locations, there is a great deal of empty space. An important requirement of a detailed atomic theory, which we will get to in Chapter 8, is that it must explain not only the repetition of properties, but also why there is so much empty space in the table.

Again, as in Mendeleev's table, the vertical columns are called **groups.** However, there is not uniform agreement among chemists on how they should be

Recall that the symbol Z stands for atomic number.

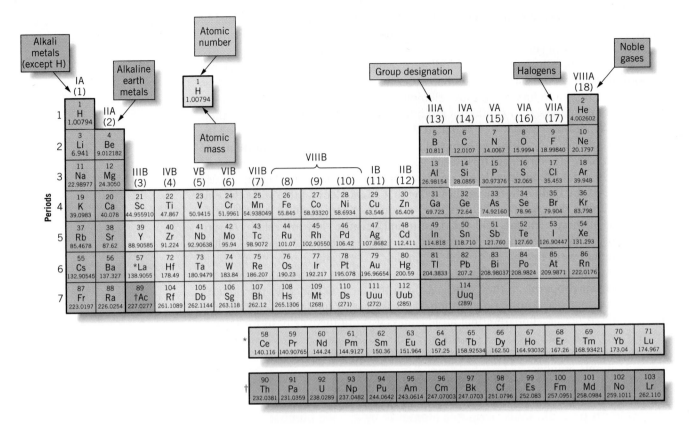

FIGURE 1.15 *The modern periodic table.*

numbered. In the past, the groups were labeled with Roman numerals and divided into A groups and B groups, separated by the three short columns headed by Fe, Co, and Ni (Group VIIIB), as indicated in Figure 1.15. This corresponds quite closely to Mendeleev's original designations and is preferred by many chemists in North America. However, another version of the table, popular in Europe, has the first seven groups from left to right labeled A, followed by the three short columns headed by Fe, Co, and Ni, and then the next seven groups labeled B. In an attempt to standardize the table, the International Union of Pure and Applied Chemistry (the IUPAC), an international body of scientists responsible for setting standards in chemistry, officially adopted a third system in which the groups are simply numbered sequentially from left to right using Arabic numerals. Thus, Group IA in the North American table is Group 1 in the IUPAC

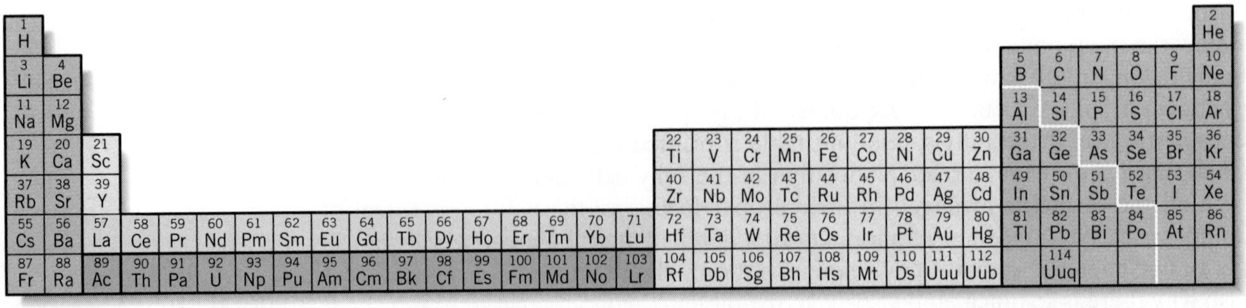

FIGURE 1.16 *Extended form of the periodic table.* The two long rows of elements below the main body of the table in Figure 1.15 are placed in their proper places in this table.

table, and Group VIIA in the North American table is Group 17 in the IUPAC table. In Figure 1.15 and on the inside front cover of the book, we have used both the North American labels as well as those preferred by the IUPAC. Because of the lack of uniform agreement among chemists on how the groups should be specified, we will use the A-group/B-group designations in Figure 1.15 when we wish to specify a particular group.

As we have already noted, the elements in a given group bear similarities to each other. Because of such similarities, groups are sometimes referred to as **families of elements.** The elements in the longer columns (the A groups) are known as the **representative elements** or **main group elements.** Those that fall into the B groups in the center of the table are called **transition elements.** The elements in the two long rows below the main body of the table are the **inner transition elements,** and each row is named after the element that it follows in the main body of the table. Thus, elements 58–71 are called the **lanthanide elements** because they follow lanthanum ($Z = 57$), and elements 90–103 are called the **actinide elements** because they follow actinium ($Z = 89$).

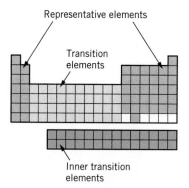

Some of the groups have acquired common names. For example, except for hydrogen, the Group IA elements are metals. They form compounds with oxygen that dissolve in water to give solutions that are strongly alkaline, or caustic. As a result, they are called the **alkali metals** or simply the *alkalis.* The Group IIA elements are also metals. Their oxygen compounds are alkaline, too, but many compounds of the Group IIA elements are unable to dissolve in water and are found in deposits in the ground. Because of their properties and where they occur in nature, the Group IIA elements became known as the **alkaline earth metals.**

On the right side of the table, in Group VIIIA, are the **noble gases.** They used to be called the inert gases until it was discovered that the heavier members of the group show a small degree of chemical reactivity. The term *noble* is used when we wish to suggest a very limited degree of chemical reactivity. Gold, for instance, is often referred to as a noble metal because so few chemicals are capable of reacting with it.

Finally, the elements of Group VIIA are called the **halogens,** derived from the Greek word meaning "sea" or "salt." Chlorine (Cl), for example, is found in familiar table salt, a compound that accounts in large measure for the salty taste of seawater.

PRACTICE EXERCISE 8: Circle the correct choices.

(a) Representative elements are: K, Cr, Pr, Ar, Al.
(b) A halogen is: Na, Fe, O, Cl, Cu.
(c) An alkaline earth metal is: Rb, Ba, La, As, Kr.
(d) A noble gas is: H, Ne, F, S, N.
(e) An alkali metal is: Zn, Ag, Br, Ca, Li.
(f) An inner transition element is: Ce, Pb, Ru, Xe, Mg.

Elements can be classified as metals, nonmetals, or metalloids

The periodic table organizes all sorts of chemical and physical information about the elements and their compounds. It allows us to study systematically the way properties vary with an element's position within the table and, in turn, makes the similarities and differences among the elements easier to understand and remember.

Even a casual inspection of samples of the elements reveals that some are familiar metals and that others, equally familiar, are not metals. Most of us are already familiar with metals such as lead, iron, or gold and nonmetals such as oxygen or nitrogen. A closer look at the nonmetallic elements, though, reveals that some of them, silicon and arsenic to name two, have properties that lie between

Periodic table and trends in the metallic character of the elements.

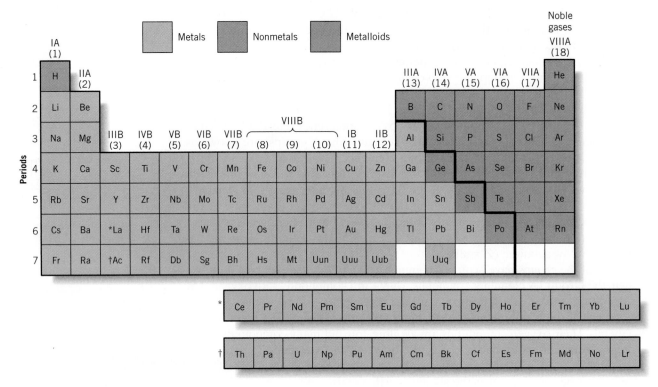

FIGURE 1.17 *Distribution of metals, nonmetals, and metalloids among the elements in the periodic table.*

Notice that the metalloids are grouped around the bold stair-step line that is drawn diagonally from boron (B) to astatine (At).

those of true metals and true nonmetals. These elements are called **metalloids.** Division of the elements into the categories of metals, nonmetals, and metalloids is not even, however (see Figure 1.17). Most elements are metals, slightly over a dozen are nonmetals, and only a handful are metalloids.

Metals

You probably know a metal when you see one. Metals tend to have a shine so unique that it's called a *metallic luster.* For example, the silvery sheen of the freshly exposed surface of sodium in Figure 1.18 would most likely lead you to identify sodium as a metal even if you had never seen or heard of it before. We also know that metals conduct electricity. Few of us would hold an iron nail in our hand and poke it into an electrical outlet. In addition, we know that metals conduct heat very well. On a cool day, metals always feel colder to the touch than do neighboring nonmetallic objects because metals conduct heat away from your hand very rapidly. Nonmetals seem less cold because they can't conduct heat away as quickly and therefore their surfaces warm up faster.

Thin lead sheets are used for sound deadening because the easily deformed lead absorbs the sound vibrations.

Other properties that metals possess, to varying degrees, are **malleability**—the ability to be hammered or rolled into thin sheets—and **ductility**—the ability to be drawn into wire. The ability of a blacksmith to make horseshoes from a bar of iron (Figure 1.19) depends on the malleability of iron and steel, and the manufacture of electrical wire (Figure 1.20) is based on the ductility of copper.

Hardness is another physical property that we usually think of for metals. Some, such as chromium or iron, are indeed quite hard, but others, like copper and lead, are rather soft. The alkali metals such as sodium (Figure 1.18) are so soft they can be cut with a knife, but they are also so chemically reactive that we rarely get to see them as free elements.

The statue of Prometheus that stands above the skating rink in Rockefeller Center in New York City is covered by a thin decorative layer of gold leaf.

All the metallic elements, except mercury, are solids at room temperature (Figure 1.21). Mercury's low freezing point (-39 °C) and fairly high boiling

One of the most malleable of metals is gold. It can be hammered into sheets so thin that they are only about 1/280,000 of an inch thick, which corresponds to a thickness of only several hundred atoms. (The sheets are so thin that some light can pass through.) Thin gold sheets are sold as "gold leaf" and are used for applying gold in a variety of decorative ways. For example, gold lettering is applied to doors by first painting the letters using a varnish and then, while the varnish is still tacky, applying the gold. Gold leaf is also used as a decorative finish on various objects, such as the statue of Prometheus that overlooks the skating rink at Rockefeller Center in New York City.

point (357 °C) make it useful as a fluid in thermometers. Most of the other metals have much higher melting points, and some are used primarily because of this. Tungsten, for example, has the highest melting point of any metal (3400 °C, or 6150 °F), which explains its use as filaments that glow white-hot in electric light bulbs.

The chemical properties of metals vary tremendously. Some, such as gold and platinum, are very unreactive toward almost all chemical agents. This property, plus their natural beauty and rarity, makes them highly prized for use in jewelry. Other metals, however, are so reactive that few people except chemists and chemistry students ever get to see them in their "free" states. For instance, the metal sodium reacts very quickly with oxygen or moisture in the air, and its bright metallic surface tarnishes almost immediately. In contrast, compounds of sodium, such as table salt and baking soda, are quite stable and very common.

We use the term *free element* to mean an element that is not chemically combined with any other element.

Nonmetals

Substances such as plastics, wood, and glass that lack the properties of metals are said to be *nonmetallic,* and an element that has nonmetallic properties is called a **nonmetal.** Most often, we encounter the nonmetals in the form of compounds or mixtures of compounds. There are some nonmetals, however, that are very important to us in their elemental forms. The air we breathe, for instance, contains mostly nitrogen and oxygen. Both are gaseous, colorless, and odorless nonmetals. Since we can't see, taste, or smell them, however, it's difficult to experience their existence. (Although if you step into an atmosphere without oxygen, your body will soon tell you that something is missing!) Probably the most commonly *observed* nonmetallic element is carbon. We find it as the graphite in pencils, as coal, and as the charcoal used for barbecues. It also occurs in a more valuable form as diamond (Figure 1.22). Although diamond and graphite differ in appearance, each is a form of elemental carbon.

Many of the nonmetals are solids at room temperature and atmospheric pressure, while many others are gases. Photographs of some of the nonmetallic elements appear in Figure 1.23. Their properties are almost completely opposite those of metals. Each of these elements lacks the characteristic appearance of a metal. They are poor conductors of heat and, with the exception of the graphite form of carbon, are also poor conductors of electricity. The electrical conductivity of graphite appears to be an accident of molecular structure, since the structures of metals and graphite are completely different.

The nonmetallic elements lack the malleability and ductility of metals. A lump of sulfur crumbles when hammered and breaks apart when pulled on. Diamond cutters rely on the brittle nature of carbon when they split a gem-quality stone by carefully striking a quick blow with a sharp blade.

As with metals, nonmetals exhibit a broad range of chemical reactivity. Fluorine, for instance, is extremely reactive. It reacts readily with almost all the other elements. At the other extreme is helium, the gas used to inflate children's balloons and the Goodyear blimp. This element does not react with anything, a fact that

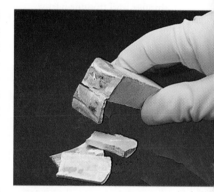

FIGURE 1.18 *Sodium is a metal.* The freshly exposed surface of a bar of sodium reveals its shiny metallic luster. The metal reacts quickly with moisture and oxygen to form a white coating. Sodium's high reactivity makes it dangerous to touch with bare skin.

FIGURE 1.19 *Malleability of iron.* A blacksmith uses the malleability of hot iron to fashion horseshoes from an iron bar.

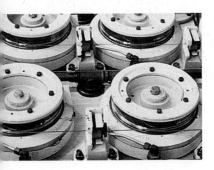

chemists find useful when they want to provide a totally *inert* (unreactive) atmosphere inside some apparatus.

Metalloids

The properties of metalloids lie between those of metals and nonmetals. This shouldn't surprise us since the metalloids are located between the metals and the nonmetals in the periodic table. In most respects, metalloids behave as nonmetals, both chemically and physically. However, in their most important physical property, electrical conductivity, they somewhat resemble metals. Metalloids tend to be **semiconductors;** they conduct electricity, but not nearly so well as metals. This property, particularly as found in silicon and germanium, is responsible for the remarkable progress made during the last four decades in the field of solid-state electronics. The operation of every computer, hi-fi stereo system, TV receiver, VCR, CD player, and AM-FM radio relies on transistors made from semiconductors. Perhaps the most amazing advance of all has been the fantastic reduction in the size of electronic components that semiconductors have allowed (Figure 1.24). To it, we owe the development of small and versatile cell phones, TV cameras, compact disc players, hand-held calculators, and microcomputers. The heart of these devices is a microcircuit embedded in the surface of a tiny silicon chip.

FIGURE 1.20 *The ductility of copper.* This property allows the metal to be drawn into wire. Here copper wire passes through one die after another as it is drawn into thinner and thinner wire.

FIGURE 1.21 *Mercury.* The metal mercury (once known as quicksilver) is a liquid at room temperature, unlike other metals, which are solids.

Trends within the periodic table help us remember facts

The occurrence of the metalloids between the metals and the nonmetals is our first example of trends in properties within the periodic table. We will frequently see that as we move from position to position across a period or down a group, chemical and physical properties change in a more or less regular way. There are few abrupt changes in the characteristics of the elements as we scan across a period or down a group. The location of the metalloids can be seen, then, as an example of the gradual transition between metallic and nonmetallic properties. From left to right across Period 3, we go from aluminum, an element that has every appearance of a metal, to silicon, a semiconductor, to phosphorus, an element with clearly nonmetallic properties. A similar gradual change is seen going down Group IVA. Carbon is certainly a nonmetal, silicon and germanium are metalloids, and tin and lead are metals. Trends such as these are useful to spot because they help us remember properties.

FIGURE 1.22 *Diamonds.* Gems such as these are simply another form of the element carbon.

FIGURE 1.23 *Some nonmetallic elements.* In the bottle on the left is dark-red liquid bromine, which vaporizes easily to give a deeply colored orange vapor. Pale green chlorine fills the round flask in the center. Solid iodine lines the bottom of the flask on the right and gives off a violet vapor. Powdered red phosphorus occupies the dish in front of the flask of chlorine, and black powdered graphite is in the watch glass. Also shown are lumps of yellow sulfur.

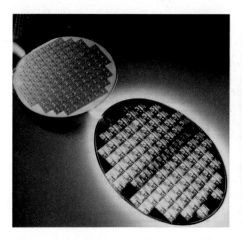

Figure 1.24 *Modern electronic circuits rely on the semiconductor properties of silicon.* The silicon wafer shown here contains more electronic components (10 billion) than there are people on our entire planet (about 6 billion)!

SUMMARY

Chemistry and Its Place among the Sciences Chemistry is a science that studies the properties and composition of **matter,** which is anything that has **mass** and occupies space. Mass is a measure of the amount of matter in an object and is commonly measured in units of **grams** (symbol **g**). Some knowledge of chemistry is needed by all scientists because all tangible things, living or not, are composed of chemicals. Chemists and scientists in other disciplines often study the same things; they simply view them from different perspectives. Today, as in the past, chemistry plays a crucial role in fulfilling the needs of society.

Scientific Method This is the general procedure by which science advances. **Observations** are made, and the **empirical facts** or **data** that are collected from many experiments are often summarized in **scientific laws,** which frequently are expressed in equation form. Scientists formulate mental images or models of nature to explain observed behavior. Models begin as **hypotheses,** and those that survive repeated testing become known as **theories.** Although the general pattern in the development of science is the cycle of observation–explanation–observation–explanation . . . , many discoveries are made accidentally by people who have learned to be observant through scientific training. The **atomic theory** proposes that matter is made of tiny particles **(atoms)** that combine to form more complex substances. **Molecules** are particles that contain two or more atoms.

Properties of Materials When studying matter we are concerned with its physical and chemical properties. **Physical properties** can be measured without changing the chemical makeup of the sample. Physical changes are changes that take place without altering the chemical composition of the sample. A **chemical property** describes a **chemical change** and relates to how substances change into other substances in **chemical reactions. Intensive properties** are independent of sample size; **extensive properties** depend on sample size. For identification purposes, intensive properties are more useful than extensive properties because they don't depend on the size of the sample.

Types of Materials Solid, liquid, and **gas** are the most common **states of matter.** Their properties can be related to the different ways the individual atomic-size particles are organized. In a solid, they are tightly packed and cannot easily move; in a liquid, they are less tightly packed and can move past each other; and in a gas, they are widely spaced with much empty space between them.

An **element** is a substance that cannot be decomposed into something simpler by a chemical reaction. Each element has an internationally agreed-upon **chemical symbol.** Symbols are used in place of the name of an element and also to represent an atom of the element. Elements combine in fixed proportions to form **compounds.** Elements and compounds are **pure substances** that may be combined in *varying* proportions to give **mixtures.** A one-phase mixture is called a **solution** and is **homogeneous.** If a mixture consists of two or more **phases** it is **heterogeneous.** Formation or separation of a mixture into its components can be accomplished by a physical change; formation or decomposition of a compound takes place by a chemical change.

Elements and Atoms When accurate masses of all the reactants and products in a reaction are measured and compared, no observable changes in mass accompany chemical reactions (the **law of conservation of mass**). The mass ratios of the elements in any compound are constant regardless of the source of the compound or how it is prepared (the **law of definite proportions**). **Dalton's atomic theory** explained the laws of chemical combination by proposing that matter consists of indestructible atoms with masses that do not change during chemical reactions. During a chemical reaction, atoms may change partners, but they are neither created nor destroyed. After Dalton had proposed his theory, it was discovered that whenever two elements form more than one compound, the different masses of one element that combine with a fixed mass of the other are in a ratio of small whole numbers (the **law of multiple proportions**). Using modern instruments such as the scanning tunneling microscope, scientists are able to "see" atoms on the surfaces of solids.

An element's **atomic mass (atomic weight)** is the relative mass of its atoms on a scale in which atoms of carbon-12 have a mass of exactly 12 u **(atomic mass units).** Most elements occur in nature as uniform mixtures of a small number of **isotopes,** whose masses differ slightly. However, all isotopes of an element have very nearly identical chemical properties, and the percentages of the isotopes that make up an element are generally so constant throughout the world that we can say that the average mass of their atoms is a constant.

Atomic Structure Atoms can be split into **subatomic particles,** such as **electrons, protons,** and **neutrons. Nucleons** are particles that make up the atomic **nucleus** and include the protons, each of which carries a single unit of **positive charge** (charge = 1+) and neutrons (no charge). The number of protons is called the **atomic number (Z)** of the element. Each element has a different atomic number. The electrons, each with a unit of **negative charge** (charge = 1−) are found outside the nucleus; their number equals the atomic number in a neutral atom. Isotopes of an element have identical atomic numbers but different numbers of neutrons. In more modern terms, an **element** can be defined as a substance whose atoms all have the same number of protons in their nuclei.

The Periodic Table The search for similarities and differences among the properties of the elements led Mendeleev to discover that when the elements are placed in (approximate) order of increasing atomic mass, similar properties recur at regular, repeating intervals. In the modern **periodic table** the elements are arranged in rows, called **periods,** in order of increasing atomic number. The rows are stacked so that elements in the columns, called **groups** or **families,** have similar chemical and physical properties. The A-group elements (IUPAC Groups 1, 2, and 13–18) are called **representative elements;** the B-group elements (IUPAC Groups 3–12) are called **transition elements.** The two long rows of **inner transition elements** located below the main body of the table consist of the **lanthanides,** which follow La ($Z = 57$), and the **actinides,** which follow Ac ($Z = 89$). Certain groups are given family names: Group IA (Group 1), except for hydrogen, are the **alkali metals** (the alkalis); Group IIA (Group 2), the **alkaline earth metals;** Group VIIA (Group 17), the **halogens;** Group VIIIA (Group 18), the **noble gases.**

Metals, Nonmetals, and Metalloids Most elements are **metals;** they occupy the lower left-hand region of the periodic table (to the left of a line drawn approximately from boron, B, to astatine, At). **Nonmetals** are found in the upper right-hand region of the table. **Metalloids** occupy a narrow band between the metals and nonmetals.

Metals exhibit a **metallic luster,** tend to be **ductile** and **malleable,** and conduct electricity. Nonmetals tend to be brittle, lack metallic luster, and are nonconductors of electricity. Many nonmetals are gases. Bromine (a nonmetal) and mercury (a metal) are the two elements that are liquids at ordinary room temperature. **Metalloids** have properties intermediate between those of metals and nonmetals and are **semiconductors** of electricity.

TOOLS ▸ **YOU HAVE LEARNED**

The table below lists the concepts that you've learned in this chapter that can be applied as tools in solving problems. Study each one carefully so that you know what each is used for. When faced with solving a problem, recall what each tool does and consider whether it will be helpful in finding a solution. This will aid you in selecting the tools you need. If necessary, refer to this table when working on the Thinking-It-Through problems and the Review Exercises that follow.

TOOL	HOW IT WORKS
Law of definite proportions (page 13)	If we know the mass ratio of the elements in one sample of a compound, we know the ratio will be the same in a different sample of the same compound.
Law of conservation of mass (page 13)	The total mass of chemicals present before a reaction starts equals the total mass after the reaction is finished. We can use this law to check whether we have accounted for all the substances formed in a reaction.
Periodic Table (page 30)	From an element's position in the periodic table, we can tell whether it's a metal, nonmetal, or metalloid.

THINKING IT THROUGH

The first step in solving a problem is to *analyze* the information available and plan what has to be done to obtain the answer. This is the most difficult part, because once you've figured out what you need to do, working through the *solution* is usually relatively easy. The goal for each of the following problems is to encourage you to separate the *analysis* and *solution* parts of problem solving. Therefore, for each of these problems we just ask you to think about the information required and to figure out the necessary steps needed to find the answer. We are *not* asking you to find the answer itself; instead, just assemble the information needed to obtain it, state what additional data (if any) are needed, and describe how you would use the data to answer the question.

The problems are divided into two groups. Those in Level 2 are more challenging than those in Level 1 and provide an opportunity to really hone your problem-solving skills. Detailed answers to these problems can be found on the web site.

Need extra help? Visit the Brady/ Senese web site at www.wiley.com/ college/brady

 ON-LINE HELP

Level 1 Problems

1. Aluminum combines with oxygen to form a compound called aluminum oxide, which is the principal constituent of the gem ruby. In an experiment, a chemist found that 15.0 g of aluminum combined with 13.3 g of oxygen. If this experiment were repeated with 25.0 g of aluminum and 13.3 g of oxygen, explain how you would determine the number of grams of aluminum oxide that would be formed. Which of the tools described in this chapter are needed to solve this problem?

2. How many times heavier than an atom of ^{12}C is the average atom of iron? (Explain how you can obtain the answer.)

3. Suppose the atomic mass unit had been defined as $\frac{1}{10}$ of the mass of an atom of phosphorus. Describe how you would calculate the atomic mass of carbon on this scale.

4. How do we find the chemical symbol for an atom that has a nucleus containing 45 protons and 58 neutrons?

5. How can you tell whether the element that has a nucleus which contains 49 protons is a metal, a nonmetal, or a metalloid?

6. Suppose you learned that scientists had discovered a new element with an atomic number of 117. How could you determine which group it belongs to in the periodic table?

Level 2 Problems

7. Carbon forms a compound with element X in which there are four atoms of X for each atom of carbon. A sample of this compound was analyzed and found to contain 1.50 g of C and 39.95 g of X. Explain how you would calculate the atomic mass of X.

8. In an experiment, it was found that 3.50 g of phosphorus combines with 12.0 g of chlorine to form a compound in which there are three chlorine atoms for each phosphorus atom. Explain how you would calculate the number of grams of chlorine that would combine with 8.50 g of phosphorus to form this same compound. Which of the tools described in this chapter are needed to solve this problem?

9. The element phosphorus burns easily in air and is used to produce some of the effects in fireworks. When phosphorus burns it is found that 2.40 g of phosphorus combines with 2.89 g of oxygen to form a compound. If this combustion were repeated using 5.60 g of phosphorus, explain how you would calculate (a) the number of grams of oxygen that would react, and (b) the number of grams of the compound that would be formed. Which of the tools described in this chapter are needed to solve this problem?

10. Arsenic forms two compounds with sulfur. In 6.00 g of one of these compounds, there were 3.62 g of As. In 6.00 g of the other compound, there were 2.89 g of As. Explain in detail how you would show that these compounds exhibit the law of multiple proportions.

11. When magnesium burns in oxygen, a brilliant light is given off, so the reaction is often used in fireworks displays. In an experiment, 1.0 g of magnesium was combined with 2.0 g of oxygen and after the reaction was complete, 1.6 g of a white solid, magnesium oxide, was collected and no magnesium metal remained. Explain how you would determine whether all of the oxygen reacted and, if any remained, how you could calculate the amount. Which of the tools described in this chapter are needed to solve this problem?

REVIEW QUESTIONS

Introduction

1.1 After some thought, give two reasons why a course in chemistry will benefit *you* in the pursuit of your particular major.

1.2 What does the science of chemistry seek to study?

1.3 Define *matter*. Which of the following are examples of matter? (a) air, (b) a pencil, (c) a cheese sandwich, (d) a squirrel, (e) your mother.

1.4 Look around your room and list ten items you see that are made of synthetic materials, that is, materials not found in nature.

1.5 In answering the preceding question, what have you learned about the contribution of chemistry to modern life?

Scientific Method

1.6 What steps are involved in the scientific method?

1.7 What is the function of a laboratory?

1.8 Define (a) data, (b) hypothesis, (c) law, and (d) theoretical model.

1.9 What role does luck play in the advancement of science?

Properties of Substances

1.10 What is *a physical property*? What is *a chemical property*? What is the chief distinction between physical and chemical properties? Define the terms *intensive property* and *extensive property*. Give two examples of each.

1.11 How does a physical change differ from a chemical change? When a solid melts, is the change a physical or a chemical change?

1.12 "A sample of calcium (an electrically conducting white metal that is shiny, relatively soft, melts at 850 °C, and boils at 1440 °C) was placed into liquid water that was at 25 °C. The calcium reacted slowly with the water to give bubbles of gaseous hydrogen and a solution of the substance calcium hydroxide." In this description, what physical properties and what chemical properties are described?

1.13 What is a chemical reaction?

1.14 Lye is a common name for a substance called sodium hydroxide. Muriatic acid is the common name for hydrochloric acid. Either of these chemicals can cause severe burns if left in contact with the skin, but when water solutions of them are mixed in just the right proportions, the resulting solution contains only sodium chloride. From this description, how do you know that a chemical reaction occurs between sodium hydroxide and hydrochloric acid?

1.15 If you swallow a water solution of baking soda, a gas (carbon dioxide) forms in your stomach, which causes you to burp. Has a chemical reaction occurred? Explain your answer.

1.16 In places like Saudi Arabia, freshwater is scarce and is recovered from seawater. When seawater is boiled, the water evaporates and the steam can be condensed to give pure water that people can drink. If all the water is evaporated, solid salt is left behind. Are the changes described here chemical or physical?

1.17 Suppose you were told that behind a screen there were two samples of liquid, one of them water and the other gasoline. You are told that Sample 1 occupies 3 fluid ounces and Sample 2 occupies 7 fluid ounces.
(a) What kind of property (intensive or extensive) is volume?
(b) Can you use the information given to you in this question to determine which sample is water and which is gasoline? (Explain.)

1.18 Name two intensive properties that you *could* use to distinguish between water and gasoline. Give one chemical property you could use.

1.19 Many reference books list physical properties that can aid in the identification of substances. One such book is the *Handbook of Chemistry and Physics,* which no doubt is available in your school library.
(a) Use the Table of Physical Constants of Inorganic Compounds in the *Handbook of Chemistry and Physics* to tabulate the melting points, boiling points, and colors of cadmium iodide, lead iodide, and bismuth tribromide.
(b) Which of these tabulated properties could you use to distinguish among these three substances?
(c) A chemist who was asked to analyze a sample was able to isolate a yellow solid from an experiment. This substance was found to melt at 402 °C and boil at 954 °C. Which of the chemicals described in part (a) of this question could the chemist have obtained from the experiment?

Types of Materials

1.20 What are the three states of matter?

1.21 How does the atomic model explain the differences in the properties of solids, liquids, and gases?

Elements, Compounds, and Mixtures

1.22 Define (a) element, (b) compound, (c) mixture, (d) homogeneous, (e) heterogeneous, (f) phase, (g) solution.

1.23 What kind of change (chemical or physical) is needed to separate a compound into its elements?

1.24 What kind of change (chemical or physical) is needed to separate a mixture into its components?

Chemical Symbols

1.25 What is the chemical symbol for each of the following elements? (a) chlorine, (b) sulfur, (c) iron, (d) silver, (e) sodium, (f) phosphorus, (g) iodine, (h) copper, (i) mercury, (j) calcium.

1.26 What is the name of each of the following elements? (a) K, (b) Zn, (c) Si, (d) Sn, (e) Mn, (f) Mg, (g) Ni, (h) Al, (i) C, (j) N.

1.27 Describe two ways that chemical symbols are used.

Laws of Chemical Combination and Dalton's Theory

1.28 Name and state the two laws of chemical combination discussed in this chapter.

1.29 Which postulate of Dalton's theory is based on the law of conservation of mass? Which is based on the law of definite proportions?

1.30 Why didn't the existence of isotopes affect the apparent validity of the atomic theory?

1.31 In your own words, describe how Dalton's theory explains the law of conservation of mass and the law of definite proportions.

1.32 Which of the laws of chemical combination is used to define the term *compound*?

Atomic Masses and Atomic Structure

1.33 Write the symbol for the isotope that forms the basis of the atomic mass scale. What is the mass of this atom expressed in atomic mass units?

1.34 What are the names, symbols, electrical charges, and masses (expressed in u) of the three subatomic particles introduced in this chapter?

1.35 Where in an atom is nearly all of its mass concentrated? Explain your answer in terms of the particles that contribute to this mass.

1.36 What is a *nucleon*? Which ones have we studied?

1.37 Define the terms *atomic number* and *mass number*.

1.38 In terms of the structures of atoms, how are isotopes of the same element alike? How do they differ?

1.39 Name one property that atoms of two *different* elements might possibly have in common.

1.40 Consider the symbol $_b^a X$, where X stands for the chemical symbol for an element. What information is given in locations (a) a and (b) b?

1.41 Write the symbols of the isotopes that contain the following. (Use the table of atomic masses and numbers printed inside the front cover for additional information, as needed.)
(a) An isotope of iodine whose atoms have 78 neutrons.
(b) An isotope of strontium whose atoms have 52 neutrons.
(c) An isotope of cesium whose atoms have 82 neutrons.
(d) An isotope of fluorine whose atoms have 9 neutrons.

The Periodic Table

1.42 In the compounds formed by Li, Na, K, Rb, and Cs with chlorine, how many atoms of Cl are there per atom of the metal? In the compounds formed by Be, Mg, Ca, Sr, and Ba with chlorine, how many atoms of Cl are there per atom of metal? How did this kind of information lead Mendeleev to develop his periodic table?

1.43 On what basis did Mendeleev construct his periodic table? On what basis are the elements arranged in the modern periodic table?

1.44 In the periodic table, what is a "period"? What is a "group"?

1.45 Why did Mendeleev leave gaps in his periodic table?

1.46 In the text, we identified two places in the modern periodic table where the atomic mass order was reversed. Using the table on the inside front cover, locate two other places where this occurs.

1.47 Which is better related to the chemistry of an element, its mass number or its atomic number? Give a brief explanation in terms of the basis for the periodic table.

1.48 On the basis of their positions in the periodic table, why is it not surprising that strontium-90, a dangerous radioactive isotope of strontium, replaces calcium in newly formed bones?

1.49 In the refining of copper, sizable amounts of silver and gold are recovered. Why is this not surprising?

1.50 Why would you reasonably expect cadmium to be a contaminant in zinc but not in silver?

1.51 Make a rough sketch of the periodic table and mark off those areas where you would find (a) the representative elements, (b) the transition elements, and (c) the inner transition elements.

1.52 What group numbers are used to designate the representative elements following (a) the North American system for designating groups and (b) the IUPAC system?

1.53 Supply the IUPAC group numbers that correspond to the following North American designations: (a) Group IA, (b) Group VIIA, (c) Group IIIB, (d) Group IB, (e) Group IVA.

1.54 Based on discussions in this chapter, explain why it is unlikely that scientists will discover a new element, never before observed, having an atomic mass of approximately 73.

1.55 Which of the following is
(a) an alkali metal: Ca, Cu, In, Li, S?
(b) a halogen: Ce, Hg, Si, O, I?
(c) a transition element: Pb, W, Ca, Cs, P?
(d) a noble gas: Xe, Se, H, Sr, Zr?
(e) a lanthanide element: Th, Sm, Ba, F, Sb?
(f) an actinide element: Ho, Mn, Pu, At, Na?
(g) an alkaline earth metal: Mg, Fe, K, Cl, Ni?

Physical Properties of Metals, Nonmetals, and Metalloids

1.56 Name five physical properties that we usually observe for metals.

1.57 Why is mercury used in thermometers? Why is tungsten used in light bulbs?

1.58 What property of metals allows them to be drawn into wire?

1.59 Gold can be hammered into sheets so thin that some light can pass through them. What property of gold allows such thin sheets to be made?

1.60 Only two metals are colored (the rest are "white," like iron or lead). You have surely seen both of them. Which metals are they?

1.61 Which nonmetals occur as monatomic gases (gases whose particles consist of single atoms)?

1.62 Which two elements exist as liquids at room temperature and pressure?

1.63 Which physical property of metalloids distinguishes them from metals and nonmetals?

1.64 Sketch the shape of the periodic table and mark off those areas where we find (a) metals, (b) nonmetals, and (c) metalloids.

1.65 Which metals can you think of that are commonly used to make jewelry? Why isn't iron used to make jewelry?

REVIEW PROBLEMS

Answers to problems whose numbers are printed in color are given in Appendix B. More challenging questions are marked with asterisks. **ILW** = Interactive LearningWare solution is available at *www.wiley.com/college/brady.*

Laws of Chemical Combination

1.66 Laughing gas is a compound formed from nitrogen and oxygen in which there are 1.75 g of nitrogen for every 1.00 g of oxygen. The compositions of several nitrogen–oxygen compounds follow. Which of these is laughing gas?
(a) 6.35 g nitrogen, 7.26 g oxygen
(b) 4.63 g nitrogen, 10.58 g oxygen
(c) 8.84 g nitrogen, 5.05 g oxygen
(d) 9.62 g nitrogen, 16.5 g oxygen
(e) 14.3 g nitrogen, 40.9 g oxygen

1.67 One of the substances used to melt ice on sidewalks and roadways in cold climates is calcium chloride. In this compound calcium and chlorine are combined in a ratio of 1.00 g of calcium to 1.77 g of chlorine. Which of the following calcium–chlorine mixtures will give calcium chloride with no calcium or chlorine left over after the reaction is complete?
(a) 3.65 g calcium, 4.13 g chlorine
(b) 0.856 g calcium, 1.56 g chlorine
(c) 2.45 g calcium, 4.57 g chlorine
(d) 1.35 g calcium, 2.39 g chlorine
(e) 5.64 g calcium, 9.12 g chlorine

1.68 Ammonia is composed of hydrogen and nitrogen in a ratio of 9.33 g of nitrogen to 2.00 g of hydrogen. If a sample of ammonia contains 6.28 g of hydrogen, how many grams of nitrogen does it contain?

1.69 A compound of phosphorus and chlorine used in the manufacture of a flame retardant treatment for fabrics contains 1.20 grams of phosphorus for every 4.12 g of chlorine. Suppose a sample of this compound contains 6.22 g of chlorine. How many grams of phosphorus does it contain?

1.70 Refer to the data about ammonia in Problem 1.68. If 4.56 g of nitrogen combined completely with hydrogen to form ammonia, how many grams of ammonia would be formed?

1.71 Refer to the data about the phosphorus–chlorine compound in Problem 1.69. If 12.5 g of phosphorus combined completely with chlorine to form this compound, how many grams of the compound would be formed?

1.72 Molecules of a certain compound of nitrogen and oxygen contain one atom each of N and O. In this compound there are 1.143 g of oxygen for each 1.000 g of nitrogen. Molecules of a different compound of nitrogen and oxygen contain one atom of N and two atoms of O. How many grams of oxygen would be combined with each 1.000 g of nitrogen in the second compound?

1.73 Tin forms two compounds with chlorine. In one of them (compound 1), there are two Cl atoms for each Sn atom; in the other (compound 2), there are four Cl atoms for each Sn atom. When combined with the same mass of tin, what would be the ratio of the masses of chlorine in the two compounds? In compound 1, 0.597 g of chlorine is combined with each 1.000 g of tin. How many grams of chlorine would be combined with 1.000 g of tin in compound 2?

Atomic Masses and Isotopes

1.74 The mass in grams of 1 atomic mass unit is $1.6605402 \times 10^{-24}$ g. Use this value to calculate the mass in grams of one atom of carbon-12.

1.75 Use the mass in grams of the atomic mass unit given in the preceding problem to calculate the average mass of one atom of sodium.

1.76 The chemical substance in natural gas is a compound called methane. Its molecules are composed of carbon and hydrogen and each molecule contains four atoms of hydrogen and one atom of carbon. In this compound, 0.33597 g of hydrogen is combined with 1.0000 g of carbon-12. Use this information to calculate the atomic mass of the element hydrogen.

1.77 A certain element X forms a compound with oxygen in which there are two atoms of X for every three atoms of O. In this compound, 1.125 g of X are combined with 1.000 g of oxygen. Use the average atomic mass of oxygen to calculate the average atomic mass of X. Use your calculated atomic mass to identify the element X.

1.78 If an atom of carbon-12 had been assigned a relative mass of 24.0000 u, what would be the average atomic mass of hydrogen relative to this mass?

1.79 One atom of ^{109}Ag has a mass that is 9.0754 times that of a ^{12}C atom. What is the atomic mass of this isotope of silver expressed in atomic mass units?

Atomic Structure

ILW 1.80 Naturally occurring copper is composed of 69.17% of ^{63}Cu, with an atomic mass of 62.9396 u, and 30.83% of ^{65}Cu, with an atomic mass of 64.9278 u. Use these data to calculate the average atomic mass of copper.

1.81 Naturally occurring magnesium (one of the elements in milk of magnesia) is composed of 78.99% of ^{24}Mg (atomic mass, 23.9850 u), 10.00% of ^{25}Mg (atomic mass, 24.9858 u), and 11.01% of ^{26}Mg (atomic mass, 25.9826 u). Use these data to calculate the average atomic mass of magnesium.

ILW 1.82 Give the numbers of neutrons, protons, and electrons in the atoms of each of the following isotopes. (Use the table of atomic masses and numbers printed inside the front cover for additional information, as needed.)
(a) radium-226   (c) $^{206}_{82}$Pb
(b) carbon-14   (d) $^{23}_{11}$Na

1.83 Give the numbers of electrons, protons, and neutrons in the atoms of each of the following isotopes. (As necessary, consult the table of atomic masses and numbers printed inside the front cover.)
(a) cesium-137   (c) $^{238}_{92}$U
(b) iodine-131   (d) $^{197}_{79}$Au

ADDITIONAL EXERCISES

1.84 An atom of an element has 25 protons in its nucleus.
(a) Is the element a metal, a nonmetal, or a metalloid?
(b) On the basis of the average atomic mass, write the symbol for the element's most abundant isotope.
(c) How many neutrons are in the isotope you described in part b?
(d) How many electrons are in atoms of this element?
(e) How many times heavier than ^{12}C is the average atom of this element?

*__1.85__ Elements X and Y form a compound in which there is one atom of X for every four atoms of Y. When these elements react, it is found that 1.00 g of X combines with 5.07 g of Y. When 1.00 g of X combines with 1.14 g of O, it forms a compound containing two atoms of O for each atom of X. Calculate the atomic mass of Y.

1.86 An iron nail is composed of four isotopes with the percentage abundances and atomic masses given in the following table. Calculate the average atomic mass of iron.

Isotope	Percentage Abundance	Atomic Mass (u)
^{54}Fe	5.80	53.9396
^{56}Fe	91.72	55.9349
^{57}Fe	2.20	56.9354
^{58}Fe	0.28	57.9333

*__1.87__ Bromine (shown in Figure 1.23, page 32) is a dark red liquid that vaporizes easily and is very corrosive to the skin. It is used commercially as a bleach for fibers and silk. Naturally occurring bromine is composed of two isotopes, ^{79}Br, with a mass of 78.9183 u, and ^{81}Br, with a mass of 80.9163 u. Use this information and the average atomic mass of bromine given in the table on the inside front cover of the book to calculate the percentage abundances of these two isotopes.

*__1.88__ Rust contains an iron–oxygen compound in which there are three oxygen atoms for each two iron atoms. In this compound, the iron to oxygen mass ratio is 2.325 g Fe to 1.000 g O. Another compound of iron and oxygen contains these elements in the ratio of 2.616 g Fe to 1.000 g O. What is the ratio of iron to oxygen atoms in this other iron–oxygen compound?

1.89 One atomic mass unit has a mass of $1.6605402 \times 10^{-24}$ g. Calculate the mass, in grams, of one atom of magnesium. What is the mass of one atom of iron, expressed in grams? Use these two answers to determine how many atoms of Mg are in 24.305 g of magnesium and how many atoms of Fe are in 55.847 g of iron. Compare your answers. What conclusions can you draw from the results of these calculations? Without actually performing any calculations, how many atoms do you think would be in 40.078 g of calcium?

CHAPTER 2 | Compounds and Chemical Reactions

The anti-tumor agent taxol, with activity against a number of leukemias and solid tumors in the breast, ovary, brain, and lung, was originally isolated from the bark of the Pacific yew tree, shown in this photo. Chemists successfully determined the structure of the taxol molecule and later were able to synthesize the compound in the laboratory. Taxol is now widely used, which may not have been possible if the yew had remained the only source of this valuable medicine. The nature of chemical compounds is the subject of this chapter.

THIS CHAPTER IN CONTEXT Now that you've learned some basic concepts about atoms and elements, we turn our attention to chemical compounds, substances formed from two or more elements. When elements combine to form a compound, they do so in two broad, general ways, either by the sharing of electrons between atoms or by the transfer of one or more electrons from one atom to another. Electron sharing produces discrete, electrically neutral particles that we call **molecules,** whereas electron transfer produces electrically charged particles that we call **ions.** Compounds composed of molecules are called *molecular compounds,* and those composed of ions are called *ionic compounds.*

This chapter has three principal goals. The first is to teach you about the kinds of elements that form these different classes of compounds and some of the properties we generally associate with these types of substances. The second is to introduce you to *chemical nomenclature*—the system used to name chemical compounds. Being able to describe compounds by name is essential for communication among scientists. Therefore, we urge you to make the effort to learn how to name compounds and how to translate chemical names into chemical formulas. Finally, our third goal is to give you an understanding of the concept of energy and how it is so intimately related to chemical change.

As with the preceding chapter, you may already be familiar with some of the topics we discuss here. Nevertheless, be sure to study them thoroughly. By doing so you begin to build your store of factual knowledge, a process that will continue in Chapters 5 and 6.

2.1 ▶ Elements combine to form compounds

A property possessed by nearly every element is the ability to combine with other elements to form compounds, although not all combinations appear to be possible. For example, iron reacts with oxygen to form a compound that we commonly call rust, but no compound is formed between sodium and iron.

In Chapter 1 we noted that during chemical reactions, the properties of the substances change, often dramatically, when a chemical reaction takes place. This is certainly true in the reactions of elements to form compounds. An example is the reaction between hydrogen and oxygen to form ordinary water.

At room temperature, both hydrogen and oxygen are clear, colorless gases. When they are mixed and ignited, these elements combine explosively to form the familiar compound water, which, of course, is a liquid at room temperature. As with nearly all chemical reactions, the properties of the substances present prior to the reaction differ quite a lot from the properties of those present afterwards.

The reaction between hydrogen and oxygen is not one we would expect to encounter in our daily lives, but it does have applications. When hydrogen and

FIGURE 2.1 *Liquid hydrogen and oxygen serve as fuel for the Space Shuttle.* Almost invisible points of blue flame come from the main engines of the Space Shuttle, which consume hydrogen and oxygen in the formation of water.

oxygen are cooled to sufficiently low temperatures, they condense to form liquids that serve as the fuel for the main rocket engines of the Space Shuttle (Figure 2.1). Hydrogen has also been used to power nonpolluting vehicles, in which its reaction with oxygen (from air) yields an exhaust containing only water vapor.

Chemical formulas specify the composition of a substance

To describe chemical substances, both elements and compounds, we commonly use **chemical formulas,** in which chemical symbols are used to represent atoms of the elements that are present. For a **free element** (*one that is not combined with another element in a compound*), we often simply use the chemical symbol. Thus, the element sodium is represented by its symbol, Na, which is interpreted to mean one atom of sodium.

Except for the noble gases, all the free nonmetallic elements exist as molecules that contain two or more atoms. Many of those we encounter frequently occur as **diatomic molecules** (molecules composed of two atoms each). Among them are the gases hydrogen, oxygen, nitrogen, and the halogens (fluorine, chlorine, bromine, and iodine). We represent these elements with chemical formulas in which subscripts indicate the number of atoms in a molecule. Thus, the formula for molecular hydrogen is H_2 and those for oxygen, nitrogen, and chlorine are O_2, N_2, and Cl_2. (See Figure 2.2.) The elements that occur as diatomic molecules are listed in the table in the margin. This would be a good time to learn them because you will come upon them often throughout the course. Other

Elements That Occur Naturally as Diatomic Molecules

Hydrogen	H_2	Flourine	F_2
Nitrogen	N_2	Chlorine	Cl_2
Oxygen	O_2	Bromine	Br_2
		Iodine	I_2

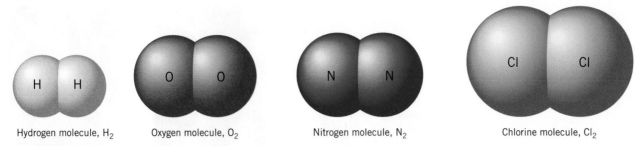

FIGURE 2.2 *Models that depict the diatomic molecules of hydrogen, oxygen, nitrogen, and chlorine.* Each contains two atoms per molecule; their different sizes reflect differences in the sizes of the atoms that make up the molecules. The atoms are shaded by color to indicate the element (hydrogen, white; oxygen, red; nitrogen, blue; and chlorine, green).

nonmetals have their atoms arranged in even more complex combinations. Elemental sulfur, for example, contains molecules of S_8, and one form of phosphorus has molecules of P_4.

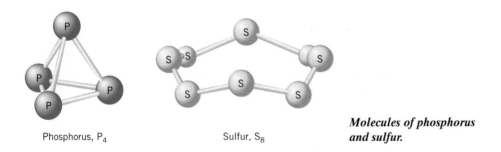

Phosphorus, P_4 Sulfur, S_8 *Molecules of phosphorus and sulfur.*

Compounds have formulas that describe their compositions

Just as chemical symbols can be used as shorthand notations for the names of elements, a chemical formula is a shorthand way of writing the name for a compound. However, *the most important characteristic of a compound's formula is that it specifies the composition of the substance.*

    In the formula of a compound, each element present is identified by its chemical symbol. Table salt, for example, has the chemical formula NaCl, which indicates it is composed of the elements sodium (Na) and chlorine (Cl). When more than one atom of an element is present, the number of atoms is given by a subscript. For instance, the iron oxide in rust has the formula Fe_2O_3, which tells us that the compound is composed of iron (Fe) and oxygen (O) and that in this compound there are two atoms of iron for every three atoms of oxygen. When no subscript is written, we assume it to be 1, so in NaCl we find one atom of sodium (Na) for each atom of chlorine (Cl). Similarly, the formula H_2O tells us that in water there are two hydrogen atoms for every one oxygen atom, and the formula for chloroform, $CHCl_3$, indicates that one atom of carbon, one atom of hydrogen, and three atoms of chlorine have combined (Figure 2.3).

    For more complicated compounds, we sometimes find formulas that contain parentheses. An example is the formula for urea, $CO(NH_2)_2$, which tells us that the group of atoms within the parentheses, NH_2, occurs twice. (The formula for urea also could be written as CON_2H_4, but there are good reasons for writing certain formulas with parentheses, as you will see later.)

Hydrates are crystals that contain water in fixed proportions

Certain compounds form crystals that contain water molecules. An example is ordinary plaster—the material often used to coat the interior walls of buildings.

Interpreting a chemical formula

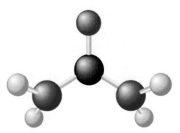

A molecule of urea, $CO(NH_2)_2$. Colors used to represent atoms are carbon, black; oxygen, red; hydrogen, white; nitrogen, blue.

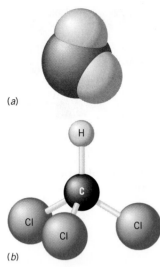

(a)

(b)

FIGURE **2.3** *Molecules of water and chloroform.* (*a*) A "space-filling" model of H_2O, (*b*) a "ball-and-stick" model of $CHCl_3$. (A standard color scheme is used to represent the different elements: carbon, black; hydrogen, white; and chlorine, green.)

FIGURE **2.4** *Water can be driven from hydrates by heating.* (*a*) Blue crystals of copper sulfate pentahydrate, $CuSO_4 \cdot 5H_2O$, about to be heated. (*b*) The hydrate readily loses water when heated. The white solid that forms is pure $CuSO_4$.

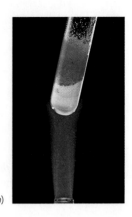

(a)

(b)

When all the water is removed, the solid is said to be **anhydrous,** meaning *without water.*

Plaster consists of crystals of calcium sulfate, $CaSO_4$, that contain two molecules of water for each $CaSO_4$. These water molecules are not held very tightly and can be driven off by heating the crystals. The dried crystals absorb water again if exposed to moisture, and the amount of water absorbed always gives crystals in which the H_2O-to-$CaSO_4$ ratio is 2 to 1. Compounds whose crystals contain water molecules in fixed ratios are quite common and are called **hydrates.** The formula for this hydrate of calcium sulfate is written $CaSO_4 \cdot 2H_2O$ to show that there are two molecules of water per $CaSO_4$. The raised dot is used to indicate that the water molecules are not bound too tightly in the crystal and can be removed.

Sometimes the *dehydration* (removal of water) of hydrate crystals produces changes in color. An example is copper sulfate, which is sometimes used as an agricultural fungicide. Copper sulfate forms blue crystals, with the formula $CuSO_4 \cdot 5H_2O$, in which there are five water molecules for each $CuSO_4$. When these blue crystals are heated, most of the water is driven off and the solid that remains, now nearly pure $CuSO_4$, is almost white (Figure 2.4). If left exposed to the air, the $CuSO_4$ will absorb moisture and form blue $CuSO_4 \cdot 5H_2O$ again.

Counting atoms in formulas is a necessary skill

Counting the number of atoms of the elements in a chemical formula is an operation you will have to perform many times, so let's look at an example.

CHEMISTRY IN PRACTICE The tendency of certain compounds to form hydrates has practical applications. For example, calcium chloride, $CaCl_2$, forms a number of hydrates, including $CaCl_2 \cdot 6H_2O$, and the *anhydrous* compound (the compound without the water) will absorb water from the air to form hydrates. Products such as the one pictured are available in hardware stores and contain porous bags of $CaCl_2$. When placed in a damp basement or closet, the $CaCl_2$ absorbs moisture and reduces the humidity, slowing the growth of mildew.

How many atoms of each element are represented by the formulas (a) $Al_2(SO_4)_3$ and (b) $CoCl_2 \cdot 6H_2O$?

ANALYSIS: This is not a difficult problem, but we do need to remember what subscripts tell us. We also must recall that a quantity within parentheses is repeated a number of times equal to the subscript that follows and a raised dot in a formula indicates the substance is a hydrate in which the number preceding H_2O specifies how many water molecules are present.

SOLUTION: (a) Here we must recognize that all the atoms within the parentheses occur three times.

Subscript 3 indicates three SO_4 units.

$$Al_2(SO_4)_3$$

Each SO_4 contains one S and four O atoms, so three of them contain three S and twelve O atoms. The subscript for Al tells us there are two Al atoms. Therefore, the formula $Al_2(SO_4)_3$ shows

$$2\,Al \qquad 3\,S \qquad 12\,O$$

(b) This is a formula for a hydrate, as indicated by the raised dot. It contains six water molecules, each with two H and one O, for every $CoCl_2$.

The 6 indicates there are six molecules of H_2O.

$$CoCl_2 \cdot 6H_2O$$

Dot indicates the compound is a hydrate.

Therefore, the formula $CoCl_2 \cdot 6H_2O$ represents

$$1\,Co \qquad 2\,Cl \qquad 12\,H \qquad 6\,O$$

Are the Answers Reasonable?
The only way to check the answer here is to perform a recount.

PRACTICE EXERCISE 1: How many atoms of each element are expressed by the formulas (a) $NiCl_2$, (b) $FeSO_4$, (c) $Ca_3(PO_4)_2$, (d) $Co(NO_3)_2 \cdot 6H_2O$?

EXAMPLE 2.1

Counting Atoms in Formulas

2.2 ▶ Chemical equations describe what happens in chemical reactions

A **chemical equation** *describes what happens when a chemical reaction occurs.* It uses chemical formulas to provide a before-and-after picture of the chemical substances involved. Consider, for example, the reaction between hydrogen and oxygen to give water. The chemical equation that describes this reaction is

$$2H_2 + O_2 \longrightarrow 2H_2O$$

The two substances that appear to the left of the arrow are the **reactants;** they are the substances present before the reaction begins. To the right of the arrow we find the formula for the **product** of the reaction, water. In this example, only one substance is formed in the reaction, so there is only one product. As we will see, however, in most chemical reactions there is more than one product. The products are the substances that are formed and that exist after the reaction is over. The arrow means "reacts to yield." Thus, this equation tells us that *hydrogen and oxygen react to yield water.*

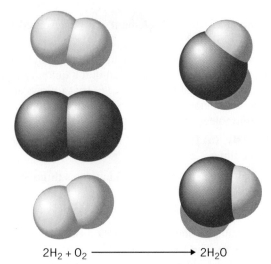

FIGURE 2.5 *The reaction between molecules of hydrogen and oxygen.* The reaction between two molecules of hydrogen and one molecule of oxygen gives two molecules of water.

$$2H_2 + O_2 \longrightarrow 2H_2O$$

Coefficients are written in front of formulas to balance an equation

In the equation for the reaction of hydrogen and oxygen, you'll notice that the number 2 precedes the formulas of hydrogen and water. Numbers in front of the formulas are called **coefficients,** and they indicate the number of molecules of each kind among the reactants and products. Thus, $2H_2$ means two molecules of H_2, and $2H_2O$ means two molecules of H_2O. When no number is written, the coefficient is assumed to be 1 (so the coefficient of O_2 equals 1).

Coefficients are needed to have the equation conform to the law of conservation of mass, as illustrated in Figure 2.5. Because atoms cannot be created or destroyed in a chemical reaction, we must have the same number of atoms of each kind present before and after the reaction (i.e., on both sides of the arrow). When this condition is met, we say the equation is **balanced.** Another example is the equation for the combustion of butane, C_4H_{10}, the fluid in disposable cigarette lighters (Figure 2.6).

FIGURE 2.6 *The combustion of butane, C_4H_{10}.* The products are carbon dioxide and water vapor.

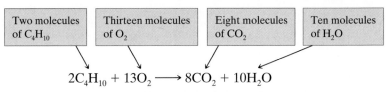

Two molecules of C_4H_{10}	Thirteen molecules of O_2	Eight molecules of CO_2	Ten molecules of H_2O

$$2C_4H_{10} + 13O_2 \longrightarrow 8CO_2 + 10H_2O$$

The 2 before the C_4H_{10} tells us that two molecules of butane react. This involves a total of 8 carbon atoms and 20 hydrogen atoms, as we see in Figure 2.7. Notice we have multiplied the numbers of atoms of C and H in one molecule of C_4H_{10} by the coefficient 2. On the right we find 8 molecules of CO_2, which contain a total of 8 carbon atoms. Similarly, 10 water molecules contain 20 hydrogen atoms. Finally, we

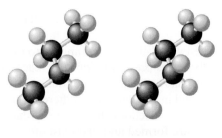

FIGURE 2.7 *Understanding coefficients in an equation.* The expression $2C_4H_{10}$ describes two molecules of butane, each of which contains 4 carbon and 10 hydrogen atoms. This gives a total of 8 carbon and 20 hydrogen atoms.

Two molecules of butane contain 8 atoms of C and 20 atoms of H can count 26 oxygen atoms on both sides of the equation. You will learn to balance equations such as this in Chapter 4.

The states of the reactants and products can be specified in a chemical equation

In a chemical equation we sometimes find it useful to specify the physical states of the reactants and products, that is, whether they are solids, liquids, or gases. This is done by writing *s* for solid, *l* for liquid, or *g* for gas in parentheses after the chemical formulas. For example, the equation for the combustion of the carbon in a charcoal briquette can be written as

$$C(s) + O_2(g) \longrightarrow CO_2(g)$$

At times, we will also find it useful to indicate that a particular substance is dissolved in water. We do this by writing *aq*, meaning *aqueous* solution, in parentheses after the formula. For instance, the reaction between stomach acid (an aqueous solution of HCl) and the active ingredient in Tums®, $CaCO_3$, is

$$2HCl(aq) + CaCO_3(s) \longrightarrow CaCl_2(aq) + H_2O(l) + CO_2(g)$$

Remember:
s = solid
l = liquid
g = gas

PRACTICE EXERCISE 2: How many atoms of each element appear on each side of the arrow in the following equation?

$$Mg(OH)_2 + 2HCl \longrightarrow MgCl_2 + 2H_2O$$

PRACTICE EXERCISE 3: Rewrite the equation in Practice Exercise 2 to show that $Mg(OH)_2$ is a solid, HCl and $MgCl_2$ are dissolved in water, and H_2O is a liquid.

2.3 ▶ Energy is an important part of chemical change

When chemical reactions occur, *energy* is almost always absorbed or given off (as heat or light, for instance). For example, the explosive reaction between hydrogen and oxygen in the main engines of the Space Shuttle, shown in Figure 2.1 (page 42), produces light, heat, and the expanding gases that help lift the space vehicle from its launch pad. Similarly, the combustion reaction between gasoline and oxygen liberates energy in the form of heat that we harness to power vehicles. In fact, one of the practical uses of chemical reactions is to satisfy the energy needs of society.

Because chemical reactions involve an absorption or release of energy, the study of energy is an integral part of the study of chemistry.

The concept of energy is more difficult to grasp than that of matter because energy is intangible; you can't hold it in your hand to study it and you can't put it in a bottle. **Energy** *is something an object has if the object is able to do work*. It can be possessed by an object in two different ways, as kinetic energy and as potential energy.

Kinetic energy *is the energy an object has when it is moving*. It depends on the object's mass and velocity; the larger its mass and the greater its velocity, the more kinetic energy it has. Thus, a heavy cement truck moving at 50 mph has more kinetic energy than a car moving at the same speed, and a car at 50 mph has more kinetic energy than a car traveling at 25 mph.

Work is done by an object when it causes something to move. A moving car has energy because it can move another car in a collision.

A simple equation relates kinetic energy (K.E.) to mass and velocity[1]:

$$K.E. = \tfrac{1}{2}mv^2 \tag{2.1}$$

where *m* is the mass and *v* is the velocity.

[1]Many equations will be numbered to make it easy to refer to them later on in the book or in class discussions.

Unlike kinetic energy, there is no single, simple equation that can be used to calculate the amount of potential energy an object has.

Potential energy *is energy an object has that can be changed to kinetic energy; it can be thought of as **stored energy.*** For example, when you wind an alarm clock, you transfer energy to a spring. The spring holds this stored energy (potential energy) and gradually releases it, in the form of kinetic energy, to make the clock's mechanism work. Water high in the mountains also has potential energy because of its attraction by gravity. When the water falls to a lower altitude, we can harness the energy that's released to turn turbines that generate electricity or operate machinery. Chemicals also possess potential energy, which is sometimes called **chemical energy.** This is energy stored in chemicals that can be liberated during chemical reactions. For example, the foods we eat have potential energy that is changed to kinetic energy by the process we call *metabolism.* This kinetic energy ultimately manifests itself as movements of muscles and as body heat.

Energy cannot be created or destroyed

One of the most important facts about energy is that *it cannot be created or destroyed; it can only be changed from one form to another.* This fact was established as the result of many experiments and observations and is known today as the **law of conservation of energy.** (Recall that when we say in science that something is "conserved," we mean that it is unchanged or remains constant.) You observe this law whenever you toss something—a ball, for instance—into the air. You give the ball some initial amount of kinetic energy when you throw it. As it rises, its potential energy increases. Because energy cannot come from nothing, the rise in potential energy comes at the expense of the ball's kinetic energy. Therefore, the ball's $\frac{1}{2}mv^2$ becomes smaller, and because the mass of the ball cannot change, the velocity (v) becomes less—the ball slows down. When all the kinetic energy has changed to potential energy the ball can go no higher; it has stopped moving and its potential energy is at a maximum. The ball then begins to fall, and its potential energy is changed back to kinetic energy.

To be *conserved* means to be kept from being lost.

Because the ball's mass can't change, the decrease in kinetic energy must be accompanied by a decrease in the ball's velocity.

Heat and temperature are not the same

In any object, the atoms and molecules are not stationary; they are constantly moving and colliding with each other. Because of these motions, the particles have varying amounts of kinetic energy, and the terms *heat* and *temperature* are related to this energy. Specifically, the temperature of an object is proportional to the average kinetic energy of its particles—the higher the average kinetic energy, the higher the temperature. What this means is that when the temperature of an object is raised, the molecules move faster. (Recall that K.E. $= \frac{1}{2}mv^2$. Increasing the average kinetic energy doesn't increase the masses of the atoms, so it must increase their speeds.)

The difference between heat and temperature. Both fires could be at the same temperature if the *average* kinetic energies of the atoms in the flames are the same. The forest fire has many more atoms with high kinetic energies, so its *total* kinetic energy is greater than that of the match flame. This gives the forest fire a greater potential for transferring heat to cooler objects, so we think of it as containing more heat.

Heat is energy (also called **thermal energy**) that is transferred between objects caused by differences in their temperatures, and, as you know, heat always passes spontaneously from the object with the higher temperature to the one that is colder. This energy transfer continues until both objects come to the same temperature. All forms of energy (potential and kinetic) can be transformed to heat energy. For example, when you "step on the brakes" to stop your car, the kinetic energy of the car is changed to heat energy by friction between the brake shoes and the wheels.

The way we interpret the transfer of heat is as a transfer of kinetic energy between atoms of the two objects. It takes place in a manner that's similar to the transfer of kinetic energy between balls on a billiard table. Here we may see a fast-moving ball strike a slow one, causing the slow one to speed up and the fast one to slow down. In this collision, kinetic energy is transferred from the ball that slows to the ball that goes faster. Similarly, when a hot object is placed in contact with a cold one, the faster atoms of the hot object collide with and lose kinetic energy to the slower atoms of the cold object. This decreases the average kinetic energy of the particles of the hot object, causing its temperature to drop. At the same time, the average kinetic energy of the particles in the cold object is raised, causing the temperature of the cold object to rise. Eventually, the average kinetic energies of the atoms in both objects become the same and the objects reach the same temperature.

Energy can also be transferred as light, which we will discuss later in the book.

Energy changes usually accompany chemical changes

The concepts discussed in the preceding paragraphs form the foundation of the *kinetic molecular theory of matter*. This is one of science's most important theories, and we will discuss it in much greater detail in Chapter 7. For now, however, let's see how these concepts allow us to explain how changes in chemical energy that accompany chemical reactions make themselves known.

Recall that chemical energy is the potential energy stored in chemicals.

In the propulsion engines of the Space Shuttle, the reaction of hydrogen with oxygen produces very hot water vapor, which expands with great force to produce the thrust that helps propel the Shuttle into orbit. What is the origin of the heat that appears in this reaction?

When hydrogen burns with oxygen, the water vapor formed has a much higher temperature than the starting materials, which means that the average kinetic energy of the atoms has increased during the reaction. But where has this increase in kinetic energy come from? Because kinetic energy cannot come from nothing, it must have come from the potential energy (chemical energy) of the hydrogen and oxygen. Therefore, we are able to conclude that during this reaction there is a drop in potential energy and that the water vapor formed in the reaction must have a lower potential energy than the hydrogen and oxygen.

The rise in temperature occurs because there is a lowering of the chemical energy during the course of the reaction.

As you can see, the analysis of temperature changes in chemical reactions can tell us about the potential energy changes that occur. And as we shall see later, the changes in potential energy that accompany reactions also reveal a great deal about the chemicals themselves.

> **PRACTICE EXERCISE 4:** Some instant cold packs sold in pharmacies contain ammonium nitrate inside a pouch of water. When this compound mixes and dissolves in the water, the mixture becomes cold. (a) How do the kinetic energies of the ammonium nitrate and water change? (b) How do the potential energies of the ammonium nitrate and water change?

2.4 ▶ Molecular compounds contain neutral particles called molecules

The concept of molecules dates to the time of Dalton's atomic theory, part of which was that atoms of elements combine in fixed numerical ratios to form "molecules" of a compound. By our modern definition, *a **molecule** is an electrically neutral particle consisting of two or more atoms*. Accordingly, the term *molecule* applies to many elements, such as H_2 and O_2, as well as to compounds.

Experimental evidence exists for molecules

One phenomenon that points to the existence of molecules is called **Brownian motion** [named after Robert Brown (1773–1858), the Scottish botanist who first observed it]. When very small particles such as tiny grains of pollen are suspended in a liquid and observed under a microscope, the tiny particles are seen to be constantly jumping and jiggling about. It appears as though they are continually being knocked to and fro by collisions with something. An explanation is that this "something" is *molecules* of the liquid. The microscopic particles are constantly bombarded by molecules of the liquid, but because the suspended particles are so small, the collisions are not occurring equally on all sides. The unequal numbers of collisions cause the lightweight particles to jerk about.

There is additional evidence for molecules, and today scientists accept the existence of molecules as fact. Looking more closely, within molecules atoms are held to each other by attractions called **chemical bonds,** which are electrical in nature. As mentioned, they arise from the sharing of electrons between one atom and another. We will discuss such bonds at considerable length in Chapters 9 and 10. What is important to know about molecules now is that *the group of atoms that make up a molecule move about together and behave as a single particle,* just as the various parts that make up a car move about as one unit. The chemical formulas that we write to describe the compositions of molecules are called **molecular formulas,** which specify the actual numbers of atoms of each kind that make up a single molecule.

Molecular compounds form when nonmetals combine

Carbon monoxide is a poisonous gas found in the exhaust of automobiles.

As a general rule, *molecular compounds are formed when nonmetallic elements combine*. For example, you learned that H_2 and O_2 combine to form molecules of water. Similarly, carbon and oxygen combine to form either carbon monoxide, CO, or carbon dioxide, CO_2. (Both are gases that are formed in various amounts as products in the combustion of fuels such as gasoline and charcoal.) Although molecular compounds can be formed by the direct combination of elements, often they are the products of reactions between compounds. You will encounter many such reactions in your study of chemistry.

Although there are relatively few nonmetals, the number of molecular substances formed by them is huge. This is because of the variety of ways in which they combine as well as the varying degrees of complexity of their molecules. Variety and complexity reach a maximum with compounds in which carbon is combined with a handful of other elements, such as hydrogen, oxygen, and nitrogen. There are so many of these compounds, in fact, that their study encompasses the chemical specialties of organic chemistry and biochemistry.

Molecules vary in size from small to very large. Some contain as few as two atoms (diatomic molecules). Most molecules are more complex, however, and contain more atoms. Molecules of water (H_2O), for example, have 3 atoms, and those of ordinary table sugar ($C_{12}H_{22}O_{11}$) have 45. There also are molecules that are very large, such as those that occur in plastics and in living organisms, some of which contain millions of atoms.

At this early stage, we can only begin to look for signs of order among the vast number of nonmetal–nonmetal compounds. To give you a taste of the subject, we will look briefly at some simple compounds that the nonmetals form with hydrogen as well as some simple compounds of carbon.

TABLE 2.1	SIMPLE HYDROGEN COMPOUNDS OF THE NONMETALLIC ELEMENTS			
	Group			
Period	IVA	VA	VIA	VIIA
2	CH_4	NH_3	H_2O	HF
3	SiH_4	PH_3	H_2S	HCl
4	GeH_4	AsH_3	H_2Se	HBr
5		SbH_3	H_2Te	HI

Hydrogen forms compounds with many nonmetals

Compounds that elements form with hydrogen are often called *hydrides*, and the formulas of the simple hydrides of the nonmetals are given in Table 2.1.[2] These compounds provide an opportunity to observe how we can use the periodic table as an aid in remembering factual information, in this case, the formulas of the hydrides. Notice that the number of hydrogen atoms combined with the nonmetal atom equals *the number of spaces to the right that we have to move in the periodic table to get to a noble gas.* (You will learn *why* this is so in Chapter 9, but for now we can just use the periodic table to help us remember the formulas.)

Periodic table and formulas for nonmetal hydrides

Two steps, so oxygen combines with two hydrogens to give H_2O.

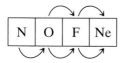

Three steps, so nitrogen combines with three hydrogens to give NH_3.

Also note in Table 2.1 that the formulas of the simple hydrides are similar for nonmetals within a given group of the periodic table. If you know the formula for the hydride of the top member of the group, then you know the formulas of all of them in that group.

Many of the nonmetals form more complex compounds with hydrogen, but we will not discuss them here.

We live in a three-dimensional world, and this is reflected in the three-dimensional shapes of molecules. The shapes of the simple nonmetal hydrides of nitrogen, oxygen, and fluorine are illustrated in Figure 2.8.

Compounds of carbon form the basis for organic chemistry

Among all the elements, carbon is unique in the variety of compounds it forms with elements such as hydrogen, oxygen, and nitrogen. As a consequence, the number and complexity of such compounds is enormous, and their study consti-

Our goal at this time is to acquaint you with some of the important kinds of organic compounds we encounter regularly, so our discussion here is brief.

Ammonia, NH_3

Water, H_2O

Hydrogen fluoride, HF

FIGURE 2.8 *Nonmetal hydrides of nitrogen, oxygen, and fluorine.*

[2]Table 2.1 shows how the formulas are normally written. The order in which the hydrogens appear in the formula is not of concern to us now. Instead, we are interested in the *number* of hydrogens that combine with a given nonmetal.

TABLE 2.2	Hydrocarbons Belonging to the Alkane Series	
Compound	Name	Boiling Point (°C)
CH_4	Methane[a]	−161.5
C_2H_6	Ethane[a]	−88.6
C_3H_8	Propane[a]	−42.1
C_4H_{10}	Butane[a]	−0.5
C_5H_{12}	Pentane	36.1
C_6H_{14}	Hexane	68.7

[a]Gases at room temperature (25 °C) and atmospheric pressure.

tutes the major specialty called **organic chemistry.** The term *organic* here comes from an early belief that these compounds could only be made by living organisms. We now know this isn't true, but the name organic chemistry persists nonetheless.

Organic compounds are around us everywhere and we will frequently use such substances as examples in our discussions. Therefore, it will be helpful if you can begin to learn some of them now.

The study of organic chemistry begins with **hydrocarbons** (compounds of carbon and hydrogen). The simplest hydrocarbon is methane, CH_4, which is a member of a series of hydrocarbons with the general formula C_nH_{2n+2}, where n is a whole number. The first six members of this series, called the **alkane** series, are given in Table 2.2 along with their boiling points. Notice that as the molecules become larger, their boiling points increase. Molecules of methane, ethane, and propane are illustrated in Figure 2.9.

The alkanes are common substances. They are the principal constituents of petroleum, from which most of our useful fuels are produced. Methane itself is the major component of natural gas. Gas-fired barbecues use propane as a fuel, and butane is the fuel in inexpensive cigarette lighters.[3] Hydrocarbons with higher boiling points are found in gasoline, kerosene, paint thinners, diesel fuel, and even candle wax.

The chemically correct names for ethylene and acetylene are ethene and ethyne, respectively. Naming organic compounds is discussed in Chapter 25.

Alkanes are not the only class of hydrocarbons. For example, there are three two-carbon hydrocarbons. In addition to ethane, C_2H_6, there are ethylene, C_2H_4 (from which polyethylene is made), and acetylene, C_2H_2 (the fuel used in *acetylene* torches).

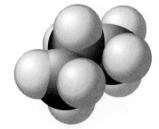

Methane, CH_4 Ethane, C_2H_6 Propane, C_3H_8

Figure 2.9 *The first three members of the alkane series of hydrocarbons.* White atoms represent hydrogen; black atoms represent carbon in these space-filling models that demonstrate the shapes of the molecules.

[3]Propane and butane are gases when they're at the pressure of the air around us, but become liquids when compressed. When you purchase these substances, they are liquids with pressurized gas above them. The gas can be drawn off and used by opening a valve to the container.

Methane   Methanol

FIGURE 2.10 *Relationship between an alkane and an alcohol.* The alcohol methanol is derived from methane by replacing one H by OH. (Color code: carbon is black, hydrogen is white, oxygen is red.)

The hydrocarbons serve as the foundation for organic chemistry. Derived from them are various other classes of organic compounds. An example is the class of compounds called alcohols, in which the atoms OH replace a hydrogen in the hydrocarbon. Thus, *methanol,* CH_3OH (also called *methyl alcohol*), is related to methane, CH_4, by removing one H and replacing it with OH (Figure 2.10). Methanol is used as a fuel and as a raw material for making other organic chemicals. Another familiar alcohol is *ethanol* (also called *ethyl alcohol*), C_2H_5OH. Ethanol, known as grain alcohol because it is obtained from the fermentation of grains, is in alcoholic beverages. It is also mixed with gasoline to produce a fuel mixture known as gasohol.

Methanol is also known as wood alcohol. It is quite poisonous. Ethanol in high doses is also a poison.

Alcohols constitute just one class of compound derived from hydrocarbons. We will discuss some others after you've learned more about how atoms bond to each other and about the structures of molecules.

PRACTICE EXERCISE 5: What is the formula of the alkane hydrocarbon having ten carbon atoms.

PRACTICE EXERCISE 6: On the basis of the discussions in this section, what are the formulas of (a) propanol and (b) butanol?

2.5 ▶ Naming molecular compounds follows a system

In conversation, chemists rarely use formulas to describe compounds. Instead, names are used. For example, you already know that water is the name for the compound having the formula H_2O and that sodium chloride is the name of NaCl.

At one time there was no uniform procedure for assigning names to compounds, and those who discovered compounds used whatever method they wished. Without some sort of system, however, remembering names for the rapidly increasing number of compounds soon became impossible. The search for a solution led chemists around the world to agree on a systematic method for naming substances. By using this method we are able to write the correct formula for any particular compound given the correct name, and vice versa.

In this section we discuss the **nomenclature** (naming) of simple molecular inorganic compounds. In general, **inorganic compounds** are substances that would *not* be considered to be derived from hydrocarbons such as methane (CH_4), ethane (C_2H_6), and other carbon–hydrogen compounds. As we noted earlier, the hydrocarbons and compounds that can be thought of as coming from them are called *organic compounds.* We will have more to say about naming organic compounds later.

Even if we exclude organic compounds, the number and variety of molecular substances is quite enormous. To introduce you to the naming of them, we will restrict ourselves to **binary compounds**—*compounds composed of two different elements.*[4]

Rules for naming molecular compounds

Binary compounds composed of two nonmetals are named using Greek prefixes

Our goal is to be able to translate a chemical formula into a name that contains information which would enable someone else, just looking at the name, to reconstruct the formula. For a binary molecular compound, therefore, we must indicate which two elements are present and the number of atoms of each in a molecule of the substance.

To identify the first element in a formula, we just specify its English name. Thus, for HCl the first word in the name is *hydrogen,* and for PCl_5 the first word is *phosphorus.* To identify the second element, we append the suffix -ide to the stem of the element's English name. Here are some examples:

Element	Stem	Name as Second Element
Oxygen	ox-	oxide
Sulfur	sulf-	sulfide
Nitrogen	nitr-	nitride
Phosphorus	phosph-	phosphide
Fluorine	fluor-	fluoride
Chlorine	chlor-	chloride
Bromine	brom-	bromide
Iodine	iod-	iodide

To form the name of the compound, we place the two parts of the name one after another. Therefore, the name of HCl is hydrogen chloride. However, to name PCl_5, we need a way to specify the number of Cl atoms bound to the phosphorus in the molecule. This is done using the following Greek prefixes.

mono-	= 1 (often omitted)		hexa-	= 6
di-	= 2		hepta-	= 7
tri-	= 3		octa-	= 8
tetra-	= 4		nona-	= 9
penta-	= 5		deca-	= 10

To name PCl_5, therefore, we add the prefix penta- to chloride to give the name phosphorus pentachloride. Notice how easily this allows us to translate the name back into the formula.

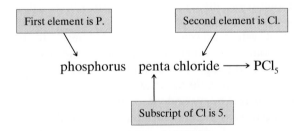

----

[4]It's important to note the different meanings of *binary* and *diatomic*. A binary substance is composed of *two different elements;* a diatomic substance is composed of molecules containing *two atoms*. Thus, CO_2 and CH_4 are binary compounds but they are not diatomic. The elements H_2 and O_2 are diatomic but they are not binary because in each molecule both atoms are of the same element. Finally, CO and HCl are binary compounds and they are diatomic; each molecule contains atoms of two different elements and each contains only two atoms.

The prefix mono- is used when we want to emphasize that only one atom of a particular element is present. For instance, carbon forms two compounds with oxygen, CO and CO_2. To clearly distinguish between them, the first is called *carbon monoxide* (one of the "o"s is omitted to make the name easier to pronounce) and the second is called *carbon dioxide*.

As indicated, the prefix mono- is often omitted from a name. Therefore, in general, if there is no prefix before the name of an element, we take it to mean there is only one atom of that element in the molecule. An exception to this is the names of binary compounds of nonmetals with hydrogen, for example, *hydrogen sulfide*. The name tells us the compound contains the two elements hydrogen and sulfur. We don't have to be told how many hydrogens are in the molecule because, as you learned earlier, we can use the periodic table to determine the number of hydrogen atoms in molecules of the simple nonmetal hydrides. Sulfur is in Group VIA, so to get to the noble gas column we have to move two steps to the right; the number of hydrogens combined with the atom of sulfur is two. The formula for hydrogen sulfide is therefore H_2S.

(a) What is the name of $AsCl_3$? (b) What is the formula for dinitrogen tetraoxide?

ANALYSIS: (a) In naming compounds, the first step is to determine what type of compound is involved. Looking at the periodic table, we see that $AsCl_3$ is made up of two nonmetals, so we anticipate that it is a molecular compound. Once we've done this, we apply the procedure described earlier.

(b) To write the formula from the name, we convert the prefixes to numbers and apply them as subscripts to the chemical symbols of the elements.

SOLUTION: (a) In $AsCl_3$, As is the symbol for arsenic and, of course, Cl is the symbol for chlorine. The first word in the name is just arsenic and the second will contain chloride with an appropriate prefix to indicate number. There are three Cl atoms, so the prefix is tri-. Therefore, the name of the compound is arsenic trichloride.

(b) As we did earlier for phosphorus pentachloride, we convert the prefixes to numbers and apply them as subscripts.

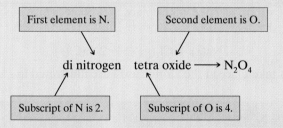

Are the Answers Reasonable?
To feel comfortable with the answers, be sure to double check for careless errors. That's about all we can do with this kind of problem.

PRACTICE EXERCISE 7: Name the following compounds, using Greek prefixes when needed: (a) PCl_3, (b) SO_2, (c) Cl_2O_7.

PRACTICE EXERCISE 8: Write formulas for the following compounds: (a) arsenic pentachloride, (b) sulfur hexachloride, (c) disulfur dichloride.

> **EXAMPLE 2.2**
> **Naming Compounds and Writing Formulas**

After we've discussed ionic compounds, this first step in the analysis will be particularly important, because different rules apply depending on the type of compound being named.

Sometimes we drop the *a* before an *o* for ease of pronunciation. N_2O_4 would then be named dinitrogen tetroxide.

Common names exist for many substances

Not every compound is named according to the systematic procedure described above. Many familiar substances were discovered long before a systematic method for naming them had been developed, and they acquired common names that are so well known that no attempt has been made to rename them. For example, following the scheme described earlier we might expect that H_2O would have the name hydrogen oxide (or even dihydrogen monoxide). Although this isn't wrong, the common name water is so well known that it is always used. Another example is ammonia, NH_3, whose odor you have no doubt experienced while using household ammonia solutions or the glass cleaner Windex®. Common names are used for the other hydrides of the nonmetals in Group VA as well. The compound PH_3 is called phosphine and AsH_3 is called arsine.

Common names are also used for very complex substances. An example is sucrose, which is the chemical name for table sugar, $C_{12}H_{22}O_{11}$. The structure of this compound is pretty complex, and its name assigned following the systematic method is equally complex. It is much easier to say the simple name sucrose, and be understood, than to struggle with the cumbersome systematic name for this common compound.

2.6 ▶ Ionic compounds are composed of charged particles called ions

Experimental evidence exists for ions in compounds

Table salt is an interesting compound. If the solid is heated to a high temperature, so that it melts, the resulting liquid is able to conduct electricity. Molecular compounds such as water, however, are very poor conductors of electricity. These observations suggest that the particles in liquid salt are quite different from those in liquid water. In fact, they suggest that salt is composed of electrically charged particles, called **ions,** rather than neutral molecules.

Formation of an ionic compound from its elements involves electron transfer

Under appropriate conditions, atoms are able to transfer electrons between them when they react. This is what happens, for example, when the metal sodium combines with the nonmetal chlorine. As shown in Figure 2.11, sodium is a typical shiny metal and chlorine is a pale green gas. When a piece of heated sodium is thrust into the chlorine, a vigorous reaction takes place yielding a white powder, salt (NaCl). The equation for the reaction is

$$2Na(s) + Cl_2(g) \longrightarrow 2NaCl(s)$$

The changes that take place at the atomic level are illustrated in Figure 2.12.

(a)

(b)

(c)

FIGURE 2.11 *Sodium reacts with chlorine to give the ionic compound sodium chloride.* (*a*) Freshly cut sodium has a shiny metallic surface. The metal reacts with oxygen and moisture, so it cannot be touched with bare fingers. (*b*) Chlorine is a pale green gas. (*c*) When a small piece of sodium is melted in a metal spoon and thrust into the flask of chlorine, it burns brightly as the two elements react to form sodium chloride. The smoke coming from the flask is composed of fine crystals of salt.

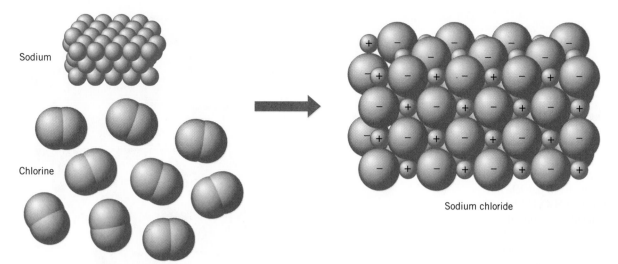

FIGURE 2.12 *The reaction of sodium with chlorine viewed at the atomic level.* Electrically neutral atoms and molecules react to yield positive and negative ions, which are held to each other by electrostatic attractions (attractions between opposite electrical charges).

The formation of the ions in salt results from the transfer of electrons between the reacting atoms. Specifically, each sodium atom gives up one electron to a chlorine atom, which thereby gains one. We can diagram the changes in equation form by using the symbol e^- to stand for an electron.

$$\overset{\overset{\displaystyle e^-}{\frown}}{Na + Cl} \longrightarrow Na^+ + Cl^-$$

The electrically charged particles formed in this reaction are a sodium ion (Na^+) and a chloride ion (Cl^-). The sodium ion has a positive 1+ charge, indicated by the superscript plus sign, because it now has one more proton in its nucleus than there are electrons outside. Similarly, by gaining one electron the chlorine atom has added one more negative charge, so the chloride ion has a single negative charge indicated by the minus sign. Solid sodium chloride is composed of these charged sodium and chloride ions and is said to be an **ionic compound.**

As a general rule, *ionic compounds are formed when metals react with nonmetals.* In the electron transfer, however, not all atoms gain or lose just one electron; some gain or lose more. For example, when calcium atoms react, they lose two electrons to form Ca^{2+} ions, and when oxygen atoms form ions, they each gain two electrons to give O^{2-} ions. (We will have to wait until a later chapter to study the reasons why certain atoms gain or lose one electron each, whereas other atoms gain or lose two or more electrons.)

Figure 2.13 compares the structures of water and sodium chloride and demonstrates an important difference between molecular and ionic compounds. For water, it is safe to say that two hydrogen atoms "belong" to each oxygen atom in a particle having the formula H_2O. However, in NaCl it is impossible to say that a particular Na^+ ion belongs to a particular Cl^- ion. The ions in a crystal of NaCl are simply packed in the most efficient way, so that positive ions and negative ions can be as close to each other as possible. In this way, the attractions between oppositely charged ions, which are responsible for holding the compound together, can be as strong as possible.

Because molecules don't exist in ionic compounds, the subscripts in their formulas are always chosen to specify the smallest whole-number ratio of the ions. This is why the formula of sodium chloride is given as NaCl rather than Na_2Cl_2 or Na_3Cl_3. Although the smallest unit of an ionic compound can't be called a molecule, the idea of "smallest unit" is still quite often useful. Therefore, we take the smallest unit

Here we are concentrating on what happens to the individual atoms, so we have not shown chlorine as diatomic Cl_2 molecules.

A neutral sodium atom has 11 protons and 11 electrons; a sodium ion has 11 protons and 10 electrons, so it carries a unit positive charge. A neutral chlorine atom has 17 protons and 17 electrons; a chloride ion has 17 protons and 18 electrons, so it carries a unit negative charge.

Notice that the charges on the ions are omitted when writing formulas for compounds. This is because compounds are electrically neutral overall.

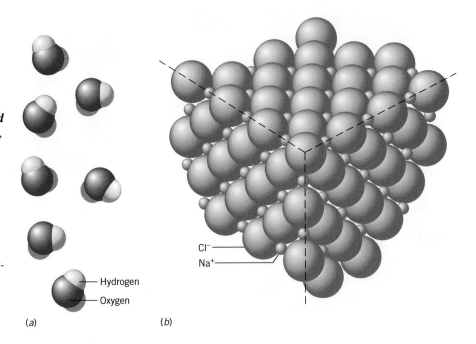

FIGURE 2.13 *Molecular and ionic substances.* (*a*) In water, there are discrete molecules that each consist of one atom of oxygen and two atoms of hydrogen. Each particle has the formula H_2O. (*b*) In sodium chloride, ions are packed in the most efficient way. Each Na^+ is surrounded by six Cl^-, and each Cl^- is surrounded by six Na^+. Because individual molecules do not exist, we simply specify the ratio of ions as NaCl.

Hydrogen
Oxygen

Cl^-
Na^+

(*a*) (*b*)

of an ionic compound to be whatever is represented in its formula and call this unit a **formula unit.** Thus, one formula unit of NaCl consists of one Na^+ and one Cl^-, whereas one formula unit of the ionic compound $CaCl_2$ consists of one Ca^{2+} and two Cl^- ions. (In a broader sense, we can use the term *formula unit* to refer to whatever is represented by a formula. Sometimes the formula specifies a set of ions, as in NaCl; sometimes it is a molecule, as in O_2 or H_2O; sometimes it can be just an ion, as in Cl^- or Ca^{2+}; and sometimes it might be just an atom, as in Na.)

PRACTICE EXERCISE 9: For each of the following atoms or ions, give the number of protons and the number of electrons in one particle. (a) an O atom, (b) an O^{2-} ion, (c) an Al^{3+} ion, (d) an Al atom.

2.7 ▶ The formulas of many ionic compounds can be predicted

In the preceding section we noted that metals combine with nonmetals to form ionic compounds. In such reactions, metal atoms lose one or more electrons to become positively charged ions and nonmetal atoms gain one or more electrons to become negatively charged ions. In referring to these particles, we will frequently call a positively charged ion a **cation** (pronounced *CAT-i-on*) and a negatively charged ion an **anion** (pronounced *AN-i-on*).[5] Thus, solid NaCl is composed of sodium cations and chloride anions.

◤TOOLS▶
Periodic table and formulas for ions

Ions formed by representative metals and nonmetals can be remembered using the periodic table

In Section 2.3 you saw that the periodic table can be helpful in remembering the formulas of the nonmetal hydrides. It can also help us remember the kinds of ions formed by many of the representative elements (elements in the A groups of the

---

[5]The names *cation* and *anion* come from the way the ions behave when electrically charged metal plates called *electrodes* are dipped into a solution that contains them. We will discuss this in detail in Chapter 21.

TABLE 2.3	SOME IONS FORMED FROM THE REPRESENTATIVE ELEMENTS					
			Group Number			
IA	IIA	IIIA	IVA	VA	VIA	VIIA
Li^+	Be^{2+}		C^{4-}	N^{3-}	O^{2-}	F^-
Na^+	Mg^{2+}	Al^{3+}	Si^{4-}	P^{3-}	S^{2-}	Cl^-
K^+	Ca^{2+}				Se^{2-}	Br^-
Rb^+	Sr^{2+}				Te^{2-}	I^-
Cs^+	Ba^{2+}					

periodic table). For example, except for hydrogen, the neutral atoms of the Group IA elements always lose one electron each when they react, thereby becoming ions with a charge of 1+. Similarly, atoms of the Group IIA elements always lose two electrons when they react, so these elements always form ions with a charge of 2+. In Group IIIA, the only important positive ion we need consider now is that of aluminum, Al^{3+}; an aluminum atom loses three electrons when it reacts to form this ion.

Positive ions are formed by metals.

All these ions are listed in Table 2.3. *Notice that the number of positive charges on each of the cations is the same as the group number when we use the North American numbering of groups in the periodic table.* Thus, sodium is in Group IA and forms an ion with a 1+ charge, barium (Ba) is in Group IIA and forms an ion with a 2+ charge, and aluminum is in Group IIIA and forms an ion with a 3+ charge. Although this generalization doesn't work for all the metallic elements (it doesn't work for the transition elements, for instance), it does help us remember what happens to the metallic elements of Groups IA and IIA and aluminum when they react.

Among the nonmetals on the right side of the periodic table we also find some useful generalizations. For example, when they combine with metals, the halogens (Group VIIA) form ions with a 1− charge and the nonmetals in Group VIA form ions with a 2− charge. Notice that *the number of negative charges on the anion is equal to the number of spaces to the right that we have to move in the periodic table to get to a noble gas.*

Negative ions are formed by the nonmetals when they combine with metals.

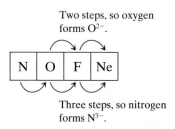

Two steps, so oxygen forms O^{2-}.

| N | O | F | Ne |

Three steps, so nitrogen forms N^{3-}.

You've probably also noticed that the number of negative charges on an anion is the same as the number of hydrogens in the simple nonmetal hydride of the element.

Writing formulas for ionic compounds follows certain rules

All chemical compounds are electrically neutral, so the ions in an ionic compound always occur in a ratio such that the total positive charge is equal to the total negative charge. This is why the formula for sodium chloride is NaCl; the l-to-l ratio of Na^+ to Cl^- gives electrical neutrality. In addition, as we've already mentioned, discrete molecules do not exist in ionic compounds, so we always use the set of smallest whole numbers that specifies the correct ratio of the ions. The following, therefore, are the rules we use in writing the formulas of ionic compounds.

A substance is electrically neutral, with a net charge of zero, if the total positive charge equals the total negative charge.

TOOLS

Rules for writing formulas of ionic compounds

Rules for Writing Formulas of Ionic Compounds

1. The positive ion is given first in the formula. (This isn't required by nature, but it is a custom we always follow.)

2. The subscripts in the formula must produce an electrically neutral formula unit. (Nature *does* require electrical neutrality.)

3. The subscripts should be the set of smallest whole numbers possible.

4. The charges on the ions are not included in the finished formula for the substance.

EXAMPLE 2.3

Writing Formulas for Ionic Compounds

Write the formulas for the ionic compounds formed from (a) Ba and S, (b) Al and Cl, and (c) Al and O.

ANALYSIS: To write the formula correctly, we have to apply the rules above, so it's important to know them. First, we need to figure out the formulas of the ions. Because we're working with representative elements, we can use the periodic table to deduce the charges.

SOLUTION: (a) The element Ba is in Group IIA, so the charge on its ion is 2+. Sulfur is in Group VIA, so its ion has a charge of 2−. Therefore, the ions are Ba^{2+} and S^{2-}. Since the charges are equal but opposite, a 1-to-1 ratio will give a neutral formula unit. Therefore, the formula is BaS. Notice that we have *not* included the charges on the ions in the finished formula.

(b) By using the periodic table, the ions of these elements are Al^{3+} and Cl^-. We can obtain a neutral formula unit by combining one Al^{3+} with three Cl^-. (The charge on Cl is 1−; the 1 is understood.)

$$1(3+) + 3(1-) = 0$$

The formula is $AlCl_3$.

(c) For these elements, the ions are Al^{3+} and O^{2-}. In the formula we seek, there must be the same number of positive charges as negative charges. This number must be a whole-number multiple of both 3 and 2. The smallest number that satisfies this condition is 6, so there must be two Al^{3+} and three O^{2-} in the formula.

$$
\begin{array}{ll}
2Al^{3+} & 2(3+) = 6+ \\
3O^{2-} & \underline{3(2-) = 6-} \\
& \text{sum} = 0
\end{array}
$$

The formula is Al_2O_3.

Are the Answers Reasonable?

In writing a formula, there are two things to check. First, be sure you've correctly written the formulas of the ions. (This is where students make a lot of mistakes.) Then check that you've combined them in a ratio that gives electrical neutrality.

PRACTICE EXERCISE 10: Write formulas for ionic compounds formed from (a) Na and F, (b) Na and O, (c) Mg and F, and (d) Al and C.

There is another rather simple way to obtain the formulas of the compounds in Example 2.3. The procedure is to make the subscript for one ion equal to the number of charges on the other. For example, for Al^{3+} and Cl^-, we can write

which gives Al_1Cl_3, or simply $AlCl_3$.

For the ions Al^{3+} and O^{2-}, we write

$$Al^{③+} \quad\longleftrightarrow\quad O^{②-}$$

This gives the formula Al_2O_3. (Notice that we've been careful to omit the charges in writing the formula for the compound.)

If you are not careful, you can be fooled in following this procedure. For example, if you apply this method to the compound formed from the ions Ba^{2+} and S^{2-}, it gives the formula Ba_2S_2. However, by convention, we always choose the smallest whole-number ratio of ions (Rule 3). Notice in Ba_2S_2 that both subscripts are divisible by 2. Therefore, to obtain the correct formula, we reduce the subscripts to the set of smallest whole numbers, which gives BaS. Exercising appropriate care, you might go back and try this method on Practice Exercise 10.

Many of our most important chemicals are ionic compounds. We have mentioned NaCl, common table salt, and $CaCl_2$, which is a substance often used to melt ice on walkways in the winter. Other examples are sodium fluoride, NaF, used by dentists to give fluoride treatments to teeth, and calcium oxide, CaO, an important ingredient in cement.

Transition and post-transition metals form more than one cation

The transition elements are located in the center of the periodic table, from Group IIIB on the left to Group IIB on the right. All of them lie to the left of the metalloids, and they all are metals. Included here are some of our most familiar metals, including iron, chromium, copper, silver, and gold.

Most of the transition metals are much less reactive than the metals of Groups IA and IIA, but when they react they also transfer electrons to nonmetal atoms to form ionic compounds. However, the charges on the ions of the transition metals do not follow as straightforward a pattern as do those of the alkali and alkaline earth metals. One of the characteristic features of the transition metals is the ability of many of them to form more than one positive ion. Iron, for example, can form two different ions, Fe^{2+} and Fe^{3+}. This means that iron can form more than one compound with a given nonmetal. For example, with chloride ion, Cl^-, iron forms two compounds, with the formulas $FeCl_2$ and $FeCl_3$. With oxygen, we find the compounds FeO and Fe_2O_3. As usual, we see that the formulas contain the ions in a ratio that gives electrical neutrality. Some of the most common ions of the transition metals are given in Table 2.4. Notice that one of the ions of mercury is diatomic Hg_2^{2+}. It consists of two Hg^+ ions joined by the same kind of bond found in

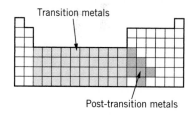

Transition metals

Post-transition metals

Distribution of transition and post-transition metals in the periodic table.

TABLE 2.4	IONS OF SOME TRANSITION METALS AND POST-TRANSITION METALS		
Transition Metals			
Chromium	Cr^{2+}, Cr^{3+}	Zinc	Zn^{2+}
Manganese	Mn^{2+}, Mn^{3+}	Silver	Ag^+
Iron	Fe^{2+}, Fe^{3+}	Cadmium	Cd^{2+}
Cobalt	Co^{2+}, Co^{3+}	Gold	Au^+, Au^{3+}
Nickel	Ni^{2+}	Mercury	Hg_2^{2+}, Hg^{2+}
Copper	Cu^+, Cu^{2+}		
Post-transition Metals			
Tin	Sn^{2+}, Sn^{4+}		
Lead	Pb^{2+}, Pb^{4+}		
Bismuth	Bi^{3+}		

molecular substances. The simple Hg^+ ion does not exist. You should be sure to learn the ions in Table 2.4.

> **PRACTICE EXERCISE 11:** Write formulas for the chlorides and oxides formed by (a) chromium and (b) copper.

The prefix post *means "after."*

The **post-transition metals** are those metals that occur in the periodic table immediately following a row of transition metals. The two most common and important ones are tin (Sn) and lead (Pb). These post-transition metals are quite different from the metals that precede the transition metals. One of the most significant differences is their ability to form two different ions and therefore two different compounds with a given nonmetal. For example, tin forms two oxides, SnO and SnO_2. Lead also forms two oxides that have similar formulas, PbO and PbO_2. The ions that these metals form are also included in Table 2.4.

A substance is **diatomic** if it is composed of molecules that contain only two atoms. It is a **binary compound** if it contains two different elements, regardless of the number of each. Thus, BrCl is a binary compound and is also diatomic; CH_4 is a binary compound but is not diatomic.

In general, polyatomic ions are not formed by the direct combination of elements. They are the products of reactions between compounds.

Many compounds contain ions composed of more than one element

The metal compounds that we have discussed so far have been *binary compounds*—compounds formed from *two* different elements. There are many other ionic compounds that contain more than two elements. These substances usually contain **polyatomic ions,** which are ions that are themselves composed of two or more atoms linked by the same kinds of bonds that hold molecules together. Polyatomic ions differ from molecules, however, in that they contain either too many or too few electrons to make them electrically neutral. Table 2.5 lists some important polyatomic ions. The formulas of compounds formed from them are determined in the same way as are those of binary ionic compounds; the ratio of the ions must be such that the formula unit is electrically neutral, and the set of smallest whole-number subscripts is used.

TABLE 2.5	FORMULAS AND NAMES OF SOME POLYATOMIC IONS		
Ion	Name (Alternate Name in Parentheses)	Ion	Name (Alternate Name in Parentheses)
NH_4^+	ammonium ion	CO_3^{2-}	carbonate ion
H_3O^+	hydronium ion[a]	HCO_3^-	hydrogen carbonate ion (bicarbonate ion)[b]
OH^-	hydroxide ion	SO_3^{2-}	sulfite ion
CN^-	cyanide ion	HSO_3^-	hydrogen sulfite ion (bisulfite ion)[b]
NO_2^-	nitrite ion	SO_4^{2-}	sulfate ion
NO_3^-	nitrate ion	HSO_4^-	hydrogen sulfate ion (bisulfate ion)[b]
ClO^-	hypochlorite ion (often written OCl^-)	SCN^-	thiocyanate ion
ClO_2^-	chlorite ion	$S_2O_3^{2-}$	thiosulfate ion
ClO_3^-	chlorate ion	CrO_4^{2-}	chromate ion
ClO_4^-	perchlorate ion	$Cr_2O_7^{2-}$	dichromate ion
MnO_4^-	permanganate ion	PO_4^{3-}	phosphate ion
$C_2H_3O_2^-$	acetate ion	HPO_4^{2-}	monohydrogen phosphate ion
$C_2O_4^{2-}$	oxalate ion	$H_2PO_4^-$	dihydrogen phosphate ion

[a]You will only encounter this ion in aqueous solutions.
[b]You will often see and hear the alternate names for these ions.

One of the minerals responsible for the strength of bones is the ionic compound calcium phosphate, which is formed from Ca^{2+} and PO_4^{3-}. Write the formula for this compound.

ANALYSIS: We already have the formulas of the ions involved, so we only need to remember the rules we use to assemble them into the correct formula of the compound.

SOLUTION: As before, we write the positive ion first and then make the number of charges on one ion equal to the subscript for the other.

$$Ca^{②+} \qquad PO_4^{③-}$$

The formula is written with parentheses to show that the PO_4^{3-} ion occurs two times in the formula unit.

$$Ca_3(PO_4)_2$$

Is the Answer Reasonable?
Double-check to see that electrical neutrality is achieved for the compound.

PRACTICE EXERCISE 12: Write the formula for the ionic compound formed from (a) Na^+ and CO_3^{2-}, (b) NH_4^+ and SO_4^{2-}, (c) potassium ion and acetate ion, (d) strontium ion and nitrate ion, and (e) Fe^{3+} and acetate ion.

> **EXAMPLE 2.4**
>
> **Formulas That Contain Polyatomic Ions**

Polyatomic ions are found in a large number of very important compounds. Examples include $CaSO_4$ (calcium sulfate, in plaster of Paris), $NaHCO_3$ (sodium bicarbonate, also called baking soda), $NaOCl$ (sodium hypochlorite, in liquid laundry bleach), $NaNO_2$ (sodium nitrite, a meat preservative), $MgSO_4$ (magnesium sulfate, also known as Epsom salts), and $NH_4H_2PO_4$ (ammonium dihydrogen phosphate, a fertilizer).

2.8 ▶ Naming ionic compounds also follows a system

In naming ionic compounds, our goal is the same as in naming molecular substances — we want a name that someone else could use to reconstruct the formula. The system we use here, however, is somewhat different than for molecular compounds.

Rules for naming ionic compounds

Binary compounds containing a metal and a nonmetal

For ionic compounds, the name of the cation is given first, followed by the name of the anion. This is the same as the sequence in which the ions appear in the formula. If the metal in the compound forms only one cation, such as Na^+ or Ca^{2+}, the cation is specified by just giving the English name of the metal. The anion in a binary compound is formed from the nonmetal and its name is created by adding the suffix -ide to the stem of the name for the nonmetal. A familiar example is NaCl, sodium chloride. Table 2.6 lists some common **monatomic** (one-atom) negative ions and their names. (Notice that the names for the anions of the nonmetals are the same as the names of the nonmetals in binary molecular compounds of two nonmetals.) It is

TABLE 2.6	MONATOMIC NEGATIVE IONS						
H^-	hydride	N^{3-}	nitride	O^{2-}	oxide	F^-	fluoride
C^{4-}	carbide	P^{3-}	phosphide	S^{2-}	sulfide	Cl^-	chloride
Si^{4-}	silicide	As^{3-}	arsenide	Se^{2-}	selenide	Br^-	bromide
				Te^{2-}	telluride	I^-	iodide

also useful to know that the -ide suffix is usually used only for monatomic ions, with just two common exceptions—hydroxide ion (OH^-) and cyanide ion (CN^-).[6]

To form the name of an ionic compound, we simply specify the names of the cation and anion. *It is not necessary to use prefixes to specify the number of cations and anions.* The reason is that once we know what the ions are, we can assemble the formula correctly just by taking them in a ratio that gives electrical neutrality. Consider, for example, the compound named magnesium nitride. The cation in the compound is magnesium and the anion is nitride. Magnesium is in Group IIA, so the magnesium ion has a charge of 2+ and its symbol is Mg^{2+}. The nitride ion is formed from a nitrogen atom, and from the position of this element in the periodic table, the anion is N^{3-}. Using the number of charges on one ion as the subscript of the other, the formula for magnesium nitride is Mg_3N_2.

> To keep the name as simple as possible, we give the minimum amount of information necessary to be able to reconstruct the formula. To write the formula of an ionic compound, we only need the formulas of the ions.

EXAMPLE 2.5
Naming Compounds and Writing Formulas

(a) What is the name of $SrBr_2$? (b) What is the formula for aluminum selenide?

ANALYSIS: The first step in naming a compound is to determine the type of compound it is. As you've already seen, there are slightly different rules for ionic and molecular substances. What clues do we have that these are ionic?
(a) The element Sr is a metal and Br is a nonmetal. Compounds of a metal and nonmetal are ionic, so we use the rules for naming ionic compounds.
(b) Aluminum is a metal. The only kind of compound of metals that we've discussed are ionic, so we'll proceed on that assumption. The -ide ending of selenide suggests the anion is composed of a single atom of a nonmetal. The only one that begins with the letters "selen-" is selenium, Se.

SOLUTION: (a) The compound is composed of the ions Sr^{2+} (an element from Group IIA) and Br^- (an element from Group VIIA). The cation simply takes the name of the metal, which is strontium. The anion's name is derived from bromine by replacing -ine with -ide; it is the bromide ion. The name of the compound is strontium bromide.
(b) The name tells us that the cation is the aluminum ion, Al^{3+}. The anion is formed from selenium (Group VIA), and its formula is Se^{2-}. The correct formula must represent an electrically neutral formula unit. Using the number of charges on one ion as the subscript of the other, the formula is Al_2Se_3.

Are the Answers Reasonable?
There's not much to check here. We can review the analysis and check to be sure we've applied the correct rules, which we have.

When a metal can form more than one ion, the charge is indicated with a Roman numeral

Many of the transition metals and post-transition metals are able to form more than one positive ion. Iron, a typical example, forms ions with either a 2+ or a 3+ charge (Fe^{2+} or Fe^{3+}). Compounds that contain these different iron ions have different formulas, so in their names it is necessary to specify which iron ion is present.

The currently preferred method for naming ions of metals that can have more than one charge in compounds is called the **Stock system.** Here we use the English name followed, *without a space,* by the numerical value of the charge written as a Roman numeral in parentheses.[7]

> Alfred Stock (1876–1946), a German inorganic chemist, was one of the first scientists to warn the public of the dangers of mercury poisoning.

---

[6]If the name of a compound ends in -ide and it isn't either a hydroxide or a cyanide, you can feel confident the substance is a binary compound.

[7]Compounds containing lead(IV), Pb^{4+}, are not very common, so compounds of Pb^{2+} are often named without using the Stock system. Thus, $PbCl_2$ is usually just called lead chloride, and lead iodide would be taken to be PbI_2. Similarly, silver and nickel are almost always found in compounds as Ag^+ and Ni^{2+}, respectively. Therefore, AgCl and $NiCl_2$ are almost always called simply silver chloride and nickel chloride.

Fe^{2+}	iron(II)	$FeCl_2$	iron(II) chloride
Fe^{3+}	iron(III)	$FeCl_3$	iron(III) chloride
Cr^{2+}	chromium(II)	CrS	chromium(II) sulfide
Cr^{3+}	chromium(III)	Cr_2S_3	chromium(III) sulfide

Remember that *the Roman numeral equals the positive charge on the metal ion*; it is not necessarily a subscript in the formula. For example, copper forms two oxides, one containing the Cu^+ ion and the other containing the Cu^{2+} ion. Their formulas are Cu_2O and CuO, and their names are as follows[8]:

Cu^+	copper(I)	Cu_2O	copper(I) oxide
Cu^{2+}	copper(II)	CuO	copper(II) oxide

These copper compounds illustrate that in deriving the formula from the name, you must figure out the formula from the ionic charges, as discussed in Section 2.7 and illustrated in the preceding example.

The compound $MnCl_2$ has a number of commercial uses, including disinfecting, in the manufacture of batteries, and in purifying natural gas. What is the name of the compound?

ANALYSIS: The first step is to determine the kind of compound involved so we can apply the appropriate rules. The compound here is made up of a metal and a nonmetal, so we use the rules for naming ionic compounds.

To name the cation, we need to know whether the metal is one that forms more than one positive ion. Manganese (Mn) is a transition element, and transition elements often do form more than one cation, so we should apply the Stock method. To do this, we need to determine the charge on the manganese cation. We can figure this out because the sum of the charges on the manganese and chlorine ions must equal zero and because chlorine forms only one negative ion.

SOLUTION: The anion of chlorine (the chloride ion) is Cl^-, so the total charge supplied by two Cl^- is $2-$. Therefore, for $MnCl_2$ to be electrically neutral, the Mn ion must carry a charge of $2+$. The cation is named as manganese(II), and the name of the compound is manganese(II) chloride.

Is the Answer Reasonable?
Performing a quick check of the arithmetic assures us we've got the correct charges on the ions. Everything appears to be okay.

EXAMPLE 2.6

Naming Compounds and Writing Formulas

What is the formula for cobalt(III) fluoride?

ANALYSIS: To answer this question, we first need to determine the formulas of the two ions. Then we assemble them into a chemical formula, being sure to achieve an electrically neutral formula unit.

SOLUTION: Cobalt(III) corresponds to Co^{3+}. The fluoride ion is F^-. To obtain an electrically neutral substance, we must have three F^- ions for each Co^{3+} ion, so the formula is CoF_3.

Is the Answer Reasonable?
We can check to see that we have the correct formulas of the ions and that we've combined them to achieve an electrically neutral formula unit. This will tell us we've obtained the correct answer.

EXAMPLE 2.7

Naming Compounds and Writing Formulas

---

[8]For some metals, such as copper and lead, one of their ions is much more commonly found in compounds than any of their others. For example, most common copper compounds contain Cu^{2+} and most common lead compounds contain Pb^{2+}. For compounds of these metals, if the charge is not indicated by a Roman numeral, we assume the ion present has a $2+$ charge. Thus, it is not unusual to find $PbCl_2$ called lead chloride, or for $CuCl_2$ to be called copper chloride.

FACETS OF CHEMISTRY 2.1

Another System for Naming Ionic Compounds

The Stock system, which we use to name compounds of metals able to form more than one positive ion, is a relatively recent development in inorganic nomenclature. A slightly different method existed before it.

In the older system, the suffix -ous is used to specify the ion with the lower charge and the suffix -ic is used to specify the ion with the higher charge. With this method, we also use the Latin stem for elements whose symbols are derived from their Latin names. Some examples are

Fe^{2+}	ferrous ion	$FeCl_2$	ferrous chloride
Fe^{3+}	ferric ion	$FeCl_3$	ferric chloride
Cu^+	cuprous ion	$CuCl$	cuprous chloride
Cu^{2+}	cupric ion	$CuCl_2$	cupric chloride

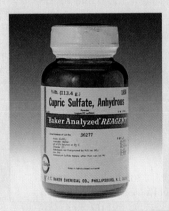

The older system of nomenclature is still found on the labels of many laboratory chemicals. This bottle contains copper(II) sulfate, which according to the older system of nomenclature is called cupric sulfate.

One difficulty with this method is that it does not specify what the charges on the metal ions are, so it becomes necessary to memorize the ions formed by the metals. Additional examples are given in the table below. Notice that mercury is an exception; we use the English stem when naming its ions.

Even though the Stock system is now preferred, some chemical companies still label bottles of chemicals using the old system. These old names also appear in the older scientific literature, which still holds much excellent data. This means that you may need to learn both systems.

Cr^{2+}	chromous	Mn^{2+}	manganous
Cr^{3+}	chromic	Mn^{3+}	manganic
Hg_2^{2+}	mercurous	Sn^{2+}	stannous
Hg^{2+}	mercuric	Sn^{4+}	stannic
Fe^{2+}	ferrous	Co^{2+}	cobaltous
Fe^{3+}	ferric	Co^{3+}	cobaltic
Pb^{2+}	plumbous	Au^+	aurous
Pb^{4+}	plumbic	Au^{3+}	auric

PRACTICE EXERCISE 13: Name the compounds K_2S, Mg_3P_2, $NiCl_2$, and Fe_2O_3. Use the Stock system where appropriate.

PRACTICE EXERCISE 14: Write formulas for (a) aluminum sulfide, (b) strontium fluoride, (c) titanium(IV) oxide, and (d) gold(III) oxide.

Names of ionic compounds that contain polyatomic ions

It is important that you learn the formulas (including charges) and the names of the polyatomic ions in Table 2.5. You will encounter them frequently throughout your chemistry course.

The extension of the nomenclature system to include ionic compounds containing polyatomic ions is straightforward. Most of the polyatomic ions listed in Table 2.5 are anions and their names are used as the second word in the name of the compound. For example, Na_2SO_4 contains the sulfate ion, SO_4^{2-}, and is called sodium sulfate. Similarly, $Cr(NO_3)_3$ contains the nitrate ion, NO_3^-. Chromium is a transition element, and in this compound its charge must be $3+$ to balance the negative charges of three NO_3^- ions. Therefore, $Cr(NO_3)_3$ is called chromium(III) nitrate.

Among the ions in Table 2.5, the only cation that forms compounds which can be isolated is ammonium ion, NH_4^+. It forms compounds such as NH_4Cl (ammonium chloride) and $(NH_4)_2SO_4$ (ammonium sulfate). Notice that the latter compound is composed of two polyatomic ions.

What is the name of $Mg(ClO_4)_2$, a compound used commercially for removing moisture from gases?

ANALYSIS: To answer this question, it is essential that you recognize that the compound contains a polyatomic ion. If you've learned the contents of Table 2.5, you know that ClO_4 is the formula (without the charge) of the perchlorate ion. With the charge, the ion's formula is ClO_4^-. We also have to decide whether we need to apply the Stock system in naming the metal Mg.

SOLUTION: Magnesium is in Group IIA and forms only the ion Mg^{2+}. Therefore, we don't need to use the Stock system in naming the cation; it is named simply as "magnesium." The anion is perchlorate, so the name of $Mg(ClO_4)_2$ is magnesium perchlorate.

Is the Answer Reasonable?
We can check to be sure we've named the anion correctly, and we have. The metal is magnesium, which only forms Mg^{2+}. Therefore, the answer seems to be correct.

PRACTICE EXERCISE 15: What are the names of (a) Li_2CO_3 and (b) $Fe(OH)_3$?

PRACTICE EXERCISE 16: Write the formulas for (a) potassium chlorate and (b) nickel(II) phosphate.

> **EXAMPLE 2.8**
> **Naming Compounds and Writing Formulas**

Hydrates are named using Greek prefixes

Earlier we discussed compounds called hydrates, such as $CuSO_4 \cdot 5H_2O$. Usually, hydrates are ionic compounds whose crystals contain water molecules in fixed proportions relative to the ionic substance. To name them, we provide two pieces of information—the name of the ionic compound and the number of water molecules in the formula. The number of water molecules is specified using the Greek prefixes mentioned earlier (mono-, di-, tri-, etc.), which precede the word *hydrate*. Thus, $CuSO_4 \cdot 5H_2O$ is named as "copper sulfate *pentahydrate*." Similarly, $CaSO_4 \cdot 2H_2O$ is named "calcium sulfate dihydrate," and $FeCl_3 \cdot 6H_2O$ is "iron(III) chloride hexahydrate."[9]

Naming a compound requires that we select the rules that apply

In this chapter we've discussed how to name two classes of compounds, molecular and ionic, and you saw that slightly different rules apply to each. To name chemical compounds successfully, therefore, it is necessary to determine which set of rules applies. This is not too difficult now, but as we study more kinds of substances, additional rules will be introduced and the problem of selecting the correct set will become more complex. Therefore, while things are still relatively simple, let's look at a systematic approach (Figure 2.14) that will lead you to the correct rules. The following example illustrates how to use the figure, and when working on the Review Questions, you may want to refer to Figure 2.14 until you are able to develop the skills that will enable you to work without it.

What is the name of (a) $CrCl_3$, (b) P_4S_3, and (c) NH_4NO_3?

ANALYSIS: For each compound, we will begin at the top of Figure 2.14 and proceed through the decision processes to arrive at the way to name the compound. As you read through the solution, be sure to refer to Figure 2.14 so you can see how we are led to the rules we need to apply.

> **EXAMPLE 2.9**
> **Applying the Rules for Naming Compounds**

[9]Chemical suppliers (who do not always follow current rules of nomenclature) sometimes indicate the number of water molecules using a number and a dash. For example, one supplier lists $Ca(NO_3)_2 \cdot 4H_2O$ as "Calcium nitrate, 4-hydrate."

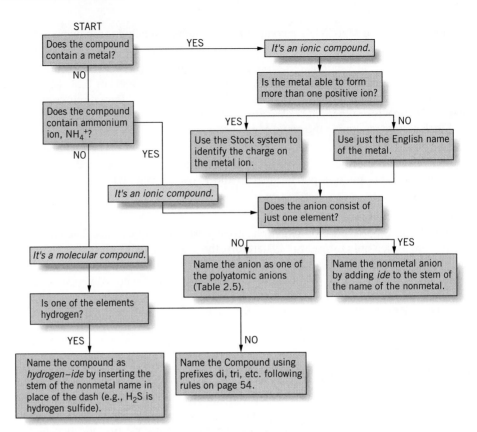

FIGURE 2.14 *Flowchart for naming molecular and ionic compounds.*

SOLUTION: (a) Starting at the top of Figure 2.14, we first determine that the compound contains a metal (Cr), so it's an ionic compound. Next, we see that the metal is a transition element, and chromium is one of those that forms more than one cation, so we have to apply the Stock method. To do this, we need to know what the charge is on the metal ion. We can figure this out using the charge on the anion and the fact that the compound must be electrically neutral overall. The anion is formed from chlorine, so its charge must be $1-$ (the anion is Cl^-). Therefore, the metal ion must be Cr^{3+}; we name the metal as *chromium(III)*. Next, we see that there is only one nonmetallic element in the compound, Cl, so the name of the anion ends in -ide; it's the *chloride* ion. The compound $CrCl_3$ is therefore named *chromium(III) chloride*.

(b) Once again, we start at the top. First we determine that the compound doesn't contain a metal. It also doesn't contain NH_4, so the compound is molecular. It doesn't contain hydrogen, so we are led to the decision that we must use Greek prefixes to specify the numbers of atoms of each element. Applying the procedure on page 54, the name of the compound P_4S_3 is *tetraphosphorus trisulfide*.

(c) We begin at the top. Studying the formula, we see that it does not contain the symbol for a metal, so we proceed down the left side of the figure. The formula does contain NH_4, which indicates the compound contains the ammonium ion, NH_4^+ (it's an ionic compound). The rest of the formula is NO_3, which consists of more than one atom. This suggests the polyatomic anion NO_3^- (*nitrate* ion). The name of the compound NH_4NO_3 is *ammonium nitrate*.

Are the Answers Reasonable?

To check the answers in a problem of this kind, review the decision processes that led you to the names. In part b, you can check to be sure you've calculated the charge on the chromium ion correctly. Also, check to be sure you've used the correct names of any polyatomic ions. Doing these things will show you've named the compounds correctly.

PRACTICE EXERCISE 17: The compound I_2O_5 is used in respirators, where it serves to react with highly toxic carbon monoxide to give the much less toxic gas carbon dioxide. What is the name of I_2O_5?

PRACTICE EXERCISE 18: The compound $Cr(C_2H_3O_2)_3$ is used in the tanning of leather. What is the name for this compound?

2.9 ▶ Molecular and ionic compounds have characteristic properties

Molecular and ionic substances have a variety of physical properties that are strongly influenced by the attractive forces that exist between the particles present within them. Four such properties, for example, are hardness, brittleness (tendency to shatter into smaller particles when struck), melting point, and electrical conductivity.

Hardness and brittleness are influenced by attractions between particles

Molecular substances, such as paraffin wax or car wax, tend to be soft and easily crushed or deformed. The reason can be seen if we examine the internal structures of these substances. Within individual molecules the atoms are held to each other very strongly by chemical bonds, but between neighboring molecules the attractions are very weak. Because the molecules are not attracted strongly to each other, they are able to slide past each other easily, and a molecular substance is easily deformed.

Compared to molecular compounds, ionic substances such as sodium chloride are much harder and tend to be brittle. These characteristics are caused by the nature of the attractive and repulsive forces between ions. Ions of opposite charge strongly attract each other, while ions of the same charge strongly repel. Therefore, within an ionic compound the ions arrange themselves so that the attractions between oppositely charged ions are at a maximum and the repulsions between like-charged ions are at a minimum. Overall, the attractions outweigh the repulsions and the particles are held together, so the solid tends to be hard.

The forces between ions can also be used to explain the brittle nature of many ionic compounds. For example, when struck by a hammer, a salt crystal shatters into many small pieces. A view at the atomic level reveals how this could occur (Figure 2.15). The slight movement of a layer of ions within an ionic crystal suddenly places ions of the *same* charge next to each other, and for that instant there are large repulsive forces that split the solid.

Melting points are influenced by attractions between particles

In general, molecular compounds composed of small molecules (molecules with relatively few atoms) tend to have low melting points, whereas ionic compounds composed of similarly sized ions have much higher melting points. Water and table salt illustrate the point. Water is a molecular substance with a low melting point (mp = 0 °C), whereas salt is an ionic compound that melts at over 800 °C. The differences in melting points are also related to the strengths of the attractions between particles in these substances. To understand this, we need to look at what's involved in causing a solid to melt.

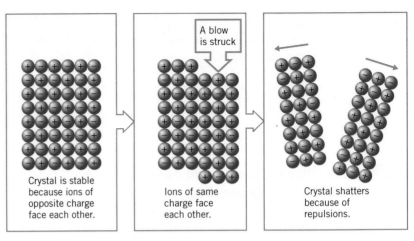

FIGURE 2.15 *An ionic crystal shatters when struck.* In the photo on the left, we see what happens when a crystal of sodium chloride is struck with a hammer. In the microview, we see that striking the crystal causes some of the layers to shift. This can bring ions of like charge face-to face. The repulsions between the ions can then force parts of the crystal apart, causing the crystal to shatter.

In a solid, the particles jiggle around within a cage of other particles, held in place by attractions between the particles. As the temperature of the solid is raised, the average kinetic energy of the particles increases and the particles bounce around more violently. Eventually, the violent motions allow the particles to overcome the attractions and the solid melts.

The molecules of molecular substances such as water and candle wax experience weak attractions to their neighbors, so they don't have to bounce around inside their crystals very violently to overcome the attractive forces and become a liquid. As a result, molecular compounds have low melting points.

In a crystal of an ionic compound such as salt, the attractions are very strong between the oppositely charged ions. For ions to overcome the attractions, they must fly around very violently, which means they have to be heated to a very high temperature. As a result, the melting point of an ionic compound is quite high. This is why nearly all ionic compounds are solids at room temperature and tend to have high melting points. In fact, some ionic substances melt only at extremely high temperatures. For instance, aluminum oxide, Al_2O_3, melts at about 2000 °C, and for this reason it is made into special bricks that are used to line the inside walls of furnaces.

Electrical conductivity requires charges that can move

Recall that we described the electrical conductivity of melted sodium chloride as evidence for the existence of ions.

In the solid state, ionic compounds do not conduct electricity. This is because electrical conductivity requires the movement of electrical charges, and in the solid state the attractive forces prevent the movement of ions through the crystal. When the solid is melted, however, the ions become free to move about and the liquid conducts electricity well. This can be demonstrated experimentally with the help of the apparatus depicted in Figure 2.16, which consists of a pair of metal electrodes that can be dipped into a container holding a substance whose electrical conductivity we wish to test. One of the electrodes is wired to an electric light bulb that will glow if electricity is able to pass between the two electrodes. When this apparatus is used to test the electrical conductivity of solid salt crystals, the bulb fails to light. However, when a flame is applied, the bulb lights brightly as soon as the salt melts.

Molecular substances differ from ionic compounds in their electrical characteristics. Molecules are uncharged particles, so they do not conduct electricity in the solid state or when melted. An example is pure water, which as we noted earlier, does not conduct electricity.

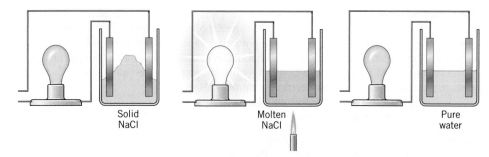

Solid NaCl

Molten NaCl

Pure water

FIGURE 2.16 *An apparatus to test for electrical conductivity.* The electrodes are dipped into the substance to be tested. If the light bulb glows when electricity is applied, the sample is an electrical conductor. Here we see that solid sodium chloride does not conduct electricity, but when the solid is melted it does conduct. Liquid water, a molecular compound, is not a conductor of electricity because it does not contain electrically charged particles.

PRACTICE EXERCISE 19: A certain element X forms a compound with chlorine that has the formula XCl_2. The compound melts at $-78\ °C$ to yield a liquid that doesn't conduct electricity. Is X a metal or a nonmetal? Explain your reasoning.

SUMMARY

Reactions of Elements to Form Compounds Nearly every element has the ability to form compounds, although not all combinations are possible. When a chemical reaction takes place, the properties of the substances present before the reaction disappear and are replaced by the properties of the substances formed in the reaction. Chemical symbols are used to write **chemical formulas,** both for **free elements** that occur as molecules (e.g., **diatomic molecules** of elements such as H_2, O_2, N_2, Cl_2) and for chemical compounds. Subscripts are used to specify how many atoms of each element are present. Some compounds form solids called **hydrates,** which contain water molecules in definite proportions. Heating a hydrate usually drives off the water.

Chemical Equations A **chemical equation** presents a before-and-after description of a chemical reaction. When **balanced,** an equation contains **coefficients** that makes the number of atoms of each kind the same among the **reactants** and the **products.** In this way, the equation conforms to the law of conservation of mass. The physical states of the reactants and products can be specified in an equation by placing the following symbols within parentheses following the chemical formulas: *s*, *l*, *g*, and *aq*, which stand for solid, liquid, gas, and aqueous solution (dissolved in water), respectively.

Energy and Chemical Change Chemical reactions are almost always accompanied by an absorption or release of energy. **Energy** is the capacity to do work. An object can have energy as **kinetic energy** (because of its motion) or as **potential energy** (stored energy). The potential energy stored in chemicals that can be released in chemical reactions is sometimes called **chemical energy.** The **law of conservation of energy** states that energy can be neither created nor destroyed, but only changed from one form to another. All forms of energy can be converted to heat. Heat and temperature are different: **heat** is energy that's transferred between objects because of a difference in their temperatures. **Temperature** is a measure of the *average* kinetic energy of the atoms in an object. When the temperature rises during a chemical reaction, the average kinetic energy of the chemicals increases and their potential energy (chemical energy) decreases.

Molecules and Molecular Compounds **Molecules** are electrically neutral particles consisting of two or more atoms. The erratic movements of microscopic particles suspended in a liquid **(Brownian motion)** can be interpreted as caused by collisions with molecules of the liquid. Molecules are held together by **chemical bonds** that arise from the sharing of electrons between atoms. Formulas we write for molecules are **molecular formulas.**

Molecular compounds are formed when nonmetals combine with each other. The simple nonmetal hydrides have formulas that can be remembered by the position of the nonmetal in the periodic table. **Organic compounds** are **hydrocarbons** or compounds considered to be derived from hydrocarbons by replacing H atoms with other atoms.

Naming Molecular Compounds The system of **nomenclature** for **binary** molecular **inorganic compounds** uses a set of Greek prefixes to specify the numbers of atoms of each kind in the formula of the compound. The first element in the formula is specified by its English name; the second element takes the suffix -ide, which is added to the stem of the English name. For simple nonmetal hydrides, it is not necessary to specify the number of hydrogens in the formula. Many familiar substances as well as very complex molecules are usually identified by **common names.**

Ions and Ionic Compounds Binary ionic compounds are formed when metals react with nonmetals. In the reaction, electrons are transferred from a metal to a nonmetal. The metal atom becomes a positive ion (a **cation**); the nonmetal atom becomes a negative ion (an **anion**). The formula of an ionic compound specifies the smallest whole-number ratio of the ions. The smallest unit of an ionic compound is called a **formula unit,** which specifies the smallest whole-number ratio of the ions that produces electrical neutrality. Many ionic compounds also contain **polyatomic ions**—ions that are composed of two or more atoms.

Naming Ionic Compounds In naming an ionic compound, the cation is specified first, followed by the anion. Metal cations take the English name of the element, and when more than one positive ion can be formed by the metal, the **Stock system** is used to identify the amount of positive charge on the cation. This is done by placing a Roman numeral equal to the positive charge in parentheses follow-

ing the name of the metal. Simple **monatomic** anions are formed by nonmetals and their names are formed by adding the suffix -ide to the stem of the nonmetal's name. Only two common polyatomic anions (cyanide and hydroxide) end in the suffix -ide.

Properties of Molecular and Ionic Compounds Molecular compounds tend to be soft, have low melting points, and do not conduct electricity in either the solid or liquid states. Ionic compounds tend to be relatively hard and shatter into small pieces when struck. They do not conduct electricity when solid, but when an ionic compound is melted, the liquid does conduct electricity. The differences in behavior between molecular and ionic substances can be related to the differing strengths of the attractions between the particles of which they're composed.

TOOLS ► **YOU HAVE LEARNED**

The following table lists the concepts that you've learned in this chapter which can be applied as tools in solving problems. Study each one carefully so that you know what each is used for. When faced with solving a problem, recall what each tool does and consider whether it will be helpful in finding a solution. This will aid you in selecting the tools you need. If necessary, refer to this table when working on the Thinking-It-Through problems and the Review Exercises that follow.

TOOL	HOW IT WORKS
Chemical formula (page 43)	Subscripts in a formula specify the number of atoms of each element in one formula unit of the substance. This gives us *atom ratios* that we will find useful when we deal with the compositions of compounds in Chapter 4.
Rules for naming molecular compounds (page 54)	They allow us to write the name of a molecular compound given its chemical formula and to write the formula given the name.
Rules for writing formulas of ionic compounds (page 60)	The rules permit us to write correct chemical formulas for ionic compounds. You will need to learn to use the periodic table (see below) to remember the charges on the cations and anions of the representative metals and nonmetals. You also should learn the ions formed by the transition and post-transition metals in Table 2.4, and it is essential that you learn the names and formulas (including charges) of the polyatomic ions in Table 2.5.
Rules for naming ionic compounds (page 63)	These rules allow us to write the name of an ionic compound given its chemical formula and to write the formula given the name.
Periodic Table (pages 51 and 58)	From a nonmetal's position in the periodic table we can write the formula of its simple hydride. For the nonmetals and the metals in Groups IA and IIA, we can use the elements' positions in the periodic table to obtain the charges on their ions.

THINKING IT THROUGH

Need extra help? **Visit the Brady/ Senese web site at** www.wiley.com/ college/brady

ON-LINE HELP

The goal for the following problems is not to find the answers themselves, but rather to assemble the information needed to solve them and explain how you would use the information to find the answers. The problems in Level 2 are more challenging than those in Level 1 and may contain more data than are required, in which case you are also asked to identify the unnecessary data. Detailed answers to the Thinking-It-Through problems can be found on the web site.

Level 1 Problems

1. Describe how you would determine the chemical formula for the compound formed from the element with atomic number 56 and the element with atomic number 35.

2. A certain compound formed by two elements is a gas at room temperature. When cooled to a very low temperature, it liquefies and then freezes to form a solid that is soft and easily crushed. How can you tell whether this compound is more likely to be molecular or ionic? What other property would be helpful in your decision?

3. In naming a compound, the first step is to determine whether the substance is ionic or molecular. What *two* clues would lead you to believe that P_4O_{10} is molecular?

4. How do we know that strontium selenide is an ionic compound?

5. How do we know that strontium selenate is not a binary compound?

Level 2 Problems

6. In an experiment, it was found that 3.50 g of phosphorus combines with 12.0 g of chlorine to form 15.5 g of the compound phosphorus trichloride, a substance used in making iridescent metallic coatings. Compared to one atom of phosphorus, how many times heavier is one atom of chlorine? (Provide the Analysis step in solving this problem without using atomic mass data found in the periodic table.)

7. A certain ionic compound formed from elements A and B has the formula AB_2. In a sample of this compound, the mass of B is 4.0 times the mass of A. Describe how you would use this information to determine the identities of elements A and B.

REVIEW QUESTIONS

Chemical Formulas

2.1 What are two ways to interpret a chemical symbol?

2.2 What is the difference between an atom and a molecule?

2.3 If a substance consists of molecules, must it be a compound, or can it be an element?

2.4 Write the formulas and names of the nonmetallic elements that exist in nature as diatomic molecules.

2.5 Epsom salts is a hydrate of magnesium sulfate, $MgSO_4 \cdot 7H_2O$. What is the formula of the substance that remains when Epsom salts is completely dehydrated?

Chemical Equations

2.6 What do we mean when we say a chemical equation is *balanced*? Why do we balance chemical equations?

2.7 For a chemical reaction, what do we mean by the term *reactants*? What do we mean by the term *products*?

2.8 What are the numbers called that are written in front of the formulas in a balanced chemical equation?

2.9 The combustion of a thin wire of magnesium metal (Mg) in an atmosphere of pure oxygen produces the brilliant light of a flashbulb, once commonly used in photography. After the reaction, a thin film of magnesium oxide is seen on the inside of the bulb. The equation for the reaction is

$$2Mg + O_2 \longrightarrow 2MgO$$

(a) State in words how this equation is read.
(b) Give the formula(s) of the reactants.
(c) Give the formula(s) of the products.
(d) Rewrite the equation to show that Mg and MgO are solids and O_2 is a gas.

2.10 The chemical equation for the combustion of octane (C_8H_{18}), a component of gasoline, is

$$2C_8H_{18} + 25O_2 \longrightarrow 16CO_2 + 18H_2O$$

Rewrite the equation so that it specifies gasoline and water as liquids and oxygen and carbon dioxide (CO_2) as gases.

Energy and Chemical Change

2.11 Give definitions for (a) energy and (b) work.

2.12 Define (a) *kinetic energy* and (b) *potential energy.* What two things determine how much kinetic energy an object possesses?

2.13 State the equation used to calculate an object's kinetic energy. Define the symbols used in the equation.

2.14 If a car increases its speed from 30 mph to 60 mph, by what factor does the kinetic energy of the car increase?

2.15 What is *chemical energy*?

2.16 How do *temperature* and *heat* differ?

2.17 A quart of boiling water will cause a more severe burn if it's spilled on you than just a drop of boiling water. Explain this in terms of the amounts of kinetic energy possessed by the water molecules in the two samples.

2.18 Which kind of energy is *chemical energy,* kinetic or potential?

2.19 State the law of conservation of energy. When a ball rolls down a hill, what happens to its potential energy? What happens to its kinetic energy? Are these observations consistent with the law of conservation of energy? (Explain.)

2.20 When a car brakes to a stop from a speed of 30 mph, it's observed that the brakes become hot. Explain this in terms of the kinetic energy the moving car possessed.

2.21 How is heat transferred from a hot object to a cold object?

2.22 If during a chemical reaction there is a reduction in the total amount of chemical energy, what happens to the temperature of the reacting chemicals? Why?

2.23 Some instant cold packs sold in pharmacies contain a bag of water inside a pouch of ammonium nitrate, NH_4NO_3. When the bag of water is broken and the two substances mixed, the ammonium nitrate dissolves and the solution becomes quite cold. What happens to the average kinetic energy of the substances in the cold pack? What happens to the chemical energy of the water and ammonium nitrate?

2.24 When gasoline burns, it reacts with oxygen in the air and forms hot gases consisting of carbon dioxide and water vapor. How does the potential energy of the gasoline and oxygen compare with the potential energy of the carbon dioxide and water vapor?

2.25 What is *thermal energy*?

Molecular Compounds of Nonmetals

2.26 Which are the only elements that exist as free, individual atoms when not chemically combined with other elements?

2.27 Write chemical formulas for the elements that normally exist in nature as diatomic molecules.

2.28 Which kind of elements normally combine to form molecular compounds?

2.29 Without referring to Table 2.1, but using the periodic table, write chemical formulas for the simplest hydrogen compounds of (a) carbon, (b) nitrogen, (c) tellurium, and (d) iodine.

2.30 The simplest hydrogen compound of phosphorus is phosphine, a highly flammable and poisonous compound with an odor of decaying fish. What is the formula for phosphine?

2.31 Astatine, a radioactive member of the halogen family, forms a compound with hydrogen. Predict its chemical formula.

2.32 Under appropriate conditions, tin can be made to form a simple molecular compound with hydrogen. Predict its formula.

2.33 Write the chemical formulas for (a) methane, (b) ethane, (c) propane, and (d) butane. Give one practical use for each of these hydrocarbons.

2.34 What are the formulas of (a) methanol and (b) ethanol?

2.35 On the basis of the molecular structure of methanol (Figure 2.10), predict the molecular structure of ethanol.

2.36 What is the formula for the alkane that has ten carbon atoms?

2.37 Candle wax is a mixture of hydrocarbons, one of which is an alkane with 23 carbon atoms. What is the formula for this hydrocarbon?

2.38 The formula for a compound is correctly given as $C_6H_{12}O_6$. State two reasons why we expect this to be a molecular compound, rather than an ionic compound.

Ionic Compounds

2.39 Describe what kind of event must occur (involving electrons) if the atoms of two different elements are to react to form (a) an ionic compound or (b) a molecular compound.

2.40 With what kind of elements do metals react?

2.41 With what kind of elements do nonmetals react?

2.42 Why are nonmetals found in more compounds than are metals, even though there are fewer nonmetals than metals?

2.43 In what major way do compounds formed between two nonmetals differ from those formed between a metal and a nonmetal?

2.44 What holds an ionic compound together? Can we identify individual molecules in an ionic compound?

2.45 What is an ion? How does it differ from an atom or a molecule?

2.46 Why do we use the term *formula unit* for ionic compounds instead of the term *molecule*?

2.47 Most compounds of aluminum are ionic, but a few are molecular. How do we know that Al_2Cl_6 is molecular?

2.48 Consider the sodium atom and the sodium ion.
(a) Write the chemical symbol of each.
(b) Do these particles have the same number of nuclei?
(c) Do they have the same number of protons?
(d) Could they have different numbers of neutrons?
(e) Do they have the same number of electrons?

2.49 Define *cation* and *anion*.

2.50 When calcium reacts with chlorine, each calcium atom loses two electrons and each chlorine atom gains one electron. What are the formulas of the ions created in this reaction?

2.51 How many electrons has a titanium atom lost if it has formed the ion Ti^{4+}? What are the total numbers of protons and electrons in a Ti^{4+} ion?

2.52 If an atom gains an electron to become an ion, what kind of electrical charge does the ion have?

2.53 How many electrons has a nitrogen atom gained if it has formed the ion N^{3-}? How many protons and electrons are in an N^{3-} ion?

2.54 What four rules do we apply when we write the formulas for ionic compounds?

2.55 What is wrong with the formula $RbCl_3$? What is wrong with the formula SNa_2?

2.56 A student wrote the formula for an ionic compound of titanium as Ti_2O_4. What is wrong with this formula? What should the formula be?

2.57 Which are the post-transition metals? Write their symbols. Write the symbols for the ions formed by the post-transition metals.

2.58 What are the formulas of the ions formed by (a) iron, (b) cobalt, (c) mercury, (d) chromium, (e) tin, and (f) copper?

2.59 Which of the following formulas are incorrect? (a) NaO_2, (b) $RbCl$, (c) K_2S, (d) Al_2Cl_3, (e) MgO_2.

2.60 What are the formulas (including charges) for (a) cyanide ion, (b) ammonium ion, (c) nitrate ion, (d) sulfite ion, (e) chlorate ion and (f) sulfate ion?

2.61 What are the formulas (including charges) for (a) hypochlorite ion, (b) bisulfate ion, (c) phosphate ion, (d) dihydrogen phosphate ion, (e) permanganate ion, and (f) oxalate ion?

2.62 What are the names of the following ions? (a) $Cr_2O_7^{2-}$, (b) OH^-, (c) $C_2H_3O_2^-$, (d) CO_3^{2-}, (e) CN^-, and (f) ClO_4^-.

2.63 From what you have learned in Section 2.7, write correct balanced equations for the reactions between (a) calcium and chlorine, (b) magnesium and oxygen, (c) aluminum and oxygen, and (d) sodium and sulfur.

Properties of Ionic and Molecular Compounds

2.64 What is the chief reason that ionic and molecular compounds have such different physical properties?

2.65 Why does molten KCl conduct electricity even though solid KCl does not?

2.66 Why are ionic compounds so brittle?

2.67 Why do most ionic compounds have high melting points? Naphthalene (moth flakes) consists of soft crystals that melt at a relatively low temperature of 80 °C. Does this substance have characteristics of an ionic or a molecular compound?

Naming Compounds

2.68 What is the difference between a binary compound and one that is diatomic? Give examples that illustrate this difference.

2.69 In naming the compounds discussed in this chapter, why is it important to know whether a compound is molecular or ionic?

2.70 In naming ionic compounds of the transition elements, why is it essential to know the charge on the anion?

REVIEW PROBLEMS

Answers to problems whose numbers are printed in color are given in Appendix B. More challenging questions are marked with asterisks.

Chemical Formulas

2.71 How many atoms of each element are represented in each of the following formulas? (a) $K_2C_2O_4$, (b) H_2SO_3, (c) $C_{12}H_{26}$, (d) $HC_2H_3O_2$, (e) $(NH_4)_2HPO_4$.

2.72 How many atoms of each kind are represented in the following formulas? (a) Na_3PO_4, (b) $Ca(H_2PO_4)_2$, (c) C_4H_{10}, (d) $Fe_3(AsO_4)_2$, (e) $C_3H_5(OH)_3$.

2.73 How many atoms of each kind are represented in the following formulas? (a) $Ni(ClO_4)_2$, (b) $CuCO_3$, (c) $K_2Cr_2O_7$, (d) CH_3CO_2H, (e) $(NH_4)_2HPO_4$.

2.74 How many atoms of each kind are represented in the following formulas? (a) $CH_3CH_2CO_2C_3H_7$, (b) $MgSO_4 \cdot 7H_2O$, (c) $KAl(SO_4)_2 \cdot 12H_2O$, (d) $Cu(NO_3)_2$, (e) $(CH_3)_3COH$.

2.75 The compound $Cr(C_2H_3O_2)_3$ is used in the tanning of leather. How many atoms of each element are given in this formula?

2.76 Asbestos, a known cancer-causing agent, has as a typical formula $Ca_3Mg_5(Si_4O_{11})_2(OH)_2$. How many atoms of each element are given in this formula?

Chemical Equations

2.77 How many atoms of each element are represented in each of the following expressions? (a) $3N_2O$, (b) $4NaHCO_3$, (c) $2CuSO_4 \cdot 5H_2O$

2.78 How many atoms of each element are represented in each of the following expressions? (a) $7CH_3CO_2H$, (b) $2(NH_2)_2CO$, (c) $5K_2Cr_2O_7$.

2.79 Consider the balanced equation

$$2Fe(NO_3)_3 + 3Na_2CO_3 \longrightarrow Fe_2(CO_3)_3 + 6NaNO_3$$

(a) How many atoms of Na are on each side of the equation?
(b) How many atoms of C are on each side of the equation?
(c) How many atoms of O are on each side of the equation?

2.80 Consider the balanced equation for the combustion of octane, a component of gasoline:

$$2C_8H_{18} + 25O_2 \longrightarrow 16CO_2 + 18H_2O$$

(a) How many atoms of C are on each side of the equation?
(b) How many atoms of H are on each side of the equation?
(c) How many atoms of O are on each side of the equation?

Ionic Compounds

2.81 Use the periodic table, but not Table 2.3, to write the symbols for the ions of (a) K, (b) Br, (c) Mg, (d) S, and (e) Al.

2.82 Use the periodic table, but not Table 2.3, to write the symbols for ions of (a) barium, (b) oxygen, (c) fluorine, (d) strontium, and (e) rubidium.

2.83 Write formulas for ionic compounds formed between (a) Na and Br, (b) K and I, (c) Ba and O, (d) Mg and Br, and (e) Ba and F.

2.84 Write the formulas for the ionic compounds formed by the following transition metals with the chloride ion, Cl^-: (a) chromium, (b) iron, (c) manganese, (d) copper, and (e) zinc.

2.85 Write formulas for the ionic compounds formed from (a) K^+ and nitrate ion, (b) Ca^{2+} and acetate ion, (c) ammonium ion and Cl^-, (d) Fe^{3+} and carbonate ion, and (e) Mg^{2+} and phosphate ion.

2.86 Write formulas for the ionic compounds formed from (a) Zn^{2+} and hydroxide ion, (b) Ag^+ and chromate ion, (c) Ba^{2+} and sulfite ion, (d) Rb^+ and sulfate ion, and (e) Li^+ and bicarbonate ion.

2.87 Write formulas for two compounds formed between O^{2-} and (a) lead, (b) tin, (c) manganese, (d) iron, and (e) copper.

2.88 Write formulas for the ionic compounds formed from Cl^- and (a) cadmium ion, (b) silver ion, (c) zinc ion, and (d) nickel ion.

Naming Compounds

2.89 Name the following molecular compounds:
(a) SiO_2 (b) XeF_4 (c) P_4O_{10} (d) Cl_2O_7.

2.90 Name the following molecular compounds:
(a) ClF_3 (b) S_2Cl_2 (c) N_2O_5 (d) $AsCl_5$.

2.91 Name the following ionic compounds: (a) CaS, (b) $AlBr_3$, (c) Na_3P, (d) Ba_3As_2, (e) Rb_2S.

2.92 Name the following ionic compounds: (a) NaF, (b) Mg_2C, (c) Li_3N, (d) Al_2O_3, (e) K_2Se.

2.93 Name the following ionic compounds using the Stock system.
(a) FeS (b) CuO (c) SnO_2 (d) $CoCl_2 \cdot 6H_2O$

2.94 Name the following ionic compounds using the Stock system.
(a) Mn_2O_3 (b) Hg_2Cl_2 (c) PbS (d) $CrCl_3 \cdot 4H_2O$

2.95 Name the following. If necessary, refer to Table 2.5 on page 62.
(a) $NaNO_2$ (c) $MgSO_4 \cdot 7H_2O$
(b) $KMnO_4$ (d) $KSCN$

2.96 Name the following. If necessary, refer to Table 2.5 on page 62.
(a) K_3PO_4 (c) $Fe_2(CO_3)_3$
(b) $NH_4C_2H_3O_2$ (d) $Na_2S_2O_3 \cdot 5H_2O$

2.97 Identify each of the following as molecular or ionic and give its name.
(a) $CrCl_2$ (f) P_4O_6
(b) S_2Cl_2 (g) $CaSO_3$
(c) $NH_4C_2H_3O_2$ (h) $AgCN$
(d) SO_3 (i) $ZnBr_2$
(e) KIO_3 (j) H_2Se

2.98 Identify each of the following as molecular or ionic and give its name.
(a) $V(NO_3)_3$ (f) K_2CrO_4
(b) $Co(C_2H_3O_2)_2$ (g) $Fe(OH)_2$
(c) Au_2S_3 (h) I_2O_4
(d) Au_2S (i) I_4O_9
(e) $GeBr_4$ (j) P_4Se_3

2.99 Write formulas for the following.
(a) sodium monohydrogen phosphate
(b) lithium selenide
(c) chromium(III) acetate
(d) disulfur decafluoride
(e) nickel(II) cyanide
(f) iron(III) oxide
(g) antimony pentafluoride

2.100 Write formulas for the following.
(a) dialuminum hexachloride
(b) tetraarsenic decaoxide
(c) magnesium hydroxide
(d) copper(II) bisulfate
(e) ammonium thiocyanate
(f) potassium thiosulfate
(g) diiodine pentaoxide

2.101 Write formulas for the following.
(a) ammonium sulfide
(b) chromium(III) sulfate hexahydrate
(c) silicon tetrafluoride
(d) molybdenum(IV) sulfide
(e) tin(IV) chloride
(f) hydrogen selenide
(g) tetraphosphorus heptasulfide

2.102 Write formulas for the following.
(a) mercury(II) acetate
(b) barium hydrogen sulfite
(c) boron trichloride
(d) calcium phosphide
(e) magnesium dihydrogen phosphate
(f) calcium oxalate
(g) xenon tetrafluoride

2.103 The compounds Se_2S_6 and Se_2S_4 have been shown to be antidandruff agents. What are their names?

2.104 The compound P_2S_5 is used to manufacture safety matches. What is the name of this compound?

ADDITIONAL EXERCISES

2.105 What are the formulas for mercury(I) nitrate dihydrate and mercury(II) nitrate monohydrate?

2.106 Consider the following substances: Cl_2, CaO, HBr, $CuCl_2$, AsH_3, $NaNO_3$, and NO_2.
(a) Which are molecular?
(b) Which are ionic?
(c) Which are binary substances?
(d) Which are diatomic?
(e) Which is a triatomic molecule?
(f) In which do we find only electron sharing?

(g) In which do we find only attractions between ions?
(h) In which do we find both electron sharing and attractions between ions?

2.107 Using the old system of nomenclature, write the names for the following compounds.
(a) gold(III) nitrate
(b) copper(II) sulfate
(c) lead(IV) oxide
(d) mercury(I) chloride
(e) mercury(II) chloride

(f) cobalt(II) hydroxide
(g) tin(II) chloride
(h) tin(IV) sulfide
(i) gold(III) sulfate

2.108 Write the names of the following compounds using the Stock system of nomenclature. Also write their formulas.
(a) chromous chloride
(b) manganous sulfate
(c) plumbous acetate

(d) cupric bromide
(e) cuprous iodide
(f) ferrous sulfate
(g) mercurous nitrate

2.109 A student needed a sample of $Fe(NO_3)_3 \cdot 9H_2O$, but when she went to the latest edition of the catalog of a major chemical supplier, she could not find it listed alphabetically under "iron." Knowing that suppliers of laboratory chemicals still often list chemicals under their older names, what name should she look for in the catalog?

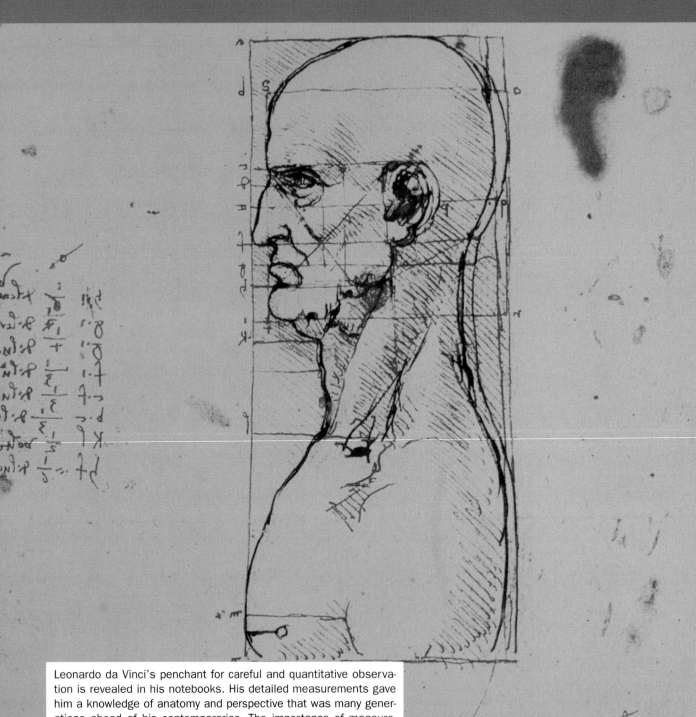

Leonardo da Vinci's penchant for careful and quantitative observation is revealed in his notebooks. His detailed measurements gave him a knowledge of anatomy and perspective that was many generations ahead of his contemporaries. The importance of measurements and the units we use to express them are the subject of this chapter.

THIS CHAPTER IN CONTEXT Every time you check the speedometer on your car's dashboard, glance at your watch, or step onto a bathroom scale, you are making a numerical observation, or **measurement.** Measurements are essential in chemistry—and in any scientific endeavor. To use them correctly in making decisions and in performing calculations, you must recognize that measurements are not the same as numbers.

In this chapter we will explore issues that affect how scientific measurements are reported and handled. We'll describe the units commonly used in science and especially in the laboratory, and discuss ways to recognize, estimate, and report the errors that are always present in measurements. We also outline strategies for converting units and using units in problem solving. These are strategies we will use often throughout the book. We conclude with a discussion of density and specific gravity—useful properties, derived from measurements, that can help identify substances.

3.1 ▶ Measurements are quantitative observations

A choking cloud of pollution has blanketed much of South Asia in recent years. The so-called "Asian Brown Cloud" is 3 kilometers (2 miles) thick and stretches from China to Pakistan, covering an area that is home to about half the world's population (Figure 3.1). A study conducted by the United Nations in 2002 links the cloud with an epidemic of respiratory disease that kills hundreds of thousands of people each year. The cloud blocks sunlight so effectively that crops in the region are beginning to fail and rainfall patterns have been altered.

The effects of the cloud are intensifying and spreading. Governments cannot afford to ignore the problem. Nor can they afford to take a trial-and-error approach to solving the problem. Any effective action is likely to be difficult and costly, and may have severe economic consequences. Leaders will demand reliable predictions of the costs and effectiveness of any proposed solution.

Scientific theories that explain how and why the cloud formed in the first place will be crucial in making those predictions. To build such a theory, scientists will have to answer the following questions:

- *What* substances are present in the cloud?
- *What* are the potential sources of pollution?
- *What* substances are present in each potential source?
- *How much* of each substance is produced?
- *How much* of a contribution does a pollution source make to the cloud?

FIGURE 3.1 *The Asian Brown Cloud is causing climatic change and an epidemic of respiratory disease in South Asia.*

Notice that some questions are qualitative (they ask "what") and some are quantitative (they ask "how much"). Both types of observations must be made to find the answers.

Qualitative (nonnumerical) observations will determine what substances are present and what the potential sources of pollution are. For example, one can observe that pumping a sample of the cloud through an air filter blackens the filter, and one can conclude that part of the cloud consists of particles that are larger than the holes in the filter. Qualitative observations can suggest what factors are important in understanding a phenomenon, so they're valuable as guides for further investigation.

Quantitative (numerical) observations must be made to match concentrations of pollutants from a potential source with those in the cloud and to determine the extent to which a pollution source contributes to the cloud. For example, one can match concentrations of trace elements in particles on the filter with those obtained from particles filtered from various pollution sources to determine whether those sources contribute significantly to the cloud. Quantitative observations are also called **measurements.** Measurements are crucial in scientific problem solving because they provide the level of detail necessary for testing and refining hypotheses.

Measurements are not numbers

Measurements involve numbers, but they differ from the numbers used in mathematics in two crucial ways.

First, measurements always involve a comparison. When you say that a person is 6 feet tall, you're really saying that the person is six times taller than a reference object that is 1 foot high. The foot is a **unit;** you measure the person's height by comparing it with an object like a ruler or a tape measure that is a

---

CHEMISTRY IN PRACTICE The importance of including units with measurements was dramatically emphasized by the loss of NASA's Mars Climate Orbiter in 1999. Ground controllers thought the probe was reporting thrust in newtons, the SI unit for force, but the spacecraft's navigational software was actually reporting thrust in pounds, an English unit. The error caused engineers to steer the $125 million spacecraft on a course that took it deep into the Martian atmosphere, where it burned up.

known number of feet long. Both the number and the unit are essential parts of the measurement. *Writing down a measurement without a unit is a common and serious mistake.*

The second important difference is that measurements always involve uncertainty. Numbers in mathematics are **exact.** There are exactly 12 inches in a foot, exactly 2.54 centimeters in an inch, exactly 14 letters in the word *sesquipedalian*. Exact numbers appear in definitions, or come from a direct count of some sort.[1]

Measurements, however, are always **inexact.** The act of measurement involves estimation of one sort or another, and both the observer and the instruments used to make the measurement have inherent physical limitations. As a result, measurements always include some uncertainty, which can be minimized but never entirely eliminated. *Any decision or conclusion based on measurements must consider the error in the measurements.* We'll explore this point further in Sections 3.2 and 3.3.

> Numbers are exact. Measurements are always inexact.

3.2 ▶ Measurements always include units

SI units are standard in science

A standard system of units is essential if measurements are to be made consistently. Metric systems of units have long been used in most countries. Metric units have two major advantages over older units. First, they are used by every industrialized nation on Earth. This helps ensure compatibility of goods such as tools and machine parts manufactured in different countries, and makes international commerce easier. Second, converting to larger or smaller units within the metric system can be done simply by moving a decimal point, because metric units are related to each other by simple multiples of 10.

Scientists working in many different countries quickly adopted the metric system to share their data with colleagues abroad. However, many different metric units were in use, and converting among these units was inconvenient. In 1960, a simplification of the original metric system was adopted by the General Conference on Weights and Measures (an international body). It is called the **International System of Units,** abbreviated **SI** from the French name, *Le Système International d'Unités*. The SI is now the dominant system of units in science and engineering, although there is still some usage of older metric units.

All SI units are built from seven base units

The SI has as its foundation a set of **base units** (Table 3.1) for seven measured quantities. For now, we will focus on the base units for length, mass, time, and temperature. The unit for amount of substance, the mole, will be discussed in Chapter 4. The unit for electrical current, the ampere, will be discussed briefly when we study electrochemistry in Chapter 21. The unit for luminous intensity, the candela, is not important in chemistry.

The SI units for *any* physical quantity can be built from these seven base units. For example, there is no SI base unit for area, but we know that to calculate the area of a rectangular room we multiply its length by its width. Therefore, the *unit* for area is derived by multiplying the *unit* for length by the *unit* for width. Length is a base measured quantity in the SI system. The base unit of length (or width) is the **meter (m).**

---

[1]Very large counts often do have some uncertainty in them because of inherent flaws in the counting process or because the count fluctuates. For example, the number of human beings in North America would be considered a measurement because it cannot be determined exactly at the present time.

TABLE 3.1	THE SI BASE UNITS	
Measurement	Unit	Symbol
Length	meter	m
Mass	kilogram	kg
Time	second	s
Electric current	ampere	A
Temperature	kelvin	K
Amount of substance	mole	mol
Luminous intensity	candela	cd

$$\text{length} \times \text{width} = \text{area}$$

$$(\text{meter}) \times (\text{meter}) = (\text{meter})^2$$

$$\text{m} \times \text{m} = \text{m}^2$$

To derive a unit for a quantity, you must first express it in terms of simpler quantities.

The SI **derived unit** for area is therefore m^2 (read as *meters squared,* or *square meter*). Table 3.2 lists several SI derived units that are used in chemistry.

In deriving SI units, we employ a very important concept that we will use repeatedly throughout this book when we perform calculations: *units undergo the same kinds of mathematical operations that numbers do.* We will see how this fact can be used to convert measurements from one unit to another in Section 3.4.

EXAMPLE 3.1
Deriving SI Units

Linear momentum is a measure of the "push" a moving object has, equal to the object's mass times its velocity. What is the SI derived unit for linear momentum?

ANALYSIS: To derive a unit for a quantity we must first express it in terms of simpler quantities. We know that linear momentum is mass times velocity. Therefore, the SI unit for linear momentum will be the SI unit for mass times the SI unit for velocity. The SI unit for mass is the **kilogram (kg).** Velocity is distance traveled (length) per unit time, so it has derived SI units of meters per second, m/s. Multiplying these units should give the derived unit for linear momentum.

TABLE 3.2	SOME DERIVED SI UNITS COMMONLY USED IN CHEMISTRY		
Measurement	Expression in Terms of Simpler Quantities	Unit and Symbol	Expression in Terms of SI Base Units
Area	length × width	square meter	m^2
Volume	length × width × height	cubic meter	m^3
Speed, velocity	distance/time	meter per second	m/s
Acceleration	velocity/time	meter per second squared	m/s^2
Density	mass/volume	kilogram per cubic meter	kg/m^3
Specific volume	volume/mass	cubic meter per kilogram	m^3/kg
Force	mass × acceleration	newton, N	$1\ N = 1\ kg\ m/s^2$
Pressure	force/area	pascal, Pa	$1\ Pa = 1\ N/m^2$ $= 1\ kg\ m^{-1}\ s^{-2}$
Energy	force × length	joule, J	$1\ J = 1\ Nm$ $= 1\ kg\ m^2/s^2$
Power	energy/time	watt, W	$1\ W = 1\ J/s$ $= 1\ kg\ m^2/s^3$

SOLUTION:

$$mass \times velocity = linear\ momentum$$

$$mass \times length\ /\ time = linear\ momentum$$

$$kilogram \times meter\ /\ second = kilogram\ meter\ /\ second$$

$$kg \times m/s = kg\ m/s$$

Is the Answer Reasonable?
The derived unit for linear momentum should be the product of units for mass and velocity, and this is obviously true.

PRACTICE EXERCISE 1: (a) The kinetic energy of an object of mass m moving at velocity v is $\frac{1}{2}mv^2$. What are the derived SI units of kinetic energy? (b) The gravitational potential energy of an object of mass m at a height h above some reference point is mgh, where g is the acceleration due to gravity (about 9.8 m/s^2). Show that the derived SI units of gravitational potential energy are the same as the units for kinetic energy.

FIGURE 3.2 *The international standard kilogram, made of a platinum–iridium alloy, which is kept at the International Bureau of Weights and Measures in France.* Other nations such as the United States maintain their own standard masses that have been carefully calibrated against this international standard.

Most of the base units are defined in terms of reproducible physical phenomena. For instance, the meter is defined as exactly the distance light travels in a vacuum in 1/299,792,458 of a second. Everyone has access to this standard because light and a vacuum are available to all. However, the SI base unit for mass is currently defined as the mass of a carefully preserved platinum–iridium alloy block stored at the International Bureau of Weights and Measures in France (Figure 3.2). This block serves indirectly as the calibrating standard for all "weights" used for scales and balances in the world. Obviously, damage or loss of the block could cause great difficulties. Researchers are currently working to redefine the kilogram in terms of a reproducible physical phenomenon.

The SI base units are defined in terms of physical phenomena that are accessible to all, rather than in terms of artifacts. The kilogram is an exception.

Scientists are working on a method of accurately counting atoms whose masses are accurately known. Their goal is to develop a new definition of the kilogram that doesn't rely on an object that can be stolen, lost, or destroyed.

Exponential notation and powers of 10 are reviewed on the web site at www.wiley.com/college/brady.

Construct SI units of any convenient size using decimal multipliers

When making measurements, we sometimes find that the basic units are either too large or too small. For instance, the meter (slightly larger than a yard) is not convenient for expressing the sizes of very small objects such as bacteria. Nor is it convenient for expressing the very large distances between planets or stars. The SI solves this problem by adjusting the size of the basic units with **decimal multipliers.** Table 3.3 lists the most commonly used decimal multipliers.

When the name of a unit is preceded by one of these prefixes, the size of the unit is modified by the corresponding decimal multiplier. For instance, the prefix *kilo* indicates a multiplying factor of 10^3, or 1000. Therefore, a kilometer is a unit of length equal to 1000 meters. The symbol for kilometer (km) is formed by applying the symbol meaning kilo (k) as a prefix to the symbol for meter (m). Thus, 1 km = 1000 m (or alternatively, 1 km = 10^3 m). Similarly a decimeter (dm) is one-tenth of a meter, so 1 dm = 0.1 m (1 dm = 10^{-1} m).

The symbols and multipliers listed in colored, boldface type in Table 3.3 are the ones most commonly encountered in chemistry.

A virus is 20 nm long. Objects that are smaller than about 2×10^{-7} m cannot be seen under an optical microscope. Can viruses be seen under an optical microscope?

ANALYSIS: First, we should state the goal of the problem. We want to know whether the virus is larger or smaller than 2×10^{-7} m.

The size of the virus has been given as 20 nm. To achieve our goal, we have to convert the size of the virus to meters, or convert 2×10^{-7} m to nm. Let's convert the size of the virus, 20 nm, to meters.

EXAMPLE 3.2
Using SI Prefixes

| TABLE 3.3 | DECIMAL MULTIPLIERS THAT SERVE AS SI PREFIXES | | | |

SI Prefixes

Prefix	Meaning	Symbol	Multiplication Factor (fraction)	Multiplication Factor (power of 10)
exa		E		10^{18}
peta		P		10^{15}
tera		T		10^{12}
giga		g		10^{9}
mega	millions of	M	1000000	10^{6}
kilo	thousands of	k	1000	10^{3}
hecto		h		10^{2}
deka		da		10^{1}
deci	tenths of	d	0.1	10^{-1}
centi	hundredths of	c	0.01	10^{-2}
milli	thousandths of	m	0.001	10^{-3}
micro	millionths of	μ	0.000001	10^{-6}
nano	billionths of	n	0.000000001	10^{-9}
pico	trillionths of	p	0.000000000001	10^{-12}
femto		f		10^{-15}
atto		a		10^{-18}

From Table 3.3, "n" is the abbreviation for *nano,* which means that the virus is 20 nanometers long. If we write this number in exponential notation, we will be able to compare it to the size of the smallest object visible under the microscope.

SOLUTION: We can replace the "n" in 20 nm with "$\times 10^{-9}$" to express the length of the virus in meters:

$$20 \, nm = 20 \times 10^{-9} \, m = 2 \times 10^{-8} \, m$$

where we have moved the decimal point to express the number in exponential notation. The length 2×10^{-8} m is smaller than 2×10^{-7} m, so the virus will not be visible.

Is the Answer Reasonable?
We can check our result by converting the size of the smallest visible object (2×10^{-7} m) to nanometers so that it can be compared directly with the size of the virus. Now 1 nanometer is 1×10^{-9} m, so the number of nanometers in 2×10^{-7} m should be

$$2 \times 10^{-7} \, m \times \left(\frac{1 \, nm}{10^{-9} \, m} \right)$$

or 200. The virus is 20 nm long, and the smallest object that can be seen under the microscope is 200 nm, so again we conclude that the virus will not be visible.

PRACTICE EXERCISE 2: Give the abbreviation for (a) microgram, (b) micrometer, (c) nanosecond. How many meters are in (a) 1 nm, (b) 1 cm, (c) 1 pm? What symbol is used to represent (a) 10^{-2} g, (b) 10^{6} m (c) 10^{-6} s?

Non-SI units are still in common use

Some older metric units that are not part of the SI system are still used in the laboratory and in the scientific literature. Some of these units are listed in Table 3.4; others will be introduced as needed in upcoming chapters.

TABLE 3.4	SOME NON-SI METRIC UNITS COMMONLY USED IN CHEMISTRY		
Measurement	Name	Symbol	Value in SI Units
Length	ångström	Å	$1\ \text{Å} = 0.1$ nm $= 10^{-10}$ m
Mass	atomic mass unit	u	$1\ \text{u} = 1.66054 \times 10^{-27}$ kg, approximately
	metric ton	t	$1\ \text{t} = 10^3$ kg
Time	minute	min	1 min $= 60$ s
	hour	h	1 h $= 60$ min $= 3600$ s
Temperature	degree Celsius	°C	Add 273.15 to Kelvin temperature
Volume	liter	L	$1\ \text{L} = 1000\ \text{cm}^3$

355ml   454g   1 liter
454g

FIGURE 3.3 *Metric units are becoming commonplace on many consumer products.*

The United States is the only large nation still using the *English system* of units, which measures distance in inches, feet, and miles; volume in ounces, quarts, and gallons; and mass in ounces and pounds. However, a gradual transition to metric units is occurring. Beverages, food packages, tools, and machine parts are often labeled in metric units (Figure 3.3). Common conversions between the English system and the SI are given in Table 3.5 and inside the rear cover of the book.[2]

Learn a few common units used in laboratory measurements

The most common measurements you will make in the laboratory will be those of length, volume, mass, and temperature.

Units in laboratory measurements

Length

The SI base unit for length, the **meter (m),** is too large for most laboratory purposes. More convenient units are the **centimeter (cm)** and the **millimeter (mm).** They are related to the meter as follows:

$$1\ \text{cm} = 10^{-2}\ \text{m} = 0.01\ \text{m}$$

$$1\ \text{mm} = 10^{-3}\ \text{m} = 0.001\ \text{m}$$

TABLE 3.5	SOME USEFUL CONVERSIONS	
Measurement	English to Metric	Metric to English
Length	1 in. = 2.54 cm 1 yd = 0.9144 m 1 mile = 1.609 km	1 m = 39.37 in. 1 km = 0.6215 mile
Mass	1 lb = 453.6 g 1 oz = 28.35 g	1 kg = 2.205 lb
Volume	1 gal = 3.785 L 1 qt = 946.4 mL 1 oz (fluid) = 29.6 mL	1 L = 1.057 qt

[2]Originally, these conversions were established by measurement. For example, if a metric ruler is used to measure the length of an inch, it is found that 1 in. equals 2.54 cm. Later, to avoid confusion about the accuracy of such measurements, it was agreed that these relationships would be taken to be exact. For instance, 1 in. is now defined as *exactly* 2.54 cm. Exact relationships also exist for the other quantities, but for simplicity many have been rounded off. For example, 1 lb = 453.59237 g, *exactly*.

An older non-SI unit called the **ångström (Å)** is often used to describe atomic and molecular-scale objects:

$$1 \text{ Å} = 0.1 \text{ nm} = 10^{-10} \text{ m}$$

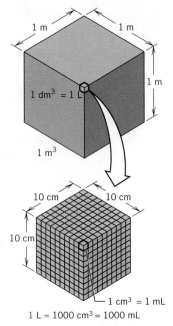

Comparison among the cubic meter, the liter (cubic decimeter), and the milliliter (cubic centimeter).

It is also useful to know the relationships

$$1 \text{ m} = 100 \text{ cm} = 1000 \text{ mm}$$

$$1 \text{ cm} = 10 \text{ mm}$$

To convert these units to English units, remember that 1 in. equals 2.54 cm, *exactly.*

Volume

Volume is a derived unit with dimensions of (length)^3. For example, the volume of a box is calculated by multiplying its length times its width times its height. With these dimensions expressed in meters, the derived SI unit for volume is the **cubic meter, m^3.**

In chemistry, measurements of volume usually arise when we measure amounts of liquids. The traditional metric unit of volume used for this is the **liter (L).** One liter is exactly 1000 cubic centimeters.

$$1 \text{ L} = 1000 \text{ cm}^3$$

The liter is too large to conveniently express most volumes measured in the lab. The glassware we normally use, such as that illustrated in Figure 3.4, is marked in **milliliters (mL).**[3]

$$1 \text{ L} = 1000 \text{ mL}$$

1 mL is exactly the same as 1 cm^3. One often finds cm^3 abbreviated as "cc" (especially in medical applications), although the SI frowns on this symbol.

Mass

In the SI, the base unit for the measurement of mass is the **kilogram (kg),** although the **gram (g)** is a more conveniently sized unit for most laboratory measurements.

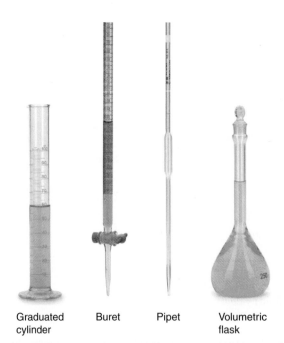

FIGURE 3.4 *Common laboratory glassware used for measuring volumes includes graduated cylinders, volumetric flasks, and a buret.*

Graduated cylinder   Buret   Pipet   Volumetric flask

---

[3]Use of the abbreviations L for liter and mL for milliliter is rather recent. Confusion between the printed letter l and the number 1 prompted the change from l for liter and ml for milliliter. You may encounter the abbreviation ml in other books or on older laboratory glassware.

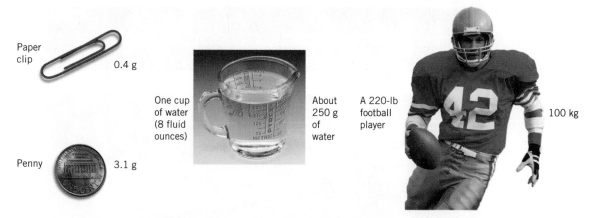

FIGURE 3.5 *Masses of several common objects in metric and English units.*

One gram, of course, is $\frac{1}{1000}$ of a kilogram (1 kg = 1000 g, so 1 g must equal 0.001 kg). One milliliter of water has a mass of about 1 gram. One liter of water has a mass of about 1 kilogram. Figure 3.5 shows the masses of several common objects in both English and metric units.

Mass is measured by comparing the weight of a sample with the weights of known standard masses. The apparatus used is called a **balance** (some examples are shown in Figure 3.6). For the balance in Figure 3.6*a*, we would place our sample on the left pan and then add standard masses to the other. When the weight of the sample and the total weight of the standards are in balance (when they match), their masses are then equal.

The kilogram is the SI base unit for mass, not the gram. The kilogram is the only SI base unit whose name has a decimal prefix.

The difference between mass and weight was discussed in Chapter 1.

Temperature

Temperature is usually measured with a thermometer. A typical laboratory thermometer consists of a long glass tube with a very thin bore connected to a reservoir containing a liquid, usually mercury (Figure 3.7*a*) or alcohol. As the temperature rises, the liquid in the reservoir expands and its height in the column increases. Electronic thermometers are also widely used in undergraduate laboratories

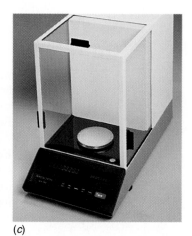

(a)   (b)   (c)

FIGURE 3.6 *Typical laboratory balances.* (*a*) A traditional two-pan analytical balance capable of measurements to the nearest 0.0001 g. (*b*) A modern top-loading balance capable of mass measurements to the nearest 0.001 g (fitted with a cover to reduce the effects of air currents and thereby improve precision). (*c*) A modern analytical balance capable of measurements to the nearest 0.0001 g.

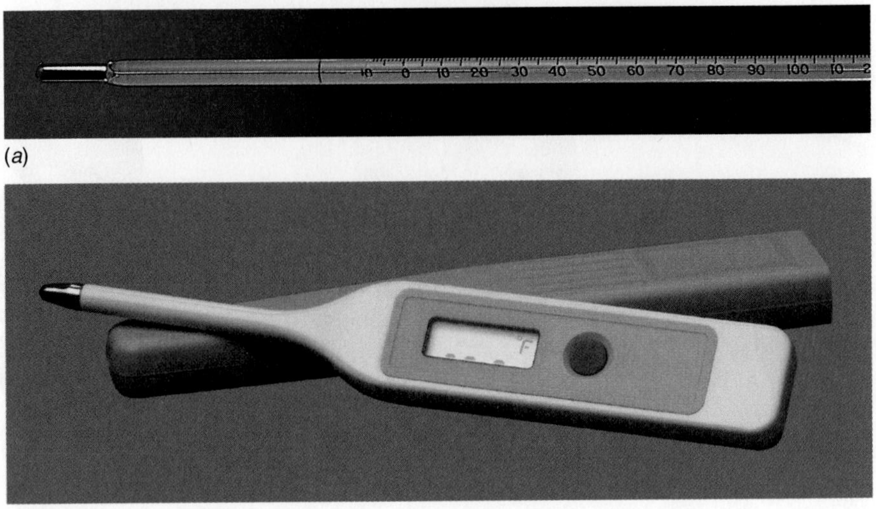

Figure 3.7 *Typical laboratory thermometers.* (*a*) A traditional mercury thermometer. (*b*) An electronic thermometer.

(*a*)

(*b*)

In chemistry, reference data is commonly tabulated at 25 °C, which is close to room temperature. Biologists often carry out their experiments at 37 °C because that is normal human body temperature.

(Figure 3.7*b*). An electronic thermometer works by sensing changes in the electrical properties of the metal probe, which are sensitive to temperature.

Thermometers are graduated in *degrees*. A degree corresponds to a change in temperature, a fixed fraction of the distance between reference points on a fixed scale.

Two temperature scales are in common use on thermometers. Both scales use as reference points the temperature at which water freezes[4] and the temperature at which it boils. On the **Celsius scale** water freezes at 0 °C and boils at 100 °C. (See Figure 3.8.) This scale is used throughout the world and is the usual scale used on laboratory thermometers. The **Fahrenheit scale** is a nonmetric temperature scale that is still used in the United States. On the Fahrenheit scale, water freezes at 32 °F and boils at 212 °F.

As you can see in Figure 3.8, on the Celsius scale there are 100 degree units between the freezing and boiling points of water, while on the Fahrenheit scale this same temperature range is spanned by 180 degree units. Therefore, each Celsius degree is nearly twice as large as a Fahrenheit degree (actually, 5 Celsius degrees is the same as 9 Fahrenheit degrees). If it is necessary to convert between these temperature scales, we can use the equation

TOOLS
Celsius–Fahrenheit conversions

$$t_F = \left(\frac{9\,{}^\circ F}{5\,{}^\circ C} \right) t_C + 32\,{}^\circ F \qquad (3.1)$$

where t_F is the Fahrenheit temperature and t_C is the Celsius temperature. Notice how °C "cancels out" in the equation, to leave only °F. The 32 °F is added to account for the fact that the freezing point of water (0 °C) occurs at 32 °F on the Fahrenheit scale.

The SI unit of temperature is the **kelvin (K),** which is the degree unit on the **Kelvin temperature scale.** Notice that the temperature unit is K, not °K (the degree symbol, °, is omitted). Also notice that the name of the unit, kelvin, is not capitalized. Equations that include temperature as a variable sometimes take on a simpler form when Kelvin temperatures are used. Temperatures must be converted to the

---

[4]Water freezes and ice melts at the same temperature. A mixture of ice and water will maintain a constant temperature of 32 °F or 0 °C. If heat is added, some ice melts; if heat is removed, some liquid water freezes, but the temperature doesn't change. The steadiness of the temperature of an ice–water mixture at 0 °C makes it convenient for calibrating thermometers.

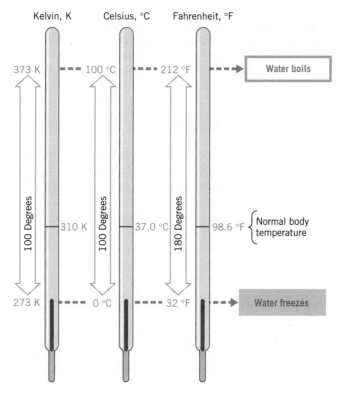

FIGURE 3.8 *Comparison among Kelvin, Celsius, and Fahrenheit temperature scales.*

Kelvin scale to correctly use these equations. We will encounter this situation many times throughout the book.

Figure 3.8 shows how the Kelvin, Celsius, and Fahrenheit temperature scales relate to each other. Notice that the kelvin is *exactly* the same size as the Celsius degree. *The only difference between these two temperature scales is the zero point.* The zero point on the Kelvin scale is called **absolute zero** and corresponds to nature's coldest temperature. It is 273.15 degree units below the zero point on the Celsius scale, which means that 0 °C equals 273.15 K, and 0 K equals −273.15 °C. Thermometers are never marked with the Kelvin scale, so when we need to express a temperature in kelvins, we have to do some arithmetic. The relationship between the Kelvin temperature, T_K, and the Celsius temperature, t_C, is

$$T_K = (t_C + 273.15 \,^\circ\text{C}) \left(\frac{1\,\text{K}}{1\,^\circ\text{C}} \right) \tag{3.2}$$

Notice how the fraction on the end of the equation causes °C to "cancel out," leaving only kelvins (K). When we are given Celsius temperatures rounded to the nearest degree, we can round 273.15 to the nearest degree as well, and use the following equation:

$$T_K = (t_C + 273 \,^\circ\text{C}) \left(\frac{1\,\text{K}}{1\,^\circ\text{C}} \right) \tag{3.3}$$

Notice that the name of the temperature scale, the Kelvin scale, is capitalized, but the name of the unit, the kelvin, is not. However, the symbol for the kelvin is the capital letter K.

◀ **TOOLS**

Celsius–Kelvin conversions

We will use a capital T to stand for the Kelvin temperature and a lowercase t (as in t_C) to stand for the Celsius temperature.

Thermal pollution, the release of large amounts of heat into rivers and other bodies of water, is a serious problem near power plants and can affect the survival of some species of fish. For example, trout will die if the temperature of the water rises above approximately 25 °C. (a) What is this temperature in °F? (b) What is this temperature in kelvins?

EXAMPLE 3.3

Converting among Temperature Scales

ANALYSIS: Both parts of the problem here deal with temperature conversions. The critical link in this problem will be relationships that connect temperatures on one scale to those on another. Let's write them.

$$\text{Equation 3.1} \qquad t_F = \left(\frac{9\,^\circ\text{F}}{5\,^\circ\text{C}}\right) t_C + 32\,^\circ\text{F}$$

$$\text{Equation 3.3} \qquad T_K = (t_C + 273\,^\circ\text{C})\left(\frac{1\,\text{K}}{1\,^\circ\text{C}}\right)$$

Equation 3.1 relates Fahrenheit temperatures to Celsius temperatures, so this is the relationship we need to answer part (a). Equation 3.3 relates Kelvin temperatures to Celsius temperatures, which is what we need for part (b). Now that we have what we need, the rest follows.

SOLUTION TO (a): We substitute the value of the Celsius temperature (25 °C) for t_C.

$$t_F = \left(\frac{9\,^\circ\text{F}}{5\,^\circ\text{C}}\right) t_C + 32\,^\circ\text{F}$$

$$= \left(\frac{9\,^\circ\text{F}}{5\,^\circ\text{C}}\right)(25\,^\circ\text{C}) + 32\,^\circ\text{F}$$

$$= 77\,^\circ\text{F}$$

Therefore, 25 °C = 77 °F. (Notice that we have canceled the unit °C in the equation above. As noted earlier, units behave the same as numbers do in calculations.)

SOLUTION TO (b): Once again, we have a simple substitution. Since $t_C = 25\,^\circ\text{C}$, the Kelvin temperature is

$$T_K = (t_C + 273\,^\circ\text{C})\left(\frac{1\,\text{K}}{1\,^\circ\text{C}}\right)$$

$$= (25\,^\circ\text{C} + 273\,^\circ\text{C})\left(\frac{1\,\text{K}}{1\,^\circ\text{C}}\right)$$

$$= 298\,\text{K}$$

Thus, 25 °C = 298 K.

Are the Answers Reasonable?

Before leaving a problem, it is always wise to examine the answer to see whether it makes sense. Ask yourself—"Is the answer too large, or too small?" Judging the answers to such questions serves as a check on the arithmetic as well as on the method of obtaining the answer and can help you find obvious errors. Let's look at how we might estimate the magnitude of the answers to both parts (a) and (b).

For part (a), we know that a Fahrenheit degree is about half the size of a Celsius degree, so 25 Celsius degrees should be about 50 Fahrenheit degrees. The positive value for the Celsius temperature tells us we have a temperature *above* the freezing point of water. Since water freezes at 32 °F, the Fahrenheit temperature should be approximately 32 °F + 50 °F = 82 °F. The answer of 77 °F is quite close.

For part (b), we recall that 0 °C = 273 K. A temperature above 0 °C must be higher than 273 K. Our calculation, therefore, appears to be correct.

PRACTICE EXERCISE 3: What Celsius temperature corresponds to 50 °F? What kelvin temperature corresponds to 68 °F (expressed to the nearest whole kelvin unit)?

3.3 Measurements always contain some uncertainty

We noted in Section 3.1 that measurements are inexact. The difference between a measurement and the "true" value we are trying to measure is called the **error** in the measurement. Don't confuse errors with mistakes. **Mistakes** are due to some preventable blunder made during the measurement. **Errors** are due to inherent limitations in the measurement procedure. Error arises because our ability to make measurements is limited by the instruments and procedures we use to make the measurement. If we identify the sources of error in a measurement, we may be able to make them smaller—but we will never be able to eliminate them entirely.

Errors arise when estimating readings from a scale

Consider reading the same temperature from each of the thermometers in Figure 3.9. The marks on the left thermometer are 1 degree apart. We can be certain that the temperature is more than 24 degrees and less than 25 degrees. We can get a better estimate of the temperature by estimating another digit between the finest markings on the scale. The fluid column falls about three-tenths of the way between the marks for 24 and 25 degrees on the left thermometer, so we can record the measurement as 24.3 °C.

It is reasonable to record the measurement this way, but it would be foolish to say that the temperature is *exactly* 24.3 °C. The last digit is only an estimate, and the left thermometer might be read as 24.2 °C by one observer or 24.4 °C by another. By convention in science, *all digits in a measurement up to and including the first estimated digit are recorded.* The digits recorded according to this convention are called **significant figures** (or **significant digits**).

When reading a scale, the last digit corresponds to one-tenth of the smallest scale division. The right thermometer reading can be written as 24.32 °C, with four significant digits, because its scale has marks that are one-tenth of a degree apart. A reading like 24 °C, with two significant digits, would be appropriate for a thermometer with marks that are 10 °C apart. Notice that the significant figures convention requires the last digit to be the one read from between the marks. If a reading appears to be exactly on a mark on the scale, a zero must be written. For example, if the fluid seemed to be exactly on the 24 °C mark on a thermometer with marks 1 °C apart, the measurement would have to be written as 24.0 °C, not 24 °C, to show that the thermometer can be read to the nearest one-tenth of a degree.

We would have more confidence in temperatures read from the thermometer on the right in Figure 3.9 because it has more digits and a smaller amount of uncertainty. *The reliability of a piece of data is indicated by the number of digits used to represent it.* This is so important we have a special terminology to describe numbers that come from measurement.

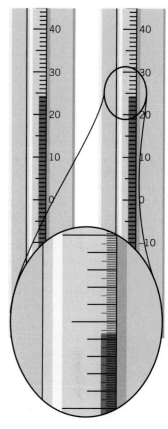

FIGURE 3.9 *Thermometers with different scales give readings with different precision.* The thermometer on the left has marks that are 1 degree apart, allowing the temperature to be estimated to the nearest tenth of a degree. The thermometer on the right has marks every 0.1 °C. This scale permits estimation of the hundredths place.

 TOOLS

Rules for counting significant figures

Digits that result from measurement such that only the digit farthest to the right is not known with certainty are called **significant figures.**

The number of significant figures in a measurement is equal to the number of digits known for sure *plus* one that is estimated.

The error inherent in reading a scale can be decreased by using an instrument with scale markings that are closer together. You can minimize the error in your measurements this way, but you can never eliminate it entirely, because at some point the scale marks will be so close together that the scale will become unreadable.

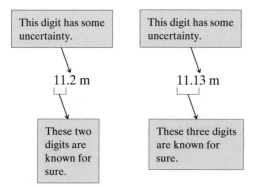

This digit has some uncertainty.

11.2 m

These two digits are known for sure.

This digit has some uncertainty.

11.13 m

These three digits are known for sure.

The first measurement of 11.2 m has three significant figures; the second, more precise value of 11.13 m has four significant figures.

Errors arise from uncontrolled conditions

It's important to realize that the error associated with reading scales is just one of many errors you'll encounter when you make a measurement. Uncontrollably changing conditions at the time of the measurement can cause errors that are more important than scale-reading errors. For example, if you are measuring a length of wire with a ruler, you may not be holding the wire perfectly straight every time. The zero mark of the ruler may not exactly line up with the wire's end every time. Materials expand when they get warmer and contract when they get cooler, so the length of the wire will change very slightly when the temperature of the room fluctuates.

Errors arise from poorly defined measurements

Ambiguities in the procedure used to take the measurement can also lead to error. For example, consider weighing yourself on a digital bathroom scale. You may think that because you can read the weight as (say) 150.0 pounds from the readout that the measurement is good to four significant figures. However, you can weigh yourself in many different ways: wet or dry, naked or clothed, standing on two legs or on one, with the scale on a hard floor or on a carpet. These weights may differ by a few pounds from each other, whereas the measurement scale implies that repeated weights should differ by only a few tenths of a pound. Which weight is correct? Would you have suspected that your readings would vary so much if you had weighed yourself only once?

Sometimes it isn't the procedure, but the measured quantity itself that is ambiguous. Suppose you want to measure the thickness of an American penny. If you measure the thickness along the rim, you'll get a different result than if you measure the thickness through Lincoln's beard, and a different result again if you measure through his bow tie. There is no single, true thickness of the coin. The variation in these measurements is a property of the coin itself.

Some errors can be detected by repeating a measurement

You will not be able to detect many of these errors if you perform only one measurement. You must repeat the measurement several times to see how consistent the values are. The values will seldom be identical if you are using a sensitive instrument to make the measurements. Some values are higher, and others are lower. When we take a large number of measurements, the values usually cluster around some central value. For example, suppose ten different people read the same thermometer. The following results might be obtained, where for convenience, we've arranged them in order of increasing size:

24.1 °C	24.3 °C
24.2 °C	24.3 °C
24.2 °C	24.4 °C
24.3 °C	24.4 °C
24.3 °C	24.5 °C

There is uncertainty in the tenths place, since the measurements range from 24.1 °C to 24.5 °C. The measurements cluster around 24.3 °C. We can estimate the central value quite simply by reporting the **average,** or **mean,** of the series of measurements. This is done by summing the measurements and then dividing by the number of measurements we made. The average temperature for our data set is 24.3 °C.

Making repeated measurements is tedious. How many measurements must we make before we can stop? The answer depends on how much confidence we want

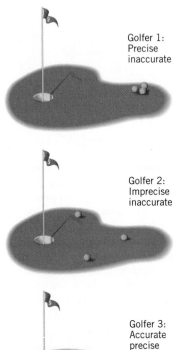

Golfer 1:
Precise
inaccurate

Golfer 2:
Imprecise
inaccurate

Golfer 3:
Accurate
precise

FIGURE 3.10 *The difference between precision and accuracy in the game of golf.* Golfer 1 hits shots that are precise (because they are tightly grouped), but the accuracy is poor because the balls are not near the target (the "true" value). Golfer 2 needs help. His shots are neither precise nor accurate. Golfer 3 wins the prize with shots that are precise (tightly grouped) and accurate (in the hole).

in the average we report. The more measurements we make, the more confident we can be that the average is close to the central value that all the measurements are grouped around.

Accuracy is correctness; precision is reproducibility

Repeated measurements can provide an estimate of the uncertainty in the measurements, and can give us valuable clues about the nature of the errors that are occurring. One can evaluate the error in a measurement by asking two critical questions about it:

- How **accurate** is the measurement, that is, how close is it to the true or correct value?[5]
- How **precise** is the measurement, that is, how close is it to the average in a series of repeated measurements?

Notice that the two questions are not equivalent, because the average doesn't always correspond to the true or correct value. A practical example of how accuracy and precision differ is illustrated in Figure 3.10.

For measurements to be **accurate,** the measuring device must be carefully calibrated. This means that it must be adjusted to give correct values when some standard reference is used with the device. For example, to calibrate an electronic balance, a known reference mass is placed on the balance and a calibration routine within the balance is initiated. Once calibrated, the balance will give correct readings, the accuracy of which is determined by the accuracy of the standard mass used. Standard reference masses (also called *weights*) can be purchased from scientific supply companies.

Precision is how closely repeated measurements of the same quantity come to each other. As we've seen, this is determined partially by our ability to read the measuring scale on the instrument and partially by uncontrolled changes while we perform the measurement.

We usually assume that a very precise measurement is also of high accuracy. We can be wrong, however, if our instruments are improperly calibrated. For example, the improperly marked ruler in Figure 3.11 might yield measurements that vary by a hundredth of a centimeter (±0.01 cm), but all the measurements would be too large by 1 cm, a case of good precision but poor accuracy.

> *Calibration* means adjusting an instrument so that it gives correct readings.

> Calibration can eliminate error that affects accuracy, but it does not eliminate errors that affect precision. Errors affecting precision (called **random errors,** or noise) can be minimized by averaging over a large number of measurements, but they can never be entirely eliminated.

How accurate would measurements be with this ruler?

FIGURE 3.11 *An improperly marked ruler.* This improperly marked ruler will yield measurements that are each wrong by one whole unit. The measurements might be precise, but the accuracy would be very poor.

PRACTICE EXERCISE 4: Four workers measure the mass of a 10.000 g mass on several different kitchen balances.

Worker A:	10.022 g, 9.976 g, 10.008 g
Worker B:	9.836 g, 10.033 g, 9.723 g
Worker C:	10.230 g, 10.231 g, 10.232 g
Worker D:	9.632 g, 9.835 g, 9.926 g

Which set of data has the best precision? Which has the best accuracy?

3.4 ▶ Measurements are written using the significant figures convention

In general, for a given measurement, *the greater the number of significant figures, the greater is the degree of precision.* If we write measurements to the proper number of significant digits, then we can give people who will read and use those

---

[5]The closeness of the average of a series of measurements to the true value is called **trueness.** One uses the word *accuracy* when talking about individual measurements, and the word *trueness* when talking about the average measurement.

measurements some idea of how precise they are. This is extremely important if decisions are being made on the basis of those measurements. Consider, for example, the following statement:

> *The Florida Secretary of State, Katherine Harris, certified that 2,912,790 people voted for George W. Bush and 2,912,253 people voted for Al Gore.*

If these numbers were exact, there is no doubt which of the candidates received more votes. However, there is some uncertainty in the numbers because recounting the ballots changed the numbers. If the numbers were uncertain in the ones or tens place, the winning candidate is still easily agreed upon. However, if the numbers were uncertain in the hundreds, thousands, or ten-thousands place, it becomes difficult to choose the winner of the election with confidence.

When are zeros significant digits?

Usually, it is simple to determine the number of significant figures in a measurement; we just count the digits. Thus 3.25 has three significant figures and 56.205 has five of them. When zeros come at the beginning or the end of a number, however, they sometimes cause confusion.

Zeros to the right of a decimal point are always counted as significant. Thus, 4.500 m has four significant figures because the zeros would not be written unless those digits were known to be zeros.

Zeros to the left of the first nonzero digit are never counted as significant. For instance, a length of 2.3 mm is the same as 0.0023 m. Since we are dealing with the same measured value, its number of significant figures cannot change when we change the units. Both quantities have two significant figures.

Zeros on the end of a number without a decimal point are assumed not to be significant. For example, suppose you were told that a protest march was attended by 45,000 people. If this was just a rough estimate, it might be uncertain by as much as several thousand, in which case the value 45,000 represents just two significant figures, since the "5" is the uncertain digit. None of the zeros would then count as significant figures. On the other hand, suppose the protesters were carefully counted using an aerial photograph so that the count could be reported to be 45,000 — give or take about 100 people. In this case, the value represents 45,000 ± 100 protesters and contains three significant figures, with the uncertain digit being the zero in the hundreds place. So a simple statement such as "there were 45,000 people attending the march" is ambiguous. We can't tell how many significant digits the number has from the number alone. We can be sure that the nonzero digits are significant, though. The best we can do is to say, "45,000 has *at least* two significant figures."

Scientific notation is reviewed on the web site at www.wiley.com/college/brady.

We can avoid confusion by using scientific notation when we report a measurement. For example, if we want to report the number of protesters as 45,000 give or take a thousand, we can write the rough estimate as 4.5×10^4. The 4.5 shows the number of significant figures and the 10^4 tells us the location of the decimal. The value obtained from the aerial photograph count, on the other hand, can be expressed as 4.50×10^4. This time the 4.50 shows three significant figures and an uncertainty of ±100 people.

$$4.5 \times 10^4 \qquad \text{two significant figures}$$

$$4.50 \times 10^4 \qquad \text{three significant figures}$$

Notice that when measurements are expressed in scientific notation to the correct number of significant digits, the number of digits written is the same no matter what units are used to express the measurement:

2.00 in.	three significant figures
= 5.08 cm	three significant figures
= 5.08×10^{-2} m	three significant figures
= 5.08×10^{1} mm	three significant figures
= 5.08×10^{7} nm	three significant figures

The number of significant digits in a measurement is the same no matter what units it is converted to, as long as approximations are not made during the conversion.

Measurements limit the precision of results calculated from them

When several measurements are obtained in an experiment they are usually combined in some way to calculate a desired quantity. For example, to determine the area of a rectangular carpet we require two measurements, length and width, which are then multiplied to give the answer we want. If one of these measurements is very precise and the other is not, we can't expect too much precision in the calculated area; the large uncertainty in the less precise measurement carries through to give a large uncertainty in the area. Therefore, to get some idea of how precise the area really is, we need a way to take into account the precision of the various values used in the calculation. To make sure this happens, we follow certain rules according to the kinds of arithmetic being performed.

The assignment of significant figures to more complex operations, such as taking logarithms and antilogarithms of measurements, is discussed on the textbook web site at www.wiley.com/college/brady.

Multiplication and division

For multiplication and division, the number of significant figures in the answer should not be greater than the number of significant figures in the least precise measurement. Let's look at a typical problem involving some measured quantities.

Significant figures in multiplication and division

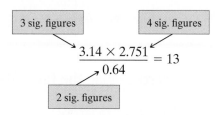

The answer to this calculation displayed on a calculator is 13.497093.[6] However, the rule says that the answer should have only as many significant figures as the least precise factor. Because the least precise factor, 0.64, has only two significant figures, the answer should have only two. The correct answer, 13, is then obtained by rounding off the calculator answer.[7]

Addition and subtraction

For addition and subtraction, the answer should have the same number of decimal places as the quantity with the fewest number of decimal places. As an example, consider the following addition of measured quantities.

Significant figures in addition and subtraction

---

[6]Calculators usually give too many significant figures. An exception is when the answer has zeros at the right that are significant figures. For example, an answer of 1.200 would be displayed on most calculators as 1.2. If the zeros belong in the answer, be sure to write them down.

[7]When we wish to round off a number at a certain point, we simply drop the digits that follow if the first of them is less than 5. Thus, 8.1634 rounds to 8.16, if we wish to have only two decimal places. If the first digit after the point of round off is larger than 5, or if it is 5 followed by other nonzero digits, then we add 1 to the preceding digit. Thus 8.167 and 8.1653 both round to 8.17. Finally, when the digit after the point of round off is a 5 and no other digits follow the 5, then we drop the 5 if the preceding digit is even and add 1 if it is odd. Thus, 8.165 rounds to 8.16 and 8.175 rounds to 8.18.

$$
\begin{array}{r}
3.247 \\
41.36 \\
+125.2 \\
\hline
169.8
\end{array}
$$

This number has only 1 decimal place.

The answer has been rounded to 1 decimal place.

In this calculation, the digits beneath the 6 and the 7 are unknown; they could be anything. (They're not necessarily zeros because if we *knew* they were zeros, then zeros would have been written there.) Adding an unknown digit to the 6 or 7 will give an answer that's also unknown, so for this sum we are not justified in writing digits in the second and third places after the decimal point. Therefore, we round the answer to the nearest tenth.

Exact numbers

Large counts (such as populations of cities or numbers of atoms in a sample of matter) are often subject to error and so have a limited number of significant digits.

We have already noted that numbers are exact, whereas measurements are not. Numbers that come from definitions, such as 12 in. = 1 ft, and those that come from a direct count, such as the number of people in a room, have no uncertainty and we can assume that they have an infinite number of significant figures. Ignore exact numbers when applying the rules described above.

PRACTICE EXERCISE 5: Perform the following calculations involving measurements and round the results so that they are written to the correct number of significant figures:

(a) 21.0233 g + 21.0 g
(b) 32.02 mL − 2.0 mL
(c) 54.183 g − 0.0278 g
(d) 10.0 g + 1.03 g + 0.243 g

(e) 10.0324 g/11.7 mL
(f) $43.4 \text{ in.} \times \dfrac{1 \text{ ft}}{12 \text{ in.}}$
(g) 1.03 m × 2.074 m × 3.9 m

3.5 ▶ Units can be converted using the factor-label method

In the example problems we have solved so far, we took two steps to arrive at a solution. The first is assembling the necessary information, and the second is using the information correctly to obtain the answer. Our goal is to teach you both, but our principal focus in this section is on a system called the **factor-label method** (also called **dimensional analysis**), which scientists use to help them perform the correct arithmetic to solve a problem.

TOOLS

Factor-label method

In the factor-label method, we treat a numerical problem as one involving a conversion of units from one kind to another. To do this, we use one or more *conversion factors* to change the units of the given quantity to the units of the answer.

(given quantity) × (conversion factor) = (desired quantity)

To construct a valid conversion factor, the relationship between the units must be true. For example, the statement

3 ft = 41 in.

is false. Although you might make a conversion factor out of it, any answers you would calculate are sure to be incorrect. *Correct answers require correct relationships between units.*

A **conversion factor** *is a fraction formed from a **valid** relationship or equality between units and is used to switch from one system of measurement and units to another.* To illustrate, suppose we want to express a person's height of 72.0 in. in centimeters. To do this, we need the relationship between the inch and the centimeter. We can obtain this from Table 1.3.

$$2.54 \text{ cm} = 1 \text{ in. (exactly)} \tag{3.4}$$

If we divide both sides of this equation by 1 in., we obtain a conversion factor.

$$\frac{2.54 \text{ cm}}{1 \text{ in.}} = \frac{1 \text{ in.}}{1 \text{ in.}} = 1$$

Notice that we have canceled the units from both the numerator and denominator of the center fraction. *Units behave just as numbers do in mathematical operations,* which is a key part of the factor-label method. This leaves the first fraction equaling 1. Let's see what happens if we multiply 72.0 in., the height that we mentioned, by this fraction.

$$72.0 \, \cancel{\text{in.}} \times \frac{2.54 \text{ cm}}{1 \, \cancel{\text{in.}}} = 183 \text{ cm}$$

$$\left(\begin{array}{c}\text{given}\\\text{quantity}\end{array}\right) \times \left(\begin{array}{c}\text{conversion}\\\text{factor}\end{array}\right) = \left(\begin{array}{c}\text{desired}\\\text{quantity}\end{array}\right)$$

The relationship between the inch and the centimeter is exact, so the numbers in 1 in. = 2.54 cm have an infinite number of significant figures.

Because we have multiplied 72.0 in. by something that is equal to 1, we know we haven't changed the magnitude of the person's height. We have, however, changed the units. Notice that we have canceled the unit inches. The only unit left is centimeters, which is the unit we want for the answer. The result, therefore, is the person's height in centimeters.

One of the benefits of the factor-label method is that it lets you know when you have done the *wrong* arithmetic. From the relationship in Equation 3.4, we can actually construct two conversion factors:

$$\frac{2.54 \text{ cm}}{1 \text{ in.}} \quad \text{and} \quad \frac{1 \text{ in.}}{2.54 \text{ cm}}$$

We used the first one correctly, but what would have happened if we had used the second by mistake?

$$72.0 \text{ in.} \times \frac{1 \text{ in.}}{2.54 \text{ cm}} = 28.3 \text{ in.}^2/\text{cm}$$

In this case, none of the units cancel. We get units of in.²/cm because inches times inches is inches squared. Even though our calculator may be very good at arithmetic, we've got the wrong answer. *The factor-label method lets us know we have the wrong answer because the units are wrong!*

We will use the factor-label method extensively throughout this book to aid us in setting up the proper arithmetic in problems. In fact, we will see that it also helps us assemble the information we need to solve the problem. The following examples illustrate the method.

Convert 3.25 m to millimeters (mm).

ANALYSIS: We can clearly show what is known and what is unknown in the problem by representing the question as an equation. We can write the given quantity (with its units) on the left and the *units* of the desired answer on the right.

$$3.25 \text{ m} = ? \text{ mm}$$

This problem is very similar to the one we solved in Example 3.2. We can solve it using the factor-label method. We need a conversion factor that relates the unit meter to the unit millimeter. From the table of decimal multipliers, the prefix milli- means "$\times 10^{-3}$," so we can write

$$1 \text{ mm} = 10^{-3} \text{ m}$$

Notice that this relationship connects the units given to the units desired.

We now have all the information we need to solve the problem.

SOLUTION: From the relationship above, we can make two conversion factors.

$$\frac{1 \text{ mm}}{10^{-3} \text{ m}} \quad \text{and} \quad \frac{10^{-3} \text{ m}}{1 \text{ mm}}$$

We know we have to cancel the unit meter, so we need to multiply by a conversion factor with this unit in the denominator. Therefore, we select the one on the left. This gives

EXAMPLE 3.4

Applying the Factor-Label Method

$$3.25 \text{ m} \times \frac{1 \text{ mm}}{10^{-3} \text{ m}} = 3.25 \times 10^3 \text{ mm}$$

Notice we have expressed the answer to three significant figures because that is how many there are in the given quantity, 3.25 m.

Is the Answer Reasonable?
We know that millimeters are much smaller than meters, so 3.25 m must represent a lot of millimeters. Our answer, therefore, makes sense.

EXAMPLE 3.5

Applying the Factor-Label Method

A liter, which is slightly larger than a quart, is defined as 1 cubic decimeter (1 dm^3). How many liters are there in 1 cubic meter (1 m^3)?

ANALYSIS: Let's begin once again by stating the problem in equation form.

$$1 \text{ m}^3 = ? \text{ L}$$

Next, we assemble the tools. What relationships do we know that relate these various units? We are given the relationship between liters and cubic decimeters.

$$1 \text{ L} = 1 \text{ dm}^3 \tag{3.5}$$

From the table of decimal multipliers, we also know the relationship between decimeters and meters,

$$1 \text{ dm} = 0.1 \text{ m}$$

but we need a relationship between cubic units. Since units undergo the same kinds of operations numbers do, we simply cube each side of this equation (being careful to cube *both* the numbers and the units).

$$(1 \text{ dm})^3 = (0.1 \text{ m})^3$$

$$1 \text{ dm}^3 = 0.001 \text{ m}^3 \tag{3.6}$$

Notice how Equations 3.5 and 3.6 provide a path from the given units to those we seek. Such a path is always a necessary condition when we apply the factor-label method.

Equation 3.5

$$\text{m}^3 \longrightarrow \text{dm}^3 \longrightarrow \text{L}$$

Equation 3.6

Now we are ready to solve the problem.

SOLUTION: The first step is to eliminate the units m^3. We use Equation 3.6.

$$1 \text{ m}^3 \times \frac{1 \text{ dm}^3}{0.001 \text{ m}^3} = 1000 \text{ dm}^3$$

Then we use Equation 3.5 to take us from dm^3 to L.

$$1000 \text{ dm}^3 \times \frac{1 \text{ L}}{1 \text{ dm}^3} = 1000 \text{ L}$$

Thus, 1 m^3 = 1000 L.

Usually, when a problem involves the use of two or more conversion factors, they can be "strung together" to avoid having to compute intermediate results. For example, this problem can be set up as follows:

$$1 \text{ m}^3 \times \frac{1 \text{ dm}^3}{0.001 \text{ m}^3} \times \frac{1 \text{ L}}{1 \text{ dm}^3} = 1000 \text{ L}$$

Is the Answer Reasonable?
One liter is about a quart. A cubic meter is about a cubic yard. Therefore, we expect a large number of liters in a cubic meter, so our answer seems reasonable. (Notice here that in our analysis we have approximated the quantities in the calculation in units of quarts and cubic yards, which may be more familiar than liters and m^3 if you've been raised in the United States. We get a feel for the approximate magnitude of the answer using our familiar units and then relate this to the actual units of the problem.)

Some mountain climbers are susceptible to high-altitude pulmonary edema (HAPE), a life-threatening condition that causes fluid retention in the lungs. It can develop when a person climbs rapidly to heights greater than 2500 meters (2.5×10^3 m). What is this distance expressed in feet?

EXAMPLE 3.6

Applying the Factor-Label Method

ANALYSIS: The problem can be stated as

$$2500 \text{ m} = ? \text{ ft}$$

We are converting a metric unit of length (the meter) into an English unit of length (the foot). The critical link between the two will be a metric-to-English length conversion. One of several sets of relationships we can use is

$$1 \text{ cm} = 10^{-2} \text{ m} \quad \text{(from Table 3.3)}$$

$$1 \text{ in.} = 2.54 \text{ cm} \quad \text{(from Table 3.5)}$$

$$1 \text{ ft} = 12. \text{ in.}$$

Notice how they provide a path from meters to centimeters to inches to feet.

SOLUTION: Now we apply the factor-label method by eliminating unwanted units to bring us to the units of the answer.

$$2.5 \times 10^3 \text{ m} \times \frac{1 \text{ cm}}{10^{-2} \text{ m}} \times \frac{1 \text{ in.}}{2.54 \text{ cm}} \times \frac{1 \text{ ft}}{12 \text{ in.}} = 8.2 \times 10^3 \text{ ft}$$

Notice that if we were to stop after the first conversion factor, the units of the answer would be centimeters; if we stop after the second, the units would be inches, and after the third we get feet—the units we want. This time the answer has been rounded to two significant figures because that's how many there were in the measured distance. Notice that the numbers 12 and 2.54 do not affect the number of significant figures in the answer because they are exact numbers derived from definitions.

This is not the only way we could have solved this problem. Other sets of conversion factors could have been chosen. For example, we could have used

$$1 \text{ yd} = 0.9144 \text{ m}$$

$$3 \text{ ft} = 1 \text{ yd}$$

Then the problem would have been set up as

$$2500 \text{ m} \times \frac{1 \text{ yd}}{0.9144 \text{ m}} \times \frac{3 \text{ ft}}{1 \text{ yd}} = 8200 \text{ ft}$$

Many problems that you meet, just like this one, have more than one path to the answer. There isn't necessarily any *one* correct way to set up the solution. *The important thing is for you to be able to reason your way through a problem and find some set of relationships that can take you from the given information to the answer*. The factor-label method can help you search for these relationships if you keep in mind the units that must be eliminated by cancellation.

Is the Answer Reasonable?
A meter is slightly longer than a yard, so let's approximate the given distance, 2500 m, as 2500 yd. In 2500 yd, there are $3 \times 2500 = 7500$ ft. Since the meter is a bit

longer than a yard, our answer should be a bit longer than 7500 ft, so the answer of 8200 ft seems to be reasonable.

PRACTICE EXERCISE 6: Use the factor-label method to perform the following conversions: (a) 3.00 yd to inches; (b) 1.25 km to centimeters; (c) 3.27 mm to feet; (d) 20.2 miles/gallon to kilometers/liter.

3.6 ▶ Density is a useful intensive property

Recall that an extensive property is one that depends on sample size.

One of the interesting things about extensive properties is that if you take the ratio of two of them, the resulting quantity is usually independent of sample size. In effect, the sample size cancels out and the calculated quantity becomes an intensive property. One useful property obtained this way is **density,** *which is defined as the ratio of an object's mass to its volume.* Using the symbols d for density, m for mass, and V for volume, we can express this mathematically as

 TOOLS
Density

$$d = \frac{m}{V} \tag{3.7}$$

Notice that to determine an object's density we make two measurements, mass and volume.

EXAMPLE 3.7
Calculating Density

A sample of blood completely fills an 8.20 cm³ vial. The empty vial has a mass of 10.30 g. The vial has a mass of 18.91 g after being filled with blood. What is the density of blood?

ANALYSIS: This problem asks you to connect the mass and volume of blood with its density. The critical link between these quantities is the definition of density, given by Equation 3.7.

SOLUTION: We need the density of blood, so we need the mass of blood and the volume of blood. The blood completely fills the vial, so the volume is 8.20 cm³. The mass of the blood is the difference between the masses of the full and empty vials:

mass of blood = 18.91 g − 10.30 g = 8.61 g

To determine the density we simply take the ratio of mass to volume.

$$\text{density} = \frac{m}{V}$$

$$= \frac{8.61 \text{ g}}{8.20 \text{ cm}^3}$$

$$= 1.05 \text{ g/cm}^3$$

This could also be written as

$$\text{density} = 1.05 \text{ g/mL}$$

because 1 cm³ = 1 mL.

Is the Answer Reasonable?
In calculating the density, we are dividing 8.61 by 8.20, a number that is slightly smaller. The answer should therefore be slightly larger than 1. Our answer of 1.05 seems reasonable. (Shortly you will learn that water has a density of about 1 g/cm³, and the density of blood is not that different from the density of water. A density of 1.05 g/cm³ seems very reasonable.)

TABLE 3.6	DENSITIES OF SOME COMMON SUBSTANCES IN g/cm³ AT ROOM TEMPERATURE
Water	1.00
Aluminum	2.70
Iron	7.86
Silver	10.5
Gold	19.3
Glass	2.2
Air	0.0012

TABLE 3.7	DENSITY OF WATER AS A FUNCTION OF TEMPERATURE
Temperature (°C)	Density (g/cm³)
10	0.999700
15	0.999099
20	0.998203
25	0.997044
30	0.995646

Each pure substance has its own characteristic density (Table 3.6). Gold, for instance, is much more dense than iron. Each cubic centimeter of gold has a mass of 19.3 g, so its density is 19.3 g/cm³. By comparison, the density of water is 1.00 g/cm³ and the density of air at room temperature is about 0.0012 g/cm³.

There is more mass in 1 cm³ of gold than in 1 cm³ of iron.

Most substances, such as the mercury in the bulb of a thermometer, expand slightly when they are heated. The same amount of matter occupies a larger volume at a higher temperature, so the amount of matter packed into each cubic centimeter is less. Therefore, density usually decreases slightly with increasing temperature.[8] For solids and liquids the size of this change is small, as you can see from the data for water in Table 3.7. When only two or three significant figures are required, we can often ignore the variation of density with temperature.

Although the density of water varies slightly with temperature, it is useful to remember the value 1.00 g/cm³. It can be used if the water is near room temperature and only three (or fewer) significant figures are required.

 **CHEMISTRY IN PRACTICE** The densities of metals are an important physical property to be considered when designing parts for aircraft and spacecraft such as the Space Shuttle. The metals aluminum and titanium are used extensively because they provide significant strength with minimal weight. Titanium is particularly useful because it is as strong as steel but about 40% lighter (for a given volume of metal). Titanium is 60% heavier than aluminum but twice as strong, and its melting point is approximately 1000 °C higher.

Use density to relate a material's mass to its volume

One reason why density is a useful property is that it provides a way to convert between the mass and volume of a substance. In Example 3.7, we found that blood has a density of 1.05 g/cm³. This density defines a relationship, which we will call an **equivalence,** between the amount of mass and its volume. In words, we could express this by saying that 1.05 g of blood are equivalent to 1.00 cm³ of blood, or alternatively, that 1.00 cm³ of blood is equivalent to 1.05 g of blood. We express this relationship symbolically as

$$1.05 \text{ g blood} \Leftrightarrow 1.00 \text{ cm}^3 \text{ blood}$$

where we have used the symbol ⇔ to mean "is equivalent to." (We can't really use an equal sign in this expression because grams can't *equal* cubic centimeters; one is a unit of mass and the other is a unit of volume.)

---

[8]Liquid water behaves oddly. Its maximum density is at 4 °C, so when water at 0 °C is warmed, its density increases until the temperature reaches 4 °C. As the temperature is increased further the density of water gradually decreases.

In setting up calculations by the factor-label method, an equivalence can be used to construct conversion factors just as equalities can. From the equivalence we have just written, we can form two conversion factors.

$$\frac{1.05 \text{ g blood}}{1.00 \text{ cm}^3 \text{ blood}} \quad \text{and} \quad \frac{1.00 \text{ cm}^3 \text{ blood}}{1.05 \text{ g blood}}$$

The following example illustrates how we use density in calculations.

EXAMPLE 3.8
Calculations Using Density

Seawater has a density of about 1.03 g/mL. (a) What mass of seawater would fill a sampling vessel to a volume of 225 mL? (b) What is the volume, in milliliters, of 45.0 g of seawater?

ANALYSIS: For both parts of this problem, we are relating the mass of a material to its volume. Density is the critical link that we need between these two quantities. For seawater, the density tells us that

$$1.00 \text{ mL seawater} \Longleftrightarrow 1.03 \text{ g seawater}$$

From this relationship, we can construct two conversion factors.

$$\frac{1.03 \text{ g seawater}}{1.00 \text{ mL seawater}} \quad \text{and} \quad \frac{1.00 \text{ mL seawater}}{1.03 \text{ g seawater}}$$

SOLUTION TO (a): The question can be restated as 225 mL seawater = ? g seawater. We need to eliminate the unit *mL seawater*, so we choose the conversion factor on the left.

$$225 \text{ mL seawater} \times \frac{1.03 \text{ g seawater}}{1.00 \text{ mL seawater}} = 232 \text{ g seawater}$$

Thus, 225 mL of the seawater has a mass of 232 g.

SOLUTION TO (b): The question is 45.0 g seawater = ? mL seawater. This time we need to eliminate the unit *g seawater*, so we use the conversion factor on the right.

$$45.0 \text{ g seawater} \times \frac{1.00 \text{ mL seawater}}{1.03 \text{ g seawater}} = 43.7 \text{ mL seawater}$$

Thus, 45.0 g of seawater has a volume of 43.7 mL.

Are the Answers Reasonable?
Notice that the density tells us that 1 mL of seawater has a mass of slightly more than 1 g. So for part (a), we might expect that 225 mL of seawater should have a mass slightly more than 225 g. Our answer, 232 g, is reasonable. For part (b), 45 g of seawater should have a volume not too far from 45 mL, so our answer of 43.7 mL is the right size.

PRACTICE EXERCISE 7: An ocean-dwelling dinosaur has an estimated body volume of 1.38×10^6 cm^3. The animal's live mass was estimated at 1.24×10^6 g. What is its density?

PRACTICE EXERCISE 8: The density of plutonium is 19.82 g/cm^3. What mass of plutonium would fit in a suitcase with a volume of 250,000 cm^3?

Specific gravity is a unitless connection between mass and volume

The numerical value for the density of a substance depends on the units used for mass and volume. For example, if we express mass in grams and volume in milliliters, the density of water is 1.00 g/mL. However, if mass is given in pounds and volume in gallons, the density of water is 8.34 lb/gal; and if mass is in pounds and volume is in cubic feet, water's density is 62.4 lb/ft³. In a similar way, the densities of other substances have different numerical values for different units. One way to avoid having to tabulate densities in all sorts of different units is to tabulate specific gravities instead. *The* **specific gravity** *of a substance is defined as the ratio of the density of the substance to the density of water.*

$$\text{sp. gr.} = \frac{d_{substance}}{d_{water}} \qquad (3.8)$$

Specific gravity

The specific gravity tells us how much denser than water a substance is. For instance, if the specific gravity of a substance is 2.00, then it is twice as dense as water; if its specific gravity is 0.50, then it is only half as dense as water. This means that if we know the density of water in a particular set of units, we can multiply it by a substance's specific gravity to obtain the substance's density in these same units. Rearranging Equation 3.8 gives

$$d_{substance} = (\text{sp. gr.})_{substance} \times d_{water} \qquad (3.9)$$

Let's look at an example.

Methanol, a liquid fuel that can be made from coal, has a specific gravity of 0.792. What is the density of methanol in units of g/mL, lb/gal, and lb/ft³?

EXAMPLE 3.9
Using Specific Gravity

ANALYSIS: The question asks us to relate specific gravity to density, so Equation 3.8 (or its equivalent, Equation 3.9) is the tool we need to get the answer. We also need some data—the density of water in each of the requested units. Ordinarily we would look for them in a table of densities of water, although this time the necessary data were given in the preceding discussion.

$$d_{water} = 1.00 \text{ g/mL}$$
$$= 8.34 \text{ lb/gal}$$
$$= 62.4 \text{ lb/ft}^3$$

SOLUTION: Now that we have all the necessary tools and data, we bring them together to obtain the answers.

$$d_{methanol} = (\text{sp. gr.})_{methanol} \times d_{water}$$
$$= 0.792 \times 1.00 \text{ g/mL}$$
$$= 0.792 \text{ g/mL}$$

Similarly, for the other units,

$$d_{methanol} = 0.792 \times 8.34 \text{ lb/gal}$$
$$= 6.61 \text{ lb/gal}$$

and

$$d_{methanol} = 0.792 \times 62.4 \text{ lb/ft}^3$$
$$= 49.4 \text{ lb/ft}^3$$

Notice that when the density is expressed in g/mL, it is numerically the same as the specific gravity.

Are the Answers Reasonable?
If the specific gravity is less than 1.00, the density of the substance is less than that of water. With a specific gravity of 0.792, the density of the methanol must there-

fore be smaller than that of water. Notice that in each case, we have obtained an answer that is smaller than the density of water. Our answers, therefore, seem to be correct.

PRACTICE EXERCISE 9: The density of aluminum is 2.70 g/mL. What is its specific gravity? What is the density of aluminum in units of lb/ft³?

PRACTICE EXERCISE 10: Ethyl acetate is a clear, colorless solvent having a fruity odor that is often used in the manufacture of plastics. Its specific gravity is 0.902. What is its density in both g/mL and lb/gal?

Conclusions must be drawn from reliable measurements

We saw in Chapter 1 that substances can be identified by their properties. If we are to rely on properties such as density for identification of substances, it is very important that our measurements be reliable. We must have some idea of what the measurement's accuracy and precision are.

The importance of accuracy is obvious. If we have no confidence that our measured values are close to the true values, we certainly cannot trust any conclusions that are based on the data we have collected.

Precision of measurements can be equally important. For example, suppose we had a gold wedding ring and we wanted to determine whether or not the gold was 24 carat. We could determine the mass of the ring, and then its volume, and compute the density of the ring. We could then compare our experimental density with the density of 24 carat gold (which is 19.3 g/mL). Suppose the ring had a volume of 1.0 mL and the ring had a mass of 18 g, as measured using a graduated cup measure and a kitchen scale. The density of the ring would then be 18 g/mL, to the correct number of significant figures. Could we conclude that the ring was made of 24 carat gold? We know the density to only two significant figures. The experimental density could be as low as 17 g/mL or as high as 19 g/mL, which means the ring could be 24 carat gold—or it could be 22 carat gold (which has a density of around 17.7 to 17.8 g/mL) or maybe even 18 carat gold (which has a density up to 16.9 g/mL).

Suppose we now measure the mass of the ring with a laboratory balance capable of measurements to the nearest ±0.001 g and obtain a mass of 18.153 g. We measure the volume using volumetric glassware and find a volume of 1.03 mL. The density is 17.6 g/mL to the correct number of significant figures. The difference between this density and the density of 24 carat gold is 19.3 g/mL − 17.6 g/mL = 1.7 g/mL. This is considerably larger than the uncertainty in the experimental density (which is about ±0.1 g/mL). We can be reasonably confident that the ring is not 24 carat gold, and in fact the measurements point towards the ring being composed of 22 carat gold.

To trust conclusions drawn from measurements, we must be sure the measurements are accurate and that they are of sufficient precision to be meaningful. This is a key consideration in designing experiments.

SUMMARY

Units of Measurement **Qualitative observations** lack numerical information, whereas **quantitative observations** require numerical measurements. The units used for measurements are based on the set of seven **SI base units,** which can be combined to give various **derived units.** These all can be scaled to various sizes by applying **decimal multiplying factors.** In the laboratory we routinely measure length, volume, mass, and temperature. Convenient units for these purposes are respectively **centimeters** or **millimeters, liters** or **milliliters, kilograms** or **grams,** and **kelvins** or **degrees Celsius.**

Significant Figures The **precision** of a measured quantity is revealed by the number of **significant figures** that it contains, which equals the number of digits known with certainty plus the first one that possesses some uncertainty. Measured values are **precise** if they contain many significant figures and therefore differ from each other by small amounts. A measurement is **accurate** if its value lies very close to the true value. When measurements are combined by multiplication or division, the answer should not contain more significant figures than the least precise factor. When addition or subtraction is used, the answer is rounded to the same number of decimal places as the quantity having the fewest number of decimal places. **Exact numbers** do not enter into determining the number of significant figures in a calculated quantity because they have no uncertainty. **Scientific notation** is useful for writing large or small numbers in compact form and for expressing unambiguously the number of significant figures in a number.

Factor-Label Method The **factor-label method** is based on the ability of units to undergo the same mathematical operations as numbers. **Conversion factors** are constructed from *valid relationships* between units. These relationships can be either equalities or **equivalencies** between units. Unit cancellation serves as a guide to the use of conversion factors and aids us in correctly setting up the arithmetic for a problem.

Density and Specific Gravity Density is a useful intensive property that is defined as the ratio of a sample's mass to its volume. Besides serving as a means for identifying substances, density is a conversion factor that relates mass to volume. **Specific gravity** is the ratio of a sample's density to that of water. The numerical values of specific gravity and density are the same if density is expressed in the units g/mL.

TOOLS ▸ YOU HAVE LEARNED

The table below lists the concepts that you've learned in this chapter that can be applied as tools in solving problems. Study each one carefully so that you know what each is used for. When faced with solving a problem, recall what each tool does and consider whether it will be helpful in finding a solution. This will aid you in selecting the tools you need. If necessary, refer to this table when working on the Thinking-It-Through problems and the Review Exercises that follow.

TOOL	HOW IT WORKS
SI prefixes (Table 3.3, page 84)	We use these prefixes to create larger and smaller units. The prefixes enable us to form conversion factors for converting between differently sized units.
Units in laboratory measurements (page 85)	Use these to convert between commonly used laboratory units of measurement.
Temperature conversions (pages 88 and 89)	We use these to convert among temperature units. Be especially sure you can convert between Celsius and Kelvin temperatures, because that will be needed most in this course.
Rules for counting significant figures (page 91)	Use these to determine the number of significant figures in a number.
Rules for arithmetic and significant figures (page 95)	Use them to round answers to the correct number of significant figures.
Factor-label method (page 96)	We set up the arithmetic in numerical problems by assembling necessary relationships into conversion factors and applying them to obtain the desired units of the answer.
Density (page 100)	We calculate the density of a substance from measurements of mass and volume. Use density to convert between mass and volume for a substance.
Specific gravity (page 103)	This provides a way to convert density from one set of units to a different set of units.

THINKING IT THROUGH

Need extra help?
Visit the Brady/
Senese web site at
www.wiley.com/
college/brady

ON-LINE HELP

The goal for the following problems is not to find the answers themselves, but rather to assemble the information needed to solve them and explain how you would use the information to find the answers. The problems in Level 2 are more challenging than those in Level 1 and may contain more data than are required, in which case you are also asked to identify the unnecessary data. Detailed answers to the Thinking-It-Through problems can be found on the web site.

Level 1 Problems

1. The following ask you to perform unit conversions. Show how you would accomplish these conversions, using appropriate conversion factors and abbreviations for the units involved.
(a) How many nanometers are there in 14.6 cm?
(b) Suppose a projectile is fired at a velocity of 1350 m/s. What is this speed expressed in kilometers per hour?
(c) A substance has a density of 8.85 g/cm³. What is this density expressed in units of $\mu g/\mu m^3$?
(d) How many cubic millimeters are there in 4.20 cm³?
(e) A person is 5 ft 4 in. tall. What is this height expressed in centimeters?

2. Criticize the following statements.
(a) One inch is approximately 2.54 centimeters.
(b) The earth is 5,232,921,474 years old.
(c) Qualitative observations are not as useful as quantitative observations.
(d) Significant digits are those digits that are meaningful.
(e) Counting always gives exact numbers.
(f) The U.S. Census can determine the exact population of the United States.
(g) Errors can always be avoided.
(h) Estimation is a fancy word for guessing.

3. Normal human body temperature is 98.6 °F. What is this temperature expressed in kelvins? (Show how you would do the calculation.)

4. In 1 second, light travels a distance of 2.9979×10^8 m. What is the speed of light expressed in *miles per hour*? (List the relationships needed and set up the equation.)

5. A cylindrical metal bar has a diameter of 0.753 cm and a length of 2.33 cm. It has a mass of 8.423 g. Explain how you would use the concept of specific gravity to calculate the density of the metal in the units lb/ft³. Set up the calculation.

6. What is the volume (in cubic centimeters) of a diamond that has a mass of 3.5 g? The density of diamond is 3.51 g/cm³. (Set up the calculation.)

7. For the preceding question, explain how you determine the number of significant figures that should be reported in the answer.

8. Jet fuel has a specific gravity of 0.75. Show how you would calculate the mass in pounds of 3.45 gallons of jet fuel.

Level 2 Problems

9. Seawater has a specific gravity of 1.03. What is the mass in pounds of 146 in.³ of seawater? (Explain how you derive the necessary relationships and set up the calculation.)

10. Because of the serious consequences of lead poisoning, the Federal Centers for Disease Control in Atlanta has set a threshold of concern for lead levels in children's blood. This threshold was based on a study that suggested that lead levels in blood as low as *10 micrograms of lead per deciliter of blood* can result in subtle effects of lead toxicity. Suppose a child had a lead level in her blood of 2.5×10^{-4} grams of lead per liter of blood. Is this person in danger of exhibiting the effects of lead poisoning? Explain how you would solve the problem.

11. Gold has a density of 19.31 g/cm³. How many grams of gold are required to provide a gold coating 0.50 mm thick on a ball bearing having a diameter of 0.700 in.? Set up the calculation.

12. Kerosene has a specific gravity of 0.965. What is its density in units of pounds per cubic meter? Set up the calculation.

13. It is possible to separate molecules based on size using a specially prepared membrane made of polycarbonate, a synthetic polymer. By chemically depositing gold from solution onto the polymer, it is possible to shrink the size of the pores in a controlled fashion and thus to design filters for molecules of various sizes. Suppose a polycarbonate membrane has pores 30 nm in diameter. How would you calculate the weight of gold required to reduce the size of a pore from 30 nm to 0.6 nm? What other information would you need to carry out the calculation?

REVIEW QUESTIONS

SI Units

3.1 Why must measurements always be written with a unit?

3.2 What does the abbreviation SI stand for?

3.3 Which SI base unit is defined in terms of a physical object?

3.4 What is the only SI base unit that includes a decimal prefix?

3.5 Newton's second law states that force is equal to mass times acceleration. What is the SI derived unit for force? (The unit is called the newton, abbreviated N.)

3.6 What is the meaning of each of the following prefixes? (a) *centi-*, (b) *milli-*, (c) *kilo-*, (d) *micro-*, (e) *nano-*, (f) *pico-*, (g) *mega-*.

3.7 What abbreviation is used for each of the prefixes named in Question 3.6?

3.8 What units are most useful in the laboratory for measuring (a) length, (b) volume, and (c) mass?

3.9 How is mass measured? What is the difference between a balance and a spring scale found in a market?

3.10 What reference points do we use in calibrating the scale of a thermometer? What temperature on the Celsius scale do we assign to each of these reference points?

3.11 In each pair, which is larger: (a) A Fahrenheit degree or a Celsius degree? (b) A Celsius degree or a kelvin? (c) A Fahrenheit degree or a kelvin?

Significant Figures; the Factor-Label Method

3.12 Define the term *significant figures*.

3.13 What is the difference between *accuracy* and *precision*?

3.14 A concentration of $1.5\ \mu g$ of a toxin is found in a milliliter sample of a patient's blood. How high might the concentration actually have been? Suppose that the measurement had been written as $1.50\ \mu g/mL$ of blood. How high might the concentration have been in this case?

3.15 Suppose someone suggested using the fraction 3 yd/1 ft as a conversion factor to change a length expressed in feet to its equivalent in yards. What is wrong with this conversion factor? Can we construct a valid conversion factor relating centimeters to meters from the equation 1 cm = 1000 m? Explain your answer.

3.16 In 1 h there are 3600 s. By what conversion factor would you multiply 250 s to convert it to hours? By what conversion factor would you multiply 3.84 h to convert it to seconds?

3.17 If you were to convert the measured length 4.165 ft to yards by multiplying by the conversion factor (1 yd/3 ft), how many significant figures should the answer contain? Why?

REVIEW PROBLEMS

Answers to problems whose numbers are printed in color are given in Appendix B. More challenging questions are marked with asterisks. **ILW** = Interactive LearningWare solution is available at *www.wiley.com/college/brady*.

SI Prefixes

3.18 What number should replace the question mark in each of the following?
(a) 1 cm = ? m (d) 1 dm = ? m
(b) 1 km = ? m (e) 1 g = ? kg
(c) 1 m = ? pm (f) 1 cg = ? g

3.19 What numbers should replace the question marks below?
(a) 1 nm = ? m (c) 1 kg = ? g (e) 1 mg = ? g
(b) $1\ \mu g$ = ? g (d) 1 Mg = ? g (f) 1 dg = ? g

Temperature Conversions

3.20 Perform the following conversions:
(a) 50 °C to °F (d) 49 °F to °C
(b) 10 °C to °F (e) 60 °C to K
(c) 25.5 °F to °C (f) −30 °C to K

3.21 Perform the following conversions:
(a) 96 °F to °C (d) 273 K to °C
(b) −6 °F to °C (e) 299 K to °C
(c) −55 °C to °F (f) 40 °C to K

3.22 A healthy dog has a temperature ranging from 37.2 to 39.2 °C. Is a dog with a temperature of 103.5 °F within normal range?

3.23 A recipe calls for an oven preheated to 200 to 225 °C. If the oven is set to 325 °F, is it within this range?

3.24 Diamonds are completely consumed by fires with temperatures of about 1500 °F. What is this temperature in °C?

3.25 The coldest permanently inhabited place on earth is the Siberian village of Oymyakon in Russia. In 1964 the temperature reached a shivering −96 °F! What is this temperature in °C?

3.26 Estimates of the temperature at the core of the sun range from 10 megakelvins to 25 megakelvins. What is this range in °C and °F?

3.27 Natural gas is mostly methane, a substance that boils at a temperature of 111 K. What is its boiling point in °C and °F?

3.28 Helium has the lowest boiling point of any liquid. It boils at 4 K. What is its boiling point in °C?

3.29 The atomic bomb detonated over Hiroshima, Japan, at the end of World War II raised the temperature on the ground below to about 6000 K. Is this hot enough to melt concrete? (Concrete melts at 2000 °C.)

3.30 Liquid nitrogen is used as a commercial refrigerant to flash freeze foods. Nitrogen boils at 77 K. Oxygen boils at −297 °F. Is oxygen a liquid or a gas at the boiling point of nitrogen?

3.31 The air within a lightning bolt is heated to a peak temperature of around 55,000 °F. The surface of the sun has a temperature of about 6000 K. Which has a higher temperature, air within a lightning bolt or the surface of the sun?

Significant Figures

3.32 How many significant figures do the following measured quantities have?
(a) 37.53 cm (d) 0.00024 kg
(b) 37.240 cm (e) 0.07080 m
(c) 202.0 g (f) 2400 mL

3.33 How many significant figures do the following measured quantities have?
(a) 0.0230 g   (d) 614.00 mg
(b) 105.303 m   (e) 10 L
(c) 0.007 kg   (f) 3.8105 mm

3.34 How many significant figures do the following measured quantities have?
(a) 1.0230 kg   (d) 27.300 g
(b) 3.0200 m   (e) 0.04320 mm
(c) 0.0030 L   (f) 200 miles

3.35 How many significant figures do the following measured quantities have?
(a) 2.303 ng   (d) 9200 g
(b) 0.04030 kg   (e) 0.0101 dL
(c) 20030 mL   (f) 420 yd

3.36 Perform the following arithmetic and round off the answers to the correct number of significant figures:
(a) 0.0023 m × 315 m
(b) 84.25 kg − 0.01075 kg
(c) (84.45 g − 94.45 g)/(31.4 mL − 9.9 mL)
(d) (23.4 g + 102.4 g + 0.003 g)/(6.478 mL)
(e) (313.44 cm − 209.1 cm) × 8.2234 cm

3.37 Perform the following arithmetic and round off the answers to the correct number of significant figures:
(a) 3.58 g/1.739 mL
(b) 4.02 mL + 0.001 mL
(c) (22.4 g − 8.3 g)/(1.142 mL − 0.002 mL)
(d) (1.345 g + 0.022 g)/(13.36 mL − 8.4115 mL)
(e) (74.335 m − 74.332 m)/(4.75 s × 1.114 s)

Review of Scientific Notation (See the web site)

3.38 Express the following numbers in scientific notation. Assume three significant figures in each number.
(a) 4340   (c) 0.003287   (e) 0.00000800
(b) 32,000,000   (d) 42,000   (f) 324,300

3.39 Express the following numbers in scientific notation. Assume, in this problem, that only the nonzero digits are significant figures:
(a) 489   (c) 82,300   (e) 2.43
(b) 0.00375   (d) 0.01225   (f) 27,320

3.40 Write the following numbers in standard, nonexponential form:
(a) 3.1×10^5 (d) 4.4×10^{-12}
(b) 4.35×10^{-6} (e) 35.6×10^{-7}
(c) 3.9×10^3 (f) 8.8×10^8

3.41 Write the following numbers in standard, nonexponential form:
(a) 5.27×10^{-8} (c) 43.5×10^{-9} (e) 4.0000×10^7
(b) 7.12×10^5 (d) 2.35×10^{-2} (f) 37.2×10^2

3.42 Perform the following arithmetic and express the answers in scientific notation:
(a) $(4.0 \times 10^7) - (2.1 \times 10^5)$
(b) $(3.0 \times 10^{-2}) + (3.21 \times 10^{-5})$
(c) $(2.1 \times 10^7) \times (2.1 \times 10^5)$
(d) $(3.0 \times 10^4) \times (8.2 \times 10^{-5})$
(e) $(9.10 \times 10^{12})/(2.0 \times 10^{-3})^2$

3.43 Perform the following arithmetic and express the answers in scientific notation:

(a) $(3.0 \times 10^4) \times (2.1 \times 10^5)$
(b) $(8.0 \times 10^{12})/(2.0 \times 10^{-3})^2$
(c) $(6.7 \times 10^{-5}) \times (8.2 \times 10^{-5})$
(d) $(1.4 \times 10^5) - (3.0 \times 10^4)$
(e) $(3.3 \times 10^{-4}) + (2.52 \times 10^{-2})$

Unit Conversions by the Factor-Label Method

3.44 Perform the following conversions:
(a) 32.0 dm to km   (d) 137.5 mL to L
(b) 8.2 mg to µg   (e) 0.025 L to mL
(c) 75.3 mg to kg   (f) 342 pm to dm

3.45 Perform the following conversions:
(a) 92 dL to mL   (d) 230 km to m
(b) 22 ng to µg   (e) 87.3 cm to km
(c) 83 pL to nL   (f) 238 mm to nm

3.46 Perform the following conversions; express your answers in scientific notation:
(a) 230 km to cm   (d) 430 µL to mL
(b) 423 kg to mg   (e) 27 ng to kg
(c) 423 kg to Mg   (f) 730 nL to kL

3.47 Perform the following conversions; express your answers in scientific notation:
(a) 183 nm to cm   (d) 33 dm to mm
(b) 3.55 g to dg   (e) 0.55 dm to km
(c) 6.22 km to nm   (f) 53.8 ng to pg

3.48 Perform the following conversions. If necessary, refer to Tables 3.4 and 3.5.
(a) 36 in. to cm   (d) 1 cup (8 oz) to mL
(b) 5.0 lb to kg   (e) 55 mi/h to km/h
(c) 3.0 qt to mL   (f) 50.0 mi to km

3.49 Perform the following conversions. If necessary, refer to Tables 3.4 and 3.5.
(a) 250 mL to qt   (d) 1.75 L to fluid oz
(b) 3.0 ft to m   (e) 35 km/h to mi/h
(c) 1.62 kg to lb   (f) 80.0 km to mi

3.50 Cola is often sold in 12 oz cans. How many milliliters are in each can?

3.51 Milk is often sold in 32 oz cartons. How many milliliters are in each carton?

3.52 An adult human being has a blood volume of approximately 4700 mL. How many quarts of blood is this?

3.53 How many fluid ounces are in a 2.00 L bottle of cola?

3.54 A metric ton is exactly 1000 kg. How many pounds is this?

3.55 A "short ton" is exactly 2000 lb. How many kilograms is this?

3.56 The human stomach can expand to hold up to 4.2 qt of food. A pistachio nut has a volume of about 0.9 mL. Use this information to estimate the maximum number of pistachios that can be eaten in one sitting.

3.57 In the movie *Cool Hand Luke* (1967), Luke wagers that he can eat 50 eggs in 1 h. The prisoners and guards bet against him, saying, "Fifty eggs gotta weigh a good six pounds. A man's gut can't hold that." A chewed, peeled chicken egg has a volume of approximately 53 mL. If

Luke's stomach has a volume of 4.2 qt, does he have any chance of winning the bet?

ILW 3.58 The winds in a hurricane can reach almost 200 mph. What is this speed in meters per second?

3.59 A bullet is fired at a speed of 2435 ft/s. What is this speed expressed in kilometers per hour?

3.60 In the United States, a "long ton" is 2240 lb. What is this mass expressed in metric tons (l metric ton = 1000 kg)?

3.61 If a person weighs 75 kg, what is the person's weight in pounds?

3.62 If a person is 6 ft 2 in. tall, what is the person's height in centimeters?

3.63 If a person weighs 131 lbs 12 oz, what is the person's weight in kilograms?

3.64 Perform the following conversions:
(a) 8.4 ft^2 to cm^2 (b) 223 mi^2 to km^2 (c) 231 ft^3 to cm^3

3.65 Perform the following conversions:
(a) 2.4 yd^2 to m^2 (b) 8.3 $in.^2$ to mm^2 (c) 9.1 ft^3 to L

3.66 Perform the following conversions:
(a) 1.0 $in.^2$ to m^2 (b) 3.7 mi^2 to km^2 (c) 144 $in.^3$ to mL

3.67 Perform the following conversions:
(a) 9.3×10^2 mm^2 to $in.^2$ (c) 314 $in.^2$ to m^2
(b) 8.6 m^3 to ft^3

3.68 Single-layer DVD disks scan at a speed of about 3.5 m/s. What is this speed in miles per hour?

3.69 In the United States, the speed limit on some highways is 65 mph. What is this speed expressed in kilometers per hour?

Density and Specific Gravity

3.70 A sample of kerosene weighs 36.4 g. Its volume was measured to be 45.6 mL. What is the density of the kerosene?

3.71 A block of magnesium has a mass of 14.3 g and a volume of 8.46 cm^3. What is the density of magnesium in g/cm^3?

3.72 A proton weighs about 1.66×10^{-24} g and has a diameter of approximately 10^{-15} m. What is the density of a proton in g/cm^3?

3.73 The earth has a mass of approximately 5.98×10^{24} kg and a radius of 6.38×10^6 m. What is the density of the earth in g/cm^3?

3.74 Acetone, the solvent in nail polish remover, has a density of 0.791 g/mL. What is the volume of 25.0 g of acetone?

3.75 A glass apparatus contains 26.223 g of water when filled at 25 °C. At this temperature, water has a density of 0.99704 g/mL. What is the volume of this apparatus?

3.76 Chloroform, a chemical once used as an anesthetic, has a density of 1.492 g/mL. What is the mass in grams of 185 mL of chloroform?

3.77 Gasoline has a density of about 0.65 g/mL. How much does 34 L (approximately 18 gal) weigh in kilograms? In pounds?

ILW 3.78 A graduated cylinder was filled with water to the 15.0 mL mark and weighed on a balance. Its mass was 27.35 g. An object made of silver was placed in the cylinder and completely submerged in the water. The water level rose to 18.3 mL. When reweighed, the cylinder, water, and silver object had a total mass of 62.00 g. Calculate the density of silver.

3.79 Titanium is a metal used to make golf clubs. A rectangular bar of this metal measuring 1.84 cm × 2.24 cm × 2.44 cm was found to have a mass of 45.7 g. What is the density of titanium?

3.80 Ethyl ether has been used as an anesthetic. Its density is 0.715 g/mL. What is its specific gravity?

3.81 Propylene glycol is used as a food additive and in cosmetics. Its density is 8.65 lb/gal. Calculate the specific gravity of propylene glycol.

3.82 Trichloroethylene is a common dry-cleaning solvent. Its specific gravity is 1.47 measured at 20 °C. What is the mass in grams of a liter of this solvent?

3.83 The Space Shuttle uses liquid hydrogen as its fuel. The external fuel tank used during takeoff carries 227,641 lb of hydrogen with a volume of 385,265 gal. Use the concept of specific gravity to calculate the density of liquid hydrogen in units of g/mL. (Express your answer to three significant figures.)

ILW 3.84 Gold has a specific gravity of 19.3. Calculate the mass, in pounds, of 1 cubic foot of gold.

3.85 From the data in Problem 3.83, you can calculate that liquid hydrogen has a specific gravity of 0.0708. How many cubic feet of liquid hydrogen weigh 565 lb?

ADDITIONAL EXERCISES

3.86 During a hit-and-run accident, a red car strikes a white car and leaves the scene. A chip of red paint is found in the dent left on the white car. Chips of white paint are found on the fender of the red car. What qualitative observations could be made to determine whether the paint chips are consistent with the paints on the vehicles? What quantitative observations could be made? Which will be more persuasive in court? Why?

3.87 You are the science reporter for a daily newspaper and your editor asks you to write a story based on a report in the scientific literature. The report states that analysis of the sediments in Hausberg Tarn (elevation 4350 m) on the side of Mount Kenya (elevation 4600–4700 m) shows that the average temperature of the water rose by 40 °C between 350 B.C. and A.D. 450. Your editor tells you that she wants all the data expressed in the English system of units. Make the appropriate conversions.

3.88 An astronomy website states that neutron stars have a density of 100 million tons per cm^3. The site does not specify whether "tons" means metric tons (1 metric ton = 1000 kg) or English tons (1 English ton = 2000 lb). How many grams would 1 teaspoon of a neutron star weigh, if the density were in metric tons per cm^3? How many grams would the teaspoon weigh if the density were in English tons per cm^3? (One teaspoon is approximately 4.93 mL.)

3.89 Distances between stars are often expressed in units of *light-years*. A light-year is the distance that light travels during 1 year. Use the factor-label method to calculate the number of miles in one light-year, given that light travels at a speed of 3.0×10^8 meters per second. [*Hint:* Begin with *1 year* (365 days) and then use conversion factors to change that to a distance traveled by light.]

3.90 A pycnometer is a glass apparatus used for accurately determining the density of a liquid. When dry and empty, a certain pycnometer had a mass of 27.314 g. When filled with distilled water at 25.0 °C, it weighed 36.842 g. When filled with chloroform (a liquid once used as an anesthetic before its toxic properties were known), the apparatus weighed 41.428 g. At 25.0 °C, the density of water is 0.99704 g/mL. (a) What is the volume of the pycnometer? (b) What is the density of chloroform?

3.91 Radio waves travel at the speed of light, 3.0×10^8 meters per second. If you were to broadcast a question to an astronaut on the moon, which is 239,000 miles from earth, what is the minimum time that you would have to wait to receive a reply?

3.92 Suppose you have a job in which you earn $4.50 for each 30 min that you work.
(a) Express this information in the form of an equivalence between dollars and minutes worked.
(b) Use the equivalence defined in (a) to calculate the number of dollars earned in 1 h 45 min.
(c) Use the equivalence defined in (a) to calculate the number of minutes you would have to work to earn $17.35.

3.93 When an object floats in water, it displaces a volume of water that has a weight equal to the weight of the object. If a ship has a weight of 4255 tons, how many cubic feet of seawater will it displace? Seawater has a specific gravity of 1.025 and 1 ton = 2000 lb.

3.94 Aerogel or "solid smoke" is a novel material that is made of silicon dioxide, like glass, but is a thousand times less dense than glass because it is extremely porous. Material scientists at NASA's Jet Propulsion Laboratory created the lightest aerogel ever in 2002, with a density of 0.00011 lb/in^3. The material was used for thermal insulation in the 2003 *Mars Exploration Rover.* If the maximum space for insulation in the spacecraft's hull is 2510 cm^3, what mass (in grams) will the aerogel insulation add to the spacecraft?

Aerogel.

3.95 Carbon tetrachloride and water do not dissolve in each other. When mixed, they form two separate layers.

Carbon tetrachloride has a density of 13.3 lb/gal. Which liquid will float on top of the other when the two are mixed?

3.96 If you were given only the masses of samples of water and methanol, or if you were given only the volume of each liquid, you could not tell them apart. However, if you were given both their masses *and* their volumes, you could determine which was the methanol and which was the water. How?

3.97 A liquid known to be either ethanol (ethyl alcohol) or methanol (methyl alcohol) was found to have a density of 0.798 ± 0.001 g/mL. Consult the *Handbook of Chemistry and Physics* to determine which liquid it is. What other measurements could help to confirm the identity of the liquid?

3.98 An unknown liquid was found to have a density of 69.22 lb/ft^3. The density of ethylene glycol (the liquid used in antifreeze) is 1.1088 g/mL. Is the unknown liquid ethylene glycol?

3.99 The density of propylene glycol, a substance used as a nontoxic antifreeze, is 1.040 g/mL at 20 °C. What is the density of this substance at 20 °C expressed in units of (a) grams per liter and (b) kilograms per cubic meter?

3.100 The density of isopropyl alcohol, used in rubbing alcohol, is 785 kg/m^3 at 20 °C. What is the density of this alcohol at 20 °C expressed in units of (a) pounds per gallon and (b) grams per liter?

3.101 When an object is heated to a high temperature, it glows and gives off light. The color balance of this light depends on the temperature of the glowing object. Photographic lighting is described, in terms of its color balance, as a temperature in kelvins. For example, a certain electronic flash gives a color balance (called *color temperature*) rated at 5800 K. What is this temperature expressed in °C?

*3.102 There exists a single temperature at which the value reported in °F is numerically the same as the value reported in °C. What is this temperature?

*3.103 In the text, the Kelvin scale of temperature is defined as an absolute scale in which one Kelvin degree unit is the same size as one Celsius degree unit. A second absolute temperature scale exists called the Rankine scale. On this scale, one Rankine degree unit (°R) is the same size as one Fahrenheit degree unit. (a) What is the only temperature at which the Kelvin and Rankine scales possess the same numerical value? Explain your answer. (b) What is the boiling point of water expressed in °R?

3.104 An old cooking adage says that "a pint is a pound, the whole world round." Is the adage accurate for water?

*3.105 Density measurements can also be used to analyze mixtures. For example, the density of solid sand (without airspaces) is about 2.84 g/mL. The density of gold is 19.3 g/mL. If a 1.00 kg sample of sand containing some gold has a density of 3.10 g/mL (without airspaces), what is the percentage by mass of gold in the sample?

The Mole: Connecting the Macroscopic and Molecular Worlds

If we know how much one piece of candy weighs, we can count large numbers of them by weighing them on a scale such as this. We just multiply the weight of one piece by the total number we want, so we know how much they must weigh altogether. We use a similar concept in chemistry when we measure large numbers of atoms and molecules for use in experiments, as you learn in this chapter.

THIS CHAPTER IN CONTEXT In the last chapter, we saw that questions that ask "how much?" are crucial in practical applications of chemistry. For example, if a manufacturer wishes to stay in business, questions like these must be answered:

- How much of a raw material is required to produce a desired amount of a commercial product?
- How much of one material would be required to completely use up another in the manufacturing process?
- If many raw materials are required, how much of each should be used so that nothing is wasted?
- How does the predicted amount of product compare with the actual amount?

To answer any of these questions, the manufacturer must be able to *quantitatively* relate amounts of raw materials with amounts of product. This usually involves scaling up the recipe for making one item to make any desired number of items. Consider the manufacture of gingerbread men, shown in Figure 4.1. On a small scale, one knows that each gingerbread man requires ten currants for eyes and buttons. On a larger scale, currants must be ordered by the ton. To run a factory, one must be able to connect the small-scale manufacture of an item with large-scale production.

In chemistry, the essential link between the small-scale world of atoms and molecules and large-scale mass measurements is the **mole.** In this chapter, we'll see how the mole can be used to quantitatively relate amounts of substances embedded in compounds or involved in chemical reactions. The part of chemistry that deals with these quantitative relationships is called **stoichiometry** (stoy-kee-AH-meh-tree), which loosely translates as "the

FIGURE 4.1 *Manufacture on small and large scales.* Making a single gingerbread man requires ten currants for eyes and buttons, one-tenth of a cup of spiced cookie dough, and a teaspoon of glacé icing. To manufacture gingerbread men by the million, you will have to order raw materials by the ton. To scale up the recipe, you will need to know the masses of raw materials required to manufacture some fixed number of gingerbread men.

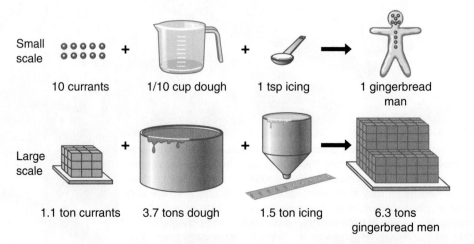

measure of the elements." We have two goals in this chapter. First, we will use the mole concept to quantitatively link numbers of atoms or molecules with mass measurements. Then we will use this link to solve practical problems involving amounts of substances that are produced or consumed in chemical reactions.

4.1 ▶ Use large-scale measurements to count tiny objects

Counting a pocketful of coins is easy. But how would you count a bucketful of coins? A truckload of coins? The more coins you have to count, the less practical a direct count becomes.

We can make a reasonable estimate of the number of coins if we know their mass. Suppose we know that a jar contains 10.0 kg of coins. Let's use the factor-label method to decide what additional information we will need to estimate the number of coins in the jar.

$$10.0 \text{ kg} \times \left(\frac{? \text{ coins}}{? \text{ kg}} \right) = ? \text{ coins}$$

If we knew the number of coins in 1 kilogram, we could fill in the conversion factor and complete the calculation. But counting a kilogram of coins is tedious. Suppose we count out 32 coins and find that they have a mass of 99.0 grams. This measurement gives us a gram-to-coin conversion factor. If we can convert kilograms to grams, and then grams to coins, we have solved the problem.

$$10.0 \text{ kg} \times \left(\frac{1000 \text{ g}}{1 \text{ kg}} \right) \times \left(\frac{32 \text{ coins}}{99.0 \text{ g}} \right) = 3.23 \times 10^3 \text{ coins}$$

We have avoided directly counting the coins by looking for a "link" between a large-scale measurement (kilograms of coins, in this case) and a small-scale quantity (number of coins).

Notice that we write 3.23×10^3 coins and not 3232 coins because the least precise measurement we've used has only three significant digits. Our estimate of the number of coins is uncertain in the tens place. To count the coins more precisely, we need to make more precise mass measurements.

Count atoms in a mass of element using atomic masses

We can use the same trick to count the number of atoms in a sample of an element. Instead of a link between mass and coins, we need a link between mass and atoms. The critical link is the atomic mass, which gives the mass of a single atom.

How many atoms of copper are in 10.0 kg of copper?

ANALYSIS: We can treat "atoms of copper" and "kg of copper" as units and search for relationships that link them.

$$10.0 \text{ kg Cu} \Longleftrightarrow ? \text{ atoms Cu}$$

We need a link between mass and number of atoms. As we have noted above, atomic masses give the mass of a single atom. The atomic mass of copper is 63.55, so one atom of copper has a mass of 63.55 atomic mass units (u). We can use atomic mass units as a stepping-stone between kilograms and atoms.

$$\boxed{63.55 \text{ u} = 1 \text{ atom Cu}}$$

$$10.0 \text{ kg} \longrightarrow \text{u} \longrightarrow \textbf{atoms Cu}$$

All we need to complete the calculation is a conversion factor from kilograms of copper (kg Cu) to atomic mass units (u Cu). The table of fundamental physical constants on the inside back cover of the book gives the conversion factor as

EXAMPLE 4.1

Counting Atoms in a Sample of Element (Using Atomic Mass, Without Moles)

Recall from Chapter 1 that an atomic mass unit is defined as exactly one-twelfth the mass of a carbon-12 atom. The atomic masses on the periodic table are unitless because they are relative to the mass of 1 atomic mass unit.

$$1 \text{ u} = 1.661 \times 10^{-27} \text{ kg}$$

where we have rounded the constant to four figures (our measurement, 10.0 kg, limits us to three figures). Our overall roadmap is

$$\boxed{63.55 \text{ u} = 1 \text{ atom Cu}}$$

$$\textbf{10.0 kg} \longrightarrow \textbf{u} \longrightarrow \textbf{atoms Cu}$$

$$\boxed{1.661 \times 10^{-27} \text{ kg} = 1 \text{ u}}$$

SOLUTION: Build conversion factors from the linking relationships. The first conversion factor must cancel kilograms to give atomic mass units. The second must cancel atomic mass units to give atoms.

$$10.0 \text{ kg Cu} \times \left(\frac{1 \text{ u Cu}}{1.661 \times 10^{-27} \text{ kg Cu}} \right) \times \left(\frac{1 \text{ atom Cu}}{63.55 \text{ u Cu}} \right) = 9.48 \times 10^{25} \text{ atoms Cu}$$

Is the Answer Reasonable?
We would expect the answer to be large, since atoms are small, but this is an unimaginably large number! Let's check to see that the order of magnitude is correct by rounding off the masses to the nearest integer:

$$\frac{10}{2 \times 10^{-27} \times 60} = 8 \times 10^{25}$$

which agrees reasonably well with our previous answer.

PRACTICE EXERCISE 1: How many atoms of gold are in a 15.0 g ring made of pure gold?

The periodic table and table of atomic masses on the inside front cover of this book give some atomic masses to six or more significant figures. Most of the numerical problems in this and later chapters involve data with only three or four significant figures. As a rule, we will round atomic masses so that they have one more significant figure than required by the data in the problem so the atomic mass data does not contribute much uncertainty to the final result.

Mass measurements can be used to count molecules using molecular masses

We can count the number of molecules in a sample of a molecular compound by the same procedure, but we need to have a **molecular mass** rather than an atomic mass. The molecular mass is simply the sum of atomic masses of the atoms in the compound's formula. For example, to five significant digits, the molecular mass of water, H_2O, is twice the atomic mass of hydrogen (1.008) plus the atomic mass of oxygen (15.999), or 18.015.

Many chemists use the terms *molecular weight* and *atomic weight* for molecular mass and atomic mass.

Recall that ionic compounds don't contain molecules; they are composed of ions, so it isn't strictly correct to refer to the "molecular mass" of an ionic compound. We shall refer to the mass of a formula unit as a **formula mass.** Formula masses are calculated just as molecular masses are. They are simply the sum of the atomic masses of the atoms shown in the formula. Consider the following example.

The smell of the sea comes from a gas called dimethyl sulfide, $(CH_3)_2S$, produced by dying phytoplankton, tiny plants that are the base of the ocean's food chain. What is the molecular mass of dimethyl sulfide, written to five significant figures?

ANALYSIS: The molecular mass is just the sum of the atomic masses. Remember that a subscript *outside* the parentheses in the formula is a multiplier for all the atoms inside.

SOLUTION: Using atomic masses of 12.011 for C, 1.008 for H, and 32.066 for S,

$$(CH_3)_2S = 2\,C + 6H + 1S$$

$$\text{molecular mass of } (CH_3)_2S = (2 \times 12.011) + (6 \times 1.008) + (1 \times 32.066)$$

$$= 24.022 + 6.048 + 32.066$$

$$= 62.136$$

Like atomic masses, molecular masses are sometimes written without units because it is understood that they are *relative* to the mass of 1 atomic mass unit. We can state that the mass of one molecule of $(CH_3)_2S$ is 62.136 atomic mass units (62.136 u), or we can say that the molecular mass of $(CH_3)_2S$ is 62.136.

PRACTICE EXERCISE 2: What is the molecular mass of a molecule of *n*-octane, $CH_3(CH_2)_6CH_3$?

PRACTICE EXERCISE 3: What is the molecular mass of a molecule of heroin, $C_{21}H_{23}NO_5$?

EXAMPLE 4.2

Computing the Molecular Mass from a Chemical Formula

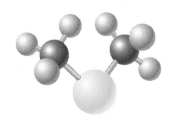

Dimethyl sulfide, $(CH_3)_2S$, is responsible for the smell of the sea.

The stench of $(CH_3)_2S$ is so powerful that it can easily be detected in very small amounts. As little as 1.0 µg of $(CH_3)_2S$ dissolved in a liter of sea air is detectable. How many molecules of $(CH_3)_2S$ are in 1.0 µg of $(CH_3)_2S$?

ANALYSIS: We can treat "molecules of $(CH_3)_2S$" and "µg of $(CH_3)_2S$" as units and search for relationships that link them.

$$1.0 \text{ µg } (CH_3)_2S \Leftrightarrow ? \text{ molecules } (CH_3)_2S$$

We need a link between mass and molecules. As we have noted, molecular masses give the mass of a single molecule. The molecular mass of $(CH_3)_2S$ was calculated in Example 4.2 as 62.136, so one molecule of $(CH_3)_2S$ has a mass of 62.136 atomic mass units (u). We can use atomic mass units as an intermediate "stepping-stone" between grams and molecules.

$$\boxed{62.136 \text{ u} = 1 \text{ molecule}}$$

$$\mathbf{1.0 \text{ µg} \longrightarrow u \longrightarrow molecules}$$

where we have dropped the formula $(CH_3)_2S$ after the units because $(CH_3)_2S$ is the only substance involved in this problem. Now we need a relationship that links grams and atomic mass units. The table of fundamental physical constants on the inside back cover of the book gives the conversion factor as

$$1 \text{ u} = 1.66 \times 10^{-27} \text{ kg}$$

so we'll have to add kg as an intermediate stepping-stone in our calculation. Our overall roadmap is now

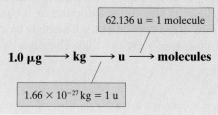

EXAMPLE 4.3

Counting the Number of Molecules in a Sample of Compound (Using Molecular Masses, Without Moles)

We now need to find relationships to link micrograms (μg) with kilograms (kg). The necessary relationships (from Chapter 3) are

$$1 \text{ kg} = 10^3 \text{ g}$$

$$1 \text{ μg} = 10^{-6} \text{ g}$$

which tells us that we should insert grams (g) as an intermediate step between micrograms and kilograms. Our complete roadmap is

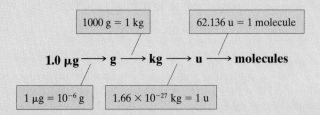

SOLUTION: All we have to do now to solve the problem is build conversion factors from the linking relationships we have listed above. The first conversion factor must cancel grams to give kilograms, so kilograms go on top and grams on bottom. The second must cancel kilograms to give atomic mass units, and the third must cancel atomic mass units to give molecules.

$$1.0 \text{ μg} \times \left(\frac{10^{-6} \text{ g}}{1 \text{ μg}} \right) \times \left(\frac{1 \text{ kg}}{1000 \text{ g}} \right) \times \left(\frac{1 \text{ u}}{1.66 \times 10^{-27} \text{ kg}} \right) \times \left(\frac{1 \text{ molecule}}{62.136 \text{ u}} \right)$$

$$= 9.7 \times 10^{15} \text{ molecules}$$

Is the Answer Reasonable?
We expect the answer to be large, since molecules are small. We can check to see that the order of magnitude is correct by multiplying the powers of 10 associated with each number:

$$\frac{10^{-6}}{10^3 \times 10^{-27} \times 10^2} = 10^{16}$$

which agrees reasonably well with our previous answer. (Notice the final power of 10 in the denominator is 10^2 because 1.66×62.136 is closer to 10^2 than 10^1.)

PRACTICE EXERCISE 4: How many molecules of heroin are in 0.010 g of heroin? The formula is $C_{21}H_{23}NO_5$.

4.2 ▶ **The mole conveniently links mass to number of atoms or molecules**

So far we have used atomic mass units to link measurements with atoms or molecules. The problem with this approach is that we must always do a tedious conversion from grams or kilograms to atomic mass units. Atomic mass units are very small. A unit that was closer in size to the kilogram or the gram would allow us to perform these conversions more conveniently. Chemists have defined a unit called the **mole** (abbreviated **mol**) to do exactly that. The mole is defined within the SI system as follows:

Mole is a Latin word with several meanings, including a shapeless mass, a large number, or trouble or difficulty.

> *One mole of any substance contains the same number of formula units as the number of atoms in exactly 12 g of carbon-12.*

We defined an atomic mass unit as exactly one-twelfth the mass of a carbon-12 atom. The definition of mole is therefore closely related to the definition of the atomic mass unit. This leads us to a more immediately useful definition of the mole:

> **One mole of a substance has a mass in grams that is numerically equal to its formula mass.**

Molar mass from atomic mass, formula mass, or molecular mass

The mass of one mole of a substance is sometimes referred to as its **molar mass.** The number of grams in a mole is different from substance to substance because different substances have different formula masses.

Why is this definition of the mole so useful? Let's calculate the number of molecules in 1 mol of water. Water has a formula mass of 18.0 (to three significant digits), so

$$18.0 \text{ g } H_2O \Leftrightarrow 1 \text{ mole } H_2O$$

$$18.0 \text{ u } H_2O \Leftrightarrow 1 \text{ molecule } H_2O$$

Using the relationships we used to solve Example 4.3, we can convert 18.0 g of water to molecules of water:

$$18.0 \text{ g } H_2O \times \left(\frac{1 \text{ kg } H_2O}{1000 \text{ g } H_2O} \right) \times \left(\frac{1 \text{ u } H_2O}{1.66 \times 10^{-27} \text{ kg } H_2O} \right) \times \left(\frac{1 \text{ molecule } H_2O}{18.0 \text{ u } H_2O} \right)$$

$$= 6.02 \times 10^{23} \text{ molecules } H_2O$$

Not impressed? *Try the same calculation with any other substance and you will get exactly the same answer.* For example, let's calculate the number of atoms in 1 mol of helium. Helium has an average atomic mass of 4.00, so

$$4.00 \text{ g He} \Leftrightarrow 1 \text{ mole He}$$

$$4.00 \text{ u He} \Leftrightarrow 1 \text{ atom He}$$

$$4.00 \text{ g He} \times \left(\frac{1 \text{ kg He}}{1000 \text{ g He}} \right) \times \left(\frac{1 \text{ u He}}{1.66 \times 10^{-27} \text{ kg He}} \right) \times \left(\frac{1 \text{ atom He}}{4.00 \text{ u He}} \right)$$

$$= 6.02 \times 10^{23} \text{ atoms He}$$

Notice that the formula mass always cancels out of the calculation, and the number of atoms depends only on the definition of the atomic mass unit.

> **One mole of any substance contains the same number of formula units (about 6.02×10^{23} formula units).**

Avogadro's number

The number of grams in a mole changes from substance to substance. However, the number of particles in a mole is the same for all substances. This confuses many students. Think of a mole as a "chemist's dozen." Just as 12 eggs is a dozen eggs, 6.02×10^{23} eggs is a mole of eggs. A dozen hen's eggs has a different mass than a dozen ostrich eggs, but in each case, you have 12 eggs. Similarly, a mole of water has a different mass than a mole of plutonium, but in each case, you have 6.02×10^{23} molecules.

The number of particles in a mole is known as **Avogadro's number** or **Avogadro's constant.** Avogadro's number is unimaginably large. For example, 6.02×10^{23} grains of sand would cover the state of Texas to a depth of 15 meters (50 feet). Yet because atoms are so tiny, Avogadro's number of atoms or molecules can usually fit into the palm of your hand (see Figures 4.2 and 4.3).

When we refer to a mole of something, it is always important to identify clearly what that "something" is. For example, the expression "1 mole of oxygen" is ambiguous. It could mean either 1 mole of oxygen atoms or 1 mole of oxygen molecules. To avoid such confusion, we usually associate a chemical formula with the unit mol. There is no ambiguity, therefore, if we write "1 mol O" to mean "1 mole of oxygen atoms" and "1 mol O_2" to mean "1 mole of oxygen molecules."

Avogadro's number was named for Amedeo Avogadro (1776–1856), an Italian scientist who was one of the pioneers of stoichiometry.

Jean Baptiste Perrin was the first person to use the term *Avogadro's number* for the number of particles in a mole and was one of the first to accurately estimate the size of the number from studies of Brownian motion (see Chapter 2).

FIGURE 4.2 *Moles of elements.* Each sample of these elements contains the same number of atoms, Avogadro's number.

FIGURE 4.3 *Moles of compounds.* One mole of four different compounds. Each sample contains the identical number of formula units or molecules, Avogadro's number.

Convert mass to moles using molar masses

We can easily convert grams of an element or compound into moles using the molar mass, as shown in the following examples.

EXAMPLE 4.4

Converting Grams of Element to Moles of Element

NASA's Cassini–Huygens space probe used plutonium dioxide pellets as fuel. There was some concern that if the craft burned up in earth's atmosphere, it would disperse small inhalable particles of plutonium dioxide that could cause lung cancer. To estimate the effects of radiation on the structural integrity of the pellets, we must first calculate the number of moles of Pu-238 in a single fuel pellet. If each pellet contained 0.880 g of Pu-238, how many moles of Pu-238 were in the fuel pellet?

ANALYSIS: The problem can be stated as follows:

$$0.880 \text{ g Pu-238} \Leftrightarrow \text{? mol Pu-238}$$

In other words, 0.880 g of Pu-238 is equivalent to how many moles of Pu-238? *The critical link between moles and mass is the molar mass.* The atomic mass gives the molar mass of an element. Note that this is an isotope of plutonium, so we'll use the isotope mass rather than the average atomic mass given on the periodic table. For Pu-238, then, the molar mass is approximately 238 g. We can write

$$238 \text{ g Pu-238} \Leftrightarrow 1 \text{ mol Pu-238}$$

Now two possible conversion factors become clear; one is the tool we need for the grams-to-moles calculation.

$$\frac{238 \text{ g Pu-238}}{1 \text{ mol Pu-238}} \quad \text{and} \quad \frac{1 \text{ mol Pu-238}}{238 \text{ g Pu-238}}$$

If we multiply the given, 0.880 g of Pu-238, by the second conversion factor, grams will cancel to yield moles of Pu-238.

SOLUTION: Draw in the cancel lines yourself.

$$0.880 \text{ g Pu-238} \times \frac{1 \text{ mol Pu-238}}{238 \text{ g Pu-238}} = 3.70 \times 10^{-3} \text{ mol Pu-238}$$

In other words,

$$0.880 \text{ g Pu-238} \Longleftrightarrow 3.70 \times 10^{-3} \text{ mol Pu-238}$$

If you are writing in your own cancel lines, *be sure to show them canceling the full unit.* For example, cancel "g Pu-238," not just "g."

Is the Answer Reasonable?

- One mole of Pu-238 would weigh 238 g.
- One-thousandth of a mole would weigh one-thousandth of 238 g, or 0.238 g.
- Our answer was about four-thousandths of a mole, which should be 4×0.24 g or about 0.92 g. This isn't far from the mass of Pu-238 we started with (0.88 g).

PRACTICE EXERCISE 5: How many moles of sulfur are present in 35.6 g of sulfur?

The human body loses about 0.010 mol of potassium daily in the urine, with larger losses in the stool. How many grams of potassium are lost each day in the urine?

ANALYSIS: We can restate the question as follows.

$$0.010 \text{ mol K} \Longleftrightarrow ? \text{ g K}$$

We need to relate moles of an element with grams of an element. *The critical link between moles and grams is molar mass.* Since this is an element, the atomic mass gives the molar mass. The periodic table gives the atomic mass of potassium as 39.0983, which we will round to 39.1. We can then write the critical link between moles and grams as

$$1 \text{ mol K} \Longleftrightarrow 39.1 \text{ g K}$$

Now we can use this to form a conversion factor to solve the problem.

SOLUTION: The given is 0.010 mol K, so we form the conversion factor so that mol K will cancel:

$$0.010 \text{ mol K} \times \frac{39.1 \text{ g K}}{1 \text{ mol K}} = 0.39 \text{ g K}$$

Is the Answer Reasonable?
One mole of potassium has a mass of 39.1 g. A hundredth of a mole (0.010 mol) should have a mass that is one-hundredth of 39.1 g. Dividing by 100 is the same as bumping the decimal point two places to the left, so 0.391 g should be the mass of a hundredth of a mole of potassium. This is what we obtained above (although we rounded to two significant digits).

PRACTICE EXERCISE 6: How many grams of silver are in 0.263 mol of Ag?

EXAMPLE 4.5
Converting Moles of Element to Grams of Element

Bone and dental implants are often coated with calcium phosphate, $Ca_3(PO_4)_2$, to permit bone to bond with the implant surface. If a coating procedure can deposit 0.115 mol of pure $Ca_3(PO_4)_2$ on an implant, how many grams does the coating weigh?

ANALYSIS: We can restate the problem as follows:

$$0.115 \text{ mol Ca}_3(PO_4)_2 \Longleftrightarrow ? \text{ g Ca}_3(PO_4)_2$$

EXAMPLE 4.6
Converting Moles of Compound to Grams of Compound

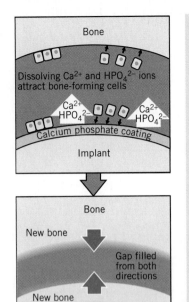

Some surfaces on bone implants are coated with calcium phosphate to permit bone to actually bond to the surface.

Avogadro's number is a **critical link** between moles of substance and particles of substance in a stoichiometry problem. If a problem does not mention atoms or molecules at all, you don't need to use Avogadro's number in the calculation!

The critical link between moles and grams of a substance is the molar mass. Since $Ca_3(PO_4)_2$ is an ionic compound, we can obtain the molar mass from the formula mass. Adding the atomic masses of three calcium atoms, two phosphorus atoms, and eight oxygen atoms gives a formula mass of 310.18 g/mol. Thus,

$$1 \text{ mol } Ca_3(PO_4)_2 \Leftrightarrow 310.18 \text{ g } Ca_3(PO_4)_2$$

SOLUTION: We have 0.115 mol $Ca_3(PO_4)_2$. Multiply it by a conversion factor constructed from the formula mass of $Ca_3(PO_4)_2$, arranged so that mol $Ca_3(PO_4)_2$ cancels, leaving g $Ca_3(PO_4)_2$:

$$0.115 \text{ mol } Ca_3(PO_4)_2 \times \frac{310.18 \text{ g } Ca_3(PO_4)_2}{1 \text{ mol } Ca_3(PO_4)_2} = 35.7 \text{ g } Ca_3(PO_4)_2$$

Is the Answer Reasonable?

Does the *size* of the answer make sense? Yes. The coating contains a little over one-tenth of a mole of $Ca_3(PO_4)_2$. A tenth of 310 is 31 g; our answer of 35.7 g is a bit more than 31 g, so it makes sense.

PRACTICE EXERCISE 7: How many grams of sodium carbonate, Na_2CO_3, must be measured to obtain 0.125 mol of Na_2CO_3?

PRACTICE EXERCISE 8: A sample of 45.8 g of H_2SO_4 contains how many moles of H_2SO_4?

Convert moles to number of molecules or atoms using Avogadro's number

Sometimes it is necessary to know the number of particles (atoms or molecules) in a sample. As we've seen, 1 mol of any substance contains Avogadro's number of particles of that substance. *Avogadro's number links moles and atoms, or moles and molecules.* It provides an easier way to link mass and atoms (or mass and molecules) than the atomic mass unit conversions we used in the previous section.

EXAMPLE 4.7
Converting Mass of an Element to Number of Atoms

Plutonium-239 is a radioactive isotope used primarily in nuclear weapons. Each atom of plutonium-239 can emit alpha particles as it decays. A single alpha particle can damage cells in a way that converts them into cancer cells. Plutonium in inhaled particles is particularly dangerous. How many atoms are in a particle of plutonium that contains 1.00×10^{-6} g (1.00 μg) of plutonium-239?

ANALYSIS: As usual, begin by expressing the problem in the form of an equation. For simplicity, we'll write "Pu" in units, rather than "^{239}Pu."

$$1.00 \times 10^{-6} \text{ g Pu} \Leftrightarrow ? \text{ atom Pu}$$

We are converting a macroscopic unit (g Pu) into microscopic particles (atoms). *The critical link in a problem that converts macroscopic amounts (such as grams or moles) to microscopic particles is Avogadro's number.* Since we are dealing with plutonium specifically, we can write

$$1 \text{ mol Pu} \Leftrightarrow 6.02 \times 10^{23} \text{ atom Pu}$$

We can now go from the unit "mol Pu" to "atom Pu." This suggests that moles will be an intermediate stepping-stone in our calculation:

$$1.00 \times 10^{-6} \text{ g Pu} \longrightarrow \text{mol Pu} \longrightarrow \text{atom Pu}$$

$$\boxed{1 \text{ mol Pu} = 6.02 \times 10^{23} \text{ atom Pu}}$$

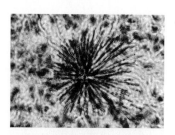

Now we need a relationship to link "g Pu" to "mol Pu." *The link between grams and moles is the molar mass.* Remember that we're dealing specifically with plutonium-239 in this problem. For this isotope, the molar mass is approximately 239 g:

$$1 \text{ mol Pu} \Leftrightarrow 239 \text{ g Pu}$$

where we have rounded the mass off to three significant digits. Our overall strategy is

The black star shows the tracks of alpha particles emitted from individual atoms of plutonium in a particle lodged in lung tissue.

$$\boxed{239 \text{ g Pu} = 1 \text{ mol Pu}}$$

$$1.00 \times 10^{-6} \text{ g Pu} \longrightarrow \text{mol Pu} \longrightarrow \text{atom Pu}$$

$$\boxed{1 \text{ mol Pu} = 6.02 \times 10^{23} \text{ atom Pu}}$$

We now have a complete set of relationships linking "g Pu" to "atom Pu."

SOLUTION: As usual, we assemble conversion factors from the linking relationships so that the units cancel correctly.

$$1.00 \times 10^{-6} \text{ g Pu} \times \left(\frac{1 \text{ mol Pu}}{239 \text{ g Pu}} \right) \times \left(\frac{6.02 \times 10^{23} \text{ atom Pu}}{1 \text{ mol Pu}} \right) = 2.52 \times 10^{15} \text{ atom Pu}$$

Thus, a 1 μg sample of plutonium-239, a microscopic speck, contains over 2,520,000,000,000,000 atoms, each of which can produce an alpha particle that may damage a lung cell.

Is the Answer Reasonable?
Let's check to see that the magnitude of the answer is reasonable. One mole or 239 g of plutonium-239 would contain 6.02×10^{23} atoms. A microgram (10^{-6} g) is about eight orders of magnitude smaller than 239 g. So we would expect the number of atoms in a microgram to be about eight orders of magnitude smaller (i.e., smaller by a factor of 10^8) than 6.02×10^{23}. In fact, $10^{23} / 10^8 = 10^{15}$, which is the order of magnitude we obtained.

PRACTICE EXERCISE 9: How many atoms are in 1.00×10^{-9} g (1.00 ng) of lead?

4.3 ▶ Chemical formulas relate amounts of substances in a compound

Consider the chemical formula for water, H_2O:

- One molecule of water contains two H atoms and one O atom.
- Two molecules of water contain four H atoms and two O atoms.
- A dozen molecules of water contain two dozen H atoms and one dozen O atoms.
- A mole of molecules of water contains 2 mol of H atoms and 1 mol of O atoms.

Whether we're dealing with atoms, dozens of atoms, or moles of atoms, the chemical formula tells us that the ratio of H atoms to O atoms is always 2 to 1.

Subscripts in a formula

> **Within chemical compounds, moles of atoms always combine in the same ratio as the individual atoms themselves.**

This fact lets us prepare mole-to-mole conversion factors involving elements in compounds as we need them. For example, in the formula P_4O_{10}, the subscripts mean that there are 4 mol of P for every 10 mol of O in this compound. We can relate P and O within the compound using the following conversion factors:

$$\frac{4 \text{ mol P}}{10 \text{ mol O}} \quad \text{or} \quad \frac{10 \text{ mol O}}{4 \text{ mol P}}$$

The formula P_4O_{10} also implies other equivalencies, each with its two associated conversion factors:

$$1 \text{ mol } P_4O_{10} \Longleftrightarrow 4 \text{ mol P} \quad \text{or} \quad \frac{1 \text{ mol } P_4O_{10}}{4 \text{ mol P}} \quad \text{and} \quad \frac{4 \text{ mol P}}{1 \text{ mol } P_4O_{10}}$$

$$1 \text{ mol } P_4O_{10} \Longleftrightarrow 10 \text{ mol O} \quad \text{or} \quad \frac{1 \text{ mol } P_4O_{10}}{10 \text{ mol O}} \quad \text{and} \quad \frac{10 \text{ mol O}}{1 \text{ mol } P_4O_{10}}$$

The reaction of phosphorus with oxygen that gives P_4O_{10} produces a brilliant light, often used in fireworks displays.

EXAMPLE 4.8
Relating Moles of Elements Within Compounds

The mineral pyrite (FeS_2) occurs naturally in coal beds. Mining the coal exposes it to air and water. This triggers a sequence of chemical reactions that release sulfur, S, into the environment as sulfuric acid, H_2SO_4. The sulfuric acid in rain runoff from the mine can devastate aquatic ecosystems around the mining operation. If every ton of soft coal mined ultimately results in the decomposition of 11 mol of FeS_2, how many moles of sulfuric acid would be released into the environment?

ANALYSIS: The problem can be restated as

$$11 \text{ mol } FeS_2 \Longleftrightarrow ? \text{ mol } H_2SO_4$$

The critical link in this problem will involve a mole-to-mole ratio, because we are dealing with two different substances. The problem states that FeS_2 releases sulfur into the environment as H_2SO_4. That suggests that moles of sulfur should be used as a link between moles of FeS_2 and moles of H_2SO_4. One mol of FeS_2 contains 2 mol of S. One mol of H_2SO_4 contains 1 mol of S. Here is our overall strategy:

$$\boxed{1 \text{ mol } FeS_2 \Longleftrightarrow 2 \text{ mol S}}$$

$$\textbf{11 mol FeS}_2 \longrightarrow \textbf{mol S} \longrightarrow \textbf{mol H}_2\textbf{SO}_4$$

$$\boxed{1 \text{ mol S} \Longleftrightarrow 1 \text{ mol } H_2SO_4}$$

SOLUTION: Starting with 11 mol FeS_2, we arrange the mole-to-mole conversion factors so that everything cancels but the desired units (mol H_2SO_4):

$$11 \text{ mol } FeS_2 \times \frac{2 \text{ mol S}}{1 \text{ mol } FeS_2} \times \frac{1 \text{ mol } H_2SO_4}{1 \text{ mol S}} = 22 \text{ mol } H_2SO_4$$

We leave two significant figures in the answer because the least precise measurement, 11 mol FeS_2, has two significant figures. (The 2 and the 1 in the conversion factors, which come from formula subscripts, are *exact* numbers of moles, and so they have an infinite number of significant figures.)

Is the Answer Reasonable?

There are 2 mol of sulfur for every mole of FeS_2. Therefore, 11 mol of FeS_2 should release 22 mol of sulfur. The number of moles of H_2SO_4 equals the number of moles of sulfur, because there is 1 mol of S for every mole of H_2SO_4.

PRACTICE EXERCISE 10: How many moles of nitrogen atoms are combined with 8.60 mol of oxygen atoms in dinitrogen pentoxide, N_2O_5?

One common use of stoichiometry in the lab occurs when we must relate the masses of two raw materials that are needed to make a compound.

Chlorophyll, the green pigment in leaves, has the formula $C_{55}H_{72}MgN_4O_5$. If 0.0011 g of Mg is available to a plant cell for chlorophyll synthesis, how many grams of carbon will be required to completely use up the magnesium?

ANALYSIS: Let's begin, as usual, by restating the problem as follows:

$$0.0011 \text{ g Mg} \Leftrightarrow ? \text{ g C} \quad \text{(for chlorophyll only)}$$

A mole-to-mole ratio is the critical link in problems that convert an amount of one substance into an amount of another. The formula of chlorophyll, $C_{55}H_{72}MgN_4O_5$, relates Mg to C within the compound. Using the formula's subscripts, the critical link between Mg and C is

$$1 \text{ mol Mg} \Leftrightarrow 55 \text{ mol C}$$

We know we'll need to use this relationship to solve the problem. Let's drop it into the middle of our roadmap:

$$\textbf{0.0011 g Mg} \longrightarrow \textbf{mol Mg} \longrightarrow \textbf{mol C} \longrightarrow \textbf{g C}$$

$$\boxed{1 \text{ mol Mg} \Leftrightarrow 55 \text{ mol C}}$$

By placing the critical link between our given information (0.0011 g Mg) and our desired units (g C) on the roadmap, we've cut a difficult problem into two simpler ones. All we have to do now to complete the flow of units is relate grams of Mg to moles of Mg, and moles of C to grams of C. The atomic mass of C links moles of C to grams of C, and the atomic mass of Mg links moles of Mg to grams of Mg. Rounding them to three significant figures, we can write

$$1 \text{ mol Mg} \Leftrightarrow 24.3 \text{ g Mg}$$

$$1 \text{ mol C} \Leftrightarrow 12.0 \text{ g C}$$

Our complete roadmap for the problem is

$$\boxed{24.3 \text{ g Mg} = 1 \text{ mol Mg}} \qquad \boxed{1 \text{ mol C} = 12.0 \text{ g C}}$$

$$\textbf{0.0011 g Mg} \longrightarrow \textbf{mol Mg} \longrightarrow \textbf{mol C} \longrightarrow \textbf{g C}$$

$$\boxed{1 \text{ mol Mg} \Leftrightarrow 55 \text{ mol C}}$$

SOLUTION: We now set up the solution by forming conversion factors so the units cancel (add the cancel marks yourself):

$$0.0011 \text{ g Mg} \times \left(\frac{1 \text{ mol Mg}}{24.3 \text{ g Mg}} \right) \times \left(\frac{55 \text{ mol C}}{1 \text{ mol Mg}} \right) \times \left(\frac{12.0 \text{ g C}}{1 \text{ mol C}} \right) = 0.030 \text{ g C}$$

EXAMPLE 4.9

Calculating Amounts of One Element from Amounts of Another in a Compound

A plant cell must supply 0.030 g C for every 0.0011 g Mg to completely use up the magnesium in the synthesis of chlorophyll.

Is the Answer Reasonable?

One gram of magnesium is 1/24.3 or about 0.04 mol Mg. We need 55 mol C for every mol Mg, so we require 55 × 0.04 or a little more than 2 mol C. One mole of carbon has a mass of 12.0 g, so a little more than 2 mol of carbon would weigh a little more than 24 g. We expect that the mass of carbon required to be a little more than 24 times the mass of magnesium we started out with.

Our answer was 0.030 g C from 0.0011 g Mg, so the mass of carbon we got is about 30 times the mass of the magnesium we started with. This seems reasonable.

PRACTICE EXERCISE 11: How many grams of iron are needed to combine with 25.6 g of O to make Fe_2O_3?

When extracting elements from their compounds, we often want to know how much element we can get from a given sample of compound. In the following example, we'll use a chemical formula to estimate the amount of copper that can be extracted from a copper sulfide ore.

EXAMPLE 4.10

Calculating the Mass of an Element in a Sample of a Compound

One of the forms in which copper (Cu) occurs in nature is copper(I) sulfide, Cu_2S. How many grams of copper metal can theoretically be obtained from 10.0 g of this compound?

ANALYSIS: We can restate the problem as follows:

$$10.0 \text{ g } Cu_2S \Longleftrightarrow ? \text{ g } Cu$$

The critical link between the mass of a compound and the mass of an element within it is the mole-to-mole ratio. We can obtain the mole-to-mole ratio using the subscripts in the formula:

$$1 \text{ mol } Cu_2S \Longleftrightarrow 2 \text{ mol } Cu$$

Our strategy could be (1) convert 10.0 g of Cu_2S to mol Cu_2S, using the molecular mass of Cu_2S; (2) convert mol Cu_2S to mol Cu, using the mole-to-mole ratio; (3) convert mol Cu to g Cu, using the atomic mass of Cu.

But there is a simpler way to do the problem. We can directly convert g Cu_2S into g Cu using intermediate results from our calculation of the formula mass of Cu_2S. We'll use the atomic masses of Cu and S and translate them into grams per mole for each element. Thus, in 1 mol of Cu_2S, we have

$$2 \text{ mol Cu:} \qquad 2 \text{ mol Cu} \times \frac{63.55 \text{ g Cu}}{1 \text{ mol Cu}} = 127.1 \text{ g Cu}$$

$$1 \text{ mol S:} \qquad 1 \text{ mol S} \times \frac{32.07 \text{ g S}}{1 \text{ mol S}} = 32.07 \text{ g S}$$

$$\overline{\qquad\qquad\qquad 1 \text{ mol } Cu_2S = 159.2 \text{ g } Cu_2S}$$

Thus, in 159.2 g of Cu_2S there is 127.1 g of Cu. We can express this as a mass equivalency for Cu_2S, namely,

$$159.2 \text{ g } Cu_2S \Longleftrightarrow 127.1 \text{ g } Cu$$

From it we can fashion a conversion factor to calculate how many grams of Cu are in only 10.0 g Cu_2S.

SOLUTION: The conversion factor we need is

$$\frac{127.1 \text{ g Cu}}{159.2 \text{ g } Cu_2S}$$

To calculate the mass of Cu in 10.00 g Cu_2S, we simply do the following:

$$10.0 \text{ g Cu}_2\text{S} \times \frac{127.1 \text{ g Cu}}{159.2 \text{ g Cu}_2\text{S}} = 7.98 \text{ g Cu}$$

Thus, in 10.0 g Cu_2S there is 7.98 g Cu.

Is the Answer Reasonable?
Notice that because of the high atomic mass of Cu and the low atomic mass of S, most of a Cu_2S sample is copper, and 7.98 g is "most of" 10.0 g.

PRACTICE EXERCISE 12: How many grams of iron are in a 15.0 g sample of iron(III) oxide, Fe_2O_3, an important iron ore called hematite?

4.4 ▶ Chemical formulas can be determined from experimental mass measurements

In pharmaceutical research, chemists often synthesize entirely new compounds, or isolate new compounds from plant and animal tissues. They must then determine the formula and structure of the new compound. This is usually accomplished using mass spectroscopy, which gives an experimental value for the molecular mass. The compound can also be decomposed chemically to find the masses of elements within a given amount of compound. Let's see how experimental mass measurements can be used to determine the compound's formula.

Percentage composition can be used to match possible formulas

The usual form for describing the relative masses of the elements in a compound is a list of *percentages by mass* called the compound's **percentage composition.** The **percentage by mass** of an element is the number of grams of the element present in 100 g of the compound. In general, a percentage by mass is found by using the following equation.

$$\% \text{ by mass of element} = \frac{\text{mass of element}}{\text{mass of whole sample}} \times 100\% \qquad (4.1)$$

Percentage composition

A sample of a liquid with a mass of 8.657 g was decomposed into its elements and gave 5.217 g of carbon, 0.9620 g of hydrogen, and 2.478 g of oxygen. What is the percentage composition of this compound?

ANALYSIS: We must apply Equation 4.1 for each element. The "mass of whole sample" here is 8.657 g, so we take each element in turn and perform the calculations.

SOLUTION:

EXAMPLE 4.11
Calculating a Percentage Composition from Chemical Analysis

For C: $\dfrac{5.217 \text{ g C}}{8.657 \text{ g compound}} \times 100\% = 60.26\% \text{ C}$

For H: $\dfrac{0.9620 \text{ g H}}{8.657 \text{ g compound}} \times 100\% = 11.11\% \text{ H}$

For O: $\dfrac{2.478 \text{ g O}}{8.657 \text{ g compound}} \times 100\% = 28.62\% \text{ O}$

Sum of percentages: 99.99%

One of the useful things about a percentage composition is that it tells us the mass of each of the elements in 100 g of the substance. For example, the results in this problem tell us that in 100.00 g of the liquid there are 60.26 g of carbon, 11.11 g of hydrogen, and 28.62 g of oxygen.

Is the Answer Reasonable?
The "check" is that the percentages must add up to 100%, allowing for small differences caused by rounding.

PRACTICE EXERCISE 13: From 0.5462 g of a compound there was isolated 0.1417 g of nitrogen and 0.4045 g of oxygen. What is the percentage composition of this compound? Are any other elements present?

Experimental percentage compositions can help identify an unknown compound

Elements can combine in many different ways. Nitrogen and oxygen, for example, form all of the following compounds: N_2O, NO, NO_2, N_2O_3, N_2O_4, and N_2O_5. To identify an unknown sample of a compound of nitrogen and oxygen, one might compare the percentage composition found by experiment with the calculated percentages for each possible formula. Which formula, for example, fits the percentage composition calculated in Practice Exercise 13? A strategy for matching empirical formulas with mass percentages is outlined in the following example.

EXAMPLE 4.12

Calculating a Theoretical Percentage Composition from a Chemical Formula

Do the mass percentages of 25.94% N and 74.06% O match the formula N_2O_5?

ANALYSIS: To calculate the theoretical percentages by mass of N and O in N_2O_5, we need the masses of N and O in a specific sample of N_2O_5. *If we choose 1 mol of the given compound to be this sample, calculating the rest of the data will be simpler.*

SOLUTION: We know that 1 mol of N_2O_5 must contain 2 mol N and 5 mol O. The corresponding number of grams of N and O are found as follows:

$$2\text{ N:} \qquad 2\text{ mol N} \times \frac{14.01\text{ g N}}{1\text{ mol N}} = 28.02\text{ g N}$$

$$5\text{ O:} \qquad 5\text{ mol O} \times \frac{16.00\text{ g O}}{1\text{ mol O}} = 80.00\text{ g O}$$

$$\overline{\qquad\qquad\qquad 1\text{ mol } N_2O_5 = 108.02\text{ g } N_2O_5}$$

Now we can calculate the percentages.

$$\text{For \% N:} \qquad \frac{28.02\text{ g}}{108.02\text{ g}} \times 100\% = 25.94\%\text{ N in } N_2O_5$$

$$\text{For \% O:} \qquad \frac{80.00\text{ g}}{108.02\text{ g}} \times 100\% = 74.06\%\text{ O in } N_2O_5$$

Thus, the experimental values do match the theoretical percentages for the formula N_2O_5.

PRACTICE EXERCISE 14: Calculate the theoretical percentage composition of N_2O_4.

An empirical formula can be determined from the masses of the different elements in a sample of a compound

The compound that forms when phosphorus burns in oxygen consists of molecules with the formula P_4O_{10}. When a formula gives the composition of one *molecule,* it is called a **molecular formula.** Notice, however, that both the subscripts 4 and 10 are divisible by 2, so the *smallest* numbers that tell us the *ratio* of P to O are 2 and 5. A simpler (but less informative) formula that expresses this ratio is P_2O_5. This is sometimes called the *simplest formula* for the compound. It is also called the **empirical formula** because it can be obtained from an experimental analysis of the compound.

To obtain an empirical formula experimentally, we need to determine the number of grams of each element in a sample of the compound. *This is the critical link in solving empirical formula problems.* Grams are then converted to moles, from which we obtain the mole ratios of the elements. Because the ratio by moles is the same as the ratio by atoms, we are able to construct the empirical formula.

The next four examples illustrate how we can calculate empirical formulas. We will then look at what additional data are required to obtain a compound's molecular formula.

A sample of a tin and chlorine compound with a mass of 2.57 g was found to contain 1.17 g of tin. What is the compound's empirical formula?

EXAMPLE 4.13

Calculating an Empirical Formula from Mass Data

ANALYSIS: The subscripts in an empirical formula give the relative number of moles of elements in a compound. If we can find the *mole* ratio of Sn to Cl, we will have the empirical formula. The first step, therefore, is to convert the numbers of grams of Sn and Cl to the numbers of moles of Sn and Cl. Then we convert these numbers into their simplest *whole-number ratio.*

The problem did not give the mass of chlorine in the 2.57 g sample. But there are only two elements present in the compound, tin and chlorine. We know the mass of the tin, and we know the total mass of compound. Therefore, the mass of Cl is simply the difference between 2.57 g and the mass of Sn present in the sample, namely, 1.17 g of Sn.

SOLUTION: First, we find the mass of Cl in 2.57 g of compound:

$$\text{mass of Cl} = 2.57 \text{ g compound} - 1.17 \text{ g Sn} = 1.40 \text{ g Cl}$$

Now we use the atomic masses to convert the mass data for tin and chlorine into moles.

$$1.17 \text{ g Sn} \times \frac{1 \text{ mol Sn}}{118.7 \text{ g Sn}} = 0.00986 \text{ mol Sn}$$

$$1.40 \text{ g Cl} \times \frac{1 \text{ mol Cl}}{35.45 \text{ g Cl}} = 0.0395 \text{ mol Cl}$$

We could now write a formula: $Sn_{0.00986}Cl_{0.0395}$, which does express the mole ratio, but we want whole numbers as subscripts so the formula can also express the proper atom ratio. To change the subscripts in $Sn_{0.00986}Cl_{0.0395}$ into a set of whole numbers *that are in the same ratio,* we divide each by a common divisor. The easiest common divisor is the smaller number in the set. *This is always the way to begin the search for whole-number subscripts; pick the smallest number of the set as the divisor.* It's guaranteed to make at least one subscript a whole number, namely, 1. Here, we divide both numbers by 0.00986.

$$Sn_{\frac{0.00986}{0.00986}} Cl_{\frac{0.0395}{0.00986}} = Sn_{1.00}Cl_{4.01}$$

If a calculated subscript (written to the proper number of significant figures) differs from a whole number by less than a few units in the last digit, we can safely round to the nearest whole number. We may round 4.01 to 4, so the empirical formula is $SnCl_4$.

Is the Answer Reasonable?

If $SnCl_4$ is the right formula, the compound should contain 1 mol of tin for every 4 mol of chlorine, or 1×118.7 g of tin for every 4×35.45, or 142 g of chlorine. Thus, the mass of chlorine in the compound should be about 20% more than the mass of tin, if the formula is correct. This is consistent with the mass of tin given and the mass of chlorine we calculated.

PRACTICE EXERCISE 15: A 1.525 g sample of a compound between nitrogen and oxygen contains 0.712 g of nitrogen. Calculate its empirical formula.

Sometimes our strategy of using the lowest common divisor does not give whole numbers. Let's see how to handle such a situation.

EXAMPLE 4.14

Calculating an Empirical Formula from Mass Composition

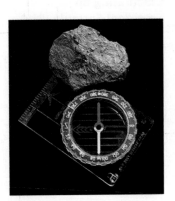

The mineral magnetite, like any magnet, is able to affect the orientation of a compass needle.

One of the compounds of iron and oxygen, "black iron oxide," occurs naturally in the mineral magnetite. When a 2.448 g sample was analyzed it was found to have 1.771 g of Fe. Calculate the empirical formula of this compound.

ANALYSIS: To calculate the empirical formula, we need to know the masses of both iron and oxygen, but we've only been given the mass of iron. When we calculate the mass of O in the sample by difference, we'll have the masses in grams of both elements, but we need the numbers of *moles* for a formula. So we'll use the atomic masses of Fe and O as tools for converting their respective masses to numbers of moles. Then we'll write a trial formula and see whether we can adjust the coefficients to their smallest whole numbers by the strategy learned in Example 4.13.

SOLUTION: The mass of O is, as we said, found by difference.

$$2.448 \text{ g compound} - 1.771 \text{ g Fe} = 0.677 \text{ g O}$$

The moles of Fe and O in the sample can now be calculated.

$$1.771 \text{ g Fe} \times \frac{1 \text{ mol Fe}}{55.845 \text{ g Fe}} = 0.03171 \text{ mol Fe}$$

$$0.677 \text{ g O} \times \frac{1 \text{ mol O}}{16.00 \text{ g O}} = 0.0423 \text{ mol O}$$

These results let us write the formula as $Fe_{0.03171}O_{0.0423}$.

Our first effort to change the ratio of 0.03171 to 0.0423 into whole numbers is to divide both by the smaller, 0.03171.

$$Fe_{\frac{0.03171}{0.03171}} O_{\frac{0.0423}{0.03171}} = Fe_{1.000}O_{1.33}$$

This time the procedure seems to fail. Given the fact that our number of significant figures is three, we can be sure that the digits 1.3 in 1.33 are known with certainty. Therefore the subscript for O, 1.33, is much too far from a whole number to round off and retain the precision allowed. In a *mole* sense, the ratio of 1 to 1.33 is correct; we just have not yet found a way to state the ratio in whole numbers. Let's look at a strategy that will give us whole-number subscripts.

Let's multiply each subscript in $Fe_{1.000}O_{1.33}$ by a whole number, 2. *This does not change the ratio;* it changes only the size of the numbers used to state it.

$$Fe_{(1.000 \times 2)}O_{(1.33 \times 2)} = Fe_{2.000}O_{2.66}$$

This didn't work either; 2.66 is also too far from a whole number (based on the allowed precision) to be rounded off. Let's try using 3 instead of 2 *on the first ratio of 1.000 to 1.33.*

$$Fe_{(1.000 \times 3)}O_{(1.33 \times 3)} = Fe_{3.000}O_{3.99}$$

We are now justified in rounding; 3.99 is acceptably close to 4. The empirical formula of the oxide of iron is Fe_3O_4.

PRACTICE EXERCISE 16: A 2.012 g sample of a compound of nitrogen and oxygen has 0.522 g of nitrogen. Calculate its empirical formula.

Empirical formulas can be determined from mass percentages

Only rarely is it possible to obtain the masses of every element in a compound by the use of just one weighed sample. Two or more analyses carried out on different samples are needed. For example, suppose an analyst is given a compound known to consist exclusively of calcium, chlorine, and oxygen. The mass of calcium in one weighed sample and the mass of chlorine in another sample would be determined in separate experiments. Then the mass data for calcium and chlorine would be converted to percentages by mass *so that the data from different samples relate to the same sample size, namely, 100 g of the compound.* The percentage of oxygen would be calculated by difference because %Ca + %Cl + %O = 100%. Each mass percentage represents a certain number of grams of the element, which is next converted into the corresponding number of moles of the element. The mole proportions are converted to whole numbers in the way we just studied, giving us the subscripts for the empirical formula. Let's see how this works.

A white powder used in paints, enamels, and ceramics has the following percentage composition: Ba, 69.6%; C, 6.09%; and O, 24.3%. What is its empirical formula?

EXAMPLE 4.15

Calculating an Empirical Formula from Percentage Composition

ANALYSIS: The percentages are *numerically* equal to masses if we choose a 100 g sample of the compound. So we have to convert masses to moles using the atomic masses of the elements. Then we proceed as before.

SOLUTION: A 100 g sample of the compound would contain 69.6 g Ba, 6.09 g C, and 24.3 g O. Let's convert these to moles.

$$Ba: \quad 69.6 \text{ g Ba} \times \frac{1 \text{ mol Ba}}{137.3 \text{ g Ba}} = 0.507 \text{ mol Ba}$$

$$C: \quad 6.09 \text{ g C} \times \frac{1 \text{ mol C}}{12.01 \text{ g C}} = 0.507 \text{ mol C}$$

$$O: \quad 24.3 \text{ g O} \times \frac{1 \text{ mol O}}{16.00 \text{ g O}} = 1.52 \text{ mol O}$$

Our preliminary empirical formula is then

$$Ba_{0.507}C_{0.507}O_{1.52}$$

We next divide each subscript by the smallest, 0.507.

$$Ba_{\frac{0.507}{0.507}} C_{\frac{0.507}{0.507}} O_{\frac{1.52}{0.507}} = Ba_{1.00}C_{1.00}O_{3.00}$$

The subscripts are whole numbers, so the empirical formula is $BaCO_3$, representing barium carbonate.

Does the Answer Make Sense?

If the formula is correct, there are 3 mol of oxygen for every 1 mol of carbon in the compound. So for every 1 mol (12 g) of carbon in the compound, there should be 3 mol (48 g) of oxygen. The carbon-to-oxygen mass ratio is 12:48, or 1:4. This is consistent with the ratio of carbon and oxygen percentages (6.09:24.3 ~ 1:4).

PRACTICE EXERCISE 17: A white solid used to whiten paper has the following percentage composition: Na, 32.4%; S, 22.6%. The unanalyzed element is oxygen. What is the compound's empirical formula?

Empirical formulas can be determined from indirect analyses

In practice, a compound is seldom broken down completely to its *elements* in a quantitative analysis. Instead, the compound is changed into other *compounds.* The reactions separate the elements by capturing each one entirely (quantitatively) in a *separate* compound *whose formula is known.*

In the following example, we illustrate an indirect analysis of a compound made entirely of carbon, hydrogen, and oxygen. Such compounds burn completely in pure oxygen—the reaction is called *combustion*—and the sole products are carbon dioxide and water. (This particular kind of indirect analysis is sometimes called a *combustion analysis.*) The complete combustion of methyl alcohol (CH_3OH), for example, occurs according to the following equation:

$$2CH_3OH + 3O_2 \longrightarrow 2CO_2 + 4H_2O$$

The carbon dioxide and water can be separated and are individually weighed. Notice that all of the carbon atoms in the original compound end up among the CO_2 molecules and all of the hydrogen atoms are in H_2O molecules. In this way at least two of the original elements, C and H, are entirely separated.

We will calculate the mass of carbon in the CO_2 collected, which equals the mass of carbon in the original sample. Similarly, we will calculate the mass of hydrogen in the H_2O collected, which equals the mass of hydrogen in the original sample. When added together, the mass of C and mass of H are less than the total mass of the sample because part of the sample is composed of oxygen. By subtracting the sum of the C and H masses from the original sample weight, we can obtain the mass of oxygen in the sample of the compound.

EXAMPLE 4.16

Empirical Formula from Indirect Analysis

A 0.5438 g sample of a liquid consisting of only C, H, and O was burned in pure oxygen, and 1.039 g of CO_2 and 0.6369 g of H_2O were obtained. What is the empirical formula of the compound?

ANALYSIS: There are two parts to this problem. For the first part, we will find the number of grams of C in the CO_2 and the number of grams of H in the H_2O. (This kind of calculation was illustrated in Example 4.10.) These values represent the number of grams of C and H in the original sample. Adding them together and subtracting the sum from the mass of the original sample will give us the mass of oxygen in the sample. In short, we have the following series of calculations:

$$\text{grams } CO_2 \longrightarrow \text{grams C}$$

$$\text{grams } H_2O \longrightarrow \text{grams H}$$

We find the mass of oxygen by difference.

$$0.5438 \text{ g sample} - (\text{g C} + \text{g H}) = \text{g O}$$

In the second half of the solution, we use the masses of C, H, and O to calculate the empirical formula, as in Example 4.14.

SOLUTION: First we find the number of grams of C in the CO_2 and of H in the H_2O. In 1 mol of CO_2 (44.009 g) there are 12.011 g of C. Therefore, in 1.039 g of CO_2 we have

$$1.039 \text{ g } CO_2 \times \frac{12.011 \text{ g C}}{44.009 \text{ g } CO_2} = 0.2836 \text{ g C}$$

In 1 mol of H_2O (18.015 g) there are 2.0158 g of H. For the number of grams of H in 0.6369 g of H_2O,

$$0.6369 \text{ g } H_2O \times \frac{2.0158 \text{ g H}}{18.015 \text{ g } H_2O} = 0.07127 \text{ g H}$$

The total mass of C and H is therefore the sum of these two quantities.

$$\text{total mass of C and H} = 0.2836 \text{ g C} + 0.07127 \text{ g H} = 0.3549 \text{ g}$$

The difference between this total and the 0.5438 g in the original sample is the mass of oxygen (the only other element).

$$\text{mass of O} = 0.5438 \text{ g} - 0.3549 \text{ g} = 0.1889 \text{ g O}$$

Now we can convert the masses of the elements to an empirical formula.

For C: $\quad 0.2836 \text{ g C} \times \dfrac{1 \text{ mol C}}{12.011 \text{ g C}} = 0.02361 \text{ mol C}$

For H: $\quad 0.07127 \text{ g H} \times \dfrac{1 \text{ mol H}}{1.008 \text{ g H}} = 0.07070 \text{ mol H}$

For O: $\quad 0.1889 \text{ g O} \times \dfrac{1 \text{ mol O}}{15.999 \text{ g O}} = 0.01181 \text{ mol O}$

Our preliminary empirical formula is thus $C_{0.02361}H_{0.07070}O_{0.01181}$. We divide all of these subscripts by the smallest number, 0.01181.

$$C_{\frac{0.02361}{0.01181}} H_{\frac{0.07070}{0.01181}} O_{\frac{0.01181}{0.01181}} = C_{1.999}H_{5.987}O_1$$

The results are acceptably close to C_2H_6O, the answer.

Does the Answer Make Sense?
First, let's check the mass of carbon and hydrogen in the carbon dioxide and water. One mole of CO_2 contains about 12 g of carbon and has a mass of about 44 g, so the computed mass of carbon should be 12/44, or a little more than a quarter, of the mass of carbon dioxide. The mass of carbon we calculated was 0.2836 g C, which is a little more than a quarter of the mass of the CO_2 (1.039 g).

Similarly, 1 mol of H_2O contains 2 g of hydrogen and has a mass of 18 g, so the mass of hydrogen we computed should have been 2/18, or one-ninth, the mass of the water. So it makes sense that 0.6369 g H_2O would contain 0.07127 g H.

We can quickly check the subscripts in the empirical formula against the mass ratios to see if they are consistent. For example, a mole of compound contains 2 mol of carbon for every 6 mol of hydrogen. A mole of compound contains about 2 mol C × (12 g C/1 mol C) = 24 g C and about 6 mol H × (1 g H/1 mol H) = 6 g H. The mass of carbon in the compound should be about four times the mass of hydrogen, if the empirical formula is correct. We calculated that there were 0.2836 g C and 0.07127 g H in the sample, which is very close to being the 4:1 carbon-to-hydrogen mass ratio we expect.

PRACTICE EXERCISE 18: The combustion of a 5.048 g sample of a compound of C, H, and O gave 7.406 g CO_2 and 4.027 g H_2O. Calculate the empirical formula of the compound.

As we said earlier, more than one sample of a substance must be analyzed whenever more than one reaction is necessary to separate the elements. This is also true in indirect analyses. For example, if a compound contains C, H, N, and O, combustion converts the C and H to CO_2 and H_2O, which are separated by special techniques and weighed. The mass of C in the CO_2 sample and the mass of H in the H_2O sample are then calculated in the usual way. A second reaction with a different sample of the compound can be used to obtain the nitrogen, either as N_2 or as NH_3. Because different-size samples are used, the masses of C and H from one sample and the mass of N from another cannot be used to calculate the empirical formula. So we convert the masses of the elements found in their respective samples into percentages of the element by mass in the compound. We then add up the percentages, subtract from 100 to get the percentage of O, and then calculate the empirical formula from the percentage composition as described earlier.

Determine molecular formulas from empirical formulas and molecular masses

The empirical formula is all that we can get for ionic compounds. For molecular compounds, however, chemists prefer *molecular* formulas because they give the number of atoms of each type in a molecule, rather than just the number of moles of elements in a mole of compound as the empirical formula does.

Sometimes an empirical formula and a molecular formula are the same. Two examples are H_2O and NH_3. Usually, however, the subscripts of a molecular formula are whole-number multiples of those in the empirical formula. The subscripts of the molecular formula P_4O_{10}, for example, are each two times those in the empirical formula, P_2O_5, as you saw earlier. The molecular mass of P_4O_{10} is likewise two times the formula mass of P_2O_5. This observation provides us with a way to find out the molecular formula for a compound provided we have a way of determining experimentally the molecular mass of the compound. If the experimental molecular mass *equals* the calculated empirical formula mass, the empirical formula itself is also a molecular formula. Otherwise, the experimental molecular mass will be some whole-number multiple of the value calculated from the empirical formula. Whatever the whole number is, it's a common multiplier for the subscripts of the empirical formula.

EXAMPLE 4.17

Determining a Molecular Formula from an Empirical Formula and a Molecular Mass

Styrene, the raw material for polystyrene foam plastics, has an empirical formula of CH. Its molecular mass is 104. What is its molecular formula?

ANALYSIS: The molecular mass of styrene, 104, is some simple multiple of the formula mass calculated for its empirical formula, CH. So we have to compute the latter, and then see how many times larger 104 is.

SOLUTION: For CH, the formula mass is

$$12.01 + 1.008 = 13.02$$

To find how much larger 104 is than 13.02, we divide.

$$\frac{104}{13.02} = 7.99$$

Rounding this to 8, we see that 104 is 8 times larger than 13.02, so the correct molecular formula of styrene must have subscripts eight times those in CH. Styrene, therefore, is C_8H_8.

Does the Answer Make Sense?

The molecular mass of C_8H_8 is approximately $(8 \times 12) + (8 \times 1) = 104$, which is consistent with the molecular mass we started with.

PRACTICE EXERCISE 19: The empirical formula of hydrazine is NH_2, and its molecular mass is 32.0. What is its molecular formula?

4.5 ▶ Chemical equations link amounts of substances in a reaction

So far we have focused on relationships between elements within a single compound. We have seen that the critical link between substances within a compound is the mole-to-mole ratio obtained from the compound's formula. In this section, we'll see that the same techniques can be used to relate substances involved in a chemical reaction. *The critical link between substances involved in a reaction is a mole-to-mole ratio obtained from the chemical equation that describes the reaction.*

Balanced chemical equation

To see how chemical equations can be used to obtain mole-to-mole relationships, consider the equation that describes the burning of octane (C_8H_{18}) in oxygen (O_2) to give carbon dioxide and steam:

$$2C_8H_{18}(l) + 25O_2(g) \longrightarrow 16CO_2(g) + 18H_2O(g)$$

This equation can be interpreted on a *microscopic* (molecular) scale as follows:

> *For every 2 molecules of liquid octane that react with 25 molecules of oxygen gas, 16 molecules of carbon dioxide gas and 18 molecules of steam are produced.*

This statement immediately suggests many equivalency relationships that can be used to build conversion factors in stoichiometry problems:

$$2 \text{ molecules } C_8H_{18} \Leftrightarrow 25 \text{ molecules } O_2$$
$$2 \text{ molecules } C_8H_{18} \Leftrightarrow 16 \text{ molecules } CO_2$$
$$2 \text{ molecules } C_8H_{18} \Leftrightarrow 18 \text{ molecules } H_2O$$
$$25 \text{ molecules } O_2 \Leftrightarrow 16 \text{ molecules } CO_2$$
$$25 \text{ molecules } O_2 \Leftrightarrow 18 \text{ molecules } H_2O$$
$$16 \text{ molecules } CO_2 \Leftrightarrow 18 \text{ molecules } H_2O$$

The chemical equation gives relative amounts of molecules of each type that participate in the reaction. It does *not* mean that 2 octane molecules actually collide with 25 O_2 molecules. The reaction occurs in many steps, which the chemical equation does not show.

Any of these microscopic relationships can be scaled up to the macroscopic level by multiplying both sides of the equivalency by Avogadro's number, which effectively allows us to replace "molecules" with "moles":

$$2 \text{ mol } C_8H_{18} \Leftrightarrow 25 \text{ mol } O_2$$
$$2 \text{ mol } C_8H_{18} \Leftrightarrow 16 \text{ mol } CO_2$$
$$2 \text{ mol } C_8H_{18} \Leftrightarrow 18 \text{ mol } H_2O$$
$$25 \text{ mol } O_2 \Leftrightarrow 16 \text{ mol } CO_2$$
$$25 \text{ mol } O_2 \Leftrightarrow 18 \text{ mol } H_2O$$
$$16 \text{ mol } CO_2 \Leftrightarrow 18 \text{ mol } H_2O$$

We can interpret the equation on a macroscopic (mole) scale as follows:

> *Two moles of liquid octane react with 25 mol of oxygen gas to produce 16 mol of carbon dioxide gas and 18 mol of steam.*

To use these equivalencies in a stoichiometry problem, the equation must be correct. Specifically, the equation must be **balanced.** That means that every atom found in the reactants must also be found somewhere in the products. You must always check to see whether this is so for a given equation before you can use the coefficients in the equation to build equivalencies and conversion factors.

We'll show you how to detect and correct an unbalanced equation in the next section.

First, let's see how mole-to-mole relationships obtained from a chemical equation can be used to convert moles of one substance to moles of another when both substances are involved in a chemical reaction.

EXAMPLE 4.18
Stoichiometry of Chemical Reactions

Whenever a problem asks you to convert an amount of one substance into an amount of a different substance, the **critical link** in the problem is usually a mole-to-mole relationship between the two substances.

How many moles of sodium phosphate, Na_3PO_4, can be made from 0.240 mol of NaOH by the following reaction?

$$3NaOH(aq) + H_3PO_4(aq) \longrightarrow Na_3PO_4(aq) + 3H_2O$$

ANALYSIS: The question asks us to relate amounts of two different substances. *A mole-to-mole relationship is usually the critical link between two different substances in a stoichiometry problem.* The balanced equation is our tool because it gives the coefficients we need. From the coefficients, we know the following:

$$3 \text{ mol NaOH} \Longleftrightarrow 1 \text{ mol } Na_3PO_4$$

This enables us to prepare the conversion factor that we need.

SOLUTION: We convert 0.240 mol NaOH to the number of moles of Na_3PO_4 equivalent to it in the reaction as follows:

$$0.240 \text{ mol NaOH} \times \frac{1 \text{ mol } Na_3PO_4}{3 \text{ mol NaOH}} = 0.0800 \text{ mol } Na_3PO_4$$

Thus, in this reaction,

$$0.240 \text{ mol NaOH} \Longleftrightarrow 0.0800 \text{ mol } Na_3PO_4$$

We can make 0.0800 mol Na_3PO_4 from 0.240 mol NaOH.

Is the Answer Reasonable?
The equation tells us that 3 mol NaOH $\Longleftrightarrow$ 1 mol Na_3PO_4, so the actual number of moles of Na_3PO_4 (0.0800 mol) should be one-third the actual number of moles of NaOH (0.240 mol), and it is.

PRACTICE EXERCISE 20: How many moles of sulfuric acid, H_2SO_4, are needed to react with 0.366 mol of NaOH by the following reaction?

$$2NaOH(aq) + H_2SO_4(aq) \longrightarrow Na_2SO_4(aq) + 2H_2O$$

PRACTICE EXERCISE 21: If 0.575 mol of CO_2 is produced by the combustion of propane, C_3H_8, how many moles of oxygen are consumed? The balanced equation is as follows (omitting physical states):

$$C_3H_8 + 5O_2 \longrightarrow 3CO_2 + 4H_2O$$

Mole-to-mole ratios link masses of different substances in chemical reactions

We often need to relate grams of one substance with grams of another in a chemical reaction. For example, glucose ($C_6H_{12}O_6$) is one of the body's primary energy sources. The body combines glucose and oxygen to give carbon dioxide and water. The balanced equation for the overall reaction is

$$C_6H_{12}O_6(aq) + 6O_2(aq) \longrightarrow 6CO_2(aq) + 6H_2O(l)$$

How many grams of oxygen must the body take in to completely process 1.00 g of glucose? The problem can be expressed as

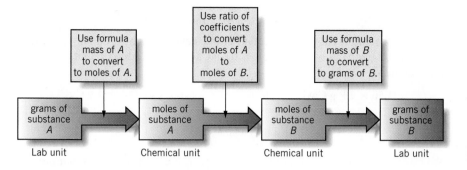

FIGURE 4.4 *The flow of the calculation for solving stoichiometry problems.* The flow applies to the calculation of the required or the expected mass of *any* reactant or product.

$$1.00 \text{ g } C_6H_{12}O_6 \Longleftrightarrow ? \text{ g } O_2$$

The first thing we should notice about this problem is that we're relating *two different substances* in a reaction. The critical link between the substances is the mole-to-mole relationship between glucose and O_2 given by the chemical equation. In this case, the equation tells us that

$$1 \text{ mol } C_6H_{12}O_6 \Longleftrightarrow 6 \text{ mol } O_2$$

If we insert that mole-to-mole conversion between our starting point (1.00 g $C_6H_{12}O_6$) and the desired quantity (g O_2) we have cut the problem into three simple pieces:

$$\boxed{1 \text{ mol } C_6H_{12}O_6 \Longleftrightarrow 6 \text{ mol } O_2}$$

$$\textbf{1.00 g } C_6H_{12}O_6 \longrightarrow \textbf{mol } C_6H_{12}O_6 \longrightarrow \textbf{mol } O_2 \longrightarrow \textbf{g } O_2$$

We convert grams of $C_6H_{12}O_6$ to mol $C_6H_{12}O_6$ (using the molecular mass of $C_6H_{12}O_6$). We then convert mol $C_6H_{12}O_6$ to mol O_2 using the equivalence relationship from the balanced equation. Finally, we convert mol O_2 into g O_2 using the molecular mass of O_2.

Figure 4.4 outlines this flow for *any* stoichiometry problem that relates reactant or product masses. If we know the *balanced equation* for a reaction and the *mass* of any reactant or product, we can calculate the required or expected mass of *any* other substance in the equation. Example 4.19 shows how it works.

Portland cement is a mixture of the oxides of calcium, aluminum, and silicon. The raw material for its calcium oxide is calcium carbonate, which occurs as the chief component of a natural rock, limestone. When calcium carbonate is strongly heated, it decomposes by the following reaction. One product, CO_2, is driven off to leave the desired CaO as the only other product.

EXAMPLE 4.19

Stoichiometric Mass Calculations

$$CaCO_3(s) \xrightarrow{\text{heat}} CaO(s) + CO_2(g)$$

A chemistry student is to prepare 1.50×10^2 g of CaO in order to test a particular "recipe" for portland cement. How many grams of $CaCO_3$ should be used, assuming that all will be converted?

ANALYSIS: Let's begin by restating the problem.

$$1.50 \times 10^2 \text{ g CaO} \Longleftrightarrow ? \text{ g } CaCO_3$$

In problems that convert an amount of one substance to an amount of a different substance in a chemical reaction, the critical link is a mole-to-mole conversion factor. From the balanced chemical equation, we have

$$1 \text{ mol CaO} \Leftrightarrow 1 \text{ mol CaCO}_3$$

We can immediately cut the problem into three easy pieces by dropping the critical link between our given information (1.50×10^2 g CaO) and our desired units (g CaCO$_3$) on the roadmap:

$$\textbf{1.50} \times \textbf{10}^2 \textbf{ g CaO} \longrightarrow \textbf{mol CaO} \longrightarrow \textbf{mol CaCO}_3 \longrightarrow \textbf{g CaCO}_3$$

$$\boxed{1 \text{ mol CaO} \Leftrightarrow 1 \text{ mol CaCO}_3}$$

We must convert g CaO to mol CaO. *The critical link between mass and moles is the molar mass:*

$$56.08 \text{ g CaO} \Leftrightarrow 1 \text{ mol CaO}$$

We must also convert mol CaCO$_3$ to g CaCO$_3$. Again, the critical link between mass and moles is the molar mass:

$$100.09 \text{ g CaCO}_3 \Leftrightarrow 1 \text{ mol CaCO}_3$$

Putting this all together, our overall strategy will be as follows:

$$\boxed{56.08 \text{ g CaO} = 1 \text{ mol CaO}} \qquad \boxed{1 \text{ mol CaCO}_3 = 100.09 \text{ g CaCO}_3}$$

$$\textbf{1.50} \times \textbf{10}^2 \textbf{ g CaO} \longrightarrow \textbf{mol CaO} \longrightarrow \textbf{mol CaCO}_3 \longrightarrow \textbf{g CaCO}_3$$

$$\boxed{1 \text{ mol CaO} \Leftrightarrow 1 \text{ mol CaCO}_3}$$

SOLUTION: We assemble conversion factors so the units cancel correctly:

$$1.50 \times 10^2 \text{ g CaO} \times \frac{1 \text{ mol CaO}}{56.08 \text{ g CaO}} \times \frac{1 \text{ mol CaCO}_3}{1 \text{ mol CaO}} \times \frac{100.09 \text{ g CaCO}_3}{1 \text{ mol CaCO}_3}$$

$$= 268 \text{ g CaCO}_3$$

Notice how the calculation flows from grams of CaO to moles of CaO, then to moles of CaCO$_3$ (using the equation), and finally to grams of CaCO$_3$. We cannot emphasize too much that *the key step in all calculations of reaction stoichiometry is the use of the balanced equation.*

Is the Answer Reasonable?
Does the *size* of the answer, which is numerically larger than the grams of CaO to be made, make sense? It does, because the formula mass of CaCO$_3$ is larger than that of CaO.

EXAMPLE 4.20

Stoichiometric Mass Calculations

One of the most spectacular reactions of aluminum, the thermite reaction, is with iron oxide, Fe$_2$O$_3$, by which metallic iron is made. So much heat is generated that the iron forms in the liquid state (Figure 4.5). The equation is

$$2\text{Al}(s) + \text{Fe}_2\text{O}_3(s) \longrightarrow \text{Al}_2\text{O}_3(s) + 2\text{Fe}(l)$$

A certain welding operation, used over and over, requires that each time at least 86.0 g of Fe be produced. What is the minimum mass in grams of Fe$_2$O$_3$ that must be used for each operation? Also calculate how many grams of aluminum are needed.

ANALYSIS: First we restate the problem:

$$86.0 \text{ g Fe} \Leftrightarrow ? \text{ g Fe}_2\text{O}_3$$

Remember that all problems in reaction stoichiometry must be solved at the mole level because an equation's coefficients disclose *mole* ratios, not mass ratios. So we have to convert the number of grams of Fe to moles. Then we can use the mole-to-mole relationship given by the coefficients in the balanced equation,

$$1 \text{ mol Fe}_2\text{O}_3 \Leftrightarrow 2 \text{ mol Fe}$$

to see how many *moles* of Fe_2O_3 are needed. We finally convert this answer into grams of Fe_2O_3. The other calculations follow the same pattern.

SOLUTION: We'll set up the first calculation as a chain. Draw all cancel lines.

$$86.0 \text{ g Fe} \times \frac{1 \text{ mol Fe}}{55.85 \text{ g Fe}} \times \frac{1 \text{ mol Fe}_2O_3}{2 \text{ mol Fe}} \times \frac{159.70 \text{ g Fe}_2O_3}{1 \text{ mol Fe}_2O_3} = 123 \text{ g Fe}_2O_3$$

$$\text{grams Fe} \longrightarrow \quad \text{moles Fe} \quad \longrightarrow \quad \text{moles Fe}_2O_3 \quad \longrightarrow \quad \text{grams Fe}_2O_3$$

A minimum of 123 g of Fe_2O_3 are required to make 86.0 g of Fe.

Next, we calculate the number of grams of Al needed, but we know that we must first find the number of *moles* of Al required. Only from this can the grams of Al be calculated. The relevant mole-to-mole relationship, again using the balanced equation, is

$$2 \text{ mol Al} \Longleftrightarrow 2 \text{ mol Fe}$$

Expressing this in the smallest whole numbers, we have

$$1 \text{ mol Al} \Longleftrightarrow 1 \text{ mol Fe}$$

Employing another chain calculation to find the mass of Al needed to make 86.0 g of Fe, we have (using 26.98 as the atomic mass of Al)

$$86.0 \text{ g Fe} \times \frac{1 \text{ mol Fe}}{55.85 \text{ g Fe}} \times \frac{1 \text{ mol Al}}{1 \text{ mol Fe}} \times \frac{26.98 \text{ g Al}}{1 \text{ mol Al}} = 41.5 \text{ g Al}$$

$$\text{grams Fe} \longrightarrow \quad \text{moles Fe} \quad \longrightarrow \quad \text{moles Al} \quad \longrightarrow \quad \text{grams Al}$$

Is the Answer Reasonable?
Think about the size of the answer, 41.5 g of Al, in relationship to the following: the mass of Fe needed, the mole-to-mole relationship in this reaction of Fe and Al (1 mol:1 mol), and the relative atomic masses of Fe (55.85) and Al (26.98). Does the answer make sense? (Should it take *less* mass of Al than the mass of Fe made?)

PRACTICE EXERCISE 22: How many grams of aluminum oxide are also produced by the reaction described in Example 4.20?

FIGURE 4.5 *The thermite reaction.* Pictured here is a device for making white hot iron by the reaction of aluminum with iron oxide and letting the molten iron run down into a mold between the ends of two steel railroad rails. The rails are thereby welded together.

4.6 ▶ Chemical equations cannot create or destroy atoms

We learned in Chapter 2 that a *chemical equation* is a shorthand, quantitative description of a chemical reaction. An equation is *balanced* when all atoms present among the reactants are also somewhere among the products. As we learned, coefficients, the numbers in front of formulas, are multiplier numbers for their respective formulas, and the values of the coefficients determine whether an equation is balanced.

Balancing Equations

Always approach the balancing of an equation as a two-step process.

> **STEP 1** *Write the unbalanced "equation."* Organize the formulas in the pattern of an equation with plus signs and an arrow. Use *correct* formulas. (You learned to write many of them in Chapter 2, but until we have studied more chemistry, you will usually be given formulas.)
>
> **STEP 2** *Adjust the coefficients to get equal numbers of each kind of atom on both sides of the arrow.*

When doing step 2, make no changes in the formulas, either in the atomic symbols or their subscripts. If you do, the equation will involve different substances from those intended. You may still be able to balance it, but the equation will not be for the reaction you want.

Zinc metal reacts with hydrochloric acid.

We'll begin with simple equations that can be balanced easily by inspection. An example is the reaction of zinc metal with hydrochloric acid (margin photo). First, we need the correct formulas, and this time we'll include the physical states because they are different. The reactants are zinc, $Zn(s)$, and hydrochloric acid, $HCl(aq)$. We also need formulas for the products. Zn changes to a water-soluble compound, zinc chloride, $ZnCl_2(aq)$, and hydrogen gas, $H_2(g)$, bubbles out as the other product. (Recall that hydrogen occurs naturally as a *diatomic molecule,* not as atoms.)

STEP 1 Write an unbalanced equation.

$$Zn(s) + HCl(aq) \longrightarrow ZnCl_2(aq) + H_2(g) \qquad \text{(unbalanced)}$$

STEP 2 Adjust the coefficients to get equal numbers of each kind of atom on both sides of the arrow.

There is no simple set of rules for adjusting coefficients. Experience is the greatest help, and experience has taught chemists that the following guidelines often get to the solution most directly when they are applied in the order given.

Some Guidelines for Balancing Equations

1. Balance elements other than H and O first.
2. Balance as a group those polyatomic ions that appear unchanged on both sides of the arrow.
3. Balance separately those elements that appear somewhere by themselves (e.g., Zn in our example).

Using the guidelines given here, we'll look at Cl first in our example. Because there are two Cl to the right of the arrow but only one to the left, we put a 2 in front of the HCl on the left side. The result is

$$Zn(s) + 2HCl(aq) \longrightarrow ZnCl_2(aq) + H_2(g)$$

Everything is now balanced. On each side we find 1 Zn, 2 H, and 2 Cl. Equations that can be balanced by inspection do not get much harder than this.

One complication is that an infinite number of *balanced* equations can be written for any given reaction! We might, for example, have adjusted the coefficients so that our equation came out as follows:

$$2Zn(s) + 4HCl(aq) \longrightarrow 2ZnCl_2(aq) + 2H_2(g)$$

This equation is also balanced. For simplicity, we usually prefer the *smallest* whole-number coefficients when writing balanced equations.

EXAMPLE 4.21
Writing a Balanced Equation

Sodium hydroxide, NaOH, and phosphoric acid, H_3PO_4, react as aqueous solutions to give sodium phosphate, Na_3PO_4, and water. The sodium phosphate remains in solution. Write the balanced equation for this reaction.

ANALYSIS: First, write an unbalanced equation that includes the reactant formulas on the left-hand side and the product formulas on the right. Then adjust stoichiometric coefficients (*never subscripts!*) until there is the same number of each type of atom on the left and right sides of the equation.

SOLUTION: We include the designation (aq) for all substances dissolved in water (except H_2O itself; we'll not give it any designation when it is in its liquid state).

$$NaOH(aq) + H_3PO_4(aq) \longrightarrow Na_3PO_4(aq) + H_2O \qquad \text{(unbalanced)}$$

There are several things not in balance, but our guidelines suggest that we work with Na first rather than with O, H, or PO_4. There are 3 Na on the right side, so we put a 3 in front of NaOH on the left, as a trial.

$$3NaOH(aq) + H_3PO_4(aq) \longrightarrow Na_3PO_4(aq) + H_2O \qquad \text{(unbalanced)}$$

Now the Na are in balance. The unit PO_4 is balanced also. Not counting the PO_4, we have on the left 3 O and 3 H in 3NaOH plus 3 H in H_3PO_4, for a net of 3 O and 6 H on the left. On the right, in H_2O, we have 1 O and 2 H. The ratio of 3 O to 6 H on the left is equivalent to the ratio of 1 O to 2 H on the right, so we write the multiplier (coefficient) 3 in front of H_2O.

$$3NaOH(aq) + H_3PO_4(aq) \longrightarrow Na_3PO_4(aq) + 3H_2O \qquad \text{(balanced)}$$

We now have a balanced equation.

Is the Answer Reasonable?
On each side we have 3 Na, 1 PO_4, 6 H, and 3 O besides those in PO_4, and it is obvious that our coefficients cannot be reduced to smaller whole numbers.

PRACTICE EXERCISE 23: When aqueous solutions of calcium chloride, $CaCl_2$, and potassium phosphate, K_3PO_4, are mixed, a reaction occurs in which solid calcium phosphate, $Ca_3(PO_4)_2$, separates from the solution. The other product is KCl(aq). Write the balanced equation.

The strategy in Example 4.21 of balancing whole unchanged units of atoms, like "PO_4," is extremely useful. These are usually polyatomic ions that go through reactions unaffected. If we leave them alone, we have less atom counting to do.

4.7 ▶ The reactant in shortest supply limits the amount of product

We've seen that balanced chemical equations can tell us how to mix reactants together in just the right proportions to get a certain amount of product. For example, ethanol, C_2H_5OH, is prepared industrially as follows:

$$\underset{\text{ethylene}}{C_2H_4} + H_2O \longrightarrow \underset{\text{ethanol}}{C_2H_5OH}$$

The equation tells us that 1 mol of ethylene will react with 1 mol of water to give 1 mol of ethanol. We can also interpret the equation on a molecular level: Every molecule of ethylene that reacts requires one molecule of water to produce one molecule of ethanol:

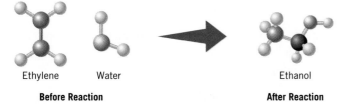

Ethylene   Water   Ethanol

Before Reaction   **After Reaction**

If we have three molecules of ethylene reacting with three molecules of water, three ethanol molecules are produced:

Notice that in both the "before" and "after" views of the reaction, the numbers of carbon, hydrogen, and oxygen atoms are the same.

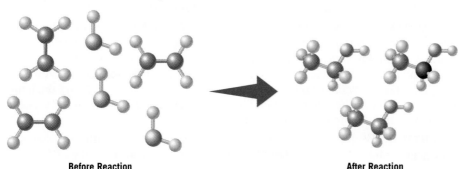

Before Reaction   **After Reaction**

What happens if we mix 3 molecules of ethylene with 5 molecules of water? The ethylene will be completely used up before all the water is, and the product will contain two unreacted water molecules:

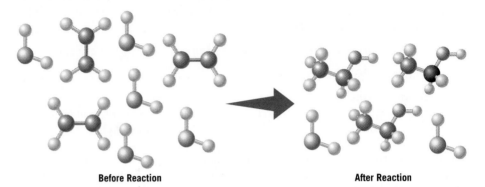

Before Reaction   **After Reaction**

We don't have enough ethylene to use up all the water. The excess water remains after the reaction stops. This situation can be a problem in the manufacture of chemicals because not only do we waste one of our reactants (water, in this case) but we also obtain a product that is contaminated with unused reactant.

In this reaction mixture, ethylene is called the **limiting reactant** because it limits the amount of product (ethanol) that forms. The water is called an **excess reactant,** because we have more of it than is needed to completely consume all the ethylene.

To predict the amount of product we'll actually obtain in a reaction, we need to know which of the reactants is the limiting reactant. In the example above, we saw that we needed only 3 H_2O molecules to react with 3 C_2H_4 molecules, but we had 5 H_2O molecules, so H_2O is present in excess and C_2H_4 is the limiting reactant. We could also have reasoned that 5 molecules of H_2O would require 5 molecules of C_2H_4, and since we have only 3 molecules of C_2H_4, it must be the limiting reactant.

Once we have identified the limiting reactant, it is possible to compute the amount of product that will actually form and the amount of excess reactant that will be left over after the reaction stops.

Example 4.22 shows how to solve a typical limiting reactant problem when the amounts of the reactants are given in mass units.

EXAMPLE 4.22

Limiting Reactant

Gold(III) hydroxide, $Au(OH)_3$, is used for electroplating gold onto other metals. It can be made by the following reaction:

$$2KAuCl_4(aq) + 3Na_2CO_3(aq) + 3H_2O \longrightarrow$$

$$2Au(OH)_3(aq) + 6NaCl(aq) + 2KCl (aq) + 3CO_2(g)$$

To prepare a fresh supply of $Au(OH)_3$, a chemist at an electroplating plant has mixed 20.00 g of $KAuCl_4$ with 25.00 g of Na_2CO_3 (both dissolved in a large excess of water). What is the maximum number of grams of $Au(OH)_3$ that can form?

ANALYSIS: The clue that tells us to *begin* the solution by figuring out what is the limiting reactant is that *the quantities of two reactants are given.* Once we determine which reactant limits the product, we use its mass in our calculation.

To find the limiting reactant, we arbitrarily pick one of the reactants ($KAuCl_4$ or Na_2CO_3) and calculate whether it would all be used up. If so, we've found the limiting reactant. If not, the other reactant limits. (We were told that water is in excess, so we know that it does not limit the reaction.)

SOLUTION: We'll choose $KAuCl_4$ as the reactant to work with and calculate how many grams of Na_2CO_3, *should* be provided to react with 20.00 g of $KAuCl_4$.

The formula masses are 377.88 for $KAuCl_4$, and 105.99 for Na_2CO_3. We'll set up a chain calculation using the following mole-to-mole relationship:

$$2 \text{ mol KAuCl}_4 \Leftrightarrow 3 \text{ mol Na}_2\text{CO}_3$$

$$20.00 \text{ g KAuCl}_4 \times \frac{1 \text{ mol KAuCl}_4}{377.88 \text{ g KAuCl}_4} \times \frac{3 \text{ mol Na}_2\text{CO}_3}{2 \text{ mol KAuCl}_4} \times \frac{105.99 \text{ g Na}_2\text{CO}_3}{1 \text{ mol Na}_2\text{CO}_3}$$

grams $KAuCl_4$ $\longrightarrow$ moles $KAuCl_4$ $\longrightarrow$ moles Na_2CO_3 $\longrightarrow$ grams Na_2CO_3

$$= 8.415 \text{ g Na}_2\text{CO}_3$$

We find that 20.00 g of $KAuCl_4$ needs 8.415 g of Na_2CO_3. The 25.00 g of Na_2CO_3 taken is therefore more than enough to let the $KAuCl_4$ react completely. $KAuCl_4$ is limiting.

Had we tested Na_2CO_3 instead of $KAuCl_4$ as the limiting reactant we would have calculated how many grams of $KAuCl_4$ are required by 25.00 g of Na_2CO_3.

$$25.00 \text{ g Na}_2\text{CO}_3 \times \frac{1 \text{ mol Na}_2\text{CO}_3}{105.99 \text{ g Na}_2\text{CO}_3} \times \frac{2 \text{ mol KAuCl}_4}{3 \text{ mol Na}_2\text{CO}_3} \times \frac{377.88 \text{ g KAuCl}_4}{1 \text{ mol KAuCl}_4}$$

grams Na_2CO_3 $\longrightarrow$ moles Na_2CO_3 $\longrightarrow$ moles $KAuCl_4$ $\longrightarrow$ grams $KAuCl_4$

$$= 59.42 \text{ g KAuCl}_4$$

The given mass of Na_2CO_3 would require much more $KAuCl_4$ than provided, so we confirm that $KAuCl_4$ is the limiting reactant.

From here on, we have a routine calculation going from mass of reactant, $KAuCl_4$, to mass of product, $Au(OH)_3$. We know from the equation's coefficients that

$$1 \text{ mol KAuCl}_4 \Leftrightarrow 1 \text{ mol Au(OH)}_3$$

Using this, plus formula masses, we do the following chain calculation:

$$20.00 \text{ g KAuCl}_4 \times \frac{1 \text{ mol KAuCl}_4}{377.88 \text{ g KAuCl}_4} \times \frac{1 \text{ mol Au(OH)}_3}{1 \text{ mol KAuCl}_4} \times \frac{247.99 \text{ g Au(OH)}_3}{1 \text{ mol Au(OH)}_3}$$

grams $KAuCl_4$ $\longrightarrow$ moles $KAuCl_4$ $\longrightarrow$ moles $Au(OH)_3$ $\longrightarrow$ grams $Au(OH)_3$

$$= 13.13 \text{ g Au(OH)}_3$$

Thus from 20.00 g of $KAuCl_4$ we can make a maximum of 13.13 g of $Au(OH)_3$.

In this synthesis, some of the initial 25.00 g of Na_2CO_3 is left over; 20.00 g of $KAuCl_4$ requires only 8.415 g of Na_2CO_3 out of 25.00 g Na_2CO_3. The difference, $(25.00 \text{ g} - 8.415 \text{ g}) = 16.58$ g of Na_2CO_3, remains unreacted. It is possible that the chemist used an excess to ensure that every last bit of the very expensive $KAuCl_4$ would be changed to $Au(OH)_3$.

The computed amount of product is always based on the limiting reactant.

Are the Answers Reasonable?

The "head check" has to be done at each step. We'll illustrate just the first. Is it reasonable that around 8 g of Na_2CO_3 would be equivalent to 20 g of $KAuCl_4$ in the reaction whose equation is given? Yes. Notice that the formula mass of Na_2CO_3 is between $\frac{1}{3}$ and $\frac{1}{4}$ the size of the formula mass of $KAuCl_4$. So on this score alone, it's reasonable that the calculated mass of Na_2CO_3 be quite a bit less than the mass of $KAuCl_4$ taken. Of course, the equation calls for 3 mol of Na_2CO_3 to 2 mol of $KAuCl_4$, so this ratio calls for more Na_2CO_3 than thus far suggested in our "head check." To correct for this, we multiply $\frac{1}{4}$ by $\frac{3}{2}$ to give $\frac{3}{8}$, and we multiply $\frac{3}{8}$ by 20 g (of $KAuCl_4$) to give us a little less than the more precisely calculated mass of Na_2CO_3. So, the answer is reasonable.

PRACTICE EXERCISE 24: In an industrial process for making nitric acid, the first step is the reaction of ammonia with oxygen at high temperature in the presence of a platinum gauze. Nitrogen monoxide forms as follows:

$$4NH_3 + 5O_2 \longrightarrow 4NO + 6H_2O$$

How many grams of nitrogen monoxide can form if a mixture of 30.00 g of NH_3 and 40.00 g of O_2 is taken initially?

4.8 ▶ **The predicted amount of product is not always obtained experimentally**

In most experiments designed for chemical synthesis, the amount of a product actually isolated falls short of the calculated maximum amount. Losses occur for several reasons. Some are mechanical, such as materials sticking to glassware. In some reactions, losses occur by the evaporation of a volatile product. In others, a product is a solid that separates from the solution as it forms because it is largely insoluble. The solid is removed by filtration. What stays in solution, although relatively small, contributes to some loss of product.

One of the common causes of obtaining less than the stoichiometric amount of a product is the occurrence of a **competing reaction.** It produces a **by-product,** a substance made by a reaction that competes with the **main reaction.** The synthesis of phosphorus trichloride, for example, gives some phosphorus pentachloride as well, because PCl_3 can react further with Cl_2.

$$\text{Main reaction:} \qquad 2P(s) + 3Cl_2(g) \longrightarrow 2PCl_3(l)$$

$$\text{Competing reaction:} \qquad PCl_3(l) + Cl_2(g) \longrightarrow PCl_5(s)$$

The competition is between newly formed PCl_3 and still unreacted phosphorus for still unchanged chlorine.

Some reactions run in both the forward and reverse directions. For example, in a solution containing lead(II) ions and chloride ions, both of these reactions occur:

$$Pb^{+2}(aq) + 2\,Cl^-(aq) \longrightarrow PbCl_2(s)$$

$$PbCl_2(s) \longrightarrow Pb^{+2}(aq) + 2Cl^-(aq)$$

Obviously, any calculation of the amount of $PbCl_2$ from the amount of Pb^{+2} added to excess chloride will fail unless it takes both reactions into account. Reactions of this sort are said to be **reversible.** We will discuss reversible reactions in some detail in later chapters.

The **actual yield** of desired product is simply how much is isolated, stated in mass units or moles. The **theoretical yield** of the product is what must be obtained if no losses occur. When less than the theoretical yield of product is obtained, chemists generally calculate the *percentage yield* of product to describe how well the preparation went. The **percentage yield** is the actual yield calculated as a percentage of the theoretical yield.

Theoretical yield is a calculated quantity. The actual yield is a measured quantity; in general, it cannot be calculated.

🏺 **TOOLS** ▶

Theoretical, actual, and percentage yields

$$\text{percentage yield} = \frac{\text{actual yield}}{\text{theoretical yield}} \times 100\%$$

Both the actual and theoretical yields must be in the same units, of course.

It is important to realize that the actual yield is an experimentally determined quantity. It usually cannot be calculated. The theoretical yield is always a calculated quantity based on a chemical equation and the amounts of the reactants available.

Let's now work an example that combines the determination of the limiting reactant with a calculation of percentage yield.

EXAMPLE 4.23

Calculating a Percentage Yield

A chemist set up a synthesis of phosphorus trichloride by mixing 12.0 g P with 35.0 g Cl_2 and obtained 42.4 g of PCl_3. Calculate the percentage yield of this compound. The equation for the main reaction is, as we noted earlier,

$$2P(s) + 3Cl_2(g) \longrightarrow 2PCl_3(l)$$

ANALYSIS: Notice that the masses of *both* reactants are given, so we know that we have a limiting reactant problem. We must first figure out which reactant, P or Cl_2, is the limiting reactant, because we have to base the calculation of the theoretical yield of PCl_3 on it.

SOLUTION: In any limiting reactant problem, we can arbitrarily pick one reactant and do a calculation to see whether it can be entirely used up. We'll choose phosphorus and see whether there is enough to react with 35.0 g of chlorine. The following calculation gives us the answer. Draw in the cancel lines yourself.

$$12.0 \text{ g P} \times \frac{1 \text{ mol P}}{30.97 \text{ g P}} \times \frac{3 \text{ mol Cl}_2}{2 \text{ mol P}} \times \frac{70.90 \text{ g Cl}_2}{1 \text{ mol Cl}_2} = 41.2 \text{ g Cl}_2$$

Thus, with 35.0 g of Cl_2 provided but 41.2 g of Cl_2 needed, there is not enough Cl_2 for the 12.0 g of P. The Cl_2 can be all used up, so Cl_2 is the limiting reactant. We therefore base the calculation of the theoretical yield of PCl_3 on Cl_2.

To find the *theoretical yield* of PCl_3, we calculate how many grams of PCl_3 could be made from 35.0 g of Cl_2 if everything went perfectly according to the equation given.

$$35.0 \text{ g Cl}_2 \times \frac{1 \text{ mol Cl}_2}{70.90 \text{ g Cl}_2} \times \frac{2 \text{ mol PCl}_3}{3 \text{ mol Cl}_2} \times \frac{137.32 \text{ g PCl}_3}{1 \text{ mol PCl}_3} = 45.2 \text{ g PCl}_3$$

$$\text{grams Cl}_2 \longrightarrow \text{moles Cl}_2 \longrightarrow \text{moles PCl}_3 \longrightarrow \text{grams PCl}_3$$

The actual yield was 42.4 g of PCl_3, not 45.2 g, so the percentage yield is calculated as follows:

$$\text{percentage yield} = \frac{42.4 \text{ g PCl}_3}{45.2 \text{ g PCl}_3} \times 100\% = 93.8\%$$

Thus 94.8% of the theoretical yield of PCl_3 was obtained.

Is the Answer Reasonable?
The obvious check is that the calculated or theoretical yield can never be *less* than the actual yield. So on this score, our answer is reasonable. But doing a "head calculation" beyond the obvious in a problem like this is difficult for most people, even when data are vigorously rounded. Doing chain calculations in which both numbers and their full units are given (for example, 1 mol Cl_2, not just 1 mol), so that full units can be canceled, should help you avoid the most common errors.

PRACTICE EXERCISE 25: Ethanol, C_2H_5OH, can be converted to acetic acid (the acid in vinegar), $HC_2H_3O_2$, by the action of sodium dichromate in aqueous sulfuric acid according to the following equation:

$$3C_2H_5OH(aq) + 2Na_2Cr_2O_7(aq) + 8H_2SO_4(aq) \longrightarrow$$

$$3HC_2H_3O_2(aq) + 2Cr_2(SO_4)_3(aq) + 2Na_2SO_4(aq) + 11H_2O$$

In one experiment, 24.0 g of C_2H_5OH, 90.0 g of $Na_2Cr_2O_7$, and a known excess of sulfuric acid were mixed, and 26.6 g of acetic acid ($HC_2H_3O_2$) were isolated. Calculate the percentage yield of $HC_2H_3O_2$.

SUMMARY

Mole Concept and Formula Mass or Molecular Mass In the SI *definition*, 1 **mole** of any substance is an amount with the same number, **Avogadro's number** (6.022×10^{23}), of atoms, molecules, or formula units as there are atoms in 12 g (exactly) of carbon-12. The sum of the atomic masses of all of the atoms appearing in a chemical formula gives the **formula mass** or the **molecular mass.**

formula (or molecular) mass (g) $\Leftrightarrow$ 1 mol of substance

Like an atomic mass, a formula or molecular mass is a tool for grams-to-moles or moles-to-grams conversions.

Chemical Formulas The actual composition of a molecule is given by its **molecular formula.** An **empirical formula** gives the ratio of atoms, but in the smallest whole numbers, and it is generally the *only* formula we write for ionic compounds. In the case of a molecular compound, the molecular mass is equal either to that calculated from the compound's empirical formula or to a simple multiple of it. When the two calculated masses are the same, then the molecular formula is the same as the empirical formula.

An empirical formula can be found from the masses of the elements obtained by the quantitative analysis of a known sample of the compound or it can be calculated from the **percentage composition.** The percentage of an element in a compound is the same as the number of grams of the element in 100 g of the compound. If there is $A\%$ of X in X_xZ_z, then 100.00 g of X_xZ_z contains A g of X. From the grams of X and Z, the moles of X and Z can be calculated. When these are adjusted to their corresponding whole-number ratios, the subscripts in the empirical formula are obtained.

Formula Stoichiometry A chemical formula is a tool for stoichiometric calculations, because its subscripts tell us the mole ratios in which the various elements are combined. In X_xZ_z, the mole-to-mole relationship between X and Z is simply

$$z \text{ mol } Z \Longleftrightarrow x \text{ mol } X$$

The conversion factors available from this equivalency are

$$\frac{z \text{ mol } Z}{x \text{ mol } X} \quad \text{and} \quad \frac{x \text{ mol } X}{z \text{ mol } Z}$$

Balanced Equations and Reaction Stoichiometry A balanced equation is a tool for reaction stoichiometry because its coefficients disclose the stoichiometric equivalencies.

When balancing an equation, only the coefficients can be adjusted, never the subscripts. In a generalized reaction,

$$xA + mW \longrightarrow yB + nZ$$

we know that

$$x \text{ mol } A \Longleftrightarrow m \text{ mol } W$$

$$x \text{ mol } A \Longleftrightarrow y \text{ mol } B$$

$$x \text{ mol } A \Longleftrightarrow n \text{ mol } Z$$

and so on.

All problems of reaction stoichiometry must be solved at the mole level of the substances involved. If grams are given, they must first be changed to moles.

Yields of Products A reactant taken in a quantity less than required by another reactant, as determined by the reaction's stoichiometry, is called the **limiting reactant.** The **theoretical yield** of a product can be no more than permitted by the limiting reactant. Sometimes **competing reactions** (side reactions) producing **by-products** reduce the **actual yield.** The ratio of the actual to the theoretical yields, expressed as a percentage, is the **percentage yield.**

◢ TOOLS ▷ YOU HAVE LEARNED

The table below lists the concepts that you've learned in this chapter that can be applied as tools in solving problems. Study each one carefully so that you know what each is used for. When faced with solving a problem, recall what each tool does and consider whether it will be helpful in finding a solution. This will aid you in selecting the tools you need. If necessary, refer to this table when working on the Thinking-It-Through problems and the Review Exercises that follow.

TOOL	HOW IT WORKS
Avogadro's number (page 117)	Relates macroscopic lab-sized quantities (e.g., moles) to numbers of individual atomic-sized particles such as atoms, molecules, or ions.
Chemical formula (page 122)	Use subscripts in a formula to establish atom ratios and mole ratios between the elements in the substance.
Atomic mass (page 117)	Use it to form a conversion factor to calculate mass from moles of an element, or moles from the mass of an element.
Formula mass; molecular mass (page 117)	Use it to form a conversion factor to calculate mass from moles of a compound, or moles from the mass of a compound.
Percentage composition (page 125)	We use it to represent the composition of a compound and as the basis for computing the empirical formula. Comparing experimental and theoretical percentage compositions can help establish the identity of a compound.
Balanced chemical equation (page 133)	Use the coefficients to establish stoichiometric equivalencies that relate moles of one substance to moles of another in a chemical reaction. The coefficients also establish the ratios by formula units among the reactants and products.
Theoretical, actual, and percentage yields (page 142)	We use it to estimate the efficiency of a reaction. Remember that the theoretical yield is calculated from the limiting reactant using the balanced chemical equation.

THINKING IT THROUGH

The goal for the following problems is not to find the answers themselves, but rather to assemble the information needed to solve them and explain how you would use the information to find the answers. The problems in Level 2 are more challenging than those in Level 1 and may contain more data than are required, in which case you are also asked to identify the unnecessary data. Detailed answers to the Thinking-It-Through problems can be found on the web site.

Need extra help?
Visit the Brady/
Senese web site at
www.wiley.com/
college/brady

Level 1 Problems

1. Show how you would answer the following questions: How many moles of carbon are combined with 0.60 mol of hydrogen atoms in the rocket fuel dimethylhydrazine, $(CH_3)_2N_2H_2$? How many moles of H_2 gas could be made from all the hydrogen in 0.56 mol of $(CH_3)_2N_2H_2$?

2. Sulfur and oxygen atoms are combined in sulfur trioxide, SO_3. How many atoms of S react with 0.400 mol of O_2 molecules to make molecules of SO_3? Set up the calculation.

3. The following information applies to a certain compound of carbon and hydrogen: In 4 mol of the compound there are 12 mol of C; in 6 molecules of the compound, there are 48 atoms of H. Describe how you would use this information to obtain the empirical formula for the compound.

4. Chloroform, a chemical once used as an anesthetic, has the formula $CHCl_3$. Set up the solutions to the following questions: (a) How many molecules of $CHCl_3$ would have to be decomposed to give 18 atoms of Cl? (b) How many moles of $CHCl_3$ would have to be decomposed to give enough chlorine atoms to make 0.36 mol of Cl_2?

5. The compound 2-furylmethanethiol, C_5H_6OS, helps give coffee its distinctive aroma. How many micrograms of sulfur are in 1.00 μg of 2-furylmethanethiol? Set up the solution following the factor-label method. Be sure to show cancellation of units.

6. A 2.05 g sample of a certain compound of tin and chlorine was found to contain 1.11 g of chlorine. Explain how you would use this information to obtain the empirical formula for the compound.

7. Concerning a certain compound of sodium, sulfur, and oxygen, you are given the following stoichiometric equivalencies:

$$4 \text{ mol Na} \Leftrightarrow 10 \text{ mol O}$$

$$5 \text{ mol O} \Leftrightarrow 2 \text{ mol S}$$

Explain how you would determine the empirical formula. Set up the calculation to determine the number of moles of sulfur combined with 2.7 mol Na.

8. How many grams of H_2 can be obtained from the electrolysis of 10.0 g of H_2O?

9. Explain how you would calculate the atomic mass of an element M if 36.0 g of M combine with 24.0 g of oxygen to give a compound with the formula MO.

10. How many grams of CuO can be made from a piece of copper wire weighing 0.3433 g?

11. A 1.015 g sample of a binary compound of calcium and bromine was dissolved in water and treated with $Na_2C_2O_4$, which caused CaC_2O_4 to form. The CaC_2O_4 was recovered from the reaction mixture and was found to weigh 0.650 g. Outline the steps needed to calculate the percentage of calcium in the original calcium–bromine compound. Explain how you would use the data given here to calculate the empirical formula for the calcium–bromine compound. Show relevant conversion factors.

12. A sample of the hydrate $CuSO_4 \cdot 5H_2O$ has a mass of 12.5 g. If it is completely dehydrated, how many grams of $CuSO_4$ would remain? How many moles of water would be released during the dehydration process? Show how you would set up the calculations.

13. Calcium hydroxide and hydrobromic acid react as follows:

$$Ca(OH)_2(s) + 2HBr(aq) \longrightarrow CaBr_2(aq) + 2H_2O$$

Describe in detail how you would answer the following:
(a) How many grams of HBr are needed to produce 0.125 mol of $CaBr_2$?
(b) If the reaction is carried out by combining 1.24 g of $Ca(OH)_2$ with 2.65 g of HBr, what is the theoretical yield of $CaBr_2$ expressed in grams?
(c) For part (b), how many grams of which reactant remain after the reaction is complete?

Level 2 Problems

14. A sample of iron ore in which the iron occurs in the mineral Fe_3O_4 is found to contain 12.5% (by mass) of iron. Describe how you would calculate the percentage by mass of Fe_3O_4 in the ore.

15. A white powder was known to be either Na_3PO_4, Na_2HPO_4, or NaH_2PO_4. A sample of it was analyzed and found to contain 21.82% phosphorus by mass. Explain how you would obtain the percentage by mass of sodium in the sample.

16. If compound X and compound Y react with each other in a 1-to-1 ratio by moles, and if 25.0 g of X combine exactly with 29.0 g of Y, which compound has the larger formula mass? Explain your answer.

17. Al_2O_3 reacts with HI as follows:

$$Al_2O_3(s) + 6HI(aq) \longrightarrow 2AlI_3(aq) + 3H_2O$$

A mixture containing 16.0 g of Al_2O_3 and 140 g of HI produces 127 g of AlI_3. Explain in detail how you would determine the number of grams of which reactant were present in excess in the original reaction mixture.

18. A 1.000 g sample of a mixture of NaCl and $CaCl_2$ was treated in solution with an excess of $AgNO_3$, which yielded a precipitate of AgCl. The AgCl weighed 2.53 g. Describe in detail how the percentage by mass of NaCl in the sample can be calculated.

19. How many distinct empirical formulas are shown by the following models for compounds formed between elements A and B? Explain. (Element A is represented by a black sphere and element B by a light gray sphere.)

20. Molecules containing A and B react to form AB as shown below. Based on the equations and the contents of the boxes labeled "Initial," sketch for each reaction the molecular models of what is present after the reaction is over. (In both cases, the species B exists as B_2. In reaction 1, A is monatomic; in reaction 2, A exists as diatomic molecules A_2.)

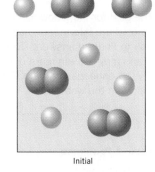

Initial

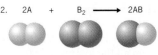

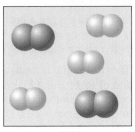

Initial

REVIEW QUESTIONS

4.1 How would you estimate the number of atoms in a gram of iron, using atomic mass units?

4.2 What is the definition of the mole?

4.3 Why are moles used when all stoichiometry problems could be done using only atomic mass units?

4.4 Which contains more molecules: 2.5 mol of H_2O, or 2.5 mol of H_2?

4.5 How many moles of iron atoms are in 1 mol of Fe_2O_3? How many iron atoms are in 1 mol of Fe_2O_3?

4.6 Write mole-to-mole conversion factors that exist between the elements in each of the following compounds:
(a) SO_2 (b) As_2O_3 (c) PbN_6 (d) Al_2Cl_6

4.7 Write a mole-to-mole conversion factor that relates the elements in each of the following compounds:
(a) Mn_3O_4 (b) Sb_2S_5 (c) P_4O_6 (d) Hg_2Cl_2

4.8 What information is required to convert grams of a substance into moles of that same substance?

4.9 Why is the expression "1.0 mol of oxygen" ambiguous? Why doesn't a similar ambiguity exist in the expression "64 g of oxygen"?

4.10 The atomic mass of aluminum is 26.98. What specific conversion factors does this value make available for relating a mass of aluminum (in grams) and a quantity of aluminum given in moles?

4.11 In general, what fundamental information, obtained from experimental measurements, is required to calculate the empirical formula of a compound?

4.12 Why is the changing of subscripts not allowed when balancing a chemical equation?

4.13 What information is required to convert grams of a substance into molecules of that same substance?

4.14 When given the *unbalanced* equation

$$Na(s) + Cl_2(g) \longrightarrow NaCl(s)$$

and asked to balance it, student A wrote

$$Na(s) + Cl_2(g) \longrightarrow NaCl_2(s)$$

and student B wrote

$$2Na(s) + Cl_2(g) \longrightarrow 2NaCl(s)$$

(a) Both equations are balanced, but which student is correct?
(b) Explain why the other student's answer is incorrect.

4.15 Give a step-by-step procedure for estimating the number grams of A required to completely react with 10 mol of B, given the following information:
   i. A and B react to form A_5B_2.
   ii. A has a molecular mass of 100.0.
   iii. B has a molecular mass of 200.0.
   iv. There are 6.02×10^{23} molecules of A in a mole of A.
Which of these pieces of information weren't needed?

4.16 If two substances react completely in a 1-to-1 ratio *both* by mass and by moles, what must be true about these substances?

4.17 What information is required to determine how many grams of sulfur would react with a gram of arsenic?

4.18 A mixture of 0.020 mol of Mg and 0.020 mol of Cl_2 reacted completely to form $MgCl_2$ according to the equation

$$Mg + Cl_2 \longrightarrow MgCl_2$$

(a) What information describes the *stoichiometry* of this reaction?

(b) What information gives the *scale* of the reaction?

4.19 In a report to a supervisor, a chemist described an experiment in the following way: "0.0800 mol of H_2O_2 decomposed into 0.0800 mol of H_2O and 0.0400 mol of O_2." Express the chemistry and stoichiometry of this reaction by a conventional chemical equation.

4.20 How would you count the number of N_2 molecules that could be produced after the explosion of 1.00 kg of NH_4NO_3?

4.21 How would Avogadro's number change if the atomic mass unit were to be redefined as 2×10^{-27} kg, exactly?

REVIEW PROBLEMS

Answers to problems whose numbers are printed in color are given in Appendix B. More challenging questions are marked with asterisks. **ILW** = Interactive LearningWare solution is available at *www.wiley.com/college/brady*.

The Mole Concept and Stoichiometric Equivalencies

4.22 In what smallest whole-number ratio must N and O atoms combine to make dinitrogen tetroxide, N_2O_4? What is the mole ratio of the elements in this compound?

4.23 In what atom ratio are the elements present in methane, CH_4 (the chief component of natural gas)? In what mole ratio are the atoms of the elements present in this compound?

4.24 How many moles of sodium atoms correspond to 1.56×10^{21} atoms of sodium?

4.25 How many moles of nitrogen molecules correspond to 1.80×10^{24} molecules of N_2?

4.26 Sucrose (table sugar) has the formula $C_{12}H_{22}O_{11}$. In this compound, what is the
(a) atom ratio of C to H? (c) atom ratio of H to O?
(b) mole ratio of C to O? (d) mole ratio of H to O?

4.27 Nail polish remover is usually the volatile liquid acetone, $(CH_3)_2CO$. In this compound, what is the
(a) atom ratio of C to O? (c) atom ratio of C to H?
(b) mole ratio of C to O? (d) mole ratio of C to H?

4.28 How many moles of Al atoms are needed to combine with 1.58 mol of O atoms to make aluminum oxide, Al_2O_3?

4.29 How many moles of vanadium atoms, V, are needed to combine with 0.565 mol of O atoms to make vanadium pentoxide, V_2O_5?

4.30 How many moles of Al are in 2.16 mol of Al_2O_3?

4.31 How many moles of O atoms are in 4.25 mol of calcium carbonate, $CaCO_3$, the chief constituent of seashells?

4.32 Aluminum sulfate, $Al_2(SO_4)_3$, is a compound used in sewage treatment plants.
(a) Construct a pair of conversion factors that relate moles of aluminum to moles of sulfur for this compound.
(b) Construct a pair of conversion factors that relate moles of sulfur to moles of $Al_2(SO_4)_3$.
(c) How many moles of Al are in a sample of this compound if the sample also contains 0.900 mol S?
(d) How many moles of S are in 1.16 mol $Al_2(SO_4)_3$?

4.33 Magnetite is a magnetic iron ore. Its formula is Fe_3O_4.
(a) Construct a pair of conversion factors that relate moles of Fe to moles of Fe_3O_4.
(b) Construct a pair of conversion factors that relate moles of Fe to moles of O in Fe_3O_4.
(c) How many moles of Fe are in 2.75 mol of Fe_3O_4?
(d) If this compound could be prepared from Fe_2O_3 and O_2, how many moles of Fe_2O_3 would be needed to prepare 4.50 mol Fe_3O_4?

4.34 How many moles of H_2 and N_2 can be formed by the decomposition of 0.145 mol of ammonia, NH_3?

4.35 How many moles of O_2 are needed to combine with 0.225 mol Al to give Al_2O_3?

ILW 4.36 How many moles of UF_6 would have to be decomposed to provide enough fluorine to prepare 1.25 mol of CF_4? (Assume sufficient carbon is available.)

4.37 How many moles of Fe_3O_4 are required to supply enough iron to prepare 0.260 mol Fe_2O_3? (Assume sufficient oxygen is available.)

4.38 How many atoms of carbon are combined with 4.13 moles of H in a sample of the compound propane, C_3H_8? (Propane is used as the fuel in gas barbecues.)

4.39 How many atoms of hydrogen are found in 2.31 mol of propane, C_3H_8?

4.40 What is the total number of C, H, and O atoms in 0.260 mol of glucose, $C_6H_{12}O_6$?

4.41 What is the total number of N, H, and O atoms in 0.356 mol of ammonium nitrate, NH_4NO_3, an important fertilizer?

Measuring Moles of Elements and Compounds

4.42 How many atoms are in 6.00 g of carbon-12?

4.43 How many atoms are in 1.50 mol of carbon-12? How many grams does this much carbon-12 weigh?

4.44 What is the mass of 1.00 mol of each of the following elements? (a) sodium (b) sulfur (c) chlorine

4.45 What is the mass of 1.00 mol of each of the following elements? (a) nitrogen (b) iron (c) magnesium

4.46 Determine the mass in grams of each of the following:
(a) 1.35 mol Fe   (b) 24.5 mol O   (c) 0.876 mol Ca

4.47 Determine the mass in grams of each of the following:
(a) 0.546 mol S   (b) 3.29 mol N   (c) 8.11 mol Al

4.48 What is the mass, in grams, of 2.00×10^{12} atoms of potassium?

4.49 What is the mass, in grams, of 4.00×10^{17} atoms of sodium?

4.50 How many moles of nickel are in 17.7 g of Ni?

4.51 How many moles of chromium are in 85.7 g of Cr?

4.52 Calculate the formula mass of each of the following and round your answer to the nearest 0.1 u:
(a) $NaHCO_3$ (c) $(NH_4)_2CO_3$ (e) $CuSO_4 \cdot 5H_2O$
(b) $K_2Cr_2O_7$ (d) $Al_2(SO_4)_3$

4.53 Calculate the formula mass of each of the following and round your answer to the nearest 0.1 u:
(a) $Ca(NO_3)_2$ (d) $Fe_4[Fe(CN)_6]_3$
(b) $Pb(C_2H_5)_4$ (e) $Na_2SO_4 \cdot 10H_2O$
(c) $Mg_3(PO_4)_2$

4.54 Calculate the mass in grams of the following:
(a) 1.25 mol $Ca_3(PO_4)_2$ (c) 0.600 mol C_4H_{10}
(b) 0.625 mol $Fe(NO_3)_3$ (d) 1.45 mol $(NH_4)_2CO_3$

4.55 What is the mass in grams of the following?
(a) 0.754 mol $ZnCl_2$ (c) 0.322 mol $POCl_3$
(b) 0.194 mol KIO_3 (d) 4.31×10^{-3} mol $(NH_4)_2HPO_4$

4.56 Calculate the number of moles of each compound in the following samples:
(a) 21.5 g $CaCO_3$ (c) 16.8 g $Sr(NO_3)_2$
(b) 1.56 g NH_3 (d) 6.98 μg Na_2CrO_4

4.57 Calculate the number of moles of each compound in the following samples:
(a) 9.36 g $Ca(OH)_2$ (c) 4.29 g H_2O_2
(b) 38.2 g $PbSO_4$ (d) 4.65 mg $NaAuCl_4$

ILW 4.58 One sample of CaC_2 contains 0.150 mol of carbon. How many moles and how many grams of calcium are also in the sample? [Calcium carbide, CaC_2, was once used to make signal flares for ships. Water dripped onto CaC_2 reacts to give acetylene (C_2H_2), which burns brightly.]

4.59 How many moles of iodine, I, are in 0.500 mol of $Ca(IO_3)_2$? How many grams of calcium iodate are needed to supply this much iodine? [Iodized salt contains a trace amount of calcium iodate, $Ca(IO_3)_2$, to help prevent a thyroid condition called goiter.]

4.60 How many moles of nitrogen, N, are in 0.650 mol of $(NH_4)_2CO_3$? How many grams of this compound supply this much nitrogen?

4.61 How many moles of nitrogen, N, are in 0.556 mol of NH_4NO_3? How many grams of NH_4NO_3 supply this much nitrogen?

4.62 How many kilograms of a fertilizer made of pure $(NH_4)_2CO_3$ would be required to supply 1 kg of nitrogen to the soil?

4.63 How many kilograms of a fertilizer made of pure P_2O_5 would be required to supply 1 kg of phosphorus to the soil?

Percentage Composition

4.64 Which has a higher percentage of oxygen: morphine ($C_{17}H_{19}NO_3$) or heroin ($C_{21}H_{23}NO_5$)?

4.65 Which has a higher percentage of nitrogen: carbamazepine ($C_{15}H_{12}N_2O$) or carbetapentane ($C_{20}H_{31}NO_3$)?

4.66 Which has a higher percentage of chlorine: Freon-12 (CCl_2F_2) or Freon 141b ($C_2H_3Cl_2F$)?

4.67 Which has a higher percentage of fluorine: Freon-12 (CCl_2F_2) or Freon 113 ($C_2Cl_3F_3$)?

4.68 Calculate the percentage composition by mass for each of the following: (a) NaH_2PO_4, (b) $NH_4H_2PO_4$, (c) $(CH_3)_2CO$, (d) $CaSO_4$, (e) $CaSO_4 \cdot 2H_2O$.

4.69 Calculate the percentage composition by mass of each of the following: (a) $(CH_3)_2N_2H_2$, (b) $CaCO_3$, (c) $Fe(NO_3)_3$, (d) C_3H_8, (e) $Al_2(SO_4)_3$.

4.70 It was found that 2.35 g of a compound of phosphorus and chlorine contained 0.539 g of phosphorus. What are the percentages by mass of phosphorus and chlorine in this compound?

4.71 An analysis revealed that 5.67 g of a compound of nitrogen and oxygen contained 1.47 g of nitrogen. What are the percentages by mass of nitrogen and oxygen in this compound?

4.72 Phencyclidine ("angel dust") is $C_{17}H_{25}N$. A sample suspected of being this illicit drug was found to have a percentage composition of 84.71% C, 10.42% H, and 5.61% N. Do these data acceptably match the theoretical data for phencyclidine?

4.73 The hallucinogenic drug LSD has the molecular formula $C_{20}H_{25}N_3O$. One suspected sample contained 74.07% C, 7.95% H, and 9.99% N.
(a) What is the percentage O in the sample?
(b) Are these data consistent for LSD?

4.74 How many grams of O are combined with 7.14×10^{21} atoms of N in the compound N_2O_5?

4.75 How many grams of C are combined with 4.25×10^{23} atoms of H in the compound C_5H_{12}?

Empirical Formulas

4.76 Write empirical formulas for the following compounds:
(a) S_2Cl_2, (b) $C_6H_{12}O_6$, (c) NH_3, (d) As_2O_6, (e) H_2O_2.

4.77 What are the empirical formulas of the following compounds? (a) $C_2H_4(OH)_2$, (b) $H_2S_2O_8$, (c) C_4H_{10}, (d) B_2H_6, (e) C_2H_5OH.

4.78 Quantitative analysis of a sample of sodium pertechnetate with a mass of 0.896 g found 0.111 g of sodium and 0.477 g of technetium. The remainder was oxygen. Calculate the empirical formula of sodium pertechnetate. (Radioactive sodium pertechnetate is used as a brain-scanning agent in medicine.)

4.79 A sample of Freon, a gas once used as a propellant in aerosol cans, was found to contain 0.423 g of C, 2.50 g Cl, and 1.34 g F. What is the empirical formula of this compound?

4.80 A dry-cleaning fluid composed of only carbon and chlorine was found to be composed of 14.5% C and 85.5% Cl (by mass). What is the empirical formula of this compound?

4.81 One compound of mercury with a formula mass of 519 contains 77.26% Hg, 9.25% C, and 1.17% H (with the balance being O). Calculate the empirical and molecular formulas.

4.82 A substance was found to be composed of 22.9% Na, 21.5% B, and 55.7% O. What is the empirical formula of this compound?

4.83 Vanillin, a compound used as a flavoring agent in food products, has the following percentage composition: 64.2% C, 5.26% H, and 31.6% O. What is the empirical formula of vanillin?

ILW 4.84 When 0.684 g of an organic compound containing only carbon, hydrogen, and oxygen was burned in oxygen, 1.312 g CO_2 and 0.805 g H_2O were obtained. What is the empirical formula of the compound?

4.85 Methyl ethyl ketone (often abbreviated MEK) is a powerful solvent with many commercial uses. A sample of this compound (which contains only C, H, and O) weighing 0.822 g was burned in oxygen to give 2.01 g CO_2 and 0.827 g H_2O. What is the empirical formula for MEK?

4.86 When 6.853 mg of a sex hormone was burned in a combustion analysis, 19.73 mg of CO_2 and 6.391 mg of H_2O were obtained. What is the empirical formula of the compound?

4.87 When a sample of a compound in the vitamin D family was burned in a combustion analysis, 5.983 mg of the compound gave 18.490 mg of CO_2 and 6.232 mg of H_2O. What is the empirical formula of the compound?

Molecular Formulas

4.88 The following are empirical formulas and the masses per mole for three compounds. What are their molecular formulas?
(a) NaS_2O_3; 270.4 g/mol   (c) C_2HCl; 181.4 g/mol
(b) C_3H_2Cl; 147.0 g/mol

4.89 The following are empirical formulas and the masses per mole for three compounds. What are their molecular formulas?
(a) Na_2SiO_3; 732.6 g/mol   (c) CH_3O; 62.1 g/mol
(b) $NaPO_3$; 305.9 g/mol

4.90 The compound described in Problem 4.86 was found to have a molecular mass of 290. What is its molecular formula?

4.91 The compound described in Problem 4.87 was found to have a molecular mass of 399. What is the molecular formula of this compound?

ILW 4.92 A sample of a compound of mercury and bromine with a mass of 0.389 g was found to contain 0.111 g bromine. Its molecular mass was found to be 561. What are its empirical and molecular formulas?

4.93 A 0.6662 g sample of "antimonal saffron" was found to contain 0.4017 g of antimony. The remainder was sulfur. The molecular mass of this compound is 404. What are the empirical and molecular formulas of this pigment? (This compound is a red pigment used in painting.)

4.94 A sample of a compound of C, H, N, and O, with a mass of 0.6216 g was found to contain 0.1735 g C, 0.01455 g H, and 0.2024 g N. Its molecular mass is 129. Calculate its empirical and molecular formulas.

4.95 Strychnine, a deadly poison, has a molecular mass of 334 and a percentage composition of 75.42% C, 6.63% H, 8.38% N, and the balance oxygen. Calculate the empirical and molecular formulas of strychnine.

Balancing Chemical Equations

4.96 How many moles of hydrogen are part of the expression "$2Ba(OH)_2 \cdot 8H_2O$," taken from a balanced equation?

4.97 How many moles of oxygen are part of the expression "$3Ca_3(PO_4)_2$," taken from a balanced equation?

4.98 Write the equation that expresses in acceptable chemical shorthand the following statement: "Iron can be made to react with molecular oxygen (O_2) to give iron oxide with the formula Fe_2O_3."

4.99 The conversion of one air pollutant, NO, produced in vehicle engines, into another, NO_2, occurs when NO reacts with molecular oxygen in the air. Write the balanced equation for this reaction.

4.100 Balance the following equations:
(a) $Ca(OH)_2 + HCl \rightarrow CaCl_2 + H_2O$
(b) $AgNO_3 + CaCl_2 \rightarrow Ca(NO_3)_2 + AgCl$
(c) $Fe_2O_3 + C \rightarrow Fe + CO_2$
(d) $NaHCO_3 + H_2SO_4 \rightarrow Na_2SO_4 + H_2O + CO_2$
(e) $C_4H_{10} + O_2 \rightarrow CO_2 + H_2O$

4.101 Balance the following equations:
(a) $SO_2 + O_2 \rightarrow SO_3$
(b) $P_4O_{10} + H_2O \rightarrow H_3PO_4$
(c) $Pb(NO_3)_2 + Na_2SO_4 \rightarrow PbSO_4 + NaNO_3$
(d) $Fe_2O_3 + H_2 \rightarrow Fe + H_2O$
(e) $Al + H_2SO_4 \rightarrow Al_2(SO_4)_3 + H_2$

4.102 Balance the following equations:
(a) $Mg(OH)_2 + HBr \rightarrow MgBr_2 + H_2O$
(b) $HCl + Ca(OH)_2 \rightarrow CaCl_2 + H_2O$
(c) $Al_2O_3 + H_2SO_4 \rightarrow Al_2(SO_4)_3 + H_2O$
(d) $KHCO_3 + H_3PO_4 \rightarrow K_2HPO_4 + H_2O + CO_2$
(e) $C_9H_{20} + O_2 \rightarrow CO_2 + H_2O$

4.103 Balance the following equations:
(a) $CaO + HNO_3 \rightarrow Ca(NO_3)_2 + H_2O$
(b) $Na_2CO_3 + Mg(NO_3)_2 \rightarrow MgCO_3 + NaNO_3$
(c) $(NH_4)_3PO_4 + NaOH \rightarrow Na_3PO_4 + NH_3 + H_2O$
(d) $LiHCO_3 + H_2SO_4 \rightarrow Li_2SO_4 + H_2O + CO_2$
(e) $C_4H_{10}O + O_2 \rightarrow CO_2 + H_2O$

Stoichiometry Based on Chemical Equations

4.104 Chlorine is used by textile manufacturers to bleach cloth. Excess chlorine is destroyed by its reaction with sodium thiosulfate, $Na_2S_2O_3$, as follows:

$$Na_2S_2O_3(aq) + 4Cl_2(g) + 5H_2O \longrightarrow$$
$$2NaHSO_4(aq) + 8HCl(aq)$$

(a) How many moles of $Na_2S_2O_3$ are needed to react with 0.12 mol of Cl_2?
(b) How many moles of HCl can form from 0.12 mol of Cl_2?
(c) How many moles of H_2O are required for the reaction of 0.12 mol of Cl_2?
(d) How many moles of H_2O react if 0.24 mol HCl is formed?

4.105 The octane in gasoline burns according to the following equation:

$$2C_8H_{18} + 25O_2 \longrightarrow 16CO_2 + 18H_2O$$

(a) How many moles of O_2 are needed to react fully with 6 mol of octane?

(b) How many moles of CO_2 can form from 0.5 mol of octane?

(c) How many moles of water are produced by the combustion of 8 mol of octane?

(d) If this reaction is used to synthesize 6.00 mol of CO_2, how many moles of oxygen are needed? How many moles of octane?

4.106 The following reaction is used to extract gold from pretreated gold ore:

$$2Au(CN)_2^-(aq) + Zn(s) \longrightarrow 2Au(s) + Zn(CN)_4^{-2}(aq)$$

(a) How many grams of Zn are needed to react with 0.11 mol of $Au(CN)_2^-$?

(b) How many grams of Au can form from 0.11 mol of $Au(CN)_2^-$?

(c) How many grams of $Au(CN)_2^-$ are required for the reaction of 0.11 mol of Zn?

4.107 Propane burns according to the following equation:

$$C_3H_8 + 5O_2 \longrightarrow 3CO_2 + 4H_2O$$

(a) How many grams of O_2 are needed to react fully with 3 mol of propane?

(b) How many grams of CO_2 can form from 0.1 mol of propane?

(c) How many grams of water are produced by the combustion of 4 mol of propane?

4.108 The incandescent white of a fireworks display is caused by the reaction of phosphorus with O_2 to give P_4O_{10}.

(a) Write the balanced chemical equation for the reaction.

(b) How many grams of O_2 are needed to combine with 6.85 g of P?

(c) How many grams of P_4O_{10} can be made from 8.00 g of O_2?

(d) How many grams of P are needed to make 7.46 g of P_4O_{10}?

4.109 The combustion of butane, C_4H_{10}, produces carbon dioxide and water. When one sample of C_4H_{10} was burned, 4.46 g of water was formed.

(a) Write the balanced chemical equation for the reaction.

(b) How many grams of butane were burned?

(c) How many grams of O_2 were consumed?

(d) How many grams of CO_2 were formed?

ILW 4.110 In *dilute* nitric acid, HNO_3, copper metal dissolves according to the following equation:

$$3Cu(s) + 8HNO_3(aq) \longrightarrow$$
$$3Cu(NO_3)_2(aq) + 2NO(g) + 4H_2O$$

How many grams of HNO_3 are needed to dissolve 11.45 g of Cu according to this equation?

4.111 The reaction of hydrazine, N_2H_4, with hydrogen peroxide, H_2O_2, has been used in rocket engines. One way these compounds react is described by the equation

$$N_2H_4 + 7H_2O_2 \longrightarrow 2HNO_3 + 8H_2O$$

According to this equation, how many grams of H_2O_2 are needed to react completely with 852 g of N_2H_4?

4.112 Oxygen gas can be produced in the laboratory by decomposition of hydrogen peroxide (H_2O_2):

$$2H_2O_2 \longrightarrow 2H_2O + O_2(g)$$

How many kilograms of O_2 can be produced from 1.0 kg of H_2O_2?

4.113 Oxygen gas can be produced in the laboratory by decomposition of potassium chlorate ($KClO_3$):

$$2KClO_3 \longrightarrow 2KCl + 3O_2(g)$$

How many kilograms of O_2 can be produced from 1.0 kg of $KClO_3$?

Limiting Reactant Calculations

4.114 The reaction of powdered aluminum and iron(III) oxide,

$$2Al + Fe_2O_3 \longrightarrow Al_2O_3 + 2Fe$$

produces so much heat the iron that forms is molten. Because of this, railroads use the reaction to provide molten steel to weld steel rails together when laying track. Suppose that in one batch of reactants 4.20 mol of Al was mixed with 1.75 mol of Fe_2O_3.

(a) Which reactant, if either, was the limiting reactant?

(b) Calculate the number of grams of iron that can be formed from this mixture of reactants.

4.115 Ethanol (C_2H_5OH) is synthesized for industrial use by the following reaction, carried out at very high pressure:

$$C_2H_4(g) + H_2O(g) \longrightarrow C_2H_5OH(l)$$

What is the maximum amount of ethanol that can be produced when 1.0 kg of ethylene (C_2H_4) and 0.010 kg of steam are placed into the reaction vessel?

ILW 4.116 Silver nitrate, $AgNO_3$, reacts with iron(III) chloride, $FeCl_3$, to give silver chloride, $AgCl$, and iron(III) nitrate, $Fe(NO_3)_3$. A solution containing 18.0 g of $AgNO_3$ was mixed with a solution containing 32.4 g of $FeCl_3$. How many grams of which reactant remains after the reaction is over?

4.117 Chlorine dioxide, ClO_2, has been used as a disinfectant in air-conditioning systems. It reacts with water according to the equation

$$6ClO_2 + 3H_2O \longrightarrow 5HClO_3 + HCl$$

If 142.0 g of ClO_2 are mixed with 38.0 g of H_2O, how many grams of which reactant remain if the reaction is complete?

4.118 Some of the acid in acid rain is produced by the following reaction:

$$3NO_2(g) + H_2O(l) \longrightarrow 2HNO_3(aq) + NO(g)$$

If a falling raindrop weighing 0.050 g comes into contact with 1.0 mg of $NO_2(g)$, how much HNO_3 can be produced?

4.119 Phosphorus pentachloride reacts with water to give phosphoric acid and hydrogen chloride according to the following equation:

$$PCl_5 + 4H_2O \longrightarrow H_3PO_4 + 5HCl$$

In one experiment, 0.360 mol of PCl_5 was slowly added to 2.88 mol of water.
(a) Which reactant, if either, was the limiting reactant?
(b) How many grams of HCl were formed in the reaction?

Theoretical Yield and Percentage Yield

4.120 Barium sulfate, $BaSO_4$, is made by the following reaction:

$$Ba(NO_3)_2(aq) + Na_2SO_4(aq) \longrightarrow BaSO_4(s) + 2NaNO_3(aq)$$

An experiment was begun with 75.00 g of $Ba(NO_3)_2$ and an excess of Na_2SO_4. After collecting and drying the product, 64.45 g of $BaSO_4$ were obtained. Calculate the theoretical yield and percentage yield of $BaSO_4$.

4.121 The Solvay process for the manufacture of sodium carbonate begins by passing ammonia and carbon dioxide through a solution of sodium chloride to make sodium bicarbonate and ammonium chloride. The equation for the overall reaction is

$$H_2O + NaCl + NH_3 + CO_2 \longrightarrow NH_4Cl + NaHCO_3$$

In the next step, sodium bicarbonate is heated to give sodium carbonate and two gases, carbon dioxide and steam.

$$2NaHCO_3 \longrightarrow Na_2CO_3 + CO_2 + H_2O$$

What is the theoretical yield of sodium carbonate, expressed in grams, if 120 g NaCl were used in the first reaction? If 85.4 g of Na_2CO_3 were obtained, what was the percentage yield?

ILW 4.122 Aluminum sulfate can be made by the following reaction:

$$2AlCl_3(aq) + 3H_2SO_4(aq) \longrightarrow Al_2(SO_4)_3(aq) + 6HCl(aq)$$

It is quite soluble in water, so to isolate it the solution has to be evaporated to dryness. This drives off the volatile HCl, but the residual solid has to be heated to a little over 200 °C to drive off all of the water. In one experiment, 25.0 g of $AlCl_3$ were mixed with 30.0 g of H_2SO_4. Eventually, 28.46 g of pure $Al_2(SO_4)_3$ were isolated. Calculate the percentage yield.

4.123 The combustion of methyl alcohol in an abundant excess of oxygen follows the equation

$$2CH_3OH + 3O_2 \longrightarrow 2CO_2 + 4H_2O$$

When 6.40 g of CH_3OH were mixed with 10.2 g of O_2 and ignited, 6.12 g of CO_2 were obtained. What was the percentage yield of CO_2?

***4.124** The potassium salt of benzoic acid, potassium benzoate ($KC_7H_5O_2$), can be made by the action of potassium permanganate on toluene (C_7H_8) as follows:

$$C_7H_8 + 2KMnO_4 \longrightarrow KC_7H_5O_2 + 2MnO_2 + KOH + H_2O$$

If the yield of potassium benzoate cannot realistically be expected to be more than 71%, what is the minimum number of grams of toluene needed to achieve this yield while producing 11.5 g of potassium benzoate?

***4.125** Manganese trifluoride, MnF_3, can be prepared by the following reaction:

$$2MnI_2(s) + 13F_2(g) \longrightarrow 2MnF_3(s) + 4IF_5(l)$$

What is the minimum number of grams of F_2 that must be used to react with 12.0 g of MnI_2 if the overall yield of MnF_3 is no more than 75% of theory?

ADDITIONAL EXERCISES

4.126 Mercury is an environmental pollutant because it can be converted by certain bacteria into the very poisonous substance methyl mercury, $(CH_3)_2Hg$. This compound ends up in the food chain and accumulates in the tissues of aquatic organisms, particularly fish, which renders them unsafe to eat. It is estimated that in the United States 263 tons of mercury are released into the atmosphere each year. If only 1.0% of this mercury is eventually changed to $(CH_3)_2Hg$, how many pounds of this compound are formed annually?

4.127 The poisonous compound methyl mercury, $(CH_3)_2Hg$, can be formed from mercury that's released into the environment.
(a) What is the formula mass of $(CH_3)_2Hg$?
(b) How many grams of $(CH_3)_2Hg$ can form from 25.0 g of Hg?

4.128 A superconductor is a substance that is able to conduct electricity without resistance, a property that is very desirable in the construction of large electromagnets. Metals have this property if cooled to temperatures a few degrees above absolute zero, but this requires the use of expensive liquid helium (boiling point, 4 K). Scientists have discovered materials that become superconductors at higher temperatures, but they are ceramics. Their brittle nature has so far prevented them from being made into long wires. A recently discovered compound of magnesium and boron, which consists of 52.9% Mg and 47.1% B, shows special promise as a high-temperature superconductor because it is inexpensive to make and can be fabricated into wire relatively easily. What is the formula of this compound?

4.129 A newspaper story describing the local celebration of Mole Day on October 23 (selected for Avogadro's number, 6.02×10^{23}) attempted to give the readers a sense of the size of the number by stating that a mole of M&Ms would be equal to 18 tractor trailers full. Assuming that an M&M occupies a volume of about 0.5 cm^3, calculate the dimensions of a cube required to hold 1 mol of M&Ms. Would 18 tractor trailers be sufficient?

4.130 Suppose you had 1 mol of pennies and that you were going to spend 500 million dollars (5.00×10^8 dollars) each and every second until you spent your entire fortune. How many years would it take you to spend all this cash? (Assume 1 year = 365 days.)

4.131 A 0.1246 g sample of a compound of chromium and chlorine was dissolved in water. All of the chloride ion was then captured by silver ion in the form of AgCl. A mass of 0.3383 g of AgCl was obtained. Calculate the empirical formula of the compound of Cr and Cl.

*__4.132__ A compound of Ca, C, N, and S was subjected to quantitative analysis and formula mass determination, and the following data were obtained. A 0.250 g sample was mixed with Na_2CO_3 to convert all of the Ca to 0.160 g of $CaCO_3$. A 0.115 g sample of the compound was carried through a series of reactions until all of its S was changed to 0.344 g of $BaSO_4$. A 0.712 g sample was processed to liberate all of its N as NH_3, and 0.155 g NH_3 was obtained. The formula mass was found to be 156. Determine the empirical and molecular formulas of this compound.

4.133 Ammonium nitrate will detonate if ignited in the presence of certain impurities. The equation for this reaction at a high temperature is

$$2NH_4NO_3(s) \xrightarrow{>300\,°C} 2N_2(g) + O_2(g) + 4H_2O(g)$$

Notice that all of the products are gases and so must occupy a vastly greater volume than the solid reactant.
(a) How many moles of *all* gases are produced from 1 mol of NH_4NO_3?
(b) If 1.00 ton of NH_4NO_3 exploded according to this equation, how many moles of *all* gases would be produced? (1 ton = 2000 lb.)
(In April, 1995, a terrorist used a rental truck full of ammonium nitrate, NH_4NO_3, to blow the nine-story Murrah Federal Building in Oklahoma in half; 168 lives were lost.)

4.134 A lawn fertilizer is rated as 6.00% nitrogen, meaning 6.00 g of N in 100 g of fertilizer. The nitrogen is present in the form of urea, $(NH_2)_2CO$. How many grams of urea are present in 100 g of the fertilizer to supply the rated amount of nitrogen?

4.135 Based solely on the amount of available carbon, how many grams of sodium oxalate, $Na_2C_2O_4$, could be obtained from 125 g of C_6H_6? (Assume that no loss of carbon occurs in any of the reactions needed to produce the $Na_2C_2O_4$.)

4.136 According to NASA, the Space Shuttle's external fuel tank for the main propulsion system carries 1,361,936 lb of liquid oxygen and 227,641 lb of liquid hydrogen. During takeoff, these chemicals are consumed as they react to form water. If the reaction is continued until all of one reactant is gone, how many pounds of which reactant are left over?

*__4.137__ For a research project, a student decided to test the effect of the lead(II) ion (Pb^{2+}) on the ability of salmon eggs to hatch. This ion was obtainable from the water-soluble salt, lead(II) nitrate, $Pb(NO_3)_2$, which the student decided to make by the following reaction. (The desired product was to be isolated by the slow evaporation of the water.)

$$PbO(s) + 2HNO_3(aq) \longrightarrow Pb(NO_3)_2(aq) + H_2O$$

Losses of product for various reasons were expected, and a yield of 86.0% was expected. In order to have 5.00 g of product at this yield, how many grams of PbO should be taken? (Assume that sufficient nitric acid, HNO_3, would be used.)

4.138 Chlorine atoms cause chain reactions in the stratosphere that destroy ozone that protects the Earth's surface from ultraviolet radiation. The chlorine atoms come from chlorofluorocarbons, compounds that contain carbon, fluorine, and chlorine, which were used for many years as refrigerants. One of these compounds is Freon-12, CF_2Cl_2. If a sample contains 1.0×10^{-9} g of Cl, how many grams of F should be present if all of the F and Cl atoms in the sample came from CF_2Cl_2 molecules?

TEST OF FACTS AND CONCEPTS

Many of the fundamental concepts and problem-solving skills that were developed in the preceding chapters will carry forward into the rest of this book. Therefore, we recommend that you pause here to see how well you have grasped the concepts, how familiar you are with important terms, and how able you are at working chemical problems. Don't be discouraged if some of the problems seem to be difficult at first. Most students of chemistry have experienced the frustration of initially not remembering what had only one or two weeks earlier seemed clear and straightforward. But if such frustrations take you back to a review, you will also experience what so many others have—what was once learned well (but since forgotten) comes back very quickly.

Some of the problems here require data or other information found in tables throughout this book, including those inside the covers. Freely use these tables as needed. For problems that require mathematical solutions, we recommend that you first assemble the necessary information in the form of equivalencies and then use them to set up appropriate conversion factors needed to obtain the answers.

1 A rectangular box was found to be 24.6 cm wide, 0.35140 m high, and 7,424 mm deep.
(a) How many significant figures are in each measurement?
(b) Calculate the volume of the box in units of cm^3. Be sure to express your answer to the correct number of significant figures.
(c) Use the answer in (b) to calculate the volume of the box in cubic feet.
(d) Suppose the box was solid and composed entirely of zinc, which has a specific gravity of 7.140. What would be the mass of the box in kilograms?

2 What is the difference between an atom and a molecule? What is the difference between a molecule and a mole?

3 If a 10 g sample of element X contains twice as many atoms as a 10 g sample of element Y, how does the atomic mass of X compare with the atomic mass of Y?

4 How did Dalton's atomic theory account for the law of conservation of mass? How did it explain the law of definite proportions?

5 If atom A has the same number of neutrons as atom B, must A and B be atoms of the same element? Explain.

6 Construct a conversion factor that would enable you to convert a volume of $3.14\ ft^3$ into cubic centimeters (cm^3).

7 The atoms of an isotope of plutonium, Pu, each contain 94 protons, 110 neutrons, and 94 electrons. Write a symbol for this element that incorporates its mass number and atomic number. Write the symbol for a different isotope of plutonium.

8 An atom of an isotope of nickel has a mass number of 60. How many protons, neutrons, and electrons are in this atom?

9 A solution was found to contain particles consisting of 12 neutrons, 10 electrons, and 11 protons. Write the chemical symbol for this particle, consulting the periodic table as needed.

10 Classify each of the following either as a substance or as a particle of which a substance can consist.
(a) Ion
(b) Mixture
(c) Isotope
(d) Atom
(e) Compound
(f) Molecule
(g) Element
(h) Nucleus

11 For the following, is it possible to have a visible sample of each? If not, explain.
(a) An isotope of iron
(b) An atom of iron
(c) A molecule of water
(d) A mole of water
(e) An ion of sodium
(f) A formula unit of sodium chloride

12 Make a sketch of the general shape of the modern periodic table and mark off those areas where we find the metals, metalloids, and nonmetals.

13 Which of the following elements would most likely be found together in nature: Ca, Hf, Sn, Cu, Zr?

14 Match an element on the left with a description on the right.

Calcium	Halogen
Iron	Noble gas
Helium	Alkali metal
Gadolinium	Alkaline earth metal
Iodine	Transition metal
Sodium	Inner transition metal

15 Define *ductile* and *malleable*.

16 Which metal is a liquid at room temperature? Which metal has the highest melting point?

17 What is the most important property that distinguishes a metalloid from a metal or a nonmetal?

18 Give the symbols of the post-transition metals.

19 Give chemical formulas for the following.
(a) potassium nitrate
(b) calcium carbonate
(c) cobalt(II) phosphate
(d) magnesium sulfite
(e) iron(III) bromide
(f) magnesium nitride
(g) aluminum selenide
(h) copper(II) perchlorate
(i) bromine pentafluoride
(j) dinitrogen pentaoxide
(k) strontium acetate
(l) ammonium dichromate
(m) copper(I) sulfide

20 Give chemical names for the following.
(a) $NaClO_3$
(b) $Ca_3(PO_4)_2$
(c) $NaMnO_4$
(d) AlP
(e) ICl_3
(f) PCl_3
(g) K_2CrO_4
(h) $Ca(CN)_2$
(i) $MnCl_2$
(j) $NaNO_2$
(k) $Fe(NO_3)_2$

21 A certain compound is hard, brittle, has a high melting point, and conducts electricity when melted but not when solid. What kind of compound is this, ionic or molecular?

22 Why do we always write empirical formulas for ionic compounds?

23 Which of the following are binary substances: Al_2O_3, Cl_2, MgO, NO_2, $NaClO_4$?

24 The formula mass of a substance is 60.2. What is the mass in grams of one of its molecules?

25 A sample of a compound with a mass of 204 g consists of 1.00×10^{23} molecules. What is its molecular mass?

26 Calculate the formula mass of $Fe_4[Fe(CN)_6]_3$.

27 How many grams of copper(II) nitrate trihydrate, $Cu(NO_3)_2 \cdot 3H_2O$, are present in 0.118 mol of this compound?

28 How many moles of nickel(II) iodide hexahydrate, $NiI_2 \cdot 6H_2O$, are in a sample of 15.7 g of this compound?

29 A sample of 0.5866 g of nicotine was analyzed and found to consist of 0.4343 g C, 0.05103 g H, and 0.1013 g N. Calculate the percentage composition of nicotine.

30 A compound of potassium had the following percentage composition: K, 37.56%; H, 1.940%; P, 29.79%. The rest was oxygen. Calculate the empirical formula of this compound (arranging the atomic symbols in the order K H P O.)

31 How many molecules of ethyl alcohol, C_2H_5OH, are in 1.00 fluid ounce of the liquid? The density of ethyl alcohol is 0.798 g/mL (1 oz = 29.6 mL)

32 What volume in liters is occupied by a sample of ethylene glycol, $C_2H_6O_2$, that consists of 5.00×10^{24} molecules. The density of ethylene glycol is 1.11 g/mL.

33 If 2.56 g of chlorine, Cl_2, are to be used to prepare dichlorine heptoxide, Cl_2O_7, how many moles and how many grams of oxygen, O_2, are needed?

34 Balance the following equations.
(a) $Fe_2O_3 + HNO_3 \rightarrow Fe(NO_3)_3 + H_2O$
(b) $C_{21}H_{30}O_2 + O_2 \rightarrow CO_2 + H_2O$

35 How many moles of nitric acid, HNO_3, are needed to react with 2.56 mol of Cu in the following reaction?

$$3Cu + 8HNO_3 \longrightarrow 3Cu(NO_3)_2 + 2NO + 4H_2O$$

36 Under the right conditions, ammonia can be converted to nitrogen monoxide, NO, by the following reaction.

$$4NH_3 + 5O_2 \longrightarrow 4NO + 6H_2O$$

How many moles and how many grams of O_2 are needed to react with 56.8 g of ammonia by this reaction?

37 Dolomite is a mineral consisting of calcium carbonate ($CaCO_3$) and magnesium carbonate ($MgCO_3$). When dolomite is strongly heated, its carbonates decompose to their oxides (CaO and MgO) and carbon dioxide is expelled.
(a) Write the separate equations for these decompositions of calcium carbonate and magnesium carbonate.
(b) When a dolomite sample with a mass of 5.78 g was heated strongly, the residue had a mass of 3.02 g. Calculate the masses in grams and the percentages of calcium carbonate and magnesium carbonate in this sample of dolomite.

38 Adipic acid, $C_6H_{10}O_4$, is a raw material for making nylon, and it can be prepared in the laboratory by the following reaction between cyclohexene, C_6H_{10}, and sodium dichromate, $Na_2Cr_2O_7$, in sulfuric acid, H_2SO_4.

$$3C_6H_{10}(l) + 4Na_2Cr_2O_7(aq) + 16H_2SO_4(aq) \longrightarrow$$
$$3C_6H_{10}O_4(s) + 4Cr_2(SO_4)_3(aq) + 4Na_2SO_4(aq) + 16H_2O$$

There are side reactions. These plus losses of product during its purification reduce the overall yield. A typical yield of purified adipic acid is 68.6%.
(a) To prepare 12.5 g of adipic acid in 68.6% yield requires how many grams of cyclohexene?
(b) The only available supply of sodium dichromate is its dihydrate, $Na_2Cr_2O_7 \cdot 2H_2O$. (Since the reaction occurs in an aqueous medium, the water in the dihydrate causes no problems, but it does contribute to the mass of what is taken of this reactant.) How many grams of this dihydrate are also required in the preparation of 12.5 g of adipic acid in a yield of 68.6%?

39 One of the ores of iron is hematite, Fe_2O_3, mixed with other rock. One sample of this ore is 31.4% hematite. How many tons of this ore are needed to make 1.00 ton of iron if the percentage recovery of iron from the ore is 91.5% (1 ton = 2000 lb)?

40 Gold occurs in the ocean in a range of concentration of 0.1 to 2 mg of gold per ton of seawater. Near one coastal city the gold concentration in the ocean is 1.5 mg/ton.
(a) How many tons of seawater have to be processed to obtain 1.0 troy ounce of gold if the recovery is 65% successful? (The troy ounce, 31.1 g, is the standard "ounce" in the gold trade.)
(b) If gold can be sold for $455 per troy ounce, what is the breakeven point in the dollar-cost per ton of processed seawater for extracting gold from the ocean at this location?

41 *C.I. Pigment Yellow 45* ("sideran yellow") is a pigment used in ceramics, glass, and enamel. When analyzed, a 2.164 g sample of this substance was found to contain 0.5259 g of Fe and 0.7345 g of Cr. The remainder was oxygen. Calculate the empirical formula of this pigment. What additional data are needed to calculate the molecular mass of this compound?

42 When 6.584 g of one of the hydrates of sodium sulfate (Na_2SO_4) was heated so as to drive off all of its water of hydration, the residue of anhydrous sodium sulfate had a mass of 2.889 g. What is the formula of the hydrate?

43 In an earlier problem we described the reaction of ammonia with oxygen to form nitrogen monoxide, NO.

$$4NH_3 + 5O_2 \longrightarrow 4NO + 6H_2O$$

How many moles and how many grams of NO could be formed from a mixture of 45.0 g of NH_3 and 58.0 g of O_2? How many grams of which reactant would remain unreacted?

Reactions Between Ions in Aqueous Solutions

If you eat this sandwich, you will probably end up with indigestion! To ease the pain, you might take an antacid such as sodium bicarbonate (commonly called baking soda or bicarbonate of soda). The result will be a burp or two. The reaction of sodium bicarbonate with stomach acid is typical of the reactions studied in this chapter.

THIS CHAPTER IN CONTEXT As you might expect, when two or more reactants are involved in a chemical reaction, their particles—atoms, ions, or molecules—must make physical contact. In other words, the particles need freedom of motion, which exists in gases and liquids but not in solids. Thus, whenever possible, reactions are carried out with all of the reactants in one fluid phase, liquid or gas. When possible, a liquid is used to dissolve solid reactants so their particles can move about.

In this chapter, our goal is to teach you what happens when ionic substances dissolve in water, the nature of the chemical reactions they undergo, and the products that form. We will also introduce you to another important class of compounds called acids and bases, and we will examine the reactions of these substances in aqueous solutions. And finally, we will extend the principles of stoichiometry that you learned in Chapter 4 to deal quantitatively with chemical reactions in solution.

5.1 ▶ Special terminology applies to solutions

Before we get to the meat of our subject, it's important that you know the terms we use to describe solutions and the substances dissolved in them. We begin, therefore, with some definitions.

When water is a component of a solution, it is usually considered to be the solvent even when it is present in small amounts.

A **solution** is a homogeneous mixture in which the molecules or ions of the components freely intermingle (see Figure 5.1). When a solution forms, at least two substances are involved. One is called the *solvent* and all of the others are called *solutes*. The **solvent** is usually taken to be the component present in largest amount and is the medium into which the solutes are mixed or dissolved. Water is a typical and very common solvent, but the solvent can actually be in any physical state—a solid, liquid, or gas. Unless stated otherwise, we will assume that any solutions we mention are *aqueous solutions,* so that liquid water is understood to be the solvent.

A **solute** is any substance dissolved in the solvent. It might be a gas, like the carbon dioxide dissolved in carbonated beverages. Some solutes are liquids, like ethylene glycol dissolved in water to protect a vehicle's radiator against freezing. Solids, of course, can be solutes, like the sugar dissolved in lemonade or the salt dissolved in seawater.

To describe the composition of a solution, we often give a ratio called a **concentration,** the *ratio* of the amount of solute either to the amount of solvent or to the amount of solution. The quantities may be in any units we desire, but grams are often used. For example, we might have a solution of salt in water in which the concentration is 2 g salt/100 g of water, in other words, a solute-to-*solvent* ratio. Often, the

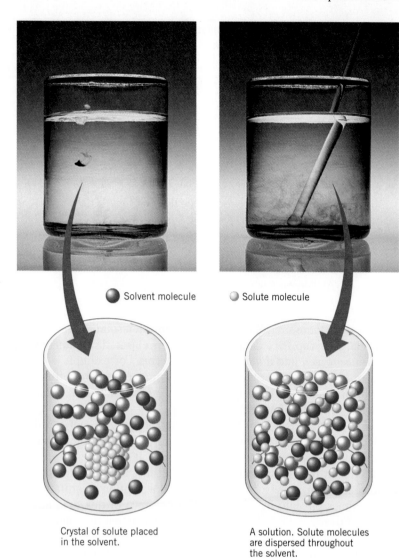

Solvent molecule   Solute molecule

Crystal of solute placed in the solvent.

A solution. Solute molecules are dispersed throughout the solvent.

FIGURE 5.1 *Formation of a solution of iodine molecules in ethyl alcohol to form "tincture of iodine," once widely used as an antiseptic.* In the first photo, a crystal of iodine, I_2, is on its way to the bottom of the beaker and is already beginning to dissolve, the purplish iodine crystal forming a reddish-brown solution. In the hugely enlarged view beneath the photo, we see the iodine molecules still bound in a crystal. The second photo shows how stirring the mixture helps the iodine molecules to disperse in the solvent, as visualized in the molecular view below the photo.

concentration of salt in seawater is given as 3 g salt/100 g seawater, which is a solute-to-*solution* ratio. When the ratio is like this—namely, number of grams of solute per 100 g of *solution*—we call the ratio the solution's **percentage concentration.**

The *relative* amounts of solute and solvent are often loosely given without specifying actual quantities. In a **dilute solution,** for example, the ratio of solute to solvent is small, sometimes very small (Figure 5.2). A few crystals of salt in a glass of water

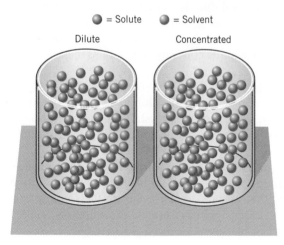

= Solute   = Solvent

Dilute   Concentrated

FIGURE 5.2 *Dilute and concentrated solutions.* The dilute solution on the left has less solute per unit volume than the more concentrated solution on the right.

TABLE 5.1	SOLUBILITIES OF SOME COMMON COMPOUNDS IN WATER	
Substance	Formula	Solubility (g/100 g water)[a]
Ammonium chloride	NH_4Cl	29.7 (0 °C)
Boric acid	H_3BO_3	6.35 (30 °C)
Calcium carbonate	$CaCO_3$	0.0015 (25 °C)
Calcium chloride	$CaCl_2$	74.5 (20 °C)
Copper sulfide	CuS	3.3×10^{-5} (18 °C)
Lead sulfate	$PbSO_4$	4.3×10^{-3} (25 °C)
Sodium hydroxide	NaOH	42 (0 °C)
		347 (100 °C)
Sodium chloride	NaCl	35.7 (0 °C)
		39.12 (100 °C)

[a]At the temperature given in parentheses.

make a very dilute salt solution. In a **concentrated solution,** the ratio of solute to solvent is large. Syrup, for example, is a very concentrated solution of sugar in water.

Concentrated and *dilute* are relative terms. For example, a solution of 100 g of sugar in 100 mL of water is concentrated compared to one with just 10 g of sugar in 100 mL of water, but the latter solution is more concentrated than one that has 1 g of sugar in 100 mL of water.

Usually there is a limit to the amount of a solute that can dissolve in a given amount of solvent. For example, at 20 °C only 36.0 g of sodium chloride is able to dissolve in 100 g of water. If more solute is added, it simply rests at the bottom of the solution. We say this is a **saturated solution** because at its present temperature it cannot dissolve any more solute. To make sure that a solution is saturated, extra solid solute is sometimes put into its container. The presence of this solid at the bottom of the container over a period of time is visible evidence that the solution is saturated.

The **solubility** of a solute is usually described by the number of grams that dissolve in 100 g of *solvent* at a given temperature to make a saturated solution. The temperature must be specified because solubilities vary with temperature. Table 5.1 shows that some saturated aqueous solutions, like sodium hydroxide, are concentrated, but others, like copper sulfide, are extremely dilute.

A solution that contains less solute than required for saturation is called an **unsaturated solution.** It is able to dissolve more solute.

Usually, the solubility of a solute increases as the temperature of the solution increases. In such cases, more solute can be dissolved by heating the mixture. If the temperature of such a warmer, saturated solution is subsequently lowered, the additional solute should separate from the solution, and indeed, this tends to happen spontaneously. However, sometimes the solute doesn't separate, leaving us with a **supersaturated solution,** a solution that actually contains more solute than required for saturation at a given temperature.

Supersaturated solutions are unstable. They can only be prepared if there are no traces of undissolved solute left in contact with the solution. If even a tiny crystal of the solute is present or is added, the extra solute crystallizes. Figure 5.3 shows what happens when a tiny crystal of sodium acetate is added to a supersaturated solution of this compound in water.

Sometimes, when a reaction is carried out in a solution, one of the products that forms has a low solubility in the solvent. As this substance forms, the solution becomes highly supersaturated and the substance separates from the solution as a solid, which we call a **precipitate.** A reaction that produces a precipitate is called a **precipitation reaction.**

FIGURE 5.3 *Crystallization.* When a small seed crystal of sodium acetate is added to a supersaturated solution of the compound, excess solute crystallizes rapidly until the solution is just saturated. The crystallization shown in this sequence took less than 10 seconds!

5.2 ▶ Ionic compounds conduct electricity when dissolved in water

Water is an amazing substance. Composed of just three atoms, it is one of the most common compounds on earth, and its ability to dissolve so many different kinds of materials is responsible, to a large degree, for the evolution of life as we know it on our planet. Our bodies are composed of approximately 60% water; it is the major component of the fluids in and around our cells and serves to transport nutrients throughout our systems.

One property of water that sets it apart from other liquids is its ability to dissolve many ionic compounds. How water accomplishes this will be discussed in detail in Chapter 14. In this chapter, we will examine what happens to ionic compounds when they dissolve in water and the kinds of reactions that take place in solutions of ionic substances.

In Chapter 2 you learned that pure water is a very poor conductor of electricity. This is because water consists of uncharged molecules that are incapable of transporting electrical charge. However, when an ionic compound such as $CuSO_4$ or NaCl is dissolved in the water, an electrically conducting solution is formed. This electrical conductivity is demonstrated in Figure 5.4a for a solution of copper sulfate.

Solutes such as $CuSO_4$ or NaCl, which yield electrically conducting aqueous solutions, are called **electrolytes.** Their ability to conduct electricity suggests the presence of electrically charged particles that are able to move through the solution. The generally accepted reason is that when an ionic compound dissolves in water, the ions separate from each other and enter the solution as more or less independent

Solutions of electrolytes conduct electricity in a way that's different from metals. This is discussed more completely in Chapter 21.

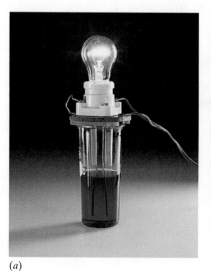

(a)   (b)

FIGURE 5.4 *Electrical conductivity of solutions of electrolytes versus nonelectrolytes.* (*a*) The copper sulfate solution is a strong conductor of electricity, and $CuSO_4$ is a strong electrolyte. (*b*) Neither sugar nor water is an electrolyte, and this sugar solution is a nonconductor.

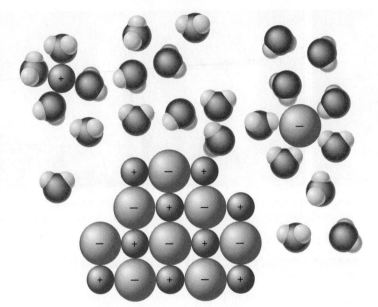

FIGURE 5.5 *Dissociation of an ionic compound as it dissolves in water.* Ions separate from the solid and become surrounded by molecules of water. The ions are said to be hydrated. In the solution, the ions are able to move freely and the solution is able to conduct electricity.

particles that are surrounded by molecules of the solvent. This change is called the **dissociation** of the ionic compound, and it is illustrated in Figure 5.5. In general, *we will assume that the dissociation of an ionic compound is complete* and that the solution contains no undissociated formula units of the salt. Thus, an aqueous solution of NaCl is really a solution that contains Na^+ and Cl^- ions, with no undissociated NaCl "molecules" in the solution. Because these solutions contain so many ions, they are strong conductors of electricity, and salts are said to be **strong electrolytes.**

Keep in mind that a strong electrolyte is 100% dissociated in aqueous solution.

Many ionic compounds have low solubilities in water. An example is AgBr, the light-sensitive compound in most photographic film. Although only a tiny amount of this compound dissolves in water, all of it that does dissolve is completely dissociated. However, because of the extremely low solubility, the number of ions in the solution is extremely small and the solution doesn't conduct electricity well. Nevertheless, it is still convenient to think of AgBr as a strong electrolyte because it serves to remind us that salts are completely dissociated in aqueous solution.

Ethylene glycol, $C_2H_4(OH)_2$, is a type of alcohol. Other alcohols, such as ethanol and methanol, are also nonelectrolytes.

Aqueous solutions of most molecular compounds do not conduct electricity, and such solutes are called **nonelectrolytes.** Examples are sugar (Figure 5.4*b*) and ethylene glycol (the solute in antifreeze solutions). Both of these solutes consist of molecules that stay intact when they dissolve. They simply intermingle with water molecules when their solutions form (Figure 5.6). The solution contains no electrically charged particles, so it is unable to conduct electricity.

FIGURE 5.6 *Formation of an aqueous solution of a non-electrolyte.* When a nonelectrolyte dissolves in water, the molecules of solute separate from each other and mingle with the water molecules. The solute molecules stay intact and do not dissociate into smaller particles.

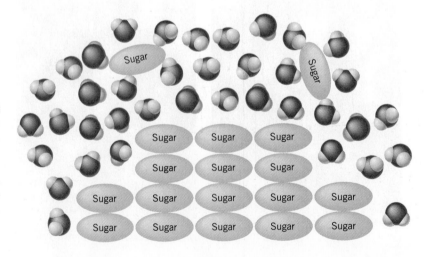

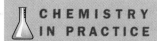 **CHEMISTRY IN PRACTICE** Electroencephalograms (EEGs) and electrocardiograms (EKGs) are diagnostic procedures for evaluating the conditions of the brain and the heart, respectively. They depend on the presence of strong electrolytes in all body fluids because these solutes must carry weak electrical currents used in the tests. An electrolyte paste is applied to those skin spots to which electrodes will be taped to provide a conducting connection into and through the skin.

Equations for dissociation reactions show the ions

A convenient way to describe the dissociation of an ionic compound is with a chemical equation.

$$NaCl(s) \longrightarrow Na^+(aq) + Cl^-(aq)$$

We use the symbol *aq* after a charged particle to mean that it is dissolved in water and surrounded by water molecules. When a solute particle is surrounded by water molecules, we say it is **hydrated.** By writing the formulas of the ions separately, we mean that they are essentially independent of each other in the solution.

We can write a similar equation for the dissociation of calcium chloride, $CaCl_2$.

$$CaCl_2(s) \longrightarrow Ca^{2+}(aq) + 2Cl^-(aq)$$

This shows that each formula unit of $CaCl_2(s)$ releases three ions, one $Ca^{2+}(aq)$ and two $Cl^-(aq)$. The symbols (*s*) and (*aq*) once again indicate that the change is the dissociation of the ionic compound. Quite often, however, these symbols are omitted. When the context makes it clear that the system is aqueous, these symbols can be "understood." You should not be disturbed, therefore, when you see an equation such as

$$CaCl_2 \longrightarrow Ca^{2+} + 2Cl^-$$

Unless something is said to the contrary, it means

$$CaCl_2(s) \longrightarrow Ca^{2+}(aq) + 2Cl^-(aq)$$

Polyatomic ions generally remain intact as dissociation occurs. When sodium sulfate, Na_2SO_4, dissolves in water, for example, the solution contains both sodium ions and intact sulfate ions, released as follows:

$$Na_2SO_4(s) \longrightarrow 2Na^+(aq) + SO_4^{2-}(aq)$$

To write an equation such as this correctly, you must know both the formulas and the charges of the polyatomic ions. If necessary, refer to Table 2.5 (page 62) for a review of the formulas of polyatomic ions.

Ammonium sulfate, $(NH_4)_2SO_4$, is used as a fertilizer to supply nitrogen to crops. Write the equation for the dissociation of this compound when it dissolves in water.

ANALYSIS: To write the equation correctly, we need to know the formulas of the ions that make up the compound. In this case, the cation is NH_4^+ and the anion is SO_4^{2-}. The formula of the compound tells us there are *two* NH_4^+ ions for each SO_4^{2-} ion. We have to be sure to indicate this in the equation.

SOLUTION: We write the formula for the solid on the left of the equation and indicate its state by (*s*). The ions are written on the right side of the equation and are shown to be in aqueous solution by the symbol (*aq*) following their formulas.

$$(NH_4)_2SO_4(s) \longrightarrow 2NH_4^+(aq) + SO_4^{2-}(aq)$$

The subscript 2 becomes the coefficient for NH_4^+.

EXAMPLE 5.1

Writing the Equation for the Dissociation of an Ionic Compound

Is the Answer Reasonable?
There are two things to check when writing equations such as this. First, be sure you have the correct formulas for the ions, including their charges. Second, be sure

you've indicated the number of ions of each kind that come from one formula unit when the compound dissociates. Performing these checks here confirms we've solved the problem correctly.

PRACTICE EXERCISE 1: Write equations that show what happens when the following solid ionic compounds dissolve in water: (a) $MgCl_2$, (b) $Al(NO_3)_3$, and (c) Na_2CO_3.

5.3 ▶ Equations for ionic reactions can be written in different ways

It is not unusual for ionic compounds to react with each other when their aqueous solutions are combined. An example is the reaction of lead nitrate, $Pb(NO_3)_2$, with potassium iodide, KI, to give a bright yellow precipitate of lead iodide (Figure 5.7).

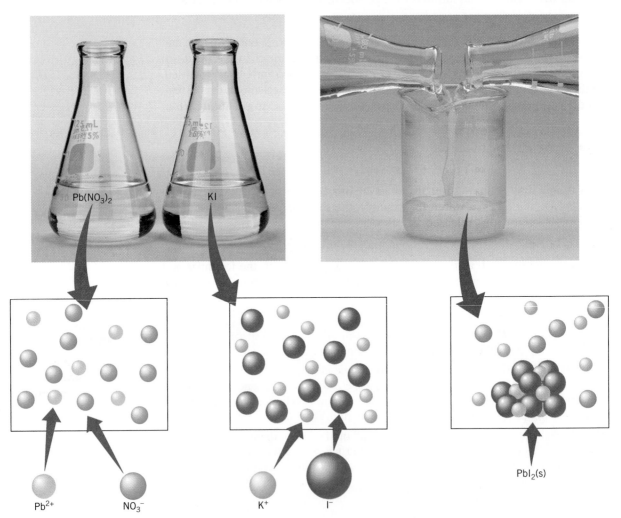

FIGURE 5.7 *The reaction of Pb(NO₃)₂ with KI.* On the left are flasks containing solutions of lead nitrate and potassium iodide. These solutes exist as separated ions in their respective solutions. On the right, we observe that when the solutions of the ions are combined, there is an immediate reaction as the Pb^{2+} ions join with the I^- ions to give a precipitate of small crystals of solid, yellow PbI_2. The reaction is so rapid that the yellow color develops where the two streams of liquid come together. If the $Pb(NO_3)_2$ and KI are combined in a 1-to-2 mole ratio, the solution would now contain only K^+ and NO_3^- ions (the ions of KNO_3).

The chemical equation for the reaction is

$$Pb(NO_3)_2(aq) + 2KI(aq) \longrightarrow PbI_2(s) + 2KNO_3(aq) \qquad (5.1)$$

where we have noted the insolubility of PbI_2 by writing (s) following its formula. Equation 5.1 is perfectly satisfactory for performing stoichiometric calculations of the type you learned in Chapter 4. However, let's take a closer look at the reaction to examine other ways that we might write the chemical equation.

Lead nitrate and potassium iodide are both ionic compounds. When they dissolve in water, they undergo complete dissociation, so their aqueous solutions actually contain ions, not "molecules," of $Pb(NO_3)_2$ and KI (Figure 5.7a). When the two solutions are mixed, the ions mingle. If no reaction were to take place, we simply would have a mixture of all four ions. However, it turns out that PbI_2 has a very low solubility in water, so the large concentrations of Pb^{2+} and I^- ions in the mixture are exactly what would be present in a highly *supersaturated* solution of PbI_2. As you learned earlier, supersaturated solutions are unstable and tend to deposit crystals of the dissolved solute, and this is exactly what happens here; lead iodide crystals form and we observe the formation of a precipitate. This is illustrated in Figure 5.7b. Notice that after the precipitation of PbI_2, the solution contains only the ions of the other product, KNO_3. The products of the reaction can be separated from each other by filtering the solid from the rest of the mixture, as pictured in Figure 5.8.

We now see that the reaction between $Pb(NO_3)_2$ and KI is really a *reaction between ions,* and can be appropriately referred to as an **ionic reaction.** We can also see that the equation given above (Equation 5.1) does not describe what actually takes place in the reaction mixture. (Chemists refer to an equation such as 5.1 as a **molecular equation** because all the formulas are written with the ions together, as if the substances in solution consist of neutral molecules.) A more accurate representation of the reaction is given by the **ionic equation,** *in which all soluble strong electrolytes are written in "dissociated" form.* To write the ionic equation for the reaction of $Pb(NO_3)_2$ with KI, we write the formulas of both reactants and the soluble product, KNO_3, in dissociated form. This gives

$$Pb(NO_3)_2(aq) \qquad + \qquad 2KI(aq) \qquad \longrightarrow \qquad PbI_2(s) + \quad 2KNO_3(aq)$$

$$Pb^{2+}(aq) + 2NO_3^-(aq) + 2K^+(aq) + 2I^-(aq) \longrightarrow PbI_2(s) + 2K^+(aq) + 2NO_3^-(aq)$$

Notice that we have *not* separated PbI_2 into its ions in this equation. This is because after the reaction is over, the Pb^{2+} and I^- ions are no longer able to move independently. They are trapped in the insoluble product, PbI_2.

The ionic equation for this reaction gives a much better representation of what actually takes place when the solutions of reactants are combined. It also lets us see that the reaction in solution really takes place between the Pb^{2+} and I^- ions; these are the ions that come together to form the product. The other ions, K^+ and NO_3^-, are unchanged by the reaction; they are present before the reaction begins and are present afterward. *Ions that do not actually take part in a reaction are sometimes called* **spectator ions;** *in a sense, they just "stand by and watch the action."*

To emphasize the actual reaction that occurs, we can write the **net ionic equation,** *which is obtained by eliminating spectator ions from the ionic equation.* Let's cross out the K^+ and NO_3^- ions.

$$Pb^{2+}(aq) + \cancel{2NO_3^-(aq)} + \cancel{2K^+(aq)} + 2I^-(aq) \longrightarrow PbI_2(s) + \cancel{2K^+(aq)} + \cancel{2NO_3^-(aq)}$$

What remains is the net ionic equation,

$$Pb^{2+}(aq) + 2I^-(aq) \longrightarrow PbI_2(s)$$

Notice how it calls our attention to the ions that are actually participating in the reaction as well as the change that occurs.

FIGURE 5.8 *Separating a precipitate from a solution by filtration.* The reaction mixture is passed through filter paper held in a funnel. The precipitate is caught by the filter paper while the clear solution passes through.

The net ionic equation is especially useful because it permits us to *generalize*. It tells us that if we combine *any* solution that contains Pb^{2+} with *any* other solution that contains I^-, we ought to expect a precipitate of PbI_2. And this is exactly what happens! For example, another soluble lead compound is lead acetate, $Pb(C_2H_3O_2)_2$, and another soluble iodide is sodium iodide, NaI. If aqueous solutions of these substances are prepared and then mixed, a yellow precipitate of PbI_2 forms immediately (Figure 5.9). Example 5.2 demonstrates how we write the molecular, ionic, and net ionic equations for the reaction.

EXAMPLE 5.2

Writing Molecular, Ionic, and Net Ionic Equations

If necessary, review Section 2.8, which discusses naming ionic compounds.

FIGURE 5.9 *Another reaction that forms lead iodide.* The net ionic equation tells us that any soluble lead compound will react with any soluble iodide compound to give lead iodide. This prediction is born out here as a precipitate of lead iodide is formed when a solution of sodium iodide is added to a solution of lead acetate.

Write the molecular, ionic, and net ionic equations for the reaction of aqueous solutions of lead acetate and sodium iodide, which yields a precipitate of lead iodide and leaves the compound sodium acetate in solution.

ANALYSIS: To write a chemical equation, we must begin with the correct formulas of the reactants and products. Only the names of the reactants and products are given, so we have to translate them into chemical formulas, which requires a knowledge of chemical nomenclature. Following the rules we discussed in Chapter 2, we have

Reactants		Products	
lead acetate	$Pb(C_2H_3O_2)_2$	lead iodide	PbI_2
sodium iodide	NaI	sodium acetate	$NaC_2H_3O_2$

Now that we have the formulas, we can arrange them to form the molecular equation, which we can then balance. To obtain the ionic equation, we will write soluble ionic compounds in dissociated form and the formula of the precipitate in "molecular" form. Finally, we look for spectator ions and eliminate them from the ionic equation to obtain the net ionic equation.

SOLUTION:
The Molecular Equation
Once we have the chemical formulas, we assemble them into the molecular equation.

$$Pb(C_2H_3O_2)_2(aq) + NaI(aq) \longrightarrow PbI_2(s) + NaC_2H_3O_2(aq)$$

Notice that we've indicated which are in solution and which is a precipitate. (This is important because we will need the information to write the ionic equation.) Next, we balance the equation by placing coefficients of 2 in front of the formulas NaI and $NaC_2H_3O_2$.

$$Pb(C_2H_3O_2)_2(aq) + 2NaI(aq) \longrightarrow PbI_2(s) + 2NaC_2H_3O_2(aq)$$

This is the balanced molecular equation.

Ionic Equation
To write the ionic equation, we keep in mind that all soluble salts are completely dissociated in water and that the formulas of precipitates are written in "molecular" form. We are also careful to use the subscripts and coefficients in the molecular equation to properly obtain the coefficients of the ions in the ionic equation.

$$Pb(C_2H_3O_2)_2(aq) \qquad\qquad 2NaI(aq)$$

$$Pb^{2+}(aq) + 2C_2H_3O_2^{-}(aq) + 2Na^+(aq) + 2I^-(aq) \longrightarrow$$

$$2NaC_2H_3O_2(aq)$$

$$PbI_2(s) + 2Na^+(aq) + 2C_2H_3O_2^{-}(aq)$$

This is the ionic equation. Notice that to write the ionic equation properly it is necessary to know both the formulas and charges of the ions.

Net Ionic Equation

We obtain this from the ionic equation by eliminating spectator ions. The spectator ions in this reaction are Na^+ and $C_2H_3O_2^-$ (they're the same on both sides of the arrow). Lets cross them out.

$$Pb^{2+}(aq) + \cancel{2C_2H_3O_2^-(aq)} + \cancel{2Na^+(aq)} + 2I^-(aq) \longrightarrow$$
$$PbI_2(s) + \cancel{2Na^+(aq)} + \cancel{2C_2H_3O_2^-(aq)}$$

What's left is the net ionic equation.

$$Pb^{2+}(aq) + 2I^-(aq) \longrightarrow PbI_2(s)$$

Notice this is the same net ionic equation as in the reaction of lead nitrate with potassium iodide.

Are the Answers Reasonable?

When you look back over a problem such as this, things to ask yourself are: (1) "Have I written the correct formulas for the reactants and products?" (2) "Is the molecular equation balanced correctly?" (3) "Have I divided the soluble ionic compounds into their ions correctly, being careful to properly apply the subscripts of the ions and the coefficients in the molecular equation?" and (4) "Have I identified and eliminated the correct ions from the ionic equation to obtain the net ionic equation?" If each of these questions can be answered in the affirmative, as they can here, the problem has been solved correctly.

PRACTICE EXERCISE 2: Write molecular, ionic, and net ionic equations for the reaction of aqueous solutions of cadmium chloride and sodium sulfide to give a precipitate of cadmium sulfide and a solution of sodium chloride.

In a balanced ionic or net ionic equation both atoms and charge must balance

In the ionic and net ionic equations we've written, not only are the atoms in balance but so is the net electrical charge, which is the same on both sides of the equation. Thus, in the ionic equation for the reaction of lead nitrate with potassium iodide, the sum of the charges of the ions on the left (Pb^{2+}, $2NO_3^-$, $2K^+$, and $2I^-$) is zero, which matches the sum of the charges on all of the formulas of the products (PbI_2, $2K^+$, and $2NO_3^-$).[1] In the net ionic equation the charges on both sides are also the same: on the left we have Pb^{2+} and $2I^-$, with a net charge of zero, and on the right we have PbI_2, also with a charge of zero. We now have an additional requirement for an ionic equation or net ionic equation to be balanced: *the net electrical charge on both sides of the equation must be the same.*

Criteria for Balanced Ionic and Net Ionic Equations

1. **Material balance.** There must be the same number of atoms of each kind on both sides of the arrow.

2. **Electrical balance.** The *net* electrical charge on the left must equal the *net* electrical charge on the right (although this charge does not necessarily have to be zero).

 TOOLS

Criteria for a balanced ionic equation

Summary

In this section we've examined three ways to write an equation for a reaction between solutions of electrolytes. Is one better than another? Not really. The molecular equation is often best for planning an experiment and performing stoichiometric

---

[1] There is no charge written for the formula of a compound such as PbI_2, so when we add up charges, we take the charge on PbI_2 to be zero.

calculations. The ionic equation is useful when we want to provide a better description of what actually takes place in the solution during the reaction. And finally, the net ionic equation is useful because it focuses our attention on the chemical changes that take place during the reaction and it helps us generalize the reaction so that we can select other sets of reactants that give the same net chemical change.

5.4 ▶ Reactions that produce precipitates can be predicted

The reaction between $Pb(NO_3)_2$ and KI, which we studied in the preceding section, is just one example of a large class of ionic reactions in which cations and anions change partners. The technical term we use to describe these reactions is **metathesis,** but they are also sometimes called **double replacement reactions.** (In the formation of the products PbI_2 and KNO_3, the I^- replaces NO_3^- in the lead compound and NO_3^- replaces I^- in the potassium compound.) Metathesis reactions in which a precipitate forms are sometimes called **precipitation reactions.**

For precipitation reactions, if we know which products are insoluble, we can predict the correct net ionic equation. Fortunately, there is a set of rules that we can use to determine, in many cases, whether an ionic compound is soluble or insoluble. These **solubility rules** are given in Table 5.2. First, let's look at some examples

---

CHEMISTRY IN PRACTICE

As you know, when someone picks up an object with their bare hands, they leave fingerprints on it. Crime investigators call them *latent* fingerprints, which means they cannot be seen unless developed in some way. (*Latent* means "hidden.") There's a number of ways latent fingerprints can be made visible, including dusting with a fine powder, as seen on TV crime shows. Another method makes use of an ionic reaction of the kind discussed in this section.

When you touch something, among the substances present in the fingerprint you leave behind is NaCl. This comes from the small amount of sweat present on your fingers. To develop a fingerprint, a dilute solution of silver nitrate is sprayed on the object of interest. As the sodium chloride in the fingerprint dissolves in the solution, the silver nitrate reacts with it to give a precipitate of silver chloride, AgCl.

$$NaCl(aq) + AgNO_3(aq) \longrightarrow AgCl(s) + NaNO_3(aq)$$

Silver chloride is white, but when exposed to sunlight it decomposes to give a black deposit of metallic silver, which reveals the lines and swirls of the fingerprint.

TOOLS

Solubility rules

TABLE 5.2	**SOLUBILITY RULES FOR IONIC COMPOUNDS IN WATER**

Soluble Compounds

1. All compounds of the alkali metals (Group IA) are soluble.
2. All salts containing NH_4^+, NO_3^-, ClO_4^-, ClO_3^-, and $C_2H_3O_2^-$ are soluble.
3. All chlorides, bromides, and iodides (salts containing Cl^-, Br^-, or I^-) are soluble *except* when combined with Ag^+, Pb^{2+}, and Hg_2^{2+} (note the subscript "2").
4. All sulfates (salts containing SO_4^{2-}) are soluble *except* those of Pb^{2+}, Ca^{2+}, Sr^{2+}, Hg_2^{2+}, and Ba^{2+}.

Insoluble Compounds

5. All metal hydroxides (ionic compounds containing OH^-) and all metal oxides (ionic compounds containing O^{2-}) are insoluble *except* those of Group IA and of Ca^{2+}, Sr^{2+}, and Ba^{2+}.

    When metal oxides do dissolve, they react with water to form hydroxides. The oxide ion, O^{2-}, does not exist in water. For example,

$$Na_2O(s) + H_2O \longrightarrow 2NaOH(aq)$$

6. All salts that contain PO_4^{3-}, CO_3^{2-}, SO_3^{2-}, and S^{2-} are insoluble, *except* those of Group IA and NH_4^+.

FACETS OF CHEMISTRY 5.1

Boiler Scale and Hard Water

Precipitation reactions occur around us all the time and we hardly ever take notice until they cause a problem. One common problem is caused by **hard water**—groundwater that contains the "hardness ions," Ca^{2+}, Mg^{2+}, Fe^{2+}, or Fe^{3+}, in concentrations high enough to form precipitates with ordinary soap. Soap normally consists of the sodium salts of organic acids derived from animal fats or oils (so-called *fatty acids*). An example is sodium stearate, $NaC_{17}H_{35}O_2$. The negative ion of the soap forms an insoluble "scum" with hardness ions, which reduces the effectiveness of the soap for removing dirt and grease.

Hardness ions can be removed from water in a number of ways. One way is to add hydrated sodium carbonate, $Na_2CO_3 \cdot 10H_2O$, often called *washing soda*, to the water. The carbonate ion forms insoluble precipitates with the hardness ions; an example is $CaCO_3$.

$$Ca^{2+}(aq) + CO_3^{2-}(aq) \longrightarrow CaCO_3(s)$$

Once precipitated, the hardness ions are not available to interfere with the soap.

Another problem when the hard water of a particular locality is rich in bicarbonate ion is the precipitation of insoluble carbonates on the inner walls of hot water pipes. When solutions containing HCO_3^- are heated, the ion decomposes as follows.

$$2HCO_3^-(aq) \longrightarrow H_2O + CO_2(g) + CO_3^{2-}(aq)$$

Like most gases, carbon dioxide becomes less soluble as the temperature is raised, so CO_2 is driven from the hot solution and the HCO_3^- is gradually converted to CO_3^{2-}. As the carbonate ions form, they are able to precipitate the hardness ions. This precipitate, which sticks to the inner walls of pipes and hot water boilers, is called *boiler scale*. In locations that have high concentrations of Ca^{2+} and HCO_3^- in the water supply, *boiler scale* is a very serious problem, as illustrated in the accompanying photograph.

Boiler scale built up on the inside of a water pipe.

to see how we apply them, and then we can see how the rules can be used to predict precipitation reactions.

The solubility rules lets us predict solubilities of many salts

To make them easier to remember, the rules given in Table 5.2 are divided into two categories. The first includes compounds that are soluble, with some exceptions. The second describes compounds that are generally insoluble, with some exceptions. Some examples will help clarify their use.

Rule 1 states that all compounds of the alkali metals are soluble in water. This means that you can expect *any* compound containing Na^+ or K^+, or any of the Group IA metal ions, *regardless of the anion,* to be soluble. If one of the reactants in a metathesis is Na_3PO_4, you now know from Rule 1 that it is soluble. Therefore, you would write it in *dissociated* form in the ionic equation. Similarly, Rule 6 states, in part, that all carbonate compounds are *insoluble* except those of the alkali metals and the ammonium ion. Thus, if one of the products in a metathesis reaction is $CaCO_3$, you'd expect it to be insoluble, because the cation is not an alkali metal or NH_4^+. Therefore, you would write its formula in undissociated form when you construct the ionic equation. (Facets of Chemistry 5.1 describes how the insolubility of $CaCO_3$ bears on the softening of "hard water.")

Solubilities are used to predict precipitation reactions

Now let's look at an example that illustrates how we use the solubility rules to predict reactions in which a precipitate forms.

EXAMPLE 5.3
Predicting Reactions and Writing Their Equations

Predict the reaction that will occur when aqueous solutions of $Pb(NO_3)_2$ and $Fe_2(SO_4)_3$ are mixed. Write molecular, ionic, and net ionic equations for it.

ANALYSIS: We know the molecular equation will take the form

$$Pb(NO_3)_2 + Fe_2(SO_4)_3 \longrightarrow$$

To complete the equation we have to determine the makeup of the products. We begin, therefore, by predicting what a double replacement (metathesis) might produce. Then we proceed to expand the molecular equation into an ionic equation, and finally we drop spectator ions to obtain the net ionic equation. To do this we need to know solubilities, and here we apply the solubility rules.

SOLUTION: Our reactants are $Pb(NO_3)_2$ and $Fe_2(SO_4)_3$, which contain the ions Pb^{2+} and NO_3^-, and Fe^{3+} and SO_4^{2-}, respectively. To write the formulas of the products, we interchange anions. We combine Pb^{2+} with SO_4^{2-}, and for electrical neutrality, we must use one ion of each. Therefore, we write $PbSO_4$ as one possible product. For the other product, we combine Fe^{3+} with NO_3^-. Electrical neutrality now demands that we use *three* NO_3^- to *one* Fe^{3+} to make $Fe(NO_3)_3$. The correct formulas of the products, then, are $PbSO_4$ and $Fe(NO_3)_3$. The unbalanced molecular equation at this point is

$$Pb(NO_3)_2 + Fe_2(SO_4)_3 \longrightarrow PbSO_4 + Fe(NO_3)_3 \quad \text{(unbalanced)}$$

Always write equations in two steps: first write correct formulas for the reactants and products, then adjust the coefficients to balance the equation.

Next, let's determine solubilities. The reactants are ionic compounds and we are told that they are in solution, so we know they are water soluble. Solubility Rules 2 and 4 tell us this also. For the products, we find that Rule 2 says that all nitrates are soluble, so $Fe(NO_3)_3$ is soluble; Rule 4 tells us that the sulfate of Pb^{2+} is *insoluble*. This means that a precipitate of $PbSO_4$ will form. Writing (*aq*) and (*s*) following appropriate formulas, the unbalanced molecular equation is

$$Fe_2(SO_4)_3(aq) + Pb(NO_3)_2(aq) \longrightarrow Fe(NO_3)_3(aq) + PbSO_4(s) \quad \text{(unbalanced)}$$

When it is balanced, we obtain the *molecular equation*.

$$Fe_2(SO_4)_3(aq) + 3Pb(NO_3)_2(aq) \longrightarrow 2Fe(NO_3)_3(aq) + 3PbSO_4(s)$$

Next, we expand this to give the ionic equation in which soluble compounds are written in dissociated (separated) form as ions and insoluble compounds are written in "molecular" form. We can place the following labels beneath the formulas of the molecular equation:

$$Fe_2(SO_4)_3(aq) + 3Pb(NO_3)_2(aq) \longrightarrow 2Fe(NO_3)_3(aq) + 3PbSO_4(s)$$

soluble	soluble	soluble	insoluble
(Rule 4)	(Rule 2)	(Rule 2)	(Rule 4)

We now separate the ions of the water-soluble species to obtain the *ionic equation*. Once again, we are careful to apply the subscripts of the ions and the coefficients.

$$2Fe^{3+}(aq) + 3SO_4^{2-}(aq) + 3Pb^{2+}(aq) + 6NO_3^-(aq) \longrightarrow$$
$$2Fe^{3+}(aq) + 6NO_3^-(aq) + 3PbSO_4(s)$$

By removing spectator ions (Fe^{3+} and NO_3^-), we obtain

$$3Pb^{2+}(aq) + 3SO_4^{2-}(aq) \longrightarrow 3PbSO_4(s)$$

Finally, we reduce the coefficients to give us the correct *net ionic equation*.

$$Pb^{2+}(aq) + SO_4^{2-}(aq) \longrightarrow PbSO_4(s)$$

Is the Answer Reasonable?

One of the main things we have to check in solving a problem such as this is that we've written the correct formulas of the products. For example, in this problem some students might be tempted (without thinking) to write $Pb(SO_4)_2$ and $Fe_2(NO_3)_3$ or even $Pb(SO_4)_3$ and $Fe_2(NO_3)_2$. *This is a common error.* Always be careful to figure out the charges of the ions that must be combined in the formula. Then take the ions in a ratio that gives an electrically neutral formula unit.

Once we're sure the formulas of the products are right, we check that we've applied the solubility rules correctly, which we have. Then we check that we've properly balanced the equation (we have), that we've correctly divided the soluble compounds into their ions (we have), and that we've eliminated the spectator ions to obtain the net ionic equation (we have).

PRACTICE EXERCISE 3: Predict the reaction that occurs on mixing the following solutions. Write molecular, ionic, and net ionic equations for the reactions that take place. (a) $AgNO_3$ and NH_4Cl (b) Na_2S and $Pb(C_2H_3O_2)_2$

5.5 ▶ Acids and bases are classes of compounds with special properties

The classification of substances according to their properties is the way scientists organize knowledge to make chemical behavior easier to remember and analyze. We've already discussed ionic and covalent compounds as classifications of chemical substances. Acids and bases constitute another class and include some of our most familiar chemicals as well as many important laboratory reagents (Figure 5.10). The vinegar in a salad dressing, the sour juice of a lemon, the gastric juice that aids digestion in the stomach, vitamin C, and the liquid in the battery of an automobile are similar in at least one respect—they all contain acids. The white crystals of lye in certain drain cleaners, the white substance that makes milk of magnesia opaque, and household ammonia are all bases.

Usually, we use the term *reagent* to mean a chemical commonly kept on hand to be used in chemical reactions.

Acids and bases have commonly recognized properties

There are some general properties that are common to aqueous solutions of acids and bases. For example, **acids** generally have a tart (sour) taste. You know this from having tasted vinegar and lemon juice. However, taste is *never* used as a laboratory

FIGURE 5.10 *Many common substances are acids or bases.*

FIGURE 5.11 *An acid–base indicator.* Litmus paper, a strip of paper impregnated with the dye litmus, becomes blue in aqueous ammonia (a base) and pink in lemon juice (which contains citric acid).

Acids and bases should be treated with respect because of their potential for causing bodily injury if spilled on the skin. If you spill an acid or base on yourself in the lab, be sure to notify your instructor at once.

Arrhenius was nearly failed by his examining committee for proposing such a bizarre idea that electrically charged particles exist in solution. In 1903, this view won him the Nobel Prize.

test for acids; some acids are poisonous and others, such as sulfuric acid in battery fluid, are extremely corrosive to animal tissue. (In general, *never taste chemicals in the laboratory!*)

Acids (and bases) also affect the colors of certain natural dye substances called *acid–base indicators.* An example is litmus (Figure 5.11), which has a pink or red color in an acidic solution and a blue color in a basic solution.[2]

Substances classified as acids also corrode many metals. (Perhaps you have seen this effect following the spill of battery acid somewhere on a metal surface under the hood of a car.)

Although some acids can be dangerous, many acids are essential to our well-being. Ascorbic acid, for example, is also known as vitamin C and is a necessary part of our diet. And, of course, acids such as citric acid and vinegar flavor our foods and make them more enjoyable.

Bases also have some properties in common. For example, aqueous solutions of **bases** have a somewhat bitter taste, and they turn litmus blue. They also have a soapy "feel." (Actually, bases change oils and fat in your skin into soap, which is why they feel slippery.) Like acids, bases pose varying risks. The base in milk of magnesia, $Mg(OH)_2$, can be taken internally as a laxative or as something to soothe the stomach when too much gastric acid (hydrochloric acid) is present. But another base, sodium hydroxide (lye) is *extremely* hazardous, particularly to the eyes but even to the skin. (The hazards posed by acids and bases are good reasons for wearing eye protective devices whenever you observe or do laboratory work.)

Arrhenius provided a useful definition of acids and bases

The first comprehensive theory concerning acids, bases, and electrically conducting solutions appeared in 1884 in the Ph.D. thesis of a Swedish chemist, Svante Arrhenius. He proposed that ions form directly when salts dissolve in water, which was a radical notion at the time. This is why *all* aqueous solutions of ionic compounds conduct electricity, according to Arrhenius. Arrhenius also theorized that all acids release hydrogen ions, H^+, in water, and all bases release hydroxide ions, OH^-. This explains why acids have many common properties and why bases behave similarly.

One of the most important properties of acids and bases is their reaction with each other, a reaction referred to as **neutralization.** For example, if solutions of hydrochloric acid, $HCl(aq)$, and sodium hydroxide, $NaOH(aq)$, are mixed in a 1-to-1 ratio by moles, the resulting solution has no effect on litmus paper because the following reaction occurs:

$$HCl(aq) + NaOH(aq) \longrightarrow NaCl(aq) + H_2O$$

The acidic and basic properties of the solutes disappear; the solution of the products is neither acidic nor basic. We say that an *acid–base neutralization* has occurred. According to Arrhenius, acid–base neutralization is simply the combination of a hydrogen ion with a hydroxide ion to produce a water molecule, thus making H^+ ions and OH^- ions disappear.

In general, the reaction of an acid with a base produces water and an ionic compound. In the reaction above, the compound is sodium chloride, or salt. This reaction is so general, in fact, that *we use the word **salt** to mean any ionic compound that doesn't contain either hydroxide ion, OH^-, or oxide ion, O^{2-}.* (Ionic compounds that contain OH^- or O^{2-} are bases.)

[2]Litmus paper, commonly found among the items in a locker in the general chemistry lab, consists of strips of absorbent paper that have been soaked in a solution of litmus and dried. Red litmus paper is used to test if a solution is basic. A basic solution turns red litmus blue. To test if the solution is acidic, blue litmus paper is used. Acidic solutions turn blue litmus red.

Acids and bases undergo ionization reactions in water

Today we know that hydrogen ions, H^+, can only exist in water when they are attached to something else. When they attach themselves to water molecules, they form hydronium ions, H_3O^+. We view H_3O^+ as a species that carries H^+ in water. However, for the sake of convenience, we often use the term *hydrogen ion* as a substitute for *hydronium ion,* and in many equations, we use $H^+(aq)$ to stand for $H_3O^+(aq)$. In fact, whenever you see the symbol $H^+(aq)$, you should realize that we are actually referring to $H_3O^+(aq)$.

For most purposes, we find that the following modified versions of Arrhenius's definitions work satisfactorily when we deal with aqueous solutions.

Even the formula H_3O^+ is something of a simplification. In water the H^+ ion is associated with more than one molecule of water, but we use the formula H_3O^+ as a simple representation.

Arrhenius Definition of Acids and Bases

An **acid** is a substance that reacts with water to produce hydronium ion, H_3O^+.

A **base** is a substance that produces hydroxide ion in water.

Substances that are acids

In general, **acids** are molecular substances that react with water to produce ions, one of which is the hydronium ion. For example, pure hydrogen chloride is a gas and is molecular, not ionic. When it dissolves in water, however, the following chemical reaction occurs, which produces ions that did not pre-exist in $HCl(g)$. (See Figure 5.12.)

$$HCl(g) + H_2O \longrightarrow H_3O^+(aq) + Cl^-(aq)$$

A reaction like this in which ions form where none existed before is called an **ionization reaction.** One result of such a reaction in water is a solution that conducts electricity, so acids can be classified as electrolytes.

Similar ionization reactions occur for other acids as well. For example, nitric acid reacts with water according to the equation

$$HNO_3(l) + H_2O \longrightarrow H_3O^+(aq) + NO_3^-(aq)$$

Notice that in the ionization reactions of HCl and HNO_3 described above, the hydronium ion is formed by the transfer of an H^+ ion from the acid molecule to the water molecule. The particle left after loss of the H^+ is an anion. We might represent this in general terms by the equation

$$\text{acid molecule} + H_2O \longrightarrow H_3O^+ + \text{anion}$$

If gaseous HCl is cooled to about $-85\ °C$, it condenses to a liquid that doesn't conduct electricity. No ions are present in pure liquid HCl.

Acids are electrolytes that yield ions by reaction with water.

HCl H_2O Cl^- H_3O^+

FIGURE 5.12 *Ionization of HCl in water.* Collisions between HCl molecules and water molecules lead to a transfer of H^+ from HCl to H_2O, giving Cl^- and H_3O^+ as products.

Acetic acid molecule
$HC_2H_3O_2$

Only this H comes off as H^+.

Acetate ion
$C_2H_3O_2^-$

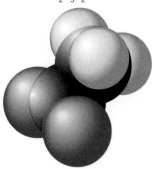

FIGURE 5.13 *Acetic acid and acetate ion.* The structures of acetic acid and acetate ion are illustrated here. In acetic acid, only the hydrogen attached to an oxygen can come off as H^+.

If we represent the acid molecule by the general formula HA, this becomes

$$HA + H_2O \longrightarrow H_3O^+ + A^-$$

As noted earlier, the "active ingredient" in the hydronium ion is H^+, which is why $H^+(aq)$ is often used in place of $H_3O^+(aq)$ in equations. Using this simplification, the ionization of HCl and HNO_3 in water can be represented as

$$HCl(g) \xrightarrow{H_2O} H^+(aq) + Cl^-(aq)$$

and

$$HNO_3(l) \xrightarrow{H_2O} H^+(aq) + NO_3^-(aq)$$

Sometimes an acid also contains hydrogen atoms that are not able to transfer to water molecules. An example is acetic acid, $HC_2H_3O_2$, the acid that gives vinegar its sour taste. This acid reacts with water as follows:

$$HC_2H_3O_2(l) + H_2O \longrightarrow H_3O^+(aq) + C_2H_3O_2^-(aq)$$

Notice that only the hydrogen written first in the formula is able to transfer to H_2O to give hydronium ions. The structures of the acetic acid molecule and the acetate ions are shown in Figure 5.13, with the hydrogen that can be lost by the acetic acid molecule indicated.

The molecules HCl, HNO_3, and $HC_2H_3O_2$ are said to be **monoprotic acids** because they are capable of furnishing only *one* H^+ per molecule of acid. **Polyprotic acids** can furnish more than one H^+ per molecule of acid and undergo reactions similar to those of HCl and HNO_3, except that the loss of H^+ by the acid occurs in two or more steps. Thus, the ionization of sulfuric acid, a **diprotic acid,** takes place by two successive steps.

$$H_2SO_4(aq) + H_2O \longrightarrow H_3O^+(aq) + HSO_4^-(aq)$$
$$HSO_4^-(aq) + H_2O \longrightarrow H_3O^+(aq) + SO_4^{2-}(aq)$$

EXAMPLE 5.4

Writing Equations for Ionization Reactions of Acids

Hydrogen ions that are able to be transferred to water molecules to form hydronium ions are usually written first in the formula for the acid.

Phosphoric acid, H_3PO_4, is a **triprotic acid** found in some soft drinks such as Coca-Cola®, where it adds a touch of tartness to the beverage. Write equations for its stepwise ionization in water.

ANALYSIS: We are told that H_3PO_4 is a triprotic acid, which is also indicated by the three hydrogens at the beginning of the formula. Because there are three hydrogens to come off the molecule, we expect there to be three steps in the ionization. Each step removes one H^+, and we can use that knowledge to deduce the formulas of the products. Let's line them up so we can see the progression.

$$H_3PO_4 \xrightarrow{-H^+} H_2PO_4^- \xrightarrow{-H^+} HPO_4^{2-} \xrightarrow{-H^+} PO_4^{3-}$$

Notice that loss of H^+ decreases the number of hydrogens by one and increases the negative charge by one unit. Also, the product of one step serves as the reactant in the next step.

SOLUTION: The first step is the reaction of H_3PO_4 with water to give H_3O^+ and $H_2PO_4^-$.

$$H_3PO_4(aq) + H_2O \longrightarrow H_3O^+(aq) + H_2PO_4^-(aq)$$

The second and third steps are similar to the first.

$$H_2PO_4^-(aq) + H_2O \longrightarrow H_3O^+(aq) + HPO_4^{2-}(aq)$$
$$HPO_4^{2-}(aq) + H_2O \longrightarrow H_3O^+(aq) + PO_4^{3-}(aq)$$

Is the Answer Reasonable?

Check to see whether the equations are balanced in terms of atoms and charge. If any mistakes were made, something would be out of balance and we would discover the error. In this case, all the equations are balanced, so we can feel confident we've written them correctly.

PRACTICE EXERCISE 4: Write the equation for the ionization of $HCHO_2$ (formic acid) in water. Formic acid is used industrially to remove hair from animal skins prior to tanning.

PRACTICE EXERCISE 5: Write equations for the stepwise ionization in water of citric acid, $H_3C_6H_5O_7$, the acid in citrus fruits.

Nonmetal oxides can be acids The acids we've discussed so far have been molecules that contain hydrogen atoms that are able to be transferred to water molecules. Another class of compounds that fits our description of acids is *nonmetal oxides*. Here we have oxides such as SO_3, CO_2, and N_2O_5 whose aqueous solutions contain H_3O^+ and turn litmus red. These oxides are called **acidic anhydrides,** where *anhydride* means "without water." They react with water to form molecular acids containing hydrogen, which are then able to undergo reaction with water to yield H_3O^+.

$$SO_3(g) + H_2O \longrightarrow H_2SO_4(aq) \qquad \text{sulfuric acid}$$

$$N_2O_5(g) + H_2O \longrightarrow 2HNO_3(aq) \qquad \text{nitric acid}$$

$$CO_2(g) + H_2O \longrightarrow H_2CO_3(aq) \qquad \text{carbonic acid}$$

Although carbonic acid is too unstable to be isolated as a pure compound, its solutions in water are quite common. Carbon dioxide from the atmosphere dissolves in rainwater and the waters of lakes and streams, for example, where it exists partly as carbonic acid and its ions (HCO_3^- and CO_3^{2-}). This makes these waters naturally slightly acidic. Carbonic acid is also present in carbonated beverages.

Not all nonmetal oxides are acidic anhydrides, only those that are able to react with water. For example, carbon monoxide doesn't react with water, so its solutions in water are not acidic; carbon monoxide, therefore, is not classified as an acidic anhydride.

Substances that are bases

Bases fall into two categories: ionic compounds that contain OH^- or O^{2-} and molecular compounds that react with water to give hydroxide ions. Because solutions of bases contain ions, they conduct electricity. Therefore, bases are electrolytes.

Ionic bases include metal hydroxides, such as $NaOH$ and $Ca(OH)_2$. When dissolved in water, they dissociate just like other soluble ionic compounds.

$$NaOH(s) \longrightarrow Na^+(aq) + OH^-(aq)$$

$$Ca(OH)_2(s) \longrightarrow Ca^{2+}(aq) + 2OH^-(aq)$$

Soluble metal oxides are **basic anhydrides** because they react with water to form the hydroxide ion as one of the products. Calcium oxide is typical.

$$CaO(s) + H_2O \longrightarrow Ca(OH)_2(aq)$$

This reaction occurs when water is added to dry cement or concrete because calcium oxide, or "quicklime," is an ingredient in these materials. In this case it is the oxide ion, O^{2-}, that actually forms the OH^-.

$$O^{2-} + H_2O \longrightarrow 2OH^-$$

$Ca(OH)_2$ is not highly soluble in water. It could be described as slightly soluble or partially soluble.

Continual contact of your hands with fresh Portland cement can lead to irritation because the mixture is quite basic.

FIGURE 5.14 *Ionization of ammonia in water.* Collisions between NH_3 molecules and water molecules lead to a transfer of H^+ from H_2O to NH_3, giving NH_4^+ and OH^- ions.

Molecular bases The most common molecular base is the gas ammonia, NH_3, which dissolves in water and reacts to give a basic solution. (See Figure 5.14.)

$$NH_3(aq) + H_2O \longrightarrow NH_4^+(aq) + OH^-(aq)$$

This is also an ionization reaction because ions have been formed where none previously existed. Notice that when a molecular base reacts with water, an H^+ is lost by the water molecule and gained by the base. One product is a cation that has one more H and one more positive charge than the reactant base. Loss of H^+ by the water gives the other product, the OH^- ion, which is why the solution is basic. We might represent this by the general equation

$$\text{base} + H_2O \longrightarrow \text{base}H^+ + OH^-$$

If we represent the base by the symbol B, this becomes

$$B + H_2O \longrightarrow BH^+ + OH^-$$

EXAMPLE 5.5

Writing the Equation for the Ionization of a Weak Base

Dimethylamine, $(CH_3)_2NH$, is used as an attractive for boll weevils so they can be destroyed. This insect has caused more than 14 billion dollars in losses in cotton yields in the United States since it arrived from Mexico in 1892. The compound is a base in water. Write an equation for its ionization.

ANALYSIS: The reactants in the equation are $(CH_3)_2NH$ and H_2O. To write the equation, we need to write the formulas of the products. When a base reacts with water, it takes an H^+ from H_2O, leaving OH^- behind.

SOLUTION: When an H^+ is added to $(CH_3)_2NH$, the product will be $(CH_3)_2NH_2^+$. Therefore, the equation for the reaction is

$$(CH_3)_2NH(aq) + H_2O \longrightarrow (CH_3)_2NH_2^+(aq) + OH^-(aq)$$

Is the Answer Reasonable?
Compare the equation we've written with the general equation for reaction of a base with water. Notice that the formula for the product has one more H and a positive charge. Also, notice that the water has become OH^- when it loses H^+. The equation is therefore correct.

PRACTICE EXERCISE 6: Hydroxylamine, $HONH_2$, is a base in water. Write an equation for its reaction with the solvent.

5.6 ▶ Naming acids and bases follows a system

Although at first there seems to be little order in the naming of acids, there are patterns that help organize names of acids and the anions that come from them when the acids are neutralized.

Hydrogen compounds of nonmetals can be acids

The binary compounds of hydrogen with many of the nonmetals are acidic, and in their water solutions they are referred to as **binary acids.** Some examples are HCl, HBr, and H_2S. In naming these substances as acids, we add the prefix *hydro-* and the suffix *-ic* to the stem of the nonmetal name, followed by the word *acid*. For example, water solutions of hydrogen chloride and hydrogen sulfide are named as follows:

Name of the Molecular Compound		**Name of the Binary Acid**	
$HCl(g)$	hydrogen chloride	$HCl(aq)$	*hydrochloric acid*
$H_2S(g)$	hydrogen sulfide	$H_2S(aq)$	*hydrosulfuric acid*

Notice that the gaseous molecular substances are named in the usual way as binary compounds. *It is their aqueous solutions that are named as acids.*

When an acid is neutralized, the salt that is produced contains the anion formed by removing a hydrogen ion, H^+, from the acid molecule. Thus, HCl yields salts containing the chloride ion, Cl^-. Similarly, HBr gives salts containing the bromide ion, Br^-. In general, then, neutralization of a binary acid yields the simple anion of the nonmetal.

PRACTICE EXERCISE 7: Name the water solutions of the following acids: HF, HBr. Name the sodium salts formed by neutralizing these acids with NaOH.

Oxoacids contain hydrogen, oxygen, and another element

Acids that contain hydrogen, oxygen, plus another element are called **oxoacids.** Examples are H_2SO_4 and HNO_3. These acids do not take the prefix *hydro-*. Many nonmetals form two or more oxoacids that differ in the number of oxygen atoms in their formulas, and they are named according to the number of oxygens. When there are two oxoacids, the one with the larger number of oxygens takes the suffix *-ic* and the one with the fewer number of oxygens takes the suffix *-ous*.

H_2SO_4	sulfur*ic acid*	HNO_3	nitr*ic acid*
H_2SO_3	sulfur*ous acid*	HNO_2	nitr*ous acid*

The halogens can occur in as many as four different oxoacids. The oxoacid with the most oxygens has the prefix *per-*, and the one with the least has the prefix *hypo-*.

HClO	*hypo*chlor*ous acid* (usually written HOCl)	$HClO_3$	chlor*ic acid*
$HClO_2$	chlor*ous acid*	$HClO_4$	*per*chlor*ic acid*

The neutralization of oxoacids produces negative polyatomic ions. There is a very simple relationship between the name of the polyatomic ion and that of its parent acid.

1. *-ic* acids give *-ate* anions: HNO_3 (nitr*ic acid*) → NO_3^- (nitr*ate* ion)
2. *-ous* acids give *-ite* anions: H_2SO_3 (sulfur*ous acid*) → SO_3^{2-} (sulf*ite* ion)

In naming polyatomic anions, the prefixes *per-* and *hypo-* carry over from the name of the parent acid. Thus perchloric acid, $HClO_4$, gives perchlorate ion, ClO_4^-, and hypochlorous acid, HClO, gives hypochlorite ion, ClO^-.

This relationship between name of the acid and name of the anion carries over to other acids that end in the suffix *-ic*. For example, acet*ic* acid gives the anion acet*ate*, and citric acid gives the anion citr*ate*.

Bromine forms four oxoacids, similar to those of chlorine. What is the name of the acid $HBrO_2$ and what is the name of the salt $NaBrO_3$?

ANALYSIS AND SOLUTION: Let's review the acids formed by chlorine and then reason by analogy. For chlorine we have

EXAMPLE 5.6

Naming Acids and Their Salts

HClO	hypochlorous acid
$HClO_2$	chlorous acid
$HClO_3$	chloric acid
$HClO_4$	perchloric acid

The acid $HBrO_2$ is similar to chlorous acid, so to name it we will use the stem of the element name bromine (brom-) in place of chlor-. Therefore, the name of $HBrO_2$ is *bromous acid.*

To find the name of $NaBrO_3$, let's begin by asking "What acid would give this salt by neutralization?" Neutralization involves removing an H^+ from the acid molecule and replacing it with a cation, in this case Na^+. Therefore, the salt $NaBrO_3$ would be obtained by neutralizing the acid $HBrO_3$. This acid has one more oxygen than bromous acid, $HBrO_2$, so it would have the ending *-ic.* Thus, $HBrO_3$ is named bromic acid. Neutralizing an acid that has a name that ends in *-ic* gives an anion with a name that ends in *-ate,* so the anion BrO_3^- is the bromate ion. Therefore, the salt $NaBrO_3$ is *sodium bromate.*

Are the Answers Reasonable?

There's really not much we can do to check the answers here. For the salt, if $HClO_3$ is chloric acid, then it seems reasonable that $HBrO_3$ would be bromic acid, which would mean that BrO_3^- is the bromate ion and $NaBrO_3$ is sodium bromate.

PRACTICE EXERCISE 8: The formula for arsenic acid is H_3AsO_4. What is the name of the salt Na_3AsO_4?

PRACTICE EXERCISE 9: Formic acid is $HCHO_2$. What is the name of the salt $Ca(CHO_2)_2$?

Acid salts can be formed by polyprotic acids

Monoprotic acids such as HCl and $HC_2H_3O_2$ have only one hydrogen that can be removed by neutralization and these acids form only one anion. However, polyprotic acids can be neutralized stepwise and the neutralization can be halted before all the hydrogens have been removed. For example, if sulfuric acid is combined with sodium hydroxide in a ratio of 1 mol of acid to 2 mol of base, then complete neutralization of the acid occurs and the SO_4^{2-} ion is formed.

$$H_2SO_4(aq) + 2NaOH(aq) \longrightarrow Na_2SO_4(aq) + 2H_2O$$

However, if the acid and base are combined in a 1-to-1 mole ratio, only half of the available hydrogens are neutralized.

$$H_2SO_4(aq) + NaOH(aq) \longrightarrow NaHSO_4(aq) + H_2O$$

The salt $NaHSO_4$, which can be isolated as crystals by evaporating the reaction mixture, is referred to as an **acid salt** because it contains the HSO_4^- ion, which is still capable of furnishing additional H^+.

In naming acid salts formed from ions such as HSO_4^-, we specify the number of hydrogens that can still be neutralized if the salt were to be treated with additional base. For example,

$NaHSO_4$	sodium hydrogen sulfate
NaH_2PO_4	sodium dihydrogen phosphate

For acid salts of diprotic acids (acids capable of releasing two H^+), the prefix *bi-* is still often used.

$NaHCO_3$	sodium bicarbonate or sodium hydrogen carbonate

Many acid salts have useful applications. As its active ingredient, this familiar product contains sodium hydrogen sulfate (sodium bisulfate), which the manufacturer calls "sodium acid sulfate."

Notice that the prefix *bi-* does *not* mean "two"; it means that there is an acidic hydrogen in the compound.

PRACTICE EXERCISE 10: Write molecular equations for the stepwise neutralization of H_3PO_4 by NaOH.

PRACTICE EXERCISE 11: What is the formula for sodium bisulfite? What is the chemically correct name for this compound?

Bases are named as hydroxides or molecules

The naming of bases is much less complicated than the naming of acids. Metal compounds that contain the ions OH^- or O^{2-}, such as NaOH and Na_2O, are ionic and are named just like any other ionic compound. Thus, NaOH is sodium hydroxide and Na_2O is sodium oxide.

Molecular bases such as NH_3 (ammonia) are specified by just giving the name of the molecule.[3] Another example is the base methylamine, CH_3NH_2, which behaves just like ammonia in water. In other words, the base extracts H^+ from H_2O to give a cation and hydroxide ion.

$$CH_3NH_2(aq) + H_2O \rightleftharpoons CH_3NH_3^+(aq) + OH^-(aq)$$

5.7 Acids and bases are classified as strong or weak

Sodium chloride and calcium chloride are both examples of **strong electrolytes**— electrolytes that break up essentially 100% into ions in water. No "molecules" of either NaCl or $CaCl_2$ are detectable in their aqueous solutions; they exist entirely as ions. This behavior is typical of all ionic compounds.

Hydrochloric acid is also a strong electrolyte. Its ionization in water is essentially complete, and no molecules of HCl can be detected in its solutions. As a result, solutions of hydrochloric acid are highly acidic and this acid is called a strong acid. In general, *acids that are strong electrolytes are called* **strong acids.** Actually, there are very few strong acids. The list below gives the most common ones. You will find it useful to memorize this list.

All strong acids are strong electrolytes.

Strong Acids

$HClO_4(aq)$	perchloric acid
$HCl(aq)$	hydrochloric acid
$HBr(aq)$	hydrobromic acid
$HI(aq)$	hydroiodic acid[4]
$HNO_3(aq)$	nitric acid
$H_2SO_4(aq)$	sulfuric acid

TOOLS

Strong acids

Metal hydroxides are ionic compounds, so they are also strong electrolytes. Those that are soluble are the hydroxides of Group IA and the hydroxides of calcium, strontium, and barium of Group IIA. Solutions of these compounds are strongly basic, so these substances are considered to be **strong bases.** (See below.)

[3]Solutions of ammonia are sometimes called *ammonium hydroxide,* although there is no evidence that the species NH_4OH actually exists.

[4]Sometimes the first "o" in the name of HI(aq) is dropped for ease of pronunciation to give *hydriodic acid.*

Strong Bases (Soluble Metal Hydroxides)

Group IA		*Group IIA*	
LiOH	lithium hydroxide		
NaOH	sodium hydroxide		
KOH	potassium hydroxide	$Ca(OH)_2$	calcium hydroxide
RbOH	rubidium hydroxide	$Sr(OH)_2$	strontium hydroxide
CsOH	cesium hydroxide	$Ba(OH)_2$	barium hydroxide

The hydroxides of other metals have very low solubilities in water. They are strong electrolytes in the sense that the small amounts of them that dissolve in solution are completely dissociated. However, because of their low solubility in water, their solutions are very weakly basic, so we have not included them in the list above.

Weak acids and bases are weak electrolytes

Most acids are not completely ionized in water. For instance, a solution of acetic acid, $HC_2H_3O_2$, is a relatively poor conductor of electricity compared to a solution of HCl with the same concentration (Figure 5.15), so acetic acid is classified as a **weak electrolyte.** This is because in the acetic acid solution only a small fraction of the molecules of the acid actually exist as H_3O^+ and $C_2H_3O_2^-$ ions. (Actually, less than 0.5% of the acid is ionized in a solution that contains 1 mol of the acid in a

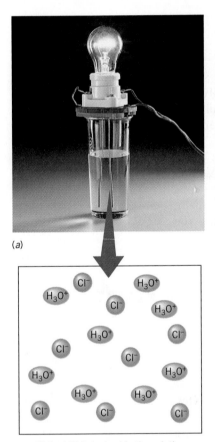

(a)

All the HCl is ionized in the solution, so there are many ions present.

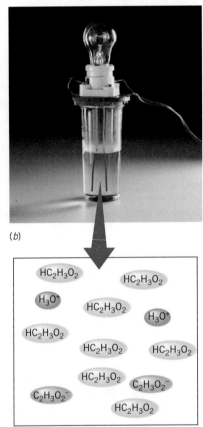

(b)

Only a small fraction of the acetic acid is ionized, so there are few ions to conduct electricity. Most of the acetic acid is present as neutral molecules of $HC_2H_3O_2$.

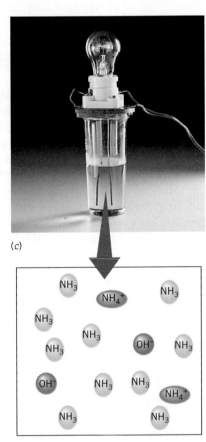

(c)

Only a small fraction of the ammonia is ionized, so few ions are present to conduct electricity. Most of the ammonia is present as neutral molecules of NH_3.

FIGURE 5.15 *Electrical conductivity of solutions of strong and weak acids and bases at equal concentrations.* (*a*) HCl is 100% ionized and is a strong conductor, enabling the light to glow brightly. (*b*) $HC_2H_3O_2$ is a weaker conductor than HCl because the extent of its ionization is far less, so the light is dimmer. (*c*) NH_3 also is a weaker conductor than HCl because the extent of its ionization is low, and the light remains dim.

liter of solution.) The rest of the acetic acid is present as molecules of $HC_2H_3O_2$. For now, let's represent this weak ionization as follows:

$$HC_2H_3O_2(aq) + H_2O \xrightarrow{\text{small percentage}} H_3O^+(aq) + C_2H_3O_2^-(aq)$$

Solutions of acetic acid do not have high concentrations of H_3O^+ and are not strongly acidic. Acetic acid is therefore said to be a **weak acid.** All weak acids are weak electrolytes. Other examples are carbonic acid, H_2CO_3, and nitrous acid, HNO_2. In fact, *if an acid is not one of the strong acids listed above, you can assume it to be a weak acid.*

Molecular bases, such as ammonia, are also weak electrolytes and have a low percentage ionization. They are classified as **weak bases.** (See Figure 5.15c.) In a solution of ammonia, only a small fraction of the solute is ionized to give NH_4^+ and OH^-. Most of the base is present as ammonia molecules.

$$NH_3(aq) + H_2O \xrightarrow{\text{small percentage}} NH_4^+(aq) + OH^-(aq)$$

It is interesting to note that *virtually all molecular bases are weak bases.*

Let's briefly summarize the results of our discussion.

> Weak acids and bases are weak electrolytes.
> Strong acids and bases are strong electrolytes.

Formic acid, $HCHO_2$, is not among the list of strong acids, so we would conclude it is a weak acid. This acid is in the venom of stinging insects, such as fire ants.

Strong bases are ionic metal hydroxides, such as NaOH. Weak bases are molecular bases, such as NH_3.

Dynamic equilibria exist in solutions of weak acids and bases

A question that might have occurred to you during the discussion of weak acids and bases is, "If some of the solute molecules in aqueous ammonia or aqueous acetic acid can react with water, why can't all of them?" Actually, all do have the *potential* to react, but we meet here examples of a very important chemical phenomenon—*dynamic equilibrium*. Let's go inside an acetic acid solution to see what's happening.

As soon as acetic acid is mixed with water, solute and solvent molecules start to collide with each other, as illustrated in the first reaction in Figure 5.16. Because

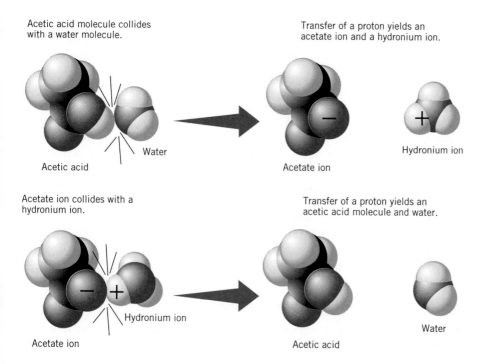

Acetic acid molecule collides with a water molecule.

Transfer of a proton yields an acetate ion and a hydronium ion.

Water

Acetic acid

Acetate ion

Hydronium ion

Acetate ion collides with a hydronium ion.

Transfer of a proton yields an acetic acid molecule and water.

Hydronium ion

Acetate ion

Acetic acid

Water

FIGURE 5.16 *Equilibrium in a solution of acetic acid.* Two opposing reactions take place simultaneously in a solution of acetic acid. Molecules of acid collide with molecules of water and form acetate ions and H_3O^+ ions. Meanwhile, acetate ions collide with H_3O^+ ions to give acetic acid molecules and water molecules. (The usual colors are used: white = H, red = O, black = C.)

For now, consider *rate* to mean the number of events per unit of volume per unit of time, like collisions per cm³ per second.

of the chemical nature of acetic acid, however, only a small percentage of such collisions produces ions. As a result, the rate (or speed) at which the ions form is quite small compared with the rate of collision. There is a *low* probability of the following reaction occurring. When it happens, we call it the **forward reaction** in the acetic acid–water system.

$$HC_2H_3O_2(aq) + H_2O \xrightarrow[\text{low probability}]{\text{small percentage}} H_3O^+(aq) + C_2H_3O_2^-(aq)$$

The probability of ions forming or disappearing as a result of collisions varies from solute to solute.

Once H_3O^+ and $C_2H_3O_2^-$ ions form, they wander around in the solution and experience collisions with solvent molecules. Occasionally, of course, an H_3O^+ ion and a $C_2H_3O_2^-$ ion meet and collide, as shown in the second reaction in Figure 5.16. Because of the chemical nature of these ions, there is a *high* probability that such a collision transfers H^+ from H_3O^+ back to $C_2H_3O_2^-$. We also represent this reaction by an equation.

$$H_3O^+(aq) + C_2H_3O_2^-(aq) \xrightarrow[\text{high probability}]{\text{high percentage}} HC_2H_3O_2(aq) + H_2O$$

which we can write in a way in which the arrow goes from right to left.

$$HC_2H_3O_2(aq) + H_2O \xleftarrow[\text{high probability}]{\text{high percentage}} H_3O^+(aq) + C_2H_3O_2^-(aq)$$

For this juggler, balls are going up at the same rate as they are coming down, so the number of balls in the air stays constant. A somewhat similar situation exists in a dynamic equilibrium, where products form at the same rate as they disappear and the amount of products stays constant.

The reaction read from left to right is the forward reaction, and the one read in the opposite direction is the reverse reaction.

We call this the **reverse reaction** because it tends to undo the forward reaction.

Let us go over the dynamics of this again. Just after adding $HC_2H_3O_2$ to water, the concentrations of H_3O^+ and $C_2H_3O_2^-$ build up because of the ionization reaction, the *forward reaction*. As their concentration increases, however, the ions naturally encounter each other more frequently. So the *reverse reaction* occurs with increasing frequency. Eventually the concentrations of the ions become large enough *to make the rate of the reverse reaction equal the rate of the forward reaction*. At this point, ions disappear as rapidly as they form and their concentrations remain constant from this moment on. The entire chemical system consisting of H_2O, $HC_2H_3O_2$, H_3O^+, and $C_2H_3O_2^-$ has reached a state of balance called **chemical equilibrium.** It is also said to be a **dynamic equilibrium** because the forward and reverse reactions don't cease; they continue to occur even after the concentrations have stopped changing.

We indicate a dynamic equilibrium by using double arrows ($\rightleftharpoons$) in the chemical equation. The ionization of acetic acid is thus represented as follows:

$$HC_2H_3O_2(aq) + H_2O \rightleftharpoons H_3O^+(aq) + C_2H_3O_2^-(aq)$$

In describing an equilibrium such as this, we will often talk about the *position of equilibrium*. By this we mean the extent to which the forward reaction proceeds toward completion. If very little of the products are present at equilibrium, the forward reaction has not gone far toward completion and we say "the position of equilibrium lies to the left," toward the reactants. On the other hand, if large amounts of the products are present at equilibrium, we say "the position of equilibrium lies to the right."

For any weak electrolyte, only a small percentage of the solute is actually ionized at any instant after equilibrium is reached, so the position of equilibrium lies to the left. To call acetic acid a *weak* acid, for example, is just another way of saying that the forward reaction in this equilibrium is far from completion.

Dynamic equilibria exist in solutions of weak bases

Weak bases undergo dynamic equilibria in solutions just as weak acids do. In a solution of ammonia, a weak base, NH_3 molecules collide with H_2O molecules, and some

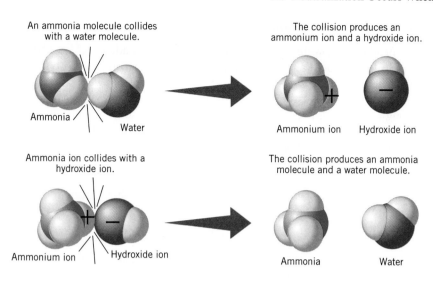

An ammonia molecule collides with a water molecule.

Ammonia / Water

The collision produces an ammonium ion and a hydroxide ion.

Ammonium ion   Hydroxide ion

Ammonia ion collides with a hydroxide ion.

Ammonium ion / Hydroxide ion

The collision produces an ammonia molecule and a water molecule.

Ammonia   Water

FIGURE 5.17 *Equilibrium in a solution of the weak base ammonia.* Collisions between water and ammonia molecules produce ammonium and hydroxide ions. The reverse process, which involves collisions between ammonium ions and hydroxide ions, removes ions from the solution and forms ammonia and water molecules.

of these collisions produce NH_4^+ and OH^- ions (Figure 5.17). The reverse reaction occurs when NH_4^+ and OH^- ions collide and give NH_3 and H_2O molecules. When the rate at which the ions form equals the rate at which they disappear, dynamic equilibrium is reached. The chemical equation that expresses the equilibrium is

$$NH_3(aq) + H_2O \rightleftharpoons NH_4^+(aq) + OH^-(aq)$$

Strong acids do not participate in equilibria because they are fully ionized

With molecular compounds that are strong electrolytes, the tendency of the forward ionization reaction to occur is very high, while the tendency of the reverse reaction to occur is extremely small. In aqueous HCl, for example, there is little tendency during a collision between Cl^- and H_3O^+ for neutral molecules of HCl and H_2O to form. As a result, the reverse reaction has essentially no tendency to occur, so in a very brief time all of the HCl molecules dissolved in water are converted to ions—the acid becomes 100% ionized. For this reason, *we do not use double arrows in describing what happens when HCl(g) or any other strong electrolyte undergoes ionization or dissociation.*

PRACTICE EXERCISE 12: Nitrous acid, HNO_2, is a weak acid thought to be responsible for certain cancers of the intestinal system. Write the chemical equation that represents the weak acid equilibrium for this substance.

PRACTICE EXERCISE 13: Methylamine, CH_3NH_2, is a fishy smelling weak base found in herring brine. Write the equation for the weak base equilibrium for this compound.

5.8 ▶ Neutralization occurs when acids and bases react

Acid–base neutralizations occur whenever a solution of an acid is mixed with a solution of a base. These reactions yield a salt as one of the products, with water usually being another product. The nature of the net ionic equation, however, differs depending on whether the acid and base are strong or weak.

Neutralization of a strong acid by a strong base gives water and a salt

Nitric acid and potassium hydroxide are examples of a strong acid and a strong base. The molecular equation for their reaction is

$$HNO_3(aq) + KOH(aq) \longrightarrow KNO_3(aq) + H_2O$$

The products are water and a salt, KNO_3, which is made of K^+ and NO_3^- ions.

Both nitric acid and potassium hydroxide are strong electrolytes, which means that when we write the ionic equation for the reaction, we write their formulas in dissociated form. On the right side of the equation only one product, KNO_3, is a strong electrolyte. The other product, water, is present in solution as molecules, so when we write the ionic equation only the KNO_3 is written in dissociated form. This gives the ionic equation,

$$H^+(aq) + NO_3^-(aq) + K^+(aq) + OH^-(aq) \longrightarrow K^+(aq) + NO_3^-(aq) + H_2O$$

The spectator ions are K^+ and NO_3^-. Removing them from the ionic equation gives the net ionic equation,

$$H^+(aq) + OH^-(aq) \longrightarrow H_2O$$

This net ionic equation applies to any reaction of a strong acid with a strong base, provided the salt formed in the reaction is water soluble.

PRACTICE EXERCISE 14: Write the molecular, ionic, and net ionic equations for the neutralization of $HCl(aq)$ by $Ca(OH)_2(aq)$.

Neutralization when one reactant is a weak acid or base

Let's look at the ionic and net ionic equations when acetic acid ($HC_2H_3O_2$), a weak acid, and sodium hydroxide, a strong base, react to give a salt, sodium acetate ($NaC_2H_3O_2$), and water. The molecular equation is

$$HC_2H_3O_2(aq) + NaOH(aq) \longrightarrow NaC_2H_3O_2(aq) + H_2O$$

In writing an ionic equation, we always write the formulas of weak electrolytes in "molecular form."

In this reaction, only two substances are strong electrolytes, NaOH and $NaC_2H_3O_2$. Acetic acid is a weak electrolyte and is present in solution mostly in the form of molecules. Therefore, when we write the ionic equation, we do not express the acid in ionic form. We simply write the formula for molecular acetic acid. The ionic equation for this reaction is therefore

$$HC_2H_3O_2(aq) + Na^+(aq) + OH^-(aq) \longrightarrow Na^+(aq) + C_2H_3O_2^-(aq) + H_2O$$

This time there is only one spectator ion, Na^+, and when we eliminate it from the equation, we obtain the net ionic equation,

$$HC_2H_3O_2(aq) + OH^-(aq) \longrightarrow C_2H_3O_2^-(aq) + H_2O$$

The net ionic equation tells us this time that the reaction in the solution is actually one between hydroxide ions and molecules of acetic acid (Figure 5.18).

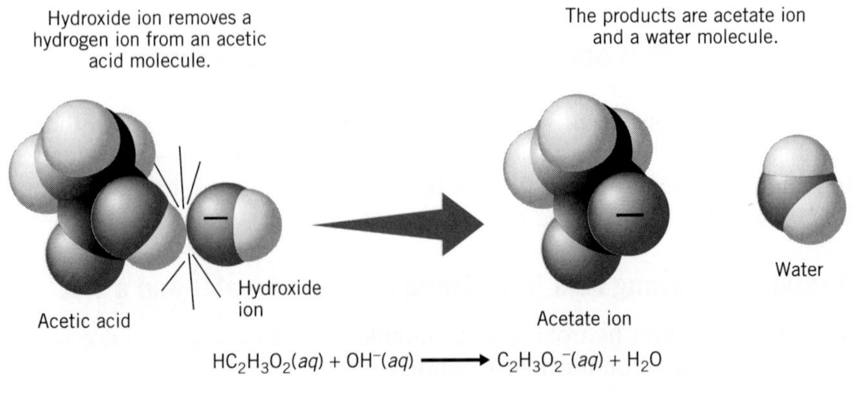

Hydroxide ion removes a hydrogen ion from an acetic acid molecule.

The products are acetate ion and a water molecule.

Acetic acid

Hydroxide ion

Acetate ion

Water

$$HC_2H_3O_2(aq) + OH^-(aq) \longrightarrow C_2H_3O_2^-(aq) + H_2O$$

FIGURE 5.18 *Net reaction of acetic acid with a strong base.* The neutralization of acetic acid by hydroxide ion occurs primarily by the removal of H^+ from acetic acid molecules by OH^- ions.

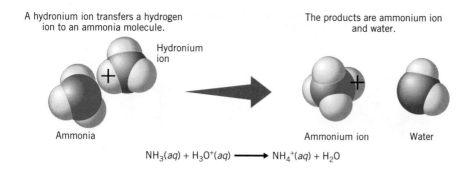

A hydronium ion transfers a hydrogen ion to an ammonia molecule.

The products are ammonium ion and water.

Hydronium ion

Ammonia

Ammonium ion   Water

$NH_3(aq) + H_3O^+(aq) \longrightarrow NH_4^+(aq) + H_2O$

FIGURE 5.19 *Reaction of ammonia with a strong acid.* The reaction occurs primarily by the direct attack of H_3O^+ on NH_3 molecules. Transfer of a proton to the ammonia molecule produces an ammonium ion and a water molecule.

A slightly different situation exists in the reaction of hydrochloric acid with aqueous ammonia, a weak base, to give the salt ammonium chloride (NH_4Cl). The molecular equation for the reaction is

$$NH_3(aq) + HCl(aq) \longrightarrow NH_4Cl(aq)$$

Notice that water does not appear as a product in the molecular equation. This is because the molecular base does not contain hydroxide ions. A solution of ammonia is basic because of the equilibrium

$$NH_3(aq) + H_2O \rightleftharpoons NH_4^+(aq) + OH^-(aq)$$

but the reaction occurs to a very small extent (less than 1% in a dilute solution); therefore, in a solution of ammonia nearly all of the base is present as molecules. In the neutralization reaction, therefore, ammonia is the principal reactant.

Water does appear as a product if we write the equation to include hydronium ion. The net ionic equation for the neutralization, illustrated in Figure 5.19, is then

$$NH_3(aq) + H_3O^+(aq) \longrightarrow NH_4^+(aq) + H_2O$$

PRACTICE EXERCISE 15: Write molecular, ionic, and net ionic equations for the reaction of (a) HCl with KOH, (b) $HCHO_2$ with LiOH, and (c) N_2H_4 with HCl.

Neutralization when the acid and base are both weak

Solutions of acetic acid and of ammonia are both poor conductors of electricity. But if their solutions are combined, the mixture becomes a very good conductor of electricity. This rather astonishing observation can be explained if we write the ionic and net ionic equations for the reaction.

The molecular equation is

$$HC_2H_3O_2(aq) + NH_3(aq) \longrightarrow NH_4C_2H_3O_2(aq)$$

For the ionic equation, the only substance to be written in dissociated form is the product, $NH_4C_2H_3O_2$. The ionic equation is therefore

$$HC_2H_3O_2(aq) + NH_3(aq) \longrightarrow NH_4^+(aq) + C_2H_3O_2^-(aq)$$

There are no spectator ions, so this is also the net ionic equation, which tells us that the reaction takes place primarily by the direct attack of ammonia on acetic acid molecules (Figure 5.20).

Notice that before the reaction begins, there are no strong electrolytes present, so the solutions conduct weakly. The product of the reaction, however, is a salt and it is a strong electrolyte, so after the acid and base react, the solution is able to conduct electricity well.

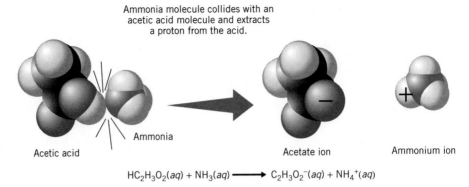

Ammonia molecule collides with an acetic acid molecule and extracts a proton from the acid.

Acetic acid   Ammonia   Acetate ion   Ammonium ion

$$HC_2H_3O_2(aq) + NH_3(aq) \longrightarrow C_2H_3O_2^-(aq) + NH_4^+(aq)$$

FIGURE 5.20 *Reaction of acetic acid with ammonia.* The collision of an ammonia molecule with an acetic acid molecule leads to a transfer of H$^+$ from the acetic acid to ammonia and the formation of ions.

> **PRACTICE EXERCISE 16:** Write molecular, ionic, and net ionic equations for the reaction of the weak base methylamine, CH_3NH_2, with formic acid, $HCHO_2$ (a weak acid).

Acids react with insoluble hydroxides and oxides

The tendency for water to be formed in a neutralization reaction is so strong that it serves to drive reactions of acids (both strong and weak) with insoluble hydroxides and oxides. An example is the reaction of magnesium hydroxide with hydrochloric acid, pictured in Figure 5.21. This is the reaction that occurs when milk of magnesia neutralizes stomach acid.

Magnesium hydroxide has a very low solubility in water and is the solid that makes milk of magnesia white. The molecular equation for its reaction with $HCl(aq)$ is

$$Mg(OH)_2(s) + 2HCl(aq) \longrightarrow MgCl_2(aq) + 2H_2O$$

When we write the ionic equation, we do not write the $Mg(OH)_2$ in dissociated form because it is insoluble. Magnesium chloride, on the other hand, is water soluble and is fully dissociated. (Notice in Figure 5.21 that the reaction mixture is not cloudy after the reaction is over, indicating that all the solid has dissolved. This tells us that the $MgCl_2$ is soluble in water.) The ionic equation is

$$Mg(OH)_2(s) + 2H^+(aq) + 2Cl^-(aq) \longrightarrow Mg^{2+}(aq) + 2Cl^-(aq) + 2H_2O$$

The net ionic equation is obtained by eliminating the Cl^- spectator ions.

$$Mg(OH)_2(s) + 2H^+(aq) \longrightarrow Mg^{2+}(aq) + 2H_2O$$

Many metal oxides are similarly dissolved by acids. For example, before steel can be given a protective coating of zinc (a process called *galvanizing*), rust must first be stripped from the steel. This is usually done by dissolving the rust, mostly Fe_2O_3, in hydrochloric acid. The molecular, ionic, and net ionic equations are

$$Fe_2O_3(s) + 6HCl(aq) \longrightarrow 2FeCl_3(aq) + 3H_2O$$
$$Fe_2O_3(s) + 6H^+(aq) + 6Cl^-(aq) \longrightarrow 2Fe^{3+}(aq) + 6Cl^-(aq) + 3H_2O$$
$$Fe_2O_3(s) + 6H^+(aq) \longrightarrow 2Fe^{3+}(aq) + 3H_2O$$

FIGURE 5.21 *Hydrochloric acid is neutralized by milk of magnesia.* A solution of hydrochloric acid is added to a beaker containing milk of magnesia. The thick white solid in milk of magnesia is magnesium hydroxide, $Mg(OH)_2$, which is able to neutralize the acid. The mixture is clear where some of the solid $Mg(OH)_2$ has already reacted and dissolved.

> **PRACTICE EXERCISE 17:** Write molecular, ionic, and net ionic equations for the reaction of $Al(OH)_3$ with HCl. (Aluminum hydroxide is an ingredient in the antacid Digel®.)

5.9 ▶ Gases are formed in some metathesis reactions

Sometimes a product of a metathesis reaction is a substance that normally is a gas at room temperature and is not very soluble in water. As a rule, these substances are formed when certain salts react with an acid, or in some cases, a base.

Sulfides and cyanides form gases when they react with acids

Hydrogen sulfide, H_2S, is the compound that gives rotten eggs their foul odor. It is a weak electrolyte and is formed when a strong acid such as HCl is added to a metal sulfide such as sodium sulfide, Na_2S. Hydrogen sulfide has a low solubility in water, so it tends to bubble out of the solution as it forms and thereby escapes as a gas. Once it has escaped, there is no possible way for it to participate in any sort of reverse reaction, so its departure drives the reaction to completion. (The departure of H_2S removes ions, $2H^+$ and S^{2-}.) The molecular, ionic, and net ionic equations for the reaction are as follows:

H_2S can be detected by odor when its concentration is only 0.15 ppb (parts per billion) or 21 μg H_2S/m^3 air.

$$2HCl(aq) + Na_2S(aq) \longrightarrow 2NaCl(aq) + H_2S(g)$$

$$2H^+(aq) + 2Cl^-(aq) + 2Na^+(aq) + S^{2-}(aq) \longrightarrow 2Na^+(aq) + 2Cl^-(aq) + H_2S(g)$$

$$2H^+(aq) + S^{2-}(aq) \longrightarrow H_2S(g)$$

Hydrogen cyanide, an extremely poisonous gas,[5] is formed in similar reactions when an acid such as sulfuric acid is added to a cyanide salt such as KCN. The molecular, ionic, and net ionic equations are

$$H_2SO_4(aq) + 2KCN(aq) \longrightarrow 2HCN(g) + K_2SO_4(aq)$$

$$2H^+(aq) + SO_4^{2-}(aq) + 2K^+(aq) + 2CN^-(aq) \longrightarrow 2HCN(g) + 2K^+(aq) + SO_4^{2-}(aq)$$

$$H^+(aq) + CN^-(aq) \longrightarrow HCN(g)$$

Acids react with carbonates, bicarbonates, sulfites, and bisulfites to give gases

Carbonic acid (H_2CO_3) forms when an acid reacts with either a bicarbonate or a carbonate. For example, consider the reaction of sodium bicarbonate ($NaHCO_3$) with hydrochloric acid, which is pictured in Figure 5.22. As the sodium bicarbonate solution is added to the hydrochloric acid, bubbles of carbon dioxide are released. This is the same reaction that occurs if you take sodium bicarbonate to soothe an upset stomach. Stomach acid is HCl and its reaction with the $NaHCO_3$ both neutralizes the acid and produces CO_2 gas (burp!). The molecular equation for the reaction is

Notice that to obtain the net ionic equation after canceling spectator ions, we've divided each of the coefficients by 2 to obtain the smallest set of whole numbers.

$$HCl(aq) + NaHCO_3(aq) \longrightarrow NaCl(aq) + H_2CO_3(aq)$$

Earlier we mentioned that carbonic acid is too unstable to be isolated in pure form. When it forms in appreciable amounts as a product in a metathesis reaction, it decomposes into its anhydride, the gas CO_2, and water. Carbon dioxide is only slightly soluble in water, so most of the CO_2 bubbles out of the solution. The decomposition reaction is

$$H_2CO_3(aq) \longrightarrow H_2O + CO_2(g)$$

Therefore, the overall molecular equation for the reaction is

$$HCl(aq) + NaHCO_3(aq) \longrightarrow NaCl(aq) + H_2O + CO_2(g)$$

FIGURE 5.22 *The reaction of sodium bicarbonate with hydrochloric acid.* The bubbles contain the gas carbon dioxide.

[5]Hydrogen cyanide is the lethal gas used in executions in the "gas chamber," where it is produced by adding potassium cyanide pellets to sulfuric acid. Most states with capital punishment statutes have abandoned this method of execution in favor of lethal injection administered under general anesthesia.

The ionic equation is

$$H^+(aq) + Cl^-(aq) + Na^+(aq) + HCO_3^-(aq) \longrightarrow Na^+(aq) + Cl^-(aq) + H_2O + CO_2(g)$$

It is this reaction, occurring in blood as it moves through the lungs, by which CO_2 is released for removal from the body as we breathe.

and the net ionic equation is

$$H^+(aq) + HCO_3^-(aq) \longrightarrow H_2O + CO_2(g)$$

Similar results are obtained if we begin with a carbonate instead of a bicarbonate. In this case, hydrogen ions combine with carbonate ions to give H_2CO_3, which subsequently decomposed to water and carbon dioxide.

$$2H^+(aq) + CO_3^{2-}(aq) \longrightarrow H_2CO_3(aq) \longrightarrow H_2O + CO_2(g)$$

The net reaction is

$$2H^+(aq) + CO_3^{2-}(aq) \longrightarrow H_2O + CO_2(g)$$

CHEMISTRY IN PRACTICE

Medications that fizz when water is added, like Alka-Seltzer, rely on the reaction of a bicarbonate with an acid. Without the water, the two key fizz-causing reactants in Alka-Seltzer, citric acid and sodium bicarbonate, are both in the solid state, so the reactants cannot mix. (Aspirin is another ingredient and serves as a pain reliever.) When a tablet is dropped into water, the solids dissolve and the reactant ions gain the freedom to find each other. They react and liberate gaseous carbon dioxide, which provides the fizz. The formula for citric acid is $H_3C_6H_5O_7$ and the molecular equation for the reaction is

$$H_3C_6H_5O_7(aq) + 3NaHCO_3(aq) \longrightarrow$$
$$3CO_2(g) + Na_3C_6H_5O_7(aq) + 3H_2O$$

Acids react with insoluble carbonates to give CO_2

The release of CO_2 by the reaction of a carbonate with an acid is such a strong driving force for reaction that it enables insoluble carbonates to dissolve in acids (strong and weak). The reaction of limestone, $CaCO_3$, with hydrochloric acid is shown in Figure 5.23. The molecular, ionic, and net ionic equations for the reaction are as follows:

$$CaCO_3(s) + 2HCl(aq) \longrightarrow CaCl_2(aq) + CO_2(g) + H_2O$$
$$CaCO_3(s) + 2H^+(aq) + 2Cl^-(aq) \longrightarrow Ca^{2+}(aq) + 2Cl^-(aq) + CO_2(g) + H_2O$$
$$CaCO_3(s) + 2H^+(aq) \longrightarrow Ca^{2+}(aq) + CO_2(g) + H_2O$$

Acids react with sulfites to give SO_2

Another unstable weak acid that decomposes when formed in large amounts is sulfurous acid, H_2SO_3. The reaction is

$$H_2SO_3(aq) \longrightarrow H_2O + SO_2(g)$$

Whenever a solution contains the ions necessary to form H_2SO_3, we can expect a release of the gas SO_2. For example, the net ionic equations for the reaction of bisulfite and sulfite ions with a strong acid are

$$HSO_3^-(aq) + H^+(aq) \longrightarrow H_2O + SO_2(g)$$
$$SO_3^{2-}(aq) + 2H^+(aq) \longrightarrow H_2O + SO_2(g)$$

Bases react with ammonium salts to give NH_3

In writing an equation for a metathesis reaction between NH_4Cl and $NaOH$, one of the products we are tempted to write is NH_4OH—so-called "ammonium hydroxide."

$$NaOH + NH_4Cl \longrightarrow NaCl + NH_4OH$$
(by exchanging cations among anions)

Carbon dioxide forms when solid sodium bicarbonate and citric acid, two ingredients of Alka-Seltzer, dissolve in water and react.

TABLE 5.3	GASES FORMED IN METATHESIS REACTIONS	
Gas	Formed by Reaction of Acids with:	Equation for Formation[a]
H_2S	Sulfides	$2H^+ + S^{2-} \rightarrow H_2S$
HCN	Cyanides	$H^+ + CN^- \rightarrow HCN$
CO_2	Carbonates	$2H^+ + CO_3^{2-} \rightarrow (H_2CO_3) \rightarrow H_2O + CO_2$
	Bicarbonates (hydrogen carbonates)	$H^+ + HCO_3^- \rightarrow (H_2CO_3) \rightarrow H_2O + CO_2$
SO_2	Sulfites	$2H^+ + SO_3^{2-} \rightarrow (H_2SO_3) \rightarrow H_2O + SO_2$
	Bisulfites (hydrogen sulfites)	$H^+ + HSO_3^- \rightarrow (H_2SO_3) \rightarrow H_2O + SO_2$
Gas	Formed by Reaction of Bases with:	Equation for Formation
NH_3	Ammonium salts[b]	$NH_4^+ + OH^- \rightarrow NH_3 + H_2O$

Gases from metathesis reactions

[a]Formulas in parentheses are of unstable compounds that break down according to the continuation of the sequence.

[b]In writing a metathesis reaction, you may be tempted sometimes to write NH_4OH as a formula for "ammonium hydroxide." This compound does not exist. In water, it is nothing more than a solution of NH_3.

However, NH_4OH has never been detected, either in pure form or in an aqueous solution. Instead, "ammonium hydroxide" is simply a solution of NH_3 in water. (Note the equivalence in terms of atoms: $NH_4OH \Leftrightarrow NH_3 + H_2O$.) Therefore, in place of NH_4OH, we write $NH_3 + H_2O$.

$$NaOH(aq) + NH_4Cl(aq) \longrightarrow NaCl(aq) + NH_3(aq) + H_2O$$

The net ionic equation, after eliminating Na^+ and Cl^-, is

$$NH_4^+(aq) + OH^-(aq) \longrightarrow NH_3(g) + H_2O$$

This equation shows that *any* ammonium salt will react with a strong base to yield ammonia. Even though ammonia is very soluble in water, some escapes as a gas and its odor can be easily detected over the reaction mixture.

Table 5.3 contains a summary of the substances that are released as gases in metathesis reactions. You should learn these reactions so you can use them in writing ionic and net ionic equations.

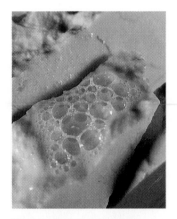

FIGURE 5.23 *Limestone reacts with acid.* Bubbles of CO_2 are formed in the reaction of limestone ($CaCO_3$) with hydrochloric acid.

5.10 ▶ Predicting metathesis reactions—A summary

In the preceding sections, we've described several kinds of metathesis reactions that occur in aqueous solutions. For each of these, we were able to write a net ionic equation. In fact, if there were no reaction, no net ionic equation would exist. Consider, for example, what happens when we mix solutions of sodium chloride and potassium nitrate. Anticipating a reaction, we might assemble a molecular equation by exchanging cations between the anions.

$$NaCl + KNO_3 \longrightarrow KCl + NaNO_3$$

The solubility rules tell us that both "reactants" and both "products" are soluble, so in the ionic equation we write all of them in dissociated form.

$$Na^+(aq) + Cl^-(aq) + K^+(aq) + NO_3^-(aq) \longrightarrow K^+(aq) + Cl^-(aq) + Na^+(aq) + NO_3^-(aq)$$

FACETS OF CHEMISTRY 5.2

Painful Precipitates — Kidney Stones

Each year, more than a million people in the United States are hospitalized because of very painful kidney stone attacks. A kidney stone is a hard mass developed from crystals that separate from the urine and build up on the inner surfaces of the kidney. The formation of the stones is caused primarily by the buildup of Ca^{2+}, $C_2O_4^{2-}$, and PO_4^{3-} ions in the urine. When the concentrations of these ions become large enough, the urine becomes supersaturated with respect to calcium oxalate and/or calcium phosphate and precipitates begin to form (70% to 80% of all kidney stones are made up of calcium oxalate and phosphate). If the crystals remain tiny enough, they can travel through the urinary tract and pass out of the body in the urine without being noticed. Sometimes, however, they continue to grow without being passed and can cause intense pain if they become stuck in the urinary tract.

Kidney stones don't all look alike. Their color depends on what substances are mixed with the inorganic precipi-

tates (e.g., proteins or blood). Most are yellow or brown, as seen in the accompanying photo, but they can be tan, gold, or even black. Stones can be round, jagged, or even have branches. They vary in size from mere specks to pebbles to stones as big as golf balls!

A calcium oxalate kidney stone. Kidney stones such as this can be extremely painful.

Notice that both sides of the equation are the same except for the order in which the ions are written. If we cancel spectator ions, there is nothing left. *There is no net ionic equation, because there is no net reaction.* The two solutions combined simply gives a mixture of all four ions.

We now see that *the existence or nonexistence of a net ionic equation reveals whether or not an ionic reaction occurs between a pair of reactants.* With this in mind, we can review those factors that we know will yield a net ionic equation, and therefore a reaction.

A net ionic equation exists when a precipitate forms from soluble reactants

A net ionic equation will exist if a precipitate is formed by mixing two soluble salts. If both products are soluble strong electrolytes, a net ionic equation will not exist and no reaction occurs.

A net ionic equation exists when water or a weak electrolyte forms

Earlier in this chapter we discussed acid–base neutralization in which acids and bases react to form water and a salt. In these reactions it is the formation of water that is the driving force. In fact, formation of water in acid–base neutralizations is such a strong driving force for reaction that strong and weak acids will even react with *insoluble* oxides and hydroxides.

The formation of water in a reaction is a special case of a more general driving force for ionic reactions—the formation of a weak electrolyte from reactants that are strong electrolytes. Consider the reaction between solutions of hydrochloric acid, HCl, and sodium formate, $NaCHO_2$. The molecular equation, written as a metathesis, is

$$HCl(aq) + NaCHO_2(aq) \longrightarrow HCHO_2(aq) + NaCl(aq)$$

The reactants are both strong electrolytes; HCl is a strong acid and $NaCHO_2$ is a salt. Among the products, we recognize NaCl as a salt, which is also a strong electrolyte.

The acid $HCHO_2$ is not on our list of strong acids, so we can expect that it is a weak acid, and therefore a weak electrolyte. The ionic equation for the reaction, then, is

$$H^+(aq) + Cl^-(aq) + Na^+(aq) + CHO_2^-(aq) \longrightarrow HCHO_2(aq) + Na^+(aq) + Cl^-(aq)$$

Removing spectator ions gives the net ionic equation,

$$H^+(aq) + CHO_2^-(aq) \longrightarrow HCHO_2(aq)$$

A net ionic equation exists when a gas is formed

When a gas is formed in a reaction, it leaves the solution, and by doing so serves as the driving force for the reaction. This permits even insoluble compounds such as calcium carbonate to dissolve in acids. For example, the net ionic equation for the reaction of $CaCO_3$ with HCl is

$$CaCO_3(s) + 2H^+(aq) \longrightarrow Ca^{2+}(aq) + CO_2(g) + H_2O$$

Notice, once again, the existence of a net ionic equation for the reaction.

Summary

The factors we've discussed above can be summarized as follows:

> A net ionic equation will exist when
>
> 1. A precipitate forms from a solution of soluble reactants.
> 2. Water forms in the reaction of an acid and a base.
> 3. A weak electrolyte forms from a solution of strong electrolytes.
> 4. A gas is formed that escapes from the reaction mixture.

To predict whether an ionic reaction will occur between a pair of reactants, we proceed as follows:

- We begin by writing a molecular equation in the form of a metathesis reaction, exchanging anions between the two cations.
- We next translate the molecular equation into an ionic equation.
- We cancel any spectator ions.

If a net ionic equation remains after removing the spectator ions, we conclude that a net reaction *does* occur. On the other hand, if there is nothing left after canceling spectator ions, *there is no net reaction*.

Let's look at several more examples that illustrate how we use these criteria to establish whether a reaction occurs between potential reactants.

What reaction, if any, occurs between $Ba(OH)_2$ and $(NH_4)_2SO_4$ in an aqueous system?

ANALYSIS: To answer a question like this, we proceed in several steps. First, we consider the possible products of a metathesis reaction and write an equation. Then, for each substance in the "reaction," we ask ourselves, "Is it a weak acid? Does it decompose to give a gas? Is it a soluble or insoluble salt?" The answers to these questions allow us to write the ionic equation. Finally, we eliminate spectator ions to determine the net ionic equation. If no net ionic equation exists, no reaction occurs.

SOLUTION: We begin by exchanging cations between the two reactants, which gives as possible products $BaSO_4$ and NH_4OH. The balanced preliminary equation is

$$Ba(OH)_2 + (NH_4)_2SO_4 \longrightarrow BaSO_4 + 2NH_4OH$$

EXAMPLE 5.7

Predicting a Reaction

Next, we scan the equation looking for acids, either weak or strong. The formulas for acids would begin with H, so none of the compounds here are acids. Next, we look for any of those substances that are, or can produce, gases (namely, H_2S, HCN, H_2CO_3, H_2SO_3, and NH_4OH). Success! We find NH_4OH, which we know should really be $NH_3(g) + H_2O$. Let's revise the balanced molecular equation and then apply the solubility rules.

> Notice that we've changed NH_4OH into NH_3 and H_2O and added coefficients.

$$\underset{\substack{\text{soluble}\\ \text{(Rule 5)}}}{Ba(OH)_2} + \underset{\substack{\text{soluble}\\ \text{(Rule 2)}}}{(NH_4)_2SO_4} \longrightarrow \underset{\substack{\text{insoluble}\\ \text{(Rule 4)}}}{BaSO_4} + 2NH_3(g) + 2H_2O$$

Now we can write the ionic equation, expressing soluble ionic compounds in dissociated form.

$$Ba^{2+}(aq) + 2OH^-(aq) + 2NH_4^+(aq) + SO_4^{2-}(aq) \longrightarrow BaSO_4(s) + 2NH_3(g) + 2H_2O$$

Notice that there are no ions that appear the same on both sides, so there are no spectator ions to eliminate. Therefore, the ionic equation is also the net ionic equation!

Is the Answer Reasonable?

There are some common errors that people make in working problems of this kind, so it is important to double-check. First, *proceed carefully.* Be sure you've written the formulas of the products correctly. (If you need review, you might look at Example 5.3 on page 166.) Look for weak acids. (You need to know the list of strong ones; if an acid isn't on the list, it's a weak acid.) Look for gases or substances that decompose into gases. (Be sure you've studied Table 5.3.). Check for insoluble compounds. (You need to know the solubility rules in Table 5.2.) If you've learned what is expected of you, and checked each step, it is likely your result is correct.

EXAMPLE 5.8
Predicting Reactions and Writing Their Equations

What reaction (if any) occurs when solutions of ammonium carbonate, $(NH_4)_2CO_3$, and propionic acid, $HC_3H_5O_2$, are mixed?

ANALYSIS: As before, we write a potential metathesis equation in molecular form. Then we examine the reactants and products to see if any are weak electrolytes or substances that give gases. We also look for soluble or insoluble ionic compounds. Then we form the ionic equation and search for spectator ions, which we eliminate to obtain the net ionic equation.

SOLUTION: We begin by constructing a molecular equation, treating the reaction as a metathesis. For the acid, we take the cation to be H^+ and the anion to be $C_3H_5O_2^-$. Therefore, exchanging cations between the two anions we can obtain the following balanced molecular equation:

$$(NH_4)_2CO_3 + 2HC_3H_5O_2 \longrightarrow 2NH_4C_3H_5O_2 + H_2CO_3$$

In the statement of the problem we are told that we are working with a *solution* of $HC_3H_5O_2$, so we know it's soluble. Also, it is not on the list of strong acids, so we expect that it is a weak acid; we will write it in molecular form when we form the ionic equation.

Next, we recognize that H_2CO_3 decomposes into $CO_2(g)$ and H_2O. Let's rewrite the molecular equation taking this into account.

$$(NH_4)_2CO_3 + 2HC_3H_5O_2 \longrightarrow 2NH_4C_3H_5O_2 + CO_2(g) + H_2O$$

Next, we need to determine which of the ionic substances are soluble. The solubility rules tell us that all ammonium salts are soluble, and we know that all salts are strong electrolytes. Therefore, we will write $(NH_4)_2CO_3$ and $NH_4C_3H_5O_2$ in dissociated form. Now we are ready to expand the molecular equation into the ionic equation.

$$2NH_4^+(aq) + CO_3^{2-}(aq) + 2HC_3H_5O_2(aq) \longrightarrow$$
$$2NH_4^+(aq) + 2C_3H_5O_2^-(aq) + CO_2(g) + H_2O$$

The only spectator ion is NH_4^+. Dropping this gives the net ionic equation.

$$CO_3^{2-}(aq) + 2HC_3H_5O_2(aq) \longrightarrow 2C_3H_5O_2^-(aq) + CO_2(g) + H_2O$$

Is the Answer Reasonable?
We perform the same checks here as in the preceding example, and they tell us our answer is correct.

What reaction (if any) occurs in water between KNO_3 and NH_4Cl?

ANALYSIS: Let's proceed by writing molecular, ionic, and net ionic equations as in the preceding example.

SOLUTION: First we write the molecular equation, being sure to construct correct formulas for the products.

$$KNO_3 + NH_4Cl \longrightarrow KCl + NH_4NO_3$$

Looking over the substances in the equation, we don't find any that are weak acids or that decompose to give gases. Next, we check solubilities.

Solubility Rule 2 tells us that both KNO_3 and NH_4Cl are soluble. By solubility Rules 1 and 2, both products are also soluble in water. The anticipated molecular equation is therefore

$$\underset{\text{soluble}}{KNO_3(aq)} + \underset{\text{soluble}}{NH_4Cl(aq)} \longrightarrow \underset{\text{soluble}}{KCl(aq)} + \underset{\text{soluble}}{NH_4NO_3(aq)}$$

and the ionic equation is

$$K^+(aq) + NO_3^-(aq) + NH_4^+(aq) + Cl^-(aq) \longrightarrow$$
$$K^+(aq) + Cl^-(aq) + NH_4^+(aq) + NO_3^-(aq)$$

Notice that the right side of the equation is the same as the left side except for the order in which the ions are written. When we eliminate spectator ions, everything goes. There is no net ionic equation, which means there is no net reaction.

Is the Answer Reasonable?
Once again, we perform the same checks here as in Example 5.7, and they tell us our answer is right.

PRACTICE EXERCISE 18: Predict whether a reaction will occur in aqueous solution between the following pairs of substances. If a reaction does occur, derive the net ionic equation. (a) $KCHO_2$ and HCl, (b) $CuCO_3$ and $HC_2H_3O_2$, (c) $Ca(C_2H_3O_2)_2$ and $AgNO_3$, and (d) $NaOH$ and $NiCl_2$.

EXAMPLE 5.9

Predicting Reactions and Writing Their Equations

5.11 ▶ The composition of a solution is described by its concentration

Earlier we noted that we use the term *concentration* to describe the composition of a solution, either as the ratio of solute to solvent or as the ratio of solute to solution. Percentage concentration (grams of solute per 100 g of solution) was given as an example. When we have to deal with the stoichiometry of reactions in solution, however, percentage concentration is not a convenient way to express concentrations of solutes. Instead, we express the amount of solute in moles and the amount of solution in liters.

Molar concentration (molarity) is a ratio of moles to liters of solution

The **molar concentration,** or **molarity** (abbreviated M), of a solution is defined as *the number of moles of solute per liter of solution*. It is a ratio of the moles of solute to the volume of the solution expressed in liters.

Molar concentration

$$\text{molarity } (M) = \frac{\text{moles of solute}}{\text{liters of solution}} \qquad (5.2)$$

Thus, a solution that contains 0.100 mol of NaCl in 1.00 L has a molarity of 0.100 M, and we would refer to this solution as 0.100 *molar* NaCl or as 0.100 M NaCl. The same concentration would result if we dissolved 0.0100 mol of NaCl in 0.100 L (100 mL) of solution, because the *ratio* of moles of solute to volume of solution is the same.

$$\frac{0.100 \text{ mol NaCl}}{1.00 \text{ L NaCl soln}} = \frac{0.0100 \text{ mol NaCl}}{0.100 \text{ L NaCl soln}} = 0.100 \text{ } M \text{ NaCl}$$

Molarity is particularly useful because it lets us obtain a given number of moles of a substance simply by measuring a volume of a previously prepared solution, and this measurement is quick and easy to do in the lab. If we had a stock supply of 0.100 M NaCl solution, for example, and we needed 0.100 mol of NaCl for a reaction, we would simply measure out 1.00 L of the solution because in this volume there is 0.100 mol of NaCl.

We need not (and seldom do) work with whole liters of solutions. It's the *ratio* of the number of moles of solute to the number of liters of solution, not the total volume, that matters. Suppose, for example, that we dissolved 0.0200 mol of sodium chromate, Na_2CrO_4, in a total final volume of 0.250 L (250 mL). The molarity of the solution would be found by the following simple calculation using Equation 5.2:

$$\text{molarity} = \frac{0.0200 \text{ mol } Na_2CrO_4}{0.250 \text{ L } Na_2CrO_4 \text{ soln}} = 0.0800 \text{ } M \text{ } Na_2CrO_4$$

A *volumetric flask* is a narrow-necked bottle having an etched mark high on its neck. When filled to the mark, the flask contains the volume given by the flask's label. Figure 5.24 shows how to use a 250 mL volumetric flask to prepare a solution of known molarity.

(*a*)   (*b*)   (*c*)   (*d*)   (*e*)

FIGURE 5.24 *The preparation of a solution having a known molarity.* (*a*) A 250 mL volumetric flask, one of a number of sizes available for preparing solutions. When filled to the line etched around its neck, this flask contains exactly 250 mL of solution. The flask here already contains a weighed amount of solute. (*b*) Water is being added. (*c*) The solute is brought completely into solution before the level is brought up to the narrow neck of the flask. (*d*) More water is added to bring the level of the solution to the etched line. (*e*) The flask is stoppered and then inverted several times to mix its contents thoroughly.

Molarity is a conversion factor relating moles of solute and volume of a solution

Whenever we have to deal with a problem that involves an amount of a chemical and a volume of a solution of that substance, the critical link in solving the problem is the molarity.

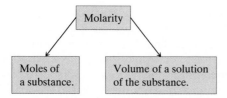

The reason is that molarity provides the conversion factors we need to convert between moles and volume (either in liters or milliliters). Consider, for example, a solution labeled 0.100 M NaCl. The unit M always translates to mean "moles per liter," so we can write

$$0.100\ M\ \text{NaCl} = \frac{0.100\ \text{mol NaCl}}{1.00\ \text{L soln}}$$

This gives us an equivalence relationship between "mol NaCl" and "L soln" that we can use to form two conversion factors.[6]

$$0.100\ \text{mol NaCl} \Longleftrightarrow 1.00\ \text{L soln}$$

$$\frac{0.100\ \text{mol NaCl}}{1.00\ \text{L NaCl soln}} \qquad \frac{1.00\ \text{L NaCl soln}}{0.100\ \text{mol NaCl}}$$

To study the effect of dissolved salt on the rusting of an iron sample, a student prepared a solution of NaCl by dissolving 1.461 g of NaCl in a 250.0 mL volumetric flask. What is the molarity of this solution?

ANALYSIS: To solve a problem such as this, it is imperative that you know the definition: molarity is the ratio of *moles of solute* to *liters of solution*. To calculate this ratio, we need two pieces of information, moles of solute and the volume of the solution, expressed in liters. Then, we just arrange them as a ratio.

$$\text{molarity} = \frac{?\ \text{mol NaCl}}{?\ \text{L soln}}$$

To obtain the numerator, we have to convert 1.461 g of NaCl to moles of NaCl. We also have to convert 250.0 mL into liters. Then we simply divide the number of moles by the number of liters to find the molarity.

SOLUTION: The number of moles of NaCl is found using the formula mass of NaCl, 58.443.

$$1.461\ \text{g NaCl} \times \frac{1\ \text{mol NaCl}}{58.443\ \text{g NaCl}} = 0.02500\ \text{mol NaCl}$$

EXAMPLE 5.10

Calculating the Molarity of a Solution

---

[6]Some students find it easier to translate the "1.00 L" part of these factors into the equivalent 1000 mL here rather than convert between liters and milliliters at some other stage of the calculation. Factors such as the two above, therefore, can be rewritten as follows whenever it is convenient. (Remember that "1000" in the following is regarded as having an infinite number of significant figures because, standing as it does for 1 L, it is part of the definition of molarity and is an exact number.)

$$\frac{0.100\ \text{mol NaCl}}{1000\ \text{mL NaCl soln}} \qquad \text{and} \qquad \frac{1000\ \text{mL NaCl soln}}{0.100\ \text{mol NaCl}}$$

If necessary, practice converting between liters and milliliters. It's a task you will have to perform frequently.

To find the volume of the solution in liters, we move the decimal three places to the left, so 250.0 mL equals 0.2500 L.

The ratio of moles to liters, therefore, is

$$\frac{0.02500 \text{ mol NaCl}}{0.2500 \text{ L}} = 0.1000 \ M \text{ NaCl}$$

Is the Answer Reasonable?

Let's use our answer to do a rough calculation of the amount of NaCl in the solution. If our answer is right, we should find a value not too far from the amount given in the problem (1.461 g). If we round the formula mass of NaCl to 60 and use 0.1 *M* as an approximate concentration, then 1 L of the solution contains 0.1 mol of NaCl, or approximately 6 g of NaCl (one-tenth of 60 g). But 250 mL is only one-fourth of a liter, so the mass of NaCl will be approximately one-fourth of 6 g, or about 1.5 g. This is pretty close to the amount that was given in the problem, so our answer is probably correct.

PRACTICE EXERCISE 19: As part of a study to determine if just the chloride ion in NaCl accelerates the rusting of iron, a student decided to see if iron rusted as rapidly when exposed to a solution of Na_2SO_4. This solution was made by dissolving 3.550 g of Na_2SO_4 in water using a 100.0 mL volumetric flask to adjust the final volume. What is the molarity of this solution?

EXAMPLE 5.11
Using Molar Concentrations

How many milliliters of 0.250 *M* NaCl solution must be measured to obtain 0.100 mol of NaCl?

ANALYSIS: We can restate the problem as follows:

$$0.100 \text{ mol NaCl} \Longleftrightarrow ? \text{ mL soln}$$

To relate moles and volume, the tool we use is the molarity.

$$0.250 \ M \text{ NaCl} = \frac{0.250 \text{ mol NaCl}}{1 \text{ L NaCl soln}}$$

The fraction on the right relates moles of NaCl to liters of the solution, which we can express as an equivalence.

$$0.250 \text{ mol NaCl} \Longleftrightarrow 1 \text{ L NaCl soln}$$

The equivalence allows us to construct two conversion factors.

$$\frac{0.250 \text{ mol NaCl}}{1 \text{ L NaCl soln}} \quad \text{and} \quad \frac{1.00 \text{ L NaCl soln}}{0.250 \text{ mol NaCl}}$$

To obtain the solution, we select the one that will allow us to cancel the unit "mol NaCl."

SOLUTION: We operate with the second factor on 0.100 mol NaCl.

$$0.100 \text{ mol NaCl} \times \frac{1.00 \text{ L NaCl soln}}{0.250 \text{ mol NaCl}} = 0.400 \text{ L of } 0.250 \ M \text{ NaCl}$$

Because 0.400 L corresponds to 400 mL, 400 mL of 0.250 *M* NaCl provides 0.100 mol of NaCl.

Is the Answer Reasonable?

One liter provides 0.250 mol, so we need somewhat less than half of a liter (1000 mL) to obtain just 0.100 mol. The answer, 400 mL, is reasonable.

PRACTICE EXERCISE 20: To continue the experiment on rusting (Practice Exercise 19), a student decided to see how rapidly iron rusted when chloride ion but not sodium ion was present. How many milliliters of 0.150 *M* KCl solution must be taken to provide 0.0500 mol of KCl?

In the laboratory, we often must prepare a solution with a specific molarity. Usually, we select the volume of solution that we'll make and then calculate how much solute must be used. Example 5.12 illustrates the kind of calculation required and also demonstrates a useful relationship. *Whenever you know **both** the volume and the molarity of a solution, you can calculate the number of moles of solute.* This is because the product of molarity and volume (expressed in liters) equals the moles of solute.

$$\text{molarity} \times \text{volume (L)} = \text{moles of solute} \qquad (5.3)$$

TOOLS

Molarity × volume (L) = moles

$$\frac{\text{mol solute}}{\text{L soln}} \times \text{L soln} = \text{mol solute}$$

This equation is a useful tool for solving problems in solution stoichiometry.

Strontium nitrate, $Sr(NO_3)_2$, is used in fireworks to produce brilliant red colors. Suppose a chemist needs to prepare 250 mL of 0.100 M $Sr(NO_3)_2$ solution. How many grams of strontium nitrate are required?

EXAMPLE 5.12

Preparing a Solution with a Known Molarity

ANALYSIS: The key to solving this problem is realizing that we need to determine how much of the solute will be in the solution after it's prepared. Because we know both the volume and molarity of the final solution, we can calculate the number of moles of $Sr(NO_3)_2$ that will be in it. Once we know the number of moles of $Sr(NO_3)_2$, we can calculate the mass using the formula mass of the salt.

SOLUTION: You've learned that the product of molarity and volume equals moles of solute. The stated molarity, 0.100 M $Sr(NO_3)_2$, translates to

$$0.100 \ M \ Sr(NO_3)_2 = \frac{0.100 \text{ mol } Sr(NO_3)_2}{1.00 \text{ L } Sr(NO_3)_2 \text{ soln}}$$

The volume 250 mL converts to 0.250 L. Therefore, multiplying the molarity by the volume in liters takes the following form:

$$\underset{\text{molarity}}{\underbrace{\frac{0.100 \text{ mol } Sr(NO_3)_2}{1.00 \text{ L } Sr(NO_3)_2 \text{ soln}}}} \times \underset{\text{volume (L)}}{\underbrace{0.250 \text{ L } Sr(NO_3)_2 \text{ soln}}} = \underset{\text{moles of solute}}{\underbrace{0.0250 \text{ mol } Sr(NO_3)_2}}$$

Finally, we convert from moles to grams using the formula mass of $Sr(NO_3)_2$, which is 211.62.

$$0.0250 \text{ mol } Sr(NO_3)_2 \times \frac{211.62 \text{ g } Sr(NO_3)_2}{1 \text{ mol } Sr(NO_3)_2} = 5.29 \text{ g } Sr(NO_3)_2$$

Thus, to prepare the solution, 5.29 g of $Sr(NO_3)_2$ would be weighed out, placed in a 250.0 mL volumetric flask, and then dissolved in water. More water would be added to make the solution reach the flask's etched mark.

We could also have set this up as a chain calculation as follows, with the conversion factors strung together:

$$0.250 \text{ L } Sr(NO_3)_2 \text{ soln} \times \frac{0.100 \text{ mol } Sr(NO_3)_2}{1.00 \text{ L } Sr(NO_3)_2 \text{ soln}} \times \frac{211.62 \text{ g } Sr(NO_3)_2}{1 \text{ mol } Sr(NO_3)_2}$$

$$= 5.29 \text{ g } Sr(NO_3)_2$$

Is the Answer Reasonable?
If we were working with a full liter of this solution, it would contain 0.1 mol of $Sr(NO_3)_2$. The formula mass of the salt is 211.62 g mol^{-1}, so 0.1 mol is slightly more than 20 g. However, we are working with just a quarter of a liter (250 mL), so the amount of $Sr(NO_3)_2$ needed is slightly more than a quarter of 20 g, or 5 g. The answer, 5.29 g, is close to this, so it makes sense. (The more you try to work these rough

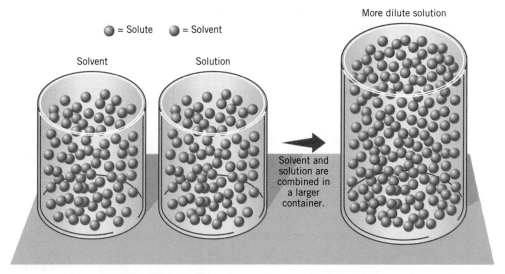

FIGURE 5.25 *Diluting a solution.* When solvent is added to a solution, the solute particles become more spread out and the solution becomes more dilute. The concentration of the solute in the solution becomes smaller, even though the *total* amount of solute remains constant.

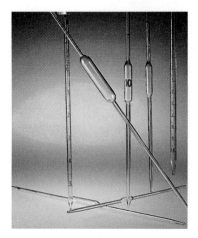

FIGURE 5.26 *Volumetric glassware.* Shown here is a selection of volumetric pipets, each marked for a particular volume and each with an etched line to designate where the filling of the pipet should stop.

calculations in your head, the quicker and surer you will become at understanding and solving stoichiometry problems.)

PRACTICE EXERCISE 21: How many grams of $AgNO_3$ are needed to prepare 250 mL of 0.0125 *M* $AgNO_3$ solution?

Diluting a solution reduces the concentration

It is not always necessary to begin with a solute in pure form to prepare a solution with a known molarity. The solute can already be dissolved in a solution of relatively high concentration, and this can be *diluted* to make a solution of the desired lower concentration. Dilution is accomplished by adding more solvent to the solution, which spreads the solute through a larger volume and causes the concentration (the amount per unit volume) to decrease (Figure 5.25). The choice of apparatus used depends on the precision required. If high precision is necessary, pipets (Figure 5.26) and volumetric flasks are used. If we can do with less precision, we might use graduated cylinders instead.

Because the process of diluting a solution involves distributing a given amount of solute throughout a larger volume, the amount of solute remains constant. This means that the product of molarity and volume, which equals the moles of solute, must be the same for both the concentrated and diluted solution.

$$\left(\begin{array}{c}\text{volume of}\\\text{dilute solution}\\\textit{to be prepared}\end{array}\right) \times M_{\text{dilute}} = \left(\begin{array}{c}\text{volume of}\\\text{concentrated solution}\\\textit{to be used}\end{array}\right) \times M_{\text{concd}}$$

moles of solute in the dilute solution   moles of solute in concentrated solution

TOOLS

Diluting solutions of known molarity

Or,

$$V_{\text{dil}} \cdot M_{\text{dil}} = V_{\text{concd}} \cdot M_{\text{concd}} \tag{5.4}$$

Any units can be used for volume in Equation 5.4 provided that the volume units are the same on both sides of the equation. We thus normally solve dilution problems using *milliliters* directly in Equation 5.4.

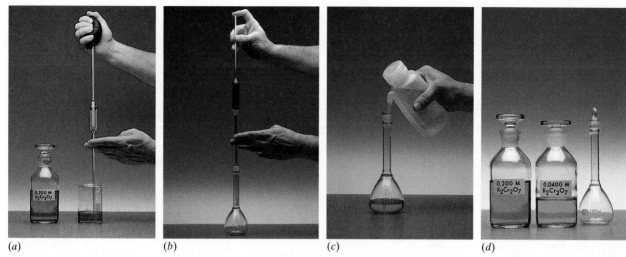

(a) (b) (c) (d)

FIGURE 5.27 *Preparing a solution by dilution.* (*a*) The calculated volume of the more concentrated solution is withdrawn from the stock solution by means of a volumetric pipet. (*b*) The solution is allowed to drain entirely from the pipet into the volumetric flask. (*c*) Water is added to the flask, the contents are mixed, and the final volume is brought up to the etched mark on the narrow neck of the flask. (*d*) The new solution is put into a labeled container.

How can we prepare 100 mL of 0.0400 M $K_2Cr_2O_7$ from 0.200 M $K_2Cr_2O_7$?

ANALYSIS: This is the way such a question comes up in the lab, but what it is really asking is "How many milliliters of 0.200 M $K_2Cr_2O_7$ (the more concentrated solution) must be diluted to give a solution with a final volume of 100 mL and a final molarity of 0.0400 M?" Once we see the question this way, we realize that Equation 5.4 applies.

SOLUTION: It's a good idea to assemble the data first, noting what is missing (and therefore what has to be calculated).

$$V_{dil} = 100 \text{ mL} \qquad M_{dil} = 0.0400 \ M$$
$$V_{concd} = ? \qquad M_{concd} = 0.200 \ M$$

Next, we use Equation 5.4.

$$100 \text{ mL} \times 0.0400 \ M = V_{concd} \times 0.200 \ M$$

Solving for V_{concd} gives

$$V_{concd} = \frac{100 \text{ mL} \times 0.0400 \ M}{0.200 \ M}$$
$$= 20.0 \text{ mL}$$

Therefore, the answer to the question as asked is: we would withdraw 20.0 mL of 0.200 M $K_2Cr_2O_7$, place it in a 100 mL volumetric flask, and then add water until the final volume is 100 mL. (See Figure 5.27.)

Is the Answer Reasonable?
Think about the *magnitude* of the dilution, from 0.2 M to 0.04 M. Notice that the concentrated solution is five times as concentrated as the dilute solution (5 × 0.04 = 0.2). To reduce the concentration by a factor of 5 requires that we increase the volume by a factor of 5, and we see that 100 mL is 5 times 20 mL. The answer appears to be correct.

PRACTICE EXERCISE 22: How could 100 mL of 0.125 M H_2SO_4 solution be made from 0.500 M H_2SO_4 solution?

EXAMPLE 5.13
Preparing a Solution of Known Molarity by Dilution

5.12 ▶ **Molarity is used for problems in solution stoichiometry**

When we deal quantitatively with reactions in solution, we often work with volumes of solutions and molarity.

One of the solids present in photographic film is silver bromide, AgBr. This compound is quite insoluble in water, and one way to prepare it is to mix solutions of two water-soluble compounds, silver nitrate and calcium bromide. Suppose we wished to prepare AgBr by the following precipitation reaction:

$$2AgNO_3(aq) + CaBr_2(aq) \longrightarrow 2AgBr(s) + Ca(NO_3)_2(aq)$$

How many milliliters of 0.125 M CaBr$_2$ solution must be used to react with the solute in 50.0 mL of 0.115 M AgNO$_3$?

ANALYSIS: As you learned in the last chapter, when we have a stoichiometry problem dealing with a chemical reaction, the tool that relates the amounts of the substances is their coefficients in the equation. For this problem, therefore, we can write

$$2 \text{ mol AgNO}_3 \Leftrightarrow 1 \text{ mol CaBr}_2$$

However, we're not given moles directly. Instead we have molarities and the volume of the AgNO$_3$ solution. The critical link in solving the problem is recognizing that the molarity and volume of the AgNO$_3$ solution provide a path to finding the number of moles of AgNO$_3$.

Knowing that we will use molarity as a tool in working the problem, let's outline the path to the answer. We can calculate the moles of AgNO$_3$ by multiplying the volume and molarity of the AgNO$_3$ solution. We then use the coefficients in the equation to translate to moles of CaBr$_2$. Finally, we use the molarity of the CaBr$_2$ solution as a conversion factor to find the volume of the solution needed. The calculation flow will look like the following:

Silver bromide (AgBr) precipitates when solutions of calcium bromide and silver nitrate are mixed.

$$\text{mol AgNO}_3 \xrightarrow{\times \frac{1 \text{ mol CaBr}_2}{2 \text{ mol AgNO}_3}} \text{mol CaBr}_2$$

$$V_{\text{AgNO}_3} \times M_{\text{AgNO}_3} \nearrow \qquad \qquad \searrow \times \frac{1.00 \text{ L CaBr}_2 \text{ soln}}{0.125 \text{ mol CaBr}_2}$$

AgNO$_3$ soln   **CaBr$_2$ soln**

Given: 50.0 mL of   *To find:*   ? mL of
          0.115 M AgNO$_3$ 0.125 M CaBr$_2$

SOLUTION: First, we find the moles of AgNO$_3$ taken. Changing 50.0 mL to 0.050 L,

$$0.0500 \text{ L AgNO}_3 \text{ soln} \times \underset{\text{volume (L)}}{\Bigg|} \times \frac{0.115 \text{ mol AgNO}_3}{1.00 \text{ L AgNO}_3 \text{ soln}} \underset{\text{molarity}}{\Bigg|} = 5.75 \times 10^{-3} \text{ mol AgNO}_3$$

Next, we use the coefficients of the equation to calculate the amount of CaBr$_2$ required.

$$5.75 \times 10^{-3} \text{ mol AgNO}_3 \times \frac{1 \text{ mol CaBr}_2}{2 \text{ mol AgNO}_3} = 2.88 \times 10^{-3} \text{ mol CaBr}_2$$

Finally we calculate the volume (mL) of 0.125 M CaBr$_2$ that contains this many moles of CaBr$_2$. Here we use the fact that the molarity of the CaBr$_2$ solution, 0.125 M, gives two possible conversion factors:

$$\frac{0.125 \text{ mol CaBr}_2}{1.00 \text{ L CaBr}_2 \text{ soln}} \quad \text{and} \quad \frac{1.00 \text{ L CaBr}_2 \text{ soln}}{0.125 \text{ mol CaBr}_2}$$

We use the one that cancels the unit "mol CaBr$_2$."

$$2.88 \times 10^{-3} \text{ mol CaBr}_2 \times \frac{1.00 \text{ L CaBr}_2 \text{ soln}}{0.125 \text{ mol CaBr}_2} = 0.0230 \text{ L CaBr}_2 \text{ soln}$$

Thus 0.0230 L, or 23.0 mL, of 0.125 M CaBr$_2$ has enough solute to combine with the AgNO$_3$ in 50.0 mL of 0.115 M AgNO$_3$.

Is the Answer Reasonable?
The molarities of the two solutions are about the same, but only 1 mol of $CaBr_2$ is needed for each 2 mol $AgNO_3$. Therefore, the volume of $CaBr_2$ solution needed (23.0 mL) should be about half the volume of $AgNO_3$ solution taken (50.0 mL), which it is.

PRACTICE EXERCISE 23: How many milliliters of 0.124 *M* NaOH contain enough NaOH to react with the H_2SO_4 in 15.4 mL of 0.108 *M* H_2SO_4 according to the following equation?

$$2NaOH(aq) + H_2SO_4(aq) \longrightarrow Na_2SO_4(aq) + 2H_2O$$

Net ionic equations can be used in stoichiometric calculations

In the preceding problem, we worked with a molecular equation in solving a stoichiometry problem. Ionic and net ionic equations can also be used, but this also requires that we work with the concentrations of the ions in solution.

Calculating concentrations of ions in solutions of electrolytes

The concentrations of the ions in a solution of an electrolyte are obtained from the formula and molar concentration of the solute. For example, suppose we are working with a solution labeled "0.10 *M* $CaCl_2$." In 1.0 L of this solution there is dissolved 0.10 mol of $CaCl_2$, which is fully dissociated into Ca^{2+} and Cl^- ions.

$$CaCl_2 \longrightarrow Ca^{2+} + 2Cl^-$$

From the stoichiometry of the dissociation, we see that 1 mol Ca^{2+} and 2 mol Cl^- are formed from each 1 mol $CaCl_2$. Therefore, 0.10 mol $CaCl_2$ will yield 0.10 mol Ca^{2+} and 0.20 mol Cl^-. In 0.10 *M* $CaCl_2$, then, the concentration of Ca^{2+} is 0.10 *M* and the concentration of Cl^- is 0.20 *M*. *Notice that the concentration of a particular ion equals the concentration of the salt multiplied by the number of ions of that kind in one formula unit of the salt.*

The solution doesn't actually contain any $CaCl_2$, even though this is the solute used to prepare the solution. Instead, the solution contains Ca^{2+} and Cl^- ions.

What are the molar concentrations of the ions in 0.20 *M* $Al_2(SO_4)_3$?

ANALYSIS: The concentrations of the ions are determined by the stoichiometry of the salt. Therefore, we determine the number of ions of each kind formed from one formula unit of $Al_2(SO_4)_3$. These values are then used along with the given concentration of the salt to calculate the ion concentrations.

SOLUTION: When $Al_2(SO_4)_3$ dissolves, it dissociates as follows:

$$Al_2(SO_4)_3(s) \longrightarrow 2Al^{3+}(aq) + 3SO_4^{2-}(aq)$$

Each formula unit of $Al_2(SO_4)_3$ yields two Al^{3+} ions and three SO_4^{2-} ions. Therefore, 0.20 mol $Al_2(SO_4)_3$ yields 0.40 mol Al^{3+} and 0.60 mol SO_4^{2-}, and we conclude that the solution contains 0.40 *M* Al^{3+} and 0.60 *M* SO_4^{2-}.

EXAMPLE 5.15
Calculating the Concentrations of Ions in a Solution

Is the Answer Reasonable?
The answers here have been obtained by simple mole reasoning. We could have found the answers in a more formal manner using the factor-label method. We can start with the given concentration and proceed as follows:

$$\frac{0.20 \text{ mol } Al_2(SO_4)_3}{1.0 \text{ L soln}} \times \frac{2 \text{ mol } Al^{3+}}{1 \text{ mol } Al_2(SO_4)_3} = \frac{0.40 \text{ mol } Al^{3+}}{1.0 \text{ L soln}} = 0.40 \text{ } M \text{ } Al^{3+}$$

$$\frac{0.20 \text{ mol } Al_2(SO_4)_3}{1.0 \text{ L soln}} \times \frac{3 \text{ mol } SO_4^{2-}}{1 \text{ mol } Al_2(SO_4)_3} = \frac{0.60 \text{ mol } SO_4^{2-}}{1.0 \text{ L soln}} = 0.60 \text{ } M \text{ } SO_4^{2-}$$

Study both methods. With just a little practice, you will have little difficulty with the reasoning approach that we used first.

EXAMPLE 5.16

Calculating the Concentration of a Salt from the Concentration of One of Its Ions

A student found that the sulfate ion concentration in a solution of $Al_2(SO_4)_3$ was 0.90 M. What was the concentration of $Al_2(SO_4)_3$ in the solution?

ANALYSIS: Once again, we use the formula of the salt to determine the number of ions released when it dissociates. This time we use the information to work backward to find the salt concentration.

SOLUTION: Let's set up the problem using the factor-label method to be sure of our procedure. We will use the fact that 1 mol $Al_2(SO_4)_3$ yields 3 mol SO_4^{2-} in solution.

$$1 \text{ mol } Al_2(SO_4)_3 \Longleftrightarrow 3 \text{ mol } SO_4^{2-}$$

Therefore,

$$\frac{0.90 \text{ mol } SO_4^{2-}}{1.0 \text{ L soln}} \times \frac{1 \text{ mol } Al_2(SO_4)_3}{3 \text{ mol } SO_4^{2-}} = \frac{0.30 \text{ mol } Al_2(SO_4)_3}{1.0 \text{ L soln}} = 0.30 \text{ } M \text{ } Al_2(SO_4)_3$$

The concentration of $Al_2(SO_4)_3$ is 0.30 M.

Is the Answer Reasonable?
We'll use the reasoning approach to check our answer. We know that 1 mol $Al_2(SO_4)_3$ yields 3 mol SO_4^{2-} in solution. Therefore, the number of moles of $Al_2(SO_4)_3$ is only one-third the number of moles of SO_4^{2-}. So the concentration of $Al_2(SO_4)_3$ must be one-third of 0.90 M, or 0.30 M.

PRACTICE EXERCISE 24: What are the molar concentrations of the ions in 0.40 M $FeCl_3$?

PRACTICE EXERCISE 25: In a solution of Na_3PO_4, the PO_4^{3-} concentration was determined to be 0.250 M. What was the sodium ion concentration in the solution?

Stoichiometry calculations using net ionic equations

You have seen that a net ionic equation is convenient for focusing on the net chemical change in an ionic reaction. Let's study some examples that illustrate how such equations can be used in stoichiometric calculations.

EXAMPLE 5.17

Stoichiometric Calculations Using a Net Ionic Equation

How many milliliters of 0.100 M $AgNO_3$ solution are needed to react completely with 25.0 mL of 0.400 M $CaCl_2$ solution? The net ionic equation for the reaction is

$$Ag^+(aq) + Cl^-(aq) \longrightarrow AgCl(s)$$

ANALYSIS: The problem refers to solutions of salts with formulas expressed in "molecular form," but the equation describes a reaction between the ions of the salts. Therefore, to use the chemical equation, we will need the concentrations of the ions. The tools we need to solve the problem, then, are the formulas of the salts (to find the concentrations of the ions) and the coefficients of the equation (to relate amounts of Ag^+ and Cl^-).

The first thing we will do is calculate the concentrations of the ions in the solutions being mixed. Then, using the volume and molarity of the Cl^- solution, we calculate the moles of Cl^- available. This value equals the moles of Ag^+ that react, because Ag^+ and Cl^- combine in a 1-to-1 mole ratio. Once we have the number of moles of Ag^+ that react, we use the molarity of the Ag^+ solution to determine the volume of the 0.100 M $AgNO_3$ solution needed.

SOLUTION: We begin by finding the concentrations of the ions in the reacting solutions:

0.100 M $AgNO_3$ contains 0.100 M Ag^+ and 0.100 M NO_3^-

0.400 M $CaCl_2$ contains 0.400 M Ca^{2+} and 0.800 M Cl^-

We're only interested in the Ag^+ and Cl^-; the Ca^{2+} and NO_3^- are spectator ions and are not involved in the reaction. For our purposes, then, the solution concentrations are 0.100 M Ag^+ and 0.800 M Cl^-.

Having these values, we can now restate the problem: how many milliliters of 0.100 M Ag^+ solution are needed to react completely with 25.0 mL of 0.800 M Cl^- solution?

$$25.0 \text{ mL Cl}^- \text{ soln} \Leftrightarrow ? \text{ mL Ag}^+ \text{ soln}$$

We have the volume and molarity of the Cl^- solution, so we calculate the number of moles of Cl^- available for reaction.

$$0.0250 \text{ L Cl}^- \text{ soln} \times \frac{0.800 \text{ mol Cl}^-}{1.00 \text{ L Cl}^- \text{ soln}} = 0.0200 \text{ mol Cl}^-$$

(Notice we have been careful to express the volume of the chloride solution in liters so that the units cancel correctly.)

Next, we use the coefficients of the equation to find the number of moles of Ag^+ that react.

$$0.0200 \text{ mol Cl}^- \times \frac{1 \text{ mol Ag}^+}{1 \text{ mol Cl}^-} = 0.0200 \text{ mol Ag}^+$$

Now we can calculate the volume of the Ag^+ solution using its molarity as a conversion factor. As we've done earlier, we use the factor that makes the units cancel correctly.

$$0.0200 \text{ mol Ag}^+ \times \frac{1.00 \text{ L Ag}^+ \text{ soln}}{0.100 \text{ mol Ag}^+} = 0.200 \text{ L Ag}^+ \text{ soln}$$

Our calculations tell us that we must use 0.200 L or 200 mL of the $AgNO_3$ solution. We could also have used the following chain calculation, of course:

$$0.0250 \text{ L Cl}^- \text{ soln} \times \frac{0.800 \text{ mol Cl}^-}{1.00 \text{ L Cl}^- \text{ soln}} \times \frac{1 \text{ mol Ag}^+}{1 \text{ mol Cl}^-} \times \frac{1.00 \text{ L Ag}^+ \text{ soln}}{0.100 \text{ mol Ag}^+}$$

$$= 0.200 \text{ L Ag}^+ \text{ soln}$$

Is the Answer Reasonable?

The silver ion concentration is one-eighth as large as the chloride ion concentration. Since the ions react one-for-one, we will need eight times as much silver ion solution as chloride solution. Eight times the amount of chloride solution, 25 mL, is 200 mL, which is the answer we obtained. Therefore, the answer appears to be correct.

PRACTICE EXERCISE 26: How many milliliters of 0.500 M KOH are needed to react completely with 60.0 mL of 0.250 M $FeCl_2$ solution to precipitate $Fe(OH)_2$?

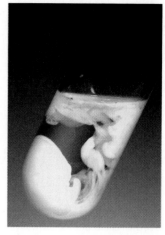

A solution of $AgNO_3$ is added to a solution of $CaCl_2$, producing a precipitate of AgCl.

Simple reasoning works well here. Since the coefficients of Ag^+ and Cl^- are the same, the numbers of moles that react must be equal.

Predicting the yield of an ionic reaction can involve a limiting reactant

In Chapter 4 you learned how to recognize and solve limiting reactant problems—problems in which amounts of both reactants are given in the statement of the problem. Similar problems can be encountered when working with net ionic equations, as illustrated in the following example.

Milk of magnesia is a suspension of $Mg(OH)_2$ in water. It can be made by adding a base to a solution containing Mg^{2+}. Suppose that 40.0 mL of 0.200 M NaOH solution is added to 25.0 mL of 0.300 M $MgCl_2$ solution. What mass of $Mg(OH)_2$ will be formed, and what will be the concentrations of the ions in the solution after the reaction is complete? The net ionic equation for the reaction is

$$Mg^{2+}(aq) + 2OH^-(aq) \longrightarrow Mg(OH)_2(s)$$

EXAMPLE 5.18

Calculation Involving the Stoichiometry of an Ionic Reaction

Milk of magnesia is a remedy for acid indigestion.

Actually, the reaction occurs as the solutions are being poured together. For purposes of the calculations, however, we may pretend that the solutions can be mixed first and then allowed to react. The initial and final states will still be the same.

The number of moles of chloride ion equals $2 \times (7.50 \times 10^{-3}$ mol).

ANALYSIS: The first thing we should notice here is that we've been given the volume and molarity for *both* solutions. By now, you should realize that volume and molarity give us moles, so in effect, we have been given the number of moles of each reactant. This means we have a limiting reactant problem (similar to the kind we studied on page 140).

To answer all the parts of the problem, the best approach is to begin by tabulating the number of *moles* of each ion supplied by their respective solutions. Then we'll use the net ionic equation to see which ions are removed and calculate which, if any, is limiting. Then we can go after the concentrations of the remaining ions and the mass of $Mg(OH)_2$ that forms.

SOLUTION: Let's begin by determining the number of moles of NaOH and $MgCl_2$ supplied by the volumes of their solutions. The conversion factors are taken from their molarities: 0.200 *M* NaOH and 0.300 *M* $MgCl_2$.

$$0.0400 \ \text{L NaOH soln} \times \frac{0.200 \ \text{mol NaOH}}{1.00 \ \text{L NaOH soln}} = 8.00 \times 10^{-3} \ \text{mol NaOH}$$

$$0.0250 \ \text{L MgCl}_2 \ \text{soln} \times \frac{0.300 \ \text{mol MgCl}_2}{1.00 \ \text{L MgCl}_2 \ \text{soln}} = 7.50 \times 10^{-3} \ \text{mol MgCl}_2$$

From this information, we obtain the number of moles of each ion present *before* any reaction occurs. In doing this, notice we take into account that 1 mol of $MgCl_2$ gives 2 mol Cl^-.

Moles of Ions before Reaction

Mg^{2+}	7.50×10^{-3} mol	Cl^-	15.0×10^{-3} mol
Na^+	8.00×10^{-3} mol	OH^-	8.00×10^{-3} mol

Now we refer to the net ionic equation, where we see that only Mg^{2+} and OH^- react. From the coefficients of the equation,

$$1 \ \text{mol} \ Mg^{2+} \Leftrightarrow 2 \ \text{mol} \ OH^-$$

This means that 7.50×10^{-3} mol Mg^{2+} (the amount of Mg^{2+} *available*) would require 15.0×10^{-3} mol OH^-. But we have only 8.00×10^{-3} mol OH^-. Insufficient OH^- is available to react with all of the Mg^{2+}, so OH^- must be the limiting reactant. Therefore, all of the OH^- will be used up and some Mg^{2+} will be unreacted. The amount of Mg^{2+} that *does* react to form $Mg(OH)_2$ can be found as follows:

$$8.00 \times 10^{-3} \ \text{mol OH}^- \times \frac{1 \ \text{mol} \ Mg^{2+}}{2 \ \text{mol OH}^-} = 4.00 \times 10^{-3} \ \text{mol} \ Mg^{2+}$$

(This amount reacts.)

Now we can tabulate the number of moles of each ion left in the mixture *after* the reaction is complete. For Mg^{2+}, the amount remaining equals the initial amount minus the amount that reacts.

Moles of Ions after Reaction

Mg^{2+}	$(7.50 \times 10^{-3}$ mol$) - (4.00 \times 10^{-3}$ mol$) = 3.50 \times 10^{-3}$ mol
Cl^-	15.0×10^{-3} mol (no change)
Na^+	8.00×10^{-3} mol (no change)
OH^-	0.00 mol (all used up)

Let's assemble all these numbers into one table to make them easier to understand.

Ion	Moles Present before Reaction	Moles That React	Moles Present after Reaction
Mg^{2+}	7.50×10^{-3} mol	4.00×10^{-3} mol	3.50×10^{-3} mol
Cl^-	15.0×10^{-3} mol	0.00 mol	15.0×10^{-3} mol
Na^+	8.00×10^{-3} mol	0.00 mol	8.00×10^{-3} mol
OH^-	8.00×10^{-3} mol	8.00×10^{-3} mol	0.00 mol

The problem asks for the concentrations of the ions in the final reaction mixture, so we must now divide each of the number of moles in the last column by the *total volume of the final solution* (40.0 mL + 25.0 mL = 65.0 mL). This volume must be expressed in liters (0.0650 L). For example, for Mg^{2+}, its concentration is

$$\frac{3.50 \times 10^{-3} \text{ mol } Mg^{2+}}{0.0650 \text{ L soln}} = 0.0538 \text{ } M \text{ } Mg^{2+}$$

Performing similar calculations for the other ions gives the following:

Concentrations of Ions after Reaction

Mg^{2+}	0.0538 *M*	Cl^-	0.231 *M*
Na^+	0.123 *M*	OH^-	0.00 *M*

Now let's turn our attention to the mass of $Mg(OH)_2$ that's formed. If 4.00×10^{-3} mol Mg^{2+} reacts, then 4.00×10^{-3} mol $Mg(OH)_2$ must be produced. The formula mass of $Mg(OH)_2$ is 58.32 g mol^{-1}, so the mass of $Mg(OH)_2$ formed is

$$4.00 \times 10^{-3} \text{ mol } Mg(OH)_2 \times \frac{58.32 \text{ g } Mg(OH)_2}{1 \text{ mol } Mg(OH)_2} = 0.233 \text{ g } Mg(OH)_2$$

The amount of $Mg(OH)_2$ formed is 0.233 g.

Stoichiometrically,
1 mol Mg^{2+} ⇔ 1 mol $Mg(OH)_2$

Is the Answer Reasonable?
This was a fairly complex problem and there are no simple calculations we can make to verify our answers. However, there are a couple of key steps we might double-check. First, we can check that we've identified the limiting reactant correctly. In the calculation above, we selected Mg^{2+} and determined if there were enough OH^- with which it could react. As a check, let's look at OH^- to see if there's enough Mg^{2+}.

We begin with 8×10^{-3} mol of OH^-, which requires half that amount of Mg^{2+}(4×10^{-3} mol Mg^{2+}) to react completely. The amount of Mg^{2+} available (7.5×10^{-3} mol) is more than enough, so Mg^{2+} will be left over and OH^- is limiting. Our analysis confirms that we've selected the correct limiting reactant.

Another place to be careful is in the calculation of the final concentrations of the ions in the reaction mixture. Whenever we add one aqueous solution to another, the mixture will have a final volume that's essentially the sum of the volumes of the two solutions combined.[7] When calculating the concentrations of anything in the final mixture, we have to use the final *combined* volume of the mixture. We see that we've taken this into account, so we can feel confident we've done this part of the calculation correctly.

PRACTICE EXERCISE 27: How many moles of $BaSO_4$ will form if 20.0 mL of 0.600 *M* $BaCl_2$ is mixed with 30.0 mL of 0.500 *M* $MgSO_4$? What will the concentrations of each ion be in the final reaction mixture?

5.13 ▶ **Chemical analysis and titration are applications of solution stoichiometry**

Chemical analyses fall into two categories. In a **qualitative analysis,** we simply determine which substances are present in a sample without measuring their amounts. In a **quantitative analysis,** our goal is to measure the amounts of the various substances in a sample.

When chemical reactions are used in a quantitative analysis, a useful strategy is to capture *all* of a desired chemical species in a compound with a known formula. From the amount of this compound obtained, we can determine how much of the desired chemical species was present in the original sample. The calculations re-

---

[7]For dilute aqueous solutions, it is generally safe to assume that volumes are additive. Mixing 100 mL of one solution with 200 mL of another will give a final mixture with a volume of 300 mL.

quired for these kinds of problems are not new; they are simply applications of the stoichiometric calculations you've already learned.

A certain insecticide is a compound known to contain carbon, hydrogen, and chlorine. Reactions were carried out on a 1.340 g sample of the compound that converted all of its chlorine to chloride ion dissolved in water. This aqueous solution was treated with an excess amount of $AgNO_3$ solution, and the AgCl precipitate was collected and weighed. Its mass was 2.709 g. What was the percentage by mass of Cl in the original insecticide sample?

ANALYSIS: The critical link in solving this problem is understanding that all the chlorine in the AgCl that was collected originated in the insecticide sample. The strategy, therefore, is to determine the mass of Cl in 2.709 g of AgCl. We then assume that this is the mass of chlorine in the original 1.340 g sample of insecticide and calculate the percentage Cl as follows:

$$\% \text{ Cl by mass} = \frac{\text{mass of Cl in sample}}{\text{mass of sample}} \times 100\%$$

SOLUTION: To find the mass of Cl in the AgCl, we employ the data used to calculate the formula mass of AgCl.

You learned to perform this kind of calculation on page 114 in Chapter 4.

Element	Moles	Mass
Ag	1 mol	1×107.87 g $= 107.87$ g Ag
Cl	1 mol	1×35.45 g $= 35.45$ g Cl
	Total	143.32 g AgCl

These data give us an equivalence between the masses of AgCl and Cl.

$$35.45 \text{ g Cl} \Leftrightarrow 143.32 \text{ g AgCl}$$

With this information, we can now calculate the mass of Cl in 2.709 g AgCl.

$$2.709 \text{ g AgCl} \times \frac{35.45 \text{ g Cl}}{143.32 \text{ g AgCl}} = 0.6701 \text{ g Cl}$$

The mass of Cl in the AgCl precipitate is 0.6701 g. We've assumed all of it came from the original insecticide sample, so the mass of Cl in the sample was also 0.6701 g. Therefore, we use this value to calculate the percentage Cl in the sample.

$$\% \text{ Cl} = \frac{0.6701 \text{ g Cl}}{1.340 \text{ g sample}} \times 100\%$$

$$= 50.01\%$$

The insecticide was 50.01% Cl by mass.

Is the Answer Reasonable?
Let's look at the formula masses of AgCl and Cl (143.3 and 35.5, respectively). If we round these off to 140 and 35, we can say that AgCl is approximately 25% Cl by mass ($35 \div 140 = 0.25$). Approximately 3 g of AgCl was obtained, and 25% of 3 g equals 0.75 g of Cl, which is close to the value we obtained. The original sample weighed 1.34 g, which is nearly twice 0.75 g (the mass of Cl), so an answer of approximately 50% Cl by mass seems reasonable.

PRACTICE EXERCISE 28: A sample of a mixture containing $CaCl_2$ and $MgCl_2$ weighed 2.000 g. The sample was dissolved in water and H_2SO_4 was added until the precipitation of $CaSO_4$ was complete. The $CaSO_4$ was filtered, dried completely, and weighed. A total of 0.736 g of $CaSO_4$ was obtained.

(a) How many moles of Ca^{2+} were in the $CaSO_4$?

(b) How many moles of Ca^{2+} were in the original 2.000 g sample?

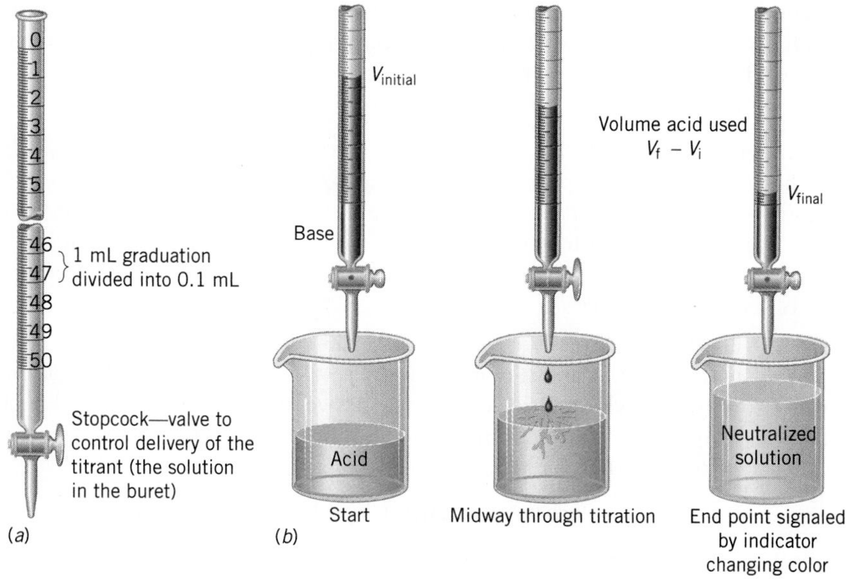

FIGURE **5.28** *Titration.* (*a*) A buret. (*b*) The titration of an acid by a base in which an acid–base indicator is used to signal the end point, which is the point at which all of the acid has been neutralized and addition of the base is halted.

(c) How many moles of $CaCl_2$ were in the 2.000 g sample?

(d) How many grams of $CaCl_2$ were in the 2.000 g sample?

(e) What was the percentage by mass of $CaCl_2$ in the original mixture?

Titration is an analytical procedure

Titration is an important laboratory procedure used in performing chemical analyses. The apparatus is shown in Figure 5.28. The long tube is called a **buret,** and it is marked for volumes, usually in increments of 0.10 mL. In a typical titration, a solution containing one reactant is placed in the receiving flask. Carefully measured volumes of a solution of the other reactant are then added from the buret. (One of the two solutions is of a precisely known concentration and is called a **standard solution**). This addition is continued until something (usually a visual effect, like a color change) signals that the two reactants have been combined in just the right proportions to give a complete reaction.

The valve at the bottom of the buret is called a **stopcock,** and it permits the analyst to control the amount of **titrant** (the solution in the buret) that is delivered to the receiving flask. This permits the addition of the titrant to be stopped once the reaction is complete.

The titration procedure is used for many kinds of reactions.

Acid–base titrations are useful in chemical analyses

Titrations are often used for acid–base reactions. Prior to the start of an acid–base titration, a drop or two of an *indicator* solution is added to the solution in the receiving flask. An **acid–base indicator** is a dye that has one color in an acidic solution and a different color in a basic solution. We have already mentioned litmus. Another, phenolphthalein, is more commonly used in acid–base titrations. It is colorless in acid and pink in base. (The theory of acid–base indicators is discussed in Chapter 18.)

During an acid–base titration, the analyst adds titrant slowly, watching for a change in the indicator color, which would signal that the solution has changed from acidic to basic (or basic to acidic, depending on the nature of the reactants in the buret and flask). Usually this color change is very abrupt, and occurs with the addition of only one final drop of the titrant just as the end of the reaction is reached. When the color change is observed, the **end point** has been reached. At this time the addition of titrant is stopped and the total volume of the titrant that's been added to the receiving flask is recorded.

EXAMPLE 5.20

Calculation Involving Acid–Base Titration

A solution of HCl is titrated with a solution of NaOH using phenolphthalein as the acid–base indicator. The pink color that phenolphthalein has in a basic solution can be seen where a drop of the NaOH solution has entered the HCl solution, to which a few drops of the indicator had been added.

A student prepares a solution of hydrochloric acid that is approximately 0.1 M and wishes to determine its precise concentration. A 25.00 mL portion of the HCl solution is transferred to a flask, and after a few drops of indicator are added, the HCl solution is titrated with 0.0775 M NaOH solution. The titration requires exactly 37.46 mL of the standard NaOH solution to reach the end point. What is the molarity of the HCl solution?

ANALYSIS: This is really a straightforward stoichiometry calculation involving a chemical reaction. The first step in solving it is to write the balanced equation, which will give us the stoichiometric equivalency between HCl and NaOH. The reaction is an acid–base neutralization, so the product is a salt plus water. Following the procedures developed earlier, the reaction is

$$HCl(aq) + NaOH(aq) \longrightarrow NaCl(aq) + H_2O$$

From the volume and molarity of the NaOH solution used, we can calculate the number of moles of NaOH dispensed from the buret. The coefficients of the equation are then used to find the number of moles of HCl that were in the receiving flask. Finally, we use the definition of molarity as a ratio of moles to volume to calculate the molarity of the HCl solution. The path to the answer can be diagrammed as follows:

$$\text{mL of NaOH soln} \xrightarrow[\text{NaOH soln}]{\text{molarity of}} \text{mol NaOH} \xrightarrow[\text{equation}]{\text{coefficients of}} \text{mol HCl} \xrightarrow[\text{of molarity}]{\text{definition}} \text{molarity of HCl}$$

SOLUTION: From the molarity and volume of the NaOH solution, we calculate the number of moles of NaOH consumed in the titration.

$$0.03746 \; \cancel{\text{L NaOH soln}} \times \frac{0.0775 \text{ mol NaOH}}{1.00 \; \cancel{\text{L NaOH soln}}} = 2.90 \times 10^{-3} \text{ mol NaOH}$$

The coefficients in the equation tell us that NaOH and HCl react in a 1-to-1 mole ratio,

$$2.90 \times 10^{-3} \text{ mol NaOH} \times \frac{1 \text{ mol HCl}}{1 \text{ mol NaOH}} = 2.90 \times 10^{-3} \text{ mol HCl}$$

so in this titration, 2.90×10^{-3} mol HCl was in the flask. To calculate the molarity of the HCl, we simply apply the definition of molarity. We take the ratio of the number of moles of HCl that reacted (2.90×10^{-3} mol HCl) to the volume (in liters) of the HCl solution used (25.00 mL, or 0.02500 L).

$$\text{molarity of HCl soln} = \frac{2.90 \times 10^{-3} \text{ mol HCl}}{0.02500 \text{ L HCl soln}}$$

$$= 0.116 \; M \text{ HCl}$$

The molarity of the hydrochloric acid is 0.116 M.

Is the Answer Reasonable?
If the concentrations of the NaOH and HCl were the same, the volumes used would have been equal. However, the volume of the NaOH solution used is larger than the volume of HCl solution. This must mean that the HCl solution is more concentrated than the NaOH solution. The value we obtained, 0.116 M, is larger than 0.0775 M, so our answer makes sense.

PRACTICE EXERCISE 29: In a titration, a sample of H_2SO_4 solution having a volume of 15.00 mL required 36.42 mL of 0.147 M NaOH solution for *complete* neutralization. What is the molarity of the H_2SO_4 solution?

PRACTICE EXERCISE 30: "Stomach acid" is hydrochloric acid. A sample of gastric juice having a volume of 5.00 mL required 11.00 mL of 0.0100 *M* KOH solution for neutralization in a titration. What was the molar concentration of HCl in this fluid? If we assume a density of 1.00 g mL^{-1} for the fluid, what was the percentage by weight of HCl?

SUMMARY

Solution Vocabulary A **solution** is a homogeneous mixture in which one or more **solutes** are dissolved in a **solvent.** A solution may be **dilute** or **concentrated,** depending on the amount of solute dissolved in a given amount of solvent. **Concentration** (e.g., **percentage concentration**) is a ratio of the amount of solute to either the amount of solvent or the amount of solution. The amount of solute required to give a **saturated** solution at a given temperature is called the solute's **solubility. Unsaturated** solutions will dissolve more solute, but **supersaturated** solutions are unstable and tend to give a **precipitate.**

Molar Concentration and Dilution The **molar concentration (molarity)** of a solution is the number of moles of solute per liter of solution. Molarity provides two conversion factors relating moles of solute and the volume of the solution.

$$\frac{\text{mol solute}}{1 \text{ L soln}} \quad \text{and} \quad \frac{1 \text{ L soln}}{\text{mol solute}}$$

Concentrated solutions of known molarity can be diluted quantitatively using volumetric glassware such as pipets and volumetric flasks. The following relationship applies:

$$V_{dil} \cdot M_{dil} = V_{concd} \cdot M_{concd}$$

Electrolytes Substances that **dissociate** or **ionize** in water to produce cations and anions are **electrolytes;** those that do not are called **nonelectrolytes.** Electrolytes include salts and metal hydroxides as well as molecular acids and bases that ionize by reaction with water. In water, ionic compounds are completely dissociated into ions and are **strong electrolytes.**

Metathesis Reactions Metathesis or **double replacement** reactions occur between the ions and are called **ionic reactions.** Solutions of soluble strong electrolytes often yield an insoluble product, which appears as a **precipitate.** Equations for these reactions can be written in three different ways. In **molecular equations,** complete formulas for all reactants and products are used. In an **ionic equation,** soluble strong electrolytes are written in dissociated (ionized) form; "molecular" formulas are used for solids and weak electrolytes. A **net ionic equation** is obtained by eliminating **spectator ions** from the ionic equation, and such an equation allows us to identify other combinations of reactants that give the same net reaction. An ionic or net ionic equation is balanced only if both atoms *and* net charge are balanced.

Acids and Bases as Electrolytes The modern version of the **Arrhenius definition of acids and bases** is that an **acid** is a substance that produces hydronium ions, H_3O^+, when dissolved in water, and a **base** produces hydroxide ions, OH^-, when dissolved in water.

The oxides of nonmetals are generally **acidic anhydrides** and react with water to give acids. Metal oxides are usually **basic anhydrides** because they tend to react with water to give metal hydroxides or bases.

Strong acids and bases are also strong electrolytes. **Weak acids and bases** are **weak electrolytes,** which are incompletely ionized in water. In a solution of a weak electrolyte there is a chemical equilibrium between the nonionized molecules of the solute and the ions formed by the reaction of the solute with water.

Acid–Base Neutralization and Ionic Reactions Acids react with bases in neutralization reactions to produce a salt and water. Acids react with insoluble oxides and hydroxides to form water and the corresponding salt. Most acid–base neutralization reactions can be viewed as a type of metathesis reaction in which one product is water.

Reactions That Produce Gases Some metathesis reactions produce gaseous products with low solubility in water or compounds that are unstable and decompose to give slightly soluble gases. In particular, acids react with soluble and insoluble carbonates and with bicarbonates to yield carbon dioxide, a gaseous decomposition product of carbonic acid, H_2CO_3.

Predicting Metathesis Reactions A metathesis reaction will occur if there is a net ionic equation. Factors that lead to the existence of a net ionic equation are formation of a precipitate from soluble products, formation of water in an acid–base neutralization, formation of a gas, and formation of a weak electrolyte from soluble strong electrolytes. You should learn the **solubility rules** (Table 5.2), and remember that all salts are strong electrolytes. Remember that all strong acids and bases are strong electrolytes, too. Everything else is a weak electrolyte or a nonelectrolyte. Be sure to learn the reactions that produce gases in metathesis reactions, which are found in Table 5.3.

Solution Stoichiometry Molarity is a useful concentration unit for working problems in the stoichiometry of reactions in solution. In ionic reactions, the concentrations of the ions in a solution of a salt can be derived from the molar concentration of the salt, taking into account the number of ions formed per formula unit of the salt.

Figure 5.29 gives an overview of the various paths through stoichiometry problems that involve chemical reac-

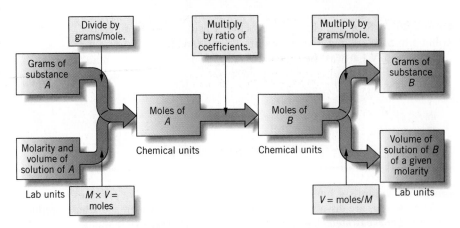

FIGURE 5.29 *Paths for working stoichiometry problems involving chemical reactions.* The critical link is the conversion between moles of one substance and moles of another using the coefficients of the balanced equation. We can obtain moles from more than one laboratory unit, namely, from grams or from volumes of solutions of known molarities. Similarly, we can present the answer either in moles, grams, or a volume of solution of known molarity.

tions. All funnel through the coefficients of the equation as the means to convert from moles of one substance to moles of another. The starting and finishing quantities can be moles, grams, or volumes of solutions of known molarity.

Titration is a technique used to make quantitative measurements of the amounts of solutions needed to obtain a complete reaction. In an acid–base titration, the **end point** is normally detected visually using an **acid–base indicator.** A color change indicates complete reaction, at which time addition of **titrant** is stopped and the volume added is recorded.

TOOLS ▶ YOU HAVE LEARNED

The table below lists the concepts that you've learned in this chapter that can be applied as tools in solving problems. Study each one carefully so that you know what each is used for. When faced with solving a problem, recall what each tool does and consider whether it will be helpful in finding a solution. This will aid you in selecting the tools you need. If necessary, refer to this table when working on the Thinking-It-Through problems and the Review Exercises that follow.

TOOL	HOW IT WORKS
List of strong acids (page 175)	When you encounter the formula for an acid and need to know whether the acid is strong or weak, you use this list. If the acid is on the list, it's a strong acid; if not, it's probably a weak acid. Strong acids are written in dissociated form in an ionic equation.
Criteria for balanced ionic equations (page 163)	If both the charge and number of atoms are the same on both sides of an equation, the equation is balanced.
Solubility rules (page 164)	You use the rules to tell if a given salt is soluble in water. Knowing that soluble salts dissociate fully in water enables you to write ionic equations correctly. The rules help you predict the course of metathesis reactions.
Types of substances that react with either acids or bases to give gases (page 185)	This list enables you to predict when to expect H_2S, HCN, CO_2, SO_2, or NH_3 to form in a metathesis reaction.
Molarity (pages 190, 193, and 194)	Molarity is the critical link in relating moles of solute to volume of solution. We use it to construct conversion factors to use in calculations.

THINKING IT THROUGH

The goal for the following problems is not to find the answers themselves, but rather to assemble the information needed to solve them and explain how you would use the information to find the answers. The problems in Level 2 are more challenging than those in Level 1 and may contain more data than are required, in which case you are also asked to identify the unnecessary data. Detailed answers to the Thinking-It-Through problems can be found on the web site.

Need extra help?
ON-LINE HELP | Visit the Brady/ Senese web site at **www.wiley.com/ college/brady**

Level 1 Problems

1. You are planning an experiment that requires the presence of the hydroxide ion at a concentration of at least 0.2 *M*. Which of the following could you use as the solute: NaOH, KOH, Mg(OH)$_2$? Describe the information you need to know to answer the question.

2. Describe the calculations necessary to determine the number of grams of CuSO$_4$·5H$_2$O required to prepare 250 mL of 0.20 *M* CuSO$_4$ solution.

3. Describe how you would calculate the number of grams of Na$_3$PO$_4$ in 20.0 mL of an aqueous solution of this compound if the sodium ion concentration is 0.160 *M*.

4. If 25 mL of water is added to 150 mL of 0.20 *M* KNO$_3$, what is the final molarity of the solute? Describe the calculation involved.

5. As the first step in answering this question, write both a molecular and a net ionic equation that describe the reaction of solid MgCO$_3$ with an aqueous solution of hydrobromic acid. Suppose that you mix 100.0 mL of 0.1250 *M* HBr with a 10.54 g sample of solid MgCO$_3$ in a large beaker. Explain the calculations required to answer the following questions: How many grams of the MgCO$_3$ dissolve? How many grams of the MgCO$_3$ remain after the reaction is over? What are the concentrations of the ions that are present in the solution after the reaction is over?

6. An unknown compound was either NaCl, NaBr, or NaI. When 0.5646 g of the compound was dissolved in water, it took 22.1 mL of 0.248 *M* AgNO$_3$ to combine with all of the halide ion in the solution. Describe how you would use this information to determine the identity of the unknown compound.

7. Consider two solutions, one of them containing Fe(NO$_3$)$_2$ and the other containing Fe$_2$(SO$_4$)$_3$. In the first solution, the NO$_3^-$ concentration is 0.240 *M*; in the other, the SO$_4^{2-}$

concentration is 0.210 *M*. How would you determine which solution has the higher concentration of iron?

Level 2 Problems

8. In an experiment, 20.0 mL of 0.25 *M* NaCl (formula mass = 58.44) will be added to 35.0 mL of 0.12 *M* AgNO$_3$ (formula mass = 169.87). Explain how you would determine the mass of AgCl that will be formed in this reaction mixture.

9. A sample of Cr$_2$(SO$_4$)$_3$ weighing 0.500 g was dissolved in 150 mL of water and treated with an excess of 0.25 *M* NaOH solution, causing a precipitate to form. The precipitate was collected by filtration. Describe the calculations required to determine the number of milliliters of 0.400 *M* HNO$_3$ needed to dissolve this precipitate.

10. A 0.100 g sample of an acid with the molecular formula C$_6$H$_8$O$_6$ was dissolved in 125 mL of water and titrated with 0.150 *M* NaOH. A volume of 7.58 mL of the base was required to completely neutralize the acid. Is the acid a monoprotic acid or a diprotic acid? (Describe the calculation involved.)

11. A mixture of two monoprotic acids—lactic acid, HC$_3$H$_5$O$_3$ (found in sour milk) and caproic acid, HC$_6$H$_{11}$O$_2$ (a foul-smelling substance found in excretions from male goats)—was titrated with 0.0500 *M* NaOH. A 0.1000 g sample of the mixture dissolved in a total volume of 225 mL of solution required 20.4 mL of the base. Explain in detail how you would calculate the mass in grams of each acid in the sample.

12. Lead chloride, PbCl$_2$, is a white solid that is slightly soluble in water at room temperature. If some crystals of PbCl$_2$ are added to a beaker of water, which of the following diagrams best represents the mixture after some PbCl$_2$ has dissolved, viewed at a submicroscopic level? Explain what is wrong with the other four diagrams. (Black spheres represent Pb and green spheres represent Cl.)

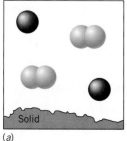
(a)

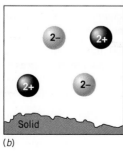
(b)

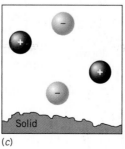
(c)

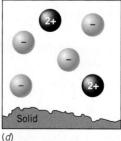
(d)

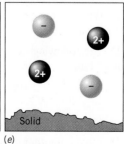
(e)

REVIEW QUESTIONS

Solution Terminology

5.1 Define the following: (a) *solvent*, (b) *solute*, (c) *concentration*.

5.2 Define the following: (a) *concentrated*, (b) *dilute*, (c) *saturated*, (d) *unsaturated*, (e) *supersaturated*.

5.3 What is meant by the term *solubility*?

5.4 Why are chemical reactions often carried out using solutions.

5.5 Define *percentage concentration* and *molar concentration*.

5.6 If water is added to a solution of sugar, what happens to the average distance between sugar molecules?

5.7 Describe what will happen if a crystal of sugar is added to (a) a saturated sugar solution, (b) a supersaturated solution of sugar, and (c) an unsaturated solution of sugar.

5.8 What is the meaning of the term *precipitate*? What condition must exist for a precipitate to form spontaneously in a solution?

Electrolytes

5.9 What is an electrolyte? What is a nonelectrolyte?

5.10 Why is an electrolyte able to conduct electricity while a nonelectrolyte cannot? What does it mean when we say that an ion is "hydrated"?

5.11 Define *dissociation* as it applies to ionic compounds that dissolve in water.

5.12 Experimentally, what is the observable difference between a strong electrolyte and a weak electrolyte?

5.13 If a substance is a weak electrolyte, what does this mean in terms of the tendency of the ions to react to re-form the molecular compound? How does this compare with strong electrolytes?

Ionic Reactions

5.14 What are the differences among molecular, ionic, and net ionic equations? What are spectator ions?

5.15 The following equation shows the formation of cobalt(II) hydroxide, a compound used to improve the drying properties of lithographic inks:

$$Co^{2+} + 2Cl^- + 2Na^+ + 2OH^- \longrightarrow$$
$$Co(OH)_2(s) + 2Na^+ + 2Cl^-$$

Which are the spectator ions? Write the net ionic equation.

5.16 What two conditions must be fulfilled by a balanced ionic equation?

5.17 The following equation is not balanced. How do we know? Find the errors and fix them.

$$3Co^{3+} + 2HPO_4^{2-} \longrightarrow Co_3(PO_4)_2 + 2H^+$$

Metathesis Reactions

5.18 What is another name for *metathesis reaction*?

5.19 In the reaction of lead nitrate with potassium iodide, why does the lead iodide that's formed in the reaction separate from the solution as a precipitate?

5.20 Write equations for the dissociation of the following in water: (a) $CaCl_2$, (b) $(NH_4)_2SO_4$, (c) $NaC_2H_3O_2$.

5.21 If a solution of trisodium phosphate, Na_3PO_4, is poured into seawater, precipitates of calcium phosphate and magnesium phosphate are formed. (Magnesium and calcium ions are among the principal ions found in seawater.) Write net ionic equations for these reactions.

5.22 Washing soda is $Na_2CO_3 \cdot 10H_2O$. Explain, using chemical equations, how this substance is able to remove Ca^{2+} ions from "hard water."

5.23 Sodium oxalate reacts with calcium chloride to yield insoluble calcium oxalate (a compound used in ceramic glazes) and sodium chloride. Express the preceding statement in the form of a metathesis reaction.

Acids, Bases, and Their Reactions

5.24 Give two general properties of an acid. Give two general properties of a base.

5.25 If you believed a solution was basic, which color litmus paper (blue or pink) would you use to test the solution to see if you were correct? What would you observe if you've selected correctly? Why would the other color litmus paper not lead to a conclusive result?

5.26 How did Arrhenius define an acid and a base?

5.27 What is the difference between *dissociation* and *ionization* as applied to the discussion of electrolytes?

5.28 Which of the following undergo dissociation in water? Which undergo ionization? (a) NaOH, (b) HNO_3, (c) NH_3, (d) H_2SO_4

5.29 Write the formulas for the strong acids discussed in Section 5.7. (If you must peek, they are listed on page 175.)

5.30 Which of the following would yield an acidic solution when they react with water? Which would give a basic solution? (a) P_4O_{10}, (b) K_2O, (c) SeO_3, (d) Cl_2O_7?

5.31 What is a *dynamic equilibrium*? Using acetic acid as an example, describe why all the $HC_2H_3O_2$ molecules are not ionized in water.

5.32 Why don't we use double arrows in the equation for the reaction of a strong acid with water?

5.33 How would the electrical conductivity of a solution of $Ba(OH)_2$ change as a solution of H_2SO_4 is added slowly to it? Use a net ionic equation to justify your answer.

5.34 Methylamine, CH_3NH_2, reacts with hydronium ion in very much the same manner as ammonia.

$$CH_3NH_2(aq) + H_3O^+(aq) \longrightarrow CH_3NH_3^+(aq) + H_2O$$

On the basis of what you have learned so far in this course, sketch the molecular structures of CH_3NH_2 and $CH_3NH_3^+$ (the methylammonium ion).

5.35 In general, what are the products when a strong acid neutralizes a strong base?

5.36 With which of the following will the weak acid $HCHO_2$ react? Where there is a reaction, write the formulas of the products.
(a) KOH   (b) MgO   (c) $CuCO_3$ (d) NH_3

Nomenclature of Acids and Bases

5.37 Name the following:
(a) $H_2Se(g)$
(b) $H_2Se(aq)$

5.38 Iodine, like chlorine, forms several acids. What are the names of the following?
(a) HIO_4 (c) HIO_2 (e) HI
(b) HIO_3 (d) HIO

5.39 For the acids in the preceding question, (a) write the formulas and (b) name the ions formed by removing a hydrogen ion (H^+) from each acid.

5.40 Write the formula for (a) chromic acid, (b) carbonic acid, and (c) oxalic acid. (*Hint:* Check the table of polyatomic ions.)

5.41 Name the following acid salts:
(a) $NaHCO_3$
(b) KH_2PO_4
(c) $(NH_4)_2HPO_4$

5.42 Write the formulas for all the acid salts that could be formed from the reaction of NaOH with the acid H_3PO_4.

5.43 Name the following oxoacids and give the names and formulas of the salts formed from them by neutralization with NaOH: (a) HOCl, (b) HIO_2, (c) $HBrO_3$, (d) $HClO_4$.

5.44 The formula for the arsenate ion is AsO_4^{3-}. What is the formula for arsenous acid?

5.45 Butyric acid, $HC_4H_7O_2$, gives rancid butter its bad odor. What is the name of the salt $NaC_4H_7O_2$?

5.46 Calcium propionate, $Ca(C_3H_5O_2)_2$, is used in baked foods as a preservative and to prevent the growth of mold. What is the name of the acid $HC_3H_5O_2$?

Reactions That Produce Gases

5.47 Suppose you suspected that a certain solution contained ammonium ions. What simple chemical test could you perform that would tell you whether your suspicion was correct?

5.48 What gas is formed if HCl is added to (a) $NaHCO_3$, (b) Na_2S, (c) K_2SO_3?

Predicting Ionic Reactions

5.49 What factors lead to the existence of a net ionic equation in a reaction between ions?

5.50 Complete the following equations and then determine the "driving force" in each reaction:
(a) $CuCl_2 + Na_2CO_3 \rightarrow$
(b) $K_2SO_3 + H_2SO_4 \rightarrow$
(c) $NaClO + HCl \rightarrow$
(d) $Ba(OH)_2 + HNO_3 \rightarrow$

5.51 Silver bromide is "insoluble." What does this mean about the concentrations of Ag^+ and Br^- in a saturated solution of AgBr? Explain why a precipitate of AgBr forms when solutions of the soluble salts $AgNO_3$ and NaBr are mixed.

Molarity and Dilution

5.52 A solution is labeled 0.150 *M* KOH. Write the two conversion factors that can be constructed from this concentration.

5.53 What is the definition of molarity? Show that the ratio of millimoles (mmol) to milliliters (mL) is equivalent to the ratio of moles to liters.

5.54 When a solution labeled 0.50 *M* HNO_3 is diluted with water to give 0.25 *M* HNO_3, what happens to the number of moles of HNO_3 in the solution?

5.55 Two solutions, A and B, are labeled "0.10 *M* $CaCl_2$" and "0.20 *M* $CaCl_2$," respectively. Both solutions contain the same number of moles of $CaCl_2$. If solution A has a volume of 50 mL, what is the volume of solution B?

5.56 When the units molarity and liter are multiplied, what are the resulting units?

Chemical Analyses and Titrations

5.57 What is the difference between a qualitative analysis and a quantitative analysis?

5.58 Describe each of the following: (a) buret, (b) titration, (c) titrant, and (d) end point.

5.59 What is the function of an indicator in a titration? What color is phenolphthalein in (a) an acidic solution and (b) a basic solution?

REVIEW PROBLEMS

Answers to problems whose numbers are printed in color are given in Appendix B. Challenging problems are marked with asterisks. **ILW** = Interactive LearningWare solution is available at *www.wiley.com/college/brady*.

Electrolytes

5.60 Write equations for the dissociation of the following ionic compounds in water: (a) LiCl, (b) $BaCl_2$, (c) $Al(C_2H_3O_2)_3$, (d) $(NH_4)_2CO_3$, (e) $FeCl_3$.

5.61 Write equations for the dissociation of the following ionic compounds in water: (a) $CuSO_4$, (b) $Al_2(SO_4)_3$, (c) $CrCl_3$, (d) $(NH_4)_2HPO_4$, (e) $KMnO_4$.

Metathesis Reactions That Produce Precipitates

5.62 Write ionic and net ionic equations for these reactions.
(a) $(NH_4)_2CO_3(aq) + MgCl_2(aq) \rightarrow$
$$2NH_4Cl(aq) + MgCO_3(s)$$
(b) $CuCl_2(aq) + 2NaOH(aq) \rightarrow Cu(OH)_2(s) + 2NaCl(aq)$
(c) $3FeSO_4(aq) + 2Na_3PO_4(aq) \rightarrow$
$$Fe_3(PO_4)_2(s) + 3Na_2SO_4(aq)$$
(d) $2AgC_2H_3O_2(aq) + NiCl_2(aq) \rightarrow$
$$2AgCl(s) + Ni(C_2H_3O_2)_2(aq)$$

5.63 Write balanced ionic and net ionic equations for these reactions.
(a) $CuSO_4(aq) + BaCl_2(aq) \rightarrow CuCl_2(aq) + BaSO_4(s)$
(b) $Fe(NO_3)_3(aq) + LiOH(aq) \rightarrow$
$$LiNO_3(aq) + Fe(OH)_3(s)$$
(c) $Na_3PO_4(aq) + CaCl_2(aq) \rightarrow Ca_3(PO_4)_2(s) + NaCl(aq)$
(d) $Na_2S(aq) + AgC_2H_3O_2(aq) \rightarrow$
$$NaC_2H_3O_2(aq) + Ag_2S(s)$$

5.64 Aqueous solutions of sodium sulfide, Na_2S, and copper(II) nitrate, $Cu(NO_3)_2$, are mixed. A precipitate of copper(II) sulfide, CuS, forms at once. Left behind is a solution of sodium nitrate, $NaNO_3$. Write the molecular, ionic, and net ionic equations for this reaction.

5.65 If an aqueous solution of iron(III) sulfate (a compound used in dyeing textiles and also for etching aluminum) is mixed with a solution of barium chloride, a precipitate of barium sulfate forms and the solution that remains contains iron(III) chloride. Write the molecular, ionic, and net ionic equations for this reaction.

5.66 Silver bromide is the chief light-sensitive substance used in the manufacture of photographic film. It can be prepared by mixing solutions of $AgNO_3$ and NaBr. Taking into account that $NaNO_3$ is soluble, write molecular, ionic, and net ionic equations for this reaction.

5.67 Trisodium phosphate, Na_3PO_4, is a useful cleaning agent, but it must be handled with care because its solutions are quite caustic. If a solution of Na_3PO_4 is added to one containing a calcium salt such as $CaCl_2$, a precipitate of calcium phosphate is formed. Write molecular, ionic, and net ionic equations for the reaction.

Acids and Bases as Electrolytes

5.68 Pure $HClO_4$ is molecular. In water it is a strong acid. Write an equation for its ionization in water.

5.69 HBr is a molecular substance that is a strong acid in water. Write an equation for its ionization in water.

5.70 Hydrazine is a toxic substance that can form when household ammonia is mixed with a bleach such as Clorox. Its formula is N_2H_4, and it is a weak base. Write a chemical equation showing its reaction with water.

5.71 Methylamine, CH_3NH_2, is a weak base with a fishy odor. Write a chemical equation showing its reaction with water.

5.72 Nitrous acid, HNO_2, is a weak acid that can form when sodium nitrite, a meat preservative reacts with stomach acid (HCl). Write an equation showing the ionization of HNO_2 in water.

5.73 Formic acid, $HCHO_2$, is a weak acid responsible for the sting of certain red ants. Write an equation showing its reaction with water.

5.74 Carbonic acid, H_2CO_3, is a weak diprotic acid formed in rainwater as it passes through the atmosphere and dissolves carbon dioxide. Write chemical equations for the equilibria involved in the stepwise ionization of H_2CO_3 in water.

5.75 Phosphoric acid, H_3PO_4, is a weak acid found in some soft drinks. It undergoes ionization in three steps. Write chemical equations for the equilibria involved in each of these reactions.

Acid–Base Neutralization

5.76 Complete and balance the following equations. For each, write the molecular, ionic, and net ionic equations. (All the products are soluble in water.)
(a) $Ca(OH)_2(aq) + HNO_3(aq) \rightarrow$
(b) $Al_2O_3(s) + HCl(aq) \rightarrow$
(c) $Zn(OH)_2(s) + H_2SO_4(aq) \rightarrow$

5.77 Complete and balance the following equations. For each, write the molecular, ionic, and net ionic equations. (All the products are soluble in water.)
(a) $HC_2H_3O_2(aq) + Mg(OH)_2(s) \rightarrow$
(b) $HClO_4(aq) + NH_3(aq) \rightarrow$
(c) $H_2CO_3(aq) + NH_3(aq) \rightarrow$

Ionic Reactions That Produce Gases

5.78 Write balanced net ionic equations for the reaction between
(a) $HNO_3(aq) + K_2CO_3(aq)$
(b) $Ca(OH)_2(aq) + NH_4NO_3(aq)$

5.79 Write balanced net ionic equations for the reaction between
(a) $H_2SO_4(aq) + NaHSO_3(aq)$
(b) $HNO_3(aq) + (NH_4)_2CO_3(aq)$

Predicting Metathesis Reactions

5.80 Explain why the following reactions take place:
(a) $CrCl_3 + 3NaOH \rightarrow Cr(OH)_3 + 3NaCl$
(b) $ZnO + 2HBr \rightarrow ZnBr_2 + H_2O$

5.81 Explain why the following reactions take place:
(a) $MnCO_3 + H_2SO_4 \rightarrow MnSO_4 + H_2O + CO_2$
(b) $Na_2C_2O_4 + 2HNO_3 \rightarrow 2NaNO_3 + H_2C_2O_4$

5.82 Use the solubility rules to decide which of the following compounds are *soluble* in water:
(a) $Ca(NO_3)_2$ (c) $Ni(OH)_2$ (e) $BaSO_4$
(b) $FeCl_2$ (d) $AgNO_3$ (f) $CuCO_3$

5.83 Predict which of the following compounds are *soluble* in water:
(a) $HgBr_2$ (c) Hg_2Br_2 (e) PbI_2
(b) $Sr(NO_3)_2$ (d) $(NH_4)_3PO_4$ (f) $Pb(C_2H_3O_2)_2$

5.84 Use the solubility rules to decide which of the following compounds are *insoluble* in water:
(a) AgCl   (d) $Ca_3(PO_4)_2$
(b) $Cr_2(SO_4)_3$ (e) $Al(C_2H_3O_2)_3$
(c) $(NH_4)_2CO_3$ (f) ZnO

5.85 Predict which of the following compounds are *insoluble* in water:
(a) $Ca(OH)_2$ (d) CsOH   (g) K_3PO_4
(b) NiS   (e) $MgSO_3$ (h) $Mg(ClO_4)_2$
(c) $PbCl_2$ (f) Fe_2O_3 (i) $CaCO_3$

5.86 Complete and balance the molecular, ionic, and net ionic equations for the following reactions:
(a) $HNO_3 + Cr(OH)_3 \rightarrow$
(b) $HClO_4 + NaOH \rightarrow$
(c) $Cu(OH)_2 + HC_2H_3O_2 \rightarrow$
(d) $ZnO + H_2SO_4 \rightarrow$

5.87 Complete and balance molecular, ionic, and net ionic equations for the following reactions:
(a) $NaHSO_3 + HBr \rightarrow$
(b) $(NH_4)_2CO_3 + NaOH \rightarrow$
(c) $(NH_4)_2CO_3 + Ba(OH)_2 \rightarrow$
(d) $FeS + HCl \rightarrow$

ILW 5.88 Complete and balance the following equations, and then write the ionic and net ionic equations. If all ions cancel, indicate that no reaction (N.R.) takes place.
(a) $Na_2SO_3 + Ba(NO_3)_2 \rightarrow$
(b) $HCHO_2 + K_2CO_3 \rightarrow$
(c) $NH_4Br + Pb(C_2H_3O_2)_2 \rightarrow$
(d) $NH_4ClO_4 + Cu(NO_3)_2 \rightarrow$

5.89 Complete and balance the following equations, and then write the ionic and net ionic equations. If all ions cancel, indicate that no reaction (N.R.) takes place.
(a) $(NH_4)_2S + NaOH \rightarrow$
(b) $Cr_2(SO_4)_3 + K_2CO_3 \rightarrow$
(c) $AgNO_3 + Cr(C_2H_3O_2)_3 \rightarrow$
(d) $Sr(OH)_2 + MgCl_2 \rightarrow$

*****5.90** Choose reactants that would yield the following net ionic equations. Write molecular equations for each.
(a) $HCO_3^- + H^+ \rightarrow H_2O + CO_2(g)$
(b) $Fe^{2+} + 2OH^- \rightarrow Fe(OH)_2(s)$
(c) $Ba^{2+} + SO_3^{2-} \rightarrow BaSO_3(s)$
(d) $2Ag^+ + S^{2-} \rightarrow Ag_2S(s)$
(e) $ZnO(s) + 2H^+ \rightarrow Zn^{2+} + H_2O$

*****5.91** Suppose that you wished to prepare copper(II) carbonate by a precipitation reaction involving Cu^{2+} and CO_3^{2-}. Which of the following pairs of reactants could you use as solutes?
(a) $Cu(OH)_2 + Na_2CO_3$ (d) $CuCl_2 + K_2CO_3$
(b) $CuSO_4 + (NH_4)_2CO_3$ (e) $CuS + NiCO_3$
(c) $Cu(NO_3)_2 + CaCO_3$

Molar Concentration

5.92 A solution is labeled 0.25 M HCl. Construct two conversion factors that relate moles of HCl to the volume of solution expressed in liters.

5.93 A solution is labeled 0.75 M $CuSO_4$. Construct two conversion factors that relate moles of $CuSO_4$ to the volume of solution, expressed in liters.

5.94 Calculate the molarity of a solution prepared by dissolving
(a) 4.00 g of NaOH in 100.0 mL of solution.
(b) 16.0 g of $CaCl_2$ in 250.0 mL of solution.
(c) 14.0 g of KOH in 75.0 mL of solution.
(d) 6.75 g of $H_2C_2O_4$ in 500 mL of solution.

5.95 Calculate the molarity of a solution that contains
(a) 3.60 g of H_2SO_4 in 450 mL of solution.
(b) 2.0×10^{-3} mol $Fe(NO_3)_2$ in 12.0 mL of solution.
(c) 1.65 mol HCl in 2.16 L of solution.
(d) 18.0 g HCl in 0.375 L of solution.

ILW 5.96 Calculate the number of grams of each solute needed to make each of the following solutions.
(a) 125 mL of 0.200 M NaCl
(b) 250 mL of 0.360 M $C_6H_{12}O_6$ (glucose)
(c) 250 mL of 0.250 M H_2SO_4

5.97 How many grams of solute are needed to make each of the following solutions?
(a) 250 mL of 0.100 M K_2SO_4
(b) 100 mL of 0.250 M K_2CO_3
(c) 500 mL of 0.400 M KOH

Dilution of Solutions

5.98 If 25.0 mL of 0.56 M H_2SO_4 is diluted to a volume of 125 mL, what is the molarity of the resulting solution?

5.99 A 150 mL sample of 0.45 M HNO_3 is diluted to 450 mL. What is the molarity of the resulting solution?

ILW 5.100 To what volume must 25.0 mL of 18.0 M H_2SO_4 be diluted to produce 1.50 M H_2SO_4?

5.101 To what volume must 50.0 mL of 1.50 M HCl be diluted to produce 0.200 M HCl?

5.102 How many milliliters of water must be added to 150 mL of 2.5 M KOH to give a 1.0 M solution? (Assume volumes are additive.)

5.103 How many milliliters of water must be added to 120 mL of 1.50 M HCl to give 1.00 M HCl?

Concentrations of Ions in Solutions of Electrolytes

5.104 Calculate the number of moles of each of the ions in the following solutions:
(a) 35.0 mL of 1.25 M KOH
(b) 32.3 mL of 0.45 M $CaCl_2$
(c) 50.0 mL of 0.40 M $AlCl_3$

5.105 Calculate the number of moles of each of the ions in the following solutions:
(a) 18.5 mL of 0.40 M $(NH_4)_2CO_3$
(b) 30.0 mL of 0.35 M $Al_2(SO_4)_3$
(c) 60.0 mL of 0.12 M $Zn(C_2H_3O_2)_2$

5.106 Calculate the concentrations of each of the ions in
(a) 0.25 M $Cr(NO_3)_2$, (b) 0.10 M $CuSO_4$, (c) 0.16 M Na_3PO_4, (d) 0.075 M $Al_2(SO_4)_3$.

5.107 Calculate the concentrations of each of the ions in
(a) 0.060 M $Ca(OH)_2$, (b) 0.15 M $FeCl_3$, (c) 0.22 M $Cr_2(SO_4)_3$, (d) 0.60 M $(NH_4)_2SO_4$.

5.108 A solution contains only the solute Na_3PO_4. What is the molar concentration of this salt if the sodium concentration is 0.21 M?

5.109 A solution contains $Fe_2(SO_4)_3$ as the only solute. The Fe^{3+} concentration in the solution is 0.48 M. What is the molar concentration of the sulfate ion?

5.110 In a solution of $Al_2(SO_4)_3$, the Al^{3+} concentration is 0.12 M. How many grams of $Al_2(SO_4)_3$ are in 50.0 mL of this solution?

5.111 In a solution of $NiCl_2$, the Cl^- concentration is 0.055 M. How many grams of $NiCl_2$ are in 250 mL of this solution?

Solution Stoichiometry

5.112 What is the molarity of an aqueous solution of potassium hydroxide if 21.34 mL is exactly neutralized by 20.78 mL of 0.116 M HCl by the following reaction?

$$KOH(aq) + HCl(aq) \longrightarrow KCl(aq) + H_2O$$

5.113 What is the molarity of an aqueous sulfuric acid solution if 12.88 mL is neutralized by 26.04 mL of 0.1024 M NaOH? The reaction is

$$2NaOH(aq) + H_2SO_4(aq) \longrightarrow Na_2SO_4(aq) + 2H_2O$$

5.114 How many milliliters of 0.25 M $NiCl_2$ solution are needed to react completely with 20.0 mL of 0.15 M Na_2CO_3 solution? How many grams of $NiCO_3$ will be formed? The reaction is

$$Na_2CO_3(aq) + NiCl_2(aq) \longrightarrow NiCO_3(s) + 2NaCl(aq)$$

5.115 Aluminum sulfate, $Al_2(SO_4)_3$, is used in water treatment to remove fine particles suspended in the water. When made basic, a gel-like precipitate forms that removes the fine particles as it settles. In an experiment, a student planned to react $Al_2(SO_4)_3$ with $Ba(OH)_2$. How many grams of $Al_2(SO_4)_3$ are needed to react with 85.0 mL of 0.0500 M $Ba(OH)_2$? The reaction is

$$Al_2(SO_4)_3(aq) + 3Ba(OH)_2(aq) \longrightarrow$$
$$2Al(OH)_3(s) + 3BaSO_4(s)$$

ILW 5.116 How many milliliters of 0.100 M NaOH are needed to completely neutralize 25.0 mL of 0.250 M H_3PO_4? The reaction is

$$3NaOH(aq) + H_3PO_4(aq) \longrightarrow Na_3PO_4(aq) + 3H_2O$$

5.117 How many grams of baking soda, $NaHCO_3$, are needed to react with 162 mL of stomach acid having an HCl concentration of 0.052 M? The reaction is

$$NaHCO_3(aq) + HCl(aq) \longrightarrow NaCl(aq) + CO_2(g) + H_2O$$

5.118 What is the molarity of an aqueous solution of barium hydroxide if 21.34 mL is exactly neutralized by 20.78 mL of 0.116 M HCl?

5.119 What is the molarity of an aqueous phosphoric acid solution if 12.88 mL is neutralized completely by 26.04 mL of 0.1024 M NaOH?

5.120 How many milliliters of 0.150 M $FeCl_3$ solution are needed to react completely with 20.0 mL of 0.0450 M $AgNO_3$ solution? How many grams of AgCl will be formed? The net ionic equation for the reaction is

$$Ag^+(aq) + Cl^-(aq) \longrightarrow AgCl(s)$$

5.121 How many grams of cobalt(II) chloride are needed to react completely with 60.0 mL of 0.200 M KOH solution? The net ionic equation for the reaction is

$$Co^{2+}(aq) + 2OH^-(aq) \longrightarrow Co(OH)_2(s)$$

5.122 How many milliliters of 0.0500 M $Ba(OH)_2$ are needed to completely neutralize 35.0 mL of 0.125 M H_3PO_4?

5.123 How many grams of baking soda, $NaHCO_3$, are needed to completely neutralize 145 mL of "battery acid," which is H_2SO_4, if the concentration of the acid is 0.0364 M?

ILW 5.124 Consider the reaction of aluminum chloride with silver acetate. How many milliliters of 0.250 M $AlCl_3$ would be needed to react completely with 20.0 mL of 0.500 M $AgC_2H_3O_2$ solution? The net ionic equation for the reaction is

$$Ag^+(aq) + Cl^-(aq) \longrightarrow AgCl(s)$$

5.125 How many milliliters of ammonium sulfate solution having a concentration of 0.250 M are needed to react completely with 50.0 mL of 1.00 M NaOH solution? The net ionic equation for the reaction is

$$NH_4^+(aq) + OH^-(aq) \longrightarrow NH_3(g) + H_2O$$

*5.126 Suppose that 4.00 g of solid Fe_2O_3 is added to 25.0 mL of 0.500 M HCl solution. What will the concentration of the Fe^{3+} be when all the HCl has reacted? What mass of Fe_2O_3 will not have reacted?

*5.127 Suppose 3.50 g of solid $Mg(OH)_2$ is added to 30.0 mL of 0.500 M H_2SO_4 solution. What will the concentration of Mg^{2+} be when all of the acid has been neutralized? How many grams of $Mg(OH)_2$ will not have dissolved?

*5.128 Suppose that 25.0 mL of 0.440 M NaCl is added to 25.0 mL of 0.320 M $AgNO_3$.
(a) How many moles of AgCl would precipitate?
(b) What would be the concentrations of each of the ions in the reaction mixture after the reaction?

*5.129 A mixture is prepared by adding 25.0 mL of 0.185 M Na_3PO_4 to 34.0 mL of 0.140 M $Ca(NO_3)_2$.
(a) What mass of $Ca_3(PO_4)_2$ will be formed?
(b) What will be the concentrations of each of the ions in the mixture after reaction?

Titrations and Chemical Analyses

5.130 A certain lead ore contains the compound $PbCO_3$. A sample of the ore weighing 1.526 g was treated with nitric acid, which dissolved the $PbCO_3$. The resulting solution was filtered from undissolved rock and treated with Na_2SO_4. This gave a precipitate of $PbSO_4$, which was dried and found to weigh 1.081 g. What is the percentage by mass of lead in the ore? (Assume all the lead is precipitated as $PbSO_4$.)

5.131 An ore of barium contains $BaCO_3$. A 1.542 g sample of the ore was treated with HCl to dissolve the $BaCO_3$. The resulting solution was filtered to remove insoluble material and then treated with H_2SO_4 to precipitate $BaSO_4$. The precipitate was filtered, dried, and found to weigh 1.159 g. What is the percentage by mass of barium in the ore? (Assume all the barium is precipitated as $BaSO_4$.)

5.132 In a titration, 23.25 mL of 0.105 M NaOH was needed to react with 21.45 mL of HCl solution. What is the molarity of the acid?

5.133 A 12.5 mL sample of vinegar (containing acetic acid, $HC_2H_3O_2$) was titrated using 0.504 M NaOH solution. The titration required 20.65 mL of the base.
(a) What was the molar concentration of acetic acid in the vinegar?

(b) Assuming the density of the vinegar to be 1.0 g mL^{-1}, what was the percentage (by mass) of acetic acid in the vinegar?

ILW 5.134 Lactic acid, $HC_3H_5O_3$, is a monoprotic acid that forms when milk sours. An 18.5 mL sample of a solution of lactic acid required 17.25 mL of 0.155 M NaOH to reach an end point in a titration. How many moles of lactic acid were in the sample?

5.135 Ascorbic acid (vitamin C) is a diprotic acid having the formula $H_2C_6H_6O_6$. A sample of a vitamin supplement was analyzed by titrating a 0.1000 g sample dissolved in water with 0.0200 M NaOH. A volume of 15.20 mL of the base was required to completely neutralize the ascorbic acid. What was the percentage by mass of ascorbic acid in the sample?

5.136 Magnesium sulfate forms a hydrate known as *Epsom salts*. A student dissolved 1.24 g of this hydrate in water and added a $BaCl_2$ solution until the precipitation of $BaSO_4$ was complete. The precipitate was filtered, dried, and found to weigh 1.174 g. Determine the formula for Epsom salts.

5.137 A sample of iron chloride weighing 0.300 g was dissolved in water and the solution was treated with $AgNO_3$ solution to precipitate the chloride as AgCl. After precipitation was complete, the AgCl was filtered, dried, and found to weigh 0.678 g. Determine the empirical formula of the iron chloride.

5.138 To a mixture of NaCl and Na_2CO_3 with a mass of 1.243 g were added 50.00 mL of 0.240 M HCl (an excess of HCl). The mixture was warmed to expel all of the CO_2 and then the unreacted HCl was titrated with 0.100 M NaOH. The titration required 22.90 mL of the NaOH solution. What was the percentage by mass of NaCl in the original mixture of NaCl and Na_2CO_3?

5.139 A mixture was known to contain both KNO_3 and K_2SO_3. To 0.486 g of the mixture, dissolved in enough water to give 50.00 mL of solution, were added 50.00 mL of 0.150 M HCl (an excess of HCl). The reaction mixture was heated to drive off all of the SO_2, and then 25.00 mL of the reaction mixture was titrated with 0.100 M KOH. The titration required 13.11 mL of the KOH solution to reach an end point. What was the percentage by mass of K_2SO_3 in the original mixture of KNO_3 and K_2SO_3?

ADDITIONAL EXERCISES

5.140 Classify each of the following as a strong electrolyte, weak electrolyte, or nonelectrolyte:
(a) KCl
(b) $C_3H_5(OH)_3$ (glycerin)
(c) NaOH
(d) $C_{12}H_{22}O_{11}$ (sucrose, or table sugar)
(e) $HC_2H_3O_2$ (acetic acid)
(f) CH_3OH (methyl alcohol)
(g) H_2SO_4
(h) NH_3

5.141 Which of the following compounds, when added to water, would not yield a strongly conducting solution? Why?
(a) $CaCO_3$ (c) CH_3OH (e) NH_3
(b) $HCHO_2$ (d) $Pb(NO_3)_2$

5.142 Write the net ionic equations for the following reactions. Include the designations for the states—solid, liquid, gas, or aqueous solution. (For any substance soluble in water, assume that it is used as its aqueous solution.)
(a) $CaCO_3 + HNO_3 \rightarrow Ca(NO_3)_2 + CO_2 + H_2O$
(b) $CaCO_3 + H_2SO_4 \rightarrow CaSO_4 + CO_2 + H_2O$
(c) $FeS + HBr \rightarrow H_2S + FeBr_2$
(d) $KOH + SnCl_2 \rightarrow KCl + Sn(OH)_2$

5.143 Examine each pair of substances, determine if they will react together, and then write the molecular and net ionic equations for any reactions that you predict.
(a) $Na_2S(aq)$ and $H_2SO_4(aq)$
(b) $LiHCO_3(s)$ and $HNO_3(aq)$
(c) $(NH_4)_3PO_4(aq)$ and $KOH(aq)$
(d) $K_2SO_3(aq)$ and $HCl(aq)$
(e) $BaCO_3(s)$ and $HBr(aq)$
(f) $CaCl_2(aq)$ and $HNO_3(aq)$

*5.144 A 0.113 g sample of a mixture of sodium chloride and sodium nitrate was dissolved in water, and enough silver nitrate solution was added to precipitate all of the chloride ion. The mass of silver chloride obtained was 0.277 g. What mass percentage of the sample was sodium chloride?

*5.145 Aspirin is a monoprotic acid called *acetylsalicylic acid*. Its formula is $HC_9H_7O_4$. A certain pain reliever was analyzed for aspirin by dissolving 0.250 g of it in water and titrating it with 0.0300 M KOH solution. The titration required 29.40 mL of base. What is the percentage by weight of aspirin in the drug?

*5.146 In an experiment, 40.0 mL of 0.270 M $Ba(OH)_2$ were mixed with 25.0 mL of 0.330 M $Al_2(SO_4)_3$.
(a) Write the net ionic equation for the reaction that takes place.
(b) What is the total mass of precipitate that forms?
(c) What are the molar concentrations of the ions that remain in the solution after the reaction is complete?

*5.147 Qualitative analysis of an unknown acid found only carbon, hydrogen, and oxygen. In a quantitative analysis, a 10.46 mg sample was burned in oxygen and gave 22.17 mg CO_2 and 3.40 mg H_2O. The molecular mass was determined to be 166 g mol^{-1}. When a 0.1680 g sample of the acid was titrated with 0.1250 M NaOH, the end point was reached after 16.18 mL of the base had been added.
(a) Calculate the percentage composition of the acid.
(b) What is its empirical formula?
(c) What is its molecular formula?
(d) Is the acid mono-, di-, or triprotic?

*5.148 How many milliliters of 0.10 M HCl must be added to 50.0 mL of 0.40 M HCl to give a final solution that has a molarity of 0.25 M?

Oxidation–Reduction Reactions

A Chinese New Year celebration wouldn't be complete without the noise and smoke of firecrackers. The explosions that make firecrackers go "bang" are produced by chemical reactions that involve the transfer of electrons between reactants. Reactions of this type are quite common and are the subject of this chapter.

THIS CHAPTER IN CONTEXT In the preceding chapter you learned about some important reactions that take place in aqueous solutions. This chapter expands on that knowledge with a discussion of a class of reactions that can be viewed as involving the transfer of one or more electrons from one reactant to another. Our goal is to teach you how to recognize and analyze the changes that occur in these reactions and how to balance equations for reactions that involve electron transfer. You will also learn how to apply the principles of stoichiometry to these reactions and how to predict a type of reaction called *single replacement*. These constitute a very broad class of reactions with many practical examples that range from combustion to batteries to the metabolism of foods by our bodies. Although we will introduce you to some of them in this chapter, we will have to wait until a later time to discuss some of the others.

6.1 ▶ Oxidation–reduction reactions involve electron transfer

Among the first reactions studied by early scientists were those that involved oxygen. The combustion of fuels and the reactions of metals with oxygen to give oxides were described by the word *oxidation*. The removal of oxygen from metal oxides to give the metals in their elemental forms was described as *reduction*.

Over time, scientists came to realize that reactions involving oxygen were actually special cases of a much more general phenomenon, one in which electrons are transferred from one substance to another. Collectively, electron transfer reactions came to be called **oxidation–reduction reactions,** or simply **redox reactions.** The term **oxidation** was used to describe the loss of electrons by one reactant, and **reduction** to the gain of electrons by another. For example, the reaction between sodium and chlorine to yield sodium chloride involves a loss of electrons by sodium (*oxidation* of sodium) and a gain of electrons by chlorine (*reduction* of chlorine). We can write these changes in equation form including the electrons, which we represent by the symbol e^-.

$$Na \longrightarrow Na^+ + e^- \qquad \text{(oxidation)}$$

$$Cl_2 + 2e^- \longrightarrow 2Cl^- \qquad \text{(reduction)}$$

We say that sodium (Na) is oxidized and chlorine (Cl_2) is reduced.

Oxidation and reduction always occur together. No substance is ever oxidized unless something else is reduced, and the total number of electrons lost by one substance is always the same as the total number gained by the other. If this were not true, electrons would appear as a product of the overall reaction, and this is never observed.[1] In the reaction of sodium with chlorine, for example, the

Notice that when we write equations of this type, the electron appears as a "product" if the process is oxidation and as a "reactant" if the process is reduction.

---

[1] If electron loss didn't equal electron gain, it would also violate the law of conservation of mass, and that doesn't happen in chemical reactions.

FACETS OF CHEMISTRY 6.1

Chlorine and Laundry Bleach

Although the combustion of fuels is probably the best-known example of a practical application of redox reactions, there are other lesser-known reactions that are almost as important. Each year huge amounts of chlorine, manufactured from salt, are used in purifying municipal drinking water supplies and treating wastewater. Chlorine is a powerful oxidizing agent, and its function in water purification is to kill bacteria. It does this by oxidizing the bacteria and causing their death.

Another use for chlorine and the chlorine-containing anion ClO^- is as a bleach. Ordinary chlorine bleaches such as Clorox consist of a dilute solution of sodium hypochlorite, NaOCl. The hypochlorite ion is a strong oxidizing agent and performs its bleaching action by destroying dyes and other colored matter. In the paper and cotton processing industries, this bleaching function has been accomplished by using chlorine directly.

Today, industry is moving away from the use of chlorine as a bleach because of environmental concerns. It is feared that the products of the reaction of chlorine with organic substances may be toxic, and in fact, many well-known insecticides are chlorine-containing compounds. Even more disturbing are reports that some chlorine-containing organic compounds mimic the behavior of the hormone estrogen.

"Chlorine" bleach, NaOCl(aq), destroys dyes by oxidizing them to colorless products.

---

overall reaction is

$$2Na + Cl_2 \longrightarrow 2NaCl$$

When two sodium atoms are oxidized, two electrons are lost, which is exactly the number of electrons gained when one Cl_2 molecule is reduced.

For a redox reaction to occur, there must be some substance to accept the electrons that another substance loses. The substance that accepts the electrons is called the **oxidizing agent** because it allows the other to lose electrons and be oxidized. Similarly, the substance that supplies the electrons is called the **reducing agent** because it helps something else to be reduced. In our example above, sodium is a serving as a reducing agent when it supplies electrons to chlorine (i.e., it is the agent that enables chlorine to be reduced). In the process, sodium is oxidized. Chlorine is an oxidizing agent when it accepts electrons from the sodium, and when that happens, chlorine is reduced to chloride ion. One way to remember this is by the following summary:

> The reducing agent is the substance that is oxidized.
> The oxidizing agent is the substance that is reduced.

Redox reactions are very common. They occur in batteries, which are built so that the electrons transferred can pass through some external circuit where they are able to light a flashlight or power a laptop computer. The metabolism of foods, which supplies our bodies with energy, also occurs by a series of redox reactions. And ordinary household bleach works by oxidizing substances that stain fabrics, making them colorless or easier to remove from the fabric. (See Facets of Chemistry 6.1.)

The *oxidizing agent* causes oxidation to occur by accepting electrons (it gets reduced). The *reducing agent* causes reduction by supplying electrons (it gets oxidized).

Liquid bleach contains hypochlorite ion, OCl^-, as the oxidizing agent.

The bright light produced by the reaction between magnesium and oxygen often is used in fireworks displays, as shown in the photo in the margin. The product of the reaction is magnesium oxide, MgO, an ionic compound.

$$2Mg + O_2 \longrightarrow 2MgO$$

Which element is oxidized and which is reduced? What are the oxidizing and reducing agents?

ANALYSIS: This question asks us to apply the definitions presented above. To accomplish this, we need to know how the electrons are transferred when magnesium and oxygen react, which means we need to know exactly what happens to Mg and O_2 in the reaction. Here we can apply what we learned in Chapter 2.

Magnesium oxide is a compound of a metal and a nonmetal, so it's an ionic substance. The locations of the elements in the periodic table tell us that the ions in MgO are Mg^{2+} and O^{2-}. (You probably already knew the formulas of these ions, but we're showing how you can apply previous knowledge to the problem at hand.) Now that we have the ions, we can determine how electrons are exchanged.

SOLUTION: When a magnesium atom becomes a magnesium ion, it must lose two electrons. Because electrons are lost in this process, they appear on the right when we express it as an equation.

$$Mg \longrightarrow Mg^{2+} + 2e^-$$

By losing electrons, magnesium is oxidized. This also means that Mg must be the reducing agent.

When oxygen reacts to yield O^{2-} ions, each oxygen atom must gain two electrons, so an O_2 molecule must gain four electrons. Because electrons are gained, they appear on the left when we write the equation.

$$O_2 + 4e^- \longrightarrow 2O^{2-}$$

By gaining electrons, O_2 is reduced and must be the oxidizing agent.

Is the Answer Reasonable?
There are two things we can do to check our answers. First, we can check to be sure that we've placed electrons on the correct sides of the equations. As with ionic equations, *the number of atoms of each kind and net charge must be the same on both sides.* We see that this is true for both equations. (If we had placed the electrons on the wrong side, the charges would not balance.) By observing the locations of the electrons in the equations (on the right for oxidation; on the left for reduction), we come to the same conclusions that Mg is oxidized and O_2 is reduced.

Another check is noting that we've identified one substance as being oxidized and the other as being reduced. If we had made a mistake, we might have found that both were oxidized, or both were reduced. But that's impossible, because in every reaction in which there is oxidation, there must also be reduction.

PRACTICE EXERCISE 1: Identify the substances oxidized and reduced and the oxidizing and reducing agents in the reaction of aluminum and chlorine to form aluminum chloride.

EXAMPLE 6.1

Identifying Oxidation–Reduction

Fireworks display over Washington, D.C.

Oxidation numbers are used to follow redox changes

The reaction of magnesium with oxygen in the preceding example is clearly a redox reaction. However, not all reactions with oxygen produce ionic products. For example, sulfur reacts with oxygen to give sulfur dioxide, SO_2, which is molecular. Nevertheless, it is convenient to also view this as a redox reaction, but to do so we must make some changes in the way we define oxidation and reduction. To do this, chemists developed a bookkeeping system called **oxidation numbers** that provides a way to keep tabs on electron transfers.

The oxidation number of an element in a particular compound is assigned according to a set of rules, which are described below. For simple, monatomic ions in a compound such as NaCl, the oxidation numbers are the same as the charges on the ions, so in NaCl the oxidation number of Na^+ is +1 and the oxidation number of Cl^- is −1. The real value of oxidation numbers is that they can also be assigned to atoms in molecular compounds. In such cases, it is important to realize that the oxidation number does not actually equal a charge on an atom. To be sure to differentiate oxidation numbers from actual electrical charges, we will specify the sign *before* the number when writing oxidation numbers, and *after* the number when writing electrical charges. Thus, a sodium ion has a charge of 1+ and an oxidation number of +1.

A term that is frequently used interchangeably with *oxidation number* is **oxidation state.** In NaCl, for example, sodium has an oxidation number of +1 and is said to be "in the +1 oxidation state." Similarly, the chlorine in NaCl is said to be in the −1 oxidation state. There are times when it is convenient to specify the oxidation state of an element when its name is written out. This is done by writing the oxidation number as a Roman numeral in parentheses after the name of the element. For example, "iron(III)" means iron in the +3 oxidation state.

We can use these new terms now to redefine a redox reaction.

> A **redox reaction** is a chemical reaction in which changes in oxidation numbers occur.

We will find it easy to follow redox reactions by taking note of the changes in oxidation numbers. To do this, however, we must be able to assign oxidation numbers to atoms in a quick and simple way.

Assignment of oxidation numbers

We can use some basic knowledge learned earlier plus the set of rules below to determine the oxidation numbers of the atoms in almost any compound.

Rule 1 means that O in O_2, P in P_4, and S in S_8 all have oxidation numbers of zero.

Rules for assigning oxidation numbers

Rules for Assigning Oxidation Numbers

1. The oxidation number of any free element (an element not combined chemically with a different element) is zero, regardless of how complex its molecules are.
2. The oxidation number for any simple, monatomic ion (e.g., Na^+ or Cl^-) is equal to the charge on the ion.
3. The sum of all the oxidation numbers of the atoms in a molecule or polyatomic ion must equal the charge on the particle.
4. In its compounds, fluorine has an oxidation number of −1.
5. In its compounds, hydrogen has an oxidation number of +1.
6. In its compounds, oxygen has an oxidation number of −2.

As you will see shortly, occasionally there is a conflict between two of these rules. When this happens, *we apply the rule with the lower number and ignore the conflicting rule.*

In addition to these basic rules, there is some other chemical knowledge you will need to remember. Recall that we can use the periodic table to help us remember the charges on certain ions of the elements. For instance, all the metals in Group IA form ions with a 1+ charge, and all those in Group IIA form ions with a 2+ charge. This means that when we find sodium in a compound, we can assign it an oxidation number of +1 because its simple ion, Na^+, has a charge of 1+. (We have applied Rule 2.) Similarly, calcium in a compound exists as Ca^{2+} and has an oxidation number of +2.

In binary ionic compounds with metals, the nonmetals have oxidation numbers equal to the charges on their anions (Table 2.3 on page 59). For example, the compound Fe_2O_3 contains the oxide ion, O^{2-}, which is assigned an oxidation number of -2. Similarly, Mg_3P_2 contains the phosphide ion, P^{3-}, which has an oxidation number of -3.

The numbered rules given above usually come into play when an element is capable of having more than one oxidation state. For example, you learned in Chapter 2 that transition metals can form more than one ion. Iron, for example, forms Fe^{2+} and Fe^{3+} ions, so in an iron compound we have to use the rules to figure out which iron ion is present. Similarly, when nonmetals are combined with hydrogen and oxygen in compounds or polyatomic ions, their oxidation numbers can vary and must be calculated using the rules.

With this as background, let's look at some examples that illustrate how we apply the rules, especially when they don't explicitly cover all the atoms in a formula.

Molybdenum disulfide, MoS_2, has a structure that allows it to be used as a dry lubricant, much like graphite. What are the oxidation numbers of the atoms in MoS_2?

ANALYSIS: This is a binary compound of a metal and a nonmetal, so for the purposes of assigning oxidation numbers, we assume it to contain ions. Molybdenum is a transition metal, so there is no simple rule to tell what its ions are. However, sulfur is a nonmetal in Group VIA, and its ion is S^{2-}. We'll begin the solution with this knowledge.

SOLUTION: Because S^{2-} is a simple monatomic ion, its charge equals its oxidation number, so sulfur has an oxidation number of -2. Once we know the oxidation number of sulfur, we can use Rule 3 (the summation rule) to determine the oxidation number of molybdenum, which we will represent by x.

$$
\begin{array}{lll}
S & (2 \text{ atoms}) \times (-2) = -4 & (\text{Rule 2}) \\
\underline{Mo} & \underline{(1 \text{ atom}) \times \quad (x) = x} & \\
& \text{Sum} = 0 & (\text{Rule 3})
\end{array}
$$

The value of x must be $+4$ for the sum to be zero. Therefore, the oxidation numbers are

$$Mo = +4 \quad \text{and} \quad S = -2$$

Is the Answer Reasonable?
There's very little to check here. All we can do is review the solution to look for errors, and there don't appear to be any.

EXAMPLE 6.2
Assigning Oxidation Numbers

If necessary, you might review Table 2.3 on page 59, which summarizes the ions formed by the representative elements.

Chlorite ion, ClO_2^-, has been shown to be a potent disinfectant, and solutions of it are sometimes used to disinfect air-conditioning systems in cars. What are the oxidation numbers of chlorine and oxygen in this ion?

ANALYSIS: This is a polyatomic ion, so the rule for simple ions doesn't work. Therefore, we have to use the other rules described above. To do this, we look over the rules to see which might apply. If there are any atoms not covered by the rules, we can expect to use the summation rule (Rule 3).

SOLUTION: Examining our rules, we see we have one for oxygen but not for chlorine. Therefore, we will use the rule for oxygen and the summation rule to figure out the oxidation number of chlorine (symbolized by x). This time the sum of oxidation numbers must add up to -1 because the charge on the ClO_2^- is $1-$.

$$
\begin{array}{lll}
O & (2 \text{ atoms}) \times (-2) = -4 & (\text{Rule 6}) \\
\underline{Cl} & \underline{(1 \text{ atom}) \times \quad (x) = x} & \\
& \text{Sum} = -1 & (\text{Rule 3})
\end{array}
$$

EXAMPLE 6.3
Assigning Oxidation Numbers

For the sum to be -1, the value of x must be $+3$, which is the oxidation number of the chlorine. Therefore,

$$Cl = +3 \qquad \text{and} \qquad O = -2$$

Is the Answer Reasonable?

As before, there is no simple check to confirm the answers are correct. Just be sure the oxidation numbers add up to the charge on the substance.

EXAMPLE 6.4

Assigning Oxidation Numbers

Hydrogen peroxide destroys bacteria by oxidizing them.

Determine the oxidation numbers of the atoms in hydrogen peroxide, H_2O_2, a common antiseptic purchased in pharmacies.

ANALYSIS: The compound is molecular, so there are no ions to work with. Scanning the rules, we see we have one for hydrogen and another for oxygen. However, we begin to see we have a conflict between two of the rules. Rule 5 tells us to assign an oxidation number of $+1$ to hydrogen, and Rule 6 tells us to give oxygen an oxidation number of -2. Both of these can't be correct because the sum must be zero (Rule 3).

$$
\begin{array}{lll}
H & (2 \text{ atoms}) \times (+1) = +2 & \text{(Rule 5)} \\
O & (2 \text{ atoms}) \times (-2) = -4 & \text{(Rule 6)} \\
\hline
 & \text{Sum} \neq 0 & \text{(violates Rule 3)}
\end{array}
$$

As mentioned above, when there is a conflict between the rules, we ignore the one with the higher number that causes the conflict.

SOLUTION: Rule 6 is the one with the higher number, so we have to ignore it this time and just apply Rules 3 (the summation rule) and 5 (the rule that tells us the oxidation number of hydrogen is $+1$). Because we don't have a rule that applies to oxygen in this case, we'll represent the oxidation number of O by x.

$$
\begin{array}{lll}
H & (2 \text{ atoms}) \times (+1) = +2 & \text{(Rule 5)} \\
O & (2 \text{ atoms}) \times (x) = 2x & \\
\hline
 & \text{Sum} = 0 & \text{(Rule 3)}
\end{array}
$$

For the sum to be zero, $2x = -2$, so $x = -1$. Therefore, in this compound, oxygen has an oxidation number of -1.

$$H = +1 \qquad O = -1$$

Is the Answer Reasonable?

A conflict between the rules occurs only rarely, but when it happens, the conflict becomes apparent because the summation rule *always* applies.

EXAMPLE 6.5

Assigning Oxidation Numbers

Calcium reacts with hydrogen to form the compound calcium hydride, CaH_2. What are the oxidation numbers of the atoms in this compound?

ANALYSIS: The compound contains a metal and a nonmetal, so we can treat it as ionic. This means that if we can determine what the ions are, we can easily assign oxidation numbers (the oxidation number equals the charge on a monatomic ion).

SOLUTION: Calcium is a Group IIA metal, so we expect it to exist as Ca^{2+} in its compounds. If calcium is the cation, then hydrogen must be an anion. We haven't encountered any negative ions of hydrogen before, but we can figure out the charge because the compound is electrically neutral overall. To balance the $2+$ charge of the Ca^{2+} ion, the charge on the two hydrogen anions (hydride ions) must be $2-$. Therefore, each hydride ion must be H^-. Having the charges on the ions, we can state that the oxidation number of calcium is $+2$ and the oxidation number of hydrogen is -1.

Because we have a rule for hydrogen in the assignment of oxidation numbers, we might be tempted to try to apply it. Once again, however, there is a conflict. We know that calcium is Ca^{2+} and according to Rule 2 its oxidation number is $+2$. Rule 3 demands that the sum of the oxidation numbers be equal to the charge on the formula, which is zero, and the rule for hydrogen (Rule 5) would require that hydrogen be $+1$. But we can't apply Rule 5 because that would lead to a violation of Rule 3.

$$Ca \quad (1 \text{ atom}) \times (+2) = +2 \quad (\text{Rule 2})$$
$$\underline{H \quad (2 \text{ atoms}) \times (+1) = +2 \quad (\text{Rule 5})}$$
$$\text{Sum} \neq 0 \quad (\text{violation of Rule 3})$$

Rule 3 is higher up than Rule 5, so Rule 3 takes precedence and we must ignore Rule 5. This gives

$$Ca \quad (1 \text{ atom}) \times (+2) = +2 \quad (\text{Rule 2})$$
$$\underline{H \quad (2 \text{ atoms}) \times (x) = 2x \quad (\text{ignoring Rule 5})}$$
$$\text{Sum} = 0 \quad (\text{Rule 3})$$

For the sum to be zero, $2x$ must equal -2, so x must equal -1. The oxidation number of hydrogen is -1 in this compound.

Is the Answer Reasonable?
There's really nothing to do here. We've already solved the problem using two different approaches, so we can feel confident the answers are correct.

PRACTICE EXERCISE 2: When hydrogen peroxide is used as an antiseptic, it kills bacteria by oxidizing them. When the H_2O_2 serves as an oxidizing agent, which product might be formed from it, O_2 or H_2O? Why?

Sometimes, oxidation numbers calculated by the rules have fractional values, as illustrated in the next example.

The air bags used as safety devices in modern autos are inflated by the very rapid decomposition of the ionic compound sodium azide, NaN_3. The reaction gives elemental sodium and gaseous nitrogen. What is the average oxidation number of the nitrogen in sodium azide?

ANALYSIS: We're told that NaN_3 is ionic, so we know there is a cation and an anion. The cation is the sodium ion, Na^+, which means that the remainder of the formula unit, "N_3," must carry one unit of negative charge. However, there could not be three nitrogen anions each with a charge of $-\frac{1}{3}$ because it would mean that each nitrogen atom had acquired one-third of an electron. Whole numbers of electrons are always involved in electron transfers. Therefore, the anion must be a single particle with a negative 1 charge, namely N_3^-. For this anion, the sum of the oxidation numbers of the three nitrogens must be -1. We can use this information, now, to solve the problem.

SOLUTION: The sum of the oxidation numbers of the nitrogen must be equal to -1. If we let x stand for the oxidation number of just one of the nitrogens, then

$$(3 \text{ atoms}) \times (x) = 3x = -1$$
$$x = -\tfrac{1}{3}$$

We could have also tackled the problem just using the rules for assigning oxidation numbers. We know that sodium exists as the ion Na^+ and that the compound is neutral overall. Therefore,

EXAMPLE 6.6
Assigning Oxidation Numbers

The explosive decomposition of sodium azide releases nitrogen gas, which rapidly inflates an air bag during a crash.

$$Na \qquad (1\text{ atom}) \times (+1) = +1 \qquad (\text{Rule 2})$$

$$N \qquad (3\text{ atoms}) \times \quad (x) = 3x$$

$$\text{Sum} = 0 \qquad (\text{Rule 3})$$

The sum of the oxidation numbers of the three nitrogen atoms in this ion must add up to -1, so each nitrogen must have an oxidation number of $-\frac{1}{3}$.

Is the Answer Reasonable?
As with the preceding problem, we've solved it two ways, so we can certainly feel confident the answer is correct.

The charge on the polyatomic ion equals the sum of the oxidation numbers of its atoms.

Many compounds contain familiar polyatomic ions such as ammonium (NH_4^+), sulfate (SO_4^{2-}), nitrate (NO_3^-), and acetate ($C_2H_3O_2^-$). The charges on these ions can be considered to represent the *net oxidation number* of the atoms that make up the ion. We can use this sometimes to help assign oxidation numbers, as shown in Example 6.7.

EXAMPLE 6.7
Assigning Oxidation Numbers

Titanium dioxide, TiO_2, is a white pigment used in making paint. A now outmoded process of making TiO_2 from its ore involved $Ti(SO_4)_2$ as an intermediate. What is the oxidation number of titanium in $Ti(SO_4)_2$?

ANALYSIS: The key here is recognizing SO_4 as the sulfate ion, SO_4^{2-}. Once we have this, the rest is straightforward. Because we're only interested in the oxidation number of titanium, we use the charge on SO_4^{2-} to equal the sum of all the oxidation numbers of the atoms in the ion.

SOLUTION: The sulfate ion has a charge of $2-$, which we can take to be its net oxidation number. Then we apply the summation rule.

$$Ti \qquad (1\text{ atom}) \times \quad (x) = x$$

$$SO_4^{2-} \qquad (2\text{ ions}) \times (-2) = -4$$

$$\text{Sum} = 0 \qquad (\text{Rule 3})$$

Obviously, the oxidation number of titanium is $+4$.

Is the Answer Reasonable?
We have the sum of oxidation numbers adding up to zero, the charge on $Ti(SO_4)_2$, so our answer is okay.

PRACTICE EXERCISE 3: Assign oxidation numbers to each atom in (a) $NiCl_2$, (b) Mg_2TiO_4, (c) $K_2Cr_2O_7$, (d) HPO_4^{2-}, (e) $V(C_2H_3O_2)_3$.

PRACTICE EXERCISE 4: Iron forms a magnetic oxide with the formula Fe_3O_4 that contains both Fe^{2+} and Fe^{3+} ions. What is the *average* oxidation number of iron in this oxide?

Oxidation numbers are used to identify oxidation and reduction

Oxidation numbers can be used in several ways. One is to define oxidation and reduction in the most comprehensive manner, as follows:

Oxidation is an increase in oxidation number.
Reduction is a decrease in oxidation number.

Let's see how these definitions apply to the reaction of hydrogen with chlorine. To avoid ever confusing oxidation numbers with actual electrical charges, we will write oxidation numbers directly above the chemical symbols of the elements.

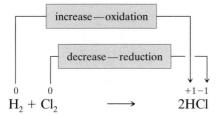

$$H_2 + Cl_2 \longrightarrow 2HCl$$

Notice we have assigned the atoms in H_2 and Cl_2 oxidation numbers of zero, in accord with Rule 1. The changes in oxidation number tell us that hydrogen is oxidized and chlorine is reduced.

Identify the substance oxidized and the substance reduced as well as the oxidizing and reducing agents in the reaction

$$2KCl + MnO_2 + 2H_2SO_4 \longrightarrow K_2SO_4 + MnSO_4 + Cl_2 + 2H_2O$$

ANALYSIS: No doubt, this is the most complex chemical equation you've encountered so far. And just looking at it, it is not even apparent that it is a redox reaction. To determine whether the reaction is redox and to give the specific answers requested, we need to apply our revised definitions of oxidation and reduction. This will tell us what is oxidized and reduced. Then we recall that the substance oxidized is the reducing agent, and the substance reduced is the oxidizing agent.

SOLUTION: To identify redox changes using oxidation numbers, we first must assign oxidation numbers to each atom on both sides of the equation. Following the rules, we get

$$\overset{+1\,-1}{2KCl} + \overset{+4\,-2}{MnO_2} + \overset{+1\,+6\,-2}{2H_2SO_4} \longrightarrow \overset{+1\,+6\,-2}{K_2SO_4} + \overset{+2\,+6\,-2}{MnSO_4} + \overset{0}{Cl_2} + \overset{+1\,-2}{2H_2O}$$

Next we look for changes, keeping in mind that an increase in oxidation number is oxidation and a decrease is reduction.

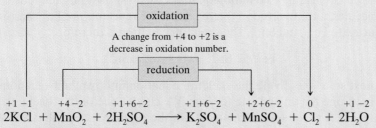

Thus the Cl in KCl is oxidized and the Mn in MnO_2 is reduced. The reducing agent is KCl and the oxidizing agent is MnO_2. (Notice that in identifying the oxidizing and reducing agents, we specify the entire formulas for the substances containing the atoms changing oxidation numbers.

Is the Answer Reasonable?
There are lots of things we could check here. That we have found changes that lead us to identify one as oxidation and the other as reduction gives us confidence that we've done the rest of the work correctly.

PRACTICE EXERCISE 5: Chlorine dioxide, ClO_2, is used to kill bacteria in the dairy industry, meat industry, and other food and beverage industry applications. It is unstable, but can be made by the following reaction:

$$Cl_2 + 2NaClO_2 \longrightarrow 2ClO_2 + 2NaCl$$

EXAMPLE 6.8

Using Oxidation Numbers to Follow Redox Reactions

Identify the substances oxidized and reduced as well as the oxidizing and reducing agents in the reaction.

PRACTICE EXERCISE 6: Identify the substances oxidized and reduced as well as the oxidizing and reducing agents in the following reaction:

$$KClO_3 + 3HNO_2 \longrightarrow KCl + 3HNO_3$$

6.2 ▶ The ion–electron method creates balanced net ionic equations for redox reactions

Many redox reactions take place in aqueous solution and many of these involve ions; they are ionic reactions. An example is the reaction of laundry bleach with substances in the wash water. The active ingredient in the bleach is hypochlorite ion, OCl^-, which is the oxidizing agent in these reactions. To study redox reactions, it is often helpful to write ionic and net ionic equations, just as we did in our analysis of metathesis reactions in Chapter 5. Balancing net ionic equations for redox reactions is especially easy if we follow a procedure called the *ion–electron method.*

The ion–electron method uses a divide and conquer approach

Equations for redox reactions are often more complex than those for metathesis reactions and can be difficult to balance by inspection. The ion–electron method is a systematic procedure for balancing redox equations.

In the **ion–electron method,** we divide the oxidation and reduction processes into individual equations called **half-reactions** that are balanced separately. Then we combine the balanced half-reactions to obtain the fully balanced net ionic equation.

To illustrate the method, let's balance the net ionic equation for the reaction in aqueous solution of iron(III) chloride, $FeCl_3$, with tin(II) chloride, $SnCl_2$, which changes the iron to Fe^{2+} and the tin to Sn^{4+}. In the reaction the chloride ion is unaffected—it's a spectator ion.

We begin by writing a **skeleton equation,** which shows only the ions (or sometimes molecules, too) involved in the reaction. The reactants are Fe^{3+} and Sn^{2+}, and the products are Fe^{2+} and Sn^{4+}. The skeleton equation is therefore

$$Fe^{3+} + Sn^{2+} \longrightarrow Fe^{2+} + Sn^{4+}$$

At first glance, the reaction appears to be balanced, but it is not. The net charge is not the same on both sides.

The next step is to divide the skeleton equation into two half-reactions. In this case, we have one for tin and one for iron. We choose one of the reactants, let's say Sn^{2+}, and write it at the left of an arrow. On the right, we write what Sn^{2+} changes to, which is Sn^{4+}. This gives us the beginnings of one half-reaction. For the second half-reaction, we write the other reactant, Fe^{3+}, on the left and the other product, Fe^{2+}, on the right.

$$Sn^{2+} \longrightarrow Sn^{4+}$$
$$Fe^{3+} \longrightarrow Fe^{2+}$$

Next, we balance the half-reactions so that each obeys *both* criteria for a balanced ionic equation: *both atoms and charge have to balance.* Obviously, the atoms are already balanced in each equation. The charge, however, is not. *To bring the charge into balance, we add electrons to whichever side of the half-reaction is more positive (or less negative).* For the first half-reaction, two electrons are added to the right, so the net charge on both sides will be 2+. In the second half-reaction, one electron is added to the left, which makes the net charge on both sides equal to 2+.

The net charge is 2+.

$$Sn^{2+} \longrightarrow Sn^{4+} + 2e^-$$

$$Fe^{3+} + e^- \longrightarrow Fe^{2+}$$

The net charge is 2+.

In any overall redox reaction, the number of electrons gained always equals the number lost (otherwise, electrons would be a reactant or a product, and they never are). For the electron gain to equal the electron loss, the second reaction has to occur twice each time the first reaction occurs once. We make this so by multiplying each of the coefficients in the second equation by 2.

$$Sn^{2+} \longrightarrow Sn^{4+} + 2e^-$$

$$2(Fe^{3+} + e^- \longrightarrow Fe^{2+})$$

This gives

$$Sn^{2+} \longrightarrow Sn^{4+} + 2e^-$$

$$2Fe^{3+} + 2e^- \longrightarrow 2Fe^{2+}$$

> We can tell Sn^{2+} is oxidized because the electrons appear on the right in its half-reaction. Similarly, we can tell Fe^{3+} is reduced because we find the electron on the left in its half-reaction.

Finally, we combine the balanced half-reactions by adding them together. The result is

$$Sn^{2+} + 2Fe^{3+} + 2e^- \longrightarrow Sn^{4+} + 2Fe^{2+} + 2e^-$$

Dropping $2e^-$ from both sides gives us the final balanced equation.

$$Sn^{2+} + 2Fe^{3+} \longrightarrow Sn^{4+} + 2Fe^{2+}$$

Notice that the charge *and* the number of atoms are now balanced.

PRACTICE EXERCISE 7: Balance the following equation by the ion–electron method:

$$Al(s) + Cu^{2+}(aq) \longrightarrow Al^{3+}(aq) + Cu(s)$$

Many redox reactions involve H⁺ and OH⁻

In many redox reactions in aqueous solutions, H^+ or OH^- ions play an important role, as do water molecules. For example, when solutions of $K_2Cr_2O_7$ and $FeSO_4$ are mixed, the acidity of the mixture decreases as dichromate ion, $Cr_2O_7^{2-}$, oxidizes Fe^{2+}. This is because the reaction uses up H^+ as a reactant and produces H_2O as a product. In other reactions, OH^- is consumed, while in still others H_2O is a reactant. Another fact is that in many cases the products (or even the reactants) of a redox reaction will differ depending on the acidity of the solution. For example, in an acidic solution MnO_4^- is reduced to Mn^{2+} ion, but in a neutral or slightly basic solution, the reduction product is insoluble MnO_2.

Because of these factors, redox reactions are generally carried out in solutions containing a substantial excess of either acid or base, so before we can apply the ion–electron method, we have to know whether the reaction occurs in an acidic or a basic solution. (This information will always be given to you in this book.)

A solution of $K_2Cr_2O_7$ being added to an acidic solution containing Fe^{2+}.

CHEMISTRY IN PRACTICE A Breathalyzer test for alcohol makes use of the oxidation of alcohol by dichromate ion. The dichromate ion is orange, but its reduction product is green Cr^{3+} ion. The breath of a person who is intoxicated contains alcohol vapor, which passes through an acidic solution containing $Cr_2O_7^{2-}$ when the person blows through the Breathalyzer device. Any alcohol in the exhaled air is oxidized, which is signaled by the appearance of green Cr^{3+}. The more alcohol the person has consumed, the greater the intensity of the green color.

H^+ and H_2O help balance redox equations for acidic solutions

As you just learned, $Cr_2O_7{}^{2-}$ reacts with Fe^{2+} in an acidic solution to give Cr^{3+} and Fe^{3+} as products. This information gives the skeleton equation,

$$Cr_2O_7{}^{2-} + Fe^{2+} \longrightarrow Cr^{3+} + Fe^{3+}$$

We then proceed through the steps described below to find the balanced equation. As you will see, the ion–electron method will tell us how H^+ and H_2O are involved in the reaction; we don't need to know this in advance.

STEP 1 *Divide the skeleton equation into half-reactions.*
We create the beginnings of two half-reactions, being sure to use the complete formulas for the ions that appear in the skeleton equation. Except for hydrogen and oxygen, the same elements must appear on both sides of a given half-reaction.

$$Cr_2O_7{}^{2-} \longrightarrow Cr^{3+}$$
$$Fe^{2+} \longrightarrow Fe^{3+}$$

STEP 2 *Balance atoms other than H and O.*
There are two Cr atoms on the left and only one on the right, so we place a coefficient of 2 in front of Cr^{3+}. The second half-reaction is already balanced in terms of atoms, so nothing need be done to it.

Many students tend to forget this step. If you do, you may end up in trouble later on.

$$Cr_2O_7{}^{2-} \longrightarrow 2Cr^{3+}$$
$$Fe^{2+} \longrightarrow Fe^{3+}$$

STEP 3 *Balance oxygen by adding H_2O to the side that needs O.*
There are seven oxygen atoms on the left of the first half-reaction and none on the right. Therefore, we add $7H_2O$ to the right side of the first half-reaction to balance the oxygens. (There is no oxygen imbalance in the second half-reaction, so there's nothing to do there.)

We use H_2O, not O or O_2, to balance oxygen atoms because H_2O is what is actually present in the solution.

$$Cr_2O_7{}^{2-} \longrightarrow 2Cr^{3+} + 7H_2O$$
$$Fe^{2+} \longrightarrow Fe^{3+}$$

The oxygen atoms now balance, but we've created an imbalance in hydrogen. That issue is addressed next.

STEP 4 *Balance hydrogen by adding H^+ to the side that needs H.*
After adding the water, we see that we've created an imbalance in the hydrogens; the first half-reaction has 14 hydrogens on the right and none on the left. To balance hydrogen, we add $14H^+$ to the left side of the half-reaction. When you do this step (or others) *be careful to write the charges on the ions.* If they are omitted, you will not obtain a balanced equation in the end.

$$14H^+ + Cr_2O_7{}^{2-} \longrightarrow 2Cr^{3+} + 7H_2O$$
$$Fe^{2+} \longrightarrow Fe^{3+}$$

Now each half-reaction is balanced in terms of atoms. Next we will balance the charge.

STEP 5 *Balance the charge by adding electrons.*
First we compute the net electrical charge on each side. For the first half-reaction, we have

$$\underbrace{14H^+ + Cr_2O_7{}^{2-}}_{\text{Net charge} = (14+) + (2-) = 12+} \longrightarrow \underbrace{2Cr^{3+} + 7H_2O}_{\text{Net charge} = 2(3+) + 0 = 6+}$$

The algebraic difference between the net charges on the two sides equals the number of electrons that must be added to the more positive (or less negative) side. In this instance, we must add $6e^-$ to the left side of the half-reaction.

$$6e^- + 14H^+ + Cr_2O_7{}^{2-} \longrightarrow 2Cr^{3+} + 7H_2O$$

This half-reaction is now complete; it is balanced in terms of both atoms and charge. (We can check this by recalculating the charge on both sides.)

To balance the other half-reaction, we add one electron to the right.

$$Fe^{2+} \longrightarrow Fe^{3+} + e^-$$

STEP 6 *Make the number of electrons gained equal to the number lost and then add the two half-reactions.*

At this point we have the two balanced half-reactions

$$6e^- + 14H^+ + Cr_2O_7{}^{2-} \longrightarrow 2Cr^{3+} + 7H_2O$$

$$Fe^{2+} \longrightarrow Fe^{3+} + e^-$$

Six electrons are gained in the first, but only one is lost in the second. Therefore, before combining the two equations we multiply all of the coefficients of the second one by 6.

$$6e^- + 14H^+ + Cr_2O_7{}^{2-} \longrightarrow 2Cr^{3+} + 7H_2O$$

$$6(Fe^{2+} \longrightarrow Fe^{3+} + e^-)$$

(Sum) $6e^- + 14H^+ + Cr_2O_7{}^{2-} + 6Fe^{2+} \longrightarrow 2Cr^{3+} + 7H_2O + 6Fe^{3+} + 6e^-$

STEP 7 *Cancel anything that is the same on both sides.*

This is the final step. Six electrons cancel from both sides to give the final balanced equation.

$$14H^+ + Cr_2O_7{}^{2-} + 6Fe^{2+} \longrightarrow 2Cr^{3+} + 7H_2O + 6Fe^{3+}$$

Notice that both the charge and the atoms balance.

In some reactions, after adding the two half-reactions you may have H_2O or H^+ on both sides—for example, $6H_2O$ on the left and $2H_2O$ on the right. Cancel as many as you can. Thus,

$$\ldots + 6H_2O \ldots \longrightarrow \ldots + 2H_2O \ldots$$

gives

$$\ldots + 4H_2O \ldots \longrightarrow \ldots$$

The following is a summary of the steps we've followed for balancing an equation for a redox reaction in an acidic solution. If you don't skip any steps and you perform them in the order given, you will always obtain a properly balanced equation.

Because we know the electrons will cancel, we really don't have to carry them down into the combined equation. We've done so here just for emphasis.

Ion–Electron Method—Acidic Solution

STEP 1 Divide the equation into two half-reactions.

STEP 2 Balance atoms other than H and O.

STEP 3 Balance O by adding H_2O.

STEP 4 Balance H by adding H^+.

STEP 5 Balance net charge by adding e^-.

STEP 6 Make e^- gain equal e^- loss; then add half-reactions.

STEP 7 Cancel anything that's the same on both sides.

 TOOLS

Ion–electron method for acidic solutions

Balance the following equation. The reaction occurs in an acidic solution.

$$MnO_4{}^- + H_2SO_3 \longrightarrow SO_4{}^{2-} + Mn^{2+}$$

ANALYSIS: In using the ion–electron method, there's not much to analyze. It's necessary to know the steps and to follow them in sequence.

SOLUTION: We follow the steps just given.

STEP 1 Divide the skeleton equation into half-reactions

EXAMPLE 6.9

Using the Ion–Electron Method

$$MnO_4^- \longrightarrow Mn^{2+}$$

$$H_2SO_3 \longrightarrow SO_4^{2-}$$

STEP 2 There is nothing to do for this step. All the atoms except H and O are already in balance.

STEP 3 Add H_2O to balance oxygens.

$$MnO_4^- \longrightarrow Mn^{2+} + 4H_2O$$

$$H_2O + H_2SO_3 \longrightarrow SO_4^{2-}$$

STEP 4 Add H^+ to balance H.

$$8H^+ + MnO_4^- \longrightarrow Mn^{2+} + 4H_2O$$

$$H_2O + H_2SO_3 \longrightarrow SO_4^{2-} + 4H^+$$

STEP 5 Balance the charge by adding electrons to the more positive side.

$$5e^- + 8H^+ + MnO_4^- \longrightarrow Mn^{2+} + 4H_2O$$

$$H_2O + H_2SO_3 \longrightarrow SO_4^{2-} + 4H^+ + 2e^-$$

STEP 6 Make electron loss equal to electron gain, then add the half-reactions.

$$2(5e^- + 8H^+ + MnO_4^- \longrightarrow Mn^{2+} + 4H_2O)$$

$$5(H_2O + H_2SO_3 \longrightarrow SO_4^{2-} + 4H^+ + 2e^-)$$

$$\overline{10e^- + 16H^+ + 2MnO_4^- + 5H_2O + 5H_2SO_3 \longrightarrow 2Mn^{2+} + 8H_2O + 5SO_4^{2-} + 20H^+ + 10e^-}$$

STEP 7 Cancel $10e^-$, $16H^+$, and $5H_2O$ from both sides. The final equation is

$$2MnO_4^- + 5H_2SO_3 \longrightarrow 2Mn^{2+} + 3H_2O + 5SO_4^{2-} + 4H^+$$

Is the Answer Reasonable?
The check is simple. Notice that each side of the equation has the same number of atoms of each element and the same net charge. This is what makes it a balanced equation and confirms that we've worked the problem correctly.

PRACTICE EXERCISE 8: The element technetium (atomic number 43) is radioactive; one of its isotopes, ^{99}Tc, is used in medicine for diagnostic imaging. The isotope is usually obtained in the form of the pertechnetate anion, TcO_4^-, but its use sometimes requires the technetium to be in a lower oxidation state. Reduction can be carried out using Sn^{2+} in an acidic solution. The skeleton equation is

$$TcO_4^- + Sn^{2+} \longrightarrow Tc^{4+} + Sn^{4+} \qquad \text{(acidic solution)}$$

Balance the equation by the ion–electron method.

PRACTICE EXERCISE 9: What is the balanced net ionic equation for the following reaction in an acidic solution?

$$Cu + NO_3^- \longrightarrow Cu^{2+} + N_2O$$

Additional steps produce a balanced equation for basic solutions

In basic solutions, the concentration of H^+ is very small; the dominant species are H_2O and OH^-. Strictly speaking, these should be used to balance the half-reactions. However, the simplest way to obtain a balanced equation for a basic solution is to first *pretend* that the solution is acidic. We balance the equation using the seven steps just described, and then we use a simple three-step procedure described below to convert the equation to the correct form for a basic solution. The conversion uses the fact that H^+ and OH^- react in a 1-to-1 ratio to give H_2O.

Additional Steps in the Ion–Electron Method for Basic Solutions

STEP 8   Add to <u>both</u> sides of the equation the same number of OH$^-$ as there are H$^+$.

STEP 9   Combine H$^+$ and OH$^-$ to form H$_2$O.

STEP 10 Cancel any H$_2$O that you can.

As an example, suppose we wanted to balance the following equation for a basic solution:

$$SO_3^{2-} + MnO_4^- \longrightarrow SO_4^{2-} + MnO_2$$

Following steps 1 through 7 for acidic solutions gives

$$2H^+ + 3SO_3^{2-} + 2MnO_4^- \longrightarrow 3SO_4^{2-} + 2MnO_2 + H_2O$$

Conversion of this equation to one appropriate for a basic solution proceeds as follows.

STEP 8 *Add to <u>both</u> sides of the equation the same number of OH$^-$ as there are H$^+$.*

The equation for acidic solution has 2H$^+$ on the left, so we add 2OH$^-$ to *each* side. This gives

$$2OH^- + 2H^+ + 3SO_3^{2-} + 2MnO_4^- \longrightarrow 3SO_4^{2-} + 2MnO_2 + H_2O + 2OH^-$$

STEP 9 *Combine H$^+$ and OH$^-$ to form H$_2$O.*

The left side has 2OH$^-$ and 2H$^+$, which become 2H$_2$O. So in place of 2OH$^-$ + 2H$^+$ we write 2H$_2$O.

$$2OH^- + 2H^+ + 3SO_3^{2-} + 2MnO_4^- \longrightarrow 3SO_4^{2-} + 2MnO_2 + H_2O + 2OH^-$$

$$2H_2O + 3SO_3^{2-} + 2MnO_4^- \longrightarrow 3SO_4^{2-} + 2MnO_2 + H_2O + 2OH^-$$

STEP 10 *Cancel any H$_2$O that you can.*

In this equation, one H$_2$O can be eliminated from both sides. The final equation, balanced for basic solution, is

$$H_2O + 3SO_3^{2-} + 2MnO_4^- \longrightarrow 3SO_4^{2-} + 2MnO_2 + 2OH^-$$

PRACTICE EXERCISE 10: Balance the following equation for a basic solution:

$$MnO_4^- + C_2O_4^{2-} \longrightarrow MnO_2 + CO_3^{2-}$$

6.3 ▶ Metals are oxidized when they react with acids

In Chapter 5 we began our discussion of acids by examining the way they ionize in water to produce solutions that contain the hydronium ion, H$_3$O$^+$. Typical reactions of acids that we studied were neutralizations, which involve reactions of the H$_3$O$^+$ ion with OH$^-$ or O^{2-}. In this section we will look at another way that acids are able to react in redox reactions.

An important property of acids is their tendency to react with certain metals. Battery acid (which is sulfuric acid, H$_2$SO$_4$) will begin to dissolve unprotected iron or steel parts of an automobile if it's spilled and not flushed away with water. A

Figure 6.1 Zinc reacts with hydrochloric acid. Bubbles of hydrogen are formed when a solution of hydrochloric acid comes in contact with metallic zinc. The same reaction takes place if hydrochloric or sulfuric acid is spilled on galvanized (zinc-coated) steel.

similar reaction occurs even faster, accompanied by vigorous bubbling, if the acid is spilled on the zinc surface of galvanized steel (see Figure 6.1).

In general, the reaction of an acid with a metal is a redox reaction in which the metal is oxidized and the acid is reduced. But in these reactions, the part of the acid that is reduced depends on the composition of the acid itself as well as on the metal.

When sulfuric acid reacts with the zinc coating on a galvanized garbage pail or a galvanized nail, the reaction is

$$Zn(s) + H_2SO_4(aq) \longrightarrow ZnSO_4(aq) + H_2(g)$$

The observed bubbling is caused by the release of hydrogen gas. If we assign oxidation numbers, we can analyze the oxidation–reduction changes that occur. In the diagram below, we use the symbol Δ (Greek letter delta) to stand for the change in oxidation number (positive, as we said, for an increase in oxidation number, negative for a decrease).

$$\Delta = (-1 \text{ per H}) \times (2\text{H}) = -2 \text{ units}$$

$$\overset{0}{Zn}(s) + \overset{+1+6-2}{H_2SO_4}(aq) \longrightarrow \overset{+2+6-2}{ZnSO_4}(aq) + \overset{0}{H_2}(g)$$

$$\Delta = +2 \text{ units per Zn}$$

The oxidation number of zinc increases from 0 to +2, so zinc is oxidized. The oxidation number of hydrogen decreases from +1 to 0, so the hydrogen of the acid is reduced.

An even clearer picture of what occurs is seen if we write the net ionic equation. Sulfuric acid, you recall, is a strong acid, so in aqueous solution we can write it in ionized form. Similarly, zinc sulfate is a salt and is therefore a strong electrolyte; we write it in fully dissociated form as well. The ionic equation for the reaction is therefore

$$Zn(s) + 2H^+(aq) + SO_4^{2-}(aq) \longrightarrow Zn^{2+}(aq) + SO_4^{2-}(aq) + H_2(g)$$

After canceling spectator ions, the net ionic equation for the reaction is

$$Zn(s) + 2H^+(aq) \longrightarrow Zn^{2+}(aq) + H_2(g)$$

Once again, assigning oxidation numbers reveals that the zinc is oxidized and the H^+ of the acid is reduced.

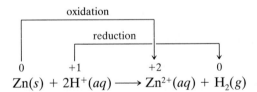

$$\overset{0}{Zn}(s) + 2\overset{+1}{H}^+(aq) \longrightarrow \overset{+2}{Zn}^{2+}(aq) + \overset{0}{H_2}(g)$$

Stated another way, *the H^+ of the acid is the oxidizing agent.*

Many metals react with acids just as zinc does—by being oxidized by hydrogen ions. In these reactions, a metal salt and gaseous hydrogen are the products.

Practice Exercise 11: Write balanced molecular, ionic, and net ionic equations for the reaction of hydrochloric acid with (a) magnesium and (b) aluminum. (Both are oxidized by hydrogen ions.)

The ease with which metals lose electrons varies considerably from one metal to another. Some metals are easily oxidized and others are not. For many metals, like iron and zinc, the hydrogen ion is a sufficiently strong oxidizing agent to do the job, so when such metals are placed in a solution of an acid they are oxidized while the hydrogen ions are reduced. Such metals are said to be *more active* than hydrogen (H_2), and they dissolve in acids like HCl and H_2SO_4 to give hydrogen gas and a salt that contains the anion of the acid. For other metals, however, hydrogen ions are not powerful enough to cause their oxidation. Copper, for example, is significantly less reactive than zinc or iron, and H^+ cannot oxidize it. If a copper penny is dropped into sulfuric or hydrochloric acid, it just sits there. No reaction occurs. Copper is an example of a metal that is *less active* than H_2.

The oxidizing power of acids depends on the nature of the acid

Hydrochloric acid contains H_3O^+ ions (which we abbreviate as H^+) and Cl^- ions. The hydrogen ion in hydrochloric acid is able to be an oxidizing agent by being reduced to H_2. However, the Cl^- ion in the solution has no tendency at all to be an oxidizing agent, because it would have to become a Cl^{2-} ion to gain an electron. This is just not feasible, so in a solution of HCl, the only oxidizing agent is H^+.

In an aqueous solution of sulfuric acid, we find a similar situation. In water this acid ionizes to give hydrogen ions and sulfate ions. Although the sulfate ion can be reduced (to SO_3^{2-}, for example), hydrogen ion is more easily reduced. Therefore, when we add a metal such as zinc to sulfuric acid, it is H^+, rather than SO_4^{2-}, that removes the electrons from Zn.

Compared with many other chemicals that we will study, the hydrogen ion in water is really a rather poor oxidizing agent, so hydrochloric acid and sulfuric acid have rather poor oxidizing abilities. For this reason, they are often called **nonoxidizing acids,** even though their hydrogen ions are able to oxidize certain metals. Actually, when we call something a nonoxidizing acid we are saying that the *anion* of the acid is a weaker oxidizing agent than H^+ (which is equivalent to saying that the anion of the acid is more difficult to reduce than H^+).

> The strongest oxidizing agent in a solution of a "nonoxidizing" acid is H^+.

Not all acids are like HCl and H_2SO_4. There are some that are called **oxidizing acids,** acids whose anions are stronger oxidizing agents than H^+. (See Table 6.1.) An example is nitric acid, HNO_3. When dissolved in water, nitric acid ionizes to give H^+ and NO_3^- ions. However, in this solution the nitrate ion is a more powerful oxidizing agent than the hydrogen ion. In the competition for electrons, therefore, it is the nitrate ion that wins, and when nitric acid reacts with a metal, it is the nitrate ion that is reduced.

Because the NO_3^- ion is a stronger oxidizing agent than H^+, it is able to oxidize metals that H^+ cannot. For example, if a copper penny is dropped into concen-

TABLE 6.1	NONOXIDIZING AND OXIDIZING ACIDS

Table of oxidizing and nonoxidizing acids

Nonoxidizing Acids
HCl(aq)
$H_2SO_4(aq)^a$
$H_3PO_4(aq)$
Most organic acids (e.g., $HC_2H_3O_2$)

Oxidizing Acids	Reduction Reaction
HNO_3	(conc.) $NO_3^- + 2H^+ + e^- \longrightarrow NO_2(g) + H_2O$
	(dilute) $NO_3^- + 4H^+ + 3e^- \longrightarrow NO(g) + 2H_2O$
	(very dilute, with strong reducing agent)
	$NO_3^- + 10H^+ + 8e^- \longrightarrow NH_4^+ + 3H_2O$
H_2SO_4	(hot, conc.) $SO_4^{2-} + 4H^+ + 2e^- \longrightarrow SO_2(g) + 2H_2O$
	(hot conc., with strong reducing agent)
	$SO_4^{2-} + 10H^+ + 8e^- \longrightarrow H_2S(g) + 4H_2O$

^{a}H_2SO_4 is a nonoxidizing acid when cold and dilute.

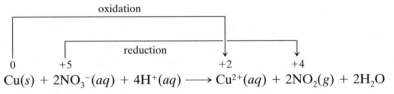

$$Cu + 4HNO_3 \longrightarrow Cu(NO_3)_2 + 2NO_2 + 2H_2O$$

FIGURE 6.2 *The reaction of copper with concentrated nitric acid.* A copper penny reacts vigorously with concentrated nitric acid, as this sequence of photographs shows. The dark red-brown vapors are nitrogen dioxide, the same gas that gives smog its characteristic color.

trated nitric acid, it reacts violently (see Figure 6.2). The reddish-brown gas is nitrogen dioxide, NO_2, which is formed by the reduction of the NO_3^- ion. The molecular equation for the reaction is

$$Cu(s) + 4HNO_3(aq) \longrightarrow Cu(NO_3)_2(aq) + 2NO_2(g) + 2H_2O$$

If we change this to a net ionic equation and then assign oxidation numbers, we can see easily which is the oxidizing agent and which is the reducing agent.

$$\overset{0}{Cu}(s) + 2\overset{+5}{N}O_3^-(aq) + 4H^+(aq) \longrightarrow \overset{+2}{Cu}^{2+}(aq) + 2\overset{+4}{N}O_2(g) + 2H_2O$$

oxidation
reduction

Notice that even though it is the nitrogen whose oxidation number is decreasing, we identify the substance that contains this nitrogen, the NO_3^- ion, as the oxidizing agent.

Nitrate ion is reduced and copper is oxidized. Therefore, nitrate ion is the oxidizing agent and copper is the reducing agent. Notice that in this reaction, *no hydrogen gas is formed.* The H^+ ions of the HNO_3 are an essential part of the reaction, but they just become part of water molecules without a change in oxidation number.

Some reactions of nitric acid as an oxidizing acid

When oxidizing acids react with metals, it is often difficult to predict the products. Reduction of the nitrate ion, for example, can produce all sorts of compounds having different oxidation states for the nitrogen, depending on the reducing power of the metal and the concentration of the acid. When *concentrated* nitric acid reacts with a metal, it often produces nitrogen dioxide, NO_2, as the reduction product. When *dilute* nitric acid is used to dissolve a metal, the product is often nitrogen monoxide (also called nitric oxide), NO, instead. Copper, for example, can display either behavior toward nitric acid. The molecular and net ionic equations for the reactions are as follows:

Concentrated HNO_3

$$Cu(s) + 4HNO_3(aq) \longrightarrow Cu(NO_3)_2(aq) + 2NO_2(g) + 2H_2O$$
$$Cu(s) + 4H^+(aq) + 2NO_3^-(aq) \longrightarrow Cu^{2+}(aq) + 2NO_2(g) + 2H_2O$$

Dilute HNO_3

$$3Cu(s) + 8HNO_3(aq) \longrightarrow 3Cu(NO_3)_2(aq) + 2NO(g) + 4H_2O$$
$$3Cu(s) + 8H^+(aq) + 2NO_3^-(aq) \longrightarrow 3Cu^{2+}(aq) + 2NO(g) + 4H_2O$$

If very dilute nitric acid reacts with a metal that is a particularly strong reducing agent, like zinc, the nitrogen can be reduced all the way down to the -3 oxidation state that it has in NH_4^+ (or NH_3). The net ionic equation is

$$4Zn(s) + 10H^+(aq) + NO_3^-(aq) \longrightarrow 4Zn^{2+}(aq) + NH_4^+(aq) + 3H_2O$$

The nitrate ion in the presence of hydrogen ions makes nitric acid quite a powerful oxidizing acid. All metals except the very unreactive ones, such as platinum and gold, are attacked by it. Nitric acid also does a good job of oxidizing organic substances, so it is wise to be especially careful when working with this acid in the laboratory. Very serious accidents have occurred when inexperienced people have used concentrated nitric acid around organic substances.

Nitric acid causes severe skin burns, so be careful when you work with it in the laboratory. If you spill any on your skin, wash it off immediately and seek the help of your lab teacher.

Hot concentrated sulfuric acid is an oxidizing acid

In a dilute solution, the sulfate ion of sulfuric acid has little tendency to serve as an oxidizing agent. However, if the sulfuric acid is both concentrated and hot, it becomes a fairly potent oxidizer. For example, copper is not bothered by cool dilute H_2SO_4, but it is attacked by hot concentrated H_2SO_4 according to the following equation

$$Cu + 2H_2SO_4(\text{hot, conc.}) \longrightarrow CuSO_4 + SO_2 + 2H_2O$$

Because of this oxidizing ability, hot concentrated sulfuric acid can be very dangerous. The liquid is viscous and can stick to the skin, causing severe burns.

6.4 ▶ A more active metal will displace a less active one from its compounds

The formation of hydrogen gas in the reaction of a metal with an acid is a special case of a more general phenomenon—one element displacing (pushing out) another element from a compound by means of a redox reaction. In the case of a metal–acid reaction, it is the metal that displaces hydrogen from the acid, changing $2H^+$ to H_2.

Another reaction of this same general type occurs when one metal displaces another metal from its compounds, illustrated by the experiment shown in Figure 6.3. Here we see a brightly polished strip of metallic zinc that is dipped into a solution of copper sulfate. After the zinc is in the solution for a while, a reddish-brown deposit of metallic copper forms on the zinc, and if the solution were analyzed, we would find that it now contains zinc ions, as well as some remaining unreacted copper ions.

Figure 6.3 *The reaction of zinc with copper ion.* (*Left*) A piece of shiny zinc next to a beaker containing a copper sulfate solution. (*Center*) When the zinc is placed in the solution, copper ions are reduced to the free metal while the zinc dissolves. (*Right*) After a while, the zinc becomes coated with a red-brown layer of copper. Notice that the solution is a lighter blue than before, showing that some of the copper ions have left the solution.

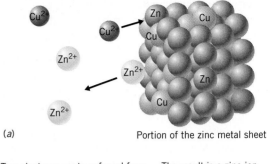

FIGURE 6.4 *The reaction of copper ions with zinc, viewed at the atomic level.* (*a*) Copper ions (blue) collide with the zinc surface where they pick up electrons from zinc atoms (gray). The zinc atoms become zinc ions (yellow) and enter the solution. The copper ions become copper atoms (red-brown) and stick to the surface of the zinc. (For clarity, the water molecules of the solution and the sulfate ions are not shown.) (*b*) A close-up view of the exchange of electrons that leads to the reaction.

(*a*) Portion of the zinc metal sheet

(*b*)

The results of this experiment can be summarized by the equation

$$Zn(s) + CuSO_4(aq) \longrightarrow Cu(s) + ZnSO_4(aq)$$

In a *double replacement* reaction such as

$$CdCl_2(aq) + Na_2S(aq) \longrightarrow$$
$$CdS(s) + 2NaCl(aq)$$

two anions (Cl$^-$ and S^{2-}) acquire different cations. In a *single replacement* reaction, *one* anion acquires a different cation. Single replacement reactions are redox reactions; double replacement reactions are not.

A reaction such as this, in which one element replaces another in a compound, is sometimes called a **single replacement reaction.**

The redox changes in the reaction above become clear if we write the net ionic equation. Both copper sulfate and zinc sulfate are soluble salts, so they are completely dissociated. Sulfate ion is a spectator ion, and the net ionic equation is

$$Zn(s) + Cu^{2+}(aq) \longrightarrow Cu(s) + Zn^{2+}(aq)$$

We see that zinc has reduced the copper ion to metallic copper, and the copper ion has oxidized metallic zinc to the zinc ion. In the process, zinc ions have taken the place of the copper ions, so a solution of copper sulfate is changed to a solution of zinc sulfate. An atomic-level view of what's happening at the surface of the zinc during the reaction is depicted in Figure 6.4.

The reaction of zinc with copper ion is quite similar to the reaction of zinc with sulfuric acid. The more "active" zinc replaces another less "active" element in a compound. Furthermore, it is easy to show that the reverse reaction doesn't occur. Nothing happens if a brightly polished piece of copper is dipped into a solution of zinc sulfate (see Figure 6.5). No matter how long the copper is in the solution, its

FIGURE 6.5 *The inability of copper to react with zinc ion.* Although metallic zinc will displace copper from a solution containing Cu^{2+} ions, metallic copper will not displace Zn^{2+} from its solutions. Here we see that the copper bar is unaffected by being dipped into a solution of zinc sulfate.

Activity series of metals

TABLE 6.2	ACTIVITY SERIES FOR SOME METALS (AND HYDROGEN)	
	Element	Oxidation Product
Least Active	Gold	Au^{3+}
	Mercury	Hg^{2+}
	Silver	Ag^+
	Copper	Cu^{2+}
	Hydrogen	H^+
	Lead	Pb^{2+}
	Tin	Sn^{2+}
	Cobalt	Co^{2+}
	Cadmium	Cd^{2+}
	Iron	Fe^{2+}
	Chromium	Cr^{3+}
	Zinc	Zn^{2+}
	Manganese	Mn^{2+}
	Aluminum	Al^{3+}
	Magnesium	Mg^{2+}
	Sodium	Na^+
	Calcium	Ca^{2+}
	Strontium	Sr^{2+}
	Barium	Ba^{2+}
	Potassium	K^+
	Rubidium	Rb^+
Most Active	Cesium	Cs^+

Increasing Ease of Oxidation of the Metal (arrow pointing downward)

surface remains untarnished. This means that the reaction of copper with zinc sulfate doesn't occur.

$$Cu(s) + ZnSO_4(aq) \longrightarrow \text{no reaction}$$

In other words, the less "active" copper is unable to displace the more "active" zinc from the solution.

The activity series arranges metals according to their ease of oxidation

Throughout the discussion in the preceding paragraph we used the word *active* to mean "easily oxidized." In other words, an element that is more easily oxidized will displace one that is less easily oxidized from its compounds. The relative ease of oxidation of two metals can be established in experiments just as simple as the ones pictured in Figures 6.3 and 6.5. After such comparisons are made for many pairs, the metals can be arranged in order of their ease of oxidation to give what is often called an **activity series** (see Table 6.2). According to the way the metals have been arranged in Table 6.2, those at the bottom are more easily oxidized (are more active) than those at the top. *This means that a given element will be displaced from its compounds by any metal below it in the table.*

Notice that we have included hydrogen in the activity series. Metals below hydrogen in the series can displace hydrogen from solutions containing H^+. These are the metals that are capable of reacting with the nonoxidizing acids discussed in the previous section. On the other hand, metals above hydrogen in the table do not react with acids having H^+ as the strongest oxidizing agent.

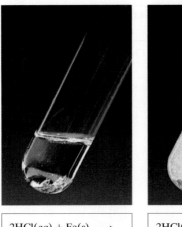  

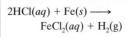

$$2HCl(aq) + Fe(s) \longrightarrow$$
$$FeCl_2(aq) + H_2(g)$$

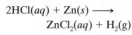

$$2HCl(aq) + Zn(s) \longrightarrow$$
$$ZnCl_2(aq) + H_2(g)$$

$$2HCl(aq) + Mg(s) \longrightarrow$$
$$MgCl_2(aq) + H_2(g)$$

Figure 6.6 *Metallic sodium reacts violently with water.* The heat of the reaction ignites the sodium metal, which can be seen burning and sending sparks from the surface of the water. In the reaction, sodium is oxidized to Na^+ and water molecules are reduced to give hydrogen gas and hydroxide ions. When the reaction is over, the solution contains sodium hydroxide.

Figure 6.7 *The relative ease of oxidation of metals parallels their rates of reaction with hydrogen ions of an acid.* The products are hydrogen gas and the metal ion in solution. All three test tubes contain $HCl(aq)$ at the same concentration. The first also contains pieces of iron; the second, pieces of zinc; and the third, pieces of magnesium. Among these three metals, the ease of oxidation increases from iron to zinc to magnesium.

Metals at the very bottom of the table are very easily oxidized and are extremely strong reducing agents. They are so reactive, in fact, that they are able to reduce the hydrogen in water. Sodium, for example, reacts vigorously with water to liberate H_2 according to the following equation (see Figure 6.6).

$$2Na(s) + 2H_2O \longrightarrow H_2(g) + 2NaOH(aq)$$

For metals below hydrogen in the activity series, an interesting parallel exists between the ease of oxidation of the metal and the speed with which it reacts with H^+. For example, in Figure 6.7, we see samples of iron, zinc, and magnesium reacting with solutions of hydrochloric acid. In each test tube the initial HCl concentration is the same, but we see that the magnesium reacts more rapidly than zinc, which reacts more rapidly than iron. You can see that the order of reactivity in Table 6.2 is the same; magnesium is more easily oxidized than zinc, which is more easily oxidized than iron.

The activity series can be used to predict reactions

The activity series in Table 6.2 permits us to make predictions of the outcome of single replacement redox reactions, as illustrated in the following examples.

EXAMPLE 6.10
Using the Activity Series

What will happen if an iron nail is dipped into a solution containing copper sulfate? If a reaction occurs, write its chemical equation.

ANALYSIS: From the discussion above, it is obvious we will need to refer to the activity series. However, if this question had been asked in another setting, how would you know what to do? What clues suggest that we have to use the activity series to find the answer?

Reading the question, we have to ask, what *could* happen? If a chemical reaction were to occur, iron would have to react with the copper sulfate. A *metal* possibly reacting with the *salt of another metal*? This is the thought that should suggest the possibility of a single replacement reaction. And you've learned how we predict such reactions. You use the locations of the potential reactants in the activity series to determine whether a reaction is possible and what the products would be.

Examining the activity series in Table 6.2, we see that iron is more easily oxidized than copper, so metallic iron will displace copper ions from a solution of copper sulfate. *A reaction will occur.* (We've answered one part of the question.) To write an equation for the reaction, we have to know the final oxidation state of the iron. In the table, this is indicated as $+2$, so the Fe atoms change to Fe^{2+} ions and pair with SO_4^{2-} ions to give $FeSO_4$. Copper(II) ions are reduced to copper atoms.

SOLUTION: Our analysis told us that a reaction *will* occur and it also gave us the products, so the equation is

$$Fe(s) + CuSO_4(aq) \longrightarrow Cu(s) + FeSO_4(aq)$$

Is the Answer Reasonable?
We can check the activity series again to be sure we've reached the correct conclusion, and we can check to be sure the equation we've written has the correct formulas and is balanced correctly. Doing this confirms that we've got the right answers.

What happens if an iron nail is dipped into a solution of aluminum sulfate?
ANALYSIS: Once again, we have to realize that we're looking for a potential single replacement reaction. Scanning the activity series, we see that aluminum metal is *more* easily oxidized than iron metal. This means that aluminum atoms would be able to displace iron ions from an iron compound. But it also means that iron atoms cannot displace aluminum ions from its compounds, and iron *atoms* plus aluminum *ions* are what we're given.

SOLUTION: If nature does not permit electrons to transfer from iron atoms to aluminum ions, we must conclude that no reaction can occur.

$$Fe(s) + Al_2(SO_4)_3(aq) \longrightarrow \text{no reaction}$$

Is the Answer Reasonable?
Checking the activity series again, we must conclude we've come to the correct answer.

PRACTICE EXERCISE 12: Write a chemical equation for the reaction that will occur, if any, when (a) aluminum metal is added to a solution of copper chloride and (b) silver metal is added to a solution of magnesium sulfate. If no reaction will occur, write "no reaction" in place of the products.

EXAMPLE 6.11
Using the Activity Series

6.5 ▶ Molecular oxygen is a powerful oxidizing agent

In 1789, the French chemist Antoine Lavoisier discovered that *combustion* involves the reaction of chemicals in various fuels, like wood and coal, not just with air but specifically with the oxygen in air. In fact, as we noted earlier, the term *oxidation* was invented to describe reactions in which one of the reactants is elemental oxygen, O_2. Oxidation originally meant the reaction of a substance with oxygen. Only much later did chemists realize that the reactions of oxygen constitute a special case of a much broader class of reactions. This realization led to the extension of the meaning of *oxidation* to cover the kinds of reactions discussed in the previous sections.

Oxygen is a plentiful chemical; it's in the air and available to anyone who wants to use it, chemist or not. Furthermore, O_2 is a very reactive oxidizing agent, so its reactions have been well studied. When substances combine with oxygen, the products are generally oxides, *molecular oxides* when the oxygen reacts with nonmetals and *ionic oxides* when the oxygen reacts with metals.

Antoine Laurent Lavoisier (1743–1794), a French chemist, is deservedly called the "father of modern chemistry." He was the first to insist on *quantitative* data from chemical research, he systematized chemical nomenclature, and his text *Traité élémentaire de chimie* was to modern chemistry what Newton's *Principia* was to the development of physics.

Organic compounds burn in oxygen

Combustion is normally taken to mean a particularly *rapid* reaction of a substance with oxygen in which both heat and light are given off. If you had to build a fire to keep warm, you no doubt would look for combustible materials like twigs, logs, or other pieces of wood to use as fuel. You know that wood burns. Experience has also taught you that certain other substances burn. When you drive a car, for example, it is probably powered by the combustion of gasoline. Wood and gasoline are examples of substances or mixtures of substances that chemists call *organic compounds*—compounds whose structures are determined primarily by the linking together of carbon atoms. When organic compounds burn, the products of the reactions are usually easy to predict.

Hydrocarbons are important fuels

Fuels such as natural gas, gasoline, kerosene, heating oil, and diesel fuel are examples of *hydrocarbons*—compounds containing only the elements carbon and hydrogen. Natural gas is composed principally of methane, CH_4. Gasoline is a mixture of hydrocarbons, the most familiar of which is octane, C_8H_{18}. Kerosene, heating oil, and diesel fuel are mixtures of hydrocarbons in which the molecules contain even more atoms of carbon and hydrogen.

When hydrocarbons burn in a *plentiful* supply of oxygen, the products of combustion are always carbon dioxide and water. Thus, methane and octane combine with oxygen according to the equations

$$CH_4 + 2O_2 \longrightarrow CO_2 + 2H_2O$$

$$2C_8H_{18} + 25O_2 \longrightarrow 16CO_2 + 18H_2O$$

Many people don't realize that water is one of the products of the combustion of hydrocarbons, even though they have seen evidence for it. Perhaps you've seen clouds of condensed water vapor coming from the exhaust pipes of automobiles on cold winter days, or you may have noticed that shortly after you first start a car, drops of water fall from the exhaust pipe. This is water that has been formed during the combustion of the gasoline. Similarly, the "smokestacks" of power stations release clouds of condensed water vapor (Figure 6.8), which is often mistaken for smoke from fires used to generate power to make electricity. Actually, many of today's power stations produce very little smoke because they burn clean natural gas instead of coal.

PRACTICE EXERCISE 13: Write a balanced equation for the combustion of butane, C_4H_{10}, in an abundant supply of oxygen. Butane is the fuel used in disposable cigarette lighters.

When there is less than an abundant supply of oxygen during the combustion of a hydrocarbon, not all of the carbon is converted to carbon dioxide. Instead,

FIGURE 6.8 *Water is a product of the combustion of hydrocarbons.* Here we see clouds of condensed water vapor coming from the stacks of an oil-fired electric-generating plant during the winter.

some of it forms carbon monoxide. Its formation is a pollution problem associated with the use of gasoline engines, as you may know.

$$2CH_4 + 3O_2 \longrightarrow 2CO + 4H_2O \qquad \text{(in a limited oxygen supply)}$$

When the oxygen supply is extremely limited, only the hydrogen of a hydrocarbon mixture is converted to the oxide (water). The carbon atoms emerge as elemental carbon. For example, when a candle burns, the fuel is a high-molecular-weight hydrocarbon (e.g., $C_{20}H_{42}$), and incomplete combustion forms tiny particles of carbon that glow brightly. If a cold surface is held in the flame, the unburned carbon deposits, as seen in Figure 6.9.

An important commercial reaction is the incomplete combustion of methane in a very limited oxygen supply, which follows the equation

$$CH_4 + O_2 \longrightarrow C + 2H_2O \qquad \text{(in a very limited oxygen supply)}$$

The carbon that forms is very finely divided and would be called *soot* by almost anyone observing the reaction. Nevertheless, such soot has considerable commercial value when collected and marketed under the name *lampblack*. This sooty form of carbon is used to manufacture inks, and much of it is used in the production of rubber tires, where it serves as a binder and a filler. When soot from incomplete combustion is released into air, its tiny particles constitute one component of air pollution, namely, *particulates,* which contribute to the haziness of smog.

This finely divided form of carbon is also called *carbon black.*

Combustion of organic compounds that contain oxygen also produces CO_2 and H_2O

Earlier we mentioned that you might choose wood to build a fire. The chief combustible ingredient in wood is cellulose, a fibrous material that gives plants their structural strength. Cellulose is composed of the elements carbon, hydrogen, and oxygen. Each cellulose molecule consists of many small, identical groups of atoms that are linked together to form a very long molecule, although the lengths of the molecules differ. For this reason we cannot specify a molecular formula for cellulose. Instead, we use the empirical formula, $C_6H_{10}O_5$, which represents the small, repeating "building block" units in large cellulose molecules. When cellulose burns, the products are also carbon dioxide and water. The only difference between its reaction and the reaction of a hydrocarbon with oxygen is that some of the oxygen in the products comes from the cellulose.

$$C_6H_{10}O_5 + 6O_2 \longrightarrow 6CO_2 + 5H_2O$$

The complete combustion of all other organic compounds containing only carbon, hydrogen, and oxygen produces the same products, CO_2 and H_2O, and follows similar equations.

The formula for cellulose can be expressed as $(C_6H_{10}O_5)_n$, which indicates that the molecule contains the $C_6H_{10}O_5$ unit repeated some large number n times.

> **PRACTICE EXERCISE 14:** Ethanol, C_2H_5OH, is now mixed with gasoline, and the mixture is sold under the name *gasohol.* Write a chemical equation for the combustion of ethanol.

FIGURE 6.9 *Incomplete combustion of a hydrocarbon.* The bright yellow color of a candle flame is caused by glowing particles of elemental carbon. Here we see that a black deposit of carbon is formed when the flame contacts a cold porcelain surface.

Burning organic compounds containing sulfur gives SO_2 as one of the products

A major pollution problem in industrialized countries is caused by the release into the atmosphere of sulfur dioxide formed by the combustion of fuels that contain sulfur or its compounds. The products of the combustion of organic compounds of sulfur are carbon dioxide, water, and sulfur dioxide. A typical reaction is

$$2C_2H_5SH + 9O_2 \longrightarrow 4CO_2 + 6H_2O + 2SO_2$$

A solution of sulfur dioxide in water is acidic, and when rain falls through polluted air it picks up SO_2 and becomes "acid rain." Some SO_2 is also oxidized to SO_3, which reacts with moisture to give H_2SO_4, making the acid rain even more acidic.

Many metals react with oxygen

Figure 6.10 *A flashbulb, before and after firing.* Fine magnesium wire in an atmosphere of oxygen fills the flashbulb at the left. After being used (right), the interior of the bulb is coated with a white film of magnesium oxide.

We don't often think of metals as undergoing combustion, but have you ever seen an old-fashioned flashbulb fired to take a photograph? The source of light is the reaction of the metal magnesium with oxygen (see Figure 6.10). A close look at a fresh flashbulb reveals a fine web of thin magnesium wire within the glass envelope. The wire is surrounded by an atmosphere of oxygen, a clear, colorless gas. When the flashbulb is used, a small electric current surges through the thin wire, causing it to become hot enough to ignite, and it burns rapidly in the oxygen atmosphere. The equation for the reaction is

$$2Mg + O_2 \longrightarrow 2MgO$$

Most metals react directly with oxygen, although not so spectacularly, and usually we refer to the reaction as **corrosion** or **tarnishing** because the oxidation products dull the shiny metal surface. Iron, for example, is oxidized fairly easily, especially in the presence of moisture. As you know, under these conditions the iron corrodes—it rusts. Rust is a form of iron(III) oxide, Fe_2O_3, that also contains an appreciable amount of absorbed water. The formula for rust is therefore normally given as $Fe_2O_3 \cdot xH_2O$ to indicate its somewhat variable composition. Although the rusting of iron is a slow reaction, the combination of iron with oxygen can be speeded up if the metal is heated to a very high temperature under a stream of pure O_2 (see Figure 6.11).

An aluminum surface, unlike that of iron, is not noticeably dulled by the reaction of aluminum with oxygen. Aluminum is a common metal found around the home in uses ranging from aluminum foil to aluminum window frames, and it surely appears shiny. Yet, aluminum is a rather easily oxidized metal, as can be seen from its position in the activity series (Table 6.2). A *freshly* exposed surface of the metal does react very quickly with oxygen and becomes coated with a very thin film of aluminum oxide, Al_2O_3, so thin that it doesn't obscure the shininess of the metal beneath. Fortunately, the oxide coating adheres very tightly to the surface of the metal and makes it very difficult for additional oxygen to combine with the aluminum. Therefore, further oxidation of aluminum occurs very slowly.

Figure 6.11 *Cutting steel with an oxyacetylene torch.* An oxygen–acetylene flame is used to heat steel until its glowing red hot. Then the acetylene is turned off and the steel is cut by a stream of pure oxygen whose reaction with the hot metal produces enough heat to melt the steel and send a shower of burning steel sparks flying.

Practice Exercise 15: The oxide formed in the reaction shown in Figure 6.11 is Fe_2O_3. Write a balanced chemical equation for the reaction.

Most nonmetals react with oxygen directly

Most nonmetals combine as readily with oxygen as do the metals, and their reactions usually occur rapidly enough to be described as combustion. To most people, the most important nonmetal combustion reaction is that of carbon because the reaction is a source of heat. Coal and charcoal, for example, are common carbon fuels. Coal is used worldwide in large amounts to generate electricity, and charcoal is a popular fuel for broiling hamburgers. If plenty of oxygen is available, the combustion of carbon gives CO_2, but when the supply of O_2 is limited, some CO forms as well. Manufacturers that package charcoal briquettes, therefore, print a warning on the bag that the charcoal shouldn't be used indoors for cooking or heating.

Sulfur is another nonmetal that burns readily in oxygen. In the manufacture of sulfuric acid, the first step is the combustion of sulfur to produce sulfur dioxide. As mentioned earlier, sulfur dioxide also forms when sulfur compounds burn, and the

The label on a bag of charcoal displays a warning about carbon monoxide.

presence of sulfur and sulfur compounds as impurities in coal and petroleum is a major source of air pollution. Power plants that burn coal are making strides to remove the SO_2 from their exhausts and it is being used to make sulfuric acid. When SO_2 does enter the atmosphere, it drifts on the wind until it finally dissolves in rainwater. Then, as a dilute solution of sulfurous acid, it falls to the Earth as one of the components of acid rain. Some SO_2 is also oxidized slowly to SO_3, which gives the strong acid H_2SO_4 when it dissolves in rainwater.

6.6 ▶ Redox reactions follow the same stoichiometric principles as other reactions

In general, redox reactions are somewhat more complex than metathesis reactions. You have witnessed this in constructing balanced net ionic equations for both types of reactions. Nevertheless, stoichiometry problems involving redox reactions are approached in the same manner as those we've discussed earlier, as illustrated by the following example.

How many grams of sodium sulfite are needed to react completely with 12.4 g of potassium dichromate in an acidic solution? The products of the reaction include sulfate ion and chromium(III) ion.

EXAMPLE 6.12
Stoichiometry of a Redox Reaction

ANALYSIS: The first hurdle to overcome is writing the formulas of the compounds named in the problem. Applying what we discussed in Chapter 2, we have

$$\text{sodium sulfite} = Na_2SO_3$$

$$\text{potassium dichromate} = K_2Cr_2O_7$$

Now we can express the problem in the form of an equation.

$$12.4 \text{ g } K_2Cr_2O_7 \Longleftrightarrow ? \text{ g } Na_2SO_3$$

As you learned in Chapter 4, the critical link in solving stoichiometry problems dealing with chemical reactions is the set of coefficients provided by the balanced chemical equation. Therefore, we will have to construct an equation and balance it. We will use the ion–electron method to construct the net ionic equation. The reactants come from the two salts above, which dissociate to give the following ions:

$$Na_2SO_3 \longrightarrow 2Na^+ + SO_3^{2-}$$

$$K_2Cr_2O_7 \longrightarrow 2K^+ + Cr_2O_7^{2-}$$

After obtaining the equation, we can proceed as with other stoichiometry problems.

SOLUTION: To write the balanced equation, we need to first construct the skeleton equation, which we can then balance by the ion–electron method. On the product side of the skeleton equation we will write sulfate ion (SO_4^{2-}) and chromium(III) ion (Cr^{3+}).

$$? \longrightarrow SO_4^{2-} + Cr^{3+}$$

Examining the ions that come from the reactant salts, we expect that the reactants will be sulfite ion (SO_3^{2-}) and dichromate ion $(Cr_2O_7^{2-})$.

$$SO_3^{2-} + Cr_2O_7^{2-} \longrightarrow SO_4^{2-} + Cr^{3+}$$

Balancing the equation by the procedure described in Section 6.2 for an acidic solution gives

$$8H^+ + 3SO_3^{2-} + Cr_2O_7^{2-} \longrightarrow 3SO_4^{2-} + 2Cr^{3+} + 4H_2O$$

The stoichiometric equivalence between sulfite and dichromate ions is given by the coefficients. Therefore,

$$3 \text{ mol } SO_3^{2-} \Longleftrightarrow 1 \text{ mol } Cr_2O_7^{2-}$$

The complete formulas of the reactants are Na_2SO_3 and $K_2Cr_2O_7$, so we can represent this equivalence as

$$3 \text{ mol } Na_2SO_3 \Leftrightarrow 1 \text{ mol } K_2Cr_2O_7$$

The path to the answer can now follow the familiar routine: we use the formula mass to convert grams of $K_2Cr_2O_7$ to moles, then we use the equivalence above to find moles of Na_2SO_3, and finally we use the formula mass of Na_2SO_3 to convert to grams of Na_2SO_3.

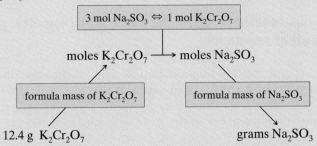

Applying the factor-label method, the solution to the problem is

$$12.4 \text{ g } K_2Cr_2O_7 \times \frac{1 \text{ mol } K_2Cr_2O_7}{294.2 \text{ g } K_2Cr_2O_7} \times \frac{3 \text{ mol } Na_2SO_3}{1 \text{ mol } K_2Cr_2O_7} \times \frac{126.1 \text{ g } Na_2SO_3}{1 \text{ mol } Na_2SO_3} = 15.9 \text{ g } Na_2SO_3$$

The amount of sodium sulfite required is 15.9 g.

Is the Answer Reasonable?
Let's do some approximate arithmetic. If the formula masses of the two compounds were the same, the amount of Na_2SO_3 needed would be three times the amount of $K_2Cr_2O_7$, or approximately 36 g of Na_2SO_3. However, the formula mass of Na_2SO_3 (126) is less than half that of $K_2Cr_2O_7$ (294), so the amount of Na_2SO_3 needed will be somewhat less than half of 36 g, or somewhat less than 18 g. Therefore, our answer, 15.9 g, seems reasonable.

PRACTICE EXERCISE 16: A researcher planned to use chlorine gas in an experiment and wished to trap excess chlorine to prevent it from escaping into the atmosphere. To accomplish this, the reaction of sodium thiosulfate ($Na_2S_2O_3$) with chlorine gas in an acidic aqueous solution to give sulfate ion and chloride ion would be used. How many grams of $Na_2S_2O_3$ are needed to trap 4.25 g of chlorine?

Redox reactions are often used in the laboratory

Because so many reactions involve oxidation and reduction, it should not be surprising that they have useful applications in the lab. An example of such an application is described in the preceding practice exercise. Some redox reactions are especially useful in chemical analyses, particularly in titrations. Unlike in acid–base titrations, however, there are no simple indicators that can be used to conveniently detect the end points in redox titrations, so we have to rely on color changes among the reactants themselves.

One of the most useful reactants for redox titrations is potassium permanganate, $KMnO_4$, especially when the reaction can be carried out in an acidic solution. Permanganate ion is a powerful oxidizing agent, so it oxidizes most substances that are capable of being oxidized. That's one reason why it is used. Especially important, though, is the fact that the MnO_4^- ion has an intense purple color and its reduction product in acidic solution is the almost colorless Mn^{2+} ion. Therefore, when a solution of $KMnO_4$ is added from a buret to a solution of a reducing agent, the chemical reaction that occurs forms a nearly colorless product. This is illus-

FIGURE 6.12 *Reduction of MnO_4^- by Fe^{2+}.* A solution of $KMnO_4$ is added to a stirred acidic solution containing Fe^{2+}. The reaction oxidizes the pale blue-green Fe^{2+} to Fe^{3+} while the MnO_4^- is reduced to the almost colorless Mn^{2+} ion. The purple color of the permanganate will continue to be destroyed until all of the Fe^{2+} has reacted. Only then will the iron-containing solution take on a pink or purple color. This ability of MnO_4^- to signal the completion of the reaction makes it especially useful in redox titrations where it serves as its own indicator.

trated in Figure 6.12, where we see a solution of $KMnO_4$ being poured into an acidic solution containing Fe^{2+}. As the $KMnO_4$ solution is added, the purple color continues to be destroyed as long as there is any reducing agent left. In a titration, after the last trace of the reducing agent has been consumed, the MnO_4^- ion in the next drop of titrant has nothing to react with, so it colors the solution pink. This signals the end of the titration. In this way, permanganate ion serves as its own indicator in redox titrations. Example 6.13 illustrates a typical analysis using $KMnO_4$ in a redox titration.

In concentrated solutions, MnO_4^- is purple, but dilute solutions of the ion appear pink.

All the iron in a 2.000 g sample of an iron ore was dissolved in an acidic solution and converted to Fe^{2+}, which was then titrated with 0.1000 *M* $KMnO_4$ solution. In the titration the iron was oxidized to Fe^{3+}. The titration required 27.45 mL of the $KMnO_4$ solution to reach the end point.

(a) How many grams of iron were in the sample?

(b) What was the percentage iron in the sample?

(c) If the iron was present in the sample as Fe_2O_3, what was the percentage by mass of Fe_2O_3 in the sample?

EXAMPLE 6.13

Redox Titrations in Chemical Analysis

ANALYSIS: The first step, of course, will be to write a balanced equation. Let's analyze the problem after that point. From the volume of the $KMnO_4$ solution and its concentration, we can determine the number of moles of titrant used. The coefficients of the equation then allow us to compute the number of moles of iron(II) that reacted. Since all the iron in the ore sample had been previously changed to iron(II), this is the number of moles of iron in the sample. Converting moles of iron to grams of iron gives the mass of iron in the sample, from which the percentage of iron can be calculated easily. Let's tackle the first two parts of the problem now, and then return to the last part afterward.

SOLUTION: The skeleton equation for the reaction is

$$Fe^{2+} + MnO_4^- \longrightarrow Fe^{3+} + Mn^{2+}$$

Balancing it by the ion–electron method for acidic solutions gives

$$5Fe^{2+} + MnO_4^- + 8H^+ \longrightarrow 5Fe^{3+} + Mn^{2+} + 4H_2O$$

The number of moles of $KMnO_4$ consumed in the reaction is calculated from the volume of the solution used in the titration and its concentration (remember: volume(L) × molarity = moles).

$$0.02745 \text{ L KMnO}_4 \times \frac{0.1000 \text{ mol KMnO}_4}{1.000 \text{ L KMnO}_4} \Leftrightarrow 0.002745 \text{ mol KMnO}_4$$

Next, we use the coefficients of the equation to calculate the number of moles of Fe^{2+} that reacted. The chemical equation tells us 5 mol of Fe^{2+} react per mole of MnO_4^- consumed. Therefore, the number of moles of Fe^{2+} that reacted is

$$0.002745 \text{ mol KMnO}_4 \times \frac{5 \text{ mol Fe}^{2+}}{1 \text{ mol KMnO}_4} \Leftrightarrow 0.01372 \text{ mol Fe}^{2+}$$

This is the number of moles of iron in the ore sample, so the mass of iron in the sample is

$$0.01372 \text{ mol Fe} \times \frac{55.845 \text{ g Fe}}{1 \text{ mol Fe}} = 0.7662 \text{ g Fe}$$

This is the answer to part (a) of the problem. Next we calculate the percentage of iron in the sample, which is the mass of iron divided by the mass of the sample, all multiplied by 100%.

$$\% \text{ Fe} = \frac{\text{mass of Fe}}{\text{mass of sample}} \times 100\%$$

Substituting gives

$$\% \text{ Fe} = \frac{0.7662 \text{ g Fe}}{2.000 \text{ g sample}} \times 100\% = 38.31\% \text{ Fe}$$

The answer to part (b) is that the sample is 38.31% iron.

Is the Answer Reasonable?
We can use some approximate arithmetic to estimate the answer. In the titration we used approximately 30 mL, or 0.030 L, of the $KMnO_4$ solution, which is 0.10 *M*. Multiplying these tells us we've used approximately 0.003 mol of $KMnO_4$. From the coefficients of the equation, five times as many moles of Fe^{2+} react, so the amount of Fe in the sample is approximately $5 \times 0.003 = 0.015$ mol. The atomic mass of Fe is about 55 g/mol, so the mass of Fe in the sample is approximately $0.015 \times 55 = 0.8$ g. Our answer (0.7662 g) is reasonable.

ANALYSIS CONTINUED: Now we can work on the last part of the question. Earlier in the problem we determined the number of moles of iron that reacted, 0.01372 mol Fe. How many moles of Fe_2O_3 would have contained this number of moles of iron? That's the critical question we have to answer. Once we know this, we can calculate the mass of the Fe_2O_3 and the percentage of Fe_2O_3 in the original sample. The critical link to solving this part of the problem is the formula of Fe_2O_3, because it relates moles of iron to moles of iron oxide.

SOLUTION CONTINUED: The chemical formula for the iron oxide gives us

$$1 \text{ mol Fe}_2O_3 \Leftrightarrow 2 \text{ mol Fe}$$

This provides the conversion factor we need to determine how many moles of Fe_2O_3 were present in the sample. Working with the number of moles of Fe,

$$0.01372 \text{ mol Fe} \times \frac{1 \text{ mol Fe}_2O_3}{2 \text{ mol Fe}} \Leftrightarrow 0.006860 \text{ mol Fe}_2O_3$$

This is the number of moles of Fe_2O_3 in the sample. The formula mass of Fe_2O_3 is 159.69 g mol^{-1}, so the mass of Fe_2O_3 in the sample was

$$0.006860 \text{ mol Fe}_2O_3 \times \frac{159.69 \text{ g Fe}_2O_3}{1 \text{ mol Fe}_2O_3} = 1.095 \text{ g Fe}_2O_3$$

Finally, the percentage of Fe_2O_3 in the sample was

$$\% \text{ Fe}_2O_3 = \frac{1.095 \text{ g Fe}_2O_3}{2.000 \text{ g sample}} \times 100\% = 54.75\% \text{ Fe}_2O_3$$

The ore sample contained 54.75% Fe_2O_3.

Is the Answer Reasonable?

We've noted that the amount of Fe in the sample is approximately 0.015 mol. The amount of Fe_2O_3 that contains this much Fe is 0.0075 mol. The formula mass of Fe_2O_3 is about 160, so the mass of Fe_2O_3 in the sample was approximately $0.0075 \times 160 = 1.2$ g, which isn't too far from the mass we obtained (1.095 g). Since 1.095 g is about half of the total sample mass of 2.000 g, the sample was approximately 50% Fe_2O_3, in agreement with the answer we obtained.

PRACTICE EXERCISE 17: A sample of a tin ore weighing 0.3000 g was dissolved in an acid solution and all the tin in the sample was changed to tin(II). In a titration, 8.08 mL of 0.0500 M $KMnO_4$ solution was required to oxidize the tin(II) to tin(IV).

(a) What is the balanced equation for the reaction in the titration?

(b) How many grams of tin were in the sample?

(c) What was the percentage by mass of tin in the sample?

(d) If the tin in the sample had been present in the compound SnO_2, what would have been the percentage by mass of SnO_2 in the sample?

SUMMARY

Oxidation–Reduction **Oxidation** is the loss of electrons or an increase in oxidation number; **reduction** is the gain of electrons or a decrease in oxidation number. Both always occur together in **redox** reactions. The substance oxidized is the **reducing agent;** the substance reduced is the **oxidizing agent. Oxidation numbers** are a bookkeeping device that we use to follow changes in redox reactions. They are assigned according to the rules on page 218. The term **oxidation state** is equivalent to oxidation number. General definitions of oxidation and reduction are the following: **oxidation** is an algebraic increase in oxidation number; **reduction** is an algebraic decrease in oxidation number.

Ion–Electron Method In a balanced redox equation, the number of electrons gained by one substance is always equal to the number lost by another substance. This fact forms the basis for the **ion–electron method** (page 227), which provides a systematic method for deriving a net ionic equation for a redox reaction in aqueous solution. According to this method, the *skeleton* net ionic equation is divided into two **half-reactions,** which are balanced separately before being recombined to give the final balanced net ionic equation. For reactions in basic solution, the equation is balanced as if it occurred in an acidic solution, and then the balanced equation is converted to its proper form for basic solution by adding an appropriate number of OH^-.

Metal-Acid Reactions In **nonoxidizing acids,** the strongest oxidizing agent is H^+ (Table 6.1). The reaction of a metal with a nonoxidizing acid gives hydrogen gas and a salt of the acid. Only metals more active than hydrogen react this way. These are metals that are located below hydrogen in the **activity series** (Table 6.2). **Oxidizing acids,** like HNO_3, contain an anion that is a stronger oxidizing agent than H^+, and they are able to oxidize many metals that nonoxidizing acids cannot.

Metal-Displacement Reactions If one metal is more easily oxidized than another, it can displace the other metal from its compounds by a redox reaction. Such reactions are sometimes called **single replacement reactions.** Atoms of the more active metal become ions; ions of the less active metal generally become atoms. In this manner, any metal in the activity series can displace any of the others above it in the series from their compounds.

Oxidations by Molecular Oxygen **Combustion,** the rapid reaction of a substance with oxygen, is accompanied by the evolution of heat and light. Combustion of a hydrocarbon in the presence of excess oxygen gives CO_2 and H_2O, two molecular oxides. When the supply of oxygen is limited, some CO also forms, and in a very limited supply of oxygen the products are H_2O and very finely divided, elemental carbon (as soot or lampblack). The combustion of organic compounds containing only carbon, hydrogen, and oxygen also gives the same products, CO_2 and H_2O. Sulfur burns to give SO_2, which also forms when sulfur-containing fuels burn. Most nonmetals also burn in oxygen to give molecular oxides.

Many metals combine with oxygen in a process often called **corrosion,** but only sometimes is the reaction rapid enough to be considered combustion. The products are ionic metal oxides.

Redox Titrations Potassium permanganate is often used in redox titrations because it is a powerful oxidizing agent and serves as its own indicator. In acidic solutions, the purple MnO_4^- ion is reduced to the nearly colorless Mn^{2+} ion.

TOOLS ▷ **YOU HAVE LEARNED**

The table below lists the concepts that you've learned in this chapter that can be applied as tools in solving problems. Study each one carefully so that you know what each is used for. When faced with solving a problem, recall what each tool does and consider whether it will be helpful in finding a solution. This will aid you in selecting the tools you need. If necessary, refer to this table when working on the Thinking-It-Through problems and the Review Exercises that follow.

TOOL	HOW IT WORKS
Rules for assigning oxidation numbers (page 218)	You use the rules to assign oxidation numbers to the atoms in a chemical formula. You use changes in oxidation numbers to identify oxidation and reduction.
Ion–electron method Follow the steps outlined on page 227 for acidic solutions, and on page 229 for basic solutions.	We use this tool when we need a balanced net ionic equation for a redox reaction in aqueous solution.
Table of oxidizing and nonoxidizing acids (page 231)	By knowing this list, you are able to identify oxidizing and nonoxidizing acids, which enables you to anticipate the products of reactions of metals with acids.
Activity series of metals (page 235)	When a question deals with the possible reaction of one metal with the salt of another, we refer to the activity series to determine the outcome.

THINKING IT THROUGH

Need extra help? Visit the Brady/ Senese web site at www.wiley.com/ college/brady

ON-LINE HELP ☞

The goal for the following problems is not to find the answers themselves, but rather to assemble the information needed to solve them and explain how you would use the information to find the answers. The problems in Level 2 are more challenging than those in Level 1 and may contain more data than are required, in which case you are also asked to identify the unnecessary data. Detailed answers to the Thinking-It-Through problems can be found on the web site.

Level 1 Problems

1. Is the reaction $HIO_4 + 2H_2O \rightarrow H_5IO_6$ a redox reaction? (Describe how you would go about answering this question.)

2. Will cadmium metal react spontaneously with Ni^{2+} ion to give Cd^{2+} and metallic nickel? (How do you determine the answer to this question?)

3. A solution contains Au^{3+} and Fe^{2+} ions. Into the solution are placed strips of metallic Au and Fe. What reaction will occur? (Explain how you would answer this question.)

4. A few minutes after you first start your car, drops of a colorless liquid can be seen falling from the exhaust pipe. After a while, this stops. Explain these observations in terms of the chemistry involved in the running of the engine and the physical changes that occur.

5. Iodate ion, IO_3^-, oxidizes $NO_2(g)$ to give iodide ion and nitrate ion as products. How many milliliters of 0.0200 M IO_3^- solution will react with 0.230 g of NO_2? (Set up the solution to the problem.)

6. A certain reaction is expected to evolve Cl_2 gas as a product, and it is desired to keep it from escaping into the lab. Therefore, the gases formed in the reaction will be trapped and passed through a solution of $Na_2S_2O_3$. The Cl_2 reacts with $S_2O_3^{2-}$ to give Cl^- and SO_4^{2-}. If 0.020 mol of Cl_2 is expected to be formed, what is the minimum volume (in milliliters) of 0.500 M $Na_2S_2O_3$ that will be required to react with all the Cl_2? (Set up the solution to the problem.)

7. How many milliliters of 0.200 M $Na_2S_2O_3$ solution are required to react completely with 0.020 mol of I_3^-? In the reaction, I_3^- is reduced to I^- and $S_2O_3^{2-}$ is oxidized to $S_4O_6^{2-}$. (Set up the solution to the problem.)

Level 2 Problems

8. A mixture is made by combining 300 mL of 0.0200 M $Na_2Cr_2O_7$ with 400 mL of 0.060 M $Fe(NO_3)_2$. Initially, the H^+ concentration in the mixture is 0.400 M. Dichromate ion oxidizes Fe^{2+} to Fe^{3+} and is reduced to Cr^{3+}. After the reaction in the mixture has ceased, how many milliliters of 0.0100 M NaOH will be required to neutralize the remaining H^+? (Set up the solution to the problem.)

9. A solution with a volume of 500.0 mL contained a mixture of SO_3^{2-} and $S_2O_3^{2-}$. A 100.0 mL portion of the solution was found to react with 80.00 mL of 0.0500 M CrO_4^{2-} in a basic solution to give CrO_2^-. The only sulfur-containing product was SO_4^{2-}. After the reaction, the solution was treated with excess 0.200 M $BaCl_2$ solution, which precipitated $BaSO_4$. This solid was filtered from the solution, dried, and found to weigh 0.9336 g. Explain in detail how you can determine the molar concentrations of SO_3^{2-} and $S_2O_3^{2-}$ in the original solution. (Set up the solution to the problem.)

10. An organic compound contains carbon, hydrogen, and sulfur. A sample of it with a mass of 1.045 g was burned in oxygen to give gaseous CO_2, H_2O, and SO_2. These gases were passed through 500.0 mL of an acidified 0.0200 M $KMnO_4$ solution, which caused the SO_2 to be oxidized to SO_4^{2-}. Only part of the available $KMnO_4$ was reduced to Mn^{2+}. Next, 50.00 mL of 0.0300 M $SnCl_2$ was added to a 50.00 mL portion of this solution, which still contained unreduced $KMnO_4$. There was more than enough added $SnCl_2$ to cause all of the remaining MnO_4^- in the 50 mL portion to be reduced to Mn^{2+}. The excess Sn^{2+} that still remained after the reaction was then titrated with 0.0100 M $KMnO_4$, using up 27.28 mL of the $KMnO_4$ solution. Show how you would calculate the percentage of sulfur in the original sample of the organic compound that had been burned.

11. A bar of copper weighing 32.00 g was dipped into 50.0 mL of 0.250 M $AgNO_3$ solution. If all the silver that deposits adheres to the copper bar, how much will the bar weigh after the reaction is complete? (Describe the calculations necessary to solve the problem. Write and balance any necessary chemical equations.)

12. You are given strips of four pure metals (A, B, C, and D) and four bottles containing 0.10 M solutions of the nitrate salts of these same four metals. Describe the experiments you could perform in order to arrange these four metals in order of increasing activity (ease of oxidation). The only other items available to you are a balance and as many beakers or test tubes as you need. You have as many strips of each metal as you need, along with emery paper to clean the metal surfaces.

REVIEW QUESTIONS

Oxidation–Reduction

6.1 Define oxidation and reduction (a) in terms of electron transfer and (b) in terms of oxidation numbers.

6.2 In the reaction $2Mg + O_2 \rightarrow 2MgO$, which substance is the oxidizing agent and which is the reducing agent? Which substance is oxidized and which is reduced?

6.3 Why must both oxidation and reduction occur simultaneously during a redox reaction? What is an oxidizing agent? What happens to it in a redox reaction? What is a reducing agent? What happens to it in a redox reaction?

6.4 In the compound As_4O_6, arsenic has an *oxidation number* of +3. What is the *oxidation state* of arsenic in this compound?

6.5 Is the following a redox reaction? Explain.

$$2NO_2 \longrightarrow N_2O_4$$

6.6 Is the following a redox reaction? Explain.

$$2CrO_4^{2-} + 2H^+ \longrightarrow Cr_2O_7^{2-} + H_2O$$

6.7 If the oxidation number of nitrogen in a certain molecule changes from +3 to −2 during a reaction, is the nitrogen oxidized or reduced? How many electrons are gained (or lost) by each nitrogen atom?

Ion–Electron Method

6.8 The following equation is not balanced. Why? Use the ion–electron method to balance it.

$$Ag + Fe^{2+} \longrightarrow Ag^+ + Fe$$

6.9 Use the ion–electron method to balance the following equation:

$$Cr^{3+} + Zn \longrightarrow Cr + Zn^{2+}$$

6.10 What are the net charges on the left and right sides of the following equations? Add electrons as necessary to make each of them a balanced half-reaction.

(a) $NO_3^- + 10H^+ \rightarrow NH_4^+ + 3H_2O$
(b) $Cl_2 + 4H_2O \rightarrow 2ClO_2^- + 8H^+$

6.11 In the preceding question, which half-reaction represents oxidation? Which represents reduction?

Reactions of Metals with Acids and the Activity Series

6.12 What is a *single replacement reaction*?

6.13 What is a nonoxidizing acid? Give two examples. What is the oxidizing agent in a nonoxidizing acid?

6.14 What is the strongest oxidizing agent in an aqueous solution of nitric acid?

6.15 If a metal is able to react with a solution of HCl, where must the metal stand relative to hydrogen in the activity series?

6.16 Where in the activity series do we find the best reducing agents? Where do we find the best oxidizing agents?

6.17 Which metals in Table 6.2 will not react with nonoxidizing acids?

6.18 Which metals in Table 6.2 will react with water? Write chemical equations for each of these reactions.

6.19 When manganese reacts with silver ion, is manganese oxidized or reduced? Is it an oxidizing agent or a reducing agent?

Oxygen as an Oxidizing Agent

6.20 Define *combustion*.

6.21 Why is "loss of electrons" described as oxidation?

6.22 What products are produced in the combustion of $C_{10}H_{22}$ (a) if there is an excess of oxygen available? (b) If there is a slightly limited oxygen supply? (c) If there is a very limited supply of oxygen?

6.23 If one of the impurities in diesel fuel has the formula C_2H_6S, what products will be formed when it burns? Write a balanced chemical equation for the reaction.

6.24 Burning ammonia in an atmosphere of oxygen produces stable N_2 molecules as one of the products. What is the other product? Write the balanced equation for the reaction.

REVIEW PROBLEMS

Answers to problems whose numbers are printed in color are given in Appendix B. More challenging problems are marked with asterisks. **ILW** = Interactive LearningWare solution is available at *www.wiley.com/college/brady.*

Oxidation–Reduction; Oxidation Numbers

6.25 For the following reactions, identify the substance oxidized, the substance reduced, the oxidizing agent, and the reducing agent:
(a) $2HNO_3 + 3H_3AsO_3 \rightarrow 2NO + 3H_3AsO_4 + H_2O$
(b) $NaI + 3HOCl \rightarrow NaIO_3 + 3HCl$
(c) $2KMnO_4 + 5H_2C_2O_4 + 3H_2SO_4 \rightarrow$
$\qquad\qquad 10CO_2 + K_2SO_4 + 2MnSO_4 + 8H_2O$
(d) $6H_2SO_4 + 2Al \rightarrow Al_2(SO_4)_3 + 3SO_2 + 6H_2O$

6.26 For the following reactions, identify the substance oxidized, the substance reduced, the oxidizing agent, and the reducing agent:
(a) $Cu + 2H_2SO_4 \rightarrow CuSO_4 + SO_2 + 2H_2O$
(b) $3SO_2 + 2HNO_3 + 2H_2O \rightarrow 3H_2SO_4 + 2NO$
(c) $5H_2SO_4 + 4Zn \rightarrow 4ZnSO_4 + H_2S + 4H_2O$
(d) $I_2 + 10HNO_3 \rightarrow 2HIO_3 + 10NO_2 + 4H_2O$

6.27 Assign oxidation numbers to the atoms indicated by boldface type: (a) **S**$^{2-}$, (b) **S**O_2, (c) **P**$_4$, (d) **P**H_3.

6.28 Assign oxidation numbers to the atoms indicated by boldface type: (a) **Cl**O_4^-, (b) **Cr**Cl_3, (c) **Sn**S_2, (d) **Au**$(NO_3)_3$.

6.29 Assign oxidation numbers to all of the atoms in the following compounds:
(a) Na_2HPO_4 (sodium dihydrogen phosphate)
(b) $BaMnO_4$ (barium manganate)
(c) $Na_2S_4O_6$ (sodium tetrathionate)
(d) ClF_3 (chlorine trifluoride)

6.30 Assign oxidation numbers to all the atoms in the following ions: (a) NO_3^-, (b) SO_3^{2-}, (c) NO^+, (d) $Cr_2O_7^{2-}$.

6.31 Assign oxidation numbers to nitrogen in the following:
(a) NO \quad (b) N_2O_5 \quad (c) NH_2OH \quad (d) NO_2 \quad (e) N_2H_4

6.32 Assign oxidation numbers to nitrogen in the following:
(a) NH_3 \quad (b) N_2O_3 \quad (c) N_2 \quad (d) NaN_3 \quad (e) HNO_2

6.33 Assign oxidation numbers to each atom in the following: (a) $NaOCl$, (b) $NaClO_2$, (c) $NaClO_3$, (d) $NaClO_4$.

6.34 Assign oxidation numbers to the elements in the following: (a) $Ca(VO_3)_2$, (b) $SnCl_4$, (c) MnO_4^{2-}, (d) MnO_2.

6.35 Assign oxidation numbers to the elements in the following: (a) PbS, (b) $TiCl_4$, (c) $Sr(IO_3)_2$, (d) Cr_2S_3.

6.36 Assign oxidation numbers to the elements in the following: (a) OF_2, (b) HOF, (c) CsO_2, (d) O_2F_2.

6.37 When chlorine is added to drinking water to kill bacteria, some of the chlorine is changed into ions by the following equilibrium:

$$Cl_2(aq) + H_2O \rightleftharpoons H^+(aq) + Cl^-(aq) + HOCl(aq)$$

In the forward reaction (the reaction going from left to right), which substance is oxidized and which is reduced? In the reverse reaction, which is the oxidizing agent and which is the reducing agent?

6.38 A pollutant in smog is nitrogen dioxide, NO_2. The gas has a reddish-brown color and is responsible for the red-brown color associated with this type of air pollution. Nitrogen dioxide is also a contributor to acid rain because when rain passes through air contaminated with NO_2, it dissolves and undergoes the following reaction:

$$3NO_2(g) + H_2O \longrightarrow NO(g) + 2H^+(aq) + 2NO_3^-(aq)$$

In this reaction, which element is reduced and which is oxidized?

Ion–Electron Method

6.39 Balance the following half-reactions occurring in an acidic solution. Indicate whether each is an oxidation or a reduction. (a) $BiO_3^- \rightarrow Bi^{3+}$, (b) $Pb^{2+} \rightarrow PbO_2$

6.40 Balance the following half-reactions occurring in an acidic solution. Indicate whether each is an oxidation or a reduction. (a) $NO_3^- \rightarrow NH_4^+$, (b) $Cl_2 \rightarrow ClO_3^-$

6.41 Balance the following half-reactions occurring in a basic solution. Indicate whether each is an oxidation or a reduction.
(a) $Fe \rightarrow Fe(OH)_2$
(b) $SO_2Cl_2 \rightarrow SO_3^{2-} + Cl^-$

6.42 Balance the following half-reactions occurring in a basic solution. Indicate whether each is an oxidation or a reduction.
(a) $Mn(OH)_2 \rightarrow MnO_4^{2-}$
(b) $H_4IO_6^- \rightarrow I_2$

ILW 6.43 Balance the following equations for reactions occurring in an acidic solution:
(a) $S_2O_3^{2-} + OCl^- \rightarrow Cl^- + S_4O_6^{2-}$
(b) $NO_3^- + Cu \rightarrow NO_2 + Cu^{2+}$
(c) $IO_3^- + AsO_3^{3-} \rightarrow I^- + AsO_4^{3-}$
(d) $SO_4^{2-} + Zn \rightarrow Zn^{2+} + SO_2$
(e) $NO_3^- + Zn \rightarrow NH_4^+ + Zn^{2+}$
(f) $Cr^{3+} + BiO_3^- \rightarrow Cr_2O_7^{2-} + Bi^{3+}$
(g) $I_2 + OCl^- \rightarrow IO_3^- + Cl^-$
(h) $Mn^{2+} + BiO_3^- \rightarrow MnO_4^- + Bi^{3+}$
(i) $H_3AsO_3 + Cr_2O_7^{2-} \rightarrow H_3AsO_4 + Cr^{3+}$
(j) $I^- + HSO_4^- \rightarrow I_2 + SO_2$

6.44 Balance these equations for reactions occurring in an acidic solution:
(a) $Sn + NO_3^- \rightarrow SnO_2 + NO$
(b) $PbO_2 + Cl^- \rightarrow PbCl_2 + Cl_2$

(c) $Ag + NO_3^- \rightarrow NO_2 + Ag^+$
(d) $Fe^{3+} + NH_3OH^+ \rightarrow Fe^{2+} + N_2O$
(e) $HNO_2 + I^- \rightarrow I_2 + NO$
(f) $C_2O_4^{2-} + HNO_2 \rightarrow CO_2 + NO$
(g) $HNO_2 + MnO_4^- \rightarrow Mn^{2+} + NO_3^-$
(h) $H_3PO_2 + Cr_2O_7^{2-} \rightarrow H_3PO_4 + Cr^{3+}$
(i) $VO_2^+ + Sn^{2+} \rightarrow VO^{2+} + Sn^{4+}$
(j) $XeF_2 + Cl^- \rightarrow Xe + F^- + Cl_2$

6.45 Balance equations for these reactions occurring in a basic solution:
(a) $CrO_4^{2-} + S^{2-} \rightarrow S + CrO_2^-$
(b) $MnO_4^- + C_2O_4^{2-} \rightarrow CO_2 + MnO_2$
(c) $ClO_3^- + N_2H_4 \rightarrow NO + Cl^-$
(d) $NiO_2 + Mn(OH)_2 \rightarrow Mn_2O_3 + Ni(OH)_2$
(e) $SO_3^{2-} + MnO_4^- \rightarrow SO_4^{2-} + MnO_2$

6.46 Balance equations for these reactions occurring in a basic solution:
(a) $CrO_2^- + S_2O_8^{2-} \rightarrow CrO_4^{2-} + SO_4^{2-}$
(b) $SO_3^{2-} + CrO_4^{2-} \rightarrow SO_4^{2-} + CrO_2^-$
(c) $O_2 + N_2H_4 \rightarrow H_2O_2 + N_2$
(d) $Fe(OH)_2 + O_2 \rightarrow Fe(OH)_3 + OH^-$
(e) $Au + CN^- + O_2 \rightarrow Au(CN)_4^- + OH^-$

6.47 Laundry bleach such as Clorox is a dilute solution of sodium hypochlorite, NaOCl. Write a balanced net ionic equation for the reaction of NaOCl with $Na_2S_2O_3$. The OCl^- is reduced to chloride ion and the $S_2O_3^{2-}$ is oxidized to sulfate ion.

6.48 Calcium oxalate is one of the minerals found in kidney stones. If a strong acid is added to calcium oxalate, the compound will dissolve and the oxalate ion will be changed to oxalic acid (a weak acid). Oxalate ion is a moderately strong reducing agent. Write a balanced net ionic equation for the oxidation of $H_2C_2O_4$ by $K_2Cr_2O_7$ in an acidic solution. The reaction yields Cr^{3+} and CO_2 among the products.

6.49 Ozone, O_3, is a very powerful oxidizing agent, and in some places ozone is used to treat water to kill bacteria and make it safe to drink. One of the problems with this method of purifying water is that if there is any bromide ion (Br^-) in the water, it becomes oxidized to bromate ion, BrO_3^-, which has shown evidence of causing cancer in test animals. Assuming that ozone is reduced to water, write a balanced chemical equation for the reaction. (Assume an acidic solution.)

6.50 Chlorine is a good bleaching agent because it is able to oxidize substances that are colored to give colorless reaction products. It is used in the pulp and paper industry as a bleach, but after it has done its work, residual chlorine must be removed. This is accomplished using sodium thiosulfate, $Na_2S_2O_3$, which reacts with the chlorine, reducing it to chloride ion. The thiosulfate ion is changed to sulfate ion, SO_4^{2-}, which is easily removed by washing with water. Write a balanced chemical equation for the reaction of chlorine with thiosulfate ion, assuming an acidic solution.

Reactions of Metals with Acids

6.51 Write balanced molecular, ionic, and net ionic equations for the reactions of the following metals with hydrochloric acid to give hydrogen plus the metal ion in solution:

(a) Manganese (gives Mn^{2+})
(b) Cadmium (gives Cd^{2+})
(c) Tin (gives Sn^{2+})
(d) Nickel (gives Ni^{2+})
(e) Chromium (gives Cr^{3+})

6.52 Write balanced molecular, ionic, and net ionic equations for the reaction of each metal in the previous problem with dilute sulfuric acid.

6.53 On the basis of the discussions in this chapter, suggest chemical equations for the oxidation of metallic silver to Ag^+ ion with (a) dilute HNO_3 and (b) concentrated HNO_3.

6.54 When hot and concentrated, sulfuric acid is a fairly strong oxidizing agent. Write a balanced net ionic equation for the oxidation of metallic copper to copper(II) ion by hot concentrated H_2SO_4, in which the sulfur is reduced to SO_2. Write a balanced molecular equation for the reaction.

6.55 For some time, potassium bromate was used in baking to improve the texture of the baked product. (Opposition to the use of bromates in food products was raised when evidence was uncovered that suggested bromate ion appeared to be a cancer-causing agent.) Bromate ion, BrO_3^-, is a good oxidizing agent, and when Sn^{2+} is added to an acidic solution of this anion, the tin is oxidized to Sn^{4+} and the bromate ion is reduced to bromide ion. Write a balanced net ionic equation for the reaction that occurs in an acidic solution.

6.56 Dichromate ion ($Cr_2O_7^{2-}$) contains chromium in the +6 oxidation state and is used in numerous industrial processes, including plating metal, anodizing aluminum, tanning leather, and manufacturing paints and explosives. In an acidic solution, dichromate ion reacts with oxalic acid to give Cr^{3+} and CO_2. Write a balanced net ionic equation for the reaction.

Single Replacement Reactions and the Activity Series

6.57 Use Table 6.2 to predict the outcome of the following reactions. If no reaction occurs, write N.R. If a reaction occurs, write a balanced chemical equation for it.
(a) $Fe + Mg^{2+} \rightarrow$ (c) $Ag^+ + Fe \rightarrow$
(b) $Cr + Pb^{2+} \rightarrow$ (d) $Ag + Au^{3+} \rightarrow$

6.58 Use Table 6.2 to predict the outcome of the following reactions. If no reaction occurs, write N.R. If a reaction occurs, write a balanced chemical equation for it.
(a) $Mn + Fe^{2+} \rightarrow$ (c) $Mg + Co^{2+} \rightarrow$
(b) $Cd + Zn^{2+} \rightarrow$ (d) $Cr + Sn^{2+} \rightarrow$

6.59 The following reactions occur spontaneously:

$$Pu + 3Tl^+ \longrightarrow Pu^{3+} + 3Tl$$

$$Ru + Pt^{2+} \longrightarrow Ru^{2+} + Pt$$

$$2Tl + Ru^{2+} \longrightarrow 2Tl^+ + Ru$$

List the metals Pu, Pt, and Tl in order of increasing ease of oxidation.

6.60 The following reactions occur spontaneously:

$$2Ga + 3Ni^{2+} \longrightarrow 3Ni + 2Ga^{3+}$$

$$3Be^{2+} + 2Pu \longrightarrow 2Pu^{3+} + 3Be$$

$$2Ga^{3+} + 3Be \longrightarrow 2Ga + 3Be^{2+}$$

List the metals Be, Ga, and Pu in order of increasing ease of oxidation.

6.61 It is found that the following reaction occurs spontaneously:

$$Ru^{2+}(aq) + Cd(s) \longrightarrow Ru(s) + Cd^{2+}(aq)$$

What reaction will occur if a mixture is prepared containing the following: $Cd(s)$, $Cd(NO_3)_2(aq)$, $Pt(s)$, $PtCl_2(aq)$? (Refer to the information in Problem 6.59.)

6.62 It is observed that when magnesium metal is dipped into a solution of beryllium chloride, some of the magnesium dissolves and beryllium metal is deposited on the surface of the magnesium. Referring to Problem 6.60, which one of the following reactions will occur spontaneously? Explain the reason for your choice.
(a) $2Ga^{3+} + 3Mg \rightarrow 3Mg^{2+} + 2Ga$
(b) $2Ga + 3Mg^{2+} \rightarrow 2Ga^{3+} + 3Mg$

Reactions of Oxygen

6.63 Write balanced chemical equations for the complete combustion (in the presence of excess oxygen) of the following:
(a) C_6H_6 (benzene, an important industrial chemical and solvent)
(b) C_3H_8 (propane, a gaseous fuel used in many stoves)
(c) $C_{21}H_{44}$ (a component of paraffin wax)

6.64 Write balanced chemical equations for the complete combustion (in the presence of excess oxygen) of the following:
(a) $C_{12}H_{26}$ (a component of kerosene)
(b) $C_{18}H_{36}$ (a component of diesel fuel)
(c) C_7H_8 (toluene, a raw material in the production of TNT)

6.65 Write balanced equations for the combustion of the hydrocarbons in Problem 6.63 in (a) a slightly limited supply of oxygen and (b) a very limited supply of oxygen.

6.66 Write balanced equations for the combustion of the hydrocarbons in Problem 6.64 in (a) a slightly limited supply of oxygen and (b) a very limited supply of oxygen.

6.67 Methanol, CH_3OH, has been suggested as an alternative to gasoline as an automotive fuel. Write a balanced chemical equation for its complete combustion.

6.68 Metabolism of carbohydrates such as glucose, $C_6H_{12}O_6$, produces the same products as complete combustion. Write a chemical equation representing the metabolism (combustion) of glucose.

Redox Reactions and Stoichiometry

6.69 Iodate ion, IO_3^-, reacts with sulfite ion to give sulfate ion and iodide ion.
(a) Write a balanced net ionic equation for the reaction.
(b) How many grams of Na_2SO_3 are needed to react with 5.00 g of $NaIO_3$?

6.70 Potable water (drinking water) should not have manganese concentrations in excess of 0.05 mg/mL. If the manganese concentration is greater than 0.1 mg/mL, it imparts a foul taste to the water and discolors laundry and porcelain surfaces. Manganese(II) ion is oxidized to permanganate

ion by bismuthate ion, BiO_3^-, in an acidic solution. In the reaction, BiO_3^- is reduced to Bi^{3+}.
(a) Write a balanced net ionic equation for the reaction.
(b) How many grams of $NaBiO_3$ are needed to oxidize the manganese in 18.5 mg of $MnSO_4$?

6.71 How many grams of copper must react to displace 12.0 g of silver from a solution of silver nitrate?

6.72 How many grams of aluminum must react to displace all the silver from 25.0 g of silver nitrate? The reaction occurs in aqueous solution.

6.73 In an acidic solution, MnO_4^- reacts with Sn^{2+} to give Mn^{2+} and Sn^{4+}.
(a) Write a balanced net ionic equation for the reaction.
(b) How many milliliters of 0.230 M $KMnO_4$ solution are needed to react completely with 40.0 mL of 0.250 M $SnCl_2$ solution?

6.74 In an acidic solution, HSO_3^- ion reacts with ClO_3^- ion to give SO_4^{2-} ion and Cl^- ion.
(a) Write a balanced net ionic equation for the reaction.
(b) How many milliliters of 0.150 M $NaClO_3$ solution are needed to react completely with 30.0 mL of 0.450 M $NaHSO_3$ solution?

6.75 Methylbromide, CH_3Br, is used in agriculture to fumigate soil to rid it of pests such as nematodes. It is injected directly into the soil, but over time it has a tendency to escape before it can undergo natural degradation to innocuous products. Soil chemists have found that ammonium thiosulfate, $(NH_4)_2S_2O_3$, a nitrogen and sulfur fertilizer, drastically reduces methylbromide emissions by causing it to degrade.

In a chemical analysis to determine the purity of a batch of commercial ammonium thiosulfate, a chemist first prepared a standard solution of iodine by the following procedure. First, 0.462 g of KIO_3 was dissolved in 100 mL of water. The solution was made acidic and treated with excess potassium iodide, which caused the following reaction to take place:

$$IO_3^- + 8I^- + 6H^+ \longrightarrow 3I_3^- + 3H_2O$$

The solution containing the I_3^- was then diluted to exactly 250 mL in a volumetric flask. Next, the chemist dissolved 0.218 g of the fertilizer in water, added starch indicator, and titrated it with the standard I_3^- solution. The reaction was

$$2S_2O_3^{2-} + I_3^- \longrightarrow S_4O_6^{2-} + 3I^-$$

The titration required 27.99 mL of the I_3^- solution.
(a) What was the molarity of the I_3^- solution used in the titration?
(b) How many grams of $(NH_4)_2S_2O_3$ were in the fertilizer sample?
(c) What was the percentage by mass of $(NH_4)_2S_2O_3$ in the fertilizer?

6.76 Sulfites are used worldwide in the wine industry as antioxidant and antimicrobial agents. However, sulfites have also been identified as causing certain allergic reactions suffered by asthmatics, and the FDA mandates that sulfites be identified on the label if they are present at levels of 10 ppm (parts per million) or higher. The analysis of sulfites in wine uses the "Ripper method," in which a standard iodine solution, prepared by the reaction of iodate and iodide ions, is used to titrate a sample of the wine. The iodine is formed in the reaction

$$IO_3^- + 5I^- + 6H^+ \longrightarrow 3I_2 + 3H_2O$$

The iodine is held in solution by adding an excess of I^-, which combines with I_2 to give I_3^-. In the titration, the SO_3^{2-} is converted to SO_2 by acidification and the reaction during the titration is

$$SO_2(aq) + I_3^-(aq) + 2H_2O \longrightarrow SO_4^{2-} + 3I^- + 4H^+$$

Starch is added to the wine sample to detect the end point, which is signaled by the formation of a dark blue color when excess iodine binds to the starch molecules. In a certain analysis, 0.0421 g of $NaIO_3$ was dissolved in dilute acid and excess NaI was added to the solution, which was then diluted to a total volume of 100.0 mL. A 50.0 mL sample of wine was then acidified and titrated with the iodine-containing solution; 2.47 mL of the iodine solution was required.
(a) What was the molarity of the iodine (actually, I_3^-) in the standard solution?
(b) How many grams of SO_2 were in the wine sample?
(c) If the density of the wine was 0.96 g/mL, what was the percentage of SO_2 in the wine?
(d) Parts per million (ppm) is calculated in a manner similar to percent (which is equivalent to *parts per hundred*).

$$ppm = \frac{\text{grams of component}}{\text{grams of sample}} \times 10^6 \text{ ppm}$$

What was the concentration of sulfite in the wine, expressed as parts per million SO_2?

ILW 6.77 A sample of a copper ore with a mass of 0.4225 g was dissolved in acid. A solution of potassium iodide was added, which caused the reaction

$$2Cu^{2+}(aq) + 5I^-(aq) \longrightarrow I_3^-(aq) + 2CuI(s)$$

The I_3^- that formed reacted quantitatively with exactly 29.96 mL of 0.02100 M $Na_2S_2O_3$ according to the following equation:

$$I_3^-(aq) + 2S_2O_3^{2-}(aq) \longrightarrow 3I^-(aq) + S_4O_6^{2-}(aq)$$

(a) What was the percentage by mass of copper in the ore?
(b) If the ore contained $CuCO_3$, what was the percentage by mass of $CuCO_3$ in the ore?

6.78 A 1.362 g sample of an iron ore that contained Fe_3O_4 was dissolved in acid and all the iron was reduced to Fe^{2+}. The solution was then acidified with H_2SO_4 and titrated with 39.42 mL of 0.0281 M $KMnO_4$, which oxidized the iron to Fe^{3+}. The net ionic equation for the reaction is

$$5Fe^{2+} + MnO_4^- + 8H^+ \longrightarrow 5Fe^{3+} + Mn^{2+} + 4H_2O$$

(a) What was the percentage by mass of iron in the ore?
(b) What was the percentage by mass of Fe_3O_4 in the ore?

6.79 Hydrogen peroxide (H_2O_2) solution can be purchased in drug stores for use as an antiseptic. A sample of such a solution weighing 1.000 g was acidified with H_2SO_4 and titrated with a 0.02000 M solution of $KMnO_4$. The net ionic equation for the reaction is

$$6H^+ + 5H_2O_2 + 2MnO_4^- \longrightarrow 5O_2 + 2Mn^{2+} + 8H_2O$$

The titration required 17.60 mL of $KMnO_4$ solution.
(a) How many grams of H_2O_2 reacted?
(b) What is the percentage by mass of the H_2O_2 in the original antiseptic solution?

6.80 Sodium nitrite, $NaNO_2$, is used as a preservative in meat products such as frankfurters and bologna. In an acidic solution, nitrite ion is converted to nitrous acid, HNO_2, which reacts with permanganate ion according to the equation

$$H^+ + 5HNO_2 + 2MnO_4^- \longrightarrow 5NO_3^- + 2Mn^{2+} + 3H_2O$$

A 1.000 g sample of a water-soluble solid containing $NaNO_2$ was dissolved in dilute H_2SO_4 and titrated with 0.01000 M $KMnO_4$ solution. The titration required 12.15 mL of the $KMnO_4$ solution. What was the percentage by mass of $NaNO_2$ in the original 1.000 g sample?

6.81 A sample of a chromium-containing alloy weighing 3.450 g was dissolved in acid, and all the chromium in the sample was oxidized to CrO_4^{2-}. It was then found that 3.18 g of Na_2SO_3 was required to reduce the CrO_4^{2-} to CrO_2^- in a basic solution, with the SO_3^{2-} being oxidized to SO_4^{2-}.
(a) Write a balanced equation for the reaction of CrO_4^{2-} with SO_3^{2-} in a basic solution.
(b) How many grams of chromium were in the alloy sample?
(c) What was the percentage by mass of chromium in the alloy?

6.82 Solder is an alloy containing the metals tin and lead. A particular sample of this alloy weighing 1.50 g was dissolved in acid. All the tin was then converted to the +2 oxidation state. Next, it was found that 0.368 g of $Na_2Cr_2O_7$ was required to oxidize the Sn^{2+} to Sn^{4+} in an acidic solution. In the reaction the chromium was reduced to Cr^{3+} ion.
(a) Write a balanced net ionic equation for the reaction between Sn^{2+} and $Cr_2O_7^{2-}$ in an acidic solution.
(b) Calculate the number of grams of tin that were in the sample of solder.
(c) What was the percentage by mass of tin in the solder?

6.83 Both calcium chloride, $CaCl_2$, and sodium chloride are used to melt ice and snow on roads in the winter. A certain company was marketing a mixture of these two compounds for this purpose. A chemist, wishing to analyze the mixture, dissolved 2.463 g of it in water and precipitated the calcium by adding sodium oxalate, $Na_2C_2O_4$.

$$Ca^{2+} + C_2O_4^{2-} \longrightarrow CaC_2O_4(s)$$

The calcium oxalate was then carefully filtered from the solution, dissolved in sulfuric acid, and titrated with 0.1000 M $KMnO_4$ solution. The reaction that occurred was

$$6H^+ + 5H_2C_2O_4 + 2MnO_4^- \longrightarrow 10CO_2 + 2Mn^{2+} + 8H_2O$$

The titration required 21.62 mL of the $KMnO_4$ solution.
(a) How many moles of $C_2O_4^{2-}$ were present in the CaC_2O_4 precipitate?
(b) How many grams of $CaCl_2$ were in the original 2.463 g sample?
(c) What was the percentage by mass of $CaCl_2$ in the sample?

6.84 A way to analyze a sample for nitrite ion is to acidify a solution containing NO_2^- and then allow the HNO_2 that is formed to react with iodide ion in the presence of excess I^-. The reaction is

$$2HNO_2 + 2H^+ + 3I^- \longrightarrow 2NO + 2H_2O + I_3^-$$

Then the I_3^- is titrated with $Na_2S_2O_3$ solution using starch as an indicator.

$$I_3^- + 2S_2O_3^{2-} \longrightarrow 3I^- + S_4O_6^{2-}$$

In a typical analysis, a 1.104 g sample that was known to contain $NaNO_2$ was treated as described above. The titration required 29.25 mL of 0.3000 M $Na_2S_2O_3$ solution to reach the end point.
(a) How many moles of I_3^- had been produced in the first reaction?
(b) How many moles of NO_2^- had been in the original 1.104 g sample?
(c) What was the percentage by mass of $NaNO_2$ in the original sample?

ADDITIONAL EXERCISES

6.85 What is the oxidation number of sulfur in the tetrathionate ion, $S_4O_6^{2-}$?

*__6.86__ In Practice Exercise 5 (page 223), some of the uses of chlorine dioxide were described along with a reaction that could be used to make ClO_2. Another reaction that is used to make this substance is

$$HCl + NaOCl + 2NaClO_2 \longrightarrow 2ClO_2 + 2NaCl + NaOH$$

Which element is oxidized? Which element is reduced? Which substance is the oxidizing agent and which is the reducing agent?

6.87 What is the average oxidation number of carbon in (a) C_2H_5OH (grain alcohol), (b) $C_{12}H_{22}O_{11}$ (sucrose—table sugar), (c) $CaCO_3$ (limestone), (d) $NaHCO_3$ (baking soda)?

6.88 The following chemical reactions are *observed to occur* in aqueous solution:

$$2Al + 3Cu^{2+} \longrightarrow 2Al^{3+} + 3Cu$$
$$2Al + 3Fe^{2+} \longrightarrow 3Fe + 2Al^{3+}$$
$$Pb^{2+} + Fe \longrightarrow Pb + Fe^{2+}$$
$$Fe + Cu^{2+} \longrightarrow Fe^{2+} + Cu$$
$$2Al + 3Pb^{2+} \longrightarrow 3Pb + 2Al^{3+}$$
$$Pb + Cu^{2+} \longrightarrow Pb^{2+} + Cu$$

Arrange the metals Al, Pb, Fe, and Cu in order of increasing ease of oxidation.

6.89 In the preceding problem, were all the experiments described actually necessary to establish the order?

6.90 According to the activity series in Table 6.2, which of the following metals react with nonoxidizing acids: (a) silver, (b) gold, (c) zinc, (d) magnesium?

6.91 In each pair below, choose the metal that would most likely react more rapidly with a nonoxidizing acid such as HCl: (a) aluminum or iron, (b) zinc or cobalt, (c) cadmium or magnesium.

6.92 In June 2002, the Department of Health and Children in Ireland began a program to distribute tablets of potassium iodate to households as part of Ireland's National Emergency Plan for Nuclear Accidents. Potassium iodate provides iodine, which when taken during a nuclear emergency, works by "topping off" the thyroid gland to prevent the uptake of radioactive iodine that might be released into the environment by a nuclear accident.

To test the potency of the tablets, a chemist dissolved one in 100 mL of water, made the solution acidic, and then added excess potassium iodide, which caused the following reaction to occur:

$$IO_3^- + 8I^- + 6H^+ \longrightarrow 3I_3^- + 3H_2O$$

The resulting solution containing I_3^- was titrated with 0.0500 M $Na_2S_2O_3$ solution, using starch indicator to detect the end point. (In the presence of iodine, starch turns dark blue. When the $S_2O_3^{2-}$ has consumed all the iodine, the solution becomes colorless.) The titration required 22.61 mL of the thiosulfate solution to reach the end point. The reaction during the titration was

$$I_3^- + 2S_2O_3^{2-} \longrightarrow 3I^- + S_4O_6^{2-}$$

How many milligrams of KIO_3 were in the tablet?

6.93 Use Table 6.2 to predict whether the following displacement reactions should occur. If no reaction occurs, write N.R. If a reaction does occur, write a balanced chemical equation for it.
(a) $Zn + Sn^{2+} \rightarrow$ (d) $Mn + Pb^{2+} \rightarrow$
(b) $Cr + H^+ \rightarrow$ (e) $Zn + Co^{2+} \rightarrow$
(c) $Pb + Cd^{2+} \rightarrow$

6.94 Sucrose, $C_{12}H_{22}O_{11}$, is ordinary table sugar. Write a balanced chemical equation representing the metabolism of sucrose. (See Review Problem 6.68.)

6.95 Write chemical equations for the reaction of oxygen with (a) zinc, (b) aluminum, (c) magnesium, (d) iron, and (e) calcium.

*__6.96__ Balance the following equations by the ion–electron method:
(a) $NBr_3 \rightarrow N_2 + Br^- + HOBr$ (basic solution)
(b) $Cl_2 \rightarrow Cl^- + ClO_3^-$ (basic solution)
(c) $H_2SeO_3 + H_2S \rightarrow S + Se$ (acidic solution)
(d) $MnO_2 + SO_3^{2-} \rightarrow Mn^{2+} + S_2O_6^{2-}$ (acidic solution)
(e) $XeO_3 + I^- \rightarrow Xe + I_2$ (acidic solution)
(f) $(CN)_2 \rightarrow CN^- + OCN^-$ (basic solution)

6.97 Lead(IV) oxide reacts with hydrochloric acid to give chlorine. The equation for the reaction is

$$PbO_2 + 4Cl^- + 4H^+ \longrightarrow PbCl_2 + 2H_2O + Cl_2$$

How many grams of PbO_2 must react to give 15.0 g of Cl_2?

*6.98 A solution contains $Ce(SO_4)_3^{2-}$ at a concentration of 0.0150 M. It was found that in a titration, 25.00 mL of this solution reacted completely with 23.44 mL of 0.032 M $FeSO_4$ solution. The reaction gave Fe^{3+} as a product in the solution. In this reaction, what is the final oxidation state of the Ce?

*6.99 A copper bar with a mass of 12.340 g is dipped into 255 mL of 0.125 M $AgNO_3$ solution. When the reaction that occurs has finally ceased, what will be the mass of unreacted copper in the bar? If all the silver that forms adheres to the copper bar, what will be the total mass of the bar after the reaction?

6.100 A solution containing 0.1244 g of $K_2C_2O_4$ was acidified, changing the $C_2O_4^{2-}$ ions to $H_2C_2O_4$. The solution was then titrated with 13.93 mL of a $KMnO_4$ solution to reach a faint pink end point. In the reaction, $H_2C_2O_4$ was oxidized to CO_2 and MnO_4^- was reduced to Mn^{2+}. What was the molarity of the $KMnO_4$ solution used in the titration?

*6.101 It was found that a 20.0 mL portion of a solution of oxalic acid, $H_2C_2O_4$, requires 6.25 mL of 0.200 M $K_2Cr_2O_7$ for complete reaction in an acidic solution. In the reaction, the oxidation product is CO_2 and the reduction product is Cr^{3+}. How many milliliters of 0.450 M NaOH are required to completely neutralize the $H_2C_2O_4$ in a separate 20.00 mL sample of the same oxalic acid solution?

Energy and Chemical Change: Breaking and Making Bonds

The kinetic energy of a spacecraft entering the atmosphere smashes molecules in the air around it into atoms. When these atoms hit the heat shield on the belly of the craft, they recombine into molecules, releasing their energy as heat. The energy changes associated with making and breaking bonds between atoms and their relationship to heat are among the topics discussed in this chapter.

THIS CHAPTER IN CONTEXT We carry out many chemical reactions not to make compounds, but to convert energy from one form to another. Combustion reactions release chemical energy in fuels as heat, which we use to heat homes, run automobile engines, and drive turbines at coal- and oil-fired power stations. Redox reactions in batteries convert chemical energy into electrical energy. Photosynthesis converts radiant energy from the sun into chemical energy, stored as sugars. In living cells, chemical reactions release chemical energy stored in foods that ultimately makes it possible for muscles to move and hearts to beat.

In this chapter we will study **thermochemistry,** the branch of chemistry that deals with energy absorbed or released by a chemical change. Thermochemistry has many practical applications, but it is also of great theoretical importance, because it is an important link between laboratory measurements (such as temperature changes) and events on the molecular level (the making and breaking of chemical bonds).

Thermochemistry is part of the science of **thermodynamics,** the study of energy transfer and energy transformation. Thermodynamics allows scientists to predict whether a proposed physical change or chemical reaction can occur under a given set of conditions. It is an essential part of chemistry (and all the natural sciences). We'll continue our study of thermodynamics in Chapter 20.

7.1 ▶ Energy is the ability to do work and supply heat

We use the word *energy* in casual conversation as a synonym for the ability to get things done. We often refer to energy as though it were a commodity, something we can buy or sell, or pour from one place to another. We speak of energy as though it were an invisible ingredient of foods and fuels.

But energy is not some sort of magical, invisible fluid that can be poured, weighed, or bottled. It is not a substance at all. Rather it's an *ability* possessed by things, the ability to do **work** and supply **heat.**

Work is motion against an opposing force. For example, lifting a weight against gravity is work. Stretching or squeezing a spring is also work (the tension of the spring provides the opposing force). Work is done when a balloon expands and pushes against the pressure of the atmosphere. We can say an object has energy if it can directly or indirectly lift a weight or stretch a spring (Figure 7.1).

Work is defined more precisely on page 275.

There are two forms of energy: energy of motion and energy of position. Let's briefly review them before we move into thermochemistry itself.

Kinetic energy is energy of motion

Consider the collision of argon atoms shown in Figure 7.2. The fast-moving atom can push and move a slower-moving atom during a collision, so it has energy. The

FIGURE 7.1 *Energy is the ability to do work.* (*a*) The crane's engine has energy because it can lift the weight. The energy comes from the controlled burning of gasoline inside the engine. (*b*) The athlete has energy because he can stretch the spring. The energy comes from the controlled breakdown of sugars and fats in his body.

(*a*)

(*b*)

faster-moving atom slows down after the collision. It has lost some of its ability to push other atoms, so it lost some of its energy in the collision. The slower-moving atom speeds up after the collision, and we say that it has gained energy. The total energy of the two atoms is the same before and after the collision. The fast-moving atom transfers some of its energy to the slower one during the collision.

Kinetic is from the Greek *kinetikos,* meaning "of motion."

All moving objects possess energy. We saw in Section 2.3 that **kinetic energy** is the energy an object has because of its motion. If we measure an object's mass (m) and velocity (v), we can calculate its kinetic energy by the equation

$$KE = \tfrac{1}{2}mv^2 \tag{7.1}$$

This equation becomes useful for predicting the masses and velocities of objects when we use it with the *law of conservation of energy,* introduced in Section 2.3. We cannot create or destroy energy. We can only change it from one form to another or transfer it from one object to another.

The law of conservation of energy is useful for discovering relationships between properties during a physical or chemical change. Often, we can write down an energy conservation relationship for a process and then relate energies to properties of the objects or substances before and after the process occurs. This gives us a relationship that shows how the properties change during the process.

Potential energy is energy due to an object's position

When an object is moved against an opposing force, its energy increases. When you lift a book off your desk, for example, you must do work. You must expend some

FIGURE 7.2 *Moving objects have kinetic energy.* (*a*) The faster-moving atom A can push other atoms during a collision, so it has energy. (*b*) After the collision, atom A is moving more slowly. It has lost some of its ability to push other atoms, so it lost some energy during the collision. Atom B is moving faster after the collision, so it has gained energy.

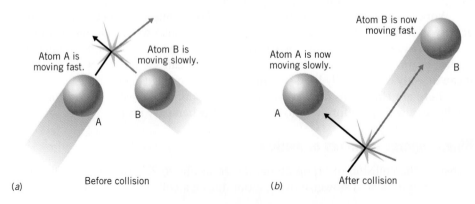

Atom A is moving fast.

Atom B is moving slowly.

Atom A is now moving slowly.

Atom B is now moving fast.

Before collision

After collision

(*a*)

(*b*)

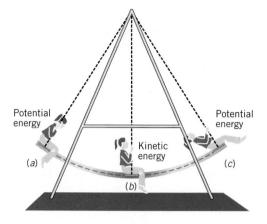

FIGURE 7.3 *Energy conversions on swing.* If friction is neglected, the total energy of the child on the swing is constant. (*a*) The child has high potential energy at the top of the swing. Because she stops momentarily at the top of the swing, her kinetic energy is zero, and her energy is all potential. (*b*) As she falls, potential energy drops. Her speed and kinetic energy increases. (*c*) At the bottom of the swing, the potential energy she had at the top of the swing has been converted into kinetic energy. Kinetic energy is at its highest and potential energy is at its lowest here. (*d*) At the top of the swing on the other side, the energy is once again entirely potential.

energy to lift the book. The energy you lose cannot simply vanish. The book gains the energy you lose. We often say that the energy is now "stored" in the book (when we actually mean that the book has gained energy). Dropping the book converts its "stored" energy into kinetic energy.

We call the energy an object has due to its position or its arrangement of parts its **potential energy,** or PE. Pushing or pulling the object against an opposing force increases its potential energy. For example, when you lift a book off a desk, you're converting the book's kinetic energy into potential energy. Dropping the book converts the potential energy back into kinetic energy. Note that the potential energy of the book depends on its position: potential energy is low when the book is on the table, and high when the book is lifted above it. Figure 7.3 shows another example.

Stretching and releasing a spring also involves conversion of potential energy into kinetic energy. You must do work to stretch the spring beyond its natural length. The more you stretch, the more work you do, and the more the potential energy of the spring rises. Releasing the spring lets it relax back to its natural length, converting its potential energy into kinetic energy in the process. Squeezing and then releasing the spring causes similar energy conversions to occur. Again, notice that the potential energy depends on position. The more we pull or push the spring away from its natural length, the higher its potential energy becomes (Figure 7.4).

Electrons in atoms continually convert potential energy into kinetic energy, and vice versa. The positively charged nucleus attracts the negatively charged elec-

It is convenient to say that potential energy is stored energy, but keep in mind that potential energy isn't some kind of substance hidden inside an object. It's the object's ability to do work or supply heat because of its position or internal arrangement.

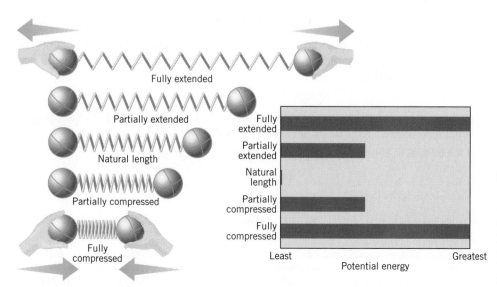

FIGURE 7.4 *The potential energy of a spring depends on its length.* Either squeezing or stretching the spring raises its potential energy. The potential energy of the spring is lowest when the spring is at its natural length.

Unlike charges attract. Like charges repel.

Nearly all chemical and physical changes involve conversion of potential energy into kinetic energy, or vice versa.

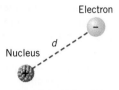

Low potential energy

Atom A

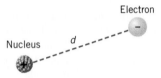

High potential energy

Atom B

The positively charged nucleus attracts the negatively charged electron. Work is required to pull the electron away from the nucleus. An electron's potential energy increases when its distance d from the nucleus increases. The electron in atom B has more potential energy than the electron in atom A.

The joule is named in honor of James Prescott Joule (1818–1889), an English physicist who determined the amount of work needed to produce a given amount of frictional heat.

   The joule is also the amount of potential energy gained when a kilogram mass is lifted slightly more than 0.1 m, or about the energy of a single heartbeat.

trons. Because of this attraction, we must do work to pull the electron away from the nucleus. The further we pull the electron away from the nucleus, the more work we must do, and the more the potential energy of the electron rises. When the electron falls back toward the nucleus, the potential energy falls and the speed and the kinetic energy of the electron increase.[1]

   We can summarize all of these examples with the following statements about gain and loss of potential energy:

- Pushing or pulling an object against an opposing force requires energy. The object's potential energy will rise.
- When the opposing force is not resisted, the object's potential energy falls.

The SI unit of energy is the joule

The SI unit of energy is called the **joule** (abbreviated **J**) and corresponds to the amount of kinetic energy possessed by a 2 kilogram object moving at a speed of 1 meter per second. We can use Equation 7.1 to derive the joule from the SI base units:

$$1\ J = \frac{1}{2}\ (2\ kg)\left(\frac{1\ m}{1\ s}\right)^2$$

$$1\ J = 1\ kg\ m^2\ s^{-2}$$

The joule is actually a rather small amount of energy and in most cases we will use the larger unit, the **kilojoule (kJ).**

$$1\ kJ = 10^3\ J$$

   Another energy unit you may be familiar with is called the **calorie (cal).** Originally, it was defined as the energy needed to raise the temperature of 1 gram of water by 1 degree Celsius.[2] With the introduction of the SI, the calorie has been redefined as follows:

$$1\ cal = 4.184\ J\quad(exactly)$$

The larger unit **kilocalorie (kcal),** which equals 1000 calories, can also be related to the kilojoule.

$$1\ kcal = 10^3\ cal$$

$$1\ kcal = 4.184\ kJ$$

   The nutritional or dietary Calorie (note the capital), Cal, is actually 1 kilocalorie.

$$1\ Cal = 1\ kcal = 4.184\ kJ$$

   While joules and kilojoules have been accepted worldwide as the standard units of energy, calories and kilocalories are still in common use, so you will need to be able to convert joules into calories, and vice versa.

EXAMPLE 7.1

Converting Joules and Calories

Children need at least 30 g of fat each day, enough to supply about 1100 kJ of energy when metabolized. How many dietary Calories does this correspond to?

ANALYSIS: This is a unit conversion problem. We can write

$$1100\ kJ \Leftrightarrow ?\ Cal$$

---

[1]Why doesn't the electron crash into the nucleus? We'll answer this question in Chapter 8!

[2]The amount of heat required to raise the temperature of 1 g of water by 1 °C varies somewhat with the initial temperature we choose, but not by much. At 25 °C, it takes 4.1796 J to raise the temperature of 1 g of water by 1 °C, which is very close to a calorie (4.184 J). To two significant figures, it takes 4.2 J to elevate the temperature of 1 g of water by 1 °C, over the whole range between its freezing and boiling points.

We've seen that one dietary Calorie is equivalent to 4.184 kJ. We therefore have a direct link between kJ and Cal and we do not need the "30 g of fat" to solve the problem.

SOLUTION: We build a conversion factor so that the given units (kJ) cancel to give the desired units (Cal):

$$1100 \text{ kJ} \times \frac{1 \text{ Cal}}{4.184 \text{ kJ}} = 2.6 \times 10^2 \text{ Cal}$$

We have rounded the result because we can safely assume no more than two significant figures in the given information.

Is the Answer Reasonable?
A Calorie is about 4 kilojoules, so the number of Calories should be about one-fourth of the number of kilojoules. One-fourth of 1000 is 250, and one-fourth of 1130 is reasonably close to the answer we obtained.

PRACTICE EXERCISE 1: A relatively inactive male needs to eat enough food to supply 54 kJ for each pound of body weight each day to maintain his current weight. How many dietary Calories is this?

A coffee cup containing 250 g of water at 25.0 °C is placed in a microwave oven and heated to 45.0 °C. How many calories of heat have been absorbed by the water? How many joules have been absorbed by the water?

EXAMPLE 7.2
Using the Definition of Calories

ANALYSIS: First, let's estimate the number of calories that would be required to raise the temperature of *just 1 gram* of water from 25.0 °C to 45.0 °C. The temperature of the water rose by 20.0 °C. If it takes 1 cal to raise the temperature of 1 g of water by 1 °C, it will take 20 cal to raise the temperature of 1 g of water by 20.0 °C. If 1 g of water absorbs 20 cal, then 250 g of water will absorb 250 times as much. Once we have the amount of heat in calories, we can convert it to joules using the relationship

$$1 \text{ cal} = 4.184 \text{ J}$$

SOLUTION: Each gram absorbs 20 cal, so 250 g absorb 250 times 20 cal, or 5.0×10^3 cal. The number of joules absorbed is

$$5.0 \times 10^3 \text{ cal} \times \frac{4.184 \text{ J}}{1 \text{ cal}} = 2.1 \times 10^4 \text{ J}$$

Is the Answer Reasonable?
Let's use the definition of calorie in a slightly different way to check the answer. If it takes 1 cal to raise the temperature of 1 g of water by 1 °C, then it will take 250 cal to raise the temperature of 250 g of water by 1 °C. We must raise the temperature 20.0 °C, not 1 °C, so it will take 20 times as many calories:

$$20.0 \text{ °C} \times \frac{250 \text{ cal}}{1 \text{ °C}} = 5.0 \times 10^3 \text{ cal}$$

which is exactly what we got before. Since there are about 4 J for every calorie, the number of joules absorbed should be $4 \times 5.0 \times 10^3$ cal, or about 2.0×10^4 J, which also is in good agreement with our previous answer.

PRACTICE EXERCISE 2: How many calories are required to raise the temperature of 50.0 g of water by 100 °C? How many joules is this?

7.2 ▶ Internal energy is the total energy of an object's molecules

In Section 2.3, we introduced **heat** as a transfer of energy that occurs between objects with different temperatures. For example, bread is often served in a basket on top of a "bread warmer," a ceramic tile that has been heated in an oven. We say that heat "flows" from the tile to the bread. We can also say that the tile is losing energy while the bread gains it.

If the basket is insulated well enough, the bread and tile will eventually come to the same temperature. At this point we say the bread and tile are in **thermal equilibrium** with each other. The temperature of the bread and tile at thermal equilibrium is less than the original temperature of the tile, but more than the original temperature of the bread.

The energy that is transferred as heat comes from an object's fund of **internal energy.** Internal energy is the sum of energies for all of the individual particles in a sample of matter. All of the particles within any object are in constant motion. For example, in a sample of air at room temperature, oxygen and nitrogen molecules travel faster than rifle bullets through the sample, constantly colliding with each other and with the walls of their container. The molecules spin as they move, and the atoms within the molecules jiggle and vibrate; these internal molecular motions also contribute to the kinetic energy of the molecule and so to the internal energy of the sample. We'll use the term **molecular kinetic energy** for the energy associated with such motions. Each particle has a certain value of molecular kinetic energy at any given moment. Molecules are continually exchanging energy with each other during collisions, but as long as the sample is isolated, the total kinetic energy of all the molecules remains constant.

Internal energy is often given the symbol E or U. In studying both chemical and physical changes, we will be interested in the *change* in internal energy that accompanies the process:

$$\Delta E = E_{\text{final}} - E_{\text{initial}}$$

For a chemical reaction, E_{final} corresponds to the internal energy of the products, so we'll write it as E_{product}. Similarly, we'll use the symbol $E_{\text{reactants}}$ for E_{initial}. So for a chemical reaction the change in internal energy is given by

$$\Delta E = E_{\text{products}} - E_{\text{reactants}} \tag{7.2}$$

Notice carefully an important convention illustrated by this equation. Changes in something like temperature (Δt) or in internal energy (ΔE) are always figured by taking "final minus initial" or "products minus reactants." This means that if a system *absorbs* energy from its surroundings during a change, its final energy is greater than its initial energy and ΔE is positive. This is what happens, for example, when photosynthesis occurs or when something in the surroundings supplies energy to charge a battery. As the system (the battery) absorbs the energy, its internal energy increases and is then available for later use elsewhere.

Temperature is related to average molecular kinetic energy

In Chapter 2 you learned that the **temperature** of a sample is a hotness/coldness property related to the *average kinetic energy of its atoms and molecules*. At first glance, temperature and internal energy may seem very similar, since both are properties of a sample related to molecular kinetic energy. But the two are quite different. Temperature is related to the *average* molecular kinetic energy, whereas internal energy is related to the *total* molecular kinetic energy.[3]

Temperature is related to average molecular kinetic energy, but it is not equal to it. Temperature is *not* an energy.

[3]When we're looking at heat flow, the total molecular kinetic energy is the part of the internal energy we're most interested in. However, internal energy is the total energy of all the particles in the object, so molecular potential energies (from forces of attraction and repulsion that operate between and within molecules) can and do make a large contribution to the internal energy. This is especially true in liquids and solids.

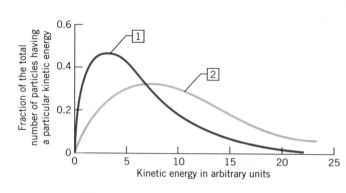

FIGURE 7.5 *The distribution of kinetic energies among gas particles.* The distribution of individual kinetic energies changes in going from a lower temperature, curve 1, to a higher temperature, curve 2. The highest point on each curve is the most probable value of kinetic energy for that temperature. It's the value of the molecular kinetic energy that we would most frequently find if we could get inside the system and observe and measure the kinetic energy of each molecule. The most probable value of molecular kinetic energy is less at the lower temperature. At the higher temperature, more molecules have high speeds and fewer molecules have low speeds, so the maximum shifts to the right and the curve flattens.

Not all of the atoms and molecules in a substance will have the same kinetic energy. The individual molecules are moving at different velocities. At some particular moment, for example, an extremely small number of *individual* molecules will have zero kinetic energy, because balanced collisions happen to cause them to stop for an instant and so have zero velocity. (When v in $\frac{1}{2}mv^2$ goes to zero, KE becomes zero.) Similarly, a small number of the molecules will have very high molecular kinetic energies because unbalanced collisions have, by chance, given them very high velocities. Between these extremes of kinetic energy there will be many molecules with intermediate amounts of kinetic energy.

Figure 7.5 shows graphs that describe the kinetic energy distributions among molecules in a sample at two different temperatures. The vertical axis represents the *fraction* of molecules with a given kinetic energy (i.e., the number of molecules with a given KE divided by the total number of molecules in the sample). Each curve in Figure 7.5 therefore describes how the fraction of molecules with a given KE varies with KE.

Notice that each curve starts out with a fraction equal to zero for zero KE. This is because the fraction of molecules that are motionless (and therefore have zero KE) at any given moment is essentially zero, regardless of the temperature. Moving to higher values of KE, we find greater fractions of molecules. For very large values of KE, the fraction drops off again because fewer molecules have very large velocities.

Each curve in Figure 7.5 has a characteristic peak or maximum, each maximum corresponding to the most frequently experienced values for molecular KE. Because the curves are not symmetrical, the *average* values of molecular KE lie slightly to the right of the maxima. *When the temperature increases, the average molecular KE increases as well.* When the cooler sample (curve 1 in Figure 7.5) is heated, the curve flattens and the maximum shifts to a higher value of average molecular KE (see curve 2). The reason the curve flattens is that the area under each curve corresponds to the sum of all the fractions, and when all the fractions are added, they must equal 1. (No matter how you cut up a pie, if you add all the fractions, you must end up with one pie!)

Internal energy is a state function

How might we measure the amount of internal energy an object has? We might try to measure the object's ability to change the temperature of another, colder object when the two are brought into contact. The more energy the original object has, the more the colder object should warm up. All we need to do is relate the temperature change to the amount of energy that has been transferred.

But what we have actually measured by this procedure is heat, not energy. *Heat is a transfer of energy due to a temperature difference.* We can measure the amount of energy that we draw out of the object, but we cannot measure the internal energy itself unless we draw it *all* off. As we'll see in the next chapter, it is not possible to do this! Fortunately, we are more interested in how internal energy *changes* in chemical reactions than we are in the absolute amount of energy that a sample of a material has. We can, for example, say that a handwarmer releases 100 J of heat, but we do not know the energy of the handwarmer before or after it supplied the heat. We do know that the final energy must be 100 J lower than the initial energy. It's rather like dumping a liter of water into the ocean. You don't know what the volume of the ocean is, but you do know that it's a liter more than it was before you came along. Heat is analogous to the liter of water (it's a volume *change* in the ocean), whereas internal energy would be analogous to the ocean's total volume, vast and impossible to measure precisely.

If we can determine heat, and not energy itself, why talk about energy at all? It turns out that *the energy of an object depends only on the object's current condition.* It doesn't matter how the object acquired that energy, or how it will lose it. This simple fact vastly simplifies thermochemical calculations.

State, as used in thermochemistry, doesn't have the same meaning as when it's used in terms such as *solid state* or *liquid state.*

We'll call a complete list of properties that specify the object's current condition the *state* of the object. In chemistry, it is usually enough to specify the object's chemical composition (substances and numbers of moles), its pressure, its temperature, and its volume to specify its state.

Any property that (like energy) depends *only* on an object's current state is called a **state function.** Pressure, temperature, and volume are examples of state functions and it's easy to understand why. A system's current temperature, for example, does not depend on what it was yesterday. Nor does it matter *how* the system acquired it, that is, the path to its current value. If it's now 25 °C, for example, we know all we can or need to know about its temperature. We do not have to specify how it got there or where it's going. Also, if the temperature were to increase, say, to 35 °C, the change in temperature, Δt, is simply the difference between the final and the initial temperatures.

The symbol Δ, pronounced "delta," denotes a *change between some initial and final state.*

$$\Delta t = t_{final} - t_{initial} \tag{7.3}$$

To make the calculation of Δt, we do not have to know what caused the temperature change—exposure to sunlight, heating by an open flame, or any other mechanism. All we need are the initial and final values. *This independence from the method or mechanism by which a change occurs is the important feature of all state functions.* As we'll see later, the advantage of recognizing that some property is a *state* function is that many calculations are then much, much easier.

Heat can be conducted by molecular collisions

The kinetic molecular theory provides a simple molecular explanation for how heat flows from one object to another.[4] In the previous section, we saw that a molecule with high kinetic energy can transfer some of its energy to other molecules when it collides with them, just as a cue ball transfers its energy to other balls on a pool table. Figure 7.6 shows how heat transfer occurs at the molecular level between a warm object, such as boiling water, and a cooler one, such as a cup. Longer arrows in this figure indicate faster-moving molecules, with greater kinetic energy. Note that the hotter objects in the figure have the higher average molecular kinetic energies.

High-energy molecules in the boiling water strike the lower-energy particles in the cup's material. These collisions give some of the molecular kinetic energy of the water molecules to the cup's molecules, which vibrate more energetically. The water molecules slow down as a result and have less average molecular kinetic energy. By this mechanism, some of the molecular kinetic energy in the hot water transfers

---

[4]There are other mechanisms for heat flow, such as convection and radiation, but we need not concern ourselves with them at this point.

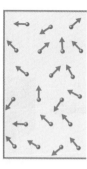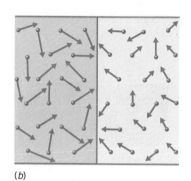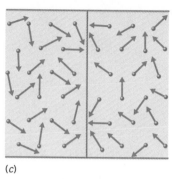

(a) (b) (c)

FIGURE 7.6 *Energy transfer from a warmer to a cooler object.* (*a*) The longer arrows on the left denote higher kinetic energies and so something warmer, like hot water just before it's poured into a cooler coffee cup. (*b*) The hot water and the cup's inner surface are now in thermal contact. Energetic water molecules are slowed down by collisions with molecules in the cup's material. Energy is transferred from the water to the material of the cup. (*c*) Thermal equilibrium is established; the temperatures of the coffee and the cup wall are now equal.

to the material of the cup. The temperature of the water decreases and the temperature of the cup increases.

7.3 ▶ Heat can be determined by measuring temperature changes

If we can devise some way of measuring the amount of heat that is absorbed or released by an object, we will be able to quantitatively study nearly any type of energy transfer. For example, if we want to measure the energy transferred by an electrical current, we can force the electric current through something with high electrical resistance, like the heating element in a toaster, and the energy transferred by the current becomes heat. But there is no instrument available that directly measures heat. We must measure temperature changes, and then use them to calculate heat.

It is very important to specify the **boundary** across which heat flows. The boundary might be visible (like the walls of a beaker) or invisible (like the boundary that separates warm air from cold air along a weather front). The boundary encloses the **system,** which is the object we are interested in studying. Everything outside the system is called the **surroundings.** The system and the surroundings together are called the **universe.**

Three types of systems are possible, depending on whether matter or energy can cross the boundary.

- **Open systems** can gain or lose mass and energy across their boundaries. The human body is an example of an open system.
- **Closed systems** can absorb or release energy, but not mass, across the boundary. The mass of a closed system is constant, no matter what happens inside. A light bulb is an example of a closed system.
- **Isolated systems** can not exchange matter or energy with their surroundings. Because energy cannot be created or destroyed, the energy of an isolated system is constant, no matter what happens inside. A stoppered thermos bottle is a good approximation of an isolated system.

Processes that occur within an isolated system are called *adiabatic,* from the Greek *a + diabatos,* not passable.

The heat an object gains or loses is directly proportional to its temperature change

We have seen that the definition of an older unit of heat, the calorie, can be used to relate the amount of heat that flows into a known mass of water with the increase in water temperature. Consider the following simple experiment:

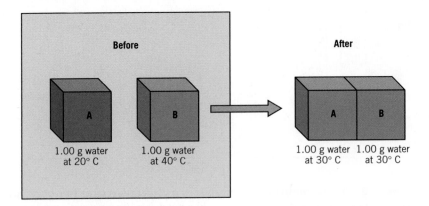

In Section 7.1, we saw that the amount of heat necessary to raise the temperature of 1 g of water by 1 °C is 1 calorie, or 4.184 joules. We can use this fact to relate the temperature change each sample undergoes with the heat it has absorbed or released.

- Cube A warmed from 20.0 °C to 30.0 °C, a rise of 10.0 °C. It takes 10.0 calories to warm 1 g of water by 10.0 °C, so we can say that cube A *gained* 10.0 calories, or 41.8 J of heat.
- Cube B cooled from 40.0 °C to 30.0 °C, a drop of 10.0 °C; 10.0 calories of heat must be released by 1 g of water to cool it by 10.0 °C, so we can say that cube B *lost* 10.0 calories, or 41.8 J of heat.

In other words, 10.0 calories (or 41.8 J) of energy was transferred from cube B to cube A.

When reporting a heat flow we must always say whether heat was gained or lost. We'll adopt the following sign convention for heat:

- If heat has been *gained* by the object, we'll write the heat as a *positive* number.
- If heat is *lost* by an object, we'll write the heat as a *negative* number.

Notice that the sign we assign to the heat that an object gains or loses is the same as the sign of the temperature change, Δt, the object undergoes, as defined in Equation 7.3.

Let's repeat this experiment with several different water temperatures. Assume that the two cubes are *insulated* from their surroundings so that heat can flow only between the cubes.

Run	Initial Temperature of A (°C)	Initial Temperature of B (°C)	Final Temperature of A & B (°C)	Heat, A (cal)	Heat, B (cal)
1	10.0	40.0	25.0	+15.0	−15.0
2	20.0	40.0	30.0	+10.0	−10.0
3	30.0	40.0	35.0	+5.0	−5.0
4	40.0	40.0	40.0	0.0	0.0
5	10.0	50.0	30.0	+20.0	−20.0
6	20.0	50.0	35.0	+15.0	−15.0
7	30.0	40.0	40.0	+10.0	−10.0

Notice the following:

- There is no heat flow in Run 4, because the temperatures of A and B are the same. *Heat does not flow between two objects of the same temperature.*

- When A and B have different temperatures, the cooler object (A) always gains heat and the warmer object (B) always loses it. *Heat flows spontaneously from the warmer object to the cooler one.*

- When the mass of A and B are the same, and when A and B are made of the same material, the final temperature is the average of the two object's initial temperatures. (We'll see in another experiment that when A and B have different masses or compositions, the final temperature can be calculated as a weighted average, not a simple average.)

- When the temperature change the object undergoes is doubled (e.g., going from Run 3 to Run 2) the heat flow is doubled too. When the temperature change is tripled (e.g., going from Run 3 to Run 1) the heat flow is tripled. When the temperature change is quadrupled (e.g., going from Run 3 to Run 5) the heat flow is quadrupled. We might conclude that *the heat gained or lost by an object is directly proportional to the temperature change it undergoes.*

A *spontaneous change* is one that continues on its own, without outside intervention. We'll see how to predict whether a change is spontaneous or not in Chapter 20.

We can write this last conclusion in the form of an equation:

$$q = C\Delta t \qquad (7.4)$$

where the heat, q, is equal to the temperature change, Δt, times some constant C. If the heat is written in joules, and the temperature change is written in °C, the units of C must be J/°C. These units tell us that C is the amount of heat the object must gain to raise its temperature by 1 °C. For this reason, C is called the **heat capacity** of a given object.

Heat capacity

Central processing chips in computers generate a tremendous amount of heat—enough to damage themselves permanently if the chip is not cooled somehow. Aluminum "heat sinks" are often attached to the chips to carry away excess heat. Suppose that a heat sink at 71.3 °C is dropped into a styrofoam cup containing 100.0 g of water at 25.0 °C. The temperature of the water rises to 27.4 °C. What is the heat capacity of the heat sink, in J/°C?

ANALYSIS: We want the heat capacity, C. We can solve Equation 7.4 for C by dividing both sides by the temperature change, Δt:

$$C = q/\Delta t$$

We must find the temperature change of the heat sink, and the amount of heat q it exchanges with the water in the cup. Let's find q first.

We can assume that the heat that is lost by the heat sink will be gained by the much cooler water, so we start by calculating the heat gained by the water:

heat lost by heat sink = −heat gained by the water

The temperature of the water rises from 25.0 °C to 27.4 °C, a rise of 2.4 °C. It would take 2.4 cal of heat to raise the temperature of each gram of water by 2.4 °C, so it would take 100.0 × 2.4 cal = +240 cal to raise the temperature of 100.0 g of water by 2.4 °C. We can convert calories to joules using the fact that 1 cal is 4.184 J.

SOLUTION: The temperature of the heat sink drops from 71.3 °C to 27.4 °C, so

$$\Delta t = t_{final} - t_{initial} = 27.4 \,°C - 71.3 \,°C = -43.9 \,°C$$

The heat gained by the water is 240 cal, or 240 cal × (4.184 J/1 cal) = 1.0×10^3 J. The heat exchanged by the heat sink will be -1.0×10^3 J (notice that the minus sign means heat was lost by the object). The heat capacity is then

$$C = \frac{q}{\Delta t} = \frac{-1.0 \times 10^3 \,J}{-43.9 \,°C} = 23 \,J/°C$$

EXAMPLE 7.3
Determining the Heat Capacity of an Object from Experimental Data

Negative heat capacities can occur in bizarre systems such as black holes, in fragmenting atomic nuclei, and in small clusters of atoms, but everyday objects always have a positive heat capacity.

Is the Answer Reasonable?

In any calculation involving energy transfer, we first check to see that all quantities have the correct signs. Heat capacities are positive for common objects.

The water gained about 1000 J and warmed up by 2.4 °C, so the water's heat capacity is 1000 J/2.4 °C, or about 400 J/°C. This is about 20 times the calculated heat capacity of the heat sink. We would expect, then, that the temperature rise of the water to be about 1/20th of the temperature drop of the heat sink, since it absorbs 20 times as much heat as the heat sink before its temperature will rise 1 degree. The temperature of the heat sink dropped by 43.9 °C, and the temperature of the water rose by 2.4 °C, which is in good agreement with this prediction.

PRACTICE EXERCISE 3: A ball bearing at 220 °C is dropped into a cup containing 250 g of water. The water warms from 25.0 to 30.0 °C. What is the heat capacity of the ball bearing, in J/°C?

Heat capacity is an extensive property

Objects with larger heat capacities can absorb a larger amount of heat without becoming much warmer. They can also release a larger amount of heat when they cool down again. This fact has many important consequences and practical applications. But why does heat capacity differ for different objects? What factors determine an object's heat capacity?

Our collision-based model of heat conduction suggests that the size of the object will be important. If energy is transferred by molecular collisions, having more molecules will mean that more energy can be transferred, so that a sample with more molecules will be able to lose and gain more heat than an equivalent sample with fewer molecules. Let's test this hypothesis.

Consider the following experiment. We will determine the heat capacity of several different masses of iron shot by measuring the temperature change when they are dropped into 50.0 g of hot water in a thermos bottle. If we know the initial temperature of the object and the water, and the final temperature of the water and the object at thermal equilibrium, we can calculate the heat capacity of the object just as we did in Example 7.3. For all runs, the initial temperature of the iron is 25.0 °C, and the initial temperature of the water is 100.0 °C.

Run	Mass of Iron (g)	Measured Final Temperature (°C)	Calculated Heat Capacity (J/°C)
1	10.0	98.4	4.56
2	20.0	96.9	9.02
3	30.0	95.5	13.4
4	40.0	94.1	17.9
5	50.0	92.7	22.6

The final temperature is the temperature of both the iron and the water at thermal equilibrium. (Can you verify that the calculated heat capacity is correct for each run?)

Notice that when the mass of iron is doubled (going from Run 1 to Run 2), the heat capacity doubles. When the mass is tripled (Run 1 to Run 3) the heat capacity triples. The data support the conclusion that *heat capacity is proportional to the mass of the sample*. Recall that any property that depends on the size of a sample is called an *extensive property*. Our experiment demonstrates that heat capacity is an extensive property.

We can write the relationship between heat capacity and mass as an equation:

$$C = ms \qquad (7.5)$$

where C is the heat capacity, m is the mass, and s is a constant called the **specific heat capacity** (or sometimes, simply the **specific heat**). If the heat capacity has units of J/°C, and if the mass is in grams, the specific heat capacity has units of (J/g °C) or J g^{-1} °C^{-1}. The units tell us that the specific heat capacity is the quantity of heat required to raise the temperature of 1 gram of a substance by 1 °C. The calorie was defined so that the specific heat of water could be written as 1 cal g^{-1} °C^{-1} (read, "one calorie per gram per degree Celsius change"). Today, specific heats are generally given in joules. Because 1 cal equals 4.184 J, the specific heat of water is 4.184 J g^{-1} °C^{-1}.

Specific heat is the heat capacity of a substance per gram. Specific heat is an *intensive property*[5]; it's an inherent thermal property and is unique for each substance (see Table 7.1). At 25 °C, iron's specific heat, for example, is 0.45 J g^{-1} °C^{-1}, much lower than that of water. Silver's specific heat, 0.235 J g^{-1} °C^{-1}, is still lower.

Water, as we see from the data in Table 7.1, has an unusually high specific heat. It takes much more heat to raise the temperature of water by a certain amount than it would take with most other substances. For example, metals have low specific heats relative to water. Notice that the specific heat of water is roughly nine times that of iron, which explains why 1 cal added to 1 g of iron raises its temperature nine times as much as when 1 cal is added to 1 g of water. In other words, the 1 cal of heat that increases the temperature of 1 g of water by 1 °C will increase the temperature of 9 g of iron that much.

Consider the experiment shown in Figure 7.7, which compares the specific heats of three different metals. Each cylinder is made of a different metal, but all three have identical masses and initial temperatures. When the hot cylinders are dropped onto the ice, the ice underneath them begins to melt. The more heat the cylinder gives up, the more the ice melts. Since all three cylinders have the same mass and undergo exactly the same temperature change, any differences in the length of the tracks they make as they melt through the ice are due to differences in specific heats. The metal with the largest specific heat can release the most heat as it cools, and so melt the most ice.

To calculate the heat absorbed or released by an object from its mass, temperature change, and specific heat, we can combine Equations 7.4 and 7.5:

$$q = ms\Delta t \tag{7.6}$$

The following example shows how this equation can be used.

The specific heat of water changes very slightly with temperature, but to two significant digits, it is 1.0 cal g^{-1} °C^{-1} (4.2 J g^{-1} °C^{-1}) over the entire range from freezing to boiling. The original definition of calorie specified a calorie as the amount of heat required to raise the temperature of 1 g of water from 14.5 °C to 15.5 °C.

The heat required to raise the temperature of 1 mol of a substance by 1 °C is called the **molar heat capacity** and has units of J mol^{-1} °C^{-1}.

TABLE 7.1	SPECIFIC HEATS
Substance	Specific Heat, J g^{-1} °C^{-1} (25 °C)
Carbon (graphite)	0.711
Copper	0.387
Ethyl alcohol	2.45
Gold	0.129
Granite	0.803
Iron	0.4498
Lead	0.128
Olive oil	2.0
Silver	0.235
Water (liquid)	4.18

TOOLS

Specific heat

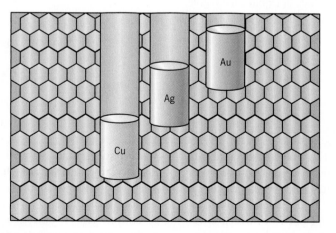

FIGURE 7.7 *Different substances have different specific heats.* Three cylinders of copper, silver, and gold with identical masses are heated to 100 °C in a pot of boiling water and then dropped onto a slab of ice. Copper has the highest specific heat and releases the most heat as it cools, so it melts more ice than the gold and silver can.

[5]This is yet another example of an intensive property that is obtained by dividing one extensive property by another. Dividing heat capacity (an extensive property) by mass (an extensive property) gives specific heat (an intensive property).

FACETS OF CHEMISTRY 7.1

Hypothermia—When the Body's "Thermal Cushion" Is Overtaxed

All of the reactions in the body, collectively called *metabolism*, have rates sensitive to temperature changes. Over the range of body temperatures, metabolic reactions generally are faster at higher temperatures and slower at lower temperatures. We will discuss here a consequence of overusing the body's thermal cushion.

Some of the critical functions sustained by the energy of metabolism are the operation of the central nervous system, including the brain, and the working of the heart. *Hypothermia* is a decrease in the temperature of the inner core of the body. As the inner core cools, rates of metabolism slow down and the symptoms given in the table progressively occur. When a newspaper states that someone died of "exposure," nearly always it is a case of hypothermia.

The onset of hypothermia is a serious medical emergency requiring prompt action by others. The victim may either not recognize the problem or be confused by it. First aid consists of getting the victim out of the wind, dried off, wrapped in warm, dry blankets, and fed warm fluids or sweet food. Contrary to the legend about the St. Bernard dogs, brandy should never be given because, once in the bloodstream, the alcohol in brandy causes the

blood capillaries to dilate. When capillaries near the cold skin suddenly release extra loads of chilled blood toward the body's core, the core cooling accelerates.

A Reference

Doris R. Kimbrough, "Heat Capacity, Body Temperature, and Hypothermia," *Journal of Chemical Education*, January 1998, page 48.

Core Temperature (°C)	Symptoms
37	Normal body temperature
35–36	Intense, uncontrollable shivering
33–35	Amnesia, confusion
30–32	Muscular rigidity, reduced shivering
26–30	Unconsciousness, erratic heartbeat
Below 26	Death by pulmonary edema and heart failure

EXAMPLE 7.4

Calculating Heat from a Temperature Change, Mass, and Specific Heat

If a gold ring with a mass of 5.50 g changes in temperature from 25.0 to 28.0 °C, how much heat has it absorbed?

ANALYSIS: The question asks us to connect the heat absorbed by the ring with its temperature change, Δt. Equations 7.4 or 7.6 could provide this connection. However, we don't know the heat capacity of the ring. We do know the mass of the ring, and the fact that it is made of gold, so we can look up the specific heat and use Equation 7.6 to solve the problem. We will use the value of the specific heat of gold given for 25 °C in Table 7.1, namely, 0.129 J g^{-1} °C^{-1}.

SOLUTION: The mass m of the ring is 5.50 g, the specific heat s is 0.129 J g^{-1} °C^{-1}, and the temperature increases from 25.0 to 28.0 °C, so Δt is 3.0 °C. Using these values in Equation 7.6 gives

$$q = ms\Delta t$$
$$= (5.50 \text{ g}) \times (0.129 \text{ J g}^{-1}\,°C^{-1}) \times (3.0\,°C)$$
$$= 2.1 \text{ J}$$

The sign of Δt will determine the sign of the heat exchanged. Δt is positive because $(t_{final} - t_{initial})$ is positive.

Thus, just 2.1 J raises the temperature of 5.50 g of gold by 3.0 °C. Because Δt is positive, so is the heat, 2.1 J. Thus the *sign* of the energy change signifies that the ring *absorbs* heat.

CHEMISTRY IN PRACTICE The adult body is about 60% water by mass, so it has a high heat capacity. This makes it relatively easy for the human body to maintain a steady temperature of 37 °C, which is vital to survival. In other words, the body can exchange considerable energy with the environment but experience only a small change in temperature.

With a substantial thermal "cushion," the body adjusts to large and sudden changes in outside temperature while experiencing very small fluctuations of its core temperature. *Hypothermia* can still happen, of course; it's an emergency brought on when the body cannot prevent its core temperature from decreasing (see Facets of Chemistry 7.1).

Is the Answer Reasonable?
If the ring had a mass of only 1 g and its temperature increased by 1 °C, we'd know from the specific heat of gold (let's round it to 0.13 J g^{-1} °C^{-1}) that the ring would absorb 0.13 J. For 3 degrees of increase, the answer would be three times as much, or 0.39 J, nearly 0.4 J. For a ring a little heavier than 5 g, the heat absorbed would be five times as much, or about 2.0 J. So our answer (2.1 J) is clearly reasonable.

PRACTICE EXERCISE 4: The temperature of 250 g of water is changed from 25.0 to 30.0 °C. How much energy was transferred into the water? Calculate your answer in joules, kilojoules, calories, and kilocalories.

7.4 ▶ Energy is absorbed or released when chemical bonds are broken or formed

Chemical energy is potential energy due to attractions between particles

We have seen that an atom's potential energy increases when its electrons are pulled farther away from its nucleus. This provides a mechanism for "storing" energy in atoms. The same is true of molecules, which can contain many nuclei and many electrons. The electrons and nuclei within molecules, being oppositely charged, *attract* each other. Electrons, however, *repel* other electrons, and atomic nuclei *repel* other atomic nuclei because like-charged particles repel. Therefore, *net attractive forces must be at work in compounds to bind their atomic nuclei and electrons together.* The term **chemical bond** is simply the name we give to such a net attractive force that exists between a pair of atomic nuclei.

Consider the following simple model of the hydrogen molecule, H$_2$:

The hydrogen molecule contains two hydrogen nuclei and two electrons. The attraction of the electrons for the nuclei is strong enough to overcome the nucleus-nucleus and electron-electron repulsions, so the molecule is held together. We say a chemical bond exists between the atoms within the molecule.

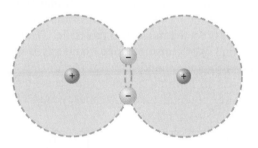

This is a highly oversimplified model of the H$_2$ molecule. We will study chemical bonds in greater detail in Chapters 9 and 10.

The H$_2$ molecule contains two hydrogen nuclei and two electrons. The attraction of the electrons for the nuclei is strong enough to overcome the nucleus–nucleus and electron–electron repulsions, so the molecule is held together. We say a chemical bond exists between the hydrogen atoms. Pulling the atoms apart will require work because we must overcome the attractions between electrons and nuclei. Thus, energy will be required to break the bond:

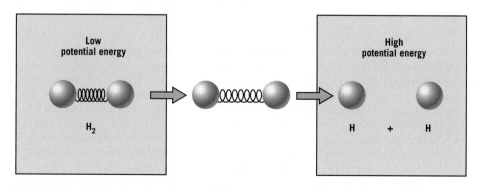

Breaking a chemical bond always requires energy.

On the other hand, if we bring together two hydrogen atoms to make an H_2 molecule, energy must be released, because we move from separated atoms with high potential energy to bonded atoms with lower potential energy:

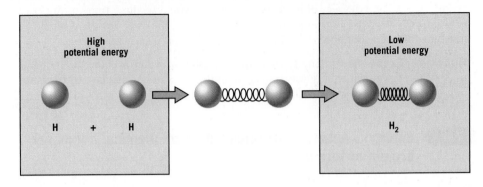

Making a chemical bond always releases energy.

When atoms collide, a bond does not just form between them spontaneously. There must be some mechanism for carrying away the energy that is released by formation of the bond, or the bond will not be able to form. Similarly, energy must be transmitted to a bond in order to break it. There must be some mechanism for supplying energy when a bond is broken. Figure 7.8 shows one way that this can happen. A third body can supply or absorb energy when a chemical bond is broken or formed between two atoms. Many other mechanisms for bond making and breaking are possible.

Chemical bonds are often modeled as springs. For example, we can model the H_2 molecule as two balls (representing the H atoms) with an interconnecting spring (representing the bond). This simple model gives a remarkably good description of many behaviors of the molecule. Stretching a spring is analogous to breaking a bond. Work must be done, so energy is required to break a bond or stretch a spring. On the other hand, allowing a stretched spring to relax to its natural length is analogous to forming a bond. The relaxed spring (and the newly formed bond) have lower potential energy, so energy must be released during bond formation.

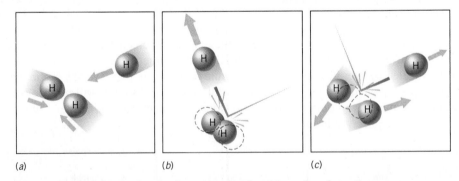

(a) (b) (c)

FIGURE 7.8 *Making a bond releases energy; breaking a bond requires energy.*
(*a*) When two hydrogen atoms approach, they attract each other. When they come close enough, a bond can begin to form. The forming bond lowers their potential energy, but their *total* energy must stay the same. This means their kinetic energy has to increase as the bond forms. Without a way to remove some of this kinetic energy the atoms will fly apart without forming a bond. (*b*) If a third atom collides with the first two at the point where they are close enough to form a bond, it can carry away the excess energy as kinetic energy and the bond forms. The average kinetic energy of the sample goes up, and so does the temperature. (*c*) The reverse happens when a bond is broken. A high kinetic energy molecule slams into two bonded atoms, and its kinetic energy drops and the potential energy of the bonded atoms rises until the bond is broken. Since the average kinetic energy of the sample is lower, the temperature goes down.

Either of these models of the chemical bond are adequate for a qualitative discussion of chemical bonds. We'll look at more sophisticated models for the chemical bond in Chapters 9 and 10. For now, the important thing to know about chemical bonds is simply that they give rise to a compound's potential energy. The potential energy that resides in chemical bonds that can be released in a chemical reaction is called **chemical energy.** Thus, the chemical energy in things like firecrackers or gasoline ultimately comes from the chemical bonds within them. The chemical energy of the gasoline in the fuel tank contributes much to the overall potential energy of a parked car.

Chemical bonds are attractions between atoms *within* a molecule. Some energy changes can also be produced by attractions *between* molecules. These "intermolecular attractions" will be discussed in Chapter 12.

The following reactions occur in the bow shock wave surrounding a spacecraft reentering Earth's atmosphere:

$$N_2(g) \longrightarrow N(g) + N(g) \tag{1}$$

$$O_2(g) \longrightarrow O(g) + O(g) \tag{2}$$

$$N_2(g) + O_2(g) \longrightarrow NO(g) + NO(g) \tag{3}$$

When the nitrogen and oxygen atoms produced by shock wave reactions hit the spacecraft's heat-shield surfaces, the following reactions occur:

$$N(g) + N(g) \longrightarrow N_2(g) \tag{4}$$

$$O(g) + O(g) \longrightarrow O_2(g) \tag{5}$$

$$N(g) + O(g) \longrightarrow NO(g) \tag{6}$$

For each of these reactions, state whether bonds are being made or broken. Which reactions will require energy? Which reactions will release energy?

ANALYSIS: A formula that contains more than one atom will always contain a chemical bond of some sort. Reactions that show a molecule breaking up into atoms will involve breaking at least one chemical bond. Conversely, reactions that show atoms combining to make molecules correspond to formation of at least one chemical bond. Breaking a bond requires an input of energy. Forming a bond releases energy.

SOLUTION: We can tabulate bonds broken and formed for each reaction as follows:

Reaction	Bonds Broken	Bonds Formed	Energy
1. $N_2(g) \rightarrow N(g) + N(g)$	1 N—N bond	none	required
2. $O_2(g) \rightarrow O(g) + O(g)$	1 O—O bond	none	required
3. $N_2(g) + O_2(g) \rightarrow$			
$\quad\quad NO(g) + NO(g)$	1 N—N bond 1 O—O bond	2 N—O bonds	more information is needed!
4. $N(g) + N(g) \rightarrow N_2(g)$	none	1 N—N bond	released
5. $O(g) + O(g) \rightarrow O_2(g)$	none	1 O—O bond	released
6. $N(g) + O(g) \rightarrow NO(g)$	none	1 N—O bond	released

EXAMPLE 7.5
Predicting Whether a Reaction Involving Atoms Will Absorb or Release Energy

Notice that all of the bond-making reactions occurring on the heat shield surface release energy. Thus, these reactions are partially responsible for the intense heat the spacecraft experiences during re-entry. Without some protection from this heat, a spacecraft with an aluminum hull, such as the Space Shuttle, will be destroyed.

Notice that in Reaction 3, two bonds are broken and two bonds are formed. The reaction will need to absorb energy to break the N—N and O—O bonds, but it will release energy when the two N—O bonds form. If the N—N and O—O bonds are collectively stronger than the N—O bonds, bond breaking will require more energy than bond formation gives back, and the overall reaction will require energy. On the other hand, if two N—O bonds together are stronger than an O—O and an N—N bond, bond formation will release more energy than bond breaking requires, so overall energy will be released. We need information on the relative strengths of the various bonds involved in the reaction before we can decide what the net balance between bond breaking or bond making is.

Is the Answer Reasonable?
There's really not much we can do here to check our answer, other than review our reasoning. It seems sound, so all appears to be okay.

PRACTICE EXERCISE 5: When an NaCl crystal is vaporized to form gaseous Na^+ and Cl^- ions, will energy have to be supplied, or will it be released?

Exothermic reactions release heat; endothermic reactions absorb heat

Chemical reactions generally involve *both* the breaking and making of chemical bonds. In most reactions, when bonds *form,* things that attract each other move closer together, which tends to decrease the potential energy of the reacting system. When bonds *break,* on the other hand, things that normally attract each other are forced apart, which increases the potential energy of the reacting system. Each reaction, therefore, has a certain net overall potential energy change, a net balance between the "costs" of breaking bonds and the "profits" from making them.

In many reactions, the products have *less* potential energy (chemical energy) than the reactants. When the methane gas burns in a Bunsen burner, for example, molecules of methane and oxygen with large amounts of chemical energy but relatively low amounts of molecular kinetic energy change into products having less chemical energy but much more molecular kinetic energy, molecules of carbon dioxide and water. Thus some chemical energy changes into molecular *kinetic* energy that can be released to the surroundings as heat, so we'll write heat as a product.

$$CH_4(g) + 2O_2(g) \longrightarrow CO_2(g) + 2H_2O(g) + heat$$

Any reaction in which heat is a product is said to be **exothermic.** Because exothermic reactions release heat, the temperature of the system will tend to rise as the reaction proceeds. If the reaction is occurring in an uninsulated vessel, the temperature of the surroundings will rise.

In some reactions the products have *more* chemical energy than the reactants. For example, plants make energy-rich molecules of glucose ($C_6H_{12}O_6$) and oxygen from carbon dioxide and water by a multistep process called *photosynthesis*. It requires a continuous supply of energy, which is supplied by portions of the sun's energy. The plant's green pigment, chlorophyll, is the solar energy absorber for photosynthesis.

$$6CO_2 + 6H_2O + solar\ energy \xrightarrow[\text{many steps}]{\text{chlorophyll}} C_6H_{12}O_6 + 6O_2$$

Photosynthesis converts solar energy, trapped by chlorophyll, into plant compounds that are rich in chemical energy.

exo means "out"
endo means "in"
therm means "heat"

Reactions that consume energy are called **endothermic.** Because endothermic reactions absorb heat, the temperature of the system will tend to drop as the reaction proceeds. If the reaction is occurring in an uninsulated vessel, the temperature of the surroundings will drop.

Energy can be released by breaking weak bonds to form stronger ones

We'll take a closer look at bonds and bond strengths in Chapters 9 and 10.

Breaking weak bonds requires relatively little energy compared to the amount of energy released when strong bonds form. This is key in understanding why burning fuels such as CH_4 produce heat. Weaker bonds between carbon and hydrogen within the fuel break, so that stronger bonds in the water and carbon dioxide molecules can form. Burning a hydrocarbon fuel produces one molecule of carbon dioxide molecule for each carbon atom and one water molecule for each pair of hydrogen atoms in the fuel. Generally, the more carbon and hydrogen a molecule of the fuel contains, the more strong bonds it can form when it is burned—and the more heat it will produce, per molecule.

Heats of reaction are measured at constant volume or at constant pressure

The amount of heat absorbed or released in a chemical reaction is called the **heat of reaction.** Heats of reaction are determined by measuring the change in temperature they cause in their surroundings, using an apparatus called a **calorimeter.** The calorimeter with a known heat capacity is often just a container for the reaction. We can calculate the heat of reaction by measuring the temperature change the reaction causes in the calorimeter. The science of using a calorimeter for determining heats of reaction is called **calorimetry.**

Calorimeter design is not standard, varying according to the kind of reaction and the precision desired. Calorimeters are usually designed to measure heats of reaction under conditions of either constant volume or constant pressure. We have constant volume conditions if we run the reaction in a closed, rigid container. Running the reaction in an open container imposes constant pressure conditions.

Pressure is an important variable in calorimetry, as we'll see in a moment. If you have ever blown up a balloon or worked with an auto or bicycle tire, you have already learned something about pressure. You know that when a tire holds air at a relatively high pressure, it is hard to dent or push in its sides because "something" is pushing back. That something is the air pressure inside the tire.

Pressure is the amount of *force* acting on a unit of area; it's the ratio of force to area.

$$\text{pressure} = \frac{\text{force}}{\text{area}}$$

You can increase the pressure in a tire by pushing or forcing air into it. Its pressure, when you quit, is whatever pressure you have added *above the initial pressure,* namely, the atmospheric pressure. We describe the air in the tire as being *compressed,* meaning that it's at a higher pressure than the atmospheric pressure.

Atmospheric pressure is the pressure exerted by the mixture of gases in the atmosphere. In English units this ratio of force to area is approximately 14.7 lb in^{-2} at sea level. The ratio does vary a bit with temperature and weather, and it varies a lot with altitude. The value of 14.7 lb in^{-2} is very close to two other common pressure units, the standard atmosphere (abbreviated **atm**), and the **bar,** which will be discussed in Chapter 11. A container that is open to the atmosphere is under a constant pressure of about 14.7 lb in^{-2}, which is approximately 1 bar or 1 atm.

We've used the symbol q for heat; the symbols q_v and q_p are often used to show heats measured at constant volume or at constant pressure, respectively. We must distinguish q_v from q_p. For reactions that involve big changes in volume (such as consumption or production of a gas), the difference between the two can be quite large.

There is no instrument that *directly* measures energy. A calorimeter does not do this; its thermometer is the only part to provide raw data, the temperature change. We calculate the *energy* change.

More precisely, 14.696 lb in^{-2} = 1.0000 atm = 1.0133 bar.

EXAMPLE 7.6

Computing Heats of Reaction Using Calorimetry

A gas-phase chemical reaction occurs inside an apparatus similar to that shown in Figure 7.9. The reaction vessel is a cylinder topped by a piston. The piston can be locked in place with a pin. For this experiment, the cylinder is immersed in an insulated bucket containing a precisely weighed amount of water. A separate experiment determined that the calorimeter (which includes the piston, cylinder, bucket, and water) has a heat capacity of 8.101 kJ/°C. We run the reaction twice, with identical amounts of reactants each time. The following data are collected.

Run	Pin Position	Initial Bucket Temperature (°C)	Final Bucket Temperature (°C)
1	*a* (piston locked)	24.00	28.91
2	*b* (piston unlocked)	27.32	31.54

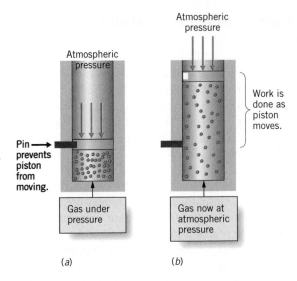

FIGURE 7.9 *Pressure–volume work.* (*a*) A gas is confined under pressure in a cylinder fitted with a piston that is held in place by a sliding pin. (*b*) When the piston is released, the gas inside the cylinder expands and pushes the piston upward against the opposing pressure of the atmosphere. As it does so, the gas does some pressure–volume work on the surroundings.

Determine q_v and q_p for this reaction.

ANALYSIS: The temperature of the water in the bucket rises, so the reaction produces heat. We'll assume that the calorimeter absorbs all of the heat the reaction produces. We can use Equation 7.4 to compute the heat absorbed from the temperature change for each run, since we are given the heat capacity of the calorimeter. Run 1 will give the heat at constant volume, q_v, since the piston cannot move. Run 2 will give the heat at constant pressure, q_p, because with the piston unlocked the entire reaction will run under atmospheric pressure.

SOLUTION: For Run 1, Equation 7.4 gives the heat absorbed by the calorimeter.

$$q = C\Delta t = (8.101 \text{ kJ/°C}) \times (28.91 \text{ °C} - 24.00 \text{ °C}) = 39.8 \text{ kJ}$$

For the calorimeter, $q = +39.8$ kJ. This heat increase occurs because the reaction releases heat, so for the reaction q must be *negative*. Therefore, we can write $q_v = -39.8$ kJ. Similarly, for Run 2,

$$q = C\Delta t = (8.101 \text{ kJ/°C}) \times (31.54 \text{ °C} - 27.32 \text{ °C}) = 34.2 \text{ kJ}$$

so $q_p = -34.2$ kJ.

Is the Answer Reasonable?
The arithmetic is straightforward, but in any calculation involving heat, always check to see that the signs of the heats in the problem makes sense. The calorimeter absorbs heat, so its heats are positive. The reaction releases heat, so its heats must be negative.

Why are q_v and q_p different for reactions that involve a significant volume change? The system in this case is the reacting mixture. If the system expands against atmospheric pressure, it is doing work. Some of the energy that would otherwise appear as heat is used up when the system pushes back the atmosphere. In the example, the work done to expand the system against atmospheric pressure is equal to the amount of "missing" heat in the constant pressure case:

$$\text{work} = (-39.8 \text{ kJ}) - (-34.2 \text{ kJ}) = -5.6 \text{ kJ}$$

The minus sign indicates that energy is leaving the system. This is called **expansion work** (or more precisely, **pressure–volume work**). A common example of pressure–volume work is the work done by the expanding gases in a cylinder of a car engine as they move a piston. Another example is the work done by expanding gases to lift a rocket from the ground. The amount of expansion work w done can be computed from atmospheric pressure and the volume change that the system undergoes[6]:

$$w = -P\Delta V \qquad (7.7)$$

P is the *opposing pressure* against which the piston pushes, and ΔV is the change in the volume of the system (the gas) during the expansion, that is, $V_{final} - V_{initial}$. When V_{final} is greater than $V_{initial}$, ΔV must be positive. This makes the expansion work negative.

Heat and work are both ways to transfer energy

In chemistry, *a minus sign on an energy transfer always means that the system loses energy.* Consider what happens whenever work can be done or heat can flow in the system shown in Figure 7.9b. If work is negative (as in an expansion) the system loses energy and the surroundings gain it. We say work is done *by* the system. If heat is negative (as in an exothermic reaction) the system loses energy and the surroundings gain it. Either event should cause a drop in internal energy. On the other hand, if work is positive (as in a compression), the system gains energy and the surroundings lose it. We say that work is done *on* the system. If heat is positive (as in an endothermic reaction), again, the system will gain energy and the surroundings will lose it. Either positive work or positive heat should cause a positive change in internal energy.

Work and heat are simply alternative ways to transfer energy. Using this sign convention, we can relate the work w and the heat q that go into the system to the internal energy change ΔE the system undergoes:

$$\Delta E = q + w \qquad (7.8)$$

In Section 7.2 we pointed out that the internal energy depended only on the current state of the system; we said that *internal energy is a state function.* This statement (together with Equation 7.8, which is a definition of internal energy change) is a statement of the **first law of thermodynamics.** The first law is one of the most subtle and most powerful principles ever devised by science. It implies that we can move energy around in various ways, but we cannot create energy, or destroy it.

ΔE is independent of how a change takes place; it depends only on where the state of the system is at the beginning and the end of the change. But the values of q and w depend on what happens *between* the initial and final states. Thus, neither q nor w is a state function. Their values depend on the *path* of the change. For example, consider the discharge of an automobile battery by two different paths (see Figure 7.10). Both paths take us between the same two states, one being the fully charged state and the other the fully discharged state. Because ΔE is a state function and because both paths have the same initial and final states, ΔE *must be the same for both paths.* But how about q and w?

---

[6]$P\Delta V$ must have units of energy if it is referred to as work. Work is accomplished when an opposing force, F, is pushed through some distance or length, L. The amount of work done is equal to the strength of the opposing force times the distance the force is moved.

$$\text{work} = F \times L$$

Because pressure is force (F) per unit area, and area is simply length squared, L^2, we can write the following equation for pressure:

$$P = \frac{F}{L^2}$$

Volume (or a volume change) has dimensions of length cubed, L^3, so pressure times volume change is

$$P\Delta V = \frac{F}{L^2} \times L^3 = F \times L$$

But, from above, $F \times L$ also equals work.

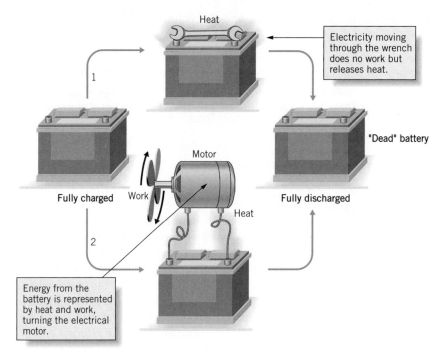

FIGURE 7.10 *Energy, heat, and work.* The complete discharge of a battery along two different paths yields the same total amount of energy, ΔE. However, if the battery is simply shorted with a heavy wrench, as shown in path 1, this energy appears entirely as heat. Path 2 gives part of the total energy as heat, but much of the energy appears as work done by the motor.

In path 1, we simply short the battery by placing a heavy wrench across the terminals. If you have ever done this, even by accident, you know how violent the result can be. Sparks fly and the wrench becomes very hot as the battery quickly discharges. Lots of heat is given off, but *the system does no work* ($w = 0$). All of ΔE appears as heat.

In path 2, we discharge the battery more slowly by using it to operate a motor. Along this path, much of the energy represented by ΔE appears as work (running the motor) and only a relatively small amount appears as heat (from the friction within the motor and the electrical resistance of the wires).

There are two vital lessons here. The first is that neither q nor w is a state function. Their values depend *entirely* on the path between the initial and final states. Yet, their sum, ΔE, as we said, *is* a state function.

Heats of combustion are determined using constant volume calorimetry

The heat produced by a combustion reaction is called a **heat of combustion.** Because combustion reactions require oxygen and produce gaseous products, we have to measure heats of combustion in a closed container. Figure 7.11 shows the apparatus that is usually used to determine heats of combustion. The instrument is called a *bomb calorimeter* because the vessel holding the reaction itself resembles a small bomb. The "bomb" has rigid walls, so the volume of the reaction mixture cannot change. The change in volume ΔV is zero when the reaction occurs. This means, of course, that $P\Delta V$ must also be zero and no expansion work is done. Therefore, w in Equation 7.7 is zero. The heat of reaction measured in a bomb calorimeter is the **heat of reaction at constant volume,** symbolized as q_v, corresponds to ΔE.

$$\Delta E = q_v$$

Food scientists determine dietary calories in foods and food ingredients by burning them in a bomb calorimeter. The reactions that break down foods in the body are complex, but they produce the same products as the combustion reaction for the food.

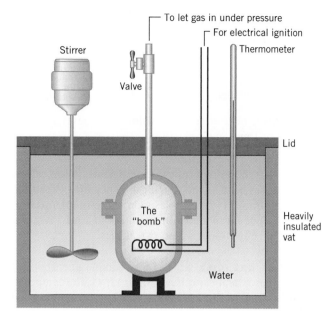

FIGURE 7.11 *A bomb calorimeter.* The water bath is usually equipped with devices for adding or removing heat from the water, thus keeping its temperature constant up to the moment when the reaction occurs in the bomb. The reaction chamber is of fixed volume, so $P\Delta V$ must equal zero for reactions in this apparatus.

EXAMPLE 7.7

Bomb Calorimetry

(a) When 1.000 g of olive oil is completely burned in pure oxygen in a bomb calorimeter like the one shown in Figure 7.11, the temperature of the water bath increases from 22.000 °C to 22.241 °C. How many dietary Calories are in olive oil, per gram? The heat capacity of the calorimeter is 9.032 kJ/°C. (b) Olive oil is almost pure glyceryl trioleate, $C_{57}H_{104}O_6$. The equation for the combustion of glyceryl trioleate is

$$C_{57}H_{104}O_6 + 80\ O_2 \longrightarrow 57CO_2 + 52H_2O$$

What is the change in internal energy, ΔE, in kJ for the combustion of 1 mol of glyceryl trioleate? Assume the olive oil burned in part *a* was pure glyceryl trioleate.

ANALYSIS: In part a, we are given the heat capacity of the calorimeter and its temperature change and asked for the heat produced by burning 1.000 g of olive oil. The calorimeter absorbs the heat released by the burning olive oil. We can use Equation 7.4 to compute the heat absorbed by the calorimeter from its temperature change and heat capacity, and put a minus sign in front of the result to get the heat released by the oil.

Since the heat capacity is in kJ, the heat released will be in kJ, too. We'll need to convert kJ to dietary Calories, which are actually kilocalories (kcal). We can use the relationship

$$1\ kJ = 4.184\ kcal$$

Part b asks for the change in internal energy, ΔE. Bomb calorimetery measures heats at constant volume. Heats at constant volume are equal to the internal energy change for the reaction. Thus the heat calculated in part a is equal to ΔE for 1 g of glyceryl trioleate. We can use the molecular mass of glyceryl trioleate to convert ΔE per gram to ΔE per mole.

SOLUTION: First, let's compute the heat absorbed by the calorimeter when 1.000 g of olive oil is burned, using Equation 7.4:

$$q(\text{calorimeter}) = C\Delta t = (9.032\ kJ/°C) \times (22.241\ °C - 22.000\ °C) = 2.18\ kJ$$

so the heat from burning 1.000 g of olive oil is −2.18 kJ. Part a asks for the heat in dietary Calories. We convert to kilocalories (kcal), which are equivalent to dietary Calories (Cal):

$$\frac{-2.18\ kJ}{1.000\ g\ oil} \times \frac{1\ kcal}{4.184\ kJ} \times \frac{1\ Cal}{1\ kcal} = -0.521\ Cal/g\ oil$$

or we can say, "0.521 dietary Calories are released when 1 gram of olive oil is burned." For part b, we'll convert the heat produced per gram to the heat produced per mole, using the molar mass of $C_{57}H_{104}O_6$, 885.4 g/mol:

$$\frac{-2.18 \text{ kJ}}{1.000 \text{ g } C_{57}H_{104}O_6} \times \frac{885.4 \text{ g } C_{57}H_{104}O_6}{1 \text{ mol } C_{57}H_{104}O_6} = -1.93 \times 10^3 \text{ kJ/mol}$$

Since this is heat at constant volume, we have $\Delta E = q_v = -1.93 \times 10^3$ kJ for combustion of 1 mol of $C_{57}H_{104}O_6$.

Is the Answer Reasonable?

Always check the signs of calculated heats first in calorimetry. The combustion reaction releases heat, so the sign must be negative. A tablespoon of butter has about 100 dietary Calories (with a tablespoon being 15 mL) so it would seem reasonable that a gram of olive oil (with a volume of about 1 mL, assuming a density near that of water) would have about 9 dietary Calories.

PRACTICE EXERCISE 6: A 1.50 g sample of carbon is burned in a bomb calorimeter that has a heat capacity of 8.930 kJ/°C. The temperature of the water jacket rises from 20.00 °C to 25.51 °C. What is ΔE for the combustion of 1 mol of carbon?

Heats of reactions in solution are determined by constant pressure calorimetry

Most reactions that are of interest to us do not occur at constant volume. Instead, they run in open containers like test tubes, beakers, and flasks, where they experience the constant pressure of the atmosphere. When reactions run under constant pressure, they may transfer energy as heat and as expansion work:

$$\Delta E = q + w$$
$$= q_p - P\Delta V \tag{7.9}$$

This is inconvenient. If we want to calculate the internal energy change for the reaction, we'll have to measure its volume change. To avoid this problem, scientists have defined a "corrected" internal energy called **enthalpy, H.** Notice that by adding $P\Delta V$ to both sides of Equation 7.8, we obtain

$$\Delta E + P\Delta V = q_p$$

Enthalpy is defined so that

$$\Delta H = q_p \tag{7.10}$$

So $\Delta H = \Delta E + P\Delta V$, and enthalpy is ultimately defined by the equation

$$H = E + PV$$

Like E, H is a state function.

Enthalpy is from the Greek en + thalpein, "to heat" or "to warm."

An **enthalpy change, ΔH,** is defined by the equation

$$\Delta H = H_{final} - H_{initial}$$

For a chemical reaction, the equation for ΔH may be rewritten as follows:

$$\Delta H = H_{products} - H_{reactants} \tag{7.11}$$

The meanings of + and − values of ΔH are the same as for ΔE. In fact, in chemistry a negative energy change always means that the system releases energy, and a positive energy change always means that the system absorbs energy.

Positive and negative values of ΔH have the same interpretation as positive and negative values of ΔE. When a system absorbs energy from its surroundings, the change is endothermic, so $H_{products}$ has to be larger than $H_{reactants}$. Therefore, *in endothermic changes ΔH is positive.*

On the other hand, if a change is exothermic, the system loses energy so $H_{products}$ must be less than $H_{reactants}$. Therefore, *in exothermic changes ΔH is negative.*

The difference between the enthalpy change and the internal energy change for a reaction is $P\Delta V$. The difference between ΔE and ΔH can be very large for reactions that produce or consume gases, because these reactions can have very large volume changes. If a reaction involves only solids and liquids, though, the values of ΔV are tiny, so ΔE and ΔH for these reactions are nearly identical.

A very simple constant pressure calorimeter, dubbed the coffee cup calorimeter, is made of two nested and capped cups made of Styrofoam, a very good insulator (Figure 7.12). A reaction occurring in such a calorimeter exchanges very little heat with the surroundings, particularly if the reaction is fast. The temperature change is rapid and easily measured. We can use Equation 7.4 to find the heat of reaction, if we have determined the heat capacity of the calorimeter and its contents before the reaction. The Styrofoam cup and the thermometer absorb only a tiny amount of heat, and we can usually ignore them in our calculations.

Usually, when objects warm up, they expand (and ΔV is positive); when objects cool, they contract (and ΔV is negative).

Research-grade calorimeters have greater accuracy and precision than the coffee cup calorimeter.

The reaction of hydrochloric acid and sodium hydroxide is very rapid and exothermic. The equation is

$$HCl(aq) + NaOH(aq) \longrightarrow NaCl(aq) + H_2O$$

In one experiment, a student placed 50.0 mL of 1.00 M HCl at 25.5 °C in a coffee cup calorimeter. To this was added 50.0 mL of 1.00 M NaOH solution also at 25.5 °C. The mixture was stirred, and the temperature quickly increased to a maximum of 32.2 °C. What is the energy evolved in joules per mole of HCl? Because the solutions are relatively dilute, we can assume that their specific heats are close to that of water, 4.18 J g^{-1} °C^{-1}. The density of 1.00 M HCl is 1.02 g mL^{-1} and that of 1.00 M NaOH is 1.04 g mL^{-1}. (We will neglect the heat lost to the Styrofoam itself, to the thermometer, and to the surrounding air.)

ANALYSIS: Before we can use Equation 7.6 to find the number of joules evolved, we have to calculate the system's total mass and the temperature change. The mass here refers to the *total* grams of the combined solutions, but we were given volumes. So we have to use their densities to calculate their masses, which we learned how to do in Chapter 3.

$$\text{mass} = \text{density} \times \text{volume}$$

For the HCl solution, the density is 1.02 g mL^{-1} and we have

$$\text{mass (HCl)} = 1.02 \text{ g mL}^{-1} \times 50.0 \text{ mL} = 51.0 \text{ g}$$

For the NaOH solution, the density is 1.04 g mL^{-1} and

$$\text{mass (NaOH)} = 1.04 \text{ g mL}^{-1} \times 50.0 \text{ mL} = 52.0 \text{ g}$$

The mass of the final solution is thus the sum, 103.0 g.

SOLUTION: The reaction changes the system's temperature by $t_{\text{final}} - t_{\text{initial}}$, so

$$\Delta t = 32.2 \text{ °C} - 25.5 \text{ °C} = 6.7 \text{ °C}$$

Now we can calculate the heat produced by the reaction using Equation 7.6.

$$\text{heat (evolved)} = \text{mass} \times \text{specific heat} \times \Delta t$$

Or

$$\text{heat evolved} = 103.0 \text{ g} \times 4.18 \text{ J g}^{-1} \text{°C}^{-1} \times 6.7 \text{ °C}$$

$$= 2.9 \times 10^3 \text{ J}$$

This is the heat evolved for the specific mixture prepared, but the problem calls for joules *per mole* of HCl. We calculate the number of moles of HCl used with the molarity-to-moles tool given by the definition of molarity. The mo-

EXAMPLE 7.8

Constant Pressure Calorimetry

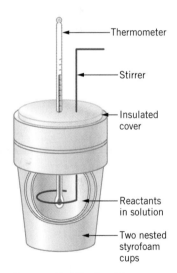

FIGURE 7.12 *A coffee cup calorimeter used to measure heats of reaction at constant pressure.*

- Thermometer
- Stirrer
- Insulated cover
- Reactants in solution
- Two nested styrofoam cups

larity of the acid is 1.00 M, so in 50.0 mL of HCl solution (0.0500 L) we have 0.0500 mol HCl.

Recall that
$M \times V(L)$ = moles

$$0.0500 \text{ L HCl soln} \times \frac{1.00 \text{ mol HCl}}{1.00 \text{ L HCl soln}} = 0.0500 \text{ mol HCl}$$

The neutralization of 0.0500 mol of acid released 2.9×10^3 J. So we should express the heat released as the following ratio:

$$\text{Energy evolved per mole of HCl} = \frac{2.9 \times 10^3 \text{ J}}{0.0500 \text{ mol HCl}} = 58 \times 10^3 \text{ J mol}^{-1}$$

$$= 58 \text{ kJ mol}^{-1}$$

Thus the energy of neutralizing HCl by NaOH is -58 kJ mol^{-1} (negative because the reaction is exothermic).

Is the Answer Reasonable?
Let's first review the logic of the steps we used. *Notice how the logic is driven by definitions, which carry specific units.* Working backward, knowing that we want units of joules per mole in the answer, we must calculate separately the number of moles of acid neutralized and the number of joules that evolved. The latter will emerge when we multiply the solution's mass (g) and specific heat (units of J g^{-1} °C^{-1}) by the degrees of temperature increase (°C).

The step in the calculation that involves the mass of the final sample is clearly easy. Because the densities are close to 1 g mL^{-1}, the masses mixed are about 50 g per solution or a total of 100 g.

The specific heat is around 4.2 J g^{-1} °C^{-1}, which means that if the final solution weighed 1 g, its heat capacity would be 4.2 J °C^{-1}. But if the mass is 100 g, then the heat capacity is 100 times as much, or 420 J °C^{-1}.

The temperature increases by about 7 °C, so when we multiply 420 J °C^{-1} by 7, we get the heat evolved, namely, about 2940 J, or about 2.9×10^3 J; this is the same as 2.9 kJ.

This much heat is associated with neutralizing 0.05 mol HCl, so the heat released per mole is 2.9 kJ divided by 0.05 mol, or 58 kJ mol^{-1}.

PRACTICE EXERCISE 7: When pure sulfuric acid dissolves in water, much heat is given off. To measure it, 175 g of water was placed in a coffee cup calorimeter and chilled to 10.0 °C. Then 4.90 g of sulfuric acid (H_2SO_4), also at 10.0 °C, was added, and the mixture was quickly stirred with a thermometer. The temperature rose rapidly to 14.9 °C. Assume that the value of the specific heat of the solution is 4.18 J g^{-1} °C^{-1} and that the solution absorbs all the heat evolved. Calculate the heat evolved in kilojoules by the formation of this solution. (Remember to use the *total* mass of the solution, the water plus the solute.) Calculate also the heat evolved *per mole* of sulfuric acid.

7.6 ▶ Thermochemical equations are chemical equations that quantitatively include heat

The amount of heat a reaction produces or absorbs depends on the number of moles of reactants we combine. It makes sense that if we burn 2 mol of carbon, we're going to get twice as much heat as we would if we had burned 1 mol. For heats of reaction to have meaning, we must describe the **system** completely. Our description must include amounts and concentrations of reactants, amounts and concentrations of products, temperature, and pressure, because all of these things can influence heats of reaction.

Unless we specify otherwise, whenever we write ΔH we mean ΔH for the *system*, not the surroundings.

Chemists have agreed to a set of **standard states** to make it easier to report and compare heats of reaction. Most thermochemical equations are written for reactants and products at a pressure of 1 **bar,** or (for substances in aqueous solution) a concentration of 1 M. A temperature of 25 °C (298 K) is often specified as well, although temperature is not part of the definition of standard states in thermochemistry.

$\Delta H°$ is the enthalpy change for a reaction with all reactants and products at standard state

The **standard heat of reaction** is the value of ΔH for a reaction occurring under standard conditions and involving the actual numbers of *moles* specified by the coefficients of the equation. To show that ΔH is for *standard* conditions, a degree sign is added to ΔH to make $\Delta H°$. The units of $\Delta H°$ are normally kilojoules.

To illustrate clearly what we mean by $\Delta H°$, let us use the reaction between gaseous nitrogen and hydrogen that produces gaseous ammonia.

$$N_2(g) + 3H_2(g) \longrightarrow 2NH_3(g)$$

When specifically 1.000 mol of N_2 and 3.000 mol of H_2 react to form 2.000 mol of NH_3 at 25 °C and 1 bar, the reaction releases 92.38 kJ. Hence, for the reaction *as given by the preceding equation,* $\Delta H° = -92.38$ kJ. Often the enthalpy change is given immediately after the equation; for example,

$$N_2(g) + 3H_2(g) \longrightarrow 2NH_3(g) \qquad \Delta H° = -92.38 \text{ kJ}$$

An equation that also shows the value of $\Delta H°$ is called a **thermochemical equation.** It always gives the physical states of the reactants and products, and *its $\Delta H°$ value is true only when the coefficients of the reactants and products are taken to mean moles of the corresponding substances.* The above equation, for example, shows a release of 92.38 kJ if *two* moles of NH_3 form. If we were to make twice as much, or 4.000 mol, of NH_3 (from 2.000 mol of N_2 and 6.000 mol of H_2), then twice as much heat (184.8 kJ) would be released. On the other hand, if only 0.5000 mol of N_2 and 1.500 mol of H_2 were to react to form only 1.000 mol of NH_3, then only half as much heat (46.19 kJ) would be released. For the various sizes of the reactions just described, for example, we would have the following thermochemical equations:

TOOLS
Thermochemical equations

$$N_2(g) + 3H_2(g) \longrightarrow 2NH_3(g) \qquad \Delta H° = -92.38 \text{ kJ}$$

$$2N_2(g) + 6H_2(g) \longrightarrow 4NH_3(g) \qquad \Delta H° = -184.8 \text{ kJ}$$

$$\tfrac{1}{2}N_2(g) + \tfrac{3}{2}H_2(g) \longrightarrow NH_3(g) \qquad \Delta H° = -46.19 \text{ kJ}$$

Because the coefficients of a thermochemical equation always mean *moles,* not molecules, we may use fractional coefficients. Normally we don't because we cannot have fractions of *molecules,* but we can have fractions of moles.

You must write down physical states for all reactants and products in thermochemical equations. The combustion of 1 mol of methane, for example, has different values of $\Delta H°$ if the water produced is in its liquid or its gaseous state.

$$CH_4(g) + 2O_2(g) \longrightarrow CO_2(g) + 2H_2O(l) \qquad \Delta H° = -890.5 \text{ kJ}$$
$$CH_4(g) + 2O_2(g) \longrightarrow CO_2(g) + 2H_2O(g) \qquad \Delta H° = -802.3 \text{ kJ}$$

(The difference in $\Delta H°$ values for these two reactions is the quantity of energy that would be released by the physical change of 2 mol of water vapor at 25 °C to 2 mol of liquid water at 25 °C.)

The following thermochemical equation is for the exothermic reaction of hydrogen and oxygen that produces water.

$$2H_2(g) + O_2(g) \longrightarrow 2H_2O(g) \qquad \Delta H° = -517.8 \text{ kJ}$$

What is the thermochemical equation for this reaction when it is conducted to produce 1.000 mol H_2O?

ANALYSIS: The given equation is for 2.000 mol of H_2O, and any changes in the coefficient for water must be made identically to all other coefficients, *as well as to the value of* $\Delta H°$.

SOLUTION: We divide everything by 2, to obtain

$$H_2(g) + \tfrac{1}{2}O_2(g) \longrightarrow H_2O(g) \qquad \Delta H° = -258.9 \text{ kJ}$$

Is the Answer Reasonable?
Compare the equation just found with the initial equation to see that the coefficients and the value of $\Delta H°$ are all divided by 2.

PRACTICE EXERCISE 8: What is the thermochemical equation for the formation of 2.500 mol of H_2O?

The great dirigible, *Hindenburg,* was filled with hydrogen, and its explosive reaction with oxygen destroyed the airship in minutes on May 6, 1937, at Lakehurst, New Jersey.

7.7 ▶ Thermochemical equations can be combined because enthalpy is a state function

Once we have the thermochemical equation for a given reaction, we can write the equation for the reverse reaction, regardless of how hard it might actually be to make it happen. For example, the thermochemical equation for the combustion of carbon in oxygen to give carbon dioxide is

$$C(s) + O_2(g) \longrightarrow CO_2(g) \qquad \Delta H° = -393.5 \text{ kJ}$$

The reverse reaction, which is extremely difficult to carry out, would be the decomposition of carbon dioxide to carbon and oxygen.

$$CO_2(g) \longrightarrow C(s) + O_2(g) \qquad \Delta H° = ?$$

Despite how hard it would be to cause this reaction, we can still know what its $\Delta H°$ *must* be. It must be +393.5 kJ, that is, equal in size but opposite in sign to $\Delta H°$ for the reaction written in the opposite direction. *The law of conservation of energy requires this remarkable result.* If the values of $\Delta H°$ for the forward and the reverse reactions were not equal but opposite in sign, then perpetual motion machines would be possible. But they are not, not even in theory. Thus, regardless of the difficulty of directly decomposing CO_2 into its elements, we can still write a thermochemical equation for it.

$$CO_2(g) \longrightarrow C(s) + O_2(g) \qquad \Delta H° = +393.5 \text{ kJ}$$

To repeat, if we know $\Delta H°$ for a given reaction, then we also know $\Delta H°$ for the reverse reaction; it has the same numerical value, but the opposite algebraic sign. This extremely useful fact makes thermochemical data available that would otherwise be impossible to measure.

Enthalpy changes depend only on initial and final states, not on the path between them

What we're leading up to is a method for combining known thermochemical equations in a way that will allow us to calculate an unknown $\Delta H°$ for some other reaction. This requires experience in other kinds of manipulations of equations. Let's revisit the combustion of carbon to see how this works and, as a bonus, see a neat demonstration that $\Delta H°$ is a state function.

We can imagine two paths leading from 1 mol each of carbon and oxygen to 1 mol of carbon dioxide.

One-Step Path

Let C and O_2 react to give CO_2 directly.

$$C(s) + O_2(g) \longrightarrow CO_2(g) \qquad \Delta H° = -393.5 \text{ kJ}$$

Two-Step Path

Let C and O_2 react to give CO, and then let CO react with O_2 to give CO_2.

$$\text{Step 1:} \quad C(s) + \tfrac{1}{2}O_2(g) \longrightarrow CO_2(g) \qquad \Delta H° = -110.5 \text{ kJ}$$
$$\text{Step 2:} \quad CO(g) + \tfrac{1}{2}O_2(g) \longrightarrow CO_2(g) \qquad \Delta H° = -283.0 \text{ kJ}$$

Overall, the two-step path consumes 1 mol each of C and O_2 to make 1 mol of CO_2, just like the one-step path. The initial and final states for the two routes to CO_2, in other words, are identical.

If $\Delta H°$ is a state function dependent only on the initial and final states and is independent of path, then the values of $\Delta H°$ for both routes should be identical. We can see that this is exactly true simply by adding the equations for the two-step path and comparing the result with the equation for the single step.

Step 1: $C(s) + \tfrac{1}{2} \quad O_2(g) \longrightarrow CO(g)$ $\Delta H° = -110.5 \text{ kJ}$

Step 2: $CO(g) + \tfrac{1}{2}O_2(g) \longrightarrow CO_2(g)$ $\Delta H° = -283.0 \text{ kJ}$

$\quad CO(g) + C(s) + O_2(g) \longrightarrow CO_2(g) + CO(g) \qquad \Delta H° = -110.5 \text{ kJ} + (-283.0 \text{ kJ})$

$$\Delta H° = -393.5 \text{ kJ}$$

The equation resulting from adding steps 1 and 2 has "$CO(g)$" appearing *identically* on opposite sides of the arrow. We can cancel them to obtain the net equation. Such cancellation is permitted only when both the formula and the physical state of a species are identically given on opposite sides of the arrow. The net thermochemical equation for the two-step process, therefore, is

$$C(s) + O_2(g) \longrightarrow CO_2(g) \qquad \Delta H° = -393.5 \text{ kJ}$$

The results, chemically and thermochemically, are thus identical for both routes to CO_2, demonstrating that $\Delta H°$ is a state function.

Enthalpy diagrams show alternative pathways between initial and final states

The energy relationships among the alternative pathways for the same overall reaction are clearly seen using a graphical construction called an **enthalpy diagram.** Figure 7.13 is an enthalpy diagram for the formation of CO_2 from C and O_2. Horizontal lines correspond to different *absolute* values of enthalpy, H, which we cannot know. However, an upper line represents a larger value of H than a lower line. *Changes* in enthalpy, $\Delta H°$, which we can know, are represented by the vertical distances between these lines. Arrows pointing upward represent increases in enthalpy and have positive values of ΔH. Arrows pointing downward represent decreases in enthalpy and have negative values of ΔH.

For an exothermic reaction, the reactants are always on a higher line than the products, because the reactants have more potential energy and so more enthalpy than the products. Notice in Figure 7.13 that the line for the absolute enthalpy of $C(s) + O_2(g)$, taken as the sum, is above the line for the final product, CO_2. A

TOOLS

Enthalpy diagrams

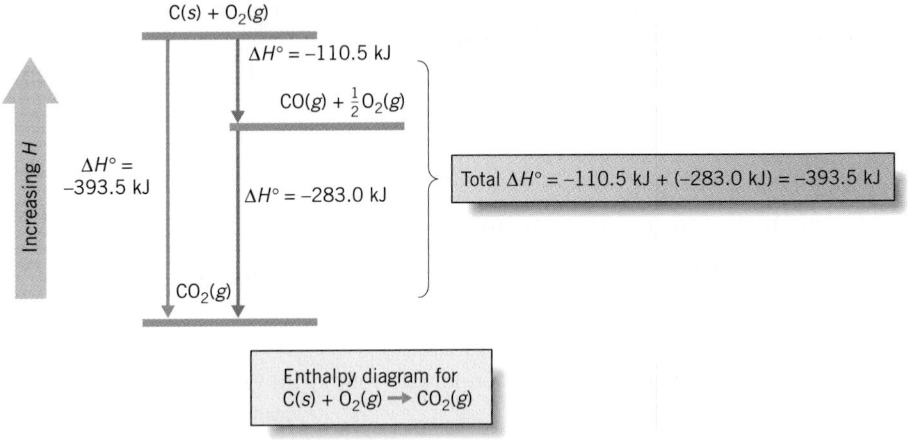

$$C(s) + O_2(g)$$

$\Delta H° = -110.5$ kJ

$$CO(g) + \tfrac{1}{2}O_2(g)$$

$\Delta H° = -393.5$ kJ

$\Delta H° = -283.0$ kJ

Total $\Delta H° = -110.5$ kJ $+ (-283.0$ kJ$) = -393.5$ kJ

Increasing H

$$CO_2(g)$$

Enthalpy diagram for
$C(s) + O_2(g) \longrightarrow CO_2(g)$

FIGURE 7.13 *An enthalpy diagram for the formation of $CO_2(g)$ from its elements by two different paths.* On the left is path 1, the direct conversion of $C(s)$ and $O_2(g)$ to $CO_2(g)$. On the right, path 2 shows two shorter, downward-pointing arrows. The first step of path 2 takes the elements to $CO(g)$, and the second step takes $CO(g)$ to $CO_2(g)$. The overall enthalpy change is identical for both paths, as it must be, because enthalpy is a state function.

horizontal line represents the *sum* of the enthalpies of all of the substances on the line, *in the physical states specified.*

Near the left side of Figure 7.13, we see a long downward arrow that connects the enthalpy level for the reactants, $C(s) + O_2(g)$, to that of the final product, $CO_2(g)$. This is for the one-step path, the direct path. On the right we have the two-step path. Here, the overall change stops at an intermediate enthalpy level corresponding to the intermediate products, $CO(g) + \tfrac{1}{2}O_2(g)$, which are those made or left over from the first step. Then step 2 occurs to give the final product. What the enthalpy diagram shows (indeed, *must* show) is that the total decrease in enthalpy is the same regardless of the path.

EXAMPLE 7.10

Preparing an Enthalpy Diagram

Hydrogen peroxide, H_2O_2, decomposes into water and oxygen by the following equation:

$$H_2O_2(l) \longrightarrow H_2O(l) + \tfrac{1}{2}O_2(g)$$

Construct an enthalpy diagram for the following two reactions of hydrogen and oxygen, and use the diagram to determine the value of $\Delta H°$ for the decomposition of hydrogen peroxide:

$$H_2(g) + O_2(g) \longrightarrow H_2O_2(l) \qquad \Delta H° = -188 \text{ kJ}$$

$$H_2(g) + \tfrac{1}{2}O_2(g) \longrightarrow H_2O(l) \qquad \Delta H° = -286 \text{ kJ}$$

ANALYSIS: The two given reactions are exothermic, so their values of $\Delta H°$ will be associated with downward-pointing arrows. The highest enthalpy level, therefore, must be for the elements themselves. The lowest level must be for the product of the reaction with the largest negative $\Delta H°$, the formation of water by the second equation. The enthalpy level for the H_2O_2, formed by a reaction with the less negative $\Delta H°$, must be in between.

SOLUTION: First, let's diagram the two reactions having known values of $\Delta H°$.

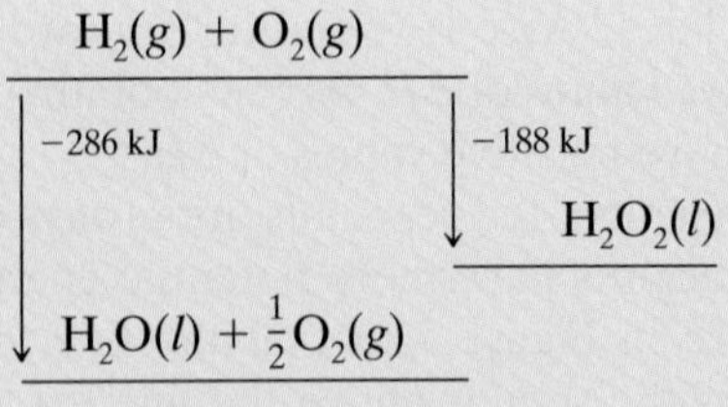

$$H_2(g) + O_2(g)$$

-286 kJ

-188 kJ

$$H_2O_2(l)$$

$$H_2O(l) + \tfrac{1}{2}O_2(g)$$

Notice that the gap on the right side corresponds exactly to the change of $H_2O_2(l)$ into $H_2O(l) + \frac{1}{2}O_2(g)$, the reaction for which $\Delta H°$ is sought. This enthalpy separation corresponds to the difference between the enthalpy levels, -286 kJ and -188 kJ. Thus for the decomposition of H_2O_2 by the equation given,

$$\Delta H° = [-286 \text{ kJ} - (-188 \text{ kJ})] = -98 \text{ kJ}$$

Our completed enthalpy diagram looks like this, showing that $\Delta H°$ for the decomposition of 1 mol of $H_2O_2(l)$ is -98 kJ.

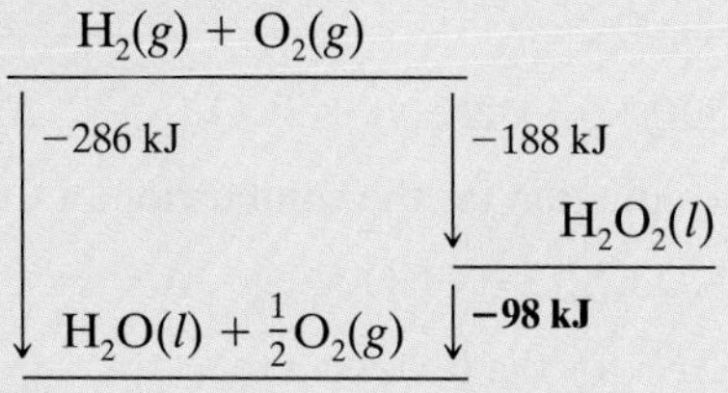

Is the Answer Reasonable?
First be sure that the chemical formulas are arrayed properly on the horizontal lines in the enthalpy diagram. The arrows pointing downward should indicate negative values of $\Delta H°$. These arrows must be consistent *in direction* with the given equations (in other words, they must point from reactants to products). Finally, notice that the total amount of energy associated with the two-step process, namely $(-188$ kJ) plus $(-98$ kJ) equals -286 kJ, the energy for the one-step process.

PRACTICE EXERCISE 9: Two oxides of copper can be made from copper by the following reactions:

$$2Cu(s) + O_2(g) \longrightarrow 2CuO(s) \qquad \Delta H° = -310 \text{ kJ}$$

$$2Cu(s) + \tfrac{1}{2}O_2(g) \longrightarrow Cu_2O(s) \qquad \Delta H° = -169 \text{ kJ}$$

Using these data, construct an enthalpy diagram that can be used to find $\Delta H°$ for the reaction, $Cu_2O(s) + \frac{1}{2}O_2(g) \longrightarrow 2CuO(s)$.

Predict any heat of reaction using Hess's law

Enthalpy diagrams, while instructive, are not necessary to calculate $\Delta H°$ for a reaction from known thermochemical equations. Using the tools for manipulating equations, we ought to be able to calculate $\Delta H°$ values simply by algebraic summing. G. H. Hess was the first to realize this, so the associated law is called **Hess's law of heat summation** or, simply, **Hess's law.**

Germain Henri Hess (1802–1850) anticipated the law of conservation of energy in the law named after him.

> **Hess's Law** The value of $\Delta H°$ for any reaction that can be written in steps equals the sum of the values of $\Delta H°$ of each of the individual steps.

The chief use of Hess's law is to calculate the enthalpy change for a reaction for which such data cannot be determined experimentally or are otherwise unavailable. Because this requires that we manipulate equations, let's recapitulate the few rules that govern these operations.

Rules for Manipulating Thermochemical Equations

1. When an equation is reversed—written in the opposite direction—the sign of $\Delta H°$ must also be reversed.[7]

We reverse an equation by interchanging reactants and products. This leaves the arrow still pointing from left to right.

[7] To illustrate, the reverse of the equation

$$C(s) + O_2(g) \longrightarrow CO_2(g) \qquad \Delta H° = -394 \text{ kJ}$$

is the following equation:

$$CO_2(g) \longrightarrow C(s) + O_2(g) \qquad \Delta H° = +394 \text{ kJ}$$

2. Formulas canceled from both sides of an equation must be for the substance in identical physical states.

3. If all the coefficients of an equation are multiplied or divided by the same factor, the value of $\Delta H°$ must likewise be multiplied or divided by that factor.

EXAMPLE 7.11

Using Hess's Law

Carbon monoxide is often used in metallurgy to remove oxygen from metal oxides and thereby give the free metal. The thermochemical equation for the reaction of CO with iron(III) oxide, Fe_2O_3, is

$$Fe_2O_3(s) + 3CO(g) \longrightarrow 2Fe(s) + 3CO_2(g) \qquad \Delta H° = -26.7 \text{ kJ}$$

Use this equation and the equation for the combustion of CO,

$$CO(g) + \tfrac{1}{2}O_2(g) \longrightarrow CO_2(g) \qquad \Delta H° = -283.0 \text{ kJ}$$

to calculate the value of $\Delta H°$ for the following reaction:

$$2Fe(s) + \tfrac{3}{2}O_2(g) \longrightarrow Fe_2O_3(s)$$

ANALYSIS: We cannot simply add the two given equations, because this will not produce the equation we want. We first have to manipulate these equations so that when we add them we will get the target equation.

SOLUTION: We can manipulate the two given equations as follows:

STEP 1 We begin by trying to get the iron atoms to come out right. The target equation must have 2Fe on the *left,* but the first equation above has 2Fe to the *right* of the arrow. To move it to the left, we must reverse the *entire* equation, remembering also to reverse the sign of $\Delta H°$. This puts Fe_2O_3 to the right of the arrow, which is where it has to be after we add our adjusted equations. After these manipulations, and reversing the sign of $\Delta H°$, we have

$$2Fe(s) + 3CO_2(g) \longrightarrow Fe_2O_3(s) + 3CO(g) \qquad \Delta H° = +26.7 \text{ kJ}$$

STEP 2 There must be $\tfrac{3}{2}O_2$ on the left, and we must be able to cancel *three* CO and *three* CO_2 when the equations are added. If we multiply the second of the equations given above by 3, we will obtain the necessary coefficients. We must also multiply the value of $\Delta H°$ of this equation by 3, because three times as many moles of substances are now involved in the reaction. When we have done this, we have

$$3CO(g) + \tfrac{3}{2}O_2(g) \longrightarrow 3CO_2(g) \qquad \Delta H° = 3 \times (-283.0 \text{ kJ}) = -849.0 \text{ kJ}$$

Let's now put our two equations together and find the answer.

$$2Fe(s) + 3CO_2(g) \longrightarrow Fe_2O_3(s) + 3CO(g) \qquad \Delta H° = +26.7 \text{ kJ}$$
$$3CO(g) + \tfrac{3}{2}O_2(g) \longrightarrow 3CO_2(g) \qquad \Delta H° = -849.0 \text{ kJ}$$

Sum: $$2Fe(s) + \tfrac{3}{2}O_2(g) \longrightarrow Fe_2O_3(s) \qquad \Delta H° = -822.3 \text{ kJ}$$

Thus the value of $\Delta H°$ for the oxidation of 2 mol Fe(s) to 1 mol $Fe_2O_3(s)$ is -822.3 kJ. (The reaction is *very* exothermic.)

Is the Answer Reasonable?
There is no quick "head check." But for each step, double-check that you have heeded the rules for manipulating thermochemical equations.

PRACTICE EXERCISE 10: Ethanol, C_2H_5OH, is made industrially by the reaction of water with ethylene, C_2H_4. Calculate the value of $\Delta H°$ for the reaction

$$C_2H_4(g) + H_2O(l) \longrightarrow C_2H_5OH(l)$$

given the following thermochemical equations:

$$C_2H_4(g) + 3O_2(g) \longrightarrow 2CO_2(g) + 2H_2O(l) \qquad \Delta H° = -1411.1 \text{ kJ}$$
$$C_2H_5OH(l) + 3O_2(g) \longrightarrow 2CO_2(g) + 3H_2O(l) \qquad \Delta H° = -1367.1 \text{ kJ}$$

7.8 ▶ **Tabulated standard heats of reaction can be used to predict any heat of reaction using Hess's law**

Enormous databases of thermochemical equations have been compiled to allow the calculation of any heat of reaction, using Hess's law. The most frequently tabulated reactions are combustion reactions, phase changes, and formation reactions. We'll discuss the enthalpy changes that accompany phase changes in Chapter 12.

The **standard heat of combustion, ΔH_c°,** of a substance is the amount of heat released when 1 mol of a **fuel** substance is completely burned in pure oxygen gas, with all reactants and products brought to 25 °C and 1 bar of pressure. All carbon in the fuel becomes carbon dioxide gas, and all the fuel's hydrogen becomes liquid water. *Combustion reactions are always exothermic, so ΔH_c° is always negative.*

How many moles of carbon dioxide gas are produced by a gas-fired power plant for every 1.0 MJ (megajoule) of energy it produces? The plant burns methane, $CH_4(g)$, for which ΔH_c° is -890 kJ/mol.

EXAMPLE 7.12

Writing an Equation for a Standard Heat of Combustion

ANALYSIS: We need to link moles of carbon dioxide with megajoules of heat produced. *The critical link between amount of substance in a reaction and the heat of reaction is a balanced thermochemical equation.* This is a combustion reaction. The reactants will be the fuel, $CH_4(g)$, and oxygen, $O_2(g)$. The products will be carbon dioxide gas (because the fuel contains carbon) and liquid water (because the fuel contains hydrogen). We'll need to balance the following equation:

$$CH_4(g) + ?\,O_2(g) \longrightarrow ?\,CO_2(g) + ?\,H_2O(l) \qquad \Delta H^\circ = -890 \text{ kJ}$$

The coefficient in front of $CO_2(g)$ will be the number of moles of CO_2 produced when 890 kJ of heat are released. We can write the critical link in this problem as an equivalency relation,

$$?\text{ mol } CO_2(g) \Longleftrightarrow 890 \text{ kJ released}$$

and build a conversion factor to convert megajoules of heat into moles of $CO_2(g)$. To link megajoules with kilojoules, we'll need to remember the meaning of the SI prefixes *kilo* and *mega*.

$$1 \text{ MJ} = 10^6 \text{ J}$$

$$1 \text{ kJ} = 10^3 \text{ J}$$

These three relations are all we need to link megajoules of heat with moles of $CO_2(g)$.

SOLUTION: We balance the equation, remembering that to write ΔH° as ΔH_c°, we have to write the equation for 1 mol of fuel (1 mol CH_4, in this case):

$$CH_4(g) + 2O_2(g) \longrightarrow CO_2(g) + 2H_2O(l) \qquad \Delta H^\circ = -890 \text{ kJ}$$

so there is 1 mol CO_2 released for every 890 kJ of heat that is released. The number of mol of CO_2 released for production of 1 MJ of energy is

$$1.0 \text{ MJ} \times \frac{10^6 \text{ J}}{1 \text{ MJ}} \times \frac{1 \text{ kJ}}{10^3 \text{ J}} \times \frac{1 \text{ mol } CO_2}{890 \text{ kJ}} = 1.1 \text{ mol } CO_2$$

Is the Answer Reasonable?
We can see that the equation has been correctly balanced for 1 mol CH_4; each side has one C atom, four H atoms, and four O atoms. We can also see that it makes sense that only slightly more than 1 mol of CO_2 is produced with 1 MJ (1000 kJ) of heat, because 1000 kJ is only slightly more than 890 kJ, the amount of heat produced with 1 mol CO_2.

PRACTICE EXERCISE 11: *N*-octane, $C_8H_{18}(l)$, has a standard heat of combustion of 5450.5 kJ/mol. A 15 gallon automobile fuel tank could hold about 480 moles of *n*-octane. How much heat could be produced by burning a full tank of *n*-octane?

Older thermochemical data used a standard pressure of 1 atm, not 1 bar. Because 1 atm = 1.01325 bar, the new definition made little difference in tabulated heats of reaction.

The **standard enthalpy of formation, ΔH_f°,** of a substance, also called its **standard heat of formation,** is the amount of heat absorbed or evolved when specifically *1 mole* of the substance is formed at 25 °C and 1 bar from its elements in their *standard states.* An element is in its **standard state** when it is at 25 °C and 1 bar and in its most stable form and physical state (solid, liquid, or gas). Oxygen, for example, is in its standard state only as a gas at 25 °C and 1 bar and only as O_2 molecules, not as O atoms or O_3 (ozone) molecules. Carbon must be in the form of graphite, not diamond, to be in its standard state, because the graphite form of carbon is the more stable form under standard conditions.

The values of ΔH_f° are taken from available reference sources.

Standard enthalpies of formation for a variety of substances are given in Table 7.2, and a more extensive table of standard heats of formation can be found in Appendix C. Notice in particular that **all values of ΔH_f° for the elements in their standard states are zero.** (Forming an element *from itself,* of course, would yield no change in enthalpy.) In most tables, values of ΔH_f° for the elements are not included for this reason.

It is important to remember the meaning of the subscript f in the symbol ΔH_f°. It is applied to a value of ΔH° only when *1 mol* of the substance is formed *from its elements in their standard states.* Consider, for example, the following four thermochemical equations and their corresponding values of ΔH°:

$$H_2(g) + \tfrac{1}{2}O_2(g) \longrightarrow H_2O(l) \qquad \Delta H_f^\circ = -285.9 \text{ kJ/mol}$$

$$2H_2(g) + O_2(g) \longrightarrow 2H_2O(l) \qquad \Delta H^\circ = -571.8 \text{ kJ}$$

$$CO(g) + \tfrac{1}{2}O_2(g) \longrightarrow CO_2(g) \qquad \Delta H^\circ = -283.0 \text{ kJ}$$

$$2H(g) + O(g) \longrightarrow H_2O(l) \qquad \Delta H^\circ = -971.1 \text{ kJ}$$

TABLE 7.2	STANDARD ENTHALPIES OF FORMATION OF TYPICAL SUBSTANCES				
Substance	ΔH_f° (kJ mol^{-1})	Substance	ΔH_f° (kJ mol^{-1})	Substance	ΔH_f° (kJ mol^{-1})
Ag(s)	0	CaO(s)	−635.5	KCl(s)	−435.89
AgBr(s)	−100.4	Ca(OH)$_2$(s)	−986.59	K$_2$SO$_4$(s)	−1433.7
AgCl(s)	−127.0	CaSO$_4$(s)	−1432.7	N$_2$(g)	0
Al(s)	0	CaSO$_4 \cdot \frac{1}{2}$H$_2$O(s)	−1575.2	NH$_3$(g)	−46.19
Al$_2$O$_3$(s)	−1669.8	CaSO$_4 \cdot$2H$_2$O(s)	−2021.1	NH$_4$Cl(s)	−315.4
C(s, graphite)	0	Cl$_2$(g)	0	NO(g)	90.37
CO(g)	−110.5	Fe(s)	0	NO$_2$(g)	33.8
CO$_2$(g)	−393.5	Fe$_2$O$_3$(s)	−822.2	N$_2$O(g)	81.57
CH$_4$(g)	−74.848	H$_2$O(g)	−241.8	N$_2$O$_4$(g)	9.67
CH$_3$Cl(g)	−82.0	H$_2$O(l)	−285.9	N$_2$O$_5$(g)	11
CH$_3$I(g)	14.2	H$_2$(g)	0	Na(s)	0
CH$_3$OH(l)	−238.6	H$_2$O$_2$(l)	−187.6	NaHCO$_3$(s)	−947.7
CO(NH$_2$)$_2$(s) (urea)	−333.19	HBr(g)	−36	Na$_2$CO$_3$(s)	−1131
CO(NH$_2$)$_2$(aq)	−391.2	HCl(g)	−92.30	NaCl(s)	−411.0
C$_2$H$_2$(g)	226.75	HI(g)	26.6	NaOH(s)	−426.8
C$_2$H$_4$(g)	52.284	HNO$_3$(l)	−173.2	Na$_2$SO$_4$(s)	−1384.5
C$_2$H$_6$(g)	−84.667	H$_2$SO$_4$(l)	−811.32	O$_2$(g)	0
C$_2$H$_5$OH(l)	−277.63	HC$_2$H$_3$O$_2$(l)	−487.0	Pb(s)	0
Ca(s)	0	Hg(l)	0	PbO(s)	−219.2
CaBr$_2$(s)	−682.8	Hg(g)	60.84	S(s)	0
CaCO$_3$(s)	−1207	I$_2$(s)	0	SO$_2$(g)	−296.9
CaCl$_2$(s)	−795.0	K(s)	0	SO$_3$(g)	−395.2

Only in the first equation is $\Delta H°$ given the subscript f. It is the only reaction that satisfies both of the conditions specified above for standard enthalpies of formation. The second equation shows the formation of *2* mol of water, not one. The third involves a *compound* as one of the reactants. The fourth involves the elements as *atoms,* which are *not standard states* for these elements. Also notice that the units of $\Delta H_f°$ are kilojoules *per mole,* not just kilojoules, because the value is for the formation of 1 mol of the compound (from its elements). We can obtain the enthalpy of formation of 2 mol of water ($\Delta H°$ for the second equation) simply by multiplying the $\Delta H_f°$ value for 1 mol of H_2O by the factor 2 mol $H_2O(l)$.

$$\left(\frac{-285.9 \text{ kJ}}{\text{mol } H_2O(l)}\right) \times 2 \text{ mol } H_2O(l) = -571.8 \text{ kJ}$$

EXAMPLE 7.13

Writing an Equation for a Standard Heat of Formation

What equation must be used to represent the formation of nitric acid, $HNO_3(l)$, when we want to include its value of $\Delta H_f°$?

ANALYSIS: The equation must show only *1 mol* of the product. We begin with its formula and take whatever fractions of moles of the elements are needed to make it. We also remember to include the physical states. Table 7.2 gives the value of $\Delta H_f°$ for $HNO_3(l)$, -173.2 kJ mol^{-1}.

SOLUTION: The three elements, H, N, and O, all occur as diatomic molecules in the gaseous state, so the following fractions of moles supply exactly enough to make 1 mol of HNO_3:

$$\tfrac{1}{2}H_2(g) + \tfrac{1}{2}N_2(g) + \tfrac{3}{2}O_2(g) \longrightarrow HNO_3(l) \qquad \Delta H_f° = -173.2 \text{ kJ mol}^{-1}$$

Is the Answer Reasonable?
The answer correctly shows only 1 mol of HNO_3, and this governs the coefficients for the reactants. So simply check if everything is balanced.

PRACTICE EXERCISE 12: Write the thermochemical equation that would be used to represent the standard heat of formation of sodium bicarbonate, $NaHCO_3(s)$.

Standard enthalpies of formation are useful because they provide a convenient method for applying Hess's law without having to manipulate thermochemical equations. This is possible because, as we will demonstrate, the net $\Delta H°_{reaction}$ equals the sum of the heats of formation of the products minus the sum of the heats of formation of the reactants, each $\Delta H_f°$ value multiplied by the appropriate coefficient given by the thermochemical equation. In other words, we can express Hess's law in the form of the following, known as the **Hess's law equation:**

$$\Delta H°_{reaction} = \left[\begin{matrix} \text{Sum of } \Delta H_f° \text{ of all} \\ \text{of the products} \end{matrix}\right] - \left[\begin{matrix} \text{Sum of } \Delta H_f° \text{ of all} \\ \text{of the reactants} \end{matrix}\right] \qquad (7.12)$$

TOOLS
Hess's law equation

Let's now demonstrate that Equation 7.12 works. Consider the reaction given by the following equation:

$$SO_3(g) \longrightarrow SO_2(g) + \tfrac{1}{2}O_2(g) \qquad \Delta H° = ?$$

We wish to calculate the heat of reaction using standard enthalpies of formation.

If we use the first method we learned, namely, the manipulation of thermochemical equations, we would need to imagine a path from the reactant to the products that involves first decomposing SO_3 into its elements in their standard states and then recombining the elements to form the products. This path is shown in Figure 7.14. The first step, whose enthalpy change is indicated as $\Delta H_1°$,

We can use either heats of combustion or heats of formation for the reactants and products in this form of Hess's law, but don't mix them. Use heats of combustion for all reactants and all products, *or* use heats of formation for all reactants and all products.

FIGURE **7.14** *Enthalpy diagram.* The reaction diagrammed is

$$SO_3(g) \longrightarrow SO_2(g) + \tfrac{1}{2}O_2(g)$$

The path of the reaction in this diagram involves the elements in their standard states as the intermediate state. Here we see the reactant being decomposed into its elements (longer upward-pointing arrow); then the elements are recombined to form the products (short downward-pointing arrow). The difference in the lengths of these two arrows is proportional to the net enthalpy change (shorter upward-pointing arrow).

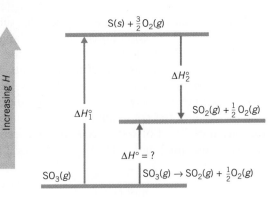

corresponds to the decomposition of SO_3 into sulfur and oxygen. This is just the *reverse* of the formation of SO_3, so we can write

$$\Delta H_1^\circ = -\Delta H_{fSO_3(g)}^\circ$$

The minus sign is written because when we reverse a process, we change the sign of its ΔH.

The second step in Figure 7.14, whose enthalpy change is indicated as ΔH_2°, is for the formation of SO_2 plus half a mole of O_2 from sulfur and oxygen. Therefore, we can write

$$\Delta H_2^\circ = \Delta H_{fSO_2(g)}^\circ + \tfrac{1}{2}\Delta H_{fO_2(g)}^\circ$$

The sum of these two steps gives the net change we want, so the sum of ΔH_1° and ΔH_2° must equal the desired ΔH°.

$$\Delta H^\circ = \Delta H_1^\circ + \Delta H_2^\circ$$

By substitution

$$\Delta H^\circ = [-\Delta H_{fSO_3(g)}^\circ] + [\Delta H_{fSO_2(g)}^\circ + \tfrac{1}{2}\Delta H_{fO_2(g)}^\circ]$$

This can be rewritten as

$$\Delta H^\circ = [\Delta H_{fSO_2(g)}^\circ + \tfrac{1}{2}\Delta H_{fO_2(g)}^\circ] + [-\Delta H_{fSO_3(g)}^\circ]$$

The change of sign gives

$$\Delta H^\circ = [\Delta H_{fSO_2(g)}^\circ + \tfrac{1}{2}\Delta H_{fO_2(g)}^\circ] - [\Delta H_{fSO_3(g)}^\circ]$$

Notice carefully the net result. $\Delta H_{reaction}^\circ$ equals the sum of the heats of formation of the products minus the sum of the heats of formation of the reactant(s), each multiplied by the appropriate coefficient. So we could have obtained the identical result by using the Hess's law equation, Equation 7.12, directly, instead of by the more laborious method of manipulating thermochemical equations.

EXAMPLE 7.14

Using Hess's Law and Standard Enthalpies of Formation

Some chefs keep baking soda, $NaHCO_3$, handy to put out grease fires. When thrown on the fire, baking soda partly smothers the fire and the heat decomposes it to give CO_2, which further smothers the flame. The equation for the decomposition of $NaHCO_3$ is

$$2NaHCO_3(s) \longrightarrow Na_2CO_3(s) + H_2O(l) + CO_2(g)$$

Use the data in Table 7.2 to calculate the ΔH° for this reaction in kilojoules.

ANALYSIS: Hess's law equation (Equation 7.12) is now our basic tool for computing values of $\Delta H°$. So we calculate the values of $\Delta H°$ for the products, taken as a set, and the values of $\Delta H°$ for the reactants, also taken as a set. Then we subtract the latter from the former to calculate $\Delta H°$. We compute values of $\Delta H°$ using $\Delta H_f°$ data from Table 7.2 and the coefficients of the chemical equation.

SOLUTION:

$$\Delta H° = [1 \text{ mol Na}_2\text{CO}_3(s) \times \Delta H°_{f\text{Na}_2\text{CO}_3(s)} + 1 \text{ mol H}_2\text{O}(l) \times \Delta H°_{f\text{H}_2\text{O}(l)}$$
$$+ 1 \text{ mol CO}_2(g) \times \Delta H°_{f\text{CO}_2(g)}] - [2 \text{ mol NaHCO}_3(s) \times \Delta H°_{f\text{NaHCO}_3(s)}]$$

We now use Table 7.2 to find the values of $\Delta H_f°$ for each substance in its proper physical state.

$$\Delta H° = \left[(1 \text{ mol Na}_2\text{CO}_3 \times \frac{(-1131 \text{ kJ})}{\text{mol Na}_2\text{CO}_3} + 1 \text{ mol H}_2\text{O} \times \frac{(-285.9 \text{ kJ})}{\text{mol H}_2\text{O}} \right.$$
$$\left. + 1 \text{ mol CO}_2 \times \frac{(-393.5 \text{ kJ})}{\text{mol CO}_2} \right]$$
$$- \left[2 \text{ mol NaHCO}_3 \times \frac{(-947.7 \text{ kJ})}{\text{mol NaHCO}_3} \right]$$

$$\Delta H° = (-1810 \text{ kJ}) - (-1895 \text{ kJ})$$
$$= +85 \text{ kJ}$$

Thus, under standard conditions, the reaction is endothermic by 85 kJ. (Notice that we did not have to manipulate any equations.)

Is the Answer Reasonable?
Double-check that all of the coefficients found in the given equation are correctly applied as multipliers in the solution. Be sure that the *signs* of the values of $\Delta H_f°$ have all been carefully used. Have you taken the order of subtracting specified in Equation 7.12? Keeping track of minus signs requires particular care.

PRACTICE EXERCISE 13: Calculate $\Delta H°$ for the following reactions:

(a) $2NO(g) + O_2(g) \rightarrow 2NO_2(g)$
(b) $NaOH(s) + HCl(g) \rightarrow NaCl(s) + H_2O(l)$

SUMMARY

Energy is the ability to do work or supply heat Because electrical attractions and repulsions occur within atoms, molecules, and ions, substances have **chemical energy,** a form of **potential energy.** The particles that make up matter—namely, atoms, molecules, or ions—are in constant motion, so they also possess kinetic energy, specifically, **molecular kinetic energy.**

The internal energy of a system, *E*, is the sum of its molecular kinetic and potential energies The more molecules you have, the higher the internal energy is, so internal energy is an extensive property. The values of E before the reaction ($E_{\text{reactants}}$) or after the reaction (E_{products}) cannot be known. However, the difference between the two, the *change* in internal energy or ΔE, is knowable.

$$\Delta E = E_{\text{products}} - E_{\text{reactants}}$$

Internal energy is a **state function,** that is, it only depends on the current condition or **state** of the system. Temperature is an intensive property that determines the direction of spontaneous heat flow. The Kelvin temperature of the sample is related to the *average* kinetic energy of its molecules. **Heat** is a transfer of energy from an object at a high temperature to one at a lower temperature. Heat can be conducted from one object to another by molecular collisions.

Heat can be determined by measuring temperature changes Before we measure a heat flow we must describe the boundary through which heat flows. The boundary encloses the **system** (the object we're interested in). Everything else in the universe is the system's **surroundings.** The heat flow, q is related to the temperature change Δt by

$$q = C\Delta t$$

where C is the **heat capacity** of the system, the heat needed to change the temperature of the system by 1 degree Celsius. The heat capacity for a pure substance can be computed from its mass, m, using the equation

$$C = ms$$

where s is the **specific heat** of the material, which is the heat needed to change the temperature of 1 g of a substance by 1 °C. Heat capacity is specific heat times mass. Heat capacity calculated on a per mole basis is called **molar heat capacity.** Water has an unusually high specific heat. We can compute a heat flow when we know the mass and specific heat of an object using the equation

$$q = ms\Delta t$$

Bond breaking requires energy; bond formation releases energy Chemical reactions involve the breaking of bonds and the forming of different bonds. To break a bond, energy must be supplied somehow; to form a bond, excess energy must be carried away. The net energy transferred is the sum of the bond breaking "costs" and the bond forming "profits." When some combinations of substances react, potential energy is converted into molecular kinetic energy or heat, called the **heat of reaction, q.** If the system is **adiabatic** (no heat leaves it), the internal temperature increases. Otherwise, the heat has a tendency to leave the system. Either way, the reaction is said to be **exothermic.**

In **endothermic** reactions, molecular kinetic energy of the reactants is converted into potential energy of the products. If the system is adiabatic, the reduction of the molecular kinetic energy causes the system's temperature to decrease. Otherwise, the system tends to absorb energy from the surroundings (heat "flows" in).

Heats of reaction are measured at constant volume or at constant pressure Pressure is the ratio of the force applied to the area over which it is applied. Atmospheric pressure is the pressure that air exerts because it has mass. When a gas evolves and pushes against the atmosphere or causes a piston to move in a cylinder, pressure–volume work is being done. When the volume change, ΔV, occurs at constant opposing pressure, P, the associated pressure–volume work is given by $w = -P\Delta V$. The energy expended in doing this pressure–volume work causes heats of reaction measured at constant volume (q_v) to differ numerically from heats measured at constant pressure (q_p).

The **first law of thermodynamics** says that no matter how the change in energy accompanying a reaction may be allocated between q and w, their sum is the same, ΔE.

$$\Delta E = q + w$$

The algebraic sign for q and w is negative when the system gives off heat to or does work on the surroundings. The sign is positive when the system absorbs heat or receives work energy done to it. When a system has rigid walls, no pressure–volume work can be done to or by the atmosphere. Consequently, w is zero, and the heat of reaction, now symbolized as q_v, is the change in the internal energy of the system, ΔE. Values of q_v are determined using a bomb calorimeter or a similar device.

When the system is under constant pressure (e.g., open to the atmosphere), pressure–volume work is now possible, so in this circumstance, the energy of the system is called its **enthalpy, H.**

$$H = E + PV$$

Like E, values of H cannot be known in an absolute sense, but a difference in enthalpy between reactants and products can be determined. The heat of reaction is now called the **enthalpy change, ΔH.** Under constant pressure,

$$\Delta H = \Delta E + P\Delta V$$

ΔH is a state function. Its value differs from that of ΔE by the work involved in interacting with the atmosphere when the change occurs at constant atmospheric pressure. In general, the difference between ΔE and ΔH is quite small.

Thermochemical equations are chemical equations that quantitatively include heat Exothermic reactions have negative values of ΔH; endothermic changes have positive values. A balanced chemical equation that includes both the enthalpy change and the physical states of the substances is called a **thermochemical equation.** These can be added, reversed (reversing also the sign of the enthalpy change), or multiplied by a constant multiplier (doing the same to the enthalpy change). If formulas are canceled or added, they must be of substances in identical physical states.

Thermochemical equations can be added because enthalpy is a state function Values of $\Delta H°$ can be determined by the manipulation of any combination of thermochemical equations that add up to the final net equation.

The reference conditions for thermochemistry, called the **standard conditions,** are 25 °C and 1 bar of pressure. An enthalpy change measured under these conditions is called the **standard enthalpy of reaction** or the **standard heat of reaction.** Its symbol is $\Delta H°$, where the degree sign signifies that standard conditions are involved. The value of $\Delta H°$ is a function of the amounts of substances involved in the reaction. If the number of moles of reactants is doubled, the standard heat of the reaction is twice as large. The units for $\Delta H°$ are generally joules or kilojoules.

When the enthalpy change is for the complete combustion of 1 mol of a pure substance under standard conditions in pure oxygen, $\Delta H°$ is called the **standard heat of combustion** of the compound, symbolized as $\Delta H°_c$.

When the enthalpy change is for the formation of 1 mol of a substance under standard conditions from its *elements in their standard states,* $\Delta H°$ is called the **standard heat of formation** of the compound, symbolized as $\Delta H°_f$, and it is generally given in units of kilojoules per mole (kJ mol^{-1}). **Hess's law of heat summation** is possible because enthalpy is a state function. The value of $\Delta H°$ for a reaction can be calculated as the sum of the $\Delta H°_f$ values of the products minus the sum of the $\Delta H°_f$ values of the reactants. The coefficients in the equation and the values of $\Delta H°_f$ from tables are used to calculate values of $\Delta H°_f$ for products and reactants.

TOOLS ▶ YOU HAVE LEARNED

The table below lists the tools you have learned in this chapter that are applicable to problem solving. Review them if necessary, and refer to them when working on the Thinking-It-Through problems and the Review Exercises that follow.

TOOL	HOW IT WORKS
Heat capacity and specific heat (pages 266 and 267)	We use a heat capacity and a temperature change to calculate the heat of a reaction or specific heat data and a mass together with a temperature change to calculate the heat of a reaction.
Thermochemical equations (page 281)	We can use thermochemical equations for one set of reactions to write a thermochemical equation for some net reaction.
Enthalpy diagrams (page 283)	We use them to visualize enthalpy changes in multistep reactions.
Hess's law (page 289)	This enables us to use standard heats of formation to calculate the enthalpy of a reaction.

THINKING IT THROUGH

The goal for the following problems is not to find the answers themselves, but rather to assemble the information needed to solve them and explain how you would use the information to find the answers. The problems in Level 2 are more challenging than those in Level 1 and may contain more data than are required, in which case you are also asked to identify the unnecessary data. Detailed answers to the Thinking-It-Through problems can be found on the web site.

Need extra help? ON-LINE HELP Visit the Brady/ Senese web site at www.wiley.com/ college/brady

Level 1 Problems

1. Which vehicle possesses more kinetic energy and why, vehicle X with a mass of 2000 kg and a velocity of 50 km h^{-1} or vehicle Y with a mass of 1000 kg and a velocity of 100 km h^{-1}?

2. A pacemaker delivers between 5 and 25 μJ of energy per pulse to the heart muscle. What is the minimum number of pulses that could be powered by the energy in a chocolate bar (which contains about 250 dietary calories)? (Set up the calculation.)

3. How many grams of water can be heated from 25.0 °C to 35.0 °C by the heat released from 85.0 g of iron that cools from 85.0 °C to 30.0 °C? (Set up the calculation.)

4. If you know the specific heat of an object, what else besides the temperature change has to be known before you can calculate the energy flow caused by a temperature change?

5. The heat capacity of a 4.50 kg bar of a metal is 1.44 kJ °C^{-1}. What is the specific heat of the metal expressed in the units cal g^{-1} °C^{-1}? (Set up the calculation.)

6. Criticize the following statements about heat:
(a) A kettle full of boiling water contains a lot of heat.
(b) Heat is the amount of energy an object contains.
(c) Heat will always flow from an object with high energy to an object with lower energy.
(d) When an object is heated, all of its molecules move faster.
(e) Heat is a state function.

(f) Heat flow between two objects will stop when the energies of the two objects become the same.
(g) Heat is an extensive property.

7. If you want to calculate how much energy is absorbed by a bar of iron when it has been kept in a fireplace with a blazing log fire, which of the following information is needed: the initial temperature of the iron before it went into the fire, the length of time it is in the fire, the final temperature of the iron, the mass of the iron bar, the volume of the iron bar, the temperature of the burning material, the specific heat of iron?

8. Consider the reaction,

$$C_{12}H_{22}O_{11}(s) + 12O_2(g) \longrightarrow 12CO_2(g) + 11H_2O(l)$$

How do the values of ΔE and ΔH compare?

9. Consider the thermochemical equation

$$H_2(g) + \tfrac{1}{2}O_2(g) \longrightarrow H_2O(l)$$

What is the value of $\Delta H°$ for the following reaction? (Describe the calculation involved.)

$$3H_2O(l) \longrightarrow 3H_2(g) + \tfrac{3}{2}O_2(g)$$

10. How many kilojoules of heat are evolved in the combustion of 35.5 g of methyl alcohol? The *unbalanced equation* for the reaction is

$$CH_3OH(l) + O_2(g) \longrightarrow CO_2(g) + H_2O(g)$$

(Set up the calculations. Include any data required.)

Level 2 Problems

11. A rock with a mass of 20 kg falls 30 ft to the ground and strikes a lever that tosses a 10 kg object into the air. How will the speed of the 10 kg object leaving the ground compare with the speed of the 20 kg object just as it strikes the lever? (Assume all the KE of the 20 kg object is transferred to the 10 kg object.)

12. A 2.00 kg piece of granite with a specific heat of 0.803 J g^{-1} °C^{-1} and a temperature of 95.0 °C is placed into 2.00 L of water at 22.0 °C. When the granite and water come to the same temperature, what will that temperature be? (Set up the calculation.)

13. A piece of metallic lead with a mass of 51.36 g was heated to 100.0 °C in boiling water. It was quickly dried and immersed in 10.0 g of liquid water that was at a temperature of 24.6 °C. The system (water plus lead) came to a temperature of 35.2 °C. What is the specific heat of lead? (Describe the calculations involved.)

14. Suppose a truck with a mass of 14.0 tons (1 ton = 2000 lb) is traveling at a speed of 45.0 mi/h. If the truck driver slams on the brakes, the kinetic energy of the truck is changed to heat as the brakes slow the truck to a stop. How much would the temperature of 5.00 gal of water increase if all this heat could be absorbed by the water? (Explain in detail all the calculations involved. Apply the factor-label method where appropriate.)

15. Consider the thermochemical equation

$$C_2H_2(g) + H_2(g) \longrightarrow C_2H_4(g) \qquad \Delta H° = -175.1 \text{ kJ}$$
acetylene   ethylene

Without referring to Table 7.2 or performing any calculations, determine which gas, acetylene or ethylene, has the more positive standard heat of formation. Explain your reasoning.

16. Two 1 L boxes contain identical amounts of argon gas represented in the figure above right as blue spheres. The insulating partition in the center can be lifted to expose the thermal conducting partition.
(a) If you were to make an animated movie showing how the argon atoms move inside the two boxes, what differences would you show in the motions of atoms in the left and right boxes? Explain.
(b) The insulating partition in the center is lifted to allow atoms on both sides to strike the conducting partition. Explain how you would change your movie over time to represent what happens to the argon atoms in the left and right boxes.
(c) What would the two thermometers show over time after the insulating partition had been lifted? Would they

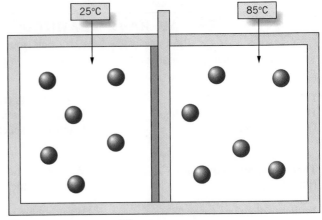

Thermal conductor   Thermal insulator

eventually stop changing values and if so, what values would each show?

17. In an experiment, 100 mL of a clear, colorless aqueous solution of a compound A is added to a beaker containing 100 mL of a clear, blue aqueous solution of a compound B. Prior to mixing both solutions were at 25 °C. After mixing the resulting solution was clear and colorless. The mixed solutions were stirred for several seconds with a thermometer. The temperature was observed to rapidly rise to a maximum value of 53 °C and then to slowly fall over a period of 30 min back to 25 °C. Explain the origins of the changes in temperature.

18. The left beaker in the figure contains a 200.0 g piece of pure copper in 100 g of boiling water. The right beaker contains 75 g of water at 10 °C. If the cold water is poured rapidly into the left beaker and the temperature of the water monitored, what will be the final temperature? Assume no heat loss to the surroundings. List the data you will need to look up to work this problem and set up the equation necessary for its solution.

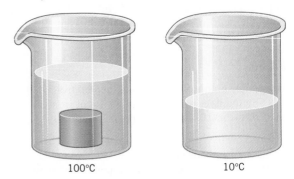

100°C   10°C

REVIEW QUESTIONS

Kinetic and Potential Energy

7.1 How are the total kinetic energy and total potential energy of atomic-sized particles related to each other in an isolated system?

7.2 What is meant by the term *chemical energy*?

7.3 In a certain chemical reaction, there is a decrease in the potential energy (chemical energy) as the reaction proceeds.
(a) How does the total kinetic energy of the particles change?
(b) How does the temperature of the reaction mixture change?

7.4 When a chemical bond forms between two colliding atoms, describe how the energy is released.

7.5 How does the potential energy change (increase, decrease, or no change) for each of the following?
(a) Two electrons come closer together.
(b) An electron and a proton become farther apart.
(c) Two atomic nuclei approach each other.
(d) A ball rolls downhill.

7.6 Describe how the potential energy of the system of atomic-sized particles changes when each of the following events occurs:
(a) The wax of a candle burns in air, giving a yellow flame. (The system consists of the wax and O_2 in the air.)
(b) Ammonium nitrate dissolves in water to produce a cooling effect.

Energy Units

7.7 How is the joule defined?

7.8 How is the calorie defined according to the SI? How was it originally defined?

7.9 How many joules are in one nutritional Calorie?

Internal Energy and the Kinetic Theory of Matter

7.10 What is meant by the term *thermal equilibrium*?

7.11 How is internal energy related to molecular kinetic and potential energy?

7.12 How is temperature different from energy?

7.13 How can the state of a system be specified?

7.14 Consider the distribution of molecular kinetic energies shown in the diagram below for a gas at 25 °C.

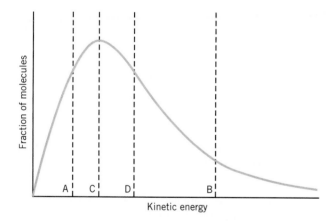

(a) Which point corresponds to the most frequently occurring (also called the *most probable*) molecular kinetic energy?
(b) Which point corresponds to the average molecular kinetic energy?
(c) If the temperature of the gas is raised to 50 °C, how will the height of the curve at *B* change?
(d) If the temperature of the gas is raised to 50 °C, how will the height of the curve at *A* change? How will the maximum height of the curve change?

7.15 Explain how heat can be conducted from one object to another.

7.16 Suppose the temperature of an object is raised from 100 °C to 200 °C by heating it with a Bunsen burner. Which of the following will be true?
(a) The average kinetic energy will increase.
(b) The total kinetic energy of all the molecules will increase.
(c) The number of fast-moving molecules will increase.
(d) The number of slow-moving molecules will increase.
(e) The chemical potential energy will decrease.

Experimental Measurement of Heat

7.17 What is the name of the thermal property whose values can have the following units?
(a) $J\,g^{-1}\,°C^{-1}$ (b) $J\,mol^{-1}\,°C^{-1}$ (c) $J\,°C^{-1}$

7.18 Which kind of substance needs more energy to undergo an increase of 5 °C, something with a *high* or with a *low* specific heat? Explain.

7.19 How do heat capacity and specific heat differ?

7.20 Which kind of substance experiences the larger increase in temperature when it absorbs 100 J, something with a *high* or a *low* specific heat?

7.21 If the specific heat values in Table 7.1 were in units of $kJ\,kg^{-1}\,K^{-1}$, would the values be *numerically* different? Explain.

7.22 What is the difference between an isolated system and a closed system?

7.23 What is the meaning of a negative value for heat?

7.24 If an amount of heat q is applied to an object, and its temperature changes by Δt, how much heat must be applied to change the temperature by $10\,\Delta t$?

7.25 What two pieces of information are necessary to determine the heat capacity of an object? What additional piece of information is necessary to determine its specific heat?

7.26 Suppose object A has twice the specific heat and twice the mass of object B. If the same amount of heat is applied to both objects, how will the temperature change of A be related to the temperature change in B?

Energy Changes in Chemical Reactions

7.27 What simple model is often used for chemical bonds?

7.28 What term do we use to describe a reaction that liberates heat to its surroundings? How does the chemical energy change during such a reaction?

7.29 What term is used to describe a reaction that absorbs heat from the surroundings? How does the chemical energy change during such a reaction?

Internal Energy and Enthalpy

7.30 What is a state function? Give two examples.

7.31 What does the internal energy, E, of a system represent in terms of the kinetic and potential energies of its particles?

7.32 Write the equation that states the first law of thermodynamics. In your own words, what does this statement mean in terms of energy exchanges between a system and its surroundings?

7.33 Which state function is given by the heat of reaction at constant volume? Which state function is given by the heat of reaction at constant pressure? Under what conditions are these heats of reaction equal?

7.34 If a system containing gases expands and pushes back a piston against a constant opposing pressure, what equation describes the work done on the system?

7.35 How is enthalpy defined?

7.36 What equation defines ΔH in general terms? How is this expressed when the system involves a chemical reaction? In general, why are chemists and biologists more interested in values of ΔH than ΔE?

7.37 If the enthalpy of a system increases by 100 kJ, what must be true about the enthalpy of the surroundings? Why?

7.38 What is the *sign* of ΔH for an exothermic change?

Enthalpy and Heats of Reaction

7.39 Why do standard reference values for temperature and pressure have to be selected when we consider and compare heats of reaction for various reactions? What are the values for the standard temperature and standard pressure?

7.40 What distinguishes a *thermochemical* equation from an ordinary chemical equation?

7.41 In a thermochemical equation, what do the coefficients represent in terms of the amounts of the reactants and products?

7.42 Why are fractional coefficients permitted in a balanced thermochemical equation? If a thermochemical equation has a coefficient of $\frac{1}{2}$ for a formula, what does it signify?

Hess's Law

7.43 What fundamental fact about ΔH makes Hess's law possible?

7.44 What *two* conditions must be met by a thermochemical equation so that its standard enthalpy change can be given the symbol ΔH_f°?

7.45 What is Hess's law of heat summation expressed in terms of standard heats of formation? What important property of enthalpy makes Hess's law possible?

7.46 Write Hess's law in terms of standard heats of combustion.

REVIEW PROBLEMS

Answers to problems whose numbers are printed in color are given in Appendix B. More challenging problems are marked with asterisks. **ILW** = Interactive LearningWare solution is available at *www.wiley.com/college/brady*.

Energy Units and Conversions

7.47 What is the kinetic energy, in kilojoules, of a 2150 kg automobile traveling at a speed of 80.0 km/h (approximately 50 mph)?

7.48 What is the kinetic energy, in kilojoules, of a cement truck with a mass of 2.04×10^4 kg traveling at 80.0 km/h?

7.49 Perform the following conversions: (a) 347 kJ to kcal, (b) 308 kcal to kJ.

7.50 Perform the following conversions: (a) 26 J to cal, (b) 78 cal to J.

7.51 A 100 W lightbulb requires 100 J of energy per second. How many calories would be required to light the bulb for 1 h? How many dietary calories is this?

7.52 A human being continually produces about 70 W (70 J per second) of heat. How many calories would be produced in one day? How many dietary calories is this?

First Law of Thermodynamics

7.53 If a system does 45 J of work and receives 28 J of heat, what is the value of ΔE for this change?

7.54 If a system absorbs 48 J of heat and does 22 J of work, what is the value of ΔE for this change?

7.55 An automobile engine converts heat into work via a cycle. The cycle must finish exactly where it started, so the energy at the start of the cycle must be exactly the same as the energy at the end of the cycle. If the engine is to do 100 J of work per cycle, how much heat must it absorb?

7.56 If the engine in the previous problem absorbs 250 J of heat per cycle, how much work can it do per cycle?

Thermal Properties, Measuring Energy Changes

7.57 How much heat in kilojoules must be removed from 175 g of water to lower its temperature from 25.0 to 15.0 °C (which would be like cooling a glass of lemonade)?

7.58 How much heat in kilojoules is needed to bring 1.0 kg of water from 25 to 99 °C (comparable to making four cups of coffee)?

7.59 How many joules are needed to increase the temperature of 15.0 g of Fe from 20.0 to 40.0 °C?

7.60 The addition of 250 J to 30.0 g of copper initially at 22 °C will change its temperature to what final value?

7.61 A 5.00 g mass of a metal was heated to 100 °C and then plunged into 100 g of water at 24.0 °C. The temperature of the resulting mixture became 28.0 °C.
(a) How many joules did the water absorb?
(b) How many joules did the metal lose?
(c) What is the heat capacity of the metal sample?
(d) What is the specific heat of the metal?

7.62 A sample of copper was heated to 120 °C and then thrust into 200 g of water at 25.00 °C. The temperature of the mixture became 26.50 °C.

(a) How much heat in joules was absorbed by the water?

(b) The copper sample lost how many joules?

(c) What was the mass in grams of the copper sample?

7.63 Fat tissue is 85% fat and 15% water. The complete breakdown of the fat itself converts it to CO_2 and H_2O, which releases 9.0 kcal per gram of fat in the tissue.

(a) How many kilocalories are released by the loss of 1.0 lb of fat *tissue* in a weight-reduction program?

(b) Running 8.0 mi h^{-1} expends about 5.0×10^2 kcal h^{-1} of extra energy. How far does a person have to run to burn off 1.0 lb of fat *tissue* by this means alone?

7.64 A well-nourished person adds about 0.50 lb of fat tissue for each 3.5×10^2 kcal of food energy taken in over and above that needed. Suppose you decided to reduce your weight simply by omitting oils and butter but keeping every other aspect of your diet and your activities the same. How many days would be needed to lose 1.0 kg of fat *tissue* by this strategy alone? The caloric content of dietary fats and oils is 9.0 kcal g^{-1}. Suppose that you have been eating 0.25 lb of salad oil and butter per day.

7.65 Calculate the molar heat capacity of iron in J mol^{-1} °C^{-1}. Its specific heat is 0.4498 J g^{-1} °C^{-1}.

7.66 What is the molar heat capacity of ethyl alcohol, C_2H_5OH, in units of J mol^{-1} °C^{-1}, if its specific heat is 0.586 cal g^{-1} °C^{-1}?

Calorimetry

7.67 A vat of 4.54 kg of water underwent a decrease in temperature from 60.25 to 58.65 °C. How much energy in kilojoules left the water? (For this range of temperature, use a value of 4.18 J g^{-1} °C^{-1} for the specific heat of water.)

7.68 A container filled with 2.46 kg of water underwent a temperature change from 25.24 °C to 27.31 °C. How much heat, measured in kilojoules, did the water absorb?

7.69 Nitric acid, HNO_3, reacts with potassium hydroxide, KOH, as follows:

$$HNO_3(aq) + KOH(aq) \longrightarrow KNO_3(aq) + H_2O(l)$$

A student placed 55.0 mL of 1.3 M HNO_3 in a coffee cup calorimeter, noted that the temperature was 23.5 °C, and added 55.0 mL of 1.3 M KOH, also at 23.5 °C. The mixture was stirred quickly with a thermometer, and its temperature rose to 31.8 °C. Calculate the heat of reaction in joules. Assume that the specific heats of all solutions are 4.18 J g^{-1} °C^{-1} and that all densities are 1.00 g mL^{-1}. Calculate the heat of reaction per mole of acid (in units of kJ mol^{-1}).

7.70 A dilute solution of hydrochloric acid with a mass of 610.29 g and containing 0.33183 mol of HCl was exactly neutralized in a calorimeter by the sodium hydroxide in 615.31 g of a comparably dilute solution. The temperature increased from 16.784 to 20.610 °C. The specific heat of the HCl solution was 4.031 J g^{-1} °C^{-1}; that of the NaOH solution was 4.046 J g^{-1} °C^{-1}. The heat capacity of the calorimeter was 77.99 J °C^{-1}. Use these data to calculate the heat evolved by the following reaction:

$$HCl(aq) + NaOH(aq) \longrightarrow NaCl(aq) + H_2O(l)$$

What is the heat of neutralization per mole of HCl? Assume that the original solutions made independent contributions to the total heat capacity of the system following their mixing.

7.71 A 1.00 mol sample of propane, a gas used for cooking in many rural areas, was placed in a bomb calorimeter with excess oxygen and ignited. The reaction was

$$C_3H_8(g) + 5O_2(g) \longrightarrow 3CO_2(g) + 4H_2O(l)$$

The initial temperature of the calorimeter was 25.000 °C and its total heat capacity was 97.1 kJ °C^{-1}. The reaction raised the temperature of the calorimeter to 27.282 °C.

(a) How many joules were liberated in this reaction?

(b) What is the heat of reaction of propane with oxygen expressed in kilojoules per mole of C_3H_8 burned?

7.72 Toluene, C_7H_8, is used in the manufacture of explosives such as TNT (trinitrotoluene). A 1.500 g sample of liquid toluene was placed in a bomb calorimeter along with excess oxygen. When the combustion of the toluene was initiated, the temperature of the calorimeter rose from 25.000 °C to 26.413 °C. The products of the combustion were $CO_2(g)$ and $H_2O(l)$, and the heat capacity of the calorimeter was 45.06 kJ °C^{-1}. The reaction was

$$C_7H_8(l) + 9O_2(g) \longrightarrow 7CO_2(g) + 4H_2O(l)$$

(a) How many joules were liberated by the reaction?

(b) How many joules would be liberated under similar conditions if 1.00 mol of toluene were burned?

Enthalpy Changes and Heats of Reaction

7.73 One thermochemical equation for the reaction of carbon monoxide with oxygen is

$$3CO(g) + \tfrac{3}{2}O_2(g) \longrightarrow 3CO_2(g) \qquad \Delta H° = -849 \text{ kJ}$$

(a) Write the thermochemical equation for the reaction using 2 mol of CO.

(b) What is $\Delta H°$ for the formation of 1 mol of CO_2 by this reaction?

7.74 Ammonia reacts with oxygen as follows:

$$4NH_3(g) + 7O_2(g) \longrightarrow 4NO_2(g) + 6H_2O(g)$$
$$\Delta H° = -1132 \text{ kJ}$$

(a) Calculate the enthalpy change for the combustion of 1 mol of NH_3.

(b) Write the thermochemical equation for the reaction in which 1 mol of H_2O is formed.

7.75 Aluminum and iron(III) oxide, Fe_2O_3, react to form aluminum oxide, Al_2O_3, and iron. For each mole of aluminum used, 426.9 kJ of energy is released under standard conditions. Write the thermochemical equation that shows the consumption of 4 mol of Al. (All of the substances are solids.)

7.76 Liquid benzene, C_6H_6, burns in oxygen to give carbon dioxide gas and liquid water (when all products are returned to 25 °C and 1 bar of pressure). The combustion of 1.00 mol of benzene liberates 3271 kJ. Write the thermochemical equation for the combustion of 3.00 mol of liquid benzene.

7.77 Magnesium burns in air to produce a bright light and is often used in fireworks displays. The combustion of magnesium follows the thermochemical equation

$$2Mg(s) + O_2(g) \longrightarrow 2MgO(s) \qquad \Delta H° = -1203 \text{ kJ}$$

How much heat (in kilojoules) is liberated by the combustion of 6.54 g of magnesium?

7.78 Methanol is the fuel in "canned heat" containers (e.g., Sterno) that are used to heat foods at cocktail parties. The combustion of methanol follows the thermochemical equation

$$2CH_3OH(l) + 3O_2(g) \longrightarrow 2CO_2(g) + 4H_2O(g)$$
$$\Delta H° = -1199 \text{ kJ}$$

How many kilojoules are liberated by the combustion of 46.0 g of methanol?

Hess's Law

7.79 Construct an enthalpy diagram that shows the enthalpy changes for a one-step conversion of germanium, $Ge(s)$, into $GeO_2(s)$, the dioxide. On the same diagram, show the two-step process, first to the monoxide, $GeO(s)$, and then its conversion to the dioxide. The relevant thermochemical equations are the following:

$$Ge(s) + \tfrac{1}{2}O_2(g) \longrightarrow GeO(s) \qquad \Delta H° = -255 \text{ kJ}$$
$$Ge(s) + O_2(g) \longrightarrow GeO_2(s) \qquad \Delta H° = -534.7 \text{ kJ}$$

Using this diagram, determine $\Delta H°$ for the following reaction:

$$GeO(s) + \tfrac{1}{2}O_2(g) \longrightarrow GeO_2(s)$$

7.80 Construct an enthalpy diagram for the formation of $NO_2(g)$ from its elements by two pathways: first, from its elements and, second, by a two-step process, also from the elements. The relevant thermochemical equations are

$$\tfrac{1}{2}N_2 + O_2(g) \longrightarrow NO_2(g) \qquad \Delta H° = +33.8 \text{ kJ}$$
$$\tfrac{1}{2}N_2(g) + \tfrac{1}{2}O_2(g) \longrightarrow NO(g) \qquad \Delta H° = +90.37 \text{ kJ}$$
$$NO(g) + \tfrac{1}{2}O_2(g) \longrightarrow NO_2(g) \qquad \Delta H° = \text{?}$$

Be sure to note the signs of the values of $\Delta H°$ associated with arrows pointing up or down. Using the diagram, determine the value of $\Delta H°$ for the third equation.

7.81 Show how the equations

$$N_2O_4(g) \longrightarrow 2NO_2(g) \qquad \Delta H° = +57.93 \text{ kJ}$$
$$2NO(g) + O_2(g) \longrightarrow 2NO_2(g) \qquad \Delta H° = -113.14 \text{ kJ}$$

can be manipulated to give $\Delta H°$ for the following reaction:

$$2NO(g) + O_2(g) \longrightarrow N_2O_4(g)$$

7.82 We can generate hydrogen chloride by heating a mixture of sulfuric acid and potassium chloride according to the equation:

$$2KCl(s) + H_2SO_4(l) \longrightarrow 2HCl(g) + K_2SO_4(s)$$

Calculate $\Delta H°$ in kilojoules for this reaction from the following thermochemical equations:

$$HCl(g) + KOH(s) \longrightarrow KCl(s) + H_2O(l)$$
$$\Delta H° = -203.6 \text{ kJ}$$

$$H_2SO_4(l) + 2KOH(s) \longrightarrow K_2SO_4(s) + 2H_2O(l)$$
$$\Delta H° = -342.4 \text{ kJ}$$

7.83 Calculate $\Delta H°$ in kilojoules for the following reaction, the preparation of the unstable acid nitrous acid, HNO_2:

$$HCl(g) + NaNO_2(s) \longrightarrow HNO_2(l) + NaCl(s)$$

Use the following thermochemical equations:

$$2NaCl(s) + H_2O(l) \longrightarrow 2HCl(g) + Na_2O(s)$$
$$\Delta H° = +507.31 \text{ kJ}$$

$$NO(g) + NO_2(g) + Na_2O(s) \longrightarrow 2NaNO_2(s)$$
$$\Delta H° = -427.14 \text{ kJ}$$

$$NO(g) + NO_2(g) \longrightarrow N_2O(g) + O_2(g)$$
$$\Delta H° = -42.68 \text{ kJ}$$

$$2HNO_2(l) \longrightarrow N_2O(g) + O_2(g) + H_2O(l)$$
$$\Delta H° = +34.35 \text{ kJ}$$

7.84 Barium oxide reacts with sulfuric acid as follows:

$$BaO(s) + H_2SO_4(l) \longrightarrow BaSO_4(s) + H_2O(l)$$

Calculate $\Delta H°$ in kilojoules for this reaction. The following thermochemical equations can be used:

$$SO_3(g) + H_2O(l) \longrightarrow H_2SO_4(l) \qquad \Delta H° = -78.2 \text{ kJ}$$
$$BaO(s) + SO_3(g) \longrightarrow BaSO_4(s) \qquad \Delta H° = -213 \text{ kJ}$$

7.85 Copper metal can be obtained by heating copper oxide, CuO, in the presence of carbon monoxide, CO, according to the following reaction:

$$CuO(s) + CO(g) \longrightarrow Cu(s) + CO_2(g)$$

Calculate $\Delta H°$ in kilojoules, using the following thermochemical equations:

$$2CO(g) + O_2(g) \longrightarrow 2CO_2(g) \qquad \Delta H° = -566.1 \text{ kJ}$$
$$2Cu(s) + O_2(g) \longrightarrow 2CuO(s) \qquad \Delta H° = -310.5 \text{ kJ}$$

7.86 Calcium hydroxide reacts with hydrochloric acid by the following equation:

$$Ca(OH)_2(aq) + 2HCl(aq) \longrightarrow CaCl_2(aq) + 2H_2O(l)$$

Calculate $\Delta H°$ in kilojoules for this reaction, using the following equations as needed:

$$CaO(s) + 2HCl(aq) \longrightarrow CaCl_2(aq) + H_2O(l)$$
$$\Delta H° = -186 \text{ kJ}$$

$$CaO(s) + H_2O(l) \longrightarrow Ca(OH)_2(s) \qquad \Delta H° = -65.1 \text{ kJ}$$

$$Ca(OH)_2(s) \xrightarrow[\text{dissolving in water}]{} Ca(OH)_2(aq)$$
$$\Delta H° = -12.6 \text{ kJ}$$

7.87 Given the following thermochemical equations,

$$CaO(s) + Cl_2(g) \longrightarrow CaOCl_2(s) \qquad \Delta H° = -110.9 \text{ kJ}$$

$$H_2O(l) + CaOCl_2(s) + 2NaBr(s) \longrightarrow$$
$$2NaCl(s) + Ca(OH)_2(s) + Br_2(l) \qquad \Delta H° = -60.2 \text{ kJ}$$

$$Ca(OH)_2(s) \longrightarrow CaO(s) + H_2O(l) \qquad \Delta H° = +65.1 \text{ kJ}$$

calculate the value of $\Delta H°$ (in kilojoules) for the reaction

$$\tfrac{1}{2}Cl_2(g) + NaBr(s) \longrightarrow NaCl(s) + \tfrac{1}{2}Br_2(l)$$

7.88 Given the following thermochemical equations,

$$2Cu(s) + S(s) \longrightarrow Cu_2S(s) \qquad \Delta H° = -79.5 \text{ kJ}$$

$$S(s) + O_2(g) \longrightarrow SO_2(g) \qquad \Delta H° = -297 \text{ kJ}$$

$$Cu_2S(s) + 2O_2(g) \longrightarrow 2CuO(s) + SO_2(g)$$
$$\Delta H° = -527.5 \text{ kJ}$$

calculate the standard enthalpy of formation (in kilojoules per mole) of $CuO(s)$.

7.89 Given the following thermochemical equations,

$$4NH_3(g) + 7O_2(g) \longrightarrow 4NO_2(g) + 6H_2O(g)$$
$$\Delta H° = -1132 \text{ kJ}$$

$$6NO_2(g) + 8NH_3(g) \longrightarrow 7N_2(g) + 12H_2O(g)$$
$$\Delta H° = -2740 \text{ kJ}$$

calculate the value of $\Delta H°$ (in kilojoules) for the reaction

$$4NH_3(g) + 3O_2(g) \longrightarrow 2N_2(g) + 6H_2O(g)$$

7.90 Given the following thermochemical equations,

$$3Mg(s) + 2NH_3(g) \longrightarrow Mg_3N_2(s) + 3H_2(g)$$
$$\Delta H° = -371 \text{ kJ}$$

$$\tfrac{1}{2}N_2(g) + \tfrac{3}{2}H_2(g) \longrightarrow NH_3(g) \qquad \Delta H° = -46 \text{ kJ}$$

calculate $\Delta H°$ (in kilojoules) for the following reaction:

$$3Mg(s) + N_2(g) \longrightarrow Mg_3N_2(s)$$

7.91 Use the following thermochemical equations,

$$8Mg(s) + Mg(NO_3)_2(s) \longrightarrow Mg_3N_2(s) + 6MgO(s)$$
$$\Delta H° = -3884 \text{ kJ}$$

$$Mg_3N_2(s) \longrightarrow 3Mg(s) + N_2(g) \qquad \Delta H° = +463 \text{ kJ}$$

$$2MgO(s) \longrightarrow 2Mg(s) + O_2(g) \qquad \Delta H° = +1203 \text{ kJ}$$

to calculate the standard heat of formation (in kilojoules per mole) of $Mg(NO_3)_2(s)$.

7.92 Given the following thermochemical equations,

$$2H_2(g) + O_2(g) \longrightarrow 2H_2O(l) \qquad \Delta H° = -571.5 \text{ kJ}$$

$$N_2O_5(g) + H_2O(l) \longrightarrow 2HNO_3(l) \qquad \Delta H° = -76.6 \text{ kJ}$$

$$\tfrac{1}{2}N_2(g) + \tfrac{3}{2}O_2(g) + \tfrac{1}{2}H_2(g) \longrightarrow$$
$$HNO_3(l) \qquad \Delta H° = -174 \text{ kJ}$$

calculate $\Delta H°$ for the reaction

$$2N_2(g) + 5O_2(g) \longrightarrow 2N_2O_5(g)$$

Hess's Law and Standard Heats of Formation

7.93 Which of the following equations has a value of $\Delta H°$ that would properly be labeled as $\Delta H_f°$?
(a) $CaCO_3(s) \rightarrow CaO(s) + CO_2(g)$
(b) $Ca(s) + \tfrac{1}{2}O_2(g) \rightarrow CaO(s)$
(c) $2Cu(s) + O_2(g) \rightarrow 2CuO(s)$

7.94 Which of the following equations has a value of $\Delta H°$ that would properly be labeled as $\Delta H_f°$?
(a) $2Fe(s) + O_2(g) \rightarrow 2FeO(s)$
(b) $SO_2(g) + \tfrac{1}{2}O_2(g) \rightarrow SO_3(g)$
(c) $N_2(g) + \tfrac{5}{2}O_2(g) \rightarrow N_2O_5(g)$

7.95 Write the thermochemical equations, including values of $\Delta H_f°$ in kilojoules per mole (from Table 7.2), for the formation of each of the following compounds from their elements, everything in standard states:
(a) $HC_2H_3O_2(l)$, acetic acid
(b) $NaHCO_3(s)$, sodium bicarbonate
(c) $CaSO_4 \cdot 2H_2O(s)$, gypsum

7.96 Write the thermochemical equations, including values of $\Delta H_f°$ in kilojoules per mole (from Table 7.2), for the formation of each of the following compounds from their elements, everything in standard states:
(a) $CO(NH_2)_2(s)$, urea
(b) $CaSO_4 \cdot \tfrac{1}{2}H_2O(s)$, plaster of Paris
(c) $CH_3OH(l)$, methyl alcohol

7.97 Using data in Table 7.2, calculate $\Delta H°$ in kilojoules for the following reactions:
(a) $2H_2O_2(l) \rightarrow 2H_2O(l) + O_2(g)$
(b) $HCl(g) + NaOH(s) \rightarrow NaCl(s) + H_2O(l)$

7.98 Using data in Table 7.2, calculate $\Delta H°$ in kilojoules for the following reactions:
(a) $CH_4(g) + Cl_2(g) \rightarrow CH_3Cl(g) + HCl(g)$
(b) $2NH_3(g) + CO_2(g) \rightarrow CO(NH_2)_2(s) + H_2O(l)$

7.99 Write the thermochemical equations that would be associated with the standard heats of formation of the following compounds:
(a) $HCl(g)$
(b) $NH_4Cl(s)$

7.100 Write the thermochemical equations that would be associated with the standard heats of formation of the following compounds:
(a) $HC_2H_3O_2(l)$
(b) $Na_2CO_3(s)$

7.101 The enthalpy change for the combustion of *1* mol of a compound under standard conditions is called the *standard heat of combustion*, and its symbol is $\Delta H°_{\text{combustion}}$. The value for sucrose, $C_{12}H_{22}O_{11}$, is $-5.65 \times 10^3 \text{ kJ mol}^{-1}$. Write the thermochemical equation for the combustion of 1 mol of sucrose and calculate the value of $\Delta H_f°$ for this compound. The sole products of combustion are $CO_2(g)$ and $H_2O(l)$.

7.102 The thermochemical equation for the combustion of acetylene gas, $C_2H_2(g)$, is

$$2C_2H_2(g) + 5O_2(g) \longrightarrow 4CO_2(g) + 2H_2O(l)$$
$$\Delta H° = -2599.3 \text{ kJ}$$

Using data in Table 7.2, determine the value of $\Delta H_f°$ for acetylene gas.

ADDITIONAL EXERCISES

7.103 A body of water with a mass of 750 g changed in temperature from 25.50 to 19.50 °C.
(a) What would have to be done to cause such a change?
(b) How much energy (in kilojoules) is involved in this change?

*7.104 A bar of hot iron with a mass of 1.000 kg and a temperature of 100.00 °C was plunged into an insulated vat of water. The mass of the water was 2.000 kg, and its initial temperature was 25.00 °C. What was the temperature of the resulting system when it stabilized?

7.105 Sulfur trioxide reacts with water to produce sulfuric acid according to the following equation. Calculate $\Delta H°$ for the reaction.

$$SO_3(g) + H_2O(l) \longrightarrow H_2SO_4(l)$$

7.106 Iron metal can be obtained from iron ore, which we can assume for this problem to be Fe_2O_3, by its reaction with hot carbon.

$$2Fe_2O_3(s) + 3C(s) \longrightarrow 4Fe(s) + 3CO_2(g)$$

What is $\Delta H°$ (in kilojoules) for this reaction? Is the reaction exothermic or endothermic?

7.107 In the recovery of iron from iron ore, the reduction of the ore is actually accomplished by reactions involving carbon monoxide. Use the following thermochemical equations,

$$Fe_2O_3(s) + 3CO(g) \longrightarrow 2Fe(s) + 3CO_2(g) \quad \Delta H° = -28 \text{ kJ}$$

$$3Fe_2O_3(s) + CO(g) \longrightarrow 2Fe_3O_4(s) + CO_2(g) \quad \Delta H° = -59 \text{ kJ}$$

$$Fe_3O_4(s) + CO(g) \longrightarrow 3FeO(s) + CO_2(g) \quad \Delta H° = +38 \text{ kJ}$$

to calculate $\Delta H°$ for the reaction

$$FeO(s) + CO(g) \longrightarrow Fe(s) + CO_2(g)$$

7.108 Use the results of Problem 7.107 and data in Table 7.2 to calculate the value of $\Delta H_f°$ for FeO. Express the answer in units of kilojoules per mole.

7.109 Phosphorus burns in air to give tetraphosphorus decaoxide.

$$4P(s) + 5O_2(g) \longrightarrow P_4O_{10}(s) \quad \Delta H° = -3062 \text{ kJ}$$

The product combines with water to give phosphoric acid, H_3PO_4.

$$P_4O_{10}(s) + 6H_2O(l) \longrightarrow 4H_3PO_4(l) \quad \Delta H° = -257.2 \text{ kJ}$$

Using these equations (and any others in the chapter, as needed), write the thermochemical equation for the formation of 1 mol of $H_3PO_4(l)$ from the elements. Include $\Delta H_f°$.

7.110 The amino acid glycine, $C_2H_5NO_2$, is one of the compounds used by the body to make proteins. The equation for its combustion is

$$4C_2H_5NO_2(s) + 9O_2(g) \longrightarrow 8CO_2(g) + 10H_2O(l) + 2N_2(g)$$

For each mole of glycine that burns, 973.49 kJ of heat is liberated. Use this information plus values of $\Delta H_f°$ for the products of combustion to calculate $\Delta H_f°$ for glycine.

7.111 Is the following reaction exothermic or endothermic? Calculate $\Delta H°$ for the reaction and then answer the question.

$$2Ag(s) + Zn(NO_3)_2(aq) \longrightarrow Zn(s) + 2AgNO_3(aq)$$

The following thermochemical equations can be used:

$$Cu(NO_3)_2(aq) + Zn(s) \longrightarrow Zn(NO_3)_2(aq) + Cu(s)$$
$$\Delta H° = -258 \text{ kJ}$$

$$2AgNO_3(aq) + Cu(s) \longrightarrow Cu(NO_3)_2(aq) + 2Ag(s)$$
$$\Delta H° = -106 \text{ kJ}$$

7.112 Calculate the $\Delta H°$ for the following reaction using the given thermochemical equations and others, as needed, that use data tabulated in this chapter or in Appendix C:

$$LiOH(aq) + HCl(aq) \longrightarrow LiCl(aq) + H_2O(l)$$

$$Li(s) + \tfrac{1}{2}O_2(g) + \tfrac{1}{2}H_2(g) \longrightarrow LiOH(s) \quad \Delta H° = -487.0 \text{ kJ}$$

$$2Li(s) + Cl_2(g) \longrightarrow 2LiCl(s) \quad \Delta H° = -815.0 \text{ kJ}$$

$$LiOH(s) \xrightarrow[\text{dissolving in water}]{} LiOH(aq) \quad \Delta H° = -19.2 \text{ kJ}$$

$$HCl(g) \xrightarrow[\text{dissolving in water}]{} HCl(aq) \quad \Delta H° = -77.0 \text{ kJ}$$

$$LiCl(s) \xrightarrow[\text{dissolving in water}]{} LiCl(aq) \quad \Delta H° = -36.0 \text{ kJ}$$

7.113 The value of $\Delta H_f°$ for HBr(g) was first evaluated using the following standard enthalpy values obtained experimentally. Use these data to calculate the value of $\Delta H_f°$ for HBr(g).

$$Cl_2(g) + 2KBr(aq) \longrightarrow Br_2(aq) + 2KCl(aq)$$
$$\Delta H° = -96.2 \text{ kJ}$$

$$H_2(g) + Cl_2(g) \longrightarrow 2HCl(g) \quad \Delta H° = -184 \text{ kJ}$$

$$HCl(aq) + KOH(aq) \longrightarrow KCl(aq) + H_2O(l)$$
$$\Delta H° = -57.3 \text{ kJ}$$

$$HBr(aq) + KOH(aq) \longrightarrow KBr(aq) + H_2O(l)$$
$$\Delta H° = -57.3 \text{ kJ}$$

$$HCl(g) \xrightarrow[\text{dissolving in water}]{} HCl(aq) \quad \Delta H° = -77.0 \text{ kJ}$$

$$Br_2(g) \xrightarrow[\text{dissolving in water}]{} Br_2(aq) \quad \Delta H° = -4.2 \text{ kJ}$$

$$HBr(g) \xrightarrow[\text{dissolving in water}]{} HBr(aq) \quad \Delta H° = -79.9 \text{ kJ}$$

*7.114 Acetylene, C_2H_2, is a gas commonly used in welding. It is formed in the reaction of calcium carbide, CaC_2, with water. Given the thermochemical equations below, calculate the value of $\Delta H_f°$ for acetylene in units of kilojoules per mole:

$$CaO(s) + H_2O(l) \longrightarrow Ca(OH)_2(s) \quad \Delta H° = -65.3 \text{ kJ}$$

$$CaO(s) + 3C(s) \longrightarrow CaC_2(s) + CO(g) \quad \Delta H° = +462.3 \text{ kJ}$$

$$CaCO_3(s) \longrightarrow CaO(s) + CO_2(g) \qquad \Delta H° = +178 \text{ kJ}$$

$$CaC_2(s) + 2H_2O(l) \longrightarrow Ca(OH)_2(s) + C_2H_2(g)$$
$$\Delta H° = -126 \text{ kJ}$$

$$2C(s) + O_2(g) \longrightarrow CO_2(g) \qquad \Delta H° = -220 \text{ kJ}$$

$$2H_2O(l) \longrightarrow 2H_2(g) + O_2(g) \qquad \Delta H° = +572 \text{ kJ}$$

7.115 The reaction for the metabolism of sucrose, $C_{12}H_{22}O_{11}$, is the same as for its combustion in oxygen to yield $CO_2(g)$ and $H_2O(l)$. The standard heat of formation of sucrose is $-2230 \text{ kJ mol}^{-1}$. Use data in Table 7.2 to compute the amount of energy (in kilojoules) released by metabolizing 1 oz (28.4 g) of sucrose.

*__7.116__ For ethanol, C_2H_5OH, which is mixed with gasoline to make the fuel gasohol, $\Delta H_f° = -277.63 \text{ kJ/mol}$. Calculate the number of kilojoules released by burning completely 1 gal of ethanol. The density of ethanol is 0.787 g cm^{-3}. Use data in Table 7.2 to help in the computation.

7.117 Consider the following thermochemical equations:

(1) $CH_3OH(l) + O_2(g) \longrightarrow HCHO_2(l) + H_2O(l)$
$$\Delta H° = -411 \text{ kJ}$$

(2) $CO(g) + 2H_2(g) \longrightarrow CH_3OH(l) \qquad \Delta H° = -128 \text{ kJ}$

(3) $HCHO_2(l) \longrightarrow CO(g) + H_2O(l) \qquad \Delta H° = -33 \text{ kJ}$

Suppose Equation 1 is reversed and divided by 2, Equations 2 and 3 are multiplied by $\frac{1}{2}$, and then the three adjusted equations are added. What is the net reaction, and what is the value of $\Delta H°$ for the net reaction?

7.118 Chlorofluoromethanes (CFMs) are carbon compounds of chlorine and fluorine and are also known as

Freons. Examples are Freon-11 ($CFCl_3$) and Freon-12 (CF_2Cl_2), which have been used as aerosol propellants. Freons have also been used in refrigeration and air-conditioning systems. It is feared that as these Freons escape into the atmosphere they will lead to a significant depletion of ozone from the upper atmosphere where ozone protects the earth's inhabitants from harmful ultraviolet radiation. In the stratosphere CFMs absorb high-energy radiation from the sun and split off chlorine atoms that hasten the decomposition of ozone, O_3. Possible reactions are

(1) $O_3(g) + Cl(g) \longrightarrow O_2(g) + ClO(g) \quad \Delta H° = -126 \text{ kJ}$

(2) $ClO(g) + O(g) \longrightarrow Cl(g) + O_2(g) \qquad \Delta H° = -268 \text{ kJ}$

(3) $O_3(g) + O(g) \longrightarrow 2O_2(g)$

The O atoms in Equation 2 come from the breaking apart of O_2 molecules caused by radiation from the sun. Use Equations 1 and 2 to calculate the value of $\Delta H°$ (in kilojoules) for Equation 3, the net reaction for the removal of O_3 from the atmosphere.

7.119 Given the following thermochemical equations,

$$2Cu(s) + O_2(g) \longrightarrow 2CuO(s) \qquad \Delta H° = -155 \text{ kJ}$$

$$Cu(s) + S(s) \longrightarrow CuS(s) \qquad \Delta H° = -53.1 \text{ kJ}$$

$$S(s) + O_2(g) \longrightarrow SO_2(g) \qquad \Delta H° = -297 \text{ kJ}$$

$$4CuS(s) + 2CuO(s) \longrightarrow 3Cu_2S(s) + SO_2(g)$$
$$\Delta H° = -13.1 \text{ kJ}$$

calculate $\Delta H°$ (in kilojoules) for the reaction

$$CuS(s) + Cu(s) \longrightarrow Cu_2S(s)$$

We pause again to allow you to test your understanding of concepts, your knowledge of scientific terms, and your skills at solving chemistry problems. Read through the following questions carefully, and answer each as fully as possible. When necessary, review topics you are uncertain of. If you can answer these questions correctly, you are ready to go on to the next group of chapters.

1 What is the difference between a strong electrolyte and a weak electrolyte? Formic acid, $HCHO_2$, is a weak acid. Write a chemical equation showing its reaction with water.

2 Methylamine, CH_3NH_2, is a weak base. Write a chemical equation showing its reaction with water.

3 Write molecular, ionic, and net ionic equations for the reaction that occurs when a solution containing hydrochloric acid is added to a solution of the weak base, methylamine (CH_3NH_2).

4 According to the solubility rules, which of the following salts would be classified as soluble?
(a) $Ca_3(PO_4)_2$ (f) $Au(ClO_4)_3$ (k) $ZnSO_4$
(b) $Ni(OH)_2$ (g) $Cu(C_2H_3O_2)_2$ (l) Na_2S
(c) $(NH_4)_2HPO_4$ (h) $AgBr$ (m) $CoCO_3$
(d) $SnCl_2$ (i) KOH (n) $BaSO_3$
(e) $Sr(NO_3)_2$ (j) Hg_2Cl_2 (o) MnS

5 Write molecular, ionic, and net ionic equations for any reactions that would occur between the following pairs of compounds. If no reaction, write "N.R."
(a) $CuCl_2(aq)$ and $(NH_4)_2CO_3(aq)$
(b) $HCl(aq)$ and $MgCO_3(s)$
(c) $ZnCl_2(aq)$ and $AgC_2H_3O_2(aq)$
(d) $HClO_4(aq)$ and $NaCHO_2(aq)$
(e) $MnO(s)$ and $H_2SO_4(aq)$
(f) $FeS(s)$ and $HCl(aq)$

6 Write a chemical equation for the complete neutralization of H_3PO_4 by NaOH.

7 Which of the following oxides are acidic and which are basic: P_4O_6, Na_2O, SeO_3, CaO, PbO, SO_2?

8 Write the formulas of any acid salts that could be formed by the reaction of the following acids with potassium hydroxide.
(a) sulfurous acid   (d) phosphoric acid
(b) nitric acid   (e) carbonic acid
(c) hypochlorous acid

9 Name the following: (a) HIO_3, (b) $HOBr$, (c) HNO_2, (d) $Ca(H_2PO_4)_2$, (e) $Fe(HSO_4)_3$.

10 Write formulas for the following:
(a) bromous acid   (c) lithium hydrogen sulfate
(b) hypoiodous acid   (d) bromic acid

11 How many milliliters of 0.200 M $BaCl_2$ must be added to 27.0 mL of 0.600 M Na_2SO_4 to give a complete reaction between their solutes?

12 What mass of $Mg(OH)_2$ will be formed when 30.0 mL of 0.200 M $MgCl_2$ solution is mixed with 25.0 mL of 0.420 M NaOH solution? What will be the molar concentrations of the ions remaining in solution?

13 How many milliliters of 6.00 M HNO_3 must be added to 200 mL of water to give 0.150 M HNO_3?

14 A certain toilet cleaner uses $NaHSO_4$ as its active ingredient. In an analysis, 0.500 g of the cleaner was dissolved in 30.0 mL of distilled water and required 24.60 mL of 0.105 M NaOH for complete neutralization in a titration. The net ionic equation for the reaction is

$$HSO_4^- + OH^- \longrightarrow H_2O + SO_4^{2-}$$

What was the percentage by weight of $NaHSO_4$ in the cleaner?

15 A volume of 28.50 mL of a freshly prepared solution of KOH was required to titrate 50.00 mL of 0.0922 M HCl solution. What was the molarity of the KOH solution?

16 How many milliliters of concentrated sulfuric acid (18.0 M) are needed to prepare 125 mL of 0.144 M H_2SO_4?

17 The density of concentrated phosphoric acid solution is 1.689 g solution/mL solution at 20 °C. It contains 144 g H_3PO_4 per 1.00×10^2 mL of solution.
(a) Calculate the molar concentration of H_3PO_4 in this solution.
(b) Calculate the number of grams of this solution required to hold 50.0 g H_3PO_4.

18 A mixture consists of lithium carbonate (Li_2CO_3) and potassium carbonate (K_2CO_3). These react with hydrochloric acid as follows.

$$Li_2CO_3(s) + 2HCl(aq) \longrightarrow 2LiCl(aq) + H_2O + CO_2(g)$$

$$K_2CO_3(s) + 2HCl(aq) \longrightarrow 2KCl(aq) + H_2O + CO_2(g)$$

When 4.43 g of this mixture was analyzed, it consumed 53.2 mL of 1.48 M HCl. Calculate the number of grams of each carbonate and their percentages.

19 One way to prepare iodine is to react sodium iodate, $NaIO_3$, with hydroiodic acid, HI. The following reaction occurs.

$$NaIO_3 + 6HI \longrightarrow 3I_2 + NaI + 3H_2O$$

Calculate the number of moles and the number of grams of iodine that can be made this way from 16.4 g of $NaIO_3$.

20 Assign oxidation numbers to the atoms in the following formulas: (a) As_4, (b) $HClO_2$, (c) $MnCl_2$, (d) $V_2(SO_3)_3$.

21 For the following unbalanced equations, write the reactants and products in the form they should appear in an ionic equation. Then write balanced net ionic equations by applying the ion-electron method.
(a) $K_2Cr_2O_7 + HCl \rightarrow KCl + Cl_2 + H_2O + CrCl_3$
(b) $KOH + SO_2(aq) + KMnO_4 \rightarrow K_2SO_4 + MnO_2 + H_2O$

22 Balance the following equations by the ion–electron method for *acidic solutions*.
(a) $Cr_2O_7^{2-} + Br^- \rightarrow Br_2 + Cr^{3+}$
(b) $H_3AsO_3 + MnO_4^- \rightarrow H_2AsO_4^- + Mn^{2+}$

23 Balance the following equations by the ion–electron method for *basic solutions*.
(a) $I^- + CrO_4^{2-} \rightarrow CrO_2^- + IO_3^-$
(b) $SO_2 + MnO_4^- \rightarrow MnO_2 + SO_4^{2-}$

24 In the previous two questions, identify the oxidizing agents and reducing agents.

25 Complete and balance the following equations if a reaction occurs.
(a) $Sn(s) + HCl(aq) \rightarrow$ (c) $Zn(s) + Cu^{2+}(aq) \rightarrow$
(b) $Cu(s) + HNO_3(concd) \rightarrow$ (d) $Ag(s) + Cu^{2+}(aq) \rightarrow$

26 Write a balanced chemical equation for the combustion of cetane, $C_{16}H_{34}$, a hydrocarbon present in diesel fuel, (a) in the presence of excess oxygen, (b) in a somewhat limited supply of oxygen, and (c) in a severely limited supply of oxygen.

27 Stearic acid, $C_{17}H_{35}CO_2H$, is derived from animal fat. Write a balanced chemical equation for the combustion of stearic acid in an abundant supply of oxygen.

28 Methanethiol, CH_3SH, is a foul-smelling gas produced in the intestinal tract by bacteria acting on albumin in the absence of air. Write a chemical equation for the combustion of CH_3SH in an excess supply of oxygen.

29 Write molecular equations for the reaction of O_2 with (a) magnesium, (b) aluminum, (c) phosphorus, (d) sulfur.

30 *Bordeaux mixture* is traditionally prepared by mixing copper sulfate and calcium hydroxide in water. The resulting suspension of copper hydroxide is sprayed on trees and shrubs to fight fungus diseases. This fungicide is also available in commercial preparations. In an analysis of one such product, a sample weighing 0.238 g was dissolved in hydrochloric acid. Excess KI solution was then added and the iodine that was formed was titrated with 0.01669 M $Na_2S_2O_3$ solution using starch as an indicator. The titration required 28.62 mL of the thiosulfate solution. What was the percentage by weight of copper in the sample of Bordeaux mixture?

31 Suppose we have a system that consists of 225 g of pure water at 25.00 °C and 1 atm of pressure. What value do we use for the heat capacity of this system? What value do we use for the specific heat of the water? (For the energy part of the unit, use the joule.)

32 Why isn't the value of ΔH_f° equal to zero for ozone, O_3?

33 The specific heat of helium is 5.19 J/g °C and that of nitrogen is 1.04 J/g °C. How many joules can one mole of each gas absorb when its temperature increases by 1.00 °C?

34 When 0.6484 g of cetyl palmitate, $C_{32}H_{64}O_2$ (a fruit wax), was burned in a bomb calorimeter with a heat capacity of 11.99 kJ/°C, the temperature of the calorimeter rose from 24.518 °C to 26.746 °C. Calculate the molar heat of combustion of cetyl palmitate in kJ/mol.

35 When we say that the value of ΔH for a chemical reaction is a *state function*, what do we mean?

36 The combustion of methane (the chief component of natural gas) follows the equation

$$CH_4(g) + 2O_2(g) \longrightarrow CO_2(g) + 2H_2O(g)$$

ΔH° for this reaction is -802.3 kJ. How many grams of methane must be burned to provide enough heat to raise the temperature of 250 mL of water from 25.0 °C to 50.0 °C?

37 Label the following thermal properties as intensive or extensive.
(a) specific heat   (d) ΔH°
(b) heat capacity   (e) molar heat capacity
(c) ΔH_f°

38 What is the definition of *internal energy*? State the first law of thermodynamics.

39 What is *pressure-volume work*? Give the equation that could be used to calculate it.

40 The change in the internal energy, ΔE, is equal to the "heat of reaction at constant volume." Explain the reason for this.

41 Suppose that a gas is produced in an exothermic chemical reaction between two reactants in an aqueous solution. Which quantity will have the larger magnitude, ΔE or ΔH?

42 The thermochemical equation for the combustion reaction of half a mole of carbon monoxide is as follows.

$$\tfrac{1}{2}CO(g) + \tfrac{1}{4}O_2(g) \longrightarrow \tfrac{1}{2}CO_2(g) \qquad \Delta H^\circ = -141.49 \text{ kJ}$$

Write the thermochemical equation for
(a) The combustion of 2 mol of $CO(g)$.
(b) The decomposition of 1 mol of $CO_2(g)$ to $O_2(g)$ and $CO(g)$.

43 The standard heat of combustion of eicosane, $C_{20}H_{42}(s)$, a typical component of candle wax, is 1.332×10^4 kJ/mol, when it burns in pure oxygen, and the products are cooled to 25 °C. The only products are $CO_2(g)$ and $H_2O(l)$. Calculate the value of the standard heat of formation of eicosane (in kJ/mol) and write the corresponding thermochemical equation.

44 Using data in Table 7.2 calculate values for the standard heats of reaction (in kilojoules) for the following reactions.
(a) $H_2SO_4(l) \rightarrow SO_3(g) + H_2O(l)$
(b) $C_2H_6(g) \rightarrow C_2H_4(g) + H_2(g)$

45 Calculate the standard heat of formation of calcium carbide, $CaC_2(s)$, in kJ/mol using the following thermochemical equations.

$Ca(s) + 2H_2O(l) \rightarrow Ca(OH)_2(s) + H_2(g)$ $\Delta H^\circ = -414.79$ kJ

$2C(s) + O_2(g) \rightarrow 2CO(g)$ $\Delta H^\circ = -221.0$ kJ

$CaO(s) + H_2O(l) \rightarrow Ca(OH)_2(s)$ $\Delta H^\circ = -65.19$ kJ

$2H_2(g) + O_2(g) \rightarrow 2H_2O(l)$ $\Delta H^\circ = -571.8$ kJ

$CaO(s) + 3C(s) \rightarrow CaC_2(s) + CO(g)$ $\Delta H^\circ = +462.3$ kJ

CHAPTER 8

The Quantum Mechanical Atom

Compact discs and DVDs are possible because information recorded on them can be read by tiny light sources called lasers. The laser emits highly monochromatic light (light of a single color), which is reflected from the surface of the CD or DVD, producing a flickering beam containing the information that's ultimately translated into sound and/or picture. The light from a laser is produced by electrons undergoing energy changes within atoms. Such energy changes are related to the electronic structure of atoms, which is the principal topic of this chapter.

THIS CHAPTER IN CONTEXT A good model is able to explain known facts and make useful predictions. In previous chapters, we've used a simple model of the atom as a collection of smaller particles—protons, neutrons, and electrons—to explain mass relationships between elements and to explain why isotopes exist. We have said that the number of protons in an atom's nucleus distinguishes it from atoms of other elements. For example, only sodium has 11 protons (and so, 11 electrons) in its atoms. Take away a proton and an electron, and you have neon, a colorless, odorless gas. Add a proton and an electron and you have magnesium, a lightweight metal that is much less reactive than sodium. In some way, then, the number of protons, neutrons, and electrons in an atom determines its properties.

But how can viewing the atom as a core nucleus surrounded by electrons explain why one element is different from another? Why are some elements metals, while others are nonmetals? Why do metals form positive ions, while nonmetals form negative ones? Why do the properties of the elements repeat when they are arranged in order of increasing atomic number? And why do elements combine in certain ratios with other elements? Why, for example, is water's formula H_2O and not H_3O or HO? Our simple model of the atom cannot answer these questions.

In Chapter 1, we saw that experimental evidence indicates that atoms have a tiny, dense, positively charged core, the nucleus, surrounded by negatively charged electrons that fill the remaining volume of the atom. Electrons must move within this volume; otherwise, they would not be able to resist the electrostatic attraction of the nucleus, and the atom would collapse. However, a vibrating charge, such as an orbiting electron, creates ripples in the electric and magnetic fields around it. The creation of these ripples, or electromagnetic waves, should cause the electron to lose energy. As the electron's energy decreases, it should spiral inwards toward the nucleus. Calculations based on classical physics predict that after emitting a brief, intense burst of radiation, the electron should immediately crash into the nucleus (Figure 8.1). That simply doesn't happen. This **collapsing atom paradox** was only one sign that classical physics was severely flawed. An avalanche of experimental evidence convinced scientists that electrons do not behave at all like objects in the everyday world. An entirely new physics would be needed to explain how electrons behaved in atoms.

New experiments led to the startling realization that electrons were not simply tiny, negatively charged particles. While electrons do behave like particles in some experiments, they behave like waves in others. This **wave/particle duality** has no real analogy in everyday life. The physics of objects that exhibit wave/particle duality is called **quantum mechanics,** or **quantum theory.** Quantum theory is a cornerstone of modern chemistry.

This simple picture of the atom makes a nice corporate logo, but the idea of an atom with electrons orbiting a nucleus as planets orbit a sun was discarded nearly a century ago.

When experiment and theory do not agree, the theory must be modified or discarded. Classical physics could not resolve the collapsing atom paradox. It also could not correctly predict the heat capacity of cold solids, the periodicity of the elements, or crucial characteristics of the radiation emitted by hot objects or hot samples of the elements.

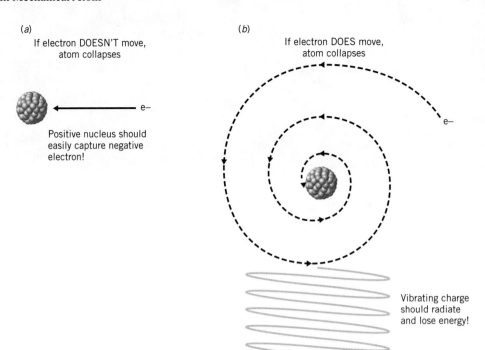

(a)

If electron DOESN'T move, atom collapses

e–

Positive nucleus should easily capture negative electron!

(b)

If electron DOES move, atom collapses

e–

Vibrating charge should radiate and lose energy!

FIGURE 8.1 *The collapsing atom paradox.* (*a*) If the electron does not move, the positively charged nucleus will attract and capture it—and the atom will collapse. (*b*) If the electron moves around the nucleus, it should radiate energy as an electromagnetic wave. The electron spirals into the nucleus, and the atom collapses.

Quantum theory can be used to explain why atoms are stable, why things have the colors they do, why the periodic table has the shape it has, why chemical bonds form, and why some molecules form but others don't. Quantum theory played a key role in the rapid advances in electronics and communications technology that took place in the last century. In this century, it has profound importance in the development of new computing technologies and in nanotechnology, an emerging field concerned with the design and construction of molecular machines.

We'll discuss nanotechnology further in Chapter 13.

Both electromagnetic radiation and electrons exhibit wave/particle duality. To understand the experimental basis for the quantum theory, we have to begin our discussion with radiation.

> ## 8.1 ▶ Electromagnetic radiation can be described as a wave or as a stream of photons

Electromagnetic radiation can be described as a wave

You've learned that objects can have energy in only two ways, as kinetic energy and as potential energy. You also learned that energy can be transferred between things, and in Chapter 7 our principal focus was on the transfer of heat. Energy can also be transferred between things as light or radiation. This is a very important form of energy in chemistry. For example, many chemical systems emit visible light as they react (see Figure 8.2).

Many experiments show that radiation carries energy through space by means of **waves.** Waves are an oscillation that moves outward from a disturbance (think of ripples moving away from a pebble dropped into a pond). In the case of radiation, the disturbance can be a vibrating electric charge. When the charge jiggles, it produces a pulse in the electric field around it. As the electric field pulses, it creates a pulse in the magnetic field. The magnetic field pulse gives rise to yet another electric field pulse further away from the disturbance. The process continues, with a pulse in one field giving rise to a pulse in the other, and the resulting train of pulses

Electricity and magnetism are closely related to each other. A moving charge creates an electric current, which in turn creates a magnetic field around it. This is the fundamental idea behind electric motors. A moving magnetic field creates an electric field or current. This is the idea behind electrical generators and turbines.

(a)

(b)

(c)

FIGURE 8.2 *Light is given off in a variety of chemical reactions.* (*a*) Combustion. (*b*) Cyalume® light sticks. (*c*) A lightning bug.

in the electric and magnetic fields is called an **electromagnetic wave.** The wave ripples away from the source at almost unimaginable speeds.

The higher the vibrating charge bounces, the greater the height or **amplitude** of the peaks will be. Amplitude affects the intensity or brightness of the radiation. Figure 8.3 shows how the amplitude or intensity of the wave varies with time and with distance as the wave travels through space. In Figure 8.3*a*, we see two complete oscillations or *cycles* of the wave during a 1 second interval. The number of cycles per second is called the **frequency** of the electromagnetic radiation, and its symbol is ν (the Greek letter *nu*, pronounced "new").

The concept of frequency extends beyond just electromagnetic radiation to other events that recur at regular intervals. For example, you go to school 5 days *per week,* or you pay your bills once *each month.* Each of these statements describes how frequently an event occurs, and they have in common the notion of *per unit of time,* or $\frac{1}{time}$. In the SI, the unit of time is the second (s), so frequency is given the unit "per second," which is $\frac{1}{second}$, or (second)$^{-1}$. This unit is given the special name **hertz (Hz).**

$$1 \text{ Hz} = 1 \text{ s}^{-1}$$

As electromagnetic radiation moves away from its source, the positions of maximum and minimum amplitude (peaks and troughs) are regularly spaced. The peak-to-peak distance is called the radiation's **wavelength,** symbolized by λ (the Greek letter *lambda*). (See Figure 8.2*b*.) Because wavelength is a distance, it has distance units (e.g., meters).

Electromagnetic waves don't need a medium to travel through, as water and sound waves do. They can cross empty space. The speed of the electromagnetic wave in a vacuum is the same no matter how the radiation is created (about 3.00×10^8 m/s).

The SI symbol for the second is s.

$$s^{-1} = \frac{1}{s}$$

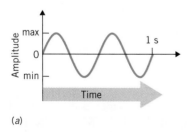
(a)

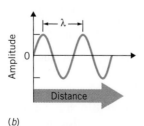
(b)

FIGURE 8.3 *Two views of electromagnetic radiation.* (*a*) The frequency, ν, of a light wave is the number of complete oscillations each second. Here two cycles span a 1 second time interval, so the frequency is two cycles per second, or 2 Hz. (*b*) Electromagnetic radiation frozen in time. This curve shows how the amplitude varies along the direction of travel. The distance between two peaks is the wavelength, λ, of the electromagnetic radiation.

For any wave, the product of its wavelength and its frequency equals the speed of the wave.

The speed of light is one of our most carefully measured constants, because the meter is defined in terms of it. The precise value of the speed of light in a vacuum is 2.99792458×10^8 m/s, and a meter is defined as exactly the distance traveled by light in 1/299792458 of a second.

Wavelength-frequency relationship

If we multiply the wavelength by frequency, the result is the speed of the wave. We can see this if we analyze the units.

$$\text{meters} \times \frac{1}{\text{second}} = \frac{\text{meters}}{\text{second}} = \text{speed}$$

In SI units

$$\text{m} \times \frac{1}{\text{s}} = \frac{\text{m}}{\text{s}} = \text{m s}^{-1}$$

The speed of electromagnetic radiation in a vacuum is a constant and is commonly called the *speed of light*. Its value to three significant figures is 3.00×10^8 m/s (or m s^{-1}). This important physical constant is given the symbol c.

$$c = 3.00 \times 10^8 \text{ m s}^{-1}$$

From the preceding discussion we obtain a very important relationship that allows us to convert between λ and ν.

$$\lambda \times \nu = c = 3.00 \times 10^8 \text{ m s}^{-1} \tag{8.1}$$

EXAMPLE 8.1

Calculating Frequency from Wavelength

Mycobacterium tuberculosis, the organism that causes tuberculosis, can be completely destroyed by irradiation with ultraviolet light with a wavelength of 254 nm. What is the frequency of this radiation?

ANALYSIS: To convert between wavelength and frequency we use Equation 8.1. However, we must be careful about the units.

SOLUTION: To calculate the frequency, we solve Equation 8.1 for ν.

$$\nu = \frac{c}{\lambda}$$

Next, we substitute for c (3.00×10^8 m s^{-1}) and for the wavelength. However, to cancel units correctly, we must have the wavelength in meters. Recall from Chapter 1 that nm means nanometer and the prefix nano implies the factor "$\times 10^{-9}$":

$$1 \text{ nm} = 10^{-9} \text{ m}$$

Therefore, 254 nm equals 254×10^{-9} m. Substituting gives

$$\nu = \frac{3.00 \times 10^8 \text{ m s}^{-1}}{254 \times 10^{-9} \text{ m}}$$

$$= 1.18 \times 10^{15} \text{ s}^{-1}$$

$$= 1.18 \times 10^{15} \text{ Hz}$$

Does the Answer Make Sense?
If the frequency is about 10^{15} Hz, the speed of light should be

$$(254 \times 10^{-9} \text{ m}) \times (10^{15} \text{ s}^{-1}) = 254 \times 10^6 \text{ m s}^{-1} \approx 3 \times 10^8 \text{ m s}^{-1}$$

which is correct to one significant digit.

EXAMPLE 8.2

Calculating Wavelength from Frequency

Radio station WGBB on Long Island, New York, broadcasts its AM signal, a form of electromagnetic radiation, at a frequency of 1240 kHz. What is the wavelength of these radio waves expressed in meters?

ANALYSIS and SOLUTION: Once again we are relating wavelength and frequency, so we must use Equation 8.1. This time we solve Equation 8.1 for the wavelength.

$$\lambda = \frac{c}{\nu}$$

Now we must cancel the unit s^{-1} in 3.00×10^8 m s^{-1}. The prefix "k" in kHz means kilo and stands for "$\times 10^3$" and Hz means s^{-1}.

$$1 \text{ kHz} = 10^3 \text{ Hz}$$

$$1 \text{ Hz} = 1 \text{ s}^{-1}$$

The frequency is therefore 1240×10^3 s^{-1}. Substituting gives

$$\lambda = \frac{3.00 \times 10^8 \text{ m s}^{-1}}{1240 \times 10^3 \text{ s}^{-1}}$$

$$= 242 \text{ m}$$

Does the Answer Make Sense?
If this wavelength is correct, we should be able to multiply it by the original frequency (in Hz) and get the speed of light, as we did in the previous example. We also should be able to divide the speed of light by the wavelength to get back the frequency. To two figures,

$$\frac{3.0 \times 10^8 \text{ m/s}}{2.4 \times 10^2 \text{ m}} = 1.2 \times 10^6 \text{ Hz} = 1200 \text{ kHz}$$

which is reasonably close to the original frequency of 1240 kHz.

PRACTICE EXERCISE 1: The most intense radiation emitted by the Earth has a wavelength of about 10.0 μm. What is the frequency of this radiation in hertz?

PRACTICE EXERCISE 2: An FM radio station in West Palm Beach, Florida, broadcasts electromagnetic radiation at a frequency of 104.3 MHz (megahertz). What is the wavelength of these radio waves, expressed in meters?

Electromagnetic waves are categorized by frequency

Electromagnetic radiation comes in a broad range of frequencies called the **electromagnetic spectrum,** illustrated in Figure 8.4. Some portions of the spectrum have popular names. For example, radio waves are electromagnetic radiations having very low frequencies (and therefore very long wavelengths). Microwaves, which also have low frequencies, are emitted by radar instruments such as those the police use to monitor the speeds of cars. In microwave ovens, similar radiation is used to heat water in foods, causing the food to cook quickly. Infrared radiation is emitted by hot objects and consists of the range of frequencies that can make molecules of most substances vibrate internally. You can't see infrared radiation, but you can feel how your body absorbs it by holding your hand near a hot radiator; the absorbed radiation makes your hand warm. Gamma rays are at the high-frequency end of the electromagnetic spectrum. They are produced by certain elements that are radioactive. X rays are very much like gamma rays, but they are usually made by special equipment. Both X rays and gamma rays penetrate living things easily.

Most of the time, you are bombarded with electromagnetic radiations from all portions of the electromagnetic spectrum. Radio and TV signals pass through you; you feel infrared radiation when you sense the warmth of a radiator; X rays and gamma rays fall on you from space; and light from a lamp reflects into your eyes from the page you're reading. Of all these radiations, your eyes are able to sense only a very narrow band of wavelengths ranging from about 400 to 700 nm. This band is called the **visible spectrum** and consists of all the colors you can see, from red through orange, yellow, green, blue, and violet. White light is composed of all these colors in roughly equal amounts, and it can be separated into them by focusing a beam of white light through a prism, which spreads the various wavelengths

Remember that there is an inverse relationship between wavelength and frequency. The lower the frequency, the longer the wavelength.

The speed of electromagnetic radiation was computed to be about 3×10^8 m/s. The fact that the same speed was determined experimentally for light supported the hypothesis that light was a form of electromagnetic radiation.

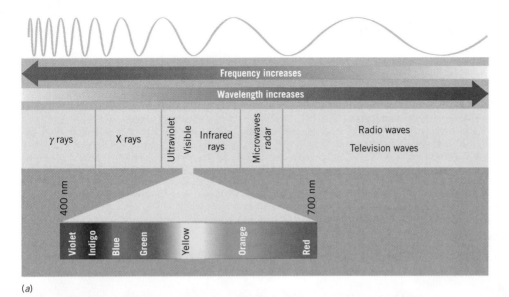

(a)

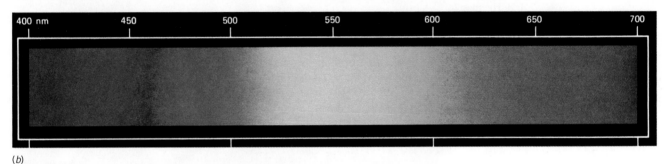

(b)

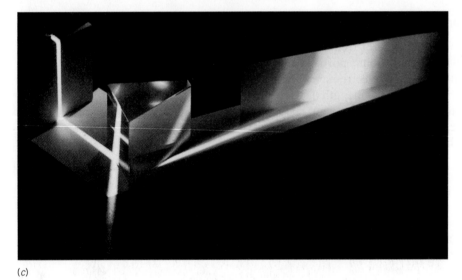

(c)

FIGURE 8.4 *The electromagnetic spectrum.* (*a*) The electromagnetic spectrum is divided into regions according to the wavelengths of the radiation. (*b*) The visible spectrum is composed of wavelengths that range from about 400 to 700 nm. (*c*) The production of a visible spectrum by splitting white light into its rainbow of colors.

apart. This is illustrated in Figure 8.4*b*. A photograph showing the production of a visible spectrum is given in Figure 8.4*c*.

The way substances absorb electromagnetic radiation often can help us characterize them. For example, each substance absorbs a uniquely different set of infrared frequencies. A plot of the wavelengths absorbed versus the intensities of absorption is called an *infrared absorption spectrum*. It can be used to identify a compound, because each infrared spectrum is as unique as a set of fingerprints. (See Figure 8.5.) Many substances absorb visible and ultraviolet radiations in unique ways, too, and they have visible and ultraviolet spectra (Figure 8.6).

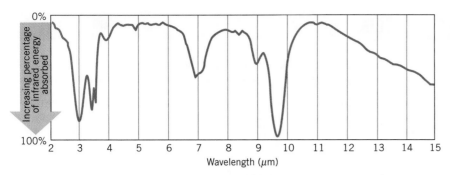

FIGURE 8.5 *Infrared absorption spectrum of methyl alcohol (also called wood alcohol), the fuel in "canned heat" products such as Sterno.* In an infrared spectrum, the usual practice is to show the amount of light absorbed increasing from top to bottom in the graph. Thus, there is a peak in the percentage of light absorbed at about 3 μm. (Spectrum courtesy Sadtler Research Laboratories, Inc., Philadelphia, Pa.)

Electromagnetic radiation can be viewed as a stream of photons

When an electromagnetic wave passes an object, the oscillating electric and magnetic fields may interact with it, as ocean waves interact with a buoy in a harbor. A tiny charged particle placed in the path of the wave will be yanked back and forth by the changing electric and magnetic fields. For example, when a radio wave strikes an antenna, electrons within the antenna begin to bounce up and down, creating an alternating current that can be detected and decoded electronically. Because the wave exerts a force on the antenna's electrons and moves them through a distance, work is done. Thus, as energy is lost by the source of the wave (the radio transmitter), energy is gained by the electrons in the antenna.

A series of groundbreaking experiments showed that classical physics does not correctly describe energy transfer by electromagnetic radiation. In 1900, a German physicist named Max Planck (1858–1947) proposed that electromagnetic radiation can be viewed as a stream of tiny packets or **quanta** of energy that were later called **photons.** Each photon travels at the speed of light. Planck proposed, and Albert

The energy of one photon is called one **quantum** of energy.

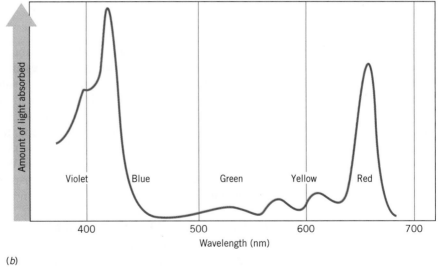

(*a*)   (*b*)

FIGURE 8.6 *Absorption of light by chlorophyll.* (*a*) Chlorophyll is the green pigment plants use to harvest solar energy for photosynthesis. (*b*) In this visible absorption spectrum of chlorophyll, the percentage of light absorbed increases from bottom to top. Thus, there is a peak in the light absorbed at about 420 nm and another at about 660 nm. This means the pigment strongly absorbs blue-violet and red light. The green color we *see* is the light that's *not* absorbed. It's composed of the wavelengths of visible light that are reflected. (Our eyes are most sensitive to green, so we don't notice the yellow components of the reflected light.)

FACETS OF CHEMISTRY 8.1

Photoelectricity and Its Applications

One of the earliest clues to the relationship between the frequency of light and its energy was the discovery of the photoelectric effect. In the latter part of the nineteenth century, it was found that certain metals acquired a positive charge when they were illuminated by light. Apparently, light is capable of kicking electrons out of the surface of the metal.

When this phenomenon was studied in detail, it was discovered that electrons could only be made to leave a metal's surface if the frequency of the incident radiation was above some minimum value, which was named the *threshold frequency*. This threshold frequency differs for different metals, depending on how tightly the metal atom holds onto electrons. Above the threshold frequency, the kinetic energy of the emitted electron increases with increasing frequency of the light. Interestingly, however, its kinetic energy does not depend on the intensity of the light. In fact, if the frequency of the light is below the minimum frequency, no electrons are observed at all, no matter how bright the light is. To physicists of that time, this was very perplexing because they believed the energy of light was related to its brightness. The explanation of the phenomenon was finally given by Albert Einstein in the form of a very simple equation.

$$KE = h\nu - w$$

where KE is the kinetic energy of the electron that is emitted, $h\nu$ is the energy of the photon of frequency ν, and w is the minimum energy needed to eject the electron from the metal's surface. Stated another way, part of the energy of the photon is needed just to get the electron off the surface of the metal. This amount is w. Any energy left over $(h\nu - w)$ appears as the electron's kinetic energy.

Besides its important theoretical implications, the photoelectric effect has many practical applications. For example, automatic "electric eye" door openers use this phenomenon by sensing the interruption of a light beam caused by the person wishing to use the door. The phenomenon is also responsible for photoconduction by certain substances that are used in light meters in cameras and other devices. The production of sound in motion pictures was first made possible by incorporating a strip along the edge of the film (called the *sound track*) that causes the light passing through it to fluctuate in intensity according to the frequency of the sound that's been recorded. A photocell converts this light to a varying electric current that is amplified and played through speakers in the theater. Even the sensitivity of photographic film to light is related to the release of photoelectrons within tiny grains of silver bromide that are suspended in a coating on the surface of the film.

Einstein (1879–1955) confirmed, that *the energy of a photon of electromagnetic radiation is proportional to the radiation's frequency,* not to its intensity or brightness as had been believed up to that time. (See Facets of Chemistry 8.1.)

TOOLS

Energy of a photon

The value of Planck's constant is 6.626×10^{-34} J s. It has units of energy (joules) multiplied by time (seconds).

$$\text{energy of a photon} = E = h\nu \qquad (8.2)$$

In this expression, h is a proportionality constant that we now call **Planck's constant.** Note that Equation 8.2 relates two representations of electromagnetic radiation. The right-hand side of the equation deals with a property of particles (energy per photon); the left-hand side deals with a property of waves (the frequency). Quantum theory unites the two representations, so we can use whichever representation of electromagnetic radiation is convenient for describing experimental results. For example, in describing the photoelectric effect (Facets of Chemistry 8.1), we represent radiation as a stream of particles. When describing how radiation can bend around small obstacles and fan out after passing through pinholes, we represent radiation as a wave phenomenon. Radiation is not a stream of particles, or a wave—it's something else entirely, something unlike anything we have ever encountered in everyday life. It is something that can behave like either particles or waves, depending on the experiment.

Planck's and Einstein's discovery was really quite surprising. If a particular event requiring energy, such as photosynthesis in green plants, is initiated by the absorption of light, it is the frequency of the light that is important, not its intensity or brightness. This makes sense when we view light as a stream of photons, since "high frequency" is associated with higher energy of the photons, while higher intensity is associated with greater numbers of photons. But it makes no sense at all when we view radiation as a wave phenomenon.

Brighter light delivers more photons; higher-frequency light delivers more energetic photons.

The idea that electromagnetic radiation can be represented as either a stream of photons or a wave is a cornerstone of the quantum theory. Physicists were able to use the concept of photons to understand many experimental results that classical physics simply had no explanation for. The success of the quantum theory in describing radiation paved the way for a second startling realization: that electrons, like radiation, could be represented as either waves or particles. We now turn our attention to the first experimental evidence that led to our modern quantum mechanical model of atomic structure: the existence of discrete lines in atomic spectra.

8.2 ▶ Atomic line spectra are experimental evidence that electrons in atoms have quantized energies

The visible spectrum described in Figure 8.4 is called a **continuous spectrum** because it contains a continuous unbroken distribution of light of *all* colors. It is formed when the light from the sun, or any other object that's been heated to a very high temperature (such as the filament in an electric light bulb), is split by a prism and displayed on a screen. A rainbow after a summer shower is a continuous spectrum that most people have seen. In this case, tiny water droplets in the air spread out the colors contained in sunlight.

A rather different kind of spectrum is observed if we examine the light that is given off when an *electric discharge,* or spark, passes through a gas such as hydrogen. The electric discharge is an electric current that *excites,* or energizes, the atoms of the gas. More specifically, the electric current excites and gives energy to the electrons in the atom. The atoms then emit the absorbed energy in the form of light as the electrons return to a lower energy state. When a narrow beam of this light is passed through a prism, as shown in Figure 8.7, we do *not* see a continuous spectrum. Instead, only a few colors are observed, displayed as a series of individual lines. This series of lines is called the element's **atomic spectrum** or **emission spectrum.** Figure 8.8 shows the visible portions of the atomic spectra of two common elements, sodium and hydrogen, and how they compare with a continuous spectrum. Notice that the spectra of these elements are quite different. In fact, each element has its own unique atomic spectrum that is as characteristic as a fingerprint.

Atoms of an element can also be excited by adding them to the flame of a Bunsen burner.

An emission spectrum is also called a *line spectrum* because the light corresponding to the individual emissions appears as lines on the screen.

A simple pattern of lines in the spectrum of hydrogen suggests a simple explanation for atomic spectra

The first success in explaining atomic spectra quantitatively came with the study of the spectrum of hydrogen. This is the simplest element, since its atoms have only one electron, and it produces the simplest spectrum with the fewest lines.

The atomic spectrum of hydrogen actually consists of several series of lines. One series is in the visible region of the electromagnetic spectrum and is shown in

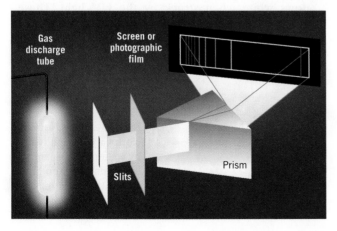

FIGURE 8.7 *Production and observation of an atomic spectrum.* Light emitted by excited atoms is formed into a narrow beam and passed through a prism, which divides the light into relatively few narrow beams with frequencies that are characteristic of the particular element that's emitting the light. When these beams fall on a screen, a series of lines is observed, which is why the spectrum is also called a *line spectrum.*

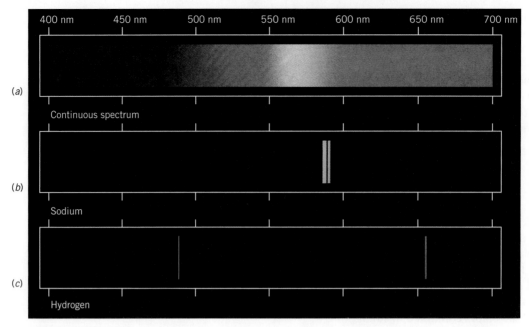

FIGURE 8.8 *Continuous and atomic emission spectra.* (*a*) The continuous visible spectrum produced by the sun or an incandescent lamp. (*b*) The atomic spectrum emission produced by sodium. The emission spectrum of sodium actually contains more than 90 lines in the visible region. The two brightest lines are shown here. All the others are less than 1% as bright as these. (*c*) The atomic spectrum (line spectrum) produced by hydrogen. There are only four lines in this visible spectrum. They vary in brightness by only a factor of 5, so they are all shown.

CHEMISTRY IN PRACTICE At one time, most of the street lighting in towns and cities was provided by incandescent lamps, in which a tungsten filament is heated white-hot by an electric current. Unfortunately, much of the light from this kind of lamp is infrared radiation, which we cannot see. As a result, only a relatively small fraction of the electrical energy used to operate the lamp actually results in visible light. Modern streetlights consist of high-intensity sodium or mercury vapor lamps in which the light is produced by passing an electric discharge through the vapors of these metals. The electric current excites the atoms, which then emit their characteristic atomic spectra. In these lamps, most of the electrical energy is converted to light in the visible region of the spectrum, so they are much more energy-efficient (and cost-efficient) than incandescent lamps. Sodium emits intense light at a wavelength of 589 nm, which is yellow. The golden glow of streetlights in scenes like that shown in the photograph at the right is from this emission line of sodium, which is produced in high-pressure sodium vapor lamps. There is another, much more yellow light emitted by low-pressure sodium lamps that you may also have seen. Both kinds of sodium lamps are very efficient, and almost all communities now use this kind of lighting to save money on the costs of electricity. In some places, the somewhat less efficient mercury vapor lamps are still used. These lamps give a bluish-white light.

Fluorescent lamps also depend on an atomic emission spectrum as the primary source of light. In these lamps an electric discharge is passed through mercury vapor, which emits some of its light in the ultraviolet region of the spectrum. The inner wall of the fluorescent tube is coated with a phosphor that glows white when struck by the UV light. These lamps are also highly efficient and convert most of the electrical energy they consume into visible light.

Park Avenue in New York City is brightly lit by sodium vapor lamps in this photo taken during the Christmas season.

Figure 8.8. Another series is in the ultraviolet region, and the rest are in the infrared. In 1885, J. J. Balmer found an equation that was able to give the wavelengths of the lines in the visible portion of the spectrum. This was soon extended to a more general equation, called the **Rydberg equation,** that could be used to calculate the wavelengths of *all* the spectral lines of hydrogen.

Rydberg equation:

$$\frac{1}{\lambda} = R_H\left(\frac{1}{n_1^2} - \frac{1}{n_2^2}\right)$$

The symbol λ stands for the wavelength, R_H is a constant (109,678 cm^{-1}), and n_1 and n_2 are variables whose values are whole numbers that range from 1 to ∞. The only restriction is that the value of n_2 must be larger than n_1. (This assures that the calculated wavelength has a positive value.) Thus, if $n_1 = 1$, acceptable values of n_2 are 2, 3, 4, . . . , ∞. The Rydberg constant, R_H, is an *empirical constant,* which means its value was chosen so that the equation gives values for λ that match the ones determined experimentally. The use of the Rydberg equation is straightforward, as illustrated in the following example.

EXAMPLE 8.3
Calculating the Wavelength of a Line in the Hydrogen Spectrum

The lines in the visible portion of the hydrogen spectrum are called the *Balmer series,* for which $n_1 = 2$ in the Rydberg equation. Calculate, to four significant figures, the wavelength in nanometers of the spectral line in this series for which $n_2 = 4$.

ANALYSIS and SOLUTION: To solve this problem, we substitute values into the Rydberg equation, which will give us $1/\lambda$. Taking the reciprocal will then give the wavelength. As usual, we must be careful with the units.

Substituting $n_1 = 2$ and $n_2 = 4$ into the Rydberg equation gives

$$\frac{1}{\lambda} = 109{,}678 \text{ cm}^{-1}\left(\frac{1}{2^2} - \frac{1}{4^2}\right)$$

$$= 109{,}678 \text{ cm}^{-1}\left(\frac{1}{4} - \frac{1}{16}\right)$$

$$= 109{,}678 \text{ cm}^{-1}(0.2500 - 0.0625)$$

$$= 109{,}678 \text{ cm}^{-1}(0.1875)$$

$$= 2.056 \times 10^4 \text{ cm}^{-1}$$

Taking the reciprocal gives the wavelength in centimeters.

$$\lambda = \frac{1}{2.056 \times 10^4 \text{ cm}^{-1}}$$

$$= 4.864 \times 10^{-5} \text{ cm}$$

Finally, we convert to nanometers.

$$\lambda = 4.864 \times 10^{-5} \text{ cm} \times \frac{10^{-2} \text{ m}}{1 \text{ cm}} \times \frac{1 \text{ nm}}{10^{-9} \text{ m}}$$

$$= 486.4 \text{ nm}$$

Is the Answer Reasonable?
Besides double-checking the arithmetic, we can check the answer against the experimental spectrum of hydrogen in Figure 8.8. This wavelength corresponds to the turquoise line in the hydrogen spectrum.

PRACTICE EXERCISE 3: Calculate the wavelength in nanometers of the spectral line in the visible spectrum of hydrogen for which $n_1 = 2$ and $n_2 = 3$. What color is this line?

The discovery of the Rydberg equation was both exciting and perplexing. The fact that the wavelength of any line in the hydrogen spectrum can be calculated by a simple equation involving just one constant and the reciprocals of the squares of two whole numbers is remarkable. What is there about the behavior of the electron in the atom that could account for such simplicity?

The energy of electrons in atoms is quantized

$E = h\nu$

$h = 6.626 \times 10^{-34}\,\text{J s}$

$E = (6.626 \times 10^{-34}\,\text{J s})$
$\times (4.567 \times 10^{14}\,\text{s}^{-1})$
$= 3.026 \times 10^{-19}\,\text{J}$

Earlier you saw that there is a simple relationship between the frequency of light and its energy, $E = h\nu$. Because excited atoms emit light of only certain specific frequencies, it must be true that only certain characteristic energy changes are able to take place within the atoms. For instance, in the spectrum of hydrogen there is a red line (see Figure 8.8) that has a wavelength of 656.4 nm and a frequency of 4.567×10^{14} Hz. As shown in the margin, the energy of a photon of this light is 3.026×10^{-19} J. Whenever a hydrogen atom emits red light, the frequency of the light is always precisely 4.567×10^{14} Hz and the energy of the atom decreases by *exactly* 3.026×10^{-19} J, never more and never less. Atomic spectra, then, tell us that *when an excited atom loses energy, not just any arbitrary amount can be lost.* The same is true if the atom gains energy.

How is it that atoms of a given element always undergo exactly the same specific energy changes? The answer seems to be that in an atom an electron can have only certain definite amounts of energy and no others. We say that the electron is restricted to certain **energy levels** and that the energy of the electron is **quantized.**

The energy of an electron in an atom might be compared to the energy of the tortoise in the zoo exhibit shown in Figure 8.9b. The tortoise trapped inside the zoo exhibit can only be "stable" on one of the ledges, so it has certain specific amounts of potential energy as determined by the "energy levels" of the various ledges. If the tortoise is raised to a higher ledge, its potential energy is increased. When it drops to a lower ledge, its potential energy decreases. If the tortoise tries to occupy heights between the ledges, it immediately falls to the lower ledge. Therefore, the energy changes for the tortoise are restricted to the differences in potential energy between the ledges.

The potential energy of the tortoise at rest is quantized.

So it is with an electron in an atom. The electron can only have energies corresponding to the set of electron energy levels in the atom. When the atom is supplied with energy (by an electric discharge, for example), an electron is raised from a low-energy level to a higher one. When the electron drops back, energy equal to the difference between the two levels is released and emitted as a photon. Because only certain energy jumps can occur, only certain frequencies of light can appear in the spectrum.

The existence of specific energy levels in atoms, as implied by atomic spectra, forms the foundation of all theories about electronic structure. Any model of the atom that attempts to describe the positions or motions of electrons must also account for atomic spectra.

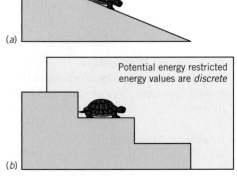

Any potential energy allowed
energy values are *continuous*

(a)

Potential energy restricted
energy values are *discrete*

(b)

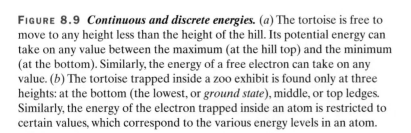

FIGURE 8.9 *Continuous and discrete energies.* (*a*) The tortoise is free to move to any height less than the height of the hill. Its potential energy can take on any value between the maximum (at the hill top) and the minimum (at the bottom). Similarly, the energy of a free electron can take on any value. (*b*) The tortoise trapped inside a zoo exhibit is found only at three heights: at the bottom (the lowest, or *ground state*), middle, or top ledges. Similarly, the energy of the electron trapped inside an atom is restricted to certain values, which correspond to the various energy levels in an atom.

The Bohr model explains the simple pattern of lines seen in the spectrum of hydrogen

The first theoretical model of the hydrogen atom that successfully accounted for the Rydberg equation was proposed in 1913 by Niels Bohr (1885–1962), a Danish physicist. In his model, Bohr likened the electron moving around the nucleus to a planet circling the sun. He suggested that the electron moves around the nucleus along fixed paths, or orbits. His model broke with the classical laws of physics by placing restrictions on the sizes of the orbits and the energy that the electron could have in a given orbit. This ultimately led Bohr to an equation that described the energy of the electron in the atom. The equation includes a number of physical constants such as the mass of the electron, its charge, and Planck's constant. It also contains an integer, n, that Bohr called a **quantum number.** Each of the orbits is identified by its value of n. When all the constants are combined, Bohr's equation becomes

$$E = \frac{-b}{n^2} \tag{8.3}$$

where E is the energy of the electron and b is the combined constant (its value is 2.18×10^{-18} J). The allowed values of n are whole numbers that range from 1 to ∞ (i.e., n could equal 1, 2, 3, 4, . . . , ∞). From this equation the energy of the electron in any particular orbit could be calculated.

Because of the negative sign in Equation 8.3, the lowest (most negative) energy value occurs when $n = 1$, which corresponds to the *first Bohr orbit.* The lowest energy state of an atom is the most stable one and is called the **ground state.** For hydrogen, the ground state occurs when its electron has $n = 1$. According to Bohr's theory, this orbit brings the electron closest to the nucleus. Conversely, an atom with $n = \infty$ would correspond to an "unbound" electron that had escaped from the nucleus. Such an electron has an energy of zero in Bohr's theory. The negative sign in Equation 8.3 ensures that any electron with a finite value of n has a lower energy than an unbound electron. Thus, energy is released when a free electron is bound to a proton to form a hydrogen atom.

When a hydrogen atom absorbs energy, as it does when an electric discharge passes through it, the electron is raised from the orbit having $n = 1$ to a higher orbit, to $n = 2$ or $n = 3$ or even higher. These higher orbits are less stable than the lower ones, so the electron quickly drops to a lower orbit. When this happens, energy is emitted in the form of light (see Figure 8.10). Since the energy of the electron in a given orbit is fixed, a drop from one particular orbit to another—say, from $n = 2$ to $n = 1$—always releases the same amount of energy, and the frequency of the light emitted because of this change is always precisely the same.

The success of Bohr's theory was in its ability to account for the Rydberg equation. When the atom emits a photon, an electron drops from a higher initial

Niels Bohr won the 1922 Nobel Prize in physics for his work on atomic structure.

Classical physical laws, such as those discovered by Issac Newton, place no restrictions on the sizes or energies of orbits.

Bohr's equation for the energy actually is

$$E = -\frac{2\pi^2 m e^4}{n^2 h^2}$$

where m is the mass of the electron, e is the charge on the electron, n is the quantum number, and h is Planck's constant. Therefore, in Equation 8.3,

$$b = \frac{2\pi^2 m e^4}{h^2} = 2.18 \times 10^{-18} \text{ J}$$

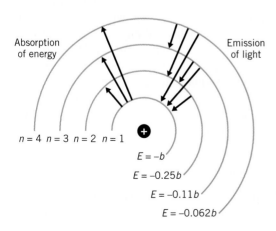

Absorption of energy

Emission of light

$n = 4$ $n = 3$ $n = 2$ $n = 1$

$E = -b$

$E = -0.25b$

$E = -0.11b$

$E = -0.062b$

Figure 8.10 *Absorption of energy and emission of light by the hydrogen atom.* When the atom absorbs energy, the electron is raised to a higher energy level. When the electron falls to a lower energy level, light of a particular energy and frequency is emitted.

energy E_h to a lower final energy E_l. If the initial quantum number of the electron is n_h and the final quantum number is n_l, then the energy change, calculated as a positive quantity, is

$$\Delta E = E_h - E_l$$

$$= \left(\frac{-b}{n_h{}^2}\right) - \left(\frac{-b}{n_l{}^2}\right)$$

This can be rearranged to give

$$\Delta E = b\left(\frac{1}{n_l{}^2} - \frac{1}{n_h{}^2}\right) \qquad \text{with } n_h > n_l$$

By combining Equations 8.1 and 8.2, the relationship between the energy "ΔE" of a photon and its wavelength λ is

$$\Delta E = \frac{hc}{\lambda} = hc\left(\frac{1}{\lambda}\right)$$

Substituting and solving for $1/\lambda$ give

$$\frac{1}{\lambda} = \frac{b}{hc}\left(\frac{1}{n_l{}^2} - \frac{1}{n_h{}^2}\right) \qquad \text{with } n_h > n_l$$

Notice how closely this equation derived from Bohr's theory matches the Rydberg equation, which was obtained solely from the experimentally measured atomic spectrum of hydrogen. Equally satisfying is that the combination of constants, b/hc, has a value of 109,730 cm^{-1}, which differs by only 0.05% from the experimentally derived value of R_H in the Rydberg equation.

The Bohr model fails for atoms with more than one electron

Bohr's model of the atom was both a success and a failure. By calculating the energy changes that occur between energy levels, Bohr was able to account for the Rydberg equation and, therefore, for the atomic spectrum of hydrogen. However, the theory was not able to explain quantitatively the spectra of atoms with more than one electron, and all attempts to modify the theory to make it work met with failure. Gradually, it became clear that Bohr's picture of the atom was flawed and that another theory would have to be found. Nevertheless, the concepts of quantum numbers and fixed energy levels were important steps forward.

8.3 ▶ **Electron waves in atoms are called orbitals**

All the objects that had been studied by scientists until the time of Bohr were large and massive in comparison with the electron, so no one had detected the limits of classical physics.

De Broglie was awarded a Nobel Prize in 1929.

Bohr's efforts to develop a theory of electronic structure were doomed from the very beginning because the classical laws of physics—those known in his day—simply do not apply to objects as small as the electron. Classical physics fails for atomic particles because matter is not really as our physical senses perceive it. When bound inside an atom, electrons behave not like solid particles, but instead like waves. This idea was proposed in 1924 by a young French graduate student, Louis de Broglie.

In Section 8.1 you learned that light waves are characterized by their wavelengths and their frequencies. The same is true of matter waves. De Broglie suggested that the wavelength of a matter wave, λ, is given by the equation

$$\lambda = \frac{h}{mv} \tag{8.4}$$

where h is Planck's constant, m is the particle's mass, and v is its velocity. Notice that this equation allows us to connect a wave property, wavelength, with particle

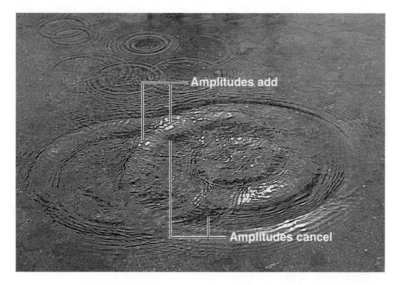

FIGURE 8.11 *Diffraction of water waves on the surface of a pond.* As the waves cross, the amplitudes increase where the waves are in phase and cancel where they are out of phase.

properties, mass and velocity. We may describe the electron either as a particle or a wave, and the de Broglie relationship provides a link between the two descriptions.

When first encountered, the concept of a particle of matter behaving as a wave rather than as a solid object is difficult to comprehend. This book certainly seems solid enough, especially if you drop it on your toe! The reason for the book's apparent solidity is that in de Broglie's equation (Equation 8.4) the mass appears in the denominator. This means that heavy objects have extremely short wavelengths. The peaks of the matter waves for heavy objects are so close together that the wave properties go unnoticed and can't even be measured experimentally. But tiny particles with very small masses have much longer wavelengths, so their wave properties become an important part of their overall behavior.

Perhaps by now you've begun to wonder if there is any way to *prove* that matter has wave properties. Actually, these properties can be demonstrated by a phenomenon that you have probably witnessed. When raindrops fall on a quiet pond, ripples spread out from where the drops strike the water, as shown in Figure 8.11. When two sets of ripples cross, there are places where the waves are *in phase*,

Gigantic waves, called *rogue waves*, with heights up to 100 ft have been observed in the ocean and are believed to be formed when a number of wave sets moving across the sea become in phase simultaneously.

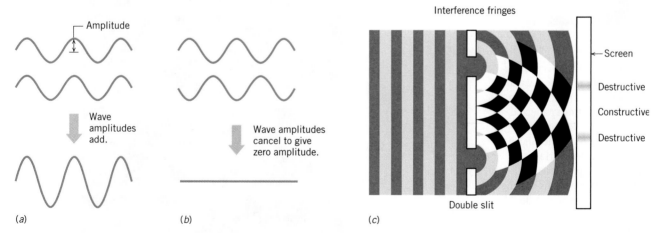

FIGURE 8.12 *Constructive and destructive interference.* (*a*) Waves in phase produce constructive interference and an increase in intensity. (*b*) Waves out of phase produce destructive interference and yield cancellation of intensity. (*c*) Light waves passing through two pinholes fan out and interfere with each other, producing an interference pattern characteristic of waves.

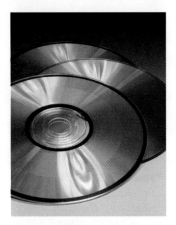

FIGURE 8.13 *Diffraction of light from a compact disc.* Colored interference fringes are produced by the diffraction of reflected light from the closely spaced grooves on the surface of a compact disc.

which means that the peak of one wave coincides with the peak of the other. At these points the intensities of the waves add and the height of the water is equal to the sum of the heights of the two crossing waves. At other places the crossing waves are *out of phase,* which means the peak of one wave occurs at the trough of the other. In these places the intensities of the waves cancel. This reinforcement and cancellation of wave intensities, referred to, respectively, as *constructive* and *destructive interference,* is a phenomenon called **diffraction.** It is examined more closely in Figure 8.12. Note how diffraction creates characteristic **interference fringes** when waves pass through adjacent pinholes or reflect off closely spaced grooves. You have seen interference fringes yourself if you've ever noticed the rainbow of colors that shine from the surface of a compact disc (Figure 8.13). When white light that contains radiations of all wavelengths is reflected from the closely spaced "grooves" on the CD, it is divided into many individual light beams. The light waves in these beams experience interference with each other, and for a given angle between the incoming and reflected light, all colors cancel except for one, which reinforces. Because we only see colors that reinforce, as the angle changes, so do the colors of the light we see.

Diffraction is a phenomenon that can only be explained as a property of waves, and we have seen how it can be demonstrated with water waves and light waves. Experiments can also be done to show that electrons, protons, and neutrons experience diffraction, which demonstrates their wave nature (see Figure 8.14). In fact, electron diffraction is the principle on which the electron microscope is based.

Bound electrons have quantized energies because they behave like standing waves

Before we can discuss how electron waves behave in atoms, we need to know a little more about waves in general. There are basically two kinds of waves, *traveling waves* and *standing waves*. On a lake or ocean the wind produces waves whose

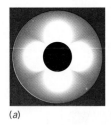

(a)

FIGURE 8.14 *Experimental evidence of the wave behavior of electrons.* (*a*) An electron diffraction pattern collected by reflecting a beam of electrons from crystalline silicon (Semiconductor Surface Physics Group, Queens University). (*b*) Electrons passing *one at a time* through a double slit. Each spot shows an electron impact on a detector. As more and more electrons are passed through the slits, interference fringes are observed.

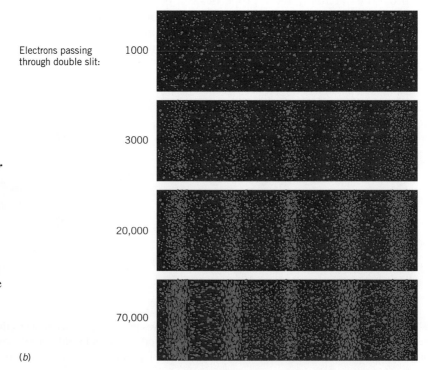

Electrons passing through double slit:

1000

3000

20,000

70,000

(b)

crests and troughs move across the water's surface, as shown in Figure 8.15. The water moves up and down while the crests and troughs travel horizontally in the direction of the wind. These are examples of **traveling waves.**

A more important kind of wave for us is the standing wave. An example is the vibrating string of a guitar. When the string is plucked, its center vibrates up and down while the ends, of course, remain fixed. The crest, or point of maximum amplitude of the wave, occurs at one position. At the ends of the string are points of zero amplitude, called **nodes,** and their positions are also fixed. A **standing wave,** then, is one in which the crests and nodes do not change position. One of the interesting things about standing waves is that they lead naturally to "quantum numbers." Let's see how this works using the guitar as an example.

As you know, many notes can be played on a guitar string by shortening its effective length with a finger placed at frets along the neck of the instrument. But even without shortening the string, we can play a variety of notes. For instance, if the string is touched momentarily at its midpoint at the same time it is plucked, the string vibrates as shown in Figure 8.16 and produces a tone an octave higher. The wave that produces this higher tone has a wavelength exactly half of that formed when the untouched string is plucked. In Figure 8.16 we see that other wavelengths are possible, too, and each gives a different note.

If you examine Figure 8.16, you will see that there are some restrictions on the wavelengths that can exist. Not just any wavelength is possible because the nodes at either end of the string are in fixed positions. The only waves that can occur are those for which a half-wavelength is repeated *exactly* a whole number of times. Expressed another way, the length of the string is a whole-number multiple of half-wavelengths. In a mathematical form we could write this as

$$L = n\left(\frac{\lambda}{2}\right)$$

where L is the length of the string, λ is a wavelength (therefore, $\lambda/2$ is half the wavelength), and n is an integer. Rearranging this to solve for the wavelength gives

$$\lambda = \frac{2L}{n} \qquad (8.5)$$

We see that the waves that are possible are determined quite naturally by a set of whole numbers (similar to quantum numbers).

We are now in a position to demonstrate how quantum theory unites wave and particle descriptions to build a simple but accurate model of a bound electron. Let's look at an electron that is confined to a wire of length L. To keep things simple, let's assume that the wire is infinitely thin, so that the electron can only move in straight lines along the wire. The wire is clamped in place at either end, and its ends cannot move up or down.

First, let's consider a classical particle model: the "bead on a wire" model shown in Figure 8.17a. The bead can slide in either direction along the wire, like a bead on an abacus. If the electron's mass is m and its velocity is v, its kinetic energy is given by

$$E = \frac{1}{2}mv^2$$

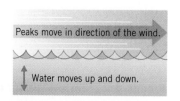

Peaks move in direction of the wind.

Water moves up and down.

FIGURE 8.15 *Traveling waves.*

The wavelength of this wave is actually twice the length of the string.

Notes played on a guitar rely on standing waves. The ends of the strings correspond to nodes of the standing waves. Different notes can be played by shortening the effective lengths of the strings with fingers placed along the neck of the instrument.

Notes played this way are called *harmonics.*

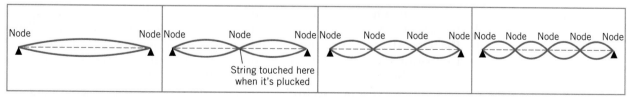

FIGURE 8.16 *Standing waves on a guitar string.*

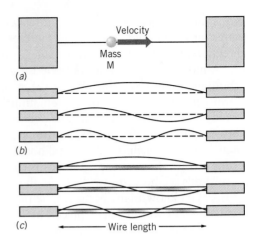

FIGURE 8.17 *Three models of an electron on an infinitely thin wire of length L.* (*a*) a classical model of the electron as a bead that can slide along the wire. Any energy is possible, even zero, and the exact position and velocity of the bead can be known simultaneously. (*b*) a classical model of the electron as a standing wave on a wire. An integer number of peaks and troughs (*n*) is required. The wavelength is restricted to values given by Equation 8.5. (*c*) a quantum mechanical model of the electron on a wire obtained by uniting model (*a*) with model (*b*), using the de Broglie relationship (Equation 8.4). The yellow areas indicate most probable positions for the electron. Energy is quantized and never zero.

The bead can have any velocity, even zero, so the energy E can have any value, even zero. No position on the wire is any more favorable than any other, and the bead is equally likely to be found anywhere on the wire. There is no reason why the bead's position and velocity cannot be known simultaneously.

Now consider a classical wave along the wire, Figure 8.17*b*. It is exactly like the guitar string we looked at in Figure 8.16. The ends of the wire are clamped in place, so there *must* be a whole number of peaks and troughs along the wire. The wavelength is restricted to values calculated by Equation 8.5. We can see that the quantum number, n, is just the number of peaks and troughs along the wave. It has integer values (1, 2, 3, . . .) because you can't have half a peak or half a trough.

Now let's use the de Broglie relation, Equation 8.4, to unite these two classical models. Our goal is to derive an expression for the energy of an electron trapped on the wire. Notice that to calculate the kinetic energy of the bead on a wire, we need to know its velocity, a particle property. The de Broglie relation lets us relate particle velocities with wavelengths. Rearranging Equation 8.4, we have

$$v = \frac{h}{m\lambda}$$

and inserting this equation for kinetic energy gives

$$E = \frac{1}{2}m\left(\frac{h}{m\lambda}\right)^2$$

$$= \frac{h^2}{2m\lambda^2}$$

This equation will give us the energy of the electron from its wavelength. If we substitute in Equation 8.5, which gives the wavelength of the standing wave in terms of the wire length L and the quantum number n, we have

$$E = \frac{n^2h^2}{8mL^2} \tag{8.6}$$

This equation has a number of profound implications. The fact that the electron's energy depends on an integer, n, means that *only certain energy states are allowed*. The allowed states are plotted on the energy-level diagram shown in Figure 8.18. The lowest value of n is 1, so the lowest energy level (the ground state) is $E = h^2/8mL^2$. Energies lower than this are not allowed, so the energy cannot be zero! This indicates that the electron will always have some residual kinetic energy. The electron is never at rest. This is true for the electron trapped in a wire and it is also

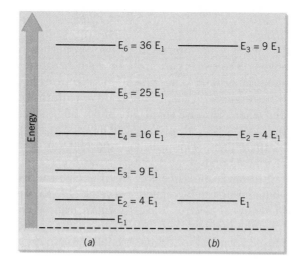

FIGURE 8.18 *Energy-level diagram for the electron-on-a-wire model.* (*a*) A long wire, with $L = 2$ nm. (*b*) A short wire, with $L = 1$ nm. Notice how the energy levels become more closely spaced when the electron has more room to move. Also notice that the energy in the ground state is not zero.

true for an electron trapped in an atom. Thus, *quantum theory resolves the collapsing atom paradox.*

Note that the spacing between energy levels is proportional to $1/L^2$. This means that when the wire is made longer, the energy levels become more closely spaced. In general, *the more room an electron has to move in, the smaller the spacings between its energy levels.* Chemical reactions sometimes change the way that electrons are confined in molecules. This causes changes in the wavelengths of light the reacting mixture absorbs. This is why color changes sometimes occur during chemical change.

The wave that corresponds to the electron is called a **wave function.** The wave function is usually represented by the symbol ψ (Greek letter *psi*). The wave function can be used to describe the shape of the electron wave and its energy. The wave function is not an oscillation of the wire, like a guitar wave, nor is it an electromagnetic wave. The wave's amplitude at any given point can be related to the probability of finding the electron there.

The electron waves shown in Figure 8.17*c* show that unlike the bead-on-a-wire model, the electron is more likely to be found at some places on the wire than others. For the ground state, with $n = 1$, the electron is most likely to be found in the center of the wire. Where the amplitude is zero—for example, at the ends of the wire or the center of the wire in the $n = 2$ state—there is a zero probability of finding the electron! Points where the amplitude of the electron wave is zero are called *nodes.* Notice that the higher the quantum number n, the more nodes the electron wave has, and from Equation 8.6, the more energy the electron has. It is generally true that *the more nodes an electron wave has, the higher its energy.*

The probability of finding an electron at a given point is proportional to the amplitude of the electron wave squared. Thus, peaks and troughs in the electron wave indicate places where there is the greatest buildup of negative charge.

Electron waves in atoms are called *orbitals*

In 1926 Erwin Schrödinger (1887–1961), an Austrian physicist, became the first scientist to successfully apply the concept of the wave nature of matter to an explanation of electronic structure. His work and the theory that developed from it are highly mathematical. Fortunately, we need only a qualitative understanding of electronic structure, and the main points of the theory can be understood without all the math.

Schrödinger developed an equation that can be solved to give wave functions and energy levels for electrons trapped inside atoms. Wave functions for electrons in atoms are called **orbitals.** Not all of the energies of the waves are different, but most are. *Energy changes within an atom are simply the result of an*

Schrödinger won a Nobel Prize in 1933 for his work. The equation he developed that gives electronic wave functions and energies is known as *Schrödinger's equation.* The equation is extremely difficult to solve. Even an approximate solution of the equation for large molecules can require hours or days of supercomputer time.

electron changing from a wave pattern with one energy to a wave pattern with a different energy.

Electron waves are described by the term *orbital* to differentiate them from the notion of *orbits*, which was part of the Bohr model of the atom.

We will be interested in two properties of orbitals, their energies and their shapes. Their energies are important because when an atom is in its most stable state (its **ground state**), the atom's electrons have waveforms with the lowest possible energies. The shapes of the wave patterns (i.e., where their amplitudes are large and where they are small) are important because the theory tells us that the amplitude of a wave at any particular place is related to the likelihood of finding the electron there. This will be important when we study how and why atoms form chemical bonds to each other.

"Most stable" almost always means "lowest energy."

In much the same way that the characteristics of a wave on a one-dimensional guitar string can be related to a single integer, wave mechanics tells us that the three-dimensional electron waves (orbitals) can be characterized by a set of *three* integer quantum numbers, n, ℓ, and m_ℓ. In discussing the energies of the orbitals, it is usually most convenient to sort the orbitals into groups according to these quantum numbers.

The principal quantum number, n

The term *shell* comes from an early notion that atoms could be thought of as similar to onions, with the electrons being arranged in layers around the nucleus.

The quantum number n is called the **principal quantum number,** and all orbitals that have the same value of n are said to be in the same **shell.** The values of n can range from $n = 1$ to $n = \infty$. The shell with $n = 1$ is called the *first shell,* the shell with $n = 2$ is the *second shell,* and so forth. The various shells are also sometimes identified by letters, beginning (for no significant reason) with K for the first shell ($n = 1$), L for the second shell ($n = 2$), and so on.

The principal quantum number is related to the size of the electron wave (i.e., how far the wave effectively extends from the nucleus). The higher the value of n, the larger is the electron's average distance from the nucleus. This quantum number is also related to the energy of the orbital. As n increases, the energies of the orbitals also increase.

Bohr was fortunate to have used the element hydrogen to develop his model of the atom. If he had chosen a different element, his model would not have worked.

Bohr's theory took into account only the principal quantum number n. His theory worked fine for hydrogen because hydrogen just happens to be the one element in which all orbitals having the same value of n also have the same energy. Bohr's theory failed for atoms other than hydrogen, however, because when the atom has more than one electron, orbitals with the same value of n can have different energies.

The secondary quantum number, ℓ

ℓ is also called the *azimuthal quantum number* and the *orbital angular momentum number.*

The **secondary quantum number, ℓ,** divides the shells into smaller groups of orbitals called **subshells.** The value of n determines which values of ℓ are allowed. For a given n, ℓ can range from $\ell = 0$ to $\ell = (n - 1)$. Thus, when $n = 1$, $(n - 1) = 0$, so the only value of ℓ that's allowed is zero. This means that when $n = 1$, there is only one subshell (the shell and subshell are really identical). When $n = 2$, ℓ can have values of 0 or 1. (The maximum value of $\ell = n - 1 = 2 - 1 = 1$.) This means that when $n = 2$, there are two subshells. One has $n = 2$ and $\ell = 0$, and the other has $n = 2$ and $\ell = 1$. The relationship between n and the allowed values of ℓ are summarized in the table in the margin.

Subshells could be identified by their value of ℓ. However, to avoid confusing numerical values of n with those of ℓ, a letter code is normally used to specify the value of ℓ.

Relationship between n and ℓ

Value of n	Values of ℓ
1	0
2	0, 1
3	0, 1, 2
4	0, 1, 2, 3
5	0, 1, 2, 3, 4
n	0, 1, 2, . . . , $(n - 1)$

Value of ℓ	0	1	2	3	4	5	. . .
Letter designation	s	p	d	f	g	h	. . .

The number of subshells in a given shell equals the value of n for that shell. For example, when $n = 3$, there are three subshells.

To designate a particular subshell, we write the value of its principal quantum number followed by the letter code for the subshell. For example, the subshell with $n = 2$ and $\ell = 1$ is the $2p$ subshell; the subshell with $n = 4$ and $\ell = 0$ is the $4s$

subshell. Notice that because of the relationship between n and ℓ, every shell has an s subshell ($1s$, $2s$, $3s$, etc.). All the shells except the first have a p subshell ($2p$, $3p$, $4p$, etc.). All but the first and second shells have a d subshell ($3d$, $4d$, etc.), and so forth.

PRACTICE EXERCISE 4: What subshells would be found in the shells with $n = 3$ and $n = 4$?

The secondary quantum number determines the shape of the orbital, which we will examine more closely later. Except for the special case of hydrogen, which has only one electron, the value of ℓ also affects the energy. This means that in atoms with two or more electrons, the subshells within a given shell differ slightly in energy, with the energy of the subshell increasing with increasing ℓ. Therefore, within a given shell, the s subshell is lowest in energy, p is the next lowest, followed by d, then f, and so on. For example,

$$4s < 4p < 4d < 4f$$
— increasing energy →

The magnetic quantum number, m_ℓ

The third quantum number, m_ℓ, is known as the **magnetic quantum number.** It divides the subshells into individual orbitals, and its values are related to the way the individual orbitals are oriented relative to each other in space. As with ℓ, there are restrictions as to the possible values of m_ℓ, which can range from $+\ell$ to $-\ell$. When $\ell = 0$, m_ℓ can have only the value 0 because $+0$ and -0 are the same. An s subshell, then, has just a single orbital. When $\ell = 1$, the possible values of m_ℓ are $+1, 0$, and -1. A p subshell therefore has three orbitals: one with $\ell = 1$ and $m_\ell = 1$, another with $\ell = 1$ and $m_\ell = 0$, and a third with $\ell = 1$ and $m_\ell = -1$. Similarly, we find that a d subshell has five orbitals and an f subshell has seven orbitals. The numbers of orbitals in the subshells are easy to remember because they follow a simple arithmetic progression.

$$
\begin{array}{ccccc}
s & p & d & f & \ldots \\
1 & 3 & 5 & 7 & \ldots
\end{array}
$$

Spectroscopists used m_ℓ to explain additional lines that appear in atomic spectra when atoms emit light while in a magnetic field, which explains how this quantum number got its name.

PRACTICE EXERCISE 5: How many orbitals are there in a g subshell?

The whole picture

The relationships among all three quantum numbers are summarized in Table 8.1. In addition, the relative energies of the subshells in an atom containing two or

TABLE 8.1	SUMMARY OF RELATIONSHIPS AMONG THE QUANTUM NUMBERS n, ℓ, AND m_ℓ			
Value of n	Value of ℓ	Values of m_ℓ	Subshell	Number of Orbitals
1	0	0	$1s$	1
2	0	0	$2s$	1
	1	$-1, 0, 1$	$2p$	3
3	0	0	$3s$	1
	1	$-1, 0, 1$	$3p$	3
	2	$-2, -1, 0, 1, 2$	$3d$	5
4	0	0	$4s$	1
	1	$-1, 0, 1$	$4p$	3
	2	$-2, -1, 0, 1, 2$	$4d$	5
	3	$-3, -2, -1, 0, 1, 2, 3$	$4f$	7

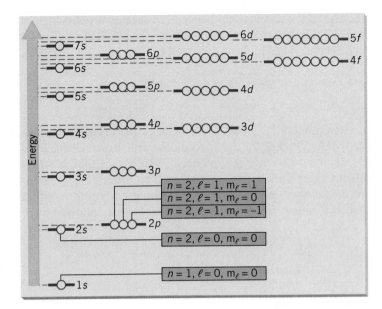

Figure 8.19 *Approximate energy-level diagram for atoms with two or more electrons.* The quantum numbers associated with the orbitals in the first two shells are also shown.

more electrons are depicted in Figure 8.19. Several important features should be noted. First, observe that each orbital on this energy diagram is indicated by a separate circle—one for an *s* subshell, three for a *p* subshell, and so forth. Second, notice that all the orbitals of a given subshell have the *same* energy. Third, note that, in going upward on the energy scale, the spacing between successive shells decreases as the number of subshells increases. This leads to the overlapping of shells having different values of *n*. For instance, the 4*s* subshell is lower in energy than the 3*d* subshell, 5*s* is lower than 4*d*, and 6*s* is lower than 5*d*. In addition, the 4*f* subshell is below the 5*d* subshell and 5*f* is below 6*d*.

We will see shortly that Figure 8.19 is very useful for predicting the electronic structures of atoms. Before discussing this, however, we must study another very important property of the electron, a property called *spin*. Electron spin gives rise to a fourth quantum number.

8.4 ▶ **Electron spin affects the distribution of electrons among orbitals in atoms**

In an atom, an electron can take on many different energies and wave shapes, each of which is called an orbital that is identified by a set of values for *n*, ℓ, and m_ℓ. When the electron wave possesses a given set of *n*, ℓ, and m_ℓ, we say the electron "occupies the orbital" with that set of quantum numbers.

Earlier it was stated that an atom is in its most stable state (its ground state) when its electrons have the lowest possible energies. This occurs when the electrons "occupy" the lowest energy orbitals that are available. But what determines how the electrons "fill" these orbitals? Fortunately, there are some simple rules that can help. These govern both the maximum number of electrons that can be in a particular orbital and how orbitals with the same energy become filled. One important factor that influences the distribution of electrons is the phenomenon known as *electron spin*.

Electrons behave like tiny charges that can spin in one of two directions

When a beam of atoms with an odd number of electrons is passed through an uneven magnetic field, the beam is split in two, as shown in Figure 8.20. The splitting occurs because the electrons within the atoms interact with the magnetic field in two different ways. The electrons behave like tiny magnets, and they are attracted to one or the other of the poles depending on their orientation. This can be explained by imagining that an electron spins around its axis, like a toy top. Recall that a moving charge creates a moving electric field, which in turn creates a magnetic field. The

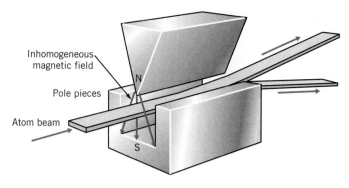

Inhomogeneous magnetic field

Pole pieces

Atom beam

FIGURE 8.20 *The discovery of electron spin.* In this classic experiment by Stern and Gerlach, a beam of atoms with an odd number of electrons is passed through an uneven magnetic field, created by magnet pole faces of different shapes. The beam splits in two, indicating that the electrons in the atoms behave as tiny magnets, which are attracted to one or the other of the poles depending on their orientation. "Electron spin" was proposed to account for the two possible orientations for the electron's magnetic field.

spinning electrical charge of the electron creates its own magnetic field. This **electron spin** could occur in two possible directions, which accounts for the two beams.

Electron spin gives us a fourth quantum number for the electron, called the **spin quantum number, m_s,** which can take on two possible values: $m_s = +\frac{1}{2}$ or $m_s = -\frac{1}{2}$, corresponding to the two beams in Figure 8.20. The actual values of m_s and the reason they are not integers aren't very important to us, but the fact that there are *only* two values is very significant.

Electrons don't actually spin. We've seen that electrons are not simply particles. But the magnetic properties of electrons are just what we'd see if the electron were a spinning charged particle, and it is useful to picture the electron as spinning.

No two electrons in an atom have identical sets of quantum numbers

In 1925 an Austrian physicist, Wolfgang Pauli (1900–1958), expressed the importance of electron spin in determining electronic structure. The **Pauli exclusion principle** states that *no two electrons in the same atom can have identical values for all four of their quantum numbers.* To understand the significance of this, suppose two electrons were to occupy the 1s orbital of an atom. Each electron would have $n = 1$, $\ell = 0$, and $m_\ell = 0$. Since these three quantum numbers are the same for both electrons, the exclusion principle requires that their fourth quantum numbers (their spin quantum numbers) be different; one electron must have $m_s = +\frac{1}{2}$ and the other, $m_s = -\frac{1}{2}$. No more than two electrons can occupy the 1s orbital of the atom simultaneously because there are only two possible values of m_s. Thus the Pauli exclusion principle is really telling us that *the maximum number of electrons in any orbital is two,* and that *when two electrons are in the same orbital, they must have opposite spins.*

The limit of two electrons per orbital also limits the maximum electron populations of the shells and subshells. For the subshells we have

Pauli received the 1945 Nobel Prize in physics for his discovery of the exclusion principle.

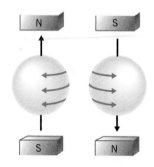

The electron can spin in either of two directions in the presence of an external magnetic field.

Subshell	Number of Orbitals	Maximum Number of Electrons
s	1	2
p	3	6
d	5	10
f	7	14

The maximum electron population per shell is shown below.

Shell	Subshells	Maximum Shell Population	
1	1s	2	
2	2s 2p	8	(2 + 6)
3	3s 3p 3d	18	(2 + 6 + 10)
4	4s 4p 4d 4f	32	(2 + 6 + 10 + 14)

Remember that a shell is a group of orbitals with the same value of *n*. A subshell is a group of orbitals with the same values of *n* and ℓ.

In general, the maximum electron population of a shell is $2n^2$.

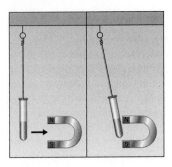

A paramagnetic substance is attracted to a magnetic field.

Diamagnetic substances are actually weakly repelled by a magnetic field.

Atoms with unpaired electrons are weakly attracted to magnets

We have seen that when two electrons occupy the same orbital they must have different values of m_s. When this occurs, we say that the spins of the electrons are *paired*, or simply that the electrons are *paired*. Such pairing leads to the cancellation of the magnetic effects of the electrons because the north pole of one electron magnet is opposite the south pole of the other. Atoms with more electrons that spin in one direction than in the other are said to contain *unpaired* electrons. For these atoms, the magnetic effects do not cancel and the atoms themselves become tiny magnets that can be attracted to an external magnetic field. This weak attraction of a substance containing unpaired electrons to a magnet is called **paramagnetism.** Substances in which all the electrons are paired are not attracted to a magnet and are said to be **diamagnetic.**

Paramagnetism and diamagnetism are measurable properties that provide experimental verification of the presence or absence of unpaired electrons in substances. In addition, the quantitative measurement of the strength of the attraction of a paramagnetic substance toward a magnetic field permits the calculation of the number of unpaired electrons in its atoms, molecules, or ions.

8.5 ▶ The ground state electron configuration is the lowest energy distribution of electrons among orbitals

The distribution of electrons among the orbitals of an atom is called the atom's **electronic structure** or **electron configuration.** This is something very useful to know about an element because the arrangement of electrons in the outer parts of an atom, which is determined by its electron configuration, controls the chemical properties of the element.

We shall be interested in the ground state electron configurations of the elements. This is the configuration that yields the lowest energy for an atom and can be predicted for many of the elements by the use of the energy-level diagram in Figure 8.19 and application of the Pauli exclusion principle. To see how we go about this, let's begin with the simplest atom of all, hydrogen.

Hydrogen has an atomic number, Z, equal to 1, so a neutral hydrogen atom has one electron. In its ground state this electron occupies the lowest energy orbital that's available, which is the $1s$ orbital. To indicate symbolically the electron configuration we list the subshells that contain electrons and indicate their electron populations by appropriate superscripts. Thus the electron configuration of hydrogen is written as

$$\text{H} \qquad 1s^1$$

Another way of expressing electron configurations that we will sometimes find useful is the **orbital diagram.** In it, a circle will represent each orbital and arrows will be used to indicate the individual electrons, head up for spin in one direction and head down for spin in the other. The orbital diagram for hydrogen is simply

$$\text{H} \qquad \textcircled{\uparrow}$$
$$1s$$

It doesn't matter whether the arrow points up or down. The energy of the electron is the same whether it has one spin or the other. For consistency, when an orbital is half-filled, we will show the electron with an arrow that points up.

Electronic configurations are built up by filling lowest energy orbitals first

To arrive at the electron configuration of an atom of another element, we imagine that we begin with a hydrogen atom and then add one proton after another (plus whatever neutrons are also needed) until we obtain the nucleus of the atom of interest. As we proceed, we also add electrons, one at a time to the lowest available orbital, until we have added enough electrons to give the neutral atom of the element in question. This imaginary process for obtaining the electronic structure of an atom is known as the **aufbau principle,** the word *aufbau* being German for *building up.* Let's look at the way this works for helium, which has $Z = 2$. This atom has two electrons, both of which are permitted to occupy the $1s$ orbital. The electron configuration of helium can therefore be written as

$$\text{He} \qquad 1s^2 \qquad \text{or} \qquad \text{He} \quad \text{⬆⬇} \atop 1s$$

Notice that the orbital diagram shows that both electrons in the $1s$ orbital are paired.

We can proceed in the same fashion to predict successfully the electron configurations of most of the elements in the periodic table. For example, the next two elements in the table are lithium, Li ($Z = 3$), and beryllium, Be ($Z = 4$), which have three and four electrons, respectively. For each of these, the first two electrons enter the $1s$ orbital with their spins paired. The Pauli exclusion principle tells us that the $1s$ subshell is filled with two electrons, and Figure 8.19 shows that the orbital of next lowest energy is the $2s$, which can also hold up to two electrons. Therefore, the third electron of lithium and the third and fourth electrons of beryllium enter the $2s$. We can represent the electronic structures of lithium and beryllium as

$$\text{Li} \qquad 1s^2 2s^1 \qquad \text{or} \qquad \text{Li} \quad \underset{1s}{\text{⬆⬇}} \ \underset{2s}{\text{⬆}}$$

$$\text{Be} \qquad 1s^2 2s^2 \qquad \text{or} \qquad \text{Be} \quad \underset{1s}{\text{⬆⬇}} \ \underset{2s}{\text{⬆⬇}}$$

After beryllium comes boron, B ($Z = 5$). Referring to Figure 8.19, we see that the first four electrons of this atom complete the $1s$ and $2s$ subshells, so the fifth electron must be placed into the $2p$ subshell.

$$\text{B} \qquad 1s^2 2s^2 2p^1$$

In the orbital diagram for boron, the fifth electron can be put into any one of the $2p$ orbitals—which one doesn't matter because they are all of equal energy.

$$\text{B} \quad \underset{1s}{\text{⬆⬇}} \ \underset{2s}{\text{⬆⬇}} \ \underset{2p}{\text{⬆○○}}$$

Notice, however, that when we give this orbital diagram we show *all* of the orbitals of the $2p$ subshell even though two of them are empty.

Next we come to carbon, which has six electrons. As before, the first four electrons complete the $1s$ and $2s$ orbitals. The remaining two electrons go in the $2p$ subshell to give

$$\text{C} \qquad 1s^2 2s^2 2p^2$$

Now, however, to write the orbital diagram we have to make a decision as to where to put the two p electrons. (At this point you may have an unprintable suggestion! But try to bear up. It's really not all that bad.) To make this decision, we apply **Hund's rule,** which states that *when electrons are placed in a set of orbitals of equal energy, they are spread out as much as possible to give as few paired electrons as possible.* Both theory and experiment have shown that if we follow this rule, we obtain the electron configuration with the lowest energy. For carbon, it

Spread them out and then line them up.

means that the two *p* electrons are in separate orbitals and their spins are in the same direction.[1]

C   ⥮ ⥮ ⥮   ⥮ ⥮ ⥮
1s   2s   2p

It doesn't matter which two orbitals are shown as occupied. Any of these are okay for the ground state of carbon.

2p

Applying the Pauli exclusion principle and Hund's rule, we can now complete the electron configurations and orbital diagrams for the rest of the elements of the second period.

		1s	2s	2p
N	$1s^2 2s^2 2p^3$	⥮	⥮	↑↑↑
O	$1s^2 2s^2 2p^4$	⥮	⥮	⥮↑↑
F	$1s^2 2s^2 2p^5$	⥮	⥮	⥮⥮↑
Ne	$1s^2 2s^2 2p^6$	⥮	⥮	⥮⥮⥮

We can continue to predict electron configurations in this way, using Figure 8.19 as a guide to tell us which subshells become occupied and in what order. For instance, after completing the 2p subshell at neon, Figure 8.19 predicts that the next two electrons enter the 3s, followed by the filling of the 3p. Then we find the 4s lower in energy than the 3d, so it is filled first. Next, the 3d is completed before we go on to fill the 4p, and so forth.

You might be thinking that you'll have to consult Figure 8.19 whenever you need to write down the ground state electron configuration of an element. However, as you will see in the next section, all the information contained in this figure is also contained in the periodic table.

8.6 ▶ Electron configurations explain the structure of the periodic table

In Chapter 1 you learned that when Mendeleev constructed his periodic table, elements with similar chemical properties were arranged in vertical columns called *groups*. Later work led to the expanded version of the periodic table we use today. The basic structure of this table is one of the strongest empirical supports for the quantum theory, which we have been using to predict electron configurations, and it also permits us to use the periodic table itself as a device for predicting electron configurations.

This would be an amazing co-incidence if the theory were wrong!

Consider, for example, the way the table is laid out (Figure 8.21). On the left there is a block of *two* columns of elements, on the right there is a block of *six* columns, in the center there is a block of *ten* columns, and below the table there are two rows consisting of *fourteen* elements each. These numbers—2, 6, 10, and 14— are *precisely* the numbers of electrons that the quantum theory tells us can occupy *s*, *p*, *d*, and *f* subshells, respectively. In fact, *we can use the structure of the periodic table to predict the filling order of the subshells when we write the electron configuration of an element.*

Periodic table and electron configurations

---

[1] Hund's rule gives us the *lowest* energy (ground state) distribution of electrons among the orbitals. However, configurations such as

C   ⥮ ⥮ ↑↓○
C   ⥮ ⥮ ⥮○○
1s   2s   2p

are not impossible, it is just that neither of them corresponds to the lowest energy distribution of electrons in the carbon atom.

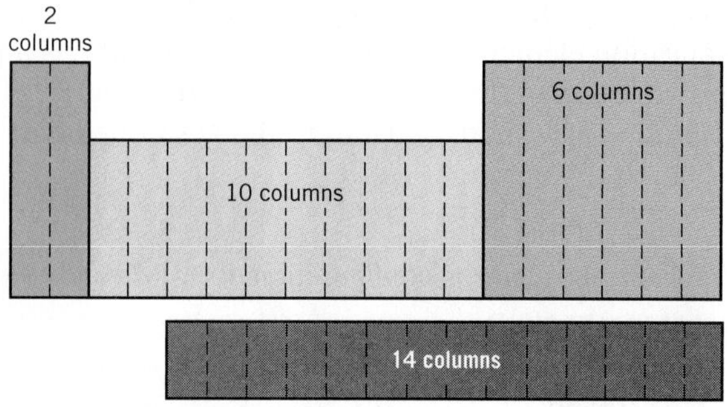

FIGURE 8.21 *The overall structure of the periodic table.* The table is naturally divided into regions of 2, 6, 10, and 14 columns, which are the numbers of electrons that can occupy *s*, *p*, *d*, and *f* subshells.

To use the periodic table to predict electron configurations, we follow the aufbau principle as before. We start with hydrogen and then move through the table row after row until we reach the element of interest, noting as we go along which regions of the table we pass through. For example, consider the element calcium. Using Figure 8.21, we obtain the electron configuration

$$\text{Ca} \qquad 1s^2 2s^2 2p^6 3s^2 3p^6 4s^2$$

To see how this configuration relates to the periodic table, refer to the inside front cover of the book. Notice that the first period has just two elements, H and He. Starting with hydrogen and passing through this period, two electrons are added to the atom. These enter the $1s$ subshell. Next we move across the second period, where the first two elements, Li and Be, are in the block of two columns. As you saw above, the last electron to enter each of these atoms goes into the $2s$ subshell. Then we enter the block of six columns, and as we move across this region in the second row we gradually fill the $2p$ subshell. Now we go to the third period where we first pass through the block of two columns, filling the $3s$ subshell, and then through the block of six columns, filling the $3p$. Finally, we move to the fourth period. To get to calcium we step through two elements in the two-column block, filling the $4s$ subshell.

Notice that every time we move across the block of two columns we add electrons to an s subshell that has a principal quantum number equal to the period number (second period, $2s$; third period, $3s$; fourth period, $4s$). Also note that every time we move across the block of six columns we add electrons to a p subshell that has a principal quantum number equal to the period number (second period, $2p$; third period, $3p$). This is summarized in Figure 8.22.

Figure 8.22 also includes the subshells that are filled when we cross through rows of transition elements and through rows of inner transition elements. For the transition elements, we simply subtract 1 from the period number to find the principal quantum number of the d subshell that's filled. For instance, as we move through the transition elements of *Period 4,* we fill the $3d$ subshell, through those in

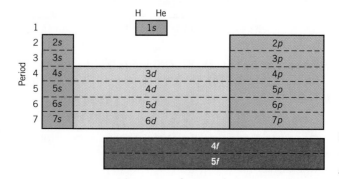

FIGURE 8.22 *Subshells that become filled as we cross periods.*

Period 5 we fill the 4d, and through those in *Period 6* we fill the 5d. For the inner transition elements, we fill an f subshell that has a principal quantum number that is two less than the period number of the element. Thus, as we cross the lanthanides in Period 6, we add electrons to the 4f subshell.

EXAMPLE 8.4 **Predicting Electron** **Configurations**	What is the electron configuration of (a) Mn and (b) Bi?

ANALYSIS: We use the periodic table as a tool to tell us which subshells become filled. It is best if you can do this without referring to Figure 8.22.

SOLUTION: (a) To get to manganese, we cross the following regions of the table, with the results indicated.

Period 1   Fill the 1s subshell.
Period 2   Fill the 2s and 2p subshells.
Period 3   Fill the 3s and 3p subshells.
Period 4   Fill the 4s and then move five places in the 3d region.

The electron configuration of Mn is therefore

$$\text{Mn} \qquad 1s^2 2s^2 2p^6 3s^2 3p^6 4s^2 3d^5$$

Often it is desirable to group all subshells of the same shell together. For manganese, this gives

$$\text{Mn} \qquad 1s^2 2s^2 2p^6 3s^2 3p^6 3d^5 4s^2$$

We'll see later that writing the configuration in this way is convenient when working out ground state electronic configurations for ions.
(b) To get to bismuth, we fill the following subshells:

Period 1   Fill the 1s.
Period 2   Fill the 2s and 2p.
Period 3   Fill the 3s and 3p.
Period 4   Fill the 4s, 3d, and 4p.
Period 5   Fill the 5s, 4d, and 5p.
Period 6   Fill the 6s, 4f, 5d, and then add three electrons to the 6p.

This gives

$$\text{Bi} \qquad 1s^2 2s^2 2p^6 3s^2 3p^6 4s^2 3d^{10} 4p^6 5s^2 4d^{10} 5p^6 6s^2 4f^{14} 5d^{10} 6p^3$$

Grouping subshells with the same value of n gives

$$\text{Bi} \qquad 1s^2 2s^2 2p^6 3s^2 3p^6 3d^{10} 4s^2 4p^6 4d^{10} 4f^{14} 5s^2 5p^6 5d^{10} 6s^2 6p^3$$

Again, either of these configurations is correct.

Are the Answers Reasonable?
There's not much to check here. One thing we can do is to count the total number of electrons represented in each electron configuration. We find they add up to 25 for manganese (as they should) and to 83 for bismuth (as they should).

PRACTICE EXERCISE 6: Use the periodic table to predict the electron configurations of (a) Mg, (b) Ge, (c) Cd, and (d) Gd. Group subshells of the same shell together.

PRACTICE EXERCISE 7: Draw orbital diagrams for (a) Na, (b) S, and (c) Fe.

The periodic recurrence of properties matches the recurrence of outer shell configurations

You learned that Mendeleev constructed the periodic table by placing elements with similar properties in the same group. We are now ready to understand the reason for these similarities in terms of the electronic structures of atoms.

When we consider the chemical reactions of atoms, our attention is usually focused on the distribution of electrons in the **outer shell** of the atom (the occupied shell with the largest value of n). This is because the **outer electrons** (those in the outer shell) are the ones that are exposed to other atoms when the atoms react. The inner electrons of an atom, called the **core electrons,** are buried deep within the atom and normally do not play a role when chemical bonds are formed. It seems reasonable, therefore, that elements with similar properties should have similar outer shell electron configurations. This is, in fact, exactly what we observe. For example, let's look at the alkali metals of Group IA. Going by our rules, we obtain the following configurations:

Li	$1s^2 \mathbf{2s^1}$
Na	$1s^2 2s^2 2p^6 \mathbf{3s^1}$
K	$1s^2 2s^2 2p^6 3s^2 3p^6 \mathbf{4s^1}$
Rb	$1s^2 2s^2 2p^6 3s^2 3p^6 3d^{10} 4s^2 4p^6 \mathbf{5s^1}$
Cs	$1s^2 2s^2 2p^6 3s^2 3p^6 3d^{10} 4s^2 4p^6 4d^{10} 5s^2 5p^6 \mathbf{6s^1}$

Each of these elements has just one outer shell electron that is in an s subshell. We know that when they react, the alkali metals each lose one electron to form ions with a charge of $1+$. For each, the electron that is lost is this outer s electron, and the electron configuration of the ion that is formed is the same as that of the preceding noble gas.

Li$^+$	$1s^2$		He	$1s^2$
Na$^+$	$1s^2 2s^2 2p^6$		Ne	$1s^2 2s^2 2p^6$
K$^+$	$1s^2 2s^2 2p^6 3s^2 3p^6$		Ar	$1s^2 2s^2 2p^6 3s^2 3p^6$
		etc.		

If you write the electron configurations of the members of any of the groups in the periodic table, you will find the same kind of similarity among the configurations of the outer shell electrons. The differences are in the value of the principal quantum number of these outer electrons.

Electron configurations can be abbreviated using noble gas core configurations

Because we are interested primarily in the electrons of the outer shell, we often write electron configurations in an abbreviated, or shorthand, form. To illustrate, let's consider the elements sodium and magnesium. Their full electron configurations are

$$\text{Na} \qquad 1s^2 2s^2 2p^6 3s^1$$

$$\text{Mg} \qquad 1s^2 2s^2 2p^6 3s^2$$

The outer electrons of both atoms are in the $3s$ subshell and each has the core configuration $1s^2 2s^2 2p^6$, which is the same as that of the noble gas neon.

To write the *shorthand configuration* for an element we indicate what the core is by placing in brackets the symbol of the noble gas whose electron configuration is the same as the core configuration. This is followed by the configuration of the outer electrons for the particular element. The noble gas used is almost always the one that occurs at the end of the period preceding the period containing the element whose configuration we wish to represent. Thus, for sodium and magnesium we write

$$Na \quad [Ne]\, 3s^1$$

$$Mg \quad [Ne]\, 3s^2$$

Even with the transition elements, we need only concern ourselves with the outermost shell and the d subshell just below. For example, iron has the configuration

$$Fe \quad 1s^2 2s^2 2p^6 3s^2 3p^6 3d^6 4s^2$$

Only the $4s$ and $3d$ electrons play a role in the chemistry of iron. In general, the electrons below the outer s and d subshells of a transition element—the *core* electrons—are relatively unimportant. In every case these core electrons have the electron configuration of a noble gas. Therefore, for iron the core is $1s^2 2s^2 2p^6 3s^2 3p^6$, which is the same as the electron configuration of argon. The abbreviated configuration of iron is therefore written as

$$Fe \quad [Ar]\, 3d^6 4s^2$$

EXAMPLE 8.5

Writing Shorthand Electron Configurations

What is the shorthand electron configuration of manganese? Draw the orbital diagram for manganese using this shorthand configuration.

ANALYSIS: In a shorthand configuration, we represent the core electrons using the symbol for the noble gas found at the end of the preceding period. This symbol is placed in brackets. Then we use the periodic table to determine the population of the orbitals beyond the noble gas core.

SOLUTION: Manganese is in Period 4. The preceding noble gas is argon, Ar, so to write the abbreviated configuration we write the symbol for argon in brackets followed by the electron configuration that exists beyond the argon core. We can obtain this by noting that to get to Mn in Period 4, we first cross the "s region" by adding two electrons to the $4s$ subshell, and then go five steps into the "d region," where we add five electrons to the $3d$ subshell. Therefore, the shorthand electron configuration for Mn is

$$Mn \quad [Ar]\, 4s^2 3d^5$$

Placing the electrons that are in the highest shell farthest to the right gives

$$Mn \quad [Ar]\, 3d^5 4s^2$$

To draw the orbital diagram, we simply distribute the electrons in the $3d$ and $4s$ orbitals following Hund's rule. This gives

$$Mn \quad [Ar] \quad \uparrow\,\uparrow\,\uparrow\,\uparrow\,\uparrow \quad \uparrow\downarrow$$
$$ 3d 4s$$

Notice that each of the $3d$ orbitals is half-filled. The atom contains five unpaired electrons and is paramagnetic.

Are the Answers Reasonable?
We can double check to be sure we've accounted for all the electrons. To get to manganese in Period 4, we have to move 7 spaces, and that's the total number of s and d electrons shown in the configuration. We can also check to be sure we've followed Hund's rule, and we have.

PRACTICE EXERCISE 8: Write shorthand configurations and abbreviated orbital diagrams for (a) P and (b) Sn. Where appropriate, place the electrons that are in the highest shell farthest to the right. How many unpaired electrons does each of these atoms have?

Chemical properties depend on valence shell electron configurations

For the representative elements (those in the longer columns), the only electrons that are normally important in controlling chemical properties are the ones in the outer shell. This outer shell is known as the **valence shell,** and it is always the occupied shell with the largest value of n. The electrons in the valence shell are called **valence electrons.** (The term *valence* comes from the study of chemical bonding and relates to the combining capacity of an element, but that's not important here.)

For the representative elements it is very easy to determine the electron configuration of the valence shell by using the periodic table. *The valence shell always consists of just the s and p subshells that we encounter crossing the period that contains the element in question.* Thus, to determine the valence shell configuration of sulfur, a Period 3 element, we note that to reach sulfur in Period 3 we place two electrons into the $3s$ and four electrons into the $3p$. The valence shell configuration of sulfur is therefore

$$\text{S} \qquad 3s^2 3p^4$$

EXAMPLE 8.6

Writing Valence Shell Configurations

Predict the electron configuration of the valence shell of arsenic ($Z = 33$).

ANALYSIS: For the valence shell configuration, we are only interested in s and p subshell electrons. The principal quantum number for them is the same as the period number.

SOLUTION: To reach arsenic in Period 4, we add electrons to the $4s$, $3d$, and $4p$ subshells. But the $3d$ is not part of the fourth shell and therefore not part of the valence shell, so all we need be concerned with are the electrons in the $4s$ and $4p$ subshells. This gives us the valence shell configuration of arsenic,

$$\text{As} \qquad 4s^2 4p^3$$

Is the Answer Reasonable?
The total number of valence electrons (i.e., those in valence shell orbitals) equals the group number. Arsenic is in Group V, and that's how many electrons we have shown in the answer, so it is correct.

PRACTICE EXERCISE 9: What is the valence shell electron configuration of (a) Se, (b) Sn, and (c) I?

Configurations for transition and rare earth elements are sometimes unexpected

The rules you've learned for predicting electron configurations work most of the time, but not always. Appendix A gives the electron configurations of all of the elements as determined experimentally. Close examination reveals that there are quite a few exceptions to the rules. Some of these exceptions are important to us because they occur with common elements.

Two important exceptions are for chromium and copper. Following the rules, we would expect the configurations to be

$$\text{Cr} \qquad [\text{Ar}]\, 3d^4 4s^2$$

$$\text{Cu} \qquad [\text{Ar}]\, 3d^9 4s^2$$

However, the actual electron configurations, determined experimentally, are

$$\text{Cr} \qquad [\text{Ar}]\, 3d^5 4s^1$$

$$\text{Cu} \qquad [\text{Ar}]\, 3d^{10} 4s^1$$

The corresponding orbital diagrams are

Cr   [Ar]   ↑ ↑ ↑ ↑ ↑   ↑

Cu   [Ar]   ↑↓ ↑↓ ↑↓ ↑↓ ↑↓   ↑

                    $3d$ $4s$

Notice that for chromium, an electron is "borrowed" from the $4s$ subshell to give a $3d$ subshell that is exactly half-filled. For copper the $4s$ electron is borrowed to give a completely filled $3d$ subshell. A similar thing happens with silver and gold, which have filled $4d$ and $5d$ subshells, respectively.

Ag   [Kr] $4d^{10}5s^1$

Au   [Xe] $4f^{14}5d^{10}6s^1$

Apparently, half-filled and filled subshells (particularly the latter) have some special stability that makes such borrowing energetically favorable. This subtle but nevertheless important phenomenon affects not only the ground state configurations of atoms but also the relative stabilities of some of the ions formed by the transition elements. Similar irregularities occur among the lanthanide and the actinide elements.

8.7 ▶ Nodes in atomic orbitals affect their energies and their shapes

The uncertainty principle is often stated mathematically as

$$\Delta x = \frac{h}{4\pi m}\left(\frac{1}{\Delta v}\right)$$

where Δx is the minimum uncertainty in the particle's location, h is Planck's constant, m is the mass of the particle, and Δv is the minimum uncertainty in the particle's velocity. Notice that we can generally measure the particle's location more precisely if the particle is heavier. Notice also that the greater the uncertainty in the velocity, the smaller the uncertainty in the particle's location.

To picture what electrons are doing within the atom, we are faced with imagining an object that behaves like a particle in some experiments and like a wave in others. There is nothing in our worldly experience that is comparable. It is easier to think about the electron as a particle, and fortunately we can still think of the electron as a particle in the usual sense by speaking in terms of the statistical probability of the electron being found at a particular place. We can then use quantum mechanics to mathematically connect the particle and wave representations of the electron. Even though we may have trouble imagining an object that can be represented both ways, mathematics describes its behavior very accurately.

Describing the electron's position in terms of statistical probability is based on more than simple convenience. The German physicist Werner Heisenberg showed mathematically that it is impossible to measure with complete precision both a particle's velocity and position at the same instant. To measure an electron's position or velocity, we have to bounce another particle off it. Thus, the very act of making the measurement changes the electron's position and velocity. We cannot determine both exact position and exact velocity simultaneously, no matter how cleverly we make the measurements. This was Heisenberg's famous **uncertainty principle.** The theoretical limitations on measuring speed and position are not significant for large objects. However, for small particles such as the electron, these limitations prevent us from ever knowing or predicting where in an atom an electron will be at a particular instant, so we speak of probabilities instead.

The amplitude of an electron wave is described by a **wave function,** which is usually given the symbol ψ (the Greek letter *psi*). The probability of finding the electron in a given location is given by ψ^2.

Wave mechanics views the probability of finding an electron at a given point in space as equal to the square of the amplitude of the electron wave (given by the square of the wave function, ψ^2) at that point. It seems quite reasonable to relate probability to amplitude, or intensity, because where a wave is intense its presence is strongly felt. The amplitude is squared because, mathematically, the amplitude can be either positive or negative, but probability only makes sense if it is positive. Squaring the amplitude assures us that the probabilities will be positive. We need not be very concerned about this point, however.

The notion of electron probability leads to two very important and frequently used concepts. One is that an electron behaves as if it were spread out around the nucleus in a sort of **electron cloud.** Figure 8.23a is a *dot-density diagram* that illustrates the way the probability of finding the electron varies in space for a $1s$ orbital.

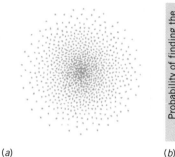
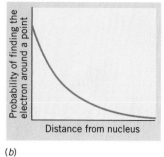

(a)   (b)

FIGURE 8.23 *Electron distribution in a 1s orbital.* (*a*) A dot-density diagram that illustrates the electron probability distribution for a 1s electron. (*b*) A graph that shows how the probability of finding the electron around a given point, ψ^2, decreases as the distance from the nucleus increases.

In those places where the dot density is large (i.e., where there are large numbers of dots per unit volume), the amplitude of the wave is large and the probability of finding the electron is also large. Figure 8.23*b* shows how the electron probability for a 1s orbital varies as we move away from the nucleus. As you might expect, the probability of finding the electron close to the nucleus is large and decreases with increasing distance from the nucleus.

The other important concept that stems from the notion that the electron probability varies from place to place is **electron density,** which relates to how much of the electron's charge is packed into a given volume. In regions of high probability there is a high concentration of electrical charge (and mass) and the electron density is large; in regions of low probability, the electron density is small.

Remember that an electron confined to a tiny space no longer behaves much like a particle. It's more like a cloud of negative charge. Like clouds made of water vapor, the density of the cloud varies from place to place. In some places the cloud is dense; in others the cloud is thinner and may be entirely absent. This is a useful picture to keep in mind as you try to visualize the shapes of atomic orbitals.

Shapes and sizes of *s* and *p* orbitals

In looking at the way the electron density distributes itself in atomic orbitals, we are interested in three things—the *shape* of the orbital, its *size,* and its *orientation* in space relative to other orbitals.

The electron density in an orbital doesn't end abruptly at some particular distance from the nucleus. It gradually fades away. Therefore, to define the size and shape of an orbital, it is useful to picture some imaginary surface enclosing, say, 90% of the electron density of the orbital and on which the probability of finding the electron is everywhere the same. For the 1s orbital in Figure 8.23, we find that if we go out a given distance from the nucleus in *any* direction, the probability of finding the electron is the same. This means that all the points of equal probability lie on the surface of a sphere, so we can say that the shape of the orbital is spherical. In fact, all *s* orbitals are spherical. As suggested earlier, their sizes increase with increasing *n*. This is illustrated in Figure 8.24. Notice that beginning with the 2s

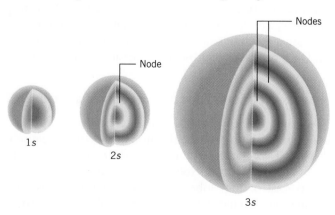

FIGURE 8.24 *Size variations among s orbitals.* The orbitals become larger as the principal quantum number, *n*, becomes larger.

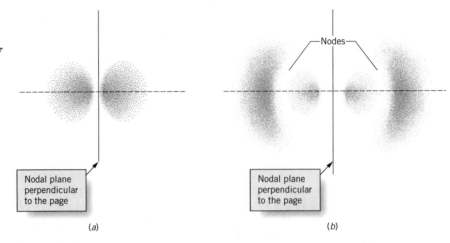

FIGURE 8.25 *Distribution of electron density in p orbitals.* (*a*) Dot-density diagram that represents a cross section of the probability distribution in a 2*p* orbital. There is a nodal plane between the two lobes of the orbital. (*b*) Cross section of a 3*p* orbital. Note the nodes in the electron density that are in addition to the nodal plane passing through the nucleus.

Nodes

Nodal plane perpendicular to the page

(*a*)

Nodal plane perpendicular to the page

(*b*)

(*a*)

(*b*)

FIGURE 8.26 *Representations of the shapes of p orbitals.* (*a*) Shape of a surface of constant probability for a 2*p* orbital. (*b*) A simplified representation of a *p* orbital that emphasizes the directional nature of the orbital.

orbital, there are certain places where the electron density drops to zero. These are the nodes of the electron wave. It is interesting that electron waves have nodes just like the waves on a guitar string. For electron waves, however, the nodes consist of imaginary *surfaces* on which the electron density is zero.

The *p* orbitals are quite different from *s* orbitals, as shown in Figure 8.25. Notice that the electron density is equally distributed in two regions on opposite sides of the nucleus. Figure 8.25*a* illustrates the two "lobes" of *one* 2*p* orbital. Between the lobes is a *nodal plane*—an imaginary flat surface on which every point has an electron density of zero. Because of this nodal plane, electrons in *p* orbitals cannot reach or "penetrate" to the nucleus as well as electrons in *s* orbitals can. This is why *p* orbitals have slightly higher energies than *s* orbitals in the same shell, for all atoms but hydrogen.

The size of the *p* orbitals also increases with increasing *n* as illustrated by the cross section of a 3*p* orbital in Figure 8.25*b*. The 3*p* and higher *p* orbitals have additional nodes besides the nodal plane that passes through the nucleus.

Figure 8.26*a* illustrates the shape of a surface of constant probability for a 2*p* orbital. Often chemists will simplify this shape by drawing two "balloons" connected at the nucleus and pointing in opposite directions, as shown in Figure 8.26*b*. Both representations emphasize the point that a *p* orbital has two equal-sized lobes that extend in opposite directions along a line that passes through the nucleus.

Orientations of orbitals in a *p* subshell

As you've learned, a *p* subshell consists of three orbitals of equal energy. Wave mechanics tells us that the lines along which the orbitals have their maximum electron densities are oriented at 90° angles to each other, corresponding to the axes of an imaginary *xyz* coordinate system (Figure 8.27). For convenience in referring to the individual *p* orbitals they are often labeled according to the axis along which they lie. The *p* orbital concentrated along the *x* axis is labeled p_x, and so forth.

Shapes and orientations of *d* orbitals in a *d* subshell

The shapes of the *d* orbitals, illustrated in Figure 8.28, are a bit more complex than are those of the *p* orbitals. Because of this, and because there are five orbitals in a *d* subshell, we haven't attempted to draw all of them at the same time on the same set of coordinate axes. Notice that four of the five *d* orbitals have the same shape and consist of four lobes of electron density. These orbitals differ only in their orientations around the nucleus (their labels come from the mathematics of wave

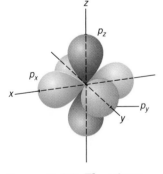

FIGURE 8.27 *The orientations of the three p orbitals in a p subshell.* Because the directions of maximum electron density lie along lines that are mutually perpendicular, like the axes of an *xyz* coordinate system, it is convenient to label the orbitals p_x, p_y, and p_z.

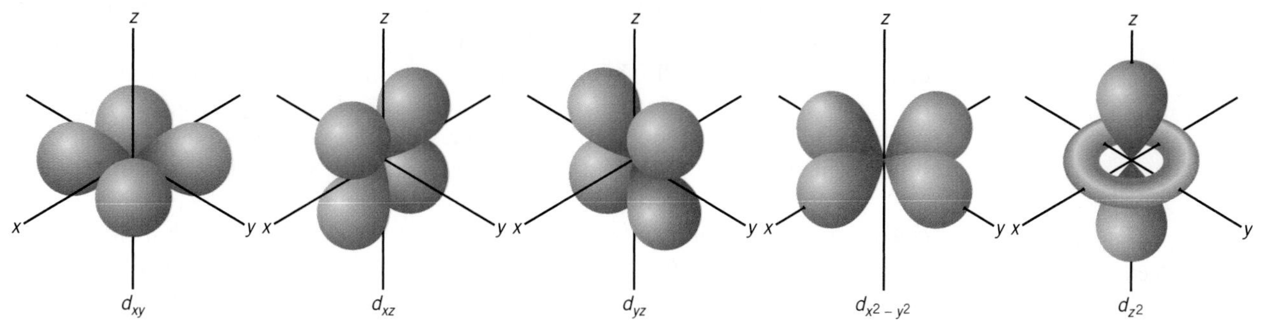

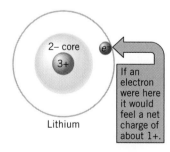

d_{xy} d_{xz} d_{yz} $d_{x^2-y^2}$ d_{z^2}

FIGURE 8.28 *The shapes and directional properties of the five d orbitals of a d subshell.*

mechanics). The fifth d orbital, labeled d_{z^2}, has two lobes that point in opposite directions along the z axis plus a doughnut-shaped ring of electron density around the center that lies in the $x-y$ plane. We will see that the d orbitals are important in the formation of chemical bonds in certain molecules and that their shapes and orientations are important in understanding the properties of the transition metals, which will be discussed in Chapter 23.

The f orbitals are even more complex than the d orbitals, but we will have no need to discuss their shapes.

8.8 ▶ Atomic properties correlate with an atom's electron configuration

There are many chemical and physical properties that vary in a more or less systematic way according to an element's position in the periodic table. For example, in Chapter 1 we noted that the metallic character of the elements increases from top to bottom in a group and decreases from left to right across a period. In this section we discuss several physical properties of the elements that have an important influence on chemical properties. We will see how these properties correlate with an atom's electron configuration, and because electron configuration is also related to the location of an element in the periodic table, we will study their periodic variations as well.

The inner electrons partially shield the outer electrons from the nucleus, so the outer electrons "feel" only a fraction of the full nuclear charge.

Effective nuclear charge is the positive charge "felt" by outer electrons

Many of an atom's properties are determined by the amount of positive charge felt by the atom's outer electrons. Except for hydrogen, this positive charge is always *less* than the full nuclear charge, because the negative charge of the electrons in inner shells partially offsets, or "neutralizes," the positive charge of the nucleus.

To gain a better understanding of this, consider the element lithium, which has the electron configuration $1s^2 2s^1$. The core electrons ($1s^2$), which lie beneath the valence shell ($2s^1$), are tightly packed around the nucleus and for the most part lie between the nucleus and the electron in the outer shell. This core has a charge of $2-$ and it surrounds a nucleus that has a charge of $3+$. When the outer $2s$ electron "looks toward" the center of the atom, it "sees" the $3+$ charge of the nucleus reduced to only about $1+$ because of the intervening $2-$ charge of the core. In other words, the $2-$ charge of the core effectively neutralizes two of the positive charges of the nucleus, so the net charge that the outer electron feels, which we call the **effective nuclear charge,** is only about $1+$. This is illustrated in an overly simplified way in Figure 8.29.

Although electrons in inner shells shield the electrons in outer shells quite effectively from the nuclear charge, electrons in the *same* shell are much less effective at shielding each other. For example, in the element beryllium ($1s^2 2s^2$)

FIGURE 8.29 *Effective nuclear charge.* If the $2-$ charge of the $1s^2$ core of lithium were 100% effective at shielding the $2s$ electron from the nucleus, the valence electron would feel an effective nuclear charge of only about $1+$.

each of the electrons in the outer $2s$ orbital is shielded quite well from the nuclear charge by the inner $1s^2$ core, but one $2s$ electron doesn't shield the other $2s$ electron very well at all. This is because electrons in the same shell are at about the same average distance from the nucleus, and in attempting to stay away from each other they only spend a very small amount of time one below the other, which is what's needed to provide shielding. Since electrons in the same shell hardly shield each other at all from the nuclear charge, *the effective nuclear charge felt by the outer electrons is determined primarily by the difference between the charge on the nucleus and the charge on the core.* With this as background, let's examine some properties controlled by the effective nuclear charge.

> An electron spends very little time between the nucleus and another electron in the same shell, so it shields that other electron poorly.

Atomic and ionic sizes increase with increasing *n* and decreasing effective nuclear charge

The wave nature of the electron makes it difficult to define exactly what we mean by the "size" of an atom or ion. As we've seen, the electron cloud doesn't simply stop at some particular distance from the nucleus; instead it gradually fades away. Nevertheless, atoms and ions do behave in many ways as though they have characteristic sizes. For example, in a whole host of hydrocarbons, ranging from methane (CH_4, natural gas) to octane (C_8H_{18}, in gasoline) to many others, the distance between the nuclei of carbon and hydrogen atoms is virtually the same. This would suggest that carbon and hydrogen have the same relative sizes in each of these compounds.

> The C—H distance in most hydrocarbons is about 110 pm (110×10^{-12} m).

Experimental measurements reveal that the diameters of atoms range from about 1.4×10^{-10} to 5.7×10^{-10} m. Their radii, which is the usual way that size is specified, range from about 7.0×10^{-11} to 2.9×10^{-10} m. Such small numbers are difficult to comprehend. A million carbon atoms placed side by side in a line would extend a little less than 0.2 mm, or about the diameter of the period at the end of this sentence.

The sizes of atoms and ions are rarely expressed in meters because the numbers are so cumbersome. Instead, a unit is chosen that makes the values easier to comprehend. A unit that scientists have traditionally used is called the **angstrom** (symbolized **Å**), which is defined as

> The angstrom is named after Anders Jonas Ångström (1814–1874), a Swedish physicist who was the first to measure the wavelengths of the four most prominent lines of the hydrogen spectrum.

$$1 \text{ Å} = 1 \times 10^{-10} \text{ m}$$

However, the angstrom is not an SI unit, and in many current scientific journals, atomic dimensions are given in picometers, or sometimes in nanometers (1 pm = 10^{-12} m and 1 nm = 10^{-9} m). In this book, we will normally express atomic dimensions in picometers, but because much of the scientific literature has these quantities in angstroms, you may someday find it useful to remember the conversions:

$$1 \text{ Å} = 100 \text{ pm}$$

$$1 \text{ Å} = 0.1 \text{ nm}$$

Atomic size varies periodically

The variations in atomic radii within the periodic table are illustrated in Figure 8.30. Here we see that atoms generally become larger going from top to bottom in a group, and they become smaller going from left to right across a period. To understand these variations we must consider two factors. One is the value of the principal quantum number of the valence electrons, and the other is the effective nuclear charge felt by the valence electrons.

Going from top to bottom within a group, the effective nuclear charge felt by the outer electrons remains nearly constant, while the principal quantum number of the valence shell increases. For example, consider the elements of Group IA. For

TOOLS

Periodic trends in atomic size

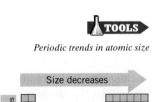

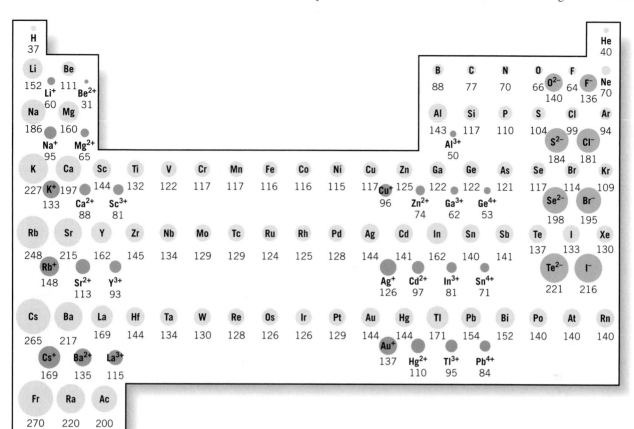

FIGURE 8.30 *Variation in atomic and ionic radii in the periodic table.* Values are in picometers.

lithium, the valence shell configuration is $2s^1$; for sodium, it is $3s^1$; for potassium, it is $4s^1$; and so forth. For each of these elements, the core has a negative charge that is 1 less than the nuclear charge, so the valence electron of each experiences a nearly constant effective nuclear charge of about 1+. However, as we descend the group, the value of n for the valence shell increases, and as you learned earlier, the larger the value of n, the larger is the orbital. Therefore, the atoms become larger as we go down a group simply because the orbitals containing the valence electrons become larger. This same argument applies whether the valence shell orbitals are s or p.

Moving from left to right across a period, electrons are added to the same shell. The orbitals holding the valence electrons all have the *same* value of n. In this case we have to examine the variation in the effective nuclear charge felt by the valence electrons.

As we move from left to right across a period, the nuclear charge increases, the outer shells of the atoms become more populated, but the inner core remains the same. For example, from lithium to fluorine the nuclear charge increases from 3+ to 9+. The core ($1s^2$) stays the same, however. As a result, the outer electrons feel an increase in positive charge (i.e., effective nuclear charge) that causes them to be drawn inward, and thereby causes the sizes of the atoms to decrease.

Across a row of transition elements or inner transition elements, the size variations are less pronounced than among the representative elements. This is because the outer shell configuration remains essentially the same while an inner shell is filled. From atomic numbers 21 to 30, for example, the outer electrons occupy the $4s$ subshell while the underlying $3d$ subshell is gradually completed. The amount of shielding provided by the addition of electrons to this inner $3d$ level is greater than the amount of shielding that would occur if the electrons were added to the outer shell, so the effective nuclear charge felt by the outer electrons increases more gradually. As a result, the decrease in size with increasing atomic number is also more gradual.

Trends in ionic size

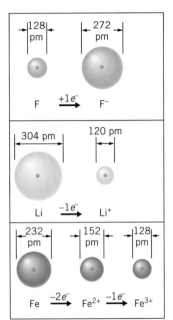

FIGURE 8.31 *Changes in size when atoms gain or lose electrons to form ions.* Adding electrons leads to an increase in the size of the particle, as illustrated for fluorine. Removing electrons leads to a decrease in the size of the particle, as shown for lithium and iron.

Ion sizes parallel atom sizes, but anions are bigger and cations are smaller

Figure 8.30 also illustrates how sizes of the ions compare with those of the neutral atoms. As you can see, when atoms gain or lose electrons to form ions, rather significant size changes take place. The reasons are easy to understand and remember.

When electrons are added to an atom, the mutual repulsions between them increase. This causes the electrons to push apart and occupy a larger volume. Therefore, *negative ions are always larger than the atoms from which they are formed* (Figure 8.31).

When electrons are removed from an atom, the electron–electron repulsions decrease, which allows the remaining electrons to be pulled closer together around the nucleus. Therefore, *positive ions are always smaller than the atoms from which they are formed.* This is also illustrated in Figure 8.31 for the elements lithium and iron. For lithium, removal of the outer $2s$ electron completely empties the valence shell and exposes the smaller $1s^2$ core. When a metal is able to form more than one positive ion, the sizes of the ions decrease as the amount of positive charge on the ion increases. To form the Fe^{2+} ion, an iron atom loses its outer $4s$ electrons. To form the Fe^{3+} ion, an additional electron is lost from the $3d$ subshell that lies beneath the $4s$. Comparing sizes, we see that the radius of an iron atom is 116 pm, whereas the radius of the Fe^{2+} ion is 76 pm. Removing yet another electron to give Fe^{3+} decreases electron–electron repulsions in the d subshell and gives the Fe^{3+} ion a radius of 64 pm.

PRACTICE EXERCISE 10: Use the periodic table to choose the largest atom or ion in each set: (a) Ge, Te, Se, Sn; (b) C, F, Br, Ga; (c) Cr, Cr^{2+}, Cr^{3+}; (d) O, O^{2-}, S, S^{2-}.

Ionization Energy

The **ionization energy** (abbreviated **IE**) *is the energy required to remove an electron from an isolated, gaseous atom or ion in its ground state.* For an atom of an element X, it is the increase in potential energy associated with the change

$$X(g) \longrightarrow X^+(g) + e^-$$

In effect, the ionization energy is a measure of how much work is required to pull an electron from an atom, so it reflects how tightly the electron is held by the atom. Usually, the ionization energy is expressed in units of kilojoules per mole (kJ/mol), so we can also view it as the energy needed to remove 1 mol of electrons from 1 mol of gaseous atoms.

Table 8.2 gives the ionization energies of the first 12 elements. As you can see, atoms with more than one electron have more than one ionization energy. These correspond to the stepwise removal of electrons, one after the other. Lithium, for example, has three ionization energies because it has three electrons. Removing the outer $2s$ electrons from 1 mol of isolated lithium atoms to give 1 mol of gaseous lithium ions, Li^+, requires 520 kJ; so the *first ionization energy* of lithium is 520 kJ/mol. The second IE of lithium is 7297 kJ/mol and corresponds to the process

$$Li^+(g) \longrightarrow Li^{2+}(g) + e^-$$

This involves the removal of an electron from the now-exposed $1s$ core of lithium. Removal of the third (and last) electron requires the third IE, which is 11,810 kJ/mol. In general, successive ionization energies always increase because each subsequent electron is being pulled away from an increasingly more positive ion, and that requires more work.

Ionization energies are additive. For example,

$Li(g) \rightarrow Li^+(g) + e^-$

$\qquad\qquad IE_1 = 520 \text{ kJ}$

$Li^+(g) \rightarrow Li^{2+}(g) + e^-$

$\qquad\qquad IE_2 = 7297 \text{ kJ}$

$Li(g) \rightarrow Li^{2+}(g) + 2e^-$

$\qquad IE \text{ total} = IE_1 + IE_2$

$\qquad\qquad = 7817 \text{ kJ}$

| TABLE 8.2 | **SUCCESSIVE IONIZATION ENERGIES IN kJ/mol FOR HYDROGEN THROUGH MAGNESIUM** | | | | | | |

Note the sharp increase in ionization energy when crossing the "staircase," indicating that the last of the valence electrons has been removed.

	1st	2nd	3rd	4th	5th	6th	7th	8th
H	1312							
He	2372	5250						
Li	520	7297	11,810					
Be	899	1757	14,845	21,000				
B	800	2426	3659	25,020	32,820			
C	1086	2352	4619	6221	37,820	47,260		
N	1402	2855	4576	7473	9442	53,250	64,340	
O	1314	3388	5296	7467	10,987	13,320	71,320	84,070
F	1680	3375	6045	8408	11,020	15,160	17,860	92,010
Ne	2080	3963	6130	9361	12,180	15,240	—	—
Na	496	4563	6913	9541	13,350	16,600	20,113	25,666
Mg	737	1450	7731	10,545	13,627	17,995	21,700	25,662

Larger atoms have lower ionization energies

Within the periodic table there are trends in the way IE varies that are useful to know and to which we will refer in later discussions. We can see these by examining a graph that shows how the first ionization energy varies with an element's position in the table, which is shown in Figure 8.32. Notice that the elements with the largest ionization energies are the nonmetals in the upper right of the periodic table and that those with the smallest ionization energies are the metals in the lower left of the table. In general, then, the following trends are observed:

> Ionization energy generally increases from bottom to top within a group and increases from left to right within a period.

TOOLS

Periodic trends in ionization energy

The same factors that affect atomic size also affect ionization energy. As the value of n increases going down a group, the orbitals become larger and the outer electrons are farther from the nucleus. Electrons farther from the nucleus are bound less tightly, so IE decreases from top to bottom. Of course, this is just the same as saying that it increases from bottom to top.

As you can see, there is a gradual overall increase in IE as we move from left to right across a period, although the horizontal variation of IE is somewhat irregular (see Facets of Chemistry 8.2). The reason for the overall trend is the increase in effective nuclear charge felt by the valence electrons as we move across a period. As we've seen, this draws the valence electrons closer to the nucleus and leads to a decrease in atomic size as we move from left to right. But the increasing effective nuclear charge also causes the valence electrons to be held more tightly, which makes it more difficult to remove them.

The results of these trends place elements with the largest IE in the upper right-hand corner of the periodic table. It is very difficult to cause these atoms to lose electrons. In the lower left-hand corner of the table are elements that have loosely held valence electrons. These elements form positive ions relatively easily, as you learned in Chapter 2.

It is often helpful to remember that the trends in IE are just the opposite of the trends in atomic size within the periodic table: when size increases, IE decreases.

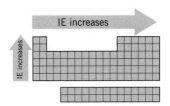

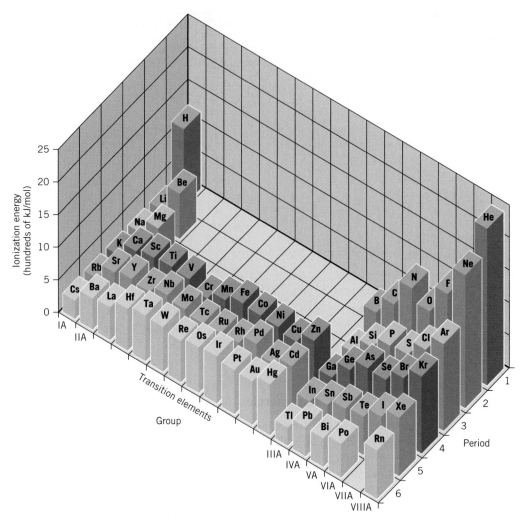

FIGURE 8.32 *Variation in first ionization energy with location in the periodic table.* Elements with the largest ionization energies are in the upper right of the periodic table. Those with the smallest ionization energies are at the lower left.

Noble gas configurations are extremely stable

Table 8.2 shows that, for a given element, successive ionization energies increase gradually until the valence shell is emptied. Then a very much larger increase in IE occurs as the core is broken into. This is illustrated graphically in Figure 8.33 for the Period 2 elements lithium through fluorine. For lithium, we see that the first electron (the 2s electron) is removed rather easily, but the second and third electrons, which come from the 1s core, are much more difficult to dislodge. For beryllium, the large jump in IE occurs after two electrons (the two 2s electrons) are removed. In fact, for all of these elements, the big jump in IE happens when the core is broken into.

The data displayed in Figure 8.33 suggest that although it may be moderately difficult to empty the valence shell of an atom, it is *extremely* difficult to break into the noble gas configuration of the core electrons. As you will learn, this is one of the factors that influences the number of positive charges on ions formed by the representative metals.

PRACTICE EXERCISE 11: Use the periodic table to select the atom with the largest IE: (a) Na, Sr, Be, Rb; (b) B, Al, C, Si.

FACETS OF CHEMISTRY 8.2

Irregularities in the Periodic Variations in Ionization Energy and Electron Affinity

The variation in first ionization energy across a period is not a smooth one, as seen in the graph in Figure 1 for the elements in Period 2. The first irregularity occurs between Be and B, where the IE increases from Li to Be but then decreases from Be to B. This happens because there is a change in the nature of the subshell from which the electron is being removed. For Li and Be, the electron is removed from the $2s$ subshell, but at boron the first electron comes from the higher-energy $2p$ subshell where it is not bound so tightly.

Another irregularity occurs between nitrogen and oxygen. For nitrogen, the electron that's removed comes from a singly occupied orbital. For oxygen, the electron is taken from an orbital that already contains an electron. We can diagram this as follows:

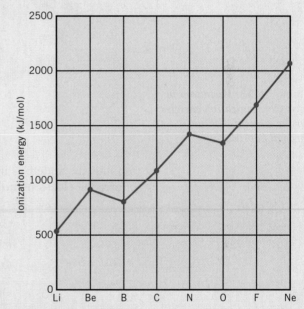

For oxygen, repulsions between the two electrons in the p orbital that's about to lose an electron help the electron leave. This "help" is absent for the electron that's about to leave the p orbital of nitrogen. As a result, it is not as difficult to remove one electron from an oxygen atom as it is to remove one electron from a nitrogen atom.

As with ionization energy, there are irregularities in the periodic trends for electron affinity. For example, the Group IIA elements have little tendency to acquire electrons because their outer shell s orbitals are filled. The incoming electron must enter a higher-energy p orbital. We also see that the EA for elements in Group VA are either endothermic or only slightly exothermic. This is because the incoming electron must enter an orbital already occupied by an electron.

One of the most interesting irregularities occurs between Periods 2 and 3 among the nonmetals. In any group, the element in Period 2 has a less exothermic electron affinity than the element below it. The reason seems to be the small size of the nonmetal atoms of Period 2, which are among the smallest elements in the periodic table. Repulsions between the many electrons in the small valence shells of these atoms leads to a lower-than-expected attraction for an incoming electron and a less exothermic electron affinity than the element below in Period 3.

FIGURE 1 Variation in IE for the Period 2 elements Li through F.

Electron affinity

The **electron affinity** (abbreviated **EA**) *is the potential energy change associated with the addition of an electron to a gaseous atom or ion in its ground state.* For an element X, it is the change in potential energy associated with the process

$$X(g) + e^- \longrightarrow X^-(g)$$

As with ionization energy, electron affinities are usually expressed in units of kilojoules per mole, so we can also view the IE as the energy change associated with adding 1 mol of electrons to 1 mol of gaseous atoms or ions.

For nearly all the elements, the addition of one electron to the neutral atom is exothermic, and the EA is given as a negative value. This is because the incoming electron experiences an attraction to the nucleus that causes the potential energy to be lowered as the electron approaches the atom. However, when a second elec-

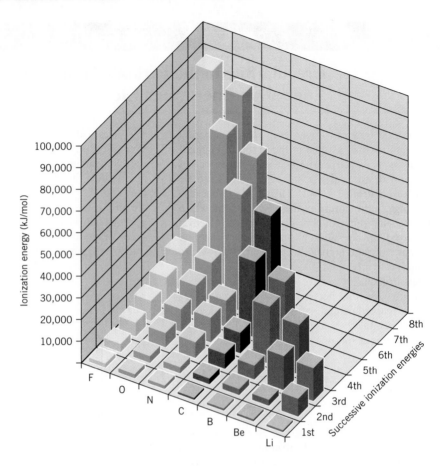

FIGURE 8.33 *Variations in successive ionization energies for the elements lithium through fluorine.*

tron must be added, as in the formation of the oxide ion, O^{2-}, work must be done to force the electron into an already negative ion.

Change	EA (kJ/mol)
$O(g) + e^- \rightarrow O^-(g)$	-141
$O^-(g) + e^- \rightarrow O^{2-}(g)$	$+844$
$O(g) + 2e^- \rightarrow O^{2-}(g)$	$+703$ (net)

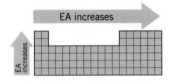

FIGURE 8.34 *Variation of electron affinity (as an exothermic quantity) within the periodic table.*

Notice that more energy is absorbed adding an electron to the O^- ion than is released by adding an electron to the O atom. Overall, the formation of an isolated oxide ion leads to a net increase in potential energy (so we say its formation is *endothermic*). The same applies to the formation of *any* negative ion with a charge larger than $1-$.

The electron affinities of the representative elements are given in Table 8.3, and we see that periodic trends in electron affinity roughly parallel those for ionization energy. (See Facets of Chemistry 8.2 for a discussion of some of the irregularities in the trends.)

TOOLS

Periodic trends in electron affinity

> Although there are some irregularities, overall, the electron affinities of the elements become more *exothermic* going from left to right across a period and from bottom to top in a group (Figure 8.34).

This shouldn't be surprising, because a valence shell that loses electrons easily (low IE) will have little attraction for additional electrons (small EA). On the other hand, a valence shell that holds its electrons tightly will also tend to bind an additional electron tightly.

TABLE 8.3	ELECTRON AFFINITIES OF THE REPRESENTATIVE ELEMENTS (kJ/mol)						
IA	IIA	IIIA	IVA	VA	VIA	VIIA
H −73							
Li −60	Be +238	B −27	C −122	N ~ +9	O −141	F −328
Na −53	Mg +230	Al −44	Si −134	P −72	S −200	Cl −348
K −48	Ca +155	Ga −30	Ge −120	As −77	Se −195	Br −325
Rb −47	Sr +167	In −30	Sn −121	Sb −101	Te −190	I −295
Cs −45	Ba +50	Tl −30	Pb −110	Bi −110	Po −183	At −270

SUMMARY

Electromagnetic Energy Electromagnetic energy, or light energy, travels through space at a constant speed of 3.00×10^8 m s^{-1} in the form of waves. The **wavelength**, λ, and **frequency**, ν, of the wave are related by the equation $\lambda\nu = c$, where c is the **speed of light**. The SI unit for frequency is the **hertz (Hz**, 1 Hz = 1 s^{-1}). Light also behaves as if it consists of small packets of energy, called **photons** or **quanta**. The energy delivered by a photon is proportional to the frequency of the light and is given by the equation $E = h\nu$, where h is **Planck's constant**. White light is composed of all the frequencies visible to the eye and can be split into a **continuous spectrum**. Visible light represents only a small portion of the entire **electromagnetic spectrum**, which also includes **X rays, ultraviolet, infrared, radio,** and **TV** waves, and **microwaves**.

Atomic Spectra The occurrence of **line spectra** tells us that atoms can emit energy only in discrete amounts and suggests that the energy of the electron is **quantized;** that is, the electron is restricted to certain specific **energy levels** in an atom. Niels Bohr recognized this and, although his theory was later shown to be incorrect, he was the first to propose a model that was able to account for the **Rydberg equation**. Bohr was the first to introduce the idea of **quantum numbers**.

Matter Waves The wave behavior of electrons and other tiny particles, which can be demonstrated by **diffraction** experiments, was suggested by de Broglie. Schrödinger applied wave theory to the atom and launched the theory we call **wave mechanics** or **quantum mechanics**. This theory tells us that electron waves in atoms are **standing waves** whose crests and **nodes** are stationary. Each standing wave, or **orbital**, is characterized by three quantum numbers, n, ℓ, and m_ℓ (**principal, secondary,** and **magnetic quantum numbers,** repectively). **Shells** are designated by n (which can range from 1 to ∞),

subshells by ℓ (which can range from 0 to $n − 1$), and orbitals within subshells by m_ℓ (which can range from $-\ell$ to $+\ell$).

Electron Configurations The electron has magnetic properties that are explained in terms of spin. The **spin quantum number**, m_s, can have values of $+\frac{1}{2}$ or $-\frac{1}{2}$. The **Pauli exclusion principle** limits orbitals to a maximum population of two electrons with **paired spins**. Substances with unpaired electrons are **paramagnetic** and are attracted weakly to a magnetic field. Substances with only paired electrons are **diamagnetic** and are slightly repelled by a magnetic field. The **electron configuration** of an element in its **ground state** is obtained by filling orbitals beginning with the 1s subshell and following the Pauli exclusion principle and **Hund's rule** (which states that electrons spread out as much as possible in orbitals of equal energy). The periodic table serves as a guide in predicting electron configurations. **Abbreviated configurations** show subshell populations outside a noble gas **core**. **Valence shell configurations** show the populations of subshells in the **outer shell** of an atom of the representative elements. Sometimes we represent electron configurations using **orbital diagrams**. Unexpected configurations occur for chromium and copper because of the extra stability of half-filled and filled subshells.

Orbital Shapes The **Heisenberg uncertainty principle** says we cannot know exactly the position and velocity of an electron both at the same instant. Consequently, wave mechanics describes the probable locations of electrons in atoms. In each orbital the electron is conveniently viewed as an **electron cloud** with a varying **electron density**. All s orbitals are spherical; each p orbital consists of two lobes with a **nodal plane** between them. A p subshell has three p orbitals whose axes are mutually perpendicular and point along the x, y, and z axes of an imaginary coordinate system centered at the nucleus. Four of the five d orbitals in a d subshell have the same shape, with four lobes of electron density each. The fifth has two lobes of electron density pointing in opposite directions along the z axis and a ring of electron density in the $x-y$ plane.

Atomic Properties The amount of positive charge felt by the valence electrons of an atom is the **effective nuclear charge.** This is less than the actual nuclear charge because core electrons partially shield the valence electrons from the full positive charge of the nucleus. **Atomic radii** depend on the value of n of the valence shell orbitals and the effective nuclear charge experienced by the valence electrons. These radii are expressed in units of picometers or nanometers, or an older unit called the **angstrom (Å)**; 1 Å = 100 pm = 0.1 nm. Atomic radii decrease from left to right in a period and from bottom to top in a group in the periodic table. Negative ions are larger than the atoms from which they are formed; positive ions are smaller than the atoms from which they are formed.

Ionization energy (IE) is the energy needed to remove an electron from an isolated gaseous atom, molecule, or ion in its ground state; it is endothermic. The first ionization energies of the elements increase from left to right in a group and from bottom to top in a period. (Irregularities occur in a period when the nature of the orbital from which the electron is removed changes and when the electron removed is first taken from a doubly occupied p orbital.) Successive ionization energies become larger, but there is a very large jump when the next electron must come from the noble gas core beneath the valence shell.

Electron affinity (EA) is the potential energy change associated with the addition of an electron to a gaseous atom or ion in its ground state. For atoms, the first EA is usually exothermic. When more than one electron is added to an atom, the overall potential energy change is endothermic. In general, electron affinity becomes more exothermic from left to right in a period and from bottom to top in a group. (However, the EA of second-period nonmetals is less exothermic than for the nonmetals of the third period. Irregularities across a period occur when the electron being added must enter the next higher-energy subshell and when it must enter a half-filled p subshell.)

TOOLS ▶ YOU HAVE LEARNED

The table below lists the tools you have learned in this chapter. Notice that only two of them are related to numerical calculations. The others are conceptual tools that we use in analyzing properties of substances in terms of the underlying structure of matter. Review all these tools and refer to them, if necessary, when working on the Thinking-It-Through problems and the Review Exercises that follow.

TOOL	HOW IT WORKS
Wavelength–frequency relationship (page 306)	We need this to convert between wavelength and frequency.
Energy of a photon (page 310)	This equation is used to calculate the energy carried by a photon of frequency ν. Also, ν can be calculated if E is known.
Periodic table (page 328)	We will use the periodic table as a tool for many purposes. In this chapter, you learned to use the periodic table as an aid in writing electron configurations of the elements and as a tool to correlate an element's location in the table to properties such as atomic radius, ionization energy, and electron affinity.
Periodic trends in atomic and ionic size (page 340)	We are able to compare the sizes of atoms and ions.
Periodic trends in ionization energy (page 341)	We are able to compare the ease with which atoms of the elements lose electrons.
Periodic trends in electron affinity (page 344)	We are able to compare the tendency of atoms or ions to gain electrons.

THINKING IT THROUGH

Need extra help?
Visit the Brady/
Senese web site at
www.wiley.com/
college/brady

The goal for the following problems is not to find the answers themselves, but rather to assemble the information needed to solve them and explain how you would use the information to find the answers. The problems in Level 2 are more challenging than those in Level 1 and may contain more data than are required, in which case you are also asked to identify the unnecessary data. Detailed answers to the Thinking-It-Through problems can be found on the web site.

Level 1 Problems

1. An FM radio station broadcasts at a frequency of 101.9 MHz. What is the wavelength of these radio waves expressed in meters? How many peaks of these waves pass a given point each second? (Set up the calculation.)

2. The human ear does not respond to sounds with frequencies higher than 20 kHz. Sound is not electromagnetic radiation; it travels at a speed of about 330 m s^{-1} through air. What is the distance between peaks of sound waves of this frequency, in meters? (Show how you would set up the calculation.)

3. Ultraviolet (UV) radiation can cause sunburn and has been implicated as a causative factor in the formation of certain skin cancers. A typical photon of UV light has a wavelength of 150 nm. How can you calculate the energy of a photon of this light? How many times larger is this than the energy of a photon emitted by a high-voltage wire carrying 60 Hz alternating current?

4. For a hydrogen atom, explain how to calculate the frequency of the photon that is emitted when the electron drops from an energy level that has $n = 6$ to $n = 2$. Describe how you would calculate the wavelength of this light. How can you determine what color the light is?

5. Atoms of which of the following elements are paramagnetic? (a) Mg, (b) Zn, (c) As, (d) Ar, (e) Ag

6. The atoms of which element have valence electrons that experience the greater effective nuclear charge, silicon or sulfur? Explain the reasoning required to answer this question.

7. State the specific relationship that allows you to determine
(a) which element has the larger atoms, Ge or S.
(b) which element is likely to have the larger first ionization energy, phosphorus or sulfur.
(c) which element is likely to have the more exothermic electron affinity, phosphorus or sulfur.

Level 2 Problems

8. Explain in detail how to calculate the ionization energy of the hydrogen atom from information provided in this chapter.

9. Magnesium forms the ion Mg^{2+} when a magnesium atom loses two electrons. Show how you would calculate the energy required to remove two electrons from a mole of magnesium atoms. How much water (in grams) could have its temperature raised from room temperature (25 °C) to the boiling point (100 °C) by the energy needed to remove two electrons from a mole of magnesium atoms? Set up the calculation.

10. Explain how one can easily determine the ionization energy of the fluoride ion, F$^-$. Is this ionization energy endothermic or exothermic?

11. How can you determine the electron affinity of an Al^{3+} ion?

12. How can you show that $Na(g) + Cl(g)$ is more stable (of lower energy) than $Na^+(g) + Cl^-(g)$, assuming infinite distance of separation between atoms or ions? The potential energy of the Na^+ and Cl^- ions is inversely proportional to the distance between the ions and is given by the equation

$$E = \frac{z^+ z^-}{4\pi d \epsilon_0^2}$$

where z^+ and z^- are the charges on the cation and anion, respectively, in units of coulombs, d is the distance between the ions in meters, and ϵ_0 is a constant which equals 8.85×10^{-12} C^2 m^{-1} J^{-1}. Describe how you would use this equation to determine the distance below which the energy of the Na^+ and Cl^- ions is lower than the energy of the neutral atoms.

13. What wavelength of light would be required to eject an electron from an atom of sodium? Set up the calculation.

REVIEW QUESTIONS

Electromagnetic Radiation

8.1 In general terms, why do we call light *electromagnetic radiation*?

8.2 In general, what does the term *frequency* imply? What is meant by the term *frequency of light*? What symbol is used for it, and what is the SI unit (and symbol) for frequency?

8.3 What is meant by the term *wavelength* of light? What symbol is used for it?

8.4 Sketch a picture of a wave and label its wavelength and its amplitude.

8.5 Which property of light waves affects the brightness of the light? Which affects the color of the light? Which affects the energy of the light?

8.6 Arrange the following regions of the electromagnetic spectrum in order of increasing wavelength (i.e., shortest wavelength → longest wavelength): microwave, TV, X-ray, ultraviolet, visible, infrared, gamma rays.

8.7 What wavelength range is covered by the *visible spectrum*?

8.8 Arrange the following colors of visible light in order of increasing wavelength: orange, green, blue, yellow, violet, red.

8.9 Write the equation that relates the wavelength and frequency of a light wave.

8.10 How is the frequency of a particular type of radiation related to the energy associated with it? (Give an equation, defining all symbols.)

8.11 What is a photon?

8.12 Show that the energy of a photon is given by the equation $E = hc/\lambda$.

8.13 Examine each of the following pairs and state which of the two has the higher *energy*:
(a) microwaves and infrared
(b) visible light and infrared
(c) ultraviolet light and X rays
(d) visible light and ultraviolet light

8.14 What is a quantum of energy?

Atomic Spectra

8.15 What is an atomic spectrum? How does it differ from a continuous spectrum?

8.16 What fundamental fact is implied by the existence of atomic spectra?

Bohr Atom and the Hydrogen Spectrum

8.17 Describe Niels Bohr's model of the structure of the hydrogen atom.

8.18 In qualitative terms, how did Bohr's model account for the atomic spectrum of hydrogen?

8.19 What is the term used to describe the lowest energy state of an atom?

8.20 In what way was Bohr's theory a success? How was it a failure?

Wave Nature of Matter

8.21 How does the behavior of very small particles differ from that of the larger, more massive objects that we meet in everyday life? Why don't we notice this same behavior for the larger, more massive objects?

8.22 Describe the phenomenon called *diffraction*. How can this be used to demonstrate that de Broglie's theory was correct?

8.23 What experiment could you perform to determine whether a beam was behaving as a wave or as a stream of particles?

8.24 What is *wave/particle duality*?

8.25 What is the difference between a *traveling wave* and a *standing wave*?

8.26 What is the collapsing atom paradox?

8.27 How does quantum mechanics resolve the collapsing atom paradox?

Electron Waves in Atoms

8.28 What are the names used to refer to the theories that apply the matter–wave concept to electrons in atoms?

8.29 What is the term used to describe a particular waveform of a standing wave for an electron?

8.30 What are the two properties of orbitals in which we are most interested? Why?

Quantum Numbers

8.31 What are the allowed values of the principal quantum number?

8.32 What is the value for n for (a) the K shell and (b) the M shell?

8.33 Why does every shell contain an s subshell?

8.34 How many orbitals are found in (a) an s subshell, (b) a p subshell, (c) a d subshell, and (d) an f subshell?

8.35 If the value of m_ℓ for an electron in an atom is 2, could another electron in the same subshell have $m_\ell = -3$?

8.36 Suppose an electron in an atom has the following set of quantum numbers: $n = 2$, $\ell = 1$, $m_\ell = 1$, $m_s = +\frac{1}{2}$. What set of quantum numbers is impossible for another electron in this same atom?

Electron Spin

8.37 What physical property of electrons leads us to propose that they spin like a toy top?

8.38 What is the name of the magnetic property exhibited by atoms that contain unpaired electrons?

8.39 What is the Pauli exclusion principle? What effect does it have on the populating of orbitals by electrons?

8.40 What are the possible values of the spin quantum number?

Electron Configuration of Atoms

8.41 What do we mean by the term *electronic structure*?

8.42 Within any given shell, how do the energies of the s, p, d, and f subshells compare?

8.43 What fact about the energies of subshells was responsible for the apparent success of Bohr's theory about electronic structure?

8.44 How do the energies of the orbitals belonging to a given subshell compare?

8.45 Give the electron configurations of the elements in Period 2 of the periodic table.

8.46 Give the correct electron configurations of (a) Cr and (b) Cu.

8.47 What is the correct electron configuration of silver?

8.48 How are the electron configurations of the elements in a given group similar? Illustrate your answer by writing shorthand configurations for the elements in Group VIA.

8.49 Define the terms *valence shell* and *valence electrons*.

Shapes of Atomic Orbitals

8.50 Why do we use probabilities when we discuss the position of an electron in the space surrounding the nucleus of an atom?

8.51 Sketch the approximate shape of (a) a $1s$ orbital and (b) a $2p$ orbital.

8.52 How does the size of a given type of orbital vary with n?

8.53 How are the p orbitals of a given p subshell oriented relative to each other?

8.54 What is a *nodal plane*?

8.55 How do nodes affect the energy of an orbital?

8.56 How many nodal planes does a p orbital have? How many does a d orbital have?

8.57 On appropriate coordinate axes, sketch the shape of the following d orbitals: (a) d_{xy}, (b) $d_{x^2-y^2}$, (c) d_{z^2}.

Atomic and Ionic Size

8.58 What is the meaning of *effective nuclear charge*? How does the effective nuclear charge felt by the outer electrons vary going down a group? How does it change as we go from left to right across a period?

8.59 In what region of the periodic table are the largest atoms found? Where are the smallest atoms found?

8.60 Going from left to right in the periodic table, why are the size changes among the transition elements more gradual than those among the representative elements?

Ionization Energy

8.61 Define *ionization energy.* Why are ionization energies of atoms and positive ions endothermic quantities?

8.62 For oxygen, write an equation for the change associated with (a) its first ionization energy and (b) its third ionization energy.

8.63 Explain why ionization energy increases from left to right in a period and decreases from top to bottom in a group.

8.64 Why is an atom's second ionization energy always larger than its first ionization energy?

8.65 Why is the fifth ionization energy of carbon so much larger than its fourth?

8.66 Why is the first ionization energy of aluminum less than the first ionization energy of magnesium?

8.67 Why does phosphorus have a larger first ionization energy than sulfur?

Electron Affinity

8.68 Define *electron affinity.*

8.69 For sulfur, write an equation for the change associated with (a) its first electron affinity and (b) its second electron affinity. How should they compare?

8.70 Why does Cl have a more exothermic electron affinity than F? Why does Br have a less exothermic electron affinity than Cl?

8.71 Why is the second electron affinity of an atom always endothermic?

8.72 How is electron affinity related to effective nuclear charge? On this basis, explain the relative magnitudes of the electron affinities of oxygen and fluorine.

REVIEW PROBLEMS

Answers to problems whose numbers are printed in color are given in Appendix B. More challenging problems are marked with asterisks. **ILW** = Interactive LearningWare solution is available at *www.wiley.com/college/brady.*

Electromagnetic Radiation

8.73 What is the frequency in hertz of blue light having a wavelength of 430 nm?

8.74 Ultraviolet light with a wavelength of more than 280 nm has little germicidal value. What is the frequency that corresponds to this wavelength?

8.75 A certain substance strongly absorbs infrared light having a wavelength of 6.85 μm. What is the frequency of this light in hertz?

8.76 The sun emits many wavelengths of light. The brightest light is emitted at about 0.48 μm. What frequency does this correspond to?

8.77 Ozone protects Earth's inhabitants from the harmful effects of ultraviolet light arriving from the sun. This shielding is a maximum for UV light having a wavelength of 295 nm. What is the frequency in hertz of this light?

8.78 The meter is defined as the length of the path light travels in a vacuum during the time interval of 1/299,792,458 of a second. The standards body recommends use of light from a helium–neon laser for realizing the meter. The light from the laser has a wavelength of 632.99139822 nm. What is the frequency of this light, in hertz?

8.79 In New York City, radio station WCBS broadcasts its FM signal at a frequency of 101.1 megahertz (MHz). What is the wavelength of this signal in meters?

8.80 Sodium vapor lamps are often used in residential street lighting. They give off a yellow light having a frequency of 5.09×10^{14} Hz. What is the wavelength of this light in nanometers?

8.81 There has been some concern in recent times about possible hazards to people who live very close to high-voltage electric power lines. The electricity in these wires oscillates at a frequency of 60 Hz, which is the frequency of any electromagnetic radiation that they emit. What is the wavelength of this radiation in meters? What is it in kilometers?

8.82 An X-ray beam has a frequency of 1.50×10^{18} Hz. What is the wavelength of this light in nanometers and in picometers?

8.83 Calculate the energy in joules of a photon of red light having a frequency of 4.0×10^{14} Hz. What is the energy of 1 mol of these photons?

8.84 Calculate the energy in joules of a photon of green light having a wavelength of 560 nm.

Atomic Spectra

8.85 In the spectrum of hydrogen, there is a line with a wavelength of 410.3 nm. (a) What color is this line? (b) What is its frequency? (c) What is the energy of each of its photons?

8.86 In the spectrum of sodium, there is a line with a wavelength of 589 nm. (a) What color is this line? (b) What is its frequency? (c) What is the energy of each of its photons?

8.87 Use the Rydberg equation to calculate the wavelength in nanometers of the spectral line of hydrogen for which $n_2 = 6$ and $n_1 = 3$. Would we be expected to see the light corresponding to this spectral line? Explain your answer.

8.88 Use the Rydberg equation to calculate the wavelength in nanometers of the spectral line of hydrogen for which $n_2 = 5$ and $n_1 = 2$. Would we be expected to see the light corresponding to this spectral line? Explain your answer.

8.89 Calculate the wavelength of the spectral line produced in the hydrogen spectrum when an electron falls from the tenth Bohr orbit to the fourth. In which region of the electromagnetic spectrum (UV, visible, or infrared) is the line?

8.90 Calculate the energy in joules and the wavelength in nanometers of the spectral line produced in the hydrogen spectrum when an electron falls from the fourth Bohr orbit to the first. In which region of the electromagnetic spectrum (UV, visible, or infrared) is the line?

Quantum Numbers

8.91 What is the letter code for a subshell with (a) $\ell = 1$ and (b) $\ell = 3$?

8.92 What is the value of ℓ for (a) an f orbital and (b) a d orbital?

8.93 Give the values of n and ℓ for the following subshells: (a) $3s$, (b) $5d$.

8.94 Give the values of n and ℓ for the following subshells: (a) $4p$, (b) $6f$.

8.95 For the shell with $n = 6$, what are the possible values of ℓ?

8.96 In a particular shell, the largest value of ℓ is 7. What is the value of n for this shell?

8.97 What are the possible values of m_ℓ for a subshell with (a) $\ell = 1$ and (b) $\ell = 3$?

8.98 If the value of ℓ for an electron in an atom is 5, what are the possible values of m_ℓ that this electron could have?

8.99 If the value of m_ℓ for an electron in an atom is -4, what is the smallest value of ℓ that the electron could have? What is the smallest value of n that the electron could have?

8.100 How many orbitals are there in an h subshell ($\ell = 5$)? What are their values of m_ℓ?

8.101 Give the complete set of quantum numbers for all of the electrons that could populate the $2p$ subshell of an atom.

8.102 Give the complete set of quantum numbers for all of the electrons that could populate the $3d$ subshell of an atom.

8.103 In an antimony atom, how many electrons have $\ell = 1$? How many electrons have $\ell = 2$ in an antimony atom?

8.104 In an atom of barium, how many electrons have (a) $\ell = 0$ and (b) $m_\ell = 1$?

Electron Configuration of Atoms

8.105 Predict the electron configurations of (a) S, (b) K, (c) Ti, and (d) Sn.

8.106 Predict the electron configurations of (a) As, (b) Cl, (c) Ni, and (d) Si.

8.107 Which of the following atoms in their ground states are expected to be paramagnetic: (a) Mn, (b) As, (c) S, (d) Sr, (e) Ar?

8.108 Which of the following atoms in their ground states are expected to be diamagnetic: (a) Ba, (b) Se, (c) Zn, (d) Si?

ILW **8.109** How many unpaired electrons would be found in the ground state of (a) Mg, (b) P, and (c) V?

8.110 How many unpaired electrons would be found in the ground state of (a) Cs, (b) S, and (c) Ni?

8.111 Write the shorthand electron configurations for (a) Ni, (b) Cs, (c) Ge, (d) Br, and (e) Bi.

8.112 Write the shorthand electron configurations for (a) Al, (b) Se, (c) Ba, (d) Sb, and (e) Gd.

8.113 Draw complete orbital diagrams for (a) Mg and (b) Ti.

8.114 Draw complete orbital diagrams for (a) As and (b) Ni.

8.115 Draw orbital diagrams for the shorthand configurations of (a) Ni, (b) Cs, (c) Ge, and (d) Br.

8.116 Draw orbital diagrams for the shorthand configurations of (a) Al, (b) Se, (c) Ba, and (d) Sb.

8.117 What is the value of n for the valence shells of (a) Sn, (b) K, (c) Br, and (d) Bi?

8.118 What is the value of n for the valence shells of (a) Al, (b) Se, (c) Ba, and (d) Sb?

8.119 Give the configuration of the valence shell for (a) Na, (b) Al, (c) Ge, and (d) P.

8.120 Give the configuration of the valence shell for (a) Mg, (b) Br, (c) Ga, and (d) Pb.

8.121 Draw the orbital diagram for the valence shell for (a) Na, (b) Al, (c) Ge, and (d) P.

8.122 Draw the orbital diagram for the valence shell of (a) Mg, (b) Br, (c) Ga, and (d) Pb.

Atomic Properties

8.123 If the core electrons were 100% effective at shielding the valence electrons from the nuclear charge and the valence electrons provided no shielding for each other, what would be the effective nuclear charge felt by a valence electron in (a) Na, (b) S, (c) Cl?

8.124 If the core electrons were 100% effective at shielding the valence electrons from the nuclear charge and the valence electrons provided no shielding for each other, what would be the effective nuclear charge felt by a valence electron in (a) Mg, (b) Si, (c) Br?

8.125 Choose the larger atom in each pair: (a) Na or Si; (b) P or Sb.

8.126 Choose the larger atom in each pair: (a) Al or Cl; (b) Al or In.

8.127 Choose the largest atom among the following: Ge, As, Sn, Sb.

8.128 Place the following in order of increasing size: N^{3-}, Mg^{2+}, Na^+, Ne, F^-, O^{2-}.

8.129 Choose the larger particle in each pair: (a) Na or Na^+; (b) Co^{3+} or Co^{2+}; (c) Cl or Cl^-.

8.130 Choose the larger particle in each pair: (a) S or S^{2-}; (b) Al^{3+} or Al; (c) Au^+ or Au^{3+}.

8.131 Choose the atom with the larger ionization energy in each pair: (a) B or C; (b) O or S; (c) Cl or As.

8.132 Choose the atom with the larger ionization energy in each pair: (a) Li or Cs; (b) Al or Si; (c) F or O.

8.133 Choose the atom with the more exothermic electron affinity in each pair: (a) Cl or Br; (b) Se or Br.

8.134 Choose the atom with the more exothermic electron affinity in each pair: (a) P or As; (b) Si or Ga.

8.135 Use the periodic table to select the element in the following list for which there is the largest difference between the second and third ionization energies: Na, Mg, Al, Si, P, Se, Cl.

8.136 Use the periodic table to select the element in the following list for which there is the largest difference between the fourth and fifth ionization energies: Na, Mg, Al, Si, P, Se, Cl.

ADDITIONAL EXERCISES

8.137 The human ear is sensitive to sound ranging from 20 to 20,000 Hz. The speed of sound is 330 m/s in air and 1500 m/s under water. What is the longest and the shortest wavelength that can be heard (a) in air and (b) under water?

8.138 Microwaves are used to heat food in microwave ovens. The microwave radiation is absorbed by moisture in the food. This heats the water, and as the water becomes hot, so does the food. How many photons having a wavelength of 3.00 mm would have to be absorbed by 1.00 g of water to raise its temperature by 1.00 °C?

8.139 In the spectrum of hydrogen, there is a line with a wavelength of 410.3 nm. Use the Rydberg equation to calculate the value of n for the higher-energy Bohr orbit involved in the emission of this light. Assume the value of n for the lower energy orbit equals 2.

8.140 Calculate the wavelength in nanometers of the shortest wavelength of light emitted by a hydrogen atom.

8.141 Which of the following electronic transitions could lead to the emission of light from an atom?

$$1s \longrightarrow 4p \longrightarrow 3d \longrightarrow 5f \longrightarrow 4d \longrightarrow 2p$$

8.142 A neon sign is a gas discharge tube in which electrons traveling from the cathode to the anode collide with neon atoms in the tube and knock electrons off of them. As electrons return to the neon ions and drop to lower energy levels, light is given off. How fast would an electron have to be moving to eject an electron from an atom of neon, which has a first ionization energy equal to 2080 kJ mol^{-1}?

*8.143 How many grams of water could have its temperature raised by 5.0 °C by a mole of photons that have a wavelength of (a) 600 nm and (b) 300 nm?

*8.144 It has been found that when the chemical bond between chlorine atoms in Cl_2 is formed, 328 kJ is released per mole of Cl_2 formed. What is the wavelength of light that would be required to break chemical bonds between chlorine atoms?

8.145 Calculate the wavelengths of the lines in the spectrum of hydrogen that result when an electron falls from a Bohr orbit with (a) $n = 5$ to $n = 1$, (b) $n = 4$ to $n = 2$, and (c) $n = 6$ to $n = 4$. In which regions of the electromagnetic spectrum are these lines?

8.146 What, if anything, is wrong with the following electron configurations for atoms in their ground states?
(a) $1s^2 2s^1 2p^3$ (c) $1s^2 2s^2 2p^4$
(b) $[Kr]3d^7 4s^2$ (d) $[Xe]4f^{14} 5d^8 6s^1$

8.147 Suppose students gave the following orbital diagrams for the 2s and 2p subshells in the ground state of an atom. What, if anything, is wrong with them? Are any of these electron distributions impossible?

(a) ○ ⬆⬇ ⬆⬇ ⬆⬇   (c) ⬆ ⬆ ⬆ ⬆

(b) ⬆ ⬇ ○ ○   (d) ⬆⬆ ⬆ ⬆ ○

8.148 How many electrons are in p orbitals in an atom of germanium?

8.149 What are the quantum numbers of the electrons that are lost by an atom of iron when it forms the ion Fe^{2+}?

*8.150 The removal of an electron from the hydrogen atom corresponds to raising the electron to the Bohr orbit that has $n = \infty$. On the basis of this statement, calculate the ionization energy of hydrogen in units of (a) joules per atom and (b) kilojoules per mole.

8.151 Use orbital diagrams to illustrate what happens when an oxygen atom gains two electrons. On the basis of what you have learned about electron affinities and electron configurations, why is it extremely difficult to place a third electron on the oxygen atom?

8.152 From the data available in this chapter, determine the ionization energy of (a) F^-, (b) O^-, and (c) O^{2-}. Are any of these energies exothermic?

8.153 For an oxygen atom, which requires more energy, the addition of two electrons or the removal of one electron?

Chemical Bonding: General Concepts

Serena Williams returns a volley from Italy's Francesca Schiavone at Wimbledon in June, 2002. The high-tech fibers in her tennis racket that make the shot possible are held together by covalent bonds, which is one of the topics we discuss in this chapter.

THIS CHAPTER IN CONTEXT In Chapter 2 you learned that we can classify substances into two broad categories, ionic and molecular. Ionic compounds, such as ordinary table salt, consist of electrically charged particles (ions) that bind to each other by electrostatic forces of attraction. We also commented in Chapter 2 that in molecular substances, such as water, the atoms are held to each other by sharing electrons. Now that you've learned about the electronic structures of atoms, we can explore these attractions, which we call **chemical bonds,** in greater depth. Our goal is to gain some insight into the reasons certain combinations of atoms prefer electron transfer and the formation of ions (leading to *ionic bonding*) while other combinations bind by electron sharing (leading to *covalent bonding*).

As with electronic structure, models of chemical bonding have also evolved, and in this chapter our goal is to introduce you to relatively simple theories. Although more complicated theories exist (some of which we will explore in Chapter 10), the basic concepts you will study in this chapter still find many useful applications in modern chemical thought.

9.1 ▶ Electron transfer leads to the formation of ionic compounds

In Chapter 2 you learned that ionic compounds are formed when metals react with nonmetals. Among the examples discussed was sodium chloride, table salt. You learned that when this compound is formed from its elements, each sodium atom loses one electron to form a sodium ion, Na^+, and each chlorine atom gains one electron to become a chloride ion, Cl^-.

$$Na \longrightarrow Na^+ + e^-$$

$$Cl + e^- \longrightarrow Cl^-$$

Once formed, these ions become tightly packed together, as illustrated in the margin, because their opposite charges attract. *This attraction between positive and negative ions in an ionic compound is what we call an* **ionic bond.**

The reason Na^+ and Cl^- ions attract each other is easy to understand. But *why* are electrons transferred between these and other atoms? *Why* does sodium form Na^+ and not Na^- or Na^{2+}? And *why* does chlorine form Cl^- instead of Cl^+ or Cl^{2-}? To answer such questions we have to consider factors that are related to the potential energy of the system of reactants and products. This is because *the dominant factor favoring the formation of a stable ionic compound from its elements is a net lowering of the potential energy.* In other words, the overall reaction must be exothermic.

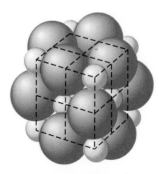

Packing of ions in NaCl.

Keep in mind the relationship between potential energy changes and endothermic and exothermic processes:

endothermic ⟺ increase in PE

exothermic ⟺ decrease in PE

The lattice energy enables ionic compounds to form

Let's begin by examining the energy change associated with the exchange of electrons between sodium atoms and chlorine atoms. If we deal with a collection of gaseous sodium atoms and chlorine atoms, we can use the ionization energy and electron affinity of sodium and chlorine, respectively. Working on a mole basis, we have

$$Na(g) \longrightarrow Na^+(g) + e^- \qquad +495.4 \text{ kJ} \qquad \text{(IE of sodium)}$$

$$Cl(g) + e^- \longrightarrow Cl^-(g) \qquad \underline{-348.8 \text{ kJ}} \qquad \text{(EA of chlorine)}$$

$$\text{Net} \quad +146.6 \text{ kJ}$$

What this calculation tells us is that forming gaseous sodium and chloride ions from gaseous sodium and chlorine atoms requires a substantial increase in the potential energy. In fact, if the IE and EA were the only energy changes involved, ionic sodium chloride would not form from sodium and chlorine. So where does the stability of the compound come from? The answer is seen if we examine a quantity called the *lattice energy.*

In the calculation above, we looked at the formation of gaseous ions, but salt is not a gas; it's a solid in which the ions are packed together in a way that maximizes the attractions between oppositely charged ions. Imagine, now, pulling these ions away from each other to form a gas of ions. This process would require a lot of work and would lead to a large increase in the potential energies of the ions. The **lattice energy** *is the energy required to completely separate the ions in 1 mol of a compound from each other to form a cloud of gaseous ions.* (The term *lattice* comes from the description of the organized packing of the ions in the solid.) For sodium chloride, the process associated with the lattice energy is pictured in Figure 9.1 and in equation form can be represented as

$$NaCl(s) \longrightarrow Na^+(g) + Cl^-(g)$$

The name *lattice energy* comes from the word *lattice,* which is used to describe the regular pattern of ions or atoms in a crystal. Lattice energies are endothermic and so are given positive signs.

The energy associated with this change (the *lattice energy* of sodium chloride) has been both measured and calculated to be $+787.0 \text{ kJ mol}^{-1}$. The positive sign means that it takes 787.0 kJ to separate the ions of 1 mol of NaCl. *It also means that if we bring together a mole of Na^+ and Cl^- ions from the gaseous state into 1 mol of crystalline NaCl, 787 kJ will be released.* If we now include the lattice energy (with its sign reversed because we're changing the direction of the change) along with the IE and EA, we have

$$Na(g) \longrightarrow Na^+(g) + e^- \qquad +495.4 \text{ kJ} \qquad \text{(IE of sodium)}$$

$$Cl(g) + e^- \longrightarrow Cl^-(g) \qquad -348.8 \text{ kJ} \qquad \text{(EA of chlorine)}$$

$$Na^+(g) + Cl^-(g) \longrightarrow NaCl(s) \qquad \underline{-787.0 \text{ kJ}} \qquad \text{(−lattice energy)}$$

$$\text{Net} \quad -640.4 \text{ kJ}$$

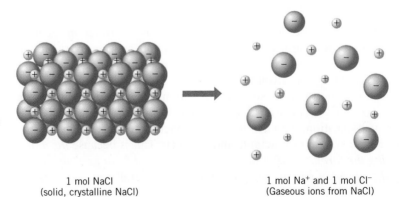

FIGURE 9.1 *Lattice energy of NaCl.* The lattice energy is equal to the amount of energy needed to separate the ions in 1 mol of an ionic compound. For NaCl, the process requires converting a mole of crystalline NaCl into 2 mol of ions (1 mol Na^+ and 1 mol Cl^-). The amount of energy absorbed equals 787 kJ.

1 mol NaCl
(solid, crystalline NaCl)

1 mol Na^+ and 1 mol Cl^-
(Gaseous ions from NaCl)

Thus, the release of energy equivalent to the lattice energy provides a large net lowering of the potential energy as solid NaCl is formed. We can also say that it is the lattice energy that provides the stabilization necessary for the formation of NaCl. Without it, the compound could not exist.

At this point, you may be wondering about our starting point in these energy calculations, gaseous sodium atoms and chlorine atoms. In nature, sodium is a solid metal and chlorine consists of gaseous Cl_2 molecules. A complete analysis of the energy changes has to take this into account, and we do this in Facets of Chemistry 9.1. Our overall conclusion doesn't change, however. *For any ionic compound, the chief stabilizing influence is the lattice energy, which when released is large enough to overcome the net energy input required to form the ions from the elements.*

Lattice energy depends on ionic size and charge

The lattice energies of some ionic compounds are given in Table 9.1. As you can see, they are all very large endothermic quantities. Their magnitudes depend on a number of factors, including the sizes of the ions and their charges.[1] In general, as the ions become smaller, the lattice energy increases; smaller ions allow the charges to get closer together, which makes them more difficult to pull apart. For example, the lattice energy for LiCl is larger than that for NaCl, reflecting the smaller size of the lithium ion.

The lattice energy also becomes larger as the amount of charge on the ions increases, because more highly charged ions attract each other more strongly. Thus, salts of Ca^{2+} have larger lattice energies than comparable salts of Na^+, and those containing Al^{3+} have even larger lattice energies.

Energy factors determine that metals form cations and nonmetals form anions

Having discussed the energy factors involved in the formation of an ionic compound, we can now understand why metals tend to form positive ions and nonmetals tend to form negative ions. At the left of the periodic table are the metals—elements with

For an ionic compound to be formed from its elements, the exothermic lattice energy must be larger than the endothermic combination of factors involved in the formation of the ions themselves, which primarily involve the IE of the metal and the EA of the nonmetal.

TABLE 9.1	LATTICE ENERGIES OF SOME IONIC COMPOUNDS	
Compound	Ions	Lattice Energy (kJ mol^{-1})
LiCl	Li^+ and Cl^-	845
NaCl	Na^+ and Cl^-	778
KCl	K^+ and Cl^-	709
LiF	Li^+ and F^-	1033
$CaCl_2$	Ca^{2+} and Cl^-	2258
$AlCl_3$	Al^{3+} and Cl^-	5492
CaO	Ca^{2+} and O^{2-}	3401
Al_2O_3	Al^{3+} and O^{2-}	15,916

[1]The potential energy of two particles with charges q_1 and q_2 separated by a distance r is

$$E = \frac{q_1 q_2}{kr}$$

where k is a proportionality constant. In an ionic solid, q_1 and q_2 have opposite signs, so E is calculated to be a negative quantity. When q_1 and q_2 become separated, r approaches infinity and E approaches zero. Therefore, the more negative the value of E, the more energy is needed to make E equal to zero and the more energy is needed to separate the ions. In a rough way, the size of the lattice energy parallels the magnitude of E. When the charges q_1 and q_2 become larger, E becomes more negative and the lattice energy becomes larger. Similarly, when r becomes smaller, corresponding to smaller ions, E also becomes more negative and the lattice energy becomes larger.

Calculating the Lattice Energy

In the main body of the text, we described the pivotal role played by the lattice energy in the formation of an ionic compound. But how can we possibly measure the amount of energy required to completely vaporize an ionic compound to give gaseous ions? Actually, we can't measure the lattice energy directly, but we can use Hess's law and some other experimental data to calculate the lattice energy indirectly.

In Chapter 7 you learned that the enthalpy change for a process is the same regardless of the path we follow from start to finish. With this in mind, we can construct a set of alternate paths from the free elements to the solid ionic compound. This is called a Born–Haber cycle after the scientists who were the first to use it to calculate lattice energies, and is shown in the accompanying figure for the formation of sodium chloride. You may recognize the figure as an enthalpy diagram.

We begin with the free elements, sodium and chlorine. The direct path at the bottom left has as its enthalpy change the heat of formation of NaCl, ΔH_f°.

$$Na(s) + \tfrac{1}{2}Cl_2(g) \longrightarrow NaCl(s) \qquad \Delta H_f^\circ = -411.3 \text{ kJ}$$

The alternate path is divided into a number of steps. The first two steps, both of which have ΔH° values that can be measured experimentally, are endothermic. They change $Na(s)$ and $Cl_2(g)$ into gaseous atoms, $Na(g)$ and $Cl(g)$. The next two steps change these atoms to ions, first by the endothermic ionization energy (IE) of Na followed by the exothermic electron affinity (EA) of Cl. This brings us to the gaseous ions, $Na^+(g)$ and $Cl^-(g)$. Notice that at this point, if we add all the energy changes, the ions are at a considerably higher energy than the reactants. If these were the only energy terms involved in the formation of NaCl, the heat of formation would be endothermic and the compound would be unstable; it could not be formed by direct combination of the elements.

The last step on the right finally brings us to solid NaCl and corresponds to the negative of the lattice energy. (Remember, the lattice energy is defined as the energy needed to *separate* the ions; in the last step, we are *bringing the ions together* to form the solid.) To make the net energy changes the same along both paths, the energy released when the ions condense to form the solid must equal -787.0 kJ. Therefore, the calculated lattice energy of NaCl must be $+787.0$ kJ mol^{-1}.

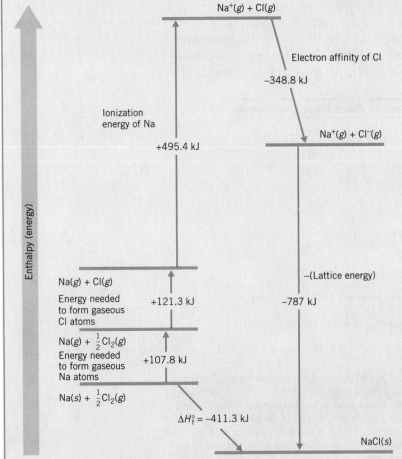

Born–Haber cycle for sodium chloride. The path at the lower left labeled ΔH_f° leads directly to NaCl(s). The upper path involves the formation of gaseous atoms from the elements, then the formation of gaseous ions from the atoms, and finally the condensation of Na^+ and Cl^- ions to give solid NaCl. The final step releases energy equivalent to the lattice energy. Both the upper and lower paths lead from the elements to solid NaCl and yield the same net energy change.

small ionization energies and electron affinities. Relatively little energy is needed to remove electrons from them to produce positive ions. At the upper right of the periodic table are the nonmetals—elements with large ionization energies and generally exothermic electron affinities. It is quite difficult to remove electrons from these elements, but sizable amounts of energy are released when they gain electrons. On an energy basis, therefore, it is least "expensive" to form a cation from a metal and an anion from a nonmetal, so it is relatively easy for the energy-lowering effect of the lattice energy to exceed the net energy-raising effect of the ionization energy and electron affinity. In fact, metals combine with nonmetals to form ionic compounds simply because ionic bonding is favored energetically over other types whenever atoms with small ionization energies combine with atoms that have large exothermic electron affinities.

The stability of the noble gas configuration can control which ions are possible

Earlier we raised the question about why sodium forms Na^+ and chlorine forms Cl^-. We're now able to find some answers by studying how the electronic structures of the elements affect the kinds of ions they form. Let's begin by examining what happens when sodium loses an electron. The electron configuration of Na is

$$Na \quad 1s^22s^22p^63s^1$$

The electron that is lost is the one least tightly held, which is the single outer $3s$ electron. The electronic structure of the Na^+ ion, then, is

$$Na^+ \quad 1s^22s^22p^6$$

Notice that this is identical to the electron configuration of the noble gas neon. We say the Na^+ ion has achieved a *noble gas configuration*.

The removal of the first electron from Na does not require much energy because the first ionization energy of sodium is relatively small. Therefore, an input of energy equal to the first ionization energy can be easily recovered by the release of energy equivalent to the lattice energy when an ionic compound containing Na^+ is formed.

Removal of a second electron from sodium is *very* difficult because it involves breaking into the $2s^22p^6$ core. Forming the Na^{2+} ion is therefore very endothermic, as we can see by adding the first and second ionization energies.

$$Na(g) \longrightarrow Na^+(g) + e^- \qquad \text{1st IE} = \quad 496 \text{ kJ mol}^{-1}$$
$$Na^+(g) \longrightarrow Na^{2+}(g) + e^- \qquad \text{2nd IE} = 4563 \text{ kJ mol}^{-1} \longleftarrow$$
$$\text{Total} \quad 5059 \text{ kJ mol}^{-1}$$

This value is so large because the electron removed is from the noble gas core beneath the outer shell of sodium.

Notice how large is the amount of energy needed to break into the neon core of the Na^+ ion to remove the second electron. Even though a compound such as $NaCl_2$ would have a larger lattice energy than NaCl (e.g., see the lattice energy for $CaCl_2$), it is not large enough to make the formation of the compound exothermic. As a result, $NaCl_2$ cannot form. Similar situations exist for other compounds of sodium, so when sodium forms a cation, electron loss stops once the Na^+ ion is formed and a noble gas electron configuration is reached.

A somewhat similar situation exists for other metals, too. Consider calcium, for example. You learned in Chapter 2 that this metal forms ions with a 2+ charge. This means that when a calcium atom reacts, it loses its two outermost electrons (giving it the same electron configuration as argon).

$$\text{Ca} \quad 1s^22s^22p^63s^23p^64s^2$$
$$\text{Ca}^{2+} \quad 1s^22s^22p^63s^23p^6$$

For calcium:

> 1st IE = 590 kJ/mol
>
> 2nd IE = 1146 kJ/mol
>
> 3rd IE = 4940 kJ/mol

The two $4s$ electrons of Ca are not held too tightly, so the total amount of energy that must be invested to remove them (the sum of the first and second ionization energies) can be recovered by the release of energy equivalent to the large lattice energy of a Ca^{2+} compound. However, the removal of yet another electron from calcium to form Ca^{3+} requires breaking into the noble gas core. As in the case of sodium, a tremendous amount of energy is needed to accomplish this, much more than can be regained by the release of the lattice energy of a Ca^{3+} compound. Therefore, a calcium atom loses just two electrons when it reacts.

In the case of sodium and calcium, we find that the large ionization energy of the noble gas core just below their outer shells limits the number of electrons they lose, so the ions that are formed have noble gas electron configurations.

Because it is so difficult to break into the noble gas core, it's convenient to think of the noble gas core as being very stable.

Nonmetals also tend to have noble gas configurations when they form anions. For example, when a chlorine atom reacts, it gains one electron. For chlorine we have

$$\text{Cl} \quad 1s^22s^22p^63s^23p^5$$

and when an electron is gained, its configuration becomes

$$\text{Cl}^- \quad 1s^22s^22p^63s^23p^6$$

At this point, electron gain ceases, because if another electron were to be added, it would have to enter an orbital in the next higher shell. With oxygen, a similar situation exists. The formation of the oxide ion, O^{2-}, is endothermic, as you learned in the previous chapter.

$$\text{O } (1s^22s^22p^4) + 2e^- \longrightarrow O^{2-}(1s^22s^22p^6) \qquad \text{EA(net)} = +703 \text{ kJ/mol}$$

However, this energy input is modest and can be overcome without much difficulty by the release of the large lattice energies of ionic metal oxides. But we never observe the formation of O^{3-} because, once again, the third electron would have to enter an orbital in the next higher shell, and this is *very* energetically unfavorable.

The octet rule can sometimes provide a guide to the ions formed by elements

Electron configurations of ions of the representative elements

In the preceding discussions you learned that a balance of energy factors causes many atoms to form ions that have a noble gas electron configuration. Historically, this is expressed in the form of a generalization: *when they form ions, atoms of most of the representative elements tend to gain or lose electrons until they have obtained a configuration identical to that of the nearest noble gas.* Because all the noble gases except helium have outer shells with eight electrons, this rule has become known as the **octet rule,** which can be stated as follows: *atoms tend to gain or lose electrons until they have achieved an outer shell that contains an* **octet of electrons** *(eight electrons).* Sodium and calcium achieve an octet by emptying their valence shells; chlorine and oxygen achieve an octet by gaining enough electrons to reach a noble gas configuration.

Many cations do not obey the octet rule

The octet rule, as applied to ionic compounds, really works well only for the cations of the Group IA and IIA metals from Period 3 down and for the anions of the nonmetals. It doesn't work for lithium and beryllium, which form ions with the same electron configuration as helium ($1s^2$). It also doesn't work for hydrogen, which is known to form an ion H^- (also with a helium configuration) when it reacts with some very reactive metals. For these three exceptions, however, the ions do have electron configurations matching that of a noble gas; the outer shell just doesn't contain eight electrons.

The most significant breakdown of the octet rule occurs with the transition metals and post-transition metals (the metals that follow a row of transition metals). To obtain the correct electron configuration of the cations of these metals, we apply the following rules:

1. The first electrons to be lost by an atom or ion are *always* those from the shell with the largest value of n (i.e., the outer shell).

2. As electrons are removed from a given shell, they come from the highest-energy occupied subshell first, before any are removed from a lower-energy subshell. Within a given shell, the energies of the subshells vary as follows: $s < p < d < f$. This means that f is emptied before d, which is emptied before p, which is emptied before s.

Let's look at two examples that illustrate these rules.

Tin (a post-transition metal) forms two ions, Sn^{2+} and Sn^{4+}. The electron configurations are

$$Sn \quad [Kr]\, 4d^{10}5s^25p^2$$

$$Sn^{2+} \quad [Kr]\, 4d^{10}5s^2$$

$$Sn^{4+} \quad [Kr]\, 4d^{10}$$

Notice that the Sn^{2+} ion is formed by the loss of the higher-energy $5p$ electrons first. Then, further loss of the two $5s$ electrons gives the Sn^{4+} ion. However, neither of these ions has a noble gas configuration.

For the transition elements, the first electrons lost are the s electrons of the outer shell. Then, if additional electrons are lost, they come from the underlying d subshell. An example is iron, which forms the ions Fe^{2+} and Fe^{3+}. The element iron has the electron configuration

$$Fe \quad [Ar]\, 3d^64s^2$$

Iron loses its $4s$ electrons fairly easily to give Fe^{2+}, which has the electron configuration[2]

$$Fe^{2+} \quad [Ar]\, 3d^6$$

The Fe^{3+} ion results when another electron is removed, this time from the $3d$ subshell.

$$Fe^{3+} \quad [Ar]\, 3d^5$$

Iron is able to form Fe^{3+} because the $3d$ subshell is close in energy to the $4s$, so it is not very difficult to remove the third electron. Notice once again that the first electrons to be removed come from the shell with the largest value of n (the $4s$ subshell). Then, after this shell is emptied, the next electrons are removed from the shell below.

Because so many of the transition elements are able to form ions in a way similar to that of iron, the ability to form more than one positive ion is usually cited as one of the characteristic properties of the transition elements. Frequently, one of the ions formed has a 2+ charge, which arises from the loss of the two outer s electrons. Ions with larger positive charges result when additional d electrons are lost.

Electron configurations of ions of transition and post-transition metals

Applying these rules to the metals of Groups IA and IIA also gives the correct electron configurations.

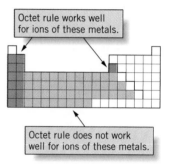

Octet rule works well for ions of these metals.

Octet rule does not work well for ions of these metals.

---

[2]In an iron atom, the $3d$ subshell is actually *lower* in energy than the $4s$ subshell, so the first electrons to be lost are those from the *higher-energy* $4s$ subshell. This energy ordering appears to violate the principles we used to derive electron configurations by the aufbau method. Although the method described in Chapter 8 will almost always yield correct ground-state electron configurations for neutral atoms, the actual ordering of the energy levels in a transition metal atom places the d subshell lower than the s subshell. The reasons for this apparent contradiction are complex and related to the way the energies of the subshells shift as we increase nuclear charge and numbers of electrons going from one element to the next in the periodic table. Further explanations are beyond the scope of this textbook, however.

Unfortunately, there is no simple way to predict exactly which ions can form for a given transition metal, nor is it simple to predict their relative stabilities with respect to oxidation or reduction.

EXAMPLE 9.1 **Writing Electron Configurations of Ions**	How do the electron configurations change (a) when a nitrogen atom forms the N^{3-} ion and (b) when an antimony atom forms the Sb^{3+} ion?

ANALYSIS: For the nonmetals, you've learned that the octet rule does work, so the ion that is formed by nitrogen will have a noble gas configuration.

Antimony is a post-transition element, so we don't expect its cation to obey the octet rule. We do have rules to handle the situation, however. When a cation is formed, electrons are removed first from the outer shell of the atom (the shell with the largest value of the principal quantum number, n). Within a given shell, electrons are always removed first from the subshell highest in energy.

SOLUTION: (a) The electron configuration for nitrogen is

$$N \quad [He]\, 2s^2 2p^3$$

To form N^{3-}, three electrons are gained. These enter the $2p$ subshell because it is the lowest available energy level. Filling the $2p$ subshell completes the octet; the configuration for the ion is therefore

$$N^{3-} \quad [He]\, 2s^2 2p^6$$

(b) Let's begin with the ground state electron configuration for antimony.

$$Sb \quad [Kr]\, 4d^{10} 5s^2 5p^3$$

To form the Sb^{3+} ion, three electrons must be removed. These will come from outer shell, which has $n = 5$. Recall that within this shell, the energies of the subshells increase in the order $s < p < d < f$. Therefore, the $5p$ subshell is higher in energy than the $5s$, so all three electrons are removed from the $5p$. This gives

$$Sb^{3+} \quad [Kr]\, 4d^{10} 5s^2$$

Are the Answers Reasonable?

In Chapter 2 you learned to use the periodic table to figure out the charges on anions of the nonmetals. For nitrogen, we would take three steps to the right to get to the nearest noble gas, neon. The electron configuration we obtained for N^{3-} is that of neon, so our answer should be correct.

For antimony, we had to remove three electrons, which completely emptied the $5p$ subshell. That's also good news, because ions do not tend to have partially filled s or p subshells (although partially filled d subshells are not uncommon for the transition metals). If we had taken the electrons from any other subshells, the Sb^{3+} ion would have had a partially filled $5p$ subshell.

EXAMPLE 9.2 **Writing Electron Configurations of Ions**	What is the electron configuration of the V^{3+} ion? Give the orbital diagram for the ion.

ANALYSIS: We will begin with the electron configuration of the neutral atom and then remove three electrons to obtain the electron configuration of the ion. We have to keep in mind that the electrons are lost first from the occupied shell with highest n.

SOLUTION: The electron configuration of vanadium is

$$V \quad \underbrace{1s^2 2s^2 2p^6 3s^2 3p^6}_{\text{Argon core}} 3d^3 4s^2$$

Notice that we've written the configuration showing the outer shell $4s$ electrons farthest to the right. To form the V^{3+} cation, three electrons must be removed from the neutral atom. The first two come from the $4s$ subshell and the third comes from the $3d$. This means we won't have to take any from the $3s$ or $3p$ subshells, so the argon core will remain intact. Therefore, let's rewrite the electron configuration in abbreviated form.

$$V \quad [Ar]\, 3d^3 4s^2$$

Removing the three electrons gives

$$V^{3+} \quad [Ar]\, 3d^2$$

To form the orbital diagram, we show all five orbitals of the $3d$ subshell and then spread the two electrons out with spins unpaired. This gives

$$V^{3+} \quad [Ar]\; \overset{\displaystyle 3d}{\textcircled{$\uparrow$}\textcircled{$\uparrow$}\bigcirc\bigcirc\bigcirc}$$

Is the Answer Reasonable?

First, we check that we've written the correct electron configuration of vanadium, which we have. (A quick count of the electrons gives 23, which is the atomic number of vanadium.) We've also taken electrons away from the atom following the rules, so the electron configuration of the ion seems okay. Finally, we remembered to show all five orbitals of the $3d$ subshell, even though only two of them are occupied.

PRACTICE EXERCISE 1: How do the electron configurations change when a chromium atom forms the following ions: (a) Cr^{2+}, (b) Cr^{3+}, (c) Cr^{6+}?

PRACTICE EXERCISE 2: How are the electron configurations of S^{2-} and Cl^- related?

9.2 ▶ Lewis symbols help keep track of valence electrons

In the last section you saw how the valence shells of atoms change when electrons are transferred during the formation of ions. We will soon see that many atoms share their valence electrons with each other when they form covalent bonds. In these discussions of bonding it is useful to be able to keep track of valence electrons. To help us do this, we use a simple bookkeeping device called *Lewis symbols*, named after their inventor, a famous American chemist, G. N. Lewis (1875–1946).

To draw the **Lewis symbol** for an element, we write its chemical symbol surrounded by a number of dots (or some other similar mark), which represent the atom's valence electrons. For example, the element lithium, which has one valence electron in its $2s$ subshell, has the Lewis symbol

Li·

In fact, each element in Group IA has a similar Lewis symbol, because each has only one valence electron. The Lewis symbols for all of the Group IA metals are

Li·   Na·   K·   Rb·   Cs·

The Lewis symbols for the eight A-group elements of Period 2 are[3]

Group	IA	IIA	IIIA	IVA	VA	VIA	VIIA	VIIIA
Symbol	Li·	·Be·	·B·	·C·	·N:	·O:	·F:	:Ne:

The elements below each of these in their respective groups have identical Lewis symbols except, of course, for the chemical symbol of the element. Notice that

Gilbert N. Lewis, chemistry professor at the University of California, helped develop theories of chemical bonding. In 1916, he proposed that atoms form bonds by sharing pairs of electrons between them.

Lewis symbols

---

[3]For beryllium, boron, and carbon, the number of unpaired electrons in the Lewis symbol doesn't agree with the number predicted from the atom's electron configuration. Boron, for example, has two electrons paired in its $2s$ orbital and a third electron in one of its $2p$ orbitals; therefore, there is actually only one unpaired electron in a boron atom. The Lewis symbols are drawn as shown, however, because when beryllium, boron, and carbon form bonds, they *behave* as if they have two, three, and four unpaired electrons, respectively. We will discuss this further in Chapter 10.

This is one of the apcantages of the North American convention for numbering groups in the periodic table.

when an atom has more than four valence electrons, the additional electrons are shown to be paired with others. Also notice that *for the representative elements, the group number is equal to the number of valence electrons* when the North American convention for numbering groups in the periodic table is followed.

EXAMPLE 9.3
Writing Lewis Symbols

What is the Lewis symbol for arsenic?

ANALYSIS: We need to know the number of valence electrons, which we can obtain from the group number. Then we distribute the electrons (dots) around the chemical symbol.

SOLUTION: The symbol for arsenic is As and we find it in Group VA. The element therefore has five valence electrons. The first four are placed around the symbol for arsenic as follows:

$$\cdot \overset{\textstyle\cdot}{\underset{\textstyle\cdot}{As}} \cdot$$

The fifth electron is paired with one of the first four. This gives

$$\cdot \overset{\textstyle\cdot}{\underset{\textstyle\cdot}{As}} \colon$$

The location of the fifth electron doesn't really matter, so equally valid Lewis symbols are

$$\cdot \overset{\textstyle\cdot\cdot}{As} \cdot \quad \text{or} \quad \colon \overset{\textstyle\cdot}{As} \cdot \quad \text{or} \quad \cdot \overset{\textstyle\cdot}{\underset{\textstyle\cdot\cdot}{As}} \cdot$$

Is the Answer Reasonable?
There's not much to check here. Have we got the correct chemical symbol? Yes. Do we have the right number of dots? Yes.

PRACTICE EXERCISE 3: Write Lewis symbols for (a) Se, (b) I, and (c) Ca.

Although we will use Lewis symbols mostly to follow the fate of valence electrons in covalent bonds, they can also be used to describe what happens during the formation of ions. For example, when a sodium atom reacts with a chlorine atom, the electron transfer can be depicted as

$$Na \cdot + \cdot \overset{\textstyle\cdot\cdot}{\underset{\textstyle\cdot\cdot}{Cl}} \colon \longrightarrow Na^+ + \left[\colon \overset{\textstyle\cdot\cdot}{\underset{\textstyle\cdot\cdot}{Cl}} \colon \right]^-$$

The valence shell of the sodium atom is emptied, so no dots remain. The outer shell of chlorine, which formerly had seven electrons, gains one to give a total of eight. The brackets are drawn around the chloride ion to show that all eight electrons are the exclusive property of the Cl^- ion. We can diagram a similar reaction between calcium and chlorine atoms.

$$\colon \overset{\textstyle\cdot\cdot}{\underset{\textstyle\cdot\cdot}{Cl}} \cdot \quad Ca \quad \cdot \overset{\textstyle\cdot\cdot}{\underset{\textstyle\cdot\cdot}{Cl}} \colon \longrightarrow Ca^{2+} + 2 \left[\colon \overset{\textstyle\cdot\cdot}{\underset{\textstyle\cdot\cdot}{Cl}} \colon \right]^-$$

EXAMPLE 9.4
Using Lewis Symbols

Use Lewis symbols to diagram the reaction that occurs between sodium and oxygen atoms to give Na^+ and O^{2-} ions.

ANALYSIS: For electrical neutrality, the formula will be Na_2O, so we know we will use two sodium atoms and one oxygen atom. The sodiums will lose one electron each to give Na^+ and the oxygen will gain two electrons to give O^{2-}.

SOLUTION: First let's draw the Lewis symbols for Na and O.

$$Na \cdot \quad \cdot \overset{\textstyle\cdot\cdot}{O} \colon$$

It takes two electrons to complete the octet around oxygen. Each Na supplies one. Therefore,

$$Na\overset{\curvearrowright}{\,\cdot}\ddot{O}:\overset{\curvearrowleft}{\,}Na \longrightarrow 2Na^+ + \left[:\ddot{O}:\right]^{2-}$$

Notice that we have put brackets around the oxide ion.

Is the Answer Reasonable?
We have accounted for all the valence electrons (an important check), the net charge is the same on both sides of the arrow (the equation is balanced), and we've placed the brackets around the oxide ion to emphasize that the octet belongs exclusively to that ion.

PRACTICE EXERCISE 4: Diagram the reaction between magnesium and oxygen atoms to give Mg^{2+} and O^{2-} ions.

9.3 ▶ Covalent bonds are formed by electron sharing

Most of the substances we encounter in our daily lives are not ionic. Rather than existing as collections of electrically charged particles (ions), they occur as electrically neutral combinations of atoms that we call *molecules*. Water, as you already know, consists of molecules made from two hydrogen atoms and one oxygen atom, and the formula for one particle of this compound is H_2O. Most substances consist of much larger molecules. For example, you've learned that the formula for table sugar is $C_{12}H_{22}O_{11}$.

Forming a covalent bond lowers the potential energy

Earlier we saw that for ionic bonding to occur, the energy-lowering effect of the lattice energy must be greater than the combined energy-raising effects of the ionization energy (IE) and electron affinity (EA). Many times this is not possible, particularly when the ionization energies of all the atoms involved are large. This happens, for example, when nonmetals combine with each other to form molecules. In such cases, nature uses a different way to lower the energy—electron sharing.

Let's look at what happens when two hydrogen atoms join to form an H_2 molecule (Figure 9.2). As the two atoms approach each other, the electron of each atom begins to feel the attraction of both nuclei. This causes the electron density around each nucleus to shift toward the region between the two atoms. Therefore, as the distance between the nuclei decreases, there is an increase in the probability of finding either electron near either nucleus. In effect, as the molecule is formed, each of the hydrogen atoms in the H_2 molecule acquires a share of two electrons.

When the electron density shifts to the region between the two hydrogen atoms, it attracts both nuclei and pulls them together. Being of the same charge,

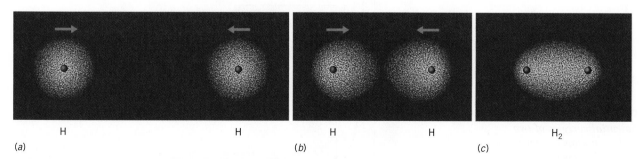

| H | H | H | H | H_2 |
| (a) | | (b) | | (c) |

FIGURE 9.2 *Formation of a covalent bond between two hydrogen atoms.* (*a*) Two H atoms separated by a large distance. (*b*) As the atoms approach each other, their electron densities are pulled into the region between the two nuclei. (*c*) In the H_2 molecule, the electron density is concentrated between the nuclei. Both electrons in the bond are distributed over both nuclei.

however, the two nuclei also repel each other, as do the two electrons. In the molecule that forms, therefore, the atoms are held at a distance at which all these attractions and repulsions are balanced. Overall, the nuclei are kept from separating, and the net force of attraction produced by the sharing of the pair of electrons is called a **covalent bond.**

Every covalent bond is characterized by two quantities: the average distance between the nuclei held together by the bond, and the amount of energy needed to separate the two atoms to produce neutral atoms again. In the hydrogen molecule, the attractive forces pull the nuclei to a distance of 75 pm, and this distance is called the **bond length** (or sometimes, the **bond distance**). Because a covalent bond holds atoms together, work must be done (energy must be supplied) to separate them. When the bond is *formed,* an equivalent amount of energy is released as the potential energies of the atoms are lowered. The amount of energy released when the bond is formed (or the amount of energy needed to "break" the bond) is called the **bond energy.**

> As the distance between the nuclei and the electron cloud that lies between them decreases, the potential energy decreases.

Figure 9.3 shows how the potential energy changes when two hydrogen atoms come together to form H_2. We see that the minimum potential energy occurs at a bond distance of 75 pm and that 1 mol of hydrogen molecules is more stable than 2 mol of hydrogen atoms by 435 kJ. In other words, the bond energy of H_2 is 435 kJ/mol.

Pairing of electrons occurs when a covalent bond forms

Before joining, each of the separate hydrogen atoms has one electron in a $1s$ orbital. When these electrons are shared, the $1s$ orbital of each atom is, in a sense, filled. Because the electrons now share the same space, they become paired as required by the Pauli exclusion principle; that is, m_s is $+\frac{1}{2}$ for one of the electrons and $-\frac{1}{2}$ for the other. In general, we almost always find that the electrons involved become paired when atoms form covalent bonds. In fact, a covalent bond is sometimes referred to as an **electron pair bond.**

> In Chapter 8 you learned that when two electrons occupy the same orbital, and therefore share the same space, their spins must be paired. The pairing of electrons is an important part of the formation of a covalent bond.

Lewis symbols are often used to keep track of electrons in covalent bonds. The electrons that are shared between two atoms are shown as a pair of dots placed between the symbols for the bonded atoms. The formation of H_2 from hydrogen atoms, for example, can be depicted as

$$H\cdot + H\cdot \longrightarrow H\!:\!H$$

Because the electrons are shared, each H atom is considered to have two electrons.

(Colored circles emphasize that two electrons can be counted around each of the H atoms.)

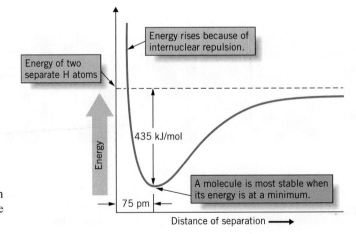

FIGURE 9.3 *Changes in the potential energies of two hydrogen atoms as they form H_2.* The energy of the molecule reaches a minimum when there is a balance between the attractions and repulsions.

FACETS OF CHEMISTRY 9.2

Sunlight and Skin Cancer

The ability of light to provide the energy for chemical reactions enables life to exist on our planet. Green plants absorb sunlight and, with the help of chlorophyll, convert carbon dioxide and water into carbohydrates (e.g., sugars and cellulose), which are essential constituents of the food chain. However, not all effects of sunlight are so beneficial.

As you know, light packs energy that's proportional to its frequency, and if the photons that are absorbed by a substance have enough energy, they can rupture chemical bonds and initiate chemical reactions. Light that is able to do this has frequencies in the UV region of the electromagnetic spectrum, and the sunlight bombarding the Earth contains substantial amounts of UV radiation. Fortunately, a layer of ozone (O_3) in the stratosphere, a region of the atmosphere extending from about 45 to 55 km altitude, absorbs most of the incoming UV, protecting life on

Dawn of a new day brings the risk of skin cancer to those particularly susceptible. Fortunately, understanding the risk allows us to protect ourselves with clothing and sunblock creams.

the surface. However, some UV radiation does get through, and the part of the spectrum of most concern is called "UV-B," with wavelengths between 280 and 320 nm.

What makes UV-B so dangerous is its ability to affect the DNA in our cells. (The structure of DNA and its replication is discussed in Chapter 25.) Absorption of UV radiation causes constituents of the DNA, called *pyrimidine bases,* to undergo reactions that form bonds between them. This causes transcription errors when the DNA replicates during cell division, giving rise to genetic mutations that can lead to skin cancers. These skin cancers fall into three classes—basal cell carcinomas, squamous cell carcinomas, and melanomas (the last being the most dangerous). Recent estimates indicate that in the United States there were about 500,000 cases of the first, 100,000 cases of the second, and 27,600 cases of the third. It has also been estimated that more than 90% of the skin cancers are due to absorption of UV-B radiation.

In recent years, concern has grown over the apparent depletion of the ozone layer in the stratosphere caused by the release of gases called *chlorofluorocarbons* (CFCs), which have been widely used in refrigerators and air conditioners. Some scientists have estimated a substantial increase in the rate of skin cancer caused by increased amounts of UV-B reaching the Earth's surface due to this ozone depletion.

For simplicity, the electron pair in a covalent bond is usually represented as a single dash. Thus, the hydrogen molecule is represented as

$$H\!-\!H$$

A formula such as this, which is drawn with Lewis symbols, is called a **Lewis formula** or **Lewis structure.** It is also called a **structural formula** because it shows which atoms are present in the molecule *and* how they are attached to each other.

Covalent bonding often follows the octet rule

You have seen that when a nonmetal atom forms an anion, electrons are gained until the s and p subshells of its valence shell are completed. This tendency to finish with a completed valence shell, usually consisting of eight electrons, also influences the number of electrons an atom tends to acquire by sharing, and it thereby controls the number of covalent bonds that an atom forms.

Hydrogen, with just one electron in its $1s$ orbital, can complete its valence shell by obtaining a share of just one electron from another atom. When this other atom is hydrogen, the H_2 molecule is formed. Because hydrogen obtains a stable valence shell configuration when it shares just one pair of electrons with another atom, a hydrogen atom forms only one covalent bond.

Many atoms form covalent bonds by sharing enough electrons to give them complete s and p subshells in their outer shells. This is the noble gas configuration mentioned earlier and is the basis of the octet rule described in Section 9.1. As applied to covalent bonding, the **octet rule** can be stated as follows: *When atoms form*

As you will see, it is useful to remember that hydrogen atoms only form one covalent bond.

covalent bonds, they tend to share sufficient electrons so as to achieve an outer shell having eight electrons.

Often, the octet rule can be used to explain the number of covalent bonds an atom forms. This number normally equals the number of electrons the atom must acquire to have a total of eight (an octet) in its outer shell. For instance, the halogens (Group VIIA) all have seven valence electrons. The Lewis symbol for a typical member of this group, chlorine, is

$$\cdot \overset{\cdot \cdot}{\underset{\cdot \cdot}{Cl}} :$$

We can see that only one electron is needed to complete its octet. Of course, chlorine can actually gain this electron and become a chloride ion. This is what it does when it forms an ionic compound such as sodium chloride (NaCl). But when chlorine combines with another nonmetal, the complete transfer of an electron is not energetically favorable. Therefore, in forming such compounds as HCl or Cl_2, chlorine gets the one electron it needs by forming a covalent bond.

$$H\cdot \; + \; \cdot \overset{\cdot \cdot}{\underset{\cdot \cdot}{Cl}} : \; \longrightarrow \; H : \overset{\cdot \cdot}{\underset{\cdot \cdot}{Cl}} :$$

$$: \overset{\cdot \cdot}{\underset{\cdot \cdot}{Cl}} \cdot \; + \; \cdot \overset{\cdot \cdot}{\underset{\cdot \cdot}{Cl}} : \; \longrightarrow \; : \overset{\cdot \cdot}{\underset{\cdot \cdot}{Cl}} : \overset{\cdot \cdot}{\underset{\cdot \cdot}{Cl}} :$$

The HCl and Cl_2 molecules can also be represented using dashes for the bonds.

$$H - \overset{\cdot \cdot}{\underset{\cdot \cdot}{Cl}} : \quad \text{and} \quad : \overset{\cdot \cdot}{\underset{\cdot \cdot}{Cl}} - \overset{\cdot \cdot}{\underset{\cdot \cdot}{Cl}} :$$

There are many nonmetals that form more than one covalent bond. For example, the three most important elements in biochemical systems are carbon, nitrogen, and oxygen.

$$\cdot \overset{\cdot}{\underset{\cdot}{C}} \cdot \qquad \cdot \overset{\cdot}{\underset{\cdot \cdot}{N}} \cdot \qquad \cdot \overset{\cdot \cdot}{\underset{\cdot \cdot}{O}} :$$

The simplest hydrogen compounds of these elements are methane, CH_4, ammonia, NH_3, and water, H_2O. Their Lewis structures are

$$
\begin{array}{ccc}
H & H & H \\
H : \overset{}{C} : H & H : \overset{\cdot \cdot}{N} : H & H : \overset{\cdot \cdot}{\underset{\cdot \cdot}{O}} : \\
\overset{}{H} & & \\
\end{array}
$$

$$
\begin{array}{ccc}
\text{or} & \text{or} & \text{or} \\
\end{array}
$$

$$
\begin{array}{ccc}
H & H & H \\
| & | & | \\
H - \overset{|}{\underset{|}{C}} - H & H - \overset{|}{\underset{\cdot \cdot}{N}} - H & H - \overset{}{\underset{\cdot \cdot}{O}} : \\
\overset{|}{H} & & \\
\text{methane} & \text{ammonia} & \text{water} \\
\end{array}
$$

It is useful to remember that *in most of their covalently bonded compounds, the number of covalent bonds formed by carbon, nitrogen, and oxygen are four, three, and two, respectively.*

Multiple bonds consist of two or more pairs of electrons

The bond produced by the sharing of *one* pair of electrons between two atoms is called a **single bond.** So far, these have been the only kind we've discussed. There are, however, many molecules in which more than a single pair of electrons are shared between two atoms. For example, we can diagram the formation of the bonds in CO_2 as follows:

$$: \overset{\cdot \cdot}{\underset{\cdot \cdot}{O}} \leftrightarrow \cdot \overset{\cdot}{C} \cdot \leftrightarrow \cdot \overset{\cdot \cdot}{\underset{\cdot \cdot}{O}} : \; \longrightarrow \; : \overset{\cdot \cdot}{\underset{}{O}} : : \overset{}{C} : : \overset{\cdot \cdot}{\underset{}{O}} :$$

The central carbon atom shares two of its electrons with each of the oxygen atoms, and each oxygen shares two electrons with carbon. The result is the formation of two **double bonds.** Notice that in the Lewis formula, both of the shared electron pairs are placed between the symbols for the two atoms joined by the double bond. Once again, if we circle the valence shell electrons that "belong" to each atom, we see that each has an octet.

8 electrons

The Lewis structure for CO_2, using dashes, is

$$:\ddot{O}=C=\ddot{O}:$$

Sometimes three pairs of electrons are shared between two atoms. The most abundant gas in the atmosphere, nitrogen, occurs in the form of diatomic molecules, N_2. As we've just seen, the Lewis symbol for nitrogen is

$$\cdot\ddot{N}:$$

and each nitrogen atom needs three electrons to complete its octet. When the N_2 molecule is formed, each of the nitrogen atoms shares three electrons with the other.

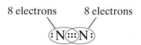

The result is called a **triple bond.** Again, notice that we place all three electron pairs of the bond between the two atoms. We count all of these electrons as though they belong to both of the atoms. Each nitrogen therefore has an octet.

8 electrons   8 electrons

$$(:N:::N:)$$

The triple bond is usually represented by three dashes, so the bonding in the N_2 molecule is normally shown as

$$:N\equiv N:$$

9.4 ▶ **Carbon compounds illustrate the variety of structures possible with covalent bonds**

Many of the substances we encounter on a daily basis can be classified as organic compounds. These include the foods we eat, the fabrics we wear, the medicines to cure our ills, the fuels that power vehicles, and many common solvents used in industry and around the home. We will frequently use organic compounds as examples in later discussions, so it will be helpful to know about some of the different types of organic compounds that exist and the kinds of structures that they have. In this section we will examine some of the ways carbon atoms combine with other atoms.

In general, organic compounds are held together by covalent bonds. In these substances carbon atoms form bonds to other atoms, such as hydrogen, oxygen, and nitrogen, to form a variety of classes of organic compounds. In Chapter 2 you learned that such substances can be considered to be derived from hydrocarbons— compounds in which the basic molecular "backbones" are composed of carbon

The locations of the unshared pairs of electrons around the oxygen are unimportant. Two equally valid Lewis structures for CO_2 are

$$:\ddot{O}=C=\ddot{O}: \quad \text{and} \quad \ddot{O}=C=\ddot{O}$$

A more comprehensive discussion of organic compounds is found in Chapter 25. In this section we look at some simple ways carbon atoms combine with other atoms to form certain important classes of organic substances that we encounter frequently.

atoms linked to one another in a chainlike fashion. (Hydrocarbons themselves are the principal constituents of petroleum.)

One of the chief features of organic compounds is the tendency of carbon to complete its octet by forming four covalent bonds. For example, in the alkane series of hydrocarbons (which we described briefly on page 52) all the bonds are single bonds, yielding compounds with the general formula C_nH_{2n+2}, where n is an integer. The structures of the first three alkanes (methane, ethane, and propane) are

These structures can be written in a condensed form as

CH_4

CH_3CH_3

$CH_3CH_2CH_3$

$$
\underset{\text{methane}}{\overset{\displaystyle H}{H-\underset{\displaystyle H}{C}-H}}
\qquad
\underset{\text{ethane}}{\overset{\displaystyle H \quad H}{H-\underset{\displaystyle H \quad H}{C-C}-H}}
\qquad
\underset{\text{propane}}{\overset{\displaystyle H \quad H \quad H}{H-\underset{\displaystyle H \quad H \quad H}{C-C-C}-H}}
$$

The shapes of their molecules are illustrated in Figure 2.9 on page 52.

When more than four carbon atoms are present, matters become more complex because there is more than one way to arrange the atoms. For example, butane has the formula C_4H_{10}, but there are two ways to arrange the carbon atoms. They occur in compounds commonly called butane and isobutane.

$$
\underset{\substack{\text{butane}\\ C_4H_{10}\\ \text{bp} = -0.5\,^\circ C}}{\overset{\displaystyle H \quad H \quad H \quad H}{H-\underset{\displaystyle H \quad H \quad H \quad H}{C-C-C-C}-H}}
\qquad
\underset{\substack{\text{isobutane}\\ C_4H_{10}\\ \text{bp} = -11.7\,^\circ C}}{\overset{\displaystyle H-\overset{\displaystyle H}{\underset{\displaystyle H}{C}}-H}{H-\underset{\displaystyle H \quad H \quad H}{\overset{\displaystyle H \quad\quad\ H}{C-C-C}}-H}}
$$

Butane and isobutane are said to be isomers of each other. In condensed form, we can write their structures as

$CH_3CH_2CH_2CH_3$

$\underset{CH_3CHCH_3}{\overset{CH_3}{|}}$

Even though they have the same molecular formula, they are actually different compounds with different properties, as you can see from the boiling points listed below their structures. The ability of atoms to arrange themselves in more than one way to give different compounds that have the same molecular formula is called *isomerism*, and is discussed more fully in Chapters 23 and 25. The existence of *isomers* is one of the reasons there are so many organic compounds. For example, the formula $C_{20}H_{42}$ is found for 366,319 distinct compounds that differ only in the way the carbon atoms are attached to each other!

Carbon can also complete its octet by forming double or triple bonds. The structures of ethylene, C_2H_4, and acetylene, C_2H_2, are shown below. Their proper IUPAC names are given in parentheses.[4]

$$
\underset{\substack{\text{ethylene}\\ \text{(ethene)}}}{\overset{\displaystyle H \quad H}{H-\underset{}{C=C}-H}}
\qquad
\underset{\substack{\text{acetylene}\\ \text{(ethyne)}}}{H-C\equiv C-H}
$$

Each of these is the first member of a class of hydrocarbons. Molecules that contain one carbon–carbon double bond (such as ethylene) are called **alkenes** and have the general formula C_nH_{2n}. Those that contain one carbon–carbon triple bond (such as acetylene) are called **alkynes** and have the general formula C_nH_{2n-2}.

---

[4]In the IUPAC system for naming organic compounds, *meth-*, *eth-*, *prop-*, and *but-* indicate carbon chains of 1, 2, 3, and 4 carbon atoms, respectively. Organic nomenclature is discussed more fully in Chapter 25.

Many organic compounds also contain oxygen and nitrogen

Most organic compounds contain elements in addition to carbon and hydrogen. As we mentioned in Chapter 2, it is convenient to consider such compounds to be derived from hydrocarbons by replacing one or more hydrogens by other groups of atoms. Such compounds can be divided into various families according to the nature of the groups attached to the parent hydrocarbon fragment. Some such families are summarized in Table 9.2.

Alcohols

In Chapter 2 it was noted that alcohols are organic compounds in which one of the hydrogen atoms of a hydrocarbon is replaced by OH. The family name for these compounds is *alcohol.* Examples are methyl alcohol (methanol) and ethyl alcohol (ethanol), which have the structures

Methyl alcohol (methanol) is the fuel in this can of Sterno.

$$
\begin{array}{cc}
\overset{\displaystyle H}{\underset{\displaystyle H}{H-\overset{|}{\underset{|}{C}}-\ddot{\overset{..}{O}}-H}} & \overset{\displaystyle H \quad H}{\underset{\displaystyle H \quad H}{H-\overset{|}{\underset{|}{C}}-\overset{|}{\underset{|}{C}}-\ddot{\overset{..}{O}}-H}} \\
\text{methyl alcohol} & \text{ethyl alcohol} \\
\text{(methanol)} & \text{(ethanol)}
\end{array}
$$

Some condensed formulas that we might write for these are CH_3OH and CH_3CH_2OH, or CH_3—OH and CH_3CH_2—OH. Methyl alcohol is used as a solvent and a fuel; ethyl alcohol is found in alcoholic beverages and is blended with gasoline in some states to yield a fuel called *gasohol.*

Ketones

In alcohols, the oxygen forms two single bonds to complete its octet, just as in water. But oxygen can also form double bonds, as you saw for CO_2. One family of

TABLE 9.2	SOME FAMILIES OF OXYGEN- AND NITROGEN-CONTAINING ORGANIC COMPOUNDS	
Family Name	General Formula[a]	Example
Alcohols	R—OH	CH_3—OH methyl alcohol
Aldehydes	$R-\overset{O}{\overset{\|}{C}}-H$	$CH_3-\overset{O}{\overset{\|}{C}}-H$ acetaldehyde
Ketones	$R-\overset{O}{\overset{\|}{C}}-R$	$CH_3-\overset{O}{\overset{\|}{C}}-CH_3$ acetone
Acids	$R-\overset{O}{\overset{\|}{C}}-OH$	$CH_3-\overset{O}{\overset{\|}{C}}-OH$ acetic acid
Amines	$R-NH_2$	CH_3-NH_2 methylamine
	R—NH—R	
	$R-\underset{\underset{\displaystyle R}{\|}}{N}-R$	

[a]R stands for a hydrocarbon fragment such as CH_3—, or CH_3CH_2—.

compounds in which a doubly bonded oxygen replaces a pair of hydrogen atoms is called *ketones*. The simplest example is acetone, a solvent in nail polish remover.

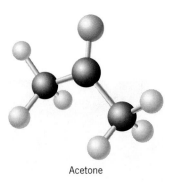

Acetone

$$H—\overset{\displaystyle H}{\underset{\displaystyle H}{\overset{|}{\underset{|}{C}}}}—\overset{:\overset{\displaystyle \cdot\cdot}{O}:}{\overset{\|}{C}}—\overset{\displaystyle H}{\underset{\displaystyle H}{\overset{|}{\underset{|}{C}}}}—H \qquad \text{or} \qquad CH_3—\overset{:O:}{\overset{\|}{C}}—CH_3$$

acetone
(propanone)

Ketones are found in many useful solvents that dissolve various plastics. An example is methyl ethyl ketone.

$$CH_3—\overset{:O:}{\overset{\|}{C}}—CH_2—CH_3$$

methyl ethyl ketone

Aldehydes

Notice that in ketones the carbon bonded to the oxygen is also attached to *two* other carbon atoms. If at least one of the atoms attached to the C=O group (called a *carbonyl group,* pronounced *car-bon-EEL*) is a hydrogen, a different family of compounds is formed called *aldehydes*. Examples are formaldehyde (used to preserve biological specimens, for embalming, and to make plastics) and acetaldehyde (used in the manufacture of perfumes, dyes, plastics, and other products).

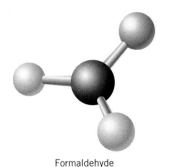

Formaldehyde

$$H—\overset{:O:}{\overset{\|}{C}}—H \qquad CH_3—\overset{:O:}{\overset{\|}{C}}—H$$

formaldehyde   acetaldehyde

Organic acids

Another important family of oxygen-containing organic compounds is the organic acids, also called *carboxylic acids*. An example is acetic acid.

$$H—\overset{\displaystyle H}{\underset{\displaystyle H}{\overset{|}{\underset{|}{C}}}}—\overset{:O:}{\overset{\|}{C}}—\overset{\cdot\cdot}{O}—H \qquad \text{or} \qquad CH_3—\overset{:O:}{\overset{\|}{C}}—\overset{\cdot\cdot}{O}—H$$

acetic acid

Notice that the acid has both a doubly bonded oxygen and an OH group attached 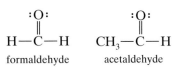 to the end carbon atom. In general, molecules that contain the $—\overset{O}{\overset{\|}{C}}—O—H$ group (called a *carboxyl group*) are weak acids. When they react with water, the hydrogen that's transferred to the water to give H_3O^+ is the one bonded to the oxygen. For example, in Chapter 5 (page 177) we described the equilibrium reaction of acetic acid with water to form hydronium ion and acetate ion.

$$H—\overset{\displaystyle H}{\underset{\displaystyle H}{\overset{|}{\underset{|}{C}}}}—\overset{:O:}{\overset{\|}{C}}—\overset{\cdot\cdot}{O}—H + H_2O \rightleftharpoons H_3O^+ + \left[H—\overset{\displaystyle H}{\underset{\displaystyle H}{\overset{|}{\underset{|}{C}}}}—\overset{:O:}{\overset{\|}{C}}—\overset{\cdot\cdot}{\underset{\cdot\cdot}{O}}: \right]^-$$

acetic acid   acetate ion

Many of the acids we will use as examples in future discussions will be organic acids. Almost all of them are weak acids and in water they participate in the same kind of ionization equilibrium as acetic acid.

Amines

Nitrogen atoms need three electrons to complete an octet and in most of its compounds, nitrogen forms three bonds. The common nitrogen-containing organic compounds can be imagined as being derived from ammonia by replacing one or more of the hydrogens of NH_3 with hydrocarbon groups. They're called *amines,* and an example is methylamine, CH_3NH_2.

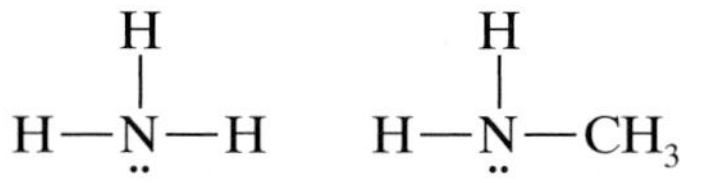

H   H
| |
H—N—H   H—N—CH_3

ammonia   methylamine

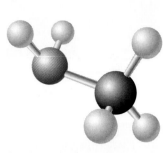

Methylamine

Amines are strong-smelling compounds and often have a "fishy" odor. Like ammonia, they're basic.

$$CH_3NH_2(aq) + H_2O \rightleftharpoons CH_3NH_3^+(aq) + OH^-(aq)$$

In reactions such as this, the hydrogen ion always becomes attached to the nitrogen of the amine. How this bond is formed will be discussed in Section 9.10.

Amino Acids

Some molecules incorporate atom groups that belong to two or more classes of compounds. An example from biology is a series of compounds called *amino acids,* which contain the amine group ($—NH_2$) as well as a carboxyl group ($—CO_2H$). The simplest of these is the amino acid glycine,

Amino acids combine in long chains to form the proteins in our bodies.

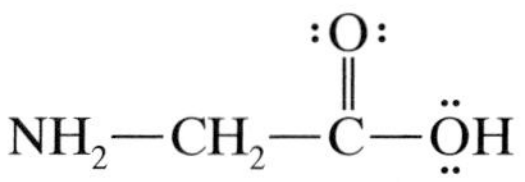

:O:
||
NH_2—CH_2—C—ÖH

glycine

We have just touched the surface of the subject of organic chemistry here, but we have introduced you to some of the kinds of organic compounds you will encounter in chapters ahead.

Glycine (an amino acid)

PRACTICE EXERCISE 5: Match the structural formulas on the left with the correct names of the families of organic compounds to which they belong.

O
||
R—C—H   amine

H
|
R—N—H   alcohol

O
||
R—C—OH   ketone

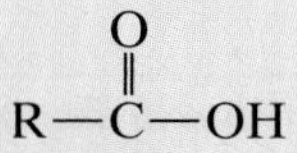

O
||
R—C—R   aldehyde

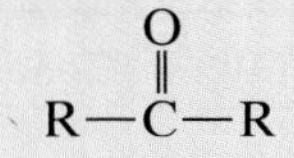

R—OH   acid

9.5 ▶ **Covalent bonds can have partial charges at opposite ends**

When two identical atoms form a covalent bond, as in H_2 or Cl_2, each has an equal share of the bond's electron pair. The electron density at both ends of the bond is the same, because the electrons are equally attracted to both nuclei. However,

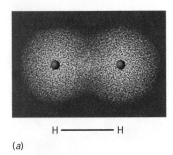

H ——— H

(a)

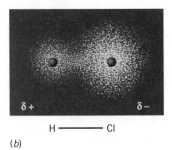

δ+   δ−

H ——— Cl

(b)

FIGURE 9.4 *Nonpolar and polar covalent bonds.* (*a*) The electron density of the electron pair in the bond is spread evenly between the two H atoms in H_2, which gives a nonpolar covalent bond. (*b*) In HCl, the electron density of the bond is pulled more tightly around the Cl end of the molecule, causing that end of the bond to become slightly negative. At the same time, the opposite end of the bond becomes slightly positive. The result is a polar covalent bond.

when different kinds of atoms combine, as in HCl, one nucleus usually attracts the electrons in the bond more strongly than the other.

The result of unequal attractions for the bonding electrons is an unbalanced distribution of electron density within the bond. For example, it has been found that in a bond, a chlorine atom attracts electrons more strongly than does a hydrogen atom. In the HCl molecule, therefore, the electron cloud is pulled more tightly around the Cl, and that end of the molecule experiences a slight buildup of negative charge. The electron density that shifts toward the chlorine is removed from the hydrogen, which causes the hydrogen end to acquire a slight positive charge. These charges are less than full 1+ and 1− charges and are called **partial charges,** which are usually indicated by the lowercase Greek letter delta, δ (see Figure 9.4). Partial charges can also be indicated on Lewis structures. For example,

$$H\!-\!\ddot{\underset{..}{Cl}}:$$
$$\delta+ \quad \delta-$$

A bond that carries partial positive and negative charges on opposite ends is called a **polar covalent bond,** or often simply a **polar bond** (the word *covalent* is understood). The term *polar* comes from the notion of *poles* of equal but opposite charge at either end of the bond. Because *two poles* of electric charge are involved, the bond is said to be an **electric dipole.**

The polar bond in HCl causes the molecule as a whole to have opposite charges on either end, so the HCl molecule as a whole is an electric dipole. We say that HCl is a **polar molecule.** The magnitude of its polarity is expressed quantitatively by its **dipole moment** (symbol μ), which is equal to the amount of charge on either end of the molecule, q, multiplied by the distance between the charges, r.

$$\mu = q \times r \qquad (9.1)$$

By separate experiments, it is possible to measure both μ and r (which corresponds to the bond length in a diatomic molecule such as HCl). Table 9.3 lists the dipole moments and bond lengths for some diatomic molecules. The dipole moments are reported in *debye* units (symbol D), where $1\text{ D} = 3.34 \times 10^{-30}$ C m (coulomb meter).

Knowledge of μ and r allows calculation of the amount of charge on opposite ends of the dipole. For HCl, such calculations show that q equals 0.17 electronic charge units, which means the hydrogen carries a charge of $+0.17e^-$ and the chlorine a charge of $-0.17e^-$.

Electronegativity expresses an atom's attraction for electrons in a bond

The degree to which a covalent bond is polar depends on the difference in the abilities of the bonded atoms to attract electrons. The greater the difference, the more polar the bond is and the more the electron density is shifted toward the atom that attracts electrons more.

TABLE 9.3	DIPOLE MOMENTS AND BOND LENGTHS FOR SOME DIATOMIC MOLECULES[a]	
Compound	Dipole Moment (D)	Bond Length (pm)
HF	1.83	91.7
HCl	1.09	127
HBr	0.82	141
HI	0.45	161
CO	0.11	113
NO	0.16	115

[a]Source: National Institute of Standards and Technology.

The term that we use to describe the relative attraction of an atom for the electrons in a bond is called the **electronegativity** of the atom. In HCl, for example, chlorine is *more electronegative* than hydrogen. The electron pair of the covalent bond spends more of its time around the more electronegative atom, which is why that end of the bond acquires a partial negative charge.

The first scientist to develop a set of numerical values for electronegativity was Linus Pauling (1901–1994). He found that polar bonds had a greater bond energy than would be expected if the opposite ends of the bonds were electrically neutral. Pauling reasoned that the attraction between the partial charges on opposite ends added to the stability of the bond. By measuring the extra bond energy, he was able to develop a scale of electronegativities for the elements. Other scientists have used different approaches to measuring electronegativities, with similar results.

A set of numerical values for the electronegativities of the elements is shown in Figure 9.5. These data are useful because the *difference* in electronegativity provides an estimate of the degree of polarity of a bond. In addition, the relative magnitudes of the electronegativities indicate which end of the bond carries the partial negative charge. For instance, fluorine is more electronegative than chlorine. Therefore, we expect an HF molecule to be more polar than an HCl molecule. (This is confirmed by the larger dipole moment of the HF molecule.) In addition, hydrogen is less electronegative than either fluorine or chlorine, so in both of these molecules the hydrogen bears the partial positive charge.

Linus Pauling (1901–1994) contributed greatly to our understanding of chemical bonding. He was the winner of two Nobel Prizes, in 1954, for Chemistry and in 1962, for Peace.

The noble gases are assigned electronegativities of zero and are omitted from the table.

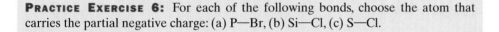

$$H\!-\!\ddot{F}\!: \qquad H\!-\!\ddot{C}l\!:$$
$$\ \ \delta+ \ \ \delta- \qquad\ \delta+ \ \ \delta-$$

PRACTICE EXERCISE 6: For each of the following bonds, choose the atom that carries the partial negative charge: (a) P—Br, (b) Si—Cl, (c) S—Cl.

By examining electronegativity values and their differences we find that there is no sharp dividing line between ionic and covalent bonding. Ionic bonding and *nonpolar covalent bonding* simply represent the two extremes. A bond is mostly ionic when the difference in electronegativity between two atoms is very large; the more electronegative atom acquires essentially complete control of the bonding

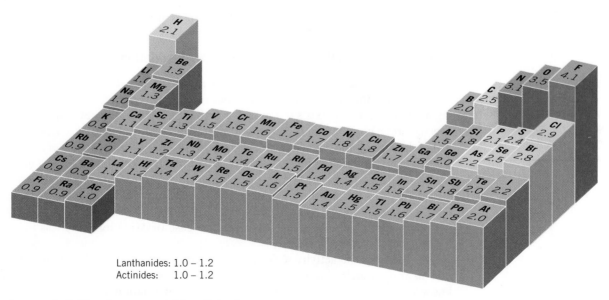

Lanthanides: 1.0 – 1.2
Actinides: 1.0 – 1.2

FIGURE 9.5 *The electronegativities of the elements.*

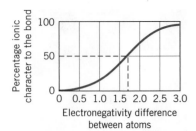

FIGURE 9.6 *Variation in the percentage ionic character of a bond with electronegativity difference.* The bond becomes about 50% ionic when the electronegativity difference equals 1.7, which means that the atoms in the bond carry a partial charge of approximately ±0.5 units.

Periodic trends in electronegativity

It is found that the electronegativity is proportional to the average of the ionization energy and the electron affinity of an element.

electrons. In a **nonpolar covalent bond,** there is no difference in electronegativity, so the pair of bonding electrons is shared equally.

$$Cs^+ \left[:\overset{..}{\underset{..}{F}}: \right]^- \qquad :\overset{..}{\underset{..}{F}}:\overset{..}{\underset{..}{F}}:$$

"bonding pair" held   bonding pair
exclusively by fluorine   shared equally

The degree to which the bond is polar, which we might think of as the amount of **ionic character** of the bond, varies in a continuous way with changes in the electronegativity difference (Figure 9.6). The bond becomes more than 50% ionic when the electronegativity difference exceeds approximately 1.7.

Electronegativity follows trends within the periodic table

Figure 9.5 also illustrates the trends in electronegativity within the periodic table; *electronegativity increases from bottom to top in a group and from left to right in a period.* Notice that the trends follow those for ionization energy. This is because an atom that has a small IE will give away an electron more easily than an atom with a large IE, just as an atom with a small electronegativity will lose its share of an electron pair more readily than an atom with a large electronegativity.

Elements located in the same region of the table (for example, the nonmetals) have similar electronegativities, which means that if they form bonds with each other, the electronegativity differences will be small and the bonds will be more covalent than ionic. On the other hand, if elements from widely separated regions of the table combine, large electronegativity differences occur and the bonds will be predominantly ionic. This is what happens, for example, when an element from Group IA or Group IIA reacts with a nonmetal from the upper right-hand corner of the periodic table.

9.6 ▶ The reactivities of metals and nonmetals can be related to their electronegativities

In addition to enabling us to determine the polarity of covalent bonds, electronegativities can also be related to chemical properties, particularly those that involve an atom's tendency to gain or lose electrons. Thus, there are parallels between an element's electronegativity and its **reactivity**—its tendency to undergo redox reactions. In this section we will examine some of these trends.

Reactivities of metals relate to their ease of oxidation

Reactivity refers in general to the tendency of a substance to react with something. The *reactivity of a metal* refers to its tendency specifically to undergo *oxidation.*

When we raise questions about the *reactivity* of a metal, we are concerned with how easily the metal is oxidized. This is because in nearly every compound containing a metal, the metal exists in a positive oxidation state, and when a free metal reacts to form a compound, it loses one or more electrons and is oxidized. As a result, a metal like sodium, which is very easily oxidized, is said to be very reactive, whereas a metal like platinum, which is very difficult to oxidize, is said to be unreactive.

There are several ways to compare how easily metals are oxidized. In Chapter 6 we saw that by comparing the abilities of metals to displace each other from compounds we are able to establish their relative ease of oxidation. This was the basis for the activity series (Table 6.2).

As useful as the activity series is in predicting the outcome of certain redox reactions, it is difficult to remember in detail. Furthermore, often it is sufficient just to know approximately where an element stands in relation to others in a broad range of reactivity. This is where the periodic table can be especially useful to us

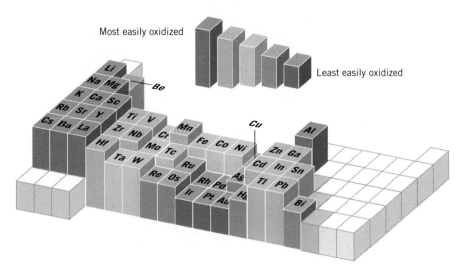

FIGURE 9.7 *The variation of the ease of oxidation of metals with position in the periodic table.*

once again, because there are trends and variations in reactivity within the periodic table that are simple to identify and remember.

Figure 9.7 illustrates how the ease of oxidation (the reactivity) of metals varies in the periodic table. In general, these trends roughly follow the variations in electronegativity, with the metal being less easily oxidized as its electronegativity increases. You might expect this, because electronegativity is a measure of how strongly the atom of an element attracts electrons when combining with an atom of a different element. The more strongly the atom attracts electrons, the more difficult it is to oxidize. This relationship between reactivity and electronegativity is only approximate, however, because many other factors affect the stability of the compounds that are formed.

In Figure 9.7, we see that the metals that are most easily oxidized are found at the far left in the periodic table. The metals in Group IA, for example, are so easily oxidized that all of them react with water to liberate hydrogen. Because of their reactivity toward moisture and oxygen, they have no useful applications that require exposure to the atmosphere, so we rarely encounter them as free metals. The same is true of the heavier metals in Group IIA, calcium through barium. These elements also react with water to liberate hydrogen.

You've learned how the electronegativities of the elements vary within the periodic table, increasing as we move from left to right in a period. This places metals with the lowest electronegativities at the left of the table—in other words, in Groups IA and IIA. It should be no surprise, therefore, that these elements are the easiest to oxidize. You also learned that electronegativity decreases as we go down a group. This explains why the heavier elements in Group IIA are easier to oxidize than those at the top of the group.

In Figure 9.7 we can also locate the metals that are the most difficult to oxidize. They occur for the most part among the heavier transition elements in the center of the periodic table, where we find the very unreactive elements platinum and gold—metals used to make fine jewelry. Their bright luster and lack of any tendency to corrode in air or water combine to make them particularly attractive for this purpose. This same lack of reactivity also is responsible for their industrial uses. Gold, for example, is used to coat the electrical contacts in low-voltage circuits found in microcomputers, because even small amounts of corrosion on more reactive metals would be sufficient to impede the flow of electricity so much as to make the devices unreliable.

The oxidizing power of nonmetals is related to their electronegativities

The reactivity of a metal is determined by its ease of oxidation and therefore its ability to serve as a reducing agent. *For a nonmetal, reactivity is usually gauged by*

The funeral mask of Tut'ankhamun, almost 3,300 years old, shows no sign of age. It was made of gold with glass eyes and lapis lazuli eyebrows and eyelashes.

its ability to serve as an oxidizing agent. This ability also varies according to the element's electronegativity. Nonmetals with high electronegativities have strong tendencies to acquire electrons and are therefore strong oxidizing agents. In parallel with changes in electronegativities in the periodic table, *the oxidizing abilities of nonmetals increase from left to right across a period and from bottom to top in a group.* Thus, the most powerful oxidizing agent is fluorine, followed closely by oxygen, both in the upper right-hand corner of the periodic table.

Single replacement reactions occur among the nonmetals, just as with the metals (which you studied in Chapter 6). For example, heating a metal sulfide in oxygen causes the sulfur to be replaced by oxygen. The displaced sulfur then combines with additional oxygen to give sulfur dioxide. The equation for a typical reaction is

$$CuS(s) + \tfrac{3}{2}O_2(g) \longrightarrow CuO(s) + SO_2(g)$$

Displacement reactions are especially evident among the halogens, where a particular halogen, as an element, will oxidize the *anion* of any halogen below it in Group VIIA. Thus, F_2 will oxidize Cl^-, Br^-, and I^-. However, Cl_2 will only oxidize Br^- and I^-, and Br_2 will only oxidize I^-. Thus, the reactions in the margin are observed.

Fluorine:

$$F_2 + 2Cl^- \longrightarrow 2F^- + Cl_2$$
$$F_2 + 2Br^- \longrightarrow 2F^- + Br_2$$
$$F_2 + 2I^- \longrightarrow 2F^- + I_2$$

Chlorine:

$$Cl_2 + 2Br^- \longrightarrow 2Cl^- + Br_2$$
$$Cl_2 + 2I^- \longrightarrow 2Cl^- + I_2$$

Bromine:

$$Br_2 + 2I^- \longrightarrow 2Br^- + I_2$$

9.7 ▶ Drawing Lewis structures is a necessary skill

Lewis structures are very useful in chemistry because they give us a relatively simple way to describe the structures of molecules. As a result, much chemical reasoning is based on them. Also, as you will learn in the next chapter, we can use the Lewis structure to make reasonably accurate predictions about the shape of a molecule.

In Sections 9.3 and 9.4 you saw Lewis structures for a variety of molecules that obey the octet rule. Examples included CO_2, Cl_2, N_2, and a variety of compounds of carbon. The octet rule is not always obeyed, however. For instance, there are some molecules in which one or more atoms must have more than an octet in the valence shell. Examples are PCl_5 and SF_6, whose Lewis structures are

Lewis structures just describe which atoms are bonded to each other and the kinds of bonds involved. Thus, the Lewis structure for water can be drawn as H—Ö—H , but it does not mean the water molecule is linear, with all the atoms in a straight line. Actually, water isn't linear; the two O—H bonds form an angle of about 104°.

SF_6 is used as a gaseous insulator in high-voltage electrical equipment.

In these molecules the formation of more than four bonds to the central atom requires that the central atom have a share of more than eight electrons. With the exception of Period 2 elements like carbon and nitrogen, most nonmetals can have more than an octet of electrons in the outer shell.

There are also some molecules (but not many) in which the central atom behaves as though it has less than an octet. The most common examples involve compounds of beryllium and boron.

$$\cdot Be \cdot + 2 \cdot \ddot{C}l : \longrightarrow : \ddot{C}l - Be - \ddot{C}l :$$

four electrons around Be

$$\dot{B} \cdot + 3 \cdot \ddot{C}l : \longrightarrow : \ddot{C}l - B - \ddot{C}l :$$

six electrons around B

Although Be and B sometimes have less than an octet, the elements in Period 2 never exceed an octet. The reason is that their valence shells, having $n = 2$, can hold a maximum of only 8 electrons. (This explains why the octet rule works so well for atoms of carbon, nitrogen, and oxygen.) However, elements in periods below Period 2, such as phosphorus and sulfur, sometimes do exceed an octet, because their valence shells can hold more than 8 electrons. For example, the valence shell for elements in Period 3, for which $n = 3$, can hold a maximum of 18 electrons, and the valence shell for Period 4 elements, which have s, p, d, and f subshells, can hold as many as 32 electrons.

A method for drawing lewis structures

We have used the Lewis structures of a variety of compounds in our discussions of covalent bonds. Such structures are also drawn for polyatomic ions, which are held together by covalent bonds, too. We now turn our attention to how these structures can be obtained in a systematic way.

The method of writing Lewis structures can be broken down into a number of steps, which are summarized in Figure 9.8. The first step is to decide which atoms are bonded to each other so that we know where to put the dots or dashes. This is not always a simple matter. Many times the formula suggests the way the atoms are arranged because the central atom, which is usually the least electronegative one, is usually written first. (Hydrogen is an exception; it forms only one ordinary covalent bond, so even if it is the least electronegative element in the compound, we would not choose it to be a central atom.) Examples are CO_2 and ClO_4^-, which have the following *skeletal structures* (i.e., arrangements of atoms):

Sometimes, obtaining the skeletal structure is not quite so simple, especially when more than two elements are present. Some generalizations are possible, however. For example, the skeletal structure of nitric acid, HNO_3, is

$$\begin{matrix} & & O & \\ H & O & N & O \end{matrix} \quad \text{(correct)}$$

rather than one of the following:

$$\begin{matrix} O & & & & & & & & \\ O & N & O & \quad \text{or} \quad & H & O & O & N & O \\ H & & & & & & & & \end{matrix} \quad \text{(incorrect)}$$

Nitric acid is an oxoacid (Section 5.6), and it happens that the hydrogen atoms that can be released from molecules of oxoacids are always bonded to oxygen atoms, which are in turn bonded to the third nonmetal atom. Therefore, recognizing HNO_3 as the formula of an oxoacid allows us to predict that the three oxygen atoms are bonded to the nitrogen, and the hydrogen is bonded to one of the oxygens. (Also, remember that hydrogen forms only one bond, so we wouldn't choose it as the central atom.)

There are times when no reasonable basis can be found for choosing a particular skeletal structure. If you must make a guess, choose the most symmetrical arrangement of atoms, because it has the greatest chance of being correct.

> **PRACTICE EXERCISE 7:** Predict reasonable skeletal structures for SO_2, NO_3^-, $HClO_3$, and H_3PO_4.

After you've decided on the skeletal structure, the next step is to count all of the *valence electrons* to find out how many dots must appear in the final formula. Using the periodic table, locate the groups in which the elements in the formula occur to de-

TOOLS

Method for drawing Lewis structures

The valence shell of hydrogen contains only the 1s subshell, which can hold a maximum of two electrons. This means hydrogen can have a share of only two electrons and can form just one covalent bond.

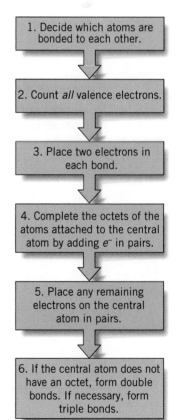

FIGURE 9.8 *Summary of steps in writing a Lewis structure.* If you follow these steps, you will obtain a Lewis structure in which the octet rule is obeyed by the maximum number of atoms.

termine the number of valence electrons contributed by each atom. If the structure you wish to draw is that of an ion, *add one additional valence electron for each negative charge or remove a valence electron for each positive charge.* Some examples are

SO_3	Sulfur (Group VIA) contributes $6e^-$.	$1 \times 6 =$	$6e^-$
	Each oxygen (Group VIA) contributes $6e^-$.	$3 \times 6 =$	$18e^-$
		Total	$24e^-$
ClO_4^-	Chlorine (Group VIIA) contributes $7e^-$.	$1 \times 7 =$	$7e^-$
	Each oxygen (Group VIA) contributes $6e^-$.	$4 \times 6 =$	$24e^-$
	Add $1e^-$ for the $1-$ charge.		$+1e^-$
		Total	$32e^-$
NH_4^+	Nitrogen (Group VA) contributes $5e^-$.	$1 \times 5 =$	$5e^-$
	Each hydrogen (Group IA) contributes $1e^-$.	$4 \times 1 =$	$4e^-$
	Subtract $1e^-$ for the $1+$ charge.		$-1e^-$
		Total	$8e^-$

Practice Exercise 8: How many valence electrons should appear in the Lewis structures of SO_2, PO_4^{3-}, and NO^+?

After we have determined the number of valence electrons, we place them into the skeletal structure in pairs following the steps outlined in Figure 9.8. Let's look at some examples of how we go about this.

EXAMPLE 9.5
Drawing Lewis Structures

What is the Lewis structure of the chloric acid molecule, $HClO_3$?

ANALYSIS: The first step is to select a reasonable skeletal structure. Because the substance is an oxoacid, we can expect the hydrogen to be bonded to an oxygen, which in turn is bonded to the chlorine. The other two oxygens would also be bonded to the chlorine. This gives

$$O$$
$$H \quad O \quad Cl \quad O$$

After this, we follow the procedure outlined in Figure 9.8.

SOLUTION: The total number of valence electrons is 26 ($1e^-$ from H, $6e^-$ from each O, and $7e^-$ from Cl). To distribute the electrons, we start by placing a pair of electrons in each bond, because we know that there must be at least one pair of electrons between each pair of atoms.

$$O$$
$$H : O : \ddot{C}l : O$$

This has used $8e^-$, so we still have $18e^-$ to go. Next, we work on the atoms surrounding the chlorine (which is the central atom in this structure). No additional electrons are needed around the H because $2e^-$ are all that can occupy its valence shell. Therefore, we next complete the octets of the oxygens.

$$: \ddot{O} :$$
$$H : \ddot{O} : \ddot{C}l : \ddot{O} :$$

We have now used a total of $24e^-$, so there are two electrons left. "Leftover" electrons are always placed on the central atom (the Cl atom, in this case). This gives

$$: \ddot{O} :$$
$$H : \ddot{O} : \ddot{C}l : \ddot{O} :$$

which we can also write as follows, using dashes for the electron pairs in the bonds.

$$\overset{\displaystyle :\!\ddot{O}\!:}{\underset{\displaystyle }{H-\ddot{O}-\overset{\displaystyle |}{Cl}-\ddot{O}\!:}}$$

The chlorine and the three oxygens have octets, and the valence shell of hydrogen is complete with $2e^-$, so we are finished.

Is the Answer Reasonable?

The most common error is to have either too many or too few valence electrons in the structure, so that's always the best place to begin your check. Doing this will confirm that the number of e^- is correct.

Draw the Lewis structure for the SO_3 molecule.

ANALYSIS: Sulfur is written first in the formula, so we expect it to be the central atom, surrounded by three oxygen atoms. This gives the skeletal structure

$$\begin{array}{c} O \\ O \quad S \quad O \end{array}$$

From here we proceed to count valence electrons and follow the appropriate steps in entering them into the structure.

SOLUTION: The total number of electrons in the formula is 24 ($6e^-$ from the sulfur plus $6e^-$ from each oxygen). We begin to distribute the electrons by placing a pair in each bond. This gives

$$\begin{array}{c} O \\ O\!:\!\ddot{S}\!:\!O \end{array}$$

We have used $6e^-$, so there are $18e^-$ left. We next complete the octets around the oxygens, which uses the remaining electrons.

$$\begin{array}{c} :\!\ddot{O}\!: \\ :\!\ddot{O}\!:\!\ddot{S}\!:\!\ddot{O}\!: \end{array}$$

At this point all the electrons have been placed into the structure, but we see that the sulfur still lacks an octet. We cannot simply add more dots because the total must be 24. Therefore, according to the last step of the procedure in Figure 9.8, we have to create a multiple bond. To do this we move a pair of electrons that we have shown to belong solely to an oxygen into a sulfur–oxygen bond so that it can be counted as belonging to both the oxygen *and* the sulfur. In other words, we place a double bond between sulfur and one of the oxygens. It doesn't matter which oxygen we choose for this honor.

$$\begin{array}{c} :\!\ddot{O}\!: \\ :\!\ddot{O}\!\curvearrowleft\!\ddot{S}\!:\!\ddot{O}\!: \end{array} \text{ gives } \begin{array}{c} :\!\ddot{O}\!: \\ :\!\ddot{O}\!:\!:\!\ddot{S}\!:\!\ddot{O}\!: \end{array} \text{ or } \begin{array}{c} :\!\ddot{O}\!: \\ :\!\ddot{O}\!=\!\overset{\displaystyle |}{S}\!-\!\ddot{O}\!: \end{array}$$

Notice that each atom has an octet.

Is the Answer Reasonable?

The key step in completing the structure is recognizing what we have to do to obtain an octet around the sulfur. We have to add more electrons to the valence shell of sulfur, but without removing them from any of the oxygen atoms. By forming the double bond, we accomplish this. A quick check also confirms that we've placed exactly the correct number of valence electrons into the structure.

What is the Lewis structure of CO?

ANALYSIS: There's really not much to analyze here. There are just two atoms, so they must be next to each other. Therefore, the skeletal structure is simply

$$C \quad O$$

EXAMPLE 9.6

Drawing Lewis Structures

EXAMPLE 9.7

Drawing Lewis Structures

Now we follow the usual procedure.

SOLUTION: The total number of valence electrons is ten ($4e^-$ from carbon plus $6e^-$ from oxygen). We begin distributing them by placing a pair into the bond.

$$C:O$$

Now we try to complete the octets with the remaining eight electrons. To do this, we arbitrarily select one atom, say carbon, to be the "central atom" and complete the valence shell of the other. The two electrons that remain are then placed on the carbon to give

$$:C:\ddot{O}:$$

Carbon has only four electrons at this point, so we have to give it four more, but without removing any from the oxygen. We can do this by moving two unshared pairs of electrons that are on the oxygen into the bond. This gives a triple bond.

$$:C \!\overset{\frown}{\underset{\smile}{:}}\! \ddot{O}: \quad \text{gives} \quad :C:::O: \quad \text{or} \quad :C \equiv O:$$

Is the Answer Reasonable?

All the valence electrons are accounted for and both C and O have octets, so the structure is correct.

Carbon monoxide is one of the few molecules in which carbon has less than four bonds and in which oxygen has more than two.

EXAMPLE 9.8
Drawing Lewis Structures

What is the Lewis structure for the ion IF_4^-?

ANALYSIS: We can anticipate that iodine will be the central atom, so our skeletal structure is

$$
\begin{array}{c}
F \\
F \quad I \quad F \\
F
\end{array}
$$

Next, we count valence electrons, remembering to add an extra electron to account for the negative charge. Then we distribute the electrons in pairs following the usual procedure.

SOLUTION: The iodine and fluorine atoms are in Group VIIA and each contribute 7 electrons, for a total of $35e^-$. The negative charge requires one additional electron, to give a total of $36e^-$.

First we place $2e^-$ into each bond, and then we complete the octets of the fluorine atoms. This uses 32 electrons.

$$
\begin{array}{c}
:\ddot{F}: \\
:\ddot{F}:\ddot{I}:\ddot{F}: \\
:\ddot{F}:
\end{array}
\quad \text{or} \quad
\begin{array}{c}
:\ddot{F}: \\
| \\
:\ddot{F}\!-\!I\!-\!\ddot{F}: \\
| \\
:\ddot{F}:
\end{array}
$$

There are four electrons left, and according to step 5 in Figure 9.8 they are placed on the central atom as *pairs* of electrons. This gives

$$
\begin{array}{c}
:\ddot{F}: \\
| \\
:\ddot{F}\!-\!\overset{\cdot\cdot}{\underset{\cdot\cdot}{I}}\!-\!\ddot{F}: \\
| \\
:\ddot{F}:
\end{array}
$$

The last step is to add brackets around the formula and write the charge outside as a superscript.

$$
\left[
\begin{array}{c}
:\ddot{F}: \\
| \\
:\ddot{F}\!-\!\overset{\cdot\cdot}{\underset{\cdot\cdot}{I}}\!-\!\ddot{F}: \\
| \\
:\ddot{F}:
\end{array}
\right]^-
$$

Is the Answer Reasonable?
We can recount the valence electrons, which tells us we have the right number of them, and all are in the Lewis structure. Each fluorine atom has an octet, which is proper. Notice that we have placed the "leftover" electrons onto the central atom. This gives iodine more than an octet, but that's okay because iodine is not a Period 2 element.

PRACTICE EXERCISE 9: Draw Lewis structures for OF_2, NH_4^+, SO_2, NO_3^-, ClF_3, and $HClO_4$.

9.8 Formal charges help select correct Lewis structures

Lewis structures are meant to describe how atoms share electrons in chemical bonds. Such descriptions are theoretical explanations or predictions that relate to the forces that hold molecules and polyatomic ions together. But, as you learned in Chapter 1, a theory is only as good as the observations on which it is based, so to have confidence in a theory about chemical bonding, we need to have a way to check it. We need experimental observations that relate to the description of bonding.

Two properties that are related to the number of electron pairs shared between two atoms are **bond length,** the distance between the nuclei of the bonded atoms, and **bond energy,** the energy required to separate the bonded atoms to give neutral particles. For example, we mentioned in Section 9.3 that measurements have shown the H_2 molecule has a bond length of 75 pm and a bond energy of 435 kJ/mol, which means that it takes 435 kJ to break the bonds of 1 mol of H_2 molecules to give 2 mol of hydrogen atoms.

For bonds between the same elements, the bond length and bond energy depend on the **bond order,** which is defined as *the number of pairs of electrons shared between two atoms.* The bond order is a measure of the amount of electron density in the bond, and the greater the electron density, the more tightly the nuclei are held and the more closely they are drawn together. This is illustrated by the data in Table 9.4, which gives typical bond lengths and bond energies for single, double, and triple bonds between carbon atoms. In summary:

> As the bond order increases, the bond length decreases and the bond energy increases, provided we are comparing bonds between the same elements.

With this as background, let's examine the Lewis structure of sulfuric acid, drawn following the rules given in the preceding section.

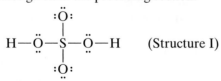

(Structure I)

It obeys the octet rule, and there doesn't seem to be any need to attempt to write any other structures for it. But a problem arises if we compare the predicted bond

Bond length = 75 pm

The hydrogen molecule has a bond length of 75 pm.

A single bond has a bond order of 1; a double bond, a bond order of 2; and a triple bond, a bond order of 3.

Correlation between bond properties and bond order

Bond	Bond Length (pm)	Bond Energy (kJ/mol)
TABLE 9.4	**AVERAGE BOND LENGTHS AND BOND ENERGIES MEASURED FOR CARBON–CARBON BONDS**	
C—C	154	348
C=C	134	615
C≡C	120	812

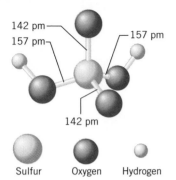

Sulfur   Oxygen   Hydrogen

FIGURE 9.9 *The structure of sulfuric acid in the vapor state.* Notice the difference in the sulfur–oxygen bond lengths.

lengths with those found experimentally. In our Lewis structure, all four sulfur–oxygen bonds are single bonds, which means they should have about the same bond lengths. However, experimentally it has been found that the bonds are not of equal length, as illustrated in Figure 9.9. The S—O bonds are shorter than the S—OH bonds, which means they must have a larger bond order. Therefore, we need to modify our Lewis structure to make it conform to reality.

Because sulfur is in Period 3, its valence shell has 3*s*, 3*p*, and 3*d* subshells, which together can accommodate more than eight electrons. Therefore, we are able to have more than four bonds to sulfur and we are able to increase the bond order in the S—O bonds by moving electron pairs to create sulfur–oxygen double bonds as shown below.

$$H—\ddot{O}—S—\ddot{O}—H \quad \text{gives} \quad H—\ddot{O}—S—\ddot{O}—H \quad \text{(Structure II)}$$

Now we have a Lewis structure that better fits experimental observations because the sulfur–oxygen double bonds are expected to be shorter than the sulfur–oxygen single bonds. Because this second Lewis structure agrees better with the actual structure of the molecule, it is the *preferred* Lewis structure, even though it violates the octet rule.

Formal charges

Are there any criteria that we could have applied that would have allowed us to predict that the second Lewis structure for H_2SO_4 is better than the one with only single bonds, even though it seems to violate the octet rule unnecessarily? To answer this question, let's take a closer look at the two Lewis structures we've drawn.

In Structure I, there are only single bonds between the sulfur and oxygen atoms. If the electrons in the bonds are shared equally by S and O, then each atom "owns" half of the electron pair, or the equivalent of one electron. In other words, the four single bonds place the equivalent of four electrons in the valence shell of the sulfur. A single atom of sulfur by itself, however, has six valence electrons. This means that in Structure I, the sulfur has two electrons *less* than it does as just an isolated atom. Thus, at least in a bookkeeping sense, it would appear that if sulfur obeyed the octet rule in H_2SO_4, it would have a charge of 2+. This *apparent* charge on the sulfur atom is called its **formal charge.**

The actual charges on the atoms in a molecule are determined by the relative electronegativities of the atoms.

Notice that in defining formal charge, we've stressed the word *apparent*. *The formal charge arises because of the bookkeeping we've done and should not be confused with whatever the actual charge is on an atom in the molecule.* (The situation is somewhat similar to the oxidation numbers you learned to assign in Chapter 6, which are artificial charges assigned according to a set of rules.)

Although formal charges don't correspond to actual charges, they are useful nonetheless, so let's study how we calculate them. The **formal charge** *on an atom is obtained by determining the number of valence electrons assigned to it in the Lewis structure and then subtracting that from the number of valence electrons in an isolated atom of the element.* In doing this calculation, electrons in bonds are divided equally between the two atoms, while unshared electrons are assigned exclusively to the atom on which they reside. For example, for Structure I (page 381), we have

> The electrons in a bond are divided equally between the two atoms.

> Unshared electrons belong exclusively to one atom.

$$H—\ddot{O}—S—\ddot{O}—H$$

To make the calculation of formal charge easier, we can use the following equation:

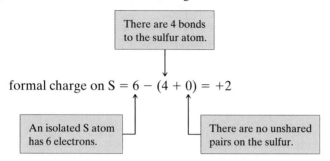

Calculated number of electrons in the valence shell of the atom in the Lewis structure.

$$\begin{bmatrix} \text{formal} \\ \text{charge} \end{bmatrix} = \begin{bmatrix} \text{number of } e^- \text{ in valence} \\ \text{shell of the isolated atom} \end{bmatrix} - \begin{bmatrix} \text{number of bonds} \\ \text{to the atom} \end{bmatrix} + \begin{bmatrix} \text{number of} \\ \text{unshared } e^- \end{bmatrix}$$ (9.2)

TOOLS

Formal charges

For example, for the sulfur in Structure I, we get

There are 4 bonds to the sulfur atom.

formal charge on S = 6 − (4 + 0) = +2

An isolated S atom has 6 electrons.

There are no unshared pairs on the sulfur.

Let's also calculate the formal charges on the hydrogen and oxygen atoms in Structure I. An isolated H atom has one electron. In Structure I each H has one bond and no unshared electrons. Therefore,

formal charge on H = 1 − (1 + 0) = 0

The hydrogens in this structure have no formal charge.

6 − (4 bonds + 0 unshared) = +2   1 − (1 bond + 0 unshared) = 0

$$\ddot{\text{O}}\text{:}$$
$$\text{H}—\ddot{\text{O}}—\text{S}—\ddot{\text{O}}—\text{H} \quad \text{(Structure I)}$$
$$\text{:}\ddot{\text{O}}\text{:}$$

In Structure I, we also see that there are two kinds of oxygen to consider. An isolated oxygen atom has six electrons, so we have, for the oxygens also bonded to hydrogen,

formal charge = 6 − (2 + 4) = 0

and for the oxygens not bonded to hydrogen,

formal charge = 6 − (1 + 6) = −1

$$\text{:}\ddot{\text{O}}\text{:}$$
$$\text{H}—\ddot{\text{O}}—\text{S}—\ddot{\text{O}}—\text{H} \quad \text{(Structure I)}$$
$$\text{:}\ddot{\text{O}}\text{:}$$

6 − (2 bonds + 4 unshared) = 0   6 − (1 bond + 6 unshared) = −1

Nonzero formal charges are indicated in a Lewis structure by placing them in circles alongside the atoms, as shown below.

$$\text{:}\ddot{\text{O}}\text{:}^{\ominus}$$
$$\text{H}—\ddot{\text{O}}—\text{S}^{2+}\ddot{\text{O}}—\text{H}$$
$$\text{:}\ddot{\text{O}}\text{:}_{\ominus}$$

Notice that the sum of the formal charges in the molecule adds up to zero. It is useful to remember that, in general, *the sum of the formal charges in any Lewis structure adds up to the charge on the species.*

Now let's look at the formal charges in Structure II. For sulfur we have

$$\text{formal charge on S} = 6 - (6 + 0) = 0$$

so the sulfur has no formal charge. The hydrogens and the oxygens that are also bonded to H are the same in this structure as before, so they have no formal charges. And finally, the oxygens that are not bonded to hydrogen have

$$\text{formal charge} = 6 - (2 + 4) = 0$$

These oxygens also have no formal charges.

Now let's compare the two structures side by side.

Imagine changing the one with the double bonds to the one with the single bonds. According to the formal charges, this involves creating two pairs of positive–negative charge from something electrically neutral. Stated another way, it involves separating negative charges from positive charges, and this requires an increase in the potential energy. Our conclusion is that the singly bonded structure on the right has a higher potential energy than the one with the double bonds. In general, the lower the potential energy of a molecule, the more stable it is. Therefore, the lower energy structure with the double bonds is, in principle, the more stable structure, so it is preferred over the one with only single bonds. This now gives us a rule that we can use in selecting the best Lewis structures for a molecule or ion:

When several Lewis structures are possible, those with the smallest formal charges are the most stable and are preferred.

EXAMPLE 9.9

Selecting Lewis Structures Based on Formal Charges

A student drew three Lewis structures for the nitric acid molecule:

Which one is preferred?

ANALYSIS: When we have to select among several Lewis structures to find the best one, the critical link is the assignment of formal charges. As a rule, the structure with the fewest formal charges will be the best structure. An exception to this is when an atom in the structure is assigned more electrons than its valence shell can actually hold. Such a structure must be eliminated from consideration.

SOLUTION: Except for hydrogen, all the atoms in the molecule are from Period 2, and therefore can have a maximum of eight electrons in their valence shells. (Period 2 elements *never* exceed an octet because their valence shells have only *s* and *p* subshells and can accommodate a maximum of eight electrons.) Scanning the structures, we see that I and II show octets around both N and O. However, the nitrogen in Structure III has five bonds to it, which require ten electrons. Therefore, this structure is not acceptable and can be eliminated. Our choice is then between Structures I and II. Let's now calculate formal charges on the atoms in each of them.

Structure I:

formal charge on N = $5 - (4 + 0) = +1$

formal charge on O attached to H = $6 - (3 + 2) = +1$

formal charge on O not attached to H = $6 - (1 + 6) = -1$

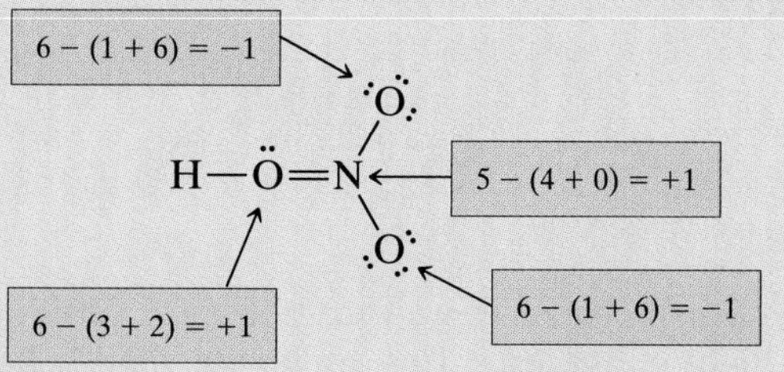

Structure II:

formal charge on N = $+1$

formal charge on O attached to H = $6 - (2 + 4) = 0$

formal charge on singly bonded O not attached to H = -1

formal charge on doubly bonded O = $6 - (2 + 4) = 0$

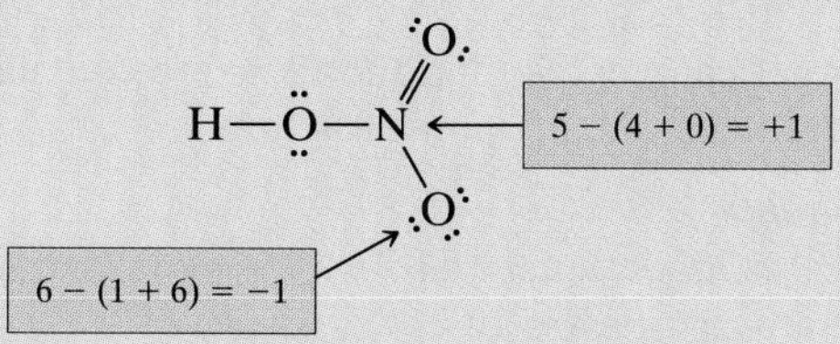

Next, we place the formal charges on the atoms in the structures.

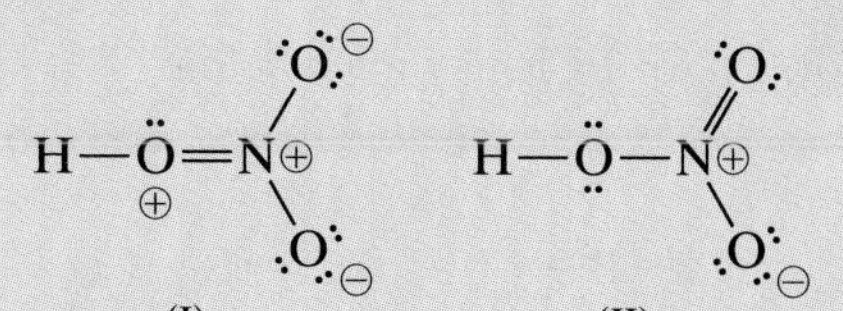

Because Structure II has fewer formal charges than Structure I, it is the lower-energy, preferred Lewis structure for HNO_3.

Is the Answer Reasonable?
One simple check we can do is to add up the formal charges in each structure. The sum must equal the net charge on the particle, which is zero for HNO_3. Adding formal charges gives zero for each structure, so we can be confident we've assigned them correctly. This gives us confidence in our answer, too.

In Structure III, the formal charges are zero on each of the atoms, but this cannot be the "preferred structure" because the nitrogen atom has too many electrons in its valence shell.

What is the preferred Lewis structure for the molecule $POCl_3$, in which phosphorus is bonded to one oxygen and three chlorine atoms?

ANALYSIS: First we must draw a Lewis structure following the rules in the previous section. Then we can assign formal charges. If there is a way to reduce these formal charges by moving electrons, we will do so to obtain a better Lewis structure.

SOLUTION: By following the rules from the previous section, we obtain the following Lewis structure:

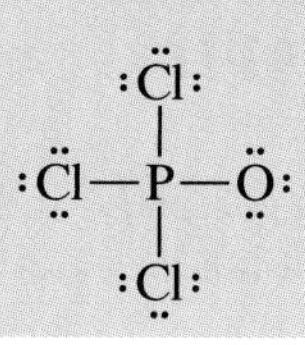

EXAMPLE 9.10

Selecting Lewis Structures Based on Formal Charges

Now we calculate the formal charges.

	Phosphorus	Oxygen	Chlorine
Formal charge	$5 - (4 + 0) = +1$	$6 - (1 + 6) = -1$	$7 - (1 + 6) = 0$

Adding formal charges to the Lewis structure gives

$$
\begin{array}{c}
:\ddot{C}l: \\
| \\
:\ddot{C}l \!-\! \overset{\oplus}{P} \!-\! \ddot{O}\overset{\ominus}{:} \\
| \\
:\ddot{C}l:
\end{array}
$$

We can reduce the formal charges if we can shift one electron from the "negative" oxygen to the "positive" phosphorus. This can be accomplished if we convert one of the unshared pairs of electrons on the oxygen into a bond between P and O (which we are permitted to do because phosphorus is a Period 3 element; its valence shell can hold more than eight electrons).

$$
\begin{array}{c}
:\ddot{C}l: \\
| \\
:\ddot{C}l \!-\! P \!-\! \ddot{O}: \\
| \\
:\ddot{C}l:
\end{array}
\quad \text{gives} \quad
\begin{array}{c}
:\ddot{C}l: \\
| \\
:\ddot{C}l \!-\! P \!=\! \ddot{O}: \\
| \\
:\ddot{C}l:
\end{array}
$$

If we recalculate formal charges, we find the new one contains no formal charges. Therefore, it is preferred over the previous structure.

Is the Answer Reasonable?
The sum of the formal charges equals zero, which is the charge on the molecule, so we appear to have a satisfactory structure.

EXAMPLE 9.11
Selecting Lewis Structures Based on Formal Charges

Two structures can be drawn for BCl_3, as shown below

$$
\begin{array}{c}
:\ddot{C}l: \\
| \\
:\ddot{C}l \!-\! B \!-\! \ddot{C}l: \\
\text{(I)}
\end{array}
\qquad
\begin{array}{c}
:\ddot{C}l: \\
| \\
:\ddot{C}l \!=\! B \!-\! \ddot{C}l: \\
\text{(II)}
\end{array}
$$

Why is the one that violates the octet rule preferred?

ANALYSIS: We're asked to select between Lewis structures, which tells us that we have to consider formal charges. We'll assign them and then see if we can answer the question.

SOLUTION: Assigning formal charges gives

$$
\text{(I)} \quad
\begin{array}{c}
:\ddot{C}l: \\
| \\
:\ddot{C}l \!-\! B \!-\! \ddot{C}l:
\end{array}
\qquad
\text{(II)} \quad
\begin{array}{c}
:\ddot{C}l: \\
| \\
:\overset{\oplus}{\ddot{C}l} \!=\! B \!-\! \ddot{C}l: \\
\ominus
\end{array}
$$

In Structure I, all the formal charges are zero. In Structure II, two of the atoms have formal charges, so this alone would argue in favor of Structure I. There is another argument in favor as well. Notice that the formal charges in Structure II place the positive charge on the more electronegative chlorine atom and the negative charge on the less electronegative boron. If charges could form in this molecule, they certainly would not be expected to form in this way. Therefore, there are two factors that make the structure with the double bond unfavorable, so we usually write the Lewis structure for BCl_3 as shown in (I).

Is the Answer Reasonable?
We've assigned the formal charges correctly, and our reasoning seems sound, so we appear to have answered the question adequately.

PRACTICE EXERCISE 10: Assign formal charges to the atoms in the following Lewis structures:

(a) $:\ddot{N}-N\equiv O:$ (b) $\left[\ddot{S}=C=\ddot{N}\right]^{-}$

PRACTICE EXERCISE 11: Select the preferred Lewis structures for (a) SO_2, (b) $HClO_3$, and (c) H_3PO_4.

9.9 ▶ Resonance applies when a single Lewis structure fails

There are some molecules and ions for which we cannot write Lewis structures that agree with experimental measurements of bond length and bond energy. One example is the formate ion, CHO_2^-, which is produced by neutralizing formic acid, $HCHO_2$ (the substance that causes the stinging sensation in bites from fire ants). The skeletal structure for this ion is

O

H   C

O

and, following the usual steps, we would write its Lewis structure as

$$\left[H-C\overset{\ddot{O}:}{\underset{\ddot{O}:}{\diagup}} \right]^{-}$$

This structure suggests that one carbon–oxygen bond should be longer than the other, but experiment shows that they are identical. In fact, the C—O bond lengths are about halfway between the expected values for a single bond and a double bond. The Lewis structure doesn't match the experimental evidence, and there's no way to write one that does. It would require showing all of the electrons in pairs and, at the same time, showing 1.5 pairs of electrons per bond.

The way we get around problems like this is through the use of a concept called **resonance.** We view the actual structure of the molecule or ion, which we cannot draw satisfactorily, as a composite, or average, of a number of Lewis structures that we can draw. For example, for formate we write

$$\left[H-C\overset{\ddot{O}:}{\underset{\ddot{O}:}{\diagup}} \right]^{-} \longleftrightarrow \left[H-C\overset{\ddot{O}:}{\underset{\ddot{O}:}{\diagdown}} \right]^{-}$$

where we have simply shifted electrons around in going from one structure to the other. The bond between the carbon and a particular oxygen is depicted as a single bond in one structure and as a double bond in the other. The average of these is 1.5 bonds (halfway between a single and a double bond), which is in agreement with the experimental bond lengths. These two Lewis structures are called **resonance structures** or **contributing structures,** and the actual structure of the ion is said to be a **resonance hybrid** of the two resonance structures that we've drawn. The double-ended arrow is used to show that we are drawing resonance structures and implies that the true hybrid structure is a composite of the two resonance structures.

The term *resonance* is often misleading to the beginning student. The word itself suggests that the actual structure flip-flops back and forth between the two

Formic acid is an organic acid. It has the structure

$$H-\overset{\overset{\textstyle :\ddot{O}}{\|}}{C}-\ddot{O}-H$$

The name *formic acid* comes from *formica,* the Latin word for ant. The one shown here is a fire ant.

No atoms have been moved; the electrons have just been redistributed.

structures shown. This is *not* the case! A mule, which is the *hybrid* offspring of a donkey and a horse, isn't a donkey one minute and a horse the next! Although it may have characteristics of both parents, a mule is a mule. A *resonance hybrid* also has characteristics of its "parents," but it never has the exact structure of any of them.

TOOLS

Determining resonance structures

There is a simple way to determine when resonance should be applied to Lewis structures. If you find that you must move electrons to create one or more double bonds while following the procedures developed in the previous sections, the number of resonance structures is equal to the number of equivalent choices for the locations of the double bonds. For example, in drawing the Lewis structure for the NO_3^- ion, we reach the stage

$$\ddot{O}\text{—}N(\ddot{O})(\ddot{O})$$

A double bond must be created to give the nitrogen an octet. Since it can be placed in any one of three locations, there are three resonance structures for this ion.

The three oxygens in NO_3^- are said to be equivalent; that is, they are all alike in their chemical environment. Each oxygen is bonded to a nitrogen atom that's attached to two other oxygen atoms.

$$\left[\ddot{O}\text{—}N(\ddot{O})(\ddot{O})\right]^- \longleftrightarrow \left[\ddot{O}\text{—}N(\ddot{O})(\ddot{O})\right]^- \longleftrightarrow \left[\ddot{O}\text{—}N(\ddot{O})(\ddot{O})\right]^-$$

Notice that each structure is the same, except for the location of the double bond.

In the nitrate ion, the extra bond that moves around from one structure to another is divided among all three bond locations. Therefore, the average bond order in the N—O bonds is expected to be $1\frac{1}{3}$, or 1.33.

EXAMPLE 9.12
Drawing Resonance Structures

A preservative used in some frankfurters and other cured meats is sodium nitrite, $NaNO_2$. Does resonance apply to the Lewis structure for the nitrite ion, NO_2^-? If so, draw the resonance structures and calculate the average bond order of the N—O bonds.

ANALYSIS: To determine whether we have to invoke resonance, we have to form the Lewis structure. If we reach a point where we have a choice as to where to place a double bond, the number of equivalent positions for the double bond equals the number of resonance structures that we draw.

SOLUTION: Following the usual procedure for drawing Lewis structures, we reach the following stage after all the electrons have been placed:

$$:\ddot{O}\text{—}\ddot{N}\text{—}\ddot{O}:$$

There is less than an octet around the nitrogen, so we have to move one of the unshared electron pairs of an oxygen into one of the N—O bonds to form a N=O double bond. There are two locations for the double bond, and both are equivalent. (This means that both oxygens have the same environment; each is attached to an "N—O" group of atoms.) Therefore, there are two choices for the location of the double bond and the answer to the first question is that *resonance does apply to this ion.*

The number of resonance structures equals the number of choices, which is two. The two structures are

$$\left[:O\text{=}N\text{—}\ddot{O}:\right]^- \longleftrightarrow \left[:\ddot{O}\text{—}N\text{=}O:\right]^-$$

(Notice that we've drawn brackets around the ions and placed the charge as a superscript.)

We have distributed the extra bond (the one used to form the double bond) over two locations, so each location shares half of it. For each NO bond position, the half-bond is added to the single bond already there, so the NO bond order is $1\frac{1}{2}$.

Is the Answer Reasonable?

Here are some questions we can ask, and if the answers are "yes," then we've solved the problem correctly: Have we counted valence electrons correctly? Have we followed the rules for placing electron pairs into the skeletal structure? Have we correctly determined the number of equivalent positions for the double bond? Have we computed the average bond order correctly?

Use formal charges to show that resonance applies to the preferred Lewis structure for the sulfite ion, SO_3^{2-}. Draw the resonance structures and determine the average bond order of the S—O bonds.

ANALYSIS: Following our usual procedure, we obtain the Lewis structure

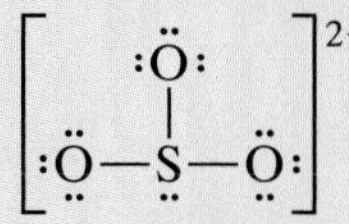

All of the valence electrons have been placed into the structure and we have octets around all of the atoms, so it doesn't seem that we need the concept of resonance. However, the question refers to the "preferred" structure, which suggests that we are going to have to assign formal charges and determine what the preferred structure is. Then we can decide whether the concept of resonance will apply.

SOLUTION: When we assign formal charges, we get

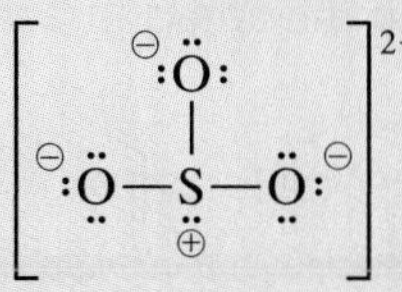

We can obtain a better Lewis structure if we can reduce the number of formal charges. This can be accomplished by moving an unshared pair from one of the oxygens into an S—O bond, thereby forming a double bond. Let's do this using the oxygen at the left.

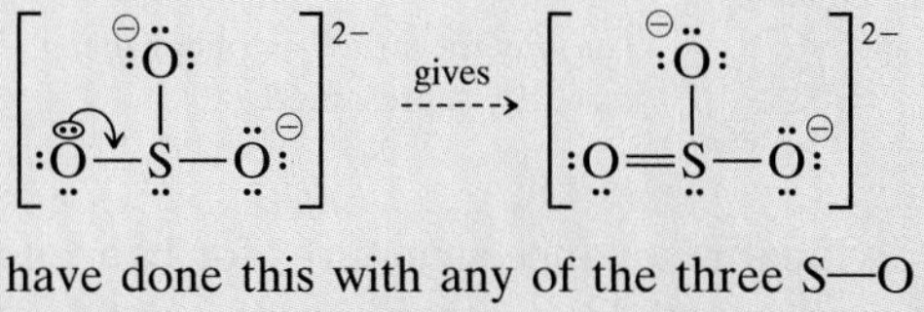

However, we could have done this with any of the three S—O bonds, so there are three choices for the location of the double bond. Therefore, there are three resonance structures.

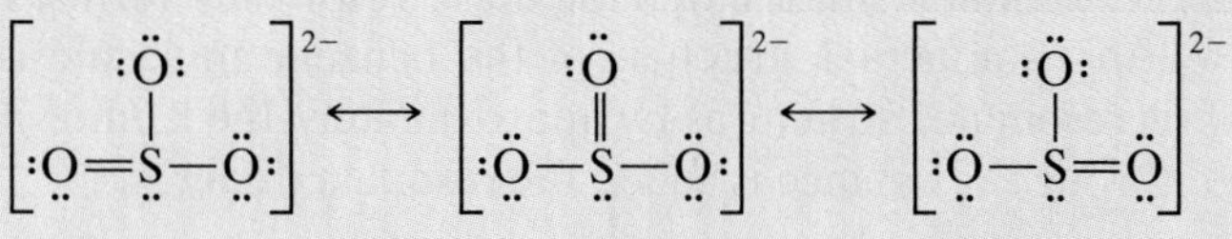

As with the nitrate ion, we expect an average bond order of 1.33.

Is the Answer Reasonable?

If we can answer "yes" to the following questions, the problem is solved correctly: Have we counted valence electrons correctly? Have we properly placed the electrons into the skeletal structure? Do the formal charges we've calculated add up to the charge on the SO_3^{2-} ion? Have we correctly determined the number of equivalent positions for the double bond? Have we computed the average bond order correctly?

EXAMPLE 9.13

Drawing Resonance Structures

PRACTICE EXERCISE 12: Draw the resonance structures for HCO_3^- .

PRACTICE EXERCISE 13: Determine the preferred Lewis structure for the PO_4^{3-} ion and if appropriate, draw resonance structures.

FIGURE 9.10 *Benzene.* The molecule has a planar hexagonal structure.

Resonance increases the stability of molecules and ions

One of the benefits that a molecule or ion derives from existing as a resonance hybrid is that its total energy is lower than any one of its resonance structures. A particularly important example of this occurs with the compound benzene, C_6H_6. This is a planar, hexagonal, ring-shaped molecule (Figure 9.10) with a basic structure that appears in many important organic molecules, ranging from plastics to amino acids.

Two resonance structures are drawn for benzene.

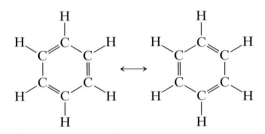

These are usually represented as hexagons with dashes showing the locations of the double bonds. It is assumed that at each apex of the hexagon there is a carbon bonded to a hydrogen as well as to the adjacent carbon atoms.

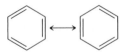

Polystyrene plastic. This common polymer contains benzene rings joined to alternating carbon atoms in a long hydrocarbon chain. When a gas is blown into melted polystyrene, the lightweight product called *Styrofoam* is formed. Styrofoam is used to make a wide variety of packaging products for consumer goods.

Usually, the actual structure of benzene (that of its resonance hybrid) is represented as a hexagon with a circle in the center. This is intended to show that the electron density of the three extra bonds is evenly distributed around the ring.

 The way the structure of benzene is usually represented.

Although the individual resonance structures for benzene show double bonds, the molecule does not react like other organic molecules that have true carbon–carbon double bonds. The reason appears to be that the resonance hybrid is considerably more stable than either of these resonance forms. In fact, it has been calculated that the actual structure of the benzene molecule is more stable than either of its resonance structures by approximately 146 kJ/mol. This extra stability achieved through resonance is called the **resonance energy.**

9.10 ▶ Both electrons in a coordinate covalent bond come from the same atom

When an ammonia molecule is formed from a nitrogen atom and three hydrogen atoms, each hydrogen shares an electron with one of the electrons of the nitrogen. By this process, three bonds are formed by nitrogen and an octet of electrons is achieved around the central atom.

$$3H\cdot + \cdot \ddot{N}: \longrightarrow H-\underset{\underset{H}{|}}{\overset{\overset{H}{|}}{N}}:$$

One might expect that at this point, no further bonds to nitrogen can be made. However, when NH_3 is placed into an acidic solution, it picks up a hydrogen ion, H^+, and becomes NH_4^+, in which there are four bonds between nitrogen and hydrogen. To understand how this can happen, let's keep tabs on all of the electrons by using red dots for the hydrogen electrons and black dots for nitrogen electrons.

$$H:\underset{\underset{H}{..}}{\overset{\overset{H}{..}}{N}}: + H^+ \longrightarrow \left[H:\underset{\underset{H}{..}}{\overset{\overset{H}{..}}{N}}:H\right]^+$$

All electrons are alike, of course. We are using different colors for them so we can see where the electrons in the bond came from.

Three of the N—H bonds are formed in the usual way, with each atom contributing one electron toward the pair that's shared. Notice, however, that the fourth bond is created differently, with both electrons in the bond being contributed by the nitrogen.

The reason this fourth bond can form is that the H^+ ion has an empty $1s$ orbital in its valence shell, which can accommodate both of the electrons from the nitrogen. The result is that when the H^+ becomes bonded to the nitrogen of NH_3, the nitrogen donates both of the electrons to the bond. *This type of bond, in which both electrons of the shared pair come from just one of the two atoms, is called a* **coordinate covalent bond.**

It is important to understand that even though we make a distinction as to the *origin* of the electrons, once the bond is formed it is really the same as any other covalent bond. In other words, we can't tell where the electrons in the bond came from after the bond has been formed. In NH_4^+, for instance, all four of the N—H bonds are identical once they've been formed, and no distinction is made among them. We usually write the structure of NH_4^+ as

The notion of coordinate covalent bonds is used to explain how bonds are formed, rather than what the bonds are like after they have formed.

$$\left[H-\underset{\underset{H}{|}}{\overset{\overset{H}{|}}{N}}-H\right]^+$$

Earlier we mentioned that when amines such as CH_3NH_2 react with water, the hydrogen ion they acquire becomes bonded to the nitrogen of the amine. The reaction is exactly analogous to the addition of a hydrogen ion to ammonia.

$$CH_3-\underset{\underset{H}{|}}{\overset{\overset{H}{|}}{N}}: + H^+ \longrightarrow \left[CH_3-\underset{\underset{H}{|}}{\overset{\overset{H}{|}}{N}}-H\right]^+$$

The concept of a coordinate covalent bond is sometimes useful when we are attempting to understand what happens to atoms in a chemical reaction. For example, it is known that if ammonia is added to boron trichloride, an exothermic reaction takes place and the compound NH_3BCl_3 is formed, in which there is a boron–nitrogen bond. Using Lewis structures, we can diagram this reaction as follows:

$$H-\underset{\underset{H}{|}}{\overset{\overset{H}{|}}{N}}: + \underset{\underset{:\ddot{Cl}:}{|}}{\overset{\overset{:\ddot{Cl}:}{|}}{B}}-\ddot{Cl}: \longrightarrow H-\underset{\underset{H}{|}}{\overset{\overset{H}{|}}{N}}:\underset{\underset{:\ddot{Cl}:}{|}}{\overset{\overset{:\ddot{Cl}:}{|}}{B}}-\ddot{Cl}:$$

In the reaction, a boron atom with less than an octet achieves a complete valence shell by sharing a pair of electrons that originates on the nitrogen atom. We might say that "the boron forms a coordinate covalent bond with the nitrogen of the ammonia molecule."

An arrow sometimes is used to represent the donated pair of electrons in a coordinate covalent bond. The direction of the arrow indicates the direction in which the electron pair is donated, in this case from the nitrogen to the boron.

Compounds like BCl₃NH₃, which are formed by simply joining two smaller molecules, are sometimes called **addition compounds.**

We will find the notion of coordinate covalent bonds useful later in the book when we describe certain kinds of acid–base reactions as well as certain compounds of transition metal ions.

PRACTICE EXERCISE 14: Use Lewis structures to explain how the reaction between hydroxide ion and hydrogen ion involves the formation of a coordinate covalent bond.

SUMMARY

Ionic Bonding In ionic compounds, the forces of attraction between positive and negative ions are called **ionic bonds.** The formation of ionic compounds by electron transfer is favored when atoms of low ionization energy react with atoms of high electron affinity. The chief stabilizing influence in the formation of ionic compounds is the release of the **lattice energy,** which is the energy required to completely separate the ions of an ionic compound. When atoms of the elements in Groups IA and IIA as well as the nonmetals form ions, they usually gain or lose enough electrons to achieve a noble gas electron configuration. Transition elements lose their outer s electrons first, followed by loss of d electrons from the shell below the outer shell. Post-transition metals lose electrons from their outer p subshell first, followed by electrons from the outer s subshell.

Covalent Bonding Electron sharing between atoms occurs when electron transfer is energetically too "expensive." Shared electrons attract the positive nuclei, and this leads to a lowering of the potential energy of the atoms as a covalent bond forms. Electrons generally become paired when they are shared. An atom tends to share enough electrons to complete its valence shell. Except for hydrogen, the valence shell usually holds eight electrons, which forms the basis of the octet rule. The **octet rule** states that atoms of the representative elements tend to acquire eight electrons in their outer shells when they form bonds. **Single, double,** and **triple bonds** involve the sharing of one, two, and three pairs of electrons, respectively, between two atoms. Boron and beryllium often have less than an octet in their compounds. Atoms of the elements of Period 2 cannot have more than an octet because their outer shells can hold only eight electrons. Elements in Period 3 and below can exceed an octet if they form more than four bonds.

Bond energy (the energy needed to separate the bonded atoms) and **bond length** (the distance between the nuclei of the atoms connected by the bond) are two experimentally measurable quantities that can be related to the number of pairs of electrons in the bond. For bonds between atoms of the same elements, bond energy increases and bond length decreases as the **bond order** increases.

Organic Compounds Carbon normally forms four covalent bonds, either to atoms of other elements or to other carbon atoms. Oxygen-containing organic compounds include **alcohols** (in which an O—H group is bonded to carbon), **ketones** (in which a **carbonyl group,** C=O, is bonded to two carbon atoms), **aldehydes** (in which the carbonyl group is attached to at least one hydrogen), and **carboxylic acids** (organic acids), which contain the **carboxyl group. Amines** are nitrogen-containing compounds derived from ammonia in which a hydrogen of NH₃ is replaced by a hydrocarbon group. **Amino acids** contain both a carboxyl group and an amine group.

Electronegativity and Polar Bonds The attraction an atom has for the electrons in a bond is called the atom's **electronegativity.** When atoms of different electronegativities form a bond, the electrons are shared unequally and the bond is **polar,** with **partial positive** and **partial negative** charges at opposite ends. This causes the bond to be an electric **dipole.** In a **polar molecule,** such as HCl, the product of the charge at either end multiplied by the distance between the charges gives the **dipole moment, μ.** When the two atoms have the same electronegativity, the bond is **nonpolar.** The extent of polarity of the bond depends on the electronegativity difference between the two bonded atoms. When the electronegativity difference is very large, ionic bonding results. A bond is approximately 50% ionic when the electronegativity difference is 1.7.

In the periodic table, electronegativity increases from left to right across a period and from bottom to top in a group.

Reactivity and Electronegativity The **reactivity** of metals is related to the ease with which they are oxidized (lose electrons); for nonmetals it is related to the ease with which they are reduced (gain electrons). Metals with low electronegativities lose electrons easily, are good reducing agents, and tend to be very reactive. The most reactive metals are located in Groups IA and IIA, and their ease of oxidation increases going down the group. For nonmetals, the higher the electronegativity, the stronger is their ability to serve as oxidizing agents. The strongest oxidizing agent is fluorine. Among the halogens, oxidizing strength decreases from fluorine to iodine.

Lewis Symbols and Lewis Structures Lewis symbols are a bookkeeping device used to keep track of valence electrons in ionic and covalent bonds. The **Lewis symbol** of an element consists of the element's chemical symbol surrounded by a number of dots equal to the number of valence electrons. In the **Lewis structure** for an ionic compound, the Lewis symbol for the anion is enclosed in brackets (with the charge written outside) to show that all the electrons belong entirely to the ion. The Lewis structure for a molecule or polyatomic ion uses pairs of dots between chemical symbols to represent shared pairs of electrons. The electron pairs in covalent bonds usually are represented by dashes; one dash equals two electrons. The following procedure is used to draw the Lewis structure: (1) decide on the skeletal structure (remember that the least electronegative atom is usually the central atom and is usually first in the formula); (2) count all the valence electrons, taking into account the charge, if any; (3) place a pair of electrons in each bond; (4) complete the octets of atoms other than the central atom (but remember that hydrogen can only have two electrons); (5) place any leftover electrons on the central atom in pairs; (6) if the central atom still has *less than* an octet, move electron pairs to make double or triple bonds.

Formal Charges The **formal charge** assigned to an atom in a Lewis structure (which usually differs from the actual charge on the atom) is calculated as the difference between the number of valence electrons of an isolated atom of the element and the number of electrons that "belong" to the atom because of its bonds to other atoms and its unshared valence electrons. The sum of the formal charges always equals the net charge on the molecule or ion. The most stable (lowest energy) Lewis structure for a molecule or ion is the one with the fewest formal charges. This is usually the preferred Lewis structure for the particle.

Resonance Two or more atoms in a molecule or polyatomic ion are *chemically equivalent* if they are attached to the same kinds of atoms or groups of atoms. Bonds to chemically equivalent atoms must be the same; they must have the same bond length and the same bond energy, which means they must involve the sharing of the same number of electron pairs. Sometimes the Lewis structures we draw suggest that the bonds to chemically equivalent atoms are not the same. Typically, this occurs when it is necessary to form multiple bonds during the drawing of a Lewis structure. When alternatives exist for the location of a multiple bond among two or more equivalent atoms, then each possible Lewis structure is actually a **resonance structure** or **contributing structure,** and we draw them all. In drawing resonance structures, the relative locations of the nuclei must be identical in all. Remember that none of the resonance structures corresponds to a real molecule, but their composite—the **resonance hybrid**—does approximate the actual structure of the molecule or ion.

Coordinate Covalent Bonding For bookkeeping purposes, we sometimes single out a covalent bond whose electron pair originated from one of the two bonded atoms. An arrow is sometimes used to indicate the donated pair of electrons. Once formed, a coordinate covalent bond is no different from any other covalent bond.

TOOLS YOU HAVE LEARNED

The table below lists the concepts that you've learned in this chapter that can be applied as tools in solving problems. Study each one carefully so that you know what each is used for. When faced with solving a problem, recall what each tool does and consider whether it will be helpful in finding a solution. This will aid you in selecting the tools you need. If necessary, refer to this table when working on the Thinking-It-Through problems and the Review Exercises that follow.

TOOL	HOW IT WORKS
Rules for electron configurations of ions (pages 358 and 359)	The rules enable us to write correct electron configurations for cations and anions.
Lewis symbols (page 361)	Lewis symbols are a bookkeeping device that we use to keep track of valence electrons in atoms and ions. You should learn to use the periodic table to construct the Lewis symbol.
Periodic trends in electronegativity (page 374)	They allow us to use the locations of elements in the periodic table to estimate the degree of polarity of bonds and to estimate which of two atoms in a bond is the most electronegative.

Method for drawing Lewis structures (page 377)	The method described yields structures in which the maximum number of atoms obey the octet rule.
Correlation between bond properties and bond order (page 381)	The correlations allow us to compare experimental properties that are related to covalent bonds with those predicted by theory.
Formal charges (page 383)	By assigning formal charges, we can select the best Lewis structure for a molecule or polyatomic ion.
Method for determining resonance structures (page 388)	By distributing multiple bonds over equivalent atoms in a molecule, we obtain a better description of the bonding in the molecule. We also know that when resonance structures can be drawn, a molecule or ion will be more stable than any of the individual resonance structures.

THINKING IT THROUGH

Need extra help?
Vist the Brady/
Senese web site at ON-LINE HELP
www.wiley.com/
college/brady

The goal for the following problems is not to find the answers themselves, but rather to assemble the information needed to solve them and explain how you would use the information to find the answers. The problems in Level 2 are more challenging than those in Level 1 and may contain more data than are required, in which case you are also asked to identify the unnecessary data. Detailed answers to the Thinking-It-Through problems can be found on the web site.

Level 1 Problems

1. Describe how you would determine how much energy would be absorbed or given off in the formation of the gaseous ions that could eventually come together to form 1 mol of solid Na_2O.

2. Describe how we could use Lewis symbols to follow the formation of K_2S from potassium and sulfur atoms.

3. Explain how you would figure out the changes in electron configuration that accompany the following:

$$Pb \longrightarrow Pb^{2+} \longrightarrow Pb^{4+}$$

4. Why is K_2S ionic but H_2S molecular? What factors must you consider in answering this question?

5. How would you determine which molecule is likely to have the more polar bonds, PCl_3 or SCl_2? In these molecules, how can you tell which atoms carry the partial positive and partial negative charges?

Level 2 Problems

6. How many double bonds are in the Lewis structure for $HBrO_3$? Explain in detail how you would answer this question.

7. What is the preferred Lewis structure for H_3AsO_4? In detail, explain what knowledge you must have to answer this question.

8. Are the following Lewis structures considered to be resonance structures? Explain. How can you predict which is the more likely structure for $POCl_3$?

9. How can you make predictions about the relative S—O bond lengths in the SO_3^{2-} and SO_4^{2-} ions?

10. What wavelength of light, if absorbed by a hydrogen molecule, would cause the molecule to split into two atoms of hydrogen? Set up the calculation. (The data required are available in this and previous chapters.)

11. What wavelength of light, if absorbed by a hydrogen molecule at 25 °C, could cause the molecule to split into the ions H^+ and H^-? Set up the calculations. (The data required are available in this and previous chapters.)

12. Dipole moments are measured in units of debyes (D), where $1\ D = 3.34 \times 10^{-30}\ C \cdot m$ (coulomb $\times$ meter). The dipole moment of HF is 1.83 D and the bond length is 917 pm. Explain how you would calculate the amount of charge on the hydrogen and the fluorine atoms in the HF molecule. (Set up the calculation so the units work out to coulombs.)

13. By ranking the strengths of the attractive forces between the cations and anions in the compounds below, arrange them in order of increasing lattice energies. Be sure to consider both the charges and the sizes of the ions.

NaCl   MgO   KBr   CaO

14. Use Figures 9.5 and 9.6 to compare the polarities of the bonds in the following compounds (some of which are hypothetical): CO_2, CS_2, CSe_2, CTe_2, MgO, MgS, MgSe, MgTe. Is it reasonable to claim that binary compounds between metals and nonmetals are always ionic and those between two nonmetals are always covalent? Explain.

REVIEW QUESTIONS

Ionic Bonding

9.1 What must be true about the change in the total potential energy of a collection of atoms for a stable compound to be formed from the elements?

9.2 What is an *ionic bond*?

9.3 How is the tendency to form ionic bonds related to the IE and EA of the atoms involved?

9.4 Define the term *lattice energy*. In what ways does the lattice energy contribute to the stability of ionic compounds?

9.5 Magnesium forms the ion Mg^{2+}, but not the ion Mg^{3+}. Why?

9.6 Why doesn't chlorine form the ion Cl^{2-}?

9.7 Why do many of the transition elements in Period 4 form ions with a 2+ charge?

9.8 If we were to compare the first, second, third, and fourth ionization energies of aluminum, between which pair of successive ionization energies would there be the largest difference? (Refer to the periodic table in answering this question.)

9.9 In each of the following pairs of compounds, which would have the larger lattice energy: (a) CaO or Al_2O_3, (b) BeO or SrO, (c) NaCl or NaBr?

Lewis Symbols

9.10 The Lewis symbol for an atom only accounts for electrons in the valence shell of the atom. Why are we not concerned with the other electrons?

9.11 What relationship exists between the number of dots in a Lewis symbol and the location of the element in the periodic table?

9.12 Which of these Lewis symbols is incorrect?

(a) $:\overset{..}{\underset{.}{O}}:$ (b) $\cdot\overset{..}{\underset{..}{Cl}}\cdot$ (c) $:\overset{..}{\underset{.}{Ne}}:$ (d) $:\overset{..}{\underset{.}{Sb}}:$

Electron Sharing

9.13 Describe what happens to the electron density around two hydrogen atoms as they come together to form an H_2 molecule.

9.14 In terms of the potential energy change, why doesn't ionic bonding occur when two nonmetals react with each other?

9.15 Is the formation of a covalent bond endothermic or exothermic?

9.16 What happens to the energy of two hydrogen atoms as they approach each other? What happens to the spins of the electrons?

9.17 Why is a covalent bond sometimes called an *electron pair bond*?

9.18 What holds the two nuclei together in a covalent bond?

9.19 What factors control the bond length in a covalent bond?

Covalent Bonding and the Octet Rule

9.20 What is the *octet rule*? What is responsible for it?

9.21 How many covalent bonds are normally formed by (a) hydrogen, (b) carbon, (c) oxygen, (d) nitrogen, and (e) chlorine?

9.22 Why do Period 2 elements never form more than four covalent bonds? Why are Period 3 elements able to exceed an octet when they form bonds?

9.23 Define (a) *single bond*, (b) *double bond,* and (c) *triple bond.*

9.24 What is a structural formula?

9.25 $H-C\equiv N:$ is the Lewis structure for hydrogen cyanide. Draw circles enclosing electrons to show that carbon and nitrogen obey the octet rule.

9.26 Why doesn't hydrogen obey the octet rule? How many covalent bonds does a hydrogen atom form?

Compounds of Carbon

9.27 What special "talent" does carbon have that allows it to form so many compounds?

9.28 Sketch the Lewis structures for (a) methane, (b) ethane, and (c) propane.

9.29 Draw the structure for a hydrocarbon that has a chain of five carbon atoms linked by single bonds. How many hydrogen atoms does the molecule have? What is the molecular formula for the compound?

9.30 How many *different* molecules have the formula C_5H_{12}? Sketch their structures.

9.31 What is a carbonyl group? In which classes of organic molecules that we've studied do we find a carbonyl group?

9.32 Match the compounds on the left with the family types on the right.

$$CH_3-CH_2-\overset{\overset{\displaystyle O}{\|}}{C}-OH \qquad \text{hydrocarbon}$$

$$CH_3-\overset{\overset{\displaystyle O}{\|}}{C}-CH_2-CH_3 \qquad \text{aldehyde}$$

$$CH_3-CH_2-\overset{\overset{\displaystyle OH}{|}}{CH}-CH_3 \qquad \text{amine}$$

$$CH_3-CH_2-\overset{\overset{\displaystyle O}{\|}}{C}-H \qquad \text{acid}$$

$$CH_3-\overset{\overset{\displaystyle H}{|}}{N}-CH_2-CH_3 \qquad \text{alcohol}$$

$$CH_3-CH_2-\underset{\underset{\displaystyle CH_3}{|}}{CH}-CH_3 \qquad \text{ketone}$$

9.33 Write a chemical equation for the ionization in water of the acid in Question 9.32. Is the acid strong or weak?

9.34 Write a chemical equation for the reaction of the amine in Question 9.32 with water.

9.35 Write the net ionic equation for the reaction of the acid and amine in Question 9.32 with each other.

9.36 Draw the structures of (a) ethylene and (b) acetylene.

Polar Bonds and Electronegativity

9.37 What is a polar covalent bond?

9.38 Define *dipole moment* in the form of an equation. What is the value of the debye (with appropriate units)?

9.39 Define *electronegativity*.

9.40 On what basis did Pauling develop his scale of electronegativities?

9.41 Which element has the highest electronegativity? Which is the second most electronegative element?

9.42 Which elements are assigned electronegativities of zero? Why?

9.43 Among the following bonds, which are more ionic than covalent: (a) Si—O, (b) Ba—O, (c) Se—Cl, (d) K—Br?

9.44 If an element has a low electronegativity, is it likely to be a metal or a nonmetal? Explain your answer.

9.45 Use the data in Table 9.3 to explain why we know that the electronegativity of nitrogen is greater than the electronegativity of carbon.

Electronegativity and the Reactivities of the Elements

9.46 When we say that aluminum is more reactive than iron, which property of these elements are we describing?

9.47 In what groups in the periodic table are the most reactive metals found? Where do we find the least reactive metals?

9.48 How is the electronegativity of a metal related to its reactivity?

9.49 Arrange the following metals in their approximate order of reactivity (most reactive first, least reactive last) based on their locations in the periodic table: (a) iridium, (b) silver, (c) calcium, (d) iron.

9.50 Complete and balance equations for the following. If no reaction occurs, write "N.R."
(a) $KCl + Br_2 \rightarrow$ (d) $CaBr_2 + Cl_2 \rightarrow$
(b) $NaI + Cl_2 \rightarrow$ (e) $AlBr_3 + F_2 \rightarrow$
(c) $KCl + F_2 \rightarrow$ (f) $ZnBr_2 + I_2 \rightarrow$

9.51 In each pair, choose the better oxidizing agent.
(a) O_2 or F_2 (c) Br_2 or I_2 (e) Se_8 or Cl_2
(b) As_4 or P_4 (d) P_4 or S_8 (f) As_4 or S_8

Failure of the Octet Rule

9.52 How many electrons are in the valence shells of (a) Be in $BeCl_2$, (b) B in BCl_3, and (c) H in H_2O?

9.53 What is the minimum number of electrons that would be expected to be in the valence shell of As in $AsCl_5$?

9.54 Nitrogen and arsenic are in the same group in the periodic table. Arsenic forms both $AsCl_3$ and $AsCl_5$, but with chlorine, nitrogen only forms NCl_3. On the basis of the electronic stuctures of N and As, explain why this is so.

Bond Length and Bond Energy

9.55 Define *bond length* and *bond energy*.

9.56 Define *bond order*. How are bond energy and bond length related to bond order? Why are there these relationships?

9.57 The energy required to break the H—Cl bond to give H^+ and Cl^- ions would not be called the H—Cl bond energy. Why?

Formal Charge

9.58 What is the definition of *formal charge*?

9.59 How are formal charges used to select the best Lewis structure for a molecule? What is the basis for this method of selection?

9.60 What are the formal charges on the atoms in the HCl molecule? What are the actual charges on the atoms in this molecule? (*Hint:* See Section 9.5.) Are formal charges the same as actual charges?

Resonance

9.61 Why is the concept of resonance needed?

9.62 What is a resonance hybrid? How does it differ from the resonance structures drawn for a molecule?

9.63 Draw the resonance structures of the benzene molecule. Why is it more stable than one would expect if the ring contained three carbon–carbon double bonds.

9.64 Polystyrene plastic consists of a long chain of carbon atoms in which every other carbon is attached to a benzene ring. The ring is attached by replacing a hydrogen of benzene with a bond to the carbon chain. In the chain, carbons not attached to other carbons are bonded to hydrogen atoms. Sketch a portion of a polystyrene molecule that contains five benzene rings.

Coordinate Covalent Bonds

9.65 What is a *coordinate covalent bond*?

9.66 Once formed, how (if at all) does a coordinate covalent bond differ from an ordinary covalent bond?

9.67 BCl_3 has an incomplete valence shell. Explain how it could form a coordinate covalent bond with a water molecule.

REVIEW PROBLEMS

Answers to problems whose numbers are printed in color are given in Appendix B. More challenging problems are marked with asterisks. **ILW** = Interactive LearningWare solution is available at *www.wiley.com/college.brady*.

Electron Configurations of Ions

9.68 Explain what happens to the electron configurations of Mg and Br when they react to form the compound $MgBr_2$.

9.69 Describe what happens to the electron configurations of lithium and nitrogen when they react to form the compound Li_3N.

ILW **9.70** What are the electron configurations of the Pb^{2+} and Pb^{4+} ions?

9.71 What are the electron configurations of the Bi^{3+} and Bi^{5+} ions?

9.72 Write the abbreviated electron configuration of the Mn^{3+} ion. How many unpaired electrons does the ion contain?

9.73 Write the abbreviated electron configuration of the Co^{3+} ion. How many unpaired electrons does the ion contain?

Lewis Symbols

9.74 Write Lewis symbols for the following atoms: (a) Si, (b) Sb, (c) Ba, (d) Al, (e) S.

9.75 Write Lewis symbols for the following atoms: (a) K, (b) Ge, (c) As, (d) Br, (e) Se.

9.76 Write Lewis symbols for the following ions: (a) K^+, (b) Al^{3+}, (c) S^{2-}, (d) Si^{4-}, (e) Mg^{2+}.

9.77 Write Lewis symbols for the following ions: (a) Br^-, (b) Se^{2-}, (c) Li^+, (d) C^{4-}, (e) As^{3-}.

9.78 Use Lewis symbols to diagram the reactions between (a) Ca and Br, (b) Al and O, and (c) K and S.

9.79 Use Lewis symbols to diagram the reactions between (a) Mg and S, (b) Mg and Cl, and (c) Mg and N.

Dipole Moments

9.80 Use the data in Table 9.3 (page 372) to calculate the amount of charge on the oxygen and nitrogen in the nitrogen monoxide molecule, expressed in electronic charge units ($e = 1.60 \times 10^{-19}$ C). Which atom carries the positive charge?

9.81 The molecule BrF has a dipole moment of 1.42 D and a bond length of 176 pm. Calculate the charge on the ends of the molecule, expressed in electronic charge units ($e = 1.60 \times 10^{-19}$ C). Which atom carries the positive charge?

Bond Energy

9.82 How many grams of water could be raised from 25 °C (room temperature) to 100 °C (the boiling point of water) by the amount of energy released in the formation of 1 mol of H_2 from hydrogen atoms? The bond energy of H_2 is 435 kJ/mol.

9.83 How much energy, in joules, is required to break the bond in *one* chlorine molecule? The bond energy of Cl_2 is 242.6 kJ/mol.

9.84 The reason there is danger in exposure to high-energy radiation (e.g., ultraviolet and X rays) is that the radiation can rupture chemical bonds. In some cases, cancer can be caused by it. A carbon–carbon single bond has a bond energy of approximately 348 kJ per mole. What wavelength of light is required to provide sufficient energy to break the C—C bond? In which region of the electromagnetic spectrum is this wavelength located?

9.85 A mixture of H_2 and Cl_2 is stable, but a bright flash of light passing through it can cause the mixture to explode. The light causes Cl_2 molecules to split into Cl atoms, which are highly reactive. What wavelength of light is necessary to cause the Cl_2 molecules to split? The bond energy of Cl_2 is 242.6 kJ per mole.

Covalent Bonds and the Octet Rule

9.86 Use Lewis structures to diagram the formation of (a) Br_2, (b) H_2O, and (c) NH_3 from neutral atoms.

9.87 Chlorine tends to form only one covalent bond because it needs just one electron to complete its octet. What are the Lewis structures for the simplest compound formed by chlorine with (a) nitrogen, (b) carbon, (c) sulfur, and (d) bromine?

9.88 Use the octet rule to predict the formula of the simplest compound formed from hydrogen and (a) Se, (b) As, and (c) Si. (Remember, however, that the valence shell of hydrogen can hold only two electrons.)

9.89 What would be the formula for the simplest compound formed from (a) P and Cl, (b) C and F, and (c) I and Cl?

Electronegativity

9.90 Use Figure 9.5 to choose the atom in each of the following bonds that carries the partial positive charge: (a) N—S, (b) Si—I, (c) N—Br, (d) C—Cl.

9.91 Use Figure 9.5 to choose the atom in each of the following bonds that carries the partial negative charge: (a) Hg—I, (b) P—I, (c) Si—F, (d) Mg—N.

9.92 Which of the bonds in Problem 9.90 is the most polar?

9.93 Which of the bonds in Problem 9.91 is the least polar?

Drawing Lewis Structures

9.94 What are the expected skeletal structures for (a) $SiCl_4$, (b) PF_3, (c) PH_3, and (d) SCl_2?

9.95 What are the expected skeletal structures for (a) HIO_3, (b) H_2CO_3, (c) HCO_3^-, and (d) PCl_4^+?

9.96 How many dots must appear in the Lewis structures of (a) $SiCl_4$, (b) PF_3, (c) PH_3, and (d) SCl_2?

9.97 How many dots must appear in the Lewis structures of (a) HIO_3, (b) H_2CO_3, (c) HCO_3^-, and (d) PCl_4^+?

ILW **9.98** Draw Lewis structures for (a) $AsCl_4^+$, (b) ClO_2^-, (c) HNO_2, and (d) XeF_2.

9.99 Draw Lewis structures for (a) TeF_4, (b) ClF_5, (c) PF_6^-, and (d) XeF_4.

9.100 Draw Lewis structures for (a) $SiCl_4$, (b) PF_3, (c) PH_3, and (d) SCl_2.

9.101 Draw Lewis structures for (a) HIO_3, (b) H_2CO_3, (c) HCO_3^-, and (d) PCl_4^+.

9.102 Draw Lewis structures for (a) CS_2 and (b) CN^-.

9.103 Draw Lewis structures for (a) SeO_3 and (b) SeO_2.

9.104 Draw Lewis structures for (a) AsH_3, (b) $HClO_2$, (c) H_2SeO_3, and (d) H_3AsO_4.

9.105 Draw Lewis structures for (a) NO^+, (b) NO_2^-, (c) $SbCl_6^-$, and (d) IO_3^-.

9.106 Draw the Lewis structure for (a) CH_2O (the central atom is carbon, which is attached to two hydrogens and an oxygen), and (b) $SOCl_2$ (the central atom is sulfur, which is attached to an oxygen and two chlorines).

9.107 Draw Lewis structures for (a) $GeCl_4$, (b) CO_3^{2-}, (c) PO_4^{3-}, and (d) O_2^{2-}.

Formal Charge

9.108 Assign formal charges to each atom in the following structures:

(a) $H\!-\!\overset{..}{\underset{..}{O}}\!-\!\overset{..}{\underset{..}{Cl}}\!-\!\overset{..}{\underset{..}{O}}:$ (c) $\overset{..}{O}\!=\!\overset{..}{S}\!-\!\overset{..}{\underset{..}{O}}:$

(b) $\overset{..}{O}\!=\!\overset{\displaystyle :\overset{..}{O}:}{\underset{|}{S}}\!-\!\overset{..}{\underset{..}{O}}:$

9.109 Assign formal charges to each atom in the following structures:

(a) $:\!\overset{..}{\underset{..}{Cl}}\!-\!\overset{\displaystyle :O:}{\underset{\displaystyle \|}{N}}\!-\!\overset{..}{\underset{..}{O}}:$ (c) $:\!\overset{..}{\underset{..}{F}}\!-\!\overset{\displaystyle :\overset{..}{O}:}{\underset{\displaystyle \underset{..}{O}:}{S}}\!-\!\overset{..}{\underset{..}{F}}:$

(b) $:\!\overset{..}{\underset{..}{F}}\!-\!\overset{\displaystyle :\overset{..}{F}:}{\underset{\displaystyle :\underset{..}{F}:}{N}}\!-\!\overset{..}{\underset{..}{O}}:$

ILW 9.110 Draw the Lewis structure for $HClO_4$ according to the rules in Section 9.7. Assign formal charges to each atom in the formula. Determine the preferred Lewis structure for this compound.

9.111 Draw the Lewis structure for $SOCl_2$ (sulfur bonded to two Cl and one O). Assign formal charges to each atom. Determine the preferred Lewis structure for this molecule.

9.112 Below are two structures for $BeCl_2$. Give two reasons why the one on the left is the preferred structure.

$:\!\overset{..}{\underset{..}{Cl}}\!-\!Be\!-\!\overset{..}{\underset{..}{Cl}}:$ $:\!\overset{..}{\underset{..}{Cl}}\!-\!Be\!=\!\overset{..}{\underset{..}{Cl}}:$

9.113 The following are two Lewis structures that can be drawn for phosgene, a substance that has been used as a war gas:

$:\!\overset{..}{\underset{..}{Cl}}\!-\!\overset{\displaystyle :\overset{..}{O}:}{\underset{\displaystyle |}{C}}\!=\!\overset{..}{\underset{..}{Cl}}:$ $:\!\overset{..}{\underset{..}{Cl}}\!-\!\overset{\displaystyle \overset{..}{O}:}{\underset{\displaystyle \|}{C}}\!-\!\overset{..}{\underset{..}{Cl}}:$

Which is the better Lewis structure? Why?

Resonance

9.114 Draw the resonance structures for CO_3^{2-}. Calculate the average C—O bond order.

ILW 9.115 Draw all the resonance structures for the N_2O_4 molecule and determine the average N—O bond order. The skeletal structure of the molecule is

$$\begin{array}{ccc} O & & O \\ & N \quad N & \\ O & & O \end{array}$$

9.116 How should the N—O bond lengths compare in the NO_3^- and NO_2^- ions?

9.117 Arrange the following in order of increasing C—O bond length: CO, CO_3^{2-}, CO_2, HCO_2^- (formate ion, page 387).

9.118 The Lewis structure of CO_2 was given as

$$\overset{..}{O}\!=\!C\!=\!\overset{..}{O}$$

but two other resonance structures can also be drawn for it. What are they? On the basis of formal charges, why are they not preferred structures?

9.119 Write Lewis structures for the nitrate ion, NO_3^-, and for nitric acid, HNO_3. How many equivalent oxygens are there in each structure? Compare the number of resonance structures for each.

9.120 Following the rules in Figure 9.8, draw the Lewis structure for the sulfate ion. Assign formal charges to each atom in this structure. Reduce formal charges to obtain the best Lewis structure for the ion. Draw the resonance structures for the ion. What is the average sulfur–oxygen bond order in the ion?

9.121 Use formal charges to establish the preferred Lewis structures for the ClO_3^- and ClO_4^- ions. Draw resonance structures for both ions and determine the average Cl—O bond order in each. Which of these ions would be expected to have the shorter Cl—O bond length?

Coordinate Covalent Bonds

9.122 Use Lewis structures to show that the hydronium ion, H_3O^+, can be considered to be formed by the creation of a coordinate covalent bond between H_2O and H^+.

9.123 Use Lewis structures to show that the reaction

$$BF_3 + F^- \longrightarrow BF_4^-$$

involves the formation of a coordinate covalent bond.

ADDITIONAL EXERCISES

9.124 Use data from the tables of ionization energies and electron affinities on pages 341 and 345 to calculate the energy changes for the following reactions:

$$Na(g) + Cl(g) \longrightarrow Na^+(g) + Cl^-(g)$$
$$Na(g) + 2Cl(g) \longrightarrow Na^{2+}(g) + 2Cl^-(g)$$

Approximately how many times larger would the lattice energy of $NaCl_2$ have to be compared to the lattice energy of NaCl for $NaCl_2$ to be more stable than NaCl?

9.125 Changing 1 mol of Mg(*s*) and 1/2 mol of $O_2(g)$ to gaseous atoms requires a total of approximately 150 kJ of energy. The first and second ionization energies of magnesium are 737 and 1450 kJ/mol; the first and second electron affinities of oxygen are −141 and +844 kJ/mol, and the standard heat of formation of MgO(*s*) is −602 kJ/mol. Construct an enthalpy diagram similar to the one in Facets of Chemistry 9.1 and use it to calculate the lattice energy of magnesium oxide. How does the lattice energy of MgO compare with that of NaCl? What might account for the difference?

***9.126** Use an enthalpy diagram to calculate the lattice energy of $CaCl_2$ from the following information: energy needed to vaporize 1 mol of Ca(*s*) is 192 kJ; for calcium, 1st IE = 589.5 kJ mol^{-1}, 2nd IE = 1146 kJ mol^{-1}; electron affinity of Cl is −348 kJ mol^{-1}; bond energy of Cl_2 is 242.6 kJ per mole of Cl—Cl bonds; standard heat of formation of $CaCl_2$ is −795 kJ mol^{-1}.

***9.127** Use an enthalpy diagram and the following data to calculate the electron affinity of bromine: standard heat of formation of NaBr is −360 kJ mol^{-1}; energy needed to vaporize 1 mol of $Br_2(l)$ to give $Br_2(g)$ is 31 kJ mol^{-1}; the first ionization energy of Na is 495.4 kJ mol^{-1}; the bond energy of Br_2 is 192 kJ per mole of Br—Br bonds; the lattice energy of NaBr is −743.3 kJ mol^{-1}.

9.128 In many ways, tin(IV) chloride behaves more like a covalent molecular species than like a typical ionic chloride. Draw the Lewis structure for the tin(IV) chloride molecule.

9.129 In each pair, choose the one with more polar bonds. (Use the periodic table to answer the question.)
(a) PCl_3 or $AsCl_3$ (c) $SiCl_4$ or SCl_2
(b) SF_2 or GeF_4 (d) SrO or SnO

9.130 How many electrons are in the outer shell of the Zn^{2+} ion?

9.131 The Lewis structure for carbonic acid (formed when CO_2 dissolves in water) is usually given as

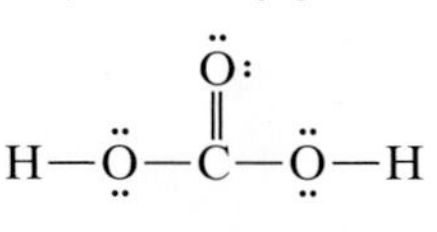

What is wrong with the following structures?

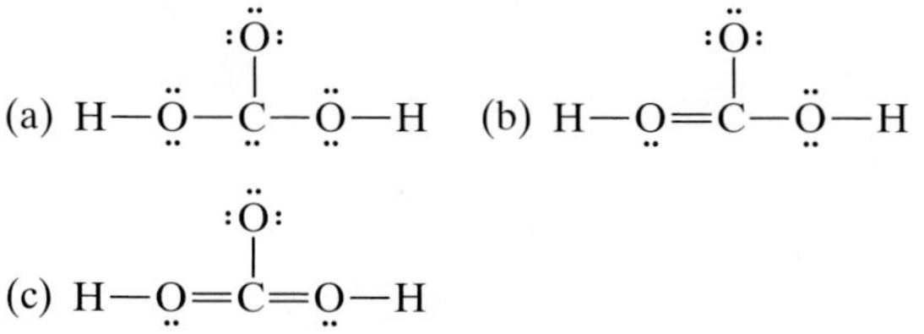

9.132 Assign formal charges to all the atoms in the following Lewis structure of hydrazoic acid, HN_3:

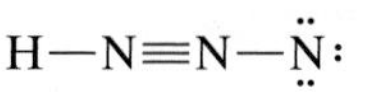

Suggest a lower energy resonance structure for this molecule.

9.133 Assign formal charges to all the atoms in the Lewis structure

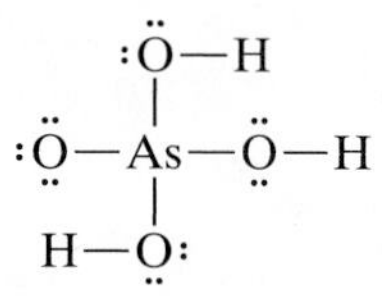

Suggest a lower energy Lewis structure for this molecule.

9.134 The inflation of an "air bag" when a car experiences a collision occurs by the explosive decomposition of sodium azide, NaN_3, which yields nitrogen gas that inflates the bag. The following resonance structures can be drawn for the azide ion, N_3^-. Identify the best and worst of them.

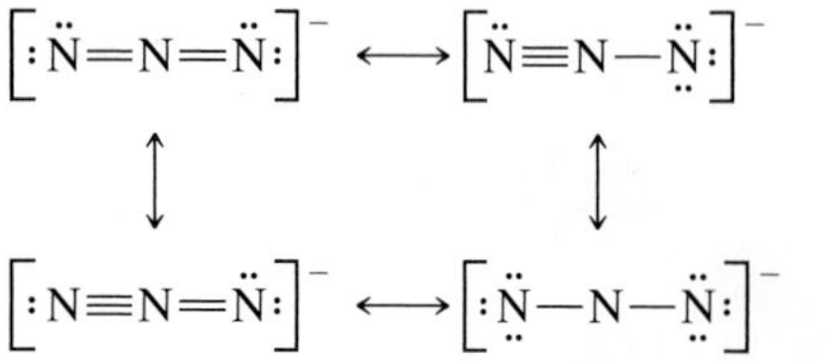

9.135 How should the sulfur oxygen bond lengths compare for the species SO_3, SO_2, SO_3^{2-}, and SO_4^{2-}?

9.136 What is the most reasonable Lewis structure for S_2Cl_2?

***9.137** There are two acids that have the formula HCNO. Which of the following skeletal structures are most likely for them? Justify your answer.

H C O N	H N O C	H O C N
H C N O	H N C O	H O N C

***9.138** What wavelength of light, if absorbed by a hydrogen molecule, could cause the molecule to split into the ions H^+ and H^-? Set up the calculations. (The data required are available in this and previous chapters.)

9.139 In the vapor state, ion pairs of KF can be identified. The dipole moment of such a pair is measured to be 8.59 D and the K—F bond length is found to be 217 pm. Is the K—F bond 100% ionic? If not, what percent of full 1+ and 1− charges do the K and F atoms carry, respectively?

9.140 The drug amphetamine is a powerful stimulant. Its structure, shown below, is similar to that of adrenaline (a powerful hormone), which probably accounts for amphetamine's effects in the body. Write the structure of the product formed when amphetamine reacts with acid (i.e., hydronium ion).

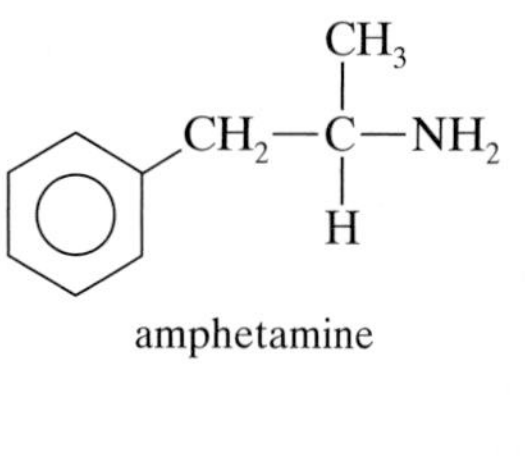

amphetamine

CHAPTER 10 | Chemical Bonding and Molecular Structure

Small differences in the shapes of objects can make a large difference in the way they function. A key operates a lock because of the way their shapes fit together, and a small change in the shape of the key will cause it to no longer open the lock. In a similar way, living things recognize molecules by the way one fits onto another, as illustrated on the following page. Shape is a very important property of molecules, and in this chapter we will explore the kinds of shapes molecules have, how to predict their shapes, and how modern theories of bonding explain molecular shape.

THIS CHAPTER IN CONTEXT If you've ever fumbled with a ring of nearly identical keys, you understand that a subtle difference in an object's shape can make a big difference in the way it functions. Living things recognize molecules in much the same way that locks "recognize" keys—by shape. Molecular shape also affects properties of substances in nonliving systems, so an understanding of molecular shape is a critical part of understanding chemistry.

In the previous chapter, you learned to draw Lewis structures to show how bonds in molecules connect atoms. In this chapter, one of our goals will be to use these two-dimensional maps to predict *molecular geometry,* the three dimensional arrangements of atoms within molecules. From this we can then predict *molecular polarity,* the imbalance of charge within a molecule.

Our second goal is to introduce two sophisticated theories that go beyond Lewis structures: the *valence bond theory* and *molecular orbital theory.* Valence bond theory views chemical bonds as the result of *overlapping atomic orbitals* on bonded atoms. Molecular orbital theory uses *mathematically combined atomic orbitals* to describe the electronic structure of a molecule. Both theories can predict molecular shape, polarity, and properties for many molecules, and provide the detailed understanding of molecular electronic structure that is needed to understand how molecular structure relates to biological activity.

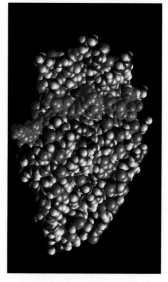

Biochemical reactions are enabled by substances called enzymes, which bind to reacting molecules. In this computer-generated image, we see a complex type of sugar molecule (green) bound to an enzyme called *lysozyme* in chicken egg white. This is an enzyme that destroys bacterial cell walls, and for it to work, there must be a very precise "lock-and-key" fit between the enzyme and the sugar molecule. Such an exact fit is controlled by the three-dimensional shapes of the molecules involved.

10.1 ▶ Molecular shapes are built from five basic arrangements

If there are only two atoms in a molecule, there is no doubt about their arrangement. One is just alongside the other. But when the molecule contains three or more atoms, it may have many different shapes. These shapes can be quite complicated, but most are built from just five basic geometrical structures. We'll describe these basic shapes on the pages ahead. As you study these structures, you should try hard to visualize them in three dimensions. You should also learn how to sketch them in a way that conveys the three-dimensional information.

Linear molecules

In a **linear molecule** the atoms lie in a straight line. All diatomic molecules are linear. When three atoms are present, the angle formed by the covalent bonds, which we call the **bond angle,** equals 180°.

A linear molecule

Basic molecular shapes

Planar triangular molecules

In a **planar triangular molecule,** a central atom sits in the center of a triangle formed by three other atoms.

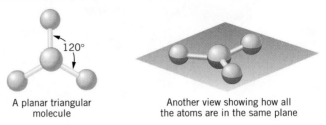

A planar triangular
molecule

Another view showing how all
the atoms are in the same plane

All four atoms lie in the same plane. The bond angles are all equal to 120°.

Tetrahedral molecules

A **tetrahedron** is a four-sided geometric figure shaped like a pyramid with triangular faces. In a **tetrahedral molecule,** the central atom lies in the center of a tetrahedron formed by four other atoms.

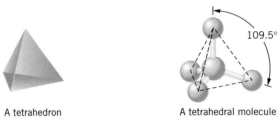

A tetrahedron

A tetrahedral molecule

All the bond angles in a tetrahedral molecule are the same and are equal to 109.5°.

This is the first molecular shape that you've encountered that can't be set down entirely on a two-dimensional surface, so you may need some practice in drawing the structure. It is really quite simple, as illustrated in Figure 10.1.

Trigonal bipyramidal molecules

A **trigonal bipyramid** consists of two *trigonal pyramids* (pyramids with triangular faces) that share a common base.

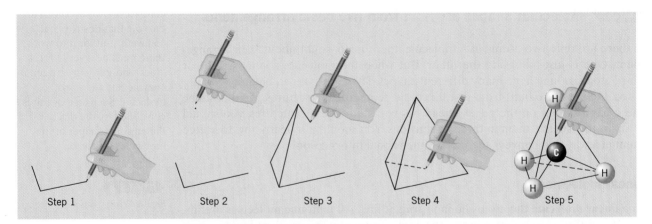

Step 1   Step 2   Step 3   Step 4   Step 5

FIGURE 10.1 *Steps in representing a tetrahedral molecule. Step 1:* Draw a check mark. *Step 2:* Select a point above the check mark at a height about equal to the width of the check mark. *Step 3:* Draw three lines that go from the point to the ends and corner of the check mark. These define the front two triangular faces of the tetrahedron. *Step 4:* Draw a dotted line to define the back edge of the tetrahedron. *Step 5:* To draw a tetrahedral molecule, place an atom in the center (a carbon atom in this example) and place an atom at each corner of the tetrahedron (H atoms in this example). Finally, draw "bonds" from the atoms at the corners to the atom in the center.

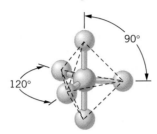

A trigonal bipyramid

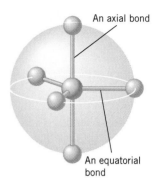

An axial bond

An equatorial bond

Axial and equatorial bonds in a trigonal bipyramidal molecule.

In a **trigonal bipyramidal molecule,** the central atom is located in the middle of the triangular plane shared by the upper and lower trigonal pyramids and is bonded to five atoms that are at the vertices of the figure.

90°

120°

A trigonal bipyramidal molecule

In this structure, not all the bonds are equivalent. If we imagine the trigonal bipyramid centered inside a sphere similar to Earth, as illustrated in the margin, the atoms in the triangular plane are located around the equator. The bonds to these atoms are called **equatorial bonds.** The angle between two equatorial bonds is 120°. The two vertical bonds pointing along the north and south axis of the sphere are 180° apart and are called **axial bonds.** The bond angle between an axial bond and an equatorial bond is 90°.

A simplified representation of a trigonal bipyramid is illustrated in the margin. The equatorial triangular plane is sketched as it would look tilted backward, so we're looking at it from its edge. The axial bonds are represented as lines pointing up and down. To add more three-dimensional character, notice that the bond pointing down appears to be partially hidden by the triangular plane in the center.

Simplified representation of a trigonal bipyramid.

Octahedral molecules

An **octahedron** is an eight-sided figure, which you might think of as two *square pyramids* sharing a common square base. The octahedron has only six vertices, and in an **octahedral molecule** we find an atom in the center of the octahedron bonded to six other atoms at these vertices.

An octahedron

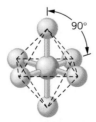

90°

An octahedral molecule

All the bonds in an octahedral molecule are equivalent, with angles between adjacent bonds equal to 90°.

A simplified representation of an octahedron is shown in the margin. The square plane in the center of the octahedron, when drawn in perspective and viewed from its edge, looks like a parallelogram. The bonds to the top and bottom

Simplified representation of an octahedron.

of the octahedron are shown as vertical lines. Once again, the bond pointing down is drawn so it appears to be partially hidden by the square plane in the center.

10.2 ▶ Molecular shapes are predicted using the VSEPR model

VSEPR Theory

A useful theoretical model should explain known facts, and it should be capable of making accurate predictions. The **valence shell electron pair repulsion model** (called the *VSEPR model*) is remarkably successful at both of these and is also conceptually simple. The model is based on the idea that groups of electrons repel each other and will position themselves as far away from each other as possible within a molecule. An **electron domain** is a region in space where electrons can be found. There are two types of domains:

An **electron domain** is a bond, lone pair, or unpaired electron. VSEPR stands for "valence shell electron pair repulsion." The name is unfortunate because we need to consider repulsions between *electron domains* rather than just electron pairs when using the VSEPR theory.

- **Bonding domains** contain electron pairs that are involved in bonds *between two atoms*. A double bond (containing four electrons) will occupy more space than a single bond (with only two electrons) but all electrons in a bond occupy the same region in space and so they all belong to the same bond domain. All of the electrons within a given single, double, or triple bond are considered to be in the same bonding domain.

- **Nonbonding domains** contain valence electrons that are associated with *a single atom*. A nonbonding domain is either an unshared pair of valence electrons (called a **lone pair**) or an unpaired electron (found in molecules with an odd number of valence electrons).

EXAMPLE 10.1
Recognizing Electron Domains

Identify all of the bonding and nonbonding domains in the acetone molecule, CH_3COCH_3, which was described on page 370.

ANALYSIS: We will begin with the Lewis structure for CH_3COCH_3. Each bond or multiple bond in the molecule is a bonding domain; each lone pair or unpaired electron is a nonbonding domain.

SOLUTION: The Lewis structure of CH_3COCH_3 is

$$
\begin{array}{ccccc}
H & & :\!O\!: & & H \\
| & & \| & & | \\
H\!-\!C & \!-\! & C & \!-\! & C\!-\!H \\
| & & & & | \\
H & & & & H
\end{array}
$$

Each single or double bond corresponds to a bonding domain (shown in orange) and each lone pair corresponds to a nonbonding domain (shown in blue), so there are nine bonding domains and two nonbonding domains.

Is the Answer Reasonable?
Check the Lewis structure. Notice that the double bond between carbon and oxygen counts as a single bonding domain, not two.

PRACTICE EXERCISE 1: Identify all the bonding and nonbonding domains in the carbonic acid molecule, H_2CO_3.

The VSEPR model is based on the notion that *electron domains keep as far away as possible from each other* while at the same time staying as close as possible to the central atom, as shown in Figure 10.2. Let's look at an example that illustrates how this simple concept enables us to predict the shape of a molecule.

Consider the $BeCl_2$ molecule. We've drawn its Lewis structure as

$$:\!\ddot{C}l\!-\!Be\!-\!\ddot{C}l\!:$$

But how are these atoms arranged? Is $BeCl_2$ linear or is it nonlinear; that is, do the atoms lie in a straight line, or do they form some angle less than 180°?

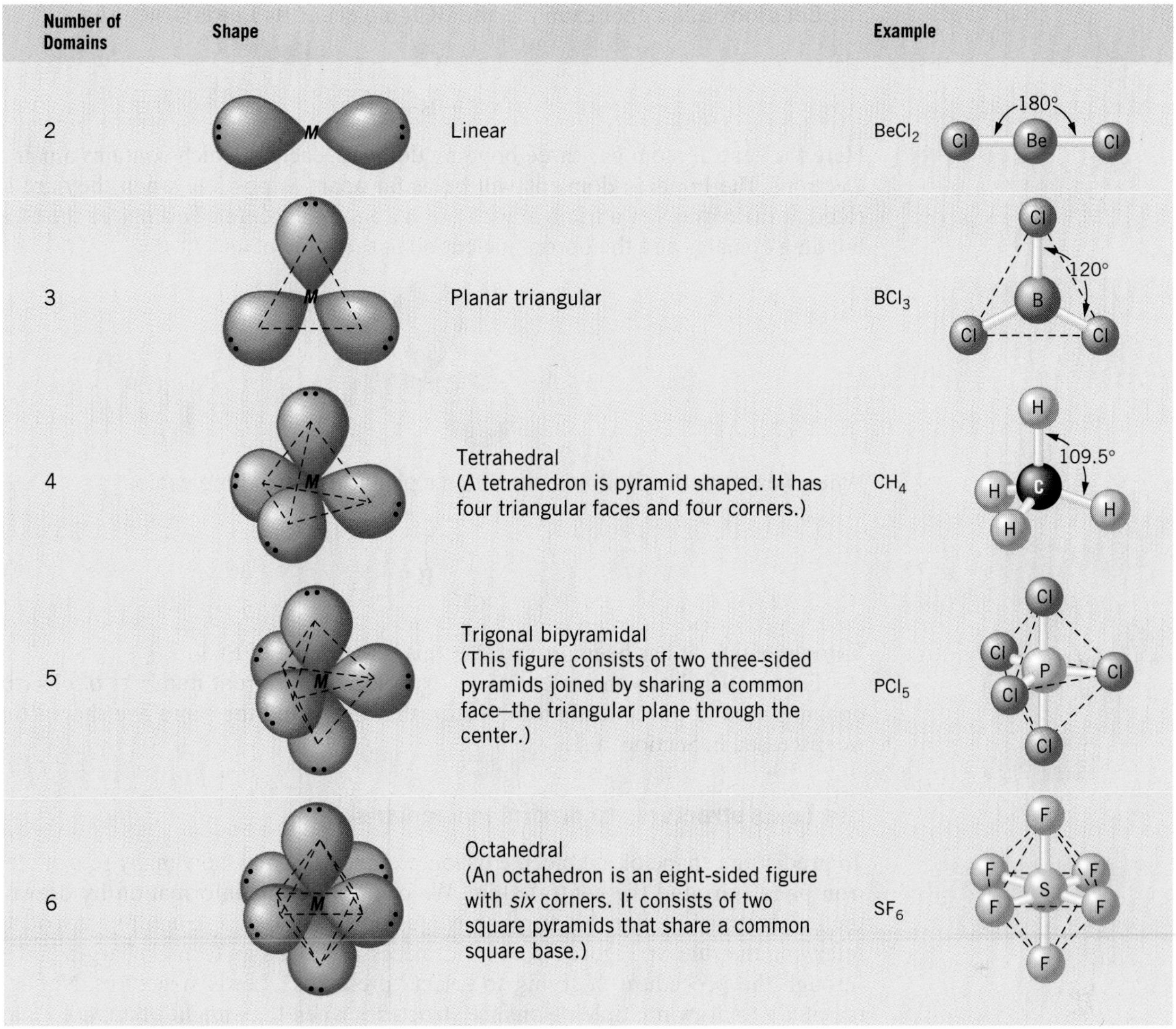
Number of Domains
Shape
Example
2
Linear
$BeCl_2$
180°
Cl
Be
Cl
3
Planar triangular
BCl_3
120°
B
4
Tetrahedral
(A tetrahedron is pyramid shaped. It has four triangular faces and four corners.)
CH_4
109.5°
H
C
5
Trigonal bipyramidal
(This figure consists of two three-sided pyramids joined by sharing a common face—the triangular plane through the center.)
PCl_5
P
6
Octahedral
(An octahedron is an eight-sided figure with *six* corners. It consists of two square pyramids that share a common square base.)
SF_6
F
S
M

FIGURE 10.2 ***Shapes expected for different numbers of electron domains around a central atom, M.*** Each lobe represents an electron domain.

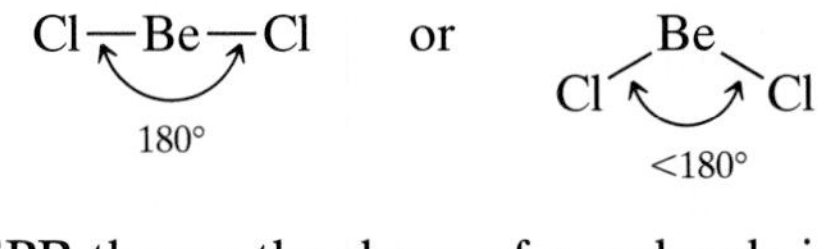
Cl
Be
Cl
180°
or
<180°

According to VSEPR theory, the shape of a molecule is determined by the tendency of electron domains to keep as far away from each other as possible. For $BeCl_2$, there are two bonding domains around the beryllium atom. The domains will seek positions as far apart as possible, which occurs when the domains are on opposite sides of the nucleus. We can represent this as

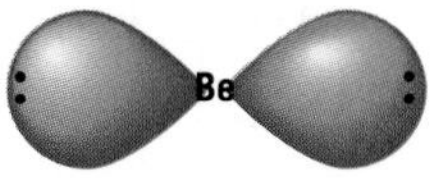
Be

to suggest the approximate locations of the electron clouds of the valence shell electron pairs. In order for the electrons to be in the Be—Cl bonds, the Cl atoms must be placed where the electrons are; the result is that we predict that a $BeCl_2$ molecule should be linear.

Cl—Be—Cl

In fact, this is the shape of $BeCl_2$ molecules in the vapor state.

Let's look at another example, the BCl_3 molecule. Its Lewis structure is

$$:\ddot{C}l:$$
$$|$$
$$:\ddot{C}l-B-\ddot{C}l:$$

Here the central atom has three bonding domains, each of which contains a pair of electrons. The bonding domains will be as far apart as possible when they are located at the corners of a triangle with the boron in the center. This places the three bonding domains and the boron nucleus all in the same plane

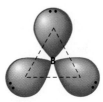

When we attach the Cl atoms, we obtain a planar triangular molecule.

$$Cl$$
$$|$$
$$B$$
$$Cl^{\diagup} \quad ^{\diagdown}Cl$$

Experimentally, it has been proved that this is the shape of BCl_3.

Figure 10.2 shows the orientations expected for different numbers of electron domains around the central atom. Notice that they yield the same five shapes that we discussed in Section 10.1.

Use Lewis structures to predict molecular shape

To predict the shape of a molecule or ion, we need to know how many sets of electron pairs surround the central atom. We can obtain this information by drawing the Lewis structure. For this purpose, we just need the Lewis structure constructed following the rules in Figure 9.8. It is not necessary to assign formal charges and go through the procedure of trying to select "preferred" Lewis structures. Nor is it necessary to draw multiple resonance structures when they might otherwise be appropriate. Just one simple Lewis structure is all we need.

EXAMPLE 10.2

Predicting Molecular Shapes

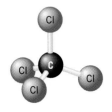

Carbon tetrachloride was once used as a cleaning fluid until it was discovered that it causes liver damage if absorbed by the body. What is the shape of the CCl_4 molecule?

ANALYSIS: Draw the Lewis structure for CCl_4. Then count the number of domains around the central atom. Finally, choose a shape that allows the domains around the central atom to be placed as far apart as possible.

SOLUTION: Following the rules, the Lewis structure of CCl_4 is

$$:\ddot{C}l:$$
$$|$$
$$:\ddot{C}l-C-\ddot{C}l:$$
$$|$$
$$:\ddot{C}l:$$

There are four bonds, and so four bonding domains around the carbon. The domains can be placed farthest apart when they are arranged tetrahedrally, so the molecule is expected to be tetrahedral. (This is, in fact, its structure.)

Is the Answer Reasonable?
Check the Lewis structure. Each atom has an octet, and the total number of valence electrons for four single bonds (8) plus twelve lone pairs (24) is 32. This matches the

total number of valence electrons for four chlorine atoms (28) plus one carbon atom (4). Finally, we check to see that we've chosen the correct shape by comparing our results with Figure 10.2.

PRACTICE EXERCISE 2: What shape is expected for the $SbCl_5$ molecule?

Nonbonding domains affect the shape of a molecule

In the molecules that we've considered so far, all of the domains around the central atom have been bonding domains. This isn't always the case, though. Some molecules have a central atom with one or more nonbonding domains, consisting of unshared electron pairs (lone pairs) or unpaired valence electrons. These nonbonding domains affect the geometry of the molecule. An example is $SnCl_2$.

$SnCl_2$ is another molecule that behaves as though it has less than an octet around the central atom.

$$:\ddot{Cl}-\ddot{Sn}-\ddot{Cl}:$$

There are three domains around the tin atom, two bonding domains plus a nonbonding domain (the lone pair). As in BCl_3, the domains are farthest apart at the corners of a triangle.

Placing the two chlorine atoms where two of the domains are gives

We can't describe this molecule as triangular, even though that is how the domains are arranged. *Molecular shape describes the arrangement of atoms, not the arrangement of domains.* Therefore, we describe the shape of the $SnCl_2$ molecule as being **nonlinear** or **bent** or **V-shaped.**

Notice, now, that when there are three domains around the central atom, *two different molecular shapes are possible.* If all three are bonding domains, as in BCl_3, a molecule with a planar triangular shape is formed. If one of the domains is nonbonding, as in $SnCl_2$, the arrangement of the atoms in the molecule is said to be nonlinear. The predicted shapes of both, however, are *derived* by first noting the triangular arrangement of domains around the central atom and *then* adding the necessary number of atoms.

We use the orientations of the domains in the valence shell of the central atom to determine how the atoms in the structure are arranged. The arrangement of the atoms is what we mean by the *shape of the molecule.*

Molecules with four domains have shapes derived from a tetrahedron

The most common type of molecule or ion has four electron pairs (an octet) in the valence shell of the central atom. When these electron pairs are used to form four bonds, as in methane (CH_4), the resulting molecule is tetrahedral, as we've seen. There are many examples, however, where nonbonding domains are present. For example,

$$H-\underset{\underset{H}{|}}{\overset{..}{N}}-H \qquad H-\overset{..}{\underset{..}{O}}-H$$

one lone pair   two lone pairs

Number of Bonding Domains	Number of Nonbonding Domains	Structure	
4	0	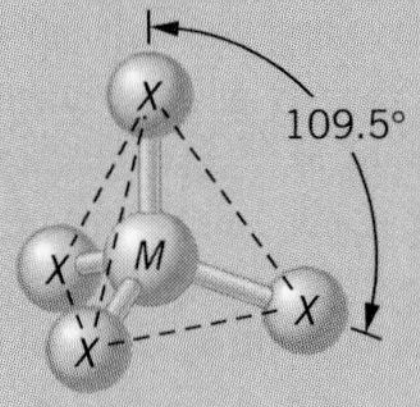	**Tetrahedral** (example, CH_4) all bond angles are 109.5°
3	1	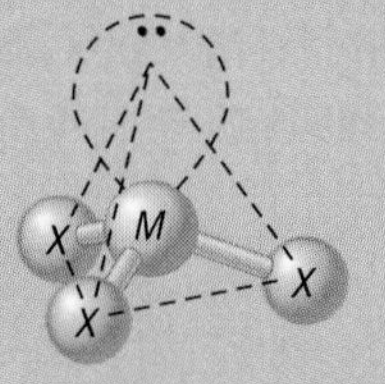	**Trigonal pyramidal** (pyramid shaped) (example, NH_3)
2	2	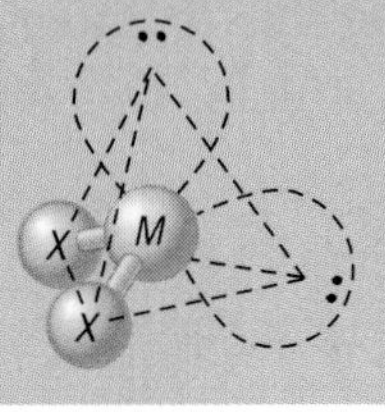	**Nonlinear, bent** (example, H_2O)

FIGURE **10.3** *Molecular shapes with four domains around the central atom.* The molecules MX_4, MX_3, and MX_2 shown here all have four domains arranged tetrahedrally around the central atom. The shapes of the molecules and their descriptions are derived from the way the *X* atoms are arranged around *M*, ignoring the nonbonding domains.

Figure 10.3 shows how the nonbonding domains affect the shapes of molecules of this type.

With one nonbonding domain, the central atom is at the top of a pyramid with three atoms at the corners of the triangular base. The resulting structure is said to be **trigonal pyramidal.** Notice once again that in specifying the shape of the molecule, we ignore the nonbonding domain. Even though the *domains* are arranged in a tetrahedron, the *atoms* are arranged in a trigonal pyramid.

When there are two nonbonding domains in the tetrahedron, the three atoms (the central atom plus the two atoms bonded to it) do not lie in a straight line, so the structure is **nonlinear.**

If a molecule contains three atoms, they either lie in a straight line and the molecule is linear, or they are not in a straight line and the molecule is nonlinear.

Molecules with five domains have shapes derived from a trigonal bipyramid

When five domains are present around the central atom, they are directed toward the vertices of a trigonal bipyramid. Molecules such as PCl_5 have this geometry, as we saw in Figure 10.2, and are said to be **trigonal bipyramidal.**

The trigonal bipyramid differs from the other basic shapes in that all positions are not equivalent. Positions on the "poles" of the central atom (called **axial positions**) each have three close neighbors at angles of 90° and one at a much larger angle of 180°. Positions around the "equator" (called **equatorial positions**) each have two neighbors at 90° and two at a much larger angle of 120°. The nearest neighbor interactions matter the most. So by counting 90° interactions, we can see that equatorial positions are less crowded than axial positions.

Electron domains arrange themselves so that they are as far apart as possible. Nonbonding domains are the largest, and they always move into the least crowded position around the central atom. In a trigonal bipyramid, the equatorial positions are less crowded than the axial positions.

The difference between the crowded axial and roomier equatorial positions is important in molecules that contain nonbonding domains. Nonbonding domains take up more room than bonding domains. Because nonbonding domains are attracted to a single nucleus, they are free to spread out as much as possible around that nucleus. However, bonding domains are attracted to two nuclei at once (one at each end of the bond), so they are smaller than nonbonding domains. *The larger*

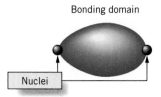

Bonding domain

Nuclei

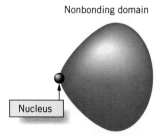

Nonbonding domain

Nucleus

Relative sizes of bonding and nonbonding domains.

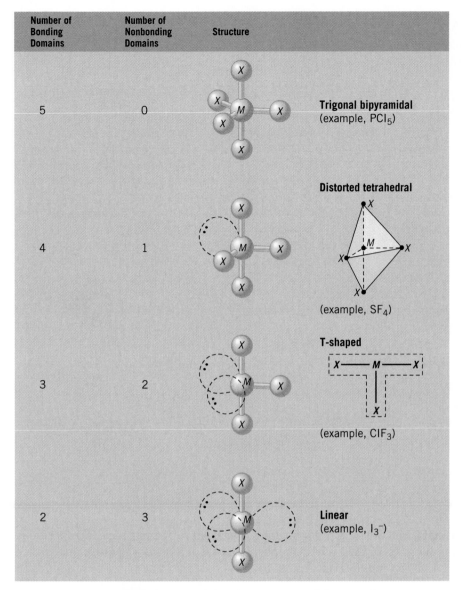

Number of Bonding Domains	Number of Nonbonding Domains	Structure	
5	0		**Trigonal bipyramidal** (example, PCl_5)
4	1		**Distorted tetrahedral** (example, SF_4)
3	2		**T-shaped** (example, ClF_3)
2	3		**Linear** (example, I_3^-)

FIGURE 10.4 *Molecular shapes with five domains around the central atom.* Four different molecular structures are possible, depending on the number of nonbonding domains around the central atom *M*.

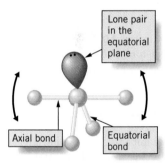

Lone pair in the equatorial plane

Axial bond

Equatorial bond

The distorted tetrahedron is sometimes described as a **seesaw structure.** The origin of this description can be seen if we tip the structure over so it stands on the two atoms in the equatorial plane, with the nonbonding domain pointing up.

nonbonding domains occupy the least crowded positions on the central atom, while the smaller bonding domains occupy whatever positions are left.

Figure 10.4 shows the kinds of geometries that we find for different numbers of nonbonding domains in the trigonal bipyramid. When there is only one nonbonding domain, as in SF_4, the structure is described as a **distorted tetrahedron** or **seesaw** in which the central atom lies along one edge of a four-sided figure. With two nonbonding domains in the equatorial plane, the molecule is **T-shaped** (shaped like the letter T), and when there are three nonbonding domains in the equatorial plane, the molecule is **linear.**

Molecules with six domains have shapes derived from an octahedron

Finally, we come to those molecules or ions that have six domains around the central atom. When all are in bonds, as in SF_6, the molecule is octahedral. When one nonbonding domain is present the molecule or ion has the shape of a **square**

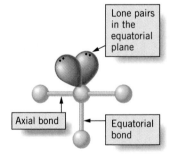

Lone pairs in the equatorial plane

Axial bond

Equatorial bond

A molecule with two nonbonding domains in the equatorial plane of the trigonal bipyramid. When tipped over, the molecule looks like a T, so it is called **T-shaped.**

Number of Bonding Domains	Number of Nonbonding Domains	Structure	
6	0		**Octahedral** (example, SF_6) all bond angles are 90°
5	1		**Square pyramidal** (example, BrF_5)
4	2		**Square planar** (example, XeF_4)

FIGURE 10.5 *Molecular shapes with six domains around the central atom.* Although more are theoretically possible, only three different molecular shapes are observed, depending on the number of nonbonding domains around the central atom.

No common molecule or ion with six domains around the central atom has more than two nonbonding domains.

pyramid, and when two nonbonding domains are present they take positions on opposite sides of the nucleus and the molecule or ion has a **square planar** structure. These shapes are shown in Figure 10.5.

EXAMPLE 10.3 **Predicting the Shapes of Molecules and Ions**	Do we expect the ClO_2^- ion to be linear? **ANALYSIS:** First, draw the Lewis structure for the ClO_2^- ion. Then count the number of domains around the central atom to choose the shape that places the domains as far away from each other as possible. We sketch this shape and then attach the two oxygens. Finally, we ignore any nonbonding domains to arrive at the description of the shape of the ion. In other words, in this last step we note how the *atoms* are arranged, not how the domains are arranged.

SOLUTION: Following our procedure, we obtain

$$\left[\ddot{\text{O}}\!-\!\ddot{\text{Cl}}\!-\!\ddot{\text{O}}\!:\right]^-$$

There are four domains around the chlorine: two bonding and two nonbonding. Four domains (according to the theory) are always arranged tetrahedrally. This gives

Now we add the two oxygens. It doesn't matter which locations in the tetrahedron we choose because all the bond angles are equal.

The O—Cl—O angle is less than 180°, so the ion is expected to be nonlinear.

Is the Answer Reasonable?
Check the Lewis structure. Each atom has an octet, and the total number of valence electrons for two single bonds (4) plus eight lone pairs (16) is 20. This matches the total number of valence electrons for one chlorine atom (7) plus two oxygen atoms (12), plus one more electron for the ion's 1− charge. Finally, we check to see that we've chosen the correct shape by comparing our results with Figure 10.3.

PRACTICE EXERCISE 3: What shape is expected for the I_3^- ion?

For a molecule with three atoms, there are only two ways they can be arranged, either in a straight line (linear) or in a nonlinear arrangement.

Xenon is one of the noble gases, and is generally quite unreactive. In fact, it was long believed that all the noble gases were totally unable to form compounds. It came as quite a surprise, therefore, when it was discovered that some compounds could be made. One of these is xenon difluoride, XeF_2. What would you expect the geometry of XeF_2 to be, linear or nonlinear?

EXAMPLE 10.4
Predicting the Shapes of Molecules and Ions

ANALYSIS: First, draw the compound's Lewis structure. Then count domains around the central atom to determine which of the basic shapes forms the basis for the structure. Sketch the structure and attach the fluorine atoms. Finally, ignore any nonbonding domains and observe how the atoms are arranged to describe the shape of the molecule.

SOLUTION: The Lewis structure of XeF_2 is

$$:\overset{..}{\underset{..}{F}}\!-\!\overset{..}{Xe}\!-\!\overset{..}{\underset{..}{F}}:$$

Next we count domains around xenon; there are five of them, three nonbonding and two bonding domains. When there are five domains, they are arranged in a trigonal bipyramid.

Now we must add the fluorine atoms. In a trigonal bipyramid, the nonbonding domains always occur in the equatorial plane through the center, so the fluorines go on the top and bottom. This gives

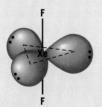

The three atoms, F—Xe—F, are arranged in a straight line, so the molecule is linear.

Is the Answer Reasonable?

Check the Lewis structure. Each fluorine atom has an octet. The xenon atom has an expanded octet, but that is not unusual for atoms in the third period or below. The total number of valence electrons for two single bonds (4) plus nine lone pairs (18) is 22. This matches the total number of valence electrons for one xenon atom (8) plus two fluorine atoms (14). Finally, we check to see that we've chosen the correct shape by comparing our results with Figure 10.4.

PRACTICE EXERCISE 4: The first known compound of the noble gas argon is HArF. What shape is expected for the HArF molecule? (Ar is the central atom.)

EXAMPLE 10.5 **Predicting the Shapes of Molecules and Ions**	What is the probable geometry of the $XeOF_4$ molecule, which contains an oxygen atom and four fluorine atoms each bonded to xenon? **ANALYSIS:** Once again, we must draw the Lewis structure and then count the number of domains around the central atom to determine the shape that places the domains as far apart as possible. We sketch the shape and attach the four fluorines and the oxygen atom.

SOLUTION: The Lewis structure is

$$\ddot{:}\overset{\displaystyle :\ddot{F}:}{\underset{\displaystyle :\ddot{O}\quad\ddot{F}:}{\overset{\displaystyle \ddot{F}\quad|\quad\ddot{F}\ddot{:}}{Xe}}}$$

There are six electron domains around xenon, and they must be arranged octahedrally.

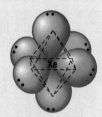

Next we attach the oxygen and four fluorines. The most symmetrical structure is

which we would describe as a square pyramid. Although the oxygen might, in principle, be placed in one of the positions at the corners of the square base (in place of one of the fluorines), the structure shown is the actual structure of the molecule.

Is the Answer Reasonable?

Check the Lewis structure. The oxygen and fluorine atoms have octets. As in the previous example, the xenon atom has an expanded octet, but this is not unusual. The total number of valence electrons for five single bonds (10) plus sixteen lone pairs (32) is 42. This matches the total number of valence electrons for one xenon atom (8) plus four fluorine atoms (28) plus one oxygen atom (6). Finally, we check to see that we've chosen the correct shape by comparing our results with Figure 10.5.

PRACTICE EXERCISE 5: What shape is expected for the XeF_4 molecule?

A bonding domain can be a single, double, or triple bond

Our discussion to this point has dealt only with the shapes of molecules or ions having single bonds. Luckily, the presence of double or triple bonds does not complicate matters at all. In a double bond, both electron pairs must stay together between the two atoms; they can't wander off to different locations in the valence shell. This is also true for the three pairs of electrons in a triple bond. *A double bond or a triple bond counts as one bonding domain.* For example, the Lewis formula for CO_2 is

$$:\ddot{O}{=}C{=}\ddot{O}:$$

The carbon atom has just two bonding domains around it, one for each double bond. The domains are located on opposite sides of the nucleus and a linear molecule is formed. Similarly, we would predict the following shapes for NO_2^- and NO_3^-.

We only need to consider one of the resonance structures to reach a conclusion about molecular structure.

nonlinear (bent)   planar triangular

The Lewis structure for the very poisonous gas hydrogen cyanide, HCN, is

$$H{-}C{\equiv}N:$$

Is the HCN molecule linear or nonlinear?

ANALYSIS: There are two bonding domains around the carbon, one for the triple bond with nitrogen and one for the single bond with hydrogen.

SOLUTION: We expect the two bonds to locate themselves 180° apart, yielding a linear HCN molecule.

Is the Answer Reasonable?
Check the Lewis structure. The carbon and nitrogen atoms have octets. The total number of valence electrons for one single bond (2), one triple bond (6), and one lone pair (2) is 10. That matches the total number of valence electrons for one hydrogen atom (1), one carbon atom (4), and one nitrogen atom (5).

PRACTICE EXERCISE 6: Predict the shapes of SO_3^{2-}, CO_3^{2-}, XeO_4, and OF_2.

EXAMPLE 10.6

Predicting the Shapes of Molecules and Ions

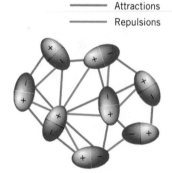

——— Attractions
——— Repulsions

FIGURE 10.6 *Attractions between dipoles.* Molecules tend to orient themselves so that the positive end of one molecular dipole is near the negative end of another.

10.3 ▶ Polar molecules are asymmetric

One of the main reasons we are concerned about whether a molecule is polar or not is because many physical properties, such as melting point and boiling point, are affected by it. This is because polar molecules attract each other. The positive end of one polar molecule attracts the negative end of another, as illustrated in Figure 10.6. The strength of the attraction depends both on the amount of charge on either end of the molecule and on the distance between the charges; in other words, it depends on the molecule's dipole moment.

The dipole moment of a molecule can be determined experimentally, and when this is done, an interesting observation is made. There are many molecules that have no dipole moment even though they contain bonds that are polar. Stated differently, they are nonpolar molecules, even though they have polar bonds. The reason for this can be seen if we examine the key role that molecular structure plays in determining molecular polarity.

Let's begin with the H—Cl molecule, which has only two atoms and therefore only one bond. As you've learned, this bond is polar because the two atoms differ

The polar HCl molecule has partial positive and negative charges on opposite ends, with the hydrogen carrying the positive charge.

H—Cl
$\delta+$ $\delta-$

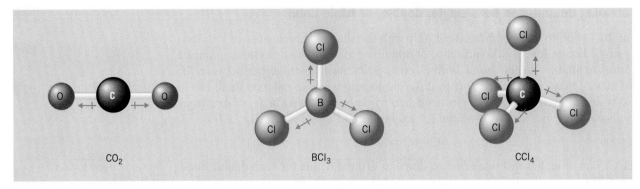

FIGURE 10.7 *In symmetric molecules such as these, the bond dipoles cancel to give nonpolar molecules.*

Bond dipoles can be treated as vectors, and the polarity of a molecule is predicted by taking the vector sum of the bond dipoles. If you've studied vectors before, this may help your understanding.

The nonbonding domains do contribute to the overall polarity of the molecule. However, because nonbonding domains are usually spread more thinly than bonding domains, the contribution is small. We can usually classify the molecule as polar or nonpolar by considering just the symmetry of the bond dipoles.

in electronegativity. This means that the atoms at opposite ends of the bond carry partial charges of opposite sign. A molecule with equal but opposite charges on opposite ends is polar, so HCl is a polar molecule. In fact, any molecule composed of just two atoms that differ in electronegativity must be polar.

For molecules that contain more than two atoms, we have to consider the combined effects of all the polar bonds. Sometimes, when all the atoms attached to the central atom are the same, the effects of the individual polar bonds cancel and the molecule as a whole is nonpolar. Some examples are shown in Figure 10.7. In this figure the dipoles associated with the bonds themselves—the **bond dipoles**—are shown as arrows crossed at one end, $\longmapsto$. The arrowhead indicates the negative end of the bond dipole and the crossed end corresponds to the positive end.

CO_2 is a symmetric molecule. Both bonds are identical, so each bond dipole is of the same magnitude. Because CO_2 is a linear molecule, these bond dipoles point in opposite directions and work against each other. The net result is that their effects cancel, and CO_2 is a nonpolar molecule. Although it isn't so easy to visualize, the same thing also happens in BCl_3 and CCl_4. In each of these molecules, the influence of one bond dipole is canceled by the effects of the others.

Perhaps you've noticed that the structures of the molecules in Figure 10.7 correspond to three of the basic shapes that we used to derive the shapes of molecules. Molecules with the remaining two structures, trigonal bipyramidal and octahedral, also are nonpolar if all the atoms attached to the central atom are the same. All of the basic shapes are "balanced," or **symmetric,** if all of the domains and groups attached to them are identical. Examples are shown in Figure 10.8. The trigonal

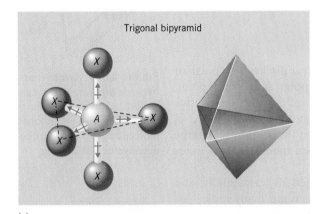

(a)

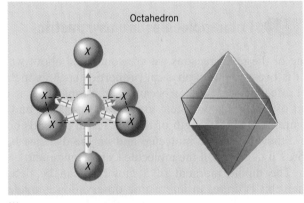

(b)

FIGURE 10.8 *Cancellation of bond dipoles in trigonal bipyramidal and octahedral molecules.* (*a*) A trigonal bipyramidal molecule, AX_5, in which the central atom A is bonded to five identical atoms X. The set of three bond dipoles in the triangular plane in the center (in blue) cancel, as do the linear set of dipoles (red). Overall, the molecule is nonpolar. (*b*) An octahedral molecule AX_6, in which the central atom is bonded to six identical atoms. This molecule contains three linear sets of bond dipoles. Cancellation occurs for each set, so the molecule is nonpolar overall.

bipyramidal structure can be viewed as a planar triangular set of atoms (shown in blue) plus a pair of atoms arranged linearly (shown in red). All the bond dipoles in the planar triangle cancel, as do the two dipoles of the bonds arranged linearly, so the molecule is nonpolar overall. Similarly, we can look at the octahedral molecule as consisting of three linear sets of bond dipoles. Cancellation of bond dipoles occurs in each set, so overall the octahedral molecule is nonpolar.

If all the atoms attached to the central atom are not the same, or if there are nonbonding domains attached the central atom, the molecule is usually polar. For example, in $CHCl_3$, one of the atoms in the tetrahedral structure is different from the others. The C—H bond is less polar than the C—Cl bonds, and the bond dipoles do not cancel (Figure 10.9). An "unbalanced" structure such as this is said to be **asymmetric.**

Two familiar molecules that have nonbonding domains on their central atoms are shown in Figure 10.10. Here the bond dipoles are oriented in such a way that their effects do not cancel. In water, for example, each bond dipole points partially in the same direction, toward the oxygen atom. As a result, the bond dipoles partially add to give a net dipole moment for the molecule. The same thing happens in ammonia, where three bond dipoles point partially in the same direction and add to give a polar NH_3 molecule.

Not every structure that contains nonbonding domains on the central atom produces polar molecules. The following are two exceptions.

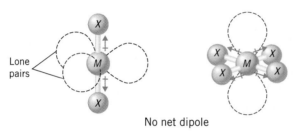

No net dipole

In the first case, we have a pair of bond dipoles arranged linearly, just as in CO_2. In the second, the bonded atoms lie at the corners of a square, which can be viewed as two linear sets of bond dipoles. If the atoms attached to the central atom are the same, cancellation of bond dipoles is bound to occur and produce nonpolar molecules. This means that molecules such as linear XeF_2 and square planar XeF_4 are nonpolar.

In Summary

A molecule will be nonpolar if (a) the bonds are nonpolar or (b) there are no lone pairs in the valence shell of the central atom and all the atoms attached to the central atom are the same.

A molecule in which the central atom has lone pairs of electrons will usually be polar, with the two exceptions described above.

On the basis of the preceding discussions, let's see how we can predict whether molecules are expected to be polar or nonpolar.

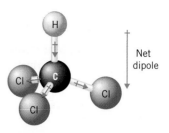

FIGURE 10.9 *Bond dipoles in the chloroform molecule, $CHCl_3$.* Because C is slightly more electronegative than H, the C—H bond dipole points toward the carbon. The small C—H bond dipole actually adds to the effects of the C—Cl bond dipoles. All the bond dipoles are additive and this causes $CHCl_3$ to be a polar molecule.

TOOLS

Molecular shape and molecular polarity

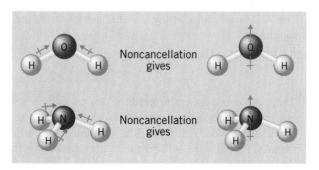

FIGURE 10.10 *When nonbonding domains occur on the central atom, the bond dipoles usually do not cancel, and polar molecules result.*

EXAMPLE 10.7
Predicting Molecular Polarity

Do we expect the PCl_3 molecule to be polar or nonpolar?

ANALYSIS: First, we need to know whether the bonds in the molecule are polar. If not, the molecule will be nonpolar regardless of its structure. If the bonds are polar, then we need to determine its structure. On the basis of the structure, we can then decide whether the bond dipoles cancel.

SOLUTION: The electronegativities of the atoms (P = 2.1, Cl = 2.9) tell us that the individual P—Cl bonds will be polar. Therefore, to predict whether or not the molecule is polar, we need to know its shape. First we draw the Lewis structure following our usual procedure.

$$\overset{\displaystyle :\ddot{\text{Cl}}:}{\underset{\displaystyle}{|}}$$
$$:\ddot{\text{Cl}}-\overset{}{\underset{..}{\text{P}}}-\ddot{\text{Cl}}:$$

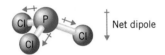
Net dipole

There are four domains around the phosphorus, so they should be arranged tetrahedrally. This means that the PCl_3 molecule should have a trigonal pyramidal shape, as shown in the margin. Because of the structure, the bond dipoles do not cancel, and we expect the molecule to be polar.

Is the Answer Reasonable?
There is not much we can do to check the answer other than to carefully check the Lewis structure, as we have done in previous examples.

EXAMPLE 10.8
Predicting Molecular Polarity

Would you expect the molecule SO_3 to be polar or nonpolar?

ANALYSIS: Oxygen is more electronegative than sulfur, so we expect the S—O bonds to be polar. We need to examine the VSEPR structure to decide whether the bond dipoles cancel.

SOLUTION: The simple Lewis structure for SO_3 is

$$\overset{\displaystyle :\ddot{\text{O}}:}{\underset{\displaystyle}{|}}$$
$$:\ddot{\text{O}}=\text{S}-\ddot{\text{O}}:$$

with two more equivalent resonance structures possible. There are three domains around the central sulfur atom, so the molecule should have a planar triangular shape. We saw in Figure 10.7 that such a molecule is nonpolar when all the attached atoms are the same, because the bond dipoles cancel.

Is the Answer Reasonable?
Again, our answer is based on the Lewis structure, which should be carefully checked. Each atom here has an octet, and the total number of valence electrons for the bonds and lone pairs shown is 24. This matches the total number of valence electrons for three oxygen atoms and one sulfur atom.

EXAMPLE 10.9
Predicting Molecular Polarity

Would you expect the molecule HCN to be polar or nonpolar?

ANALYSIS: The molecule has polar bonds because carbon is slightly more electronegative than hydrogen and nitrogen is slightly more electronegative than carbon. We should look at the VSEPR model of the molecule to see whether the bond dipoles cancel.

SOLUTION: The Lewis structure of HCN is

$$\text{H}-\text{C}\equiv\text{N}:$$

There are two domains around the central carbon atom, so we expect a linear shape. However, the two bond dipoles do *not* cancel. One reason is that they are not of equal magnitude, which we know because the difference in electronegativity between C and H is 0.4, whereas the difference in electronegativity between C and N is

0.6. The other reason is that both bond dipoles point in the same direction, from the atom of low electronegativity to the one of high electronegativity. This is illustrated below.

Bond dipoles

Net dipole

Notice that the bond dipoles add to give a net dipole moment for the molecule.

Is the Answer Reasonable?
Again there is not much we can do to check the answer other than to carefully check the Lewis structure, and compare the molecular shape we have chosen with those shown in Figure 10.2.

PRACTICE EXERCISE 7: Which of the following molecules would you expect to be polar: (a) SF_6, (b) SO_2, (c) $BrCl$, (d) AsH_3, (e) CF_2Cl_2?

10.4 ▶ Valence bond theory explains bonding as an overlap of orbitals

So far, we have taken a very simple view of covalent bonding based primarily on the use of Lewis structures for electron bookkeeping. Lewis structures, however, tell us nothing about *why* covalent bonds are formed or *how* electrons manage to be shared between atoms. Nor does the VSEPR theory, as useful and accurate as it generally is at predicting molecular geometry, explain *why* electrons group themselves into domains as they do. Thus, as helpful as these simple models are, we must look beyond them if we wish to understand more fully the covalent bond and the factors that determine molecular geometry.

Modern theories of bonding are based on the fact that electrons behave like standing waves when confined to atoms. Atomic orbitals are mathematical representations of these waves in atoms. In molecules, we can explain how atoms form a covalent bond if we can understand how the orbitals of the bonded atoms interact with each other.

There are fundamentally two theories of covalent bonding that have evolved, and in many ways they complement one another. They are called the **valence bond theory** (or **VB theory,** for short) and the **molecular orbital theory (MO theory).** They differ principally in the way they construct a theoretical model of the bonding in a molecule. The valence bond theory imagines individual atoms, each with its own orbitals and electrons, coming together to form the covalent bonds of the molecule. On the other hand, the molecular orbital theory doesn't concern itself with *how* the molecule is formed. It just views a molecule as a collection of positively charged nuclei surrounded in some way by electrons that occupy a set of *molecular orbitals,* in much the same way that the electrons in an atom occupy *atomic orbitals.* (In a sense, this theory would look at an atom as if it were a special case— a molecule having only one positive center, instead of many.)

In their simplest forms, the VB and MO theories appear to be rather different. However, it has been found that both theories can be extended and refined to give the same results. This is as it should be, of course, since they both attempt to explain the same set of facts—the structures and shapes of molecules and the strengths of chemical bonds. We will examine both theories in an elementary way, but the greater emphasis will be on the VB theory because it is easier to understand.

According to VB theory, *a bond between two atoms is formed when a **pair of electrons** with their **spins paired** is shared by two **overlapping** atomic orbitals, one*

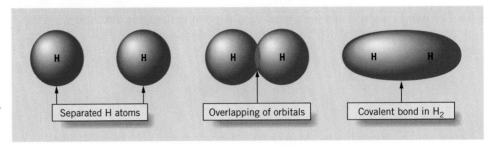

FIGURE 10.11 *The formation of the hydrogen molecule according to valence bond theory.*

Remember, just *two electrons* with *paired spins* can be shared between two overlapping atomic orbitals.

orbital from each of the atoms joined by the bond. By **overlap of orbitals** we mean that portions of two atomic orbitals from different atoms share the same space.

An important part of the theory, as suggested by the bold italic type above, is that only *one* pair of electrons, with paired spins, can be shared by two overlapping orbitals. This electron pair becomes concentrated in the region of overlap and helps "cement" the nuclei together, so the amount that the potential energy is lowered when the bond is formed is determined in part by the extent to which the orbitals overlap. Therefore, *atoms tend to position themselves so that the maximum amount of orbital overlap occurs because this yields the minimum potential energy and therefore the strongest bonds.* As we will see, this is one of the major factors that control the shapes of molecules.

The way VB theory views the formation of a hydrogen molecule is shown in Figure 10.11. As the two atoms approach each other, their $1s$ orbitals overlap and the electron pair spreads out over both orbitals, thereby giving the H—H bond. The description of the bond in this molecule provided by VB theory is essentially the same as that discussed in Section 9.3.

Now let's look at the HF molecule, which is a bit more complex than H_2. Following the usual rules we can write its Lewis structure as

$$H—\ddot{\underset{..}{F}}:$$

and we can diagram the formation of the bond as

$$H\cdot + \cdot\ddot{\underset{..}{F}}: \longrightarrow H—\ddot{\underset{..}{F}}:$$

Our Lewis symbols suggest that the H—F bond is formed by the pairing of electrons, one from hydrogen and one from fluorine. To explain this according to VB theory, we must have two half-filled orbitals, one from each atom, that can be joined by overlap. (They must be half-filled, because we can't place more than two electrons into the bond.) To see clearly what must happen, it is best to look at the orbital diagrams of the valence shells of hydrogen and fluorine.

$$H \quad \boxed{\uparrow} \\ \quad\quad 1s$$

$$F \quad \boxed{\uparrow\downarrow} \quad \boxed{\uparrow\downarrow}\boxed{\uparrow\downarrow}\boxed{\uparrow} \\ \quad\quad 2s \quad\quad\quad 2p$$

The requirements for bond formation are met by overlapping the half-filled $1s$ orbital of hydrogen with the half-filled $2p$ orbital of fluorine; there are then two orbitals plus two electrons whose spins can adjust so they are paired. The formation of the bond is illustrated in Figure 10.12.

The overlap of orbitals provides a means for sharing electrons, thereby allowing each atom to complete its valence shell. It is sometimes convenient to indicate this using orbital diagrams. For example, the diagram below shows how the fluorine atom completes its $2p$ subshell by acquiring a share of an electron from hydrogen.

$$F \text{ (in HF)} \quad \boxed{\uparrow\downarrow} \quad \boxed{\uparrow\downarrow}\boxed{\uparrow\downarrow}\boxed{\uparrow\downarrow} \quad \text{(Colored arrow is the H electron.)} \\ \quad\quad\quad\quad 2s \quad\quad\quad 2p$$

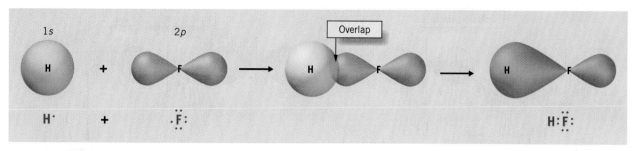

FIGURE 10.12 *The formation of the hydrogen fluoride molecule according to valence bond theory.* Only the half-filled $2p$ orbital of fluorine is shown.

Notice that in both the Lewis and VB descriptions of the formation of the H—F bond, the atoms' valence shells are completed.

Let's look now at a still more complex molecule, hydrogen sulfide, H_2S. Experiments have shown that this is a nonlinear molecule in which the H—S—H bond angle is about $92°$.

H_2S is the compound that gives rotten eggs their foul odor.

$$\underset{92°}{\overset{\displaystyle S}{H\diagdown\diagup H}}$$

Valence bond theory explains this structure as follows.

The orbital diagram for the valence shell of sulfur is

$$S \quad \underset{3s}{⇅} \quad \underset{3p}{⇅\,↑\,↑}$$

Sulfur has two $3p$ orbitals that each contain only one electron. Each of these can overlap with the $1s$ orbital of a hydrogen atom, as shown in Figure 10.13. This overlap completes the $2p$ subshell of sulfur because each hydrogen provides one electron.

$$S \text{ (in } H_2S\text{)} \quad \underset{3s}{⇅} \quad \underset{3p}{⇅\,⇅\,⇅} \qquad \text{(Colored arrows are H electrons.)}$$

In Figure 10.13, notice that when the $1s$ orbital of a hydrogen atom overlaps with a p orbital of sulfur, the best overlap occurs when the hydrogen atom lies along the axis of the p orbital. Because p orbitals are oriented at $90°$ angles to each other, the H—S bonds are expected to be at this angle, too. Therefore, the predicted bond angle is $90°$. This is very close to the actual bond angle of $92°$ found by experiment.

Two orbitals from different atoms never overlap simultaneously with opposite ends of the same p orbital.

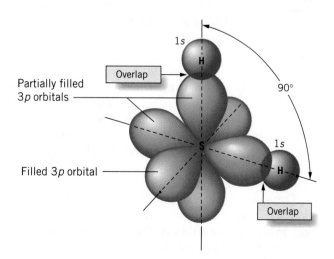

FIGURE 10.13 *Bonding in H_2S.* We expect the hydrogen $1s$ orbitals to position themselves so that they can best overlap with the two partially filled $3p$ orbitals of sulfur, which gives a predicted bond angle of $90°$. The experimentally measured bond angle of $92°$ is very close to the predicted angle.

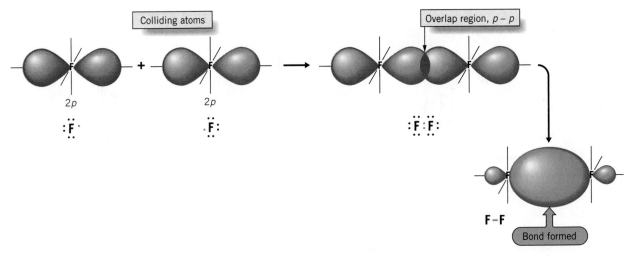

FIGURE 10.14 *Bonding in the fluorine molecule according to valence bond theory.* The two completely filled *p* orbitals on each fluorine atom are omitted, for clarity.

Thus, the VB theory requirement for maximum overlap quite nicely explains the geometry of the hydrogen sulfide molecule.

The overlap of *p* orbitals with each other is also possible. For example, according to VB theory the bonding in the fluorine molecule, F_2, occurs by the overlap of two $2p$ orbitals, as shown in Figure 10.14. The formation of the other diatomic molecules of the halogens, all of which are held together by single bonds, could be similarly described.

> **PRACTICE EXERCISE 8:** Use the principles of VB theory to explain the bonding in HCl. Give the orbital diagram for chlorine in the HCl molecule and indicate the orbital that shares the electron with one from hydrogen. Sketch the orbital overlap that gives rise to the H—Cl bond.
>
> **PRACTICE EXERCISE 9:** The phosphine molecule, PH_3, has a trigonal pyramidal shape with H—P—H bond angles equal to 93.7°. Give the orbital diagram for phosphorus in the PH_3 molecule and indicate the orbitals that share electrons with those from hydrogen. On a set of *xyz* coordinate axes, sketch the orbital overlaps that give rise to the P—H bonds.

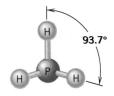

10.5 ▶ Hybrid orbitals are used to explain experimental molecular geometries

The approach that we've taken so far has worked well with some simple molecules. Their shapes are explained quite nicely by the overlap of simple atomic orbitals. However, there are many molecules with shapes and bond angles that fail to fit the model that we have developed. For example, methane, CH_4, has a shape that the VSEPR model predicts (correctly) to be tetrahedral. The H—C—H bond angles in this molecule are 109.5°. But no simple atomic orbitals are oriented at this angle with respect to each other. Therefore, to explain the bonds in molecules such as CH_4, we must study the way atomic orbitals *of the same atom* can be mixed with each other when bonds are formed.

Orbitals are the basic building blocks of valence bonds. To realistically describe bonds, we sometimes must blend two or more of these basic building blocks to form **hybrid atomic orbitals.** These new orbitals have new shapes and new directional properties, and they can be overlapped to give structures that have realistic

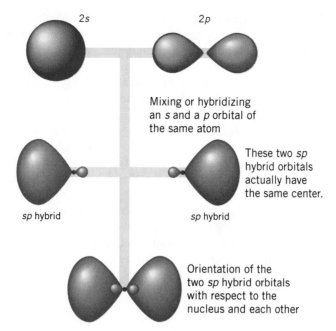

FIGURE 10.15 *Formation of sp hybrid orbitals.* Mixing of the 2s and 2p atomic orbitals produces a pair of *sp* hybrid orbitals. The large lobes of these orbitals point in opposite directions.

bond angles. It's important to realize that hybrid orbitals are part of the valence bond *theory*. They can't be directly observed in an experiment. We use them to describe molecular structures that have been determined experimentally.

Mixing an *s* orbital and a *p* orbital forms one kind of hybrid atomic orbital. This creates *two* new orbitals called **sp hybrid orbitals.** The label *sp* identifies the kinds of pure atomic orbitals from which the hybrids are formed, in this case, one *s* and one *p* orbital. The shapes and directional properties of the *sp* hybrid orbitals are illustrated in Figure 10.15. Notice that each of the hybrid orbitals has the same shape; each has one large lobe and another much smaller one. The large lobe extends farther from the nucleus than either the *s* or *p* orbital from which the hybrid was formed. This allows the hybrid orbital to overlap more effectively with an orbital on another atom when a bond is formed. Therefore, hybrid orbitals form stronger, more stable bonds than would be possible if just simple atomic orbitals were used. In fact, the strength of these bonds is in good agreement with bond strengths that are determined experimentally.

Another point to notice in Figure 10.15 is that the large lobes of the two *sp* hybrid orbitals point in opposite directions; that is, they are 180° apart. If bonds are formed by the overlap of these hybrids with orbitals of other atoms, the other atoms will occupy positions on opposite sides of this central atom. Let's look at a specific example, the linear beryllium hydride molecule, BeH_2, as it would be formed in the gas phase.[1]

The orbital diagram for the valence shell of beryllium is

Be ⬆⬇ ○○○
    2s   2p

Notice that the 2s orbital is filled and the three 2p orbitals are empty. For bonds to form at a 180° angle between beryllium and the two hydrogen atoms, two conditions must be met:

1. The two orbitals that beryllium uses to form the Be—H bonds must point in opposite directions, and

2. Each of the beryllium orbitals must contain only one electron.

In general, the greater the overlap of two orbitals, the stronger is the bond. At a given distance between nuclei, the greater "reach" of a hybrid orbital gives better overlap than either an *s* or *p* orbital.

[1]In the solid state, BeH_2 has a complex structure not consisting of simple BeH_2 molecules.

The reason for the first requirement is obvious; the overlap of Be and H orbitals must give the correct experimentally measured bond angle. The reason for the second is that each bond must contain just two electrons. Since a hydrogen 1s orbital supplies one electron, the beryllium orbital with which it overlaps must also contain one electron. A filled Be orbital won't do, because overlap of a half-filled hydrogen 1s with a filled orbital on Be would produce a "bond" with three electrons in it—a situation that's forbidden by VB theory. An empty Be orbital won't do either, because overlap with a hydrogen 1s orbital would give a bond with only one electron in it. Although such a bond isn't forbidden, it would be much weaker than a bond with two electrons, so electron pair bonds are definitely preferred. The net effect of all this is that when the Be—H bonds form, the electrons of the beryllium atom become unpaired.

It takes quite a bit of energy to move one of the electrons from the 2s to the 2p orbital, but this energy can be recovered by the strong bonds that the hybrid orbitals form. When the orbitals mix (hybridize), two identical sp hybrid orbitals form. Both sp orbitals have the same energy, so it is much easier to realize a configuration with two unpaired electrons:

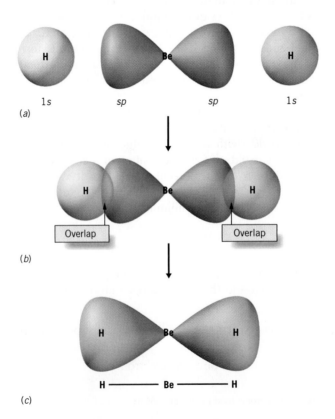

Now the 1s orbitals of the hydrogen atoms can overlap with the sp hybrids of beryllium as shown in Figure 10.16. Because the two sp hybrid orbitals of beryllium

FIGURE 10.16 Bonding in BeH₂ according to valence bond theory. Only the larger lobe of each sp hybrid orbital is shown. (a) The two hydrogen 1s orbitals approach the pair of sp hybrid orbitals of beryllium. (b) Overlap of the hydrogen 1s orbitals with the sp hybrid orbitals. (c) A representation of the distribution of electron density in the two Be—H bonds after they have been formed.

are identical in shape, the two Be—H bonds are identical except for the directions in which they point, and we say that the bonds are *equivalent*. Since the bonds point in opposite directions, the linear geometry of the molecule is also explained. The orbital diagram for beryllium in this molecule is

These *sp* hybrid orbitals are said to be *equivalent* because they have the same size, shape, and energy.

Be (in BeH₂) ⊕⊕ ◯◯ (Colored arrows are H electrons.)
  sp unhybridized
  2*p* orbitals

Even if we had not known the shape of the BeH₂ molecule, we could have obtained the same bonding picture by applying the VSEPR model first. The Lewis structure for BeH₂ is H:Be:H, with two bonding domains around the central atom, so the molecule is linear. Once the shape is known, we can apply the VB theory to explain the bonding in terms of orbital overlaps. Thus, the VB and VSEPR theories complement each other well. VSEPR theory allows us to predict geometry in a simple way, and once the geometry is known, it is relatively easy to analyze the bonding in terms of VB theory.

Other hybrids formed from *s* and *p* orbitals

Combining an *s* orbital with two or three *p* orbitals can also form hybrid orbitals. When two *p* orbitals are mixed with an *s* orbital, a set of *three sp² hybrid orbitals* is formed. When three *p* orbitals are mixed with an *s* orbital, a set of *four sp³ hybrid orbitals* is formed. The superscript 2 or 3 indicates the number of *p* orbitals in the mix. Notice that the number of hybrid orbitals in a set equals the number of atomic orbitals used to form them.

There is a conservation of orbitals during hybrid orbital formation. The number of hybrid orbitals of a given kind is always equal to the number of atomic orbitals that are mixed.

The three *sp²* hybrids lie in the same plane and point to the corners of a triangle. The four *sp³* hybrids point to the corners of a tetrahedron. Their directional properties are summarized in Figure 10.17. The following examples illustrate how they can be used to explain the bonding in molecules.

The BCl₃ molecule has a planar triangular shape. How is this explained in terms of valence bond theory?

ANALYSIS: Since we know the shape of the molecule, we can anticipate the kinds of hybrid orbitals used to form the bonds. A planar triangular shape is consistent

EXAMPLE 10.10
Explaining Bonding with Hybrid Orbitals

Orientations of hybrid orbitals

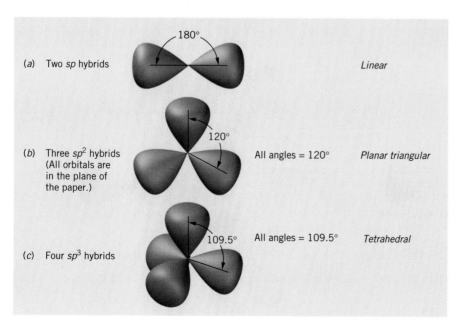

(a) Two *sp* hybrids — 180° — Linear

(b) Three *sp²* hybrids (All orbitals are in the plane of the paper.) — 120° — All angles = 120° — Planar triangular

(c) Four *sp³* hybrids — 109.5° — All angles = 109.5° — Tetrahedral

FIGURE 10.17 *Directional properties of hybrid orbitals formed from s and p atomic orbitals.* (*a*) *sp* hybrid orbitals oriented at 180° to each other. (*b*) *sp²* hybrid orbitals formed from an *s* orbital and two *p* orbitals. The angle between them is 120°. (*c*) *sp³* hybrid orbitals formed from an *s* orbital and three *p* orbitals. The angle between them is 109.5°.

with sp^2 hybridization in the valence shell of the central atom, so we proceed from that point to determine whether the electronic structures of the atoms involved fit with the use of these kinds of orbitals.

SOLUTION: First, let's examine the orbital diagram for the valence shell of boron.

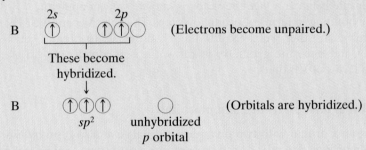

To form the three bonds to chlorine, boron needs three half-filled orbitals. These can be obtained by unpairing the electrons in the 2s orbital and placing one of them in the 2p. Then the 2s orbital and two of the 2p orbitals can be combined to give the set of sp^2 hybrids.

Now let's look at the valence shell of chlorine.

The half-filled 3p orbital of each chlorine atom can overlap with a hybrid sp^2 orbital of boron to give three B—Cl bonds.

Figure 10.18 illustrates the overlap of the orbitals to give the bonding in the molecule.

Is the Answer Reasonable?
We have already seen that the Lewis structure of BCl_3 is

$$:\overset{..}{\underset{}{Cl}}:$$
$$|$$
$$:\overset{..}{\underset{..}{Cl}}—B—\overset{..}{\underset{..}{Cl}}:$$

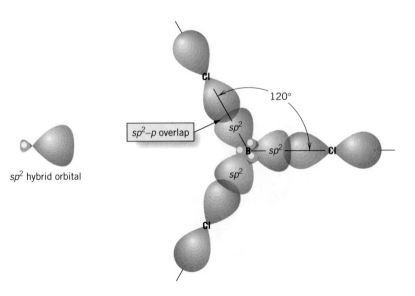

FIGURE 10.18 *Valence bond description of the bonding in BCl₃.* Each B—Cl bond is formed by the overlap of a half-filled p orbital of chlorine with an sp^2 hybrid orbital of boron. (Only the half-filled p orbital of each chlorine is shown.)

with three bonding domains around the central atom. The VSEPR model predicts a planar triangular molecule, in agreement with the sp^2 hybridization on the boron atom shown in Figure 10.18. Notice that there is one hybrid orbital around the central atom in the VB model for each electron domain around the central atom in the VSEPR model. Our VB model shows how those domains can be built from atomic orbitals on the boron and chlorine atoms.

Methane, CH_4, is a tetrahedral molecule. How is this explained in terms of valence bond theory?

ANALYSIS: A tetrahedral shape is consistent with sp^3 hybridization in the valence shell of the central atom. We'll need to hybridize one s and three p orbitals on the central atom to form sp^3 hybrids.

SOLUTION: Let's examine the valence shell of carbon.

$$ C \quad \boxed{\uparrow\downarrow}_{2s} \quad \boxed{\uparrow}\boxed{\uparrow}\boxed{\,}_{2p} $$

To form four C—H bonds, we need four half-filled orbitals. Unpairing the electrons in the $2s$ and moving one to the vacant $2p$ orbital satisfies this requirement. Then we can hybridize all the orbitals to give the desired sp^3 set.

$$ C \quad \underbrace{\overset{2s}{\boxed{\uparrow}} \quad \overset{2p}{\boxed{\uparrow}\boxed{\uparrow}\boxed{\uparrow}}} \quad \text{(Electrons become unpaired.)} $$

These become
hybridized.
↓

$$ C \quad \boxed{\uparrow}\boxed{\uparrow}\boxed{\uparrow}\boxed{\uparrow} $$
sp^3 hybrids

Then we form the four bonds to hydrogen $1s$ orbitals.

$$ C\ (\text{in } CH_4) \quad \boxed{\uparrow\downarrow}\boxed{\uparrow\downarrow}\boxed{\uparrow\downarrow}\boxed{\uparrow\downarrow} \quad \text{(Colored arrows are H electrons.)} $$
sp^3

This is illustrated in the margin.

Is the Answer Reasonable?
The positions of the hydrogen atoms around the carbon give the correct tetrahedral shape for the molecule. The Lewis structure for CH_4 shows four electron domains around the carbon, which is consistent with the idea of four sp^3 orbitals overlapping with hydrogen $1s$ orbitals to form four valence bonds.

EXAMPLE 10.11
Explaining Bonding with Hybrid Orbitals

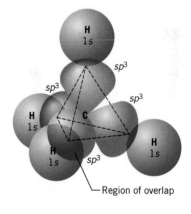

Region of overlap

In methane, carbon forms four single bonds with hydrogen atoms by using sp^3 hybrid orbitals. Carbon uses these same kinds of orbitals in all of its compounds in which it is bonded to four other atoms by single bonds. This makes the tetrahedral orientation of atoms around carbon one of the primary structural features of organic compounds, and organic chemists routinely think in terms of "tetrahedral carbon," shown in Figure 10.19.

In the alkane hydrocarbons, carbon atoms are bonded to other carbon atoms. An example is ethane, C_2H_6.

$$ \begin{array}{c} \;\;\;H\;\;\;H \\ \;\;\;|\;\;\;\;| \\ H-C-C-H \\ \;\;\;|\;\;\;\;| \\ \;\;\;H\;\;\;H \end{array} $$
ethane

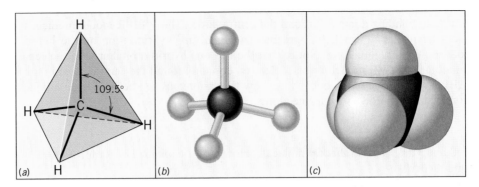

FIGURE 10.19 *The "tetrahedral carbon."* (*a*) The heavy lines are the axes of the bonds. (*b*) A ball-and-stick model of the CH₄ molecule. (*c*) A scale model of CH₄ that indicates the relative volumes occupied by the electron clouds.

The VB theory describes the bond between carbons in ethane as an overlap of sp^3 hybrid orbitals (Figure 10.20). Spinning the —CH₃ groups doesn't affect the overlap of these orbitals, so the groups can spin relatively freely around the C—C bond, like wheels spinning on an axle. Many different relative orientations of the hydrogen atoms are possible. These different relative orientations are called **conformations.** With complex molecules, the number of possible conformations is enormous. For example, Figure 10.21 illustrates three of the enormous number of possible conformations of the pentane molecule, C₅H₁₂, one of the low-molecular-weight organic compounds in gasoline.

$$\text{H}-\underset{\underset{\text{H}}{|}}{\overset{\overset{\text{H}}{|}}{\text{C}}}-\underset{\underset{\text{H}}{|}}{\overset{\overset{\text{H}}{|}}{\text{C}}}-\underset{\underset{\text{H}}{|}}{\overset{\overset{\text{H}}{|}}{\text{C}}}-\underset{\underset{\text{H}}{|}}{\overset{\overset{\text{H}}{|}}{\text{C}}}-\underset{\underset{\text{H}}{|}}{\overset{\overset{\text{H}}{|}}{\text{C}}}-\text{H}$$

pentane

Hybridization for five or more domains around an atom involves *d* orbitals

Earlier we saw that certain molecules have atoms that must violate the octet rule because they form more than four bonds. In these cases, the atom must reach

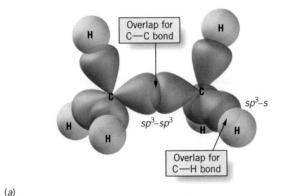

(*a*)

FIGURE 10.20 *The bonds in the ethane molecule.* (*a*) Overlap of orbitals. (*b*) The degree of overlap of the sp^3 orbitals in the carbon–carbon bond is not appreciably affected by the rotation of the two CH₃— groups relative to each other around the bond.

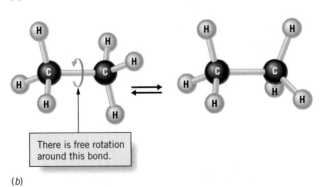

(*b*)

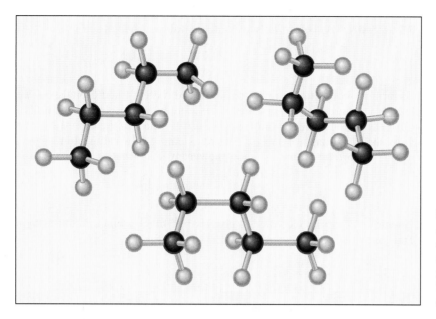

FIGURE **10.21** *Three of the many conformations of the atoms in the pentane molecule, C_5H_{12}.* Free rotation around single bonds makes these different conformations possible.

beyond its *s* and *p* valence shell orbitals to form sufficient half-filled orbitals for bonding. This is because the *s* and *p* orbitals can be mixed to form a maximum of only four hybrid orbitals. When five or more hybrid orbitals are needed, *d* orbitals are brought into the mix. The two most common kinds of hybrid orbitals involving *d* orbitals are **sp^3d** and **sp^3d^2 hybrid orbitals.** Their directional properties are illustrated in Figure 10.22. Notice that the sp^3d hybrids point toward the corners of a trigonal bipyramid and the sp^3d^2 hybrids point toward the corners of an octahedron.

The sulfur hexafluoride molecule has an octahedral shape. Describe the bonding in this molecule in terms of valence bond theory.

ANALYSIS: An octahedral structure suggests the use of sp^3d^2 hybrid orbitals, so we need to find some *d* orbitals to form the hybrids.

EXAMPLE 10.12

Explaining Bonding with Hybrid Orbitals

Orientations of hybrid orbitals

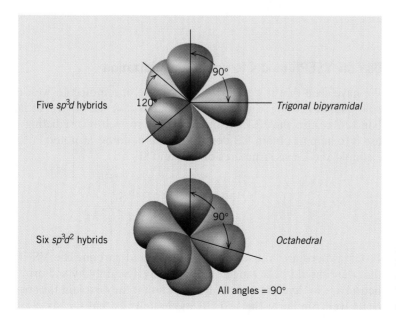

Five sp^3d hybrids — 120° — 90° — Trigonal bipyramidal

Six sp^3d^2 hybrids — 90° — Octahedral

All angles = 90°

FIGURE **10.22** *Orientations of hybrid orbitals that involve d atomic orbitals.* (*a*) sp^3d hybrid orbitals, formed by mixing an *s* orbital, three *p* orbitals, and a *d* orbital. The orbitals point to the corners of a trigonal bipyramid. (*b*) sp^3d^2 hybrid orbitals, formed by mixing an *s* orbital, three *p* orbitals, and two *d* orbitals. The orbitals point to the corners of an octahedron.

SOLUTION: As before, let's examine the valence shell of sulfur.

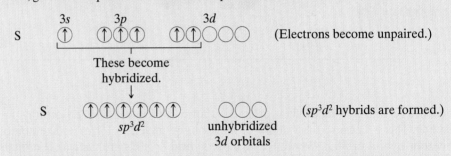

To form six bonds to fluorine atoms we need six half-filled orbitals, but we see only four orbitals altogether. An isolated sulfur atom has electrons only in its $3s$ and $3p$ subshells, so these are the only ones we usually show in the orbital diagram. But the third shell also has a d subshell, which is empty in a sulfur atom. Therefore, let's rewrite the orbital diagram to show the vacant $3d$ subshell.

Unpairing all of the electrons to give six half-filled orbitals, followed by hybridization, gives the required set of half-filled sp^3d^2 orbitals.

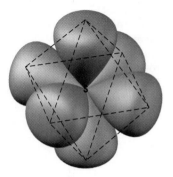

sp³d² hybrid orbitals of sulfur in SF₆

Finally, the six S—F bonds are formed by overlap of the half-filled $2p$ orbitals of fluorine with these half-filled sp^3d^2 hybrids.

Is the Answer Reasonable?

The Lewis structure for SF_6 shows six bonding domains around the central sulfur atom, which is consistent with the formation of six valence bonds in our VB model of the molecule.

PRACTICE EXERCISE 10: Use valence bond theory to describe the bonding in the molecule $AsCl_5$, which has a trigonal bipyramidal shape.

Use the VSEPR model to predict hybridization

We have seen that if we know the structure of a molecule, we can make a reasonable guess as to the kind of hybrid orbitals that the central atom uses to form its bonds. Because the VSEPR model works so well in predicting geometry, we can use it to help us obtain VB descriptions of bonding, as noted in Example 10.10. For example, the Lewis structures of CH_4 and SF_6 are

In CH_4, there are four electron pairs around carbon. The VSEPR model tells us that they should be arranged tetrahedrally. The only hybrid orbitals that are tetrahedral are sp^3 hybrids, and we have seen that they explain the structure of this molecule well. Similarly, the VSEPR model tells us that the six electron pairs around sulfur should be arranged octahedrally. The only octahedrally oriented hybrids are sp^3d^2, so the sulfur in the SF_6 molecule must use these hybrids.

PRACTICE EXERCISE 11: What kind of hybrid orbitals are expected to be used by the central atom in (a) SiH_4 and (b) PCl_5?

Hybrid orbitals can be used to describe molecules with nonbonding domains

Methane is a tetrahedral molecule with sp^3 hybridization of the orbitals of carbon and H—C—H bond angles that are each equal to 109.5°. In ammonia, NH_3, the H—N—H bond angles are 107°, and in water the H—O—H bond angle is 104.5°. Both NH_3 and H_2O have H—X—H bond angles that are close to the bond angles expected for a molecule whose central atom has sp^3 hybrids. The use of sp^3 hybrids by oxygen and nitrogen, therefore, is often used to explain the geometry of H_2O and NH_3.

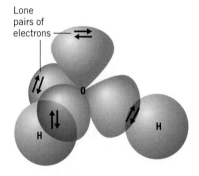

Lone pairs of electrons

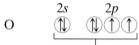

Hybridization and bond formation occur.

O (in H_2O) (Colored arrows are H electrons.)
sp^3

Hybridization and bond formation occur.

N (in NH_3) (Colored arrows are H electrons.)
sp^3

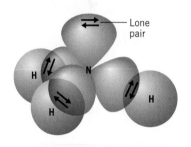

Lone pair

According to these descriptions, not all of the hybrid orbitals of the central atom must be used for bonding. Lone pairs of electrons can be accommodated in them too. In fact, putting the lone pair on the nitrogen in an sp^3 hybrid orbital gives a geometry that agrees well with the experimentally determined structure of the ammonia molecule.

Use valence bond theory to explain the bonding in the SF_4 molecule.

ANALYSIS: First, we will have to draw a Lewis structure to determine how many electron domains we have around the central atom in the molecule. We can then use VSEPR theory to predict the molecule's shape. That will allow us to choose a hybridization for the central atom. We can then look at the atomic orbitals on the central atom to see how this hybridization can be realized.

SOLUTION: The Lewis structure of SF_4 is

EXAMPLE 10.13

Explaining Bonding with Hybrid Orbitals

The VSEPR theory predicts that the electron pairs around the sulfur should be in a trigonal bipyramidal arrangement, and the only hybrids that fit this geometry are sp^3d. To see how they are formed, we look at the valence shell of sulfur, including the vacant $3d$ subshell.

S

$3s$ $3p$ $3d$

To form the four bonds to fluorine atoms, we need four half-filled orbitals, so we un-pair the electrons in one of the filled orbitals. This gives

S

Next, we form the hybrid orbitals. In doing this, we use all the valence shell orbitals that have electrons in them.

S

These become hybridized.

S

sp^3d unhybridized 3d orbitals

Now, four S—F bonds can be formed by overlap of half-filled 2p orbitals of fluorine with the sp^3d hybrid orbitals of sulfur.

S (in SF_4)

sp^3d unhybridized 3d orbitals   (Colored arrows are F electrons.)

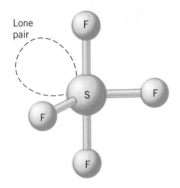

Lone pair

The structure of SF_4.

Is the Answer Reasonable?

We should check to see that the Lewis structure is correct, since everything we have done depends on it. The structure shows 8 valence electrons in bonds and 26 electrons in lone pairs, for a total of 34 valence electrons. That agrees with the total valence electrons contributed by a sulfur atom (6) and four fluorine atoms (28). The VB model shows five valence bonds around the sulfur, which correspond to the five electron domains around the sulfur in the Lewis structure.

PRACTICE EXERCISE 12: What kind of hybrid orbitals would we expect the central atom to use for bonding in (a) PCl_3 and (b) ClF_3?

Hybrid orbitals can be used to explain the formation of coordinate covalent bonds

In Section 9.10 we defined a coordinate covalent bond as one in which both of the shared electrons are provided by just one of the joined atoms. For example, boron trifluoride, BF_3, can combine with an additional fluoride ion to form the tetrafluoroborate ion, BF_4^-, according to the equation

$$BF_3 + F^- \longrightarrow BF_4^-$$

tetrafluoroborate ion

We can diagram this reaction as follows:

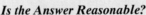

As we mentioned previously, the coordinate covalent bond is really no different from any other covalent bond once it has been formed. The distinction between them is made *only* for bookkeeping purposes. One place where such bookkeeping is useful is in keeping track of the orbitals and electrons used when atoms bond together.

The VB theory requirements for bond formation—two overlapping orbitals sharing two paired electrons—can be satisfied in two ways. One, as we have seen, is by the overlapping of two half-filled orbitals. This gives an "ordinary" covalent bond. The other is the overlapping of one filled orbital with one empty orbital. The atom with the filled orbital donates the shared pair of electrons, and a coordinate covalent bond is formed.

The structure of the BF_4^- ion can therefore be explained as follows. First we examine the orbital diagram for boron.

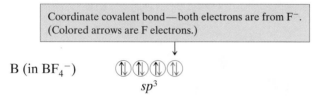

To form four bonds, we need four hybrid orbitals. Since VSEPR theory predicts that the ion will have a tetrahedral shape, the boron will use sp^3 hybrids. Notice we spread the electrons out over the hybrid orbitals as much as possible.

B

sp^3

Boron forms three ordinary covalent bonds with fluorine atoms plus one coordinate covalent bond with a fluoride ion.

> Coordinate covalent bond—both electrons are from F⁻.
> (Colored arrows are F electrons.)

B (in BF_4^-)

sp^3

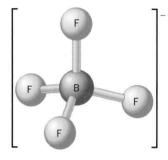

The structure of BF_4^-.

PRACTICE EXERCISE 13: What hybrid orbitals are used by phosphorus in PCl_6^-? Draw the orbital diagram for phosphorus in PCl_6^-. What is the shape of the PCl_6^- ion?

10.6 Hybrid orbitals can be used to explain multiple bonds

The types of orbital overlap that we have described so far produce bonds in which the electron density is concentrated most heavily *between* the nuclei of the two atoms along an imaginary line that joins their centers. Any bond of this kind, whether formed from the overlap of *s* orbitals, *p* orbitals, or hybrid orbitals (Figure 10.23), is called a **sigma bond** (or **σ bond**).

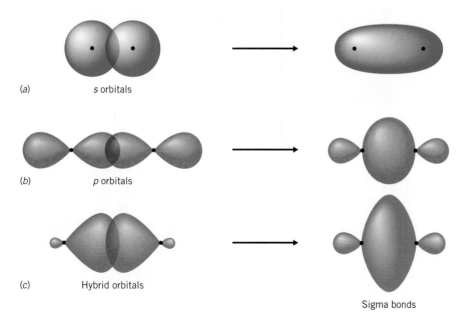

(a) s orbitals

(b) p orbitals

(c) Hybrid orbitals

Sigma bonds

FIGURE 10.23 *Formation of σ bonds.* Sigma bonds concentrate electron density along the line between the two atoms joined by the bond. (*a*) From the overlap of *s* orbitals. (*b*) From the end-to-end overlap of *p* orbitals. (*c*) From the overlap of hybrid orbitals.

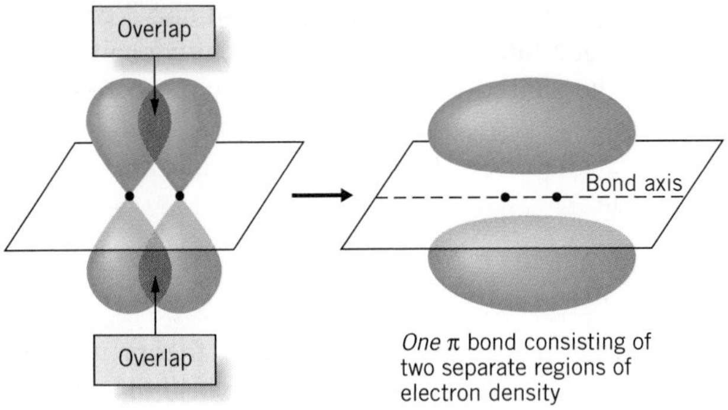

FIGURE 10.24 *Formation of a π bond.* Two *p* orbitals overlap sideways instead of end-to-end. The electron density is concentrated in two regions on opposite sides of the bond axis.

Another way that *p* orbitals can overlap is shown in Figure 10.24. This produces a bond in which the electron density is divided between two separate regions that lie on opposite sides of an imaginary line joining the two nuclei. This kind of bond is called a **pi bond** (or **π bond**). Notice that a π bond, like a *p* orbital, consists of two parts, and each part makes up just half of the π bond; it takes *both* of them to equal *one* π bond.

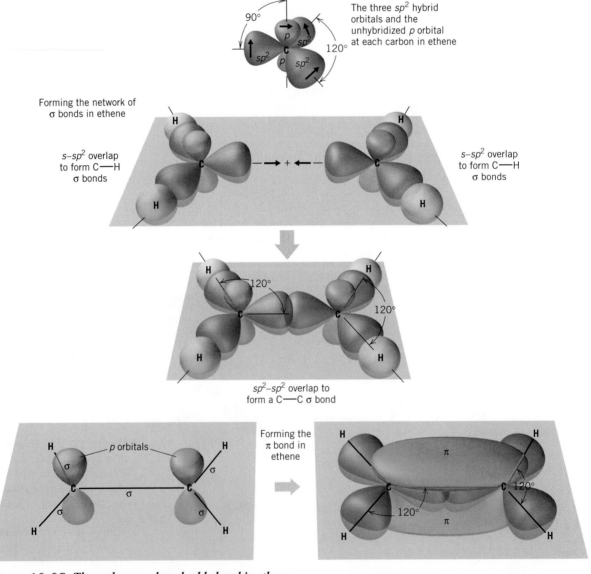

FIGURE 10.25 *The carbon–carbon double bond in ethene.*

The formation of π bonds allows atoms to form double and triple bonds. To see how this occurs, let's look at the bonding in the compound ethene, C_2H_4, which has the Lewis structure

$$
\begin{array}{ccc}
H & & H \\
\diagdown & & \diagup \\
& C = C & \\
\diagup & & \diagdown \\
H & & H
\end{array}
$$

ethene

Ethene is also called ethylene. Polyethylene, a common plastic, is made from C_2H_4.

The molecule is planar and each carbon atom lies in the center of a triangle surrounded by three other atoms (two H and one C atom). As we've seen, this structure suggests that carbon uses sp^2 hybrid orbitals to form its bonds. Therefore, let's look at the distribution of electrons among the orbitals that carbon has available in its valence shell, assuming sp^2 hybridization.

C $\underset{2s}{\uparrow}$ $\underset{2p}{\uparrow \uparrow \uparrow}$ (Electrons become unpaired for bond formation.)

These become hybridized.

↓

C $\underset{\substack{sp^2 \\ \text{(used to form} \\ \sigma \text{ bonds)}}}{\uparrow \uparrow \uparrow}$ $\underset{\substack{\text{unhybridized} \\ 2p \text{ orbital}}}{\uparrow}$ (sp^2 hybrids are formed.)

Restricted rotation around double bonds has important consequences in organic chemistry and biochemistry.

Notice that the carbon atom has an unpaired electron in an unhybridized $2p$ orbital. This p orbital is oriented perpendicular to the triangular plane of the sp^2 hybrid orbitals, as shown in Figure 10.25. Now we can see how the molecule goes together.

The basic framework of the molecule is determined by the formation of σ bonds. Each carbon uses two of its sp^2 hybrids to form σ bonds to hydrogen atoms. The third sp^2 hybrid on each carbon is used to form a σ bond between the two carbon atoms, thereby accounting for one of the two bonds of the double bond. Finally, the remaining unhybridized $2p$ orbitals, one from each carbon atom, overlap to produce a π bond, which accounts for the second bond of the double bond.

Notice how well this description of bonding fits with the observed (and predicted) structure of the molecule. The use of sp^2 hybrids by carbon makes available the unpaired electrons in the unhybridized $2p$ orbitals, which in turn makes it possible for the carbon atoms to form the extra bond between them.

This explanation also accounts for one of the most important properties of double bonds: rotation of one portion of the molecule relative to the rest around the axis of the double bond occurs only with great difficulty. The reason for this is illustrated in Figure 10.26. As one CH_2 group is rotated relative to the other around the carbon–carbon bond, the unhybridized p orbitals become misaligned and can no longer overlap effectively. To rotate around a double bond, we must destroy the π bond. This requires more energy than is normally available to molecules at room temperature. As a result, rotation around the axis of a double bond usually doesn't take place.

In almost every instance, a double bond consists of a σ bond and a π bond. Another example is the compound formaldehyde (the substance used as a preservative for biological specimens and as an embalming fluid). The Lewis structure of this compound is

$$
\begin{array}{cc}
H & \\
\diagdown & \\
& C = \ddot{\underset{\cdot\cdot}{O}} \\
\diagup & \\
H &
\end{array}
$$

formaldehyde

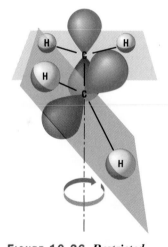

FIGURE 10.26 *Restricted rotation around a double bond.* If the CH_2 group closest to us were to rotate relative to the one at the rear, the unhybridized p orbitals would become misaligned, as shown here. This would destroy the overlap and break the π bond. Bond breaking requires a lot of energy, more than is available to the molecule through the normal bending and stretching of its bonds at room temperature. Because of this, rotation about the double bond axis is hindered or "restricted."

As with ethene, the carbon forms sp^2 hybrids, leaving an unpaired electron in an unhybridized p orbital.

C   ⬆⬆⬆   ⬆

sp^2 (unhybridized $2p$ orbital)

The oxygen can also form sp^2 hybrids, with electron pairs in two of them and an unpaired electron in the third. This means that the remaining unhybridized p orbital also has an unpaired electron.

O   ⬆⬇   ⬆⬇⬆⬆   (ground state of oxygen)

$2s$ $2p$

These form sp^2 hybrids.
↓

O   ⬆⬇⬆⬇⬆   ⬆

sp^2 (unhybridized $2p$ orbital)

The two filled sp^2 hybrids on the oxygen become lone pairs on the oxygen atom in the molecule.

Figure 10.27 shows how the carbon, hydrogen, and oxygen atoms come together to form the molecule. As before, the basic framework of the molecule is formed by the σ bonds. These determine the molecular shape. The carbon–oxygen double bond also contains a π bond formed by the overlap of the unhybridized p orbitals.

Now let's look at a molecule containing a triple bond. An example is ethyne, also known as acetylene, C_2H_2 (a gas used as a fuel for welding torches).

H—C≡C—H

ethyne (acetylene)

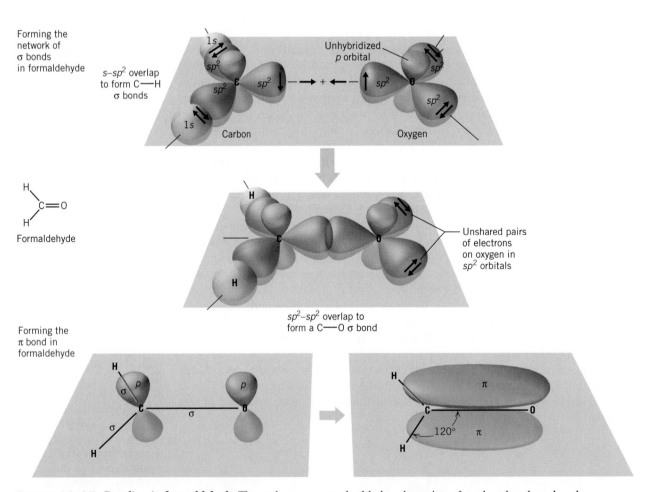

FIGURE 10.27 *Bonding in formaldehyde.* The carbon–oxygen double bond consists of a σ bond and a π bond.

In the linear acetylene molecule, each carbon needs two hybrid orbitals to form two σ bonds—one to a hydrogen atom and one to the other carbon atom. These can be provided by mixing the $2s$ and one of the $2p$ orbitals to form sp hybrids. To help us visualize the bonding, we will imagine that there is an xyz coordinate system centered at each carbon atom and that it is the $2p_z$ orbital that becomes mixed in the hybrid orbitals.

We label the orbitals p_x, p_y, and p_z just for convenience; they are really all equivalent.

C $\quad$ $2s$ $\quad$ $2p_z$ $2p_y$ $2p_x$
$\quad$ ↑ $\quad$ ↑ ↑ ↑ $\quad$ (Electrons become unpaired for bonding.)

These become hybridized.

↓

C $\quad$ ↑↑ $\quad$ ↑ ↑
$\quad$ sp $\quad$ $2p_x$ $2p_y$ $\quad$ (unhybridized)

Figure 10.28 shows how the molecule is formed. The sp orbitals point in opposite directions and are used to form the σ bonds. The unhybridized $2p_x$ and $2p_y$ orbitals

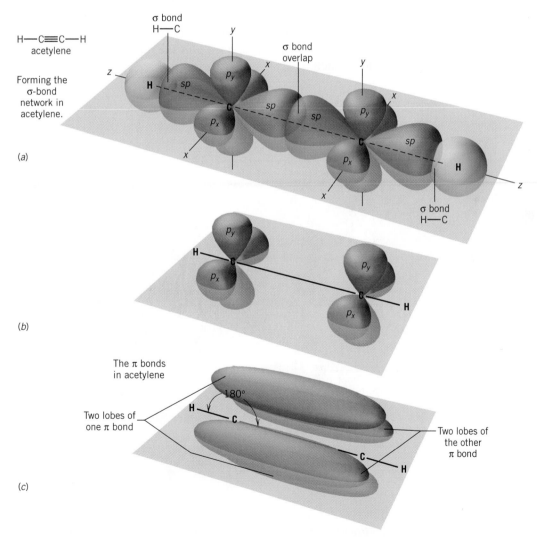

FIGURE 10.28 *The carbon–carbon triple bond in acetylene.* (*a*) The sp hybrid orbitals on each carbon atom are used to form the σ bonds. (*b*) Sideways overlap of unhybridized $2p_x$ and $2p_y$ orbitals of the carbon atoms produces two π bonds. (*c*) The two π bonds in acetylene after they've formed.

are perpendicular to the C—C bond axis and overlap sideways to form two separate π bonds that surround the C—C σ bond. Notice that we now have three pairs of electrons in three bonds—one σ bond and two π bonds—whose electron densities are concentrated in different places. The three electron pairs therefore manage to avoid each other as much as possible. Also notice that the use of sp hybrid orbitals for the σ bonds allows us to explain the linear arrangement of atoms in the molecule.

Similar descriptions can be used to explain the bonding in other molecules that have triple bonds. Figure 10.29, for example, shows how the nitrogen molecule, N_2, is formed. In it, too, the triple bond is composed of one σ bond and two π bonds.

A brief summary

On the basis of the preceding discussion, we can make some observations that are helpful in applying the valence bond theory to a variety of molecules.

1. The basic molecular framework of a molecule is determined by the arrangement of its σ bonds.

2. Hybrid orbitals are used by an atom to form its σ bonds and to hold lone pairs of electrons.

3. The number of hybrid orbitals needed by an atom in a structure equals the number of atoms to which it is bonded *plus* the number of lone pairs of electrons in its valence shell.

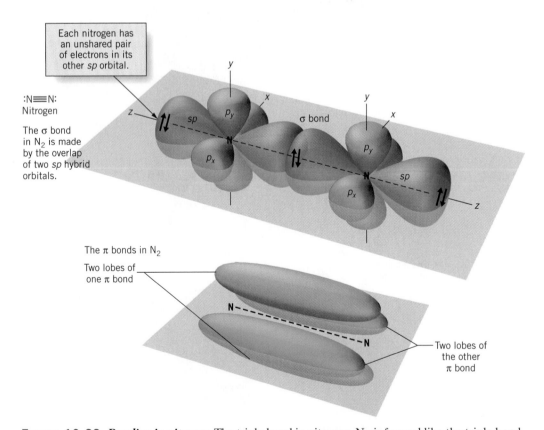

FIGURE 10.29 *Bonding in nitrogen.* The triple bond in nitrogen, N_2, is formed like the triple bond in acetylene. A σ bond is formed by overlap of sp hybrid orbitals. The two unhybridized $2p$ orbitals on each nitrogen atom overlap to give the two π bonds. On each nitrogen, there is a lone pair of electrons in the sp hybrid orbital that's not used to form the σ bond.

4. When there is a double bond in a molecule, it consists of one σ bond and one π bond.

5. When there is a triple bond in a molecule, it consists of one σ bond and two π bonds.

10.7 ▶ Molecular orbital theory explains bonding as constructive interference of atomic orbitals

Molecular orbital theory takes the view that a molecule is similar to an atom in one important respect. Both have energy levels that correspond to various orbitals that can be populated by electrons. In atoms, these orbitals are called *atomic orbitals;* in molecules, they are called **molecular orbitals.** (We shall frequently call them MOs.)

In most cases, the actual shapes and energies of molecular orbitals cannot be determined exactly. Nevertheless, theoreticians have found that reasonably good estimates of their shapes and energies can be obtained by combining the electron waves corresponding to the atomic orbitals of the atoms that make up the molecule. In forming molecular orbitals, these waves interact by constructive and destructive interference just like other waves that we've seen. Their intensities are either added or subtracted when the atomic orbitals overlap. The way this occurs can be seen if we look at the overlap of a pair of $1s$ orbitals from two atoms in a molecule like H_2 (Figure 10.30). The *two* $1s$ orbitals combine when the molecule is formed to give *two* MOs. In one MO, the intensities of the electron waves add together between the nuclei, which gives a buildup of electron density that helps hold the nuclei near each other. Such an MO is said to be a **bonding molecular orbital.** *Electrons in bonding MOs tend to stabilize a molecule.*

In the other MO, cancellation of the electron waves reduces the electron density between the nuclei. With less electron density between the nuclei, the nuclei repel each other more strongly, so this MO is called an **antibonding molecular orbital.** *Antibonding MOs tend to destabilize a molecule when occupied by electrons.*

Both the bonding and antibonding MOs formed by the overlap of s orbitals have their maximum electron density on an imaginary line that passes through the two nuclei. Earlier, we called bonds that have this property sigma bonds. Molecular orbitals like this are also designated as sigma (σ). An asterisk is used to indicate the MOs that are antibonding, and a subscript is written to report which atomic orbitals make up the MO. For example, the bonding MO formed by the overlap of $1s$ orbitals is symbolized as σ_{1s} and the antibonding MO is written as σ_{1s}^*.

The number of MOs formed is always equal to the number of atomic orbitals that are combined.

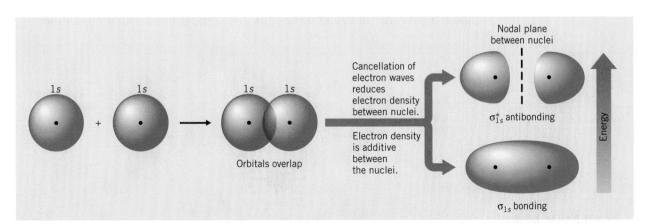

FIGURE 10.30 *Interaction of 1s atomic orbitals to produce bonding and antibonding molecular orbitals.* These are σ-type orbitals because the electron density is concentrated along the imaginary line that passes through both nuclei.

Bonding MOs are lower in energy than antibonding MOs formed from the same atomic orbitals. This is also depicted in Figure 10.30. When electrons populate molecular orbitals, they fill the lower-energy, bonding MOs first. The rules that apply to filling MOs are the same as those for filling atomic orbitals: *electrons spread out over orbitals of equal energy (Hund's rule) and two electrons can only occupy the same orbital if their spins are paired.*

Molecular orbital theory can explain why some molecules exist, and others do not

Let's see how molecular orbital theory can be used to account for the existence of certain molecules, as well as the nonexistence of others. Figure 10.31 is an MO energy-level diagram for H_2. The energies of the separate $1s$ atomic orbitals are indicated at the left and right; those of the molecular orbitals are shown in the center. The H_2 molecule has two electrons, and both can be placed in the σ_{1s} orbital. The shape of this bonding orbital, shown in Figure 10.30, should be familiar. It's the same as the shape of the electron cloud that we described using the valence bond theory.

Next, let's consider what happens when two helium atoms come together. Why can't a stable molecule of He_2 be formed? Figure 10.32 is the energy diagram for He_2. Notice that both bonding and antibonding orbitals are filled. In situations such as this there is a net destabilization because the antibonding MO is raised in energy more than the bonding MO is lowered, relative to the orbitals of the separated atoms. This means the total energy of He_2 is larger than that of two separate He atoms, so the "molecule" is unstable and immediately comes apart.

In general, the effects of *antibonding electrons* (those in antibonding MOs) cancel the effects of an equal number of bonding electrons, and molecules with equal numbers of bonding and antibonding electrons are unstable. If we remove an antibonding electron from He_2 to give He_2^+, there is a net excess of bonding electrons, and the ion should be capable of existence. In fact, the emission spectrum of He_2^+ can be observed when an electric discharge is passed through a helium-filled tube, which shows that He_2^+ is present during the electric discharge. However, the ion is not very stable and cannot be isolated.

He_2^+ has two bonding electrons and one antibonding electron.

Bond order is the difference in number of electron pairs in bonding and antibonding orbitals

The concept of bond order was introduced in Chapter 9. Recall that it was defined as the number of pairs of electrons shared between two atoms. Thus, the sharing of one pair gives a single bond with a bond order of 1, two pairs give a double bond and a bond order of 2, and three pairs give a triple bond with a bond order of 3.

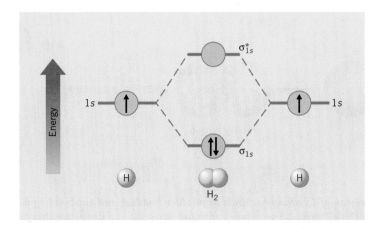

Figure 10.31 *Molecular orbital energy-level diagram for H_2.*

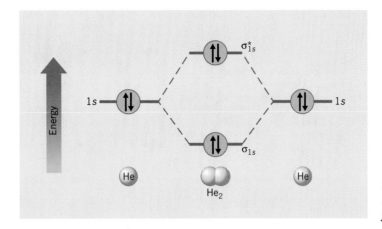

FIGURE 10.32 *Molecular orbital energy-level diagram for He$_2$.*

To translate the MO description into these terms, we compute the bond order as follows:

$$\text{bond order} = \frac{(\text{number of bonding } e^-) - (\text{number of antibonding } e^-)}{2}$$

For the H$_2$ molecule, we have

$$\text{bond order} = \frac{2 - 0}{2} = 1$$

A bond order of 1 corresponds to a single bond. For He$_2$ we have

$$\text{bond order} = \frac{2 - 2}{2} = 0$$

A bond order of zero means no bond exists, so the He$_2$ molecule is unable to exist. However, the He$_2^+$ ion does form, and its calculated bond order is

$$\text{bond order} = \frac{2 - 1}{2} = 0.5$$

Notice that the bond order does not have to be a whole number.

Molecular orbital theory successfully predicts the properties of second period diatomics

The outer shell of a Period 2 element (Li through Ne) consists of 2s and 2p sub-shells. When atoms of this period bond to each other, the atomic orbitals of these subshells interact strongly to produce molecular orbitals. The 2s orbitals, for example, overlap to form σ_{2s} and σ_{2s}^* molecular orbitals having essentially the same shapes as the σ_{1s} and σ_{1s}^* MOs, respectively. Figure 10.33 shows the shapes of the bonding and antibonding MOs produced when the 2p orbitals overlap. If we label those that point toward each other as 2p_z, a set of bonding and antibonding MOs are formed that we can label as σ_{2p_z} and $\sigma_{2p_z}^*$. The 2p_x and 2p_y orbitals, which are perpendicular to the 2p_z orbitals, overlap sideways to give π-type molecular orbitals. They are labeled π_{2p_x} and $\pi_{2p_x}^*$, and π_{2p_y} and $\pi_{2p_y}^*$.

The z axis is taken as the imaginary line between the two bonded nuclei.

The approximate relative energies of the MOs formed from the second shell atomic orbitals are shown in Figure 10.34. Notice that from Li to N, the energies of the π_{2p_x} and π_{2p_y} orbitals are lower than the energy of the σ_{2p_z}. Then from O to Ne, the energies of the two levels are reversed.

Using Figure 10.34, we can predict the electronic structures of diatomic molecules of Period 2. These *MO electron configurations* are obtained using the same rules that are applied to the filling of atomic orbitals in atoms.

 TOOLS

Molecular orbital energy diagram

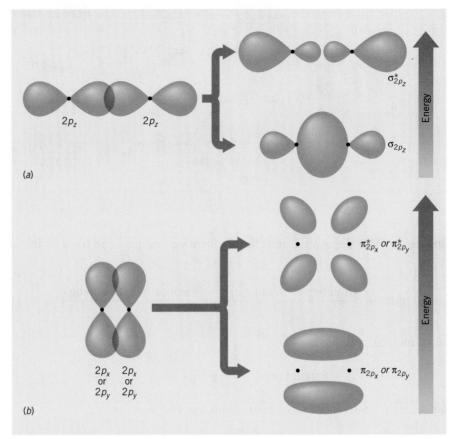

FIGURE 10.33 *Formation of molecular orbitals by the overlap of p orbitals.* (*a*) Two p_z orbitals that point at each other give bonding and antibonding σ-type MOs. (*b*) Perpendicular to the $2p_z$ orbitals are $2p_x$ and $2p_y$ orbitals that overlap to give two sets of bonding and antibonding π-type MOs.

1. Electrons fill the lowest-energy orbitals that are available.
2. No more than two electrons, with spins paired, can occupy any orbital.
3. Electrons spread out as much as possible, with spins unpaired, over orbitals that have the same energy.

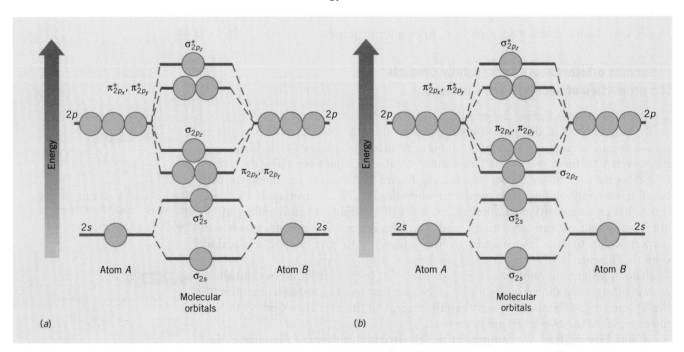

FIGURE 10.34 *Approximate relative energies of molecular orbitals in second period diatomic molecules:* (*a*) Li_2 through N_2, (*b*) O_2 through Ne_2.

Applying these rules to the valence electrons of Period 2 atoms gives the MO electron configurations shown in Table 10.1. Let's see how well MO theory performs by examining data that are available for these molecules.

According to Table 10.1, MO theory predicts that molecules of Be_2 and Ne_2 should not exist at all because they have bond orders of zero. In beryllium vapor and in gaseous neon, no evidence of Be_2 or Ne_2 has ever been found. MO theory also predicts that diatomic molecules of the other Period 2 elements should exist because they all have bond orders greater than zero. These molecules have, in fact, been observed. Although lithium, boron, and carbon are complex solids under ordinary conditions, they can be vaporized. In the vapor, molecules of Li_2, B_2, and C_2 can be detected. Nitrogen, oxygen, and fluorine, as you know, are gaseous elements that exist as N_2, O_2, and F_2.

In Table 10.1, we also see that the predicted bond order increases from boron to carbon to nitrogen and then decreases from nitrogen to oxygen to fluorine. As the bond order increases, the *net* number of bonding electrons increases, which means that more electron density is concentrated between the nuclei. This greater concentration of negative charge binds the nuclei more tightly and therefore gives a stronger bond. The attraction between the increased electron density and the positive nuclei also draws the nuclei closer to the center of the bond, thereby decreasing the bond length. The *experimentally measured* bond energies and bond lengths given in Table 10.1 follow these predictions quite nicely.

Molecular orbital theory is particularly successful in explaining the electronic structure of the oxygen molecule. Experiments show that O_2 is paramagnetic; the molecule contains two unpaired electrons. In addition, the bond length in O_2 is about what is expected for an oxygen–oxygen double bond. These data are not explained by valence bond theory. For example, if we write a Lewis structure for O_2 that shows a double bond and also obeys the octet rule, all the electrons appear in pairs.

> :Ö::Ö: (not acceptable based on experimental
>   evidence because all electrons are paired)

On the other hand, if we show the unpaired electrons, the structure has only a single bond and doesn't obey the octet rule.

> :Ö:Ö: (not acceptable based on experimental
>   evidence because of the O—O single bond)

Molecular oxygen is attracted weakly by a magnet.

Although MO theory handles easily the things that VB theory has trouble with, MO theory loses the simplicity of VB theory. For even quite simple molecules, MO theory is too complicated to make predictions without extensive calculations.

TABLE 10.1   MOLECULAR ORBITAL POPULATIONS AND BOND ORDERS FOR PERIOD 2 DIATOMIC MOLECULES[a]

	Li_2	Be_2	B_2	C_2	N_2		O_2	F_2	Ne_2
Number of Bonding Electrons	2	2	4	6	8		8	8	8
Number of Antibonding Electrons	0	2	2	2	2		4	6	8
Bond Order	1	0	1	2	3		2	1	0
Bond Energy (kJ/mol)	110	—	300	612	953		501	129	—
Bond Length (pm)	267	—	158	124	109		121	144	—

(Energy level diagrams for each molecule show the molecular orbitals $\sigma_{2p_z}^*$, $\pi_{2p_x}^*, \pi_{2p_y}^*$, σ_{2p_z}, π_{2p_x}, π_{2p_y}, σ_{2s}^*, σ_{2s} for Li_2 through N_2; and $\sigma_{2p_z}^*$, $\pi_{2p_x}^*, \pi_{2p_y}^*$, π_{2p_x}, π_{2p_y}, σ_{2p_z}, σ_{2s}^*, σ_{2s} for O_2, F_2, Ne_2.)

[a]Although the order of the energy levels corresponding to the σ_{2p_z} and the π bonding MOs become reversed at oxygen, either sequence would yield the same result—a triple bond for N_2, a double bond for O_2, and a single bond for F_2.

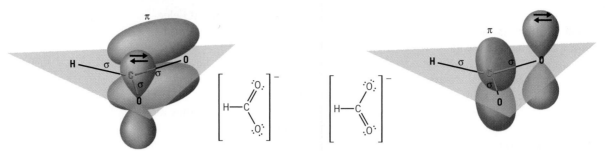

FIGURE 10.35 *Valence bond descriptions of the resonance structures of the formate ion, CHO$_2^-$.* Only the *p* orbitals of oxygen that can form π bonds to carbon are shown. The ion is planar because carbon uses *sp^2* hybrid orbitals to form the σ-bond framework.

With MO theory, we don't have any of these difficulties. By applying Hund's rule, the two electrons in the π^* orbitals of O_2 spread out over these orbitals with their spins unpaired because both orbitals have the same energy. The electrons in the two antibonding π^* orbitals cancel the effects of two electrons in the two bonding π orbitals, so the net bond order is 2 and the bond is effectively a double bond.

> **PRACTICE EXERCISE 14:** The MO energy-level diagram for the nitric oxide molecule, NO, is essentially the same as that shown in Table 10.1 for O_2. Indicate which MOs are populated in NO and calculate the bond order for the molecule.

10.8 ▶ Molecular orbital theory uses delocalized orbitals to describe molecules with resonance structures

One of the least satisfying aspects of the way valence bond theory explains chemical bonding is the need to write resonance structures for certain molecules and ions. In Chapter 9 we discussed this in terms of Lewis structures, which we have since learned can be equated to shorthand notations for valence bond descriptions of molecules. As an example, let's look at the formate ion, CHO_2^-, which we described in Chapter 9 as a resonance hybrid of the following two structures:

In terms of the overlap of orbitals, VB theory would picture the formate ion as shown in Figure 10.35. The σ-bond framework is formed by overlap of the *sp^2* hybrid orbitals of carbon with the 1*s* orbital of hydrogen and the *p* orbitals of the oxygen atoms. As you can see in Figure 10.35, the π bond can be formed with either oxygen atom, which is how we obtain the two resonance structures.

Molecular orbital theory avoids the problem of resonance by recognizing that electron pairs can sometimes be shared among overlapping orbitals from three or more atoms. For the CHO_2^- ion, it allows *three p* orbitals (one from carbon and one each from the two oxygen atoms) to overlap simultaneously to form one large π-type molecular orbital that spreads over all three nuclei, as shown in Figure 10.36. Since the π electrons in this MO are not required to stay "localized"

FIGURE 10.36 *Delocalized pi bond in CHO$_2^-$.* Molecular orbital theory allows the electron pair in the π-type bond to be spread out or delocalized over all three atoms of the —CO$_2^-$ unit in CHO_2^-.

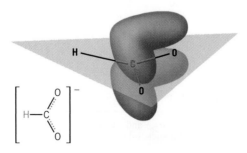

between just two nuclei, we say that the bond is **delocalized.** Delocalized π-type molecular orbitals permit a single description of the electronic structure of molecules and ions.

The delocalized nature of π bonds that extend over three or more nuclei can be indicated with dotted lines rather than dashes when Lewis-type structural formulas are drawn. The formate ion, for example, can be drawn as

$$\left[H-C\overset{\displaystyle O}{\underset{\displaystyle O}{\cdots}} \right]^{-}$$

Delocalized π bonds are quite common in many kinds of molecules and ions. Some other examples are

$$\left[O\overset{\displaystyle O}{\underset{\displaystyle}{=}C}O \right]^{2-} \quad \left[O\overset{\displaystyle O}{\underset{\displaystyle}{=}N}O \right]^{-} \quad \left[O\overset{\displaystyle \ddot{N}}{=}O \right]^{-}$$

$$CO_3^{2-} \qquad\qquad NO_3^{-} \qquad\qquad NO_2^{-}$$

An important example from the realm of organic chemistry is benzene, C_6H_6. As you learned in Chapter 9, this molecule has a ring structure whose resonance structures can be written as

In benzene, the σ-bond framework requires that the carbon atoms use sp^2 hybrid orbitals because each carbon forms three σ bonds. This leaves each carbon atom with a half-filled unhybridized p orbital perpendicular to the plane of the ring. These p orbitals overlap to give a delocalized π-electron cloud that looks like two doughnuts stacked one on top of the other, with the σ-bond framework sandwiched between them (Figure 10.37). The delocalized nature of the π electrons is the reason we usually represent the structure of benzene as

In benzene

One of the special characteristics of delocalized bonds, like those found in CO_3^{2-}, NO_3^{-}, NO_2^{-}, and C_6H_6, is that they make a molecule or ion more stable than it would be if it had localized bonds. In Chapter 9 this was described in terms of resonance energy. In the molecular orbital theory, we no longer speak of resonance; instead, we refer to the electrons as being delocalized. The extra stability that is associated with this delocalization is therefore described, in the language of MO theory, as the **delocalization energy.**

*A **localized bond** is one in which the electrons spend all their time shared between just two atoms. The electrons in a delocalized bond become shared among more than two atoms.*

The bonding in each of these ions or molecules is explained by just one structure in the MO theory.

Benzene is an important industrial solvent, but it is also quite toxic.

*Functionally, the terms **resonance energy** and **delocalization energy** are the same; they just come from different approaches to bonding theory.*

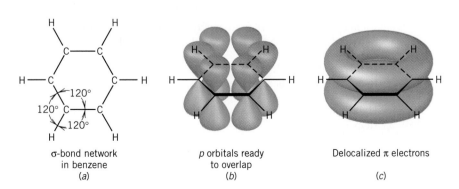

FIGURE 10.37 *Benzene.*
(*a*) The σ-bond framework. All atoms lie in the same plane. (*b*) The unhybridized p orbitals at each carbon prior to side-to-side overlap. (*c*) The double doughnut-shaped electron cloud formed by the delocalized π electrons.

σ-bond network in benzene
(*a*)

p orbitals ready to overlap
(*b*)

Delocalized π electrons
(*c*)

SUMMARY

Molecular Shapes and VSEPR Theory The structures of most molecules can be described in terms of one or another of five basic geometries: **linear, planar triangular, tetrahedral, trigonal bipyramidal,** and **octahedral.** The **VSEPR theory** predicts molecular geometry by assuming that **electron domains**—regions of space that contain bonding electrons, unpaired valence electrons, or lone pairs—stay as far apart as possible from each other, while staying as close as possible to the central atom. Figures 10.2 to 10.5 illustrate the structures obtained with different numbers of groups of electrons in the valence shell of the central atom in a molecule or ion and with different numbers of lone pairs and attached atoms. The correct shape of a molecule or polyatomic ion can usually be predicted from the Lewis structure.

Molecular Shape and Molecular Polarity A molecule that contains identical atoms attached to a central atom will be nonpolar if there are no lone pairs of electrons in the central atom's valence shell. It will be polar if lone pairs are present, except in two cases: (1) when there are three lone pairs and two attached atoms, and (2) when there are two lone pairs and four attached atoms. If all the atoms attached to the central atom are not alike, the molecule will usually be polar.

Valence Bond (VB) Theory According to VB theory, a covalent bond is formed between two atoms when an atomic orbital on one atom **overlaps** with an atomic orbital on the other and a pair of electrons with paired spins is shared between the overlapping orbitals. In general, the better the overlap of the orbitals, the stronger the bond. A given atomic orbital can only overlap with one other orbital on a different atom, so a given atomic orbital can only form one bond with an orbital on one other atom.

Hybrid Atomic Orbitals **Hybrid orbitals** are formed by mixing pure s, p, and/or d orbitals. Hybrid orbitals overlap better with other orbitals than the pure atomic orbitals from which they are formed, so bonds formed by hybrid orbitals are stronger than those formed by ordinary atomic orbitals. **Sigma bonds** (σ bonds) are formed by the following kinds of orbital overlap: s–s, s–p, end-to-end p–p, and overlap of hybrid orbitals. Sigma bonds allow free rotation around the bond axis. The side-by-side overlap of p orbitals produces a **pi bond** (π bond). Pi bonds do not permit free rotation around the bond axis because rotation around the axis of a π bond involves bond breaking. In complex molecules, the basic molecular framework is built with σ bonds. A double bond consists of one σ bond and one π bond. A triple bond consists of one σ bond and two π bonds.

Molecular Orbital (MO) Theory This theory begins with the supposition that molecules are similar to atoms, except they have more than one positive center. They are treated as collections of nuclei and electrons, with the electrons of the molecule distributed among **molecular orbitals** of different energies. Molecular orbitals can spread over two or more nuclei and can be considered to be formed by the constructive and destructive interference of the overlapping electron waves corresponding to the atomic orbitals of the atoms in the molecule. **Bonding MOs** concentrate electron density between nuclei; **antibonding MOs** remove electron density from between nuclei. The rules for the filling of MOs are the same as those for atomic orbitals. The ability of MO theory to describe **delocalized orbitals** avoids the need for resonance theory. Delocalization of bonds leads to a lowering of the energy by an amount called the **delocalization energy** and produces more stable molecular structures.

TOOLS ▶ **YOU HAVE LEARNED**

The table below lists the tools you have learned in this chapter that are applicable to problem solving. Review them if necessary, and refer to them when working on the Thinking-It-Through problems and the Review Exercises that follow.

TOOL	HOW IT WORKS
Basic molecular shapes (page 401)	Knowing how to draw them enables you to sketch the shapes of most molecules.
VSEPR theory (page 404)	This theory enables you to predict the shape of a molecule or polyatomic ion when its Lewis structure is known. It also enables you to determine the kind of hybrid orbitals used by the central atom in a molecule or polyatomic ion.
Molecular shape (page 415)	In this chapter you learned how we can use the shape of a molecule as a tool to determine whether the molecule is polar.
Orientations of hybrid orbitals (pages 423 and 427)	Hybrid orbitals allow us to explain the bonding in molecules based on their shapes.
Molecular orbital energy diagram (page 439)	The energy diagram allows us to explain the bonding in Period 2 diatomic molecules.

THINKING IT THROUGH

The goal for the following problems is not to find the answers themselves, but rather to assemble the information needed to solve them and explain how you would use the information to find the answers. The problems in Level 2 are more challenging than those in Level 1 and may contain more data than are required, in which case you are also asked to identify the unnecessary data. Detailed answers to the Thinking-It-Through problems can be found on the web site.

Need extra help?
ON-LINE HELP Visit the Brady/
Senese web site at
**www.wiley.com/
college/brady**

Level 1 Problems

1. The BrO_3^- ion is pyramidal. Without drawing the Lewis structure for the ion, explain how we know that there is at least one lone pair of electrons on the bromine atom.

2. The molecule SF_2 is nonlinear, whereas the molecule XeF_2 is linear. Without drawing their Lewis structures, what differences could there be between the valence shells of S and Xe in these compounds?

3. The molecule SF_5Cl is octahedral. Is it polar or nonpolar? Explain.

4. Antimony forms a compound with hydrogen that is called *stibine*. Its formula is SbH_3 and the H—Sb—H bond angles are 91.3°. Which kinds of orbitals does Sb most likely use to form the Sb—H bonds, pure p orbitals or hybrid orbitals? Explain your reasoning.

5. Describe in detail how you would predict whether the SF_2 molecule is polar or nonpolar.

6. The sequence of energy levels in the molecular orbital energy level diagram for nitrogen monoxide, NO, is essentially the same as for O_2. How would you determine the average bond order in the NO molecule?

7. How could you predict whether or not the NO molecule is paramagnetic?

Level 2 Problems

8. The molecule XCl_3 is pyramidal. In which group in the periodic table is element X found? If the molecule were planar triangular, in which group would X be found? If the molecule were T-shaped, in which group would X be found. Why is it unlikely that element X is in Group VI?

9. Consider the molecule $COCl_2$, whose Lewis structure is

$$\ddot{O}=C\begin{matrix} \ddot{C}\ddot{l} \\ \\ \ddot{C}\ddot{l} \end{matrix}$$

VSEPR theory predicts this is a planar triangular molecule with bond angles of 120°. However, in this molecule the O—C—Cl bond angles are not the same as the Cl—C—Cl bond angle. Suggest two reasons why all the bond angles are not the same.

REVIEW QUESTIONS

Shapes of Molecules

10.1 Sketch the following molecular shapes and give the various bond angles in the structure: (a) planar triangular, (b) tetrahedral, (c) octahedral.

10.2 Sketch the following molecular shapes and give the bond angles in the structure: (a) linear, (b) trigonal bipyramidal.

VSEPR Theory

10.3 What is the underlying principle on which the VSEPR theory is based?

10.4 What is an electron domain?

10.5 How many bonding domains and how many nonbonding domains are there in a molecule of formaldehyde, HCHO?

10.6 What arrangements of domains around an atom are expected when there are (a) three domains, (b) six domains, (c) four domains, and (d) five domains?

10.7 What molecular shapes are expected when the valence shell of the central atom has (a) three single bonds and one lone pair, (b) four single bonds and two lone pairs, and (c) a double bond and one lone pair? Give the name for each structure and also sketch its shape.

Predicting Molecular Polarity

10.8 Why is it useful to know the polarities of molecules?

10.9 How do we indicate a bond dipole when we draw the structure of a molecule?

10.10 Are all asymmetric molecules polar?

10.11 What condition must be met if a molecule having polar bonds is to be nonpolar?

10.12 Use a drawing to show why the SO_2 molecule is polar.

Modern Bonding Theories

10.13 What is the theoretical basis of both valence bond (VB) theory and molecular orbital (MO) theory?

10.14 What shortcomings of Lewis structures and VSEPR theory do VB and MO theories attempt to overcome?

10.15 What is the main difference in the way VB and MO theories view the bonds in a molecule?

Valence Bond Theory

10.16 What is meant by orbital overlap?

10.17 If an atom uses p orbitals to form bonds to two other atoms, what is the expected bond angle?

10.18 What are the principal postulates of the valence bond theory?

10.19 Use sketches of orbitals to describe how the covalent bond in H_2 is formed.

10.20 Use sketches of orbitals to describe how VB theory would explain the formation of the H—Br bond in hydrogen bromide.

Hybrid Orbitals

10.21 What term is used to describe the mixing of atomic orbitals of the same atom?

10.22 Why do atoms usually prefer to use hybrid orbitals for bonding rather than plain atomic orbitals?

10.23 Sketch figures that illustrate the directional properties of the following hybrid orbitals: (a) sp, (b) sp^2, (c) sp^3, (d) sp^3d, (e) sp^3d^2.

10.24 Why do Period 2 elements never use sp^3d or sp^3d^2 hybrid orbitals for bond formation?

10.25 What relationship is there, if any, between Lewis structures and the valence bond descriptions of molecules?

10.26 How can the VSEPR model be used to predict the hybridization of an atom in a molecule?

10.27 Using orbital diagrams, describe how sp^3 hybridization occurs in each atom: (a) carbon, (b) nitrogen, (c) oxygen. If these elements use sp^3 hybrid orbitals to form bonds, how many lone pairs of electrons would be found on each?

10.28 Sketch the way the orbitals overlap to form the bonds in each of the following: (a) CH_4, (b) NH_3, (c) H_2O. (Assume the central atom uses hybrid orbitals.)

10.29 We explained the bond angles of $107°$ in NH_3 by using sp^3 hybridization of the central nitrogen atom. Had the original unhybridized p orbitals of the nitrogen been used to overlap with $1s$ orbitals of each hydrogen, what would have been the H—N—H bond angles? Explain.

10.30 If the central oxygen in the water molecule did not use sp^3 hybridized orbitals (or orbitals of any other kind of hybridization), what would be the expected bond angle in H_2O (assuming no angle-spreading force)?

10.31 Using sketches of orbitals and orbital diagrams, describe sp^2 hybridization at (a) boron and (b) carbon.

10.32 What two basic shapes have hybridizations that include d orbitals?

Coordinate Covalent Bonds and VB Theory

10.33 The ammonia molecule, NH_3, can combine with a hydrogen ion, H^+ (which has an empty $1s$ orbital), to form the ammonium ion, NH_4^+. (This is how ammonia can neutralize acid and therefore function as a base.) Sketch the geometry of the ammonium ion, indicating the bond angles.

10.34 Use orbital diagrams to show the formation of a coordinate covalent bond in H_3O^+ when H_2O reacts with H^+.

Multiple Bonds and Hybrid Orbitals

10.35 How do σ and π bonds differ?

10.36 Why can free rotation occur easily around a σ-bond axis but not around a π-bond axis?

10.37 Using sketches, describe the bonds and bond angles in ethylene, C_2H_4.

10.38 Sketch the way the bonds form in acetylene, C_2H_2.

10.39 How does VB theory treat the benzene molecule? (Draw sketches describing the orbital overlaps and the bond angles.)

Molecular Orbital Theory

10.40 Why is the higher-energy MO in H_2 called an *antibonding* orbital?

10.41 Using a sketch, describe the two lowest-energy MOs of H_2 and their relationship to their parent atomic orbitals.

10.42 Explain why He_2 does not exist but H_2 does.

10.43 How does MO theory account for the paramagnetism of O_2?

10.44 On the basis of MO theory, explain why Li_2 molecules can exist but Be_2 molecules cannot. Could the ion Be_2^+ exist?

10.45 What are the bond orders in (a) O_2^+, (b) O_2^-, and (c) C_2^+?

10.46 What relationship is there between bond order and bond energy?

10.47 Sketch the shapes of the π_{2p_y} and $\pi_{2p_y}^*$ MOs.

10.48 What is a delocalized MO?

10.49 Use a Lewis-type structure to indicate delocalized bonding in the nitrate ion.

10.50 What problem encountered by VB theory does MO theory avoid by delocalized bonding?

10.51 Draw the representation of the benzene molecule that indicates its delocalized π system.

10.52 What effect does delocalization have on the stability of the electronic structure of a molecule?

10.53 What is delocalization energy? How is it related to resonance energy?

REVIEW PROBLEMS

Answers to problems whose numbers are printed in color are given in Appendix B. More challenging questions are marked with asterisks. **ILW** = Interactive LearningWare solution is available at *www.wiley.com/college/brady.*

10.54 Predict the shapes of (a) NH_2^-, (b) CO_3^{2-}, (c) IF_3, (d) Br_3^-, and (e) GaH_3.

10.55 Predict the shapes of (a) SF_3^+, (b) NO_3^-, (c) SO_4^{2-}, (d) O_3, and (e) N_2O.

ILW 10.56 Predict the shapes of (a) FCl_2^+, (b) AsF_5, (c) AsF_3, (d) SbH_3, and (e) SeO_2.

10.57 Predict the shapes of (a) TeF_4, (b) $SbCl_6^-$, (c) NO_2^-, (d) PCl_4^+, and (e) PO_4^{3-}.

10.58 Predict the shapes of (a) IO_4^-, (b) ICl_4^-, (c) TeF_6, (d) SiO_4^{4-}, and (e) ICl_2^-.

10.59 Predict the shapes of (a) CS_2, (b) BrF_4^-, (c) ICl_3, (d) ClO_3^-, and (e) SeO_3.

10.60 Acetylene, a gas used in welding torches, has the Lewis structure $H—C\equiv C—H$. What would you expect the H—C—C bond angle to be in this molecule?

10.61 Ethylene, a gas used to ripen tomatoes artificially, has the Lewis structure

$$\begin{array}{ccc} H & & H \\ | & & | \\ H—C & = & C—H \end{array}$$

What would you expect the H—C—H and H—C—C bond angles to be in this molecule? (*Caution:* Don't be fooled by the way the structure is drawn here.)

10.62 Which of the following compounds contain non-bonding domains: (a) BrF_3, (b) IF_4^+, (c) H_2Se, (d) CBr_4, (e) N_2H_4?

10.63 Which of the following compounds contain non-bonding domains: (a) $HOBr$, (b) NH_4^+, (c) BrF_5, (d) PH_3, (e) TeF_4?

10.64 Predict the bond angle for each of the following molecules: (a) OCl_2, (b) H_2O, (c) SO_2, (d) I_3^-, (e) NH_2^-.

10.65 Predict the bond angle for each of the following molecules: (a) $HOCl$, (b) PH_2^-, (c) OCN^-, (d) O_3, (e) SnF_2.

Predicting Molecular Polarity

ILW 10.66 Which of the following molecules would be expected to be polar: (a) HBr, (b) $POCl_3$, (c) CH_2O, (d) $SnCl_4$, (e) $SbCl_5$?

10.67 Which of the following molecules would be expected to be polar: (a) PBr_3, (b) SO_3, (c) $AsCl_3$, (d) ClF_3, (e) BCl_3?

10.68 Which of the following would be expected to be polar: (a) $ClNO$, (b) XeF_3^+, (c) $SeBr_4$, (d) NO, (e) NO_2?

10.69 Which of the following would be expected to be polar: (a) H_2S, (b) BeH_2, (c) SCN^-, (d) CN^-, (e) $BrCl_3$?

10.70 Explain why SF_6 is nonpolar, but SF_5Br is polar.

10.71 Explain why CH_3Cl is polar, but CCl_4 is not.

Valence Bond Theory

10.72 Hydrogen selenide is one of nature's most foul-smelling substances. Molecules of H_2Se have H—Se—H bond angles very close to 90°. How would VB theory explain the bonding in H_2Se? Use sketches of orbitals to show how the bonds are formed. Illustrate with appropriate orbital diagrams as well.

10.73 Use sketches of orbitals to show how VB theory explains the bonding in the F_2 molecule. Illustrate with appropriate orbital diagrams as well.

Hybrid Orbitals

10.74 Use orbital diagrams to explain how the $BeCl_2$ molecule is formed. What kind of hybrid orbitals does beryllium use in $BeCl_2$?

10.75 Use orbital diagrams to describe the bonding in (a) $SnCl_4$ and (b) $SbCl_5$. Be sure to indicate hybrid orbital formation.

10.76 Draw Lewis structures for the following and use the geometry predicted by the VSEPR model to determine what kind of hybrid orbitals the central atom uses in bond formation: (a) ClO_3^-, (b) SO_3, (c) OF_2.

10.77 Draw Lewis structures for the following and use the geometry predicted by the VSEPR model to determine what kind of hybrid orbitals the central atom uses in bond formation: (a) $SbCl_6^-$, (b) $BrCl_3$, (c) XeF_4.

10.78 Use the VSEPR model to help you describe the bonding in the following molecules according to VB theory: (a) $AsCl_3$, (b) ClF_3.

10.79 Use the VSEPR model to help you describe the bonding in the following molecules according to VB theory: (a) $SbCl_5$, (b) $SeCl_2$.

Coordinate Covalent Bonds and VB Theory

10.80 Use orbital diagrams to show that the bonding in SbF_6^- involves the formation of a coordinate covalent bond.

10.81 What kind of hybrid orbitals are used by tin in $SnCl_6^{2-}$? Draw the orbital diagram for Sn in $SnCl_6^{2-}$. What is the geometry of $SnCl_6^{2-}$?

Multiple Bonding and Valence Bond Theory

10.82 A nitrogen atom can undergo sp^2 hybridization when it becomes part of a carbon–nitrogen double bond, as in $H_2C=NH$.
(a) Using a sketch, show the electron configuration of sp^2 hybridized nitrogen just before the overlapping occurs to make this double bond.
(b) Using sketches (and the analogy to the double bond in C_2H_4), describe the two bonds of the carbon–nitrogen double bond.
(c) Describe the geometry of $H_2C=NH$ (using a sketch that shows all expected bond angles).

10.83 A nitrogen atom, can undergo *sp* hybridization and then become joined to carbon by a triple bond to give the structural unit —C≡N:. This triple bond consists of one σ bond and two π bonds.

(a) Write the orbital diagram for *sp* hybridized nitrogen as it would look before any bonds form.

(b) Using the carbon–carbon triple bond as the analogy, and drawing pictures to show which atomic orbitals overlap with which, show how the three bonds of the triple bond in —C≡N: form.

(c) Again using sketches, describe all the bonds in hydrogen cyanide, H—C≡N:.

(d) What is the likeliest H—C—N bond angle in HCN?

10.84 Tetrachloroethylene, a common dry-cleaning solvent, has the formula C_2Cl_4. Its structure is

Use the electron domain and VB theories to describe the bonding in this molecule. What are the expected bond angles in the molecule?

10.85 Phosgene, $COCl_2$, was used as a war gas during World War I. It reacts with moisture in the lungs of its victims to form CO_2 and gaseous HCl, which cause the lungs to fill with fluid. Phosgene is a simple molecule with the structure

Describe the bonding in this molecule using VB theory.

10.86 What kind of hybrid orbitals do the numbered atoms use in the following molecule?

10.87 What kinds of bonds (σ or π) are found in the numbered bonds in the following molecule?

Molecular Orbital Theory

ILW **10.88** Use the MO energy diagram to predict which in each pair has the greater bond energy: (a) O_2 or O_2^+, (b) O_2 or O_2^-, (c) N_2 or N_2^+.

10.89 Assume that in the NO molecule the molecular orbital energy level sequence is similar to that for O_2. What happens to the NO bond length when an electron is removed from NO to give NO^+?

10.90 Which of the following molecules is paramagnetic: (a) O_2^+, (b) O_2, (c) O_2^-, (d) NO, (e) N_2?

10.91 What is the bond order for each of the following molecules: (a) CO, (b) CN, (c) OH, (d) BN, (e) NO?

ADDITIONAL EXERCISES

10.92 Formaldehyde has the Lewis structure

$$H—C=\ddot{O}$$

with H above the C.

What would you predict its shape to be?

10.93 Describe the changes in molecular geometry that take place during the following reactions:
(a) $BF_3 + F^- \rightarrow BF_4^-$
(b) $PCl_5 + Cl^- \rightarrow PCl_6^-$
(c) $ICl_3 + Cl^- \rightarrow ICl_4^-$
(d) $2PCl_3 + Cl_2 \rightarrow PCl_5$
(e) $C_2H_2 + H_2 \rightarrow C_2H_4$

10.94 Cyclopropane is a triangular molecule with C—C—C bond angles of 60°. Explain why the σ bonds joining carbon atoms in cyclopropane are weaker than the carbon–carbon σ bonds in the noncyclic propane.

cyclopropane   propane

10.95 What facts about boron strongly suggested the need to consider *sp²* hybridization of its second shell atomic orbitals?

10.96 Phosphorus trifluoride, PF_3, has F—P—F bond angles of 97.8°.

(a) How would VB theory use hybrid orbitals to explain these data?

(b) How would VB theory use unhybridized orbitals to account for these data?

(c) Do either of these models work very well?

10.97 A six-membered ring of carbons can hold a double bond but not a triple bond. Explain.

cyclohexene (exists)   cyclohexyne (unknown)

*****10.98** There exists a hydrocarbon called butadiene, which has the molecular formula C_4H_6 and the structure

The C=C bond lengths are 134 pm (about what is expected for a carbon–carbon double bond), but the C—C bond length in this molecule is 147 pm, which is shorter than a normal C—C single bond. The molecule is planar (i.e., all the atoms lie in the same plane).

(a) What kind of hybrid orbitals do the carbon atoms use in this molecule to form the carbon–carbon bonds?

(b) Between which pairs of carbon atoms do we expect to find sideways overlap of p orbitals (i.e., π-type p–p overlap)?

(c) On the basis of your answer to part b, do you expect to find localized or delocalized π bonding in the carbon chain in this molecule?

(d) Based on your answer to part c, explain why the center carbon–carbon bond is shorter than a carbon–carbon single bond?

10.99 Draw the structure of the molecule CCl_2F_2 (carbon is the central atom). Is the molecule polar or nonpolar? Explain.

*__10.100__ *The more electronegative are the atoms bonded to the central atom, the less are the repulsions between the electron pairs in the bonds.* On the basis of this statement, predict the most probable structure for the molecule PCl_3F_2. Do we expect the molecule to be polar or nonpolar?

*__10.101__ *A lone pair of electrons in the valence shell of an atom has a larger effective volume than a bonding electron pair. Lone pairs therefore repel other electron pairs more strongly than do bonding pairs.* On the basis of these statements, describe how the bond angles in TeF_4 and BrF_4^- deviate from those found in a trigonal bipyramid and an octahedron, respectively. Sketch the molecular shapes of TeF_4 and BrF_4^- and indicate these deviations on your drawing.

*__10.102__ *The two electron pairs in a double bond repel other electron pairs more than the single pair of electrons in a single bond.* On the basis of this statement, which bond angles should be larger in SO_2Cl_2, the O—S—O bond angles or the Cl—S—Cl bond angles? In the molecule, sulfur is bonded to two oxygen atoms and two chlorine atoms. (*Hint:* Assign formal charges and work with the best Lewis structure for the molecule.)

*__10.103__ *A hybrid orbital does not distribute electron density symmetrically around the nucleus of an atom. Therefore, a lone pair in a hybrid orbital contributes to the overall polar-

ity of a molecule.* On the basis of these statements and the fact that NH_3 is a very polar molecule and NF_3 is a nearly nonpolar molecule, justify the notion that the lone pair of electrons in each of these molecules is held in an sp^3 hybrid orbital.

10.104 In a certain molecule, a p orbital overlaps with a d orbital as shown below. What kind of bond is formed, σ or π? Explain your choice.

10.105 If we take the internuclear axis in a diatomic molecule to be the z axis, what kind of p orbital (p_x, p_y, or p_z) on one atom would have to overlap with a d_{xz}, orbital on the other atom to give a pi bond?

10.106 The peroxynitrite ion, $OONO^-$, is a potent toxin formed in cells affected by diseases such as diabetes or atherosclerosis. Peroxynitrite ion can oxidize and destroy biomolecules crucial for the survival of the cell.

(a) Estimate the O—O—N and O—N—O bond angles in peroxynitrite ion.

(b) What is the hybridization of the N atom in peroxynitrite ion?

(c) Suggest why the peroxynitrite ion is expected to be much less stable than the nitrate ion, NO_3^-.

10.107 A *neurotransmitter* is a chemical substance released by one nerve cell to communicate with others. The first neurotransmitter ever discovered was acetylcholine:

$$\begin{array}{c}
\text{O} \\
\parallel \\
\text{H}_3\text{C}-\text{C} \\
\diagdown \text{O} \\
\mid \\
\text{H}_2\text{C} \\
\diagdown \text{CH}_2 \\
\mid \\
\text{H}_3\text{C}-\overset{+}{\text{N}}-\text{CH}_3 \\
\mid \\
\text{CH}_3
\end{array}$$

(a) Predict the C—N—C bond angle.

(b) Predict the O—C—O bond angle.

(c) What is the hybridization of the nitrogen atom?

TEST OF FACTS AND CONCEPTS

Once again we permit you to test your understanding of concepts, your knowledge of scientific terms, and your problem solving skills. Read through the following questions carefully, and answer each as fully as possible. When necessary, review topics that give you difficulty. When you are able to answer these questions correctly, you are ready to study the next group of chapters.

1 What are the three principle particles that make up the atom? On the atomic mass scale, what are their approximate masses? What are their electrical charges?

2 A beam of green light has a wavelength of 500 nm. What is the frequency of this light? What is the energy, in joules, of one photon of this light? What is the energy, in joules, of one mole of photons of this light? Would blue light have more or less energy per photon than this light?

3 Arrange the following kinds of electromagnetic radiation in order of increasing frequency: X rays, blue light, radio waves, gamma rays, microwaves, red light, infrared light, ultraviolet light.

4 What is a continuous spectrum? How does it differ from an atomic spectrum?

5 What experimental evidence is there that matter has wave-like properties?

6 What is the difference between a traveling wave and a standing wave? What is a node?

7 How is the energy of an electron related to the number of nodes in its electron wave?

8 What is a wave function? What Greek letter is usually used to represent a wave function? What word do we use to refer to an electron wave in an atom?

9 What are the quantum numbers of the electrons in the valence shells of (a) sulfur, (b) strontium, (c) lead, (d) bromine, (e) boron?

10 If a given shell has $n = 4$, which kinds of subshells (s, p, etc.) does it have? What is the maximum number of electrons that could populate this shell?

11 Use the periodic table to predict the electron configurations of (a) tin, (b) germanium, (c) silicon, (d) lead, and (e) nickel.

12 Give the electron configurations of the ions (a) Pb^{2+}, (b) Pb^{4+}, (c) S^{2-}, (d) Fe^{3+}, and (e) Zn^{2+}.

13 What causes an atom, molecule, or ion to be paramagnetic? Which of the ions in the preceding question are paramagnetic? What term describes the magnetic properties of the others?

14 Give the shorthand electron configurations of (a) Ni, (b) Cr, (c) Sr, (d) Sb, and (e) Po.

15 Define *ionization energy* and *electron affinity*. In terms of these properties, which kinds of elements tend to react to form ionic compounds?

16 In general, the second ionization energy of an atom is larger than the first, the third is larger than the second, and so on. Why?

17 Which of the following elements has the largest difference between its second and third ionization energy? Explain your choice.
(a) Li   (b) Be   (c) B   (d) C

18 Which of the following processes are endothermic?
(a) $P^-(g) + e^- \rightarrow P^{2-}(g)$ (c) $Cl(g) + e^- \rightarrow Cl^-(g)$
(b) $Fe^{3+}(g) + e^- \rightarrow Fe^{2+}(g)$ (d) $S(g) + 2e^- \rightarrow S^{2-}(g)$

19 Sketch the shape of (a) an s orbital, (b) a p orbital, (c) the $3d_{xz}$ orbital.

20 What is meant by the term *electron density*?

21 Give orbital diagrams for the valence shells of selenium and thallium.

22 Which ion would be larger: (a) Fe^{2+} or Fe^{3+}, (b) O^- or O^{2-}?

23 Which of the following pairs of elements would be expected to form ionic compounds: (a) Br and F, (b) H and P, (c) Ca and F?

24 Use Lewis symbols to diagram the reaction of calcium with sulfur to form CaS.

25 Draw Lewis structures for (a) SbH_3, (b) IF_3, (c) $HClO_2$, (d) C_2^{2-}, (e) AsF_5, (f) O_2^{2-}, (g) HCO_3^-, (h) TeF_6, (i) HNO_3.

26 Use the VSEPR theory to predict the shapes of (a) $SbCl_3$, (b) IF_5, (c) AsH_3, (d) BrF_2, (e) OF_2?

27 What kinds of hybrid orbitals are used by the central atom in each of the species in the preceding question?

28 Referring to your answers to questions 25 and 26, which of the following molecules would be nonpolar: SbH_3, IF_3, AsF_5, $SbCl_3$, OF_2?

29 The oxalate ion has the following arrangement of atoms.

$$O \qquad O$$
$$C \quad C$$
$$O \qquad O$$

Draw all of its resonance structures.

30 What is meant by the term *overlap of orbitals*?

31 What are *sigma bonds*? What are *pi bonds*? How are sigma and pi bonds used to explain the formation of double and triple bonds?

32 Some resonance structures that can be drawn for carbon dioxide are shown below.

$$:\ddot{O}=C=\ddot{O}: \qquad :O\equiv C-\ddot{O}: \qquad :\ddot{O}-C\equiv O:$$
$$\quad I \qquad\qquad\quad II \qquad\qquad\quad III$$

Assign formal charges to the atoms in these structures. Explain why Structure I is the preferred structure.

33 Ozone, O_3, consists of a chain of three oxygen atoms.
(a) Draw the two resonance structures for ozone that obey the octet rule.
(b) Based on your answer to (a), is the molecule linear or nonlinear?
(c) Assign formal charges to the atoms in the resonance structures you have drawn in part (a).
(d) On the basis of your answers to (b) and (c), explain why ozone is a polar molecule even though it is composed of three atoms that have identical electronegativities.

34 Why, on the basis of formal charges and relative electronegativities, is it more reasonable to expect the structure of $POCl_3$ to be the one on the left rather than the one on the right?

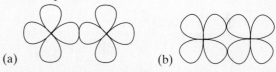

Is either of these the "best" Lewis structure that can be drawn for this molecule?

35 A certain element X was found to form three compounds with chlorine having the formulas XCl_2, XCl_4, and XCl_6. One of its oxides has the formula XO_3, and X reacts with sodium to form the compound Na_2X.
(a) Is X a metal or a nonmetal?
(b) In which group in the periodic table is X located?
(c) In which periods in the periodic table could X possibly be located?
(d) Draw Lewis structures for XCl_2, XCl_4, XCl_6, and XO_3. (Where possible, follow the octet rule.) Which has multiple bonding?
(e) What do we expect the molecular structures of XCl_2, XCl_4, XCl_6, and XO_3 to be? Which are polar molecules?
(f) The element X also forms the oxide XO_2. Draw a Lewis structure for XO_2 that obeys the octet rule.
(g) Assign formal charges to the atoms in the Lewis structures for XO_2 and XO_3 drawn for parts d and f.
(h) What kinds of hybrid orbitals would X use for bonding in XCl_4 and XCl_6?
(i) If X were to form a compound with aluminum, what would be its formula?
(j) Which compound of X would have the more ionic bonds, Na_2X or MgX?
(k) If X were in Period 5, what would be the electron configuration of its valence shell?

36 Where in the periodic table are the very reactive metals located? Where are the least reactive ones located?

37 What are bonding and antibonding molecular orbitals? How do they differ in shape and energy?

38 Describe how molecular orbital theory explains the bonding in the oxygen molecule.

39 What is a delocalized molecular orbital? How does molecular orbital theory avoid the concept of resonance?

40 Predict the shape of the following molecules and ions: (a) PF_3, (b) PF_4^+, (c) PF_6^-, (d) PF_5.

41 Which of the substances in the preceding question have a net dipole moment?

42 Consider the following statements: (1) Fe^{2+} is easily oxidized to Fe^{3+}, and (2) Mn^{2+} is difficult to oxidize to Mn^{3+}. On the basis of the electron configurations of the ions, explain the difference in ease of oxidation.

43 For each of the following pairs of compounds, which has the larger lattice energy: (a) MgO or NaCl, (b) MgO or BeO, (c) NaI or NaF, (d) MgO or CaS? Explain your choices.

44 The melting point of Al_2O_3 is much higher than the melting point of NaCl. On the basis of lattice energies, explain why this is so.

45 Why is the change in atomic size, going from one element to the next in a period, smaller among the transition elements than among the representative elements?

46 Which kind of bond, σ or π, is produced by the overlap of d orbitals pictured below:

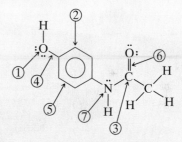

47 The following is the chemical structure of acetaminophen, the pain killer in Tylenol®.

(a) What kinds of hybrid orbitals are used by atoms numbered 1, 2, and 3?
(b) How many sigma and pi bonds are in each of the bonds numbered 4 and 6?
(c) What is the bond order of bonds numbered 4, 5, and 6?
(d) Identify the carbonyl group in the molecule.
(e) Identify the alcohol group in the molecule.
(f) Identify the amine group in the molecule.
(g) What is the geometry of electron domains around atoms labeled 1, 2, 3, and 7?
(h) What is the molecular formula of the compound?

Properties of Gases

Hot gases expand, and the expansion of hot gases formed by the combustion of jet fuel provides the thrust that enables this passenger plane to take off. The way the volume of a gas is affected by its temperature is one of the properties of gases you will study in this chapter.

THIS CHAPTER IN CONTEXT In the preceding chapters we've discussed the chemical properties of a variety of different substances. We've also studied the kinds of forces (chemical bonds) that hold molecular and ionic substances together. In fact, it is the nature of the chemical bonds that dictate chemical properties. With this chapter we begin a systematic study of the *physical properties* of materials, as well as the factors that govern the behavior of gases, liquids, and solids. We study gases first because they are the easiest to understand and their behavior will help explain some of the properties of liquids and solids in the next chapter.

Many common substances are gases at room temperature, including the entire mix of elements and compounds that make up Earth's atmosphere. Because you've spent your entire existence surrounded by air, you are already familiar with many of the properties that gases have. Our goal in this chapter is to refine this understanding in terms of the physical laws that govern the way gases behave. You will also learn how this behavior is interpreted in terms of the way we view gases at a molecular level. In our discussions we will describe how the kinetic molecular theory, first introduced to you in Chapter 7, provides an explanation of the gas laws. Finally, you will learn how a close examination of gas properties furnishes clues about molecular size and the attractions that exist between molecules.

11.1 ▶ Familiar properties of gases can be explained at the molecular level

For a long time, early scientists didn't recognize the existence of gases as examples of matter. Of course, we now understand that gases are composed of chemical substances that exist in one of the three common states of matter. The reason for the early confusion is that the physical properties of gases differ so much from liquids and solids. Consider water, for example. We can see and feel it as a liquid, but it seems to disappear when it evaporates and surrounds us as water vapor. With this in mind, let's examine some of the properties of gaseous substances to look for clues that suggest the nature of gases when viewed at a molecular level.

The most common gas familiar to people is the air that surrounds us. (Actually, air is a mixture, but that doesn't matter much when it comes to the physical properties of gases.) Because you've grown up surrounded by gas, you already are aware of many properties that gases have. Let's look at two of them.

- You can wave your hand through air with little resistance. (Compare that with waving your hand through a tub filled with water.)
- The air in a bottle has little weight to it, so if a bottle of air is submerged under water and released, it quickly bobs to the surface.

Because air has so little weight for a given volume, it makes things filled with air float, much to the relief of the occupants of an inflatable life raft.

Air is roughly 21% O_2 and 79% N_2, but it has traces of several other gases.

Aerosol cans carry a warning about subjecting them to high temperatures because the internal pressure can become large enough to cause the can to explode.

Both of these observations suggest that a given volume of air doesn't have much matter in it. (We can express this by saying that air has a low density.) What else do you know about gases?

- Gases can be compressed. Inflating a tire involves pushing more and more air into the same container (the tire). This behavior is a lot different from liquids; you can't squeeze more water into an already filled bottle.
- Gases exert a pressure. Whenever you inflate a balloon you have an experience with gas pressure, and the "feel" of a balloon suggests that the pressure acts equally in all directions.
- The pressure of a gas depends on *how much* gas is confined. The *more air* you pump into a tire, the greater the pressure.
- Gases fill completely any container into which they're placed. You've never heard of half a bottle of air. If you put air in a container, it expands and fills the container's entire volume. (This is certainly a lot different than the behavior of liquids and solids.)
- Gases mix freely and quickly with each other. You've experienced this when you've smelled the cologne of someone passing by. The vapors of the person's cologne mix with and spread through the air.
- The pressure of a gas rises when its temperature is increased. That's why the warning "Do Not Incinerate" is printed on aerosol cans. A sealed can, if made too hot, is in danger of exploding from the increased pressure.[1]

Properties suggest a molecular model

The simple qualitative observations about gases described above suggest what gases must be like when viewed at a molecular level (Figure 11.1). The fact that there's so little matter in a given volume suggests that there is a lot of space between the individual molecules, especially when compared to liquids or solids. This would also explain why gases can be so easily compressed—squeezing a gas simply removes some of the empty space.

It also seems reasonable to believe that the molecules of a gas are moving around fairly rapidly. How else could we explain the travel of the molecules of a cologne so quickly through the air? Furthermore, if gas molecules didn't move, gravity would cause them to settle to the bottom of a container (which they don't do). And if gas molecules are moving, some must be colliding with the walls of the container, and the force of these tiny collisions would explain the pressure a gas exerts. It also explains why adding more gas increases the pressure; the more gas in the container, the more collisions with the walls, and the higher the pressure.

Finally, the fact that gas pressure rises with increasing temperature suggests that the molecules move faster with increasing temperature, because faster molecules would exert greater forces when they collide with the walls.

Perhaps by now you recognize this simple model of a gas as the basis for the kinetic-molecular theory discussed in Section 7.2. Keep this model in mind as we continue our discussion of gas properties and gas laws. We will come back to the model later in the chapter when we will use it to explain in a semiquantitative way the laws that govern gas behavior.

Container wall

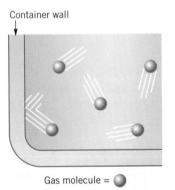

Gas molecule =

FIGURE **11.1** *A gas viewed at the molecular level.* Simple qualitative observations of the properties of gases leads us to conclude that a gas is composed of widely spaced molecules that are in constant motion. Collisions of molecules with the walls produce tiny forces that, when taken all together, are responsible for the gas pressure.

---

[1] An amusing newspaper report, "The Awful Day of the Ham," told on herself by Donna Tabbert Long, described how a peaceful Sunday was punctuated by a huge explosion that shook the apartment building. When her husband asked her how the ham dinner was coming, she reported that when she last looked in the oven the ham (in its unopened can) was "puffing up nicely." Her husband reached the kitchen door when the explosion occurred and soon reported, in a quiet voice tinged with awe, "There's ham hanging all over in there." (*Minneapolis Tribune,* August 2, 1980, page 5B.)

11.2 ▶ Pressure is a measured property of gases

Most people are familiar with the concept of pressure, although it is often confused with force. As we discussed in Chapter 7, **pressure** *is force per unit area*, calculated by dividing the force by the area over which the force acts.

$$\text{pressure} = \frac{\text{force}}{\text{area}}$$

Earth exerts a gravitational force on everything with mass on it or near it. What we call the *weight* of an object, like a book, is simply our measure of the gravitational force acting on it. Hence, we use "weight" for "force" in the equation for pressure when the gravitational force is involved.

The distinction between force and pressure is important. The weight of the books in a backpack causes less shoulder pain, for example, when the pack has wide straps rather than narrow straps of string. The force (the weight) is the same, but the pressure, or force *per unit area,* on your shoulders is much less when the weight is distributed over wide straps. The *ratio* of force to area, the pressure, is less, and you can feel the difference.

The pressure of the atmosphere is measured with a barometer

Earth's gravity pulls on the air mass of the atmosphere, causing it to cover the Earth's surface like an invisible blanket. The molecules in the air collide with any object the air contacts, and by doing so, produce a pressure we call the *atmospheric pressure.*

At any particular location on Earth, the atmospheric pressure acts equally in all directions—up, down, and sideways. In fact, it presses against our bodies with a surprising amount of force, but we don't really feel it because the fluids in our bodies push back with equal pressure. We can observe atmospheric pressure, however, if we pump the air from a collapsible container such as the can in Figure 11.2.

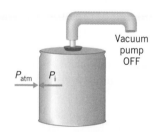

(a)　　　　　(b)

FIGURE 11.2 *The effect of an unbalanced pressure.*
(*a*) The pressure inside the can, P_{inside}, is the same as the atmospheric pressure outside, P_{atm}. The pressures are balanced; $P_{\text{inside}} = P_{\text{atm}}$.
(*b*) When a vacuum pump reduces the pressure inside the can, P_{inside} is made less than P_{atm}, and the unbalanced outside pressure quickly and violently makes the can collapse.

Before air is removed, the walls of the can experience atmospheric pressure equally inside and out. When some air is pumped out, however, the pressure inside decreases, making the atmospheric pressure outside of the can greater than the pressure inside. The net inward pressure is sufficiently great to make the can crumple.

To measure atmospheric pressure we use a device called a **barometer.** The simplest type is the *Torricelli barometer* (Figure 11.3), which consists of a glass tube sealed at one end, 80 cm or more in length.[2] To set up the apparatus, the tube is filled with mercury, capped, and then inverted with its capped end immersed in a dish of mercury. When the cap is removed, some mercury runs out, but not all.[3] Atmospheric pressure, pushing on the surface of the mercury in the dish, holds most of the mercury in the tube. Opposing the atmosphere is the downward pressure caused by the weight of the mercury still inside the tube. When the two pressures become equal, no more mercury can run out, but a space inside the tube above the mercury level has been created having essentially no atmosphere; it's a *vacuum.*

The height of the mercury column, measured from the surface of the mercury in the dish, is directly proportional to atmospheric pressure. On days when the atmospheric pressure is high, more mercury is forced from the dish into the tube and the height of the column increases. When the atmospheric pressure drops, during an approaching storm, for example, some mercury flows out of the tube and the height of the column decreases. Most people live where this height fluctuates between 730 and 760 mm.

Mercury, a shiny, metallic element, is a liquid above −39 °C.

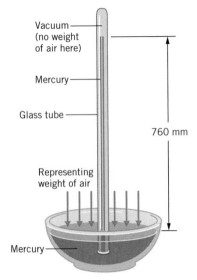

Vacuum
(no weight
of air here)

Mercury

Glass tube

760 mm

Representing
weight of air

Mercury

FIGURE 11.3 *The mercurial or Torricelli barometer.*

In English units, one atmosphere of pressure is 14.7 lb/in[2]. This means that at sea level, each square inch of your body is experiencing a force of nearly 15 pounds.

Units of pressure include the pascal, atmosphere, and torr

The height of the mercury column in a barometer varies with altitude. For example, on top of Mount Everest, Earth's highest peak, the air is so "thin" that the atmospheric pressure can maintain a height of only about 250 mm of mercury. At sea level, the height fluctuates around a value of 760 mm. Some days it's a little higher, some a little lower, depending on the weather. The average pressure at sea level has long been used by scientists as a standard unit of pressure. The **standard atmosphere (atm)** was defined as the pressure needed to support a column of mercury 760 mm high measured at 0 °C.[4]

In the SI, the unit of pressure is the **pascal,** symbolized **Pa.** In SI units, the pascal is the ratio of force in *newtons* (N, the SI unit of force) to area in meters squared,

$$1 \text{ Pa} = \frac{1 \text{ N}}{1 \text{ m}^2} = 1 \text{ N m}^{-2}$$

It's a very small pressure; 1 Pa is approximately the pressure exerted by the weight of a lemon spread over an area of 1 m[2].

To bring the standard atmosphere unit in line with other SI units, it has been redefined in terms of the pascal as follows:

$$1 \text{ atm} = 101,325 \text{ Pa} \qquad \text{(exactly)}$$

---

[2]In 1643 Evangelista Torricelli, an Italian mathematician, suggested an experiment, later performed by a colleague, that demonstrated that atmospheric pressure determines the height to which a fluid will rise in a tube inverted over the same liquid. This concept led to the development of the Torricelli barometer, which is named in his honor.

[3]Today, mercury is kept as much as possible in closed containers. Although atoms of mercury do not readily escape into the gaseous state, mercury vapor is a dangerous poison.

[4]Because any metal, including mercury, expands or contracts as the temperature increases or decreases, the height of the mercury column varies with temperature (just like in a thermometer). Therefore, the definition of the standard atmosphere required that the temperature at which the mercury height is measured be specified.

In Canada and many other countries, pascals or kilopascals (kPa) are the preferred unit of pressure. For instance, tire pressures as well as barometric pressures in weather reports are given in kilopascals (1 atm is about 101 kPa). A unit of pressure related to the pascal is the **bar,** which is defined as 100 kPa. Consequently, 1 bar is slightly smaller than 1 standard atmosphere (1 bar = 0.9868 atm). You may have heard the **millibar** unit (1 mb = 10^{-3} bar) used in weather reports describing pressures inside storms such as hurricanes. For example, Hurricane Andrew, which devastated south Florida in August of 1992, had a central pressure of 922 mb, the third lowest of the twentieth century. [The lowest atmospheric pressure at sea level ever observed in the Western Hemisphere was 887 mb (0.876 atm) during Hurricane Gilbert, September 1988.]

For ordinary laboratory work, the pascal (or kilopascal) is not a conveniently measured unit. As you will see, we often use tubes filled with mercury to measure pressures, so it is convenient to use a unit of pressure called the **torr** (named after Torricelli). The torr is defined as 1/760th of 1 atm.

$$1 \text{ torr} = \frac{1}{760} \text{ atm}$$

$$1 \text{ atm} = 760 \text{ torr} \qquad \text{(exactly)}$$

The torr is very close to the pressure that is able to support a column of mercury 1 mm high. In fact, the *millimeter of mercury* (abbreviated mm Hg) is often itself used as a pressure unit. Except when the most exacting measurements are being made, it is safe to use the relationship

$$1 \text{ torr} = 1 \text{ mm Hg}$$

The torr and atm are the units that we will use throughout this book, but those going into medicine in the United States will often see "mm Hg" used in connection with the gases of respiration or of anesthesia.

> **PRACTICE EXERCISE 1:** In a comment above, it was reported that the lowest barometric pressure ever recorded at sea level in the Western Hemisphere was 887 mb during Hurricane Gilbert in 1988. What was the pressure in torr?

Manometers measure the pressures of trapped gas samples

Gases used as reactants or formed as products in chemical reactions are kept from escaping by employing closed glassware. To measure pressures inside such vessels, a **manometer** is used. Two types are common, open-end and closed-end.

The **open-end manometer** consists of a U-tube partly filled with a liquid, usually mercury (see Figure 11.4). One end of the U-tube's arm is open to the atmosphere. The other end is exposed to the entrapped gas.

When the two mercury levels are equal, as shown in Figure 11.4a, the pressure of the gas, P_{gas}, equals atmospheric pressure, P_{atm}. Measuring the atmospheric pressure with a barometer then gives the pressure of the trapped gas.

When the gas pressure is *greater* than that of the atmosphere, the gas pushes the mercury higher in the open tube, as we see in Figure 11.4b. The difference in the heights of the two mercury levels, measured in millimeters of mercury, equals the difference between the pressure of the gas and the pressure of the atmosphere. We will call this difference P_{Hg}, and because it's been measured in millimeters, it is the pressure difference in units of torr. Because the gas pressure is higher than that of the atmosphere, we have to *add* P_{Hg} to P_{atm} to calculate the pressure of the gas.

$$P_{gas} = P_{atm} + P_{Hg} \qquad \text{(used when } P_{gas} > P_{atm}) \qquad (11.1)$$

Figure 11.4c illustrates how the apparatus would look when the pressure of the trapped gas is *less* than atmospheric pressure. The higher outside pressure is able to

Below are values of the standard atmosphere (atm) expressed in different pressure units. Studying the table will give you a feel for the sizes of the different units.

> 760 torr
> 101,325 Pa
> 101.325 kPa
> 1.013 bar
> 1013 mb
> 14.7 lb in.$^{-2}$

The last image taken by the National Weather Service radar in Miami, Florida, before the radar transmitter was blown off the roof at 4:35 A.M. on August 24, 1992. It turned out to be the most expensive natural disaster in U.S. history.

Advantages of mercury over other liquids are its low reactivity, its low melting point, and particularly its very high density, which permits short manometer tubes.

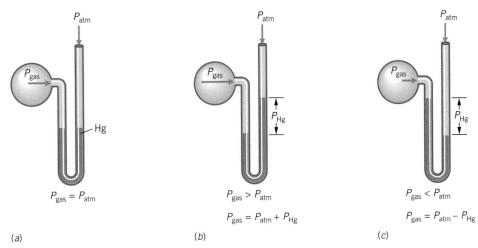

FIGURE 11.4 An open-end manometer. (*a*) The pressure of the entrapped gas, P_{gas}, equals P_{atm}, the atmospheric pressure. (Note that the arrows representing each pressure are here of *equal* length). (*b*) The entrapped gas pressure exceeds that of the atmosphere by an amount equal to the value of P_{Hg} when expressed in the unit of millimeters of mercury or mm Hg (but remember that 1 mm Hg = 1 torr). In other words, $P_{gas} = P_{atm} + P_{Hg}$. (*c*) The gas pressure is less than that of the atmosphere by an amount equal to P_{Hg}, so $P_{gas} = P_{atm} - P_{Hg}$.

force mercury in the direction of the gas bulb, so the level open to the atmosphere is now lower than that exposed to the gas. Now to find the gas pressure we must do a *subtraction*.

$$P_{gas} = P_{atm} - P_{Hg} \qquad \text{(used when } P_{gas} < P_{atm}) \qquad (11.2)$$

EXAMPLE 11.1

Measuring the Pressure of a Gas Using a Manometer

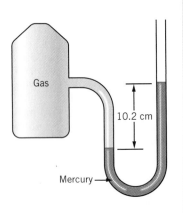

A student collected a gas in an apparatus connected to an open-end manometer, as illustrated in the figure in the margin. The difference in the heights of the mercury in the two columns was 10.2 cm and the atmospheric pressure was measured to be 756 torr. What was the pressure of the gas in the apparatus?

ANALYSIS: From the preceding discussion, we know we will use the atmospheric pressure and either add to it or subtract from it the differences in heights of the mercury column, expressed in pressure units of torr. But which should we do? In a problem of this type, it is best not to try to memorize equations such as 11.1 or 11.2. Instead, we'll use some common sense (something that will help a lot in working problems involving gases).

When we look at the diagram of the apparatus, we see that the mercury is pushed up into the arm of the manometer that's open to the air. Common sense tells us that the pressure of the gas inside must be larger than the pressure of the air outside. Therefore, we will add the pressure difference to the atmospheric pressure.

Finally, before we can do the arithmetic, we must be sure the pressure difference is calculated in torr. We know that 1 torr is equivalent to 1 mm Hg, but the difference in heights was measured in centimeters. Therefore, we have to convert 10.2 cm to millimeters.

SOLUTION: You probably know that 1 cm equals 10 mm. If so, that makes the conversion of the pressure difference easy:

$$10.2 \text{ cm Hg} = 102 \text{ mm Hg}$$

If you weren't sure of the conversion, you could always go back to the definitions of the SI prefixes.

$$1 \text{ cm} = 10^{-2} \text{ m}$$

$$1 \text{ mm} = 10^{-3} \text{ m}$$

$$10.2 \text{ cm} \times \frac{10^{-2} \text{ m}}{1 \text{ cm}} \times \frac{1 \text{ mm}}{10^{-3} \text{ m}} = 102 \text{ mm}$$

Either way, we find the difference in pressures to be 102 mm Hg, which is equivalent to 102 torr.

To find the gas pressure, we add 102 torr to the atmospheric pressure.

$$P_{\text{gas}} = P_{\text{atm}} + P_{\text{Hg}}$$

$$= 756 \text{ torr} + 102 \text{ torr}$$

$$= 858 \text{ torr}$$

The gas pressure is 858 torr.

Is the Answer Reasonable?
The arithmetic is so simple there's not much chance of error here. But we can double-check that we've done the right *kind* of arithmetic (adding or subtracting). Look at the apparatus again. If the gas pressure were lower than atmospheric pressure, it would appear as though the gas were sucking the mercury higher. That's not what we see here. It almost looks like the gas is trying to blow the mercury out of the manometer, so the gas pressure is higher than atmospheric. That's exactly what we've found in our calculation, so we can feel confident we've solved the problem correctly.

PRACTICE EXERCISE 2: In another experiment, it was found that the mercury level in the arm of the manometer attached to the container of gas was 8.7 cm higher than in the arm open to the air. What was the pressure of the gas?

A closed-end manometer avoids the need to measure atmospheric pressure

A **closed-end manometer** (Figure 11.5) is made by sealing the arm that will be farthest from the gas sample and then filling the closed arm completely with mercury. When the gas pressures to be measured are expected to be small, the filled arm can be made short, making the entire apparatus compact. In this design, the mercury is held to the top of the closed arm when the open arm of the manometer is exposed to the atmosphere. When connected to a gas at a low pressure, however, the mercury level in the sealed arm will drop, leaving a vacuum above it. The pressure of the gas can then be measured just by reading the difference in heights of the mercury in the two arms, P_{Hg}. No separate measurement of the atmospheric pressure is required.

Liquids other than mercury can be used in manometers

Mercury is so dense (13.6 g mL^{-1}) that when a gas pressure differs very little from atmospheric pressure, the difference in mercury levels in a manometer is very small and difficult to measure. The precision of measurement therefore suffers. By using a liquid less dense than mercury, the difference in liquid levels is magnified. For example, a pressure difference of 1 torr would show up as a height difference of 1 mm if mercury is used, whereas the height difference would be 13.6 mm if water ($d = 1.0 \text{ g mL}^{-1}$) is used in the manometer. The reason is that it takes a larger volume of the less dense water to have the same weight (and

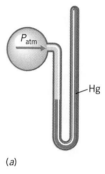

(a)

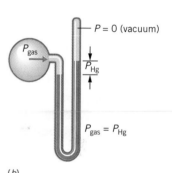

(b)

FIGURE 11.5 *A closed-end manometer for measuring gas pressures less than 1 atm or 760 torr.* (*a*) When constructed, the tube is fully evacuated and then mercury is allowed to enter the tube so that the closed arm fills completely. (*b*) When the tube is connected to a bulb containing a gas at low pressure, the gas pressure is insufficient to keep the mercury level at the top of the closed arm, and mercury flows out of that arm and into the other arm. The mercury comes to rest when the difference in mercury levels, P_{Hg}, represents the pressure of the confined gas: $P_{\text{gas}} = P_{\text{Hg}}$.

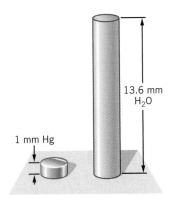

13.6 mm
H₂O

1 mm Hg

FIGURE 11.6 Columns of mercury and water that exert the same pressure. Here we see two columns of liquid, one of mercury and the other of water, both with the same diameter. Because the mercury is 13.6 times as dense as water, the 1 mm high column of mercury weighs just as much as the water column. This means that both columns exert the same pressure (1 torr) on the surface beneath.

therefore exert the same pressure) as the more dense mercury (see Figure 11.6). The net result is that *for a given difference in pressure, the difference in heights between the two levels is inversely proportional to the density of the liquid used in the manometer.* Using *h* for the height difference and *d* for the density, we can express this mathematically as

$$h \propto \frac{1}{d}$$

This inverse proportionality is most often used when we compare the heights of two liquids, *A* and *B*, with different densities used to measure the same gas pressure. The differences in heights, h_A and h_B, are related to the densities, d_A and d_B, as follows.

$$\frac{h_B}{h_A} = \frac{d_A}{d_B}$$

If we know the densities and have measured the height difference h_A, we can calculate the height difference h_B by solving the equation for h_B.

$$h_B = h_A \times \frac{d_A}{d_B} \tag{11.3}$$

We usually use Equation 11.3 when we measure the pressure of a gas with a manometer filled with a liquid other than mercury. In this case, we cannot use the height difference directly to calculate the pressure of the gas. We first have to calculate the height of a mercury column that would exert the *same* pressure as the height difference for the liquid. The calculation is illustrated in Example 11.2.

EXAMPLE 11.2

Using a Fluid Other Than Mercury in a Manometer

A liquid with a density of 0.854 g mL⁻¹ was used in an open-end manometer to measure a gas pressure that was slightly greater than atmospheric pressure (which was 745.0 mm Hg at the time of measurement). In the manometer, the liquid level was 17.5 mm higher in the open arm than in the arm nearest the gas sample (similar to the situation in Figure 11.4b). What was the gas pressure in torr?

ANALYSIS: First, looking at Figure 11.4b, common sense tells us that we will have to add a pressure to the atmospheric pressure to find the pressure of the gas. But the pressure we add must have the unit torr, and the unit torr corresponds to height differences measured in millimeters of *mercury*. Therefore, the critical link in solving this problem is realizing that we have to convert millimeters of liquid (*d* = 0.854 g mL⁻¹) to millimeters of mercury (*d* = 13.6 g mL⁻¹). The tool to accomplish this is Equation 11.3, where liquid *B* is mercury and liquid *A* is the one with a density of 0.854 g mL⁻¹.

SOLUTION: All the hard work has been done. We just have to substitute into Equation 11.3.

$$h_{Hg} = 17.5 \text{ mm} \times \frac{0.854 \text{ g mL}^{-1}}{13.6 \text{ g mL}^{-1}} = 1.10 \text{ mm Hg}$$

The height of the corresponding column of mercury, 1.10 mm, translates into a pressure of 1.10 torr. To find the gas pressure, we add this to the atmospheric pressure, 745.0 torr.

$$P_{gas} = 745.0 \text{ torr} + 1.10 \text{ torr}$$

$$= 746.1 \text{ torr}$$

Is the Answer Reasonable?
To check the answer, we need to remember the inverse proportionality between density and the height of the column required to exert a given pressure. Mercury is much more dense that the liquid in the manometer, so the height of the mer-

cury column needed to exert the same pressure will be much less. That's just what we found in our calculation, so that part of the calculation seems okay. The calculated pressure of the gas is larger than atmospheric, so our final answer makes sense, too.

PRACTICE EXERCISE 3: If water were used instead of mercury to construct a Torricelli barometer (Figure 11.3), at least how long would the tube have to be? Express the answer in millimeters and in feet.

11.3 ▶ The gas laws summarize experimental observations

Earlier, we examined some properties of gases with which you're familiar. Our discussion was only qualitative, however, and now that we've discussed pressure and its units, we are ready to examine gas behavior quantitatively.

There are four variables that affect the properties of a gas—pressure, volume, temperature, and the amount of gas. In this section, we will study situations in which the amount of gas (measured either by grams or moles) remains constant and observe how gas samples respond to changes in pressure, volume, and temperature.

At constant temperature for a fixed amount of gas, the volume is inversely proportional to the pressure

When you inflate a bicycle tire with a hand pump, you squeeze air into a smaller volume and increase its pressure. Packing molecules into a smaller space causes an increased number of collisions with the walls, and because these collisions are responsible for the pressure, the pressure increases (Figure 11.7).

Robert Boyle, an English scientist (1627–1691), performed experiments to determine quantitatively how the volume of a fixed amount of gas varies with pressure. Because the volume is also affected by the temperature, he held that variable constant. When mercury was added to a J-shaped tube containing a trapped gas (at constant temperature), the increasing pressure caused the volume of the trapped gas to decrease, as illustrated in Figure 11.8. A graph of typical data collected in this kind of experiment is shown in Figure 11.9, and demonstrates that *the volume*

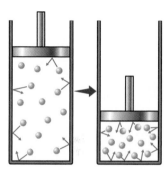

FIGURE 11.7 *Compressing a gas increases its pressure.* A molecular view of what happens when a gas is squeezed into a smaller volume. By crowding the molecules together, the number of collisions with a given area of the walls increases, which causes the pressure to rise.

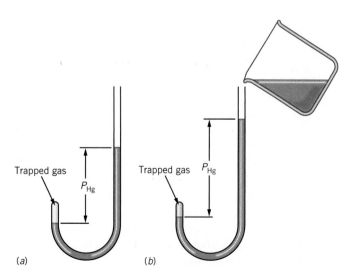

Trapped gas P_{Hg}

Trapped gas P_{Hg}

(a)   (b)

FIGURE 11.8 *Studying the effect of pressure on the volume of a gas.* (*a*) Air is trapped in the end of the J-shaped tube by mercury. (*b*) As more mercury is added, the pressure on the trapped gas increases and the volume decreases.

FIGURE **11.9** *The variation of volume with pressure at constant temperature for a fixed amount of gas.* (*a*) A typical graph of volume versus pressure, showing that as the pressure increases, the volume decreases. (*b*) A straight line is obtained when volume is plotted against $\dfrac{1}{P}$, which shows that $V \propto \dfrac{1}{P}$.

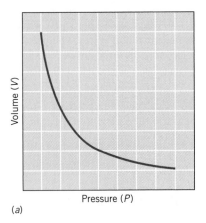

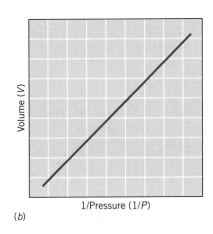

of a given amount of gas held at constant temperature varies inversely with the applied pressure. Mathematically, this can be expressed as

$$V \propto \frac{1}{P} \qquad \text{(temperature and amount of gas held constant)}$$

where V is the volume and P is the pressure. This relationship between pressure and volume is now called **Boyle's law** or the **pressure–volume law.**

In the expression above, the proportionality sign, $\propto$, can be removed by introducing a proportionality constant, C.

$$V = \frac{1}{P} \times C$$

Rearranging gives

$$PV = C$$

This equation tells us, for example, that if at constant temperature the gas pressure is doubled, the gas volume must be cut in half so that the product $P \times V$ doesn't change.

What is remarkable about Boyle's discovery, is that *this relationship is essentially the same for all gases at temperatures and pressures usually found in the laboratory.* As we will see later, neither liquids nor solids have a comparably general law.

An ideal gas would obey Boyle's law perfectly

When very precise measurements are made, Boyle's law doesn't quite work. No gas obeys Boyle's law exactly over a wide range of temperatures and pressures, although most gases come very close to it at temperatures and pressures normally encountered in the lab. It's when a gas is under high pressure (10 atm or more) or at a low temperature (below 200 K) or some combination of these *and is therefore on the verge of changing to a liquid* that the gas fits Boyle's law most poorly. We'll see why later in the chapter.

Although real gases do not *exactly* obey Boyle's law or any of the other gas laws that we'll study, it is often useful to imagine a hypothetical gas that would. We call such a hypothetical gas an *ideal gas*. An **ideal gas** would obey the gas laws exactly over all temperatures and pressures. A real gas behaves more and more like an ideal gas as its pressure decreases and its temperature increases. Most gases we work with in the lab can be treated as ideal gases unless we're dealing with extremely precise measurements.

At constant pressure for a fixed amount of gas, the volume increases with increasing temperature

In 1787 a French chemist and mathematician named Jacques Alexander Charles became interested in hot-air ballooning, which at the time was becoming popular in France. His new interest led him to study what happens to the volume of a sample of gas when the temperature is varied, keeping the pressure constant.

When data from experiments such as his are plotted, a graph like that shown in Figure 11.10 is obtained. Here the volume of the gas is plotted against the temperature in degrees Celsius. The colored points correspond to typical data, and the lines are drawn to most closely fit the data. Each line represents data collected for a different sample. Because all gases eventually condense if cooled sufficiently, the solid portions of the lines correspond to temperatures at which measurements are possible; at lower temperatures the gas condenses. However, if the lines are extrapolated (reasonably extended) back to a point where the volume of the gas would become zero if it didn't condense, all the lines meet at the same temperature, −273.15 °C. Especially significant is the fact that this exact same behavior is exhibited by all gases; when plots of volume versus temperature are extrapolated to zero volume, the temperature axis is always crossed at −273.15 °C. This point represents the temperature at which all gases, if they did not condense, would have a volume of zero and below which they would have a negative volume. Negative volumes are impossible, of course, so it was reasoned that −273.15 °C must be nature's coldest temperature, and it was called **absolute zero.**

As you learned earlier, absolute zero corresponds to the zero point on the Kelvin temperature scale, and to obtain a Kelvin temperature, we add 273.15 °C to the Celsius temperature.[5]

$$T_K = T_{°C} + 273.15$$

For most purposes, we will need only three significant figures, so we can use the following approximate relationship.

$$T_K = T_{°C} + 273$$

Jacques Alexandre César Charles (1746–1823), a French scientist, had a keen interest in hot-air balloons. He was the first to inflate a balloon with hydrogen.

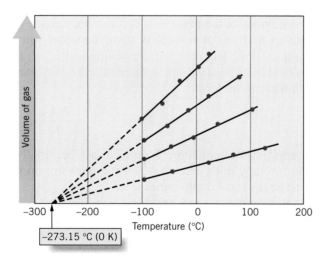

Volume of gas

−300   −200   −100   0   100   200
Temperature (°C)

−273.15 °C (0 K)

FIGURE 11.10 *Charles' law plots.* Each line shows how the gas volume changes with temperature for a different size sample of the same gas.

[5]In Chapter 3 we presented this equation as

$$T_K = T_{°C} \left(\frac{1\ K}{1\ °C} \right) + 273.15\ K$$

to emphasize unit cancelation. Operationally, however, we just add 273.15 to the Celsius temperature to obtain the Kelvin temperature (or we just add 273 if three significant figures are sufficient).

The straight lines in Figure 11.10 suggest that at constant pressure, the volume of a gas is directly proportional to its temperature, provided the temperature is expressed in kelvins. This became known as **Charles' law** (or the **temperature–volume law**) and is expressed mathematically as

$$V \propto T \qquad \text{(pressure and amount of gas held constant)}$$

The value of C' depends on the size and pressure of the gas sample.

Using a constant of proportionality, C', we can write

$$V = C'T \qquad \text{(pressure and amount of gas held constant)}$$

At constant volume for a fixed amount of gas, the pressure is proportional to the absolute temperature

Joseph Louis Gay-Lussac (1778–1850), a French scientist, was a codiscoverer of the element boron.

During the winter of 1801–1802, the French scientist Joseph Louis Gay-Lussac, a skilled and diligent experimentalist, studied how the pressure and temperature of a fixed amount of gas at constant volume are related. (Such conditions exist, for example, when a gas is confined in a vessel with rigid walls, like an aerosol can.) The relationship that he firmly established, called **Gay-Lussac's law** or the **pressure–temperature law,** states that *the pressure of a fixed amount of gas held at constant volume is directly proportional to the Kelvin temperature.* Thus,

$$P \propto T \qquad \text{(volume and amount of gas held constant)}$$

We're using different symbols for the various gas law constants because they are different for each law.

Using still another constant of proportionality, Gay-Lussac's law becomes

$$P = C''T \qquad \text{(volume and amount of gas held constant)}$$

The combined gas law brings together the other gas laws we've studied

The three gas laws we've just examined can be brought together into a single equation known as the **combined gas law,** which states that *the ratio PV/T is a constant for a fixed amount of gas.*

$$\frac{PV}{T} = \text{constant} \qquad \text{(for a fixed amount of gas)}$$

Usually, we use the combined gas law in problems where we know some given set of conditions of temperature, pressure, and volume (for a fixed amount of gas) and wish to find out how one of these variables will change when the others are changed. If we label the initial conditions of P, V, and T with the subscript 1 and the final conditions with the subscript 2, the combined gas law can be written in the following useful form:

Combined gas law

$$\frac{P_1 V_1}{T_1} = \frac{P_2 V_2}{T_2} \tag{11.4}$$

In applying this equation, T must always be in kelvins. The pressure and volume can have any units, but whatever the units are on one side of the equation, they must be the same on the other side.

Equation 11.4 is the combined gas law in the mathematical form most useful for calculations.

It is simple to show that Equation 11.4 contains each of the other gas laws as special cases. Boyle's law, for example, applies when the temperature is constant. Under these conditions, T_1 equals T_2 and temperature cancels from the equation. This leaves us with

$$P_1 V_1 = P_2 V_2 \qquad \text{(when } T_1 \text{ equals } T_2)$$

which is one way to write Boyle's law. Similarly, under conditions of constant pressure, P_1 equals P_2 and the pressure cancels, so Equation 11.4 reduces to

$$V_1/T_1 = V_2/T_2 \qquad \text{(when } P_1 \text{ equals } P_2)$$

This, of course, is another way of writing Charles' law. Under the constant volume conditions required by Gay-Lussac's law, V_1 equals V_2, and Equation 11.4 reduces to

$$P_1/T_1 = P_2/T_2 \qquad \text{(when } V_1 \text{ equals } V_2\text{)}$$

Anesthetic gas is normally given to a patient when the room temperature is 20 °C and the patient's body temperature is 37 °C. What would this temperature change do to 1.60 L of gas if the pressure remains constant?

ANALYSIS: In working a problem about gases, the first step is determining which equation applies to the situation. To use the combined gas law, the problem must deal with changes in the variables P, V, and/or T for a fixed amount of gas. Although not explicitly stated, that's the condition we have here. We can infer a fixed amount of gas from the way the question is asked: "what would this *temperature change* do to *1.60 L of gas*. . . . ?"

To solve the problem, we begin with the combined gas law,

$$\frac{P_1 V_1}{T_1} = \frac{P_2 V_2}{T_2}$$

and then eliminate any variable that doesn't change. Because the pressure is constant, we eliminate P_1 and P_2, which leaves

$$\frac{V_1}{T_1} = \frac{V_2}{T_2}$$

The procedure from this point is straightforward. We will assemble the data for volume and temperature, being careful to change temperatures from degrees Celsius to kelvins. Then we can substitute into the equation and solve for the unknown quantity.

SOLUTION: Let's collect the data, changing degrees Celsius to kelvins. For the variables, "1" refers to the initial conditions and "2" refers to the final ones.

	Initial (1)	Final (2)
V	1.60 L	?
T	293 K (20 + 273)	310 K (37 + 273)

Next, we insert the data into the equation.

$$\frac{1.60 \text{ L}}{293 \text{ K}} = \frac{V_2}{310 \text{ K}}$$

Solving for V_2 gives[6]

$$V_2 = 1.60 \text{ L} \times \frac{310 \text{ K}}{293 \text{ K}}$$

$$= 1.69 \text{ L}$$

The volume increases by 0.09 L, about 90 mL. Although this is only a 6% change, it is a factor that anesthesiologists have to consider in administering anesthetic gases.

---

[6]Multiplying both sides of the equation by 310 K gives

$$V_2 = \frac{1.60 \text{ L} \times 310 \text{ K}}{293 \text{ K}}$$

We've rewritten this to isolate the ratio of temperatures. Notice that temperature units cancel, so we have the correct units for the answer.

Is the Answer Reasonable?

The most important thing we can do to check our answer is to see if the *direction* of the change (from initial to final) is correct. We have a gas sample for which the temperature is increasing. We know gases expand when heated, so the final volume should be larger than the initial volume, and that's what we found. This tells us we've solved the equation correctly for V_2. Second, if we look at the volume-raising ratio of Kelvin temperatures, 310 K/293 K, we see that it is not much larger than 1, so the volume increase will not be very large, and it isn't. Our answer seems to be okay.

PRACTICE EXERCISE 4: A sample of nitrogen has a volume of 880 mL and a pressure of 740 torr. What pressure will change the volume to 870 mL at the same temperature?

EXAMPLE 11.4

Using the Combined Gas Law

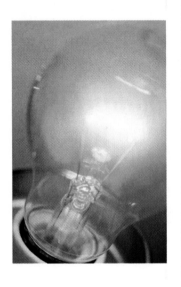

An ordinary incandescent light bulb contains a tungsten filament that becomes white-hot (about 2500 °C) when electricity is passed through it. To prevent the filament from oxidizing, the bulb is filled to a low pressure with the inert (unreactive) gas argon. Suppose a 12.0 L cylinder containing compressed argon at a pressure of 57.8 atm measured at 25 °C is to be used to fill electric light bulbs, each with a volume of 158 mL, to a pressure of 3.00 torr at 20 °C. How many of these light bulbs could be filled by the argon in the cylinder?

ANALYSIS: What do we need to know to figure out how many light bulbs can be filled? If we knew the total volume of gas with a pressure of 3.00 torr at 20 °C, we could just divide by the volume of one light bulb; the result is the number of light bulbs that can be filled. So, our main problem is determining what volume the argon in the cylinder will occupy when its pressure is reduced to 3.00 torr and its temperature is lowered to 20 °C. It's a calculation that will require the combined gas law.

The initial conditions will be 12.0 L of argon at 57.8 atm and 25 °C. The final pressure and temperature will be 3.00 torr and 20 °C. To apply the ideal gas law, the pressure units will have to be the same, so we can convert 57.8 atm to torr. The temperatures will have to be in kelvins, so we add 273 to the Celsius temperatures. We're now ready to perform the math.

SOLUTION: We'll begin by converting 57.8 atm to torr. We know that 1 atm = 760 torr, so

$$57.8 \text{ atm} \times \frac{760 \text{ torr}}{1 \text{ atm}} = 4.39 \times 10^4 \text{ torr}$$

Now we can assemble the data, again using "1" for the initial conditions and "2" for the final ones.

	Initial (1)	Final (2)
P	4.39×10^4 torr	3.00 torr
V	12.0 L	?
T	298 K (25 + 273)	293 K (20 + 273)

Now we have to use the combined gas law.

$$\frac{P_1 V_1}{T_1} = \frac{P_2 V_2}{T_2}$$

To obtain the final volume, we solve for V_2. We'll do this and then rearrange the equation slightly.

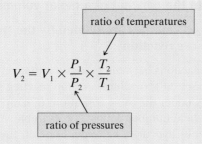

Notice that for V_2 to have the same units as V_1, the units in numerator and denominator of both ratios must cancel. That's the reason we had to convert atmospheres to torr for P_1. Let's now substitute values from our table of data.

$$V_2 = 12.0 \text{ L} \times \frac{4.39 \times 10^4 \text{ torr}}{3.00 \text{ torr}} \times \frac{293 \text{ K}}{298 \text{ K}}$$

$$= 1.73 \times 10^5 \text{ L}$$

Before we can divide by the volume of one light bulb (158 mL), we have to convert the total volume to milliliters.

$$1.73 \times 10^5 \text{ L} \times \frac{1000 \text{ mL}}{1 \text{ L}} = 1.73 \times 10^8 \text{ mL}$$

The volume per light bulb gives us the relationship

$$1 \text{ light bulb} \Leftrightarrow 158 \text{ mL argon}$$

which we can use to find the number of light bulbs that can be filled.

$$1.73 \times 10^8 \text{ mL argon} \times \frac{1 \text{ light bulb}}{158 \text{ mL argon}} = 1.09 \times 10^6 \text{ light bulbs}$$

That's 1.09 million light bulbs! As you can see, there's not much argon in each one.

Is the Answer Reasonable?

Do the sizes of the numbers make sense, and have we substituted correctly into the combined gas law? The first calculation was converting from atm to torr, and we expect the number of torr to be a lot larger than the number of atmospheres, so that calculation seems to be okay. We've also converted from Celsius to Kelvin temperatures, which is important to check, too.

To determine whether we've solved the combined gas law properly, we check to see if the pressure and temperature ratios move the volume in the right direction. Going from the initial to final conditions, the pressure *decreases* from 4.39×10^4 torr to 3.00 torr, so the volume should *increase* a lot. The pressure ratio we used is much larger than 1, so multiplying by it should increase the volume, which agrees with what we expect. Next, look at the temperature change; a *drop in temperature* should tend to *decrease* the volume, so we should be multiplying by a ratio that's less than 1. The ratio 293/298 is less than 1, so that ratio is correct, too.

After applying the combined gas law, the volume was in liters. Did we convert to milliliters correctly? A quick check there tells us we have. Finally, we divided the total volume of argon by the volume of one bulb. The calculation was $(1.73 \times 10^8) \div (1.58 \times 10^2)$, which gives an answer slightly larger than 1×10^6, and that agrees with our answer.

Notice that we use our knowledge of how gases behave and some common sense to determine whether we've done the correct arithmetic. If you learn to do this, you can catch your mistakes.

PRACTICE EXERCISE 5: What will be the final pressure of a sample of nitrogen with a volume of 950 m³ at 745 torr and 25.0 °C if it is heated to 60.0 °C and given a final volume of 1150 m³?

Avogadro's law leads to the concept of a molar volume of a gas

At constant *T* and *P*, gas volumes are related by the stoichiometric coefficients in a gas phase reaction

When scientists studied reactions between gases quantitatively, they made an interesting discovery. If they measured the volumes of the reacting gases, as well as the volumes of gaseous products, under the same conditions of temperature and pressure, the volumes are in simple whole-number ratios. For example, hydrogen gas reacts with chlorine gas to give gaseous hydrogen chloride. Beneath the names in the following equation are the measured volumes with which these gases interact, *provided that the volumes are compared under identical conditions of temperature and pressure:*

$$\text{hydrogen} + \text{chlorine} \longrightarrow \text{hydrogen chloride}$$

| 1 volume | 1 volume | 2 volumes |

What this means is that if we were to use 1.0 L of hydrogen, it would react with 1.0 L of chlorine and produce 2.0 L of hydrogen chloride. If we were to use 10.0 L of hydrogen, all the other volumes would be multiplied by 10 as well.

Similar simple, whole-number ratios by volume are observed when hydrogen combines with oxygen to give water, which is a gas above 100 °C.

$$\text{hydrogen} + \text{oxygen} \longrightarrow \text{water (gaseous)}$$

| 2 volumes | 1 volume | 2 volumes |

Notice that the reacting volumes, *measured under identical temperatures and pressures,* are in ratios of simple, whole numbers.[7]

Observations such as those above led Gay-Lussac to formulate his **law of combining volumes,** which states that *when gases react at the same temperature and pressure, their combining volumes are in ratios of simple, whole numbers.* Much later, we learned that these "simple, whole numbers" are the coefficients of the equations for the reactions.

At a given *T* and *P*, equal volumes of gas contain the same number of molecules

Amedeo Avogadro (1776–1856), an Italian scientist, helped to put chemistry on a quantitative basis.

When Amedeo Avogadro thought about how the volume ratios in the reactions of gases are in *whole* numbers, he was driven to conclude that equal volumes of gases must have identical numbers of molecules (at the same *T* and *P*). Today, we know that "equal numbers of *molecules*" is the same as "equal numbers of *moles*," so Avogadro's insight, now called **Avogadro's principle,** is expressed as follows. *When measured at the same temperature and pressure, equal volumes of gases contain equal numbers of moles* (see Figure 11.11).

We will use Avogadro's principle in Section 11.6 to work problems dealing with the stoichiometry of reactions between gases.

In modern terms, Avogadro's principle means that in reactions of gases, their volumes (at the same *T* and *P*) and their numbers of moles must be in the same ratio. In the equation for the reaction of hydrogen and chlorine, for example, the coefficients are all in the same 1:1:2 ratio as are the numbers of volumes and the numbers of moles.

$$H_2(g) \quad + \quad Cl_2(g) \longrightarrow \quad 2HCl(g)$$

Coefficients	1	1	2
Volumes	1 volume	1 volume	2 volumes (experimental)
Molecules (or moles)	1	1	2 (Avogadro's principle)

[7]The great French chemist Antoine Laurent Lavoisier (1743–1794) was the first to observe the volume relationships of this particular reaction. In his 1789 textbook, *Elements of Chemistry,* he wrote that the formation of water from hydrogen and oxygen requires that two volumes of hydrogen be used for every volume of oxygen. Lavoisier was unable to extend the study of this behavior of hydrogen and oxygen to other gas reactions because he was beheaded during the French Revolution. (See Michael Laing, *The Journal of Chemical Education,* February 1998, page 177.)

	He (at. mass, 4.003)	Ne (at. mass, 20.18)	Ar (at. mass, 39.95)
Number of grams	4.003 g	20.18 g	39.95 g
Number of moles	1.000 mol	1.000 mol	1.000 mol
Number of atoms of gas	6.022×10^{23} atoms	6.022×10^{23} atoms	6.022×10^{23} atoms
Pressure	1.00 atm	1.00 atm	1.00 atm
Temperature	273 K	273 K	273 K
Volume	22.4 L	22.4 L	22.4 L
Density	0.179 g L^{-1}	0.901 g L^{-1}	1.78 g L^{-1}

FIGURE 11.11 *Avogadro's principle.* When measured at the same temperature and pressure, equal volumes of gases contain equal numbers of moles. Shown here are data for three Group VIIIA elements, helium, neon, and argon, all under identical temperatures and pressures. Notice that 1 mol of each gas, or Avogadro's number of gaseous atoms, occupies the same volume. The reason why the numbers of grams and densities differ is that the masses of the individual atoms are not the same.

A corollary to Avogadro's principle is that *the volume of a gas is directly proportional to its number of moles, n.*

$$V \propto n \quad \text{(at constant } T \text{ and } P\text{)}$$

Avogadro's principle was a remarkable advance in our understanding of gases. His insight enabled chemists for the first time to determine the formulas of gaseous elements.[8]

The standard molar volume is the volume occupied by one mole of gas at STP

Avogadro's principle implies that the volume occupied by 1 mol of *any* gas—its *molar volume*—must be identical for all gases under the same conditions of pressure and temperature. To compare the molar volumes of different gases, scientists agreed to use 1 atm and 273.15 K (0 °C) as the **standard conditions of temperature and pressure,** or **STP,** for short. If we measure the molar volumes for a variety of gases at STP, we find that the values fluctuate somewhat because the gases are not "ideal." Some typical values are shown in Table 11.1 and if we were to examine the data for many gases, we would find an average of around 22.4 L per mole. This

TABLE 11.1	MOLAR VOLUMES OF SOME GASES AT STP	
Gas	Molar Formula	Volume (L)
Helium	He	22.398
Argon	Ar	22.401
Hydrogen	H_2	22.410
Nitrogen	N_2	22.413
Oxygen	O_2	22.414
Carbon dioxide	CO_2	22.414

[8]Suppose that hydrogen chloride, for example, is correctly formulated as HCl, not as H_2Cl_2 or H_3Cl_3 or higher, and certainly not as $H_{0.5}Cl_{0.5}$. Then the only way that *two* volumes of hydrogen chloride could come from just *one* volume of hydrogen and *one* of chlorine is if each particle of hydrogen and each of chlorine were to consist of *two* atoms of H and Cl, respectively, H_2 and Cl_2. If these particles were single-atom particles, H and Cl, then one volume of H and one volume of Cl could give only *one* volume of HCl, not two. Of course, if the initial assumption were incorrect so that hydrogen chloride is, say, H_2Cl_2 instead of HCl, then hydrogen would be H_4 and chlorine would be Cl_4. The extension to larger subscripts works in the same way.

value is taken to be the molar volume of an *ideal gas* at STP and is now called the **standard molar volume** of a gas.

For an ideal gas at STP:

$$1 \text{ mol gas} \Longleftrightarrow 22.4 \text{ L gas}$$

11.4 ▶ The ideal gas law relates *P*, *V*, *T*, and the number of moles of gas, *n*

In our discussion of the combined gas law, we noted that the ratio *PV/T* equals a constant for a fixed amount of gas. However, the value of this "constant" is actually proportional to the number of moles of gas, *n*, in the sample.[9]

To create an equation even more general than the combined gas law, therefore, we can write

$$\frac{PV}{T} \propto n$$

We can replace the proportionality symbol, ∝, with an equal sign by including another proportionality constant.

$$\frac{PV}{T} = n \times \text{constant}$$

This new constant is given the symbol **R** and is called the **universal gas constant.** We can now write the combined gas law in a still more general form called the **ideal gas law.**

$$\frac{PV}{T} = nR$$

An ideal gas would obey this law exactly over all ranges of the gas variables. The equation, sometimes called the **equation of state for an ideal gas,** is usually rearranged and written as follows:

In some references, it is called the universal gas law.

 TOOLS

Ideal gas law

If n, P, and T in Equation 11.5 are known, for example, then V can have only one value.

Ideal Gas Law (Equation of State for an Ideal Gas)

$$PV = nRT \tag{11.5}$$

Equation 11.5 tells us how the four important variables for a gas—*P*, *V*, *n*, and *T*—are related. If we know the values of three, we can calculate the fourth. In fact, Equation 11.5 tells us that if values for three of the four variables are fixed for a given gas, *the fourth can only have one value.* We can define the *state* of a given gas simply by specifying any three of the four variables.

To use the ideal gas law, we have to know the value of the universal gas constant, *R*. To calculate it, we use the standard conditions of pressure and temperature, and we use the standard molar volume, which sets *n* equal to 1 mol, the number of moles of the sample. We still have to decide what units to use for pressure and volume, and the value of *R* differs with these choices. Our choices for this chapter are to express volumes in *liters* and pressures in *atmospheres.* Thus for one

[9]In the problems we worked earlier, we were able to use the combined gas law expressed as

$$\frac{P_1V_1}{T_1} = \frac{P_2V_2}{T_2}$$

because the amount of gas remained fixed.

molar volume at STP, $n = 1$ mol, $V = 22.4$ L, $P = 1.00$ atm, and $T = 273$ K. Using these values let's calculate R as follows:

$$R = \frac{PV}{nT} = \frac{(1.00 \text{ atm})(22.4 \text{ L})}{(1.00 \text{ mol})(273 \text{ K})}$$

$$= 0.0821 \frac{\text{atm L}}{\text{mol K}}$$

Or, arranging the units in a commonly used order,

$$R = 0.0821 \text{ L atm mol}^{-1} \text{ K}^{-1}$$

More precise measurements give $R = 0.082057$ L atm mol^{-1} K^{-1}.

To use this value of R in working problems, we have to be sure to express volumes in liters and pressures in atmospheres. And, of course, temperatures must be expressed in kelvins.

In Example 11.4, we described filling a 158 mL light bulb with argon at a temperature of 20 °C and a pressure of 3.00 torr. How many grams of argon are in the light bulb under these conditions?

EXAMPLE 11.5

Using the Ideal Gas Law

ANALYSIS: The critical link in solving this and many other problems involving gases is recognizing that whenever we have P, V, and T for a gas, we have enough information to calculate the number of moles of gas. The tool to accomplish this is the ideal gas law equation, $PV = nRT$. We will solve it for n and substitute values for P, V, R, and T. Once we have the number of moles of argon, a simple moles-to-grams conversion using the atomic mass will give us the mass of Ar in grams.

In performing the calculation, we will have to be sure to convert the volume to liters, pressure to atmospheres, and the temperature to kelvins so the units will cancel correctly.

SOLUTION: To use $R = 0.0821$ L atm mol^{-1} K^{-1}, we must have V in liters, P in atmospheres, and T in kelvins. Gathering the data and making the necessary unit conversions as we go, we have

$P = 3.95 \times 10^{-3}$ atm   From: 3.00 torr $\times \dfrac{1 \text{ atm}}{760 \text{ torr}}$

$V = 0.158$ L   From: 158 mL

$T = 293$ K   From: (20 °C + 273)

Solving the ideal gas law for n gives us

$$n = \frac{PV}{RT}$$

Substituting the proper values of P, V, R, and T into this equation gives

$$n = \frac{(3.95 \times 10^{-3} \text{ atm})(0.158 \text{ L})}{(0.0821 \text{ L atm mol}^{-1} \text{ K}^{-1})(293 \text{ K})}$$

$$= 2.59 \times 10^{-5} \text{ mol of Ar}$$

The atomic mass of Ar is 39.95, so 1 mol Ar = 39.95 g Ar. A conversion factor made from this relationship lets us convert "mol Ar" into "g Ar."

$$2.59 \times 10^{-5} \text{ mol Ar} \times \frac{39.95 \text{ g Ar}}{1 \text{ mol Ar}} = 1.04 \times 10^{-3} \text{ g Ar}$$

Thus, the light bulb contains only about one-thousandth of a gram of argon.

Is the Answer Reasonable?
Approximate arithmetic would not be easy in this problem, but we can get some idea of whether our answers seem reasonable based on some common sense. First,

the pressure of the gas, 3.00 torr, is very low and we expect it to be a small fraction of 1 atm. The value we obtained, 3.95×10^{-3} atm, seems reasonable. We can also expect that the amount of argon in the bulb will also be quite small, so our final answer also makes sense. As a final check, we can look to be sure the units cancel correctly, and we see that they do. (Whenever you're working a problem where there is unit cancellation, be sure to check to be sure the units do cancel as they are supposed to.)

PRACTICE EXERCISE 6: How many grams of argon were in the 12.0 L cylinder of argon used to fill the light bulbs described in Example 11.4? The pressure of the argon was 57.8 atm and the temperature was 25 °C.

Molecular mass can be calculated from measurements of *P, V, T,* and mass

When a chemist makes a new compound, its molecular mass is usually determined to help establish its chemical identity. In general, *to determine the molecular mass of a compound experimentally, we need to find two pieces of information about a given sample—the mass of the sample and the number of moles of the substance in the sample.* (This is the critical link in working *any* problem involving the experimental determination of molecular mass.) Once we have mass and moles for the same sample, we simply divide the number of grams by the number of moles to find the molecular mass. For instance, if we had a sample weighing 6.40 g and found that it also contained 0.100 mol of the substance, the molecular mass would be

When the molecular mass of a substance is expressed in units of grams per mole, the quantity is often called the *molar mass* and has the units g mol^{-1}.

$$\frac{6.40 \text{ g}}{0.100 \text{ mol}} = 64.0 \text{ g mol}^{-1}$$

If the compound is a gas, its molecular mass can be found using experimental values of pressure, volume, temperature, and sample mass. The *P, V, T* data allow us to calculate the number of moles using the ideal gas law (as in Example 11.5). Once we know the number of moles of gas and the mass of the gas sample, the molecular mass is obtained by taking the ratio of *grams to moles*.

EXAMPLE 11.6
Determining the Molecular Mass of a Gas

As part of a rock analysis, a student added hydrochloric acid to a rock sample and observed a fizzing action, indicating a gas was being evolved (see the figure in the margin). The student collected a sample of the gas in a 0.220 L gas bulb until its pressure reached 0.757 atm at a temperature of 25.0 °C. The sample weighed 0.299 g. What is the molecular mass of the gas? What kind of mineral probably exists in the rock?

ANALYSIS: The strategy for finding the molecular mass was described above. We use the *P, V, T* data to calculate the number of moles of gas in the sample. To do this, we solve the ideal gas law for *n* and then substitute values, being sure the units will cancel correctly. Then we divide the mass by the number of moles to find the molecular mass.

SOLUTION: Pressure is already in atmospheres and the volume is in liters, but we must convert degrees Celsius into kelvins. Gathering our data, we have

$$P = 0.757 \text{ atm} \qquad V = 0.220 \text{ L} \qquad T = 298 \text{ K} \quad (25.0 \text{ °C} + 273)$$

Next, we solve the ideal gas law ($PV = nRT$) for the number of moles, *n*.

$$n = \frac{PV}{RT}$$

Hydrochloric acid reacting with a rock sample.

Now we can substitute the data for *P, V,* and *T,* along with the value of *R*. This gives

$$n = \frac{(0.757 \text{ atm})(0.220 \text{ L})}{(0.0821 \text{ L atm mol}^{-1} \text{ K}^{-1})(298 \text{ K})}$$

$$= 6.81 \times 10^{-3} \text{ mol}$$

The molecular mass is obtained from the ratio of grams to moles.

$$\text{molecular mass} = \frac{0.299 \text{ g}}{6.81 \times 10^{-3} \text{ mol}} = 43.9 \text{ g mol}^{-1}$$

We now know the measured molecular mass is 43.9, but what gas could this be? Looking back on our discussions in Chapter 5, what gases do we know are given off when a substance reacts with acids? On page 164 we find some options; the gas might be H_2S, HCN, CO_2, or SO_2. Using atomic masses to calculate their molecular masses we get

H_2S	34 g mol^{-1}	CO_2	44 g mol^{-1}
HCN	27 g mol^{-1}	SO_2	64 g mol^{-1}

The only gas with a molecular mass close to 43.9 is CO_2, and that gas would be evolved if we treat a carbonate with an acid. The rock probably contains a carbonate mineral. (Limestone and marble are examples of such minerals.)

Is the Answer Reasonable?
We don't have to do much mental arithmetic to gain some confidence in our answer. The volume of the gas sample (0.220 L) is a bit less than 1/100 of 22.4 L, the volume of 1 mol at STP. If the gas were at STP, then 0.299 g would be 1/100 of a mole, so an entire mole would weigh about 30 g (rounded from 29.9 g). Although 30 g per mole is not very close to 43.9 g per mole, we do appear to have the decimal in the right place. In addition, the fact that the answer (43.9 g mol^{-1}) agrees so well with the molecular mass of one of the gases formed when substances react with acids makes all the puzzle pieces fit. Viewed in total, therefore, we can feel confident in our answers.

PRACTICE EXERCISE 7: The label on a cylinder of a noble gas became illegible, so a student allowed some of the gas to flow into a gas bulb with a volume of 300.0 mL until the pressure was 685 torr. The sample now weighed 1.45 g; its temperature was 27.0 °C. What is the molecular mass of this gas? Which of the Group VIIIA gases (the noble gases) was it?

Gas densities depend on molecular masses

Because 1 mol of any gas occupies the same volume at a particular pressure and temperature, the mass contained in that volume depends on the molecular mass of the gas. Consider, for example, 1 mol samples of O_2 and CO_2 at STP (Figure 11.12). Each sample occupies a volume of 22.4 L. The oxygen sample has a mass of 32.0 g while the carbon dioxide sample has a mass of 44.0 g. If we calculate the densities of the gases, we find the density of CO_2 is larger than that of O_2.

Recall from Chapter 3 that density is the ratio of mass to volume. To make the numbers easier to comprehend, gas densities are usually expressed in grams per liter.

$$d_{O_2} = \frac{32.0 \text{ g}}{22.4 \text{ L}} = 1.43 \text{ g L}^{-1} \qquad d_{CO_2} = \frac{44.0 \text{ g}}{22.4 \text{ L}} = 1.96 \text{ g L}^{-1}$$

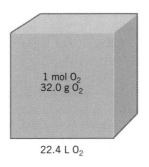

1 mol O_2
32.0 g O_2

22.4 L O_2

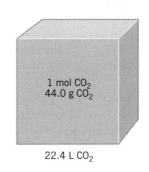

1 mol CO_2
44.0 g CO_2

22.4 L CO_2

FIGURE 11.12 *One mole samples of O_2 and CO_2 at STP.* Each sample occupies 22.4 L, but the O_2 weighs 32.0 g, whereas the CO_2 weighs 44.0 g. The CO_2 has more mass per unit volume than the O_2 and has the higher density.

Because the volume of a gas is affected by temperature and pressure, the density of a gas changes as these variables change. Gases become less dense as their temperatures rise, which is why hot-air balloons are able to float; the less dense hot air in the balloon floats in the more dense cool air that surrounds it. Gases also become more dense as their pressures increase because increasing the pressure packs more molecules into the same space. To calculate the density of a gas at conditions other than STP, we use the ideal gas law, as illustrated in Example 11.7.

EXAMPLE 11.7

Calculating the Density of a Gas

One procedure used to separate the isotopes of uranium to obtain material to construct a nuclear weapon employs a uranium compound with the formula UF_6. The compound boils at about 56 °C, so at 100 °C it is a gas. What is the density of UF_6 at 100 °C if the pressure of the gas is 740 torr? (Assume the gas contains the mix of uranium isotopes commonly found in nature.)

ANALYSIS: In our discussion above, you learned that if the gas were at STP, we could calculate the density by dividing the mass of 1 mol of the gas by its molar volume, 22.4 L. We can do a similar calculation here if we first calculate the molar volume at 100 °C and 740 torr. To do this, we can use the ideal gas law, setting n equal to 1.00 mol and converting the pressure to atmospheres and temperature to kelvins.

SOLUTION: The pressure is 740 torr, so to convert to atmospheres, we use the relationship between the atmosphere and torr.

$$740 \text{ torr} \times \frac{1 \text{ atm}}{760 \text{ torr}} = 0.974 \text{ atm}$$

The temperature is 100 °C, which converts to 373 K, and we said we would use $n = 1.00$ mol. To find the molar volume, we solve the ideal gas law for V.

$$V = \frac{nRT}{P}$$

Now we substitute values.

$$V = \frac{1.00 \text{ mol} \times 0.0821 \text{ L atm mol}^{-1} \text{ K}^{-1} \times 373 \text{ K}}{0.974 \text{ atm}} = 31.4 \text{ L}$$

At these conditions of temperature and pressure, the molar volume is 31.4 L per mole. To calculate the density, we need one more piece of data, the molecular mass of UF_6. Adding up the atomic masses, we get a molecular mass of 352.0 g mol^{-1}. The density is therefore

$$d_{UF_6} = \frac{352.0 \text{ g}}{31.4 \text{ L}} = 11.2 \text{ g L}^{-1}$$

Is the Answer Reasonable?
At STP, the density would be equal to 352 g ÷ 22.4 L = 15.7 g L^{-1}. The pressure is almost 1 atm, but the temperature is quite a bit larger than 273 K. Gases expand when heated, so a liter of the gas at the higher temperature will have less UF_6 in it. That means the density will be lower at the higher temperature, and our answer agrees with this analysis. It is probably correct.

PRACTICE EXERCISE 8: Sulfur dioxide is a gas that's been used in commercial refrigeration but not in residential refrigeration, because it is toxic. If your refrigerator used SO_2, you could be injured if it developed a leak. What is the density of SO_2 gas measured at −20 °C and a pressure of 96.5 kPa?

FACETS OF CHEMISTRY 11.1

Gas Densities and Exploding Boats

A pleasant summer afternoon turns to tragedy as a boat explodes in flames when the owner attempts to start the engine after refueling the vessel with gasoline. Unfortunately, such accidents are not uncommon and can be avoided by pulling fresh air through the engine compartment of the boat with an exhaust fan prior to starting the engine. Another precaution that's recommended is to keep the craft's doors and windows closed during refueling. But why, besides its flammability, does gasoline pose such a threat on boats? Refueling your car is no way near as dangerous even though you're using the same fuel.

The answer has to do with the density of gasoline vapor relative to the density of air. Gas densities affect how gases behave when one is added to another. Although all gases will completely mix eventually, when a dense gas is first added to a less dense gas, it will sink. This is similar to placing a high-density "heavy" solid such as iron into a low-density liquid such as water; the high-density solid sinks.

Gasoline is composed of hydrocarbons such as C_7H_{16} and C_8H_{18}, which have relatively high molecular masses (100 g mol^{-1} for C_7H_{16} and 114 g mol^{-1} for C_8H_{18}). Because the density of a gas is proportional to its molecular mass, gasoline vapor is about three times more dense than air (which has an average molecular mass of about 29 g mol^{-1}). If some gasoline spills during refueling, vapors of the fuel will settle into the lowest parts of the boat, which is where the engine is located. When the engine is started, a spark can cause the fuel to ignite explosively. If this happens, the boat is often a total loss and people are injured or even killed. The problem can be avoided by excluding vapors from the inside of the boat (keeping doors and windows closed during refueling) and by exhausting any vapors with a safely wired fan before attempting to start the engine. It's also a good idea to do a "sniff" check, just to be sure no gasoline fumes have accumulated.

When you fill the gas tank of your car, the hazards are minimized because any gasoline vapors tend to drift away, rather than being confined to the engine space. As a result, there is very little danger of an explosion when you start your car.

Failure to follow safe practices during refueling can lead to tragedy. It is likely that this accident could have been avoided if gasoline vapors had been exhausted from the engine compartment before the motors were started.

Gas densities can be used to calculate molecular masses

One of the ways we can use the density of a gas is to determine the molecular mass. To do this, we also need to know the temperature and pressure at which the density was measured. Example 11.8 illustrates the reasoning and calculation involved.[10]

[10]From the ideal gas law we can derive an equation from which we could calculate the molar mass directly from the density. If we let the mass of gas equal *g*, we could calculate the number of moles, *n*, by the ratio

$$n = \frac{g}{\text{molar mass}}$$

Substituting into the ideal gas law gives

$$PV = nRT = \frac{gRT}{\text{molar mass}}$$

Solving for molar mass, we have

$$\text{molar mass} = \frac{gRT}{PV} = \left(\frac{g}{V}\right) \times \frac{RT}{P}$$

The quantity *g*/*V* is the ratio of mass to volume, which is the density *d*, so making this substitution gives

$$\text{molar mass} = \frac{dRT}{P}$$

This equation could also be used to solve the problem in Example 11.8 by substituting values for *d*, *R*, *T*, and *P*.

EXAMPLE 11.8

Calculating the Molecular Mass from Gas Density

A liquid sold under the name Perclene is used as a dry-cleaning solvent. It has an empirical formula CCl_2 and a boiling point of 121 °C. When vaporized, the gaseous compound has a density of 4.93 g L^{-1} at 785 torr and 150 °C. What is the molecular mass of the compound and what is its molecular formula?

ANALYSIS: Earlier you saw that for gases, we can calculate the molecular mass if we have P, V, and T data for a weighed sample of the substance. We use the P, V, T data with the ideal gas law to calculate n, and then divide the number of grams by the number of moles. But, in this problem we seem to be given values only for P and T. How will we obtain the other data we need (V and grams of gas)? The answer comes from what the density tells us. A value of 4.93 g L^{-1} translates to 4.93 g per liter, which means that if we had 1.00 L of the gas, it would have a mass of 4.93 g. Thus, the density gives the other two data items we need.

$$1.00 \text{ L gas} \Leftrightarrow 4.93 \text{ g gas}$$

After we've calculated the molecular mass, we find the molecular formula following the procedure discussed in Chapter 4. We divide the molecular mass by the empirical formula mass to find the factor by which we multiply the subscripts in the empirical formula to obtain the molecular formula.

SOLUTION: To calculate the number of moles of gas in 1.00 L, we use the ideal gas law, making unit conversions as needed. The data are

$$T = 423 \text{ K} \qquad \text{From } (150 \text{ °C} + 273)$$

$$P = 1.03 \text{ atm} \qquad \text{From } 785 \text{ torr} \times \frac{1 \text{ atm}}{760 \text{ torr}}$$

$$V = 1.00 \text{ L}$$

Solving the ideal gas law for n and substituting values, we have

$$n = \frac{PV}{RT} = \frac{(1.03 \text{ atm})(1.00 \text{ L})}{(0.0821 \text{ L atm mol}^{-1} \text{ K}^{-1})(423 \text{ K})}$$

$$= 2.97 \times 10^{-2} \text{ mol}$$

Now we can calculate the molecular mass by dividing the mass, 4.93 g, by the number of moles, 2.97×10^{-2} mol.

$$\frac{4.93 \text{ g}}{2.97 \times 10^{-2} \text{ mol}} = 166 \text{ g mol}^{-1}$$

The molecular mass of the compound is thus 166 g mol^{-1}.

The empirical formula mass that we calculate from the empirical formula, CCl_2, is 82.9. We now divide the molecular mass by this value to see how many times CCl_2 occurs in the molecular formula.

$$\frac{166}{82.9} = 2.00$$

To find the molecular formula, we multiply the subscripts of the empirical formula by 2.

$$\text{molecular formula} = C_{1 \times 2}Cl_{2 \times 2} = C_2Cl_4$$

(This is the formula for a compound commonly called *tetrachloroethylene*, which is indeed used as a dry-cleaning fluid.)

Are the Answers Reasonable?

Sometimes we don't have to do any arithmetic to see that an answer is almost surely correct. The fact that the molecular mass we calculated from the gas density is evenly divisible by the empirical formula mass suggests we've worked the problem correctly.

PRACTICE EXERCISE 9: A gaseous compound of phosphorus and fluorine with an empirical formula of PF_2 has a density of 5.60 g L^{-1} at 23.0 °C and 750 torr. Calculate its molecular mass and its molecular formula.

11.5 ▶ In a mixture each gas exerts its own partial pressure

So far in our discussions we've dealt with only pure gases. Gas mixtures, however, are not at all uncommon. In fact, we live our lives surrounded by a mixture of gases we call air. In general, gas mixtures obey the same laws as pure gases, so Boyle's law applies equally to both pure oxygen and to air. There are times, though, when we must be concerned with the composition of a gas mixture, such as when we are concerned with a pollutant in the atmosphere. In these cases, the variables affected by the composition of a gas mixture are the numbers of moles of each component and the contribution each component makes to the total observed pressure. Because gases mix completely, all the components of the mixture occupy the same volume—that of the container holding them. Furthermore, the temperature of each gaseous component is the same as the temperature of the entire mixture. Therefore, in a gas mixture, each of the components have the same volume and the same temperature.

The composition of a mixture can be expressed in mole fractions and mole percents

One of the useful ways of describing the composition of a mixture is in terms of the *mole fractions* of the components. The **mole fraction** is *the ratio of the number of moles of a given component to the total number of moles of all components.* Expressed mathematically, the mole fraction of substance A in a mixture of A, B, C, . . . , Z substances is

The concept of mole fraction applies to any uniform mixture in any physical state—gas, liquid, or solid.

$$X_A = \frac{n_A}{n_A + n_B + n_C + n_D + \cdots + n_Z} \qquad (11.6)$$

 TOOLS
Mole fractions

where X_A is the mole fraction of component A, and $n_A, n_B, n_C, . . . , n_Z$ are the numbers of moles of each component, A, B, C, . . . , Z, respectively. The sum of all mole fractions for a mixture must always equal 1.

You can see in Equation 11.6 both numerator and denominator have the same units (moles), so they cancel. As a result, a mole fraction has no units. Nevertheless, always remember the definition: a mole fraction stands for the ratio of *moles* of one component to the total number of *moles* of all components.

Sometimes the mole fraction composition of a mixture is expressed on a percentage basis; we call it a **mole percent (mol%).** The mole percent is obtained by multiplying the mole fraction by 100 mol%.

What is the mole fraction (and mole percent) of each component of a mixture composed of 0.200 mol O_2 and 0.500 mol N_2?

ANALYSIS: There's not much to analyze here. We simply calculate the mole fraction according to the definition. Mole percent, as described above, is mole fraction multiplied by 100 mol%.

SOLUTION: Let's begin with oxygen.

$$X_{O_2} = \frac{\text{moles of } O_2}{\text{moles of } O_2 + \text{moles of } N_2}$$

$$= \frac{0.200 \text{ mol}}{0.200 \text{ mol} + 0.500 \text{ mol}}$$

$$= \frac{0.200 \text{ mol}}{0.700 \text{ mol}} = 0.286$$

Similarly, for N_2 we have

$$X_{N_2} = \frac{0.500 \text{ mol}}{0.200 \text{ mol} + 0.500 \text{ mol}} = 0.714$$

EXAMPLE 11.9
Calculating Mole Fractions

Container wall

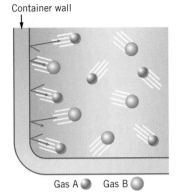

Gas A ● Gas B ○

FIGURE 11.13 *Partial pressures viewed at the molecular level.* In a mixture of two gases, *A* and *B*, both collide with the walls of the container and thereby contribute their partial pressures to the total pressure.

The air we breathe is a common mixture of nonreacting gases. At least under ordinary temperatures and pressures, nitrogen and oxygen do not react.

Notice that the sum of the mole fractions (0.286 + 0.714) equals 1.00. In fact, we could have obtained the mole fraction of nitrogen with less calculation by subtracting the mole fraction of oxygen from 1.00.

$$X_{O_2} + X_{N_2} = 1.00$$

$$X_{N_2} = 1.00 - X_{O_2}$$

$$= 1.00 - 0.286 = 0.714$$

Finally, we calculate the mole percents as

$$\text{mol\% } O_2 = 0.286 \times 100 \text{ mol\%} = 28.6 \text{ mol\%}$$

$$\text{mol\% } N_2 = 0.714 \times 100 \text{ mol\%} = 71.4 \text{ mol\%}$$

Are the Answers Reasonable?
The check in this kind of calculation is to make sure the mole fractions add up to 1 and that the mole percents add up to 100. They do, so the answers seem to be correct (especially because we calculated the mole fractions independently for both O_2 and N_2).

Each gas in a mixture exerts its own pressure that contributes to the total pressure

In a mixture of nonreacting gases such as the air around us, each gas contributes to the total pressure in proportion to the fraction (by moles) in which it is present (see Figure 11.13). This contribution to the total pressure is called the **partial pressure** of the gas. It is the pressure the gas would exert if it were the only gas in a container of the same size at the same temperature.

The general symbol we will use for the partial pressure of gas *A* is P_A. For a particular gas, the formula of the gas may be put into the subscript, as in P_{O_2}. What John Dalton discovered about partial pressures is now called **Dalton's law of partial pressures:** *the total pressure of a mixture of gases is the sum of their individual partial pressures.* In equation form, the law is

Dalton's law of partial pressures

$$P_{\text{total}} = P_A + P_B + P_C + \cdots \tag{11.7}$$

In dry CO_2-free air at STP, for example, P_{O_2} is 159.12 torr, P_{N_2} is 593.44 torr, and P_{Ar} is 7.10 torr. These partial pressures add up to 759.66 torr, just 0.34 torr less than 760 torr or 1.00 atm. The remaining 0.34 torr is contributed by several trace gases, including other noble gases.

EXAMPLE 11.10

Using Dalton's Law of Partial Pressures

Suppose you want to fill a pressurized tank having a volume of 4.00 L with oxygen-enriched air for use in diving, and you want the tank to contain 50.0 g of O_2 and 150 g of N_2. What will the total gas pressure have to be at 25 °C?

ANALYSIS: We're dealing with a mixture of gases and their pressures, which is the clue that tells us that our tool for solving the problem is Dalton's law of partial pressures. If we calculate the pressure that each gas would have if it were *alone* in the tank, these will be their partial pressures in the mixture. We can then add the partial pressures to find the answer. To find the individual partial pressures, we apply the ideal gas law using a temperature of 298 K (for 25 °C) and a volume of 4.00 L. To use this law, however, we must first convert the amounts of the gases in grams to numbers of moles, a routine grams-to-moles conversion.

SOLUTION:

For moles of oxygen, $\quad n = 50.0 \text{ g } O_2 \times \dfrac{1 \text{ mol } O_2}{32.00 \text{ g } O_2} = 1.56 \text{ mol } O_2$

For moles of nitrogen, $\quad n = 150 \text{ g } N_2 \times \dfrac{1 \text{ mol } N_2}{28.00 \text{ g } N_2} = 5.36 \text{ mol } N_2$

Next, we calculate the partial pressures. For each component,

$$P_{component} = \frac{nRT}{V}$$

where V is the volume of the tank.

$$P_{O_2} = \frac{1.56\ \text{mol} \times 0.0821\ \text{L atm mol}^{-1}\ \text{K}^{-1} \times 298\ \text{K}}{4.00\ \text{L}}$$

$$= 9.54\ \text{atm}$$

$$P_{N_2} = \frac{5.36\ \text{mol} \times 0.0821\ \text{L atm mol}^{-1}\ \text{K}^{-1} \times 298\ \text{K}}{4.00\ \text{L}}$$

$$= 32.8\ \text{atm}$$

For the total pressure, using Dalton's law, we have

$$P_{total} = 9.54\ \text{atm} + 32.8\ \text{atm}$$

$$= 42.3\ \text{atm}$$

Is the Answer Reasonable?

Once you've seen that the *total* number of moles of gases is about 7 mol, you'll know that this many moles would occupy 7 mol × 22.4 L/mol, or close to 160 L at STP. Compressing a gas from 160 L to 4 L, or 1/40th as much, would require increasing the pressure from 1 atm to 40 times as much, which is rather close to our answer.

PRACTICE EXERCISE 10: How many grams of oxygen are present at 25 °C in a 5.00 L tank of oxygen-enriched air under a total pressure of 30.0 atm when the only other gas is nitrogen at a partial pressure of 15.0 atm?

Collecting gases over water

When gases that do not react with water are prepared in the laboratory, they can be trapped over water by an apparatus like that shown in Figure 11.14. Because of the way the gas is collected, it is saturated with water vapor. Water vapor in a mixture of gases has a partial pressure like that of any other gas.

The vapor present in the space above *any* liquid always contains some of the liquid's vapor, which exerts its own pressure, called the liquid's **vapor pressure.** Its value for any given substance depends only on the temperature. The vapor pressures of water at different temperatures, for example, are given in Table 11.2.

When we collect a (wet) gas over water, we usually want to know how much *dry* gas this corresponds to, and we can use data for the wet gas to find out. From Dalton's law we can write

Even the mercury in a barometer has a tiny vapor pressure—0.0012 torr at 20 °C—which is much too small to affect readings of barometers and the manometers studied in this chapter.

$$P_{total} = P_{gas} + P_{water\ vapor}$$

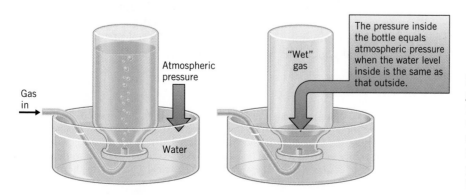

Gas in

Atmospheric pressure

Water

"Wet" gas

The pressure inside the bottle equals atmospheric pressure when the water level inside is the same as that outside.

FIGURE 11.14 *Collecting a gas over water.* As the gas bubbles through the water, water vapor goes into the gas, so the total pressure inside the bottle includes the partial pressure of the water vapor at the temperature of the water.

TABLE 11.2	VAPOR PRESSURE OF WATER AT VARIOUS TEMPERATURES		
Temperature (°C)	Vapor Pressure (torr)	Temperature (°C)	Vapor Pressure (torr)
0	4.579	50	92.51
5	6.543	55	118.0
10	9.209	60	149.4
15	12.79	65	187.5
20	17.54	70	233.7
25	23.76	75	289.1
30	31.82	80	355.1
35	42.18	85	433.6
37[a]	47.07	90	527.8
40	55.32	95	633.0
45	71.88	100	760.0

[a]Human body temperature.

Solving for P_{gas}, we have

$$P_{gas} = P_{total} - P_{water\ vapor}$$

We find values for $P_{water\ vapor}$ at various temperatures from Table 11.2. If we have adjusted the height of the collecting jar so the water level inside matches that outside, the total pressure of the trapped gas equals the atmospheric pressure, so the value for P_{total} is obtained from the laboratory barometer. We thus calculate P_{gas}, which is the pressure that the gas would exert if it were dry and inside the same volume that was used to collect it.[11]

EXAMPLE 11.11

Collection of a Gas over Water

A sample of oxygen is collected over water at 20 °C and a pressure of 738 torr. Its volume is 310 mL. (a) What is the partial pressure of the oxygen? (b) What would be its volume when dry at STP?

ANALYSIS: There are two parts to this problem, so before we get started, let's determine which of the gas laws will be our tools for solving them. It seems pretty clear that for part (a) we will need Dalton's law of partial pressures. We have O_2 collected over water, so we know the container holds both O_2 and $H_2O(g)$. For part (b), we have to determine how the volume of the O_2 collected will change when the conditions change to those of STP. We're not changing the amount of O_2. The amount of O_2 is fixed (constant); it just undergoes changes in P, V, and T. To work problems of this kind, our tool is the combined gas law.

To solve part (a), we will have to subtract the vapor pressure of water from the total pressure of the gas mixture (738 torr). That's all there is to it.

To solve part (b), we substitute into the combined gas law that's been solved for V_2. Our starting conditions will be 310 mL, 20 °C, and the pressure obtained in part (a). The final conditions will have a temperature of 0 °C and a pressure of 760 torr.

---

[11]If the water levels are not the same inside the flask and outside, a correction has to be calculated and applied to the room pressure to obtain the true pressure in the flask. For example, if the water level is higher inside the flask than outside, the pressure in the flask is lower than atmospheric pressure. The difference in levels is in millimeters of *water*, so this has to be converted to the equivalent in millimeters of mercury, using the approach of Example 11.2, before the room pressure is corrected.

SOLUTION: To calculate the partial pressure of the oxygen, we use Dalton's law. We will need the vapor pressure of water at 20 °C, which we find in Table 11.2 to be 17.5 torr.

$$P_{O_2} = P_{total} - P_{water\ vapor}$$

$$= 738\ torr - 17.5\ torr = 720\ torr$$

The answer to part (a) is that the partial pressure of O_2 is 720 torr.
   For part (b), we'll begin by assembling the data.

	Initial (1)	Final (2)
P	720 torr (which is P_{O_2})	760 torr (standard pressure)
V	310 mL	?
T	293 K (20.0 °C + 273)	273 K (standard temperature)

We use these in the combined gas law equation:

$$\frac{P_1 V_1}{T_1} = \frac{P_2 V_2}{T_2}$$

Solving for V_2 and rearranging the equation a bit, we have

$$V_2 = V_1 \times \left(\frac{P_1}{P_2}\right) \times \left(\frac{T_2}{T_1}\right)$$

Now we can substitute values and calculate V_2.

$$V_2 = 310\ mL \times \left(\frac{720\ torr}{760\ torr}\right) \times \left(\frac{273\ K}{293\ K}\right)$$

$$= 274\ mL$$

Thus, when the water vapor is removed from the gas sample, the dry oxygen will occupy a volume of 274 mL at STP.

Are the Answers Reasonable?
We know the pressure of the dry O_2 will be less than that of the wet gas, so the answer to part (a) seems reasonable. To check part (b), we can see if the pressure and temperature ratios move the volume in the right direction. The pressure is increasing (720 torr → 760 torr), which should tend to lower the volume. The pressure ratio above will do that. The temperature change (293 K → 273 K) should also lower the volume, and once again, the temperature ratio above will have that effect. Our answer to part (b) is probably okay.

PRACTICE EXERCISE 11: Suppose you prepared a sample of nitrogen and collected it over water at 15 °C at a total pressure of 745 torr and a volume of 310 mL. Find the partial pressure of the nitrogen and the volume it would occupy at STP.

Mole fractions of gases are related to partial pressures

Partial pressure data can be used to calculate the mole fractions of individual gases in a gas mixture because the number of *moles* of each gas is directly proportional to its partial pressure. We can demonstrate this as follows. The partial pressure, P_A, for any one gas A in a gas mixture with a total volume V at a temperature T is found by the ideal gas law equation, $PV = nRT$. So to calculate the number of moles of A present, we have

$$n_A = \frac{P_A V}{RT}$$

For any particular gas mixture at a given temperature, the values of V, R, and T are all constants, making the ratio V/RT a constant, too. We can therefore simplify the previous equation by using C to stand for V/RT. In other words, we can write

$$n_A = P_A C$$

The result is the same as saying that *the number of moles of a gas in a mixture of gases is directly proportional to the partial pressure of the gas*. The constant C is the same for all gases in the mixture. So by using different letters to identify individual gases, we can let $P_B C$ stand for n_B, $P_C C$ stand for n_C, and so on in Equation 11.6. Thus,

$$X_A = \frac{P_A C}{P_A C + P_B C + P_C C + \cdots + P_Z C}$$

The constant, C, can be factored out and canceled, so

$$X_A = \frac{P_A}{P_A + P_B + P_C + \cdots + P_Z}$$

The denominator is the sum of the partial pressures of all the gases in the mixture, but this sum equals the total pressure of the mixture (Dalton's law of partial pressures). Therefore, the previous equation simplifies to

$$X_A = \frac{P_A}{P_{total}} \tag{11.8}$$

Thus, the mole fraction of a gas in a gas mixture is simply the ratio of its partial pressure to the total pressure. Equation 11.8 also gives us a simple way to calculate the partial pressure of a gas in a gas mixture when we know its mole fraction.

EXAMPLE 11.12

Using Partial Pressure Data to Calculate Mole Fractions

What are the mole fractions and mole percents of nitrogen and oxygen in air when their partial pressures are 160 torr for oxygen and 600 torr for nitrogen? Assume no other gases are present.

ANALYSIS: This calls for Equation 11.8, which gives us the mole fraction of a gas from its partial pressure and the total pressure. But we must first calculate P_{total}, which is just the sum of the partial pressures (Equation 11.7). When we have P_{total}, we can apply Equation 11.8 to each gas.

SOLUTION: First we find the total pressure.

$$P_{total} = 600 \text{ torr} + 160 \text{ torr} = 760 \text{ torr}$$

We use Equation 11.8 first for the mole fraction of N_2.

$$X_{N_2} = \frac{P_{N_2}}{P_{total}} = \frac{600 \text{ torr}}{760 \text{ torr}}$$

$$= 0.789, \text{ or } 78.9 \text{ mol\% } N_2$$

We do the same for oxygen.

$$X_{O_2} = \frac{P_{O_2}}{P_{total}} = \frac{160 \text{ torr}}{760 \text{ torr}}$$

$$= 0.211, \text{ or } 21.1 \text{ mol\% } O_2$$

Are the Answers Reasonable?
The mole fractions must add up to 1 and the mole percents must add up to 100 mol%, and they do.

PRACTICE EXERCISE 12: What is the mole fraction and the mole percent of oxygen in exhaled air if P_{O_2} is 116 torr and P_{total} is 760 torr?

Partial pressures can be calculated from mole fractions

As we said, if we can calculate mole fractions from partial pressures, it must be equally simple to calculate partial pressures from mole fractions. A mountain climber, for example, might be interested in this kind of calculation before working at high altitudes, wondering uneasily about the partial pressure of oxygen. We breathe easily at or near sea level because our lungs and the hemoglobin concentration of our blood are adapted to the partial pressure of oxygen in air, roughly 160 torr, at sea level. At the summit of Mount Everest (elevation 8848 m), however, the total atmospheric pressure is only about 250 torr. *The mole fractions of oxygen and nitrogen in air are essentially constant at all altitudes in the lower atmosphere*, so the mole fraction of oxygen, 0.211, is the same at sea level as at the top of Mount Everest. For any gas in air, what changes with altitude is not the mole fraction but the partial pressure, because proportionately fewer numbers of moles of the gas are present in a given volume. Thus, the value of P_{O_2} at any altitude is some fraction, the mole fraction, of the atmospheric pressure there. So for oxygen on top of Mount Everest, the value of P_{O_2} must be the mole fraction of O_2 times the atmospheric pressure there.

$$P_{O_2} = X_{O_2} P_{total}$$

$$P_{O_2} = 0.211 \times 250 \text{ torr} = 52.8 \text{ torr} \qquad \text{(at Everest's summit)}$$

11.6 Gas volumes are used in solving stoichiometry problems

For reactions involving gases, Avogadro's principle lets us use a new kind of stoichiometric equivalency, one between *volumes* of gases. Earlier, for example, we noted the following reaction and its gas volume relationships:

$$2H_2(g) + O_2(g) \longrightarrow 2H_2O(g)$$

2 volumes   1 volume   2 volumes

Provided we are dealing with gas volumes measured at the same temperature and pressure, we can write the following stoichiometric equivalencies:

2 volumes $H_2(g) \Leftrightarrow 1$ volume $O_2(g)$ just as   2 mol $H_2 \Leftrightarrow 1$ mol O_2

2 volumes $H_2(g) \Leftrightarrow 2$ volumes $H_2O(g)$ just as   2 mol $H_2 \Leftrightarrow 2$ mol H_2O

1 volume $O_2(g) \Leftrightarrow 2$ volumes $H_2O(g)$ just as   1 mol $O_2 \Leftrightarrow 2$ mol H_2O

Relationships such as these can greatly simplify stoichiometry problems, as we see in the next example.

The recognition that equivalencies in gas *volumes* are numerically the same as those for numbers of moles of gas in reactions involving gases simplifies many calculations.

How many liters of hydrogen, measured at STP, are needed to combine exactly with 1.50 L of nitrogen, also measured at STP, to form ammonia?

ANALYSIS: The critical link in solving this problem is simply the fact that the gases are at the same temperature and pressure. That makes the ratios by volume the same as the ratio by moles, which means that the ratios by volume are the same as the ratios of the coefficients of the balanced equation.

We can restate the problem as

$$1.50 \text{ L N}_2 \Leftrightarrow ? \text{ L H}_2$$

We're dealing with a chemical reaction, so we need the chemical equation.

$$3H_2(g) + N_2(g) \longrightarrow 2NH_3(g)$$

Avogadro's principle gives us the equivalency we need to perform the calculation.

$$3 \text{ volumes H}_2 \Leftrightarrow 1 \text{ volume N}_2$$

EXAMPLE 11.13

Stoichiometry of Reactions of Gases

SOLUTION: The volume of hydrogen is given in liters, so we need to express the volume equivalency in those units as well.

$$3 \text{ L } H_2 \Longleftrightarrow 1 \text{ L } N_2$$

We then multiply the given, 1.50 L N_2, by a conversion factor made from the equivalency above.

$$\text{volume of } H_2 = 1.50 \text{ L } N_2 \times \frac{3 \text{ L } H_2}{1 \text{ L } N_2}$$

$$= 4.50 \text{ L } H_2$$

Is the Answer Reasonable?
The volume of H_2 needed is three times the volume of N_2, and 3 × 1.5 equals 4.5, so the answer is correct. Remember, however, that the simplicity of this problem arises because the volumes are at the same temperature and pressure.

PRACTICE EXERCISE 13: Methane burns according to the following equation:

$$CH_4(g) + 2O_2(g) \longrightarrow CO_2(g) + 2H_2O(g)$$

The combustion of 4.50 L of CH_4 consumes how many liters of O_2, both volumes measured at 25 °C and 740 torr?

In the preceding example, volumes were measured at the same temperature and pressure, but this is not always the case, as illustrated in the following example.

EXAMPLE 11.14

Stoichiometry Calculations Involving Gases

Nitrogen monoxide, a pollutant released by automobile engines, is oxidized by molecular oxygen to give the reddish-brown gas nitrogen dioxide, which gives smog its characteristic color. The equation is

$$2NO(g) + O_2(g) \longrightarrow 2NO_2(g)$$

How many milliliters of O_2, measured at 20 °C and 755 torr, are needed to react with 180 mL of NO, measured at 45 °C and 720 torr?

ANALYSIS: Once again, we have a stoichiometry problem, but it's more complicated than the preceding problem because the gases are not at the same temperature and pressure. There are two ways we can proceed.

Path 1

We could use the ideal gas law and the P, V, and T for NO to find moles of NO. Then the coefficients of the equation would be used to find moles of O_2. Finally, we would use the ideal gas law and the P, T, and moles of O_2 to find the volume of O_2.

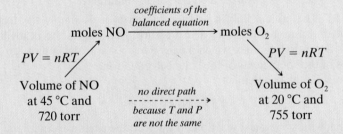

Path 2

We could use the combined gas law to find what volume the NO would occupy if it were at the same temperature and pressure as the O_2. Then we would use the coefficients of the equation to find the volume of O_2.

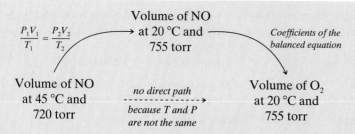

Both path 1 and path 2 will work, but path 2 looks easier, so let's proceed that way.

SOLUTION: The first step is to use the combined gas law applied to the given volume of NO, so let's set up the data as usual.

	Initial (1)	Final (2)
P	720 torr	755 torr
V	180 mL	?
T	318 K (45 °C + 273)	293 K (20 °C + 273)

Solving the combined gas law for V_2 gives

$$V_2 = V_1 \times \left(\frac{P_1}{P_2} \right) \times \left(\frac{T_2}{T_1} \right)$$

Next we substitute values.

$$V_2 = 180 \text{ mL} \times \left(\frac{720 \text{ torr}}{755 \text{ torr}} \right) \times \left(\frac{293 \text{ K}}{318 \text{ K}} \right)$$

$$= 158 \text{ mL NO}$$

Now we have the volume of NO *at the same temperature and pressure as the oxygen.* This lets us use the coefficients of the equation to establish the equivalency

$$2 \text{ mL NO} \Longleftrightarrow 1 \text{ mL O}_2$$

and apply it to find the volume of O_2 required for the reaction.

$$158 \text{ mL NO } \times \frac{1 \text{ mL O}_2}{2 \text{ mL NO}} = 79.0 \text{ mL O}_2$$

Is the Answer Reasonable?
We've selected the shorter method for solving the problem, which reduces the opportunity for error. For the first part of the calculation, we can check to see whether the pressure and temperature ratios move the volume in the right direction. The pressure is increasing (702 torr → 755 torr), so that should have a lowering effect on the volume. The pressure ratio is smaller than 1, so it is having the proper effect. The temperature is dropping (318 K → 293 K), so the change should also have a volume-lowering effect. The temperature ratio is smaller than 1, so it is also having the correct effect. The volume of NO at 20 °C and 755 torr is probably correct.

The check of the second part of the calculation is simple. According to the equation, the volume of O_2 required should be half the volume of NO, which it is, so the final answer seems to be okay.

PRACTICE EXERCISE 14: Solve the preceding problem using the method that involves the ideal gas law to show that the same answer is obtained by both methods.

Not all stoichiometry problems requiring the use of the gas laws are reactions between gases, as illustrated in Example 11.15.

An important chemical reaction in the manufacture of Portland cement is the high-temperature decomposition of calcium carbonate to give calcium oxide and carbon dioxide.

$$CaCO_3(s) \longrightarrow CO_2(g) + CaO(s)$$

Suppose a 1.25 g sample of $CaCO_3$ is decomposed by heating. How many milliliters of CO_2 gas will be evolved if it will be measured at 740 torr and 25 °C?

ANALYSIS: The balanced equation tells us

$$1 \text{ mol } CaCO_3 \Leftrightarrow 1 \text{ mol } CO_2$$

Once we calculate the *moles* of $CaCO_3$, we can use this value for n in the ideal gas law equation to find the gas volume.

SOLUTION: The formula mass of $CaCO_3$ is 100.1, so

$$\text{moles of } CaCO_3 = 1.25 \text{ g } CaCO_3 \times \frac{1 \text{ mol } CaCO_3}{100.1 \text{ g } CaCO_3}$$

$$= 1.25 \times 10^{-2} \text{ mol } CaCO_3$$

Because $1 \text{ mol } CaCO_3 \Leftrightarrow 1 \text{ mol } CO_2$,

$$n = 1.25 \times 10^{-2} \text{ mol } CO_2$$

Before we use n in the ideal gas law equation, we must convert the given pressure and temperature into the units required by R.

$$P = 740 \text{ torr} \times \frac{1 \text{ atm}}{760 \text{ torr}} = 0.974 \text{ atm} \qquad T = 298 \text{ K} \quad (25\,°\text{C} + 273)$$

By rearranging the ideal gas law equation, we obtain

$$V = \frac{nRT}{P}$$

$$= \frac{(1.25 \times 10^{-2} \text{ mol})(0.0821 \text{ L atm mol}^{-1}\text{K}^{-1})(298 \text{ K})}{0.974 \text{ atm}}$$

$$= 0.314 \text{ L} = 314 \text{ mL}$$

The reaction will yield 314 mL of CO_2 at the conditions specified.

Is the Answer Reasonable?
Let's use some approximate arithmetic to see if the answer is in the right ballpark. The reaction generates somewhat more than 0.0100 mol of CO_2. At STP, 0.0100 mol would occupy 1/100th of the standard molar volume, or 0.224 L. We expect somewhat more than this, so 0.314 L is probably okay. At least we can feel confident the decimal is in the right place.

PRACTICE EXERCISE 15: In one lab, the gas-collecting apparatus used a gas bulb with a volume of 250 mL. How many grams of $Na_2CO_3(s)$ reacting with hydrochloric acid would be needed to prepare enough $CO_2(g)$ to fill this bulb to a pressure of 738 torr at a temperature of 23 °C? The equation is

$$Na_2CO_3(s) + 2HCl(aq) \longrightarrow 2NaCl(aq) + CO_2(g) + H_2O$$

11.7 ▶ Effusion and diffusion in gases lead to Graham's law

If you've ever had a pet that's had an encounter with a skunk, you've learned first-hand about diffusion! **Diffusion** is the spontaneous intermingling of the molecules of one gas, like those of skunk "perfume," with molecules of another gas, like the air in a room (see Figure 11.15a). **Effusion,** on the other hand, is the gradual move-

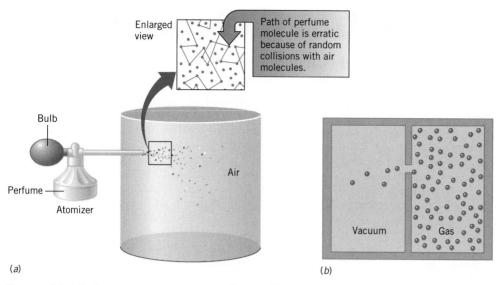

FIGURE 11.15 *Spontaneous movements of gases.* (*a*) Diffusion. (*b*) Effusion.

ment of gas molecules through a very tiny hole into a vacuum (Figure 11.15*b*). The rates at which both of these processes occur depend on the speeds of gas molecules; the faster the molecules move, the more rapidly diffusion and effusion occur.

The Scottish chemist Thomas Graham (1805–1869), studied the rates of diffusion and effusion of a variety of gases through porous clay pots and through small apertures. Comparing different gases at the same temperature and pressure, Graham found that their rates of effusion were inversely proportional to the square roots of their densities. This relationship is now called **Graham's law.**

$$\text{effusion rate} \propto \frac{1}{\sqrt{d}} \quad \text{(when compared at the same } T \text{ and } P\text{)}$$

Graham's law is usually used in comparing the rates of effusion of different gases, so the proportionality constant can be eliminated and an equation can be formed by taking the ratio of effusion rates.

$$\frac{\text{effusion rate } (A)}{\text{effusion rate } (B)} = \frac{\sqrt{d_B}}{\sqrt{d_A}} = \sqrt{\frac{d_B}{d_A}} \tag{11.9}$$

By taking the ratio, the proportionality constant cancels from numerator and denominator.

Earlier you saw that the density of a gas is directly proportional to its molecular mass. Therefore, we can re-express Equation 11.9 as follows:

$$\frac{\text{effusion rate } (A)}{\text{effusion rate } (B)} = \sqrt{\frac{d_B}{d_A}} = \sqrt{\frac{M_B}{M_A}} \tag{11.10}$$

Graham's law of effusion

where M_A and M_B are the molecular masses of gases A and B.

Graham's law works for diffusion as well as effusion. It tells us that gases with low molecular masses will effuse (and diffuse) more rapidly than gases with high molecular masses. Thus, hydrogen with a molecular mass of 2 will effuse more rapidly than methane, CH_4, with a molecular mass of 16.

At a given temperature and pressure, which effuses more rapidly and by what factor, ammonia or hydrogen chloride?

ANALYSIS: A gas effusion problem requires the use of Graham's law, which says that the gas with the smaller molecular mass will effuse more rapidly. So we need to find the molecular masses. Then the relative rates of effusion are found by using them in Equation 11.10.

EXAMPLE 11.16

Using Graham's Law

FACETS OF CHEMISTRY 11.2

Effusion and Nuclear Energy

The fuel used in almost all nuclear reactors is uranium, but only one of its naturally occurring isotopes, ^{235}U, can be easily split to yield energy. Unfortunately, this isotope is present in a very low concentration (about 0.72%) in naturally occurring uranium. Most of the element as it is mined consists of the more abundant isotope ^{238}U. Therefore, before uranium can be fabricated into fuel elements, it must be enriched to a ^{235}U concentration of about 2 to 5%. Enrichment requires that the isotopes be separated, at least to some degree.

Separating the uranium isotopes is not feasible by chemical means because the chemical properties of both isotopes are essentially identical. Instead, a method is re-quired that is based on the very small difference in the masses of the isotopes. As it happens, uranium forms a compound with fluorine, UF_6, that is easily vaporized at a relatively low temperature. The UF_6 gas thus formed con-sists of two kinds of molecules, $^{235}UF_6$ and $^{238}UF_6$, with molecular masses of 349 and 352, respectively. Because of their different masses, their rates of effusion are slightly different; $^{235}UF_6$ effuses 1.0043 times faster than $^{238}UF_6$. Although the differnce is small, it is sufficient to enable enrichment, provided the effusion is carried out over and over again enough times. In fact, it takes over 1400 sepa-rate diffusion chambers arranged one after another to achieve the necessary level of enrichment.

SOLUTION: The molecular masses are 17.03 for NH_3 and 36.46 for HCl, so we know that NH_3, with its smaller molecular mass, effuses more rapidly than HCl. The ratio of the effusion rates is given by

$$\frac{\text{effusion rate }(NH_3)}{\text{effusion rate }(HCl)} = \sqrt{\frac{M_{HCl}}{M_{NH_3}}}$$

$$= \sqrt{\frac{36.46}{17.03}} = 1.463$$

We can rearrange the result as

$$\text{effusion rate }(NH_3) = 1.463 \times \text{effusion rate }(HCl)$$

Thus, ammonia effuses 1.463 times more rapidly than HCl under the same condi-tions.

Is the Answer Reasonable?
The only quick check is to be sure that the arithmetic tells us ammonia effuses more rapidly than the HCl, and that's what our result tells us. In fact, the result of this cal-culation is confirmed by the experiment shown in Figure 11.16.

PRACTICE EXERCISE 16: The hydrogen halide gases all have the same general formula, HX, where X can be Cl, Br, or I. If HCl(g) effuses 1.88 times more rapidly than one of the others, which hydrogen halide is the other, HBr or HI?

11.8 ▶ **The kinetic-molecular theory explains the gas laws**

Back in the nineteenth century, scientists who already knew the gas laws couldn't help but ask, "How do gases 'work'?" They wondered what had to be true about all gases to explain their conformity to a common set of gas laws. The **kinetic theory of gases** was the answer. We introduced some of its ideas in Chapter 7, and at the be-ginning of this chapter we described a number of observations that suggest what gases must be like when viewed at a molecular level. Let's look more closely now at the kinetic theory to see how well it explains the behavior of gases.

The strategy for answering the question about the nature of gases began with postulating a model of an ideal gas. Then the laws of physics and statistics were

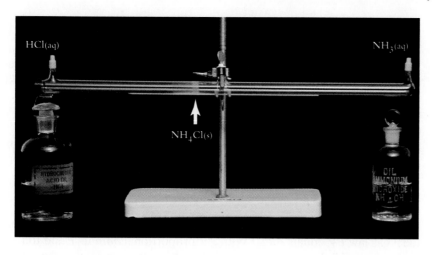

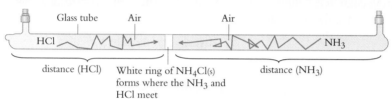

FIGURE 11.16 *Rates of diffusion of ammonia and hydrogen chloride.* Concentrated aqueous NH_3 and HCl were introduced at opposite ends of the glass tube. The two gases, released from their solutions, diffused toward each other through the air in the tube. Where they met, they reacted to form solid NH_4Cl, which deposited on the wall of the tube. From the location of the NH_4Cl, we can see that the NH_3 had moved farther down the tube than the HCl, which demonstrates that NH_3 diffuses faster in air than HCl.

applied to see whether the observed gas laws could be predicted from the model. The results were splendidly successful.

Postulates of the Kinetic Theory of Gases

1. A gas consists of an extremely large number of very tiny particles that are in constant, random motion.

2. The gas particles themselves occupy a net volume so small in relation to the volume of their container that their contribution to the total volume can be ignored.

3. The particles often collide in perfectly elastic collisions[12] with themselves and with the walls of the container, and they move in straight lines between collisions, neither attracting nor repelling each other.

The particles are assumed to be so small that they have no dimensions at all.

In summary, the model pictures an ideal gas as a collection of supersmall, constantly moving billiard balls that continually bounce off each other and the walls of their container and so exert a net pressure on the walls (as described in Figure 11.1, page 452). The gas particles are assumed to be so small that their individual volumes can be ignored, so an ideal gas is effectively all empty space.

The gas laws are predicted by the kinetic theory

According to the model, gases are mostly empty space. As we noted earlier, this explains why gases, unlike liquids and solids, can be compressed so much (squeezed to smaller volumes). It also explains why we have gas laws for gases, and *the same laws for all gases,* but not comparable laws for liquids or solids. The chemical identity of the gas does not matter, because gas molecules do not touch each other except when they collide, and there are extremely weak interactions, if any, between them.

We cannot go over the mathematical details, but we can describe some of the ways in which the model of an ideal gas used in the kinetic theory helps us understand the gas laws.

---

[12]In *perfectly elastic* collisions, no energy is lost by friction as the colliding objects deform momentarily.

Lower pressure

(a)

Higher pressure

(b)

FIGURE 11.17 *The kinetic theory and the pressure–volume law (Boyle's law).* When the gas volume is made smaller in going from (a) to (b), the frequency of the collisions per unit area of the container's walls increases. Therefore, the pressure increases.

Kinetic theory relates temperature to average kinetic energy

The greatest triumph of the kinetic theory came with its explanation of gas temperature, which we discussed in Section 7.2. What the calculations showed was that the product of gas pressure and volume, PV, is proportional to the average kinetic energy of the gas molecules.

$$PV \propto \text{average molecular KE}$$

But from the experimental study of gases, culminating in the equation of state for an ideal gas, we have another term to which PV is proportional, namely, the Kelvin temperature of the gas.

$$PV \propto T$$

(We know what the proportionality constant here is, namely, nR, because by the ideal gas law PV equals nRT.) With PV proportional *both* to T and to the "average molecular KE," then it must be true that the temperature of a gas is proportional to the average molecular KE.

$$T \propto \text{average molecular KE}$$

Kinetic theory explains the pressure–volume law (Boyle's law)

Using the model of an ideal gas, physicists were able to demonstrate that gas pressure is the net effect of innumerable collisions made by gas particles with the walls of the container. Let's imagine that one wall of a gas container is a movable piston that we can push in (or pull out) and so change the volume (see Figure 11.17). If we make the volume smaller without changing the temperature, no change occurs in the average kinetic energy of the molecules. But now there would be more gas particles per unit volume. If we reduce the volume by one-half, for example, we double the number of molecules per unit volume. This would double the number of collisions per second with a unit area and therefore double the pressure. Thus, cutting the volume in half forces the pressure to double, which is exactly what Boyle discovered:

$$P \propto \frac{1}{V} \quad \text{or} \quad V \propto \frac{1}{P}$$

Kinetic theory explains the pressure–temperature law (Gay-Lussac's law)

The kinetic theory, as we've said, tells us that an increase in gas temperature increases the average velocity of gas particles. At higher velocities, the particles must strike the container's walls more frequently and with greater force. But at constant volume, the *area* being struck is still the same, so the force per unit area—the pressure—must therefore increase (see Figure 11.18). In this way the kinetic theory explains how the pressure of a fixed amount of gas is proportional to temperature (at constant volume), which is the pressure–temperature law of Gay-Lussac.

Kinetic theory explains the temperature–volume law (Charles' law)

The kinetic theory tells us that increasing the temperature of a gas increases the average kinetic energy of its particles, which tends to increase the pressure of the gas. The only way we can keep P constant (a condition of Charles' law) is to have the gas expand (see Figure 11.19). Therefore, a gas expands with increasing T in order to keep P constant, which is another way of saying that V is proportional to T at constant P. Thus, the kinetic theory explains Charles' law.

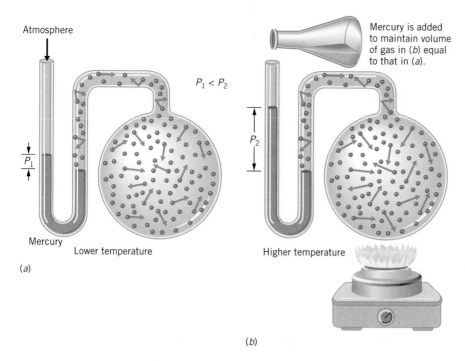

FIGURE 11.18 *The kinetic theory and the pressure–temperature law (Gay-Lussac's law).* (*a*) The gas exerts only a small pressure, P_1, over and above the atmospheric pressure. (*b*) Raising the gas temperature raises the pressure to P_2, with the extra mercury being added to keep the gas from expanding its volume. At the higher temperature the gas particles move with greater average velocity (as suggested by the longer motion arrows), so they strike the walls with greater force and cause the pressure to increase. To prevent the gas from expanding, more mercury has to be added. The size of the pressure increase is shown by the greater difference between the heights of the mercury columns.

Kinetic theory explains Dalton's law of partial pressure

The law of partial pressures is actually evidence for that part of the third postulate in the kinetic theory that pictures particles of an ideal gas moving in straight lines between collisions, neither attracting nor repelling each other (see Figure 11.20). By not interacting with each other, the molecules act *independently,* so each gas behaves as though it were alone in the container. Only if the particles of each gas do act independently can the partial pressures of the gases add up in a simple way to give the total pressure.

Kinetic theory explains Graham's law of effusion

The key conditions of Graham's law are that the rates of effusion of two gases with different molecular masses must be compared at the same pressure and tempera-

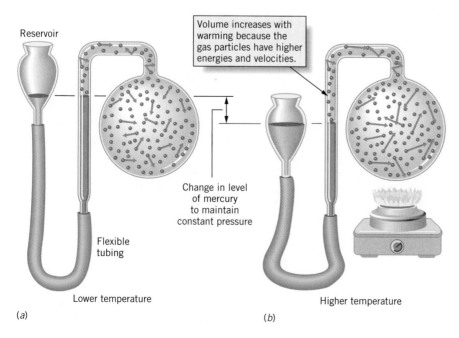

FIGURE 11.19 *The kinetic theory and the temperature–volume law (Charles' law).* The pressure is the same in both (*a*) and (*b*), as indicated by the mercury levels. At the higher temperature (*b*), the gas particles have a higher average kinetic energy and velocity, and therefore strike the walls with greater force. To hold the pressure constant (a condition of Charles' law), the gas has to be given more room by letting some mercury flow out of the tube and back into the reservoir.

FIGURE 11.20 *Gas molecules act independently when they neither attract nor repel each other.* Gas molecules would not travel in straight lines if they attracted each other as in (*a*) or repelled each other as in (*b*). This would influence the length of time between collisions with the walls and would therefore affect the collision frequency with the walls. In turn, this would affect the pressure each gas would exert. Only if the molecules traveled in straight lines with no attractions or repulsions, as in (*c*), would their individual pressures not be influenced by near misses or by collisions between the molecules.

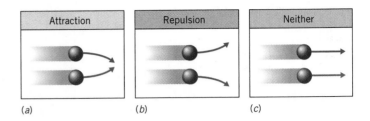

ture and under conditions where the gas molecules do not hinder each other. When two gases have the same temperature, their particles have identical average molecular kinetic energies. Using subscripts 1 and 2 to identify two gases with molecules having different masses m_1 and m_2, we can write that at a given temperature

$$\overline{KE_1} = \overline{KE_2}$$

where the bar over KE signifies "average."

For a single molecule, its kinetic energy is $KE = \frac{1}{2}mv^2$. For a large collection of molecules of the same substance, the average kinetic energy is $\overline{KE} = \frac{1}{2}m\overline{v^2}$, where $\overline{v^2}$ is the *average of the velocities squared* (called the *mean square* velocities). We have not extended the "average" notation (the bar) over the mass because all the molecules of a given substance have the same mass (the average of their masses is just the mass).

Once again comparing two gases, 1 and 2, we take $\overline{v_1^2}$ and $\overline{v_2^2}$ to be the average of the velocities squared of their molecules. If both gases are at the same temperature, we have

$$\overline{KE_1} = \frac{1}{2}m_1\overline{v_1^2} = \frac{1}{2}m_2\overline{v_2^2} = \overline{KE_2}$$

Now let's rearrange the previous equation to get the ratio of $\overline{v^2}$ terms.

$$\frac{\overline{v_1^2}}{\overline{v_2^2}} = \frac{m_2}{m_1}$$

Next, we'll take the square root of both sides. When we do this, we obtain a ratio of quantities called the **root mean square** (abbreviated **rms**) speeds, which we will represent as $(\overline{v_1})_{rms}$ and $(\overline{v_2})_{rms}$.

$$\frac{(\overline{v_1})_{rms}}{(\overline{v_2})_{rms}} = \sqrt{\frac{m_2}{m_1}}$$

The rms speed, $\overline{v_{rms}}$, is not actually the same as the average speed of the gas molecules, but instead represents the speed of a molecule that would have the average kinetic energy. (The difference is subtle, and the two averages do not differ by much, as noted in the margin.)

For any substance, the mass of an individual molecule is proportional to the molecular mass. Representing a molecular mass of a gas by M, we can restate this as $m \propto M$. The proportionality constant is the same for all gases. (It's in grams per atomic mass unit when we express m in atomic mass units.) When we take a ratio of two molecular masses, the constant cancels anyway, so we can write

$$\frac{(\overline{v_1})_{rms}}{(\overline{v_2})_{rms}} = \sqrt{\frac{m_2}{m_1}} = \sqrt{\frac{M_2}{M_1}}$$

Notice that the preceding equation tells us that the rms speed of the molecules is inversely proportional to the square root of the molecular mass. This means that *at*

Suppose we have two molecules with velocities of 6 and 10 m s^{-1}. The average speed is $\frac{1}{2}(6 + 10) = 8$ m s^{-1}. The rms speed is obtained by squaring each speed, averaging the squared values, and then taking the square root of the result. Thus,

$$\overline{v}_{rms} = \sqrt{\tfrac{1}{2}(6^2 + 10^2)} = 8.2 \text{ m s}^{-1}$$

a given temperature, molecules of a gas with a high molecular mass move more slowly, on average, than molecules of a gas with a low molecular mass.

As you might expect, fast-moving molecules will find an opening in the wall of a container more often than slow-moving molecules, so they will effuse faster. Therefore, the rate of effusion of a gas is proportional to the average speed of its molecules, and therefore, it is also proportional to $1/\sqrt{M}$.

$$\text{effusion rate} \propto \overline{v_{rms}} \propto \frac{1}{\sqrt{M}}$$

Let's use k as the proportionality constant. This gives

$$\text{effusion rate} \propto \frac{k}{\sqrt{M}}$$

Comparing two gases, 1 and 2, and taking a ratio of effusion rates to cause k to cancel, we have

$$\frac{\text{effusion rate (gas 1)}}{\text{effusion rate (gas 2)}} = \sqrt{\frac{M_2}{M_1}}$$

This is the way we expressed Graham's law in Equation 11.10. Thus, still another gas law supports the model of an ideal gas.

Kinetic theory predicts an absolute zero

The kinetic theory found that the temperature is proportional to the average kinetic energy of the molecules.

$$T \propto \text{average molecular KE} \propto \tfrac{1}{2}m\overline{v^2}$$

If the average molecular KE becomes zero, the temperature must also become zero. But mass (m) cannot become zero, so the only way that the average molecular KE can be zero is if v goes to zero. A particle cannot move any slower that it does at a dead standstill, so if the particles stop moving entirely, the substance is as cold as anything can get. It's at absolute zero.[13]

11.9 ▶ Real gases don't obey the ideal gas law perfectly

According to the ideal gas law, the ratio PV/T equals a product of two constants, nR. But, experimentally, for real gases PV/T is actually not quite a constant. When we use experimental values of P, V, and T for a real gas, such as O_2, to plot actual values of PV/T as a function of P, we get the curve shown in Figure 11.21. The *horizontal* line at $PV/T = 1$ in Figure 11.21 is what we should see if PV/T were truly constant over all values of P, as it would be for an ideal gas.

A real gas, like oxygen, deviates from ideal behavior for two important reasons. First, the model of an ideal gas assumes that gas molecules are infinitesimally small—that the individual molecules have no volume. But real molecules do take up some space. (If all of the kinetic motions of the gas molecules ceased and the molecules settled, you could imagine the net space that the molecules would occupy in and of themselves.) Second, in an ideal gas there would be no attractions between molecules, but in a real gas molecules do experience weak attractions toward each other.

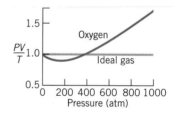

FIGURE 11.21 *Deviation from the ideal gas law.* A graph of PV/T versus P for an ideal gas is like plotting a constant versus P. The graph must be a straight line, as shown, because the ideal gas law equation tells us that $PV/T = nR$ (a product of constants). The same plot for oxygen, a real gas, is not a straight line, showing that O_2 is not "ideal."

[13]Actually, even at 0 K, there must be some slight motion. It's required by the Heisenberg uncertainty principle, which says (in one form) that it's impossible to know precisely both the speed and the location of a particle simultaneously. (If one knows the speed, then there's uncertainty in the location, for example.) If the molecules were actually dead still at absolute zero, there would be no uncertainty in their speed. But then the uncertainty in their *position* would be infinitely great. We would not know where they were! But we do know; they're in this or that container. Thus some uncertainty in speed must exist to have less uncertainty in position and so locate the sample!

At room temperature and atmospheric pressure, most gases behave nearly like an ideal gas, also for two reasons. First, the space between the molecules is so large that the volume occupied by the molecules themselves is insignificant. By doubling the pressure, we are able to squeeze the gas into very nearly half the volume. Second, the molecules are moving so rapidly and are so far apart that the attractions between them are hardly felt. As a result, the gas behaves almost as though there are no attractions.

Deviations from ideal behavior are felt most when the gas is at very high pressure and when the temperature is low. Raising the pressure can reduce only the empty space between the molecules, not the volume of the individual particles themselves. At high pressure, the space taken by the molecules themselves is a significant part of the total volume, so doubling the pressure cannot halve the total volume. As a result, the actual volume of a real gas is larger than expected for an ideal gas, and the ratio PV/T is larger than if the gas were ideal. We see this for O_2 at the right side of the graph in Figure 11.21.

The attractive forces between molecules reveal themselves by causing the pressure of a real gas to be slightly lower than that expected for an ideal gas. The attractions cause the paths of the molecules to bend whenever they pass near each other (Figure 11.22). Because the molecules are not traveling in straight lines, as they would in an ideal gas, they have to travel further between collisions with the walls. As a result, the molecules of a real gas don't strike the walls as frequently as they would if the gas were ideal, and this reduced frequency of collision translates to a reduced pressure. Thus, the ratio PV/T is *less* than that for an ideal gas, particularly where the problem of particle volume is least, at lower pressures. The curve for O_2 in Figure 11.21, therefore, dips at lower pressures.

The van der Waals equation corrects for deviations from ideal behavior

J. D. van der Waals (1837–1923), a Dutch scientist, won the 1910 Nobel Prize in physics.

Many attempts have been made to modify the equation of state of an ideal gas to get an equation that better fits the experimental data for individual real gases. One of the more successful efforts was that of J. D. van der Waals. He found ways to correct the measured values of P and V to give better fits of the data to the general gas law equation. The result of his derivation is called the *van der Waals equation of state for a real gas.* Let's take a brief look at how van der Waals made corrections to measured values of P and V to obtain expressions that fit the ideal gas law.

FIGURE 11.22 *The effect of attractive forces on the pressure of a real gas.* (*a*) In an ideal gas, the molecules would travel in straight lines. (*b*) In a real gas, the paths curve as one molecule passes close to another because the molecules attract each other. Asterisks indicate the points at which molecules come close to each other. Because of the curved paths, molecules of a real gas take longer to reach the walls between collisions, which reduces the collision frequency and so slightly lowers the pressure.

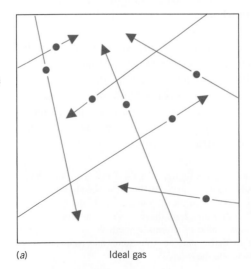

(*a*)   Ideal gas

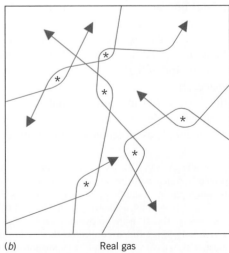

(*b*)   Real gas

As you know, if a gas were ideal, it would obey the equation

$$P_{\text{ideal}}V_{\text{ideal}} = nRT$$

But for a real gas, using the measured pressure, P_{meas}, and measured volume, V_{meas},

$$P_{\text{meas}}V_{\text{meas}} \neq nRT$$

The reason is that P_{meas} is smaller than P_{ideal} (as a result of attractive forces between real gas molecules) and that V_{meas} is larger than V_{ideal} (because real molecules do take up some space). Therefore, to get the pressure and volume to obey the ideal gas law, we have to *add* something to the measured pressure and *subtract* something from the measured volume. That's exactly what van der Waals did. Here's his equation.

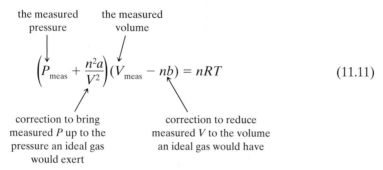

$$\left(P_{\text{meas}} + \frac{n^2 a}{V^2}\right)\left(V_{\text{meas}} - nb\right) = nRT \tag{11.11}$$

The constants a and b are called *van der Waals constants* (see Table 11.3). They are determined for each real gas by carefully measuring P, V, and T under varying conditions. Then trial calculations are made to figure out what values of the constants give the best matches between the observed data and the van der Waals equation.

Notice that the constant a involves a correction to the pressure term of the ideal gas law, so the size of a would indicate something about attractions between molecules. Larger values of a mean stronger attractive forces between molecules. Thus, the most easily liquefied substances, like water and ethyl alcohol, have the largest values of the van der Waals constant a, suggesting relatively strong attractive forces between their molecules.

The constant b helps to correct for the volume occupied by the molecules themselves, so the size of b would indicate something about the sizes of particles in the gas. Larger values of b mean larger molecular sizes. Looking at data for the noble gases in Table 11.3, we see that as the atoms become larger from helium through xenon, the values of b become also larger. In the next chapter we'll continue the study of factors that control the physical state of a substance, particularly attractive forces and their origins.

TABLE 11.3	VAN DER WAALS CONSTANTS				
Substance	a $(\text{L}^2\,\text{atm}\,\text{mol}^{-2})$	b $(\text{L}\,\text{mol}^{-1})$	Substance	a $(\text{L}^2\,\text{atm}\,\text{mol}^{-2})$	b $(\text{L}\,\text{mol}^{-1})$
Noble Gases			*Other Gases*		
Helium, He	0.03421	0.02370	Hydrogen, H_2	0.02444	0.02661
Neon, Ne	0.2107	0.01709	Oxygen, O_2	1.360	0.03183
Argon, Ar	1.345	0.03219	Nitrogen, N_2	1.390	0.03913
Krypton, Kr	2.318	0.03978	Methane, CH_4	2.253	0.04278
Xenon, Xe	4.194	0.05105	Carbon dioxide, CO_2	3.592	0.04267
			Ammonia, NH_3	4.170	0.03707
			Water, H_2O	5.464	0.03049
			Ethyl alcohol, C_2H_5OH	12.02	0.08407

SUMMARY

Barometers, Manometers, and Pressure Units Atmospheric pressure is measured with a **barometer** in which a pressure of 1 **standard atmosphere** (1 **atm**) will support a column of mercury 760 mm high. This is a pressure of 760 **torr.** By definition, 1 atm = 101,325 **pascals** (**Pa**) and 1 **bar** = 100 kPa. **Manometers,** both open-end and closed-end, are used to measure the pressure of trapped gases. The height of fluid in a manometer or barometer supported by a given pressure is inversely proportional to the density of the fluid. In comparing two liquids A and B, the equation

$$\frac{h_B}{h_A} = \frac{d_A}{d_B}$$

relates the heights of the fluid columns, h, to the fluid densities, d.

Gas Laws An **ideal gas** is a hypothetical gas that obeys the gas laws exactly over all ranges of pressure and temperature. Real gases exhibit ideal gas behavior most closely at low pressures and high temperatures, which are conditions remote from those that liquefy a gas.

Boyle's Law (Pressure–Volume Law). For a fixed amount of gas at constant temperature, volume varies inversely with pressure. $V \propto 1/P$.

Charles' Law (Temperature–Volume Law). For a fixed amount of gas at constant pressure, volume varies directly with the Kelvin temperature. $V \propto T$.

Gay-Lussac's Law (Temperature–Pressure Law). For a fixed amount of gas at constant volume, pressure varies directly with the Kelvin temperature. $P \propto T$.

Avogadro's Principle. Equal volumes of gases contain equal numbers of moles when compared at the same temperature and pressure. At **STP,** 1 mol of an ideal gas occupies a volume of 22.4 L.

Combined Gas Law. PV divided by T for a given gas sample is a constant. $PV/T = C$.

Ideal Gas Law. $PV = nRT$. When P is in atm and V is in L, the value of R is 0.0821 L atm mol^{-1} K^{-1} (T being, as usual, in kelvins).

Gay-Lussac's Law of Combining Volumes. When measured at the same temperature and pressure, the volumes of gases consumed and produced in chemical reactions are in the same ratios as their coefficients.

Mole Fraction. The mole fraction X_A of a substance A equals the ratio of the number of moles of A, n_A, to the total number of moles n_{total} of all the components of a mixture.

$$X_A = \frac{n_A}{n_{total}}$$

Dalton's Law of Partial Pressures. The total pressure of a mixture of gases is the sum of the partial pressures of the individual gases.

$$P_{total} = P_A + P_B + P_C + \cdots$$

In terms of mole fractions, $P_A = X_A P_{total}$ and $X_A = P_A/P_{total}$.

Graham's Law of Effusion. The rate of effusion of a gas varies inversely with the square root of its density (or the square root of its molecular mass) at constant pressure and temperature. Comparing different gases at the same temperature and pressure,

$$\frac{\text{effusion rate } (A)}{\text{effusion rate } (B)} = \sqrt{\frac{d_B}{d_A}} = \sqrt{\frac{M_B}{M_A}}$$

Kinetic Theory of Gases An ideal gas consists of a large number of particles, with individual volumes close to zero, that are in constant, chaotic, random motion, traveling in straight lines with no attractions or repulsions between them. When the laws of physics and statistics are applied to this model, and the results compared with the ideal gas law, the Kelvin temperature of a gas is found to be proportional to the average kinetic energy of the gas particles. Pressure is the result of forces of collision of the particles with the container's walls.

Real Gases Because individual gas particles do have real volumes and because small forces of attraction do exist between them, real gases do not exactly obey the gas laws. The van der Waals equation of state for a real gas makes corrections for the volume of the gas molecules and for the attractive force between gas molecules. The van der Waals constant a provides a measure of the attractive forces between molecules, whereas the constant b gives a measure of the relative size of the gas molecules.

Reaction Stoichiometry: A Summary With the study in this chapter of the stoichiometry of reactions involving gases, we have completed our study of the tools needed for the calculations of all variations of reaction stoichiometry. The critical link in all such calculations is the set of coefficients given by the balanced equation, which provides the stoichiometric equivalencies needed to convert from the

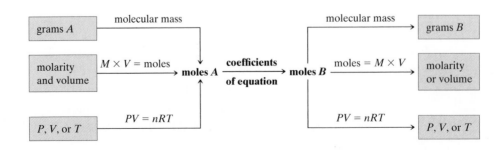

number of moles of one substance into the numbers of moles of any of the others in the reaction. To use the coefficients requires that all the calculations must funnel through *moles*. Whether we start with grams of some compound in a reaction, or the molarity of its solution plus a volume, or *P-V-T* data for a gas in the reaction, *we must get the essential calculation into moles*. After applying the coefficients, we can then move back to any other kind of unit we wish.

The flow chart on page 494 summarizes what we have been doing.

The labels on the arrows of the flowchart suggest the basic tools. Formula masses or molecular masses get us from grams to moles or from moles to grams. Molarity and volume data move us from concentration to moles or back. With *P-V-T* data we can find moles or, knowing moles of a gas, we can find any one of *P*, *V*, or *T*, given the other two.

TOOLS ➤ YOU HAVE LEARNED

The table below lists the concepts that you've learned in this chapter that can be applied as tools in solving problems. Study each one carefully so that you know what each is used for. When faced with solving a problem, recall what each tool does and consider whether it will be helpful in finding a solution. This will aid you in selecting the tools you need. If necessary, refer to this table when working on the Thinking-It-Through problems and the Review Exercises that follow.

TOOL	HOW IT WORKS
Combined gas law (page 462)	This law applies when the amount of gas is constant and we are asked how one of the variables (P, V, or T) changes when we change two of the others. When the amount of gas and one of the variables (P, V, or T) are constant, the problem reduces to one involving Boyle's, Charles', or Gay-Lussac's law.
Ideal gas law (page 468)	This law applies when any three of the four variables P, V, T, or n, are known and we wish to calculate the value of the fourth.
Dalton's law of partial pressures (page 476)	We use this law to calculate the partial pressure of one gas in a mixture of gases. This requires the total pressure and either the partial pressures of the other gases or their mole fractions. If the partial pressures are known, their sum is the total pressure. When a gas is collected over water, this law is used to obtain the partial pressure of the collected gas in the "wet" gas mixture.
Mole fractions (page 475)	Given the composition of a gas mixture, we can calculate the mole fraction of a component. The mole fraction can then be used to find the partial pressure of the component given the total pressure. If the total pressure and partial pressure of a component are known, we can calculate the mole fraction of the component.
Graham's law of effusion (page 485)	This law allows us to calculate relative rates of effusion of gases. It also allows us to calculate molecular masses from relative rates of effusion.

THINKING IT THROUGH

The goal for the following problems is not to find the answers themselves, but rather to assemble the information needed to solve them and explain how you would use the information to find the answers. The problems in Level 2 are more challenging than those in Level 1 and may contain more data than are required, in which case you are also asked to identify the unnecessary data. Detailed answers to the Thinking-It-Through problems can be found on the web site.

Need extra help?
ON-LINE HELP | Visit the Brady/
Senese web site at
www.wiley.com/
college/brady

Level 1 Problems

1. The plunger of a bicycle tire pump is to be pushed in so that the pressure of the trapped air changes from 734 torr to 2.50 atm. The air hose has been closed off; no air can escape. No temperature change occurs. If the initial volume of air is 175 mL, explain how you would calculate the final volume. (Set up the calculations.)

2. At 70 °C a sample of a gas has a volume of 550 mL. If you wished to reduce its volume to 500 mL, to what temperature must you cool it? Assume no change in pressure. (Set up the calculation.)

3. A sample of (dry) methane, CH_4, was collected at an atmospheric pressure of 1.00 atm and a temperature of 25 °C in a sealed glass flask that the manufacturer said could withstand an internal pressure of 1.50 atm. Assuming the claim to be accurate, describe how you would determine whether this flask could be safely heated to 50 °C.

4. A glass vessel with a volume of 1.50 L was filled with 1.68 g of dry nitrogen at 24 °C and attached to a closed-end mercury manometer. The mercury in the arm of the manometer attached to the glass vessel was measured to be 18 mm higher than that in the arm open to the atmosphere. What was the atmospheric pressure? (Explain the calculations required to answer the question.)

5. Hydrochloric acid was added to a sample of magnesium. The reaction

$$2HCl(aq) + Mg(s) \longrightarrow MgCl_2(aq) + H_2(g)$$

produced hydrogen gas that had a volume of 37.6 mL and a pressure of 746 torr when it was collected over water at 20 °C. How many grams of magnesium were consumed in the reaction? (Describe the calculations in detail.)

6. A gas sample weighing 1.67 g has a volume of 276 mL at 589 torr and 28 °C. What is the molecular mass of the gas?

Level 2 Problems

7. A 1.56 g sample of gas was collected over water at 25 °C and a pressure of 745 torr in a 275 mL container. What is the density of a dry sample of this gas at 45 °C and 770 torr? (Describe the calculations in detail.)

8. A small amount of hydrogen gas was burned in a 1.000 L container initially filled with 0.2500 g of dry O_2. Some O_2 remained after the reaction was complete. The container and its contents were returned to 25 °C and the pressure in the container was measured to be 57.48 torr. Describe in detail how you would calculate how many grams of hydrogen had been burned. (*Hint:* Some of the water produced is liquid and some is gas.)

9. A 20.0 mL sample of dilute, aqueous hydrogen peroxide was placed in a 1.00 L sealed vessel. The pressure inside the flask was determined to be 740 torr. The system was maintained at 25 °C as the hydrogen peroxide decomposed according to the following equation:

$$2H_2O_2(aq) \longrightarrow O_2(g) + 2H_2O(l)$$

After all of the hydrogen peroxide had thus decomposed, the pressure was measured again and found to be 907 torr. Set up the calculations you would use to determine the number of grams of H_2O_2 present in the initial sample. (Ignore the very small *change* in the gas space available in the flask caused by the formation of some additional water. Assume that oxygen is insoluble in water.) Is it necessary to do a Dalton's law calculation to correct for the vapor pressure of water? Explain.

10. A cylinder with a movable piston contains helium gas over water at 35 °C. The piston is lowered until the volume of the gas is halved. The pressure inside the cylinder rose to 840 torr. What was the initial pressure inside the cylinder before the piston was moved? (Describe the necessary calculations in detail.)

11. In a chemical analysis, the nitrogen in a 0.465 g sample of a compound containing C, H, N, and O was converted to N_2 gas. The gas had a volume of 76.6 mL at 25 °C and 752 torr. In a separate analysis, the compound was determined to also

contain 32.0% C and 6.67% H. What is the empirical formula of the compound? (Explain the calculations in detail.)

12. Both CO_2 and SF_6 are colorless and odorless gases that are chemically quite stable in the atmosphere. If you planned to manufacture tennis balls and wished them to retain their "bounce" for the longest time, which gas would you use to give internal pressure to the balls? Explain. Assume that costs are not what matters the most, only quality.

13. Use kinetic molecular theory to explain the increase in size of a weather balloon as it rises from ground level into the upper atmosphere, keeping in mind that the pressure and temperature both decrease with increasing altitude. Consider separately the effects upon volume of changing the pressure and temperature. Which effect is dominant, and how do you know this?

14. A 1 L bulb is filled with equal parts of hydrogen gas and oxygen gas. Hydrogen atoms are depicted as light gray spheres and oxygen atoms as red spheres. Which of the following figures best represents the contents of the bulb. Explain why you selected one figure and why you rejected each of the others.

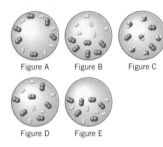

Figure A   Figure B   Figure C

Figure D   Figure E

15. At 25 °C the distribution of molecular speeds for argon atoms is as shown in the figure. Note that the figure is not symmetric about its maximum value but is steeper on the low-speed side.

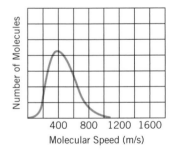

(a) What does the area under the curve represent?

(b) Sketch two additional curves to represent the distribution of molecular speeds for argon, one at a lower temperature and one at a higher temperature. Be sure to maintain in your other curves the property identified in part a.

16. Samples of NO gas and O_2 gas each at 1.0 atm pressure are separated by a closed stopcock as shown in the figure. Both NO and O_2 are colorless gases. At 25 °C, these gases react rapidly according to the equation $2NO + O_2 \rightarrow 2NO_2$. The product of this reaction is a brown gas.

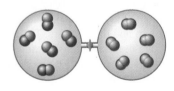

(a) What would you expect to see when the stopcock is turned to the open position? (Both bulbs do not instantaneously become brown.)

(b) What concept developed in this chapter explains the development of color?

(c) If the reaction were to go to completion, show by means of a sketch how the system would appear in terms of the molecules present.

REVIEW QUESTIONS

Concept of Pressure; Manometers and Barometers

11.1 Define *pressure*. Explain why the height of the mercury in a barometer is independent of the cross-sectional area of the tube.

11.2 If you get jabbed by a pencil, why does it hurt so much more if it's with the sharp point rather than the eraser? Explain in terms of the concepts of force and pressure.

11.3 Write expressions that could be used to form conversion factors to convert between
(a) kilopascal and atm   (d) torr and pascal
(b) torr and mm Hg   (e) bar and pascal
(c) torr and atm   (f) bar and atm

11.4 At 20 °C the density of mercury is 13.6 g mL^{-1} and that of water is 1.00 g mL^{-1}. At 20 °C, the vapor pressure of mercury is 0.0012 torr and that of water is 18 torr. Give and explain two reasons why water would be an inconvenient fluid to use in a Torricelli barometer.

11.5 What is the advantage of using a closed-end manometer, rather than an open-end one, when measuring the pressure of a trapped gas?

11.6 Why wouldn't water be a good liquid to use in a closed-end manometer?

Relationships for a Fixed Amount of Gas

11.7 Express the following gas laws in equation form: (a) temperature–volume law (Charles' law), (b) temperature–pressure law (Gay-Lussac's law), (c) pressure–volume law (Boyle's law), (d) combined gas law.

11.8 Which of the four important variables in the study of the physical properties of gases are assumed to be held constant in each of the following laws: (a) Boyle's law, (b) Charles' law, (c) Gay-Lussac's law, (d) combined gas law?

11.9 What is meant by an *ideal gas*? Under what conditions does a real gas behave most like an ideal gas?

Ideal Gas Law

11.10 State the ideal gas law in the form of an equation. What is the value of the gas constant in units of L atm mol^{-1} K^{-1}?

11.11 Using O_2 as an example, carefully distinguish between 1 *molecule*, 1 *mole*, 1 *molar mass*, and 1 *molar volume*.

11.12 At STP how many molecules of H_2 are in 22.4 L?

Dalton's Law and Graham's Law

11.13 State Dalton's law of partial pressures in the form of an equation.

11.14 Define *mole fraction*. How is the partial pressure of a gas related to its mole fraction and the total pressure?

11.15 Consider the diagrams below that illustrate three mixtures of gases *A* and *B*. If the total pressure of the mixture is 1.00 atm, which of the drawings corresponds to a mixture in which the partial pressure of *A* equals 0.600 atm? What are the partial pressures of *A* in the other mixtures? What are the partial pressures of *B*?

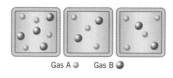

Gas A ●   Gas B ●

11.16 What is the difference between *diffusion* and *effusion*? State Graham's law in the form of an equation.

11.17 Show that the molecular mass of a gas, *M*, is directly proportional to its density, *d*.

Kinetic Theory of Gases

11.18 What model of a gas was proposed by the kinetic theory of gases?

11.19 To what properties of an ideal gas is its temperature proportional?
(a) According to the kinetic theory
(b) According to the combined gas law

11.20 What aspects of the kinetic theory of gases explain why two gases, given the freedom to mix, will always mix *entirely*?

11.21 If the molecules of a gas at constant volume are somehow given a lower average kinetic energy, what two measurable properties of the gas will change and in what direction?

11.22 Explain *how* raising the temperature of a gas causes it to expand at constant pressure. (Describe how the model of an ideal gas connects the increase in temperature to the gas expansion.)

11.23 Explain in terms of the kinetic theory *how* raising the temperature of a confined gas makes its pressure increase.

11.24 How does the kinetic theory explain the existence of an absolute zero, 0 K?

11.25 What is meant by "root mean square speed"? How is it related to the kinetic energies of the molecules of a gas?

11.26 Which of the following gases has the largest value of $\overline{v_{rms}}$ at 25 °C: (a) N_2, (b) CO_2, (c) NH_3, or (d) HBr?

11.27 In your own words, explain why a gas with a low molecular mass effuses faster than a gas with a high molecular mass.

11.28 How would you expect the rate of effusion of a gas to depend on (a) the pressure of the gas, and (b) the temperature of the gas?

11.29 In what way does Dalton's law of partial pressures provide evidence that the molecules in a gas move and behave independently of each other?

11.30 What postulates of the kinetic theory are not strictly true, and why?

11.31 If a certain gas *A* has a larger value of the van der Waals constant *b* than gas *B*, what does this suggest about the molecules of gas *A*?

11.32 A small value for the van der Waals constant *a* suggests something about the molecules of the gas. What?

11.33 Which of the molecules at the right has the larger value of the van der Waals constant *b*? Explain your choice.

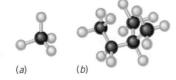

(a)   (b)

11.34 Under the same conditions of *T* and *V*,

why is the pressure of a real gas less than the pressure the gas would exert if it were ideal? At a given *T* and *P*, why is the volume of a real gas larger than it would be if the gas were ideal?

11.35 Suppose we have a mixture of helium and argon. On average, which atoms are moving faster at 25 °C, and why?

REVIEW PROBLEMS

Answers to problems whose numbers are printed in color are given in Appendix B. More challenging problems are marked with asterisks. **ILW** = Interactive LearningWare solution is available at *www.wiley.com/college/brady*.

Pressure Unit Conversions

11.36 Carry out the following unit conversions: (a) 1.26 atm to torr, (b) 740 torr to atm, (c) 738 torr to mm Hg, (d) 1.45×10^3 Pa to torr.

11.37 Carry out the following unit conversions: (a) 0.625 atm to torr, (b) 825 torr to atm, (c) 62 mm Hg to torr, (d) 1.22 kPa to bar.

11.38 What is the pressure in torr of each of the following?
(a) 0.329 atm (summit of Mt. Everest, the world's highest mountain)
(b) 0.460 atm (summit of Mt. Denali, the highest mountain in the United States)

11.39 What is the pressure in atm of each of the following? (These are the values of the pressures exerted individually by N_2, O_2, and CO_2, respectively, in typical inhaled air.)
(a) 595 torr   (b) 160 torr   (c) 0.300 torr

Manometers and Barometers

11.40 An open-end manometer containing mercury was connected to a vessel holding a gas at a pressure of 720 torr. The atmospheric pressure was 765 torr. Sketch a diagram of the apparatus showing the relative heights of the mercury in the two arms of the manometer. What is the difference in the heights of the mercury expressed in centimeters?

11.41 An open-end manometer containing mercury was connected to a vessel holding a gas at a pressure of 820 torr. The atmospheric pressure was 750 torr. Sketch a diagram of the apparatus showing the relative heights of the mercury in the two arms of the manometer. What is the difference in the heights of the mercury expressed in centimeters?

11.42 An open-end mercury manometer was connected to a flask containing a gas at an unknown pressure. The mercury in the arm open to the atmosphere was 65 mm higher than the mercury in the arm connected to the flask. The atmospheric pressure was 748 torr. What was the pressure of the gas in the flask (in torr)?

11.43 An open-end mercury manometer was connected to a flask containing a gas at an unknown pressure. The mercury in the arm open to the atmosphere was 82 mm lower than the mercury in the arm connected to the flask. The at-

mospheric pressure was 752 torr. What was the pressure of the gas in the flask (in torr)?

11.44 A student carried out a reaction using a bulb connected to an open-end mercury manometer on a day when the pressure was 746 torr. Before the reaction, the level of the mercury in both arms was at a height of 12.50 cm, measured by a meter stick placed between the two arms of the U-shaped tube. After the reaction, the level in the arm connected to the bulb was at 8.50 cm. Did the reaction produce or consume gases? What was the final pressure in the bulb (in torr)?

11.45 In another experiment, using the same apparatus as described in the previous problem, the mercury level in the arm nearest the bulb stood at 10.20 cm after the reaction. Before the reaction both levels were at 12.50 cm. The pressure in the lab that day was 741 torr. Did the reaction produce or consume gases? What was the final pressure in the bulb (in torr)?

11.46 Suppose that in a closed-end manometer the mercury in the closed arm was 12.5 cm higher than the mercury in the arm connected to a vessel containing a gas. What is the pressure of the gas expressed in torr?

11.47 Suppose a gas is in a vessel connected to both an open-end and a closed-end manometer. The difference in heights of the mercury in the closed-end manometer was 236 mm, while in the open-end manometer the mercury level in the arm open to the atmosphere was 512 mm below the level in the arm connected to the vessel. Calculate the atmospheric pressure. (It may help to sketch the apparatus.)

11.48 Suppose you were to construct a barometer using a fluid with a density of 1.22 g mL^{-1}. How high would the liquid level be in this barometer if the atmospheric pressure was 755 torr? (Mercury has a density of 13.6 g mL^{-1}.)

11.49 A suction pump works by creating a vacuum into which some fluid will rise. If the pump were able to create a perfect vacuum (i.e., giving a pressure equal to 0 torr), what is the maximum height, in meters, that water could be raised if the atmospheric pressure is 765 torr? Suppose that you work for a construction company in Florida and are told to use a suction pump to remove water from a pit. The water level is over 35 ft below the site where you must place the pump. Can *any* suction pump be expected to raise water to this height at sea level? (The density of water is 1.00 g mL^{-1}; that of mercury is 13.6 g mL^{-1}.)

Gas Laws for a Fixed Amount of Gas

11.50 A gas has a volume of 255 mL at 725 torr. What volume will the gas occupy at 365 torr if the temperature of the gas doesn't change?

11.51 A bicycle pump has a barrel that is 75.0 cm long (about 30 in.). If air is drawn into the pump at a pressure of 1.00 atm during the upstroke, how long must the downstroke be, in centimeters, to raise the pressure of the air to 5.50 atm (approximately the pressure in the tire of a ten-speed bike)? Assume no change in the temperature of the air.

11.52 A gas has a volume of 3.86 L at a temperature of 45 °C. What will the volume of the gas be if its temperature is raised to 80 °C while its pressure is kept constant?

11.53 A balloon has a volume of 2.50 L indoors at a temperature of 22 °C. If the balloon is taken outdoors on a cold day when the air temperature is −15 °C (5 °F), what will its volume be in liters? Assume constant air pressure within the balloon.

11.54 A sample of a gas has a pressure of 850 torr at 285 °C. To what Celsius temperature must the gas be heated to double its pressure if there is no change in the volume of the gas?

11.55 Before taking a trip, you check the air in a tire of your automobile and find it has a pressure of 45 lb in.$^{-2}$ on a day when the air temperature is 10 °C (50 °F). After traveling some distance, you find that the temperature of the air in the tire has risen to 43 °C (approximately 110 °F). What is the air pressure in the tire at this higher temperature, expressed in units of lb in.$^{-2}$?

ILW 11.56 A sample of helium at a pressure of 740 torr and in a volume of 2.58 L was heated from 24.0 to 75.0 °C. The volume of the container expanded to 2.81 L. What was the final pressure (in torr) of the helium?

11.57 When a sample of neon with a volume of 648 mL and a pressure of 0.985 atm was heated from 16.0 to 63.0 °C, its volume became 689 mL. What was its final pressure (in atm)?

11.58 What must be the new volume of a sample of nitrogen (in L) if 2.68 L at 745 torr and 24.0 °C is heated to 375.0 °C under conditions that let the pressure change to 760 torr?

11.59 When 280 mL of oxygen at 741 torr and 18.0 °C was warmed to 33.0 °C, the pressure became 760 torr. What was the final volume (in mL)?

11.60 A sample of argon with a volume of 6.18 L, a pressure of 761 torr, and a temperature of 20.0 °C expanded to a volume of 9.45 L and a pressure of 373 torr. What was its final temperature in °C?

11.61 A sample of a refrigeration gas in a volume of 455 mL, at a pressure of 1.51 atm, and at a temperature of 25.0 °C was compressed into a volume of 220 mL with a pressure of 2.00 atm. To what temperature (in °C) did it have to change?

Ideal Gas Law

11.62 What would be the value of the gas constant R in units of $mL\ torr\ mol^{-1}\ K^{-1}$?

11.63 The SI generally uses its base units to compute constants involving derived units. The SI unit for volume, for example, is the cubic meter, called the *stere*, because the meter is the base unit of length. We learned about the SI unit of pressure, the pascal, in this chapter. The temperature unit is the kelvin. Calculate the value of the gas constant in the SI units $m^3\ Pa\ mol^{-1}\ K^{-1}$.

11.64 What volume in liters does 0.136 g of O_2 occupy at 20.0 °C and 748 torr?

11.65 What volume in liters does 1.67 g of N_2 occupy at 22.0 °C and 756 torr?

11.66 What pressure (in torr) is exerted by 10.0 g of O_2 in a 2.50 L container at a temperature of 27 °C?

11.67 If 12.0 g of water is converted to steam in a 3.60 L pressure cooker held at a temperature of 108 °C, what pressure would be produced?

11.68 A sample of carbon dioxide has a volume of 26.5 mL at 20.0 °C and 624 torr. How many grams of CO_2 are in the sample?

11.69 Methane is formed in landfills by the action of certain bacteria on buried organic matter. If a sample of methane collected from a landfill has a volume of 250 mL at 750 torr and 27 °C, how many grams of methane are in the sample?

11.70 A sample of 4.18 mol of H_2 at 18.0 °C occupies a volume of 24.0 L. Under what pressure, in atmospheres, is this sample?

11.71 If a steel cylinder with a volume of 1.60 L contains 10.0 mol of oxygen, under what pressure (in atm) is the oxygen if the temperature is 25.0 °C?

11.72 To three significant figures, calculate the density in g L^{-1} of the following gases at STP: (a) C_2H_6 (ethane), (b) N_2, (c) Cl_2, (d) Ar.

11.73 To three significant figures, calculate the density in g L^{-1} of the following gases at STP: (a) Ne, (b) O_2, (c) CH_4 (methane), (d) CF_4.

11.74 What density (in g L^{-1}) does oxygen have at 24.0 °C and 742 torr?

11.75 At 748.0 torr and 20.65 °C, what is the density of argon (in g L^{-1})?

ILW 11.76 A chemist isolated a gas in a glass bulb with a volume of 255 mL at a temperature of 25.0 °C and a pressure (in the bulb) of 10.0 torr. The gas weighed 12.1 mg. What is the molecular mass of this gas?

11.77 Boron forms a variety of unusual compounds with hydrogen. A chemist isolated 6.3 mg of one of the boron hydrides in a glass bulb with a volume of 385 mL at 25.0 °C and a bulb pressure of 11 torr.
(a) What is the molecular mass of this hydride?
(b) Which of the following is likely to be its molecular formula, BH_3, B_2H_6, or B_4H_{10}?

11.78 At 22.0 °C and a pressure of 755 torr, a gas was found to have a density of 1.13 g L^{-1}. Calculate its molecular mass.

11.79 A gas was found to have a density of 0.08747 mg mL^{-1} at 17.0 °C and a pressure of 760 torr. What is its molecular mass? Can you tell what the gas most likely is?

Stoichiometry of Reactions of Gases

11.80 How many liters of F_2 at STP are needed to react with 4.00 L of H_2, also at STP, in the following reaction?

$$H_2(g) + F_2(g) \longrightarrow 2HF(g)$$

11.81 In the Haber process for the synthesis of ammonia,

$$N_2(g) + 3H_2(g) \longrightarrow 2NH_3(g)$$

how many liters of N_2 are needed to react completely with 45.0 L of H_2, if the volumes of both gases are measured at STP?

11.82 How many milliliters of oxygen are required to react completely with 175 mL of C_4H_{10} if the volumes of both gases are measured at the same temperature and pressure? The reaction is

$$2C_4H_{10}(g) + 13O_2(g) \longrightarrow 8CO_2(g) + 10H_2O(g)$$

11.83 How many milliliters of O_2 are consumed in the complete combustion of a sample of hexane, C_6H_{14}, if the reaction produces 855 mL of CO_2? Assume all gas volumes are measured at the same temperature and pressure. The reaction is

$$2C_6H_{14}(g) + 19O_2(g) \longrightarrow 12CO_2(g) + 14H_2O(g)$$

ILW 11.84 Propylene, C_3H_6, reacts with hydrogen under pressure to give propane, C_3H_8:

$$C_3H_6(g) + H_2(g) \longrightarrow C_3H_8(g)$$

How many liters of hydrogen (at 740 torr and 24 °C) react with 18.0 g of propylene?

11.85 Nitric acid is formed when NO_2 is dissolved in water:

$$3NO_2(g) + H_2O(l) \longrightarrow 2HNO_3(aq) + NO(g)$$

How many milliliters of NO_2 at 25 °C and 752 torr are needed to form 12.0 g of HNO_3?

11.86 How many milliliters of O_2 measured at 27 °C and 654 torr are needed to react completely with 16.8 mL of CH_4 measured at 35 °C and 725 torr?

11.87 How many milliliters of H_2O vapor, measured at 318 °C and 735 torr, are formed when 33.6 mL of NH_3 at 825 torr and 127 °C react with oxygen according to the following equation?

$$4NH_3(g) + 3O_2(g) \longrightarrow 2N_2(g) + 6H_2O(g)$$

11.88 Calculate the maximum number of milliliters of CO_2, at 745 torr and 27 °C, that could be formed in the combustion of carbon monoxide if 300 mL of CO at 683 torr and 25 °C is mixed with 150 mL of O_2 at 715 torr and 125 °C.

11.89 A mixture of ammonia and oxygen is prepared by combining 300 mL of NH_3 (measured at 750 torr and 28 °C) with 220 mL of O_2 (measured at 780 torr and 50 °C). How many milliliters of N_2 (measured at 740 torr and 100 °C) could be formed if the following reaction occurs?

$$4NH_3(g) + 3O_2(g) \longrightarrow 2N_2(g) + 6H_2O(g)$$

Dalton's Law of Partial Pressures

11.90 A 1.00 L container was filled by pumping into it 1.00 L of N_2 at 200 torr, 1.00 L of O_2 at 150 torr, and 1.00 L

of He at 300 torr. All volumes and pressures were measured at the same temperature. What was the total pressure inside the container after the mixture was made?

11.91 A mixture of N_2, O_2, and CO_2 has a total pressure of 740 torr. In this mixture the partial pressure of N_2 is 120 torr and the partial pressure of O_2 is 400 torr. What is the partial pressure of the CO_2?

ILW 11.92 A 22.4 L container at 0 °C contains 0.30 mol N_2, 0.20 mol O_2, 0.40 mol He, and 0.10 mol CO_2. What are the partial pressures of each of the gases?

11.93 A 0.200 mol sample of a mixture of N_2 and CO_2 with a total pressure of 840 torr is exposed to solid CaO, which reacts with CO_2 according to the equation

$$CaO(s) + CO_2(g) \longrightarrow CaCO_3(s)$$

After reaction was complete, the pressure of the gas had dropped to 320 torr. How many moles of CO_2 were in the original mixture?

11.94 A sample of carbon monoxide was prepared and collected over water at a temperature of 20 °C and a total pressure of 754 torr. It occupied a volume of 268 mL. Calculate the partial pressure of the CO in torr as well as its dry volume (in mL) under a pressure of 1.00 atm and 20 °C.

11.95 A sample of hydrogen was prepared and collected over water at 25 °C and a total pressure of 742 torr. It occupied a volume of 288 mL. Calculate its partial pressure (in torr) and what its dry volume would be (in mL) under a pressure of 1.00 atm.

11.96 What volume of "wet" methane would you have to collect at 20.0 °C and 742 torr to be sure that the sample contains 244 mL of dry methane (also at 742 torr)?

11.97 What volume of "wet" oxygen would you have to collect if you need the equivalent of 275 mL of dry oxygen at 1.00 atm? (The atmospheric pressure in the lab is 746 torr.) The oxygen is to be collected over water at 15.0 °C.

11.98 What are the mole percents of the components of air inside the lungs when they have the following partial pressures? For N_2, 570 torr; O_2, 103 torr; CO_2, 40 torr; and water vapor, 47 torr.

11.99 Assuming no other components are present, calculate the mole percents of oxygen and nitrogen in the air at the following places:
(a) At the top of Mt. Everest (elevation 8.8 km) on a day when $P_{N_2} = 197$ torr and $P_{O_2} = 53$ torr
(b) At sea level and 1 atm pressure, with $P_{N_2} = 600$ torr and $P_{O_2} = 160$ torr
(c) Comparing the answers calculated for parts a and b, what has to be the reason why it is hard to breathe without supplemental oxygen at high altitudes?

Graham's Law

11.100 Under conditions in which the density of CO_2 is 1.96 g L^{-1} and that of N_2 is 1.25 g L^{-1}, which gas will effuse more rapidly? What will be the ratio of the rates of effusion of N_2 to CO_2?

11.101 Arrange the following gases in order of increasing rate of diffusion at 25 °C: Cl_2, C_2H_4, SO_2.

11.102 Uranium hexafluoride is a white solid that readily passes directly into the vapor state. (Its vapor pressure at 20.0 °C is 120 torr.) A trace of the uranium in this compound—about 0.7%—is uranium-235, which can be used in a nuclear power plant. The rest of the uranium is essentially uranium-238, and its presence interferes with these applications for uranium-235. Gas effusion of UF_6 can be used to separate the fluoride made from uranium-235 and the fluoride made from uranium-238. Which hexafluoride effuses more rapidly? By how much? (You can check your answer by reading Facets of Chemistry 11.2 on page 486.)

11.103 An unknown gas X effuses 1.65 times faster than C_3H_8. What is the molecular mass of gas X?

ADDITIONAL EXERCISES

11.104 One of the oldest units for atmospheric pressure is lb in.$^{-2}$ (pounds per square inch, or psi). Calculate the numerical value of the standard atmosphere in these units to three significant figures. Calculate the mass in pounds of a uniform column of water 33.9 ft high having an area of 1.00 in.^{2} at its base. (Use the following data: density of mercury = 13.6 g mL^{-1}; density of water = 1.00 g mL^{-1}; 1 mL = 1 cm^3; 1 lb = 454 g; 1 in. = 2.54 cm.)

*__11.105__ A typical automobile has a weight of approximately 3500 lb. If the vehicle is to be equipped with tires, each of which will contact the pavement with a "footprint" that is 6.0 in. wide by 3.2 in. long, what must the gauge pressure of the air be in each tire? (Gauge pressure is the amount that the gas pressure exceeds atmospheric pressure. Assume that atmospheric pressure is 14.7 lb in.$^{-2}$.)

11.106 Suppose you were planning to move a house by transporting it on a large trailer. The house has an estimated weight of 45.6 tons (1 ton = 2000 lb). The trailer is expected to weigh 8.3 tons. Each wheel of the trailer will have tires inflated to a gauge pressure of 85 psi (which is actually 85 psi above atmospheric pressure). If the area of contact between a tire and the pavement can be no larger than 100 in.^{2} (10 in. × 10 in.), what is the minimum number of wheels the trailer must have? (Remember, tires are mounted in multiples of two on a trailer. Assume that atmospheric pressure is 14.7 psi.)

*__11.107__ The motion picture *Titanic* described the tragedy of the collision of the ocean liner of the same name with an iceberg in the North Atlantic. The ship sank soon after the collision on April 14, 1912, and now rests on the seafloor at a depth of 12,468 ft. Recently, the wreck was explored by the research vessel *Nautile,* which has successfully recovered a variety of items from the debris field surrounding the sunken ship. Calculate the pressure in atmospheres and pounds per square inch exerted on the hull of the *Nautile* as it explores the seabed surrounding the *Titanic.* (Seawater has a density of approximately 1.025 g mL^{-1}; mercury has a density of 13.6 g mL^{-1}; 1 atm = 14.7 lb in.$^{-2}$.)

*__11.108__ Two flasks (which we will refer to as Flask 1 and Flask 2) are connected to each other by a U-shaped tube filled with an oil having a density of 0.826 g mL^{-1}. The oil level in the arm connected to Flask 2 is 16.24 cm higher than in the arm connected to Flask 1. Flask 1 is also connected to an open-end mercury manometer. The mercury level in the arm open to the atmosphere is 12.26 cm higher than the level in the arm connected to Flask 1. The atmospheric pressure is 0.827 atm. What is the pressure of the gas in Flask 2 expressed in torr?

*__11.109__ A bubble of air escaping from a diver's mask rises from a depth of 100 ft to the surface where the pressure is 1.00 atm. Initially, the bubble has a volume of 10.0 mL. Assuming none of the air dissolves in the water, how many times larger is the bubble just as it reaches the surface. Use your answer to explain why scuba divers constantly exhale as they slowly rise from a deep dive. (The density of seawater is approximately 1.025 g mL^{-1}; the density of mercury is 13.6 g mL^{-1}.)

*__11.110__ In a diesel engine, the fuel is ignited when it is injected into hot compressed air, heated by the compression itself. In a typical high-speed diesel engine, the chamber in the cylinder has a diameter of 10.7 cm and a length of 13.4 cm. On compression, the length of the chamber is shortened by 12.7 cm (a "5-inch stroke"). The compression of the air changes its pressure from 1.00 to 34.0 atm. The temperature of the air before compression is 364 K. As a result of the compression, what will be the final air temperature (in K and °C) just before the fuel injection?

*__11.111__ Early one cool (60.0 °F) morning you start on a bike ride with the atmospheric pressure at 14.7 lb in.$^{-2}$ and the tire gauge pressure at 50.0 lb in.$^{-2}$. (Gauge pressure is the amount that the pressure exceeds atmospheric pressure.) By late afternoon, the air had warmed up considerably, and this plus the heat generated by tire friction sent the temperature inside the tire to 104 °F. What will the tire gauge now read, assuming that the volume of the air in the tire and the atmospheric pressure have not changed?

11.112 The range of temperatures over which an automobile tire must be able to function is roughly −50 to 120 °F. If a tire is filled to a gauge pressure of 35 lb in.$^{-2}$ at −50 °F (on a cold day in Alaska, for example), what will be the gauge pressure in the tire (in the same pressure units) on a hot day in Death Valley when the temperature is 120 °F? (Assume that the volume of the tire does not change.)

11.113 Chlorine reacts with sulfite ion to give sulfate ion and chloride ion. How many milliliters of Cl_2 gas measured at 25 °C and 734 torr are required to react with all the $SO_3{}^{2-}$ in 50.0 mL of 0.200 M Na_2SO_3 solution?

11.114 Ammonia effuses at a rate that is 2.93 times larger than that of an unknown gas at the same T and P. What is the molecular mass of the unknown?

11.115 A common laboratory preparation of hydrogen on a small scale uses the reaction of zinc with hydrochloric acid. Zinc chloride is the other product.
(a) Write the balanced equation for the reaction.
(b) If 12.0 L of H_2 at 760 torr and 20.0 °C is wanted, how many grams of zinc are needed, in theory?

(c) If the acid is available as 8.00 M HCl, what is the minimum volume of this solution (in milliliters) required to produce the amount of H_2 described in part b?

*11.116 In an experiment designed to prepare a small amount of hydrogen by the method described in the preceding problem, a student was limited to using a gas-collecting bottle with a maximum capacity of 335 mL. The method involved collecting the hydrogen over water. What are the minimum number of grams of Zn and the minimum number of milliliters of 6.00 M HCl needed to produce the *wet* hydrogen that can exactly fit this collecting bottle at 740 torr and 25.0 °C?

11.117 Carbon dioxide can be made in the lab by the reaction of hydrochloric acid with calcium carbonate.

$$CaCO_3(s) + 2HCl(aq) \longrightarrow CaCl_2(aq) + H_2O(l) + CO_2(g)$$

How many milliliters of dry CO_2 at 20.0 °C and 745 torr can be prepared from a mixture of 12.3 g of $CaCO_3$ and 185 mL of 0.250 M HCl?

11.118 A mixture was prepared in a 500 mL reaction vessel from 300 mL of O_2 (measured at 25 °C and 740 torr) and 400 mL of H_2 (measured at 45 °C and 1250 torr). The mixture was ignited and the H_2 and O_2 reacted to form water. What was the final pressure inside the reaction vessel after the reaction was over if the temperature was held at 120 °C?

11.119 A student collected 18.45 mL of H_2 over water at 24 °C. The water level inside the collection apparatus was 8.5 cm higher than the water level outside. The barometric pressure was 746 torr. How many grams of zinc had to react with HCl(aq) to produce the H_2 that was collected?

11.120 A mixture of gases is prepared from 87.5 g of O_2 and 12.7 g of H_2. After the reaction of O_2 and H_2 is complete, what is the total pressure of the mixture if its temperature is 160 °C and its volume is 12.0 L? What are the partial pressures of the gases remaining in the mixture?

11.121 In many countries, the fertilizer ammonia is made using methane as the source of hydrogen (and energy). The overall process—and there are several steps—is by the following equation, where N_4O is used as an approximate formula for air, another raw material:

$$7CH_4 + 10H_2O + 4N_4O \longrightarrow 16NH_3 + 7CO_2$$

Methane is sold in units of tcf, where 1 tcf = 1×10^3 ft³ = 28.3×10^3 L at STP.
(a) What volume of NH_3 (in liters at STP) can be made by this process from 1.00 tcf of methane (also at STP)?
(b) One thousand cubic feet (1.00 tcf) of methane at STP is equivalent to how many kilograms of ammonia in this reaction?

11.122 A sample of an unknown gas with a mass of 3.620 g was made to decompose into 2.172 g of O_2 and 1.448 g of S. Prior to the decomposition, this sample occupied a volume of 1120 mL at 750 torr and 25.0 °C.
(a) What is the percentage composition of the elements in this gas?
(b) What is the empirical formula of the gas?
(c) What is its molecular formula?

*11.123 A sample of a new antimalarial drug with a mass of 0.2394 g was made to undergo a series of reactions that changed all of the nitrogen in the compound into N_2. This gas had a volume of 18.90 mL when collected over water at 23.80 °C and a pressure of 746.0 torr. At 23.80 °C, the vapor pressure of water is 22.110 torr.
(a) Calculate the percentage of nitrogen in the sample.
(b) When 6.478 mg of the compound was burned in pure oxygen, 17.57 mg of CO_2 and 4.319 mg of H_2O were obtained. What are the percentages of C and H in this compound? Assuming that any undetermined element is oxygen, write an empirical formula for the compound.
(c) The molecular mass of the compound was found to be 324. What is its molecular formula?

*11.124 In one analytical procedure for determining the percentage of nitrogen in unknown compounds, weighed samples are made to decompose to N_2, which is collected over water at known temperatures and pressures. The volumes of N_2 are then translated into grams and then into percentages.
(a) Show that the following equation can be used to calculate the percentage of nitrogen in a sample having a mass of W grams when the N_2 has a volume of V mL and is collected over water at t_c °C at a total pressure of P torr. The vapor pressure of water occurs in the equation as $P^\circ_{H_2O}$.

$$\text{percentage N} = 0.04489 \times \frac{V(P - P^\circ_{H_2O})}{W(273 + t_c)}$$

(b) Use this equation to calculate the percentage of nitrogen in the sample described in the preceding problem.

11.125 The "rotten-egg" odor is caused by hydrogen sulfide, H_2S. Most people can detect it at a concentration of 0.15 ppb (parts per billion), meaning 0.15 L of H_2S in 10^9 L of space. A typical student lab is 40 × 20 × 8 ft.
(a) At STP, how many liters of H_2S could be present in a typical lab to have a concentration of 0.15 ppb?
(b) How many milliliters of 0.100 M Na_2S would be needed to generate the amount of H_2S in part a by the following reaction with hydrochloric acid?

$$Na_2S(aq) + 2HCl(aq) \longrightarrow H_2S(g) + 2NaCl(aq)$$

Intermolecular Attractions and the Properties of Liquids and Solids

Geckos have sticky feet. They can rapidly run up walls and along ceilings, and they can cling to almost anything with just one foot. The gecko's secret is the thousands of tiny hairs that cover the bottom of each of its feet. The hairs are so small and so numerous that when they come in contact with a surface, huge numbers of weak attractions between the molecules in the hairs and molecules in the surface enable the gecko to hold on tightly. The gecko releases its grip by curling its toes, peeling the hairs away from the surface. In this chapter we will explore the kinds of attractions that exist between molecules and how these attractions affect the physical properties of liquids and solids.

THIS CHAPTER IN CONTEXT In the preceding chapter we studied the physical properties of gases, and one of our observations was that all gases behave pretty much alike, regardless of their chemical makeup. This is especially so at low pressures and high temperatures, which allows us to use a set of gas laws to describe the behavior of *any* gas. However, when we compare substances in their liquid or solid states (their *condensed states*), the situation is quite different. When a substances is a liquid or a solid, its particles are packed closely together and the forces between them, which we call **intermolecular forces,** are quite strong. Chemical composition and molecular structure play an important role in determining the strengths of such forces, and this causes different substances to behave quite differently from each other when they are liquids or solids.

In this chapter we focus most of our attention on the properties of liquids, which are almost always molecular substances at room temperature. We will reserve a detailed discussion of the solid state for Chapter 13. We begin our study by looking at the basic differences among the states of matter in terms of both common observable properties and the way the states of matter differ at the molecular level. In this chapter we will also examine the different kinds and relative strengths of intermolecular forces. You will learn how they are related to molecular composition and structure and how intermolecular forces influence a variety of familiar physical properties of liquids, such as boiling points and ease of evaporation. And by studying the energy changes associated with changes of states (for example, evaporation or condensation), you will become familiar with the forces that affect weather on our planet and how we can control our local environment through air conditioning and refrigeration.

12.1 | Gases, liquids, and solids differ because intermolecular forces depend on the distances between molecules

There are differences among gases, liquids, and solids that are immediately obvious and familiar to everyone. For example, any gas will expand to fill whatever volume is available to it, even if it has to mix with other gases to do so. That's why the mouth-watering aroma of a fine cooked meal permeates your home while it's being prepared. Liquids and solids, however, retain a constant volume when transferred from one container to another. A solid, such as an ice cube, also keeps its shape, but a liquid such as soda conforms to the shape of whatever bottle or glass we put it in.

In Chapter 11 you learned that gases are easily compressed. Liquids and solids, on the other hand, are nearly *incompressible,* which means their volumes change very little when they are subjected to high pressures. Properties such as the

Observable Properties

Gases are easily compressed, but expand spontaneously to fill whatever container they are in.

Liquids retain their volume when placed into a container, but conform to the shape of the container. They are fluid and are able to flow. Liquids are nearly incompressible

Solids keep their shape and volume *and* are virtually incompressible. They often have crystalline shapes, which are discussed in Chapter 13.

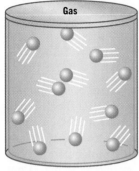 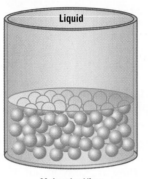 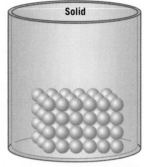

Molecular View

Widely spaced molecules with much empty space between them. Motion is random and with very weak attractions between the molecules.

Molecules tightly packed but with little order. They are able to move past each other with little difficulty. Intermolecular attractive forces are relatively strong.

Molecules tightly packed and highly ordered. Very strong attractions between molecules hold them in place so they are effectively locked in position.

FIGURE 12.1 *General properties of gases, liquids, and solids.* Properties can be understood in terms of how tightly the molecules are packed together and the strengths of the intermolecular attractions between them.

ones we've described can be understood in terms of the way the particles are distributed in the three states of matter, which is illustrated in Figure 12.1.

Intermolecular forces depend on distance

Earlier we noted that laws comparable to the gas laws do not exist for liquids and solids because the properties of substance in these condensed states are so strongly influenced by intermolecular forces. A question we might ask is "Why, for a given substance, are these attractions so strong in liquids and solids, but almost negligible in gases?"

If you have ever played with magnets you know that the attractive force between them rapidly becomes weaker as the distance between the magnets increases. Thus, when two magnets are near each other, the attraction can be quite strong, but if they are far apart hardly any attraction is felt at all. The intermolecular attractions are similarly affected by the distance between the molecules, rapidly becoming weaker as the distance between the molecules increases.

In gases, the molecules are so far apart that the attractive forces are almost negligible, so any differences between the attractive forces are so small that they hardly matter at all. As a result, chemical composition has little effect on the properties of a gas. But in a liquid or a solid, the molecules are close together and the attractions are strong. Differences among these attractions caused by differences in chemical makeup are greatly amplified, so the properties of liquids and solids depend quite heavily on chemical composition.

The closer two molecules are, the more strongly they attract each other.

12.2 ▶ Intermolecular attractions involve electrical charges

Intermolecular attractions are much weaker than chemical bonds

As we noted in the preceding discussion, most of the *physical* properties of substances are controlled by the strengths of intermolecular attractions. Before we study how these forces arise, it is important to understand that the attractions *between* molecules (the ***inter*molecular forces**) are always much weaker than the attractions between atoms *within* molecules (***intra*molecular** forces, which are the *chemical bonds* that hold molecules together). In a molecule of HCl, for example,

The strengths of intramolecular forces (chemical bonds) determine chemical properties; the strengths of intermolecular attractions determine physical properties.

FIGURE 12.2 *Attractions within and between hydrogen chloride molecules.* Strong *intramolecular* attractions (chemical bonds) exist between H and Cl atoms within HCl molecules. These attractions control the chemical properties of HCl. Weaker *intermolecular* attractions exist between neighboring HCl molecules. The intermolecular attractions control the physical properties of this substance.

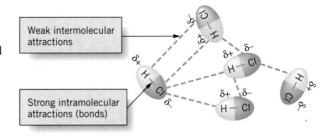

Weak intermolecular attractions

Strong intramolecular attractions (bonds)

the H and Cl atoms are held very tightly to each other by a covalent bond, and it is the strength of this bond that affects the *chemical properties* of HCl. The strength of the chemical bond also keeps the molecule intact as it moves about. When a particular chlorine atom moves, the hydrogen atom bonded to it is forced to follow along (see Figure 12.2). Attractions between neighboring HCl molecules, in contrast, are much weaker. In fact, they are only about 4% as strong as the covalent bond in HCl. These weaker attractions are what determine the *physical properties* of liquid and solid HCl.

From the preceding discussion, we can see that to understand and explain the physical properties of liquids and solids, we need to learn about the kinds and relative strengths of intermolecular attractions. *These forces all arise from attractions between opposite charges and are classified according to how they originate.* Collectively, they are called **van der Waals forces,** after J. D. van der Waals, who studied the nonideal behavior of real gases.

> The force between two electrical charges is called the *electrostatic force*. Other forces in nature are the magnetic, gravitational, nuclear strong, and nuclear weak forces.

Dipole–dipole attractions

> **TOOLS**
>
> *Intermolecular attractions: dipole–dipole forces*
>
> $\delta+$ $\delta-$
>
> H—Cl

In Chapter 9 you saw that the HCl molecule is an example of a *polar molecule,* one with a partial positive charge at one end and a partial negative charge at the other. Because unlike charges attract, polar molecules tend to line up so that the positive end of one dipole is near the negative end of another. Thermal energy (molecular kinetic energy), however, causes the molecules to collide and become disoriented, so the alignment isn't perfect. Nevertheless, there is still a net attraction between polar molecules (see Figure 12.3). We call this kind of intermolecular attractive force a **dipole–dipole attraction.** It is generally a much weaker force than a covalent bond, being only about 1% as strong. Dipole–dipole attractions fall off rapidly with distance, with the energy required to separate a pair of dipoles being proportional to $1/d^3$, where d is the distance between the dipoles.

There are two principal reasons for the relative weakness of dipole–dipole attractions. One is that the charges associated with dipoles are only *partial* charges, not full charges. The other reason, which we have already discussed, is that at ordinary temperatures (e.g., room temperature) collisions between molecules cause the dipoles to be somewhat misaligned, thereby reducing the effectiveness of the attractions.

Attractions (--) are greater than repulsions (--), so the molecules feel a net attraction to each other.

FIGURE 12.3 *Dipole–dipole attractions.* Attractions between polar molecules occur because the molecules tend to align themselves so that opposite charges are near each other and like charges are as far apart as possible. The alignment is not perfect because the molecules are constantly moving and colliding.

Hydrogen bonds

A particularly important kind of dipole–dipole attraction occurs when hydrogen is covalently bonded to a very small, highly electronegative atom (principally, fluorine, oxygen, and nitrogen). In this situation, unusually strong dipole–dipole attractions are often observed, for which there are several reasons. First, F—H, O—H, and N—H bonds are very polar. Because of the large electronegativity differences, the ends of the bond dipoles carry substantial amounts of positive and negative charge. Second, the charges are highly concentrated because of the small sizes of the atoms involved. And third, the positive end of one dipole can get quite close to the negative end of another, also because of the small sizes of the atoms. All of these factors combine to produce an exceptionally strong dipole–dipole attraction that is given the special name **hydrogen bond.** Typically, a hydrogen bond is about five to ten times stronger than other dipole–dipole attractions. Many molecules in biological systems, like proteins and nucleic acids, contain N—H and O—H bonds, and hydrogen bonding in these substances is one factor that determines their overall structures and biological functions.

Hydrogen bonds in water

The strength of the hydrogen bond gives water some very unusual and special properties. Most substances become more dense when they change from a liquid to a solid. Not so with water. In liquid water, the molecules experience hydrogen bonds that continually break and re-form as the molecules move around (Figure 12.4*b*). As water begins to freeze, however, the molecules become locked in place, and each water molecule participates in four hydrogen bonds (Figure 12.4*c*). The resulting structure occupies a larger volume than the same amount of liquid water, so ice is less dense than the liquid. Because of this, ice cubes and icebergs float in the more dense liquid (much to the distress of the people on the *Titanic*). The expansion of freezing water is also the reason why we must add antifreeze to a car's cooling system when the temperature is expected to drop below 0 °C in the winter. The expanding ice is capable of cracking the engine block. Ice formation is also responsible for erosion, causing rocks to split where water has seeped into cracks. And in northern cities, freezing water breaks up pavement, creating potholes in the streets.

Intermolecular attractions: hydrogen bonding

Relative Electronegativities

F	4.1
O	3.5
N	3.1
H	2.1

Hydrogen bonding in ice causes the molecules to be farther apart than in liquid water. This causes the ice to be less dense than water, which is why icebergs like this one float.

Density (*d*) is the ratio of mass (*m*) to volume (*v*):

$$d = \frac{m}{v}$$

When *v* increases while *m* remains constant, *d* decreases.

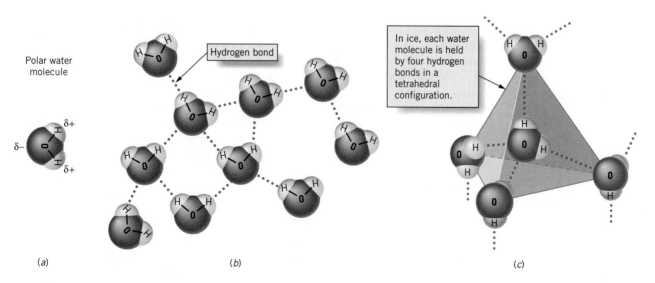

Polar water molecule

Hydrogen bond

In ice, each water molecule is held by four hydrogen bonds in a tetrahedral configuration.

(a)   (b)   (c)

Figure 12.4 *Hydrogen bonding in water.* (*a*) The polar water molecule. (*b*) Hydrogen bonding produces strong attractions between water molecules in the liquid. (*c*) Hydrogen bonding (dotted lines) between water molecules in ice, where each water molecule is held by four hydrogen bonds in a tetrahedral configuration.

London forces

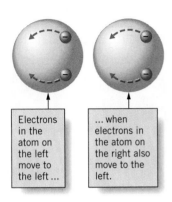

| Electrons in the atom on the left move to the left ... | ... when electrons in the atom on the right also move to the left. |

The attractions between polar molecules such as HCl and SO_2 are fairly easy to understand. But attractive forces occur even between the particles of nonpolar substances, such as the atoms of the noble gases and the nonpolar molecules of Cl_2 and CH_4. These nonpolar substances can also be condensed to liquids and even to solids if they are cooled to low-enough temperatures. Therefore, attractions between their particles, although weak, must exist to cause them to cling together. Electronegativity differences cannot be the origin of these attractions, so what is?

In 1930 Fritz London, a German physicist, explained how the particles in even nonpolar substances can experience intermolecular attractions. He noted that in any atom or molecule the electrons are constantly moving. If we could examine such motions in two neighboring particles, we would find that the movement of electrons in one influences the movement of electrons in the other. This is because electrons repel each other and tend to stay as far apart as they can. Therefore, as an electron of one particle gets near the other particle, electrons on the second particle are pushed away. This happens continually as the electrons move around, so to some extent, the electron density in both particles flickers back and forth in a synchronous fashion. This is illustrated in Figure 12.5, which depicts a series of instantaneous "frozen" views of the electron density. Notice that *at any given moment the electron density of a particle can be unsymmetrical,* with more negative charge on one side than on the other. For that particular instant, the particle is a dipole, and we call it a *momentary dipole* or **instantaneous dipole.**

As an instantaneous dipole forms in one particle, it causes the electron density in its neighbor to become unsymmetrical, too. As a result, this second particle also becomes a dipole. We call it an **induced dipole** because it is caused by, or *induced* by, the formation of the first dipole. Because of the way the dipoles are formed, they always have the positive end of one near the negative end of the other, so there is a dipole–dipole attraction between them. It is a very short-lived attraction, however, because the electrons keep moving; the dipoles vanish as quickly as they form. But, in another moment, the dipoles will reappear in a different orientation and there will be another brief dipole–dipole attraction. In this way the short-lived dipoles cause momentary tugs between the particles. When averaged over a period of time, there is a net, overall attraction. It tends to be relatively weak, however, because the attractive forces are only "turned on" part of the time.

The momentary dipole–dipole attractions that we've just discussed are called *instantaneous dipole-induced dipole attractions,* to distinguish them from the kind of permanent dipole–dipole attractions that exist without interruption in polar substances like HCl. They are also called **London dispersion forces** (or simply **London forces** or **dispersion forces**).

London forces exist between all molecules and ions. Although they are the only kind of attraction possible between nonpolar molecules, London forces also contribute significantly to the total intermolecular attraction between polar molecules, where they are present in addition to the regular dipole–dipole attractions. London forces even occur between oppositely charged ions, but their effects are relatively weak compared to ionic attractions. London forces contribute little to the net overall attractions between ions and are often ignored.

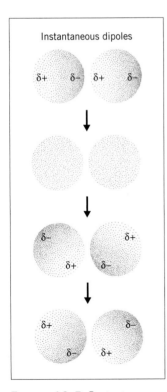

Instantaneous dipoles

FIGURE 12.5 *Instantaneous "frozen" views of the electron density in two neighboring particles.* Attractions exist between the instantaneous dipoles while they exist.

The strengths of London forces

To compare the strengths of intermolecular attractions, a property we can use is boiling point. As we will explain in more detail later in this chapter, the higher the boiling point, the stronger are the attractions between molecules in the liquid.

The strengths of London forces are found to depend chiefly on three factors. One is the **polarizability** of the electron cloud of a particle, which is a measure of the ease with which the electron cloud is distorted, and thus is a measure of the

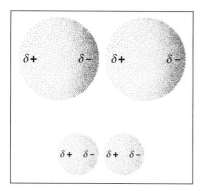

FIGURE 12.6 *Effect of molecular size on the strengths of London dispersion forces.* A large electron cloud is more easily deformed than a small one, so in a large molecule the charges on opposite ends of an instantaneous dipole are larger than in a small molecule. Large molecules therefore experience stronger London forces than small molecules.

ease with which the instantaneous and induced dipoles can form. In general, *as the volume of the electron cloud increases, its polarizability also increases.* When an electron cloud is large, the outer electrons are generally not held very tightly by the nucleus (or nuclei, if the particle is a molecule). This causes the electron cloud to be "mushy" and rather easily deformed, so instantaneous dipoles and induced dipoles form without much difficulty (see Figure 12.6). As a result, particles with large electron clouds experience stronger London forces than do similar particles with small electron clouds.

The effects of size can be seen if we compare the boiling points of the halogens or the noble gases (see Table 12.1). Going down Group VIIIA, for example, the noble gas atoms become larger and the boiling points increase, reflecting increasingly stronger intermolecular attractions (stronger London forces). A similar trend is found among the nonpolar halogen molecules in Group VIIA.

The effects of size are also seen if we examine the values of the van der Waals constant *a* for the noble gases (Table 11.3 on page 493). Recall that this constant is proportional to the attractive forces between molecules in a "real" gas, and the data in Table 11.3 show that the attractive forces increase as we go down the group of noble gases from helium to xenon. The increase in the strengths of the attractive forces parallels the increase in the sizes of the atoms.

Another factor that affects the strengths of London forces is the number of atoms in a molecule. For molecules containing the same elements, London forces increase with the number of atoms, as illustrated by the hydrocarbons (see Table 12.2). Their molecules consist of chains of carbon atoms with hydrogen atoms bonded to them all along the chain. If we compare two hydrocarbon molecules of different chain length—for example, C_3H_8 and C_6H_{14}—the one having the longer chain also has the higher boiling point. This means that the molecule with the longer chain length experiences the stronger intermolecular attractive forces. They are strong between the longer molecules because there are more places along their lengths where instantaneous dipoles can develop and lead to London attractions to other molecules (see Figure 12.7). Even if the strength of attraction at each location is about the same, the *total* attraction experienced between the longer C_6H_{14}

London forces decrease very rapidly as the distance between particles increases. The energy required to separate particles held by London forces varies as $1/d^6$, where d is the distance between the particles.

$$
\begin{array}{cc}
 & CH_3 \\
 & | \\
 & CH_2 \\
 & | \\
 & CH_2 \\
CH_3 & | \\
| & CH_2 \\
CH_2 & | \\
| & CH_2 \\
CH_3 & | \\
\text{propane} & CH_3 \\
C_3H_8 & \text{hexane} \\
 & C_6H_{14}
\end{array}
$$

TABLE 12.1	BOILING POINTS OF THE HALOGENS AND NOBLE GASES		
Group VIIA	Boiling Point (°C)	Group VIIIA	Boiling Point (°C)
F_2	−188.1	He	−268.6
Cl_2	−34.6	Ne	−245.9
Br_2	58.8	Ar	−185.7
I_2	184.4	Kr	−152.3
		Xe	−107.1
		Rn	−61.8

TABLE 12.2	BOILING POINTS OF SOME HYDROCARBONS[a]
Molecular Formula	Boiling Point at 1 atm (°C)
CH_4	−161.5
C_2H_6	−88.6
C_3H_8	−42.1
C_4H_{10}	−0.5
C_5H_{12}	36.1
C_6H_{14}	68.7
⋮	⋮
$C_{10}H_{22}$	174.1
⋮	⋮
$C_{22}H_{46}$	327

[a]The molecules of each hydrocarbon in this table have carbon chains of the type C—C—C—C—etc.; that is, one carbon follows another.

The effect of large numbers of atoms on the total strengths of London forces can be compared to the bond between loop and hook layers of the familiar product Velcro. Each loop-to-hook attachment is not very strong, but when large numbers of them are involved, the overall bond between Velcro layers is quite strong.

molecules is greater than that felt between shorter C_3H_8 molecules. In Table 12.2 you can see that some of the hydrocarbons have boiling points that are quite high, which shows that the cumulative effects of London forces can lead to very strong attractions. We will see how this affects the properties of some important polymers in Chapter 13.

The third factor that affects the strengths of London forces is molecular shape. Even with molecules that have the same number of the same kinds of atoms, the shapes of the molecules can have an influence. For example, compare the molecules shown in Figure 12.8. Both have the same molecular formula, C_5H_{12}. However, the more compact neopentane molecule, $(CH_3)_4C$, has a lower boiling point than the long chainlike *n*-pentane molecule, $CH_3CH_2CH_2CH_2CH_3$. Presumably, because of the compact shape of the $(CH_3)_4C$ molecule, the individual hydrogens on neighboring molecules cannot interact with each other as well as those on the chainlike molecule.

London forces and dipole–dipole forces compared

At first thought, we might expect that London forces are quite weak compared to dipole–dipole attractions. However, for many substances the London forces are as strong as or stronger than any dipole–dipole forces also present. For example, consider HCl (bp −84.9 °C) and HBr (bp −67.0 °C). Molecules of HCl are more polar

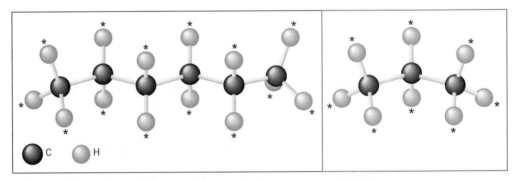

FIGURE 12.7 *The number of atoms in a molecule affects London forces.* The C_6H_{14} molecule, left, has more sites (indicated by asterisks, *) along its chain where it can be attracted to other molecules nearby than does the shorter C_3H_8 molecule, right. As a result, the boiling point of C_6H_{14} (hexane, 68.7 °C) is higher than that of C_3H_8 (propane, −42.1 °C).

neopentane, (CH₃)₄C   n-pentane, CH₃CH₂CH₂CH₂CH₃
bp = 9.5 °C   bp = 36.1 °C

FIGURE 12.8 *Molecular shape affects the strengths of London forces.* Shown are two molecules with the formula C_5H_{12}. Not all hydrogen atoms can be seen in these space-filling models. The neopentane molecule, $(CH_3)_4C$, has a more compact shape than the *n*-pentane molecule, $CH_3CH_2CH_2CH_2CH_3$. In the more compact structure, the H atoms cannot interact with those on neighboring molecules as well as the H atoms in the long chainlike structure, so overall the intermolecular attractions are weaker between the more compact molecules.

than those of HBr because Cl is more electronegative than Br. This means that the dipole–dipole attractions in HCl(l) are stronger than those in HBr(l). But their boiling points tell us that the intermolecular attractions in HBr(l) are stronger than those in HCl(l). This could only be true if the London forces in HBr are a lot stronger than in HCl, so that they can compensate for the weaker dipole–dipole attractions.

London forces are expected to increase going from HCl to HBr because Br is larger than Cl and is more polarizable. The fact that the increased strengths of the London forces are able to dominate over the changes in dipole–dipole attractions reveals that London forces contribute significantly to the overall intermolecular attractions in these substances. In fact, one source suggests that in HCl(l), the dipole–dipole attractions account for only about 20% of the total intermolecular attractions and that London forces account for about 80%.

Ion–dipole and ion–induced dipole forces of attraction

Ions are able to interact with the charged ends of polar molecules to give **ion–dipole attractions.** This occurs in water, for example, when ionic compounds dissolve to give hydrated ions. Cations become surrounded by water molecules that are oriented with the negative ends of their dipoles pointing toward the cation. Similarly, anions attract the positive ends of water dipoles. This is illustrated in Figure 12.9. These same interactions can persist into the solid state as well. As you learned earlier, some salts crystallize as hydrates in which water molecules are held in fixed stoichiometric ratios within the solid. Many aluminum salts, for example, crystallize as hexahydrates. In $AlCl_3 \cdot 6H_2O$, for instance, we find the water molecules surrounding the aluminum ion at the vertices of an octahedron, as illustrated in Figure 12.10. They are held there by ion–dipole attractions.

The creation of a dipole in a neighbor is not restricted to instantaneous irregularities in the electron densities of neutral atoms and molecules. Ions, with full positive and negative charges, also induce dipoles in particles nearby (like molecules of a solvent, or even other ions), leading to **ion–induced dipole** attractions. These can be quite strong because the charge on the ion does not flicker on and off like the instantaneous charges responsible for ordinary London dispersion forces.

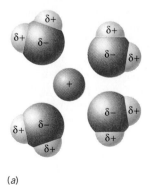

(a)

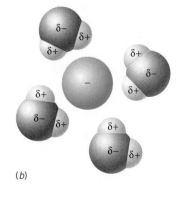

(b)

FIGURE 12.9 *Ion–dipole attractions.* Here we see the attractions between water molecules and positive and negative ions. (*a*) The negative ends of water dipoles surround a cation and are attracted to the ion. (*b*) The positive ends of water molecules surround an anion, which gives a net attraction.

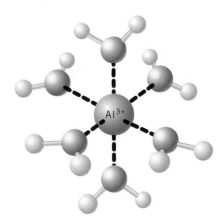

Figure 12.10 *Ion–dipole attractions hold water molecules in a hydrate.* Water molecules are arranged at the vertices of an octahedron around an aluminum ion in $AlCl_3 \cdot 6H_2O$.

Estimating the effects of intermolecular forces

In this section we have described a number of different types of intermolecular attractive forces and the kinds of substances in which they occur (see the summary in Table 12.3). With this knowledge, you should now be able to make some estimate of the nature and relative strengths of intermolecular attractions if you know the molecular structure of a substance. This will enable you to understand and sometimes predict how the physical properties of different substances compare. For example, we've already mentioned that boiling point is a property that depends on the strengths of intermolecular attractions. By being able to compare intermolecular forces in different substances, we can sometimes predict how their boiling points compare. This is illustrated in Example 12.1.

EXAMPLE 12.1
Using Relative Attractive Forces to Predict Properties

Below are structural formulas of ethanol (ethyl alcohol) and propylene glycol (a compound used as a nontoxic antifreeze). Which of these compounds would be expected to have the higher boiling point?

$$
\underset{\text{ethanol}}{
\begin{array}{c}
\ \ \text{H}\ \ \text{H} \\
\ \ \ \ | \ \ \ \ | \\
\text{H--C--C--OH} \\
\ \ \ \ | \ \ \ \ | \\
\ \ \text{H}\ \ \text{H}
\end{array}}
\qquad
\underset{\text{propylene glycol}}{
\begin{array}{c}
\ \ \text{H}\ \ \text{H}\ \ \text{H} \\
\ \ \ \ | \ \ \ \ | \ \ \ \ | \\
\text{H--C--C--C--OH} \\
\ \ \ \ | \ \ \ \ | \ \ \ \ | \\
\ \ \text{H}\ \ \text{OH H}
\end{array}}
$$

ANALYSIS: We know that boiling points are related to the strengths of intermolecular attractions—the stronger the attractions, the higher the boiling point. Therefore, if we can determine which compound has the stronger intermolecular attractions, we can answer the question. Let's decide which kinds of attractions are present and then try to determine their relative strengths.

SOLUTION: We know that both substances will experience London forces, because they are present between *all* molecules. London forces become stronger as molecules become larger, so the London forces should be stronger in propylene glycol.

Looking at the structures, we see that both contain —OH groups (one in ethanol and two in propylene glycol). This means we can expect that there will be hydrogen bonding in both liquids. Because there are more —OH groups per molecule in propylene glycol than in ethanol, we might reasonably expect that there are more opportunities for the ethylene glycol molecules to participate in hydrogen bonding. This would make the hydrogen bonding forces greater in propylene glycol.

Our analysis tells us that both kinds of attractions are stronger in propylene glycol than in ethanol, so propylene glycol should have the higher boiling point.

Is the Answer Reasonable?
There's not much we can do to check our answer other than to review the reasoning, which is sound. (We could also check a reference book, where we would find that the

TABLE 12.3	SUMMARY OF INTERMOLECULAR ATTRACTIONS	
Intermolecular Attraction	Types of Substances That Exhibit Attraction	Strength Relative to a Covalent Bond
Dipole–dipole attractions	Occurs between molecules that have permanent dipoles (i.e., polar molecules).	1%–5%
Hydrogen bonding	Occurs when molecules contain N—H and O—H bonds.	5%–10%
London dispersion forces	All atoms, molecules, and ions experience these kinds of attractions. They are present in all substances.	Depends on sizes and shapes of molecules. For large molecules, the cumulative effect of many weak attractions can lead to a large net attraction.
Ion–dipole attractions	Occurs when ions interact with polar molecules.	~10%; depends on ion charge and polarity of molecule.
Ion–induced dipole attractions	Occurs when an ion creates a dipole in a neighboring particle, which may be a molecule or another ion.	Variable, depending on the charge on the ion and the polarizability of its neighbor.

boiling point of ethanol is 78.5 °C and the boiling point of propylene glycol is 188.2 °C!)

PRACTICE EXERCISE 1: Propylamine and trimethylamine have the same molecular formula, C_3H_9N, but quite different structures, as shown below. Which of these substances is expected to have the higher boiling point? Why?

$$CH_3—CH_2—CH_2—NH_2 \qquad \begin{array}{c} CH_3 \\ | \\ H_3C—N—CH_3 \end{array}$$
propylamine   trimethylamine

12.3 ▶ Intermolecular forces and tightness of packing affect the properties of liquids and solids

Earlier we briefly described some properties of liquids and solids. We continue here with a more in-depth discussion, and we'll start by examining two properties that depend mostly on how tightly packed the molecules are, namely, *compressibility* and *diffusion*. Other properties depend much more on the strengths of intermolecular attractive forces, properties such as *retention of volume or shape, surface tension,* the ability of a liquid to *wet* a surface, the *viscosity* of a liquid, and a solid's or liquid's *tendency to evaporate.*

Compressibility and diffusion depend primarily on tightness of packing

Compressibility

The **compressibility** of a substance is a measure of the ability of the substance to be forced into a smaller volume. Gases are highly compressible because the molecules are far apart and the gas is mostly empty space. Earlier we noted that in a liquid or solid most of the space is taken up by the molecules, and that there is very little empty space into which to crowd other molecules. As a result, it is very difficult to compress liquids or solids to a smaller volume by applying pressure, so we say that these states of matter are nearly **incompressible.** This is a property that we often make use of. When you "step on the brakes" of a car, for example, you rely on the incompressibility of the brake fluid to transmit the pressure you apply with your foot to the brake shoes on the wheels. If some air should get into the brake lines,

Hydraulic machinery such as this earthmover uses the incompressibility of liquids to transmit forces that accomplish work.

you're in big trouble, because when you apply pressure to the brakes it simply compresses the air and doesn't force the brake shoes to stop the car. A similar application of the incompressibility of liquids is the foundation of the engineering science of *hydraulics,* which uses fluids to transmit forces that lift or move heavy objects.

Diffusion

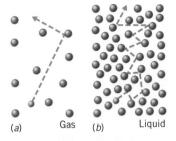

FIGURE 12.11 *Diffusion in a gas and a liquid viewed at the molecular level.* (*a*) Diffusion in a gas is rapid because relatively few collisions occur between widely spaced molecules. (*b*) Diffusion in a liquid is slow because of many collisions between closely spaced particles.

Diffusion occurs much more rapidly in gases than in liquids, and hardly at all in solids. In gases, molecules diffuse rapidly because they travel relatively long distances between collisions, as illustrated in Figure 12.11. In liquids, however, a given molecule suffers many collisions as it moves about, so it takes longer to move from place to place and diffusion is much slower. Diffusion in solids is almost nonexistent at room temperature because the particles of a solid are held tightly in place. At high temperatures, though, the particles of a solid sometimes have enough kinetic energy to jiggle their way past each other, and diffusion can occur slowly. Such high-temperature solid-state diffusion is used to make electronic devices, like transistors (described in Section 13.4). Solid-state diffusion allows for small, carefully controlled amounts of some impurity—arsenic, for example—to be added to a semiconductor such as silicon or germanium. This process, called "doping," allows the conductivity of these materials to be modified in desirable ways.

Most physical properties depend primarily on the strengths of intermolecular attractions

Retention of volume and shape

Volume and shape are physical properties we discussed earlier when we compared gases, liquids, and solids. The underlying reasons for the differences are simply intermolecular attractive forces.

In gases, intermolecular attractions are too weak to prevent the molecules from moving apart to fill the entire container, so a gas will conform to the shape and volume of whatever container it's placed in. In liquids and solids, however, the attractions are much stronger and are able to hold the particles closely together,

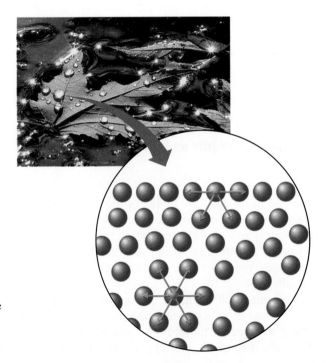

FIGURE 12.12 *Surface tension and intermolecular attractions.* In water, as in other liquids, molecules at the surface are surrounded by fewer molecules than those below the surface. As a result, surface molecules experience fewer attractions than molecules within the liquid.

thereby preventing liquids and solids from expanding in volume in the manner of a gas. As a result, liquids and solids keep the same volume regardless of the size of their container. In a solid, the attractions are even stronger than in a liquid. They hold the particles more or less rigidly in place, so a solid retains its shape when moved from one container to another.

Surface tension

A property that is especially evident for liquids is the phenomenon of *surface tension,* a property related to the tendency of a liquid to seek a shape that yields the minimum surface area. For a given volume, the shape with the minimum surface area is a sphere—it's a principle of solid geometry. This is why raindrops tend to be little spheres.

To understand surface tension, we need to examine why molecules would prefer to be within a liquid rather than at its surface. Let's begin by examining the environment surrounding molecules in these two conditions (see Figure 12.12). We see that a molecule that's *within* the liquid is surrounded by densely packed molecules on all sides, whereas one at the *surface* has neighbors beside and below it, but none above. As a result, a surface molecule is attracted to fewer neighbors than one within the liquid. With this in mind, let's imagine how we might change an interior molecule to one at the surface. To accomplish this, we would have to pull away some of the surrounding molecules. Because there are intermolecular attractions, removing neighbors requires work, so there's an increase in potential energy involved. This leads to the conclusion that *a molecule at the surface has a higher potential energy than a molecule in the bulk of the liquid.*

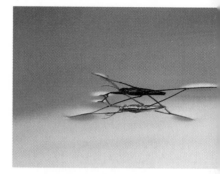

An insect called a *waterstrider,* shown here, is able to walk on water because of the liquid's surface tension, which causes the water to behave as though it has a skin that resists piercing by the insect's legs.

What does potential energy have to do with surface tension? In general, a system becomes more stable when its potential energy decreases. That's why a yardstick standing on end falls over; by doing so its potential energy drops and it reaches a lower-energy, more stable position. For a liquid, reducing its surface area (and thereby reducing the number of molecules at the surface) lowers its potential energy, and the lowest energy is achieved when the liquid has the smallest surface area possible (namely, a spherical shape). In more accurate terms, then, *the* **surface tension** of a liquid is proportional to the energy needed to expand its surface area.

The tendency of a liquid to spontaneously acquire a minimum surface area explains many common observations. For example, surface tension causes the sharp edges of glass tubing to become rounded when the glass is softened in a flame, an operation called "fire polishing." Surface tension is also what allows us to fill a water glass above the rim, giving the surface a rounded appearance (Figure 12.13). The surface behaves as if it has a thin, invisible "skin" that lets the water in the glass pile up, trying to assume a spherical shape. Gravity, of course, works in opposition, tending to pull the water down. If too much water is added to the glass, the gravitational force finally wins and the skin breaks; the water overflows. If you push on the surface of a liquid, it resists expansion and pushes back, so the surface "skin" appears to resist penetration. This is what enables certain insects to "walk on water," as illustrated in the photo in the margin.

Another consequence of surface tension is that it enables grains of moist sand to stick together and allows children to build sand castles. As illustrated in Figure 12.14, separating sand particles that are coated with a film of water requires increasing the surface area of the water. The liquid resists this and thereby serves as a "glue" holding the sand together. (If you completely submerge sand in water, the "glue" is lost, because there's no longer a surface area to be affected by separating the sand particles.)

Surface tension is a property that varies with the strengths of intermolecular attractions. Liquids with strong intermolecular attractive forces have large differences in potential energy between their interior and surface molecules, so expanding the surface requires a large increase in energy. The generalization, then, is that *liquids*

FIGURE 12.13 *Surface tension in a liquid.* Surface tension allows a glass to be filled with water above the rim.

FIGURE 12.14 *Surface tension holds moist grains of sand together.* If the sand particles are pulled apart, the surface area of the water film increases, which would lead to an increase in potential energy. As a result, the sand particles resist being pulled apart, and therefore seem to stick together.

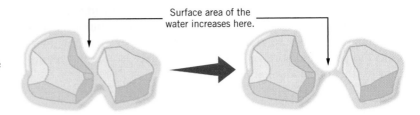

Surface area of the water increases here.

with strong intermolecular attractions have large surface tensions. Not surprisingly, water's surface tension is among the highest known (comparisons being at the same temperature); its intermolecular forces are hydrogen bonds, the strongest kind of dipole–dipole attraction. In fact, the surface tension of water is roughly three times that of gasoline, which consists of relatively nonpolar hydrocarbon molecules able to experience only London forces.

Because of strong hydrogen bonding, the surface tension of water is roughly two to three times larger than the surface tension of any common organic solvent.

Wetting of a surface by a liquid

A property we associate with liquids, especially water, is their ability to wet things. **Wetting** is the spreading of a liquid across a surface to form a thin film. Water wets clean glass, such as the windshield of a car, by forming a thin film over the surface of the glass (see Figure 12.15*a*). Water won't wet a greasy windshield, however. Instead, on greasy glass water forms tiny beads (see Figure 12.15*b*).

For wetting to occur, the intermolecular attractive forces between the liquid and the surface must be of about the same strength as the forces within the liquid itself. Such a rough equality of forces exists when water touches clean glass. This is because the glass surface contains lots of oxygen atoms. The water molecules can form hydrogen bonds to the surface oxygens nearly as well as they can form hydrogen bonds to each other. As a result, part of the energy needed to expand the water's surface area when wetting occurs is recovered by the formation of hydrogen bonds to the glass surface.

Glass is characterized by a vast network of silicon–oxygen bonds.

When the glass is coated by a film of oil or grease, however, the conditions are quite different (Figure 12.15*b*). The surface now is actually no longer glass, but oil and grease instead. Their molecules are relatively nonpolar, so their attractions to other molecules (including water) are largely limited to London forces. These are weak compared with hydrogen bonds, so the attractions *within* liquid water are much stronger than the attractions *between* water molecules and the greasy surface. The weak water-to-grease London forces can't overcome the hydrogen bonding within liquid water, so the water doesn't spread out; it forms beads instead.

One of the reasons why detergents are used for such chores as doing laundry or washing floors is that detergents contain chemicals called **surfactants** that drastically lower the surface tension of water. This makes the water "wetter," which allows the detergent solution to spread more easily across the surface to be cleaned.

When a liquid has a low surface tension, like gasoline, we know that it has weak intermolecular attractions, and such a liquid easily wets solid surfaces. The weak attractions between molecules in gasoline, for example, are readily replaced by attractions to almost any surface, so gasoline easily spreads to a thin film. If you've ever spilled a little gasoline, you have experienced firsthand that it doesn't bead.

Viscosity

As everybody knows, syrup flows less readily or is more resistant to flow than water (both at the same temperature). Flowing is a change in the *form* of the liquid, and such resistance to a change in form is called the liquid's **viscosity.** We say that syrup is more *viscous* than water. The concept of viscosity is not confined to liquids, however, although it is with liquids that the property is most commonly associated.

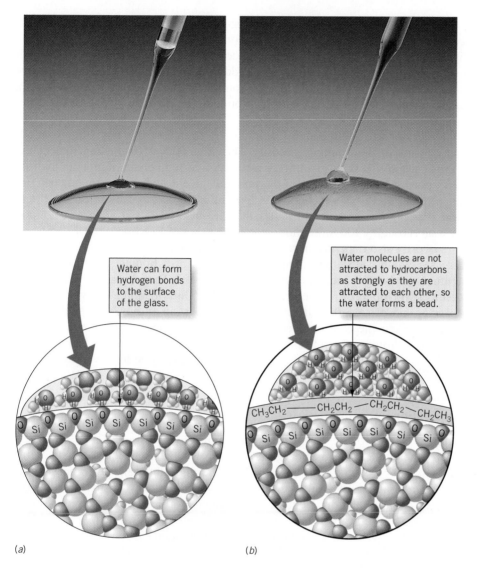

Water can form hydrogen bonds to the surface of the glass.

Water molecules are not attracted to hydrocarbons as strongly as they are attracted to each other, so the water forms a bead.

(a)

(b)

FIGURE 12.15 *Intermolecular attractions affect the ability of water to wet a surface.* (*a*) Water wets a clean glass surface because the surface contains many oxygen atoms to which water molecules can form hydrogen bonds. (*b*) If the surface has a layer of grease, to which water molecules are only weakly attracted, the water doesn't wet it. The water resists spreading and forms a bead instead.

Solid things, even rock, also yield to forces acting to change their shapes, but normally do so only gradually and imperceptibly. Gases also have viscosity, but they respond almost instantly to form-changing forces.

Viscosity has been called the *internal friction* of a material. It is influenced both by intermolecular attractions and by molecular shape and size. For molecules of similar size, we find that as the strengths of the intermolecular attractions increase, so does the viscosity. For example, consider acetone (nail polish remover) and ethylene glycol (automotive antifreeze), each of which contains ten atoms.

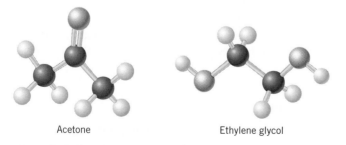

Acetone

Ethylene glycol

If you've ever spilled a little acetone, you know that it flows very easily. In contrast, ethylene glycol has an "oily" thickness to it and flows more slowly. Ethylene glycol is more viscous than acetone, and looking at the molecular structures, it's easy to see why. Acetone contains a polar carbonyl group ($>$C$=$O), so it experiences dipole–dipole attractions as well as London forces. Ethylene glycol, on the other hand, contains two —OH groups, so in addition to London forces, ethylene glycol molecules also participate in hydrogen bonding (a much stronger interaction than dipole–dipole forces). Strong hydrogen bonding in ethylene glycol makes it more viscous than acetone.

Molecular size and the ability of molecules to tangle with each other is another major factor in determining viscosity. The long, floppy, entangling molecules in heavy machine oil (almost entirely a mixture of long-chain, nonpolar hydrocarbons), plus the London forces in the material, give it a viscosity roughly 600 times that of water at 15 °C. Vegetable oils, like the olive oil or corn oil used to prepare salad dressings, consist of molecules that are also large but generally nonpolar. Olive oil is roughly 100 times more viscous than water.

Viscosity also depends on temperature; as the temperature drops, the viscosity increases. When water, for example, is cooled from its boiling point to room temperature, its viscosity increases by over a factor of 3. The increase in viscosity with cooling is why operators of vehicles use a "light," thin (meaning less viscous) motor oil during subzero weather. It's much easier to start an automobile at subzero temperatures when the cold crankcase oil flows readily and is not thick and syrupy. During very hot months, a "heavy" (more viscous) motor oil is used so that it cannot slip as readily between cylinders and pistons and so be lost even as it is doing its lubricating work.

> Recall that the sum of the London forces experienced by the many atoms in a large molecule can far outweigh dipole–dipole forces and even the hydrogen bonding between small molecules.

> As the temperature drops, molecules move more slowly and intermolecular forces become more effective at restraining flow.

Evaporation and sublimation

> A change of state is also called a *phase change*.

Finally, we come to one of the most important physical properties of liquids and solids—their tendency to undergo a change of state from liquid to gas or from solid to gas. For liquids, the change is called **evaporation,** and everyone has seen liquids evaporate; it's what happens to water, for example, when streets, wet from a rain shower, gradually dry. Solids can also change directly to the gaseous state by evaporation without going through the liquid state, except that for a solid-to-gas change of state we use a special term, **sublimation.** Solid carbon dioxide, commonly called *dry ice,* is an example. It is "dry" ice because it doesn't melt; instead, at atmospheric pressure it *sublimes;* it changes directly to gaseous CO_2. Naphthalene, the ingredient in some brands of moth flakes, is a substance that can sublime and seemingly disappear.

To understand evaporation and sublimation, we have to examine the motions of molecules. As you know by now, molecules do not stand still in liquids and solids. They bounce around against their neighbors, and at a given temperature, there is *exactly the same* distribution of kinetic energies in a liquid or a solid as there is in a gas. This means that Figure 7.5 on page 261 applies to liquids and solids as well as gases. Some molecules have low kinetic energies and move slowly, while others with higher kinetic energies move faster. A small fraction of the molecules have very large kinetic energies and therefore very high velocities. If one of these high-velocity molecules is at the surface and is moving outward fast enough, it can escape the attractions of its neighbors and enter the vapor state. When this happens, we say the molecule has left by evaporation (or sublimation, if the substance is a solid).

Naphthalene sublimes when heated and the vapor condenses directly to a solid when it encounters a cool surface. Here, beautiful flaky naphthalene crystals have been formed on the bottom of a flask containing ice water.

Evaporation and cooling

One of the things we notice about the evaporation of a liquid is that it produces a cooling effect. Have you ever come out of the water after swimming and been chilled by a breeze? The evaporation of water from your body produced this effect. In fact, our bodies use the evaporation of water to maintain a constant body temperature.

During warm weather or vigorous exercise we perspire, and the evaporation of our perspiration cools our skin. Evaporative cooling also occurs within the lungs, which puts water vapor into position to be exhaled, carrying heat out of the body with it. You've probably also used the cooling effect caused by evaporation by blowing gently on the surface of a hot bowl of soup or a cup of coffee. The stream of air stirs the liquid, bringing more hot liquid to the surface to be cooled as it evaporates.

We can see why liquids become cool during evaporation by examining Figure 12.16, which illustrates the kinetic energy distribution in a liquid at a particular temperature. A marker along the horizontal axis shows the minimum kinetic energy needed by a molecule to escape the attractions of its neighbors. Only molecules with kinetic energies equal to or greater than the minimum can leave the liquid. Others with less kinetic energy may begin to leave, but before they can escape they slow to a stop and then fall back. The situation is somewhat like attempting to launch a spaceship using a cannon. If the velocity of the projectile is not very great, it will not rise very far before it slows to a halt and falls to Earth. However, if it is going fast enough (if it has its "escape velocity"), it can overcome Earth's gravity and leave.

Figure 12.16 also allows us to understand why liquids cool as they evaporate. Notice in Figure 12.16 that the minimum kinetic energy needed to escape is much larger than the average kinetic energy, which means that when molecules evaporate they carry with them large amounts of kinetic energy. As a result, the average kinetic energy of the molecules left behind decreases. (You might think of this as being similar to removing people taller than 6 ft from a large class of students. When this is done, the average height of those who are left is less.) Because the Kelvin temperature of the remaining liquid is directly proportional to the now lower average kinetic energy, the temperature is lower; in other words, evaporation causes the liquid that remains to be cooler.

The rate of evaporation depends on surface area, temperature, and strengths of intermolecular attractions

One of the important things we are going to be concerned about later in this chapter is the *rate of evaporation* of a liquid. There are several factors that control this. You are probably already aware of one of them—the surface area of the liquid. Because evaporation occurs from the liquid's surface and not from within, it makes sense that as the surface area is increased, more molecules are able to escape and the liquid evaporates more quickly. For liquids having the same surface area, the rate of evaporation depends on two factors, namely, temperature and the strengths of intermolecular attractions. Let's examine each of them separately.

When a canteen of water is covered with a water-absorbing fabric that is kept wet, the evaporation of the water helps to keep the water in the canteen cool.

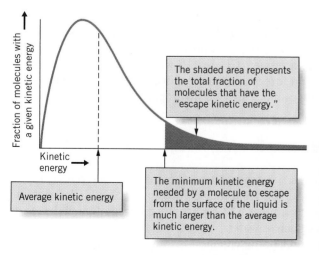

The shaded area represents the total fraction of molecules that have the "escape kinetic energy."

Average kinetic energy

The minimum kinetic energy needed by a molecule to escape from the surface of the liquid is much larger than the average kinetic energy.

FIGURE 12.16 *Cooling of a liquid by evaporation.* Molecules that are able to escape from the liquid have kinetic energies larger than the average. When they leave, the average kinetic energy of the molecules left behind is less, so the temperature is lower.

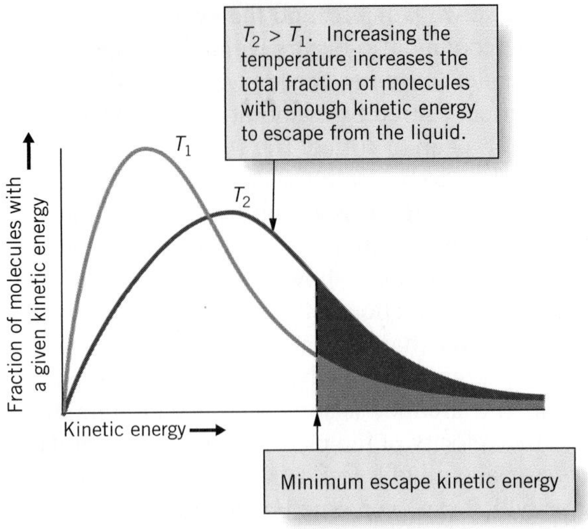

FIGURE 12.17 *Effect of increasing the temperature on the rate of evaporation of a liquid.* At the higher temperature, the total fraction of molecules with enough kinetic energy to escape is larger, so the rate of evaporation is larger.

To understand how temperature and intermolecular forces affect the rate of evaporation, we must compare evaporation rates from the same-size surface area. In this discussion, therefore, "rate of evaporation" means "rate of evaporation *per unit surface area.*"

Effect of temperature on the rate of evaporation

Intermolecular forces and the rate of evaporation

The influence of temperature on evaporation rate is no surprise; you already know that hot water evaporates faster than cold water. The underlying reason is brought out in Figure 12.17, which shows the kinetic energy distributions for the *same* liquid at two temperatures. Notice two important features of the figure. First, the same minimum kinetic energy is needed for the escape of molecules at both temperatures. This minimum is determined by the kinds of attractive forces between the molecules and is independent of temperature. Second, the shaded area of the curve represents the *total* fraction of molecules having kinetic energies equal to or greater than the minimum. At the higher temperature, the total fraction is larger, which means that at the higher temperature a greater total fraction has the ability to evaporate. As you might expect, when more molecules have the needed energy, more evaporate in a unit of time. Therefore, *the rate of evaporation per unit surface area of a given liquid is greater at a higher temperature.*

The effect of intermolecular attractions on evaporation rate can be seen by studying Figure 12.18. Here we have kinetic energy distributions for two *different* liquids—call them *A* and *B*—both at the same temperature. In liquid *A*, the attractive forces are weak; they might be of the London type, for example. As we see, the minimum kinetic energy needed by *A* molecules to escape is not very large because they are not attracted very strongly to each other. In liquid *B*, the intermolecular attractive forces are much stronger; they might be hydrogen bonds, for instance. Molecules of *B*, therefore, are held more tightly to each other at the liquid's surface and must have a higher kinetic energy to evaporate. As you can see from the figure, the total fraction of molecules with enough energy to evaporate is greater for *A* than for *B*, which means that *A* evaporates faster than *B*. In general, then, *the weaker the intermolecular attractive forces, the faster is the rate of evaporation at a given temperature.* You are probably also aware of this phenomenon. At room temperature, for example, nail polish remover [acetone, $(CH_3)_2CO$], whose molecules experience weak dipole–dipole and London forces of attraction, evaporates faster than water, whose molecules feel the effects of much stronger hydrogen bonds.

12.4 ▶ Changes of state lead to dynamic equilibria

A **change of state** occurs when a substance is transformed from one physical state to another. The evaporation of a liquid and the sublimation of a solid, described in the preceding section, are two examples. Others are the melting of a solid such as ice and the freezing of a liquid such as water.

One of the important features about changes of state is that, at any particular

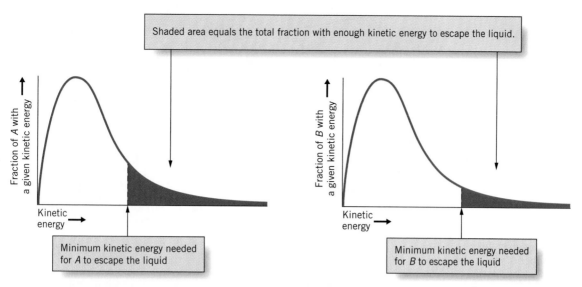

FIGURE 12.18 *Kinetic energy distribution in two different liquids, A and B, at the same temperature.* The minimum kinetic energy required by molecules of *A* to escape is less than for *B* because the intermolecular attractions in *A* are weaker than in *B*. This causes *A* to evaporate faster than *B*.

temperature, they always tend toward a condition of *dynamic equilibrium.* We introduced the concept of dynamic equilibrium on page 177 with an example of a system at chemical equilibrium. The same general principles apply to a physical equilibrium, such as that between a liquid and its vapor. Let's examine in some detail what happens when a liquid evaporates in a sealed container and what is involved when equilibrium is reached.

If a liquid is placed in an empty container, it immediately begins to evaporate, and molecules of the substance begin to collect in the space above the liquid (see Figure 12.19*a*). As they fly around in the vapor, the molecules collide with each other, with the walls of the container, and with the surface of the liquid itself. Those that strike the liquid's surface tend to stick because their kinetic energies become scattered among the surface molecules. It is somewhat like throwing a ping-pong ball into a large box of ping-pong balls. The incoming ball knocks others around, and by giving them kinetic energy it loses some of its own. Because its kinetic energy has been reduced, there is a high probability that the incoming ball won't bounce out.

The change of a vapor to its liquid state is called **condensation;** we say that the vapor *condenses* when it changes to a liquid. The rate at which vapor molecules collide with a liquid's surface and condense depends on the concentration of molecules in the vapor. When there are only a few in a given volume, the number of collisions per second with a unit area of the liquid's surface is small, and the rate of condensation is slow. When there are many molecules per unit of volume in the vapor, many such collisions occur each second with a unit of surface area, so the rate of condensation is higher.

In chemistry (unless otherwise indicated), when we use the term *equilibrium,* we always mean *dynamic equilibrium.*

Evaporation

increases number of
molecules in vapor

| molecules in the liquid state | | molecules in the gaseous state |

Condensation

decreases number of
molecules in vapor

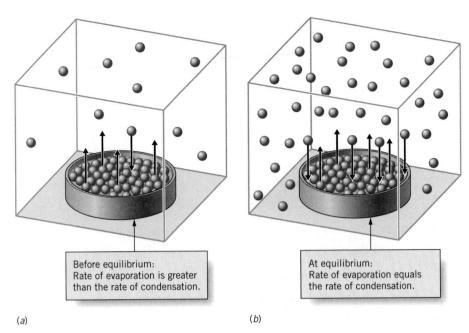

FIGURE **12.19** *Evaporation of a liquid into a sealed container.* (*a*) The liquid has just begun to evaporate into the container. The rate of evaporation is greater than the rate of condensation. (*b*) A dynamic equilibrium is reached when the rate of evaporation equals the rate of condensation. In a given time period, the number of molecules entering the vapor equals the number that leave, so there is no net change in the number of gaseous molecules.

Before equilibrium:
Rate of evaporation is greater than the rate of condensation.

At equilibrium:
Rate of evaporation equals the rate of condensation.

(*a*)   (*b*)

Let's go back now to when the liquid was first introduced into the empty container. With virtually none of its molecules yet in the vapor state, evaporation is occurring faster than condensation, so the number of molecules in the vapor is increasing. But as molecules accumulate in the vapor, the condensation rate increases, and it continues to increase until the rates of condensation and evaporation become the same (Figure 12.19*b*). From that moment on, the number of molecules in the vapor will remain constant, because over a given period of time the number that enters the vapor is the same as the number that leaves. At this point we have a condition of *dynamic equilibrium,* one in which two opposing effects, evaporation and condensation, are occurring at equal rates so that their effects cancel. It is an *equilibrium* because there is no apparent change; the *number* of vapor molecules remains constant as does the number of liquid molecules. It is a *dynamic* equilibrium because activity has not ceased. Molecules continue to evaporate and condense; they just do so at equal rates.

Similar equilibria are also reached in melting and sublimation. At a temperature called the **melting point,** a solid begins to change to a liquid as heat is added. At this temperature a dynamic equilibrium can exist between molecules in the solid and those in the liquid. Molecules leave the solid and enter the liquid at the same rate as molecules leave the liquid and join the solid (Figure 12.20). As long as no heat is added or removed from such a solid–liquid equilibrium mixture, melting and freezing occur at equal rates. For sublimation, the situation is exactly the same

FIGURE **12.20** *Dynamic equilibrium at the melting point of a solid.* As long as no heat is added or removed, melting (red arrows) and freezing (black arrows) occur at equal rates and the number of particles in the solid remains constant.

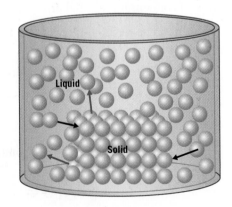

Liquid

Solid

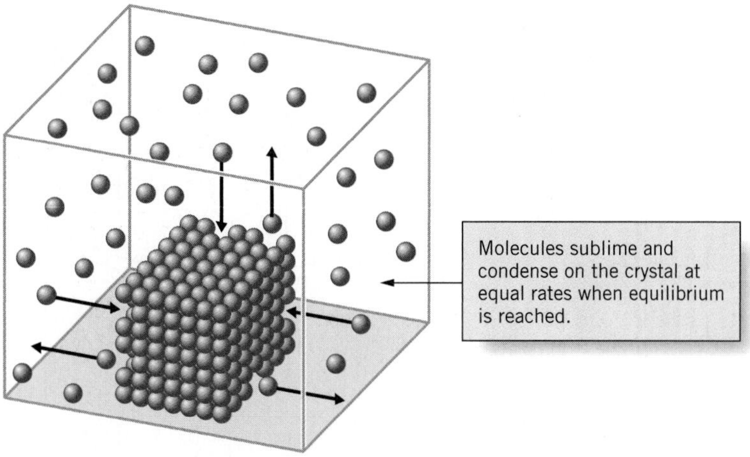

Molecules sublime and condense on the crystal at equal rates when equilibrium is reached.

FIGURE 12.21 *A solid–vapor equilibrium.* Molecules evaporate from the solid at the same rate as molecules condense from the vapor.

as in the evaporation of a liquid into a sealed container (see Figure 12.21). After a few moments, the rates of sublimation and condensation become the same and equilibrium is established. All these various equilibria have very profound effects on the properties of solids and liquids, as we shall soon see.

12.5 ▶ Vapor pressures of liquids and solids are controlled by temperature and intermolecular attractions

When a liquid evaporates, the molecules that enter the vapor exert a pressure called the **vapor pressure.** From the very moment a liquid begins to evaporate into the vapor space above it, there is a vapor pressure. If the evaporation is taking place inside a sealed container, this pressure grows until finally equilibrium is reached. Once the rates of evaporation and condensation become equal, the concentration of molecules in the vapor remains constant and the vapor exerts a constant pressure. This final pressure is called the **equilibrium vapor pressure** of the liquid. In general, when we refer to the *vapor pressure,* we really mean the equilibrium vapor pressure.

A vapor–liquid equilibrium is possible only in a closed container. When the container is open, vapor molecules drift away and the liquid might completely evaporate.

We can measure the vapor pressure of a liquid with an apparatus like that shown in Figure 12.22. Initially, both sides of the apparatus are open to the atmosphere, so the pressure in both arms of the manometer is the same. A small amount of liquid is then added to the flask from the funnel and the left side of the apparatus is sealed immediately. As some of the liquid evaporates, the pressure in the flask rises, forcing the fluid in the left side of the manometer downward. The increase in pressure, as measured by the difference in the levels in the manometer, is equal to the pressure exerted by the vapor, namely, the vapor pressure of the liquid.

Factors that affect the equilibrium vapor pressure

Factors that affect the vapor pressure

Figure 12.23 shows plots of equilibrium vapor pressure versus temperature for a few liquids. From these graphs we see that both a liquid's temperature and its chemical composition are the major factors affecting its vapor pressure. Once we have selected a particular liquid, however, only the temperature matters. The reason is that the vapor pressure of a given liquid is a function solely of its rate of evaporation *per unit area of the liquid's surface.* When this rate is large, a large concentration of molecules in the vapor state is necessary to establish equilibrium, which is another way of saying that the vapor pressure is relatively high when the evaporation rate is high. As the temperature of a given liquid increases, so does its rate of evaporation and so does its equilibrium vapor pressure.

As chemical composition changes in going from one liquid to another, the strengths of intermolecular attractions change. If the attractions increase, the rates

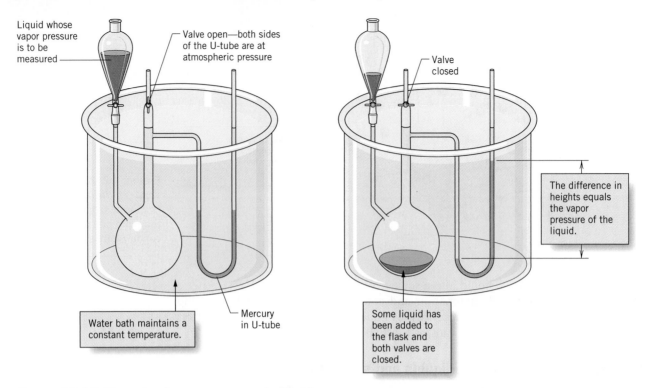

Figure 12.22 *Measuring the vapor pressure of a liquid.*

of evaporation at a given temperature decrease, and the vapor pressures decrease. These data on relative vapor pressures tell us that of the four liquids in Figure 12.23, intermolecular attractions are strongest in propylene glycol, next strongest in acetic acid, third strongest in water and weakest in ether. Thus, *we can use vapor pressures as indications of relative strengths of the attractive forces in liquids.*

A liquid with a high vapor pressure at a given temperature is said to be *volatile*.

Factors that do not affect the vapor pressure

An important fact about vapor pressure is that *its magnitude doesn't depend on the total surface area of the liquid, nor on the volume of the liquid in the container, nor on the volume of the container itself, just as long as some liquid remains when equilibrium is reached.* The reason is that none of these factors affects the rate of evaporation *per unit surface area.*

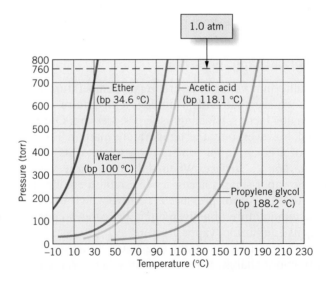

Figure 12.23 *Variation of vapor pressure with temperature for some common liquids.*

Increasing the *total* surface area does increase the *total* rate of evaporation, but the larger area is also available for condensation, so the rate at which molecules return to the liquid also increases. The rates of both evaporation and condensation are thus affected equally, and no change occurs to the equilibrium vapor pressure.

Adding more liquid to the container can't affect the equilibrium either, because evaporation occurs from the *surface*. Having more molecules in the bulk of the liquid does not change what is going on at the surface.

To understand why the vapor pressure doesn't depend on the *size* of the vapor space, consider a liquid in equilibrium with its vapor in a cylinder with a movable piston, as illustrated in Figure 12.24*a*. Withdrawing the piston (Figure 12.24*b*) increases the volume of the vapor space; as the vapor expands the pressure it exerts becomes less, so there's a momentary drop in the pressure. The molecules of the vapor, being more spread out now, no longer strike the surface as frequently, so the rate of condensation has also decreased. The rate of evaporation hasn't changed, however, so for a moment the system is not at equilibrium and the substance is evaporating faster than it is condensing (Figure 12.24*b*). This condition prevails, changing more liquid into vapor, until the concentration of molecules in the vapor has risen enough to make the condensation rate again equal to the evaporation rate (Figure 12.24*c*). At this point the vapor pressure has returned to its original value. Therefore, the net result of expanding the space above the liquid is to change more liquid into vapor, but it does not affect the equilibrium vapor pressure. Similarly, we expect that reducing the volume of the vapor space above the liquid will also not affect the equilibrium vapor pressure.

> Recall that when the volume of a gas increases at constant temperature, its pressure decreases.

> **PRACTICE EXERCISE 2:** Suppose a liquid is in equilibrium with its vapor in a piston-cylinder apparatus like that just described. If the piston is pushed in a short way and the system is allowed to return to equilibrium, what will happen to the *total number* of molecules in the liquid and the vapor?

Vapor pressures of solids

Solids have vapor pressures just as liquids do. In a crystal, the particles are not stationary, as we have said. They are constantly jiggling around, bumping into their neighbors. At a given temperature there is a distribution of kinetic energies, so some particles jiggle about slowly while others bounce around with a great deal of molecular kinetic energy. Some particles at the surface have large enough kinetic

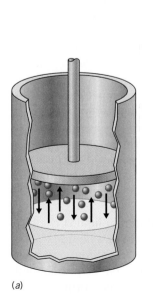 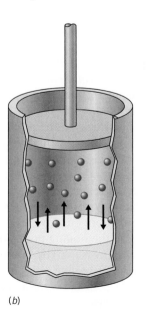 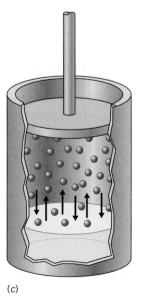

(*a*) (*b*) (*c*)

FIGURE 12.24 *Effect of a volume change on the vapor pressure of a liquid.* (*a*) Equilibrium exists between liquid and vapor. (*b*) The volume is increased, which upsets the equilibrium and causes the pressure to drop. Because the concentration of molecules in the vapor has dropped, the rate of condensation is now less than the rate of evaporation, which hasn't changed. (*c*) After more liquid has evaporated, equilibrium is restored and the rates of condensation and evaporation are again equal and the vapor pressure has returned to its initial value.

In many solids, such as NaCl, the attractive forces are so strong that virtually no particles have enough kinetic energy to escape at room temperature, so essentially no evaporation occurs. Their vapor pressures at room temperature are virtually zero.

energies to break away from their neighbors and enter the vapor state. When particles in the vapor collide with the crystal, they can be recaptured, so condensation can occur too. Eventually, the concentration of particles in the vapor reaches a point where the rate of sublimation equals the rate of condensation, and a dynamic equilibrium is established. The pressure of the vapor that is in equilibrium with the solid is called the **equilibrium vapor pressure of the solid.** As with liquids, this equilibrium vapor pressure is usually referred to simply as the vapor pressure. Like that of a liquid, the vapor pressure of a solid is determined by the strengths of the attractive forces between the particles and by the temperature.

12.6 ▶ Boiling occurs when a liquid's vapor pressure equals atmospheric pressure

We know the water in this pot of vegetables is boiling because we can see bubbles of steam rising to the surface. Because the temperature of boiling water cannot rise above its boiling point, foods cooked in water cannot become too hot and burn.

On the top of Mt. Everest, the world's tallest peak, water boils at only 69 °C.

If you were asked to check whether a pot of water was boiling, what would you look for? The answer, of course, is *bubbles.* When a liquid boils, large bubbles usually form at many places on the inner surface of the container and rise to the top. If you were to place a thermometer into the boiling water, you would find that the temperature remains constant, regardless of how you adjust the flame under the pot. A hotter flame just makes the water bubble faster, but it doesn't raise the temperature. *Any pure liquid remains at a constant temperature while it is boiling,* a temperature that's called the liquid's **boiling point.**

If you measure the boiling point of water in Philadelphia, New York, or any place else that is nearly at sea level, your thermometer will read 100 °C or very close to it. However, if you try this experiment in Denver, Colorado, you will find that water boils at about 95 °C. Denver, at a mile above sea level, has a lower atmospheric pressure, so we find that the boiling point depends on the atmospheric pressure.

These observations raise some interesting questions. Why do liquids boil? And why does the boiling point depend on the pressure of the atmosphere? The answers become apparent when we realize that inside the bubbles of a boiling liquid is the *liquid's vapor,* not air. When water boils, the bubbles contain water vapor (steam); when alcohol boils, the bubbles contain alcohol vapor. As a bubble grows, liquid evaporates into it, and the pressure of the vapor pushes the liquid aside, making the size of the bubble increase (see Figure 12.25). Opposing the bubble's internal vapor pressure, however, is the pressure of the atmosphere pushing down on the top of the liquid, attempting to collapse the bubble. The only way the bubble can exist and grow is for the vapor pressure within it to equal (maybe just slightly exceed) the pressure exerted by the atmosphere. In other words, bubbles of vapor cannot even form until the temperature of the liquid rises to a point at which the liquid's vapor pressure equals the atmospheric pressure. Thus, in scientific terms, the **boiling point** is defined as *the temperature at which the vapor pressure of the liquid is equal to the prevailing atmospheric pressure.*

Now we can easily understand why water boils at a lower temperature in Denver than it does in New York City. Because the atmospheric pressure is lower in Denver, the water there doesn't have to be heated to as high a temperature to make its vapor pressure equal to the atmospheric pressure. The lower temperature of boiling water at places with high altitudes, like Denver, makes it necessary to cook foods longer. At the other extreme, a pressure cooker is a device that increases the pressure over the boiling water and thereby raises the boiling point. At the higher temperature, foods cook more quickly.

To make it possible to compare the boiling points of different liquids, chemists have chosen 1 atm as the reference pressure. The boiling point of a liquid at 1 atm is called its **normal boiling point.** (If a boiling point is reported without also mentioning the pressure at which it was measured, we assume it to be the normal boiling point.) Notice in Figure 12.23, page 524, that we can find the normal boiling

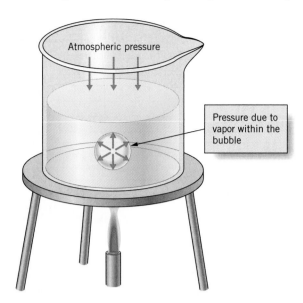

Atmospheric pressure

Pressure due to vapor within the bubble

FIGURE 12.25 *A liquid at its boiling point.* The pressure of the vapor within a bubble in a boiling liquid pushes the liquid aside against the opposing pressure of the atmosphere. Bubbles can't form unless the vapor pressure of the liquid is at least equal to the pressure of the atmosphere.

points of ether, water, acetic acid, and propylene glycol by noting the temperatures at which their vapor pressure curves cross the 1 atm pressure line.

Earlier we mentioned that the boiling point is a property whose value depends on the strengths of the intermolecular attractions in a liquid. When the attractive forces are strong, the liquid has a low vapor pressure at a given temperature, so it must be heated to a high temperature to bring its vapor pressure up to atmospheric pressure. High boiling points therefore result from strong intermolecular attractions, so we often use normal boiling point data to assess relative intermolecular attractions among different liquids. (In fact, we did this in solving Example 12.1.)

TOOLS

Boiling points of substances

The effects of intermolecular attractions on boiling point are easily seen by examining Figure 12.26, which gives the plots of the boiling points versus period numbers for some families of binary hydrogen compounds. Notice, first, the gradual increase in boiling point for the hydrogen compounds of the Group IVA elements (CH_4 through GeH_4). These compounds are composed of nonpolar tetrahedral molecules. The boiling points increase from CH_4 to GeH_4 simply because the molecules become larger and their electron clouds become more polarizable, which leads to an increase in the strengths of the London forces.

When we look at the hydrogen compounds of the other nonmetals, we find the same trend from Period 3 through Period 5. Thus, for three compounds of the Group VA series (PH_3, AsH_3, and SbH_3), there is a gradual increase in boiling point, corresponding again to the increasing strengths of London forces. Similar increases occur for the three Group VIA compounds (H_2S, H_2Se, and H_2Te) and for the three Group VIIA compounds (HCl, HBr, and HI). Significantly, however, the Period 2 members of each of these series (NH_3, H_2O, and HF) have much higher boiling points than might otherwise be expected. The reason is that each is involved in hydrogen bonding, which is a much stronger attraction than London forces.

One of the most interesting and far-reaching consequences of hydrogen bonding is that it causes water to be a liquid, rather than a gas, at temperatures near 25 °C. If it were not for hydrogen bonding, water would have a boiling point somewhere near −80 °C and could not exist as a liquid except at still lower temperatures. At such low temperatures it is unlikely that life as we know it could have developed.

PRACTICE EXERCISE 3: The atmospheric pressure at the top of Mt. McKinley in Alaska, 3.85 miles above sea level, is 330 torr. Use Figure 12.23 to estimate the boiling point of water at the top of this mountain.

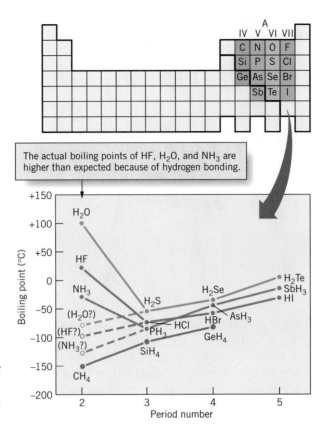

FIGURE 12.26 *Effects of intermolecular attractions on boiling point.* Boiling points of the hydrogen compounds of elements of Groups IVA, VA, VIA, and VIIA of the periodic table.

12.7 ▶ Energy changes occur during changes of state

When a liquid or solid evaporates or a solid melts, there are increases in the distances between the particles of the substance. Particles that normally attract each other are forced apart, increasing their potential energies. Such energy changes affect our daily lives in many ways, especially the energy changes associated with the changes in state of water, changes which even control the weather on our planet. To study these energy changes, let's begin by examining how the temperature of a substance varies as it is heated.

Heating curves and cooling curves

Figure 12.27*a* illustrates the way the temperature of a substance changes as we add heat to it *at a constant rate,* starting with the solid and finishing with the gaseous state of the substance. The graph is sometimes called a **heating curve** for the substance.

A heating curve can be used to measure accurately the melting point and boiling point of a liquid.

When a solid or liquid is heated, the volume expands only slightly, so there are only small changes in the average distance between the particles. This means that very small changes in potential energy take place, so almost all the heat added goes to increasing the kinetic energy.

Notice that as heat is added to the solid, the temperature increases until the solid begins to melt. The temperature then remains constant as long as both solid and liquid phases coexist. When all the solid has melted, the continued addition of heat to what is now a liquid causes the temperature to climb once again. This continues until the boiling point is reached, where the temperature levels off again. As we noted, the temperature of the boiling liquid stays the same until all of it has boiled away and changed to a gas. Then, finally, more heat just raises the temperature of the gas. Let's analyze these changes in terms of changes in the kinetic and potential energies of the particles.

First, let's look at the portions of the graph that slope upward. These occur where we are increasing the temperature of the solid, the liquid, or the gas phases. Because temperature is related to average kinetic energy, nearly all of the heat we add in these regions of the heating curve goes to increasing the average kinetic energies of the particles. In other words, the added heat makes the particles go faster and collide with each other with more force.

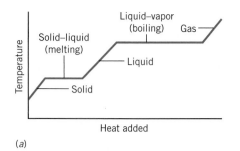
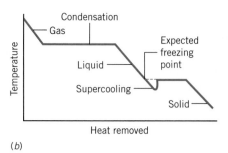

(a) (b)

FIGURE 12.27 *Heating and cooling curves.* (*a*) A heating curve observed when heat is added to a substance at a constant rate. The temperatures corresponding to the flat portions of the curve occur at the melting point and boiling point. (*b*) A cooling curve observed when heat is removed from a substance at a constant rate. Condensation of vapor to a liquid occurs at the same temperature as the liquid boils. Supercooling is seen here as the temperature of the liquid dips below its freezing point (the same temperature as its melting point). Once a tiny crystal forms, the temperature rises to the freezing point.

In those portions of the heating curve where the temperature remains constant, the average kinetic energy of the particles is not changing. This means that all the heat being added must go to increase the *potential energies* of the particles. During melting, the particles held rigidly in the solid begin to separate slightly as they form the mobile liquid phase. This separation of the particles is accompanied by a potential energy increase, which is what we measure by following the quantity of heat input during the melting process. During boiling, an even greater increase occurs in the distances between the molecules. Here they go from the relatively tight packing in the liquid to the widely spaced distribution of molecules in the gas. This gives rise to an even larger increase in the potential energy, which we see as a longer flat region on the heating curve during the boiling of the liquid.

The opposite of a heating curve is a **cooling curve** (see Figure 12.27*b*). Here we start with a gas and gradually cool it—remove heat from it at a constant rate—until we have reached a solid. The cooling curve looks very much like the opposite of a heating curve, except for what often happens when the temperature of the liquid approaches the freezing point. We might expect that when the freezing point is reached, removal of more heat would cause the immediate formation of some solid. Often, however, the liquid continues to cool *below its freezing point*, a phenomenon called **supercooling.**

Supercooling can happen when the freezing point is reached because the molecules of the liquid are still in a jumbled disordered arrangement, not in the ordered structure characteristic of a solid (as we'll describe in the next chapter). While the molecules move about, waiting for a few to form the beginnings of a crystal, the temperature continues to drop because heat is continually being removed. Finally, a tiny crystal forms and other molecules quickly join it, losing potential energy, which is changed to kinetic energy. The sudden increase in molecular kinetic energy raises the temperature to the freezing point, and then freezing continues normally as more heat is taken away.

Supercooling of a vapor is also possible. Condensation of supercooled water vapor onto solid surfaces leads to dew in warm weather and frost in freezing weather.

Molar heats of fusion, vaporization, and sublimation

Enthalpy changes in phase changes

Because phase changes occur at constant temperature and pressure, the potential energy changes associated with melting and vaporization can be expressed as enthalpy changes. Usually, enthalpy changes are expressed on a "per mole" basis and are given special names to identify the kind of change involved. For example, using the word **fusion** instead of *melting,* the **molar heat of fusion,** ΔH_{fusion}, is *the heat absorbed by one mole of a solid when it melts to give a liquid at the same temperature and pressure.* Similarly, the **molar heat of vaporization,** $\Delta H_{\text{vaporization}}$, is *the heat absorbed when one mole of a liquid is changed to one mole of vapor at a constant tem-*

The concept of enthalpy was introduced in Chapter 7.

These are also called *enthalpies* of fusion, vaporization, and sublimation.

One of the bits of information we like to know about the weather is the **relative humidity.** On humid days we know we are likely to feel uncomfortable, and when the humidity is very low, it can bother our breathing and dry our eyes. Relative humidity is a measure of how nearly saturated air is with moisture. If the air contains as much moisture as it can hold, the pressure of the water vapor will be the same as the equilibrium vapor pressure. Wet clothes won't dry under these conditions, and neither will perspiration, so we are likely to feel uncomfortable if we exercise under these conditions.

If the air is less than saturated with water vapor, the pressure of the water vapor will be below the equilibrium vapor pressure. For example, at 10 °C

(50 °F), the equilibrium vapor pressure of water is 9.2 torr (see Table 11.2, page 478). If the water vapor in air at this temperature only exerts a partial pressure of 4.6 torr, the percent relative humidity is 50%:

% relative humidity =

$$\frac{\text{partial pressure of } H_2O \text{ vapor}}{\text{equilib. vapor pressure of } H_2O} \times 100\%$$

$$= \frac{4.6 \text{ torr}}{9.2 \text{ torr}} \times 100\% = 50\%$$

The lower the percent relative humidity, the more rapidly water will evaporate into the air and the faster wet things will dry. We will also feel more comfortable when we exercise, because perspiration will evaporate more quickly and produce a greater cooling effect on our bodies.

Fusion means melting. The thin metal band in an electrical fuse becomes hot as electricity passes through it. It protects a circuit by melting if too much current is drawn. On the right we see a fuse that has done its job.

Intermolecular attractions and heats of vaporization and sublimation

perature and pressure. Finally, the **molar heat of sublimation, $\Delta H_{sublimation}$,** *is the heat absorbed by one mole of a solid when it sublimes to give one mole of vapor, once again at a constant temperature and pressure.* The values of ΔH for fusion, vaporization, and sublimation are all positive because the phase change in each case is endothermic, being accompanied by a net increase in potential energy.

You have probably made use of these enthalpy changes at one time or another. For example, you've added ice to a drink to keep it cool because as the ice melts, it absorbs heat (its heat of fusion). Your body uses the heat of vaporization of water to cool itself through the evaporation of perspiration. During the summer, ice cream trucks carry dry ice because the sublimation of CO_2 absorbs its heat of sublimation and keeps the ice cream cold. Refrigerators and air conditioners work on the basis of making the heat unwanted in one space function as the heat of vaporization for the system's operating fluid (see Facets of Chemistry 12.1).

Energy changes and intermolecular attractions

For a given substance, $\Delta H_{vaporization}$ is larger than ΔH_{fusion}. This is because the particles separate to greater distances during vaporization than during melting, so the changes in potential energy are greater. For the same substance, $\Delta H_{sublimation}$ is even larger than $\Delta H_{vaporization}$ because the attractive forces in the solid are larger than those in the liquid. As you would expect, overcoming these stronger forces requires more energy.

When a liquid evaporates or a solid sublimes, the particles go from a situation

Solar energy is the major driving force for weather on our planet, and much of it is captured as the heat of vaporization of water. Sunlight shining on the oceans heats the water and provides the energy to cause water to evaporate in large amounts. When moisture-laden air rises, it cools and the water vapor condenses to form clouds. At the same time, the energy released during condensation (i.e., the heat of vaporization of the water) causes air currents to move and creates the wind and lightning associated with thunderstorms. Over land, the collision of warm moist air with cold air produces strong storms, dangerous lightning, and sometimes torna-

does. All of the energy for these weather events comes from the heat of vaporization of water being released when water vapor condenses to liquid water.

Over the oceans, large storms such as hurricanes rely on a continued supply of warm moist air produced by rapid evaporation of water from tropical waters. Continual condensation in the high clouds forms rain and supplies the energy needed to feed the storm's winds. When such storms encounter cooler waters, they often die out, because cold water evaporates slowly. When a hurricane encounters a large land mass, it drops huge amounts of rain but rapidly loses strength because it can no longer be fed by evaporating ocean waters.

FACETS OF CHEMISTRY 12.1

Refrigeration and Air Conditioning

The fact that the evaporation of a liquid is endothermic and the condensation of a gas is exothermic forms the basis for the operation of refrigerators and air conditioners. These devices use a gas that can be liquefied at room temperature—ammonia or a halogenated hydrocarbon—to pump heat from one place to another.

In a refrigerator, a compressor squeezes the gas into a small volume, which causes the pressure to increase. At the same time, the work done compressing the gas becomes stored as heat—the gas becomes warm. The warm gas is then circulated through cooling coils that are usually located on the back of the refrigerator. (Did you ever notice that these coils are warm?) When the gas cools under pressure, it liquefies when its temperature drops below a certain value called the gas's *critical temperature*. (For a further discussion of critical temperature, see page 537.)

Condensation of the gas is exothermic and releases the substance's heat of vaporization, which is also given off to the surroundings. After being cooled, the pressurized liquid moves back into the refrigerator where it passes through a nozzle into a region of low pressure. At this low pressure, the liquid evaporates, and heat equal to the heat of vaporization is absorbed, which causes the inside of the refrigerator to cool. Then the gas is sucked back into the compressor, and the cycle is repeated.

Through condensation and evaporation, the compressor uses the gas as the fluid in a heat pump that absorbs heat from inside the refrigerator and dumps it outside. An air conditioner works in essentially the same way, except that a room becomes the equivalent of the inside of a refrigerator, and the heat pump dumps the heat into the air outside.

---

in which the attractive forces are very strong to one in which the attractive forces are so small they can almost be ignored. Therefore, the values of $\Delta H_{\text{vaporization}}$ and $\Delta H_{\text{sublimation}}$ give us directly the energy needed to separate molecules from each other. We can examine such values to obtain reliable comparisons of the strengths of intermolecular attractions. (How heats of vaporization are determined is described in Facets of Chemistry 12.2.)

In Table 12.4, notice that the heats of vaporization of water and ammonia are very large, which is just what we would expect for hydrogen-bonded substances. By comparison, CH_4, a nonpolar substance composed of atoms of similar size, has a very small heat of vaporization. Note also that polar substances such as HCl and SO_2 have fairly large heats of vaporization compared with nonpolar substances. For example, compare HCl with Cl_2. Even though Cl_2 contains two relatively large atoms, and therefore would be expected to have larger London forces than HCl, the latter substance has the larger $\Delta H_{\text{vaporization}}$. This must be due to dipole–dipole attractions between polar HCl molecules—attractions that are absent in nonpolar Cl_2.

The stronger the attractions, the more the potential energy will increase when the molecules become separated, and the larger will be the value of ΔH.

The Lewis structure of SO_2 tells us that the molecule is nonlinear, and this means that it must be polar.

TABLE 12.4	SOME TYPICAL HEATS OF VAPORIZATION	
Substance	$\Delta H_{\text{vaporization}}$ (kJ/mol)	Type of Attractive Force
H_2O	+43.9	Hydrogen bonding and London
NH_3	+21.7	Hydrogen bonding and London
HCl	+15.6	Dipole–dipole and London
SO_2	+24.3	Dipole–dipole and London
F_2	+5.9	London
Cl_2	+10.0	London
Br_2	+15.0	London
I_2	+22.0	London
CH_4	+8.16	London
C_2H_6	+15.1	London
C_3H_8	+16.9	London
C_6H_{14}	+30.1	London

FACETS OF CHEMISTRY 12.2

Determining Heats of Vaporization

The way the vapor pressure varies with temperature, which was described in Section 12.5, depends on the heat of vaporization of a substance. The relationship, however, is not a simple proportionality. Instead, it involves *natural logarithms,* which are logarithms to the base e, a nonrepeating decimal number (an irrational number) having the value 2.71828. . . . The natural logarithm of a quantity x, symbolized as $\ln x$, is the power to which e must be raised to give x. (More information about natural logarithms can be found on our web site: www.wiley.com/college/brady.)

Rudolf Clausius (1822–1888), a German physicist, and Benoit Clapeyron (1799–1864), a French engineer, used the principles of thermodynamics (a subject that's discussed in Chapter 20) to derive the following equation that relates the vapor pressure, heat of vaporization (ΔH_{vap}), and temperature:

$$\ln P = \frac{-\Delta H_{vap}}{RT} + C \qquad (1)$$

The quantity $\ln P$ is the natural logarithm of the vapor pressure, R is the gas constant expressed in energy units ($R = 8.314$ J mol^{-1} K^{-1}), T is the absolute temperature, and C is a constant. Scientists call this the Clausius–Clapeyron equation.

The Clausius–Clapeyron equation provides a convenient graphical method for determining heats of vaporization from experimentally measured vapor pressure–temperature data. To see this, let's rewrite the equation as follows.

$$\ln P = \left(\frac{-\Delta H_{vap}}{R}\right)\frac{1}{T} + C$$

Recall from algebra that a straight line is represented by the general equation

$$y = mx + b$$

where x and y are variables, m is the slope, and b is the intercept of the line with the y axis. In this case, we can make the substitutions

$$y = \ln P \qquad x = \frac{1}{T} \qquad m = \left(\frac{-\Delta H_{vap}}{R}\right) \qquad b = C$$

Therefore, we have

$$\ln P = \left(\frac{-\Delta H_{vap}}{R}\right)\frac{1}{T} + C$$
$$\Updownarrow \qquad \qquad \Updownarrow \quad \Updownarrow \quad \Updownarrow$$
$$y = \qquad m \quad x + b$$

Thus, a graph of $\ln P$ versus $1/T$ should give a straight line that has a slope equal to $-\Delta H_{vap}/R$. Such straight-line relationships are illustrated in Figure 1, in which experimental data are plotted for water, acetone, and ethanol. From the graphs in Figure 1, the calculated values of ΔH_{vap} are as follows: for water, 43.9 kJ mol^{-1}; for acetone, 32.0 kJ mol^{-1}; and for ethanol, 40.5 kJ mol^{-1}.

Using Equation 1 above, a "two point" form of the Clausius–Clapeyron equation can be derived that can be used to calculate ΔH_{vap} if the vapor pressure is known at two different temperatures. This equation is

$$\ln \frac{P_1}{P_2} = \frac{\Delta H_{vap}}{R}\left(\frac{1}{T_2} - \frac{1}{T_1}\right) \qquad (2)$$

If we know the value of the heat of vaporization, Equation 2 can also be used to calculate the vapor pressure at some particular temperature (say P_2 at a temperature T_2) if we already know the vapor pressure P_1 at a temperature T_1.

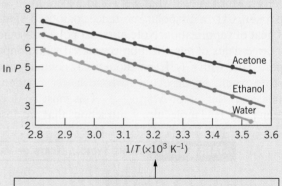

The numbers along the horizontal axis are equal to $1/T$ values multiplied by 1000 to make the axis easier to label.

FIGURE 1 A graph showing plots of $\ln P$ versus $1/T$ for acetone, ethanol, and water.

Substances with large heats of vaporization also tend to have high boiling points. Both properties are affected the same way by the strengths of intermolecular attractions.

Heats of vaporization also reflect the factors that control the strengths of London forces. For example, the data in Table 12.4 show the effect of chain length on the intermolecular attractions between hydrocarbons; as the chain length increases from one carbon in CH_4 to six carbons in C_6H_{14}, the heat of vaporization also increases. This is in agreement with the expected increase in the total strengths of the London forces. In our discussion of London forces, we also mentioned that their strengths increase as the electron clouds of the particles become larger, and we used the boiling points of the halogens to illustrate this. The heats of vaporization of the halogens in Table 12.4 are in agreement with this, too.

12.8 ▶ Changes in a dynamic equilibrium can be analyzed using Le Châtelier's principle

Throughout this chapter, we have studied various dynamic equilibria. One example was the equilibrium that exists between a liquid and its vapor in a closed container. You learned that when the temperature of the liquid is increased in this system, its vapor pressure also increases. Let's briefly review why this occurs.

Initially, the liquid is in equilibrium with its vapor, which exerts a certain pressure. When the temperature is increased, equilibrium no longer exists because evaporation occurs more rapidly than condensation. Eventually, as the concentration of molecules in the vapor increases, *the system reaches a new equilibrium* in which there is more vapor and a little less liquid. The greater concentration of molecules in the vapor causes a larger pressure.

> In the new equilibrium, there is less liquid and more vapor, but there is *equilibrium* nonetheless.

The way a liquid's equilibrium vapor pressure responds to a temperature change is an example of a general phenomenon. *Whenever a dynamic equilibrium is upset by some disturbance, the system changes in a way that will, if possible, bring the system back to equilibrium again.* It's also important to understand that in the process of regaining equilibrium, the system undergoes a net change. Thus, when the temperature of a liquid is raised, there is some net conversion of liquid into vapor as the system returns to equilibrium. When the new equilibrium is reached, the amounts of liquid and vapor are not the same as they were before.

Throughout the remainder of this book, we will deal with many kinds of equilibria, both chemical and physical. It would be very time-consuming and sometimes very difficult to carry out a detailed analysis each time we wish to know the effects of some disturbance on an equilibrium system. Fortunately, there is a relatively simple and fast method for predicting the effect of a disturbance, one based on a principle proposed in 1888 by a brilliant French chemist, Henry Le Châtelier (1850–1936).

Le Châtelier's Principle

When a dynamic equilibrium in a system is upset by a disturbance, the system responds in a *direction* that tends to counteract the disturbance and, if possible, restore equilibrium.

Le Châtelier's principle

Let's see how we can apply Le Châtelier's principle to a liquid–vapor equilibrium that is subjected to a temperature increase. We cannot increase a temperature, of course, without adding heat. Thus, *the addition of heat is really the disturbing influence when a temperature is increased.* So let's incorporate "heat" as a member of the equation used to represent the liquid–vapor equilibrium.

$$\text{heat} + \text{liquid} \rightleftharpoons \text{vapor} \tag{12.1}$$

Recall that we use double arrows, $\rightleftharpoons$, to indicate a dynamic equilibrium in an equation. They imply opposing changes happening at equal rates. Evaporation is endothermic, so the heat is placed on the left side of Equation 12.1 to show that heat is absorbed by the liquid when it changes to the vapor and that heat is released when the vapor condenses to a liquid.

Le Châtelier's principle tells us that when we add heat to raise the temperature of the equilibrium system, the system will try to adjust in a way that absorbs some of the added heat. This can happen if some liquid evaporates, because vaporization is endothermic. When liquid evaporates, the amount of vapor increases and causes the pressure to rise. Thus, we have reached the correct conclusion in a very simple way, namely, that heating a liquid must increase its vapor pressure.

Recall that we often use the term **position of equilibrium** to refer to the relative amounts of the substances on opposite sides of the double arrows in an equilibrium expression such as Equation 12.1. Thus, we can think of how a disturbance affects the position of equilibrium. For example, increasing the temperature increases the amount of vapor and decreases the amount of liquid, and we say *the position of equilibrium has shifted;* in this case, it has shifted in the direction of the vapor, or it has *shifted to the right.* In using Le Châtelier's principle, it is often convenient to think of a disturbance as "shifting the position of equilibrium" in one direction or another in the equilibrium equation.

PRACTICE EXERCISE 4: Use Le Châtelier's principle to predict how a temperature increase will affect the vapor pressure of a solid. (*Hint:* solid + heat ⇌ vapor.)

12.9 ▶ Phase diagrams graphically represent pressure–temperature relationships

TOOLS

Phase diagram

Sometimes it is useful to know under what combinations of temperature and pressure a substance will be a liquid, a solid, or a gas, or the conditions of temperature and pressure that produce an equilibrium between any two phases. A simple way to determine this is to use a **phase diagram**—a graphical representation of the pressure–temperature relationships that apply to the equilibria between the phases of the substance.

Figure 12.28 is the phase diagram for water. On it, there are three lines that intersect at a common point. Points on these lines correspond to temperatures and pressures at which equilibria between phases can exist. For example, line *AB* is the vapor pressure curve for the solid (ice). Every point on this line gives a temperature and a pressure at which ice and its vapor are able to coexist in equilibrium. For instance, at −10 °C ice has an equilibrium vapor pressure of 2.15 torr. This kind of information would be very useful to know if you wanted to design a system for making freeze-dried foods. It tells you that ice will sublime at −10 °C if a vacuum pump can reduce the pressure of water vapor above the ice to less than 2.15 torr. The lower pressure disturbs the ice–vapor equilibrium by removing water vapor, and ice sublimes in a vain attempt to replace the water vapor and reestablish the equilibrium.

Line *BD* is the vapor pressure curve for liquid water. It gives the temperatures and pressures at which the liquid and vapor are able to coexist in equilibrium. Notice that when the temperature is 100 °C, the vapor pressure is 760 torr. Therefore, this diagram also tells us that water boils at 100 °C when the pressure is 1 atm (760 torr), because that is the temperature at which the vapor pressure equals 1 atm. At the top of Mt. McKinley, in Alaska, the atmospheric pressure is only about 330 torr. The phase diagram tells us that the boiling point of water there will be 78 °C.

Campers use freeze-dried foods because they are easily reconstituted by adding water.

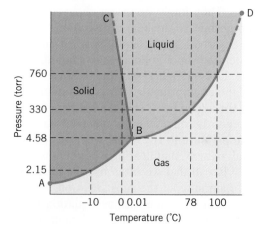

FIGURE 12.28 *The phase diagram for water, distorted to emphasize certain features.* Temperatures and pressures corresponding to the dashed lines on the diagram are referred to in the text discussion.

The solid–vapor equilibrium line, *AB*, and the liquid–vapor line, *BD*, intersect at a common point, *B*. Because this point is on both lines, there is equilibrium among all three phases at the same time.

$$\text{vapor}$$
$$\nearrow\swarrow \qquad \nwarrow\searrow$$
$$\text{liquid} \rightleftharpoons \text{solid}$$

The temperature and pressure at which this triple equilibrium occurs define the **triple point** of the substance. For water, the triple point occurs at 0.01 °C and 4.58 torr. Every known chemical substance except helium has its own characteristic triple point, which is controlled by the balance of intermolecular forces in the solid, liquid, and vapor.

In the SI, the triple point of water is used to define the Kelvin temperature of 273.16 K.

Line *BC*, which extends upward from the triple point, is the solid–liquid equilibrium line or *melting point line*. It gives temperatures and pressures at which the solid and the liquid are able to be in equilibrium. At the triple point, the melting of ice occurs at +0.01 °C (and 4.58 torr); at 760 torr, melting occurs very slightly lower, at 0 °C. Thus, we can tell that *increasing the pressure on ice lowers its melting point*.

The decrease in the melting point of ice that occurs when there is an increase in pressure can be predicted using Le Châtelier's principle and the knowledge that liquid water is more dense than ice. Let's reflect again on the equilibrium between ice and liquid water at 0 °C and 1 atm.

$$H_2O(s) \rightleftharpoons H_2O(l)$$

Imagine that the equilibrium is established in an apparatus like that shown in Figure 12.29 (and, again, keep in mind that ice is less dense than liquid water, which is why it is shown floating on top of the liquid). If the piston is forced in slightly, the pressure increases, and this serves to disturb the equilibrium. According to Le Châtelier's principle, the system should respond in a way that tends to reduce the pressure. This can happen if some of the ice melts, so that the ice–liquid mixture won't require as much space. Then the molecules won't push as hard against each other, and the pressure will drop. Thus, a pressure-increasing disturbance to the system favors a volume-decreasing change, which corresponds to the melting of some ice.

Next, suppose we have ice at a pressure just below the solid–liquid line, *BC*. If, *at constant temperature*, we raise the pressure to a point just above the line, the ice will melt and become a liquid. This could only happen if the melting point becomes lower as the pressure is raised.

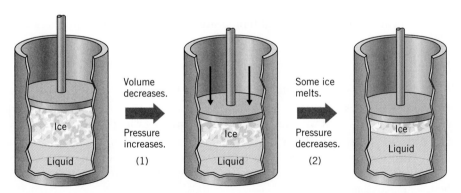

FIGURE 12.29 *The effect of pressure on the equilibrium,* H_2O*(solid)* $\rightleftharpoons$ H_2O *(liquid).* (1) Pushing down on the piston decreases the volume of both the ice and liquid water by a small amount and increases the pressure. (2) Some of the ice melts, producing the more dense liquid. As the total volume of ice and liquid water decreases, the pressure drops in accordance with Le Châtelier's principle.

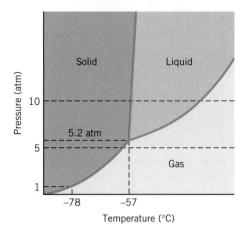

Figure 12.30 *The phase diagram for carbon dioxide.*

Water is very unusual. Almost all other substances have melting points that increase with increasing pressure. Consider, for example, the phase diagram for carbon dioxide (see Figure 12.30). For CO_2 the solid–liquid line slants to the right (it slanted to the left for water). Also notice that carbon dioxide has a triple point that's above 1 atm. At atmospheric pressure, the only equilibrium that can be established is between solid carbon dioxide and its vapor. At a pressure of 1 atm this equilibrium occurs at a temperature of -78 °C. This is the temperature of dry ice, which sublimes at atmospheric pressure at -78 °C.

Single phase regions can be identified in a phase diagram

Besides specifying phase equilibria, the three intersecting lines on a phase diagram serve to define regions of temperature and pressure at which only a single phase can exist. For example, between lines *BC* and *BD* in Figure 12.28 are temperatures and pressures at which water exists as a liquid without being in equilibrium with either vapor or ice. At 760 torr, water is a liquid anywhere between 0 °C and 100 °C. For instance, we are told by the diagram that we can't have ice with a temperature of 25 °C if the pressure is 760 torr (which, of course, you already knew; ice never has a temperature of 25 °C.) The diagram also says that we can't have water vapor with a pressure of 760 torr when the temperature is 25 °C (which, again, you already knew; the temperature has to be taken to 100 °C for the vapor pressure to reach 760 torr). Instead, we are told by the phase diagram that the *only* phase for pure water at 25 °C and 1 atm is the liquid. Below 0 °C at 760 torr, water is a solid; above 100 °C at 760 torr, water is a vapor. On the phase diagram for water, the phases that can exist in the different temperature–pressure regions are marked.

EXAMPLE 12.2	What phase would we expect for water at 0 °C and 4.58 torr?
Interpreting a Phase Diagram	**ANALYSIS:** The words, "What phase . . . ," as well as the specified temperature and pressure suggest that we refer to the phase diagram of water (Figure 12.28).

SOLUTION: First we find 0 °C on the temperature axis of the phase diagram of water. Then we move upward until we intersect a line corresponding to 4.58 torr. This intersection occurs in the "solid" region of the diagram. At 0 °C and 4.58 torr, then, water exists as a solid.

Is the Answer Reasonable?
We've seen that the freezing point of water increases slightly when we lower the pressure, so below 1 atm, water should still be a solid at 0 °C. That agrees with the answer we obtained from the phase diagram.

What phase changes occur if water at 0 °C is gradually compressed from a pressure of 2.15 torr to 800 torr?

EXAMPLE 12.3
Interpreting a Phase Diagram

ANALYSIS: Asking about "what phase changes occur" suggests once again that we use the phase diagram of water (Figure 12.28).

SOLUTION: According to the phase diagram, at 0 °C and 2.15 torr, water exists as a gas (water vapor). As the vapor is compressed, we move upward along the 0 °C line until we encounter the solid–vapor line. Here, an equilibrium will exist as compression gradually transforms the gas into solid ice. Once all the vapor has frozen, further compression raises the pressure and we continue the climb along the 0 °C line until we next encounter the solid–liquid line at 760 torr. As further compression takes place, the solid will gradually melt. After all the ice has melted, the pressure will continue to climb while the water remains a liquid. At 800 torr and 0 °C, the water will be liquid.

Is the Answer Reasonable?
There's not too much we can do to check all this except to take a fresh look at the phase diagram. We do expect that above 760 torr, the melting point of ice will be less than 0 °C, so at 0 °C and 800 torr we can anticipate that water will be a liquid.

PRACTICE EXERCISE 5: What phase changes will occur if water at −20 °C and 2.15 torr is heated to 50 °C under constant pressure?

PRACTICE EXERCISE 6: What phase will water be in if it is at a pressure of 330 torr and a temperature of 50 °C?

There is one final aspect of the phase diagram that has to be mentioned. For water (Figure 12.28), the vapor pressure line for the liquid, which begins at point *B*, terminates at point *D*, which is known as the **critical point.** The temperature and pressure at *D* are called the **critical temperature, T_c,** and **critical pressure, P_c.** Above the critical temperature, a distinct liquid phase cannot exist, *regardless of the pressure.*

Above the critical temperature, a substance cannot be liquefied by the application of pressure.

Figure 12.31 illustrates what happens to a substance as it approaches its critical point. In Figure 12.31*a*, we see a liquid in a container with some vapor above it. We can distinguish between the two phases because they have different densities, which causes them to bend light differently. This allows us to see the interface, or surface, between the more dense liquid and the less dense vapor. If this liquid is now heated, two things happen. First, more liquid evaporates. This causes an increase in the number of molecules per cubic centimeter of vapor which, in turn, causes the density of the vapor to increase. Second, the liquid expands (just like mercury does in a thermometer). This means that a given mass of liquid occupies more volume, so its density decreases. As the temperature of the liquid and vapor continue to increase, the vapor density rises and the liquid density falls; they approach each other. Eventually the densities become equal, and a separate liquid phase no longer exists; everything is the same (see Figure 12.31*b*). The highest tem-

This interface between liquid and gas is called the **meniscus.**

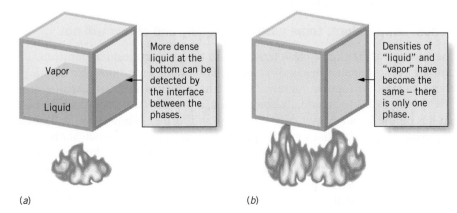

(a)   (b)

FIGURE 12.31 *Changes that are observed when a liquid is heated in a sealed container.* (*a*) Below the critical temperature. (*b*) Above the critical temperature.

FACETS OF CHEMISTRY 12.3

Decaffeinated Coffee and Supercritical Carbon Dioxide

Many people prefer to avoid caffeine, yet still enjoy their cup of coffee. For them, decaffeinated coffee is just the thing. To satisfy this demand, coffee producers remove caffeine from the coffee beans before roasting them. Several methods have been used, some of which use solvents such as methylene chloride (CH_2Cl_2) or ethyl acetate ($CH_3CO_2C_2H_5$) to dissolve the caffeine. Even though only trace amounts of these solvents remain after the coffee beans are dried, there are those who would prefer not to have any such chemicals in their coffee. And that's where carbon dioxide comes into the picture.

It turns out that supercritical carbon dioxide is an excellent solvent for many organic substances, including caffeine. To make it, gaseous CO_2 is heated to a temperature above its critical temperature of 31 °C, typically to about 80 °C. It is then compressed to about 200 atm. This gives it a density near that of a liquid, but with some properties of a gas. The fluid has a very low viscosity and readily penetrates coffee beans that have been softened with steam, drawing out the water and caffeine. After several hours, the CO_2 has removed as much as 97% of the caffeine, and the fluid containing the water and caffeine is then drawn

off. When the pressure of the supercritical CO_2 solution is reduced, the CO_2 turns to a gas and the water and caffeine separate. The caffeine is recovered and sold to beverage or pharmaceutical companies. Meanwhile, the pressure over the coffee beans is also reduced and the beans are warmed to about 120 °C, causing residual CO_2 to evaporate. Because CO_2 is not a toxic gas, any small traces of CO_2 that remain are totally harmless.

Decaffeination of coffee is not the only use of supercritical CO_2. It is also used to extract the essential flavor ingredients in spices and herbs for use in a variety of products. As with coffee, using supercritical CO_2 as a solvent completely avoids any potential harm that might be caused by small residual amounts of other solvents.

perature at which a liquid phase still exists is the critical temperature, and the pressure of the vapor at this temperature is the critical pressure. A substance that has a temperature above its critical temperature and a density near its liquid density is described as a **supercritical fluid.** Supercritical fluids have some unique properties that make them excellent solvents, and one that is particularly useful is supercritical carbon dioxide, which is discussed in Facets of Chemistry 12.3.

The values of the critical temperature and critical pressure are unique for every chemical substance and are controlled by the intermolecular attractions (see Table 12.5). Notice that liquids with strong intermolecular attractions, like water, tend to have high critical temperatures. Under pressure, the strong attractions between the molecules are able to hold them together in a liquid state even when the molecules are jiggling about violently at an elevated temperature. In contrast, substances with weak intermolecular attractions, such as methane and helium, have low critical temperatures. For these substances, even the small amounts of kinetic energy possessed by the molecules at low temperatures is sufficient to overcome the intermolecular attractions and prevent the molecules from sticking together as a liquid, despite being held close together under high pressure.

At room temperature, some gases will liquefy and others will not

On a very hot day, when the temperature is in the 90s, a filled CO_2 fire extinguisher won't give the sensation that it's filled with a liquid. At such temperatures, the CO_2 is in a supercritical state and no separate liquid phase exists.

When a gaseous substance has a temperature below its critical temperature, it is capable of being liquefied by compressing it. For example, carbon dioxide is a gas at room temperature (approximately 25 °C). This is below its critical temperature of 31 °C. If the $CO_2(g)$ is gradually compressed, a pressure will eventually be reached that lies on the liquid–vapor curve for CO_2, and further compression will cause the CO_2 to liquefy. In fact, that's what happens when a CO_2 fire extinguisher is filled; the CO_2 that's pumped in is a liquid under a high pressure. If you shake a filled CO_2 fire extinguisher, you can feel the liquid sloshing around inside, provided the temperature of the fire extinguisher is below 31 °C (88 °F). When the fire extinguisher is used, a valve releases the pressurized CO_2, which rushes out to extinguish the fire.

Refrigerators and air-conditioning systems, which are described in Facets of Chemistry 12.1 on page 531, use condensable gases to transfer heat from one place to another. The transfer of heat depends on releasing the heat of vaporization during condensation and absorbing the heat of vaporization during evaporation. For this to work, the gases must have high critical temperatures, because the condensation step must take place under pressure at or near room temperature. Among the first gases used for refrigeration were ammonia (T_c = 132.5 °C) and sulfur dioxide (T_c = 157.8 °C). However, these gases are toxic and pose a threat if used in home refrigerators (although they are still used in commercial refrigeration systems). In the 1930s the refrigerant CCl_2F_2 (called Freon-12) was discovered, which is nontoxic and also has a high T_c. This compound and other chlorofluorocarbons (CFCs) were used extensively in both refrigerators and air-conditioning systems. They are now being replaced either by hydrocarbon mixtures, CF_3CH_2F, or other compounds that don't contain chlorine because of the harmful effects that CFCs have on the Earth's protective layer of ozone in the upper atmosphere.

Gases such as O_2 and N_2, which have critical temperatures far below 0 °C, can never be liquids at room temperature. When they are compressed, they simply become high-pressure gases. To make liquid N_2 or O_2, the gases must be made very cold as well as be compressed to high pressures.

TABLE 12.5	SOME CRITICAL TEMPERATURES AND PRESSURES	
Compound	T_c (°C)	P_c (atm)
Water	374.1	217.7
Ammonia	132.5	112.5
Carbon dioxide	31	72.9
Ethane (C_2H_6)	32.2	48.2
Methane (CH_4)	−82.1	45.8
Helium	−267.8	2.3

SUMMARY

Physical Properties: Gases, Liquids, and Solids A gas expands to fill the entire volume of its container. Liquids and solids retain a constant volume if transferred from one container to another. Solids also retain a constant shape. These characteristics are related to how tightly packed the particles are and on the relative strengths of the intermolecular attractions in the different states of matter. Most physical properties depend primarily on intermolecular attractions. In gases, these attractions are weak because the molecules are so far apart. They are much stronger in liquids and solids, where the particles are packed together tightly.

Intermolecular Attractions Polar molecules attract each other by **dipole–dipole attractions,** which arise because the positive end of one dipole attracts the negative end of another. Nonpolar molecules are attracted to each other by **London dispersion forces,** which are **instantaneous dipole–induced dipole attractions.** London forces are present between all particles, including atoms, polar and nonpolar molecules, and ions. Among different substances, London forces increase with an increase in size of a particle's electron cloud; they also increase with increasing chain length among molecules such as the hydrocarbons. Compact molecules experience weaker London forces than similar long-chain molecules. For large molecules, the cumulative effect of large numbers of weak London force interactions can be quite strong and outweigh other intermolecular attractions. **Hydrogen bonding,** a special case of dipole–dipole attractions, occurs between molecules in which hydrogen is covalently bonded to a small, very electronegative atom—principally, nitrogen, oxygen, or fluorine. Hydrogen bonding is much stronger than the other types of intermolecular attractions. **Ion–dipole attractions** occur when ions interact with polar substances. **Ion–induced dipole attractions** result when an ion creates a dipole in a neighboring molecule or ion.

General Properties of Liquids and Solids Properties that depend mostly on closeness of packing of particles are

compressibility (or the opposite, incompressibility) and **diffusion.** Diffusion is slow in liquids and almost nonexistent in solids at room temperature. Properties that depend mostly on the strengths of intermolecular attractions are **retention of volume and shape, surface tension,** and **ease of evaporation. Surface tension** is related to the energy needed to expand a liquid's surface area. A liquid can **wet** a surface if its molecules are attracted to the surface about as strongly as they are attracted to each other. **Evaporation** of liquids and solids is endothermic and produces a cooling effect. The overall rate of evaporation increases with increasing surface area. The rate of evaporation from a given surface area of a liquid increases with increasing temperature and with decreasing intermolecular attractions. Evaporation of a solid is called **sublimation.**

Changes of State Changes from one physical state to another, such as melting, vaporization, or sublimation, can occur as dynamic equilibria. In a **dynamic equilibrium,** opposing processes occur continually at equal rates, so over time there is no apparent change in the composition of the system. For liquids and solids, equilibria are established when vaporization occurs in a sealed container. A solid is in equilibrium with its liquid at the melting point.

Vapor Pressures When the rates of evaporation and condensation of a liquid are equal, the vapor above the liquid exerts a pressure called the **equilibrium vapor pressure** (or more commonly, just the **vapor pressure**). The vapor pressure is controlled by the rate of evaporation *per unit surface area.* When the intermolecular attractive forces are large, the rate of evaporation is small and the vapor pressure is small. Vapor pressure increases with increasing temperature because the rate of evaporation increases as the temperature rises. The vapor pressure is independent of the *total* surface area of the liquid because as the surface area increases, the rates of evaporation and condensation are both increased equally; there is no change in the concentration of molecules in the vapor, so there is no change in the vapor pressure. Solids have vapor pressures just as liquids do.

Boiling Point A substance boils when its vapor pressure equals the prevailing atmospheric pressure. The **normal boiling point** of a liquid is the temperature at which its vapor pressure equals 1 atm. Substances with high boiling points have strong intermolecular attractions.

Energy Changes Associated with Changes of State On a **heating curve,** flat portions correspond to phase changes in which the heat added changes the potential energies of the particles without changing their average kinetic energy. On a **cooling curve, supercooling** sometimes occurs when the temperature of the liquid drops below the freezing point of the substance. The enthalpy changes for melting, vaporization of a liquid, and sublimation are the **molar heat of fusion,** ΔH_{fusion}, the **molar heat of vaporization,** $\Delta H_{\text{vaporization}}$, and the **molar heat of sublimation,** $\Delta H_{\text{sublimation}}$, respectively. They are all endothermic and are related in size as follows: $\Delta H_{\text{fusion}} < \Delta H_{\text{vaporization}} < \Delta H_{\text{sublimation}}$. The sizes of these enthalpy changes are large for substances with strong intermolecular attractive forces.

Le Châtelier's Principle When the equilibrium in a system is upset by a disturbance, the system changes in a direction that minimizes the disturbance and, if possible, brings the system back to equilibrium. By this principle, we find that raising the temperature favors an endothermic change. Decreasing the volume favors a change toward a more dense phase.

Phase Diagrams Temperatures and pressures at which equilibria can exist between phases are given graphically in a **phase diagram.** The three equilibrium lines intersect at the **triple point.** The liquid–vapor line terminates at the **critical point.** At the **critical temperature,** a liquid has a vapor pressure equal to its **critical pressure.** Above the critical temperature a liquid phase cannot be formed; the single phase that exists is called a **supercritical fluid.** The equilibrium lines also divide a phase diagram into temperature–pressure regions in which a substance can exist in just a single phase. Water is different from most substances in that its melting point decreases with increasing pressure.

Summary Relating Intermolecular Attractions and Properties For liquids, as the strengths of intermolecular attractions *increase*:

- Surface tension *increases.*
- Viscosity *increases.*
- Rate of evaporation *decreases.*
- Equilibrium vapor pressure *decreases.*
- Boiling point *increases.*
- Heat of vaporization *increases.*
- Critical temperature *increases.*

TOOLS ▶ **YOU HAVE LEARNED**

The table below lists the concepts you've learned in this chapter that can be applied as tools in solving problems. Study each one carefully so that you know what each is used for. When faced with solving a problem, recall what each tool does and consider whether it will be helpful in finding a solution. This will aid you in selecting the tools you need. If necessary, refer to this table when working on the Thinking-It-Through problems and the Review Exercises that follow.

TOOL	HOW IT WORKS
Relationship between intermolecular forces and molecular structure (pages 506, 507, and 508)	From molecular structure, we can determine whether a molecule is polar or not and whether it has N—H or O—H bonds. This lets us predict and compare the strengths of intermolecular attractions. You should be able to identify when dipole–dipole, London, and hydrogen bonding occurs. See the summary table on page 513.
Factors that affect rates of evaporation and vapor pressure (pages 520 and 523)	They allow us to compare the relative rates of evaporation based on temperature and the strengths of intermolecular forces in substances. They also allow us to compare the strengths of intermolecular forces based on the relative magnitudes of vapor pressures at a given temperature.
Boiling points of substances (page 527)	They allow us to compare the strengths of intermolecular forces in substances based on their boiling points.
Enthalpy changes during phase changes (page 529)	They allow us to compare the strengths of intermolecular forces in substances based on relative values of $\Delta H_{\text{vaporization}}$ and $\Delta H_{\text{sublimation}}$.
Le Châtelier's principle (page 533)	Le Châtelier's principle enables us to predict the direction in which the position of equilibrium is shifted when a dynamic equilibrium is upset by a disturbance. You should be able to predict how the position of equilibrium between phases is affected by temperature and pressure changes.
Phase diagram (page 534)	We use a phase diagram to identify temperatures and pressures at which equilibrium can exist between phases of a substance and to identify conditions under which only a single phase can exist.

THINKING IT THROUGH

Need extra help?
Visit the Brady/
Senese web site at
www.wiley.com/
college/brady

The goal for the following problems is not to find the answers themselves, but rather to assemble the information needed to solve them and explain how you would use the information to find the answers. The problems in Level 2 are more challenging than those in Level 1 and may contain more data than are required, in which case you are also asked to identify the unnecessary data. Detailed answers to the Thinking-It-Through problems can be found on the web site.

Level 1 Problems

1. When hot coffee in an insulated cup gradually cools, the cooler liquid at the surface circulates to the bottom as fresh hot liquid rises to the top. Why?

2. Vigorous exercise on a hot, muggy day produces more discomfort than the same exercise on an equally hot but dry day. Why?

3. If you fill a can halfway with gasoline and then seal the container, you will find that when the container is reopened some time later there is a noticeable hiss, as gases escape from the can. Why is there an increase in the pressure inside the can even if its temperature hasn't changed?

4. London forces are generally weaker than hydrogen bonds, yet water boils at 100 °C, whereas $C_{20}H_{42}$ (a component of candle wax) boils at 343 °C. Why?

5. At a given temperature, which of the liquids below is likely to have the greater vapor pressure?

$(CH_3CH_2)_2O$ $(CH_3CH_2)_2NH$

6. A flask is partially filled with water and fitted with a stopcock. With the stopcock open, the water is brought to a boil. After a few minutes heating is stopped and the stopcock is closed. The flask is allowed to cool for a short time and then ice water is poured over the top of it. This causes the water to boil vigorously! Why?

7. Why are indoor relative humidities generally very low in Minnesota in winter, even when outside the humidity is 100% and it's snowing?

8. Why do clouds form when the humid air of a weather system known as a *warm front* encounters the cool, relatively dry air of a *cold front*?

9. When a liquid is spread over a larger surface area, it evaporates more rapidly. Why, then, does the surface area of a liquid in a sealed container not influence the equilibrium vapor pressure of the liquid?

10. Some liquid water is added to a large copper vessel. After waiting for a time period sufficient for the water to have reached equilibrium with its vapor it was discovered that the vapor pressure of the water was less than the equilibrium vapor pressure. What is the most probable reason for this?

11. When one mole of a liquid freezes, less energy is given off than when one mole of it condenses from a vapor to a liquid. Why?

12. Which of the following is likely to have the larger molar heat of vaporization?

13. Which of the following is likely to have the larger molar heat of vaporization?

14. Room temperature is above hydrogen's critical temperature. If H_2 at a pressure of 1 atm is compressed at room temperature, will a pressure eventually be reached at which it will condense to a liquid?

Level 2 Problems

15. At a given temperature, which of the liquids below is likely to have the higher vapor pressure?

methylene chloride   chloroform

16. If a gas such as Freon-12, CCl_2F_2, is allowed to expand freely into a vacuum, it cools slightly. Why does this cooling prove there are attractive forces between the CCl_2F_2 molecules?

17. Melting point is sometimes used as an indication of the extent of covalent bonding in a compound—the higher the melting point, the more ionic the substance. On this basis, oxides of metals seem to become less ionic as the charge on the metal ion increases. Thus, Cr_2O_3 has a melting point of 2266 °C, whereas CrO_3 has a melting point of only 196 °C. The explanation often given is similar in some respects to explanations of the variations in the strengths of certain intermolecular attractions given in this chapter. Provide an explanation for the greater degree of electron sharing in CrO_3 as compared with Cr_2O_3.

REVIEW QUESTIONS

Comparisons among the States of Matter

12.1 Under what conditions would we expect gases to obey the gas laws best?

12.2 Why aren't there liquid laws that are comparable to the gas laws?

12.3 Why is the behavior of a gas affected very little by its chemical composition?

12.4 Why are the intermolecular attractive forces stronger in liquids and solids than they are in gases?

12.5 Compare the behavior of gases, liquids, and solids when they are transferred from one container to another.

12.6 At a molecular level, how do gases, liquids, and solids differ?

12.7 For a given substance, how do the intermolecular attractive forces compare in its gaseous, liquid, and solid states?

Intermolecular Attractions

12.8 In general, how do the strengths of intermolecular attractions compare with the strengths of chemical bonds?

12.9 What kinds of attractive forces (inter- or intramolecular) are responsible for chemical properties? What kind are responsible for physical properties?

12.10 Describe *dipole–dipole attractions*.

12.11 What are *London forces*? How are they affected by the sizes of the atoms in a molecule? How are they affected by the number of atoms in a molecule? How are they affected by the shape of a molecule?

12.12 Define *polarizability*. How does this property affect the strengths of London forces?

12.13 What are *hydrogen bonds*?

12.14 Which nonmetals, besides hydrogen, are most often involved in hydrogen bonding? Why these and not others?

12.15 Which is expected to have the higher boiling point, C_8H_{18} or C_4H_{10}? Explain your choice.

12.16 Ethanol and dimethyl ether have the same molecular formula, C_2H_6O. Ethanol boils at 78.4 °C, whereas dimethyl ether boils at −23.7 °C. Their structural formulas are

$$CH_3CH_2OH \qquad CH_3OCH_3$$
ethanol   dimethyl ether

Explain why the boiling point of the ether is so much lower than the boiling point of ethanol.

12.17 How do the strengths of covalent bonds and dipole–dipole attractions compare? How do the strengths of ordinary dipole–dipole attractions compare with the strengths of hydrogen bonds?

12.18 Explain why London forces are called *instantaneous dipole–induced dipole forces*. What effect does the polarizability of atoms have on the strengths of London forces?

12.19 What are *ion–induced dipole* attractions? How are the

strengths of these forces affected by the amount of charge on the cation, assuming the cations to be of equal size?

12.20 In which compound are the ion–induced dipole attractions stronger, CaO or CaS?

General Properties of Liquids and Solids

12.21 Name two physical properties of liquids and solids that are controlled primarily by how tightly packed the particles are. Name three that are controlled mostly by the strengths of the intermolecular attractions.

12.22 Why does diffusion occur more slowly in liquids than in gases? Why does diffusion occur extremely slowly in solids?

12.23 Suppose you have a container holding ice and liquid water, both at 0 °C. How do the average kinetic energies of the molecules compare in these two physical states of water?

12.24 Why are liquids and solids so difficult to compress?

12.25 On the basis of kinetic theory, would you expect the rate of diffusion in a liquid to increase or decrease as the temperature is increased? Explain your answer.

12.26 What is *surface tension*? Why do molecules at the surface of a liquid behave differently from those within the interior?

12.27 What kinds of observable effects are produced by the surface tension of a liquid?

12.28 What relationship is there between surface tension and the intermolecular attractions in the liquid?

12.29 Which liquid is expected to have the larger surface tension at a given temperature, CCl_4 or H_2O? Explain your answer.

12.30 What does *wetting* of a surface mean? What is a *surfactant*? What is its purpose and how does it function?

12.31 Polyethylene plastic consists of long chains of carbon atoms, each of which is also bonded to hydrogens, as shown below:

Water forms beads when placed on a polyethylene surface. Why?

12.32 The structural formula for glycerol is

Would you expect this liquid to wet glass surfaces? Explain your answer.

12.33 Water forms beads on a nonpolar surface, yet a nonpolar liquid such as gasoline does not form beads on a polar surface such as glass. Why?

12.34 On the basis of what happens on a molecular level, why does evaporation lower the temperature of a liquid?

12.35 On the basis of the distribution of kinetic energies of the molecules of a liquid, explain why increasing the liquid's temperature increases the rate of evaporation.

12.36 How is the rate of evaporation of a liquid affected by increasing the surface area of the liquid? How is the rate of evaporation affected by the strengths of intermolecular attractive forces?

12.37 During the cold winter months, snow often disappears gradually without melting. How is this possible? What is the name of the process responsible for this phenomenon?

12.38 How is freeze-drying accomplished? What are its advantages?

12.39 Why do puddles of water evaporate more quickly in the summer than in the winter?

Changes of State and Equilibrium

12.40 What terms do we use to describe the following changes of state: (a) solid → gas, (b) liquid → gas, (c) gas → liquid, (d) solid → liquid, (e) liquid → solid?

12.41 When a molecule escapes from the surface of a liquid by evaporation, it has a kinetic energy that's much larger than the average KE. Why is it likely that after being in the vapor for a while its kinetic energy will be much less? If this molecule collides with the surface of the liquid, is it likely to bounce out again?

12.42 Why does a molecule of a vapor that collides with the surface of a liquid tend to be captured by the liquid, even if the incoming molecule has a large kinetic energy?

12.43 When an equilibrium is established in the evaporation of a liquid into a sealed container, we refer to it as a *dynamic equilibrium*. Why?

12.44 Viewed at a molecular level, what is happening when a dynamic equilibrium is established between the liquid and solid forms of a substance? What is the temperature called at which there is an equilibrium between a liquid and a solid?

12.45 Is it possible to establish an equilibrium between a solid and its vapor?

12.46 Explain what would happen if moist air contacts a surface that has a temperature that's below the freezing point of water.

Vapor Pressure

12.47 Define *equilibrium vapor pressure*. Why do we call the equilibrium involved a *dynamic equilibrium*?

12.48 Explain why changing the volume of a container in which there is a liquid–vapor equilibrium has no effect on the equilibrium vapor pressure.

12.49 Why doesn't a change in the surface area of a liquid cause a change in the equilibrium vapor pressure?

12.50 What effect does increasing the temperature have on the equilibrium vapor pressure of a liquid? Why?

12.51 Why does humid air at 30 °C contain more moisture per cubic meter than humid air at 25 °C?

12.52 Why does rain often form when humid air is forced to rise over a mountain range?

12.53 Why does moisture condense on the outside of a cool glass of water in the summertime?

12.54 Why do we feel more uncomfortable in humid air at 90 °F than in dry air at 90 °F?

12.55 Why does the air in a heated building in the winter have such a low humidity?

Boiling Points of Liquids

12.56 Define *boiling point* and *normal boiling point*.

12.57 Why does the boiling point vary with atmospheric pressure?

12.58 Mt. Kilimanjaro in Tanzania is the tallest peak in Africa (19,340 ft). The normal barometric pressure at the top of this mountain is about 345 torr. At what Celsius temperature would water be expected to boil there? (See Figure 12.23.)

12.59 Why is the boiling point a useful property with which to identify liquids?

12.60 Explain how a pressure cooker works.

12.61 When liquid ethanol begins to boil, what is present inside the bubbles that form?

12.62 The radiator cap of an automobile engine is designed to maintain a pressure of approximately 15 lb/in.^{2} above normal atmospheric pressure. How does this help prevent the engine from "boiling over" in hot weather?

12.63 Butane, C_4H_{10}, has a boiling point of -0.5 °C (which is 31 °F). Despite this, liquid butane can be seen sloshing about inside a typical butane lighter, even at room temperature. Why isn't the butane boiling inside the lighter at room temperature?

12.64 Why does H_2S have a lower boiling point than H_2Se? Why does H_2O have a much higher boiling point than H_2S?

12.65 An H—F bond is more polar than an O—H bond, so HF forms stronger hydrogen bonds than H_2O. Nevertheless, HF has a lower boiling point than H_2O. Explain why this is so.

Energy Changes That Accompany Changes of State

****12.66** At the right is a cooling curve for 1 mol of a substance.

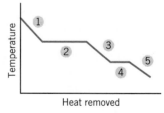

(a) On which portions of this graph do we find the average kinetic energy of the molecules of the substance changing?

(b) On which portions of this graph is the amount of heat removed related primarily to a lowering of the potential energy of the molecules?

(c) Which portion of the graph corresponds to the release of the heat of vaporization?

(d) Which portion of the graph corresponds to the release of the heat of fusion?

(e) Which is larger, the heat of fusion or the heat of vaporization?

(f) On the graph, indicate the melting point of the solid.

(g) On the graph, indicate the boiling point of the liquid.

(h) On the drawing, indicate how supercooling of the liquid would affect the graph.

12.67 What are the names and symbols given to the enthalpy changes involved in (a) the conversion of 1 mol of liquid to 1 mol of vapor, (b) the conversion of 1 mol of solid to 1 mol of vapor, and (c) the conversion of 1 mol of solid to 1 mol of liquid?

12.68 Why is $\Delta H_{vaporization}$ larger than ΔH_{fusion}? How does $\Delta H_{sublimation}$ compare with $\Delta H_{vaporization}$? Explain your answer.

12.69 Would the "heat of condensation," $\Delta H_{condensation}$, be exothermic or endothermic?

12.70 Hurricanes can travel for thousands of miles over warm water, but they rapidly lose their strength when they move over a large land mass or over cold water. Why?

12.71 Rain forms when water vapor condenses in clouds. Explain why this is the source of energy for producing winds in storms.

12.72 Ethanol (grain alcohol) has a molar heat of vaporization of 39.3 kJ/mol. Ethyl acetate, a common solvent, has a molar heat of vaporization of 32.5 kJ/mol. Which of these substances has the larger intermolecular attractions?

12.73 A burn caused by steam is much more serious than one caused by the same amount of boiling water. Why?

12.74 Arrange the following substances in order of their increasing values of $\Delta H_{vaporization}$: (a) HF, (b) CH_4, (c) CF_4, (d) HCl.

Le Châtelier's Principle

12.75 State Le Châtelier's principle in your own words.

12.76 What do we mean by the *position of equilibrium*?

12.77 Use Le Châtelier's principle to predict the effect of adding heat in the equilibrium: solid + heat $\rightleftharpoons$ liquid.

12.78 Use Le Châtelier's principle to explain why lowering the temperature lowers the vapor pressure of a solid.

Phase Diagrams

12.79 For most substances, the solid is more dense than the liquid. Use Le Châtelier's principle to explain why the melting point of such substances should *increase* with increasing pressure. Sketch the phase diagram for such a substance, being sure to have the solid–liquid equilibrium line slope in the correct direction.

12.80 Define *critical temperature* and *critical pressure*.

12.81 What is a supercritical fluid? Why is supercritical CO_2 used to decaffeinate coffee?

12.82 For refrigeration systems, why is it important for the gas that's used to have a high critical temperature? Why is it important that the boiling point of the substance be low?

12.83 What phases of a substance are in equilibrium at the triple point?

12.84 Why doesn't CO_2 have a normal boiling point?

12.85 At room temperature, hydrogen can be compressed to very high pressures without liquefying. On the other hand, butane becomes a liquid at high pressure (at room temperature). What does this tell us about the critical temperatures of hydrogen and butane?

REVIEW PROBLEMS

Answers to problems whose numbers are printed in color are given in Appendix B. More challenging problems are marked with asterisks.

Intermolecular Attractions and Molecular Structure

12.86 Which liquid evaporates faster at 25 °C, diethyl ether (an anesthetic) or butanol (a solvent used in the preparation of shellac and varnishes)? Both have the molecular formula $C_4H_{10}O$, but their structural formulas are different, as shown below.

$$CH_3CH_2CH_2CH_2OH \qquad CH_3CH_2{-}O{-}CH_2CH_3$$
$$\text{butanol} \qquad\qquad \text{diethyl ether}$$

12.87 Which compound should have the higher vapor pressure at 25 °C, butanol or diethyl ether? Which should have the higher boiling point?

12.88 What kinds of intermolecular attractive forces (dipole–dipole, London, hydrogen bonding) are present in the following substances?

(a) HF (b) PCl_3 (c) SF_6 (d) SO_2

12.89 What kinds of intermolecular attractive forces are present in the following substances?

(a) $CH_3{-}\overset{\overset{\displaystyle O}{\|}}{C}{-}OH$ (b) H_2S (c) SO_3 (d) CH_3NH_2

12.90 Consider the compounds $CHCl_3$ (chloroform, an important solvent that was once used as an anesthetic) and $CHBr_3$ (bromoform, which has been used as a sedative). Compare the strengths of their dipole–dipole attractions and the strengths of their London forces. Their boiling points are 61 °C and 149 °C, respectively. For these compounds, which kinds of attractive forces (dipole–dipole or London) are more important in determining their boiling points? Justify your answer.

12.91 Carbon dioxide does not liquefy at atmospheric pressure, but instead forms a solid that sublimes at −78 °C. Nitrogen dioxide forms a liquid that boils at 21 °C at atmospheric pressure. How do these data support the statement that CO_2 is a linear molecule whereas NO_2 is nonlinear?

12.92 Which should have the higher boiling point, ethanol (CH_3CH_2OH, found in alcoholic beverages) or ethanethiol (CH_3CH_2SH, a foul-smelling liquid found in the urine of rabbits that have feasted on cabbage)?

12.93 How do the strengths of London forces compare in $CO_2(l)$ and $CS_2(l)$? Which of these is expected to have the higher boiling point? (Check your answer by referring to the *Handbook of Chemistry and Physics,* which is available in your school library.)

12.94 Below are the vapor pressures of some relatively common chemicals measured at 20 °C. Arrange these substances in order of increasing intermolecular attractive forces.

Benzene, C_6H_6	80 torr
Acetic acid, $HC_2H_3O_2$	11.7 torr
Acetone, C_3H_6O	184.8 torr
Diethyl ether, $C_4H_{10}O$	442.2 torr
Water	17.5 torr

12.95 The boiling points of some common substances are given here. Arrange these substances in order of increasing strengths of intermolecular attractions.

Ethanol, C_2H_5OH	78.4 °C
Ethylene glycol, $C_2H_4(OH)_2$	197.2 °C
Water	100 °C
Diethyl ether, $C_4H_{10}O$	34.5 °C

Vapor Pressure

12.96 On a certain day the air temperature was 25 °C and the pressure exerted by the water vapor in the air was 18.42 torr. What was the percent relative humidity?

12.97 What is the pressure exerted by the water vapor in air at 30 °C if its relative humidity is 65%?

Energy Changes That Accompany Changes of State

12.98 The molar heat of vaporization of water at 25 °C is +43.9 kJ/mol. How many kilojoules of heat would be required to vaporize 125 mL (0.125 kg) of water?

12.99 The molar heat of vaporization of acetone, C_3H_6O, is 30.3 kJ/mol at its boiling point. How many kilojoules of heat would be liberated by the condensation of 5.00 g of acetone?

12.100 Suppose 45.0 g of water at 85 °C is added to 105.0 g

of ice at 0 °C. The molar heat of fusion of water is 6.01 kJ/mol, and the specific heat of water is 4.18 J/g °C. On the basis of these data, (a) what will be the final temperature of the mixture and (b) how many grams of ice will melt?

12.101 A cube of solid benzene (C_6H_6) at its melting point and weighing 10.0 g is placed in 10.0 g of water at 30 °C. Given that the heat of fusion of benzene is 9.92 kJ/mol, to what temperature will the water have cooled by the time all of the benzene has melted?

12.102 Which would be expected to have a higher molar heat of vaporization, water (bp 100 °C) or ethanol (bp 78.4 °C)? Explain your answer.

12.103 Which has the higher molar heat of fusion, water (bp 100 °C) or ethylene glycol (bp 197.2 °C)? Explain your answer.

Phase Diagrams

12.104 Sketch the phase diagram for a substance that has a triple point at −15.0 °C and 0.30 atm, melts at −10.0 °C at 1 atm, and has a normal boiling point of 90 °C.

12.105 Based on the phase diagram of the preceding problem, below what pressure will the substance undergo sublimation? How does the density of the liquid compare with the density of the solid?

12.106 According to Figure 12.30, what phase(s) should exist for CO_2 at (a) −60 °C and 6 atm, (b) −60 °C and 2 atm, (c) −40 °C and 10 atm, and (d) −57 °C and 5.2 atm?

12.107 Looking at the phase diagram for CO_2 (Figure 12.30), how can we tell that solid CO_2 is more dense than liquid CO_2?

12.108 Describe what happens when the pressure on CO_2 is gradually raised from 1 atm to 20 atm at a constant temperature of −56 °C. What happens if the pressure is increased from 1 atm to 20 atm at a constant temperature of −58 °C?

12.109 Describe what happens when CO_2 at 2 atm is warmed from −80 °C to 0 °C at constant pressure.

ADDITIONAL EXERCISES

12.110 Make a list of *all* of the inter- and intramolecular attractive forces that exist in solid Na_2SO_3.

12.111 Calculate the mass of water vapor present in 10.0 L of air at 20 °C if the relative humidity is 75%.

12.112 Should acetone molecules be attracted to water molecules more strongly than to other acetone molecules? Explain your answer. The structure of acetone is shown below.

$$
\begin{array}{c}
\quad\ \ \text{H}\quad\ \text{O}\quad\ \text{H} \\
\quad\ \ | \qquad || \qquad | \\
\text{H}-\text{C}-\text{C}-\text{C}-\text{H} \\
\quad\ \ | \qquad\qquad | \\
\quad\ \ \text{H}\qquad\quad\ \text{H}
\end{array}
$$

*__**12.113** The intermolecular forces of attraction in ethylene glycol are much stronger than in liquid acetone. Therefore, more energy is absorbed when a given amount of the glycol

is converted to a liquid than when the same amount of acetone is changed to a liquid. Yet, if you spill acetone on your hand it produces a much greater cooling effect than if you spill an equivalent amount of the glycol on you hand. Why?

$$
\begin{array}{cc}
\begin{array}{c}
\ \ \text{H}\ \ \ \text{H} \\
\ \ |\ \ \ \ \ | \\
\text{HO}-\text{C}-\text{C}-\text{OH} \\
\ \ |\ \ \ \ \ | \\
\ \ \text{H}\ \ \ \text{H} \\
\text{ethylene glycol}
\end{array}
&
\begin{array}{c}
\ \ \text{H}\ \ \ \text{O}\ \ \ \text{H} \\
\ \ |\ \ \ \ ||\ \ \ \ | \\
\text{H}-\text{C}-\text{C}-\text{C}-\text{H} \\
\ \ |\ \ \ \ \ \ \ \ \ | \\
\ \ \text{H}\ \ \ \ \ \ \text{H} \\
\text{acetone}
\end{array}
\end{array}
$$

12.114 Acetic acid has a heat of fusion of 10.8 kJ/mol and a heat of vaporization of 24.3 kJ/mol.

$$HC_2H_3O_2(s) \longrightarrow HC_2H_3O_2(l) \qquad \Delta H_{fusion} = 10.8 \text{ kJ/mol}$$

$$HC_2H_3O_2(l) \longrightarrow HC_2H_3O_2(g) \qquad \Delta H_{vaporization} = 24.3 \text{ kJ/mol}$$

Use Hess's law to estimate the value for the heat of sublimation of acetic acid, in kilojoules per mole.

*12.115 When warm, moist air sweeps in from the ocean and rises over a mountain range, it expands and cools. Explain how this cooling is related to the attractive forces between gas molecules. Why does this cause rain to form? When the air drops down the far side of the range, its pressure rises as it is compressed. Explain why this causes the air temperature to rise. How does the humidity of this air compare with the air that originally came in off the ocean? Now, explain why the coast of California is lush farmland, whereas valleys (such as Death Valley) that lie to the east of the tall Sierra Nevada mountains are arid and dry.

*12.116 If you shake a CO_2 fire extinguisher on a cool day, you can feel the liquid CO_2 sloshing around inside. However, if you shake the same fire extinguisher on a very hot summer day, you will not feel the liquid sloshing around. What is responsible for this difference?

12.117 The critical temperature of acetone is 236 °C, whereas the critical temperature of propane, C_3H_8, is 97 °C. Why are they so different?

Measuring the Heat of Vaporization (Facets of Chemistry 12.2)

12.118 Ethyl acetate, a solvent used to decaffeinate coffee and to dissolve certain plastics, has a vapor pressure of 72.8 torr at 20 °C and a vapor pressure of 186.2 torr at 40 °C. Estimate the value of ΔH_{vap} for this solvent in units of kilojoules per mole.

12.119 Isopropyl alcohol, used in rubbing alcohol, has a heat of vaporization of 42.09 kJ/mol and a vapor pressure of 31.6 torr at 20 °C. What is the vapor pressure of this alcohol at 60 °C?

12.120 The vapor pressure of liquid nitrogen is 28.1 torr at −216 °C and 47.0 torr at −213 °C. Calculate the heat of vaporization of N_2 in units of kJ/mol. Estimate the normal boiling point of liquid nitrogen.

12.121 Diethyl ether, a liquid that has been used as an anesthetic, has a vapor pressure of 185 torr at 0 °C. At 10 °C its vapor pressure is 0.384 atm. What is the heat of vaporization of diethyl ether in kJ/mol?

12.122 The heat of vaporization of toluene (used to make trinitrotoluene, TNT) is 38.0 kJ/mol. The vapor pressure of toluene is 20.0 torr at 18.4 °C. What will its vapor pressure be at 25 °C?

12.123 Referring to the data in the preceding problem, determine the temperature at which toluene will have a vapor pressure of 40 torr.

12.124 Isopropyl alcohol (rubbing alcohol) has a heat of vaporization of 42.09 kJ/mol and a vapor pressure of 31.6 torr at 20 °C. Estimate the normal boiling point of isopropyl alcohol.

12.125 Propane, the fuel used in gas barbecues, has the following vapor pressures:

Temperature (°C)	Vapor pressure (torr)
−92.4	40
−87.0	60
−79.6	100
−61.2	300
−47.3	600

Graphically determine the value of ΔH_{vap} for propane in units of kJ/mol.

*12.126 At 25 °C, liquid A has a vapor pressure of 100 torr while liquid B has a vapor pressure of 200 torr. The heat of vaporization of liquid A is 32.0 kJ/mol and that of liquid B is 18.0 kJ/mol. At what Celsius temperature will A and B have the same vapor pressure?

Structures, Properties, and Applications of Solids

Skis are available today that contain high-tech "smart" ceramics that are able to detect vibrations in the skis. These "smart materials" then respond in a way to reduce the vibrations, which enables the skier to maintain better control. This is just one example of a large number of applications of the modern materials discussed in this chapter.

THIS CHAPTER IN CONTEXT Most of the materials in the world around us are solids. Some are hard and some are soft; some are flexible and others are rigid; some conduct electricity and others are insulators. In fact, it is this wide variety of properties that accounts for the myriad of uses of solid substances.

Solids fall into two broad categories. *Crystalline solids* have highly regular features that reflect a great deal of the internal order among the particles within them. *Amorphous solids,* on the other hand, lack the high degree of order found in crystals. Useful materials fall into both categories, as you will learn in this chapter.

Until relatively recently, the number of materials available for use by humans was small, being limited to rocks and minerals from the earth and to plant and animal products such as wood and leather. Chemical science has since expanded this number tremendously through the discovery of synthetic polymers, metal alloys, specialized ceramics, and specially prepared solids with unusual electronic properties. Our goal in this chapter is to take a close look at some of these materials so you are able to gain insight into their structures, how they work, and their modern applications.

FIGURE 13.1 *Crystals of table salt.* The size of the tiny cubic sodium chloride crystals can be seen in comparison with a penny.

A high degree of regularity is the principal feature that makes solids different from liquids. A liquid lacks this long-range repetition of structure because the particles in a liquid are jumbled and disorganized as they move about.

13.1 ▶ Crystalline solids have an ordered internal structure

When many substances freeze, or when they separate as a solid from a solution, they tend to form crystals that have highly regular features. For example, Figure 13.1 is a photograph of crystals of one of our most familiar chemicals, sodium chloride—ordinary table salt. Notice that each particle is very nearly a perfect little cube. You might think that the manufacturers went to a lot of trouble and expense to make such uniformly shaped crystals. Actually, they could hardly avoid it. Whenever a solution of NaCl is evaporated, the crystals that form have edges that intersect at 90° angles. Thus, cubes are the norm for NaCl, not the exception.

Crystals in general tend to have flat surfaces that meet at angles that are characteristic of the substance. The regularity of these surface features reflects the high degree of order among the particles that lie within the crystal. This is true whether the particles are atoms, molecules, or ions.

Crystal structures are described by lattices and unit cells

Any repetitive pattern has a symmetrical aspect about it, whether it be the stacking of bricks in a wall, a wallpaper design, or the orderly packing of particles in a crystal (Figure 13.2). For example, we can easily recognize certain repeating distances between the elements of the pattern, and we can see that the lines along which the elements of the pattern repeat are at certain angles to each other.

Symmetry in some repetitive patterns

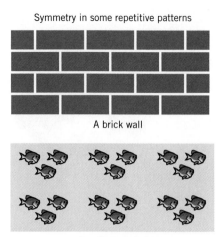

A brick wall

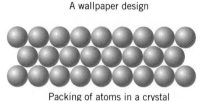

A wallpaper design

Packing of atoms in a crystal

FIGURE 13.2 *Symmetry among repetitive patterns.* A brick wall, a wallpaper design, and particles arranged in a crystal each show a repeating pattern of structural units. The pattern can be described by the distances between the repeating units and the angles along which the repetition of structure occurs.

To concentrate on the symmetrical features of a repeating structure, it is convenient to describe it in terms of a set of points that have the same repeat distances as the structure, arranged along lines oriented at the same angles. Such a pattern of points is called a **lattice,** and when we apply it to describe the packing of particles in a solid, we often call it a **crystal lattice.**

In a crystal, the number of particles is enormous. If you could imagine being at the center of even the tiniest crystal, you would find that the particles go on as far as you can see in every direction. Describing the positions of all these particles or their lattice points is impossible and, fortunately, unnecessary. All we need to do is describe the repeating unit of the lattice, which we call the *unit cell.* To see this, and to gain an insight into the usefulness of the lattice concept, let's begin in two dimensions.

In Figure 13.3 we see a two-dimensional square lattice. It's a *square lattice* because the lattice points lie at the corners of squares. The repeating unit of the lattice, its **unit cell,** is indicated in the drawing. If we began with this unit cell, we

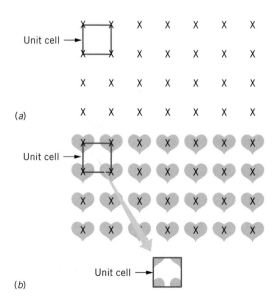

FIGURE 13.3 *A two-dimensional lattice.* (*a*) A simple square lattice, for which the unit cell is a square with lattice points at the corners. (*b*) A wallpaper pattern formed by associating a design element (pink heart) with each lattice point. The x centered on each heart design element corresponds to a lattice point. Notice that the unit cell contains portions of a heart at each corner. (The rest of each heart lies in adjacent unit cells.) In both (*a*) and (*b*), all the information needed to generate the entire array is present in the unit cell, which is repeated over and over.

FIGURE 13.4 *A three-dimensional simple cubic lattice.* (*a*) A simple cubic unit cell showing the locations of the lattice points. (*b*) A portion of a simple cubic lattice built by stacking simple cubic unit cells. (*c*) A substance that forms crystals with a simple cubic lattice. In the unit cell for this substance, identical atoms have their nuclei at the lattice points. Notice that only a portion of each atom lies within this particular unit cell. The rest of each atom is located in adjacent unit cells.

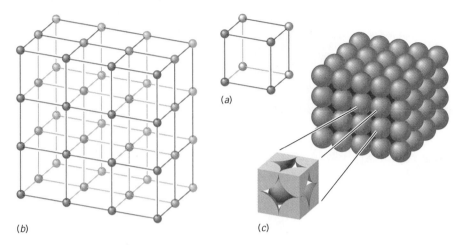

could produce the entire lattice by moving it repeatedly left and right and up and down by distances equal to its edge length. In this sense, all the properties of a lattice are contained in the properties of its unit cell.

An important fact about lattices is that the same lattice can be used to describe many different designs or structures. For example, in Figure 13.3*b*, we see a design formed by associating a pink heart with each lattice point. Using a square lattice, we could form any number of designs just by using different design elements (for example, a rose or a diamond) or by changing the lengths of the edges of the unit cell. *The only requirement is that the same design element must be associated with each lattice point.* In other words, if there is a rose at one lattice point, then there must be a rose at all the other lattice points.

Extending the lattice concept to three dimensions is straightforward. Illustrated in Figure 13.4 is a **simple cubic** (also called a **primitive cubic**) lattice, the simplest and most symmetrical three-dimensional lattice. Its unit cell, the **simple cubic unit cell,** is a cube with lattice points only at its eight corners. Figure 13.4*c* shows the packing of atoms in a substance that crystallizes in a simple cubic lattice as well as the unit cell for that substance. Notice that when the unit cell is "carved out" of the crystal, we find only part of an atom (one-eighth of an atom, actually) at each corner. The rest of each atom resides in adjacent unit cells. Because the unit cell has eight corners, if we put all the corner pieces together we would obtain one complete atom. Thus, we conclude that this unit cell contains just one atom.

$$8 \text{ corners} \times \frac{1/8 \text{ atom}}{\text{corner}} = 1 \text{ atom}$$

As with the two-dimensional lattice, we could use the same simple cubic lattice to describe the structures of many different substances. The *sizes* of the unit cells would vary because the sizes of atoms vary, but the essential symmetry of the stacking would be the same in them all. This fact about lattices makes it possible to describe limitless numbers of different compounds with just a limited set of three-dimensional lattices. In fact, it has been shown mathematically that there are only 14 different three-dimensional lattices possible, which means that all the chemical substances that can exist must form crystals with one or another of these 14 lattice types.

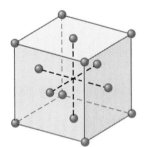

FIGURE 13.5 *A face-centered cubic unit cell.* Lattice points are found at each of the eight corners and in the center of each face.

There are three cubic lattices

TOOLS

Cubic unit cells

In addition to simple cubic, two other cubic lattices are possible: face-centered cubic and body-centered cubic. The **face-centered cubic** (abbreviated **fcc**) **unit cell** has lattice points (and therefore, identical particles) at each of its eight corners plus another

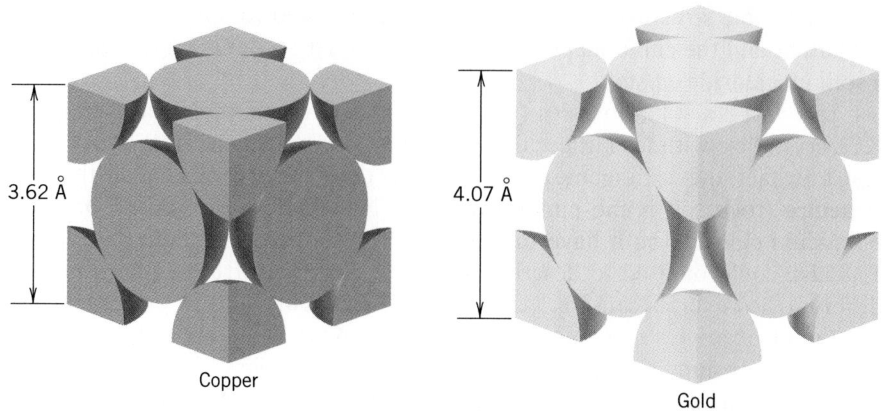

Copper

Gold

FIGURE **13.6** *Unit cells for copper and gold.* These metals both crystallize in a face-centered cubic structure with similar face-centered cubic unit cells. The atoms are arranged in the same way, but their unit cells have edges of different lengths because the atoms are of different sizes ($1 \text{ Å} = 1 \times 10^{-10}$ m).

in the center of each face, as shown in Figure 13.5. Many common metals—copper, silver, gold, aluminum, and lead, for example—form crystals that have face-centered cubic lattices. Each of these metals has the same *kind* of lattice, but the *sizes* of their unit cells differ because the sizes of the atoms differ (see Figure 13.6).

The **body-centered cubic (bcc) unit cell** has lattice points at each corner plus one in the center of the cell, as illustrated in Figure 13.7. The body-centered cubic lattice is also common among a number of metals—examples are chromium, iron, and platinum. Again, these are substances with the same *kind* of lattice, but the dimensions of the lattices vary because of the *different sizes* of the particular atoms.

Not all unit cells are cubic. Some have edges of different lengths or edges that intersect at angles other than 90°. Although you should be aware of the existence of other unit cells and lattices, we will limit the remainder of our discussion to cubic lattices and their unit cells.

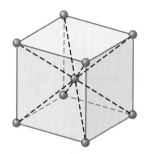

FIGURE **13.7** *A body-centered cubic unit cell.* Lattice points are located at each of the eight corners and in the center of the unit cell.

Many compounds crystallize with cubic lattices

We have seen that a number of metals have cubic lattices. The same is true for many compounds. Figure 13.8, for example, is a view of a portion of a sodium chloride crystal. The Cl^- ions (green) are shown at the lattice points that correspond to a face-centered cubic unit cell. The smaller gray spheres represent Na^+ ions. Notice that they fill the spaces between the Cl^- ions. If we look at the locations of identical particles (Cl^-, for example) we find them at lattice points that describe a face-

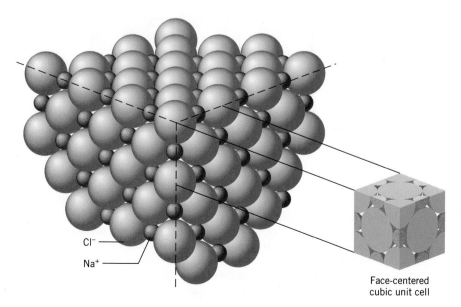

Cl^-

Na^+

Face-centered cubic unit cell

FIGURE **13.8** *The packing of ions in a sodium chloride crystal.* Chloride ions are shown here to be associated with the lattice points of a face-centered cubic unit cell, with the sodium ions placed between the chloride ions.

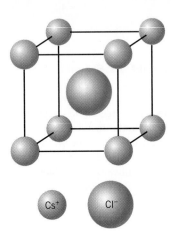

FIGURE 13.9 *The unit cell for cesium chloride, CsCl.* The chloride ion is located in the center of the unit cell. The ions are not shown full-size to make it easier to see their locations in the unit cell.

centered cubic structure. Thus, sodium chloride is said to have a face-centered cubic lattice, and the cubic shape of this lattice is what accounts for the cubic shape of a sodium chloride crystal.

Many of the alkali halides (Group IA–VIIA compounds), such as NaBr and KCl, crystallize with fcc lattices that have the same arrangement of ions as found in NaCl. In fact, this arrangement of ions is so common that it's called the **rock salt structure** (rock salt is the mineral name of NaCl). Because sodium bromide and potassium chloride both have the same kind of lattice as sodium chloride, Figure 13.8 also could be used to describe their unit cells. The *sizes* of their unit cells are different, however, because K^+ is a larger ion than Na^+, and Br^- is larger than Cl^-.

Other examples of cubic unit cells are shown in Figures 13.9 and 13.10. The structure of cesium chloride in Figure 13.9 is simple cubic, although at first glance it may appear to be body-centered. This is because in a crystal lattice, identical chemical units must be at each lattice point. In CsCl, Cs^+ ions are found at the corners, but not in the center, so the Cs^+ ions describe a simple cubic unit cell.

Both zinc sulfide and calcium fluoride in Figure 13.10 have face-centered cubic unit cells that differ from that for sodium chloride, which illustrate once again how the same basic kind of lattice can be used to describe a variety of chemical structures.

Stoichiometry affects the packing of atoms in a unit cell

At this point, you may wonder why a compound crystallizes with a particular structure. Although this is a complex issue, at least one factor is the stoichiometry of the substance. Because the crystal is made up of a huge number of identical unit cells, the stoichiometry within the unit cell must match the overall stoichiometry of the compound. Let's see how this applies to sodium chloride.

EXAMPLE 13.1

Counting Atoms or Ions in a Unit Cell

How many sodium and chloride ions are there in the unit cell of sodium chloride?

ANALYSIS: To answer this question, we have to look closely at the unit cell of sodium chloride. The critical link is realizing that when the unit cell is carved out of the crystal, it encloses *parts of ions,* so we have to determine how many *whole* sodium and chloride ions can be constructed from the pieces within a given unit cell.

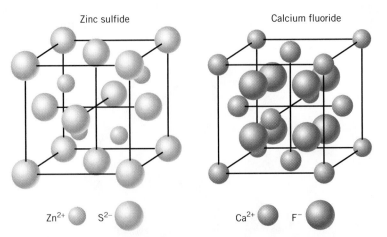

FIGURE 13.10 *Crystal structures based on the face-centered cubic lattice.* Both zinc sulfide, ZnS, and calcium fluoride, CaF$_2$, have crystal structures that fit a face-centered cubic lattice. In ZnS, the sulfide ions are shown at the fcc lattice sites with the four zinc ions entirely within the unit cell. In CaF$_2$, the calcium ions are at the lattice points with the eight fluorides entirely within the unit cell. *Note:* The ions are not shown full-size to make it easier to see their locations in the unit cells.

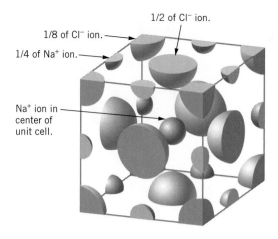

1/2 of Cl⁻ ion.

1/8 of Cl⁻ ion.

1/4 of Na⁺ ion.

Na⁺ ion in center of unit cell.

FIGURE 13.11 *An exploded view of the unit cell of sodium chloride.*

SOLUTION: Let's look at an "exploded" view of the NaCl unit cell, shown in Figure 13.11. We see that we have parts of chloride ions at the corners and in the centers of each face. Let's add the parts.

For chloride:

$$8 \text{ corners} \times \tfrac{1}{8} \text{ Cl}^- \text{ per corner} = 1 \text{ Cl}^-$$

$$4 \text{ faces} \times \tfrac{1}{2} \text{ Cl}^- \text{ per face} = 3 \text{ Cl}^-$$

$$\text{Total} = 4 \text{ Cl}^-$$

For the sodium ions, we have parts along each of the 12 edges plus one whole Na^+ ion in the center of the unit cell. Let's add them.

For sodium:

$$12 \text{ edges} \times \tfrac{1}{4} \text{ Na}^+ \text{ per edge} = 3 \text{ Na}^+$$

$$1 \text{ Na}^+ \text{ in the center} = 1 \text{ Na}^+$$

$$\text{Total} = 4 \text{ Na}^+$$

Thus, in one unit cell, there are four chloride ions and four sodium ions.

Is the Answer Reasonable?
The ratio of the ions is 4 to 4, which is the same as 1 to 1. That's the ratio of the ions in NaCl, so the answer is correct.

PRACTICE EXERCISE 1: What is the ratio of the ions in the unit cell of cesium chloride, which is shown in Figure 13.9?

The calculation in Example 13.1 shows why NaCl can have the crystal structure it does; the unit cell has the proper ratio of cations to anions. It also shows why a compound such as $CaCl_2$ could *not* crystallize with the same kind of unit cell as NaCl. The sodium chloride structure demands a 1 to 1 ratio of cation to anion, so it could not be used by $CaCl_2$ (which has a 1 to 2 cation-to-anion ratio).

Efficiency of packing also can affect how atoms are arranged in a solid

For many substances, particularly metals, the type of crystal structure formed is controlled by maximizing the number of neighbors that surround a given atom. The more neighbors an atom has, the greater are the number of interatomic attractions and the greater is the energy lowering when the solid forms. Structures that achieve the maximum density of packing are known as **closest-packed structures,** and there are two of them that are only slightly different. To visualize how these are produced, let's look at ways to pack spheres of identical size.

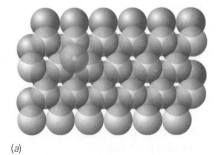

(a)

(b)

Figure 13.12 *Packing of spheres.* (*a*) One layer of closely packed spheres. (*b*) A second layer is started by placing a sphere (colored red) in a depression formed between three spheres in the first layer. (*c*) The second layer of spheres is shown slightly transparent so we can see how the atoms are stacked over the first layer.

(c)

Figure 13.12*a* shows a layer of blue spheres packed as tightly as possible. Notice that each sphere is touched by six others in this layer. Next, we place a second layer on top of the first. We've colored the spheres in this layer red to distinguish them from those in the first layer. We begin with one red sphere, which rests in a depression formed by three atoms in the first layer (Figure 13.12*b*). As others are added, they too fit into depressions in the first layer, until we have a second layer looking just like the first, but shifted slightly and arranged as shown in Figure 13.12*c*. Notice that we've made the red spheres slightly transparent so we can see where they sit above the first layer.

When we add the third layer, which we will show as green spheres, there are two different ways to begin. The choice we make will affect the way the third layer is stacked relative to the others. One way to start is by placing a green sphere over a depression between red spheres in the second layer that is *also* over a depression between blue spheres in the first layer. This is shown in Figure 13.13*a*. The second choice is to place the green sphere so that it is directly over an atom in the first layer (Figure 13.13*b*). As the third layer is built up, two slightly different patterns emerge according to how the third layer was begun, as illustrated in Figure 13.14. The packing in Figure 13.14*a* is called **cubic closest packing,** abbreviated **ccp.** This is because when viewed from a different perspective we find the atoms are located at positions corresponding to a face-centered cubic lattice. The stacking of layers in Figure 13.14*b* is called **hexagonal closest packing,** or **hcp.** The lattice corresponding to this packing is a hexagonal lattice.

In the hcp structure, the layers alternate in an A-B-A-B . . . pattern, where A stands for the orientations of the first, third, fifth, etc. layers, and B stands for the

Figure 13.13 *Adding a third layer of spheres.* Two options exist for starting the third layer. (*a*) The first atom is placed over a hole in both the first *and* second layer. (*b*) The first atom is placed over a hole in the second layer that is directly over an atom in the first layer.

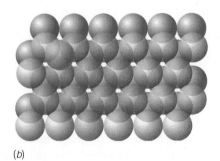

(a)

(b)

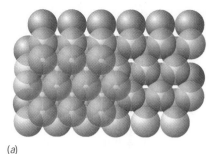 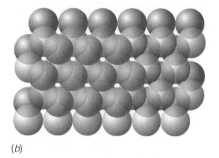

(a)   (b)

FIGURE 13.14 *Closest-packed structures.* (*a*) Cubic closest packing of spheres. (*b*) Hexagonal closest packing of spheres. Notice that there are subtle differences between the two modes of packing.

orientations of the second, fourth, sixth, etc. layers. Thus, spheres in the third layer are directly above those in the first, spheres in the fifth layer are directly above those in the third, and so on. In the ccp structure, there is an A-B-C-A-B-C . . . pattern. The first layer is oriented like the fourth, the second like the fifth, and the third like the sixth.

Both the ccp and hcp structures yield very efficient packing of identically sized atoms. In both structures, each atom is in contact with 12 neighboring atoms—6 atoms in its own layer, 3 atoms in the layer below, and 3 atoms in the layer above. Metals that crystallize with the ccp structure include copper, silver, gold, aluminum, and lead. Metals with the hcp structure include titanium, zinc, cadmium, and magnesium.

Not all solids are crystalline

If a cubic salt crystal is broken, the pieces still have flat faces that intersect at 90° angles. If you shatter a piece of glass, on the other hand, the pieces often have surfaces that are not flat. Instead, they tend to be smooth and curved (see Figure 13.15). This behavior illustrates a major difference between crystalline solids, such as NaCl, and noncrystalline solids, also called **amorphous solids,** such as glass.

The word *amorphous* is derived from the Greek word *amorphos,* which means "without form." Amorphous solids do not have the kinds of long-range repetitive internal structures that are found in crystals. In some ways their structures, being jumbled, are more like liquids than solids. Examples of amorphous solids are ordinary glass and many polymers of the kind we will discuss later in this chapter. In fact, the word **glass** is often used as a general term to refer to any amorphous solid.

As suggested in Figure 13.15, substances that form amorphous solids often consist of long, chainlike molecules that are intertwined in the liquid state somewhat like long strands of cooked spaghetti. To form a crystal from the melted material, these long molecules would have to become untangled and line up in specific patterns. But as the liquid cools, the molecules slow down. Unless the liquid is cooled extremely slowly, the molecular motion decreases too rapidly for the untangling to take place, and the substance solidifies with the molecules still intertwined. As a result, amorphous solids are sometimes described as **supercooled liquids,** a term suggesting the kind of structural disorder found in liquids.

Within the solid, long molecules become tangled. Solid lacks the long-range order found in crystals.

FIGURE 13.15 *Glass is noncrystalline solid.* When glass breaks, the pieces have sharp edges, but their surfaces are not flat planes. This is because in an amorphous solid like glass, long molecules (much simplified here for clarity) are tangled and disorganized, so there is no long-range order characteristic of a crystal.

13.2 ▶ X-ray diffraction is used to study crystal structures

When atoms are bathed in X rays, they absorb some of the radiation and then emit it again in all directions. In effect, each atom becomes a tiny X-ray source. If we look at radiation from two such atoms (see Figure 13.16), we find that the X rays emitted are in phase in some directions but out of phase in others. In Chapter 8 you learned that constructive (in-phase) and destructive (out-of-phase) interferences create a phenomenon called *diffraction.* X-ray diffraction by crystals has enabled many scientists to win Nobel prizes by determining the structures of extremely complex compounds in a particularly elegant way.

In a crystal, there are enormous numbers of atoms, but *they are evenly spaced throughout the lattice.* When the crystal is bathed in X rays, intense beams are diffracted because of constructive interference, and they appear only in specific directions. In other directions, no X rays appear because of destructive interference. When the X rays coming from the crystal fall on photographic film, the diffracted beams form a **diffraction pattern** (see Figure 13.17). The film is darkened only where the X rays strike.[1]

In 1913, the British physicist William Henry Bragg and his son William Lawrence Bragg discovered that just a few variables control the appearance of an X-ray diffraction pattern. These are shown in Figure 13.18, which illustrates the conditions necessary to obtain constructive interference of the X rays from successive layers of atoms (planes of atoms) in a crystal. A beam of X rays having a wavelength λ strikes the layers at an angle θ. Constructive interference causes an intense diffracted beam to emerge at the same angle θ. The Braggs derived an equation, now called the **Bragg equation,** relating λ, θ, and the distance between the planes of atoms, d,

Bragg equation

$$n\lambda = 2d \sin \theta \qquad (13.1)$$

where n is a whole number. The Bragg equation is the basic tool used by scientists in the study of solid structures. Let's briefly see how they use it.

In any crystal, many different sets of planes can be passed through the atoms. Figure 13.19 illustrates this idea in two dimensions for a simple pattern of points. When a crystal produces an X-ray diffraction pattern, many diffracted beams are produced because of the diffraction from the many sets of planes. The physical geometry of the apparatus used to record the diffraction pattern allows the measurement of the angles at which the diffracted beams emerge from each distinct set of planes. Knowing the wavelength of the X rays, the values of n (which can be determined), and the measured angles, θ, the distances between planes of atoms, d, can be computed. The next step is to use the calculated interplanar distances to work backward to deduce where the atoms in the crystal must be located so that layers of atoms are indeed separated by these distances. If this sounds like a difficult task, it is! Some sophisticated mathematics as well as computers are needed to accomplish it. The efforts, however, are well rewarded because the calculations give the locations of atoms within the unit cell and the distances between them. This information, plus a lot of chemical "common sense," is used by chemists to arrive at the shapes and sizes of the molecules in the crystal. Example 13.2 (page 558) illustrates how such data is used.

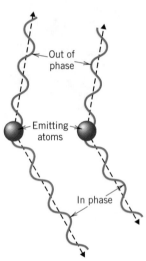

FIGURE 13.16 *Diffraction of X rays from atoms in a crystal.* X rays emitted from atoms are in phase in some directions and out of phase in other directions.

FIGURE 13.17 *X-ray diffraction.* (*a*) The production of an X-ray diffraction pattern. (*b*) An X-ray diffraction pattern produced by sodium chloride recorded on photographic film.

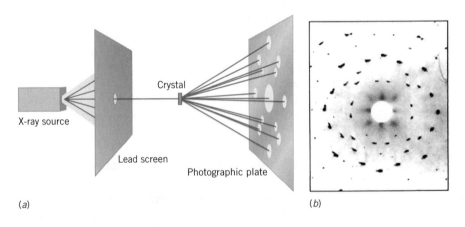

(*a*)

(*b*)

[1]Modern X-ray diffraction instruments use electronic devices to detect and measure the angles and intensities of the diffracted X rays.

FACETS OF CHEMISTRY 13.1

X-Ray Diffraction and Biochemistry

Biochemists today often use X-ray diffraction to determine the structures of enzymes and other large molecules produced by living systems. The most famous example of this was the experimental determination of the molecular structure of DNA. As you may know, DNA is found in the nuclei of cells and serves to carry an organism's genetic information. In 1953, using X-ray diffraction photographs of DNA fibers obtained by Rosalind Franklin and Maurice Wilkins (see Figure 1), James Watson and Francis Crick came to the conclusion that the DNA structure consists of the now-famous double helix (see Figure 2). We will examine this important structure in more detail in Chapter 25. Watson, Crick, and Wilkins shared the 1962 Nobel prize in physiology or medicine for their discovery.

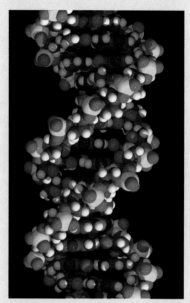

FIGURE 2 *A computer-generated model of the DNA double helix.* The structure of DNA is discussed in Chapter 25.

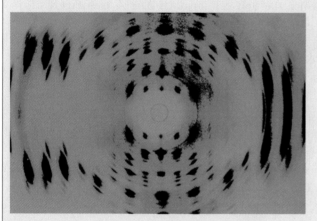

FIGURE 1 *X-ray diffraction photograph of DNA.*

X-ray diffraction has had a profound impact on the study of biochemical molecules. For example, the general shape of DNA molecules, the chemicals of genes, was deduced by using X-ray diffraction (see Facets of Chemistry 13.1). Today, X-ray diffraction continues to be one of the tools used by biochemists to determine the structures of complex proteins and enzymes.

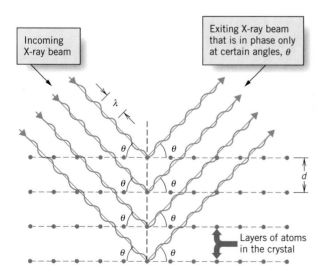

Incoming X-ray beam

Exiting X-ray beam that is in phase only at certain angles, θ

Layers of atoms in the crystal

FIGURE 13.18 *Diffraction of X rays from successive layers of atoms in a crystal.* The layers of atoms are separated by a distance d. The X rays of wavelength λ enter and emerge at an angle θ relative to the layers of the atoms. For the emerging beam of X rays to have any intensity, the condition $n\lambda = 2d \sin \theta$ must be fulfilled, where n is a whole number.

FIGURE **13.19** *A two-dimensional pattern of points with many possible sets of parallel lines.* In a three-dimensional crystal lattice there are many sets of parallel planes.

EXAMPLE 13.2

Using Crystal Structure Data to Calculate Atomic Sizes

X-ray diffraction measurements reveal that copper crystallizes with a face-centered cubic lattice in which the unit cell length is 3.62 Å (see Figure 13.6). What is the radius of a copper atom expressed in angstroms and in picometers?

ANALYSIS: In Figure 13.6, we see that copper atoms are in contact along a diagonal (the dashed line below) that runs from one corner of a face to another corner.

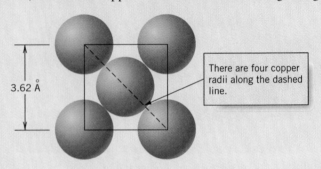

There are four copper radii along the dashed line.

By geometry, we can calculate the length of this diagonal, which equals four times the radius of a copper atom. Once we calculate the radius in angstrom units we can convert to picometers using the relationships

$$1 \text{ Å} = 1 \times 10^{-10} \text{ m}$$

$$1 \text{ pm} = 1 \times 10^{-12} \text{ m}$$

SOLUTION: From geometry, the length of the diagonal is $\sqrt{2}$ times the length of the edge of the unit cell.

$$\text{diagonal} = \sqrt{2} \times (3.62 \text{ Å}) = 5.12 \text{ Å}$$

If we call the radius of the copper atom r_{Cu}, then the diagonal equals $4 \times r_{Cu}$. Therefore,

$$4 \times r_{Cu} = 5.12 \text{ Å}$$

$$r_{Cu} = 1.28 \text{ Å}$$

The calculated radius of the copper atom is 1.28 Å.
Next we convert this to picometers.

$$1.28 \text{ Å} \times \frac{1 \times 10^{-10} \text{ m}}{1 \text{ Å}} \times \frac{1 \text{ pm}}{1 \times 10^{-12} \text{ m}} = 128 \text{ pm}$$

Is the Answer Reasonable?
It's difficult to get an intuitive feel for the sizes of atoms, so we should be careful to check the calculation. The length of the diagonal seems about right; it is longer than the edge of the unit cell. If we look again at the diagram above, we can see that along the diagonal there are four copper radii. The rest of the arithmetic is okay, so our answer is correct.

13.3 ▶ Physical properties are related to crystal types

You know from personal experience that solids exhibit a wide range of properties. Some solids, like diamond, are very hard. Others, such as naphthalene (moth flakes) or ice, are soft by comparison and are easily crushed. Some solids, like salt crystals or iron, have high melting points, whereas others, such as candle wax, melt at low temperatures. Some solids conduct electricity well, but others are nonconducting.

The physical properties just described depend on the kinds of particles in the solid as well as on the strengths of attractive forces holding the solid together. Even though we can't make exact predictions about such properties, some generalizations do exist. In discussing them, it is convenient to divide crystals into four types: ionic, molecular, covalent, and metallic.

◀ **TOOLS**

Properties of crystal types

Ionic crystals have cations and anions at lattice sites

We already discussed some of the properties of **ionic crystals** in Chapter 2. They are relatively hard, have high melting points, and are brittle. Such properties reflect strong attractive forces between ions of opposite charge as well as repulsions that occur when ions of like charge are near each other. You also learned that ionic compounds do not conduct electricity in their solid states, but in their molten states they conduct electricity well. This behavior is consistent with the mobility of ions in the liquid state but not in the solid state.

Molecular crystals have neutral molecules at lattice sites

Molecular crystals are solids in which the lattice sites are occupied either by atoms (as in solid argon or krypton) or by molecules (as in solid CO_2, SO_2, or H_2O). If the molecules of such solids are relatively small, the crystals tend to be soft and have low melting points because the particles in the solid experience relatively weak intermolecular attractions. In crystals of argon, for example, the attractive forces are exclusively London forces. In SO_2, which is composed of polar molecules, there are dipole–dipole attractions as well as London forces. And in water crystals (ice) the molecules are held in place primarily by strong hydrogen bonds. (When the molecules in a solid are large, as in the polymers we will discuss in Section 13.5, the physical properties can be much different from those of solids containing small molecules.)

Molecular crystals are soft because little effort is needed to separate the particles or cause them to move past each other.

Covalent crystals have atoms at lattice sites covalently bonded to other atoms

Covalent crystals are solids in which lattice positions are occupied by atoms that are covalently bonded to other atoms at neighboring lattice sites. The result is a crystal that is essentially one gigantic molecule. These solids are sometimes called **network solids** because of the interlocking network of covalent bonds extending throughout the crystal in all directions. A typical example is diamond (see Figure 13.20). Covalent crystals tend to be very hard and to have very high melting points because of the strong attractions between covalently bonded atoms. Other examples of covalent crystals are quartz (SiO_2, found in some types of sand) and silicon carbide (SiC, a common abrasive used in sandpaper).

Metallic crystals have cations at lattice sites surrounded by mobile electrons

Metallic crystals have properties that are quite different from those of the other three types. Metallic crystals conduct heat and electricity well, and they have the luster characteristically associated with metals. A number of different models have been developed to explain metallic crystals. One of the simplest models views the

FIGURE 13.20 *The structure of diamond.* Each carbon atom is covalently bonded to four others at the corners of a tetrahedron. This is just a tiny portion of a diamond, of course; the structure extends throughout the entire diamond crystal.

TABLE 13.1		TYPES OF CRYSTALS		
Crystal Type	Particles Occupying Lattice Sites	Type of Attractive Force	Typical Examples	Typical Properties
Ionic	Positive and negative ions	Attractions between ions of opposite charge	NaCl, CaCl$_2$, NaNO$_3$	Relatively hard; brittle; high melting points; nonconductors of electricity as solids, but conduct when melted
Molecular	Atoms or molecules	Dipole–dipole attractions, London forces, hydrogen bonding	HCl, SO$_2$, N$_2$, Ar, CH$_4$, H$_2$O	Soft; low melting points; nonconductors of electricity in both solid and liquid states
Covalent (network)	Atoms	Covalent bonds between atoms	Diamond, SiC (silicon carbide), SiO$_2$ (sand, quartz)	Very hard; very high melting points; nonconductors of electricity
Metallic	Positive ions	Attractions between positive ions and an electron cloud that extends throughout the crystal	Cu, Ag, Fe, Na, Hg	Range from very hard to very soft; melting points range from high to low; conduct electricity in both solid and liquid states; have characteristic luster

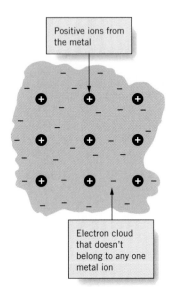

Positive ions from the metal

Electron cloud that doesn't belong to any one metal ion

FIGURE 13.21 *The "electron sea" model of a metallic crystal.* In this highly simplified view of a metallic solid, metal atoms lose valence electrons to the solid as a whole and exist as positive ions surrounded by a mobile "sea" of electrons.

lattice positions of a metallic crystal as being occupied by *positive ions,* which consist of the nuclei plus core electrons of the metal atoms. Surrounding these ions is a "cloud" of electrons formed by the valence electrons, which extends throughout the entire solid (see Figure 13.21). The electrons in this cloud belong to no single positive ion, but rather to the crystal as a whole. Because the electrons aren't localized on any one atom, they are free to move easily, which accounts for the high electrical conductivity of metals. By their movement, the electrons can also transmit kinetic energy rapidly through the solid, so metals are also good conductors of heat. This model explains the luster of metals, too. When light shines on the metal, the loosely held electrons vibrate easily and readily re-emit the light with essentially the same frequency and intensity.

It is not possible to make many simple generalizations about the melting points of metals. We've seen before that some, like tungsten, have very high melting points. Others, such as sodium (mp = 97.8 °C) and mercury, which is a liquid at room temperature, have quite low melting points. To some extent, the melting point depends on the charge of the positive ions in the metallic crystal. The Group IA metals have just one valence electron to lose to the lattice. Their cores are cations with a 1+ charge and are only weakly attracted to the "electron sea" that surrounds them. Atoms of the Group IIA metals, however, each lose two electrons to the metallic lattice and form ions with a 2+ charge. These more highly charged ions are attracted more strongly to the surrounding electron sea, so the Group IIA metals have higher melting points than their neighbors in Group IA. For example, magnesium melts at 650 °C, whereas sodium, also in Period 3, melts at only 98 °C. Those metals with very high melting points, like tungsten, must have very strong attractions between their atoms, which suggests that there probably is some covalent bonding between them as well.

The different ways of classifying crystals and a summary of their general properties are given in Table 13.1.

EXAMPLE 13.3

Identifying Crystal Types from Physical Properties

The metal osmium, Os, forms an oxide with the formula OsO$_4$. The soft crystals of OsO$_4$ melt at 40 °C, and the resulting liquid does not conduct electricity. To which crystal type does solid OsO$_4$ probably belong?

ANALYSIS: You might be tempted to suggest that the compound is ionic simply because it is formed from a metal and a nonmetal. However, the properties of the compound are inconsistent with it being ionic. Therefore, we have to consider that there may be exceptions to the generalization discussed earlier about metal–nonmetal compounds. If so, what do the properties of OsO_4 suggest about its crystal type?

SOLUTION: The characteristics of the OsO_4 crystals—softness and low melting point—suggest that solid OsO_4 is a molecular solid and that it contains molecules of OsO_4. This is further supported by the fact that liquid OsO_4 does not conduct electricity, which is evidence for the lack of ions in the liquid.

Is the Answer Reasonable?
There's not much we can do to check ourselves here except to review our analysis.

PRACTICE EXERCISE 2: Boron nitride, which has the empirical formula BN, melts at 2730 °C and is almost as hard as a diamond. What is the probable crystal type for this compound?

PRACTICE EXERCISE 3: Crystals of elemental sulfur are easily crushed and melt at 113 °C to give a clear yellow liquid that does not conduct electricity. What is the probable crystal type for solid sulfur?

13.4 ▶ Band theory explains the electronic structures of solids

In the preceding section, we described a very simple model to explain some of the properties of metals. However, the model fails to explain *why* metal atoms lose their valence electrons to the lattice as a whole, and it makes no attempt to account for the properties of metalloids (which are semiconductors) or why nonmetallic substances are electrical insulators. The need for a more sophisticated model led to a theory of bonding in solids called the **band theory.**

According to band theory, an **energy band** in a solid is composed of a very large number of closely spaced energy levels that are formed by combining atomic orbitals of similar energy from each of the atoms within the substance. For example, in sodium the $1s$ atomic orbitals, one from each atom, combine to form a single $1s$ band. The number of energy levels in the band equals the number of $1s$ orbitals supplied by the entire collection of sodium atoms. The same thing occurs with the $2s$, $2p$, etc., orbitals, so that we also have $2s$, $2p$, etc., bands within the solid.

Figure 13.22 illustrates the energy bands in solid sodium. Notice that the electron density in the $1s$, $2s$, and $2p$ bands does not extend far from each individual nucleus, so these bands produce effectively localized energy levels in the solid. However, the $3s$ band (which is formed by overlap of the valence shell orbitals of the sodium atoms) is delocalized and extends continuously through the solid. The same applies to energy bands formed by higher-energy orbitals.

Sodium atoms have filled $1s$, $2s$, and $2p$ orbitals, so the corresponding bands in the solid are also filled. The $3s$ orbital of sodium, however, is only half-filled, which leads to a half-filled $3s$ band. The $3p$ and higher-energy bands in sodium are completely empty. When a voltage is applied across sodium, electrons in filled bands cannot move through the solid because orbitals in the same band on neighboring atoms are already filled and cannot accept an additional electron. In the $3s$ band, which is half-filled, an electron can hop from atom to atom with ease, and this allows sodium to conduct electricity well.

We refer to the band containing the outer shell (valence shell) electrons as the **valence band.** Any band that is either vacant or partially filled and uninterrupted throughout the lattice is called a **conduction band,** because electrons in it are able to move through the solid and thereby serve to carry electricity.

In metallic sodium the valence band and conduction band are the same, so sodium is a good conductor. In magnesium the $3s$ valence band is filled and therefore cannot be used to transport electrons. However, the vacant $3p$ conduction

Delocalized orbitals were discussed in Chapter 10.

In terms of the band theory, the electrons in the "electron sea" model correspond to the electrons in the conduction band.

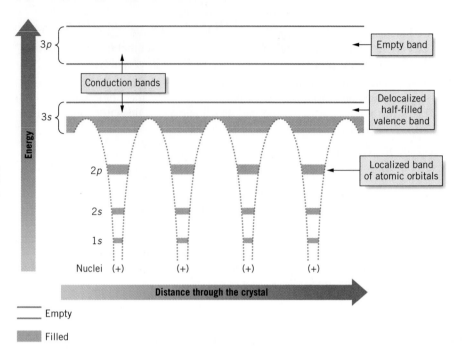

FIGURE 13.22 *Energy bands in solid sodium.* The 1s, 2s, and 2p bands do not extend far in either direction from the nucleus, so the electrons in them are localized around each nucleus and can't move through the crystal. The delocalized 3s valence band extends throughout the entire solid, as do bands such as the 3p band formed from higher-energy orbitals of the sodium atoms.

band actually overlaps the valence band and can easily be populated by electrons when voltage is applied (Figure 13.23a). This permits magnesium to be a conductor.

In an insulator such as glass, diamond, or rubber, all the valence electrons are used to form covalent bonds, so all the orbitals of the valence band are filled and cannot contribute to electrical conductivity. In addition, the energy separation, or **band gap,** between the filled valence band and the nearest conduction band (empty band) is large. As a result, electrons cannot populate the conduction band, so these substances are unable to conduct electricity (Figure 13.23b).

In a semiconductor such as silicon or germanium, the valence band is also filled, but the band gap between the filled valence band and the nearest conduction band is small (Figure 13.23c). At room temperature, thermal energy possessed by the electrons is sufficient to promote some electrons to the conduction band and a small degree of electrical conductivity is observed. One of the interesting properties of semiconductors is that their electrical conductivity increases with increasing temperature. This is because as the temperature rises, the number of electrons with

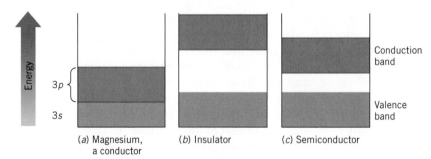

FIGURE 13.23 *Energy bands in different types of solids.* (*a*) In magnesium, a good electrical conductor, the empty 3p conduction band overlaps the filled 3s valence band and provides a way for this metal to conduct electricity. (*b*) In an insulator, the energy gap between the filled valence band and the empty conduction band prevents electrons from populating the conduction band. (*c*) In a semiconductor, there is a small band gap and thermal energy can promote some electrons from the filled valence band to the empty conduction band. This enables the solid to conduct electricity weakly.

enough energy to populate the conduction band also increases. Photons can also provide the energy needed to promote electrons to the conduction band. This happens, for example, with photoconductors such as cadmium sulfide.

Transistors and other electronic devices employ semiconductors

One of the most significant discoveries of the twentieth century was the way that the electrical characteristics of semiconductors can be modified by the controlled introduction of carefully selected impurities. This led to the discovery of transistors, which made possible all the marvelous electronic devices we now take for granted, such as portable CD players, electronic cameras, cell phones, and microcomputers. In fact, almost everything we do today is influenced in some way by electronic circuits etched into tiny silicon chips.

In a semiconductor such as silicon, all the valence electrons are used to form covalent bonds to other atoms. If a small amount of a Group IIIA element such as boron is added to silicon, it can replace silicon atoms in the structure of the solid. (We say the silicon has been **doped** with boron.) However, for each boron added, one of the covalent bonds in the structure will be deficient in electrons because boron has only three valence electrons instead of the four that silicon has. Under an applied voltage, an electron from a neighboring atom can move to fill this deficiency and thereby leave a positive "hole" behind. This "hole" can then be filled by an electron from a neighboring atom, which creates a new "hole" on that atom. As this happens again and again, the positive "hole" migrates through the solid. The net result is an electrical conduction of a positive charge. Because of the positive nature of the moving charge carrier, the substance is said to be a **p-type semiconductor.**

If the impurity added to the silicon is a Group VA element such as arsenic, it has one more electron in its valence shell than silicon. When the bonds are formed in the solid, there will be an electron left over that's not used in bonding. The extra electrons supplied by the impurity can enter the conduction band and move through the solid under an applied voltage, and because the moving charge now consists of negatively charged electrons, the solid is said to be an **n-type semiconductor.**

Transistors are made from n- and p-type semiconductors and can be formed directly on the surface of a silicon chip, which has made possible the microcircuits in computers and calculators. Another interesting application is in solar batteries, which are easier to understand.

A silicon **solar battery** (also called a **solar cell**) is composed of a silicon wafer doped with arsenic (giving an n-type semiconductor) over which is placed a thin layer of silicon doped with boron (a p-type semiconductor). This is illustrated in Figure 13.24. In the dark there is an equilibrium between electrons and holes at

Recall that at any given temperature there is a distribution of kinetic energies among the particles of a substance. *Thermal energy* is kinetic energy a particle possesses because of its temperature. The higher the temperature, the larger is the thermal energy.

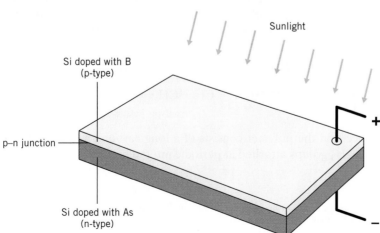

Sunlight

Si doped with B (p-type)

p–n junction

Si doped with As (n-type)

+

–

Figure 13.24 *Construction of a silicon solar battery.*

A handheld solar calculator. The solar panel is located above the display.

the interface between the two layers, which is called a **p-n junction.** Some electrons from the n-type layer diffuse into the holes in the p-layer and are trapped. This leaves some positive holes in the n-layer. Equilibrium is achieved when the positive holes in the n-layer prevent further movement of electrons into the p-layer.

When light falls on the surface of the cell, the equilibrium is upset. Energy is absorbed, which permits electrons that were trapped in the p-layer to return to the n-layer. As these electrons move across the p-n junction into the n-layer, other electrons leave the n-layer through the wire, pass through the electrical circuit, and enter the p-layer. Thus an electric current flows when light falls on the cell and the external circuit is completed. This electric current can be used to run a motor, power a handheld calculator, or perform whatever other task we wish.

13.5 ▶ Polymers are composed of many repeating molecular units

Figure 13.25 Synthetics in recreation. Lightweight synthetic fabrics have made it easier to pack strong tents into remote regions (here, the Wasatch Mountains of Utah).

Nearly all of the compounds that we have studied so far have relatively low molecular masses. Both in nature and in the world of synthetics, however, many substances consist of **macromolecules** made up of hundreds or even thousands of atoms. Synthetics made of macromolecules are examples of how chemists have been able to take very ordinary substances in nature, like coal, oil, air, and water, and make new materials, never seen before, with useful applications. Recording tapes, skis, composites in recreational vehicles and their tires, backpacking gear, and all sorts of other recreational equipment derive strength from macromolecules (Figure 13.25). Materials that are made into thread and cloth consist of macromolecules, and when paints cure, some of their molecules change into other molecules of enormous sizes.

In nature, substances with macromolecules are almost everywhere you look. Trees and anything made of wood, for example, derive their strength from lignins and cellulose, both consisting of enormous molecules that overlap and intertwine. The proteins and the DNA found in our bodies are also huge molecules. Macromolecules of biological origin will be discussed in Chapter 25; in this chapter we will look at synthetic macromolecules and some of their uses.

Polymers have structural order

Figure 13.26 Artificial turf on a football field is made of polypropylene fibers.

Some macromolecular substances have more structural order than others. A **polymer,** for example, is a macromolecular substance all of whose molecules have a small characteristic structural feature that repeats itself over and over. An example is polypropylene, a polymer with many uses, including dishwasher-safe food containers, indoor-outdoor carpeting, and artificial turf (Figure 13.26). The molecules of polypropylene have the following system.

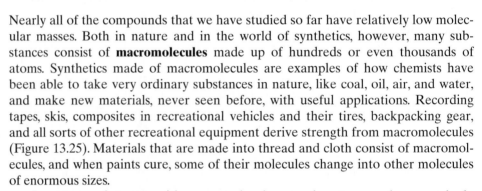

polypropylene

Notice that the polymer consists of a long carbon chain (the polymer's **backbone**) with CH_3 groups attached at periodic intervals.[2]

---

[2]The drawing of the molecule shows one way the CH_3 groups can be oriented relative to the backbone and each other. Other orientations are also possible, and they affect the physical properties of the polymer.

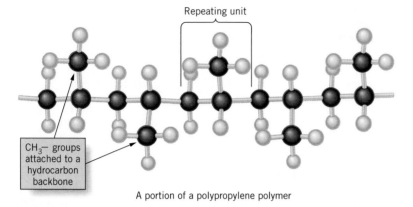

Repeating unit

CH$_3$— groups attached to a hydrocarbon backbone

A portion of a polypropylene polymer

If you study this structure, you can see that one structural unit occurs repeatedly (actually thousands of times). In fact, the structure of a polymer is usually represented by the use of only its repeating unit, enclosed in parentheses, with a subscript *n* standing for several thousand units.

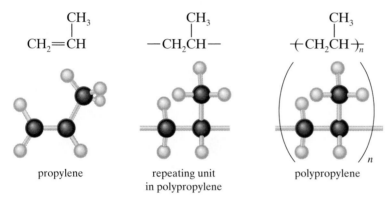

propylene   repeating unit in polypropylene   polypropylene

The value of *n* is not a constant for every molecule in a given polymer sample. A polymer, therefore, does not consist of molecules identical in *size,* just identical in *kind;* they have the same repeating unit. Notice that despite the *-ene* ending (which is used in naming compounds with double bonds), "polypropylene" has no double bonds. The polymer is named after its starting material.

The repeating unit of a polymer is contributed by a chemical raw material called a **monomer.** Thus propylene is the monomer for polypropylene. The reaction that makes a polymer out of a monomer is called **polymerization,** and the verb is "to polymerize." Most (but not all) useful polymers are formed from monomers that are considered to be organic compounds.

Some polymers form by addition of monomer units

There are basically two ways that monomers become joined to form polymers. One path involves the simple addition of one monomer unit to another, a process that continues over and over until a very long chain of monomer units is produced. Polymers formed by this process are called **addition polymers.** Polypropylene, discussed above, is an example. A simpler example is *polyethylene,* formed from ethylene ($CH_2=CH_2$) monomer units. Under the right conditions and with the aid of a substance called an *initiator,* a pair of electrons in the carbon–carbon double bond of ethylene becomes unpaired. The initiator binds to one carbon, leaving an unpaired electron on the other carbon. The result is a very reactive substance called a *free radical,* which can attack the double bond of another ethylene. In the attack, one of the electron pairs of the double bond becomes unpaired. One of the electrons becomes shared with the unpaired electron of the free radical, forming a bond that joins the two hydrocarbon units together. The other unpaired electron moves to the end of the chain, as follows.

Reactions forming addition polymers

As the free radical approaches the ethylene molecule . . .   . . . electrons in one of the bonds unpair and move to opposite ends of the molecule.

$$\boxed{initiator} - \underset{\underset{H}{|}}{\overset{\overset{H}{|}}{C}} - \underset{\underset{H}{|}}{\overset{\overset{H}{|}}{C}} \cdot \quad \overset{H}{\underset{H}{C}} = \overset{H}{\underset{H}{C}} \quad \longrightarrow \quad \boxed{initiator} - \underset{\underset{H}{|}}{\overset{\overset{H}{|}}{C}} - \underset{\underset{H}{|}}{\overset{\overset{H}{|}}{C}} - \underset{\underset{H}{|}}{\overset{\overset{H}{|}}{C}} - \underset{\underset{H}{|}}{\overset{\overset{H}{|}}{C}} \cdot$$

A bond forms between the two hydrocarbon units and the unpaired electron moves to the end of the chain.

This process is repeated over and over as a long hydrocarbon chain grows. Eventually the chain becomes terminated and the result is a polyethylene molecule. The molecule is so large that the initiator, which is still present at one end of the chain, is an insignificant part of the whole, so in writing the structure of the polymer, the initiator is generally ignored.

etc. — polyethylene chain —etc.

polyethylene

unit derived from ethylene

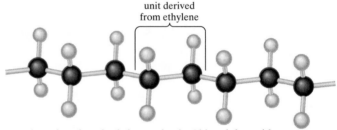

Actually, after ethylene has polymerized, the repeating unit in polyethylene is simply CH_2.

A portion of a polyethylene molecule. Although formed from a monomer with the formula C_2H_4, the actual repeating unit is CH_2.

The process by which the polymer forms has a significant influence on its ultimate structure. For example, the least expensive method for making polyethylene leads to **branching,** which means that polymer chains grow off the main backbone of the molecule as it grows longer (Figure 13.27). Other, more expensive procedures produce molecules without branching, and as we will discuss later, this has a significant effect on the properties of the polymer.

In addition to polyethylene and polypropylene, another very common addition polymer, called **polystyrene,** is formed by polymerizing styrene,

FIGURE 13.27 *Branching of the polymer chain in polyethylene.* Segments of polyethylene grow off the main polymer backbone.

H H
| |
C=C—H

styrene (C_6H_5—CH=CH$_2$)

$\left[\!\!\begin{array}{c}CH—CH_2\end{array}\!\!\right]_n$

polystyrene

FIGURE 13.28 *Styrofoam.* Made of polystyrene, Styrofoam is widely used as an insulation in construction.

Notice that the monomer is similar to propylene, but with a benzene ring in place of the methyl group (CH$_3$—). Therefore, in the polymer we find C_6H_5— attached to every other atom in the hydrocarbon backbone. Polystyrene has almost as many uses as polyethylene. It's used to make clear plastic drinking glasses, molded car parts, and housings of things like computers and kitchen appliances. Sometimes a gas, like carbon dioxide, is blown through molten polystyrene as it is molded into articles. As the hot liquid congeals, tiny pockets of gas are trapped, and the product is a foamed plastic, like the familiar Styrofoam cups or insulation materials (Figure 13.28).

Hundreds of substances similar to ethylene, propylene, and styrene, as well as their halogen derivatives, have been tested as monomers, and Table 13.2 gives some examples.

Use the information in Table 13.2 to write the structure of the polymer poly(vinyl chloride) showing three repeat units. Write the general formula for this polymer.

ANALYSIS: The polymer we're dealing with is an addition polymer, so we can anticipate that the entire CH$_2$=CHCl molecule will repeat over and over. The polymer will be formed by opening the double bond, as we saw for ethylene.

SOLUTION: When the double bond opens, bonds to two other monomer units will form.

H Cl
| |
other monomer unit—C—C—other monomer unit
| |
H H

We need to attach three repeating units to have the answer to the problem.

H Cl H Cl H Cl
| | | | | |
—C—C—C—C—C—C—
| | | | | |
H H H H H H

repeating unit

The general formula for the polymer should indicate the repeating unit occurring *n* times.

$\left(\!\!\begin{array}{c}H \;\; Cl \\ | \;\; | \\ C—C \\ | \;\; | \\ H \;\; H\end{array}\!\!\right)_n$

EXAMPLE 13.4
Writing the Formula for an Addition Polymer

A "halogen derivative" is a compound in which one or more halogen atoms substitute for hydrogens in a parent molecule. Thus, CH$_3$Cl is a halogen derivative of CH$_4$.

Is the Answer Reasonable?
There's not much to check here, other than to be sure that the repeating unit has the same molecular formula as the monomer, which it does.

PRACTICE EXERCISE 4: Write a formula showing three repeating units of the monomer in the polymer Teflon.

TABLE 13.2	SOME ADDITION POLYMERS FORMED FROM COMPOUNDS RELATED TO ETHYLENE, $CH_2{=}CH_2$	
Polymer	Monomer	Uses
Polyethylene	$CH_2{=}CH_2$	Grocery bags, bottles, children's toys, bulletproof vests
Polypropylene	$\begin{array}{c}CH_3\\ \vert\\ CH_2{=}CH\end{array}$	Dishwasher-safe plastic kitchenware, indoor-outdoor carpeting, rope
Polystyrene	$\bigcirc{-}CH{=}CH_2$	Plastic cups, toys, housings for kitchen appliances, Styrofoam insulation
Poly(vinyl chloride) (PVC)	$CH_2{=}CHCl$	Insulation, credit cards, vinyl siding for houses, bottles, plastic pipe
Poly(tetrafluoroethylene) (Teflon)	$F_2C{=}CF_2$	Nonstick surfaces on cookware, valves
Poly(vinyl acetate) (PVA)	$CH_2{=}C\overset{H}{\underset{O{-}C{=}O{-}CH_3}{\diagup}}$	Latex paint, coatings, glue, molded items
Poly(methyl methacrylate) (Lucite)	$CH_2{=}C\overset{CH_3}{\underset{C{=}O{-}O{-}CH_3}{\diagdown}}$	Shatter-resistant windows, coatings, acrylic paints, molded items

Polymers can be formed by condensation reactions

TOOLS

Reactions forming condensation polymers

The second way that monomer units can combine to form a polymer is by a process called **condensation,** in which a small molecule is eliminated when the two monomer units become joined. In a simplified way, we can diagram this as follows.

$$A{-}A{-}A{-}A{-}A{-}A{-}OH \quad H{-}B{-}B{-}B{-}B{-}B{-}B \longrightarrow H_2O$$

$$A{-}A{-}A{-}A{-}A{-}A{-}B{-}B{-}B{-}B{-}B{-}B + H_2O$$

In this example, an OH group from one molecule combines with an H from another, forming a water molecule. At the same time, the two molecules A and B become joined by a covalent bond. If this can be made to happen at both ends of A and B, long chains are formed and a **condensation polymer** is the result.

The two most familiar condensation polymers are nylon and polyesters. Nylon is formed by combining two different compounds, so it's considered a **copolymer.** The first nylon to be manufactured is called **nylon 6,6** because it forms by combining two compounds each with six carbon atoms.

$$HO-\overset{O}{\underset{\|}{C}}-CH_2CH_2CH_2CH_2-\overset{O}{\underset{\|}{C}}\boxed{-OH \quad H}+\overset{H}{\underset{|}{N}}-CH_2CH_2CH_2CH_2CH_2CH_2-\overset{H}{\underset{|}{N}}-H$$

adipic acid   hexamethylene diamine

$$HO-\overset{O}{\underset{\|}{C}}-CH_2CH_2CH_2CH_2-\underset{\underbrace{\qquad}_{amide\ bond}}{\overset{O}{\underset{\|}{C}}-\overset{H}{\underset{|}{N}}}-CH_2CH_2CH_2CH_2CH_2CH_2-\overset{H}{\underset{|}{N}}-H$$

Adipic acid is an organic acid; notice that it contains two carboxyl groups (it's called a dicarboxylic acid). The other compound is an amine (diamine, actually, because it has two amine groups). By the elimination of water, the two molecules become joined by a linkage called an *amide bond*. This same linkage is found in proteins, including silk (a fiber nylon was invented to replace) and proteins found in our bodies.

The molecule above with the amide bond still has a carboxyl group on one end and an amine group on the other, so further condensation can occur, ultimately leading to the formation of nylon 6,6.

$$\left(\underset{\underbrace{\qquad\qquad}_{6\ carbon\ atoms}}{\overset{O}{\underset{\|}{C}}-CH_2CH_2CH_2CH_2-\overset{O}{\underset{\|}{C}}-\overset{H}{\underset{|}{N}}}-\underset{\underbrace{\qquad\qquad}_{6\ carbon\ atoms}}{CH_2CH_2CH_2CH_2CH_2CH_2}-\overset{H}{\underset{|}{N}}\right)_n$$

Nylon 6,6

Nylon was invented in 1940 and became popular as a substitute for silk in women's stockings. It forms strong elastic fibers and is used to make fishing line as well as fibers found in all sorts of clothing and many other products.

An example of a polyester is shown below.

$$\left(-O-\overset{O}{\underset{\|}{C}}-\bigcirc-\overset{\overset{ester\ group}{\boxed{\overset{O}{\underset{\|}{C}}-O}}}{}+CH_2-CH_2\right)_n$$

terephthalate group   ethylene group

poly(ethylene terephthalate)—also known as PET

It is a copolymer made by condensation polymerization, the first step of which is

$$H_3C-O-\overset{O}{\underset{\|}{C}}-\bigcirc-\overset{O}{\underset{\|}{C}}\boxed{-O-CH_3 \quad H}O-CH_2-CH_2-OH$$

dimethyl terephthalate   ethylene glycol

$$H_3C-O-\overset{O}{\underset{\|}{C}}-\bigcirc-\overset{O}{\underset{\|}{C}}-O-CH_2-CH_2-OH + CH_3OH$$

Notice that this time the small molecule that's displaced is CH_3OH, methyl alcohol. Continued polymerization ultimately leads to the PET polymer shown above. You probably have heard of this polymer because it also goes by the name *Dacron*.

A variety of starting materials can be used to form different polyesters with a range of properties. Their uses include fibers for fabrics, shatterproof

Dacron. The exceptional wet strength and lightness of Dacron are just the right properties for sails.

plastic bottles for soft drinks, Mylar for making recording tapes and balloons that don't easily deflate, and shatterproof windows and eyeglasses. See Facets of Chemistry 13.2.

Cross-linking between polymer strands gives increased strength

When natural rubber latex was first discovered, it wasn't particularly useful. You couldn't make rubber tires out of it because when it got hot, the rubber became sticky and would melt; when it became cold, it became brittle and hard. In 1839, Charles Goodyear (of tire fame) discovered that adding sulfur to latex, and then heating it, drastically altered the properties of the rubber. He called his new product **vulcanized rubber.**

What happens when sulfur reacts with latex is that groups of sulfur atoms form bridges, called **cross-links,** between strands of the latex polymer (known technically as polyisoprene). This is shown in Figure 13.29. By linking the strands of polyisoprene together, they are no longer able to slip by each other when hot, so the rubber doesn't melt. The increased strength also prevents the polymer from becoming brittle and easily broken when cold. Cross-linking also gives the polymer a

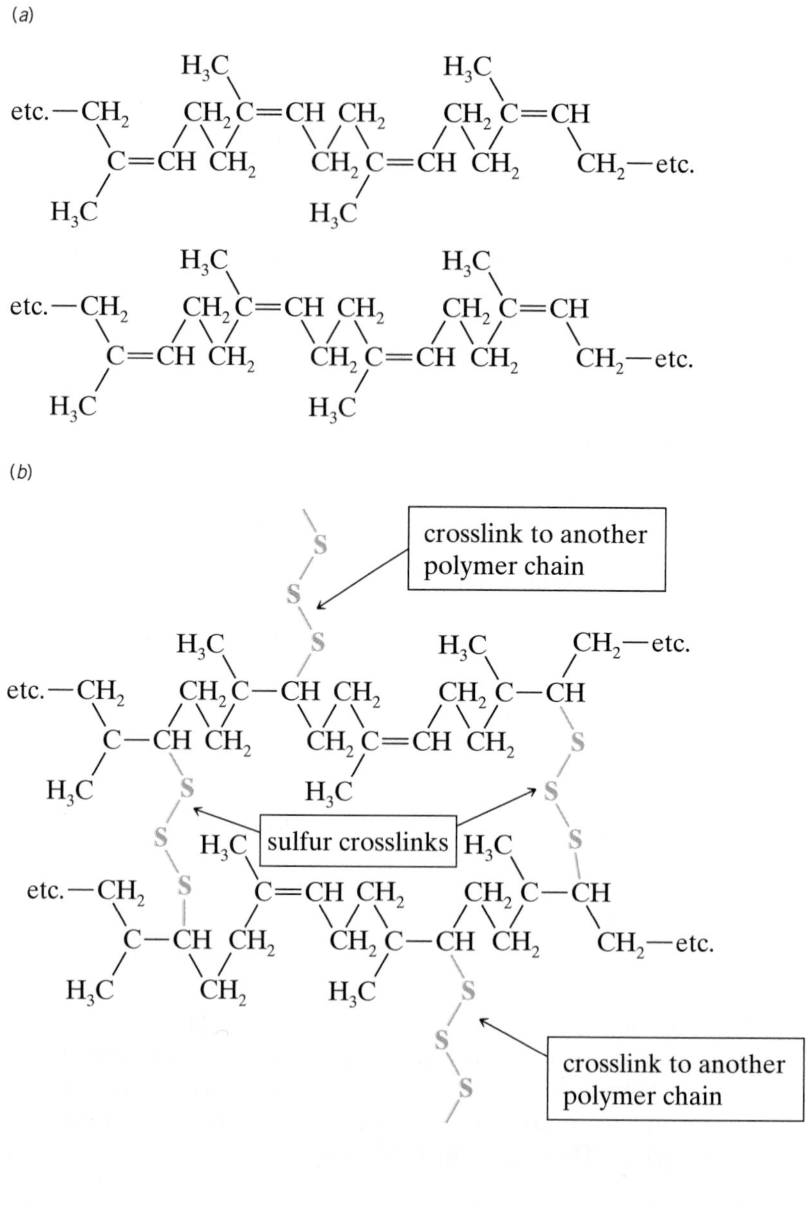

Notice that the isoprene in Figure 13.29 has lots of bends and turns in it. When you stretch rubber, you tend to straighten out these polymer strands, but when you release it, the strands snap back to their original shapes.

FIGURE **13.29** *Crosslinking of polymer chains in rubber by reaction with sulfur.* (*a*) Two strands of polyisoprene molecules. (*b*) Sulfur reacts by opening double bonds in the polymer molecules and forming bridges between adjacent strands.

FACETS OF CHEMISTRY 13.2

Bioplastics and Biodegradable Polymers

Almost all the polymers we find incorporated in fabrics and consumer plastics are made from materials derived from petroleum. This leads to two major problems. First, there is only a finite amount of petroleum available on Earth, so when we manufacture plastics, we are consuming a precious commodity that can't be renewed. Second, synthetic polymers and plastics are not readily attacked by microorganisms, so when placed in landfills or scattered into the environment, they do not degrade and pose a continuing pollution problem. But changes are in the wind.

DuPont, one of the world's leading chemical producers, is manufacturing a polyester polymer called *Sorona,* a versatile polymer with properties that make it excellent for apparel and upholstery fabrics (Figure 1). It is made by condensation polymerization of propanediol with terephthalic acid.

DuPont has been obtaining the propanediol from petroleum sources, but has developed a genetically engineered microbe to produce it by fermentation of corn sugar. The terephthalic acid, however, is still made from petroleum.

Cargill Dow's *NatureWorks* is a completely bio-based polymer made by polymerization of lactic acid, which is also derived by fermentation of corn sugar. The reaction is

The NatureWorks polymer uses no petroleum products and is being used to make fibers for apparel where it is combined with cotton fibers. Other fabric blends for apparel include wool and silk. Clear cold drink cups are also being made with the polymer, which is biodegradable. It's expected that use of the polymer will spread to a variety of packaging materials, too (see Figure 2).

Metabolix, Inc. in Cambridge, Massachusetts, has found a way to avoid the chemical synthesis step in making polymers by using genetic engineering to develop microbes that convert plant matter directly into polymers that are similar to NatureWorks. Depending on the microbe used, a variety of polymers can be produced with the general formula

Different polymer properties are obtained depending on the value of x. At the present time, production of such polymers is at the pilot plant stage.

FIGURE 1 *Sorona fibers in these fabrics have superior stretch recovery, soft touch, and good dye and print capabilities.*

FIGURE 2 *The plastic film used to package these golf balls is made from NatureWorks polylactide polymer.*

(*a*) Two branched polymer chains become twisted with little order.

(*b*) Polymer chains are able to align, producing a tightly packed structure with a large degree of crystallinity.

FIGURE 13.30 *Amorphous and crystalline polyethylene.* (*a*) When branching occurs in LDPE, the polymer strands are not able to become aligned and an amorphous product results. (*b*) Linear HDPE has a high degree of crystallinity, which makes for excellent strong fibers.

A new house under construction is wrapped with a Tyvek® fabric to prevent water and air intrusion, thereby lowering heating and cooling costs.

HDPE molecules contain approximately 30,000 CH_2 units linked end to end.

These molecules contain between 200,000 and 400,000 CH_2 units attached end to end!

"memory," enabling it to snap back to its original shape when released after being stretched (a property you've experienced using rubber bands). The amount of sulfur used to vulcanize the latex also affects the properties of the finished product. If only a small amount of sulfur is used, the polymer is elastic. But as the amount of sulfur increases, so does the amount of cross-linking, and the product becomes harder and less resilient.

Cross-linking provides strength to a variety of polymers you may be familiar with. Formica (used in kitchen countertops), epoxy resins (in epoxy glues), and polycarbonates (which have a polyester-like structure) made strong for use in shatterproof eyeglasses are examples. Even the material used to make soft contact lenses is a cross-linked polymer that's capable of absorbing lots of water.

Polymer crystallinity affects physical properties

Beyond chemical stability, physical properties are the features of polymers most sought after. Desirable properties of Teflon, for example, are its chemical inertness and its slipperiness toward just about anything. Nylon isn't eaten by moths—a chemical property, in fact—but its superior strength and its ability to be made into fibers and fabrics of great beauty are what make nylon valuable. Dacron (a polyester) does not mildew, like cotton, and when made into fibers for a sail, it is superior to cotton. Dacron has greater strength with lower mass than cotton, and Dacron fibers do not stretch as much.

In many ways, the physical properties of a polymer are related to how the individual polymer strands are able to pack in the solid. For example, earlier we noted that the least expensive way of making polyethylene yields a product that has branching. The branches prevent the molecules from lining up in an orderly fashion, so the molecules twist and intertwine to give an essentially amorphous solid (which means it lacks the kind of order found in crystalline solids). See Figure 13.30*a*. This amorphous product is called *low density polyethylene (LDPE);* the polymer molecules have a relatively low molecular mass and the solid has little structural strength. LDPE is the kind of polyethylene used to make the plastic bags grocery stores use to pack your purchases.

Following different methods, polyethylene can be made to form without branching and with molecular masses ranging from 200,000 to 500,000 u. This polymer is called *linear* polyethylene or *high-density polyethylene, HDPE.* In HDPE, the polymer strands are able to line up alongside each other to produce a large degree of order (and therefore, crystallinity), as illustrated in Figure 13.30*b*. This enables the molecules to form fibers easily, and because the molecules are large and packed so well, the London forces between them are very strong. The result is a strong, tough fiber. DuPont's Tyvek®, for example, is made from thin crystalline HDPE polyethylene fibers randomly oriented and pressed together into a material resembling paper. It is lightweight, strong, and resists water, tears, punctures, and abrasion. Federal Express has been using it for years for envelopes, and builders use it to wrap new construction to prevent water and air intrusion, thereby lowering heating and cooling costs. Tyvek is also used to make limited-use protective clothing for use in hazardous environments.

Under the right conditions, linear polyethylene molecules can be made extremely long, yielding *ultrahigh-molecular-weight polyethylene (UHMWPE)* with molecules having molecular masses of three to six million. The fibers produced from this polymer are so strong that they are used to make bulletproof vests! Honeywell is producing an oriented polyethylene polymer they call Spectra, which forms flexible fibers that can be woven into a strong, cut-resistant fabric. It is used to make thin, lightweight liners for surgical gloves that resist cuts by scalpels, industrial work gloves, and even sails for sailboats (see Figure 13.31). Mixed with other plastics, it can be molded into strong rigid forms such as helmets for military or sporting applications.

FIGURE 13.31 *Spectra fibers.* Strong, lightweight sails made of Spectra fibers propelled Brad Van Liew to victory in the 2002–2003 "Around Alone" around-the-world yacht race. Van Liew sailed 31,094 miles in seven months to win the race.

Other good fiber-forming polymers also have long molecules with shapes that permit strong interactions between the individual polymer strands. Nylon, for example, possesses polar carbonyl groups, $>C=O$, and N—H bonds that form strong hydrogen bonds between the individual molecules.

Three strands of nylon 6,6 bound to each other by hydrogen bonds.

Kevlar, another type of nylon, also forms strong hydrogen bonds between polymer molecules and is very crystalline. Its strong fibers are also used to make bulletproof vests. Because the fibers are so strong, they are also used to make thin yet strong hulls of racing boats. This lightweight construction improves speed and performance without sacrificing safety.

Inorganic polymers have backbones composed of atoms other than carbon

The polymers we've discussed so far would be classified as organic polymers because their main polymer chains (the polymer's backbone) are composed primarily of carbon atoms. Typically, such polymers tend to become brittle when cold and deteriorate when very hot. They also tend to be flammable, swell in organic solvents, and are usually not compatible with living tissue. (Only a few organic polymers can be used for medical implants.) With inorganic polymers, whose molecular backbones consist of atoms other than carbon, such problems are often minimized.

High-speed racing boats often are built with hulls made of Kevlar. Because of the polymer's high strength, the hulls can be made thin, thereby reducing weight and increasing speed.

Silicone polymers have a silicon–oxygen backbone

Among the synthetic inorganic polymers that have received the most widespread use are the **silicones.** These polymers contain an alternating silicon–oxygen backbone and have organic groups attached to the backbone at each silicon atom. Usually, these are methyl groups, —CH_3. For example, the structure of a typical "linear" silicone polymer is illustrated below.

$$CH_3-\underset{\underset{CH_3}{|}}{\overset{\overset{CH_3}{|}}{Si}}-O\left(\underset{\underset{CH_3}{|}}{\overset{\overset{CH_3}{|}}{Si}}-O\right)_n\underset{\underset{CH_3}{|}}{\overset{\overset{CH_3}{|}}{Si}}-O-\underset{\underset{CH_3}{|}}{\overset{\overset{CH_3}{|}}{Si}}-CH_3$$

a silicone polymer chain

Silicone polymers are formed by condensation polymerization, which begins with the *hydrolysis* (reaction with water) of such compounds as $(CH_3)_2SiCl_2$ and $(CH_3)_3SiCl$. As shown below, this gives Si—OH bonds that subsequently expel the components of water and form Si—O—Si linkages. Thus the first step in the polymerization of $(CH_3)_2SiCl_2$ is

$$Cl-\underset{\underset{CH_3}{|}}{\overset{\overset{CH_3}{|}}{Si}}-Cl + 2H_2O \longrightarrow HO-\underset{\underset{CH_3}{|}}{\overset{\overset{CH_3}{|}}{Si}}-OH + 2HCl$$

which is followed immediately by the expulsion of two hydrogens and an oxygen from between pairs of silicon atoms to give the polymer.

$$CH_3-\underset{\underset{CH_3}{|}}{\overset{\overset{CH_3}{|}}{Si}}-O\boxed{H \quad HO}\underset{\underset{CH_3}{|}}{\overset{\overset{CH_3}{|}}{Si}}-O\boxed{H \quad HO}\underset{\underset{CH_3}{|}}{\overset{\overset{CH_3}{|}}{Si}}-O\boxed{H \quad HO}\underset{\underset{CH_3}{|}}{\overset{\overset{CH_3}{|}}{Si}}-OH \ldots \text{etc.}$$

$$CH_3-\underset{\underset{CH_3}{|}}{\overset{\overset{CH_3}{|}}{Si}}-O-\underset{\underset{CH_3}{|}}{\overset{\overset{CH_3}{|}}{Si}}-O-\underset{\underset{CH_3}{|}}{\overset{\overset{CH_3}{|}}{Si}}-O-\underset{\underset{CH_3}{|}}{\overset{\overset{CH_3}{|}}{Si}}-O-\ldots$$

methylsilicone polymer

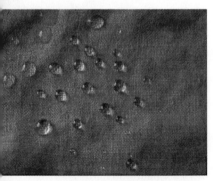

FIGURE 13.32 *Silicone polymers.* Water forms beads on a fabric made waterproof by treatment with a silicone polymer.

Depending on the chain length and the degree to which chains are cross-linked to each other, the silicones are oils, greases, or rubbery solids.

Silicone oils repel water and water-based products. For this reason, they have been used in polishes for furniture and cars and as waterproofing treatments for fabrics and leather (Figure 13.32). Silicone greases are used as permanent lubricants in clocks and ball bearings. Silicone resins are used in some paints because of their resistance to peeling and high temperatures. Silicone rubber has excellent electrical insulating properties and is used to coat electrical wire and in electric motors. These rubbery materials also retain their flexibility over wide temperature ranges and can be used in place of natural rubber for both high- and low-temperature applications. Silicones are even found in medical preparations. The polymer with the structure illustrated above is known medicinally as simethicone and is a common antigas agent found in antacid products.

13.6 ▶ Liquid crystals have properties of both liquids and crystals

As you've learned, molecules in a crystalline solid are arranged in a highly ordered rigid pattern, whereas those in a liquid are jumbled and disordered and able to move past each other. However, some substances behave in some ways as both liquids and crystals when they are just above their melting points. They have liquid properties because they are fluid, but the molecules are not arranged altogether randomly. To describe these substances the term **liquid crystal** was coined.

In general, molecules that can form a liquid crystal phase tend to be long, thin molecules with a central rigid region (so the molecules can hold their shapes without twisting). They have a rodlike shape and they also possess strong dipoles that help them align in the liquid state. The first such substance discovered to have these properties is cholesteryl benzoate, a molecule that has a structure derived from the molecule cholesterol.

Long straight-chain hydrocarbon molecules such as $C_{15}H_{32}$ are not good candidates for liquid crystals because rotation around the C—C single bonds allows the molecules to bend and twist, making it difficult for them to maintain any alignment in the fluid phase.

cholesteryl benzoate

A number of different liquid crystalline phases have been observed. The least ordered of them is called the **nematic phase.** As shown in Figure 13.33b, molecules in this state have a tendency to line up parallel to each other in one direction but are still able to move past each other up, down, and sideways.

A more ordered liquid crystalline phase is called the **smectic phase,** illustrated in Figure 13.33c. In this phase, the rodlike molecules are arranged parallel to each other and approximately in layers, which are able to slide over each other. The word *smectic* is derived from the Greek word for soap, and the thick, slippery substance often found at the bottom of a soap dish is, in fact, a type of smectic liquid crystal. More than one type of smectic phase is observed, and Figure 13.33d shows another one called **smectic C.** Here, the axes of the molecules tilt relative to a line perpendicular to the plane of a layer.

A particularly interesting and useful liquid crystalline phase is formed by molecules similar to cholesteryl benzoate. It is called a **cholesteric phase,** and is illustrated in Figure 13.34. As suggested in Figure 13.34a, molecules align as in a nematic phase, but in thin layers. In the figure we see that going down through the layers, the orientations of the molecules rotate, producing twisted stacks of molecules.

Liquid phase
(a)

Nematic phase
(b)

Smectic phase
(c)

Smectic C phase
(d)

FIGURE 13.33 *Liquid and liquid crystal phases.* (*a*) Molecules in a liquid phase are without any order. (*b*) Molecules in a nematic liquid phase are ordered in one direction. (*c*) Molecules in a smectic liquid phase are oriented within layers. (*d*) The axes of the molecules in a smectic C phase are not perpendicular to the planes of the layers.

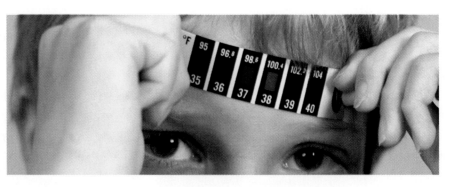

Thin layer of molecules arranged in a nematic-like pattern.

In this example, there is a counter clockwise twist going from top to bottom in the stack of molecules.

The orientations of molecules in the layers rotate from one layer to the next, producing a twist going down one layer after another.

FIGURE 13.34 *Structure of a cholesteric liquid crystal.* (*a*) Molecules align as in a nematic liquid crystal, but in very thin layers. Going down, the orientation of each layer is rotated relative to the layer above. (*b*) Viewed from the side, the rodlike molecules are arranged in twisted stacks from top to bottom.

(*a*)   (*b*)

Among the unique properties of cholesteric liquid crystals is their ability to reflect light that has a wavelength equal to the distance required for the twist to make one full turn. This distance changes with temperature, so the color of the reflected light also changes with temperature. As a result, it is possible to make a liquid crystal thermometer that displays the temperature of its environment by its reflected color (Figure 13.35).

Liquid crystal displays rely on polarized light

Several different liquid crystalline phases have been used to form electronic liquid crystal (LC) displays for devices such as digital watches, calculators, and screens for laptop computers. All of them rely on the interaction of polarized light with the liquid crystalline substance.

Light is *electromagnetic radiation* that possesses both electric and magnetic components that behave like vectors. These vectors oscillate in directions perpendicular to the direction in which the light wave is traveling (Figure 13.36*a*). In ordinary light, the oscillations of the electric and magnetic fields of the photons are oriented randomly around the direction of the light beam. In *plane-polarized light,* all the vibrations occur in the same plane (Figure 13.36*b*). Ordinary light can be polarized in several ways. One is to pass it through a special film of plastic, as in a pair of Polaroid sunglasses. This has the effect of filtering out all the vibrations except those that are in one plane (Figure 13.36*b*).

In the LC display ordinarily found on a wristwatch, a cholesteric liquid crystal is sandwiched between two thin layers of glass that have transparent electrodes de-

This modern stop watch uses a liquid crystal display.

FIGURE 13.35 *Liquid crystal thermometer.* The spacings between layers of molecules in the liquid crystal change with temperature, causing the color of the reflected light to also change. The color can be correlated with the temperature and used as a thermometer.

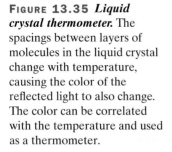

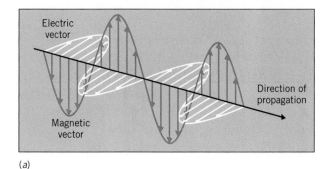

(a)

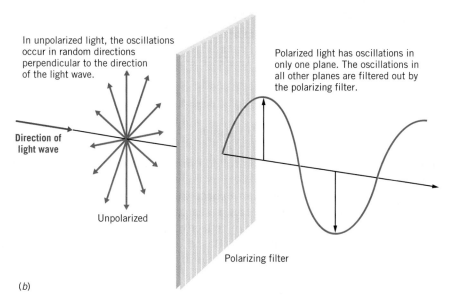

In unpolarized light, the oscillations occur in random directions perpendicular to the direction of the light wave.

Direction of light wave

Unpolarized

Polarized light has oscillations in only one plane. The oscillations in all other planes are filtered out by the polarizing filter.

Polarizing filter

(b)

FIGURE 13.36 *Unpolarized and polarized light.* (*a*) Light possesses electric and magnetic components that oscillate perpendicular to the direction of propagation of the light wave. (*b*) In unpolarized light, the electromagnetic oscillations of the photons are oriented at random angles around the axis of propagation of the light wave. The polarizing filter has the effect of preventing oscillations from passing through unless they are in one particular plane. The result is called *plane-polarized light.*

posited on their surfaces. The shapes and locations of the electrodes are arranged so that when activated they display numbers, letters, or other characters. Polarizing filters are then added to the top and bottom of the sandwich, but arranged so that the directions of polarization are at 90° to each other (we say the polarizers are "crossed"). Finally, a mirror is placed below the bottom polarizer.

Figure 13.37 shows what happens to light when it passes through the LC sandwich while the voltage to the electrodes is turned off. As light passes through the first filter, it becomes polarized. If the liquid crystal weren't present, it would be blocked by the second filter. But with the cholesteric liquid crystal between the two polarizers, the polarized light coming through the first filter is rotated as it goes through the helical structure of the liquid crystal. When the light reaches the next filter, its plane of polarization has been rotated sufficiently so that it is able to pass through. The light is then reflected off the mirror and travels back through the liquid crystal, is rotated again, and comes back out. This is seen as the characteristic gray background color of the display.

When a charge is placed on the electrodes, the molecules of the liquid crystal leave their helical orientations and become aligned with the electric field. This causes them to lose the ability to rotate the polarized light, so now when the light reaches the second filter, it is blocked and no light passes through. Instead, it is absorbed and the display appears black. By activating various electrode segments, the device is able to display different characters and numbers, which appear black against the gray background.

In today's laptop computers, color LC displays use other liquid crystalline materials with more rapid response times, but the basis of operation is essentially the same. A white light coming out through the display is filtered to give colored pixels

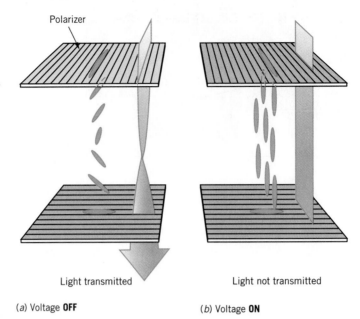

Polarizer

Light transmitted

Light not transmitted

(*a*) Voltage **OFF**

(*b*) Voltage **ON**

FIGURE 13.37 *Operation of a typical liquid display.*
(*a*) With the voltage off, light passing through the first filter is polarized. It is then rotated by the helical stacking of molecules in the liquid crystal, so when it reaches the second filter, it is able to pass through.
(*b*) With the voltage on, the helical packing of the liquid crystal molecules is disrupted and the light is not rotated after being polarized. As a result, it is blocked from passing through the second filter.

(tiny picture elements) that are controlled using polarizing filters in the manner described above.

Looking to the future, we can expect to see more lightweight and energy-efficient flexible LC displays. In fact, researchers at the Liquid Crystal Institute have been working on a new type of reflective LC display that could become the "digital ink" in electronic publishing. Based on a cholesteric liquid crystal, these displays can be switched from black to color by application of a small electric field. In fact, they can be made to do this without using polarizing filters, so they are much brighter and have better contrast than current LC reflective displays.

13.7 ▶ **Modern ceramics have applications far beyond porcelain and pottery**

Ceramic materials have a long history, dating to prehistoric times. Examples of pottery about 13,000 years old have been found in several parts of the world. Today, ceramics include common inorganic building materials such as brick, cement, and glass. We find ceramics around the home as porcelain dinnerware, tiles, sinks, toilets, and artistic pottery and figurines. We also find ceramics in places we wouldn't expect, such as cell phones and inside diesel engines.

Manufactured **ceramics,** such as building materials and ceramic objects we find around our home, are made from inorganic minerals such as clay, silica (sand), and other silicates (compounds containing anions composed of silicon and oxygen), which are taken from the earth's crust. Such materials have been produced for hundreds or even thousands of years. In recent times, an entirely new set of materials, generally referred to as *advanced ceramics,* have been prepared by chemists in laboratories and have high-tech applications. We will focus our discussions mainly on these.

Ceramics have high melting points and are very hard

There are some properties that ceramics have in common, and there are others that can be tailored by controlling the ceramic's composition and method of preparation. For example, almost all ceramic materials are very hard and have very high melting points. Table 13.3 contains a list of some advanced ceramic materials and their properties, along with the properties of some metals in common use. The hardest substance known is diamond, and as you can see from the data, some ceramics have hardnesses that approach diamond. Silicon carbide, which is relatively

TABLE 13.3	SOME PROPERTIES OF CERAMIC MATERIALS, DIAMOND, AND TYPICAL METALS			
Ceramic Material	Hardness[a] (GPa)	Melting Point (°C)	Elastic Modulus[b] (GPa)	Density (g cm^{-3})
Titanium carbide, TiC	28	3067	470	4.93
Titanium nitride, TiN	21	2950	590	5.40
Titanium dioxide, TiO$_2$ (Titania)	11	1867	205	4.25
Zirconium carbide, ZrC	25	3445	400	6.63
Zirconium oxide, ZrO$_2$ (Zirconia)	12	2677	190	5.76
Aluminum nitride, AlN	12	2250	350	3.36
Aluminum oxide, Al$_2$O$_3$ (Alumina)	21	2047	400	3.98
Beryllium oxide, BeO	15	2550	390	3.03
Tungsten carbide, WC	23	2776	720	15.72
Silicon carbide, SiC	26	2760	480	3.22
Boron nitride, BN	50	2730	660	3.48
Diamond, C	80	3800	910	3.52
Stainless steel	2.35	1420	200	7.89
Aluminum	1.51	658	69	2.70

[a]Pressure required to dent the material. The larger the value, the harder the material.

[b]Force per unit cross-sectional area required to elongate (stretch) the solid. The larger the value, the stronger the material and the more it resists stretching.

inexpensive to make in bulk, has long been used as an abrasive in sandpaper and grinding wheels.

Notice that all of the ceramics are harder than iron and steel and also have lower densities and very high melting points. Their high strength and relatively low density make them useful for applications in space.

Because of their high melting points, we often find ceramics used as *refractories* (heat-resistant materials) for lining furnaces and rocket engine exhausts. The exterior of the Space Shuttle (Figure 13.38) is coated with protective ceramic tiles that serve as thermal "heat" shields during re-entry from space. When given a glassy porcelain surface, ceramics are impervious to water, but under the right conditions

FIGURE **13.38** *Heat shield of the Space Shuttle.* Individual heat-shielding ceramic tiles are visible on the surface of the Space Shuttle Challenger. The tiles are made of a low-density porous silica ceramic derived from common sand, SiO$_2$. They are able to withstand very high temperatures and have a very low thermal conductivity, preventing heat transfer to the body of the shuttle. The photo was taken shortly after the spacecraft was completed. In 1986, the Challenger was lost in a tragic accident shortly after lift-off. Damage to the tiles during liftoff is believed to be responsible for the loss of the Columbia and its crew in February 2003 when the spacecraft broke apart during reentry into the Earth's atmosphere.

they can be made to be porous and can be used as filters in applications where high temperatures would consume other materials.

Most ceramics do not conduct electricity, so ceramics are used as insulators for high-voltage power lines, automobile spark plugs, and in TV sets. However, some ceramic materials become excellent conductors of electricity when cooled to very low temperatures. There is a whole class of these materials that are superconductors with many potential applications. We will say more about them later.

Materials used to make ceramics have large lattice energies

If you study Table 13.3, you will see that the compounds used to form ceramics generally contain metals with large positive oxidation states combined with small nonmetals (e.g., O, N, and C) with large negative oxidation states. If these compounds were purely ionic, they would have very large lattice energies because of the high charges and small radii of the ions involved. Large lattice energies require very large thermal energies to overcome, which explains why these substances have such high melting points. The high charges on cation and anion would also produce very strong electrostatic attractions and make it very difficult to pull the ions apart. This would explain their hardness.

Actually, the compounds involved possess substantial covalent bonding between the atoms. Small cations with high charges tend to pull electron density from highly charged anions into the region between the ions, causing the bonds to become substantially covalent. (This topic is discussed in some depth in Section 23.3.) Furthermore, elements such as oxygen, nitrogen, and carbon are able to form covalent bridges between two or more atoms, enabling the formation of network-type solids. Therefore, covalent bonding between the atoms also contributes substantially to both the strength and high melting points of ceramic materials.

How ceramic materials are formed into products

Several methods are used to form ceramics from their raw materials. In many cases, the components of the ceramic are first pulverized to give a very fine powder. The powder is then mixed with water to form a slurry that can be poured into a mold where it sets up into a solid form that has little structural strength. (Alternatively, the fine powder is mixed with a binder and pressed into the desired shape.) The newly formed object is next placed in a kiln where it is heated to a high temperature, often over 1000 °C. At these high temperatures the fine particles stick together by a process called *sintering,* which produces the finished ceramic. Although sintering is sometimes accompanied by partial melting of the fine particles, often the particles remain entirely solid during the process.

There are some problems with making ceramics by the traditional methods described above. Because it is difficult to produce uniform and very small particle sizes by grinding, ceramics made from materials prepared this way often contain small cracks and voids, which adversely affect physical properties such as strength. In addition, the chemical composition of the ceramic cannot be easily and reproducibly controlled by mixing various powdered components.

The sol-gel process gives control to the formation of ceramic materials

Sol-gel process

For some types of ceramics, the problems of particle size and uniformity can be avoided by using a method called the **sol-gel process.** The chemistry involved resembles the formation of condensation polymers, which we discussed in Section 13.5. The starting materials are metal salts or compounds in which a metal or metalloid (e.g., Si) is bonded to some number of alkoxide groups. An *alkoxide* is an anion formed by removing a hydrogen ion from an alcohol.[3]

$$C_2H_5OH \xrightarrow{-H^+} C_2H_5O^-$$

ethanol   ethoxide ion

Alcohols have very little tendency to lose H^+ ions, and as a result, alkoxide ions have a *very strong* affinity for H^+. In fact, when placed in water they react by removing H^+ from H_2O to give the alcohol and hydroxide ion.

$$C_2H_5O^- + H_2O \longrightarrow C_2H_5OH + OH^-$$

This reaction forms the basis for the start of a sequence of reactions that ultimately yields the ceramic material. Let's use zirconium ethoxide, $Zr(C_2H_5O)_4$, to illustrate the process.

The sequence of reactions begins with the gradual addition of water to an alcohol solution of $Zr(C_2H_5O)_4$, which causes ethoxide to react to form ethanol and be replaced by hydroxide ions. This reaction is termed **hydrolysis** because it involves a reaction with water.

$$Zr(C_2H_5O)_4 + 4H_2O \longrightarrow Zr(OH)_4 + 4C_2H_5OH$$

The next reaction involves condensation, in which the components of H_2O are lost and a Zr—O—Zr bridge is formed.

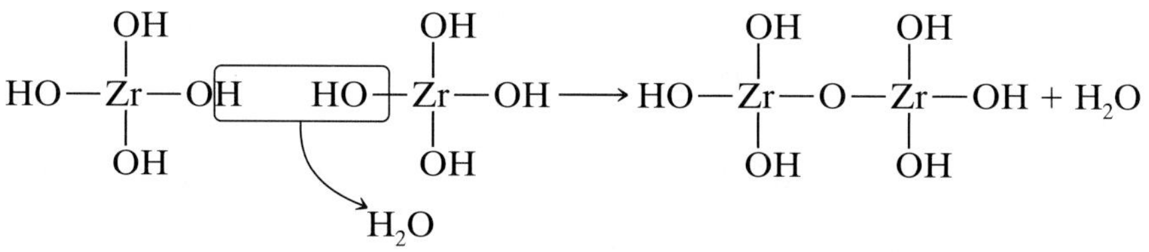

Continued polymerization leads to a network of zirconium atoms bridged by oxygens, which produces extremely small insoluble particles that are essentially oxides with many residual hydroxide ions. These particles are suspended in the alcohol solvent and have a gel-like quality.

Once the sol-gel suspension is formed, it can be used in a number of ways, as indicated in Figure 13.39. It can be deposited on a surface by dip coating to yield very thin ceramic coatings. It can be cast into a mold to produce a semisolid gelatinlike material (wet gel). The wet gel can be dried by evaporation of the solvent to give a porous solid called a **xerogel**. Further heating causes the porous structure of the xerogel to collapse and form a dense ceramic or glass with a uniform structure. If the solvent is removed from the wet gel under supercritical conditions (at a temperature above the critical temperature of the solvent), a very porous and extremely low density solid called an **aerogel** is formed (see the photo on page 110). By adjusting the viscosity of the gel suspension, ceramic fibers can be formed. And by precipitation, ultrafine and uniform ceramic powders are formed. What is amazing is that all these different forms can be made from the same material depending on how the suspension is handled.

The sol-gel process is not suitable for all ceramic materials and a variety of other methods are used for their preparation. For example, silicon carbide, a very desirable ceramic because of its very high strength and hardness, can be formed into fibers by thermal decomposition of polymers containing silicon and carbon in

Typical metals used in the sol-gel process are Si, Ti, Zr, Al, Sn, and Ce.

Notice how similar this reaction is to the formation of condensation polymers such as nylon and polyesters.

Xerogels have very large internal surface areas because of their fine porous structure. Some have been found to be useful catalysts that help other reactions take place.

[3]An **alkyl** group is a hydrocarbon fragment derived from an alkane by removal of a hydrogen atom. For example, the methyl group, CH_3—, is derived from methane, CH_4, and is an alkyl group. If we represent an alcohol by the general formula R—OH, where R represents an alkyl group, then the general formula of an alkoxide ion is R—O^-.

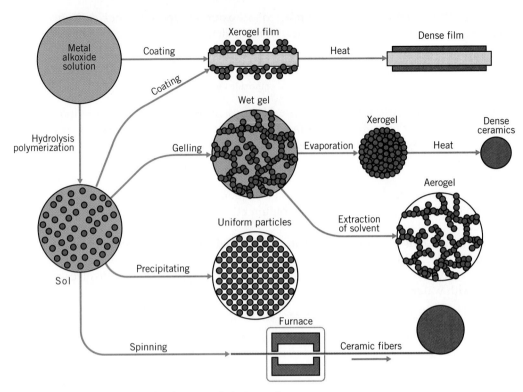

FIGURE 13.39 *Sol-gel technologies and their products.*

the right stoichiometric ratio. One such substance is poly(methylsilane), whose structure is

$$
\begin{array}{c}
\overset{\displaystyle CH_3}{\underset{\displaystyle H}{|}} \left(\overset{\displaystyle CH_3}{\underset{\displaystyle H}{|}} \right) \overset{\displaystyle CH_3}{\underset{\displaystyle H}{|}} \overset{\displaystyle CH_3}{\underset{\displaystyle H}{|}} \\
-Si - \left(Si \right)_n Si - Si-
\end{array}
$$

Heating this polymer to high temperatures drives off two H_2 molecules per repeating unit, accompanied by cross-linking between adjacent polymer chains to give fibers with the formula SiC.

Applications of advanced ceramics vary widely

High-tech ceramics are relatively new materials and new applications for them are continually being found. It is interesting to note how varied their uses are. Here are a few examples.

Thin ceramic films, deposited by the sol-gel process or other methods, are found on optical surfaces where they serve as antireflective coatings and as filters for lighting. Tools such as drill bits are given thin coatings of titanium nitride (TiN) to make them more wear resistant (Figure 13.40).

Zirconia (ZrO_2) is used to make ceramic golf spikes for golf shoes, as well as portions of hip-joint replacement parts for medicine, and knives that stay sharp much longer than steel. All these applications are possible because of the hardness of the ZrO_2 ceramic.

Boron nitride, BN, powder is composed of flat, platelike crystals that can easily slide over one another, and one of its uses is in cosmetics, where it gives a silky texture (Figure 13.41). Another boron-containing ceramic, boron carbide, is used along with Kevlar polymer to make bulletproof vests. When a bullet strikes the vest, it is shattered by the ceramic, which absorbs most of the kinetic energy, with residual energy being absorbed by the Kevlar backing.

FIGURE 13.40 *Titanium nitride.* A drill bit with a thin golden coating of titanium nitride, TiN, will retain its sharpness longer than steel drill bits.

Silicon nitride, Si_3N_4, is used to make engine components for diesel engines because it is extremely hard and wear resistant, has a high stiffness with low density, and can withstand extremely high temperatures and harsh chemical environments.

Piezoelectric ceramics have the property that they produce an electric potential when their shape is deformed. Conversely, they deform when an electric potential is applied to them. A company makes "smart skis" that incorporate piezoelectric devices that use both properties. Vibrations in the skis are detected by the potential developed when they deform. A potential is then applied to cancel the vibration. This "smart materials" technology was first developed to dampen vibrations in optical components of the "Star Wars" weapons system being developed to protect the United States from missile attack.

Smart materials possess both sensing and action capability. They adaptively respond to changing stimuli.

High-temperature superconductors are ceramics

At ordinary temperatures, metals like copper have a small but significant resistance to the flow of electricity, converting a percentage of the electrical energy to heat. This is a major problem in the transmission of electrical power over long distances because of the energy losses that occur. To minimize this, very high voltages are used, but energy loss still occurs.

A **superconductor** is a material in a state in which it offers no resistance to the flow of electricity. In addition, a superconductor repels a surrounding magnetic field. Prior to 1987, only a few materials, nearly all of them metals and metal alloys, could be put into such a state and then only if their temperatures were lowered to just a few degrees above absolute zero. To accomplish this, they would have to be immersed in liquid helium (boiling point 4.2 K). Because of the high cost of this refrigerant, few practical applications of superconductivity were possible.

The temperature at and below which a material is superconducting is given the symbol T_c. In 1986 it was discovered that a previously known ceramic material with the formula $LaBa_2CuO_4$ has a value of T_c equal to 30 K. Keeping the material in a superconducting state at 30 K, however, still requires refrigeration technology too expensive and too cumbersome to permit any important practical applications. To make superconductivity practical, materials had to be discovered that could become superconducting at temperatures much higher than

H. Kamerlingh-Onnes, a Dutch physicist, won the 1913 Nobel prize in physics for his discovery that mercury is superconducting at 4.1 K in liquid helium (boiling point 4.2 K).

FIGURE 13.41 *Ceramics in cosmetics.* Elisabeth Arden markets a collection of cosmetics that use boron nitride, BN, as an ingredient.

FACETS OF CHEMISTRY 13.3

Composite Materials

Long ago, primitive people learned that mixing straw with mud produced better bricks than pure mud alone. Their building materials were early examples of **composites,** in which properties of different materials work together to give a product that is better (in this example, stronger) than any of its components. Within the composite, the materials retain their individual identities; they do not dissolve in each other or react chemically.

These days, composite materials are everywhere. Concrete, for example, is a mixture of sand, cement, and larger stones or gravel, and it is often reinforced by steel bars. The cement holds it all together, but the stones and steel combine to give a building material that is inexpensive and very strong. Composites also exist naturally. Wood, for example, is a composite consisting of long fibers of cellulose held together by a much weaker substance called *lignin.* Cotton, which is also composed of cellulose, is much weaker than wood because it lacks the binding power of the lignin.

With the development of polymer plastics, designers had a new material at their disposal that could easily be formed into complex shapes. But the plastic lacked strength and could crack easily. Then it was found that by incorporating fibers of various kinds into the plastic, a strong and resilient combination was formed that's now called **fiber-reinforced polymer** (**FRP,** for short). The most familiar example is fiberglass, in which thin, flexible glass threads are embedded in a polyester or epoxy polymer. About 65% of all composites produced today are fiberglass and are used for boat hulls, surfboards, sporting goods, swimming pool linings, and car bodies.

There are many applications of composites in construction. For example, deck surfaces for bridges (Figure 1) are being made from FRP that are strong, rigid, will not corrode, and can be quickly installed (which offsets the higher initial costs of the materials). FRP is also being

FIGURE 2 *A prefabricated house with an exterior made entirely of a ceramic fiber-reinforced composite.*

used to repair and support deteriorating concrete structures such as old bridges and highway overpasses.

With the development of high-tech ceramic materials, new composites are being made that have some exceptional properties and applications. For example, a company in Ohio is using a composite of very thin ($\sim10^{-6}$ m) ceramic fibers mixed with a durable polymer resin to build prefabricated segments of houses that can be quickly assembled to give a finished structure at a reasonable cost. What is unusual is that the entire exterior of the structure (Figure 2), including the roof, gutters, stone foundation, and clapboard siding, is molded from the ceramic composite material. The ceramic composite is strong, fire resistant, and is expected to have a life span of more than 150 years.

Applications in the automotive industry include a new brake disk being used by Porsche (Figure 3), which is made of a fiber-reinforced ceramic material that is lightweight, able to withstand extremely high temperatures, and is both hard and fracture resistant. Its high abrasion resistance is expected to give a useful life of more than 185,000 miles! (With brakes like that, you may never ever need a brake job.)

FIGURE 1 *Installation of an FRP composite bridge deck.* A special adhesive is used to attach it to the steel beams of the bridge.

FIGURE 3 *Porsche Ceramic Composite Brake (PCCB) disk, which Porsche is using on its 911 Turbos made in Zuffenhausen, Germany.*

the boiling point of liquid helium. Specifically, a material with T_c above the boiling point of liquid nitrogen, 77.4 K, was sought. The refrigeration technology for producing and handling liquid nitrogen is well known and relatively simple, and liquid nitrogen costs about as much as milk or beer. Today, materials that have values of T_c above the boiling point of nitrogen are called *high-temperature superconductors*.

In 1987, a ceramic mixed metal oxide was discovered with T_c equal to 93 K. The compound, $YBa_2Cu_3O_7$, is sometimes called the "1-2-3 compound," because of the 1:2:3 ratio of Y, Ba, and Cu. The significant structural feature of this substance is the layering of copper and oxygen atoms (see Figure 13.42). The 1-2-3 compound is just one of a small family of the mixed oxides of these elements that are superconducting above 77 K.

Experimentally, the ceramic superconductors are easy to make. All it takes are the parent oxides and a high-temperature furnace. Teams around the world worked arduously to find materials that pushed T_c to still higher values. A mixed oxide of thallium, calcium, barium, and copper—discovered in 1988—was found that had zero electrical resistance at 107 K. An IBM team (San Jose, California), working with the same elements, found a material for which $T_c = 122$ K. Since then, materials with values of T_c above 130 K (at atmospheric pressure) have been found.

Applications

As you can imagine, the chief problem with ceramic superconductors is that they are brittle. But during the 1990s scientists and engineers successfully made wirelike materials that could be used to build electrical cables (see Figure 13.43). At the start of the new millennium, Detroit Edison Company was using these cables to replace copper cables at its Frisbie power substation. The new cables carry up to three times the amount of electricity as the old ones. This enables a large expansion in electrical service without having to lay down more copper cables. More recently, patents have been granted for inexpensive methods to produce round, continuous superconducting wire that promises to further reduce the cost of implementing this technology.

Among the applications of superconducting cables is the development of superconducting coils for the storage of standby electricity. In the absence of electrical resistance in superconducting cables, the electrons would just keep on circulating (in theory, forever). The electrical energy could then be tapped to smooth

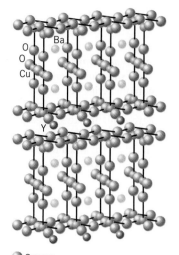

Oxygen
Copper
Barium
Yttrium

FIGURE 13.42 *The 1-2-3 superconducting ceramic, $YBa_2Cu_3O_7$.* The charge is transported through the copper–oxygen layers.

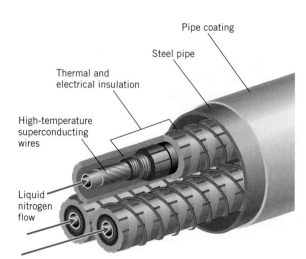

Pipe coating
Steel pipe
Thermal and electrical insulation
High-temperature superconducting wires
Liquid nitrogen flow

FIGURE 13.43 *Superconducting electrical cables can be made from a bismuth–strontium–calcium–copper–oxygen ceramic incorporating small amounts of lead.* The ceramic flexes somewhat like thin glass filaments. The "wires" are protected by a silver alloy. (*Source:* American Superconductor/John Schneidman.)

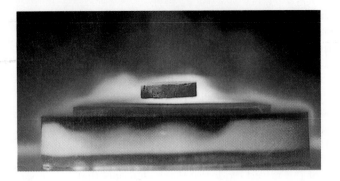

FIGURE 13.44 *Levitation of a magnet above a superconductor.* When a superconducting ceramic is cooled below T_c, it becomes superconducting and repels a magnetic field. That repulsion is sufficient to hold a magnet suspended above it.

out blips in electrical transmission that mess up sensitive applications, like computer uses.

One of the most fascinating properties of materials in their superconducting states is that they can be levitated by a magnetic field.[4] Besides zero electrical resistance, a substance in its superconducting state permits no magnetic field within itself. A weak magnetic field is actually *repelled* by a superconductor, as we mentioned earlier, and this is what makes the levitation of superconductors possible (Figure 13.44). Someday we might levitate entire trains and send them at high speed with no more frictional resistance than offered to airplanes by the surrounding air. Prototypes of such "mag-lev" trains are currently being tested by Germany and Japan.

13.8 ▶ Nanotechnology deals with controlling structure at the molecular level

Atoms and small molecules have dimensions of the order of several tenths of a nanometer. For example, the diameter of a carbon atom is about 0.15 nm. Therefore, when we examine matter at the nanometer level, we are looking at very small structures with dimensions of perhaps tens to hundreds of atoms. **Nanotechnology** deals with using such small-scale objects and the special properties that accompany them to develop useful applications. Ultimately, the goal of nanotechnology (also sometimes called **molecular nanotechnology**) is to be able to build materials from the atom up. Such technology doesn't quite exist yet, but scientists are beginning to make progress in that direction. This section, therefore, is kind of a progress report that will give you some feeling of where materials science is now and where it's heading—sort of a glimpse at the future.

There are several reasons why there is so much interest in nanotechnology. One is because the properties of materials are related to their structures. By controlling structures at the atomic and molecular level, we can (in principle) tailor materials to have specific properties. Driving much of the research in this area is the continuing efforts by computer and electronics designers to produce ever smaller circuits. The reductions in size achieved through traditional methods are near their limit, so new ways to achieve smaller circuits and smaller electrical devices are being sought.

Studying nanometer-scale objects requires visualizing and manipulating very tiny structures

What has enabled scientists to begin the exploration of the nanoworld is the development of tools that allow them to see and sometimes manipulate individual atoms and molecules. We've already discussed one of these important devices, the scanning tunneling microscope (STM), in Chapter 1 when we discussed experimental

---

[4]*Levitation* means "floating on air;" the verb is "to levitate."

evidence for atoms (see page 17). This instrument, which can only be used with electrically conducting samples, enables the imaging of individual atoms. What is very interesting is that it can also be used to move atoms around on a surface. To illustrate this, scientists have arranged atoms to spell out words (Figure 13.45). Although writing words with atoms doesn't have much practical use, it demonstrates that one of the required capabilities for working with substances at the molecular level is achievable.

To study nonconducting samples, a device called an **atomic force microscope (AFM)** can be used. Figure 13.46 illustrates its basic principles. A very sharp stylus (sort of like a phonograph needle) is moved across the surface of the sample under study. Intermolecular forces between the tip of the probe and the surface molecules cause the probe to flex as it follows the ups and downs of the bumps that are the individual molecules and atoms. A mirrored surface attached to the probe reflects a laser beam at angles proportional to the amount of deflection of the probe. A sensor picks up the signal from the laser and translates it into data that can be analyzed by a computer to give three-dimensional images of the sample's surface. A typical image produced by an AFM is shown in Figure 13.47.

FIGURE 13.45 *Atoms of iron on copper.* Scientists at IBM used an STM instrument to manipulate iron atoms into the Kanji letters for "atom." The literal translation is something like "original child."

Discoveries and progress in nanoscale materials

Among the new materials discovered through research in nanotechnology are carbon nanotubes, very thin tubes composed of rolled-up sheets of carbon atoms. Until the mid 1980s, only two forms of carbon were known, diamond and graphite. The structure of diamond, with its three-dimensional network lattice, was described earlier on page 559 and is shown in Figure 13.48*a*. **Graphite** is quite different from diamond and consists of layers of carbon atoms, each of which consists of many hexagonal "benzene-like" rings fused together in a structure reminiscent of chicken wire.

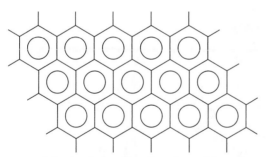

A fragment of a carbon layer in graphite.

In graphite, these sheets are stacked one on top of another, as shown in Figure 13.48*b*.

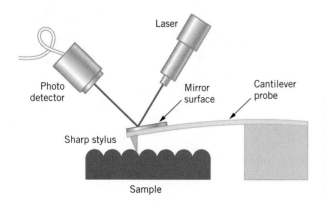

FIGURE 13.46 *An atomic force microscope (AFM).* A sharp stylus attached to the end of a cantilever probe rides up and down over the surface features of the sample. A laser beam, reflected off a mirrored surface at the end of the probe, changes angle as the probe moves up and down. A photo detector reads these changes and sends the information to a computer, which translates the data into an image.

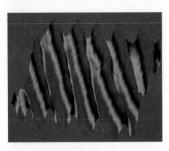

In 1985, a new form of carbon was discovered that consists of tiny balls of carbon atoms, the simplest of which has the formula C_{60} (Figure 13.48c). They were named **fullerenes** and the C_{60} molecule itself was named **buckminsterfullerene** (nickname **buckyball**) in honor of R. Buckminster Fuller, the designer of a type of structure called a geodesic dome. The bonds between carbon atoms in the buckyball are arranged in a pattern of five- and six-member rings arranged like the seams in a soccer ball as well as the structural elements of the geodesic dome.

Carbon nanotubes, discovered in 1991, are another form of carbon that is related to the fullerenes. They are formed, along with fullerenes, when an electric arc is passed between carbon electrodes. The nanotubes consist of tubular carbon molecules that we can visualize as rolled-up sheets of graphite (with hexagonal rings of carbon atoms). The tubes are capped at each end with half of a spherical fullerene molecule, so a short tube would have a shape like a frankfurter. A portion of a carbon nanotube is illustrated in Figure 13.48d. The first nanotubes to be discovered were multiwalled, meaning that they were tubes within tubes. Since then, single-wall nanotubes have been made with diameters of about 1 nanometer and lengths that can be as much as millions of times greater than their widths. Studying the properties and applications of these materials is what nanotechnology is all about.

What has sparked so much interest in nanotubes is their unusual properties. Single-walled tubes have high strengths and low densities compared to other materials. For example, weight-for-weight they're about 100 times stronger than stainless steel and about 40 times stronger than carbon fibers used to make tennis rackets and shafts for golf clubs. Because they are so tiny and so strong, carbon nanotubes have been attached to electrodes at the end of a glass thread where, with an applied voltage, they pinch together and behave as nanosized tweezers!

When formed, carbon nanotubes have surfaces free of other substances, so they are able to align closely to each other where strong London forces can bind

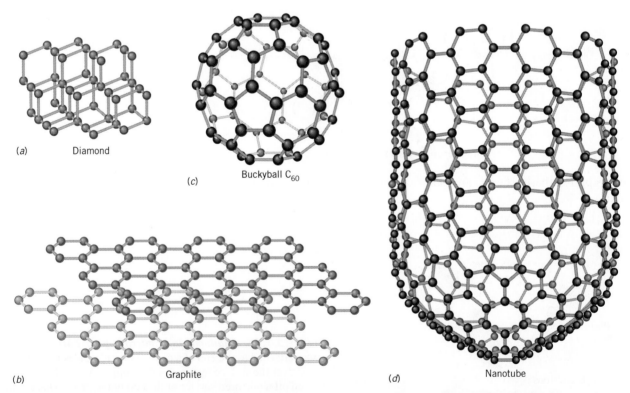

FIGURE **13.48** *Molecular forms of carbon.* (*a*) Diamond. (*b*) Graphite. (*c*) Buckminsterfullerene, or "Buckyball," C_{60}. (*d*) A portion of a carbon nanotube showing one closed end.

FIGURE 13.49 *Ceramic fibers of alumina,* Al_2O_3. The fine fibers shown in this image, which was formed by an electron microscope, are about 2 to 4 nm in diameter, which is less than 1/10,000th of the thickness of a human hair. (The reference scale in the lower left corner is 100 nm long.)

them tightly together. Chinese researchers reported in 2002 that they were able to form nanotube yarn up to 30 cm long by drawing bundles of nanotubes from a layer of these structures they had deposited on a surface. This raises the prospect of forming very light high-strength fibers for use in space applications.

Depending on structural characteristics, carbon nanotubes either conduct electricity like a metal or they are semiconductors. They are also excellent conductors of heat, so researchers are studying them as means of removing heat from tiny computer chips. Their electrical properties are being studied as possible nanosized wires and semiconductors in electronic devices.

In the preceding discussion we've focused entirely on carbon. However, other substances have also been observed to form nanotubes. Molybdenum sulfide, MoS_2, assembles in nanotube forms with diameters as small as 0.96 nm and are considered candidates for separating gases such as mixtures of H_2 and O_2. The smaller H_2 molecules would pass through the tiny pores while the larger O_2 molecules would not.

Boron nitride, BN, is another material that has been made to form nanotubes. It's a substances that in larger forms is a strong ceramic. Although the properties of these nanotubes have not yet been thoroughly studied, it is expected that the material will be a semiconductor and will certainly be more resistant to oxidation than carbon nanotubes.

Nanotubes are not the only small-scale structures of interest to researchers. For example, tiny rods composed of crystalline barium titanate, $BaTiO_3$, and strontium titanate, $SrTiO_3$, have been made that can be electrically polarized in either of two directions. This is an attractive property because it makes the materials candidates for very small-scale data storage devices for computers.

The two states of polarization can represent binary numbers of 0 and 1 and so can be used to store digital information.

Alumina, a ceramic material, is produced in a fiber form in which the individual alumina "whiskers" have diameters of 2 to 4 nm and lengths of hundreds of nanometers (Figure 13.49). Anticipated applications include using mats of the fibers to make ultrafine filters with pores smaller than 5 nm capable of filtering viruses and bacteria.

In conclusion . . .

Our discussion here has only scratched the surface of this emerging technology. It is a field advancing so rapidly that what you've read here is probably already "ancient history," even though we've incorporated some of the most recent developments at the time of writing. There is little doubt among scientists that nanotechnology will have a profound influence on our lives in times to come.

SUMMARY

Crystalline Solids Crystalline solids have very regular features that are determined by the highly ordered arrangements of particles within them, which can be described in terms of repeating three-dimensional arrays of points called **lattices.** The simplest portion of a lattice is its **unit cell.** Many structures can be described by the same lattice by associating different units (atoms, molecules, or ions) to lattice points and by changing the dimensions of the unit cell. Three cubic unit cells are possible—**simple cubic, face-centered cubic,** and **body-centered cubic.** Sodium chloride and many other alkali metal halides crystallize in the **rock salt structure,** which contains four formula units per unit cell. Two modes of closest packing of atoms are **cubic closest packing (ccp)** and **hexagonal closest packing (hcp).** The ccp structure has an A-B-C-A-B-C . . . alternating stacking of layers of spheres; the hcp structure has an A-B-A-B . . . stacking of layers. **Amorphous** solids lack the internal structure of crystalline solids. Glass is an amorphous solid and is sometimes called a **supercooled liquid.**

X-Ray Diffraction Information about crystal structures is obtained experimentally from **X-ray diffraction patterns** produced by constructive and destructive interference of X rays scattered by atoms. Distances between planes of atoms in a crystal can be calculated by the Bragg equation, $n\lambda = 2d \sin \theta$, where n is a whole number, λ is the wavelength of the X rays, d is the distance between planes of atoms producing the diffracted beam, and θ is the angle at which the diffracted X-ray beam emerges relative to the planes of atoms producing the diffracted beam.

Crystal Types Crystals can be divided into four general types: **ionic, molecular, covalent,** and **metallic.** Their properties depend on the kinds of particles within the lattice and on the attractions between the particles, as summarized in Table 13.1.

Band Theory of Solids In solids, atomic orbitals of the atoms combine to yield **energy bands** that consist of many energy levels. The **valence band** is formed by orbitals of the valence shells of the atoms. A **conduction band** is a partially filled or empty band. In an **electrical conductor** the conduction band is either partially filled or is empty and overlaps a filled band. In an **insulator**, the **band gap** between the filled valence band and the empty conduction band is large, so no electrons populate the conduction band. In a **semiconductor,** the band gap between the filled valence band and the conduction band is small and thermal energy can promote some electrons to the conduction band. Silicon becomes a **p-type semiconductor,** in which the charge is carried by positive **"holes,"** if it is **doped** with a Group IIIA element such as boron. It becomes an **n-type semiconductor,** in which the charge is carried by electrons, if it is doped with a Group VA element such as arsenic.

Polymers Polymers, which are **macromolecules,** are made up of a very large number of atoms in which a small characteristic feature repeats over and over many times.

Polypropylene, an **addition polymer** that consists of a hydrocarbon **backbone** with a methyl group, CH_3, attached to every other carbon, is formed by **polymerization** of the **monomer** propylene. Polyethylene and polystyrene, which are also addition polymers, are formed from ethylene and styrene, respectively. **Condensation polymers** are formed by elimination of a small molecule such as H_2O and CH_3OH from two monomer units accompanied by the formation of a covalent bond between the monomers. **Nylons** and **polyesters** are **copolymers** because they are formed from two different monomers. **Silicone** polymers have a backbone composed of alternating silicon and oxygen atoms with organic groups (usually CH_3) attached to the silicon.

Cross-linking occurs when bridging groups of atoms link polymer chains together. Latex (polyisoprene) can be cross-linked by heating it with sulfur to give **vulcanized rubber.** Polymerization of ethylene can lead to **branching,** which produces an amorphous polymer called low-density polyethylene (LDPE). High-density polyethylene (HDPE) and ultrahigh-molecular-weight polyethylene (UHMWPE) are not branched and are more crystalline, which makes them stronger. Nylon's properties are affected by hydrogen bonding between polymer strands.

Liquid Crystals Certain substances whose molecules are long and thin with a rigid central region have properties just above their melting points that resemble both liquids and crystals. They are in a **liquid crystalline** phase when they exhibit these properties. In a **nematic phase,** molecules are aligned parallel to each other but are otherwise able to move in all directions. In a **smectic phase,** molecules are aligned along their long axes as well as in layers. In a **smectic C phase,** the long axes of the molecules are tilted relative to a line perpendicular to the plane of a layer. In a **cholesteric phase,** molecules are in thin layers aligned as in a nematic phase, but going down through layers the orientations of the molecules rotate, producing twisted stacks of molecules. Cholesteric liquid crystals reflect light that has a wavelength equal to the distance required for the twist to make one full turn. The twisted layers in a cholesteric crystal rotate **polarized light,** a property used in LC displays.

Ceramic Materials Ceramics, in general, are very hard and have very high melting points. Common ceramic materials are made from inorganic minerals obtained from the Earth's crust. They are ground to fine powders, formed into shapes, and fired at high temperatures where *sintering* occurs, causing the fine particles to stick together to give a rigid solid.

Materials used to make advanced ceramics generally contain a metal or metalloid in a high oxidation state combined with a small, highly charged nonmetal anion. Large lattice energies and significant covalent bonding cause the materials to be hard and high melting. The **sol-gel process** uses the reaction of water with metal **alkoxides** to form extremely small, gel-like particles suspended in alcohol that can be used in a variety of ways to give coatings, **xerogels, aerogels,** and ultrafine powders for making ceramics. Thermal decomposition of polymethylsilane can give silicon carbide fibers.

Ceramic **superconductors** have been found that lose all electrical resistance at temperatures above 77 K, which can be achieved by cooling with liquid nitrogen. The 1-2-3 compound $YBa_2Cu_3O_7$ features a layering of copper and oxygen atoms through which the conduction occurs. Applications of superconductors are expected to include electric power storage and magnetically levitated (mag-lev) trains. The latter is possible because superconductors repel magnetic fields.

Nanotechnology Nanotechnology (also called **molecular nanotechnology**) deals with using the unique properties of nanometer-scale structures to develop useful applications. The scanning tunneling microscope (STM) is used to image electrically conducting surfaces at the atomic scale and can manipulate individual atoms on a surface. The atomic force microscope (AFM) uses an extremely sharp stylus to follow the contours of a surface, which need not be electrically conducting. **Carbon nanotubes** are formed along with **fullerenes** (such as **buckminsterfullerene** or **"buckyballs"**) when an electric arc is passed between carbon electrodes. They consist of rolled-up sheets of **graphite** (a form of carbon composed of sheets of hexagonal rings of carbon atoms stuck together) and are capped at each end by half of a spherical buckyball. The nanotubes are very strong, can be electrical conductors or semiconductors, and are good conductors of heat. They can be made to form fibers. A variety of ceramiclike materials also form nanoscale fibers and other structures of interest.

TOOLS ▶ YOU HAVE LEARNED

The table below lists the concepts you've learned in this chapter that can be applied as tools in solving problems. Study each one carefully so that you know what each is used for. When faced with solving a problem, recall what each tool does and consider whether it will be helpful in finding a solution. This will aid you in selecting the tools you need. If necessary, refer to this table when working on the Thinking-It-Through problems and the Review Exercises that follow.

TOOL	HOW IT WORKS
Bragg equation (page 556)	Using the wavelength of X rays and angles at which X rays are diffracted from a crystal, the distances between planes of atoms can be calculated.
Unit cell structures for simple cubic, face-centered cubic, and body-centered cubic lattices (page 550)	By knowing the arrangements of atoms in these unit cells, we can use the dimensions of the unit cell to calculate atomic radii and other properties.
Properties of crystal types (page 559)	By examining certain physical properties of a solid (hardness, melting point, electrical conductivity in the solid and liquid state), we can often predict the nature of the particles that occupy lattice sites in the solid and the kinds of attractive forces between them.
Polymerization reactions forming addition polymers (page 565)	From the structure of a monomer, we can predict the structure of the polymer.
Polymerization reactions forming condensation polymers (page 568)	From the structures of the monomers, we can predict the structure of the polymer.
Hydrolysis reactions in the sol-gel process (page 580)	We can use structural formulas to write reactions for the formation of oxygen bridges between metals and nonmetals as part of the creation of ceramic materials.

THINKING IT THROUGH

The goal for the following problems is not to find the answers themselves, but rather to assemble the information needed to solve them and explain how you would use the information to find the answers. The problems in Level 2 are more challenging than those in Level 1 and may contain more data than are required, in which case you are also asked to identify the unnecessary data. Detailed answers to the Thinking-It-Through problems can be found on the web site.

 ON-LINE HELP

Need extra help? Visit the Brady/Senese web site at www.wiley.com/college/brady

---

Level 1 Problems

1. Describe how you would calculate the number of unit cells in 26.0 g of chromium, which crystallizes with a body-centered cubic unit cell.

2. If a compound with the formula A_2B_3 crystallizes with a unit cell that contains four atoms of A, how many atoms of B must be in the unit cell. Explain the reasoning involved.

3. To form a two-dimensional lattice from its unit cell, the unit cell is moved repeatedly a distance equal to its edge length along directions parallel to its edges. Each move generates a new unit cell with a new set of lattice points. Which of the following cannot be unit cells in a two-dimensional lattice? Explain your reasoning.

(a)   (b)   (c)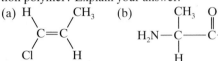

4. Which of the following is most likely to form a condensation polymer? Explain your answer.

5. Light observed after it passes through a polarizing filter is much less intense (bright) than light passing through ordinary glass. Explain why this is so.

6. Consider the hydrocarbon illustrated by the drawing below. Why isn't this compound as likely to form a liquid crystal phase as cholesteryl benzoate (page 575)?

7. Why wouldn't we expect $TiBr_3$ to form a type of ceramic material?

Level 2 Problems

8. Calcium oxide crystallizes with the same structure as NaCl, which has a density of 2.17 g cm^{-3}. The radius of a Ca^{2+} ion is approximately 99 pm and that of an O^{2-} ion is approximately 140 pm. Using these values, describe how you would estimate the density of CaO.

9. The unit cell for gold is described in Figure 13.6. Describe how you would use the information in this figure to calculate the atomic radius of a gold atom.

10. For a substance that crystallizes in a simple cubic structure, describe how you would calculate the percentage of unoccupied space (i.e., the space not actually occupied by the atoms themselves).

11. Approximately how many carbon nanotubes with a diameter of 1.4 nm would have to be bundled together to have the same cross-sectional area as a human hair, which has a diameter of 200 μm? Describe how you would perform the calculation.

12. Suppose a beam of X rays, diffracted from a crystal of gold at 25 °C, was observed at an angle of 18.5° relative to the planes of atoms defining the faces of the unit cell. Explain how you would calculate the largest value that the wavelength of the X rays could have.

REVIEW QUESTIONS

Crystalline Solids and X-Ray Diffraction

13.1 What is the difference between a crystalline solid and an amorphous solid?

13.2 What surface features do crystals have that suggest a high degree of order among the particles within them?

13.3 What is a *lattice*? What is a *unit cell*?

13.4 What relationship is there between a crystal lattice and its unit cell?

13.5 The diagrams below illustrate two typical arrangements of paving bricks in a patio or driveway. Sketch the unit cells that correspond to these two patterns of bricks.

(a)

(b)

13.6 Below is illustrated the way the atoms of two different elements are packed in a certain solid. The nuclei occupy positions at the corners of a cube. Is this cube the unit cell for this substance? Explain your answer.

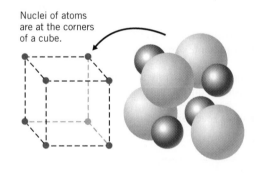

Nuclei of atoms are at the corners of a cube.

13.7 Describe simple cubic, face-centered cubic, and body-centered cubic unit cells. (Make a sketch of each.)

13.8 Make a sketch of a layer of sodium ions and chloride ions in a NaCl crystal. Indicate how the ions are arranged in a face-centered cubic pattern, regardless of whether we place lattice points at the Cl$^-$ ions or Na$^+$ ions.

13.9 How do the crystal structures of copper and gold differ? In what way are they similar? On the basis of the locations of the elements in the periodic table, what kind of crystal structure would you expect for silver?

13.10 What kind of lattice does zinc sulfide have?

13.11 What kind of lattice does calcium fluoride have?

13.12 Only 14 different kinds of crystal lattices are possible. How can this be true, considering the fact that there are millions of different chemical compounds that are able to form crystals?

13.13 Write the Bragg equation and define the symbols.

13.14 Atomic dimensions are usually given either in picometers (an SI unit) or angstroms (1 Å = 10^{-10} m). How many picometers are there in 1 angstrom? Write a conversion factor relating nanometers and angstroms.

13.15 Explain, in general terms, how an X-ray diffraction pattern of a crystal and the Bragg equation provide information that allows chemists to figure out the structures of molecules.

13.16 Why can't $CaCl_2$ or $AlCl_3$ form crystals with the same structure as NaCl?

Crystal Types

13.17 What kinds of particles are located at the lattice sites in a metallic crystal?

13.18 What kinds of attractive forces exist between particles in (a) molecular crystals, (b) ionic crystals, and (c) covalent crystals?

13.19 Why are covalent crystals sometimes called *network solids*?

13.20 Magnesium chloride is ionic. What general physical properties are expected of its crystals?

Amorphous Solids

13.21 What does the word *amorphous* mean?

13.22 What is an amorphous solid? Compare what happens when crystalline and amorphous solids are broken into pieces.

Band Theory of Solids

13.23 On the basis of the band theory of solids, how do conductors, insulators, and semiconductors differ?

13.24 Define the terms (a) *valence band* and (b) *conduction band*.

13.25 Why does the electrical conductivity of a semiconductor increase with increasing temperature?

13.26 In calcium, why can't electrical conduction take place by movement of electrons through the 2s energy band? How does calcium conduct electricity?

13.27 What is a p-type semiconductor? What is an n-type semiconductor?

13.28 Name two elements that would make germanium a p-type semiconductor when added in small amounts.

13.29 Name two elements that would make silicon an n-type semiconductor when added in small amounts.

13.30 How does a solar cell work?

Polymers

13.31 What is a *macromolecule*? Name two naturally occurring macromolecular substances.

13.32 What is a *polymer*? Are all macromolecules polymers?

13.33 What is a *monomer*? Draw structures for the monomers used to make (a) polypropylene, (b) poly(tetrafluoroethylene), and (c) poly(vinyl chloride).

13.34 What do we mean by the term *polymer backbone*?

13.35 What is the repeating unit in polypropylene? Write the formula for the polypropylene polymer. Give three uses for polypropylene.

13.36 How do propylene and the repeating unit in polypropylene differ?

13.37 What is the difference between an *addition polymer* and a *condensation polymer*?

13.38 Write the structure of polystyrene showing three of the repeating units. What are three uses for polystyrene plastic?

13.39 What is a *copolymer*?

13.40 Write the structural formula for (a) nylon 6,6 and (b) poly(ethylene terephthalate).

13.41 The structural formula for Kevlar is

Identify the amide bond in this polymer.

13.42 Polycarbonate polymers are polyesters. An example is

Identify the structural feature that makes it a polyester. Identify the structural feature that makes it a poly*carbonate*.

13.43 What is meant by the term *branching* as applied to polymers? How does branching affect the properties of low-density polyethylene?

13.44 What is *cross-linking*? How does it affect the properties of a polymer?

13.45 What is vulcanized rubber?

13.46 How does polymer crystallinity affect the physical properties of the polymer?

13.47 Why doesn't low-density polyethylene form strong crystalline polymer fibers?

13.48 Why does nylon form strong fibers?

13.49 Show how hydrogen bonding can bind Kevlar polymer chains together. (See Question 13.41.)

13.50 What are some applications of crystalline polymers such as HDPE, UHMWPE, and Kevlar?

13.51 Structurally, how do inorganic polymers differ from organic polymers?

13.52 Sketch a portion of the molecular backbone of the silicone polymers.

13.53 What is the function of $(CH_3)_3SiCl$ in the reaction mixture used to prepare a silicone polymer? What would happen if some CH_3SiCl_3 were included?

13.54 What is the nature of the polymer found in the anti-gas medication called simethicone?

Liquid Crystals

13.55 What is the origin of the term *liquid crystal*?

13.56 Which of the following molecules is most likely to form a liquid crystalline phase? Justify your selection.

(a)

$-CH_2-CH-CH_2-CH_2-CH_2-CH_3$

CH_2-CH_3

(b)

$N{\equiv}C-$ $-CH_2-CH_2-CH_2-CH_2-CH_3$

(c)

H_3C

CH_3

$C-CH_3$

CH_3

H_3C

13.57 Describe how molecules are arranged in the following liquid crystalline phases: (a) nematic, (b) smectic, (c) cholesteric.

13.58 Why does the color of reflected light from a cholesteric liquid crystal depend on the temperature?

13.59 Why might a smectic liquid crystal have a slippery feel?

13.60 What is plane-polarized light? How does it differ from unpolarized light?

13.61 Explain why crossed polarizers do not permit the passage of light.

13.62 Explain how the application of an electric field is able to cause a liquid crystal display to display an image.

Ceramic Materials

13.63 Name five common applications of ceramic materials found around the home.

13.64 What are two physical properties that make ceramics useful materials?

13.65 What is a *refractory*? Give two uses for refractories.

13.66 Why are ceramic materials used to make supports for high-voltage electrical transmission cables?

13.67 Which of the following pairs of ions might likely be used to make a useful ceramic?
(a) Cr^{3+} and N^{3-} (c) Cu^{2+} and S^{2-}
(b) Be^{2+} and C^{4-} (d) Y^{3+} and O^{2-}

13.68 How are ceramic materials formed into ceramic products?

13.69 What is *sintering*? How does it affect the physical properties of ceramic materials?

13.70 How does the sol-gel process resemble the formation of condensation polymers?

13.71 What is an *alkoxide*? What kind of reaction is called *hydrolysis*?

13.72 Use structural formulas to describe what happens when water is added to a solution of $Si(OC_2H_5)_4$ in alcohol.

13.73 What is a *xerogel*? How can one be made? What happens to a xerogel if it is heated to high temperature?

13.74 What is an *aerogel*? How is one formed? What are its unusual properties?

13.75 Write an equation that shows what happens when poly(methylsilane) is heated to high temperature.

13.76 Use specific examples to describe four applications of "advanced ceramics."

13.77 What is a superconductor? Why is it very expensive to use typical metallic superconductors in applications?

13.78 Suppose a substance was found that was an excellent superconductor, with a T_c of -175 °C. Would this material be of economic interest? Explain your answer.

13.79 Write the formula for a "1-2-3 ceramic superconductor."

13.80 Why is it possible to levitate a strong magnet above a substance that's in its superconducting state?

Nanotechnology

13.81 What does the SI prefix *nano* stand for? Why the interest in *nanotechnology* rather than *femtotechnology* or *millitechnology*?

13.82 What is the ultimate goal of molecular nanotechnology?

13.83 Describe how a scanning tunneling microscope works. Could it be used to investigate the surface of a ceramic object?

13.84 Describe how an atomic force microscope works. Could it be used to investigate the surface of a ceramic object?

13.85 Describe the structures of diamond, graphite, and buckminsterfullerene. What kind of hybrid orbitals does carbon use in these forms of carbon?

13.86 Describe the structure of a single-walled carbon nanotube. Give three reasons why there is so much interest in carbon nanotubes.

13.87 How are the structures of graphite and carbon nanotubes related?

13.88 How does the diameter of a carbon nanotube compare with the diameter of a human hair?

13.89 Give the formulas of two other substances that have been observed to form nanotubes.

13.90 Magnetic media are able to store digital information by representing the binary digits 0 and 1 by the magnetic states (either magnetized on unmagnetized) of small regions in the material making up the media surface. Why are $BaTiO_3$ and $SrTiO_3$ "nanorods" attractive candidates for data storage devices.

13.91 Alumina (a ceramic) can form fine fibers scientists have called "whiskers." What is the chemical composition of alumina?

REVIEW PROBLEMS

Answers to problems whose numbers are printed in color are given in Appendix B. More challenging questions are marked with asterisks. **ILW** = Interactive LearningWare solution is available at *www.wiley.com/college/brady*.

Crystalline Solids and X-Ray Diffraction

13.92 How many zinc and sulfide ions are present in the unit cell of zinc sulfide? (See Figure 13.10.)

13.93 How many copper atoms are within the face-centered cubic unit cell of copper? (*Hint:* See Figure 13.6 and add up all the *parts* of atoms in the fcc unit cell.)

ILW 13.94 The atomic radius of nickel is 1.24 Å. Nickel crystallizes in a face-centered cubic lattice. What is the length of the edge of the unit cell expressed in angstroms and in picometers?

13.95 Silver forms face-centered cubic crystals. The atomic radius of a silver atom is 144 pm. Draw the face of a unit cell with the nuclei of the silver atoms at the lattice points. The atoms are in contact along the diagonal. Calculate the length of an edge of this unit cell.

13.96 Potassium ions have a radius of 133 pm, and bromide ions have a radius of 195 pm. The crystal structure of potassium bromide is the same as for sodium chloride. Estimate the length of the edge of the unit cell in potassium bromide.

13.97 The unit cell edge in sodium chloride has a length of 564.0 pm. The sodium ion has a radius of 95 pm. What is the *diameter* of a chloride ion?

13.98 Calculate the angles at which X rays of wavelength 229 pm will be observed to be defracted from crystal planes spaced (a) 1000 pm apart and (b) 250 pm apart. Assume $n = 1$ for both calculations.

13.99 Calculate the interplanar spacings (in picometers) that correspond to defracted beams of X rays at $\theta = 20.0°$, $27.4°$, and $35.8°$, if the X rays have a wavelength of 141 pm. Assume that $n = 1$.

13.100 Aluminum crystallizes with a face-centered cubic structure. If the aluminum atom has an atomic radius of 1.43 Å, what is the length of the edge of the unit cell of aluminum?

13.101 Chromium, used to protect and beautify other metals, crystallizes with a body-centered cubic lattice in which the chromium atoms are in contact along the body diagonal of the unit cell. (The body diagonal starts at one corner and then runs through the center of the cell to the opposite corner.) The edge of the unit cell has a length of 2.884 Å. What is the radius of a chromium atom in angstroms?

13.102 Cesium chloride, CsCl, crystallizes with a cubic unit cell of edge length 412.3 pm. The density of CsCl is 3.99 g cm^{-3}. Show that the unit cell cannot be body-centered cubic or face-centered cubic.

13.103 Metallic sodium crystallizes with a body-centered cubic unit cell. The element has a density of 0.97 g cm^{-3}. What is the length of an edge of the unit cell? What is the radius of a sodium atom?

13.104 Cesium chloride forms a simple cubic lattice in which Cs$^+$ ions are at the corners and a Cl$^-$ ion is in the center (see Figure 13.9). The cation–anion contact occurs along the body diagonal of the unit cell (see Problem 13.101). The length of the edge of the unit cell is 412.3 pm. The Cl$^-$ ion has a radius of 181 pm. Calculate the radius of the Cs$^+$ ion.

13.105 Rubidium chloride has the rock salt structure. Cations and anions are in contact along the edge of the unit cell, which is 658 pm long. The radius of the chloride ion is 181 pm. What is the radius of the Rb$^+$ ion?

Crystal Types

13.106 Tin(IV) chloride, SnCl$_4$, has soft crystals with a melting point of -30.2 °C. The liquid is nonconducting. What type of crystal is formed by SnCl$_4$?

13.107 Elemental boron is a semiconductor, is very hard, and has a melting point of about 2250 °C. What type of crystal is formed by boron?

13.108 Gallium crystals are shiny and conduct electricity. Gallium melts at 29.8 °C. What type of crystal is formed by gallium?

13.109 Titanium(IV) bromide forms soft orange-yellow crystals that melt at 39 °C to give a liquid that doesn't conduct electricity. The liquid boils at 230 °C. What type of crystals does TiBr$_4$ form?

13.110 Columbium is another name for one of the elements. This element is shiny, soft, and ductile. It melts at 2468 °C, and the solid conducts electricity. What kind of solid does columbium form?

13.111 Elemental phosphorus consists of soft, white, "waxy" crystals that are easily crushed and melt at 44 °C. The solid does not conduct electricity. What type of crystal does phosphorus form?

13.112 Indicate which type of crystal (ionic, molecular, covalent, metallic) each of the following would form when it solidifies: (a) Br$_2$, (b) LiF, (c) MgO, (d) Mo, (e) Si, (f) PH$_3$, (g) NaOH.

13.113 Indicate which type of crystal (ionic, molecular, covalent, metallic) each of the following would form when it solidifies: (a) O$_2$, (b) H$_2$S, (c) Pt, (d) KCl, (e) Ge, (f) Al$_2$(SO$_4$)$_3$, (g) Ne.

Polymers

13.114 The structure of vinyl acetate is shown below. This compound forms an addition polymer in the same way as ethylene and propylene. Draw a segment of the poly(vinyl acetate) polymer that contains three of the repeating units.

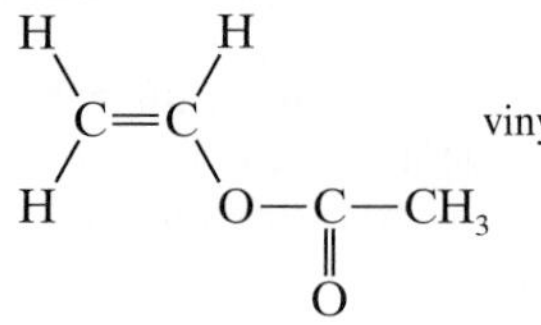

vinyl acetate

13.115 If the following two compounds polymerized in the same way that Dacron forms, what would be the repeating unit of their polymer?

$$
HOCCH_2CH_2COH \qquad HOCH_2CH_2OH
$$

13.116 Kodel, a polyester made from the following monomers, is used to make fibers for weaving crease-resistant fabrics.

$$
HOCH_2-\text{⬡}-CH_2OH \qquad HOC-\text{◯}-COH
$$

monomers for Kodel

What is the structure of the repeating unit in Kodel?

13.117 If the following two compounds polymerized in the same way that nylon forms, what would be the repeating unit in the polymer?

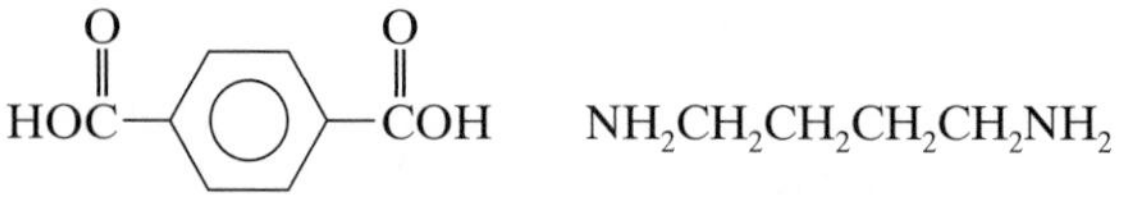

$$
HOC-\text{◯}-COH \qquad NH_2CH_2CH_2CH_2CH_2NH_2
$$

ADDITIONAL EXERCISES

13.118 Below is an illustration of a "herring-bone" pattern for patio pavers. Identify the unit cell for this pattern of bricks.

*13.119 Gold crystallizes in a face-centered cubic lattice. The edge of the unit cell has a length of 407.86 pm. The density of gold is 19.31 g/cm³. Use these data and the atomic mass of gold to calculate the value of Avogadro's number.

13.120 Gold crystallizes with a face-centered cubic unit cell with an edge length of 407.86 pm. Calculate the atomic radius of gold in units of picometers.

*13.121 Calculate the amount of empty space (in pm³) in simple cubic, body-centered cubic, and face-centered cubic unit cells if the lattice points are occupied by identical atoms with a diameter of 1.00 pm. Which of these structures gives the most efficient packing of atoms?

*13.122 Lithium bromide has the rock salt structure in which Br⁻ ions, centered at the lattice points, are in contact with each other. Calculate the ionic radii of Br⁻ and Li⁺ in picometers, given that the edge length of the unit cell equals 550 pm. The value of the ionic radius of Li, r_{Li^+}, found in tables is 60 pm. Explain why the value you obtained for r_{Li^+} is larger than the tabulated value.

13.123 Silver has an atomic radius of 144 pm. What would be the density of silver in g cm⁻³ if it were to crystallize in (a) a simple cubic lattice, (b) a body-centered cubic lattice, and (c) a face-centered cubic lattice? The actual density of silver is 10.6 g cm⁻³. Which cubic lattice does silver have?

*13.124 Potassium chloride crystallizes with the rock salt structure. When bathed in X rays, the layers of atoms corresponding to the surfaces of the unit cell produce a diffracted beam of X rays (λ = 1.54Å) at an angle of 6.97°. Calculate the density of KCl.

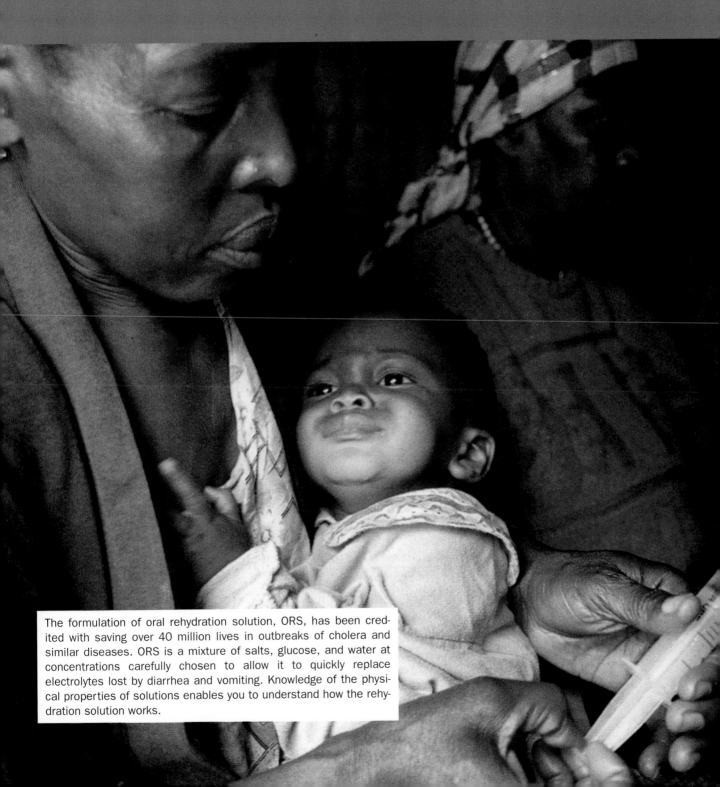

The formulation of oral rehydration solution, ORS, has been credited with saving over 40 million lives in outbreaks of cholera and similar diseases. ORS is a mixture of salts, glucose, and water at concentrations carefully chosen to allow it to quickly replace electrolytes lost by diarrhea and vomiting. Knowledge of the physical properties of solutions enables you to understand how the rehydration solution works.

Homogeneous means "everywhere alike." In a homogeneous mixture, composition, density, and other properties are the same throughout the sample.

THIS CHAPTER IN CONTEXT *Solutions* are homogeneous mixtures. The most abundant component of the mixture is called the *solvent* (Section 5.1). The other substances in the mixture are called *solutes.* We've already seen that many reactions occur easily in solution (Chapter 5). In Sections 14.1 and 14.2, we'll apply what we've learned about intermolecular forces (Chapters 11 and 12) as well as certain principles of thermochemistry (Chapter 7) to study the formation of solutions. We'll see how temperature and pressure affect the *solubility* of a substance in Sections 14.3 and 14.4. In Section 14.5, we'll introduce several important temperature-independent measures of concentration.

In Sections 14.6 through 14.9, we will study *colligative properties,* physical properties of solutions that depend more on the concentration of a solution than the chemical identity of the solute. We'll see that colligative properties can be extremely useful in determining the molecular masses of large molecules, and we will look at several biological applications.

In Section 14.8 we'll study *osmosis,* a flow of material through a membrane that is permeable to some substances (such as water) and not to others (such as salts). This phenomenon is of tremendous importance in biology and medicine. It is the operating principle behind kidney dialysis, which removes small waste molecules from the blood without loss of critically needed protein molecules. It is a key to understanding the action of diarrheal diseases, which kill about three and a half million children under the age of five each year. The deaths result from massive loss of fluid due to unbalanced salt concentrations across membranes in the lower digestive tract. "Oral rehydration solutions" containing precisely adjusted concentrations of sugar and electrolytes can equalize these salt concentrations and prevent these deaths in most cases. The development of oral rehydration solutions has been called one of the most important medical advances of the last century.

Section 14.10 deals with *colloidal dispersions* and *suspensions,* solutions with particle sizes that range from the size of large molecules to the size of biological cells. Colloidal dispersions are of supreme importance in industry, where they are applied to manufacture paints, inks, cosmetics, medicines, liquid crystals, and foams.

14.1 ▶ Substances mix spontaneously when there is no energy barrier to mixing

To understand the effect of solute and solvent on solubility, we need a detailed study of the *process* by which solutions form. We'll begin by taking a closer look at how gases spontaneously mix with or dissolve in each other.

Imagine a container holding two unmixed gases separated by a removable panel (see Figure 14.1*a*). When the panel is slid away (Figure 14.1*b*) the process of mixing begins *spontaneously* because of the random motions of the molecules. The gases mingle with no outside help. Once they have formed a homogeneous solution, their molecules will never spontaneously separate from one another and re-form the original, unmixed state. The unmixed state is so unlikely statistically once the panel has been removed that we can say "never" with confidence.

The spontaneous mixing of gases illustrates one of nature's strong "driving forces" for change. *A system, left to itself, will tend toward the most probable state.*[1] At the instant we remove the panel (Figure 14.1), the container holds two separate gas samples, in contact but unmixed. This represents a highly improbable state, like lines of boys and girls before a sixth-grade dance. The natural motions of the molecules make this state highly improbable once the panel is removed. Now the vastly more probable distribution is one in which the molecules are thoroughly mixed.

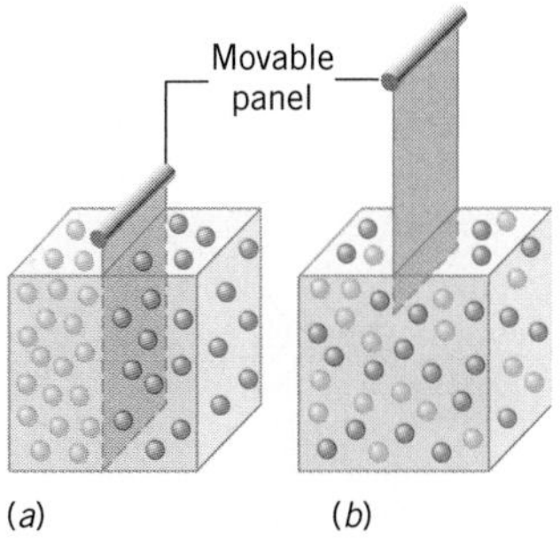

The drive to attain the most probable state accounts for the formation of gaseous solutions. We can also use it to explain the formation of liquid solutions whenever the attractive forces between molecules in the solution are all much the same. Sometimes, though, attractions between like molecules are stronger than those between unlike molecules. This imbalance of attractive forces can block the natural drive toward the more probable mixed state.

Attractions between solute and solvent molecules are another driving force behind the formation of a solution

Attractive forces between particles do not affect the formation of gaseous solutions much, because such forces are very weak when particles are as far apart as they are in gases.

Attractive forces are very important, however, to the formation of liquid solutions. Such forces hold solute and solvent particles close together (see Figure 14.2). This works against getting them to intermingle. Yet ions and polar molecules must separate from each other if intermingling is to occur and a solution is to form. Success in preparing a nongaseous solution depends not only on the fact that the mixed state is highly probable but also on the strengths of the intermolecular attractive forces in *both solvent and solute.*

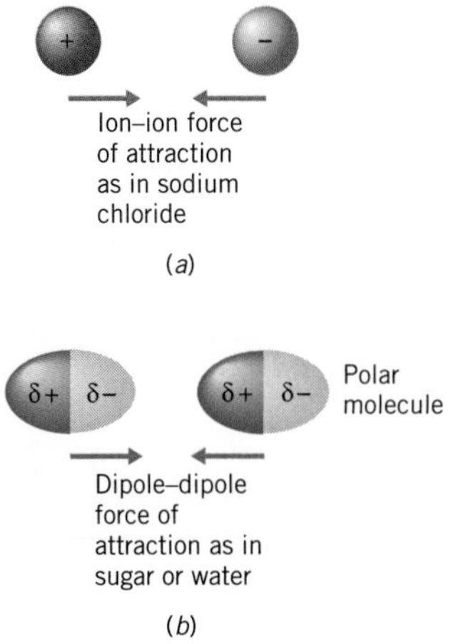

Like dissolves like

If we mix ethanol (the alcohol present in alcoholic beverages) and water, they completely blend or dissolve in any proportion we choose; we say that the two are completely *miscible.* Benzene, C_6H_6, on the other hand, is a liquid that is virtually insoluble in water; the two are *immiscible.* To understand this different behavior, let's look closely at what happens in each case when the liquids are combined. Overall, we know that the molecules of solute and solvent must be pushed apart to make room for those of the other if a solution is to form.

FIGURE 14.1 *Mixing of gases.* When two gases, initially in separate compartments (*a*), suddenly find themselves in the same container (*b*), they mix spontaneously.

Movable panel

(*a*)   (*b*)

Ion–ion force of attraction as in sodium chloride

(*a*)

$\delta+$ $\delta-$ $\delta+$ $\delta-$ Polar molecule

Dipole–dipole force of attraction as in sugar or water

(*b*)

FIGURE 14.2 *Opposite charges attract.* (*a*) Interionic force of attraction. (*b*) Intermolecular force of attraction.

[1] In Chapter 20 this driving force will be called *entropy.*

H—C—C—O—H (ethanol structure with δ− δ+ labels) and benzene structure

ethanol
(a polar OH group)

benzene
(no polar group)

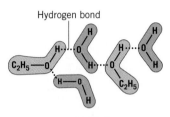

Hydrogen bond

FIGURE 14.3 *Hydrogen bonds in aqueous ethanol.* Ethanol molecules form hydrogen bonds (indicated by ···) to water molecules.

Ethanol has a molecular structure that includes a polar O—H group, so its molecules can form hydrogen bonds with each other and with water molecules, which also have O—H groups (see Figure 14.3). Water molecules can attract alcohol molecules almost as strongly as they attract each other. The forces that must be overcome when water molecules are pushed apart to make room for alcohol molecules are compensated by similar attractions to the incoming alcohol molecules. Alcohol molecules moving into the water are surrounded by water molecules that attract them and impede them from moving out of the water again. Similarly, water molecules moving into the alcohol are caged by alcohol molecules. Nature's strong tendency toward the more probable mixed state of a solution is free to work, and water and alcohol are miscible.

Quite a different situation occurs if we try to dissolve benzene in water. At room temperature and pressure, the two liquids do not mix. Benzene molecules are nonpolar. They are attracted to each other only by London forces, which are relatively weak. Benzene molecules cannot hydrogen bond, and so are not attracted to water molecules very strongly. To accommodate the benzene molecules in a true solution, the water molecules would have to move apart to make room for them. This would require energy, because strong hydrogen bonds between the water molecules would have to be broken. The energy required to do this would not be balanced by attractions to the incoming benzene molecules. Without the input of energy, the strong attractions between water molecules block nature's tendency toward the more probable mixed state.[2]

It is also easy to see why a solution of water in benzene cannot be made. Suppose that we did manage to disperse water molecules in benzene. As they move about, the water molecules would occasionally encounter each other. Because they attract each other so much more strongly than they attract benzene molecules, water molecules would stick together at each such encounter by hydrogen bonds. This would continue to happen until all the water was in a separate phase. A solution of water in benzene would not be stable.

Although benzene cannot dissolve in water, it does dissolve quite well in nonpolar liquids, like carbon tetrachloride, CCl_4. The forces of attraction between CCl_4 molecules are about as weak as those between benzene molecules. The energy required to open holes in the benzene for CCl_4 molecules to occupy is small and compensated by the energy released when weak attractions form between CCl_4 and benzene molecules. There are no obstacles to nature's drive toward the more probable mixed state, and the solution readily forms.

In summary, we see that when the strengths of intermolecular attractions are *similar* in solute and solvent, solutions can form as illustrated by the easy formation of alcohol–water and benzene–carbon tetrachloride solutions (Figure 14.4). This is what is behind a rule of thumb that chemists widely use, namely the **"like dissolves like" rule:** when solute and solvent have molecules "like" each other in

TOOLS

"Like dissolves like" rule

---

[2]The full explanation for the immiscibility of benzene (or oil) and water is complicated. It requires more background in thermodynamics (Chapter 20) than we have thus far been able to provide. For a discussion with leading references, however, see T. P. Silverstein, *Journal of Chemical Education,* January 1998, page 116.

(a)

(b)

FIGURE 14.4 *Like dissolves like.* (*a*) Ethanol is soluble in water because the mixed ethanol–water attractions are not significantly weaker than the water–water or ethanol–ethanol attractions. (*b*) Benzene is not soluble in water because the mixed benzene–water attraction is much weaker than the water–water and benzene–benzene attractions. The stronger attractions in the pure liquids constitute an energy barrier to mixing.

polarity, they tend to form a solution. When solute and solvent molecules are quite different in polarity, solutions of any appreciable concentration do not form. The rule has long enabled chemists to use chemical composition and molecular structure to predict the likelihood of two substances dissolving in each other.

The solubility of solids depends on the relative strengths of intermolecular attractions

In moving to solutions of solids in liquids, the basic principles remain the same. We'll look first at what happens when sodium chloride, a crystalline salt, dissolves in water.

Figure 14.5 depicts a section of a crystal of NaCl in contact with water. The dipoles of water molecules orient themselves so that the negative ends of some point toward Na^+ ions and the positive ends of others point at Cl^- ions. In other

---

CHEMISTRY IN PRACTICE If you've ever removed a butter, chocolate, or grease stain using a solvent like naphtha, kerosene, or turpentine, you've applied the "like dissolves like" rule. Nonpolar substances like grease or butter dissolve easily in nonpolar solvents like naphtha, but only with great difficulty in water, a polar solvent. You can often remove a stain by scrubbing with a solvent that is similar to the stain. In Section 14.10, we'll see that the "like dissolves like" rule can be used to explain how soaps and detergents allow nonpolar stains to be washed out with water.

The "like dissolves like" rule is helpful in understanding how contaminants are able to persist in the environment. A pollutant that is water soluble is easily diluted by a river, lake, or ocean. However, nonpolar substances tend to concentrate in fat tissue, which is nonpolar. Once these substances dissolve in fat within an animal or human body, they tend to stay there for many years. This gives them a chance to accumulate to dangerous levels over time within the body. Carnivores acquire even higher concentrations by eating contaminated fat from organisms lower on the food chain.

The "like dissolves like" rule is of critical importance in developing new drugs to treat brain and neurological disorders. The *blood–brain barrier* is a hydrocarbon-rich curtain that isolates the brain from the bloodstream. Any drug that affects the brain must cross this barrier. Small, nonpolar molecules such as caffeine or nicotine easily dissolve in the nonpolar barrier and make their way into the brain. Unfortunately, many new drugs that might be useful in curing diseases of the central nervous system are not able to cross the barrier, and so they cannot be administered orally or by injection into the bloodstream.

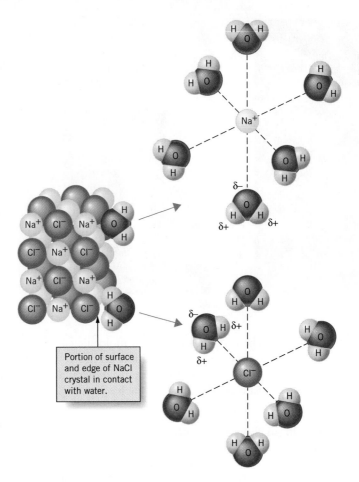

FIGURE 14.5 *Hydration of ions.* Hydration involves a complex redirection of forces of attraction and repulsion. Before this solution forms, water molecules are attracted only to each other, and Na⁺ and Cl⁻ ions have only each other in the crystal to be attracted to. In the solution, the ions have water molecules to take the places of their oppositely charged counterparts; in addition, water molecules find ions more attractive than even other water molecules.

Portion of surface and edge of NaCl crystal in contact with water.

Water molecules collide everywhere along the crystal surface, but *successful* collisions—those that dislodge ions—are more likely to occur at corners and edges.

We're clearly at the borderline between chemical and physical changes when we place the binding of water molecules to sodium or chloride ions in the realm of physical changes.

words, *ion–dipole* attractions occur that tend to tug and pull ions from the crystal. At the corners and edges of the crystal, ions are held by fewer neighbors within the solid and so are more readily dislodged than those elsewhere on the crystal's surface. As water molecules dislodge these ions, new corners and edges are exposed, and the crystal continues to dissolve.

As they become free, the ions become completely surrounded by water molecules (also shown in Figure 14.5). The phenomenon is called the **hydration** of ions. The *general* term for the surrounding of a solute particle by solvent molecules is **solvation,** so hydration is just a special case of solvation. Ionic compounds are able to dissolve in water when the attractions between water dipoles and ions overcome the attractions of the ions for each other within the crystal.

Similar events explain why solids composed of polar molecules, like those of sugar, dissolve in water (see Figure 14.6). Attractions between the solvent and solute dipoles help to dislodge molecules from the crystal and bring them into solution. Again we see that "like dissolves like"; a polar solute dissolves in a polar solvent.

The same reasoning explains why nonpolar solids like wax are soluble in nonpolar solvents such as benzene. Wax is a solid mixture of long-chain hydrocarbons, held together by London forces. The attractions between the molecules of the solvent (benzene) and the solute (wax) are also London forces, of comparable strength. The natural drive toward the more probable mixed state can proceed.

When intermolecular attractive forces within solute and solvent are sufficiently different, the two do not form a solution. For example, ionic solids or very polar molecular solids (like sugar) are insoluble in nonpolar solvents such as benzene, gasoline, or lighter fluid. The molecules of these solvents, all hydrocarbons, are unable to attract ions or very polar molecules with enough force to overcome the

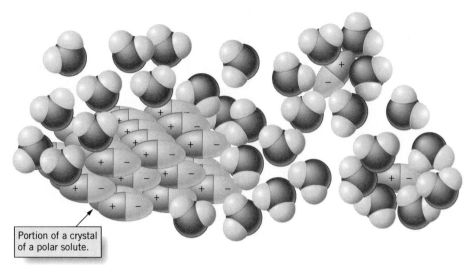

FIGURE **14.6** *Hydration of a polar molecule.* A polar solid dissolves in water because its molecules are attracted to the very polar water molecules. The solute molecules experience hydration when they become surrounded by molecules of water. Notice that the water molecules orient themselves so their positive ends are near the negative ends of the solute dipoles and their negative ends are near the positive ends of the solute dipoles.

Portion of a crystal of a polar solute.

much stronger attractions that the ions or polar molecules experience within their own crystals. Nature's drive toward the more probable mixed state can be blocked by strong interionic or intermolecular forces.

14.2 ▸ Enthalpy of solution comes from unbalanced intermolecular attractions

Because intermolecular attractive forces are important when liquids and solids are involved, the formation of a solution is inevitably associated with energy exchanges. The total energy absorbed or released when a solute dissolves in a solvent at constant pressure to make a solution is called the *molar enthalpy of solution,* or usually just the **heat of solution, ΔH_{soln}.**

Energy is required to separate the particles of solute and also those of the solvent and make them spread out to make room for each other. This step is *endothermic,* because we must overcome the attractions between molecules to spread the particles out. But once the particles come back together as a *solution,* the attractive forces between approaching solute and solvent particles yield a decrease in the system's potential energy, and this is an *exothermic* change. The enthalpy of solution, ΔH_{soln}, is simply the net result of these two opposing enthalpy contributions.

In Chapter 7 you learned that enthalpy is a *state function.* So the magnitude of a *change* in enthalpy, like ΔH_{soln}, does not depend on *how* the system goes from one state to another. We may devise any route or path we please for the *process* of forming a solution as long as we go from the same initial state—separated samples of solute and solvent—to the same final state, the solution. To look at this more closely, let us follow the potential energy changes that occur when a solid dissolves in a liquid.

$$\Delta H_{soln} = H_{soln} - H_{solvent} - H_{solute}$$

The enthalpy of solution for a solid is the lattice energy plus the solvation energy

We can model the solution of a solid in a liquid as a two-step process, as shown in the enthalpy diagram in Figure 14.7. Overall, the two steps will take us from the solid solute and liquid solvent to the final solution, but you'll see that they do not represent the way a solution is actually made in the lab. In the first step, we imagine that the solid separates into its individual particles; in effect, we imagine that the solid is vaporized. In the second step, we imagine that the gaseous solute particles

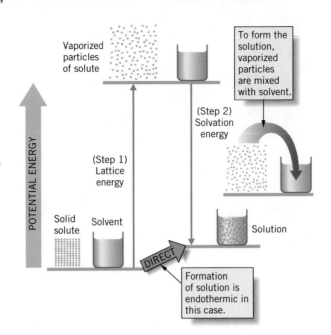

Figure 14.7 *Enthalpy diagram for a solid dissolving in a liquid.* In the real world, the solution is formed directly as indicated by the purple arrow. We can analyze the energy change by imagining the two separate steps, because enthalpy changes are functions of state and are independent of path. The energy change along the direct path is the algebraic sum of Step 1 and Step 2.

ΔH is positive for an endothermic change, so the arrow for Step 1 in the figure points upward, the positive direction.

enter the solvent where they become solvated. As we said, we imagine this particular process to let us use existing experimental data, being permitted to do so only because the enthalpy of solution is a state function.

As you can see from Figure 14.7, the first step requires an input of energy. The particles in the solid attract each other, so energy must be supplied to separate them. The amount of energy required is called the *lattice energy.* Recall from Section 9.1 that the lattice energy is the energy required to separate 1 mol of a crystalline compound into gaseous particles.[3] For ionic compounds, the gaseous particles are ions; for molecular compounds, they are molecules. For example, the lattice energy of potassium iodide, KI, is given by ΔH in the equation

$$KI(s) \longrightarrow K^+(g) + I^-(g) \qquad \Delta H = +632 \text{ kJ}$$

In the second step, gaseous solute particles enter the solvent and are solvated. The potential energy of the system *decreases,* because solute and solvent particles attract each other. This step must be exothermic. The enthalpy change when gaseous solute particles (obtained from 1 mol of solute) are dissolved in a solvent is called the **solvation energy.** If the solvent is water, the solvation energy can also be called the **hydration energy.** For example, the hydration energy of KI is given by ΔH in the equation

$$K^+(g) + I^-(g) \longrightarrow K^+(aq) + I^-(aq) \qquad \Delta H = -619 \text{ kJ}$$

The magnitude of ΔH_{soln} depends somewhat on the final concentration of the solution being made. With more solute particles, more energy is required to break up the lattice, but at higher concentrations there are fewer water molecules available for solvation, and the second step releases less energy.

The *enthalpy of solution* is the difference between the energy required for Step 1 and the energy released in Step 2. It is the enthalpy change when 1 mol of a crystalline substance is dissolved in a solvent. For example, the enthalpy of solution of KI calculated from the data given above is given by the equation

$$KI(s) \longrightarrow K^+(aq) + I^-(aq) \qquad \Delta H_{\text{soln}} = +13 \text{ kJ}$$

When the energy required for Step 1 exceeds the energy released in Step 2, the solution forms endothermically. But when more energy is released in Step 2 than is needed for Step 1, the solution forms exothermically.

---

[3]Note that a few texts and thermochemical databases define *lattice energy* as the energy released when a mole of a crystalline solid forms from its gaseous component ions or molecules, so that the lattice energy they report is always negative (exothermic). If you use such data, simply remove the minus sign to obtain a lattice energy that is consistent with our discussion.

TABLE 14.1	LATTICE ENERGIES, HYDRATION ENERGIES, AND HEATS OF SOLUTION FOR SOME GROUP IA METAL HALIDES			

| | | | ΔH_{soln} | |
Compound	Lattice Energy (kJ mol^{-1})	Hydration Energy (kJ mol^{-1})	Calculated[a] ΔH_{soln} (kJ mol^{-1})	Measured[b] ΔH_{soln} (kJ mol^{-1})
LiCl	+833	−883	−50	−37.0
NaCl	+766	−770	−4	+3.9
KCl	+690	−686	+4	+17.2
LiBr	+787	−854	−67	−49.0
NaBr	+728	−741	−13	−0.602
KBr	+665	−657	+8	+19.9
KI	+632	−619	+13	+20.33

[a]Calculated ΔH_{soln} = lattice energy + hydration energy.
[b]Heats of solution refer to the formation of extremely dilute solutions.

We can test this analysis by comparing the experimental heats of solution of some salts in water with those calculated from lattice and hydration energies (Table 14.1). The calculations are described in Figures 14.8 and 14.9, using enthalpy diagrams for the formation of aqueous solutions of two salts, KI and NaBr.

The agreement between calculated and measured values in Table 14.1 is not particularly impressive. This is partly because lattice and hydration energies are not precisely known and partly because the model used in our analysis is evidently too simple. (The model, for example, does not consider nature's tendency toward the most probable state, which is a major factor in the formation of solutions.) Notice, however, that when "theory" predicts relatively large heats of solution, the experimental values are also relatively large, and that both values have the same

Small percentage errors in very large numbers can cause huge percentage changes in the *differences* between such numbers.

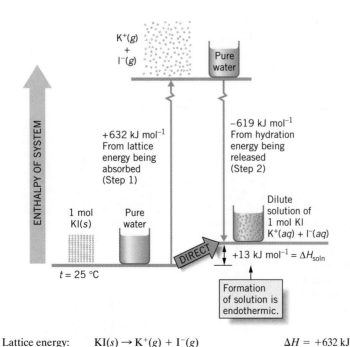

FIGURE 14.8 *Enthalpy of solution.* The formation of aqueous potassium iodide.

Lattice energy:	KI(s) → K⁺(g) + I⁻(g)	$\Delta H = +632$ kJ
Hydration energy:	K⁺(g) + I⁻(g) → K⁺(aq) + I⁻(aq)	$\Delta H = -619$ kJ
Total:	KI(s) → K⁺(aq) + I⁻(aq)	$\Delta H_{soln} = +13$ kJ

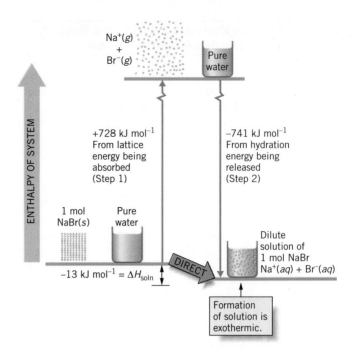

FIGURE 14.9 *Enthalpy of solution.* The formation of aqueous sodium bromide.

Lattice energy:	$NaBr(s) \rightarrow Na^+(g) + Br^-(g)$	$\Delta H = +728$ kJ
Hydration energy:	$Na^+(g) + Br^-(g) \rightarrow Na^+(aq) + Br^-(aq)$	$\Delta H = -741$ kJ
Total:	$NaBr(s) \rightarrow Na^+(aq) + Br^-(aq)$	$\Delta H_{soln} = -13$ kJ

sign (except for NaCl). Notice also that the changes in values show the same trends when we compare the three chloride salts—LiCl, NaCl, and KCl—or the three bromide salts—LiBr, NaBr, and KBr.

Solution of a liquid in another liquid can be modeled as a three-step process

To consider heats of solution when liquids dissolve in liquids, we will work with a three-step path to go from the initial to the final state (see Figure 14.10). We will designate one liquid as the solute and the other as the solvent.

FIGURE 14.10 *Enthalpy of solution.* To analyze the enthalpy change for the formation of a solution of two liquids, we can imagine the hypothetical steps shown here. **Step 1.** The molecules of the liquid designated as the solute move apart slightly to make room for the solvent molecules, an endothermic process. **Step 2.** The molecules of the solvent are made to take up a larger volume to make room for the solute molecules, which is also an endothermic change. **Step 3.** The expanded samples of solute and solvent come together to form the solution. Because the molecules attract each other, this step is exothermic.

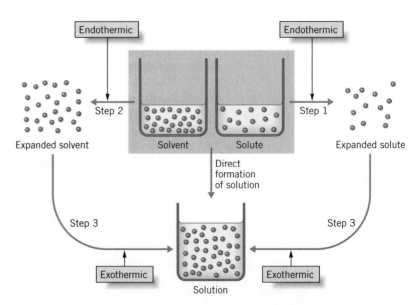

1. *Expand the solute liquid.* First, we imagine that the molecules of one liquid are moved apart just far enough to make room for molecules of the other liquid. Because we have to overcome forces of attraction, this step increases the system's potential energy and so is *endothermic.*

2. *Expand the solvent liquid.* The second step is like the first, but is done to the other liquid (solvent). On an enthalpy diagram (Figure 14.11) we have climbed two energy steps and have both the solvent and the solute in their slightly "expanded" conditions.

3. *Mix the expanded liquids.* The third step lets nature's drive for the most probable mixed state take its course, as the molecules of expanded solvent and solute come together and intermingle. Because the molecules of the two liquids now experience mutual forces of attraction, the system's potential energy *decreases,* and Step 3 is exothermic. The value of ΔH_{soln} will, again, be the net energy change for these steps.

Ideal solutions

The enthalpy diagram in Figure 14.11 shows the case when the sum of the energy inputs for Steps 1 and 2 is equal to the energy released in Step 3, so the overall value of ΔH_{soln} is zero. This is very nearly the case when we make a solution of benzene and carbon tetrachloride. Attractive forces between molecules of benzene are almost exactly the same as those between molecules of CCl_4, or between molecules of benzene and those of CCl_4. If all such intermolecular forces were identical, the net ΔH_{soln} would be exactly zero, and the resulting solution would be called an **ideal solution.** Be sure to notice the difference between an *ideal solution* and an *ideal gas.* In an ideal gas, there are no attractive forces. In an ideal solution, there are attractive forces, but they are all the same.

Acetone and water form a solution exothermically (see Figure 14.12*a*). With these liquids, the third step releases more energy than the sum of the first two chiefly because molecules of water and acetone attract each other more strongly than acetone molecules attract each other.

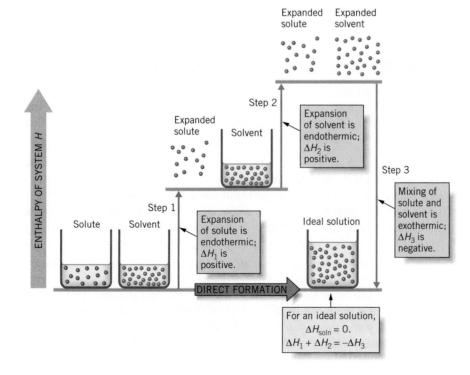

FIGURE 14.11 *Enthalpy changes in the formation of an ideal solution.* The three-step and the direct-formation paths both start and end at the same place with the same enthalpy outcome. The sum of the positive ΔH values for the two endothermic Steps, 1 and 2, numerically equals the negative ΔH value for the exothermic Step 3. The net ΔH for the formation of an ideal solution is therefore zero.

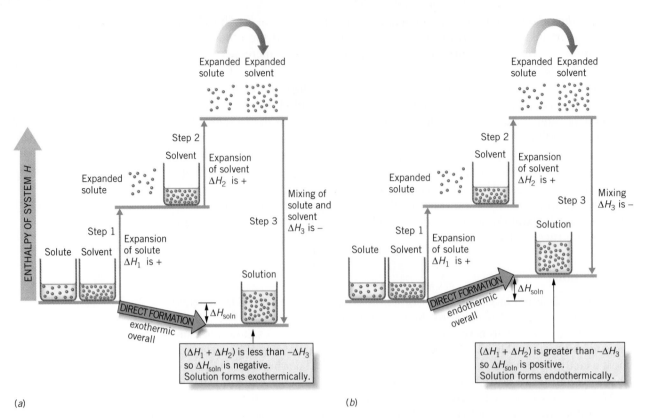

FIGURE 14.12 *Enthalpy changes when a nonideal solution forms.* Because ΔH_{soln} is a state function, the multistep paths and the direct-formation paths have identical enthalpy outcomes. (*a*) The process is exothermic. (*b*) The process is endothermic.

Ethanol and hexane form a solution endothermically (see Figure 14.12*b*). In this case, the release of energy in the third step is not enough to compensate for the energy demands of Steps 1 and 2, and the solution cools as it forms. Ethanol molecules attract each other more strongly than they can attract hexane molecules. Hexane molecules cannot push their way into ethanol without breaking up some of the hydrogen bonding between ethanol molecules.

$$CH_3CH_2CH_2CH_2CH_2CH_3$$
hexane

Gas solubility can be understood using a simple molecular model

Unlike solid and liquid solutes, only very weak attractions exist between gas molecules, so the energy required to "expand the solute" is negligible. The heat absorbed or released when a gas dissolves in a liquid has essentially two contributions, as shown in Figures 14.13:

1. *Energy is absorbed to open "pockets" in the solvent that can hold gas molecules.* The solvent must be expanded slightly to accommodate the molecules of the gas. This requires a small energy input since attractions between solvent molecules must be overcome. Water is a special case—it already contains open holes in its network of loose hydrogen bonds around room temperature. For water, very little energy is required to create pockets that can hold gas molecules.

2. *Energy is released when gas molecules are popped into these pockets.* Intermolecular attractions between the gas molecules and the surrounding solvent molecules lower the total energy, and energy is released as heat. The stronger the attractions are, the more heat is released. Water is capable of forming hydrogen bonds with some gases, such as NH_3, whereas organic solvents often can't. A larger amount of heat is released when such a gas molecule is placed in the pocket in water than in organic solvents.

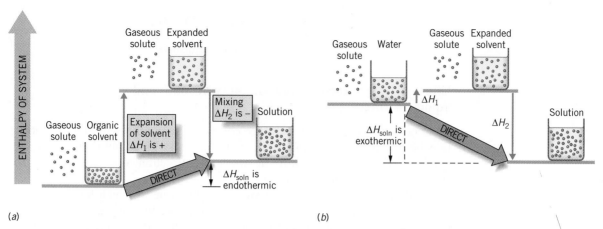

(a)   (b)

FIGURE **14.13** *A molecular model of gas solubility.* (*a*) A gas dissolves in an organic solvent. Energy is absorbed to open "pockets" in the solvent that can hold the gas molecules. In the second step, energy is released when the gas molecules enter the pockets where they are attracted to the solvent molecules. Here, the solution process is shown to be endothermic. (*b*) Around room temperature, water's loose network of hydrogen bonds already contains pockets that can accommodate gas molecules, so little energy is needed to prepare the solvent to accept the gas. In the second step, energy is released as the gas molecules take their place in the pockets where they experience attractions to the water molecules. In this case, the solution process is exothermic.

Heats of solution for gases in organic solvents are often endothermic because the energy required to open up pockets is greater than the energy released by attractions formed between the gas and solvent molecules.

Heats of solution for gases in water are often exothermic because water already contains pockets to hold the gas molecules, and energy is released when water and gas molecules attract each other.

14.3 ▶ A substance's solubility changes with temperature

By "solubility" we mean the mass of solute that forms a *saturated* solution with a given mass of solvent at a specified temperature. The units often are grams of solute per 100 g of solvent. Notice that solubility always refers to a *saturated* solution. If we add more solute, the extra solute will remain as a separate phase in contact with the solution. There will be a coming and going of solute particles between the dissolved and undissolved phases, of course, because in a saturated solution we have equilibrium between the two phases (see Figure 14.14). We can write this equilibrium as shown on the next page.

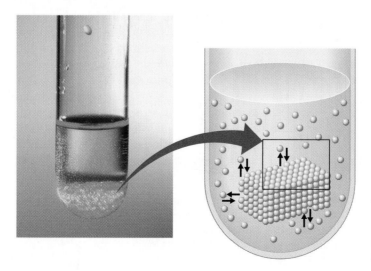

FIGURE **14.14** *Saturated solution.* In a saturated solution, a dynamic equilibrium exists between the undissolved solute and the solute in the solution.

$$\text{solute}_{\text{undissolved}} \rightleftharpoons \text{solute}_{\text{dissolved}}$$
(Solute contacts the   (Solute is in a
saturated solution.)   saturated solution.)

Effect of temperature on the solubility of solids and liquids in other liquids

As long as we keep the temperature constant, the equilibrium in a saturated solution holds. Heating or cooling the solution shifts its equilibrium. As we saw in Section 12.8, equilibrium shifts in whichever direction most directly absorbs the "stress" (Le Châtelier's principle).

Let's consider the more common situation, one in which heating an initially saturated solution causes more solute to dissolve. We may even place "energy" within the equilibrium expression, putting it on the left side, the undissolved solute side, because heat is absorbed when solute dissolves.

$$\text{solute}_{\text{undissolved}} + \text{energy} \rightleftharpoons \text{solute}_{\text{dissolved}}$$
(Solute is in contact with a
saturated solution.)

The stress of heating the solution causes the equilibrium to shift to the *right*. This shift consumes the energy being added, and *it also causes more solute to dissolve.* The amount that dissolves depends on the particular solute and the solvent. Compare in Figure 14.15, for example, the widely different responses to increasing temperature in the solubilities of ammonium nitrate and sodium chloride in water.

Some solutes become *less* soluble with increasing temperature, for example, cerium(III) sulfate, $Ce_2(SO_4)_3$ (Figure 14.15). Energy must be *released* from a saturated solution of cerium(III) sulfate to make more solute dissolve. For its equilibrium equation, we must show "energy" on the right side because heat is released when more of the solute dissolves into a saturated solution.

$$\text{solute}_{\text{undissolved}} \rightleftharpoons \text{solute}_{\text{dissolved}} + \text{energy}$$

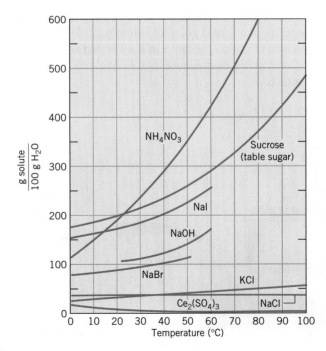

FIGURE 14.15 *Solubility in water versus temperature for several substances.* Most substances become more soluble when the temperature of the solution is increased, but the amount of this increased solubility varies considerably.

TABLE 14.2	SOLUBILITIES OF COMMON GASES IN WATER[a]			
	Temperature			
Gas	0 °C	20 °C	50 °C	100 °C
Nitrogen, N_2	0.0029	0.0019	0.0012	0
Oxygen, O_2	0.0069	0.0043	0.0027	0
Carbon dioxide, CO_2	0.335	0.169	0.076	0
Sulfur dioxide, SO_2	22.8	10.6	4.3	1.8[b]
Ammonia, NH_3	89.9	51.8	28.4	7.4[c]

[a]Solubilities are in grams of solute per 100 g of water when the gaseous space over the liquid is saturated with the gas and the total pressure is 1 atm.

[b]Solubility at 90 °C.

[c]Solubility at 96 °C.

When we *increase* the temperature of this system, the equilibrium absorbs the stress, the added energy, by shifting to the left, and some of the dissolved solute comes out of solution. Such systems are not common.

Effect of temperature on the solubility of a gas in a liquid

Table 14.2 gives data for the solubilities of several common gases in water at different temperatures, but all under 1 atm of pressure. In water, gases are usually more soluble at colder temperatures. But the solubility of gases, like other solubilities, can increase or decrease with temperature, depending on the gas and the solvent. For example, the solubilities of H_2, N_2, CO, He, and Ne actually rise with rising temperature in common organic solvents like carbon tetrachloride, toluene, and acetone.

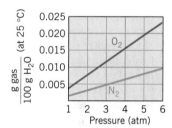

FIGURE 14.16 *Solubility in water versus pressure for two gases.* The amount of gas that dissolves increases as the pressure is raised.

14.4 ▶ Gases become more soluble at higher pressures

The solubility of a gas in a liquid is affected not only by temperature but also by pressure. The gases in air, for example, chiefly nitrogen and oxygen, are not very soluble in water under ordinary pressures, but at elevated pressures both become increasingly soluble (see Figure 14.16). Facets of Chemistry 14.1 describes the dangerous implications of these changes to those who must work under higher-than-normal air pressures.

The relevant equilibrium is

$$\text{gas} + \text{solvent} \rightleftharpoons \text{solution} \qquad (14.1)$$

Suppose we now increase the pressure on the system and so reduce the volume available to the gas. Le Châtelier's principle says that the system will change in a way to counteract this "stress," that is, the equilibrium will shift so that the pressure is reduced. To relieve the stress, the amount of gas must decrease. Some gas molecules must leave the gas phase and dissolve in the solution. Increasing the pressure of the gas shifts Equilibrium 14.1 to the right, and the solubility of the gas increases.

If we make the pressure on the solution *lower*, the equilibrium must shift to the left, and some dissolved gas leaves the solution. Every time you pop open a carbonated beverage, dissolved carbon dioxide gas fizzes out in response to the sudden lowering of pressure.

To see how pressure affects gas solubility on a molecular level, imagine a closed container with a movable wall and partly filled with a solution of some gas

Bottled carbonated beverages fizz when the bottle is opened because the sudden drop in pressure causes a sudden drop in gas solubility.

The Bends—A Potentially Lethal Consequence of Henry's Law

When people, like deep-sea divers, work where the air pressure is much above normal, they have to be careful to return slowly to normal atmospheric pressure (see Figure 1). Otherwise, they face the danger of the "bends"—severe pain in joints and muscles, fainting, possible deafness, paralysis, and death. Each 10 m of water depth adds roughly 1 atm to the pressure. Also at risk are workers who dig deep tunnels, where higher-than-normal air pressure is maintained to keep water out.

The bends can develop because the solubilities of both nitrogen and oxygen in blood are higher under higher pressure, as Henry's law says. Once the blood is enriched in these gases it must not be allowed to lose them suddenly. If microbubbles of nitrogen or oxygen appear at blood capillaries, they will block the flow of blood. Such a loss is particularly painful at joints, and any reduction in blood flow to the brain can be extremely serious.

For each atmosphere of pressure above the normal, about 20 minutes of careful decompression is usually recommended. This allows time for the respiratory system to gather and expel excess nitrogen. Excess oxygen is mostly used up by normal metabolism.

FIGURE 1 If deep sea divers return too rapidly to atmospheric pressure they will suffer the "bends."

in a liquid (see Figure 14.17). On the left, we begin with equilibrium; gas molecules come and go at equal rates between the dissolved and undissolved states (Figure 14.17a). Gas molecules enter the solution at a rate proportional to the frequency with which they collide with the surface of the solution. This frequency increases when we increase the pressure of the gas (Figure 14.17b) because the gas molecules are squeezed closer together. More of them now collide with a given surface area of the solution in each second of time. Because liquids and liquid solutions are incompressible, the increased pressure alone has no effect on the frequency with which gas molecules can leave the solution. That means that increasing the pressure favors the forward reaction in Equation 14.1 over the reverse reaction.

As more gas dissolves, however, the forward rate steadily slows and the reverse rate increases. It increases because at a higher concentration of dissolved gas, there are simply more gas molecules in solution at each unit of surface area. Their frequency of escape is proportional to this concentration. But the forward gas-dissolving reaction dominates until enough additional gas has dissolved to equalize the opposing rates. We again have equilibrium (Figure 14.17c); in terms of what species are where, it is not the original equilibrium system, but a new one, because more gas molecules are in solution.

FIGURE 14.17 *How pressure increases the solubility of a gas in a liquid.* (*a*) At some specific pressure, equilibrium exists between the vapor phase and the solution. (*b*) An increase in pressure puts stress on the equilibrium. More gas molecules are dissolving than are leaving the solution. (*c*) More gas has dissolved and equilibrium is restored.

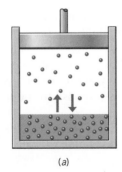

(a)

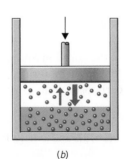

(b)

(c)

For gases *that do not react with the solvent,* there is a simple relationship between gas pressure and gas solubility—the **pressure–solubility law,** usually called **Henry's law,** after William Henry.

William Henry (1775–1836) was an English chemist and physician.

Henry's Law (Pressure–Solubility Law)

The concentration of a gas in a liquid at any given temperature is directly proportional to the partial pressure of the gas over the solution.

$$C_{gas} = k_H P_{gas} \qquad (T \text{ is constant})$$

TOOLS

Henry's law

In the equation, C_{gas} is the concentration of the gas and P_{gas} is the partial pressure of the gas above the solution. The proportionality constant, k_H, called the *Henry's law constant,* is unique to each gas. (Notice that constant temperature is assumed.) The equation is true only at low concentrations and pressures and, as we said, for gases that do not react with the solvent.

An alternate (and commonly used) expression of Henry's law is

$$\frac{C_1}{P_1} = \frac{C_2}{P_2} \tag{14.2}$$

where C_1 and P_1 refer to initial conditions and C_2 and P_2 to final conditions.

Equation 14.2 is true because the two ratios, C_1/P_1 and C_2/P_2, equal the same Henry's law constant, k_H.

At 20 °C the solubility of N_2 in water is 0.0150 g L^{-1} when the partial pressure of nitrogen is 580 torr. What will be the solubility of N_2 in water at 20 °C when its partial pressure is 800 torr?

ANALYSIS: This problem deals with the effect of gas pressure on gas solubility, so Henry's law applies. We use this law in its form given by Equation 14.2 because it lets us avoid having to know or calculate the Henry's law constant.

SOLUTION: Let's gather the data first.

$$C_1 = 0.0150 \text{ g } L^{-1} \qquad C_2 = ?$$
$$P_1 = 580 \text{ torr} \qquad P_2 = 800 \text{ torr}$$

Using Equation 14.2, we have

$$\frac{0.0150 \text{ g } L^{-1}}{580 \text{ torr}} = \frac{C_2}{800 \text{ torr}}$$

Solving for C_2,

$$C_2 = 0.0207 \text{ g } L^{-1}$$

The solubility under the higher pressure is 0.0207 g L^{-1}.

Is the Answer Reasonable?
In relationship to the initial concentration, the size of the answer makes sense because Henry's law tells us to expect a greater solubility at the higher pressure.

PRACTICE EXERCISE 1: How many grams of nitrogen and oxygen are dissolved in 100 g of water at 20 °C when the water is saturated with air? At 1 atm pressure, the solubility of oxygen in water is 0.00430 g O_2/100 g H_2O, and the solubility of nitrogen in water is 0.00190 g N_2/100 g H_2O. In pure, dry air, P_{N_2} equals 593 torr and P_{O_2} equals 159 torr.

EXAMPLE 14.1

Using Henry's Law

Gases with polar bonds in their molecules are more soluble in water

The gases sulfur dioxide, ammonia, and, to a lesser extent, carbon dioxide are far more soluble in water than are oxygen or nitrogen (see Table 14.2). Part of the reason is that SO_2, NH_3, and CO_2 molecules have polar bonds and sites of partial charge that attract water molecules, forming hydrogen bonds to help hold the gases in solution. Ammonia molecules, in addition, not only can accept hydrogen bonds from water (O—H···N) but also can donate them through their N—H bonds (N—H···O).

The more soluble gases also react with water to some extent as the following chemical equilibria show:

$$CO_2(aq) + H_2O \rightleftharpoons H_2CO_3(aq) \rightleftharpoons H^+(aq) + HCO_3^-(aq)$$

$$SO_2(aq) + H_2O \rightleftharpoons H^+(aq) + HSO_3^-(aq)$$

$$NH_3(aq) + H_2O \rightleftharpoons NH_4^+(aq) + OH^-(aq)$$

The forward reactions contribute to the higher concentrations of the gases in solution, as compared to gases such as O_2 and N_2 that do not react with water at all. Gaseous sulfur trioxide is very soluble in water because it reacts quantitatively (i.e., completely) with water to form sulfuric acid.[4]

$$SO_3(g) + H_2O \longrightarrow H_2SO_4(aq)$$

14.5 ▶ Molarity changes with temperature; molality, mass percentages, and mole fractions do not

For the stoichiometry of chemical reactions in solution, *molar concentration* or *molarity,* mol/L, is a convenient unit of concentration. Molarity lets us to measure out moles of a solute simply by measuring the volume of solution. Unfortunately, solution volumes change slightly with temperature. Most solutions expand when heated and contract when cooled. This means that molarity will also change somewhat with temperature. For example, a saline solution prepared at room temperature will have a larger volume and a lower molarity when it is warmed to body temperature.

To describe and compare the *physical properties* of solutions, temperature-insensitive expressions for concentration work best. We have already studied one, the *mole fraction,* or its near relative, the *mole percent* (see Section 11.6). Two other temperature-insensitive concentration units that are in common use are mass fractions (and mass percents) and molal concentrations.

> In a mixture, the mole fraction of component A, X_A, is given by
> $$X_A = \frac{\text{moles of } A}{\text{total moles of all components}}$$

Percent concentrations

Mass percents

Concentrations of solutions are often expressed as a **percent by mass** (sometimes called a *percent by weight*), which gives grams of solute per 100 grams of solution. Percent by mass is indicated by the symbol % (w/w), where the "w" stands for "weight." To calculate a percent by mass from solute and solution masses, we can use the following formula:

$$\text{percent by mass} = \frac{\text{mass of solute}}{\text{mass of solution}} \times 100\%$$

For example, a solution labeled "0.85% (w/w) NaCl" is one in which the ratio of solute to solution is 0.85 g of NaCl to 100 g of NaCl solution.

[4]The "concentrated sulfuric acid" of commerce has a concentration of 93 to 98% H_2SO_4, or roughly 18 *M*. This solution takes up water avidly and very exothermically, removing moisture even out of humid air itself (and so must be kept in stoppered bottles). Because it is dense, oily, sticky, and highly corrosive, it must be handled with extreme care. Safety goggles and gloves must be worn when dispensing concentrated sulfuric acid. When making a more dilute solution, *always pour (slowly with stirring) the concentrated acid into the water.* If water is poured onto concentrated sulfuric acid, it can layer on the acid's surface, because the density of the acid is so much higher than water's (1.8 g mL^{-1} vs. 1.0 g mL^{-1} for water). At the interface, such intense heat can be generated that the steam thereby created could explode from the container, spattering acid around.

The "(w/w)" is sometimes omitted. This is unfortunate, because there are other ways to define percent concentration. Clinical laboratories sometimes report concentrations as **percent by mass/volume,** using the symbol "% (w/v)":

$$\text{percent by mass/volume} = \frac{\text{mass of solute (g)}}{\text{volume of solution (mL)}} \times 100\%$$

For example, a solution that is 4% (w/v) $SrCl_2$ contains 4 g of $SrCl_2$ dissolved in 100 mL of solution. Notice that with a percent by mass, we can use any units for the mass of the solute and mass of the solution, as long as they are the same for both. With percent by mass/volume, we *must* use grams for the mass of solute and milliliters for the volume of solution.

If a percent concentration does not include a (w/v) or (w/w) designation, we assume that it is a percent by mass. The two are very similar for dilute aqueous solutions. One milliliter of solution has a mass of about 1 g, so the percent by mass is roughly equal to the percent by mass/volume. However, when the solvent is not water, or when the solution is concentrated, the two percentages can be significantly different.

Percentages are simply parts per hundred. Other expressions of concentration are *parts per million* (ppm) and *parts per billion* (ppb), where 1 ppm equals 1 g of component in 10^6 g of the mixture and 1 ppb equals 1 g of component in 10^9 g of the mixture.

Seawater is typically 3.5% sea salt and has a density of about 1.03 g/mL. How many grams of sea salt would be needed to prepare enough seawater solution to completely fill a 62.5 L aquarium?

EXAMPLE 14.2

Using Percent Concentrations

ANALYSIS: We need the number of grams of sea salt in 62.5 L of seawater solution. We can write the problem as

$$62.5 \text{ L solution} \Leftrightarrow ? \text{ g sea salt}$$

To solve the problem, we'll need a relationship that links seawater and sea salt. The percent concentration is the link we need. If we assume that the percent is a percent by mass, we can write "3.5% sea salt" as follows:

$$3.5 \text{ g sea salt} \Leftrightarrow 100 \text{ g solution}$$

We now need a link between grams of solution and liters of solution. Density relates the mass of a substance to its volume, so we can write a density of 1.03 g/mL as

$$1.03 \text{ g solution} \Leftrightarrow 1.00 \text{ mL solution}$$

If we convert 62.5 L solution into mL solution, we can use this relationship to obtain grams of solution. We can then use the percent to convert grams of solution into grams of sea salt.

SOLUTION: Writing the conversion factors so that the units cancel properly, we have

$$62.5 \text{ L solution} \times \frac{1000 \text{ mL solution}}{1 \text{ L solution}} \times \frac{1.03 \text{ g solution}}{1 \text{ mL solution}} \times \frac{3.5 \text{ g sea salt}}{100 \text{ g solution}}$$

$$= 2.25 \times 10^3 \text{ g sea salt}$$

Is the Answer Reasonable?
If seawater is about 4% sea salt, 100 g of seawater should contain about 4 g of sea salt, so the mass of the seawater should be about 100/4 or 25 times the mass of the sea salt. If we have 2 kg of sea salt, we should have about 2 kg × 25, or 50 kg of seawater. Since a gram of seawater has a volume of about 1 mL, a kilogram of seawater should have a volume of about a liter. So 2 kg of sea salt should make about 50 L of seawater. This is reasonably close to the volume of the aquarium, so the answer should be correct.

Molal concentration

Molal concentration

The number of moles of solute per kilogram of *solvent* is called the **molal concentration** or the **molality** of a solution. The usual symbol for molality is ***m.***

Be sure to notice that molality is defined per kilogram of *solvent,* not kilogram of solution.

We can, of course, obtain any mass we need by taking the equivalent in volume, using the solvent's density to calculate the volume.

$$\text{molality} = m = \frac{\text{mol of solute}}{\text{kg of solvent}}$$

For example, if we dissolve 0.5000 mol of sugar in 1000 g (1.000 kg) of water, we have a 0.5000 *m* solution of sugar. We would not need a volumetric flask to prepare the solution, because we weigh the solvent. Some important physical properties of a solution are related in a simple way to its molality, as we'll soon see.

Don't confuse *molarity* and *molality.*

$$\text{molality} = m = \frac{\text{mol of solute}}{\text{kg of } solvent} \qquad \text{molarity} = M = \frac{\text{mol of solute}}{\text{L of } solution}$$

As we pointed out at the beginning of this section, the molarity of a solution changes slightly with temperature, but molality does not. Molality is more convenient in experiments involving temperature changes.

When water is the solvent, a solution's molarity approaches its molality as the solution becomes more dilute. In very dilute solutions, 1 L of *solution* is nearly 1 L of water, which has a mass close to 1 kg. Now the ratio of moles/liter (molarity) is very nearly the same as the ratio of moles/kilogram (molality).

EXAMPLE 14.3

Calculation to Prepare a Solution of a Given Molality

An experiment calls for a 0.150 *m* solution of sodium chloride in water. How many grams of NaCl would have to be dissolved in 500.0 g of water to prepare a solution of this molality?

ANALYSIS: *Molality* means moles of solute per kilogram of solvent. The concentration 0.150 *m* means that the solution must contain 0.150 mol of NaCl for every 1 kg (1000 g) of water. We can write

$$0.150 \text{ mol NaCl} \Longleftrightarrow 1000 \text{ g H}_2\text{O}$$

To calculate the number of moles of NaCl we need for 500.0 g of H$_2$O, we use this relationship as a conversion factor. We can then use the molar mass of NaCl to convert moles of NaCl into grams of NaCl.

SOLUTION:

$$500.0 \text{ g H}_2\text{O} \times \frac{0.150 \text{ mol NaCl}}{1000 \text{ g H}_2\text{O}} \times \frac{58.44 \text{ g NaCl}}{1 \text{ mol NaCl}} = 4.38 \text{ g NaCl}$$

When 4.38 g of NaCl is dissolved in 500 g of H$_2$O, the concentration is 0.150 *m* NaCl.

Is the Answer Reasonable?
We'll round the formula mass of NaCl to 60, so 0.1 mol is 6 g and 0.2 mol is 12 g. We also notice that 0.150 *m* is halfway between 0.1 and 0.2 *m.* So for a whole kilogram of water, we'd need halfway between 6 g (0.1 mol) and 12 g (0.2 mol) of NaCl. For half

as much water, or 500 g of water, we cut these limits in two, so we'd need halfway between 3 and 6 g of NaCl. Our answer is in this range.

PRACTICE EXERCISE 4: Water freezes at a lower temperature when it contains solutes. To study the effect of methanol on the freezing point of water, we might begin by preparing a series of solutions of known molalities. Calculate the number of grams of methanol (CH_3OH) needed to prepare a 0.250 m solution, using 2000 g of water.

PRACTICE EXERCISE 5: If you prepare a solution by dissolving 4.00 g of NaOH in 250 g of water, what is the molality of the solution?

Conversions among concentration units

We often need to relate concentrations expressed in different units. The following example shows that all we need to convert concentrations are the definitions of the units.

What is the molality of 10.0% (w/w) aqueous NaCl?

ANALYSIS: First, recall the definitions of molality and percent by mass. The question asks us to go from

$$\frac{10.0 \text{ g NaCl}}{100.0 \text{ g NaCl soln}} \quad \text{to} \quad \frac{? \text{ mol NaCl}}{1 \text{ kg water}}$$

We have two conversions to perform:

- In the numerator, we must convert 10.0 g of NaCl into moles of NaCl. The two can be linked using the molar mass of NaCl, 58.44 g mol^{-1}.
- In the denominator, we must convert 100.0 grams of NaCl *solution* to kilograms of *water*. If we subtract the mass of NaCl (10.0 g) from the mass of the NaCl solution (100.0 g), we see that 100.0 g of NaCl solution contains 90.0 g of water. We can write

$$100.0 \text{ g NaCl soln} \Leftrightarrow 90.0 \text{ g water}$$

We can then convert grams of water to kilograms.

SOLUTION: We write the linking relationships as conversion factors, arranging them so that units cancel properly:

$$\frac{10.0 \text{ g NaCl}}{100.0 \text{ g NaCl soln}} \times \frac{1 \text{ mol NaCl}}{58.44 \text{ g NaCl}} \times \frac{100.0 \text{ g NaCl soln}}{90.0 \text{ g water}} \times \frac{1000 \text{ g water}}{1 \text{ kg water}}$$

$$= 1.90 \text{ } m \text{ NaCl}$$

A 10.0% NaCl solution is also 1.90 molal.

Is the Answer Reasonable?
If you're satisfied that you've calculated the moles of NaCl correctly (about 1/6th mol or about 0.17 mol), then notice that this is dissolved in 10% *less* than 0.1 kg water (90% of 0.1 kg is 0.09 kg). The ratio of 0.17 mol to 0.1 kg is the same as the ratio of 1.7 mol to 1 kg, which aligns the ratio with the definition of molality. So the molality should be about 10% *more* than 1.7, which the answer is.

PRACTICE EXERCISE 6: A certain sample of concentrated hydrochloric acid is 37.0% HCl. Calculate the molality of this solution.

EXAMPLE 14.4
Finding Molality from Mass Percent

Another kind of calculation is to find the molarity of a solution from its mass percent. This cannot be done without the density of the solution, as the next example shows.

EXAMPLE 14.5
**Finding Molarity
from Mass Percent**

A certain supply of concentrated hydrochloric acid has a concentration of 36.0% HCl. The density of the solution is 1.19 g mL^{-1}. Calculate the molar concentration of HCl.

ANALYSIS: Recall the definitions of percent by mass and molarity. We must perform the following conversion:

$$\frac{36.0 \text{ g HCl}}{100 \text{ g HCl soln}} \quad \text{to} \quad \frac{? \text{ mol HCl}}{1 \text{ L HCl soln}}$$

We have two conversions to perform:

- The numerators show that we need a grams-to-moles conversion for the solute, HCl. The link between grams and moles is the molar mass (36.46 g mol^{-1} for HCl).
- The denominators tell us that we must carry out a grams-to-liters conversion for the HCl solution, which is why we need the solution's density. Remember from Chapter 3 that *density is the link between a substance's mass and volume.* We can write

$$1.19 \text{ g HCl soln} \Leftrightarrow 1.00 \text{ mL HCl soln}$$

To complete the conversion we'll have to change milliliters to liters.

When the density of a *solution* is given, "g/mL" always means the ratio of grams of *solution* to milliliters of *solution.*

SOLUTION: We can do the conversions in either order. Here, we've converted the numerator first:

$$\frac{36.0 \text{ g HCl}}{100 \text{ g HCl soln}} \times \frac{1 \text{ mol HCl}}{36.46 \text{ g HCl}} \times \frac{1.19 \text{ g HCl soln}}{1.00 \text{ mL HCl soln}} \times \frac{1000 \text{ mL HCl soln}}{1 \text{ L HCl soln}}$$

$$= 11.8 \text{ } M \text{ HCl}$$

So 36.0% HCl is also 11.8 *M* HCl.

Is the Answer Reasonable?
With HCl having a molar mass of approximately 36 g mol^{-1}, the 36% solution contains about 1 mol of HCl in 100 g of solution. If the density of the solution were the same as water (1 g/mL), then the solution would have 1 mol HCl in 100 mL, which is equivalent to 10 mol HCl in 1000 mL. This would give a molarity of 10 *M*, which isn't too far from our answer, so the answer is probably correct. However, let's look a little more closely at the data. Notice that the density of the solution is about 1.2 g/mL, which is about 20% *more* than pure water. This would make the volume of 1000 g of solution about 20% *less* than 1000 mL. Because the volume appears in the denominator of the molarity calculation, a smaller volume would lead to a larger molarity. Therefore, we would expect the molarity to be about 20% *more* than 10 *M*, or about 12 *M*. That's *very* close to our answer, so we can be confident it's correct.

PRACTICE EXERCISE 7: Hydrobromic acid can be purchased as 40.0% HBr. The density of this solution is 1.38 g mL^{-1}. What is the molar concentration of HBr in this solution?

14.6 ▶ **Substances have lower vapor pressures in solution**

After the Greek *kolligativ,* "depending on number and not on nature."

The physical properties of solutions to be studied in this and succeeding sections are called **colligative properties,** because they depend mostly on the relative *populations* of particles in mixtures, not on their chemical identities. In this section, we examine the effects of solutes on the vapor pressures of solvents in liquid solutions.

All liquid solutions of *nonvolatile solutes* (solutes that can't evaporate from a solution) have lower vapor pressures than their pure solvents. When the solution is dilute (and, always remember, the solute is nonvolatile), a simple law, **Raoult's law,** says that the vapor pressure of the solution, P_{soln}, equals the product of the mole

Francois Marie Raoult (1830–1901) was a French scientist.

fraction of the solvent, $X_{solvent}$, and its vapor pressure when pure, $P°_{solvent}$. In equation form, Raoult's law is expressed as follows.

Raoult's Law equation

$$P_{solution} = X_{solvent} \, P°_{solvent}$$

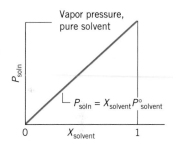

◄ TOOLS ⬛

Raoult's law

Because of the form of this equation, a plot of $P_{solution}$ versus $X_{solvent}$ should be *linear* at all concentrations when the system obeys Raoult's law (see Figure 14.18).

Notice that the mole fraction in Raoult's law refers to the solvent, not the solute. Usually we're more interested in the effect of the *solute's* mole fraction concentration, X_{solute}, on the vapor pressure. We can show that the change in vapor pressure, or ΔP, is directly proportional to the mole fraction of solute, X_{solute}, as follows.[5]

$$\Delta P = X_{solute} \, P°_{solvent} \qquad (14.3)$$

The change in vapor pressure equals the mole fraction of the solute times the solvent's vapor pressure when pure.

Carbon tetrachloride has a vapor pressure of 100 torr at 23 °C. This solvent can dissolve candle wax, which is essentially nonvolatile. Although candle wax is a mixture, we can take its molecular formula to be $C_{22}H_{46}$ (molecular mass 311). What is the vapor pressure at 23 °C of a solution prepared by dissolving 10.0 g of wax in 40.0 g of CCl_4 (molecular mass 154)?

ANALYSIS: This problem involves the vapor pressure of a solution and so we need Equation 14.3. We need the mole fraction of the *solute* to use the equation, so we must first calculate the total number of moles of solute and solvent.

SOLUTION:

For CCl_4 $\qquad$ $40.0 \text{ g } CCl_4 \times \dfrac{1 \text{ mol } CCl_4}{154 \text{ g } CCl_4} = 0.260 \text{ mol } CCl_4$

For $C_{22}H_{46}$ $\qquad$ $10.0 \text{ g } C_{22}H_{46} \times \dfrac{1 \text{ mol } C_{22}H_{46}}{311 \text{ g } C_{22}H_{46}} = 0.0322 \text{ mol } C_{22}H_{46}$

The total number of moles = 0.292 mol

Now, we can calculate the mole fraction of the solute, $C_{22}H_{46}$.

$$X_{C_{22}H_{46}} = \frac{0.0322 \text{ mol}}{0.292 \text{ mol}} = 0.110$$

The amount of the lowering of the vapor pressure of the solute, ΔP, will be this particular mole fraction, 0.110, times the vapor pressure of pure CCl_4 (100 torr), calculated by Equation 14.3.

$$\Delta P = 0.110 \times 100 \text{ torr} = 11.0 \text{ torr}$$

EXAMPLE 14.6

Calculating with Raoult's Law

FIGURE 14.18 *A Raoult's law plot.* When the vapor pressure of a solution is plotted against the mole fraction of the solvent, the result is a straight line.

---

[5]To derive Equation 14.3, note that the value of ΔP is simply the following difference:

$$\Delta P = (P°_{solvent} - P_{solution})$$

The mole fractions for our two-component system, $X_{solvent}$ and X_{solute}, must add up to 1; it's the nature of mole fractions.

$$X_{solvent} = 1 - X_{solute}$$

We now insert this expression for $X_{solvent}$ into the Raoult's law equation.

$$P_{solution} = X_{solvent} \, P°_{solvent}$$

$$P_{solution} = (1 - X_{solute})P°_{solvent} = P°_{solvent} - X_{solute} \, P°_{solvent}$$

So, by rearranging terms,

$$X_{solute} \, P°_{solvent} = (P°_{solvent} - P_{solution}) = \Delta P$$

This result can be rearranged again to give Equation 14.3.

The presence of the wax in the CCl_4 lowers the vapor pressure of the CCl_4 by 11.0 torr (from 100 to 89 torr).

Is the Answer Reasonable?

Raoult's law implies that *the vapor pressure of a solvent in a solution must be less than that of the pure solvent.* Our answer does show a lowering of the vapor pressure of CCl_4 in the wax solution. Now let's apply the Raoult's law equation directly. We must multiply the mole fraction of the *solvent* by the vapor pressure of the solvent. The mole fraction of the solvent must be 1 minus 0.11 (the other mole fraction), or 0.89. Taking 0.89 times 100 torr gives 89 torr, which is what we calculated.

PRACTICE EXERCISE 8: Dibutyl phthalate, $C_{16}H_{22}O_4$ (molecular mass 278), is an oil sometimes used to soften plastic articles. Its vapor pressure is negligible around room temperature. What is the vapor pressure, at 20 °C, of a solution of 20.0 g of dibutyl phthalate in 50.0 g of octane, C_8H_{18} (molecular mass 114)? The vapor pressure of pure octane at 20 °C is 10.5 torr.

How does a nonvolatile solute lower the vapor pressure of the solvent in a solution? Consider the following analogy. Imagine a crowded stadium, filled to capacity. The stadium has separate restrooms for men and women. Women will be entering and leaving the women's restroom at a certain rate. The rate will depend on the number of women in the stadium. If 50% of the people in the stadium are women, we expect the women's restroom to be twice as busy as it would be if only 25% of the people were women. In other words, increasing the concentration of *men* in the stadium lowers the amount of traffic in and out of the women's room. Increasing the number of men in the stadium also lowers the number of women occupying the women's restroom, on average.

In our analogy, the stadium and the women's restroom represent the solution and vapor phases. The men and women represent nonvolatile solute and solvent molecules, respectively. Adding a nonvolatile solute to a solution will lower the evaporation rate of the solvent, just as increasing the fraction of men in the stadium will decrease traffic to the women's restroom. Increasing the concentration of solute lowers the number of solvent molecules in the vapor phase, on average, just as increasing the fraction of men in the stadium lowers the number of women occupying the women's room. If the number of solvent molecules in the vapor phase is lowered, the partial pressure of the solvent above the solution is lowered, as well. The effect is shown in Figure 14.19.

Only ideal solutions obey Raoult's law. The vapor pressures of components in real solutions are sometimes higher or sometimes lower than Raoult's law would predict. These differences between ideal and real solution behavior are quite useful. They give us a simple experimental way to compare the relative strengths of attractions between molecules in solution. When the attractions between unlike molecules are strongest, the experimental vapor pressure will be lower than calculated with Raoult's law. Conversely, when the attractions between identical molecules are strongest, the experimental vapor pressure is higher than Raoult's law would predict.

FIGURE 14.19 *Effect of a nonvolatile solute on the vapor pressure of a solvent.* (*a*) Equilibrium between a pure solvent and its vapor. With a high number of solvent molecules in the liquid phase, the rate of evaporation and condensation is relatively high. (*b*) In the solution, some of the solvent molecules have been replaced with solute molecules. There are fewer solvent molecules available to evaporate from solution. The evaporation rate is lower. When equilibrium is established, there are fewer molecules in the vapor. The vapor pressure of the solution is less than that of the pure solvent.

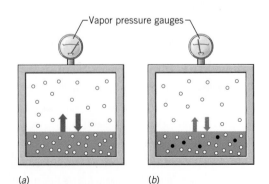

(a)　　(b)

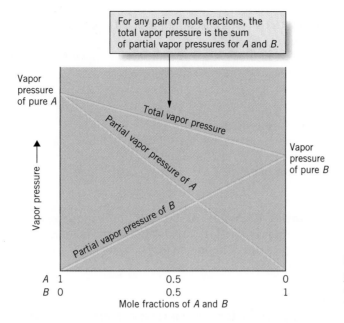

For any pair of mole fractions, the total vapor pressure is the sum of partial vapor pressures for A and B.

FIGURE 14.20 *The vapor pressure of an ideal, two-component solution of volatile compounds.*

How can we predict the vapor pressure of a solution that contains more than one volatile component?

When two (or more) components of a liquid solution can evaporate, the vapor contains molecules of each. Each volatile component contributes its own partial pressure to the total pressure. By Raoult's law, the partial pressure of a particular component is directly proportional to the component's mole fraction in the solution. By Dalton's law of partial pressures, the total vapor pressure will be the sum of the partial pressures. To calculate these partial pressures, we use the Raoult's law equation for each component. When component A is present in a mole fraction of X_A, its partial pressure (P_A) is this fraction of its vapor pressure when pure, namely, P_A°.

$$P_A = X_A P_A^\circ$$

And, by the same argument, P_B, the partial pressure of component B, is

$$P_B = X_B P_B^\circ$$

Remember, P_A and P_B here are the *partial* pressures as calculated by Raoult's law.

The total pressure of the solution of liquids A and B is then, by Dalton's law of partial pressures, the sum of P_A and P_B.

$$P_{\text{total}} = X_A P_A^\circ + X_B P_B^\circ$$

Notice that this equation contains the Raoult's law equation as a special case. If one component, say, component B, is nonvolatile, it has no vapor pressure (P_B° is zero) so the $X_B P_B^\circ$ term drops out, leaving the Raoult's law equation.

Figure 14.20 shows how the vapor pressure of an ideal, two-component solution of volatile compounds changes with composition.

Acetone is a solvent for both water and molecular liquids that do not dissolve in water, like benzene. At 20 °C, acetone has a vapor pressure of 162 torr. The vapor pressure of water at 20 °C is 17.5 torr. Assuming that the mixture obeys Raoult's law, what would be the vapor pressure of a solution of acetone and water with 50.0 mol% of each?

ANALYSIS: To find P_{total} we need to calculate the individual partial pressures and then add them.

EXAMPLE 14.7

Calculating the Vapor Pressure of a Solution of Two Volatile Liquids

SOLUTION: A concentration of 50.0 mol% corresponds to a mole fraction of 0.500, so

$$P_{acetone} = 0.500 \times 162 \text{ torr} = 81.0 \text{ torr}$$

$$P_{water} = 0.500 \times 17.5 \text{ torr} = \underline{8.75 \text{ torr}}$$

$$P_{total} = 89.8 \text{ torr}$$

Is the Answer Reasonable?
The vapor pressure of the solution (89.8 torr) has to be much higher than that of pure water (17.5 torr), because of the volatile acetone, but much less than that of pure acetone (162 torr), because of the high mole fraction of water. The answer seems to be okay.

PRACTICE EXERCISE 9: At 20 °C, the vapor pressure of cyclohexane, a hydrocarbon solvent, is 66.9 torr and that of toluene (another solvent) is 21.1 torr. What is the vapor pressure of a solution of the two at 20 °C when each is present at a mole fraction of 0.500?

14.7 ▶ Solutions have lower freezing points and higher boiling points than pure solvents

Solutes influence more than the vapor pressures of solutions. They also affect boiling and freezing points. The boiling point of a solution of a nonvolatile solute, for example, is higher than that of the solvent itself, and the freezing point of the solution is lower than the pure solvent's freezing point. For aqueous solutions, we can understand why by returning to the phase diagram for water, first studied in Section 12.9.

Aqueous solutions have vapor pressures that are everywhere lower than the vapor pressure of pure water

Figure 14.21*a* shows the phase diagram for pure water by the curves in blue. Remember that the curves are plots of the values of pressures and temperatures at which the various phases are able to be *in equilibrium*. The *triple point* occurs where all three phases are able to exist in equilibrium simultaneously. Two vertical dashed lines in Figure 14.21*a* extend from particular points on the solid–liquid and the vapor–liquid curves, points that correspond to a pressure of 760 torr (1 atm). These vertical lines intersect the temperature axis at the *normal* freezing point and the *normal* boiling point of pure water.

The word *normal* in normal boiling or freezing point means that the system is under 1 atm of pressure when boiling or freezing.

One technology for obtaining salt-free water from the ocean is to freeze seawater, separate the ice crystals (pure water, when washed free of brine), and let the ice melt.

Now let's replace the pure water in the phase diagram by an aqueous solution of a nonvolatile solute (see Figure 14.21*b*). *The solute stays entirely in the liquid phase.* None of its molecules are in the vapor phase, which is pure water vapor. None are in the solid (ice) phase, because the structure of the ice crystals does not admit alien molecules. How does the presence of solute in the liquid affect the phase diagram?

- Neither ice nor water vapor contains solute molecules, so we expect the ice–water vapor boundary line to be unaffected.
- The solute makes the vapor pressure of the solution less than that of pure water. The liquid–gas boundary line for a solution will be consistently lower than the liquid–gas boundary line for pure water, at every temperature. We can read the boiling point of pure water from the blue line and the boiling point of the solution from the red line. At any given pressure, the solution boils at a higher temperature than pure water. For example, at 1 atm (760 torr), pure water boils at 100 °C, and the aqueous solution always boils at a slightly higher temperature than that. The boiling point of the solution

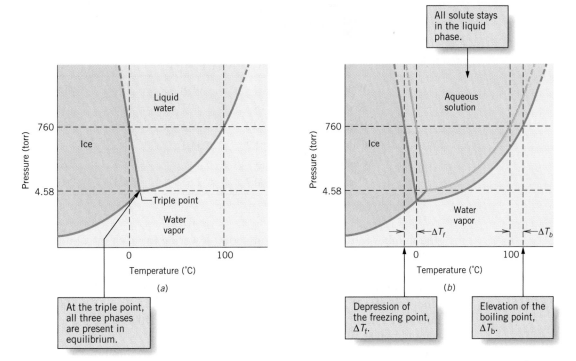

FIGURE 14.21 *Phase diagrams for water and an aqueous solution (not drawn to scale).* (*a*) Phase diagram for pure water. (*b*) Phase diagram for an aqueous solution of a nonvolatile solute.

minus the boiling point of pure water is called the **boiling point elevation.** It is given the symbol ΔT_b.

- Lowering the liquid–gas boundary line forces the triple point to a lower value. The line defining the equilibrium between the solid and liquid phases, which slants upward from the triple point, is pushed to the left. At any given pressure, the freezing point of the solution (on the red line) is lower than the freezing point of pure water (on the blue line). For example, at 760 torr, water freezes at 0 °C and the aqueous solution freezes at a temperature below 0 °C. The freezing point of pure water minus the freezing point of the solution is called the **freezing point depression.** It is given the symbol ΔT_f.

Solutions freeze at lower temperatures and boil at higher temperatures than the pure solvent

Freezing point depression and boiling point elevation are both *colligative properties.* The magnitudes of ΔT_f and ΔT_b are proportional to the relative populations of solute and solvent molecules. Molality (*m*) is the preferred concentration expression (rather than mole fraction or percent by mass) because of the resulting simplicity of the equations relating ΔT to concentration. The following equations work reasonably well only for dilute solutions, however.

$$\Delta T_f = K_f m \tag{14.4}$$

$$\Delta T_b = K_b m \tag{14.5}$$

TOOLS

Freezing point depression; boiling point elevation

K_f and K_b are the proportionality constants and are called, respectively, the **molal freezing point depression constant** and the **molal boiling point elevation constant.** The values of both K_f and K_b are characteristic of each solvent (see Table 14.3). The units of each constant are $°C\ m^{-1}$; the value of K_f for a given solvent corresponds to

	bp (°C)	K_b (°C m^{-1})	mp (°C)	K_f (°C m^{-1})
TABLE 14.3 MOLAL BOILING POINT ELEVATION AND FREEZING POINT DEPRESSION CONSTANTS				
Solvent				
Water	100	0.51	0	1.86
Acetic acid	118.3	3.07	16.6	3.57
Benzene	80.2	2.53	5.45	5.07
Chloroform	61.2	3.63	—	—
Camphor	—	—	178.4	37.7
Cyclohexane	80.7	2.69	6.5	20.0

the number of degrees of freezing point lowering for each molal unit of concentration. The K_f for water is 1.86 °C m^{-1}, and water freezes normally at 0 °C. A 1.00 m solution in water freezes at 1.86 °C *below* the normal freezing point, or at -1.86 °C. A 2.00 m solution should freeze at -3.72 °C. (We say "should," but systems are seldom this ideal.) Similarly, because K_b for water is 0.51 °C m^{-1}, a 1.00 m aqueous solution at 1 atm pressure boils at $(100 + 0.51)$ °C or 100.51 °C, and a 2.00 m solution should boil at 101.2 °C.

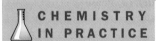

CHEMISTRY IN PRACTICE Water expands when it freezes, enough to crack boulders, concrete, or vehicle radiators. To prevent the latter, motorists dissolve commercial, nonvolatile, permanent-type ethylene glycol antifreezes in water to use as the radiator fluid. The "antifreeze" concentration is made sufficiently high to have a freezing point substantially below that of pure water. In the most northerly tier of inland states as well as in Canada, motorists typically ask for protection to -40 °C, which is approximately achieved with a 50:50 solution, by volume, of the antifreeze.

The ethylene glycol not only lowers the freezing point of the coolant, it also raises the boiling point. In a modern automobile, the radiator solution is kept under pressure, which further raises the boiling point. Both factors combine to allow the engine to run warmer than if plain water were in the radiator, so the chance that the radiator will "boil over" is decreased.

EXAMPLE 14.8

Estimating a Freezing Point Using a Colligative Property

Estimate the freezing point of a solution made from 10.00 g of urea, $CO(NH_2)_2$ (molar mass 60.06 g mol^{-1}), in 100.0 g of water.

ANALYSIS: Equation 14.4 relates concentration to freezing point depression. To use the equation, we must first calculate the molality of the solution.

SOLUTION: For the number of moles of urea,

$$10.00 \text{ g } CO(NH_2)_2 \times \frac{1 \text{ mol } CO(NH_2)_2}{60.06 \text{ g } CO(NH_2)_2} = 0.1665 \text{ mol } CO(NH_2)_2$$

This is the number of moles of $CO(NH_2)_2$ in 100.0 g or 0.1000 kg of water, so the molality is given by

$$\text{molality} = \frac{0.1665 \text{ mol}}{0.1000 \text{ kg}} = 1.665 \ m$$

For our estimate we must next use Equation 14.4. Table 14.3 says that K_f for water is 1.86 °C m^{-1}.

$$\Delta T_f = K_f m = (1.86 \text{ °C } m^{-1})(1.665 \ m)$$

$$= 3.10 \text{ °C}$$

The solution should freeze at 3.10 °C below 0 °C, or at -3.10 °C.

Is the Answer Reasonable?

For every unit of molality, the freezing point must be depressed by about 2 °C. The molality of this solution is between 1 m and 2 m, so we expect the freezing point depression to be between about 2 °C and 4 °C. It is.

PRACTICE EXERCISE 10: At what temperature will a 10% aqueous solution of sugar ($C_{12}H_{22}O_{11}$) boil?

Freezing point depression and boiling point elevation can be used to determine molar masses

We have described the properties of freezing point depression and boiling point elevation as *colligative;* that is, they depend on the relative *numbers* of particles, not on their kinds. Because the effects are proportional to molal concentrations, experimental values of ΔT_f or ΔT_b can be useful for calculating the molar masses of unknown solutes. We'll see how this works in the next example.

A solution made by dissolving 5.65 g of an unknown molecular compound in 110.0 g of benzene froze at 4.39 °C. What is the molar mass of the solute?

ANALYSIS: We can use Equation 14.4 to calculate the number of moles of solute in the given solution. Then we can divide grams of solute by moles of solute to find the molar mass.

SOLUTION: Table 14.3 tells us that the melting point of benzene is 5.45 °C and that the value of K_f for benzene is 5.07 °C m^{-1}. The amount of freezing point depression is

$$\Delta T_f = 5.45 \,°C - 4.39 \,°C = 1.06 \,°C$$

We now use Equation 14.4 to find the molality of the solution.

$$\Delta T_f = K_f m$$

$$\text{molality} = \frac{\Delta T_f}{K_f}$$

$$= \frac{1.06 \,°C}{5.07 \,°C \, m^{-1}} = 0.209 \, m$$

This means that for every kilogram of benzene in the solution, there are 0.209 mol of solute. But we have only 110.0 g or 0.1100 kg of benzene. So the actual number of moles of solute in the given solution is found by

$$0.1100 \text{ kg benzene} \times \frac{0.209 \text{ mol solute}}{1 \text{ kg benzene}} = 0.0230 \text{ mol solute}$$

We can now obtain the molar mass. There are 5.65 g of solute per 0.0230 mol of solute:

$$\frac{5.65 \text{ g}}{0.0230 \text{ mol}} = 246 \text{ g mol}^{-1}$$

The mass of 1 mol of the solute is 246 g.

Is the Answer Reasonable?

Once we're satisfied that the amount of freezing point depression is about 1 °C, we can concentrate on the implication of the molal freezing point depression constant for benzene; we'll round it to 5 °C per molal unit. So a depression that is 1/5th as much must mean that the molal concentration must also be (roughly) 1/5th as much, or about 0.2 mol per kilogram of solvent. In about 1/10th as much solvent, 0.1 kg benzene, there would be 1/10th as much solute, or 0.02 mol. But this amount comes

EXAMPLE 14.9

Calculating a Molar Mass from Freezing Point Depression Data

from a mass of solute between 5 and 6 g. The ratio of grams to moles (the molar mass) is then 5 or 6 divided by 0.02, or between 250 and 300 g mol^{-1}. The answer is close to that range, not within it, because our estimate of the number of moles of solute is low by at least 10% (0.02 versus 0.023).

PRACTICE EXERCISE 11: A solution made by dissolving 3.46 g of an unknown compound in 85.0 g of benzene froze at 4.13 °C. What is the molar mass of the compound?

14.8 ▶ Osmosis is flow of material through a semipermeable membrane due to unequal concentrations

In living things, membranes of various kinds keep mixtures and solutions organized and separated. Yet some substances have to be able to pass through membranes in order that nutrients and products of chemical work can be distributed correctly. These membranes, in other words, must have a selective *permeability*. They must keep some substances from going through while letting others pass. Such membranes are said to be *semipermeable*.

The degree of permeability varies with the kind of membrane. Cellophane, for example, is permeable to water and small solute particles—ions or molecules—but impermeable to very large molecules, like those of starch or proteins. Special membranes can even be prepared that are permeable only to water, not to any solutes.

Depending on the kind of membrane separating solutions of different concentration, two similar phenomena, *dialysis* and *osmosis,* can be observed. Both are functions of the relative populations of the particles of the dissolved materials, so they are colligative properties.

When a membrane is able to let both water and *small* solute particles through, such as the membranes in living systems, the process is called **dialysis,** and the membrane is called a *dialyzing membrane.* It does not permit huge molecules through, like those of proteins and starch.

Osmosis

When a semipermeable membrane will let only solvent molecules get through, this movement is called *osmosis,* and the special membrane needed to observe it is called an *osmotic membrane.* Such membranes are rare, but they can be made.

The *direction* in which osmosis occurs, namely, the net movement of solvent from the more dilute solution (or pure solvent) into the more concentrated solution, may be explained with the help of Figure 14.22 as follows. Water molecules move freely across the membrane. The concentration of water in the aqueous solution (*B*) is lower than the concentration in the pure water (*A*). More water molecules are available to cross the membrane from *A* to *B*, so the flow of water molecules *into* the solution is somewhat greater than the flow *out* of the solution. In Figure 14.22*b*, the net flow of water into the solution has visibly increased the volume of *B*. The phenomenon is **osmosis,** a net shift of solvent (only) through the membrane from the side less concentrated in *solute* to the side more concentrated.

Osmotic pressure

If we push on the solution, we can force water back through the membrane (from *B* to *A*). The weight of the rising fluid column in Figure 14.22*b* provides a push or opposing pressure that eventually stops osmosis. If we apply further pressure, as illustrated in Figure 14.22*c*, we can force enough water back through the membrane to restore the system to its original condition. The exact opposing pressure needed to prevent any osmotic flow *when one of the liquids is pure solvent* is called the **osmotic pressure** of the solution.

Latin *permeare*, "to go through."

Artificial kidney machines use dialyzing membranes to help remove the smaller molecules of wastes from the blood while letting the blood retain its larger protein molecules.

The direction of osmosis is here traditionally defined with reference to the concentration of the *solute*, not the solvent.

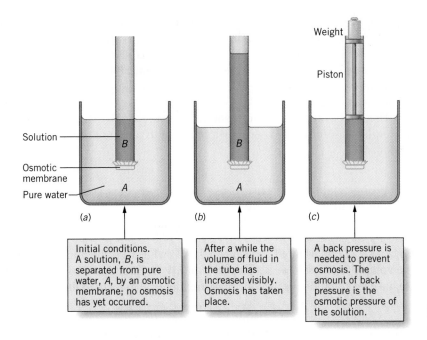

FIGURE **14.22** *Osmosis and osmotic pressure.*

Notice that the term "osmotic pressure" uses the word *pressure* in a novel way. A solution does not "have" a special pressure called osmotic pressure apart from osmosis. What the solution has is a *concentration* that can generate the occurrence of osmosis and the associated osmotic pressure in a special circumstance. Then, in proportion to the solution's concentration, it requires a specific back pressure to prevent osmosis. By *exceeding* this back pressure, osmosis can be reversed. *Reverse osmosis* is widely used to purify seawater.

It might be helpful in understanding osmosis to refer to Figure 14.23, where we have a beaker of water in an enclosed space *already saturated in water vapor,* and we have just added a beaker of an aqueous solution of a nonvolatile solute. This upsets equilibria. Water molecules tend to escape into the vapor phase from each beaker, but their rate of escape from the pure water (larger upward arrow) is greater than their rate of escape from the solution (smaller upward arrow). This must be so because the vapor pressure of pure water is greater than that of the solution. The rate at which water molecules return to the solution from the saturated vapor state is not diminished by the presence of the solute. At the solution–vapor interface, then, more water molecules return to the solution than leave it, and this takes water from the vapor state. To keep the vapor phase saturated, more of the pure water must evaporate. The net effect is that water moves from the beaker of pure water into the vapor

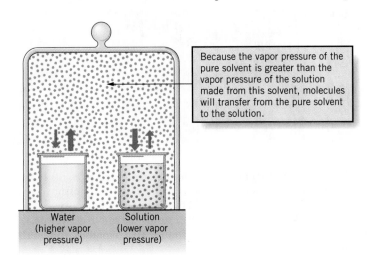

FIGURE **14.23** *Principles at work in osmosis.* Here, beakers instead of a semipermeable membrane separate the pure solvent from its solution, one containing a nonvolatile solute.

space and then down into the solution. In osmosis, an osmotic membrane instead of beakers separates the liquids, but the same principles are at work.

It is not hard to imagine that the higher the concentration of the *solution* is, the more water will transfer to it (referring to Figure 14.23). In other words, the osmotic pressure of a solution is proportional to the relative concentrations of its solute and solvent.

The symbol that's used for osmotic pressure is the Greek capital pi, Π. In a dilute aqueous solution, Π is proportional both to temperature, T, and molar concentration, M:

$$\Pi \propto MT$$

The proportionality constant turns out to be the gas constant, R, so for a *dilute* aqueous solution we can write

$$\Pi = MRT \tag{14.6}$$

Of course, M equals mol/L, which equals n/V, where n is the number of moles and V is the volume in liters. If we replace M with n/V in Equation 14.6, and rearrange terms, we have an equation for osmotic pressure identical in form to the ideal gas law.

Osmotic pressure

$$\Pi V = nRT \tag{14.7}$$

Equation 14.7 is the *van't Hoff equation for osmotic pressure.*

Osmotic pressure is of tremendous importance in biology and medicine. Cells are surrounded with membranes that restrict the flow of salts but allow water to pass through freely. To maintain a constant amount of water, the osmotic pressure of solutions on either side of the cell membrane must be identical. For an example, a solution that is 0.9% NaCl by mass has the same osmotic pressure as the contents of red blood cells, and red blood cells bathed in this solution can maintain their normal water content. The solution is said to be *isotonic* with red blood cells. Blood plasma is an isotonic solution.

If the cell is placed in a solution with a salt concentration higher than the concentration within the cell, osmosis causes water to flow out of the cell. Such a solution is said to be *hypertonic*. The cell shrinks and dehydrates, and eventually dies. This process kills freshwater fish and plants that are washed out to sea.

On the other hand, water will flow into the cell if it is placed into a solution with an osmotic pressure that is much lower than the osmotic pressure of the cell's contents. Such a solution is called a *hypotonic* solution. A cell placed in distilled water, for example, will swell and burst. If you've ever tried to put in a pair of contact lenses with tap water instead of an isotonic saline solution, you've experienced cell damage from a hypotonic solution. The effects of isotonic, hypotonic, and hypertonic solutions on a cell are shown in Figure 14.24.

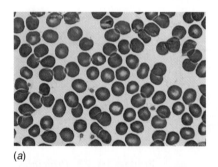

(a)

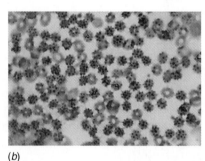

(b)

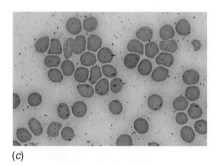

(c)

FIGURE 14.24 *Effects of isotonic, hypertonic, and hypotonic solutions on red blood cells.* (*a*) In an isotonic solution (0.85% NaCl by mass), solutions on either side of the membrane have the same osmotic pressure, and there is no flow of water across the cell membrane. (*b*) In this hypertonic solution (5.0% NaCl by mass), water flows from areas of lower salt concentration (inside the cell) to higher concentration (the hypertonic solution), causing the cell to dehydrate. (*c*) In this hypotonic solution (0.1% NaCl by mass), water flows from areas of lower salt concentration (the hypotonic solution) to higher concentration (inside the cell). The cell swells and bursts.

Obviously, the measurement of osmotic pressure can be very important in preparing solutions that will be used to culture tissues or to administer medicines intravenously. Osmotic pressures can be measured by an instrument called an *osmometer,* illustrated and explained in Figure 14.25. Osmotic pressures can be very high, even in dilute solutions, as Example 14.9 shows.

A very dilute solution, 0.00100 *M* sugar in water, is separated from pure water by an osmotic membrane. What osmotic pressure in torr develops at 25 °C or 298 K?

ANALYSIS: Equation 14.6 applies; recall that $R = 0.0821$ L atm mol^{-1} K^{-1}.

SOLUTION:

$$\Pi = (0.00100 \text{ mol L}^{-1})(0.0821 \text{ L atm mol}^{-1} \text{ K}^{-1})(298 \text{ K})$$

$$= 0.0245 \text{ atm}$$

In torr we have

$$\Pi = 0.0245 \text{ atm} \times \frac{760 \text{ torr}}{1 \text{ atm}} = 18.6 \text{ torr}$$

The osmotic pressure of 0.00100 *M* sugar in water is 18.6 torr.

EXAMPLE 14.10

Calculating an Osmotic Pressure

Is the Answer Reasonable?
A 1 *M* solution should have an osmotic pressure of *RT.* Rounding the gas law constant to 0.08 and the temperature to 300, the osmotic pressure of a 1 *M* solution should be about 24 atm. The osmotic pressure of a 0.001 *M* solution should be 1/1000 of this, or about 0.024 atm, which is very close to what we got before we converted to torr.

PRACTICE EXERCISE 12: What is the osmotic pressure in torr of a 0.0115 *M* glucose solution at body temperature (37 °C)?

In the preceding example, you saw that a 0.0010 *M* solution of sugar has an osmotic pressure of 18.6 torr. This pressure would support a column of the solution roughly 25 cm, or 10 in., high. If the solution had been 100 times as concentrated, 0.100 *M* sugar—still relatively dilute—the height of the column supported would be roughly 25 m, or over 80 ft!

Molar mass can be estimated from osmotic pressure measurements

An osmotic pressure measurement taken of a dilute solution can be used to determine the molar concentration (see Equation 14.6) and so the molecular mass of a molecular solute. Determination of molar mass by osmotic pressure is much more sensitive than determination by freezing point depression or boiling point elevation. The following example shows how experimental measurements of osmotic pressure can be used to determine molar mass.

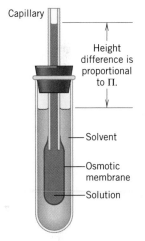

Capillary

Height difference is proportional to Π.

Solvent

Osmotic membrane

Solution

FIGURE 14.25 *Simple osmometer.* When solvent moves into the solution by osmosis, the level of the solution in the capillary rises. The height reached can be related to the osmotic pressure of the solution.

CHEMISTRY IN PRACTICE

Osmotic pressure is a factor (not the only factor) in the rise of sap in trees. Water, of course, can move into a plant from the ground only if the groundwater is more dilute in solutes than the plant sap. If this is not true, as in the overzealous application of fertilizer or the accidental spillage of salt onto the soil, the plant does not get the water it needs. It may even lose water to the soil, wilt, and die.

When agricultural fields are heavily irrigated but poorly drained, minerals in the soil tend to migrate toward the surface and stay there. They can make the growing zone of the soil too salty for crops. Much acreage of agricultural land has been lost to soil salination caused by irrigation without effective drainage. About a quarter of the land in Iraq with the potential for irrigation, for example, in what was once called the "fertile crescent," contains too much salt for crops. Some historians believe that the weakening and collapse of the Sumerian civilization, located in what is now southern Iraq, may have been hastened not just by attacks of outsiders but also by soil salination and subsequent famines.

EXAMPLE 14.11

Calculating Molecular Mass from Osmotic Pressure

An aqueous solution with a volume of 100 mL and containing 0.122 g of an unknown molecular compound has an osmotic pressure of 16.0 torr at 20.0 °C. What is the molar mass of the solute?

ANALYSIS: A molar mass is the ratio of grams to moles. We're given the number of *grams* of the solute, so we need to find the number of *moles* equivalent to this mass in order to compute the ratio.

To find the number of *moles* of solute we must first use Equation 14.6 to find the moles-per-liter concentration of the solution, its *molarity*. Remember that to use the gas constant ($R = 0.0821$ L atm mol^{-1} K^{-1}), we need P in atm and T in kelvins. Then we can use the given volume of the solution (in liters) and the molarity to calculate the number of moles of solute that are represented by 0.122 g of solute. Finally, we can calculate the molar mass by dividing grams of solute by moles of solute.

SOLUTION: First, the solution's molarity, M, is calculated using the osmotic pressure. This pressure (Π), 16.0 torr, corresponds to 0.0211 atm. The temperature, 20.0 °C, corresponds to 293 K (273 + 20.0). Using Equation 14.6,

$$16.0 \text{ torr} \times \frac{1 \text{ atm}}{760 \text{ torr}} = 0.0211 \text{ atm}$$

$$\Pi = MRT$$

$$0.0211 \text{ atm} = (M)(0.0821 \text{ L atm mol}^{-1}\text{ K}^{-1})(293 \text{ K})$$

$$M = 8.77 \times 10^{-4} \frac{\text{mol solute}}{\text{L soln}}$$

The 0.122 g of solute is in 100 mL or 0.100 L of solution, not in a whole liter. So the number of moles of solute corresponding to 0.122 g is found by

$$\text{moles of solute} = (0.100 \text{ L soln})\left(8.77 \times 10^{-4} \frac{\text{mol solute}}{\text{L soln}}\right)$$

$$= 8.77 \times 10^{-5} \text{ mol}$$

The molar mass of the solute is the number of grams of solute per mole of solute:

$$\text{molar mass} = \frac{0.122 \text{ g}}{8.77 \times 10^{-5} \text{ mol}} = 1.39 \times 10^{3} \text{ g mol}^{-1}$$

Is the Answer Reasonable?
First, about all you can do is double-check the calculation of the number of moles, which turns out to be a little less than 10×10^{-5} mol, or 10^{-4} mol. Dividing the mass, roughly 0.12 g, by 10^{-4} mol gives 0.12×10^{4} g per mol, or 1.2×10^{3} g per mol, which is very roughly what we found.

PRACTICE EXERCISE 13: A solution of a carbohydrate prepared by dissolving 72.4 mg in 100 mL of solution has an osmotic pressure of 25.0 torr at 25.0 °C. What is the molecular mass of the compound?

14.9 ▶ **Ionic solutes affect colligative properties differently than nonionic solutes**

The molal freezing point depression constant for water is 1.86 °C m^{-1}. So you might think that a 1.00 m solution of NaCl would freeze at −1.86 °C. Instead, it freezes at −3.37 °C. This greater depression of the freezing point by the salt, which is almost twice 1.86 °C, is not hard to understand if we remember that colligative properties depend on the concentrations of *particles*. We know that NaCl(s) dissociates into ions in water.

$$\text{NaCl}(s) \longrightarrow \text{Na}^{+}(aq) + \text{Cl}^{-}(aq)$$

If the ions are truly separated, 1.00 m NaCl actually has a concentration in dissolved solute particles of 2.00 m, twice the given molal concentration. Theoretically, a 1.00 m NaCl should freeze at $2 \times (-1.86 \ °C)$ or $-3.72 \ °C$. (Why it actually freezes a little higher than this, at $-3.37 \ °C$, will be discussed shortly.)

If we made up a solution of 1.00 m $(NH_4)_2SO_4$, we would have to consider the following dissociation:

$$(NH_4)_2SO_4(s) \longrightarrow 2NH_4^+(aq) + SO_4^{2-}(aq)$$

One mole of $(NH_4)_2SO_4$ can give a total of 3 mol of ions (2 mol of NH_4^+ and 1 mol of SO_4^{2-}). We would expect the freezing point of a 1 m solution of $(NH_4)_2SO_4$ to be $3 \times (-1.86 \ °C) = -5.58 \ °C$.

When we want to roughly *estimate* a colligative property of a solution of an electrolyte, we recalculate the solution's molality using an assumption about the way the solute dissociates or ionizes.

We say "estimate" because the calculations can only give rough values when the simple assumption of 100% dissociation is used.

Estimate the freezing point of aqueous 0.106 m $MgCl_2$, assuming that it dissociates completely.

ANALYSIS: The equation that relates a change in freezing temperature to molality is Equation 14.4.

$$\Delta T_f = K_f m$$

From Table 14.3, K_f for water is 1.86 $°C \ m^{-1}$. We cannot simply use 0.106 m as the molarity because $MgCl_2$ is an ionic compound that dissociates in water. We must use the total molality of ions in the equation.

SOLUTION: When $MgCl_2$ dissolves in water it breaks up as follows:

$$MgCl_2(s) \longrightarrow Mg^{2+}(aq) + 2Cl^-(aq)$$

Because 1 mol of $MgCl_2$ gives 3 mol of ions, the effective (assumed) molality of what we have called 0.106 m $MgCl_2$ is three times as great.

$$\text{Effective molality} = (3)(0.106 \ m) = 0.318 \ m$$

Now we can use Equation 14.4.

$$\Delta T_f = (1.86 \ °C \ m^{-1})(0.318 \ m)$$
$$= 0.591 \ °C$$

The freezing point is depressed below 0.000 $°C$ by 0.591 $°C$, so we calculate that this solution freezes at $-0.591 \ °C$.

Is the Answer Reasonable?
The molality, as recalculated, is roughly 0.3, so 3/10th of 1.86 (call it 2) is 0.6, which, after we add the unit, $°C$, and subtract from 0 $°C$ gives $-0.6 \ °C$, which is close to the answer.

PRACTICE EXERCISE 14: Calculate the freezing point of aqueous 0.237 m LiCl on the assumption that it is 100% dissociated. Calculate what its freezing point would be if the percent dissociation were 0%.

EXAMPLE 14.12
Estimating the Freezing Point of a Salt Solution

Percent ionization can be estimated from colligative property measurements

Experiments show that neither the 1.00 m NaCl nor the 1.00 m $(NH_4)_2SO_4$ solution described earlier in this section freezes quite as low as calculated. Our assumption that an electrolyte separates *completely* into its ions is incorrect. Some oppositely charged ions exist as very closely associated pairs, called *ion pairs,* which behave as single "molecules." Clusters larger than two ions probably also exist. The formation of ion pairs and clusters makes the actual *particle* concentration in a 1.00 m NaCl solution somewhat less than 2.00 m. As a result, the freezing point depression of 1.00 m NaCl is not quite as large as calculated on the basis of 100% dissociation.

"Association" is the opposite of dissociation. It is the coming together of particles to form larger particles.

Jacobus van't Hoff (1852–1911), a Dutch scientist, won the first Nobel prize in chemistry (1901).

As solutions of electrolytes are made more and more *dilute,* the observed and calculated freezing points come closer and closer together. At greater dilutions, the association of ions is less and less a complication because the ions can be farther apart. So the solutes behave more and more as if they were 100% separated into their ions at ever higher dilutions.

Chemists compare the degrees of dissociation of electrolytes at different dilutions by a quantity called the **van't Hoff factor, *i*.** It is the ratio of the observed freezing point depression to the value calculated on the assumption that the solute dissolves as a nonelectrolyte.

$$i = \frac{(\Delta T_f)_{\text{measured}}}{(\Delta T_f)_{\text{calcd as a nonelectrolyte}}}$$

The hypothetical van't Hoff factor, i, is 2 for NaCl, KCl, and $MgSO_4$, which break up into two ions on 100% dissociation. For K_2SO_4, the theoretical value of i is 3 because one K_2SO_4 unit gives three ions. The actual van't Hoff factors for several electrolytes at different dilutions are given in Table 14.4. Notice that with decreasing concentration (that is, at higher dilutions) the experimental van't Hoff factors agree better with their corresponding hypothetical van't Hoff factors.

The increase in the percentage of complete dissociation that comes with greater dilution is not the same for all salts. In going from concentrations of 0.1 to 0.001 m, the increase in percentage dissociation of KCl, as measured by the change in i, is only about 7%. But for K_2SO_4, the increase for the same dilution is about 22%, a difference caused by the anion, SO_4^{2-}. It has twice the charge as the anion in KCl and so the SO_4^{2-} ion attracts K^+ more strongly than can Cl^-. Hence, letting an ion of 2− charge and an ion of 1+ charge get farther apart by dilution has a greater effect on their acting independently than giving ions of 1− and 1+ charge more room. When *both* cation and anion are doubly charged, the improvement in percent dissociation with dilution is even greater. We can see from Table 14.4 that there is an almost 50% increase in the value of i for $MgSO_4$ as we go from a 0.1 to a 0.001 m solution.

Estimation of percent dissociation from freezing point depression data

In 1.00 m aqueous acetic acid (a weak acid), the following equilibrium exists:

$$HC_2H_3O_2(aq) \rightleftharpoons H^+(aq) + C_2H_3O_2^-(aq)$$

The solution freezes at −1.90 °C, which is only a little lower than it would be expected to freeze (−1.86 °C) if no ionization occurred. Evidently some ionization, but not much, has happened. We can estimate the percent ionization by the following calculation.

We first use the data to calculate the *apparent molality* of the solution, that is, the molality of all dissolved species, $HC_2H_3O_2 + C_2H_3O_2^- + H^+$. We again use

TABLE 14.4	VAN'T HOFF FACTORS VERSUS CONCENTRATION			
	van't Hoff Factor, *i*			
	Molal Concentration (mol salt/kg water)			If 100% Dissociation Occurred
Salt	0.1	0.01	0.001	
NaCl	1.87	1.94	1.97	2.00
KCl	1.85	1.94	1.98	2.00
K_2SO_4	2.32	2.70	2.84	3.00
$MgSO_4$	1.21	1.53	1.82	2.00

Equation 14.4 ($\Delta T_f = K_f m$), only now letting m stand for the apparent molality; K_f is 1.86 °C m^{-1}.

$$\text{apparent molality} = \frac{\Delta T_f}{K_f} = \frac{1.90 \, °C}{1.86 \, °C \, m^{-1}}$$

$$= 1.02 \, m$$

If there is 1.02 mol of all solute species in 1 kg of solvent, we have 0.02 mol of solute species more than we started with, because we began with 1.00 mol of acetic acid. Since we get just one *extra* particle for each acetic acid molecule that ionizes, the extra 0.02 mol of particles must have come from the ionization of 0.02 mol of acetic acid. So the percent ionization is

$$\text{percent ionization} = \frac{\text{mol of acid ionized}}{\text{mol of acid available}} \times 100\%$$

$$= \frac{0.02}{1.00} \times 100\% = 2\%$$

In other words, the percent ionization in 1.00 m acetic acid is estimated to be 2% by this procedure. (Other kinds of measurements offer better precision, and the percent ionization of acetic acid at this concentration is actually less than 1%.)

Colligative properties provide evidence for clustering of solute particles

Some molecular solutes produce *weaker* colligative effects than their molal concentrations would lead us to predict. These unexpectedly weak colligative properties are often evidence that solute molecules are clustering or **associating** in solution. For example, when dissolved in benzene, benzoic acid molecules associate as **dimers.** They are held together by hydrogen bonds, indicated by the dotted lines in the following:

Di- signifies "two," so a dimer is the result of the combination of two single molecules.

benzoic acid   benzoic acid dimer

C_6H_5- means

Because of association, the depression of the freezing point of a 1.00 m solution of benzoic acid in benzene is only about one-half the calculated value. By forming a dimer, benzoic acid has an effective molecular mass that is twice as much as normally calculated. The larger effective molecular mass reduces the molal concentration of particles by half, and the effect on the freezing point depression is reduced by one-half.

14.10 ▶ Colloids are molecular assemblies of submicrometer dimensions, suspended in a solvent

As we mentioned at the beginning of this chapter, the *sizes* of particles, quite apart from their relative numbers or identities, affect the physical properties of homogeneous mixtures. If we mix powdered sand with water, for example, and shake the mixture constantly, it continues to be somewhat homogeneous, but the sand particles soon settle when we stop shaking it. Nothing like this occurs when we dissolve salt or sugar in water. Small ions and molecules do not settle out of solution. Their weights are too small for gravity to overcome the mixing influence of their kinetic motions.

The shaken mixture of sand and water is an example of a **suspension.** To form suspensions, particles must have at least one dimension larger than about 1000 nm

	TABLE 14.5	COLLOIDAL SYSTEMS	
Type	Dispersed Phase[a]	Dispersing Medium[b]	Common Examples
Foam	Gas	Liquid	Soapsuds, whipped cream
Solid foam	Gas	Solid	Pumice, marshmallow
Liquid aerosol	Liquid	Gas	Mist, fog, clouds, liquid air pollutants
Emulsion	Liquid	Liquid	Cream, mayonnaise, milk
Solid emulsion	Liquid	Solid	Butter, cheese
Smoke	Solid	Gas	Dust, particulates in smog, aerogels
Sol	Solid	Liquid	Starch in water, paint, jellies[c]
Solid sol	Solid	Solid	Alloys, pearls, opals

[a]The colloidal particles constitute the *dispersed phase.*

[b]The continuous matter into which the colloidal particles are dispersed is called the *dispersing medium.*

[c]Sols that become semisolid or semirigid, like gelatin dessert and fruit jellies, are called *gels.*

(1 μm). Suspensions that don't settle at once, like powdered sand in water, can nearly always be separated by filtration using ordinary filter paper or an ordinary laboratory centrifuge. The high-speed rotor of a centrifuge creates a strong gravitation-like force sufficient to pull the suspended material down through the fluid.

A **colloidal dispersion**—often simply called a **colloid**—is a mixture in which the dispersed particles have at least one dimension in the range of 1 to 1000 nm (see Table 14.5). Colloidally dispersed particles are generally too small to be trapped by ordinary filter paper.

The tendency of colloidal dispersions in a fluid state not to separate is aided by the collisions that the dispersed particles experience from the constantly moving molecules of the "solvent." In Chapter 2, we saw that tiny particles such as pollen grains move erratically in water due to uneven bombardment by water molecules. This erratic movement of colloidally dispersed particles is called **Brownian movement** after a Scottish botanist, Robert Brown (1773–1858). He first noticed it when he used a microscope to study pollen grains in water. Particles in suspension, of course, experience the same bumping around, but they are too large to be kept in suspension without outside help, such as by stirring or shaking the mixture.

Colloidal dispersions that do eventually separate are those in which the dispersed particles, over time, grow too large. For example, you can make a colloidal dispersion of olive oil in vinegar—a salad dressing—by vigorously shaking the mixture. Shaking breaks at least some of the oil into microdroplets having sizes in the right range for colloidal dispersions. The microdroplets, however, eventually coalesce and the oil separates. Evidently, to prepare a *stable* colloidal dispersion, we must not only make the dispersed particles initially small enough but must also keep them from joining together.

The dispersed particles will not coalesce if they carry the same kind of electrical charge, either all positive or all negative (see Figure 14.26*a*). (Ions of the opposite charge are in the solvent, keeping the whole system electrically neutral.) Some of the most stable dispersions form when the surfaces of their colloidal particles have preferentially attracted ions of just one kind of charge from a dissolved salt. The dispersed particles of most **sols,** which are colloidal dispersions of solids in a fluid, are examples. Alternatively, dispersions may form when extremely large, like-charged ions, such as those of proteins, are involved (see Figure 14.26*b*).

Colloidal dispersions of one liquid in another are called **emulsions.** They are often relatively stable provided that a third component called an **emulsifying**

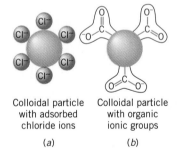

Colloidal particle with adsorbed chloride ions

(*a*)

Colloidal particle with organic ionic groups

(*b*)

FIGURE 14.26 *Stabilizing colloidal dispersions.* Colloidal particles often bear electrical charges that stabilize the dispersion, preventing the particles from joining together. (*a*) The colloidal particle has attracted chloride ions to itself. (*b*) A particle whose extremely large molecules carry negatively charged groups. In both cases, these colloidal particles cannot join together, grow larger, and eventually separate from the dispersing medium.

FIGURE **14.27** *An oil-in-water emulsion.*

agent is also present. Mayonnaise, for example, is an oil-in-water emulsion (see Figure 14.27) of an edible oil in a dilute solution of an edible organic acid. The emulsifying agent is lecithin, a component of egg yolk. Its molecules act to give an electrically charged surface to each microdroplet of the oil, which keeps the microdroplets from coalescing. In homogenized milk, microparticles of butterfat are coated with the milk protein casein. Water-in-oil emulsions are also possible (Figure 14.28). Margarine is an example in which a soybean product is the emulsifying agent.

Starch has very large molecules and forms a colloidal dispersion in water. Usually, colloidal starch mixtures have a milky look or can even be opaque, because the colloidal particles are large enough to reflect and scatter visible light. Even when a beam of light is focused on a starch dispersion so dilute as to look as clear as water, the path of the beam is revealed by the light scattered to the side (see Figure 14.29). Light scattering by colloidal dispersions is called the **Tyndall effect,** after John Tyndall (1820–1893), a British scientist. Solutes in true solutions, however, involve species too small to scatter light, so solutions do not give the Tyndall effect.

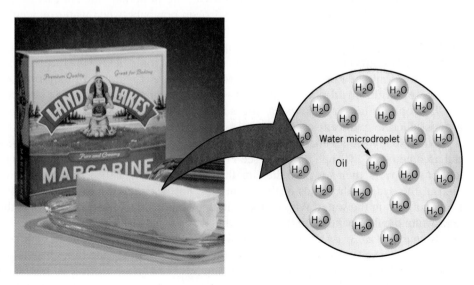

FIGURE **14.28** *A water-in-oil emulsion.*

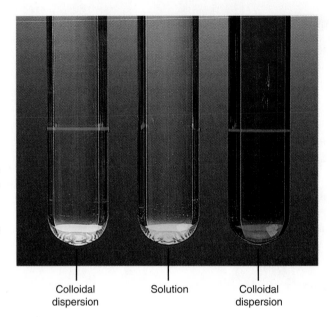

Figure 14.29 *The Tyndall effect, light scattering by colloidal dispersions.* The tube on the left contains a colloidal starch dispersion, and the tube on the right has a colloidal dispersion of Fe_2O_3 in water. The middle tube has a solution of Na_2CrO_4, a colored solute. The red laser light is partly scattered in the two colloidal dispersions so it can be seen, but it passes through the middle solution unchanged.

Colloidal dispersion · Solution · Colloidal dispersion

How soaps and detergents work

The "like dissolves like" rule helps explain how soaps and detergents work. Oily and greasy matter constitute the "glue" that binds soil to fabrics or skin. Hot water alone does a poor job of removing such dirt-binding materials, but even a small amount of detergent makes a cleansing difference.

All detergents have a common feature, whether they are ionic or molecular. Their structures include a long, nonpolar, hydrocarbon "tail" holding a very polar or ionic "head." A typical system in ordinary soap, for example, has the following anion. (The charge-balancing cation is Na^+ in most soaps.)

In synthetic detergents, the hydrophilic group is not

$$-\overset{\overset{O}{\|}}{C}-O^- \quad \text{but} \quad -\overset{\overset{O}{\|}}{\underset{\underset{O}{\|}}{S}}-O^-.$$

$$CH_3CH_2CH_2CH_2CH_2CH_2CH_2CH_2CH_2CH_2CH_2CH_2CH_2CH_2CH_2CH_2CH_2C\overset{\overset{O}{\|}}{O}^-$$

hydrocarbon tail · anionic head

Like all hydrocarbons, the tail of this molecule is nonpolar. The tail slips easily into nonpolar solvents like oil and grease. However, the nonpolar tail can't hydrogen bond with water. The network of strong hydrogen bonds between water molecules tends to prevent the tail from mixing freely with water. Because the tail appears to avoid water, we say that the tail is **hydrophobic** (water fearing). The entire ion is dragged into solution in hot water only because of its ionic head, which can be hydrated. The ionic head is said to be **hydrophilic** (water loving). To minimize contacts between the hydrophobic tails and water and to maximize contacts between the hydrophilic heads and water, the soap's anions gather into colloidal-sized groups called soap **micelles** (see Figure 14.30).

When soap micelles encounter a greasy film, their anions find something else that can accommodate hydrophobic tails (see Figure 14.31*a*). The tails of innumerable anions of soap work their way into the film; the tails, in a sense, dissolve in the grease or oil. But the hydrophilic heads of the anions will not be dragged into the grease; the heads stay out in the water. With a little agitating help from the washing machine, the greasy film breaks up into countless tiny droplets *each pincushioned by projecting, negatively charged heads*, as shown in Figure 14.31*b*. Because the droplets, now colloidally dispersed, all have the same kind of charge, they repel each other and cannot again coalesce. For all practical purposes, the greasy film has been dissolved, and we send it on its way to be destroyed by bacteria at wastewater treatment plants or in the ground.

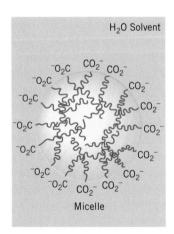

H_2O Solvent

Micelle

Figure 14.30 *Soap micelles.* The formation of soap micelles is driven by the water-avoiding properties of the hydrophobic tails of soap units and the water-attracting properties of their hydrophilic heads.

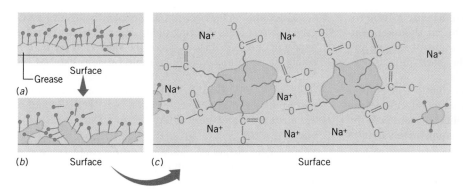

Figure 14.31 *How soap loosens grease that might be coating a cloth surface.* (*a*) The hydrophobic tails of soap anions work their way into the grease layer. The hydrophilic heads (represented by the solid dots) stay exposed to water. (*b*) The grease layer breaks up. (*c*) In an enlarged view, microdroplets of grease form, each pincushioned and made like-charged by soap anions. The micro-droplets are now easily poured out with the wash water.

SUMMARY

Solutions In **solutions,** the dissolved particles have dimensions on the order of 0.05 to 0.25 nm (sometimes up to 1 nm). Solutions are fully homogeneous. They cannot be separated by filtration, and they remain physically stable indefinitely. (They do not display the *Tyndall effect* or *Brownian motion,* either.)

When liquid solutions form by physical means, generally polar or ionic solutes dissolve well in polar solvents, like water. Nonpolar, molecular solutes dissolve well in nonpolar solvents. These observations are behind the **"like dissolves like" rule.** Nature's driving force for establishing the more statistically probable mixed state plus intermolecular attractions are the major factors in the formation of a solution. When both solute and solvent are nonpolar, nature's tendency toward the more probable mixed state dominates because intermolecular attractive forces are weak. Ion–dipole attractions and the **solvation** or **hydration** of dissolved species are major factors in forming solutions of ionic or polar solutes in something polar, like water.

Heats of Solution The molar enthalpy of solution, also called the **heat of solution,** is the net of the **lattice energy** and the **solvation energy** (or, when water is the solvent, the **hydration energy**). The lattice energy is the increase in the potential energy of the system required to separate the molecules or ions of the solute from each other. The solvation energy corresponds to the potential energy lowering that occurs in the system from the subsequent attractions of these particles to the solvent molecules as intermingling occurs.

When liquids dissolve in liquids, an **ideal solution** forms if the potential energy increase needed to separate the molecules equals the energy lowering as the separated molecules come together. Usually, some net energy exchange occurs, however, and few solutions are ideal.

When a gas dissolves in a gas, the energy cost to separate the particles is virtually zero, because they're already separated. They remain separated in the gas mixture, so the net enthalpy of solution is very small if not zero. When a gas dissolves in an organic solvent, the solution process is often endothermic. When a gas dissolves in water, the solution process is often exothermic.

Pressure and Gas Solubility At pressures not too much different from atmospheric pressure, the solubility of a gas in a liquid is directly proportional to the partial pressure of the gas—**Henry's law.**

Concentration Expressions To study and use colligative properties, a solution's concentration ideally is expressed either in mole fractions or as a **molal concentration,** or **molality.** Such expressions provide a temperature-independent description of a concentration that, in the final analysis, gives a ratio of solute to solvent particles. The **molality** of a solution, *m,* is the ratio of the number of moles of solute to the kilograms of *solvent* (not solution, but solvent).

Mass fractions and **mass percents** are concentration expressions used often for reagents when direct information about number of moles of solutes is unimportant.

Colligative Properties Colligative properties are those that depend on the ratio of the particles of the solute to the molecules of the solvent. These properties include the **lowering of the vapor pressure,** the **depression of a freezing point,** the **elevation of a boiling point,** and the **osmotic pressure** of a solution.

According to **Raoult's law,** the vapor pressure of a solution of a nonvolatile (molecular) solute is the vapor pressure of the pure solvent times the mole fraction of the solvent. An alternative expression of this law is that the change in vapor pressure caused by the solute equals the mole fraction of the solute times the pure solvent's vapor pressure. When the components of a solution are volatile liquids, then Raoult's law calculations find the *partial pressures* of the vapors of the individual liquids. The sum of the partial pressures equals the total vapor pressure of the solution.

An **ideal solution** is one that would obey Raoult's law exactly and in which all attractions between molecules are equal. An ideal solution has $\Delta H_{soln} = 0$. Solutions of liquids that form exothermically usually display negative deviations from Raoult's law. Those that form endothermically have positive deviations.

In proportion to its molal concentration, a solute causes a **freezing point depression** and a **boiling point elevation.** The proportionality constants are the *freezing point depression constant* and the *boiling point elevation constant,* and they differ from solvent to solvent. Freezing point and boiling point data from a solution made from known masses of both solute and solvent can be used to calculate the molecular mass of a solute.

When a solution is separated from the pure solvent (or from a less concentrated solution) by an *osmotic membrane,* **osmosis** occurs, which is the net flow of solvent into the more concentrated solution. The back pressure required to prevent osmosis is called the solution's **osmotic pressure,** which, in a dilute aqueous solution, is proportional to the product of the Kelvin temperature and the molar concentration. The proportionality constant is *R*, the ideal gas constant, and the equation relating these variables is $\Pi = MRT$.

When the membrane separating two different solutions is a *dialyzing membrane,* it permits **dialysis,** the passage not only of solvent molecules but also of small solute molecules or ions. Only very large molecules are denied passage.

Colligative Properties of Electrolytes Because an electrolyte releases more ions in solution than indicated by the molal (or molar) concentration, a solution of an electrolyte has more pronounced colligative properties than a solution of a molecular compound at the same molality. The dissociation of a strong electrolyte approaches 100% in very dilute solutions, particularly when both ions are singly charged. Weak electrolytes provide solutions more like those of nonelectrolytes.

The ratio of the value of a particular colligative property, like ΔT_f for a freezing point depression, to the value expected under no dissociation is the **van't Hoff factor, *i*.** A solute whose formula unit breaks into two ions, like NaCl or $MgSO_4$, would have a van't Hoff factor of 2 if it were 100% dissociated. Observed van't Hoff factors approach those corresponding to 100% dissociation only as the solutions are made more and more dilute.

Other Homogeneous Mixtures Relative particle sizes and homogeneity distinguish among the three simplest kinds of mixtures—**suspensions, colloidal dispersions,** and **solutions.** Suspensions have particles with at least one dimension larger than 1000 nm, they are not homogeneous, and they settle out spontaneously. They can be separated by filtration.

Colloidal Dispersions In colloidal dispersions, the dispersed particles have one or more dimensions in the range of 1 to 1000 nm. They are considered to be homogeneous, but we're at a borderline. They separate more slowly than suspensions in response to gravity, but many can be kept indefinitely if the dispersed particles bear like charges or are coated with an emulsifying agent. Colloidal particles can scatter light **(Tyndall effect),** can display **Brownian motion,** and cannot ordinarily be separated by filtration.

TOOLS **YOU HAVE LEARNED**

The table below lists the tools you have learned in this chapter that are applicable to problem solving. Review them if necessary, and refer to them when working on the Thinking-It-Through problems and the Review Exercises that follow.

TOOL	HOW IT WORKS
"Like dissolves like" rule (page 600)	Using chemical composition and structure, the rule predicts whether two substances can form a solution.
Henry's law (page 613) $C_{gas} = k_H P_{gas}$	Calculates the solubility of a gas at a given pressure from its solubility at another pressure.
Mass fraction; mass percent (page 614) $\%(w/w) = \dfrac{\text{mass of solute}}{\text{mass of solution}} \times 100\%$	Calculates the mass of a solution that delivers a given mass of solute.
Molal concentration (page 616) $\text{molality} = \dfrac{\text{moles of solute}}{\text{kg of solvent}}$	Provides a temperature-independent concentration expression for use with colligative properties.
Raoult's law (page 619) $P_{solution} = X_{solvent} P°_{solvent}$	Calculates the effect of a solute on the vapor pressure of a solution.
Equations for freezing point depression and boiling point elevation (page 623) $\Delta T_f = K_f m$ $\Delta T_b = K_b m$	Enables us to estimate freezing and boiling points or molecular masses. Enables us to estimate percent dissociation of a weak electrolyte (page 632).
Equation for osmotic pressure (page 628) $\Pi V = nRT$	We use this equation to estimate the osmotic pressure of a solution or to calculate molecular masses from osmotic pressure and concentration data (page 630).

THINKING IT THROUGH

The goal for the following problems is not to find the answers themselves, but rather to assemble the information needed to solve them and explain how you would use the information to find the answers. The problems in Level 2 are more challenging than those in Level 1 and may contain more data than are required, in which case you are also asked to identify the unnecessary data. Detailed answers to the Thinking-It-Through problems can be found on the web site.

Need extra help?
Visit the Brady/
Senese web site at
www.wiley.com/
college/brady

ON-LINE HELP

Level 1 Problems

1. The solubility of a gas in water at 20 °C is 0.0176 g L^{-1} at 681 torr. How would you calculate its solubility at 0.989 atm at 20 °C?

2. A solution of sulfuric acid is 10.0% H_2SO_4. How many grams of this solution are needed to provide 0.100 mol of H_2SO_4? Explain the calculation. If the question had asked for the number of *milliliters* of the solution instead of the number of grams, what additional data about the solution would be needed, and how would they be used?

3. For an experiment involving osmosis, you need to know the molality of a 12.5% (w/w) solution of sugar ($C_{12}H_{22}O_{11}$) in water. How can this be calculated?

4. The mole fraction of NaCl in an aqueous solution (with no other component) is 0.0100. How would you calculate the mass percent of NaCl in the solution?

5. A solution of KNO_3 has a concentration that is 0.750 *m*. How would you calculate the mole fraction of KNO_3 in the solution?

6. The stockroom has a solution that is 0.125 *m* NH_4Cl. Describe how you would calculate the amount of this solution you would need to supply 0.575 mol of NH_4Cl. If you want to measure the amount of the solution in milliliters, what additional data would you need?

7. A 50.00% H_2SO_4 solution has a specific gravity at 20 °C of 1.3977. Describe how you would calculate the molarity of the solution.

8. A solution of glucose ($C_6H_{12}O_6$, a nonvolatile solute) in 100 g of water has a vapor pressure of 20.134 torr at a temperature at which the vapor pressure of pure water is 23.756 torr. Describe how you would calculate the number of grams of glucose in the solution.

9. The cooling system of an automobile usually contains a solution of antifreeze prepared by mixing equal volumes of ethylene glycol, $C_2H_4(OH)_2$, and water. The density of ethylene glycol is 1.113 g mL^{-1}. Describe how you would calculate the freezing point of the mixture. How would you calculate the boiling point of the solution?

10. Describe how you would calculate the vapor pressure of a solution of tetrachloroethylene in methyl acetate at 40.0 °C in which the mole fraction of tetrachloroethylene is 0.045. At 40 °C, the vapor pressure of tetrachloroethylene is 40.0 torr and that of methyl acetate is 400 torr.

11. What is the approximate osmotic pressure of a 0.0125 *M* solution of $CaCl_2$ in water at 20 °C? Explain the calculation.

Level 2 Problems

12. If you add 75.0 g of water to 150 g of a 10.0% solution of NaCl, what is the percent concentration of the new solution?

13. A solution of $AlCl_3$ in water has a freezing point of −4.50 °C. Assuming complete dissociation of the solute and ideal solution behavior, describe how you would calculate the vapor pressure of this solution at 35 °C. What data, not given in the problem, would you need to obtain the answer?

14. A solution of glycerol, a nonvolatile solute, in 150 g of water has a vapor pressure of 26.36 torr at 30 °C. At this temperature, pure water has a vapor pressure of 31.82 torr. Explain how you would calculate the amount of water that would have to be added to this solution to raise the vapor pressure to 29.00 torr at 30 °C.

15. A solution of a nonvolatile solute in 150 g of water has a freezing point of −0.835 °C. An additional 1.20 g of the solute was added to the solution, and the freezing point dropped to −1.045 °C. How many grams of the solute had been dissolved in the original solution? Explain the calculation in detail.

16. The figure below shows a sealed vessel containing liquid water in equilibrium with water vapor. Liquid water is shown as pale blue with water vapor represented by individual water molecules. In the vessel on the right, green spheres represent a nonvolatile solute. (Only solute molecules on the surface layer of water are shown.) The vapor pressure in the vessel on the right, which contains the nonvolatile solute, is observed to be less than for pure water in the vessel on the left.

(a) Assume that the solute molecules on the surface of the solution on the right occupy 10% of the surface area. If the water in the vessel on the left was transferred to a container with 10% less cross-sectional area, would the vapor pressure be reduced to the same value as the vapor pressure in the vessel on the right? Explain.

(b) Use these models to explain why the vapor pressure in the vessel on the right is less than the vapor pressure of the pure water in the vessel shown on the left. Consider the relative rates of evaporation and condensation in the two vessels and the concept of dynamic equilibria.

17. Acetic acid found in vinegar, CH_3CO_2H, is a weak acid that is partially dissociated in aqueous solution:

$$CH_3CO_2H(aq) \rightleftharpoons CH_3CO_2^-(aq) + H^+(aq)$$

(a) Based upon the partial dissociation of a weak acid would you expect the van't Hoff factor for acetic acid to equal 1.0, 3.0, or be some number larger than 1.0 but much smaller than 2.0? Explain.

(b) From a freezing point depression experiment using a nonpolar solvent, the van't Hoff factor for acetic acid is found to be very close to 0.5. Explain this observation. Use Lewis structures to justify your explanation.

18. Make a sketch to represent solid NaCl and another to show NaCl(*aq*). Have your sketch of the solid show the three-dimensional structure of NaCl. In both sketches show the fundamental species that comprise NaCl. In the sketch for NaCl(*aq*), use water molecules to distinguish the fundamental species in aqueous solution from those in the solid. What intermediate step between your two sketches should you draw if you wished to analyze the role of lattice energy and hydration energy in determining the enthalpy of solution of NaCl? Draw a sketch for the species involved in this intermediate step and use the three sketches to explain how lattice and hydration energies determine the enthalpy of solution.

REVIEW QUESTIONS

Why Solutions Form

14.1 Name three kinds of *homogeneous* mixtures. In which are the particle sizes the smallest?

14.2 Why do two gases spontaneously mix when they are brought into contact?

14.3 When substances form liquid solutions, what two factors are involved in determining the solubility of the solute in the solvent?

14.4 Methanol, CH_3—O—H, and water are miscible in all proportions. What does this mean? Explain how the O—H unit in methanol contributes to this.

14.5 Gasoline and water are immiscible. What does this mean? Explain why they are immiscible in terms of structural features of their molecules and forces of attraction between them.

14.6 Explain how ion–dipole forces help to bring potassium chloride into solution in water.

14.7 What does hydration refer to, and what kinds of substances, electrolytes or nonelectrolytes, does it tend to assist most in dissolving in water?

14.8 What kinds of substances are more soluble in water, hydrophilic compounds or hydrophobic compounds?

14.9 Explain why potassium chloride will not dissolve in carbon tetrachloride, CCl_4.

14.10 Iodine, I_2, is very slightly soluble in water but dissolves readily in carbon tetrachloride. Why won't it dissolve very much in water and what makes so easy the formation of the solution of iodine in carbon tetrachloride?

14.11 Iodine dissolves far better in ethanol (to form "tincture of iodine," an old antiseptic) than in water. What does this tell us about the alcohol molecules, as compared with water molecules?

Enthalpy of Solution

14.12 How do we define *enthalpy of solution*? Give both a narrative definition and an equation.

14.13 No solution actually forms by the two-step process that we used as a model for the formation of a solution of some ionic compound in water. When we are interested in the molar enthalpy of solution, however, we can get away with this model. Explain. What makes attractive the particular two steps we used for the purpose of estimating ΔH_{soln}?

14.14 The value of ΔH_{soln} for a soluble compound is, say, $+26 \text{ kJ mol}^{-1}$, and a nearly saturated solution is prepared in an insulated container (e.g., a coffee cup calorimeter). Will the system's temperature increase or decrease as the solute dissolves?

14.15 Referring to the preceding question, which value for this compound would be numerically larger, its lattice energy or its hydration energy?

14.16 Which would be expected to have the larger hydration energy, Al^{3+} or Li^+? Why? (Both ions are about the same size.)

14.17 Suggest a reason why the value of ΔH_{soln} for a gas such as CO_2, dissolving in water, is negative.

14.18 The value of ΔH_{soln} for the formation of an acetone–water solution is negative. Explain this in general terms that discuss intermolecular forces of attraction.

14.19 The value of ΔH_{soln} for the formation of an ethanol–hexane solution is positive. Explain this in general terms that involve intermolecular forces of attraction.

14.20 How can Le Châtelier's principle be applied to explain how the addition of heat causes undissolved solute to go into solution (in the cases of most saturated solutions)?

14.21 If the value of ΔH_{soln} for the formation of a mixture of two liquids A and B is zero, what does this imply about the relative strengths of A-A, B-B, and A-B intermolecular attractions?

Temperature and Solubility

14.22 If a saturated solution of NH_4NO_3 at 70 °C is cooled to 10 °C, how many grams of solute will separate if the quantity of the solvent is 100 g? (Use data in Figure 14.15.)

14.23 A hot concentrated solution in which equal masses of NaI and NaBr are dissolved is slowly cooled from 50 °C. Which salt precipitates first? (Use data in Figure 14.15.)

14.24 Fishermen know that on hot summer days, the largest fish will be found in deep sinks in lake bottoms, where the water is coolest. Use the temperature dependence of oxygen solubility in water to explain why.

Pressure and Solubility

14.25 What is Henry's law?

14.26 Mountain streams often contain fewer living things than equivalent streams at sea level. Give one reason why this might be true in terms of oxygen solubilities at different pressures.

14.27 What makes ammonia or carbon dioxide so much more soluble in water than nitrogen?

14.28 Sulfur dioxide is an air pollutant wherever sulfur-containing fuels have been used. Rainfall that washes sulfur dioxide out of the air and onto the land is called acid rain. How does sulfur dioxide contribute to acid rain? Write an equation as part of your answer.

14.29 Why does a bottled carbonated beverage fizz when you take the cap off?

Expressions of Concentration

14.30 Write the definition for each of the following concentration units: mole fraction, mole percent, molality, percent by mass.

14.31 How does the molality of a solution vary with increasing temperature? How does the molarity of a solution vary with increasing temperature?

14.32 Suppose a 1.0 *m* solution of a solute is made using a solvent with a density of 1.15 g/mL. Will the molarity of this solution be numerically larger or smaller than 1.0? Explain.

Lowering of the Vapor Pressure

14.33 What specific fact about a physical property of a solution must be true to call it a colligative property?

14.34 Viewed at a molecular level, what causes a solution with a nonvolatile solute to have a lower vapor pressure than the solvent at the same temperature?

14.35 How is Raoult's law used when we want to calculate the total vapor pressure of a liquid solution made up of two *volatile* liquids?

14.36 What kinds of data would have to be obtained to find out if a binary solution of two miscible liquids is almost exactly an ideal solution?

14.37 When octane is mixed with methanol, the vapor pressure of the octane over the solution is higher than what we would calculate using Raoult's law. Why? Explain the discrepancy in terms of intermolecular attractions.

Freezing Point Depression and Boiling Point Elevation

14.38 When an aqueous solution of sodium chloride starts to freeze, why don't the ice crystals contain ions of the salt?

14.39 Explain why a nonvolatile solute dissolved in water makes the system have (a) a higher boiling point than water and (b) a lower freezing point than water.

Dialysis and Osmosis

14.40 Why do we call dialyzing and osmotic membranes *semipermeable*? What is the opposite of *permeable*?

14.41 What is the key difference between dialyzing and osmotic membranes?

14.42 What is meant by *dialysis*?

14.43 At a molecular level, explain why in osmosis there is a net migration of solvent from the side of the membrane less concentrated in solute to the side more concentrated in solute.

14.44 Two glucose solutions of unequal molarity are separated by an osmotic membrane. Which solution will *lose* water, the one with the higher or the lower molarity?

14.45 Which aqueous solution has the higher osmotic pressure, 10% glucose, $C_6H_{12}O_6$, or 10% sucrose, $C_{12}H_{22}O_{11}$? (Both are molecular compounds.)

14.46 When a solid is *associated* in a solution, what does this mean? What difference does it make to expected colligative properties?

14.47 What is the difference between a *hypertonic* solution and a *hypotonic* solution?

Colligative Properties of Electrolytes

14.48 Why are colligative properties of solutions of ionic compounds usually more pronounced than those of solutions of molecular compounds of the same molalities?

14.49 What is the van't Hoff factor? What is its expected value for all nondissociating molecular solutes? If its measured value is slightly larger than 1.0, what does this suggest about the solute? What is suggested by a van't Hoff factor of approximately 0.5?

14.50 Which aqueous solution, if either, is likely to have the lower freezing point, 10% NaCl or 10% NaI?

14.51 Which aqueous solution, if either, is likely to have the higher boiling point, 0.50 *m* NaI or 0.50 *m* Na_2CO_3?

Kinds of Mixtures

14.52 What is a homogeneous mixture?

14.53 What single feature most distinguishes suspensions, colloidal dispersions, and solutions from each other?

14.54 Blood cells are about 5–6 micrometers in diameter. Would blood be considered a solution, a colloidal dispersion, or a suspension?

Colloidal Dispersions

14.55 List three factors that can contribute to the stability of a colloidal dispersion.

14.56 What general name is given to a colloidal dispersion of one liquid in another?

14.57 What is an emulsifying agent? Give an example.

14.58 What is the Tyndall effect, and why do colloidal dispersions but not solutions show it?

14.59 What causes Brownian movement in fluid colloidal dispersions?

14.60 What is a sol, and how can one often be stabilized?

14.61 What simple test could be used to tell if a clear, aqueous fluid contains colloidally dispersed particles?

14.62 Soap, as a solute in water, spontaneously forms micelles. What are micelles and what is the driving force for their formation?

REVIEW PROBLEMS

Answers to problems whose numbers are printed in color are given in Appendix B. More challenging problems are marked with asterisks. **ILW** = Interactive LearningWare solution is available at *www.wiley.com/college/brady*.

Enthalpy of Solution

14.63 Consider the formation of a solution of aqueous potassium chloride. Write the thermochemical equations for (a) the conversion of solid KCl into its gaseous ions and (b) the subsequent formation of the solution by hydration of the ions. The lattice energy of KCl is -690 kJ mol^{-1}, and the hydration energy of the ions is -686 kJ mol^{-1}. Calculate the enthalpy of solution of KCl in kJ mol^{-1}.

14.64 Referring to Table 14.1, suppose the lattice energy of KCl were just 2% greater than the value given. What then would be the calculated ΔH_{soln} (in kJ mol^{-1})? How would this new value compare with that given in the table? By what percentage is the new value different from the value in the table? (The point of this calculation is to show that small percentage errors in two large numbers, such as lattice energy and hydration energy, can cause large percentage changes in the absolute difference between them.)

14.65 If the enthalpy of solution of an ionic compound in water is $+14$ kJ mol^{-1}, and the lattice energy is -630 kJ mol^{-1}, estimate the hydration energy of the ions in the compound.

14.66 If the enthalpy of solution of an ionic compound in water is -50 kJ mol^{-1}, and the hydration energy is -890 kJ mol^{-1}, estimate the lattice energy of the ionic compound.

Henry's Law

14.67 The solubility of methane, the chief component of natural gas, in water at 20 °C and 1.0 atm pressure is 0.025 g L^{-1}. What is its solubility in water at 1.5 atm and 20 °C?

14.68 At 740 torr and 20 °C, nitrogen has a solubility in water of 0.018 g L^{-1}. At 620 torr and 20 °C, its solubility is 0.015 g L^{-1}. Show that nitrogen obeys Henry's law.

14.69 If the solubility of a gas in water is 0.010 g L^{-1} at 25 °C with the partial pressure of the gas over the solution at 1.0 atm, predict the solubility of the gas at the same temperature but at double the pressure.

14.70 100.0 mL of water shaken with oxygen gas at 1.0 atm will dissolve 0.0039 g O$_2$. Estimate the Henry's law constant for oxygen gas in water.

Expressions of Concentration

14.71 What is the molality of NaCl in a solution that is 3.000 M NaCl, with a density of 1.07 g mL^{-1}?

14.72 A solution of acetic acid, CH$_3$COOH, has a concentration of 0.143 M and a density of 1.00 g mL^{-1}. What is the molality of this solution?

14.73 What is the molal concentration of glucose, C$_6$H$_{12}$O$_6$, a sugar found in many fruits, in a solution made by dissolving 24.0 g of glucose in 1.00 kg of water? What is the mole fraction of glucose in the solution? What is the mass percent of glucose in the solution?

14.74 If you dissolved 11.5 g of NaCl in 1.00 kg of water, what would be its molal concentration? What are the mass percent NaCl and the mole percent NaCl in the solution? The volume of this solution is virtually identical to the original volume of the 1.00 kg of water. What is the molar concentration of NaCl in this solution? What would have to be true about any solvent for one of its dilute solutions to have essentially the same molar and molal concentrations?

14.75 A solution of ethanol, CH$_3$CH$_2$OH, in water has a concentration of 1.25 m. Calculate the mass percent of ethanol in the solution.

14.76 A solution of NaCl in water has a concentration of 19.5%. Calculate the molality of the solution.

14.77 A solution of NH$_3$ in water is at a concentration of 5.00% by mass. Calculate the mole percent NH$_3$ in the solution. What is the molal concentration of the NH$_3$?

14.78 An aqueous solution of isopropyl alcohol, C$_3$H$_8$O, rubbing alcohol, has a mole fraction of alcohol equal to 0.250. What is the percent by mass of alcohol in the solution? What is the molality of the alcohol?

ILW 14.79 Sodium nitrate, NaNO$_3$, is sometimes added to tobacco to improve its burning characteristics. An aqueous solution of NaNO$_3$ has a concentration of 0.363 m. Its density is 1.0185 g mL^{-1}. Calculate the molar concentration of NaNO$_3$ and the mass percent of NaNO$_3$ in the solution. What is the mole fraction of NaNO$_3$ in the solution?

14.80 In an aqueous solution of sulfuric acid, the concentration is 1.89 mol% of acid. The density of the solution is 1.0645 g mL^{-1}. Calculate the following: (a) the molal concentration of H$_2$SO$_4$, (b) the mass percent of the acid, and (c) the molarity of the solution.

Raoult's Law

14.81 At 25 °C, the vapor pressure of water is 23.8 torr. What is the vapor pressure of a solution prepared by dissolving 65.0 g of C$_6$H$_{12}$O$_6$ (a nonvolatile solute) in 150 g of water? (Assume the solution is ideal.)

14.82 The vapor pressure of water at 20 °C is 17.5 torr. A 20% solution of the nonvolatile solute ethylene glycol, C$_2$H$_4$(OH)$_2$, in water is prepared. Estimate the vapor pressure of the solution.

ILW 14.83 At 25 °C, the vapor pressures of benzene (C$_6$H$_6$) and toluene (C$_7$H$_8$) are 93.4 and 26.9 torr, respectively. A solution made by mixing 60.0 g of benzene and 40.0 g of toluene is prepared. At what applied pressure, in torr, will this solution boil?

14.84 Pentane (C$_5$H$_{12}$) and heptane (C$_7$H$_{16}$) are two hydrocarbon liquids present in gasoline. At 20 °C, the vapor pressure of pentane is 420 torr and the vapor pressure of heptane is 36.0 torr. What will be the total vapor pressure (in torr) of a solution prepared by mixing equal masses of the two liquids?

*14.85 Benzene and toluene help get good engine performance from lead-free gasoline. At 40 °C, the vapor pressure of benzene is 180 torr and that of toluene is 60 torr. Suppose you wished to prepare a solution of these liquids that will have a total vapor pressure of 96 torr at 40 °C. What must be the mol% concentrations of each in the solution?

*14.86 The vapor pressure of pure methanol, CH_3OH, at 30 °C is 160 torr. How many grams of the nonvolatile solute glycerol, $C_3H_5(OH)_3$, must be added to 100 g of methanol to obtain a solution with a vapor pressure of 140 torr?

14.87 A solution containing 8.3 g of a nonvolatile, nondissociating substance dissolved in 1 mol of chloroform, $CHCl_3$, has a vapor pressure of 511 torr. The vapor pressure of pure $CHCl_3$ at the same temperature is 526 torr. Calculate (a) the mole fraction of the solute, (b) the number of moles of solute in the solution, and (c) the molecular mass of the solute.

14.88 At 21.0 °C, a solution of 18.26 g of a nonvolatile, nonpolar compound in 33.25 g of ethyl bromide, C_2H_5Br, had a vapor pressure of 336.0 torr. The vapor pressure of pure ethyl bromide at this temperature is 400.0 torr. What is the molecular mass of the compound?

Freezing Point Depression and Boiling Point Elevation

14.89 Calculate what would be the estimated boiling point of 2.00 m sugar in water. What would be its estimated freezing point? (It's largely the sugar in ice cream that makes it difficult to keep ice cream frozen hard.)

14.90 Glycerol, $C_3H_5(OH)_3$ (molecular mass 92), is essentially a nonvolatile liquid that is very soluble in water. A solution is made by dissolving 46.0 g of glycerol in 250 g of water. Calculate the approximate freezing point of the solution and the boiling point of the solution (at 1 atm).

14.91 How many grams of sucrose ($C_{12}H_{22}O_{11}$) are needed to lower the freezing point of 100 g of water by 3.00 °C?

14.92 What will be the boiling point of the solution described in the preceding problem?

14.93 A solution of 12.00 g of an unknown nondissociating compound dissolved in 200.0 g of benzene freezes at 3.45 °C. Calculate the molecular mass of the unknown.

14.94 A solution of 14 g of a nonvolatile, nondissociating compound in 1.0 kg of benzene boils at 81.7 °C. Calculate the molecular mass of the unknown.

ILW 14.95 What are the molecular mass and molecular formula of a nondissociating molecular compound whose empirical formula is C_4H_2N if 3.84 g of the compound in 500 g of benzene gives a freezing point depression of 0.307 °C?

14.96 Benzene reacts with hot concentrated nitric acid dissolved in sulfuric acid to give chiefly nitrobenzene, $C_6H_5NO_2$. A by-product is often obtained, which consists of 42.86% C, 2.40% H, and 16.67% N (by mass). The boiling point of a solution of 5.5 g of the by-product in 45 g of benzene was 1.84 °C higher than that of benzene. (a) Calculate the empirical formula of the by-product. (b) Calculate a molecular mass of the by-product and determine its molecular formula.

Osmotic Pressure

ILW 14.97 (a) Show that the following equation is true.

$$\text{molar mass of solute} = \frac{(\text{grams of solute})RT}{\Pi V}$$

(b) An aqueous solution of a compound with a very high molecular mass was prepared in a concentration of 2.0 g L^{-1} at 25 °C. Its osmotic pressure was 0.021 torr. Calculate the molecular mass of the compound.

14.98 A saturated solution of 0.400 g of a polypeptide in 1.00 L of an aqueous solution has an osmotic pressure of 3.74 torr at 27 °C. What is the approximate molecular mass of the polypeptide?

Colligative Properties of Electrolyte Solutions

14.99 The vapor pressure of water at 20 °C is 17.5 torr. What would be the vapor pressure at 20 °C of a solution made by dissolving 10.0 g of NaCl in 100 g of water? (Assume complete dissociation of the solute and an ideal solution.)

14.100 How many grams of $AlCl_3$ would have to be dissolved in 150 mL of water to give a solution that has a vapor pressure of 38.7 torr at 35 °C? Assume complete dissociation of the solute and ideal solution behavior. (At 35 °C, the vapor pressure of pure water is 42.2 torr.)

14.101 What is the osmotic pressure, in torr, of a 2.0% solution of NaCl in water when the temperature of the solution is 25 °C?

14.102 Below are the concentrations of the most abundant ions in seawater.

Ion	Molality	Ion	Molality
Chloride	0.566	Calcium	0.011
Sodium	0.486	Potassium	0.011
Magnesium	0.055	Bicarbonate	0.002
Sulfate	0.029		

Use these data to estimate the osmotic pressure of seawater at 25 °C in units of atm. What is the minimum pressure in atm needed to desalinate seawater by reverse osmosis?

14.103 What is the expected freezing point of a 0.20 m solution of $CaCl_2$? (Assume complete dissociation.)

14.104 The freezing point of a 0.10 m solution of mercury(I) nitrate is approximately −0.27 °C. Show that these data suggest that the formula of the mercury(I) ion is Hg_2^{2+}.

14.105 A 1.00 m aqueous solution of HF freezes at −1.91 °C. According to these data, what is the percent ionization of HF in the solution?

14.106 An aqueous solution of a weak electrolyte, HX, with a concentration of 0.125 m has a freezing point of −0.261 °C. What is the percent ionization of the compound (to two significant figures)?

Interionic Attractions and Colligative Properties

14.107 The van't Hoff factor for the solute in 0.100 m $NiSO_4$ is 1.19. What would this factor be if the solution behaved as if it were 100% dissociated?

14.108 What is the expected van't Hoff factor for K_2SO_4 in an aqueous solution, assuming 100% dissociation?

14.109 A 0.118 m solution of LiCl has a freezing point of $-0.415\ °C$. What is the van't Hoff factor for this solute at this concentration?

14.110 What is the approximate osmotic pressure of a 0.118 m solution of LiCl at 10 °C? Express the answer in torr. (Use the data in the preceding problem.)

ADDITIONAL EXERCISES

*14.111 As discussed in Facets of Chemistry 14.1, the "bends" is a medical emergency that a diver faces if he or she rises too quickly to the surface from a deep dive. The origin of the problem is seen in the calculations in this problem. At 37 °C (normal body temperature), the solubility of N_2 in water is 0.015 g L^{-1} when its pressure over the solution is 1 atm. Air is approximately 78 mol% N_2. How many moles of N_2 are dissolved per liter of blood (essentially an aqueous solution) when a diver inhales air at a pressure of 1 atm? How many moles of N_2 dissolve per liter of blood when the diver is submerged to a depth of approximately 100 ft, where the total pressure of the air being breathed is 4 atm? If the diver suddenly surfaces, how many milliliters of N_2 gas, in the form of tiny bubbles, are released into the bloodstream from each liter of blood (at 37 °C and 1 atm)?

14.112 Deep-sea divers sometimes substitute helium for nitrogen in the air that they carry in their tanks to prevent the "bends," because helium is much less soluble than nitrogen in blood. Use the simple molecular model of gas solubility discussed in Section 14.2 to explain why helium is relatively insoluble in water.

14.113 Use the simple molecular model of gas solubility discussed in Section 14.2 to explain why N_2 gas solubility in water decreases as the temperature of the solution rises from room temperature to about 70 °C, and then begins to increase.

14.114 The vapor pressure of a mixture of 400 g of carbon tetrachloride and 43.3 g of an unknown compound is 137 torr at 30 °C. The vapor pressure of pure carbon tetrachloride at 30 °C is 143 torr, while that of the pure unknown is 85 torr. What is the approximate molecular mass of the unknown?

*14.115 An experiment calls for the use of the dichromate ion, $Cr_2O_7^{2-}$, in sulfuric acid as an oxidizing agent for isopropyl alcohol, C_3H_8O. The chief product is acetone, C_3H_6O, which forms according to the following equation:

$$3C_3H_8O + Na_2Cr_2O_7 + 4H_2SO_4 \longrightarrow$$
$$3C_3H_6O + Cr_2(SO_4)_3 + Na_2SO_4 + 7H_2O$$

(a) The oxidizing agent is available only as sodium dichromate dihydrate. What is the minimum number of grams of sodium dichromate dihydrate needed to oxidize 21.4 g of isopropyl alcohol according to the balanced equation?

(b) The amount of acetone actually isolated was 12.4 g. Calculate the percentage yield of acetone.

(c) The reaction produces a volatile by-product. When a sample of it with a mass of 8.654 mg was burned in oxygen, it was converted into 22.368 mg of carbon dioxide and 10.655 mg of water, the sole products. (Assume that any unaccounted-for element is oxygen.) Calculate the percentage composition of the by-product and determine its empirical formula.

(d) A solution prepared by dissolving 1.338 g of the by-product in 115.0 g of benzene had a freezing point of 4.87 °C. Calculate the molecular mass of the by-product and write its molecular formula.

14.116 What is the osmotic pressure in torr of a 0.010 M aqueous solution of a molecular compound at 25 °C?

*14.117 Ethylene glycol, $C_2H_6O_2$, is used in some antifreeze mixtures. Protection against freezing to as low as $-40\ °F$ is sought.

(a) How many moles of solute are needed per kilogram of water to ensure this protection?

(b) The density of ethylene glycol is 1.11 g mL^{-1}. To how many milliliters of solute does your answer to part a correspond?

(c) Calculate the number of quarts of ethylene glycol that should be mixed with each quart of water to get the desired protection.

14.118 The osmotic pressure of a dilute solution of a slightly soluble polymer in water was measured using the osmometer in Figure 14.25. The difference in the heights of the liquid levels was determined to be 1.26 cm at 25 °C. Assume the solution has a density of 1.00 g mL^{-1}. (a) What is the osmotic pressure of the solution in torr? (b) What is the molarity of the solution? (c) At what temperature would the solution be expected to freeze? (d) On the basis of the results of these calculations, explain why freezing point depression cannot be used to determine the molecular masses of compounds composed of very large molecules.

14.119 A solution of ethanol, C_2H_5OH, in water has a concentration of 4.613 mol L^{-1}. At 20 °C, its density is 0.9677 g mL^{-1}. Calculate the following: (a) the molality of the solution and (b) the percent concentration of the alcohol. The density of ethanol is 0.7893 g mL^{-1} and the density of water is 0.9982 g mL^{-1} at 20 °C.

14.120 Consider an aqueous 1.00 m solution of Na_3PO_4, a compound with useful detergent properties.

(a) Calculate the boiling point of the solution on the assumption that it does not ionize at all in solution.

(b) Do the same calculation assuming that the van't Hoff factor for Na_3PO_4 reflects 100% dissociation into its ions.

(c) The 1.00 m solution boils at 100.183 °C at 1 atm. Calculate the van't Hoff factor for the solute in this solution.

14.121 In an aqueous solution of KNO_3, the concentration is 0.9159 mol% of the salt. The solution's density is 1.0489 g mL^{-1}. Calculate (a) the molal concentration of KNO_3, (b) the percent (w/w) of KNO_3, and (c) the molarity of the KNO_3 in the solution.

Here is another chance for you to test your understanding of concepts, your knowledge of scientific terms, and your skills at problem solving. Read through the following questions carefully, and answer each as fully as possible. Review topics when necessary. When you are able to answer these questions correctly, you are ready to go on to the next group of chapters.

1 A 15.5 L sample of neon at 25.0 °C and a pressure of 748 torr is kept at 25.0 °C as it is allowed to expand to a final volume of 25.4 L. What is the final pressure?

2 An 8.95 L sample of nitrogen at 25.0 °C and 1.00 atm is compressed to a volume of 0.895 L and a pressure of 5.56 atm. What must its final temperature be?

3 A sample of oxygen-enriched air with a volume of 12.5 L at 25.0 °C and 1.00 atm consists of 45.0% (v/v) oxygen and 55.0% (v/v) nitrogen. What are the partial pressures of oxygen and nitrogen (in torr) in this sample after it has been warmed to a temperature of 37.0 °C and is still at a final volume of 12.5 L?

4 If a gas in a cylinder pushes back a piston against a constant opposing pressure of 3.0×10^5 pascals and undergoes a volume change of 0.50 m^3, how much work will the gas do, expressed in joules?

5 What is the formula mass of a gaseous element if 6.45 g occupies 1.92 L at 745 torr and 25.0 °C? Which element is it?

6 What is the formula mass of a gaseous element if at room temperature it effuses through a pinhole 2.16 times as rapidly as xenon? Which element is it?

7 Which has a higher value of the van der Waals constant a, a gas whose molecules are polar or one whose molecules are nonpolar? Explain.

8 How many milliliters of dry CO_2, measured at STP, could be evolved in the reaction between 20.0 mL of 0.100 M $NaHCO_3$ and 30.0 mL of 0.0800 M HCl?

9 How many milliliters of Cl_2 gas, measured at 25 °C and 740 torr, are needed to react with 10.0 mL of 0.10 M NaI if the I$^-$ is oxidized to IO_3^- and Cl_2 is reduced to Cl$^-$?

10 Potassium hypobromite, KOBr, converts ammonia to nitrogen by the following reaction.

$$3KOBr + 2NH_3 \longrightarrow N_2 + 3KBr + 3H_2O$$

To prepare 475 mL of dry N_2, when measured at 24.0 °C and 738 torr, what is the minimum number of grams of KOBr required?

11 Hydrogen peroxide, H_2O_2, is decomposed by potassium permanganate according to the following reaction.

$$5H_2O_2 + 2KMnO_4 + 3H_2SO_4 \longrightarrow$$
$$5O_2 + 2MnSO_4 + K_2SO_4 + 8H_2O$$

What is the minimum number of milliliters of 0.125 M $KMnO_4$ required to prepare 375 mL of dry O_2 when the gas volume is measured at 22.0 °C and 738 torr?

12 One way to make chlorine is to let manganese dioxide, MnO_2, react with hydrochloric acid according to the following equation.

$$4HCl + MnO_2 \longrightarrow Cl_2 + MnCl_2 + 2H_2O$$

What is the minimum volume (in mL) of 6.44 M HCl needed to prepare 525 mL of dry chlorine when the gas is obtained at 24.0 °C and 742 torr?

13 A sample of 248 mL of wet nitrogen gas was collected over water at a total gas pressure of 736 torr and a temperature of 21.0 °C. (The vapor pressure of water at 21.0 °C is 18.7 torr.) The nitrogen was produced by the reaction of sulfamic acid, HNH_2SO_3, with 425 mL of a solution of sodium nitrite according to the following equation.

$$NaNO_2 + HNH_2SO_3 \longrightarrow N_2 + NaHSO_4 + H_2O$$

Calculate what must have been the molar concentration of the sodium nitrite.

14 Consider the molecule $POCl_3$, in which phosphorus is the central atom and is bonded to an oxygen atom and three chlorine atoms.
(a) Draw the Lewis structure of $POCl_3$ and predict its geometry.
(b) Is the molecule polar or nonpolar? Explain.
(c) What kinds of attractive forces would be present between $POCl_3$ molecules in the liquid?

15 What kinds of attractive forces, including chemical bonds, would be present between the particles in
(a) $H_2O(l)$ (c) $CH_3OH(l)$ (e) $NaCl(s)$
(b) $CCl_4(l)$ (d) $BrCl(l)$ (f) $Na_2SO_4(s)$

16 What is a *dynamic equilibrium*? In terms of Le Châtelier's principle and the "equation"

$$\text{liquid} + \text{heat} \rightleftharpoons \text{vapor}$$

explain why raising the temperature of a liquid increases the liquid's equilibrium vapor pressure.

17 Can a solid have a vapor pressure? How would the vapor pressure of a solid vary with temperature?

18 Trimethylamine $(CH_3)_3N$, is a substance responsible in part for the smell of fish. It has a boiling point of 3.5 °C and a molecular weight of 59.1. Dimethylamine, $(CH_3)_2NH$, has a similar odor and boils at a slightly higher temperature, 7 °C, even though it has a somewhat lower molecular mass (45.1). How can this be explained in terms of the kinds of attractive forces between their molecules?

19 Methanol, CH_3OH, commonly known as wood alcohol, has a boiling point of 64.7 °C. Methylamine, a fishy-smelling chemical found in herring brine, has a boiling point of −6.3 °C. Ethane, a hydrocarbon present in petroleum, has a boiling point of −88 °C.

$$H-\overset{\overset{\displaystyle H}{|}}{\underset{\underset{\displaystyle H}{|}}{C}}-O-H \qquad H-\overset{\overset{\displaystyle H}{|}}{\underset{\underset{\displaystyle H}{|}}{C}}-\overset{\overset{\displaystyle H}{|}}{N}-H \qquad H-\overset{\overset{\displaystyle H}{|}}{\underset{\underset{\displaystyle H}{|}}{C}}-\overset{\overset{\displaystyle H}{|}}{\underset{\underset{\displaystyle H}{|}}{C}}-H$$

methanol   methylamine   ethane
b.p. 64.7 °C   b.p. −6.3 °C   b.p. −88 °C

Each has nearly the same molecular mass. Account for the large differences in their boiling points in terms of the attractive forces between their molecules.

20 Based on what you've learned in these chapters, explain:
(a) Why a breeze cools you when you're perspiring.
(b) Why droplets of water form on the outside of a glass of cold soda on a warm, humid day.
(c) Why you feel more uncomfortable on a warm, humid day than on a warm, dry day.
(d) The origin of the energy in a violent thunderstorm.
(e) Why clouds form as warm, moist air flows over a mountain range.

21 Make sketches of (a) a face-centered cubic unit cell, (b) a body-centered cubic unit cell, and (c) a simple cubic unit cell. Which type of unit cell does NaCl have?

22 Aluminum has a density of 2.70 g cm^{-3} and crystallizes in a face-centered cubic lattice. Use these and other data to calculate the atomic radius of an aluminum atom.

23 What is the difference between the closest packed structures identified as ccp and hcp? In each of these structures, how many atoms are in contact with any given atom?

24 A certain compound has the formula MCl_2. Crystals of the compound melt at 772 °C and give a liquid that is electrically conducting. What kind of crystal does this compound form?

25 Silicon dioxide, SiO_2, forms very hard crystals that melt at 1610 °C to yield a liquid that does not conduct electricity. What crystal type does SiO_2 form?

26 Sketch the phase diagram for a substance that has a triple point at 25 °C and 100 torr, a normal boiling point of 150 °C, and a melting point at 1 atm of 27 °C. Is the solid more dense or less dense than the liquid? Where on the curve would the critical temperature and critical pressure be? What phase would exist at 30 °C and 10.0 torr?

27 Draw diagrams that illustrate the band structures in typical conductors, nonconductors, and semiconductors.

28 If the element arsenic, As, is added to germanium, which kind of semiconductor (n or p) will result? Explain your answer.

29 How many grams of 4.00% (w/w) solution of KOH in water are needed to neutralize completely the acid in 10.0 mL 0.256 M H_2SO_4?

30 In general, what structural characteristics do molecules have that are able to form liquid crystal phases?

31 What effect does a cholesteric liquid crystal have on polarized light that passes through it?

32 In general, what characteristics are possessed by the ions in compounds that form ceramics?

33 Describe the following: (a) sintering, (b) xerogel, (c) areogel.

34 Use $Si(C_2H_5O)_4$ to illustrate how the sol-gel process works.

35 What is a carbon nanotube? Describe its structure. Describe two properties of carbon nanotubes.

36 Calculate the molar concentration of 15.00% (w/w) Na_2CO_3 solution at 20.0 °C given that its density is 1.160 g/mL.

37 The solubility of pure oxygen in water in 20.0 °C and 760 torr is 4.30×10^{-2} g O_2 per liter of H_2O. When air is in contact with water and the air pressure is 585 torr at 20 °C, how many grams of oxygen from the air dissolve in 1.00 L of water? The average concentration of oxygen in the air is 21.1% (vol/vol).

38 Compound A is a white solid with a high melting point. When it melts, it conducts electricity. In which solvent is it likely to be more soluble, water or gasoline? Explain.

39 Compound XY is an ionic compound that dissociates as it dissolves in water. The lattice energy of XY is −600 kJ/mol. The hydration energy of its ions is −610 kJ/mol.
(a) Write the thermochemical equations for the two steps in the formation of a solution of XY in water.
(b) Write the sum of these two equations in the form of a thermochemical equation, showing the net ΔH.
(c) Draw an enthalpy diagram for the formation of this solution.

40 A 0.270 M KOH solution has a density of 1.01 g/mL. Calculate the percent concentration (w/w) of KOH.

41 At 20 °C a 40.00% (v/v) solution of ethyl alcohol, C_2H_5OH, in water, has a density of 0.9369 g/mL. The density of pure ethyl alcohol at this temperature is 0.7907 g/mL and of water is 0.9982 g/mL.
(a) Calculate the molar concentration and the molal concentration of C_2H_5OH in this solution.
(b) Calculate the concentration of C_2H_5OH in this solution in mole fractions and mole percents.
(c) The vapor pressure of ethyl alcohol at 20 °C is 41.0 torr and of water is 17.5 torr. If the 40.00% (v/v) solution were ideal, what would be the vapor pressure of each component over the solution?

42 Estimate the boiling point of 1.0 molal $Al(NO_3)_3$, assuming that it dissociates entirely into Al^{3+} and NO_3^- ions in solution.

43 Squalene is an oil found chiefly in shark liver oil but also present in low concentrations in olive oil, wheat germ oil, and yeast. A qualitative analysis disclosed that its molecules consist entirely of carbon and hydrogen. When a sample of squalene with a mass of 0.5680 g was burned in pure oxygen, there was obtained 1.8260 g of carbon dioxide and 0.6230 g of water.
(a) Calculate the empirical formula of squalene.
(b) When 0.1268 g of squalene was dissolved in 10.50 g of molten camphor, the freezing point of this solution was 177.3 °C. (The melting point of pure camphor is 178.4 °C, and its molal freezing point depression constant is 37.5 °C kg-camphor^{-1}. Calculate the formula mass of squalene and determine its molecular formula.

Kinetics: The Study of Rates of Reaction

Lightsticks glow more brightly when they are warm rather than cold because the light-producing chemical reaction proceeds faster at the higher temperature. Temperature is one of several factors studied in this chapter that affect the speeds at which chemical reactions occur.

Egyptian-American chemist Ahmed H. Zewail won the 1999 Nobel Prize in Chemistry for developing ultrafast laser techniques that can record the motion of atoms in a molecule during a chemical reaction.

THIS CHAPTER IN CONTEXT Chemical reactions run at many different speeds. Reactions like the rusting of iron, the ripening of fruit, the weathering of stone, the spoiling of milk, or the breakdown of plastics in the environment take place very slowly. Other reactions, like the combustion of gasoline or the explosion of gunpowder, occur very quickly indeed. The fastest reactions are complete in under a picosecond (10^{-12} s). Newly developed high-speed "cameras" use ultrashort laser flashes to monitor these reactions at the atomic level, detecting motions of atoms and molecules over times on the order of femtoseconds (10^{-15} s). Slow-motion replay movies of chemical reactions taken with such cameras are being used in the design of nanoscale machines and electronics and to study biochemical processes occurring in living cells.

Reaction speeds are influenced (and so, controlled) by many different variables, such as temperature, pressure, and concentration. For example, milk doesn't spoil as quickly when it is kept cold. Concentrated acid reacts much more quickly with magnesium ribbon than dilute acid does. Nitrogen and oxygen in air do not react at any significant rate at room temperature and pressure, but at the high temperature and pressure found in the cylinder of an automobile engine, these two elements in air do react somewhat, giving gases that contribute both to smog and to the greenhouse effect. Similarly, when you have a fever, the oxygen-requiring reactions of metabolism run faster, stepping up the body's demand for oxygen. This puts a strain on the heart as it must now pump faster to deliver the oxygen and to remove waste carbon dioxide.

Chemical kinetics is the study of reaction rates. On a practical level, kinetics is concerned with factors that affect speeds of reactions and in how reaction speeds can be controlled. This is essential in industry, where synthetic reactions must take place at controlled speeds. If a reaction takes weeks or months to occur, it may not be economically feasible; on the other hand, if the reaction occurs too quickly or uncontrollably, it may not be safe to carry out. Fine control over reaction rates allows manufacturers to improve productivity and hold down operating costs. On a theoretical level, reaction rates give valuable clues about how a reaction takes place step-by-step at the molecular level. Understanding the reaction at this level of detail often allows even finer control of the reaction's speed and suggests ways to modify the reaction to produce new types of products or to improve the reaction's yield by preventing undesirable side reactions from occurring.

15.1 ▶ The rate of a reaction is the change in reactant or product concentrations with time

The **rate of reaction** for a given chemical change is the speed with which its reactants disappear and its products form. The speed is measured by the amount of products produced or reactants consumed per unit time. Usually this is done by monitoring the concentrations of the reactants or products over time, as the reaction runs (see Figure 15.1).

At a more fundamental level, a study of the rate of a reaction often gives detailed information about *how* reactants change into products. A balanced equation generally describes only a net overall change. Usually, however, the net change is the result of a series of simple reactions that are not at all evident from the equation. Consider, for example, the combustion of propane, C_3H_8.

$$C_3H_8(g) + 5O_2(g) \longrightarrow 3CO_2(g) + 4H_2O(g)$$

Anyone who has ever played billiards knows that this reaction simply cannot occur in a single, simultaneous collision between one propane molecule and five oxygen molecules. Just getting only three balls to come together with but one "click" on a flat, two-dimensional surface is extremely improbable. How unlikely it must be, then, for the *simultaneous* collision in three-dimensional space of six reactant molecules. Instead, the combustion of propane proceeds very rapidly by a series of much more probable steps, involving colliding chemical species of fleeting existence. *The series of individual steps that add up to the overall observed reaction is called the* **mechanism of the reaction.** Information about reaction mechanisms is one of the dividends paid by the study of rates.

15.2 ▶ Five factors affect reaction rates

Before we take up the quantitative aspects of reaction rates, let's look qualitatively at factors that can make a reaction run faster or slower. There are five principal factors that influence reaction rates.

1. Chemical nature of the reactants
2. Ability of the reactants to come in contact with each other

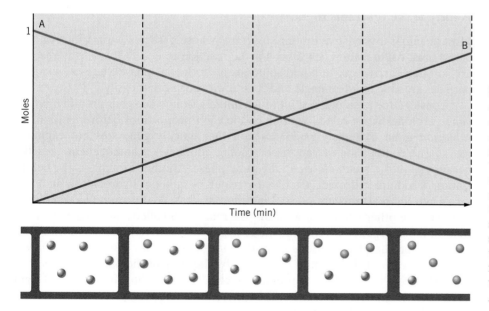

FIGURE 15.1 *Reaction rates are measured by monitoring concentration changes over time.* The progress of the reaction A → B. Note that the number of A molecules (in red) decreases with time, as the number of B molecules (in blue) increases. The steeper the concentration versus time curves are, the faster the rate of reaction is. The filmstrip represents the relative numbers of molecules of A and B at each time.

(a)   (b)

FIGURE 15.2 *The chemical nature of the reactants affects reaction rates.* (*a*) Sodium loses electrons easily, so it reacts quickly with water. (*b*) Potassium loses electrons even more easily than sodium, so its reaction with water is explosively fast.

3. Concentrations of the reactants

4. Temperature

5. Availability of rate-accelerating agents called *catalysts*

Chemical nature of the reactants

Bonds break and new bonds form during reactions. The most fundamental differences among reaction rates, therefore, lie in the reactants themselves, in the inherent tendencies of their atoms, molecules, or ions to undergo changes in chemical bonds. Some reactions are fast by nature and others are slow (Figure 15.2). Because sodium atoms by nature lose electrons so easily, for example, a freshly exposed surface of metallic sodium tarnishes almost instantly when exposed to air and moisture. Under identical conditions, potassium also reacts with air and moisture, but the reaction is much faster because potassium atoms lose electrons more easily than sodium.

Ability of the reactants to meet

Most reactions involve two or more reactants whose particles (atoms, ions, or molecules) must collide with each other for the reaction to occur. This is why reactions are so often carried out in liquid solutions or in the gas phase, states in which the particles are able to intermingle and collide with each other easily.

Consider, for example, that although *liquid* gasoline does burn rapidly, gasoline *vapor* explodes when mixed with air in the right proportions before ignition. (An *explosion* is an extremely rapid reaction that quickly generates hot expanding gases.) The combustion of vaporized gasoline illustrates a **homogeneous reaction,** because all of the reactants are in the same phase. Another example is the neutralization of sodium hydroxide, dissolved in water, by aqueous hydrochloric acid.

When the reactants are present in different phases—for example, when one is a gas and the other a liquid or a solid—the reaction is called a **heterogeneous reaction.** In a heterogeneous reaction, the reactants are able to meet only at the interface between the phases, so *the area of contact between the phases determines the rate of the reaction.* This area is controlled by the sizes of the particles of the reactants. By pulverizing a solid, the total surface area can be hugely increased (Figure 15.3). This maximizes contact between the atoms, ions, or molecules in the solid state with those in a different phase.

Cube 1 cm on edge

1 cm

Total surface area = 6 cm^2

Dividing into cubes 0.01 cm on an edge gives 1,000,000 cubes.

0.01 cm

Total surface area of all cubes = 600 cm^2

FIGURE 15.3 *Effect of crushing a solid.* When a single solid is subdivided into much smaller pieces, the total surface area on all of the pieces becomes very large.

⚗️ **CHEMISTRY IN PRACTICE** If you have ever tried to light a campfire, you've learned that holding a match to finely divided kindling gets the fire going far more rapidly than trying to ignite a log directly. The kindling, with its greater surface area, allows for more contact with oxygen and a faster reaction.

When grain dust particles are extremely small, dry, and mixed with air, grain elevators have been known to explode. Although the reaction is heterogeneous, the contact between the two reactants, the dust and oxygen, is at a maximum, so an accidental spark sets off an explosion that destroys the elevator. Similar explosions caused by coal dust in underground coal mines have killed many miners.

The same situation is deliberately created in the operation of a coal-fueled power plant, where the coal is first pulverized and then blown with air into the combustion chamber. The combustion is extremely rapid and virtually 100% complete, thus delivering the maximum amount of the chemical energy in coal to the production of electricity.

Particle size and reaction rate. An explosion in this grain elevator in New Orleans, Louisiana, killed 35 people in December 1977.

Although heterogeneous reactions such as those just illustrated are obviously important, they are very complex and difficult to analyze. In this chapter, therefore, we'll focus mostly on homogeneous systems.

Concentrations of the reactants

The rates of both homogeneous and heterogeneous reactions are affected by the concentrations of the reactants. For example, wood burns relatively quickly in air but extremely rapidly in pure oxygen. Someone has estimated that if air were 30% oxygen instead of 20%, it would not be possible to put out forest fires. Even red-hot steel wool, which only sputters and glows in air, bursts into flame when thrust into pure oxygen (see Figure 15.4).

Temperature of the system

Almost all chemical reactions occur faster at higher temperatures than they do at lower temperatures. You may have noticed, for example, that insects move more slowly when the air is cool. An insect is a cold-blooded creature, which means that its body temperature is determined by the temperature of its surroundings. As the air cools, insects cool, and so the rates of their chemical metabolism slow down, making insects sluggish.

Presence of catalysts

Catalysts are substances that increase the rates of chemical reactions without being used up. It's amazing, but *catalysts affect every moment of our lives.* That's because the enzymes that direct our body chemistry are all catalysts. So are many of the substances used by the chemical industry to make gasoline, plastics, fertilizers, and other products that have become virtual necessities in our lives. We will discuss how catalysts affect reaction rates later in Section 15.9.

With this qualitative overview, let's now delve into the quantitative aspects of kinetics.

FIGURE 15.4 *Effect of concentration on rate.* Steel wool, after being heated to redness in a flame, burns spectacularly when dropped into pure oxygen.

15.3 ▶ Rates of reaction are measured by monitoring change in concentration over time

A **rate** in any field of study is always expressed as a ratio in which a unit of time is in the denominator. Suppose, for example, that you have a job with a pay rate of ten dollars per hour. Because *per* can be translated as *divided by,* your pay rate can be written as a fraction (abbreviating hour as hr).

$$\text{rate of pay} = \frac{10 \text{ dollars}}{1 \text{ hr}}$$

$\frac{1}{x} = x^{-1}.$

The fraction $\frac{1}{hr}$ can also be written as hr^{-1}, so your pay rate can also be given as

$$\text{rate of pay} = 10 \text{ dollars hr}^{-1}$$

When chemical reactions occur, the concentrations of reactants decrease as they are used up, while the concentrations of the products increase as they form. So one way to describe a reaction's rate is to pick one species in the reaction's equation and describe its change in concentration per unit of time. The result is the rate of the reaction *with respect to that species.* Remembering that we always take "final minus initial," the rate of reaction with respect, say, to species X is

Once again, we use the symbol Δ to mean a change.

$$\text{rate with respect to } X = \frac{(\text{conc. of } X \text{ at time } t_2 - \text{conc. of } X \text{ at time } t_1)}{(t_2 - t_1)}$$

$$= \frac{\Delta(\text{conc. of } X)}{\Delta t}$$

Molarity (mol/L) is normally the concentration unit, and the second (s) is the most often used unit of time. Therefore, the units for reaction rates are most frequently the following:

$$\frac{\text{mol/L}}{\text{s}}$$

Because 1/L and 1/s can also be written as L^{-1} and s^{-1}, the units for a reaction rate can be expressed as **mol L^{-1} s^{-1}.** For instance, if the concentration of one product of a reaction increases by 0.50 mol/L each second, the rate of its formation is 0.50 mol L^{-1} s^{-1}. Similarly, if the concentration of a reactant decreases by 0.20 mol/L per second, its rate of reaction is 0.20 mol L^{-1} s^{-1}.

By convention, reaction rate is reported as a positive value whether something increases or decreases in concentration.

Rates and coefficients

When we know the value of a reaction rate with respect to one species, the coefficients of the reaction's balanced equation may be used to find the rates with respect to the other species. For example, in the combustion of propane described earlier,

$$C_3H_8(g) + 5O_2(g) \longrightarrow 3CO_2(g) + 4H_2O(g)$$

5 mol of O_2 *must* be consumed per unit of time for each mole of C_3H_8 used in the same time. Therefore, in this reaction oxygen *must* react five times faster than propane in units of mol L^{-1} s^{-1}. Similarly, CO_2 forms three times faster than C_3H_8 reacts, and H_2O four times faster. The magnitudes of the rates relative to each other are thus in the same relationship as the coefficients in the balanced equation.

Butane, C_4H_{10}, burns in oxygen to give CO_2 and H_2O according to the equation

$$2C_4H_{10}(g) + 13O_2(g) \longrightarrow 8CO_2(g) + 10H_2O(g)$$

If the butane concentration is decreasing at a rate of 0.20 mol L^{-1} s^{-1}, what is the rate at which the oxygen concentration is decreasing, and what are the rates at which the product concentrations are increasing?

ANALYSIS: We need to relate the rates in terms of oxygen and the products to the given rate in terms of butane. The chemical equation links amounts of these substances to the amount of butane. The magnitudes of the rates relative to each other are in the same relationship as the coefficients in the balanced equation.

SOLUTION: For oxygen,

$$\frac{0.20 \text{ mol } C_4H_{10}}{\text{L s}} \times \frac{13 \text{ mol } O_2}{2 \text{ mol } C_4H_{10}} = \frac{1.3 \text{ mol } O_2}{\text{L s}}$$

Oxygen is reacting at a rate of 1.3 mol L^{-1} s^{-1}. For CO_2 and H_2O, we have similar calculations.

$$\frac{0.20 \text{ mol } C_4H_{10}}{\text{L s}} \times \frac{8 \text{ mol } CO_2}{2 \text{ mol } C_4H_{10}} = \frac{0.80 \text{ mol } CO_2}{\text{L s}}$$

$$\frac{0.20 \text{ mol } C_4H_{10}}{\text{L s}} \times \frac{10 \text{ mol } H_2O}{2 \text{ mol } C_4H_{10}} = \frac{1.0 \text{ mol } H_2O}{\text{L s}}$$

Therefore,

$$\text{rate of formation of } CO_2 = 0.80 \text{ mol } L^{-1} s^{-1}$$

$$\text{rate of formation of } H_2O = 1.0 \text{ mol } L^{-1} s^{-1}$$

Are the Answers Reasonable?
If what we've calculated is correct, then the ratio of the numerical values of the last two rates, namely, 0.80 to 1.0, should check out to be the same as the ratio of the corresponding coefficients in the chemical equation, namely, 8 to 10 (the same as 0.8 to 1.0). The ratios match, so we can be confident in our answers.

PRACTICE EXERCISE 1: Hydrogen sulfide burns in oxygen to form sulfur dioxide and water.

$$2H_2S(g) + 3O_2(g) \longrightarrow 2SO_2(g) + 2H_2O(g)$$

If sulfur dioxide is being formed at a rate of 0.30 mol L^{-1} s^{-1}, what are the rates of disappearance of hydrogen sulfide and oxygen?

EXAMPLE 15.1
Relationships of Rates within a Reaction

Butane is lighter fluid.

Because the rates of reaction of reactants and products are all related, it doesn't matter which species we pick to follow concentration changes over time. For example, to study the decomposition of hydrogen iodide, HI, into H_2 and I_2,

$$2HI(g) \longrightarrow H_2(g) + I_2(g)$$

it is easiest to monitor the I_2 concentration because it is the only colored substance in the reaction. As the reaction proceeds, purple iodine vapor forms, and there are instruments that allow us to relate the intensity of the color to the iodine concentration. Then, once we know the rate of formation of iodine, we also know the rate of formation of hydrogen. It's the same because the coefficients of H_2 and I_2 are the same. And the rate of disappearance of HI, which has a coefficient of 2 in the equation, is twice as fast as the rate of formation of I_2.

Most reactions slow down as reactants are used up

A reaction rate is generally not constant throughout the reaction but commonly changes as the reactants are used up. This is because the rate usually depends on the concentrations of the reactants, and these change as the reaction proceeds. For

TABLE 15.1	DATA, AT 508 °C, FOR THE REACTION $2HI(g) \longrightarrow H_2(g) + I_2(g)$		
Concentration of HI (mol/L)	Time (s)	Concentration of HI (mol/L)	Time (s)
0.100	0	0.0387	200
0.0716	50	0.0336	250
0.0558	100	0.0296	300
0.0457	150	0.0265	350

example, Table 15.1 contains data for the decomposition of hydrogen iodide at a temperature of 508 °C. The data, which show the changes in molar HI concentration over time, are plotted in Figure 15.5. Notice that the molar HI concentration drops fairly rapidly during the first 50 s of the reaction, which means that the initial rate is relatively fast. However, later, in the interval between 300 s and 350 s, the concentration changes by only a small amount, so the rate has slowed considerably. Thus, the steepness of the curve at any moment reflects the rate of the reaction; the steeper the curve, the higher the rate.

See the web site for a discussion of slope.

"Reaction rate" refers to the instantaneous rate, unless we say otherwise.

The rate at which the HI is being consumed at any particular moment is called the *instantaneous rate*. The instantaneous rate can be determined from the slope (or tangent) of the curve measured at the time we have chosen. The slope, which can be read off the graph, is the ratio (expressed positively) of the change in concentration to the change in time. In Figure 15.5, for example, the rate of the decomposition of hydrogen iodide is determined for a time 100 s from the start of the reaction. After the tangent to the curve is drawn, we measure a concentration change (a decrease of 0.027 mol/L) and the time change (110 s) from the graph. Because the rate is based on a *decreasing* concentration, we use a minus sign for the *equation* that describes this rate so that the rate itself will be a positive quantity. We use square brackets to signify concentrations specifically in moles per liter; [HI] thus means the molar concentration of HI.

$$\text{rate}_{\text{with respect to HI}} = -\left\{ \frac{[HI]_{\text{final}} - [HI]_{\text{initial}}}{t_{\text{final}} - t_{\text{initial}}} \right\} = -\left\{ \frac{-0.027 \text{ mol/L}}{110 \text{ s}} \right\}$$

$$\text{rate}_{\text{with respect to HI}} = 2.5 \times 10^{-4} \text{ mol L}^{-1} \text{ s}^{-1}$$

Thus, at this moment in the reaction, the rate with respect to HI is 2.5×10^{-4} mol L^{-1} s^{-1}. In the following example, we'll use this technique to obtain the *initial instantaneous rate* of the reaction, that is, the instantaneous rate of reaction at time zero.

FIGURE 15.5 *Effect of time on concentration.* The data for this plot of the change in the concentration of HI with time for the reaction

$$2HI(g) \longrightarrow H_2(g) + I_2(g)$$

at 508 °C are taken from Table 15.1. The slope is negative because we're measuring the *disappearance* of HI. But when its value is used as a rate of reaction, we express the rate as positive, as we do all rates of reaction.

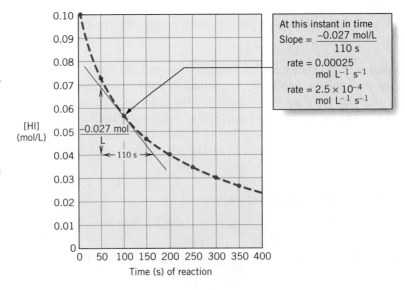

What is the initial rate of the reaction

$$2HI(g) \longrightarrow H_2(g) + I_2(g)$$

at 508 °C, with respect to HI?

ANALYSIS: The question asks about the initial rate, which is the instantaneous rate of the reaction at time zero. Using Figure 15.5, we can draw a tangent line to the curve showing HI concentration as a function of time at time zero. The instantaneous rate will be the slope of the tangent line. Remember that the slope of a line can be calculated from the coordinates of any two points (x_1, y_1) and (x_2, y_2) using the equation

$$slope = \frac{y_2 - y_1}{x_2 - x_1}$$

SOLUTION: Remembering that a tangent line to a curve touches the curve at only one point, we can draw the line as follows:

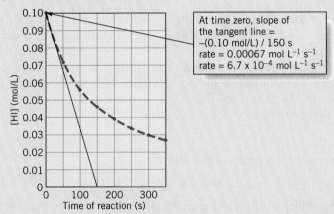

At time zero, slope of
the tangent line =
–(0.10 mol/L) / 150 s
rate = 0.00067 mol L^{-1} s^{-1}
rate = 6.7 x 10^{-4} mol L^{-1} s^{-1}

To more precisely determine the slope of the tangent line, we should choose two points that are as far apart as possible. The point on the curve (0 s, 0.10 mol/L) and the intersection of the tangent line with the time axis (150 s, 0 mol/L) are widely separated:

$$slope = \frac{0.10 \text{ mol/L} - 0.00 \text{ mol/L}}{0 \text{ s} - 150 \text{ s}} = -6.7 \times 10^{-4} \text{ mol L}^{-1} \text{s}^{-1}$$

The slope is negative because the concentration of HI is decreasing as time increases. Rates are positive quantities, so we can report the initial rate of reaction as 6.7×10^{-4} mol L^{-1} s^{-1}. Since the tangent line might be drawn a number of different ways, the time difference is uncertain by more than 10 s, and the rate can be reported simply as 7×10^{-4} mol L^{-1} s^{-1}.

Are the Answers Reasonable?
The instantaneous rate at time zero ought to be slightly larger than the average rate between zero and 50 s:

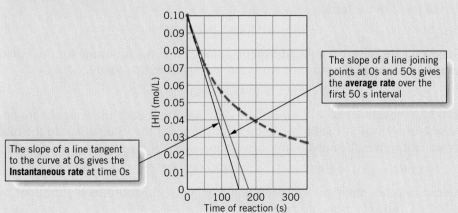

The slope of a line joining
points at 0s and 50s gives
the **average rate** over the
first 50 s interval

The slope of a line tangent
to the curve at 0s gives the
Instantaneous rate at time 0s

EXAMPLE 15.2
**Estimating the Initial
Rate of a Reaction**

The average rate is the slope of a chord connecting two points on a concentration versus time graph. The instantaneous rate is the slope of a tangent line at a single point. The average and instantaneous rates are quite different from each other.

We can compute the average rate directly from data in Table 15.1:

$$\text{slope} = \frac{0.0716 \text{ mol/L} - 0.100 \text{ mol/L}}{50 \text{ s} - 0 \text{ s}} = -5.7 \times 10^{-4} \text{ mol L}^{-1} \text{s}^{-1}$$

so the average rate from 0 to 50 s is 5.7×10^{-4} mol L^{-1} s^{-1}. As expected, this is slightly less than the instantaneous rate at time zero, 7×10^{-4} mol L^{-1} s^{-1}.

PRACTICE EXERCISE 2: Use the graph in Figure 15.5 to estimate the rate of reaction with respect to HI at 250 s after the start of the reaction.

15.4 ▶ Rate laws give reaction rate as a function of reactant concentrations

Thus far we have focused on a rate with respect to *one* component of a reaction. We'll now broaden our focus to consider a rate expression that includes all reactants.

The rate of a homogeneous reaction at any instant is proportional to the product of the molar concentrations of the reactants, each molarity raised to some power or exponent that has to be found by experiment. Let's consider a chemical reaction with an equation of the following form:

$$A + B \longrightarrow \text{products}$$

Its rate of reaction can be expressed as follows:

Rate law of a reaction

$$\text{rate} \propto [A]^m [B]^n \tag{15.1}$$

As we said, the values of the exponents *n* and *m* are found by experiment, which we'll go into shortly.

Rate laws

The proportionality symbol, $\propto$, in Equation 15.1 can be replaced by an equal sign if we introduce a proportionality constant, *k*, which is called the **rate constant** for the reaction. This gives Equation 15.2.

$$\text{rate} = k [A]^m [B]^n \tag{15.2}$$

The value of *k* depends on the particular reaction being studied as well as the temperature at which the reaction occurs.

Equation 15.2 is called the **rate law** for the reaction of *A* with *B*. Once we have found values for *k*, *n*, and *m*, the rate law allows us to calculate the rate of the reaction at any set of known values of concentrations. Consider, for example, the following reaction:

$$H_2SeO_3 + 6I^- + 4H^+ \longrightarrow Se + 2I_3^- + 3H_2O$$

Its rate law is of the form

$$\text{rate} = k [H_2SeO_3]^x [I^-]^y [H^+]^z$$

The exponents have been found experimentally to be the following for the initial rate of this reaction (i.e., the rate when the reactants are first combined):

The units of the rate constant are generally such that the calculated rate will have the units mol L^{-1} s^{-1}.

$$x = 1, \qquad y = 3, \qquad \text{and} \qquad z = 2$$

At 0 °C, *k* equals 5.0×10^5 L^5 mol^{-5} s^{-1}. (We have to specify the temperature because *k* varies with it.) Substituting the exponents and the value of *k* into the rate law equation gives the rate law for the reaction.

When the exponent is equal to 1, as it is for [H$_2$SeO$_3$], it is "understood" and usually omitted.

$$\text{rate} = (5.0 \times 10^5 \text{ L}^5 \text{ mol}^{-5} \text{ s}^{-1}) [H_2SeO_3] [I^-]^3 [H^+]^2 \quad \text{(at 0 °C)}$$

We can calculate the rate of the reaction at 0 °C for any set of concentrations of H$_2$SeO$_3$, I$^-$, and H$^+$ using this rate law.

In the stratosphere, molecular oxygen (O_2) can be broken into two oxygen atoms by ultraviolet radiation from the sun. When one of these oxygen atoms strikes an ozone (O_3) molecule in the stratosphere, the ozone molecule is destroyed and two oxygen molecules are created:

$$O(g) + O_3(g) \longrightarrow 2O_2(g)$$

This reaction is part of the natural cycle of ozone destruction and creation in the stratosphere. What is the rate of ozone destruction *for this reaction alone* at an altitude of 25 km, if the rate law for the reaction is

$$\text{rate} = 4.15 \times 10^5 \, \text{L mol}^{-1}\,\text{s}^{-1}\,[O_3][O]$$

and the reactant concentrations at 25 km are the following: $[O_3] = 1.2 \times 10^{-8} \, M$ and $[O] = 1.7 \times 10^{-14} \, M$?

ANALYSIS: Because we already know the rate law, the answer to this question is merely a matter of substituting the given molar concentrations into this law.

SOLUTION: To see how the units work out, let's write all of the concentration values as well as the rate constant's units in fraction form.

$$\text{rate} = \frac{4.15 \times 10^5 \, \text{L}}{\text{mol s}} \times \left(\frac{1.2 \times 10^{-8} \, \text{mol}}{\text{L}} \right) \times \left(\frac{1.7 \times 10^{-14} \, \text{mol}}{\text{L}} \right)$$

Performing the arithmetic and canceling the units, we see that

$$\text{rate} = \frac{8.5 \times 10^{-17} \, \text{mol}}{\text{L s}} = 8.5 \times 10^{-17} \, \text{mol L}^{-1}\text{s}^{-1}$$

Is the Answer Reasonable?
There's obviously no simple check. Multiplying the powers of 10 for the rate constant and the concentrations together reassures us that the rate is of the correct order of magnitude, and we can see that at least the answer has the correct units for a reaction rate.

PRACTICE EXERCISE 3: The rate law for the decomposition of HI to I_2 and H_2 is

$$\text{rate} = k \, [HI]^2$$

At 508 °C, the rate of the reaction of HI was found to be 2.5×10^{-4} mol L^{-1} s^{-1} when the HI concentration was 0.0558 M (see Figure 15.5). (a) What is the value of k? (b) What are the units of k?

You cannot predict the rate law for a reaction from the overall balanced equation for the reaction

Although a rate law's exponents are generally unrelated to the chemical equation's coefficients, they sometimes are the same by coincidence, as is the case in the decomposition of hydrogen iodide.

$$2HI(g) \longrightarrow H_2(g) + I_2(g)$$

The rate law, as we've said, is

$$\text{rate} = k \, [HI]^2$$

The exponent of [HI] in the rate law, namely 2, happens to match the coefficient of HI in the overall chemical equation, but *there is no way we could have predicted this match without experimental data.* Therefore, *never* simply assume the exponents and the coefficients are the same—it's a trap that many students fall into.

An exponent in a rate law is called the **order of the reaction**[1] with respect to the corresponding reactant. For instance, the decomposition of gaseous N_2O_5 into NO_2 and O_2,

$$2N_2O_5 \longrightarrow 4NO_2 + O_2$$

has the rate law

$$\text{rate} = k\,[N_2O_5]$$

The exponent of $[N_2O_5]$ is 1, so the reaction rate is said to be *first order* in N_2O_5. The rate law for the decomposition of HI has an exponent of 2 for the HI concentration, so its reaction rate is *second order* in HI. The rate law

$$\text{rate} = k[H_2SeO_3]\,[I^-]^3\,[H^+]^2$$

describes a reaction rate that is first order with respect to H_2SeO_3, third order with respect to I^-, and second order with respect to H^+. The overall order for this reaction is $1 + 3 + 2 = 6$.

The **overall order of a reaction** is the sum of the orders with respect to each reactant in the rate law. The decomposition of N_2O_5 is a first-order reaction, and the decomposition of HI is a second-order reaction.

The exponents in a rate law are usually small whole numbers, but fractional and negative exponents are occasionally found. A negative exponent means that the concentration term really belongs in the denominator, which means that as the concentration of the species increases, the rate of reaction decreases.

There are even *zero-order* reactions. They have reaction rates that are independent of the concentration of any reactant. Zero-order reactions usually involve a small amount of a catalyst that is saturated with reactants. This is rather like the situation in a crowded supermarket with only a single checkout lane open. It doesn't matter how many people join the line; the line will move at the same rate no matter how many people are standing in it. An example of a zero-order reaction is the elimination of ethyl alcohol in the liver. Regardless of the blood alcohol level, the rate of alcohol removal by the body is constant, because the number of available catalyst molecules present in the liver is constant. Another zero-order reaction is the decomposition of gaseous ammonia into H_2 and N_2 on a hot platinum surface. The rate at which ammonia decomposes is the same, regardless of its concentration in the gas. The rate law for a zero-order reaction is simply

$$\text{rate} = k$$

where the rate constant k has units of mol L^{-1} s^{-1}. The rate constant depends on the amount, quality, and available surface area of the catalyst. For example, forcing the ammonia through hot platinum powder (with a high surface area) would cause it to decompose faster than simply passing it over a hot platinum surface.

PRACTICE EXERCISE 4: The following reaction

$$BrO_3^- + 3SO_3^{2-} \longrightarrow Br^- + 3SO_4^{2-}$$

has the rate law

$$\text{rate} = k\,[BrO_3^-]\,[SO_3^{2-}]$$

What is the order of the reaction with respect to each reactant? What is the overall order of the reaction?

---

[1] The reason for describing the *order* of a reaction is to take advantage of a great convenience, namely, the mathematics involved in the treatment of the data is the same for all reactions having the same order. We will not go into this very deeply, but you should be familiar with this terminology; it's often used to describe the effects of concentration on reaction rates.

The order of a reaction must be determined experimentally

We've mentioned several times that the exponents in the rate law of an overall re-action must be determined experimentally. *This is the only way to know for sure what the exponents are.* To determine the exponents, we study how changes in con-centration affect the rate of the reaction. For example, consider again the following hypothetical reaction:

$$A + B \longrightarrow \text{products}$$

Suppose, further, that the data in Table 15.2 have been obtained in a series of five experiments. We know the form of the rate law for the reaction will be

$$\text{rate} = k\,[A]^n\,[B]^m$$

The values of n and m can be discovered by looking for patterns in the rate data given in the table. *One of the easiest ways to reveal patterns in data is to form ratios of results using different sets of conditions.* Because this technique is quite generally useful, let's look at how it is applied to the problem of finding the rate law expo-nents in some detail.

For experiments 1, 2, and 3 in Table 15.2, the concentration of B has been held constant at 0.10 M. Any change in the rate for these first three experiments must be due to the change in $[A]$. The rate law tells us that when the concentration of B is held constant, the rate must be proportional to $[A]^n$, so if we take the ratio of rate laws for experiments 2 and 1, we obtain

$$\frac{\text{rate}_2}{\text{rate}_1} = \left(\frac{[A]_2}{[A]_1}\right)^n$$

For experiments 1 and 2,

$$\frac{\text{rate}_2}{\text{rate}_1} = \frac{0.40 \text{ mol L}^{-1}\,\text{s}^{-1}}{0.20 \text{ mol L}^{-1}\,\text{s}^{-1}} = 2$$

and

$$\frac{[A]_2}{[A]_1} = \frac{0.20 \text{ mol L}^{-1}}{0.10 \text{ mol L}^{-1}} = 2$$

so doubling $[A]$ in going from experiment 1 to experiment 2 doubles the rate, and the relationship reduces to $2.0 = 2.0^n$. For each unique combination of experiments 1, 2, and 3, we have

$$2.0 = 2.0^n \quad \text{(for experiments 2 and 1)}$$
$$3.0 = 3.0^n \quad \text{(for experiments 3 and 1)}$$
$$1.5 = 1.5^n \quad \text{(for experiments 3 and 2)}$$

The only value of n that makes all of these equations true is $n = 1$. The reaction must be first order with respect to A.

TABLE 15.2	CONCENTRATION-RATE DATA FOR THE HYPOTHETICAL REACTION $A + B \rightarrow$ PRODUCTS		
	Initial Concentrations		
Experiment	$[A]$ (mol L^{-1})	$[B]$ (mol L^{-1})	Initial Rate of Formation of Products (mol L^{-1} s^{-1})
1	0.10	0.10	0.20
2	0.20	0.10	0.40
3	0.30	0.10	0.60
4	0.30	0.20	2.40
5	0.30	0.30	5.40

In the final three experiments, the concentration of B changes while the concentration of A is held constant. This time it is the concentration of B that affects the rate. Taking the ratio of rate laws for experiments 4 and 3, we have

$$\frac{\text{rate}_4}{\text{rate}_3} = \left(\frac{[B]_4}{[B]_3}\right)^m$$

For each unique combination of experiments 3, 4, and 5, we have

$$4.0 = 2.0^m \qquad \text{(for experiments 4 and 3)}$$
$$9.0 = 3^m \qquad \text{(for experiments 5 and 3)}$$
$$2.25 = 1.5^m \qquad \text{(for experiments 5 and 4)}$$

The only value of m that makes all of these equations true is $m = 2$. The reaction must be second order with respect to B.

Having determined the exponents for the concentration terms, we now know that the rate law for the reaction must be

$$\text{rate} = k\,[A]^1\,[B]^2$$

To calculate the value of k, we need only substitute rate and concentration data into the rate law for any one of the sets of data.

$$k = \frac{\text{rate}}{[A]^1\,[B]^2}$$

Using the data from the first set in Table 15.2,

$$k = \frac{0.20\ \text{mol L}^{-1}\,\text{s}^{-1}}{(0.10\ \text{mol L}^{-1})(0.10\ \text{mol L}^{-1})^2}$$

$$= \frac{0.20\ \text{mol L}^{-1}\,\text{s}^{-1}}{0.0010\ \text{mol}^3\,\text{L}^{-3}}$$

After canceling such units as we can, the value of k with the net units is

$$k = 2.0 \times 10^2\ \text{L}^2\,\text{mol}^{-2}\,\text{s}^{-1}$$

PRACTICE EXERCISE 5: Use the data from the other four experiments (Table 15.2) to calculate k for this reaction. What do you notice about the values of k?

Table 15.3 summarizes the reasoning used to determine the order with respect to each reactant from experimental data.

TABLE 15.3 RELATIONSHIP BETWEEN THE ORDER OF A REACTION AND CHANGES IN CONCENTRATION AND RATE

Factor by Which the Concentration Is Changed	Factor by Which the Rate Changes	Exponent on the Concentration Term in the Rate Law
2	Rate	0
3	is	0
4	unchanged	0
2	$2 = 2^1$	1
3	$3 = 3^1$	1
4	$4 = 4^1$	1
2	$4 = 2^2$	2
3	$9 = 3^2$	2
4	$16 = 4^2$	2
2	$8 = 2^3$	3
3	$27 = 3^3$	3
4	$64 = 4^3$	3

Sulfuryl chloride, SO_2Cl_2, is used to manufacture the antiseptic chlorophenol. The following data were collected on the decomposition of SO_2Cl_2 at a certain temperature.

$$SO_2Cl_2(g) \longrightarrow SO_2(g) + Cl_2(g)$$

Initial Concentration of SO_2Cl_2 (mol L^{-1})	Initial Rate of Formation of SO_2 (mol L^{-1} s^{-1})
0.100	2.2×10^{-6}
0.200	4.4×10^{-6}
0.300	6.6×10^{-6}

What are the rate law and the value of the rate constant for this reaction?

ANALYSIS: The first step is to write the general form of the expected rate law so we can see which exponents have to be determined. Then we study the data to see how the rate changes when the concentration is changed by a certain factor.

SOLUTION: We expect the rate law to have the form

$$\text{rate} = k\,[SO_2Cl_2]^x$$

Let's examine the data from the first two experiments. Notice that when we double the concentration from 0.100 M to 0.200 M, the initial rate doubles (from 2.2×10^{-6} mol L^{-1} s^{-1} to 4.4×10^{-6} mol L^{-1} s^{-1}). If we look at the first and third, we see that when the concentration triples (from 0.100 M to 0.300 M), the rate also triples (from 2.2×10^{-6} mol L^{-1} s^{-1} to 6.6×10^{-6} mol L^{-1} s^{-1}). This behavior tells us that the reaction must be first order in the SO_2Cl_2 concentration. The rate law is therefore

$$\text{rate} = k\,[SO_2Cl_2]^1$$

To evaluate k, we can use any of the three sets of data. Choosing the first,

$$k = \frac{\text{rate}}{[SO_2Cl_2]^1}$$

$$= \frac{2.2 \times 10^{-6}\ \text{mol}\ L^{-1}\,s^{-1}}{0.100\ \text{mol}\ L^{-1}}$$

$$= 2.2 \times 10^{-5}\ s^{-1}$$

Is the Answer Reasonable?
We should get the same value of k by picking any other pair of values. With the last pair of data, at an initial molar concentration of SO_2Cl_2 of 0.300 mol L^{-1} and an initial rate of 6.6×10^{-6} mol L^{-1} s^{-1}, we calculate k again to be 2.2×10^{-5} s^{-1}.

Isoprene forms a dimer called dipentene. (A *dimer* is a compound formed by joining two simpler molecules.)

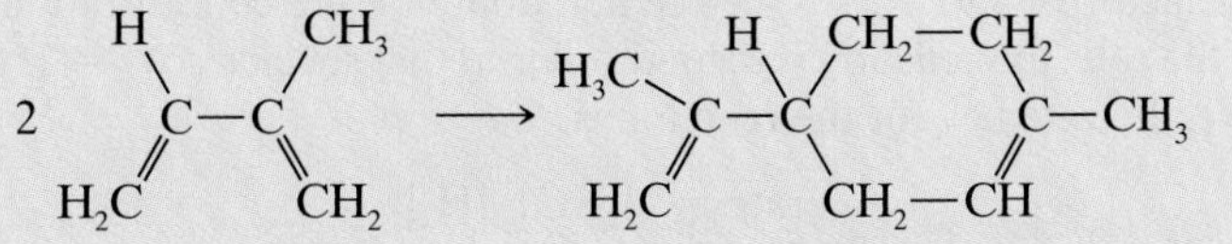

What is the rate law for this reaction, given the following data for the initial rate of formation of dipentene?

EXAMPLE 15.4

Determining the Exponents of a Rate Law

Sulfuryl chloride is a dense liquid with a pungent odor and is corrosive to the skin and lungs.

We could also use experiments 2 and 3. From the second to the third, the rate increases by the same factor, 1.5, as the concentration, so by these data, too, the reaction must be first order.

EXAMPLE 15.5

Determining the Exponents of a Rate Law

Isoprene is used industrially to make a product identical with natural rubber.

Initial Concentration of Isoprene (mol L^{-1})	Initial Rate of Formation of Dipentene (mol L^{-1} s^{-1})
0.50	1.98
1.50	17.8

ANALYSIS: With only one reactant, we expect the rate law to be of the form

$$rate = k\,[isoprene]^x$$

To discover x, we must compare the results of the two experiments.

SOLUTION: From the first to the second experiment, we see that the isoprene concentration increases by a factor of 3. The *rate* of the reaction, however, increases by a factor of (17.8/1.98) = 8.99. This is very nearly a factor of 9, which is 3^2. Therefore, the value of x is 2 and the rate law is

$$rate = k\,[isoprene]^2$$

Is the Answer Reasonable?
Given so few data, we don't have a way to try another calculation.

EXAMPLE 15.6

Determining the Exponents of a Rate Law

The following data were measured for the reduction of nitric oxide with hydrogen.

$$2NO(g) + 2H_2(g) \longrightarrow N_2(g) + 2H_2O(g)$$

Initial Concentrations (mol L^{-1})		Initial Rate of Formation of H$_2$O (mol L^{-1} s^{-1})
[NO]	[H$_2$]	
0.10	0.10	1.23×10^{-3}
0.10	0.20	2.46×10^{-3}
0.20	0.10	4.92×10^{-3}

Nitric oxide is what forms to some extent from the air's nitrogen and oxygen in operating vehicle engine cylinders where the temperature and pressure are both high.

What is the rate law for the reaction?

ANALYSIS: This time we have two reactants. To see how their concentrations affect the rate we must vary only one concentration at a time. Therefore, we choose two experiments in which the concentration of one reactant doesn't change and examine the effect of a change in the concentration of the other reactant. Then we repeat the procedure for the second reactant.

SOLUTION: We expect the rate law to have the form

$$rate = k\,[NO]^n\,[H_2]^m$$

Let's look at the first two experiments. Here the concentration of NO remains the same, so the rate is being affected by the change in the H$_2$ concentration. When we double the H$_2$ concentration, the rate doubles, so the reaction is first order with respect to H$_2$. This means $m = 1$.

Next, we need to pick two experiments in which the H$_2$ concentration doesn't change. Working with the first and third, we see that [NO] doubles and the rate increases by a factor of 4.92/1.23 = 4.00. When doubling the concentration of a species quadruples the rate, the reaction is second order in that species, so $n = 2$.

Therefore, the rate law for the reaction is

$$rate = k\,[NO]^2\,[H_2]$$

When the NO in vehicle exhaust reaches the outside air, it quickly oxidizes to nitrogen dioxide, NO$_2$, a poison and the chief cause of the reddish color of smog.

Is the Answer Reasonable?
The only data that we haven't used as a pair are the data for the second and third reactions. The value of [NO] increases by 2 in going from the second to the third set of

data, so this should multiply the rate by 2^2 or 4, if we've found the right exponents. But the value for $[H_2]$ halves at the same time, so this should take a rate that is otherwise four times as large and cut it by a factor of $(1/2)^1$, or in half. The net effect, then, is to make the rate of the third reaction two times as large as that of the second reaction, which is the observed rate change.

PRACTICE EXERCISE 6: Ordinary sucrose (table sugar) reacts with water in an acidic solution to produce two simpler sugars, glucose and fructose, that have the same molecular formulas.

$$C_{12}H_{22}O_{11} + H_2O \longrightarrow C_6H_{12}O_6 + C_6H_{12}O_6$$
sucrose   glucose   fructose

In a series of experiments, the following data were obtained:

Initial Sucrose Concentration (mol L^{-1})	Rate of Formation of Glucose (mol L^{-1} s^{-1})
0.10	6.17×10^{-5}
0.20	1.23×10^{-4}
0.50	3.09×10^{-4}

What is the order of the reaction with respect to sucrose?

PRACTICE EXERCISE 7: A certain reaction has the following equation: $A + B \longrightarrow C + D$. Experiments yielded the following results:

Initial Concentrations (mol L^{-1})		Initial Rate of Formation of C (mol L^{-1} s^{-1})
$[A]$	$[B]$	
0.40	0.30	1.0×10^{-4}
0.80	0.30	4.0×10^{-4}
0.80	0.60	1.6×10^{-3}

What is the rate law for the reaction?

15.5 ▶ Integrated rate laws give concentration as a function of time

The rate law tells us how the speed of a reaction varies with the concentrations of the reactants. Often, however, we wish to have more information about the status of a reaction than how fast it is going. For instance, if we were manufacturing some chemical in a large reaction vessel, we might want to know what the concentrations of the reactants and products are at some specified time after the reaction has started so we could decide whether it is time to collect the products. We might like to know how long it would take for the reactant concentrations to drop below some minimum optimum values so we could replenish them. Obtaining information of this kind requires some sort of expression that relates concentration to time.

The relationship between the concentration of a reactant and time can be derived from a rate law using calculus. By summing or "integrating" the instantaneous rates of a reaction from the start of the reaction until some specified time t, we can obtain *integrated rate laws* that quantitatively give concentration as a function of time. The form of the integrated rate law depends on the order of the reaction. The mathematical expressions that relate concentration and time in complex reactions can be complicated, so we will concentrate on using integrated rate laws for a few simple first- and second-order reactions.

The natural logarithm of concentration is related linearly with time for first-order reactions

A reaction that is, overall, first order has a rate law of the type

$$\text{rate} = k\,[A]$$

It can be shown that the following equation is the relationship between the concentration of A and time:

Integrated rate law, first-order reactions

Logarithms, both common (base 10) and natural (base e), are discussed on our web site: www.wiley.com/college/brady. Natural logarithms (ln) are related as follows to common logarithms (log):

$$\ln x = 2.303 \log x$$

$$\ln \frac{[A]_0}{[A]_t} = kt \qquad (15.3)$$

Where $[A]_0$ is the initial concentration of A ($t = 0$), and $[A]_t$ is the molar concentration of A at some time t after the start of the reaction. The equation involves a natural logarithm, ln (see Facets of Chemistry 12.2, page 532). The natural logarithm of a particular ratio of concentrations equals the product of the rate constant and time.

We can take the antilogarithm of both sides of Equation 15.3 and rearrange it to obtain the concentration at time t directly as a function of time:

$$[A]_t = [A]_0\, e^{-kt} \qquad (15.3a)$$

where e is the base of the system of natural logarithms ($e = 2.718 \ldots$). Equation 15.3a shows that the concentration of A decays (decreases) exponentially with time. Calculations can use either form of Equation 15.3, as illustrated by the following examples.

EXAMPLE 15.7
Concentration–Time Calculations for First-Order Reactions

N_2O_5, an acid anhydride, is changed to nitric acid by a reaction with water.

$$N_2O_5 + H_2O \longrightarrow 2HNO_3$$

Dinitrogen pentoxide, N_2O_5, is not very stable. In the gas phase or dissolved in a nonaqueous solvent, like carbon tetrachloride, it decomposes by a first-order reaction into N_2O_4 and O_2.

$$2N_2O_5 \longrightarrow 2N_2O_4 + O_2$$

The rate law is

$$\text{rate} = k\,[N_2O_5]$$

At 45 °C, the rate constant for the reaction in carbon tetrachloride is $6.22 \times 10^{-4}\,\text{s}^{-1}$. If the initial concentration of N_2O_5 in the solution is 0.100 M, how many minutes will it take for the concentration to drop to 0.0100 M?

ANALYSIS: To solve the problem, we simply substitute quantities into Equation 15.3 and solve for t.

SOLUTION: First, let's assemble the data.

$$[N_2O_5]_0 = 0.100\ M \qquad\qquad [N_2O_5]_t = 0.0100\ M$$
$$k = 6.22 \times 10^{-4}\,\text{s}^{-1} \qquad\qquad t = ?$$

Substituting,

$$\ln\left(\frac{0.100\ M}{0.0100\ M}\right) = (6.22 \times 10^{-4}\,\text{s}^{-1})t$$

$$\ln(10.0) = (6.22 \times 10^{-4}\,\text{s}^{-1})t$$

Using a scientific calculator to find the natural logarithm of 10.0,[2]

$$2.303 = (6.22 \times 10^{-4}\,\text{s}^{-1})t$$

---

[2]There are special rules for significant figures for logarithms. When taking a logarithm of a quantity, the number of digits written *after the decimal point* equals the number of significant figures in the quantity. Here, 10.0 has three significant figures, so the natural logarithm of 10.0, which is 2.303, has three digits after the decimal.

Solving for t,

$$t = \frac{2.303}{6.22 \times 10^{-4} \text{ s}^{-1}}$$

$$= 3.70 \times 10^3 \text{ s}$$

This is the time in seconds. Dividing by 60 gives the time in minutes, 61.7 min.

Is the Answer Reasonable?

There's no simple check other than to redo the calculation. We can substitute the time we obtained, along with the given initial concentration, into Equation 15.3a and see whether we get back the given concentration at time t:

$$[A]_t = [A]_0 \, e^{-kt} = (0.100 \, M) \times e^{(-6.22 \times 10^{-4} \text{s}^{-1} \times 3.70 \times 10^3 \text{s})}$$

$$= 0.0100 \, M$$

which is exactly what we started with.

On most scientific calculators there is an e^x key (or some equivalent). First calculate the value of the exponent, change its sign, and then press this key to obtain the value of e^{-kt}.

EXAMPLE 15.8

Concentration–Time Calculations for First-Order Reactions

If the initial concentration of N_2O_5 in a carbon tetrachloride solution at 45 °C is 0.500 M (see Example 15.7), what will its concentration be after exactly 1 hr?

ANALYSIS: This time we have to solve for an unknown concentration, which is within the logarithm expression. The easiest way to do this is to first solve for the *ratio* of the concentrations. Once this ratio is known, we can then substitute the known concentration and calculate the unknown one. Remember that the unit of k involves seconds, not hours, so we must convert the given 1 hr into seconds (1 hr = 3600 s).

SOLUTION: Let's begin by listing the data.

$$[N_2O_5]_0 = 0.500 \, M \qquad\qquad [N_2O_5]_t = ? \, M$$

$$k = 6.22 \times 10^{-4} \text{ s}^{-1} \qquad\qquad t = 3600 \text{ s}$$

Now we solve for the concentration ratio:

$$\ln\left(\frac{[N_2O_5]_0}{[N_2O_5]_t}\right) = (6.22 \times 10^{-4} \text{ s}^{-1}) \times 3600 \text{ s}$$

$$= 2.24$$

To take the antilogarithm (antiln), we use a pocket calculator to find e raised to the 2.24 power:

$$\text{antiln}\left[\ln\left(\frac{[N_2O_5]_0}{[N_2O_5]_t}\right)\right] = \frac{[N_2O_5]_0}{[N_2O_5]_t}$$

$$\text{antiln}(2.24) = e^{2.24}$$

$$= 9.4$$

This means that

$$\frac{[N_2O_5]_0}{[N_2O_5]_t} = 9.4$$

Now we can substitute the known concentration, $[N_2O_5]_0 = 0.500 \, M$. This gives

$$\frac{0.500 \, M}{[N_2O_5]_t} = 9.4$$

Solving for $[N_2O_5]_t$ gives

$$[N_2O_5]_t = \frac{0.500 \, M}{9.4}$$

$$= 0.053 \, M$$

After 1 hr, the concentration of N_2O_5 will have dropped to 0.053 M.

Is the Answer Reasonable?
At least the final concentration of N_2O_5 is *less* than its initial concentration. You'd know that you made a huge mistake if the final concentration were greater; reactants are used up by reactions. Other than this, there is no simple check except to redo the calculations.

PRACTICE EXERCISE 8: In Practice Exercise 6, the reaction of sucrose with water in an acidic solution was described.

$$C_{12}H_{22}O_{11} + H_2O \longrightarrow C_6H_{12}O_6 + C_6H_{12}O_6$$
sucrose   glucose   fructose

The reaction is first order with a rate constant of 6.2×10^{-5} s^{-1} at 35 °C, when the H^+ concentration is 0.10 M. Suppose, in an experiment, the initial sucrose concentration was 0.40 M. (a) What will its concentration be after exactly 2 hr? (b) How many minutes will it take for the concentration of sucrose to drop to 0.30 M?

The logarithmic relationship between concentration and time for a first-order reaction (Equation 15.3) provides an elegant way to determine accurately the rate constant by graphical means. Using the properties of logarithms,[3] Equation 15.3 can be rewritten in a form that corresponds to the equation for a straight line.

> The equation for a straight line is usually written
> $$y = mx + b$$
> where x and y are variables, m is the slope, and b is the intercept of the line with the y axis.

$$\ln[A]_t = -kt + \ln[A]_0$$

$$\downarrow \qquad \updownarrow \quad \downarrow$$
$$y \quad = \quad mx + b$$

A plot of the values of $\ln[A]_t$ (vertical axis) versus values of t (horizontal axis) should give a straight line that has a slope equal to $-k$. Such a plot is illustrated in Figure 15.6 for the decomposition of N_2O_5 into N_2O_4 and O_2 in the solvent carbon tetrachloride.

The reciprocal of the concentration is related linearly to time for second-order reactions

For simplicity, we will only consider a second-order reaction with a rate law of the following type:

$$\text{rate} = k\,[B]^2$$

The relationship between concentration and time for a reaction with such a rate law is given by Equation 15.4, an equation that is quite different from that for a first-order reaction.

The slope of this line is the negative of the rate constant.

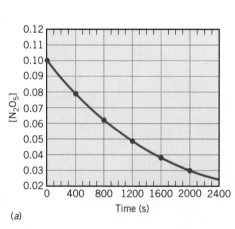

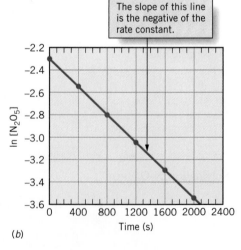

FIGURE 15.6 *The decomposition of N_2O_5.* (*a*) A graph of concentration versus time for the decomposition at 45 °C. (*b*) A straight line is obtained if the logarithm of the concentration is plotted versus time. The slope of this line equals the negative of the rate constant for the reaction.

---

[3]The logarithm of a quotient, $\ln\dfrac{a}{b}$, can be written as the difference, $\ln a - \ln b$.

$$\frac{1}{[B]_t} - \frac{1}{[B]_0} = kt \qquad (15.4)$$

Integrated rate law, second-order reactions

$[B]_0$ is the initial concentration of B, and $[B]_t$ is the concentration at time t. The next example illustrates how Equation 15.4 is applied to calculations.

Nitrosyl chloride, NOCl, decomposes slowly to NO and Cl_2.

$$2NOCl \longrightarrow 2NO + Cl_2$$

The rate law shows that the rate is second order in NOCl.

$$rate = k\,[NOCl]^2$$

The rate constant k equals 0.020 L mol^{-1} s^{-1} at a certain temperature. If the initial concentration of NOCl in a closed reaction vessel is 0.050 M, what will the concentration be after 30 min?

ANALYSIS: We're given a rate law and so can see that it is for a second-order reaction and has the simple form to which our study is limited. We must calculate $[NOCl]_t$, the molar concentration of NOCl, after 30 min (1800 s) using Equation 15.4.

SOLUTION: Let's begin by tabulating the data.

$$[NOCl]_0 = 0.050\ M \qquad\qquad [NOCl]_t = ?\ M$$

$$k = 0.020\ \text{L mol}^{-1}\,\text{s}^{-1} \qquad\qquad t = 1800\ \text{s}$$

The equation we wish to substitute into is

$$\frac{1}{[NOCl]_t} - \frac{1}{[NOCl]_0} = kt$$

Making the substitutions gives

$$\frac{1}{[NOCl]_t} - \frac{1}{0.050\ \text{mol L}^{-1}} = (0.020\ \text{L mol}^{-1}\,\text{s}^{-1}) \times (1800\ \text{s})$$

Solving for $1/[NOCl]_t$ gives

$$\frac{1}{[NOCl]_t} - 20\ \text{L mol}^{-1} = 36\ \text{L mol}^{-1}$$

$$\frac{1}{[NOCl]_t} = 56\ \text{L mol}^{-1}$$

Taking the reciprocals of both sides gives us the value of [NOCl].

$$[NOCl]_t = \frac{1}{56\ \text{L mol}^{-1}}$$

$$= 0.018\ \text{mol L}^{-1}$$

$$= 0.018\ M$$

The molar concentration of NOCl has decreased from 0.050 M to 0.018 M after 30 min.

Is the Answer Reasonable?
The concentration of NOCl has decreased, as it must, but there is, again, no simple check other than to redo the calculation.

PRACTICE EXERCISE 9: For the reaction in the preceding example, determine how many minutes it would take for the NOCl concentration to drop from 0.040 M to 0.010 M.

EXAMPLE 15.9

Concentration–Time Calculations for Second-Order Reactions

NOCl is a corrosive, reddish-yellow gas that is intensely irritating to the eyes and skin.

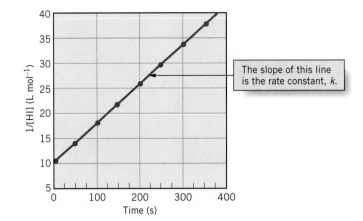

FIGURE 15.7 *Second-order kinetics.* A graph of 1/[HI] versus time for the data in Table 15.1.

The rate constant k for a second-order reaction, one with a rate law following Equation 15.4, can be determined graphically by a method similar to that used for a first-order reaction. We can rearrange Equation 15.4 so that it corresponds to an equation for a straight line.

$$\frac{1}{[B]_t} = kt + \frac{1}{[B]_0}$$

$$y = mx + b$$

When a reaction is second order, then, a plot of $1/[B]_t$ versus t should yield a straight line having a slope k. This is illustrated in Figure 15.7 for the decomposition of HI, using data in Table 15.1.

First-order reactions have constant half-lives; second-order reactions don't

Half-lives

Fast reactions are characterized by large values of k and small values of $t_{1/2}$.

The *half-life* of a reactant is a convenient way to describe how fast it reacts, particularly for an overall first-order process. A reactant's **half-life, $t_{1/2}$,** is the amount of time required for half of the reactant to disappear. A rapid reaction has a short half-life because half of the reactant disappears quickly. The equations for half-lives depend on the order of the reaction.

Half-lives of first-order reactions

When a reaction, overall, is first order, the half-life of the reactant can be obtained from Equation 15.3 (page 662) by setting $[A]_t$ equal to one-half of $[A]_0$.

$$[A]_t = \tfrac{1}{2}[A]_0$$

Substituting $\tfrac{1}{2}[A]_0$ for $[A]_t$ and $t_{1/2}$ for t in Equation 15.3, we have

$$\ln \frac{[A]_0}{\tfrac{1}{2}[A]_0} = kt_{1/2}$$

Noting that the left-hand side of the equation simplifies to ln 2, and solving the equation for $t_{1/2}$, we have

Because ln 2 equals 0.693, Equation 15.5 is sometimes written

$$t_{1/2} = \frac{0.693}{k}$$

$$t_{1/2} = \frac{\ln 2}{k} \tag{15.5}$$

Because k is a constant for a given reaction, the half-life is also a constant for any particular first-order reaction (at any given temperature). Remarkably, in other

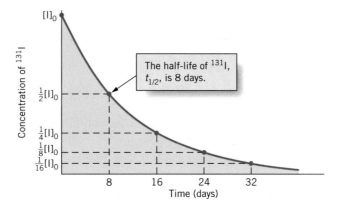

FIGURE 15.8 *First-order radioactive decay of iodine-131.* The initial concentration of the isotope is represented by $[I]_0$.

words, *the half-life of a first-order reaction is not affected by the initial concentration of the reactant.* This can be illustrated by one of the most common first-order events in nature, the change that radioactive isotopes undergo during radioactive "decay." In fact, you have probably heard the term *half-life* used in reference to the life spans of radioactive substances.

Iodine-131, an unstable, radioactive isotope of iodine, undergoes a nuclear reaction whereby it emits a type of radiation and changes into a stable isotope of xenon.[4] The intensity of the radiation decreases, or *decays*, with time (see Figure 15.8). Notice that the time it takes for the first half of the ^{131}I to disappear is 8 days. Then, during the next 8 days half of the remaining ^{131}I disappears, and so on. Regardless of the initial amount, it takes 8 days for half of that amount of ^{131}I to disappear, which means that the half-life of ^{131}I is a constant.

Half-lives of second-order reactions

The half-life of a second-order reaction *does* depend on initial reactant concentrations. We can see this by examining Figure 15.5 (page 652), which follows the decomposition of gaseous HI, a second-order reaction. The reaction begins with a hydrogen iodide concentration of 0.10 M. After 125 s, the concentration of HI drops to 0.050 M, so 125 s is the observed half-life when the initial concentration of HI is 0.050 M. If we then take 0.050 M as the next "initial" concentration, we find that it takes 250 s (at a *total* elapsed time of 375 s) to drop to 0.025 M. If we cut the initial concentration in half, from 0.10 M to 0.05 M, the half-life doubles, from 125 to 250 s.

It can be shown for a second-order reaction of the type we're studying, like the decomposition of HI, that the half-life is inversely proportional to the initial concentration of the reactant. The half-life is related to the rate constant by Equation 15.6.

$$t_{1/2} = \frac{1}{k \times (\text{initial concentration of reactant})} \qquad (15.6)$$

This equation comes from Equation 15.4 by letting $[B]_t = \frac{1}{2}[B]_0$.

One of the most dangerous radioisotopes released in nuclear explosions or accidents is iodine-131, which concentrates in the thyroid gland. The half-life of radioactive iodine-131 is 8.0 days. What fraction of the initial iodine-131 would be present in a patient after 24.0 days if none of it were eliminated through natural body processes?

ANALYSIS: Having learned that iodine-131 is a radioactive isotope and so decays by a first order process, we know that the half-life is constant and does not depend on the ^{131}I concentration.

EXAMPLE 15.10
Half-Life Calculations

---

[4]Iodine-131 is used in the diagnosis of thyroid disorders. The thyroid gland is a small organ located just below the "Adam's apple" and astride the windpipe. It uses iodide ion to make a hormone, so when a patient is given a dose of $^{131}I^-$ mixed with nonradioactive I^-, both ions are taken up by the thyroid gland. The change (temporary) in radioactivity of the gland is a measure of thyroid activity.

The fraction remaining after n half-lives is $\left(\dfrac{1}{2}\right)^n$, or simply $\dfrac{1}{2^n}$.

SOLUTION: A period of 24.0 days is exactly three half-lives. If we take the fraction initially present to be 1, we can set up a table

Half-life	0	1	2	3
Fraction	1	$\dfrac{1}{2}$	$\dfrac{1}{4}$	$\dfrac{1}{8}$

Half of the iodine-131 is lost in the first half-life, half of that disappears in the second half-life, and so on. Therefore, the fraction remaining after three half-lives is $\frac{1}{8}$.

Is the Answer Reasonable?
We could also have solved the problem using the integrated first-order rate law, Equation 15.3. We'll need the first-order rate constant k, which we can obtain from the half-life by rearranging Equation 15.5:

$$k = \frac{\ln 2}{t_{1/2}} = \frac{0.693}{8.0 \text{ days}} = 0.0866 \text{ day}^{-1}$$

Then we can use Equation 15.3 to compute the fraction $\dfrac{[A]_0}{[A]_t}$:

$$\ln \frac{[A]_0}{[A]_t} = kt = (0.0866 \text{ day}^{-1})(24.0 \text{ day}) = 2.08$$

Taking the antilogarithm of both sides, we have

$$\frac{[A]_0}{[A]_t} = e^{2.08} = 8.0$$

The initial concentration, $[A]_0$, is 8.0 times as large as the concentration after 24.0 days. The fraction remaining after 24.0 days is $\frac{1}{8}$, which is exactly what we obtained much more simply above.

EXAMPLE 15.11
Half-Life Calculations

The reaction $2HI(g) \rightarrow H_2(g) + I_2(g)$ has the rate law, rate $= k \, [HI]^2$, with $k = 0.079$ L mol^{-1} s^{-1} at 508 °C. What is the half-life for this reaction at this temperature when the initial HI concentration is 0.050 M?

ANALYSIS: The rate law tells us that the reaction is second order. To calculate the half-life, we need to use Equation 15.6.

SOLUTION: The initial concentration is 0.050 mol L^{-1}; $k = 0.079$ L mol^{-1} s^{-1}. Substituting these values into Equation 15.6 gives

$$t_{1/2} = \frac{1}{(0.079 \text{ L mol}^{-1} \text{ s}^{-1})(0.050 \text{ mol L}^{-1})}$$

$$= 250 \text{ s (rounded)}$$

Is the Answer Reasonable?
Other than checking our units and arithmetic, there is no simple way to check this problem.

PRACTICE EXERCISE 10: In Practice Exercise 6, the reaction of sucrose with water was found to be first order with respect to sucrose. The rate constant under the conditions of the experiments was 6.17×10^{-4} s^{-1}. Calculate the value of $t_{1/2}$ for this reaction in minutes. How many minutes would it take for three-quarters of the sucrose to react? (*Hint:* What fraction of the sucrose remains?)

PRACTICE EXERCISE 11: Suppose that the value of $t_{1/2}$ for a certain reaction was found to be independent of the initial concentration of the reactants. What could you say about the order of the reaction?

15.6 ▶ **Reaction rate theories explain experimental rate laws in terms of molecular collisions**

In Section 15.2 we mentioned that nearly all reactions proceed faster at higher temperatures. As a rule, the reaction rate increases by a factor of about 2 or 3 for each 10 °C increase in temperature, although the actual amount of increase differs from one reaction to another. Temperature evidently has a strong effect on reaction rate. To understand why, we need to develop some theoretical models that explain our observations. One of the simplest models is called *collision theory.*

Reaction rate is related to the number of effective collisions per second between reactant molecules

The basic postulate of **collision theory** is that the rate of a reaction is proportional to the number of *effective* collisions per second among the reactant molecules. An *effective collision* is one that actually gives product molecules. Anything that can increase the frequency of effective collisions should, therefore, increase the rate.

The kinetic theory provides insights for reaction rate theory.

One of the several factors that influences the number of effective collisions per second is *concentration.* As reactant concentrations increase, the number of collisions per second of all types, including effective collisions, cannot help but increase. We'll return to the significance of concentration in Section 15.8.

Not *every* collision between reactant molecules actually results in a chemical change. We know this because the reactant atoms or molecules in a gas or a liquid undergo an enormous number of collisions per second with each other. If every collision were effective, all reactions would be over in an instant. *Only a very small fraction of all the collisions can really lead to a net change.* Why is this so?

At the start of the reaction described by Figure 15.5, only about one of every billion billion (10^{18}) collisions leads to a net chemical reaction. In each of the other collisions, the reactant molecules just bounce off each other.

The importance of molecular orientation

In most reactions, when two reactant molecules collide they must be oriented correctly for a reaction to occur. For example, the reaction represented by the following equation:

$$2NO_2Cl \longrightarrow 2NO_2 + Cl_2$$

appears to proceed by a two-step mechanism. One step involves the collision of an NO_2Cl molecule with a chlorine atom.

$$NO_2Cl + Cl \longrightarrow NO_2 + Cl_2$$

The orientation of the NO_2Cl molecule when hit by the Cl atom is important (see Figure 15.9). The poor orientation shown in Figure 15.9a cannot result in the formation of Cl_2 because the two Cl atoms are not being brought close enough together for a new Cl—Cl bond to form as an N—Cl bond breaks. Figure 15.9b shows the necessary orientation if the collision of NO_2Cl and Cl is to effectively lead to products.

The importance of molecular kinetic energy

Not all collisions, even those correctly oriented, are energetic enough to result in products, and this is the major reason that only a small percentage of all collisions actually lead to chemical change. The colliding particles must carry into the collision a certain minimum combined molecular kinetic energy, called the **activation energy, E_a.** In a successful collision, activation energy changes over to potential energy as the particles hit each other and chemical bonds become reorganized into those of the products. For most chemical reactions, the activation energy is quite large, and only a small fraction of all well-oriented, colliding molecules have it.

We can understand the existence of activation energy by studying in detail what actually takes place during a collision. For old bonds to break and new bonds to form, the atomic nuclei within the colliding particles must get close enough together. The molecules on a collision course must, therefore, be moving with a combined

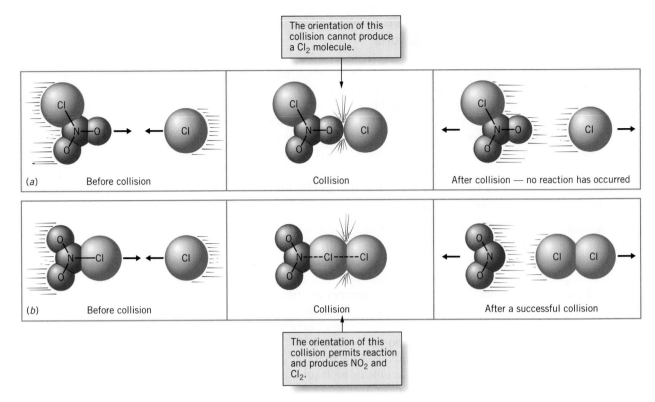

The orientation of this collision cannot produce a Cl₂ molecule.

(a)   Before collision   Collision   After collision — no reaction has occurred

(b)   Before collision   Collision   After a successful collision

The orientation of this collision permits reaction and produces NO₂ and Cl₂.

FIGURE 15.9 *The importance of molecular orientation during a collision in a reaction.* The key step in the decomposition of NO_2Cl to NO_2 and Cl_2 is the collision of a Cl atom with a NO_2Cl molecule. (*a*) A poorly oriented collision. (*b*) An effectively oriented collision.

kinetic energy great enough to overcome the natural repulsions between electron clouds. Otherwise, the molecules simply veer away or bounce apart. Only fast molecules with large kinetic energies can collide with enough collision energy to enable their nuclei and electrons to overcome repulsions and thereby reach the positions required for the bond breaking and bond making that the chemical change demands.

How temperature greatly affects rates

With the concept of activation energy, we can now explain why the rate of a reaction increases so much with increasing temperature. We'll use the two plots in Figure 15.10, each plot corresponding to a different temperature for the same mixture of reactants. Each curve is a plot of the different *fractions* of all collisions (vertical axis) that have particular values of kinetic energy of collision (horizontal axis). (The total area under a curve then represents the total number of collisions,

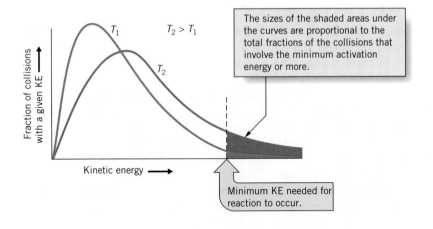

The sizes of the shaded areas under the curves are proportional to the total fractions of the collisions that involve the minimum activation energy or more.

$T_2 > T_1$

Minimum KE needed for reaction to occur.

FIGURE 15.10 *Kinetic energy distributions for a reaction mixture at two different temperatures.*

because all of the fractions must add up to this total.) Notice what happens to the plots when the temperature is increased; the maximum point shifts to the right and the curve flattens somewhat. However, *a modest increase in temperature generally does not affect the reaction's activation energy.* Within reason, the activation energy of a reaction is not affected by a change in temperature. In other words, as the curve flattens and shifts to the right with an increase in temperature, the value of E_a stays the same.

The shaded areas under the curves in Figure 15.10 represent the sum of all those fractions of the total collisions that equal or exceed the activation energy. This sum—we could call it the *reacting fraction*—is relatively much greater at the higher temperature than at the lower temperature because a significant fraction of the curve shifts beyond the activation energy in even a modest change to a higher temperature. In other words, at the higher temperature, a much greater fraction of the collisions occurring each second results in a chemical change, so the reactants disappear faster at the higher temperature.

The transition state is the arrangement of atoms at the top of the activation energy "hill"

Transition state theory is used to explain in detail what happens when reactant molecules come together in a collision. Most often, those in a head-on collision slow down, stop, and then fly apart unchanged. When a collision does cause a reaction, the particles that separate are those of the products. Regardless of what happens to them, however, as the molecules on a collision course slow down, their total kinetic energy decreases as it changes into potential energy (PE). It's like the momentary disappearance of the kinetic energy of a tennis ball when it hits the racket. In the deformed racket and ball, this energy becomes potential energy, which soon changes back to kinetic energy as the ball takes off in a new direction.

The law of conservation of energy requires that the total energy, the sum of KE and PE, be constant in a collision.

To visualize the relationship between activation energy and the development of total potential energy we sometimes use a *potential-energy diagram* (see Figure 15.11). The vertical axis represents changes in *potential* energy as the kinetic energy of the colliding particles changes over to this form. The horizontal axis is called the **reaction coordinate,** and it represents the extent to which the reactants have changed to the products. It helps us follow the path taken by the reaction as reactant molecules come together and change into product molecules. Activation energy appears as a potential energy "hill" or barrier between the reactants and products. Only colliding molecules, properly oriented, that can deliver kinetic energy into potential energy at least as large as E_a are able to climb over the hill and produce products.

We can use a potential energy diagram to follow the progress of both an unsuccessful and a successful collision (see Figure 15.12). As two reactant molecules

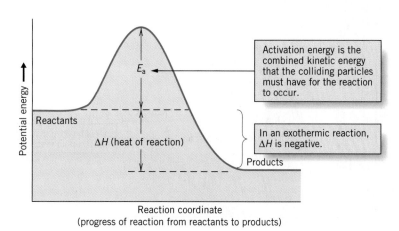

FIGURE **15.11** *Potential-energy diagram for an exothermic reaction.*

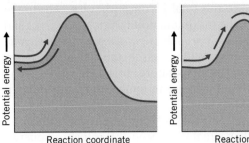

Potential energy

Reaction coordinate

(*a*) An unsuccessful collision; the colliding molecules separate unchanged.

Potential energy

Reaction coordinate

(*b*) A successful collision; the activation energy barrier is crossed and the products are formed.

FIGURE 15.12 *The difference between an unsuccessful and a successful collision.*

collide, we say that they begin to climb the potential-energy barrier as they slow down and experience the conversion of their kinetic energy into potential energy. But if their combined initial kinetic energies are equivalent to a potential energy that is less than E_a, the molecules are unable to reach the top of the hill (Figure 15.12*a*). Instead, they fall back toward the reactants. They bounce apart chemically unchanged with their original total kinetic energy; no net reaction has occurred. On the other hand, if the combined kinetic energy of the colliding molecules equals or exceeds E_a, and if the molecules are oriented properly, they are able to pass over the activation energy barrier and form product molecules (Figure 15.12*b*).

Heats of reaction, energies of activation, and potential-energy diagrams

A reaction's potential-energy diagram, such as that of Figure 15.11, helps us to visualize the *heat of reaction, ΔH,* a concept introduced in Chapter 7. It's the difference between the potential energy of the products and the potential energy of the reactants. Figure 15.11 is for an *exothermic* reaction because the products have a *lower* potential energy than the reactants. In such a system, the net decrease in potential energy appears as an increase in the molecular kinetic energy of the emerging product molecules. The temperature of the system increases during an exothermic reaction because the average molecular kinetic energy of the system increases.

A potential-energy diagram for an endothermic reaction is shown in Figure 15.13. Now the products have a *higher* potential energy than the reactants and, in terms of the heat of reaction, a net input of energy is needed to form the products. Endothermic reactions produce a cooling effect as they proceed because there is a net conversion of molecular kinetic energy to potential energy. As the *total* molecular kinetic energy decreases, the *average* molecular kinetic energy decreases as well, and the temperature drops.

Notice that E_a for an endothermic process is invariably greater than (or it might be equal to) the heat of reaction. If ΔH is both positive and *high, E_a* must

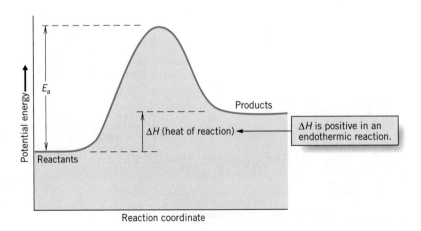

Potential energy

E_a

Products

ΔH (heat of reaction)

ΔH is positive in an endothermic reaction.

Reactants

Reaction coordinate

FIGURE 15.13 *A potential-energy diagram for an endothermic reaction.*

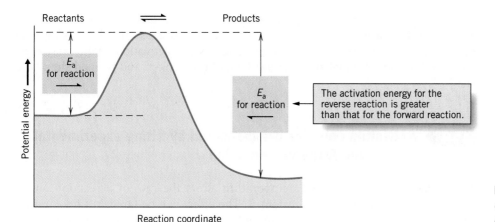

Reactants ⇌ Products

E_a for reaction →

E_a for reaction ←

The activation energy for the reverse reaction is greater than that for the forward reaction.

Potential energy

Reaction coordinate

FIGURE 15.14 *Activation energy barrier for the forward and reverse reactions.*

also be high, making such reactions very slow. However, for an exothermic reaction (ΔH is negative) we cannot tell from ΔH how large E_a is. It could be high, making for a slow reaction despite its being exothermic. If E_a is low, the reaction would be rapid and all its heat would appear quickly.

In Chapter 7 we saw that when the direction of a reaction is reversed, the sign given to the enthalpy change, ΔH, is reversed. In other words, a reaction that is exothermic in the forward direction *must* be endothermic in the reverse direction, and vice versa. This might seem to suggest that reactions are generally reversible. Many are, but if we look again at the energy diagram for a reaction that is exothermic in the forward direction (Figure 15.11), it is obvious that in the opposite direction the reaction is endothermic *and must have a significantly higher activation energy* than the forward reaction. What differs most for the forward and reverse directions is the relative height of the activation energy barrier (see Figure 15.14).

One of the main reasons for studying activation energies is that they provide information about what actually occurs during an effective collision. For example, in Figure 15.9*b* on page 670, we described a way that NO_2Cl could react successfully with a Cl atom during a collision. During this collision, there is a moment when the N—Cl bond is partially broken and the new Cl—Cl bond is partially formed. This brief moment during a successful collision is called the reaction's **transition state.** The potential energy of the transition state corresponds to the high point on the potential-energy diagram (see Figure 15.15). The unstable chemical species that momentarily exists at this instant, $O_2N\cdots Cl\cdots Cl$, with its partially formed and partially broken bonds, is called the **activated complex.**

The size of the activation energy tells us about the relative importance of bond breaking and bond making during the formation of the activated complex. A very

FIGURE 15.15 *Transition state and the activated complex.* Formation of an activated complex in the reaction between NO_2Cl and Cl.

$NO_2Cl + Cl \longrightarrow NO_2 + Cl_2$

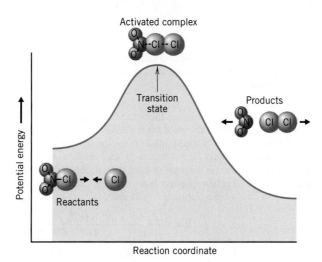

Activated complex

Transition state

Products

Reactants

Potential energy

Reaction coordinate

high activation energy suggests, for instance, that bond *breaking* contributes very heavily to the formation of the activated complex because bond breaking is an energy-absorbing process. On the other hand, a small activation energy may mean that bonds of about equal strength are being both broken and formed simultaneously.

15.7 ▶ Activation energies are measured by fitting experimental data to the Arrhenius equation

As we have said, changing the temperature alters the rate of a reaction and so changes the value of the rate constant, k. The effect can be large, but how much k changes with temperature depends on the magnitude of the activation energy.

The activation energy is related to the rate constant by a relationship discovered in 1889 by Svante Arrhenius, whose name you may recall from our discussion of electrolytes and acids and bases in Chapter 5. The usual form of the **Arrhenius equation** is

Arrhenius equation

$$k = A\, e^{-E_a/RT} \tag{15.7}$$

where k is the rate constant, e is the base of the natural logarithm system, and T is the Kelvin temperature. A is a proportionality constant sometimes called the **frequency factor** or the **pre-exponential factor**. R is the gas constant, which we'll express in our study of kinetics in energy units; namely, R equals 8.314 J mol^{-1} K^{-1}.[5]

The activation energy can be determined graphically

Equation 15.7 is normally used in its logarithmic form. If we take the natural logarithm of both sides, we obtain

$$\ln k = \ln A - E_a/RT$$

Let's rewrite the equation as

$$\ln k = \ln A - (E_a/R) \times (1/T) \tag{15.8}$$

We know that the rate constant k varies with the temperature T, which also means that the quantity $\ln k$ varies with the quantity $(1/T)$. These two quantities, namely, $\ln k$ and $1/T$, are variables, so Equation 15.8 is in the form of an equation for a straight line.

$$
\begin{array}{ccccccc}
\ln k & = & \ln A & + & (-E_a/R) & \times & (1/T) \\
\updownarrow & & \updownarrow & & \updownarrow & & \updownarrow \\
y & = & b & + & m & & x
\end{array}
$$

---

[5]The units of R given here are actually SI units, namely, the joule (J), the mole (n), and the kelvin (K). To calculate R in these units we need to go back to the defining equation for the universal gas law, rearranging terms.

$$R = PV/nT$$

In Chapter 11 we learned that the standard conditions of pressure and temperature are 1 atm and 273.15 K; we expressed the standard molar volume in liters, namely, 22.414 L. But 1 atm equals 1.01325 $\times$ 10^5 N m^{-2}, where N is the SI unit of force, the newton, and m is the meter, the SI unit of length. So m^2 is area given in SI units. From Chapter 11, the ratio of force to area given by N m^{-2} is called the *pascal*, Pa, and that force times distance, or N m, defines one unit of energy in the SI and is called the *joule*, J. In the SI, volume must be expressed as m^3, to employ the SI unit of length to define volume, and 1 L equals 10^{-3} m^3. So now we can calculate R in SI units.

$$R = \frac{(1.01325 \times 10^5 \text{ N m}^{-2}) \times (22.414 \times 10^{-3}\text{ m}^3)}{(1 \text{ mol} \times 273.15 \text{ K})}$$

$$= 8.314 \text{ N m mol}^{-1}\text{ K}^{-1} = 8.314 \text{ J mol}^{-1}\text{ K}^{-1}$$

To determine the activation energy, we can make a graph of $\ln k$ versus $1/T$, measure the slope of the line, and then use the relationship

$$\text{slope} = -E_a/R$$

to calculate E_a. Example 15.12 illustrates how this is done.

Consider again the decomposition of NO_2 into NO and O_2. The equation is

$$2NO_2(g) \longrightarrow 2NO(g) + O_2(g)$$

The following data were collected for the reaction:

Rate Constant, k ($L\,mol^{-1}\,s^{-1}$)	Temperature (°C)
7.8	400
10	410
14	420
18	430
24	440

Determine the activation energy for the reaction in kilojoules per mole.

ANALYSIS: Equation 15.8 applies, but the use of rate data to determine the activation energy graphically requires that we plot $\ln k$, not k, versus the *reciprocal* of the *Kelvin* temperature, so we have to convert the given data into $\ln k$ and $1/T$ before we can make the graph.

SOLUTION: To illustrate, using the first set of data, the conversions are

$$\ln k = \ln (7.8) = 2.05$$

$$\frac{1}{T} = \frac{1}{(400 + 273)\,K} = \frac{1}{673\,K} = 1.486 \times 10^{-3}\,K^{-1}$$

We are carrying extra "significant figures" for the purpose of graphing the data. The remaining conversions give the table in the margin.

Then we plot $\ln k$ versus $1/T$ as shown in the figure below.

EXAMPLE 15.12
Determining Energy of Activation Graphically

$\ln k$	$1/T$ (K^{-1})
2.05	1.486×10^{-3}
2.30	1.464×10^{-3}
2.64	1.443×10^{-3}
2.89	1.422×10^{-3}
3.18	1.403×10^{-3}

There is a statistical method for computing the slope of the straight line that best fits the data. Ordinarily, this is what chemists would use.

$\Delta(1/T) = (1.460 \times 10^{-3}) - (1.410 \times 10^{-3})$
$= 5.0 \times 10^{-5}$

Δx

Slope $= \Delta y/\Delta x$
$= \dfrac{y_2 - y_1}{x_2 - x_1}$

$\Delta \ln k = 2.36 - 3.06$
$= -0.70$

Δy

(1/T) [multiply all values by 10^{-3}]

The slope of the curve is obtained as the ratio

$$\text{slope} = \frac{\Delta(\ln k)}{\Delta(1/T)} = \frac{-0.70}{5.0 \times 10^{-5}\ \text{K}^{-1}}$$

$$= -1.4 \times 10^4\ \text{K} = -E_a/R$$

After changing signs and solving for E_a we have

$$E_a = (8.314\ \text{J mol}^{-1}\ \text{K}^{-1})(1.4 \times 10^4\ \text{K})$$

$$= 1.2 \times 10^5\ \text{J mol}^{-1}$$

$$= 1.2 \times 10^2\ \text{kJ mol}^{-1}$$

Is the Answer Reasonable?
The activation energy is positive and has the correct units. Otherwise, there's no simple check.

The activation energy can be calculated from rate constants measured at two temperatures

In obtaining an activation energy, sometimes we may not wish to go to the bother of graphing data, or we may have only two rate constants measured at two temperatures. In such cases the activation energy can be obtained algebraically. The following relationship can be derived from Equation 15.8:

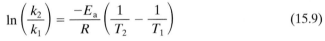

$$\ln\left(\frac{k_2}{k_1}\right) = \frac{-E_a}{R}\left(\frac{1}{T_2} - \frac{1}{T_1}\right) \tag{15.9}$$

TOOLS

Arrhenius equation, alternate form

The following example illustrates the use of this equation.

EXAMPLE 15.13

Calculating an Energy of Activation from Two Rate Constants

The decomposition of HI has rate constants $k = 0.079\ \text{L mol}^{-1}\ \text{s}^{-1}$ at 508 °C and $k = 0.24\ \text{L mol}^{-1}\ \text{s}^{-1}$ at 540 °C. What is the activation energy of this reaction in kJ mol^{-1}?

ANALYSIS: When there are only two data sets, the simplest way to estimate E_a is to solve Equation 15.9.

SOLUTION: Let's begin by organizing the data; a small table will be helpful. Choose one of the rate constants as k_1 (it doesn't matter which one) and then fill in the table.

	k (L mol^{-1} s^{-1})	T (K)
1	0.079	508 + 273 = 781
2	0.24	540 + 273 = 813

We also have $R = 8.314\ \text{J mol}^{-1}\ \text{K}^{-1}$. Substituting into Equation 15.9 gives

$$\ln\left(\frac{0.24\ \text{L mol}^{-1}\ \text{s}^{-1}}{0.079\ \text{L mol}^{-1}\ \text{s}^{-1}}\right) = \frac{-E_a}{8.314\ \text{J mol}^{-1}\ \text{K}^{-1}}\left(\frac{1}{813\ \text{K}} - \frac{1}{781\ \text{K}}\right)$$

$$\ln(3.0) = \frac{-E_a}{8.314\ \text{J mol}^{-1}\ \text{K}^{-1}}(0.00123\ \text{K}^{-1} - 0.00128\ \text{K}^{-1})$$

$$1.10 = \frac{-E_a}{8.314\ \text{J mol}^{-1}}(-0.000050)$$

Multiplying both sides by 8.314 J mol^{-1} gives

$$9.15\ \text{J mol}^{-1} = E_a(0.000050)$$

Solving for E_a, we have

$$E_a = 1.8 \times 10^5 \text{ J mol}^{-1}$$

Converting to kilojoules

$$E_a = 1.8 \times 10^2 \text{ kJ mol}^{-1}$$

Is the Answer Reasonable?
There's no simple check.

The reaction $2NO_2 \rightarrow 2NO + O_2$ has an activation energy of 111 kJ/mol. At 400 °C, $k = 7.8$ L mol^{-1} s^{-1}. What is the value of k at 430 °C?

ANALYSIS: This time we know the activation energy and k at one temperature. We will need to use Equation 15.9 to obtain k at the other temperature. Since the logarithm term contains the ratio of the rate constants, we will solve for the value of this ratio, substitute the known value of k, and then solve for the unknown k.

SOLUTION: Let's begin by writing Equation 15.9.

$$\ln\left(\frac{k_2}{k_1}\right) = \frac{-E_a}{R}\left(\frac{1}{T_2} - \frac{1}{T_1}\right)$$

Organizing the data gives us the following table:

	k (L mol^{-1} s^{-1})	T (K)
1	7.8	400 + 273 = 673 K
2	?	430 + 273 = 703 K

We must use $R = 8.314$ J mol^{-1} K^{-1} and express E_a in joules ($E_a = 1.11 \times 10^5$ J mol^{-1}). Next, we substitute values into the right side of the equation and solve for $\ln(k_2/k_1)$.

$$\ln\left(\frac{k_2}{k_1}\right) = \frac{-1.11 \times 10^5 \text{ J mol}^{-1}}{8.314 \text{ J mol}^{-1} \text{ K}^{-1}}\left(\frac{1}{703 \text{ K}} - \frac{1}{673 \text{ K}}\right)$$

$$= (-1.34 \times 10^4 \text{ K})(-6.34 \times 10^{-5} \text{ K}^{-1})$$

Therefore,

$$\ln\left(\frac{k_2}{k_1}\right) = 0.850$$

Taking the antilog gives the ratio of k_2 to k_1.

$$\frac{k_2}{k_1} = e^{0.850} = 2.34$$

Solving for k_2,

$$k_2 = 2.34k_1$$

Substituting the value of k_1 from the data table gives

$$k_2 = 2.34 (7.8 \text{ L mol}^{-1} \text{ s}^{-1})$$

$$= 18 \text{ L mol}^{-1} \text{ s}^{-1}$$

Is the Answer Reasonable?
Although no simple check exists, we have at least found that the value of k for the higher temperature is greater than it is for the lower temperature, as it should be.

EXAMPLE 15.14
Calculating the Rate Constant at a Particular Temperature

PRACTICE EXERCISE 12: The reaction $CH_3I + HI \rightarrow CH_4 + I_2$ was observed to have rate constants $k = 3.2$ L mol^{-1} s^{-1} at 350 °C and $k = 23$ L mol^{-1} s^{-1} at 400 °C. (a) What is the value of E_a expressed in kJ/mol? (b) What would be the rate constant at 300 °C?

15.8 ▶ Experimental rate laws can be used to support or reject proposed mechanisms for a reaction

At the beginning of this chapter we mentioned that most reactions do not occur in a single step. Instead, the net overall reaction is the result of a series of simple reactions, each of which is called an *elementary process*. An **elementary process** is a reaction whose rate law can be written from its own chemical equation, using its coefficients as exponents for the concentration terms without requiring experiments to determine the exponents. The entire series of elementary processes is called the reaction's **mechanism.** For most reactions, the individual elementary processes cannot actually be observed; instead, we only see the net reaction. Therefore, the mechanism a chemist writes is really a *theory* about what occurs step-by-step as the reactants are changed to the products.

What makes an elementary process "elementary"? Remember that the exponents of the concentration terms for the rate law of an *overall* reaction bear no *necessary* relationship to the coefficients in the overall balanced equation. As we keep emphasizing, the exponents in the *overall* rate law must be determined experimentally. So what makes an elementary process "elementary" is that a simple relationship between coefficients and exponents *does exist,* as we are about to see.

Because the individual steps in a mechanism sometimes cannot be observed directly, devising a mechanism for a reaction requires some ingenuity. However, we can immediately tell whether a proposed mechanism is feasible. *The overall rate law derived from the mechanism must agree with the observed rate law for the overall reaction.*

The rate law for an elementary process can be predicted from the chemical equation for the process

Let's see how the exponents of the concentration terms in a rate law for an elementary process can be devised. Consider the following example of an elementary process that involves collisions between two identical molecules leading directly to the products shown:

$$2NO_2 \longrightarrow NO_3 + NO \tag{15.10}$$

$$\text{rate} = k\,[NO_2]^x$$

There are twice as many NO_2 molecules, each of which collides twice as often, so the total number of collisions per second doubles, i.e., it increases by a factor of 4.

How can we predict the value of the exponent x? If the NO_2 concentration were doubled, there would be *twice* as many individual NO_2 molecules and *each* would have *twice* as many neighbors with which to collide. The number of NO_2-to-NO_2 collisions per second would therefore be increased by a factor of 4, resulting in an increase in the rate by a factor of 4, which is 2^2. Earlier we saw that when the doubling of a concentration leads to a fourfold increase in the rate, the concentration of that reactant is raised to the second power in the rate law. Thus, if Equation 15.10 represents an elementary process, its rate law should be

$$\text{rate} = k\,[NO_2]^2$$

The exponent in the rate law for the elementary process is the same as the coefficient in the chemical equation. If we found, experimentally, that the exponent in this rate law was *not* equal to 2, we would know that the reaction is not really an elementary process.

> The exponents in the rate law for an elementary process are equal to the coefficients of the reactants in the chemical equation for that elementary process.

Remember that this rule applies only to *elementary processes*. If all we know is the balanced equation for the overall reaction, the only way we can find the exponents of the rate equation is by doing experiments.

The rate law for the slowest step in a mechanism should agree with the experimental rate law

How does the ability to predict the rate law of an elementary process help chemists predict reaction mechanisms? To answer this question, let's look at two reactions and what is believed to be their mechanisms. (There are many other, more complicated systems, and Facets of Chemistry 15.1 describes one type, the free-radical chain reaction, that is particularly important.)

First, consider the gaseous reaction

$$2NO_2Cl \longrightarrow 2NO_2 + Cl_2 \tag{15.11}$$

Experimentally, the rate is first order in NO_2Cl, so the rate law is

$$\text{rate} = k\,[NO_2Cl]$$

The first question we might ask is, Could the overall reaction (Equation 15.11) occur in a single step by the collision of two NO_2Cl molecules? The answer is no, because then it would be an elementary process and the rate law predicted for it would include a squared term, $[NO_2Cl]^2$. But the experimental rate law is first order in NO_2Cl. So the predicted and experimental rate laws don't agree, and we must look further to find the mechanism of the reaction.

On the basis of chemical intuition and other information that we won't discuss here, chemists believe the actual mechanism of the reaction in Equation 15.11 is the following two-step sequence of elementary processes:

$$NO_2Cl \longrightarrow NO_2 + Cl$$
$$NO_2Cl + Cl \longrightarrow NO_2 + Cl_2$$

The Cl atom formed here is called a *reactive intermediate*. We never actually observe the Cl because it reacts so quickly.

Notice that when the two reactions are added the intermediate, Cl, drops out and we obtain the net overall reaction given in Equation 15.11. *Being able to add the elementary processes and thus to obtain the overall reaction is another major test of a mechanism.*

In any multistep mechanism, one step is usually much slower than the others. In this mechanism, for example, it is believed that the first step is slow and that once a Cl atom forms, it reacts very rapidly with another NO_2Cl molecule to give the final products.

The final products of a multistep reaction cannot appear faster than the products of the slow step, so the slow step in a mechanism is called the **rate-determining step** or the **rate-limiting step.** In the two-step mechanism above, then, the first reaction is the rate-determining step because the final products can't be formed faster than the rate at which Cl atoms form.

The rate-determining step is similar to a slow worker on an assembly line. The production rate depends on how quickly the slow worker works, regardless of how fast the other workers are. The factors that control the speed of the rate-determining step therefore also control the overall rate of the reaction. This means that *the rate law for the rate-determining step is directly related to the rate law for the overall reaction.*

Because the rate-determining step is an elementary process, we can predict its rate law from the coefficients of its reactants. The coefficient of NO_2Cl in its relatively slow breakdown to NO_2 and Cl is 1. Therefore, the rate law predicted for the first step is

$$\text{rate} = k\,[NO_2Cl]$$

Notice that the predicted rate law derived for the two-step mechanism agrees with the experimentally measured rate law. Although this doesn't *prove* that the mechanism is correct, it does provide considerable support for it. From the standpoint of kinetics, therefore, the mechanism is reasonable.

Free Radicals, Explosions, Octane Ratings, and Aging

A **free radical** is a very reactive species that contains one or more unpaired electrons. Examples are chlorine atoms formed when a Cl_2 molecule absorbs a photon (light) of the appropriate energy:

$$Cl_2 + \text{light energy } (h\nu) \longrightarrow 2Cl\cdot$$

(A dot placed next to the symbol of an atom or molecule represents an unpaired electron and indicates that the particle is a free radical.) The reason free radicals are so reactive is because of the tendency of electrons to become paired through the formation of either ions or covalent bonds.

Free radicals are important in many gaseous reactions, including those responsible for the production of photochemical smog in urban areas. Reactions involving free radicals have useful applications, too. As we saw in Chapter 13, many plastics are made by polymerization reactions that take place by mechanisms that involve free radicals. In addition, free radicals play a part in one of the most important processes in the petroleum industry, *thermal cracking*. This reaction is used to break C—C and C—H bonds in long-chain hydrocarbons to produce the smaller molecules that give gasoline a higher octane rating. An example is the formation of free radicals in the thermal cracking reaction of butane. When butane is heated to 700–800 °C, one of the major reactions that occurs is

$$CH_3-CH_2\!:\!CH_2-CH_3 \xrightarrow{\text{heat}} CH_3CH_2\cdot + CH_3CH_2\cdot$$

The central C—C bond of butane is shown here as a pair of dots, :, rather than the usual dash. When the bond is broken, the electron pair is divided between the two free radicals that are formed. This reaction produces two ethyl radicals, $CH_3CH_2\cdot$.

Free radical reactions tend to have high initial activation energies because chemical bonds must be broken to form the radicals. Once the free radicals are formed, however, reactions in which they are involved tend to be very rapid.

Free-Radical Chain Reactions

In many cases, a free radical reacts with a reactant molecule to give a product molecule plus another free radical. Reactions that involve such a step are called **chain reactions.**

Many explosive reactions are chain reactions involving free-radical mechanisms. One of the most studied reactions of this type is the formation of water from hydrogen and oxygen. The elementary processes involved can be described according to their roles in the mechanism.

The reaction begins with an **initiation step** that gives free radicals.

$$H_2 + O_2 \xrightarrow{\text{hot surface}} 2OH\cdot \quad \text{(initiation)}$$

The chain continues with a **propagation step,** which produces the product plus another free radical.

$$OH\cdot + H_2 \longrightarrow H_2O + H\cdot \quad \text{(propagation)}$$

The reaction of H_2 and O_2 is explosive because the mechanism also contains **branching steps.**

$$\left.\begin{array}{l} H\cdot + O_2 \longrightarrow OH\cdot + O\cdot \\ O\cdot + H_2 \longrightarrow OH\cdot + H\cdot \end{array}\right] \text{(branching)}$$

Thus, the reaction of one $H\cdot$ with O_2 leads to the net production of two $OH\cdot$ plus an $O\cdot$. Every time an $H\cdot$ reacts with oxygen, then, there is an increase in the number of free radicals in the system. The free-radical concentration grows rapidly, and the reaction rate becomes explosively fast.

Chain mechanisms also contain termination steps, which remove free radicals from the system. In the reaction of H_2 and O_2, the wall of the reaction vessel serves to remove $H\cdot$, which tends to halt the chain process.

$$2H\cdot \xrightarrow{\text{wall}} H_2$$

Free Radicals and Aging

Direct experimental evidence also exists for the presence of free radicals in functioning biological systems. These highly reactive species play many roles, but one of the most interesting is their apparent involvement in the aging process. One theory suggests that free radicals attack pro-

FIGURE 1 People exposed to sunlight over long periods, like this woman from Nepal (a small country between India and Tibet), tend to develop wrinkles because ultraviolet radiation causes changes in their skin. This is particularly evident when compared to the smooth skin of the child that she's holding.

tein molecules in collagen. Collagen is composed of long strands of fibers of proteins and is found throughout the body, especially in the flexible tissues of the lungs, skin, muscles, and blood vessels. Attack by free radicals seems to lead to cross-linking between these fibers, which stiffens them and makes them less flexible. The most readily observable result of this is the stiffening and hardening of the skin that accompanies aging or too much sunbathing (see Figure 1).

Free radicals also seem to affect fats (lipids) within the body by promoting the oxidation and deactivation of enzymes that are normally involved in the metabolism of lipids. Over a long period, the accumulated damage gradu-

ally reduces the efficiency with which the cell carries out its activities. Interestingly, vitamin E appears to be a natural free-radical inhibitor. Diets that are deficient in vitamin E produce effects resembling radiation damage and aging.

Still another theory of aging suggests that free radicals attack the DNA in the nuclei of cells. DNA is the substance that is responsible for directing the chemical activities required for the successful functioning of the cell. Reactions with free radicals cause a gradual accumulation of errors in the DNA, which reduce the efficiency of the cell and can lead, ultimately, to malfunction and cell death.

The second reaction mechanism that we will study is that of the following gas phase reaction:

$$2NO + 2H_2 \longrightarrow N_2 + 2H_2O \qquad (15.12)$$

The experimentally determined rate law is

$$rate = k\,[NO]^2\,[H_2] \qquad (experimental)$$

We can quickly tell from this rate law that Equation 15.12 could *not* itself be an elementary process. If it were, the exponent for $[H_2]$ would have to be 2. Obviously, a mechanism involving two or more steps must be involved.

A chemically reasonable mechanism that yields the correct form for the rate law consists of the following two steps:

$$2NO + H_2 \longrightarrow N_2O + H_2O \qquad (slow)$$
$$N_2O + H_2 \longrightarrow N_2 + H_2O \qquad (fast)$$

Dinitrogen monoxide or nitrous oxide is the "laughing gas" used as an anesthetic in medicine and dentistry.

One test of the mechanism, as we said, is that the two equations must add to give the correct overall equation, and they do. Further, the chemistry of the second step has actually been observed in separate experiments. N_2O is a known compound, and it does react with H_2 to give N_2 and H_2O. Another test of the mechanism involves the coefficients of NO and H_2 in the predicted rate law for the first step, the supposed rate-determining step.

$$rate = k\,[NO]^2\,[H_2] \qquad (predicted)$$

This rate equation does match the experimental rate law, but there is still a serious flaw in the proposed mechanism. If the postulated slow step actually describes an elementary process, it would involve the simultaneous collision between three molecules, two NO and one H_2. A three-way collision is so unlikely that if it were really involved in the mechanism, the overall reaction would be extremely slow. Reaction mechanisms seldom include elementary processes that involve more than two-body, or **bimolecular, collisions.**

Chemists believe the reaction in Equation 15.12 proceeds by the following three-step sequence of bimolecular elementary processes:

$$2NO \rightleftharpoons N_2O_2 \qquad (fast)$$
$$N_2O_2 + H_2 \longrightarrow N_2O + H_2O \qquad (slow)$$
$$N_2O + H_2 \longrightarrow N_2 + H_2O \qquad (fast)$$

In this mechanism the first step is proposed to be a rapidly established equilibrium in which the unstable intermediate N_2O_2 forms in the forward reaction and then quickly decomposes into NO by the reverse reaction. The rate-determining step is the reaction of N_2O_2 with H_2 to give N_2O and a water molecule. The third step is

Recall that in a dynamic equilibrium forward and reverse reactions occur at equal rates.

the reaction mentioned above. Once again, notice that the three steps add to give the net overall change.

Since the second step is rate determining, the rate law for the reaction should match the rate law for this step. We predict this to be

$$\text{rate} = k\,[N_2O_2]\,[H_2] \qquad (15.13)$$

The occurrence of N_2O_2 in the proposed mechanism can only be surmised. The compound is never present at a detectable concentration because, as supposed, it's too unstable.

However, the experimental rate law does not contain the species N_2O_2. Therefore, we must find a way to express the concentration of N_2O_2 in terms of the reactants in the overall reaction. To do this, let's look closely at the first step of the mechanism, which we view as a reversible reaction.

The rate in the forward direction, in which NO is the reactant, is

$$\text{rate (forward)} = k_f\,[NO]^2$$

The rate of the reverse reaction, in which N_2O_2 is the reactant, is

$$\text{rate (reverse)} = k_r\,[N_2O_2]$$

If we view this as a dynamic equilibrium, then the rates of the forward and reverse reactions are equal, which means that

$$k_f\,[NO]^2 = k_r\,[N_2O_2] \qquad (15.14)$$

Since we would like to eliminate N_2O_2 from the rate law in Equation 15.13, let's solve Equation 15.14 for $[N_2O_2]$.

$$[N_2O_2] = \frac{k_f}{k_r}\,[NO]^2$$

Substituting into the rate law in Equation 15.13 yields

$$\text{rate} = k\left(\frac{k_f}{k_r}\right)[NO]^2\,[H_2]$$

Combining all the constants into one (k') gives

$$\text{rate} = k'\,[NO]^2\,[H_2]$$

Now the rate law derived from the mechanism matches the rate law obtained experimentally. The three-step mechanism does appear to be reasonable on the basis of kinetics.

The procedure we have worked through here applies to many reactions that proceed by mechanisms involving sequential steps. Steps that precede the rate-determining step are considered to be rapidly established equilibria involving unstable intermediates.

Summary

Although chemists may devise other experiments to help prove or disprove the correctness of a mechanism, one of the strongest pieces of evidence is the experimentally measured rate law for the overall reaction. No matter how reasonable a particular mechanism may appear, if its elementary processes cannot yield a predicted rate law that matches the experimental one, the mechanism is wrong and must be discarded.

PRACTICE EXERCISE 13: Ozone, O_3, reacts with nitric oxide, NO, to form nitrogen dioxide and oxygen.

$$NO + O_3 \longrightarrow NO_2 + O_2$$

This is one of the reactions involved in the formation of photochemical smog. If this reaction occurs in a single step, what is the expected rate law for the reaction?

PRACTICE EXERCISE 14: The mechanism for the decomposition of NO_2Cl is

$$NO_2Cl \longrightarrow NO_2 + Cl$$

$$NO_2Cl + Cl \longrightarrow NO_2 + Cl_2$$

What would the predicted rate law be if the second step in the mechanism were the rate-determining step?

15.9 ▶ Catalysts change reaction rates by providing alternative paths between reactants and products

A **catalyst** is a substance that changes the rate of a chemical reaction without itself being used up. In other words, all of the catalyst added at the start of a reaction is present chemically unchanged after the reaction has gone to completion. The action caused by a catalyst is called **catalysis.** Broadly speaking, there are two kinds of catalysts. *Positive catalysts* speed up reactions, and *negative catalysts,* usually called *inhibitors,* slow reactions down. After this, when we use "catalyst" we'll mean positive catalyst, the usual connotation.

Although the catalyst is not part of the overall reaction, it does participate by changing the mechanism of the reaction. The catalyst provides a path to the products that has a rate-determining step with a lower activation energy than that of the uncatalyzed reaction (see Figure 15.16). Because the activation energy along this new route is smaller, a greater fraction of the collisions of the reactant molecules have the minimum energy needed to react, so the reaction proceeds faster.

Catalysts can be divided into two groups—**homogeneous catalysts,** which exist in the same phase as the reactants, and **heterogeneous catalysts,** which exist in a separate phase.

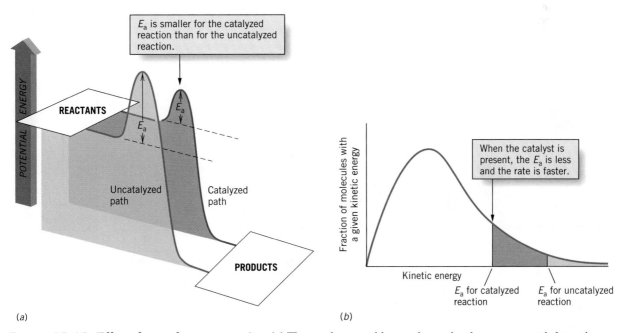

FIGURE 15.16 *Effect of a catalyst on a reaction.* (*a*) The catalyst provides an alternative, low-energy path from the reactants to the products. (*b*) A larger fraction of molecules have sufficient energy to react when the catalyzed path is available.

Homogeneous catalysts are in the same phase as the reactants

In the modern process for making sulfuric acid, the *contact process,* vanadium(V) oxide, V_2O_5, is the catalyst for the oxidation of sulfur dioxide to sulfur trioxide.

An example of homogeneous catalysis is found in the *lead chamber process* for manufacturing sulfuric acid. To make sulfuric acid by this process, sulfur is burned to give SO_2, which is then oxidized to SO_3. The SO_3 is dissolved in water as it forms to give H_2SO_4.

$$S + O_2 \longrightarrow SO_2$$

$$SO_2 + \tfrac{1}{2}O_2 \longrightarrow SO_3$$

$$SO_3 + H_2O \longrightarrow H_2SO_4$$

Unassisted, the second reaction, oxidation of SO_2 to SO_3, occurs slowly. In the lead chamber process, the SO_2 is combined with a mixture of NO, NO_2, air, and steam in large lead-lined reaction chambers. The NO_2 readily oxidizes the SO_2 to give NO and SO_3. The NO is then reoxidized to NO_2 by oxygen.

The NO_2 is regenerated in the second reaction and so is recycled over and over. Thus, only small amounts of it are needed in the reaction mixture to do an effective catalytic job.

$$NO_2 + SO_2 \longrightarrow NO + SO_3$$

$$NO + \tfrac{1}{2}O_2 \longrightarrow NO_2$$

The NO_2 serves as a catalyst by being an oxygen carrier and by providing a low-energy path for the oxidation of SO_2 to SO_3. Notice, as must be true for any catalyst, the NO_2 is regenerated; it has not been permanently changed.

Heterogeneous catalysts are in a separate phase from the reactants

Adsorption means that molecules bind to a surface.

A heterogeneous catalyst is commonly a solid, and it usually functions by promoting a reaction on its surface. One or more of the reactant molecules are adsorbed onto the surface of the catalyst, where an interaction with the surface increases their reactivity. An example is the synthesis of ammonia from hydrogen and nitrogen by the Haber process.

$$3H_2 + N_2 \longrightarrow 2NH_3$$

The reaction takes place on the surface of an iron catalyst that contains traces of aluminum and potassium oxides. It is thought that hydrogen molecules and nitro-

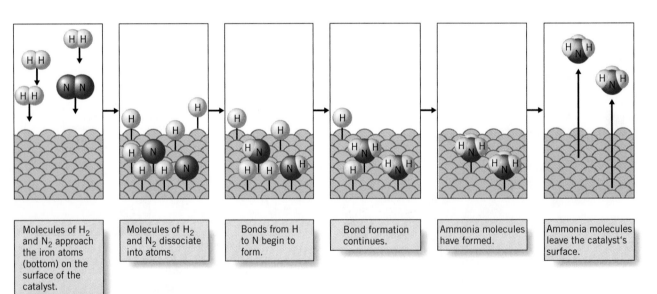

| Molecules of H_2 and N_2 approach the iron atoms (bottom) on the surface of the catalyst. | Molecules of H_2 and N_2 dissociate into atoms. | Bonds from H to N begin to form. | Bond formation continues. | Ammonia molecules have formed. | Ammonia molecules leave the catalyst's surface. |

FIGURE 15.17 *The Haber process.* Catalytic formation of ammonia molecules from hydrogen and nitrogen on the surface of a catalyst.

(a)

(b)

FIGURE 15.18 *Catalysts are very important in the petroleum industry.* (*a*) Catalytic cracking towers at a Standard Oil refinery. (*b*) A variety of catalysts are available as beads, powders, or in other forms for various refinery operations.

gen molecules dissociate while being held on the catalytic surface. The hydrogen atoms then combine with the nitrogen atoms to form ammonia. Finally, the completed ammonia molecule breaks away, freeing the surface of the catalyst for further reaction. This sequence of steps is illustrated in Figure 15.17.

Heterogeneous catalysts are used in many important commercial processes. The petroleum industry uses heterogeneous catalysts to crack hydrocarbons into smaller fragments and then re-form them into the useful components of gasoline (see Figure 15.18). The availability of such catalysts allows refineries to produce gasoline, jet fuel, or heating oil from crude oil in any ratio necessary to meet the demands of the marketplace.

A vehicle that uses unleaded gasoline is equipped with a catalytic converter (Figure 15.19) designed to lower the concentrations of exhaust gas pollutants, such as carbon monoxide, unburned hydrocarbons, and nitrogen oxides. Air is introduced into the exhaust stream that then passes over a catalyst that adsorbs CO, NO, and O_2. The NO dissociates into N and O atoms, and the O_2 also dissociates into atoms. Pairing of nitrogen atoms then produces N_2, and oxidation of CO by oxygen atoms produces CO_2. Unburned hydrocarbons are also oxidized to CO_2 and H_2O. The catalysts in catalytic converters are deactivated or "poisoned" by lead-based octane boosters like tetraethyl lead, $[Pb(C_2H_5)_4]$, so "leaded" gasoline cannot be legally used in vehicles anymore.

The poisoning of catalysts is also a major problem in many industrial processes. Methyl alcohol (methanol, CH_3OH), for example, is a promising fuel that can be made from coal and steam by the reaction

$$C \text{ (from coal)} + H_2O \longrightarrow CO + H_2$$

followed by

$$CO + 2H_2 \longrightarrow CH_3OH$$

A catalyst for the second step is copper(I) ion held in solid solution with zinc oxide. However, traces of sulfur, a contaminant in coal, must be avoided because sulfur reacts with the catalyst and destroys its catalytic activity.

In living systems, complex protein-based molecules called *enzymes* catalyze almost every reaction that occurs in living cells. Enzymes contain a specially shaped area called an "active site" that lowers the energy of the transition state of the reaction being catalyzed. Poisons often work by distorting or blocking the active sites.

FIGURE 15.19 *A modern catalytic converter of the type used in about 80% of new cars.* Part of the converter has been cut away to reveal the porous ceramic material that serves as the support for the catalyst.

We'll study enzymes further in Section 25.9.

SUMMARY

Reaction Rates The speeds or **rates** of reaction are controlled by five factors: (1) the nature of the reactants, (2) the ability of reactants to meet, (3) the concentrations of the reactants, (4) the temperature, and (5) the presence of catalysts. The rates of **heterogeneous reactions** are determined largely by the area of contact between the phases; the rates of **homogeneous reactions** are determined by the concentrations of the reactants. The rate is measured by monitoring the change in reactant or product concentrations with time.

$$\text{rate} = \Delta(\text{concentration})/\Delta(\text{time})$$

In any chemical reaction, the rates of formation of products and the rates of disappearance of reactants are related by the coefficients of the balanced overall chemical equation.

Rate Laws The **rate law** for a reaction relates the reaction rate to the molar concentrations of the reactants. The rate is proportional to the product of the molar concentrations of the reactants, each raised to an appropriate power. These exponents must be determined by experiments in which the concentrations are varied and the effects of the variations on the rate are measured. The proportionality constant, k, is called the **rate constant.** Its value depends on temperature but not on the concentrations of the reactants. The sum of the exponents in the rate law is the **order** (or overall order) of the reaction.

Concentration and Time Equations exist that relate the concentration of a reactant at a given time t to the initial concentration and the rate constant. The time required for half of a reactant to disappear is the **half-life, $t_{1/2}$.** For a first-order reaction, the half-life is a constant that depends only on the rate constant for the reaction; it is independent of the initial concentration. The half-life for a second-order reaction is inversely proportional both to the initial concentration of the reactant and to the rate constant.

Theories of Reaction Rate According to **collision theory,** the rate of a reaction depends on the number of **effective collisions** per second of the reactant particles, which is only an extremely small fraction of the total number of collisions per second. This fraction is small partly because collisions usually do not result in products unless the reactant molecules are suitably oriented. The major reason, however, why the fraction of successful collisions is small is that the colliding molecules must jointly possess a minimum molecular kinetic energy, called the **activation energy, E_a.** As the

temperature increases, a larger fraction of the collisions has this necessary energy, making more collisions effective each second and the reaction faster.

Transition state theory visualizes how the energies of molecules and the orientations of their nuclei interact as they collide. In this theory, the energy of activation is viewed as an energy barrier on the reaction's potential-energy diagram. The *heat of reaction* is the net potential-energy difference between the reactants and the products. In reversible reactions, the values of E_a for both the forward and reverse reactions can be identified on an energy diagram. The species at the high point on an energy diagram is the **activated complex** and is said to be in the **transition state.**

Determining the Activation Energy The **Arrhenius equation** lets us see how changes in activation energy and temperature affect a rate constant. The Arrhenius equation also lets us determine E_a either graphically or by a calculation using the appropriate form of the Arrhenius equation. The calculation requires two rate constants determined at two temperatures. The graphical method uses more values of rate constants at more temperatures and thus usually yields more accurate results. The activation energy and the rate constant at one temperature can be used to calculate the rate constant at another temperature.

Reaction Mechanisms The detailed sequence of elementary processes that lead to the net chemical change is the **mechanism** of the reaction. Since intermediates usually cannot be detected, the mechanism is a theory. Support for a mechanism comes from matching the predicted rate law for the mechanism with the rate law obtained from experimental data. For the **rate-determining step** or for any **elementary process,** the corresponding rate law has exponents equal to the coefficients in the balanced equation for the elementary process.

Catalysts are substances that change a reaction rate but are not consumed by the reaction. Negative catalysts inhibit reactions. Positive catalysts provide alternative paths for reactions for which at least one step has a smaller activation energy than the uncatalyzed reaction. **Homogeneous catalysts** are in the same phase as the reactants. **Heterogeneous catalysts** provide a path of lower activation energy by having a surface on which the reactants are adsorbed and react. Catalysts in living systems are called *enzymes.*

TOOLS ▶ YOU HAVE LEARNED

The table below lists the tools you have learned in this chapter that are applicable to problem solving. Review them if necessary, and refer to them when working on the Thinking-It-Through problems and the Review Exercises that follow.

TOOL	HOW IT WORKS
Rate law of a reaction (page 654)	Uses data on how concentration affects rate to evaluate the exponents of the concentration terms in a rate law, determines the overall order of a reaction from the exponents of the concentration terms in its rate law, and calculates a rate constant.
Integrated first-order rate law (page 662)	For a first-order reaction with known k, use this equation to calculate the concentration of a reactant at some specified time after the start of the reaction, or the time required for the concentration to drop to some specified value.
Integrated second-order rate law (page 665)	For a second-order reaction with known k, use this equation to calculate the concentration of a reactant at some specified time after the start of the reaction, or the time required for the concentration to drop to some specified value.
Half-lives (page 666)	Calculates a half-life from experimental data and reaction order, or calculates the rate of disappearance of a reactant.
Arrhenius equation (pages 674 and 676)	Determines an activation energy graphically and interrelates rate constants, activation energy, and temperature.

THINKING IT THROUGH

The goal for the following problems is not to find the answers themselves, but rather to assemble the information needed to solve them and explain how you would use the information to find the answers. The problems in Level 2 are more challenging than those in Level 1 and may contain more data than are required, in which case you are also asked to identify the unnecessary data. Detailed answers to the Thinking-It-Through problems can be found on the web site.

 ON-LINE HELP | **Need extra help?** Visit the Brady/ Senese web site at www.wiley.com/ college/brady

Level 1 Problems

1. Consider the following reaction:

$$C_3H_8(g) + 5O_2(g) \longrightarrow 3CO_2(g) + 4H_2O(g)$$

If at a given moment C_3H_8 is reacting at a rate of 0.400 mol L^{-1} s^{-1}, what are the rates of formation of CO_2 and H_2O? What is the rate at which O_2 is reacting? (Explain how you would obtain the answers.)

2. Suppose the following data were collected for the reaction

$$2A \longrightarrow B + 2C$$

Concentration of A (mol L^{-1})	Time (min)
0.2000	0
0.1902	10
0.1721	20
0.1482	30
0.1213	40
0.0945	50
0.0700	60

How would you determine the rate at which A is reacting at $t = 25$ min? Explain how you would determine the rates at which B and C are being formed at $t = 25$ min.

3. The reaction $I^-(aq) + OCl^-(aq) \rightarrow IO^-(aq) + Cl^-(aq)$ occurs in a basic solution. When the $[OH^-] = 1.00$ M, the reaction is first order with respect to each reactant and has a rate constant of 60.0 L mol^{-1} s^{-1}. If $[I^-] = 0.0200$ M and $[OCl^-] = 0.0300$ M at a certain time, what is the rate at which OI^- is being formed? (Describe how you would find the answer. Provide any necessary equations.)

4. Carbon-14 is a radioactive isotope that decays by the nuclear reaction

$$^{14}_{6}C \longrightarrow {}^{14}_{7}N + {}_{-1}^{0}e$$

It is a first-order process with a half-life of 5770 years. What is the rate constant for the decay? Explain how you would determine the number of years that it will take for 1/100th of the ^{14}C in a sample of naturally occurring carbon (a mixture of carbon isotopes, including ^{14}C) to decay.

5. The reaction $2NO_2 \rightarrow 2NO + O_2$ has the rate constant $k = 0.63$ L mol^{-1} s^{-1} at 600 K. If the initial concentration of NO_2

in a reaction vessel is 0.050 M, how would you determine how long it will take for the concentration to drop to 0.020 M?

6. For the reaction described in the preceding question, how can you calculate the number of molecules of NO_2 that react each second when the NO_2 concentration is $3.0 \times 10^{-3}\ M$?

Level 2 Problems

7. The following data were collected at a certain temperature for the reaction

$$2N_2O_5(g) \longrightarrow 4NO_2(g) + O_2(g)$$

Experiment	$[N_2O_5]$ (mol L^{-1})	Rate of Formation of O_2 (mol L^{-1} s^{-1})
1	0.050	0.084
2	0.075	0.126

What is the order of the reaction with respect to N_2O_5? How would you obtain the value of the rate constant at this temperature?

8. For the reaction

$$2N_2O_5(g) \longrightarrow 4NO_2(g) + O_2(g)$$

it was observed at a particular moment that the pressure in a container of these gases was increasing at the rate of 20.0 torr per second. What additional information would you need to calculate the rate at which the N_2O_5 was reacting in units of mol L^{-1} s^{-1}? Explain how you would perform the calculation.

9. At high temperatures, collisions between argon atoms and oxygen molecules can lead to the formation of oxygen atoms:

$$Ar + O_2 \longrightarrow Ar + O + O$$

At 5000 K, the rate constant for this reaction is 5.49×10^6 L mol^{-1} s^{-1}. At 10,000 K, $k = 9.86 \times 10^8$ L mol^{-1} s^{-1}. Describe the calculation required to estimate the activation energy for the reaction.

10. The rate law for the following reaction,

$$F_2(g) + 2NO_2(g) \longrightarrow 2NO_2F(g)$$

as found by experiment, is

$$\text{rate} = k\,[NO_2]\,[F_2]$$

The mechanism of the reaction is believed to consist of the following elementary processes:

$$F_2 + NO_2 \longrightarrow NO_2F + F \quad \text{(step 1)}$$

$$F + NO_2 \longrightarrow NO_2F \quad \text{(step 2)}$$

(a) Derive the rate law that would be expected if step 1 is the rate-determining step.
(b) What would the rate law be if step 2 is the rate-determining step?
(c) Which step is rate determining, according to the experimental rate law?

11. The five figures represent relative concentrations of the species A (red spheres), B (blue spheres) and AB (red-blue

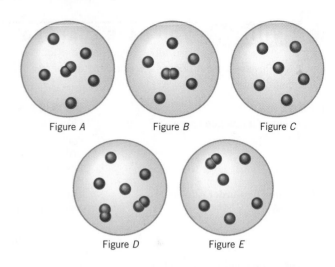

Figure A   Figure B   Figure C

Figure D   Figure E

joined spheres), which form reactants and products for the reaction:

$$A + B \longrightarrow AB$$

For each of the following rate laws identify the figure with the fastest rate and the slowest rate.
(a) rate $= k\,[A]^2$
(b) rate $= k\,[A][B]$
(c) rate $= k\,[A][B]^2$

12. The figure below shows the potential-energy reaction coordinate diagram for a chemical reaction that can occur via two reaction pathways.

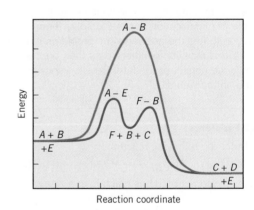

(a) What fundamental kinetic concept does this diagram illustrate?
(b) What is the equation for the overall reaction?
(c) What is the mechanism for this reaction using the higher energy pathway?
(d) What is the mechanism for this reaction using the lower energy pathway?
(e) On the figure above, identify (1) the activation energy for the reaction for the higher energy path, (2) the activation energy for the reaction for the lower energy path, and (3) ΔH for the reaction.
(f) Identify each of the species A to F as a reactant, intermediate, product, or catalyst.

REVIEW QUESTIONS

Factors That Affect Reaction Rate

15.1 What does *rate of reaction* mean in qualitative terms?

15.2 Give an example from everyday experience of (a) a very fast reaction, (b) a moderately fast reaction, and (c) a slow reaction.

15.3 In terms of reaction rates, what is an explosion?

15.4 From an economic point of view, why would industrial corporations want to know about the factors that affect the rate of a reaction?

15.5 Suppose we compared two reactions, one requiring the simultaneous collision of three molecules and the other requiring a collision between two molecules. From the standpoint of statistics, and all other factors being equal, which reaction should be faster? Explain your answer.

15.6 List the five factors that affect the rates of chemical reactions.

15.7 How does an instantaneous rate of reaction differ from an average rate of reaction?

15.8 Explain how the initial instantaneous rate of reaction can be determined from experimental concentration versus time data.

15.9 What is a *homogeneous reaction*? Give an example.

15.10 What is a *heterogeneous reaction*? Give an example.

15.11 Why are chemical reactions usually carried out in solution?

15.12 What is the major factor that affects the rate of a heterogeneous reaction?

15.13 How does particle size affect the rate of a heterogeneous reaction? Why?

15.14 The rate of hardening of epoxy glue depends on the amount of hardener that is mixed into the glue. What factor affecting reaction rates does this illustrate?

15.15 A Polaroid instant photograph develops faster if it's kept warm than if it is exposed to cold. Why?

15.16 What is a *catalyst*?

15.17 In cool weather, the number of chirps per minute from crickets diminishes. How can this be explained?

15.18 On the basis of what you learned in Chapter 12, why do foods cook faster in a pressure cooker than in an open pot of boiling water?

15.19 Persons who have been submerged in very cold water and who are believed to have drowned sometimes can be revived. On the other hand, persons who have been submerged in warmer water for the same length of time have died. Explain this in terms of factors that affect the rates of chemical reactions.

Concentration and Rate; Rate Laws

15.20 What are the units of reaction rate?

15.21 What is a *rate law*? What is the proportionality constant called?

15.22 What is meant by the *order* of a reaction?

15.23 What are the units of the rate constant for (a) a first-order reaction, (b) a second-order reaction, and (c) a zero-order reaction?

15.24 How must the exponents in a rate law be determined?

15.25 How does the dependence of reaction rate on concentration differ between a zero-order and a first-order reaction?

15.26 Is there any way of using the coefficients in the balanced overall equation for a reaction to predict with certainty what the exponents are in the rate law?

15.27 If the concentration of a reactant is doubled and the reaction rate is unchanged, what must be the order of the reaction with respect to that reactant?

15.28 If the concentration of a reactant is doubled and the reaction rate doubles, what must be the order of the reaction with respect to that reactant?

15.29 If the concentration of a reactant is doubled, by what factor will the rate increase if the reaction is second order with respect to that reactant?

15.30 In an experiment, the concentration of a reactant was tripled. The rate increased by a factor of 27. What is the order of the reaction with respect to that reactant?

15.31 Biological reactions usually involve the interaction of an enzyme with a *substrate*, the substance that actually undergoes the chemical change. In many cases, the rate of reaction depends on the concentration of the enzyme but is independent of the substrate concentration. What is the order of the reaction with respect to the substrate in such instances?

15.32 A reaction has the following rate law:

$$\text{rate} = k[A]^2[B][C]$$

What are the units of the rate constant, k?

Concentration and Time; Half-Lives

15.33 Give the equations that relate concentration to time for (a) a first-order reaction and (b) a second-order reaction.

15.34 What is meant by the term *half-life*?

15.35 How is the half-life of a first-order reaction affected by the initial concentration of the reactant?

15.36 How is the half-life of a second-order reaction affected by the initial reactant concentration?

15.37 Derive the equations for $t_{1/2}$ for first- and second-order reactions from Equations 15.5 and 15.6, respectively.

15.38 The integrated rate law for a zero-order reaction is

$$[A]_t - [A]_0 = kt$$

Derive an equation for the half-life of a zero-order reaction.

Effect of Temperature on Rate

15.39 What is the basic postulate of collision theory?

15.40 What two factors influence the effectiveness of molecular collisions in producing chemical change?

15.41 In terms of the kinetic theory, why does an increase in temperature increase the reaction rate?

15.42 Draw the potential-energy diagram for an endothermic reaction. Indicate on the diagram the activation energy for both the forward and reverse reactions. Also indicate the heat of reaction.

15.43 Explain, in terms of the law of conservation of energy, why an endothermic reaction leads to a cooling of the reaction mixture (provided heat cannot enter from outside the system).

15.44 Define the terms *transition state* and *activated complex*.

15.45 Draw a potential-energy diagram for an exothermic reaction and indicate on the diagram the location of the transition state.

15.46 Suppose a certain slow reaction is found to have a very small activation energy. What does this suggest about the importance of molecular orientation in the formation of the activated complex?

15.47 The decomposition of carbon dioxide,

$$CO_2 \longrightarrow CO + O$$

has a very large activation energy of approximately 460 kJ/mol. Explain why this is consistent with a mechanism that involves the breaking of a $C{=}O$ bond.

15.48 State the Arrhenius equation (which relates the rate constant to temperature and the activation energy) and define the symbols.

Reaction Mechanisms

15.49 What is the definition of an *elementary process*? How are elementary processes related to the mechanism of a reaction?

15.50 What is a *rate-determining step*?

15.51 In what way is the rate law for a reaction related to the rate-determining step?

15.52 A reaction has the following mechanism:

$$2NO \longrightarrow N_2O_2$$
$$N_2O_2 + H_2 \longrightarrow N_2O + H_2O$$
$$N_2O + H_2 \longrightarrow N_2 + H_2O$$

What is the net overall change that occurs in this reaction?

15.53 If the experimental rate law for the mechanism in the previous question is second order in NO and first order in H_2, which of the reaction steps is the rate-determining step?

15.54 If the reaction $NO_2 + CO \rightarrow NO + CO_2$ occurs by a one-step collision process, what would be the expected rate law for the reaction? The actual rate law is rate $= k\,[NO_2]^2$. Could the reaction actually occur by a one-step collision between NO_2 and CO? Explain.

15.55 Oxidation of NO to NO_2—one of the reactions in the production of smog—appears to involve carbon monoxide. A possible mechanism is

$$CO + {\cdot}OH \longrightarrow CO_2 + H{\cdot}$$
$$H{\cdot} + O_2 \longrightarrow HOO{\cdot}$$
$$HOO{\cdot} + NO \longrightarrow {\cdot}OH + NO_2$$

(The formulas with dots represent extremely reactive species with unpaired electrons and are called *free radicals*.) Write the net chemical equation for the reaction.

15.56 Show that the following two mechanisms give the same net overall reaction:

Mechanism 1

$$OCl^- + H_2O \longrightarrow HOCl + OH^-$$
$$HOCl + I^- \longrightarrow HOI + Cl^-$$
$$HOI + OH^- \longrightarrow H_2O + OI^-$$

Mechanism 2

$$OCl^- + H_2O \longrightarrow HOCl + OH^-$$
$$I^- + HOCl \longrightarrow ICl + OH^-$$
$$ICl + 2OH^- \longrightarrow OI^- + Cl^- + H_2O$$

15.57 The experimental rate law for the reaction $NO_2 + CO \rightarrow CO_2 + NO$ is rate $= k\,[NO_2]^2$. If the mechanism is

$$2NO_2 \longrightarrow NO_3 + NO \quad \text{(slow)}$$
$$NO_3 + CO \longrightarrow NO_2 + CO_2 \quad \text{(fast)}$$

show that the predicted rate law is the same as the experimental rate law.

15.58 Enzymes are biological catalysts. One proposed mechanism for enzyme catalysis of the reaction $A \rightarrow B$ is

$$E + A \longrightarrow EA \quad \text{(step 1)}$$
$$EA \longrightarrow E + B \quad \text{(step 2)}$$

where E is the enzyme. What is the overall rate law for this reaction if step 1 is the slower step? What is the overall rate law if step 2 is the slower step?

Catalysts

15.59 How does a catalyst increase the rate of a chemical reaction?

15.60 What is a *homogeneous catalyst*? How does it function, in general terms?

15.61 What is a *heterogeneous catalyst*? How does it function?

15.62 What is the difference in meaning between the terms *adsorption* and *absorption*? (If necessary, use a dictionary.) Which one applies to heterogeneous catalysts?

15.63 What does the catalytic converter do in the exhaust system of an automobile? Why should leaded gasoline not be used in cars equipped with catalytic converters?

REVIEW PROBLEMS

Answers to problems whose numbers are printed in color are given in Appendix B. More challenging problems are marked with asterisks. **ILW** = Interactive LearningWare solution is available at *www.wiley.com/college/brady*.

Measuring Rates of Reaction

15.64 The following data were collected at a certain temperature for the decomposition of sulfuryl chloride, SO_2Cl_2, a chemical used in a variety of organic syntheses.

$$SO_2Cl_2 \longrightarrow SO_2 + Cl_2$$

Time (min)	$[SO_2Cl_2]$ (mol L^{-1})	Time (min)	$[SO_2Cl_2]$ (mol L^{-1})
0	0.1000	600	0.0453
100	0.0876	700	0.0397
200	0.0768	800	0.0348
300	0.0673	900	0.0305
400	0.0590	1000	0.0267
500	0.0517	1100	0.0234

Make a graph of concentration versus time and determine the rate of formation of SO_2 at $t = 200$ min and $t = 600$ min.

15.65 The following data were collected for the decomposition of acetaldehyde, CH_3CHO (used in the manufacture of a variety of chemicals including perfumes, dyes, and plastics), into methane and carbon monoxide. The data were collected at a temperature of 530 °C.

$$CH_3CHO \longrightarrow CH_4 + CO$$

$[CH_3CHO]$ (mol L^{-1})	Time (s)	$[CH_3CHO]$ (mol L^{-1})	Time (s)
0.200	0	0.070	120
0.153	20	0.063	140
0.124	40	0.058	160
0.104	60	0.053	180
0.090	80	0.049	200
0.079	100		

Make a graph of concentration versus time and determine the rate of reaction of CH_3CHO after 60 s and after 120 s.

15.66 In the reaction $3H_2 + N_2 \rightarrow 2NH_3$, how does the rate of disappearance of hydrogen compare to the rate of disappearance of nitrogen? How does the rate of appearance of NH_3 compare to the rate of disappearance of nitrogen?

15.67 For the reaction $2A + B \rightarrow 3C$, it was found that the rate of disappearance of B was 0.30 mol L^{-1} s^{-1}. What were the rate of disappearance of A and the rate of appearance of C?

15.68 In the combustion of hexane (a low-boiling component of gasoline),

$$2C_6H_{14}(g) + 19O_2(g) \longrightarrow 12CO_2(g) + 14H_2O(g)$$

it was found that the rate of reaction of C_6H_{14} was 1.20 mol L^{-1} s^{-1}.
(a) What was the rate of reaction of O_2?
(b) What was the rate of formation of CO_2?
(c) What was the rate of formation of H_2O?

15.69 At a certain moment in the reaction

$$2N_2O_5 \longrightarrow 4NO_2 + O_2$$

N_2O_5 is decomposing at a rate of 2.5×10^{-6} mol L^{-1} s^{-1}. What are the rates of formation of NO_2 and O_2?

Rate Laws for Reactions

15.70 Estimate the rate of the reaction

$$H_2SeO_3 + 6I^- + 4H^+ \longrightarrow Se + 2I_3^- + 3H_2O$$

given that the rate law for the reaction (at 0 °C) is

$$\text{rate} = (5.0 \times 10^5 \text{ L}^5 \text{ mol}^{-5} \text{ s}^{-1}) [H_2SeO_3] [I^-]^3 [H^+]^2$$

and the reactant concentrations are $[H_2SeO_3] = 2.0 \times 10^{-2}$ M, $[I^-] = 2.0 \times 10^{-3}$ M, and $[H^+] = 1.0 \times 10^{-3}$ M.

15.71 Estimate the rate of the reaction

$$H^+(aq) + OH^-(aq) \longrightarrow H_2O(l)$$

given that the rate law for the reaction is

$$\text{rate} = (1.3 \times 10^{11} \text{ L mol}^{-1} \text{ s}^{-1}) [OH^-][H^+]$$

for neutral water where $[H^+] = 1.0 \times 10^{-7}$ M and $[OH^-] = 1.0 \times 10^{-7}$ M.

15.72 The oxidation of NO (released in small amounts in the exhaust of automobiles) produces the brownish-red gas NO_2, which is a component of urban air pollution.

$$2NO + O_2 \longrightarrow 2NO_2$$

The rate law for the reaction is rate $= k[NO]^2 [O_2]$. At 25 °C, $k = 7.1 \times 10^9$ L^2 mol^{-2} s^{-1}. What would be the rate of the reaction if $[NO] = 0.0010$ mol L^{-1} and $[O_2] = 0.034$ mol L^{-1}?

15.73 The rate law for the decomposition of N_2O_5 is

$$\text{rate} = k[N_2O_5].$$

If $k = 1.0 \times 10^{-5}$ s^{-1}, what is the reaction rate when the N_2O_5 concentration is 0.0010 mol L^{-1}?

15.74 For the reaction

$$2HCrO_4^- + 3HSO_3^- + 5H^+ \longrightarrow 2Cr^{3+} + 3SO_4^{2-} + 5H_2O$$

the rate law is rate $= k [HCrO_4^-] [HSO_3^-]^2 [H^+]$.
(a) What is the order of the reaction with respect to each reactant?
(b) What is the overall order of the reaction?

15.75 A key elementary process in the destruction of stratospheric ozone from nitrogen oxides in jet exhaust is the reaction

$$NO + O_3 \longrightarrow NO_2 + O_2$$

The rate law is rate $= k\,[NO][O_3]$.
(a) What is the order with respect to each reactant?
(b) What is the overall order of the reaction?

15.76 The following data were collected for the reaction

$$M + N \longrightarrow P + Q$$

Initial Concentrations (mol L^{-1})		Initial Rate of Reaction (mol L^{-1} s^{-1})
[M]	[N]	
0.010	0.010	2.5×10^{-3}
0.020	0.010	5.0×10^{-3}
0.020	0.030	4.5×10^{-2}

What is the rate law for the reaction? What is the value of the rate constant (with correct units)?

15.77 Cyclopropane, C_3H_6, is a gas used as a general anesthetic. It undergoes a slow molecular rearrangement to propylene.

cyclopropane   propylene

At a certain temperature, the following data were obtained relating concentration and rate.

Initial Concentration of C_3H_6 (mol L^{-1})	Rate of Formation of Propylene (mol L^{-1} s^{-1})
0.050	2.95×10^{-5}
0.100	5.90×10^{-5}
0.150	8.85×10^{-5}

What is the rate law for the reaction? What is the value of the rate constant, with correct units?

15.78 The reaction of iodide ion with hypochlorite ion, OCl^- (the active ingredient in a "chlorine bleach" such as Clorox), follows the equation $OCl^- + I^- \rightarrow OI^- + Cl^-$. It is a rapid reaction that gives the following rate data.

Initial Concentrations (mol L^{-1})		Rate of Formation of Cl$^-$ (mol L^{-1} s^{-1})
[OCl$^-$]	[I$^-$]	
1.7×10^{-3}	1.7×10^{-3}	1.75×10^{4}
3.4×10^{-3}	1.7×10^{-3}	3.50×10^{4}
1.7×10^{-3}	3.4×10^{-3}	3.50×10^{4}

What is the rate law for the reaction? Determine the value of the rate constant with its correct units.

15.79 The formation of small amounts of nitric oxide, NO, in automobile engines is the first step in the formation of smog. As noted earlier, nitric oxide is readily oxidized to nitrogen dioxide by the reaction

$$2NO(g) + O_2(g) \longrightarrow 2NO_2(g)$$

The following data were collected in a study of the rate of this reaction.

Initial Concentrations (mol L^{-1})		Rate of Formation of NO$_2$ (mol L^{-1} s^{-1})
[O$_2$]	[NO]	
0.0010	0.0010	7.10
0.0040	0.0010	28.4
0.0040	0.0030	255.6

What is the rate law for the reaction? What is the rate constant with its correct units?

ILW 15.80 At a certain temperature the following data were collected for the reaction $2ICl + H_2 \rightarrow I_2 + 2HCl$.

Initial Concentrations (mol L^{-1})		Initial Rate of Formation of I$_2$ (mol L^{-1} s^{-1})
[ICl]	[H$_2$]	
0.10	0.10	0.0015
0.20	0.10	0.0030
0.10	0.0500	0.00075

Determine the rate law and the rate constant (with correct units) for the reaction.

15.81 The following data were obtained for the reaction of $(CH_3)_3CBr$ with hydroxide ion at 55 °C.

$$(CH_3)_3CBr + OH^- \longrightarrow (CH_3)_3COH + Br^-$$

Initial Concentrations (mol L^{-1})		Initial Rate of Formation of $(CH_3)_3COH$ (mol L^{-1} s^{-1})
[(CH$_3$)$_3$CBr]	[OH$^-$]	
0.10	0.10	1.0×10^{-3}
0.20	0.10	2.0×10^{-3}
0.30	0.10	3.0×10^{-3}
0.10	0.20	1.0×10^{-3}
0.10	0.30	1.0×10^{-3}

What is the rate law for the reaction? What is the value of the rate constant (with correct units) at this temperature?

Concentration and Time

15.82 Data for the decomposition of SO_2Cl_2 according to the equation $SO_2Cl_2(g) \rightarrow SO_2(g) + Cl_2(g)$ were given in Problem 15.64. Show graphically that these data fit a first-order rate law. Graphically determine the rate constant for the reaction.

15.83 For the data in Problem 15.65, decide graphically whether the reaction is first or second order. Determine the rate constant for the reaction described in that problem.

ILW 15.84 The decomposition of SO_2Cl_2 described in Problem 15.64 has a first-order rate constant $k = 2.2 \times 10^{-5}$ s^{-1} at 320 °C. If the initial SO_2Cl_2 concentration in a container is 0.0040 M, what will its concentration be (a) after 1.00 hour, (b) after 1.00 day?

15.85 If it takes 75.0 min for the concentration of a reactant to drop to 20% of its initial value in a first-order reaction, what is the rate constant for the reaction in the units min^{-1}?

15.86 The concentration of a drug in the body is often expressed in units of milligrams per kilogram of body weight. The initial dose of a drug in an animal was 25.0 mg/kg body weight. After 2.00 hours, this concentration had dropped to 15.0 mg/kg body weight. If the drug is eliminated metabolically by a first-order process, what is the rate constant for the process in units of min^{-1}?

15.87 In the preceding problem, what must the initial dose of the drug be in order for the drug concentration 3.00 hr afterward to be 5.0 mg/kg body weight?

15.88 The decomposition of hydrogen iodide follows the equation $2HI(g) \rightarrow H_2(g) + I_2(g)$. The reaction is second order and has a rate constant equal to 1.6×10^{-3} L mol^{-1} s^{-1} at 700 °C. If the initial concentration of HI in a container is 3.4×10^{-2} M, how many minutes will it take for the concentration to be reduced to 8.0×10^{-4} M?

15.89 The second-order rate constant for the decomposition of HI at 700 °C was given in the preceding problem. At 2.5×10^3 min after a particular experiment had begun, the HI concentration was equal to 4.5×10^{-4} mol L^{-1}. What was the initial molar concentration of HI in the reaction vessel?

Half-Lives

15.90 The half-life of a certain first-order reaction is 15 min. What fraction of the original reactant concentration will remain after 2.0 hr?

15.91 Strontium-90 has a half-life of 28 years. How long will it take for all of the strontium-90 presently on Earth to be reduced to 1/32nd of its present amount?

15.92 Using the graph from Problem 15.64, determine the time required for the SO_2Cl_2 concentration to drop from 0.100 mol L^{-1} to 0.050 mol L^{-1}. How long does it take for the concentration to drop from 0.050 mol L^{-1} to 0.025 mol L^{-1}? What is the order of this reaction? (*Hint:* How is the half-life related to concentration?)

15.93 Using the graph from Problem 15.65, determine how long it takes for the CH_3CHO concentration to decrease from 0.200 mol L^{-1} to 0.100 mol L^{-1}. How long does it take the concentration to drop from 0.100 mol L^{-1} to 0.050 mol L^{-1}? What is the order of this reaction? (*Hint:* How is the half-life related to concentration?)

15.94 A certain first-order reaction has a rate constant $k = 1.6 \times 10^{-3}$ s^{-1}. What is the half-life for this reaction?

15.95 The decomposition of NOCl, the compound that gives a yellow-orange color to aqua regia (a mixture of concentrated HCl and HNO_3 that's able to dissolve gold and platinum) follows the reaction $2NOCl \rightarrow 2NO + Cl_2$. It is a second-order reaction with $k = 6.7 \times 10^{-4}$ L mol^{-1} s^{-1} at 400 K. What is the half-life of this reaction if the initial concentration of NOCl is 0.20 mol L^{-1}?

Calculations Involving the Activation Energy

15.96 The following data were collected for a reaction.

Rate Constant (L mol^{-1} s^{-1})	Temperature (°C)
2.88×10^{-4}	320
4.87×10^{-4}	340
7.96×10^{-4}	360
1.26×10^{-3}	380
1.94×10^{-3}	400

Determine the activation energy for the reaction in kJ/mol both graphically and by calculation using Equation 15.9. For the calculation of E_a, use the first and last sets of data in the table above.

15.97 Rate constants were measured at various temperatures for the reaction $HI(g) + CH_3I(g) \rightarrow CH_4(g) + I_2(g)$. The following data were obtained.

Rate Constant (L mol^{-1} s^{-1})	Temperature (°C)
1.91×10^{-2}	205
2.74×10^{-2}	210
3.90×10^{-2}	215
5.51×10^{-2}	220
7.73×10^{-2}	225
1.08×10^{-1}	230

Determine the activation energy in kJ/mol both graphically and by calculation using Equation 15.9. For the calculation of E_a, use the first and last sets of data in the table above.

ILW **15.98** The decomposition of NOCl, $2NOCl \rightarrow 2NO + Cl_2$, has $k = 9.3 \times 10^{-5}$ L mol^{-1} s^{-1} at 100 °C and $k = 1.0 \times 10^{-3}$ L mol^{-1} s^{-1} at 130 °C. What is E_a for this reaction in kJ mol^{-1}? Use the data at 100 °C to calculate the frequency factor.

15.99 The conversion of cyclopropane, an anesthetic, to propylene (see Problem 15.77) has a rate constant $k = 1.3 \times 10^{-6}$ s^{-1} at 400 °C and $k = 1.1 \times 10^{-5}$ s^{-1} at 430 °C.
(a) What is the activation energy in kJ/mol?
(b) What is the value of the frequency factor, A, for this reaction?
(c) What is the rate constant for the reaction at 350 °C?

15.100 The reaction of CO_2 with water to form carbonic acid, $CO_2(aq) + H_2O \rightarrow H_2CO_3(aq)$, has $k = 3.75 \times 10^{-2}$ s^{-1} at 25 °C and $k = 2.1 \times 10^{-3}$ s^{-1} at 0 °C. What is the activation energy for this reaction in kJ/mol?

15.101 If a reaction has $k = 3.0 \times 10^{-4}$ s^{-1} at 25 °C and an activation energy of 100.0 kJ/mol, what will the value of k be at 50 °C?

15.102 The decomposition of N_2O_5 has an activation energy of 103 kJ/mol and a frequency factor of 4.3×10^{13} s^{-1}. What is the rate constant for this decomposition at (a) 20 °C and (b) 100 °C?

15.103 At 35 °C, the rate constant for the reaction

$$C_{12}H_{22}O_{11} + H_2O \longrightarrow C_6H_{12}O_6 + C_6H_{12}O_6$$

$$\text{sucrose} \qquad\qquad\qquad \text{glucose} \quad\;\; \text{fructose}$$

is $k = 6.2 \times 10^{-5}$ s^{-1}. The activation energy for the reaction is 108 kJ mol^{-1}. What is the rate constant for the reaction at 45 °C?

ADDITIONAL EXERCISES

5.104 For the reaction and data given in Problem 15.65, make a graph of concentration versus time for the *formation* of CH_4. What are the rates of formation of CH_4 at $t = 40$ s and $t = 100$ s?

15.105 The age of wine can be determined by measuring the trace amount of radioactive tritium, ^3H, present in a sample. Tritium is formed from hydrogen in water vapor in the upper atmosphere by cosmic bombardment, so all naturally occurring water contains a small amount of this isotope. Once the water is in a bottle of wine, however, the formation of additional tritium from the water is negligible, so the tritium initially present gradually diminishes by a first-order radioactive decay with a half-life of 12.5 years. If a bottle of wine is found to have a tritium concentration that is 0.100 that of freshly bottled wine (i.e., $[^3H]_t = 0.100 [^3H]_0$), what is the age of the wine?

15.106 Carbon-14 dating can be used to estimate the age of formerly living materials because the uptake of carbon-14 from carbon dioxide in the atmosphere stops once the organism dies. If tissue samples from a mummy contain about 81.0% of the carbon-14 expected in living tissue, how old is the mummy? The half-life for decay of carbon-14 is 5730 years.

15.107 One of the reactions that occurs in polluted air in urban areas is $2NO_2(g) + O_3(g) \rightarrow N_2O_5(g) + O_2(g)$. It is believed that a species with the formula NO_3 is involved in the mechanism, and the observed rate law for the overall reaction is rate $= k[NO_2][O_3]$. Propose a mechanism for this reaction that includes the species NO_3 and is consistent with the observed rate law.

$*$**15.108** Suppose a reaction occurs with the mechanism

(1) $2A \rightleftharpoons A_2$ (fast)

(2) $A_2 + E \longrightarrow B + C$ (slow)

in which the first step is a very rapid reversible reaction that can be considered to be essentially an equilibrium (forward and reverse reactions occurring at the same rate) and the second is a slow step.
(a) Write the rate law for the forward reaction in step 1.
(b) Write the rate law for the reverse reaction in step 1.
(c) Write the rate law for the rate-determining step.
(d) What is the chemical equation for the net reaction that occurs in this chemical change?
(e) Use the results of parts a and b to rewrite the rate law of the rate-determining step in terms of the concentrations of the reactants in the overall balanced equation for the reaction.

$*$**15.109** A reaction that has the stoichiometry

$2A + B \longrightarrow$ products

was found to yield the following data.

Initial Concentrations (mol L^{-1})		Initial Rate of Reaction
[A]	[B]	of A (mol L^{-1} s^{-1})
0.020	0.030	0.0150
0.025	0.030	0.0188
0.025	0.040	0.0334

(a) What is the rate law for the reaction?
(b) What is the rate constant for the reaction with its correct units?

15.110 The decomposition of urea, $(NH_2)_2CO$, in 0.10 M HCl follows the equation

$$(NH_2)_2CO(aq) + 2H^+(aq) + H_2O \longrightarrow 2NH_4^+(aq) + CO_2(g)$$

At 60 °C, $k = 5.84 \times 10^{-6}$ min^{-1} and at 70 °C, $k = 2.25 \times 10^{-5}$ min^{-1}. If this reaction is run at 80 °C starting with a urea concentration of 0.0020 M, how many minutes will it take for the urea concentration to drop to 0.0012 M?

15.111 Show that for a reaction that obeys the general rate law

$$\text{rate} = k[A]^n$$

a graph of log(rate) versus log$[A]$ should yield a straight line with a slope equal to the order of the reaction. For the reaction in Review Exercise 15.64, measure the rate of the reaction at $t = 150$, 300, 450, and 600 s. Then graph log (rate) versus log $[SO_2Cl_2]$ and determine the order of the reaction with respect to SO_2Cl_2.

15.112 The bonding in O_2 was discussed in Section 10.7 as was the bonding in N_2. Molecular nitrogen is very unreactive, whereas molecular oxygen is very reactive. On the basis of what you have learned in this chapter, what fact about O_2 is responsible, at least in part, for its reactivity? What might account for the low degree of reactivity of molecular nitrogen?

15.113 It was mentioned that the rates of many reactions approximately double for each 10 °C rise in temperature. Assuming a starting temperature of 25 °C, what would the activation energy be, in kJ mol^{-1}, if the rate of a reaction were to be twice as large at 35 °C?

15.114 The development of a photographic image on film is a process controlled by the kinetics of the reduction of silver halide by a developer. The time required for development at a particular temperature is inversely proportional to the rate constant for the process. Below are published data on development times for Kodak's Tri-X film using Kodak D-76 developer. From these data, estimate the activation energy (in units of kJ mol^{-1}) for the development process. Also estimate the development time at 15 °C.

Temperature (°C)	Development Time (minutes)
18	10
20	9
21	8
22	7
24	6

15.115 The rate at which crickets chirp depends on the ambient temperature, because crickets are cold-blooded insects whose body temperature follows the temperature of their environment. It has been found that the Celsius temperature can be estimated by counting the number of chirps in 8 seconds and then adding 4. In other words, $t_C = $ (number of chirps in 8 seconds) + 4.

(a) Calculate the number of chirps in 8 seconds for temperatures of 20, 25, 30, and 35 °C.

(b) The number of chirps per unit of time is directly proportional to the rate constant for a biochemical reaction involved in the cricket's chirp. On the basis of this assumption, make a graph of ln(chirps in 8 s) versus $(1/T)$. Calculate the activation energy for the biochemical reaction involved.

(c) How many chirps would a cricket make in 8 seconds at a temperature of 40 °C?

*15.116 The cooking of an egg involves the denaturation of a protein called albumen. The time required to achieve a particular degree of denaturation is inversely proportional to the rate constant for the process. This reaction has a high activation energy, $E_a = 418$ kJ mol^{-1}. Calculate how long it would take to cook a traditional three-minute egg on top of Mt. McKinley in Alaska on a day when the atmospheric pressure there is 355 torr.

*15.117 In the upper atmosphere, a layer of ozone, O_3, shields Earth from harmful ultraviolet radiation. The ozone is generated by the reactions

$$O_2 \xrightarrow{hv} 2O$$

$$O + O_2 \longrightarrow O_3$$

The migration of chlorofluorocarbon aerosol propellants and refrigerant fluids such as Freon 12 (CCl_2F_2) into the stratosphere would at least partially destroy the ozone shield by the reactions

$$CCl_2F_2 \longrightarrow CClF_2 + Cl$$

$$Cl + O_3 \longrightarrow ClO + O_2$$

$$ClO + O \longrightarrow Cl + O_2$$

Read Facets of Chemistry 15.1 and then explain how the second and third reactions constitute a chain reaction that could destroy huge numbers of O_3 molecules as well as O atoms that might otherwise react with O_2 to replenish the ozone.

*15.118 The following question is based on Facets of Chemistry 15.1. The reaction of hydrogen and bromine appears to follow the mechanism.

$$Br_2 \xrightarrow{hv} 2Br \cdot$$

$$Br \cdot + H_2 \longrightarrow HBr + H \cdot$$

$$H \cdot + Br_2 \longrightarrow HBr + Br \cdot$$

$$2Br \cdot \longrightarrow Br_2$$

(a) Identify the initiation step in the mechanism.
(b) Identify any propagation steps.
(c) Identify the termination step.

The mechanism also contains the reaction

$$H \cdot + HBr \longrightarrow H_2 + Br \cdot$$

How does this reaction affect the rate of formation of HBr?

Chemical Equilibrium— General Concepts

Coral is made of calcium carbonate. Rising ocean temperatures and rising carbon dioxide levels in the Earth's oceans have shifted the equilibrium between solid and dissolved calcium carbonate on many coral reefs. The way changes in concentrations affect the position of equilibrium is one of the topics you will study in this chapter.

THIS CHAPTER IN CONTEXT In Chapter 15, we were concerned with *how fast* chemical reactions occur. We saw that some chemical reactions slow down as the reactants are consumed and the products form. After a time, the concentrations stop changing, and we are often left with a mixture that contains both reactants and products. In this and the following three chapters, we'll be concerned with *how far* the reaction has gone once concentrations stop changing.

Although no further change in concentration occurs after the reaction has run for a sufficient time, the reaction hasn't stopped. The forward and reverse reactions are simply running at the same rates, so that we observe no net change. The system is in a state of *dynamic chemical equilibrium*. We introduced dynamic equilibrium in Section 5.7, and have already studied many specific examples of dynamic equilibria in Chapters 12, 14, and 15. In a liquid–vapor equilibrium, for example, two opposing processes, evaporation and condensation, run at the same rates, so there is no net change in the amount of liquid or vapor. In a chemical equilibrium, such as the ionization of a weak acid, the acid dissociates into ions at the same rate that the ions recombine to form the acid, so no net change in concentration is observed.

The reaction read from left to right is the *forward reaction.* The reaction read from right to left is the *reverse reaction.*

Our goals in this chapter are to quantitatively predict the concentrations of reactants and products in a mixture at equilibrium, and to understand factors that affect the composition of a system at equilibrium. Knowing these factors gives us some control over the outcome of a reaction. For example, we'll see why the formation of the pollutant nitric oxide (NO) by reaction of N_2 and O_2 in a gasoline engine can be reduced, in a predictable way, by lowering the combustion temperature within the engine. The concept of equilibrium is crucial in understanding living systems, because concentrations of substances within living cells must be carefully controlled. The concepts introduced in this chapter are fundamental in understanding many current environmental problems, including global warming, acid rain, and stratospheric ozone depletion. We'll use the basic principles of equilibrium developed in this chapter to understand acid–base equilibria in Chapter 17 and solubility equilibria in Chapter 18. In Chapter 20, we will tie the concept of chemical equilibrium with the energetics of chemical change and the drive toward the most probable state.

16.1 ▶ Dynamic equilibrium is achieved when the rates of two opposing processes are equal

We have seen that acetic acid ($HC_2H_3O_2$) rapidly establishes equilibrium in water:

$$HC_2H_3O_2(aq) + H_2O \rightleftharpoons H_3O^+(aq) + C_2H_3O_2^-(aq)$$

At equilibrium, $HC_2H_3O_2$ and H_2O are reacting at the same rate as H_3O^+ and $C_2H_3O_2^-$. How is this dynamic equilibrium established?

When acetic acid is first added to the water, the forward reaction (ionization) occurs rapidly as molecules of $HC_2H_3O_2$ and H_2O collide. But as the number of $HC_2H_3O_2$ molecules decreases, the forward reaction slows. Just the opposite effect is found for the reverse reaction. As the concentration of ions increases, the ions encounter each other more frequently, and the reverse reaction speeds up. After some time, the reverse process will have to wait for the forward process to produce more ions. The forward process will have to wait for the reverse process to produce more acetic acid molecules. This means that eventually *the rates of the forward and reverse reactions become equal,* so the concentrations of the ions cease to change. At this point the system has reached a state of dynamic equilibrium.

> It is an *equilibrium* because the concentrations don't change. It is *dynamic* because the opposing reactions never cease.

The reaction of acetic acid with water is just one example of a general phenomenon. For almost any reaction, the concentrations change quickly at first. Eventually the concentrations approach steady equilibrium values, as shown in Figure 16.1 for the decomposition of $N_2O_4(g)$ into $NO_2(g)$. These concentrations will never change, unless the system is disturbed in some way. Remember the forward and reverse reactions are still running; even after equilibrium has been reached, both the forward and reverse reactions continue to occur. That is what the double arrows ($\rightleftharpoons$) represent in the equilibrium equation.

> Sometimes an equal sign (=) is used in place of the double arrows.

Most chemical systems reach a state of dynamic equilibrium, given enough time. Sometimes, though, this equilibrium is extremely difficult (or even impossible) to detect, because in some reactions the amounts of either the reactants or products present at equilibrium are virtually zero. For instance, when a strong acid such as HCl is dissolved in water, it reacts so completely to produce H_3O^+ and Cl^- that no detectable amount of HCl molecules remains; we say that the HCl is completely ionized.

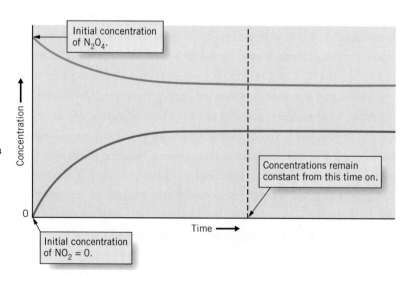

FIGURE **16.1** *The approach to equilibrium.* In the decomposition of $N_2O_4(g)$ into $NO_2(g)$, $N_2O_4(g) \rightleftharpoons 2NO_2(g)$, the concentrations of the N_2O_4 and NO_2 change relatively fast at first. As time passes, the concentrations change more and more slowly. When equilibrium is reached, the concentrations of N_2O_4 and NO_2 no longer change with time; they remain constant.

16.2 ▶ Closed systems reach the same equilibrium concentrations whether we start with reactants or with products

The composition of the equilibrium mixture is found to be independent of whether we begin the reaction from the "reactant side" or the "product side." In the equilibrium between acetic acid and water, for example, it doesn't matter whether we start by combining a mole of acetic acid with a mole of water, or whether we start with a mole of acetate ion and a mole of hydronium ion. The final amounts of acetic acid, acetate ion, water, and hydronium ion will be the same at equilibrium in either case.

Let's consider some experiments that we might perform at 25 °C on the gaseous decomposition of N_2O_4 into NO_2 shown in Figure 16.1.

$$N_2O_4(g) \rightleftharpoons 2NO_2(g)$$

It's easy to detect changes in the position of equilibrium in this reaction. N_2O_4 is colorless, and NO_2 is brown. When the reaction runs in the forward direction, there will be more NO_2 and the mixture will be browner; when the reaction runs in reverse, the mixture will become more and more transparent.

Suppose we set up the two experiments shown in Figure 16.2. In the first 1 liter flask we place 0.0350 mol N_2O_4. Since no NO_2 is present, some N_2O_4 must decompose for the mixture to reach equilibrium, so the reaction will proceed in the forward direction (i.e., from left to right). When equilibrium is reached, we find the concentration of N_2O_4 has dropped to 0.0292 mol L^{-1} and the concentration of NO_2 has become 0.0116 mol L^{-1}.

In the second 1 liter flask we place 0.0700 mol of NO_2 (*precisely* the amount of NO_2 that would form if 0.0350 mol of N_2O_4—the amount placed in the first flask—decomposed completely). In this second flask there is no N_2O_4 present initially, so NO_2 molecules must combine, following the reverse reaction (right to left) shown above, to give enough N_2O_4 for equilibrium. When we measure the concentrations at equilibrium in the second flask, we find, once again, 0.0292 mol L^{-1} of N_2O_4 and 0.0116 mol L^{-1} of NO_2.

We see here that the same equilibrium composition is reached whether we begin with pure NO_2 or pure N_2O_4, as long as the *total* amount of nitrogen and oxygen to be divided between these two substances is the same. Similar observations apply to other chemical systems as well, which leads to the following generalization.

> For a given *overall* system composition, we always reach the same equilibrium concentrations whether equilibrium is approached from the forward or reverse direction.

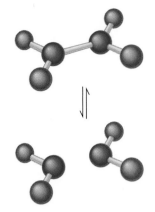

The equilibrium between N_2O_4 and NO_2.

Stoichiometrically, 0.0700 mol of NO_2 could be formed from 0.0350 mol of N_2O_4, since the ratio of NO_2 to N_2O_4 is 2:1.

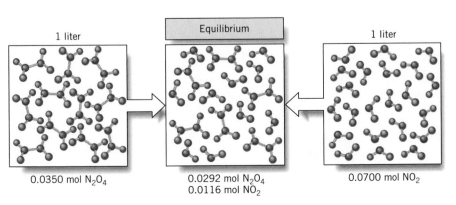

1 liter

0.0350 mol N_2O_4

Equilibrium

0.0292 mol N_2O_4
0.0116 mol NO_2

1 liter

0.0700 mol NO_2

FIGURE 16.2 *Reaction reversibility for the equilibrium $N_2O_4(g) \rightleftharpoons 2NO_2(g)$.* The same equilibrium composition is reached from either the forward or reverse direction, provided the *overall* system composition is the same. Because pure NO_2 is brown and pure N_2O_4 is colorless, the amber color of the equilibrium mixture indicates that both species are present at equilibrium.

16.3 ▶ A law relating equilibrium concentrations can be derived from the balanced chemical equation for a reaction

For any chemical system at equilibrium, there exists a simple relationship among the molar concentrations of the reactants and products. To see this, let's consider the gaseous reaction of hydrogen with iodine to form hydrogen iodide. The equation for the reaction is

$$H_2(g) + I_2(g) \rightleftharpoons 2HI(g)$$

Figure 16.3 shows the results of several experiments measuring equilibrium amounts of each gas, starting with different amounts of the reactants and product in a 10.0 L reaction vessel. When equilibrium is reached, the amounts of H_2, I_2, and HI are different for each experiment, as are their molar concentrations. This isn't particularly surprising, but what is amazing is that the relationship among the concentrations is very simple and can actually be predicted (once we've learned how) from the balanced equation for the reaction.

The molar concentrations are obtained by dividing the number of moles of each substance by the volume, 10.0 L.

For each experiment in Figure 16.3, if we square the molar concentration of HI at equilibrium and then divide this by the product of the equilibrium molar concentrations of H_2 and I_2, we obtain the same numerical value. This is shown in Table 16.1 where we have once again used square brackets around formulas as symbols for molar concentrations.

The fraction used to calculate the values in the last column of Table 16.1,

$$\frac{[HI]^2}{[H_2]\,[I_2]}$$

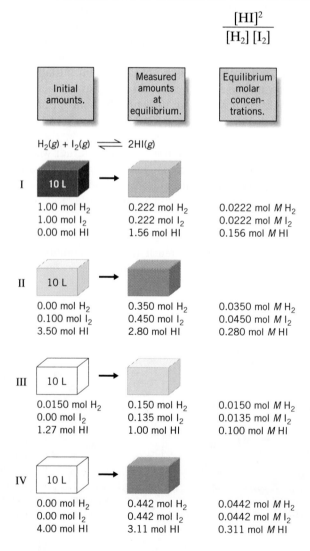

FIGURE 16.3 *Four experiments to study the equilibrium among H_2, I_2, and HI gases.* Different amounts of the reactants and product are placed in a 10.0 L reaction vessel at 440 °C where the gases established the equilibrium: $H_2(g) + I_2(g) \rightleftharpoons 2HI(g)$. When equilibrium is reached, different amounts of reactants and products remain in each experiment, which gives different equilibrium concentrations.

TABLE 16.1	EQUILIBRIUM CONCENTRATIONS AND THE MASS ACTION EXPRESSION			
	Equilibrium Concentrations (mol L^{-1})			$\dfrac{[HI]^2}{[H_2][I_2]}$
Experiment	$[H_2]$	$[I_2]$	$[HI]$	
I	0.0222	0.0222	0.156	$(0.156)^2/(0.0222)(0.0222) = 49.4$
II	0.0350	0.0450	0.280	$(0.280)^2/(0.0350)(0.0450) = 49.8$
III	0.0150	0.0135	0.100	$(0.100)^2/(0.0150)(0.0135) = 49.4$
IV	0.0442	0.0442	0.311	$(0.311)^2/(0.0442)(0.0442) = 49.5$
				Average $= 49.5$

is called the **mass action expression.**[1] The origin of this term isn't important; just consider it a name we use to refer to this fraction. The numerical value of the mass action expression is called the **reaction quotient** and is often symbolized by the letter Q. For this reaction, we can write

$$Q = \frac{[HI]^2}{[H_2][I_2]}$$

for any set of H_2, I_2, and HI concentrations, whether we are at equilibrium or not.

In Table 16.1, notice that when H_2, I_2, and HI are in dynamic equilibrium at 440 °C, the reaction quotient is equal to essentially the same constant value of 49.5. In fact, if we repeated the experiments in Figure 16.3 over and over again starting with different amounts of H_2, I_2, and HI, we would still obtain the same reaction quotient, provided the systems had reached equilibrium and the temperature was 440 °C. Therefore, for this reaction at equilibrium we can write

$$\frac{[HI]^2}{[H_2][I_2]} = 49.5 \quad \text{(at equilibrium at 440 °C)} \quad (16.1)$$

This relationship is called the **equilibrium law** for the system. Significantly, it tells us that for a mixture of these three gases to be at equilibrium at 440 °C, the value of the mass action expression (the reaction quotient) must equal 49.5. If the reaction quotient has any other value, then the gases are not in equilibrium at this temperature. The constant 49.5, which characterizes this equilibrium system, is called the **equilibrium constant.** The equilibrium constant is usually symbolized by K_c (the subscript c because we write the mass action expression using molar concentrations). Thus, we can state the equilibrium law as follows:

$$\frac{[HI]^2}{[H_2][I_2]} = K_c = 49.5 \quad \text{(at 440 °C)} \quad (16.2)$$

Equation 16.2 can be called the *equilibrium condition* for the reaction of H_2, I_2, and HI at 440 °C.

It is often useful to think of an equilibrium law such as Equation 16.2 as a *condition* that must be met for equilibrium to exist.

For chemical equilibrium to exist in a reaction mixture, the reaction quotient Q must be equal to the equilibrium constant, K_c.

---

[1]The mass action expression is derived using thermodynamics, which we'll discuss in Chapter 20. Technically, each concentration in the mass action expression should be divided by its value at standard state before inserting it into the expression. This makes the value of any mass action expression unitless. *For substances in solution,* the standard state is an effective molar concentration of 1 *M*, so we would have to divide each concentration by 1 *M* (1 mol L^{-1}), which makes no difference in the numerical value of the mass action expression as long as all concentrations are molarities. *For gases,* the standard state is 1 bar, so we would have to divide each gas concentration by its concentration at that pressure, or each gas partial pressure by 1 bar. For simplicity, we'll leave the standard state values out of our mass action expressions. Later we'll see that these values are lumped into the numerical value of the expression.

In general, it is necessary to specify the temperature when giving a value of K_c, because K_c changes when the temperature changes. For example, for the reaction $CH_4(g) + H_2O(g) \rightleftharpoons CO(g) + 3H_2(g)$,

$K_c = 1.78 \times 10^{-3}$ at 800 °C

$K_c = 4.68 \times 10^{-2}$ at 1000 °C

$K_c = 5.67$ at 1500 °C

Notice that even though we cannot predict the *rate law* from the balanced overall equation, we can predict the *equilibrium law*.

As you've probably noticed, we have repeatedly mentioned the temperature when referring to the value of K_c. This is because the value of the equilibrium constant changes when the temperature changes. Thus, if we had performed the experiments in Figure 16.3 at a temperature other than 440 °C, we would have obtained a different value for K_c.

An important fact about the mass action expression and the equilibrium law is that it can *always* be predicted from the balanced chemical equation for the reaction. For example, for the general chemical equation

$$dD + eE \rightleftharpoons fF + gG$$

where D, E, F, and G represent chemical formulas and d, e, f, and g are their coefficients, the mass action expression is

$$\frac{[F]^f [G]^g}{[D]^d [E]^e}$$

The exponents in the mass action expression are the same as the stoichiometric coefficients in the balanced equation. The condition for equilibrium in this reaction is given by the equation

$$\frac{[F]^f [G]^g}{[D]^d [E]^e} = K_c$$

where the only concentrations that satisfy the equation are *equilibrium concentrations.*

Notice that in writing the mass action expression the molar concentrations of the products are always placed in the numerator and those of the reactants appear in the denominator. Also note that after being raised to appropriate powers the concentration terms are *multiplied,* not added.

TOOLS

Constructing the equilibrium law

| **EXAMPLE 16.1**

 Writing the Equilibrium Law | Most of the hydrogen produced in the United States is derived from methane (CH_4) in natural gas, using the forward reaction of the equilibrium $$CH_4(g) + H_2O(g) \rightleftharpoons CO(g) + 3H_2(g)$$ What is the equilibrium law for this reaction?

 ANALYSIS: The equilibrium law sets the mass action expression equal to the equilibrium constant. To form the mass action expression, we place the concentrations of the products in the numerator and the concentrations of the reactants in the denominator. The coefficients in the equation become exponents on the concentrations.

 SOLUTION: The equilibrium law is $$\frac{[CO][H_2]^3}{[CH_4][H_2O]} = K_c$$

 Is the Answer Reasonable?
 Check to see that products are on top of the fraction bar and that reactants are on the bottom. Also check the superscripts and be sure that they are the same as those in the balanced chemical equation. Notice that we omit writing the exponent when it is equal to 1. |

PRACTICE EXERCISE 1: Write the equilibrium law for each of the following:
(a) $2H_2(g) + O_2(g) \rightleftharpoons 2H_2O(g)$
(b) $CH_4(g) + 2O_2(g) \rightleftharpoons CO_2(g) + 2H_2O(g)$

The rule that we always write the concentrations of the products in the numerator of the mass action expression and the concentrations of the reactants in the denominator is not required by nature. It is simply a convention chemists have agreed on. Certainly, if the mass action expression is equal to a constant,

$$\frac{[HI]^2}{[H_2][I_2]} = K_c$$

its reciprocal is also equal to a constant (let's call it K_c'),

$$\frac{[H_2][I_2]}{[HI]^2} = \frac{1}{K_c} = K_c'$$

As we mentioned earlier, if we have the chemical equation for the equilibrium, we can *always* construct the correct mass action expression from it. For example, suppose we're told that at a particular temperature $K_c = 10.0$ for the reaction

$$2NO_2(g) \rightleftharpoons N_2O_4(g)$$

From the chemical equation, we can write the correct mass action expression and the correct equilibrium law.

$$K_c = \frac{[N_2O_4]}{[NO_2]^2} = 10.0$$

The balanced chemical equation contains all the information we need to write the equilibrium law.

Equilibrium laws can be combined, scaled, and reversed

Sometimes it is useful to be able to combine chemical equilibria to obtain the equation for some other reaction of interest. In doing this, we perform various operations such as reversing an equation, multiplying the coefficients by some factor, and adding the equations to give the desired equation. In our discussion of thermochemistry, you learned how such manipulations affect ΔH values. Some different rules apply to changes in the mass action expressions and equilibrium constants.

 **TOOLS**

Manipulating equilibrium equations

Changing the direction of an equilibrium

When the direction of an equation is reversed, the new equilibrium constant is the reciprocal of the original. You have just seen this in the discussion above. As another example, when we reverse the equilibrium

$$PCl_3 + Cl_2 \rightleftharpoons PCl_5 \qquad K_c = \frac{[PCl_5]}{[PCl_3][Cl_2]}$$

we obtain

$$PCl_5 \rightleftharpoons PCl_3 + Cl_2 \qquad K_c' = \frac{[PCl_3][Cl_2]}{[PCl_5]}$$

The mass action expression for the second reaction is the reciprocal of that for the first, so K_c' equals $1/K_c$.

Multiplying the coefficients by a factor

When the coefficients in an equation are multiplied by a factor, the equilibrium constant is raised to a power equal to that factor. For example, suppose we multiply the coefficients of the equation

$$PCl_3 + Cl_2 \rightleftharpoons PCl_5 \qquad K_c = \frac{[PCl_5]}{[PCl_3][Cl_2]}$$

by 2. This gives

$$2PCl_3 + 2Cl_2 \rightleftharpoons 2PCl_5 \qquad K_c'' = \frac{[PCl_5]^2}{[PCl_3]^2[Cl_2]^2}$$

Comparing mass action expressions, we see that $K_c'' = K_c^2$.

Adding chemical equilibria

When chemical equilibria are added, their equilibrium constants are multiplied. For example, suppose we add the following two equations:

We have numbered the equilibrium constants just to distinguish one from the other.

$$2N_2 + O_2 \rightleftharpoons 2N_2O \qquad K_{c1} = \frac{[N_2O]^2}{[N_2]^2[O_2]}$$

$$2N_2O + 3O_2 \rightleftharpoons 4NO_2 \qquad K_{c2} = \frac{[NO_2]^4}{[N_2O]^2[O_2]^3}$$

$$2N_2 + 4O_2 \rightleftharpoons 4NO_2 \qquad K_{c3} = \frac{[NO_2]^4}{[N_2]^2[O_2]^4}$$

If we multiply the mass action expression for K_{c1} by that for K_{c2}, we obtain the mass action expression for K_{c3}.

$$\frac{[\cancel{N_2O}]^2}{[N_2]^2[O_2]} \times \frac{[NO_2]^4}{[\cancel{N_2O}]^2[O_2]^3} = \frac{[NO_2]^4}{[N_2]^2[O_2]^4}$$

Therefore, $K_{c1} \times K_{c2} = K_{c3}$.

PRACTICE EXERCISE 2: At 25 °C, $K_c = 7.0 \times 10^{25}$ for the reaction

$$2SO_2(g) + O_2(g) \rightleftharpoons 2SO_3(g)$$

What is the value of K_c for the reaction $SO_3(g) \rightleftharpoons SO_2(g) + \frac{1}{2}O_2(g)$?

PRACTICE EXERCISE 3: At 25 °C, the following reactions have the equilibrium constants noted to the right of their equations.

$$2CO(g) + O_2(g) \rightleftharpoons 2CO_2(g) \qquad K_c = 3.3 \times 10^{91}$$

$$2H_2(g) + O_2(g) \rightleftharpoons 2H_2O(g) \qquad K_c = 9.1 \times 10^{80}$$

Use these data to calculate K_c for the reaction

$$H_2O(g) + CO(g) \rightleftharpoons CO_2(g) + H_2(g)$$

16.4 ▶ Equilibrium laws for gaseous reactions can be written in terms of concentrations or pressures

When all the reactants and products are gases, we can formulate mass action expressions in terms of partial pressures as well as molar concentrations. This is possible because the molar concentration of a gas is proportional to its partial pressure. This comes from the ideal gas law,

$$PV = nRT$$

Solving for P gives

$$P = \left(\frac{n}{V}\right)RT$$

If you double the molar concentration of a gas without changing its temperature or volume, you double its pressure.

The quantity n/V has units of mol/L and is simply the molar concentration. Therefore, we can write

$$P = (\text{molar concentration}) \times RT \qquad (16.3)$$

This equation applies whether the gas is by itself in a container or part of a mixture. In the case of a gas mixture, P is the partial pressure of the gas.

The relationship expressed in Equation 16.3 lets us write the mass action expression for reactions between gases either in terms of molarities or partial pressures. However, when we make a switch we can't expect the numerical values of the equilibrium constants to be the same, so we use two different symbols for K. When molar concentrations are used, we use the symbol K_c. When partial pressures are used, then K_P is the symbol. For example, the equilibrium law for the reaction of nitrogen with hydrogen to form ammonia

$$N_2(g) + 3H_2(g) \rightleftharpoons 2NH_3(g)$$

can be written in either of the following two ways:

$$\frac{[NH_3]^2}{[N_2][H_2]^3} = K_c \qquad \begin{pmatrix} \text{because molar concentrations} \\ \text{are used in the mass action} \\ \text{expression} \end{pmatrix}$$

$$\frac{P_{NH_3}^2}{P_{N_2}P_{H_2}^3} = K_P \qquad \begin{pmatrix} \text{because partial pressures are} \\ \text{used in the mass action} \\ \text{expression} \end{pmatrix}$$

The equilibrium molar concentrations can be used to calculate K_c, whereas the equilibrium partial pressures can be used to calculate K_P. We will discuss how to convert between K_c and K_P in Section 16.6.

EXAMPLE 16.2
Writing Expressions for K_P

Most of the world's supply of methanol, CH_3OH, is produced by the following reaction:

$$CO(g) + 2H_2(g) \rightleftharpoons CH_3OH(g)$$

Write the expression for K_P for this equilibrium.

ANALYSIS: For K_P we use partial pressures in the mass action expression. We put the equilibrium partial pressures of the products in the numerator and the equilibrium partial pressures of the reactants in the denominator. The coefficients in the equation become exponents on the pressures.

SOLUTION: The expression for K_P for this reaction is

$$K_P = \frac{P_{CH_3OH}}{P_{CO}\,P_{H_2}^2}$$

Is the Answer Reasonable?
Check to see that the mass action expression gives products over reactants, and not the other way around. Check each superscript against the coefficients in the balanced chemical equation.

PRACTICE EXERCISE 4: Using partial pressures, write the equilibrium law for the reaction

$$H_2(g) + I_2(g) \rightleftharpoons 2HI(g)$$

16.5 **A large K means a product-rich equilibrium mixture; a small K means a reactant-rich mixture at equilibrium**

Whether we work with K_P or K_c, a bonus of always writing the mass action expression with the product concentrations in the numerator is that the size of the equilibrium constant gives us a measure of how far the reaction proceeds toward completion when equilibrium is reached. For example, the reaction

$$2H_2(g) + O_2(g) \rightleftharpoons 2H_2O(g)$$

has $K_c = 9.1 \times 10^{80}$ at 25 °C. This means that when there is an equilibrium between these gases,

$$K_c = \frac{[H_2O]^2}{[H_2]^2[O_2]} = \frac{9.1 \times 10^{80}}{1}$$

Actually, you would need about 200,000 L of water vapor at 25 °C just to find one molecule of O_2 and two molecules of H_2.

By writing K_c as a fraction, $(9.1 \times 10^{80})/1$, we see that the numerator of the mass action expression must be enormous compared with the denominator, which means that the concentration of H_2O has to be enormous in comparison to the concentrations of H_2 and O_2. At equilibrium, therefore, most of the hydrogen and oxygen atoms in the system are found in the H_2O and very few are present in H_2 and O_2. The enormous value of K_c tells us that the reaction between H_2 and O_2 goes to completion.

The reaction between N_2 and O_2 to give NO

$$N_2(g) + O_2(g) \rightleftharpoons 2NO(g)$$

has a very small equilibrium constant; $K_c = 4.8 \times 10^{-31}$ at 25 °C. The equilibrium law for this reaction is

$$\frac{[NO]^2}{[N_2][O_2]} = 4.8 \times 10^{-31}$$

Since $10^{-31} = 1/10^{31}$, we can write this as

$$\frac{[NO]^2}{[N_2][O_2]} = \frac{4.8}{10^{31}}$$

In air at 25 °C, the equilibrium concentration of NO *should be* about 10^{-17} mol/L. It is usually higher because NO is formed in various reactions, such as those responsible for air pollution caused by automobiles.

Here the denominator is huge compared with the numerator, so the concentrations of N_2 and O_2 must be very much larger than the concentration of NO. This means that in a mixture of N_2 and O_2 at this temperature, the amount of NO that is formed is negligible. The reaction hardly proceeds at all toward completion before equilibrium is reached.

The relationship between the equilibrium constant and the position of equilibrium can thus be summarized as follows:

TOOLS
Significance of the magnitude of K

When K is very large	The reaction proceeds far toward completion. The position of equilibrium lies far toward the products.
When $K \approx 1$	The concentrations of reactants and products are nearly the same at equilibrium. The position of equilibrium lies approximately midway between reactants and products.
When K is very small	Extremely small amounts of products are formed. The position of equilibrium lies far toward the reactants.

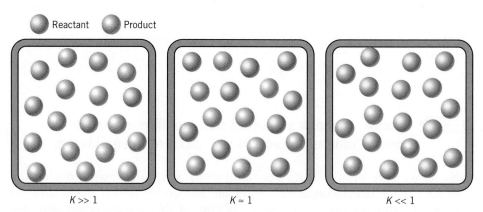

The magnitude of K and the position of equilibrium. For the reaction

$$\text{reactant} \rightleftharpoons \text{product}$$

a large amount of product and very little reactant are in the reaction mixture at equilibrium when K is very large ($K \gg 1$), so we say *the position of equilibrium lies to the right.* When $K \approx 1$, approximately equal amounts of reactant and product are present at equilibrium. When $K \ll 1$, the reaction mixture contains a large amount of reactant at equilibrium and very little product, so we say *the position of equilibrium lies to the left.*

Notice that we have omitted the subscript for K in this summary. The same qualitative predictions about the extent of reaction apply whether we use K_P or K_c.

One of the ways that we can use equilibrium constants is to compare the extents to which two or more reactions proceed to completion. Take care in making such comparisons, however, because unless the K's are greatly different, the comparison is valid only if both reactions have the same number of reactant and product molecules appearing in their balanced chemical equations.

PRACTICE EXERCISE 5: Which of the following reactions will tend to proceed farthest toward completion?
(a) $H_2(g) + Br_2(g) \rightleftharpoons 2HBr(g)$ $K_c = 1.4 \times 10^{-21}$
(b) $2NO(g) \rightleftharpoons N_2(g) + O_2(g)$ $K_c = 2.1 \times 10^{30}$
(c) $2BrCl \rightleftharpoons Br_2 + Cl_2$ (in CCl_4 solution) $K_c = 0.145$

16.6 ▶ A simple expression relates K_P and K_c

For some reactions K_P is equal to K_c, but for many others the two constants have different values. It is therefore desirable to have a way to calculate one from the other. Converting between K_P and K_c uses the relationship between partial pressure and molarity described in Section 16.4. Equation 16.3 can be used to change K_P to K_c by substituting

$$(molar\ concentration) \times RT$$

for the partial pressure of each gas in the mass action expression for K_P. Similarly, K_c can be changed to K_P by solving Equation 16.3 for the molar concentrations and then substituting the result, P/RT, into the appropriate expression for K_c. This sounds like a lot of work, and it is. Fortunately, there is a general equation, which can be derived from these relationships, that we can use to make these conversions simply.

$PV = nRT$
Solving for the concentration of the gas, n/V, gives

$$\frac{n}{V} = \frac{P}{RT}$$

$$K_P = K_c(RT)^{\Delta n_g} \tag{16.4}$$

 TOOLS

Converting between K_P and K_c

In this equation, the value of Δn_g is equal to the change in the *number of moles of gas* in going from the reactants to the products.

$$\Delta n_g = (moles\ of\ gaseous\ products) - (moles\ of\ gaseous\ reactants)$$

Δn_g is calculated from the coefficients of the equation, taking them to stand for moles.

We use the coefficients of the balanced equation for the reaction to calculate the numerical value of Δn_g. For example, the equation

$$N_2(g) + 3H_2(g) \rightleftharpoons 2NH_3(g) \tag{16.5}$$

tells us that 2 mol of NH_3 are formed when 1 mol of N_2 and 3 mol of H_2 react. In other words, 2 mol of gaseous product are formed from a total of 4 mol of gaseous reactants. That's a decrease of 2 mol of gas, so Δn_g for this reaction equals -2.

For some reactions, the value of Δn_g is equal to zero. An example is the decomposition of HI.

$$2HI(g) \rightleftharpoons H_2(g) + I_2(g)$$

Notice that if we take the coefficients to mean moles, there are 2 mol of gas on each side of the equation. This means that $\Delta n_g = 0$. Since (RT) raised to the zero power is equal to 1, $K_P = K_c$.

EXAMPLE 16.3

Converting Between K_P and K_c

At 500 °C, the reaction between N_2 and H_2 to form ammonia

$$N_2(g) + 3H_2(g) \rightleftharpoons 2NH_3(g)$$

has $K_c = 6.0 \times 10^{-2}$. What is the numerical value of K_P for this reaction?

ANALYSIS: The equation that we wish to use is

$$K_P = K_c (RT)^{\Delta n_g}$$

In the discussion above, we saw that $\Delta n_g = -2$ for this reaction. All we need now are appropriate values of R and T. The temperature, T, must be expressed in kelvins. (When used to stand for temperature, a capital letter T in an equation always means the absolute temperature.) Next we must choose an appropriate value for R. Referring back to Equation 16.3, if the partial pressures are expressed in atm and the concentration in mol L^{-1}, the only value of R that is consistent with these units is $R = 0.0821$ L atm $mol^{-1} K^{-1}$, and this is the *only* value of R that can be used in Equation 16.4.

SOLUTION: Assembling the data, then, we have

$$K_c = 6.0 \times 10^{-2} \qquad\qquad \Delta n_g = -2$$

$$T = (500 + 273) \text{ K} = 773 \text{ K} \qquad R = 0.0821 \text{ L atm mol}^{-1} \text{ K}^{-1}$$

Substituting these into the equation for K_P gives

$$K_P = (6.0 \times 10^{-2}) \times [(0.0821) \times (773)]^{-2}$$

$$= (6.0 \times 10^{-2})/(63.5)^2$$

$$= 1.5 \times 10^{-5}$$

In this case, K_P has a numerical value quite different from that of K_c.

Is the Answer Reasonable?

In working these problems, check to be sure you have used the correct value for R and that the temperature is expressed in kelvins. Notice that if Δn_g is negative, as it is in this reaction, the expression $(RT)^{\Delta n_g}$ will be less than one. That should make K_P smaller than K_c, which is consistent with our results.

EXAMPLE 16.4

Converting Between K_P and K_c

At 25 °C, K_P for the reaction

$$N_2O_4(g) \rightleftharpoons 2NO_2(g)$$

has a value of 0.140. Calculate the value of K_c.

ANALYSIS: Once again, the equation we need is

$$K_P = K_c (RT)^{\Delta n_g}$$

This time, $\Delta n_g = 2 - 1 = +1$.

SOLUTION: Tabulating the data, we have:

$$K_P = 0.140 \qquad \Delta n_g = +1$$

$$T = 298 \text{ K} \qquad R = 0.0821 \text{ L atm mol}^{-1} \text{ K}^{-1}$$

Solving the equation for K_c gives

$$K_c = \frac{K_P}{(RT)^{\Delta n_g}}$$

Substituting values into this equation yields

$$K_c = \frac{0.140}{[(0.0821) \times (298)]^1}$$

$$= 5.72 \times 10^{-3}$$

Once again, there is a substantial difference between the values of K_P and K_c.

Is the Answer Reasonable?
Examining Equation 16.4 we see that when Δn_g is positive, the value of K_P will be *larger* than K_c because K_c is multiplied by (RT) raised to a positive power. (The opposite is true if Δn_g is negative, as we saw in Example 16.3.) In this example, Δn_g is positive and the given value of K_P (0.140) is larger than the calculated value of K_c (5.72×10^{-3}).

PRACTICE EXERCISE 6: Methanol, CH_3OH, is a promising fuel that can be synthesized from carbon monoxide and hydrogen according to the reaction

$$CO(g) + 2H_2(g) \rightleftharpoons CH_3OH(g)$$

For this reaction at 200 °C, $K_P = 3.8 \times 10^{-2}$. Do you expect K_P to be larger or smaller than K_c? Calculate the value of K_c at this temperature.

PRACTICE EXERCISE 7: Nitrous oxide, N_2O, is a gas used as an anesthetic; it is sometimes called "laughing gas." This compound has a strong tendency to decompose into nitrogen and oxygen following the equation

$$2N_2O(g) \rightleftharpoons 2N_2(g) + O_2(g)$$

but the reaction is so slow that the gas appears to be stable at room temperature (25 °C). The decomposition reaction has $K_c = 7.3 \times 10^{34}$. What is the value of K_P for this reaction at 25 °C?

16.7 ▶ Heterogeneous equilibria involve reaction mixtures with more than one phase

In a **homogeneous reaction**—or a **homogeneous equilibrium**—all of the reactants and products are in the same phase. Equilibria among gases are homogeneous because all gases mix freely with each other, so a single phase exists. There are also many equilibria in which reactants and products are dissolved in the same liquid phase.

When more than one phase exists in a reaction mixture, we call it a **heterogeneous reaction.** A common example is the combustion of wood, in which a solid fuel reacts with gaseous oxygen. Another is the thermal decomposition of sodium bicarbonate (baking soda), which occurs when the compound is sprinkled on a fire.

$$2NaHCO_3(s) \longrightarrow Na_2CO_3(s) + H_2O(g) + CO_2(g)$$

Safety-minded cooks keep a box of baking soda nearby because this reaction makes it an excellent fire extinguisher for burning fats or oil. The fire is smothered by the products of the reaction.

Heterogeneous reactions are able to reach equilibrium just as homogeneous reactions can. If $NaHCO_3$ is placed in a sealed container so that no CO_2 or H_2O can escape, the gases and solids come to a heterogeneous equilibrium.

$$2NaHCO_3(s) \rightleftharpoons Na_2CO_3(s) + H_2O(g) + CO_2(g)$$

Following our usual procedure, we can write the equilibrium law as

$$\frac{[Na_2CO_3(s)][H_2O(g)][CO_2(g)]}{[NaHCO_3(s)]^2} = K$$

However, the equilibrium law for reactions involving pure liquids and solids can be written in an even simpler form. This is because the concentration of a pure liquid or solid is unchangeable at a given temperature. *For any pure liquid or solid, the ratio of amount of substance to volume of substance is a constant.* For example, if we had a 1 mol crystal of $NaHCO_3$, it would occupy a volume of 38.9 cm³. Two moles of $NaHCO_3$ would occupy twice this volume, 77.8 cm³ (Figure 16.4), but the *ratio* of

1 mol $NaHCO_3$
38.9 cm³

$$\text{Molarity} = \frac{1 \text{ mol } NaHCO_3}{0.0389 \text{ L}}$$
$$= 25.7 \text{ mol/L}$$

2 mol $NaHCO_3$
77.8 cm³

$$\text{Molarity} = \frac{2 \text{ mol } NaHCO_3}{0.0778 \text{ L}}$$
$$= 25.7 \text{ mol/L}$$

FIGURE 16.4 *The concentration of a substance in the solid state is a constant.* Doubling the number of moles also doubles the volume, but the *ratio* of moles to volume remains the same.

moles to liters (i.e., the molar concentration) remains the same. For $NaHCO_3$, the concentration of the substance in the solid is

$$\frac{1 \text{ mol}}{0.0389 \text{ L}} = \frac{2 \text{ mol}}{0.0778 \text{ L}} = 25.7 \text{ mol L}^{-1}$$

This is the concentration of $NaHCO_3$ in the solid, regardless of the size of the solid sample. In other words, the concentration of $NaHCO_3$ is constant, provided that some of it is present in the reaction mixture.

Similar reasoning shows that the concentration of Na_2CO_3 in pure solid Na_2CO_3 is a constant, too. This means that the equilibrium law now has three constants, K plus two of the concentration terms. It makes sense to combine all of the numerical constants together.

$$[H_2O(g)][CO_2(g)] = \frac{K[NaHCO_3(s)]^2}{[Na_2CO_3(s)]} = K_c$$

> The equilibrium law for a heterogeneous reaction is written without concentration terms for pure solids or liquids.

Equilibrium constants that are given in tables represent all of the constants combined.[2]

<table>
<tr><td>

EXAMPLE 16.5

Writing the Equilibrium Law for a Heterogeneous Reaction

</td><td>

The air pollutant sulfur dioxide can be removed from a gas mixture by passing the gases over calcium oxide. The equation is

$$CaO(s) + SO_2(g) \rightleftharpoons CaSO_3(s)$$

Write the equilibrium law for K_c for this reaction.

ANALYSIS: The concentrations of the two solids, CaO and $CaSO_3$, are incorporated into the equilibrium constant K_c for the reaction. The only concentration that should appear in the mass action expression is that of SO_2.

SOLUTION: Therefore, the equilibrium law is simply

$$\frac{1}{[SO_2]} = K_c$$

Is the Answer Reasonable?
There is no easy way to check this result without testing it experimentally, but this equilibrium law is typical for reactions that involve solids or liquids. Pure liquids, pure solids, and the solvents in dilute solutions are built into the equilibrium constant and do not appear in the mass action expression.

PRACTICE EXERCISE 8: Write the equilibrium law for the following heterogeneous reactions:
(a) $2Hg(l) + Cl_2(g) \rightleftharpoons Hg_2Cl_2(s)$
(b) $NH_3(g) + HCl(g) \rightleftharpoons NH_4Cl(s)$
(c) $Na(s) + H_2O(l) \rightleftharpoons NaOH(aq) + H_2(g)$
(d) Dissolving solid Ag_2CrO_4 in water: $Ag_2CrO_4(s) \rightleftharpoons 2Ag^+(aq) + CrO_4^{2-}(aq)$
(e) $CaCO_3(s) + H_2O(l) + CO_2(aq) \rightleftharpoons Ca^{+2}(aq) + 2HCO_3^-(aq)$

</td></tr>
</table>

[2]Thermodynamics handles heterogeneous equilibria in a more elegant way by expressing the mass action expression in terms of "effective concentrations," or **activities.** In doing this, thermodynamics *defines* the *activity* of any pure liquid or solid as equal to 1, which means terms involving such substances drop out of the mass action expression.

16.8 When a system at equilibrium is stressed, it reacts to relieve the stress

In Section 16.9 you will see that it is possible to perform calculations that tell us what the composition of an equilibrium system is. However, many times we really don't need to know exactly what the equilibrium concentrations are. Instead, we may want to know what actions we should take to control the relative amounts of the reactants or products at equilibrium. For instance, if we were designing gasoline engines, we would like to know what could be done to minimize the formation of nitrogen oxide pollutants. (See Facets of Chemistry 16.1.) Or, if we were preparing ammonia, NH_3, by the reaction of N_2 with H_2, we might want to know how to maximize the yield of NH_3.

Le Châtelier's principle, introduced in Chapter 12, provides us with the means for making qualitative predictions about changes in chemical equilibria. It does this in much the same way that it allows us to predict the effects of outside influences on equilibria that involve physical changes, such as liquid–vapor equilibria. Recall that **Le Châtelier's principle** states that *if an outside influence upsets an equilibrium, the system undergoes a change in a direction that counteracts the disturbing influence and, if possible, returns the system to equilibrium.* Let's examine what kinds of "outside influences" can affect chemical equilibria.

When we say an equilibrium system is "stressed," we mean it is subjected to a disturbance that tends to upset the equilibrium.

 TOOLS

Le Châtelier's principle

Adding a reactant or product

If we add or remove a reactant or product, its concentration changes. This changes the value of Q so that it is no longer equal to K. Equilibrium has been upset. The system restores equilibrium by changing the concentrations so that Q once again equals K. The chemical reaction runs either to the right or left, depending on whether Q needs to increase or decrease to make it equal to the value of K. Le Châtelier's principle lets us predict which way the reaction goes.

As an example, let's study the equilibrium between two ions of copper.

$$\underset{\text{blue}}{Cu(H_2O)_4^{2+}(aq)} + 4Cl^-(aq) \rightleftharpoons \underset{\text{yellow}}{CuCl_4^{2-}(aq)} + 4H_2O(l)$$

As noted, $Cu(H_2O)_4^{2+}$ is blue and $CuCl_4^{2-}$ is yellow. Mixtures of the two have an intermediate color and therefore appear blue-green, as illustrated in Figure 16.5, center.

Suppose we add chloride ion to an equilibrium mixture of these copper ions. The system can remove some of it by reacting it with $Cu(H_2O)_4^{2+}$. This gives more $CuCl_4^{2-}$ (Figure 16.5, right), and we say that the equilibrium has "shifted to the right" or "shifted toward the products." In this new position of equilibrium, there is less $Cu(H_2O)_4^{2+}$ and more $CuCl_4^{2-}$ and uncombined H_2O. There is also more Cl^-, because not all that we add reacts. In other words, in this new position of equilibrium, *all* the concentrations have changed in a way that causes Q to become equal to K_c. Similarly, the position of equilibrium is shifted to the left when we add water to the mixture (Figure 16.5, left). The system is able to get rid of some of the H_2O by reaction with $CuCl_4^{2-}$, so more of the blue $Cu(H_2O)_4^{2+}$ is formed.

If we were able to remove a reactant or product, the position of equilibrium would also be changed. For example, if we add Ag^+ to a solution that contains both copper ions in equilibrium, we see an enhancement of the blue color.

$$\underset{\text{blue}}{Cu(H_2O)_4^{2+}(aq)} + 4Cl^-(aq) \rightleftharpoons \underset{\text{yellow}}{CuCl_4^{2-}(aq)} + 4H_2O$$

$$\Downarrow +Ag^+$$

$$Ag^+(aq) + Cl^-(aq) \rightleftharpoons AgCl(s)$$

$Cu(H_2O)_4^{2+}$ and $CuCl_4^{2-}$ are called **complex ions.** Some of the interesting properties and applications of complex ions of metals are discussed in Chapter 23.

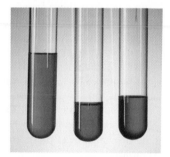

FIGURE 16.5 *The effect of concentration changes on the position of equilibrium.* The solution in the center contains a mixture of blue $Cu(H_2O)_4^{2+}$ and yellow $CuCl_4^{2-}$, so it has a blue-green color. At the right, some of the same solution after the addition of concentrated HCl. It has a more pronounced green color because the equilibrium is shifted toward $CuCl_4^{2-}$. At the left is some of the original solution after the addition of water. It is blue because the equilibrium has shifted toward $Cu(H_2O)_4^{2+}$.

FACETS OF CHEMISTRY 16.1

Air Pollution and Le Châtelier's Principle

As everyone knows, one of the most serious causes of air pollution is the automobile. All sorts of obnoxious chemicals are present in the exhaust gases leaving the engine, and various methods have been devised to control the amounts of these emissions that enter the atmosphere. For instance, almost all cars today are equipped with catalytic converters (see Chapter 15). These devices mix air with the exhaust gases and promote oxidation of unburned fuel and carbon monoxide to carbon dioxide. They also cause the decomposition of nitrogen oxides to elemental nitrogen and oxygen. Catalytic converters are expensive, however, and other methods to accomplish the same goals have also been studied.

When air is drawn into a car's engine, both N_2 and O_2 are present. During combustion of the gasoline, oxygen reacts with the hydrocarbons in the fuel to produce CO_2, CO, and H_2O. However, N_2 and O_2 can also form NO.

$$N_2(g) + O_2(g) \rightleftharpoons 2NO(g)$$

At room temperature, K_c for this reaction is 4.8×10^{-31}. Its small value tells us that the equilibrium concentration of NO should be very small. Therefore, we don't find N_2 reacting with O_2 under ordinary conditions, so the atmosphere is quite stable.

The reaction of N_2 and O_2 to form NO is endothermic. Le Châtelier's principle tells us that at high temperatures, such as those found in the cylinders of a gasoline or diesel engine during combustion, this equilibrium should be shifted to the right, so at high temperatures some NO does form. Unfortunately, when the exhaust leaves the engine, it cools so rapidly that the NO can't decompose. The reaction rate at the lower temperature becomes too slow. The result is that some NO is present among the exhaust gases. Once in the atmosphere, this NO becomes oxidized to NO_2, which is responsible for the brownish haze often associated with severe air pollution.

One way to reduce the amount of NO pollution of the atmosphere is to reduce the amount of NO that's formed in automobile engines. Since the extent to which the reaction proceeds toward the formation of NO increases as the temperature is raised, the amount of NO

that's formed can be reduced by simply running the combustion reaction at a lower temperature. One method that has been used to accomplish this is to lower the compression ratio of the engine. This is the ratio of the volume of the cylinder when the piston is at the bottom of its stroke divided by the volume after the piston has compressed the air-fuel mixture. At high compression ratios, the air-fuel mixture is heated to a high temperature before it's ignited. After combustion, the gases are very hot, which favors the production of NO. Lowering the compression ratio lowers the maximum combustion temperature which decreases the tendency for NO to be formed. Unfortunately, lowering the compression ratio also lowers the efficiency of the engine, which makes for poorer fuel economy.

Another method for controlling NO emissions that has been experimented with involves mixing water with the air-fuel mixture. Some of the heat from the combustion is absorbed by the water vapor, so the mixture of exhaust gases doesn't get as hot as it would otherwise. At these lower temperatures, the concentration of NO in the exhaust is greatly reduced.

FIGURE 1 At times, air pollution from auto exhaust emissions poses a health hazard. This can be particularly severe in large cities like Los Angeles where heavy traffic and atmospheric conditions can combine to produce dangerous levels of smog.

As the Ag^+ reacts with Cl^- to form insoluble AgCl in the second reaction, the equilibrium in the first reaction shifts to the left to replace some of the Cl^-, which also produces more blue $Cu(H_2O)_4^{2+}$. The position of equilibrium shifts to replace the substance that is removed.

Equilibrium shifts to remove reactants or products that have been added.
Equilibrium shifts to replace reactants or products that have been removed.

Changing the volume in gaseous reactions

Changing the volume of a mixture of reacting gases changes their molar concentrations and their partial pressures, so we expect it to have some effect on the position of equilibrium. Let's consider the equilibrium

$$3H_2(g) + N_2(g) \rightleftharpoons 2NH_3(g)$$

If we reduce the volume of the reaction mixture, we expect the pressure to increase. The system can oppose the pressure change if it is able to reduce the number of molecules of gas, because fewer molecules of gas exert a lower pressure. If the reaction proceeds to the right, two NH_3 molecules appear when four molecules (one N_2 and three H_2) disappear. Therefore, this equilibrium is shifted to the right when the volume of the reaction mixture is reduced.

Now let's look at the equilibrium

$$H_2(g) + I_2(g) \rightleftharpoons 2HI(g)$$

If this reaction proceeds in either direction, there is no change in the number of molecules of gas. This reaction, then, cannot respond to pressure changes, so changing the volume of the reaction vessel has virtually no effect on the equilibrium.

The simplest way to analyze the effects of a volume change on an equilibrium system is to count the number of molecules of gaseous substances on both sides of the equation.

> Reducing the volume of a gaseous reaction mixture causes the reaction to decrease the number of molecules of gas, if it can.

As a final note here, *moderate pressure changes have negligible effect on reactions involving only liquids or solids.* Substances in these states are virtually incompressible, and reactions involving them have no way to counteract pressure changes.

Recall from Chapter 11 that according to Boyle's law, volume is inversely proportional to pressure (when the temperature and number of moles of gas are held constant).

Changing the temperature

If the system is heated, the reaction will shift in the direction that absorbs heat. If the reaction is endothermic, it will run in the forward direction to remove the added energy. For example, the melting of ice is endothermic:

$$H_2O(s) \rightleftharpoons H_2O(l) \qquad \Delta H° = +6 \text{ kJ (at 0 °C)}$$

If you cup your hands around a glass of ice and water at 0 °C, the heat you've added to the system shifts the equilibrium to the right, and some of the ice melts. If we include energy as a reactant in the equation, we could write

$$\text{energy} + H_2O(s) \rightleftharpoons H_2O(l)$$

Adding a reactant drives the reaction in the forward direction, with energy behaving like any other reactant in this case.

Removing some amount of reactant should shift the reaction to the left. We expect that removing energy from the system would have the same effect. Putting the glass in the freezer causes some of the water to freeze. Some of the energy is removed, so the equilibrium shifts towards the reactant side of the equation.

The same arguments can be used for equations that describe chemical changes. For instance, the reaction to produce NH_3 from N_2 and H_2 is exothermic; heat is released when NH_3 molecules are formed ($\Delta H_f° = -46.19$ kJ/mol from Table 7.2). Since the reaction produces energy, we can include energy as a product in the equilibrium equation

$$3H_2(g) + N_2(g) \rightleftharpoons 2NH_3(g) + \text{energy}$$

In Chapter 12, we saw that freezing is exothermic and melting is endothermic.

This same kind of analysis was used in Chapter 14 to predict how solubility changes with temperature.

It is easy to predict what will happen when the reaction mixture is heated or cooled. Heating the mixture adds energy, so the system will counter by shifting to the left. The added energy will be absorbed, and at the same time NH_3 will decompose to produce more H_2 and N_2. Cooling the mixture removes energy, so the equilibrium shifts to the right. H_2 and N_2 will react to produce more NH_3, and energy will be released to compensate for energy removed by cooling.

> Increasing the temperature shifts a reaction in a direction that produces an endothermic (heat-absorbing) change.
>
> Decreasing the temperature shifts a reaction in a direction that produces an exothermic (heat-releasing) change.

When heat is added to an equilibrium mixture, it is added to all of the substances present (reactants *and* products). As the system returns to equilibrium, the net reaction that occurs is the one that is endothermic.

This gives us a way to experimentally determine whether a reaction is exothermic or endothermic, without using a thermometer. Earlier we described the equilibrium involving complex ions of copper. The effect of temperature on this equilibrium is demonstrated in Figure 16.6. We can use the observed color change to determine which way the equilibrium shifts when the system is heated. It shifts toward the products when heated, so we know the reaction must be endothermic.

The effect of temperature on an equilibrium system differs from the effects of other stresses we've discussed in one important way. Changes in concentrations or volume can shift the position of equilibrium, but they do not change the equilibrium constant. But changing the temperature changes the value of the mass action expression at equilibrium, which means that K has changed.

Temperature is the *only* factor that can change K for a given reaction.

How does K change when temperature changes? The enthalpy change of the reaction is the critical factor. Let's look once again at the equilibrium law for the industrial synthesis of ammonia.

$$3H_2(g) + N_2(g) \rightleftharpoons 2NH_3(g) \qquad \Delta H° = -92 \text{ kJ}$$

$$\frac{[NH_3]^2}{[N_2][H_2]^3} = K_c$$

We'll see how to quantitatively relate K and temperature in Chapter 20.

Because the reaction is exothermic, increasing the temperature shifts the equilibrium to the left. The concentration of NH_3 decreases while the concentrations of N_2 and H_2 increase. Therefore, the numerator of the mass action expression becomes *smaller* and the denominator becomes *larger*. This gives a smaller reaction quotient and therefore a smaller value of K_c.

FIGURE 16.6 *The effect of temperature on the equilibrium* $Cu(H_2O)_4^{2+} + 4Cl^- \rightleftharpoons CuCl_4^{2-} + 4H_2O$. In the center, an equilibrium mixture of the two complexes. When the solution is cooled in ice (left), the equilibrium shifts toward the blue $Cu(H_2O)_4^{2+}$. When heated in boiling water (right), the equilibrium shifts toward $CuCl_4^{2-}$. This behavior indicates that the reac ɪdothermic in the forward direction.

Increasing the temperature of an exothermic reaction makes its equilibrium constant smaller.

Increasing the temperature of an endothermic reaction makes its equilibrium constant larger.

Catalysts have no effect on the position of equilibrium

Recall that catalysts are substances that affect the speeds of chemical reactions without actually being used up. However, catalysts do not affect the position of equilibrium in a system. The reason is that a catalyst affects both the forward and reverse reactions equally. Both are speeded up to the same degree, so adding a catalyst to a system has no net effect on the system's equilibrium composition. The catalyst's only effect is to bring the system to equilibrium faster.

Catalysts don't appear in the chemical equation for a reaction, so they don't appear in the mass action expression, either.

Adding an inert gas at constant volume has no effect on the position of equilibrium

A change in volume is not the only way to change the pressure in an equilibrium system of gaseous reactants and products. The pressure can also be changed by keeping the volume the same and adding another gas. If this gas cannot react with any of the gases already present (i.e., if the added gas is *inert* toward the substances in equilibrium), the concentrations of the reactants and products won't change. The concentrations will continue to satisfy the equilibrium law and the reaction quotients will continue to equal K_c, so there will be no change in the position of equilibrium.

The reaction $N_2O_4(g) \rightleftharpoons 2NO_2(g)$ is endothermic, with $\Delta H° = +56.9$ kJ. How will the amount of NO_2 at equilibrium be affected by (a) adding N_2O_4, (b) lowering the pressure by increasing the volume of the container, (c) raising the temperature, and (d) adding a catalyst to the system? Which of these changes will alter the value of K_c?

EXAMPLE 16.6

Application of Le Châtelier's Principle

ANALYSIS: We are applying various types of stresses to an equilibrium mixture of NO_2 and N_2O_4. We can apply Le Châtelier's principle in each case. The equilibrium will shift to the right or left to counteract the applied stress.

SOLUTION:
(a) Adding N_2O_4 will cause the equilibrium to shift to the right—in a direction that will consume some of the added N_2O_4. The amount of NO_2 will increase.

(b) When the pressure in the system drops, the system responds by producing more molecules of gas, which will tend to raise the pressure and partially offset the change. Since more gas molecules are formed if some N_2O_4 decomposes, the amount of NO_2 at equilibrium will increase.

(c) Because the reaction is endothermic, we write the equation showing heat as a reactant

$$\text{heat} + N_2O_4(g) \rightleftharpoons 2NO_2(g)$$

Raising the temperature is accomplished by adding heat, so the system will respond by absorbing heat. This means that the equilibrium will shift to the right and the amount of NO_2 at equilibrium will increase.

(d) A catalyst causes a reaction to reach equilibrium more quickly, but it has no effect on the position of chemical equilibrium. Therefore, the amount of NO_2 at equilibrium will not be affected.

Finally, the *only* change that alters K is the temperature change. Raising the temperature (adding heat) will increase K_c for this endothermic reaction.

Is the Answer Reasonable?

Compare each of our answers with examples given in this section to confirm that they are logically consistent.

PRACTICE EXERCISE 9: Consider the equilibrium $PCl_3(g) + Cl_2(g) \rightleftharpoons PCl_5(g)$, for which $\Delta H° = -88$ kJ. How will the amount of Cl_2 at equilibrium be affected by (a) adding PCl_3, (b) adding PCl_5, (c) raising the temperature, and (d) decreasing the volume of the container? How (if at all) will each of these changes affect K_P for the reaction?

CHEMISTRY IN PRACTICE The production of ammonia from nitrogen and hydrogen is an essential process in the chemical industry. Without inorganic fertilizers produced from ammonia, agriculture could not support the world's current population. Ammonia is also the basis for the manufacture of nearly all nitrogen-containing chemicals.

In the **Haber process,** heated nitrogen and hydrogen are passed over a catalyst prepared from finely divided iron mixed with K_2O, Al_2O_3, SiO_2, and MgO to produce the ammonia in a sealed reactor:

$$N_2(g) + 3H_2(g) \rightleftharpoons 2NH_3(g) \qquad \Delta H° = -92 \text{ kJ}$$

Le Châtelier's principle can be used to maximize the yield of ammonia by careful control of the pressure and temperature within the reactor. First, if we increase the pressure, the equilibrium will shift toward the products to reduce the number of moles of gas in the reactor. Haber process reactors have been run at pressures of hundreds of atmospheres. High-pressure reactors are expensive, though, and recent refinements in technology have allowed pressures to be lowered to 40–80 atmospheres.

We can also remove ammonia from the reactor as it is formed. The system will then have to produce more ammonia to maintain equilibrium. This is usually accomplished by passing the gases over a refrigeration coil. Ammonia liquefies at -33 °C, but nitrogen and hydrogen have much lower boiling points and remain gaseous. The liquid ammonia is drained away and the remaining gases are then recycled back into the reactor.

The reaction is exothermic, so lowering the temperature should shift the equilibrium toward the products. But what effect will lowering the temperature have on the rate of the reaction? We saw in Chapter 15 that most reactions slow down at lower temperatures. Unfortunately, the Haber process is no exception. Lowering the temperature beyond a certain point causes the reaction to run too slowly. Economically, it is better to obtain a moderate yield quickly at higher temperature than it is to obtain a high yield very slowly at a lower temperature. Most modern reactors run at temperatures of 250–400 °C.

16.9 ▶ Equilibrium concentrations can be used to predict equilibrium constants, and vice versa

You have seen that the magnitude of an equilibrium constant gives us some feel for the extent to which the reaction proceeds at equilibrium. Sometimes, however, it is necessary to have more than merely a qualitative knowledge of equilibrium concentrations. This requires that we be able to use the equilibrium law for purposes of calculation.

Equilibrium calculations for gaseous reactions can be performed using either K_P or K_c, but for reactions in solution we must use K_c. Whether we deal with concentrations or partial pressures, however, the same basic principles apply.

Overall, we can divide equilibrium calculations into two main categories:

1. Calculating equilibrium constants from known equilibrium concentrations or partial pressures.

2. Calculating one or more equilibrium concentrations or partial pressures using the known value of K_c or K_P.

Calculating K_c from equilibrium concentrations

One way to determine the value of K_c is to carry out the reaction, measure the concentrations of reactants and products after equilibrium has been reached, and then use the equilibrium values in the equilibrium law to compute K_c. As an example, let's look again at the decomposition of N_2O_4.

$$N_2O_4(g) \rightleftharpoons 2NO_2(g)$$

In Section 16.2, we saw that if 0.0350 mol of N_2O_4 is placed into a 1 liter flask at 25 °C, the concentrations of N_2O_4 and NO_2 at equilibrium are

$$[N_2O_4] = 0.0292 \text{ mol/L}$$

$$[NO_2] = 0.0116 \text{ mol/L}$$

To calculate K_c for this reaction, we substitute the equilibrium concentrations into the mass action expression of the equilibrium law.

$$\frac{[NO_2]^2}{[N_2O_4]} = K_c$$

$$\frac{(0.0116)^2}{(0.0292)} = K_c$$

Performing the arithmetic gives

$$K_c = 4.61 \times 10^{-3}$$

Although calculating an equilibrium constant this way is easy, sometimes we have to do a little work to figure out what all the concentrations are, as shown in Example 16.7.

At a certain temperature, a mixture of H_2 and I_2 was prepared by placing 0.200 mol of H_2 and 0.200 mol of I_2 into a 2.00 liter flask. After a period of time the equilibrium

$$H_2(g) + I_2(g) \rightleftharpoons 2HI(g)$$

was established. The purple color of the I_2 vapor was used to monitor the reaction, and from the decreased intensity of the purple color it was determined that, at equilibrium, the I_2 concentration had dropped to 0.020 mol L^{-1}. What is the value of K_c for this reaction at this temperature?

ANALYSIS: The first step in any equilibrium problem is to write the balanced chemical equation and the related equilibrium law. The equation is already given, and the equilibrium law corresponding to it is

$$\frac{[HI]^2}{[H_2][I_2]} = K_c$$

To calculate the value of K_c we must substitute the *equilibrium concentrations* of H_2, I_2, and HI into the mass action expression. But what are they? We have been given only one directly, the value of $[I_2]$. To obtain the others, we have to do some reasoning based on the chemical equation.

The system starts with a set of initial concentrations, and to reach equilibrium a chemical change occurs. There is no HI present initially, so the reaction above must proceed to the right, because there has to be some HI present for the system to be at equilibrium. This change will increase the HI concentration and decrease the concentrations of H_2 and I_2. If we can determine what the *changes* are, we can figure out the equilibrium concentrations. *In fact, the key to solving almost all equilibrium problems is determining how the concentrations change as the system comes to equilibrium.*

EXAMPLE 16.7

Calculating K_c from Equilibrium Concentrations

Concentration table

To help us in our analysis we will construct a **concentration table** beneath the chemical equation. We will do this for most of the equilibrium problems we encounter from now on, so let's examine the general form of the table.

In the first row we write the initial *molar concentrations* of the reactants and products. (All entries in the table will be *molar concentrations,* not *moles.* These are the only quantities that satisfy the equilibrium law.) Then, in the second row we enter the changes in concentration, using a positive sign if the concentration is increasing and a negative sign if it is decreasing. Finally, adding the changes to the initial concentrations gives the equilibrium concentrations.

$$\left(\begin{array}{c}\text{equilibrium}\\\text{concentration}\end{array}\right) = \left(\begin{array}{c}\text{initial}\\\text{concentration}\end{array}\right) + \left(\begin{array}{c}\text{change in}\\\text{concentration}\end{array}\right)$$

*To set up the concentration table, it is essential that you realize that the changes in concentration **must be in the same ratio as the coefficients of the balanced equation.*** This is because the changes are caused by the chemical reaction proceeding in one direction or the other. In this problem, the coefficients of H_2 and I_2 are the same, so as the reaction proceeds to the right, the concentrations of H_2 and I_2 must decrease by the *same* amount. Similarly, because the coefficient of HI is twice that of H_2 or I_2, the concentration of HI must increase by *twice* as much. Now let's look at the finished concentration table, how the individual entries are obtained, and the solution of the problem.

SOLUTION: To construct the concentration table, we begin by entering the data given in the statement of the problem. These are shown in colored type. The values in regular type are then derived as described below.

	$H_2(g)$	$+$	$I_2(g)$	$\rightleftharpoons$	$2HI(g)$
Initial concentrations (*M*)	0.100		0.100		0.000
Changes in concentrations (*M*)	−0.080		−0.080		+2(0.080)
Equilibrium concentrations (*M*)	0.020		0.020		0.160

Initial Concentrations

In any system, the initial concentrations are controlled by the person doing the experiment.

Notice that we have calculated the ratio of moles to liters for both the H_2 and I_2, (0.200 mol/2.00 L) = 0.100 *M*. Because no HI was placed into the reaction mixture, its initial concentration has been set to zero.

Changes in Concentrations

The changes in concentrations are controlled by the stoichiometry of the reaction.

If we can find one of the changes, we can calculate the others from it.

We have been given both the initial and equilibrium concentrations of I_2, so by difference we can calculate the change for I_2 (−0.080 *M*). The other changes are then calculated from the mole ratios specified in the chemical equation. As noted above, because H_2 and I_2 have the same coefficients (i.e., 1), their changes are equal. The coefficient of HI is 2, so its change must be twice that of I_2. Reactant concentrations decrease, so the change is negative, while product concentrations increase, so the change in HI concentration is positive.

Equilibrium Concentrations

For H_2 and HI, we just add the change to the initial value.

Now we can substitute the equilibrium concentrations into the mass action expression and calculate K_c.

$$K_c = \frac{(0.160)^2}{(0.020)(0.020)}$$

$$= 64$$

Is the Answer Reasonable?

First, carefully examine the equilibrium law. As always, products must be placed in the numerator and reactants in the denominator. Be sure that each exponent corresponds to the correct coefficient in the balanced chemical equation for the reaction.

Now let's check the concentration table. Notice that the changes for both reactants have the same sign, which is opposite that of the product. *This relationship among the signs of the changes is always true and can serve as a useful check when you construct a concentration table.*

Finally, check to make sure that each of the equilibrium concentrations is properly placed in the mass action expression. The equilibrium constant we obtained is greater than one. That implies that a significant amount of product should be present at equilibrium, and examination of the concentration table shows that this is so.

The Concentration Table—A Summary

Let's review some key points that apply not only to this problem but to others that deal with equilibrium calculations.

1. The only values that we can substitute into the mass action expression in the equilibrium law are equilibrium concentrations—the values that appear in the last row of the table.

2. When we enter initial concentrations into the table, they should be in units of moles per liter (mol L^{-1}). The initial concentrations are those present in the reaction mixture when it's prepared; we imagine that no reaction occurs until everything is mixed.

By expressing concentrations in moles per liter (mol L^{-1}), we follow what happens to reactants and products in each liter of solution.

3. The changes in concentrations *always* occur in the same ratio as the coefficients in the balanced equation. For example, if we were dealing with the equilibrium

$$3H_2(g) + N_2(g) \rightleftharpoons 2NH_3(g)$$

and found that the $N_2(g)$ concentration decreases by 0.10 M during the approach to equilibrium, the entries in the "change" row would be as follows:

$3H_2(g)$	$+$	$N_2(g)$	$\rightleftharpoons$	$2NH_3(g)$
change in concentration: $-3 \times (0.10\ M)$		$-1 \times (0.10\ M)$		$+2 \times (0.10\ M)$
$\downarrow$		$\downarrow$		$\downarrow$
$-0.30\ M$		$-0.10\ M$		$+0.20\ M$

4. In constructing the "change" row, be sure the reactant concentrations all change in the same direction and that the product concentrations all change in the opposite direction. If the concentrations of the reactants decrease, all the entries for the reactants in the "change" row should have a minus sign and all the entries for the products should be positive.

If the initial concentration of some reactant is zero, its change must be positive (an increase) because the final concentration can't be negative.

Keep these ideas in mind as we construct the concentration tables for other equilibrium problems.

PRACTICE EXERCISE 10: An equilibrium was established for the reaction

$$CO(g) + H_2O(g) \rightleftharpoons CO_2(g) + H_2(g)$$

at 500 °C. (This is an industrially important reaction for the preparation of hydrogen.) At equilibrium, the following concentrations were found in the reaction vessel: $[CO] = 0.180\ M$, $[H_2O] = 0.0411\ M$, $[CO_2] = 0.150\ M$, and $[H_2] = 0.200\ M$. What is the value of K_c for this reaction?

PRACTICE EXERCISE 11: In a particular experiment, it was found that when $O_2(g)$ and $CO(g)$ were mixed and reacted according to the equation

$$2CO(g) + O_2(g) \rightleftharpoons 2CO_2(g)$$

the O_2 concentration decreased by 0.030 mol L^{-1} when the reaction reached equilibrium. How had the concentrations of CO and CO_2 changed?

PRACTICE EXERCISE 12: A student placed 0.200 mol of $PCl_3(g)$ and 0.100 mol of $Cl_2(g)$ into a 1.00 liter container at 250 °C. After the reaction

$$PCl_3(g) + Cl_2(g) \rightleftharpoons PCl_5(g)$$

came to equilibrium it was found that the flask contained 0.120 mol of PCl_3.
(a) What were the initial concentrations of the reactants and product?
(b) By how much had the concentrations changed when the reaction reached equilibrium?
(c) What were the equilibrium concentrations?
(d) What is the value of K_c for this reaction at this temperature?

Equilibrium concentrations can be calculated using K_c

In the simplest calculation of this type, all but one of the equilibrium concentrations are known, as illustrated in Example 16.8.

EXAMPLE 16.8

Using K_c to Calculate Concentrations at Equilibrium

The reversible reaction

$$CH_4(g) + H_2O(g) \rightleftharpoons CO(g) + 3H_2(g)$$

has been used as a commercial source of hydrogen. At 1500 °C, an equilibrium mixture of these gases was found to have the following concentrations: $[CO] = 0.300\ M$, $[H_2] = 0.800\ M$, and $[CH_4] = 0.400\ M$. At 1500 °C, $K_c = 5.67$ for this reaction. What was the equilibrium concentration of $H_2O(g)$ in this mixture?

ANALYSIS: The first step, once we have the chemical equation for the equilibrium, is to write the equilibrium law for the reaction.

$$K_c = \frac{[CO]\,[H_2]^3}{[CH_4]\,[H_2O]}$$

The equilibrium constant and all of the equilibrium concentrations except that for H_2O are known, so we solve the equation for H_2O.

SOLUTION: Solving the equation for H_2O gives

$$[H_2O] = \frac{[CO]\,[H_2]^3}{[CH_4]\,K_c}$$

Substituting the given equilibrium concentrations and the equilibrium constant into this equation, we have

$$[H_2O] = \frac{(0.300)(0.800)^3}{(0.400)(5.67)}$$

$$= \frac{0.154}{2.27}$$

$$= 0.0678\ M$$

Is the Answer Reasonable?
We can always check problems of this kind by substituting all of the equilibrium concentrations into the mass action expression to see if we get the original equilibrium constant back. We have

$$K_c = \frac{[CO]\,[H_2]^3}{[CH_4]\,[H_2O]} = \frac{(0.300)(0.800)^3}{(0.400)(0.0678)} = 5.66$$

Rounding off the H_2O concentration caused a slight error in K_c but it is consistent with the equilibrium constant given in the problem.

PRACTICE EXERCISE 13: Ethyl acetate, $CH_3CO_2C_2H_5$, is an important solvent used in lacquers, adhesives, the manufacture of plastics, and even as a food flavoring. It is produced from acetic acid and ethanol by the reaction

$$CH_3CO_2H(l) + C_2H_5OH(l) \rightleftharpoons CH_3CO_2C_2H_5(l) + H_2O(l)$$
acetic acid   ethanol

At 25 °C, K_c = 4.10 for this reaction. In a reaction mixture, the following equilibrium concentrations were observed: $[CH_3CO_2H]$ = 0.210 M, $[H_2O]$ = 0.00850 M, and $[CH_3CO_2C_2H_5]$ = 0.910 M. What was the concentration of C_2H_5OH in the mixture?

Acetic acid has the structure

H   O
| ||
H—C—C—O—H
|
H

where the hydrogen released by ionization is indicated in red. The formula of the acid is written either as CH_3CO_2H (which emphasizes its molecular structure) or as $HC_2H_3O_2$ (which emphasizes that the acid is monoprotic).

Equilibrium concentrations can be calculated using K_c and initial concentrations

A more complex type of calculation involves the use of initial concentrations and K_c to compute equilibrium concentrations. Although some of these problems can be so complicated that a computer is needed to solve them, we can learn the general principles involved by working on simple calculations. Even these, however, require a little applied algebra. This is where the concentration table can be very helpful.

The reaction

$$CO(g) + H_2O(g) \rightleftharpoons CO_2(g) + H_2(g)$$

has K_c = 4.06 at 500 °C. If 0.100 mol of CO and 0.100 mol of $H_2O(g)$ are placed in a 1.00 liter reaction vessel at this temperature, what are the concentrations of the reactants and products when the system reaches equilibrium?

ANALYSIS: The key to solving this kind of problem is recognizing that at equilibrium the mass action expression must equal K_c.

$$\frac{[CO_2][H_2]}{[CO][H_2O]} = 4.06 \quad \text{(at equilibrium)}$$

We must find values for the concentrations that satisfy this condition.

Because we don't know what these concentrations are, we represent them algebraically as unknowns. This is where the concentration table is very helpful. To build the table, we need quantities to enter into the "initial concentrations," "changes in concentrations," and "equilibrium concentrations" rows.

Initial Concentrations
The initial concentrations of CO and H_2O are each 0.100 mol/1.00 L = 0.100 M. Since no CO_2 or H_2 is initially placed into the reaction vessel, their initial concentrations both are zero.

Changes in Concentration
Some CO_2 and H_2 must form for the reaction to reach equilibrium. This also means that some CO and H_2O must react. But how much? If we knew the answer, we could calculate the equilibrium concentrations. Therefore, the changes in concentration are our unknown quantities.

Let us allow x to be equal to the number of moles per liter of CO that react. The change in the concentration of CO is then $-x$ (it is negative because the change decreases the CO concentration). Because CO and H_2O react in a 1:1 mole ratio, the change in the H_2O concentration is also $-x$. Since 1 mol each of CO_2 and H_2 is formed from 1 mol of CO, the CO_2 and H_2 concentrations each increase by x (their changes are $+x$).

EXAMPLE 16.9

Using K_c to Calculate Equilibrium Concentrations

We could just as easil· chosen x to be the nu· mol/L of H_2O that · the number of mol/ or H_2 that forms. T¹ ing special ab chosen CO to de·

Equilibrium Concentrations

We obtain the equilibrium concentrations as

$$\left(\begin{array}{c}\text{equilibrium}\\\text{concentration}\end{array}\right) = \left(\begin{array}{c}\text{initial}\\\text{concentration}\end{array}\right) + \left(\begin{array}{c}\text{change in}\\\text{concentration}\end{array}\right)$$

SOLUTION: Here is the completed concentration table.

	$CO(g)$	$+$	$H_2O(g)$	$\rightleftharpoons$	$CO_2(g)$	$+$	$H_2(g)$
Initial concentrations (M)	0.100		0.100		0.0		0.0
Changes in concentrations (M)	$-x$		$-x$		$+x$		$+x$
Equilibrium concentrations (M)	$0.100 - x$		$0.100 - x$		x		x

The coefficients of x can be the same as the coefficients in the balanced chemical equation. This would assure us that the changes are in the same ratio as the coefficients in the equation.

The equilibrium concentration values must satisfy the equation given by the equilibrium law.

Note that the last line in the table tells us that the equilibrium CO and H_2O concentrations are equal to the number of moles per liter that were present initially minus the number of moles per liter that react. The equilibrium concentrations of CO_2 and H_2 equal the number of moles per liter of each that forms, since no CO_2 or H_2 is present initially.

Next, we substitute the quantities from the "equilibrium concentrations" row into the mass action expression and solve for x.

$$\frac{(x)(x)}{(0.100 - x)(0.100 - x)} = 4.06$$

which we can write as

$$\frac{x^2}{(0.100 - x)^2} = 4.06$$

In this problem we can solve the equation for x most easily by taking the square root of both sides.

$$\frac{x}{(0.100 - x)} = \sqrt{4.06} = 2.01$$

Clearing fractions gives

$$x = 2.01(0.100 - x)$$
$$x = 0.201 - 2.01x$$

Collecting terms in x gives

$$x + 2.01x = 0.201$$
$$3.01x = 0.201$$
$$x = 0.0668$$

Now that we know the value of x, we can calculate the equilibrium concentrations from the last row of the table.

$$[CO] = 0.100 - x = 0.100 - 0.0668 = 0.033\ M$$
$$[H_2O] = 0.100 - x = 0.100 - 0.0668 = 0.033\ M$$
$$[CO_2] = x = 0.0668\ M$$
$$[H_2] = x = 0.0668\ M$$

Does the Answer Make Sense?

First, we should check to see that all concentrations are positive numbers. They are. One way to check the answers is to substitute the equilibrium concentrations we've found into the mass action expression and evaluate the reaction quotient. If our answers are correct, Q should equal K_c. Let's do this.

$$Q = \frac{(0.0668)^2}{(0.033)^2} = 4.1$$

Rounding K_c to two significant figures gives 4.1, so the calculated concentrations satisfy the equilibrium law.

In the preceding example it was stated that the reaction

$$CO(g) + H_2O(g) \rightleftharpoons CO_2(g) + H_2(g)$$

has $K_c = 4.06$ at 500 °C. Suppose 0.0600 mol each of CO and H_2O is mixed with 0.100 mol each of CO_2 and H_2 in a 1.00 L reaction vessel. What will the concentrations of all the substances be when the mixture reaches equilibrium at this temperature?

ANALYSIS: We will proceed in much the same way as in the preceding example. However, this time determining the algebraic signs of x will not be quite so simple because none of the initial concentrations is zero. The best way to determine the algebraic signs is to use the initial concentrations to calculate the initial reaction quotient. Then we can compare Q to K_c, and then by reasoning we will figure out which way the reaction must proceed to make Q equal to K_c.

SOLUTION: The equilibrium law for the reaction is

$$\frac{[CO_2][H_2]}{[CO][H_2O]} = K_c$$

Let's use the initial concentrations, shown in the first row of the concentration table below, to determine the initial value of the reaction quotient.

$$Q_{initial} = \frac{(0.100)(0.100)}{(0.0600)(0.0600)} = 2.78 < K_c$$

As indicated, $Q_{initial}$ is less than K_c, so the system is not at equilibrium. To reach equilibrium Q must become larger, which requires an increase in the concentrations of CO_2 and H_2 as the reaction proceeds. This means that for CO_2 and H_2, the change must be positive, and for CO and H_2O, the change must be negative.

Here is the completed concentration table.

	CO(g)	+	H₂O(g)	⇌	CO₂(g)	+	H₂(g)
Initial concentrations (*M*)	0.0600		0.0600		0.100		0.100
Change in concentrations (*M*)	$-x$		$-x$		$+x$		$+x$
Equilibrium concentrations (*M*)	$0.0600 - x$		$0.0600 - x$		$0.100 + x$		$0.100 + x$

Substituting equilibrium quantities into the mass action expression in the equilibrium law gives us

$$\frac{(0.100 + x)^2}{(0.0600 - x)^2} = 4.06$$

Taking the square root of both sides yields

$$\frac{0.100 + x}{0.0600 - x} = 2.01$$

EXAMPLE 16.10

Using K_c to Calculate Equilibrium Concentrations

To solve for x we first multiply each side by $0.0600 - x$ to obtain

$$0.100 + x = 2.01(0.0600 - x)$$

$$0.100 + x = 0.121 - 2.01x$$

Collecting terms in x to one side and the constants to the other gives

$$x + 2.01x = 0.121 - 0.100$$

$$3.01x = 0.021$$

$$x = 0.0070$$

Now we can calculate the equilibrium concentrations:

$$[CO] = [H_2O] = (0.0600 - x) = 0.0600 - 0.0070 = 0.0530 \ M$$

$$[CO_2] = [H_2] = (0.100 + x) = 0.100 + 0.0070 = 0.107 \ M$$

Is the Answer Reasonable?
As a check, let's evaluate the reaction quotient using the calculated equilibrium concentrations.

$$Q = \frac{(0.107)^2}{(0.0530)^2} = 4.08$$

This is acceptably close to the value of K_c. (That it is not *exactly* equal to K_c is because of the rounding of answers during the calculations.)

In each of the preceding two examples, we were able to simplify the solution by taking the square root of both sides of the algebraic equation obtained by substituting equilibrium concentrations into the mass action expression. Such simplifications are not always possible, however, as illustrated in the next example.

EXAMPLE 16.11

Using K_c to Calculate Equilibrium Concentrations

At a certain temperature, $K_c = 4.50$ for the reaction

$$N_2O_4(g) \rightleftharpoons 2NO_2(g)$$

If 0.300 mol of N_2O_4 is placed into a 2.00 L container at this temperature, what will be the equilibrium concentrations of both gases?

ANALYSIS: As in the preceding example, at equilibrium the mass action expression must be equal to K_c.

$$\frac{[NO_2]^2}{[N_2O_4]} = 4.50$$

We will need to find algebraic expressions for the equilibrium concentrations and substitute them into the mass action expression. To obtain these, we set up the concentration table for the reaction.

Initial Concentrations
The initial concentration of N_2O_4 is 0.300 mol/2.00 L = 0.150 M. Since no NO_2 was placed in the reaction vessel, its initial concentration is 0.000 M.

Changes in Concentrations
There is no NO_2 in the reaction mixture, so we know its concentration must increase. This means the N_2O_4 concentration must decrease as some of the NO_2 is formed. Let's allow x to be the number of moles per liter of N_2O_4 that reacts, so the change in the N_2O_4 concentration is $-x$. Because of the stoichiometry of the reaction, the NO_2 concentration must increase by $2x$, so its change in concentration is $+2x$.

Equilibrium Concentrations
As before, we add the change to the initial concentration in each column to obtain expressions for the equilibrium concentrations.

SOLUTION: Here is the concentration table.

	$N_2O_4(g)$	$\rightleftharpoons$	$2NO_2(g)$
Initial concentrations (M)	0.150		0.000
Changes in concentrations (M)	$-x$		$+2x$
Equilibrium concentrations (M)	$0.150 - x$		$2x$

Notice that we have chosen the coefficients of x to be the same as the coefficients in the balanced chemical equation.

Now we substitute the equilibrium quantities into the mass action expression.

$$\frac{(2x)^2}{(0.150 - x)} = 4.50$$

or

$$\frac{4x^2}{(0.150 - x)} = 4.50 \tag{16.6}$$

This time the left side of the equation is not a perfect square, so we cannot just take the square root of both sides as in Example 16.10. However, because the equation involve terms in x^2, x, and a constant, we can use the quadratic formula. Recall that for a quadratic equation of the form $ax^2 + bx + c = 0$,

$$x = \frac{-b \pm \sqrt{b^2 - 4ac}}{2a}$$

Expanding Equation 16.6 above gives

$$4x^2 = 4.50 \, (0.150 - x)$$

$$= 0.675 - 4.50x$$

Arranging terms in the standard order gives

$$4x^2 + 4.50x - 0.675 = 0$$

Therefore, the quantities we will substitute into the quadratic formula are as follows: $a = 4$, $b = 4.50$, and $c = -0.675$. Making these substitutions gives

$$x = \frac{-4.50 \pm \sqrt{(4.50)^2 - 4(4)(-0.675)}}{2(4)}$$

$$= \frac{-4.50 \pm \sqrt{31.05}}{8}$$

$$= \frac{-4.50 \pm 5.57}{8}$$

Because of the $\pm$ term, there are two values of x that satisfy the equation, $x = 0.134$ and $x = -1.26$. However, only the first value, $x = 0.134$, makes any sense chemically. Using this value, the equilibrium concentrations are

$$[N_2O_4] = 0.150 - 0.134 = 0.016 \, M$$

$$[NO_2] = 2(0.134) = 0.268 \, M$$

Notice that if we had used the negative root, -1.26, the equilibrium concentration of NO_2 would be negative. Negative concentrations are impossible, so $x = -1.26$ is not acceptable *for chemical reasons*. In general, whenever you use the quadratic equation in a chemical calculation, one root will be satisfactory and the other will lead to answers that are nonsense.

Is the Answer Reasonable?

Once again, we can evaluate the reaction quotient using the calculated equilibrium values. When we do this, we obtain $Q = 4.49$, which is acceptably close to the value of K_c given.

Equilibrium problems can be much more complex than the one we have just discussed. However, you can often make assumptions that simplify the problem so that an approximate solution can be obtained.

PRACTICE EXERCISE 14: During an experiment, 0.200 mol of H_2 and 0.200 mol of I_2 were placed into a 1.00 L vessel where the reaction

$$H_2(g) + I_2(g) \rightleftharpoons 2HI(g)$$

came to equilibrium. For this reaction, $K_c = 49.5$ at the temperature of the experiment. What were the equilibrium concentrations of H_2, I_2, and HI?

Equilibrium calculations when K_c is very small

Many chemical reactions have equilibrium constants that are either very large or very small. For example, most weak acids have very small values for K_c. Therefore, only very tiny amounts of products form when these weak acids react with water.

When the K_c for a reaction is very small, the equilibrium calculations can usually be simplified considerably, as shown in the next example.

EXAMPLE 16.12

Simplifying Equilibrium Calculations for Reactions with Small K_c

Hydrogen, a potential fuel, is found in great abundance in water. Before the hydrogen can be used as a fuel, however, it must be separated from the oxygen; the water must be split into H_2 and O_2. One possibility is thermal decomposition, but this requires very high temperatures. Even at 1000 °C, $K_c = 7.3 \times 10^{-18}$ for the reaction

$$2H_2O(g) \rightleftharpoons 2H_2(g) + O_2(g)$$

If at 1000 °C the H_2O concentration in a reaction vessel is set initially at 0.100 M, what will the H_2 concentration be when the reaction reaches equilibrium?

ANALYSIS: We know that at equilibrium

$$\frac{[H_2]^2[O_2]}{[H_2O]^2} = 7.3 \times 10^{-18}$$

We now set up the concentration table.

Initial Concentrations
The initial concentration of H_2O is 0.100 M; those of H_2 and O_2 are both 0.0 M.

Changes in Concentrations
We know the changes must be in the same ratio as the coefficients in the balanced equation, so we place x's in this row with coefficients equal to those in the chemical equation. Because there are no products present initially, their changes must be positive and the change for the water must be negative.

Once we've set up the concentration table, we can substitute equilibrium concentrations into the equilibrium law. The equilibrium concentrations will be expressions that contain the unknown quantity x. We can solve the equilibrium law for x, and use the concentration table to relate x to the H_2 concentration that we're trying to calculate.

SOLUTION: The completed concentration table is

	$2H_2O(g)$	$\rightleftharpoons$	$2H_2(g)$	$+$	$O_2(g)$
Initial concentrations (M)	0.100		0.0		0.0
Changes in concentrations (M)	$-2x$		$+2x$		$+x$
Equilibrium concentrations (M)	$0.100 - 2x$		$2x$		x

When we substitute the equilibrium quantities into the mass action expression we get

$$\frac{(2x)^2 x}{(0.100 - 2x)^2} = 7.3 \times 10^{-18}$$

or

$$\frac{4x^3}{(0.100 - 2x)^2} = 7.3 \times 10^{-18}$$

$(2x)^2 x = (4x^2)x = 4x^3$

This is a *cubic equation* (one term involves x^3) and can be rather difficult to solve unless we can simplify it. In this instance we are able to do so because the very small value of K_c tells us that hardly any of the H_2O will decompose. Whatever the actual value of x, we know that it is going to be very small. This means that $2x$ will also be small, so when this tiny value is subtracted from 0.100, the result will still be very, very close to 0.100. We will make the assumption, then, that the term in the denominator will be essentially unchanged from 0.100 by subtracting $2x$ from it; that is, we will assume that $0.100 - 2x \approx 0.100$.

Even before we solve the problem, we know that hardly any H_2 and O_2 will be formed, because K_c is so small.

This assumption greatly simplifies the math. We now have

$$\frac{4x^3}{(0.100)^2} = 7.3 \times 10^{-18}$$

$$4x^3 = (0.0100)(7.3 \times 10^{-18}) = 7.3 \times 10^{-20}$$

$$x^3 = 1.8 \times 10^{-20}$$

$$x = \sqrt[3]{1.8 \times 10^{-20}}$$

$$= 2.6 \times 10^{-7}$$

Notice that the value of x that we've obtained is indeed very small. If we double it and subtract the answer from 0.100, we still get 0.100 when we round to the correct number of significant figures (the third decimal place).

$$0.100 - 2x = 0.100 - 2(2.6 \times 10^{-7}) = 0.09999948$$

$$= 0.100 \text{ (rounded correctly)}$$

This check verifies that our assumption was valid. Finally, we have to obtain the H_2 concentration. Our table gives

$$[H_2] = 2x$$

Always be sure to check your assumptions when solving a problem of this kind.

Therefore,

$$[H_2] = 2(2.6 \times 10^{-7}) = 5.2 \times 10^{-7} \, M$$

Is the Answer Reasonable?

The equilibrium constant is very small for this reaction (7.3×10^{-18}), so we expect the amount of product present at equilibrium to be small as well. The very low concentration of H_2 seems reasonable. The value of x was positive, which means that there was a decrease in reactant concentrations and an increase in product concentrations, as we would expect if only reactants were originally present. Finally, we can check the calculation by seeing if the mass action expression is equal to the equilibrium constant:

$$\frac{[H_2]^2[O_2]}{[H_2O]^2} = \frac{(5.2 \times 10^{-7})^2(2.6 \times 10^{-7})}{0.100^2} = 7.0 \times 10^{-18}$$

which differs from the original equilibrium constant only because the value of x was rounded off.

The simplifying assumption made in the preceding example is valid because a very small number is subtracted from a much larger one. We could also have neglected x (or $2x$) if it were a very small number that was being added to a much larger one. Remember that you can only neglect an x that's *added* or *subtracted;* you can never drop an x that occurs as a multiplying or dividing factor. Some examples are

Simplifying assumptions when K is very small

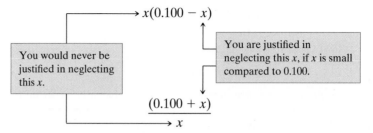

As a rule of thumb, you can expect that these simplifying assumptions will be valid if the concentration from which x is subtracted, or to which x is added, is at least 1000 times greater than K. For instance, in the preceding example, $2x$ was subtracted from 0.100. Since 0.100 is much larger than $1000 \times (7.3 \times 10^{-18})$ we expect the assumption $0.100 - 2x \approx 0.100$ to be valid. However, even though the simplifying assumption is *expected* to be valid, always check to see if it really is after finishing the calculation. If the assumption proves invalid, then some other way to solve the algebra must be found.

> **PRACTICE EXERCISE 15:** In air at 25 °C and 1 atm, the N_2 concentration is 0.033 M and the O_2 concentration is 0.00810 M. The reaction
>
> $$N_2(g) + O_2(g) \rightleftharpoons 2NO(g)$$
>
> has $K_c = 4.8 \times 10^{-31}$ at 25 °C. Taking the N_2 and O_2 concentrations given above as initial values, calculate the equilibrium NO concentration that should exist in our atmosphere from this reaction at 25 °C.

SUMMARY

Dynamic Equilibrium When the forward and reverse reactions in a chemical system occur at equal rates, a dynamic equilibrium exists and the concentrations of the reactants and products remain constant. For a given overall chemical composition, the amounts of reactants and products that are present at equilibrium are the same regardless of whether the equilibrium is approached from the direction of pure "reactants," pure "products," or any mixture of them. (In a chemical equilibrium, the terms *reactants* and *products* do not have the usual significance because the reaction is proceeding in both directions simultaneously. Instead, we use *reactants* and *products* simply to identify the substances on the left- and right-hand sides of the equation for the equilibrium.)

The Equilibrium Law The **mass action expression** is a fraction. The concentrations of the products, raised to powers equal to their coefficients in the chemical equation, are multiplied together in the numerator. The denominator is constructed in the same way from the concentrations of the reactants raised to powers equal to their coefficients. The numerical value of the mass action expression is the **reaction quotient, Q.** At equilibrium, the reaction quotient is equal to the **equilibrium constant, K_c.** If partial pressures of gases are used in the mass action expression, K_P is obtained. The magnitude of the equilibrium constant is roughly pro-

portional to the extent to which the reaction proceeds to completion when equilibrium is reached.

Manipulating Equilibrium Equations When an equation is multiplied by a factor n to obtain a new equation, we raise its K to the power n to obtain the K for the new equation. When two equations are added, we multiply their K's to obtain the new K. When an equation is reversed, we take the reciprocal of its K to obtain the new K.

Relating K_P to K_c The values of K_P and K_c are only equal if the same number of moles of gas are represented on both sides of the chemical equation. When the numbers of moles of gas are different, K_P is related to K_c by the equation $K_P = K_c (RT)^{\Delta n_g}$. Remember to use $R = 0.0821$ L atm mol^{-1} K^{-1} and T = absolute temperature. Also, be careful to calculate Δn_g as the difference between the number of moles of *gaseous* products and the number of moles of *gaseous* reactants in the balanced equation.

Heterogeneous Equilibria An equilibrium involving substances in more than one phase is a **heterogeneous equilibrium.** The mass action expression for a heterogeneous equilibrium omits concentration terms for pure liquids and/or pure solids.

Le Châtelier's Principle This principle states that *when an equilibrium is upset, a chemical change occurs in a direction that opposes the disturbing influence and brings the system to equilibrium again.* Adding a reactant or a product causes a reaction to occur that uses up part of what has been added. Removing a reactant or a product causes a reaction that replaces part of what has been removed. Increasing the pressure (by reducing the volume) drives a reaction in the direction of the fewer number of moles of gas. Pressure changes have virtually no effect on equilibria involving only solids and liquids. Raising the temperature causes an equilibrium to shift in an endothermic direction. The value of K increases with increasing temperature for reactions that are endothermic in the forward direction. A change in temperature is the only factor that changes K. Addition of a catalyst has no effect on an equilibrium.

Equilibrium Calculations The initial concentrations in a chemical system are controlled by the person who combines the chemicals at the start of the reaction. The changes in concentration are determined by the stoichiometry of the reaction. Only equilibrium concentrations satisfy the equilibrium law. When these are used, the mass action expression is equal to K_c. When a change in concentration is expected to be very small compared to the initial concentration, the change may be neglected and the algebraic equation derived from the equilibrium law can be simplified. In general, this simplification is valid if the initial concentration is at least 1000 times larger than K.

TOOLS ▶ YOU HAVE LEARNED

In this chapter you have learned the following tools that we apply to various aspects of solving problems dealing with chemical equilibria.

TOOL	HOW IT WORKS
Mass action expression and the equilibrium law (page 702)	Defines the relationship that exists between the equilibrium concentrations or partial pressures of the reactants and products in a chemical reaction. Be sure you know how to form the mass action expression from the balanced chemical equation. Writing the equilibrium law is one of the first steps in solving equilibrium problems.
Manipulation of equilibrium equations (page 703)	Permits you to determine K for a particular reaction by manipulating or combining one or more different chemical equations and their associated equilibrium constants. Be sure you know the rules that apply for the various operations.
Magnitude of the equilibrium constant (page 706)	Ascertains qualitatively the extent to which a reaction proceeds toward completion when equilibrium is reached. The size of K determines whether you are able to apply simplifying assumptions in solving problems.
$K_P = K_c (RT)^{\Delta n_g}$ (page 707)	Converts between K_P and K_c. Remember that Δn_g is the change in the number of moles of gas, which is computed using the coefficients of gaseous products and reactants.
Le Châtelier's principle (page 711)	Analyzes the effects of disturbances on the position of equilibrium.
Concentration table (page 718)	Organizes the approach to solving problems dealing with chemical equilibria. It incorporates initial concentrations, changes in concentrations, and equilibrium concentration expressions.
Simplifying assumptions (page 728)	We use these to simplify the algebra when working equilibrium problems for reactions for which K is very small.

THINKING IT THROUGH

Need extra help?
Visit the Brady/
Senese web site at
ON-LINE HELP
www.wiley.com/
college/brady

The goal for the following problems is not to find the answers themselves, but rather to assemble the information needed to solve them and explain how you would use the information to find the answers. The problems in Level 2 are more challenging than those in Level 1 and may contain more data than are required, in which case you are also asked to identify the unnecessary data. Detailed answers to the Thinking-It-Through problems can be found on the web site.

Level 1 Problems

1. Suppose the following data were collected at 25 °C for the system

$$NO_2(g) + NO(g) \rightleftharpoons N_2O_3(g)$$

Equilibrium Concentrations		
$[NO_2]$	$[NO]$	$[N_2O_3]$
0.020 M	0.033 M	8.6×10^{-3} M
0.040 M	0.0085 M	4.4×10^{-3} M
0.015 M	0.012 M	2.3×10^{-3} M

Explain in detail how you would use these data to show that they obey the principles of the equilibrium law.

2. The reaction $NO_2(g) + NO(g) \rightleftharpoons N_2O(g) + O_2(g)$ has $K_c = 0.914$ at a certain temperature. What manipulations are necessary to determine, at this same temperature, the value of K_c for the following reaction?

$$2N_2O(g) + 2O_2(g) \rightleftharpoons 2NO_2(g) + 2NO(g)$$

3. In the preceding question, is there sufficient information to calculate the value of K_P for the reaction below? If not, what additional data are needed?

$$NO_2(g) + NO(g) \rightleftharpoons N_2O(g) + O_2(g)$$

4. The reaction

$$NO_2(g) + NO(g) \rightleftharpoons N_2O(g) + O_2(g)$$

has $\Delta H° = -42.9$ kJ. List the kinds of changes that would alter the amount of NO_2 present in the system at equilibrium. What could be done to this system *without* changing the amount of NO_2 at equilibrium?

5. At 100 °C, $K_c = 0.135$ for the reaction

$$3H_2(g) + N_2(g) \rightleftharpoons 2NH_3(g)$$

In a reaction mixture at equilibrium at this temperature, $[NH_3] = 0.030$ M and $[N_2] = 0.50$ M. Explain how you can calculate the molar concentration of $H_2(g)$. (Set up the calculation.)

6. In the reaction mixture described in the preceding question, explain how you can calculate the partial pressures of each of the gases in the mixture. (Set up the calculation.)

7. How would you use the results obtained in the preceding questions to calculate the value of K_P for the following reaction?

$$3H_2(g) + N_2(g) \rightleftharpoons 2NH_3(g)$$

8. The reaction $N_2O_3(g) \rightleftharpoons NO(g) + NO_2(g)$ is to be stud-

ied at a certain temperature. A mixture was prepared in which the initial concentrations of NO and NO_2 were 0.40 M and 0.60 M, respectively. When equilibrium was reached, the concentration of N_2O_3 was measured to be 0.13 M. What were the equilibrium concentrations of NO and NO_2? (Explain each step in the solution to the problem.)

Level 2 Problems

9. One function of the catalysts in an automotive catalytic converter is to promote the decomposition of nitrogen oxides into nitrogen and oxygen. Given that the reaction of $N_2(g)$ with $O_2(g)$ to form NO(g) is endothermic, explain how the functioning of a catalytic converter is related to chemical equilibrium and Le Châtelier's principle.

10. The following are equilibrium constants calculated for the reaction

$$3H_2(g) + N_2(g) \rightleftharpoons 2NH_3(g)$$

t (°C)	K_P	K_c
25	7.11×10^5	4.26×10^8
300	4.61×10^{-9}	1.02×10^{-5}
400	2.62×10^{-10}	8.00×10^{-7}

Explain how you can determine in a simple way how the amount of NH_3 at equilibrium varies as the temperature of the reaction mixture is increased.

11. Suppose we begin the reaction

$$2H_2(g) + O_2(g) \rightleftharpoons 2H_2O(g) \qquad K_c = 1.4 \times 10^{17}$$

with the following initial concentrations: $[H_2] = 0.020$ M, $[O_2] = 0.030$ M, and $[H_2O] = 0.00$ M. If we let x equal the number of moles per liter of O_2 that react, why are we not justified in neglecting x in expressing the equilibrium concentrations of H_2 and O_2?

12. In the preceding question, how could the initial concentrations be recalculated so as to make the simplifying assumption valid when the calculation is performed? (*Hint:* The position of equilibrium does not depend on the direction from which equilibrium is approached.)

13. For the equilibrium

$$3NO_2(g) \rightleftharpoons N_2O_5(g) + NO(g)$$

$K_c = 1.0 \times 10^{-11}$. If a 4.00 L container initially holds 0.20 mol of NO_2, how many moles of N_2O_5 will be present when this system reaches equilibrium? (Describe each step in the solution to the problem.)

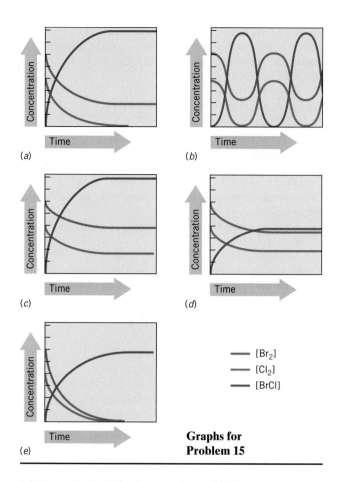

(a) *(b)*

(c) *(d)*

(e)

— [Br₂]
— [Cl₂]
— [BrCl]

Graphs for Problem 15

14. To study the following reaction at 20 °C:

$$NO(g) + NO_2(g) + H_2O(g) \rightleftharpoons 2HNO_2(g)$$

a mixture of $NO(g)$, $NO_2(g)$, and $H_2O(g)$ was prepared in a 10.0 L glass bulb. For NO, NO₂, and HNO₂, the initial concentrations were as follows: $[NO] = [NO_2] = 2.59 \times 10^{-3}$ *M* and $[HNO_2] = 0.0$ *M*. The initial partial pressure of $H_2O(g)$ was 17.5 torr. When equilibrium was reached, the HNO₂ concentration was 4.0×10^{-4} *M*. Explain in detail how to calculate the equilibrium constant, K_c, for this reaction.

15. In CCl₄ solution Cl₂ reacts with Br₂ to form BrCl according to the reaction

$$Br_2 + Cl_2 \rightleftharpoons 2BrCl \qquad K_c = 2$$

If the initial concentrations were $[Br_2] = 0.6$ *M*, $[Cl_2] = 0.4$ *M*, and $[BrCl] = 0.0$ *M*, which of the concentration versus time graphs on the left could represent this reaction? Explain why you rejected each of the other four graphs.

16. Two 1.00 L bulbs are filled with 0.500 atm of $F_2(g)$ and $PF_3(g)$, as illustrated in the figure below. At a particular temperature, $K_P = 4.0$ for the reaction of these gases to form $PF_5(g)$:

$$F_2(g) + PF_3(g) \rightleftharpoons PF_5(g)$$

(a) The stopcock between the bulbs is opened and the pressure falls. Make a sketch of this apparatus showing the gas composition once the pressure is stable.
(b) List any chemical reactions that continue to occur once the pressure is stable.
(c) Suppose all the gas in the left bulb is forced into the bulb on the right and then the stopcock is closed. Make a second sketch to show the composition of the gas mixture after equilibrium is reached. Comment on how the composition of the one-bulb system differs from that for the two-bulb system.

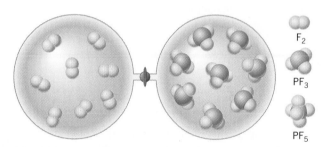

F₂

PF₃

PF₅

REVIEW QUESTIONS

General

16.1 Sketch a graph showing how the concentrations of the reactants and products of a typical chemical reaction vary with time during the course of the reaction. Assume no products are present at the start of the reaction.

16.2 What is meant when we say that chemical reactions are *reversible*?

Mass Action Expression, K_P and K_c

16.3 What is an *equilibrium law*?

16.4 How is the term *reaction quotient* defined?

16.5 Under what conditions does the reaction quotient equal K_c?

16.6 When a chemical equation and its equilibrium constant are given, why is it not necessary to also specify the form of the mass action expression?

16.7 At 225 °C, $K_P = 6.3 \times 10^{-3}$ for the reaction

$$CO(g) + 2H_2(g) \rightleftharpoons CH_3OH(g)$$

Would we expect this reaction to go nearly to completion?

16.8 Here are some reactions and their equilibrium constants:
(a) $2CH_4(g) \rightleftharpoons C_2H_6(g) + H_2(g)$ $\qquad K_c = 9.5 \times 10^{-13}$
(b) $CH_3OH(g) + H_2(g) \rightleftharpoons CH_4(g) + H_2O(g)$
$\qquad\qquad\qquad\qquad\qquad\qquad K_c = 3.6 \times 10^{20}$
(c) $H_2(g) + Br_2(g) \rightleftharpoons 2HBr(g)$ $\qquad K_c = 2.0 \times 10^9$

Arrange these reactions in order of their increasing tendency to go toward completion.

Converting between K_P and K_c

16.9 State the equation relating K_P to K_c and define all terms. Which is the only value of R that can be properly used in this equation?

16.10 Use the ideal gas law to show that the partial pressure of a gas is directly proportional to its molar concentration. What is the proportionality constant?

Heterogeneous Equilibria

16.11 What is the difference between a *heterogeneous equilibrium* and a *homogeneous equilibrium*?

16.12 Why do we omit the concentrations of pure liquids and solids from the mass action expressions of heterogeneous reactions?

Le Châtelier's Principle

16.13 State Le Châtelier's principle in your own words.

16.14 How will the equilibrium

$$\text{heat} + CH_4(g) + 2H_2S(g) \rightleftharpoons CS_2(g) + 4H_2(g)$$

be affected by the following?
(a) The addition of $CH_4(g)$
(b) The addition of $H_2(g)$
(c) The removal of $CS_2(g)$
(d) A decrease in the volume of the container
(e) An increase in temperature

16.15 The reaction $CO(g) + 2H_2(g) \rightleftharpoons CH_3OH(g)$ has $\Delta H° = -18$ kJ. How will the amount of CH_3OH present at equilibrium be affected by the following?
(a) Adding $CO(g)$
(b) Removing $H_2(g)$
(c) Decreasing the volume of the container
(d) Adding a catalyst
(e) Increasing the temperature

16.16 Consider the equilibrium

$$N_2O(g) + NO_2(g) \rightleftharpoons 3NO(g) \qquad \Delta H° = +155.7 \text{ kJ}$$

In which direction will this equilibrium be shifted by the following changes?
(a) Addition of N_2O
(b) Removal of NO_2
(c) Addition of NO
(d) Increasing the temperature of the reaction mixture
(e) Addition of helium gas to the reaction mixture at constant volume
(f) Decreasing the volume of the container at constant temperature

16.17 In Review Questions 16.15 and 16.16, which change(s) will alter the value of K_c?

16.18 Consider the equilibrium $2NO(g) + Cl_2(g) \rightleftharpoons 2NOCl(g)$ for which $\Delta H° = -77.07$ kJ. How will the amount of Cl_2 at equilibrium be affected by the following?
(a) Removal of $NO(g)$
(b) Addition of $NOCl(g)$
(c) Raising the temperature
(d) Decreasing the volume of the container

REVIEW PROBLEMS

Answers to problems whose numbers are printed in color are given in Appendix B. More challenging problems are marked with asterisks. **ILW** = Interactive LearningWare solution is available at *www.wiley.com/college/brady*.

Equilibrium Laws for K_P and K_c

16.19 Write the equilibrium law for each of the following reactions in terms of molar concentrations:
(a) $2PCl_3(g) + O_2(g) \rightleftharpoons 2POCl_3(g)$
(b) $2SO_3(g) \rightleftharpoons 2SO_2(g) + O_2(g)$
(c) $N_2H_4(g) + 2O_2(g) \rightleftharpoons 2NO(g) + 2H_2O(g)$
(d) $N_2H_4(g) + 6H_2O_2(g) \rightleftharpoons 2NO_2(g) + 8H_2O(g)$
(e) $SOCl_2(g) + H_2O(g) \rightleftharpoons SO_2(g) + 2HCl(g)$

16.20 Write the equilibrium law for each of the following gaseous reactions in terms of molar concentrations:
(a) $3Cl_2(g) + NH_3(g) \rightleftharpoons NCl_3(g) + 3HCl(g)$
(b) $PCl_3(g) + PBr_3(g) \rightleftharpoons PCl_2Br(g) + PClBr_2(g)$
(c) $NO(g) + NO_2(g) + H_2O(g) \rightleftharpoons 2HNO_2(g)$
(d) $H_2O(g) + Cl_2O(g) \rightleftharpoons 2HOCl(g)$
(e) $Br_2(g) + 5F_2(g) \rightleftharpoons 2BrF_5(g)$

16.21 Write the equilibrium law for the reactions in Problem 16.19 in terms of partial pressures.

16.22 Write the equilibrium law for the reactions in Problem 16.20 in terms of partial pressures.

16.23 Write the equilibrium law for each of the following reactions in aqueous solution:
(a) $Ag^+(aq) + 2NH_3(aq) \rightleftharpoons Ag(NH_3)_2^+(aq)$
(b) $Cd^{2+}(aq) + 4SCN^-(aq) \rightleftharpoons Cd(SCN)_4^{2-}(aq)$

16.24 Write the equilibrium law for each of the following reactions in aqueous solution:
(a) $HClO(aq) + H_2O \rightleftharpoons H_3O^+(aq) + ClO^-(aq)$
(b) $CO_3^{2-}(aq) + HSO_4^-(aq) \rightleftharpoons HCO_3^-(aq) + SO_4^{2-}(aq)$

Manipulating Equilibrium Equations

16.25 At 25 °C, $K_c = 1 \times 10^{-85}$ for the reaction

$$7IO_3^- + 9H_2O + 7H^+ \rightleftharpoons I_2 + 5H_5IO_6$$

What is the value of K_c for the following reaction?

$$I_2 + 5H_5IO_6 \rightleftharpoons 7IO_3^- + 9H_2O + 7H^+$$

16.26 Use the following equilibria:

$$2CH_4(g) \rightleftharpoons C_2H_6(g) + H_2(g) \qquad K_c = 9.5 \times 10^{-13}$$

$$CH_4(g) + H_2O(g) \rightleftharpoons CH_3OH(g) + H_2(g)$$

$$K_c = 2.8 \times 10^{-21}$$

to calculate K_c for the reaction

$$2CH_3OH(g) + H_2(g) \rightleftharpoons C_2H_6(g) + 2H_2O(g)$$

16.27 Write the equilibrium law for each of the following reactions in terms of molar concentrations:
(a) $H_2(g) + Cl_2(g) \rightleftharpoons 2HCl(g)$
(b) $\frac{1}{2}H_2(g) + \frac{1}{2}Cl_2(g) \rightleftharpoons HCl(g)$

How does K_c for reaction (a) compare with K_c for reaction (b)?

16.28 Write the equilibrium law for the reaction

$$2HCl(g) \rightleftharpoons H_2(g) + Cl_2(g)$$

How does K_c for this reaction compare with K_c for reaction (a) in the preceding problem?

Converting between K_P and K_c

16.29 A 345 mL container holds NH_3 at a pressure of 745 torr and a temperature of 45 °C. What is the molar concentration of ammonia in the container?

16.30 In a certain container at 145 °C the concentration of water vapor is 0.0200 M. What is the partial pressure of H_2O in the container?

16.31 For which of the following reactions does $K_P = K_c$?
(a) $2H_2(g) + C_2H_2(g) \rightleftharpoons C_2H_6(g)$
(b) $N_2(g) + O_2(g) \rightleftharpoons 2NO(g)$
(c) $2NO(g) + O_2(g) \rightleftharpoons 2NO_2(g)$

16.32 For which of the following reactions does $K_P = K_c$?
(a) $CO_2(g) + H_2(g) \rightleftharpoons CO(g) + H_2O(g)$
(b) $PCl_3(g) + Cl_2(g) \rightleftharpoons PCl_5(g)$
(c) $N_2O_4(g) \rightleftharpoons 2NO_2(g)$

ILW 16.33 The reaction $CO(g) + 2H_2(g) \rightleftharpoons CH_3OH(g)$ has $K_P = 6.3 \times 10^{-3}$ at 225 °C. What is the value of K_c at this temperature?

16.34 The reaction $HCO_2H(g) \rightleftharpoons CO(g) + H_2O(g)$ has $K_P = 1.6 \times 10^6$ at 400 °C. What is the value of K_c for this reaction at this temperature?

16.35 The reaction $N_2O(g) + NO_2(g) \rightleftharpoons 3NO(g)$ has $K_c = 4.2 \times 10^{-4}$ at 500 °C. What is the value of K_P at this temperature?

16.36 One possible way of removing NO from the exhaust of a gasoline engine is to cause it to react with CO in the presence of a suitable catalyst.

$$2NO(g) + 2CO(g) \rightleftharpoons N_2(g) + 2CO_2(g)$$

At 300 °C, this reaction has $K_c = 2.2 \times 10^{59}$. What is K_P at 300 °C?

16.37 At 773 °C, the reaction $CO(g) + 2H_2(g) \rightleftharpoons CH_3OH(g)$ has $K_c = 0.40$. What is K_P at this temperature?

16.38 The reaction $COCl_2(g) \rightleftharpoons CO(g) + Cl_2(g)$ has $K_P = 4.6 \times 10^{-2}$ at 395 °C. What is K_c at this temperature?

Heterogeneous Equilibria

16.39 Calculate the molar concentration of water in (a) 18.0 mL of H_2O, (b) 100.0 mL of H_2O, and (c) 1.00 L of H_2O. Assume that the density of water is 1.00 g/mL.

16.40 The density of sodium chloride is 2.164 g cm^{-3}. What is the molar concentration of NaCl in a 12.0 cm^3 sample of pure NaCl? What is the molar concentration of NaCl in a 25.0 g sample of pure NaCl?

16.41 Write the equilibrium law corresponding to K_c for each of the following heterogeneous reactions:
(a) $2C(s) + O_2(g) \rightleftharpoons 2CO(g)$
(b) $2NaHSO_3(s) \rightleftharpoons Na_2SO_3(s) + H_2O(g) + SO_2(g)$
(c) $2C(s) + 2H_2O(g) \rightleftharpoons CH_4(g) + CO_2(g)$
(d) $CaCO_3(s) + 2HF(g) \rightleftharpoons CaF_2(s) + H_2O(g) + CO_2(g)$
(e) $CuSO_4 \cdot 5H_2O(s) \rightleftharpoons CuSO_4(s) + 5H_2O(g)$

16.42 Write the equilibrium law corresponding to K_c for each of the following heterogeneous reactions:
(a) $CaCO_3(s) + SO_2(g) \rightleftharpoons CaSO_3(s) + CO_2(g)$
(b) $AgCl(s) + Br^-(aq) \rightleftharpoons AgBr(s) + Cl^-(aq)$
(c) $Cu(OH)_2(s) \rightleftharpoons Cu^{2+}(aq) + 2OH^-(aq)$
(d) $Mg(OH)_2(s) \rightleftharpoons MgO(s) + H_2O(g)$
(e) $3CuO(s) + 2NH_3(g) \rightleftharpoons 3Cu(s) + N_2(g) + 3H_2O(g)$

16.43 The heterogeneous reaction $2HCl(g) + I_2(s) \rightleftharpoons 2HI(g) + Cl_2(g)$ has $K_c = 1.6 \times 10^{-34}$ at 25 °C. Suppose 0.100 mol of HCl and solid I_2 are placed in a 1.00 L container. What will be the equilibrium concentrations of HI and Cl_2 in the container?

16.44 At 25 °C, $K_c = 360$ for the reaction

$$AgCl(s) + Br^-(aq) \rightleftharpoons AgBr(s) + Cl^-(aq)$$

If solid AgCl is added to a solution containing 0.10 M Br$^-$, what will be the equilibrium concentrations of Br$^-$ and Cl$^-$?

Equilibrium Calculations

16.45 At a certain temperature, $K_c = 0.18$ for the equilibrium

$$PCl_3(g) + Cl_2(g) \rightleftharpoons PCl_5(g)$$

Suppose a reaction vessel at this temperature contained these three gases at the following concentrations: [PCl_3] = 0.0420 M, [Cl_2] = 0.0240 M, [PCl_5] = 0.00500 M.
(a) Is the system in a state of equilibrium?
(b) If not, which direction will the reaction have to proceed to get to equilibrium?

16.46 At 460 °C, the reaction

$$SO_2(g) + NO_2(g) \rightleftharpoons NO(g) + SO_3(g)$$

has $K_c = 85.0$. A reaction flask at 460 °C contains these gases at the following concentrations: [SO_2] = 0.00250 M, [NO_2] = 0.00350 M, [NO] = 0.0250 M, [SO_3] = 0.0400 M.
(a) Is the reaction at equilibrium?
(b) If not, which way will the reaction have to proceed to arrive at equilibrium?

16.47 At a certain temperature, the reaction

$$CO(g) + 2H_2(g) \rightleftharpoons CH_3OH(g)$$

has $K_c = 0.500$. If a reaction mixture at equilibrium contains 0.180 M CO and 0.220 M H_2, what is the concentration of CH_3OH?

16.48 $K_c = 64$ for the reaction

$$N_2(g) + 3H_2(g) \rightleftharpoons 2NH_3(g)$$

at a certain temperature. Suppose it was found that an equilibrium mixture of these gases contained 0.360 M NH_3 and 0.0192 M N_2. What was the concentration of H_2 in the mixture?

16.49 At 773 °C, a mixture of $CO(g)$, $H_2(g)$, and $CH_3OH(g)$ was allowed to come to equilibrium. The following equilibrium concentrations were then measured: $[CO] = 0.105$ M, $[H_2] = 0.250$ M, $[CH_3OH] = 0.00261$ M. Calculate K_c for the reaction

$$CO(g) + 2H_2(g) \rightleftharpoons CH_3OH(g)$$

16.50 Ethylene, C_2H_4, and water react under appropriate conditions to give ethanol. The reaction is

$$C_2H_4(g) + H_2O(g) \rightleftharpoons C_2H_5OH(g)$$

An equilibrium mixture of these gases at a certain temperature had the following concentrations: $[C_2H_4] = 0.0148$ M, $[H_2O] = 0.0336$ M, $[C_2H_5OH] = 0.180$ M. What is the value of K_c?

ILW 16.51 At high temperature, 2.00 mol of HBr was placed in a 4.00 L container where it decomposed to give the equilibrium

$$2HBr(g) \rightleftharpoons H_2(g) + Br_2(g)$$

At equilibrium, the concentration of Br_2 was measured to be 0.0955 M. What is K_c for this reaction at this temperature?

16.52 A 0.050 mol sample of formaldehyde vapor, CH_2O, was placed in a heated 500 mL vessel and some of it decomposed. The reaction is

$$CH_2O(g) \rightleftharpoons H_2(g) + CO(g)$$

At equilibrium, the $CH_2O(g)$ concentration was 0.066 mol L^{-1}. Calculate the value of K_c for this reaction.

16.53 The reaction $NO_2(g) + NO(g) \rightleftharpoons N_2O(g) + O_2(g)$ reached equilibrium at a certain high temperature. Originally, the reaction vessel contained the following initial concentrations: $[N_2O] = 0.184$ M, $[O_2] = 0.377$ M, $[NO_2] = 0.0560$ M, and $[NO] = 0.294$ M. The concentration of the NO_2, the only colored gas in the mixture, was monitored by following the intensity of the color. At equilibrium, the NO_2 concentration had become 0.118 M. What is the value of K_c for this reaction at this temperature?

16.54 At 25 °C, 0.0560 mol O_2 and 0.020 mol N_2O were placed in a 1.00 L container where the following equilibrium was then established:

$$2N_2O(g) + 3O_2(g) \rightleftharpoons 4NO_2(g)$$

At equilibrium, the NO_2 concentration was 0.020 M. What is the value of K_c for this reaction?

16.55 At 25 °C, $K_c = 0.145$ for the following reaction in the solvent CCl_4:

$$2BrCl \rightleftharpoons Br_2 + Cl_2$$

If the initial concentration of BrCl in the solution is 0.050 M, what will the equilibrium concentrations of Br_2 and Cl_2 be?

16.56 At 25 °C, $K_c = 0.145$ for the following reaction in the solvent CCl_4:

$$2BrCl \rightleftharpoons Br_2 + Cl_2$$

If the initial concentrations of Br_2 and Cl_2 are each 0.0250 M, what will their equilibrium concentrations be?

16.57 The equilibrium constant, K_c, for the reaction

$$SO_3(g) + NO(g) \rightleftharpoons NO_2(g) + SO_2(g)$$

was found to be 0.500 at a certain temperature. If 0.240 mol of SO_3 and 0.240 mol of NO are placed in a 2.00 L container and allowed to react, what will be the equilibrium concentration of each gas?

16.58 For the reaction in the preceding problem, a reaction mixture is prepared in which 0.120 mol NO_2 and 0.120 mol of SO_2 are placed in a 1.00 L vessel. After the system reaches equilibrium, what will be the equilibrium concentrations of all four gases? How do these equilibrium values compare to those calculated in Problem 16.57? Account for your observation.

16.59 At a certain temperature, the reaction

$$CO(g) + H_2O(g) \rightleftharpoons CO_2(g) + H_2(g)$$

has $K_c = 0.400$. Exactly 1.00 mol of each gas was placed in a 100 L vessel and the mixture underwent reaction. What was the equilibrium concentration of each gas?

16.60 At 25 °C, $K_c = 0.145$ for the following reaction in the solvent CCl_4:

$$2BrCl \rightleftharpoons Br_2 + Cl_2$$

If the initial concentrations of each substance in a solution are 0.0400 M, what will their equilibrium concentrations be?

ILW 16.61 The reaction $2HCl(g) \rightleftharpoons H_2(g) + Cl_2(g)$ has $K_c = 3.2 \times 10^{-34}$ at 25 °C. If a reaction vessel contains initially 0.0500 mol L^{-1} of HCl which then reacts to reach equilibrium, what will be the concentrations of H_2 and Cl_2?

16.62 At 200 °C, $K_c = 1.4 \times 10^{-10}$ for the reaction

$$N_2O(g) + 2NO_2(g) \rightleftharpoons 3NO(g)$$

If 0.200 mol of N_2O and 0.400 mol NO_2 are placed in a 4.00 L container, what would the NO concentration be if this equilibrium were established?

ILW 16.63 At 2000 °C, the decomposition of CO_2,

$$2CO_2(g) \rightleftharpoons 2CO(g) + O_2(g)$$

has $K_c = 6.4 \times 10^{-7}$. If a 1.00 L container holding 1.0 × 10^{-2} mol of CO_2 is heated to 2000 °C, what will be the concentration of CO at equilibrium?

16.64 At 500 °C, the decomposition of water into hydrogen and oxygen,

$$2H_2O(g) \rightleftharpoons 2H_2(g) + O_2(g)$$

has $K_c = 6.0 \times 10^{-28}$. How many moles of H_2 and O_2 are present at equilibrium in a 5.00 L reaction vessel at this temperature if the container originally held 0.015 mol H_2O?

***16.65** At a certain temperature, $K_c = 0.18$ for the equilibrium

$$PCl_3(g) + Cl_2(g) \rightleftharpoons PCl_5(g)$$

If 0.026 mol of PCl_5 is placed in a 2.00 L vessel at this temperature, what will the concentration of PCl_3 be at equilibrium?

*16.66 At 460 °C, the reaction

$$SO_2(g) + NO_2(g) \rightleftharpoons NO(g) + SO_3(g)$$

has $K_c = 85.0$. Suppose 0.100 mol of SO_2, 0.0600 mol of NO_2, 0.0800 mol of NO, and 0.120 mol of SO_3 are placed in a 10.0 L container at this temperature. What will the concentrations of all the gases be when the system reaches equilibrium?

*16.67 At a certain temperature, $K_c = 0.500$ for the reaction

$$SO_3(g) + NO(g) \rightleftharpoons NO_2(g) + SO_2(g)$$

If 0.100 mol SO_3 and 0.200 mol NO are placed in a 2.00 L container and allowed to come to equilibrium, what will the NO_2 and SO_2 concentrations be?

*16.68 At 25 °C, $K_c = 0.145$ for the following reaction in the solvent CCl_4:

$$2BrCl \rightleftharpoons Br_2 + Cl_2$$

A solution was prepared with the following initial concentrations: [BrCl] = 0.0400 M, $[Br_2]$ = 0.0300 M, and $[Cl_2]$ = 0.0200 M. What will their equilibrium concentrations be?

*16.69 At a certain temperature, $K_c = 4.3 \times 10^5$ for the reaction

$$HCO_2H(g) \rightleftharpoons CO(g) + H_2O(g)$$

If 0.200 mol of $HCHO_2$ is placed in a 1.00 L vessel, what will be the concentrations of CO and H_2O when the system reaches equilibrium?

*16.70 The reaction $H_2(g) + Br_2(g) \rightleftharpoons 2HBr(g)$ has $K_c = 2.0 \times 10^9$ at 25 °C. If 0.100 mol of H_2 and 0.200 mol of Br_2 are placed in a 10.0 L container, what will all the equilibrium concentrations be at 25 °C?

ADDITIONAL EXERCISES

16.71 The reaction $N_2O_4(g) \rightleftharpoons 2NO_2(g)$ has $K_P = 0.140$ at 25 °C. In a reaction vessel containing these gases in equilibrium at this temperature, the partial pressure of N_2O_4 was 0.250 atm.
(a) What was the partial pressure of the NO_2 in the reaction mixture?
(b) What was the total pressure of the mixture of gases?

16.72 The following reaction in aqueous solution has $K_c = 1 \times 10^{-85}$ at a temperature of 25 °C.

$$7IO_3^- + 9H_2O + 7H^+ \rightleftharpoons I_2 + 5H_5IO_6$$

What is the equilibrium law for this reaction?

*16.73 At a certain temperature, $K_c = 0.914$ for the reaction

$$NO_2(g) + NO(g) \rightleftharpoons N_2O(g) + O_2(g)$$

A mixture was prepared containing 0.200 mol NO_2, 0.300 mol NO, 0.150 mol N_2O, and 0.250 mol O_2 in a 4.00 L container. What will be the equilibrium concentrations of each gas?

*16.74 At 400 °C, $K_c = 2.9 \times 10^4$ for the reaction

$$HCHO_2(g) \rightleftharpoons CO(g) + H_2O(g)$$

A mixture was prepared with the following initial concentrations: [CO] = 0.20 M, $[H_2O]$ = 0.30 M. No formic acid, $HCHO_2$, was initially present. What was the equilibrium concentration of $HCHO_2$?

*16.75 At 27 °C, $K_P = 1.5 \times 10^{18}$ for the reaction

$$3NO(g) \rightleftharpoons N_2O(g) + NO_2(g)$$

If 0.030 mol of NO were placed in a 1.00 L vessel and this equilibrium were established, what would be the equilibrium concentrations of NO, N_2O, and NO_2?

16.76 Consider the equilibrium

$$2NaHSO_3(s) \rightleftharpoons Na_2SO_3(s) + H_2O(g) + SO_2(g)$$

How will the position of equilibrium be affected by the following?
(a) Adding $NaHSO_3$ to the reaction vessel
(b) Removing Na_2SO_3 from the reaction vessel
(c) Adding H_2O to the reaction vessel
(d) Increasing the volume of the reaction vessel

*16.77 For the reaction below, $K_P = 1.6 \times 10^6$ at 400 °C.

$$HCHO_2(g) \rightleftharpoons CO(g) + H_2O(g)$$

A mixture of CO(g) and $H_2O(l)$ was prepared in a 2.00 L reaction vessel at 25 °C in which the pressure of CO was 0.177 atm. The mixture also contained 0.391 g of H_2O. The vessel was sealed and heated to 400 °C. When equilibrium was reached, what was the partial pressure of the $HCHO_2$?

*16.78 At a certain temperature, $K_c = 0.914$ for the reaction

$$NO_2(g) + NO(g) \rightleftharpoons N_2O(g) + O_2(g)$$

A mixture was prepared containing 0.200 mol of each gas in a 5.00 L container. What will be the equilibrium concentration of each gas? How will the concentrations then change if 0.050 mol of NO_2 is added to the equilibrium mixture?

*16.79 At a certain temperature, $K_c = 0.914$ for the reaction

$$NO_2(g) + NO(g) \rightleftharpoons N_2O(g) + O_2(g)$$

Equal amounts of NO and NO_2 are to be placed in a 5.00 L container until the N_2O concentration at equilibrium is 0.050 M. How many moles of NO and NO_2 must be placed in the container?

Acids and Bases:
A Second Look

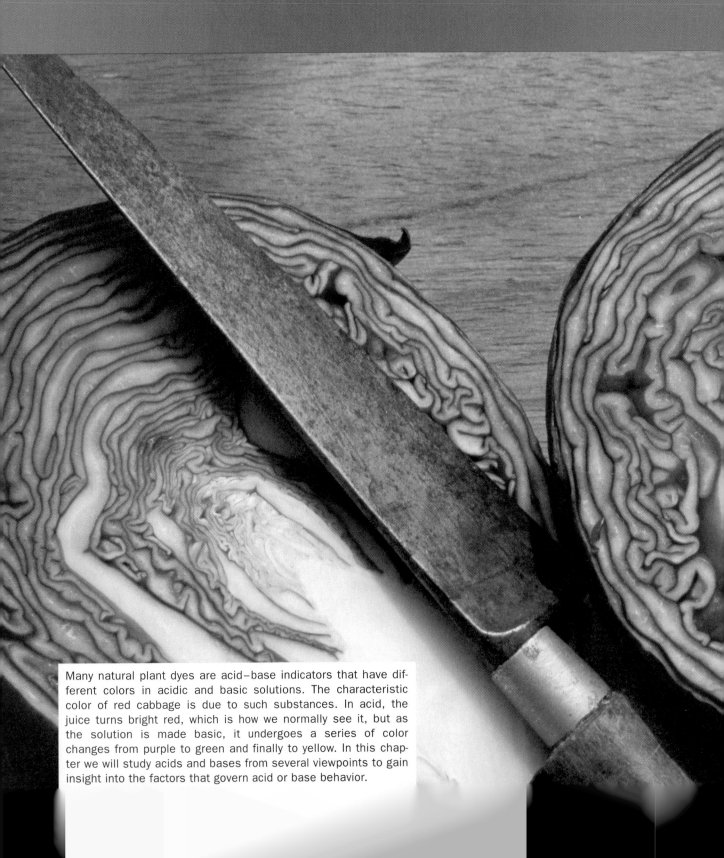

Many natural plant dyes are acid–base indicators that have different colors in acidic and basic solutions. The characteristic color of red cabbage is due to such substances. In acid, the juice turns bright red, which is how we normally see it, but as the solution is made basic, it undergoes a series of color changes from purple to green and finally to yellow. In this chapter we will study acids and bases from several viewpoints to gain insight into the factors that govern acid or base behavior.

THIS CHAPTER IN CONTEXT In Chapter 5 we introduced you to compounds that we call acids and bases. You learned that many common substances, from household products such as vinegar and ammonia, to biologically important compounds such as amino acids, are conveniently classified into one or the other of these categories. The significant property that makes such classifications useful is that *acids react with bases*. In fact, this is such a useful relationship that the acid–base concept has been expanded far beyond the limited Arrhenius definition we discussed in Chapter 5. We now return to acid–base chemistry to learn about these broader and often more useful views. At the same time, we will also study how trends in the strengths of acids and bases correlate with the periodic table.

17.1 ▶ Brønsted–Lowry acids and bases exchange protons

In our earlier discussion, an acid was described as a substance that produces H_3O^+ in water, whereas a base gives OH^-. An acid–base neutralization, according to Arrhenius, is a reaction in which an acid and a base combine to produce water and a salt. However, many reactions resemble neutralizations without involving H_3O^+, OH^-, or even H_2O. For example, when open bottles of concentrated hydrochloric acid and concentrated aqueous ammonia are placed side by side, a white cloud forms when the vapors from the two bottles mix (see Figure 17.1). The cloud consists of tiny crystals of ammonium chloride that form when ammonia and hydrogen chloride gases, escaping from the open bottles, mix in air and react.

$$NH_3(g) + HCl(g) \longrightarrow NH_4Cl(s)$$

What's interesting is that this is the same net reaction that occurs when an aqueous solution of ammonia (a base) is neutralized by an aqueous solution of hydrogen chloride (an acid). Yet, the gaseous reaction doesn't fit the description of an acid–base neutralization according to the Arrhenius definition because there's no water involved.

If we look at both the aqueous and gaseous reactions, they do have something in common. Both involve the transfer of a *proton* (a hydrogen ion, H^+) from one particle to another.[1] In water, where HCl is completely ionized, the transfer is from H_3O^+ to NH_3, as we discussed on page 181. The ionic equation is

$$NH_3(aq) + H_3O^+(aq) + Cl^-(aq) \longrightarrow \underbrace{NH_4^+(aq) + Cl^-(aq)}_{\text{the ions of } NH_4Cl} + H_2O$$

In the gas phase, the proton is transferred directly from the HCl molecule to the NH_3 molecule.

FIGURE 17.1 *The reaction of gaseous HCl with gaseous NH₃.* As each gas escapes from its concentrated aqueous solution and mingles with the other, a cloud of microcrystals of NH₄Cl forms above the bottles.

Chemists often use the terms *proton* and *hydrogen ion* interchangeably in discussions of acid–base chemistry.

---

[1] When the single electron is removed from a hydrogen atom, what remains is just the nucleus of the atom, which is a proton. Therefore, a hydrogen ion, H^+, consists of a proton, and the terms *proton* and *hydrogen ion* are often used interchangeably.

Proton (H^+) transfers
from Cl to N.

As ammonium ions and chloride ions form, they attract each other, gather, and settle as crystals of ammonium chloride.

The Brønsted–Lowry concept views acid–base reactions as proton transfers

The gaseous reaction between NH_3 and HCl has the look and feel of an acid–base reaction, so it makes sense to alter the definitions so it can be treated as one. Johannes Brønsted (1879–1947), a Danish chemist, and Thomas Lowry (1874–1936), a British scientist, realized that the important event in most acid–base reactions is simply the transfer of a proton from one species to another. They proposed, therefore, that *acids* be redefined as species that donate protons and *bases* as species that accept protons. The heart of the *Brønsted–Lowry concept of acids and bases* is that *acid–base reactions are proton transfer reactions.* The definitions are therefore very simple.

> **Brønsted–Lowry Definitions of Acids and Bases[2]**
>
> An **acid** is a proton donor.
>
> A **base** is a proton acceptor.

Accordingly, hydrogen chloride is an acid because when it reacts with ammonia, HCl molecules donate protons to NH_3 molecules. Similarly, ammonia is a base because NH_3 molecules accept protons.

Even when water is the solvent, chemists use the Brønsted–Lowry definitions more often than those of Arrhenius. Thus, the reaction between hydrogen chloride and water to form hydronium ion (H_3O^+) and chloride ion (Cl^-), which is another proton transfer reaction, is clearly a Brønsted–Lowry acid–base reaction. Molecules of HCl are the acid in this reaction, and water molecules are the base. HCl molecules collide with water molecules and protons transfer during the collisions.

| water | hydrogen chloride | collision "complex" of proper orientation and energy for proton transfer | hydronium ion | chloride ion |

Conjugate acids and bases

Under the Brønsted–Lowry view, it is useful to consider any acid–base reaction as a chemical equilibrium, having both a forward and a reverse reaction. We first encountered a chemical equilibrium in a discussion of weak acids on page 178. Let's examine it again in the light of the Brønsted–Lowry definitions, using formic acid, $HCHO_2$, as an example. Formic acid is a weak acid, so we represent its ionization

---

[2]Although Brønsted and Lowry are both credited with defining acids and bases in terms of proton transfer, Brønsted carried the concepts further. For the sake of brevity, we will often use the terms *Brønsted acid* and *Brønsted base* when referring to the substances involved in proton transfer reactions.

as a chemical equilibrium in which water is not just a solvent but also a chemical reactant, a proton acceptor.[3]

$$HCHO_2(aq) + H_2O \rightleftharpoons H_3O^+(aq) + CHO_2^-(aq)$$

In the forward reaction, a formic acid molecule donates a proton to the water molecule and changes to a formate ion, CHO_2^- (see Figure 17.2a). Thus, $HCHO_2$ behaves as a Brønsted acid, a proton donor. Because water accepts this proton from $HCHO_2$, water behaves as a Brønsted base, a proton acceptor.

Now let's look at the reverse reaction (see Figure 17.2b). In it, H_3O^+ behaves as a Brønsted acid because it donates a proton to the CHO_2^- ion. The CHO_2^- ion behaves as a Brønsted base by accepting the proton.

The equilibrium involving $HCHO_2$, H_2O, H_3O^+, and CHO_2^- is typical of proton transfer equilibria in general, in that we can identify *two* acids (e.g., $HCHO_2$ and H_3O^+) and *two* bases (e.g., H_2O and CHO_2^-). Notice that in the aqueous formic acid equilibrium, the acid on the right of the arrows (H_3O^+) is formed from the base on the left (H_2O), and the base on the right (CHO_2^-) is formed from the acid on the left ($HCHO_2$).

Two substances, such as H_3O^+ and H_2O, that differ from each other by only *one* proton, are referred to as a **conjugate acid–base pair.** Thus, H_3O^+ and H_2O are such a pair. One member of the pair is called the **conjugate acid** because it is the proton donor of the two. The other member is the **conjugate base,** because it is the pair's proton acceptor. We say that *H_3O^+ is the conjugate acid of H_2O, and H_2O is the conjugate base of H_3O^+.* Notice that *the acid member of the pair has one more H^+ than the base member.*

The pair $HCHO_2$ and CHO_2^- is the other conjugate acid–base pair in the aqueous formic acid equilibrium. $HCHO_2$ has one more H^+ than CHO_2^-, so the conjugate acid of CHO_2^- is $HCHO_2$; the conjugate base of $HCHO_2$ is CHO_2^-. One way to highlight the two members of a conjugate acid–base pair in an equilibrium equation is to connect them by a line.

formic acid ($HCHO_2$)

Only the H in red is available in an acid–base reaction.

formate ion (CHO_2^-)

$$\overbrace{HCHO_2 \ + \ H_2O}^{\text{conjugate pair}} \rightleftharpoons \underbrace{H_3O^+ \ + \ CHO_2^-}$$
$$\quad \text{acid} \qquad \text{base} \qquad \text{acid} \qquad \text{base}$$

conjugate pair

Notice how the conjugate acid always has one more H^+ than the conjugate base.

In any Brønsted–Lowry acid–base equilibrium, there are invariably *two* conjugate acid–base pairs. It is important that you learn how to pick them out of an equation by inspection and to write them from formulas.

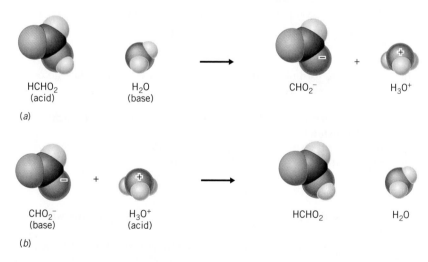

Figure 17.2 *Brønsted acids and bases in aqueous formic acid.* (*a*) Formic acid transfers a proton to a water molecule. $HCHO_2$ is the acid and H_2O is the base. (*b*) When hydronium ion transfers a proton to the CHO_2^- ion, H_3O^+ is the acid and CHO_2^- is the base.

[3]Just as we have represented acetic acid as $HC_2H_3O_2$, placing the ionizable or acidic hydrogen first (as we do in HCl or HNO_3), so we give the formula of formic acid as $HCHO_2$ and the formula of the formate ion as CHO_2^-. As the margin comment shows, however, the acidic hydrogen in formic acid resides on O, not on C. Many chemists prefer to write formic acid as HCO_2H.

EXAMPLE 17.1

Determining the Conjugate Bases of Brønsted Acids

What are the conjugate bases of nitric acid, HNO_3, and the hydrogen sulfate ion, HSO_4^-?

ANALYSIS: The members of a conjugate acid–base pair differ by one H^+, with the member having the greater number of hydrogens being the acid. We are asked to find the conjugate base of something, so that "something" must be the acid member of the pair. To find the formula of the base, we remove one H^+ from the acid. This is equivalent to removing one hydrogen from the acid and decreasing its positive charge by one unit (or, increasing the negative charge by one unit).

SOLUTION: Removing one H^+ (both the atom and the charge) from HNO_3 leaves NO_3^-. The nitrate ion, NO_3^-, is thus the conjugate base of HNO_3. Deleting an H^+ from HSO_4^- leaves its conjugate base, SO_4^{2-}. (Notice that the charge goes from $1-$ to $2-$ because we've removed the positively charged H^+.)

Is the Answer Reasonable?
As a check, we can quickly compare the two formulas in each pair.

$$HNO_3 \qquad NO_3^-$$
$$HSO_4^- \qquad SO_4^{2-}$$

In each case, the formula on the right has one less H^+ than the one on the left, so it is the conjugate base. We've answered the question correctly.

PRACTICE EXERCISE 1: Write the formula of the conjugate base for each of the following Brønsted acids:
(a) H_2O (c) HNO_2 (e) $H_2PO_4^-$ (g) H_2
(b) HI (d) H_3PO_4 (f) HPO_4^{2-} (h) NH_4^+

EXAMPLE 17.2

Determining the Conjugate Acids of Brønsted Bases

What are the formulas of the conjugate acids of OH^- and PO_4^{3-}?

ANALYSIS: This time we have to find the conjugate acid of something, so the "something" in each case must be a base. To find the conjugate acid we have to add one H^+ to the formula of the given base. This amounts to adding an H and increasing the positive charge by one unit (or decreasing the negative charge by one unit).

SOLUTION: Adding H^+ to OH^- gives H_2O, the conjugate acid of the hydroxide ion. Adding H^+ to the formula PO_4^{3-} gives HPO_4^{2-}, the conjugate acid of PO_4^{3-}. Notice how we have adjusted the electrical charge in each case.

Is the Answer Reasonable?
As before, we can compare the two formulas in each pair.

$$OH^- \qquad H_2O$$
$$PO_4^{3-} \qquad HPO_4^{2-}$$

In each case, the formula on the right has one more H^+ than the one on the left, so it is the conjugate acid. We've answered the question correctly.

PRACTICE EXERCISE 2: Write the formula of the conjugate acid for each of the following Brønsted bases:
(a) HO_2^- (c) CO_3^{2-} (e) NH_2^- (g) $H_2PO_4^-$
(b) SO_4^{2-} (d) CN^- (f) NH_3 (h) HPO_4^{2-}

EXAMPLE 17.3

Identifying Conjugate Acid–Base Pairs in a Brønsted–Lowry Acid–Base Reaction

The anion of sodium hydrogen sulfate, HSO_4^-, reacts as follows with the phosphate ion, PO_4^{3-}:

$$HSO_4^-(aq) + PO_4^{3-}(aq) \longrightarrow SO_4^{2-}(aq) + HPO_4^{2-}(aq)$$

Identify the two conjugate acid–base pairs.

ANALYSIS: There are two things to look for in identifying the conjugate acid–base pairs in an equation. One is that the members of a conjugate pair are alike except for the number of hydrogens and charge. The second is that the members of each pair must be on opposite sides of the arrow. In each pair, of course, the acid is the one with the greater number of hydrogens.

SOLUTION: Two of the formulas in the equation contain "PO_4," so they must belong to the same conjugate pair. The one with the greater number of hydrogens, HPO_4^{2-}, must be the Brønsted acid, and the other, PO_4^{3-}, must be the Brønsted base. Therefore, one conjugate acid–base pair is HPO_4^{2-} and PO_4^{3-}. The other two ions, HSO_4^- and SO_4^{2-}, belong to the second conjugate acid–base pair; HSO_4^- is the conjugate acid and SO_4^{2-} is the conjugate base.

Sodium hydrogen sulfate (also called sodium bisulfate) is used in the manufacture of certain kinds of cement and to clean oxide coatings from metals.

$$\overset{\text{conjugate pair}}{\overbrace{HSO_4^-(aq) + \underset{}{PO_4^{3-}(aq)} \longrightarrow \underset{}{SO_4^{2-}(aq)} + HPO_4^{2-}(aq)}}$$

$$\underset{\text{acid}}{} \quad \underset{\text{base}}{\phantom{PO_4^{3-}}} \quad \underset{\text{base}}{\phantom{SO_4^{2-}}} \quad \underset{\text{acid}}{\phantom{HPO_4^{2-}}}$$

$$\underset{\text{conjugate pair}}{\underbrace{\phantom{PO_4^{3-}(aq) \longrightarrow SO_4^{2-}(aq)}}}$$

Is the Answer Reasonable?

A check satisfies us that we have fulfilled the requirements that each conjugate pair has one member on one side of the arrow and the other member on the opposite side of the arrow and that the members of each pair differ from each other by one (and *only* one) H^+.

PRACTICE EXERCISE 3: One kind of baking powder contains sodium bicarbonate and calcium dihydrogen phosphate. When water is added, a reaction occurs by the following net ionic equation:

$$HCO_3^-(aq) + H_2PO_4^-(aq) \longrightarrow H_2CO_3(aq) + HPO_4^{2-}(aq)$$

Identify the two Brønsted acids and the two Brønsted bases in this reaction. (The H_2CO_3 decomposes to release CO_2, which causes the cake batter to rise.)

PRACTICE EXERCISE 4: When some of the strong cleaning agent "trisodium phosphate" is mixed with household vinegar, which contains acetic acid, the following equilibrium is one of the many that are established. (The position of equilibrium lies to the right.) Identify the pairs of conjugate acids and bases.

$$PO_4^{3-}(aq) + HC_2H_3O_2(aq) \rightleftharpoons HPO_4^{2-}(aq) + C_2H_3O_2^-(aq)$$

Amphoteric substances can behave as either acids or bases

Some molecules or ions are able to function either as an acid or as a base, depending on the kind of substance mixed with them. For example, in its reaction with hydrogen chloride, water behaves as a *base* because it *accepts* a proton from the HCl molecule.

$$\underset{\text{base}}{H_2O} + \underset{\text{acid}}{HCl(g)} \longrightarrow H_3O^+(aq) + Cl^-(aq)$$

On the other hand, water behaves as an *acid* when it reacts with the weak base ammonia.

$$\underset{\text{acid}}{H_2O} + \underset{\text{base}}{NH_3(aq)} \rightleftharpoons NH_4^+(aq) + OH^-(aq)$$

Here, H_2O *donates* a proton to NH_3 in the forward reaction.

A substance that can be either an acid or a base depending on the other substance present is said to be **amphoteric.** Another term is **amphiprotic,** to stress that the *proton* donating or accepting ability is of central concern.

From the Greek *amphoteros*, "partly one and partly the other."

Amphoteric or amphiprotic substances may be either molecules or ions. For example, anions of acid salts, such as the bicarbonate ion of baking soda, are amphoteric. The HCO_3^- ion can either donate a proton to a base or accept a proton from an acid. Thus, toward the hydroxide ion, the bicarbonate ion is an acid; it donates its proton to OH^-.

$$\underset{\text{acid}}{HCO_3^-(aq)} + \underset{\text{base}}{OH^-(aq)} \longrightarrow CO_3^{2-}(aq) + H_2O$$

Toward hydronium ion, however, HCO_3^- is a base; it accepts a proton from H_3O^+.

$$\underset{\text{base}}{HCO_3^-(aq)} + \underset{\text{acid}}{H_3O^+(aq)} \longrightarrow H_2CO_3(aq) + H_2O$$

[Recall that H_2CO_3 (carbonic acid) almost entirely decomposes to $CO_2(g)$ and water as it forms.]

PRACTICE EXERCISE 5: The anion of sodium monohydrogen phosphate, Na_2HPO_4, is amphoteric. Using H_3O^+ and OH^-, write net ionic equations that illustrate this property.

17.2 ▶ Strengths of Brønsted acids and bases follow periodic trends

Acid and base strengths are measured relative to a standard

When we speak of the strength of a Brønsted acid, we are referring to its ability to donate a proton to a base. We measure this by determining the extent to which the reaction of the acid with the base proceeds toward completion—the more complete the reaction, the stronger the acid. To compare the strengths of a series of acids, we have to select some reference base, and because we are interested most in reactions in aqueous media, that base is usually water (although other reference bases could also be chosen).

In Chapter 5 we discussed strong and weak acids from the Arrhenius point of view, and much of what we said there applies when the same acids are studied using the Brønsted–Lowry concept. Thus, acids such as HCl and HNO_3 react completely with water to give H_3O^+ because they are strong proton donors. Hence, we classify them as *strong Brønsted acids*. On the other hand, acids such as HNO_2 (nitrous acid) and $HC_2H_3O_2$ (acetic acid) are much weaker proton donors. Their reactions with water are far from complete and we classify them as *weak acids*.

In a similar manner, the relative strengths of Brønsted *bases* are assigned according to their abilities to accept and bind protons. Once again, to compare strengths, we have to choose a standard acid. Because water is amphiprotic, it can also serve as the standard acid. Substances that are powerful proton acceptors, such as the oxide ion, react completely and are considered to be *strong Brønsted bases*.

$$O^{2-} + H_2O \xrightarrow{100\%} 2OH^-$$

Weaker proton acceptors, such as ammonia, undergo incomplete reactions with water, and we classify them *weak bases*.

Hydronium ion and hydroxide ion are the strongest acid and base that can exist in the presence of water

Both HCl and HNO_3 are very powerful proton donors. They are so powerful, in fact, that when placed into water they react completely, losing their protons to water molecules to yield H_3O^+ ions. Representing them by the general formula HA, we have

$$\underset{\text{acid}}{HA} + \underset{\text{base}}{H_2O} \xrightarrow{100\%} \underset{\text{acid}}{H_3O^+} + \underset{\text{base}}{A^-}$$

Because both reactions go to completion, we really can't tell which of the two, HCl or HNO_3, is actually the better proton donor (stronger acid). This would require a reference base less willing than water to accept protons. In water, both HCl and HNO_3 are converted quantitatively to another acid, H_3O^+. The conclusion, therefore, is that *H_3O^+ is the strongest acid we will ever find in an aqueous solution,* because stronger acids react completely with water to give H_3O^+.

A similar conclusion is reached regarding hydroxide ion. We noted that the strong Brønsted base O^{2-} reacts completely with water to give OH^-. Another very powerful proton acceptor is the amide ion, NH_2^-, which also reacts completely with water.

$$NH_2^- + H_2O \xrightarrow{100\%} NH_3 + OH^-$$
$$\text{base} \qquad \text{acid} \qquad\qquad \text{acid} \qquad \text{base}$$

Using water as the reference acid, we can't tell which is the better proton acceptor, O^{2-} or NH_2^-, because both react completely, being replaced by another base, OH^-. Therefore, we can say that *OH^- is the strongest base we will ever find in an aqueous solution,* because stronger bases react completely with water to give OH^-.

Comparing the acid–base strengths of conjugate pairs

As we noted earlier, the chemical equation for the equilibrium present in an aqueous solution of a weak acid actually shows two acids. One is almost always stronger than the other, and the position of the equilibrium tells us which of the two acids is stronger. Let's see how this works using the acetic acid equilibrium, in which the position of equilibrium lies to the left.

$$HC_2H_3O_2(aq) + H_2O \rightleftharpoons H_3O^+(aq) + C_2H_3O_2^-(aq)$$
$$\text{acid} \qquad\qquad \text{base} \qquad\quad \text{acid} \qquad\qquad \text{base}$$

The two Brønsted acids in this equilibrium are $HC_2H_3O_2$ and H_3O^+, and it's helpful to think of them as competing with each other in donating protons to acceptors. The fact that nearly all potential protons stay on the $HC_2H_3O_2$ molecules, and only a relative few spend their time on the H_3O^+ ions, means that the *hydronium ion is a better proton donor than the acetic acid molecule.* Thus, the hydronium ion is a stronger Brønsted acid than acetic acid, and we inferred this relative acidity from the position of equilibrium.

The acetic acid equilibrium also has two bases, $C_2H_3O_2^-$ and H_2O. Both compete for any available protons. But at equilibrium, most of the protons originally carried by acetic acid are still found on $HC_2H_3O_2$ molecules; relatively few are joined to H_2O in the form of H_3O^+ ions. This means that *acetate ions must be more effective than water molecules at obtaining and holding protons from proton donors.* This is the same as saying that the acetate ion is a stronger base than the water molecule. So this illustrates a relative basicity that we are able to infer from the position of the acetic acid equilibrium.

Notice what our discussion of the two conjugate pairs in the acetic acid equilibrium has brought out. Both the weaker of the two acids and the weaker of the two bases are found on the same side of the equation, which is the side favored by the position of equilibrium.

> *The position of an acid–base equilibrium favors the weaker acid and base.*

$$HC_2H_3O_2(aq) + H_2O \rightleftharpoons H_3O^+(aq) + C_2H_3O_2^-(aq)$$
$$\text{weaker acid} \qquad \text{weaker base} \qquad \text{stronger acid} \qquad \text{stronger base}$$

← *Position of equilibrium lies to the left, in favor of the weaker acid and base.*

It is the chemical nature of acetic acid not to ionize much.

The concepts we've just developed allow us now to understand better one of the driving forces for metathesis reactions discussed in Chapter 5, namely, the formation of a weak electrolyte from a solution of strong electrolytes.

Consider what happens when solutions of hydrochloric acid and sodium acetate are mixed. At the moment of mixing, the solution contains high concentrations of H_3O^+ and $C_2H_3O_2^-$ ions. But we've just seen that the hydronium ion (a strong acid) will donate a proton to the acetate ion to make acetic acid (a weak acid). Therefore, to reach equilibrium in the net ionic equation below, there will be a *net reaction* of the ions on the left to form the molecules on the right.

When aqueous solutions of hydrochloric acid and sodium acetate are mixed, the vinegary odor of acetic acid can be soon noticed.

$$H_3O^+(aq) + C_2H_3O_2^-(aq) \longrightarrow HC_2H_3O_2(aq) + H_2O$$

| stronger acid | stronger base | weaker acid | weaker base |

> *Net reaction will occur from left to right.* ➡

This analysis provides us with another way of stating the previous generalization.

> *Stronger acids and bases tend to react with each other to produce their weaker conjugates.*

You can see that once we know the relative strengths of acids and bases, we can predict reactions.

There's a reciprocal relationship within a conjugate acid–base pair

One aid in predicting the relative strengths of acids and bases is the existence of a reciprocal relationship.

> *The stronger a Brønsted acid is, the weaker is its conjugate base.*

To illustrate, recall that HCl(g) is a very strong Brønsted acid; it's 100% ionized in a dilute aqueous solution.

$$HCl + H_2O \xrightarrow{100\%} H_3O^+(aq) + Cl^-(aq)$$

As we explained in Section 5.7, we don't write double equilibrium arrows for the ionization of a strong acid. By not doing so with HCl is another way of saying that the chloride ion, the conjugate base of HCl, must be a very weak Brønsted base. Even in the presence of H_3O^+, a very strong proton donor, chloride ions aren't able to win protons. So HCl, the strong acid, has a particularly weak conjugate base, Cl^-.

There's a matching reciprocal relationship.

> *The weaker a Brønsted acid is, the stronger is its conjugate base.*

Consider, for example, the conjugate pair, OH^- and O^{2-}. The hydroxide ion is the conjugate acid and the oxide ion is the conjugate base. But the hydroxide ion must be an *extremely* weak Brønsted acid; in fact, we've known it so far only as a base. Given the extraordinary weakness of OH^- as an acid, its conjugate base, the oxide ion, must be an exceptionally strong base. And as you've already learned, oxide ion is such a strong base that its reaction with water is 100% complete. That's why we don't write double equilibrium arrows in the equation for the reaction.

$$O^{2-} + H_2O \xrightarrow{100\%} OH^- + OH^-$$

| base | acid | acid | base |

The very strong base O^{2-} has
a very weak conjugate acid, OH^-.

In the reaction below, will the position of equilibrium lie to the left or the right, given the fact that acetic acid is known to be a stronger acid than the hydrogen sulfite ion?

$$HSO_3^-(aq) + C_2H_3O_2^-(aq) \rightleftharpoons HC_2H_3O_2(aq) + SO_3^{2-}(aq)$$

ANALYSIS: We just learned that the position of an acid–base equilibrium favors the weaker acid and base. That's the critical link we need to solve this problem. We have to identify which acid and base make up the weaker set. When we do this, we'll have discovered which substances make up the stronger set, and we can then predict where the position of equilibrium will lie.

SOLUTION: We'll write the equilibrium equation using the given fact about the relative strengths of acetic acid and the hydrogen sulfite ion to start writing labels.

$$HSO_3^-(aq) + C_2H_3O_2^-(aq) \rightleftharpoons HC_2H_3O_2(aq) + SO_3^{2-}(aq)$$
weaker acid   stronger acid

Now we'll use the reciprocal relationships to label the two bases. The stronger acid must have the weaker conjugate base; the weaker acid must have the stronger conjugate base.

$$HSO_3^-(aq) + C_2H_3O_2^-(aq) \rightleftharpoons HC_2H_3O_2(aq) + SO_3^{2-}(aq)$$
weaker acid   weaker base   stronger acid   stronger base

Finally, the answer flows from the fact that the position of an acid–base equilibrium favors the weaker acid and base. So here the position of equilibrium lies to the left.

Is the Answer Reasonable?
There are two things we can check. First, both of the weaker conjugates should be on the same side of the equation. They are, so that suggests we've made the correct assignments. Second, the reaction will proceed farther in the direction of the weaker acid and base, so that places the position of equilibrium on the left, which agrees with our answer.

PRACTICE EXERCISE 6: Given the fact that HSO_4^- is a stronger acid than HPO_4^{2-}, determine whether the substances on the left of the arrows or those on the right are favored in the following equilibrium:

$$HSO_4^-(aq) + PO_4^{3-}(aq) \rightleftharpoons SO_4^{2-}(aq) + HPO_4^{2-}(aq)$$

EXAMPLE 17.4

Using Reciprocal Relationships to Predict Equilibrium Positions

In the preceding example, you saw that if we know the relative strengths of Brønsted acids, we can predict the position of an equilibrium involving them. There are considerable variations in the strengths of Brønsted acids, and the periodic table can sometimes help us assign relative strengths (and to remember the assignments). If we know how to do this for *acids,* we'll have also done this for their conjugate bases, because of the reciprocal relationships discussed above.

Periodic trends exist in the strengths of binary acids

Many of the binary compounds between hydrogen and nonmetals, which we may represent by HX, H_2X, H_3X, etc., are acidic and are called **binary acids.** HCl is a common example, but Table 17.1 lists the others that are acids in water. The strong acids are marked by asterisks.

The relative strengths of binary acids correlate with the periodic table in two ways.

◄ **TOOLS** ⬛

Periodic table and strengths of binary acids

The strengths of the binary acids increase from left to right within the same period.

As we go left to right within a period, the increase in electronegativities causes the corresponding H—X bonds to become more polar, making the partial positive

Electronegativity was discussed in Section 9.5.

TABLE 17.1	**ACIDIC BINARY COMPOUNDS OF HYDROGEN AND NONMETALS**[a]		
Group VIA		**Group VIIA**	
(H$_2$O)		HF	Hydrofluoric acid
H$_2$S	Hydrosulfuric acid	*HCl	Hydrochloric acid
H$_2$Se	Hydroselenic acid	*HBr	Hydrobromic acid
H$_2$Te	Hydrotelluric acid	*HI	Hydroiodic acid

[a]The *names* are for the aqueous solutions of these compounds. Strong acids are marked with asterisks.

charge ($\delta+$) on H greater. As the $\delta+$ on H becomes larger, it becomes easier for the hydrogen to separate as H$^+$, so the molecule becomes a better proton donor. For example, as we go from S to Cl in Period 3, the electronegativity increases and we find that HCl is a stronger acid than H$_2$S. A similar increase in electronegativity occurs going left to right in Period 2, from O to F, and HF is a stronger acid than H$_2$O.

The second correlation is as follows:

> The strengths of binary acids increase from top to bottom within the same group.

Among the binary acids of the halogens, for example, the following is the order of relative acidity[4]:

$$HF < HCl < HBr < HI$$

Thus, HF is the weakest acid in the series, and HI is the strongest.

The identical trend occurs in the series of the binary acids of Group VIA elements, having formulas of the general type H$_2$X. The farther X is from the top of the group, the stronger H$_2$X is as an acid. Thus, S lies below O, and H$_2$S is a stronger acid than H$_2$O.

These trends in acidity are opposite what we would expect on the basis of trends in electronegativities, which tell us that the H—F bond is more polar than the H—I bond and that the O—H bond is more polar than the H—S bond. Evidently, the relative electronegativity of X is just one factor affecting the acidity of an H—X bond. Another is the strength of the H—X bond.

In analyzing the changes associated with the donation of a proton by an acid, one of the most important factors to be considered, besides the polarity of the bond, is the strength of the H—X bond. The breaking of this bond is essential for the hydrogen to become separated as an H$^+$ ion, so anything that contributes to variations in bond strength will also impact variations in acid strength.

In general, the bond strength is determined to a significant degree by the sizes of the atoms involved. Small atoms form strong covalent bonds, whereas large atoms form much weaker bonds. Moving horizontally within a period, atomic size varies relatively little, so the strengths of the H—X bonds are nearly the same. As a result, the most significant influence on acid strength is variations in the polarity of H—X bonds. Within a group, however, there is a relatively large increase in atomic size from one element to the next, which means there is a rapid decrease in the strength of the H—X bonds as we descend a group. Apparently, this decrease in

---

[4]When we compare acid strengths, we compare the abilities of different acids to protonate a particular base. As we noted earlier, for strong acids such as HCl, HBr, and HI, water is too strong a proton acceptor to permit us to see differences among their proton-donating abilities. All three of these acids are completely ionized in water and so appear to be of equal strength, a phenomenon called the *leveling effect* (the differences are obscured, or *leveled* out). To compare the acidities of these acids, a solvent that is a weaker proton acceptor than water (HF or HC$_2$H$_3$O$_2$, for example) has to be used.

bond strength is enough to more than compensate for the decrease in the polarity of the H—X bonds, and the molecules become better able to release protons as we go down a group. The net effect of the two opposing factors, therefore, is an increase in the strengths of the binary acids as we go from top to bottom in the group.

PRACTICE EXERCISE 7: Using *only* the periodic table, choose the stronger acid of each pair: (a) H_2Se or HBr, (b) H_2Se or H_2Te, (c) CH_3OH or CH_3SH.

Trends exist in the strengths of oxoacids

Acids composed of hydrogen, oxygen, and some other element are called **oxoacids** (see Table 17.2). Those that are strong acids in water are marked in the table by asterisks.

A feature common to the structures of all oxoacids is the presence of O—H groups bonded to some central atom. For example, the structures of two oxoacids of the Group VIA elements are

$$
\begin{array}{cc}
\overset{\displaystyle O}{\underset{\displaystyle O}{\overset{\|}{\underset{\|}{\text{H—O—S—O—H}}}}} & \overset{\displaystyle O}{\underset{\displaystyle O}{\overset{\|}{\underset{\|}{\text{H—O—Se—O—H}}}}} \\
H_2SO_4 & H_2SeO_4 \\
\text{sulfuric acid} & \text{selenic acid}
\end{array}
$$

When an oxoacid ionizes, the hydrogen that's lost as an H^+ comes from the same kind of bond in every instance, specifically, an O—H bond. The "acidity" of such a hydrogen, meaning the ease with which it's released as H^+, is determined by how the group of atoms attached to the oxygen affects the polarity of the O—H bond. If this group of atoms makes the O—H bond more polar, it will cause the H to come off more easily as H^+ and thereby increase the acidity of the molecule.

TABLE 17.2	SOME OXOACIDS OF NONMETALS AND METALLOIDS[a]						
Group IVA		**Group VA**		**Group VIA**		**Group VIIA**	
H_2CO_3	Carbonic acid	*HNO_3	Nitric acid			HFO	Hypofluorous acid
		HNO_2	Nitrous acid				
		H_3PO_4	Phosphoric acid	*H_2SO_4	Sulfuric acid	*$HClO_4$	Perchloric acid
		H_3PO_3	Phosphorous acid[b]	H_2SO_3	Sulfurous acid[c]	*$HClO_3$	Chloric acid
						$HClO_2$	Chlorous acid
						HClO	Hypochlorous acid
		H_3AsO_4	Arsenic acid	*H_2SeO_4	Selenic acid	*$HBrO_4$	Perbromic acid[d]
		H_3AsO_3	Arsenous acid	H_2SeO_3	Selenous acid	*$HBrO_3$	Bromic acid
						HIO_4	Periodic acid
						$(H_5IO_6)^e$	
						HIO_3	Iodic acid

[a]Strong acids are marked with asterisks.

[b]Phosphorous acid, despite its formula, is only a diprotic acid.

[c]Hypothetical. An aqueous solution actually contains just dissolved sulfur dioxide, $SO_2(aq)$.

[d]Pure perbromic acid is unstable; a dihydrate is known.

[e]H_5IO_6 is formed from $HIO_4 + 2H_2O$.

$$\overset{\delta-}{G}\!-\!\overset{}{O}\!-\!\overset{\delta+}{H}$$

> If the group of atoms, G, attached to the O—H group is able to draw electron density from the O atom, the O will pull electron density from the O—H bond, thereby making the bond more polar.

It turns out that there are two principal factors that determine how the polarity of the O—H bond is affected. One is the electronegativity of the central atom in the oxoacid and the other is the number of oxygens attached to the central atom.

The electronegativity of the central atom affects the acidity of an oxoacid

To study the effects of the electronegativity of the central atom, we must compare oxoacids having the same number of oxygens. When we do this, we find that as the electronegativity of the central atom increases, the oxoacid becomes a better proton donor (i.e., a stronger acid). The following diagram illustrates the effect.

$$-\overset{|}{\underset{|}{X}}\!-\!\overset{\delta-}{O}\!-\!\overset{\delta+}{H}$$

> As the electronegativity of X increases, electron density is drawn away from O, which draws electron density away from the O—H bond. This makes the bond more polar and makes the molecule a better proton donor.

Because electronegativity increases from bottom to top within a group and from left to right within a period, we can make the following generalization.

Periodic table and strengths of oxoacids

> When the central atoms of oxoacids hold the same number of oxygen atoms, the acid strength increases from bottom to top within a group and from left to right within a period.

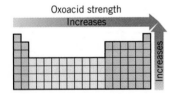

Oxoacid strength Increases

In Group VIA, for example, H_2SO_4 is a stronger acid than H_2SeO_4 because sulfur is more electronegative than selenium. Similarly, among the halogens, acid strength increases for acids with the formula HXO_4 as follows:

$$HIO_4 < HBrO_4 < HClO_4$$

Going from left to right within Period 3, we can compare the acids H_3PO_4, H_2SO_4, and $HClO_4$, where we find the following order of acidities:

$$H_3PO_4 < H_2SO_4 < HClO_4$$

PRACTICE EXERCISE 8: Which is the stronger acid, $HClO_3$ or $HBrO_3$?

The number of oxygens affects the acidity of an oxoacid

To examine how the number of oxygens attached to the central atom affects the acidity of a molecule, we must compare acids that have the same central atom. When we do this, we find that as the number of *lone oxygens* increases, the oxoacid becomes a better proton donor. (A *lone oxygen* is one that is bonded only to the central atom and not to a hydrogen.) Thus, comparing HNO_3 with HNO_2, we find that HNO_3 is the stronger acid. To understand why, let's look at their molecular structures.

$$\overset{\delta-}{\ddot{O}}=\ddot{N}-\ddot{O}-H \quad < \quad$$

nitrous acid   nitric acid

Oxygen, as you know, is a very electronegative element and has a strong tendency to pull electron density away from any atom to which it is attached. In an oxoacid, lone oxygens pull electron density away from the central atom, which increases the central atom's ability to draw electron density away from the O—H bond. It's as though the lone oxygens make the central atom more electronegative. It makes sense, therefore, that the more lone oxygens there are attached to a central atom, the more polar will be the O—H bonds of the acid and the stronger will be the acid. Thus, in HNO_3 the two lone oxygens produce a greater effect than the one lone oxygen in HNO_2, so HNO_3 is the stronger acid.

Similar effects are seen among other oxoacids, as well. For example, H_2SeO_4 is a stronger acid than H_2SeO_3 because there are two lone oxygens attached to the selenium in H_2SeO_4 while there's only one in H_2SeO_3.

H_2SeO_4 H_2SeO_3
selenic acid   >   selenous acid

Among the halogens, we find the same trend in acid strengths. For the oxoacids of chlorine, for instance, we find this trend.

$$HClO < HClO_2 < HClO_3 < HClO_4$$

Comparing their structures we have,

HClO $HClO_2$ $HClO_3$ $HClO_4$

This leads to another generalization.

For a given central atom, the acid strength of an oxoacid increases with the number of oxygens held by the central atom.

The ability of lone oxygens to affect acid strength extends to organic compounds as well. For example, compare the molecules below.

The molecule on the left is ethanol (ethyl alcohol). In water it is not acidic at all. Replacing the two hydrogens on the carbon adjacent to the OH group with an oxygen, however, yields acetic acid, the molecule on the right. The greater ability of oxygen to pull electron density from the carbon produces a greater polarity of the

Usually, the formula for hypochlorous acid is written HOCl. We've written it HClO here to make it easier to follow the trend in acid strengths among the oxoacids of chlorine.

O—H bond, which is one factor that causes acetic acid to be a better proton donor than ethanol.

Dispersal of negative charge to the lone oxygens affects the basicity of the anion of an oxoacid

In the ionization of an acid, H*A*, the strength of the acid is analyzed in terms of the position of equilibrium in the reaction

$$HA + H_2O \rightleftharpoons H_3O^+ + A^-$$

The stronger the acid, the farther to the right is the position of equilibrium. This position of equilibrium is determined by two factors. One is the ability of the acid to push protons onto water molecules (the strength of H*A* as a proton donor), and the other is the willingness of A^- to accept protons from H_3O^+ (the strength of A^- as a proton acceptor).

For oxoacids, the lone oxygens play a part in determining the basicity of the anion formed in the ionization reaction. Consider the acids H_2SO_4 and H_3PO_4.

As you can see in their structures, H_2SO_4 has two lone oxygens and H_3PO_4 has only one. Following the ionizations of their first protons, three lone oxygens are present in the resulting HSO_4^- anion but only two are in the $H_2PO_4^-$ anion.

In oxoanions such as these, the *lone* oxygens carry most of the negative charge, not the oxygens still bonded to hydrogens. This charge actually spreads over the lone oxygens. In HSO_4^-, therefore, each oxygen carries a charge of about $\frac{1}{3}-$. This is not a *formal* charge obtained by the rules on page 383. We're only saying that the charge of $1-$ is spread more or less evenly over three oxygens to give each a charge of $\frac{1}{3}-$. By the same reasoning, in $H_2PO_4^-$ each of the two lone oxygens carries a charge of about $\frac{1}{2}-$. The smaller negative charge on the lone oxygens in HSO_4^- makes this ion less able to attract H^+ ions from H_3O^+, so HSO_4^- is a weaker base than $H_2PO_4^-$. In other words, HSO_4^- cannot become H_2SO_4 again as readily as $H_2PO_4^-$ can become H_3PO_4. There are thus *two* factors that make H_2SO_4 more fully ionized than H_3PO_4 in water. One is the greater tendency of H_2SO_4 to lose H^+ and the other is the lesser tendency of HSO_4^- to recapture H^+. So if we compare solutions of H_2SO_4 and H_3PO_4 of equal molar concentrations, there will be a larger percentage of HSO_4^- ions than of $H_2PO_4^-$ ions. Sulfuric acid, to put it another way, will be more fully ionized than phosphoric acid. Of course, this is how we define acid strength—in terms of the *percentage* of the acid molecules that become ionized in solution. So sulfuric acid is a stronger acid than phosphoric acid.

The HSO_4^- ion is also a stronger acid than the $H_2PO_4^-$ ion.

PRACTICE EXERCISE 9: In each pair, select the stronger acid: (a) HIO_3 or HIO_4, (b) H_2TeO_3 or H_2TeO_4, (c) H_3AsO_3 or H_3AsO_4.

17.3 ▶ Lewis acids and bases involve coordinate covalent bonds

Earlier you saw that the Brønsted–Lowry definitions extended the Arrhenius acid–base concept to deal with reactions in which water was not involved as a re- actant or product, yet had all the appearances of acid–base reactions. There are also reactions not involving proton transfer that beg to be treated as acid–base re- actions, too. For example, if gaseous SO_3 is passed over solid CaO, a reaction occurs in which $CaSO_4$ forms.

$$CaO(s) + SO_3(g) \longrightarrow CaSO_4(s) \qquad (17.1)$$

If these reactants are dissolved in water first, they react to form $Ca(OH)_2$ and H_2SO_4, and when their solutions are mixed the following reaction takes place:

$$Ca(OH)_2(aq) + H_2SO_4(aq) \longrightarrow CaSO_4(s) + 2H_2O \qquad (17.2)$$

The same two initial reactants, CaO and SO_3, form the same ultimate product, $CaSO_4$. In one case, water is never involved at all. In the other, water first combines with the CaO and SO_3, and is then liberated again. It certainly seems that if Reac- tion 17.2 is an acid–base reaction, we should be able to consider Reaction 17.1 to be an acid–base reaction, too. But there are no protons being transferred, so our definitions require further generalizations to be able to encompass situations such as we have just described.

When a proton transfers to a Brønsted base, the base provides the pair of elec- trons needed to form the covalent bond to H. The proton accepts this pair as the bond forms. The general nature of these facts was noticed by G. N. Lewis, after whom Lewis symbols are named (Section 9.2), so he proposed very broad defini- tions of acids and bases, definitions that would include both the Arrhenius and Brønsted kinds but would embrace other systems as well. **Lewis acids and bases** are defined as follows.

Lewis Definitions of Acids and Bases

1. An **acid** is any ionic or molecular species that can accept a pair of electrons in the formation of a coordinate covalent bond.

2. A **base** is any ionic or molecular species that can donate a pair of electrons in the formation of a coordinate covalent bond.

3. Neutralization is the formation of a coordinate covalent bond between the donor (base) and the acceptor (acid).

Remember, a coordinate cova- lent bond is just like any other covalent bond once it has formed. By using this term, we are following the origin of the electron pair that forms the bond.

Examples of Lewis acid–base reactions

The reaction between BF_3 and NH_3 illustrates a Lewis acid–base neutralization. The reaction is exothermic because a bond is formed between N and B, with the ni- trogen donating an electron pair and the boron accepting it.

A similar reaction was de- scribed in Chapter 9 between BCl_3 and ammonia.

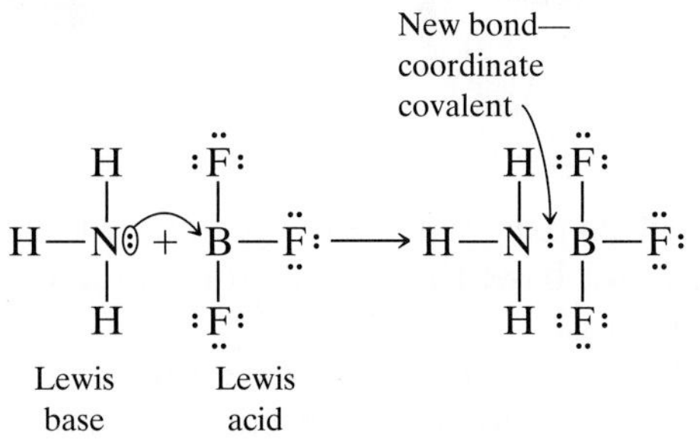

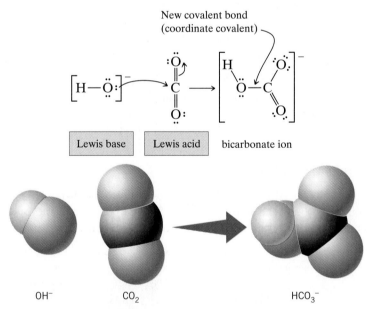

NH₃   BF₃   NH₃BF₃

Note that as the bond forms from ammonia to boron trifluoride, the geometry around the boron changes from planar triangular to tetrahedral, which is what is expected with four bonds to the boron.

Compounds like BF₃NH₃, which are formed by simply joining two smaller molecules, are called **addition compounds.**

The ammonia molecule thus acts as a *Lewis base*. The boron atom in BF₃, having only six electrons in its valence shell and needing two more to achieve an octet, accepts the pair of electrons from the ammonia molecule. Hence, BF₃ is functioning as a *Lewis acid*.

As this example illustrates, Lewis bases are substances that have completed valence shells *and unshared pairs of electrons* (e.g., NH_3, H_2O, and O^{2-}). A Lewis acid, on the other hand, can be a substance with an incomplete valence shell, such as BF_3 or H^+.

A substance can also be a Lewis acid even when it has a central atom with a complete valence shell. This works when the central atom has a double bond that, by the shifting of an electron pair to an adjacent atom, can make room for an incoming pair of electrons from a Lewis base. Carbon dioxide is an example. When carbon dioxide is bubbled into aqueous sodium hydroxide, the gas is instantly trapped as the bicarbonate ion.

$$CO_2(g) + OH^-(aq) \longrightarrow HCO_3^-(aq)$$

Lewis acid–base theory represents the movement of electrons in this reaction as follows.

New covalent bond
(coordinate covalent)

Lewis base   Lewis acid   bicarbonate ion

OH⁻   CO₂   HCO₃⁻

The donation of an electron pair *from* the oxygen of the OH^- ion produces a bond, so the OH^- ion is the Lewis base. The carbon atom of the CO_2 accepts the electron pair, so CO_2 is the Lewis acid.

Lewis acids can also be substances that have valence shells capable of holding more electrons. For example, consider the reaction of sulfur dioxide as a Lewis acid with oxide ion as a Lewis base to make the sulfite ion. This reaction occurs when gaseous sulfur dioxide, an acidic anhydride, mingles with solid calcium oxide, a basic anhydride, to give calcium sulfite, $CaSO_3$.

$$SO_2(g) + CaO(s) \longrightarrow CaSO_3(s)$$

Let's see how electrons relocate as the sulfite ion forms. We use one of the two resonance structures of SO_2 and one of the three such structures for the sulfite ion.

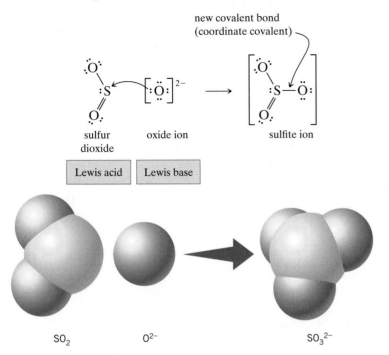

The two very electronegative oxygens attached to the sulfur in SO_2 give the sulfur a substantial positive partial charge, which induces the formation of the coordinate covalent bond from the oxide ion to the sulfur. In this case, sulfur can accommodate more than an octet in its valence shell, so relocation of electron pairs is not necessary.

A summary of the kinds of substances that behave as Lewis acids and bases is located in Table 17.3. Study it and then work on the Practice Exercise below.

PRACTICE EXERCISE 10: (a) Is the fluoride ion more likely to behave as a Lewis acid or a Lewis base? Explain. (b) Is the $BeCl_2$ molecule more likely to behave as a Lewis acid or a Lewis base? Explain. (c) Is the SO_3 molecule more likely to behave as a Lewis acid or a Lewis base? Explain.

TABLE 17.3	TYPES OF SUBSTANCES THAT ARE LEWIS ACIDS AND BASES

Lewis Acids

Molecules or ions with incomplete valence shells (e.g., BF_3, H^+).

Molecules or ions with complete valence shells, but with multiple bonds that can be shifted to make room for more electrons (e.g., CO_2).

Molecules or ions that have central atoms capable of holding additional electrons (usually, atoms of elements in Period 3 and below; e.g., SO_2).

Lewis Bases

Molecules or ions that have unshared pairs of electrons and that have complete valence shells (e.g., O^{2-}, NH_3).

A Brønsted acid–base reaction is an exchange of a Lewis acid between two Lewis bases

The Lewis concept includes Brønsted acids and bases as special cases. What takes place during a Brønsted–Lowry acid–base reaction is that the proton, a Lewis acid, moves from one Lewis base to another. As an example, let's look at the reaction between hydronium ion and ammonia. The chemical equation is

$$H_3O^+ + NH_3 \longrightarrow H_2O + NH_4^+$$

Here we view the hydronium ion as the "neutralization" product of the Lewis base H_2O with the Lewis acid H^+. In the reaction, the H^+ is transferred from the weaker Lewis base (H_2O) to the stronger Lewis base (NH_3). To emphasize that it is the proton that is transferring, we might write the equation as

$$H_2O\!-\!H^+ + NH_3 \longrightarrow H_2O + H^+\!\!-\!NH_3$$

We can diagram this as follows using Lewis structures:

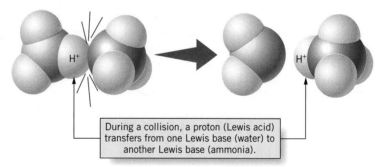

H$^+$ (a Lewis acid) transfers from the oxygen of the hydronium ion to the nitrogen of the ammonia molecule.

The molecular view of the reaction is

During a collision, a proton (Lewis acid) transfers from one Lewis base (water) to another Lewis base (ammonia).

In general, all Brønsted–Lowry acid–base reactions follow this pattern, and the position of equilibrium favors the proton being attached to the stronger Lewis base.

In analyzing proton transfer reactions, it is wise to stick to one interpretation or the other (Brønsted–Lowry or Lewis). Usually we will use the Brønsted–Lowry approach because it is useful to think in terms of conjugate acid–base pairs. However, if you switch to the Lewis interpretation, don't try to apply Brønsted–Lowry terminology at the same time; it won't work and you'll only get confused.

17.4 ▶ Elements and their oxides demonstrate acid–base properties

As you have probably noticed, the elements most likely to form acids of any type are the nonmetals in the upper right-hand corner of the periodic table. The elements most likely to form basic hydroxides are similarly grouped in one general location in the table, among the metals, particularly those in Groups IA (alkali metals) and IIA (alkaline earth metals) along the left side. Thus, elements can themselves be classified according to their abilities to be involved in acids or bases.

In general, the experimental basis for classifying the elements according to their abilities to form acids or bases depends on how their *oxides* behave toward water. In general, *most metal oxides react with water to form bases, and nonmetal oxides react with water to form acids.* Typical metal oxides that give hydroxides in water are sodium oxide, Na_2O, and calcium oxide, CaO.

$$Na_2O + H_2O \longrightarrow 2NaOH \qquad \text{sodium hydroxide}$$

$$CaO + H_2O \longrightarrow Ca(OH)_2 \qquad \text{calcium hydroxide}$$

In Section 5.5 we learned that metal oxides like Na_2O and CaO are called *basic anhydrides* ("anhydride" means "without water") because they can neutralize acids and their aqueous mixtures test basic to litmus. The reactions of metal oxides with water are really reactions of their oxide ions, which win H^+ from molecules of H_2O, leaving ions of OH^-.

Metal oxides are solids at room temperature, and the oxide ions (O^{2-}) in many metal oxides are so tightly bound in the crystals that when they are stirred with water the oxide ions are unable to take H^+ ions from surrounding water molecules. When a strong acid is present in the surrounding medium, however, then even some otherwise water-insoluble metal oxides dissolve. Iron(III) oxide, for example, dissolves in water by reacting as follows with acid:

The reactions of insoluble metal oxides with acids were introduced in Section 5.8.

$$Fe_2O_3(s) + 6H^+(aq) \longrightarrow 2Fe^{3+}(aq) + 3H_2O$$

This is a common method for removing rust from iron in industrial processes.

Hydrated metal ions can behave as weak acids

Water molecules, because of the partial negative sites on their O atoms, are attracted to metal ions in water, with H_2O acting as a Lewis base and the metal ion as a Lewis acid. When water molecules gather about a cation, they form a species called a *hydrated ion* (see Section 5.2). Hydrated metal ions tend to be Brønsted acids because of the equilibrium shown below. For simplicity, the equation represents a metal ion as a *mono*hydrate, namely, $M(H_2O)^{n+}$ with a net positive charge of $n+$ (n being 1, 2, or 3, depending on M).

The force of attraction between an ion and water molecules is often strong enough to persist when the solvent water evaporates. Solid hydrates crystallize from solution, for example, $CuSO_4 \cdot 5H_2O$ (page 44).

$$M(H_2O)^{n+} + H_2O \rightleftharpoons MOH^{(n-1)+} + H_3O^+$$

In other words, hydrated metal ions tend to be proton donors in water. Let's see why. The reason will be, by now, a familiar argument.

The positive charge on the central metal ion attracts the water molecule, holds it by its O atom, and draws electron density from the O—H bonds.

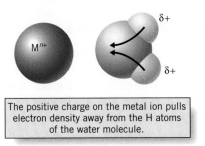

The positive charge on the metal ion pulls electron density away from the H atoms of the water molecule.

This intensifies the partial positive charge on H and weakens the O—H bond with respect to the transfer of H^+ to another water molecule in the formation of a hydronium ion, which we can illustrate as follows:

$$M^{n+} :\ddot{O}: \quad \overset{\delta+}{H} :\ddot{O}: \quad \overset{H}{\diagdown} \quad \longrightarrow \quad \left[M-\ddot{O} \overset{H}{\diagdown} \right]^{(n-1)+} + \left[H-\ddot{O}: \overset{H}{\diagdown} \right]^{+}$$

> Electron density is reduced in the O—H bonds of water (indicated by the curved arrows) by the positive charge of the metal ion, thereby increasing the partial positive charge on the hydrogens. This promotes the transfer of H$^+$ to a water molecule.

The degree to which metal ions produce acidic solutions depends on two principal factors. One is the amount of charge on the cation and the other is the cation's size. As you might expect, increasing the charge on the cation increases the metal ion's ability to draw electron density toward itself and away from the O—H bond, which thereby favors the release of H$^+$. This means that highly charged metal ions ought to produce more acidic solutions than ions of low charge, and that is generally the case.

The reason the size of the cation also affects its acidity is that when the cation is small, the positive charge is highly concentrated. A highly concentrated positive charge is better able to pull electrons from an O—H bond than a positive charge that is more spread out. Therefore, for a given positive charge, the smaller the cation, the more acidic are its solutions.

Both size and amount of charge can be considered together by referring to a metal ion's *positive charge density,* the ratio of the positive charge to the volume of the cation (its ionic volume).

$$\text{charge density} = \frac{\text{ionic charge}}{\text{ionic volume}}$$

The higher the positive charge density is, the more effective the metal ion is at drawing electron density from the O—H bond and the more acidic is the hydrated cation.

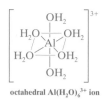

octahedral Al(H$_2$O)$_6^{3+}$ ion

Very small cations with large positive charges have large positive charge densities and tend to be quite acidic. An example is the hydrated aluminum ion, Al^{3+}. The hexahydrate, Al(H$_2$O)$_6^{3+}$, is one of several of this cation's hydrated forms that are present in an aqueous solution of an aluminum salt. This ion is acidic in water because of the equilibrium

$$[Al(H_2O)_6]^{3+}(aq) + H_2O \rightleftharpoons [Al(H_2O)_5(OH)]^{2+}(aq) + H_3O^+(aq)$$

The equilibrium, while not actually *strongly* favoring the products, does produce enough hydronium ion so that a 0.1 M solution of AlCl$_3$ in water has about the same concentration of hydronium ions as a 0.1 M solution of acetic acid, roughly 1×10^{-3} M.

Periodic trends in the acidity of metal ions

Within the periodic table, atomic size increases down a group and decreases from left to right in a period. Cation sizes follow these same trends, so within a given group, the cation of the metal at the top of the group has the smallest volume and the largest charge density. Therefore, hydrated metal ions at the top of a group in the periodic table are the most acidic within the group.

The cations of the Group IA metals (Li$^+$, Na$^+$, K$^+$, Rb$^+$, or Cs$^+$), with charges of just 1+, have little tendency to increase the H$_3$O$^+$ concentration in an aqueous solution. Within Group IIA, the Be^{2+} cation is very small and has sufficient charge density to cause the hydrated ion to be a weak acid. The other cations of Group IIA (Mg^{2+}, Ca^{2+}, Sr^{2+}, Ba^{2+}) have charge densities that become progressively smaller as we go down the group. Although their hydrated ions all generate some hydronium ion in water, the amount is negligible.

Some transition metal ions are also acidic, especially those with charges of 3+. For example, solutions containing salts of Fe^{3+} and Cr^{3+} tend to be acidic because their ions in solution exist as $Fe(H_2O)_6^{3+}$ and $Cr(H_2O)_6^{3+}$, respectively, and undergo the same ionization reaction as does the $Al(H_2O)_6^{3+}$ ion discussed above.

Acid–base properties of metal oxides are influenced by the oxidation number of the metal

Not all metal oxides are basic. As the oxidation number (or charge) on a metal ion increases, the metal ion becomes more acidic; it becomes a better electron pair acceptor. For metal hydrates, we've seen that this causes electron density to be pulled from the O—H bonds of water molecules, causing the hydrate itself to become a weak proton donor. The increasing acidity of metal ions with increasing charge also affects the basicity of their oxides.

When the positive charge on a metal is small, the oxide tends to be basic, as we've seen for oxides such as Na_2O and CaO. With ions having a 3+ charge, the oxides are less basic and begin to take on acidic properties as well; they become amphoteric. (Recall that a substance is *amphoteric* if it is capable of reacting as either an acid or a base.) Aluminum oxide is an example; it is able to react with both acids and bases. It has basic properties when it dissolves in acid.

$$Al_2O_3(s) + 6H^+(aq) \longrightarrow 2Al^{3+}(aq) + 3H_2O$$

As noted earlier, the hydrated aluminum ion has six water molecules surrounding it, so in an acidic solution the aluminum exists primarily as $Al(H_2O)_6^{3+}$.

Aluminum oxide exhibits acidic properties when it dissolves in a base. One way to write the equation for the reaction is

$$Al_2O_3(s) + 2OH^-(aq) \longrightarrow 2AlO_2^-(aq) + H_2O$$

Actually, in basic solution the formula of the aluminum-containing species is more complex than this and is better approximated by $Al(H_2O)_2(OH)_4^-$. Note that the difference between the two formulas is just the number of water molecules involved in the formation of the ion.

$$Al(H_2O)_2(OH)_4^- \rightleftharpoons AlO_2^- + 4H_2O$$

However we write the formula for the aluminum-containing ion, it is an anion, not a cation.

When the metal is in a very high oxidation state, the oxide becomes acidic. Chromium(VI) oxide, CrO_3, is an example. When dissolved in water, the resulting solution is quite acidic and is called *chromic acid*. One of the principal species in the solution is H_2CrO_4, which is a strong acid that is more than 95% ionized. The acid forms salts containing the chromate ion, CrO_4^{2-}.

Nonmetal oxides form acids when they react with water

Nonmetal oxides are usually *acidic anhydrides;* those that react with water give acidic solutions. Typical examples of the formation of acids from nonmetal oxides are the following reactions:

Not all nonmetal oxides react with water. Carbon monoxide is an example of one that does not.

$$SO_3(g) + H_2O \longrightarrow H_2SO_4(aq) \qquad \text{sulfuric acid}$$

$$N_2O_5(g) + H_2O \longrightarrow 2HNO_3(aq) \qquad \text{nitric acid}$$

$$CO_2(g) + H_2O \longrightarrow H_2CO_3(aq) \qquad \text{carbonic acid}$$

17.5 ▶ pH is a measure of the acidity of a solution

There are literally thousands of weak acids and bases, and they vary widely in how weak they are. Both the acetic acid in vinegar and the carbonic acid in pressurized soda water, for example, are classified as weak. Yet carbonic acid is only about 3% as strong an acid as acetic acid. To study such differences quantitatively, we need to explore acid–base equilibria in greater depth. This requires that we first discuss a particularly important equilibrium that exists in *all* aqueous solutions, namely, the ionization of water itself.

Water is a very weak electrolyte

Using sensitive instruments, pure water is observed to weakly conduct electricity, indicating the presence of very small concentrations of ions. They arise from the very slight self-ionization, or *autoionization,* of water itself, represented by the following equilibrium equation:

We've omitted the usual (*aq*) following the symbols for ions in water for the sake of simplicity.

$$H_2O + H_2O \rightleftharpoons H_3O^+ + OH^-$$

The forward reaction requires a collision of two H_2O molecules.

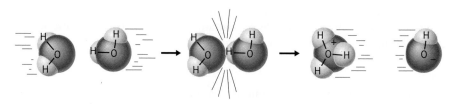

Its equilibrium law, following the procedures developed in Chapter 16, is

The density of water is 1.00 g mL^{-1} or 1.00×10^3 g L^{-1}, so the molar concentration of water (at 18.0 g mol^{-1}) is 55.6 mol L^{-1}.

$$\frac{[H_3O^+][OH^-]}{[H_2O]^2} = K_c$$

In pure water, and even in dilute aqueous solutions, the molar concentration of water is essentially a constant, with a value of 55.6 *M*. Therefore, the term $[H_2O]^2$ in the denominator is a constant that we can combine with K_c.

$$[H_3O^+][OH^-] = K_c \times [H_2O]^2$$

The product of the two constants, $K_c \times [H_2O]^2$, must also be a constant. Because of the importance of the autoionization equilibrium, it is given the special symbol **K_w** and is called the **ion product constant of water.**

$$[H_3O^+][OH^-] = K_w$$

Often, for convenience, we omit the water molecule that carries the hydrogen ion and write H^+ in place of H_3O^+. The equilibrium equation for the autoionization of water then simplifies as follows:

$$H_2O \rightleftharpoons H^+ + OH^-$$

The equation for K_w based on this is likewise simplified.

Ion product constant of water

$$[H^+][OH^-] = K_w \tag{17.3}$$

In pure water, the concentrations of H^+ and OH^- produced by the autoionization are equal because the ions are formed in equal numbers. It's been found that the concentrations have the following values at 25 °C:

$$[H^+] = [OH^-] = 1.0 \times 10^{-7} \text{ mol } L^{-1}$$

Therefore, at 25 °C,

$$K_w = (1.0 \times 10^{-7}) \times (1.0 \times 10^{-7})$$

$$K_w = 1.0 \times 10^{-14} \qquad \text{(at 25 °C)} \qquad\qquad (17.4)$$

As with other equilibrium constants, the value of K_w varies with temperature (see Table 17.4). But, for simplicity, we will generally deal with systems at 25 °C, so we'll usually not specify the temperature each time. The value of K_w at 25 °C ($K_w = 1.0 \times 10^{-14}$) is so important that it should be learned (memorized).

Solutes can affect [H⁺] and [OH⁻], but they can't alter K_w

Water's autoionization takes place in *any* aqueous solution, but because of the effects of other solutes, the molar concentrations of H^+ and OH^- may not be equal. Nevertheless, their product, K_w, is the same. Thus, although Equations 17.3 and 17.4 were derived for pure water, they also apply to dilute aqueous solutions. The significance of this must be emphasized. *In any aqueous solution, the product of* [H⁺] *and* [OH⁻] *equals* K_w, *although the two molar concentrations may not actually equal each other.*

Criteria for acidic, basic, and neutral solutions

One of the consequences of the autoionization of water is that *in any aqueous solution, there are always both H_3O^+ and OH^- ions, regardless of what solutes are present.* This means that in a solution of the acid HCl there is some OH^-, and in a solution of the base NaOH there is some H_3O^+. So what criteria do we use to determine whether a solution is neutral, acidic, or basic?

A *neutral solution* is one in which the molar concentrations of H_3O^+ and OH^- are equal; neither ion is present in a greater concentration than the other. An *acidic solution* is one in which some solute has made the molar concentration of H_3O^+ greater than that of OH^-. On the other hand, a *basic solution* exists when the molar concentration of OH^- exceeds that of H_3O^+. We therefore define acidic and basic solutions in terms of the *relative* molarities of H_3O^+ and OH^-.

Because of the autoionization of water, even the most acidic solution has some OH^- and even the most basic solution has some H_3O^+.

Neutral solution	$[H_3O^+] = [OH^-]$
Acidic solution	$[H_3O^+] > [OH^-]$
Basic solution	$[H_3O^+] < [OH^-]$

In a sample of blood at 25 °C, $[H^+] = 4.6 \times 10^{-8}$ *M*. Find the molar concentration of OH^-, and decide if the sample is acidic, basic, or neutral.

ANALYSIS: To reach a decision, we need to know how [H⁺] and [OH⁻] compare. The critical link is the relationship between the concentrations of [H⁺] and [OH⁻]. We know that the values of [H⁺] and [OH⁻] are *always* related to each other and to K_w at 25 °C as follows:

$$1.0 \times 10^{-14} = [H^+][OH^-]$$

EXAMPLE 17.5

Finding [H⁺] from [OH⁻] or Finding [OH⁻] from [H⁺]

TABLE 17.4	K_w AT VARIOUS TEMPERATURES		
Temperature (°C)	K_w	Temperature (°C)	K_w
0	1.5×10^{-15}	30	1.5×10^{-14}
10	3.0×10^{-15}	40	3.0×10^{-14}
20	6.8×10^{-15}	50	5.5×10^{-14}
25	1.0×10^{-14}	60	9.5×10^{-14}

If we know one concentration, we can *always* find the other.

SOLUTION: We substitute the given value of [H$^+$] into this equation and solve for [OH$^-$].

$$1.0 \times 10^{-14} = (4.6 \times 10^{-8})\,[\text{OH}^-]$$

Solving for [OH$^-$], and remembering that its units are mol L^{-1} or M,

$$[\text{OH}^-] = \frac{1.0 \times 10^{-14}}{4.6 \times 10^{-8}}\,M$$

$$= 2.2 \times 10^{-7}\,M$$

When we compare [H$^+$] equaling $4.6 \times 10^{-8}\,M$ with [OH$^-$] equaling $2.2 \times 10^{-7}\,M$, we see that [OH$^-$] > [H$^+$]. Our answer, then, is that the blood is slightly basic.

Are the Answers Reasonable?
We know $K_w = 1.0 \times 10^{-14} = $ [H$^+$][OH$^-$]. So one check is to note that if [H$^+$] is slightly *less* than 1×10^{-7}, [OH$^-$] will have to be slightly *more* than 1×10^{-7}, as it is. (Be careful in comparing numbers when the exponent on 10 is negative; a value of 10^{-7} is larger than 10^{-8}.)

PRACTICE EXERCISE 11: An aqueous solution of sodium bicarbonate, NaHCO$_3$, has a molar concentration of hydroxide ion of $7.8 \times 10^{-6}\,M$. What is the molar concentration of hydrogen ion? Is the solution acidic, basic, or neutral?

The pH concept provides a logarithmic scale of acidity

In most solutions of weak acids and bases, the molar concentrations of H$^+$ and OH$^-$ are very small, like those in Example 17.5. When you tried to compare the two values in that example, namely, $4.6 \times 10^{-8}\,M$ and $2.2 \times 10^{-7}\,M$, you had to look in four places, two before the multiplication symbol and the two negative exponents. There's an easier approach involving only one number. A Danish chemist, S. P. L. Sørenson (1868–1939), suggested it.

To make comparisons of small values of [H$^+$] easier, Sørenson defined a quantity that he called the **pH** of the solution as follows:

pH equation

$$\text{pH} = -\log\,[\text{H}^+] \qquad (17.5)^5$$

The properties of logarithms let us rearrange Equation 17.5 as follows:

$$[\text{H}^+] = 10^{-\text{pH}} \qquad (17.6)$$

Equation 17.5 can be used to calculate the pH of a solution if its molar concentration of H$^+$ is known. On the other hand, if the pH is known and we wish to calculate the molar concentration of hydrogen ion, we apply Equation 17.6. We will illustrate these calculations with examples shortly.

The logarithmic definition of pH, Equation 17.5, has proved to be so useful that it has been adapted to quantities other than [H$^+$]. Thus, for any quantity X, we may define a term, pX, in the following way:

Defining equation for pX

$$\text{p}X = -\log\,X \qquad (17.7)$$

For example, to express small concentrations of hydroxide ion, we can define the **pOH** of a solution as

$$\text{pOH} = -\log\,[\text{OH}^-]$$

[5]In this equation and similar ones, like Equation 17.7, the logarithm of only the numerical part of the bracketed term is taken. The physical units must be mol L^{-1}, but they are set aside for the calculation.

Similarly, for K_w we can define **pK_w** as follows.

$$pK_w = -\log K_w$$

The numerical value of pK_w at 25 °C equals $-\log (1.0 \times 10^{-14})$ or $-(-14.00)$. Thus,

$$pK_w = 14.00 \quad \text{(at 25 °C)}$$

A useful relationship among pH, pOH, and pK_w can be derived from Equation 17.3 that defines K_w.

$$[H^+][OH^-] = K_w \quad \text{(Equation 17.3)}$$

By the properties of logarithms,

$$\log [H^+] + \log [OH^-] = \log K_w$$

We next multiply the terms on both sides by -1.

$$-\log [H^+] + -\log [OH^-] = -\log K_w$$

But each of these terms is in the form of pX (Equation 17.7), so

$$pH + pOH = pK_w = 14.00 \quad \text{(at 25 °C)} \quad (17.8)$$

◄ TOOLS

Relationship of pH to pOH

This tells us that in an aqueous solution of any solute at 25 °C, the sum of pH and pOH is 14.00.

Acidic, basic, and neutral solutions are identified by their pH

One meaning attached to pH is that it is a measure of the acidity of a solution. Hence, we may define *acidic, basic,* and *neutral* in terms of pH values. In pure water, or in any solution that is *neutral,*

$$[H^+] = [OH^-] = 1.0 \times 10^{-7} M$$

Therefore, by Equation 17.5, *the pH of a neutral solution at 25 °C is 7.00.*[6]

An *acidic solution* is one in which $[H^+]$ is larger than $10^{-7} M$ and so has a pH *less* than 7.00. Thus, *as a solution's acidity increases, its pH decreases.*

A *basic solution* is one in which the value of $[H^+]$ is less than $10^{-7} M$ and so has a pH that is *greater* than 7.00. *As a solution's acidity decreases, its pH increases.* These simple relationships may be summarized as follows. At 25 °C:

pH < 7.00	**Acidic solution**
pH > 7.00	**Basic solution**
pH = 7.00	**Neutral solution**

The pH values for some common substances are given on a pH scale in Figure 17.3.

One of the deceptive features of pH, particularly to those just learning the concept, is that a hydrogen ion concentration multiplies by 10 for each decrease in pH by only one unit. For example, by Equation 17.5, at pH 6 the hydrogen ion concentration is 1×10^{-6} mol L^{-1}. At pH 5, it's 1×10^{-5} mol L^{-1}, or 10 times greater. That's a major difference.

The pH concept is almost never used when a pH value would be a negative number, namely, with solutions having hydrogen ion concentrations greater than $1\ M$. For example, in a solution where the value of $[H^+]$ is $2\ M$, the "2" is a simple enough number; we can't simplify it further by using its corresponding pH value.

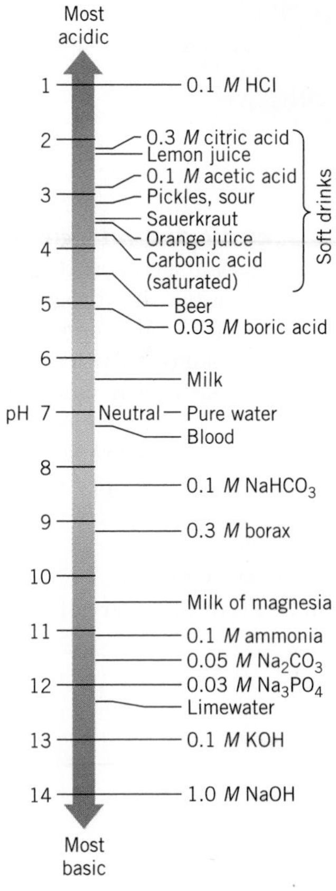

FIGURE 17.3 *The pH scale.*

[6]Recall that the rule for significant figures in logarithms is that *the number of decimal places in the logarithm of a number equals the number of significant figures in the number.* For example, 3.2 × 10⁻⁵ has just two significant figures. The logarithm of this number, displayed on a pocket calculator, is −4.494850022. We write the logarithm, when correctly rounded, as −4.49.

The pH concept is used not only in chemistry but also throughout the life sciences, because the processes of life are very sensitive to the pH of the fluids associated with them. Small changes in pH matter. The fingerlings of game fish, for example, generally cannot live in water with a pH less than about 5.5. So when sulfur oxides or nitrogen oxides wash into lakes from acid rain, fish habitats can be ruined.

If the pH of your blood is not right, your ability to breathe can be fatally impaired. If the pH of your blood goes lower than 7.2, for example, or higher than 7.7, you are experiencing a serious medical emergency. In untreated diabetes, the blood pH is on a downward move, from about 7.35 toward 7.2; in altitude sickness, it is moving upward, from 7.35 to 7.9. Unchecked hysterics, diarrhea, or certain drug overdoses also upset the pH of the blood. And consider that you can change the pH of a liter of pure water from 7 to 4 just by adding only one drop of concentrated hydrochloric acid to it. That's not enough to make the water taste sour. Fortunately, the blood has solutes that help to prevent such huge swings in its pH.

When $[H^+]$ is 2.00 M, the pH, calculated by Equation 17.5, would be $-\log (2.00)$ or -0.301. There is nothing wrong with negative pH values, but they offer no advantage over the actual value of $[H^+]$.

pH can be measured or estimated

One of the remarkable things about pH is that it can be easily measured using an instrument called a *pH meter* (Figure 17.4). An electrode system sensitive to the hydrogen ion concentration in a solution is first dipped into a solution of known pH to calibrate the instrument. Once calibrated, the apparatus can then be used to measure the pH of any other solution simply by immersing the electrodes into it. Most modern pH meters are able to determine pH values to within ±0.01 pH units, and research-grade instruments are capable of even greater precision in the pH range of 0.0 to 14.0.

Another less precise method of obtaining the pH uses acid–base indicators. As you learned in Chapter 5, these are dyes whose colors in aqueous solution depend on the acidity of the solution. Table 17.5 gives several examples. Indicators change

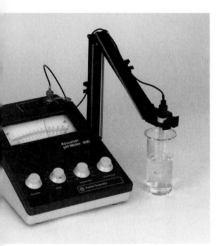

FIGURE 17.4 *A pH meter.* A special combination electrode that is sensitive to hydrogen ion concentration is dipped into the solution to be tested. After the instrument has been calibrated using a solution of known pH, the electrode is dipped into the solution to be tested and the pH is read from the meter. (How the pH electrode works is described in Facets of Chemistry 21.2.)

TABLE 17.5	COMMON ACID–BASE INDICATORS	
Indicator	Approximate pH Range over Which the Color Changes	Color Change (lower to higher pH)
Methyl green	0.2–1.8	Yellow to blue
Thymol blue	1.2–2.8	Yellow to blue
Methyl orange	3.2–4.4	Red to yellow
Ethyl red	4.0–5.8	Colorless to red
Methyl purple	4.8–5.4	Purple to green
Bromocresol purple	5.2–6.8	Yellow to purple
Bromothymol blue	6.0–7.6	Yellow to blue
Phenol red	6.4–8.2	Yellow to red/violet
Litmus	4.7–8.3	Red to blue
Cresol red	7.0–8.8	Yellow to red
Thymol blue	8.0–9.6	Yellow to blue
Phenolphthalein	8.2–10.0	Colorless to pink
Thymolphthalein	9.4–10.6	Colorless to blue
Alizarin yellow R	10.1–12.0	Yellow to red
Clayton yellow	12.2–13.2	Yellow to amber

FACETS OF CHEMISTRY 17.1

Swimming Pools, Aquariums, and Flowers

Something these have in common is the need for a proper pH. The water in swimming pools is more than just water; its a dilute mix of chemicals that prevent the growth of bacteria and stabilize the pool lining. For the pool chemistry to be properly balanced, the optimum pH range is between 7.2 to 7.6. Aquariums have to have their pH controlled because fish are very sensitive to how acidic or basic the water is. Depending on the species of fish in the tank, the pH should be between 6.0 and 7.6, and if it drifts much outside this range, the fish will die.

The pH can also affect the availability of substances that plants need to grow. A pH of from 6 to 7 is best for most plants because most nutrients are more soluble when the soil is slightly acidic than when it is neutral or slightly basic. If the soil pH is too high, metal ions such as iron, manganese, and boron that plants need will precipitate and not be available in the groundwater. A very low pH is not good either. If the pH drops to between 4 and 5, metal ions that are toxic to many plants become released as their compounds become more soluble. A low pH also inhibits the growth of certain beneficial bacteria that are needed to decompose organic matter in the soil and release nutrients, especially nitrogen.

There are chemicals that can be added to swimming pools, aquariums, and soil that will raise or lower the pH, but to know how to use them, it's necessary to measure the pH. Commercial test kits are available to enable you to do this, and they involve the use of acid–base indicators. Kits for testing the pH of pool water (Figure 1) use *phenol red* indicator, which changes color over the pH range 6.4–8.2. This places the intermediate color of the indicator right in the middle of the desired range of 7.2–7.6. To use the kit, a sample of pool water is placed into the plastic cylinder and five drops of the indicator solution are added. After shaking the mixture to ensure uniform mixing, the color is compared to the standards to gauge the pH. It can then be decided whether chemicals have to be added to either raise or lower the pH.

Similar test kits are available to test the water in an aquarium, except that the indicator used is bromothymol

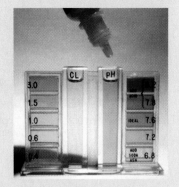

FIGURE 1 *Swimming pool test kit.* This apparatus is used to test for both pH and chlorine concentration. The color of the indicator on the right tells us the pH of the pool water is approximately 7.6, which is slightly basic.

blue, which changes color over a pH range of 6.0–7.6. This generally matches the desired pH range for the water in the aquarium. Soil test kits, such as that shown in Figure 2, use a "universal indicator," which is a mixture of indicators that enable estimation of the pH over a wider range. A sample of the soil to be tested is placed in the plastic apparatus that comes with the kit. Water is added along with a tablet of the indicator. The mixture is shaken and then the color of the water is compared with the color chart. The color that most closely matches the color of the solution gives the estimated pH.

FIGURE 2 *Testing soil pH.* The indicator reveals that the soil sample being tested has a pH of approximately 6.5, which is just slightly acidic.

color over a narrow range of pH values, and Figure 17.5 shows the colors of some indicators at opposite ends of their color change ranges. Facets of Chemistry 17.1 describes some applications of indicators to measuring pH.

pH test papers are available that are impregnated with one or more indicator dyes. To obtain a rough idea of the pH, a drop of the solution to be tested is touched to a strip of the test paper and the resulting color is compared with a color code. Some commercial test papers (e.g., Hydrion®) are impregnated with several dyes, with their vials carrying the color code (see Figure 17.6).

Litmus paper, commonly found in chemistry labs, is often used qualitatively to test whether a solution is acidic or basic. It consists of porous paper impregnated with litmus dye, made either red or blue by exposure to acid or base, and then

pH 8.2   pH 10.0
Phenolphthalein

pH 3.2   pH 4.4
Methyl orange

FIGURE 17.5 *Colors of some common acid–base indicators.*

pH 6.0   pH 7.6
Bromothymol blue

pH 4.8   pH 5.4
Methyl purple

dried. Below pH 4.7, litmus is red, and above pH 8.3, it is blue. The transition for the color change occurs over a pH range of 4.7 to 8.3 with the center of the change at about 6.5, very nearly a neutral pH. To test whether a solution is acidic, a drop of it is placed on blue litmus paper. If the dye turns pink, the solution is acidic. Similarly, to test if the solution is basic, a drop is touched to red litmus paper. If the dye turns blue, the solution is basic.

pH calculations

Let's now look at some examples of pH calculations using Equations 17.5 and 17.6. These are calculations you will be performing often in Section 17.6 and throughout Chapter 18, so be sure you become proficient at them. The first example calculates a pH from a value either of $[H^+]$ or of $[OH^-]$. Another finds the value of $[H^+]$ or $[OH^-]$ given a pH.

EXAMPLE 17.6
Calculating pH and pOH from [H⁺]

Be sure to notice that *common* logs (base 10 logs) are used in pH calculations. Don't use the *natural* log function on your scientific calculator by mistake.

Because rain washes pollutants out of the air, the lakes in many parts of the world have undergone pH changes. In a New England state, the water in one lake was found to have a hydrogen ion concentration of 3.2×10^{-5} mol L^{-1}. What are the calculated pH and pOH values of the lake's water? Is the water acidic or basic?

ANALYSIS: If we know the value of $[H^+]$, Equation 17.5 is the tool we use to find the pH. The calculation is straightforward.

SOLUTION: First, let's write the equation we will use.

$$pH = -\log [H^+] \qquad \text{(Equation 17.5)}$$

Next, we simply make the substitution for $[H^+]$.

$$pH = -\log (3.2 \times 10^{-5})$$

Using a calculator to find the logarithm of 3.2×10^{-5} gives the value -4.49 (note the negative sign). To find the pH, we change its algebraic sign.

$$pH = -(-4.49) = 4.49$$

The pH is less than 7.00, so the lake's water is acidic (too acidic, in fact, for game fish to survive).

Once we know the pH, finding the pOH is most easily done by using the fact that the sum of pH and pOH equals 14.00. Therefore, the pOH of the lake water is

$$pOH = 14.00 - 4.49$$
$$= 9.51$$

Are the Answers Reasonable?
The value of $[H^+]$ is between 1×10^{-5} and 1×10^{-4}, so the pH has to be between 5 and 4, which it is. Therefore, the pOH must be between 9 and 10, as it is.

PRACTICE EXERCISE 12: A carbonated beverage was found to have a hydrogen ion concentration of 3.67×10^{-4} mol L^{-1}. What are the calculated pH and pOH values of this beverage? Is it acidic, basic, or neutral?

FIGURE 17.6 *Using a pH test paper.* The color of this Hydrion® test strip changed to orange when a drop of the lemon juice solution in the beaker was placed on it. According to the color code, the pH of the solution is closer to 3 than to 4.

What is the pH of a sodium hydroxide solution at 25 °C in which the hydroxide ion concentration equals 0.0026 M?

ANALYSIS: There are two ways to solve this problem. We could use the K_w expression ($K_w = [H^+][OH^-] = 1 \times 10^{-14}$) and the given value of the hydroxide ion concentration to find $[H^+]$, and then calculate the pH using Equation 17.5. The second way would be to calculate the pOH from the hydroxide ion concentration ($pOH = -\log [OH^-]$) and then subtract the pOH from 14.00 to find the pH. The second path requires less effort, so let's proceed that way.

SOLUTION: First, we substitute the $[OH^-]$, which equals 0.0026 M, into the equation for pOH.

$$pOH = -\log [OH^-]$$
$$= -\log (0.0026)$$
$$= -(-2.59) = 2.59$$

Then we subtract this pOH value from 14.00 to find the pH.

$$pH = 14.00 - 2.59$$
$$= 11.41$$

Notice that the pH in this basic solution is well above 7.

Is the Answer Reasonable?
The molar concentration of hydroxide ion is between 0.001 (or 10^{-3}) and 0.01 (or 10^{-2}), so the pOH must be between 3 and 2, which it is. Thus, the pH ought to be between 11 and 12.

PRACTICE EXERCISE 13: The molarity of OH$^-$ in the water in which a soil sample has soaked overnight is 1.47×10^{-9} mol L^{-1}. What is the pH of the solution?

EXAMPLE 17.7
Calculating pH from [OH$^-$]

EXAMPLE 17.8
**Calculating [H⁺]
from pH**

"Calcareous soil" is soil rich in calcium carbonate. The pH of the moisture in such soil generally ranges from just over 7 to as high as 8.3. After one particular soil sample was soaked in water, the pH of the water was measured to be 8.14. What value of [H⁺] corresponds to a pH of 8.14? Is the soil acidic or basic?

ANALYSIS: The tool we use to calculate [H⁺] from pH is Equation 17.6.

$$[H^+] = 10^{-pH}$$

SOLUTION: We substitute the pH value, 8.14, into the equation above, which gives

$$[H^+] = 10^{-8.14}$$

We are calculating the antilogarithm of −8.14. If you need help with this, a review is available on the web site, www.wiley.com/college/brady.

Because the pH has two digits following the decimal point, we obtain two significant figures in the hydrogen ion concentration. Therefore, the answer, correctly rounded, is

$$[H^+] = 7.2 \times 10^{-9} \, M$$

Because the pH > 7 and [H⁺] < 1 × 10⁻⁷ M, the soil is basic.

Is the Answer Reasonable?
The given value of 8.14 is between 8 and 9, so at a pH of 8.14, the hydrogen ion concentration must lie between 10^{-8} and 10^{-9} mol L⁻¹. And that's what we found.

PRACTICE EXERCISE 14: Find the values of [H⁺] and [OH⁻] that correspond to each of the following values of pH. State whether each solution is acidic or basic.
(a) 2.90 (the approximate pH of lemon juice)
(b) 3.85 (the approximate pH of sauerkraut)
(c) 10.81 (the pH of milk of magnesia, a laxative)
(d) 4.11 (the pH of orange juice, on the average)
(e) 11.61 (the pH of dilute, household ammonia)

17.6 ▶ Strong acids and bases are fully dissociated in solution

Calculating the pH of dilute solutions of strong acids and bases takes into account their complete dissociation

If necessary, review the list of strong acids on page 175.

By now you know that strong acids and bases are considered to be 100% dissociated in an aqueous solution. This makes calculating the concentrations of H⁺ and OH⁻ in their solutions a relatively simple task.

When the solute in a solution is a strong monoprotic acid, such as HCl or HNO₃, we expect to obtain one mole of H⁺ for every mole of the acid in the solution. Thus, a 0.010 M solution of HCl contains 0.010 mol L⁻¹ of H⁺, and a 0.0020 M solution of HNO₃ contains 0.0020 mol L⁻¹ of H⁺. (In this chapter we will not consider strong diprotic acids, such as H₂SO₄, because only the first step in their ionization is complete. In solutions of H₂SO₄, for example, only about 10% of the HSO₄⁻ ions are further ionized to SO₄²⁻ ions and hydrogen ions.)

To calculate the pH of a solution of a strong monoprotic acid, we use the molarity of the H⁺ obtained from the stated molar concentration of the acid. Thus, the 0.010 M HCl solution mentioned above has [H⁺] = 0.010 M, from which we calculate the pH to be equal to 2.00.

For strong bases, calculating the OH⁻ concentration is similarly straightforward. A 0.050 M solution of NaOH contains 0.050 mol L⁻¹ of OH⁻ because the base is fully dissociated and each mole of NaOH releases one mole of OH⁻ when it dissociates. For bases such as Ba(OH)₂, we have to recognize that two moles of OH⁻ are released by each mole of the base.

$$Ba(OH)_2(s) \longrightarrow Ba^{2+}(aq) + 2OH^-(aq)$$

Therefore, if a solution contained 0.010 mol Ba(OH)₂ per liter, the concentration of OH⁻ would be 0.020 M. Of course, once we know the OH⁻ concentration we can calculate pOH, from which we can calculate the pH. The following example illustrates the kinds of calculations we've just described.

Calculate the values of pH, pOH, and $[OH^-]$ for the following solutions: (a) 0.020 M HCl and (b) 0.00035 M $Ba(OH)_2$.

EXAMPLE 17.9

Calculating the pH, pOH, and $[OH^-]$ for Solutions of Strong Acids or Strong Bases

ANALYSIS: The first step in calculating the pH of a solution is an examination of the solute (or solutes) that are present. There are several questions you need to ask yourself. Are they strong or weak electrolytes? Are they acids or bases? How will they affect the pH? The answers to these questions will determine how you proceed next.

In this example, we expect the solutes to be strong electrolytes because that's what we've been discussing. However, let's take a look at them. The solute in part (a) is HCl, which you should recognize as one of the strong acids. Therefore, we expect it to be 100% ionized. From each mole of HCl, we expect one mole of H^+, so we will use the molar concentration of HCl to obtain $[H^+]$, from which we can calculate the pH, pOH, and $[OH^-]$.

The solute in part (b), $Ba(OH)_2$, is a soluble metal hydroxide, which is also a strong electrolyte. From each mole of $Ba(OH)_2$, there are *two* moles of OH^- liberated. We'll use the molarity of the $Ba(OH)_2$ to calculate the hydroxide concentration. From $[OH^-]$ we can calculate pOH and then pH.

SOLUTION: (a) In 0.020 M HCl, $[H^+]$ = 0.020 M. Therefore,

$$pH = -\log (0.020)$$
$$= 1.70$$

Thus in 0.020 M HCl, the pH is 1.70. The pOH is $(14.00 - 1.70) = 12.30$. To find $[OH^-]$, we can use this value of pOH.

$$[OH^-] = 10^{-pOH}$$
$$= 10^{-12.30}$$
$$= 5.0 \times 10^{-13} M$$

Notice how much smaller $[OH^-]$ is in this acidic solution than it is in pure water. This makes sense, because the solution is quite acidic.

(b) As noted, for each mole of $Ba(OH)_2$ we obtain 2 mol of OH^-. Therefore,

$$[OH^-] = 2 \times 0.00035 M$$
$$= 0.00070 M$$
$$pOH = -\log (0.00070)$$
$$= 3.15$$

Thus the pOH of this solution is 3.15, and the pH = $(14.00 - 3.15) = 10.85$.

Are the Answers Reasonable?
In (a), the molarity of H^+ is between 0.01 M and 0.1 M, or between 10^{-2} M and 10^{-1} M. So the pH must be between 1 and 2, as we found. In (b), the molarity of OH^- is between 0.0001 and 0.001, or between 10^{-4} and 10^{-3}. Hence, the pOH must be between 3 and 4, which it is.

PRACTICE EXERCISE 15: Calculate $[H^+]$ and the pH in 0.0050 M NaOH.

PRACTICE EXERCISE 16: Rhododendrons are shrubs that produce beautiful flowers in the springtime. They only grow well in soil that has a pH that is 5.5 or slightly lower. What is the hydrogen ion concentration in the soil moisture if the pH is 5.5?

Acidic or basic solutes suppress the ionization of water

In the preceding calculations, we have made a critical and correct assumption, namely, that the autoionization of water contributes negligibly to the total $[H^+]$ in a solution of an acid and to the total $[OH^-]$ in a solution of a base.[7] Let's take a closer look at this.

---

[7] The autoionization of water cannot be neglected when working with very dilute solutions of acids and bases.

In a solution of an acid, there are actually *two* sources of H^+. One is from the ionization of the acid solute itself and the other is from the autoionization of water. Thus,

$$[H^+]_{total} = [H^+]_{from\ solute} + [H^+]_{from\ H_2O}$$

A similar equation applies to the total OH^- concentration in a solution of a base.

$[OH^-]_{total} = $
$\quad [OH^-]_{from\ solute} + [OH^-]_{from\ H_2O}$

Except in very dilute solutions of acids, the amount of H^+ contributed by the water ($[H^+]_{from\ H_2O}$) is small compared to the amount of H^+ contributed by the solute ($[H^+]_{from\ solute}$). For instance, in Example 17.9 we saw that in 0.020 M HCl the molarity of OH^- was 5.0×10^{-13} M. The only source of OH^- in this acidic solution is from the autoionization of water, and the amounts of OH^- and H^+ *formed by the autoionization of water* must be equal. Therefore, $[H^+]_{from\ H_2O}$ also equals 5.0×10^{-13} M. If we now look at the total $[H^+]$ for this solution, we have

$$[H^+]_{total} = 0.020\ M + 5.0 \times 10^{-13}M$$
$$\text{(from HCl)}\quad \text{(from } H_2O)$$

$$= 0.020\ M \qquad \text{(rounded correctly)}$$

In any solution of an acid, the autoionization of water is suppressed by the H^+ furnished by the solute. It's simply an example of Le Châtelier's principle. If we look at the autoionization reaction, we can see that if some H^+ is provided by an external source (an acidic solute, for example), the position of equilibrium will be shifted to the left.

$$H_2O \rightleftharpoons H^+ + OH^-$$

Adding H^+ from a solute causes the position of equilibrium to shift to the left.

Only in very dilute solutions of acids or bases (10^{-6} M or less) does this assumption fail.

As the results of our calculation have shown, the concentrations of H^+ and OH^- *from the autoionization reaction* are reduced well below their values in a neutral solution (1.0×10^{-7} M). Therefore, except for *very* dilute solutions (10^{-6} M or less), we will assume that all of the H^+ in the solution of an acid comes from the solute. Similarly, *we'll assume that in a solution of a base, all the OH^- comes from the dissociation of the solute.*

SUMMARY

Brønsted–Lowry Acids and Bases A **Brønsted–Lowry acid** (or, more simply, a **Brønsted acid**) is a proton donor; a **Brønsted base** is a proton acceptor. According to the Brønsted–Lowry approach, an acid–base reaction is a proton transfer event. In an equilibrium involving a Brønsted acid and base, there are two **conjugate acid–base pairs.** The members of any given pair differ from each other by only one H^+. A substance that can be either an acid or a base, depending on the nature of the other reactant, is **amphoteric** or, with emphasis on proton transfer reactions, **amphiprotic.**

Lewis Acids and Bases A **Lewis acid** accepts a pair of electrons from a **Lewis base** in the formation of a coordinate covalent bond. Lewis bases often have filled valence shells and must have at least one unshared electron pair. Lewis acids have an incomplete valence shell that can accept an electron pair, have double bonds that allow electron pairs to be moved to make room for an incoming electron pair from a Lewis base, or have valence shells that can accept more than an octet of electrons.

We may summarize the three views of acids and bases as follows.

Arrhenius view   Acids give H^+ in water; bases provide OH^- in water.

Brønsted view   Acids are proton donors; bases are proton acceptors.

Lewis view   Acids are electron pair acceptors; bases are electron pair donors.

Relative Acidities and the Periodic Table **Binary acids** contain only hydrogen and another nonmetal. Their strengths increase from top to bottom within a group and left to right across a period. **Oxoacids,** which contain oxygen atoms in addition to hydrogen and another element, increase in strength as the number of oxygen atoms on the same central atom increases. Oxoacids having the same number of oxygens generally increase in strength as the central atom moves from bottom to top within a group and from left to right across a period.

Acid–Base Properties of the Elements and Their Oxides Oxides of metals are basic anhydrides when the charge on the ion is small. Those of the Groups IA and IIA metals neutralize acids and tend to react with water to form metal hydroxides. The hydrates of metal ions tend to be proton donors when the positive charge density of the metal ion is itself sufficiently high, as it is when the ion has a 3+ charge. The small beryllium ion, Be^{2+}, forms a weakly acidic hydrated ion. Metal oxides become more acidic as the oxidation number of the metal becomes larger. Aluminum oxide is amphoteric, dissolving in both acids and bases. Chromium(VI) oxide is acidic, forming chromic acid when it dissolves in water.

The Autoionization of Water and the pH Concept Water reacts with itself to produce small amounts of H_3O^+ (usually abbreviated H^+) and OH^- ions. The concentrations of these ions both in pure water *and* in dilute aqueous solutions are related by the expression

$$[H^+][OH^-] = K_w = 1.0 \times 10^{-14} \quad \text{(at 25 °C)}$$

K_w is the **ion-product constant of water.** In pure water,

$$[H^+] = [OH^-] = 1.0 \times 10^{-7}\ M$$

The **pH** of a solution is a measure of the acidity of a solution and is normally measured with a pH meter. As the pH decreases, the acidity increases. The defining equation for pH is $pH = -\log [H^+]$. In exponential form, this relationship between $[H^+]$ and pH is given by $[H^+] = 10^{-pH}$.

Comparable expressions can be used to describe low OH^- ion concentrations in terms of **pOH** values: $pOH = -\log [OH^-]$ and $[OH^-] = 10^{-pOH}$. At 25 °C, $pH + pOH = 14.00$. A solution is acidic if its pH is less than 7.00 and basic if its pH is greater than 7.00. A neutral solution has a pH of 7.00.

Solutions of Strong Acids and Strong Bases In calculating the pH of solutions of strong acids and bases, we assume that they are 100% ionized. The autoionization of water contributes a negligible amount to the $[H^+]$ in a solution of an acid. It also contributes a negligible amount to the $[OH^-]$ in a solution of a base.

TOOLS ▶ YOU HAVE LEARNED

The table below lists the concepts that you've learned in this chapter that can be applied as tools in solving problems. Study each one carefully so that you know what each is used for. When faced with solving a problem, recall what each tool does and consider whether it will be helpful in finding a solution. This will aid you in selecting the tools you need. If necessary, refer to this table when working on the Thinking-It-Through problems and the Review Exercises that follow.

TOOL	HOW IT WORKS
Periodic trends in strengths of binary acids (page 745)	You can use the trends to predict the relative acidities of X—H bonds, both for the binary hydrides themselves and for molecules that contain X—H bonds.
Periodic trends in strengths of oxoacids (page 748)	You use the trends to predict the relative acidities of oxoacids according to the nature of the central nonmetal as well as the number of oxygens attached to a given nonmetal. The principles involved also let you compare acidities of compounds containing different electronegative elements.
Ion-product constant for water $[H^+][OH^-] = K_w$ (page 758)	Use this equation to calculate $[H^+]$ if you know $[OH^-]$, and vice versa. Be sure you have learned the value of K_w.
Defining equations for pX $pX = -\log X$ and $X = 10^{-pX}$ (page 760)	Use the equations of this type to calculate pH and pOH from $[H^+]$ and $[OH^-]$, respectively. You also use them to calculate $[H^+]$ and $[OH^-]$ from pH and pOH, respectively.
Relationship between pH and pOH: $pH + pOH = 14.00$ (page 761)	This relationship is used to calculate pH if you know pOH, and vice versa.

THINKING IT THROUGH

The goal for the following problems is not to find the answers themselves, but rather to assemble the information needed to solve them and explain how you would use the information to find the answers. The problems in Level 2 are more challenging than those in Level 1 and may contain more data than are required, in which case you are also asked to identify the unnecessary data. Detailed answers to the Thinking-It-Through problems can be found on the web site.

Need extra help? Visit the Brady/Senese web site at www.wiley.com/college/brady

Level 1 Problems

1. How do you obtain the formula for the conjugate base of CH_4? How can you tell whether this base is strong or weak?

2. How can you tell whether the following are amphoteric: $H_3O^+, H_2O, HSO_4^-, NO_3^-$?

3. Can the ammonia molecule serve as a Lewis acid? Can it be a Lewis base? Explain the basis for your answer.

4. The hydration of a metal ion when it is placed into water can be viewed as a Lewis acid–base reaction. Explain, using Al^{3+} and $Al(H_2O)_6^{3+}$ as examples.

Level 2 Problems

5. Consider a metal ion, M^{n+}, in contact with n hydroxide ions. When the charge on the metal ion is small, the compound is basic. When the charge on the metal ion is large, the compound is acidic. From what you have learned in this chapter, explain this observation.

6. The reaction

$$N_2H_5^+ + NH_3 \longrightarrow N_2H_4 + NH_4^+$$

can be viewed as one in which one Lewis base displaces another from a compound. Among the reactants, which is the Lewis base? Which is the Lewis acid?

7. A 250 mL volumetric flask holds a 0.035 M solution of an acid. The pH of the solution was measured to be 3.5. Is the solute a strong acid or not? How can you tell?

8. In our calculations of the H^+ concentration in a solution of a strong acid, we ignored the small contribution made by the ionization of water. Suppose a solution of HCl had a concentration of 1.0×10^{-6} M. Show how you would calculate the molarity of H^+ in this solution taking into account the autoionization of water.

9. The pH of a 0.50 M solution of formic acid, $HCHO_2$, is 2.02. How could you tell that this is not a strong acid? Explain how you would use this information to calculate the percentage of the formic acid that has ionized in the solution.

10. Describe in detail how you would calculate the number of grams of solid NaOH that must be added to 150 mL of 0.0100 M HCl to give a solution with a pH of 3.00.

REVIEW QUESTIONS

Brønsted–Lowry Acids and Bases

17.1 How is a Brønsted acid defined? How is a Brønsted base defined?

17.2 How are the formulas of the members of a conjugate acid–base pair related to each other? Within the pair, how can you tell which is the acid?

17.3 Is H_2SO_4 the conjugate acid of SO_4^{2-}? Explain your answer.

17.4 What is meant by the term *amphoteric*? Give two chemical equations that illustrate the amphoteric nature of water.

17.5 Define the term *amphiprotic*.

Trends in Acid–Base Strengths

17.6 Within the periodic table, how do the strengths of the binary acids vary from left to right across a period? How do they vary from top to bottom within a group?

17.7 Astatine, atomic number 85, is radioactive and does not occur in appreciable amounts in nature. On the basis of what you have learned in this chapter, answer the following:
(a) How would the acidity of HAt compare to HI?
(b) How would the acidity of $HAtO_3$ compare with $HBrO_3$?

17.8 Explain why nitric acid is a stronger acid than nitrous acid.

$$\text{HO—NO}_2 \qquad \text{HO—NO}$$
$$\text{nitric acid} \qquad \text{nitrous acid}$$

17.9 Explain why H_2S is a stronger acid than H_2O.

17.10 Which is the stronger Brønsted base, $CH_3CH_2O^-$ or $CH_3CH_2S^-$? What is the basis for your selection?

17.11 Explain why $HClO_4$ is a stronger acid than H_2SeO_4.

17.12 Explain why the NO_2^- ion is a stronger base than the SO_3^{2-} ion.

17.13 The position of equilibrium in the equation below lies far to the left. Identify the conjugate acid–base pairs. Which of the two acids is stronger?

$$HOCl(aq) + H_2O \rightleftharpoons H_3O^+(aq) + OCl^-(aq)$$

17.14 Consider the following: CO_3^{2-} is a weaker base than hydroxide ion, and HCO_3^- is a stronger acid than water. In the equation below, would the position of equilibrium lie to the left or to the right? Justify your answer.

$$CO_3^{2-}(aq) + H_2O \rightleftharpoons HCO_3^-(aq) + OH^-(aq)$$

17.15 Acetic acid, $HC_2H_3O_2$, is a weaker acid than nitrous acid, HNO_2. How do the strengths of the bases $C_2H_3O_2^-$ and NO_2^- compare?

17.16 Nitric acid, HNO_3, is a very strong acid. It is 100% ionized in water. In the reaction below, would the position of equilibrium lie to the left or to the right?

$$NO_3^-(aq) + H_2O \rightleftharpoons HNO_3(aq) + OH^-(aq)$$

17.17 $HClO_4$ is a stronger proton donor than HNO_3, but in water both acids appear to be of equal strength; they are both 100% ionized. Why is this so? What solvent property would be necessary in order to distinguish between the acidities of these two Brønsted acids?

17.18 Formic acid, $HCHO_2$, and acetic acid, $HC_2H_3O_2$, are classified as weak acids, but in water $HCHO_2$ is more fully ionized than $HC_2H_3O_2$. However, if we use liquid ammonia as a solvent for these acids, they both appear to be of equal strengths; both are 100% ionized in liquid ammonia. Explain why this is so.

17.19 Which of the following molecules is expected to be the stronger Brønsted acid? Why?

17.20 In which of the molecules below is the hydrogen printed in color the more acidic. Explain your choice.

Lewis Acids and Bases

17.21 Define *Lewis acid* and *Lewis base*.

17.22 Explain why the addition of a proton to a water molecule to give H_3O^+ is a Lewis acid–base reaction.

17.23 Methylamine has the formula CH_3NH_2 and the structure

Use Lewis structures to illustrate the reaction of methylamine with boron trifluoride, BF_3.

17.24 Use Lewis structures to show the Lewis acid–base reaction between CO_2 and H_2O to give H_2CO_3. Identify the Lewis acid and the Lewis base in the reaction.

17.25 Explain why the oxide ion, O^{2-}, can function as a Lewis base but not as a Lewis acid.

17.26 The molecule SbF_5 is able to function as a Lewis acid. Explain why it is able to be a Lewis acid.

17.27 In the reaction of calcium with oxygen to form calcium oxide, each calcium gives a pair of electrons to an oxygen atom. Why isn't this viewed as a Lewis acid–base reaction?

17.28 Boric acid is very poisonous and is used in ant bait (to kill ant colonies) and to poison cockroaches. It is a weak acid with a formula often written as H_3BO_3, although it is better written as $B(OH)_3$. It functions not as a Brønsted acid, but as a Lewis acid. Using Lewis structures, show how $B(OH)_3$ can bind to a water molecule and cause the resulting product to be a weak Brønsted acid.

Acid–Base Properties of the Elements and Their Oxides

17.29 Suppose that a new element was discovered. Based on the discussions in this chapter, what properties (both physical and chemical) might be used to classify the element as a metal or a nonmetal?

17.30 If the oxide of an element dissolves in water to give an acidic solution, is the element more likely to be a metal or a nonmetal?

17.31 Many chromium salts crystallize as hydrates containing the ion $Cr(H_2O)_6^{3+}$. Solutions of these salts tend to be acidic. Explain why.

17.32 Which ion is expected to give the more acidic solution, Fe^{2+} or Fe^{3+}? Why?

17.33 Ions of the alkali metals have little effect on the acidity of a solution. Why?

17.34 What acid is formed when the following oxides react with water: (a) SO_3, (b) CO_2, (c) P_4O_{10}?

17.35 Consider the following oxides: CrO, Cr_2O_3, CrO_3.
(a) Which is most acidic?
(b) Which is most basic?
(c) Which is most likely to be amphoteric?

17.36 Write equations for the reaction of Al_2O_3 with (a) a strong acid and (b) a strong base.

Ionization of Water and the pH Concept

17.37 Write the chemical equation for the autoionization of water and the equilibrium law for K_w.

17.38 How are acidic, basic, and neutral solutions in water defined (a) in terms of $[H^+]$ and $[OH^-]$ and (b) in terms of pH?

17.39 At 25 °C, how are the pH and pOH of a solution related to each other?

REVIEW PROBLEMS

Answers to problems whose numbers are printed in color are given in Appendix B. More challenging problems are marked with asterisks. **ILW** = Interactive LearningWare solution is available at *www.wiley.com/college/brady*.

Brønsted Acids and Bases

17.40 Write the formula for the conjugate acid of each of the following:
(a) F^- (d) O_2^{2-}
(b) N_2H_4 (e) $HCrO_4^-$
(c) C_5H_5N

17.41 Write the formula for the conjugate base of each of the following:
(a) NH_2OH (d) H_5IO_6
(b) HSO_3^- (e) HNO_2
(c) HCN

17.42 Identify the conjugate acid–base pairs in the following reactions:
(a) $HNO_3 + N_2H_4 \rightleftharpoons NO_3^- + N_2H_5^+$
(b) $NH_3 + N_2H_5^+ \rightleftharpoons NH_4^+ + N_2H_4$
(c) $H_2PO_4^- + CO_3^{2-} \rightleftharpoons HPO_4^{2-} + HCO_3^-$
(d) $HIO_3 + HC_2O_4^- \rightleftharpoons IO_3^- + H_2C_2O_4$

17.43 Identify the conjugate acid–base pairs in the following reactions:
(a) $HSO_4^- + SO_3^{2-} \rightleftharpoons HSO_3^- + SO_4^{2-}$
(b) $S^{2-} + H_2O \rightleftharpoons HS^- + OH^-$
(c) $CN^- + H_3O^+ \rightleftharpoons HCN + H_2O$
(d) $H_2Se + H_2O \rightleftharpoons HSe^- + H_3O^+$

Trends in Acid–Base Strengths

17.44 Choose the stronger acid: (a) H_2S or H_2Se; (b) H_2Te or HI; (c) PH_3 or NH_3. Give your reasons.

17.45 Choose the stronger acid: (a) HBr or HCl; (b) H_2O or HF; (c) H_2S or HBr. Give your reasons.

17.46 Choose the stronger acid: (a) HIO_3 or HIO_4; (b) H_3AsO_4 or H_3AsO_3. Give your reasons.

17.47 Choose the stronger acid: (a) HOCl or $HClO_2$; (b) H_2SeO_4 or H_2SeO_3. Give your reasons.

17.48 Choose the stronger acid: (a) H_3AsO_4 or H_3PO_4; (b) H_2CO_3 or HNO_3; (c) H_2SeO_4 or $HClO_4$. Give your reasons.

17.49 Choose the stronger acid: (a) $HClO_3$ or HIO_3; (b) HIO_2 or $HClO_3$; (c) H_2SeO_3 or $HBrO_4$. Give your reasons.

Lewis Acids and Bases

17.50 Use Lewis symbols to diagram the reaction

$$NH_2^- + H^+ \longrightarrow NH_3$$

Identify the Lewis acid and Lewis base in the reaction.

17.51 Use Lewis symbols to diagram the reaction

$$BF_3 + F^- \longrightarrow BF_4^-$$

Identify the Lewis acid and Lewis base in the reaction.

17.52 Aluminum chloride, $AlCl_3$, forms molecules with itself with the formula Al_2Cl_6. Its structure is

Use Lewis structures to show how the reaction $2AlCl_3 \rightarrow Al_2Cl_6$ is a Lewis acid–base reaction.

17.53 Beryllium chloride, $BeCl_2$, exists in the solid as a polymer, consisting of long chains of $BeCl_2$ units arranged as indicated below. The formula of the chain can be represented as $(BeCl_2)_n$, where n is a large number. Use Lewis structures to show how the reaction $nBeCl_2 \rightarrow (BeCl_2)_n$ is a Lewis acid–base reaction.

17.54 Use Lewis structures to diagram the reaction

$$CO_2 + O^{2-} \longrightarrow CO_3^{2-}$$

Identify the Lewis acid and Lewis base in this reaction.

17.55 Use Lewis structures to diagram the reaction

$$CO_2 + H_2O \longrightarrow H_2CO_3$$

Identify the Lewis acid and Lewis base in this reaction.

17.56 Use Lewis structures to show how the following reaction can be viewed as the displacement of one Lewis base by another Lewis base from a Lewis acid. Identify the two Lewis bases and the Lewis acid.

$$NH_2^- + H_2O \longrightarrow NH_3 + OH^-$$

17.57 Use Lewis structures to show how the following reaction involves the transfer of a Lewis base from one Lewis acid to another. Identify the two Lewis acids and the Lewis base.

$$CO_3^{2-} + SO_2 \longrightarrow CO_2 + SO_3^{2-}$$

Autoionization of Water and pH

17.58 At the temperature of the human body, 37 °C, the value of K_w is 2.4×10^{-14}. Calculate $[H^+]$, $[OH^-]$, pH, and pOH of pure water at this temperature. What is the relation between pH, pOH, and K_w at this temperature? Is water neutral at this temperature?

17.59 Deuterium oxide, D_2O, ionizes like water. At 20 °C its K_w, or ion product constant, analogous to that of water, is 8.9×10^{-16}. Calculate $[D^+]$ and $[OD^-]$ in deuterium oxide at 20 °C. Calculate also the pD and the pOD.

17.60 Calculate the H^+ concentration in each of the following solutions in which the hydroxide ion concentrations are (a) 0.0024 M (c) 5.6×10^{-9} M (b) 1.4×10^{-5} M (d) 4.2×10^{-13} M

17.61 Calculate the OH^- concentration in each of the following solutions in which the H^+ concentrations are (a) 3.5×10^{-8} M (c) 2.5×10^{-13} M (b) 0.0065 M (d) 7.5×10^{-5} M

17.62 Calculate the pH of each solution in Problem 17.60.

17.63 Calculate the pH of each solution in Problem 17.61.

17.64 A certain brand of beer had a H^+ concentration equal to 1.9×10^{-5} mol L^{-1}. What is the pH of the beer?

17.65 A soft drink was put on the market with $[H^+] = 1.4 \times 10^{-5}$ mol L^{-1}. What is its pH?

17.66 Calculate the molar concentrations of H^+ and OH^- in solutions that have the following pH values: (a) 3.14   (b) 2.78   (c) 9.25   (d) 13.24   (e) 5.70

17.67 Calculate the molar concentrations of H^+ and OH^- in solutions that have the following pOH values: (a) 8.26   (b) 10.25   (c) 4.65   (d) 6.18   (e) 9.70

17.68 The interaction of water droplets in rain with carbon dioxide that is naturally present in the atmosphere causes rainwater to be slightly acidic because CO_2 is an acidic anhydride. As a result, pure clean rain has a pH of about 5.6. What are the hydrogen ion and hydroxide ion concentrations in this rainwater?

17.69 "Acid rain" forms when rain falls through air polluted by oxides of sulfur and nitrogen, which dissolve to form acids such as H_2SO_3, H_2SO_4, and HNO_3. Trees and plants are affected if the acid rain has a pH of 3.5 or lower. What is the hydrogen ion concentration in acid rain that has a pH of 3.42?

Solutions of Strong Acids and Bases

ILW 17.70 What is the concentration of H^+ in 0.030 M HNO_3? What is the pH of the solution? What is the OH^- concentration in the solution?

17.71 What is the concentration of H^+ in 0.0025 M $HClO_4$? What is the pH of the solution? What is the OH^- concentration in the solution?

ILW 17.72 A sodium hydroxide solution is prepared by dissolving 6.0 g NaOH in 1.00 L of solution. What is the molar concentration of OH^- in the solution? What is the pOH and the pH of the solution? What is the hydrogen ion concentration in the solution?

17.73 A solution was made by dissolving 0.837 g $Ba(OH)_2$ in 100 mL final volume. What is the molar concentration of OH^- in the solution? What are the pOH and the pH? What is the hydrogen ion concentration in the solution?

17.74 A solution of $Ca(OH)_2$ has a measured pH of 11.60. What is the molar concentration of the $Ca(OH)_2$ in the solution?

17.75 A solution of HCl has a pH of 2.50. How many grams of HCl are there in 250 mL of this solution?

17.76 How many milliliters of 0.100 M KOH are needed to completely neutralize the HCl in 300 mL of a hydrochloric acid solution that has a pH of 2.25?

17.77 It was found that 20.20 mL of an HNO_3 solution are needed to react completely with 300 mL of a LiOH solution that has a pH of 12.05. What is the molarity of the HNO_3 solution?

17.78 In a 0.0020 M solution of NaOH, how many moles per liter of OH^- come from the ionization of water?

17.79 In a certain solution of HCl, the ionization of water contributes 3.4×10^{-11} moles per liter to the H^+ concentration. What is the total H^+ concentration in the solution?

ADDITIONAL EXERCISES

17.80 What is the formula of the conjugate acid of dimethylamine, $(CH_3)_2NH$? What is the formula of its conjugate base?

17.81 Suppose 10.0 mL of HCl gas at 25 °C and 734 torr is bubbled into 250 mL of pure water. What will be the pH of the resulting solution, assuming all the HCl dissolves in the water?

17.82 Suppose the HCl described in the preceding problem is bubbled into 200 mL of a solution of NaOH that has a pH of 10.50. What will the pH of the resulting solution be?

17.83 Write equations that illustrate the amphiprotic nature of the bicarbonate ion.

17.84 Hydrogen peroxide is a stronger Brønsted acid than water. (a) Explain why this is so. (b) Is an aqueous solution of hydrogen peroxide acidic or basic?

*17.85 How would you expect the degree of ionization of $HClO_3$ to compare in the solvents $H_2O(l)$ and $HF(l)$? The reactions are

$$HClO_3 + H_2O \rightleftharpoons H_3O^+ + ClO_3^-$$

$$HClO_3 + HF \rightleftharpoons H_2F^+ + ClO_3^-$$

Justify your answer.

17.86 Hydrazine, N_2H_4, is a weaker Brønsted base than ammonia. In the following reaction, would the position of equilibrium lie to the left or to the right? Justify your answer.

$$N_2H_5^+ + NH_3 \rightleftharpoons N_2H_4 + NH_4^+$$

17.87 Identify the two Brønsted acids and two Brønsted bases in the reaction

$$NH_2OH + CH_3NH_3^+ \rightleftharpoons NH_3OH^+ + CH_3NH_2$$

17.88 In the reaction in the preceding exercise, the position of equilibrium lies to the left. Identify the stronger acid in each of the conjugate pairs in the reaction.

*17.89 Suppose 38.0 mL of 0.00200 M HCl is added to 40.0 mL of 0.00180 M NaOH. What will be the pH of the final mixture?

*17.90 What is the pH of a 1.0×10^{-7} M solution of HCl?

17.91 At normal body temperature (37 °C), $K_w = 2.42 \times 10^{-14}$. What is the pH of a neutral aqueous solution at body temperature?

*17.92 How many milliliters of 0.10 M NaOH must be added to 200 mL of 0.010 M HCl to give a mixture with a pH of 3.00?

17.93 Milk of magnesia is a suspension of magnesium hydroxide in water. Although $Mg(OH)_2$ is relatively insoluble, a small amount does dissolve in the water, which makes the mixture slightly basic and gives it a pH of 10.08. How many grams of $Mg(OH)_2$ are actually dissolved in 100 mL of milk of magnesia?

*17.94 A 1.0 M solution of acetic acid has a pH of 2.37. What percentage of the acetic acid is ionized in the solution?

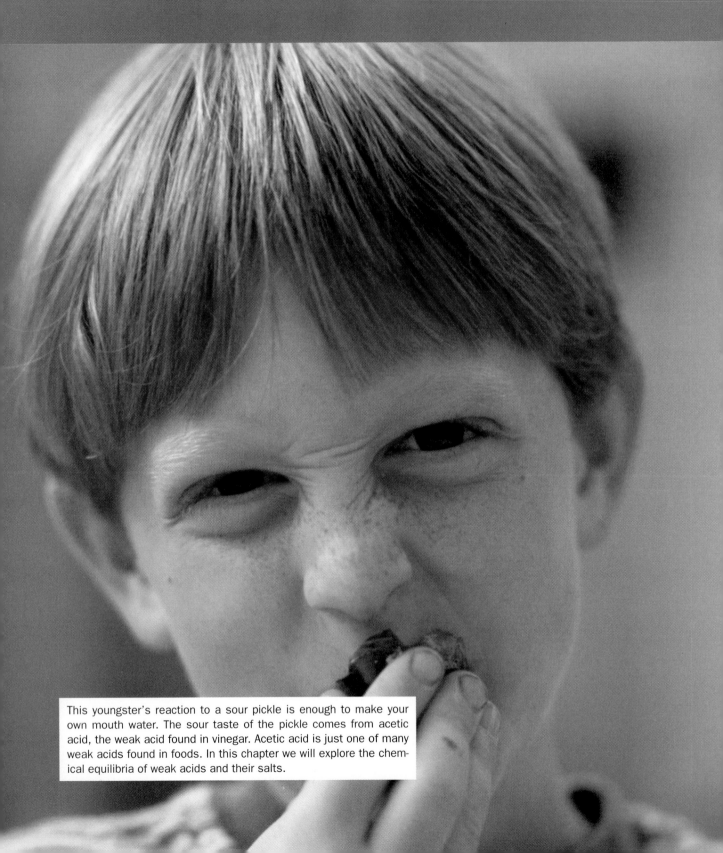

This youngster's reaction to a sour pickle is enough to make your own mouth water. The sour taste of the pickle comes from acetic acid, the weak acid found in vinegar. Acetic acid is just one of many weak acids found in foods. In this chapter we will explore the chemical equilibria of weak acids and their salts.

THIS CHAPTER IN CONTEXT In the preceding chapters we've discussed the general principles of chemical equilibrium and how to deal with calculating the hydrogen and hydroxide ion concentrations and pH of solutions of strong acids and bases. Our goal in this chapter is to bring these concepts together as we examine the chemical equilibria that exist in solutions of weak acids and bases. There are literally thousands of such substances, many of which are found in the environment or in biological systems as a consequence of the natural acidity or basicity of biological molecules. The principles we develop here have applications, therefore, not only in traditional chemistry labs but also in labs that focus on environmentally and biochemically related problems.

18.1 ▶ Ionization constants can be defined for weak acids and bases

As you've learned, weak acids and bases are incompletely ionized in water and exist in solution as molecules that are in equilibria with the ions formed by their reactions with water. To deal quantitatively with these equilibria it is essential that you be able to write correct chemical equations for the equilibrium reactions, from which you can then obtain the corresponding correct equilibrium laws. Fortunately, there is a pattern that applies to the way these substances react. *Once you've learned how to write the correct equation for one acid or base, you can write the correct equation for any other.*

A weak acid reacts with water to give its conjugate base and hydronium ion

In aqueous solutions, all weak acids behave the same way. They are Brønsted acids and, therefore, proton donors. Some examples are $HC_2H_3O_2$, HSO_4^-, and NH_4^+. In water, these participate in the following equilibria.

$$HC_2H_3O_2 + H_2O \rightleftharpoons H_3O^+ + C_2H_3O_2^-$$

$$HSO_4^- + H_2O \rightleftharpoons H_3O^+ + SO_4^{2-}$$

$$NH_4^+ + H_2O \rightleftharpoons H_3O^+ + NH_3$$

Notice that in each case, the acid reacts with water to give H_3O^+ and the corresponding conjugate base. We can represent these reactions in a general way using HA to stand for the formula of the acid.

$$HA + H_2O \rightleftharpoons H_3O^+ + A^- \qquad (18.1)$$

As you can see from the equations above, HA does not have to be electrically neutral; it can be a molecule such as $HC_2H_3O_2$, a negative ion such as HSO_4^-, or a

Acids do not have to be molecular. Many ions are also proton donors.

TOOLS

General equation for the ionization of a weak acid

positive ion such as NH_4^+. (Of course, the actual charge on the conjugate base will then depend on the charge on the parent acid.)

Following the procedure developed in Chapter 16, we can also write a general equation for the equilibrium law for Reaction 18.1, using K_c' to stand for the equilibrium constant.

$$K_c' = \frac{[H_3O^+][A^-]}{[HA][H_2O]}$$

In our discussion of the autoionization of water in Chapter 17, we noted that in dilute aqueous solutions $[H_2O]$ can be considered a constant, so it can be combined with K_c' to give a new equilibrium constant. Doing this gives

$$K_c' \times [H_2O] = \frac{[H_3O^+][A^-]}{[HA]} = K_a$$

*Some call K_a the **acid dissociation constant**.*

The new constant $\boldsymbol{K_a}$ is called an **acid ionization constant.** Abbreviating H_3O^+ as H^+, the equation for the ionization of the acid can be simplified as

$$HA \rightleftharpoons H^+ + A^-$$

from which the expression for K_a is obtained directly.

$$K_a = \frac{[H^+][A^-]}{[HA]} \tag{18.2}$$

You should learn how to write the chemical equation for the ionization of a weak acid and be able to write the equilibrium law corresponding to its K_a. This is illustrated in Example 18.1.

EXAMPLE 18.1

Writing the K_a Expression for a Weak Acid

Nitrites in meat. The meat products shown here contain nitrite ion as a preservative.

Nitrous acid, HNO_2, is a weak acid that's formed in the stomach when nitrite food preservatives encounter stomach acid. There has been some concern that this acid may form carcinogenic products by reacting with proteins. Write the chemical equation for the equilibrium ionization of HNO_2 in water and the appropriate K_a expression.

ANALYSIS: To solve this problem, we have to think about what happens when the acid reacts with water so we can construct the chemical equation. Once we have the equation, we use it to construct the equilibrium law, which will consist of the mass action expression set equal to K_a.

SOLUTION: When HNO_2 reacts with water, the products will be hydronium ion, H_3O^+, and the conjugate base. The conjugate base of HNO_2 is NO_2^-, so the equation for the ionization reaction is

$$HNO_2 + H_2O \rightleftharpoons H_3O^+ + NO_2^-$$

In the K_a expression, we leave out the H_2O that appears on the left

$$K_a = \frac{[H_3O^+][NO_2^-]}{[HNO_2]}$$

For simplicity, we usually represent H_3O^+ as H^+, so we can write the expression as

$$K_a = \frac{[H^+][NO_2^-]}{[HNO_2]}$$

Alternatively, we could have written the simplified equation for the ionization reaction, from which the K_a expression above is obtained directly.

$$HNO_2 \rightleftharpoons H^+ + NO_2^-$$

Is the Answer Reasonable?
Notice that the form of the K_a expression matches Equation 18.2, so we are confident we have solved the problem correctly.

PRACTICE EXERCISE 1: For each of the following acids, write the equation for its ionization in water and the appropriate expression for K_a: (a) $HCHO_2$, (b) $(CH_3)_2NH_2^+$, (c) $H_2PO_4^-$.

For weak acids, values of K_a are usually quite small and can be conveniently represented in a logarithmic form similar to pH. Thus, we can define the **pK_a** of an acid as

$$pK_a = -\log K_a$$

Also, $K_a = 10^{-pK_a}$

The strength of a weak acid is determined by its value of K_a; the larger the K_a, the stronger and more fully ionized the acid. Because of the negative sign in the defining equation for pK_a, the stronger the acid, the *smaller* is its value of pK_a. The values of K_a and pK_a for some typical weak acids are given in Table 18.1. A more complete list is located in Appendix C.

The values of K_a for strong acids are very large and are not tabulated. For many strong acids, K_a values have not been measured.

A certain acid was found to have a pK_a equal to 4.88. Is this acid stronger or weaker than acetic acid? What is the value of K_a for the acid?

EXAMPLE 18.2

Interpreting pK_a and Finding K_a

ANALYSIS: We can compare the acid strengths by comparing their pK_a values. The larger the pK_a, the weaker the acid. Finding K_a from pK_a involves the same kind of calculation as finding $[H^+]$ from pH, which you learned to do in Chapter 17.

SOLUTION: From Table 18.1, the pK_a of acetic acid equals 4.74. The acid referred to in the problem has pK_a equal to 4.88. Because the acid has a larger pK_a than acetic acid, it is a weaker acid.

TABLE 18.1	K_a AND pK_a VALUES FOR WEAK MONOPROTIC ACIDS AT 25 °C		
Name of Acid	Formula	K_a	pK_a
Iodic acid	HIO_3	1.7×10^{-1}	0.23
Chloroacetic acid	$HC_2H_2O_2Cl$	1.36×10^{-3}	2.87
Nitrous acid	HNO_2	7.1×10^{-4}	3.15
Hydrofluoric acid	HF	6.8×10^{-4}	3.17
Cyanic acid	$HOCN$	3.5×10^{-4}	3.46
Formic acid	$HCHO_2$	1.8×10^{-4}	3.74
Barbituric acid	$HC_4H_3N_2O_3$	9.8×10^{-5}	4.01
Butanoic acid	$HC_4H_7O_2$	1.52×10^{-5}	4.82
Acetic acid	$HC_2H_3O_2$	1.8×10^{-5}	4.74
Propanoic acid	$HC_3H_5O_2$	1.34×10^{-5}	4.87
Hydrazoic acid	HN_3	1.8×10^{-5}	4.74
Hypochlorous acid	$HOCl$	3.0×10^{-8}	7.52
Hydrogen cyanide (*aq*)	HCN	6.2×10^{-10}	9.21
Phenol	HC_6H_5O	1.3×10^{-10}	9.89
Hydrogen peroxide	H_2O_2	1.8×10^{-12}	11.74

To find K_a from pK_a, we use an equation similar to that for finding $[H^+]$ from pH, namely,

$$K_a = 10^{-pK_a}$$

Substituting,

$$K_a = 10^{-4.88} = 1.3 \times 10^{-5}$$

Is the Answer Reasonable?

As a quick check, we can compare the K_a values. For acetic acid, $K_a = 1.8 \times 10^{-5}$. This is larger than 1.3×10^{-5}, which tells us that acetic acid is the stronger acid. That agrees with our conclusion based on the pK_a values.

PRACTICE EXERCISE 2: Two acids, HA and HB, have pK_a values of 3.16 and 4.14, respectively. Which is the stronger acid? What are the K_a values for these acids?

A weak base reacts with water to give its conjugate acid and hydroxide ion

As with weak acids, all weak bases behave in a similar manner in water. They are weak Brønsted bases and are therefore proton acceptors. Examples are ammonia, NH_3, and acetate ion, $C_2H_3O_2^-$. Their reactions with water are

$$NH_3 + H_2O \rightleftharpoons NH_4^+ + OH^-$$

$$C_2H_3O_2^- + H_2O \rightleftharpoons HC_2H_3O_2 + OH^-$$

Notice that in each instance, the base reacts with water to give OH^- and the corresponding conjugate acid. We can also represent these reactions by a general equation. If we represent the base by the symbol B, the reaction is

General equation for the ionization of a weak base

$$B + H_2O \rightleftharpoons BH^+ + OH^- \tag{18.3}$$

This yields the equilibrium law

$$K_c' = \frac{[BH^+][OH^-]}{[B][H_2O]}$$

K_b *is also called the **base dissociation constant**.*

As with solutions of weak acids, the quantity $[H_2O]$ in the denominator is effectively a constant that can be combined with K_c' to give a new constant that we call the **base ionization constant, K_b.**

$$K_b = \frac{[BH^+][OH^-]}{[B]} \tag{18.4}$$

EXAMPLE 18.3

Writing the K_b Expression for a Weak Base

Hydrazine, N_2H_4, is a weak base. It is a poisonous substance that's sometimes formed when a "chlorine bleach," which contains hypochlorite ion, is added to an aqueous solution of ammonia. Write the equation for the reaction of hydrazine with water and write the expression for its K_b.

ANALYSIS: To solve this problem, we have to think about what happens when the base reacts with water so we can correctly construct the chemical equation. Once we have the equation, we use it to construct the equilibrium law, which will consist of the mass action expression set equal to K_b.

SOLUTION: When N_2H_4 reacts with water, the products will be OH^- and the conjugate acid. The conjugate acid of N_2H_4 is $N_2H_5^+$, which we obtain by adding a proton (H^+) to N_2H_4. Therefore, the equation for the ionization reaction in water is[1]

$$N_2H_4 + H_2O \rightleftharpoons N_2H_5^+ + OH^-$$

In the expression for K_b we omit the H_2O, so the equilibrium law is

$$K_b = \frac{[N_2H_5^+][OH^-]}{[N_2H_4]}$$

Is the Answer Reasonable?
Comparing our answer to Equations 18.3 and 18.4, we see they fit the correct pattern for a weak base, so our answers are correct.

Hydrazine is a rocket fuel and is used in the manufacture of semiconductor devices.

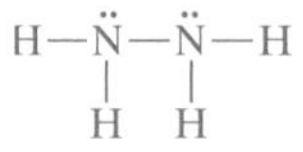

hydrazine
N_2H_4

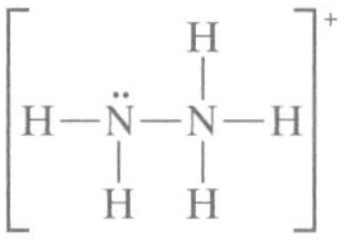

hydrazinium ion
$N_2H_5^+$

Solutions of chlorine bleach such as Clorox contain the hypochlorite ion, OCl^-, which is a weak base. Write the chemical equation for the reaction of OCl^- with water and the appropriate expression for K_b for this anion.

ANALYSIS: This is similar to the preceding problem. We have to be sure we have the correct formulas for the reactants and products of the equation.

SOLUTION: The reaction of hypochlorite ion as a base will yield its conjugate acid plus OH^-. We obtain the formula for the conjugate acid of OCl^- by adding one H^+ to the anion, which gives $HOCl$. Therefore, the equation for the equilibrium is

$$OCl^- + H_2O \rightleftharpoons HOCl + OH^-$$

As before, we omit H_2O from the expression for K_b.

$$K_b = \frac{[HOCl][OH^-]}{[OCl^-]}$$

Is the Answer Reasonable?
Once again, we can compare our answers to Equations 18.3 and 18.4. The answers are okay.

PRACTICE EXERCISE 3: For each of the following bases, write the equation for its ionization in water and the appropriate expression for K_b. (*Hint*: See footnote 1.)

(a) $(CH_3)_3N$ (trimethylamine) (b) SO_3^{2-} (sulfite ion) (c) NH_2OH (hydroxylamine)

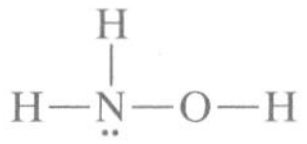

EXAMPLE 18.4

Writing the K_b Expression for a Weak Base

hydroxylamine, NH_2OH

---

[1]When a proton is accepted by a weak base that contains a nitrogen atom, such as NH_3, N_2H_4, or $(CH_3)_2NH$, the proton binds to a lone pair of electrons on a nitrogen atom of the base. For reaction with water, we can diagram the reactions of these bases as

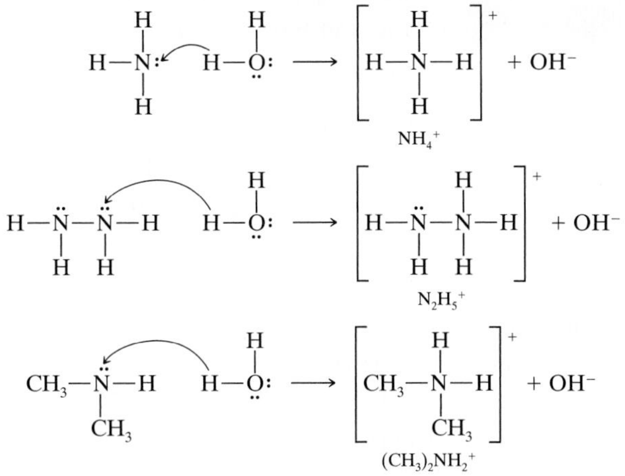

Because the K_b values for weak bases are usually small numbers, the same kind of logarithmic notation is often used to represent their equilibrium constants. Thus, **pK_b** is defined as

Also, $K_b = 10^{-pK_b}$

$$pK_b = -\log K_b$$

Table 18.2 lists some molecular bases and their corresponding values of K_b and pK_b. A more complete list is located in Appendix E.

The product of K_a and K_b equals K_w for an acid–base conjugate pair

Formica is Latin for "ant."

Formic acid, $HCHO_2$ (a substance partly responsible for the sting of a fire ant), is a typical weak acid that ionizes according to the equation

$$HCHO_2 + H_2O \rightleftharpoons H_3O^+ + CHO_2^-$$

As you've seen, we write its K_a expression as

$$K_a = \frac{[H^+][CHO_2^-]}{[HCHO_2]}$$

formic acid, $HCHO_2$
(Acidic hydrogen shown in red.)

The conjugate base of formic acid is the formate ion, CHO_2^-, and when a solute that contains this ion (e.g., $NaCHO_2$) is dissolved in water, the solution is slightly basic. In other words, the formate ion is a weak base in water and participates in the equilibrium

$$CHO_2^- + H_2O \rightleftharpoons HCHO_2 + OH^-$$

formate ion, CHO_2^-

The K_b expression for this base is

$$K_b = \frac{[HCHO_2][OH^-]}{[CHO_2^-]}$$

There is an important relationship between the equilibrium constants for this acid–base pair: the product of K_a times K_b equals K_w. We can see this by multiplying the mass action expressions.

$$K_a \times K_b = \frac{[H^+][\cancel{CHO_2^-}]}{[\cancel{HCHO_2}]} \times \frac{[\cancel{HCHO_2}][OH^-]}{[\cancel{CHO_2^-}]} = [H^+][OH^-] = K_w$$

In fact, this same relationship exists for *any* acid–base conjugate pair.

TOOLS

Relationship between K_a and K_b

For *any* acid–base conjugate pair,

$$K_a \times K_b = K_w \qquad (18.5)$$

TABLE 18.2	K_b AND pK_b VALUES FOR WEAK MOLECULAR BASES AT 25 °C		
Name of Base	Formula	K_b	pK_b
Butylamine	$C_4H_9NH_2$	5.9×10^{-4}	3.23
Methylamine	CH_3NH_2	4.4×10^{-4}	3.36
Ammonia	NH_3	1.8×10^{-5}	4.74
Hydrazine	N_2H_4	1.7×10^{-6}	5.77
Strychnine	$C_{21}H_{22}N_2O_2$	1.0×10^{-6}	6.00
Morphine	$C_{17}H_{19}NO_3$	7.5×10^{-7}	6.13
Hydroxylamine	$HONH_2$	6.6×10^{-9}	8.18
Pyridine	C_5H_5N	1.7×10^{-9}	8.82
Aniline	$C_6H_5NH_2$	4.4×10^{-10}	9.36

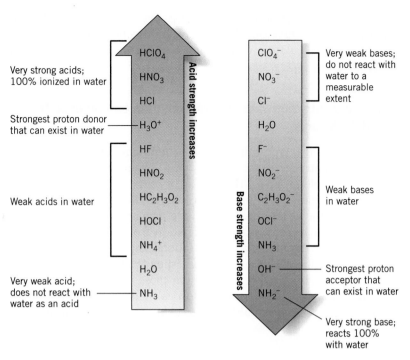

FIGURE 18.1 *The relative strengths of conjugate acid–base pairs.* The stronger the acid is, the weaker is its conjugate base. The weaker the acid is, the stronger is its conjugate base. Very strong acids are 100% ionized and their conjugate bases do not react with water to any measurable extent.

Another useful relationship, which can be derived by taking the negative logarithm of each side of Equation 18.5, is

$$pK_a + pK_b = pK_w = 14.00 \qquad \text{(at 25 °C)} \qquad (18.6)$$

There are some important consequences of the relationship expressed in Equation 18.5. One is that it is not necessary to tabulate both K_a and K_b for the members of an acid–base pair; if one K is known, the other can be calculated. For example, the K_a for $HCHO_2$ and the K_b for NH_3 will be found in most tables of acid–base equilibrium constants, but these tables usually will not contain the equilibrium constants for the ions that are the conjugates. Thus, tables usually will not contain the K_b for CHO_2^- or the K_a for NH_4^+. If we need them for a calculation, we can calculate them using Equation 18.5.

$-\log (K_a \times K_b) = -\log K_w$

$(-\log K_a) + (-\log K_b) =$
$\qquad\qquad\qquad -\log K_w$

$pK_a + pK_b = pK_w$

Tables of ionization constants usually give values for only the molecular member of an acid–base pair.

PRACTICE EXERCISE 4: The value of K_a for $HCHO_2$ is 1.8×10^{-4}. What is the value of K_b for the CHO_2^- ion?

Another interesting and useful observation is that *there is an inverse relationship between the strengths of the acid and base members of a conjugate pair*. This is illustrated graphically in Figure 18.1. Because the product of K_a and K_b is a constant, the larger the value of K_a is, the smaller is the value of K_b. In other words, *the stronger the conjugate acid, the weaker is its conjugate base* (a fact that we noted in Chapter 17 in our discussion of the strengths of Brønsted acids and bases). We will say more about this relationship in Section 18.3.

18.2 ▶ Calculations can involve finding or using K_a and K_b

Our goal in this section is to develop a general strategy for dealing quantitatively with the equilibria of weak acids and bases in water. Generally, these calculations fall into two categories. The first involves calculating the value of K_a or K_b from the initial concentration of the acid or base and the measured pH of the solution (or some other data about the equilibrium composition of the solution). The second involves calculating equilibrium concentrations given K_a or K_b and initial concentrations.

Initial concentrations and equilibrium data can be used to calculate K_a or K_b

In problems of this type, our *first* goal is to obtain numerical values for *all* the equilibrium concentrations that are needed to evaluate the mass action expression in the definition of K_a or K_b. (The reason, of course, is that at equilibrium the reaction quotient, Q, equals the equilibrium constant.) We are usually given the molar concentration of the acid or base as it would appear on the label of a bottle containing the solution. Also provided is information from which we can obtain directly at least one of the equilibrium concentrations. Thus, we might be given the measured pH of the solution, which provides an estimate of the equilibrium concentration of H^+. (Equilibrium is achieved rapidly in these solutions, so when we measure the pH, the value obtained can be used to calculate the equilibrium concentration of H^+.) Alternatively, we might be given the **percentage ionization** of the acid or base, which we define as follows:

Recall that the reaction quotient, Q, is the numerical value of the mass action expression for the reaction mixture.

Percent ionization

$$\text{percentage ionization} = \frac{\text{moles ionized per liter}}{\text{moles available per liter}} \times 100\% \qquad (18.7)$$

Let's look at some examples that illustrate how to determine K_a and K_b from the kind of data mentioned.

EXAMPLE 18.5

Calculating K_a and pK_a from pH

Sauerkraut contains lactic acid. For some, nothing goes better on a hot dog than sauerkraut and mustard. Lactic acid gives sauerkraut its sour taste.

In the equilibrium law:
$[H^+]$ means $[H^+]_{\text{at equilibrium}}$.
$[C_3H_5O_3^-]$ means
  $[C_3H_5O_3^-]_{\text{at equilibrium}}$.
$[HC_3H_5O_3]$ means
  $[HC_3H_5O_3]_{\text{at equilibrium}}$.

Lactic acid ($HC_3H_5O_3$), which is present in sour milk, also gives sauerkraut its tartness. It is a monoprotic acid. In a 0.100 M solution of lactic acid, the pH is 2.44 at 25 °C. Calculate the K_a and pK_a for lactic acid at this temperature.

ANALYSIS: In any problem involving the ionization of a weak acid or base, the first step is to write the chemical equation. From the equation we can write the correct equilibrium law and perform any necessary stoichiometric reasoning. You've learned how to write equations for such reactions, so all we need to know is whether the solute is an acid or a base. We're told its an acid, so we'll begin by writing the equilibrium equation, and then the K_a expression.

We're given the pH of the solution, so we can use this to calculate the H^+ concentration. This will be the *equilibrium* $[H^+]$; as noted above, equilibrium is reached rapidly in solutions of acids and bases, so when a pH is measured, it is the pH at equilibrium. Once we know the $[H^+]$, we will perform some reasoning using a concentration table of the type we developed in Chapter 16 to figure out the rest of the equilibrium concentrations. After we know them all, we will substitute them into the mass action expression to calculate K_a. (It's important to remember that values that satisfy the K_a expression are *equilibrium* values only.)

SOLUTION: We know the solute is an acid, and we know the general equation for the ionization of an acid, which we apply to the solute in question.

$$HC_3H_5O_3 \rightleftharpoons H^+ + C_3H_5O_3^- \qquad K_a = \frac{[H^+][C_3H_5O_3^-]}{[HC_3H_5O_3]}$$

In this problem, the only solute is the weak acid, so the only source of the ions, H^+ and $C_3H_5O_3^-$, is the ionization of the acid. (Remember, in a solution of an acid, it is safe to ignore the small amount of H^+ contributed by the autoionization of water.) As a result, the equilibrium concentration of H^+ must be identical to that of $C_3H_5O_3^-$ because the ionization reaction gives these ions in a 1:1 ratio. However, we do not have the values of either $[H^+]$ or $[C_3H_5O_3^-]$; all we have is the pH of the

0.100 M solution. We must start by finding [H$^+$] from pH. This gives us the *equilibrium* value of [H$^+$]. Then we'll put together the concentration table.

First we'll convert pH into [H$^+$].

$$[H^+] = 10^{-2.44}$$

$$= 3.6 \times 10^{-3}\ M$$

$$= 0.0036\ M$$

Because [C$_3$H$_5$O$_3^-$] = [H$^+$], we now know that [C$_3$H$_5$O$_3^-$] = 0.0036 M. With these data we can now set up the concentration table. Study the "Notes" below the table to be sure you understand how the numbers in the table are obtained.

	HC$_3$H$_5$O$_3$ $\rightleftharpoons$	H$^+$ +	C$_3$H$_5$O$_3^-$
Initial concentrations (M)	0.100	0	0 (Note 1)
Changes in concentrations caused by the ionization (M)	-0.0036	$+0.0036$	$+0.0036$ (Note 2)
Final concentrations at equilibrium (M)	(0.100 $-$ 0.0036) = 0.096 (correctly rounded) (Note 3)	0.0036	0.0036

Note 1. The *initial* concentration of the acid is what the label tells us, 0.100 M; we take it to be the concentration before any ionization occurs. The *initial* values of [H$^+$] and [C$_3$H$_5$O$_3^-$] in the table correspond to the concentrations of the two ions *produced by solutes that are strong electrolytes*. Because no strong electrolytes are present in this solution, we set [H$^+$] and [C$_3$H$_5$O$_3^-$] equal to zero. (As we noted earlier, the *extremely* small amount of H$^+$ contributed by the self-ionization of water can be safely ignored. We will ignore it in all calculations involving acids in this chapter.)

Note 2. From the pH we obtained the equilibrium concentration of H$^+$, which equals the concentration of C$_3$H$_5$O$_3^-$. The *changes* for the ions are obtained by difference. The H$^+$ and C$_3$H$_5$O$_3^-$ concentrations both increase by 0.0036 M, so these values are positive. For every H$^+$ ion that forms by the ionization, there is one less molecule of the initial acid. The minus sign in -0.0036 in the column for HC$_3$H$_5$O$_3$ is used because the concentration of this species is *decreased* by the ionization.

Note 3. The equilibrium concentration of HC$_3$H$_5$O$_3$ is obtained by adding the change ($-0.0036\ M$) algebraically to the initial concentration.

The last row of data contains the equilibrium concentrations that we now use to calculate K_a. We simply substitute them into the K_a expression.

$$K_a = \frac{(3.6 \times 10^{-3})(3.6 \times 10^{-3})}{0.096}$$

$$= 1.4 \times 10^{-4}$$

Thus the acid ionization constant for lactic acid is 1.4×10^{-4}. To find pK_a, we take the negative logarithm of K_a.

$$pK_a = -\log K_a$$

$$= -\log (1.4 \times 10^{-4})$$

$$pK_a = 3.85$$

Is the Answer Reasonable?

Weak acids have small ionization constants, so the value we obtained for K_a seems to be reasonable. *We should also check the entries in the concentration table to be sure they are reasonable.* For example, the "changes" for the ions are both positive, meaning both of their concentrations are increasing. This is the way it should be. The changes are caused by the ionization reaction, so they both have to change in the same direction. Also, we have the concentration of the molecular acid decreasing, as it should if the ions are being formed by the ionization.

Remember, the measured pH *always* gives the equilibrium concentration of H$^+$.

EXAMPLE 18.6

Calculating K_b and pK_b from Percent Ionization

methylamine

We take 6.4% of 0.100 M to find the moles per liter of the base that has ionized.

Notice that because we know *both* the initial concentration of CH_3NH_2 and the change in this value, we need make no assumptions in computing the final concentration.

Methylamine, CH_3NH_2, is a weak base and one of several substances that give herring brine its pungent odor. In 0.100 M CH_3NH_2, only 6.4% of the base has undergone ionization. What are K_b and pK_b of methylamine?

ANALYSIS: In this problem we've been given the percentage ionization of the base. We will use this to calculate the equilibrium concentrations of the ions in the solution. After we know these, solving the problem follows the same path as in the preceding example.

SOLUTION: The first step is to write the chemical equation for the equilibrium and the equilibrium law. To write the chemical equation we need the formula for the conjugate acid of CH_3NH_2, which is $CH_3NH_3^+$ (we've added one H^+ to the nitrogen atom of the base to obtain the conjugate acid). Therefore, applying the general equation for the ionization of a weak base given on page 778, the chemical equation and equilibrium law are

$$CH_3NH_2(aq) + H_2O \rightleftharpoons CH_3NH_3^+(aq) + OH^-(aq)$$

$$K_b = \frac{[CH_3NH_3^+][OH^-]}{[CH_3NH_2]}$$

The percentage ionization tells us that 6.4% of the CH_3NH_2 has reacted. Therefore, the number of moles per liter of the base that has ionized at equilibrium in this solution is

$$\text{moles per liter of } CH_3NH_2 \text{ ionized} = 0.064 \times 0.100 \; M = 0.0064 \; M$$

This value represents the *decrease* in the concentration of CH_3NH_2, which we record in the "change" row of the concentration table as -0.0064. We can now use the change in $[CH_3NH_2]$ to determine the concentrations of the other species at equilibrium. Be sure to study the "Notes" below the table to be sure you understand where the entries in the table come from.

	H_2O	+	CH_3NH_2	$\rightleftharpoons$	$CH_3NH_3^+$	+	OH^-
Initial concentrations (M)			0.100		0		0 (Note 1)
Changes in concentrations caused by the ionization (M)			-0.0064		$+0.0064$		$+0.0064$ (Note 2)
Final concentrations at equilibrium (M)			(0.100 − 0.0064) = 0.094 (properly rounded)		0.0064		0.0064

Note 1. Once again, we record the concentrations of solute species, assuming no initial ionization. We can safely ignore the extremely small contribution to $[OH^-]$ produced by the self-ionization of water.

Note 2. The concentration of CH_3NH_2 decreases by 0.0064 M, so the concentrations of the ions each increase by this amount. (Remember: However the reactants change, the products change in the opposite direction.)

Now we can calculate K_b by substituting equilibrium quantities into the mass action expression.

$$K_b = \frac{(0.0064)(0.0064)}{(0.094)}$$

$$= 4.4 \times 10^{-4}$$

and

$$pK_b = -\log(4.4 \times 10^{-4})$$

$$= 3.36$$

Methylamine. The fishy aroma you experience when preparing to consume pickled herring is due in part to the weak base methylamine.

Is the Answer Reasonable?

First, the value of K_b is small, so that suggests we've probably worked the problem correctly. We can also check the "Change" row to be sure the algebraic signs are correct. Both initial concentrations start out at zero, so they must both increase, which they do. The change for the CH_3NH_2 is opposite in sign to that for the ions, and that's as it should be. Therefore, we appear to have set up the problem correctly.

PRACTICE EXERCISE 5: When butter turns rancid, its foul odor is mostly that of butyric acid, $HC_4C_7O_2$, a weak monoprotic acid similar to acetic acid in structure. In a 0.0100 M solution of butyric acid at 20 °C, the acid is 4.0% ionized. Calculate the K_a and pK_a of butyric acid at this temperature.

PRACTICE EXERCISE 6: Few substances are more effective in relieving intense pain than morphine. Morphine is an alkaloid—an alkali-like compound obtained from plants—and alkaloids are all weak bases. In 0.010 M morphine, the pH is 10.10. Calculate the K_b and pK_b for morphine. (You don't need to know the formula for morphine, just that it's a base. Use whatever symbol you want to write the equation.)

$$CH_3-CH_2-CH_2-\overset{\displaystyle O}{\overset{\displaystyle \|}{C}}-O-H$$
butyric acid
$HC_4H_7O_2$

Calculating equilibrium concentrations from K_a (or K_b) and initial concentrations

Acid–base equilibrium problems of this type often present a confusing picture to students. Usually, the difficulty is getting started on the right foot. Therefore, before we begin, let's take an overview of the "landscape" to develop a strategy for selecting the correct attack on the problem.

Almost any problem in which you are given a value of K_a or K_b falls into one of three categories: (1) the aqueous solution contains a weak acid as its *only* solute, (2) the solution contains a weak base as its *only* solute, or (3) the solution contains *both* a weak acid and its conjugate base. The approach we take for each of these conditions is described below and is summarized in Figure 18.2.

1. If the solution contains only a weak acid as the solute, then the problem must be solved using K_a. This means that the correct chemical equation for the problem is the ionization of the weak acid. It also means that if you are given the K_b for the acid's conjugate base, you will have to calculate the K_a in order to solve the problem.

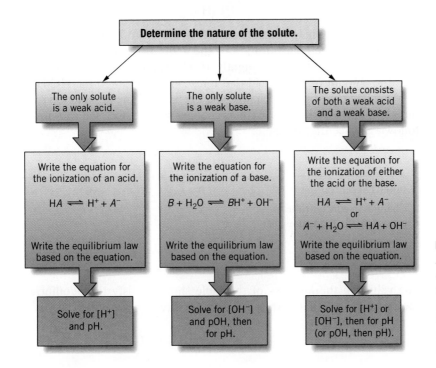

FIGURE 18.2 *Determining how to proceed in acid–base equilibrium problems.* The nature of the solute species determines how the problem is approached. Following this diagram will get you started in the right direction.

Conditions 1 and 2 apply to solutions of weak molecular acids and bases and to solutions of salts that contain an ion that is an acid or a base. Condition 3 applies to solutions called *buffers,* which are discussed in Section 18.5.

2. If the solution contains only a weak base, the problem must be solved using K_b. The correct chemical equation is for the ionization of the weak base. If the statement of the problem provides you with the K_a for the base's conjugate acid, then you must calculate the K_b to solve the problem.

3. Solutions that contain two solutes, one a weak acid and the other its conjugate base, have some special properties that we will discuss later. (Such solutions are called *buffers.*) To work problems for these kinds of mixtures, we can use *either* K_a or K_b— it doesn't matter which we use, because we will obtain the same answers either way. However, if we elect to use K_a, then the chemical equation we use must be for the ionization of the acid. If we decide to use K_b, then the correct chemical equation is for the ionization of the base. Usually, the choice of whether to use K_a or K_b is made on the basis of which constant is most readily available.

When you begin the solution of a problem, decide which of the three conditions above applies. If necessary, keep Figure 18.2 handy to guide you in your decision, so that you select the correct path to follow.

Simplifications can usually be made in acid–base equilibrium calculations

In Chapter 16 you learned that when the equilibrium constant is small, it is frequently possible to make simplifying assumptions that greatly reduce the algebraic effort in obtaining equilibrium concentrations. For many acid–base equilibrium problems such simplifications are particularly useful.

Let's consider a solution of 1.0 M acetic acid, $HC_2H_3O_2$, for which $K_a = 1.8 \times 10^{-5}$. What is involved in determining the equilibrium concentrations in the solution?

To answer this question, we begin with the chemical equation for the equilibrium. Because the only solute in the solution is the weak acid, we must use K_a and therefore write the equation for the ionization of the acid. Let's use the simplified version.

$$HC_2H_3O_2 \rightleftharpoons H^+ + C_2H_3O_2^-$$

The equilibrium law is

$$K_a = \frac{[H^+][C_2H_3O_2^-]}{[HC_2H_3O_2]} = 1.8 \times 10^{-5}$$

$$\begin{matrix} & O \\ & \| \\ CH_3 &\!\!\!\!-C-O-H \end{matrix}$$
acetic acid
$HC_2H_3O_2$

We will take the given concentration, 1.0 M, to be the initial concentration of $HC_2H_3O_2$ (i.e., the concentration of the acid before any ionization has occurred). This concentration will drop slightly as the acid ionizes and the ions form. If we let x be the amount of acetic acid that ionizes per liter, then the $HC_2H_3O_2$ concentration will decrease by x (its change will be $-x$) and the H^+ and $C_2H_3O_2^-$ concentrations will each increase by x (their changes will be $+x$). We can now construct the following concentration table.

	$HC_2H_3O_2$	$\rightleftharpoons$	H^+	$+$	$C_2H_3O_2^-$
Initial concentrations (M)	1.0		0		0
Changes in concentrations caused by the ionization (M)	$-x$		$+x$		$+x$
Concentrations at equilibrium (M)	$1.0 - x$		x		x

The initial concentrations of the ions are set equal to zero because none of them have been supplied by a solute.

The values in the last row of the table should satisfy the equilibrium law. When they are substituted into the mass action expression we obtain

$$\frac{(x)(x)}{1.0 - x} = 1.8 \times 10^{-5}$$

This equation involves a term in x^2, and it could be solved using the quadratic formula. However, this and many other similar calculations involving weak acids and bases can be simplified. Let's look at the reasoning.

The equilibrium constant, 1.8×10^{-5}, is quite small, so we can anticipate that very little of the acetic acid will be ionized at equilibrium. This means x will be very small, so we make the approximation $(1.0 - x) \approx 1.0$. Replacing $1.0 - x$ with 1.0 yields the equation

We are assuming that when x is subtracted from 1.0 and the result is rounded to the correct number of significant figures, the answer will round to 1.0.

$$\frac{x^2}{1.0} = 1.8 \times 10^{-5}$$

Solving for x gives $x = 0.0042\ M$. We see that the value of x is indeed negligible compared to $1.0\ M$ (i.e., if we subtract $0.0042\ M$ from $1.0\ M$ and round correctly, we obtain $1.0\ M$).

Notice that when we make this approximation, *the initial concentration of the acid is used as if it were the equilibrium concentration.* The approximation is valid when the equilibrium constant is small and the concentrations of the solutes are reasonably high—conditions that will apply to situations you will encounter in all but Section 18.4. In Section 18.4 we will discuss the conditions under which the approximation is not valid.

As we will discuss later, if the solute concentration is at least 400 times the value of K, we can use initial concentrations as though they are equilibrium values.

Let's look at some examples that illustrate typical acid–base equilibrium problems.

A student planned an experiment that would use $0.10\ M$ propionic acid, $HC_3H_5O_2$. Calculate the values of $[H^+]$ and pH for this solution. For propionic acid, $K_a = 1.4 \times 10^{-5}$.

ANALYSIS: First, we note that the only solute in the solution is a weak acid, so we know we will have to use K_a and write the equation for the ionization of the acid.

$$HC_3H_5O_2 \rightleftharpoons H^+ + C_3H_5O_2^-$$

$$K_a = \frac{[H^+][C_3H_5O_2^-]}{[HC_3H_5O_2]}$$

The initial concentration of the acid is $0.10\ M$, and the initial concentrations of the ions are both $0\ M$. Then we construct the concentration table.

SOLUTION: All concentrations in the table are in moles per liter.

	$HC_3H_5O_2$	$\rightleftharpoons$	H^+	$+$	$C_3H_5O_2^-$
Initial concentrations (M)	0.10		0		0
Changes in concentrations caused by the ionization (M)	$-x$		$+x$		$+x$
Final concentrations at equilibrium (M)	$(0.10 - x)$ ≈ 0.10		x		x

Notice that the equilibrium concentrations of H^+ and $C_3H_5O_2^-$ are the same; they are represented by x. Anticipating that x will be very small, we make the simplifying approximation $(0.10 - x) \approx 0.10$, so we take the equilibrium concentration of $HC_3H_5O_2$ to be $0.10\ M$. Substituting these quantities into the K_a expression gives

$$K_a = \frac{[H^+][C_3H_5O_2^-]}{[HC_3H_5O_2]} = \frac{(x)(x)}{(0.10 - x)} \approx \frac{(x)(x)}{(0.10)} = 1.4 \times 10^{-5}$$

Solving for x yields

$$x = 1.2 \times 10^{-3}$$

Because $x = [H^+]$,

$$[H^+] = 1.2 \times 10^{-3}\ M$$

EXAMPLE 18.7

Calculating the Values of $[H^+]$ and pH for a Solution of a Weak Acid from Its K_a Value

The calcium salt of propionic acid, calcium propionate, is used as a preservative in baked products.

$$CH_3-CH_2-\overset{\displaystyle O}{\overset{\displaystyle \|}{C}}-O-H$$

propionic acid
$HC_3H_5O_2$

Finally, we calculate the pH

$$pH = -\log(1.2 \times 10^{-3})$$

$$= 2.92$$

Is the Answer Reasonable?

First, we see that the calculated pH is less than 7. This tells us that the solution is acidic, which it should be for a solution of an acid. Also, the pH is higher than it would be if the acid were strong. (A 0.10 M solution of a strong acid would have $[H^+] = 0.10$ M and pH = 1.0.) If we wish to further check the accuracy of the calculation, we can substitute the calculated equilibrium concentrations into the mass action expression. If the calculated quantities are correct, the reaction quotient should equal K_a. Let's do the calculation.

$$\frac{[H^+][C_3H_5O_2^-]}{[HC_3H_5O_2]} = \frac{(x)(x)}{(0.10)} = \frac{(1.2 \times 10^{-3})^2}{0.10} = 1.4 \times 10^{-5} = K_a$$

The check works, so we know we have done the calculation correctly.

PRACTICE EXERCISE 7: Nicotinic acid, $HC_2H_4NO_2$, is a B vitamin. It is also a weak acid with $K_a = 1.4 \times 10^{-5}$. Calculate $[H^+]$ and the pH of a 0.050 M solution of $HC_2H_4NO_2$.

EXAMPLE 18.8

Calculating the pH of a Solution and the Percentage Ionization of the Solute

A solution of hydrazine, N_2H_4, has a concentration of 0.25 M. What is the pH of the solution, and what is the percentage ionization of the hydrazine? Hydrazine has $K_b = 1.7 \times 10^{-6}$.

ANALYSIS: Hydrazine must be a weak base, because we have its value of K_b (the "b" in K_b tells us this is a *base* ionization constant). Since hydrazine is the only solute in the solution, we will have to write the equation for the ionization of a weak base and then set up the K_b expression. The conjugate acid of N_2H_4 has one additional H^+, so its formula is $N_2H_5^+$. We need this formula so we can write the correct chemical equation.

SOLUTION: We begin with the chemical equation and the K_b expression.

$$N_2H_4 + H_2O \rightleftharpoons N_2H_5^+ + OH^-$$

$$K_b = \frac{[N_2H_5^+][OH^-]}{[N_2H_4]}$$

Let's once again construct the concentration table. The only sources of $N_2H_5^+$ and OH^- are the ionization of the N_2H_4, so their initial concentrations are both zero. They will form in equal amounts, so we let their changes in concentration equal $+x$. The concentration of N_2H_4 will decrease by x, so its change is $-x$.

	H_2O +	N_2H_4 $\rightleftharpoons$	$N_2H_5^+$ +	OH^-
Initial concentrations (M)		0.25	0	0
Changes in concentrations caused by the ionization (M)		$-x$	$+x$	$+x$
Final concentrations at equilibrium (M)		$(0.25 - x)$ ≈ 0.25	x	x

Because K_b is so small, $[N_2H_4] \approx 0.25$ M. (As before, we assume the initial concentration will be effectively the same as the equilibrium concentration, an approximation that we expect to be valid.) Substituting into the K_b expression,

$$\frac{(x)(x)}{0.25 - x} \approx \frac{(x)(x)}{0.25} = 1.7 \times 10^{-6}$$

Solving for x gives $x = 6.5 \times 10^{-4}$. This value represents the hydroxide ion concentration, from which we can calculate the pOH.

$$pOH = -\log (6.5 \times 10^{-4})$$

$$= 3.19$$

The pH of the solution can then be obtained from the relationship

$$pH + pOH = 14.00$$

Thus,

$$pH = 14.00 - 3.19$$

$$= 10.81$$

To calculate the percentage ionization, we need to use Equation 18.7 on page 782. This requires that we know the number of moles per liter of the base that has ionized. From the concentration table, we see that this value is also equal to x, the amount of N_2H_4 that ionizes per liter (i.e., the change in the N_2H_4 concentration). Therefore,

$$\text{percentage ionization} = \frac{6.5 \times 10^{-4}\,M}{0.25\,M} \times 100\%$$

$$= 0.26\%$$

The base is 0.26% ionized.

Is the Answer Reasonable?
First, we can quickly check to see whether the value of x and our assumed equilibrium concentration of N_2H_4 (0.25 M) satisfy the equilibrium law by substituting them into the mass action expression and comparing the result with K_b. Doing this, the value of the mass action expression we obtain is 1.7×10^{-6}, which is the same as K_b, so the value of x is correct. Because x is so small (i.e., because so little of the base is ionized at equilibrium), we expect a small percentage ionization, which agrees with the value we calculated.

PRACTICE EXERCISE 8: Pyridine, C_5H_5N, is a bad-smelling liquid for which $K_b = 1.7 \times 10^{-9}$. What is the pH of a 0.010 M aqueous solution of pyridine?

PRACTICE EXERCISE 9: Phenol is an acidic organic compound for which $K_a = 1.3 \times 10^{-10}$. What is the pH of a 0.15 M solution of phenol in water?

18.3 ▶ Salt solutions are not neutral if the ions are weak acids or bases

In Section 18.1 you saw that weak acids and bases are not limited to molecular substances. For instance, on page 775 NH_4^+ was given as an example of a weak acid, and on page 778 $C_2H_3O_2^-$ was cited as an example of a weak base. To prepare solutions of these ions, however, we cannot simply add them to water. Ions always come to us in compounds in which there is both a cation *and* an anion. Therefore, to place NH_4^+ in water, we need a salt such as NH_4Cl, and to place $C_2H_3O_2^-$ in water, we need a salt such as $NaC_2H_3O_2$.

Because a salt contains two ions, the pH of its solution can potentially be affected by either the cation or the anion, or perhaps even both. Therefore, we have to consider *both* ions as we assess the effect of a salt on the pH of a solution.

Cations can be acids

Conjugate acids of molecular bases are weak acids

The reaction of a salt with water to give either an acidic or basic solution is sometimes called **hydrolysis.** Hydrolysis means *reaction with water.*

The ammonium ion, NH_4^+, is the conjugate acid of the molecular base, NH_3. We have already learned that NH_4^+, supplied for example by NH_4Cl, is a weak acid. The equation for its reaction as an acid is

$$NH_4^+(aq) + H_2O \rightleftharpoons NH_3(aq) + H_3O^+(aq)$$

Writing this in a simplified form, along with the K_a expression, we have

$$NH_4^+(aq) \rightleftharpoons NH_3(aq) + H^+(aq) \qquad K_a = \frac{[NH_3][H^+]}{[NH_4^+]}$$

Because the K_a values for ions are seldom tabulated, we would usually expect to calculate the K_a value using the relationship $K_a \times K_b = K_w$. In Table 18.2 the K_b for NH_3 is given as 1.8×10^{-5}. Therefore,

$$K_a = \frac{K_w}{K_a} = \frac{1.0 \times 10^{-14}}{1.8 \times 10^{-5}} = 5.6 \times 10^{-10}$$

Another example is the hydrazinium ion, $N_2H_5^+$, which is also a weak acid.

$$N_2H_5^+(aq) \rightleftharpoons N_2H_4(aq) + H^+(aq) \qquad K_a = \frac{[N_2H_4][H^+]}{[N_2H_5^+]}$$

The tabulated K_b for N_2H_4 is 1.7×10^{-6}. From this, the calculated value of K_a equals 5.9×10^{-9}.

These examples illustrate that *the conjugate acids of molecular bases tend to be acidic.* Therefore,

Identification of acidic cations

> Salts that contain cations that are the conjugate acids of weak molecular bases can affect the pH of a solution. These cations are weak acids.

Metal ions with high charge densities are weak acids

It's the *charge density* (charge per unit volume) that matters, not just the size of the charge, so if the cation is very small, like Be^{2+}, the ratio of its smaller 2+ charge to its small volume can be large enough to make its hydrated ion a weak acid.

In Section 17.4 you learned that small, highly charged cations such as Al^{3+} are acidic in water because the water molecules that surround the metal ion are able to release H^+ ions rather easily. For the hydrated aluminum ion, which can be represented as $Al(H_2O)_6^{3+}$, the equilibrium can be expressed as

$$Al(H_2O)_6^{3+}(aq) + H_2O \rightleftharpoons Al(H_2O)_5(OH^-)^{2+}(aq) + H_3O^+(aq)$$

The aluminum ion is just one example. Many other cations with 3+ charges, such as Cr^{3+} and Fe^{3+}, also yield aqueous solutions that are acidic. The equilibria in such solutions can be treated by the same procedures that we've used for other weak

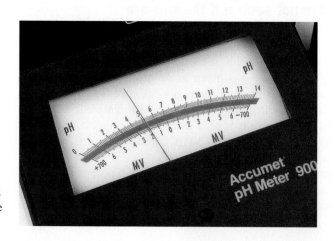

Ammonium ion as a weak acid. The measured pH of a solution of the salt NH_4NO_3 is less than 7, which indicates the solution is acidic.

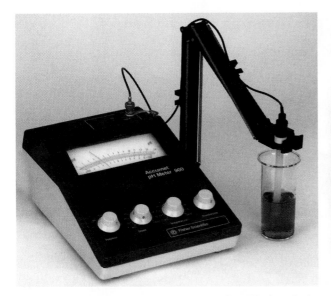
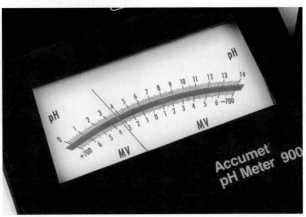

Acidity of chromium(III) ion in water. The pH of a solution of chromium(III) nitrate, which contains the blue-violet $Cr(H_2O)_6^{3+}$ ion, is distinctly acidic. The ion undergoes the same kind of equilibrium ionization as does the $Al(H_2O)_6^{3+}$ ion.

acids, but we will not discuss them further in this book. (Salts that may be of concern to us in this chapter will not contain these ions.)

Metal ions with small charges are "nonacids"

None of the singly charged cations of the Group IA metals—Li^+, Na^+, K^+, Rb^+, or Cs^+—directly affects the pH of an aqueous solution. Except for Be^{2+}, neither do any of the doubly charged cations of the Group IIA metals—Mg^{2+}, Ca^{2+}, Sr^{2+}, or Ba^{2+}. In none of these is the ratio of charge to size apparently large enough to affect the loss of protons by water molecules.

Anions can be bases

When a Brønsted acid loses a proton, its conjugate base is formed. Thus, Cl^- is the conjugate base of HCl, and $C_2H_3O_2^-$ is the conjugate base of $HC_2H_3O_2$.

Acid	Base
HCl	Cl^-
$HC_2H_3O_2$	$C_2H_3O_2^-$

Although both Cl^- and $C_2H_3O_2^-$ are bases, only the latter affects the pH of an aqueous solution. Why?

Earlier you learned that there is an inverse relationship between the strength of an acid and its conjugate base—the stronger the acid, the weaker is the conjugate base. Therefore, when an acid is *extremely strong*, as in the case of HCl or other "strong" acids that are 100% ionized (such as HNO_3), the conjugate base is *extremely weak*—too weak to affect in a measurable way the pH of a solution. Consequently, we have the following generalization.

> The anion of a strong acid is too weak a base to influence the pH of a solution.

Acetic acid is much weaker than HCl, as evidenced by its value of K_a (1.8×10^{-5}). Because acetic acid is a weak acid, its conjugate base is much stronger than

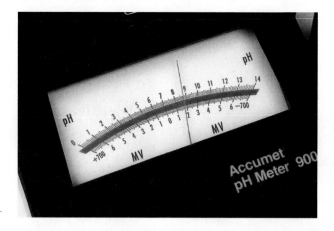

Acetate ion is a weak base in water. The pH of a solution of $KC_2H_3O_2$ is greater than 7, which shows that the solution is basic. Acetate ion reacts with water to yield small amounts of hydroxide ion.

$$C_2H_3O_2^- + H_2O \rightleftharpoons$$
$$HC_2H_3O_2 + OH^-$$

Cl^-. We can calculate the value of K_b for $C_2H_3O_2^-$ from the K_a for $HC_2H_3O_2$, also with the equation $K_a \times K_b = K_w$.

$$K_b = \frac{1.0 \times 10^{-14}}{1.8 \times 10^{-5}} = 5.6 \times 10^{-10}$$

This leads to another conclusion:

Identification of basic anions

> The anion of a weak acid is a weak base and can influence the pH of a solution. It will tend to make the solution basic.

Predicting the acid–base properties of a salt

To decide if any given salt will affect the pH of an aqueous solution, we must examine each of its ions and see what it alone might do. There are four possibilities:

1. If neither the cation nor the anion can affect the pH, the solution should be neutral.
2. If only the cation of the salt is acidic, the solution will be acidic.
3. If only the anion of the salt is basic, the solution will be basic.
4. If a salt has a cation that is acidic and an anion that is basic, the pH of the solution is determined by the *relative* strengths of the acid and base.

Let's work some examples to show how to use these generalizations.

EXAMPLE 18.9 **Predicting the Effect of a Salt on the pH of a Solution**	Sodium hypochlorite, NaOCl, is an ingredient in many common bleaching and disinfecting agents. Will a NaOCl solution be acidic, basic, or neutral? **ANALYSIS:** The solute is a salt, so we assume it to be 100% dissociated in water. $$NaOCl(s) \xrightarrow{H_2O} Na^+(aq) + OCl^-(aq)$$ To answer the question, we take each ion, in turn, and examine its effect on the acidity of the solution. **SOLUTION:** First we examine the cation, Na^+, which could potentially be acidic. However, Na^+ is an ion of a Group IA metal and is *not* acidic. Therefore, it does not affect the pH of the solution. We might say Na^+ has a "neutral" effect on the pH.

The only time an anion might be acidic is if it is from a partially neutralized polyprotic acid, e.g., HSO_4^-. Because OCl^- doesn't have a hydrogen, it can't be a Brønsted acid.

Next, we need to determine whether the anion might be basic. To decide, we need to know about the strength of its conjugate acid, which we obtain by adding H^+ to OCl^- to give HOCl. The list of strong acids is short, and HOCl is not on it, so we can expect that it is a weak acid. If HOCl is a weak acid, then OCl^- is a weak base and its presence should tend to make the solution basic.

Of the two ions in the salt, one is basic and the other is neutral. The answer to the question, therefore, is that this salt solution will be basic.

Is the Answer Reasonable?
All we can do here is check our reasoning, which appears to be sound.

PRACTICE EXERCISE 10: Is a solution of $NaNO_2$ acidic, basic, or neutral?

PRACTICE EXERCISE 11: Is a solution of KCl acidic, basic, or neutral?

PRACTICE EXERCISE 12: Is a solution of NH_4Br acidic, basic, or neutral?

What is the pH of a 0.10 *M* solution of NaOCl? For HOCl, $K_a = 3.0 \times 10^{-8}$.

ANALYSIS: Problems such as this are just like the other acid–base equilibrium problems you have learned to solve. We proceed as follows: (1) We determine the nature of the solute; is it a weak acid, a weak base, or are both a weak acid and weak base present? (2) We write the appropriate chemical equation and equilibrium law. (3) We proceed with the solution.

The solute is a salt, which is dissociated into the ions Na^+ and OCl^-. Only the latter can affect the pH. It is the conjugate *base* of HOCl, so for the purposes of problem solving, the active solute species is a weak base. On page 786 you learned that when the *only* solute is a weak base, we must use the K_b expression to solve the problem. This is where we begin the solution to the problem.

SOLUTION: We write the chemical equation for the equilibrium ionization of the base, OCl^-, and its K_b expression.

$$OCl^- + H_2O \rightleftharpoons HOCl + OH^- \qquad K_b = \frac{[HOCl][OH^-]}{[OCl^-]}$$

The data provided in the problem gives the K_a for HOCl. But we can easily calculate K_b because $K_a \times K_b = K_w$.

$$K_b = \frac{K_w}{K_a} = \frac{1.0 \times 10^{-14}}{3.0 \times 10^{-8}} = 3.3 \times 10^{-7}$$

Now let's set up the concentration table. The only source of HOCl and OH^- is the reaction of the OCl^-, so their concentrations will each increase by *x* and the concentration of OCl^- will decrease by *x*.

H_2O	+	OCl^-	$\rightleftharpoons$	HOCl	+	OH^-
Initial concentrations (*M*)		0.10		0		0
Changes in concentrations caused by the reaction (*M*)		$-x$		$+x$		$+x$
Final concentrations at equilibrium (*M*)		$(0.10 - x)$ ≈ 0.10		*x*		*x*

At equilibrium, the concentrations of HOCl and OH^- are the same, *x*.

$$[HOCl] = [OH^-] = x$$

Because K_b is so small, $[OCl^-] \approx 0.10$ *M*. (Once again, we expect *x* to be so small that we are able to use the initial concentration of the base as if it were the equilibrium concentration.) Substituting into the K_b expression gives

$$\frac{(x)(x)}{0.10 - x} \approx \frac{(x)(x)}{0.10} = 3.3 \times 10^{-7}$$

$$x = 1.8 \times 10^{-4} \ M$$

EXAMPLE 18.10
Calculating the pH of a Salt Solution

If you would have had trouble writing the correct chemical equation for this equilibrium, you should review Section 18.1.

This value of x represents the OH^- concentration, from which we can calculate the pOH and then the pH.

$$pOH = -\log(1.8 \times 10^{-4})$$

$$= 3.74$$

$$pH = 14.00 - pOH$$

$$= 14.00 - 3.74$$

$$= 10.26$$

The pH of this solution is 10.26.

Is the Answer Reasonable?

From the nature of the salt, we expect the solution to be basic. The calculated pH corresponds to a basic solution, so the answer seems to be reasonable. You've also seen that we can check the accuracy of the answer by substituting the calculated equilibrium concentrations into the mass action expression.

$$\frac{[HOCl][OH^-]}{[OCl^-]} = \frac{(x)(x)}{0.10} = \frac{(1.8 \times 10^{-4})^2}{0.10} = 3.2 \times 10^{-7}$$

The result is quite close to the value of K_b, so our answers are correct.

EXAMPLE 18.11

Calculating the pH of a Salt Solution

What is the pH of a 0.20 M solution of hydrazinium chloride, N_2H_5Cl? Hydrazine, N_2H_4, is a weak base with $K_b = 1.7 \times 10^{-6}$.

ANALYSIS: This problem is quite similar to the preceding one. Looking over the statement of the problem, we should realize that N_2H_5Cl is a salt composed of $N_2H_5^+$ (the conjugate acid of N_2H_4) and Cl^-. Since N_2H_4 is a weak base, we expect the $N_2H_5^+$ ion to be a weak acid and thereby affect the pH of the solution. On the other hand, Cl^- is the conjugate base of HCl (a strong acid) and is too weak to influence the pH. Therefore, the only active solute species is the acid, $N_2H_5^+$, which means that to solve the problem we must write the equation for the ionization of the acid and use the K_a expression.

SOLUTION: We will begin with the simplified chemical equation for the equilibrium and write the K_a expression.

$$N_2H_5^+ \rightleftharpoons H^+ + N_2H_4 \qquad K_a = \frac{[H^+][N_2H_4]}{[N_2H_5^+]}$$

The problem has given us K_b for N_2H_4, but we need K_a for $N_2H_5^+$. We obtain this by solving the equation $K_a \times K_b = K_w$ for K_a.

$$K_a = \frac{K_w}{K_b} = \frac{1.0 \times 10^{-14}}{1.7 \times 10^{-6}} = 5.9 \times 10^{-9}$$

Now we set up the concentration table. The initial concentrations of H^+ and N_2H_4 are both set equal to zero; there is no strong acid to give H^+ in the solution and no N_2H_4 is present before the reaction of water with $N_2H_5^+$. Next, we indicate that the concentration of $N_2H_5^+$ decreases by x and the concentrations of H^+ and N_2H_4 both increase by x.

	$N_2H_5^+$	$\rightleftharpoons$	H^+	+	N_2H_4
Initial concentrations (M)	0.20		0		0
Changes in concentrations caused by the ionization (M)	$-x$		$+x$		$+x$
Final concentrations at equilibrium (M)	$(0.20 - x)$ ≈ 0.20		x		x

At equilibrium, equal amounts of H^+ and N_2H_4 are present, and their equilibrium concentrations are each equal to x.

$$[H^+] = [N_2H_4] = x$$

We also assume that $[N_2H_5^+] \approx 0.20\ M$ and then substitute quantities into the mass action expression.

$$\frac{(x)(x)}{(0.20 - x)} \approx \frac{(x)(x)}{0.20} = 5.9 \times 10^{-9}$$

$$x = 3.4 \times 10^{-5}\ M$$

Since $x = [H^+]$, the pH of the solution is

$$pH = -\log (3.4 \times 10^{-5})$$

$$= 4.47$$

Is the Answer Reasonable?
The active solute species in the solution is a weak acid and the calculated pH is less than 7, so the answer seems reasonable. Check the accuracy yourself by substituting equilibrium concentrations into the mass action expression.

PRACTICE EXERCISE 13: What is the pH of a 0.10 M solution of $NaNO_2$?

PRACTICE EXERCISE 14: What is the pH of a 0.10 M solution of NH_4Br?

Solutions that contain the salt of a weak acid and a weak base

There are many salts whose ions are both able to affect the pH of the salt solution. Whether or not the salt has a net effect on the pH now depends on the relative strengths of its ions in functioning, one as an acid and the other as a base. If they are matched in their respective strengths, the salt has no net effect on pH. In ammonium acetate, for example, the ammonium ion is an acidic cation and the acetate ion is a basic anion. However, the K_a of NH_4^+ is 5.6×10^{-10} and the K_b of $C_2H_3O_2^-$ just happens to be the same, 5.6×10^{-10}. The cation tends to produce H^+ ions to the same extent that the anion tends to produce OH^-. So in aqueous ammonium acetate, $[H^+] = [OH^-]$, and the solution has a pH of 7.

Consider, now, ammonium formate, NH_4CHO_2. The formate ion, CHO_2^-, is the conjugate base of the weak acid, formic acid, so it is a Brønsted base. Its K_b is 5.6×10^{-11}. Comparing this value to the (slightly larger) K_a of the ammonium ion, 5.6×10^{-10}, we see that NH_4^+ is slightly stronger as an acid than the formate ion is as a base. As a result, a solution of ammonium formate is slightly acidic. We are not concerned here about calculating a pH, only in predicting if the solution is acidic, basic, or neutral.

If *neither* the cation nor anion is able to affect the pH, the salt solution will be neutral (provided no other acidic or basic solutes are present).

Will an aqueous solution that is 0.20 M NH_4F be acidic, basic, or neutral?

ANALYSIS: This is a salt in which the cation is a weak acid (it's the conjugate acid of a weak base, NH_3) and the anion is a weak base (it's the conjugate base of a weak acid, HF). The question, then, is, "How do the two ions compare in their abilities to affect the pH of the solution?" We have to calculate their respective K_a and K_b to compare their strengths.

SOLUTION: The K_a of NH_4^+ (calculated from the K_b for NH_3) is 5.6×10^{-10}. Similarly, the K_b of F^- is 1.5×10^{-11} (calculated from the K_a for HF, 6.8×10^{-4}). Comparing the two equilibrium constants, we see that the acid (NH_4^+) is stronger than the base (F^-), so we expect the solution to be slightly acidic.

Is the Answer Reasonable?
There's not much to check here except to be sure we've done the arithmetic correctly, and we have.

EXAMPLE 18.12

Predicting How a Salt Affects the pH of its Solution

PRACTICE EXERCISE 15: Will an aqueous solution of ammonium cyanide, NH_4CN, be acidic, basic, or neutral?

18.4 ▶ Simplifications fail for some equilibrium calculations

In the preceding two sections we used initial concentrations of solutes as though they were equilibrium concentrations when we performed calculations. This is only an approximation, as we discussed on page 787, but it is one that works most of the time. Unfortunately, it does not work in all cases, so now that you have learned the basic approach to solving equilibrium problems, we will examine those conditions under which simplifying approximations do and do not work. We will also study how to solve problems when the approximations cannot be used.

When simplifying assumptions fail

When a weak acid, HA, ionizes in water, its concentration is reduced as the ions form. If we let x represent the amount of acid that ionizes per liter, the equilibrium concentration becomes

$$[HA]_{equilib} = [HA]_{initial} - x$$

For reasons beyond the scope of this text,[2] the accuracy of acid–base equilibrium calculations is limited to about ±5%. So as long as x does not exceed ±5% of $[HA]_{initial}$, we assume its value is negligible and say that

$$[HA]_{equilib} \approx [HA]_{initial}$$

It is not difficult to show that for x to be less than or equal to ±5% of $[HA]_{initial}$, $[HA]_{initial}$ must be greater than or equal to 400 times the value of K_a.

When simplifications work

Simplifications work when $[HA]_{initial} \geq 400 \times K_a$

For the equilibrium problems in the preceding sections, this condition was fulfilled, so our simplifications were valid. We now want to study what to do when $[HA]_{initial} < 400 \times K_a$ (i.e., when we cannot justify simplifying the algebra).

The quadratic solution

When the algebraic equation obtained by substituting quantities into the equilibrium law is a quadratic equation, we can use the quadratic formula to obtain the solution. This is illustrated by the following example.

EXAMPLE 18.13

Using the Quadratic Formula in Equilibrium Problems

Chloroacetic acid, $HC_2H_2O_2Cl$, is used as an herbicide and in the manufacture of dyes and other organic chemicals. It is a weak acid with $K_a = 1.4 \times 10^{-3}$. What is the pH of a 0.010 M solution of $HC_2H_2O_2Cl$?

ANALYSIS: Before we begin the solution, we check to see whether we can use our usual simplifying approximation. We do this by calculating $400 \times K_a$ and then comparing the result to the initial concentration of the acid.

$$400 \times K_a = 400 (1.4 \times 10^{-3}) = 0.56$$

---

[2]To properly perform equilibrium calculations, we should use quantities called *activities* instead of molar concentrations. The activity of a substance is its *effective concentration,* which is affected by its electrical charge and the concentrations of other charged species in the solution. By using molar concentrations we introduce errors in the calculations that limit the accuracy we are able to obtain, so we generally carry only two significant figures. Using molar concentrations, however, makes the problems much easier to cope with mathematically.

The initial concentration of $HC_2H_2O_2Cl$ is *less than* 0.56, so we know the simplification does not work.

SOLUTION: Let's begin by writing the chemical equation and the K_a expression. (We know we must use K_a because the only solute is the acid.)

$$HC_2H_2O_2Cl \rightleftharpoons H^+ + C_2H_2O_2Cl^-$$

$$K_a = \frac{[H^+][C_2H_2O_2Cl^-]}{[HC_2H_2O_2Cl]} = 1.4 \times 10^{-3}$$

The initial concentration of the acid will be reduced by an amount x as it undergoes ionization to form the ions. From this, let's build the concentration table.

	$HC_2H_2O_2Cl$	$\rightleftharpoons$	H^+	$+$	$C_2H_2O_2Cl^-$
Initial concentrations (M)	0.010		0		0
Changes in concentrations (M)	$-x$		$+x$		$+x$
Equilibrium concentrations (M)	$(0.010 - x)$		x		x

Substituting equilibrium concentrations into the equilibrium law gives

$$\frac{(x)(x)}{(0.010 - x)} = 1.4 \times 10^{-3}$$

This time we cannot neglect x. To solve the problem, we first clear fractions by multiplying both sides by $(0.010 - x)$. This gives

$$x^2 = (0.010 - x)\, 1.4 \times 10^{-3}$$

$$x^2 = (1.4 \times 10^{-5}) - (1.4 \times 10^{-3})x$$

The general quadratic equation has the form

$$ax^2 + bx + c = 0$$

The values of x that make this equation true are related to the coefficients, a, b, and c by the quadratic formula:

$$x = \frac{-b \pm \sqrt{b^2 - 4ac}}{2a}$$

Rearranging our equation to follow the general form gives

$$x^2 + (1.4 \times 10^{-3})x - (1.4 \times 10^{-5}) = 0$$

so we make the substitutions $a = 1$, $b = 1.4 \times 10^{-3}$, and $c = -1.4 \times 10^{-5}$. Entering these into the quadratic formula gives

$$x = \frac{-1.4 \times 10^{-3} \pm \sqrt{(1.4 \times 10^{-3})^2 - 4(1)(-1.4 \times 10^{-5})}}{2(1)}$$

$$= \frac{-1.4 \times 10^{-3} \pm \sqrt{2.0 \times 10^{-6} + 5.6 \times 10^{-5}}}{2}$$

$$= \frac{-1.4 \times 10^{-3} \pm \sqrt{5.8 \times 10^{-5}}}{2}$$

$$= \frac{-1.4 \times 10^{-3} \pm 7.6 \times 10^{-3}}{2}$$

Because of the $\pm$ sign we obtain two values for x, but as you saw in Chapter 16, only one of them makes any sense. Here are the two values:

$$x = 3.1 \times 10^{-3}\ M \quad \text{and} \quad x = -4.5 \times 10^{-3}\ M$$

chloroacetic acid

We know that x cannot be negative because that would give negative concentrations for the ions (which is impossible), so we must choose the first value as the correct one. This yields the following equilibrium concentrations:

$$[H^+] = 3.1 \times 10^{-3} \, M$$

$$[C_2H_2O_2Cl^-] = 3.1 \times 10^{-3} \, M$$

$$[HC_2H_2O_2Cl] = 0.010 - 0.0031$$

$$= 0.007 \, M$$

Finally, we calculate the pH of the solution.

$$pH = -\log(3.1 \times 10^{-3})$$

$$= 2.51$$

Notice that, indeed, x is not negligible compared to the initial concentration, so the simplifying approximation would not have been valid.

Is the Answer Reasonable?
As before, a quick check can be performed by substituting the calculated equilibrium concentrations into the mass action expression.

$$\frac{(3.1 \times 10^{-3})^2}{0.007} = 1.4 \times 10^{-3}$$

The value we obtain equals K_a, so the equilibrium concentrations are correct.

Practice Exercise 16: Calculate the pH of a 0.0010 M solution of dimethylamine, $(CH_3)_2NH$, for which $K_b = 9.6 \times 10^{-4}$.

Solving by successive approximations

Method of successive approximations

If you worked your way through the discussion of the use of the quadratic equation, you'll surely be better able to appreciate the *method of successive approximations*. It's not only much faster, particularly with a scientific calculator, but just as accurate. The procedure is outlined in Figure 18.3; follow the diagram as we apply the method below.

If we were to attempt to use the simplifying approximation in the preceding example, we would obtain the following:

$$\frac{x^2}{0.010} = 1.4 \times 10^{-3}$$

for which we obtain the solution $x = 3.7 \times 10^{-3}$. We will call this our *first approximation* to a solution to the equation

$$\frac{(x)(x)}{(0.010 - x)} = 1.4 \times 10^{-3}$$

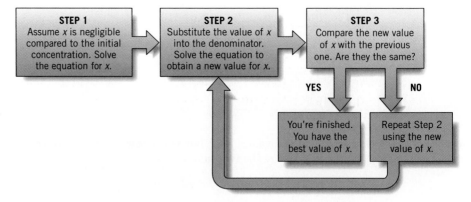

Figure 18.3 *The method of successive approximations.*
Following the steps outlined here leads to a rapid solution of equilibrium problems when the usual simplifications fail.

Let's put $x = 3.7 \times 10^{-3}$ into the term in the denominator, $(0.100 - x)$, and recalculate x. This gives us

$$\frac{x^2}{(0.010 - 0.0037)} = 1.4 \times 10^{-3}$$

or,

$$x^2 = (0.006) \times 1.4 \times 10^{-3}$$

Taking only the positive root,

$$x = 2.9 \times 10^{-3}$$

Notice that this value of x is much closer to the value calculated using the quadratic equation (which gave $x = 3.1 \times 10^{-3}$). Now we'll call $x = 2.9 \times 10^{-3}$ our *second approximation,* and repeat the process. This gives

$$\frac{x^2}{(0.010 - 0.0029)} = 1.4 \times 10^{-3}$$

$$x^2 = (0.007) \times 1.4 \times 10^{-3}$$

$$x = 3.1 \times 10^{-3}$$

This is the identical value obtained using the quadratic equation. Notice also that the *change* in x between the two approximations grew smaller. The second approximation gave a smaller correction than the first. Each succeeding approximation differs from the preceding one by smaller and smaller amounts. We stop the calculation when the difference between two approximations is insignificant. Try this approach by reworking Practice Exercise 16 and solving for $[OH^-]$ in the 0.0010 M $(CH_3)_2NH$ solution using the method of successive approximations.[3]

18.5 Buffers enable the control of pH

We all know that acids cause things made of metals, like cars or battery terminals, to corrode faster. Chefs know that adding just a little lemon juice to milk makes it curdle. If the pH of your blood were to change from what it should be, within the range of 7.35 to 7.42, either to 7.00 or to 8.00, you would die. Thus, a change in pH can promote chemical reactions, sometimes entirely unwanted. Fortunately, there are ways to protect systems against large changes in pH.

Every life-form is extremely sensitive to slight changes in pH.

By a careful choice of solutes, a solution can be prepared that will experience no more than a small change in pH even if small amounts of some strong acid or strong base is added or produced by some reaction. Such mixtures of solutes are called *buffers,* because they do just that—they buffer the system against a change in pH. The solution itself is said to be *buffered* or it is described as a *buffer solution.* There is almost no area of experimental work in chemistry, or in its applications in other fields, where the concept of buffers is not important. Buffers introduce no new concepts to what we have already studied, just a new application.

Buffers contain a weak acid and a weak base

Usually, a **buffer** consists of two solutes, one providing a weak Brønsted acid and the other a weak Brønsted base. Usually, the acid and base represent a conjugate pair. If the acid is molecular, then the conjugate base is *supplied by a soluble salt of*

---

[3]Many modern handheld calculators have "solver" functions that enable the user to solve equilibrium problems such as those discussed here without having to make approximations. You might wish to check the instruction manual for your calculator to see if it has these capabilities. You should also check with your instructor to be sure you are allowed to use this kind of calculator on an exam.

CHEMISTRY IN PRACTICE

Although most buffer systems consist of two separate species that react with H^+ or OH^-, the bicarbonate ion is an example of a single ion that is able to serve both functions. The reactions are

$$HCO_3^-(aq) + H^+(aq) \longrightarrow H_2CO_3(aq)$$
$$HCO_3^-(aq) + OH^-(aq) \longrightarrow H_2O + CO_3^{2-}(aq)$$

Because sodium bicarbonate is nontoxic and because the pH of a solution of the salt is close to 7.0, there are many practical applications of the HCO_3^- buffer. One is in controlling the pH of a swimming pool (Figure 18.4). The chemicals that are added to a swimming pool to retard the growth of bacteria can also affect the pH of the water, making it unpleasant for swimmers. Adding nontoxic $NaHCO_3$ is an inexpensive and effective way to maintain the pool's pH at an acceptable value.

FIGURE 18.4 *A practical application of a buffer.* Baking soda, which is sodium bicarbonate, is sometimes added to swimming pools to control the pH of the water.

the acid. For example, a common buffer system consists of acetic acid plus sodium acetate, with the salt's acetate ion serving as the Brønsted base. In your blood, carbonic acid (H_2CO_3, a weak diprotic acid) and the bicarbonate ion (HCO_3^-, its conjugate base) serve as one of the buffer systems used to maintain a remarkably constant pH in the face of the body's production of organic acids by metabolism. Another common buffer consists of the weakly acidic cation, NH_4^+, supplied by a salt like NH_4Cl, and its conjugate base, NH_3.

One important point about buffers is the distinction between keeping a solution at a particular pH and keeping it neutral—at a pH of 7. Although it is certainly possible to prepare a buffer to work at pH 7, buffers can be made that will work around any pH value throughout the pH scale. One topic we must study, therefore, is how to pick the weak acid and its salt—and their mole ratio—that would make a buffer good for some chosen pH.

Buffer systems do *not* protect a solution against a tide of strong acid or base, just relatively small amounts. So another topic we have to consider is the *capacity* of a buffer, which is the amount of strong acid or base it can absorb without undergoing more than a specified change in pH.

Reactions in a buffer when H^+ or OH^- is added

How a buffer works

For simplicity, we will first confine our discussion to the HA/A^- type of buffer system, like the acetic acid/sodium acetate buffer. Let's first see *how* this system can buffer a solution.

To work, a buffer must be able to neutralize either a strong acid or strong base that is added. This is precisely what the weak base and weak acid components of the buffer do. If we add extra H^+ to the buffer (from a strong acid), the weak conjugate base can react with it as follows:

$$H^+(aq) + A^-(aq) \longrightarrow HA(aq)$$

Not all the added H^+ is neutralized, so the pH is lowered a little. Soon we'll see how much it changes.

Thus, the added acid changes some of the buffer's Brønsted base, A^-, to its conjugate (weak) acid, HA. This reaction prevents a large buildup of H^+ that would otherwise be caused by the addition of the strong acid.

A similar response occurs when a strong base is added to the buffer. The OH^- from the strong base will react with some HA.

$$HA(aq) + OH^-(aq) \longrightarrow A^-(aq) + H_2O$$

Here the added OH^- changes some of the buffer's Brønsted acid, HA, into its conjugate base, A^-. This prevents a buildup of OH^-, which would otherwise cause a large change in the pH. Thus, one member of a buffer team neutralizes H^+ that might get into the solution, and the other member neutralizes OH^-. *Understanding these reactions is the critical link in working many problems having to do with buffers.*

Calculating the pH of a buffer solution

The acetic acid/acetate ion buffer system is acidic

The acetate buffer is a solution of both acetic acid and sodium acetate, which provides the acetate ion in solution. The $HC_2H_3O_2$ in the buffer neutralizes OH^- as follows:

$$HC_2H_3O_2(aq) + OH^-(aq) \longrightarrow C_2H_3O_2^-(aq) + H_2O$$

and its acetate ion neutralizes H^+ as follows:

$$C_2H_3O_2^-(aq) + H^+(aq) \longrightarrow HC_2H_3O_2(aq)$$

The following example illustrates how we can calculate the pH of such a buffer mixture.

A mixture of $HC_2H_3O_2$ and $C_2H_3O_2^-$ is called the *acetate buffer*.

EXAMPLE 18.14

Calculating the pH of a Buffer

To study the effects of a weakly acidic medium on the rate of corrosion of a metal alloy, a student prepared a buffer solution containing both 0.11 M $NaC_2H_3O_2$ and 0.090 M $HC_2H_3O_2$. What is the pH of this solution?

ANALYSIS: The buffer solution contains both the weak acid $HC_2H_3O_2$ and its conjugate base $C_2H_3O_2^-$. Earlier we noted that when both solute species are present we can use *either* K_a or K_b to perform calculations, whichever is handy. In our tables we find $K_a = 1.8 \times 10^{-5}$ for $HC_2H_3O_2$, so the simplest approach is to use the equation for the ionization of the acid. We will also be able to use the simplifying approximations developed earlier; these always work for buffers, so *we will be able to use the initial concentrations as though they are equilibrium values.*

SOLUTION: We begin with the chemical equation and the expression for K_a.

$$HC_2H_3O_2 \rightleftharpoons H^+ + C_2H_3O_2^- \qquad K_a = \frac{[H^+][C_2H_3O_2^-]}{[HC_2H_3O_2]} = 1.8 \times 10^{-5}$$

Let's set up the concentration table this time so we can proceed carefully. We will take the initial concentrations of $HC_2H_3O_2$ and $C_2H_3O_2^-$ to be the values given in the problem. There's no H^+ present from a strong acid, so we set this concentration equal to zero. If the initial concentration of H^+ is zero, its concentration must increase on the way to equilibrium, so under H^+ in the change row we enter $+x$. The other changes follow from that. Here's the completed table.

	$HC_2H_3O_2$	$\rightleftharpoons$	H^+	+	$C_2H_3O_2^-$
Initial concentrations (M)	0.090		0		0.11
Changes in concentrations (M)	$-x$		$+x$		$+x$
Equilibrium concentrations (M)	$(0.090 - x) \approx 0.090$		x		$(0.11 + x) \approx 0.11$

For buffer solutions the quantity x will be very small, so it is safe to make the simplifying approximations. What remains, then, is to substitute the quantities from the last row of the table into the K_a expression.

$$\frac{(x)(0.11 + x)}{(0.090 - x)} \approx \frac{(x)(0.11)}{(0.090)} = 1.8 \times 10^{-5}$$

Solving for x gives us

$$x = \frac{(0.090) \times 1.8 \times 10^{-5}}{(0.11)}$$

$$= 1.5 \times 10^{-5}$$

Notice how small x is compared to the initial concentrations. The simplification was valid.

Because x equals $[H^+]$, we now have $[H^+] = 1.5 \times 10^{-5}$ M. Then we calculate pH:

$$pH = -\log(1.5 \times 10^{-5})$$

$$= 4.82$$

Thus the pH of the buffer is 4.82.

Checking the Answer

We can check the answer in the usual way by substituting our calculated equilibrium values into the mass expression. Let's do it.

$$\frac{[H^+][C_2H_3O_2^-]}{[HC_2H_3O_2]} = \frac{(1.5 \times 10^{-5})(0.11)}{(0.090)} = 1.8 \times 10^{-5}$$

The reaction quotient equals K_a, so the values we've obtained are correct equilibrium concentrations.

PRACTICE EXERCISE 17: Calculate the pH of the buffer solution in the preceding example by using the K_b for $C_2H_3O_2^-$. (Be sure to write the chemical equation for the equilibrium as the reaction of $C_2H_3O_2^-$ with water. Then use the chemical equation as a guide in setting up the equilibrium expression for K_b. If you work the problem correctly, you should obtain the same answer as above.)

Simplifications are permitted in buffer calculations

There are two useful simplifications that we can use in working buffer calculations. The first is the one we made in Example 18.14:

There are some buffer systems for which these simplifications might not work. However, you will not encounter them in this text.

> We will be justified in using the *initial* concentrations of both the weak acid and its conjugate base as though they were equilibrium values.

There is a further simplification that can be made because the mass action expression contains the ratio of the molar concentrations (in units of moles per liter) of the acid and conjugate base. Let's enter these units for the acid and its conjugate base into the mass action expression. For an acid HA,

$$K_a = \frac{[H^+][A^-]}{[HA]} = \frac{[H^+](\text{mol } A^- \, \text{L}^{-1})}{(\text{mol } HA \, \text{L}^{-1})} = \frac{[H^+](\text{mol } A^-)}{(\text{mol } HA)} \qquad (18.8)$$

Notice that the units L^{-1} cancel from the numerator and denominator. This means that for a given acid–base pair, the $[H^+]$ is determined by the *mole* ratio of conjugate base to conjugate acid; we don't *have* to use molar concentrations.

> *For buffer solutions **only**,* we can use either molar concentrations or moles in the K_a (or K_b) expression to express the amounts of the members of the conjugate acid–base pair (but we must use the same units for each member of the pair).

A further consequence of the relationship derived above is that the pH of a buffer should not change if the buffer is diluted. Dilution changes the volume of a solution but it does not change the number of moles of the solutes, so their mole *ratio* remains constant and so does $[H^+]$.

The ammonia/ammonium ion buffer is basic

A solution of ammonium chloride in aqueous ammonia also provides a weak acid, NH_4^+, and its conjugate base, NH_3, and so it is a buffer. It is able to neutralize OH^- by the following reaction:

$$NH_4^+(aq) + OH^-(aq) \longrightarrow NH_3(aq) + H_2O$$

The conjugate base can neutralize H^+.

$$NH_3(aq) + H^+(aq) \longrightarrow NH_4^+(aq)$$

To study the influence of an alkaline medium on the rate of a reaction, a student prepared a buffer solution by dissolving 0.12 mol of NH_3 and 0.095 mol of NH_4Cl in 250 mL of water. What is the pH of the buffer?

ANALYSIS: The pH of the buffer is determined by the mole ratio of the members of the acid–base pair, so to calculate the pH we will be able to use the moles of NH_3 and NH_4^+ directly in the mass action expression. To set up the equilibrium law we can use either the K_a for NH_4^+ or the K_b for NH_3. Since K_b is tabulated, we will use it and write the equation for the ionization of the base.

SOLUTION: We begin with the chemical equation and the K_b expression.

$$NH_3 + H_2O \rightleftharpoons NH_4^+ + OH^- \qquad K_b = \frac{[NH_4^+][OH^-]}{[NH_3]} = 1.8 \times 10^{-5}$$

The solution contains 0.12 mol of NH_3 and 0.095 mol of NH_4^+ from the complete dissociation of the salt NH_4Cl. We can enter these quantities into the mass action expression and solve for $[OH^-]$.

$$1.8 \times 10^{-5} = \frac{(0.095)[OH^-]}{0.12}$$

solving for $[OH^-]$ gives

$$[OH^-] = 2.3 \times 10^{-5}$$

To calculate the pH, we obtain pOH and subtract it from 14.00.

$$pOH = -\log (2.3 \times 10^{-5})$$
$$= 4.64$$
$$pH = 14.00 - 4.64$$
$$= 9.36$$

Checking the Answer
By now you know that you can check the answer by substituting equilibrium concentrations into the mass action expression. Try it.

PRACTICE EXERCISE 18: Determine the pH of the buffer in the preceding example using the K_a for NH_4^+, which you can calculate from K_b for NH_3. (If you work the problem correctly, you should obtain the same answer as above.)

EXAMPLE 18.15
Calculating the pH of an Ammonia/ Ammonium Ion Buffer

The solution also contains 0.095 mol of Cl^-, of course, but this ion is not involved in the equilibrium.

Preparation of a buffer with a given pH

In experimental work, a chemist or biologist usually first decides the pH at which a particular system should be buffered and then chooses the buffer components that will best deliver this pH. This choice requires careful attention to what most affects the pH of a buffer.

Biologists also must be concerned with possible toxic side effects produced by the components of a buffer.

The factors that govern the pH of a buffer solution

We can more clearly see the two factors that dominate the pH of a buffered solution by rearranging Equation 18.8 to solve for $[H^+]$:

$$[H^+] = K_a \frac{[HA]}{[A^-]} \qquad (18.9)$$

or

$$[H^+] = K_a \frac{\text{mol } HA}{\text{mol } A^-} \qquad (18.10)$$

The first factor to affect $[H^+]$ (and therefore pH) is the K_a of the weak acid. The second is the *ratio* of the molarities (in Equation 18.9) or the ratio of moles (in

Equation 18.10). The quantities we use in these two equations, of course, are the "initial" values—either concentrations or moles. In other words, we assume their values don't change significantly as a result of ionization of the weak acid. To emphasize this, let's rewrite them as follows:

$$[H^+] = K_a \frac{[HA]_{initial}}{[A^-]_{initial}} \qquad (18.11)$$

$$[H^+] = K_a \frac{(mol\ HA)_{initial}}{(mol\ A^-)_{initial}} \qquad (18.12)$$

Notice particularly what happens if we prepare a buffer so as to make the concentrations of the two buffer components identical. Then the ratio $[HA]_{initial}/[A^-]_{initial}$ in Equation 18.11 equals 1, which means that $[H^+] = K_a$ and therefore $pH = pK_a$.

If you take a biology course, you're likely to run into a logarithmic form of Equation 18.11 called the **Henderson–Hasselbalch equation.** This is obtained by taking the negative logarithm of both sides of Equation 18.11 and rearranging the term involving the concentrations.

$$pH = pK_a + \log \frac{[A^-]_{initial}}{[HA]_{initial}} \qquad (18.13)$$

In most buffers used in the life sciences, the anion A^- comes from a salt in which the cation has a charge of $1+$, such as NaA, and the acid is monoprotic. With these as conditions, the equation is sometimes written

$$pH = pK_a + \log \frac{[salt]}{[acid]} \qquad (18.14)$$

For practice, you may wish to apply this equation to buffer problems at the end of the chapter.

Selecting the weak acid for the preparation of a buffer solution

When $[H^+] = K_a$, then $-\log [H^+] = -\log K_a$, so $pH = pK_a$.

Usually buffers are made so that the ratio $[HA]_{initial}/[A^-]_{initial}$ is not greatly different from 1. Consequently, *what mostly determines where, on the pH scale, a buffer can work best is the* pK_a *of the weak acid.* Thus, to prepare a specific buffer for use at a prechosen pH, we first select a weak acid whose pK_a is near the pH we desire. Almost never can an acid be found, however, whose pK_a *exactly* equals the pH we want. So we try for an acid whose pK_a is *close* to this pH. Then, by experimentally adjusting the ratio $[HA]_{initial}/[A^-]_{initial}$, we can make a final adjustment to get the desired pH.

The desirable range for the ratio $[HA]_{initial}/[A^-]_{initial}$ is from $\frac{10}{1}$ to $\frac{1}{10}$. Outside this range we can run into problems with the solubilities of the solutes, or we can have a buffer of low capacity. (The question of buffer *capacity* will be studied soon.) According to Equation 18.11, keeping the ratio $[HA]_{initial}/[A^-]_{initial}$ within these limits means that the $[H^+]$ will range from $10 \times K_a$ to $0.1 \times K_a$. Taking the negative logarithm of both sides, this translates into the desirable range of pH values for a buffer normally being

$$pH = pK_a \pm 1 \qquad (18.15)$$

EXAMPLE 18.16

Preparing a Buffer Solution to Have a Predetermined pH

A solution buffered at a pH of 5.00 is needed in an experiment. Can we use acetic acid and sodium acetate to make it? If so, how many moles of $NaC_2H_3O_2$ must be added to 1.0 L of a solution that contains 1.0 mol $HC_2H_3O_2$ to prepare the buffer?

ANALYSIS: There are two parts to this problem. To answer the first, we must check the pK_a of acetic acid to see if it is in the desired range of $pH = pK_a \pm 1$. If it is, we can convert pH to $[H^+]$ and then use Equation 18.12 to calculate the necessary mole ratio. Once we have this, we can proceed to calculate the number of moles of $C_2H_3O_2^-$ needed and then the number of moles of $NaC_2H_3O_2$.

SOLUTION: Because we want the pH to be 5.00, the pK_a of the selected acid should be 5.00 ± 1, meaning the pK_a should be between 4.00 and 6.00. Because $K_a = 1.8 \times 10^{-5}$ for acetic acid, $pK_a = -\log(1.8 \times 10^{-5}) = 4.74$. So the pK_a of acetic acid falls in the desired range and acetic acid can be used together with the acetate ion to make the buffer.

Next, we use Equation 18.12 to find the mole ratio of solutes.

$$[H^+] = K_a \times \frac{(\text{mol } HC_2H_3O_2)_{\text{initial}}}{(\text{mol } C_2H_3O_2^-)_{\text{initial}}}$$

First let's solve for the mole ratio.

$$\frac{(\text{mol } HC_2H_3O_2)_{\text{initial}}}{(\text{mol } C_2H_3O_2^-)_{\text{initial}}} = \frac{[H^+]}{K_a}$$

The desired pH = 5.00, so $[H^+] = 1.0 \times 10^{-5}$; also $K_a = 1.8 \times 10^{-5}$. Substituting gives

$$\frac{(\text{mol } HC_2H_3O_2)_{\text{initial}}}{(\text{mol } C_2H_3O_2^-)_{\text{initial}}} = \frac{1.0 \times 10^{-5}}{1.8 \times 10^{-5}} = 0.56$$

This is the *mole* ratio of the buffer components we want. The solution we are preparing contains 1.0 mol $HC_2H_3O_2$, so the number of moles of acetate ion required is

$$\text{moles } C_2H_3O_2^- = \frac{1.0 \text{ mol } HC_2H_3O_2}{0.56}$$

$$= 1.78 \text{ mol } C_2H_3O_2^-$$

For each mole of $NaC_2H_3O_2$ there is 1 mol of $C_2H_3O_2^-$, so to prepare the solution we need 1.78 mol $NaC_2H_3O_2$.

Is the Answer Reasonable?

A 1-to-1 mole ratio of $C_2H_3O_2^-$ to $HC_2H_3O_2$ would give pH = pK_a = 4.74. The desired pH of 5.00 is slightly more basic than 4.74, so the amount of conjugate base needed should be larger than the amount of conjugate acid. Our answer of 1.78 mol of $NaC_2H_3O_2$ appears to be reasonable.

PRACTICE EXERCISE 19: A chemist needed an aqueous buffer with a pH of 3.90. Would formic acid and its salt, sodium formate, make a good pair for this purpose? If so, what mole ratio of the acid, $HCHO_2$, to the anion of this salt, CHO_2^-, is needed? How many grams of $NaCHO_2$ would have to be added to a solution that contains 0.10 mol $HCHO_2$?

Buffer capacity describes the buffer's ability to absorb strong acid or base

If enough strong acid is added to react with all of the base component of a buffer, the mixture will no longer be able to neutralize more strong acid if it enters the solution. Similarly, if enough strong base is added to neutralize all of the acid component of the buffer, the mixture will no longer be able to react with additional strong base. In either case, the buffering ability of the mixture will have been exhausted. Thus, there is a limit to a buffer's *capacity*—how much strong acid or strong base the buffer is able to absorb before its buffering ability is essentially destroyed.

A buffer's *capacity* is determined by the magnitudes of the actual molarities of its components. Therefore, we must decide before making the buffer solution what outer limits to the change in pH can be tolerated—0.1 unit, 0.2 unit, or something else. Then, consistent with the necessary *ratio* of base to acid, we calculate the actual number of moles of each that we will dissolve in the solution. Let's do a calculation to show how effective a buffer is at resisting pH changes produced by a relatively small amount of extra acid (or base).

A buffer of low capacity might be needed if its ions and molecules could, if too concentrated, interfere with another use of the solution.

EXAMPLE 18.17

The Capacity of a Buffer to Resist Changes in pH

A researcher had available 1.00 L of an acetic acid/sodium acetate buffer that contained 1.00 M $HC_2H_3O_2$ and 1.00 M $NaC_2H_3O_2$. This entire solution was intended for an experiment in which 0.15 mol of H^+ was expected to be generated by a reaction without changing the volume of the solution. By how much will the pH of the buffer change by the addition of this much H^+? Could the buffer be used in this experiment if the maximum allowable pH change could be no larger than 0.2 unit?

ANALYSIS: To answer the question, we need to know the initial pH of the buffer and the pH after the addition of the strong acid. We use the given concentrations of the $HC_2H_3O_2$ and $C_2H_3O_2^-$ to obtain the initial pH. To find the pH after the addition of H^+, we need to determine how the concentrations of the $HC_2H_3O_2$ and $C_2H_3O_2^-$ change when the system absorbs the strong acid. We then use the new concentrations to calculate the new pH of the buffer. The difference between the new and original pH values tells us how much the pH changes.

SOLUTION: First, we calculate the pH of the buffer before addition of the H_3O^+. For acetic acid, we have seen that

$$K_a = \frac{[H^+][C_2H_3O_2^-]}{[HC_2H_3O_2]} = 1.8 \times 10^{-5}$$

The given data tell us that initially,

$$[HC_2H_3O_2]_{initial} = 1.00 \ M$$

$$[C_2H_3O_2^-]_{initial} = 1.00 \ M$$

We will assume these initial values are effectively the same as the equilibrium concentrations. Substituting them into the mass action expression lets us calculate the initial pH of the buffer.

$$\frac{[H^+](1.00)}{1.00} = 1.8 \times 10^{-5}$$

$$[H^+] = 1.8 \times 10^{-5}$$

$$pH = 4.74$$

Next, we consider the reaction that takes place as H^+ enters the solution. Recall that the conjugate base of the buffer neutralizes H^+.

$$H^+ + C_2H_3O_2^- \longrightarrow HC_2H_3O_2$$

The equilibrium constant for this reaction is 5.6×10^4, which is the reciprocal of K_a for acetic acid. Its large value tells us the reaction proceeds very far toward completion, so nearly all the added H^+ reacts with $C_2H_3O_2^-$.

The reaction is nearly complete, in the sense that virtually all the added H^+ is consumed. As an approximation, we assume complete reaction, so for each mole of H^+ added, a mole of $C_2H_3O_2^-$ is changed to a mole of $HC_2H_3O_2$. Since 0.15 mol L^{-1} of H^+ is added, the concentration of $HC_2H_3O_2$ *increases* by this amount and the concentration of $C_2H_3O_2^-$ *decreases* by this amount. Applying these changes gives the final concentrations.

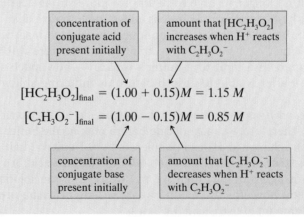

We now substitute these new values into the mass action expression to calculate the final $[H^+]$ in the solution.

$$\frac{[H^+](0.85)}{(1.15)} = 1.8 \times 10^{-5}$$

$$[H^+] = 2.4 \times 10^{-5} \text{ mol L}^{-1}$$

From this we calculate the final pH,

$$pH = 4.62$$

By adding 0.15 mol of strong acid to a liter of the buffer, its pH has gone from 4.74 to 4.62, a change of only 0.12 pH unit (which is smaller than the limit of ± 0.2 pH unit). The buffer solution, in other words, was able to absorb 0.15 mol of H^+ without exceeding the limit, even though the added H^+ consumed 15% of the available base, $C_2H_3O_2^-$.

Is the Answer Reasonable?
There's no simple way to estimate how *much* the pH will change when H^+ or OH^- is added to the buffer; however, we can anticipate the *direction* of the change. Adding H^+ to a buffer will lower the pH somewhat and adding OH^- will raise it. In this example, the buffer is absorbing H^+, so its pH should drop, and our calculations agree. If our calculated pH had been higher than the original, we might expect to find errors in the calculation of the final values of $[HC_2H_3O_2]$ and $[C_2H_3O_2^-]$.

PRACTICE EXERCISE 20: Suppose that the buffer of the preceding example is used in an experiment in which 0.11 mol of OH^- ion will be generated (with no volume change). Can the buffer handle this without having the pH change by more than 0.2 unit? Calculate the new pH.

Example 18.17 and Practice Exercise 20 demonstrate how effective a buffer can be at absorbing H^+ and OH^- with only minimal changes in pH. In many experiments it's necessary to limit the pH change to ranges smaller than 0.2 pH unit (e.g., to ± 0.1 pH unit or less). Generally, the mole quantities of buffer required to do this must be upwards of ten times greater than the expected influx of strong acid or base. Maintaining the pH within such narrow ranges is particularly important in work on biological systems where a change of even 0.1 pH unit in some internal fluid is usually lethal to a living system.

The resistance to pH changes defines the effectiveness of a buffer system

Although buffers do not have unlimited *capacities,* they still operate remarkably well in preventing wide swings in pH. The question of *capacity* is related to the question of buffer *effectiveness*. How well does a buffer system work? How well does it actually hold the pH? In Example 18.17, we added 0.15 mol of strong acid to 1.00 L of a buffer and the pH changed by 0.12 pH unit. Just think of how much the pH would change had we added this much acid to 1.00 L of pure water. The solution would now have

$$[H^+] = 0.15 \text{ mol L}^{-1}$$

and a pH of

$$pH = -\log (0.15)$$

$$= 0.82$$

With the buffer, the pH changes from 4.74 to 4.62; without the buffer the pH (starting with pure water) changes from 7.00 to 0.82. This is an enormous difference. Buffers, unless overwhelmed by the addition of excessive amounts of strong acid or base, truly prevent wide swings of pH.

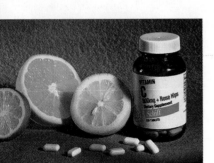

Vitamin C. Many fruits and vegetables contain ascorbic acid (vitamin C), a weak diprotic acid with the formula $H_2C_6H_6O_6$.

18.6 ▶ Polyprotic acids ionize in two or more steps

Until now our discussion of weak acids has focused entirely on equilibria involving monoprotic acids. There are, of course, many acids capable of supplying more than one H^+ per molecule. Recall that these are called *polyprotic acids.* Examples include sulfuric acid, H_2SO_4, carbonic acid, H_2CO_3, and phosphoric acid, H_3PO_4. These acids undergo ionization in a series of steps, each of which releases one proton. For weak polyprotic acids, such as H_2CO_3 and H_3PO_4, each step is an equilibrium. Even sulfuric acid, which we consider a strong acid, is not completely ionized. Loss of the first proton to yield the HSO_4^- ion is complete, but the loss of the second proton is incomplete and involves an equilibrium. In this section, we will focus our attention on weak polyprotic acids.

Let's begin with the weak diprotic acid, H_2CO_3. In water, the acid ionizes in two steps, each of which is an equilibrium that transfers an H^+ ion to a water molecule.

$$H_2CO_3 + H_2O \rightleftharpoons H_3O^+ + HCO_3^-$$
$$HCO_3^- + H_2O \rightleftharpoons H_3O^+ + CO_3^{2-}$$

As usual, we can use H^+ in place of H_3O^+ and simplify these equations to give

$$H_2CO_3 \rightleftharpoons H^+ + HCO_3^-$$
$$HCO_3^- \rightleftharpoons H^+ + CO_3^{2-}$$

Each step has its own ionization constant, K_a, which we identify as K_{a_1} for the first step and K_{a_2} for the second. For carbonic acid,

$$K_{a_1} = \frac{[H^+][HCO_3^-]}{[H_2CO_3]} = 4.3 \times 10^{-7}$$

$$K_{a_2} = \frac{[H^+][CO_3^{2-}]}{[HCO_3^-]} = 4.7 \times 10^{-11}$$

Notice that each ionization makes a contribution to the total molar concentration of H^+, and one of our goals here is to relate the K_a values and the concentration of the acid to $[H^+]$. At first glance, this seems to be a formidable task, but certain simplifications are justified that make the problem relatively simple to solve.

TABLE 18.3	ACID IONIZATION CONSTANTS FOR POLYPROTIC ACIDS				
		Acid Ionization Constant for Successive Ionizations (25 °C)			
Name	Formula	1st	2nd	3rd	
Carbonic acid	H_2CO_3	4.3×10^{-7}	4.7×10^{-11}		
Hydrogen sulfide	$H_2S(aq)$	9.5×10^{-8}	1×10^{-19}		
Phosphoric acid	H_3PO_4	7.1×10^{-3}	6.3×10^{-8}	4.5×10^{-13}	
Arsenic acid	H_3AsO_4	5.6×10^{-3}	1.7×10^{-7}	4.0×10^{-12}	
Sulfuric acid	H_2SO_4	Large	1.0×10^{-2}		
Selenic acid	H_2SeO_4	Large	1.2×10^{-2}		
Telluric acid	H_6TeO_6	2×10^{-8}	1×10^{-11}		
Sulfurous acid	H_2SO_3	1.2×10^{-2}	6.6×10^{-8}		
Selenous acid	H_2SeO_3	3.5×10^{-3}	5×10^{-8}		
Tellurous acid	H_2TeO_3	3×10^{-3}	2×10^{-8}		
Ascorbic acid (vitamin C)	$H_2C_6H_6O_6$	6.7×10^{-5}	2.7×10^{-12}		
Oxalic acid	$H_2C_2O_4$	6.5×10^{-2}	6.1×10^{-5}		
Citric acid (18 °C)	$H_3C_6H_5O_7$	7.4×10^{-4}	1.7×10^{-5}	4.0×10^{-7}	

Simplifications in calculations involving polyprotic acids

The principal factor that simplifies calculations involving many polyprotic acids is the large differences between successive ionization constants. Notice that for H_2CO_3, K_{a_1} is much larger than K_{a_2} (they differ by a factor of nearly 10,000). Similar differences between K_{a_1} and K_{a_2} are observed for many diprotic acids. One reason is that an H^+ is lost much more easily from the neutral H_2A molecule than from the HA^- ion. The stronger attraction of the opposite charges inhibits the second ionization. Typically, K_{a_1} is greater than K_{a_2} by a factor of between 10^4 and 10^5, as the data in Table 18.3 show. For a triprotic acid, such as phosphoric acid, H_3PO_4, the second acid ionization constant is similarly greater than the third.

Because K_{a_1} is so much larger than K_{a_2}, virtually all the H^+ in a solution of the acid comes from the first step in the ionization. In other words,

$$[H^+]_{total} = [H^+]_{first\ step} + [H^+]_{second\ step}$$

$$[H^+]_{first\ step} \gg [H^+]_{second\ step}$$

Therefore, we make the approximation that

$$[H^+]_{total} \approx [H^+]_{first\ step}$$

This means that as far as calculating the H^+ concentration is concerned, we can treat the acid as though it were a monoprotic acid and ignore the second step in the ionization. Example 18.18 illustrates how this is applied.

Additional K_a values of polyprotic acids are located in Appendix C.

Calculate the concentrations of all the species produced in the ionization of 0.040 M H_2CO_3 as well as the pH of the solution.

ANALYSIS: As in any equilibrium problem, we begin with the chemical equations and the K_a expressions.

$$H_2CO_3 \rightleftharpoons H^+ + HCO_3^- \quad K_{a_1} = \frac{[H^+][HCO_3^-]}{[H_2CO_3]} = 4.3 \times 10^{-7}$$

$$HCO_3^- \rightleftharpoons H^+ + CO_3^{2-} \quad K_{a_2} = \frac{[H^+][CO_3^{2-}]}{[HCO_3^-]} = 4.7 \times 10^{-11}$$

EXAMPLE 18.18

Calculating the Concentrations of Solute Species in a Solution of a Polyprotic Acid

We will want to calculate the concentrations of H^+, HCO_3^-, and CO_3^{2-} and the concentration of H_2CO_3 that remains at equilibrium. As noted in the preceding discussion, the problem is simplified considerably by assuming that the second reaction occurs to a negligible extent compared to the first. This assumption (which we will justify later in the problem) allows us to calculate the H^+ and HCO_3^- concentrations as though the solute were a monoprotic acid. This type of problem is one we've worked on in Section 18.2.

To obtain $[CO_3^{2-}]$, we will have to use the second step in the ionization. Once again, the large difference between K_{a_1} and K_{a_2} permits some simplifications. The first involves $[H^+]$, which we have already examined.

$$[H^+]_{equilib} \approx [H^+]_{first\ step}$$

The second involves $[HCO_3^-]$, for which we can write

$$[HCO_3^-]_{equilib} = [HCO_3^-]_{formed\ in\ first\ step} - [HCO_3^-]_{lost\ in\ second\ step}$$

However, because $K_{a_1} \gg K_{a_2}$, we expect that

$$[HCO_3^-]_{formed\ in\ first\ step} \gg [HCO_3^-]_{lost\ in\ second\ step}$$

Therefore, we can make the approximation that

$$[HCO_3^-]_{equilib} \approx [HCO_3^-]_{formed\ in\ first\ step}$$

SOLUTION: Treating H_2CO_3 as though it were a monoprotic acid yields a problem similar to many others we worked earlier in this chapter. We know some of the H_2CO_3 ionizes; let's call this amount x. If x moles per liter of H_2CO_3 ionize, then x moles per liter of H^+ and HCO_3^- are formed, and the concentration of H_2CO_3 is reduced by x.

	$H_2CO_3 \rightleftharpoons H^+ + HCO_3^-$		
Initial concentrations (M)	0.040	0	0
Changes in concentrations (M)	$-x$	$+x$	$+x$
Equilibrium concentrations (M)	$0.040 - x$	x	x

$$[H^+] = [HCO_3^-] = x$$

$$[H_2CO_3] = 0.040 - x$$

Substituting into the expression for K_{a_1} gives

$$\frac{(x)(x)}{(0.040 - x)} = 4.3 \times 10^{-7}$$

Because K_{a_1} is so small, we expect that $(0.040 - x) \approx 0.040$. (This is the usual approximation we made in solving problems of this kind.) The algebra then reduces to

> Notice that $0.040 > 400 \times K_{a_1}$, so it is safe to use the initial concentration, $0.040\ M$, as though it were the equilibrium concentration.

$$\frac{x^2}{0.040} = 4.3 \times 10^{-7}$$

Solving for x gives $x = 1.3 \times 10^{-4}$, so $[H^+] = [HCO_3^-] = 1.3 \times 10^{-4}\ M$. From this we can now calculate the pH.

$$\text{pH} = -\log(1.3 \times 10^{-4}) = 3.89$$

Now let's calculate the concentration of CO_3^{2-}. We obtain this from the expression for K_{a_2}.

$$K_{a_2} = \frac{[H^+][CO_3^{2-}]}{[HCO_3^-]} = 4.7 \times 10^{-11}$$

> Keep in mind that for a given solution, the values of H^+ used in the expressions for K_{a_1} and K_{a_2} are identical. At equilibrium there is only *one* equilibrium H^+ concentration.

In our analysis, we concluded that we can use the approximations

$$[H^+]_{\text{equilib}} \approx [H^+]_{\text{first step}}$$

$$[HCO_3^-]_{\text{equilib}} \approx [HCO_3^-]_{\text{first step}}$$

Substituting the values obtained above gives us

$$\frac{(1.3 \times 10^{-4})[CO_3^{2-}]}{1.3 \times 10^{-4}} = 4.7 \times 10^{-11}$$

$$[CO_3^{2-}] = 4.7 \times 10^{-11} = K_{a_2}$$

Let's summarize the results. At equilibrium, we have

$$[H_2CO_3] = 0.040\ M$$

$$[H^+] = [HCO_3^-] = 1.3 \times 10^{-4}\ M \ (\text{and pH} = 3.89)$$

$$[CO_3^{2-}] = 4.7 \times 10^{-11}\ M$$

Checking the Simplifying Assumptions
In the second step of the ionization, H^+ and CO_3^{2-} are formed in equal amounts from HCO_3^-. This means that the contribution to $[H^+]$ *by the second step* equals $[CO_3^{2-}]$. Therefore, $[H^+]_{\text{second step}} = 4.7 \times 10^{-11}\ M$. Comparing this value to the $[H^+]$ computed using K_{a_1}, we see that the second step of the ionization contributes a negligible amount to the total hydrogen ion concentration in the solution.

$$[H^+]_{\text{equilib}} = [H^+]_{\text{first step}} + [H^+]_{\text{second step}}$$

$$[H^+]_{equilib} = (1.3 \times 10^{-4} \, M) + (5.6 \times 10^{-11} \, M)$$

$$= 1.3 \times 10^{-4} \, M \text{ (correctly rounded)}$$

Thus, we were safe in treating H_2CO_3 as though it were a monoprotic acid when we computed the H^+ concentration.

Our second approximation,

$$[HCO_3^-]_{equilib} \approx [HCO_3^-]_{formed \, in \, first \, step}$$

is also valid. The concentration of HCO_3^- formed in the first step equals $1.3 \times 10^{-4} \, M$. This value is decreased by $4.7 \times 10^{-11} \, mol \, L^{-1}$ when the H^+ and CO_3^{2-} form in the second step. Thus,

$$[HCO_3^-]_{equilib} = [HCO_3^-]_{\substack{formed \, in \\ first \, step}} - [HCO_3^-]_{\substack{lost \, in \\ second \, step}}$$

$$= (1.3 \times 10^{-4} \, M) - (4.7 \times 10^{-11})$$

$$= 1.3 \times 10^{-4} \, M \text{ (correctly rounded)}$$

Because 4.7×10^{-11} is negligible compared to 1.3×10^{-4}, the equilibrium concentration of HCO_3^- equals the concentration computed using K_{a_1}.

Checking the Accuracy of the Answers

By now you know that you can always perform a quick check of the answers by substituting them into the appropriate mass action expressions:

$$\frac{(1.3 \times 10^{-4})(1.3 \times 10^{-4})}{0.040} = 4.2 \times 10^{-7} \, (\text{which is very close to } K_{a_1})$$

$$\frac{(1.3 \times 10^{-4})(4.7 \times 10^{-11})}{(1.3 \times 10^{-4})} = 4.7 \times 10^{-11} \, (\text{which equals } K_{a_2})$$

PRACTICE EXERCISE 21: Ascorbic acid (vitamin C) is a diprotic acid, $H_2C_6H_6O_6$. See Table 18.3. Calculate $[H^+]$, pH, and $[C_6H_6O_6^{2-}]$ in a 0.10 M solution of ascorbic acid.

One of the most interesting observations we can derive from the preceding example is that *in a solution that contains a polyprotic acid **as the only solute**, the concentration of the ion formed in the second step of the ionization equals K_{a_2}.*

18.7 ▶ Salts of polyprotic acids give basic solutions

You learned earlier that the pH of a salt solution is controlled by whether the salt's cation, anion, or both are able to react with water. For simplicity, the salts of polyprotic acids that we will discuss here will be limited to those containing *nonacidic* cations, such as sodium or potassium ion. In other words, we will study salts in which the anion alone is basic and thereby affects the pH of a solution.

A typical example of a salt of a polyprotic acid is sodium carbonate, Na_2CO_3. It is a salt of H_2CO_3, and the salt's carbonate ion is a Brønsted base that is responsible for *two* equilibria that furnish OH^- ion and thereby affect the pH of the solution. These equilibria and their corresponding expressions for K_b are as follows:

$$CO_3^{2-}(aq) + H_2O \rightleftharpoons HCO_3^-(aq) + OH^-(aq) \quad (18.16)$$

$$K_{b_1} = \frac{[HCO_3^-][OH^-]}{[CO_3^{2-}]}$$

$$HCO_3^-(aq) + H_2O \rightleftharpoons H_2CO_3(aq) + OH^-(aq) \quad (18.17)$$

$$K_{b_2} = \frac{[H_2CO_3][OH^-]}{[HCO_3^-]}$$

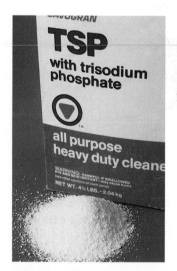

The salt Na_3PO_4 is sold under the name trisodium phosphate, or TSP. Solutions of the salt in water are quite basic and are used as an aid in cleaning grime from painted surfaces. Some states restrict the sale of TSP because the phosphate ion it contains presents a pollution problem by promoting the growth of algae in lakes.

The calculations required to determine the concentrations of the species involved in these equilibria are very much like those involving weak diprotic acids. The difference is that the chemical reactions here are those of bases instead of acids. In fact, *the simplifying assumptions in our calculations will be almost identical to those for weak polyprotic acids.*

First, let's compare the K_b values for the successive equilibria. To obtain these constants, we will use the relationship that $K_a \times K_b = K_w$ along with the values of K_{a_1} and K_{a_2} for H_2CO_3.

In general, K_b values for anions of polyprotic acids are not tabulated. When needed, they are calculated from the appropriate K_a values for the acids.

Notice that in the equilibrium in which CO_3^{2-} is the base (Equation 18.16), the conjugate acid is HCO_3^-. Therefore, to calculate K_b for CO_3^{2-} (which we called K_{b_1}), we must use the K_a for HCO_3^-, which corresponds to K_{a_2} for carbonic acid. Thus,

$$K_{b_1} = \frac{K_w}{K_{a_2}} = \frac{1.0 \times 10^{-14}}{4.7 \times 10^{-11}} = 2.1 \times 10^{-4}$$

Similarly, in the equilibrium in which HCO_3^- is the base (Equation 18.17), the conjugate acid is H_2CO_3. Therefore, to calculate K_b for HCO_3^- (which we called K_{b_2}), we must use the K_a for H_2CO_3, which is K_{a_1} for carbonic acid.

Remember: K_{b_1} comes from K_{a_2}, and K_{b_2} comes from K_{a_1}.

$$K_{b_2} = \frac{K_w}{K_{a_1}} = \frac{1.0 \times 10^{-14}}{4.3 \times 10^{-7}} = 2.3 \times 10^{-8}$$

Now we can compare the two K_b values. For CO_3^{2-}, $K_{b_1} = 2.1 \times 10^{-4}$, and for the (much) weaker base, HCO_3^-, K_{b_2} is 2.3×10^{-8}. Thus, CO_3^{2-} has a K_b nearly 10,000 times that of HCO_3^-. This means that the reaction of CO_3^{2-} with water (Equation 18.16) generates far more OH^- than the reaction of HCO_3^- (Equation 18.17). The contribution of the latter to the total pool of OH^- is relatively so small that we can safely ignore it. The simplification here is very much the same as in our treatment of the ionization of weak polyprotic acids.

$$[OH^-]_{\text{total}} = [OH^-]_{\underset{\text{first step}}{\text{formed in}}} + [OH^-]_{\underset{\text{second step}}{\text{formed in}}}$$

$$[OH^-]_{\underset{\text{first step}}{\text{formed in}}} \gg [OH^-]_{\underset{\text{second step}}{\text{formed in}}}$$

$$[OH^-]_{\text{total}} \approx [OH^-]_{\underset{\text{first step}}{\text{formed in}}}$$

Thus, if we wish to calculate the pH of a solution of a basic anion of a polyprotic acid, *we may work exclusively with K_{b_1} and ignore any further reactions.* Let's study an example of how this works.

EXAMPLE 18.19

Calculating the pH of a Solution of a Salt of a Weak Diprotic Acid

Here, $400 \times K_{b_1} = 0.084$. Because $0.15 > 0.084$, we are able to use the initial concentration of CO_3^{2-} as the equilibrium concentration.

What is the pH of a 0.15 M solution of Na_2CO_3?

ANALYSIS: We can ignore the reaction of HCO_3^- with water, so the only relevant equilibrium is

$$CO_3^{2-} + H_2O \rightleftharpoons HCO_3^- + OH^- \qquad K_{b_1} = \frac{[HCO_3^-][OH^-]}{[CO_3^{2-}]} = 2.1 \times 10^{-4}$$

(The value of K_{b_1} was calculated in the discussion preceding this example.) Also, the initial concentration of CO_3^{2-} (0.15 M) is larger than $400 \times K_{b_1}$, so we can safely use 0.15 M as the equilibrium concentration of CO_3^{2-}.

SOLUTION: Some of the CO_3^{2-} will react; we'll let this be x mol L^{-1}. The concentration table, then, is as follows.

	CO_3^{2-}	+	H_2O	$\rightleftharpoons$	HCO_3^-	+	OH^-
Initial concentrations (M)	0.15				0		0
Changes in concentrations caused by the ionization (M)	$-x$				$+x$		$+x$
Equilibrium concentrations (M)	$(0.15 - x) \approx 0.15$				x		x

Now we're ready to insert the values in the last row of the table into the equilibrium expression for the carbonate ion.

$$\frac{[HCO_3^-][OH^-]}{[CO_3^{2-}]} = \frac{(x)(x)}{0.15 - x} \approx \frac{(x)(x)}{0.15} = 2.1 \times 10^{-4}$$

$$x = 3.2 \times 10^{-3}$$

Notice that the value of x really is sufficiently smaller than 0.15 to justify our using 0.15 as a good approximation of $(0.15 - x)$. Because $x = 5.2 \times 10^{-3}$, we have

$$[OH^-] = 3.2 \times 10^{-3} \text{ mol L}^{-1}$$

$$pOH = -\log(3.2 \times 10^{-3}) = 2.49$$

Finally, we calculate the pH by subtracting pOH from 14.00.

$$pH = 14.00 - 2.49$$

$$= 11.51$$

Thus, the pH of 0.15 M Na_2CO_3 is calculated to be 11.51. Once again we see that an aqueous solution of a salt composed of a basic anion and a neutral cation is basic.

Is the Answer Reasonable?

The pH we obtained corresponds to a basic solution, which is what we expect when the anion of the salt reacts as a weak base. The answer seems reasonable.

PRACTICE EXERCISE 22: What is the pH of a 0.20 M solution of Na_2SO_3 at 25 °C? For the diprotic acid, H_2SO_3, $K_{a_1} = 1.2 \times 10^{-2}$ and $K_{a_2} = 6.6 \times 10^{-8}$.

PRACTICE EXERCISE 23: Reasoning by analogy from what you learned about solutions of weak polyprotic acids, what is the molar concentration of H_2SO_3 in a 0.010 M solution of Na_2SO_3?

Some detergents that use sodium carbonate as their caustic ingredient. Detergents are best able to dissolve grease and grime if their solutions are basic. In these products, the basic ingredient is sodium carbonate. The salt Na_2CO_3 is relatively nontoxic and the carbonate ion it contains is a relatively strong base. The carbonate ion also serves as a water softener by precipitating Ca^{2+} ions as insoluble $CaCO_3$.

18.8 ## Acid–base titrations have sharp changes in pH at the equivalence point

The *titrant* is the solution being slowly added from a buret to a solution in the receiving flask.

In Section 5.13, we studied the overall procedure for an acid–base titration and we saw how titration data can be used in various stoichiometric calculations. In performing this procedure, the titration is halted at the **end point** when a change in color of an indicator occurs. Ideally, this end point should occur at the **equivalence point,** when stoichiometrically equivalent amounts of acid and base have combined. To obtain this ideal result, selecting an appropriate indicator requires foresight. We will understand this better by studying how the pH of the solution being titrated changes with the addition of titrant.

When the pH of a solution at different stages of a titration is plotted against the volume of titrant added, we obtain a **titration curve.** The values of pH in these plots can be measured using a pH meter during the titration, or it can be calculated by the procedures studied in this chapter. We will use the calculation method, so this discussion will serve as a review.

Titration of a strong acid by a strong base

The titration of HCl(aq) with a standardized NaOH solution illustrates the titration of a strong acid by a strong base. The molecular and net ionic equations are

$$HCl(aq) + NaOH(aq) \longrightarrow NaCl(aq) + H_2O$$

$$H^+(aq) + OH^-(aq) \longrightarrow H_2O$$

An acid–base titration. Phenolphthalein is often used as an indicator to detect the end point.

Despite the high precision assumed for the molarity and the volume of the titrant, we will retain only two significant figures in the calculated pH. Precision higher than this is hard to obtain and seldom sought in actual lab work.

Let's consider what happens to the pH of a solution, initially 25.00 mL of 0.2000 M HCl, as small amounts of the titrant, 0.2000 M NaOH, are added. We will calculate the pH of the resulting solution at various stages of the titration, retaining only two significant figures, and plot the values against the volume of titrant.

Before any titrant has been added, the receiving flask contains just 0.2000 M HCl. Because this is a strong acid, we know that

$$[H^+] = [HCl] = 0.2000 \ M$$

So

$$pH = -\log (0.2000)$$
$$= 0.70 \text{ (the initial pH)}$$

The amount of HCl initially present in 25.00 mL of 0.2000 M HCl is

$$25.00 \text{ mL HCl soln} \times \frac{0.2000 \text{ mol HCl}}{1000 \text{ mL HCl soln}} = 5.000 \times 10^{-3} \text{ mol HCl}$$

Now suppose we add 10.00 mL of 0.2000 M NaOH from the buret. The amount of NaOH added is

$$10.00 \text{ mL NaOH soln} \times \frac{0.2000 \text{ mol NaOH}}{1000 \text{ mL NaOH soln}} = 2.000 \times 10^{-3} \text{ mol NaOH}$$

This base neutralizes 2.000×10^{-3} mol of HCl, so the amount of HCl remaining is

$$(5.000 \times 10^{-3} - 2.000 \times 10^{-3}) \text{ mol HCl} = 3.000 \times 10^{-3} \text{ mol HCl remaining}$$

The total volume of the solution is now $(25.00 + 10.00)$ mL $= 35.00$ mL $= 0.03500$ L, so the $[H^+]$ is calculated as

$$[H^+] = \frac{3.000 \times 10^{-3} \text{ mol}}{0.03500 \text{ L}} = 8.571 \times 10^{-2} \ M$$

The corresponding pH is 1.07.

Table 18.4 shows the results of calculating the pH of the solution in the receiving flask after the addition of further small volumes of the titrant. After 25.00 mL of 0.2000 NaOH is added, all of the HCl is neutralized and we have reached the equivalence point. At this point the solution is a mixture of Na^+ and Cl^- ions, neither of which has an effect on pH, so the solution is neutral with a pH $= 7.00$.

TABLE 18.4 | **TITRATION OF 25.00 mL OF 0.2000 M HCl WITH 0.2000 M NaOH**

Initial Volume of HCl (mL)	Initial Amount of HCl (mol)	Volume of NaOH Added (mL)	Amount of NaOH (mol)	Amount of Excess Reagent (mol)	Total Volume of Solution (mL)	Molarity of Ion in Excess (mol L^{-1})	pH
25.00	5.000×10^{-3}	0	0	5.000×10^{-3} (H^+)	25.00	0.2000 (H^+)	0.70
25.00	5.000×10^{-3}	10.00	2.000×10^{-3}	3.000×10^{-3} (H^+)	35.00	8.571×10^{-2} (H^+)	1.07
25.00	5.000×10^{-3}	20.00	4.000×10^{-3}	1.000×10^{-3} (H^+)	45.00	2.222×10^{-2} (H^+)	1.65
25.00	5.000×10^{-3}	24.00	4.800×10^{-3}	2.000×10^{-4} (H^+)	49.00	4.082×10^{-3} (H^+)	2.39
25.00	5.000×10^{-3}	24.90	4.980×10^{-3}	2.000×10^{-5} (H^+)	49.90	4.000×10^{-4} (H^+)	3.40
25.00	5.000×10^{-3}	24.99	4.998×10^{-3}	2.000×10^{-6} (H^+)	49.99	4.000×10^{-5} (H^+)	4.40
25.00	5.000×10^{-3}	25.00	5.000×10^{-3}	0	50.00	0	7.00
25.00	5.000×10^{-3}	25.01	5.002×10^{-3}	2.000×10^{-6} (OH^-)	50.01	3.999×10^{-5} (OH^-)	9.60
25.00	5.000×10^{-3}	25.10	5.020×10^{-3}	2.000×10^{-5} (OH^-)	50.10	3.992×10^{-4} (OH^-)	10.60
25.00	5.000×10^{-3}	26.00	5.200×10^{-3}	2.000×10^{-4} (OH^-)	51.00	3.922×10^{-3} (OH^-)	11.59
25.00	5.000×10^{-3}	50.00	1.000×10^{-2}	5.000×10^{-3} (OH^-)	75.00	6.667×10^{-2} (OH^-)	12.82

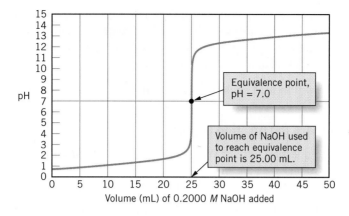

Any further addition of titrant is simply the addition of more and more OH⁻ ion to an increasing volume of solution in the flask. So after the HCl has been neutralized by 25.00 mL of NaOH solution, the pH calculation reflects only a dilute solution of NaOH.

Figure 18.5 shows a plot of the pH of the solution in the receiving flask versus the volume of the 0.2000 *M* NaOH solution added—the *titration curve* for the titration of a strong acid with a strong base. Notice how slowly the pH of the solution increases until we are almost exactly at the equivalence point, pH = 7.00. With the addition of a very small amount of base, the curve rises very sharply and then, almost as suddenly, bends to reflect just a gradual increase in pH. Notice also that the equivalence point occurs at a pH of 7.00, *which characterizes the titration of any strong monoprotic acid with any strong base*. The only ions at the equivalence point in our example, aside from the traces of those from the self-ionization of water, are Na⁺ and Cl⁻. Na⁺ is not an acidic cation, and Cl⁻ is not a basic anion, so the solution at this point is neutral with a pH of 7.00.

The equivalence point in a titration can be found using a pH meter by plotting pH versus volume of base added and noting the sharp rise in pH. If a pH meter is used, we don't have to use an indicator.

Titration of a weak acid by a strong base

The calculations for the titration of a weak acid by a strong base are a bit more complex than when both the acid and base are strong. This is because we have to consider the equilibria involving the weak acid and its conjugate base. As an example, let's do the calculations and draw a titration curve for the titration of 0.200 *M* HC₂H₃O₂ with 0.200 *M* NaOH. The molecular and net ionic equations are

$$HC_2H_3O_2(aq) + NaOH(aq) \longrightarrow NaC_2H_3O_2(aq) + H_2O$$

$$HC_2H_3O_2(aq) + OH^-(aq) \longrightarrow C_2H_3O_2^-(aq) + H_2O$$

We will start with 25.00 mL of 0.200 *M* HC₂H₃O₂ and use 0.200 *M* NaOH as the titrant. Because the concentrations of both solutions are the same, and because the acid and base react in a 1-to-1 mole ratio, we know that we will need exactly 25.00 mL of the base to reach the equivalence point. With this as background, let's look at what is involved in calculating the pH at various points along the titration curve.

1. **Before the titration begins.** The solution at this point is simply a solution of the weak acid, HC₂H₃O₂. We must use K_a to calculate the pH.

2. **During the titration but before the equivalence point.** As we add NaOH to the HC₂H₃O₂, the chemical reaction produces C₂H₃O₂⁻, so the solution contains both HC₂H₃O₂ and C₂H₃O₂⁻ (it is a buffer solution). We can use K_a to calculate the pH here as well.

3. **At the equivalence point.** Here all the HC₂H₃O₂ has reacted and the solution contains the salt NaC₂H₃O₂ (i.e., we have a mixture of the ions Na⁺ and C₂H₃O₂⁻). The anion is a weak base, so we must use K_b to calculate the pOH and then the pH.

4. After the equivalence point. The OH^- introduced by the further addition of NaOH has nothing with which to react and causes the solution to become basic. The concentration of OH^- produced by the unreacted base is used to calculate the pOH and ultimately the pH.

We will look at each of these briefly.

Before the titration begins

We worked a similar problem on page 787.

The question here is: What is the pH of 0.200 M $HC_2H_3O_2$? For acetic acid, the equilibrium is

$$HC_2H_3O_2(aq) \rightleftharpoons H^+(aq) + C_2H_3O_2^-(aq)$$

and $K_a = 1.8 \times 10^{-5}$. If we let x equal the number of moles per liter of $HC_2H_3O_2$ that ionizes, then

$$[H^+] = [C_2H_3O_2^-] = x$$

$$[HC_2H_3O_2] = 0.200\ M - x$$

Because K_a is small, we assume that x will be very small compared to 0.200 M, so $(0.200\ M - x) \approx 0.200\ M$. Substituting into the K_a expression gives,

$$K_a = \frac{[H^+][C_2H_3O_2^-]}{[HC_2H_3O_2]} = \frac{(x)(x)}{(0.200 - x)} \approx \frac{(x)(x)}{(0.200)} = 1.8 \times 10^{-5}$$

$$x^2 = 3.6 \times 10^{-6}\ \text{(rounded)}$$

$$x = [H^+] = 1.9 \times 10^{-3}$$

So, $[H^+] = 1.9 \times 10^{-3}\ M$, which yields a pH of 2.72. This is the pH before any NaOH is added.

During the titration but before the equivalence point

As we add NaOH, it reacts with $HC_2H_3O_2$, changing it to $C_2H_3O_2^-$. Therefore the solution contains both $HC_2H_3O_2$ and $C_2H_3O_2^-$, which makes it a buffer system. To calculate the pH, we need concentrations (or moles) of both $HC_2H_3O_2$ and $C_2H_3O_2^-$ in the solution. Let's calculate the pH after 10.0 mL of the base has been added.

The number of moles of NaOH added is

$$10.0\ \text{mL NaOH soln} \times \frac{0.200\ \text{mol NaOH}}{1000\ \text{mL soln}} = 2.00 \times 10^{-3}\ \text{mol NaOH}$$

The NaOH reacted with 2.00×10^{-3} mol of $HC_2H_3O_2$ to produce 2.00×10^{-3} mol of $C_2H_3O_2^-$.

$$\text{moles of } C_2H_3O_2^- = 2.00 \times 10^{-3}$$

The amount of $HC_2H_3O_2$ remaining in the solution is the difference between the amount originally present and the amount that reacted. Before the titration began, the amount of $HC_2H_3O_2$ in the solution was

$$25.0\ \text{ml } HC_2H_3O_2\ \text{soln} \times \frac{0.200\ \text{mol } HC_2H_3O_2}{1000\ \text{mL soln}} = 5.00 \times 10^{-3}\ \text{mol } HC_2H_3O_2$$

The amount of $HC_2H_3O_2$ remaining after addition of the base is therefore

$$5.00 \times 10^{-3}\ \text{mol } HC_2H_3O_2 - 2.00 \times 10^{-3}\ \text{mol } HC_2H_3O_2 = 3.00 \times 10^{-3}\ \text{mol } HC_2H_3O_2$$

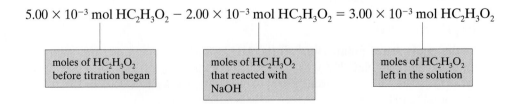

| moles of $HC_2H_3O_2$ before titration began | moles of $HC_2H_3O_2$ that reacted with NaOH | moles of $HC_2H_3O_2$ left in the solution |

The total volume after addition of the base is now $(25.0 \text{ mL} + 10.0 \text{ mL}) = 35.0 \text{ mL} = 0.0350 \text{ L}$. Therefore, the concentrations of acetate ion and acetic acid now are

$$[C_2H_3O_2^-] = \frac{2.00 \times 10^{-3} \text{ mol}}{0.0350 \text{ L}} = 5.71 \times 10^{-2} \text{ mol L}^{-1}$$

$$[HC_2H_3O_2] = \frac{3.00 \times 10^{-3} \text{ mol}}{0.0350 \text{ L}} = 8.57 \times 10^{-2} \text{ mol L}^{-1}$$

For the buffer, we can now substitute these values as though they were equilibrium concentrations and substitute them into the mass action expression to calculate $[H^+]$.

$$\frac{[H^+](5.71 \times 10^{-2})}{(8.57 \times 10^{-2})} = 1.8 \times 10^{-5}$$

Therefore,

$$[H^+] = 1.8 \times 10^{-5} \times \frac{8.57 \times 10^{-2}}{5.71 \times 10^{-2}}$$

$$= 2.7 \times 10^{-5} \, M$$

The pH calculates to be

$$pH = 4.57$$

As long as we have not reached the equivalence point—as long as unreacted acetic acid remains together with acetate ion—we can compute the pH of the solution in the receiving flask by this kind of buffer calculation.

At the equivalence point

At the equivalence point we've combined equal numbers of moles of $HC_2H_3O_2$ and NaOH. Their reaction yields $NaC_2H_3O_2$, so the solution is simply a dilute solution of sodium acetate. It no longer is a buffer. We now have to deal with the reaction of the acetate ion with water. Because $C_2H_3O_2^-$ is a base, we have to use K_b and write the chemical equation for the reaction of a base with water.

$$C_2H_3O_2^-(aq) + H_2O \rightleftharpoons HC_2H_3O_2(aq) + OH^-(aq)$$

$$K_b = \frac{[OH^-][HC_2H_3O_2]}{[C_2H_3O_2^-]} = \frac{K_w}{K_a} = 5.6 \times 10^{-10}$$

At the equivalence point, we have added 25.0 mL of 0.200 M NaOH to the original 25.0 mL of 0.200 M $HC_2H_3O_2$. The volume is now 50.0 mL = 0.0500 L. We began with 5.00×10^{-3} mol of $HC_2H_3O_2$, so we now have 5.00×10^{-3} mol of $C_2H_3O_2^-$ (in 0.0500 L). The concentration of $C_2H_3O_2^-$, therefore, is

$$[C_2H_3O_2^-] = \frac{5.00 \times 10^{-3} \text{ mol}}{0.0500 \text{ L}} = 0.100 \, M$$

To find the pH, we realize that a small amount of the $C_2H_3O_2^-$ reacts to give equal amounts of OH^- and $HC_2H_3O_2$. We let the concentrations of these equal x, so $[OH^-] = [C_2H_3O_2^-] = x$. The concentration of $C_2H_3O_2^-$ that is left is $(0.100 - x) \approx 0.100 \, M$. (As before, we expect x to be very small because K_b is so small.) Substituting into the mass action expression gives

$$\frac{[OH^-][HC_2H_3O_2]}{[C_2H_3O_2^-]} = \frac{(x)(x)}{0.100 - x} \approx \frac{(x)(x)}{0.100} = 5.6 \times 10^{-10}$$

$$x^2 = 5.6 \times 10^{-11}$$

$$x = 7.5 \times 10^{-6}$$

This means that $[OH^-] = 7.5 \times 10^{-6} \, M$, so the pOH calculates to be 5.12. Finally, the pH is $(14.00 - 5.12)$ or 8.88. Thus the pH at the equivalence point in this titration is 8.88, somewhat basic because the solution contains the Brønsted base,

It is interesting that although we describe the overall reaction in the titration as *neutralization*, at the equivalence point the solution is slightly basic, not neutral.

TABLE 18.5	Titration of 25.0 mL of 0.200 M $HC_2H_3O_2$ with 0.200 M NaOH	
Volume of Base Added (mL)	Molar Concentration of Species in Parentheses	pH
None	1.9×10^{-3} (H^+)	2.72
10.0	2.7×10^{-5} (H^+)	4.57
24.9	7.2×10^{-8} (H^+)	7.14
24.99	7.1×10^{-9} (H^+)	8.14
25.0	7.5×10^{-6} (OH^-)	8.88
25.01	4.0×10^{-5} (OH^-)	9.60
25.1	4.0×10^{-4} (OH^-)	10.60
26.0	3.9×10^{-3} (OH^-)	11.59
35.0	3.3×10^{-2} (OH^-)	12.52

$C_2H_3O_2{}^-$. *In the titration of any weak acid with a strong base, the pH at the equivalence point will be greater than 7.*

After the equivalence point

What happens if we continue to add NaOH(aq) to the solution beyond the equivalence point? Now the additional OH^- shifts the following equilibrium to the left:

$$C_2H_3O_2{}^-(aq) + H_2O \rightleftharpoons HC_2H_3O_2(aq) + OH^-(aq)$$

The production of OH^- *by this route* is thus suppressed as we add more and more NaOH solution, so effectively the only source of OH^- to affect the pH is from the base added after the equivalence point. From here on, therefore, the pH calculations become identical to those for the last half of the titration curve for HCl and NaOH. Each point is just a matter of calculating the additional number of moles of NaOH, calculating the new final volume, taking their ratio (to obtain the molar concentration), and then calculating pOH and pH.

The titration curve

Table 18.5 summarizes the titration data, and the titration curve is given in Figure 18.6. Once again, notice how the change in pH occurs slowly at first, then shoots up through the equivalence point, and finally looks just like the last half of the curve in Figure 18.5.

Practice Exercise 24: Suppose we titrate 20.0 mL of 0.100 M $HCHO_2$ with 0.100 M NaOH. Calculate the pH (a) before any base is added, (b) when half of the $HCHO_2$ has been neutralized, (c) after a total of 15.0 mL of base has been added, and (d) at the equivalence point.

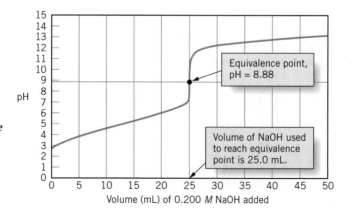

Figure 18.6 *Titration curve for titrating a weak acid with a strong base.* In this titration, we follow the pH as 0.200 M acetic acid is titrated with 0.200 M NaOH.

FIGURE 18.7 ***Titration curve for the titration of a weak base with a strong acid.*** Here we follow the pH as 0.200 *M* NH_3 is titrated with 0.200 *M* HCl.

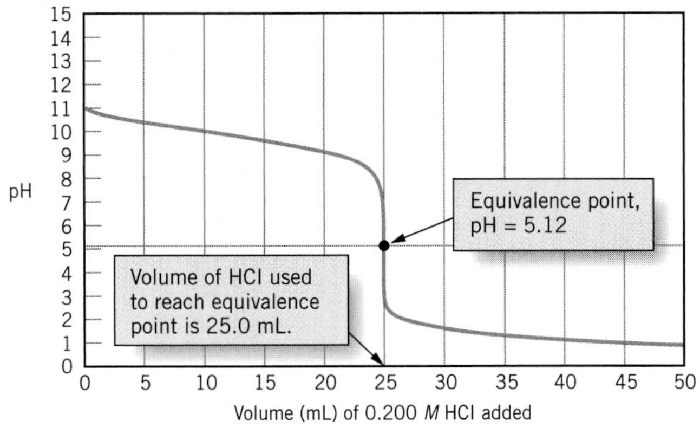

PRACTICE EXERCISE 25: Suppose 30.0 mL of 0.15 *M* NaOH is added to 50.0 mL of 0.20 *M* $HCHO_2$. What will be the pH of the resulting solution?

Titration of a weak base by a strong acid

The calculations involved here are nearly identical to those of the weak acid–strong base titration. We will review what is required but will not actually perform the calculations.

If we titrate 25.0 mL of 0.200 *M* NH_3 with 0.200 *M* HCl, the molecular and net ionic equations are

$$NH_3(aq) + HCl(aq) \longrightarrow NH_4Cl(aq)$$

$$NH_3(aq) + H^+(aq) \longrightarrow NH_4^+(aq)$$

1. **Before the titration begins.** The solution at this point is simply a solution of the weak base, NH_3. Since the only solute is a base, we must use K_b to calculate the pH.
2. **During the titration but before the equivalence point.** As we add HCl to the NH_3, the neutralization reaction produces NH_4^+, so the solution contains both NH_3 and NH_4^+ (it is a buffer solution). We can use K_b for NH_3 to calculate the pH here as well. (We could instead use K_a for NH_4^+ if it were available.)
3. **At the equivalence point.** Here all the NH_3 has reacted and the solution contains the salt NH_4Cl. The anion, Cl^-, comes from the strong acid HCl and is so weak that it cannot affect the pH. However, the cation NH_4^+ is a weak acid derived from the weak base NH_3, so we use its K_a to calculate the pH. In doing the calculation, it's important to use the total volume of the solution to calculate the concentration of the NH_4^+.
4. **After the equivalence point.** The H^+ introduced by further addition of HCl has nothing with which to react, so it causes the solution to become more and more acidic. The concentration of H^+ produced by the unreacted HCl is used directly to calculate the pH.

Figure 18.7 illustrates the titration curve for this system.

Titration curves for diprotic acids show two equivalence points

When a weak diprotic acid such as ascorbic acid (vitamin C) is titrated with a strong base, there are two protons to be neutralized and there are two equivalence points. Provided that the values of K_{a_1} and K_{a_2} differ by several powers of 10, the neutralization takes place stepwise and the resulting titration curve shows two sharp increases in pH. We won't perform the calculations here, because they're complex, but instead just look at the general shape of the titration curve, which is shown in Figure 18.8.

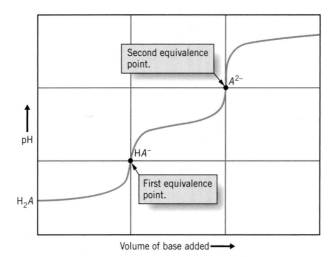

Figure 18.8 *Titration of a diprotic acid, H_2A, by a strong base.* As each equivalence point is reached, there is a sharp rise in the pH.

Acid–base indicators are weak acids

Most dyes that work as acid–base indicators are also weak acids. Therefore, let's represent an indicator by the formula H*In*. In its un-ionized state, H*In* has one color. Its conjugate base, *In*⁻, has a different color, the more strikingly different the better. In solution, the indicator is involved in a typical acid–base equilibrium:

$$HIn(aq) \rightleftharpoons H^+(aq) + In^-(aq)$$

acid form · · · · · · · · · · base form
(one color) · · · · · · · (another color)

The corresponding acid ionization constant, K_{In}, is given by

$$K_{In} = \frac{[H^+][In^-]}{[HIn]}$$

In a strongly acidic solution, when the H^+ concentration is high, the equilibrium is shifted to the left and most of the indicator exists in its "acid form." Under these conditions, the color we observe is that of H*In*. If the solution is made basic, the H^+ concentration drops and the equilibrium shifts to the right, toward *In*⁻, and the color we observe is that of the "base form" of the indicator.

The *observed* change in color for an indicator actually occurs over a range of pH values. This is because of the human eye's limited ability to discern color changes. As discussed further in Facets of Chemistry 18.1, sometimes a change of as much as 2 pH units is needed for some indicators—more for litmus—before the eye notices the color change. This is why tables of acid–base indicators, such as Table 17.5 on page 762, provide approximate pH ranges for the color changes.

How acid–base indicators work

In a typical acid-base titration, you've seen that as we pass the equivalence point, there is a sudden and large change in the pH. For example, in the titration of HCl with NaOH described earlier, the pH just one-half drop before the equivalence point (when 24.97 mL of the base has been added) is 3.92. Just one-half drop later (when 25.03 mL of base has been added) we have passed the equivalence point and the pH has risen to 10.08. This large swing in pH (from 3.92 to 10.08) causes a sudden shift in the position of equilibrium for the indicator, and we go from a condition where most of the indicator is in its acid form to a condition in which most is in the base form. This is observed visually as a change in color from that of the acid form to that of the base form.

FACETS OF CHEMISTRY 18.1

Color Ranges for Indicators

Suppose that the color of HIn is red and the color of its anion, In^-, is yellow. In terms of the eye's abilities, the ratio of $[HIn]/[In^-]$ has to be roughly at least as large as 10 to 1 for the eye to see the color as definitely red. And this ratio has to be at least as small as 1 to 10 for the eye to see the color as definitely yellow. Therefore, the color transition that the eye actually sees at the end point occurs as the ratio of $[HIn]$ to $[In^-]$ changes from 10/1 to 1/10.

We can calculate the relationship of pH to pK_a for the indicator at the start of this range as follows. We know that

$$K_{In} = \frac{[H^+][In^-]}{[HIn]}$$

Therefore, at the start of the range, we must have

$$K_{In} = [H^+] \times \frac{10}{1}$$

$$pK_{In} = pH - \log 10$$

Or, at the start of the transition we should have

$$pH = pK_{In} - 1$$

At the other end of the range, when $[HIn]/[In^-]$ is 1/10, we have

$$K_{In} = [H^+] \times \frac{1}{10}$$

$$pK_{In} = pH - \log(10^{-1})$$

So at the end of the color transition, the pH should be

$$pH = pK_{In} + 1$$

Therefore, the *range* of pH over which a definite change in color occurs is

$$pH = pK_{In} \pm 1$$

A range of ± 1 pH unit on either side of the value of pK_{In} corresponds to a span of 2 pH units over which the eye can definitely notice the color change.

Selecting the best indicator for an acid–base titration

The explanation in the preceding paragraph assumes that the *midpoint* of the color change of the indicator corresponds to the pH at the equivalence point. At this midpoint we expect that there will be equal amounts of both forms of the indicator, which means that $[In^-] = [HIn]$. If this condition is true, then

$$K_{In} = \frac{[H^+]\cancel{[In^-]}}{\cancel{[HIn]}}$$

so

$$K_{In} = [H^+]$$

Taking the negative logarithms of both sides gives us

$$-\log K_{In} = -\log [H^+]$$

or

$$pK_{In} = pH_{(\text{at the equivalence point})}$$

This important result tells us that once we know the pH that the solution should have at the equivalence point of a titration, we also know the pK_{In} that the indicator should have to function most effectively. Thus, we want an indicator whose pK_{In} equals (or is as close as possible to) the pH at the equivalence point. Phenolphthalein, for example, changes a solution from colorless to pink as the solution's pH changes over a range of 8.2 to 10.0. (See Figure 17.5, page 764.) *Phenolphthalein is therefore a nearly perfect indicator for the titration of a weak acid by a strong base,* where the pH at the equivalence point is on the basic side. (Refer back to Figure 18.6 on page 818.)

Phenolphthalein also works very well for the titration of a strong acid by a strong base. As discussed earlier, just one drop of titrant can bring the pH from below 7 to nearly 10, which spans the pH range for phenolphthalein.

Good indicators have intense colors

When we perform a titration, we want to use as little indicator as possible. The reason is that indicators are weak acids and also react with the titrant. If we were forced to use a lot of indicator, it would upset our measurements and make the titration less precise. Therefore, the best indicators are those with the most intense colors. Then, even extremely small amounts can give striking colors and color changes without consuming too much of the titrant.

SUMMARY

Acid and Base Ionization Constants A weak acid HA ionizes according to the general equation

$$HA + H_2O \rightleftharpoons H_3O^+ + A^-$$

or more simply,

$$HA \rightleftharpoons H^+ + A^-$$

The equilibrium constant is called the **acid ionization constant, K_a** (sometimes called an *acid dissociation constant*).

$$K_a = \frac{[H^+][A^-]}{[HA]}$$

A weak base B ionizes by the general equation

$$B + H_2O \rightleftharpoons BH^+ + OH^-$$

The equilibrium constant is called the **base ionization constant, K_b** (sometimes called a *base dissociation constant*).

$$K_b = \frac{[OH^-][BH^+]}{[B]}$$

The smaller the values of K_a (or K_b), the weaker are the substances as Brønsted acids (or bases).

Another way to compare the relative strengths of acids or bases is to use the negative logarithms of K_a and K_b, called **pK_a** and **pK_b**, respectively.

$$pK_a = -\log K_a \qquad pK_b = -\log K_b$$

The *smaller* the pK_a or pK_b, the *stronger* is the acid or base. For a conjugate acid–base pair,

$$K_a \times K_b = K_w$$

and

$$pK_a + pK_b = 14.00 \qquad \text{(at 25 °C)}$$

Equilibrium Calculations—Determining K_a and K_b The values of K_a and K_b can be obtained from initial concentrations of the acid or base and either the pH of the solution or the percentage ionization of the acid or base. The measured pH gives the equilibrium value for $[H^+]$. The **percentage ionization** is defined as

$$\text{percentage ionization} = \frac{\text{amount ionized}}{\text{amount available}} \times 100\%$$

Equilibrium Calculations—Determining Equilibrium Concentrations when K_a or K_b Is Known Problems fall into one of three categories: (1) the only solute is a weak acid (we must use the K_a expression), (2) the only solute is a

weak base (we must use the K_b expression), and (3) the solution contains both a weak acid and its conjugate base (we can use either K_a or K_b).

When the initial concentration of the acid (or base) is larger than 400 times the value of K_a (or K_b), it is safe to use initial concentrations of acid or base as though they were equilibrium values in the mass action expression. When this approximation cannot be used, we use either the quadratic formula or the method of successive approximations (usually the latter is faster).

Ions as Acids or Bases A solution of a salt is acidic if the cation is acidic but the anion is neutral. Metal ions with high charge densities generally are acidic, but those of Groups IA and IIA (except Be^{2+}) are not. Cations such as NH_4^+, which are the conjugate acids of weak *molecular* bases, are themselves proton donors and are acidic.

When the anion of a salt is the conjugate base of a *weak* acid, the anion is a weak base. Anions of strong acids, such as Cl^- and NO_3^-, are such weak Brønsted bases that they cannot affect the pH of a solution.

If the salt is derived from a weak acid *and* a weak base, its net effect on pH has to be determined on a case-by-case basis by determining which of the two ions is the stronger.

Buffers A solution that contains both a weak acid and a weak base (usually an acid–base conjugate pair) is called a **buffer,** because it is able to absorb $[H^+]$ from a strong acid or $[OH^-]$ from a strong base without suffering large changes in pH. For the general acid–base pair, HA and A^-, the following reactions neutralize H^+ and OH^-.

$$H^+ \text{ is added to the buffer:} \qquad A^- + H^+ \longrightarrow HA$$

$$OH^- \text{ is added to the buffer:} \qquad HA + OH^- \longrightarrow A^- + H_2O$$

The pH of a buffer is controlled by the ratio of weak acid to weak base, expressed either in terms of molarities or moles.

$$[H^+] = K_a \times \frac{[HA]}{[A^-]} = K_a \times \frac{\text{moles } HA}{\text{moles } A^-}$$

Because the $[H^+]$ is determined by the mole ratio of HA to A^-, dilution does not change the pH of a buffer. The **Henderson–Hasselbalch equation,**

$$pH = pK_a + \log \frac{[A^-]_{\text{initial}}}{[HA]_{\text{initial}}}$$

can be used to calculate the pH directly from the pK_a of the acid and the concentrations of the conjugate acid, HA, and base, A^-.

In performing buffer calculations, the usually valid simplifications are

$$[HA]_{equilibrium} = [HA]_{initial}$$

$$[A^-]_{equilibrium} = [A^-]_{initial}$$

The value of $[A^-]_{initial}$ is found from the molarity of the *salt* of the weak acid in the buffer. Once $[H^+]$ is found by this calculation, the pH is calculated.

Buffer calculations can also be performed using K_b for the weak acid component. Similar equations apply, the principal difference being that it is the OH^- concentration that is calculated. Thus, using K_b we have

$$[OH^-] = K_b \times \frac{conjugate\ base}{conjugate\ acid}$$

$$= K_b \times \frac{moles\ conjugate\ base}{moles\ conjugate\ acid}$$

Buffers are most effective when the pK_a of the acid member of the buffer pair lies within ± 1 unit of the desired pH.

Although the molar or mole *ratio* determines the pH at which a buffer works, the *amounts* of the acid and its conjugate base give the buffer its *capacity* for taking on strong acid or base with little change in pH.

Equilibria in Solutions of Polyprotic Acids A polyprotic acid has a K_a value for the ionization of each of its hydrogen ions. Successive values of K_a often differ by a factor of 10^4 to 10^5. This allows us to calculate the pH of a solution of a polyprotic acid by using just the value of K_{a_1}. If the polyprotic acid is the only solute, the anion formed in the second step of the ionization, A^{2-}, has a concentration equal to K_{a_2}.

Solutions of Salts of Polyprotic Acids The anions of weak polyprotic acids are bases that react with water in successive steps, the last of which has the molecular polyprotic acid as a product. For a diprotic acid, $K_{b_1} = K_w/K_{a_2}$ and $K_{b_2} = K_w/K_{a_1}$. Usually, $K_{b_1} \gg K_{b_2}$, so virtually all the OH^- produced in the solution comes from the first step. The pH of the solution can be calculated using just K_{b_1} and the reaction

$$A^{2-} + H_2O \rightleftharpoons HA^- + OH^-$$

Acid–Base Titrations and Indicators The pH at the equivalence point depends on the ions of the salt formed during the titration. In the titration of a strong acid by a strong base, the pH at the equivalence point is 7 because neither of the ions produced in the reaction affects the pH of an aqueous solution.

For the titration of a weak acid by a strong base, the approach toward calculation of the pH varies depending on which stage of the titration is considered. (1) Before the titration begins, the only solute is the weak acid and K_a is used. (2) After starting, but before the equivalence point, the solution is a buffer. Either K_a or K_b can be used. (3) At the equivalence point, the active solute is the conjugate base of the weak acid. K_b for the conjugate base must be used. (4) After the equivalence point, excess base determines the pH of the mixture. No equilibrium calculations are required. At the equivalence point, the pH is above 7. Similar calculations apply in the titration of a weak base by a strong acid. In such titrations, the pH at the equivalence point is below 7.

The indicator chosen for a titration should have the center of its color change near the pH at the equivalence point. For the titration of a strong or weak acid with a strong base, phenolphthalein is the best indicator. When a weak base is titrated with a strong acid, several indicators including methyl orange work satisfactorily.

TOOLS YOU HAVE LEARNED

The table below lists the concepts you've learned in this chapter that can be applied as tools in solving problems. Study each one carefully so that you know what each is used for. When faced with solving a problem, recall what each tool does and consider whether it will be helpful in finding a solution. This will aid you in selecting the tools you need. If necessary, refer to this table when working on the Thinking-It-Through problems and the Review Exercises that follow.

TOOL	HOW IT WORKS
General equation for the ionization of a weak acid (page 775)	You follow the form of the general equation to write the correct chemical equilibrium for a given weak acid, from which you construct the correct K_a expression.
General equation for the ionization of a weak base (page 778)	You follow the form of the general equation to write the correct chemical equilibrium for a given weak base, from which you construct the correct K_b expression.
$K_a \times K_b = K_w$ (page 780)	Use this equation to calculate K_a given K_b, or vice versa.
percentage ionization $= \dfrac{moles\ per\ liter\ ionized}{moles\ per\ liter\ available} \times 100\%$ (page 782)	Use this equation to calculate the percentage ionization from the initial concentration of acid (or base) and the change in the concentrations of the ions. If the percentage ionization is known, you can calculate the change in the concentration of the acid or base and then use that information to calculate the K_a (or K_b).

Identification of acidic cations and basic anions (pages 790 and 792)	We use the concepts developed here to determine whether a salt solution is acidic, basic, or neutral. This is also the first step in calculating the pH of a salt solution and, in conjunction with the next tool, to establish the correct direction in analyzing the problem.
Criterion for applying simplifications in acid–base equilibrium calculations: $[X]_{initial} > 400 \times K$ (page 796)	We use this to determine whether initial concentrations can be used as though they are equilibrium values in the mass action expression when working acid–base equilibrium problems. If the condition fails, then we have to use the quadratic equation or the method of successive approximations.
Method of successive approximations (page 798)	Use this method to quickly calculate equilibrium concentrations when the usual simplifications are inappropriate.
Reactions when H^+ or OH^- are added to a buffer (page 800)	These reactions determine how the concentrations of conjugate acid and base change when a strong acid or strong base is added to a buffer. Adding H^+ decreases $[A^-]$ and increases $[HA]$; adding OH^- decreases $[HA]$ and increases $[A^-]$.

THINKING IT THROUGH

Need extra help?
Visit the Brady/
Senese web site at
www.wiley.com/
college/brady

ON-LINE HELP

The goal for the following problems is not to find the answers themselves, but rather to assemble the information needed to solve them and explain how you would use the information to find the answers. The problems in Level 2 are more challenging than those in Level 1 and may contain more data than are required, in which case you are also asked to identify the unnecessary data. Detailed answers to the Thinking-It-Through problems can be found on the web site.

Level 1 Problems

1. Propionic acid, $HC_3H_5O_2$, is a monoprotic acid. Suppose you've prepared a 0.100 M solution of this acid. What experimental information is required to determine the value of K_a for this acid? Explain how you would use this information to calculate K_a.

2. A 0.10 M solution of a weak base is prepared. The value of K_b for the base is 2.0×10^{-8}. Describe how you would calculate the pH of the solution. Explain any simplifications you would make and how you would justify their validity.

3. A 0.10 M solution of a weak acid with $K_a = 2.5 \times 10^{-6}$ has a pH of 3.28. Is the weak acid the only solute in the solution? (Describe in detail how you would answer this question.)

4. What information do you need to actually prepare, in the laboratory, a buffer that has a pH of 5.00 using acetic acid and potassium acetate? Describe any calculations that would be necessary.

5. How would you calculate the acetate ion concentration in a solution that contains 0.0010 M HCl and 0.10 M $HC_2H_3O_2$? Describe and justify any simplifications you would make in solving the problem.

6. A 0.035 M solution of an acid has a pH of 3.5. Is the solute a strong acid or not? How can you tell?

7. Acetic acid ($HC_2H_3O_2$) and propionic acid ($HC_3H_5O_2$) have nearly identical pK_a values. If you had an aqueous solution that is 0.10 M in acetic acid and also 0.10 M in the sodium salt of the other acid, propionic acid, would this

solution function as a *buffer*? Explain. Write net ionic equations that describe its chemical behavior if a small amount of hydrochloric acid is added or if a small amount of sodium hydroxide is added.

8. Describe how you would calculate the value of K_{b_1} for the PO_4^{3-} ion.

9. In a solution of sodium sulfite, there are actually three sources of OH^- ion. Write chemical equations for the three reactions involved. Which source(s) of OH^- can be ignored, and why?

Level 2 Problems

10. A simple way to measure the pK_a of an acid is to mix equal numbers of moles of the acid and a base (such as NaOH) to form an aqueous solution and then measure the pH with a pH meter. Explain why this method avoids having to perform equilibrium calculations.

11. Suppose 25.0 mL of 0.10 M HCl is added to 50.0 mL of 0.10 M KNO_2. Explain how you would calculate the pH of the resulting mixture.

12. The value of K_w increases as the temperature of water is raised.
(a) Explain what effect increasing the temperature of pure water will have on its pH. (Will the pH increase, decrease, or stay the same?)
(b) As the temperature of pure water increases, will the water become more acidic, more basic, or remain neutral? Justify your answer.

13. A solution of dichloroacetic acid, $HC_2HO_2Cl_2$, has a concentration of 0.50 M. Describe how you would calculate the freezing point of this solution. (Since it is a dilute aqueous solution, assume that the molarity equals the molality.)

14. Suppose 250 mL of $NH_3(g)$ measured at 20 °C and 750 torr is dissolved in 100 mL of 0.050 M HCl. Describe how you would calculate the pH of the resulting solution.

15. How many milliliters of 0.0010 M HCl would have to be added to 100 mL of a 0.100 M solution of $HC_2H_3O_2$ to make the $C_2H_3O_2^-$ concentration equal 1.0×10^{-5} M? (Explain all the calculations that are necessary to obtain the answer.)

16. What is the concentration of H_2CO_3 in a 0.10 M solution of Na_2CO_3? (Explain how you would calculate the answer. In what way is this calculation similar to finding the CO_3^{2-} concentration in a solution of H_2CO_3?)

17. Show that the concentration of PO_4^{3-} in a 0.10 M solution of H_3PO_4 is not equal to K_{a_3} for this acid. Describe how you would calculate the PO_4^{3-} concentration in this solution.

18. Describe in detail how you would calculate the hydrogen ion concentration in 0.10 M H_2SO_4. (*Hint*: The H^+ concentration is *not* 0.20 M.) What data are needed to solve the problem?

19. At the right is a diagram illustrating a mixture of HF and F^- in an aqueous solution. For this mixture, does pH equal pK_a for HF? Explain. Describe how the number of HF molecules and F^- ions will change after three OH^- ions are added. How will the number of HF and F^- change if two H^+ ions are added? Explain your answers by using chemical equations.

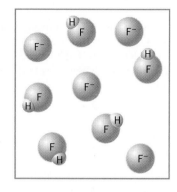

20. A student wrote the following equation for the equilibrium reaction of the weak base CH_3NH_2 with water. What, if anything, is wrong with it? Explain your answer.

$$CH_3NH_2(aq) + H_2O \rightleftharpoons CH_4NH_2^+(aq) + OH^-(aq)$$

REVIEW QUESTIONS

Acid and Base Ionization Constants: K_a, K_b, pK_a, and pK_b

18.1 Write the general equation for the ionization of a weak acid in water. Give the equilibrium law corresponding to K_a.

18.2 Write the chemical equation for the ionization of each of the following weak acids in water. (Some are polyprotic acids; for these write only the equation for the first step in the ionization.)
(a) HNO_2 (c) $HAsO_4^{2-}$
(b) H_3PO_4 (d) $(CH_3)_3NH^+$

18.3 For each of the acids in Review Question 18.2, write the appropriate K_a expression.

18.4 Write the general equation for the ionization of a weak base in water. Give the equilibrium law corresponding to K_b.

18.5 Write the chemical equation for the ionization of each of the following weak bases in water:
(a) $(CH_3)_3N$ (c) NO_2^-
(b) AsO_4^{3-} (d) $(CH_3)_2N_2H_2$

18.6 For each of the bases in Review Question 18.5, write the appropriate K_b expression.

18.7 Benzoic acid, $C_6H_5CO_2H$, is an organic acid whose sodium salt, $C_6H_5CO_2Na$, has long been used as a safe food additive to protect beverages and many foods against harmful yeasts and bacteria. The acid is monoprotic. Write the equation for its K_a.

18.8 Write the equation for the equilibrium that the benzoate ion, $C_6H_5CO_2^-$ (Review Question 18.7), would produce in water as it functions as a Brønsted base. Then write the expression for the K_b of the conjugate base of benzoic acid.

18.9 The pK_a of HCN is 9.21 and that of HF is 3.17. Which is the stronger Brønsted base, CN^- or F^-?

18.10 Write the structural formulas for the conjugate acids of the following:

(a) $CH_3{-}CH_2{-}\overset{\overset{CH_3}{|}}{\underset{\cdot\cdot}{N}}{-}H$ (c) $H{-}\overset{\cdot\cdot}{\underset{\cdot\cdot}{O}}{-}\overset{\overset{H}{|}}{\underset{\cdot\cdot}{N}}{-}H$

(b) [benzene ring]$-N{:}$

18.11 Write the structural formulas for the conjugate bases of the following:

(a) [benzene ring]$-NH_3^+$ (c) $\left[H{-}\overset{\overset{H}{|}}{\underset{\cdot\cdot}{N}}{-}\overset{\overset{H}{|}}{\underset{\underset{H}{|}}{N}}{-}H \right]^+$

(b) $\left[CH_3{-}\overset{\overset{CH_3}{|}}{\underset{\underset{CH_3}{|}}{N}}{-}H \right]^+$

18.12 How is percentage ionization defined? Write the equation.

18.13 If a weak acid is 1.2% ionized in a 0.22 M solution, what is the molar concentration of H^+ in the solution?

Acid–Base Properties of Salt Solutions

18.14 Aspirin is acetylsalicylic acid, a monoprotic acid whose K_a value is 3.27×10^{-4}. Does a solution of the sodium salt of aspirin in water test acidic, basic, or neutral? Explain.

18.15 The K_b value of the oxalate ion, $C_2O_4^{2-}$, is 1.9×10^{-10}. Is a solution of $K_2C_2O_4$ acidic, basic, or neutral? Explain.

18.16 Consider the following compounds and suppose that 0.5 M solutions are prepared of each: NaI, KF, $(NH_4)_2SO_4$, KCN, $KC_2H_3O_2$, CsNO$_3$, and KBr. Write the *formulas* of those that have solutions that are (a) acidic, (b) basic, and (c) neutral.

18.17 Will an aqueous solution of $AlCl_3$ turn litmus red or blue? Explain.

18.18 Explain why the beryllium ion is a more acidic cation than the calcium ion.

18.19 Ammonium nitrate is commonly used in fertilizer mixtures as a source of nitrogen for plant growth. What effect, if any, will this compound have on the acidity of the moisture in the ground? Explain.

18.20 A solution of hydrazinium acetate is slightly acidic. Without looking at the tables of equilibrium constants, is K_a for acetic acid larger or smaller than K_b for hydrazine? Justify your answer.

18.21 When ammonium nitrate is added to a suspension of magnesium hydroxide in water, the $Mg(OH)_2$ dissolves. Write a net ionic equation to show how this occurs.

Equilibrium Calculations when Simplifications Fail

18.22 What criterion do we use to determine whether or not we are able to use the initial concentration of an acid or base as though it were the equilibrium concentration when we calculate the pH of the solution?

18.23 For which of the following are we permitted to use the initial concentration of the acid or base to calculate the pH of the solution specified?
(a) 0.020 M $HC_2H_3O_2$ (c) 0.002 M N_2H_4
(b) 0.10 M CH_3NH_2 (d) 0.050 M $HCHO_2$

18.24 Describe the method of successive approximations.

Buffers

18.25 Write ionic equations that illustrate how each pair of compounds can serve as a buffer pair.
(a) H_2CO_3 and $NaHCO_3$ (the "carbonate" buffer in blood)
(b) NaH_2PO_4 and Na_2HPO_4 (the "phosphate" buffer inside body cells)
(c) NH_4Cl and NH_3

18.26 Which buffer would be better able to hold a steady pH on the addition of strong acid, buffer 1 or buffer 2? Explain. Buffer 1 is a solution containing 0.10 M NH_4Cl and 1 M NH_3. Buffer 2 is a solution containing 1 M NH_4Cl and 0.10 M NH_3.

18.27 Bicarbonate ion is able to act as a buffer all by itself. Write chemical equations that show how this ion reacts with (a) H^+ and (b) OH^-.

Ionization of Polyprotic Acids

18.28 Arsenic acid, H_3AsO_4, is a triprotic acid.
(a) Write the three equilibrium equations that represent the successive ionizations of this acid.

(b) The negative logarithms of the acid ionization constants for the equilibria are, respectively, 2.2, 6.9, and 11.5 at 25°C. What are the numerical values of the acid ionization constants?

(c) Which is the stronger acid, $HAsO_4^{2-}$ or HPO_4^{2-}? How can you tell?

18.29 When sulfur dioxide, an air pollutant from the burning of sulfur-containing coal or oil, dissolves in water, an acidic solution is formed that can be viewed as containing sulfurous acid, H_2SO_3.

$$H_2O + SO_2(g) \rightleftharpoons H_2SO_3(aq)$$

Write the expression for K_{a_1} and K_{a_2} for sulfurous acid.

18.30 Citric acid, found in citrus fruits, is a triprotic acid, $H_3C_6H_5O_7$. Write chemical equations for the three-step ionization of this acid in water and write the expressions for their K_a.

18.31 If water is treated as any other weak diprotic Brønsted acid, what is the equation expressing its K_{a_1}? How does this expression differ from K_w? Write the expression for K_{a_2} for water.

18.32 What simplifying assumptions do we usually make in working problems involving the ionization of polyprotic acids? Why are they usually valid? Under what conditions do they fail?

Salts of Polyprotic Acids

18.33 Write the equations for the chemical equilibria that exist in solutions of (a) Na_2SO_3, (b) Na_3PO_4, and (c) $K_2C_4H_4O_6$.

18.34 What simplifying assumptions do we usually make in working problems involving equilibria of salts of polyprotic acids? Why are they usually valid? Under what conditions do they fail?

Titrations and Acid–Base Indicators

18.35 When a formic acid solution is titrated with sodium hydroxide, will the solution be acidic, neutral, or basic at the equivalence point?

18.36 When a solution of hydrazine is titrated by hydrochloric acid, will the solution be acidic, neutral, or basic at the equivalence point?

18.37 Define the terms *equivalence point* and *end point* as they apply to an acid–base titration.

18.38 Qualitatively, describe how an acid–base indicator works. Why do we want to use a minimum amount of indicator in a titration?

18.39 Explain why ethyl red is a better indicator than phenolphthalein in the titration of dilute ammonia by dilute hydrochloric acid.

18.40 What is a good indicator for titrating potassium hydroxide with hydrobromic acid? Explain.

18.41 If you use methyl orange in the titration of $HC_2H_3O_2$ with NaOH, will the end point of the titration correspond to the equivalence point? Justify your answer.

REVIEW PROBLEMS

Answers to problems whose numbers are printed in color are given in Appendix B. More challenging problems are marked with asterisks. **ILW** = Interactive LearningWare solution is available at *www.wiley.com/college/brady.*

Acid and Base Ionization Constants: K_a, K_b, pK_a, and pK_b

18.42 The K_a for HF is 6.8×10^{-4}. What is the K_b for F⁻?

18.43 The barbiturate ion, $C_4H_3N_2O_3^-$, has $K_b = 1.0 \times 10^{-10}$. What is K_a for barbituric acid?

18.44 Hydrogen peroxide, H_2O_2, is a weak acid with $K_a = 1.8 \times 10^{-12}$. What is the value of K_b for the HO_2^- ion?

18.45 Methylamine, CH_3NH_2, resembles ammonia in odor and basicity. Its K_b is 4.4×10^{-4}. Calculate the K_a of its conjugate acid.

18.46 Lactic acid, $HC_3H_5O_3$, is responsible for the sour taste of sour milk. At 25 °C, its $K_a = 1.4 \times 10^{-4}$. What is the K_b of its conjugate base, the lactate ion, $C_3H_5O_3^-$?

18.47 Iodic acid, HIO_3, has a pK_a of 0.23. (a) What is the formula and the K_b of its conjugate base? (b) Is its conjugate base a stronger or a weaker base than the acetate ion?

Equilibrium Calculations

18.48 A 0.20 *M* solution of a weak acid, H*A*, has a pH of 3.22. What is the percentage ionization of the acid?

18.49 If a weak base is 0.030% ionized in 0.030 *M* solution, what is the pH of the solution?

18.50 What is the value of K_a for the acid in Problem 18.48?

18.51 What is the value of K_b for the base in Problem 18.49?

18.52 Periodic acid, HIO_4, is an important oxidizing agent and a moderately strong acid. In a 0.10 *M* solution, $[H^+] = 3.8 \times 10^{-2}$ mol L^{-1}. Calculate the K_a and pK_a for periodic acid.

18.53 Chloroacetic acid, $HC_2H_2O_2Cl$, is a stronger monoprotic acid than acetic acid. In a 0.10 *M* solution, the pH is 1.96. Calculate the K_a and pK_a for chloroacetic acid.

ILW 18.54 Ethylamine, $CH_3CH_2NH_2$, has a strong, pungent odor similar to that of ammonia. Like ammonia, it is a Brønsted base. A 0.10 *M* solution has a pH of 11.86. Calculate the K_b and pK_b for ethylamine. What is the percentage ionization of ethylamine in this solution?

18.55 Hydroxylamine, $HONH_2$, like ammonia, is a Brønsted base. A 0.15 *M* solution has a pH of 10.12. What is the K_b and pK_b for hydroxylamine? What is the percentage ionization of the $HONH_2$?

ILW 18.56 What are the concentrations of all the solute species in 0.150 *M* lactic acid, $HC_3H_5O_2$? What is the pH of the solution? This acid has $K_a = 1.4 \times 10^{-4}$.

18.57 What are the concentrations of all the solute species in a 1.0 *M* solution of hydrogen peroxide, H_2O_2? What is the pH of the solution? For H_2O_2, $K_a = 1.8 \times 10^{-12}$.

18.58 What is the pH of 0.15 *M* hydrazoic acid, HN_3? For HN_3, $K_a = 1.8 \times 10^{-5}$. What percentage of the HN_3 is ionized?

18.59 Phenol, also known as carbolic acid, is sometimes used as a disinfectant. What are the concentrations of all of the solute species in a 0.050 *M* solution of phenol, HC_6H_5O? What percentage of the phenol is ionized? For this acid, $K_a = 1.3 \times 10^{-10}$.

18.60 Codeine, a cough suppressant extracted from crude opium, is a weak base with a pK_b of 5.79. What will be the pH of a 0.020 *M* solution of codeine? (Use *Cod* as a symbol for codeine.)

18.61 Pyridine, C_5H_5N, is a bad-smelling liquid that is a weak base in water. Its pK_b is 8.82. What is the pH of a 0.20 *M* aqueous solution of this compound?

18.62 A solution of acetic acid has a pH of 2.54. What is the concentration of acetic acid in this solution?

18.63 How many moles of NH_3 must be dissolved in water to give 500 mL of solution with a pH of 11.22?

Acid–Base Properties of Salt Solutions

18.64 Calculate the pH of 0.20 *M* NaCN. What is the concentration of HCN in the solution?

18.65 Calculate the pH of 0.40 *M* KNO_2. What is the concentration of HNO_2 in the solution?

ILW 18.66 Calculate the pH of 0.15 *M* CH_3NH_3Cl. For methylamine, CH_3NH_2, $K_b = 4.4 \times 10^{-4}$.

18.67 Calculate the pH of 0.10 *M* hydrazinium chloride, N_2H_5Cl.

18.68 Many drugs that are natural Brønsted bases are put into aqueous solution as their much more soluble salts with strong acids. The powerful painkiller morphine, for example, is very slightly soluble in water, but morphine nitrate is quite soluble. We may represent morphine by the symbol *Mor* and its conjugate acid as H—*Mor*⁺. The pK_b of morphine is 6.13. What is the calculated pH of a 0.20 *M* solution of H—*Mor*⁺?

18.69 Quinine, an important drug in treating malaria, is a weak Brønsted base that we may represent as *Qu*. To make it more soluble in water, it is put into a solution as its conjugate acid, which we may represent as H—*Qu*Cl. What is the calculated pH of a 0.15 *M* solution of H—*Qu*⁺? Its pK_a is 8.52 at 25 °C.

18.70 A 0.18 *M* solution of the sodium salt of nicotinic acid (also known pharmaceutically as niacin) has a pH of 9.05. What is the value of K_a for nicotinic acid?

18.71 A weak base *B* forms the salt *B*HCl, composed of the ions *B*H⁺ and Cl⁻. A 0.15 *M* solution of the salt has a pH of 4.28. What is the value of K_b for the base *B*?

18.72 What percentage of $C_5H_5NH^+$ reacts with water in a 0.10 *M* solution of pyridinium chloride, C_5H_5NHCl?

18.73 Calculate the number of grams of NH_4Br that have to be dissolved in 1.00 L of water at 25 °C to have a solution with a pH of 5.16.

*18.74 Liquid chlorine bleach is really nothing more than a solution of sodium hypochlorite, NaOCl, in water. Usually, the concentration is approximately 5% NaOCl by weight. Use this information to calculate the approximate pH of a bleach solution, assuming no other solutes are in the solution except NaOCl. (Assume the bleach has a density of 1.0 g/mL.)

*18.75 The conjugate acid of a molecular base has the hypothetical formula, BH^+, which has a pK_a of 5.00. A solution of a salt of this cation, BHY, tests slightly basic. Will the conjugate acid of Y^-, HY, have a pK_a greater than 5.00 or less than 5.00? Explain.

Equilibrium Calculations when Simplifications Fail

18.76 What is the percentage ionization in a 0.15 M solution of HF? What is the pH of the solution?

18.77 What is the percentage ionization of the solute in 1.0×10^{-3} M acetic acid? What is the pH of the solution?

18.78 What is the pH of a 0.0050 M solution of sodium cyanide?

18.79 What is the pH of a 0.020 M solution of chloroacetic acid, for which $K_a = 1.36 \times 10^{-3}$?

18.80 The compound *para*-aminobenzoic acid (PABA) is a powerful sunscreening agent whose salts were once used widely in suntanning and sunscreening lotions. The parent acid, which we may symbolize as H—*Paba*, is a weak acid with a pK_a of 4.92 (at 25 °C). What is the $[H^+]$ and pH of a 0.030 M solution of this acid?

18.81 Barbituric acid, $HC_4H_3N_2O_3$ (which we will abbreviate H—*Bar*), was discovered by the Nobel Prize-winning organic chemist Adolph von Baeyer and named after his friend Barbara. It is the parent compound of widely used sleeping drugs, the barbiturates. Its pK_a is 4.01. What is the $[H^+]$ and pH of a 0.020 M solution of H—*Bar*?

Buffers

18.82 What is the pH of a solution that contains 0.15 M $HC_2H_3O_2$ and 0.25 M $C_2H_3O_2^-$? Use $K_a = 1.8 \times 10^{-5}$ for $HC_2H_3O_2$.

18.83 Rework the preceding problem using the K_b for the acetate ion. (Be sure to write the proper chemical equation and equilibrium law.)

18.84 A buffer is prepared containing 0.25 M NH_3 and 0.45 M NH_4^+. Calculate the pH of the buffer using the K_b for NH_3.

18.85 Calculate the pH of the buffer in the preceding problem using the K_a for NH_4^+.

18.86 Suppose 25.0 mL of 0.10 M HCl is added to a 250 mL portion of a buffer composed of 0.25 M NH_3 and 0.20 M NH_4Cl. By how much will the concentrations of the NH_3 and NH_4^+ ions change after the addition of the strong acid?

18.87 A student added 100 mL of 0.10 M NaOH to 250 mL of a buffer that contained 0.15 M $HC_2H_3O_2$ and 0.25 M $C_2H_3O_2^-$. By how much did the concentrations of $HC_2H_3O_2$ and $C_2H_3O_2^-$ change after the addition of the strong base?

18.88 By how much will the pH change if 0.050 mol of HCl is added to 1.00 L of the buffer in Review Problem 18.82?

18.89 By how much will the pH change if 50.0 mL of 0.10 M NaOH is added to 500 mL of the buffer in Review Problem 18.82?

18.90 By how much will the pH change if 0.020 mol of HCl is added to 1.00 L of the buffer in Review Problem 18.84?

18.91 By how much will the pH change if 75 mL of 0.10 M KOH is added to 200 mL of the buffer in Review Problem 18.84?

18.92 What mole ratio of NH_4Cl to NH_3 would buffer a solution at pH 9.25?

18.93 What mole ratio of $NaC_3H_5O_2$ to $HC_3H_5O_2$ (propanoic acid) would give a buffer with a pH of 4.15?

18.94 How many grams of sodium acetate, $NaC_2H_3O_2$, would have to be added to 1.0 L of 0.15 M acetic acid (pK_a 4.74) to make the solution a buffer for pH 5.00?

18.95 How many grams of sodium formate, $NaCHO_2$, would have to be dissolved in 1.0 L of 0.12 M formic acid (pK_a 3.74) to make the solution a buffer for pH 3.80?

18.96 How many grams of ammonium chloride would have to be dissolved in 500 mL of 0.20 M NH_3 to prepare a solution buffered at pH 10.00?

18.97 How many grams of ammonium chloride have to be dissolved in 125 mL of 0.10 M NH_3 to make it a buffer with a pH of 9.15?

ILW 18.98 Suppose 25.00 mL of 0.100 M HCl is added to an acetate buffer prepared by dissolving 0.100 mol of acetic acid and 0.110 mol of sodium acetate in water. What are the initial and final pH values? What would be the pH if the same amount of HCl solution were added to 125 mL of pure water?

18.99 How many milliliters of 0.15 M HCl would have to be added to 100 mL of the buffer described in Review Problem 18.98 to make the pH decrease by 0.05 pH unit? How many milliliters of the same HCl solution would, if added to 100 mL of pure water, make the pH decrease by 0.05 pH unit?

Solutions of Polyprotic Acids

ILW 18.100 Calculate the concentrations of all the solute species in a 0.15 M solution of ascorbic acid (vitamin C). What is the pH of the solution?

18.101 Tellurium, in the same family as sulfur, forms an acid analogous to sulfuric acid called telluric acid. It exists, however, as H_6TeO_6 (which looks like the formula H_2TeO_4 + $2H_2O$). It is a diprotic acid with $K_{a_1} = 2 \times 10^{-8}$ and $K_{a_2} = 1 \times 10^{-11}$. Calculate the concentrations of H^+, $H_5TeO_6^-$, and $H_4TeO_6^{2-}$ in a 0.25 M solution of H_6TeO_6. What is the pH of the solution?

18.102 Calculate the concentrations of all of the solute species involved in the equilibria in a 3.0 M solution of H_3PO_4. Calculate the pH of the solution.

18.103 What is the pH of a 0.50 M solution of arsenic acid, H_3AsO_4? In this solution, what are the concentrations of $H_2AsO_4^-$ and $HAsO_4^{2-}$?

18.104 Tartaric acid, $H_2C_4H_4O_6$, is a diprotic acid for which $K_{a_1} = 9.2 \times 10^{-4}$ and $K_{a_2} = 4.3 \times 10^{-5}$. Calculate the molar concentrations of H^+ and the two anions of tartaric acid in a solution that has a concentration of 0.10 M.

18.105 A 0.20 M solution of tartaric acid contains enough HCl so that its hydrogen ion concentration equals 0.030 M. What are the molar concentrations of $[HC_4H_4O_6^-]$ and $C_4H_4O_6^{2-}$ in the solution?

*18.106 Phosphorous acid, H_3PO_3, is actually a diprotic acid for which $K_{a_1} = 1.0 \times 10^{-2}$ and $K_{a_2} = 2.6 \times 10^{-7}$. What are the values of $[H^+]$, $[H_2PO_3^-]$, and $[HPO_3^{2-}]$ in a 1.0 M solution of H_3PO_3? What is the pH of the solution?

*18.107 What is the pH of a 0.20 M solution of oxalic acid, $H_2C_2O_4$?

Solutions of Salts of Polyprotic Acids

18.108 Calculate the pH of 0.12 M Na_2SO_3. What are the concentrations of HSO_3^- and H_2SO_3 in the solution?

18.109 Calculate the pH of 0.15 M K_2CO_3. What are the concentrations of HCO_3^- and H_2CO_3 in the solution?

18.110 The decahydrate of sodium carbonate, $Na_2CO_3 \cdot 10H_2O$, is known as washing soda. How many grams of $Na_2CO_3 \cdot 10H_2O$ have to be dissolved in 1.00 L of water at 25 °C to have a solution with a pH of 11.62?

18.111 Calculate the number of grams of $Na_2SO_3 \cdot 7H_2O$ that you would have to dissolve in 0.50 L of water at 25 °C to have a solution with a pH of 10.15.

18.112 Sodium citrate, $Na_3C_6H_5O_7$, is used as an anticoagulant in the collection of blood. What is the pH of a 0.10 M solution of this salt?

18.113 What is the pH of a 0.25 M solution of sodium oxalate, $Na_2C_2O_4$?

*18.114 What is the pH of a 0.50 M solution of Na_3PO_4? In this solution, what are the concentrations of HPO_4^{2-}, $H_2PO_4^-$, and H_3PO_4?

*18.115 The pH of a 0.10 M Na_2CO_3 solution is adjusted to 12.00 using a strong base. What is the concentration of CO_3^{2-} in this solution?

Acid–Base Titrations

ILW 18.116 When 50 mL of 0.10 M formic acid, $HCHO_2$, is titrated with 0.10 M sodium hydroxide, what is the pH at the equivalence point? (Be sure to take into account the change in volume during the titration.) Select a good indicator for this titration from Table 17.5.

18.117 When 25 mL of 0.10 M aqueous ammonia is titrated with 0.10 M hydrobromic acid, what is the pH at the equivalence point? Select a good indicator for this titration from Table 17.5.

18.118 What is the pH of a solution prepared by mixing 25.0 mL of 0.180 M $HC_2H_3O_2$ with 35.0 mL of 0.250 M NaOH?

18.119 What is the pH of a solution prepared by mixing exactly 30.0 mL of 0.200 M $HC_2H_3O_2$ with 15.0 mL of 0.400 M KOH?

*18.120 For the titration of 25.00 mL of 0.1000 M acetic acid with 0.1000 M NaOH, calculate the pH (a) before the addition of any NaOH solution, (b) after 10.00 mL of the base has been added, (c) after half of the $HC_2H_3O_2$ has been neutralized, and (d) at the equivalence point.

*18.121 For the titration of 25.00 mL of 0.1000 M ammonia with 0.1000 M HCl, calculate the pH (a) before the addition of any HCl solution, (b) after 10.00 mL of the acid has been added, (c) after half of the NH_3 has been neutralized, and (d) at the equivalence point.

ADDITIONAL EXERCISES

18.122 Calculate the percentage ionization of acetic acid in solutions having concentrations of 1.0 M, 0.10 M, and 0.010 M. How does the percentage ionization of a weak acid change as the acid becomes more dilute?

18.123 What is the pH of a solution that is 0.100 M in HCl and also 0.125 M in $HC_2H_3O_2$? What is the concentration of acetate ion in this solution?

18.124 A solution is prepared by mixing 300 mL of 0.500 M NH_3 and 100 mL of 0.500 M HCl. Assuming that the volumes are additive, what is the pH of the resulting mixture?

18.125 What is the pH of 0.120 M NH_4NO_3?

18.126 A solution is prepared by dissolving 15.0 g of pure $HC_2H_3O_2$ and 25.0 g of $NaC_2H_3O_2$ in 750 mL of solution (the final volume). (a) What is the pH of the solution? (b) What is the pH of the solution after 25.00 mL of 0.25 M NaOH is added? (c) What is the pH after 25.0 mL of 0.40 M HCl is added to the original solution?

*18.127 For an experiment involving what happens to the growth of a particular fungus in a slightly acidic medium, a biochemist needs 250 mL of an acetate buffer with a pH of 5.12. The buffer solution has to be able to hold the pH to

within ±0.10 pH unit of 5.12 even if 0.0100 mol of NaOH or 0.0100 mol of HCl enters the solution.
(a) What is the minimum number of grams of acetic acid and of sodium acetate dihydrate that must be used to prepare the buffer?
(b) Describe the buffer by giving its molarity in acetic acid and its molarity in sodium acetate.
(c) What is the pH of an unbuffered solution made by adding 0.0100 mol of NaOH to 250 mL of pure water?
(d) What is the pH of an unbuffered solution made by adding 0.0100 mol of HCl to 250 mL of pure water?

*18.128 An analytical procedure for separating certain components of a dietary supplement requires that they be dissolved in a buffered solution having a pH of 3.80. This buffer has to be able to handle the introduction of 5.0×10^{-3} mol of either OH^- or H^+ without changing its pH by more than 0.050 unit.
(a) Using data in Table 18.1, pick the best acid (and its sodium salt) that could be used for preparing the buffered solution.
(b) Calculate the minimum number of grams of the pure acid and of the pure salt needed to make 750 mL of the buffered solution.

(c) Describe the buffer by giving its molarity in the selected acid and its molarity in the sodium salt of the acid.

(d) If 5.0×10^{-3} mol of HCl were added to 750 mL of a solution of hydrochloric acid having a pH of 3.80, what would be the new pH?

(e) If 5.0×10^{-3} mol of NaOH were added to 750 mL of a solution of hydrochloric acid with a pH of 3.80, what would be the new pH?

18.129 Predict whether the pH of 0.120 M NH_4CN is greater than, less than, or equal to 7.00. Give your reasons.

*__18.130__ What is the approximate freezing point of a 0.50 M solution of dichloroacetic acid, $HC_2HO_2Cl_2$ ($K_a = 5.0 \times 10^{-2}$)? Assume the density of the solution is 1.0 g/mL.

*__18.131__ How many milliliters of ammonia gas measured at 25 °C and 740 torr must be dissolved in 250 mL of 0.050 M HNO_3 to give a solution with a pH of 9.26?

*__18.132__ The hydrogen sulfate ion, HSO_4^-, is a moderately strong Brønsted acid with a K_a of 1.0×10^{-2}.

(a) Write the chemical equation for the ionization of the acid and give the appropriate K_a expression.

(b) What is the value of $[H^+]$ in 0.010 M HSO_4^- (furnished by the salt, $NaHSO_4$)? Do NOT make simplifying assumptions.

(c) What is the calculated $[H^+]$ in 0.010 M HSO_4^-, obtained by using the usual simplifying assumption?

(d) How much error is introduced by incorrectly using the simplifying assumption?

18.133 Some people who take megadoses of ascorbic acid will drink a solution containing as much as 6.0 g of ascorbic acid dissolved in a glass of water. Assuming the volume to be 250 mL, calculate the pH of this solution.

18.134 What molar concentration of HCO_3^- can exist in equilibrium in arterial blood at a pH of 7.35 when the molar concentration of CO_2 is 2.2×10^{-2} mol L^{-1}? Assume that all of the CO_2 is present as H_2CO_3.

*__18.135__ Calculate the concentrations of all solute species in 0.30 M H_3PO_4. (Use the method of successive approximations for the first step in the ionization.)

*__18.136__ For the titration of 25.00 mL of 0.1000 M HCl with 0.1000 M NaOH, calculate the pH of the reaction mixture after each of the following total volumes of base have been added to the original solution. (Remember to take into account the change in total volume.) Construct a graph showing the titration curve for this experiment.

(a) 0 mL	(d) 24.99 mL	(g) 25.10 mL
(b) 10.00 mL	(e) 25.00 mL	(h) 26.00 mL
(c) 24.90 mL	(f) 25.01 mL	(i) 50.00 mL

Solubility and Simultaneous Equilibria

Creatures such as this sea scallop are able to precipitate calcium carbonate from sea water to form their hard protective shells. The tiny blue dots around the edge of its shell are eyes that enable the scallop to sense light and movement and thereby elude potential predators. In this chapter you will learn about solubility equilibria and the conditions necessary for precipitates to form.

CHAPTER OUTLINE

THIS CHAPTER IN CONTEXT By now you've learned that the principles of equilibrium govern the outcome of many of the chemical reactions that occur in the world around us. And many of these reactions involve the formation and dissolving of precipitates (a topic we introduced in Chapter 5). For example, groundwater rich in carbon dioxide dissolves deposits of calcium carbonate, producing vast underground caverns, and as the remaining calcium-containing solution gradually evaporates, stalactites and stalagmites form. Similar deposits of $CaCO_3$ form from hard water and leave hard-to-dissolve spots on glass surfaces and in pipes. Within living organisms, precipitation reactions form the hard calcium carbonate shells of clams, oysters, and coral as well as the unwanted calcium oxalate and calcium phosphate deposits we call kidney stones (Facets of Chemistry 5.2). On the other hand, dilute acids in our mouths promote the dissolving of tooth enamel, which is composed of a mineral made up of calcium phosphate and calcium hydroxide.

In this chapter we will apply the principles of equilibrium to the study of solubility. You will learn how we can calculate solubilities of "insoluble" salts in water and in other solutions, and how the formation of substances called *complex ions* can affect solubilities. The concepts we develop here can tell us when precipitates will form and when they will dissolve, and we can use these in a practical lab setting in the separation of metal ions for chemical analysis.

19.1 An insoluble salt is in equilibrium with the solution around it

The equilibrium constant for an "insoluble" salt is called the solubility product constant, K_{sp}

None of the salts we described in Chapter 5 as being insoluble are *totally* insoluble. When placed into pure water, a small amount will dissolve. For example, consider the salt AgCl, which the solubility rules tell us is "insoluble." If we place some solid AgCl in water, a very small amount dissolves, and the following equilibrium exists when the solution has become saturated.

$$AgCl(s) \rightleftharpoons Ag^+(aq) + Cl^-(aq)$$
(in a saturated solution of AgCl)

This is a heterogeneous equilibrium because it involves a solid reactant (AgCl) in equilibrium with ions in aqueous solution. Using the procedure developed in Section 16.7 (page 709), we write the equilibrium law as follows, omitting the solid from the mass action expression:

$$[Ag^+][Cl^-] = K_{sp} \tag{19.1}$$

Recall that when salts dissolve, they dissociate essentially completely, so the equilibrium is between the solid and the ions that are in the solution.

TOOLS

Solubility product expression

The equilibrium constant, K_{sp}, is called the **solubility product constant** (because the system is a *solubility* equilibrium and the constant equals a *product* of ion concentrations).

It's important that you understand the distinction between solubility and solubility product. The *solubility* of a salt is the amount of the salt that dissolves in a given amount of solvent to give a saturated solution. The *solubility product* is the product of the molar concentrations of the ions in the saturated solution, raised to appropriate powers (see below).

The solubilities of salts change with temperature, so a value of K_{sp} applies only at the temperature at which it was determined. Some typical K_{sp} values are in Table 19.1 and in Appendix C. (Insoluble oxides and sulfides make up a special group to be discussed in Section 19.2.)

Ion products

The term on the left in Equation 19.1, the product of the molar concentrations of the dissolved ions of the solute, is called the **ion product** of AgCl. At any dilution of the salt throughout the entire range of possibilities for an *unsaturated* solution, there will be varying values for the ion concentrations and, therefore, for the ion product. However, the ion product acquires a constant value, K_{sp}, in a *saturated* solution. Only then is the ion product equal to the *solubility product constant*. When a solution of AgCl is less than saturated, the value of the ion product is less than K_{sp}. Thus, we can use the numerical value of the ion product for a given solution as a test for saturation by comparing it to the value of K_{sp}.

Many salts produce more than one of a given ion when they dissociate, and this introduces exponents into the ion product expression. For example, when silver chromate, Ag_2CrO_4, precipitates (Figure 19.1), it enters into the following solubility equilibrium:

$$Ag_2CrO_4(s) \rightleftharpoons 2Ag^+(aq) + CrO_4^{2-}(aq)$$

◄**TOOLS**
Ion product of a salt

TABLE 19.1	**SOLUBILITY PRODUCT CONSTANTS**						
Type	Salt	Ions of Salt	K_{sp} (25 °C)	Type	Salt	Ions of Salt	K_{sp} (25 °C)
Halides	$CaF_2 \rightleftharpoons Ca^{2+} + 2F^-$		3.9×10^{-11}	**Carbonates**	$SrCO_3 \rightleftharpoons Sr^{2+} + CO_3^{2-}$		9.3×10^{-10}
	$PbF_2 \rightleftharpoons Pb^{2+} + 2F^-$		3.6×10^{-8}	**(cont.)**	$BaCO_3 \rightleftharpoons Ba^{2+} + CO_3^{2-}$		5.0×10^{-9}
	$AgCl \rightleftharpoons Ag^+ + Cl^-$		1.8×10^{-10}		$CoCO_3 \rightleftharpoons Co^{2+} + CO_3^{2-}$		1.0×10^{-10}
	$AgBr \rightleftharpoons Ag^+ + Br^-$		5.0×10^{-13}		$NiCO_3 \rightleftharpoons Ni^{2+} + CO_3^{2-}$		1.3×10^{-7}
	$AgI \rightleftharpoons Ag^+ + I^-$		8.3×10^{-17}		$ZnCO_3 \rightleftharpoons Zn^{2+} + CO_3^{2-}$		1.0×10^{-10}
	$PbCl_2 \rightleftharpoons Pb^{2+} + 2Cl^-$		1.7×10^{-5}	**Chromates**	$Ag_2CrO_4 \rightleftharpoons 2Ag^+ + CrO_4^{2-}$		1.2×10^{-12}
	$PbBr_2 \rightleftharpoons Pb^{2+} + 2Br^-$		2.1×10^{-6}		$PbCrO_4 \rightleftharpoons Pb^{2+} + CrO_4^{2-}$		$1.8 \times 10^{-14}(d)$
	$PbI_2 \rightleftharpoons Pb^{2+} + 2I^-$		7.9×10^{-9}	**Sulfates**	$CaSO_4 \rightleftharpoons Ca^{2+} + SO_4^{2-}$		2.4×10^{-5}
Hydroxides	$Al(OH)_3 \rightleftharpoons Al^{3+} + 3OH^-$		$3 \times 10^{-34}(a)$		$SrSO_4 \rightleftharpoons Sr^{2+} + SO_4^{2-}$		3.2×10^{-7}
	$Ca(OH)_2 \rightleftharpoons Ca^{2+} + 2OH^-$		6.5×10^{-6}		$BaSO_4 \rightleftharpoons Ba^{2+} + SO_4^{2-}$		1.1×10^{-10}
	$Fe(OH)_2 \rightleftharpoons Fe^{2+} + 2OH^-$		7.9×10^{-16}		$PbSO_4 \rightleftharpoons Pb^{2+} + SO_4^{2-}$		6.3×10^{-7}
	$Fe(OH)_3 \rightleftharpoons Fe^{3+} + 3OH^-$		1.6×10^{-39}	**Oxalates**	$CaC_2O_4 \rightleftharpoons Ca^{2+} + C_2O_4^{2-}$		2.3×10^{-9}
	$Mg(OH)_2 \rightleftharpoons Mg^{2+} + 2OH^-$		7.1×10^{-12}		$MgC_2O_4 \rightleftharpoons Mg^{2+} + C_2O_4^{2-}$		8.6×10^{-5}
	$Zn(OH)_2 \rightleftharpoons Zn^{2+} + 2OH^-$		$3.0 \times 10^{-16}(b)$		$BaC_2O_4 \rightleftharpoons Ba^{2+} + C_2O_4^{2-}$		1.2×10^{-7}
Carbonates	$Ag_2CO_3 \rightleftharpoons 2Ag^+ + CO_3^{2-}$		8.1×10^{-12}		$FeC_2O_4 \rightleftharpoons Fe^{2+} + C_2O_4^{2-}$		2.1×10^{-7}
	$MgCO_3 \rightleftharpoons Mg^{2+} + CO_3^{2-}$		3.5×10^{-8}		$PbC_2O_4 \rightleftharpoons Pb^{2+} + C_2O_4^{2-}$		2.7×10^{-11}
	$CaCO_3 \rightleftharpoons Ca^{2+} + CO_3^{2-}$		$4.5 \times 10^{-9}(c)$				

[a]Alpha form. [b]Amorphous form. [c]Calcite form. [d]At 10 °C.

FIGURE 19.1 *Silver chromate.* When sodium chromate is added to a solution of silver nitrate, deep red, "insoluble" silver chromate, Ag_2CrO_4, precipitates.

Molar solubility

When fully dissociated, each formula unit of Ag_2CrO_4 gives two Ag^+ ions and one CrO_4^{2-} ion. In the ion product of such a salt, each concentration term carries an exponent that equals the number of ions released per formula unit. So the ion product for silver chromate is $[Ag^+]^2[CrO_4^{2-}]$; the exponent of 2 appears because the formula of Ag_2CrO_4 has two Ag^+ ions per formula unit. Also notice that to obtain the correct ion product expression, you have to know the formulas of the ions that make up the salt. Thus, you have to realize that Ag_2CrO_4 is composed of Ag^+ and the polyatomic anion CrO_4^{2-}. If necessary, review the list of polyatomic ions in Table 2.5 on page 62.

> **PRACTICE EXERCISE 1:** What are the ion product expressions for the following salts, assuming they are fully dissociated in water: (a) BaC_2O_4, (b) Ag_3PO_4?

K_{sp} can be determined from molar solubilities

One way to determine the value of K_{sp} for a slightly soluble salt is to measure its solubility—how much of the salt is required to give a saturated solution in a specified amount of solution. The molar concentration of the salt in its saturated solution is called the **molar solubility;** it equals *the number of moles of salt dissolved in one liter of the saturated solution.* The molar solubility can be used to calculate the K_{sp} under the assumption that all of the salt that dissolves is 100% dissociated into the ions implied in the salt's formula.[1]

EXAMPLE 19.1

Calculating K_{sp} from Solubility Data

Silver bromide, AgBr, is the light-sensitive compound used in nearly all photographic film. The solubility of AgBr in water was measured to be 1.3×10^{-4} g L^{-1} at 25 °C. Calculate K_{sp} for AgBr at this temperature.

ANALYSIS: As usual, we begin with the chemical equation for the equilibrium, from which we construct the equilibrium law (here, the expression for K_{sp}).

$$AgBr(s) \rightleftharpoons Ag^+(aq) + Br^-(aq)$$

$$K_{sp} = [Ag^+][Br^-]$$

To calculate K_{sp}, we need the concentrations of the ions expressed in moles per liter. We will calculate these concentrations from the solubility, which tells us the number of grams of AgBr that dissolve per liter to give a saturated solution. The first step will be to convert 1.3×10^{-4} g of AgBr to moles, which will give us the molar solubility of the salt—the number of moles of AgBr dissolved per liter. From the molar solubility of the salt, we will calculate the molarities of the ions in the solution. We will proceed carefully by setting up a concentration table similar to those we've used in other equilibrium calculations.

In setting up the table, it is helpful if we imagine the formation of the saturated solution to occur stepwise. First, we will look at the composition of the solvent into which the salt will be placed. Does it contain any of the ions involved in the equilibrium? If it doesn't, the initial concentrations will be set equal to zero. However, if the solvent contains a solute that is a source of one of the ions in the equilibrium, we will use its concentration as the initial concentration of that ion.

When the salt dissolves, the concentrations of the ions increase, so the entries in the "change" row will be positive and will have values determined by the stoichiometry of the salt. We will obtain these from the molar solubility.

---

[1]This assumption works reasonably well for slightly soluble salts made up of univalent ions, like silver bromide. For simplicity, and to illustrate the nature of calculations involving solubility equilibria, we will work on the assumption that *all* salts behave as though they are 100% dissociated. This is not entirely true, especially for salts of polyvalent ions, so the accuracy of our calculations is limited. The issues involved are discussed further in Facets of Chemistry 19.1.

The entries in the last row, which are the equilibrium values, are obtained by adding the initial concentrations to the changes. Once we have the equilibrium ion concentrations, we substitute them into the ion product expression to calculate K_{sp}.

SOLUTION: First, we calculate the number of moles of AgBr dissolved per liter.

$$\text{solubility} = \frac{1.3 \times 10^{-4}\,\text{g AgBr}}{1.00\,\text{L soln}} \times \frac{1.00\,\text{mol AgBr}}{187.77\,\text{g AgBr}} = 6.9 \times 10^{-7}\,\text{mol L}^{-1}$$

Now we can begin to set up the concentration table, which is shown below. Notice first that there are no entries in the "reactant" column. This is because AgBr is a solid.

Initial Concentrations

In the first row, under the formulas of the ions, we enter the initial concentrations of Ag^+ and Br^-. *Remember, these are the concentrations of the ions present in the solvent before any of the AgBr dissolves.* Careful reading of the problem tells us that the solvent is pure water; if the solvent already contained another source of Ag^+ or Br^- (such as $AgNO_3$ or KBr), it would be stated in the problem. Because neither Ag^+ nor Br^- is present in pure water, we set the initial concentrations equal to zero.

Changes in Concentrations

In the "change" row, we enter data on how the concentrations of the $Ag^+(aq)$ and $Br^-(aq)$ change when the AgBr dissolves. Because dissolving the salt always increases these concentrations, the quantities are entered as positive values. The entries in this row are related to each other by the stoichiometry of the dissociation reaction. Thus, when AgBr dissolves, Ag^+ and Br^- ions are released in a 1:1 ratio. So, when 6.9×10^{-7} mol of AgBr dissolves (per liter), 6.9×10^{-7} mol of Ag^+ ion and 6.9×10^{-7} mol of Br^- ion go into solution. The concentrations of these species thus *change* (increase) by these amounts.

Equilibrium Concentrations

As usual, we obtain the equilibrium values by adding the "initial concentrations" to the "changes."

	AgBr(s) $\rightleftharpoons$	$Ag^+(aq)$ +	$Br^-(aq)$
Initial concentrations (M)	No entries in this column	0	0
Changes in concentrations when AgBr dissolves (M)		$+6.9 \times 10^{-7}$	$+6.9 \times 10^{-7}$
Equilibrium concentrations (M)		6.9×10^{-7}	6.9×10^{-7}

There are no entries in the column under AgBr(s) because this substance does not appear in the K_{sp} expression.

We now substitute the equilibrium ion concentrations into the K_{sp} expression.

$$K_{sp} = [Ag^+]\,[Br^-]$$
$$= (6.9 \times 10^{-7})(6.9 \times 10^{-7})$$
$$= 4.8 \times 10^{-13}$$

The K_{sp} of AgBr is thus calculated to be 4.8×10^{-13} at 25 °C.

The K_{sp} we calculated from the solubility data here differs by only 4% from the value in Table 19.1.

Is the Answer Reasonable?

We've divided 1.3×10^{-4} by a number that's approximately 200, which would give a value of about 6.5×10^{-7}, so our molar solubility seems reasonable. (Also, we know AgBr has a very low solubility in water, so we expect the molar solubility to be very small.) The reasoning involved in the change row also makes sense; the number of moles of ions formed per liter must each equal the number of moles of AgBr that dissolve. Finally, if we round 6.9×10^{-7} to 7×10^{-7} and square it, we obtain $49 \times 10^{-14} = 4.9 \times 10^{-13}$. Our answer of 4.8×10^{-13} seems to be okay.

EXAMPLE 19.2

Calculating K_{sp} from Molar Solubility Data

The molar solubility of silver chromate, Ag_2CrO_4, in water is 6.7×10^{-5} mol L^{-1} at 25 °C. What is K_{sp} for Ag_2CrO_4?

ANALYSIS: We begin with the equilibrium equation and K_{sp} expression.

$$Ag_2CrO_4(s) \rightleftharpoons 2Ag^+(aq) + CrO_4^{2-}(aq) \qquad K_{sp} = [Ag^+]^2[CrO_4^{2-}]$$

The next step is to set up the concentration table.

The solute is pure water, so neither Ag^+ nor CrO_4^{2-} is present before the salt dissolves. The initial concentrations of Ag^+ and CrO_4^{2-} are therefore zero. For the "change" row, we have to be careful to take into account the stoichiometry of the salt. When one liter of pure water dissolves 6.7×10^{-5} mol of Ag_2CrO_4, we can see by the coefficients in the equilibrium equation that 100% dissociation would produce 6.7×10^{-5} mol of CrO_4^{2-} and $2 \times (6.7 \times 10^{-5}$ mol) of Ag^+. With this information, we can fill in the "initial" and "change" rows.

SOLUTION: As usual, we add the values in the "initial" and "change" rows to obtain the equilibrium concentrations of the ions.

$Ag_2CrO_4(s)$	$\rightleftharpoons$	$2Ag^+(aq)$	+	$CrO_4^{2-}(aq)$
Initial concentrations (M)		0		0
Changes in concentrations when Ag_2CrO_4 dissolves (M)		$+[2 \times (6.7 \times 10^{-5})]$ $= +1.3 \times 10^{-4}$		$+6.7 \times 10^{-5}$
Equilibrium concentrations (M)		1.3×10^{-4}		6.7×10^{-5}

Substituting the equilibrium concentrations into the mass action expression for K_{sp} gives

$$K_{sp} = (1.3 \times 10^{-4})^2(6.7 \times 10^{-5})$$
$$= 1.1 \times 10^{-12}$$

So the K_{sp} of Ag_2CrO_4 at 25 °C is calculated to be 1.1×10^{-12}.

Is the Answer Reasonable?
The critical link in solving the problem correctly is placing the correct quantities in the "initial" and "change" rows. The Ag_2CrO_4 is dissolved in water, so the initial concentrations of the ions must be zero; these entries are okay. For the "changes," it's important to remember that the quantities are related to each other by the coefficients in the chemical equation. That means the change for Ag^+ must be twice as large as the change for CrO_4^{2-}. Studying the table, we see that we've done this correctly. We can also see that we've added the "change" to the "initial" values correctly and that we've performed the proper arithmetic in evaluating the mass action expression.

PRACTICE EXERCISE 2: The solubility of thallium(I) iodide, TlI, in water at 20 °C is 1.8×10^{-5} mol L^{-1}. Using this fact, calculate K_{sp} for TlI on the assumption that it is 100% dissociated in the solution.

PRACTICE EXERCISE 3: One liter of water is able to dissolve 2.15×10^{-3} mol of PbF_2 at 25 °C. Calculate the value of K_{sp} for PbF_2.

Now let's introduce another complication. Let's suppose that the aqueous system into which we're dissolving a slightly soluble salt is not pure water, but instead is a solution of a solute that provides one of the ions of the salt.

The molar solubility of $PbCl_2$ in a 0.10 M NaCl solution is 1.7×10^{-3} mol L^{-1} at 25 °C. Calculate the K_{sp} for $PbCl_2$.

ANALYSIS: Again, we start by writing the equation for the equilibrium and the K_{sp} expression.

$$PbCl_2(s) \rightleftharpoons Pb^{2+}(aq) + 2Cl^-(aq) \qquad K_{sp} = [Pb^{2+}][Cl^-]^2$$

In this problem, the $PbCl_2$ is being dissolved not in pure water, but in 0.10 M NaCl, which contains 0.10 M Cl^-. (It also contains 0.10 M Na^+, but that's not important here because Na^+ doesn't affect the equilibrium and so doesn't appear in the K_{sp} expression.) The initial concentration of Pb^{2+} is zero because none is in solution to begin with. The *initial* concentration of Cl^-, however, is 0.10 M. When the $PbCl_2$ dissolves in the NaCl solution, the Pb^{2+} concentration increases by 1.7×10^{-3} M and the Cl^- concentration increases by $2 \times (1.7 \times 10^{-3}$ $M)$. With these data we build the concentration table and evaluate the ion product to obtain K_{sp}.

SOLUTION:

$PbCl_2(s)$	$\rightleftharpoons$	$Pb^{2+}(aq)$	$+$	$2Cl^-(aq)$
Initial concentrations (M)		0		0.10 (because the solvent is 0.10 M NaCl)
Changes in concentrations when $PbCl_2$ dissolves (M)		$+1.7 \times 10^{-3}$		$+[2 \times (1.7 \times 10^{-3})]$ $= +3.4 \times 10^{-3}$
Equilibrium concentrations (M)		1.7×10^{-3}		$0.10 + 0.0034 = 0.10$ (rounded correctly)

Substituting the equilibrium concentrations into the K_{sp} expression gives

$$K_{sp} = (1.7 \times 10^{-3})(0.10)^2$$

$$= 1.7 \times 10^{-5}$$

The K_{sp} of $PbCl_2$ is calculated to be 1.7×10^{-5} at 25 °C.

Is the Answer Reasonable?
The solvent here is 0.10 M NaCl, so we have Cl^- in the mixture before any $PbCl_2$ dissolves. We check the "initial" row to be sure we've entered 0.10 M under Cl^-, and we have. Next, we check to be sure we've taken the stoichiometry of the equilibrium equation into account when placing values into the "change" row. We've done this correctly, too, because the change for Cl^- is twice the change for Pb^{2+}. Finally, we can check to be sure we've performed the correct arithmetic on the equilibrium concentrations. We have, so our answer should be correct.

PRACTICE EXERCISE 4: At 25 °C, the molar solubility of $CoCO_3$ in a 0.10 M Na_2CO_3 solution is 1.0×10^{-9} mol L^{-1}. Calculate the value of K_{sp} for $CoCO_3$.

PRACTICE EXERCISE 5: The molar solubility of PbF_2 in a 0.10 M $Pb(NO_3)_2$ solution at 25 °C is 3.1×10^{-4} mol L^{-1}. Calculate the value of K_{sp} for PbF_2.

EXAMPLE 19.3

Calculating K_{sp} from Molar Solubility Data

We are assuming that for each formula unit of $PbCl_2$ that dissolves, we get one Pb^{2+} ion and two Cl^- ions.

Molar solubility can be calculated from K_{sp}

Besides calculating K_{sp} from solubility information, we can also compute solubilities from values of K_{sp}. The following examples show how to make such calculations.[2]

---

[2]As noted earlier, these calculations ignore the fact that not all of the salt that dissolves is 100% dissociated into the ions implied in the salt's formula. This is particularly a problem with the salts of polyvalent ions, so the solubilities of such salts, when calculated from their K_{sp} values, must be taken as rough estimates. In fact, many calculations involving K_{sp} give values that are just estimates.

FACETS OF CHEMISTRY 19.1

Ion Pairs

Like silver bromide, calcium sulfate, $CaSO_4$, has a 1:1 ratio of ions. How well does it work, then, to take the square of the molar solubility of $CaSO_4$ to calculate the value of this salt's K_{sp}? Let's see.

The equation for the solubility product constant of $CaSO_4$ is

$$K_{sp} = [Ca^{2+}][SO_4^{2-}]$$

We know by experiment that the actual molar solubility of $CaSO_4$ at $25\,°C$ is 1.53×10^{-2} mol L^{-1}. If we assume complete dissociation, we obtain a value of 2.34×10^{-4} for the (calculated) K_{sp} of calcium sulfate. However, the ion product we calculate this way is not the true value of the actual *ion product* in the saturated solution, namely, $[Ca^{2+}][SO_4^{2-}]$, an expression that requires that we use the effective molarities of the separated ions. In a saturated solution of $CaSO_4$, the true value of K_{sp} is known to be closer to 2.4×10^{-5}, which is about one-tenth the value of K_{sp} calculated from the molar solubility. The true value of K_{sp} implies that the molarities of the separated ions are less than we assume from the molar solubility. What's wrong?

Formation of Ion Pairs

As we mentioned in Chapter 14, in a solution of a salt, the cations and anions encounter each other from time to time, and because of their opposite charges, they attract and tend to linger near each other for a while, forming **ion pairs.** When this happens, the salt behaves somewhat like it is less than 100% dissociated.

The tendency for ions to form ion pairs depends on the concentration (the higher the concentration, the greater the chance that the ions will meet) and it also depends to a large extent on the charges on the ions. With a substance such as $CaSO_4$, the divalent ions released when the salt dissociates are likely to form more stable (and longer-lasting) ion pairs than monovalent ions of a solute such as AgBr. As a result, $CaSO_4$ behaves as though it is less fully dissociated than AgBr. Therefore, the effective concentration of ions in a solution of a salt composed of divalent ions is less than in a solution of a salt made up of monovalent ions. The consequence of all this is that the K_{sp} calculated assuming complete dissociation is substantially larger than it would be if we were to take interionic attractions into account. It also means that a molar solubility calculated from K_{sp} would tend to be *smaller* than the actual solubility of the salt expressed in moles per liter.

Interionic attractions cause severe limitations in the accuracy of K_{sp} calculations. Nevertheless, such calculations help us understand in at least a semiquantitative way the principles involved in phenomena like the common ion effect, our ability to selectively precipitate metal ions as insoluble salts, and the influence of the formation of complex ions on the solubilities of salts—all of which are discussed in this chapter.

EXAMPLE 19.4 **Calculating Molar Solubility from K_{sp}**	What is the molar solubility of AgCl in pure water at $25\,°C$?

ANALYSIS: To solve this problem, we need three things:

1. The equilibrium chemical equation for a saturated solution of AgCl
2. The K_{sp} equation, which we figure out from the equilibrium equation
3. The value of K_{sp} for AgCl

The relevant equations are

$$AgCl(s) \rightleftharpoons Ag^+(aq) + Cl^-(aq) \qquad K_{sp} = [Ag^+][Cl^-] = 1.8 \times 10^{-10}$$

The value of K_{sp} is from Table 19.1. Now we can build a concentration table. First, we see that the solvent is pure water, so neither Ag^+ nor Cl^- ion is present in solution at the start; their *initial* concentrations are zero.

Next, we turn to the "change" row. If we knew what the changes were, we could calculate the equilibrium concentrations and figure out the molar solubility. But we don't know the changes, so we will have to find them algebraically. To do this, *we will define our unknown x as the molar solubility of the salt*—the number of moles of AgCl that dissolves in 1 L. Because 1 mol of AgCl yields 1 mol Ag^+ and 1 mol Cl^-, the concentration of each of these ions increases by x. Thus, our concentration table is as follows.

AgCl(s)	$\rightleftharpoons$	Ag$^+$(aq)	+	Cl$^-$(aq)
Initial concentrations (M)		0		0
Changes in concentrations when AgCl dissolves (M)		$+x$		$+x$
Equilibrium concentrations (M)		x		x

By defining x as the molar solubility, the coefficients of x in the "change" row are the same as the coefficients of Ag$^+$ and Cl$^-$ in the equation for the equilibrium.

SOLUTION: We make the substitutions into the K_{sp} expression, using equilibrium quantities:

$$(x)(x) = 1.8 \times 10^{-10}$$

$$x = 1.3 \times 10^{-5}$$

The calculated molar solubility of AgCl in water at 25 °C is 1.3×10^{-5} mol L^{-1}.

Is the Answer Reasonable?
We check our entries in the concentration table. The solvent is water, so the initial concentrations are zero. If x is the molar solubility, then the changes in the concentrations of Ag$^+$ and Cl$^-$ when the salt dissolves must also be equal to x. We can also check the algebra, which we've done correctly, so the answer seems to be okay.

Calculate the molar solubility of lead iodide, PbI$_2$, from its K_{sp} in water at 25 °C.

ANALYSIS: As usual, we begin with the equation for the equilibrium and the K_{sp} expression. We obtain K_{sp} from Table 19.1.

$$\text{PbI}_2(s) \rightleftharpoons \text{Pb}^{2+}(aq) + 2\text{I}^-(aq) \qquad K_{sp} = [\text{Pb}^{2+}][\text{I}^-]^2 = 7.9 \times 10^{-9}$$

The solvent is water, so the initial concentrations of the ions are zero. As before, we will define x as the molar solubility of the salt. By doing this, the coefficients of x in the "change" row will be identical to the coefficients of the ions in the chemical equation. This assures us that the changes in the concentrations are in the correct mole ratio.

SOLUTION: Here is the concentration table.

PbI$_2$(s)	$\rightleftharpoons$	Pb^{2+}(aq)	+	2I$^-$(aq)
Initial concentrations (M)		0		0
Changes in concentrations when PbI$_2$ dissolves (M)		$+x$		$+2x$
Equilibrium concentrations (M)		x		$2x$

Substituting equilibrium quantities into the K_{sp} expression gives

$$K_{sp} = (x)(2x)^2 = 4x^3 = 7.9 \times 10^{-9}$$

$$x^3 = 2.0 \times 10^{-9}$$

$$x = 1.3 \times 10^{-3}$$

Thus, the molar solubility of PbI$_2$ is calculated to be 1.3×10^{-3} mol L^{-1}.

Is the Answer Reasonable?
We check the entries in the table. The solvent is water, so the initial concentrations are equal to zero. By setting x equal to the moles per liter of PbI$_2$ that dissolves (i.e., the molar solubility), the coefficients of x in the "change" row have to be the same as the coefficients of Pb^{2+} and Cl$^-$ in the chemical equation. The entries in the "change" row are okay. In performing the algebra, notice that when we square $2x$, we get $4x^2$. Multiplying $4x^2$ by x gives $4x^3$. After that, the rest is straightforward.

EXAMPLE 19.5

Calculating Molar Solubility from K_{sp}

The precipitation of yellow PbI$_2$ occurs when the two colorless solutions, one with sodium iodide and the other of lead(II) nitrate, are mixed.

PRACTICE EXERCISE 6: What is the calculated molar solubility in water at 25 °C of (a) AgBr and (b) Ag_2CO_3?

FIGURE 19.2 *The common ion effect.* The test tube shown here initially held a saturated solution of NaCl, where the equilibrium $NaCl(s) \rightleftharpoons Na^+(aq) + Cl^-(aq)$ had been established. Addition of a few drops of concentrated HCl, containing a high concentration of the common ion Cl^-, forced the equilibrium to shift to the left. This caused some white crystals of solid NaCl to precipitate.

A salt is less soluble if the solvent already contains an ion of the salt

Suppose that we stir some lead chloride (a compound having a low solubility) with water long enough to establish the following equilibrium:

$$PbCl_2(s) \rightleftharpoons Pb^{2+}(aq) + 2Cl^-(aq)$$

If we now add a concentrated solution of a soluble lead compound, such as $Pb(NO_3)_2$, the increased concentration of Pb^{2+} in the $PbCl_2$ solution will drive the position of equilibrium to the left, causing some $PbCl_2$ to precipitate. The phenomenon is simply an application of Le Châtelier's principle, the net result being that $PbCl_2$ is less soluble in a solution that contains Pb^{2+} from another source than it is in pure water.

When the mixture of $PbCl_2$ and $Pb(NO_3)_2$ comes to equilibrium, there are two sources of Pb^{2+} in the solution. An ion in a solution that has been supplied from more than one solute is called a **common ion.** The addition of the common ion lowers the solubility of $PbCl_2$; it becomes less soluble in the presence of $Pb(NO_3)_2$ (or any other soluble lead salt) than it is in pure water. The same effect is produced if a solution of a soluble chloride salt is added to the saturated $PbCl_2$ solution. The added Cl^- will drive the equilibrium to the left, reducing the amount of dissolved $PbCl_2$. In this case, the common ion is Cl^-.

The lowering of the solubility of an ionic compound by the addition of a common ion is called the **common ion effect.** Figure 19.2 shows how even a relatively soluble salt, NaCl, can be forced out of its saturated solution simply by adding concentrated hydrochloric acid, which serves as a source of the common ion, Cl^-. The common ion effect can dramatically lower the solubility of a salt, as the next example demonstrates.

EXAMPLE 19.6 **Calculations Involving the Common Ion Effect**

What is the molar solubility of PbI_2 in a 0.10 M NaI solution?

ANALYSIS: As usual, we begin with the chemical equation for the equilibrium and the appropriate K_{sp} expression. The value of K_{sp} is obtained from Table 19.1.

$$PbI_2(s) \rightleftharpoons Pb^{2+}(aq) + 2I^-(aq) \qquad K_{sp} = [Pb^{2+}][I^-]^2 = 7.9 \times 10^{-9}$$

Next, we begin to assemble the concentration table. As before, we imagine that we are adding the PbI_2 to a solvent into which it dissolves. This time, however, the solvent isn't water; it's a solution of NaI, which contains one of the ions of the salt PbI_2. Therefore, our initial concentrations will not both be zero. The solvent doesn't contain any lead compound, so the initial concentration of Pb^{2+} is equal to zero, but the solvent does contain 0.10 M NaI, which is completely dissociated and yields 0.10 M Na^+ and 0.10 M I^-. The initial concentration of I^- is therefore 0.10 M. These are the values we place in the "initial" row.

Next, we let x be the molar solubility of PbI_2. When x mol of PbI_2 dissolves per liter, the concentration of Pb^{2+} changes by $+x$ and that of I^- by twice as much, or $+2x$. Finally, the equilibrium concentrations are obtained by summing the initial concentrations and the changes. Here is the concentration table.

$PbI_2(s)$	$\rightleftharpoons$	$Pb^{2+}(aq)$	$+$	$2I^-(aq)$
Initial concentrations (M)		0		0.10
Changes in concentrations when PbI_2 dissolves (M)		$+x$		$+2x$
Equilibrium concentrations (M)		x		$0.10 + 2x$

SOLUTION: Substituting equilibrium values into the K_{sp} expression gives

$$K_{sp} = (x)(0.10 + 2x)^2 = 7.9 \times 10^{-9}$$

Just a brief inspection reveals that solving this expression for x will be difficult if we cannot simplify the math. Fortunately, a simplification is possible, because the small value of K_{sp} for PbI_2 tells us that the salt has a very low solubility. This means very little of the salt will dissolve, so x (or even $2x$) will be quite small. Let's assume that $2x$ will be much smaller than 0.10. If this is so, then

$$0.10 + 2x \approx 0.10 \text{ (assuming } 2x \text{ is negligible}$$
$$\text{compared to 0.10)}$$

Substituting 0.10 M for the I^- concentration gives

$$K_{sp} = (x)(0.10)^2 = 7.9 \times 10^{-9}$$

$$x = \frac{7.9 \times 10^{-9}}{(0.10)^2}$$

$$= 7.9 \times 10^{-7} M$$

Thus, the molar solubility of PbI_2 in 0.10 M NaI solution is calculated to be 7.9×10^{-7} M.

Is the Answer Reasonable?
Check the entries in the table. The initial concentrations come from the solvent, which contains no Pb^{2+} but does contain 0.10 M I^-. By letting x equal the molar solubility, the coefficients of x in the "change" row equal the coefficients in the equation for the equilibrium. We should also check to see if our simplifying assumption is valid. Notice that $2x$, which equals 1.6×10^{-6}, is indeed vastly smaller than 0.10, just as we anticipated. (If we add 1.6×10^{-6} to 0.10 and round correctly, we obtain 0.10.)

A Mistake to Avoid

The most common mistake that students make with problems like Example 19.6 is to use the coefficient of an ion in the solubility equilibrium at the wrong moment in the calculation. The coefficient of I^- in the PbI_2 equilibrium is 2. The mistake is to use this 2 to double the *initial* concentration of I^-. However, the *initial* concentration of I^- was provided not by PbI_2 but by NaI. When 0.10 mol of NaI dissociates, it gives 0.10 mol of I^-, not 2×0.10 mol. *The coefficients in the equation for the equilibrium are only used to obtain the quantities in the "change" row.*

To avoid mistakes, it is useful to always view the formation of the final solution as a two-step process. You begin with a solvent into which the "insoluble solid" will be placed. In some problems, the solvent may be pure water, in which case the initial concentrations of the ions will be zero. In other problems, like Example 19.6, the solvent will be a *solution* that contains a common ion. When this is so, first decide what the concentration of the common ion is and enter this value into the "initial concentration" row of the table. Next, imagine that the solid is added to the solvent and a little of it dissolves. *The amount that dissolves is what gives the values in the "change" row.* These entries must be in the same ratio as the coefficients in the equilibrium, which is accomplished if we let x be the molar solubility of the salt. Then the coefficients of x will be the same as the coefficients of the ions in the chemical equation for the equilibrium.

READ THESE TWO
PARAGRAPHS!

In Example 19.5 (page 839) we found that the molar solubility of PbI_2 in *pure* water is 1.3×10^{-3} M. In water that contains 0.10 M NaI (Example 19.6), the solubility of PbI_2 is 7.9×10^{-7} M, well over a thousand times smaller. As we said, the common ion effect can cause huge reductions in the solubilities of sparingly soluble compounds.

PRACTICE EXERCISE 7: Calculate the molar solubility of AgI in 0.20 M NaI solution. Compare the answer to the calculated molar solubility of AgI in pure water.

PRACTICE EXERCISE 8: Calculate the molar solubility of $Fe(OH)_3$ in a solution where the OH^- concentration is initially 0.050 M. Assume the dissociation of $Fe(OH)_3$ is 100%.

We can use K_{sp} to determine if a precipitate will form in a solution

Sometimes in setting up an experiment, the plans tentatively include an aqueous solution in which several ions are to be dissolved (or will form). Before proceeding, we might wish to ask, "Will any combination of the desired ions, at the planned concentrations, be of a salt that is too insoluble to be in solution?" If so, we may wish to change the plans for the experiment.

For a precipitate of a salt to form in a solution, the solution must be supersaturated. If it were unsaturated, the solution would be capable of dissolving more of the salt, and if it were just saturated, any precipitation would make the solution unsaturated and cause the precipitate to redissolve. Our goal, therefore, is to determine whether the solution is supersaturated.

For a given salt, we can judge whether its solution is unsaturated, saturated, or supersaturated, by using the concentrations of the ions to compute the ion product and then compare the result to the value of K_{sp}. If the solution is saturated, the ion product will equal K_{sp}. If the solution is unsaturated, the ion concentrations will be less than in a saturated solution and when we calculate the value of the ion product, it will be smaller than K_{sp}. However, if the solution is supersaturated, it contains more solute than required for saturation and the ion concentrations will be larger than in a saturated solution. When the value of the ion product is computed, the result will be larger than K_{sp}. This can be summarized as follows.

A substance will not precipitate from a solution if its concentration is such that the solution is either unsaturated or saturated.

Precipitate will form	Ion product $> K_{sp}$	(supersaturated)
No precipitate will form	$\begin{cases} \text{Ion product} = K_{sp} \\ \text{Ion product} < K_{sp} \end{cases}$	(saturated) (unsaturated)

Let's look at some sample calculations.

EXAMPLE 19.7

Predicting Whether a Precipitate Will Form

A student wished to prepare 500 mL of a solution containing 0.0075 mol of NaCl and 0.075 mol of $Pb(NO_3)_2$. Knowing from the solubility rules that the chloride of Pb^{2+} is "insoluble," there was concern that a precipitate of $PbCl_2$ might form. Will it?

ANALYSIS: To answer this question, we will compute the ion product for $PbCl_2$ *using the concentrations of the ions in the solution to be prepared.* Only if the computed ion product is larger than K_{sp} will a precipitate be expected. To perform the calculation correctly, we need the correct form of the ion product. We can obtain this by writing the solubility equilibrium and the K_{sp} expression that applies to a saturated solution.

$$PbCl_2(s) \rightleftharpoons Pb^{2+}(aq) + 2Cl^-(aq) \qquad K_{sp} = [Pb^{2+}][Cl^-]^2$$

From Table 19.1, $K_{sp} = 1.7 \times 10^{-5}$. Also, we have to convert moles to molarity, because these are the quantities that we have to substitute into the ion product expression.

SOLUTION: The planned solution would have the following molar concentrations:

$$[Pb^{2+}] = \frac{0.075 \text{ mol}}{0.500 \text{ L}} = 0.15 \text{ } M$$

$$[Cl^-] = \frac{0.0075 \text{ mol}}{0.500 \text{ L}} = 0.015 \text{ } M$$

We use these values to compute the ion product for $PbCl_2$ in such a solution.

$$[Pb^{2+}][Cl^-]^2 = (0.15)(0.015)^2 = 3.4 \times 10^{-5}$$

But this value of the ion product is larger than the K_{sp} of $PbCl_2$, 1.7×10^{-5}. This means that a precipitate of $PbCl_2$ should form if the preparation of this solution is attempted.

Is the Answer Reasonable?

We can double-check that the molarities of the ions in the planned solution are correct. Then, we check the calculation of the ion product. All seems to be in order, so our answer is correct.

It is usually difficult to prevent the extra salt from precipitating out of a supersaturated solution. (Sodium acetate is a notable exception; supersaturated solutions of this salt are easily made.)

PRACTICE EXERCISE 9: Will a precipitate of $CaSO_4$ form in a solution if the Ca^{2+} concentration is 0.0025 M and the SO_4^{2-} concentration is 0.030 M?

PRACTICE EXERCISE 10: Will a precipitate form in a solution containing 3.4×10^{-4} M CrO_4^{2-} and 4.8×10^{-5} M Ag^+?

What possible precipitate might form by mixing 50.0 mL of 1.0×10^{-4} M NaCl with 50.0 mL of 1.0×10^{-6} M $AgNO_3$? Will it form? (Assume that the volumes are additive.)

EXAMPLE 19.8

Predicting Whether a Precipitate Will Form

ANALYSIS: In this problem we are being asked, in effect, whether a metathesis reaction will occur between NaCl and $AgNO_3$. We should be able to predict whether this *might* occur by using the solubility rules. If the solubility rules suggest a precipitate, we can then calculate the ion product for the compound using the concentrations of the ions in the final solution. If this ion product exceeds K_{sp} for the salt, then a precipitate is expected.

To calculate the ion product correctly requires that we use the concentrations of the ions *after the solutions have been mixed*. Therefore, before computing the ion product, we must first take into account that mixing the solutions dilutes each of the solutes.

SOLUTION: Let's begin by writing the equation for the potential metathesis reaction between NaCl and $AgNO_3$. We use the solubility rules to determine whether each product is soluble or insoluble.

$$NaCl(aq) + AgNO_3(aq) \longrightarrow AgCl(s) + NaNO_3(aq)$$

The solubility rules indicate that we expect a precipitate of AgCl. But are the concentrations of Ag^+ and Cl^- actually high enough?

In the original solutions, the 1.0×10^{-6} M $AgNO_3$ contains 1.0×10^{-6} M Ag^+ and the 1.0×10^{-4} M NaCl contains 1.0×10^{-4} M Cl^-. What are the concentrations of these ions after dilution? To determine these we use the equation that applies to all dilution problems involving molarity.

$$M_i V_i = M_f V_f \qquad (19.2)$$

Solving for M_f gives

This equation was developed in Section 5.11 on page 194.

$$M_f = \frac{M_i V_i}{V_f} \qquad (19.3)$$

The initial volumes of both solutions are 50.0 mL and when the two solutions are combined, the final total volume is 100 mL. Therefore,

$$[Ag^+]_{final} = \frac{(50.0 \text{ mL})(1.0 \times 10^{-6} M)}{100.0 \text{ mL}} = 5.0 \times 10^{-7} M$$

$$[Cl^-]_{final} = \frac{(50.0 \text{ mL})(1.0 \times 10^{-4} M)}{100.0 \text{ mL}} = 5.0 \times 10^{-5} M$$

The units for V_i and V_f can be any we please provided they are the same for both. The units for M, of course, are mol L^{-1}.

Now we use these to calculate the ion product for AgCl, which we can write from the dissociation reaction of the salt.

$$AgCl(s) \rightleftharpoons Ag^+(aq) + Cl^-(aq)$$

$$\text{ion product for AgCl} = [Ag^+][Cl^-]$$

Substituting the concentrations computed above gives

$$\text{ion product} = (5.0 \times 10^{-7})(5.0 \times 10^{-5}) = 2.5 \times 10^{-11}$$

In Table 19.1, the K_{sp} for AgCl is given as 1.8×10^{-10}. Notice that the ion product is *smaller* than K_{sp}, which means that a precipitate will *not* form when these solutions are poured together.

Is the Answer Reasonable?

There are several things we should check here. They include writing the equation for the metathesis, calculating the concentrations of the ions after dilution (we've doubled the volume, so the concentrations are halved), and setting up the correct ion product. All appear to be correct, so the answer should be okay.

PRACTICE EXERCISE 11: What precipitate might be expected if we pour together 100.0 mL of $1.0 \times 10^{-3} M$ Pb(NO$_3$)$_2$ and 100.0 mL of $2.0 \times 10^{-3} M$ MgSO$_4$? Will some form? (Assume that the volumes are additive.)

PRACTICE EXERCISE 12: What precipitate might be expected if we pour together 50.0 mL of 0.10 M Pb(NO$_3$)$_2$ and 20.0 mL of 0.040 M NaCl? Will some form? (Assume that the volumes are additive.)

19.2 ▶ Solubility equilibria of metal oxides and sulfides involve reactions with water

Most water-insoluble metal oxides, like Fe$_2$O$_3$, dissolve in acid. For example,

$$Fe_2O_3(s) + 6H^+(aq) \longrightarrow 2Fe^{3+}(aq) + 3H_2O$$

The acid neutralizes the oxide ion and thus releases the metal ion from the solid. Other metal oxides dissolve in water without the aid of an acid. They do so, however, by reacting with water instead of by a simple dissociation of ions that remain otherwise unchanged. Sodium oxide, for example, consists of Na$^+$ and O^{2-} ions, and it readily dissolves in water. The solution, however, does not contain the oxide ion, O^{2-}. Instead, the hydroxide ion forms. The equation for the reaction is

$$Na_2O(s) + H_2O \longrightarrow 2NaOH(aq)$$

This actually involves the reaction of oxide ions with water as the crystals of Na$_2$O break up.

$$O^{2-}(s) + H_2O \longrightarrow 2OH^-(aq)$$

The oxide ion is simply too powerful a base to exist in water at any concentration worthy of experimental note. We can understand why from the extraordinarily high (estimated) value of K_b for O^{2-}, 1×10^{22}. Thus there is no way to supply *oxide ions* to an aqueous solution in order to form an insoluble metal oxide directly. Oxide ions react with water, instead, to generate the hydroxide ion. When an insoluble metal *oxide* instead of an insoluble metal *hydroxide* does precipitate from a solution, it forms because the specific metal ion is able to react with OH^-, extract O^{2-}, and leave H^+ (or H_2O) in the solution. The silver ion, for example, precipitates as brown silver oxide, Ag_2O, when OH^- is added to aqueous silver salts (Figure 19.3).

$$2Ag^+(aq) + 2OH^-(aq) \longrightarrow Ag_2O(s) + H_2O$$

FIGURE 19.3 *Silver oxide.* A brown, mudlike precipitate of silver oxide forms as sodium hydroxide solution is added to a solution of silver nitrate.

Hydrogen sulfide as a diprotic acid

When we shift from oxygen to sulfur in Group VIA and consider metal sulfides, we find many similarities to oxides. One is that the sulfide ion, S^{2-}, like the oxide ion, is such a strong Brønsted base that it does not exist in any ordinary aqueous solution. The sulfide ion has not been detected in an aqueous solution even in the presence of 8 *M* NaOH, where one might think that the reaction

$$OH^- + HS^- \longrightarrow H_2O + S^{2-}$$

could generate some detectable S^{2-}. An 8 *M* NaOH solution is at a concentration well outside the bounds of the "ordinary." Thus, Na_2S, like Na_2O, dissolves in water *by reacting with it,* not by releasing an otherwise unchanged divalent anion, S^{2-}.

$$Na_2S(s) + H_2O \longrightarrow 2Na^+(aq) + HS^-(aq) + OH^-(aq)$$

Just as some metal oxides form by a reaction between a metal ion and OH^-, many metal sulfides form when their metal ions are exposed to HS^-. Some metal ions even react with H_2S directly to form sulfides. Simply bubbling hydrogen sulfide gas into an aqueous solution of any one of a number of metal ions—Cu^{2+}, Pb^{2+}, and Ni^{2+}, for example—causes their sulfides to precipitate. Many of these have distinctive colors (Figure 19.4) that can be used to help identify which metal ion is in solution. A typical reaction is that of Cu^{2+} with H_2S.

$$Cu^{2+}(aq) + H_2S(aq) \rightleftharpoons CuS(s) + 2H^+(aq) \qquad K = 1.7 \times 10^{15}$$

The extremely large value of the equilibrium constant tells us that the forward reaction is essentially the only reaction; the reverse reaction is very unimportant. If we turn this equilibrium around, we would have something that looks very much like an equilibrium constant for defining a solubility product. Let's look at this possibility more closely.

In the laboratory, the qualitative analysis of metal ions uses the precipitation of metal sulfides to separate some metal ions from others.

K_{spa} values relate solubilities of metal sulfides

Copper sulfide is extremely insoluble in aqueous acid judging from the value of K for the reaction of Cu^{2+} with H_2S just described. However, if we were to write the equilibrium that describes the solubility of CuS in water, we would write

$$CuS(s) \rightleftharpoons Cu^{2+}(aq) + S^{2-}(aq)$$

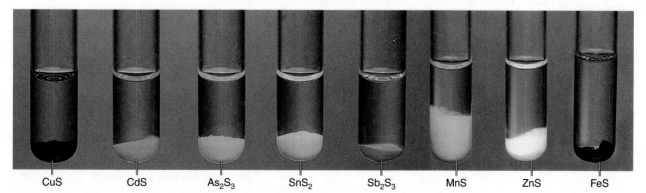

| CuS | CdS | As_2S_3 | SnS_2 | Sb_2S_3 | MnS | ZnS | FeS |

FIGURE 19.4 *The colors of some metal sulfides.*

This assumes that S^{2-} exists in water, and it doesn't, as we said. So we must instead write the equilibrium when $CuS(s)$ dissolves in water as follows in order to show which anions are actually produced:

$$CuS(s) + H_2O \rightleftharpoons Cu^{2+}(aq) + HS^-(aq) + OH^-(aq) \qquad (19.4)$$

Now the ion product is $[Cu^{2+}][HS^-][OH^-]$, and the solubility product constant for CuS is expressed by the following equation:

$$K_{sp} = [Cu^{2+}][HS^-][OH^-]$$

Several values of K_{sp} of this form for a number of metal sulfides are given in the last column in Table 19.2. Notice particularly how much the K_{sp} values vary—from 2×10^{-53} to 3×10^{-11}, a spread of a factor of 10^{42}.

Acid-insoluble sulfides

Several metal sulfides, the *acid-insoluble sulfides,* have K_{sp} values so low that they do not dissolve in acid. The cations in this group can be precipitated from the other cations simply by bubbling hydrogen sulfide into a sufficiently acidic solution that contains several metal ions. However, when the solution is acidic, we have to treat the solubility equilibria differently. In acid, HS^- and OH^- would be neutralized, leaving their conjugate acids, H_2S and H_2O. When we introduce acid ($2H^+$) into the left side of Equation 19.4, for example, to neutralize the bases on the right side (HS^- and OH^-), we obtain

$$CuS(s) + H_2O + 2H^+(aq) \rightleftharpoons Cu^{2+}(aq) + H_2S(aq) + H_2O$$

After removing the H_2O from each side, we have the net equation for the $CuS(s)$ solubility equilibrium *in dilute acid.*

$$CuS(s) + 2H^+(aq) \rightleftharpoons Cu^{2+}(aq) + H_2S(aq)$$

Acid solubility product

This changes the mass action expression for the solubility product equilibrium, which we will now call the **acid solubility product, K_{spa}.** The "a" in the subscript indicates that the medium is acidic.

$$K_{spa} = \frac{[Cu^{2+}][H_2S]}{[H^+]^2}$$

Table 19.2 also gives K_{spa} values for the metal sulfides. Notice that all K_{spa} values are about 10^{21} larger than the K_{sp} values. Metal sulfides are clearly vastly more soluble in dilute acid than in water. Yet several—the acid-insoluble sulfides—are

TABLE 19.2	METAL IONS SEPARABLE BY SELECTIVE PRECIPITATION OF SULFIDES[a]						
Metal Ion	Sulfide	K_{spa}	K_{sp}	Metal Ion	Sulfide	K_{spa}	K_{sp}
Acid-Insoluble Sulfides				*Base-Insoluble Sulfides (Acid-Soluble Sulfides)*			
Hg^{2+}	HgS (black form)	2×10^{-32}	2×10^{-53}	Zn^{2+}	α-ZnS	3×10^{-4}	3×10^{-25}
Ag^+	Ag_2S	6×10^{-30}	6×10^{-51}		β-ZnS	3×10^{-2}	3×10^{-23}
Cu^{2+}	CuS	6×10^{-16}	6×10^{-37}	Co^{2+}	CoS	5×10^{-1}	5×10^{-22}
Cd^{2+}	CdS	3×10^{-7}	3×10^{-28}	Ni^{2+}	NiS	4×10^{1}	4×10^{-20}
Pb^{2+}	PbS	3×10^{-7}	3×10^{-28}	Fe^{2+}	FeS	6×10^{2}	6×10^{-19}
Sn^{2+}	SnS	1×10^{-5}	1×10^{-26}	Mn^{2+}	MnS (pink form)	3×10^{10}	3×10^{-11}
					MnS (green form)	3×10^{7}	3×10^{-14}

[a]Data are for 25 °C. See R. J. Meyers, *J. Chem. Ed.,* Vol. 63, 1986, p. 689.

so insoluble that even the most soluble of them, SnS, barely dissolves, even in moderately concentrated acid. So there are two families of sulfides, the *acid-insoluble sulfides* and the acid-soluble ones, otherwise known as the *base-insoluble sulfides*. As we discuss in the next section, the differing solubilities of their sulfides in acid provide a means of separating the two classes of metal ions from each other.

19.3 ▶ Metal ions can be separated by selective precipitation

Selective precipitation means causing one metal ion to precipitate while holding the other in solution. Often this is possible because of the large differences in the solubilities of salts that we would generally consider to be insoluble. For example, consider AgCl and $PbCl_2$, which the solubility rules describe as being "insoluble." Their K_{sp} values are 1.8×10^{-10} for AgCl and 1.7×10^{-5} for $PbCl_2$, which gives them molar solubilities in water of 1.3×10^{-5} M for AgCl and 1.6×10^{-2} M for $PbCl_2$. In terms of molar solubilities, lead chloride is approximately 1200 times more soluble in water than AgCl. If we had a solution containing both 0.10 M Pb^{2+} and 0.10 M Ag^+ and began adding Cl^-, AgCl would precipitate first (Figure 19.5). In fact, we can calculate that before any $PbCl_2$ starts to precipitate, the concentration of Ag^+ will have been reduced to 1.6×10^{-8} M. Thus, nearly all the silver is removed from the solution without precipitating any of the lead, and the ions are effectively separated. All we need to do is find a way of adjusting the concentration of the Cl^- to achieve the separation. In this section we examine how we can control the concentration of a precipitating anion so as to achieve selectivity.

Selective precipitation of metal sulfides

The large differences in K_{spa} values between the acid-insoluble and the base-insoluble metal sulfides make it possible to separate the corresponding cations when they are in the same solution. The sulfides of the acid-insoluble cations are selectively precipitated by hydrogen sulfide from a solution kept at a pH that keeps the other cations in solution. A solution saturated in H_2S, is used, for which the molarity of H_2S is 0.1 M.

Let's work an example to show how we can calculate the pH needed to allow the selective precipitation of two metal cations as their sulfides. We will use Cu^{2+} and Ni^{2+} ions to represent a cation from each class.

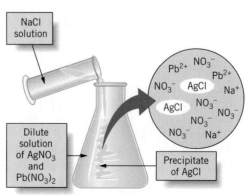

FIGURE 19.5 *Selective precipitation.* When dilute sodium chloride is added to a solution containing both Ag^+ ions and Pb^{2+} ions (both dissolved as their nitrate salts), the less soluble AgCl precipitates before the more soluble $PbCl_2$. Precipitation of the AgCl is nearly complete before any $PbCl_2$ begins to form.

Over what range of hydrogen ion concentrations (and pH) is it possible to separate Cu^{2+} from Ni^{2+} when both metal ions are present in a solution at a concentration of 0.010 M and the solution is made saturated in H_2S (where $[H_2S] = 0.1$ M)?

ANALYSIS: We must work with the chemical equilibria for the metal sulfides and their associated equilibrium expressions (the equations for their K_{spa}).

$$CuS(s) + 2H^+(aq) \rightleftharpoons Cu^{2+}(aq) + H_2S(aq) \qquad K_{spa} = \frac{[Cu^{2+}][H_2S]}{[H^+]^2} = 6 \times 10^{-16}$$

$$NiS(s) + 2H^+(aq) \rightleftharpoons Ni^{2+}(aq) + H_2S(aq) \qquad K_{spa} = \frac{[Ni^{2+}][H_2S]}{[H^+]^2} = 4 \times 10^1$$

The K_{spa} values (from Table 19.2) tell us that NiS is much more soluble in an acidic solution than CuS. Therefore, we want to make the H^+ concentration large enough to prevent NiS from precipitating and small enough that CuS does precipitate.

The problem reduces to two questions. The first is, "What hydrogen ion concentration would be needed to keep the Cu^{2+} *in solution*?" (The answer to this question will give us the *upper limit* on $[H^+]$; we would really want a lower H^+ concentration so CuS *will* precipitate.) The second question is, "What is the hydrogen ion concentration just before NiS precipitates?" The answer to this will be the *lower limit* on $[H^+]$. At any lower H^+ concentration, NiS will precipitate, so we want an H^+ concentration equal to or larger than this value. Once we know these limits, we know that any hydrogen ion concentration in between them will permit CuS to precipitate but retain Ni^{2+} in solution.

SOLUTION: We will find the upper limit first. If CuS *does not precipitate*, the Cu^{2+} concentration will be the given value, 0.010 M, so we substitute this along with the H_2S concentration (0.1 M) into the expression for K_{spa}.

$$K_{spa} = \frac{[Cu^{2+}][H_2S]}{[H^+]^2} = \frac{(0.010)(0.1)}{[H^+]^2} = 6 \times 10^{-16}$$

Now we solve for $[H^+]$.

$$[H^+]^2 = \frac{(0.010)(0.1)}{6 \times 10^{-16}} = 2 \times 10^{12}$$

$$[H^+] = 1 \times 10^6 \ M$$

If we could make $[H^+] = 1 \times 10^6$ M, we could prevent CuS from forming. However, it isn't possible to have 10^6 or a *million* moles of H^+ per liter! What the calculated $[H^+]$ tells us, therefore, is that *no matter how acidic the solution is, we cannot prevent CuS from precipitating when we saturate the solution with H_2S.* (You can now see why CuS is classed as an "acid-insoluble sulfide.")

To obtain the lower limit, we calculate the $[H^+]$ required to give an equilibrium concentration of Ni^{2+} equal to 0.010 M. If we keep the value of $[H^+]$ *equal to or larger* than this value, then NiS will be prevented from precipitating. The calculation is exactly like the one above. First, we substitute values into the K_{spa} expression.

$$K_{spa} = \frac{[Ni^{2+}][H_2S]}{[H^+]^2} = \frac{(0.010)(0.1)}{[H^+]^2} = 4 \times 10^1$$

Once again, we solve for $[H^+]$.

$$[H^+]^2 = \frac{(0.010)(0.1)}{4 \times 10^1}$$

$$[H^+] = 5 \times 10^{-3} \ M$$

$$pH = 2.3$$

If we maintain the pH of the solution of 0.010 M Cu^{2+} and 0.010 M Ni^{2+} at 2.3 or lower (more acidic), as we make the solution saturated in H_2S, virtually all the Cu^{2+} will precipitate as CuS, but all the Ni^{2+} will stay in solution.

Is the Answer Reasonable?

The K_{spa} values certainly tell us that CuS is quite insoluble in acid and that NiS is a lot more soluble, so from that standpoint the results seem reasonable. There's not much more we can do to check the answer, other than to review the entire set of calculations.

In actual experimental work involving the separation of acid-insoluble cations from base-insoluble cations, a solution more acidic than pH 2.3 (calculated in Example 19.9) is employed. A pH of about 0.5 is used, which corresponds to $[H^+] = 0.3\ M$. We can also see why NiS can be classified as a "base-insoluble sulfide." If the solution is *basic* when it is made saturated in H_2S, the pH will surely be larger than 2.3 and NiS will precipitate.

PRACTICE EXERCISE 13: Consider a solution containing Hg^{2+} and Fe^{2+}, both at molarities of 0.010 M. It is to be saturated with H_2S. Calculate the highest pH that this solution could have that would keep Fe^{2+} in solution while causing Hg^{2+} to precipitate as HgS.

Selective precipitation of metal carbonates

The principles of selective precipitation by the control of pH also apply to any system where the anion is that of a weak acid. The metal carbonates are examples, with many being quite insoluble in water (see Table 19.1). The dissociation of magnesium carbonate in water, for example, involves the following equilibrium and K_{sp} equations:

$$MgCO_3(s) \rightleftharpoons Mg^{2+}(aq) + CO_3^{2-}(aq) \qquad K_{sp} = 3.5 \times 10^{-8}$$

For strontium carbonate, the equations are

$$SrCO_3(s) \rightleftharpoons Sr^{2+}(aq) + CO_3^{2-}(aq) \qquad K_{sp} = 9.3 \times 10^{-10}$$

Would it be possible to separate the magnesium ion from the strontium ion by taking advantage of the difference in K_{sp} of their carbonates?

We can do so if we can control the carbonate ion concentration. Because the carbonate ion is a relatively strong Brønsted base, its control is available, indirectly, through adjusting the pH of the solution. This is because the hydrogen ion is one of the species in the following equilibria[3]:

$$H_2CO_3 \rightleftharpoons H^+ + HCO_3^- \qquad K_{a_1} = 4.3 \times 10^{-7}$$

$$HCO_3^- \rightleftharpoons H^+ + CO_3^{2-} \qquad K_{a_2} = 4.7 \times 10^{-11}$$

The control of the concentration of the carbonate ion by pH is like the control of the concentration of hydrogen sulfide ion, HS^-, by pH.

You can see that if we increase the hydrogen ion concentration, both equilibria will shift to the left, in accordance with Le Châtelier's principle, and this will reduce the concentration of the carbonate ion. The value of $[CO_3^{2-}]$ thus decreases with decreasing pH. On the other hand, if we decrease the hydrogen ion concentration, making the solution more basic, we will cause the two equilibria to shift to the right. The concentration of the carbonate ion thus increases with increasing pH.

[3]The situation involving aqueous carbonic acid is complicated by the presence of dissolved CO_2, which we could represent in an equation as $CO_2(aq)$. In fact, this is how most of the dissolved CO_2 exists, namely, as $CO_2(aq)$, not as $H_2CO_3(aq)$. But we may use $H_2CO_3(aq)$ as a surrogate or stand-in for $CO_2(aq)$, because the latter changes smoothly to the former on demand. The following two successive equilibria involving carbonic acid exist in a solution of aqueous CO_2:

$$CO_2(aq) + H_2O \rightleftharpoons H_2CO_3(aq) \rightleftharpoons H^+(aq) + HCO_3^-(aq)$$

The value of K_{a_1} cited here for $H_2CO_3(aq)$ is really the product of the equilibrium constants of these two equilibria.

Now let's see what kinds of calculations we must do to find the pH range within which the less soluble carbonate will precipitate while the more soluble carbonate remains dissolved.

It will be useful to combine the two hydrogen carbonate equilibria into one overall equation that relates the molar concentrations of carbonic acid, carbonate ion, and hydrogen ion. So we first add the two.

$$H_2CO_3 \rightleftharpoons H^+ + HCO_3^-$$

$$\underline{HCO_3^- \rightleftharpoons H^+ + CO_3^{2-}}$$

$$H_2CO_3 \rightleftharpoons 2H^+ + CO_3^{2-} \tag{19.5}$$

Recall from Chapter 16 that when we add two equilibria to get a third, the equilibrium constant of the latter is the product of those of the two equilibria that are combined. Thus, for Equation 19.5 the equilibrium constant, which we'll symbolize as K_a, is obtained as follows:

We can safely use Equation 19.6 *only* when two of the three concentrations are known.

$$K_a = K_{a_1} \times K_{a_2} = \frac{[H^+]^2 [CO_3^{2-}]}{[H_2CO_3]} \tag{19.6}$$

$$= (4.3 \times 10^{-7}) \times (4.7 \times 10^{-11})$$

$$= 2.0 \times 10^{-17}$$

With this K_a value we may now study how to find the pH range within which Mg^{2+} and Sr^{2+} can be separated by taking advantage of their difference in K_{sp} values.

EXAMPLE 19.10

Separation of Metal Ions by the Selective Precipitation of Their Carbonates

A solution contains magnesium nitrate and strontium nitrate, each at a concentration of 0.10 M. Carbon dioxide is to be bubbled in to make the solution saturated in $CO_2(aq)$, approximately 0.030 M. What pH range would make it possible for the carbonate of one cation to precipitate but not that of the other?

ANALYSIS: There are really two parts to this. First, what is the range in values of $[CO_3^{2-}]$ that allows one carbonate to precipitate but not the other? Second, given this range, what are the values of the solution's pH that correspond to this range in $[CO_3^{2-}]$ values?

To answer the first question, we will use the K_{sp} values of the two carbonate salts and their molar concentrations to find the CO_3^{2-} concentrations in their saturated solutions. To keep the more soluble carbonate from precipitating, the CO_3^{2-} concentration must be at or below that required for saturation. For the less soluble of the two, we must have a CO_3^{2-} concentration larger than in a saturated solution so that it will precipitate.

To answer the second question, we will use the combined K_a expression for H_2CO_3 to find the $[H^+]$ that gives the necessary CO_3^{2-} concentrations obtained in answering the first question. Having these $[H^+]$, it is then a simple matter to convert them to pH values.

SOLUTION: The range in $[CO_3^{2-}]$ values is obtained by using the K_{sp} values of the two carbonates.

$$K_{sp} = [Mg^{2+}] [CO_3^{2-}] = 3.5 \times 10^{-8}$$

$$K_{sp} = [Sr^{2+}] [CO_3^{2-}] = 9.3 \times 10^{-10}$$

On the basis of the K_{sp} values, we see that $MgCO_3$ is the more soluble of the two. Let's find the $[CO_3^{2-}]$ value required to give its saturated solution.

$$[CO_3^{2-}] = \frac{K_{sp}}{[Mg^{2+}]} = \frac{3.5 \times 10^{-8}}{0.10}$$

$$= 3.5 \times 10^{-7} \, M$$

If the value of $[CO_3^{2-}]$ *equals* 3.5×10^{-7}, the solution will be saturated in $MgCO_3$ and none will precipitate. If the value of $[CO_3^{2-}]$ goes *above* 3.5×10^{-7} M, $MgCO_3$ will start to precipitate, thereby contaminating the less soluble carbonate, $SrCO_3$. Therefore, to prevent $MgCO_3$ from precipitating requires that $[CO_3^{2-}] \leq 3.5 \times 10^{-7}$ M.

The value of $[CO_3^{2-}]$ that we must create to have Sr^{2+} precipitate must be *larger* than that in a saturated solution of $SrCO_3$. Once again, we first find the $[CO_3^{2-}]$ needed for saturation.

$$[CO_3^{2-}] = \frac{K_{sp}}{[Sr^{2+}]} = \frac{9.3 \times 10^{-10}}{0.10}$$

$$= 9.3 \times 10^{-9} \ M$$

Thus, the value of $[CO_3^{2-}]$ must be adjusted, by adjusting the pH, so that it is *larger* than 9.3×10^{-9} M to make $SrCO_3$ precipitate.

In summary, for the solution given, the ranges of carbonate ion molarities under which $SrCO_3$ will precipitate but $MgCO_3$ will not are

$$[CO_3^{2-}] > 9.3 \times 10^{-9} \ M$$

$$[CO_3^{2-}] \leq 3.5 \times 10^{-7} \ M$$

The second phase of our calculation now asks what values of $[H^+]$ correspond to the calculated limits on $[CO_3^{2-}]$. For this we will use the combined equilibrium equation (Equation 19.5) and the associated K_a expression, Equation 19.6. First we solve Equation 19.6 for the square of $[H^+]$, using the molarity of the dissolved CO_2, 0.030 M, as the value of $[H_2CO_3]$. This gives us

$$[H^+]^2 = K_a \times \frac{[H_2CO_3]}{[CO_3^{2-}]} = 2.0 \times 10^{-17} \times \frac{0.030}{[CO_3^{2-}]}$$

We will use this equation for each of the two boundary values of $[CO_3^{2-}]$ to calculate the corresponding two values of $[H^+]^2$. Once we have them, the steps to values of $[H^+]$ and pH are easy.

For magnesium carbonate. To *prevent* the precipitation of $MgCO_3$, $[CO_3^{2-}]$ must be no higher than 3.5×10^{-7} M, so $[H^+]^2$ must not be less than

$$[H^+]^2 = 2.0 \times 10^{-17} \times \frac{0.030}{3.5 \times 10^{-7}}$$

$$= 1.7 \times 10^{-12}$$

Taking the square root,

$$[H^+] = 1.3 \times 10^{-6} \ M$$

This corresponds to a pH of 5.88. At a higher (more basic) pH, magnesium carbonate precipitates. Thus, pH $\leq$ 5.88 to prevent precipitation of $MgCO_3$.

For strontium carbonate. To have a saturated solution of $SrCO_3$ with $[Sr^{2+}] = 0.10$ M, the value of $[CO_3^{2-}]$ would be 9.3×10^{-9} M, as we calculated above. This corresponds to a value of $[H^+]^2$ found as follows:

$$[H^+]^2 = 2.0 \times 10^{-17} \times \frac{0.030}{9.3 \times 10^{-9}}$$

$$= 6.5 \times 10^{-11}$$

$$[H^+] = 8.1 \times 10^{-6} \ M$$

$$pH = 5.09$$

To cause $SrCO_3$ to precipitate, the $[CO_3^{2-}]$ would have to be higher than 9.3×10^{-9} M, and that would require that $[H^+]$ be *less than* 8.1×10^{-6} M. If $[H^+]$ were less than 8.1×10^{-6} M, the pH would have to be higher than 5.09. Thus, to cause $SrCO_3$ to precipitate from the given solution, pH $>$ 5.09.

Because $CO_2(aq)$ can rapidly combine with water to give $H_2CO_3(aq)$, the value of $[H_2CO_3]$ is taken to be that of $[CO_2(aq)]$, namely, 0.030 M.

In summary, when the pH of the given solution is kept above 5.09 and less than or equal to 5.88, Sr^{2+} will precipitate as $SrCO_3$ but Mg^{2+} will remain dissolved.

Is the Answer Reasonable?
The only way to be sure of the answer is to go back over the reasoning and the calculations. However, there is a sign that the answer is probably correct. Notice that the two K_{sp} values are not vastly different—they differ by just a factor of about 40. Therefore, it's not surprising that to achieve separation we would have to keep the pH within a rather narrow range (from 5.09 to 5.88).

PRACTICE EXERCISE 14: A solution contains calcium nitrate and nickel nitrate, each at a concentration of 0.10 *M*. Carbon dioxide is to be bubbled in to make its concentration equal 0.030 *M*. What pH range would make it possible for the carbonate of one cation to precipitate but not that of the other?

19.4 ▶ Complex ions participate in equilibria in aqueous solutions

Metal ions can combine with anions or neutral molecules to form complex ions

In our previous discussions of metal-containing compounds, we left you with the impression that the only kinds of bonds in which metals are ever involved are ionic bonds. For some metals, like the alkali metals of Group IA, this is close enough to the truth to warrant no modifications. But for many other metal ions, especially those of the transition metals and the post-transition metals, it is not. This is because the ions of many of these metals are able to behave as Lewis acids (i.e., as electron pair acceptors in the formation of coordinate covalent bonds). Thus, by participating in Lewis acid–base reactions they become *covalently* bonded to other atoms. Copper(II) ion is a typical example.

In aqueous solutions of copper(II) salts, like $CuSO_4$ or $Cu(NO_3)_2$, the copper is not present as simple Cu^{2+} ions. Instead, each Cu^{2+} ion becomes bonded to four water molecules to give a pale blue ion with the formula $Cu(H_2O)_4^{2+}$ (see Figure 19.6). We call this species a **complex ion** because it is composed of a number of simpler species (i.e., it is *complex,* not simple). The chemical equation for the formation of the $Cu(H_2O)_4^{2+}$ ion is

$$Cu^{2+} + 4H_2O \longrightarrow Cu(H_2O)_4^{2+}$$

which can be diagrammed using Lewis structures as follows:

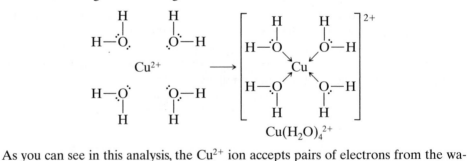

FIGURE 19.6 *The complex ion of Cu²⁺ and water.* A solution containing copper sulfate has a blue color because it contains the complex ion $Cu(H_2O)_4^{2+}$.

As you can see in this analysis, the Cu^{2+} ion accepts pairs of electrons from the water molecules, so Cu^{2+} is a Lewis acid and the water molecules are each Lewis bases.

The number of complex ions formed by metals, especially the transition metals, is enormous, and the study of the properties, reactions, structures, and bonding in complex ions like $Cu(H_2O)_4^{2+}$ has become an important specialty within chemistry. We will provide a more complete discussion of them in Chapter 23. For now, we will introduce you to some of the terminology that we use in describing these substances.

Recall that a *Lewis base* is an electron pair donor in the formation of a coordinate covalent bond.

A Lewis base that attaches itself to a metal ion is called a **ligand** (from the Latin *ligare,* meaning "to bind"). Ligands can be neutral molecules with unshared pairs of electrons (like H_2O), or they can be anions (like Cl^- or OH^-). The atom in the ligand that actually provides the electron pair is called the **donor atom,** and the metal ion is the **acceptor.** The result of combining a metal ion with one or more ligands is a *complex ion,* or simply just a **complex.** The word *complex* avoids problems when the particle formed is electrically neutral, as sometimes happens. Compounds that contain complex ions are generally referred to as **coordination compounds** because the bonds in a complex ion can be viewed as coordinate covalent bonds. Sometimes the complex itself is called a **coordination complex.**

In an aqueous solution, the formation of a complex ion is really a reaction in which water molecules are replaced by other ligands. Thus, when NH_3 is added to a solution of copper ion, the water molecules in the $Cu(H_2O)_4^{2+}$ ion are replaced, one after another, by molecules of NH_3 until the deeply blue complex $Cu(NH_3)_4^{2+}$ is formed (Figure 19.7). Each successive reaction is an equilibrium, so the entire chemical system involves many species and is quite complicated. Fortunately, when the ligand concentration is *large* relative to that of the metal ion, the concentrations of the intermediate complexes are very small and we can work only with the *overall* reaction for the formation of the final complex. Our study of complex ion equilibria will be limited to these situations. The equilibrium equation for the formation of $Cu(NH_3)_4^{2+}$, therefore, can be written as though the complex forms in one step.

$$Cu(H_2O)_4^{2+}(aq) + 4NH_3(aq) \rightleftharpoons Cu(NH_3)_4^{2+}(aq) + 4H_2O$$

We will simplify this equation for the purposes of dealing quantitatively with the equilibria by omitting the water molecules. (It's safe to do this because the concentration of H_2O in aqueous solutions is taken to be effectively a constant and need not be included in mass action expressions.) In simplified form, we write the equilibrium above as follows:

$$Cu^{2+}(aq) + 4NH_3(aq) \rightleftharpoons Cu(NH_3)_4^{2+}(aq)$$

We have two goals here: to study such equilibria themselves and to learn how they can be used to influence the solubilities of metal ion salts.

According to Le Châtelier's principle, when the concentration of ammonia is high, the position of equilibrium in this reaction is shifted far to the right, so effectively all complex ions with water molecules are changed to those with ammonia molecules.

A solution containing the lighter blue $Cu(H_2O)_4^{2+}$ complex.

Adding ammonia produces the darker blue $Cu(NH_3)_4^{2+}$ complex.

FIGURE 19.7 *The complex ion of Cu^{2+} and ammonia.* Ammonia molecules displace water molecules from $Cu(H_2O)_4^{2+}$ to give the deep blue $Cu(NH_3)_4^{2+}$ ion.

Formation constants reflect the stabilities of complex ions

Formation constants of complex ions

When the chemical equation for the equilibrium is written so that the complex ion is the product, the equilibrium constant for the reaction is called the **formation constant, K_{form}**. The equilibrium law for the formation of $Cu(NH_3)_4^{2+}$ in the presence of excess NH_3, for example, is

$$\frac{[Cu(NH_3)_4^{2+}]}{[Cu^{2+}][NH_3]^4} = K_{form}$$

Sometimes this equilibrium constant is called the **stability constant.** The larger its value, the greater is the concentration of the complex at equilibrium, and so the more stable is the complex.

Table 19.3 provides several more examples of complex ion equilibria and their associated equilibrium constants. (Additional examples are in Appendix C.) Notice that the most stable complex in the table, $Co(NH_3)_6^{3+}$, has, as you would expect, the largest value of K_{form}.

Instability constants are the inverse of formation constants

Some chemists prefer to describe the relative stabilities of complex ions differently. The *inverses* of formation constants are cited and are called **instability constants, K_{inst}**. This approach focuses attention on the *breakdown* of the complex, not its formation. Therefore, the associated equilibrium equation is written as the reverse of the formation of the complex. The equilibrium for the copper–ammonia complex would be written as follows, for example:

$$Cu(NH_3)_4^{2+}(aq) \rightleftharpoons Cu^{2+}(aq) + 4NH_3(aq)$$

The equilibrium constant for this equilibrium is called the *instability constant.*

$$K_{inst} = \frac{[Cu^{2+}][NH_3]^4}{[Cu(NH_3)_4^{2+}]} = \frac{1}{K_{form}}$$

Notice that K_{inst} is the reciprocal of K_{form}. The K_{inst} is called an *instability* constant because the larger its value is, the more *unstable* the complex is. The data in the last column of Table 19.3 show this. The least stable complex in the table, $Co(NH_3)_6^{2+}$, has the largest value of K_{inst}.

TABLE 19.3	FORMATION CONSTANTS AND INSTABILITY CONSTANTS FOR SOME COMPLEX IONS		
Ligand	Equilibrium	K_{form}	K_{inst}
NH_3	$Ag^+ + 2NH_3 \rightleftharpoons Ag(NH_3)_2^+$	1.6×10^7	6.3×10^{-8}
	$Co^{2+} + 6NH_3 \rightleftharpoons Co(NH_3)_6^{2+}$	5.0×10^4	2.0×10^{-5}
	$Co^{3+} + 6NH_3 \rightleftharpoons Co(NH_3)_6^{3+}$	4.6×10^{33}	2.2×10^{-34}
	$Cu^{2+} + 4NH_3 \rightleftharpoons Cu(NH_3)_4^{2+}$	1.1×10^{13}	9.1×10^{-14}
	$Hg^{2+} + 4NH_3 \rightleftharpoons Hg(NH_3)_4^{2+}$	1.8×10^{19}	5.6×10^{-20}
F^-	$Al^{3+} + 6F^- \rightleftharpoons AlF_6^{3-}$	1×10^{20}	1×10^{-20}
	$Sn^{4+} + 6F^- \rightleftharpoons SnF_6^{2-}$	1×10^{25}	1×10^{-25}
Cl^-	$Hg^{2+} + 4Cl^- \rightleftharpoons HgCl_4^{2-}$	5.0×10^{15}	2.0×10^{-16}
Br^-	$Hg^{2+} + 4Br^- \rightleftharpoons HgBr_4^{2-}$	1.0×10^{21}	1.0×10^{-21}
I^-	$Hg^{2+} + 4I^- \rightleftharpoons HgI_4^{2-}$	1.9×10^{30}	5.3×10^{-31}
CN^-	$Fe^{2+} + 6CN^- \rightleftharpoons Fe(CN)_6^{4-}$	1.0×10^{24}	1.0×10^{-24}
	$Fe^{3+} + 6CN^- \rightleftharpoons Fe(CN)_6^{3-}$	1.0×10^{31}	1.0×10^{-31}

No More Soap Scum—Complex Ions and Solubility

A problem that has plagued homeowners with hard water—water that contains low concentrations of divalent cations, especially Ca^{2+}—is the formation of insoluble deposits of "soap scum" as well as "hard water spots" on surfaces such as shower tiles, shower curtains, and bathtubs. These deposits form when calcium ions interact with large anions in soap to form precipitates and also when hard water that contains bicarbonate ion evaporates, causing precipitation of calcium carbonate, $CaCO_3$.

$$Ca^{2+}(aq) + 2HCO_3^-(aq) \rightarrow CaCO_3(s) + CO_2(g) + H_2O$$

A variety of consumer products are sold that contain ingredients intended to prevent these precipitates from forming, and they accomplish this by forming complex ions with calcium ions, which has the effect of increasing the solubilities of the soap scum and calcium carbonate deposits. A principal ingredient in these products is an organic compound called *ethylenediaminetetraacetic* acid (mercifully abbreviated *EDTA*). The structure of the compound, which is also abbreviated as H_4EDTA to emphasize that it contains four acidic hydrogens which are part of carboxyl groups, is

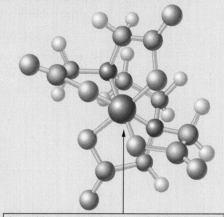

A metal ion surrounded octahedrally by all six donor atoms if H_4EDTA looses all of the acidic hydrogens.

One consumer product, called Clean Shower (Figure 1), contains H_4EDTA along with substances called *surfactants*. When the EDTA binds with calcium ions, it releases just two H^+ ions:

$$H_4EDTA(aq) + Ca^{2+}(aq) \rightarrow CaH_2EDTA(aq) + 2H^+(aq)$$

The H^+ combines with anions of the soap to form neutral organic compounds called *fatty acids* that are usually not water-soluble. The surfactants in the product, however, enable the fatty acids to dissolve, preventing formation of precipitates. The Clean Shower product is sprayed on the wet walls after taking a shower, forming the products described above. The next time you take a shower, the water washes away the soluble products, keeping soap scum from building up and keeping the shower walls clean.

H_4EDTA
Acidic hydrogens are shown in blue,
donor atoms are shown in red.

This molecule is an excellent complex-forming ligand; it contains a total of *six* donor atoms (in red) that can bind to a metal ion, enabling the ligand to wrap itself around the metal ion as illustrated at the top of the next column. (The usual colors are used to identify the various elements in the ligand.)

FIGURE 1 The product Clean Shower contains EDTA, which prevents the formation of soap scum and hard water spots in showers and on bathtubs.

Complex ion formation increases the solubility of a salt

The silver halides are extremely insoluble salts, as we've learned. The K_{sp} of AgBr at 25 °C, for example, is only 5.0×10^{-13}. In a saturated aqueous solution, the concentration of each ion of AgBr is only 7.1×10^{-7} mol L^{-1}. Suppose that we start with a saturated solution in which undissolved AgBr is present and equilibrium exists. Now suppose that we begin to add aqueous ammonia to the system. Because NH_3 molecules bind strongly to silver ions, they begin to form $Ag(NH_3)_2^+$ with the trace amount of Ag^+ ions initially in solution. The reaction is

Complex ion equilibrium: $\quad Ag^+(aq) + 2NH_3(aq) \rightleftharpoons Ag(NH_3)_2^+(aq)$

Because the forward reaction removes *uncomplexed* Ag^+ ions from solution, it upsets the solubility equilibrium:

Solubility equilibrium: $\quad AgBr(s) \rightleftharpoons Ag^+(aq) + Br^-(aq)$

By withdrawing uncomplexed Ag^+ ions, the ammonia causes the solubility equilibrium to shift to the right to generate more Ag^+ ions from $AgBr(s)$. The way the two equilibria are related can be viewed as follows.

$$AgBr(s) \rightleftharpoons Ag^+(aq) + Br^-(aq)$$

Addition of ammonia removes free silver ion from the solution, causing the solubility equilibrium to shift to the right.

$$+$$
$$2NH_3(aq)$$
$$\updownarrow$$
$$Ag(NH_3)_2^+(aq)$$

Thus, adding ammonia causes more $AgBr(s)$ to dissolve—an example of Le Châtelier's principle at work. Our example illustrates a general phenomenon:

> The solubility of a slightly soluble salt increases when one of its ions can be changed to a soluble complex ion.

To analyze what happens, lets put the two equilibria together.

Complex ion equilibrium: $\quad Ag^+(aq) + 2NH_3(aq) \rightleftharpoons Ag(NH_3)_2^+(aq)$

Solubility equilibrium: $\quad\quad\quad\quad\quad AgBr(s) \rightleftharpoons Ag^+(aq) + Br^-(aq)$

Sum of equilibria: $\quad\quad AgBr(s) + 2NH_3(aq) \rightleftharpoons Ag(NH_3)_2^+(aq) + Br^-(aq)$

The equilibrium constant for the net overall reaction is written in the usual way. The term for $[AgBr(s)]$ is omitted because it refers to a solid and so has a constant value.

$$K_c = \frac{[Ag(NH_3)_2^+][Br^-]}{[NH_3]^2}$$

Recall that when equilibria are added, their equilibrium constants are multiplied.

Because we have added two equations to obtain a third equation, K_c for this expression is the product of K_{form} and K_{sp}. The values for K_{form} for $Ag(NH_3)_2^+$ and K_{sp} for AgBr are known, so by multiplying the two, we find the overall value of K_c for the silver bromide–ammonia system.

$$K_c = (1.6 \times 10^7)(5.0 \times 10^{-13})$$
$$= 8.0 \times 10^{-6}$$

This approach thus gives us a way to calculate the solubility of a sparingly soluble salt when a substance able to form a complex with the metal ion is put into its solution. The next example shows how this works.

EXAMPLE 19.11

Calculating the Solubility of a Slightly Soluble Salt in the Presence of a Ligand

How many moles of AgBr can dissolve in 1.0 L of 1.0 M NH_3?

ANALYSIS: A few preliminaries have to be done before we can take advantage of a concentration table. We need the overall equation and its associated equation for the equilibrium constant. The overall equilibrium is

$$AgBr(s) + 2NH_3(aq) \rightleftharpoons Ag(NH_3)_2^+(aq) + Br^-(aq)$$

The equation for K_c is

$$K_c = \frac{[Ag(NH_3)_2^+][Br^-]}{[NH_3]^2} = 8.0 \times 10^{-6} \text{ (as calculated earlier)}$$

Now let's prepare the concentration table. To do this, we imagine we are adding solid AgBr to the ammonia solution.

Before any reaction takes place, the concentration of NH_3 is 1.0 M and the concentration of $Ag(NH_3)_2^+$ is zero. The concentration of Br^- is also zero, because there is none of it in the ammonia solution.

If we define x as the molar solubility of the AgBr in the solution, then the concentrations of $Ag(NH_3)_2^+$ and Br^- both increase by x and the concentration of ammonia decreases by $2x$ (because of the coefficient of NH_3 in the equation).

The equilibrium values are obtained as usual by adding the initial and change rows. However, another comment is in order here. Letting the concentration of Br^- equal that of $Ag(NH_3)_2^+$ is valid because and only because K_{form} is such a large number. Essentially *all* Ag^+ ions that do dissolve from the insoluble AgBr are changed to the complex ion. There are relatively few uncomplexed Ag^+ ions in the solution, so the numbers of $Ag(NH_3)_2^+$ and Br^- ions in the solution are very nearly the same.

SOLUTION: Here is the concentration table, built using the reasoning above.

	AgBr(s)	+	2NH₃(aq)	⇌	Ag(NH₃)₂⁺(aq)	+	Br⁻(aq)
Initial concentrations (M)			1.0		0		0
Changes in concentrations caused by NH_3 (M)			$-2x$		$+x$		$+x$
Equilibrium concentrations (M)			$(1.0 - 2x)$		x		x

We've done all the hard work. Now all we need to do is substitute the values in the last row of the concentration table into the equation for K_c.

$$K_c = \frac{(x)(x)}{(1.0 - 2x)^2} = 8.0 \times 10^{-6}$$

This can be simplified by taking the square root of both sides. This gives

$$\frac{x}{(1.0 - 2x)} = \sqrt{8.0 \times 10^{-6}} = 2.8 \times 10^{-3}$$

Solving for x gives

$$x = 2.8 \times 10^{-3}$$

Because we've defined x as the molar solubility of AgBr in the ammonia solution, we can say that 2.8×10^{-3} mol of AgBr dissolves in 1.0 L of 1.0 M NH_3. (This is not very much, of course, but in contrast, only 7.1×10^{-7} mol of AgBr dissolves in 1.0 L of pure water. Thus AgBr is nearly 4000 times more soluble in the 1.0 M NH_3 than in pure water.)

Is the Answer Reasonable?
Everything depends on the reasoning used in preparing the concentration table, particularly letting $-2x$ represent the change in the concentration of NH_3 as a result of the presence of AgBr and the formation of the complex. If you are comfortable with this, then the calculation should be straightforward.

PRACTICE EXERCISE 15: Calculate the solubility of silver chloride in 0.10 M NH_3 and compare it with its solubility in pure water. (Refer to Table 19.1 for the K_{sp} for AgCl.)

PRACTICE EXERCISE 16: How many moles of NH_3 have to be added to 1.0 L of water to dissolve 0.20 mol of AgCl? The complex ion $Ag(NH_3)_2^+$ forms.

SUMMARY

Solubility Equilibria for Salts The **ion product** of a salt is the product of the molar concentrations of its ions, each concentration raised to the power equal to the subscript of the ion in the salt's formula. At a given temperature, this *ion product* in a *saturated* solution of the salt equals a constant called the **solubility product constant,** or K_{sp}.

The **common ion effect** is the ability of an ion of a soluble salt to suppress the solubility of a sparingly soluble compound that has the same (the "common") ion.

If soluble salts are mixed together in the same solution, a cation of one and an anion of another will precipitate if the ion product exceeds the solubility product constant of the salt formed from them.

When metal oxides and sulfides dissolve in water, neither O^{2-} nor S^{2-} ions exist in the solution. Instead, these anions react with water, and OH^- or HS^- ions form.

Selective Precipitation Metal sulfides are vastly more soluble in an acidic solution than in water. To express their solubility equilibria in acid, the **acid solubility constant,** or K_{spa}, is used. Several metal sulfides have such low values of K_{spa} that they are insoluble even at low pH. These **acid-insoluble sulfides** can thus be selectively separated from the **base-insoluble sulfides** by making the solution both quite acidic as well as saturated in H_2S. By adjusting the pH, salts of other weak acids, like the carbonate salts, can also be selectively precipitated.

Complex Ions of Metals Coordination compounds contain **complex ions** (also called **complexes** or **coordination complexes**), formed from a metal ion and a number of ligands. **Ligands** are Lewis bases and may be electrically neutral or negatively charged. Water and ammonia are common neutral ligands. The equilibrium constant for the formation of a complex in the presence of an excess of ligand is called the **formation constant,** K_{form}, of the complex. The larger the value of K_{form}, the more stable is the complex. Salts whose cations form stable complexes, like Ag^+ salts whose cation forms a stable complex with ammonia, $Ag(NH_3)_2{}^+$, are made more soluble when the ligand is present. **Instability constants,** K_{inst}, are the inverses of formation constants. The larger the value of an instability constant is, the more unstable is the complex.

TOOLS ▷ YOU HAVE LEARNED

The table below lists the concepts you've learned in this chapter that can be applied as tools in solving problems. Study each one carefully so that you know what each is used for. When faced with solving a problem, recall what each tool does and consider whether it will be helpful in finding a solution. This will aid you in selecting the tools you need. If necessary, refer to this table when working on the Thinking-It-Through problems and the Review Exercises that follow.

TOOL	HOW IT WORKS
Molar solubility (page 834)	From the molar solubility we can calculate the value of K_{sp} for a sparingly soluble salt.
Solubility product constant, K_{sp} (page 832)	We use K_{sp} to calculate the molar solubility of a salt either in pure water or in a solution that contains a common ion. Comparing K_{sp} with the ion product of a potential precipitating salt permits us to decide whether a precipitate will form.
Ion product of a salt (page 833)	Compare the ion product with K_{sp} to determine whether a precipitate will form: ion product > K_{sp}, a precipitate forms ion product ≤ K_{sp} $\begin{cases} \text{a precipitate} \\ \text{does not form} \end{cases}$
Acid solubility product constant, K_{spa} (page 846)	Use K_{spa} data to calculate the solubility of a metal sulfide at a given pH. We can use K_{spa} data for two or more metal sulfides to calculate the pH at which one will selectively precipitate from a solution saturated in H_2S.
Formation constants (stability constants) of complexes (page 854)	We can use them to make judgments concerning the relative stabilities of complexes. Along with K_{sp} values, we can use formation constants of complexes to determine the solubility of a sparingly soluble salt in a solution containing a ligand able to form a complex with the cation of the salt.

THINKING IT THROUGH

The goal for the following problems is not to find the answers themselves, but rather to assemble the information needed to solve them and explain how you would use the information to find the answers. The problems in Level 2 are more challenging than those in Level 1 and may contain more data than are required, in which case you are also asked to identify the unnecessary data. Detailed answers to the Thinking-It-Through problems can be found on the web site.

Need extra help?
On-Line Help Visit the Brady/
Senese web site at
www.wiley.com/
college/brady

Level 1 Problems

1. In water, the solubility of lead(II) chloride is 0.016 M. Explain how you would use this information to calculate the value of K_{sp} for $PbCl_2$.

2. Suppose 50.0 mL each of 0.0100 M solutions of NaBr and $Pb(NO_3)_2$ are poured together. How could you determine whether a precipitate would form? Describe how you would calculate the concentrations of the ions at equilibrium.

3. Suppose you wished to control the PO_4^{3-} concentration in a solution of phosphoric acid by controlling the pH of the solution. If you assume you know the H_3PO_4 concentration, what combined equation would be useful for this purpose?

4. What is the highest concentration of Pb^{2+} that can exist in a solution of 0.10 M HCl? (Explain how you would calculate the answer.)

5. Will lead(II) bromide be less soluble in 0.10 M $Pb(C_2H_3O_2)_2$ or 0.10 M NaBr? (Describe how you would perform the calculations necessary to answer the question.)

Level 2 Problems

6. How many milliliters of 0.10 M HCl would have to be added to 100 mL of a saturated solution of $PbCl_2$ in contact with 50.0 g of solid $PbCl_2$ to reduce the Pb^{2+} concentration to 0.0050 M? (Explain the calculations in detail. Don't forget to take into account the combined volumes of the two solutions.)

7. If a solution of 0.10 M Mn^{2+} and 0.10 M Cd^{2+} is gradually made basic, what will the concentration of Cd^{2+} be when $Mn(OH)_2$ just begins to precipitate? Assume no change in the volume of the solution. (Explain how you would approach the problem and state what data you need to answer the question.)

8. A solution of $MgBr_2$ can be changed to a solution of $MgCl_2$ by adding AgCl(*s*) and stirring the mixture well. In terms of the equilibria involved, explain how this happens.

9. After 100 g of solid $CaCO_3$ was added to a slightly basic solution, the pH was measured to be 8.50. What was the molar solubility of $CaCO_3$ in this solution? (Explain the calcu-

lations involved. Take into account the reaction of CO_3^{2-} with water.)

10. Suppose that some dipositive cation, M^{2+}, is able to form a complex ion with a ligand, L, by the following equation:

$$M^{2+} + 2L \rightleftharpoons [M(L)_2]^{2+}$$

The cation also forms a sparingly soluble salt, MCl_2. In which of the following circumstances would a given quantity of ligand be more able to bring larger quantities of the salt into solution? Explain and justify the calculation involved.
(a) $K_{form} = 1 \times 10^2$ and $K_{sp} = 1 \times 10^{-15}$
(b) $K_{form} = 1 \times 10^{10}$ and $K_{sp} = 1 \times 10^{-20}$

11. Silver forms a sparingly soluble iodide salt, AgI, and it also forms a soluble iodide complex, AgI_2^-. How much potassium iodide must be added to 100 mL of water to dissolve 0.020 mol of AgI? (Explain in detail how would you obtain the answer. *Hint:* Be sure to include *all* the iodide that's added to the solution.)

12. Solid ammonium chloride was added to 250 mL of a milky suspension of insoluble $Mg(OH)_2$ and the mixture was shaken. Afterwards, it was a clear, colorless solution. Use chemical equilibria to describe what happened in the mixture to cause the $Mg(OH)_2$ to dissolve.

13. You are given a sample containing NaCl(*aq*) and NaBr(*aq*), both with concentrations of 0.020 M. Some of the figures below represent a series of snapshots of what molecular-level views of the sample would show as a 0.200 M $Pb(NO_3)_2(aq)$ solution is slowly added. In these figures, Br^- is red-brown and Cl^- is green. Also, spectator ions are not shown, nor are ions whose concentrations are much less than the other ions. The K_{sp} constants for $PbCl_2$ and $PbBr_2$ are 1.7×10^{-5} and 2.1×10^{-6}, respectively.
(a) Arrange the figures in a time sequence to show what happens as the lead nitrate solution is added, excluding any that do not "make sense."
(b) Explain why you exclude any figures that do not belong in the observed time sequence.

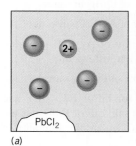

(a)

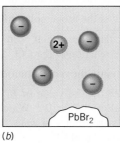

(b)

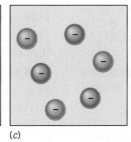

(c)

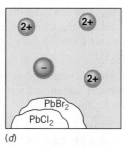

(d)

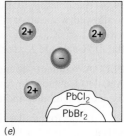

(e)

REVIEW QUESTIONS

Solubility Products

19.1 What is the difference between an *ion product* and an *ion product constant*?

19.2 Use the equilibrium below to demonstrate why the K_{sp} expression does not include the concentration of $Ba_3(PO_4)_2$ in the denominator.

$$Ba_3(PO_4)_2(s) \rightleftharpoons 3Ba^{2+}(aq) + 2PO_4^{3-}(aq)$$

19.3 What is the common ion effect? How does Le Châtelier's principle explain it? Use the solubility equilibrium for AgCl to illustrate the common ion effect.

19.4 If sodium acetate is added to a solution of acetic acid, the pH increases. Explain how this is an example of the common ion effect.

19.5 With respect to K_{sp}, what conditions must be met if a precipitate is going to form in a solution?

19.6 What limits the accuracy and reliability of solubility calculations based on K_{sp} values?

Oxides, Sulfides, and Selective Precipitations

19.7 Potassium oxide is readily soluble in water, but the resulting solution contains essentially no oxide ion. Explain, using an equation, what happens to the oxide ion.

19.8 What chemical reaction takes place when solid sodium sulfide is dissolved in water? Write the chemical equation.

19.9 Consider cobalt(II) sulfide.
(a) Write its solubility equilibrium and K_{sp} equation for a saturated solution in water. (Remember, there is no free sulfide ion in the solution.)
(b) Write its solubility equilibrium and K_{spa} equation for a saturated solution in aqueous acid.

19.10 Use Le Châtelier's principle to explain how adjusting the pH enables the control of the concentration of $C_2O_4^{2-}$ in a solution of oxalic acid, $H_2C_2O_4$.

Simultaneous and Complex Ion Equilibria

19.11 On the basis of Le Châtelier's principle, explain why the addition of solid NH_4Cl to a beaker containing solid $Mg(OH)_2$ in contact with water causes the $Mg(OH)_2$ to dissolve.

19.12 Using Le Châtelier's principle, explain how the addition of aqueous ammonia dissolves silver chloride. If HNO_3 is added after the AgCl has dissolved in the NH_3 solution, it causes AgCl to re-precipitate. Explain why.

19.13 For $PbCl_3^-$, $K_{form} = 2.5 \times 10^1$. If a solution containing this complex ion is diluted with water, $PbCl_2$ precipitates. Write the equations for the equilibria involved and use them together with Le Châtelier's principle to explain how this happens.

REVIEW PROBLEMS

Answers to problems whose numbers are printed in color are given in Appendix B. More challenging problems are marked with asterisks. **ILW** = Interactive LearningWare solution is available at *www.wiley.com/college/brady*.

Solubility Products

19.14 Write the K_{sp} expressions for each of the following compounds:
(a) CaF_2 (c) $PbSO_4$ (e) PbI_2
(b) Ag_2CO_3 (d) $Fe(OH)_3$ (f) $Cu(OH)_2$

19.15 Write the K_{sp} expressions for each of the following compounds:
(a) AgI (c) $PbCrO_4$ (e) $ZnCO_3$
(b) Ag_3PO_4 (d) $Al(OH)_3$ (f) $Zn(OH)_2$

Determining K_{sp}

19.16 Barium sulfate, $BaSO_4$, is so insoluble that it can be swallowed without significant danger, even though Ba^{2+} is toxic. At 25 °C, 1.00 L of water dissolves only 0.00245 g of $BaSO_4$. Calculate K_{sp} for $BaSO_4$.

19.17 A student found that a maximum of 0.800 g $AgC_2H_3O_2$ is able to dissolve in 100 mL of water. What is the molar solubility and the K_{sp} for this salt?

19.18 It was found that the molar solubility of $BaSO_3$ in 0.10 M $BaCl_2$ is 8.0×10^{-6} M. What is the value of K_{sp} for $BaSO_3$?

19.19 A student prepared a saturated solution of $CaCrO_4$ and found that when 156 mL of the solution was evaporated, 0.649 g of $CaCrO_4$ was left behind. What is the value of K_{sp} for this salt?

19.20 At 25 °C, the molar solubility of Ag_3PO_4 is 1.8×10^{-5} mol L^{-1}. Calculate K_{sp} for this salt.

19.21 The molar solubility of $Ba_3(PO_4)_2$ in water at 25 °C is 1.4×10^{-8} mol L^{-1}. What is the value of K_{sp} for this salt?

Using K_{sp} to Calculate Solubilities

19.22 What is the molar solubility of $PbBr_2$ in water?

19.23 What is the molar solubility of Ag_2CrO_4 in water?

19.24 Calculate the molar solubility of Ag_2CO_3 in water. (Ignore the reaction of the CO_3^{2-} ion with water.)

19.25 Calculate the molar solubility of PbI_2 in water.

19.26 At 25 °C, the value of K_{sp} for LiF is 1.7×10^{-3} and that for BaF_2 is 1.7×10^{-6}. Which salt, LiF or BaF_2, has the larger molar solubility in water? Calculate the solubility of each in units of mol L^{-1}.

19.27 At 25 °C, the value of K_{sp} for AgCN is 2.2×10^{-16} and that for $Zn(CN)_2$ is 3×10^{-16}. In terms of grams per 100 mL of solution, which salt is the more soluble in water? Do the calculations.

19.28 A salt whose formula is of the form MX has a value of K_{sp} equal to 3.2×10^{-10}. Another sparingly soluble salt, MX_3, must have what value of K_{sp} if the molar solubilities of the two salts are to be identical?

19.29 A salt having a formula of the type M_2X_3 has $K_{sp} = 2.2 \times 10^{-20}$. Another salt, M_2X, has to have what K_{sp} value if M_2X has twice the molar solubility of M_2X_3?

19.30 Calcium sulfate is found in plaster. At 25 °C, the value of K_{sp} for $CaSO_4$ is 2.4×10^{-5}. What is the calculated molar solubility of $CaSO_4$ in water?

19.31 Chalk is $CaCO_3$, and at 25 °C its $K_{sp} = 4.5 \times 10^{-9}$. What is the molar solubility of $CaCO_3$? How many grams of $CaCO_3$ dissolve in 100 mL of aqueous solution? (Ignore the reaction of CO_3^{2-} with water.)

Common Ion Effect

ILW **19.32** Copper(I) chloride has $K_{sp} = 1.9 \times 10^{-7}$. Calculate the molar solubility of CuCl in (a) pure water, (b) 0.0200 *M* HCl solution, (c) 0.200 *M* HCl solution, and (d) 0.150 *M* $CaCl_2$ solution.

19.33 Gold(III) chloride, $AuCl_3$, has $K_{sp} = 3.2 \times 10^{-25}$. Calculate the molar solubility of $AuCl_3$ in (a) pure water, (b) 0.010 *M* HCl solution, (c) 0.010 *M* $MgCl_2$ solution, and (d) 0.010 *M* $Au(NO_3)_3$ solution.

19.34 Calculate the molar solubility of $Mg(OH)_2$ in a solution that is basic with a pH of 12.50.

19.35 Calculate the molar solubility of $Al(OH)_3$ in a solution that is slightly basic with a pH of 9.50.

19.36 Calculate the molar solubility of Ag_2CrO_4 at 25 °C in (a) 0.200 *M* $AgNO_3$ and (b) 0.200 *M* Na_2CrO_4. For Ag_2CrO_4 at 25 °C, $K_{sp} = 1.2 \times 10^{-12}$.

19.37 What is the molar solubility of $Mg(OH)_2$ in 0.20 *M* NaOH? For $Mg(OH)_2$, $K_{sp} = 7.1 \times 10^{-12}$.

19.38 Calculate the molar solubility of $CaSO_4$ in 0.015 *M* $CaCl_2$.

19.39 Calculate the molar solubility of AgCl in 0.050 *M* $AlCl_3$.

19.40 How much will the percent ionization of the acetic acid change in 500 mL of 0.10 *M* $HC_2H_3O_2$ if 0.050 mol of solid $NaC_2H_3O_2$ is added? (Assume no change in the volume of the solution.) How much will the pH change?

19.41 How much will the percent ionization of the acetic acid change in 500 mL of 0.10 *M* $HC_2H_3O_2$ if 0.025 mol of gaseous HCl is dissolved in the solution. (Assume no change in volume.) How will the pH of the solution change?

*19.42 In an experiment 2.20 g of NaOH(*s*) are added to 250 mL of 0.10 *M* $FeCl_2$ solution. What mass of $Fe(OH)_2$ will be formed? What will the molar concentration of Fe^{2+} be in the final solution?

19.43 Suppose that 1.75 g of NaOH(*s*) are added to 250 mL of 0.10 *M* $NiCl_2$ solution. What mass, in grams, of $Ni(OH)_2$ will be formed? What will be the pH of the final solution? For $Ni(OH)_2$, $K_{sp} = 6 \times 10^{-16}$.

19.44 What is the molar solubility of $Fe(OH)_2$ in a buffer that has a pH of 9.50?

19.45 What is the molar solubility of $Ca(OH)_2$ in (a) 0.10 *M* $CaCl_2$ and (b) 0.10 *M* NaOH?

Precipitation

19.46 Does a precipitate of $PbCl_2$ form when 0.0150 mol of $Pb(NO_3)_2$ and 0.0120 mol of NaCl are dissolved in 1.00 L of solution?

19.47 Silver acetate, $AgC_2H_3O_2$, has $K_{sp} = 2.3 \times 10^{-3}$. Does a precipitate form when 0.015 mol of $AgNO_3$ and 0.25 mol of $Ca(C_2H_3O_2)_2$ are dissolved in a total volume of 1.00 L of solution?

ILW **19.48** Does a precipitate of $PbBr_2$ form if 50.0 mL of 0.0100 *M* $Pb(NO_3)_2$ are mixed with (a) 50.0 mL of 0.0100 *M* KBr and (b) 50.0 mL of 0.100 *M* NaBr?

19.49 Would a precipitate of silver acetate form if 22.0 mL of 0.100 *M* $AgNO_3$ were added to 45.0 mL of 0.0260 *M* $NaC_2H_3O_2$? For $AgC_2H_3O_2$, $K_{sp} = 2.3 \times 10^{-3}$.

19.50 Both AgCl and AgI are very sparingly soluble salts, but the solubility of AgI is much less than that of AgCl, as can be seen by their K_{sp} values. Suppose that a solution contains both Cl^- and I^- with $[Cl^-] = 0.050$ *M* and $[I^-] = 0.050$ *M*. If solid $AgNO_3$ is added to 1.00 L of this mixture (so that no appreciable change in volume occurs), what is the value of $[I^-]$ when AgCl first begins to precipitate?

19.51 Suppose that Na_2SO_4 is added gradually to 100 mL of a solution that contains both Ca^{2+} ion (0.15 *M*) and Sr^{2+} ion (0.15 *M*). (a) What will the Sr^{2+} concentration be (in mol L^{-1}) when $CaSO_4$ just begins to precipitate? (b) What percentage of the strontium ion has precipitated when $CaSO_4$ just begins to precipitate?

Oxides, Sulfides, and Selective Precipitations

ILW **19.52** What value of $[H^+]$ and what pH permit the selective precipitation of the sulfide of just one of the two metal ions in a solution that has a concentration of 0.010 *M* Pb^{2+} and 0.010 *M* Co^{2+}?

19.53 Can a selective separation of Mn^{2+} from Sn^{2+} be accomplished by a suitable adjustment of the pH of a solution that is 0.010 *M* in Mn^{2+}, 0.010 *M* in Sn^{2+}, and saturated in H_2S? (Assume the green form of MnS in Table 19.2.)

19.54 What range of pH would permit the selective precipitation of Cu^{2+} as $Cu(OH)_2$ from a solution that contains 0.10 *M* Cu^{2+} and 0.10 *M* Mn^{2+}? For $Mn(OH)_2$, $K_{sp} = 1.6 \times 10^{-13}$ and for $Cu(OH)_2$, $K_{sp} = 4.8 \times 10^{-20}$.

19.55 Kidney stones often contain insoluble calcium oxalate, CaC_2O_4, which has $K_{sp} = 2.3 \times 10^{-9}$. Calcium oxalate is considerably less soluble than magnesium oxalate, MgC_2O_4, which has $K_{sp} = 8.6 \times 10^{-5}$. Suppose a solution contained both Ca^{2+} and Mg^{2+} at a concentration of 0.10 *M*. What pH would be required to achieve maximum separation of these ions by precipitation of CaC_2O_4 if the solution also contains oxalic acid, $H_2C_2O_4$ at a concentration of 0.10 *M*? For $H_2C_2O_4$, $K_{a_1} = 6.5 \times 10^{-2}$ and $K_{a_2} = 6.1 \times 10^{-5}$.

Complex Ion Equilibria

19.56 Write the chemical equilibria and equilibrium laws that correspond to K_{form} for the following complexes:
(a) $CuCl_4^{2-}$ (b) AgI_2^- (c) $Cr(NH_3)_6^{3+}$

19.57 Write the chemical equilibria and equilibrium laws that correspond to K_{inst} for the following complexes:
(a) $Ag(S_2O_3)_2^{3-}$ (b) $Zn(NH_3)_4^{2+}$ (c) SnS_3^{2-}

19.58 Write equilibria that correspond to K_{inst} for each of the following complex ions and write the equations for K_{inst}:
(a) $Co(NH_3)_6^{3+}$ (b) HgI_4^{2-} (c) $Fe(CN)_6^{4-}$

19.59 Write the equilibria that are associated with the equations for K_{form} for each of the following complex ions. Write also the equations for the K_{form} of each.
(a) $Hg(NH_3)_4^{2+}$ (b) SnF_6^{2-} (c) $Fe(CN)_6^{3-}$

19.60 For $PbCl_3^-$, $K_{form} = 2.5 \times 10^1$. Use this information plus the K_{sp} for $PbCl_2$ to calculate K_c for the reaction

$$PbCl_2(s) + Cl^-(aq) \rightleftharpoons PbCl_3^-(aq)$$

19.61 The overall formation constant for $Ag(CN)_2^-$ equals 5.5×10^{18}, and the K_{sp} for $AgCN$ equals 1.6×10^{-14}. Calculate K_c for the reaction

$$AgCN(s) + CN^-(aq) \rightleftharpoons Ag(CN)_2^-(aq)$$

19.62 How many grams of solid NaCN have to be added to 1.2 L of water to dissolve 0.11 mol of $Fe(OH)_3$ in the form of $Fe(CN)_6^{3-}$? Use data as needed from Tables 19.1 and 19.3. (For simplicity, ignore the reaction of CN^- ion with water.)

19.63 In photography, unexposed silver bromide is removed from film by soaking the film in a solution of sodium thiosulfate, $Na_2S_2O_3$. Silver ion forms a soluble complex with thiosulfate ion, $S_2O_3^{2-}$, that has the formula $Ag(S_2O_3)_2^{3-}$, and formation of the complex causes the AgBr in the film to dissolve. The $Ag(S_2O_3)_2^{3-}$ complex has $K_{form} = 2.0 \times 10^{13}$. How many grams of AgBr ($K_{sp} = 5.0 \times 10^{-13}$) will dissolve in 125 mL of 1.20 M $Na_2S_2O_3$ solution?

19.64 Silver iodide is very insoluble and can be difficult to remove from glass apparatus, but it forms a relatively stable complex ion, AgI_2^- ($K_{form} = 1 \times 10^{11}$) that makes AgI fairly soluble in a solution containing I^-. When a solution containing the AgI_2^- ion is diluted with water, AgI precipitates. Explain why this happens in terms of the equilibria that are involved. How many grams of AgI will dissolve in 100 mL of 1.0 M KI solution?

19.65 The formation constant for $Ag(CN)_2^-$ equals 5.3×10^{18}. Use the data in Table 19.1 to determine the molar solubility of AgI in 0.010 M KCN solution.

19.66 The molar solubility of $Zn(OH)_2$ in 1.0 M NH_3 is 5.7×10^{-3} mol L^{-1}. Determine the value of the instability constant of the complex ion, $Zn(NH_3)_4^{2+}$. Ignore the reaction, $NH_3 + H_2O \rightleftharpoons NH_4^+ + OH^-$.

19.67 Calculate the molar solubility of $Cu(OH)_2$ in 2.0 M NH_3. (For simplicity, ignore the reaction of NH_3 as a base.)

ADDITIONAL EXERCISES

19.68 Magnesium hydroxide, $Mg(OH)_2$, found in milk of magnesia, has a solubility of 7.05×10^{-3} g L^{-1} at 25 °C. Calculate K_{sp} for $Mg(OH)_2$.

19.69 Does iron(II) sulfide dissolve in 8 M HCl? Perform the calculations that prove your answer.

19.70 As noted earlier, milk of magnesia is an aqueous suspension of $Mg(OH)_2$. If we assume that besides the $Mg(OH)_2$ the only other component is water, use K_{sp} to estimate the pH of the milk of magnesia.

***19.71** Suppose that 25.0 mL of 0.10 M HCl are added to 1.000 L of saturated $Mg(OH)_2$ in contact with more than enough $Mg(OH)_2(s)$ to react with all the HCl. After reaction has ceased, what will the molar concentration of Mg^{2+} be? What will the pH of the solution be?

***19.72** Solid $Mn(OH)_2$ is added to a solution of 0.100 M $FeCl_2$. After reaction, what will be the molar concentrations of Mn^{2+} and Fe^{2+} in the solution? What will be the pH of the solution? For $Mn(OH)_2$, $K_{sp} = 1.6 \times 10^{-13}$.

***19.73** Suppose that 50.0 mL of 0.12 M $AgNO_3$ are added to 50.0 mL of 0.048 M NaCl solution.
(a) What mass of AgCl would form?
(b) Calculate the final concentrations of all of the ions in the solution that is in contact with the precipitate.
(c) What percentage of the Ag^+ ions have precipitated?

***19.74** A sample of hard water was found to have 278 ppm Ca^{2+} ion. Into 1.00 L of this water, 1.00 g of Na_2CO_3 was dissolved. What is the new concentration of Ca^{2+} in parts per million? (Assume that the addition of Na_2CO_3 does not change the volume, and assume that the densities of the aqueous solutions involved are all 1.00 g mL^{-1}.)

***19.75** What value of $[H^+]$ and what pH would allow the selective separation of the carbonate of just one of the two metal ions in a solution that has a concentration of 0.010 M La^{3+} and 0.010 M Pb^{2+}. For $La_2(CO_3)_3$, $K_{sp} = 4.0 \times 10^{-34}$; for $PbCO_3$, $K_{sp} = 7.4 \times 10^{-14}$. A saturated solution of CO_2 in water has a concentration of H_2CO_3 equal to 3.3×10^{-2} M.

***19.76** When solid NH_4Cl is added to a suspension of $Mg(OH)_2(s)$, some of the $Mg(OH)_2$ dissolves.
(a) Write equations for *all* the chemical equilibria that exist in the solution after the addition of the NH_4Cl.
(b) Use Le Châtelier's principle to explain why adding NH_4Cl causes $Mg(OH)_2$ to dissolve.
(c) How many moles of NH_4Cl must be added to 1.0 L of a suspension of $Mg(OH)_2$ to dissolve 0.10 mol of $Mg(OH)_2$?
(d) What is the pH of the solution after the 0.10 mol of $Mg(OH)_2$ has dissolved in the solution containing the NH_4Cl?

*19.77 After solid $CaCO_3$ was added to a slightly basic solution, the pH was measured to be 8.50. What was the molar solubility of $CaCO_3$ in this solution? (See Thinking It Through, Problem 9, on page 859.)

19.78 In modern construction, walls and ceilings are constructed of "drywall," which consists of plaster sandwiched between sheets of heavy paper. Plaster is composed of calcium sulfate, $CaSO_4·2H_2O$. Suppose you had a leak in a water pipe that was dripping water on a drywall ceiling 1/2 in. thick at a rate of 2.00 L per day. Use the K_{sp} of calcium sulfate to estimate how many days it would take to dissolve a hole 1.0 cm in diameter. Assume the density of the plaster is 0.97 g cm^{-3}.

19.79 What is the molar solubility of $Fe(OH)_3$ in water? (*Hint:* You have to take into account the self-ionization of water.)

19.80 The value of K_{inst} for $SnCl_4^{2-}$ is 5.6×10^{-3}.
(a) What is the value of K_{form}?
(b) Is this complex ion more or less stable than those in Table 19.3?

*19.81 In Example 19.11 (page 857) we say, "there are relatively few uncomplexed Ag^+ ions in the solution." Calculate the molar concentration of Ag^+ ion actually left after the complex forms as described in Example 19.11.

*19.82 What are the concentrations of Pb^{2+}, Br^-, and I^- in an aqueous solution that's in contact with both PbI_2 and $PbBr_2$?

*19.83 Will a precipitate form in a solution made by dissolving 1.0 mol of $AgNO_3$ and 1.0 mol $HC_2H_3O_2$ in 1.0 L of solution? For $AgC_2H_3O_2$, $K_{sp} = 2.3 \times 10^{-3}$ and for $HC_2H_3O_2$, $K_a = 1.8 \times 10^{-5}$.

*19.84 How many moles of solid NH_4Cl must be added to 1.0 L of water in order to dissolve 0.10 mol of solid $Mg(OH)_2$? *Hint:* Consider the simultaneous equilibria

$$Mg(OH)_2(s) \rightleftharpoons Mg^{2+}(aq) + 2OH^-(aq)$$

$$NH_3(aq) + H_2O \rightleftharpoons NH_4^+(aq) + OH^-(aq)$$

*19.85 How many grams of solid sodium acetate must be added to 200 mL of a solution containing 0.200 M $AgNO_3$ and 0.10 M nitric acid to cause silver acetate to begin to precipitate? For $HC_2H_3O_2$, $K_a = 1.8 \times 10^{-5}$ and for $AgC_2H_3O_2$, $K_{sp} = 2.3 \times 10^{-3}$.

*19.86 How many grams of solid potassium fluoride must be added to 200 mL of a solution that contains 0.20 M $AgNO_3$ and 0.10 M acetic acid to cause silver acetate to begin to precipitate? For HF, $K_a = 6.5 \times 10^{-4}$; for $HC_2H_3O_2$, $K_a = 1.8 \times 10^{-5}$; for $AgC_2H_3O_2$, $K_{sp} = 2.3 \times 10^{-3}$.

*19.87 What is the molar solubility of $Mg(OH)_2$ in 0.10 M NH_3 solution? Remember that NH_3 is a weak base.

*19.88 If 100 mL of 2.0 M NH_3 is added to 400 mL of a solution containing 0.10 M Mn^{2+} and 0.10 M Sn^{2+}, what minimum number of grams of HCl would have to be added to the mixture to prevent $Mn(OH)_2$ from precipitating? For $Mn(OH)_2$, $K_{sp} = 1.6 \times 10^{-13}$. Assume that virtually all the tin is precipitated as $Sn(OH)_2$ by the reaction

$$Sn^{2+}(aq) + 2NH_3(aq) + 2H_2O \longrightarrow$$
$$Sn(OH)_2(s) + 2NH_4^+(aq)$$

*19.89 What is the molar concentration of Cu^{2+} ion in a solution prepared by mixing 0.50 mol of NH_3 and 0.050 mol of $CuSO_4$ in 1.00 L of solution? For NH_3, $K_b = 1.8 \times 10^{-5}$; for $Cu(OH)_2$, $K_{sp} = 4.8 \times 10^{-20}$; and for $Cu(NH_3)_4^{2+}$, $K_{form} = 4.8 \times 10^{12}$.

*19.90 On the basis of the K_{sp} of $Fe(OH)_3$, what would be the pH of a mixture consisting of solid $Fe(OH)_3$ mixed with pure water? (Assume 100% dissociation of the iron(III) hydroxide.)

Once again you have an opportunity to test your understanding of concepts, your knowledge of scientific terms, and your skills at solving chemistry problems. Read through the following questions carefully, and answer each as fully as possible. Review topics when necessary. When you are able to answer these questions correctly, you are ready to go on to the next group of chapters.

1 At 25 °C and $[OH^-] = 1.00\ M$, the reaction

$$I^-(aq) + OCl^-(aq) \longrightarrow OI^-(aq) + Cl^-(aq)$$

has the rate law: rate = $(0.60\ L\ mol^{-1}\ s^{-1})[I^-][OCl^-]$. Calculate the rate of the reaction when $[OH^-] = 1.00\ M$ and
(a) $[I^-] = 0.0100\ M$ and $[OCl^-] = 0.0200\ M$
(b) $[I^-] = 0.100\ M$ and $[OCl^-] = 0.0400\ M$

2 A reaction has the stoichiometry: $3A + B \rightarrow C + D$. The following data were obtained for the initial rate of formation of C at various concentrations of A and B.

Initial Concentrations		Initial Rate of Formation
$[A]$	$[B]$	of C $(mol\ L^{-1}\ s^{-1})$
0.010	0.010	2.0×10^{-4}
0.020	0.020	8.0×10^{-4}
0.020	0.010	8.0×10^{-4}

(a) What is the rate law for the reaction?
(b) What is the value of the rate constant?
(c) What is the rate at which C is formed if $[A] = 0.017\ M$ and $[B] = 0.033\ M$?

3 If the concentration of a particular reactant is doubled and the rate of the reaction is cut in half, what must be the order of the reaction with respect to that reactant?

4 Organic compounds that contain large proportions of nitrogen and oxygen tend to be unstable and are easily decomposed. Hexanitroethane, $C_2(NO_2)_6$, decomposes according to the equation

$$C_2(NO_2)_6 \longrightarrow 2NO_2 + 4NO + 2CO_2$$

The reaction in CCl_4 as a solvent is first-order with respect to $C_2(NO_2)_6$. At 70 °C, $k = 2.41 \times 10^{-6}\ s^{-1}$ and at 100 °C $k = 2.22 \times 10^{-4}\ s^{-1}$.
(a) What is the half-life of $C_2(NO_2)_6$ at 70 °C? What is the half-life at 100 °C?
(b) If 0.100 mol of $C_2(NO_2)_6$ is dissolved in CCl_4 at 70 °C to give 1.00 L of solution, what will be the $C_2(NO_2)_6$ concentration after 500 minutes?
(c) What is the value of the activation energy of this reaction, expressed in kilojoules?
(d) What is the reaction's rate constant at 120 °C?

5 The reaction $2A + 2B \rightarrow M + N$ has the rate law: rate = $k[A]^2$. At 25 °C, $k = 1.0 \times 10^{-4}\ L\ mol^{-1}\ s^{-1}$. If the initial concentrations of A and B are 0.250 M and 0.150 M, respectively
(a) What is the half-life of the reaction?
(b) What will be the concentrations of A and B after 30 minutes?

6 Define *reaction mechanism, rate-determining step,* and *elementary process.*

7 The decomposition of ozone, O_3, is believed to occur by the two-step mechanism

$O_3 \longrightarrow O_2 + O$	(slow)
$O + O_3 \longrightarrow 2O_2$	(fast)
$2O_3 \longrightarrow 3O_2$	(net reaction)

If this is the mechanism, what is the reaction's rate law?

8 Draw a diagram showing how the potential energy varies during an exothermic reaction. Identify the activation energy for both the forward and reverse reactions. Also, identify the heat of reaction.

9 How does a heterogeneous catalyst increase the rate of a chemical reaction?

10 Write the appropriate mass-action expression, using molar concentrations, for these reactions.
(a) $NO_2(g) + N_2O(g) \rightleftharpoons 3NO(g)$
(b) $CaSO_3(s) \rightleftharpoons CaO(s) + SO_2(g)$
(c) $NiCO_3(s) \rightleftharpoons Ni^{2+}(aq) + CO_3^{2-}(aq)$

11 At a certain temperature, the reaction $2HF(g) \rightleftharpoons H_2(g) + F_2(g)$ has $K_c = 1 \times 10^{-13}$. Does this reaction proceed far toward completion when equilibrium is reached? If 0.010 mol HF was placed in a 1.00 L container and the system was permitted to come to equilibrium, what would be the concentrations of H_2 and F_2 in the container?

12 At 100 °C, the reaction $2NO_2(g) \rightleftharpoons N_2O_4(g)$ has $K_P = 6.5 \times 10^{-2}$. What is the value of K_c at this temperature?

13 At 1000 °C, the reaction $NO_2(g) + SO_2(g) \rightleftharpoons NO(g) + SO_3(g)$ has $K_c = 3.60$. If 0.100 mol NO_2 and 0.100 mol SO_2 are placed in a 5.00 L container and allowed to react, what will all the concentrations be when equilibrium is reached? What will the new equilibrium concentrations be if 0.010 mol NO and 0.010 mol SO_3 are added to this original equilibrium mixture?

14 For the reaction in the preceding question, $\Delta H° = -41.8$ kJ. How will the equilibrium concentration of NO be affected if:
(a) More NO_2 is added to the container?
(b) Some SO_3 is removed from the container?
(c) The temperature of the reaction mixture is raised?
(d) Some SO_2 is removed from the mixture?
(e) The pressure of the gas mixture is lowered by expanding the volume to 10.0 L?

15 At 25 °C, the water in a natural pool of water in one of the western states was found to contain hydroxide ions at a concentration of 4.7×10^{-7} g OH^- per liter. Calculate the pH of this water and state if it is acidic, basic, or neutral.

16 Which is the stronger acid, H_3PO_3 or H_3PO_4? How can one tell without a table of weak and strong acids?

17 Which is the stronger acid, H_2S or H_2Te?

18 What are the conjugate acids of (a) HSO_3^-, (b) N_2H_4?

19 What are the conjugate bases of (a) HSO_3^-, (b) N_2H_4, (c) $C_5H_5NH^+$?

20 Identify the conjugate acid-base pairs in the reaction

$$CH_3NH_2 + NH_4^+ \rightleftharpoons CH_3NH_3^+ + NH_3$$

21 Use Lewis structures to diagram the reaction between the Lewis base OH^- and the Lewis acid SO_3.

22 X, Y, and Z are all nonmetallic elements in the same period of the periodic table where they occur, left to right, in the order given. Which would be a stronger binary acid than the binary acid of Y, the binary acid of X or the binary acid of Z? Explain.

23 The first antiseptic to be used in surgical operating rooms was phenol, C_6H_5OH, a weak acid and a potent bactericide. A 0.550 M solution of phenol in water was found to have a pH of 5.07.
(a) Write the chemical equation for the equilibrium involving C_6H_5OH in the solution.
(b) Write the equilibrium law corresponding to K_a for C_6H_5OH.
(c) Calculate the values of K_a and pK_a for phenol.
(d) Calculate the values of K_b and pK_b for the phenoxide ion, $C_6H_5O^-$.

24 The pK_a of saccharin, $HC_7H_3SO_3$, a sweetening agent, is 11.68.
(a) What is the pK_b of the saccharinate ion, $C_7H_3SO_3^-$?
(b) Does a solution of sodium saccharinate in water have a pH of 7, or is the solution acidic or basic. If the pH is not 7, calculate the pH of a 0.010 M solution of sodium saccharinate in water?

25 At 25 °C the value of K_b for codeine, a pain-killing drug, is 1.63×10^{-6}. Calculate the pH of a 0.0115 M solution of codeine in water.

26 The pK_b of methylamine, CH_3NH_2, is 3.43. Calculate the pK_a of its conjugate acid, $CH_3NH_3^+$.

27 Ascorbic acid, $H_2C_6H_6O_6$, is a diprotic acid usually known as vitamin C. For this acid, pK_{a_1} is 4.10 and pK_{a_2} is 11.79. When 125 mL of a solution of ascorbic acid was evaporated to dryness, the residue of pure ascorbic acid has a mass of 3.12 g.
(a) Calculate the molar concentration of ascorbic acid in the solution before it was evaporated.
(b) Calculate the pH of the solution and the molar concentration of the ascorbate ion, $C_6H_6O_6^{2-}$, before the solution was evaporated.

28 What ratio of molar concentrations of sodium acetate to acetic acid can buffer a solution at a pH of 4.50?

29 Write a chemical equation for the reaction that would occur in a buffer composed of $NaC_2H_3O_2$ and $HC_2H_3O_2$ if
(a) some HCl were added.
(b) some NaOH were added.

30 If 0.020 mol of NaOH were added to 500 mL of a sodium acetate-acetic acid buffer that contains 0.10 M $NaC_2H_3O_2$ and 0.15 M $HC_2H_3O_2$, by how many pH units will the pH of the buffer change?

31 When 50.00 mL of an acid with a concentration of 0.115 M (for which $pK_a = 4.87$) is titrated with 0.100 M NaOH, what is the pH at the equivalence point? What would be a good indicator for this titration?

32 Calculate the pH of a 0.050 M solution of sodium ascorbate, $NaC_6H_6O_6$. For ascorbic acid, $H_2C_6H_6O_6$, $K_{a_1} = 7.9 \times 10^{-5}$ and $K_{a_2} = 1.6 \times 10^{-12}$.

33 When 25.0 mL of 0.100 M NaOH was added to 50.0 mL of a 0.100 M solution of a weak acid, HX, the pH of the mixture reached a value of 3.56. What is the value of K_a for the weak acid?

34 How many grams of solid NaOH would have to be added to 100 mL of a 0.100 M solution of NH_4Cl to give a mixture with a pH of 9.26?

35 The molar solubility of silver chromate, Ag_2CrO_4, in water is 6.7×10^{-5} M. What is K_{sp} for Ag_2CrO_4?

36 What is the pH of a saturated solution of $Mg(OH)_2$?

37 What is the solubility of $Fe(OH)_2$ in grams per liter if the solution is buffered to a pH of 10.00?

38 Suppose 30.0 mL of 0.100 M $Pb(NO_3)_2$ is added to 20.0 mL of 0.500 M KI.
(a) How many grams of PbI_2 will be formed?
(b) What will the molar concentrations of all the ions be in the mixture after equilibrium has been reached?

39 How many moles of NH_3 must be added to 1.00 L of solution to dissolve 1.00 g of $CuCO_3$? For $CuCO_3$, $K_{sp} = 2.5 \times 10^{-10}$. Ignore hydrolysis of CO_3^{2-}, but consider the formation of the complex ion, $Cu(NH_3)_4^{2+}$.

40 Over what pH range must a solution be buffered to achieve a selective separation of the carbonates of barium, $BaCO_3$, ($K_{sp} = 5.0 \times 10^{-9}$), and lead, $PbCO_3$ ($K_{sp} = 7.4 \times 10^{-14}$)? The solution is initially 0.010 M in Ba^{2+} and 0.010 M in Pb^{2+}.

41 A solution containing 0.10 M Pb^{2+} and 0.10 M Ni^{2+} is to be saturated with H_2S. What range of pH values could this solution have so that when the procedure is completed one of the ions remains in solution while the other is precipitated as its sulfide?

42 A solution that contains 0.10 M Fe^{2+} and 0.10 M Sn^{2+} is maintained at a pH of 3.00 while H_2S is gradually added to it. What will be the concentration of Sn^{2+} in the solution when FeS just begins to precipitate?

Thermodynamics

Lava from a Hawaiian volcano reaches the cooling waters of the Pacific Ocean. Heat always flows spontaneously from the red hot lava to the cooler ocean waters. There are many more ways to distribute energy among the atoms and molecules of the lava and water together than there are ways to distribute the energy in the lava alone, so the lava loses energy and the water gains it. This analysis of heat transfer leads to the concept of entropy, which is one of the principal topics discussed in this chapter.

THIS CHAPTER IN CONTEXT When fuels such as octane (C_8H_{18}) are burned in excess oxygen, carbon dioxide and water are produced and heat is released:

$$2C_8H_{18}(l) + 25O_2(g) \longrightarrow 16CO_2(g) + 18H_2O(l)$$

Many of our current problems with fossil fuel supplies and greenhouse gas production might be solved if there were an efficient way to run this reaction *in reverse*. If the equilibrium could be shifted to the left, octane would be produced from carbon dioxide and water—and the supply of this fuel might be essentially unlimited. Unfortunately, the reaction runs overwhelmingly in the forward direction. There is no measurable tendency for carbon dioxide and water molecules to reunite on their own to produce octane molecules. We say that the reaction takes place *spontaneously* as written, because once the reaction starts, octane will keep burning until equilibrium is established. In this case, the equilibrium is so far to the right that the reaction continues until all octane or all available oxygen is used up.

Why does the reaction naturally "go" as written? All combustion reactions show the same overwhelming tendency to completion. In Chapter 16, we saw that reactions often run in both the forward and reverse directions, so the strong tendency for combustion reactions to run forward is *not* typical of all reactions. What makes this reaction different?

What determines the size of the equilibrium constant? In Chapter 16, we compared the equilibrium constant K with the reaction quotient Q to predict which way a reaction would run. When Q is less than K, the reactants are converted into products. When Q is greater than K, products are converted into reactants. The mixture of reactants and products is at equilibrium when $Q = K$. However, we do not yet understand how properties of the reactants and the products in a reaction affect the size of the equilibrium constant. Understanding the factors that affect the size of the equilibrium constant would tell us how to control conditions to improve the yield of a reaction. We might also be able to predict conditions that make it possible for the reaction to run in reverse.

How much energy must be supplied to run the reaction in the desired direction? We'll need to answer this question to determine whether or not running a reaction is economically feasible. The question is also of great importance in biology. Many of the vital reactions occurring in living cells run in reverse. The cell must supply energy to drive these reactions. The energy is released when sugars and other "foods" are broken down into carbon dioxide and water within the cell.

The equilibrium constant K is roughly 10^{1860} for the combustion of octane under standard conditions.

Thermo implies heat, *dynamics* implies movement.

If necessary, review in Chapter 7 the meanings of such terms as *system, surroundings, state, state function,* and *isothermal* and *adiabatic* changes.

Chemical thermodynamics is the study of the role of energy in chemical change and in determining the behavior of materials. It can provide answers for each of the questions we've just asked. You were introduced to some of the principles of thermodynamics in Chapter 7 when you learned to calculate enthalpy changes (heats of reaction). You should recall that enthalpy changes deal with the exchange of heat between chemical systems and their surroundings. Such transfers of heat, however, represent only one aspect of thermodynamics. Another is nature's drive toward the most probable state, which we discussed briefly in Chapter 14. We will see that this drive is the basis of one of the fundamental laws of thermodynamics and the key to understanding equilibrium in chemical systems.

20.1 ▶ Internal energy can be transferred as heat or work, but it cannot be created or destroyed

Thermodynamics is based on a few laws that summarize centuries of experimental observation. Each law is a statement about relationships among energy, heat, work, and temperature. Because the laws manifest themselves in so many different ways, and underly so many different phenomena, there are many alternative but equivalent ways to state them. For ease of reference, the laws are identified by number and are called the *first law,* the *second law,* and the *third law.*

The **first law of thermodynamics,** which was discussed in Chapter 7, states that internal energy may be transferred as heat or work but it cannot be created or destroyed. This law, you recall, serves as the foundation for Hess's law, which we used in our computations involving enthalpy changes in Chapter 7. Let's review the first law in more detail.

Recall that the **internal energy** of a system, which is given the symbol E, is the system's total energy—the sum of all the kinetic and potential energies of its particles.

$$E_{\text{system}} = (\text{KE})_{\text{system}} + (\text{PE})_{\text{system}}$$

Although we can't actually measure the system's total energy, we *can* measure energy changes, and the equation

$$\Delta E = E_{\text{final}} - E_{\text{initial}}$$

or, for a chemical reaction,

$$\Delta E = E_{\text{products}} - E_{\text{reactants}}$$

serves to relate the algebraic sign of ΔE to the direction of energy flow. Thus, ΔE is positive if energy flows into a system and negative if energy flows out.

The first law of thermodynamics considers two ways by which energy can be exchanged between a system and its surroundings. One is by the absorption or release of heat, which is given the symbol q. The other involves **work, w.** If *work is done on a system,* as in the compression of a gas, the system gains and stores energy. Conversely, if *the system does work on the surroundings,* as when a gas expands and pushes a piston, the system loses some energy by changing part of its potential energy to kinetic energy, which is transferred to the surroundings. The **first law of thermodynamics** expresses the net change in energy mathematically by the equation

$$\Delta E = q + w$$

In a steam engine, such as the one powering this locomotive, work is derived from the expansion of steam, which is produced by supplying heat to boil water. The relationships between heat and work are important factors in the design of engines of all kinds.

$\Delta E = $ (heat input)
    + (work input)

In words, this equation states that the change in the internal energy equals the sum of the energy gained as heat and the energy gained by work being done on the system. This equation also defines the algebraic signs of q and w for energy flow. Thus, when

q is (+)   Heat is absorbed by the system.

q is (−)   Heat is released by the system.

w is (+)   Work is done on the system.

w is (−)   Work is done by the system.

In Chapter 7 you learned that ΔE is a state function, which means that its value does not depend on how a change from one state to another is carried out. The opposite is true for q and w.

PRACTICE EXERCISE 1: Which of the following changes is accompanied by the most negative value of ΔE?
(a) A spring is compressed and heated.
(b) A compressed spring expands and is cooled.
(c) A spring is compressed and cooled.
(d) A compressed spring expands and is heated.

When a chemical system expands, it does pressure–volume work on the surroundings

There are two kinds of work that chemical systems can do (or have done on them) that are of concern to us. One is electrical work, which is examined in the next chapter and will be discussed there. The other is work associated with the expansion or contraction of a system under the influence of an external pressure. Such "pressure–volume" or P–V work was discussed in Chapter 7 where it was shown that this kind of work is given by the equation

$$w = -P\Delta V$$

where P is the *external pressure* on the system.

If P–V work is the only kind of work involved in a chemical change, the equation for ΔE takes the form

$$\Delta E = q + (-P\Delta V) = q - P\Delta V$$

When a reaction takes place in a container whose volume cannot change (so that $\Delta V = 0$), $\Delta E = q$, which means the entire energy change must appear as heat that's either absorbed or released. For this reason, ΔE is called the *heat at constant volume* (q_v)

$$\Delta E = q_v$$

Work being done on a gas.
When a gas is *compressed* by an external pressure, w is positive because work is done on the gas. Because V decreases when the gas is compressed, the negative sign assures that w will be positive for the compression of a gas.

PRACTICE EXERCISE 2: Molecules of an ideal gas have no intermolecular attractions and therefore undergo no change in potential energy on expansion of the gas. If the expansion is also at constant temperature, there is no change in the kinetic energy, so the isothermal (constant temperature) expansion of an ideal gas has $\Delta E = 0$. Suppose such a gas expands from a volume of 1.0 L to 12.0 L against a constant opposing pressure of 14.0 atm. In units of L atm, what are q and w for this change?

PRACTICE EXERCISE 3: If a gas is compressed under adiabatic conditions (allowing no transfer of heat to or from the surroundings) by application of an external pressure, the temperature of the gas increases. Why?

Diesel engines are used to power large trucks and other heavy equipment. In the cylinders of a diesel engine, air is compressed to very small volumes, raising the temperature to the point where fuel ignites spontaneously when injected into it.

Enthalpy is a convenient state function for studying heats of reaction under constant pressure

Rarely do we carry out reactions in containers of fixed volume. Usually, reactions take place in containers open to the atmosphere where they are exposed to a constant pressure. To study heats of reactions under constant pressure conditions, enthalpy was invented. Recall that the **enthalpy, H,** is defined by the equation

$$H = E + PV$$

At constant pressure, an **enthalpy change, ΔH,** is then

$$\Delta H = \Delta E + P\Delta V$$

Substituting for ΔE gives

$$\Delta H = (q - P\Delta V) + P\Delta V$$

$$= q_P$$

where q_P is the **heat at constant pressure.** Thus, the value of ΔH for a system equals the heat at constant pressure.

The pressure–volume work is the difference between ΔE and ΔH for a chemical reaction

In Chapter 7 we noted that ΔE and ΔH are not equal. They differ by the pressure–volume work, $-P\Delta V$. We can see this by rearranging the equation above.

$$\Delta E - \Delta H = -P\Delta V$$

The ΔV values for reactions involving only solids and liquids are very tiny, so ΔE and ΔH for these reactions are nearly the same size.

The only time ΔE and ΔH differ by a significant amount is when gases are formed or consumed in a reaction, and even then they do not differ by much. To calculate ΔE from ΔH (or ΔH from ΔE), we must have a way of calculating the pressure–volume work.

If we assume the gases in the reaction behave as ideal gases, then we can use the ideal gas law. Solved for V this is

$$V = \frac{nRT}{P}$$

A volume change can therefore be expressed as

$$\Delta V = \Delta\left(\frac{nRT}{P}\right)$$

For a change at constant pressure and temperature we can rewrite this as

$$\Delta V = \Delta n\left(\frac{RT}{P}\right)$$

Thus, when the reaction occurs, the volume change is caused by a change in the number of moles of *gas*. Of course, in chemical reactions not all reactants and products need be gases, so to be sure we compute Δn correctly, let's express the change in the number of moles of gas as Δn_{gas}. It's defined as

Remember, in calculating Δn, we only count moles of *gases* among the reactants and products.

$$\Delta n_{gas} = (n_{gas})_{products} - (n_{gas})_{reactants}$$

The $P\Delta V$ product is therefore

$$P\Delta V = P \cdot \Delta n_{gas}\left(\frac{RT}{P}\right)$$

$$= \Delta n_{gas}\, RT$$

Substituting into the equation for ΔH gives

$$\Delta H = \Delta E + \Delta n_{gas} RT \qquad (20.1)$$

 TOOLS

Converting between ΔE and ΔH

The following example illustrates how small the difference is between ΔE and ΔH.

EXAMPLE 20.1

Conversion between ΔE and ΔH

The decomposition of calcium carbonate in limestone is used industrially to make carbon dioxide.

$$CaCO_3(s) \longrightarrow CaO(s) + CO_2(g)$$

The reaction is endothermic and has $\Delta H° = +571$ kJ. What is the value of $\Delta E°$ for this reaction? What is the percentage difference between $\Delta E°$ and $\Delta H°$?

ANALYSIS: The problem asks us to convert between ΔE and ΔH, so we need to use Equation 20.1. The superscript $°$ on ΔH and ΔE tells us the temperature is 25 °C (standard temperature for measuring heats of reaction). It is also important to remember that in calculating Δn_{gas}, we count just the numbers of moles of gas.

We also need a value for R. Since we want to calculate the term ΔnRT in units of kilojoules, we must have R in appropriate energy units. The value of R we will use in this and other calculations that involve energies is found in a table of constants like the one inside the rear cover of this book. It is: $R = 8.314$ J mol^{-1} K^{-1}.

SOLUTION: The equation we need is

$$\Delta H = \Delta E + \Delta n_{gas} RT$$

Solving for ΔE and applying the superscript $°$ gives

$$\Delta E° = \Delta H° - \Delta n_{gas} RT$$

To calculate Δn_{gas} we take the coefficients in the equation to represent numbers of moles. Therefore, there is one mole of gas among the products and no moles of gas among the reactants, so

$$\Delta n_{gas} = 1 - 0 = 1$$

The temperature is 298 K, and $R = 8.314$ J mol^{-1} K^{-1}. Substituting,

$$\Delta E° = +571 \text{ kJ} - (1 \text{ mol})(8.314 \text{ J mol}^{-1} \text{ K}^{-1})(298 \text{ K})$$

$$= +571 \text{ kJ} - 2.48 \text{ kJ}$$

$$= +569 \text{ kJ}$$

The percentage difference is calculated as follows:

$$\% \text{ difference} = \frac{2.48 \text{ kJ}}{571 \text{ kJ}} \times 100\%$$

$$= 0.434\%$$

Notice how small the relative difference is between ΔE and ΔH.

Is the Answer Reasonable?
A gas is formed in the reaction, so the system must expand and push against the opposing pressure of the atmosphere when the reaction occurs at constant pressure. Energy must be supplied to accomplish this. At constant volume this expansion would not be necessary; the pressure would simply increase. Therefore, decomposing the $CaCO_3$ should require more energy at constant pressure ($\Delta H°$) than at constant volume ($\Delta E°$) by an amount equal to the work done pushing back the atmosphere, and we see that $\Delta H°$ is larger than $\Delta E°$, so the answer is reasonable.

PRACTICE EXERCISE 4: The reaction

$$CaO(s) + 2HCl(g) \longrightarrow CaCl_2(s) + H_2O(g)$$

has $\Delta H° = -217.1$ kJ. Calculate $\Delta E°$ for this reaction. What is the percentage difference between $\Delta E°$ and $\Delta H°$?

A spontaneous change is a change that continues without outside intervention

Now that we've reviewed the way thermodynamics treats energy changes, we turn our attention to one of our main goals in studying this subject—finding relationships among the factors that control whether events are spontaneous. By **spontaneous change** we mean one that occurs by itself, without continuous outside assistance. Examples are water flowing over a waterfall and the melting of ice cubes in a cold drink on a warm day. These are events that proceed on their own.

We take many spontaneous events for granted. What would you think if you dropped a book and it rose to the ceiling?

Some spontaneous changes occur very rapidly. An example is the detonation of a stick of dynamite or the exposure of photographic film. Other spontaneous events, such as the rusting of iron or the erosion of stone, occur slowly and many years may pass before a change is noticed. Still others occur at such an extremely slow rate under ordinary conditions that they appear not to be spontaneous at all. Gasoline–oxygen mixtures appear perfectly stable indefinitely at room temperature because under these conditions they react very slowly. However, if heated, their rate of reaction increases tremendously and they react explosively.

Each day we also witness events that are obviously *not* spontaneous. We may pass by a pile of bricks in the morning and later in the day find that they have become a brick wall. We know from experience that the wall didn't get there by itself. A pile of bricks becoming a brick wall is *not* spontaneous; it requires the intervention of a bricklayer. Similarly, the decomposition of water into hydrogen and oxygen is not spontaneous. We see water all the time and we know that it's stable. Nevertheless, we can cause water to decompose by passing an electric current through it in a process called *electrolysis* (Figure 20.1).

$$2H_2O(l) \xrightarrow{\text{electrolysis}} 2H_2(g) + O_2(g)$$

This decomposition will continue, however, only as long as the electric current is maintained. As soon as the supply of electricity is cut off, the decomposition ceases. This example demonstrates the difference between spontaneous and nonspontaneous changes. Once a spontaneous event begins, it has a tendency to continue

FIGURE 20.1 *The electrolysis of water produces H_2 and O_2 gases.* It is a nonspontaneous change that only continues as long as electricity is supplied.

until it is finished. A nonspontaneous event, on the other hand, can continue only as long as it receives some sort of outside assistance.

Nonspontaneous changes have another common characteristic. They are able to occur only when accompanied by some spontaneous change. For example, a bricklayer consumes food, and a series of spontaneous biochemical reactions then occur that supply the necessary muscle power to build a wall. Similarly, the non-spontaneous electrolysis of water requires some sort of spontaneous mechanical or chemical change to generate the needed electricity. In short, *all nonspontaneous events occur at the expense of spontaneous ones.* Everything that happens can be traced, either directly or indirectly, to spontaneous changes.

The driving of nonspontaneous reactions to completion by linking them to spontaneous ones is an important principle in biochemistry.

The direction of spontaneous change is often but not always in the direction of lower energy

What determines the direction of spontaneous change? Let's begin by examining some everyday events, such as those depicted in Figure 20.2. When iron rusts, heat is released, so the reaction lowers the internal energy of the system. Similarly, the chemical substances in the gasoline–oxygen mixture lose chemical energy by evolving heat as the gasoline burns to produce CO_2 and H_2O.

We might be tempted to conclude (as some nineteenth-century chemists did) that spontaneous events occur in the direction of lowest energy, so that energy lowering is a "driving force" behind spontaneous change. For example, we may argue that when we drop a book, it falls to the floor because that lowers its potential energy. Since a change that lowers the potential energy of a system can be said to be exothermic, we can state this factor another way—*exothermic changes have a tendency to proceed spontaneously.*

Most, but not all, chemical reactions that are exothermic occur spontaneously.

Yet if this is so, how do we explain the third photograph in Figure 20.2? The melting of ice at room temperature is clearly a spontaneous process. But ice absorbs heat from the surroundings as it melts. This absorbed heat gives the water from the melted ice cube a higher internal energy than the original ice had. This is an example of a spontaneous but endothermic process. There are many other examples of spontaneous endothermic processes: the evaporation of water from a lake, the expansion of carbon dioxide gas into a vacuum, or the operation of a chemical "cold pack."

In Chapter 14 we discussed the solution process, and we said that one of the principal driving forces in the formation of a solution is the fact that the mixed state is much more probable than the unmixed state. It turns out that arguments like this can be used to explain the direction of any spontaneous process. Let's look at a concrete example.

FIGURE 20.2 *Three common spontaneous events*—iron rusts, fuel burns, and an ice cube melts at room temperature.

20.3 ▶ **Spontaneous processes tend to proceed from states of low probability to states of higher probability**

In Chapter 7, you learned that when a hot object is placed in contact with a colder one, heat will flow spontaneously from the hot object to the colder one. But why? Energy can be conserved no matter what the direction of the heat flow was. And if we believed that the lowering of energy was the driving force behind spontaneous change, we might wonder why the lowering of the energy of the hot object is favored over the gain in energy that results for the colder one.

We learned how temperature affects the distribution of energies in a sample of molecules in Chapter 7.

Let's build a simple model to explain the direction of heat flow. Imagine that we have two objects made of molecules that have a low-energy ground state and a higher-energy excited state. They will actually have many different excited states, but let's keep things simple and assume that the molecules can be either "low energy" in the ground state or "high energy" in just one excited state. We'll use a blue dot to represent a low-energy molecule, and a red dot to represent a high-energy molecule. If we place three high-energy molecules in contact with three low-energy molecules, we initially have a configuration like this:

where the three molecules on the left represent an object that is "hot" and the three on the right represent a "cold" object. After the objects are placed in contact, energy can be transferred between molecules in the two objects. The total energy of the two objects must be the same before and after contact. Because our molecules can have only high-energy (red) or low-energy (blue) states, this means that the total number of red molecules must be the same before and after contact. Here are all possible distributions of energy among the six molecules, after the objects have been placed into contact:

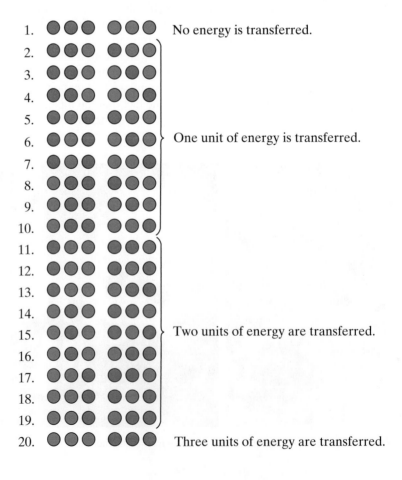

We can see four possible outcomes: 0, 1, 2, or 3 units of energy can be transferred from the hot body to the colder one. *Which is most likely to occur?*

Notice that some of the outcomes can be produced in a number of different ways. For example, items 2 through 10 represent ways energy can be distributed among the molecules after one unit of energy is transferred. *The more ways that a state can be produced, the more likely it is to occur.* We can use this fact to estimate the probability of the state occurring. There are 20 possible distributions of energy among the particles after the hot and cold objects are brought into contact. Let's assume that all of these distributions are equally probable. Then the probability of a particular outcome can be calculated as follows:

$$\text{probability of an outcome} = \frac{\left[\begin{array}{c}\text{number of ways the}\\\text{outcome can be produced}\end{array}\right]}{\left[\begin{array}{c}\text{total number of ways all}\\\text{outcomes can be produced}\end{array}\right]}$$

For example, there are 9 distributions that correspond to a transfer of one energy unit and 20 distributions in all, so the probability that one energy unit will be transferred is $9/20 = 0.45$ or 45%.

Units of Energy Transferred	Number of Equivalent Ways to Produce the Energy Transfer	Probability of Energy Transfer
0	1	$1/20 = 5\%$
1	9	$9/20 = 45\%$
2	9	$9/20 = 45\%$
3	1	$1/20 = 5\%$

In this model system, there is a 19/20 or 95% chance that some amount of energy transfer will occur. This agrees with our expectation that heat should flow spontaneously from the hot object to the colder one.

If we repeat this analysis with objects that contain more and more particles, the probability of energy transfer becomes so high that we can be quite certain that heat will flow from the hot object to the colder one. If we are dealing with hot and cold objects that contain moles of particles, the chance that no energy transfer would occur once they were placed into contact is completely negligible.

Although our model for heat transfer is very simple, it demonstrates the role of probability in determining the direction of a spontaneous process. *Spontaneous processes tend to proceed from states of low probability to states of higher probability.* The higher probability states are those that allow more options for distributing energy among the molecules, so we can also say that *spontaneous processes tend to disperse energy.*

Entropy is a measure of the number of equivalent ways to distribute energy in the system

Because statistical probability is so important in determining the outcome of chemical and physical events, thermodynamics defines a quantity, called **entropy** (symbol S), that describes the number of equivalent ways that energy can be distributed in the system. The larger the value of the entropy, the more energetically equivalent versions there are of the system and, therefore, the larger is its statistical probability.

In chemistry, we usually deal with systems that contain very large numbers of particles. It is usually impractical to count the number of ways that the particles can be arranged to produce a system with a particular energy, as we did in our simple model for energy flow between a hot and cold object. Fortunately, we will not need to do so. The entropy of the system can be related to experimental heat and temperature measurements (see Facets of Chemistry 20.1, page 883). We will also

The greater the statistical probability of a particular state, the greater is the entropy.

(a)   Two ways to count out $2 with paper money.

If energy were money, entropy would describe the number of different ways of counting it out. For example, there are only two ways of counting out $2 with American paper money, but there are five ways of counting out $2 using 50-cent and 25-cent coins. We could say that the "entropy" of a system that dealt in coins was higher than that of a system that dealt only in paper money.

(b)   Five ways to count out $2 with 50¢ and 25¢ coins.

be able to recognize increases and decreases in entropy without explicitly counting the number of ways the system can be realized.

Like enthalpy, entropy is a state function. It depends only on the state of the system, so a **change in entropy, ΔS,** is independent of the path from start to finish. As with other thermodynamic quantities, ΔS is defined as "final minus initial" or "products minus reactants." Thus

$$\Delta S = S_{final} - S_{initial}$$

or, for a chemical system,

$$\Delta S = S_{products} - S_{reactants}$$

As you can see, when S_{final} is larger than $S_{initial}$ (or when $S_{products}$ is larger than $S_{reactants}$), the value of ΔS is positive. A positive value for ΔS means an increase in the number of energy-equivalent ways the system can be produced, and we have seen that this kind of change tends to be spontaneous (see Figure 20.3). This leads to a general statement about entropy:

> Any event that is accompanied by an increase in the entropy of the system will have a *tendency* to occur spontaneously.

FIGURE 20.3 *A positive value for ΔS means an increase in the number of ways energy can be distributed among a system's molecules.* Consider the reaction $3A \rightarrow 3B$, where A molecules can take on energies that are multiples of 10 energy units, and B molecules can take on energies that are multiples of 5 units. Suppose that the total energy of the reacting mixture is 20 units. (a) There are two ways to distribute 20 units of energy among three molecules of A. (b) There are four ways to distribute 20 units of energy among three molecules of B. The entropy of B is higher than the entropy of A because there are more ways to distribute the same amount of energy in B than in A.

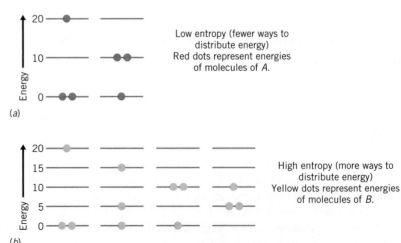

An increase in freedom of molecular motions corresponds to an increase in entropy

It is often possible to predict whether ΔS is positive or negative for a particular change. This is because several factors influence the magnitude of the entropy in predictable ways.

Predicting the sign of ΔS

Volume

For gases, the entropy increases with increasing volume, as illustrated in Figure 20.4. At the left we see a gas confined to one side of a container, separated from a vacuum by a removable partition. Let's suppose the partition could be pulled away in an instant, as shown in Figure 20.4*b*. Now we find a situation in which all the molecules of the gas are at one end of a larger container. There are many more possible ways that the total kinetic energy can be distributed if we give the molecules greater freedom of movement and spread them throughout the larger volume. That makes the configuration in Figure 20.4*b* extremely unlikely. Therefore, the gas expands spontaneously to achieve a more probable (higher entropy) particle distribution.

Temperature

The entropy is also affected by the temperature; the higher the temperature the larger is the entropy. For example, when a substance is a solid at absolute zero, its particles are essentially motionless. There is relatively little kinetic energy, and so there are few ways to distribute kinetic energy among the particles, so the entropy of the solid is relatively low (Figure 20.5*a*). If some heat is added to the solid, the kinetic energy of the particles increases along with the temperature. This causes the particles to move and vibrate within the crystal, so at a particular moment (pictured in Figure 20.5*b*) the particles are not found exactly at their lattice sites. There is more kinetic energy and more freedom of movement, so there are more ways to distribute the energy among the molecules. At a still higher temperature there is more kinetic energy and more possible ways to distribute it and the solid has a still higher entropy (Figure 20.5*c*).

Physical state

One of the major factors that affects the entropy of a system is its physical state, which is demonstrated in Figure 20.6. Suppose that the blocks represent ice, liquid water, and steam at the same temperature. There is greater freedom of molecular movement in water than in ice at the same temperature, and so there are more ways to distribute kinetic energy among the molecules of liquid water than there are in ice. Steam molecules are free to move through the entire container, so there are many more possible ways to distribute the kinetic

As you will learn soon, absolute values for S can be obtained. This is entirely different from E and H, where we can only determine differences (i.e., ΔE and ΔH).

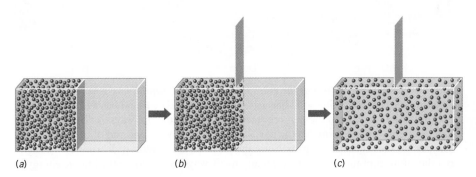

(a) (b) (c)

Figure 20.4 *The expansion of a gas into a vacuum.* (*a*) A gas in a container separated from a vacuum by a partition. (*b*) The gas at the moment the partition is removed. (*c*) The gas expands to achieve a more probable (higher entropy) particle distribution.

FIGURE 20.5 *Variation of entropy with temperature.* (*a*) At absolute zero, the atoms, represented by colored circles, rest at their equilibrium lattice positions, represented by black dots. Entropy is relatively low. (*b*) At a higher temperature, the particles vibrate about their equilibrium positions and there are more different ways to distribute kinetic energy among the molecules. Entropy is higher than in (*a*). (*c*) At a still higher temperature, vibration is more violent and at any instant the particles are found in even more arrangements. Entropy is higher than in (*b*).

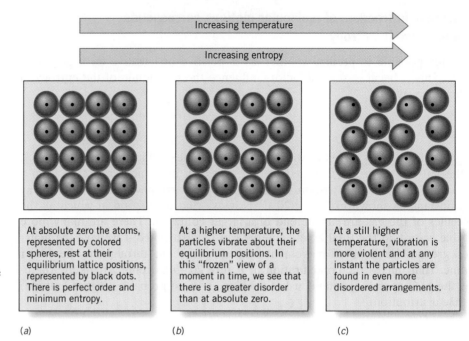

At absolute zero the atoms, represented by colored spheres, rest at their equilibrium lattice positions, represented by black dots. There is perfect order and minimum entropy.

At a higher temperature, the particles vibrate about their equilibrium positions. In this "frozen" view of a moment in time, we see that there is a greater disorder than at absolute zero.

At a still higher temperature, vibration is more violent and at any instant the particles are found in even more disordered arrangements.

(*a*)   (*b*)   (*c*)

$$S_{solid} < S_{liquid} \ll S_{gas}$$

For reactions that involve gases, we can simply calculate the change in the number of moles of gas, Δn_{gas}, on going from reactants to products. When Δn_{gas} is positive, so is the entropy change.

energy among the gas molecules than there are in liquid or solid water. In fact, any gas has such a large entropy compared with a liquid or solid that changes that produce gases from liquids or solids are almost always accompanied by increases in entropy.

The sign of the entropy change for a chemical reaction can sometimes be predicted by examining the chemical equation

When a chemical reaction produces or consumes gases, the sign of its entropy change is usually easy to predict. This is because the entropy of a gas is so much larger than that of either a liquid or solid. For example, the thermal decomposition of sodium bicarbonate produces two gases, CO_2 and H_2O.

$$2NaHCO_3(s) \xrightarrow{\text{heat}} Na_2CO_3(s) + CO_2(g) + H_2O(g)$$

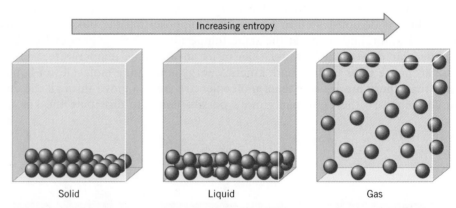

Solid   Liquid   Gas

FIGURE 20.6 *Comparison of the entropies of the solid, liquid, and gaseous states of a substance.* The crystalline solid has a very low entropy. The liquid has a higher entropy because its molecules can move more freely and there are more ways to distribute kinetic energy among them. All the particles are still found at one end of the container. The gas has the highest entropy because the particles are randomly distributed throughout the entire container, so there are many, many ways to distribute the kinetic energy.

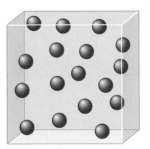
Lower entropy

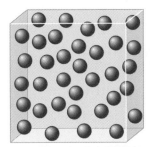
Higher entropy

FIGURE 20.7 *Entropy is affected by number of particles.* Adding additional particles to a system increases the number of ways that energy can be distributed in the system, so with all other things being equal, a reaction that produces more particles will have a positive value of ΔS.

Because the amount of gaseous products is larger than the amount of gaseous reactants, we can predict that the entropy change for the reaction is positive. On the other hand, the reaction

$$CaO(s) + SO_2(g) \longrightarrow CaSO_3(s)$$

(which can be used to remove sulfur dioxide from a gas mixture) has a negative entropy change.

Number of Particles

For chemical reactions, another major factor that affects the sign of ΔS is an increase in the total number of molecules as the reaction proceeds. When more molecules are produced during a reaction, more ways of distributing the energy among the molecules are possible. *When all other things are equal, reactions that increase the number of particles in the system tend to have a positive entropy change*, as shown in Figure 20.7.

Predict the algebraic sign of ΔS for the reactions:
(a) $2NO_2(g) \rightarrow N_2O_4(g)$
(b) $C_3H_8(g) + 5O_2(g) \rightarrow 3CO_2(g) + 4H_2O(g)$

EXAMPLE 20.2
Predicting the Sign of ΔS

ANALYSIS: As we examine the equations we look for changes in the number of moles of gas and changes in the number of particles on going from reactants to products.

SOLUTION: In reaction (a) we are forming fewer, more complex molecules (N_2O_4) from simpler ones (NO_2). Since we are forming fewer molecules, there are fewer ways to distribute energy among them, which means that the entropy must be decreasing. Therefore, ΔS must be negative.

For reaction (b), we can count the number of molecules on both sides. On the left of the equation we have six molecules; on the right there are seven. There are more ways to distribute kinetic energy among seven molecules than among six, so for reaction (b), we expect ΔS to be positive.

Are the Answers Reasonable?
We reach the same conclusion by counting the number of moles of gas on both sides of the equation. In reaction (a) 1 mol of gaseous product is formed from 2 mol of gaseous reactant. When there is a decrease in the number of moles of gas, the reaction tends to have a negative ΔS. In reaction (b), there are 6 mol of gas on the reactant side of the equation and 7 mol of gas on the product side. Because the number of moles of gas is increasing, we expect ΔS to be positive.

PRACTICE EXERCISE 5: Predict the sign of the entropy change for (a) the condensation of steam to liquid water and (b) the sublimation of a solid.

PRACTICE EXERCISE 6: Predict the sign of ΔS for the following reactions:
(a) $2SO_2(g) + O_2(g) \rightarrow 2SO_3(g)$
(b) $CO(g) + 2H_2(g) \rightarrow CH_3OH(g)$

PRACTICE EXERCISE 7: What is the expected sign of ΔS for the following reactions? Justify your answers.
(a) $2H_2(g) + O_2(g) \rightarrow 2H_2O(l)$
(b) $N_2(g) + 3H_2(g) \rightarrow 2NH_3(g)$
(c) $Ca(OH)_2(s) \xrightarrow{H_2O} Ca^{2+}(aq) + 2OH^-(aq)$

20.4 ▶ All spontaneous processes increase the total entropy of the universe

We have studied two factors, enthalpy and entropy, that affect whether or not a physical or chemical event will be spontaneous. Sometimes these factors work together. For example, the combustion of octane releases heat into the surroundings and results in a greater number of molecules among which energy can be distributed. Therefore, the enthalpy decreases and entropy increases. Since both of these favor a spontaneous change, the two factors complement one another. In other situations, the effects of enthalpy and entropy are in opposition. Such is the case, as we have seen, in the melting of ice or the evaporation of water. The endothermic nature of these changes tends to make them nonspontaneous, while the greater freedom of motion tends to make them spontaneous. In the reaction of H_2 with O_2 to form H_2O, the enthalpy and entropy changes are also in opposition. In this case, the exothermic nature of the reaction is sufficient to overcome the negative value of ΔS and cause the reaction to be spontaneous.

When enthalpy and entropy oppose one another, their relative importance in determining spontaneity is far from obvious. In addition, temperature becomes a third factor that can influence the direction in which a change is spontaneous. For example, if we attempt to raise the temperature of an ice–water slush to 25 °C, all the solid will melt. At 25 °C the change *solid → liquid* is spontaneous. On the other hand, if we attempt to cool an ice–water slush to −25 °C, freezing occurs, so at −25 °C the opposite change is spontaneous. *Thus, there are actually three factors that can influence spontaneity: the enthalpy change, the entropy change, and the temperature.* The balance among these factors comes into focus through the second law of thermodynamics.

The second law of thermodynamics states that all real processes increase the total entropy of the universe

One of the most far-reaching observations in science is incorporated into the **second law of thermodynamics,** which states, in effect, that *whenever a spontaneous event takes place in our universe, the total entropy of the universe increases* ($\Delta S_{total} > 0$). Notice that the increase in entropy that's referred to here is for the *total* entropy of the *universe* (system *plus* surroundings), not just the system alone. This means that a system's entropy can decrease, just as long as there is a larger increase in the entropy of the surroundings so that the *overall* entropy change is positive. Because everything that happens relies on spontaneous changes of some sort, the entropy of the universe is constantly rising.

Now let's examine more closely the entropy change for the universe. As we've suggested, this quantity equals the sum of the entropy change for the system plus the entropy change for the surroundings.

$$\Delta S_{total} = \Delta S_{system} + \Delta S_{surroundings}$$

It can be shown that the entropy change for the surroundings is equal to the heat transferred *to* the surroundings *from* the system, $q_{surroundings}$, divided by the Kelvin temperature, T, at which it is transferred.

$$\Delta S_{\text{surroundings}} = \frac{q_{\text{surroundings}}}{T}$$

The law of conservation of energy requires that the heat transferred to the surroundings equals the negative of the heat added to the system, so we can write

$$q_{\text{surroundings}} = -q_{\text{system}}$$

In our study of the first law of thermodynamics we saw that for changes at constant temperature and pressure, $q_{\text{system}} = \Delta H$ for the system. By substitutions, therefore, we arrive at the relationship

$$\Delta S_{\text{surroundings}} = \frac{-\Delta H_{\text{system}}}{T}$$

and the entropy change for the entire universe becomes

$$\Delta S_{\text{total}} = \Delta S_{\text{system}} - \frac{\Delta H_{\text{system}}}{T}$$

By rearranging the right side of this equation, we obtain

$$\Delta S_{\text{total}} = \frac{T\Delta S_{\text{system}} - \Delta H_{\text{system}}}{T}$$

Now let's multiply both sides of the equation by T to give

$$T\Delta S_{\text{total}} = T\Delta S_{\text{system}} - \Delta H_{\text{system}}$$

or

$$T\Delta S_{\text{total}} = -(\Delta H_{\text{system}} - T\Delta S_{\text{system}})$$

Because ΔS_{total} must be positive for a spontaneous change, the quantity in parentheses, $(\Delta H_{\text{system}} - T\Delta S_{\text{system}})$, must be negative.

> For a change to be spontaneous,
> $$\Delta H_{\text{system}} - T\Delta S_{\text{system}} < 0 \qquad (20.2)$$

Equation 20.2 gives us a way of examining the balance between ΔH, ΔS, and temperature in determining the spontaneity of an event, and it becomes convenient at this point to introduce another thermodynamic state function. It is called the **Gibbs free energy, G,** named to honor one of America's greatest scientists, Josiah Willard Gibbs (1839–1903). (It's called *free energy* because it is related, as we will see later, to the maximum energy in a change that is "free" or "available" to do useful work.) The Gibbs free energy is defined as

The origin of *free* in *free energy* is discussed in Section 20.7.

$$G = H - TS \qquad (20.3)$$

so for changes at constant T and P, this equation becomes

$$\Delta G = \Delta H - T\Delta S \qquad (20.4)$$

Because G is defined entirely in terms of state functions, it is also a state function. This means that

$$\Delta G = G_{\text{final}} - G_{\text{initial}} \qquad (20.5)$$

By comparing Equations 20.2 and 20.5, we arrive at the special importance of the free energy change:

> *At constant temperature and pressure, a change can only be spontaneous if it is accompanied by a decrease in the free energy of the system.*

ΔG as a predictor of spontaneity

Reactions that occur with a free energy decrease are sometimes said to be **exergonic.** Those that occur with a free energy increase are sometimes said to be **endergonic.**

In other words, for a change to be spontaneous, G_{final} must be less than $G_{initial}$ and ΔG must be negative. With this in mind, we can now examine how ΔH, ΔS, and T are related in determining spontaneity.

When ΔH is negative and ΔS is positive, the process will be spontaneous

The combustion of octane is accompanied by a large lowering of the potential energy as many strong bonds are formed in carbon dioxide and water from relatively weaker ones in the fuel and oxygen. There is also a large increase in entropy since the number of particles in the system is increased. For this change, ΔH is negative and ΔS is positive, both of which favor spontaneity.

$$\Delta H \text{ is negative } (-)$$
$$\Delta S \text{ is positive } (+)$$
$$\Delta G = \Delta H - T\Delta S$$
$$= (-) - [T(+)]$$

Notice that regardless of the absolute temperature, which must be a positive number, ΔG will be negative. This means that regardless of the temperature, such a change must be spontaneous. In fact, once started, fires will continue to consume all available fuel or oxygen at nearly any temperature because combustion reactions are always spontaneous (see Figure 20.8a).

When ΔH is positive and ΔS is negative, the process is not spontaneous

When a change is endothermic and is accompanied by a lowering of the entropy, both factors work against spontaneity.

$$\Delta H \text{ is positive } (+)$$
$$\Delta S \text{ is negative } (-)$$
$$\Delta G = \Delta H - T\Delta S$$
$$= (+) - [T(-)]$$

(a) (b) (c)

FIGURE 20.8 *Process spontaneity can be predicted if ΔH, ΔS, and T are known.* (a) When ΔH is negative and ΔS is positive, as in any combustion reaction, the reaction is spontaneous at any temperature. (b) When ΔH is positive and ΔS is negative, the process is not spontaneous. Wood does not spontaneously reform from ash, carbon dioxide, and water. (c) When ΔH and ΔS have the same sign, temperature determines whether the process is spontaneous or not. Water spontaneously becomes ice below 0 °C, and ice spontaneously melts into water above 0 °C.

Now, no matter what the temperature is, ΔG will be positive and the change must be nonspontaneous. An example would be carbon dioxide and water coming back together to form wood and oxygen again in a fire. If you saw such a thing happen on a film, experience would tell you that the film was being played backwards.

CHEMISTRY IN PRACTICE Plants can manufacture wood from carbon dioxide and water that they take in, but the reaction by itself is *not* spontaneous. Plant cells manufacture wood by coupling this nonspontaneous reaction to a complex series of reactions with negative values of ΔG, so that the entire chain of reactions taken together is spontaneous overall. This same trick is used to drive nonspontaneous processes forward in human cells. The high negative free energy change from the breakdown of sugar and other nutrients is coupled with the nonspontaneous synthesis of complex proteins from simple starting materials, driving cell growth and making life possible.

When ΔH and ΔS have the same sign, temperature determines spontaneity

When ΔH and ΔS have the same algebraic sign, the temperature becomes the determining factor in controlling spontaneity. If ΔH and ΔS are both positive, then

$$\Delta G = (+) - [T(+)]$$

Thus, ΔG is the difference between two positive quantities, ΔH and $T\Delta S$. This difference will only be negative if the term $T\Delta S$ is larger in magnitude than ΔH, and this will only be true when the temperature is high. In other words, *when ΔH and ΔS are both positive, the change will be spontaneous at high temperature but not at low temperature.* A familiar example is the melting of ice.

$$H_2O(s) \longrightarrow H_2O(l)$$

This is a change that is endothermic and also accompanied by an increase in entropy. We know that at high temperature (above 0 °C) melting is spontaneous, but at low temperature (below 0 °C) it is not.

For similar reasons, when ΔH and ΔS are both negative, ΔG will be negative (and the change spontaneous) only when the temperature is low.

$$\Delta G = (-) - [T(-)]$$

Only when the negative value of ΔH is larger in magnitude than the negative value of $T\Delta S$ will ΔG be negative. Such a change is only spontaneous at low temperature. An example is the freezing of water (see Figure 20.8c).

$$H_2O(l) \longrightarrow H_2O(s)$$

This is an exothermic change that is accompanied by a decrease in entropy; it is only spontaneous at low temperatures (i.e., below 0 °C).

Figure 20.9 summarizes the effects of the signs of ΔH and ΔS on ΔG, and hence on the spontaneity of physical and chemical events.

20.5 ▶ The third law of thermodynamics makes experimental measurement of absolute entropies possible

Earlier we described how the entropy of a substance depends on temperature, and we noted that at absolute zero the order within a crystal is a maximum and the entropy is a minimum. The **third law of thermodynamics** goes one step farther by stat-

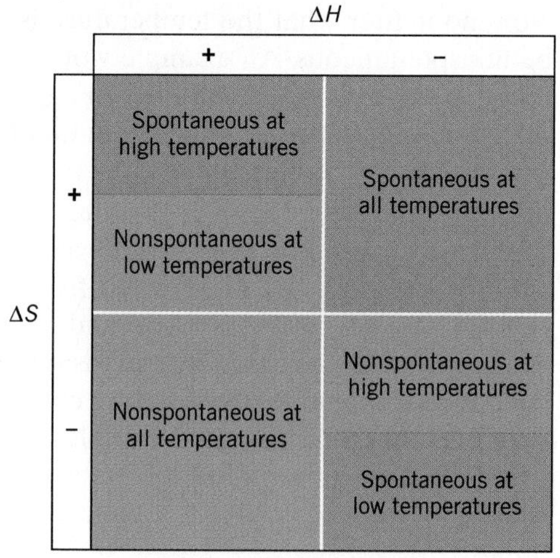

Figure 20.9 *Summary of the effects of the signs of ΔH and ΔS on spontaneity.*

ing: *at absolute zero the entropy of a perfectly ordered pure crystalline substance is zero.*

$$S = 0 \quad \text{at} \quad T = 0\,\text{K}$$

25 °C and 1 atm are the same standard conditions we used in our discussion of $\Delta H°$ in Chapter 7.

Because we know the point at which entropy has a value of zero, it is possible by *experimental measurement* and calculation to determine the total amount of entropy that a substance has at temperatures above 0 K. If the entropy of 1 mol of a substance is determined at a temperature of 298 K (25 °C) and a pressure of 1 atm, we call it the **standard entropy, $S°$.** Table 20.1 lists the standard entropies for a

TABLE 20.1	STANDARD ENTROPIES OF SOME TYPICAL SUBSTANCES AT 298.15 K				
Substance	$S°$ (J mol^{-1} K^{-1})	Substance	$S°$ (J mol^{-1} K^{-1})	Substance	$S°$ (J mol^{-1} K^{-1})
Ag(s)	42.55	CaO(s)	40	N$_2$(g)	191.5
AgCl(s)	96.2	Ca(OH)$_2$(s)	76.1	NH$_3$(g)	192.5
Al(s)	28.3	CaSO$_4$(s)	107	NH$_4$Cl(s)	94.6
Al$_2$O$_3$(s)	51.00	CaSO$_4\cdot\frac{1}{2}$H$_2$O(s)	131	NO(g)	210.6
C(s, graphite)	5.69	CaSO$_4\cdot$2H$_2$O(s)	194.0	NO$_2$(g)	240.5
CO(g)	197.9	Cl$_2$(g)	223.0	N$_2$O(g)	220.0
CO$_2$(g)	213.6	Fe(s)	27	N$_2$O$_4$(g)	304
CH$_4$(g)	186.2	Fe$_2$O$_3$(s)	90.0	Na(s)	51.0
CH$_3$Cl(g)	234.2	H$_2$(g)	130.6	Na$_2$CO$_3$(s)	136
CH$_3$OH(l)	126.8	H$_2$O(g)	188.7	NaHCO$_3$(s)	102
CO(NH$_2$)$_2$(s)	104.6	H$_2$O(l)	69.96	NaCl(s)	72.38
CO(NH$_2$)$_2$(aq)	173.8	HCl(g)	186.7	NaOH(s)	64.18
C$_2$H$_2$(g)	200.8	HNO$_3$(l)	155.6	Na$_2$SO$_4$(s)	149.4
C$_2$H$_4$(g)	219.8	H$_2$SO$_4$(l)	157	O$_2$(g)	205.0
C$_2$H$_6$(g)	229.5	HC$_2$H$_3$O$_2$(l)	160	PbO(s)	67.8
C$_8$H$_{18}$(l)	466.9	Hg(l)	76.1	S(s)	31.9
C$_2$H$_5$OH(l)	161	Hg(g)	175	SO$_2$(g)	248.5
Ca(s)	154.8	K(s)	64.18	SO$_3$(g)	256.2
CaCO$_3$(s)	92.9	KCl(s)	82.59		
CaCl$_2$(s)	114	K$_2$SO$_4$(s)	176		

Why the Units of Entropy Are Energy/Temperature

Entropy is a state function, just like enthalpy, so the value of ΔS doesn't depend on the "path" that is followed during a change. In other words, we can proceed from one state to another in any way we like and ΔS will be the same. If we choose a *reversible path* in which just a slight alteration in the system can change the direction of the process, we can measure ΔS directly. For example, at 25 °C an ice cube will melt and there is nothing we can do at that temperature to stop it. This change is *nonreversible* in the sense described above, so we couldn't use this path to measure ΔS. At 0 °C, however, it is simple to stop the melting process and reverse its direction. At 0 °C the melting of ice is a thermodynamically reversible process.

If we set up a change so that it is reversible, we can calculate the entropy change as $\Delta S = q/T$, where q is the heat added to the substance and T is the temperature at which the heat is added. The more energy that we add to the system as heat, the more possible ways there are to distribute it among the molecules. To understand this, let's use the analogy between energy and money that we used

on page 874. We said that if energy were money, entropy would describe the number of different ways there are of counting it out. Just as there are more possible ways to count out $10 than there are ways of counting out $2, there are more ways to distribute a large amount of heat among a group of molecules than there are to disperse a smaller amount of heat. This entropy increase is *directly proportional to the amount of heat added*.

Temperature also affects the size of the entropy increase. To see why, study Figure 20.6. At lower temperatures, the entropy of a material is low, so the *change* in entropy produced when a given amount of heat is added to the material will be greater than it would have been at higher temperatures. The low temperature material can disperse the additional energy in more ways than the higher temperature material can. For a given quantity of heat the entropy change is *inversely proportional to the temperature* at which the heat is added. Thus, $\Delta S = q/T$, and entropy has units of energy divided by temperature (e.g., J K^{-1}).

---

number of substances.[1] Notice that entropy has the dimensions of energy/temperature (i.e., joules per kelvin; this is explained in Facets of Chemistry 20.1).

Once we have the entropies of a variety of substances, we can calculate the standard **entropy change, $\Delta S°$,** for chemical reactions in much the same way as we calculated $\Delta H°$ in Chapter 7.

$$\Delta S° = (\text{sum of } S° \text{ of the products}) - (\text{sum of } S° \text{ of the reactants}) \quad (20.6)$$

If the reaction we are working with happens to correspond to the formation of 1 mol of a compound from its elements, then the $\Delta S°$ that we calculate can be referred to as the **standard entropy of formation, $\Delta S°_f$.** Values of $\Delta S°_f$ are not tabulated, however; if we need them for some purpose, we must calculate them from tabulated values of $S°$.

◀ **TOOLS**

Standard entropies

This is simply a Hess's law type of calculation. Note, however, that elements have nonzero $S°$ values, which must be included in the bookkeeping.

Urea (a compound found in urine) is manufactured commercially from CO_2 and NH_3. One of its uses is as a fertilizer, where it reacts slowly with water in the soil to produce ammonia and carbon dioxide. The ammonia provides a source of nitrogen for growing plants.

$$CO(NH_2)_2(aq) + H_2O(l) \longrightarrow CO_2(g) + 2NH_3(g)$$
$$\text{urea}$$

EXAMPLE 20.3

Calculating $\Delta S°$ from Standard Entropies

---

[1] In our earlier discussions of standard states (Chapter 7) we defined the *standard pressure* as 1 atm. This was the original pressure unit used by thermodynamicists. However, in the SI, the recognized unit of pressure is the pascal (Pa), not the atmosphere. After considerable discussion, the SI adopted the **bar** as the standard pressure for thermodynamic quantities: 1 bar $= 10^5$ Pa. One bar differs from 1 atmosphere by only 1.3%, and for thermodynamic quantities that we deal with in this text, their values at 1 atm and at 1 bar differ by an insignificant amount. Since most available thermodynamic data is still specified at 1 atm rather than 1 bar, we shall continue to use 1 atm for the standard pressure.

What is the standard entropy change when 1 mol of urea reacts with water?

ANALYSIS: We can use Equation 20.6 to compute the standard entropy change for the reaction. We'll need the standard entropies $S°$ of each reactant and product.

The data we need can be found in Table 20.1, and is collected in the table to the right.

Substance	$S°$ (J/mol K)
$CO(NH_2)_2(aq)$	173.8
$H_2O(l)$	69.96
$CO_2(g)$	213.6
$NH_3(g)$	192.5

SOLUTION: Applying Equation 20.6, we have

$$\Delta S° = [S°_{CO_2(g)} + 2S°_{NH_3(g)}] - [S°_{CO(NH_2)_2(aq)} + S°_{H_2O(l)}]$$

$$= \left[1 \text{ mol} \times \left(\frac{213.6 \text{ J}}{\text{mol K}}\right) + 2 \text{ mol} \times \left(\frac{192.5 \text{ J}}{\text{mol K}}\right)\right]$$

$$- \left[1 \text{ mol} \times \left(\frac{173.8 \text{ J}}{\text{mol K}}\right) + 1 \text{ mol} \times \left(\frac{69.96 \text{ J}}{\text{mol K}}\right)\right]$$

$$= (598.6 \text{ J/K}) - (243.8 \text{ J/K})$$

$$= 354.8 \text{ J/K}$$

Notice that the unit mol cancels in each term, so the units of $\Delta S°$ are joules per kelvin.

Thus, the standard entropy change for this reaction is $+354.8$ J/K (which we can also write as $+354.8$ J K^{-1}).

Is the Answer Reasonable?

In the reaction, gases are formed from liquid reactants. Since gases have much larger entropies than liquids, we expect $\Delta S°$ to be positive, which agrees with our answer.

PRACTICE EXERCISE 8: Calculate the standard entropy change, $\Delta S°$, in J K^{-1} for the following:
(a) $CaO(s) + 2HCl(g) \rightarrow CaCl_2(s) + H_2O(l)$
(b) $C_2H_4(g) + H_2(g) \rightarrow C_2H_6(g)$

Sometimes, the temperature is specified as a subscript in writing the symbol for the standard free energy change. For example, $\Delta G°$ can also be written $\Delta G°_{298}$. As you will see later, there are times when it is desirable to indicate the temperature explicitly.

Calculating $\Delta G°$ from $\Delta H°$ and $\Delta S°$

20.6 ▶ **The standard free energy change, $\Delta G°$, is ΔG at standard conditions**

When ΔG is determined at 25 °C (298 K) and 1 atm, we call it the **standard free energy change, $\Delta G°$.** There are several ways of obtaining $\Delta G°$ for a reaction. One of them is to compute $\Delta G°$ from $\Delta H°$ and $\Delta S°$.

$$\Delta G° = \Delta H° - (298.15 \text{ K})\Delta S°$$

Experimental measurement of $\Delta G°$ is also possible, but we will discuss how this is done later.

EXAMPLE 20.4
Calculating $\Delta G°$ from $\Delta H°$ and $\Delta S°$

Calculate $\Delta G°$ for the reaction of urea with water from values of $\Delta H°$ and $\Delta S°$.

$$CO(NH_2)_2(aq) + H_2O(l) \longrightarrow CO_2(g) + 2NH_3(g)$$

ANALYSIS: We can calculate $\Delta G°$ with the equation

$$\Delta G° = \Delta H° - T\Delta S°$$

The data needed to calculate $\Delta H°$ come from Table 7.2 and require a Hess's law calculation. To obtain $\Delta S°$, we normally would need to do a similar calculation with data from Table 20.1. However, we already performed this calculation in Example 20.3.

SOLUTION: First we calculate $\Delta H°$ from data in Table 7.2.

$$\Delta H° = [\Delta H°_{f\,CO_2(g)} + 2\Delta H°_{f\,NH_3(g)}] - [\Delta H°_{f\,CO(NH_2)_2(aq)} + \Delta H°_{f\,H_2O(l)}]$$

$$= \left[1\text{ mol} \times \left(\frac{-393.5\text{ kJ}}{\text{mol}} \right) + 2\text{ mol} \times \left(\frac{-46.19\text{ kJ}}{\text{mol}} \right) \right]$$

$$- \left[1\text{ mol} \times \left(\frac{-319.2\text{ kJ}}{\text{mol}} \right) + 1\text{ mol} \times \left(\frac{-285.9\text{ kJ}}{\text{mol}} \right) \right]$$

$$= (-485.9\text{ kJ}) - (-605.1\text{ kJ})$$

$$= +119.2\text{ kJ}$$

In Example 20.3 we found $\Delta S°$ to be $+354.8$ J K^{-1}. To calculate $\Delta G°$ we also need the Kelvin temperature, which we must express to at least four significant figures to match the number of significant figures in $\Delta S°$. Since standard temperature is *exactly* 25 °C, $T_K = (25.00 + 273.15)$ K $= 298.15$ K. Also, we must be careful to express $\Delta H°$ and $T\Delta S°$ in the same energy units, so we'll change the units of the entropy change to give $\Delta S° = +0.3548$ kJ K^{-1}. Substituting into the equation for $\Delta G°$,

To be exact: $T_K = t_C + 273.15$

354.8 J K^{-1} $= 0.3548$ kJ K^{-1}

$$\Delta G° = +119.2\text{ kJ} - (298.15\text{ K})(0.3548\text{ kJ K}^{-1})$$

$$= +119.2\text{ kJ} - 105.8\text{ kJ}$$

$$= +13.4\text{ kJ}$$

Therefore, for this reaction, $\Delta G° = +13.4$ kJ.

Is the Answer Reasonable?
There's not much we can do to check the reasonableness of the answer. We just need to check to be sure we've done the calculations correctly.

PRACTICE EXERCISE 9: Use the data in Table 7.2 and Table 20.1 to calculate $\Delta G°$ for the formation of iron(III) oxide (the iron oxide in rust). The equation for the reaction is

$$4Fe(s) + 3O_2(g) \longrightarrow 2Fe_2O_3(s)$$

In Section 7.8 you learned that it is useful to have tabulated standard heats of formation, $\Delta H°_f$, because they can be used with Hess's law to calculate $\Delta H°$ for many different reactions. Standard free energies of formation, $\Delta G°_f$, can be used in similar calculations to obtain $\Delta G°$.

$$\Delta G° = (\text{sum of } \Delta G°_f \text{ of products}) - (\text{sum of } \Delta G°_f \text{ of reactants}) \qquad (20.7)$$

TOOLS

Using $\Delta G°_f$ values

The $\Delta G°_f$ values for some typical substances are found in Table 20.2. Example 20.5 shows how we can use them to calculate $\Delta G°$ for a reaction.

Ethanol, C_2H_5OH (also called ethyl alcohol), is made from grain by fermentation and has been used as an additive to gasoline to produce a product called gasohol. What is $\Delta G°$ for the combustion of liquid ethanol to give $CO_2(g)$ and $H_2O(g)$?

EXAMPLE 20.5

Calculating $\Delta G°$ from $\Delta G°_f$

ANALYSIS: We can use Equation 20.7 to compute the standard Gibbs free energy change for the reaction. We'll need the standard Gibbs free energy changes of formation for each reactant and product. These data can be found in Table 20.2.

SOLUTION: First, we need the balanced equation for the reaction.

$$C_2H_5OH(l) + 3O_2(g) \longrightarrow 2CO_2(g) + 3H_2O(g)$$

Applying Equation 20.7, we have

$$\Delta G° = [2\Delta G°_{f\,CO_2(g)} + 3\Delta G°_{f\,H_2O(g)}] - [\Delta G°_{f\,C_2H_5OH(l)} + 3\Delta G°_{f\,O_2(g)}]$$

TABLE 20.2	STANDARD FREE ENERGIES OF FORMATION OF TYPICAL SUBSTANCES AT 298.15 K					

Substance	ΔG_f° (kJ mol^{-1})	Substance	ΔG_f° (kJ mol^{-1})	Substance	ΔG_f° (kJ mol^{-1})
$Ag(s)$	0	$CaO(s)$	-604.2	$N_2(g)$	0
$AgCl(s)$	-109.7	$Ca(OH)_2(s)$	-896.76	$NH_3(g)$	-16.7
$Al(s)$	0	$CaSO_4(s)$	-1320.3	$NH_4Cl(s)$	-203.9
$Al_2O_3(s)$	-1576.4	$CaSO_4 \cdot \frac{1}{2} H_2O(s)$	-1435.2	$NO(g)$	$+86.69$
$C(s, graphite)$	0	$CaSO_4 \cdot 2H_2O(s)$	-1795.7	$NO_2(g)$	$+51.84$
$CO(g)$	-137.3	$Cl_2(g)$	0	$N_2O(g)$	$+103.6$
$CO_2(g)$	-394.4	$Fe(s)$	0	$N_2O_4(g)$	$+98.28$
$CH_4(g)$	-50.79	$Fe_2O_3(s)$	-741.0	$Na(s)$	0
$CH_3Cl(g)$	-58.6	$H_2(g)$	0	$Na_2CO_3(s)$	-1048
$CH_3OH(l)$	-166.2	$H_2O(g)$	-228.6	$NaHCO_3(s)$	-851.9
$CO(NH_2)_2(s)$	-197.2	$H_2O(l)$	-237.2	$NaCl(s)$	-384.0
$CO(NH_2)_2(aq)$	-203.8	$HCl(g)$	-95.27	$NaOH(s)$	-382
$C_2H_2(g)$	$+209$	$HNO_3(l)$	-79.91	$Na_2SO_4(s)$	-1266.8
$C_2H_4(g)$	$+68.12$	$H_2SO_4(l)$	-689.9	$O_2(g)$	0
$C_2H_6(g)$	-32.9	$HC_2H_3O_2(l)$	-392.5	$PbO(s)$	-189.3
$C_2H_5OH(l)$	-174.8	$Hg(l)$	0	$S(s)$	0
$C_8H_{18}(l)$	$+17.3$	$Hg(g)$	$+31.8$	$SO_2(g)$	-300.4
$Ca(s)$	0	$K(s)$	0	$SO_3(g)$	-370.4
$CaCO_3(s)$	-1128.8	$KCl(s)$	-408.3		
$CaCl_2(s)$	-750.2	$K_2SO_4(s)$	-1316.4		

As with ΔH°, the ΔG_f° for any element in its standard state is zero. Therefore, using the data from Table 20.2,

$$\Delta G^\circ = \left[2 \text{ mol} \times \left(\frac{-394.4 \text{ kJ}}{\text{mol}}\right) + 3 \text{ mol} \times \left(\frac{-228.6 \text{ kJ}}{\text{mol}}\right)\right]$$
$$- \left[1 \text{ mol} \times \left(\frac{-174.8 \text{ kJ}}{\text{mol}}\right) + 3 \text{ mol} \times \left(\frac{0 \text{ kJ}}{\text{mol}}\right)\right]$$
$$= (-1474.6 \text{ kJ}) - (-174.8 \text{ kJ})$$
$$= -1299.8 \text{ kJ}$$

The standard free energy change for the reaction equals -1299.8 kJ.

Is the Answer Reasonable?
There's no quick way to estimate the answer, although we could calculate ΔG° from ΔH° and ΔS° following the method used in Example 20.5 for a thorough check. We know that ethanol is quite flammable under standard conditions, so we expect this combustion reaction to proceed spontaneously. The large negative standard free energy change makes sense.

PRACTICE EXERCISE 10: Calculate $\Delta G^\circ_{reaction}$ in kilojoules for the following reactions using the data in Table 20.2:
(a) $2NO(g) + O_2(g) \rightarrow 2NO_2(g)$
(b) $Ca(OH)_2(s) + 2HCl(g) \rightarrow CaCl_2(s) + 2H_2O(g)$

20.7 ▶ ΔG is the maximum amount of work that can be done by a process

One of the chief uses of spontaneous chemical reactions is the production of useful work. For example, fuels are burned in gasoline or diesel engines to power automobiles and heavy machinery, and chemical reactions in batteries start our autos and run all sorts of modern electronic gadgets, including cellular phones, beepers, and laptop computers.

When chemical reactions occur, however, their energy is not always harnessed to do work. For instance, if gasoline is burned in an open dish, the energy evolved is lost entirely as heat and no useful work is accomplished. Engineers, therefore, seek ways to capture as much energy as possible in the form of work. One of their primary goals is to maximize the efficiency with which chemical energy is converted to work and to minimize the amount of energy transferred unproductively to the environment as heat.

Scientists have discovered that the maximum conversion of chemical energy to work occurs if a reaction is carried out under conditions that are said to be thermodynamically reversible. A process is defined as **thermodynamically reversible** if its driving force is opposed by another force that is just the slightest bit weaker, so that the slightest increase in the opposing force will cause the direction of the change to be reversed. An example of a nearly reversible process is illustrated in Figure 20.10, where we have a compressed gas in a cylinder pushing against a piston that's held in place by liquid water above it. If a water molecule evaporates, the external pressure drops slightly and the gas can expand just a bit. Gradually, as one water molecule after another evaporates, the gas inside the cylinder slowly expands. At any time, however, the process can be reversed by the condensation of a water molecule.

Note that thermodynamic reversibility implies that the system is very close to equilibrium at every stage of a process. While we sometimes say that a reaction like dissociation of a weak acid in water is "reversible" simply because it runs in both the forward and backward directions, we cannot say the reaction is *thermodynamically reversible* unless the concentrations are only infinitesmally different from their equilibrium values as the reaction occurs.

Although we could obtain the maximum work by carrying out a change reversibly, a thermodynamically reversible process requires so many steps that it proceeds at an extremely slow speed. If the work cannot be done at a reasonable rate, it is of little value to us. Our goal, then, is to approach thermodynamic reversibility for maximum efficiency, but to carry out the change at a pace that will deliver work at acceptable rates.

A reversible process requires an infinite number of tiny steps. This takes forever to accomplish.

The relationship of useful work to thermodynamic reversibility was illustrated earlier (Section 7.4) in our discussion of the discharge of an automobile battery. Re-

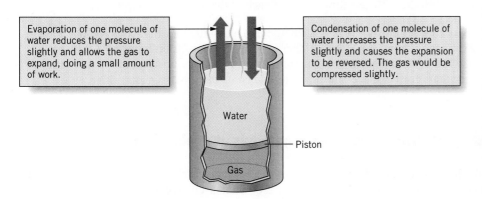

Evaporation of one molecule of water reduces the pressure slightly and allows the gas to expand, doing a small amount of work.

Condensation of one molecule of water increases the pressure slightly and causes the expansion to be reversed. The gas would be compressed slightly.

Water

Piston

Gas

Figure 20.10 *A reversible expansion of a gas.* As water molecules evaporate one at a time, the external pressure gradually decreases and the gas slowly expands. The process would be reversed if a molecule of water were to condense into the liquid. The ability of the expansion to be reversed by the slightest increase in the opposing pressure is what makes this a reversible process.

call that when the battery is shorted with a heavy wrench, no work is done and all the energy appears as heat. In this case there is nothing opposing the discharge, and it occurs in the most thermodynamically irreversible manner possible. However, when the current is passed through a small electric motor, the motor itself offers resistance to the passage of the electricity and the discharge takes place slowly. In this instance, the discharge occurs in a more nearly thermodynamically reversible manner because of the opposition provided by the motor, and a relatively large amount of the available energy appears in the form of the work accomplished by the motor.

The preceding discussion leads naturally to the question, "Is there a limit to the amount of the available energy in a reaction that can be harnessed as useful work?" The answer to this question is to be found in the Gibbs free energy.

> The maximum amount of energy produced by a reaction that can be *theoretically* harnessed as work is equal to ΔG.

This is the energy that need not be lost to the surroundings as heat and is therefore *free* to be used for work. Thus, by determining the value of ΔG, we can find out whether or not a given reaction will be an effective source of useful energy. Also, by comparing the actual amount of work derived from a given system with the ΔG values for the reactions involved, we can measure the efficiency of the system.

EXAMPLE 20.6
Calculating Maximum Work

Calculate the maximum work available, expressed in kilojoules, from the oxidation of 1 mol of octane, $C_8H_{18}(l)$, by oxygen to give $CO_2(g)$ and $H_2O(l)$ at 25 °C and 1 atm.

ANALYSIS: The maximum work is equal to ΔG for the reaction. Standard thermodynamic conditions are specified, so we need to calculate $\Delta G°$.

SOLUTION: First we need a balanced equation for the reaction. For the combustion of 1 mol of C_8H_{18} we have

$$C_8H_{18}(l) + 12\tfrac{1}{2}O_2(g) \longrightarrow 8CO_2(g) + 9H_2O(l)$$

Then we apply Equation 20.7.

$$\Delta G° = [8\,\Delta G°_{f\,CO_2(g)} + 9\,\Delta G°_{f\,H_2O(l)}] - [\Delta G°_{f\,C_8H_{18}(l)} + 12.5\,\Delta G°_{f\,O_2(g)}]$$

Referring to Table 20.2 and dropping the canceled mol units,

$$\Delta G° = [8 \times (-394.4)\text{ kJ} + 9 \times (-237.2)\text{ kJ}] - [1 \times (+17.3)\text{ kJ} + 12.5 \times (0)\text{ kJ}]$$

$$= (-5290\text{ kJ}) - (+17.3\text{ kJ})$$

$$= -5307\text{ kJ}$$

Thus, at 25 °C and 1 atm, we can expect no more than 5307 kJ of work from the oxidation of 1 mol of C_8H_{18}.

Is the Answer Reasonable?
Be sure to check the algebraic signs of each of the terms in the calculation.

PRACTICE EXERCISE 11: Calculate the maximum work that could be obtained at 25 °C and 1 atm from the oxidation of 1.00 mol of aluminum by $O_2(g)$ to give $Al_2O_3(s)$. (The combustion of aluminum in booster rockets provides part of the energy that lifts the Space Shuttle off its launching pad.)

Blastoff of the Space Shuttle.
The large negative heat of formation of Al_2O_3 provides power to the solid booster rockets that lift the Space Shuttle from its launch pad.

20.8 ▶ ΔG is zero when a system is at equilibrium

We have seen that when the value of ΔG for a given change is negative, the change occurs spontaneously. We have also seen that a change is nonspontaneous when ΔG is positive. However, when ΔG is neither positive nor negative, the change is neither spontaneous nor nonspontaneous—the system is in a state of equilibrium. This occurs when ΔG is equal to zero.

When a system is in a state of dynamic equilibrium,
$$G_{\text{products}} = G_{\text{reactants}} \quad \text{and} \quad \Delta G = 0$$

Let's again consider the freezing of water.
$$H_2O(l) \rightleftharpoons H_2O(s)$$

Below 0 °C, ΔG for this change is negative and the freezing is spontaneous. On the other hand, above 0 °C we find that ΔG is positive and freezing is nonspontaneous. When the temperature is exactly 0 °C, $\Delta G = 0$ and an ice–water mixture exists in a condition of equilibrium. As long as heat isn't added or removed from the system, neither freezing nor melting is spontaneous and the ice and liquid water can exist together indefinitely.

No work can be done by a system at equilibrium

We have identified ΔG as a quantity that specifies the amount of work that is available from a system. Since ΔG is zero at equilibrium, the amount of work available is zero also. Therefore, when a system is at equilibrium, no work can be extracted from it. As an example, let's consider again the common lead storage battery that we use to start a car.

When the battery is fully charged, there are virtually no products of the discharge reaction present. The chemical reactants, however, are present in large amounts. Therefore, the total free energy of the reactants far exceeds the total free energy of products and, since $\Delta G = G_{\text{products}} - G_{\text{reactants}}$, the ΔG of the system has a large negative value. This means that a lot of energy is available to do work. As the battery discharges, the reactants are converted to products and G_{products} gets larger while $G_{\text{reactants}}$ gets smaller; thus ΔG becomes less negative, and less energy is available to do work. Finally, the battery reaches equilibrium. The total free energies of the reactants and the products have become equal, so $G_{\text{products}} - G_{\text{reactants}} = 0$ and $\Delta G = 0$. No further work can be extracted and we say the battery is dead.

Melting points and boiling points can be estimated from ΔH and ΔS

When we have equilibrium in any system, we know that $\Delta G = 0$. For a phase change such as $H_2O(l) \rightarrow H_2O(s)$, equilibrium can only exist at one particular temperature at atmospheric pressure. In this instance, that temperature is 0 °C. Above 0 °C, only liquid water can exist, and below 0 °C all the liquid will freeze to give ice. This yields an interesting relationship between ΔH and ΔS for a phase change. Since $\Delta G = 0$,

$$\Delta G = 0 = \Delta H - T\Delta S$$

Therefore,

$$\Delta H = T\Delta S$$

and

$$\Delta S = \frac{\Delta H}{T} \tag{20.8}$$

Thus, if we know ΔH for the phase change and the temperature at which the two phases coexist, we can calculate ΔS for the phase change. Another interesting relationship that we can obtain is

$$T = \frac{\Delta H}{\Delta S} \tag{20.9}$$

Thus, if we know ΔH and ΔS, we can calculate the temperature at which equilibrium will occur.

| **EXAMPLE 20.7** **Calculating the Equilibrium Temperature for a Phase Change** | For the phase change $Br_2(l) \rightarrow Br_2(g)$, $\Delta H° = +31.0$ kJ mol^{-1} and $\Delta S° = 92.9$ J mol^{-1} K^{-1}. Assuming that ΔH and ΔS are nearly temperature independent, calculate the approximate Celsius temperature at which $Br_2(l)$ will be in equilibrium with $Br_2(g)$ at 1 atm (i.e., the normal boiling point of liquid Br_2). |

ANALYSIS: The temperature at which equilibrium exists is given by Equation 20.9,

$$T = \frac{\Delta H}{\Delta S}$$

ΔH and ΔS do not change much with changes in temperature. This is because temperature changes affect the enthalpies and entropies of both the reactants and products by about the same amount, so the differences between reactants and products stay fairly constant.

If ΔH and ΔS do not depend on temperature, then we can use $\Delta H°$ and $\Delta S°$ in this equation. That is,

$$T = \frac{\Delta H°}{\Delta S°}$$

SOLUTION: Substituting the data given in the problem,

$$T = \frac{3.10 \times 10^4 \text{ J mol}^{-1} \text{ K}^{-1}}{92.9 \text{ J mol}^{-1} \text{ K}^{-1}}$$

$$= 334 \text{ K}$$

The Celsius temperature is $334 - 273 = 61$ °C. Notice that we were careful to express $\Delta H°$ in joules, not kilojoules, so the units would cancel correctly. It is also interesting that the boiling point we calculated is quite close to the measured normal boiling point of 58.8 °C.

Is the Answer Reasonable?
The $\Delta H°$ value equals 31,000 and the $\Delta S°$ value equals approximately 100, which means the temperature should be about 310 K. Our value, 334 K, is not far from that, so the answer is reasonable.

PRACTICE EXERCISE 12: The heat of vaporization of mercury is 60.7 kJ/mol. For $Hg(l)$, $S° = 76.1$ J mol^{-1} K^{-1} and for $Hg(g)$, $S° = 175$ J mol^{-1} K^{-1}. Estimate the normal boiling point of liquid mercury.

Free energy diagrams show how free energy changes during a process

We have said that in a phase change such as $H_2O(l) \rightarrow H_2O(s)$, equilibrium can exist for a given pressure only at one particular temperature; for water at a pressure of 1 atm, this temperature is 0 °C, the freezing point. At other temperatures, the phase change proceeds entirely to completion in one direction or another. One way to gain a better understanding of this is by studying **free energy diagrams,** which depict how the free energy changes as we proceed from the "reactants" to the "products."

Figure 20.11 illustrates three different free energy diagrams for water–ice mixtures. On the left of each diagram is indicated the free energy of pure liquid water, and on the right the free energy of the pure solid. Points along the horizontal axis represent mixtures of both phases. Thus, going from left to right across a graph we are able to see how the free energy varies as a system that consists entirely of $H_2O(l)$ changes ultimately to one that consists entirely of $H_2O(s)$.

Below 0 °C, we see that the free energy of the liquid is higher than that of the solid. You've learned that a spontaneous change will occur if the free energy can decrease. Therefore, the first diagram tells us that if we start with the liquid phase, or any mixture of liquid or solid, freezing will occur until only the solid is present. This is because the free energy decreases continually until all the liquid has frozen.

Above 0 °C, we have the opposite situation. The free energy decreases in the direction of $H_2O(s) \rightarrow H_2O(l)$, and it continues to drop until all the solid has

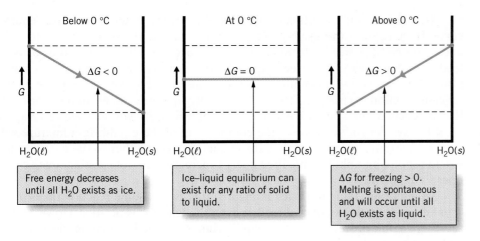

FIGURE 20.11 *Free energy diagram for conversion of $H_2O(l)$ to $H_2O(s)$.* At the left of each diagram, the system consists entirely of $H_2O(l)$. At the right is $H_2O(s)$. The horizontal axis represents the extent of conversion from $H_2O(l)$ to $H_2O(s)$.

melted. This means that if we have ice, or any mixture of ice and liquid water, it will continue to melt until only the liquid is present.

Above or below 0 °C, the system is unable to establish an equilibrium mixture of liquid and solid. Melting or freezing occurs until only one phase is present. However, at 0 °C there is no change in free energy if either melting or freezing occurs, so there is no driving force for either change. Therefore, as long as a system of ice and liquid water is insulated from warmer or colder surroundings, any particular mixture of the two phases is stable and a state of equilibrium exists.

Free energy diagrams for chemical reactions show a minimum in free energy at equilibrium

The free energy changes that occur in most chemical reactions are more complex than those in phase changes. As an example, let's study a reaction you've seen before—the decomposition of N_2O_4 into NO_2.

$$N_2O_4(g) \longrightarrow 2NO_2(g)$$

In our discussion of chemical equilibrium in Chapter 16 (page 699), we noted that equilibrium in this system can be approached from *either* direction, with the same equilibrium concentrations being achieved provided we begin with the same overall system composition.

Figure 20.12 shows the free energy diagram for the reaction. Notice that in going from reactant to product, the free energy has a minimum. It drops below that of either pure N_2O_4 or pure NO_2.

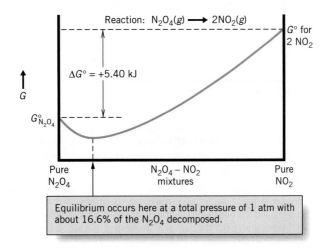

FIGURE 20.12 *Free energy diagram for the decomposition of $N_2O_4(g)$.* The minimum on the curve indicates the composition of the reaction mixture at equilibrium. Because $\Delta G°$ is positive, the position of equilibrium lies close to the reactants. Not much product will form by the time the system reaches equilibrium.

When a system moves "downhill" on its free energy curve, $G_{final} < G_{initial}$ and ΔG is negative. Changes with negative ΔG are spontaneous.

Any system will spontaneously seek the lowest point on its free energy curve. If we begin with pure $N_2O_4(g)$, the reaction will proceed from left to right and some $NO_2(g)$ will be formed, because proceeding in the direction of NO_2 leads to a lowering of the free energy. If we begin with pure $NO_2(g)$, a change also will occur. Going downhill on the free energy curve now takes place as the reverse reaction occurs [i.e., $2NO_2(g) \rightarrow N_2O_4(g)$]. Once the bottom of the "valley" is reached, the system has come to equilibrium. As you learned in Chapter 16, if the system isn't disturbed, the composition of the equilibrium mixture will remain constant. Now we see that any change (moving either to the left or right) would require an uphill climb. Free energy increases are not spontaneous, so this doesn't happen.

An important thing to notice in Figure 20.12 is that some reaction takes place spontaneously in the forward direction even though $\Delta G°$ is positive. However, the reaction doesn't proceed far before equilibrium is reached. For comparison, Figure 20.13 shows the shape of the free energy curve for a reaction with a negative $\Delta G°$. We see here that at equilibrium there has been a much greater conversion of reactants to products. Thus, $\Delta G°$ *tells us where the position of equilibrium lies between pure reactants and pure products.* When $\Delta G°$ is positive, the position of equilibrium lies close to the reactants and little reaction occurs by the time equilibrium is reached. When $\Delta G°$ is negative, the position of equilibrium lies close to the products and a large amount of products will have formed by the time equilibrium is reached.

$\Delta G°$ can be used to predict the outcome of a chemical reaction

Using $\Delta G°$ to determine the position of equilibrium

In general, the value of $\Delta G°$ for most reactions is much larger numerically than the $\Delta G°$ for the $N_2O_4 \rightarrow NO_2$ reaction. In addition, the extent to which a reaction proceeds is very sensitive to the size of $\Delta G°$. If the $\Delta G°$ value for a reaction is reasonably large—about 20 kJ or more—almost no observable reaction will occur when $\Delta G°$ is positive. On the other hand, the reaction will go almost to completion if $\Delta G°$ is both large and negative.[2] From a practical standpoint, then, *the size and sign of $\Delta G°$ serve as indicators of whether an observable spontaneous reaction will occur.*

Figure 20.13 *Free energy curve for a reaction having a negative $\Delta G°$.* Because $G_B°$ is less than $G_A°$, $\Delta G°$ is negative. This causes the position of equilibrium to lie far to the right, near the products. When the system reaches equilibrium, there will be a large amount of products present and little reactants.

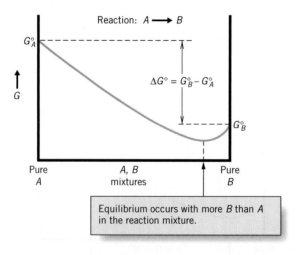

[2]To actually see a change take place, the speed of a spontaneous reaction must be reasonably fast. For example, the decomposition of the nitrogen oxides into N_2 and O_2 is thermodynamically spontaneous ($\Delta G°$ is negative), but their rates of decomposition are so slow that these substances appear to be stable and some are obnoxious air pollutants.

Would we expect to be able to observe the following reaction at 25 °C?

$$NH_4Cl(s) \longrightarrow NH_3(g) + HCl(g)$$

ANALYSIS: We will need to determine the magnitude and sign of $\Delta G°$ for the reaction. If $\Delta G°$ is reasonably large and positive, the reaction won't be observed. If it is reasonably large and negative, we can expect to see the reaction go nearly to completion.

SOLUTION: First let's calculate $\Delta G°$ for the reaction using the data in Table 20.2. The procedure is the same as that discussed earlier.

$$\Delta G° = [\Delta G°_{f\,NH_3(g)} + \Delta G°_{f\,HCl(g)}] - [\Delta G°_{f\,NH_4Cl(s)}]$$

$$= [(-16.7\text{ kJ}) + (-95.27\text{ kJ})] - [-203.9\text{ kJ}]$$

$$= +91.9\text{ kJ}$$

Because $\Delta G°$ is large and positive, we only expect to observe the spontaneous formation of extremely small amounts of products at this temperature.

Is the Answer Reasonable?
Be sure to check the algebraic signs of each term in the calculation.

PRACTICE EXERCISE 13: Use the data in Table 20.2 to determine whether the reaction

$$SO_2(g) + O_2(g) \longrightarrow SO_3(g)$$

should "occur spontaneously" at 25 °C.

PRACTICE EXERCISE 14: Use the data in Table 20.2 to determine whether we should expect to see the formation of $CaCO_3(s)$ in the following reaction at 25 °C:

$$CaCl_2(s) + H_2O(g) + CO_2(g) \longrightarrow CaCO_3(s) + 2HCl(g)$$

EXAMPLE 20.8
Using $\Delta G°$ as a Predictor of the Outcome of a Reaction

The position of equilibrium changes with temperature because ΔG changes with temperature

So far, we have confined our discussion of the relationship of free energy and equilibrium to a special case, 25 °C. But what about other temperatures? Equilibria certainly can exist at temperatures other than 25 °C, and in Chapter 16 you learned how to apply Le Châtelier's principle to predicting the way temperature affects the position of equilibrium. Now let's see how thermodynamics deals with this.

You've learned that at 25 °C the position of equilibrium is determined by the difference between the free energy of pure products and the free energy of pure reactants. This difference is given by $\Delta G°_{298}$, where we have now used the subscript 298 to indicate the temperature, 298 K. $\Delta G°_{298}$ is defined as

298 K = 25 °C

$$\Delta G°_{298} = (G°_{products})_{298} - (G°_{reactants})_{298}$$

At temperatures other than 25 °C, it is still the difference between the free energies of the products and reactants that determines the position of equilibrium. We might write this as $\Delta G°_T$. Thus, at a temperature other than 25 °C (298 K), we have

$$\Delta G°_T = (G°_{products})_T - (G°_{reactants})_T$$

where $(G°_{products})_T$ and $(G°_{reactants})_T$ are the total free energies of the pure products and reactants, respectively, at this other temperature.

Next, we must find a way to compute $\Delta G°_T$. Earlier we saw that $\Delta G°$ can be obtained from the equation

$$\Delta G° = \Delta H° - (298\text{ K})\Delta S°$$

At a different temperature, T, the equation becomes

$$\Delta G°_T = \Delta H°_T - T\Delta S°_T$$

Now we seem to be getting closer to our goal. If we can compute or estimate the values of ΔH_T° and ΔS_T° for a reaction, we have solved our problem.

The magnitudes of ΔH and ΔS are relatively insensitive to temperature changes. However, ΔG is very temperature sensitive because

$$\Delta G = \Delta H - T\Delta S.$$

The size of ΔG_T° obviously depends very strongly on the temperature—the equation above has temperature as one of its variables. However, the magnitudes of the ΔH and ΔS for a reaction are relatively insensitive to the temperature. This is because the enthalpies and entropies of *both* the reactants and products increase about equally with increasing temperature, so their differences, ΔH and ΔS, remain nearly the same. As a result, we can use ΔH_{298}° and ΔS_{298}° as reasonable approximations of ΔH_T° and ΔS_T°. This allows us to rewrite the equation for ΔG_T° as

$$\Delta G_T^\circ \approx \Delta H_{298}^\circ - T\Delta S_{298}^\circ \qquad (20.10)$$

 TOOLS

Calculating ΔG° at temperatures other than 25 °C

The following examples illustrate how this equation is useful.

EXAMPLE 20.9

ΔG° at Temperatures Other than 25 °C

Earlier we saw that at 25 °C the value of ΔG° for the reaction

$$N_2O_4(g) \longrightarrow 2NO_2(g)$$

has a value of +5.40 kJ. What is the approximate value of ΔG_T° for this reaction at 100 °C?

ANALYSIS: We can use Equation 20.10 to estimate ΔG_T° but we will first need values of ΔH° and ΔS°. The ΔH° for the reaction can be calculated from the data in Table 7.2 on page 288. Here we find the following standard heats of formation:

$$N_2O_4(g) \qquad \Delta H_f^\circ = +9.67 \text{ kJ/mol}$$

$$NO_2(g) \qquad \Delta H_f^\circ = +33.8 \text{ kJ/mol}$$

We combine these by a Hess's law calculation to compute ΔH° for the reaction. Next, we compute ΔS° for the reaction using Equation 20.6. We can find absolute entropy data for each reactant and product in Table 20.1.

$$N_2O_4(g) \qquad S^\circ = 304 \text{ J/mol K}$$

$$NO_2(g) \qquad S^\circ = 240.5 \text{ J/mol K}$$

SOLUTION: We first compute the standard enthalpy change for the reaction using Hess's Law:

$$\Delta H^\circ = [2\,\Delta H_{f\,NO_2(g)}^\circ] - [\Delta H_{f\,N_2O_4(g)}^\circ]$$

$$= \left[2 \text{ mol} \times \left(\frac{33.8 \text{ kJ}}{\text{mol}}\right)\right] - \left[1 \text{ mol} \times \left(\frac{9.67 \text{ kJ}}{\text{mol}}\right)\right]$$

$$= +57.9 \text{ kJ}$$

Next, we use Equation 20.6 to compute ΔS°:

$$\Delta S^\circ = \left[2 \text{ mol} \times \left(\frac{240.5 \text{ J}}{\text{mol K}}\right)\right] - \left[1 \text{ mol} \times \left(\frac{304 \text{ J}}{\text{mol K}}\right)\right]$$

$$= +177 \text{ J K}^{-1} \text{ or } 0.177 \text{ kJ K}^{-1}$$

The temperature is 100 °C, which is 373 K, so we can call the free energy change ΔG_{373}°. Substituting into Equation 20.10 using $T = 373$ K, we get

$$\Delta G_{373}^\circ \approx (+57.9 \text{ kJ}) - (373 \text{ K})(0.177 \text{ kJ K}^{-1})$$

$$\approx -8.1 \text{ kJ}$$

Notice that at this higher temperature, the sign of ΔG_T° has become negative.

Is the Answer Reasonable?
Once again, check to be sure the signs are correct for the various terms in the calculation.

Using the results of Example 20.9, sketch a free energy diagram for the decomposition of N_2O_4 into NO_2 at 100 °C. For this reaction, how does the position of equilibrium change as the temperature is increased?

ANALYSIS: Since ΔG_T° is negative for the reaction at 100 °C, the G° of the reactants must be higher than the G° of the products. Therefore, when we draw the free energy curve, the minimum must lie closer to the products than to the reactants.

SOLUTION: The diagram for the reaction at 100 °C is shown in the margin. Comparing this diagram to the one in Figure 20.12, we see that at the higher temperature the minimum lies closer to the products. This means that at the higher temperature the decomposition will proceed further toward completion when equilibrium has been reached.

Is the Answer Reasonable?
The reaction is endothermic ($\Delta H^\circ = +57.9$ kJ). Le Châtelier's principle tells us that the position of equilibrium in an endothermic reaction is shifted toward the products when the temperature is increased. This is in agreement with the conclusion we reached in this problem.

PRACTICE EXERCISE 15: Use the data in Table 20.2 to determine ΔG_{298}° for the reaction

$$2NaHCO_3(s) \longrightarrow Na_2CO_3(s) + CO_2(g) + H_2O(g)$$

Then calculate the approximate value for ΔG° for the reaction at 200 °C using the data in Tables 7.2 and 20.1. How does the position of equilibrium for this reaction change as the temperature is increased?

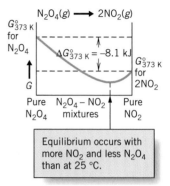

EXAMPLE 20.10
Position of Equilibrium as a Function of Temperature

$N_2O_4(g) \longrightarrow 2NO_2(g)$

$G_{373\ K}^\circ$ for N_2O_4

$\Delta G_{373\ K}^\circ = -8.1$ kJ

$G_{373\ K}^\circ$ for $2NO_2$

G

Pure N_2O_4 $N_2O_4 - NO_2$ mixtures   Pure NO_2

Equilibrium occurs with more NO_2 and less N_2O_4 than at 25 °C.

20.9 Equilibrium constants can be estimated from standard free energy changes

In the preceding discussion, you learned in a qualitative way that the position of equilibrium in a reaction is determined by the sign and magnitude of ΔG°. You also learned that the direction in which a reaction proceeds depends on where the system composition stands relative to the minimum on the free energy curve. Thus, the reaction will proceed spontaneously in the forward direction only if it will lead to a lowering of the free energy (i.e., if ΔG is negative).

Quantitatively, the relationship between ΔG and ΔG° is expressed by the following equation, which we will not attempt to justify:

$$\Delta G = \Delta G^\circ + RT \ln Q \tag{20.11}$$

Here R is the gas constant in appropriate energy units (e.g., 8.314 J mol^{-1} K^{-1}), T is the Kelvin temperature, and $\ln Q$ is the natural logarithm of the reaction quotient. For gaseous reactions, Q is calculated using partial pressures expressed in atm;[3] for reactions in solution, Q is calculated from molar concentrations. Equation 20.11 allows us to predict the direction of the spontaneous change in a reaction mixture if we know ΔG° and the composition of the mixture, as illustrated in Example 20.11.

 TOOLS

Relating the reaction quotient to ΔG

Recall that in Chapter 16 we defined Q as the *reaction quotient*—the numerical value of the mass action expression.

[3]Strictly speaking, we should express pressures in atm or bar depending on which was used to obtain the ΔG° data. For simplicity, we will use the more familiar pressure unit atm in all of our calculations. Because the atm and bar differ by only about 1 percent, any errors that might be introduced will be very small.

EXAMPLE 20.11

Determining the Direction of a Spontaneous Reaction

In Chapter 16, you learned that you can also predict the direction of a reaction by comparing Q with K (page 723).

The reaction $2NO_2(g) \rightleftharpoons N_2O_4(g)$ has $\Delta G^\circ_{298} = -5.40$ kJ per mole of N_2O_4. In a reaction mixture, the partial pressure of NO_2 is 0.25 atm and the partial pressure of N_2O_4 is 0.60 atm. In which direction must this reaction proceed to reach equilibrium?

ANALYSIS: Since we know that reactions proceed spontaneously *toward* equilibrium, we are really being asked to determine whether the reaction will proceed spontaneously in the forward or reverse direction. We can use Equation 20.11 to calculate ΔG for the forward reaction. If ΔG is negative, then the forward reaction is spontaneous. However, if the calculated ΔG is positive, the forward reaction is nonspontaneous and it is really the reverse reaction that is spontaneous.

SOLUTION: First, we need the correct form for the mass action expression so we can calculate Q correctly. Expressed in terms of partial pressures, the mass action expression is

$$\frac{P_{N_2O_4}}{P^2_{NO_2}}$$

Therefore, the equation we will use is

$$\Delta G = \Delta G^\circ + RT \ln \left(\frac{P_{N_2O_4}}{P^2_{NO_2}} \right)$$

Next, let's assemble the data:

$$\Delta G^\circ_{298} = -5.40 \text{ kJ mol}^{-1} \qquad\qquad T = 298 \text{ K}$$
$$= -5.40 \times 10^3 \text{ J mol}^{-1} \qquad P_{N_2O_4} = 0.60 \text{ atm}$$
$$R = 8.314 \text{ J mol}^{-1} \text{ K}^{-1} \qquad P_{NO_2} = 0.25 \text{ atm}$$

Notice we have changed the energy units of ΔG° to joules so they will be compatible with those calculated using R. Substituting quantities gives[4]

$$\Delta G = -5.40 \times 10^3 \text{ J mol}^{-1} + (8.314 \text{ J mol}^{-1} \text{ K}^{-1}) (298 \text{ K}) \ln \left[\frac{0.60 \text{ atm}}{(0.25 \text{ atm})^2} \right]$$

$$= -5.40 \times 10^3 \text{ J mol}^{-1} + (8.314 \text{ J mol}^{-1} \text{ K}^{-1}) (298 \text{ K}) (2.26)$$

$$= -5.40 \times 10^3 \text{ J mol}^{-1} + 5.60 \times 10^3 \text{ J mol}^{-1}$$

$$= +2.0 \times 10^2 \text{ J mol}^{-1}$$

Since ΔG is positive, the forward reaction is nonspontaneous. The reverse reaction is the one that will occur, and to reach equilibrium, some N_2O_4 will have to decompose.

Is the Answer Reasonable?

There is no simple check. However, we can check to be sure the energy units in both terms on the right are the same. Notice that here we have changed the units for ΔG° to joules to match those of R. Also notice that the temperature is expressed in kelvins, to match the temperature units in R.

PRACTICE EXERCISE 16: In which direction will the reaction described in the preceding example proceed to reach equilibrium if the partial pressure of NO_2 is 0.60 atm and the partial pressure of N_2O_4 is 0.25 atm?

---

[4]As we noted in the footnote on page 662, when taking the logarithm of a quantity, the number of digits *after the decimal point* should equal the number of significant figures in the quantity. Since 0.60 atm and 0.25 atm both have two significant figures, the logarithm of the quantity in square brackets (2.26) is rounded to give two digits after the decimal point.

The thermodynamic equilibrium constant K can be computed from $\Delta G°$

Earlier you learned that when a system reaches equilibrium the free energy of the products equals the free energy of the reactants and ΔG equals zero. We also know that at equilibrium the reaction quotient equals the equilibrium constant.

$$\text{At equilibrium} \quad \begin{cases} \Delta G = 0 \\ Q = K \end{cases}$$

If we substitute these into Equation 20.11, we obtain

$$0 = \Delta G° + RT \ln K$$

which can be rearranged to give

$$\Delta G° = -RT \ln K \qquad (20.12)$$

TOOLS

Determining thermodynamic equilibrium constants

The equilibrium constant K calculated from this equation is often called the **thermodynamic equilibrium constant** and corresponds to K_p for reactions involving gases (with partial pressures expressed in atm) and to K_c for reactions in solution (with concentrations expressed in mol L^{-1}).[5]

Equation 20.12 is useful because it permits us to determine equilibrium constants from either measured or calculated values of $\Delta G°$. As you know, $\Delta G°$ can be determined by a Hess's law type of calculation from tabulated values of $\Delta G_f°$. It also allows us to obtain values of $\Delta G°$ from measured equilibrium constants.

The brownish haze associated with air pollution is caused by nitrogen dioxide, NO_2, a red-brown gas. Nitric oxide, NO, is formed in auto engines and some of it escapes into the air, where it is oxidized to NO_2 by oxygen.

$$2NO(g) + O_2(g) \rightleftharpoons 2NO_2(g)$$

The value of K_p for this reaction is 1.7×10^{12} at 25.00 °C. What is $\Delta G°$ for the reaction, expressed in joules per mole? In kilojoules per mole?

ANALYSIS: We can compute $\Delta G°$ from the given value of K_p using the following relationship:

$$\Delta G° = -RT \ln K_p$$

We'll need the following data. Note that the temperature is given to the nearest hundredth of a degree, so we'll use the equation $T_K = t_C + 273.15$ to compute the Kelvin temperature.

$$R = 8.314 \text{ J mol}^{-1} \text{ K}^{-1}$$

$$T = 298.15 \text{ K}$$

$$K_p = 1.7 \times 10^{12}$$

SOLUTION: Substituting these values into the equation, we have

$$\Delta G° = -(8.314 \text{ J mol}^{-1} \text{ K}^{-1} \times 298.15 \text{ K}) \ln (1.7 \times 10^{12})$$

$$= -(8.314 \text{ J mol}^{-1} \text{ K}^{-1} \times 298.15 \text{ K}) \times (28.16)$$

$$= -6.980 \times 10^4 \text{ J mol}^{-1} \text{ (to four significant figures)}$$

Expressed in kilojoules, $\Delta G° = -69.80 \text{ kJ mol}^{-1}$.

EXAMPLE 20.12

Thermodynamic Equilibrium Constants

The units in this calculation give $\Delta G°$ on a per mole (mol^{-1}) basis. This reminds us that we are viewing the coefficients of the reactants and products as representing moles, rather than some other-sized quantity.

---

[5]The thermodynamic equilibrium constant requires that gases *always* be included as partial pressures in mass action expressions. The pressure units must be standard pressure units (atmospheres or bars, depending on which convention was used for collecting the $\Delta G°$ data). For a heterogeneous equilibrium involving a gas and a substance in solution, the mass action expression will mix partial pressures (for the gas) with molarities (for the dissolved substance).

Is the Answer Reasonable?

The value of K_p tells us that the position of equilibrium lies far to the right, which means $\Delta G°$ must be large and negative. Therefore, the answer, -69.80 kJ mol^{-1}, seems reasonable.

EXAMPLE 20.13
Thermodynamic Equilibrium Constants

To make the units cancel correctly, we have added the unit mol^{-1} to the value of $\Delta G°$. This just emphasizes that the amounts of reactants and products are specified in mole-sized quantities.

Sulfur dioxide, which is sometimes present in polluted air, reacts with oxygen when it passes over the catalyst in automobile catalytic converters. The product is the very acidic oxide SO_3.

$$2SO_2(g) + O_2(g) \rightleftharpoons 2SO_3(g)$$

For this reaction, $\Delta G° = -1.40 \times 10^2$ kJ mol^{-1} at 25 °C. What is the value of K_p?

ANALYSIS: Once again we use the equation

$$\Delta G° = -RT \ln K_p$$

Our data are

$$R = 8.314 \text{ J mol}^{-1} \text{ K}^{-1}$$

$$T = 298 \text{ K}$$

$$\Delta G° = -1.40 \times 10^2 \text{ kJ mol}^{-1}$$

$$= -1.40 \times 10^5 \text{ J mol}^{-1}$$

To calculate K_p, let's first solve for $\ln K_p$.

$$\ln K_p = \frac{-\Delta G°}{RT}$$

SOLUTION: Substituting values gives

$$\ln K_p = \frac{-(-1.40 \times 10^5 \text{ J mol}^{-1})}{(8.314 \text{ J mol}^{-1} \text{ K}^{-1})(298 \text{ K})}$$

$$= +56.5$$

To calculate K_p, we take the antilogarithm,[6]

$$K_p = e^{56.5}$$

$$= 3 \times 10^{24}$$

Notice that we have expressed the answer to only one significant figure. As discussed earlier, when taking a logarithm, the number of digits written after the decimal place equals the number of significant figures in the number. Conversely, the number of significant figures in the antilogarithm equals the number of digits after the decimal in the logarithm.

Is the Answer Reasonable?

The value of $\Delta G°$ is large and negative, so the position of equilibrium should favor the products. The large value of K_p is therefore reasonable.

PRACTICE EXERCISE 17: The reaction $N_2(g) + 3H_2(g) \rightleftharpoons 2NH_3(g)$ has $K_p = 6.9 \times 10^5$ at 25.0 °C. Calculate $\Delta G°$ for this reaction in units of kilojoules.

PRACTICE EXERCISE 18: The reaction $H_2(g) + I_2(g) \rightleftharpoons 2HI(g)$ has $\Delta G° = +3.3$ kJ mol^{-1} at 25.0 °C. What is the value of K_p at this temperature?

---

[6]The rule is that the number of digits after the decimal point in a quantity equals the number of significant figures in its antilogarithm. Since 56.5 has only one digit past the decimal point, its antilogarithm has only one significant figure.

Thermodynamic equilibrium constants at temperatures other than 25 °C can be computed from ΔG_T°

Earlier we discussed the way temperature affects the position of equilibrium in a reaction. At 25 °C, the relative proportions of reactants and products at equilibrium are determined by ΔG_{298}°, and we have just seen that we can calculate the value of the equilibrium constant at 25 °C from ΔG_{298}°. As the temperature moves away from 25 °C, the position of equilibrium also changes because of changes in the value of ΔG_T°. Therefore, ΔG_T° can be used to calculate K at temperatures other than 25 °C by the same methods used in Examples 20.12 and 20.13.

Nitrous oxide, N_2O, is an anesthetic known as laughing gas because it sometimes relieves patients of their inhibitions. The decomposition of nitrous oxide has $K_p = 1.8 \times 10^{36}$ at 25 °C. The equation is

$$2N_2O(g) \rightleftharpoons 2N_2(g) + O_2(g)$$

For this reaction, $\Delta H^\circ = -163$ kJ and $\Delta S^\circ = +148$ J K^{-1}. What is the approximate value for K_p for this reaction at 40 °C?

EXAMPLE 20.14
Calculating K at Temperatures Other Than 25 °C

ANALYSIS: To apply Equation 20.12, we need to have the value of ΔG° at 40 °C (313 K) which we can represent as ΔG_{313}°. We can estimate ΔG_{313}° using the values of ΔH° and ΔS° measured at 25 °C.

$$\Delta G_{313}^\circ \approx \Delta H_{298}^\circ - (313 \text{ K})\Delta S_{298}^\circ$$

Recall that this is permitted because ΔH° and ΔS° do not change much with temperature.

SOLUTION: Substituting the values of ΔH° and ΔS° provided in the problem gives

$$\Delta G_{313}^\circ \approx -1.63 \times 10^5 \text{ J} - (313 \text{ K})(+148 \text{ J K}^{-1})$$

Notice that we've converted kilojoules to joules. Performing the arithmetic gives

$$\Delta G_{313}^\circ \approx -2.09 \times 10^5 \text{ J (rounded)}$$

The next step is to use this value of ΔG_{313}° to compute K_p with Equation 20.12. First, let's solve for ln K_p.

$$\ln K_p = \frac{-\Delta G_{313}^\circ}{RT}$$

Substituting, with $R = 8.314$ J mol^{-1} K^{-1}, $T = 313$ K, and expressing ΔG_{313}° in joules on a per mole basis gives

$$\ln K_p = \frac{2.09 \times 10^5 \text{ J mol}^{-1}}{(8.314 \text{ J mol}^{-1} \text{ K}^{-1})(313 \text{ K})}$$

$$= +80.3$$

Taking the antilogarithm,

$$K_p = e^{80.3}$$

$$= 7 \times 10^{34}$$

Notice that at this higher temperature, N_2O is actually slightly more stable than at 25 °C, as reflected in the slightly smaller value for the equilibrium constant for its decomposition.

Is the Answer Reasonable?

In Chapter 16 you learned that Le Châtelier's principle predicts that when a reaction is exothermic, as this one is, an increase in temperature decreases the size of the equilibrium constant. The smaller value of K_p at the higher temperature here is consistent with Le Châtelier's principle, so the answer seems reasonable.

PRACTICE EXERCISE 19: The reaction $N_2(g) + 3H_2(g) \rightleftharpoons 2NH_3(g)$ has a standard enthalpy of reaction of -92.4 kJ and a standard entropy of reaction equal to -198.3 J K^{-1}. Estimate the value of K_p for this reaction at 50 °C.

20.10 ▶ Bond energies can be estimated from reaction enthalpy changes

You have seen how thermodynamic data allow us to predict the spontaneity of chemical reactions as well as the nature of chemical systems at equilibrium. In addition to these useful and important benefits of the study of thermodynamics, there is a bonus. By studying heats of reaction, and heats of formation in particular, we can obtain fundamental information about the chemical bonds in the substances that react, because the origin of the energy changes in chemical reactions is changes in bond energies.

> Bond strength relates to bond energy. A strong bond is one with a large bond energy.

Recall that the **bond energy** is the amount of energy needed to break a chemical bond to give electrically neutral fragments. It is a useful quantity to know in the study of chemical properties, because during chemical reactions, bonds within the reactants are broken and new ones are formed as the products appear. The first step—bond breaking—is one of the factors that controls the reactivity of substances. Elemental nitrogen, for example, has a very low degree of reactivity, which is generally attributed to the very strong triple bond in N_2. Reactions that involve the breaking of this bond in a single step simply do not occur. When N_2 does react, it is by a stepwise involvement of its three bonds, one at a time.

> The heats of formation of the nitrogen oxides are endothermic because of the large bond energy of the N_2 molecule.

Bond energies can be measured or calculated using Hess's law

> The emission spectrum of a molecule gives information about the energy levels in the molecule, just as the emission spectrum of an atom gives information about atomic energy levels.

The bond energies of simple diatomic molecules such as H_2, O_2, and Cl_2 are usually measured *spectroscopically*. A flame or an electric spark is used to excite (energize) the molecules, causing them to emit light. An analysis of the spectrum of emitted light allows scientists to compute the amount of energy needed to break the bond.

For more complex molecules, thermochemical data can be used to calculate bond energies in a Hess's law kind of calculation. We will use the standard heat of formation of methane to illustrate how this is accomplished. However, before we can attempt such a calculation, we must first define a thermochemical quantity that we will call the **atomization energy,** symbolized ΔH_{atom}. This is the amount of energy needed to rupture all the chemical bonds in 1 mol of gaseous molecules to give gaseous atoms as products. For example, the atomization of methane is

> The kind of bond breaking described here divides the electrons of the bond equally between the two atoms. It could be symbolized as
>
> A:B $\longrightarrow$ A· + ·B

$$CH_4(g) \longrightarrow C(g) + 4H(g)$$

> Atomization energies are always endothermic. The formation of a molecule from its *gaseous* atoms is always exothermic because bond formation is exothermic.

and the enthalpy change for the process is ΔH_{atom}. For this particular molecule, ΔH_{atom} corresponds to the total amount of energy needed to break all the C—H bonds in 1 mol of CH_4; therefore, division of ΔH_{atom} by 4 would give the average C—H bond energy in methane, expressed in kJ mol^{-1}.

Figure 20.14 shows how we can use the standard heat of formation, ΔH_f°, to calculate the atomization energy. Across the bottom we have the chemical equation

FIGURE 20.14 *Two paths for the formation of methane from its elements in their standard states.* As described in the text, steps 1, 2, and 3 of the upper path involve the formation of gaseous atoms of the elements and the formation of the bonds in CH_4. The lower path corresponds to the direct combination of the elements in their standard states to give CH_4. Because ΔH is a state function, the sum of the enthalpy changes along the upper path must equal the enthalpy change for the lower path (ΔH_f°).

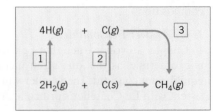

for the formation of CH_4 from its elements. The enthalpy change for this reaction, of course, is ΔH_f°. In this figure we also can see an alternative three-step path that leads to $CH_4(g)$. One step is the breaking of H—H bonds in the H_2 molecules to give gaseous hydrogen atoms, another is the vaporization of carbon to give gaseous carbon atoms, and the third is the combination of the gaseous atoms to form CH_4 molecules. These changes are labeled 1, 2, and 3 in the figure.

Since ΔH is a state function, the net enthalpy change from one state to another is the same regardless of the path that we follow. This means that the sum of the enthalpy changes along the upper path must be the same as the enthalpy change along the lower path, ΔH_f°. Perhaps this can be more easily seen in Hess's law terms if we write the changes along the upper path in the form of thermochemical equations.

Steps 1 and 2 have enthalpy changes that are called *standard heats of formation of gaseous atoms*. Values for these quantities have been measured for many of the elements, and some are given in Table 20.3. Step 3 is the opposite of atomization, and its enthalpy change will therefore be the negative of ΔH_{atom} (recall that if we reverse a reaction, we change the sign of its ΔH).

A more complete table of standard heats of formation of gaseous atoms is located in Appendix C.

$$
\begin{array}{lll}
\text{(Step 1)} & 2H_2(g) \longrightarrow 4H(g) & \Delta H_1^\circ = 4\Delta H_{f\,H(g)}^\circ \\
\text{(Step 2)} & C(s) \longrightarrow C(g) & \Delta H_2^\circ = \Delta H_{f\,C(g)}^\circ \\
\text{(Step 3)} & \underline{4H(g) + C(g) \longrightarrow CH_4(g)} & \underline{\Delta H_3^\circ = -\Delta H_{atom}} \\
& 2H_2(g) + C(s) \longrightarrow CH_4(g) & \Delta H^\circ = \Delta H_{f\,CH_4(g)}^\circ
\end{array}
$$

Notice that by adding the first three equations, we get the equation for the formation of CH_4 from its elements in their standard states. This means that adding the ΔH° values of the first three equations should give ΔH_f° for CH_4.

When the equations are added, $C(g)$ and $4H(g)$ cancel.

$$\Delta H_1^\circ + \Delta H_2^\circ + \Delta H_3^\circ = \Delta H_{f\,CH_4(g)}^\circ$$

Let's substitute for ΔH_1°, ΔH_2°, and ΔH_3°, and then solve for ΔH_{atom}. First, we substitute for the ΔH° quantities.

$$4\Delta H_{f\,H(g)}^\circ + \Delta H_{f\,C(g)}^\circ + (-\Delta H_{atom}) = \Delta H_{f\,CH_4(g)}^\circ$$

Next, we solve for $(-\Delta H_{atom})$.

$$-\Delta H_{atom} = \Delta H_{f\,CH_4(g)}^\circ - 4\Delta H_{f\,H(g)}^\circ - \Delta H_{f\,C(g)}^\circ$$

Changing signs and rearranging the right side of the equation just a bit gives

$$\Delta H_{atom} = 4\Delta H_{f\,H(g)}^\circ + \Delta H_{f\,C(g)}^\circ - \Delta H_{f\,CH_4(g)}^\circ$$

TABLE 20.3	STANDARD HEATS OF FORMATION OF SOME GASEOUS ATOMS FROM THE ELEMENTS IN THEIR STANDARD STATES		
Atom	ΔH_f° per Mole of Atoms (kJ mol^{-1})^a	Atom	ΔH_f° per Mole of Atoms (kJ mol^{-1})^a
H	217.89	S	276.98
Li	161.5	F	79.14
Be	324.3	Si	450
B	560	P	332.2
C	716.67	Cl	121.47
N	472.68	Br	112.38
O	249.17	I	107.48

^{a}All values are positive because forming the gaseous atoms from the elements involves bond breaking and is endothermic.

Now all we need are values for the ΔH_f°'s on the right side. From Table 20.3 we obtain $\Delta H_{f\,H(g)}^\circ$ and $\Delta H_{f\,C(g)}^\circ$, and the value of $\Delta H_{f\,CH_4(g)}^\circ$ is obtained from Table 7.2. We will round these to the nearest 0.1 kJ/mol.

$$\Delta H_{f\,H(g)}^\circ = +217.9 \text{ kJ/mol}$$

$$\Delta H_{f\,C(g)}^\circ = +716.7 \text{ kJ/mol}$$

$$\Delta H_{f\,CH_4(g)}^\circ = -74.8 \text{ kJ/mol}$$

Substituting these values gives

$$\Delta H_{\text{atom}} = 1663.1 \text{ kJ/mol}$$

and division by 4 gives an estimate of the average C—H bond energy in this molecule.

$$\text{bond energy} = \frac{1663.1 \text{ kJ/mol}}{4}$$

$$= 415.8 \text{ kJ/mol of C—H bonds}$$

This value is quite close to the one in Table 20.4, which is an average of the C—H bond energies in many different compounds. The other bond energies in Table 20.4 are also based on thermochemical data and were obtained by similar calculations.

Bond energies can be used to estimate heats of formation

An amazing thing about many covalent bond energies is that they are very nearly the same in many different compounds. This suggests, for example, that a C—H bond is very nearly the same in CH_4 as it is in a large number of other compounds that contain this kind of bond.

Because the bond energy doesn't vary much from compound to compound, we can use tabulated bond energies to estimate the heats of formation of substances. For example, let's calculate the standard heat of formation of methyl alcohol vapor, $CH_3OH(g)$. The structural formula for methanol is

$$
\begin{array}{c}
\text{H} \\
| \\
\text{H—C—O—H} \\
| \\
\text{H}
\end{array}
$$

To perform this calculation, we set up two paths from the elements to the compound, as shown in Figure 20.15. The lower path has an enthalpy change corresponding to $\Delta H_{f\,CH_3OH(g)}^\circ$, while the upper path takes us to the gaseous elements and then through the energy released when the bonds in the molecule are formed. This latter energy can be computed from the bond energies in Table 20.4. As before, the

TABLE 20.4	SOME AVERAGE BOND ENERGIES				
Bond	Bond Energy (kJ mol^{-1})	Bond	Bond Energy (kJ mol^{-1})	Bond	Bond Energy (kJ mol^{-1})
C—C	348	C=O	743	H—Br	366
C=C	612	C—F	484	H—I	299
C≡C	960	C—Cl	338	H—N	388
C—H	412	C—Br	276	H—O	463
C—N	305	C—I	238	H—S	338
C=N	613	H—H	436	H—Si	376
C≡N	890	H—F	565		
C—O	360	H—Cl	431		

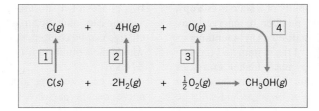

Figure 20.15 *Two paths for the formation of methyl alcohol vapor from its elements in their standard states.* The numbered paths are referred to in the discussion.

sum of the energy changes along the upper path must be the same as the energy change along the lower path, and this permits us to compute $\Delta H^\circ_{f\,CH_3OH(g)}$.

Steps 1, 2, and 3 involve the formation of the gaseous atoms from the elements, and their enthalpy changes are obtained from Table 20.3.

$$\Delta H^\circ_1 = \Delta H^\circ_{f\,C(g)} = 1\text{ mol} \times 716.7\text{ kJ/mol} = 716.7\text{ kJ}$$

$$\Delta H^\circ_2 = 4\Delta H^\circ_{f\,H(g)} = 4\text{ mol} \times (217.9\text{ kJ/mol}) = 871.6\text{ kJ}$$

$$\Delta H^\circ_3 = \Delta H^\circ_{f\,O(g)} = 1\text{ mol} \times (249.2\text{ kJ/mol}) = 249.2\text{ kJ}$$

The sum of these values, $+1837.5$ kJ, is the total energy input (the net ΔH°) for the first three steps.

The formation of the CH_3OH molecule from the gaseous atoms is exothermic because energy is always released when atoms become joined by covalent bonds. In this molecule we can count three C—H bonds, one C—O bond, and one O—H bond. Their formation releases energy equal to their bond energies, which we obtain from Table 20.4.

Bond	Energy (kJ)
3(C—H)	$3 \times (412\text{ kJ/mol}) = 1236$
C—O	360
O—H	463

Adding these together gives a total of 2059 kJ. Therefore, ΔH°_4 is -2059 kJ (because it is exothermic). Now we can compute the total enthalpy change for the upper path.

$$\Delta H^\circ = (+1837.5\text{ kJ}) + (-2059\text{ kJ})$$

$$= -222\text{ kJ}$$

The value just calculated should be equal to ΔH°_f for $CH_3OH(g)$. For comparison, it has been found experimentally that ΔH°_f for this molecule (in the vapor state) is -201 kJ/mol. At first glance, the agreement doesn't seem very good, but on a relative basis the calculated value (-222 kJ) differs from the experimental one by only about 10%.

SUMMARY

First Law of Thermodynamics The change in the **internal energy** of a system, ΔE, equals the sum of the heat absorbed by the system, q, and the work done on the system, w. ΔE is a state function, but q and w are not. The values of q and w depend on how the change takes place. For pressure–volume work, $w = -P\Delta V$, where $\Delta V = V_{final} - V_{initial}$. The heat at constant volume, q_v, is equal to ΔE, whereas the heat at constant pressure, q_p, is equal to ΔH. The value of ΔH differs from ΔE by the work expended in pushing back the atmosphere when the change occurs at constant atmospheric pressure. In general, the difference between ΔE and

ΔH is quite small. For a chemical reaction, $\Delta H = \Delta E + \Delta n_{gas}RT$, where Δn_{gas} is the change in the number of moles of *gas* on going from reactants to products.

Spontaneity A spontaneous change occurs without outside assistance, whereas a nonspontaneous change requires continuous help and can occur only if it is accompanied by and linked to some other spontaneous event.

Spontaneity is associated with statistical probability— spontaneous processes tend to proceed from lower probability states to those of higher probability. The thermodynamic quantity associated with the probability of a state is **entropy, S.** Entropy is a measure of the number of energetically equivalent ways a state can be realized. Any spontaneous change increases the total entropy of the system and the surroundings. In general, gases have much higher entropies than liquids, which have somewhat higher entropies than solids. Entropy increases with volume for a gas and with the temperature. During a chemical reaction, the entropy tends to increase if the number of molecules increases.

Second Law of Thermodynamics This law states that the entropy of the universe increases whenever a spontaneous change occurs.

Gibbs Free Energy The Gibbs free energy change, ΔG, allows us to determine the combined effects of enthalpy and entropy changes on the spontaneity of a chemical or physical change. A change is spontaneous only if the free energy of the system decreases (ΔG is negative). When ΔH and ΔS have the same algebraic sign, the temperature becomes the critical factor in determining spontaneity.

Third Law of Thermodynamics The entropy of a pure crystalline substance is equal to zero at absolute zero (0 K). Because we know where the zero point is on the entropy scale, it is possible to measure absolute amounts of entropy possessed by substances. **Standard entropies, $S°$,** are calculated for 25 °C and 1 atm (Table 20.1) and can be used to calculate $\Delta S°$ for chemical reactions.

$$\Delta S° = (\text{sum of } S° \text{ of products}) - (\text{sum of } S° \text{ of reactants})$$

Standard Free Energy Changes When ΔG is measured at 25 °C and 1 atm, it is the **standard free energy change, $\Delta G°$.** As with enthalpy changes, the **standard free energies of formation, $\Delta G_f°$** (Table 20.2) can be used to obtain $\Delta G°$ for chemical reactions by a Hess's law type of calculation.

$$\Delta G° = (\text{sum of } \Delta G_f° \text{ of products}) - (\text{sum of } \Delta G_f° \text{ of reactants})$$

For any system, the value of ΔG is equal to the maximum amount of energy that can be obtained in the form of useful work. This maximum work can be obtained only if the change takes place reversibly. All real changes are irreversible and we always obtain less work than is theoretically available; the rest is lost as heat.

Free Energy and Equilibrium When a system reaches equilibrium, $\Delta G = 0$ and no useful work can be obtained from it. At any particular pressure, an equilibrium between two phases of a substance (e.g., liquid ⇌ solid, or solid ⇌ vapor) can only occur at one temperature. The entropy change can be computed as $\Delta S = \Delta H/T$. The temperature at which the equilibrium occurs can be calculated from $T = \Delta H/\Delta S$.

In chemical reactions, a minimum on the free energy curve occurs partway between pure reactants and pure products. This minimum can be approached from either the reactants or products, and the composition of the equilibrium mixture is determined by where the minimum lies along the reactant → product axis; when it lies close to the products, the proportion of products to reactants is large and the reaction goes far toward completion.

When a reaction has a value of $\Delta G°$ that is both large and negative, it will appear to occur spontaneously because a lot of products will be formed by the time equilibrium is reached. If $\Delta G°$ is large and positive, it may be difficult to observe any reaction at all because only tiny amounts of products will be formed. The sign and magnitude of $\Delta G°$ can therefore be used to predict the *apparent* spontaneity of a chemical reaction.

At a temperature other than 25 °C, the value of $\Delta G_T°$ can be estimated using $\Delta H°$ and $\Delta S°$ as approximations of $\Delta H_T°$ and $\Delta S_T°$. The equation is $\Delta G_T° \approx \Delta H_{298\text{ K}}° - T\Delta S_{298\text{ K}}°$. Computing $\Delta G_T°$ allows us to see how temperature affects the position of equilibrium in a chemical reaction.

Thermodynamic Equilibrium Constants The spontaneity of a reaction is determined by ΔG (how the free energy changes with a change in concentration). This is related to the standard free energy change, $\Delta G°$, by the equation $\Delta G = \Delta G° + RT \ln Q$, where Q is the reaction quotient for the system. At equilibrium, $\Delta G° = -RT \ln K$, where $K = K_p$ for gaseous reactions and $K = K_c$ for reactions in solution. For temperatures other than 25 °C, we can estimate $\Delta G_T°$ by the equation $\Delta G_T° \approx \Delta H_{298}° - T\Delta S_{298}°$ and then use it to calculate K.

Bond Energies and Heats of Reaction The bond energy equals the amount of energy needed to break a bond to give neutral fragments. The sum of all the bond energies in a molecule is the **atomization energy, ΔH_{atom},** and, on a mole basis, it corresponds to the energy needed to break 1 mol of molecules into individual atoms. The heat of formation of a gaseous compound equals the sum of the energies needed to form atoms of the elements that are found in the substance plus the negative of the atomization energy.

TOOLS ▶ **YOU HAVE LEARNED**

The table below lists the tools that you have learned in this chapter. You will need them to answer chemistry questions. Review them if necessary, and refer to the tools when working on the Thinking-It-Through problems and the Review Exercises that follow.

TOOL	HOW IT WORKS
$\Delta H = \Delta E + \Delta n_{gas} RT$ (page 869)	Converts between ΔE and ΔH for a reaction.
Standard entropies (page 883)	They can be used to calculate the value of $\Delta S°$ for a reaction.
Predicting the sign of ΔS (page 875)	Determines whether the entropy change favors spontaneity.

Sign of ΔG (page 880)	Determines whether or not a change is spontaneous.
$\Delta G° = \Delta H° - (298\ K)\Delta S°$ **(page 884)**	Calculates $\Delta G°$ from $\Delta H°$ and $\Delta S°$ values.
Standard free energies of formation (page 885)	Calculates $\Delta G°$ for a reaction from tabulated values of $\Delta G_f°$.
Value of $\Delta G°$ for a reaction (page 892)	Determines qualitatively whether or not a significant amount of products will form.
$\Delta G_T° \approx \Delta H_{298}° - T\Delta S_{298}°$ **(page 894)**	We calculate the value of $\Delta G°$ at temperatures other than 25 °C.
$\Delta G = \Delta G° + RT\ \ln\ Q$ **(page 895)**	Relates the reaction quotient to ΔG, which enables us to determine whether a reaction is at equilibrium, and if not, the direction the reaction must proceed to reach equilibrium.
$\Delta G° = -RT\ \ln\ K$ **(page 897)**	Relates $\Delta G°$ to the equilibrium constant; K equals K_p for gaseous reactions and K_c for reactions in solution.

THINKING IT THROUGH

The goal for the following problems is not to find the answers themselves, but rather to assemble the information needed to solve them and explain how you would use the information to find the answers. The problems in Level 2 are more challenging than those in Level 1 and may contain more data than are required, in which case you are also asked to identify the unnecessary data. Detailed answers to the Thinking-It-Through problems can be found on the web site.

Need extra help?
ON-LINE HELP Visit the Brady/ Senese web site at www.wiley.com/ college/brady

Level 1 Problems

1. Describe how you would calculate the work done by a gas as it expands at constant temperature from a volume of 3.00 L and a pressure of 5.00 atm to a volume of 8.00 L. The external pressure against which the gas expands is 1.00 atm. Set up the calculation so the answer is in units of joules (1 atm = 101,325 Pa).

2. The reaction

$$2N_2O(g) \longrightarrow 2N_2(g) + O_2(g)$$

has $\Delta H° = -163.14$ kJ. What is the value of ΔE for the decomposition of 180 g of N_2O at 25 °C? If we assume that ΔH doesn't change appreciably with temperature, what is ΔE for this same reaction at 200 °C? (Set up the calculations.)

3. What factors must you consider to determine the sign of ΔS for the reaction $2N_2O(g) \rightarrow 2N_2(g) + O_2(g)$ if it occurs at constant temperature?

4. For the reaction in Question 3, what additional information do we need to determine whether this reaction should tend to proceed spontaneously in the forward direction?

5. What factors must you consider to determine the sign of ΔS for the following reaction?

$$2HI(g) \longrightarrow H_2(g) + I_2(s)$$

6. How would you calculate the value of $\Delta S_f°$ for $HI(g)$? (Set up the calculation.)

7. The reaction

$$2C_4H_{10}(g) + 13O_2(g) \longrightarrow 8CO_2(g) + 10H_2O(g)$$

has $\Delta G° = -5407$ kJ. Show how you would determine the value of $\Delta G_f°$ for $C_4H_{10}(g)$. (Set up the calculation.)

8. Explain how you would calculate the value of $\Delta G_{500\ K}°$ for the reaction in Question 7. (Set up the calculation.)

9. A 10.0 L vessel at 20 °C contains butane, $C_4H_{10}(g)$, at a pressure of 2.00 atm. What is the maximum amount of work that can be obtained by the combustion of this butane if the gas is brought to a pressure of 1 atm and the temperature is brought to 25 °C? Assume the products are also returned to this same temperature and pressure. (Set up the calculation.)

10. Given the following reactions and their values of $\Delta G°$, explain in detail how you would calculate the value of $\Delta G_f°$ for $N_2O_5(g)$. (Set up the calculation.)

$$2H_2(g) + O_2(g) \longrightarrow 2H_2O(l) \qquad \Delta G° = -474.4\ kJ$$

$$N_2O_5(g) + H_2O(l) \longrightarrow 2HNO_3(l) \qquad \Delta G° = -37.6\ kJ$$

$$\tfrac{1}{2}N_2(g) + \tfrac{3}{2}O_2(g) + \tfrac{1}{2}H_2(g) \longrightarrow HNO_3(l) \quad \Delta G° = -79.91\ kJ$$

11. How would you determine the atomization energy of the molecule whose structure is shown below?

12. Explain how you would use bond energies and the heats of formation of atoms of the elements to calculate the heat of formation of gaseous hydrogen cyanide, HCN.

Level 2 Problems

13. In the SI, pressure is expressed in pascals. One pascal is equal to one newton (N, the SI unit of force) per square meter: $1\ Pa = 1\ N/m^2$. Work is force times distance, and the joule is defined therefore as $1\ J = 1\ N \cdot m$. Show that when the pressure is expressed in pascals and volume in cubic meters that the product $P\Delta V$ yields work in units of joules.

14. When an ideal gas expands or contracts at constant temperature, $\Delta E = 0$. In terms of the definition of an ideal gas and the kinetic theory interpretation of temperature, explain why this is true.

15. An ideal gas in a cylinder fitted with a piston expands at constant temperature from a pressure of 5 atm and a volume of 6.0 L to a final volume of 12 L against a constant opposing pressure of 2.0 atm. How much heat does the gas absorb, expressed in units of L atm (liter × atm)? Set up the calculation. (*Hint:* Refer to the preceding question.)

16. The reaction in Question 5 has $\Delta H° = -53.2$ kJ. How can you determine how the sign and magnitude of $\Delta G_T°$ will be affected by a change in temperature?

17. At room temperature (25 °C), the gas ClNO is impure because it decomposes slightly according to the equation

$$2ClNO(g) \rightleftharpoons Cl_2(g) + 2NO(g)$$

The extent of decomposition is about 5%. What is the approximate value of $\Delta G_{298}°$ for this reaction at this temperature? (Explain each step in the solution to the problem.)

18. The reaction

$$N_2O(g) + O_2(g) \rightleftharpoons NO_2(g) + NO(g)$$

has $\Delta H° = -42.9$ kJ and $\Delta S° = -26.1$ J/K. Suppose 0.100 mol of N_2O and 0.100 mol of O_2 were placed in a container at 500 °C and this equilibrium were established. Explain how you would determine what percentage of the N_2O would have reacted.

REVIEW QUESTIONS

First Law of Thermodynamics

20.1 What is the origin of the name *thermodynamics*?

20.2 What kinds of energy contribute to the internal energy of a system? Why can't we measure or calculate a system's internal energy?

20.3 How is a change in the internal energy defined in terms of the initial and final internal energies?

20.4 What is the algebraic sign of ΔE for an endothermic change? Why?

20.5 State the first law of thermodynamics in words. What equation defines the change in the internal energy in terms of heat and work? Define the meaning of the symbols, including the significance of their algebraic signs.

20.6 Which quantities in the statement of the first law are state functions and which are not?

20.7 Which thermodynamic quantity corresponds to the heat at constant volume? Which corresponds to the heat at constant pressure?

20.8 What are the units of $P\Delta V$ if pressure is expressed in pascals and volume is expressed in cubic meters?

20.9 How are ΔE and ΔH related to each other?

20.10 When are ΔE and ΔH numerically equal to each other?

20.11 If there is a decrease in the number of moles of gas during an exothermic chemical reaction, which is numerically larger, ΔE or ΔH? Why?

Spontaneous Change

20.12 What is a *spontaneous change*?

20.13 List five changes that you have encountered recently that occurred spontaneously. List five changes that are nonspontaneous that you have caused to occur.

20.14 Which of the items that you listed in Question 20.13 are exothermic (leading to a lowering of the potential energy) and which are endothermic (accompanied by an increase in potential energy)?

20.15 At constant pressure, what role does the enthalpy change play in determining the spontaneity of an event?

20.16 How can one estimate the probability of a state of the system?

20.17 How do the probabilities of the initial and final states in a process affect the spontaneity of the process?

Entropy

20.18 An instant cold pack purchased in a pharmacy contains a packet of solid ammonium nitrate surrounded by a pouch of water. When the packet of NH_4NO_3 is broken, the solid dissolves in water and a cooling of the mixture occurs because the solution process for NH_4NO_3 in water is endothermic. Explain, in terms of what happens to the molecules and ions, why this mixing occurs spontaneously.

20.19 What is *entropy*?

20.20 Will the entropy change for each of the following be positive or negative?
(a) Moisture condenses on the outside of a cold glass.
(b) Raindrops form in a cloud.
(c) Gasoline vaporizes in the carburetor of an automobile engine.
(d) Air is pumped into a tire.
(e) Frost forms on the windshield of your car.
(f) Sugar dissolves in coffee.

20.21 On the basis of our definition of entropy, suggest why entropy is a state function.

20.22 State the second law of thermodynamics.

20.23 In animated cartoons, visual effects are often created (for amusement) that show events that ordinarily don't occur in real life because they are accompanied by huge entropy decreases. Can you think of an example of this? Explain why there is an entropy decrease in your example.

20.24 How can a process have a negative entropy change for the system and yet still be spontaneous?

Third Law of Thermodynamics and Standard Entropies

20.25 What is the third law of thermodynamics?

20.26 Would you expect the entropy of an alloy (a solution of two metals) to be zero at 0 K? Explain your answer.

20.27 Why does entropy increase with increasing temperature?

20.28 Does glass have $S = 0$ at 0 K? Explain.

Gibbs Free Energy

20.29 What is the equation expressing the change in the Gibbs free energy for a reaction occurring at constant temperature and pressure?

20.30 In terms of the algebraic signs of ΔH and ΔS, under what circumstances will a change be spontaneous:
(a) At all temperatures?
(b) At low temperatures but not at high temperatures?
(c) At high temperatures but not at low temperatures?

20.31 Under what circumstances will a change be nonspontaneous regardless of the temperature?

20.32 Is it possible to run a combustion reaction in reverse, so that fuel and oxygen are produced from carbon dioxide and water?

Free Energy and Work

20.33 How is free energy related to useful work?

20.34 What is a thermodynamically reversible process? How is the amount of work obtained from a change related to thermodynamic reversibility?

20.35 How is the *rate* at which energy is withdrawn from a system related to the amount of that energy which can appear as useful work?

20.36 When glucose is oxidized by the body to generate energy, part of the energy is used to make molecules of ATP (adenosine triphosphate). However, of the total energy released in the oxidation of glucose, only 38% actually goes to making ATP. What happens to the rest of the energy?

20.37 Why are real, observable changes not considered to be thermodynamically reversible processes?

Free Energy and Equilibrium

20.38 In what way is free energy related to equilibrium?

20.39 Considering the fact that the formation of a bond between two atoms is exothermic and is accompanied by an entropy decrease, explain why all chemical compounds decompose into individual atoms if heated to a high enough temperature.

20.40 When a warm object is placed in contact with a cold one, they both gradually come to the same temperature. On a molecular level, explain how this is related to entropy and spontaneity.

20.41 Sketch a graph to show how the free energy changes during a phase change such as the melting of a solid.

20.42 Sketch the shape of the free energy curve for a chemical reaction that has a positive $\Delta G°$. Indicate the composition of the reaction mixture corresponding to equilibrium.

20.43 Many reactions that have large, negative values of $\Delta G°$ are not actually observed to happen at 25 °C and 1 atm. Why?

Effect of Temperature on the Free Energy Change

20.44 Why is the value of ΔG for a change so dependent on temperature?

20.45 What equation can be used to obtain an approximate value for $\Delta G_T°$ for a reaction at a temperature other than 25 °C?

Thermodynamics and Equilibrium

20.46 Suppose a reaction has a negative $\Delta H°$ and a negative $\Delta S°$. Will more or less product be present at equilibrium as the temperature is raised?

20.47 Write the equation that relates the free energy change to the value of the reaction quotient for a reaction.

20.48 How is the equilibrium constant related to the standard free energy change for a reaction? (Write the equation.)

20.49 What is a thermodynamic equilibrium constant?

20.50 What is the value of $\Delta G°$ for a reaction for which $K = 1$?

Bond Energies and Heats of Reaction

20.51 Define the term *atomization energy*.

20.52 Why are the heats of formation of gaseous atoms from their elements endothermic quantities?

20.53 The gaseous C_2 molecule has a bond energy of 602 kJ mol^{-1}. Why isn't the standard heat of formation of C(*g*) equal to half this value?

REVIEW PROBLEMS

Answers to problems whose numbers are printed in color are given in Appendix B. More challenging problems are marked with asterisks. **ILW** = Interactive LearningWare solution is available at *www.wiley.com/college/brady*.

First Law of Thermodynamics

20.54 A certain system absorbs 300 J of heat and has 700 J of work performed on it. What is the value of ΔE for the change? Is the overall change exothermic or endothermic?

20.55 The value of ΔE for a certain change is -1455 J. During the change, the system absorbs 812 J of heat. Did the system do work, or was work done on the system? How much work, expressed in joules, was involved?

20.56 Suppose that you were pumping an automobile tire with a hand pump that pushed 24.0 in.^{3} of air into the tire on each stroke. During one such stroke the opposing pressure in the tire was 30.0 lb/in.^{2} above the normal atmospheric pressure of 14.7 lb/in.^{2}. Calculate the number of joules of work accomplished during this stroke. (1 L atm = 101.325 J.)

20.57 Consider the reaction between aqueous solutions of baking soda, $NaHCO_3$, and vinegar, $HC_2H_3O_2$.

$$NaHCO_3(aq) + HC_2H_3O_2(aq) \longrightarrow$$
$$NaC_2H_3O_2(aq) + H_2O(l) + CO_2(g)$$

If this reaction occurs at atmospheric pressure ($P = 1$ atm), how much work, expressed in L atm, is done by the system in pushing back the atmosphere when 1.00 mol $NaHCO_3$ reacts at a temperature of 25 °C? (*Hint:* Review the gas laws.)

20.58 Calculate $\Delta H°$ and $\Delta E°$ for the following reactions at 25 °C. (If necessary, refer to the data in Table C.1 in Appendix C.)
(a) $3PbO(s) + 2NH_3(g) \rightarrow 3Pb(s) + N_2(g) + 3H_2O(g)$
(b) $NaOH(s) + HCl(g) \rightarrow NaCl(s) + H_2O(l)$
(c) $Al_2O_3(s) + 2Fe(s) \rightarrow Fe_2O_3(s) + 2Al(s)$
(d) $2CH_4(g) \rightarrow C_2H_6(g) + H_2(g)$

20.59 Calculate $\Delta H°$ and $\Delta E°$ for the following reactions at 25 °C. (If necessary, refer to the data in Table C.1 in Appendix C.)
(a) $2C_2H_2(g) + 5O_2(g) \rightarrow 4CO_2(g) + 2H_2O(g)$
(b) $C_2H_2(g) + 5N_2O(g) \rightarrow 2CO_2(g) + H_2O(g) + 5N_2(g)$
(c) $NH_4Cl(s) \rightarrow NH_3(g) + HCl(g)$
(d) $(CH_3)_2CO(l) + 4O_2(g) \rightarrow 3CO_2(g) + 3H_2O(g)$

Factors That Affect Spontaneity

20.60 Use the data from Table 7.2 to calculate $\Delta H°$ for the following reactions. On the basis of their values of $\Delta H°$, which are favored to occur spontaneously?
(a) $CaO(s) + CO_2(g) \rightarrow CaCO_3(s)$
(b) $C_2H_2(g) + 2H_2(g) \rightarrow C_2H_6(g)$
(c) $3CaO(s) + 2Fe(s) \rightarrow 3Ca(s) + Fe_2O_3(s)$
(d) $Ca(OH)_2(s) \rightarrow CaO(s) + H_2O(l)$
(e) $2NaCl(s) + H_2SO_4(l) \rightarrow Na_2SO_4(s) + 2HCl(g)$

20.61 Use the data from Table 7.2 to calculate $\Delta H°$ for the following reactions. On the basis of their values of $\Delta H°$, which are favored to occur spontaneously?
(a) $2C_2H_2(g) + 5O_2(g) \rightarrow 4CO_2(g) + 2H_2O(g)$
(b) $C_2H_2(g) + 5N_2O(g) \rightarrow 2CO_2(g) + H_2O(g) + 5N_2(g)$
(c) $Fe_2O_3(s) + 2Al(s) \rightarrow Al_2O_3(s) + 2Fe(s)$
(d) $NH_4Cl(s) \rightarrow NH_3(g) + HCl(g)$
(e) $Ag(s) + KCl(s) \rightarrow AgCl(s) + K(s)$

20.62 When a coin is tossed, there is a 50–50 chance of it landing either heads or tails. Suppose that you tossed four coins. On the basis of the different possible outcomes, what is the probability of all four coins coming up heads? What is the probability of an even heads–tails distribution? (Now work Problem 20.63.)

20.63 Suppose that you had two equal-volume containers sharing a common wall with a hole in it. Suppose there were four molecules in this system. What is the probability that all four would be in one container at the same time? What is the probability of finding an even distribution of molecules between the two containers? (*Hint:* If you haven't already done so, work Problem 20.62 first.) What do the results of Problems 20.62 and 20.63 suggest about why gases expand spontaneously?

20.64 Predict the algebraic sign of the entropy change for the following reactions:
(a) $PCl_3(g) + Cl_2(g) \rightarrow PCl_5(g)$
(b) $SO_2(g) + CaO(s) \rightarrow CaSO_3(s)$
(c) $CO_2(g) + H_2O(l) \rightarrow H_2CO_3(aq)$
(d) $Ni(s) + 2HCl(aq) \rightarrow H_2(g) + NiCl_2(aq)$

20.65 Predict the algebraic sign of the entropy change for the following reactions:
(a) $I_2(s) \rightarrow I_2(g)$
(b) $Br_2(g) + 3Cl_2(g) \rightarrow 2BrCl_3(g)$
(c) $NH_3(g) + HCl(g) \rightarrow NH_4Cl(s)$
(d) $CaO(s) + H_2O(l) \rightarrow Ca(OH)_2(s)$

Third Law of Thermodynamics

20.66 Calculate $\Delta S°$ for the following reactions in J K^{-1} from the data in Table 20.1. On the basis of their values of $\Delta S°$, which of these reactions are favored to occur spontaneously?
(a) $N_2(g) + 3H_2(g) \rightarrow 2NH_3(g)$
(b) $CO(g) + 2H_2(g) \rightarrow CH_3OH(l)$
(c) $2C_2H_6(g) + 7O_2(g) \rightarrow 4CO_2(g) + 6H_2O(g)$
(d) $Ca(OH)_2(s) + H_2SO_4(l) \rightarrow CaSO_4(s) + 2H_2O(l)$
(e) $S(s) + 2N_2O(g) \rightarrow SO_2(g) + 2N_2(g)$

20.67 Calculate $\Delta S°$ for the following reactions in J K^{-1}, using the data in Table 20.1.
(a) $Ag(s) + \frac{1}{2}Cl_2(g) \rightarrow AgCl(s)$
(b) $H_2(g) + \frac{1}{2}O_2(g) \rightarrow H_2O(g)$
(c) $H_2(g) + \frac{1}{2}O_2(g) \rightarrow H_2O(l)$
(d) $CaCO_3(s) + H_2SO_4(l) \rightarrow CaSO_4(s) + H_2O(g) + CO_2(g)$
(e) $NH_3(g) + HCl(g) \rightarrow NH_4Cl(s)$

20.68 Calculate $\Delta S_f°$ for the following compounds in J mol^{-1} K^{-1}:
(a) $C_2H_4(g)$ (c) $NaCl(s)$ (e) $HC_2H_3O_2(l)$
(b) $N_2O(g)$ (d) $CaSO_4 \cdot 2H_2O(s)$

20.69 Calculate $\Delta S_f°$ for the following compounds in J mol^{-1} K^{-1}:
(a) $Al_2O_3(s)$ (c) $N_2O_4(g)$ (e) $CaSO_4 \cdot \frac{1}{2}H_2O(s)$
(b) $CaCO_3(s)$ (d) $NH_4Cl(s)$

20.70 Nitrogen dioxide, NO_2, an air pollutant, dissolves in rainwater to form a dilute solution of nitric acid. The equation for the reaction is

$$3NO_2(g) + H_2O(l) \longrightarrow 2HNO_3(l) + NO(g)$$

Calculate $\Delta S°$ for this reaction in J K^{-1}.

20.71 Good wine will turn to vinegar if it is left exposed to air because the alcohol is oxidized to acetic acid. The equation for the reaction is

$$C_2H_5OH(l) + O_2(g) \longrightarrow HC_2H_3O_2(l) + H_2O(l)$$

Calculate $\Delta S°$ for this reaction in J K^{-1}.

Gibbs Free Energy

ILW 20.72 Phosgene, $COCl_2$, was used as a war gas during World War I. It reacts with the moisture in the lungs to produce HCl, which causes the lungs to fill with fluid, and CO_2, which asphyxiates the victim. Both lead ultimately to death. For $COCl_2(g)$, $S° = 284$ J/mol K and $\Delta H_f° = -223$ kJ/mol. Use this information and the data in Table 20.1 to calculate $\Delta G_f°$ for $COCl_2(g)$ in kJ mol^{-1}.

20.73 Aluminum oxidizes rather easily but forms a thin, protective coating of Al_2O_3 that prevents further oxidation of the aluminum beneath. Use the data for $\Delta H_f°$ (Table 7.2) and $S°$ to calculate $\Delta G_f°$ for $Al_2O_3(s)$ in kJ mol^{-1}.

20.74 Compute $\Delta G°$ in kJ for the following reactions, using the data in Table 20.2:
(a) $SO_3(g) + H_2O(l) \rightarrow H_2SO_4(l)$
(b) $2NH_4Cl(s) + CaO(s) \rightarrow CaCl_2(s) + H_2O(l) + 2NH_3(g)$
(c) $CaSO_4(s) + 2HCl(g) \rightarrow CaCl_2(s) + H_2SO_4(l)$
(d) $C_2H_4(g) + H_2O(g) \rightarrow C_2H_5OH(l)$
(e) $Ca(s) + 2H_2SO_4(l) \rightarrow CaSO_4(s) + SO_2(g) + 2H_2O(l)$

20.75 Compute $\Delta G°$ in kJ for the following reactions, using the data in Table 20.2:
(a) $2HCl(g) + CaO(s) \rightarrow CaCl_2(s) + H_2O(g)$
(b) $H_2SO_4(l) + 2NaCl(s) \rightarrow 2HCl(g) + Na_2SO_4(s)$
(c) $3NO_2(g) + H_2O(l) \rightarrow 2HNO_3(l) + NO(g)$
(d) $2AgCl(s) + Ca(s) \rightarrow CaCl_2(s) + 2Ag(s)$
(e) $NH_3(g) + HCl(g) \rightarrow NH_4Cl(s)$

20.76 Plaster of Paris, $CaSO_4 \cdot \frac{1}{2}H_2O(s)$, reacts with liquid water to form gypsum, $CaSO_4 \cdot 2H_2O(s)$. Write a chemical equation for the reaction and calculate $\Delta G°$ in kJ, using the data in Table 20.2.

20.77 When phosgene, the war gas described in Review Problem 20.72, reacts with water vapor, the products are $CO_2(g)$ and $HCl(g)$. Write an equation for the reaction and compute $\Delta G°$ in kJ. For $COCl_2(g)$, $\Delta G_f°$ is -210 kJ/mol.

20.78 Given the following

$$4NO(g) \longrightarrow 2N_2O(g) + O_2(g) \quad \Delta G° = -139.56 \text{ kJ}$$
$$2NO(g) + O_2(g) \longrightarrow 2NO_2(g) \quad \Delta G° = -69.70 \text{ kJ}$$

calculate $\Delta G°$ for the reaction

$$2N_2O(g) + 3O_2(g) \longrightarrow 4NO_2(g)$$

20.79 Given these reactions and their $\Delta G°$ values,

$$COCl_2(g) + 4NH_3(g) \longrightarrow CO(NH_2)_2(s) + 2NH_4Cl(s)$$
$$\Delta G° = -332.0 \text{ kJ}$$
$$COCl_2(g) + H_2O(l) \longrightarrow CO_2(g) + 2HCl(g)$$
$$\Delta G° = -141.8 \text{ kJ}$$
$$NH_3(g) + HCl(g) \longrightarrow NH_4Cl(s) \quad \Delta G° = -91.96 \text{ kJ}$$

calculate the value of $\Delta G°$ for the reaction

$$CO(NH_2)_2(s) + H_2O(l) \longrightarrow CO_2(g) + 2NH_3(g)$$

Free Energy and Work

20.80 Gasohol is a mixture of gasoline and ethanol (grain alcohol), C_2H_5OH. Calculate the maximum work that could be obtained at 25 °C and 1 atm by burning 1 mol of C_2H_5OH.

$$C_2H_5OH(l) + 3O_2(g) \longrightarrow 2CO_2(g) + 3H_2O(g)$$

20.81 What is the maximum amount of useful work that could theoretically be obtained at 25 °C and 1 atm from the combustion of 48.0 g of natural gas, $CH_4(g)$, to give $CO_2(g)$ and $H_2O(g)$?

Free Energy and Equilibrium

20.82 Chloroform, formerly used as an anesthetic and now believed to be a carcinogen (cancer-causing agent), has a heat of vaporization $\Delta H_{vaporization} = 31.4$ kJ mol^{-1}. The change, $CHCl_3(l) \rightarrow CHCl_3(g)$ has $\Delta S° = 94.2$ J mol^{-1} K^{-1}. At what temperature do we expect $CHCl_3$ to boil (i.e., at what temperature will liquid and vapor be in equilibrium at 1 atm pressure)?

20.83 For the melting of aluminum, $Al(s) \rightarrow Al(l)$, $\Delta H° = 10.0$ kJ mol^{-1} and $\Delta S° = 9.50$ J/mol K. Calculate the melting point of Al. (The actual melting point is 660 °C.)

20.84 Isooctane, an important constituent of gasoline, has a boiling point of 99.3 °C and a heat of vaporization of 37.7 kJ mol^{-1}. What is ΔS (in J mol^{-1} K^{-1}) for the vaporization of 1 mol of isooctane?

20.85 Acetone (nail polish remover) has a boiling point of 56.2 °C. The change, $(CH_3)_2CO(l) \rightarrow (CH_3)_2CO(g)$, has $\Delta H° = 31.9$ kJ mol^{-1}. What is $\Delta S°$ for this change?

Free Energy and Spontaneity of Chemical Reactions

ILW 20.86 Determine whether the following reaction will be spontaneous. (Do we expect appreciable amounts of products to form?)

$$C_2H_4(g) + 2HNO_3(l) \longrightarrow$$
$$HC_2H_3O_2(l) + H_2O(l) + NO(g) + NO_2(g)$$

20.87 Which of the following reactions (equations unbalanced) would be expected to be spontaneous at 25 °C and 1 atm?
(a) $PbO(s) + NH_3(g) \rightarrow Pb(s) + N_2(g) + H_2O(g)$
(b) $NaOH(s) + HCl(g) \rightarrow NaCl(s) + H_2O(l)$
(c) $Al_2O_3(s) + Fe(s) \rightarrow Fe_2O_3(s) + Al(s)$
(d) $2CH_4(g) \rightarrow C_2H_6(g) + H_2(g)$

20.88 Calculate the value of $\Delta G_{373}°$ in kJ for the following reaction at 100 °C, using data in Tables 7.2 and 20.1:

$$C_2H_4(g) + H_2(g) \longrightarrow C_2H_6(g)$$

20.89 Calculate the value of $\Delta G_{373}°$ in kJ for the following reaction at 100 °C, using data in Tables 7.2 and 20.1:

$$5SO_3(g) + 2NH_3(g) \longrightarrow 2NO(g) + 5SO_2(g) + 3H_2O(g)$$

Thermodynamic Equilibrium Constants

20.90 Calculate the value of the thermodynamic equilibrium constant for the following reactions at 25 °C. (Refer to the data in Appendix C.)
(a) $2PCl_3(g) + O_2(g) \rightleftharpoons 2POCl_3(g)$
(b) $2SO_3(g) \rightleftharpoons 2SO_2(g) + O_2(g)$

20.91 Calculate the value of the thermodynamic equilibrium constant for the following reactions at 25 °C. (Refer to the data in Appendix C.)
(a) $N_2H_4(g) + 2O_2(g) \rightleftharpoons 2NO(g) + 2H_2O(g)$
(b) $N_2H_4(g) + 6H_2O_2(g) \rightleftharpoons 2NO_2(g) + 8H_2O(g)$

ILW 20.92 The reaction $NO_2(g) + NO(g) \rightleftharpoons N_2O(g) + O_2(g)$ has $\Delta G^\circ_{1273} = -9.67$ kJ mol^{-1}. A 1.00 L reaction vessel at 1000 °C contains 0.0200 mol NO_2, 0.040 mol NO, 0.015 mol N_2O, and 0.0350 mol O_2. Is the reaction at equilibrium? If not, in which direction will the reaction proceed to reach equilibrium?

20.93 The reaction $CO(g) + H_2O(g) \rightleftharpoons HCHO_2(g)$ has $\Delta G^\circ_{673} = +79.8$ kJ mol^{-1}. If a mixture at 400 °C contains 0.040 mol CO, 0.022 mol H_2O, and 3.8×10^{-3} mol $HCHO_2$ in a 2.50 L container, is the reaction at equilibrium? If not, in which direction will the reaction proceed spontaneously?

20.94 A reaction that can convert coal to methane (the chief component of natural gas) is

$$C(s) + 2H_2(g) \rightleftharpoons CH_4(g)$$

for which $\Delta G^\circ = -50.79$ kJ mol^{-1}. What is the value of K_p for this reaction at 25 °C? Does this value of K_p suggest that studying this reaction as a means of methane production is worth pursuing?

20.95 One of the important reactions in living cells from which the organism draws energy is the reaction of adenosine triphosphate (ATP) with water to give adenosine diphosphate (ADP) and free phosphate ion.

$$ATP + H_2O \rightleftharpoons ADP + PO_4^{3-}$$

The value of ΔG°_T for this reaction at 37 °C (normal human body temperature) is -33 kJ mol^{-1}. Calculate the value of the equilibrium constant for the reaction at this temperature.

20.96 What is the value of the equilibrium constant for a reaction for which $\Delta G^\circ = 0$? What will happen to the composition of the system if we begin the reaction with the pure products?

20.97 Methanol, a potential replacement for gasoline as an automotive fuel, can be made from H_2 and CO by the reaction

$$CO(g) + 2H_2(g) \rightleftharpoons CH_3OH(g)$$

At 500 K, this reaction has $K_p = 6.25 \times 10^{-3}$. Calculate ΔG°_{500} for this reaction in units of kilojoules.

20.98 Refer to the data in Table 20.2 to determine ΔG° in kilojoules for the reaction $2NO(g) + 2CO(g) \rightleftharpoons N_2(g) + 2CO_2(g)$. What is the value of K_p for this reaction at 25 °C?

20.99 Use the data in Appendix C.1 to calculate ΔG°_{773} (at 500 °C) in kilojoules for the following reaction:

$$2NO(g) + 2CO(g) \rightleftharpoons N_2(g) + 2CO_2(g)$$

What is the value of K_p for the reaction at this temperature?

Bond Energies and Heats of Reaction

20.100 Use the data in Table 20.4 to compute the approximate atomization energy of NH_3.

20.101 Approximately how much energy would be released during the formation of the bonds in 1 mol of acetone molecules? Acetone, the solvent usually found in nail polish remover, has the structural formula

20.102 The standard heat of formation of ethanol vapor, $C_2H_5OH(g)$, is -235.3 kJ mol^{-1}. Use the data in Table 20.3 and the average bond energies for C—C, C—H, and O—H bonds to estimate the C—O bond energy in this molecule. The structure of the molecule is

20.103 The standard heat of formation of ethylene, $C_2H_4(g)$, is $+52.284$ kJ mol^{-1}. Calculate the C=C bond energy in this molecule.

ILW 20.104 Carbon disulfide, CS_2, has the Lewis structure $:\!S\!=\!C\!=\!S\!:$, and for $CS_2(g)$, $\Delta H^\circ_f = +115.3$ kJ mol^{-1}. Use the data in Table 20.3 to calculate the average C=S bond energy in this molecule.

20.105 Gaseous hydrogen sulfide, H_2S, has $\Delta H^\circ_f = -20.15$ kJ mol^{-1}. Use the data in Table 20.3 to calculate the average S—H bond energy in this molecule.

20.106 For $SF_6(g)$, $\Delta H^\circ_f = -1096$ kJ mol^{-1}. Use the data in Table 20.3 to calculate the average S—F bond energy in SF_6.

20.107 Use the results of the preceding problem and the data in Table C.2 to calculate the standard heat of formation of $SF_4(g)$. The measured value of ΔH°_f for $SF_4(g)$ is -718.4 kJ mol^{-1}. What is the percentage difference between your calculated value of ΔH°_f and the experimentally determined value?

20.108 Use the data in Tables 20.3 and 20.4 to estimate the standard heat of formation of acetylene, H—C≡C—H, in the gaseous state.

20.109 What would be the approximate heat of formation of CCl_4 vapor at 25 °C and 1 atm?

20.110 Which substance should have the more exothermic heat of formation, CF_4 or CCl_4?

*20.111 Would you expect the value of ΔH°_f for benzene, C_6H_6, computed from tabulated bond energies, to be very close to the experimentally measured value of ΔH°_f? Justify your answer.

ADDITIONAL EXERCISES

20.112 For each of the reactions in Problem 20.90, calculate K_p at 300 °C.

20.113 If pressure is expressed in atmospheres and volume is expressed in liters, $P\Delta V$ has units of L atm (liters × at-mospheres). In Chapter 11 you learned that 1 atm = 101,325 Pa, and in Chapter 3 you learned that 1 L = 1 dm^3. Use this information to determine the number of joules corresponding to 1 L atm.

*20.114 When an ideal gas expands at a constant temperature, $\Delta E = 0$ for the change. Why?

*20.115 When a real gas expands at a constant temperature, $\Delta E > 0$ for the change. Why?

20.116 How much work is accomplished by the following chemical reaction if it occurs inside a bomb calorimeter?

$$2C_4H_{10}(g) + 13O_2(g) \longrightarrow 8CO_2(g) + 10H_2O(g)$$

How much work is done if the reaction is carried out at 1 atm, with all reactants and products at 25 °C? [For $C_4H_{10}(g)$, $\Delta H_f^\circ = -126$ kJ mol^{-1}.]

*20.117 A cylinder fitted with a piston contains 5.00 L of a gas at a pressure of 4.00 atm. The entire apparatus is contained in a water bath to maintain a constant temperature of 25 °C. The piston is released and the gas expands until the pressure inside the cylinder equals the atmospheric pressure outside, which is 1 atm. Assume ideal gas behavior and calculate the amount of work done by the gas as it expands at constant temperature.

*20.118 The experiment described in the preceding problem is repeated, but this time a weight, which exerts a pressure of 2 atm, is placed on the piston. When the gas expands, its pressure drops to this 2 atm pressure. Then the weight is removed and the gas is allowed to expand again to a final pressure of 1 atm. Throughout both expansions the temperature of the apparatus was held at a constant 25 °C. Calculate the amount of work performed by the gas in each step. How does the combined total amount of work in this two-step expansion compare to the amount of work done by the gas in the one-step expansion described in the preceding problem?

20.119 What is the algebraic sign of ΔS for the following changes?
(a) $2Ag^+(aq) + CrO_4^{2-}(aq) \rightarrow Ag_2CrO_4(s)$
(b) $NaCl(s) \rightarrow Na^+(aq) + Cl^-(aq)$
(c) $NH_3(g) \xrightarrow{\text{H}_2\text{O}} NH_3(aq)$
(d) naphthalene$(g) \rightarrow$ naphthalene(s)
(e) A gas is cooled from 40 °C to 25 °C.
(f) A gas is compressed at constant temperature from 4.0 L to 2.0 L.

20.120 When potassium iodide dissolves in water, the mixture becomes cool. For this change, which is of a larger magnitude, ΔS or ΔH?

20.121 The reaction $Cl_2(g) + Br_2(g) \rightarrow 2BrCl(g)$ has a very small value for ΔS° ($+11.6$ J K^{-1}). Why?

20.122 For a phase change, why is there only one temperature at which there can be an equilibrium between the phases?

20.123 The enthalpy of combustion, $\Delta H^\circ_{\text{combustion}}$, of oxalic acid, $H_2C_2O_4(s)$, is -246.05 kJ mol^{-1}. Consider the following data:

Substance	ΔH_f° (kJ mol^{-1})	S° (J mol^{-1} K^{-1})
C(s)	0	5.69
CO$_2$(g)	-393.5	213.6
H$_2$(g)	0	130.6
H$_2$O(l)	-285.8	69.96
O$_2$(g)	0	205.0
H$_2$C$_2$O$_4$(s)	?	120.1

(a) Write the balanced thermochemical equation that describes the combustion of 1 mol of oxalic acid.
(b) Write the balanced thermochemical equation that describes the formation of 1 mol of oxalic acid.
(c) Use the information in the table above and the equations in parts a and b to calculate ΔH_f° for oxalic acid.
(d) Calculate ΔS_f° for oxalic acid and ΔS° for the combustion of 1 mol of oxalic acid.
(e) Calculate ΔG_f° for oxalic acid and ΔG° for the combustion of 1 mol of oxalic acid.

20.124 Many biochemical reactions have positive values for ΔG° and seemingly should not be expected to be spontaneous. They occur, however, because they are chemically coupled with other reactions that have negative values of ΔG°. An example is the set of reactions that forms the beginning part of the sequence of reactions involved in the metabolism of glucose, a sugar. Given these reactions and their corresponding ΔG° values,

glucose + phosphate $\longrightarrow$ glucose 6-phosphate + H$_2$O

$$\Delta G^\circ = +13.13 \text{ kJ}$$

ATP + H$_2$O $\longrightarrow$ ADP + phosphate $\quad \Delta G^\circ = -32.22$ kJ

calculate ΔG° for the coupled reaction

glucose + ATP $\longrightarrow$ glucose 6-phosphate + ADP

*20.125 Cars, trucks, and other machines that use gas or diesel engines for power have cooling systems. In terms of thermodynamics, what makes these cooling systems necessary?

20.126 Ethyl alcohol, C_2H_5OH, has been suggested as an alternative to gasoline as a fuel. In Example 20.5 we calculated ΔG° for combustion of 1 mol of C_2H_5OH; in Example 20.6 we calculated ΔG° for combustion of 1 mol of octane. Let's assume that gasoline has the same properties as octane (one of its constituents). The density of C_2H_5OH is 0.7893 g/mL; the density of octane, C_8H_{18}, is 0.7025 g/mL. Calculate the maximum work that could be obtained by burning 1 gal (3.78 L) each of C_2H_5OH and C_8H_{18}. On a *volume* basis, which is a better fuel?

20.127 Use the data in Table 20.3 to calculate the bond energy in the nitrogen molecule and the oxygen molecule.

20.128 The heat of vaporization of carbon tetrachloride, CCl_4, is 29.9 kJ mol^{-1}. Using this information and data in Tables 20.3 and 20.4, estimate the standard heat of formation of liquid CCl_4.

Electrochemistry

It seems almost everyone is using laptop computers today, including the Buddhist monk seen here. Advances in battery technology, which is one of the topics discussed in this chapter, have enabled portable computers to have brighter displays and operate for longer periods of time.

THIS CHAPTER IN CONTEXT Oxidation and reduction (redox) reactions occur in many chemical systems. Examples include the rusting of iron, the action of bleach on stains, and the reactions of photosynthesis in the leaves of green plants. All of these changes involve the transfer of electrons from one chemical species to another. In this chapter we will study how it is possible to separate the processes of oxidation (electron loss) and reduction (electron gain) and cause them to occur in different physical locations. When we are able to do this, we can use spontaneous redox reactions to produce electricity. And by reversing the process, we can use electricity to make nonspontaneous redox reactions happen.

Because electricity plays a role in these systems, the processes involved are described as **electrochemical changes.** The study of such changes is called **electrochemistry.**

The applications of electrochemistry are widespread. In industry, many important chemicals, including liquid bleach (sodium hypochlorite) and lye (sodium hydroxide), are manufactured by electrochemical reactions. Batteries, which produce electrical energy by means of chemical reactions, are used to power toys, flashlights, electronic calculators, laptop computers, heart pacemakers, video cameras, cell phones, and even some automobiles. In the laboratory, electrical measurements enable us to monitor chemical reactions of all sorts, even those in systems as tiny as a living cell.

Besides the practical applications, we will also see how electrical measurements and the principles of thermodynamics combine to give fundamental information about chemical reactions, such as free energy changes and equilibrium constants.

21.1 ▶ Galvanic cells use redox reactions to generate electricity

If you have silver fillings in your teeth, you may have experienced a strange and perhaps even unpleasant sensation while accidentally biting on a piece of aluminum foil. The sensation is caused by a very mild electric shock produced by a voltage difference between your metal fillings and the aluminum. In effect, you created a battery. Batteries not too much different from this serve to power all sorts of gadgets, as mentioned at the beginning of this chapter. The energy from a battery comes from a spontaneous redox reaction in which the electron transfer is forced to take place through a wire. The apparatus that provides electricity in this way is called a **galvanic cell,** after Luigi Galvani (1737–1798), an Italian anatomist who discovered that electricity can cause the contraction of muscles. (It is also called a **voltaic cell,** after another Italian scientist, Allesandro Volta, 1745–1827, whose inventions led ultimately to the development of modern batteries.)

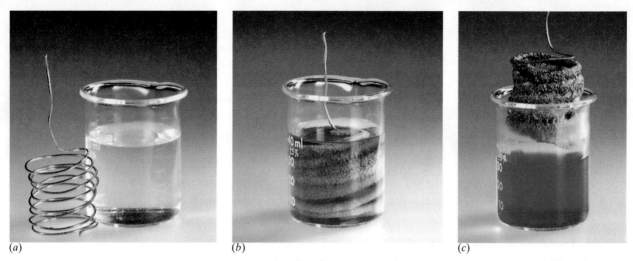

(a) (b) (c)

FIGURE 21.1 *Reaction of copper with a solution of silver nitrate.* (*a*) A coil of copper wire stands next to a beaker containing a silver nitrate solution. (*b*) When the copper wire is placed in the solution, copper dissolves, giving the solution its blue color, and metallic silver deposits as glittering crystals on the wire. (*c*) After a while, much of the copper has dissolved and nearly all of the silver has deposited as the free metal.

Setting up a galvanic cell

If a shiny piece of metallic copper is placed into a solution of silver nitrate, a spontaneous reaction occurs. Gradually, a grayish white deposit forms on the copper and the solution itself becomes pale blue as hydrated Cu^{2+} ions enter the solution (see Figure 21.1). The equation is

$$2Ag^+(aq) + Cu(s) \longrightarrow Cu^{2+}(aq) + 2Ag(s)$$

Although the reaction is exothermic, no usable energy can be harnessed from it because all the energy is dispersed as heat.

To produce *electrical* energy, the two half-reactions involved in the net reaction must be made to occur in separate containers or compartments called **half-cells.** An apparatus to accomplish this—a **galvanic cell**—is made up of two half-cells, as illustrated in Figure 21.2. On the left, a silver electrode dips into a solution of $AgNO_3$, and, on the right, a copper electrode dips into a solution of $Cu(NO_3)_2$. The two electrodes are connected by an external electrical circuit and the two solutions

A galvanic cell is composed to two half-cells connected by an external circuit and a salt bridge.

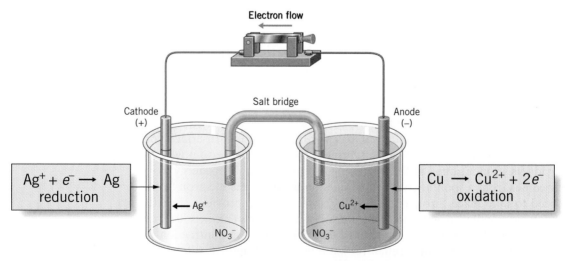

FIGURE 21.2 *A galvanic cell.* The cell consists of two half-cells where the oxidation and reduction half-reactions take place.

are connected by a *salt bridge,* the function of which will be described shortly. When the circuit is completed by closing the switch, the reduction of Ag^+ to Ag occurs spontaneously in the beaker on the left and oxidation of Cu to Cu^{2+} occurs spontaneously in the beaker on the right. The reaction that takes place in each half-cell is a *half-reaction* of the type you learned to balance by the ion–electron method in Chapter 6. In the silver half-cell, the following half-reaction occurs:

$$Ag^+(aq) + e^- \longrightarrow Ag(s) \qquad \text{(reduction)}$$

In the copper half-cell, the half-reaction is

$$Cu(s) \longrightarrow Cu^{2+}(aq) + 2e^- \qquad \text{(oxidation)}$$

When these reactions take place, electrons left behind by oxidation of the copper travel as an electric current through the external circuit to the other electrode where they are picked up by the silver ions, which are thereby reduced.

The cell reaction is the net overall reaction in the cell

The overall reaction that takes place in the galvanic cell is called the **cell reaction.** To obtain it, we add the individual electrode half-reactions together, but only after we make sure that the number of electrons gained in one half-reaction equals the number lost in the other, a requirement of every redox reaction. The procedure we use is the one described as the *ion–electron method* (Section 6.2). Thus, to obtain the cell reaction we multiply the half-reaction for the reduction of silver by 2 and then add the two half-reactions to obtain the net reaction. (Notice that $2e^-$ appears on each side, and so they cancel.)

$$
\begin{array}{ll}
2Ag^+(aq) + 2e^- \longrightarrow 2Ag(s) & \text{(reduction)} \\
\underline{Cu(s) \longrightarrow Cu^{2+}(aq) + 2e^-} & \text{(oxidation)} \\
2Ag^+(aq) + Cu(s) + \cancel{2e^-} \longrightarrow 2Ag(s) + Cu^{2+}(aq) + \cancel{2e^-} & \text{(cell reaction)}
\end{array}
$$

When electrons appear as a reactant, the process is reduction; when they appear as a product, it is oxidation.

Electrodes are named according the chemical processes that occur at them

The electrodes in electrochemical systems are identified by the names *cathode* and *anode.* The names are *always* assigned according to the nature of the chemical changes that occur at the electrodes. In any electrochemical system, *the* **cathode** *is the electrode at which reduction takes place; the* **anode** *is the electrode at which oxidation occurs.*

> The **cathode** is the electrode at which reduction (electron gain) occurs.
>
> The **anode** is the electrode at which oxidation (electron loss) occurs.

Thus, in the galvanic cell we've been discussing, the silver electrode is the cathode and the copper electrode is the anode.

Conduction of charge occurs in two ways

When the reactions take place in the copper–silver galvanic cell, positive copper ions *enter* the liquid that surrounds the anode while positive silver ions *leave* the liquid that surrounds the cathode (Figure 21.3). If these processes were to continue, the solution around the anode would quickly become positively charged because of the excess of positive ions. Similarly, the solution surrounding the cathode would acquire a negative charge because the departure of positive ions would leave an excess of negative ions. Nature does not permit large amounts of positive

Figure 21.3 *Changes that take place at the anode and cathode in the copper–silver galvanic cell.* (not drawn to scale). At the anode, Cu^{2+} ions enter the solution when copper atoms are oxidized, leaving electrons behind on the electrode. Unless Cu^{2+} ions move away from the electrode or NO_3^- ions move toward it, the solution around the electrode will become positively charged. At the cathode, Ag^+ ions leave the solution and become silver atoms by acquiring electrons from the electrode surface. Unless more silver ions move toward the cathode or negative ions move away, the solution around the electrode will become negatively charged.

Cathode

Anode

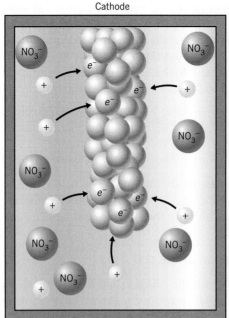

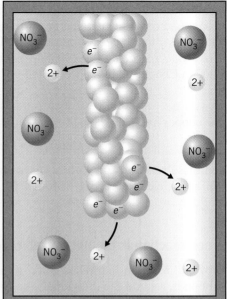

Reduction of silver ions at the cathode extracts electrons from the electrode, so the electrode becomes positively charged.

Oxidation of copper atoms at the anode leaves electrons behind on the electrode, which becomes negatively charged.

and negative charge to accumulate, so for the reactions to continue, and for electricity to continue to flow through the external circuit, a means for balancing the charges in the solutions around the electrodes must exist. To understand how this happens, let's examine how electrical charge is conducted in the cell.

In the external circuit of the cell, electrical charge is transported from one electrode to the other by the movement of *electrons* through the wires. This type of conduction is called **metallic conduction** and is how metals in general conduct electricity. In the cell, electrons always travel from the anode, where they are left behind by the oxidation process, to the cathode, where they are picked up by the substance being reduced.

In electrochemical cells there is another kind of electrical conduction that also takes place. In a solution that contains ions (or in a molten ionic compound), *electrical charge is carried through the liquid by the movement of ions, not electrons.* The transport of electrical charge by ions is called **electrolytic conduction.**

For a galvanic cell to work, the solutions in both half-cells must remain electrically neutral. This requires that ions be permitted to enter or leave the solutions. For example, when copper is oxidized, the solution surrounding the electrode becomes filled with Cu^{2+} ions, so negative ions are needed to balance their charge. Similarly, when Ag^+ ions are reduced, NO_3^- ions are left behind in the solution and positive ions are needed to maintain neutrality. The salt bridge shown in Figure 21.2 allows the movement of ions required to keep the solutions neutral.

A **salt bridge** is a tube filled with a solution of a salt composed of ions not involved in the cell reaction. Often KNO_3 or KCl are used. The tube is fitted with porous plugs at each end that prevent the solution from pouring out, but at the same time enable the solution in the salt bridge to exchange ions with the solutions in the half-cells.

During operation of the cell, negative ions can diffuse from the salt bridge into the copper half-cell, or Cu^{2+} ions can leave the solution and enter the salt bridge. Both processes help keep the copper half-cell electrically neutral. At the silver half-cell, positive ions from the salt bridge can enter or negative NO_3^- ions can leave the half-cell to keep it electrically neutral, too.

Without the salt bridge, electrical neutrality could not be maintained and no electrical current could be produced by the cell. Therefore, *electrolytic contact*—contact by means of a solution containing ions—must be maintained for the cell to function.

If we look closely at the overall movement of ions during the operation of the galvanic cell, we find that negative ions (*anions*) move away from the cathode, where they are present in excess, and *toward the anode,* where they are needed to balance the charge of the positive ions being formed. Similarly, we find that positive ions (*cations*) move away from the anode, where they are in excess, and *toward the cathode,* where they can balance the charge of the anions left in excess. In fact, the reason positive ions are called cations and negative ions are called anions is because of the nature of the electrodes toward which they move. In summary:

> Cations move in the general direction of the cathode.
>
> Anions move in the general direction of the anode.

Charges on the electrodes come from electron loss and gain

At the anode of the galvanic cell described in Figures 21.2 and 21.3, copper atoms spontaneously leave the electrode and enter the solution as Cu^{2+} ions. The electrons that are left behind give the anode a slight negative charge. (We say the anode has a *negative polarity.*) At the cathode, electrons spontaneously join Ag^+ ions to produce neutral atoms, but the effect is the same as if Ag^+ ions become part of the electrode, so the cathode acquires a slight positive charge. (The cathode has a *positive polarity.*) During the operation of the cell, the amount of positive and negative charge on the electrodes is kept small by the flow of electrons (an electric current) through the external circuit from the anode to the cathode when the circuit is complete. In fact, unless electrons can flow out of the anode and into the cathode, the chemical reactions that occur at their surfaces will cease.

The small difference in charge between the electrodes is forced by the spontaneity of the overall reaction, that is, by the favorable free energy change. Nature's tendency toward electrical neutrality prevents a large buildup of charge on the electrodes and promotes the spontaneous flow of electricity through the external circuit.

Cell notation gives a shorthand description of a galvanic cell

As a matter of convenience, chemists have devised a shorthand way of describing the makeup of a galvanic cell. For example, the copper–silver cell that we have been using in our discussion is represented as follows:

$$Cu(s)\,|\,Cu^{2+}(aq)\,\|\,Ag^+(aq)\,|\,Ag(s)$$

By convention, in **standard cell notation,** the anode half-cell is specified on the left, with the electrode material of the anode given first. In this case, the anode is copper metal, but in other galvanic cells an electrode could be an inert (unreactive) metal, like platinum, provided that the redox reaction in the half-cell is between two dissolved species (e.g., Fe^{2+} and Fe^{3+}). The single vertical bar represents a *phase boundary*—here, between the copper electrode and the solution that surrounds it. The double vertical bars represent the salt bridge, which connects the solutions in the two half-cells. On the right, the cathode half-cell is described, with the material of the cathode given last. Thus, the electrodes themselves (copper and silver) are specified at opposite ends of the cell description.

Also, notice that for each half-cell, the reactant in the redox change is given first. In the anode compartment, Cu is the reactant and is oxidized to Cu^{2+}, whereas in the cathode compartment, Ag^+ is the reactant and is reduced to Ag.

Sometimes, the oxidized and reduced forms of the reactants in a half-cell are both in solution. For example, a galvanic cell can be made using an anode composed of a zinc electrode dipping into a solution containing Zn^{2+} and a cathode composed of inert (nonreacting) platinum electrode dipping into a solution containing both Fe^{2+} and Fe^{3+}. The cell reaction is

$$2Fe^{3+}(aq) + Zn(s) \longrightarrow 2Fe^{2+}(aq) + Zn^{2+}(aq)$$

The cell notation for this galvanic cell is written as follows:

$$Zn(s) \mid Zn^{2+}(aq) \parallel Fe^{2+}(aq), Fe^{3+}(aq) \mid Pt(s)$$

where we have separated the formulas for the two iron ions by a comma. In this cell, the reduction of the Fe^{3+} to Fe^{2+} takes place at the surface of the inert platinum electrode.

EXAMPLE 21.1
Describing Galvanic Cells

The following spontaneous reaction occurs when metallic zinc is dipped into a solution of copper sulfate:

$$Zn(s) + Cu^{2+}(aq) \longrightarrow Zn^{2+}(aq) + Cu(s)$$

Describe a galvanic cell that could take advantage of this reaction. What are the half-cell reactions? What is the standard cell notation? Make a sketch of the cell and label the cathode and anode, the charges on each electrode, the direction of ion flow, and the direction of electron flow.

ANALYSIS: Answering all these questions relies on identifying the anode and cathode from the equation for the cell reaction; that's the critical link in solving all problems of this type. By definition, the anode is the electrode at which oxidation happens, and the cathode is where reduction occurs. The first step, therefore, is to determine which reactant is oxidized and which is reduced. One way to do this is to divide the cell reaction into half-reactions and balance them by adding electrons. Then, if electrons appear as a product, the half-reaction is oxidation; if the electrons appear as a reactant, the half-reaction is reduction.

SOLUTION: The balanced half-reactions are as follows:

$$Zn(s) \longrightarrow Zn^{2+}(aq) + 2e^-$$
$$Cu^{2+}(aq) + 2e^- \longrightarrow Cu(s)$$

Zinc loses electrons and is oxidized, so it is the anode. The anode half-cell is therefore a zinc electrode dipping into a solution that contains Zn^{2+} [e.g., from dissolved $Zn(NO_3)_2$ or $ZnSO_4$]. Symbolically, the anode half-cell is written with the electrode material at the left of the vertical bar and the oxidation product at the right.

$$Zn(s) \mid Zn^{2+}(aq)$$

Copper ion gains electrons and is reduced to metallic copper, so the cathode half-cell consists of a copper electrode dipping into a solution containing Cu^{2+} [e.g., from dissolved $Cu(NO_3)_2$ or $CuSO_4$]. The copper half-cell can be represented as follows, with the electrode material to the right of the vertical bar and the substance reduced on the left:

$$Cu^{2+}(aq) \mid Cu(s)$$

The standard cell notation places the zinc anode half-cell on the left and the copper cathode half-cell on the right, separated by double bars that represent the salt bridge.

$$\underset{\text{anode}}{Zn(s) \mid Zn^{2+}(aq)} \parallel \underset{\text{cathode}}{Cu^{2+}(aq) \mid Cu(s)}$$

A sketch of the cell is shown in the margin. The anode always carries a negative charge in a galvanic cell, so the zinc electrode is negative and the copper electrode is positive. Electrons in the external circuit travel from the negative electrode to the

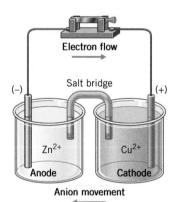

The zinc–copper cell. In the cell notation described in this Example, we indicate the anode on the left and the cathode on the right. In this drawing of the apparatus, the anode half-cell is also shown on the left, but it could just as easily be shown on the right, as in Figure 21.2. Be sure you understand that where we place the apparatus on the lab bench doesn't affect which half-cell is the anode and which is the cathode.

positive electrode (i.e., from the Zn anode to the Cu cathode). Anions move toward the anode, and cations move toward the cathode.

Are the Answers Reasonable?

All of the answers depend on determining which substance is oxidized and which is reduced, so that's what to check. Oxidation is electron loss, and Zn must lose electrons to become Zn^{2+}, so zinc is oxidized and must be the anode. If zinc is the anode, then copper must be the cathode. All the rest follows by reasoning.

PRACTICE EXERCISE 1: Sketch and label a galvanic cell that makes use of the following spontaneous redox reaction:

$$Mg(s) + Fe^{2+}(aq) \longrightarrow Mg^{2+}(aq) + Fe(s)$$

Write the half-reactions for the anode and cathode. Give the standard cell notation.

PRACTICE EXERCISE 2: Write the anode and cathode half-reactions for the following galvanic cell. Write the equation for the overall cell reaction.

$$Al(s) \mid Al^{3+}(aq) \parallel Pb^{2+}(aq) \mid Pb(s)$$

21.2 ▶ Cell potentials can be related to reduction potentials

A galvanic cell has an ability to push electrons through the external circuit. The magnitude of this ability is expressed as a **potential.** Potential is expressed in an electrical unit called the **volt (V),** which is a measure of the amount of energy, in joules, that can be delivered per SI unit of charge (the **coulomb**) as the current moves through the circuit. Thus, a current flowing under a potential of 1 volt can deliver 1 joule of energy per coulomb.

$$1 \text{ V} = 1 \text{ J/C} \tag{21.1}$$

The potential generated by a galvanic cell has also been called an **electromotive force (emf),** suggesting that it can be thought of as the force with which an electric current is pushed through a wire. Electrical current and emf are often likened to the flow of water in a hose ("current") and the pressure of the water ("emf").

The cell potential is the maximum potential produced by a galvanic cell

The voltage or potential of a galvanic cell varies with the amount of current flowing through the circuit. The *maximum* potential that a given cell can generate is called its **cell potential,** E_{cell}, and it depends on the composition of the electrodes, the concentrations of the ions in the half-cells, and the temperature. Therefore, to compare the potentials for different cells we use the **standard cell potential,** symbolized $E°_{cell}$. This is the potential of the cell when all of the ion concentrations are 1.00 M, the temperature is 25 °C, and any gases that are involved in the cell reaction are at a pressure of 1 atm.

If current is drawn from a cell, some of the cell's voltage is lost overcoming its own internal resistance, and the measured voltage isn't E_{cell}.

Cell potentials are rarely larger than a few volts. For example, the standard cell potential for the galvanic cell constructed from silver and copper electrodes shown in Figure 21.4 is only 0.46 V, and one cell in an automobile battery produces only about 2 V. Batteries that generate higher voltages, such as an automobile battery, contain a number of cells arranged in series so that their potentials are additive.

Reduction potentials are a measure of the tendency of reduction half-reactions to occur

It is useful to imagine that the measured overall cell potential arises from a competition, or "tug-of-war," between the two half-cells for electrons. Thus, we think of each half-cell as having a certain natural tendency to acquire electrons and proceed as a *reduction*. The magnitude of this tendency is expressed by the half-reaction's **reduction potential.** When measured under standard conditions, namely, 25 °C, concentrations of 1.00 M for all solutes, and a pressure of 1 atm, the reduction

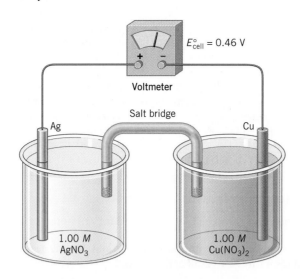

$E^\circ_{cell} = 0.46$ V

Voltmeter

Salt bridge

Ag   Cu

1.00 M
AgNO$_3$

1.00 M
Cu(NO$_3$)$_2$

FIGURE 21.4 *A cell designed to generate the standard cell potential.* The concentrations of the ions in the half-cells are 1.00 M.

Standard reduction potentials are also called **standard electrode potentials.**

potential is called the **standard reduction potential.** To represent a standard reduction potential, we will add a subscript to the symbol E° that identifies the substance undergoing reduction. Thus, the standard reduction potential for the half-reaction

$$Cu^{2+}(aq) + 2e^- \longrightarrow Cu(s)$$

is specified as $E^\circ_{Cu^{2+}}$.

The half-reaction with the larger reduction potential wins the tug-of-war and gets the electrons. The other half-reaction loses and is forced to become an oxidation.

When two half-cells are connected, the one with the larger reduction potential (the one with the greater tendency to undergo reduction) acquires electrons from the half-cell with the lower reduction potential, which is therefore forced to undergo oxidation. The measured cell potential, which is always taken to be a positive number, actually represents the magnitude of the *difference* between the reduction potential of one half-cell and the reduction potential of the other. In general, therefore,

 TOOLS

Standard reduction potentials

$$E^\circ_{cell} = \begin{pmatrix} \text{standard reduction} \\ \text{potential of the} \\ \text{substance reduced} \end{pmatrix} - \begin{pmatrix} \text{standard reduction} \\ \text{potential of the} \\ \text{substance oxidized} \end{pmatrix} \qquad (21.2)$$

Equation 21.2 will be used frequently, so be sure you know it.

As an example, let's look at the copper–silver cell. From the cell reaction,

$$2Ag^+(aq) + Cu(s) \longrightarrow 2Ag(s) + Cu^{2+}(aq)$$

we can see that silver ions are reduced and copper is oxidized. If we compare the two possible reduction half-reactions,

$$Ag^+(aq) + e^- \longrightarrow Ag(s)$$

$$Cu^{2+}(aq) + 2e^- \longrightarrow Cu(s)$$

the one for Ag$^+$ must have a greater tendency to proceed than the one for Cu^{2+}, because it is the silver ion that is actually reduced. This means that the standard reduction potential of Ag$^+$ must be algebraically larger than the standard reduction potential of Cu^{2+}. In other words, if we knew the values of $E^\circ_{Ag^+}$ and $E^\circ_{Cu^{2+}}$, we could calculate E°_{cell} with Equation 21.2 by subtracting the smaller reduction potential from the larger one.

$$E^\circ_{cell} = E^\circ_{Ag^+} - E^\circ_{Cu^{2+}}$$

Assigning standard reduction potentials requires a reference electrode

Unfortunately there is no way to measure the standard reduction potential of an isolated half-cell. All we can measure is the difference in potential produced when two half-cells are connected. Therefore, to assign values to the various standard reduction potentials, a reference electrode has been arbitrarily chosen and its standard reduction potential has been assigned a value of *exactly* 0 V. This reference electrode is called the **standard hydrogen electrode** (see Figure 21.5). Gaseous hydrogen at a pressure of 1 atm is bubbled over a platinum electrode coated with very finely divided platinum, which provides a large catalytic surface area on which the electrode reaction can occur. This electrode is surrounded by a solution whose temperature is 25 °C and in which the hydrogen ion concentration is 1.00 *M*. The half-cell reaction at the platinum surface, written as a reduction, is

$$2H^+(aq, 1.00\ M) + 2e^- \rightleftharpoons H_2(g, 1\ atm) \qquad E^\circ_{H^+} = 0\ V$$

The double arrows indicate only that the reaction is reversible, not that there is true equilibrium. Whether the half-reaction occurs as reduction or oxidation depends on the reduction potential of the half-cell with which it is paired.

Figure 21.6 illustrates the hydrogen electrode connected to a copper half-cell to form a galvanic cell. To obtain the cell reaction, we have to know what is oxidized and what is reduced, and we need to know which is the cathode and which is the anode. We can determine this by measuring the charges on the electrode, because we know that in a galvanic cell the cathode is the positive electrode and the anode is the negative electrode. When we use a voltmeter to measure the potential of the cell, we find that the copper electrode carries a positive charge and the hydrogen electrode a negative charge. Therefore, copper must be the cathode, and Cu^{2+} is reduced to Cu when the cell operates. Similarly, hydrogen must be the anode, and H_2 is oxidized to H^+. The half-reactions and cell reaction, therefore, are

$$Cu^{2+}(aq) + 2e^- \longrightarrow Cu(s) \qquad \text{(cathode)}$$
$$\underline{H_2(g) \longrightarrow 2H^+(aq) + 2e^- \qquad \text{(anode)}}$$
$$Cu^{2+}(aq) + H_2(g) \longrightarrow Cu(s) + 2H^+(aq) \qquad \text{(cell reaction)}[1]$$

Using Equation 21.2, we can express E°_{cell} in terms of $E^\circ_{Cu^{2+}}$ and $E^\circ_{H^+}$.

$$E^\circ_{cell} = E^\circ_{Cu^{2+}} - E^\circ_{H^+}$$

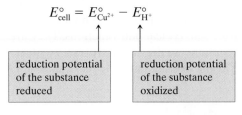

reduction potential of the substance reduced

reduction potential of the substance oxidized

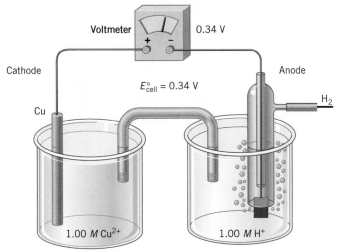

Voltmeter · 0.34 V

Cathode

Anode

$E^\circ_{cell} = 0.34\ V$

Cu

H_2

1.00 *M* Cu^{2+}

1.00 *M* H^+

The large surface area provided by the fine platinum coating on the electrode enables the electrode reaction to occur rapidly.

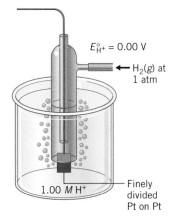

$E^\circ_{H^+} = 0.00\ V$

← $H_2(g)$ at 1 atm

1.00 *M* H^+ — Finely divided Pt on Pt

FIGURE 21.5 *The hydrogen electrode.* The half-reaction is $2H^+(aq) + 2e^- \rightleftharpoons H_2(g)$.

FIGURE 21.6 *A galvanic cell composed of copper and hydrogen half-cells.* The cell reaction is $Cu^{2+}(aq) + H_2(g) \rightarrow Cu(s) + 2H^+(aq)$.

[1]The cell notation for this cell is written as

$$Pt(s), H_2(g)\ |\ H^+(aq)\ \|\ Cu^{2+}(aq)\ |\ Cu(s)$$

The notation for the hydrogen electrode (the anode in this case) is shown at the left of the double vertical bars.

In a galvanic cell, the measured cell potential is *always* taken to be a positive value. This is important to remember.

The measured standard cell potential is 0.34 V and $E°_{H^+}$ equals 0.00 V. Therefore,

$$0.34 \text{ V} = E°_{Cu^{2+}} - 0.00 \text{ V}$$

Relative to the hydrogen electrode, then, the standard reduction potential of Cu^{2+} is +0.34 V. (We have written the value with a plus sign because some reduction potentials are negative, as we will see.)

Now let's look at a galvanic cell set up between a zinc electrode and a hydrogen electrode (see Figure 21.7). This time we find that the hydrogen electrode is positive and the zinc electrode is negative, which tells us that the hydrogen electrode is the cathode and the zinc electrode is the anode. This means that hydrogen ion is being reduced and zinc is being oxidized. The half-reactions and cell reaction are therefore

Remember, in a galvanic cell, the cathode is (+) and the anode is (−).

$$2H^+(aq) + 2e^- \longrightarrow H_2(g) \quad \text{(cathode)}$$

$$\underline{\quad\quad Zn(s) \longrightarrow Zn^{2+}(aq) + 2e^- \quad\quad} \quad \text{(anode)}$$

$$2H^+(aq) + Zn(s) \longrightarrow H_2(g) + Zn^{2+}(aq) \quad \text{(cell reaction)}^2$$

From Equation 21.2, the standard cell potential is given by

$$E°_{cell} = E°_{H^+} - E°_{Zn^{2+}}$$

Substituting into this the measured standard cell potential of 0.76 V and $E°_{H^+} = 0.00$ V, we have

$$0.76 \text{ V} = 0.00 \text{ V} - E°_{Zn^{2+}}$$

which gives

$$E°_{Zn^{2+}} = -0.76 \text{ V}$$

Notice that the reduction potential of zinc is negative. A negative reduction potential simply means that the substance is not as easily reduced as H^+. In this case, it tells us that Zn is oxidized when it is paired with the hydrogen electrode.

The standard reduction potentials of many half-reactions can be compared to that for the standard hydrogen electrode in the manner described above. Table 21.1 lists values obtained for some typical half-reactions. They are arranged in decreasing order — the half-reactions at the top have the greatest tendency to occur as reduction, while those at the bottom have the least tendency to occur as reduction.

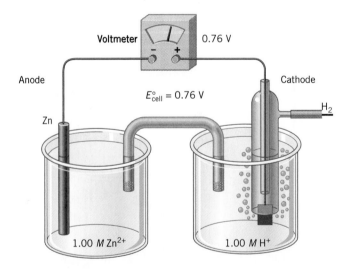

FIGURE 21.7 *A galvanic cell composed of zinc and hydrogen half-cells.* The cell reaction is $Zn(s) + 2H^+(aq) \rightarrow Zn^{2+}(aq) + H_2(g)$.

---

²This cell is represented as

$$Zn(s) \mid Zn^{2+}(aq) \parallel H^+(aq) \mid H_2(g), Pt(s)$$

This time the hydrogen electrode is the cathode and appears at the right of the double vertical bars.

TABLE 21.1	STANDARD REDUCTION POTENTIALS AT 25 °C	
	Half Reaction	$E°$ (volts)

Half Reaction	$E°$ (volts)
$F_2(g) + 2e^- \rightleftharpoons 2F^-(aq)$	+2.87
$S_2O_8^{2-}(aq) + 2e^- \rightleftharpoons 2SO_4^{2-}(aq)$	+2.01
$PbO_2(s) + HSO_4^-(aq) + 3H^+(aq) + 2e^- \rightleftharpoons PbSO_4(s) + 2H_2O$	+1.69
$2HOCl(aq) + 2H^+(aq) + 2e^- \rightleftharpoons Cl_2(g) + 2H_2O$	+1.63
$MnO_4^-(aq) + 8H^+(aq) + 5e^- \rightleftharpoons Mn^{2+}(aq) + 4H_2O$	+1.51
$PbO_2(s) + 4H^+(aq) + 2e^- \rightleftharpoons Pb^{2+}(aq) + 2H_2O$	+1.46
$BrO_3^-(aq) + 6H^+(aq) + 6e^- \rightleftharpoons Br^-(aq) + 3H_2O$	+1.44
$Au^{3+}(aq) + 3e^- \rightleftharpoons Au(s)$	+1.42
$Cl_2(g) + 2e^- \rightleftharpoons 2Cl^-(aq)$	+1.36
$O_2(g) + 4H^+(aq) + 4e^- \rightleftharpoons 2H_2O$	+1.23
$Br_2(aq) + 2e^- \rightleftharpoons 2Br^-(aq)$	+1.07
$NO_3^-(aq) + 4H^+(aq) + 3e^- \rightleftharpoons NO(g) + 2H_2O$	+0.96
$Ag^+(aq) + e^- \rightleftharpoons Ag(s)$	+0.80
$Fe^{3+}(aq) + e^- \rightleftharpoons Fe^{2+}(aq)$	+0.77
$I_2(s) + 2e^- \rightleftharpoons 2I^-(aq)$	+0.54
$NiO_2(s) + 2H_2O + 2e^- \rightleftharpoons Ni(OH)_2(s) + 2OH^-(aq)$	+0.49
$Cu^{2+}(aq) + 2e^- \rightleftharpoons Cu(s)$	+0.34
$SO_4^{2-} + 4H^+(aq) + 2e^- \rightleftharpoons H_2SO_3(aq) + H_2O$	+0.17
$AgBr(s) + e^- \rightleftharpoons Ag(s) + Br^-$	+0.07
$2H^+(aq) + 2e^- \rightleftharpoons H_2(g)$	0
$Sn^{2+}(aq) + 2e^- \rightleftharpoons Sn(s)$	−0.14
$Ni^{2+} + 2e^- \rightleftharpoons Ni(s)$	−0.25
$Co^{2+}(aq) + 2e^- \rightleftharpoons Co(s)$	−0.28
$PbSO_4(s) + H^+(aq) + 2e^- \rightleftharpoons Pb(s) + HSO_4^-(aq)$	−0.36
$Cd^{2+}(aq) + 2e^- \rightleftharpoons Cd(s)$	−0.40
$Fe^{2+}(aq) + 2e^- \rightleftharpoons Fe(s)$	−0.44
$Cr^{3+}(aq) + 3e^- \rightleftharpoons Cr(s)$	−0.74
$Zn^{2+}(aq) + 2e^- \rightleftharpoons Zn(s)$	−0.76
$2H_2O + 2e^- \rightleftharpoons H_2(g) + 2OH^-(aq)$	−0.83
$Al^{3+}(aq) + 3e^- \rightleftharpoons Al(s)$	−1.66
$Mg^{2+}(aq) + 2e^- \rightleftharpoons Mg(s)$	−2.37
$Na^+(aq) + e^- \rightleftharpoons Na(s)$	−2.71
$Ca^{2+}(aq) + 2e^- \rightleftharpoons Ca(s)$	−2.76
$K^+(aq) + e^- \rightleftharpoons K(s)$	−2.92
$Li^+(aq) + e^- \rightleftharpoons Li(s)$	−3.05

Substances located to the left of the double arrows are *oxidizing agents,* because they become reduced when the reactions proceed in the forward direction. The best oxidizing agents are those most easily reduced, and they are located at the top of the table (e.g., F_2).

Substances located to the right of the double arrows are *reducing agents;* they become oxidized when the reactions proceed from right to left. The best reducing agents are those found at the bottom of the table (e.g., Li).

We mentioned earlier that the standard cell potential of the silver–copper galvanic cell has a value of 0.46 V. The cell reaction is

$$2Ag^+(aq) + Cu(s) \longrightarrow 2Ag(s) + Cu^{2+}(aq)$$

and we have seen that the reduction potential of Cu^{2+}, $E°_{Cu^{2+}}$, is +0.34 V. What is the value of $E°_{Ag^+}$, the reduction potential of Ag^+?

ANALYSIS: Since we know the potential of the cell and one of the two reduction potentials, we can use Equation 21.2 to calculate the unknown reduction potential.

EXAMPLE 21.2

Calculating Half-Cell Potentials

This requires that we identify the substance oxidized and the substance reduced. We can do this either by dividing the cell reaction into half-reactions, or we can observe how the oxidation numbers of the reactants change. Recall that if the oxidation number increases algebraically, the substance undergoes oxidation, whereas if the oxidation number decreases, the substance is reduced.

SOLUTION: Silver changes from Ag^+ to Ag; its oxidation number decreases from $+1$ to 0, so Ag^+ is reduced. Similar reasoning tells us that copper is oxidized from Cu to Cu^{2+}. Therefore, according to Equation 21.2,

$$E^\circ_{cell} = E^\circ_{Ag^+} - E^\circ_{Cu^{2+}}$$

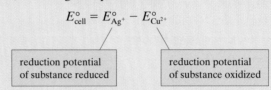

reduction potential of substance reduced   reduction potential of substance oxidized

Substituting values for E°_{cell} and $E^\circ_{Cu^{2+}}$,

$$0.46\,V = E^\circ_{Ag^+} - 0.34\,V$$

Then we solve for $E^\circ_{Ag^+}$.

$$E^\circ_{Ag^+} = 0.46\,V + 0.34\,V$$
$$= 0.80\,V$$

The standard reduction potential of silver ion is therefore $+0.80$ V.

Is the Answer Reasonable?
We know the cell potential is the difference between the two reduction potentials. The difference between $+0.80$ V and $+0.34$ V (subtracting the smaller from the larger) is 0.46 V. Our calculated reduction potential for Ag^+ appears to be correct.

PRACTICE EXERCISE 3: The galvanic cell described in Practice Exercise 1 has a standard cell potential of 1.93 V. The standard reduction potential of Fe^{2+} corresponding to the half-reaction $Fe^{2+}(aq) + 2e^- \rightleftharpoons Fe(s)$ is -0.44 V. Calculate the standard reduction potential of magnesium. Check your answer by referring to Table 21.1.

21.3 ▶ Standard reduction potentials can predict spontaneous reactions

One of the goals of chemistry is to be able to predict reactions. The half-reactions and standard reduction potentials of Table 21.1 enable us to do this for redox reactions, whether they occur in a galvanic cell or just in a container with all the chemicals combined in one reaction mixture. For example, we can use the data in the table to determine whether a reaction will occur between Cl_2 and Br^- to give Cl^- and Br_2, or whether Br_2 will react with Cl^- to give Br^- and Cl_2. We can also determine what reaction would happen if the reduced and oxidized form of one species, such as Fe and a salt of Fe^{2+}, were mixed with a similar pair of another, such as Ni and a salt of Ni^{2+}. (We can determine whether Fe would reduce Ni^{2+}, or whether Ni would reduce Fe^{2+}.) And we can use the data in Table 21.1 to calculate the standard cell potential when a given redox reaction is set up as a galvanic cell. Let's look at some examples.

Redox reactions can be predicted by comparing reduction potentials

It's easy to predict the spontaneous reaction between the substances in two half-reactions, because we know that *the half-reaction with the more positive reduction potential always takes place as written (namely, as a reduction), while the other half-reaction is forced to run in reverse (as an oxidation).*

FACETS OF CHEMISTRY 21.1

Corrosion of Iron and Cathodic Protection

A problem that has plagued humanity ever since the discovery of methods for obtaining iron and other metals from their ores has been corrosion—the reaction of a metal with substances in the environment. The rusting of iron in particular is a serious problem because iron and steel have so many uses.

The rusting of iron is a complex chemical reaction that involves both oxygen and moisture (see Figure 1). Iron won't rust in pure water that's oxygen free, and it won't rust in pure oxygen in the absence of moisture. The corrosion process is apparently electrochemical in nature, as shown in the accompanying diagram. At one place on the surface, iron becomes oxidized in the presence of water and enters solution as Fe^{2+}.

$$Fe(s) \longrightarrow Fe^{2+}(aq) + 2e^-$$

At this location the iron is acting as an anode.

The electrons that are released when the iron is oxidized travel through the metal to some other place where the iron is exposed to oxygen. This is where reduction takes place (it's a cathodic region on the metal surface), and oxygen is reduced to give hydroxide ion.

$$\tfrac{1}{2}O_2(aq) + H_2O + 2e^- \longrightarrow 2OH^-(aq)$$

The iron(II) ions that are formed at the anodic regions gradually diffuse through the water and eventually contact the hydroxide ions. This causes a precipitate of $Fe(OH)_2$ to form, which is very easily oxidized by O_2 to give $Fe(OH)_3$. This hydroxide readily loses water. In fact, complete dehydration gives the oxide,

$$2Fe(OH)_3 \longrightarrow Fe_2O_3 + 3H_2O$$

When partial dehydration of the $Fe(OH)_3$ occurs, *rust* is formed. It has a composition that lies between that of the hydroxide and that of the oxide, Fe_2O_3, and is usually referred to as a *hydrated oxide*. Its formula is generally represented as $Fe_2O_3 \cdot xH_2O$.

This mechanism for the rusting of iron explains one of the more interesting aspects of this damaging process. Perhaps you've noticed that when rusting occurs on the body of a car, the rust appears at and around a break (or a scratch) in the surface of the paint, but the damage extends under the painted surface for some distance. Apparently, the Fe^{2+} ions that are formed at the anode sites are able to diffuse rather long distances to the hole in the paint, where they finally react with air to form the rust.

Cathodic Protection

One way to prevent the rusting of iron is to coat it with another metal. This is done with "tin" cans, which are actually steel cans that have been coated with a thin layer of tin. However, if the layer of tin is scratched and the iron beneath is exposed, the corrosion is accelerated because iron has a lower reduction potential than tin; the iron be-

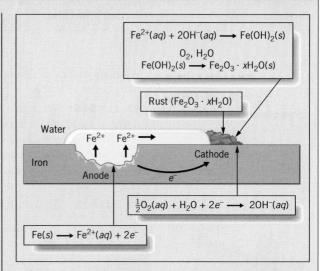

FIGURE 1 *Corrosion of iron.* Iron dissolves in anodic regions to give Fe^{2+}. Electrons travel through the metal to cathodic sites where oxygen is reduced, forming OH^-. The combination of the Fe^{2+} and OH^-, followed by air oxidation, gives rust.

comes the anode in an electrochemical cell and is easily oxidized.

Another way to prevent corrosion is called *cathodic protection*. It involves placing the iron in contact with a metal that is *more easily* oxidized. This causes iron to be a cathode and the other metal to be the anode. If corrosion occurs, iron is protected from oxidation because it is cathodic and the other metal reacts instead.

FIGURE 2 *Cathodic protection.* Before launching, a shiny new zinc anode disk is attached to the bronze rudder of this boat to provide cathodic protection. Over time, the zinc will corrode instead of the less reactive bronze. (The rudder is painted with a special blue paint to inhibit the growth of barnacles.)

Zinc is most often used to provide cathodic protection to other metals. For example, zinc sacrificial anodes can be attached to the rudder of a boat (see Figure 2). When the rudder is submerged, the zinc will gradually corrode but the metal of the rudder will not. Periodically, the anodes are replaced to provide continued protection.

Steel objects that must withstand the weather are often coated with a layer of zinc, a process called *galvanizing*. You've seen this on chain-link fences and metal garbage pails. Even if the steel is exposed through a scratch, it is prevented from being oxidized because it is in contact with a metal that is more easily oxidized.

EXAMPLE 21.3
Predicting a Spontaneous Reaction

What spontaneous reaction occurs if Cl_2 and Br_2 are added to a solution that contains both Cl^- and Br^-?

ANALYSIS: We know that in the spontaneous redox reaction, the more easily reduced substance will be the one that undergoes reduction. If we can find reduction potentials for Cl_2 and Br_2, we can compare their $E°$ values to determine which is the more easily reduced, and then use that information to write the correct "cell reaction." This is the spontaneous reaction, *whether or not it occurs in a galvanic cell.*

SOLUTION: There are two possible reduction half-reactions:

$$Cl_2(g) + 2e^- \longrightarrow 2Cl^-(aq)$$

$$Br_2(aq) + 2e^- \longrightarrow 2Br^-(aq)$$

Strictly speaking, the $E°$ values only tell us what to expect under standard conditions. However, only when $E°_{cell}$ is small can changes in the concentrations change the direction of the spontaneous reaction.

Referring to Table 21.1, we find that Cl_2 has a more positive reduction potential (+1.36 V) than does Br_2 (+1.07 V). This means Cl_2 will be reduced and the half-reaction for Br_2 will be reversed, changing to an oxidation. Therefore, the spontaneous reaction has the following half-reactions:

$$Cl_2(g) + 2e^- \longrightarrow 2Cl^-(aq) \qquad \text{(a reduction)}$$

$$2Br^-(aq) \longrightarrow Br_2(aq) + 2e^- \qquad \text{(an oxidation)}$$

The net reaction is obtained by combining the half-reactions.

$$Cl_2(g) + 2Br^-(aq) \longrightarrow Br_2(aq) + 2Cl^-(aq)$$

Is the Answer Reasonable?
We can check to be sure we've read the correct values for $E°_{Cl_2}$ and $E°_{Br_2}$ from Table 21.1, and we can check the half-reactions we used to find the equation for the net reaction. (Experimentally, chlorine does indeed oxidize bromide ion to bromine, a fact used to recover bromine from seawater and natural brine solutions.)

The reactants and products of *spontaneous* redox reactions are easy to spot when reduction potentials are listed in order of most positive to least positive (most negative), as in Table 21.1. For *any* pair of half-reactions, the one higher up in the table has the more positive reduction potential and occurs as a reduction. The other half-reaction is reversed and occurs as an oxidation. *Therefore, for a spontaneous reaction, the **reactants** are found on the left side of the higher half-reaction and on the right side of the lower half-reaction.*

EXAMPLE 21.4
Predicting the Outcome of Redox Reactions

Predict the reaction that will occur when Ni and Fe are added to a solution that contains both Ni^{2+} and Fe^{2+}.

ANALYSIS: The first question we would ask is "What *possible* reactions could occur?" We have a situation involving possible changes of ions to atoms or of atoms to ions. In other words, the system involves a possible redox reaction, and you've seen we can predict these using data in Table 21.1. One way to do this is to note the relative positions of the half-reactions when arranged as they are in Table 21.1.

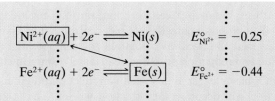

The activity series of the metals (Section 6.4), also used to predict reactions, is derived from Table 21.1.

In the table, the half-reaction higher up has the more positive (in this case, less negative) reduction potential and will occur as a reduction. As a result, the reactants in the spontaneous reaction are related by the diagonal line that slants from upper left to lower right, as illustrated above. In other words, Ni^{2+} will react with Fe. The products are the substances on the opposite sides of the half-reaction, Ni and Fe^{2+}.

SOLUTION: We've done nearly all the work in our analysis of the problem. All that's left is to write the equation. The reactants are Ni^{2+} and Fe; the products are Ni and Fe^{2+}.

$$Ni^{2+}(aq) + Fe(s) \longrightarrow Ni(s) + Fe^{2+}(aq)$$

The equation is balanced both in terms of atoms and charges, so this is the reaction that will occur in the system specified in the problem.

Is the Answer Reasonable?
Notice that we've predicted the reaction very easily using just the *positions* of the half-reactions relative to each other in the table; we really didn't have to study their reduction potentials. We could check ourselves by proceeding as in Example 21.3. In the table, we see that Ni^{2+} has a more positive (less negative) reduction potential than Fe^{2+}, so Ni^{2+} is reduced and its half-cell reaction is written just as in Table 21.1. The half-cell reaction for Fe^{2+} in Table 21.1, however, must be reversed; it is Fe that will be oxidized.

$$Ni^{2+}(aq) + 2e^- \longrightarrow Ni(s) \qquad \text{(reduction)}$$
$$\underline{Fe(s) \longrightarrow Fe^{2+}(aq) + 2e^- \qquad \text{(oxidation)}}$$
$$Ni^{2+}(aq) + Fe(s) \longrightarrow Ni(s) + Fe^{2+}(aq) \qquad \text{(net reaction)}$$

PRACTICE EXERCISE 4: Use the positions of the half-reactions in Table 21.1 to predict the spontaneous reaction when Br^-, SO_4^{2-}, H_2SO_3, and Br_2 are mixed in an acidic solution.

PRACTICE EXERCISE 5: From the positions of the respective half-reactions in Table 21.1, predict whether the following reaction is spontaneous. If it is not, write the equation for the spontaneous reaction.

$$Sn^{2+}(aq) + 2Fe^{2+}(aq) \longrightarrow Sn(s) + 2Fe^{3+}(aq)$$

Reduction potentials predict the cell reaction and cell potential of a galvanic cell

We've just seen that we can use reduction potentials to predict spontaneous redox reactions. If we intend to use these reactions in a galvanic cell, we can also predict what the standard cell potential will be, as illustrated in the next example.

A typical cell of a lead storage battery of the type used to start automobiles is constructed using electrodes made of lead and lead(IV) oxide (PbO_2) and with sulfuric acid as the electrolyte. The half-reactions and their reduction potentials in this system are

$$PbO_2(s) + 3H^+(aq) + HSO_4^-(aq) + 2e^- \rightleftharpoons PbSO_4(s) + 2H_2O \qquad E^\circ_{PbO_2} = 1.69 \text{ V}$$
$$PbSO_4(s) + H^+(aq) + 2e^- \rightleftharpoons Pb(s) + HSO_4^-(aq) \qquad E^\circ_{PbSO_4} = -0.36 \text{ V}$$

What is the cell reaction and what is the standard potential of this cell?

EXAMPLE 21.5

Predicting the Cell Reaction and Cell Potential of a Galvanic Cell

ANALYSIS: In the spontaneous cell reaction, the half-reaction with the larger (more positive) reduction potential will take place as reduction while the other half-reaction will be reversed and occur as oxidation. The cell potential is simply the difference between the two reduction potentials, calculated using Equation 21.2.

SOLUTION: PbO_2 has a larger, more positive reduction potential than $PbSO_4$, so the first half-reaction will occur in the direction written. The second must be reversed to occur as an oxidation. In the cell, therefore, the half-reactions are

Remember that half-reactions are combined following the same procedure used in the ion–electron method of balancing redox reactions (Section 6.2).

$$PbO_2(s) + 3H^+(aq) + HSO_4^-(aq) + 2e^- \longrightarrow PbSO_4(s) + 2H_2O$$

$$Pb(s) + HSO_4^-(aq) \longrightarrow PbSO_4(s) + H^+(aq) + 2e^-$$

Adding the two half-reactions and canceling electrons gives the cell reaction,

$$PbO_2(s) + Pb(s) + 2H^+(aq) + 2HSO_4^-(aq) \longrightarrow 2PbSO_4(s) + 2H_2O$$

The cell potential is obtained by using Equation 21.2.

$$E^\circ_{cell} = (E^\circ \text{ of substance reduced}) - (E^\circ \text{ of substance oxidized})$$

Since the first half-reaction occurs as a reduction and the second as an oxidation,

$$E^\circ_{cell} = E^\circ_{PbO_2} - E^\circ_{PbSO_4}$$
$$= (1.69 \text{ V}) - (-0.36 \text{ V})$$
$$= 2.05 \text{ V}$$

Are the Answers Reasonable?
The half-reactions involved in the problem are located in Table 21.1, and their relative positions tell us that PbO_2 will be reduced and that lead will be oxidized. Therefore, we've combined the half-reactions correctly and we can feel confident that we've also applied Equation 21.2 correctly.

EXAMPLE 21.6

Predicting the Cell Reaction and Cell Potential of a Galvanic Cell

What would be the cell reaction and the standard cell potential of a galvanic cell employing the following half-reactions?

$$Al^{3+}(aq) + 3e^- \rightleftharpoons Al(s) \qquad E^\circ_{Al^{3+}} = -1.66 \text{ V}$$

$$Cu^{2+}(aq) + 2e^- \rightleftharpoons Cu(s) \qquad E^\circ_{Cu^{2+}} = +0.34 \text{ V}$$

Which half-cell would be the anode?

ANALYSIS: This problem is very similar to the preceding one, so we expect to proceed in essentially the same way.

SOLUTION: The half-reaction with the more positive reduction potential will occur as a reduction; the other will occur as an oxidation. In this cell, then, Cu^{2+} is reduced and Al is oxidized. To obtain the cell reaction, we add the two half-reactions, remembering that the electrons must cancel. This means we must multiply the copper half-reaction by 3 and the aluminum half-reaction by 2.

$$3[Cu^{2+}(aq) + 2e^- \longrightarrow Cu(s)] \qquad \text{(reduction)}$$

$$2[Al(s) \longrightarrow Al^{3+}(aq) + 3e^-] \qquad \text{(oxidation)}$$

$$\overline{3Cu^{2+}(aq) + 2Al(s) \longrightarrow 3Cu(s) + 2Al^{3+}(aq)} \qquad \text{(cell reaction)}$$

The anode in the cell is aluminum because that is where oxidation takes place (by definition).

To obtain the cell potential, we substitute into Equation 21.2.

$$E^\circ_{cell} = E^\circ_{Cu^{2+}} - E^\circ_{Al^{3+}}$$
$$= (0.34 \text{ V}) - (-1.66 \text{ V})$$
$$= 2.00 \text{ V}$$

An important point to notice here is that *although we multiply the half-reactions by factors to make the electrons cancel,* **we do not multiply the reduction potentials by these factors.**[3] To obtain the cell potential, we simply subtract one reduction potential from the other.

Are the Answers Reasonable?

If we locate these half-reactions in Table 21.1, their relative positions tell us we've written the correct equation for the spontaneous reaction. It also means we've identified correctly the substances reduced and oxidized, so we've correctly applied Equation 21.2.

PRACTICE EXERCISE 6: What are the overall cell reaction and the standard cell potential of a galvanic cell employing the following half-reactions?

$$NiO_2(s) + 2H_2O + 2e^- \rightleftharpoons Ni(OH)_2(s) + 2OH^-(aq) \qquad E°_{NiO_2} = 0.49 \text{ V}$$

$$Fe(OH)_2(s) + 2e^- \rightleftharpoons Fe(s) + 2OH^-(aq) \qquad E°_{Fe(OH)_2} = -0.88 \text{ V}$$

These are the reactions in an Edison cell, a type of rechargeable storage battery.

PRACTICE EXERCISE 7: What are the overall cell reaction and the standard cell potential of a galvanic cell employing the following half-reactions?

$$Cr^{3+}(aq) + 3e^- \rightleftharpoons Cr(s) \qquad E°_{Cr^{3+}} = -0.74 \text{ V}$$

$$MnO_4^-(aq) + 8H^+(aq) + 5e^- \rightleftharpoons Mn^{2+}(aq) + 4H_2O \qquad E°_{MnO_4^-} = +1.51 \text{ V}$$

The calculated cell potential can tell us whether a reaction is spontaneous

Because we can predict the spontaneous redox reaction that will take place among a mixture of reactants, it also should be possible to predict whether or not a particular reaction, *as written,* can occur spontaneously. We can do this by calculating the cell potential that corresponds to the reaction in question and seeing if the potential is *positive.*

Standard cell potentials

In a galvanic cell, the calculated cell potential for the spontaneous reaction is always positive. If the calculated cell potential is negative, the reaction is spontaneous in the reverse direction.

For example, to obtain the cell potential for a spontaneous reaction in our previous examples, we subtracted the reduction potentials in a way that gave a positive answer. Therefore, if we compute the cell potential for a particular reaction *based on the way the equation is written* and the potential comes out positive, we know the reaction is spontaneous. If the calculated cell potential comes out negative, however, the reaction is nonspontaneous. In fact, it is really spontaneous in the opposite direction.

These generalizations apply under standard conditions: 1 *M* concentrations of all ions, 1 atm pressure for gases, and 25 °C.

Determine whether the following reactions are spontaneous as written. If a reaction is not spontaneous, write the equation for the reaction that is.

(1) $Cu(s) + 2H^+(aq) \longrightarrow Cu^{2+}(aq) + H_2(g)$

(2) $3Cu(s) + 2NO_3^-(aq) + 8H^+(aq) \longrightarrow 3Cu^{2+}(aq) + 2NO(g) + 4H_2O$

ANALYSIS: Our goal for each reaction will be to calculate the cell potential based on the reaction as written. If $E°_{cell}$ is positive, then the reaction is spontaneous.

EXAMPLE 21.7

Determining whether a Reaction is Spontaneous by Using the Calculated Cell Potential

---

[3]Reduction potentials are intensive quantities; they have the units volts, which are joules *per coulomb.* The same number of joules are available for each coulomb of charge regardless of the total number of electrons shown in the equation. Therefore, reduction potentials are never multiplied by factors before they are subtracted to give the cell potential.

However, if E°_{cell} is negative, then the reaction is not spontaneous as written and reversing the equation will give the spontaneous reaction.

To calculate E°_{cell}, we need to divide the equation into its half-reactions, find the necessary reduction potentials in Table 21.1, and then use Equation 21.2 to calculate E°_{cell}. The signs of E°_{cell} will then tell us whether the reactions are spontaneous under standard conditions.

SOLUTION: (1) The half-reactions involved in this reaction are

$$Cu(s) \longrightarrow Cu^{2+}(aq) + 2e^- \qquad \text{(oxidation)}$$

$$2H^+(aq) + 2e^- \longrightarrow H_2(g) \qquad \text{(reduction)}$$

The H^+ is reduced and Cu is oxidized, so Equation 21.2 will take the form

$$E^{\circ}_{cell} = E^{\circ}_{H^+} - E^{\circ}_{Cu^{2+}}$$

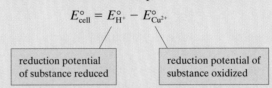

reduction potential of substance reduced

reduction potential of substance oxidized

Substituting values from Table 21.1 gives

$$E^{\circ}_{cell} = (0.00\text{ V}) - (0.34\text{ V})$$

$$= -0.34\text{ V}$$

The calculated cell potential is negative, so reaction (1) is not spontaneous in the forward direction. The spontaneous reaction is actually the reverse of (1).

$$Cu^{2+}(aq) + H_2(g) \longrightarrow Cu(s) + 2H^+(aq)$$

reaction (1) reversed

Copper doesn't dissolve in acids like HCl because the only oxidizing agent is H^+, which is a weaker oxidizing agent than Cu^{2+}.

(2) The half-reactions involved in this equation are

$$Cu(s) \longrightarrow Cu^{2+}(aq) + 2e^-$$

$$NO_3^-(aq) + 4H^+(aq) + 3e^- \longrightarrow NO(g) + 2H_2O$$

The Cu is oxidized while the NO_3^- is reduced. According to Equation 21.2,

$$E^{\circ}_{cell} = E^{\circ}_{NO_3^-} - E^{\circ}_{Cu^{2+}}$$

Substituting values from Table 21.1 gives

$$E^{\circ}_{cell} = (0.96\text{ V}) - (0.34\text{ V})$$

$$= +0.62\text{ V}$$

Because the calculated cell potential is positive, reaction (2) is spontaneous in the forward direction, as written.

Copper dissolves in HNO_3 because it contains the oxidizing agent NO_3^-.

Are the Answers Reasonable?
By noting the relative positions of the half-reactions in Table 21.1, you can confirm that we've answered the questions correctly.

PRACTICE EXERCISE 8: Which of the following reactions occur spontaneously in the forward direction?
(a) $Br_2(aq) + Cl_2(g) + 2H_2O \rightarrow 2Br^-(aq) + 2HOCl(aq) + 2H^+(aq)$
(b) $3Zn(s) + 2Cr^{3+}(aq) \rightarrow 3Zn^{2+}(aq) + 2Cr(s)$

21.4 ▶ Cell potentials are related to free energy changes

The fact that cell potentials allow us to predict the spontaneity of redox reactions is no coincidence. There is a relationship between the cell potential and the free energy change for a reaction. In Chapter 20 we saw that ΔG for a reaction is a measure of the maximum useful work that can be obtained from a chemical reaction. Specifically, the relationship is

$$-\Delta G = \text{maximum work} \qquad (21.3)$$

Determining a free energy change from a cell potential

In an electrical system, work is supplied by the electric current that is pushed along by the potential of the cell. It can be calculated from the equation

$$\text{maximum work} = n\mathscr{F}E_{\text{cell}} \qquad (21.4)$$

where n is the number of moles of electrons transferred; $\mathscr{F}$ is a constant called the **Faraday constant,** which is equal to the number of coulombs of charge equivalent to 1 mol of electrons (9.65×10^4 coulombs per mole of electrons); and E_{cell} is the potential of the cell in volts. To see that Equation 21.4 gives work (which has the units of energy) we can analyze the units. In Equation 21.1 you saw that 1 volt = 1 joule/coulomb. Therefore,

$$\text{maximum work} = (\text{mol } e^-) \times \left(\frac{\text{coulombs}}{\text{mol } e^-}\right) \times \left(\frac{\text{joule}}{\text{coulomb}}\right) = \text{joule}$$

$$\updownarrow \qquad\qquad\qquad \updownarrow \qquad\qquad\qquad \updownarrow$$

$$n \qquad\qquad\qquad \mathscr{F} \qquad\qquad\qquad E_{\text{cell}}$$

Combining Equations 21.3 and 21.4 gives us

$$\Delta G = -n\mathscr{F}E_{\text{cell}} \qquad (21.5)$$

If we are dealing with the *standard* cell potential, we can calculate the *standard* free energy change.

$$\Delta G^\circ = -n\mathscr{F}E^\circ_{\text{cell}} \qquad (21.6)$$

 TOOLS

Faraday constant

More precisely, 1 $\mathscr{F}$ = 96,485 C.

Note that when E_{cell} is positive, ΔG will be negative and the cell reaction will be spontaneous. We used this idea in the preceding section in predicting spontaneous cell reactions.

TOOLS

Free energy changes from cell potentials

Calculate ΔG° for the following reaction, given that its standard cell potential is 0.320 V at 25 °C:

$$NiO_2(s) + 2Cl^-(aq) + 4H^+(aq) \longrightarrow Cl_2(g) + Ni^{2+}(aq) + 2H_2O$$

ANALYSIS: This is a straightforward application of Equation 21.6. Taking the coefficients in the equation to stand for *moles,* 2 mol of Cl^- are oxidized to Cl_2 and 2 mol of electrons are transferred ($n = 2$ mol e^-). We will also use the Faraday constant, 1 $\mathscr{F}$ = 96,500 C/mol e^- (the SI abbreviation for coulomb is C).

$$1 \, \mathscr{F} = \frac{96,500 \text{ C}}{1 \text{ mol } e^-}$$

SOLUTION: Using Equation 21.6, we have

$$\Delta G^\circ = -(2 \text{ mol } e^-) \times \left(\frac{9.65 \times 10^4 \text{ C}}{1 \text{ mol } e^-}\right) \times \left(\frac{0.320 \text{ J}}{\text{C}}\right)$$

$$= -6.18 \times 10^4 \text{ J}$$

$$= -61.8 \text{ kJ}$$

EXAMPLE 21.8

Calculating the Standard Free Energy Change

Is the Answer Reasonable?
Let's do some approximate arithmetic. The Faraday constant equals approximately 100,000, or 10^5. The product $2 \times 0.32 = 0.64$, so ΔG° should be about 0.64×10^5, or 6.4×10^4. The answer seems to be okay.

Equilibrium constants can be calculated from $E°_{cell}$

One useful application of electrochemistry is the determination of equilibrium constants. In Chapter 20 you saw that $\Delta G°$ is related to the equilibrium constant by the expression

$$\Delta G° = -RT \ln K_c$$

where we have used K_c for the equilibrium constant because electrochemical reactions occur in solution. We've seen in this chapter that $\Delta G°$ is also related to $E°_{cell}$

$$\Delta G° = -n\mathscr{F}E°_{cell}$$

Therefore, $E°_{cell}$ and the equilibrium constant are also related. Equating the right sides of the two equations, we have

$$-n\mathscr{F}E°_{cell} = -RT \ln K_c$$

Solving for $E°_{cell}$ gives[4]

Equilibrium constants from cell potentials

$$E°_{cell} = \frac{RT}{n\mathscr{F}} \ln K_c \qquad (21.7)$$

For the units to work out correctly, the value of R must be 8.314 J mol^{-1} K^{-1}, T must be the temperature in kelvins, $\mathscr{F}$ equals 9.65×10^4 C per mole of e^-, and n equals the number of moles of electrons transferred in the reaction.

EXAMPLE 21.9

Calculating Equilibrium Constants from $E°_{cell}$

Calculate K_c for the reaction in Example 21.8.

ANALYSIS: The problem simply involves substituting values into Equation 21.7 and solving for the equilibrium constant.

SOLUTION: The reaction in Example 21.8 has $E°_{cell} = 0.320$ V and $n = 2$. The temperature is 25 °C or 298 K. Let's solve Equation 21.7 for $\ln K_c$ and then substitute values.

$$\ln K_c = \frac{E°_{cell} \, n\mathscr{F}}{RT}$$

Substituting values and using the relationship that 1 V = 1 J C^{-1},

$$\ln K_c = \frac{0.320 \text{ J C}^{-1} \times 2 \times 9.65 \times 10^4 \text{ C mol}^{-1}}{8.314 \text{ J mol}^{-1} \text{ K}^{-1} \times 298 \text{ K}}$$

$$= 24.9$$

Taking the antilogarithm,

$$K_c = e^{24.9} = 7 \times 10^{10}$$

[4]For historical reasons, Equation 21.7 is sometimes expressed in terms of common logs (base 10 logarithms). Natural and common logarithms are related by the equation

$$\ln x = 2.303 \log x$$

For reactions at 25 °C (298 K), all of the constants (R, T, and $\mathscr{F}$) can be combined with the factor 2.303 to give 0.0592 joules/coulomb. Because joules/coulomb equals volts, Equation 21.7 reduces to

$$E°_{cell} = \frac{0.0592 \text{ V}}{n} \log K_c$$

where n is the number of moles of electrons transferred in the cell reaction as it is written.

Is the Answer Reasonable?

As a rough check, we can look at the magnitude of $E°_{cell}$ and apply some simple reasoning. When $E°_{cell}$ is positive, $\Delta G°$ is negative, and in Chapter 20 you learned that when $\Delta G°$ is negative, the reaction proceeds far toward completion when equilibrium is reached. Therefore, we expect that K_c will be large, and that agrees with our answer.

A more complete check would require evaluating the fraction used to compute $\ln K_c$. First, we should check to be sure we've substituted correctly into the equation for $\ln K_c$. Next, we could do some approximate arithmetic to check the value of $\ln K_c$, but it's quicker and less error prone to check the calculation using a calculator.

PRACTICE EXERCISE 10: The calculated standard cell potential for the reaction

$$Cu^{2+}(aq) + 2Ag(s) \rightleftharpoons Cu(s) + 2Ag^+(aq)$$

is $E°_{cell} = -0.46$ V. Calculate K_c for this reaction. Does this reaction proceed very far toward completion?

21.5 ▶ Concentrations in a galvanic cell affect the cell potential

When all of the ion concentrations in a cell are 1.00 M and when the partial pressures of any gases involved in the cell reaction are 1 atm, the cell potential is equal to the standard potential. When the concentrations or pressures change, however, so does the potential. For example, in an operating cell or battery, the potential gradually drops as the reactants are used up and as the cell reaction approaches its natural equilibrium status. When it reaches equilibrium, the potential has dropped to zero—the battery is dead.

The Nernst equation defines the relationship of cell potential to ion concentrations

The effect of concentration on the cell potential can be obtained from thermodynamics. In Chapter 20, you learned that the free energy change is related to the reaction quotient Q by the equation

$$\Delta G = \Delta G° + RT \ln Q$$

Substituting for ΔG and $\Delta G°$ from Equations 21.5 and 21.6 gives

$$-n\mathscr{F}E_{cell} = -n\mathscr{F}E°_{cell} + RT \ln Q$$

Dividing both sides by $-n\mathscr{F}$ gives

$$E_{cell} = E°_{cell} - \frac{RT}{n\mathscr{F}} \ln Q \qquad (21.8)$$

TOOLS
Nernst equation

This equation is commonly known as the **Nernst equation,**[5] named after Walter Nernst, a German chemist and physicist.

In writing the Nernst equation for a galvanic cell, we will construct the mass action expression (from which we calculate Q) using molar concentrations for ions and partial pressures in atmospheres for gases.[6] Thus, for the following cell using a

---

[5]Using common logarithms instead of natural logarithms and calculating the constants for 25 °C gives another form of the Nernst equation that is sometimes used.

$$E_{cell} = E°_{cell} - \frac{0.0592 \text{ V}}{n} \log Q$$

[6]Because of interionic attractions, ions do not always behave as though their concentrations are equal to their molarities. Strictly speaking, therefore, we should use effective concentrations (called *activities*) in the mass action expression. Effective concentrations are difficult to calculate, so for simplicity we will use molarities and accept the fact that our calculations are not entirely accurate.

hydrogen electrode (with the partial pressure of H_2 not necessarily equal to 1 atm) and having the reaction

$$Cu^{2+}(aq) + H_2(g) \longrightarrow Cu(s) + 2H^+(aq)$$

the Nernst equation would be written

$$E_{cell} = E°_{cell} - \frac{RT}{n\mathscr{F}} \ln \frac{[H^+]^2}{[Cu^{2+}]P_{H_2}}$$

EXAMPLE 21.10

Calculating the Effect of Concentration on E_{cell}

Suppose a galvanic cell employs the following half-reactions:

$$Ni^{2+}(aq) + 2e^- \rightleftharpoons Ni(s) \qquad E°_{Ni^{2+}} = -0.25\ V$$

$$Cr^{3+}(aq) + 3e^- \rightleftharpoons Cr(s) \qquad E°_{Cr^{3+}} = -0.74\ V$$

Calculate the cell potential when $[Ni^{2+}] = 1.0 \times 10^{-4}\ M$ and $[Cr^{3+}] = 2.0 \times 10^{-3}\ M$.

ANALYSIS: Because the concentrations are not 1.00 M, we must use the Nernst equation, but first we need the cell reaction. We need it to determine the number of electrons transferred, n, and we need it to determine the correct form of the mass action expression from which we calculate Q. We must also note that the reacting system is heterogeneous; both solid metals and a liquid solution of their dissolved ions are involved. So we have to remember that a mass action expression does not contain concentration terms for solids, such as Ni and Cr.

SOLUTION: Nickel has the more positive (less negative) reduction potential, so its half-reaction will occur as a reduction. This means that chromium will be oxidized. Making electron gain equal to electron loss, the cell reaction is found as follows:

$$3[Ni^{2+}(aq) + 2e^- \longrightarrow Ni(s)] \qquad \text{(reduction)}$$

$$\underline{2[Cr(s) \longrightarrow Cr^{3+}(aq) + 3e^-]} \qquad \text{(oxidation)}$$

$$3Ni^{2+}(aq) + 2Cr(s) \longrightarrow 3Ni(s) + 2Cr^{3+}(aq) \qquad \text{(cell reaction)}$$

The total number of electrons transferred is six, which means $n = 6$. Now we can write the Nernst equation for the system.

$$E_{cell} = E°_{cell} - \frac{RT}{6\mathscr{F}} \ln \frac{[Cr^{3+}]^2}{[Ni^{2+}]^3}$$

The mass action expression, Q, for the cell reaction is

$$Q = \frac{[Cr^{3+}]^2}{[Ni^{2+}]^3}$$

Notice that we've constructed the mass action expression, from which we will calculate the reaction quotient, using the concentrations of the ions raised to powers equal to their coefficients in the net cell reaction, and that we have not included concentration terms for the two solids. This is the procedure we followed for heterogeneous equilibria in Chapter 16.

Next we need $E°_{cell}$. Since Ni^{2+} is reduced,

$$E°_{cell} = E°_{Ni^{2+}} - E°_{Cr^{3+}}$$

$$= (-0.25\ V) - (-0.74\ V)$$

$$= 0.49\ V$$

Now we can substitute this value for $E°_{cell}$ along with $R = 8.314\ J\ mol^{-1}\ K^{-1}$, $T = 298\ K$, $n = 6$, $\mathscr{F} = 9.65 \times 10^4\ C\ mol^{-1}$, $[Ni^{2+}] = 1.0 \times 10^{-4}\ M$, and $[Cr^{3+}] = 2.0 \times 10^{-3}\ M$ into the Nernst equation. This gives

$$E_{cell} = 0.49\ V - \frac{8.314\ J\ mol^{-1}\ K^{-1} \times 298\ K}{6 \times 9.65 \times 10^4\ C\ mol^{-1}} \ln \frac{(2.0 \times 10^{-3})^2}{(1.0 \times 10^{-4})^3}$$

$$= 0.49\ V - (0.00428\ V) \ln (4.0 \times 10^6)$$

$$= 0.49\ V - 0.0651\ V$$

$$= 0.42\ V$$

The potential of the cell is expected to be 0.42 V.

Is the Answer Reasonable?
There's no simple way to check the answer. However, there are certain critical points to look over. First, be sure you've combined the half-reactions correctly to give the balanced cell reaction, because we need the coefficients of the equation to obtain the correct superscripts in the Nernst equation. Then, be sure you've used the Kelvin temperature, $R = 8.314$ J mol^{-1} K^{-1}, and made the other substitutions correctly.

PRACTICE EXERCISE 11: In a certain zinc–copper cell,

$$Zn(s) + Cu^{2+}(aq) \longrightarrow Zn^{2+}(aq) + Cu(s)$$

the ion concentrations are $[Cu^{2+}] = 0.0100$ M and $[Zn^{2+}] = 1.0$ M. What is the cell potential at 25 °C?

Experimental cell potentials can be used to determine ion concentrations

One of the principal uses of the relationship between concentration and cell potential is in the measurement of concentrations. Experimental determination of cell potentials combined with modern developments in electronics has provided a means of monitoring and analyzing the concentrations of all sorts of substances in solution, even some that are not themselves ionic and that are not involved directly in electrochemical changes. In fact, the operation of a pH meter relies on the logarithmic relationship between hydrogen ion concentration and the potential of a special kind of electrode, as described in Facets of Chemistry 21.2. Example 21.11 illustrates how the Nernst equation is applied in determining concentrations.

To measure the concentration of Cu^{2+} in a large number of samples of water in which the copper ion concentration is expected to be quite small, an electrochemical cell was assembled that consists of a silver electrode, dipping into a 1.00 M solution of $AgNO_3$, connected by a salt bridge to a second half-cell containing a copper electrode. The copper half-cell was then filled with one water sample after another, with the cell potential being measured for each sample. In the analysis of one sample, the cell potential at 25 °C was measured to be 0.62 V. The copper electrode was observed to carry a negative charge, so it served as the anode. What was the concentration of copper ion in this sample?

EXAMPLE 21.11

Using the Nernst Equation to Determine Concentrations

ANALYSIS: In this problem, we've been given the cell potential, E_{cell}, and we can calculate E°_{cell} from the reduction potentials in Table 21.1. The unknown quantity is one of the concentration terms in the Nernst equation.

SOLUTION: The first step is to write the proper equation for the cell reaction, because we need it to compute E°_{cell} and to construct the mass action expression for use in the Nernst equation. Because copper is the anode, it is being oxidized. This also means that Ag^+ is being reduced. Therefore, the equation for the cell reaction is

$$Cu(s) + 2Ag^+(aq) \longrightarrow Cu^{2+}(aq) + 2Ag(s)$$

Two electrons are transferred, so $n = 2$ and the Nernst equation is

$$E_{cell} = E^\circ_{cell} - \frac{RT}{2\mathscr{F}} \ln \frac{[Cu^{2+}]}{[Ag^+]^2}$$

The mass action expression, Q, for this reaction is

$$Q = \frac{[Cu^{2+}]}{[Ag^+]^2}$$

FACETS OF CHEMISTRY 21.2

Measurement of pH

The Nernst equation tells us that the potential of a cell changes with the concentrations of the ions involved in the cell reaction. As noted in the conclusion to Example 21.11, one of the most useful results of this is that cell potential measurements provide a way indirectly to measure and monitor ion concentrations in aqueous solutions.

Scientists have developed a number of specialized electrodes that can be dipped into one solution after another and whose potential is affected in a reproducible way by the concentration of only one species in the solution. An example is the glass electrode of a pH meter for the measurement of pH (see Figure 1). The electrode is constructed from a hollow glass tube sealed with a special thin-walled glass membrane at the bottom. The tube is filled partway with a dilute solution of HCl, and dipping into this HCl solution is a silver wire coated with a layer of silver chloride. The potential of the electrode is controlled by the difference between the hydrogen ion concentrations inside and outside the thin glass membrane at the bottom. Since the H+ concentration inside the electrode is constant, the electrode's potential varies only with the concentration of H+ in the solution outside. In fact, because concentrations appear in the logarithm term of the Nernst equation, the cell potential is proportional to the logarithm of the H+ concentration and therefore to pH of the outside solution.

A glass electrode is always used with another reference electrode whose potential is a constant, and this gives a galvanic cell whose potential depends on the pH of the solution into which the electrodes are immersed. The measurement of the potential of this galvanic cell and the translation of the potential into pH are the job of a pH meter such as the one shown in Figure 17.4 on page 762.

As in Figure 17.4, sometimes the reference and glass electrodes are combined into one physical unit, so they appear to be just one electrode.

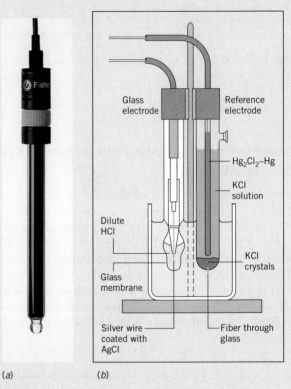

(a)   (b)

FIGURE 1 *Glass electrode.* (*a*) Photograph of a typical glass electrode. (*b*) Cutaway view of the construction of a glass electrode and a reference electrode.

The value of E°_{cell} can be obtained from the tabulated standard reduction potentials in Table 21.1. Following our usual procedure and recognizing that silver ion is reduced,

$$E^\circ_{cell} = E^\circ_{Ag^+} - E^\circ_{Cu^{2+}}$$
$$= (0.80\,V) - (0.34\,V)$$
$$= 0.46\,V$$

Now we can substitute values into the Nernst equation and solve for the concentration ratio in the mass action expression.

$$0.62\,V = 0.46\,V - \frac{8.314\,J\,mol^{-1}\,K^{-1} \times 298\,K}{2 \times 9.65 \times 10^4\,C\,mol^{-1}} \ln \frac{[Cu^{2+}]}{[Ag^+]^2}$$

Solving for $\ln([Cu^{2+}]/[Ag^+]^2)$ gives

$$\ln \frac{[Cu^{2+}]}{[Ag^+]^2} = -12$$

Taking the antilog gives us the value of the mass action expression.

$$\frac{[Cu^{2+}]}{[Ag^+]^2} = 6 \times 10^{-6}$$

Since we know that the concentration of Ag^+ is 1.00 M, we can now solve for the Cu^{2+} concentration.

$$\frac{[Cu^{2+}]}{(1.00)^2} = 6 \times 10^{-6}$$

$$[Cu^{2+}] = 6 \times 10^{-6}\ M$$

Is the Answer Reasonable?

All we can do easily is check to be sure we've written the correct chemical equation, on which all the rest of the solution to the problem rests. Be careful about algebraic signs and that you select the proper value for R and the temperature in kelvins. Also, notice that we first solved for the logarithm of the ratio of concentration terms. Then, after taking the (natural) antilogarithm, we substitute the known value for $[Ag^+]$ and solve for $[Cu^{2+}]$.

As a final point, notice that the Cu^{2+} concentration is indeed very small and that it can be obtained very easily by simply measuring the potential generated by the electrochemical cell. Determining the concentrations in many samples is also very simple—just change the water sample and measure the potential again.

The ease of such operations and the fact that they lend themselves well to automation and computer analysis make electrochemical analyses especially attractive to scientists.

PRACTICE EXERCISE 12: In the analysis of two other water samples by the procedure described in Example 21.11, cell potentials (E_{cell}) of 0.57 V and 0.82 V were obtained. Calculate the Cu^{2+} ion concentration in each of these samples.

PRACTICE EXERCISE 13: A galvanic cell was constructed by connecting a nickel electrode that was dipped into 1.20 M $NiSO_4$ solution to a chromium electrode that was dipped into a solution containing Cr^{3+} at an unknown concentration. The potential of the cell was measured to be 0.552 V, with the chromium serving as the anode. The standard cell potential for this system was determined to be 0.487 V. What was the concentration of Cr^{3+} in the solution of unknown concentration?

21.6 ▶ Batteries are practical examples of galvanic cells

One of the most familiar uses of galvanic cells is the generation of portable electrical energy.[7] In this section we will briefly examine some of the more common present-day uses of galvanic cells, popularly called *batteries*, and some promising future applications. These devices are classified as being either **primary cells** (cells not designed to be recharged; they are discarded after their energy is depleted) or **secondary cells** (cells designed for repeated use; they are able to be recharged).

The lead storage battery

The common **lead storage battery** used to start an automobile is composed of a number of secondary cells, each having a potential of about 2 V, that are connected in series so that their voltages are additive. Most automobile batteries contain six such cells and give about 12 V, but 6, 24, and 32 V batteries are also available.

A typical lead storage battery is shown in Figure 21.8. The anode of each cell is composed of a set of lead plates, the cathode consists of another set of plates that

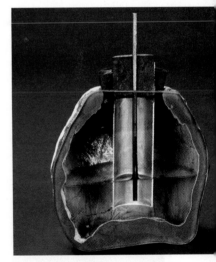

The oldest known electric battery in existence, discovered in 1938 in Baghdad, Iraq, consists of a copper tube surrounding an iron rod. If filled with an acidic liquid such as vinegar, the cell could produce a small electric current.

[7]Strictly speaking, a cell is a single electrochemical unit consisting of a cathode and an anode. A battery is a collection of cells connected in series.

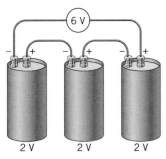

2 V   2 V   2 V

If three 2 volt cells are connected in series, their voltages are additive to provide a total of 6 volts. In today's autos, 12 volt batteries containing six cells are the norm.

FIGURE 21.8 *Lead storage battery.* A 12 volt lead storage battery, such as those used in most automobiles, consists of six cells like the one shown here. Notice that the anode and cathode each consist of several plates connected together. This allows the cell to produce the large currents necessary to start a car.

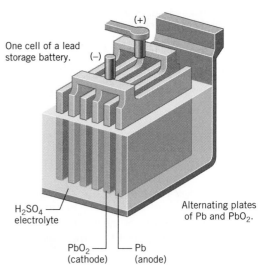

One cell of a lead storage battery.

H_2SO_4 electrolyte

PbO_2 (cathode)   Pb (anode)

Alternating plates of Pb and PbO_2.

hold a coating of PbO_2, and the electrolyte is sulfuric acid. When the battery is discharging the electrode reactions are

$$PbO_2(s) + 3H^+(aq) + HSO_4^-(aq) + 2e^- \longrightarrow PbSO_4(s) + 2H_2O \qquad \text{(cathode)}$$

$$Pb(s) + HSO_4^-(aq) \longrightarrow PbSO_4(s) + H^+(aq) + 2e^- \quad \text{(anode)}$$

The net reaction taking place in each cell is

$$PbO_2(s) + Pb(s) + \underbrace{2H^+(aq) + 2HSO_4^-(aq)}_{2H_2SO_4} \longrightarrow 2PbSO_4(s) + 2H_2O$$

As the cell discharges, the sulfuric acid concentration decreases, which provides a convenient means for checking the state of the battery. Because the density of a sulfuric acid solution decreases as its concentration drops, the concentration can be determined very simply by measuring the density with a **hydrometer,** which consists of a rubber bulb that is used to draw the battery fluid into a glass tube containing a float (see Figure 21.9). The depth to which the float sinks is inversely proportional to the density of the liquid—the deeper the float sinks, the lower is the density of the acid and the weaker is the charge on the battery. The narrow neck of the float is usually marked to indicate the state of charge of the battery.

The principal advantage of the lead storage battery is that the cell reactions that occur spontaneously during discharge can be reversed by the application of a voltage from an external source. In other words, the battery can be recharged by electrolysis. The reaction for battery recharge is

$$2PbSO_4(s) + 2H_2O \xrightarrow{\text{electrolysis}} PbO_2(s) + Pb(s) + 2H^+(aq) + 2HSO_4^-(aq)$$

Disadvantages of the lead storage battery are that it is very heavy and that its corrosive sulfuric acid can spill. The most modern lead storage batteries use a lead–calcium alloy as the anode. This reduces the need to have the individual cells vented, and the battery can be sealed, thereby preventing spillage of the electrolyte.

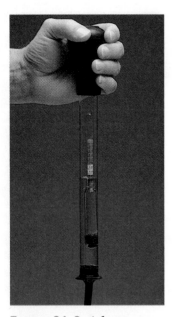

FIGURE 21.9 *A battery hydrometer.* Battery acid is drawn into the glass tube. The depth to which the float sinks is inversely proportional to the concentration of the acid and, therefore, to the state of charge of the battery.

The zinc–manganese dioxide cell

The ordinary, relatively inexpensive 1.5 V dry cell, is the **zinc–manganese dioxide cell,** or **Leclanché cell** (named after its inventor, George Leclanché). It is a primary cell used to power flashlights, tape recorders, and the like, but it is not really dry (see Figure 21.10). Its outer shell is made of zinc, which serves as the anode. The exposed outer surface at the bottom of the cell is the negative end of the battery. The

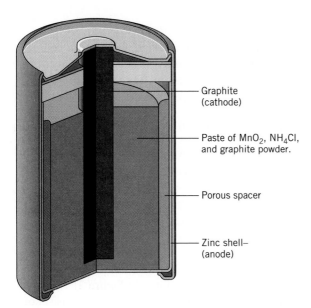

FIGURE 21.10 *A cut-away view of a zinc–manganese dioxide dry cell (Leclanché cell).*

Graphite (cathode)

Paste of MnO_2, NH_4Cl, and graphite powder.

Porous spacer

Zinc shell– (anode)

cathode—the positive terminal of the battery—consists of a carbon (graphite) rod surrounded by a moist paste of graphite powder, manganese dioxide, and ammonium chloride.

The anode reaction is simply the oxidation of zinc.

$$Zn(s) \longrightarrow Zn^{2+}(aq) + 2e^- \qquad \text{(anode)}$$

The cathode reaction is complex, and a mixture of products is formed. One of the major reactions is

$$2MnO_2(s) + 2NH_4^+(aq) + 2e^- \longrightarrow Mn_2O_3(s) + 2NH_3(aq) + H_2O \qquad \text{(cathode)}$$

The ammonia that forms at the cathode reacts with some of the Zn^{2+} produced from the anode to form a complex ion, $Zn(NH_3)_4^{2+}$. Because of the complexity of the cathode half-cell reaction, no simple overall cell reaction can be written.

The major advantages of the dry cell are its relatively low price and the fact that it normally works without leaking. One disadvantage is that the cell loses its ability to function rather rapidly under heavy current drain, a loss that occurs because ions formed in the cell reaction cannot easily diffuse away from the electrodes. If unused for a while, however, these batteries spontaneously rejuvenate somewhat as the products diffuse away from the electrodes. Another disadvantage of the zinc–carbon dry cell is that it can't be recharged. Devices that claim to recharge such cells simply drive the reaction products away from the electrodes, which allows the cell to function again. This doesn't work very many times, however. Before too long the zinc casing develops holes and the battery must be discarded.

A more popular version of the Leclanché battery uses a basic, or *alkaline* electrolyte and is called an **alkaline battery** or **alkaline dry cell.** It too uses Zn and MnO_2 as reactants, but under basic conditions (Figure 21.11). The half-cell reactions are

$$Zn(s) + 2OH^-(aq) \longrightarrow ZnO(s) + H_2O + 2e^- \qquad \text{(anode)}$$

$$\underline{2MnO_2(s) + H_2O + 2e^- \longrightarrow Mn_2O_3(s) + 2OH^-(aq)} \qquad \text{(cathode)}$$

$$Zn(s) + 2MnO_2(s) \longrightarrow ZnO(s) + Mn_2O_3(s) \qquad \text{(net cell reaction)}$$

and the voltage is about 1.54 V. It has a longer shelf life and is able to deliver higher currents for longer periods than the less expensive zinc–carbon cell. The reason is that there are no ions produced around the electrodes that have to diffuse away to maintain the flow of current.

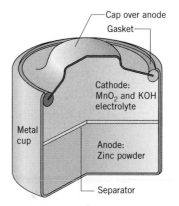

Cap over anode

Gasket

Cathode: MnO_2 and KOH electrolyte

Metal cup

Anode: Zinc powder

Separator

FIGURE 21.11 *A simplified diagram of an alkaline zinc–manganese dioxide dry cell.*

The alkaline battery is also a primary cell.

The nickel–cadmium storage cell

The **nickel–cadmium storage cell,** or **nicad battery,** is a secondary cell that produces a potential of about 1.4 V, which is slightly lower than that of the zinc–carbon cell. The electrode reactions in the cell during discharge are

$$Cd(s) + 2OH^-(aq) \longrightarrow Cd(OH)_2(s) + 2e^- \qquad \text{(anode)}$$

$$\underline{NiO_2(s) + 2H_2O + 2e^- \longrightarrow Ni(OH)_2(s) + 2OH^-(aq) \qquad \text{(cathode)}}$$

$$Cd(s) + NiO_2(s) + 2H_2O \longrightarrow Ni(OH)_2(s) + Cd(OH)_2(s) \quad \text{(net cell reaction)}$$

The nickel–cadmium battery can be recharged, in which case the anode and cathode reactions above are reversed to remake the reactants. The battery also can be sealed to prevent leakage, which is particularly important in electronic devices.

Nickel–cadmium batteries work especially well in applications such as portable power tools, CD players, and even electric cars. They have a high *energy density* (available energy per unit volume), they are able to release the energy quickly, and they can be rapidly recharged.

The construction of a modern nicad battery is illustrated in Figure 21.12. It consists of an electrode "sandwich" rolled up like a jelly roll. The cathode consists of a metal support with a porous coating of nickel that holds the NiO_2, a compound of nickel in the +4 oxidation state. The anode consists of a cadmium coating over a steel mesh. This construction produces large electrode surface areas, which allows the cell to deliver large amounts of energy rapidly. Disposal of nicad batteries poses an environmental hazard because of the toxicity of cadmium. Efforts are underway to encourage the recycling of these batteries, rather than disposal in landfills.

If a battery can supply large amounts of energy and is contained in a small package, it will have a desirably high energy density.

The silver oxide battery

This miniature battery has often been used in wristwatches, calculators, and auto-exposure cameras. It was developed to replace earlier mercury batteries, which posed an environmental hazard if not disposed of properly. The anode material in

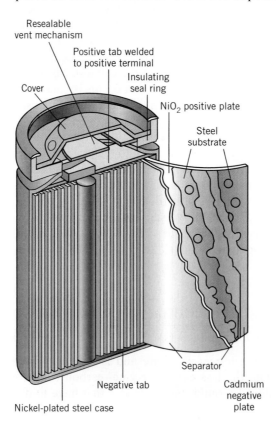

FIGURE 21.12 *Cutaway view of the construction of a nickel–cadmium battery.*

the cell is zinc, and the substance that reacts at the cathode is silver oxide, Ag_2O. The electrolyte is basic, and the half-reactions that occur at the electrodes are

$$Zn(s) + 2OH^-(aq) \longrightarrow Zn(OH)_2(s) + 2e^- \qquad \text{(anode)}$$

$$Ag_2O(s) + H_2O + 2e^- \longrightarrow 2Ag(s) + 2OH^-(aq) \qquad \text{(cathode)}$$

The overall net cell reaction is

$$Zn(s) + Ag_2O(s) + H_2O \longrightarrow Zn(OH)_2(s) + 2Ag(s)$$

As you might expect, these cells are quite expensive because they contain silver. They generate a potential of about 1.5 V.

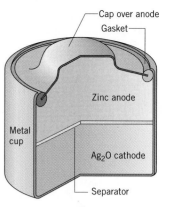

A silver oxide battery.

Modern high-performance batteries

As everyone knows, the field of electronics has undergone fantastic changes in the past decade or two. Cell phones, portable CD players, electronic cameras, laptop computers, electronic calculators, cordless tools, and even heart pacemakers have been made possible by advances in battery technology. Developers of batteries are concerned with a number of properties, including shelf life, rate of energy output, **energy density** (the ratio of the energy available to the volume of the battery), and the **specific energy** (the ratio of available energy to weight). We will discuss two types of batteries that have become popular in modern applications—nickel–metal hydride batteries and lithium batteries.

Nickel–metal hydride battery

These secondary cells, which are often referred to as Ni–MH batteries, have been used extensively in recent years to power devices such as cell phones, camcorders, and even electric vehicles. They are similar in many ways to the alkaline nickel–cadmium cells discussed above, except for the anode reactant, which is hydrogen. At first, this seems odd, because hydrogen is a gas at room temperature and atmospheric pressure. However, in the late 1960s it was discovered that some metal alloys [e.g., $LaNi_5$ (an alloy of lanthanum and nickel) and Mg_2Ni (an alloy of magnesium and nickel)] have the ability to absorb and hold substantial amounts of hydrogen and that the hydrogen could be made to participate in reversible electrochemical reactions. The term *metal hydride* has come to be used to describe the hydrogen-holding alloy.

There are compounds of hydrogen with metals such as sodium that actually contain the *hydride ion*, H^-. The metal "hydrides" described here are not of that type.

The cathode in the Ni–MH cell is NiO(OH), a compound of nickel in the +3 oxidation state, and the electrolyte is a solution of KOH. A diagram of a cylindrical cell is shown in Figure 21.13. Using the symbol MH to stand for the metal hydride, the reactions in the cell during discharge are

Anode reaction:

$$MH(s) + OH^-(aq) \longrightarrow M(s) + H_2O + e^-$$

Cathode reaction:

$$NiO(OH)(s) + H_2O + e^- \longrightarrow Ni(OH)_2(s) + OH^-(aq)$$

Cell reaction:

$$MH(s) + NiO(OH)(s) \longrightarrow Ni(OH)_2(s) + M(s) \qquad E^\circ_{cell} = 1.35 \text{ V}$$

When the cell is recharged, these reactions are reversed.

The principal advantage of the Ni–MH cell over the Ni–Cd cell is that it can store about 50% more energy in the same volume. This means, for example, that comparing cells of the same size and weight, a nickel–metal hydride cell can power a laptop computer or a cell phone about 50% longer than can a nickel–cadmium cell.

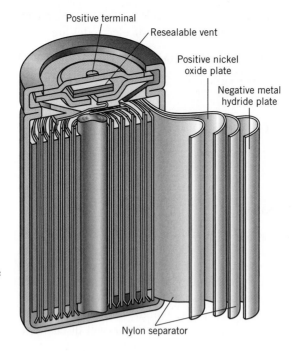

Positive terminal

Resealable vent

Positive nickel oxide plate

Negative metal hydride plate

Nylon separator

FIGURE 21.13 *Cutaway view of a nickel–metal hydride cell.* The electrode sandwich is rolled up "jelly roll" style, which yields a large effective electrode area. This enables the cell to deliver large amounts of energy quickly.

Although they have advantages, Ni–MH cells are not without problems. For example, they undergo relatively rapid self-discharge, which means that they lose their charge relatively rapidly even when not in use. Early models lost as much as 5% of their capacity per day, although refinements in their design have improved on this substantially.

Lithium batteries

If you look at the location of lithium in the table of reduction potentials (Table 21.1), you will see that it has the most negative reduction potential of any metal. This means that lithium is very easily oxidized electrochemically, and its large negative reduction potential suggests it has a lot of appeal as an anode material. Furthermore, lithium is a very lightweight metal, so a cell employing lithium as a reactant would also be lightweight.

The major problem with using lithium in a galvanic cell is the fact that lithium reacts vigorously with water to produce hydrogen gas and lithium hydroxide.

$$2\text{Li}(s) + 2\text{H}_2\text{O} \longrightarrow 2\text{LiOH}(aq) + \text{H}_2(g)$$

Therefore, to employ lithium in a galvanic cell scientists had to find a way to avoid aqueous electrolytes, which meant that a nonaqueous medium had to be used. This became possible in the 1970s with the introduction of organic solvents and solvent mixtures that were able dissolve certain lithium salts and thereby serve as electrolytes.

A number of different lithium cells have been studied, and some have achieved commercial success. They fall into two categories, primary batteries that can be used once and then discarded when fully discharged, and rechargeable cells.

One of the most common nonrechargeable cells is the **lithium–manganese dioxide battery,** which accounts for about 80% of all primary lithium cells. This cell uses a solid lithium anode and a cathode made of a heat-treated manganese(IV) oxide, MnO_2. The electrolyte is a mixture of propylene carbonate and dimethoxyethane (see structures in the margin) containing a dissolved lithium salt

propylene carbonate

dimethoxyethane

such as $LiClO_4$. The cell reactions are as follows (superscripts are the oxidation numbers of the manganese):

$$Li \longrightarrow Li^+ + e^- \qquad \text{(anode)}$$

$$Mn^{IV}O_2 + Li^+ + e^- \longrightarrow Mn^{III}O_2(Li^+) \qquad \text{(cathode)}$$

$$Li + Mn^{IV}O_2 \longrightarrow Mn^{III}O_2(Li^+) \qquad \text{(net cell reaction)}$$

This cell produces a no-load voltage of 3.4 V (which drops to about 2.8 V during use). This voltage is more than twice that of an alkaline dry cell, and because of the light weight of the lithium, it produces more than twice as much energy for a given weight. The cell also has an extremely long shelf life of about 7 yr, so the battery doesn't go dead if it isn't used. The battery comes in two basic configurations. Thin, flat, coin-shaped cells are often used for memory backup and in watches. Wound cells are configured in a manner similar to cylindrical Ni–MH cells and, because of the larger electrode area, are able to deliver more power. They are used in applications that require a higher current drain or energy pulses (e.g., photoflash).

Lithium ion cells

Attempts to make rechargeable lithium batteries containing lithium metal electrodes have been plagued with safety problems. The major problem is that upon repeated charging and discharging of the cell, the lithium tends to form long whiskers that reach across the separator and eventually short out the cell, sometimes with catastrophic results. The rapid rise in temperature would tend to ignite the organic electrolyte fluid, causing the battery to vent with flames. (That's definitely not what you want to happen with a cell phone at your ear!)

Success came with the development of the **lithium ion cell,** which doesn't use metallic lithium, but lithium ions instead. In fact, the cell's operation doesn't actually involve true oxidation and reduction. Instead, it uses the transport of Li^+ ions through the electrolyte from one electrode to the other accompanied by the transport of electrons through the external circuit to maintain charge balance. Here's how it works.

It was discovered that Li^+ ions are able to slip between layers of atoms in certain crystals (a process called **intercalation**). Graphite is one such substance. It consists of layers of fused hexagonal rings of carbon atoms (Figure 13.48, page 588). The most commonly used other compound able to intercalate Li^+ ions is $LiCoO_2$. These are the materials used to make the electrodes.

When the cell is constructed, it is in its "uncharged" state, with no Li^+ ions between layers of carbon atoms in the graphite. When the cell is charged (Figure 21.14a), Li^+ ions leave $LiCoO_2$ and travel through the electrolyte to the graphite (represented below by the formula C_6).

Initial charging: $LiCoO_2 + C_6 \longrightarrow Li_{1-x}CoO_2 + Li_xC_6$

When the cell spontaneously discharges to provide electrical power (Figure 21.14b), Li^+ ions move back through the electrolyte to the cobalt oxide while electrons move through the external circuit from the graphite electrode to the cobalt oxide electrode. If we represent the amount of Li^+ transferring by y, the discharge "reaction" is

Discharge: $Li_{1-x}CoO_2 + Li_xC_6 \longrightarrow Li_{1-x+y}CoO_2 + Li_{x-y}C_6$

Thus, the charging and discharging cycles simply sweep Li^+ ions back and forth between the two electrodes, with the electrons flowing through the external circuit to keep the charge in balance.

Two types of lithium ion cells have been developed. The one most commonly used today in mobile phones, camcorders, and laptop computers employs a liquid electrolyte (usually containing $LiPF_6$, a compound containing Li^+ and PF_6^- ions).

Pronounced in-*terca-la*-tion.

Rhymes with *percolate*.

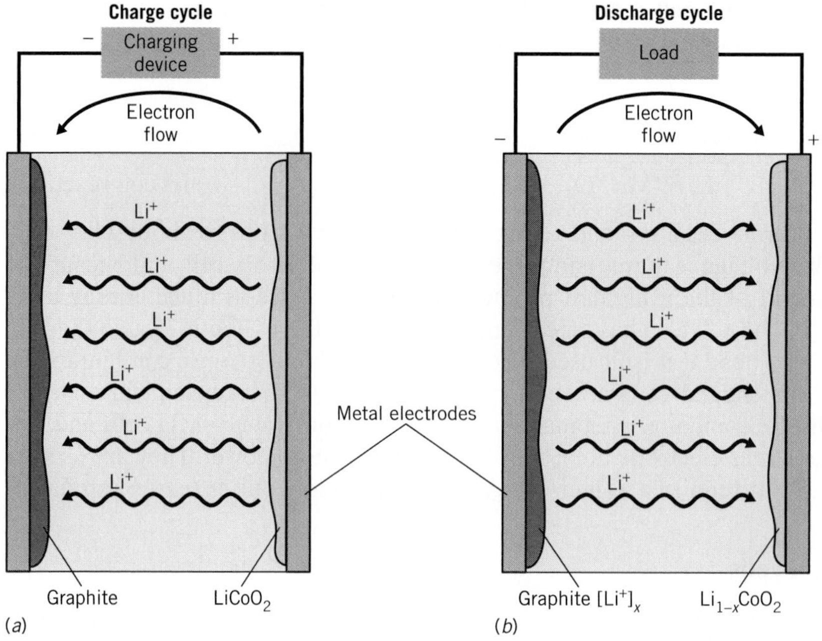

Charge cycle

Charging device

Electron flow

Li^+
Li^+
Li^+
Li^+
Li^+
Li^+

Graphite $LiCoO_2$

(a)

Discharge cycle

Load

Electron flow

Li^+
Li^+
Li^+
Li^+
Li^+
Li^+

Metal electrodes

Graphite $[Li^+]_x$ $Li_{1-x}CoO_2$

(b)

FIGURE 21.14 *Lithium ion cell.* (*a*) During the charging cycle, an external voltage forces electrons through the external circuit and causes lithium ions to travel from the $LiCoO_2$ electrode to the graphite electrode. (*b*) During discharge, the lithium ions spontaneously migrate back to the $LiCoO_2$ electrode, and electrons flow through the external circuit to balance the charge.

The cell generates about 3.7 V, which means three Ni–Cd cells connected in series would be required to form an equivalent battery pack. In addition, the lithium ion cell has twice the energy density of a standard Ni–Cd cell. The second type of cell, which is under development, uses a thin, flexible, conducting polymer film in place of the liquid electrolyte. Such cells can be made very thin (less than 1 mm) and can be made into almost any shape.

Fuel cells

The galvanic cells we've discussed until now can only produce power for a limited time because the electrode reactants are eventually depleted. Fuel cells are different; they are electrochemical cells in which the electrode reactants are supplied continuously and are able to operate without theoretical limit as long as the supply of reactants is maintained. This makes fuel cells an attractive source of power where long-term generation of electrical energy is required.

Figure 21.15 illustrates an early design of a hydrogen–oxygen **fuel cell.** The electrolyte, a hot (~200 °C) concentrated solution of potassium hydroxide in the center compartment, is in contact with two porous electrodes that contain a catalyst (usually platinum) to facilitate the electrode reactions. Gaseous hydrogen and oxygen under pressure are circulated so as to come in contact with the electrodes. At the cathode, oxygen is reduced.

$$O_2(g) + 2H_2O + 4e^- \longrightarrow 4OH^-(aq) \qquad \text{(cathode)}$$

At the anode, hydrogen is oxidized to water.

$$H_2(g) + 2OH^-(aq) \longrightarrow 2H_2O + 2e^- \qquad \text{(anode)}$$

Part of the water formed at the anode leaves as steam mixed with the circulating hydrogen gas. The net cell reaction, after making electron loss equal to electron gain, is

$$2H_2(g) + O_2(g) \longrightarrow 2H_2O$$

A major advantage of the fuel cell is that there is no electrode material to be replaced, as there is in an ordinary battery. The fuel can be fed in continuously to produce power. In fact, hydrogen–oxygen fuel cells were used in the Gemini and Apollo missions and other space programs for just this reason.

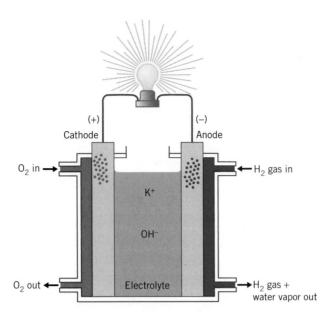

FIGURE 21.15 *A hydrogen–oxygen fuel cell.*

One reason fuel cells are so appealing is their thermodynamic efficiency. In a fuel cell, the net reaction is equivalent to combustion. The production of usable energy by the direct combustion of fuels is an extremely inefficient process. Largely because of the constraints set by fundamental thermodynamic principles, modern electric power plants are unable to harness more than about 35 to 40% of the potential energy in oil, coal, or natural gas. A gasoline or diesel engine has an efficiency of only about 25 to 30%. The rest of the energy is lost to the surroundings as heat, which is the reason why a vehicle must have an effective cooling system.

Fuel cells "burn" fuel under conditions that are much more nearly thermodynamically reversible than simple combustion. They therefore achieve much greater efficiencies—75% is quite feasible. In addition, hydrogen–oxygen fuel cells are essentially pollution-free. The only product formed by the cell is water.

Advances in fuel cell development have led to lower operating temperatures and the ability to use methanol, a liquid at room temperature, as a source of hydrogen. The hydrogen is generated from methanol vapor by a catalytic process. The net reaction is

$$CH_3OH(g) + H_2O(g) \longrightarrow CO_2(g) + 3H_3(g)$$
methanol

With some modifications, equipment used to dispense gasoline could be used to dispense methanol, thereby solving the problem of fuel distribution for fuel-cell-powered vehicles.

DaimlerChrysler recently demonstrated the feasibility of such a system by driving its NECAR 5 fuel-cell-powered automobile 3000 miles across the United States in 16 days. An onboard fuel reformer extracts hydrogen from methanol to feed the fuel cells.

The use of fuel cells is still in its infancy, but will surely grow. It has been estimated that by 2010, the world fuel cell market will exceed $23 billion.

21.7 ▶ Electrolysis uses electrical energy to cause chemical reactions

In our preceding discussions, we've examined how spontaneous redox reactions can be used to generate electrical energy. We now turn our attention to the opposite process, the use of electrical energy to force nonspontaneous redox reactions to occur. In fact, these are precisely the kinds of reactions that take place when rechargeable batteries are charged.

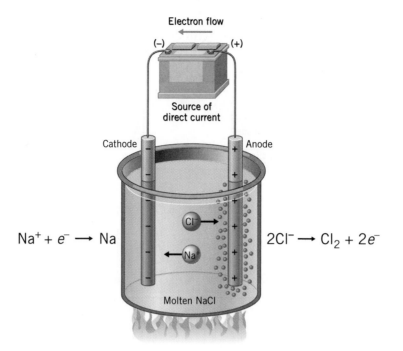

Figure 21.16 *Electrolysis of molten sodium chloride.* In this electrolysis cell, the passage of an electric current decomposes molten sodium chloride into metallic sodium and gaseous chlorine. Unless the products are kept apart, they react on contact to re-form NaCl.

To perform electrolysis, we must use direct current in which electrons move in only one direction, not in the oscillating, back-and-forth pattern of alternating current.

When electricity is passed through a molten (melted) ionic compound or through a solution of an electrolyte, a chemical reaction occurs that we call **electrolysis.** A typical electrolysis apparatus, called an **electrolysis cell** or **electrolytic cell,** is shown in Figure 21.16. This particular cell contains molten sodium chloride. (A substance undergoing electrolysis must be molten or in solution so its ions can move freely and conduction can occur.) *Inert electrodes*—electrodes that won't react with the molten NaCl—are dipped into the cell and then connected to a source of direct current (DC) electricity.

The DC source serves as an "electron pump," pulling electrons away from one electrode and pushing them through the external wiring onto the other electrode. The electrode from which electrons are removed becomes positively charged, while the other electrode becomes negatively charged. When electricity starts to flow, chemical changes begin to happen. At the positive electrode, oxidation occurs as electrons are pulled away from negatively charged chloride ions. Because of the nature of the chemical change, therefore, *the positive electrode becomes the anode.* The DC source pumps the electrons through the external electrical circuit to the negative electrode. Here, reduction takes place as the electrons are forced onto positively charged sodium ions, so *the negative electrode is the cathode.*

The chemical changes that occur at the electrodes can be described by chemical equations.

At the melting point of NaCl, 801 °C, metallic sodium is a liquid.

$$\text{Na}^+(l) + e^- \longrightarrow \text{Na}(l) \qquad \text{(cathode)}$$

$$2\text{Cl}^-(l) \longrightarrow \text{Cl}_2(g) + 2e^- \qquad \text{(anode)}$$

As in a galvanic cell, the overall reaction that takes place in the electrolysis cell is called the *cell reaction.* To obtain it, we add the individual electrode half-reactions together, making sure that the number of electrons gained in one half-reaction equals the number lost in the other.

$$2\text{Na}^+(l) + 2e^- \longrightarrow 2\text{Na}(l) \qquad \text{(cathode)}$$

$$2\text{Cl}^-(l) \longrightarrow \text{Cl}_2(g) + 2e^- \qquad \text{(anode)}$$

$$\overline{2\text{Na}^+(l) + 2\text{Cl}^-(l) + \cancel{2e^-} \longrightarrow 2\text{Na}(l) + \text{Cl}_2(g) + \cancel{2e^-}} \qquad \text{(cell reaction)}$$

As you know, table salt is quite stable. It doesn't normally decompose because the reverse reaction, the reaction of sodium and chlorine to form sodium chloride, is highly spontaneous. Therefore, we often write the word *electrolysis* above the arrow in the equation to show that electricity is the driving force for this otherwise nonspontaneous reaction.

$$2Na^+(l) + 2Cl^-(l) \xrightarrow{\text{electrolysis}} 2Na(l) + Cl_2(g)$$

Electrolytic and galvanic cells compared

In a galvanic cell, the spontaneous cell reaction deposits electrons on the anode and removes them from the cathode. As a result, the anode carries a slight negative charge and the cathode a slight positive charge. In an *electrolysis cell,* the situation is reversed. Here, the oxidation at the anode must be forced to occur, which requires that the anode be positive so it can remove electrons from the reactant at that electrode. On the other hand, the cathode must be made negative so it can force the reactant at the electrode to accept electrons.

Electrolytic Cell	Galvanic Cell
Cathode is negative (reduction).	Cathode is positive (reduction).
Anode is positive (oxidation).	Anode is negative (oxidation).

Even though the charges on the cathode and anode differ between electrolytic cells and galvanic cells, the ions in solution always move in the same direction. A cation is a positive ion that always moves away from the anode toward the cathode. In both types of cells, positive ions move toward the cathode. They are attracted there by the negative charge on the cathode in an electrolysis cell; they diffuse toward the cathode in our galvanic cell to balance the charge of negative ions left behind when ions are reduced. Similarly, anions are negative ions that move away from the cathode and toward the anode. They are attracted to the positive anode in an electrolysis cell, and they diffuse toward the anode in our galvanic cell to balance the charge of the positive ions entering the solution.

By agreement among scientists, the names *anode* and *cathode* are assigned according to the nature of the reaction taking place at the electrode. If the reaction is oxidation, the electrode is called the anode; if it's reduction, the electrode is called the cathode.

Redox must occur for conduction to continue in an electrolytic cell

The electrical conductivity of a molten salt or a solution of an electrolyte is possible only because of the reactions that take place at the surface of the electrodes. For example, when charged electrodes are dipped into molten NaCl they become surrounded by a layer of ions of the opposite charge. Let's look closely at what happens at one of the electrodes, say the anode (Figure 21.17). Here the positive charge of the electrode attracts negative Cl^- ions, which form a coating on the electrode's surface. The charge on the anode pulls electrons from the ions, causing them to be oxidized and changing them into neutral Cl atoms that join to become Cl_2 molecules. Because the molecules are neutral, they are not held by the electrode and so move away from the electrode's surface. Their places are quickly taken by negative ions from the surrounding liquid, which tends to leave the surrounding liquid positively charged. Other negative ions from farther away move toward the anode to keep the liquid there electrically neutral. In this way, negative ions gradually migrate toward the anode. By a similar process, positive ions diffuse through the liquid toward the negatively charged cathode, where they become reduced.

Recall that the terms *cation* and *anion* are derived from the directions the ions move during electrochemical reactions. Cations (positive ions) move toward the cathode and anions (negative ions) migrate toward the anode. This happens in both electrolytic and galvanic cells.

Electrolysis reactions in aqueous solutions can involve redox of water

When electrolysis is carried out in an aqueous solution, the electrode reactions are more difficult to predict because at the electrodes there are competing reactions. Not only do we have to consider the possible oxidation and reduction of the solute,

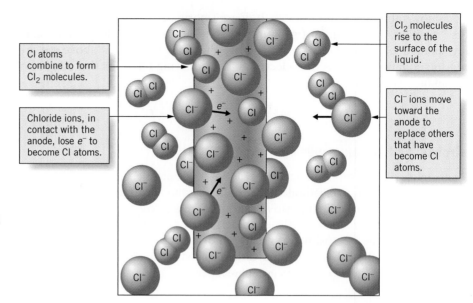

FIGURE 21.17 *A microscopic view of changes at the anode in the electrolysis of molten NaCl.* The positive charge of the electrode attracts a coating of Cl^- ions. At the surface of the electrode, electrons are pulled from the ions, yielding neutral Cl atoms, which combine to form Cl_2 molecules that move away from the electrode and eventually rise to the surface as a gas.

Cl atoms combine to form Cl_2 molecules.

Chloride ions, in contact with the anode, lose e^- to become Cl atoms.

Cl_2 molecules rise to the surface of the liquid.

Cl^- ions move toward the anode to replace others that have become Cl atoms.

but also the oxidation and reduction of water. For example, consider what happens when electrolysis is performed on a solution of potassium sulfate (Figure 21.18). The products are hydrogen and oxygen. At the cathode, water is reduced, not K^+.

$$2H_2O(l) + 2e^- \longrightarrow H_2(g) + 2OH^-(aq) \qquad \text{(cathode)}$$

At the anode, water is oxidized, not the sulfate ion.

$$2H_2O(l) \longrightarrow O_2(g) + 4H^+(aq) + 4e^- \qquad \text{(anode)}$$

Color changes of an acid–base indicator dissolved in the solution confirm that the solution becomes basic around the cathode, where OH^- is formed, and acidic around the anode, where H^+ is formed (see Figure 21.19). In addition, the gases H_2 and O_2 can be separately collected.

We can understand why these redox reactions happen if we examine reduction potential data from Table 21.1. For example, at the cathode we have the following competing reactions.

$$K^+(aq) + e^- \longrightarrow K(s) \qquad\qquad E^\circ_{K^+} = -2.92 \text{ V}$$

$$2H_2O(l) + 2e^- \longrightarrow H_2(g) + 2OH^-(aq) \qquad E^\circ_{H_2O} = -0.83 \text{ V}$$

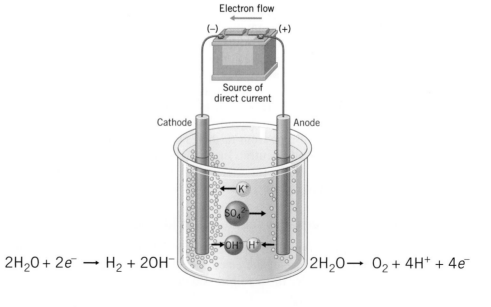

Electron flow

$(-)$ $(+)$

Source of direct current

Cathode Anode

K^+

$SO_4{}^{2-}$

OH^- H^+

FIGURE 21.18 *Electrolysis of an aqueous solution of potassium sulfate.* The products of the electrolysis are the gases hydrogen and oxygen.

$2H_2O + 2e^- \longrightarrow H_2 + 2OH^-$

$2H_2O \longrightarrow O_2 + 4H^+ + 4e^-$

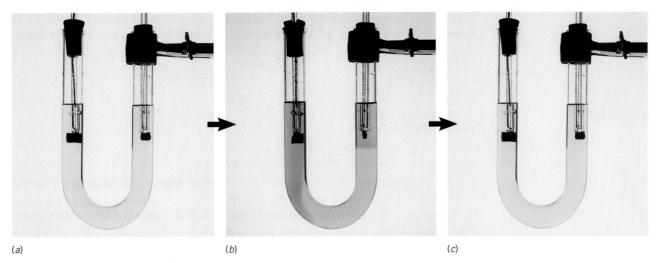

(a) (b) (c)

FIGURE 21.19 *Electrolysis of an aqueous solution of potassium sulfate in the presence of acid–base indicators.* (*a*) The initial yellow color indicates that the solution is neutral (neither acidic nor basic). (*b*) As the electrolysis proceeds, H^+ is produced at the anode (along with O_2) and causes the solution there to become pink. At the cathode, H_2 is evolved and OH^- ions are formed, which turns the solution around that electrode a bluish violet. (*c*) After the electrolysis is stopped and the solution is stirred, the color becomes yellow again as the H^+ and OH^- ions formed by the electrolysis neutralize each other.

Water has a much less negative (and therefore more positive) reduction potential than K^+, which means H_2O is much easier to reduce than K^+. As a result, when the electrolysis is performed the more easily reduced substance is reduced and we observe H_2 being formed at the cathode.

Now let's look at the anode, where we have the following possible oxidation half-reactions:

$$SO_4^{2-}(aq) \longrightarrow S_2O_8^{2-}(aq) + 2e^-$$

$$2H_2O \longrightarrow 4H^+(aq) + O_2(g) + 2e^-$$

In Table 21.1, we find them written in the opposite direction:

$$S_2O_8^{2-}(aq) + 2e^- \longrightarrow SO_4^{2-}(aq) \qquad E^{\circ}_{S_2O_8^{2-}} = +2.01 \text{ V}$$

$$O_2(g) + 4H^+(aq) + 4e^- \longrightarrow 2H_2O \qquad E^{\circ}_{O_2} = +1.23 \text{ V}$$

The E° values tell us that $S_2O_8^{2-}$ is much more easily reduced than O_2. But if $S_2O_8^{2-}$ is the *more easily reduced,* then the product, SO_4^{2-}, must be *the less easily oxidized.* Stated another way, *the half-reaction with the smaller reduction potential is more easily reversed as an oxidation.* As a result, when electrolysis is performed, water is oxidized instead of SO_4^{2-} and we observe O_2 being formed at the anode.

If the first half-reaction proceeds more easily to the right, then the second will proceed more easily to the left.

What function does potassium sulfate serve?

The overall cell reaction for the electrolysis of the K_2SO_4 solution can be obtained as before. Because the number of electrons lost has to equal the number gained, the cathode reaction must occur twice each time the anode reaction occurs once.

$$2 \times [2H_2O(l) + 2e^- \longrightarrow H_2(g) + 2OH^-(aq)] \qquad \text{(cathode)}$$

$$2H_2O(l) \longrightarrow O_2(g) + 4H^+(aq) + 4e^- \qquad \text{(anode)}$$

After adding, we combine the coefficients for water and cancel the electrons from both sides to obtain the cell reaction.

$$6H_2O(l) \longrightarrow 2H_2(g) + O_2(g) + 4H^+(aq) + 4OH^-(aq)$$

Notice that hydrogen ions and hydroxide ions are produced in equal numbers. In Figure 21.19, we see that when the solution is stirred, they combine to form water.

$$6H_2O(l) \longrightarrow 2H_2(g) + O_2(g) + 4H^+(aq) + 4OH^-(aq)$$

$$\downarrow$$

$$4H_2O$$

The net change, then, is

$$2H_2O \xrightarrow{\text{electrolysis}} 2H_2(g) + O_2(g)$$

At this point you may have begun to wonder whether the potassium sulfate serves any function, since neither K^+ nor SO_4^{2-} ions are changed by the reaction. Nevertheless, if the electrolysis is attempted with pure distilled water, nothing happens. There is no current flow, and no H_2 or O_2 forms. Apparently, the potassium sulfate must have some purpose.

The function of K_2SO_4 (or some other electrolyte) is to maintain electrical neutrality in the vicinity of the electrodes. If the K_2SO_4 were not present and the electrolysis were to occur anyway, the solution around the anode would become positively charged. It would become filled with H^+ ions, with no negative ions to balance their charge. Similarly, the solution surrounding the cathode would become negatively charged as it is filled with OH^- ions, with no nearby positive ions. The formation of positively or negatively charged solutions just doesn't happen because doing so requires too much energy, so in the absence of an electrolyte the electrode reactions cannot take place.

When K_2SO_4 is in the solution, K^+ ions can move toward the cathode and mingle with the OH^- ions as they are formed. Similarly, the SO_4^{2-} ions can move toward the anode and mingle with the H^+ ions as they are produced there. In this way, at any moment, each small region of the solution is able to contain the same number of positive and negative charges and thereby remain neutral.

Often we can use reduction potentials to predict electrolysis products

Let's look at an example of how we can use reduction potentials to anticipate the products of an electrolysis. Suppose we were planning to carry out electrolysis on an aqueous solution of copper(II) bromide, $CuBr_2$. What products should we expect to be formed? To answer this question, we need to examine the possible electrode half-reactions and their respective reduction potentials.

At the cathode, possible reactions are the reduction of copper ion and the reduction of water. The half-reactions with their reduction potentials from Table 21.1 are

$$Cu^{2+}(aq) + 2e^- \longrightarrow Cu(s) \qquad E^\circ_{Cu^{2+}} = +0.34\,V$$

$$2H_2O(l) + 2e^- \longrightarrow H_2(g) + 2OH^-(aq) \qquad E^\circ_{H_2O} = -0.83\,V$$

The much more positive reduction potential for Cu^{2+} tells us to anticipate that Cu^{2+} will be reduced at the cathode.

At the anode, possible reactions are the oxidation of Br^- and the oxidation of water. The half-reactions are

$$2Br^-(aq) \longrightarrow Br_2(aq) + 2e^-$$

$$2H_2O(l) \longrightarrow O_2(g) + 4H^+(aq) + 4e^-$$

In Table 21.1 we find these written as reductions with the corresponding E° values.

$$Br_2(aq) + 2e^- \longrightarrow 2Br^-(aq) \qquad E^\circ_{Br_2} = +1.07\,V$$

$$O_2(g) + 4H^+(aq) + 4e^- \longrightarrow 2H_2O(l) \qquad E^\circ_{O_2} = +1.23\,V$$

The data tell us that O_2 is more easily reduced than Br_2, which means that Br^- is more easily oxidized than H_2O. Therefore, we expect that Br^- will be oxidized at the anode.

In fact, our predictions are confirmed when we perform the electrolysis, as illustrated in Figure 21.20. The cathode, anode, and net cell reactions are

$$Cu^{2+}(aq) + 2e^- \longrightarrow Cu(s) \qquad \text{(cathode)}$$
$$2Br^-(aq) \longrightarrow Br_2(aq) + 2e^- \qquad \text{(anode)}$$

$$Cu^{2+}(aq) + 2Br^-(aq) \xrightarrow{\text{electrolysis}} Cu(s) + Br_2(aq) \quad \text{(net reaction)}$$

Electrolysis is planned for an aqueous solution that contains a mixture of 0.50 *M* ZnSO₄ and 0.50 *M* NiSO₄. On the basis of reduction potentials, what products are expected to be observed at the electrodes? What is the expected net cell reaction?

ANALYSIS: We need to consider the competing reactions at the cathode and the anode. At the cathode, the half-reaction with the most positive (or least negative)

> **EXAMPLE 21.12**
> **Predicting the Products in an Electrolysis Reaction**

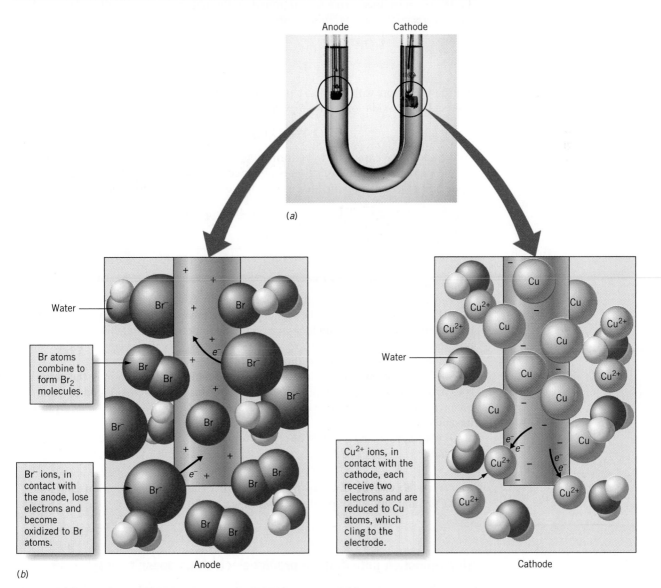

(a)

(b)

Anode · Cathode

Water

Br atoms combine to form Br₂ molecules.

Br⁻ ions, in contact with the anode, lose electrons and become oxidized to Br atoms.

Anode

Water

Cu²⁺ ions, in contact with the cathode, each receive two electrons and are reduced to Cu atoms, which cling to the electrode.

Cathode

FIGURE 21.20 *Electrolysis of an aqueous solution of copper(II) bromide.* The blue color of the CuBr₂ solution is due to copper ion [actually, Cu(H₂O)₄²⁺ ion]. At the anode on the left, bromide ions are oxidized to bromine atoms, which combine to form Br₂ molecules, imparting an orange-yellow color to the solution. At the cathode, copper ions are reduced. Here we see the copper as a black deposit building up on the electrode and flaking away to fall to the bottom of the apparatus. The diagrams below the photograph indicate what is happening at the atomic level at the electrodes. Notice that electron transfer occurs at the surface of the electrodes and that electrons do not travel through the solution.

reduction potential will be the one expected to occur. At the anode, the half-reaction with the *least positive* reduction potential is the one most easily reversed and should occur as an oxidation.

SOLUTION: At the cathode, the competing reduction reactions involve the two cations and water. The reactions and their reduction potentials are

$$Ni^{2+}(aq) + 2e^- \rightleftharpoons Ni(s) \qquad\qquad E° = -0.25 \text{ V}$$

$$Zn^{2+}(aq) + 2e^- \rightleftharpoons Zn(s) \qquad\qquad E° = -0.76 \text{ V}$$

$$2H_2O + 2e^- \rightleftharpoons H_2(g) + 2OH^-(aq) \qquad E° = -0.83 \text{ V}$$

The least negative reduction potential is that of Ni^{2+}, so we expect this ion to be reduced at the cathode and solid nickel to be formed.

At the anode, the competing oxidation reactions are for water and SO_4^{2-} ion. In Table 21.1, substances oxidized are found on the right side of the half-reactions. The two half-reactions having these as products are

$$S_2O_8^{2-}(aq) + 2e^- \rightleftharpoons 2SO_4^{2-}(aq) \qquad E° = +2.05 \text{ V}$$

$$O_2(g) + 4H^+(aq) + 4e^- \rightleftharpoons 2H_2O \qquad E° = +1.23 \text{ V}$$

The half-reaction with the least positive $E°$ (the second one here) is most easily reversed as an oxidation, so we expect the oxidation half-reaction to be

$$2H_2O \rightleftharpoons O_2(g) + 4H^+(aq) + 4e^-$$

At the anode, we expect O_2 to be formed.

The predicted net cell reaction is obtained by combining the two expected electrode half-reactions, making the electron loss equal to the electron gain.

$$2H_2O \longrightarrow O_2(g) + 4H^+(aq) + 4e^- \qquad\qquad \text{(anode)}$$

$$\underline{2 \times [Ni^{2+}(aq) + 2e^- \longrightarrow Ni(s)] \qquad\qquad\qquad\quad \text{(cathode)}}$$

$$2H_2O + 2Ni^{2+}(aq) \longrightarrow O_2(g) + 4H^+(aq) + 2Ni(s) \quad \text{(net cell reaction)}$$

Are the Answers Reasonable?

We can check the locations of the half-reactions in Table 21.1 to confirm our conclusions. For the reduction step, the higher up in the table a half-reaction is, the greater its tendency to occur as reduction. Among the competing half-reactions at the cathode, the one for Ni^{2+} is highest, so we expect that Ni^{2+} is the easiest to reduce and $Ni(s)$ should be formed at the cathode.

For the oxidation step, the lower down in the table a half-reaction is, the easier it is to reverse and cause to occur as oxidation. On this basis, the oxidation of water is easier than the oxidation of SO_4^{2-}, so we expect H_2O to be oxidized and O_2 to be formed at the anode.

Of course, we could also test our prediction by carrying out the electrolysis experimentally.

PRACTICE EXERCISE 14: In the electrolysis of an aqueous solution containing both Cd^{2+} and Sn^{2+}, what product do we expect at the cathode?

Using reduction potentials to predict electrolysis doesn't always work

Although we can use reduction potentials to predict electrolysis reactions most of the time, there are occasions when standard reduction potentials do not successfully predict electrolysis products. Sometimes concentration effects or the formation of complex ions interfere. And sometimes the electrodes themselves are the culprits. For example, in the electrolysis of aqueous NaCl using inert platinum electrodes, we find experimentally that Cl_2 is formed at the anode. Is this what we would have expected? Let's examine the reduction potentials of O_2 and Cl_2 to find out.

$$Cl_2(g) + 2e^- \rightleftharpoons 2Cl^-(aq) \qquad E° = +1.36 \text{ V}$$

$$O_2(g) + 4H^+(aq) + 4e^- \rightleftharpoons 2H_2O \qquad E° = +1.23 \text{ V}$$

Because of its less positive reduction potential, we would expect the oxygen half-reaction to be the easier to reverse (with water being oxidized to O_2). Thus, reduction potentials predict that O_2 should be formed, but experiment shows that Cl_2 is produced. The explanation for why this happens is beyond the scope of this text, but the unexpected result does teach us that we must be cautious in predicting products in electrolysis reactions solely on the basis of reduction potentials.

21.8 ▶ Stoichiometry of electrochemical reactions involves electric current and time

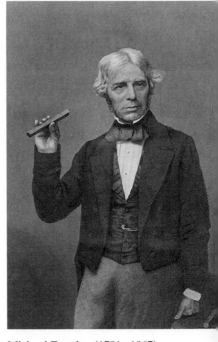

Much of the early research in electrochemistry was performed by Michael Faraday. In fact, it was Faraday who coined the terms *anode, cathode, electrode, electrolyte,* and *electrolysis*. In about 1833, Faraday discovered that the amount of chemical change that occurs during electrolysis is directly proportional to the amount of electrical charge that is passed through an electrolysis cell. For example, the reduction of copper ion at a cathode is given by the equation

Michael Faraday (1791–1867), a British scientist and both a chemist and a physicist, made key discoveries leading to electric motors, generators, and transformers.

$$Cu^{2+}(aq) + 2e^- \longrightarrow Cu(s)$$

The equation tells us that to deposit 1 mol of metallic copper requires 2 mol of electrons. Therefore, to deposit 2 mol of copper requires 4 mol of electrons, and that takes twice as much electricity. The half-reaction for an oxidation or reduction, therefore, relates the amount of chemical substance consumed or produced to the amount of electrons that the electric current must supply. To use this information, however, we must be able to relate it to electrical measurements that can be made in the laboratory.

The SI unit of electric current is the **ampere (A)** and the SI unit of charge is the **coulomb (C).** A coulomb is the amount of charge that passes by a given point in a wire when an electric current of 1 ampere flows for 1 second. This means that coulombs are the product of amperes of current multiplied by seconds. Thus

$$1 \text{ coulomb} = 1 \text{ ampere} \times 1 \text{ second}$$

$$1 \text{ C} = 1 \text{ A} \times \text{s}$$

For example, if a current of 4 A flows through a wire for 10 s, 40 C pass by a given point in the wire.

$$(4 \text{ A}) \times (10 \text{ s}) = 40 \text{ A} \times \text{s}$$

$$= 40 \text{ C}$$

As we noted earlier, it has been determined that 1 mol of electrons carries a charge of 9.65×10^4 C, which in honor of Michael Faraday is often called the Faraday constant, $\mathscr{F}$.

$$1 \text{ mol } e^- \Leftrightarrow 9.65 \times 10^4 \text{ C} \qquad \text{(to three significant figures)}$$

$$1 \, \mathscr{F} = 9.65 \times 10^4 \text{ C/mol } e^-$$

Now we have a way to relate laboratory measurements to the amount of chemical change that occurs during an electrolysis. Measuring the current in amperes and the time in seconds allows us to calculate the charge sent through the system in coulombs. From this we can get the amount of electrons (in moles), which we can then use to calculate the amount of chemical change produced. The following examples demonstrate the principles involved for electrolysis, but similar calculations also apply to reactions in galvanic cells.

How many grams of copper are deposited on the cathode of an electrolytic cell if an electric current of 2.00 A is run through a solution of $CuSO_4$ for a period of 20.0 min?

ANALYSIS: The balanced half-reaction serves as our tool for relating chemical change to amounts of electricity. The ion being reduced is Cu^{2+}, so the half-reaction is

$$Cu^{2+} + 2e^- \longrightarrow Cu$$

Therefore,

$$1 \text{ mol Cu} \Leftrightarrow 2 \text{ mol } e^-$$

The product of current (in amperes) and time (in seconds) will give us charge (in coulombs). We can relate this to the number of moles of electrons by the Faraday constant. Then, from the number of moles of electrons we calculate moles of copper, from which we calculate the mass of copper by using the atomic mass. Here's a diagram of the path to the solution.

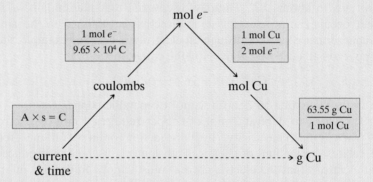

$$20.0 \text{ min } \frac{60 \text{ s}}{1 \text{ min}} = 1.20 \times 10^3 \text{ s}$$

SOLUTION: First we convert minutes to seconds; $20.0 \text{ min} = 1.20 \times 10^3 \text{ s}$. Then we multiply the current by the time to obtain the number of coulombs ($1 \text{ A} \times \text{s} = 1 \text{ C}$).

$$(1.20 \times 10^3 \text{ s}) \times (2.00 \text{ A}) = 2.40 \times 10^3 \text{ A} \times \text{s}$$

$$= 2.40 \times 10^3 \text{ C}$$

Because $1 \text{ mol } e^- \Leftrightarrow 9.65 \times 10^4 \text{ C}$,

$$2.40 \times 10^3 \text{ C} \times \frac{1 \text{ mol } e^-}{9.65 \times 10^4 \text{ C}} = 0.0249 \text{ mol } e^-$$

Next, we use the relationship between mol e^- and mol Cu from the balanced half-reaction along with the atomic mass of copper.

$$0.0249 \text{ mol } e^- \times \left(\frac{1 \text{ mol Cu}}{2 \text{ mol } e^-} \right) \times \left(\frac{63.55 \text{ g Cu}}{1 \text{ mol Cu}} \right) = 0.791 \text{ g Cu}$$

The electrolysis will deposit 0.791 g of copper on the cathode.

We could have combined all these steps in a single calculation by stringing together the various conversion factors and using the factor label method to cancel units.

$$2.00 \text{ A} \times 20.0 \text{ min} \times \frac{60 \text{ s}}{1 \text{ min}} \times \frac{1 \text{ mol } e^-}{9.65 \times 10^4 \text{ A s}} \times \frac{1 \text{ mol Cu}}{2 \text{ mol } e^-} \times \frac{63.55 \text{ g Cu}}{1 \text{ mol Cu}}$$

$$= 0.790 \text{ g Cu}$$

The small difference between the two answers is because of rounding off in the step-wise calculation.

Is the Answer Reasonable?
There is no simple check on the numerical answer, but the cancellation of units tells us we've set up the calculation correctly.

Electrolysis provides a useful way to deposit a thin metallic coating on an electrically conducting surface. The technique is called *electroplating*. How much time would it take in minutes to deposit 0.500 g of metallic nickel on a metal object using a current of 3.00 A? The nickel is reduced from the +2 oxidation state.

EXAMPLE 21.14
Calculations Related to Electrolysis

ANALYSIS: We need an equation for the reduction. Because the nickel is reduced to the free metal from the +2 state, we can write

$$Ni^{2+} + 2e^- \longrightarrow Ni(s)$$

This gives the relationship

$$1 \text{ mol Ni} \Leftrightarrow 2 \text{ mol } e^-$$

We wish to deposit 0.500 g of Ni, which we can convert to moles. Then we can calculate the number of moles of electrons required, which in turn is used with the Faraday constant to determine the number of coulombs required. Because this is the product of amperes and seconds, we can calculate the time needed to deposit the metal. The calculation can be diagrammed as follows.

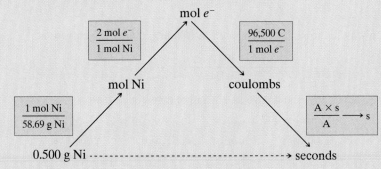

Faraday constant =
$$\frac{9.65 \times 10^4 \text{ C}}{\text{mole } e^-}$$

SOLUTION: First, we calculate the number of moles of electrons required.

$$0.500 \text{ g Ni} \times \left(\frac{1 \text{ mol Ni}}{58.69 \text{ g Ni}}\right) \times \left(\frac{2 \text{ mol } e^-}{1 \text{ mol Ni}}\right) = 0.0170 \text{ mol } e^-$$

Then we calculate the number of coulombs needed.

$$0.0170 \text{ mol } e^- \times \left(\frac{9.65 \times 10^4 \text{ C}}{1 \text{ mol } e^-}\right) = 1.64 \times 10^3 \text{ C}$$

$$= 1.64 \times 10^3 \text{ A} \times \text{s}$$

This tells us that the product of current multiplied by time equals 1.64×10^3 A × s. The current is 3.00 A. Dividing 1.64×10^3 A × s by 3.00 A gives the time required in seconds, which we then convert to minutes.

$$\left(\frac{1.64 \times 10^3 \text{ A} \times \text{s}}{3.00 \text{ A}}\right) \times \left(\frac{1 \text{ min}}{60 \text{ s}}\right) = 9.11 \text{ min}$$

We could also have combined these calculations in a single string of conversion factors.

$$0.500 \text{ g Ni} \times \frac{1 \text{ mol Ni}}{58.69 \text{ g Ni}} \times \frac{2 \text{ mol } e^-}{1 \text{ mol Ni}} \times \frac{9.65 \times 10^4 \text{ A} \text{ s}}{1 \text{ mol } e^-} \times \frac{1}{3.00 \text{ A}} \times \frac{1 \text{ min}}{60 \text{ s}}$$

$$= 9.13 \text{ min}$$

In this string of conversion factors, we have replaced 9.65×10^4 C by 9.65×10^4 A × s. Remember, 1 C = 1 A × s.

As before, the small difference between the two answers is caused by rounding in the stepwise calculation.

Is the Answer Reasonable?
As in the preceding example, the units cancel correctly, so we can be confident we've set up the problem correctly.

<table>
<tr><td>

EXAMPLE 21.15

Calculations Related to Electrolysis

</td><td>

What current is needed to deposit 0.500 g of chromium metal from a solution of Cr^{3+} in a period of 1.00 hr?

ANALYSIS: The procedure here is very similar to the preceding problem. The difference is that at the end we will find the current instead of the time.

SOLUTION: First we write the balanced half-reaction.

</td></tr>
</table>

$$Cr^{3+} + 3e^- \longrightarrow Cr$$

This gives the relationship: 1 mol Cr ⟺ 3 mol e^-, which we use as before to calculate the number of moles of electrons required.

$$0.500 \text{ g Cr} \times \left(\frac{1 \text{ mol Cr}}{52.00 \text{ g Cr}}\right) \times \left(\frac{3 \text{ mol } e^-}{1 \text{ mol Cr}}\right) = 0.0288 \text{ mol } e^-$$

Then we find the number of coulombs.

$$0.0288 \text{ mol } e^- \times \left(\frac{9.65 \times 10^4 \text{ C}}{1 \text{ mol } e^-}\right) = 2.78 \times 10^3 \text{ C}$$

$$= 2.78 \times 10^3 \text{ A} \times \text{s}$$

Since we want the metal to be deposited in 1.00 hr (3600 s), the current is

$$\frac{2.78 \times 10^3 \text{ A} \times \text{s}}{3600 \text{ s}} = 0.772 \text{ A}$$

We can also set up the calculation as a string of conversion factors.

As before, rounding in the stepwise calculation causes a slight discrepancy between the two answers.

$$0.500 \text{ g Cr} \times \frac{1 \text{ mol Cr}}{52.00 \text{ g Cr}} \times \frac{3 \text{ mol } e^-}{1 \text{ mol Cr}} \times \frac{9.65 \times 10^4 \text{ A s}}{1 \text{ mol } e^-} \times \frac{1}{3600 \text{ s}} = 0.773 \text{ A}$$

Is the Answer Reasonable?
Once again, the units cancel properly, so the problem is set up correctly.

PRACTICE EXERCISE 15: How many moles of hydroxide ion will be produced at the cathode during the electrolysis of water with a current of 4.00 A for a period of 200 s? The cathode reaction is

$$2e^- + 2H_2O \longrightarrow H_2 + 2OH^-$$

PRACTICE EXERCISE 16: How many minutes will it take for a current of 10.0 A to deposit 3.00 g of gold from a solution of $AuCl_3$?

PRACTICE EXERCISE 17: What current must be supplied to deposit 3.00 g of gold from a solution of $AuCl_3$ in 20.0 min?

PRACTICE EXERCISE 18: Suppose the solutions in the galvanic cell depicted in Figure 21.2 (page 914) have a volume of 125 mL and suppose the cell is operated for a period of 1.25 hr with a constant current of 0.100 A flowing through the external circuit. By how much will the concentration of the copper ion increase during this time period?

21.9 ▶ **Electrolysis has many industrial applications**

Besides being a useful tool in the chemistry laboratory, electrolysis has many important industrial applications. In this section we will briefly examine the chemistry of electroplating and the production of some of our most common chemicals.

Electroplating deposits metal on a surface

Electroplating, which was mentioned in Example 21.14, is a procedure in which electrolysis is used to apply a thin (generally 0.03 to 0.05 mm thick) ornamental or protective coating of one metal over another. It is a common technique for improving the appearance and durability of metal objects. For instance, a thin, shiny coating of metallic chromium is applied over steel objects to make them attractive and to prevent rusting. Silver and gold plating are applied to jewelry made from less expensive metals, and silver plate is common on eating utensils (knives, forks, spoons, etc.).

Figure 21.21 illustrates a typical apparatus used for plating silver. Silver ion in the solution is reduced at the cathode, where it is deposited as metallic silver on the object to be plated. At the anode, silver from the metal bar is oxidized, replenishing the supply of the silver ion in the solution. As time passes, silver is gradually transferred from the bar at the anode onto the object at the cathode.

The exact composition of the electroplating bath varies, depending on the metal to be deposited, and it can affect the appearance and durability of the finished surface. For example, silver deposited from a solution of silver nitrate ($AgNO_3$) does not stick to other metal surfaces very well. However, if it is deposited from a solution of silver cyanide containing $Ag(CN)_2^-$, the coating adheres well and is bright and shiny. Other metals that are electroplated from a cyanide bath are gold and cadmium. Nickel, which can also be applied as a protective coating, is plated from a nickel sulfate solution, and chromium is plated from a chromic acid (H_2CrO_4) solution.

This motorcycle sparkles with chrome plating that was deposited by electrolysis. The shiny hard coating of chromium is both decorative and a barrier to corrosion.

Aluminum is produced from aluminum oxide

Aluminum does not occur as the free metal in nature. It is even so reactive that the traditional methods for obtaining a metal from its ores do not work. Because of this, until the latter part of the nineteenth century, aluminum was so costly to isolate that only the very wealthy could afford aluminum products. Early efforts to produce aluminum by the electrolysis of a molten form of an aluminum compound were unfeasible. Its anhydrous halide salts (those with no water of hydration) are difficult to prepare and volatile—they tend to evaporate rather than melt. On the other hand, its oxide, Al_2O_3, has such a high melting point (over 2000 °C) that no practical method of melting it could be found.

Aluminum is used today as a structural metal, in alloys, and in such products as aluminum foil, electrical wire, window frames, and kitchen pots and pans.

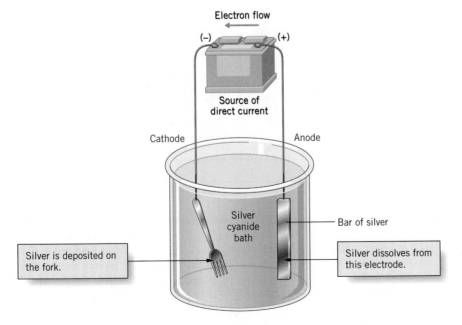

FIGURE 21.21 *Apparatus for electroplating silver.* Silver dissolves at the anode, where it is oxidized to Ag^+. Silver is deposited on the cathode (the fork), where Ag^+ is reduced.

In 1886, a young student at Oberlin College, Charles M. Hall, discovered that Al_2O_3 dissolves in the molten form of a mineral called cryolite, Na_3AlF_6, to give a conducting mixture with a relatively low melting point from which aluminum could be produced electrolytically. The process was also discovered by Paul Héroult in France at nearly the same time, and today this method for producing aluminum is usually called the **Hall–Héroult process** (see Figure 21.22). Purified aluminum oxide, which is obtained from an ore called *bauxite,* is dissolved in molten cryolite in which the oxide dissociates to give Al^{3+} and O^{2-} ions. At the cathode, aluminum ions are reduced to produce the free metal, which forms as a layer of molten aluminum below the less dense solvent. At the carbon anodes, oxide ion is oxidized to give free O_2.

A large cell can produce as much as 900 lb of aluminum per day.

$$Al^{3+} + 3e^- \longrightarrow Al(l) \qquad \text{(cathode)}$$

$$2O^{2-} \longrightarrow O_2(g) + 4e^- \qquad \text{(anode)}$$

The net cell reaction is

$$4Al^{3+} + 6O^{2-} \longrightarrow 4Al(l) + 3O_2(g)$$

The oxygen formed at the anode attacks the carbon electrodes (producing CO_2), so the electrodes must be replaced frequently.

The production of aluminum consumes enormous amounts of electrical energy and is therefore very costly, not only in terms of dollars but also in terms of energy resources. For this reason, recycling of aluminum has a high priority as we seek to minimize our use of energy.

Magnesium is produced from seawater

Metallic magnesium has a number of structural uses because of its low density ("light weight"). Around the home, for example, you may have a magnesium alloy ladder.

The major source of magnesium is seawater in which, on a mole basis, Mg^{2+} is the third most abundant ion, exceeded only by Na^+ and Cl^-. To obtain metallic magnesium, seawater is made basic, which causes Mg^{2+} ions to precipitate as $Mg(OH)_2$. The precipitate is separated by filtration and dissolved in hydrochloric acid.

$$Mg(OH)_2(s) + 2HCl(aq) \longrightarrow MgCl_2(aq) + 2H_2O$$

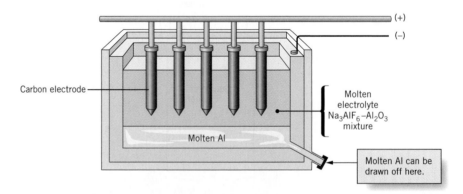

FIGURE 21.22 *Production of aluminum by electrolysis.* In the apparatus used to produce aluminum electrolytically by the Hall–Héroult process, Al_2O_3 is dissolved in molten cryolite, Na_3AlF_6. Al^{3+} is reduced to metallic Al and O^{2-} is oxidized to O_2, which reacts with the carbon anodes to give CO_2. Periodically, molten aluminum is drawn off at the bottom of the cell and additional Al_2O_3 is added to the cryolite. The carbon anodes also must be replaced from time to time as they are consumed by their reaction with O_2.

The resulting solution is evaporated to give solid $MgCl_2$, which is then melted and electrolyzed. Free magnesium is deposited at the cathode and chlorine gas is produced at the anode.

$$MgCl_2(l) \xrightarrow{\text{electrolysis}} Mg(l) + Cl_2(g)$$

The chlorine is recycled to make HCl, which is returned to the process.

Sodium is made by electrolysis of sodium chloride

Sodium is prepared by the electrolysis of molten sodium chloride (see Section 21.7). The metallic sodium and the chlorine gas that form must be kept apart or they will react violently and re-form NaCl. The **Downs cell** accomplishes this separation (see Figure 21.23).

Both sodium and chlorine are commercially important. Chlorine is used largely to manufacture plastics such as polyvinyl chloride (PVC), many solvents, and industrial chemicals. A small percentage of the annual chlorine production is used to chlorinate drinking water.

PVC is widely used to make pipes and conduits for water and sanitary systems.

Sodium has been used in the manufacture of tetraethyllead, an octane booster for gasoline that has been phased out in the United States but that is still used in many other countries. Sodium is also used in the production of sodium vapor lamps. These lamps, with their familiar bright-yellow color, have the advantage of giving off most of their energy in a portion of the spectrum that humans can see. In terms of useful light output versus energy input, they are about 15 times more efficient than ordinary incandescent lightbulbs and about 3 to 4 times more efficient than the bluish-white mercury vapor lamps once commonly used for street lighting.

Refining of copper

One of the most interesting and economically attractive applications of electrolysis is the purification or refining of metallic copper. When copper is first obtained from its ore, it is about 99% pure. The impurities—mostly silver, gold, platinum, iron, and zinc—decrease the electrical conductivity of the copper enough that even 99% pure copper must be further refined before it can be used in electrical wire.

The impure copper is used as the anode in an electrolysis cell that contains a solution of copper sulfate and sulfuric acid as the electrolyte (see Figure 21.24).

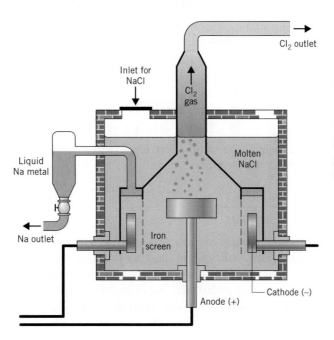

FIGURE 21.23 *Commercial electrolysis of molten sodium chloride.* Cross section of the *Downs cell* used for the electrolysis of molten sodium chloride. The cathode is a circular ring that surrounds the anode. The electrodes are separated from each other by an iron screen. During the operation of the cell, molten sodium collects at the top of the cathode compartment, from which it is periodically drained. The chlorine gas bubbles out of the anode compartment and is collected.

FIGURE 21.24 *Purification of copper by electrolysis.* Impure copper anodes dissolve and pure copper is deposited on the cathodes. Metals less easily reduced than copper remain in solution, while metals less easily oxidized than copper settle to the bottom of the apparatus as "anode mud."

Cathode (−)   Anode (+)

Pure Cu is deposited on the cathode.

Fe^{2+}
Zn^{2+}
Cu^{2+}
SO_4^{2-}
H^+

Cu and metals more easily oxidized than Cu are oxidized here and enter the solution as ions.

Impure copper

Pure copper

$CuSO_4/H_2SO_4$ electrolyte

Sludge containing silver, gold, and platinum

The cathode is a thin sheet of very pure copper. When the cell is operated at the correct voltage, only copper and impurities more easily oxidized than copper (iron and zinc) dissolve at the anode. The less active metals simply fall off the electrode and settle to the bottom of the container. At the cathode, copper ions are reduced, but the zinc ions and iron ions remain in solution because they are more difficult to reduce than copper. Gradually, the impure copper anode dissolves and the copper cathode, about 99.96% pure, grows larger. The accumulating sludge—called *anode mud*—is removed periodically, and the value of the silver, gold, and platinum recovered from it virtually pays for the entire refining operation.

Copper refining provides one fourth of the silver and one eighth of the gold produced annually in the United States.

Electrolysis of brine produces several important chemicals

One of the most important commercial electrolysis reactions is the electrolysis of concentrated aqueous sodium chloride solutions called **brine.** An apparatus that could be used for this in the laboratory is shown in Figure 21.25. At the cathode, water is much more easily reduced than sodium ion, so H_2 forms.

$$2H_2O(l) + 2e^- \longrightarrow H_2(g) + 2OH^-(aq) \qquad \text{(cathode)}$$

As we noted earlier, even though water is more easily oxidized than chloride ion, complicating factors at the electrodes actually allow chloride ion to be oxidized instead. At the anode, therefore, we observe the formation of Cl_2.

$$2Cl^-(aq) \longrightarrow Cl_2(g) + 2e^- \qquad \text{(anode)}$$

The net cell reaction is therefore

$$2Cl^-(aq) + 2H_2O(l) \longrightarrow H_2(g) + Cl_2(g) + 2OH^-(aq)$$

If we include the sodium ion, already in the solution as a spectator ion and not involved in the electrolysis directly, we can see why this is such an important reaction.

$$\underbrace{2Na^+(aq) + 2Cl^-(aq)}_{2NaCl(aq)} + 2H_2O \xrightarrow{\text{electrolysis}} H_2(g) + Cl_2(g) + \underbrace{2Na^+(aq) + 2OH^-(aq)}_{2NaOH(aq)}$$

Thus, the electrolysis converts inexpensive salt to valuable chemicals: H_2, Cl_2, and NaOH. The hydrogen is used to make other chemicals, including hydrogenated vegetable oils. The chlorine is used for the purposes mentioned earlier. Among the uses of sodium hydroxide, one of industry's most important bases, are the manufacture of soap and paper, the neutralization of acids in industrial reactions, and the purification of aluminum ores.

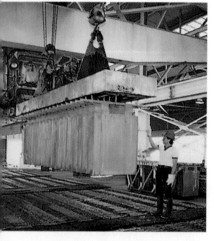

Copper refining. Copper cathodes, 99.96% pure, are pulled from the electrolytic refining tanks at Kennecott's Utah copper refinery. It takes about 28 days for the impure copper anodes to dissolve and deposit the pure metal on the cathodes.

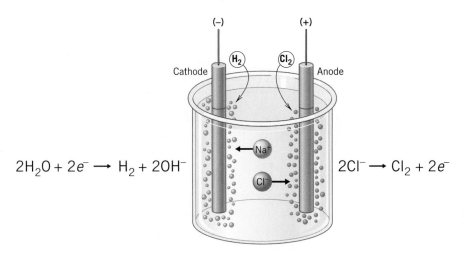

FIGURE 21.25 *The electrolysis of brine.* In the electrolysis of *brine,* a concentrated solution of sodium chloride, hydrogen gas is formed at the cathode and chlorine gas is formed at the anode. The solution becomes basic as the reaction progresses.

In the industrial electrolysis of brine to make pure NaOH, an apparatus as simple as that in Figure 21.25 cannot be used for three chief reasons. First, it is necessary to capture the H_2 and Cl_2 separately to prevent them from mixing and reacting (explosively). Second, the NaOH from the reaction is contaminated with unreacted NaCl. Third, if Cl_2 is left in the presence of NaOH, the solution becomes contaminated by hypochlorite ion (OCl^-), which forms by the reaction of Cl_2 with OH^-.

Sodium hydroxide is commonly known as *lye* or as *caustic soda.*

$$Cl_2(g) + 2OH^-(aq) \longrightarrow Cl^-(aq) + OCl^-(aq) + H_2O$$

In one manufacturing operation, however, the Cl_2 is not removed as it forms, and its reaction with hydroxide ion is used to manufacture aqueous sodium hypochlorite. For this purpose, the solution is stirred vigorously during the electrolysis so that very little Cl_2 escapes. As a result, a stirred solution of NaCl gradually changes during electrolysis to a solution of NaOCl, a dilute solution of which is sold as liquid laundry bleach (e.g., Clorox).

Most of the pure NaOH manufactured today is made in an apparatus called a **diaphragm cell.** The design varies somewhat, but Figure 21.26 illustrates its basic

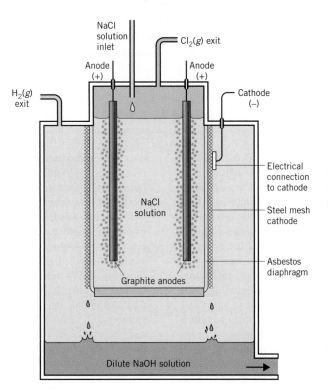

FIGURE 21.26 *A diaphragm cell used in the commercial production of NaOH by the electrolysis of aqueous NaCl.* This is a cross section of a cylindrical cell in which the NaCl solution is surrounded by an asbestos diaphragm supported by a steel mesh cathode. (From J. E. Brady and G. E. Humiston, *General Chemistry: Principles and Structure,* 4th ed. Copyright © 1986, John Wiley & Sons, New York. Used by permission.)

FIGURE 21.27 *Electrolysis of aqueous sodium chloride (brine) using a mercury cell.* At the anode, chloride ions are oxidized to chlorine, Cl_2. At the cathode, sodium ions are reduced to sodium atoms, which dissolve in the mercury. This mercury is pumped to a separate compartment, where it is exposed to water. As sodium atoms come to the surface of the mercury, they react with water to give H_2 and NaOH. (From J. E. Brady and G. E. Humiston, *General Chemistry: Principles and Structure,* 4th ed. Copyright © 1986, John Wiley & Sons, New York. Used by permission.)

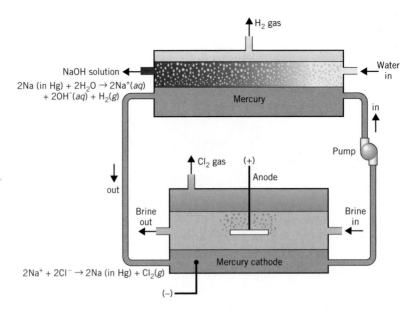

H_2 gas

NaOH solution

Water in

2Na (in Hg) + 2H$_2$O → 2Na$^+$(aq) + 2OH$^-$(aq) + H$_2$(g)

Mercury

in

Pump

Cl_2 gas

(+)

Anode

out

Brine out

Brine in

2Na$^+$ + 2Cl$^-$ → 2Na (in Hg) + Cl$_2$(g)

Mercury cathode

(−)

The mercury cell which gives pure NaOH, poses an environmental hazard because of the potential for mercury pollution. It is a process being phased out in many countries around the world.

features. The cell consists of an iron wire mesh cathode that encloses a porous asbestos shell—the diaphragm. The NaCl solution is added to the top of the cell and seeps slowly through the diaphragm. When it contacts the iron cathode, hydrogen is evolved and is pumped out of the surrounding space. The solution, now containing dilute NaOH, drips off the cell into the reservoir below. Meanwhile, within the cell, chlorine is generated at the anodes dipping into the NaCl solution. Because there is no OH$^-$ in this solution, the Cl_2 can't react to form OCl$^-$ ion and simply bubbles out of the solution and is captured.

The NaOH obtained from the diaphragm cell, although free of OCl$^-$, is still contaminated by small amounts of NaCl. The production of pure NaOH is accomplished using a **mercury cell** (see Figure 21.27). In this apparatus, mercury serves as the cathode, and the cathode reaction actually involves the reduction of sodium ions to sodium atoms, which dissolve in the liquid mercury. The mercury–sodium solution (called *sodium amalgam*) is pumped to another chamber and exposed to water free of NaCl. Here, the sodium atoms react with water to give H_2 and NaOH. The net overall chemical change is still the conversion of NaCl into H_2, Cl_2, and NaOH, but the NaOH produced in the mercury cell isn't contaminated by NaCl.

SUMMARY

Galvanic Cells A **galvanic cell** is composed of two **half-cells,** each containing an **electrode** in contact with an electrolyte reactant. A spontaneous redox reaction is thus divided into separate oxidation and reduction half-reactions, with the electron transfer occurring through an external electrical circuit. Reduction occurs at the **cathode;** oxidation occurs at the **anode.** In a galvanic cell, the cathode is positively charged and the anode is negatively charged. The half-cells must be connected electrolytically by a **salt bridge** to complete the electrical circuit, which permits electrical neutrality to be maintained, by allowing cations to move toward the cathode and anions toward the anode. The **potential** (expressed in volts) produced by a cell is equal to the **standard cell potential** when all ion concentrations are 1 *M* and the partial pressures of any gases involved equal 1 atm. The **cell potential** can be considered to be the difference be-

tween the **reduction potentials** of the half-cells. In the spontaneous reaction, the half-cell with the higher reduction potential undergoes reduction and forces the other to undergo oxidation. The reduction potentials of isolated half-cells can't be measured, but values are assigned by choosing the **hydrogen electrode** as a reference electrode; its reduction potential is assigned a value of exactly 0 V. Species more easily reduced than H$^+$ have positive reduction potentials; those less easily reduced have negative reduction potentials. Reduction potentials can be used to predict the cell reaction and to calculate $E°_{cell}$. They can also be used to predict spontaneous redox reactions not occurring in galvanic cells and to predict whether or not a given reaction is spontaneous. Often, they can be used to determine the products of electrolysis reactions.

Thermodynamics and Cell Potentials The values of $\Delta G°$ and K_c for a reaction can be calculated from $E°_{cell}$. They all involve the **faraday**, $\mathscr{F}$, a constant equal to the number of **coulombs (C)** of charge per mole of electrons ($1\ \mathscr{F} = 96{,}500$ C/mol e^-). The Nernst equation relates the cell potential to the standard cell potential and the reaction quotient. It allows the cell potential to be calculated for ion concentrations other than $1.00\ M$. The important equations in this section are

$$\Delta G° = -n\mathscr{F}E°_{cell}$$

$$E°_{cell} = \frac{RT}{n\mathscr{F}} \ln K_c$$

$$E_{cell} = E°_{cell} - \frac{RT}{n\mathscr{F}} \ln Q$$

Practical Galvanic Cells The **lead storage battery** and the **nickel–cadmium (nicad) battery** are **secondary cells** and are rechargeable. The state of charge of the lead storage battery can be tested with a **hydrometer,** which measures the density of the sulfuric acid electrolyte. The **zinc–manganese dioxide cell** (the **Leclanché cell** or common **dry cell**) and the common **alkaline battery** (which uses essentially the same reactions as the less expensive dry cell) are **primary cells** and are not rechargeable. The **zinc–silver oxide cell** is often used in small electronic devices and cameras. The rechargeable **nickel–metal hydride** (Ni–MH) battery uses hydrogen contained in a metal alloy as its anode reactant and has a higher **energy density** than the nicad battery. Primary **lithium–manganese dioxide cells** and rechargeable **lithium ion cells** produce a large cell potential and have a very large energy density. Lithium ion cells store and release energy by transferring lithium ions between electrodes where the Li^+ are **intercalated** between layers of atoms in the electrode materials. **Fuel cells,** which have high thermodynamic efficiencies, are able to provide continuous power because they consume fuel that can be fed continuously.

Electrolysis In an **electrolytic cell,** a flow of electricity causes an otherwise nonspontaneous reaction to occur. A negatively charged **cathode** causes reduction of one reactant and a positively charged **anode** causes oxidation of another. Ion movement instead of electron transport occurs in the electrolyte. The electrode reactions are determined by which species is most easily reduced and which is most easily oxidized, but in aqueous solutions complex surface effects at the electrodes can alter the natural order. In the electrolysis of water, an electrolyte must be present to maintain electrical neutrality at the electrodes.

Quantitative Aspects of Electrochemical Reactions A **faraday** ($\mathscr{F}$) is the charge carried by 1 mol of electrons, 9.65×10^4 C. The product of current (**amperes**) and time (seconds) gives coulombs. These relationships and the half-reactions that occur at the anode or cathode permit us to relate the amount of chemical change to measurements of current and time.

Applications of Electrolysis **Electroplating,** the production of aluminum and magnesium, the refining of copper, and the electrolysis of molten and aqueous sodium chloride are examples of practical applications of electrolysis.

TOOLS ▶ YOU HAVE LEARNED

The table below lists the concepts you've learned in this chapter that can be applied as tools in solving problems. Study each one carefully so that you know what each is used for. When faced with solving a problem, recall what each tool does and consider whether it will be helpful in finding a solution. This will aid you in selecting the tools you need. If necessary, refer to this table when working on the Thinking-It-Through problems and the Review Exercises that follow.

TOOL	HOW IT WORKS
Standard reduction potentials (page 920)	In a galvanic cell, the difference between two reduction potentials equals the standard cell potential. Comparing reduction potentials lets us predict the electrode reactions in electrolysis.
Standard cell potentials (page 929)	By calculating $E°_{cell}$ we can predict the spontaneity of a redox reaction. We can use $E°_{cell}$ to calculate $\Delta G°$ (page 931) and equilibrium constants (page 932). They are also needed in the Nernst equation to relate concentrations of species in galvanic cells to the cell potential.
$\Delta G° = -n\mathscr{F}E°$ (page 931)	This equation lets us calculate standard free energy changes from cell potentials, and vice versa.
$E°_{cell} = \dfrac{RT}{n\mathscr{F}} \ln K_c$ (page 932)	This equation lets us calculate equilibrium constants from cell potentials.
Nernst equation $E_{cell} = E°_{cell} - \dfrac{RT}{n\mathscr{F}} \ln Q$ (page 933)	This equation lets us calculate the cell potential from $E°_{cell}$ and concentration data; we can also calculate the concentration of a species in solution from $E°_{cell}$ and a measured value for E_{cell}.
Faraday constant $1\ \mathscr{F} = 9.65 \times 10^4$ **C/mol** e^- (page 931)	Besides being a constant in the equations above, it allows us to relate coulombs (obtained from the product of current and time) to moles of chemical change in electrochemical reactions.

THINKING IT THROUGH

The goal for the following problems is not to find the answers themselves, but rather to assemble the information needed to solve them and explain how you would use the information to find the answers. The problems in Level 2 are more challenging than those in Level 1 and may contain more data than are required, in which case you are also asked to identify the unnecessary data. Detailed answers to the Thinking-It-Through problems can be found on the web site.

Level 1 Problems

1. How can you determine whether the reaction $Au + NO_3^- + 4H^+ \rightarrow Au^{3+} + NO + 2H_2O$ will occur spontaneously?

2. In a galvanic cell established between $Fe(s)$ and $Fe^{2+}(aq)$ plus $Co(s)$ and $Co^{2+}(aq)$, how can you determine which electrode will carry the positive charge?

3. Explain how you would use data in this chapter to calculate the value of the equilibrium constant at 25 °C for the following reaction:

$$2Br^-(aq) + I_2(s) \rightleftharpoons Br_2(aq) + 2I^-(aq)$$

4. Explain how you would calculate the number of grams of Pb that will be deposited on the cathode of an electrolytic cell if 500 mL of a solution of $Pb(NO_3)_2$ is electrolyzed for a period of 30.00 min at a current of 4.00 A.

5. At a current of 2.00 A, describe in detail how you would calculate the number of minutes it would take to deposit a coating of silver on a metal object by electrolysis of $Ag(CN)_2^-$ if the silver coating is to be 0.020 mm thick and is to have an area of 25 cm². The density of silver is 10.50 g/cm³.

6. Suppose you carried out electrolysis on 250 mL of 1.00 M solution of $AgNO_3$ using a current of 1.00 A for a period of 30.0 min. Describe the calculations necessary to determine the pH of the solution after the electrolysis is over. Assume that NO_3^- ion is more difficult to oxidize than water.

Level 2 Problems

7. A current was passed through two electrolytic cells connected in series (so the same current passed through both of them). In one cell was a solution of $Cr_2(SO_4)_3$, and in the other, a solution of $AgNO_3$. If 1.45 g of Ag was deposited on the cathode in the second cell, how many grams of Cr were deposited on the cathode in the first cell? The temperature of the cell was 25 °C. Describe the calculations in detail.

8. The following cell was established:

$$Sn(s) \mid Sn^{2+}(? M) \parallel Cl^-(1.00 \times 10^{-2}M) \mid AgCl(s), Ag(s)$$

The notation on the right indicates that the cathode consists of a metallic silver electrode coated with solid silver chloride in contact with a solution containing chloride ion. The half-reaction for the cathode is

$$AgCl(s) + e^- \rightleftharpoons Ag(s) + Cl^-(aq)$$

The standard reduction potentials are $E°_{AgCl} = 0.2223$ V and $E°_{Sn^{2+}} = -0.1375$ V. The tin electrode was dipped into a

solution of Sn^{2+} at 25 °C and the potential of the cell was measured to be 0.3114 V. Describe how you would calculate the concentration of Sn^{2+} in the solution.

9. Describe how you would calculate the value of K_c at 25 °C for the reaction in Question 8.

10. Describe in detail how you would calculate the potential of a cell in which the cell reaction is

$$NiO_2(s) + 4H^+(aq) + 2Ag(s) \longrightarrow$$
$$Ni^{2+}(aq) + 2H_2O + 2Ag^+(aq)$$

if the pH is 2.50 and the half-cells contain 125 mL each of solutions containing Ni^{2+} and Ag^+, each at a concentration of 0.020 M.

11. How many milliliters of H_2 gas at 25 °C and a total pressure of 755 torr will be generated by the electrolysis of water using a KNO_3 electrolyte if a current of 1.25 A is passed through the electrolysis cell for a period of 10.00 min? Describe the calculation in detail.

12. Suppose a galvanic cell was set up at 25 °C using the reaction

$$Ni(s) + 2Ag^+(aq) \longrightarrow Ni^{2+}(aq) + 2Ag(s)$$

in which one half-cell contains a nickel electrode dipping into 125 mL of 0.100 M $NiSO_4$ and the other half-cell contains a silver electrode dipping into 125 mL of 0.100 M $AgNO_3$. If the cell supplies a constant current of 0.100 A for a period of 2.00 hr, what will the potential of the cell have become? Describe how you would perform the calculation.

13. Make a sketch of a galvanic cell that utilizes a standard hydrogen electrode and another electrode that would enable you to determine the K_{sp} for AgCl. Describe the contents of each half-cell, including concentrations of dissolved species. Explain how you would find the K_{sp} value from the measured cell potential.

14. A galvanic cell utilizing a copper electrode in 1.00 M $CuSO_4$ and a metal Z electrode in 1.00 M $Z(NO_3)_2$ is set up as shown in the figure below. In a second experiment, also shown below, a copper rod is placed in a beaker of 1.00 M $Z(NO_3)_2$ and a metal Z rod is placed in a 1.00 M $CuSO_4$ solution shown in the figure. The colorless solution of 1.00 M $Z(NO_3)_2$ is observed to develop a pale blue color, while the blue color of the 1.00 M $CuSO_4$ solution appears to remain constant.
(a) Write the chemical equation for the spontaneous cell reaction occurring in the galvanic cell.
(b) Which way do the electrons flow in the external circuit (toward or away from the copper electrode)?
(c) The salt bridge contains saturated KNO_3 solution. Toward which electrodes do the ions of the salt bridge move?

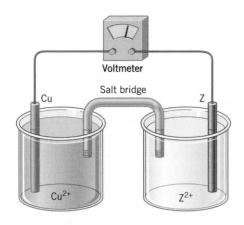 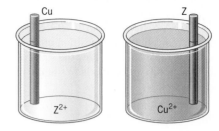

(d) In the galvanic cell, does the color of the $CuSO_4$ remain constant, fade, or become darker?

(e) Over time, will the voltage shown on the voltmeter remain constant, fall, or increase?

15. A galvanic cell was constructed at 25.0 °C by connecting a silver electrode, dipping into 1.000 *M* $AgNO_3$, to a hydrogen electrode (with the partial pressure of H_2 equal to 1.000 atm) dipping into a solution in which the hydrogen ion concentration was unknown. The potential of the cell was measured to be 1.0764 V and the silver electrode was the cathode. The standard reduction potential of Ag^+ has the value $E°_{Ag^+} = +0.7996$ V. Explain how you would calculate the pH of the solution in contact with the hydrogen electrode. Indicate the values of the constants you would use to be sure you have a sufficient number of significant figures.

REVIEW QUESTIONS

Galvanic Cells

21.1 What is a galvanic cell? What is a half-cell?

21.2 What is the function of a *salt bridge*?

21.3 In the copper–silver cell, why must the Cu^{2+} and Ag^+ solutions be kept in separate containers?

21.4 Which redox processes take place at the anode and cathode in a galvanic cell? What electrical charges do the anode and cathode carry in a galvanic cell?

21.5 In a galvanic cell, do electrons travel from anode to cathode, or from cathode to anode? Explain.

21.6 Explain how the movement of the ions relative to the electrodes is the same in both galvanic and electrolytic cells.

21.7 When magnesium metal is placed into a solution of copper sulfate, the magnesium dissolves to give Mg^{2+} and copper metal is formed. Write a net ionic equation for this reaction. Describe how you could use the reaction in a galvanic cell. Which metal, copper or magnesium, is the cathode?

21.8 Aluminum will displace tin from solution according to the equation $2Al(s) + 3Sn^{2+}(aq) \rightarrow 2Al^{3+}(aq) + 3Sn(s)$. What would be the individual half-cell reactions if this were the cell reaction in a galvanic cell? Which metal would be the anode and which the cathode?

21.9 At first glance, the following equation may appear to be balanced: $MnO_4^- + Sn^{2+} \rightarrow SnO_2 + MnO_2$. What is wrong with it?

Cell Potentials and Reduction Potentials

21.10 For a galvanic cell, what is the meaning of the term *potential*? What are its units?

21.11 What is the difference between a *cell potential* and a *standard cell potential*?

21.12 How are standard reduction potentials combined to give the standard cell potential for a spontaneous reaction?

21.13 What ratio of units gives volts? What are the units of amperes × volts × seconds?

21.14 Is it possible to measure the potential of an isolated half-cell? Explain your answer.

21.15 Describe the hydrogen electrode. What is the value of its standard reduction potential?

21.16 What do the positive and negative signs of reduction potentials tell us?

21.17 If $E°_{Cu^{2+}}$ had been chosen as the standard reference electrode and had been assigned a potential of 0.00 V, what would the reduction potential of the hydrogen electrode be relative to it?

21.18 If you set up a galvanic cell using metals not found in Table 21.1, what experimental information will tell you which is the anode and which is the cathode in the cell?

Using Standard Reduction Potentials

21.19 Compare Table 6.2 with Table 21.1. What can you say about the basis for the activity series for metals?

21.20 Make a sketch of a galvanic cell for which the cell notation is

$$Fe(s) \mid Fe^{3+}(aq) \parallel Ag^+(aq) \mid Ag(s)$$

(a) Label the anode and the cathode.

(b) Indicate the charge on each electrode.

(c) Indicate the direction of electron flow in the external circuit.

(d) Write the equation for the net cell reaction.

21.21 Make a sketch of a galvanic cell in which inert platinum electrodes are used in the half-cells for the system

$$Pt(s) \mid Fe^{2+}(aq), Fe^{3+}(aq) \parallel Br_2(aq), Br^-(aq) \mid Pt(s)$$

Label the diagram and indicate the composition of the electrolytes in the two cell compartments. Show the signs of the electrodes and label the anode and cathode. Write the equation for the net cell reaction.

Cell Potentials and Thermodynamics

21.22 Write the equation that relates the standard cell potential to the standard free energy change for a reaction.

21.23 What is the equation that relates the equilibrium constant to the cell potential?

21.24 Show how the equation that relates the equilibrium constant to the cell potential can be derived from the Nernst equation.

21.25 What is the maximum amount of work, expressed in joules, that can be obtained from the discharge of a silver oxide battery (see page 941) if 0.500 g of Ag_2O reacts? Assume the potential of the cell remains constant at 1.50 V.

The Effect of Concentration on Cell Potential

21.26 You have learned that the principles of thermodynamics allow the following equation to be derived: $\Delta G = \Delta G° + RT \ln Q$, where Q is the reaction quotient. Without referring to the text, use this equation and the relationship between ΔG and the cell potential to derive the Nernst equation.

21.27 The cell reaction during the discharge of a lead storage battery is

$$Pb(s) + PbO_2(s) + 2H^+(aq) + 2HSO_4^-(aq) \longrightarrow$$
$$2PbSO_4(s) + 2H_2O$$

The standard cell potential is 2.05 V. What is the correct form of the Nernst equation for this reaction at 25 °C?

Practical Galvanic Cells

21.28 What are the anode and cathode reactions during the discharge of a lead storage battery? How can a battery produce a potential of 12 V if the cell reaction has a standard potential of only 2 V?

21.29 What are the anode and cathode reactions during the charging of a lead storage battery?

21.30 How is a hydrometer constructed? How does it measure density? Why can a hydrometer be used to check the state of charge of a lead storage battery?

21.31 What role does the element calcium play in the construction of a modern automobile battery?

21.32 What reactions occur at the electrodes in the ordinary dry cell?

21.33 What chemical reactions take place at the electrodes in an alkaline dry cell?

21.34 Give the half-cell reactions and the cell reaction that take place in a nicad battery during discharge. What are the reactions that take place during the charging of the cell?

21.35 Describe the chemical reactions that take place in a silver oxide battery.

21.36 When used to describe the properties of batteries, what do the terms *energy density* and *specific energy* mean?

21.37 Which has the larger specific energy, a lead storage battery or a nickel–metal hydride battery? Explain.

21.38 How is hydrogen held as a reactant in a nickel–metal hydride battery? Write the chemical formula for a typical alloy used in this battery. What is the electrolyte?

21.39 What are the anode, cathode, and net cell reactions that take place in a nickel–metal hydride battery during discharge? What are the reactions when the battery is charged?

21.40 What is the principal advantage that a nickel–metal hydride battery has compared with a nickel–cadmium battery? How does their relatively rapid rate of self-discharge affect their shelf life? Would you use a Ni–MH battery in a wrist watch? Explain your answer.

21.41 Give two reasons why lithium is such an attractive anode material for use in a battery. What are the problems associated with using lithium for this purpose?

21.42 What are the electrode materials in a typical primary lithium cell? Write the equations for the anode, cathode, and cell reactions.

21.43 What problems are encountered in attempting to make reliable lithium-metal rechargeable cells?

21.44 What are the electrode materials in a typical lithium ion cell? Explain what happens when the cell is charged. Explain what happens when the cell is discharged.

21.45 Write the cathode, anode, and net cell reaction in a hydrogen–oxygen fuel cell.

21.46 What advantages do fuel cells offer over conventional means of obtaining electrical power by the combustion of fuels?

Electrolysis

21.47 What electrical charges do the anode and the cathode carry in an electrolytic cell? What does the term *inert electrode* mean?

21.48 How do *electrolytic conduction* and *metallic conduction* differ?

21.49 Why must electrolysis reactions occur at the electrodes in order for electrolytic conduction to continue?

21.50 What is the difference between a *direct current* and an *alternating current*?

21.51 Why must NaCl be melted before it is electrolyzed to give Na and Cl_2? Write the anode, cathode, and overall cell reactions for the electrolysis of molten NaCl.

21.52 Write half-reactions for the oxidation and the reduction of water.

21.53 What happens to the pH of the solution near the cathode and anode during the electrolysis of K_2SO_4? What function does K_2SO_4 serve in the electrolysis of a K_2SO_4 solution?

Stoichiometric Relationships in Electrolysis

21.54 What is a *faraday*? What relationships relate faradays to current and time measurements?

21.55 Using the same current, which will require the greater length of time, depositing 0.10 mol Cu from a Cu^{2+} solution, or depositing 0.10 mol of Cr from a Cr^{3+} solution? Explain your reasoning.

21.56 An electric current is passed through two electrolysis cells connected in series (so the same amount of current passes through each of them). One cell contains Cu^{2+} and the other contains Ag^+. In which cell will the larger number of moles of metal be deposited? Explain your answer.

21.57 An electric current is passed through two electrolysis cells connected in series (so the same amount of current passes through each of them). One cell contains Cu^{2+} and the other contains Fe^{2+}. In which cell will the greater mass of metal be deposited? Explain your answer.

Applications of Electrolysis

21.58 What is *electroplating*? Sketch an apparatus to electroplate silver.

21.59 Describe the Hall–Héroult process for producing metallic aluminum. What half-reaction occurs at the anode? What half-reaction occurs at the cathode? What is the overall cell reaction?

21.60 In the Hall–Héroult process, why must the carbon anodes be replaced frequently?

21.61 Describe by means of chemical equation how magnesium is recovered from seawater. What electrochemical reaction is involved?

21.62 How is metallic sodium produced? What are some uses of metallic sodium? Write equations for the anode and cathode reactions.

21.63 Describe the electrolytic refining of copper. What economic advantages offset the cost of electricity for this process? What chemical reactions occur at (a) the anode and (b) the cathode?

21.64 Describe the electrolysis of aqueous sodium chloride. How do the products of the electrolysis compare for stirred and unstirred reactions? Write chemical equations for the reactions that occur at the electrodes.

21.65 What are the advantages and disadvantages in the electrolysis of aqueous NaCl of (a) the diaphragm cell and (b) the mercury cell?

REVIEW PROBLEMS

Answers to problems whose numbers are printed in color are given in Appendix B. More challenging problems are marked with asterisks. **ILW** = Interactive LearningWare solution is available at *www.wiley.com/college/brady*.

Cell Notation

21.66 Write the half-reactions and the balanced cell reaction for each of the following galvanic cells:
(a) $Cd(s) \mid Cd^{2+}(aq) \parallel Au^{3+}(aq) \mid Au(s)$
(b) $Pb(s), PbSO_4(s) \mid HSO_4^-(aq) \parallel$
$\qquad\qquad H^+(aq), HSO_4^-(aq) \mid PbO_2(s), PbSO_4(s)$
(c) $Cr(s) \mid Cr^{3+}(aq) \parallel Cu^{2+}(aq) \mid Cu(s)$

21.67 Write the half-reactions and the balanced cell reaction for the following galvanic cells:
(a) $Zn(s) \mid Zn^{2+}(aq) \parallel Cr^{3+}(aq) \mid Cr(s)$
(b) $Fe(s) \mid Fe^{2+}(aq) \parallel Br_2(aq), Br^-(aq) \mid Pt(s)$
(c) $Mg(s) \mid Mg^{2+}(aq) \parallel Sn^{2+}(aq) \mid Sn(s)$

21.68 Write the cell notation for the following galvanic cells. For half-reactions in which all the reactants are in solution or are gases, assume the use of inert platinum electrodes.
(a) $Cd^{2+}(aq) + Fe(s) \rightarrow Cd(s) + Fe^{2+}(aq)$
(b) $Cl_2(g) + 2Br^-(aq) \rightarrow Br_2(aq) + 2Cl^-(aq)$
(c) $Au^{3+}(aq) + 3Ag(s) \rightarrow Au(s) + 3Ag^+(aq)$

21.69 Write the cell notation for the following galvanic cells. For half-reactions in which all the reactants are in solution or are gases, assume the use of inert platinum electrodes.
(a) $NO_3^-(aq) + 4H^+(aq) + 3Fe^{2+}(aq) \rightarrow$
$\qquad\qquad 3Fe^{3+}(aq) + NO(g) + 2H_2O$
(b) $NiO_2(s) + 4H^+(aq) + 2Ag(s) \rightarrow$
$\qquad\qquad Ni^{2+}(aq) + 2H_2O + 2Ag^+(aq)$
(c) $Mg(s) + Cd^{2+}(aq) \rightarrow Mg^{2+}(aq) + Cd(s)$

Using Reduction Potentials

21.70 For each pair of substances, use Table 21.1 to choose the better reducing agent.
(a) Sn(s) or Ag(s)   (c) Co(s) or Zn(s)
(b) $Cl^-(aq)$ or $Br^-(aq)$ (d) $I^-(aq)$ or Au(s)

21.71 For each pair of substances, use Table 21.1 to choose the better oxidizing agent.
(a) $NO_3^-(aq)$ or $MnO_4^-(aq)$ (c) $PbO_2(s)$ or $Cl_2(g)$
(b) $Au^{3+}(aq)$ or $Co^{2+}(aq)$ (d) $NiO_2(s)$ or $HOCl(aq)$

21.72 Use the data in Table 21.1 to calculate the standard cell potential for each of the following reactions:
(a) $Cd^{2+}(aq) + Fe(s) \rightarrow Cd(s) + Fe^{2+}(aq)$
(b) $Br_2(aq) + 2Cl^-(aq) \rightarrow Cl_2(g) + 2Br^-(aq)$
(c) $Au^{3+}(aq) + 3Ag(s) \rightarrow Au(s) + 3Ag^+(aq)$

21.73 Use the data in Table 21.1 to calculate the standard cell potential for each of the following reactions:
(a) $NO_3^-(aq) + 4H^+(aq) + 3Fe^{2+}(aq) \rightarrow$
$\qquad\qquad 3Fe^{3+}(aq) + NO(g) + 2H_2O$
(b) $NiO_2(s) + 4H^+(aq) + 2Ag(s) \rightarrow$
$\qquad\qquad Ni^{2+}(aq) + 2H_2O + 2Ag^+(aq)$
(c) $Mg(s) + Cd^{2+}(aq) \rightarrow Mg^{2+}(aq) + Cd(s)$

21.74 From the positions of the half-reactions in Table 21.1, determine whether the following reactions are spontaneous under standard state conditions:
(a) $2Au^{3+} + 6I^- \rightarrow 3I_2 + 2Au$
(b) $3Fe^{2+} + 2NO + 4H_2O \rightarrow 3Fe + 2NO_3^- + 8H^+$
(c) $3Ca + 2Cr^{3+} \rightarrow 2Cr + 3Ca^{2+}$

21.75 Use the data in Table 21.1 to determine which of the following reactions should occur spontaneously under standard state conditions:
(a) $Br_2 + 2Cl^- \rightarrow Cl_2 + 2Br^-$
(b) $Ni^{2+} + Fe \rightarrow Fe^{2+} + Ni$
(c) $H_2SO_3 + H_2O + Br_2 \rightarrow 4H^+ + SO_4^{2-} + 2Br^-$

21.76 Use the data in Table 21.1 to calculate the standard cell potential for the reaction

$$MnO_4^- + 8H^+ + 5Fe^{2+} \rightarrow 5Fe^{3+} + Mn^{2+} + 4H_2O$$

21.77 Use the data in Table 21.1 to calculate the standard cell potential for the reaction

$$2Ag^+ + Fe \longrightarrow 2Ag + Fe^{2+}$$

21.78 Write the cell notation for the galvanic cells formed by using the following pairs of half-reactions. Calculate their standard cell potentials. For each pair, identify the anode and cathode.
(a) $Co^{2+}(aq) + 2e^- \rightleftharpoons Co(s)$
    $Zn^{2+}(aq) + 2e^- \rightleftharpoons Zn(s)$
(b) $Ni^{2+}(aq) + 2e^- \rightleftharpoons Ni(s)$
    $Mg^{2+}(aq) + 2e^- \rightleftharpoons Mg(s)$
(c) $Au^{3+}(aq) + 3e^- \rightleftharpoons Au(s)$
    $Sn^{2+}(aq) + 2e^- \rightleftharpoons Sn(s)$

21.79 Write the cell notation for the galvanic cells formed by using the following pairs of half-reactions. For half-reactions in which all the reactants are in solution, assume the use of inert platinum electrodes. Calculate their standard cell potentials. For each pair, identify the anode and cathode.
(a) $BrO_3^-(aq) + 6H^+(aq) + 6e^- \rightleftharpoons Br^-(aq) + 3H_2O$
    $Cu^{2+}(aq) + 2e^- \rightleftharpoons Cu(s)$
(b) $Fe^{3+}(aq) + e^- \rightleftharpoons Fe^{2+}(aq)$
    $Ag^+(aq) + e^- \rightleftharpoons Ag(s)$
(c) $NO_3^-(aq) + 4H^+(aq) + 3e^- \rightleftharpoons NO(g) + 2H_2O$
    $MnO_4^-(aq) + 8H^+(aq) + 5e^- \rightleftharpoons Mn^{2+}(aq) + 4H_2O$

ILW 21.80 From the half-reactions below, determine the cell reaction and standard cell potential.

$$BrO_3^- + 6H^+ + 6e^- \rightleftharpoons Br^- + 3H_2O \qquad E°_{BrO_3^-} = 1.44\,V$$

$$I_2 + 2e^- \rightleftharpoons 2I^- \qquad E°_{I_2} = 0.54\,V$$

21.81 What is the standard cell potential and the net reaction in a galvanic cell that has the following half-reactions?

$$MnO_2 + 4H^+ + 2e^- \rightleftharpoons Mn^{2+} + 2H_2O \qquad E°_{MnO_2} = 1.23\,V$$

$$PbCl_2 + 2e^- \rightleftharpoons Pb + 2Cl^- \qquad E°_{PbCl_2} = -0.27\,V$$

21.82 What will be the spontaneous reaction among H_2SO_3, $S_2O_3^{2-}$, $HOCl$, and Cl_2? The half-reactions involved are

$$2H_2SO_3 + 2H^+ + 4e^- \rightleftharpoons S_2O_3^{2-} + 3H_2O \quad E°_{H_2SO_3} = 0.40\,V$$

$$2HOCl + 2H^+ + 2e^- \rightleftharpoons Cl_2 + 2H_2O \qquad E°_{HOCl} = 1.63\,V$$

21.83 What will be the spontaneous reaction among Br_2, I_2, Br^-, and I^-?

21.84 Will the following reaction occur spontaneously under standard state conditions?

$$SO_4^{2-} + 4H^+ + 2I^- \longrightarrow H_2SO_3 + I_2 + H_2O$$

Use $E°_{cell}$ calculated from data in Table 21.1 to answer this question.

21.85 Determine whether the reaction

$$S_2O_8^{2-} + Ni(OH)_2 + 2OH^- \longrightarrow 2SO_4^{2-} + NiO_2 + 2H_2O$$

will occur spontaneously under standard state conditions. Use $E°_{cell}$ calculated from the data below to answer the question.

$$NiO_2 + 2H_2O + 2e^- \rightleftharpoons Ni(OH)_2 + 2OH^- \quad E°_{NiO_2} = 0.49\,V$$

$$S_2O_8^{2-} + 2e^- \rightleftharpoons 2SO_4^{2-} \qquad E°_{S_2O_8^{2-}} = 2.01\,V$$

Cell Potentials and Thermodynamics

ILW 21.86 Calculate $\Delta G°$ for the following reaction *as written:*

$$2Br^- + I_2 \longrightarrow 2I^- + Br_2$$

21.87 Calculate $\Delta G°$ for the reaction

$$2MnO_4^- + 6H^+ + 5HCHO_2 \longrightarrow 2Mn^{2+} + 8H_2O + 5CO_2$$

for which $E°_{cell} = 1.69\,V$.

21.88 Given the following half-reactions and their standard reduction potentials,

$$2ClO_3^- + 12H^+ + 10e^- \rightleftharpoons Cl_2 + 6H_2O \quad E°_{ClO_3^-} = 1.47\,V$$

$$S_2O_8^{2-} + 2e^- \rightleftharpoons 2SO_4^{2-} \qquad E°_{S_2O_8^{2-}} = 2.01\,V$$

calculate (a) $E°_{cell}$, (b) $\Delta G°$ for the cell reaction, and (c) the value of K_c for the cell reaction.

21.89 Calculate K_c for the system $Ni^{2+} + Co \rightleftharpoons Ni + Co^{2+}$. Use the data in Table 21.1. Assume $T = 298\,K$.

21.90 The system $2AgI + Sn \rightleftharpoons Sn^{2+} + 2Ag + 2I^-$ has a calculated $E°_{cell} = -0.015\,V$. What is the value of K_c for this system?

21.91 Determine the value of K_c at 25 °C for the reaction

$$2H_2O + 2Cl_2 \rightleftharpoons 4H^+ + 4Cl^- + O_2$$

The Effect of Concentration on Cell Potential

ILW 21.92 The cell reaction

$$NiO_2(s) + 4H^+(aq) + 2Ag(s) \longrightarrow$$
$$Ni^{2+}(aq) + 2H_2O + 2Ag^+(aq)$$

has $E°_{cell} = 2.48\,V$. What will be the cell potential at a pH of 5.00 when the concentrations of Ni^{2+} and Ag^+ are each 0.010 M?

21.93 $E°_{cell} = 0.135\,V$ for the reaction

$$3I_2(s) + 5Cr_2O_7^{2-}(aq) + 34H^+ \longrightarrow$$
$$6IO_3^-(aq) + 10Cr^{3+}(aq) + 17H_2O$$

What is E_{cell} if $[Cr_2O_7^{2-}] = 0.010\,M$, $[H^+] = 0.10\,M$, $[IO_3^-] = 0.00010\,M$, and $[Cr^{3+}] = 0.0010\,M$?

21.94 A cell was set up having the following reaction:

$$Mg(s) + Cd^{2+}(aq) \rightarrow Mg^{2+}(aq) + Cd(s) \qquad E°_{cell} = 1.97\,V$$

The magnesium electrode was dipping into a 1.00 M solution of $MgSO_4$ and the cadmium electrode was dipping into a solution of unknown Cd^{2+} concentration. The potential of the cell was measured to be 1.54 V. What was the unknown Cd^{2+} concentration?

21.95 A silver wire coated with AgCl is sensitive to the presence of chloride ion because of the half-cell reaction

$$AgCl(s) + e^- \rightleftharpoons Ag(s) + Cl^- \qquad E°_{AgCl} = 0.2223 \text{ V}$$

A student, wishing to measure the chloride ion concentration in a number of water samples, constructed a galvanic cell using the AgCl electrode as one half-cell and a copper wire dipping into 1.00 M CuSO$_4$ solution as the other half-cell. In one analysis, the potential of the cell was measured to be 0.0925 V, with the copper half-cell serving as the cathode. What was the chloride ion concentration in the water? (Take $E°_{Cu^{2+}} = 0.3419$ V.)

*21.96 At 25 °C, a galvanic cell was set up having the following half-reactions:

$$Fe^{2+}(aq) + 2e^- \rightleftharpoons Fe(s) \qquad E°_{Fe^{2+}} = -0.447 \text{ V}$$

$$Cu^{2+}(aq) + 2e^- \rightleftharpoons Cu(s) \qquad E°_{Cu^{2+}} = +0.3419 \text{ V}$$

The copper half-cell contained 100 mL of 1.00 M CuSO$_4$. The iron half-cell contained 50.0 mL of 0.100 M FeSO$_4$. To the iron half-cell was added 50.0 mL of 0.500 M NaOH solution. The mixture was stirred and the cell potential was measured to be 1.175 V. Calculate the value of K_{sp} for Fe(OH)$_2$.

*21.97 Suppose a galvanic cell was constructed at 25 °C using a Cu/Cu^{2+} half-cell (in which the molar concentration of Cu^{2+} was 1.00 M) and a hydrogen electrode having a partial pressure of H$_2$ equal to 1 atm. The hydrogen electrode dips into a solution of unknown hydrogen ion concentration, and the two half-cells are connected by a salt bridge. The precise value of $E°_{Cu^{2+}}$ is +0.3419 V.
(a) Derive an equation for the pH of the solution with the unknown hydrogen ion concentration, expressed in terms of E_{cell} and $E°_{cell}$.
(b) If the pH of the solution were 5.15, what would be the observed potential of the cell?
(c) If the potential of the cell were 0.645 V, what would be the pH of the solution?

Quantitative Aspects of Electrochemical Reactions

21.98 How many coulombs are passed through an electrolysis cell by (a) a current of 4.00 A for 600 s? (b) A current of 10.0 A for 20.0 min? (c) A current of 1.50 A for 6.00 hr?

21.99 How many moles of electrons correspond to the answers to each part of Problem 21.98?

21.100 How many moles of electrons are required to
(a) Reduce 0.20 mol Fe^{2+} to Fe?
(b) Oxidize 0.70 mol Cl$^-$ to Cl$_2$?
(c) Reduce 1.50 mol Cr^{3+} to Cr?
(d) Oxidize 1.0×10^{-2} mol Mn^{2+} to MnO$_4^-$?

21.101 How many moles of electrons are required to (a) produce 5.00 g Mg from molten MgCl$_2$? (b) Form 41.0 g Cu from a CuSO$_4$ solution?

21.102 How many moles of Cr^{3+} would be reduced to Cr by the same amount of electricity that produces 12.0 g Ag from a solution of AgNO$_3$?

21.103 In Problem 21.102, if a current of 4.00 A was used, how many minutes would the electrolysis take?

ILW 21.104 How many grams of Fe(OH)$_2$ are produced at an iron anode when a basic solution undergoes electrolysis at a current of 8.00 A for 12.0 min?

21.105 How many grams of Cl$_2$ would be produced in the electrolysis of molten NaCl by a current of 4.25 A for 35.0 min?

ILW 21.106 How many hours would it take to produce 75.0 g of metallic chromium by the electrolytic reduction of Cr^{3+} with a current of 2.25 A?

21.107 How many hours would it take to generate 35.0 g of lead from PbSO$_4$ during the charging of a lead storage battery using a current of 1.50 A? The half-reaction is

$$PbSO_4 + H^+ + 2e^- \longrightarrow Pb + HSO_4^-$$

21.108 How many amperes would be needed to produce 60.0 g of magnesium during the electrolysis of molten MgCl$_2$ in 2.00 hr?

21.109 A large electrolysis cell that produces metallic aluminum from Al$_2$O$_3$ by the Hall–Héroult process is capable of yielding 900 lb (409 kg) of aluminum in 24 hr. What current is required?

21.110 The electrolysis of 250 mL of a brine solution (NaCl) was carried out for a period of 20.00 min with a current of 2.00 A. The resulting solution was titrated with 0.620 M HCl. How many milliliters of the HCl solution were required for the titration?

21.111 An unstirred solution of 2.00 M NaCl was electrolyzed for a period of 25.0 min and then titrated with 0.250 M HCl. The titration required 15.5 mL of the acid. What was the average current in amperes during the electrolysis?

21.112 A solution of NaCl in water was electrolyzed with a current of 2.50 A for 15.0 min. How many milliliters of Cl$_2$ gas would be formed if it was collected over water at 25 °C and a total pressure of 750 torr?

21.113 How many milliliters of dry gaseous H$_2$, measured at 20 °C and 735 torr, would be produced at the cathode in the electrolysis of dilute H$_2$SO$_4$ with a current of 0.750 A for 15.00 min?

Predicting Electrolysis Reactions

21.114 If electrolysis is carried out on an aqueous solution of aluminum sulfate, what products are expected at the electrodes? Write the equation for the net cell reaction.

21.115 If electrolysis is carried out on an aqueous solution of cadmium iodide, what products are expected at the electrodes? Write the equation for the net cell reaction.

21.116 Write the anode reaction for the electrolysis of an aqueous solution that contains (a) SO$_4^{2-}$, (b) Br$^-$, and (c) SO$_4^{2-}$ and Br$^-$.

21.117 Write the cathode reaction for the electrolysis of an aqueous solution that contains (a) K$^+$, (b) Cu^{2+}, and (c) K$^+$ and Cu^{2+}.

21.118 What products would we expect at the electrodes if a solution containing both KBr and CuSO$_4$ were electrolyzed? Write the equation for the net cell reaction.

21.119 What products would we expect at the electrodes if a solution containing both BaCl$_2$ and CuI$_2$ were electrolyzed? Write the equation for the net cell reaction.

ADDITIONAL EXERCISES

*21.120 A watt is a unit of electrical power and is equal to 1 joule per second (1 watt = 1 J s^{-1}). How many hours can a calculator drawing 5×10^{-4} watt be operated by a mercury battery having a cell potential equal to 1.34 V if a mass of 1.00 g of HgO is available at the cathode? The cell reaction is

$$HgO(s) + Zn(s) \longrightarrow ZnO(s) + Hg(l)$$

*21.121 Suppose that a galvanic cell were set up having the net cell reaction

$$Zn(s) + 2Ag^+(aq) \longrightarrow Zn^{2+}(aq) + 2Ag(s)$$

The Ag^+ and Zn^{2+} concentrations in their respective half-cells initially are 1.00 M, and each half-cell contains 100 mL of electrolyte solution. If this cell delivers current at a constant rate of 0.10 A, what will the cell potential be after 15.00 hr?

21.122 It was desired to determine the reduction potential of Pt^{2+} ion. A galvanic cell was set up in which one half-cell consisted of a Pt electrode dipping into a 0.0100 M solution of $Pt(NO_3)_2$ and the other was prepared by dipping a silver wire coated with AgCl into a 0.100 M solution of HCl. The potential of the cell was measured to be 0.778 V, and it was found that the Pt electrode carried a positive charge. Given the following half-reaction and its reduction potential,

$$AgCl(s) + e^- \rightleftharpoons Ag(s) + Cl^-(aq) \qquad E^\circ_{AgCl} = 0.2223 \text{ V}$$

calculate the standard reduction potential for the half-reaction

$$Pt^{2+}(aq) + 2e^- \rightleftharpoons Pt(s)$$

*21.123 The value of K_{sp} for AgBr is 5.0×10^{-13}. What will be the potential of a cell constructed of a standard hydrogen electrode as one half-cell and a silver wire coated with AgBr dipping into 0.10 M HBr as the other half-cell. For the Ag/AgBr electrode,

$$AgBr(s) + e^- \rightleftharpoons Ag(s) + Br^-(aq) \qquad E^\circ_{AgBr} = +0.070 \text{ V}$$

*21.124 A student set up an electrolysis apparatus and passed a current of 1.22 A through a 3 M H_2SO_4 solution for 30.0 min. The H_2 formed at the cathode was collected and found to have a volume, over water at 27 °C, of 288 mL at a total pressure of 767 torr. Use these data to calculate the charge on the electron, expressed in coulombs.

*21.125 A hydrogen electrode is immersed in a 0.10 M solution of acetic acid at 25 °C. This electrode is connected to another consisting of an iron nail dipping into 0.10 M $FeCl_2$. What will be the measured potential of this cell? Assume $P_{H_2} = 1.00$ atm.

21.126 Consider the following half-reactions and their reduction potentials:

$$MnO_4^- + 8H^+ + 5e^- \longrightarrow Mn^{2+} + 4H_2O$$

$$E^\circ_{MnO_4^-} = +1.507 \text{ V}$$

$$ClO_3^- + 6H^+ + 6e^- \longrightarrow Cl^- + 3H_2O$$

$$E^\circ_{ClO_3^-} = +1.451 \text{ V}$$

(a) What is the value of E°_{cell}?
(b) What is the value of ΔG° for this reaction?
(c) What is the value of K_c for the reaction at 25 °C?
(d) Write the Nernst equation for this reaction.

(e) What is the potential of the cell when $[MnO_4^-]$ = 0.20 M, $[Mn^{2+}]$ = 0.050 M, $[Cl^-]$ = 0.0030 M, $[ClO_3^-]$ = 0.110 M, and the pH of the solution is 4.25?

21.127 In the absence of unusual electrode effects, what should be expected to be observed at the cathode when electrolysis is performed on a solution that contains $NiSO_4$ and $CdSO_4$ at equal concentrations? (See Table 21.1.)

*21.128 What current would be required to deposit 1.00 m^2 of chrome plate having a thickness of 0.050 mm in 4.50 hr from a solution of H_2CrO_4? The density of chromium is 7.19 g cm^{-3}.

21.129 A solution containing vanadium (chemical symbol V) in an unknown oxidation state was electrolyzed with a current of 1.50 A for 30.0 min. It was found that 0.475 g of V was deposited on the cathode. What was the original oxidation state of the vanadium ion?

*21.130 Calculate the value of ΔG (in kilojoules) for a system containing the following species at the concentrations in parentheses: Mn^{2+} (0.10 M), $Cr_2O_7^{2-}$ (0.010 M), MnO_4^- (0.0010 M), Cr^{3+} (0.0010 M). The pH of the solution is 6.00. The reaction you should consider is

$$6Mn^{2+}(aq) + 5Cr_2O_7^{2-}(aq) + 22H^+(aq) \rightleftharpoons$$
$$6MnO_4^-(aq) + 10Cr^{3+}(aq) + 11H_2O(l)$$

Use data in Table 21.1 and Appendix C to determine the direction the reaction will proceed to reach equilibrium, given the starting condition described above.

21.131 A solution of NaCl is neutral, with an expected pH of 7. If electrolysis were carried out on 500 mL of a NaCl solution with a current of 0.500 A, how many seconds would it take for the pH of the solution to rise to a value of 9.00?

21.132 What masses of H_2 and O_2 in grams would have to react each second in a fuel cell at 110 °C to provide 1.00 kilowatt (kW) of power if we assume a thermodynamic efficiency of 70%? (*Hint:* Use data in Chapters 7 and 20 to compute the value of ΔG° for the reaction $H_2(g) + \frac{1}{2}O_2(g) \rightarrow H_2O(g)$ at 110 °C. 1 watt = 1 J s^{-1}.)

21.133 How much work in kilojoules is able to be accomplished by a 5.00 minute flow of electricity having a voltage of 110 V and current of 1.00 A?

*21.134 A Ag/AgCl electrode dipping into 1.00 M HCl has a standard reduction potential of +0.2223 V. The half-reaction is

$$AgCl(s) + e^- \rightleftharpoons Ag(s) + Cl^-(aq)$$

A second Ag/AgCl electrode is dipped into a solution containing Cl^- at an unknown concentration. The cell generates a potential of 0.0435 V, with the electrode in the solution of unknown concentration having a negative charge. What is the molar concentration of Cl^- in the unknown solution?

21.135 Consider the following galvanic cell:

$$Ag(s) \mid Ag^+(3.0 \times 10^{-4} \ M) \parallel$$
$$Fe^{3+}(1.1 \times 10^{-3} \ M), Fe^{2+}(0.040 \ M) \mid Pt(s)$$

Calculate the cell potential. Determine the sign of the electrodes in the cell. Write the equation for the spontaneous cell reaction.

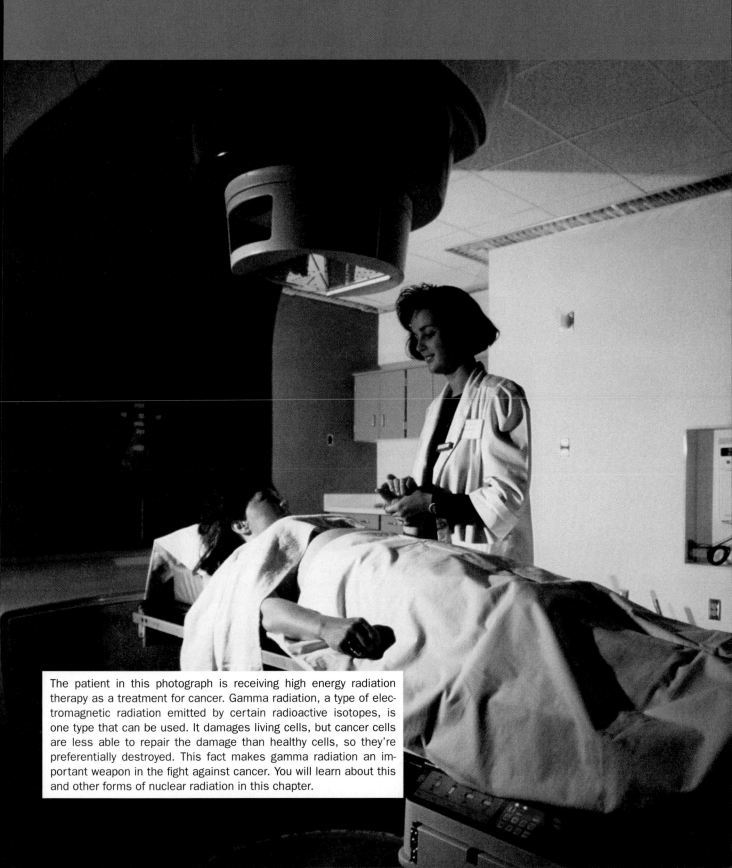

The patient in this photograph is receiving high energy radiation therapy as a treatment for cancer. Gamma radiation, a type of electromagnetic radiation emitted by certain radioactive isotopes, is one type that can be used. It damages living cells, but cancer cells are less able to repair the damage than healthy cells, so they're preferentially destroyed. This fact makes gamma radiation an important weapon in the fight against cancer. You will learn about this and other forms of nuclear radiation in this chapter.

THIS CHAPTER IN CONTEXT As in an ocean storm, nearly all of the action in chemistry occurs near the surface, among electron energy levels, and not in the depths among atomic nuclei. The nuclei of most isotopes are exceptionally stable and are completely unaffected by events high above. But there are exceptions. Many elements have one or more isotopes with unstable nuclei, resulting in properties both useful and dangerous. You will study them in this chapter.

22.1 ▶ Mass and energy are conserved in *all* of their forms

Isotopes with unstable atomic nuclei are **radioactive,** so called because they emit high-energy streams of particles or electromagnetic radiation, sometimes both at the same time. Often called **radionuclides,** radioactive isotopes undergo *nuclear reactions,* changes new to our study. Scientists exploit these reactions for many reasons, such as to study chemical changes, to conduct many kinds of analyses (including the dating of rocks and ancient objects), to diagnose or treat many diseases, and to obtain energy.

Mass has an energy equivalent

To begin our study of radionuclides we reexamine two physical laws that, until this chapter, have been assumed to be separate and independent, namely, the laws of conservation of energy and conservation of mass. They may be safely treated as distinct for chemical reactions but not for nuclear reactions. These two laws, however, are only different aspects of a deeper, more general law.

As atomic and nuclear physics developed in the early 1900s, physicists realized that the mass of a particle cannot be treated as a constant in all circumstances. The mass, m, of a particle depends on the particle's velocity, v, meaning the velocity relative to the observer. A particle's mass is related to this velocity, and to the velocity of light, c, by the following equation:

$c = 3.00 \times 10^8$ m s^{-1}, the speed of light.

$$m = \frac{m_0}{\sqrt{1 - (v/c)^2}} \qquad (22.1)$$

Notice what happens when v is zero and the particle has no velocity (relative to the observer). The ratio v/c is then zero, the whole denominator reduces to a value of 1, and Equation 22.1 becomes

$$m = m_0$$

This is why the symbol m_0 stands for the particle's *rest mass.*

Rest mass is what we measure in all lab operations, because any object, like a chemical sample, is either at rest (from our viewpoint) or is not moving extraordinarily rapidly. Only as the particle's velocity approaches the speed of light, c, does the v/c term in Equation 22.1 become important. As v approaches c, the ratio v/c approaches 1, and so $[1 - (v/c)^2]$ gets closer and closer to 0. The whole denominator, in other words, approaches a value of 0. If it actually reached 0, then m, which would be $(m_0 \div 0)$, would go to infinity. In other words, the mass, m, of the particle moving at the velocity of light would be infinitely great, a physical impossibility. This is why the speed of light is seen as an absolute upper limit on the speed that any particle can have.

Even at $v = 1000$ m s^{-1} (about 2250 mph), the denominator (unrounded) is 0.99999333, or within 7×10^{-4}% of 1.

At the velocities of everyday experience, the mass of anything calculated by Equation 22.1 equals the rest mass to several significant figures. The difference cannot be detected by weighing devices. Thus, in all of our normal work, mass appears to be conserved, and the law of conservation of mass functions this way in chemistry.

In a universe where mass changes with velocity, physicists also realized that a new understanding of energy is necessary. Mass and energy were seen to be interconvertible, just like potential and kinetic energies. What is conserved with respect to the energy of a system is *all* its forms of energy, including the system's mass calculated as an equivalent amount of energy. Alternatively, what is conserved about the mass of a system is *all* its forms, including the system's energy expressed as an equivalent amount of mass. The single, deeper law that summarizes these conclusions is now called the **law of conservation of mass–energy.**

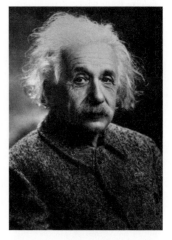

Law of Conservation of Mass–Energy
The sum of all the energy in the universe and of all the mass (expressed as an equivalent in energy) is a constant.

Albert Einstein (1879–1955); Nobel prize, 1921 (physics).

The Einstein equation quantitatively relates rest mass and energy

Albert Einstein, one of the many famous physicists of the twentieth century, was able to show that when mass converts to energy, the amount of the change in energy, ΔE, is related to the change in rest mass, Δm_0, by the following equation, now called the **Einstein equation:**

The Einstein equation is given as $E = mc^2$ in the popular press.

$$\Delta E = \Delta m_0 c^2$$

Einstein equation

Again, c is the velocity of light, 3.00×10^8 m s^{-1}.

Because the velocity of light is very large, even an enormous value of ΔE corresponds to an extremely small change in mass, Δm_0. The combustion of methane, for example, releases considerable heat per mole.

$$CH_4(g) + 2O_2(g) \longrightarrow CO_2(g) + 2H_2O(l) \qquad \Delta H° = -890 \text{ kJ}$$

The source of the 890 kJ of heat energy is a loss of mass. When calculated by the Einstein equation, the 890 kJ originates from the conversion into energy of 9.89 ng of mass, about 1×10^{-7}% of the mass of 1 mol of CH_4 and 2 mol of O_2. Such a tiny change is not detectable by laboratory balances. Although the Einstein equation has no practical application in chemistry, its importance certainly became clear in physics when atomic fission was first observed in 1939.

22.2 ▶ The energy required to break a nucleus into separate nucleons is called the nuclear binding energy

The actual mass of an atomic nucleus is always a little smaller than the sum of the rest masses of all of its *nucleons,* its protons and neutrons. The mass difference, sometimes called the **mass defect,** represents mass that changed into energy as the nucleons gathered to form the nucleus, this energy being released from the system.

The absorption of the same amount of energy by the nucleus would be required to break it apart into its nucleons again, so the energy is called the **binding energy** of the nucleus.

Nuclear binding energy

Nuclear binding energy is not energy actually possessed by the nucleus but is, instead, the energy the nucleus would have to absorb to break apart. Thus, the *higher* the binding energy, the *more stable* is the nucleus.

Nuclear binding energy of helium

The atomic mass unit, u, equals $1.6605389 \times 10^{-24}$ g.

We can calculate nuclear binding energy using the Einstein equation. Helium-4, for example, has atomic number 2, so its nucleus consists of 4 nucleons, 2 protons and 2 neutrons. The rest mass of one helium-4 nucleus is known to be 4.0015061792 u. However, the sum of the rest masses of its four separated nucleons is slightly more, 4.0318827650 u, which we can show as follows. The rest mass of an isolated proton is 1.0072764669 u and that of a neutron is 1.0086649156 u.

$$\text{For 2 protons: } 2 \times 1.0072764669 \text{ u} = 2.0145529338 \text{ u}$$

$$\text{For 2 neutrons: } 2 \times 1.0086649156 \text{ u} = \underline{2.0173298312 \text{ u}}$$

$$\text{Total rest mass of nucleons in } {}^4\text{He} = 4.0318827650 \text{ u}$$

The mass defect, the difference between the calculated and measured rest masses for the helium-4 nucleus, is 0.030375858 u, found by

$$4.0318827650 \text{ u} - 4.0015061792 \text{ u} = 0.030375858 \text{ u}$$

Using Einstein's equation, the nuclear binding energy that is equivalent to the mass defect of 0.030375858 u is found as follows, where we have to convert u (the atomic mass unit) to kilograms by the relationships 1 u = $1.6605389 \times 10^{-24}$ g, 1 kg = 1000 g, and 1 J = 1 kg m^2 s^{-2}:

$$\Delta E = \Delta mc^2 = \underbrace{(0.030375858 \text{ u}) \frac{(1.6605389 \times 10^{-24} \text{ g})}{1 \text{ u}} \frac{1 \text{ kg}}{1000 \text{ g}}}_{\Delta m \text{ (in kg)}} \underbrace{(3.00 \times 10^8 \text{ m s}^{-1})^2}_{c^2}$$

$$= 4.54 \times 10^{-12} \text{ kg m}^2 \text{ s}^{-2}$$

$$= 4.54 \times 10^{-12} \text{ J}$$

There are four nucleons in the helium-4 nucleus, so the binding energy per nucleon is $(4.54 \times 10^{-12} \text{ J})/4$ nucleons or 1.14×10^{-12} J/nucleon.

The formation of just one nucleus of helium-4 releases 4.54×10^{-12} J. If we could make Avogadro's number or 1 mol of helium-4 nuclei—the total mass would be only 4 g—the net release of energy would be

$$(6.02 \times 10^{23} \text{ nuclei}) \times (4.54 \times 10^{-12} \text{ J/nucleus}) = 2.73 \times 10^{12} \text{ J}$$

This is a huge amount of energy from forming only 4 g of helium. It could keep a 100 watt lightbulb lit for nearly 900 years!

Quite often you'll see the terms *atomic fusion* and *atomic fission* used for *nuclear fusion* and *nuclear fission.*

The formation of a nucleus from its nucleons is called **nuclear fusion,** and our calculation shows that the energy potentially available from nuclear fusion is enormous. The technological problems of achieving sustained fusion are immense, however, and practical power from fusion appears to be decades off.

Relationship between binding energy per nucleon and nuclear stability

Figure 22.1 shows a plot of binding energies per nucleon versus mass numbers for most of the elements. The curve passes through a maximum at iron-56, which means that the nuclei of iron-56 atoms are the most stable of all. The plot in Figure 22.1, however, does not have a sharp maximum. Thus, a large number of

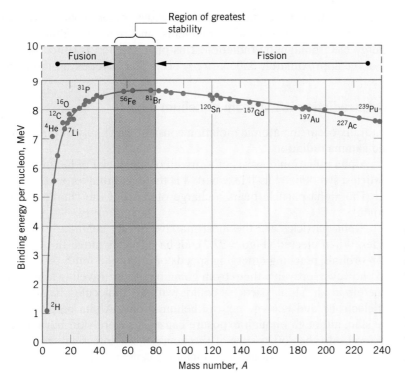

FIGURE 22.1 *Binding energies per nucleon.* The energy unit here is the megaelectron volt or MeV: 1 MeV = 10^6 eV = 1.602×10^{-13} J. (From D. Halliday and R. Resnick, *Fundamentals of Physics,* 2nd ed., Revised, 1986. John Wiley & Sons, Inc. Used by permission.)

elements with intermediate mass numbers in the broad center of the periodic table include the most stable isotopes in nature.

As we continue to follow the plot of Figure 22.1 to the highest mass numbers, the nuclei decrease in stability as the binding energies decrease. Among the heaviest atoms, therefore, we might expect to find isotopes that could change to more stable forms by breaking up, by undergoing nuclear fission. **Nuclear fission,** the subject of Section 22.8, is the spontaneous breaking apart of a nucleus to form isotopes of intermediate mass number.

22.3 ▶ Radioactivity is an emission of particles and/or electromagnetic radiation by unstable atomic nuclei

Except for hydrogen, all atomic nuclei have more than one proton. Each proton bears a positive charge, and we have learned that like charges repel. So we might well ask how can *any* nucleus be stable? Electrostatic forces of attraction and repulsion, such as the kinds present among the ions in a crystal of sodium chloride, are not the only forces at work in the nucleus, however. Protons do, indeed, repel each other electrostatically, but another force, a force of attraction called the *nuclear strong force,* also acts in the nucleus. The nuclear strong force overcomes the electrostatic force of repulsion between protons, and it binds both protons and neutrons into a nuclear package. Moreover, the neutrons, by helping to keep the protons farther apart, also lessen repulsions between protons.

Still another nuclear force is called the *weak force,* and it is involved in beta decay.

Radioactive decay

One consequence of the difference between the nuclear strong force and the electrostatic force occurs among nuclei carrying large numbers of protons but with too few intermingled neutrons to dilute the electrostatic repulsions between protons. Such nuclei are often unstable and occur among radionuclides. Their nuclei carry more energy than do other arrangements of the same nucleons. To achieve less

Adjacent neutrons experience no electrostatic repulsion between each other, only the strong force (of attraction).

energy and thus more stability, radionuclides undergo **radioactive decay,** meaning that they eject small nuclear fragments, and many simultaneously release high-energy electromagnetic radiation. This phenomenon is called **radioactivity.** About 50 of the approximately 350 naturally occurring isotopes are radioactive.

Alpha radiation is a stream of helium nuclei

Naturally occurring atomic radiation consists principally of three kinds: alpha, beta, and gamma radiation.

Alpha-emitting radionuclides are common among the elements near the end of the periodic table.

Alpha radiation consists of a stream of the nuclei of helium atoms, called **alpha particles,** symbolized as ^4_2He, where 4 is the mass number and 2 is the atomic number. The alpha particle bears a charge of 2+, but the charge is omitted from the symbol.

Alpha particles are the most massive of any commonly emitted by radionuclides. When ejected (Figure 22.2), alpha particles move through the atom's electron orbitals, reaching emerging speeds of up to one-tenth the speed of light. Their size, however, prevents them from going far. After traveling at most only a few centimeters in air, alpha particles collide with air molecules, lose kinetic energy, pick up electrons, and become neutral helium atoms. Alpha particles cannot penetrate the skin, although enough exposure causes a severe skin burn. If carried in air or on food into the soft tissues of the lungs or the intestinal tract, emitters of alpha particles can cause serious harm, including cancer.

Nuclear equations describe the decay of radioactive nuclei

To symbolize the decay of a nucleus, we construct a **nuclear equation,** which we can illustrate by the alpha decay of uranium-238 to thorium-234.

$$^{238}_{92}\text{U} \longrightarrow {}^{234}_{90}\text{Th} + {}^4_2\text{He}$$

Unlike chemical reactions, nuclear reactions produce new isotopes, so we need separate rules for balancing nuclear equations. A nuclear equation is balanced when

 **TOOLS**

Balancing nuclear equations

1. The sums of the mass numbers on each side of the arrow are equal.

2. The sums of the atomic numbers on each side are equal.

In the nuclear equation for the decay of uranium-238, the atomic numbers balance (90 + 2 = 92), and the mass numbers balance (234 + 4 = 238). Notice that electrical charges are not included, even though they are there (initially). The alpha particle, for example, has a charge of 2+. Initially, therefore, the thorium particle has as much opposite charge, 2−. These charged particles, however, eventually pick up or lose electrons either from each other or from molecules in the matter through which they travel.

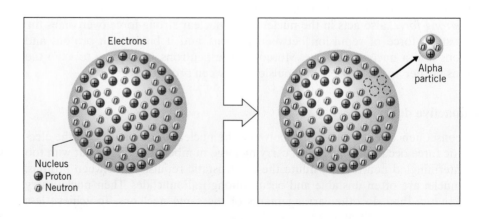

Figure 22.2 *Emission of an alpha particle from an atomic nucleus.*

Beta radiation is a stream of electrons

Naturally occurring **beta radiation** consists of a stream of electrons, which in this context are called **beta particles.** In a nuclear equation, the beta particle has the symbol $_{-1}^{0}e$, because the electron's mass number is 0 and its charge is $1-$. Hydrogen-3 (tritium) is a beta emitter that decays by the following equation:

Tritium is a synthetic radionuclide.

$$\underset{\text{tritium}}{^{3}_{1}\text{H}} \longrightarrow \underset{\text{helium-3}}{^{3}_{2}\text{He}} + \underset{\substack{\text{beta particle} \\ \text{(electron)}}}{^{0}_{-1}e} + \underset{\text{antineutrino}}{\bar{v}}$$

Both the antineutrino (to be described shortly) and the beta particle come from the atom's nucleus, not its electron shells. We do not think of them as having a prior existence in the nucleus, any more than a photon exists before its emission from an excited atom (see Figure 22.3). Both the beta particle and the antineutrino are created during the decay process in which a neutron is transformed into a proton.

$$\underset{\substack{\text{neutron} \\ \text{(in the} \\ \text{nucleus)}}}{^{1}_{0}n} \longrightarrow \underset{\substack{\text{beta particle} \\ \text{(emitted)}}}{^{0}_{-1}e} + \underset{\substack{\text{proton} \\ \text{(remains in} \\ \text{the nucleus)}}}{^{1}_{1}p} + \underset{\text{antineutrino}}{\bar{v}}$$

Unlike alpha particles, which are all emitted with the same discrete energy from a given radionuclide, beta particles emerge from a given beta emitter with a continuous spectrum of energies. Their energies vary from zero to some characteristic fixed upper limit for each radionuclide. This fact once gave nuclear physicists considerable trouble, partly because it was an apparent violation of energy conservation. To solve this problem, Wolfgang Pauli proposed in 1927 that beta emission is accompanied by yet another decay particle, this one electrically neutral and almost massless. Enrico Fermi suggested the name *neutrino* ("little neutral one"), but eventually it was named the *antineutrino,* symbolized by $\bar{v}$.

An electron is extremely small, so a beta particle is less likely to collide with the molecules of anything through which it travels. Depending on its initial kinetic energy, a beta particle can travel up to 300 cm in dry air, much farther than alpha particles. Only the highest-energy beta particles can penetrate the skin, however.

Enrico Fermi (1901–1954); Nobel prize 1938 (physics).

Gamma radiation is very high energy electromagnetic radiation

Gamma radiation, which often accompanies either alpha or beta radiation, consists of high-energy photons, given the symbol $_{0}^{0}\gamma$ or, often, simply γ in equations. Gamma radiation is extremely penetrating and is effectively blocked only by very dense materials, like lead.

The emission of gamma radiation involves transitions between energy levels *within* the nucleus. Nuclei have energy levels of their own, much as atoms have

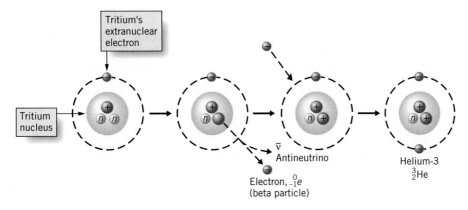

FIGURE 22.3 *Emission of a beta particle from a tritium nucleus.*

orbital energy levels. When a nucleus emits an alpha or beta particle, it sometimes is left in an excited energy state. By the emission of a gamma-ray photon, the nucleus relaxes into a more stable state.

Electron volt as an energy unit

The energy carried by a given radiation is usually described by an energy unit new to our study, the **electron volt,** abbreviated **eV**; 1 eV is the energy an electron receives when accelerated under the influence of 1 volt. It is related to the joule as follows:

$$1 \text{ eV} = 1.602 \times 10^{-19} \text{ J}$$

As you can see, the electron volt is an extremely small amount of energy, so multiples are commonly used, like the kilo-, mega-, and gigaelectron volt.

In the interconversion of mass and energy,
$$1 \text{ eV} = 1.783 \times 10^{-36} \text{ kg}$$
$$1 \text{ GeV} = 1.783 \times 10^{-24} \text{ g}$$

$$1 \text{ keV} = 10^3 \text{ eV}$$

$$1 \text{ MeV} = 10^6 \text{ eV}$$

$$1 \text{ GeV} = 10^9 \text{ eV}$$

An alpha particle radiated by radium-224 has an energy of 5 MeV. Hydrogen-3 (tritium) emits beta radiation at an energy of 0.05 to 1 MeV. The gamma radiation from cobalt-60, the radiation currently used to kill bacteria and other pests in certain foods, consists of photons with energies of 1.173 MeV and 1.332 MeV.

X rays are high-energy electromagnetic radiation

X rays used in diagnosis typically have energies of 100 keV or less.

X rays, like gamma rays, consist of high-energy electromagnetic radiation, but their energies are usually less than those of gamma radiation. Although X rays are emitted by some synthetic radionuclides, they normally are generated by special machines in which a high-energy electron beam is focused onto a metal target.

EXAMPLE 22.1

Writing a Balanced Nuclear Equation

Ions of cesium, which is in the same family as sodium, travel in the body to many of the same sites where sodium ions go.

Cesium-137, $^{137}_{55}\text{Cs}$, one of the radioactive wastes from a nuclear power plant or an atomic bomb explosion, emits beta and gamma radiation. Write the nuclear equation for the decay of cesium-137.

ANALYSIS: We start with an incomplete equation using the given information and then figure out any other data needed to complete the equation.

SOLUTION: The incomplete nuclear equation is

$$^{137}_{55}\text{Cs} \longrightarrow {}^{0}_{-1}e + {}^{0}_{0}\gamma + \underset{\underset{\text{Atomic number}}{\uparrow}}{\overset{\overset{\text{Mass number goes here.}}{\downarrow}\,\overset{\text{Atomic symbol goes here.}}{\downarrow}}{\underline{\quad}\underline{\quad}}}$$

goes here.

The atomic symbol can be obtained from the table inside the front cover after we have determined the atomic number, Z. Z is found using the fact that the atomic number (55) on the left side of the equation must equal the sum of the atomic numbers on the right side.

$$55 = -1 + 0 + Z$$

$$Z = 56$$

The periodic table tells us that element 56 is Ba (barium). To determine which isotope of barium forms, we recall that the sums of the mass numbers on either side of the equation must also be equal. Letting A be the mass number of the barium isotope,

$$137 = 0 + 0 + A$$

$$A = 137$$

The balanced nuclear equation, therefore, is

$$^{137}_{55}Cs \longrightarrow \ _{-1}^{0}e + \ _{0}^{0}\gamma + \ _{56}^{137}Ba$$

Is the Answer Reasonable?

Besides double-checking that element 56 is barium, the answer satisfies the requirements of a nuclear equation; namely, the sums of the mass numbers, 137, are the same on both sides as are the sums of the atomic numbers.

PRACTICE EXERCISE 1: Radium-226, $^{226}_{88}Ra$, is an alpha and gamma emitter. Write a balanced nuclear equation for its decay. (Marie Curie earned one of her two Nobel Prizes for isolating the element radium, which soon became widely used to treat cancer.)

PRACTICE EXERCISE 2: Write the balanced nuclear equation for the decay of strontium-90, a beta emitter. (Strontium-90 is one of the many radionuclides present in the wastes of operating nuclear power plants.)

Marie Curie (1867–1934); Nobel prizes 1903 (physics) and 1911 (chemistry).

A radioactive disintegration series is a sequence of successive nuclear reactions

Sometimes one radionuclide does not decay to a stable isotope but decays instead to another unstable radionuclide. The decay of one radionuclide after another will continue until a stable isotope forms. The sequence of such successive nuclear reactions is called a **radioactive disintegration series.** Four series occur naturally. Uranium-238 is at the head of one (Figure 22.4).

We studied half-lives in Section 15.5. One **half-life** period in nuclear science is the time it takes for a given sample of a radionuclide to decay to one-half of its initial amount. Because radioactive decay is a first-order process, the period of time taken by one half-life is independent of the initial number of nuclei. The huge variations in the half-lives of several radionuclides are shown in Table 22.1.

Some radioisotopes emit positrons or neutrons; others capture electrons

Many synthetic isotopes emit positrons, particles with the mass of an electron but a positive instead of a negative charge. A **positron** is a positive beta particle, a positive electron, and its symbol is $_{1}^{0}e$. It forms in the nucleus by the conversion of a

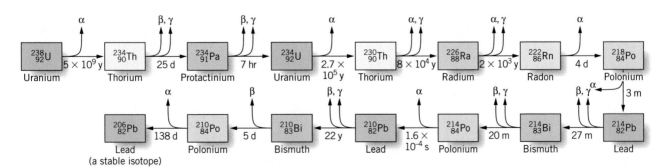

FIGURE 22.4 *The uranium-238 radioactive disintegration series.* The time given beneath each arrow is the half-life period of the preceding isotope (yr = year, m = month, d = day, hr = hour, mi = minute, and s = second).

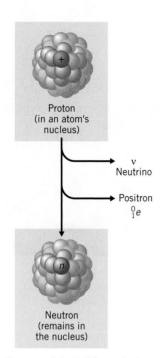

Proton
(in an atom's
nucleus)

ν
Neutrino

Positron
^0_1e

Neutron
(remains in
the nucleus)

FIGURE 22.5 *The emission of a positron replaces a proton by a neutron.*

TABLE 22.1	TYPICAL HALF-LIFE PERIODS		
Element	Isotope	Half-life	Radiations or Mode of Decay
Naturally occurring radionuclides			
Potassium	$^{40}_{19}K$	1.3×10^9 yr	beta, gamma
Tellurium	$^{123}_{52}Te$	1.2×10^{13} yr	electron capture
Neodymium	$^{144}_{60}Nd$	5×10^{15} yr	alpha
Samarium	$^{149}_{62}Sm$	4×10^{14} yr	alpha
Rhenium	$^{187}_{75}Re$	7×10^{10} yr	beta
Radon	$^{222}_{86}Rn$	3.82 day	alpha
Radium	$^{226}_{88}Ra$	1590 yr	alpha, gamma
Thorium	$^{230}_{90}Th$	8×10^4 yr	alpha, gamma
Uranium	$^{238}_{92}U$	4.51×10^9 yr	alpha
Synthetic radionuclides			
Tritium	^3_1T	12.26 yr	beta
Oxygen	$^{15}_8O$	124 s	positron
Phosphorus	$^{32}_{15}P$	14.3 day	beta
Technetium	$^{99m}_{43}Tc$	6.02 hr	gamma
Iodine	$^{131}_{53}I$	8.07 day	beta
Cesium	$^{137}_{55}Cs$	30 yr	beta
Strontium	$^{90}_{38}Sr$	28.1 yr	beta
Plutonium	$^{238}_{94}Pu$	87.8 yr	alpha
Americium	$^{243}_{95}Am$	7.37×10^3 yr	alpha

proton to a neutron (Figure 22.5). Positron emission, like beta emission, is accompanied by a chargeless and virtually massless particle, a *neutrino* (v), the counterpart of the antineutrino ($\bar{v}$) in the realm of antimatter (defined below). Cobalt-54, for example, is a positron emitter and changes to a stable isotope of iron.

$$^{54}_{27}Co \longrightarrow \ ^{54}_{26}Fe \ + \ ^0_1e \ + \ v$$

positron   neutrino

A positron, when emitted, eventually collides with an electron, and the two annihilate each other (Figure 22.6). Their masses change into the energy of two gamma-ray photons called *annihilation radiation photons,* each with an energy of 511 keV.

$$^{\ 0}_{-1}e \ + \ ^0_1e \longrightarrow 2\,^0_0\gamma$$

Because a positron destroys a particle of ordinary matter (an electron), it is called a particle of *antimatter.* To be classified as **antimatter,** a particle must have a counterpart among one of ordinary matter, and the two must annihilate each other when they collide. For example, a neutron destroys an antineutron.

Neutron emission, another kind of nuclear reaction, does not lead to an isotope of a different element. Krypton-87, for example, decays as follows to krypton-86:

$$^{87}_{36}Kr \longrightarrow \ ^{86}_{36}Kr \ + \ ^1_0n$$

neutron

Electron capture, yet another kind of nuclear reaction, is very rare among natural isotopes but common among synthetic radionuclides. Vanadium-50 nuclei, for example, can capture orbital K shell or L shell electrons, change to nuclei of stable atoms of titanium, and emit X rays and neutrinos.

The symbol $^+\beta$ is sometimes used for the positron.

The K shell is the shell with principal quantum number $n = 1$; the L shell is at $n = 2$.

$$\ce{^{50}_{23}V} + \ce{^{0}_{-1}e} \xrightarrow{\text{electron capture}} \ce{^{50}_{22}Ti} + \text{X rays} + \nu$$

The net effect of electron capture is the conversion of a proton into a neutron (Figure 22.7).

$$\underset{\substack{\text{proton} \\ \text{(in the} \\ \text{nucleus)}}}{\ce{^{1}_{1}p}} + \underset{\substack{\text{electron} \\ \text{(captured} \\ \text{from the} \\ \text{K shell)}}}{\ce{^{0}_{-1}e}} \longrightarrow \underset{\substack{\text{neutron} \\ \text{(in the} \\ \text{nucleus)}}}{\ce{^{1}_{0}n}}$$

Electron capture does not change an atom's mass number, only its atomic number. It also leaves a hole in the K or L shell, and the atom emits photons of X rays as other orbital electrons drop down to fill the hole. Moreover, the nucleus that has just captured an orbital electron may be in an excited energy state and so can emit a gamma-ray photon.

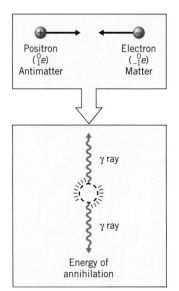

FIGURE 22.6 *Gamma radiation is produced when a positron and an electron collide.*

Band of stability

22.4 Stable isotopes fall within the "band of stability" on a plot based on numbers of protons and neutrons

When all known isotopes of each element, both stable and unstable, are arrayed on a plot according to numbers of protons and neutrons, an interesting zone can be defined (Figure 22.8). The two curved lines in the array of Figure 22.8 enclose this zone, called the **band of stability,** within which lie all stable nuclei. (No isotope above element 83, bismuth, is included in Figure 22.8 because none has a *stable* isotope.) Within the band of stability are also some unstable isotopes, because smooth lines cannot be drawn to exclude them.

Any isotope not represented anywhere on the array, inside or outside the band of stability, probably has a half-life too short to permit its detection. An isotope with 50 neutrons and 60 protons, for example, would be too unstable to justify the time and money for an attempt to make it.

Notice that the band curves slightly upward as the number of protons increases. The curvature means that the *ratio* of neutrons to protons gradually increases from 1:1, a ratio indicated by the straight line in Figure 22.8. The reason is easy to understand. More protons require more neutrons to provide a compensating nuclear strong force and to dilute electrostatic proton–proton repulsions.

Isotopes occurring above and to the left of the band of stability tend to be beta emitters. Isotopes lying below and to the right of the band are positron emitters. The isotopes with atomic numbers above 83 tend to be alpha emitters. Are there any reasons for these tendencies?

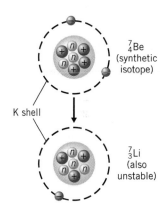

FIGURE 22.7 *Electron capture.* Electron capture is the collapse of an orbital electron into the nucleus, and this changes a proton into a neutron.

Alpha emitters

The alpha emitters, as we said, occur mostly among the radionuclides above atomic number 83. Their nuclei have too many protons, and the most efficient way to lose protons is by the loss of an alpha particle.

Beta emitters

Beta emitters are generally above the band of stability and so have neutron:proton ratios that evidently are too high. By beta decay a nucleus loses a neutron and gains a proton, thus decreasing the ratio.

The proton can also be given the symbol $\ce{^{1}_{1}H}$ in nuclear equations.

$$\ce{^{1}_{0}n} \longrightarrow \ce{^{1}_{1}p} + \ce{^{0}_{-1}e} + \bar{\nu}$$

For example, by beta decay fluorine-20 decreases its neutron:proton ratio from 11/9 to 10/10.

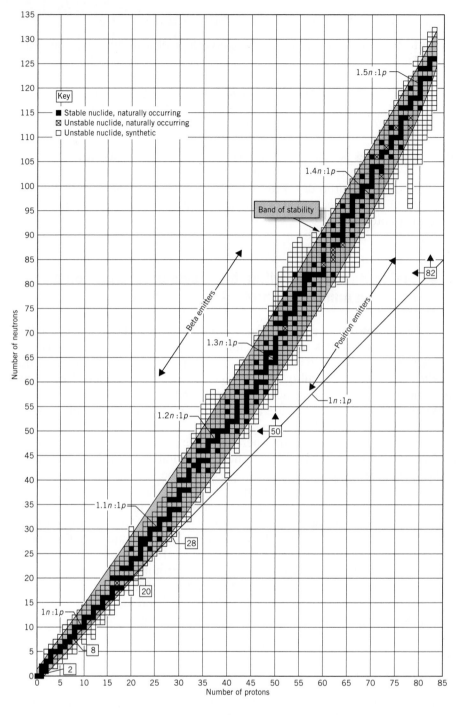

Figure 22.8 *The band of stability.*

$$^{20}_{9}\text{F} \longrightarrow ^{20}_{10}\text{Ne} + ^{0}_{-1}e + \bar{\nu}$$

$$\frac{\text{neutron}}{\text{proton}} = \frac{11}{9} \qquad \frac{10}{10}$$

The surviving nucleus, that of neon-10, is closer to the center of the band of stability. Figure 22.9, an enlargement of the fluorine part of Figure 22.8, explains this change further. Figure 22.9 also shows how the beta decay of magnesium-27 to aluminum-27 also lowers the neutron:proton ratio.

Positron emitters

In nuclei with too few neutrons to be stable, positron emission increases the neutron:proton ratio. A fluorine-17 nucleus, for example, increases its neutron:proton

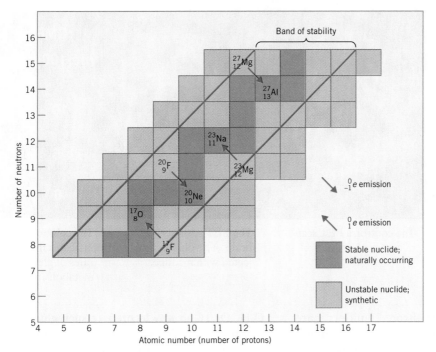

FIGURE 22.9 *An enlarged section of the band of stability.* Beta decay from magnesium-27 and fluorine-20 reduces their neutron:proton ratio and moves them closer to the band of stability. Positron decay from magnesium-23 and fluorine-17 increases this ratio and moves these nuclides closer to the band of stability, too.

ratio, improves its stability, and moves into the band of stability by emitting a positron and a neutrino and changing to oxygen-17 (see Figure 22.9).

$$^{17}_{9}F \longrightarrow\ ^{17}_{8}O +\ ^{0}_{1}e + \nu$$

$$\frac{\text{neutron}}{\text{proton}} = \frac{8}{9} \qquad \frac{9}{8}$$

Positron decay by magnesium-23 to sodium-23 produces a similarly favorable shift (also shown in Figure 22.9).

Nuclei with even numbers of protons and neutrons are likely to be stable

Nature favors even numbers for protons and neutrons, as summarized by the **odd–even rule.**

> **Odd–Even Rule**
> When the numbers of neutrons and protons in a nucleus are both even, the isotope is far more likely to be stable than when both numbers are odd.

TOOLS
Odd–even rule

Of the 264 stable isotopes, only 5 have odd numbers of both protons and neutrons, whereas 157 have even numbers of both. The rest have an odd number of one nucleon and an even number of the other. To see this, notice in Figure 22.8 how the horizontal lines with the largest numbers of dark squares (stable isotopes) most commonly correspond to even numbers of neutrons. Similarly, the vertical lines with the most dark squares most often correspond to even numbers of protons.

The odd–even rule is related to the spins of nucleons. Both protons and neutrons behave as though they spin, like orbital electrons. When two protons or two neutrons have paired spins, meaning the spins are opposite, their combined energy is less than when the spins are unpaired. Only when there are even numbers of protons and of neutrons can all spins be paired and so give the nucleus less energy and more stability. The least stable nuclei tend to be those with both an odd proton and an odd neutron.

The isotopes $^{2}_{1}H$, $^{6}_{3}Li$, $^{10}_{5}B$, $^{14}_{7}N$, and $^{138}_{57}La$ all have odd numbers of both protons and neutrons.

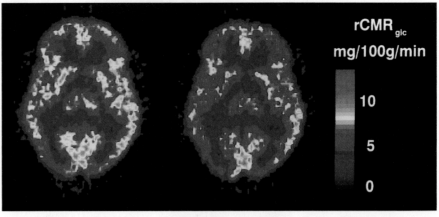

FIGURE 22.10 *Positron emission tomography (PET) in the study of brain activity.* (*Left*) Nonsmoker. (*Right*) Smoker. The PET scan reveals widespread reduction in the rate of glucose metabolism when nicotine is present. (Photos courtesy of E. D. London.)

The control, a PET scan of a normal brain.

The PET scan of the brain of a volunteer injected with nicotine.

The color code indicates the rates of glucose metabolism.

CHEMISTRY IN PRACTICE

Positron emitters are now being used in an important method for studying brain function called the PET scan, for *positron emission tomography*. The technique begins by chemically incorporating positron-emitting radionuclides into molecules, like glucose, that can be absorbed by the brain directly from the blood. It's like inserting radiation generators that act from within the brain rather than focusing X rays or gamma rays from the outside. Carbon-11, for example, is a positron emitter whose atoms can be used in place of carbon-12 atoms in glucose molecules,

$C_6H_{12}O_6$. (One way to prepare such glucose is to let a leafy vegetable, Swiss chard, use $^{11}CO_2$ to make the glucose by photosynthesis.)

A PET scan using carbon-11 glucose detects situations where glucose is not taken up normally, e.g., in manic depression, schizophrenia, and Alzheimer's disease. After the carbon-11 glucose is ingested, radiation detectors outside the body pick up the annihilation radiation of positron emission from specifically those brain sites that use glucose. PET scan technology, for example, showed that the uptake of glucose by the brains of smokers is less than that of nonsmokers (see Figure 22.10).

Isotopes with "magic numbers" are especially stable

Another rule of thumb for nuclear stability is based on *magic numbers* of nucleons. Isotopes with specific numbers of protons or neutrons, the **magic numbers,** are more stable than the rest. The magic numbers of nucleons are 2, 8, 20, 28, 50, 82, and 126, and where they fall is shown in Figure 22.8 (except for magic number 126).

When the numbers of both protons and neutrons are the same magic number, as they are in ^4_2He, $^{16}_8O$, and $^{40}_{20}Ca$, the isotope is very stable. $^{100}_{50}Sn$ also has two identical magic numbers. Although this isotope of tin is unstable, having a half-life of only several seconds, it is much more stable than nearby radionuclides, whose half-lives are in milliseconds. Thus, although tin-100 lies well outside the band of stability, it is stable enough to be observed. One stable isotope of lead, $^{208}_{82}Pb$, involves two different magic numbers, 82 protons and 126 neutrons.

Magic numbers do not cancel the need for a favorable neutron:proton ratio. An atom with 82 protons and 82 neutrons lies far outside the band of stability, and yet 82 is a magic number.

The existence of magic numbers supports the hypothesis that a nucleus has a shell structure with energy levels analogous to electron energy levels. Electron levels, as you already know, are associated with their own special numbers, those that equal the maximum number of electrons allowed in a principal energy level: 2, 8, 18, 32, 50, 72, and 98 (for principal levels 1, 2, 3, 4, 5, 6, and 7, respectively). The total numbers of electrons in the atoms of the most chemically stable elements—the noble gases—also make up a special set: 2, 10, 18, 36, 54, and 86 electrons. Thus, special sets of numbers are not unique to nuclei.

22.5 ▶ Transmutation is the change of one isotope into another

The change of one isotope into another is called **transmutation,** and radioactive decay is only one cause. Transmutation can also be forced by the bombardment of nuclei with high-energy particles, such as alpha particles from natural emitters, neutrons from atomic reactors, or protons made by stripping electrons from hydrogen. To make them better bombarding missiles, protons and alpha particles can be accelerated in an electrical field (Figure 22.11). This gives them greater energy and enables them to sweep through the target atom's orbital electrons and become buried in its nucleus. Although beta particles can be accelerated, their disadvantage is that they are repelled by a target atom's own electrons.

Transmutation occurs when compound nuclei decay

Both the energy and the mass of a bombarding particle enter the target nucleus at the moment of capture. The energy of the new nucleus, called a **compound nucleus,** quickly becomes distributed among all of the nucleons, but the nucleus is nevertheless rendered somewhat unstable. To get rid of the excess energy, a compound nucleus generally ejects something, a neutron, proton, or electron, and often emits gamma radiation as well. This leaves a new nucleus of an isotope different than the original target, so a transmutation has occurred overall.

Compound here refers only to the idea of combination, not to a chemical.

Ernest Rutherford observed the first example of artificial transmutation. When he let alpha particles pass through a chamber containing nitrogen atoms, an entirely new radiation was generated, one much more penetrating than alpha radiation. It proved to be a stream of protons (Figure 22.12). Rutherford was able to show that the protons came from the decay of the compound nuclei of fluorine-18, produced when nitrogen-14 nuclei captured bombarding alpha particles.

$$\underset{\substack{\text{alpha}\\\text{particle}}}{^4_2\text{He}} + \underset{\substack{\text{nitrogen}\\\text{nucleus}}}{^{14}_7\text{N}} \longrightarrow \underset{\substack{\text{fluorine}\\\text{(compound}\\\text{nucleus)}}}{^{18}_9\text{F*}} \longrightarrow \underset{\substack{\text{oxygen}\\\text{(a rare but}\\\text{stable iso-}\\\text{tope)}}}{^{17}_8\text{O}} + \underset{\substack{\text{proton}\\\text{(high-}\\\text{energy)}}}{^1_1p}$$

The asterisk, *, symbolizes a high-energy nucleus, a compound nucleus.

In the synthesis of alpha particles from lithium-7, protons are used as bombardment particles. The resulting compound nucleus, that of beryllium-8, splits in two.

$$\underset{\text{proton}}{^1_1p} + \underset{\text{lithium}}{^7_3\text{Li}} \longrightarrow \underset{\text{beryllium}}{^8_4\text{Be*}} \longrightarrow \underset{\substack{\text{alpha}\\\text{particles}}}{2\,^4_2\text{He}}$$

FIGURE 22.11 *Linear accelerator.* In this linear accelerator at Brookhaven National Laboratory on Long Island, New York, protons can be accelerated to just under the speed of light and given up to 33 GeV of energy before they strike their target. (1 GeV = 10^9 eV.)

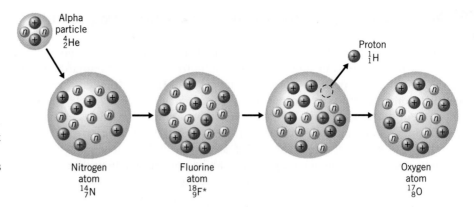

FIGURE 22.12 *Transmutation of nitrogen into oxygen.* When the nucleus of nitrogen-14 captures an alpha particle it becomes a compound nucleus of fluorine-18. This then expels a proton and becomes the nucleus of oxygen-17.

Decay modes of compound nuclei

A given compound nucleus can be made in a variety of ways. Aluminum-27, for example, forms by any of the following routes:

$$^4_2He + {}^{23}_{11}Na \longrightarrow {}^{27}_{13}Al*$$

$$^1_1p + {}^{26}_{12}Mg \longrightarrow {}^{27}_{13}Al*$$

$$\underset{\text{deuteron}}{^2_1H} + {}^{25}_{12}Mg \longrightarrow {}^{27}_{13}Al*$$

A deuteron is the nucleus of a deuterium atom, just as a proton is the nucleus of a protium (hydrogen) atom.

Once it forms, the compound nucleus, $^{27}_{13}Al*$, has no memory of how it was made. It only knows how much energy it has. Depending on this energy, different paths of decay are available, and all of the following routes have been observed. They illustrate how the synthesis of such a large number of synthetic isotopes, some stable and others unstable, has been possible.

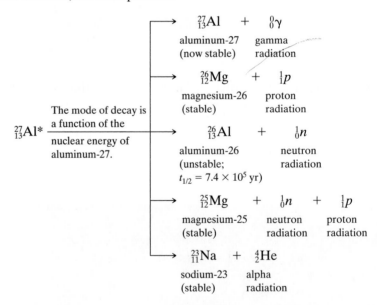

Transmutation is used to produce synthetic elements

Over a thousand isotopes have been made by transmutations. Most do not occur naturally; they appear in the band of stability (Figure 22.8) as open squares, nearly 900 in number. The naturally occurring isotopes above atomic number 83 all have very long half-lives. Others might have existed, but their half-lives probably were too short to permit them to last into our era. All of the elements from neptunium (atomic number 93) and up, the **transuranium elements,** are synthetic. Elements from atomic numbers 93 to 103 complete the actinide series of the periodic table, which starts with element 90, thorium. Beyond this series, elements 104–112 and 114 have also been made.

To make the heaviest elements, bombarding particles larger than neutrons are used, such as alpha particles or the nuclei of heavier atoms. For example, element 110—recently given the official name darmstadtium (Ds)—was made when a neutron was ejected from the compound nucleus formed by the fusion of nickel-62 and lead-208.

$$^{62}_{28}Ni + ^{208}_{82}Pb \longrightarrow ^{270}_{110}Ds \longrightarrow ^{269}_{110}Ds + ^{1}_{0}n$$

Some atoms of element 111 formed when a neutron was lost from the compound nucleus made by bombarding bismuth-209 with nickel-64. Atoms of element 112 emerged from the loss of a neutron from a compound nucleus made by bombarding lead-208 with zinc-70. The synthesis of the 287 isotope of element 114 was done by Russian scientists by bombarding plutonium-242 with calcium-48. The loss of three neutrons from the compound nucleus gave the 287 isotope, which has a half-life of about 5 s.

The study of the chemistry of the elements from 104 and higher must be done virtually one atom at a time because their half-lives are so short.

22.6 ▶ How is radiation measured?

Atomic radiation is often described as **ionizing radiation** because it creates ions by knocking electrons from molecules in the matter through which the radiation travels. The generation of ions is behind some of the devices for detecting radiation.

Geiger counter

The Geiger–Müller tube, one part of a *Geiger counter,* detects beta and gamma radiation having energy high enough to penetrate the tube's window. Inside the tube is a gas under low pressure in which ions form when radiation enters. The ions permit a pulse of electricity to flow, which activates a current amplifier, a pulse counter, and an audible click.

Scintillation counters

A *scintillation counter* (Figure 22.13) has a surface coated with a special chemical that emits a tiny flash of light when struck by a particle of radiation. These flashes can be magnified electronically and automatically counted.

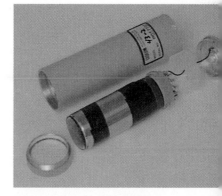

FIGURE 22.13 *Scintillation probe.* Energy received from radiations striking the phosphor at the end of this probe is processed by the photomultiplier unit and sent to an instrument (not shown) for recording.

Film dosimeters

The darkening of a photographic film exposed to radiation over a period of time is proportional to the total quantity of radiation received. *Film dosimeters* work on this principle (Figure 22.14), and people who work near radiation sources use them. Each person keeps a log of total exposure, and if a predetermined limit is exceeded, the worker must be reassigned to an unexposed workplace.

Radioactive decay is a first-order process

The **activity** of a radioactive material is the number of disintegrations per second. The SI unit of activity is the **becquerel (Bq),** and it equals one disintegration per second. A liter of air has an activity of about 0.04 Bq, due to carbon-14 in its carbon dioxide. A gram of natural uranium has an activity of about 2.6×10^4 Bq.

The **curie (Ci),** named after Marie Curie, the discoverer of radium, is an older unit, being equal to the activity of a 1.0 g sample of radium-226.

The becquerel is named after Henri Becquerel (1852–1908), the discoverer of radioactivity, who won a Nobel Prize in 1903.

$$1 \text{ Ci} = 3.7 \times 10^{10} \text{ disintegrations s}^{-1} = 3.7 \times 10^{10} \text{ Bq} \qquad (22.2)$$

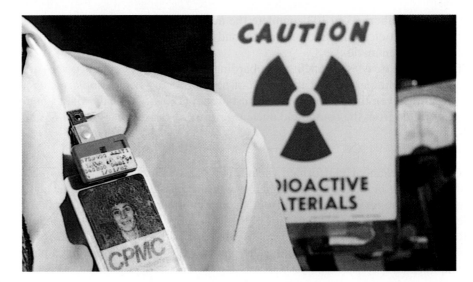

FIGURE 22.14 *Badge dosimeter.* The doses of various kinds of radiation received by the film in the dosimeter for predetermined periods of time are entered into the worker's log.

For a sufficiently large sample of a radioactive material, the activity is experimentally found to be proportional to the number of radioactive nuclei, N:

$$\text{activity} = kN$$

The constant of proportionality, k, is called the **decay constant.** The decay constant is characteristic of the particular radioactive nuclide, and it gives the activity per nuclide in the sample. Since activity is the number of disintegrations per second, and so the change in the number of nuclei per second, we can write

Law of radioactive decay

$$\text{activity} = -\frac{\Delta N}{\Delta t} = kN \tag{22.3}$$

which is called the **law of radioactive decay.**[1] The law shows that radioactive decay is a first-order kinetic process, and the decay constant is really just a first-order rate constant in terms of number of nuclei, rather than concentrations.

Recall from Chapter 15 that the half-life of a first-order reaction is given by Equation 15.5:

Half-life

$$t_{1/2} = \frac{\ln 2}{k}$$

If we know the half-life of a radioisotope, we can use this relationship to compute its decay constant and also the activity of a known mass of the radioisotope, as Example 22.2 demonstrates.

EXAMPLE 22.2

Using the Law of Radioactive Decay

Deep space probes such as NASA's Cassini spacecraft need to keep their instruments warm enough to operate effectively. Since solar power is not available in the darkness of deep space, these craft generate heat from the radioactive decay of small pellets of plutonium dioxide. Each pellet is about the size of a pencil eraser and weighs about 2.7 g. If the pellets are pure PuO_2, with the plutonium being ^{238}Pu, what is the activity of one of the fuel pellets, in becquerels?

ANALYSIS: We can use the law of radioactive decay, Equation 22.3, to compute the activity. But to use the equation, we'll need the decay constant, k, and the number of plutonium-238 atoms, N, in the fuel pellet.

From Table 22.1, the half-life of ^{238}Pu is 87.8 years. We can rearrange Equation 15.5 to compute the decay constant k from the half-life:

---

[1]Note that the minus sign is introduced to make the activity a positive number, since the change in the number of radioactive nuclei, ΔN, is negative.

$$k = \frac{\ln 2}{t_{1/2}}$$

We'll want the decay constant in terms of seconds because the becquerel is defined as disintegrations per second. We should therefore convert the half-life into seconds before substituting it into Equation 15.5.

We know that the pellet contains 2.7 g of PuO_2. To calculate N, we'll have to perform the following stoichiometric conversion:

$$2.7 \text{ g } PuO_2 \Leftrightarrow \text{? atoms Pu}$$

We are trying to convert an amount of one substance (PuO_2) into another (Pu), so the critical link we need is a mole-to-mole ratio. Our strategy is to perform the following conversions:

$$2.7 \text{ g } PuO_2 \longrightarrow \text{mol } PuO_2 \longrightarrow \text{mol Pu} \longrightarrow \text{atoms Pu}$$

We'll need the formula mass of PuO_2 to perform the first conversion. Since the plutonium is plutonium-238, the formula mass of PuO_2 is 270 g mol^{-1}. For the second conversion, the formula of PuO_2 tells us that there is 1 mol Pu in 1 mol PuO_2. For the final conversion, recall that 1 mol of atoms is 6.02×10^{23} atoms.

SOLUTION: First, let's compute the decay constant from the half-life. We must convert the half-life into seconds:

$$87.8 \text{ years} \times \frac{365 \text{ days}}{1 \text{ year}} \times \frac{24 \text{ hours}}{1 \text{ day}} \times \frac{60 \text{ min}}{1 \text{ hour}} \times \frac{60 \text{ s}}{1 \text{ min}} = 2.77 \times 10^9 \text{ s}$$

Now we can solve Equation 15.5 for the decay constant:

$$k = \frac{\ln 2}{t_{1/2}}$$

$$= \frac{0.693}{2.77 \times 10^9 \text{ s}}$$

$$= 2.50 \times 10^{-10} \text{ s}^{-1}$$

Next, we need the number of ^{238}Pu atoms in the fuel pellet:

$$2.7 \text{ g } PuO_2 \times \frac{1 \text{ mol } PuO_2}{270 \text{ g } PuO_2} \times \frac{1 \text{ mol Pu}}{1 \text{ mol } PuO_2} \times \frac{6.02 \times 10^{23} \text{ atoms Pu}}{1 \text{ mol Pu}}$$

$$= 6.0 \times 10^{21} \text{ atoms Pu}$$

From the law of radioactive decay, Equation 22.3, the activity of the fuel pellet is

$$\text{activity} = kN = 2.50 \times 10^{-10} \text{ s}^{-1} \times 6.0 \times 10^{21} \text{ atoms Pu} = 1.5 \times 10^{12} \text{ atoms Pu/s}$$

Since the becquerel is defined as the number of disintegrations per second, and each plutonium atom corresponds to one disintegration, we can report the activity as 1.5×10^{12} Bq.

Is the Answer Reasonable?

There is no easy way to check the size of the decay constant beyond checking the arithmetic. The number of Pu atoms in the pellet makes sense, because if we have 2.7 g PuO_2 and the formula mass is 270 g mol^{-1}, we have 1/100th of a mole of PuO_2 and so 1/100 of a mole of Pu. We should have (1/100) of 6.02×10^{23} atoms, or 6.02×10^{21} atoms of Pu.

The fuel pellet has about 40 times the activity of a gram of radium-226, as indicated by Equation 22.2, so the activity of ^{238}Pu per gram is about 15 times the activity of ^{226}Ra per gram.

PRACTICE EXERCISE 3: The EPA limit for radon-222 is 6 pCi (picocuries) per liter of air. How many atoms of ^{222}Rn per liter of air will produce this activity?

Absorbed dose is the energy equivalent to the quantity of radioactivity absorbed

The energy equivalent to the quantity of radiation absorbed by some material is defined in units of *absorbed dose* (or simply *dose*). The **gray (Gy)** is the SI unit of absorbed dose, and 1 gray corresponds to 1 joule of energy absorbed per kilogram of absorbing material. The **rad** is an older unit of absorbed dose, 1 rad being the absorption of 10^{-2} joule per kilogram of tissue. Thus, 1 Gy equals 100 rad. In terms of danger, if every individual in a large population received 450 rad (4.5 Gy), roughly half of the population would die in 60 days.

The gray is not a good basis for comparing the biological effects of radiation in tissue, because these effects depend not just on the energy absorbed but also on the kind of radiation and the tissue itself. The **sievert (Sv),** the SI unit of *dose equivalent,* was invented to meet this problem. If we let D stand for absorbed dose in grays, Q for quality factor (meaning relative effectiveness for causing harm in tissue), and N for any other modifying factors, then the dose equivalent, H, in sieverts is defined as the following product:

$$H = DQN$$

In other words, we multiply the absorbed dose (D) from some radiation by a factor (Q) that takes into account the biologically significant properties of the radiation and by any other factor (N) bearing on the net effect. The value of Q for alpha radiation is 20, but it is only 1.0 for beta and gamma radiation. The alpha particle has a large value of Q because of its relatively large size and charge, which enable it to damage molecules with considerable efficiency.

The **rem** is an older unit of dose equivalent, one still used in medicine. Its value is generally taken to equal 10^{-2} Sv. The U.S. government has set guidelines of 0.3 rem per week as the maximum exposure workers may receive. (For comparison, a chest X ray typically involves about 0.007 rem, or 7 mrem.)

Radiation damages living tissue

A whole body dose of 25 rem (0.25 Sv) induces noticeable changes in human blood. A set of symptoms called *radiation sickness* develops at about 100 rem, becoming severe at 200 rem. Among the symptoms are nausea, vomiting, a drop in the white cell count, diarrhea, dehydration, prostration, hemorrhaging, and loss of hair. If each person in a large population absorbed 400 rem, half would die in 60 days. A 600 rem dose would kill everyone in the group in a week. Many workers received at least 400 rem in the moments following the steam explosion that tore apart one of the nuclear reactors at the Ukraine energy park near Chernobyl in 1986.

Radiation-produced free radicals

Even small absorbed doses can be biologically harmful. The danger does not lie in the heat energy associated with the dose, which is usually very small. Rather, the harm is in the ability of ionizing radiation to create unstable ions or neutral species with odd (unpaired) electrons, species that can set off other reactions. Water, for example, can interact as follows with ionizing radiation:

$$\text{H-\overset{..}{\underset{..}{O}}-H} \xrightarrow{\text{radiation}} [\text{H-\overset{.}{\underset{..}{O}}-H}]^{+} + {}_{-1}^{0}e$$

The new cation, $[\text{H-\underset{..}{O}-H}]^{+}$, is unstable, and one breakup path is

$$[\text{H-\underset{..}{O}-H}]^{+} \longrightarrow \text{H}^{+} + :\overset{.}{\underset{..}{O}}\text{-H}$$
$$\qquad\qquad\qquad\qquad \text{proton} \quad \text{hydroxyl} \atop \text{radical}$$

The proton might pick up a stray electron to become a hydrogen atom, $\text{H} \cdot$.

The gray is named after Harold Gray, a British radiologist.

Rad stands for radiation absorbed dose.

Rem stands for roentgen equivalent for man, where the roentgen is a unit related to X-ray and gamma radiation.

The results of radiation damage appear earliest in tissues with rapidly dividing cells, like bone marrow and the gastrointestinal tract.

FACETS OF CHEMISTRY 22.1

Radon-222 in the Environment

Several years ago an engineer unintentionally but repeatedly set off radiation alarms at his workplace, a nuclear power plant in Pennsylvania. Such alarms are triggered when radiation levels surpass a predetermined value. The problem eventually was traced to his home, where the radiation level was 2700 picocuries per liter of air, compared with the level in the average house of 1 picocurie per liter. (The picocurie is 10^{-12} curie.) The chief culprit was radon-222, a natural intermediate in the uranium-238 disintegration series (Figure 22.4). (One picocurie per liter is equivalent to the decay of two radon atoms per minute.) The engineer had carried radon-222 and its decay products on his clothing from his home to the power plant.

One result of this episode was to send geologists scurrying to map the general area for concentrations of uranium-238. They found that the bedrock of the Reading Prong, a geological formation that cuts across Pennsylvania, New Jersey, New York, and up into the New England states, is relatively rich in uranium.

Radon-222 is a noble gas and is as *chemically* unreactive as argon, krypton, and the other noble gases. But this decay product of uranium-238 has an unstable nucleus. Its half-life is 4 days, and it emits both alpha and gamma radiation. Being a gas, radon-222 diffuses through rocks and soil and enters buildings. In fact, buildings can act much like chimneys and draw radon as a chimney draws smoke. Radon-222 in the air is breathed into the lungs. Because of its short half-life, some of it decays in the lungs before it is exhaled. The decay products are not gases, and several have short half-lives. When these are left in the lungs, their radiation can cause lung cancer. Scientists of the National Research Council estimate that roughly 12% of all lung cancer deaths, 15,000 to 22,000 per year (out of a total 157,000 deaths per year from lung cancer) are caused by exposure to radon. Almost 90% of these victims are also smokers, indicating that smoking and radon exposure make a particularly dangerous mix.

The Environmental Protection Agency (EPA) has recommended an upper limit of 4 picocuries per liter in the air of homes. Estimates released in the late 1980s by the EPA were that roughly 1% of the homes in the United States have radon levels that generate radiation of more than 10 picocuries per liter. Those who estimate risks present in day-to-day activities say that such a level poses less a risk to life than the risk of driving a car. Radon test kits for homeowners are available at home supply stores (see Figure 1).

FIGURE 1 Test kits for measuring radon levels in rooms can be purchased by home owners.

Both the hydrogen atom and the hydroxyl radical are examples of **free radicals,** which are neutral or charged particles having one or more unpaired electrons, often called simply *radicals.* They are chemically very reactive. What they do next depends on the other chemical species nearby, but radicals can set off a series of totally undesirable chemical reactions inside a living cell. This is what makes the injury from absorbed radiation of far greater magnitude than the energy alone could inflict. A dose of 600 rem is a lethal dose in a human, but the same dose absorbed by pure water causes the ionization of only 1 water molecule in every 36 million.

Background radiation comes from a variety of sources

The presence of radionuclides in nature makes it impossible for us to be free from all exposure to ionizing radiation. High-energy cosmic rays shower on us from the sun. They interact with the air's nitrogen molecules to produce carbon-14, a beta emitter, which enters the food chain via the foods made when carbon dioxide is used in photosynthesis (see "Carbon-14 Dating" in Section 22.7). From soil and from building stone comes the radiation of radionuclides native to Earth's crust. The top 40 cm of Earth's soil holds an average of 1 gram of radium, an alpha emitter, per square kilometer. Naturally occurring potassium-40, a beta emitter, adds its radiation wherever potassium ions are found in the body. The presence of carbon-14 and potassium-40 together produce about 5×10^5 nuclear disintegrations per minute inside an adult human. Radon gas seeps into basements from underground

Radium salts were once put into the paint used for the numerals on watch faces to make them glow in the dark.

formations (see Facets of Chemistry 22.1). In fact, a little over half of the radiation we experience, on average, is from radon-222 and its decay products.

Diagnostic X rays, both medical and dental, also expose us to ionizing radiation. All these sources produce a combined **background radiation** that averages 360 mrem per person annually in the United States. The averages are roughly 82% from natural radiation and 18% from medical sources.

Radiation can be reduced by shielding and by distance

Gamma radiation and X rays are so powerful that they are effectively shielded only by very dense materials, like lead. Otherwise, one should stay as far from a source as possible, because the intensity of radiation diminishes with the *square* of the distance. This relationship is the **inverse square law,** which can be written mathematically as follows, where d is the distance from the source:

$$\text{radiation intensity} \propto \frac{1}{d^2}$$

When the ratio is taken, the proportionality constant cancels, so we don't have to know what its value is.

When the intensity, I_1, is known at distance d_1, then the intensity I_2 at distance d_2 can be calculated by the following equation:

$$\frac{I_1}{I_2} = \frac{d_2{}^2}{d_1{}^2} \tag{22.4}$$

Inverse square law

This law applies only to a small source that radiates equally in all directions, with no intervening shields.

> **PRACTICE EXERCISE 4:** If an operator 10 m from a small source is exposed to 1.4 units of radiation, what will be the intensity of the radiation if he moves to 1.2 m from the source?

22.7 ▶ Radionuclides have many medical and analytical applications

The fact that the *chemical* properties of both the radioactive and stable isotopes of an element are the same is exploited in some uses of radionuclides. The chemical and physical properties enable scientists to get radionuclides into place in systems of interest. Then the radiation is exploited for medical or analytical purposes. Tracer analysis is an example.

Tracer analysis is used in medicine to study specific tissues in the body

In **tracer analysis,** the chemical form and properties of a radionuclide enable the system to distribute it to a particular location. The intensity of the radiation then tells something about how that site is working. In the form of the iodide ion, for example, iodine-131 is carried by the body to the thyroid gland, the only user of iodide ion in the body. The gland takes up the iodide ion to synthesize the hormone thyroxine. An underactive thyroid gland is unable to concentrate iodide ion normally and will emit less intense radiation under standard test conditions.

The tumor appears as a "hot spot" under radiological analysis.

Tracer analyses are also used to pinpoint the locations of brain tumors, which are uniquely able to concentrate the pertechnetate ion, $TcO_4{}^-$, made from technetium-99m.[2] This strong gamma emitter, which resembles the chloride ion in some respects, is one of the most widely used radionuclides in medicine.

[2] The *m* in technetium-99m means that the isotope is in a metastable form. Its nucleus is at a higher energy level than the nucleus in technetium-99, to which technetium-99m decays.

Neutron activation analysis is used to detect trace elements

A number of stable nuclei can be changed into emitters of gamma radiation by capturing neutrons, and this makes possible a procedure called **neutron activation analysis.** Neutron capture followed by gamma emission can be represented by the following equation (where A is a mass number and X is a hypothetical atomic symbol):

$$^{sA}X \quad + \quad ^{1}_{0}n \quad \longrightarrow \quad ^{(A+1)}X* \quad \longrightarrow \quad ^{(A+1)}X \quad + \quad ^{0}_{0}\gamma$$

| isotope of element X being analyzed | neutron | compound nucleus (unstable) | more stable form of a new isotope of X | gamma radiation |

A neutron-enriched compound nucleus emits gamma radiation at its own set of unique frequencies, and these sets of frequencies are now known for each isotope. (Not all isotopes, however, become gamma emitters by neutron capture.) The element can be identified by measuring the specific *frequencies* of gamma radiation emitted. The *concentration* of the element can be determined by measuring the *intensity* of the gamma radiation.

Neutron activation analysis is so sensitive that concentrations as low as $10^{-9}\%$ can be determined. A museum might have a lock of hair of some famous but long-dead person suspected of having been slowly murdered by arsenic poisoning. If so, some arsenic would be in the hair, and neutron activation analysis could find it without destroying the specimen of hair.

Radiological dating determines the age of a sample using kinetics

The determination of the age of a geological deposit or an archaeological find by the use of the radionuclides that are naturally present is called **radiological dating.** It is based partly on the premise that the half-lives of radionuclides have been constant throughout the entire geological period. This premise is supported by the finding that half-lives are insensitive to all environmental forces such as heat, pressure, magnetism, or electrical stresses.

In geological dating, a pair of isotopes is sought that are related as a "parent" to a "daughter" in a radioactive disintegration series, like the uranium-238 series (Figure 22.4). Uranium-238 (as "parent") and lead-206 (as "daughter") have thus been used as a radiological dating pair of isotopes. The half-life of uranium-238 is very long, a necessary criterion for geological dating. After the concentrations of uranium-238 and lead-206 are determined in a rock specimen, the *ratio* of the concentrations together with the half-life of uranium-238 is used to calculate the age of the rock.

Probably the most widely used isotopes for dating rock are the potassium-40/argon-40 pair. Potassium-40 is a naturally occurring radionuclide with a half-life nearly as long as that of uranium-238. One of its modes of decay is electron capture, and argon-40 forms.

For ^{238}U, $t_{1/2} = 4.51 \times 10^9$ yr

For ^{40}K, $t_{1/2} = 1.3 \times 10^9$ yr

$$^{40}_{19}\text{K} + ^{0}_{-1}e \longrightarrow ^{40}_{18}\text{Ar}$$

The argon produced by this reaction remains trapped within the crystal lattices of the rock and is freed only when the rock sample is melted. How much has accumulated is measured with a mass spectrometer (page 24), and the observed ratio of argon-40 to potassium-40, together with the half-life of the parent, permits the age of the specimen to be estimated. Because the half-lives of uranium-238 and potassium-40 are so long, samples have to be at least 300,000 years old for either of the two parent–daughter isotope pairs to work.

Carbon-14 dating

For dating organic remains, like wooden or bone objects from ancient tombs, carbon-14 analysis is used. The ratio of carbon-14 (a beta emitter) to stable carbon in

Extraodinary precautions are taken to ensure that specimens are not contaminated by more recent sources of carbon or carbon compounds.

the ancient sample is measured and compared to the ratio in recent organic remains. The data can then be used to calculate the age of the sample.

There are two approaches to carbon-14 dating. The older method, introduced by its discoverer, Willard F. Libby (Nobel Prize in chemistry, 1960), measures the *radioactivity* of a sample taken from the specimen. The radioactivity is proportional to the concentration of carbon-14.

The newer and current method of carbon-14 dating relies on a device resembling a mass spectrometer that is able to separate the atoms of carbon-14 from the other isotopes of carbon (as well as from nitrogen-14) *and count all of them,* not just the carbon-14 atoms that decay. This approach permits the use of smaller samples (0.5–5 mg versus 1–20 g for the Libby method); it works at much higher efficiencies; and it gives more precise dates. Objects of up to 70,000 years old can be dated, but the highest accuracy involves systems no older than 7000 years.

Carbon-14 has been produced in the atmosphere for many thousands of years by the action of cosmic rays on atmospheric nitrogen. Primary cosmic rays are streams of particles from the sun, mostly protons but also the nuclei of many elements (as high as atomic number 26). Their collisions with atmospheric atoms and molecules generate secondary cosmic rays, which include all subatomic particles. Secondary cosmic rays are what contribute to our background radiation, and people living at high altitudes receive more of this radiation than other people.

Neutrons of the proper energy in secondary cosmic rays can transmute nitrogen atoms into carbon-14 atoms according to the following equation (Figure 22.15):

$$\underset{\substack{\text{neutron} \\ \text{(in cosmic} \\ \text{ray)}}}{{}^{1}_{0}n} \; + \; \underset{\substack{\text{nitrogen} \\ \text{atom in the} \\ \text{atmosphere}}}{{}^{14}_{7}N} \; \longrightarrow \; \underset{\substack{\text{compound} \\ \text{nucleus}}}{{}^{15}_{7}N^{*}} \; \longrightarrow \; \underset{\substack{\text{carbon-14} \\ \text{atom (a} \\ \text{beta emitter)}}}{{}^{14}_{6}C} \; + \; \underset{\substack{\text{high-energy} \\ \text{proton}}}{{}^{1}_{1}p}$$

The newly formed carbon-14 diffuses into the lower atmosphere. It becomes oxidized to carbon dioxide and enters Earth's biosphere by means of photosynthesis. Carbon-14 thus becomes incorporated into plant substances and into the materials of animals that eat plants. As the carbon-14 decays, more is ingested by the living thing. The net effect is an overall equilibrium involving carbon-14 in the global system. As long as the plant or animal is alive, its ratio of carbon-14 atoms to carbon-12 atoms is constant.

At death, an organism's remains have as much carbon-14 as they can ever have, and they now slowly lose this carbon-14 by decay. The decay is a first-order process with a rate independent of the *number* of original carbon atoms. The ratio of carbon-14 to carbon-12, therefore, can be related to the years that have elapsed between the time of death and the time of the measurement. The critical assumption in carbon-14 dating is that the steady-state availability of carbon-14 from the atmosphere has remained largely unchanged over the period for which measurements are valid.[3]

In contemporary biological samples the ratio $^{14}C/^{12}C$ is about 1.2×10^{-12}. Thus, each fresh 1.0 g sample of biological carbon in equilibrium with the $^{14}CO_2$ of the atmosphere has a ratio of 5.8×10^{10} atoms of carbon-14 to 4.8×10^{22} atoms of carbon-12. The ratio decreases by a factor of 2 for each half-life period of ^{14}C (5730 yr).

---

[3]The available atmospheric pool of carbon-14 atoms fluctuates somewhat with the intensities of cosmic ray showers, with slow, long-term changes in Earth's magnetic field, and with the huge injections of carbon-12 into the atmosphere from the large-scale burning of coal and petroleum in the 1900s. To reduce the uncertainties in carbon-14 dating, results of the method have been corrected against dates made by tree-ring counting. For example, an uncorrected carbon-14 dating of a Viking site at L'anse aux Meadows, Newfoundland, gave a date of AD 895 ± 30. When corrected, the date of the settlement became AD 997, almost exactly the time indicated in Icelandic sagas for Leif Eriksson's landing at "Vinland," now believed to be the L'anse aux Meadows site.

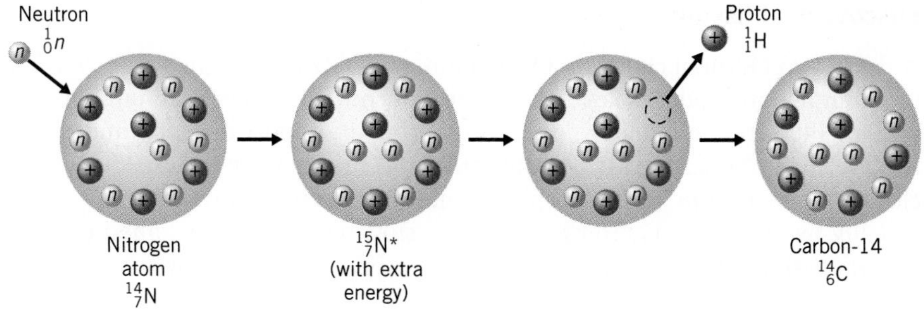

FIGURE **22.15** *Formation of carbon-14.* Carbon-14 forms in the atmosphere when neutrons in secondary cosmic radiation are captured by nitrogen-14 nuclei.

The dating of an object makes use of the fact that radioactive decay is a first-order process of the kind we studied in Chapter 15. If we let r_0 stand for the ^{14}C/^{12}C ratio at the time of death of the carbon-containing species and r_t stand for the ^{14}C/^{12}C ratio now, after the elapse of t years, we can substitute into Equation 15.3 (page 662) to obtain

$$\ln \frac{r_0}{r_t} = kt \tag{22.5}$$

where k is the rate constant for the decay (the decay constant for ^{14}C) and t is the elapsed time. We can obtain the rate constant from the half-life of ^{14}C using Equation 15.5.

$$\ln 2 = kt_{1/2}$$

Substituting 5730 yr for $t_{1/2}$ and solving for k gives $k = 1.21 \times 10^{-4}$ yr^{-1}. We can now substitute this value into Equation 22.5 to give

$$\ln \frac{r_0}{r_t} = (1.21 \times 10^{-4} \text{ yr}^{-1})\, t \tag{22.6}$$

Equation 22.6 can be used to calculate the age of a once-living object if its current ^{14}C/^{12}C ratio can be measured.

Suppose, now, that a sample of an ancient wooden object is found to have a ratio of ^{14}C to ^{12}C equal to 3.0×10^{-13}. Taking r_0 equal to 1.2×10^{-12} and substituting into Equation 22.6,

$$\ln \left(\frac{1.2 \times 10^{-12}}{3.0 \times 10^{-13}} \right) = (1.21 \times 10^{-4} \text{ yr}^{-1})\, t$$

$$\ln 4.0 = (1.21 \times 10^{-4})\, t$$

Solving for t gives an age of 1.2×10^4 years.

22.8 ▶ Nuclear fission is the breakup of a nucleus into two fragments of comparable size after capture of a slow neutron

Because of their electrical neutrality, neutrons penetrate an atom's electron cloud relatively easily and so are able to enter the nucleus. Enrico Fermi discovered in the early 1930s that even slow-moving *thermal neutrons* can be captured. (Thermal neutrons are those whose average kinetic energy puts them in thermal equilibrium with their surroundings at room temperature.)

When Fermi directed thermal neutrons at a uranium target, several different species of nuclei were produced, a result that Fermi could not explain. None had an atomic number above 82. He and other physicists did not follow up on a suggestion by Ida Noddack-Tacke, a chemist, that nuclei of much lower atomic number probably formed. The forces holding a nucleus together were considered too strong to permit such a result.

Discovery of fission

Otto Hahn received the 1944 Nobel Prize in chemistry.

In 1939, two German chemists, Otto Hahn and Fritz Strassman, finally verified that uranium fission produces several isotopes much lighter than uranium, for example, barium-140 (atomic number 56). Shortly after the report of Hahn and Strassman, two German physicists, Lise Meitner and Otto Frisch, were able to show that a nucleus of one of the rare isotopes in natural uranium, uranium-235, splits into roughly two equal parts after it captures a slow neutron. They named this breakup nuclear **fission.** The general reaction can be represented as follows:

$$^{235}_{92}U + ^{1}_{0}n \longrightarrow X + Y + b\,^{1}_{0}n$$

X and Y can be a large variety of nuclei with intermediate atomic numbers; over 30 have been identified. The coefficient b has an average value of 2.47, the average number of neutrons produced by fission events. A typical specific fission is

$$^{235}_{92}U + ^{1}_{0}n \longrightarrow ^{236}_{92}U^* \longrightarrow ^{94}_{36}Kr + ^{139}_{56}Ba + 3\,^{1}_{0}n$$

What actually undergoes fission is the compound nucleus of uranium-236. It has 144 neutrons and 92 protons for a neutron:proton ratio of roughly 1.6. Initially, the emerging krypton and barium isotopes have the same ratio, and this is much too high for them. The neutron:proton ratio for stable isotopes with 36 to 56 protons is nearer 1.2 to 1.3 (Figure 22.8). Therefore, the initially formed, neutron-rich krypton and barium nuclides promptly eject neutrons, called *secondary neutrons,* that generally have much higher energies than thermal neutrons.

An isotope that can undergo fission after neutron capture is called a **fissile isotope.** The naturally occurring fissile isotope of uranium used in reactors is uranium-235, whose abundance among the uranium isotopes today is only 0.72%. Two other fissile isotopes, uranium-233 and plutonium-239, can be made in nuclear reactors.

Nuclear chain reactions

The secondary neutrons released by fission become thermal neutrons as they are slowed by collisions with surrounding materials. They can now be captured by unchanged uranium-235 atoms. Because each fission event produces, on the average, more than two new neutrons, the potential exists for a **nuclear chain reaction** (Figure 22.16). A *chain reaction* is a self-sustaining process whereby products from one event cause one or more repetitions of the process.

If the sample of uranium-235 is small enough, the loss of neutrons to the surroundings is sufficiently rapid to prevent a chain reaction. However, at a certain *critical mass* of uranium-235, a few kilograms, this loss of neutrons is insufficient to prevent a sustained reaction. A virtually instantaneous fission of the sample ensues, in other words, an atomic bomb explosion. To trigger an atomic bomb, therefore, two subcritical masses of uranium-235 (or plutonium-239) are forced together to form a critical mass.

Energy yield from fission

1 MeV = 1.602×10^{-13} J

The binding energy per nucleon (Figure 22.1) in uranium-235 (about 7.6 MeV) is lower than the binding energies of the new nuclides (about 8.5 MeV). The net change for a single fission event can be calculated as follows (where we intend only *two* significant figures in each result):

Binding energy in krypton-94:

$$(8.5 \text{ MeV/nucleon}) \times 94 \text{ nucleons} = 800 \text{ MeV}$$

Binding energy in barium-139:

$$(8.5 \text{ MeV/nucleon}) \times 139 \text{ nucleons} = \underline{1200 \text{ MeV}}$$

Total binding energy of products: 2000 MeV

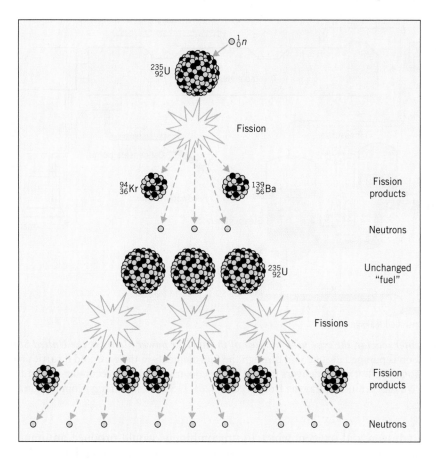

FIGURE 22.16 *Nuclear chain reaction.* Whenever the concentration of a fissile isotope is high enough (at or above the critical mass), the neutrons released by one fission can be captured by enough unchanged nuclei to cause more than one additional fission event. In civilian nuclear reactors, the fissile isotope is too dilute for this to get out of control. In addition, control rods of nonfissile materials that are able to capture excess neutrons can be inserted or withdrawn from the reactor core to make sure that the heat generated can be removed as fast as it forms.

Binding energy in uranium-235:

$$(7.6 \text{ MeV/nucleon}) \times 235 \text{ nucleons} = 1800 \text{ MeV}$$

The difference in total binding energy is (2000 MeV − 1800 MeV), or 200 MeV (3.2×10^{-11} J), which has to be taken as just a rough calculation. This is the energy released by each fission event going by the equation given. The energy produced by the fission of 1 kg (4.25 mol) of uranium-235 is calculated to be roughly 8×10^{13} J, enough to keep a 100 watt lightbulb in energy for 3000 years.

The energy available from 1 kg of uranium-235 is equivalent to the energy of 3000 tons of soft coal or 13,200 barrels of oil.

Heat from fission reactions can be used to drive electrical generators

Virtually all civilian nuclear power plants throughout the world operate on the same general principles. The heat of fission is used either directly or indirectly to increase the pressure of some gas that then drives an electrical generator.

The heart of a nuclear power plant is the *reactor,* where fission takes place in the *fuel core.* The nuclear fuel is generally uranium oxide, enriched to 2–4% in uranium-235 and formed into glasslike pellets. These are housed in long, sealed metal tubes called *cladding.* Bunches of tubes are assembled in spacers that permit a coolant to circulate around the tubes. A reactor has several such assemblies in its fuel core. The coolant carries away the heat of fission.

There is no danger of a nuclear power plant undergoing an atomic bomb explosion. An atomic bomb requires pure uranium-235 or plutonium-239. The concentration of fissile isotopes in a reactor is in the range of 2 to 4%, and much of the remainder is the common, nonfissile uranium-238. However, if the coolant fails to carry away the heat of fission, the reactor core can melt down, and the molten mass might even go through the thick walled containment vessel in which the reactor is kept. Or the high heat of the fusion might split molecules of coolant water into

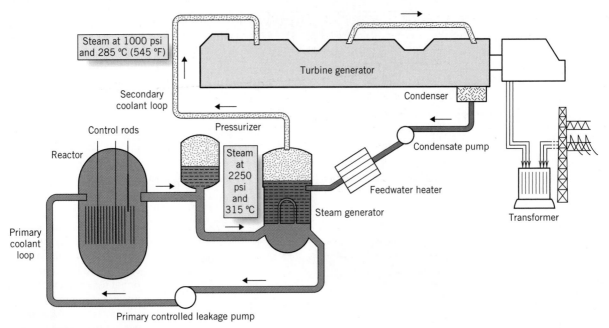

FIGURE 22.17 *Pressurized water reactor, the type used in most of the nuclear power plants in the United States.* Water in the primary coolant loop is pumped around and through the fuel elements in the core, and it carries away the heat of the nuclear chain reactions. The hot water delivers its heat to the cooler water in the secondary coolant loop, where steam is generated to drive the turbines. (Drawing from WASH-1261, U.S. Atomic Energy Commission, 1973.)

hydrogen and oxygen, which, on recombining, would produce an immense explosion. This is what happened when the reactor at the Ukrainian energy park at Chernobyl failed in 1986.

To convert secondary neutrons to thermal neutrons, the fuel core has a *moderator,* which is the coolant water itself in nearly all civilian reactors. Collisions between secondary neutrons and moderator molecules heats up the moderator. This heat energy eventually generates steam that enables an electric turbine to run. Ordinary water is a good moderator, but so are heavy water (D_2O) and graphite.

Two main types of reactors dominate civilian nuclear power, the *boiling water reactor* and the *pressurized water reactor.* Both use ordinary water as the moderator and so are sometimes called *light water reactors.* Roughly two thirds of the reactors in the United States are the pressurized water type (Figure 22.17). Such a reactor has two loops, and water circulates in both. The primary loop moves water through the reactor core, where it picks up thermal energy from fission. The water is kept in the *liquid* state by being under high pressure (hence the name *pressurized* water reactor).

The hot water in the primary loop transfers thermal energy to the secondary loop at the steam generator (Figure 22.17). This makes steam at high temperature and pressure, which is piped to the turbine. As the steam drives the turbine, the steam pressure drops. The condenser at the end of the turbine, cooled by water circulating from a river or lake or from huge cooling towers, forces a maximum pressure drop within the turbine by condensing the steam to liquid water. The returned water is then recycled to high-pressure steam. (In the boiling water reactor, there is only one coolant loop. The water heated in the reactor itself is changed to the steam that drives the turbine.)

Nuclear power plants produce several types of radioactive waste

Radioactive wastes from nuclear power plants occur as gases, liquids, and solids. The gases are mostly radionuclides of krypton and xenon but, with the exception of xenon-85 ($t_{1/2} = 10.4$ years), the gases have short half-lives and decay quickly. During decay, they must be contained, and this is one function of the cladding. Other dangerous radionuclides produced by fission include iodine-131, strontium-90, and cesium-137.

Iodine-131 must be contained because the human thyroid gland concentrates iodide ion to make the hormone thyroxine. Once the isotope is in the gland, beta radiation from iodine-131 could cause harm, possibly thyroid cancer and possibly impaired thyroid function. An effective countermeasure to iodine-131 poisoning is to take doses of ordinary iodine (as sodium iodide). Statistically, this increases the likelihood that the thyroid will take up stable iodide ion rather than the unstable anion of iodine-131.

Cesium-137 and strontium-90 also pose problems to humans. Cesium is in Group IA together with sodium, so radiating cesium-137 cations travel to some of the same places in the body where sodium ions go. Strontium is in Group IIA with calcium, so strontium-90 cations can replace calcium ions in bone tissue, sending radiation into bone marrow and possibly causing leukemia. The half-lives of cesium-137 and strontium-90, however, are relatively short.

> Cesium-137 has a half-life of 30 years; that of strontium-90 is 28.1 years. Both are beta emitters.

Some radionuclides in wastes are so long-lived that solid reactor wastes must be kept away from all human contact for dozens of centuries, longer than any nation has ever yet endured. Probably the most intensively studied procedure for making solid radioactive wastes secure is to convert them to glasslike or rocklike solids and bury them deeply within a rock stratum or in a mountain believed to be geologically stable with respect to earthquakes or volcanoes.

> The entire reactor of a power plant becomes a solid waste problem when its working life is over.

SUMMARY

The Einstein Equation Mass and energy are interconvertible. The **Einstein equation,** $\Delta E = \Delta mc^2$ (where c is the speed of light), lets us calculate one from the other. The total of the energy in the universe and all the mass calculated as an equivalent of energy is a constant, which is the **law of conservation of mass–energy.**

Nuclear Binding Energies When a nucleus forms from its nucleons, some mass changes into energy. This amount of energy, the **nuclear binding energy,** would be required to break up the nucleus again. The higher the binding energy per nucleon, the more stable is the nucleus.

Radioactivity The *electrostatic force* by which protons repel each other is overcome in the nucleus by the *nuclear strong force.* The ratio of neutrons to protons is a factor in nuclear stability. By radioactivity, the naturally occurring **radionuclides** adjust their neutron:proton ratios, lower their energies, and so become more stable by emitting **alpha** or **beta radiation,** sometimes gamma radiation as well.

The loss of an **alpha particle** leaves a nucleus with four fewer units of mass number and two fewer of atomic number. Loss of a **beta particle** leaves a nucleus with the same mass number and an atomic number one unit higher. **Gamma radiation** lets a nucleus lose some energy without a change in mass or atomic number.

Depending on the specific isotope, synthetic radionuclides emit alpha, beta, and gamma radiation. Some emit **positrons** (positive electrons) that produce gamma radiation by annihilation collisions with electrons. Other synthetic radionuclides decay by **electron capture** and emit X rays. Some radionuclides emit neutrons.

Nuclear equations are balanced when the mass numbers and atomic numbers on either side of the arrow respectively balance. The energies of emission are usually described in **electron volts** or multiples thereof.

A few very long-lived radionuclides in nature, like ^{238}U, are at the heads of **radioactive disintegration series,** which represent the successive decays of "daughter" radionuclides until a stable isotope forms.

Nuclear Stability Stable nuclides generally fall within a curving band, called the **band of stability,** when all known nuclides are plotted according to their numbers of neutrons and protons. Radionuclides that have too high neutron:proton ratios eject beta particles to adjust their ratios downward. Those with neutron:proton ratios too low generally emit positrons to change their ratios upward.

Isotopes whose nuclei consist of even numbers of both neutrons and protons are generally much more stable than those with odd numbers of both; this is the **odd–even rule.** Having all neutrons paired and all protons paired is energetically better (more stable) than having any nucleon unpaired.

Isotopes with specific numbers of protons or neutrons, the **magic numbers** of 2, 8, 20, 28, 50, 82, and 126, are generally more stable than others.

Transmutation When a bombardment particle—a proton, deuteron, alpha particle, or neutron—is captured, the resulting **compound nucleus** contains the energy of both the captured particle and its nucleons. The mode of decay of the compound nucleus is a function of its extra energy, not its extra mass. Many radionuclides have been made by these nuclear reactions, including all of the elements from 93 and higher.

Detecting and Measuring Radiations Instruments to detect and measure radiation—Geiger counters, for example—take advantage of the ability of radiation to generate ions in air or other matter.

The electron volt describes how much energy a radiation carries; $1 \text{ eV} = 1.602 \times 10^{-19} \text{ J}$.

The **curie (Ci)** and the **becquerel (Bq)**, the SI unit, describe how active a source is; $1 \text{ Ci} = 3.7 \times 10^{10} \text{ Bq}$, where $1 \text{ Bq} = 1$ disintegration/s.

The SI unit of absorbed dose, the **gray (Gy),** is used to describe how much energy is absorbed by a unit mass of absorber; $1 \text{ Gy} = 1 \text{ J/kg}$. An older unit, the **rad,** is equal to 0.01 Gy.

The **sievert**, an SI unit, and the **rem**, an older unit, are used to compare doses absorbed by different tissues and caused by different kinds of radiation. A 600 rem whole-body dose is lethal.

The normal **background radiation** causes millirem exposures per year. Naturally occurring radon, cosmic rays, radionuclides in soil and stone building materials, medical X rays, and releases from nuclear tests or from nuclear power plants all contribute to this background.

Protection against radiation can be achieved by using dense shields (e.g., lead or thick concrete), by avoiding overuse of radionuclides or X rays in medicine, and by taking advantage of the **inverse square law.** This law tells us that the intensity of radiation decreases with the square of the distance from its source.

Dating Radiological dating uses the known half-lives of naturally occurring radionuclides to date geological and archaeological objects. During life, living systems take up carbon-14. Following the organism's death, the age of the remains can be determined from the ratio of carbon-14 to carbon-12.

Fission Uranium-235, which occurs naturally, and plutonium-239, which can be made from uranium-238, are **fissile isotopes** that serve as the fuel in present-day reactors. When either isotope captures a thermal neutron, the isotope splits in one of several ways to give two smaller isotopes plus energy and more neutrons. The neutrons can generate additional fission events, enabling a nuclear chain reaction. If a critical mass of a fissile isotope is allowed to form, the **nuclear chain reaction** proceeds out of control, and the material detonates as an atomic bomb explosion.

Fission Reactors The goal of controlling the chain reaction and keeping it from becoming supercritical has resulted in various designs for commercial reactors. *Pressurized water reactors* are most commonly used; these have two loops of circulating fluids. In the primary loop, water circulates around the reactor core and absorbs the heat of fission. In the secondary loop, water accepts the heat and changes to high-pressure steam, which drives the electrical generator. If the circulation of fluid in the primary loop fails, the reactor must be shut down immediately to prevent a core meltdown and a possible breach of the reactor's containment vessel. One major problem with nuclear energy is the storage of radioactive wastes.

TOOLS ▶ **YOU HAVE LEARNED**

In this chapter you have learned the following tools that we apply to various aspects of solving problems dealing with nuclear chemistry.

TOOL	HOW IT WORKS
The Einstein equation, $\Delta E = \Delta m_0 c^2$ (page 973)	Computes the energy equivalent of an amount of mass.
Magnitude of the nuclear binding energy (page 974)	Compares the stability of nuclei.
Nuclear equations (page 976)	Quantitatively describes nuclear reactions, and relates amounts of nuclear reactants and products.
Band of stability (page 981)	Ascertains the stability of nuclei, and identifies the likely decay type from the ratio of neutrons to protons.
Odd–even rule (page 983)	Compares the stability of nuclei.
The law of radioactive decay (page 988)	Relates activities, decay constants, and the number of disintegrating atoms in a sample.
Inverse square law (page 992)	Computes radiation intensity at various distances from a radioactive source.
Half-life (page 998)	Computes decay constants and quickly estimates the time required for the number of disintegrating atoms to decay to a certain level.

REVIEW QUESTIONS

Conservation of Mass Energy

22.1 When a substance is described as *radioactive,* what do we know about it?

22.2 In chemical calculations involving chemical reactions we can regard the law of conservation of mass as a law independent of the law of conservation of energy despite Einstein's union of the two. What fact(s) make this possible?

22.3 How can we know that the speed of light is the absolute upper limit on the speed of any object?

22.4 State the following:
(a) Law of conservation of mass–energy
(b) Einstein equation

22.5 Why isn't the sum of the masses of all nucleons in one nucleus equal to the mass of the actual nucleus?

Radioactivity

22.6 Three kinds of radiation make up nearly all of the radiation observed from naturally occurring radionuclides. What are they?

22.7 Give the composition of each of the following:
(a) Alpha particle   (c) Positron
(b) Beta particle   (d) Deuteron

22.8 Why is the penetrating ability of alpha radiation less than that of beta or gamma radiation?

22.9 With respect to their formation, how do gamma rays and X rays differ?

22.10 How does electron capture generate X rays?

Nuclear Stability

22.11 What data are plotted and what criterion is used to identify the actual band in the band of stability?

22.12 Both barium-123 and barium-140 are radioactive, but which is more likely to have the *longer* half-life? Explain your answer.

22.13 Tin-112 is a stable nuclide but indium-112 is radioactive and has a very short half-life ($t_{1/2}$ = 14 min). What does tin-112 have that indium-112 does not to account for this difference in stability?

22.14 Lanthanum-139 is a stable nuclide but lanthanum-140 is unstable ($t_{1/2}$ = 40 hr). What rule of thumb concerning nuclear stability is involved?

22.15 As the atomic number increases, the neutron:proton ratio increases. What does this suggest is a factor in nuclear stability?

22.16 Radionuclides of high atomic number are far more likely to be alpha emitters than those of low atomic number. Offer an explanation for this phenomenon.

22.17 Although lead-164 has two magic numbers, 82 protons and 82 neutrons, it is unknown. Lead-208, however, is known and stable. What problem accounts for the nonexistence of lead-164?

22.18 What decay particle is emitted from a nucleus of low to intermediate atomic number but a relatively high neutron:proton ratio? How does the emission of this particle benefit the nucleus?

22.19 What decay particle is emitted from a nucleus of low to intermediate atomic number but a relatively low neutron:proton ratio? How does the emission of this particle benefit the nucleus?

22.20 What does electron capture do to the neutron:proton ratio in a nucleus, increase it, decrease it, or leave it alone? Which kinds of radionuclides are more likely to undergo this change, those above or those below the band of stability?

Transmutations

22.21 Compound nuclei form and decay almost at once. What accounts for the instability of a compound nucleus?

22.22 What explains the existence of several decay modes for the compound nucleus aluminum-27?

22.23 Rutherford theorized that a compound nucleus forms when helium nuclei hit nitrogen-14 nuclei. If this compound nucleus decayed by the loss of a neutron instead of a proton, what would be the other product?

Detecting and Measuring Radiations

22.24 What specific property of radiation is used by the Geiger counter?

22.25 Dangerous doses of radiation can actually involve very small quantities of energy. Explain.

22.26 What units, SI and common, are used to describe each of the following?
(a) The *activity* of a radioactive sample
(b) The *energy* of a particle or of a photon of radiation
(c) The quantity of energy absorbed by a given mass
(d) Dose equivalents for comparing biological effects

22.27 A sample giving 3.7×10^{10} disintegrations/s has what activity in Ci and in Bq?

22.28 Explain the necessity in health sciences for the *sievert.*

Applications of Radionuclides

22.29 Why should a radionuclide used in diagnostic work have a short half-life? If the half-life is too short, what problem arises?

22.30 An alpha emitter is not used in diagnostic work. Why?

22.31 In general terms, explain how neutron activation analysis is used and how it works.

22.32 What is one assumption in the use of the uranium:lead ratio for dating ancient geologic formations?

*22.33** If a sample used for carbon-14 dating is contaminated by air, there is a potentially serious problem with the method. What is it?

22.34 List some of the kinds of radiation that make up our background radiation.

Nuclear Energy

22.35 Why is it easier for a nucleus to capture a neutron than a proton?

22.36 What do each of the following terms mean?
(a) Thermal neutron
(b) Nuclear fission
(c) Fissile isotope

22.37 Which fissile isotope occurs in nature?

22.38 What fact about the fission of uranium-235 makes it possible for a *chain reaction* to occur?

22.39 Explain in general terms why fission generates more neutrons than needed to initiate it.

22.40 Why would there be a *subcritical mass* of a fissile isotope? (Why isn't *any* mass of uranium-235 critical?)

22.41 What purpose is served by a *moderator* in a nuclear reactor?

22.42 Why is there no possibility of an atomic bomb explosion from a nuclear power plant?

REVIEW PROBLEMS

Answers to problems whose numbers are printed in color are given in Appendix B. More challenging problems are marked with asterisks. **ILW** = Interactive LearningWare solution is available at *www.wiley.com/college/brady*.

Conservation of Mass Energy

22.43 Calculate the mass equivalent in grams of 1.00 kJ.

*__22.44__ Calculate the mass in kilograms of a 1.00 kg object when its velocity, relative to us, is (a) 3.00×10^7 m s^{-1}, (b) 2.90×10^8 m s^{-1}, and (c) 2.99×10^8 m s^{-1}. (Notice the progression of these numbers toward the velocity of light, 3.00×10^8 m s^{-1}.)

22.45 Calculate the quantity of mass in nanograms that is changed into energy when 1 mol of liquid water forms by the combustion of hydrogen, all measurements being made at 1 atm and 25 °C.

22.46 Show that the mass equivalent to the energy released by the complete combustion of 1 mol of methane (890 kJ) is 9.89 ng.

Nuclear Binding Energies

ILW 22.47 Calculate the binding energy in J/nucleon of the deuterium nucleus whose mass is 2.0135 u.

22.48 Calculate the binding energy in J/nucleon of the tritium nucleus whose mass is 3.01550 u.

Radioactivity

22.49 Complete the following nuclear equations by writing the symbols of the missing particles:
(a) $^{211}_{82}\text{Pb} \rightarrow ^{0}_{-1}e +$ _____
(b) $^{177}_{73}\text{Ta} \xrightarrow{\text{electron capture}}$ _____
(c) $^{220}_{86}\text{Rn} \rightarrow ^{4}_{2}\text{He} +$ _____
(d) $^{19}_{10}\text{Ne} \rightarrow ^{0}_{1}e +$ _____

22.50 Write the symbols of the missing particles to complete the following nuclear equations:
(a) $^{245}_{96}\text{Cm} \rightarrow ^{4}_{2}\text{He} +$ _____
(b) $^{146}_{56}\text{Ba} \rightarrow ^{0}_{-1}e +$ _____
(c) $^{58}_{29}\text{Cu} \rightarrow ^{0}_{1}e +$ _____
(d) $^{68}_{32}\text{Ge} \xrightarrow{\text{electron capture}}$ _____

22.51 Write a balanced nuclear equation for each of the following changes:
(a) Alpha emission from plutonium-242
(b) Beta emission from magnesium-28
(c) Positron emission from silicon-26
(d) Electron capture by argon-37

22.52 Write the balanced nuclear equation for each of the following nuclear reactions:
(a) Electron capture by iron-55
(b) Beta emission by potassium-42
(c) Positron emission by ruthenium-93
(d) Alpha emission by californium-251

22.53 Write the symbols, including the atomic and mass numbers, for the radionuclides that would give each of the following products:
(a) Fermium-257 by alpha emission
(b) Bismuth-211 by beta emission
(c) Neodymium-141 by positron emission
(d) Tantalum-179 by electron capture

22.54 Each of the following nuclides forms by the decay mode described. Write the symbols of the parents, giving both atomic and mass numbers.ss
(a) Rubidium-80 formed by electron capture
(b) Antimony-121 formed by beta emission
(c) Chromium-50 formed by positron emission
(d) Californium-253 formed by alpha emission

22.55 Krypton-87 decays to krypton-86. What other particle forms?

22.56 Write the symbol of the nuclide that forms from cobalt-58 when it decays by electron capture.

Nuclear Stability

22.57 If an atom of potassium-38 had the option of decaying by positron emission or beta emission, which route would it likely take, and why? Write the nuclear equation.

22.58 Suppose that an atom of argon-37 could decay by either beta emission or electron capture. Which route would it likely take, and why? Write the nuclear equation.

22.59 If we begin with 3.00 mg of iodine-131 ($t_{1/2}$ = 8.07 hr), how much remains after 6 half-life periods?

22.60 A sample of technetium-99*m* with a mass of 9.00 ng will have decayed to how much of this radionuclide after 4 half-life periods (about 1 day)?

Transmutations

22.61 When vanadium-51 captures a deuteron ($^{2}_{1}H$), what compound nucleus forms? (Write its symbol.) This particle expels a proton ($^{1}_{1}p$). Write the nuclear equation for the overall change from vanadium-51.

22.62 The alpha-particle bombardment of fluorine-19 generates sodium-22 and neutrons. Write the nuclear equation, including the intermediate compound nucleus.

22.63 Gamma ray bombardment of bromine-81 causes a transmutation in which neutrons are one product. Write the symbol of the other product.

22.64 Neutron bombardment of cadmium-115 results in neutron capture and the release of gamma radiation. Write the nuclear equation.

22.65 When manganese-55 is bombarded by protons, neutrons are released. What else forms? Write the nuclear equation.

22.66 Which nuclide forms when sodium-23 is bombarded by alpha particles and the compound nucleus emits a gamma ray photon?

22.67 The nuclei of which isotope of zinc would be needed as bombardment particles to make nuclei of element 112 from lead-208 if the intermediate compound nucleus loses a neutron?

22.68 Write the symbol of the nuclide whose nuclei would be the target for bombardment by nickel-64 nuclei to produce nuclei of $^{272}_{111}Uuu$ after the intermediate compound nucleus loses a neutron. (Uuu is the temporary symbol for element 111)

Detecting and Measuring Radiations

22.69 Suppose that a radiologist who is 2.0 m from a small, unshielded source of radiation receives 2.8 units of radiation. To reduce the exposure to 0.28 units of radiation, to what distance from the source should the radiologist move?

22.70 By what percentage should a radiation specialist increase the distance from a small, unshielded source to reduce the radiation intensity by 10.0%?

22.71 If exposure from a distance of 1.60 m gave a worker a dose of 8.4 rem, how far should the worker move away from the source to reduce the dose to 0.50 rem for the same period?

22.72 During work with a radioactive source, a worker was told that he would receive 50 mrem at a distance of 4.0 m during 30 min of work. What would be the received dose if the worker moved closer, to 0.50 m, for the same period?

Law of Radioactive Decay

22.73 Smoke detectors contain a small amount of americium-241, which has a half-life of 1.70×10^5 days. If the detector contains 0.20 mg of ^{241}Am, what is the activity, in becquerels? In microcuries?

22.74 Strontium-90 is a dangerous radioisotope present in fallout produced by nuclear weapons. ^{90}Sr has a half-life of 1.00×10^4 days. What is the activity of 1.00 g of ^{90}Sr, in becquerels? In microcuries?

22.75 Iodine-131 is a radioisotope present in radioactive fallout that targets the thyroid gland. If 1.00 mg of ^{131}I has an activity of 4.6×10^{12} Bq, what is the decay constant for ^{131}I? What is the half-life, in seconds?

22.76 A 10.0 mg sample of thallium-201 has an activity of 7.9×10^{13} Bq. What is the decay constant for ^{201}Tl? What is the half-life of ^{201}Tl, in seconds?

Applications of Radionuclides

22.77 What percentage of cesium chloride made from cesium-137 ($t_{1/2}$ = 30 yr; beta emitter) remains after 150 yr? What *chemical* product forms?

22.78 A sample of waste has a radioactivity, caused solely by strontium-90 (beta emitter, $t_{1/2}$ = 28.1 yr), of 0.245 Ci g^{-1}. How many years will it take for its activity to decrease to 1.00×10^{-6} Ci g^{-1}?

22.79 A worker in a laboratory unknowingly became exposed to a sample of radiolabeled sodium iodide made from iodine-131 (beta emitter, $t_{1/2}$ = 8.07 days). The mistake was realized 28.0 days after the accidental exposure, at which time the activity of the sample was 25.6×10^{-5} Ci/g. The safety officer needed to know how active the sample was at the time of the exposure. Calculate this value in Ci/g.

22.80 Technetium-99*m* (gamma emitter, $t_{1/2}$ = 6.02 hr) is widely used for diagnosis in medicine. A sample prepared in the early morning for use that day had an activity of 4.52×10^{-6} Ci. What will its activity be at the end of the day, that is, after 8.00 hr?

22.81 A 500 mg sample of rock was found to have 2.45×10^{-6} mol of potassium-40 ($t_{1/2}$ = 1.3×10^9 yr) and 2.45×10^{-6} mol of argon-40. How old was the rock? (What assumption is made about the origin of the argon-40?)

22.82 If a rock sample was found to contain 1.16×10^{-7} mol of argon-40, how much potassium-40 would also have to be present for the rock to be 1.3×10^9 yr old?

ILW **22.83** A tree killed by being buried under volcanic ash was found to have a ratio of carbon-14 atoms to carbon-12 atoms of 4.8×10^{-14}. How long ago did the eruption occur?

22.84 A wooden door lintel from an excavated site in Mexico would be expected to have what ratio of carbon-14 to carbon-12 atoms if the lintel is 9.0×10^3 yr old?

Nuclear Energy

22.85 Complete the following nuclear equation by supplying the symbol for the other product of the fission.

$$^{235}_{92}U + ^{1}_{0}n \longrightarrow ^{94}_{38}Sr + \underline{\quad} + 2^{1}_{0}n$$

22.86 Both products of the fission in the previous problem are unstable. According to Figures 22.8 and 22.9, what is the most likely way for each of them to decay, by alpha emission, beta emission, or by positron emission? Explain. What are some of the possible fates of the extra neutrons produced by the fission shown in the previous problem?

ADDITIONAL EXERCISES

22.87 What is the nuclear equation for each of the following changes?
(a) Beta emission from aluminum-30
(b) Alpha emission from einsteinium-252
(c) Electron capture by molybdenum-93
(d) Positron emission by phosphorus-28

*22.88 Calculate the binding energy in J/nucleon of the nucleus of an atom of iron-56. The observed mass of one *atom* is 55.9349 u. What information lets us know that no isotope has a *larger* binding energy per nucleon?

*22.89 Calculate the binding energy in J/nucleon of uranium-235. The observed mass of one *atom* is 235.0439 u.

22.90 Give the nuclear equation for each of these changes.
(a) Positron emission by carbon-10
(b) Alpha emission by curium-243
(c) Electron capture by vanadium-49
(d) Beta emission by oxygen-20

*22.91 If a positron is to be emitted spontaneously, how much more *mass* (as a minimum) must an *atom* of the parent have than an *atom* of the daughter nuclide? Explain.

22.92 There is a gain in binding energy per nucleon when light nuclei fuse to form heavier nuclei. Yet, a tritium atom and a deuterium atom, in a mixture of these isotopes, does not spontaneously fuse to give helium (and energy). Explain why not.

22.93 ^{214}Bi decays to isotope *A* by alpha emission; *A* then decays to *B* by beta emission, which decays to *C* by another beta emission. Element *C* decays to *D* by still another beta emission, and *D* decays by alpha emission to a stable isotope, *E*. What is the proper symbol of element *E*? (Contributed by Prof. W. J. Wysochansky, Georgia Southern University.)

22.94 ^{15}O decays by positron emission with a half-life of 124 s. (a) Give the proper symbol of the product of the decay. (b) How much of a 750 mg sample of ^{15}O remains after 5.0 min of decay? (Contributed by Prof. W. J. Wysochansky, Georgia Southern University.)

22.95 Alpha decay of ^{238}U forms ^{234}Th. What kind of decay from ^{234}Th produces ^{234}Ac? (Contributed by Prof. Mark Benvenuto, University of Detroit—Mercy.)

22.96 A sample of rock was found to contain 2.07×10^{-5} mol of ^{40}K and 1.15×10^{-5} mol of ^{40}Ar. If we assume that all of the ^{40}Ar came from the decay of ^{40}K, what is the age of the rock in years ($t_{1/2} = 1.3 \times 10^9$ yr for ^{40}K.)

22.97 The ^{14}C content of an ancient piece of wood was found to be one eighth of that in living trees. How many years old is this piece of wood ($t_{1/2} = 5730$ yr for ^{14}C)?

22.98 Dinitrogen trioxide, N_2O_3, is largely dissociated into NO and NO_2 in the gas phase, where there exists the equilibrium $N_2O_3 \rightleftharpoons NO + NO_2$. In an effort to determine the structure of N_2O_3, a mixture of NO and *NO_2 was prepared containing isotopically labeled N in the NO_2. After a period of time the mixture was analyzed and found to contain substantial amounts of both *NO and *NO_2. Explain how this is consistent with the structure for N_2O_3 being ONONO.

22.99 The reaction $(CH_3)_2Hg + HgI_2 \rightarrow 2CH_3HgI$ is believed to occur through a transition state with the structure

If this is so, what should be observed if CH_3HgI and *HgI_2 are mixed, where the asterisk denotes a radioactive isotope of Hg? Explain your answer.

*22.100 A large, complex piece of apparatus has built into it a cooling system containing an unknown volume of cooling liquid. It is desired to measure the volume of the coolant without draining the lines. To the coolant was added 10.0 mL of methanol whose molecules included atoms of ^{14}C and that had a specific activity of 580 counts per minute per gram (cpm/g), determined using a Geiger counter. The coolant was permitted to circulate to assure complete mixing before a samples was withdrawn that was found to have a specific activity of 29 cpm/g. Calculate the volume of coolant in the system in milliliters. The density of methanol is 0.792 g/mL, and the density of the coolant is 0.884 g/mL.

*22.101 A complex ion of chromium(III) with oxalate ion was prepared from ^{51}Cr-labeled $K_2Cr_2O_7$, having a specific activity of 843 cpm/g (counts per minute per gram), and ^{14}C-labeled oxalic acid, $H_2C_2O_4$, having a specific activity of 345 cpm/g. Chromium-51 decays by electron capture with the emission of gamma radiation, whereas ^{14}C is a pure beta emitter. Because of the characteristics of the beta and gamma detectors, each of these isotopes may be counted independently. A sample of the complex ion was observed to give a gamma count of 165 cpm and a beta count of 83 cpm. From these data, determine the number of oxalate ions bound to each Cr(III) in the complex ion. (*Hint:* For the starting materials calculate the cpm per mole of Cr and oxalate, respectively.)

22.102 Iodine-131 is used to treat Graves disease, a disease of the thyroid gland. The amount of ^{131}I used depends on the size of the gland. If the dose is 86 microcuries per gram of thyroid gland, how many grams of ^{131}I should be administered to a patient with a thyroid gland weighing 20 g?

Again you have an opportunity to test your understanding of concepts, your knowledge of scientific terms, and your skills at solving chemistry problems. Read through the following questions carefully, and answer each as fully as possible. Review topics when necessary.

1 What is meant by the term *spontaneous change*?

2 Which of these are state functions: E, H, q, S, G, w, T?

3 How is entropy related to statistical probability?

4 What would be the algebraic signs of ΔS for the following reactions?
(a) $Br_2(l) + Cl_2(g) \rightarrow 2BrCl(g)$
(b) $CaO(s) + CO_2(g) \rightarrow CaCO_3(s)$

5 Which of the following states has the greatest entropy?
(a) $2H_2O(l)$
(b) $2H_2O(s)$
(c) $2H_2(l) + O_2(g)$
(d) $2H_2(g) + O_2(g)$
(e) $4H(g) + 2O(g)$

6 Calculate $\Delta S°$ (in J/K) for the following reactions.
(a) $H_2O(l) + SO_3(g) \rightarrow H_2SO_4(l)$
(b) $2KCl(s) + H_2SO_4(l) \rightarrow K_2SO_4(s) + 2HCl(g)$
(c) $C_2H_4(g) + H_2O(g) \rightarrow C_2H_5OH(l)$

7 Calculate $\Delta G°$ in kJ for the following reactions.
(a) $CaSO_4 \cdot \frac{1}{2}H_2O(s) + \frac{3}{2}H_2O(l) \rightarrow CaSO_4 \cdot 2H_2O(s)$
(b) $CH_4(g) + Cl_2(g) \rightarrow CH_3Cl(g) + HCl(g)$
(c) $CaSO_4(s) + CO_2(g) \rightarrow CaCO_3(s) + SO_3(g)$

8 Which of the reactions in the preceding question would appear to be spontaneous?

9 Calculate $\Delta S°$, $\Delta H°$, and $\Delta G_T°$ (at 400 °C) using energy units of joules or kilojoules for these reactions.
(a) $CaSO_4 \cdot 2H_2O \rightarrow CaSO_4 \cdot \frac{1}{2}H_2O(s) + \frac{3}{2}H_2O(g)$
(b) $NaOH(s) + NH_4Cl(s) \rightarrow NaCl(s) + NH_3(g) + H_2O(g)$
(c) $SO_3(g) \rightarrow SO_2(g) + \frac{1}{2}O_2(g)$

10 For the reaction $3NO(g) \rightleftharpoons NO_2(g) + N_2O(g)$, calculate K_p at 25 °C using values of $\Delta G_f°$ from Table 20.2. Calculate the value of K_c at 25 °C for the reaction.

11 Consider the reaction

$$CH_4(g) + Cl_2(g) \rightleftharpoons CH_3Cl(g) + HCl(g).$$

(a) Calculate $\Delta G_{473}°$ for this reaction at 200 °C.
(b) What is the value of K_p for this reaction at 200 °C?
(c) What is the value of K_c for this reaction at 200 °C?

12 Glycine, one of the important amino acids, has the structure

$$
\begin{array}{c}
\;\;\;\;\;\;\;\;\; H \;\;\; O \\
\;\;\;\;\;\;\;\;\; | \;\;\;\;\; \| \\
H-N-C-C-O-H \\
\;\;\;\;\;\;\;\; | \;\;\; | \\
\;\;\;\;\;\;\;\; H \;\; H
\end{array}
$$

Calculate the atomization energy of this molecule from the data in Table 20.4.

13 Use bond energies in Table 20.4 to calculate the approximate energy that would be absorbed or given off in the formation of 25.0 g of C_2H_6 by the following reaction in the gas phase.

$$
H-C\equiv C-H + 2H_2 \longrightarrow
\begin{array}{c}
H \;\; H \\
| \;\;\; | \\
H-C-C-H \\
| \;\;\; | \\
H \;\; H
\end{array}
$$

14 Sketch a diagram of an electrolysis cell in which a concentrated solution of NaCl is undergoing electrolysis.
(a) Label the cathode and anode, including their charges.
(b) Write half-reactions for the changes taking place at the electrodes.
(c) Write a balanced equation for the net cell reaction.

15 Suppose that the electrolysis cell described in the previous question contains 250 mL of brine.
(a) What will be the pH of the solution if the electrolysis is carried out for 20.0 minutes using a current of 1.00 A?
(b) How many milliliters of H_2, measured at STP, would be evolved if the cell were operated at 5.00 A for 10.0 minutes?

16 What current would be required to deposit 0.100 g of nickel in 20.0 minutes from a solution of $NiSO_4$?

17 A large electrolysis cell can produce as much as 900 lb of aluminum in 1 day from Al_2O_3. What current is required to accomplish this?

18 Sketch a diagram of a galvanic cell consisting of a copper electrode dipping into 1.00 M $CuSO_4$ solution and an iron electrode dipping into 1.00 M $FeSO_4$ solution.
(a) Identify the cathode and the anode. Indicate the charge carried by each.
(b) Write the equation for the net cell reaction.
(c) Describe the cell by writing its standard cell notation.
(d) What is the potential of the cell?

19 What is the purpose of a salt bridge in a galvanic cell?

20 Suppose the cell described in the previous question contains 100 mL of each solution and is operated for a period of 50.0 hr at a constant current of 0.10 A. At the end of this time, what will be the concentrations of Cu^{2+} and Fe^{2+} in their respective solutions? What will the cell potential be at this point?

21 Use data from Table 21.1 to calculate the value of K_c at 25 °C for the reaction

$$O_2 + 4Br^- + 4H^+ \longrightarrow 2Br_2 + 2H_2O$$

22 A galvanic cell was assembled as follows. In one compartment, a copper electrode was immersed in a 1.00 M solution of $CuSO_4$. In the other compartment, a manganese electrode was immersed in a 1.00 M solution of $MnSO_4$. The potential of the cell was measured to be 1.52 V, with the Mn electrode as the negative electrode. What is the value of $E°$ for the following half-reaction?

$$Mn^{2+}(aq) + 2e^- \rightleftharpoons Mn(s)$$

23 Calculate E_{cell} for the reaction

$$3Cu(s) + 2NO_3^-(aq) + 8H^+(aq) \longrightarrow$$
$$2NO(g) + 4H_2O + 3Cu^{2+}(aq)$$

24 What is the value of K_c for the reaction described in the preceding question?

25 Consider a galvanic cell formed by using the half-reactions

$$NiO_2(s) + 2H_2O + 2e^- \rightleftharpoons Ni(OH)_2(s) + 2OH^-(aq)$$
$$E° = +0.49 \text{ V}$$

$$PbO_2(s) + H_2O + 2e^- \rightleftharpoons PbO(s) + 2OH^-(aq)$$
$$E° = +0.25 \text{ V}$$

(a) Write the equation for the spontaneous cell reaction.
(b) Calculate the value of $E°_{cell}$ for this system.
(c) Calculate $\Delta G°$ for the spontaneous cell reaction.

26 A galvanic cell was constructed in which one half-cell consists of a silver electrode coated with silver chloride dipping into a solution that contains chloride ion and the second half-cell consists of a nickel electrode dipping into a solution that contains Ni^{2+}. The half-cell reactions and their reduction potentials are

$$AgCl(s) + e^- \rightleftharpoons Ag(s) + Cl^-(aq) \quad E° = +0.222 \text{ V}$$
$$Ni^{2+}(aq) + 2e^- \rightleftharpoons Ni(s) \quad E° = -0.257 \text{ V}$$

(a) Write the equation for the spontaneous cell reaction.
(b) Calculate $E°_{cell}$ for the cell reaction.
(c) Write the Nernst equation for the cell.
(d) Calculate the cell potential if $[Cl^-] = 0.020 \, M$ and $[Ni^{2+}] = 0.10 \, M$.
(e) The galvanic cell was used to measure an unknown chloride ion concentration. The Ni^{2+} concentration was 0.200 M and the measured cell potential was 0.388 V. What was the chloride ion concentration?

27 Describe how a nickel-metal hydride cell works.

28 Describe how a lithium-ion cell works. What does intercalation mean?

29 What factors involving the nucleus of helium might be responsible for the large abundance of helium in the universe?

30 What is the symbol for (a) an alpha particle, (b) a beta particle, (c) a positron?

31 What is the rest mass of a particle of gamma radiation?

32 Suppose that the total mass of the reactants in a chemical reaction was 100.00000 g. How many kilojoules of energy would have to evolve from this reaction if the total mass of the products could be no greater than 99.99900 g? If all this energy were used to heat water, how many liters of water could have its temperature raised from 10 °C to 100 °C?

33 Calculate the binding energy in joules/nucleon for the nucleus of the deuterium atom, ^2_1H. The mass of an atom of deuterium, including its electron, is 2.014102 u.

34 Which nuclear force acts over the longer distance, the electrostatic force or the nuclear strong force?

35 Why is an isotope of hydrogen a better candidate for nuclear fusion than an isotope of helium?

36 When uranium-238 is bombarded by a particular bombarding particle, it can be transmuted into plutonium-239 and neutrons. Which bombarding particle accomplishes this? How many neutrons are produced for each plutonium-239 atom? What is the nuclear equation for this change?

37 If an atom of beryllium-7 decays by the capture of a K-electron, into which isotope does it change? Write the equation.

38 An atom of neodymium-144 decays by alpha emission. Write the nuclear equation.

39 Phosphorus-32 decays by beta emission. Write the nuclear equation.

40 The half-life of samarium-149, an alpha emitter, is 4×10^{14} years. The half-life of oxygen-15 is 124 seconds. Assuming that equimolar samples are compared, which is the more intensely radioactive?

41 The half-life of cesium-137 is 30 years. Of an initial 100 g sample, how much cesium-137 will remain after 300 years?

42 What are some "rules of thumb" that can be used to judge if a particular radionuclide might have a long enough half-life to warrant the effort to make it?

43 The ratio of neutrons to protons in atomic nuclei grows larger with increasing atomic number. What is an explanation for this?

44 Which is likelier to be the more stable isotope of calcium, calcium-40 or calcium-42? (What "rule of thumb" about nuclear stability is involved here?)

45 A particular compound nucleus can form from a variety of combinations of targets and bombarding particles, as in the example of aluminum. What determines how a compound nucleus breaks up?

46 Strontium-90, a beta emitter, has a half-life of 28.1 yr. If 36.2 mg of ^{90}Sr were incorporated in the bones of a growing child, how many beta particles would the child absorb from this source in 1.00 day?

47 Which kind of radiation poses a greater health risk, beta or alpha radiation? Why?

Electron Configurations of the Elements

Atomic Number			Atomic Number			Atomic Number		
1	H	$1s^1$	39	Y	$[Kr]\,5s^2\,4d^1$	77	Ir	$[Xe]\,6s^2\,4f^{14}\,5d^7$
2	He	$1s^2$	40	Zr	$[Kr]\,5s^2\,4d^2$	78	Pt	$[Xe]\,6s^1\,4f^{14}\,5d^9$
3	Li	$[He]\,2s^1$	41	Nb	$[Kr]\,5s^1\,4d^4$	79	Au	$[Xe]\,6s^1\,4f^{14}\,5d^{10}$
4	Be	$[He]\,2s^2$	42	Mo	$[Kr]\,5s^1\,4d^5$	80	Hg	$[Xe]\,6s^2\,4f^{14}\,5d^{10}$
5	B	$[He]\,2s^2\,2p^1$	43	Tc	$[Kr]\,5s^2\,4d^5$	81	Tl	$[Xe]\,6s^2\,4f^{14}\,5d^{10}\,6p^1$
6	C	$[He]\,2s^2\,2p^2$	44	Ru	$[Kr]\,5s^1\,4d^7$	82	Pb	$[Xe]\,6s^2\,4f^{14}\,5d^{10}\,6p^2$
7	N	$[He]\,2s^2\,2p^3$	45	Rh	$[Kr]\,5s^1\,4d^8$	83	Bi	$[Xe]\,6s^2\,4f^{14}\,5d^{10}\,6p^3$
8	O	$[He]\,2s^2\,2p^4$	46	Pd	$[Kr]\,4d^{10}$	84	Po	$[Xe]\,6s^2\,4f^{14}\,5d^{10}\,6p^4$
9	F	$[He]\,2s^2\,2p^5$	47	Ag	$[Kr]\,5s^1\,4d^{10}$	85	At	$[Xe]\,6s^2\,4f^{14}\,5d^{10}\,6p^5$
10	Ne	$[He]\,2s^2\,2p^6$	48	Cd	$[Kr]\,5s^2\,4d^{10}$	86	Rn	$[Xe]\,6s^2\,4f^{14}\,5d^{10}\,6p^6$
11	Na	$[Ne]\,3s^1$	49	In	$[Kr]\,5s^2\,4d^{10}\,5p^1$	87	Fr	$[Rn]\,7s^1$
12	Mg	$[Ne]\,3s^2$	50	Sn	$[Kr]\,5s^2\,4d^{10}\,5p^2$	88	Ra	$[Rn]\,7s^2$
13	Al	$[Ne]\,3s^2\,3p^1$	51	Sb	$[Kr]\,5s^2\,4d^{10}\,5p^3$	89	Ac	$[Rn]\,7s^2\,6d^1$
14	Si	$[Ne]\,3s^2\,3p^2$	52	Te	$[Kr]\,5s^2\,4d^{10}\,5p^4$	90	Th	$[Rn]\,7s^2\,6d^2$
15	P	$[Ne]\,3s^2\,3p^3$	53	I	$[Kr]\,5s^2\,4d^{10}\,5p^5$	91	Pa	$[Rn]\,7s^2\,5f^2\,6d^1$
16	S	$[Ne]\,3s^2\,3p^4$	54	Xe	$[Kr]\,5s^2\,4d^{10}\,5p^6$	92	U	$[Rn]\,7s^2\,5f^3\,6d^1$
17	Cl	$[Ne]\,3s^2\,3p^5$	55	Cs	$[Xe]\,6s^1$	93	Np	$[Rn]\,7s^2\,5f^4\,6d^1$
18	Ar	$[Ne]\,3s^2\,3p^6$	56	Ba	$[Xe]\,6s^2$	94	Pu	$[Rn]\,7s^2\,5f^6$
19	K	$[Ar]\,4s^1$	57	La	$[Xe]\,6s^2\,5d^1$	95	Am	$[Rn]\,7s^2\,5f^7$
20	Ca	$[Ar]\,4s^2$	58	Ce	$[Xe]\,6s^2\,4f^1\,5d^1$	96	Cm	$[Rn]\,7s^2\,5f^7\,6d^1$
21	Sc	$[Ar]\,4s^2\,3d^1$	59	Pr	$[Xe]\,6s^2\,4f^3$	97	Bk	$[Rn]\,7s^2\,5f^9$
22	Ti	$[Ar]\,4s^2\,3d^2$	60	Nd	$[Xe]\,6s^2\,4f^4$	98	Cf	$[Rn]\,7s^2\,5f^{10}$
23	V	$[Ar]\,4s^2\,3d^3$	61	Pm	$[Xe]\,6s^2\,4f^5$	99	Es	$[Rn]\,7s^2\,5f^{11}$
24	Cr	$[Ar]\,4s^1\,3d^5$	62	Sm	$[Xe]\,6s^2\,4f^6$	100	Fm	$[Rn]\,7s^2\,5f^{12}$
25	Mn	$[Ar]\,4s^2\,3d^5$	63	Eu	$[Xe]\,6s^2\,4f^7$	101	Md	$[Rn]\,7s^2\,5f^{13}$
26	Fe	$[Ar]\,4s^2\,3d^6$	64	Gd	$[Xe]\,6s^2\,4f^7\,5d^1$	102	No	$[Rn]\,7s^2\,5f^{14}$
27	Co	$[Ar]\,4s^2\,3d^7$	65	Tb	$[Xe]\,6s^2\,4f^9$	103	Lr	$[Rn]\,7s^2\,5f^{14}\,6d^1$
28	Ni	$[Ar]\,4s^2\,3d^8$	66	Dy	$[Xe]\,6s^2\,4f^{10}$	104	Rf	$[Rn]\,7s^2\,5f^{14}\,6d^2$
29	Cu	$[Ar]\,4s^1\,3d^{10}$	67	Ho	$[Xe]\,6s^2\,4f^{11}$	105	Db	$[Rn]\,7s^2\,5f^{14}\,6d^3$
30	Zn	$[Ar]\,4s^2\,3d^{10}$	68	Er	$[Xe]\,6s^2\,4f^{12}$	106	Sg	$[Rn]\,7s^2\,5f^{14}\,6d^4$
31	Ga	$[Ar]\,4s^2\,3d^{10}\,4p^1$	69	Tm	$[Xe]\,6s^2\,4f^{13}$	107	Bh	$[Rn]\,7s^2\,5f^{14}\,6d^5$
32	Ge	$[Ar]\,4s^2\,3d^{10}\,4p^2$	70	Yb	$[Xe]\,6s^2\,4f^{14}$	108	Hs	$[Rn]\,7s^2\,5f^{14}\,6d^6$
33	As	$[Ar]\,4s^2\,3d^{10}\,4p^3$	71	Lu	$[Xe]\,6s^2\,4f^{14}\,5d^1$	109	Mt	$[Rn]\,7s^2\,5f^{14}\,6d^7$
34	Se	$[Ar]\,4s^2\,3d^{10}\,4p^4$	72	Hf	$[Xe]\,6s^2\,4f^{14}\,5d^2$	110	Ds	$[Rn]\,7s^2\,5f^{14}\,6d^8$
35	Br	$[Ar]\,4s^2\,3d^{10}\,4p^5$	73	Ta	$[Xe]\,6s^2\,4f^{14}\,5d^3$	111	Uuu	$[Rn]\,7s^2\,5f^{14}\,6d^9$
36	Kr	$[Ar]\,4s^2\,3d^{10}\,4p^6$	74	W	$[Xe]\,6s^2\,4f^{14}\,5d^4$	112	Uub	$[Rn]\,7s^2\,5f^{14}\,6d^{10}$
37	Rb	$[Kr]\,5s^1$	75	Re	$[Xe]\,6s^2\,4f^{14}\,5d^5$	114	Uuq	$[Rn]\,7s^2\,5f^{14}\,6d^{10}\,7p^2$
38	Sr	$[Kr]\,5s^2$	76	Os	$[Xe]\,6s^2\,4f^{14}\,5d^6$			

Answers to Practice Exercises and Selected Review Problems

Chapter 1

Practice Exercises

1. (a) iron and oxygen; (b) sodium, phosphorus, and oxygen; (c) aluminum and oxygen; (d) calcium, carbon, and oxygen. **2.** 12.25 g Cd. **3.** 26.9814 u. **4.** 5.2955 times as heavy as carbon. **5.** 10.8 u. **6.** $^{240}_{94}$Pu. **7.** 17 protons, 17 electrons, and 18 neutrons. **8.** (a) K, Ar, Al; (b) Cl; (c) Ba; (d) Ne; (e) Li; (f) Ce.

Review Problems

1.66 Compound (c). **1.68** 29.3 g nitrogen. **1.70** 5.54 g ammonia. **1.72** 2.286 g oxygen. **1.74** $1.9926482 \times 10^{-23}$ g. **1.76** 1.0079 u. **1.78** 2.0158 u. **1.80** 63.55 u. **1.82** (a) 138 neutrons, 88 protons, 88 electrons; (b) 8 neutrons, 6 protons, 6 electrons; (c) 124 neutrons, 82 protons, 82 electrons; (d) 12 neutrons, 11 protons, 11 electrons.

Chapter 2

Practice Exercises

1. (a) 1 Ni, 2 Cl; (b) 1 Fe, 1 S, 4 O; (c) 3 Ca, 2 P, 8 O; (d) 1 Co, 2 N, 12 O, 12 H. **2.** 1 Mg, 2 O, 4 H, and 2 Cl (on each side). **3.** $Mg(OH)_2(s) + 2 HCl(aq) \rightarrow MgCl_2(aq) + 2 H_2O(l)$. **4.** (a) Decreases; (b) Increases. **5.** $C_{10}H_{22}$. **6.** (a) C_3H_8O; (b) $C_4H_{10}O$. **7.** (a) phosphorus trichloride; (b) sulfur dioxide; (c) dichlorine heptaoxide. **8.** (a) $AsCl_5$; (b) SCl_6; (c) S_2Cl_2. **9.** (a) 8 protons, 8 electrons; (b) 8 protons, 10 electrons; (c) 13 protons, 10 electrons; (d) 13 protons, 13 electrons. **10.** (a) NaF; (b) Na_2O; (c) MgF_2; (d) Al_4C_3. **11.** (a) $CrCl_3$ and $CrCl_2$; Cr_2O_3 and CrO; (b) CuCl, $CuCl_2$; Cu_2O and CuO. **12.** (a) Na_2CO_3; (b) $(NH_4)_2SO_4$; (c) $KC_2H_3O_2$; (d) $Sr(NO_3)_2$; (e) $Fe(C_2H_3O_2)_3$. **13.** (a) potassium sulfide; (b) magnesium phosphide; (c) nickel(II) chloride; (d) iron(III) oxide. **14.** (a) Al_2S_3; (b) SrF_2; (c) TiO_2; (d) Au_2O_3. **15.** (a) Lithium carbonate; (b) iron(III) hydroxide. **16.** (a) $KClO_3$; (b) $Ni_3(PO_4)_2$. **17.** diiodine pentaoxide. **18.** chromium(II) acetate. **19.** X is a nonmetal.

Review Problems

2.71 (a) 2 K, 2 C, 4 O; (b) 2 H, 1 S, 3 O; (c) 12 C, 26 H; (d) 4 H, 2 C, 2 O; (e) 9 H, 2 N, 1 P, 4 O. **2.73** (a) 1 Ni, 2 Cl, 8 O; (b) 1 Cu, 1 C, 3 O; (c) 2 K, 2 Cr, 7 O; (d) 2 C, 4 H, 2 O; (e) 2 N, 9 H, 1 P, 4 O. **2.75** 1 Cr, 6 C, 9 H, 6 O. **2.77** (a) 6 N, 3 O; (b) 4 Na, 4 H, 4 C, 12 O; (c) 2 Cu, 2 S, 18 O, 20 H. **2.79** (a) 6 atoms Na; (b) 3 atoms C; (c) 27 atoms O. **2.81** (a) K^+; (b) Br^-; (c) Mg^{2+}; (d) S^{2-}; (e) Al^{3+}. **2.83** (a) NaBr; (b) KI; (c) BaO; (d) $MgBr_2$; (e) BaF_2. **2.85** (a) KNO_3; (b) $Ca(C_2H_3O_2)_2$; (c) NH_4Cl; (d) $Fe_2(CO_3)_3$; (e) $Mg_3(PO_4)_2$. **2.87** (a) PbO and PbO_2; (b) SnO and SnO_2; (c) MnO and MnO_2; (d) FeO and Fe_2O_3; (e) Cu_2O and CuO. **2.89** (a) silicon dioxide; (b) xenon tetrafluoride; (c) tetraphosphorus decaoxide; (d) dichlorine heptaoxide. **2.91** (a) calcium sulfide; (b) aluminum bromide; (c) sodium phosphide; (d) barium arsenide; (e) rubidium sulfide. **2.93** (a) iron(II) sulfide; (b) copper(II) oxide; (c) tin(IV) oxide; (d) cobalt(II) chloride hexahydrate. **2.95** (a) sodium nitrite; (b) potassium permanganate; (c) magnesium sulfate heptahydrate; (d) potassium thiocyanate. **2.97** (a) ionic, chromium(II) chloride; (b) molecular, disulfur dichloride; (c) ionic, ammonium acetate; (d) molecule, sulfur trioxide; (e) ionic, potassium iodate; (f) molecular, tetraphosphorus decaoxide; (g) ionic, calcium sulfite; (h) ionic, silver cyanide; (i) ionic, zinc bromide; (j) molecular, hydrogen selenide. **2.99** (a) Na_2HPO_4; (b) Li_2Se; (c) $Cr(C_2H_3O_2)_3$; (d) S_2F_{10}; (e) $Ni(CN)_2$; (f) Fe_2O_3; (g) SbF_5. **2.101** (a) $(NH_4)_2S$; (b) $Cr_2(SO_4)_3 \cdot 6H_2O$; (c) SiF_4; (d) MoS_2; (e) $SnCl_4$; (f) H_2Se; (g) P_4S_7.

2.103 Se_2S_6 is diselenium hexasulfide; Se_2S_4 is diselenium tetrasulfide.

Chapter 3

Practice Exercises

1. (a) $m \cdot v^2$ would have units of $kg \cdot (m/s)^2 = kg \cdot m^2/s^2$; (b) mgh would have units of $kg \cdot (m/s^2) \cdot m = kg \cdot m^2/s^2$. **2.** (a) μg; (b) μm; (c) ns. (a) 1×10^{-9}; (b) 1×10^{-2}; (c) 1×10^{-12}; (a) cg; (b) Mm; (c) μs. **3.** 10 °C; 293 K.

4. The data set from Worker C has the best precision. The data set from Worker A has the best accuracy.

5. (a) 42.0 g; (b) 30.0 mL; (c) 54.155 g; (d) 11.3 g; (e) 0.857 g/mL; (f) 3.62 ft (1 and 12 are exact numbers); (g) 8.3 m³.
6. (a) 108 in.; (b) 1.25×10^5 cm; (c) 0.0107 ft; (d) 8.59 km/L.
7. 0.899 g/cm³. **8.** 5.0×10^6 g (or 5.0 Mg). **9.** 2.70; 168 lb/ft³.
10. 0.902 g/mL; 7.52 lb/gal.

Review Problems

3.18 (a) 0.01 m; (b) 1000 m; (c) 1×10^{12} pm; (d) 0.1 m; (e) 0.001 kg; (f) 0.01 g. **3.20** (a) 122 °F; (b) 50 °F; (c) −3.6 °C; (d) 9.4 °C; (e) 333K; (f) 243K. **3.22** 39.7 °C; dog has a fever. **3.24** 830 °C. **3.26** 1.0×10^7 °C to 2.5×10^7 °C; 1.8×10^7 °F to 4.5×10^7 °F. **3.28** −269 °C. **3.30** Liquid. **3.32** (a) 4 significant figures; (b) 5 significant figures; (c) 4 significant figures; (d) 2 significant figures; (e) 4 significant figures; (f) 2 significant figures. **3.34** (a) 5 significant figures; (b) 5 significant figures; (c) 2 significant figures; (d) 5 significant figures; (e) 4 significant figures; (f) 1 significant figure. **3.36** (a) 0.72 m²; (b) 84.24 kg; (c) −0.465 g/cm³; (d) 19.42 g/mL; (e) 857.7 cm². **3.38** (a) 4.34×10^3; (b) 3.20×10^7; (c) 3.29×10^{-3};

(d) 4.20×10^4; (e) 8.00×10^{-6}; (f) 3.24×10^5. **3.40** (a) 310,000;
(b) 0.000,004,35; (c) 3,900; (d) 0.000,000,000,004,4;
(e) 0.000,000,356; (f) 88,000,000. **3.42** (a) 4.0×10^7; (b) 3.0×10^{-2}; (c) 4.4×10^{12}; (d) 2.5; (e) 2.3×10^{18}. **3.44** (a) 3.20×10^{-3} km; (b) 8.2×10^3 μg; (c) 7.53×10^{-5} kg; (d) 0.1375 L;
(e) 25 mL; (f) 3.42×10^{-9} dm. **3.46** (a) 2.3×10^7 cm;
(b) 4.23×10^8 mg; (c) 4.23×10^{-1} mg; (d) 4.3×10^{-1} mL;
(e) 2.7×10^{-11} kg. **3.48** (a) 91 cm; (b) 2.3 kg; (c) 2800 mL;
(d) 200 mL; (e) 88 km/hr; (f) 80.4 km. **3.50** 360 mL.
3.52 5.0 qts. **3.54** 2205 lb. **3.56** 4,000 pistachios. **3.58** 90 m/s.
3.60 1.02 metric tons. **3.62** 190 cm. **3.64** (a) 7,800 cm^2;
(b) 580 km^2; (c) 6.54×10^6 cm^3. **3.66** (a) 0.000,65 m^2;
(b) 9.6 km^2; (c) 2,360 mL. **3.68** 7.8 mi/hr. **3.70** 0.798 g/mL.
3.72 3×10^{15} g/cm^3. **3.74** 31.6 mL. **3.76** 276 g. **3.78** 11 g/mL.
3.80 0.715. **3.82** 1470 g. **3.84** 1.20×10^3 lb.

Chapter 4

Practice Exercises

1. 4.59×10^{22} atoms Au. **2.** 114.26. **3.** 369.41. **4.** 1.6×10^{19}
molecules heroin. **5.** 1.11 mol S. **6.** 28.4 g Ag. **7.** 13.3 g.
8. 0.467 mol H_2SO_4. **9.** 2.91×10^{12} atoms Pb. **10.** 3.44 mol N
atoms. **11.** 59.5 g Fe. **12.** 10.5 g Fe. **13.** 25.94% N, 74.06% O;
no other elements present. **14.** 30.45% N, 69.55% O. **15.** NO.
16. N_2O_5. **17.** Na_2SO_4. **18.** CH_2O. **19.** N_2H_4. **20.** 0.183 mol
H_2SO_4. **21.** 0.958 mol O_2. **22.** 78.5 g Al_2O_3. **23.** $3CaCl_2(aq) + 2K_3PO_4(aq) \rightarrow Ca_3(PO_4)_2(s) + 6KCl(aq)$. **24.** 30.01 g NO.
25. 86.1%.

Review Problems

4.22 1:2. **4.24** 2.59×10^{-3} mole Na. **4.26** (a) 6:11;
(b) 12:11; (c) 2:1; (d) 2:1. **4.28** 1.05 mol Al. **4.30** 4.32 mol Al.

4.32 (a) $\left(\dfrac{2 \text{ mol Al}}{3 \text{ mol S}}\right)$ or $\left(\dfrac{3 \text{ mol S}}{2 \text{ mol Al}}\right)$; (b) $\left(\dfrac{3 \text{ mol S}}{1 \text{ mol Al}_2(SO_4)_3}\right)$
or $\left(\dfrac{1 \text{ mol Al}_2(SO_4)_3}{3 \text{ mol S}}\right)$; (c) 0.600 mol Al; (d) 3.48 mol S.

4.34 0.0725 mol N_2, 0.218 mol H_2. **4.36** 0.833 moles UF_6.
4.38 9.33×10^{23} atoms C. **4.40** 9.39×10^{23} atoms C, 1.88×10^{24} atoms H, 9.39×10^{23} atoms O. **4.42** 3.01×10^{23} atoms.
4.44 (a) 23.0 g Na; (b) 32.1 g S; (c) 35.5 g Cl. **4.46** (a) 75.4 g
Fe; (b) 392 g O; (c) 35.1 g Ca. **4.48** 1.30×10^{-10} g K.
4.50 0.302 mol Ni. **4.52** (a) 84.0 g/mol; (b) 294.2 g/mol;
(c) 96.1 g/mol; (d) 342.2 g/mol; (e) 249.7 g/mol.
4.54 (a) 388 g $Ca_3(PO_4)_2$; (b) 151 g $Fe(NO_3)_3$; (c) 34.9 g C_4H_{10};
(d) 139 g $(NH_4)_2CO_3$. **4.56** (a) 0.215 mol $CaCO_3$;
(b) 9.16×10^{-2} mol NH_3; (c) 7.94×10^{-2} mol $Sr(NO_3)_2$;
(d) 4.31×10^{-8} mol Na_2CrO_4. **4.58** 0.075 mol Ca, 3.01 g Ca.
4.60 1.30 mol N; 62.5 g $(NH_4)_2CO_3$. **4.62** 3.43 kg fertilizer.
4.64 Heroin. **4.66** Freon 141b. **4.68** (a) 19.2% Na, 1.68% H,
25.8% P, 53.3% O; (b) 12.2% N, 5.26% H, 26.9% P, 55.6% O;
(c) 62.0% C, 10.4% H, 27.6% O; (d) 29.4% Ca, 23.6% S,
47.0% O; (e) 23.3% Ca, 18.6% S, 55.7% O, 2.34% H.
4.70 22.9% P, 77.1% Cl. **4.72** Yes. **4.74** 0.474 g O.
4.76 (a) SCl; (b) CH_2O; (c) NH_3; (d) AsO_3; (e) OH.
4.78 $NaTcO_4$. **4.80** CCl_2. **4.82** $Na_2B_4O_7$. **4.84** C_2H_6O.
4.86 $C_{19}H_{30}O_2$. **4.88** (a) $Na_2S_4O_6$; (b) $C_6H_4Cl_2$; (c) $C_6H_3Cl_3$.
4.90 $C_{19}H_{30}O_2$. **4.92** Empirical formula is HgBr; molecular
formula is Hg_2Br_2. **4.94** Empirical formula is CHNO; molecular formula is $C_3H_3N_3O_3$. **4.96** 36 moles of hydrogen.

4.98 $4Fe(s) + 3O_2(g) \rightarrow 2Fe_2O_3(s)$

4.100 (a) $Ca(OH)_2 + 2HCl \rightarrow CaCl_2 + 2H_2O$
(b) $2AgNO_3 + CaCl_2 \rightarrow Ca(NO_3)_2 + 2AgCl$
(c) $2Fe_2O_3 + 3C \rightarrow 4Fe + 3CO_2$
(d) $2NaHCO_3 + H_2SO_4 \rightarrow Na_2SO_4 + 2H_2O + 2CO_2$
(e) $2C_4H_{10} + 13O_2 \rightarrow 8CO_2 + 10H_2O$

4.102 (a) $Mg(OH)_2 + 2HBr \rightarrow MgBr_2 + 2H_2O$
(b) $2HCl + Ca(OH)_2 \rightarrow CaCl_2 + 2H_2O$
(c) $Al_2O_3 + 3H_2SO_4 \rightarrow Al_2(SO_4)_3 + 3H_2O$
(d) $2KHCO_3 + H_3PO_4 \rightarrow K_2HPO_4 + 2H_2O + 2CO_2$
(e) $C_9H_{20} + 14O_2 \rightarrow 9CO_2 + 10H_2O$

4.104 (a) 0.030 mol $Na_2S_2O_3$; (b) 0.24 mol HCl; (c) 0.15 mol
H_2O; (d) 0.15 mol H_2O. **4.106** (a) 3.6 g Zn; (b) 22 g Au;
(c) 55 g $Au(CN)_2{}^-$. **4.108** (a) $4P + 5O_2 \rightarrow P_4O_{10}$; (b) 8.85 g
O_2; (c) 14.2 g P_4O_{10}; (d) 3.26 g P. **4.110** 30.31 g HNO_3.
4.112 0.47 kg O_2. **4.114** (a) Fe_2O_3; (b) 195 g Fe. **4.116** 26.7 g
$FeCl_3$. **4.118** 0.00091 g HNO_3. **4.120** Theoretical yield =
66.98 g $BaSO_4$; % yield = 96.22%. **4.122** 88.74%. *4.124 9.2
g C_7H_8.

Chapter 5

Practice Exercises

1. (a) $MgCl_2(s) \rightarrow Mg^{2+}(aq) + 2Cl^-(aq)$; (b) $Al(NO_3)_3(s) \rightarrow Al^{3+}(aq) + 3NO_3{}^-(aq)$; (c) $Na_2CO_3(s) \rightarrow 2Na^+(aq) + CO_3{}^{2-}(aq)$.

2. molecular: $CdCl_2(aq) + Na_2S(aq) \rightarrow CdS(s) + 2NaCl(aq)$
ionic: $Cd^{2+}(aq) + 2Cl^-(aq) + 2Na^+(aq) + S^{2-}(aq) \rightarrow$
$\qquad\qquad\qquad\qquad CdS(s) + 2Na^+(aq) + 2Cl^-(aq)$
net ionic: $Cd^{2+}(aq) + S^{2-}(aq) \rightarrow CdS(s)$

3. (a) molecular: $AgNO_3(aq) + NH_4Cl(aq) \rightarrow$
$\qquad\qquad\qquad\qquad AgCl(s) + NH_4NO_3(aq)$
ionic: $Ag^+(aq) + NO_3{}^-(aq) + NH_4{}^+(aq) + Cl^-(aq) \rightarrow$
$\qquad\qquad\qquad AgCl(s) + NH_4{}^+(aq) + NO_3{}^-(aq)$
net ionic: $Ag^+(aq) + Cl^-(aq) \rightarrow AgCl(s)$
(b) molecular: $Na_2S(aq) + Pb(C_2H_3O_2)_2(aq) \rightarrow$
$\qquad\qquad\qquad\qquad 2NaC_2H_3O_2(aq) + PbS(s)$
ionic: $2Na^+(aq) + S^{2-}(aq) + Pb^{2+}(aq) + 2C_2H_3O_2{}^-(aq) \rightarrow$
$\qquad\qquad 2Na^+(aq) + 2C_2H_3O_2{}^-(aq) + PbS(s)$
net ionic: $S^{2-}(aq) + Pb^{2+}(aq) \rightarrow PbS(s)$

4. $HCHO_2(aq) + H_2O \rightarrow H_3O^+(aq) + CHO_2{}^-(aq)$

5. $H_3C_6H_5O_7(s) + H_2O \rightarrow H_3O^+(aq) + H_2C_6H_5O_7{}^-(aq)$
$H_2C_6H_5O_7{}^-(aq) + H_2O \rightarrow H_3O^+(aq) + HC_6H_5O_7{}^{2-}(aq)$
$HC_6H_5O_7{}^{2-}(aq) + H_2O \rightarrow H_3O^+(aq) + C_6H_5O_7{}^{3-}(aq)$

6. $HONH_2(aq) + H_2O \rightarrow HONH_3{}^+(aq) + OH^-(aq)$

7. HF: Hydrofluoric acid, sodium salt = sodium fluoride
(NaF); HBr: Hydrobromic acid, sodium salt = sodium bromide (NaBr).

8. Sodium arsenate. **9.** Calcium formate.

10. $H_3PO_4(aq) + NaOH(aq) \rightarrow NaH_2PO_4(aq) + H_2O$
$\qquad NaH_2PO_4(aq) + NaOH(aq) \rightarrow Na_2HPO_4(aq) + H_2O$
$\qquad Na_2HPO_4(aq) + NaOH(aq) \rightarrow Na_3PO_4(aq) + H_2O$

11. $NaHSO_3$ (sodium hydrogen sulfite)

12. $HNO_2(aq) + H_2O \rightleftharpoons H_3O^+(aq) + NO_2{}^-(aq)$

13. $CH_3NH_2(aq) + H_2O \rightleftharpoons CH_3NH_3{}^+(aq) + OH^-(aq)$

14. molecular: $2HCl(aq) + Ca(OH)_2(aq) \rightarrow$
$\qquad\qquad\qquad\qquad CaCl_2(aq) + 2H_2O$

ionic: $2H^+(aq) + 2Cl^-(aq) + Ca^{2+}(aq) + 2OH^-(aq) \rightarrow$
$$Ca^{2+}(aq) + 2Cl^-(aq) + 2H_2O$$
net ionic: $2H^+(aq) + 2OH^-(aq) \rightarrow 2H_2O$

15. (a) molecular: $HCl(aq) + KOH(aq) \rightarrow H_2O + KCl(aq)$
ionic: $H^+(aq) + Cl^-(aq) + K^+(aq) + OH^-(aq) \rightarrow$
$$H_2O + K^+(aq) + Cl^-(aq)$$
net ionic: $H^+(aq) + OH^-(aq) \rightarrow H_2O$
(b) molecular: $HCHO_2(aq) + LiOH(aq) \rightarrow$
$$H_2O + LiCHO_2(aq)$$
ionic: $HCHO_2(aq) + Li^+(aq) + OH^-(aq) \rightarrow$
$$H_2O + Li^+(aq) + CHO_2^-(aq)$$
net ionic: $HCHO_2(aq) + OH^-(aq) \rightarrow H_2O + CHO_2^-(aq)$
(c) molecular: $N_2H_4(aq) + HCl(aq) \rightarrow N_2H_5Cl(aq)$
ionic: $N_2H_4(aq) + H^+(aq) + Cl^-(aq) \rightarrow N_2H_5^+(aq) + Cl^-(aq)$
net ionic: $N_2H_4(aq) + H^+(aq) \rightarrow N_2H_5^+(aq)$

16. molecular: $CH_3NH_2(aq) + HCHO_2(aq) \rightarrow$
$$CH_3NH_3CHO_2(aq)$$
ionic: $CH_3NH_2(aq) + HCHO_2(aq) \rightarrow$
$$CH_3NH_3^+(aq) + CHO_2^-(aq)$$
net ionic: $CH_3NH_2(aq) + HCHO_2(aq) \rightarrow$
$$CH_3NH_3^+(aq) + CHO_2^-(aq)$$

17. molecular: $Al(OH)_3(s) + 3HCl(aq) \rightarrow$
$$AlCl_3(aq) + 3H_2O$$
ionic: $Al(OH)_3(s) + 3H^+(aq) + 3Cl^-(aq) \rightarrow$
$$Al^{3+}(aq) + 3Cl^-(aq) + 3H_2O$$
net ionic: $Al(OH)_3(s) + 3H^+(aq) \rightarrow Al^{3+}(aq) + 3H_2O$

18. (a) Formic acid, a weak acid will form. Net ionic equation: $H^+(aq) + CHO_2^-(aq) \rightleftharpoons HCHO_2(aq)$; (b) Carbonic acid will form and it will further dissociate to water and carbon dioxide: $CuCO_3(s) + 2H^+(aq) \rightarrow 2CO_2(g) + 2H_2O + Cu^{2+}(aq)$; (c) NR; (d) Insoluble nickel hydroxide will precipitate: $Ni^{2+}(aq) + 2OH^-(aq) \rightarrow Ni(OH)_2(s)$. **19.** 0.2498 M.
20. 333 mL. **21.** 0.53 g $AgNO_3$. **22.** Mix 25.0 mL of 0.500 M H_2SO_4 with enough water to make 100 mL of total solution.
23. 26.8 mL NaOH. **24.** 0.40 M Fe^{3+}, 1.2 M Cl^-. **25.** 0.750 M Na^+. **26.** 60.0 mL KOH. **27.** 1.20×10^{-2} mol $BaSO_4$; 0.00 M Ba^{2+}, 0.480 M Cl^-, 0.300 M Mg^{2+}, 6.0×10^{-2} mol SO_4^{2-}.
28. (a) 5.41×10^{-3} mol Ca^{2+}; (b) 5.41×10^{-3} moles Ca^{2+}; (c) 5.41×10^{-3} moles $CaCl_2$; (d) 0.600 g $CaCl_2$; (e) 30.0% $CaCl_2$. **29.** 0.178 M H_2SO_4. **30.** 0.0220 M HCl; 0.0803%.

Review Problems

5.60 (a) $LiCl(s) \rightarrow Li^+(aq) + Cl^-(aq)$
(b) $BaCl_2(s) \rightarrow Ba^{2+}(aq) + 2Cl^-(aq)$
(c) $Al(C_2H_3O_2)_3(s) \rightarrow Al^{3+}(aq) + 3C_2H_3O_2^-(aq)$
(d) $(NH_4)_2CO_3(s) \rightarrow 2NH_4^+(aq) + CO_3^{2-}(aq)$
(e) $FeCl_3(s) \rightarrow Fe^{3+}(aq) + 3Cl^-(aq)$

5.62 (a) ionic: $2NH_4^+(aq) + CO_3^{2-}(aq) + Mg^{2+}(aq) +$
$$2Cl^-(aq) \rightarrow 2NH_4^+(aq) + 2Cl^-(aq) + MgCO_3(s)$$
net: $Mg^{2+}(aq) + CO_3^{2-}(aq) \rightarrow MgCO_3(s)$
(b) ionic: $Cu^{2+}(aq) + 2Cl^-(aq) + 2Na^+(aq) + 2OH^-(aq) \rightarrow$
$$Cu(OH)_2(s) + 2Na^+(aq) + 2Cl^-(aq)$$
net: $Cu^{2+}(aq) + 2OH^-(aq) \rightarrow Cu(OH)_2(s)$
(c) ionic: $3Fe^{2+}(aq) + 3SO_4^{2-}(aq) + 6Na^+(aq) + 2PO_4^{3-}(aq) \rightarrow$
$$Fe_3(PO_4)_2(s) + 6Na^+(aq) + 3SO_4^{2-}(aq)$$
net: $3Fe^{2+}(aq) + 2PO_4^{3-}(aq) \rightarrow Fe_3(PO_4)_2(s)$
(d) ionic: $2Ag^+(aq) + 2C_2H_3O_2^-(aq) + Ni^{2+}(aq) + 2Cl^-(aq) \rightarrow$
$$2AgCl(s) + Ni^{2+}(aq) + 2C_2H_3O_2^-(aq)$$
net: $2Ag^+(aq) + 2Cl^-(aq) \rightarrow 2AgCl(s)$

5.64 molecular: $Na_2S(aq) + Cu(NO_3)_2(aq) \rightarrow$
$$2NaNO_3(aq) + CuS(s)$$
ionic: $2Na^+(aq) + S^{2-}(aq) + Cu^{2+}(aq) + 2NO_3^-(aq) \rightarrow$
$$2Na^+(aq) + 2NO_3^-(aq) + CuS(s)$$
net: $S^{2-}(aq) + Cu^{2+}(aq) \rightarrow CuS(s)$

5.66 molecular: $AgNO_3(aq) + NaBr(aq) \rightarrow$
$$AgBr(s) + NaNO_3(aq)$$
ionic: $Ag^+(aq) + NO_3^-(aq) + Na^+(aq) + Br^-(aq) \rightarrow$
$$AgBr(s) + Na^+(aq) + NO_3^-(aq)$$
net: $Ag^+(aq) + Br^-(aq) \rightarrow AgBr(s)$

5.68 $HClO_4(aq) + H_2O \rightleftharpoons H_3O^+(aq) + ClO_4^-(aq)$
5.70 $N_2H_4(aq) + H_2O \rightleftharpoons N_2H_5^+(aq) + OH^-(aq)$
5.72 $HNO_2(aq) + H_2O \rightleftharpoons H_3O^+(aq) + NO_2^-(aq)$
5.74 $H_2CO_3(aq) + H_2O \rightleftharpoons H_3O^+(aq) + HCO_3^-(aq)$
$HCO_3^-(aq) + H_2O \rightleftharpoons H_3O^+(aq) + CO_3^{2-}(aq)$

5.76 (a) molecular: $Ca(OH)_2(aq) + 2HNO_3(aq) \rightarrow$
$$Ca(NO_3)_2(aq) + 2H_2O$$
ionic: $Ca^{2+}(aq) + 2OH^-(aq) + 2H^+(aq) + 2NO_3^-(aq) \rightarrow$
$$Ca^{2+}(aq) + 2NO_3^-(aq) + 2H_2O$$
net: $H^+(aq) + OH^-(aq) \rightarrow H_2O$
(b) molecular: $Al_2O_3(s) + 6HCl(aq) \rightarrow 2AlCl_3(aq) + 3H_2O$
ionic: $Al_2O_3(s) + 6H^+(aq) + 6Cl^-(aq) \rightarrow$
$$2Al^{3+}(aq) + 6Cl^-(aq) + 3H_2O$$
net: $Al_2O_3(s) + 6H^+(aq) \rightarrow 2Al^{3+}(aq) + 3H_2O$
(c) molecular: $Zn(OH)_2(s) + H_2SO_4(aq) \rightarrow ZnSO_4(aq) + 2H_2O$
ionic: $Zn(OH)_2(s) + 2H^+(aq) + SO_4^{2-}(aq) \rightarrow$
$$Zn^{2+}(aq) + SO_4^{2-}(aq) + 2H_2O$$
net: $Zn(OH)_2(s) + 2H^+(aq) \rightarrow Zn^{2+}(aq) + 2H_2O$

5.78 (a) $2H^+(aq) + CO_3^{2-}(aq) \rightarrow H_2O + CO_2(g)$
(b) $NH_4^+(aq) + OH^-(aq) \rightarrow NH_3(aq) + H_2O$

5.80 (a) formation of insoluble $Cr(OH)_3$; (b) formation of water, a weak electrolyte. **5.82** (a), (b), and (d) are soluble. **5.84** (a), (d), and (f) are insoluble.

5.86 (a) $3HNO_3(aq) + Cr(OH)_3(s) \rightarrow Cr(NO_3)_3(aq) + 3H_2O$
ionic: $3H^+(aq) + 3NO_3^-(aq) + Cr(OH)_3(s) \rightarrow$
$$Cr^{3+}(aq) + 3NO_3^-(aq) + 3H_2O$$
net: $3H^+(aq) + Cr(OH)_3(s) \rightarrow Cr^{3+}(aq) + 3H_2O$
(b) $HClO_4(aq) + NaOH(aq) \rightarrow NaClO_4(aq) + H_2O$
ionic: $H^+(aq) + ClO_4^-(aq) + Na^+(aq) + OH^-(aq) \rightarrow$
$$Na^+(aq) + ClO_4^-(aq) + H_2O$$
net: $H^+(aq) + OH^-(aq) \rightarrow H_2O$
(c) $Cu(OH)_2(s) + 2HC_2H_3O_2(aq) \rightarrow$
$$Cu(C_2H_3O_2)_2(aq) + 2H_2O$$
ionic: $Cu(OH)_2(s) + 2HC_2H_3O_2(aq) \rightarrow$
$$Cu^{2+}(aq) + 2C_2H_3O_2^-(aq) + 2H_2O$$
net: $Cu(OH)_2(s) + 2HC_2H_3O_2(aq) \rightarrow$
$$Cu^{2+}(aq) + 2C_2H_3O_2^-(aq) + 2H_2O$$
(d) $ZnO(s) + H_2SO_4(aq) \rightarrow ZnSO_4(aq) + H_2O$
ionic: $ZnO(s) + 2H^+(aq) + SO_4^{2-}(aq) \rightarrow$
$$Zn^{2+}(aq) + SO_4^{2-}(aq) + H_2O$$
net: $ZnO(s) + 2H^+(aq) \rightarrow Zn^{2+}(aq) + H_2O$

5.88 (a) $Na_2SO_3(aq) + Ba(NO_3)_2(aq) \rightarrow$
$$BaSO_3(s) + 2NaNO_3(aq)$$
ionic: $2Na^+(aq) + SO_3^{2-}(aq) + Ba^{2+}(aq) + 2NO_3^-(aq) \rightarrow$
$$BaSO_3(s) + 2Na^+(aq) + 2NO_3^-(aq)$$
net: $Ba^{2+}(aq) + SO_3^{2-}(aq) \rightarrow BaSO_3(s)$
(b) $2HCHO_2(aq) + K_2CO_3(aq) \rightarrow$
$$2KCHO_2(aq) + H_2O + CO_2(g)$$

ionic: $2HCHO_2(aq) + 2K^+(aq) + CO_3{}^{2-}(aq) \rightarrow$
$$2K^+(aq) + 2CHO_2{}^-(aq) + H_2O + CO_2(g)$$
net: $2HCHO_2(aq) + CO_3{}^{2-}(aq) \rightarrow$
$$2CHO_2{}^-(aq) + H_2O + CO_2(g)$$
(c) $2NH_4Br(aq) + Pb(C_2H_3O_2)_2(aq) \rightarrow$
$$2NH_4C_2H_3O_2(aq) + PbBr_2(s)$$
ionic: $2NH_4{}^+(aq) + 2Br^-(aq) + Pb^{2+}(aq) + 2C_2H_3O_2{}^-(aq) \rightarrow$
$$2NH_4{}^+(aq) + 2C_2H_3O_2{}^-(aq) + PbBr_2(s)$$
net: $Pb^{2+}(aq) + 2Br^-(aq) \rightarrow PbBr_2(s)$
(d) $2NH_4ClO_4(aq) + Cu(NO_3)_2(aq) \rightarrow$
$$Cu(ClO_4)_2(aq) + 2NH_4NO_3(aq)$$
ionic: $2NH_4{}^+(aq) + 2ClO_4{}^-(aq) + Cu^{2+}(aq) + 2NO_3{}^-(aq) \rightarrow$
$$Cu^{2+}(aq) + 2ClO_4{}^-(aq) + 2NO_3{}^-(aq) + 2NH_4{}^+(aq)$$
net: N.R.

*5.90 Numerous possible answers. One of many possible sets of answers would be:
(a) $NaHCO_3(aq) + HCl(aq) \rightarrow NaCl(aq) + CO_2(g) + H_2O$
(b) $FeCl_2(aq) + 2NaOH(aq) \rightarrow Fe(OH)_2(s) + 2NaCl(aq)$
(c) $Ba(NO_3)_2(aq) + K_2SO_3(aq) \rightarrow BaSO_3(s) + 2KNO_3(aq)$
(d) $2AgNO_3(aq) + Na_2S(aq) \rightarrow Ag_2S(s) + 2NaNO_3(aq)$
(e) $ZnO(s) + 2HCl(aq) \rightarrow ZnCl_2(aq) + H_2O$

5.92 $\dfrac{0.25 \text{ mol HCl}}{1 \text{ L HCl soln}}$ and $\dfrac{1 \text{ L HCl soln}}{0.25 \text{ mol HCl}}$

5.94 (a) 1.00 M; (b) 0.577 M; (c) 3.33 M; (d) 0.150 M.
5.96 (a) 1.46 g NaCl; (b) 16.2 g $C_6H_{12}O_6$; (c) 6.13 g H_2SO_4.
5.98 0.11 M H_2SO_4. **5.100** 3.00×10^2 mL. **5.102** 230 mL.
5.104 (a) 0.0438 mol OH^-, 0.0438 mol K^+; (b) 0.015 mol Ca^{2+}, 0.030 mol Cl^-; (c) 0.020 mol Al^{3+}, 0.060 mol Cl^-.
5.106 (a) 0.25 M Cr^{2+}, 0.50 M $NO_3{}^-$; (b) 0.10 M Cu^{2+}, 0.10 M $SO_4{}^{2-}$; (c) 0.48 M Na^+, 0.16 M $PO_4{}^{3-}$; (d) 0.15 M Al^{3+}, 0.23 M $SO_4{}^{2-}$. **5.108** 0.070 M Na_3PO_4. **5.110** 1.0 g $Al_2(SO_4)_3$.
5.112 0.1130 M KOH. **5.114** 12 mL $NiCl_2$; 0.36 g $NiCO_3$.
5.116 188 mL NaOH. **5.118** 0.0565 M $Ba(OH)_2$.
5.120 2.00 mL $FeCl_3$; 0.129 g AgCl. **5.122** 131 mL $Ba(OH)_2$.
5.124 13.3 mL $AlCl_3$. ***5.126** 0.167 M Fe^{3+}; 3.67 g Fe_2O_3.
*5.128 (a) 8.00×10^{-3} mol AgCl; (b) 0.060 M Cl^-; 0.160 M $NO_3{}^-$; 0.220 M Na^+. **5.130** 48.40% Pb. **5.132** 0.114 M HCl.
5.134 2.67×10^{-3} mol $HC_3H_5O_3$. **5.136** $MgSO_4 \cdot 7H_2O$.
5.138 59% NaCl.

Chapter 6

Practice Exercises

1. Aluminum is oxidized and is the reducing agent; chlorine is reduced and is the oxidizing agent. **2.** Water, since the oxidation number of oxygen drops from -1 to -2. **3.** (a) Ni $+2$; Cl -1; (b) Mg $+2$; Ti $+4$; O -2; (c) K $+1$; Cr $+6$; O -2; (d) H $+1$; P $+5$, O -2; (e) V $+3$; C 0; H $+1$; O -2. **4.** $+8/3$.
5. In this reaction, all the chlorine changes oxidation number. The simplest analysis suggests that $NaClO_2$ is oxidized and is the reducing agent, and Cl_2 is reduced and is the oxidizing agent. **6.** $KClO_3$ is reduced and HNO_2 is oxidized. This means $KClO_3$ is the oxidizing agent and HNO_2 is the reducing agent.

7. $2Al(s) + 3Cu^{2+}(aq) \rightarrow 2Al^{3+}(aq) + 3Cu(s)$
8. $3Sn^{2+} + 16H^+ + 2TcO_4{}^- \rightarrow 2Tc^{4+} + 8H_2O + 3Sn^{4+}$
9. $4Cu + 2NO_3{}^- + 10H^+ \rightarrow 4Cu^{2+} + N_2O + 5H_2O$
10. $2MnO_4{}^- + 3C_2O_4{}^{2-} + 4OH^- \rightarrow$
$$2MnO_2 + 6CO_3{}^{2-} + 2H_2O$$

11. (a) molecular: $Mg(s) + 2HCl(aq) \rightarrow MgCl_2(aq) + H_2(g)$
ionic: $Mg(s) + 2H^+(aq) + 2Cl^-(aq) \rightarrow$
$$Mg^{2+}(aq) + 2Cl^-(aq) + H_2(g)$$
net ionic: $Mg(s) + 2H^+(aq) \rightarrow Mg^{2+}(aq) + H_2(g)$
(b) molecular: $2Al(s) + 6HCl(aq) \rightarrow 2AlCl_3(aq) + 3H_2(g)$
ionic: $2Al(s) + 6H^+(aq) + 6Cl^-(aq) \rightarrow$
$$2Al^{3+}(aq) + 6Cl^-(aq) + 3H_2(g)$$
net ionic: $2Al(s) + 6H^+(aq) \rightarrow 2Al^{3+}(aq) + 3H_2(g)$
12. (a) $2Al(s) + 3Cu^{2+}(aq) \rightarrow 2Al^{3+}(aq) + 3Cu(s)$; (b) N.R.
13. $2C_4H_{10}(l) + 13O_2(g) \rightarrow 8CO_2(g) + 10H_2O(g)$
14. $C_2H_5OH(l) + 3O_2(g) \rightarrow 2CO_2(g) + 3H_2O(g)$
15. $4Fe(s) + 3O_2(g) \rightarrow 2Fe_2O_3(s)$. **16.** 2.37 g $Na_2S_2O_3$.
17. (a) $5Sn^{2+} + 2MnO_4{}^- + 16H^+ \rightarrow 5Sn^{4+} + 2Mn^{2+} + 8H_2O$; (b) 0.120 g Sn; (c) 40.0% Sn; (d) 50.7% SnO_2.

Review Problems

6.25 (a) substance reduced (and oxidizing agent): HNO_3; substance oxidized (and reducing agent): H_3AsO_3; (b) substance reduced (and oxidizing agent): HOCl; substance oxidized (and reducing agent): NaI; (c) substance reduced (and oxidizing agent): $KMnO_4$; substance oxidized (and reducing agent): $H_2C_2O_4$; (d) substance reduced (and oxidizing agent): H_2SO_4; substance oxidized (and reducing agent): Al.
6.27 (a) -2; (b) $+4$; (c) 0; (d) -3.
6.29 (a) Na: $+1$ (c) Na: $+1$
      H: $+1$ S: $+2.5$
      P: $+5$ O: -2
      O: -2
   (b) Ba: $+2$ (d) Cl: $+3$
      Mn: $+6$ F: -1
      O: -2
6.31 (a) $+2$; (b) $+5$; (c) -1; (d) $+4$; (e) -2.
6.33 (a) O: -2 (c) O: -2
      Na: $+1$ Na: $+1$
      Cl: $+1$ Cl: $+5$
   (b) O: -2 (d) O: -2
      Na: $+1$ Na: $+1$
      Cl: $+3$ Cl: $+7$
6.35 (a) S: -2 (c) O: -2
      Pb: $+2$ Sr: $+2$
               I: $+5$
   (b) Cl: -1 (d) S: -2
      Ti: $+4$ Cr: $+3$

6.37 *In the forward direction:* Cl_2 is reduced, and it is also oxidized! *In the reverse direction:* Cl^- is the reducing agent, HOCl is the oxidizing agent. **6.39** (a) $BiO_3{}^- + 6H^+ + 2e^- \rightarrow Bi^{3+} + 3H_2O$; this is a reduction of $BiO_3{}^-$; (b) $Pb^{2+} + 2H_2O \rightarrow PbO_2 + 4H^+ + 2e^-$; this is an oxidation of Pb^{2+}.
6.41 (a) $Fe + 2OH^- \rightarrow Fe(OH)_2 + 2e^-$; this is an oxidation of Fe; (b) $2e^- + 2OH^- + SO_2Cl_2 \rightarrow SO_3{}^{2-} + 2Cl^- + H_2O$; this is a reduction of SO_2Cl_2.

6.43 (a) $OCl^- + 2S_2O_3{}^{2-} + 2H^+ \rightarrow S_4O_6{}^{2-} + Cl^- + H_2O$
(b) $2NO_3{}^- + Cu + 4H^+ \rightarrow 2NO_2 + Cu^{2+} + 2H_2O$
(c) $3AsO_3{}^{3-} + IO_3{}^- \rightarrow I^- + 3AsO_4{}^{3-}$
(d) $Zn + SO_4{}^{2-} + 4H^+ \rightarrow Zn^{2+} + SO_2 + 2H_2O$
(e) $NO_3{}^- + 4Zn + 10H^+ \rightarrow 4Zn^{2+} + NH_4{}^+ + 3H_2O$
(f) $2Cr^{3+} + 3BiO_3{}^- + 4H^+ \rightarrow Cr_2O_7{}^{2-} + 3Bi^{3+} + 2H_2O$
(g) $I_2 + 5OCl^- + H_2O \rightarrow 2IO_3{}^- + 5Cl^- + 2H^+$

(h) $2Mn^{2+} + 5BiO_3^- + 14H^+ \rightarrow 2MnO_4^- + 5Bi^{3+} + 7H_2O$

(i) $3H_3AsO_3 + Cr_2O_7^{2-} + 8H^+ \rightarrow$
$$3H_3AsO_4 + 2Cr^{3+} + 4H_2O$$

(j) $2I^- + HSO_4^- + 3H^+ \rightarrow I_2 + SO_2 + 2H_2O$

6.45 (a) $2CrO_4^{2-} + 3S^{2-} + 4H_2O \rightarrow 2CrO_2^- + 3S + 8OH^-$

(b) $3C_2O_4^{2-} + 2MnO_4^- + 4H_2O \rightarrow$
$$6CO_2 + 2MnO_2 + 8OH^-$$

(c) $4ClO_3^- + 3N_2H_4 \rightarrow 4Cl^- + 6NO + 6H_2O$

(d) $NiO_2 + 2Mn(OH)_2 \rightarrow Ni(OH)_2 + Mn_2O_3 + H_2O$

(e) $3SO_3^{2-} + 2MnO_4^- + H_2O \rightarrow 3SO_4^{2-} + 2MnO_2 + 2OH^-$

6.47 $OCl^- + S_2O_3^{2-} \rightarrow 2SO_4^{2-} + Cl^-$

6.49 $O_3 + Br^- \rightarrow BrO_3^-$

6.51 (a) Molecular: $Mn(s) + 2HCl(aq) \rightarrow$
$$MnCl_2(aq) + H_2(g)$$
Ionic: $Mn(s) + 2H^+(aq) + 2Cl^-(aq) \rightarrow$
$$Mn^{2+}(aq) + 2Cl^-(aq) + H_2(g)$$
Net: $Mn(s) + 2H^+(aq) \rightarrow Mn^{2+}(aq) + H_2(g)$

(b) Molecular: $Cd(s) + 2HCl(aq) \rightarrow CdCl_2(aq) + H_2(g)$
Ionic: $Cd(s) + 2H^+(aq) + 2Cl^-(aq) \rightarrow$
$$Cd^{2+}(aq) + Cl^-(aq) + H_2(g)$$
Net: $Cd(s) + 2H^+(aq) \rightarrow Cd^{2+}(aq) + H_2(g)$

(c) Molecular: $Sn(s) + 2HCl(aq) \rightarrow SnCl_2(aq) + H_2(g)$
Ionic: $Sn(s) + 2H^+(aq) + 2Cl^-(aq) \rightarrow$
$$Sn^{2+}(aq) + 2Cl^-(aq) + H_2(g)$$
Net: $Sn(s) + 2H^+(aq) \rightarrow Sn^{2+}(aq) + H_2(g)$

(d) Molecular: $Ni(s) + 2HCl(aq) \rightarrow NiCl_2(aq) + H_2(g)$
Ionic: $Ni(s) + 2H^+(aq) + 2Cl^-(aq) \rightarrow$
$$Ni^{2+}(aq) + 2Cl^-(aq) + H_2(g)$$
Net: $Ni(s) + 2H^+(aq) \rightarrow Ni^{2+}(aq) + H_2(g)$

(e) Molecular: $2Cr(s) + 6HCl(aq) \rightarrow 2CrCl_3(aq) + 3H_2(g)$
Ionic: $2Cr(s) + 6H^+(aq) + 6Cl^-(aq) \rightarrow$
$$2Cr^{3+}(aq) + 6Cl^-(aq) + 3H_2(g)$$
Net: $2Cr(s) + 6H^+(aq) \rightarrow 2Cr^{3+}(aq) + 3H_2(g)$

6.53 (a) $3Ag(s) + 4HNO_3(aq) \rightarrow$
$$3AgNO_3(aq) + 2H_2O + NO(g)$$
(b) $Ag(s) + 2HNO_3(aq) \rightarrow AgNO_3(aq) + H_2O + NO_2(g)$

6.55 $3Sn^{2+} + 6H^+ + BrO_3^- \rightarrow Br^- + 3H_2O + 3Sn^{4+}$

6.57 (a) $Fe + Mg^{2+} \rightarrow NR$

(b) $2Cr + 3Pb^{2+} \rightarrow 3Pb + 2Cr^{3+}$

(c) $Fe + 2Ag^+ \rightarrow 2Ag + Fe^{2+}$

(d) $3Ag + Au^{3+} \rightarrow Au + 3Ag^+$

6.59 $Pu > Tl > Ru > Pt$

6.61 $Cd(s) + PtCl_2(aq) \rightarrow CdCl_2(aq) + Pt(s)$

6.63 (a) $2C_6H_6(l) + 15O_2(g) \rightarrow 12CO_2(g) + 6H_2O(g)$

(b) $C_3H_8(g) + 5O_2(g) \rightarrow 3CO_2(g) + 4H_2O(g)$

(c) $C_{21}H_{44}(s) + 32O_2(g) \rightarrow 21CO_2(g) + 22H_2O(g)$

6.65 (a) $2C_6H_6(l) + 9O_2(g) \rightarrow 12CO(g) + 6H_2O(g)$
$2C_3H_8(g) + 7O_2(g) \rightarrow 6CO(g) + 8H_2O(g)$
$2C_{21}H_{44}(s) + 43O_2(g) \rightarrow 42CO(g) + 44H_2O(g)$

(b) $2C_6H_6(l) + 3O_2(g) \rightarrow 12C(s) + 6H_2O(g)$
$C_3H_8(g) + 2O_2(g) \rightarrow 3C(s) + 4H_2O(g)$
$C_{21}H_{44}(s) + 11O_2(g) \rightarrow 21C(s) + 22H_2O(g)$

6.67 $2CH_3OH(l) + 3O_2(g) \rightarrow 2CO_2(g) + 4H_2O(g)$

6.69 (a) $IO_3^- + 3SO_3^{2-} \rightarrow I^- + 3SO_4^{2-}$; (b) 9.55 g Na_2SO_3.
6.71 3.53 g Cu. **6.73** (a) $2MnO_4^- + 5Sn^{2+} + 16H^+ \rightarrow 2Mn^{2+}$
$+ 5Sn^{4+} + 8H_2O$; (b) 17.4 mL. **6.75** (a) 0.0259 M I_3^-;
(b) 0.2150 g $(NH_4)_2S_2O_3$; (c) 98.6%. **6.77** (a) 9.463% Cu;

(b) 18.40% $CuCO_3$. **6.79** (a) 0.02994 g H_2O_2; (b) 2.994%
H_2O_2. **6.81** (a) $2CrO_4^{2-} + 3SO_3^{2-} + H_2O \rightarrow 2CrO_2^- +$
$3SO_4^{2-} + 2OH^-$; (b) 0.875 g Cr in the original alloy;
(c) 25.4% Cr. **6.83** (a) 5.405×10^{-3} mol $C_2O_4^{2-}$; (b) 0.5999 g
$CaCl_2$; (c) 24.35% $CaCl_2$.

Chapter 7

Practice Exercises

1. 13 Cal. **2.** 5.00×10^3 cal; 20,900 J. **3.** 27.5 J/°C. **4.** 5200 J;
5.2 kJ; 1200 cal; 1.2 kcal. **5.** Supplied. **6.** −394 kJ/mol C.
7. 3.7 kJ; 74 kJ/mole. **8.** $2.5H_2(g) + 1.25O_2(g) \rightarrow 2.5H_2O(g)$,
$\Delta H = -647.3$ kJ.

9.

10. $\Delta H° = -44.0$ kJ. **11.** 2.62×10^6 kJ. **12.** $Na(s) + \frac{1}{2}H_2(g) +$
$C(s) + \frac{3}{2}O_2(g) \rightarrow NaHCO_3(s), \Delta H_f° = -947.7$ kJ/mol.
13. (a) −113.1 kJ; (b) −177.8 kJ.

Review Problems

7.47 531 kJ. **7.49** (a) 82.9 kJ; (b) 1290 kJ. **7.51** 8600 cal;
about 8.6 dietary calories. **7.53** −17 J. **7.55** 100 J. **7.57** 7.32
kJ. **7.59** 135 J. **7.61** (a) 1.67×10^3 J; (b) 1.67×10^3 J; (c) 23.2
J °C^{-1}; (d) 4.64 J g^{-1} °C^{-1}. **7.63** (a) 3.5×10^3 kcal; (b) 56
miles. **7.65** 25.12 J/mol °C. **7.67** −30.4 kJ. **7.69** 53 kJ/mol.
7.71 (a) 2.22×10^5 J; (b) −222 kJ/mol. **7.73** (a) $2CO(g) +$
$O_2(g) \rightarrow 2CO_2(g), \Delta H° = -566$ kJ; (b) −283 kJ/mol.
7.75 $4Al(s) + 2Fe_2O_3(s) \rightarrow 2Al_2O_3(s) + 4Fe(s), \Delta H° =$
−1708 kJ. **7.77** −162 kJ.

7.79

The enthalpy change for the reaction is −280 kJ.

7.81 Add the second equation to the inverse of the first,
changing the sign of the first equation. **7.83** −78.61 kJ.
7.85 −127.8 kJ. **7.87** −53 kJ. **7.89** −1268 kJ. **7.91** −188 kJ.
7.93 Only (b).

7.95 (a) $2C(\text{graphite}) + 2H_2(g) + O_2(g) \rightarrow HC_2H_3O_2(l)$,
$$\Delta H_f° = -487.0 \text{ kJ}$$
(b) $Na(s) + \frac{1}{2}H_2(g) + C(\text{graphite}) + \frac{3}{2}O_2(g) \rightarrow$
$$NaHCO_3(s) \qquad \Delta H_f° = -947.7 \text{ kJ}$$
(c) $Ca(s) + 8S(s) + 3O_2(g) + 2H_2(g) \rightarrow$
$$CaSO_4·2H_2O(s) \qquad \Delta H_f° = -2021.1 \text{ kJ}$$

7.97 (a) −196.6 kJ; (b) −177.8 kJ.

7.99 (a) $\frac{1}{2}H_2(g) + \frac{1}{2}Cl_2(g) \rightarrow HCl(g) \quad \Delta H_f° = -92.30$ kJ/mol
(b) $\frac{1}{2}N_2(g) + 2H_2(g) + \frac{1}{2}Cl_2(g) \rightarrow NH_4Cl(s)$,
$$\Delta H_f° = -315.4 \text{ kJ/mol}$$

7.101 $C_{12}H_{22}O_{11}(s) + 12O_2(g) \rightarrow 12CO_2(g) + 11H_2O(\ell)$
$$\Delta H° = -5.65 \times 10^3 \text{ kJ/mol}$$
$\Delta H_f°[C_{12}H_{22}O_{11}(s)] = -2.22 \times 10^3$ kJ

Chapter 8

Practice Exercises

1. 3.00×10^{13} Hz. **2.** 2.874 m. **3.** 656.6 nm, which is red.
4. When $n = 3$: s, p and d subshells; when $n = 4$: s, p, d and f subshells. **5.** 9 orbitals.

6. (a) Mg: $1s^2 2s^2 2p^6 3s^2$
(b) Ge: $1s^2 2s^2 2p^6 3s^2 3p^6 3d^{10} 4s^2 4p^2$
(c) Cd: $1s^2 2s^2 2p^6 3s^2 3p^6 3d^{10} 4s^2 4p^6 4d^{10} 5s^2$
(d) Gd: $1s^2 2s^2 2p^6 3s^2 3p^6 3d^{10} 4s^2 4p^6 4d^{10} 4f^7 5s^2 5p^6 5d^1 6s^2$

7.

(a) Na:

$1s$ $2s$ $2p$ $3s$ $3p$ $4s$ $3d$

(b) S:

$1s$ $2s$ $2p$ $3s$ $3p$ $4s$ $3d$

(c) Fe:

$1s$ $2s$ $2p$ $3s$ $3p$ $4s$ $3d$

8.

(a) P: [Ne]$3s^2 3p^3$
[Ne]
$3s$ $3p$ (3 unpaired electrons)

(b) Sn: [Kr]$4d^{10} 5s^2 5p^2$
[Kr]
$4d$ $5s$ $5p$ (2 unpaired electrons)

9. (a) Se: $4s^2 4p^4$; (b) Sn: $5s^2 5p^2$; (c) I: $5s^2 5p^5$. **10.** (a) Sn; (b) Ga; (c) Cr; (d) S^{2-}. **11.** (a) Be; (b) C.

Review Problems

8.73 6.98×10^{14} Hz. **8.75** 4.38×10^{13} Hz. **8.77** 1.02×10^{15} Hz. **8.79** 2.98 m. **8.81** 5.0×10^6 m; 5.0×10^3 km. **8.83** 2.7×10^{-19} J; 1.6×10^5 J mol^{-1}. **8.85** (a) violet (see Figure 8.8); (b) 7.31×10^{14} s^{-1}; (c) 4.85×10^{-19} J. **8.87** 1090 nm, which is not in the visible region. (We would not see the line.) **8.89** 1.737×10^{-6} m; this is in the infrared region. **8.91** (a) p; (b) f. **8.93** (a) $n = 3, \ell = 0$; (b) $n = 5, \ell = 2$. **8.95** $\ell = 0, 1, 2, 3, 4, 5$. **8.97** (a) $m_\ell = 1, 0,$ or -1; (b) $m_\ell = 3, 2, 1, 0, -1, -2,$ or -3. **8.99** When $m_\ell = -4$, the minimum value of ℓ is 4 and the minimum value of n is 5.

8.101

n	ℓ	m_ℓ	m_s
2	1	-1	$+1/2$
2	1	-1	$-1/2$
2	1	0	$+1/2$
2	1	0	$-1/2$
2	1	$+1$	$+1/2$
2	1	$+1$	$-1/2$

8.103 21 electrons have $\ell = 1$, 20 electrons have $\ell = 2$

8.105 (a) S $1s^2 2s^2 2p^6 3s^2 3p^4$
(b) K $1s^2 2s^2 2p^6 3s^2 3p^6 4s^1$
(c) Ti $1s^2 2s^2 2p^6 3s^2 3p^6 3d^2 4s^2$
(d) Sn $1s^2 2s^2 2p^6 3s^2 3p^6 4s^2 3d^{10} 4p^6 4d^{10} 5s^2 5p^2$

8.107 (a) Mn is paramagnetic; (b) As is paramagnetic; (c) S is paramagnetic; (d) Sr is not paramagnetic; (e) Ar is not paramagnetic. **8.109** (a) zero; (b) three; (c) three.
8.111 (a) [Ar] $3d^8 4s^2$; (b) [Xe] $6s^1$; (c) [Ar] $3d^{10} 4s^2 4p^2$; (d) [Ar] $3d^{10} 4s^2 4p^5$; (e) [Xe] $4f^{14} 5d^{10} 6s^2 6p^3$.

8.113

(a) Mg: $1s$ $2s$ $2p$ $3s$ $3p$ $4s$ $3d$

(b) Ti: $1s$ $2s$ $2p$ $3s$ $3p$ $4s$ $3d$

8.115

(a) Ni: [Ar] $4s$ $3d$

(b) Cs: [Xe] $6s$

(c) Ge: [Ar] $4s$ $3d$ $4p$

(d) Br: [Ar] $4s$ $3d$ $4p$

8.117 (a) 5; (b) 4; (c) 4; (d) 6. **8.119** (a) $3s^1$; (b) $3s^2 3p^1$; (c) $4s^2 4p^2$; (d) $3s^2 3p^3$.

8.121

(a) Na: $3s$ (c) Ge: $4s$ $4p$

(b) Al: $3s$ $3p$ (d) P: $3s$ $3p$

8.123 (a) 1; (b) 6; (c) 7. **8.125** (a) Na; (b) Sb. **8.127** Sb.
8.129 (a) Na; (b) Co^{2+}; (c) Cl^-. **8.131** (a) C; (b) O; (c) Cl.
8.133 (a) Cl; (b) Br. **8.135** Mg.

Chapter 9

Practice Exercises

1. Cr: [Ar] $3d^4 4s^2$; (a) Cr^{2+}: [Ar]$3d^4$; (b) Cr^{3+}: [Ar]$3d^3$; (c) Cr^{6+}: [Ar]. **2.** The electron configurations are identical.

3. (a) :Se: (b) :I·   (c) ·Ca·

4. ·Mg·   :O: ⟶ Mg^{2+} [:O:]$^{2-}$

5.

O
‖
R—C—H aldehyde

O
‖
R—C—OH acid

H
|
R—N—H amine

O
‖
R—C—R ketone

R—O—H alcohol

6. (a) Br; (b) Cl; (c) Cl.

7.

SO_2 O S O NO_3^- O
                                    O N O

$HClO_3$ O
H O Cl O

O
H O P O H

H_3PO_4 O
              H

8. SO_2 has 18 valence electrons; PO_4^{3-} has 32 valence electrons; NO^+ has 10 valence electrons.

9.

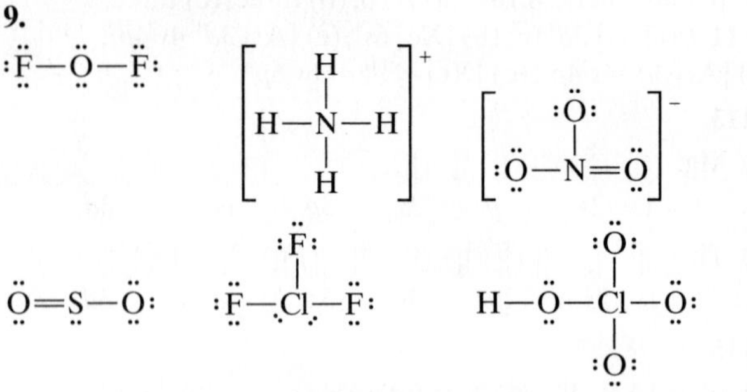

:F̈—Ö—F̈:

$$\left[\begin{array}{c} H \\ H-N-H \\ H \end{array}\right]^+$$

$$\left[:\ddot{O}-N=\ddot{O}\right]^-$$ (with :Ö: above N)

Ö=S—Ö: :F̈—C̈l—F̈:

H—Ö—Cl—Ö: (with :Ö: above Cl and :Ö: below Cl)

10.

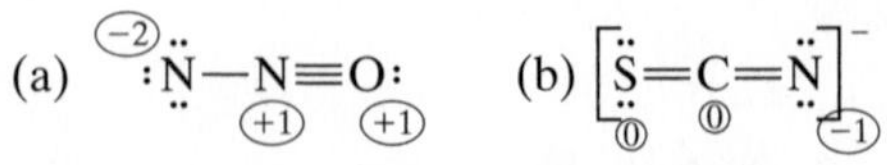

(a) :N̈—N≡O: with charges (-2) on N, (+1) on N, (+1) on O

(b) $\left[\ddot{S}=C=\ddot{N}\right]^-$ with charges 0, 0, (-1)

11.

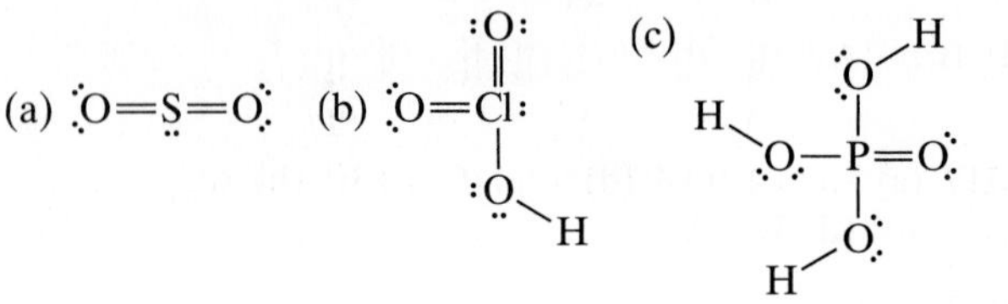

(a) :Ö=S=Ö: (b) :Ö=C̈l:

(c)

H—O—P=O with O—H groups (phosphorus structure with two O—H and one =O)

12.

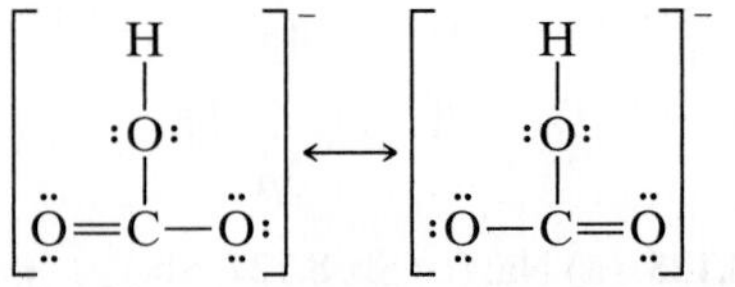

$$\left[\begin{array}{c} H \\ :O: \\ \ddot{O}=C-\ddot{O}: \end{array}\right]^- \longleftrightarrow \left[\begin{array}{c} H \\ :O: \\ :\ddot{O}-C=\ddot{O} \end{array}\right]^-$$

13.

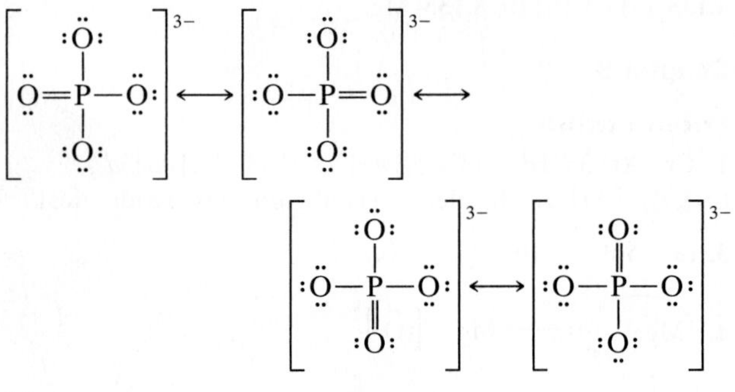

$$\left[\begin{array}{c} :\ddot{O}: \\ \ddot{O}=P-\ddot{O}: \\ :\ddot{O}: \end{array}\right]^{3-} \longleftrightarrow \left[\begin{array}{c} :\ddot{O}: \\ :\ddot{O}-P=O \\ :\ddot{O}: \end{array}\right]^{3-} \longleftrightarrow$$

$$\left[\begin{array}{c} :\ddot{O}: \\ :\ddot{O}-P-\ddot{O}: \\ :\ddot{O}: \end{array}\right]^{3-} \longleftrightarrow \left[\begin{array}{c} :O: \\ :\ddot{O}-P-\ddot{O}: \\ :\ddot{O}: \end{array}\right]^{3-}$$

14.

H—Ö:⁻ ⤻ H⁺ ⟶ H—Ö—H (arrow pointing up to)

coordinate covalent bond

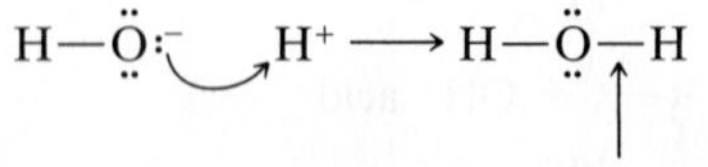

Review Problems

9.68 Magnesium loses two electrons: $Mg \rightarrow Mg^{2+} + 2e^-$; $[Ne]3s^2 \rightarrow [Ne]$. Each bromine gains an electron: $Br + e^- \rightarrow Br^-$; $[Ar]3d^{10}4s^24p^5 \rightarrow [Kr]$. **9.70** Pb^{2+}: $[Xe]4f^{14}5d^{10}6s^2$; Pb^{4+}: $[Xe]4f^{14}5d^{10}$. **9.72** Mn^{3+}: $[Ar]3d^4$ (4 unpaired electrons).

9.74

(a) ·Si·   (b) ·Sb·   (c) ·Ba·   (d) ·Al·   (e) :S̈·

9.76

(a) $[K]^+$ (b) $[Al]^{3+}$ (c) $\left[:\ddot{S}:\right]^{2-}$ (d) $\left[:\ddot{Si}:\right]^{4-}$

(e) $[Mg]^{2+}$

9.78

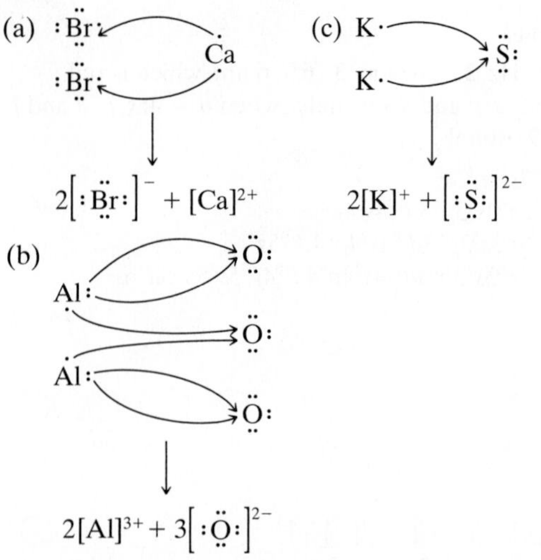

(a) :B̈r⤻ and :B̈r⤻ with Ca· ⟶ $2\left[:\ddot{B}r:\right]^- + [Ca]^{2+}$

(c) K·⤻ and K·⤻ with ·S̈: ⟶ $2[K]^+ + \left[:\ddot{S}:\right]^{2-}$

(b) Al: and Al: with →Ö: groups ⟶ $2[Al]^{3+} + 3\left[:\ddot{O}:\right]^{2-}$

9.80 +0.029 for the nitrogen atom, and −0.029 for the oxygen atom. **9.82** 1400 g. **9.84** 344 nm (this is in the ultraviolet region).

9.86

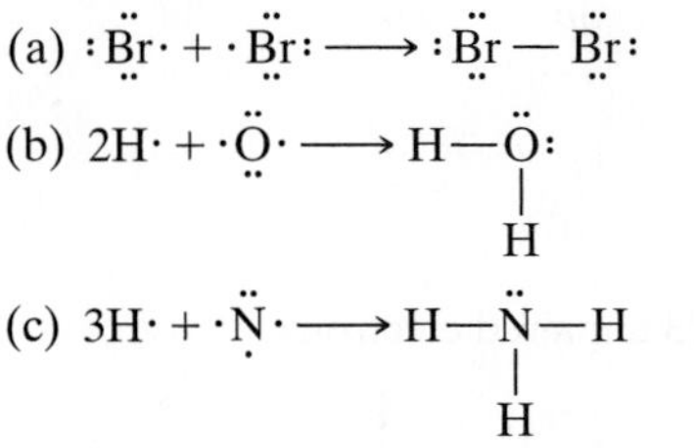

(a) :B̈r· + ·B̈r: ⟶ :B̈r—B̈r:

(b) 2H· + ·Ö· ⟶ H—Ö—H

(c) 3H· + ·N̈· ⟶ H—N̈—H with H below

9.88 (a) H_2Se; (b) H_3As; (c) SiH_4. **9.90** (a) S; (b) Si; (c) Br; (d) C. **9.92** N—S.

9.94

(a) Cl Si Cl with Cl above and Cl below   (b) F P F with F above and F below

(c) H P H with H below   (d) Cl S Cl

9.96 (a) 32; (b) 26; (c) 8; (d) 20.

9.98

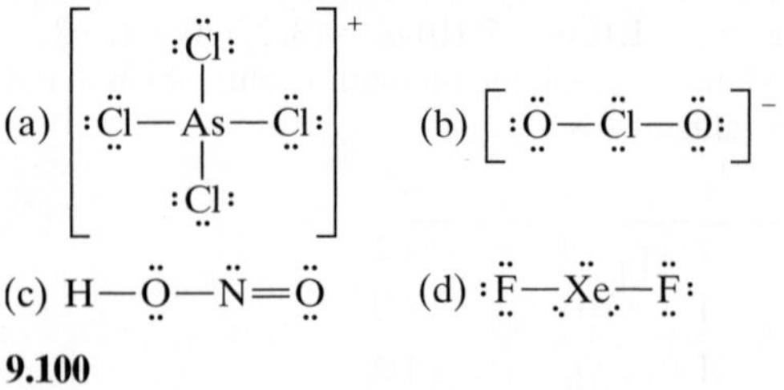

(a) $$\left[\begin{array}{c} :\ddot{C}l: \\ :\ddot{C}l-As-\ddot{C}l: \\ :\ddot{C}l: \end{array}\right]^+$$ (b) $\left[:\ddot{O}-\ddot{C}l-\ddot{O}:\right]^-$

(c) H—Ö—N=Ö   (d) :F̈—Xe—F̈:

9.100

(a) :C̈l—Si—C̈l: with :Cl: above and :Cl: below   (b) :F̈—P—F̈: with :F: below

(c) H—P—H with H below   (d) :C̈l—S̈—C̈l:

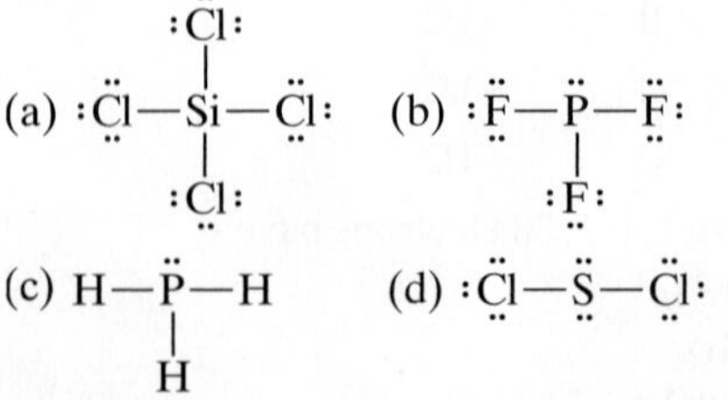

9.102 (a) :S̈=C=S̈:   (b) $\left[:C\equiv N:\right]^-$

9.104

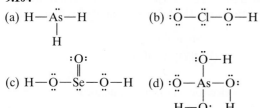

(a) H—Äs—H
       |
       H

(b) :Ö—Cl—Ö—H

(c) H—Ö—Se—Ö—H (with :O: double bonded above Se)

(d) :Ö—As—Ö: (with :Ö—H above and H—Ö: H below)

9.106

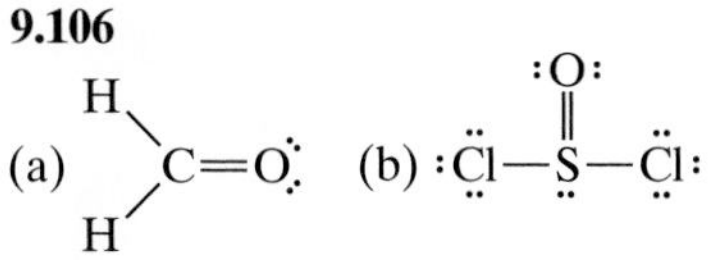

(a) H₂C=Ö: (H and H on left, C=O)

(b) :Cl—S—Cl: (with :O: double bonded above S)

9.108

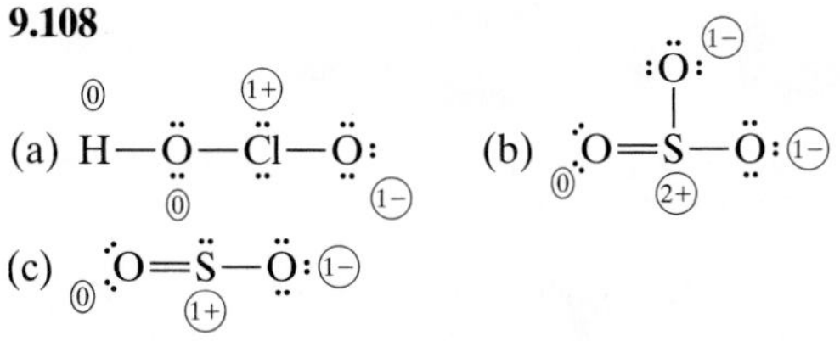

(a) H—Ö—Cl—Ö: (charges (0) over O, (1+) over Cl, (0) under O, (1−) under O)

(b) :Ö:(1−) O=S—Ö:(1−) (charges (0), (2+))

(c) :O=S—Ö:(1−) (charges (0), (1+))

9.110

:Ö:(1−)
(0)Ö—Cl—Ö:(1−)
    (3+)
H :Ö:
(0) (1−)

←→

    Ö
:Ö—Cl—Ö:(1−)
    (1+)
H Ö.

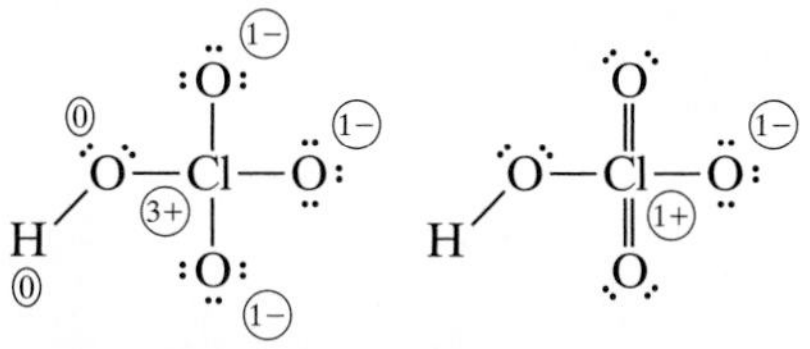

9.112 The formal charges on all of the atoms of the left structure are zero. Also, the non-preferred structure places a positive formal charge on chlorine.

9.114 The average bond order is 4/3.

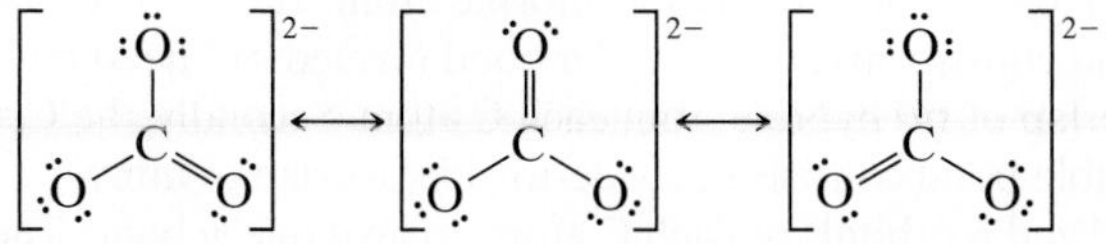

$$\left[\begin{array}{c}:Ö: \\ | \\ :O—C—O:\end{array}\right]^{2-} \leftrightarrow \left[\begin{array}{c}:O: \\ \| \\ :O—C—O.\end{array}\right]^{2-} \leftrightarrow \left[\begin{array}{c}:Ö: \\ | \\ :O—C—O.\end{array}\right]^{2-}$$

9.116 The N—O bonds in NO_2^- should be shorter than those in NO_3^-.

9.118 These are not preferred structures, because they contain formal charges.

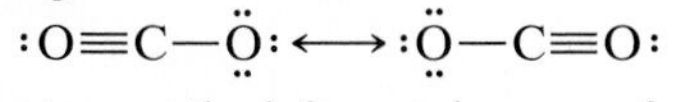

:O≡C—Ö: ←→ :Ö—C≡O:

9.120 The structure at the left contains more formal charges than the others.

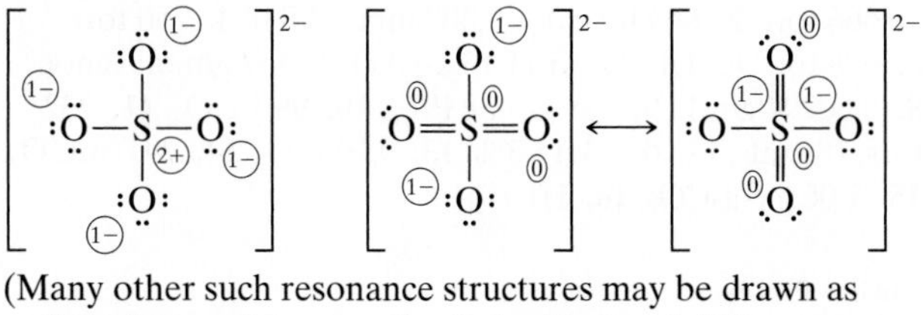

$$\left[\begin{array}{c}:Ö:(1-) \\ | \\ (1-):O—S—O:(1-) \\ (2+) \\ :O:(1-)\end{array}\right]^{2-} \leftrightarrow \left[\begin{array}{c}:Ö:(1-) \\ | \\ (0)O=S=O:(0) \\ (1-):O:(0)\end{array}\right]^{2-} \leftrightarrow \left[\begin{array}{c}(0)O \\ \| \\ (1-):O—S—O:(1-) \\ \| \\ (0)O.\end{array}\right]^{2-}$$

(Many other such resonance structures may be drawn as well, each containing two S=O double bonds.) The average bond order is 1.5.

9.122

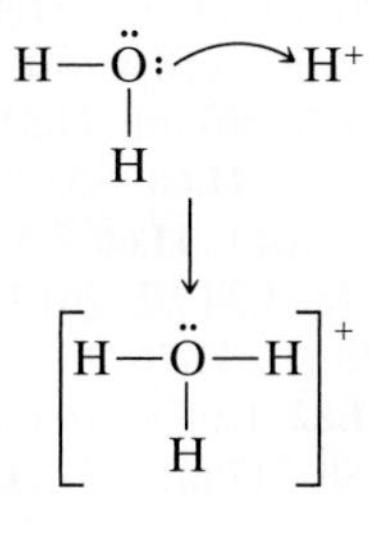

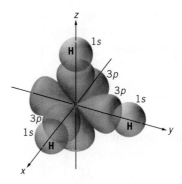

Chapter 10

Practice Exercises

1. Each bond below corresponds to a bonding domain, including the double bond, which is considered a single bonding domain. (Carbon has 3 bonding domains.)

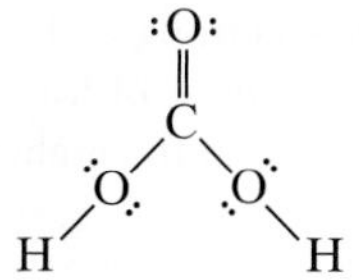

The lone pairs on the oxygens correspond to non bonding domains. **2.** Trigonal bipyramidal. **3.** Linear. **4.** Linear. **5.** Square planar

6. SO_3^{2-}: Trigonal pyramidal
CO_3^{2-}: Planar triangular shape
XeO_4: Tetrahedral
OF_2: Bent

7. (b), (c), (d), and (e)

8. The H—Cl bond is formed by the overlap of the half-filled 1s atomic orbital of a H atom with the half-filled 3p valence orbital of a Cl atom:

Cl atom in HCl (x = H electron):

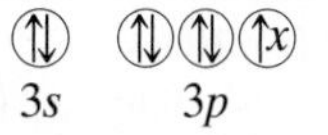

⇅ ⇅⇅⇡⤫
3s   3p

The overlap that gives rise to the H—Cl bond is that of a 1s orbital of H with a 3p orbital of Cl:

9. P atom in PH_3 (x = H electron):

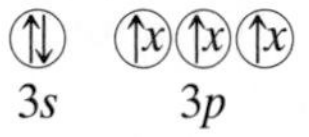

⇅ ⤫⤫⤫
3s   3p

The orbital overlap that forms the P—H bond combines a 1s orbital of hydrogen with a 3p orbital of phosphorus (Note: Only half of each p orbital is shown):

10. Each of arsenic's five sp^3d hybrid orbitals overlaps with a 3p atomic orbital of a chlorine atom to form a total of five As—Cl single bonds. Four of the 3d atomic orbitals of As remain unhybridized. **11.** (a) sp^3; (b) sp^3d. **12.** (a) sp^3; (b) sp^3d.

13. sp^3d^2, since six atoms are bonded to the central atom.

P atom in PCl_6^- (x = Cl electron):

sp^3d^2 $3d$

The ion is octahedral.

14. NO has 11 valence electrons, and the MO diagram is similar to that shown in Table 10.1 for O_2, except that one fewer electron is employed at the highest energy level

$$\sigma^*_{2p_z} \quad \text{-------} \quad \bigcirc$$
$$\pi^*_{2p_x}, \pi^*_{2p_y} \quad \text{-------} \quad (\uparrow)\bigcirc$$
$$\pi_{2p_x}, \pi_{2p_y} \quad \text{-------} \quad (\uparrow\downarrow)(\uparrow\downarrow)$$
$$\sigma_{2p_z} \quad \text{-------} \quad (\uparrow\downarrow)$$
$$\sigma^*_{2s} \quad \text{-------} \quad (\uparrow\downarrow)$$
$$\sigma_{2s} \quad \text{-------} \quad (\uparrow\downarrow)$$

The bond order is calculated to be 5/2

Review Problems

10.54 (a) bent; (b) planar triangular; (c) T-shaped; (d) linear; (e) planar triangular. **10.56** (a) nonlinear; (b) trigonal bipyramidal; (c) trigonal pyramidal; (d) trigonal pyramidal; (e) nonlinear. **10.58** (a) tetrahedral; (b) square planar; (c) octahedral; (d) tetrahedral; (e) linear. **10.60** 180°. **10.62** (a), (b), (c), and (e). **10.64** (a) 109.5°; (b) 109.5°; (c) 120°; (d) 180°; (e) 109.5°. **10.66** (a), (b), and (c). **10.68** All are polar. **10.70** In SF_6, although the individual bonds in this substance are polar bonds, the geometry of the bonds is symmetrical which serves to cause the individual dipole moments of the various bonds to cancel one another. In SF_5Br, one of the six bonds has a different polarity so the individual dipole moments of the various bonds do not cancel one another.

10.72 The 1s atomic orbitals of the hydrogen atoms overlap with the mutually perpendicular p atomic orbitals of the selenium atom.

Se atom in H_2Se (x = H electron):

$$(\uparrow\downarrow) \quad (\uparrow\downarrow)(\uparrow x)(\uparrow x)$$
$$4s \qquad 4p$$

10.74 Atomic Be: Hybridized Be:

$$(\uparrow\downarrow)\bigcirc\bigcirc\bigcirc \qquad (\uparrow x)(\uparrow x)\bigcirc\bigcirc$$
$$2s \quad 2p \qquad\qquad sp \quad 2p$$
$$(x = \text{a Cl electron})$$

10.76 (a) sp^3 hybridized:

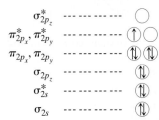

(b) sp^2 hybridized:

(c) sp^3 hybridized:

10.78 (a) There are three bonds to As and one lone pair at As, requiring As to be sp^3 hybridized. (b) There are three

atoms bonded to the central Cl atom, and it also has two lone pairs of electrons. The hybridization of Cl is thus sp^3d.

10.80   Sb in SbF_6^-:

sp^3d^2

(xx = an electron pair from the donor F^-)

10.82 (a) N in the C≡N system:

sp^2 $2p$

(b)

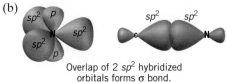

Overlap of 2 sp^2 hybridized orbitals forms σ bond.

Unhybridized p orbitals form π bond.

(c)

10.84 Each carbon atom is sp^2 hybridized, and each C—Cl bond is formed by the overlap of an sp^2 hybrid of carbon with a p atomic orbital of a chlorine atom. The C=C double bond consists first of a C—C σ bond formed by "head on" overlap of sp^2 hybrids from each C atom. Secondly, the C=C double bond consists of a side-to-side overlap of unhybridized p orbitals of each C atom, to give one π bond. The molecule is planar, and the expected bond angles are all 120°. **10.86** 1. sp^3; 2. sp; 3. sp^2; 4. sp^2. **10.88** (a) O_2^+; (b) O_2; (c) N_2. **10.90** (a)–(d).

Chapter 11

Practice Exercises

1. 666 torr. **2.** 669 torr. **3.** 10,300 mm; 33.8 ft. **4.** 750 torr. **5.** 688 torr. **6.** 1,130 g Ar (1.13 kg Ar). **7.** 132 g/mol; xenon. **8.** 0.00294 g/mL. **9.** 138 g/mol; P_2F_4. **10.** 98.1 g O_2. **11.** 732 torr; 283 mL. **12.** 0.153; 15.3%. **13.** 9.00 L O_2. **14.** 79.0 mL O_2. **15.** 1.06 g Na_2CO_3. **16.** HI.

Review Problems

11.36 (a) 958 torr; (b) 0.974 atm; (c) 738 mm Hg; (d) 10.9 torr. **11.38** (a) 250 torr; (b) 350 torr. **11.40** 4.5 cm Hg. **11.42** 813 torr. **11.44** Produce; 826 torr. **11.46** 125 torr. **11.48** 8.42×10^3 mm. **11.50** 507 mL. **11.52** 4.28 L. **11.54** 843 °C. **11.56** 796 torr. **11.58** 5.73 L. **11.60** −53 °C. **11.62** 6.24×10^4 mL·torr/mol·K. **11.64** 0.104 L. **11.66** 2,340 torr. **11.68** 0.0398 g. **11.70** 4.16 atm. **11.72** (a) 1.34 g/L; (b) 1.25 g/L; (c) 3.17 g/L; (d) 1.78 g/L. **11.74** 1.28 g/L. **11.76** 88.3 g/mol. **11.78** 27.5 g/mol. **11.80** 4.00 L. **11.82** 1.14×10^3 mL. **11.84** 10.7 L H_2. **11.86** 36.3 mL O_2. **11.88** 217 mL CO_2. **11.90** 650 torr.

11.92 $P_{N_2} = 0.30$ atm $= 228$ torr; $P_{O_2} = 0.20$ atm $= 152$ torr; $P_{He} = 0.40$ atm $= 304$ torr; $P_{CO_2} = 0.10$ atm $= 76$ torr. **11.94** 736 torr; 260 mL. **11.96** 250 mL. **11.98** 75.0% nitrogen, 13.6% oxygen, 6.2% water, 5.3% carbon dioxide. **11.100** Nitrogen effuses 1.25 as rapidly as carbon dioxide. **11.102** $^{235}UF_6$ diffuses 1.0043 times as rapidly as $^{238}UF_6$.

Chapter 12

Practice Exercises

1. Propylamine. It can form hydrogen bonds. **2.** Remains the same. **3.** About 75 °C. **4.** Increases vapor pressure. **5.** Sublimation. **6.** Liquid.

Review Problems

12.86 Diethyl ether. **12.88** London forces are possible in them all. Dipole-dipole in (a), (b), and (d). Hydrogen bonding in (a). **12.90** London forces. Since bromoform has a higher boiling point that chloroform, we must conclude that it experiences stronger intermolecular attractions than chloroform, which can only be due to London forces. **12.92** Ethanol. **12.94** ether < acetone < benzene < water < acetic acid. **12.96** 77.4%. **12.98** 305 kJ. **12.100** (a) 0 °C; (b) 47.9 g. **12.102** Water. (Higher intermolecular attractions.)

12.104

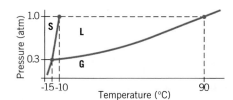

12.106 (a) solid; (b) gas; (c) liquid; (d) solid, liquid, and gas. **12.108** At -56 °C, the vapor is compressed until the liquid–vapor line is reached, at which point the vapor condenses to a liquid. As the pressure is increased further, the solid–liquid line is reached, and the liquid freezes.

At -58 °C, the gas is compressed until the solid–vapor line is reached, at which point the vapor condenses directly to a solid.

Chapter 13

Practice Exercises

1. 1 to 1. **2.** Covalent solid. **3.** Molecular solid.
4. $-(CF_2-CF_2)-(CF_2-CF_2)-(CF_2-CF_2)-$

Review Problems

13.92 $4 Zn^{2+}, 4 S^{2-}$. **13.94** 3.51 Å (351 pm). **13.96** 656 pm. **13.98** (a) $\theta = 6.57°$; (b) $\theta = 27.3°$. **13.100** 4.04 Å. **13.102** Its density does not match that for body-centered or face-centered cubic. **13.104** 176 pm. **13.106** Molecular solid. **13.108** Metallic solid. **13.110** Metallic solid. **13.112** (a) molecular; (b) ionic; (c) ionic; (d) metallic; (e) covalent; (f) molecular; (g) ionic.

13.114

13.116

Chapter 14

Practice Exercises

1. 0.00430 g of O_2 and 0.00190 g of N_2. **2.** 2.50 g NaOH, 248 g H_2O, 251 mL H_2O. **3.** 2.0×10^1 g solution. **4.** 20 g CH_3OH (rounded from 16.0 g). **5.** 0.40 m. **6.** 16.1 m. **7.** 6.82 M. **8.** 9.02 torr. **9.** 44.1 torr. **10.** 100.16 °C. **11.** 157 g/mol. **12.** 223 torr. **13.** 5.38×10^2 g/mol. **14.** 100% dissociated: -0.882 °C; 0% dissociated: -0.441 °C.

Review Problems

14.63 This is to be very much like that shown in Figures 14.8 and 14.9:

(a) $KCl(s) \rightarrow K^+(g) + Cl^-(g)$, $\Delta H = +690$ kJ mol^{-1}
(b) $K^+(g) + Cl^-(g) \rightarrow K^+(aq) + Cl^-(aq)$, $\Delta H = -686$ kJ mol^{-1}
  $KCl(s) \rightarrow K^+(aq) + Cl^-(aq)$, $\Delta H = +4$ kJ mol^{-1}

14.65 -616 kJ/mol. **14.67** 0.038 g/L. **14.69** 0.020 g/L. **14.71** 3.35 m. **14.73** 0.133 m; 2.39×10^{-3}; 2.34%. **14.75** 5.45%. **14.77** 5.28%; 3.09 m. **14.79** 0.359 M; 3.00%; 6.49×10^{-3}. **14.81** 22.8 torr. **14.83** 69.4 torr. *14.85 70% toluene and 30% benzene. **14.87** (a) 0.029; (b) 2.99×10^{-2} moles; (c) 278 g/mol. **14.89** 101 °C; -3.72 °C. **14.91** 60 g. **14.93** 152 g/mol. **14.95** 127 g/mol; $C_8H_4N_2$. **14.97** (a) The units on the right side of this equation are g/mol, which is correct. (b) 1.8×10^6 g/mol. **14.99** 16.5 torr. **14.101** 1.3×10^4 torr. **14.103** -1.1 °C. **14.105** 3%. **14.107** 2. **14.109** 1.89.

Chapter 15

Practice Exercises

1. $O_2 = 0.45$ mol L^{-1} s^{-1}; $H_2S = 0.30$ mol L^{-1} s^{-1}. **2.** A value near 1×10^{-4} mol L^{-1} s^{-1} is correct. **3.** (a) $k = 8.0 \times 10^{-2}$ L mol^{-1} s^{-1}; (b) L mol^{-1} s^{-1}. **4.** With respect to $[BrO_3^-] = 1$; with respect to $[SO_3^{2-}] = 1$; overall $= 2$. **5.** Each of the other data sets also gives the same value: $k = 2.0 \times 10^2$ L^2 mol^{-2} s^{-1}. **6.** First order. **7.** Rate $= k[A]^2[B]^2$. **8.** (a) 0.26 M; (b) 77 min. **9.** 63 min. **10.** 18.7 min; 37.4 min. **11.** The reaction is first-order. **12.** (a) 1.4×10^2 kJ/mol; (b) 0.30 L mol^{-1} s^{-1}. **13.** Rate $= k[NO][O_3]$.

14. Rate $= \dfrac{k[NO_2Cl]^2}{[NO_2]}$

Review Problems

15.64

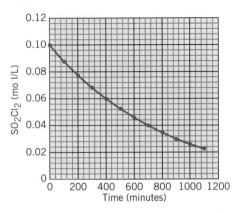

At 200 min, 1×10^{-4} M/s; at 600 minutes, 7×10^{-5} M/s.

15.66 Three times the rate; twice the rate

15.68 (a) 11.4 mol $L^{-1}\,s^{-1}$; (b) 7.20 mol $L^{-1}\,s^{-1}$; (c) 8.40 mol $L^{-1}\,s^{-1}$. **15.70** 8.0×10^{-11} mol $L^{-1}\,s^{-1}$. **15.72** 2.4×10^{2} mol $L^{-1}\,s^{-1}$. **15.74** (a) With respect to $HCrO_4^-$, the order is 1; with respect to HSO_3^-, the order is 2; with respect to H^+, the order is 1. (b) Overall order = 4. **15.76** Rate = $k[M][N]^2$; $k = 2.5 \times 10^3$ $L^2\,mol^{-2}\,s^{-1}$. **15.78** Rate = $k[OCl^-][I^-]$; $k = 6.1 \times 10^9$ L $mol^{-1}\,s^{-1}$. **15.80** Rate = $k[ICl][H_2]$; $k = 1.5 \times 10^{-1}$ L $mol^{-1}\,s^{-1}$.

15.82 A graph of ln $[SO_2Cl_2]_t$ versus t will yield a straight line if the data obeys a first-order rate law.

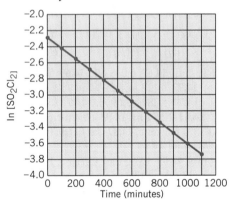

$k = 1.32 \times 10^{-3}$ min^{-1}.

15.84 (a) 3.7×10^{-3} M; (b) 6.0×10^{-4} M. **15.86** 4.26×10^{-3} min^{-1}. **15.88** 1.3×10^4 min. **15.90** $1/256^{th}$. **15.92** 500 min; another 500 minutes, first-order. **15.94** 4.3×10^2 seconds. **15.96** Graphically, 79 kJ/mol; By equation, 79.3 kJ/mol. **15.98** 99 kJ/mol; 6.6×10^9 L $mol^{-1}\,s^{-1}$. **15.100** 78 kJ/mol. **15.102** (a) $1.9 \times 10^{-5}\,s^{-1}$; (b) $1.6 \times 10^{-1}\,s^{-1}$.

Chapter 16

Practice Exercises

1. (a) $\dfrac{[H_2O]^2}{[H_2]^2[O_2]} = K_c$; (b) $\dfrac{[CO_2][H_2O]^2}{[CH_4][O_2]^2} = K_c$.

2. $K_c = 1.2 \times 10^{-13}$. **3.** $K_c = 1.9 \times 10^5$

4. $K_p = \dfrac{(P_{HI})^2}{(P_{H_2})(P_{I_2})}$

5. Reaction (b). **6.** Smaller; 57. **7.** 1.8×10^{36}

8. (a) $K_c = \dfrac{1}{[Cl_2(g)]}$; (b) $K_c = \dfrac{1}{[NH_3(g)][HCl(g)]}$;

(c) $K_c = [Na^+(aq)][OH^-(aq)][H_2(g)]$;

(d) $K_c = [Ag^+]^2[CrO_4^{2-}]$; (e) $K_c = \dfrac{[Ca^{+2}(aq)][HCO_3^-(aq)]^2}{[CO_2(aq)]}$

9. (a) Shift to the right; (b) Shift to the left; (c) Shift to the left; (d) Shift to the right. **10.** 4.06. **11.** [CO] decreases by 0.060 mol/L and $[CO_2]$ increases by 0.060 mol/L. **12.** (a) $[PCl_3] = 0.20\ M$; $[Cl_2] = 0.100\ M$; $[PCl_5] = 0.000\ M$; (b) PCl_3 and Cl_2 have decreased by 0.080 M; PCl_5 has increased by 0.080 M; (c) $[PCl_3] = 0.120\ M$; $[PCl_5] = 0.080\ M$; $[Cl_2] = 0.020\ M$; (d) 33. **13.** 8.98×10^{-3} M. **14.** $[H_2] = [I_2] = 0.044\ M$; $[HI] = 0.312\ M$. **15.** 1.1×10^{-17} M.

Review Problems

16.19 (a) $K_c = \dfrac{[POCl_3]^2}{[PCl_3]^2[O_2]}$ (d) $K_c = \dfrac{[NO_2]^2[H_2O]^8}{[N_2H_4][H_2O_2]^6}$

(b) $K_c = \dfrac{[SO_2]^2[O_2]}{[SO_3]^2}$ (e) $K_c = \dfrac{[SO_2][HCl]^2}{[SOCl_2][H_2O]}$

(c) $K_c = \dfrac{[NO]^2[H_2O]^2}{[N_2H_4][O_2]^2}$

16.21 (a) $K_p = \dfrac{(P_{POCl_3})^2}{(P_{PCl_3})^2(P_{O_2})}$ (d) $K_p = \dfrac{(P_{NO_2})^2(P_{H_2O})^8}{(P_{N_2H_4})(P_{H_2O_2})^6}$

(b) $K_p = \dfrac{(P_{SO_2})^2(P_{O_2})}{(P_{SO_3})^2}$ (e) $K_p = \dfrac{(P_{SO_2})(P_{HCl})^2}{(P_{SOCl_2})(P_{H_2O})}$

(c) $K_p = \dfrac{(P_{NO})^2(P_{H_2O})^2}{(P_{N_2H_4})(P_{O_2})^2}$

16.23 (a) $K_c = \dfrac{[Ag(NH_3)_2^+]}{[Ag^+][NH_3]^2}$; (b) $K_c = \dfrac{[Cd(SCN)_4^{2-}]}{[Cd^{2+}][SCN^-]^4}$.

16.25 1×10^{85}

16.27 (a) $K_c = \dfrac{[HCl]^2}{[H_2][Cl_2]}$; (b) $K_c = \dfrac{[HCl]}{[H_2]^{1/2}[Cl_2]^{1/2}}$;

K_c for reaction (b) is the square root of K_c for reaction (a). **16.29** 0.0375 M. **16.31** (b). **16.33** 11. **16.35** 2.7×10^{-2}. **16.37** 5.4×10^{-5}. **16.39** In each case we get approximately 55.5 M.

16.41 (a) $K_c = \dfrac{[CO]^2}{[O_2]}$ (d) $K_c = \dfrac{[H_2O][CO_2]}{[HF]^2}$

(b) $K_c = [H_2O][SO_2]$ (e) $K_c = [H_2O]^5$

(c) $K_c = \dfrac{[CH_4][CO_2]}{[H_2O]^2}$

16.43 [HI] = 1.47×10^{-12} M; $[Cl_2] = 7.37 \times 10^{-13}$ M. **16.45** (a) No; (b) Shift to the left. **16.47** 4.36×10^{-3} M. **16.49** 0.398. **16.51** 0.0955. **16.53** 0.915. **16.55** $[Br_2] = [Cl_2] = 0.011$ M. **16.57** $[NO_2] = [SO_2] = 0.0497$ mol/L; [NO] = $[SO_3] = 0.0703$ mol/L. **16.59** $[H_2] = [CO_2] = 7.7 \times 10^{-3}$ M; [CO] = $[H_2O] = 0.0123$ M. **16.61** $[H_2] = [Cl_2] = 8.9 \times 10^{-19}$ M. **16.63** 5.0×10^{-4} M. *16.65 3.0×10^{-5} M. *16.67 $[NO_2] = [SO_2] = 0.0281$ M. *16.69 [CO] = $[H_2O] = 0.200$ M.

Chapter 17

Practice Exercises

1. (a) OH^-; (b) I^-; (c) NO_2^-; (d) $H_2PO_4^-$; (e) HPO_4^{2-}; (f) PO_4^{3-}; (g) H^-; (h) NH_3. **2.** (a) H_2O_2; (b) HSO_4^-;

(c) HCO_3^-; (d) HCN; (e) NH_3; (f) NH_4^+; (g) H_3PO_4; (h) $H_2PO_4^-$. **3.** The Brønsted acids are $H_2PO_4^-(aq)$ and $H_2CO_3(aq)$. The Brønsted bases are $HCO_3^-(aq)$ and $HPO_4^{2-}(aq)$.

4.

conjugate pair

$$PO_4^{3-}(aq) + HC_2H_3O_2(aq) \rightleftharpoons HPO_4^{2-}(aq) + C_2H_3O_2^-(aq)$$

base   acid   acid   base

conjugate pair

5. $HPO_4^{2-}(aq) + OH^-(aq) \rightarrow PO_4^{3-}(aq) + H_2O$; HPO_4^{2-} acting as an acid; $HPO_4^{2-}(aq) + H_3O^+(aq) \rightarrow H_2PO_4^- + H_2O$; HPO_4^{2-} acting as a base. **6.** The substances on the right. **7.** (a) HBr; (b) H_2Te; (c) CH_3SH. **8.** $HClO_3$. **9.** (a) HIO_4; (b) H_2TeO_4; (c) H_3AsO_4. **10.** (a) Lewis bases; (b) Lewis acid; (c) Lewis acid. **11.** 1.3×10^{-9} M; basic. **12.** $pH = 3.44$, $pOH = 10.56$; acidic. **13.** 5.17

14. (a) $[H^+] = 1.3 \times 10^{-3}$ M; $[OH^-] = 7.7 \times 10^{-12}$ M; acidic (b) $[H^+] = 1.4 \times 10^{-4}$ M; $[OH^-] = 7.1 \times 10^{-11}$ M; acidic (c) $[H^+] = 1.5 \times 10^{-11}$ M; $[OH^-] = 6.7 \times 10^{-4}$ M; basic (d) $[H^+] = 7.8 \times 10^{-5}$ M; $[OH^-] = 1.3 \times 10^{-10}$ M; acidic (e) $[H^+] = 2.5 \times 10^{-12}$ M; $[OH^-] = 4.0 \times 10^{-3}$ M; basic

15. $pH = 11.70$, $[H^+] = 2.0 \times 10^{-12}$ M. **16.** 3.2×10^{-6} M.

Review Problems

17.40 (a) HF; (b) $N_2H_5^+$; (c) $C_5H_5NH^+$; (d) HO_2^-; (e) H_2CrO_4.

17.42

conjugate pair

(a) $HNO_3 + N_2H_4 \rightleftharpoons N_2H_5^+ + NO_3^-$

acid   base   acid   base

conjugate pair

conjugate pair

(b) $N_2H_5^+ + NH_3 \rightleftharpoons NH_4^+ + N_2H_4$

acid   base   acid   base

conjugate pair

conjugate pair

(c) $H_2PO_4^- + CO_3^{2-} \rightleftharpoons HCO_3^- + HPO_4^{2-}$

acid   base   acid   base

conjugate pair

conjugate pair

(d) $HIO_3 + HC_2O_4^- \rightleftharpoons H_2C_2O_4 + IO_3^-$

acid   base   acid   base

conjugate pair

17.44 (a) H_2Se, larger central atom; (b) HI, more electronegative atom; (c) PH_3, larger central atom. **17.46** (a) HIO_4, because it has more oxygen atoms; (b) H_3AsO_4, because it has more oxygen atoms. **17.48** (a) H_3PO_4, since P is more electronegative; (b) HNO_3, because N is more electronegative; (c) $HClO_4$, because Cl is more electronegative.

17.50

$$[H-\ddot{N}-H]^- + H^+ \longrightarrow H-N-H$$

17.52

17.54

17.56

17.58 $[H^+] = [OH^-] = 1.5 \times 10^{-7}$ M; $pH = 6.82$, $pOH = 6.82$, $pK_w = pH + pOH = 6.82 + 6.82 = 13.64$; water is neutral at this temperature. **17.60** (a) 4.2×10^{-12} M; (b) 7.1×10^{-10} M; (c) 1.8×10^{-6} M; (d) 2.4×10^{-2} M. **17.62** (a) 11.38; (b) 9.15; (c) 5.74; (d) 1.62. **17.64** 4.72.

17.66 (a) $[H^+] = 7.2 \times 10^{-4}$ M; $[OH^-] = 1.4 \times 10^{-11}$ M (b) $[H^+] = 1.7 \times 10^{-3}$ M; $[OH^-] = 6.0 \times 10^{-12}$ M (c) $[H^+] = 5.6 \times 10^{-10}$ M; $[OH^-] = 1.8 \times 10^{-5}$ M (d) $[H^+] = 5.8 \times 10^{-14}$ M; $[OH^-] = 1.7 \times 10^{-1}$ M (e) $[H^+] = 2.0 \times 10^{-6}$ M; $[OH^-] = 5.0 \times 10^{-9}$ M

17.68 $[H^+] = 2.5 \times 10^{-6}$ M; $[OH^-] = 4.0 \times 10^{-9}$ M. **17.70** $[H^+] = 0.030$ M, $pH = 1.5$; $[OH^-] = 3.2 \times 10^{-13}$ M. **17.72** 0.15 M OH^-; $pOH = 0.82$, $pH = 13.18$; $[H^+] = 6.6 \times 10^{-14}$ M. **17.74** 2.0×10^{-3} M $Ca(OH)_2$. **17.76** 16.9 mL KOH solution. **17.78** 5.01×10^{-12} M.

Chapter 18

Practice Exercises

1. (a) $HCHO_2 + H_2O \rightleftharpoons H_3O^+ + CHO_2^-$

$$K_a = \frac{[H_3O^+][CHO_2^-]}{[HCHO_2]}$$

(b) $(CH_3)_2NH_2^+ + H_2O \rightleftharpoons H_3O^+ + (CH_3)_2NH$

$$K_a = \frac{[H_3O^+][CH_3NH]}{[(CH_3)_2NH_2^+]}$$

(c) $H_2PO_4^- + H_2O \rightleftharpoons H_3O^+ + HPO_4^{2-}$

$$K_a = \frac{[H_3O^+][HPO_4^{2-}]}{[H_2PO_4^-]}$$

2. HA is the stronger acid. For HA: $K_a = 6.9 \times 10^{-4}$; for HB: $K_a = 7.2 \times 10^{-5}$

3. (a) $(CH_3)_3N + H_2O \rightleftharpoons (CH_3)_3NH^+ + OH^-$

$$K_b = \frac{[(CH_3)_3NH^+][OH^-]}{[(CH_3)_3N]}$$

(b) $SO_3^{2-} + H_2O \rightleftharpoons HSO_3^- + OH^-$

$$K_b = \frac{[HSO_3^-][OH^-]}{[SO_3^{2-}]}$$

(c) $NH_2OH + H_2O \rightleftharpoons NH_3OH^+ + OH^-$

$$K_b = \frac{[NH_3OH^+][OH^-]}{[NH_2OH]}$$

4. 5.6×10^{-11}. **5.** 1.7×10^{-5}; 4.78. **6.** 1.7×10^{-6}; 5.77.
7. 8.4×10^{-4} M; 3.08. **8.** 8.62. **9.** 5.36. **10.** neutral. **11.** neutral.
12. acidic. **13.** 8.07. **14.** 5.13. **15.** basic. **16.** 10.79. **17.** 4.84.
18. 9.36. **19.** Yes, 0.72 mol $HCHO_2$ for every mol CHO_2^-;
4.9 g $NaCHO_2$. **20.** Yes, 4.84. **21.** $[H^+] = 2.6 \times 10^{-3}$ M;
pH = 2.6; $[HC_6H_6O_6^-] = 2.7 \times 10^{-12}$ M. **22.** 10.24. **23.** Equal
to K_{b_2} for SO_3^{2-}. **24.** (a) 2.37; (b) 3.74; (c) 4.22; (d) 8.22. **25.** 3.66.

Review Problems

18.42 1.5×10^{-11}. **18.44** 5.6×10^{-3}. **18.46** 7.1×10^{-11}.
18.48 0.30%. **18.50** 1.8×10^{-6}. **18.52** 2×10^{-2}; $pK_a = 1.7$.
18.54 6×10^{-4}; $pK_b = 3.2$; % ionization = 7.2%.
18.56 $[HC_3H_5O_2] = 0.145$ M; $[C_3H_5O_2^-] = [H_3O^+] =$
0.0046 M, pH = 2.3. **18.58** pH = 2.78; % ionization = 1.1%.
18.60 10.26. **18.62** 0.47 M. **18.64** pH = 11.26; [HCN] = $1.8 \times$
10^{-3} M. **18.66** 5.72. **18.68** 4.29. **18.70** 1.4×10^{-5}. **18.72** 0.77%.
***18.74** pH = 10.7. **18.76** % ionization = 6.5%; pH = 2.01.
18.78 10.44. **18.80** $[H^+] = 6.0 \times 10^{-4}$ M; pH = 3.22. **18.82** 4.97.
18.84 9.00. **18.86** $\Delta[NH_4^+] = -0.01$ M; $\Delta[NH_3] = -0.03$ M.
18.88 -0.23 pH units. **18.90** Decreases by 0.06 pH units.
18.92 1 to 1. **18.94** 22 g $NaCHO_2$. **18.96** 0.96 g NH_4Cl.
18.98 Initial pH = 4.786, final pH = 4.766; 1.78 with no
buffer. **18.100** 2.50. **18.102** $[H^+] = [H_2PO_4^-] = 0.14$ M;
$[H_3PO_4] = 2.9$ M; $[HPO_4^{2-}] = 6.3 \times 10^{-8}$ M; 2.0×10^{-19} $M =$
$[PO_4^{3-}]$; pH = 0.85. **18.104** $[H^+] = [HC_4H_4O_6^-] = 0.0091$ M;
$[C_4H_4O_6^{2-}] = 4.3 \times 10^{-5}$ M. ***18.106** 0.095 $M = [H^+] =$
$[H_2PO_3^-]$; $[HPO_3^{2-}] = 2.6 \times 10^{-7}$ M; pH = 1.02. **18.108** pH =
10.13; $[HSO_3^-] = 1.3 \times 10^{-4}$ M; $[H_2SO_3] = 8.3 \times 10^{-13}$ M.
18.110 25.2 g. **18.112** 9.70. ***18.114** $[H_3PO_4] = 2.4 \times 10^{-18}$ M;
$[H_2PO_4^-] = 1.6 \times 10^{-7}$ M; $[HPO_4^{2-}] = 9.4 \times 10^{-2}$ M; pH =
12.97. **18.116** 8.23; cresol red would be a good indicator.
18.118 12.850. ***18.120** (a) 2.8724; (b) 4.5686; (c) 4.7447;
(d) 8.7236.

Chapter 19

Practice Exercises

1. (a) $K_{sp} = [Ba^{2+}][C_2O_4^{2-}]$; (b) $K_{sp} = [Ag^+]^3[PO_4^{3-}]$.
2. 3.2×10^{-10}. **3.** 3.98×10^{-8}. **4.** 1.0×10^{-10}. **5.** 3.9×10^{-8}.
6. (a) 7.1×10^{-7} M AgBr; (b) 1.3×10^{-4} M Ag_2CO_3. **7.** $4.2 \times$
10^{-16} M; 9.1×10^{-9} M in pure water. **8.** 1.3×10^{-35}. **9.** A precipitate will form. **10.** No precipitate will form. **11.** A pre-
cipitate of $PbSO_4$ is not expected. **12.** A precipitate of $PbCl_2$
is not expected. **13.** 2.9. **14.** pH = 5.40 – 6.13. **15.** $4.9 \times$
10^{-3} M; about 380 times greater than in pure water. **16.** 4.1
moles of NH_3.

Review Problems

19.14 (a) $CaF_2(s) \rightleftharpoons Ca^{2+} + 2F^-$ $K_{sp} = [Ca^{2+}][F^-]^2$
(b) $Ag_2CO_3(s) \rightleftharpoons 2Ag^+ + CO_3^{2-}$ $K_{sp} = [Ag^+]^2[CO_3^{2-}]$
(c) $PbSO_4(s) \rightleftharpoons Pb^{2+} + SO_4^{2-}$ $K_{sp} = [Pb^{2+}][SO_4^{2-}]$
(d) $Fe(OH)_3(s) \rightleftharpoons Fe^{3+} + 3OH^-$ $K_{sp} = [Fe^{3+}][OH^-]^3$
(e) $PbI_2(s) \rightleftharpoons Pb^{2+} + 2I^-$ $K_{sp} = [Pb^{2+}][I^-]^2$
(f) $Cu(OH)_2(s) \rightleftharpoons Cu^{2+} + 2OH^-$ $K_{sp} = [Cu^{2+}][OH^-]^2$

19.16 1.10×10^{-10}. **19.18** 8.0×10^{-7}. **19.20** 2.8×10^{-18}.
19.22 8.1×10^{-3} M. **19.24** 1.3×10^{-4} M. **19.26** 4.1×10^{-2} M
= molar solubility of LiF, 7.5×10^{-3} M = molar solubility of
BaF_2. **19.28** 2.8×10^{-18}. **19.30** 4.9×10^{-3} M. **19.32** (a) $4.4 \times$
10^{-4} M; (b) 9.5×10^{-6} M; (c) 9.5×10^{-7} M; (d) 6.3×10^{-7}
M. **19.34** 8.4×10^{-5} M. **19.36** (a) 3.0×10^{-11} M; (b) $1.2 \times$
10^{-6} M. **19.38** 1.5×10^{-3} M. **19.40** 1.3%; +1.8 pH units.

***19.42** 2.2 g solid $Fe(OH)_2$; 1.6×10^{-14} M. **19.44** 7.9×10^{-7}
M. **19.46** No precipitate will form. **19.48** (a) No precipitate
will form; (b) Yes, a precipitate will form. **19.50** 2.3×10^{-8}
M. **19.52** 0.045 M; 1.35. **19.54** 4.8–8.1.
19.56 (a) $Cu^{2+}(aq) + 4Cl^-(aq) \rightleftharpoons CuCl_4^{2-}(aq)$

$$K_{form} = \frac{[CuCl_4^{2-}]}{[Cu^{2+}][Cl^-]^4}$$

(b) $Ag^+(aq) + 2I^-(aq) \rightleftharpoons AgI_2^-(aq)$ $\quad K_{form} = \dfrac{[AgI_2^-]}{[Ag^+][I^-]^2}$

(c) $Cr^{3+}(aq) + 6NH_3(aq) \rightleftharpoons Cr(NH_3)_6^{3+}(aq)$

$$K_{form} = \frac{[Cr(NH_3)_6^{3+}]}{[Cr^{3+}][NH_3]^6}$$

19.58 (a) $Co(NH_3)_6^{3+}(aq) \rightleftharpoons Co^{3+}(aq) + 6NH_3(aq)$

$$K_{inst} = \frac{[Co^{3+}][NH_3]^6}{[Co(NH_3)_6^{3+}]}$$

(b) $HgI_4^{2-}(aq) \rightleftharpoons Hg^{2+}(aq) + 4I^-(aq)$ $\quad K_{inst} = \dfrac{[Hg^{2+}][I^-]^4}{[HgI_4^{2-}]}$

(c) $Fe(CN)_6^{4-}(aq) \rightleftharpoons Fe^{2+}(aq) + 6CN^-(aq)$

$$K_{inst} = \frac{[Fe^{2+}][CN^-]^6}{[Fe(CN)_6^{4-}]}$$

19.60 4.3×10^{-4}. **19.62** 450 g NaCN. **19.64** 1.9×10^{-4} g AgI.
19.66 3.9×10^{-10}.

Additional Exercises

19.68 7.09×10^{-12}. **19.70** 10.4. ***19.72** $[Fe^{2+}] = 0.001$ M,
$[Mn^{2+}] = 0.0995$ M; pH around 7. ***19.74** 7.2×10^{-2} ppm
Ca^{2+}. ***19.76** (a) $Mg(OH)_2(s) \rightleftharpoons Mg^{2+} + 2OH^-$; $NH_4^+ +$
$OH^- \rightleftharpoons NH_3 + H_2O$; (b) The NH_4^+ reacts with any OH^-
produced in the dissociation of $Mg(OH)_2$ thereby shifting
the equilibrium to the right; (c) Add 0.63 mol of NH_4Cl;
(d) 11.28. **19.78** 0.60 days, or about 14 hours. **19.80** (a) $1.8 \times$
10^2; (b) Much less stable. ***19.82** $[Pb^{2+}] = 8.40 \times 10^{-3}$ M,
$[I^-] = 9.70 \times 10^{-4}$ M and $[Br^-] = 1.58 \times 10^{-2}$ M.
***19.84** Add 0.63 mol of NH_4Cl. ***19.86** 0.78 g KF.
***19.88** 3.7 g HCl. ***19.90** About 7.

Chapter 20

Practice Exercises

1. Case (b). **2.** $w = -154$ L atm; $q = 154$ L atm. **3.** Energy is
added to the system in the form of work. **4.** -214.6 kJ; 1.14%.
5. (a) ΔS is negative; (b) ΔS is positive. **6.** (a) ΔS is negative;
(b) ΔS is negative. **7.** (a) ΔS is negative since there is a change
from a gas phase to a liquid phase. (b) ΔS is negative since
there are less gas molecules. (c) ΔS is positive since the particles go from an ordered, crystalline state to a more disordered,
aqueous state. **8.** (a) -229 J/K; (b) -120.9 J/K. **9.** 1482 kJ.
10. (a) -69.7 kJ/mol; (b) -120.1 kJ/mol. **11.** 788 kJ.
12. 341 °C. **13.** The reaction should be spontaneous.
14. We do not expect to see products. **15.** $+33$ kJ; -29 kJ.
The position of the equilibrium will be shifted more towards
products at the higher temperature. **16.** Reaction will proceed to the right. **17.** -33 kJ. **18.** 0.26. **19.** 3.84×10^4.

Review Problems

20.54 $+1000$ J, endothermic. **20.56** 121 J. **20.58** (a) $+24.58$
kJ, 19.6 kJ; (b) -178 kJ, -175 kJ; (c) $+847.6$ kJ, 847.6 kJ;
(d) $+65.029$ kJ, 65.029 kJ. **20.60** (a) -178 kJ, $\therefore$ favored;

(b) -311 kJ, $\therefore$ favored; (c) $+1084.3$ kJ, $\therefore$ not favored; (d) $+65.2$ kJ, $\therefore$ not favored; (e) $+64.2$ kJ, $\therefore$ not favored. **20.62** 1 in 16; 6 in 16. **20.64** (a) negative; (b) negative; (c) negative; (d) positive. **20.66** (a) -198.3 J/K, $\therefore$ not favored; (b) -332.3 J/K, $\therefore$ not favored; (c) $+92.6$ J/K, $\therefore$ favored; (d) $+14$ J/K, $\therefore$ favored; (e) $+159.6$ J/K, $\therefore$ favored. **20.68** (a) -52.8 J mol^{-1} K^{-1}; (b) -74.0 J mol^{-1} K^{-1}; (c) -90.1 J mol^{-1} K^{-1}; (d) -868.9 J mol^{-1} K^{-1}; (e) -318 J mol^{-1} K^{-1}. **20.70** -269.7 J/K. **20.72** -209 kJ/mol. **20.74** (a) -82.3 kJ; (b) -8.8 kJ; (c) $+70.7$ kJ; (d) -14.3 kJ; (e) -715.3 kJ. **20.76** $CaSO_4 \cdot \frac{1}{2}H_2O(s) + \frac{3}{2}H_2O(l) \rightarrow CaSO_4 \cdot 2H_2O(s)$; -4.7 kJ. **20.78** $+0.16$ kJ. **20.80** 1,299.8 kJ. **20.82** 60 °C. **20.84** 101 J mol^{-1} K^{-1}. **20.86** Yes, the reaction is spontaneous. **20.88** -91.3 kJ. **20.90** (a) 10^{263}; (b) 2.90×10^{-25}. **20.92** The system is not at equilibrium and must shift to the right. **20.94** 8.00×10^8; this is worth studying as a method for methane production. **20.96** If $\Delta G° = 0$, $K_c = 1$. The equilibrium will shift towards the reactants. **20.98** -687.6 kJ; 10^{121}. **20.100** 1.17×10^3 kJ/mol. **20.102** 517 kJ/mol. **20.104** 577.7 kJ/mol. **20.106** 308.0 kJ/mol. **20.108** 587 kJ/mol. **20.110** CF_4.

Chapter 21

Practice Exercises

1. anode: $Mg(s) \rightarrow Mg^{2+}(aq) + 2e^-$;
cathode: $Fe^{2+}(aq) + 2e^- \rightarrow Fe(s)$;
cell notation: $Mg(s) \,|\, Mg^{2+}(aq) \,\|\, Fe^{2+}(aq) \,|\, Fe(s)$

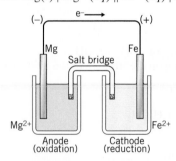

2. anode: $Al(s) \rightarrow Al^{3+}(aq) + 3e^-$;
cathode: $Pb^{2+}(aq) + 2e^- \rightarrow Pb(s)$;
overall: $3Pb^{2+}(aq) + 2Al(s) \rightarrow 2Al^{3+}(aq) + 3Pb(s)$.

3. -2.37 V; this agrees exactly with Table 21.1. **4.** $Br_2(aq) + H_2SO_3(aq) + H_2O \rightarrow 2Br^-(aq) + SO_4^{2-}(aq) + 4H^+(aq)$. **5.** The reaction will not be spontaneous in the direction shown. **6.** $NiO_2(s) + Fe(s) + 2H_2O \rightarrow Ni(OH)_2(s) + Fe(OH)_2(s)$; 1.37 V. **7.** $3MnO_4^-(aq) + 24H^+(aq) + 5Cr(s) \rightarrow 5Cr^{3+}(aq) + 3Mn^{2+}(aq) + 12H_2O$; 2.25 V. **8.** (a) nonspontaneous; (b) spontaneous. **9.** -3.72×10^2 kJ; -8.86×10^2 kJ. **10.** 2.7×10^{-16}; we do not expect much product to form. **11.** 1.04 V. **12.** 1.9×10^{-4} M; 6.9×10^{-13} M. **13.** 6.6×10^{-4} M. **14.** $Sn(s)$. **15.** 8.29×10^{-3} mol OH^-. **16.** 7.33 min. **17.** 3.67 A. **18.** $+0.0187$ M.

Review Problems

21.66 (a) anode: $Cd(s) \rightarrow Cd^{2+}(aq) + 2e^-$
cathode: $Au^{3+}(aq) + 3e^- \rightarrow Au(s)$
cell: $3Cd(s) + 2Au^{3+}(aq) \rightarrow 3Cd^{2+}(aq) + 2Au(s)$
(b) anode: $Pb(s) + HSO_4^-(aq) \rightarrow PbSO_4(s) + H^+(aq) + 2e^-$
cathode: $PbO_2(s) + 3H^+(aq) + HSO_4^-(aq) + 2e^- \rightarrow PbSO_4(s) + 2H_2O$

cell: $Pb(s) + PbO_2(s) + 2H^+(aq) + 2HSO_4^-(aq) \rightarrow 2PbSO_4(s) + 2H_2O$
(c) anode: $Cr(s) \rightarrow Cr^{3+}(aq) + 3e^-$
cathode: $Cu^{2+}(aq) + 2e^- \rightarrow Cu(s)$
cell: $2Cr(s) + 3Cu^{2+}(aq) \rightarrow 2Cr^{3+}(aq) + 3Cu(s)$

21.68 (a) $Fe(s) \,|\, Fe^{2+}(aq) \,\|\, Cd^{2+}(aq) \,|\, Cd(s)$;
(b) $Pt(s) \,|\, Br^-(aq), Br_2(g) \,\|\, Cl_2(aq), Cl^-(aq) \,|\, Pt(s)$;
(c) $Ag(s) \,|\, Ag^+(aq) \,\|\, Au^{3+}(aq) \,|\, Au(s)$. **21.70** (a) $Sn(s)$;
(b) $Br^-(aq)$; (c) $Zn(s)$; (d) $I^-(aq)$. **21.72** (a) 0.04 V;
(b) -0.29 V; (c) 0.62 V. **21.74** (a) spontaneous; (b) not spontaneous; (c) spontaneous. **21.76** 0.74 V.
21.78 (a) $Zn(s) \,|\, Zn^{2+}(aq) \,\|\, Co^{2+}(aq) \,|\, Co(s)$; 0.48 V;
(b) $Mg(s) \,|\, Mg^{2+}(aq) \,\|\, Ni^{2+}(aq) \,|\, Ni(s)$; 2.12 V;
(c) $Sn(s) \,|\, Sn^{2+}(aq) \,\|\, Au^{3+}(aq) \,|\, Au(s)$; 1.56 V.
21.80 $BrO_3^-(aq) + 6I^-(aq) + 6H^+(aq) \rightarrow 3I_2(s) + Br^-(aq) + 3H_2O$; 0.90 V. **21.82** $4HOCl(aq) + 2H^+(aq) + S_2O_3^{2-}(aq) \rightarrow 2Cl_2(g) + H_2O + 2H_2SO_3(aq)$. **21.84** Not spontaneous. **21.86** 1.0×10^2 kJ. **21.88** (a) 0.54 V; (b) -5.2×10^2 kJ; (c) 2.1×10^{91}. **21.90** 0.31. **21.92** 2.07 V. **21.94** 2.86×10^{-15} M. *21.96** 1.96×10^{-15}. **21.98** (a) 2.40×10^3 C; (b) 1.20×10^4 C; (c) 3.24×10^4 C. **21.100** (a) 0.40 mol e^-; (b) 0.70 mol e^-; (c) 4.50 mol e^-; (d) 5.0×10^{-2} mol e^-. **21.102** 0.0371 mol Cr^{3+}. **21.104** 2.68 g $Fe(OH)_2$. **21.106** 51.5 hr. **21.108** 66.2 A. **21.110** 40.2 mL HCl. **21.112** 233 mL HCl. **21.114** Cathode: $H_2(g) + 2OH^-(aq)$; anode: $O_2(g) + 4H^+(aq)$; Overall reaction: $2H_2O \rightleftharpoons 2H_2(g) + O_2(g)$. **21.116** (a) $2H_2O \rightleftharpoons O_2(g) + 4H^+(aq) + 4e^-$; (b) $2Br^- \rightleftharpoons Br_2 + 2e^-$; (c) $2Br^- \rightleftharpoons Br_2 + 2e^-$. **21.118** Cathode: $Cu(s)$; anode: Br_2; Overall reaction: $Cu^{2+} + 2Br^- \rightleftharpoons Br_2 + Cu(s)$.

Chapter 22

Practice Exercises

1. $^{226}_{88}Ra \rightarrow ^{222}_{86}Rn + ^4_2He + ^0_0\gamma$. **2.** $^{90}_{38}Sr \rightarrow ^{90}_{39}Y + ^0_{-1}e$. **3.** 1.10×10^5 atoms Rn-222. **4.** 100 units (1 sig. fig.).

Review Problems

22.43 1.11×10^{-11} g. **22.45** 3.18 ng. **22.47** 1.8×10^{-13} J per nucleon. **22.49** (a) $^{211}_{83}Bi$; (b) $^{177}_{72}Hf$; (c) $^{216}_{84}Po$; (d) $^{19}_9F$. **22.51** (a) $^{242}_{94}Pu \rightarrow ^4_2He + ^{238}_{92}U$; (b) $^{28}_{12}Mg \rightarrow ^0_{-1}e + ^{28}_{13}Al$; (c) $^{26}_{14}Si \rightarrow ^0_1e + ^{26}_{13}Al$; (d) $^{37}_{18}Ar + ^0_{-1}e \rightarrow ^{37}_{17}Cl$. **22.53** (a) $^{261}_{102}No$; (b) $^{211}_{82}Pb$; (c) $^{141}_{61}Pm$; (d) $^{179}_{74}W$. **22.55** ^1_0n. **22.57** Positron emission, because this produces a product having a higher neutron-to-proton ratio: $^{38}_{19}K \rightarrow ^0_1e + ^{38}_{18}Ar$. **22.59** 0.0469 mg. **22.61** $^{53}_{24}Cr^* ; ^{51}_{23}V + ^2_1H \rightarrow ^{53}_{24}Cr^* \rightarrow ^1_1p + ^{52}_{23}V$. **22.63** $^{80}_{35}Br$. **22.65** $^{55}_{26}Fe; ^{55}_{25}Mn + ^1_1p \rightarrow ^1_0n + ^{55}_{26}Fe$. **22.67** Zn-70. **22.69** 6.3 m. **22.71** 6.6 m. **22.73** 2.4×10^7 Bq, 6.5×10^2 μCi. **22.75** 1.0×10^{-6}, 6.9×10^5 s. **22.77** 3.2%, $BaCl_2$. **22.79** 2.84×10^{-3} Ci/g. **22.81** 1.3×10^9 years old. **22.83** 27,000 years ago. **22.85** $^{235}_{92}U + ^1_0n \rightarrow ^{94}_{38}Sr + ^{140}_{54}Xe + 2^1_0n$.

Chapter 23

Practice Exercises

1. 5570 K. **2.** $[Ag(S_2O_3)_2]^{3-}$; $(NH_4)_3[Ag(S_2O_3)_2]$. **3.** $AlCl_3 \cdot 6H_2O$; $[Al(H_2O)_6]^{3+}$. **4.** (a) potassium hexacyanoferrate(III); (b) dichlorobis(ethylenediamine)chromium(III) sulfate; **5.** (a) $[SnCl_6]^{2-}$; (b) $(NH_4)_2[Fe(CN)_4(H_2O)_2]$. **6.** (a) six; (b) six; (c) six; (d) six.

Review Problems

23.89 734 K. **23.91** (a) Bi_2O_5; (b) PbS; (c) PbI_2.

23.93 (a) SnO; (b) SnS. **23.95** (a) $CaCl_2$; (b) BeF_2.
23.97 (a) HgS; (b) Ag_2S. **23.99** (a) HgS; (b) Ag_2S.
23.101 -3, $[Fe(CN)_6]^{3-}$. **23.103** $[CoCl_2(en)_2]^+$.
23.105 (a) oxalato; (b) sulfido (thio); (c) chloro;
(d) dimethylamine. **23.107** (a) hexaamminenickel(II) ion;
(b) triamminetrichlorochromate(II) ion; (c) hexanitrocobal-
tate(III) ion; (d) diamminetetracyanomanganate(II) ion;
(e) trioxalatoferrate(III) ion or trisoxalatoferrate(III) ion.
23.109 (a) $[Fe(CN)_2(H_2O)_4]^+$; (b) $Ni(C_2O_4)(NH_3)_4$;
(c) $[Fe(CN)_4(H_2O)_2]^-$; (d) $K_3[Mn(SCN)_6]$; (e) $[CuCl_4]^{2-}$.
23.111 Six, $+2$.

23.113

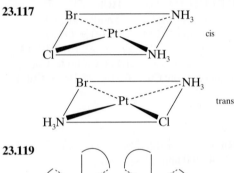

(The curved lines represent —$CH_2\overset{\displaystyle O}{\overset{\|}{C}}$— groups.)

23.115 Since both are the *cis* isomer, they are identical. One
can be superimposed on the other after simple rotation.

23.117

Br——————NH₃
    Pt   cis
Cl——————NH₃

Br——————NH₃
    Pt   trans
H₃N——————Cl

23.119

Co   Co

23.121 (a) $Cr(H_2O)_6^{3+}$; (b) $Cr(en)_3^{3+}$. **23.123** $[Cr(CN)_6]^{3-}$.
23.125 (a) The value of Δ increases down a group. Therefore, we choose: $[RuCl(NH_3)_5]^{3+}$; (b) The value of Δ increases with oxidation state of the metal. Therefore, we
choose: $[Ru(NH_3)_6]^{3+}$. **23.127** This is the one with the
strongest field ligand, since Co^{2+} is a d^7 ion: $[CoA_6]^{3+}$.
23.129 High-spin; cannot be diamagnetic.

Chapter 24

Review Problems

24.89 (a) $AlP(s) + 3H_2O \rightarrow Al(OH)_3(s) + PH_3(g)$
(b) $Mg_2C(s) + 4H_2O(l) \rightarrow CH_4(g) + 2Mg(OH)_2(s)$
(c) $FeS(s) + 2HCl(aq) \rightarrow FeCl_2(aq) + H_2S(aq)$
(d) $MgSe(s) + H_2SO_4(aq) \rightarrow H_2Se(aq) + MgSO_4(aq)$

24.91 (a) $2CO + O_2 \rightarrow 2CO_2$
(b) $2C_2H_6 + 7O_2 \rightarrow 4CO_2 + 6H_2O$
(c) $P_4O_6 + 2O_2 \rightarrow P_4O_{10}$
(d) $4NH_3 + 5O_2 \rightarrow 4NO + 6H_2O$

24.93 1.57×10^{18}; to the right; reaction kinetics are too slow to
achieve equilibrium. **24.95** -57.9 kJ; If we increase the temper-

ature, effectively adding heat (a product), Le Châtelier's Principle states that the equilibrium will shift to the left and N_2O_4
will dissociate. **24.97** (a) IO_3^- is trigonal pyramidal; (b) ClO_2^-
is bent; (c) IO_6^- is octahedral; (d) ClO_4^- is tetrahedral.

24.99 (a) trigonal pyramidal   (d) planar triangular

:Cl—P—Cl: :Br—B—Br:
    | |
  :Cl: :Br:

(b) T-shaped   (e) square planar

:F—Br—F:
   |
  :F:

$$\left[\begin{array}{c} :F: \\ :F—Br—F: \\ :F: \end{array}\right]^-$$

(c) octahedral

$$\left[\begin{array}{c} :F: \\ F—Si—F \\ :F: \end{array}\right]^{2-}$$

Chapter 25

Practice Exercises

1. (a) 3-methylhexane; (b) 4-ethyl-2,3-dimethylheptane;
(c) 5-ethyl-2,4,6-trimethyloctane.

2.

(a) $CH_3-\overset{\displaystyle O}{\overset{\|}{C}}H$ or $CH_3-\overset{\displaystyle O}{\overset{\|}{C}}OH$
(depending on the strength of the oxidizing agent)

(b) $CH_3CH_2\overset{\displaystyle O}{\overset{\|}{C}}CH_2CH_3$

Review Problems

25.82

(a) H—C—N—H   (d) H—N—O—H
(with H's)

(b) Br—C—Br   (e) H—C≡C—H
(with H's)

(c) H—C—Cl   (f) H—N—N—H
(with Cl's)

25.84 (a) alkene; (b) alcohol; (c) ester; (d) carboxylic acid;
(e) amine; (f) alcohol. **25.86** The saturated compounds are b, e,
and f. **25.88** (a) amine; (b) amine; (c) amide; (d) amine, ketone.
25.90 (a) identical; (b) identical; (c) unrelated; (d) isomers;
(e) identical; (f) identical; (g) isomers. **25.92** (a) pentane;
(b) 2-methylpentane; (c) 2,4-dimethylhexane

25.94 (a) no isomers

(b)

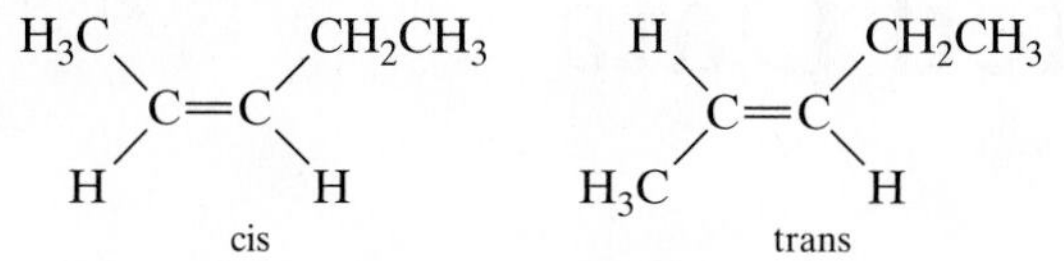

cis   trans

(c)

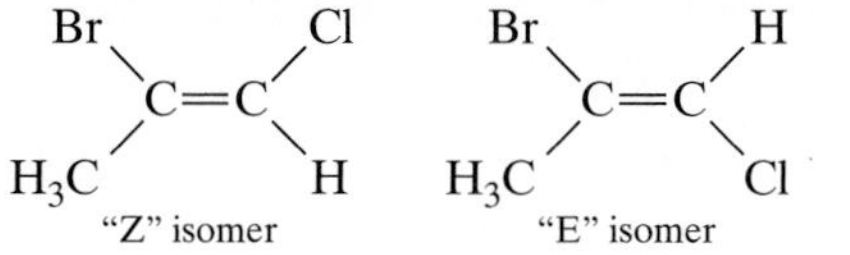

"Z" isomer   "E" isomer

Two isomers exist here, but not correctly called *cis* and *trans* isomers.

25.96 (a) CH_3CH_3; (b) $ClCH_2CH_2Cl$; (c) $BrCH_2CH_2Br$; (d) CH_3CH_2Cl; (e) CH_3CH_2Br; (f) CH_3CH_2OH.

25.98

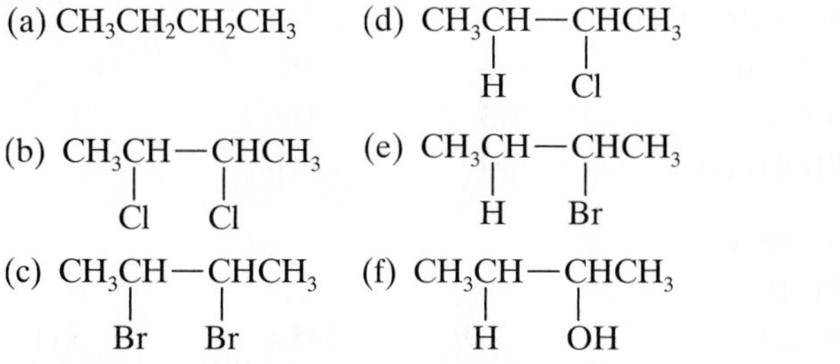

(a) $CH_3CH_2CH_2CH_3$ (d) $CH_3\underset{\underset{H}{|}}{CH}-\underset{\underset{Cl}{|}}{CH}CH_3$

(b) $CH_3\underset{\underset{Cl}{|}}{CH}-\underset{\underset{Cl}{|}}{CH}CH_3$ (e) $CH_3\underset{\underset{H}{|}}{CH}-\underset{\underset{Br}{|}}{CH}CH_3$

(c) $CH_3\underset{\underset{Br}{|}}{CH}-\underset{\underset{Br}{|}}{CH}CH_3$ (f) $CH_3\underset{\underset{H}{|}}{CH}-\underset{\underset{OH}{|}}{CH}CH_3$

25.100 This sort of reaction would disrupt the π-delocalization of the benzene ring. The subsequent loss of resonance energy would not be favorable.

25.102 CH_3OH IUPAC name = methanol; common name = methyl alcohol;

CH_3CH_2OH IUPAC name = ethanol; common name = ethyl alcohol;

$CH_3CH_2CH_2OH$ IUPAC name = 1-propanol; common name = propyl alcohol;

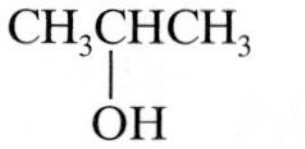

$CH_3\underset{\underset{OH}{|}}{CH}CH_3$ IUPAC name = 2-propanol; common name = isopropyl alcohol.

25.104 $CH_3CH_2CH_2-O-CH_3$ methyl propyl ether
$CH_3CH_2-O-CH_2CH_3$ diethyl ether
$(CH_3)_2CH-O-CH_3$ methyl 2-propyl ether

***25.106** (a)

(b)

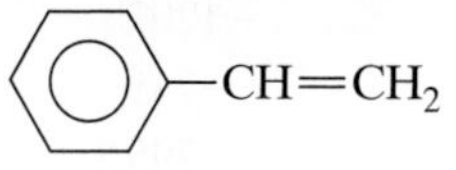

$-CH=CH_2$

(c)

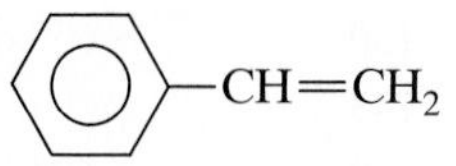

$-CH=CH_2$

25.108 (a)

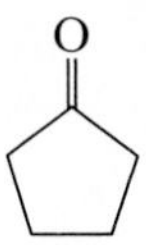

(b)

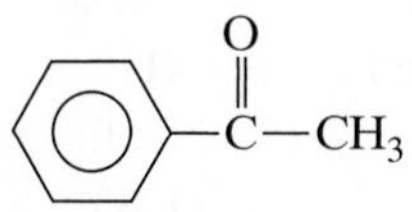

$-\overset{\overset{O}{\|}}{C}-CH_3$

(c)

$-CH_2-\overset{\overset{O}{\|}}{C}-H$

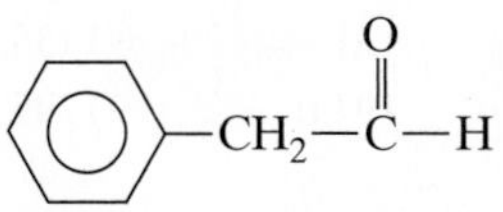

***25.110** The elimination of water can result in a C=C double bond in two locations:
CH_2=$CHCH_2CH_3$ CH_3CH=$CHCH_3$
1-butene   2-butene

25.112 The aldehyde is more easily oxidized. The product is:

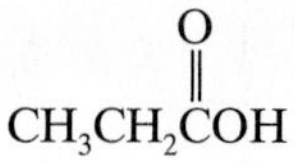

$$CH_3CH_2\overset{\overset{O}{\|}}{C}OH$$

25.114 (a) $CH_3CH_2CO_2H$; (b) $CH_3CH_2CO_2H + CH_3OH$; (c) $Na^+ + CH_3CH_2CH_2CO_2^- + H_2O$. ***25.116** $CH_3CO_2H + CH_3CH_2NHCH_2CH_3$

25.118

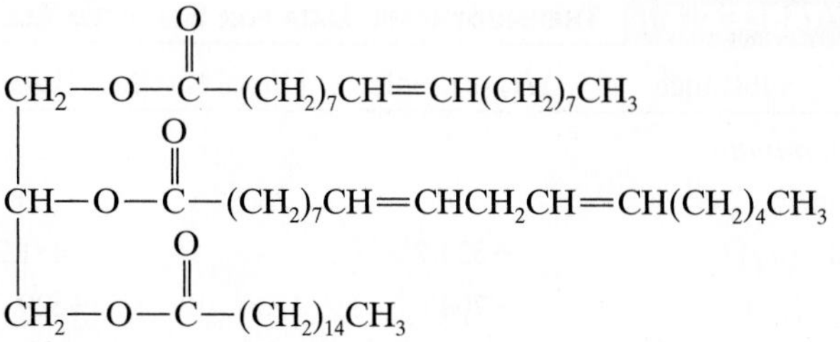

25.120

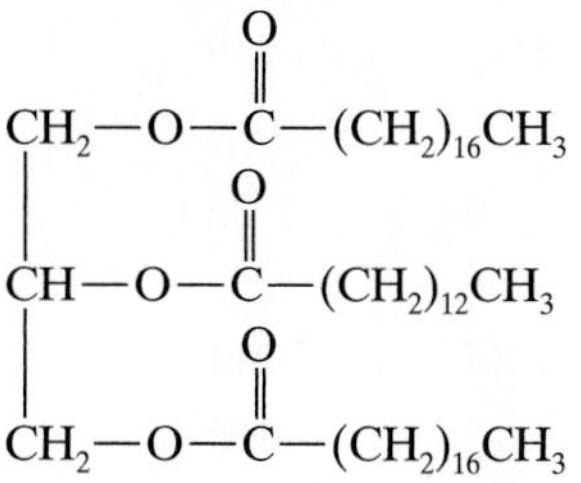

25.122 Hydrophobic sites are composed of fatty acid units. Hydrophilic sites are composed of charged units.

25.124

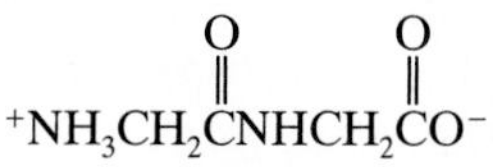

$^+NH_3CH_2\overset{\overset{O}{\|}}{C}NHCH_2\overset{\overset{O}{\|}}{C}O^-$

25.126

$^+NH_3\underset{\underset{CH_2C_6H_5}{|}}{CH}\overset{\overset{O}{\|}}{C}-NHCH_2\overset{\overset{O}{\|}}{C}O^-$ $^+NH_3CH_2\overset{\overset{O}{\|}}{C}-NH-\underset{\underset{CH_2C_6H_5}{|}}{CH}\overset{\overset{O}{\|}}{C}O^-$

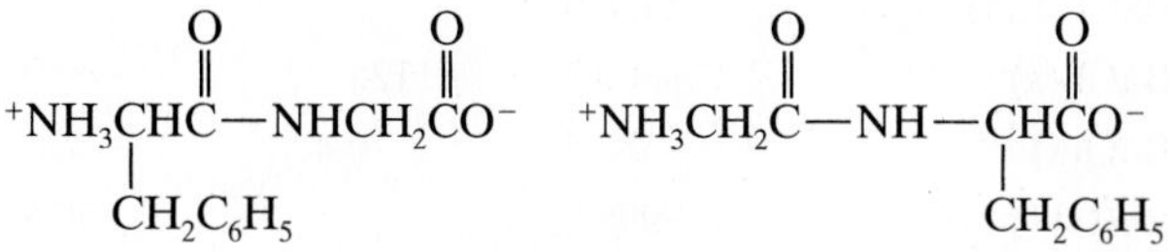

Tables of Selected Data

TABLE C.1	THERMODYNAMIC DATA FOR SELECTED ELEMENTS, COMPOUNDS, AND IONS (25 °C)						
Substance	ΔH_f° (kJ/mol)	S° (J/mol K)	ΔG_f° (kJ/mol)	Substance	ΔH_f° (kJ/mol)	S° (J/mol K)	ΔG_f° (kJ/mol)
Aluminum				*Boron* (*continued*)			
Al(s)	0	28.3	0	$B_2H_6(g)$	+36	232	+87
$Al^{3+}(aq)$	−524.7		−481.2	$B_2O_3(s)$	−1273	53.8	−1194
$AlCl_3(s)$	−704	110.7	−629	$B(OH)_3(s)$	−1094	88.8	−969
$Al_2O_3(s)$	−1676	51.0	−1576.4	*Bromine*			
$Al_2(SO_4)_3(s)$	−3441	239	−3100	$Br_2(l)$	0	152.2	0
Arsenic				$Br_2(g)$	+30.9	245.4	3.11
As(s)	0	35.1	0	$HBr(g)$	−36	198.5	53.1
$AsH_3(g)$	+66.4	223	+68.9	$Br^-(aq)$	−121.55	82.4	−103.96
$As_4O_6(s)$	−1314	214	−1153	*Cadmium*			
$As_2O_5(s)$	−925	105	−782	Cd(s)	0	51.8	0
$H_3AsO_3(aq)$	−742.2			$Cd^{2+}(aq)$	−75.90	−73.2	−77.61
$H_3AsO_4(aq)$	−902.5			$CdCl_2(s)$	−392	115	−344
Barium				CdO(s)	−258.2	54.8	−228.4
Ba(s)	0	66.9	0	CdS(s)	−162	64.9	−156
$Ba^{2+}(aq)$	−537.6	9.6	−560.8	$CdSO_4(s)$	−933.5	123	−822.6
$BaCO_3(s)$	−1219	112	−1139	*Calcium*			
$BaCrO_4(s)$	−1428.0			Ca(s)	0	41.4	0
$BaCl_2(s)$	−860.2	125	−810.8	$Ca^{2+}(aq)$	−542.83	−53.1	−553.58
BaO(s)	−553.5	70.4	−525.1	$CaCO_3(s)$	−1207	92.9	−1128.8
$Ba(OH)_2(s)$	−998.22	−8	−875.3	$CaF_2(s)$	−741	80.3	−1166
$Ba(NO_3)_2(s)$	−992	214	−795	$CaCl_2(s)$	−795.8	114	−750.2
$BaSO_4(s)$	−1465	132	−1353	$CaBr_2(s)$	−682.8	130	−663.6
Beryllium				$CaI_2(s)$	−535.9	143	
Be(s)	0	9.50	0	CaO(s)	−635.5	40	−604.2
$BeCl_2(s)$	−468.6	89.9	−426.3	$Ca(OH)_2(s)$	−986.6	76.1	−896.76
BeO(s)	−611	14	−582	$Ca_3(PO_4)_2(s)$	−4119	241	−3852
Bismuth				$CaSO_3(s)$	−1156		
Bi(s)	0	56.9	0	$CaSO_4(s)$	−1433	107	−1320.3
$BiCl_3(s)$	−379	177	−315	$CaSO_4 \cdot \frac{1}{2}H_2O(s)$	−1573	131	−1435.2
$Bi_2O_3(s)$	−576	151	−497	$CaSO_4 \cdot 2H_2O(s)$	−2020	194.0	−1795.7
Boron				*Carbon*			
B(s)	0	5.87	0	C(s, graphite)	0	5.69	0
				C(s, diamond)	+1.88	2.4	+2.9
$BCl_3(g)$	−404	290	−389	$CCl_4(l)$	−134	214.4	−65.3

(continued)

Table C.1 • **A-19**

TABLE C.1 | THERMODYNAMIC DATA FOR SELECTED ELEMENTS, COMPOUNDS, AND IONS (25 °C) (*Continued*)

Substance	ΔH_f° (kJ/mol)	S° (J/mol K)	ΔG_f° (kJ/mol)	Substance	ΔH_f° (kJ/mol)	S° (J/mol K)	ΔG_f° (kJ/mol)
Carbon (*continued*)				*Cobalt* (*continued*)			
$CO(g)$	−110	197.9	−137.3	$Co^{2+}(aq)$	−59.4	−110	−53.6
$CO_2(g)$	−394	213.6	−394.4	$CoCl_2(s)$	−325.5	106	−282.4
$CO_2(aq)$	−413.8	117.6	−385.98	$Co(NO_3)_2(s)$	−422.2	192	−230.5
$H_2CO_3(aq)$	−699.65	187.4	−623.08	$CoO(s)$	−237.9	53.0	−214.2
$HCO_3^-(aq)$	−691.99	91.2	−586.77	$CoS(s)$	−80.8	67.4	−82.8
$CO_3^{2-}(aq)$	−677.14	−56.9	−527.81				
$CS_2(l)$	+89.5	151.3	+65.3	*Copper*			
$CS_2(g)$	+117	237.7	+67.2	$Cu(s)$	0	33.15	0
$HCN(g)$	+135.1	201.7	+124.7	$Cu^{2+}(aq)$	+64.77	−99.6	+65.49
$CN^-(aq)$	+150.6	94.1	+172.4	$CuCl(s)$	−137.2	86.2	−119.87
$CH_4(g)$	−74.9	186.2	−50.79	$CuCl_2(s)$	−172	119	−131
$C_2H_2(g)$	+227	200.8	+209	$Cu_2O(s)$	−168.6	93.1	−146.0
$C_2H_4(g)$	+51.9	219.8	+68.12	$CuO(s)$	−155	42.6	−127
$C_2H_6(g)$	−84.5	229.5	−32.9	$Cu_2S(s)$	−79.5	121	−86.2
$C_3H_8(g)$	−104	269.9	−23	$CuS(s)$	−53.1	66.5	−53.6
$C_4H_{10}(g)$	−126	310.2	−17.0	$CuSO_4(s)$	−771.4	109	−661.8
$C_6H_6(l)$	+49.0	173.3	+124.3	$CuSO_4 \cdot 5H_2O(s)$	−2279.7	300.4	−1879.7
$CH_3OH(l)$	−238	126.8	−166.2	*Fluorine*			
$C_2H_5OH(l)$	−278	161	−174.8	$F_2(g)$	0	202.7	0
$HCHO_2(g)$	−363	251	+335	$F^-(aq)$	−332.6	−13.8	−278.8
$HC_2H_3O_2(l)$	−487.0	160	−392.5	$HF(g)$	−271	173.5	−273
$HCHO(g)$	−108.6	218.8	−102.5	*Gold*			
$CH_3CHO(g)$	−167	250	−129	$Au(s)$	0	47.7	0
$(CH_3)_2CO(l)$	−248.1	200.4	−155.4	$Au_2O_3(s)$	+80.8	125	+163
$C_6H_5CO_2H(s)$	−385.1	167.6	−245.3	$AuCl_3(s)$	−118	148	−48.5
$CO(NH_2)_2(s)$	−333.5	104.6	−197.2	*Hydrogen*			
$CO(NH_2)_2(aq)$	−319.2	173.8	−203.8	$H_2(g)$	0	130.6	0
$CH_2(NH_2)CO_2H(s)$	−532.9	103.5	−373.4	$H_2O(l)$	−285.9	69.96	−237.2
Chlorine				$H_2O(g)$	−241.8	188.7	−228.6
$Cl_2(g)$	0	223.0	0	$H_2O_2(l)$	−187.8	109.6	−120.3
$Cl^-(aq)$	−167.2	56.5	−131.2	$H_2Se(g)$	+76	219	+62.3
$HCl(g)$	−92.5	186.7	−95.27	$H_2Te(g)$	+154	234	+138
$HCl(aq)$	−167.2	56.5	−131.2	*Iodine*			
$HClO(aq)$	−131.3	106.8	−80.21	$I_2(s)$	0	116.1	0
Chromium				$I_2(g)$	+62.4	260.7	+19.3
$Cr(s)$	0	23.8	0	$HI(g)$	+26	206	+1.30
$Cr^{3+}(aq)$	−232			*Iron*			
$CrCl_2(s)$	−326	115	−282	$Fe(s)$	0	27	0
$CrCl_3(s)$	−563.2	126	−493.7	$Fe^{2+}(aq)$	−89.1	−137.7	−78.9
$Cr_2O_3(s)$	−1141	81.2	−1059	$Fe^{3+}(aq)$	−48.5	−315.9	−4.7
$CrO_3(s)$	−585.8	72.0	−506.2	$Fe_2O_3(s)$	−822.2	90.0	−741.0
$(NH_4)_2Cr_2O_7(s)$	−1807			$Fe_3O_4(s)$	−1118.4	146.4	−1015.4
$K_2Cr_2O_7(s)$	−2033.01			$FeS(s)$	−100.0	60.3	−100.4
Cobalt				$FeS_2(s)$	−178.2	52.9	−166.9
$Co(s)$	0	30.0	0				

TABLE C.1	THERMODYNAMIC DATA FOR SELECTED ELEMENTS, COMPOUNDS, AND IONS (25 °C) (Continued)						
Substance	ΔH_f° (kJ/mol)	S° (J/mol K)	ΔG_f° (kJ/mol)	Substance	ΔH_f° (kJ/mol)	S° (J/mol K)	ΔG_f° (kJ/mol)
Lead				*Nickel*			
$Pb(s)$	0	64.8	0	$Ni(s)$	0	30	0
$Pb^{2+}(aq)$	−1.7	10.5	−24.4	$NiCl_2(s)$	−305	97.5	−259
$PbCl_2(s)$	−359.4	136	−314.1	$NiO(s)$	−244	38	−216
$PbO(s)$	−217.3	68.7	−187.9	$NiO_2(s)$			−199
$PbO_2(s)$	−277	68.6	−219	$NiSO_4(s)$	−891.2	77.8	−773.6
$Pb(OH)_2(s)$	−515.9	88	−420.9	$NiCO_3(s)$	−664.0	91.6	−615.0
$PbS(s)$	−100	91.2	−98.7	$Ni(CO)_4(g)$	−220	399	−567.4
$PbSO_4(s)$	−920.1	149	−811.3	*Nitrogen*			
Lithium				$N_2(g)$	0	191.5	0
$Li(s)$	0	28.4	0	$NH_3(g)$	−46.0	192.5	−16.7
$Li^+(aq)$	−278.6	10.3		$NH_4^+(aq)$	−132.5	113	−79.37
$LiF(s)$	−611.7	35.7	−583.3	$N_2H_4(g)$	+95.40	238.4	+159.3
$LiCl(s)$	−408	59.29	−383.7	$N_2H_4(l)$	+50.6	121.2	+149.4
$LiBr(s)$	−350.3	66.9	−338.87	$NH_4Cl(s)$	−314.4	94.6	−203.9
$Li_2O(s)$	−596.5	37.9	−560.5	$NO(g)$	+90.4	210.6	+86.69
$Li_3N(s)$	−199	37.7	−155.4	$NO_2(g)$	+34	240.5	+51.84
Magnesium				$N_2O(g)$	+81.5	220.0	+103.6
$Mg(s)$	0	32.5	0	$N_2O_4(g)$	+9.16	304	+98.28
$Mg^{2+}(aq)$	−466.9	−138.1	−454.8	$N_2O_5(g)$	+11	356	+115
$MgCO_3(s)$	−1113	65.7	−1029	$HNO_3(l)$	−174.1	155.6	−79.9
$MgF_2(s)$	−1124	79.9	−1056	$NO_3^-(aq)$	−205.0	146.4	−108.74
$MgCl_2(s)$	−641.8	89.5	−592.5	*Oxygen*			
$MgCl_2 \cdot 2H_2O(s)$	−1280	180	−1118	$O_2(g)$	0	205.0	0
$Mg_3N_2(s)$	−463.2	87.9	−411	$O_3(g)$	+143	238.8	+163
$MgO(s)$	−601.7	26.9	−569.4	$OH^-(aq)$	−230.0	−10.75	−157.24
$Mg(OH)_2(s)$	−924.7	63.1	−833.9	*Phosphorus*			
Manganese				$P(s, white)$	0	41.09	0
$Mn(s)$	0	32.0	0	$P_4(g)$	+314.6	163.2	+278.3
$Mn^{2+}(aq)$	−223	−74.9	−228	$PCl_3(g)$	−287.0	311.8	−267.8
$MnO_4^-(aq)$	−542.7	191	−449.4	$PCl_5(g)$	−374.9	364.6	−305.0
$KMnO_4(s)$	−813.4	171.71	−713.8	$PH_3(g)$	+5.4	210.2	+12.9
$MnO(s)$	−385	60.2	−363	$P_4O_6(s)$	−1640		
$Mn_2O_3(s)$	−959.8	110	−882.0	$POCl_3(g)$	−1109.7	646.5	−1019
$MnO_2(s)$	−520.9	53.1	−466.1	$POCl_3(l)$	−1186	26.36	−1035
$Mn_3O_4(s)$	−1387	149	−1280	$P_4O_{10}(s)$	−2984	228.9	−2698
$MnSO_4(s)$	−1064	112	−956	$H_3PO_4(s)$	−1279	110.5	−1119
Mercury				*Potassium*			
$Hg(l)$	0	76.1	0	$K(s)$	0	64.18	0
$Hg(g)$	+61.32	175	+31.8	$K^+(aq)$	−252.4	102.5	−283.3
$Hg_2Cl_2(s)$	−265.2	192.5	−210.8	$KF(s)$	−567.3	66.6	−537.8
$HgCl_2(s)$	−224.3	146.0	−178.6	$KCl(s)$	−436.8	82.59	−408.3
$HgO(s)$	−90.83	70.3	−58.54	$KBr(s)$	−393.8	95.9	−380.7
$HgS(s, red)$	−58.2	82.4	−50.6	$KI(s)$	−327.9	106.3	−324.9
				$KOH(s)$	−424.8	78.9	−379.1
				$K_2O(s)$	−361	98.3	−322
				$K_2SO_4(s)$	−1433.7	176	−1316.4

Table C.2 • **A-21**

TABLE C.1	THERMODYNAMIC DATA FOR SELECTED ELEMENTS, COMPOUNDS, AND IONS (25 °C) (Continued)						
Substance	ΔH_f° (kJ/mol)	S° (J/mol K)	ΔG_f° (kJ/mol)	Substance	ΔH_f° (kJ/mol)	S° (J/mol K)	ΔG_f° (kJ/mol)
Silicon				*Sodium* (*continued*)			
Si(s)	0	19	0	Na_2SO_4(s)	−1384.49	149.49	−1266.83
SiH_4(g)	+33	205	+52.3	*Sulfur*			
SiO_2(s,alpha)	−910.0	41.8	−856	S(s, rhombic)	0	31.8	0
Silver				SO_2(g)	−297	248	−300
Ag(s)	0	42.55	0	SO_3(g)	−396	256	−370
Ag^+(aq)	+105.58	72.68	+77.11	H_2S(g)	−20.6	206	−33.6
AgCl(s)	−127.1	96.2	−109.8	H_2SO_4(l)	−813.8	157	−689.9
AgBr(s)	−100.4	107.1	−96.9	H_2SO_4(aq)	−909.3	20.1	−744.5
$AgNO_3$(s)	−124	141	−32	SF_6(g)	−1209	292	−1105
Ag_2O(s)	−31.1	121.3	−11.2	*Tin*			
Sodium				Sn(s,white)	0	51.6	0
Na(s)	0	51.0	0	Sn^{2+}(aq)	−8.8	−17	−27.2
Na^+(aq)	−240.12	59.0	−261.91	$SnCl_4$(l)	−511.3	258.6	−440.2
NaF(s)	−571	51.5	−545	SnO(s)	−285.8	56.5	−256.9
NaCl(s)	−413	72.38	−384.0	SnO_2(s)	−580.7	52.3	−519.6
NaBr(s)	−360	83.7	−349	*Zinc*			
NaI(s)	−288	91.2	−286	Zn(s)	0	41.6	0
$NaHCO_3$(s)	−947.7	102	−851.9	Zn^{2+}(aq)	−153.9	−112.1	−147.06
Na_2CO_3(s)	−1131	136	−1048	$ZnCl_2$(s)	−415.1	111	−369.4
Na_2O_2(s)	−510.9	94.6	−447.7	ZnO(s)	−348.3	43.6	−318.3
Na_2O(s)	−510	72.8	−376	ZnS(s)	−205.6	57.7	−201.3
NaOH(s)	−426.8	64.18	−382	$ZnSO_4$(s)	−982.8	120	−874.5

TABLE C.2	HEATS OF FORMATION OF GASEOUS ATOMS FROM ELEMENTS IN THEIR STANDARD STATES*		
Element	ΔH_f° (kJ mol^{-1})	Element	ΔH_f° (kJ mol^{-1})
Group IA		*Group IVA*	
H	217.89	C	716.67
Li	161.5	Si	450
Na	108.2	*Group VA*	
K	89.62	N	472.68
Rb	82.0	P	332.2
Cs	78.2	*Group VIA*	
Group IIA		O	249.17
Be	324.3	S	276.98
Mg	146.4	*Group VIIA*	
Ca	178.2	F	79.14
Sr	163.6	Cl	121.47
Ba	177.8	Br	112.38
Group IIIA		I	107.48
B	560		
Al	329.7		

*All values in this table are positive because forming the gaseous atoms from the elements is endothermic; it involves bond breaking.

TABLE C.3	AVERAGE BOND ENERGIES		
Bond	Bond Energy (kJ mol^{-1})	Bond	Bond Energy (kJ mol^{-1})
C—C	348	C—Br	276
C=C	612	C—I	238
C≡C	960	H—H	436
C—H	412	H—F	565
C—N	305	H—Cl	431
C=N	613	H—Br	366
C≡N	890	H—I	299
C—O	360	H—N	388
C=O	743	H—O	463
C—F	484	H—S	338
C—Cl	338	H—Si	376

TABLE C.4	VAPOR PRESSURE OF WATER AS A FUNCTION OF TEMPERATURE						
Temp (°C)	Vapor Pressure (Torr)	Temp (°C)	Vapor Pressure (Torr)	Temp (°C)	Vapor Pressure (Torr)	Temp (°C)	Vapor Pressure (Torr)
0	4.58	26	25.2	52	102.1	78	327.3
1	4.93	27	26.7	53	107.2	79	341.0
2	5.29	28	28.3	54	112.5	80	355.1
3	5.68	29	30.0	55	118.0	81	369.7
4	6.10	30	31.8	56	123.8	82	384.9
5	6.54	31	33.7	57	129.8	83	400.6
6	7.01	32	35.7	58	136.1	84	416.8
7	7.51	33	37.7	59	142.6	85	433.6
8	8.04	34	39.9	60	149.4	86	450.9
9	8.61	35	41.2	61	156.4	87	468.7
10	9.21	36	44.6	62	163.8	88	487.1
11	9.84	37	47.1	63	171.4	89	506.1
12	10.5	38	49.7	64	179.3	90	525.8
13	11.2	39	52.4	65	187.5	91	546.0
14	12.0	40	55.3	66	196.1	92	567.0
15	12.8	41	58.3	67	205.0	93	588.6
16	13.6	42	61.5	68	214.2	94	610.9
17	14.5	43	64.8	69	223.7	95	633.9
18	15.5	44	68.3	70	233.7	96	657.6
19	16.5	45	71.9	71	243.9	97	682.1
20	17.5	46	75.6	72	254.6	98	707.3
21	18.6	47	79.6	73	265.7	99	733.2
22	19.8	48	83.7	74	277.2	100	760.0
23	21.1	49	88.0	75	289.1		
24	22.4	50	92.5	76	301.4		
25	23.8	51	97.2	77	314.1		

Table C.5 • **A-23**

TABLE C.5	SOLUBILITY PRODUCT CONSTANTS	
Salt	Solubility Equilibrium	K_{sp}
Fluorides		
MgF_2	$MgF_2(s) \rightleftharpoons Mg^{2+}(aq) + 2F^-(aq)$	6.6×10^{-9}
CaF_2	$CaF_2(s) \rightleftharpoons Ca^{2+}(aq) + 2F^-(aq)$	3.9×10^{-11}
SrF_2	$SrF_2(s) \rightleftharpoons Sr^{2+}(aq) + 2F^-(aq)$	2.9×10^{-9}
BaF_2	$BaF_2(s) \rightleftharpoons Ba^{2+}(aq) + 2F^-(aq)$	1.7×10^{-6}
LiF	$LiF(s) \rightleftharpoons Li^+(aq) + F^-(aq)$	1.7×10^{-3}
PbF_2	$PbF_2(s) \rightleftharpoons Pb^{2+}(aq) + 2F^-(aq)$	3.6×10^{-8}
Chlorides		
CuCl	$CuCl(s) \rightleftharpoons Cu^+(aq) + Cl^-(aq)$	1.9×10^{-7}
AgCl	$AgCl(s) \rightleftharpoons Ag^+(aq) + Cl^-(aq)$	1.8×10^{-10}
Hg_2Cl_2	$Hg_2Cl_2(s) \rightleftharpoons Hg_2^{2+}(aq) + 2Cl^-(aq)$	1.2×10^{-18}
TlCl	$TlCl(s) \rightleftharpoons Tl^+(aq) + Cl^-(aq)$	1.8×10^{-4}
$PbCl_2$	$PbCl_2(s) \rightleftharpoons Pb^{2+}(aq) + 2Cl^-(aq)$	1.7×10^{-5}
$AuCl_3$	$AuCl_3(s) \rightleftharpoons Au^{3+}(aq) + 3Cl^-(aq)$	3.2×10^{-25}
Bromides		
CuBr	$CuBr(s) \rightleftharpoons Cu^+(aq) + Br^-(aq)$	5×10^{-9}
AgBr	$AgBr(s) \rightleftharpoons Ag^+(aq) + Br^-(aq)$	5.0×10^{-13}
Hg_2Br_2	$Hg_2Br_2(s) \rightleftharpoons Hg_2^{2+}(aq) + 2Br^-(aq)$	5.6×10^{-23}
$HgBr_2$	$HgBr_2(s) \rightleftharpoons Hg^{2+}(aq) + 2Br^-(aq)$	1.3×10^{-19}
$PbBr_2$	$PbBr_2(s) \rightleftharpoons Pb^{2+}(aq) + 2Br^-(aq)$	2.1×10^{-6}
Iodides		
CuI	$CuI(s) \rightleftharpoons Cu^+(aq) + I^-(aq)$	1×10^{-12}
AgI	$AgI(s) \rightleftharpoons Ag^+(aq) + I^-(aq)$	8.3×10^{-17}
Hg_2I_2	$Hg_2I_2(s) \rightleftharpoons Hg_2^{2+}(aq) + 2I^-(aq)$	4.7×10^{-29}
HgI_2	$HgI_2(s) \rightleftharpoons Hg^{2+}(aq) + 2I^-(aq)$	1.1×10^{-28}
PbI_2	$PbI_2(s) \rightleftharpoons Pb^{2+}(aq) + 2I^-(aq)$	7.9×10^{-9}
Hydroxides		
$Mg(OH)_2$	$Mg(OH)_2(s) \rightleftharpoons Mg^{2+}(aq) + 2OH^-(aq)$	7.1×10^{-12}
$Ca(OH)_2$	$Ca(OH)_2(s) \rightleftharpoons Ca^{2+}(aq) + 2OH^-(aq)$	6.5×10^{-6}
$Mn(OH)_2$	$Mn(OH)_2(s) \rightleftharpoons Mn^{2+}(aq) + 2OH^-(aq)$	1.6×10^{-13}
$Fe(OH)_2$	$Fe(OH)_2(s) \rightleftharpoons Fe^{2+}(aq) + 2OH^-(aq)$	7.9×10^{-16}
$Fe(OH)_3$	$Fe(OH)_3(s) \rightleftharpoons Fe^{3+}(aq) + 3OH^-(aq)$	1.6×10^{-39}
$Co(OH)_2$	$Co(OH)_2(s) \rightleftharpoons Co^{2+}(aq) + 2OH^-(aq)$	1×10^{-15}
$Co(OH)_3$	$Co(OH)_3(s) \rightleftharpoons Co^{3+}(aq) + 3OH^-(aq)$	3×10^{-45}
$Ni(OH)_2$	$Ni(OH)_2(s) \rightleftharpoons Ni^{2+}(aq) + 2OH^-(aq)$	6×10^{-16}
$Cu(OH)_2$	$Cu(OH)_2(s) \rightleftharpoons Cu^{2+}(aq) + 2OH^-(aq)$	4.8×10^{-20}
$V(OH)_3$	$V(OH)_3(s) \rightleftharpoons V^{3+}(aq) + 3OH^-(aq)$	4×10^{-35}
$Cr(OH)_3$	$Cr(OH)_3(s) \rightleftharpoons Cr^{3+}(aq) + 3OH^-(aq)$	2×10^{-30}
Ag_2O	$Ag_2O(s) + H_2O \rightleftharpoons 2Ag^+(aq) + 2OH^-(aq)$	1.9×10^{-8}
$Zn(OH)_2$	$Zn(OH)_2(s) \rightleftharpoons Zn^{2+}(aq) + 2OH^-(aq)$	3.0×10^{-16}
$Cd(OH)_2$	$Cd(OH)_2(s) \rightleftharpoons Cd^{2+}(aq) + 2OH^-(aq)$	5.0×10^{-15}
$Al(OH)_3$ (alpha form)	$Al(OH)_3(s) \rightleftharpoons Al^{3+}(aq) + 3OH^-(aq)$	3×10^{-34}
Cyanides		
AgCN	$AgCN(s) \rightleftharpoons Ag^+(aq) + CN^-(aq)$	2.2×10^{-16}

(continued)

TABLE C.5	SOLUBILITY PRODUCT CONSTANTS *(Continued)*	
Salt	Solubility Equilibrium	K_{sp}

Cyanides (continued)

$Zn(CN)_2$	$Zn(CN)_2(s) \rightleftharpoons Zn^{2+}(aq) + 2CN^-(aq)$	3×10^{-16}

Sulfites

$CaSO_3$	$CaSO_3(s) \rightleftharpoons Ca^{2+}(aq) + SO_3{}^{2-}(aq)$	3×10^{-7}
Ag_2SO_3	$Ag_2SO_3(s) \rightleftharpoons 2Ag^+(aq) + SO_3{}^{2-}(aq)$	1.5×10^{-14}
$BaSO_3$	$BaSO_3(s) \rightleftharpoons Ba^{2+}(aq) + SO_3{}^{2-}(aq)$	8×10^{-7}

Sulfates

$CaSO_4$	$CaSO_4(s) \rightleftharpoons Ca^{2+}(aq) + SO_4{}^{2-}(aq)$	2.4×10^{-5}
$SrSO_4$	$SrSO_4(s) \rightleftharpoons Sr^{2+}(aq) + SO_4{}^{2-}(aq)$	3.2×10^{-7}
$BaSO_4$	$BaSO_4(s) \rightleftharpoons Ba^{2+}(aq) + SO_4{}^{2-}(aq)$	1.1×10^{-10}
$RaSO_4$	$RaSO_4(s) \rightleftharpoons Ra^{2+}(aq) + SO_4{}^{2-}(aq)$	4.3×10^{-11}
Ag_2SO_4	$Ag_2SO_4(s) \rightleftharpoons 2Ag^+(aq) + SO_4{}^{2-}(aq)$	1.5×10^{-5}
Hg_2SO_4	$Hg_2SO_4(s) \rightleftharpoons Hg_2{}^{2+}(aq) + SO_4{}^{2-}(aq)$	7.4×10^{-7}
$PbSO_4$	$PbSO_4(s) \rightleftharpoons Pb^{2+}(aq) + SO_4{}^{2-}(aq)$	6.3×10^{-7}

Chromates

$BaCrO_4$	$BaCrO_4(s) \rightleftharpoons Ba^{2+}(aq) + CrO_4{}^{2-}(aq)$	2.1×10^{-10}
$CuCrO_4$	$CuCrO_4(s) \rightleftharpoons Cu^{2+}(aq) + CrO_4{}^{2-}(aq)$	3.6×10^{-6}
Ag_2CrO_4	$Ag_2CrO_4(s) \rightleftharpoons 2Ag^+(aq) + CrO_4{}^{2-}(aq)$	1.2×10^{-12}
Hg_2CrO_4	$Hg_2CrO_4(s) \rightleftharpoons Hg_2{}^{2+}(aq) + CrO_4{}^{2-}(aq)$	2.0×10^{-9}
$CaCrO_4$	$CaCrO_4(s) \rightleftharpoons Ca^{2+}(aq) + CrO_4{}^{2-}(aq)$	7.1×10^{-4}
$PbCrO_4$	$PbCrO_4(s) \rightleftharpoons Pb^{2+}(aq) + CrO_4{}^{2-}(aq)$	1.8×10^{-14}

Carbonates

$MgCO_3$	$MgCO_3(s) \rightleftharpoons Mg^{2+}(aq) + CO_3{}^{2-}(aq)$	3.5×10^{-8}
$CaCO_3$	$CaCO_3(s) \rightleftharpoons Ca^{2+}(aq) + CO_3{}^{2-}(aq)$	4.5×10^{-9}
$SrCO_3$	$SrCO_3(s) \rightleftharpoons Sr^{2+}(aq) + CO_3{}^{2-}(aq)$	9.3×10^{-10}
$BaCO_3$	$BaCO_3(s) \rightleftharpoons Ba^{2+}(aq) + CO_3{}^{2-}(aq)$	5.0×10^{-9}
$MnCO_3$	$MnCO_3(s) \rightleftharpoons Mn^{2+}(aq) + CO_3{}^{2-}(aq)$	5.0×10^{-10}
$FeCO_3$	$FeCO_3(s) \rightleftharpoons Fe^{2+}(aq) + CO_3{}^{2-}(aq)$	2.1×10^{-11}
$CoCO_3$	$CoCO_3(s) \rightleftharpoons Co^{2+}(aq) + CO_3{}^{2-}(aq)$	1.0×10^{-10}
$NiCO_3$	$NiCO_3(s) \rightleftharpoons Ni^{2+}(aq) + CO_3{}^{2-}(aq)$	1.3×10^{-7}
$CuCO_3$	$CuCO_3(s) \rightleftharpoons Cu^{2+}(aq) + CO_3{}^{2-}(aq)$	2.3×10^{-10}
Ag_2CO_3	$Ag_2CO_3(s) \rightleftharpoons 2Ag^+(aq) + CO_3{}^{2-}(aq)$	8.1×10^{-12}
Hg_2CO_3	$Hg_2CO_3(s) \rightleftharpoons Hg_2{}^{2+}(aq) + CO_3{}^{2-}(aq)$	8.9×10^{-17}
$ZnCO_3$	$ZnCO_3(s) \rightleftharpoons Zn^{2+}(aq) + CO_3{}^{2-}(aq)$	1.0×10^{-10}
$CdCO_3$	$CdCO_3(s) \rightleftharpoons Cd^{2+}(aq) + CO_3{}^{2-}(aq)$	1.8×10^{-14}
$PbCO_3$	$PbCO_3(s) \rightleftharpoons Pb^{2+}(aq) + CO_3{}^{2-}(aq)$	7.4×10^{-14}

Phosphates

$Ca_3(PO_4)_2$	$Ca_3(PO_4)_2(s) \rightleftharpoons 3Ca^{2+}(aq) + 2PO_4{}^{3-}(aq)$	2.0×10^{-29}
$Mg_3(PO_4)_2$	$Mg_3(PO_4)_2(s) \rightleftharpoons 3Mg^{2+}(aq) + 2PO_4{}^{3-}(aq)$	6.3×10^{-26}
$SrHPO_4$	$SrHPO_4(s) \rightleftharpoons Sr^{2+}(aq) + HPO_4{}^{2-}(aq)$	1.2×10^{-7}
$BaHPO_4$	$BaHPO_4(s) \rightleftharpoons Ba^{2+}(aq) + HPO_4{}^{2-}(aq)$	4.0×10^{-8}
$LaPO_4$	$LaPO_4(s) \rightleftharpoons La^{3+}(aq) + PO_4{}^{3-}(aq)$	3.7×10^{-23}
$Fe_3(PO_4)_2$	$Fe_3(PO_4)_2(s) \rightleftharpoons 3Fe^{2+}(aq) + 2PO_4{}^{3-}(aq)$	1×10^{-36}
Ag_3PO_4	$Ag_3PO_4(s) \rightleftharpoons 3Ag^+(aq) + PO_4{}^{3-}(aq)$	2.8×10^{-18}
$FePO_4$	$FePO_4(s) \rightleftharpoons Fe^{3+}(aq) + PO_4{}^{3-}(aq)$	4.0×10^{-27}

(continued)

Table C.6 • **A-25**

TABLE C.5	**SOLUBILITY PRODUCT CONSTANTS** (*Continued*)	
Salt	Solubility Equilibrium	K_{sp}

Phosphates (*continued*)

$Zn_3(PO_4)_2$	$Zn_3(PO_4)_2(s) \rightleftharpoons 3Zn^{2+}(aq) + 2PO_4^{3-}(aq)$	5×10^{-36}
$Pb_3(PO_4)_2$	$Pb_3(PO_4)_2(s) \rightleftharpoons 3Pb^{2+}(aq) + 2PO_4^{3-}(aq)$	3.0×10^{-44}
$Ba_3(PO_4)_2$	$Ba_3(PO_4)_2(s) \rightleftharpoons 3Ba^{2+}(aq) + 2PO_4^{3-}(aq)$	5.8×10^{-38}

Ferrocyanides

$Zn_2[Fe(CN)_6]$	$Zn_2[Fe(CN)_6](s) \rightleftharpoons 2Zn^{2+}(aq) + Fe(CN)_6^{4-}(aq)$	2.1×10^{-16}
$Cd_2[Fe(CN)_6]$	$Cd_2[Fe(CN)_6](s) \rightleftharpoons 2Cd^{2+}(aq) + Fe(CN)_6^{4-}(aq)$	4.2×10^{-18}
$Pb_2[Fe(CN)_6]$	$Pb_2[Fe(CN)_6](s) \rightleftharpoons 2Pb^{2+}(aq) + Fe(CN)_6^{4-}(aq)$	9.5×10^{-19}

TABLE C.6	**FORMATION CONSTANTS OF COMPLEXES (25 °C)**

Complex Ion Equilibrium	K_{form}	Complex Ion Equilibrium	K_{form}
Halide Complexes		*Cyanide Complexes* (*continued*)	
$Al^{3+} + 6F^- \rightleftharpoons [AlF_6]^{3-}$	2.5×10^4	$Cd^{2+} + 4CN^- \rightleftharpoons [Cd(CN)_4]^{2-}$	7.7×10^{16}
$Al^{3+} + 4F^- \rightleftharpoons [AlF_4]^-$	2.0×10^8	$Au^+ + 2CN^- \rightleftharpoons [Au(CN)_2]^-$	2×10^{38}
$Be^{2+} + 4F^- \rightleftharpoons [BeF_4]^{2-}$	1.3×10^{13}	*Complexes with Other Monodentate Ligands*	
$Sn^{4+} + 6F^- \rightleftharpoons [SnF_6]^{2-}$	1×10^{25}	methylamine (CH_3NH_2)	
$Cu^+ + 2Cl^- \rightleftharpoons [CuCl_2]^-$	3×10^5	$Ag^+ + 2CH_3NH_2 \rightleftharpoons [Ag(CH_3NH_2)_2]^+$	7.8×10^6
$Ag^+ + 2Cl^- \rightleftharpoons [AgCl_2]^-$	1.8×10^5	thiocyanate ion (SCN^-)	
$Pb^{2+} + 4Cl^- \rightleftharpoons [PbCl_4]^{2-}$	2.5×10^{15}	$Cd^{2+} + 4SCN^- \rightleftharpoons [Cd(SCN)_4]^{2-}$	1×10^3
$Zn^{2+} + 4Cl^- \rightleftharpoons [ZnCl_4]^{2-}$	1.6	$Cu^{2+} + 2SCN^- \rightleftharpoons [Cu(SCN)_2]$	5.6×10^3
$Hg^{2+} + 4Cl^- \rightleftharpoons [HgCl_4]^{2-}$	5.0×10^{15}	$Fe^{3+} + 3SCN^- \rightleftharpoons [Fe(SCN)_3]$	2×10^6
$Cu^+ + 2Br^- \rightleftharpoons [CuBr_2]^-$	8×10^5	$Hg^{2+} + 4SCN^- \rightleftharpoons [Hg(SCN)_4]^{2-}$	5.0×10^{21}
$Ag^+ + 2Br^- \rightleftharpoons [AgBr_2]^-$	1.7×10^7	hydroxide ion (OH^-)	
$Hg^{2+} + 4Br^- \rightleftharpoons [HgBr_4]^{2-}$	1×10^{21}	$Cu^{2+} + 4OH^- \rightleftharpoons [Cu(OH)_4]^{2-}$	1.3×10^{16}
$Cu^+ + 2I^- \rightleftharpoons [CuI_2]^-$	8×10^8	$Zn^{2+} + 4OH^- \rightleftharpoons [Zn(OH)_4]^{2-}$	2×10^{20}
$Ag^+ + 2I^- \rightleftharpoons [AgI_2]^-$	1×10^{11}	*Complexes with Bidentate Ligands**	
$Pb^{2+} + 4I^- \rightleftharpoons [PbI_4]^{2-}$	3×10^4	$Mn^{2+} + 3 en \rightleftharpoons [Mn(en)_3]^{2+}$	6.5×10^5
$Hg^{2+} + 4I^- \rightleftharpoons [HgI_4]^{2-}$	1.9×10^{30}	$Fe^{2+} + 3 en \rightleftharpoons [Fe(en)_3]^{2+}$	5.2×10^9
Ammonia Complexes		$Co^{2+} + 3 en \rightleftharpoons [Co(en)_3]^{2+}$	1.3×10^{14}
$Ag^+ + 2NH_3 \rightleftharpoons [Ag(NH_3)_2]^+$	1.6×10^7	$Co^{3+} + 3 en \rightleftharpoons [Co(en)_3]^{3+}$	4.8×10^{48}
$Zn^{2+} + 4NH_3 \rightleftharpoons [Zn(NH_3)_4]^{2+}$	7.8×10^8	$Ni^{2+} + 3 en \rightleftharpoons [Ni(en)_3]^{2+}$	4.1×10^{17}
$Cu^{2+} + 4NH_3 \rightleftharpoons [Cu(NH_3)_4]^{2+}$	1.1×10^{13}	$Cu^{2+} + 2 en \rightleftharpoons [Cu(en)_2]^{2+}$	3.5×10^{19}
$Hg^{2+} + 4NH_3 \rightleftharpoons [Hg(NH_3)_4]^{2+}$	1.8×10^{19}	$Mn^{2+} + 3 bipy \rightleftharpoons [Mn(bipy)_3]^{2+}$	1×10^6
$Co^{2+} + 6NH_3 \rightleftharpoons [Co(NH_3)_6]^{2+}$	5.0×10^4	$Fe^{2+} + 3 bipy \rightleftharpoons [Fe(bipy)_3]^{2+}$	1.6×10^{17}
$Co^{3+} + 6NH_3 \rightleftharpoons [Co(NH_3)_6]^{3+}$	4.6×10^{33}	$Ni^{2+} + 3 bipy \rightleftharpoons [Ni(bipy)_3]^{2+}$	3.0×10^{20}
$Cd^{2+} + 6NH_3 \rightleftharpoons [Cd(NH_3)_6]^{2+}$	2.6×10^5	$Co^{2+} + 3 bipy \rightleftharpoons [Co(bipy)_3]^{2+}$	8×10^{15}
$Ni^{2+} + 6NH_3 \rightleftharpoons [Ni(NH_3)_6]^{2+}$	2.0×10^8	$Mn^{2+} + 3 phen \rightleftharpoons [Mn(phen)_3]^{2+}$	2×10^{10}
Cyanide Complexes		$Fe^{2+} + 3 phen \rightleftharpoons [Fe(phen)_3]^{2+}$	1×10^{21}
$Fe^{2+} + 6CN^- \rightleftharpoons [Fe(CN)_6]^{4-}$	1.0×10^{24}	$Co^{2+} + 3 phen \rightleftharpoons [Co(phen)_3]^{2+}$	6×10^{19}
$Fe^{3+} + 6CN^- \rightleftharpoons [Fe(CN)_6]^{3-}$	1.0×10^{31}	$Ni^{2+} + 3 phen \rightleftharpoons [Ni(phen)_3]^{2+}$	2×10^{24}
$Ag^+ + 2CN^- \rightleftharpoons [Ag(CN)_2]^-$	5.3×10^{18}	$Co^{2+} + 3C_2O_4^{2-} \rightleftharpoons [Co(C_2O_4)_3]^{4-}$	4.5×10^6
$Cu^+ + 2CN^- \rightleftharpoons [Cu(CN)_2]^-$	1.0×10^{16}	$Fe^{3+} + 3C_2O_4^{2-} \rightleftharpoons [Fe(C_2O_4)_3]^{3-}$	3.3×10^{20}

(*continued*)

| TABLE C.6 | FORMATION CONSTANTS OF COMPLEXES (25 °C) (*Continued*) | | | |

Complex Ion Equilibrium	K_{form}	Complex Ion Equilibrium	K_{form}
Complexes of Other Ligands		$Ca^{2+} + 2NTA^{3-} \rightleftharpoons [Ca(NTA)_2]^{4-}$	3.2×10^{11}
$Zn^{2+} + EDTA^{4-} \rightleftharpoons [Zn(EDTA)]^{2-}$	3.8×10^{16}		
$Mg^{2+} + 2NTA^{3-} \rightleftharpoons [Mg(NTA)_2]^{4-}$	1.6×10^{10}		

*en = ethylenediamine

bipy = bipyridyl

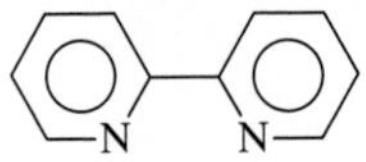

bipyridyl

phen = 1,10-phenanthroline

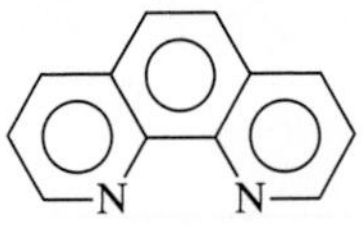

1,10-phenanthroline

$EDTA^{4-}$ = ethylenediaminetetraacetate ion

NTA^{3-} = nitrilotriacetate ion

| TABLE C.7 | IONIZATION CONSTANTS OF WEAK ACIDS AND BASES (ALTERNATIVE FORMULAS IN PARENTHESES) | | |

Monoprotic Acids	Name	K_a
$HC_2O_2Cl_3$ (Cl_3CCO_2H)	trichloriacetic acid	2.2×10^{-1}
HIO_3	iodic acid	1.69×10^{-1}
$HC_2HO_2Cl_2$ (Cl_2CHCO_2H)	dichloroacetic acid	5.0×10^{-2}
$HC_2H_2O_2Cl$ (ClH_2CCO_2H)	chloroacetic acid	1.36×10^{-3}
HNO_2	nitrous acid	7.1×10^{-4}
HF	hydrofluoric acid	6.8×10^{-4}
HOCN	cyanic acid	3.5×10^{-4}
$HCHO_2$ (HCO_2H)	formic acid	1.8×10^{-4}
$HC_3H_5O_3$ ($CH_3CH(OH)CO_2H$)	lactic acid	1.38×10^{-4}
$HC_4H_3N_2O_3$	barbituric acid	9.8×10^{-5}
$HC_7H_5O_2$ ($C_6H_5CO_2H$)	benzoic acid	6.28×10^{-5}
$HC_4H_7O_2$ ($CH_3CH_2CH_2CO_2H$)	butanoic acid	1.52×10^{-5}
HN_3	hydrazoic acid	1.8×10^{-5}
$HC_2H_3O_2$ (CH_3CO_2H)	acetic acid	1.8×10^{-5}
$HC_3H_5O_2$ ($CH_3CH_2CO_2H$)	propanoic acid	1.34×10^{-5}
HOCl	hypochlorous acid	3.0×10^{-8}
HOBr	hypobromous acid	2.1×10^{-9}
HCN	hydrocyanic acid	6.2×10^{-10}
HC_6H_5O	phenol	1.3×10^{-10}
HOI	hypoiodous acid	2.3×10^{-11}
H_2O_2	hydrogen peroxide	1.8×10^{-12}

Polyprotic Acids	Name	K_{a_1}	K_{a_2}	K_{a_3}
H_2SO_4	sulfuric acid	large	1.0×10^{-2}	
H_2CrO_4	chromic acid	5.0	1.5×10^{-6}	
$H_2C_2O_4$	oxalic acid	5.6×10^{-2}	5.4×10^{-5}	

Table C.8 • **A-27**

| TABLE C.7 | IONIZATION CONSTANTS OF WEAK ACIDS AND BASES (ALTERNATIVE FORMULAS IN PARENTHESES) (*Continued*) | | | |

Polyprotic Acids	Name	K_{a_1}	K_{a_2}	K_{a_3}
H_3PO_3	phosphorous acid	3×10^{-2}	1.6×10^{-7}	
$H_2SO_3 \, [SO_2(aq)]$	sulfurous acid	1.2×10^{-2}	6.6×10^{-8}	
H_2SeO_3	selenous acid	4.5×10^{-3}	1.1×10^{-8}	
H_2TeO_3	tellurous acid	3.3×10^{-3}	2.0×10^{-8}	
$H_2C_3H_2O_4$ $(HO_2CCH_2CO_2H)$	malonic acid	1.4×10^{-3}	2.0×10^{-6}	
$H_2C_8H_4O_4$	phthalic acid	1.1×10^{-3}	3.9×10^{-6}	
$H_2C_4H_4O_6$	tartaric acid	9.2×10^{-4}	4.3×10^{-5}	
$H_2C_6H_6O_6$	ascorbic acid	6.8×10^{-5}	2.7×10^{-12}	
H_2CO_3	carbonic acid	4.5×10^{-7}	4.7×10^{-11}	
H_3PO_4	phosphoric acid	7.1×10^{-3}	6.3×10^{-8}	4.5×10^{-13}
H_3AsO_4	arsenic acid	5.6×10^{-3}	1.7×10^{-7}	4.0×10^{-12}
$H_3C_6H_5O_7$	citric acid	7.1×10^{-4}	1.7×10^{-5}	6.3×10^{-6}

Weak Bases	Name	K_b
$(CH_3)_2NH$	dimethylamine	9.6×10^{-4}
CH_3NH_2	methylamine	4.4×10^{-4}
$CH_3CH_2NH_2$	ethylamine	4.3×10^{-4}
$(CH_3)_3N$	trimethylamine	7.4×10^{-5}
NH_3	ammonia	1.8×10^{-5}
N_2H_4	hydrazine	9.6×10^{-7}
NH_2OH	hydroxylamine	6.6×10^{-9}
C_5H_5N	pyridine	1.5×10^{-9}
$C_6H_5NH_2$	aniline	4.1×10^{-10}
PH_3	phosphine	10^{-28}

| TABLE C.8 | STANDARD REDUCTION POTENTIALS (25 °C) |

$E°$ (Volts)	Half-Cell Reaction
+2.87	$F_2(g) + 2e^- \rightleftharpoons 2F^-(aq)$
+2.08	$O_3(g) + 2H^+(aq) + 2e^- \rightleftharpoons O_2(g) + H_2O$
+2.05	$S_2O_8^{2-}(aq) + 2e^- \rightleftharpoons 2SO_4^{2-}(aq)$
+1.82	$Co^{3+}(aq) + e^- \rightleftharpoons Co^{2+}(aq)$
+1.77	$H_2O_2(aq) + 2H^+(aq) + 2e^- \rightleftharpoons 2H_2O$
+1.69	$PbO_2(s) + HSO_4^-(aq) + 3H^+(aq) + 2e^- \rightleftharpoons PbSO_4(s) + 2H_2O$
+1.63	$2HOCl(aq) + 2H^+(aq) + 2e^- \rightleftharpoons Cl_2(g) + 2H_2O$
+1.51	$Mn^{3+}(aq) + e^- \rightleftharpoons Mn^{2+}(aq)$
+1.507	$MnO_4^-(aq) + 4H^+(aq) + 3e^- \rightleftharpoons MnO_2(s) + 2H_2O$
+1.49	$MnO_4^-(aq) + 8H^+(aq) + 5e^- \rightleftharpoons Mn^{2+}(aq) + 4H_2O$
+1.46	$PbO_2(s) + 4H^+(aq) + 2e^- \rightleftharpoons Pb^{2+}(aq) + 2H_2O$
+1.44	$BrO_3^-(aq) + 6H^+(aq) + 6e^- \rightleftharpoons Br^-(aq) + 3H_2O$
+1.42	$Au^{3+}(aq) + 3e^- \rightleftharpoons Au(s)$
+1.36	$Cl_2(g) + 2e^- \rightleftharpoons 2Cl^-(aq)$
+1.33	$Cr_2O_7^{2-}(aq) + 14H^+(aq) + 6e^- \rightleftharpoons 2Cr^{3+}(aq) + 7H_2O$
+1.24	$O_3(g) + H_2O + 2e^- \rightleftharpoons O_2(g) + 2OH^-(aq)$

(*continued*)

TABLE C.8	STANDARD REDUCTION POTENTIALS (25 °C) (*Continued*)
$E°$ (Volts)	Half-Cell Reaction

$E°$ (Volts)	Half-Cell Reaction
+1.23	$MnO_2(s) + 4H^+(aq) + 2e^- \rightleftharpoons Mn^{2+}(aq) + 2H_2O$
+1.23	$O_2(g) + 4H^+(aq) + 4e^- \rightleftharpoons 2H_2O$
+1.20	$Pt^{2+}(aq) + 2e^- \rightleftharpoons Pt(s)$
+1.07	$Br_2(aq) + 2e^- \rightleftharpoons 2Br^-(aq)$
+0.96	$NO_3^-(aq) + 4H^+(aq) + 3e^- \rightleftharpoons NO(g) + 2H_2O$
+0.94	$NO_3^-(aq) + 3H^+(aq) + 2e^- \rightleftharpoons HNO_2(aq) + H_2O$
+0.91	$2Hg^{2+}(aq) + 2e^- \rightleftharpoons Hg_2^{2+}(aq)$
+0.87	$HO_2^-(aq) + H_2O + 2e^- \rightleftharpoons 3OH^-(aq)$
+0.80	$NO_3^-(aq) + 4H^+(aq) + 2e^- \rightleftharpoons 2NO_2(g) + 2H_2O$
+0.80	$Ag^+(aq) + e^- \rightleftharpoons Ag(s)$
+0.77	$Fe^{3+}(aq) + e^- \rightleftharpoons Fe^{2+}(aq)$
+0.69	$O_2(g) + 2H^+(aq) + 2e^- \rightleftharpoons H_2O_2(aq)$
+0.54	$I_2(s) + 2e^- \rightleftharpoons 2I^-(aq)$
+0.49	$NiO_2(s) + 2H_2O + 2e^- \rightleftharpoons Ni(OH)_2(s) + 2OH^-(aq)$
+0.45	$SO_2(aq) + 4H^+(aq) + 4e^- \leftrightharpoons S(s) + 2H_2O$
+0.401	$O_2(g) + 2H_2O + 4e^- \rightleftharpoons 4OH^-(aq)$
+0.34	$Cu^{2+}(aq) + 2e^- \rightleftharpoons Cu(s)$
+0.27	$Hg_2Cl_2(s) + 2e^- \rightleftharpoons 2Hg(l) + 2Cl^-(aq)$
+0.25	$PbO_2(s) + H_2O + 2e^- \rightleftharpoons PbO(s) + 2OH^-(aq)$
+0.2223	$AgCl(s) + e^- \rightleftharpoons Ag(s) + Cl^-(aq)$
+0.172	$SO_4^{2-}(aq) + 4H^+(aq) + 2e^- \rightleftharpoons H_2SO_3(aq) + H_2O$
+0.169	$S_4O_6^{2-}(aq) + 2e^- \rightleftharpoons 2S_2O_3^{2-}(aq)$
+0.16	$Cu^{2+}(aq) + e^- \rightleftharpoons Cu^+(aq)$
+0.15	$Sn^{4+}(aq) + 2e^- \rightleftharpoons Sn^{2+}(aq)$
+0.14	$S(s) + 2H^+(aq) + 2e^- \rightleftharpoons H_2S(g)$
+0.07	$AgBr(s) + e^- \rightleftharpoons Ag(s) + Br^-(aq)$
0.00	$2H^+(aq) + 2e^- \rightleftharpoons H_2(g)$
−0.13	$Pb^{2+}(aq) + 2e^- \rightleftharpoons Pb(s)$
−0.14	$Sn^{2+}(aq) + 2e^- \rightleftharpoons Sn(s)$
−0.15	$AgI(s) + e^- \rightleftharpoons Ag(s) + I^-(aq)$
−0.25	$Ni^{2+}(aq) + 2e^- \rightleftharpoons Ni(s)$
−0.28	$Co^{2+}(aq) + 2e^- \rightleftharpoons Co(s)$
−0.34	$In^{3+}(aq) + 3e^- \rightleftharpoons In(s)$
−0.34	$Tl^+(aq) + e^- \rightleftharpoons Tl(s)$
−0.36	$PbSO_4(s) + H^+(aq) + 2e^- \rightleftharpoons Pb(s) + HSO_4^-(aq)$
−0.40	$Cd^{2+}(aq) + 2e^- \rightleftharpoons Cd(s)$
−0.44	$Fe^{2+}(aq) + 2e^- \rightleftharpoons Fe(s)$
−0.56	$Ga^{3+}(aq) + 3e^- \rightleftharpoons Ga(s)$
−0.58	$PbO(s) + H_2O + 2e^- \rightleftharpoons Pb(s) + 2OH^-(aq)$
−0.74	$Cr^{3+}(aq) + 3e^- \rightleftharpoons Cr(s)$
−0.76	$Zn^{2+}(aq) + 2e^- \rightleftharpoons Zn(s)$
−0.81	$Cd(OH)_2(s) + 2e^- \rightleftharpoons Cd(s) + 2OH^-(aq)$
−0.83	$2H_2O + 2e^- \rightleftharpoons H_2(g) + 2OH^-(aq)$
−0.88	$Fe(OH)_2(s) + 2e^- \rightleftharpoons Fe(s) + 2OH^-(aq)$
−0.91	$Cr^{2+}(aq) + 2e^- \rightleftharpoons Cr(s)$

Table C.8 • **A-29**

TABLE C.8	STANDARD REDUCTION POTENTIALS (25 °C) (*Continued*)
$E°$ (Volts)	Half-Cell Reaction
−1.16	$N_2(g) + 4H_2O + 4e^- \rightleftharpoons N_2O_4(aq) + 4OH^-(aq)$
−1.18	$V^{2+}(aq) + 2e^- \rightleftharpoons V(s)$
−1.216	$ZnO_2^{2-}(aq) + 2H_2O + 2e^- \rightleftharpoons Zn(s) + 4OH^-(aq)$
−1.63	$Ti^{2+}(aq) + 2e^- \rightleftharpoons Ti(s)$
−1.66	$Al^{3+}(aq) + 3e^- \rightleftharpoons Al(s)$
−1.79	$U^{3+}(aq) + 3e^- \rightleftharpoons U(s)$
−2.02	$Sc^{3+}(aq) + 3e^- \rightleftharpoons Sc(s)$
−2.36	$La^{3+}(aq) + 3e^- \rightleftharpoons La(s)$
−2.37	$Y^{3+}(aq) + 3e^- \rightleftharpoons Y(s)$
−2.37	$Mg^{2+}(aq) + 2e^- \rightleftharpoons Mg(s)$
−2.71	$Na^+(aq) + e^- \rightleftharpoons Na(s)$
−2.76	$Ca^{2+}(aq) + 2e^- \rightleftharpoons Ca(s)$
−2.89	$Sr^{2+}(aq) + 2e^- \rightleftharpoons Sr(s)$
−2.90	$Ba^{2+}(aq) + 2e^- \rightleftharpoons Ba(s)$
−2.92	$Cs^+(aq) + e^- \rightleftharpoons Cs(s)$
−2.92	$K^+(aq) + e^- \rightleftharpoons K(s)$
−2.93	$Rb^+(aq) + e^- \rightleftharpoons Rb(s)$
−3.05	$Li^+(aq) + e^- \rightleftharpoons Li(s)$

Glossary

This Glossary has the definitions of the Key Terms that were marked in boldface throughout the chapters plus a few additional terms. The numbers in parentheses that follow the definitions are the numbers of the sections in which the glossary entries received their principal discussions.

A

Absolute Zero: 0 K, −273.15 °C. Nature's lowest temperature. (3.2, 11.3)

Acceptor: A Lewis acid; the central metal ion in a complex ion. (19.4, 23.4)

Accuracy: Freedom from error. The closeness of a measurement to the true value. (3.3)

Acid: *Arrhenius theory*—a substance that produces hydronium ions (hydrogen ions) in water. (5.4)
> *Brønsted theory*—a proton donor. (17.1)
> *Lewis theory*—an electron-pair acceptor. (17.3)

Acid-Base Indicator: A dye with one color in acid and another color in base. (5.13, 17.5, 18.8)

Acid-Base Neutralization: The reaction of an acid with a base. (5.5)

Acidic Anhydride: An oxide that reacts with water to make the solution acidic. (5.5)

Acidic Solution: An aqueous solution in which $[H^+] > [OH^-]$. (17.5)

Acid Ionization Constant (K_a):
$$K_a = \frac{[H^+][A^-]}{[HA]} \text{ for the equilibrium,}$$
$$HA \rightleftharpoons H^+ + A^- \qquad (18.1)$$

Acid Rain: Rain made acidic by dissolved sulfur and nitrogen oxides.

Acid Salt: A salt of a partially neutralized polyprotic acid, for example, $NaHSO_4$ or $NaHCO_3$. (5.6)

Acid Solubility Product: The special solubility product expression for metal sulfides in dilute acid and related to the equation for their dissolving. For a divalent metal sulfide, MS,
$$MS(s) + 2H^+(aq) \longrightarrow$$
$$M^{2+}(aq) + H_2S(aq)$$
$$K_{spa} = \frac{[M^{2+}][H_2S]}{[H^+]^2} \qquad (19.2)$$

Actinide Elements (Actinide Series): Elements 90−103. (1.7)

Activated Complex: The chemical species that exists with partly broken and partly formed bonds in the transition state. (15.6)

Activation Energy: The minimum kinetic energy that must be possessed by the reactants in order to give an effective collision (one that produces products). (15.6)

Activities: Effective concentrations which properly should be substituted into a mass action expression to satisfy the equilibrium law. The activity of a solid is defined as having a value of 1.

Activity: For a radioactive material, the number of disintegrations per second. (22.6)

Activity Series: A list of metals in their order of reactivity as reducing agents. (6.4)

Addition Compound: A molecule formed by the joining of two simpler molecules through formation of a covalent bond (usually a coordinate covalent bond). (9.10)

Addition Polymer: A polymer formed by the simple addition of one monomer unit to another, a process that continues over and over until a very long chain of monomer units is produced. (13.5)

Addition Reaction: The addition of a molecule to a double or triple bond. (25.2)

Adiabatic Change: A change within a system during which no energy enters or leaves the system.

Aerogel: A very porous and extremely low density solid formed when solvent is removed from the wet gel (made by the sol-gel process) under supercritical conditions. (13.7)

Alcohol: An organic compound whose molecules have the OH group attached to tetrahedral carbon. (25.3)

Aldehyde: An organic compound whose molecules have the group $-CH{=}O$. (25.5)

Alkali Metals: The Group IA elements (except hydrogen)—lithium, sodium, potassium, rubidium, and cesium. (1.7)

Alkaline Dry Cell: A zinc−manganese dioxide galvanic cell of 1.54 V used commonly in flashlight batteries. (21.6)

Alkaline Earth Metals: The Group IIA elements—beryllium, magnesium, calcium, strontium, barium, and radium. (1.7)

Alkalis: (a) The alkali metals. (1.7) (b) Hydroxides of the alkali metals; strong bases. (5.7)

Alkane: A hydrocarbon whose molecules have only single bonds. (2.4, 9.4, 25.2)

Alkene: A hydrocarbon whose molecules have one or more double bonds. (9.4, 25.2)

Alkoxide: An anion formed by removing a hydrogen ion from an alcohol. (13.7)

Alkyl Group: An organic group of carbon and hydrogen atoms related to an alkane but with one less hydrogen atom (e.g., CH_3-, methyl; CH_3CH_2-, ethyl). (25.2)

Alkyne: A hydrocarbon whose molecules have one or more triple bonds. (9.4, 25.2)

Allotrope: One of two or more forms of an element. (24.2)

Alpha Particle ($^4_2$He): The nucleus of a helium atom. (22.3)

Alpha Radiation: A high-velocity stream of alpha particles produced by radioactive decay. (22.3)

Alum: A double salt with the general formula $M^+M^{3+}(SO_4)_2 \cdot 12H_2O$, such as potassium alum: $KAl(SO_4)_2 \cdot 12H_2O$.

Amalgam: A solution of a metal in mercury.

Amide: An organic compound whose molecules have any one of the following groups: (25.5)

$$\underset{-CNH_2}{\overset{O}{\|}} \qquad \underset{-CNHR}{\overset{O}{\|}} \qquad \underset{-CNR_2}{\overset{O}{\|}}$$

α-Amino Acid: One of 20 monomers of polypeptides. (25.9)

Amine: An organic compound whose molecules contain the group NH_2, NHR, or NR_2. (25.4)

Amorphous Solid: A noncrystalline solid. A glass. (13.1)

Ampere (A): The SI unit for electric current; one coulomb per second. (21.8)

Amphiprotic Compound: A compound that can act either as a proton-donor or a proton-acceptor; an amphoteric compound. (17.1)

Amphoteric Compound: A compound that can react as either an acid or a base. (17.1)

Amplitude: The height of a wave, which is a measure of the wave's intensity. (8.1)

AMU: See *Atomic Mass Unit.*

Angstrom (Å): 1 Å = 10^{-10} m = 100 pm = 0.1 nm. (8.8)

Anhydrous: Without water. (2.1)

Anion: A negatively charged ion. (2.7)

Anode: The positive electrode in a gas discharge tube. The electrode at which oxidation occurs during an electrochemical change. (Facets of Chemistry 1.1, 21.1)

Antibonding Electrons: Electrons that occupy antibonding molecular orbitals. (10.7)

Antibonding Molecular Orbital: A molecular orbital that denies electron density to the space between nuclei and destabilizes a molecule when occupied by electrons. (10.7)

Anticodon: A triplet of bases on a tRNA molecule that pairs to a matching triplet—a codon—on an mRNA molecule during mRNA-directed polypeptide synthesis. (25.10)

Antimatter: Any particle annihilated by a particle of ordinary matter. (22.3)

Aqua Regia: One part concentrated nitric acid and three parts concentrated hydrochloric acid (by volume).

Aqueous Solution: A solution that has water as the solvent.

Aromatic Compound: An organic compound whose molecules have the benzene ring system. (25.2)

Arrhenius Acid: See *Acid.*

Arrhenius Base: See *Base.*

Arrhenius Equation: An equation that relates the rate constant of a reaction to the reaction's activation energy. (15.7)

Association: The joining together of molecules by hydrogen bonds. (14.9)

Asymmetric: Lacking symmetry. (10.3)

Atmosphere, Standard (Atm): 101,325 Pa. The pressure that supports a column of mercury 760 mm high at 0 °C; 760 torr. (11.2)

Atmospheric Pressure: The pressure exerted by the mixture of gases in our atmosphere. (7.5, 11.2)

Atom: A neutral particle having one nucleus; the smallest representative sample of an element. (1.2)

Atomic Force Microscope (AFM): A device for 3-dimensional imaging the surfaces of solids (conducting or nonconducting) at the scale of atoms. (13.8)

Atomic Mass: The average mass (in u) of the atoms of the isotopes of a given element as they occur naturally. (1.5)

Atomic Mass Unit (u): 1.6606×10^{-24} g; 1/12th the mass of one atom of carbon-12. Sometimes given the symbol amu. (1.5)

Atomic Number: The number of protons in a nucleus. (1.6)

Atomic Spectrum: The line spectrum produced when energized or excited atoms emit electromagnetic radiation. (8.2)

Atomic Weight: See *Atomic Mass.*

Atomization Energy (ΔH_{atom}): The energy needed to rupture all of the bonds in one mole of a substance in the gas state and produce its atoms, also in the gas state. (20.10)

Aufbau Principle: A set of rules enabling the construction of an electron structure of an atom from its atomic number. (8.5)

Avogadro's Number (Avogadro's Constant): 6.02×10^{23}; the number of particles or formula units in one mole. (4.2)

Avogadro's Principle: Equal volumes of gases contain equal numbers of molecules when they are at identical temperatures and pressures. (11.2)

Axial Bonds: Covalent bonds oriented parallel to the vertical axis in a trigonal bipyramidal molecule. (10.1)

Azimuthal Quantum Number (ℓ): The quantum number ℓ. (See also: secondary quantum number.) (8.3)

B

Backbone (polymer): The long chain of atoms in a polymer to which other groups are attached. (13.5)

Background Radiation: The atomic radiation from the natural radionuclides in the environment and from cosmic radiation. (22.6)

Balance: An apparatus for measuring mass. (3.2)

Balanced Equation: A chemical equation that has on opposites sides of the arrow the same number of each atom and the same net charge. (2.2, 4.5, 5.3)

Band Gap: The energy separation between the filled valence shell and the nearest conduction band (empty band). (13.4)

Band of Stability: The envelope that encloses just the stable nuclides in a plot of all nuclides constructed according to their numbers of neutrons versus their numbers of protons. (22.4)

Band Theory of Solids: An energy band of closely spaced energy levels exists in solids some of which levels make up a valence band and others of which make up the (partially filled or empty) conduction band. (13.4)

Bar: The standard pressure for thermodynamic quantities; 1 bar = 10^5 pascals, 1 atm = 101,325 Pa. (11.2)

Barometer: An apparatus for measuring atmospheric pressure. (11.2)

Base: *Arrhenius theory*—a substance that releases OH$^-$ ions in water. (5.5)

   Brønsted theory—a proton-acceptor. (17.1)

   Lewis theory—an electron-pair acceptor. (17.3)

Base Ionization Constant, K_b:

$K_b = \dfrac{[BH^+][OH^-]}{[B]}$ for the equilibrium,

$$B + H_2O \rightleftharpoons BH^+ + OH^- \quad (18.1)$$

Base Units: The units of the fundamental measurements of the SI. (3.2)

Basic Anhydride: An oxide that can neutralize acid or that reacts with water to give OH$^-$. (5.5)

Basic Oxygen Process: A method to convert pig iron into steel. (23.2)

Basic Solution: An aqueous solution in which [H$^+$] < [OH$^-$]. (17.5)

Battery: One or more galvanic cells arranged to serve as a practical source of electricity. (21.6)

Becquerel (Bq): 1 disintegration s^{-1}. The SI unit for the activity of a radioactive source. (22.6)

Bent Molecule (V-Shaped Molecule): A molecule that is nonlinear. (10.2)

Beta Particle ($_{-1}^{0}e$): An electron emitted by radioactive decay. (22.3)

Beta Radiation: A stream of electrons produced by radioactive decay. (22.3)

Bidentate Ligand: A ligand that has two atoms that can become simultaneously attached to the same metal ion. (23.4)

Bimolecular Collision: A collision of two molecules. (15.8)

Binary Acid: An acid with the general formula H_nX, where X is a nonmetal. (5.6)

Binary Compound: A compound composed of two different elements. (2.5)

Binding Energy, Nuclear: The energy equivalent of the difference in mass between an atomic nucleus and the sum of the masses of its nucleons. (22.2)

Biochemistry: The study of the organic substances in organisms. (25.6)

Black Phosphorus: An allotrope of phosphorus that has a layered structure. (24.2)

Blast Furnace: A structure in which iron ore is reduced to iron. (23.2)

Body-Centered Cubic (bcc) Unit Cell: A unit cell having identical atoms, molecules, or ions at the corners of a cube plus one more particle in the center of the cube. (13.1)

Boiling Point: The temperature at which the vapor pressure of the liquid equals the atmospheric pressure. (12.6)

Boiling Point Elevation: A colligative property of a solution by which the solution's boiling point is higher than that of the pure solvent. (14.7)

Bond Angle: The angle formed by two bonds that extend from the same atom. (10.1)

Bond Dipole: A dipole within a molecule associated with a specific bond. (10.3)

Bond Dissociation Energy: See *Bond Energy.*

Bond Distance: See *Bond Length.*

Bond Energy: The energy needed to break one mole of a particular bond to give electrically neutral fragments. (9.3, 20.10)

Bond Length: The distance between two nuclei that are held by a chemical bond. (9.3)

Bond Order: The number of electron pairs shared between two atoms. The *net* number of pairs of bonding electrons.

bond order = 1/2 × (No. of bonding e^- − No. of antibonding e^-). (9.8, 10.7)

Bonding Domain: A region between two atoms that contains one or more electron pairs in bonds and that influences molecular shape. (10.2)

Bonding Electrons: Electrons that occupy bonding molecular orbitals. (10.7)

Bonding Molecular Orbital: A molecular orbital that produces a buildup of electron density between nuclei and stabilizes a molecule when occupied by electrons. (10.7)

Boundary: The interface between a system and its surroundings across which energy or matter might pass. (7.3)

Boyle's Law: See *Pressure-Volume Law.*

Bragg Equation: $n\lambda = 2\,d\sin\theta$. The equation used to convert X ray diffraction data into crystal structure. (13.2)

Branched-Chain Compounds: An organic compound in whose molecules the carbon atoms do not all occur one after another in a continuous sequence. (25.2)

Branching (Polymer): The formation of side chains (branches) along the main backbone of a polymer. (13.5)

Branching Step: A step in a chain reaction that produces more chain-propagating species than it consumes. (Facets of Chemistry 15.1)

Brine: An aqueous solution of sodium chloride, often with other salts. (21.9)

Brønsted Acid: See *Acid.*

Brønsted Base: See *Base.*

Brownian Movement: The random, erratic motions of colloidally dispersed particles in a fluid. (2.4)

Buckminsterfullerene: The C_{60} molecule. Also called buckyball. (13.8)

Buckyball: See *buckminsterfullerene.*

Buffer: (a) A pair of solutes that can keep the pH of a solution almost constant if either acid or base is added. (b) A solution containing such a pair of solutes. (18.5)

Buffer Capacity: A measure of how much strong acid or strong base is needed to change the pH of a buffer by some specified amount. (18.5)

Buret: A long tube of glass usually marked in mL and 0.1 mL units and equipped with a stopcock for the controlled addition of a liquid to a receiving flask. (5.13)

By-product: The substances formed by side reactions. (4.8)

C

Calorie (cal): 4.184 J. The energy that will raise the temperature of 1.00 g of water from 14.5 to 15.5 °C. (In popular books on foods, the term *Calorie,* with a capital C, means 1000 cal or 1 kcal.) (7.1)

Calorimeter: An apparatus used in the determination of the heat of a reaction. (7.5)

Calorimetry: The science of measuring the quantities of heat that are involved in a chemical or physical change. (7.5)

Carbohydrates: Polyhydroxyaldehydes or polyhydroxyketones or substances that yield these by hydrolysis and that are obtained from plants or animals. (25.7)

Carbon Family: Group IVA in the periodic table—carbon, silicon, germanium, tin, and lead. (1.7)

Carbon Nanotube: Tubular carbon molecules that can be visualized as rolled up sheets of graphite (with hexagonal rings of carbon atoms) capped at each end by half of a spherical fullerene molecule. (13.8)

Carbonyl Group: An organic functional group consisting of a carbon atom joined to an oxygen atom by a double bond; C=O. (25.5)

Carboxylic Acids: An organic compound whose molecules have the carboxyl group CO_2H. (25.5)

Catalysis: Rate enhancement caused by a catalyst. (15.2, 15.9)

Catalyst: A substance that in relatively small proportion accelerates the rate of a reaction without being permanently chemically changed. (15.2, 15.9)

Catenation: The linking together of atoms of the same element to form chains. (24.4)

Cathode: The negative electrode in a gas discharge tube. The electrode at which reduction occurs during an electrochemical change. (21.1, Facets of Chemistry 1.1)

Cathode Ray: A stream of electrons ejected from a hot metal and accelerated toward a positively charged site in a vacuum tube. (Facets of Chemistry 1.1)

Cation: A positively charged ion. (2.7)

Ccp Structure: See *Cubic Closest Packing.*

Cell Potential, E_{cell}: The emf of a galvanic cell when no current is drawn from the cell. (21.2)

Cell Reaction: The overall chemical change that takes place in an electrolytic cell or a galvanic cell. (21.1)

Celsius Scale: A temperature scale on which water freezes at 0 °C and boils at 100 °C (at 1 atm) and that has 100 divisions called Celsius degrees between these two points. (3.2)

Centimeter (cm): 0.01 m. (3.2)

Ceramic: A very hard, strong, and high-melting inorganic substance. (13.7)

Chain Reaction: A self-sustaining change in which the products of one event cause one or more new events. (Facets of Chemistry 15.1, 22.8)

Change of State: Transformation of matter from one physical state to another. In thermochemistry, any change in a variable used to define the state of a particular system—a change in composition, pressure, volume, or temperature. (12.4)

Charge: The mixture of raw materials added to a blast furnace. (23.2)

Charge Transfer Absorption Band: An absorption band that arises from the absorption of photons, each of which supplies energy to transfer an electron from an anion to a cation. (23.3)

Charles' Law: See *Temperature-Volume Law.*

Chelate: A complex ion containing rings formed by polydentate ligands. (23.4)

Chelate Effect: The extra stability found in complexes that contain chelate rings. (23.4)

Chemical Bond: The force of electrical attraction that holds atoms together in compounds. (2.4, 9.1)

Chemical Change: A change that converts substances into other substances; a chemical reaction. (1.3)

Chemical Energy: The potential energy of chemicals that is transferred during chemical reactions. (2.3, 7.4)

Chemical Equation: A before-and-after description that uses formulas and coefficients to represent a chemical reaction. (2.2)

Chemical Equilibrium: Dynamic equilibrium in a chemical system. (5.7)

Chemical Formula: A formula written using chemical symbols and subscripts that describes the composition of a chemical compound or element. (2.1)

Chemical Kinetics: The study of rates of reaction. (15.1)

Chemical Property: The ability of a substance, either by itself or with other substances, to undergo a change into new substances. (1.3)

Chemical Reaction: A change in which new substances (products) form from starting materials (reactants). (1.3)

Chemical Symbol: A formula for an element. (1.4)

Chemistry: The study of the compositions of substances and the ways by which their properties are related to their compositions. (1.1)

Cholesteric Phase: A liquid crystal in which molecules align as in a nematic phase, but in thin layers. Going down through the layers, the orientations of the molecules rotate, producing twisted stacks of molecules. (13.6)

Chirality: The "handedness" of an object; the property of an object (like a molecule) that makes it unable to be superimposed onto a model of its own mirror image. (23.7)

Cis Isomer: A stereoisomer whose uniqueness is in having two groups on the same side of some reference plane. (23.7, 25.2)

Clausius-Clapeyron Equation: The relationship between the vapor pressure, the temperature, and the molar heat of vaporization of a substance (where C is a constant).

$$\ln P = \frac{\Delta H_{vap}}{RT} + C$$

(Facets of Chemistry 12.2)

Closed-End Manometer: See *Manometer.*

Closed System: A system that can absorb or release energy but not mass across the boundary between the system and its surroundings. (7.3)

Closest Packed Structure: A crystal structure in which atoms or molecules are packed as efficiently as possible. (13.1)

Codon: An individual unit of hereditary instruction that consists of three, side by side, side-chains on a molecule of mRNA. (25.10)

Coefficients: Numbers in front of formulas in chemical equations. (2.2)

Coinage Metals: Copper, silver, and gold.

Coke: Coal that has been strongly heated to drive off its volatile components and that is mostly carbon. (23.2)

Collapsing Atom Paradox: The paradox faced by classical physics that predicts a moving electron in an atom should emit energy and spiral into the nucleus. (8.1)

Colligative Property: A property such as vapor pressure lowering, boiling point elevation, freezing point depression, and osmotic pressure whose physical value depends only on the ratio of the numbers of moles of solute and solvent particles and not on their chemical identities. (14.6)

Collision Theory: The rate of a reaction is proportional to the number of effective collisions that occur each second between the reactants. (15.6)

Colloidal Dispersion (Colloid): A homogeneous mixture in which the particles of one or more components have at least one dimension in the range of 1 to 1000 nm—larger than those in a solution but smaller than those in a suspension. (14.10)

Combined Gas Law: See *Gas Law, Combined.*

Combustion: A rapid reaction with oxygen accompanied by a flame and the evolution of heat and light. (6.5)

Common Ion: The ion in a mixture of ionic substances that is common to the formulas of at least two. (19.1)

Common Ion Effect: The solubility of one salt is reduced by the presence of another having a common ion. (19.1)

Competing Reaction: A reaction that reduces the yield of the main product by forming by-products. (4.8)

Complementary Color: The color of the reflected or transmitted light when one component of white light is removed by absorption. (23.8)

Complex Ion (or simply a *Complex*): The combination of one or more anions or neutral molecules (ligands) with a metal ion. (19.4, 23.4)

Compound: A substance consisting of chemically combined atoms from two or more elements and present in a definite ratio. (1.4)

Compound Nucleus: An atomic nucleus carrying excess energy following its capture of some bombarding particle. (22.5)

Compressibility: Capable of undergoing a reduction in volume under increasing pressure. (12.3)

Concentrated Solution: A solution that has a large ratio of the amounts of solute to solvent. (5.1)

Concentration: The ratio of the quantity of solute to the quantity of solution (or the quantity of solvent). (See *Molal Concentration, Molar Concentration, Mole Fraction, Normality, Percent Concentration.*) (5.1)

Concentration Table: A part of the strategy for organizing data needed to make certain calculations, particularly any involving equilibria. (16.9)

Conclusion: A statement that is based on what we think about a series of observations. (1.2)

Condensation: The change of a vapor to its liquid or solid state. (12.4)

Condensation Polymer: A polymer formed from monomers by splitting out a small molecule such as H_2O or CH_3OH. (13.5)

Condensation Polymerization: The process of forming a condensation polymer. (13.5)

Conduction Band: A vacant or partially filled but uninterrupted band in the valence band region of a solid. (13.4)

Conformation: A particular relative orientation or geometric form of a flexible molecule. (10.5)

Conjugate Acid: The species in a conjugate acid-base pair that has the greater number of H^+ units. (17.1)

Conjugate Acid-Base Pair: Two substances (ions or molecules) whose formulas differ by only one H^+ unit. (17.1)

Conjugate Base: The species in a conjugate acid-base pair that has the fewer number of H^+ units. (17.1)

Conservation of Energy, Law of: See *Law of Conservation of Energy.*

Conservation of Mass-Energy, Law of: See *Law of Conservation of Mass-Energy.*

Contact Process: A method for manufacturing sulfuric acid that involves the catalytic oxidation of SO_2 to SO_3, dissolving SO_3 in H_2SO_4 to give $H_2S_2O_7$, and then addition of water to form H_2SO_4. (24.5)

Continuous Spectrum: The electromagnetic spectrum corresponding to the mixture of frequencies present in white light. (8.2)

Contributing Structure: One of a set of two or more Lewis structures used in applying the theory of resonance to the structure of a compound. A resonance structure. (9.9)

Conversion Factor: A ratio constructed from the relationship between two units such as 2.54 cm/1 in., from 1 in. = 2.54 cm. (3.5)

Cooling Curve: A graph showing how the temperature of a substance changes as heat is removed from it at a constant rate as the substance undergoes changes in its physical state. (12.7)

Coordinate Covalent Bond: A covalent bond in which both electrons originated from one of the joined atoms, but otherwise like a covalent bond in all respects. (9.10)

Coordination Compound (Coordination Complex): A complex or its salt. (19.4, 23.4)

Coordination Number: The number of donor atoms that surround a metal ion. (23.6)

Copolymer: A polymer made from two or more different monomers. (13.5)

Core Electrons: The inner electrons of an atom that are not exposed to the electrons of other atoms when chemical bonds form. (8.6)

Corrosion: The slow oxidation of metals exposed to air or water. (6.5)

Coulomb (C): The SI unit of electrical charge; the charge on 6.25×10^{18} electrons; the amount of charge that passes a fixed point of a wire conductor when a current of 1 A flows for 1 s. (21.2)

Covalent Bond: A chemical bond that results when atoms share electron pairs. (9.1)

Covalent Crystal (Network Solid): A crystal in which the lattice positions are occupied by atoms that are covalently bonded to the atoms at adjacent lattice sites. (13.3)

Critical Mass: The mass of a fissile isotope above which a self-sustaining chain reaction occurs. (22.8)

Critical Point: The point at the end of a vapor pressure versus temperature curve for a liquid and that corresponds to the critical pressure and the critical temperature. (12.9)

Critical Pressure (P_c): The vapor pressure of a substance at its critical temperature. (12.9)

Critical Temperature (T_c): The temperature above which a substance cannot exist as a liquid regardless of the pressure. (12.9)

Cross-Link: A bridge formed between polymer strands. (13.5)

Crystal Field Splitting (Δ): The difference in energy between sets of d orbitals in a complex ion. (23.8)

Crystal Field Theory: A theory that considers the effects of the polarities or the charges of the ligands in a complex ion on the energies of the d orbitals of the central metal ion. (23.8)

Crystal Lattice: The repeating symmetrical pattern of atoms, molecules, or ions that occurs in a crystal. (13.1)

Cubic Closest Packing (ccp): Efficient packing of spheres with an A-B-C-A-B-C . . . alternating stacking of layers of spheres. (13.1)

Cubic Meter (m^3): The SI derived unit of volume. (3.2)

Curie (Ci): A unit of activity for radioactive samples, equal to 3.7×10^{10} disintegrations s^{-1}. (22.6)

D

Dalton: One atomic mass unit, u.

Dalton's Atomic Theory: Matter consists of tiny, indestructible particles called atoms. All atoms of one element are identical. The atoms of different elements have different masses. Atoms combine in definite ratios by atoms when they form compounds. (1.5)

Dalton's Law of Partial Pressures: See *Partial Pressures, Law of.*

Data: The information (often in the form of physical quantities) obtained in an experiment or other experience or from references. (1.2)

Debye: Unit used to express dipole moments. $1 \text{ D} = 3.34 \times 10^{-30} \text{ C m}$ (coulomb meter). (9.5)

Decay Constant: The first order rate constant for radioactive decay. (22.6)

Decimal Multipliers: Factors—exponentials of 10 or decimals—that are used to define larger or smaller units. (3.2)

Decomposition: A chemical reaction that changes one substance into two or more simpler substances.

Dehydration: Removal of water from a substance. (2.1)

Dehydration Reaction: Formation of a carbon−carbon double bond by removal of the components of water from an alcohol. (25.3)

Deliquescent Compound: A compound able to absorb enough water from humid air to form a concentrated solution.

Delocalization Energy: The difference between the energy a substance would have if its molecules had no delocalized molecular orbitals and the energy it has because of such orbitals. (10.8)

Delocalized Molecular Orbital: A molecular orbital that spreads over more than two nuclei. (10.8)

ΔH_{fusion}: The amount of energy needed to change one mole of a solid into one mole of liquid at a constant temperature. (12.7)

$\Delta H_{sublimation}$: The amount of energy needed to change one mole of solid to one mole of vapor at a constant temperature. (12.7)

$\Delta H_{vaporization}$: The amount of energy needed to convert one mole of liquid to one mole of vapor at a constant temperature. (12.7)

Density: The ratio of an object's mass to its volume. (3.6)

Dependent Variable: The experimental variable of a pair of variable whose value is determined by the other, the independent variable.

Derived Unit: Any unit defined solely in terms of base units. (3.2)

Deuterium, ${}^2_1$H: The isotope of hydrogen with a mass number of 2.

Dialysis: The passage of small molecules and ions, but not species of a colloidal size, through a semipermeable membrane. (14.8)

Diamagnetism: The condition of not capable of being attracted to a magnet. (8.4)

Diaphragm Cell: An electrolytic cell used to manufacture sodium hydroxide by the electrolysis of aqueous sodium chloride. (21.9)

Diatomic Substance (Diatomic Molecule): A molecular substance made from two atoms. (2.1)

Diborane: The simplest hydrogen compound of boron, B_2H_6. (24.3)

Diffraction: Constructive and destructive interference by waves. (8.3, 13.2)

Diffraction Pattern: The image formed on a screen or a photographic film caused by the diffraction of electromagnetic radiation such as visible light or X rays. (8.3, 13.2)

Diffusion: The spontaneous intermingling of one substance with another. (11.7)

Dilute Solution: A solution in which the ratio of the quantities of solute to solvent is small. (5.1)

Dilution: The process whereby a concentrated solution is made more dilute. (5.11)

Dimensional Analysis: See *Factor Label Method.*

Dipole: Partial positive and partial negative charges separated by a distance. (9.5)

Dipole-Dipole Attractions: Attractions between molecules that are dipoles. (12.2)

Dipole Moment: The product of the sizes of the partial charges in a dipole multiplied by the distance between them; a measure of the polarity of a molecule. (9.5)

Diprotic Acid: An acid that can furnish two H^+ per molecule. (5.5)

Disaccharide: A carbohydrate whose molecules can be hydrolyzed to two monosaccharides. (25.7)

Dispersion Forces: Another term for London forces. (12.2)

Disproportionation: A redox reaction in which a portion of a substance is oxidized at the expense of the rest, which is reduced. (24.5)

Dissociation: The separation of preexisting ions when an ionic compound dissolves or melts. (5.2)

Distorted Tetrahedron: A description of a molecule in which the central atom is surrounded by 5 electron pairs, one of which is a lone pair of electrons. The central atom is bonded to four other atoms. The structure is also said to have a see-saw shape. (10.2)

DNA: Deoxyribonucleic acid; a nucleic acid that hydrolyzes to deoxyribose, phosphate ion, adenine, thymine, guanine, and cytosine, and that is the carrier of genes. (25.10)

DNA Double Helix: Two oppositely running strands of DNA held in a helical configuration by inter-strand hydrogen bonds. (25.10)

Donor Atom: The atom on a ligand that makes an electron pair available in the formation of a complex. (19.4, 23.4)

Doped Semiconductor: A semiconductor chip (e.g., silicon) to which a trace amount of an element (like boron or arsenic) has been added to alter the conducting properties of the semiconductor. (13.4)

Double Bond: A covalent bond formed by sharing two pairs of electrons. (9.3) A covalent bond consisting of one sigma bond and one pi bond. (10.6)

Double Replacement Reaction (Meta-thesis Reaction): A reaction of two salts in which cations and anions exchange partners (e.g., $AgNO_3 + NaCl \rightarrow AgCl + NaNO_3$). (5.4)

Double Salt: A salt consisting of two different salts combined in a definite ratio.

Downs Cell: An electrolytic cell for the industrial production of sodium. (21.9)

Ductility: A metal's ability to be drawn (or stretched) into wire. (1.7)

Dynamic Equilibrium: A condition in which two opposing processes are occurring at equal rates. (5.7)

E

Effective Collision: A collision between molecules that is capable of leading to a net chemical change. (15.6)

Effective Nuclear Charge: The net positive charge an outer electron experiences as a result of the partial screening of the full nuclear charge by core electrons. (8.8)

Effusion: The movement of a gas through a very tiny opening into a region of lower pressure. (11.7)

Effusion, Law of (Graham's Law): The rates of effusion of gases are inversely proportional to the square roots of their densities when compared at identical pressures and temperatures.

$$\text{effusion rate} \propto \frac{1}{\sqrt{d}} \ (\text{constant } P \text{ and } T)$$

where d is the gas density. (11.7)

Einstein Equation: $\Delta E = \Delta m_\text{o} c^2$ where ΔE is the energy obtained when a quantity of rest mass, Δm_o, is destroyed, or the energy lost when this quantity of mass is created. (22.1)

Elastomers: Polymers with elastic properties. (13.5)

Electric Dipole: Two poles of electric charge separated by a distance. (9.5)

Electrochemical Change: A chemical change that is caused by or that produces electricity. (21 Introduction)

Electrochemistry: The study of electrochemical changes. (21 Introduction)

Electrolysis: The production of a chemical change by the passage of electricity through a solution that contains ions or through a molten ionic compound. (21.7)

Electrolysis Cell: An apparatus for electrolysis. (21.7)

Electrolyte: A compound that conducts electricity either in solution or in the molten state. (5.2)

Electrolytic Cell: See *Electrolysis Cell.*

Electrolytic Conduction: The transport of electrical charge by ions. (21.1)

Electrolyze: To pass electricity through an electrolyte and cause a chemical change. (21.7)

Electromagnetic Energy: Energy transmitted by wavelike oscillations in the strengths of electrical and magnetic fields; light energy. (8.1)

Electromagnetic Wave (Electromagnetic Radiation): The successive series of oscillations in the strengths of electrical and magnetic fields associated with light, microwaves, gamma rays, ultraviolet rays, infrared rays and the like. (8.1)

Electromagnetic Spectrum: The distribution of frequencies of electromagnetic radiation among various types of such radiation—microwave, infrared, visible, ultraviolet, X, and gamma rays. (8.1)

Electromotive Force (emf): The voltage produced by a galvanic cell and that can make electrons move in a conductor. (21.1)

Electron (e^- or $_{-1}^0 e$): (a) A subatomic particle with a charge of 1− and mass of 0.0005486 u (9.10939×10^{-28} g) and that occurs outside an atomic nucleus. The particle that moves when an electric current flows. (1.6) (b) A beta particle. (22.3)

Electron Affinity: The energy change (usually expressed in kJ mol^{-1}) that occurs when an electron adds to an isolated gaseous atom or ion. (8.8)

Electron Capture: The capture by a nucleus of an orbital electron and that changes a proton into a neutron in the nucleus. (22.3)

Electron Cloud: Because of its wave properties, an electron's influence spreads out like a cloud around the nucleus. (8.7)

Electron Configuration: The distribution of electrons in an atom's orbitals. (8.5)

Electron Density: The concentration of the electron's charge within a given volume. (8.7)

Electron Domain: A region around an atom where one or more electron pairs are concentrated and which influences the shape of a molecule. (10.2)

Electron Domain Model: See *Valence Shell Electron-Pair Repulsion Theory.*

Electronegativity: The relative ability of an atom to attract electron density toward itself when joined to another atom by a covalent bond. (9.5)

Electronic Structure: The distribution of electrons in an atom's orbitals. (8.5)

Electron-Pair Bond: A covalent bond. (9.3)

Electron Spin: The spinning of an electron about its axis that is believed to occur because the electron behaves as a tiny magnet. (8.4)

Electron Volt (eV): The energy an electron receives when it is accelerated under the influence of 1 V and equal to 1.6×10^{-19} J. (22.3)

Electroplating: Depositing a thin metallic coating on an object by electrolysis. (21.9)

Element: A substance in which all of the atoms have the same atomic number. A substance that cannot be broken down by chemical reactions into anything that is both stable and simpler. (1.4)

Elementary Process: One of the individual steps in the mechanism of a reaction. (15.8)

Elimination Reaction: The loss of a small molecule from a larger molecule as in the elimination of water from an alcohol. (25.3)

Emission Spectrum: See *Atomic Spectrum.*

Empirical Facts: Facts discovered by performing experiments. (1.2)

Empirical Formula: A chemical formula that uses the smallest whole-number subscripts to give the proportions by atoms of the different elements present. (4.4)

Emulsifying Agent: A substance that stabilizes an emulsion. (14.10)

Emulsion: A colloidal dispersion of one liquid in another. (10.10)

Enantiomers: Stereoisomers whose molecular structures are related as an object to its mirror image but that cannot be superimposed. (23.7)

Endothermic: Descriptive of a change in which a system's internal energy increases. (7.4)

End Point: The moment in a titration when the indicator changes color and the titration is ended. (5.13, 18.8)

Energy: Something that matter possesses by virtue of an ability to do work. (2.3, 7.1)

Energy Band: A large number of closely spaced energy levels in a solid formed by combining atomic orbitals of similar energy from each of the atoms in the solid. (13.4)

Energy Density: For a galvanic cell, the ratio of the energy available to the volume of the cell. (21.6)

Energy Level: A particular energy an electron can have in an atom or a molecule. (8.2)

Enthalpy (H): The heat content of a system. (7.5, 20.1)

Enthalpy Change (ΔH): The difference in enthalpy between the initial state and the final state for some change. (7.5, 20.1)

Enthalpy Diagram: A graphical depiction of enthalpy changes following different paths from reactants to products. (7.7)

Enthalpy of Solution: The enthalpy change that accompanies the formation of a solution from the solute and the solvent. (14.2)

Entropy (S): A thermodynamic quantity related to the number of equivalent ways the energy of a system can be distributed. The greater this number, the more probable is the state and the higher is the entropy.

Enzyme: A catalyst in a living system and that consists of a protein. (25.9)

Equation of State of an Ideal Gas: See *Gas Law, Ideal.*

Equatorial Bond: A covalent bond located in the plane perpendicular to the long axis of a trigonal bipyramidal molecule. (10.1)

Equilibrium: See *Dynamic Equilibrium.*

Equilibrium Constant: The value that the mass action expression has when the system is at equilibrium. (16.3)

Equilibrium Law: The mathematical equation for a particular equilibrium system that sets the mass action expression equal to the equilibrium constant. (16.3)

Equilibrium Vapor Pressure of a Liquid: The pressure exerted by a vapor in equilibrium with its liquid state. (12.5)

Equilibrium Vapor Pressure of a Solid: The pressure exerted by a vapor in equilibrium with its solid state. (12.5)

Equivalence: A relationship between two quantities expressed in different units. (3.6)

Equivalence Point: The moment in a titration when the number of equivalents of the reactant added from a buret equals the number of equivalents of another reactant in the receiving flask. (18.8)

Error in a Measurement: The difference between a measurement and the "true" value we are trying to measure. (3.3)

Ester: An organic compound whose molecules have the ester group. (25.5)

$$-\overset{\overset{\displaystyle O}{\|}}{C}-O-C$$
Ester group

Ether: An organic compound in whose molecules two hydrocarbon groups are joined to an oxygen. (25.3)

Ethyl Group: CH_3CH_2-.

Evaporate: To change from a liquid to a vapor. (12.3)

Exact Number: A number obtained by a direct count or that results by a definition; and that is considered to have an infinite number of significant figures. (3.1)

Exon: One of a set of sections of a DNA molecule (separated by introns) that, taken together, constitute a gene. (25.10)

Exothermic: Descriptive of a change in which energy leaves a system and enters the surroundings. (7.4)

Expansion Work: See *Pressure-Volume Work*.

Exponential Notation: See *Scientific Notation*.

Extensive Property: A property of an object that is described by a physical quantity whose magnitude is proportional to the size or amount of the object (e.g., mass or volume). (1.3)

F

Face-Centered Cubic (fcc) Unit Cell: A unit cell having identical atoms, molecules, or ions at the corners of a cube and also in the center of each face of the cube. (13.1)

Factor-Label Method: A problem-solving technique that uses the correct cancellation of the units of physical quantities as a guide for the correct setting-up of the solution to the problem. (3.5)

Fahrenheit Scale: A temperature scale on which water freezes at 32 °F and boils at 212 °F (at 1 atm) and between which points there are 180 degree divisions called Fahrenheit degrees. (3.2)

Family of Elements: See *Group*.

Faraday ($\mathscr{F}$): One mole of electrons; 9.65×10^4 coulombs. (21.3)

Faraday Constant ($\mathscr{F}$): 9.65×10^4 coulomb/mol e^-. (21.3)

Fatty Acid: One of several long-chain carboxylic acids produced by the hydrolysis (digestion) of a lipid. (25.8)

First Law of Thermodynamics: A formal statement of the law of conservation of energy. $\Delta E = q + w$ (7.5, 20.1)

First Order Reaction: A reaction with a rate law in which rate $= k[A]^1$, where A is a reactant. (15.4)

Fissile Isotope: An isotope capable of undergoing fission following neutron capture. (22.8)

Fission: The breaking apart of atomic nuclei into smaller nuclei accompanied by the release of energy, and the source of energy in nuclear reactors. (22.8)

Flotation: A method for concentrating sulfide ores of copper and lead by bubbling air through a slurry of oil-coated ore particles. The sulfides, but not soil or other rock particles, stick to the rising air bubbles and collect in the foam at the surface. (23.2)

Force: Anything that can cause an object to change its motion or direction.

Formal Charge: The apparent charge on an atom in a molecule or polyatomic ion as calculated by a set of rules that generally assign a bonding pair of electrons to the more electronegative of the two atoms held by the bond. (9.8)

Formation Constant (K_{form}): The equilibrium constant for an equilibrium involving the formation of a complex ion. Also called the stability constant. (19.4)

Formula: See *Chemical Formula*.

Formula Unit: A particle that has the composition given by the chemical formula. (2.6)

Formula Mass: The sum of the atomic masses (in u) of all of the atoms represented in the chemical formula; called the *molecular mass* when applied to molecular substances; called the *molar mass* when given units of g mol^{-1}. (4.1)

Forward Reaction: In a chemical equation, the reaction as read from left to right. (5.7)

Fossil Fuels: Coal, oil, and natural gas. (25.2)

Free Element: An element that is not combined with another element in a compound. (2.1)

Free Energy: See *Gibbs Free Energy*.

Free Energy Diagram: A plot of the changes in free energy for a multicomponent system versus the composition. (20.8)

Free Radical: An atom, molecule, or ion that has one or more unpaired electrons. (13.5, Facets of Chemistry 15.1, 22.6)

Freezing Point Depression: A colligative property of a liquid solution by which the freezing point of the solution is lower than that of the pure solvent. (14.7)

Frequency (ν): The number of cycles per second of electromagnetic radiation. (8.5)

Frequency Factor: The proportionality constant, A, in the Arrhenius equation. (15.7)

Fuel Cell: An electrochemical cell in which electricity is generated from the same reactions that accompany the burning of a fuel. (21.6)

Fullerene: An allotrope of carbon made of an extended joining together of five- and six-membered rings of carbon atoms. (13.8)

Functional Group: The group of atoms of an organic molecule that enters into a characteristic set of reactions that are independent of the rest of the molecule. (25.1)

Fusion: (a) Melting. (12.7) (b) The formation of atomic nuclei by the joining together of the nuclei of lighter atoms. (22.2)

G

G: See *Gibbs Free Energy*.

ΔG: See *Gibbs Free Energy Change*.

$\Delta G°$: See *Standard Free Energy Change*.

$\Delta G_f°$: See *Standard Free Energy of Formation*.

Galvanic Cell: An electrochemical cell in which a spontaneous redox reaction produces electricity. (21.1)

Galvanizing: The coating of a metal, like iron, with another, like zinc, to protect against corrosion.

Gamma Radiation: Electromagnetic radiation with wave lengths in the range of 1 Å or less (the shortest wave lengths of the spectrum). (22.3)

Gangue: The unwanted rock and sand that is separated from an ore. (23.2)

Gas Constant, Universal (R): $R = 0.0821$ liter atm mol^{-1} K^{-1} (11.4)

Gas Law, Combined: For a given mass of gas, the product of its pressure and volume divided by its Kelvin temperature is a constant.

$$PV/T = \text{a constant.} \qquad (11.3)$$

Gas Law, Ideal: $PV = nRT$ (11.4)

Gay-Lussac's Law: See *Pressure-Temperature Law.*

Genetic Code: The correlation of codons with amino acids. (25.10)

Genetic Engineering: The use of enzymes, nucleic acids, and genes to modify an organism's genes for some purpose. (25.10)

Geometric Isomer: One of a set of isomers that differ only in geometry. (23.7, 25.2)

Geometric Isomerism: The existence of isomers whose molecules have identical atomic organizations but different geometries; cis-trans isomers. (23.7, 25.2)

Gibbs Free Energy (G): A thermodynamic quantity that relates enthalpy, (H), entropy (S), and temperature (T) by the equation,

$$G = H - TS. \qquad (20.4)$$

Gibbs Free Energy Change (ΔG): The difference given by

$$\Delta G = \Delta H - T\Delta S \qquad (20.6)$$

Glass: Any amorphous solid. (13.1)

Graham's Law: See *Effusion, Law of.*

Gram (g): 0.001 kg (3.2)

Graphite: The most stable allotrope of carbon, consisting of layers of joined six-member rings of carbon atoms. (13.8)

Gray (Gy): The SI unit of radiation-absorbed dose.

$$1\,Gy = 1\,J\,kg^{-1} \qquad (22.6)$$

Greenhouse Effect: The retention of solar energy made possible by the ability of the greenhouse gases (e.g., CO_2, CH_4, H_2O, and the chlorofluorocarbons) to absorb outgoing radiation and re-radiate some of it back to earth.

Ground State: The lowest energy state of an atom or molecule. (8.2, 8.3)

Group: A vertical column of elements in the periodic table. (1.7)

H

ΔH: See *Enthalpy Change.*

ΔH_{atom}: See *Atomization Energy.*

$\Delta H°$: See *Standard Heat of Reaction.*

$\Delta H_{combustion}$: See *Heat of Combustion.*

$\Delta H_f°$: See *Standard Heat of Formation.*

ΔH_{fusion}: See *Molar Heat of Fusion.*

ΔH_{soln}: See *Heat of Solution.*

$\Delta H_{sublimation}$: See *Molar Heat of Sublimation.*

$\Delta H_{vaporization}$: See *Molar Heat of Vaporization.*

Haber-Bosch Process: An industrial synthesis of ammonia from nitrogen and hydrogen under pressure and at a moderately high temperature in the presence of a catalyst. (16.8)

Half-Cell: That part of a galvanic cell in which either oxidation or reduction takes place. (21.1)

Half-Life ($t_{1/2}$): The time required for a reactant concentration or the mass of a radionuclide to be reduced by half. (15.5, 22.6)

Half-Reaction: A hypothetical reaction that constitutes exclusively either the oxidation or the reduction half of a redox reaction and in whose equation the correct formulas for all species taking part in the change are given together with enough electrons to give the correct electrical balance. (6.2)

Halogen Family: Group VIIA in the periodic table—fluorine, chlorine, bromine, iodine, and astatine. (1.7)

Hall-Héroult Process: A method for manufacturing aluminum by the electrolysis of aluminum oxide in molten cryolite. (21.9)

Hard Water: Water with dissolved Mg^{2+}, Ca^{2+}, Fe^{2+}, or Fe^{3+} ions at a concentration high enough (above 25 mg L^{-1}) to interfere with the use of soap.

Heat: A form of energy. (2.3, 7.1)

Heat Capacity: The quantity of heat needed to raise the temperature of an object by 1 °C. (7.3)

Heat of Combustion: The heat evolved in the combustion of a substance. (7.5)

Heat of Formation, Standard: See *Standard Heat of Formation.*

Heat of Reaction: The heat exchanged between a system and its surroundings when a chemical change occurs in the system. (7.5, 20.1)

Heat of Reaction at Constant Pressure: The heat of a reaction in an open system. (7.5, 20.1)

Heat of Reaction at Constant Volume: The heat of a reaction in a sealed vessel, like a bomb calorimeter. (7.5, 20.1)

Heat of Reaction, Standard: See *Standard Heat of Reaction.*

Heat of Solution (ΔH_{soln}): The energy exchanged between the system and its surroundings when one mole of a solute dissolves in a solvent to make a dilute solution. (14.2)

Heating Curve: A graph showing how the temperature of a substance changes as heat is added to it at a constant rate as the substance undergoes changes in its physical state. (12.7)

Henderson-Hasselbalch equation:

$$pH = pK_a + \log \frac{[A^-]_{initial}}{[HA]_{initial}} \quad \text{or}$$

$$pH = pK_a + \log \frac{[salt]}{[acid]} \qquad (18.5)$$

Henry's Law: See *Pressure-Solubility Law.*

Hertz (Hz): 1 cycle s^{-1}; the SI unit of frequency. (8.1)

Hess's Law: For any reaction that can be written in steps, the standard heat of reaction is the same as the sum of the standard heats of reaction for the steps. (7.7)

Hess's Law Equation: For the change,

$$aA + bB + \cdots \rightarrow nN + mM + \cdots$$

$$\Delta H° = \left[\begin{array}{l} \text{Sum of } \Delta H_f° \text{ of all} \\ \text{of the products.} \end{array} \right]$$
$$- \left[\begin{array}{l} \text{Sum of } \Delta H_f° \text{ of all} \\ \text{of the reactants.} \end{array} \right] \quad (7.8)$$

Heterocyclic Compound: A compound whose molecules have rings that include one or more multivalent atoms other than carbon. (25.1)

Heterogeneous Catalyst: A catalyst that is in a different phase than the reactants and onto whose surface the reactant molecules are adsorbed and where they react. (15.9)

Heterogeneous Equilibrium: An equilibrium involving more than one phase. (16.7)

Heterogeneous Mixture: A mixture that has two or more phases with different properties. (1.4)

Heterogeneous Reaction: A reaction in which not all of the chemical species are in the same phase. (15.2, 16.7)

Hexagonal Closest Packing (hcp): Efficient packing of spheres with an A-B-A-B-... alternating stacking of layers of spheres. (13.1)

High Spin Complex: A complex ion or coordination compound in which there is the maximum number of unpaired electrons. (23.8)

Homogeneous Catalyst: A catalyst that is in the same phase as the reactants. (15.9)

Homogeneous Equilibrium: An equilibrium system in which all components are in the same phase. (16.7)

Homogeneous Mixture: A mixture that has only one phase and that has uniform properties throughout. (1.4)

Homogeneous Reaction: A reaction in which all of the chemical species are in the same phase. (15.2, 16.7)

Hund's Rule: Electrons that occupy orbitals of equal energy are distributed with unpaired spins as much as possible among all such orbitals. (8.5)

Hybrid Atomic Orbitals: Orbitals formed by mixing two or more of the basic atomic orbitals of an atom and that make possible more effective overlaps with the orbitals of adjacent atoms than do ordinary atomic orbitals. (10.5)

Hydrate: A compound that contains molecules of water in a definite ratio to other components. (2.1)

Hydrated Ion: An ion surrounded by a cage of water molecules that are attracted by the charge on the ion. (5.2)

Hydration: The development in an aqueous solution of a cage of water molecules about ions or polar molecules of the solute. (14.1)

Hydration Energy: The enthalpy change associated with the hydration of gaseous ions or molecules as they dissolve in water. (14.2)

Hydride: (a) A binary compound of hydrogen. (b) A compound containing the hydride ion (H^-). (24.3)

Hydrocarbon: An organic compound whose molecules consist entirely of carbon and hydrogen atoms. (2.4, 25.2)

Hydrogen Bond: An extra strong dipole-dipole attraction between a hydrogen bound covalently to nitrogen, oxygen, or fluorine and another nitrogen, oxygen, or fluorine atom. (12.2)

Hydrogen Electrode: The standard of comparison for reduction potentials and for which $E^\circ_{H^+}$ has a value of 0.00 V (25 °C, 1 atm) when $[H^+] = 1\ M$ in the reversible half-cell reaction:

$$2H^+(aq) + 2e^- \rightleftharpoons H_2(g) \qquad (21.2)$$

Hydrolysis: A reaction with water. (13.7, 24.3)

Hydrometer: A device for measuring specific gravity. (21.6)

Hydronium Ion: H_3O^+. (5.6)

Hydrophilic Group: A polar molecular unit capable of having dipole-dipole attractions or hydrogen bonds with water molecules. (14.10, 25.8)

Hydrophobic Group: A nonpolar molecular unit with no affinity for the molecules of a polar solvent, like water. (14.10, 25.8)

Hypertonic Solution: A solution that has a higher osmotic pressure than cellular fluids. (14.8)

Hypothesis: A tentative explanation of the results of experiments. (1.2)

Hypotonic Solution: A solution that has a lower osmotic pressure than cellular fluids. (14.8)

I

Ideal Gas: A hypothetical gas that obeys the gas laws exactly. (11.3)

Ideal Gas Law: $PV = nRT$ (11.4)

Ideal Solution: A hypothetical solution that would obey the vapor pressure–concentration law (Raoult's law) exactly. (14.2)

Immiscible: Insoluble. (14.1)

Incompressible: Incapable of losing volume under increasing pressure. (12.3)

Independent Variable: The experimental variable of a pair of variables whose value is first selected and from which the value of the dependent variable then results.

Indicator: A chemical put in a solution being titrated and whose change in color signals the end point. (5.13, 17.5, 18.8)

Induced Dipole: A dipole created when the electron cloud of an atom or a molecule is distorted by a neighboring dipole or by an ion. (12.2)

Inert Gas: Any of the noble gases—Group VIIIA of the periodic table. Any gas that has virtually no tendency to react. (1.7)

Inexact Number: Any number obtained by measurement. (3.1)

Initiation Step: The step in a chain reaction that produces reactive species that can start chain propagation steps. (Facets of Chemistry 15.1)

Initiator: A substance that initiates a free radical polymerization reaction. (13.5)

Inner Transition Elements: Members of the two long rows of elements below the main body of the periodic table—elements 58–71 and elements 90–103. (1.7)

Inorganic Compound: A compound made from any elements except those compounds of carbon classified as organic compounds. (2.5)

Instability Constant (K_{inst}): The reciprocal of the formation constant for an equilibrium in which a complex ion forms. (19.4)

Instantaneous Dipole: A momentary dipole in an atom, ion, or molecule caused by the erratic movement of electrons. (12.2)

Instantaneous Rate: The rate of reaction at any particular moment during a reaction. (15.3)

Intensive Property: A property whose physical quantity is independent of the size of the sample, such as density or temperature. (1.3)

Intercalation: The insertion of small atoms or ions between layers in a crystal such as graphite. (21.6)

Interference Fringes: Pattern of light produced by waves that undergo diffraction. (8.3)

Integrated Rate Law: A rate law that relates concentration to time. (15.5)

Interhalogen Compound: A compound whose molecules are made from two different halogens. (24.7)

Intermolecular Forces (Intermolecular Attractions): Attractions *between* neighboring molecules. (12.2)

Internal Energy (E): The sum of all of the kinetic energies and potential energies of the particles within a system. (7.2, 20.1)

International System of Units (SI): The successor to the metric system of measurements that retains most of the units of the metric system and their decimal relationships but employs new reference standards. (3.2)

Intramolecular Forces: Forces of attraction within molecules; chemical bonds. (12.2)

Intron: One of a set of sections of a DNA molecule that separate the exon sections of a gene from each other. (25.10)

Inverse Square Law: The intensity of a radiation is inversely proportional to the square of the distance from its source. (22.6)

Ion: An electrically charged particle on the atomic or molecular scale of size. (2.6)

Ion Pair: A more or less loosely associated pair of ions in a solution. (14.9)

Ion-Dipole Attraction: The attraction between an ion and the charged end of a polar molecule. (12.2)

Ion-Electron Method: A method for balancing redox reactions that uses half-reactions. (6.2)

Ion-Induced Dipole Attraction: Attraction between an ion and a dipole induced in a neighboring molecule. (12.2)

Ionic Bond: The attractions between ions that hold them together in ionic compounds. (9.1)

Ionic Character: The extent to which a covalent bond has a dipole moment and is polarized. (9.5, 23.3)

Ionic Compound: A compound consisting of positive and negative ions. (2.6)

Ionic Crystal: A crystal that has ions located at the lattice points. (13.3)

Ionic Equation: A chemical equation in which soluble strong electrolytes are written in dissociated or ionized form. (5.3)

Ionic Potential: The ratio of an ion's charge to its radius $\left(\phi = \dfrac{q}{r}\right)$. (23.3)

Ionic Reaction: A chemical reaction in which ions are involved. (5.3)

Ionization Energy (IE): The energy needed to remove an electron from an isolated, gaseous atom, ion, or molecule (usually given in units of kJ mol^{-1}). (8.8)

Ionization Reaction: A reaction of chemical particles that produces ions. (5.4)

Ionizing Radiation: Any high-energy radiation—X rays, gamma rays, or radiations from radionuclides—that generates ions as it passes through matter. (22.6)

Ion Product: The mass action expression for the solubility equilibrium involving the ions of a salt and equal to the product of the molar concentrations of the ions, each concentration raised to a power that equals the number of ions obtained from one formula unit of the salt. (19.1)

Ion Product Constant of Water (K_w): $K_w = [H^+][OH^-]$. (17.5)

Ion Product of Water: $[H^+][OH^-]$. (17.5)

Isoelectronic: Particles having the same electronic structure for two or more atoms, molecules, or ions are said to be isoelectronic. (24.5)

Isolated System: A system that can not exchange matter or energy with its surroundings. (7.3)

Isomer: One of a set of compounds that have identical molecular formulas but different structures. (23.7, 25.2)

Isomerism: The existence of sets of isomers. (23.7, 25.2)

Isopropyl Group: $(CH_3)_2CH-$

Isotonic Solution: A solution that has the same osmotic pressure as cellular fluids. (14.8)

Isotopes: Atoms of the same element with different atomic masses. Atoms of the same element with different numbers of neutrons in their nuclei. (1.5, 1.6)

IUPAC Rules: The formal rules for naming substances as developed by the International Union of Pure and Applied Chemistry.

J

Joule (J): The SI unit of energy.

$$1 \text{ J} = 1 \text{ kg m}^2 \text{ s}^{-2}$$

and

$$1 \text{ J} = 4.184 \text{ cal (exactly).} \qquad (7.1)$$

K

K: See *Kelvin*.

K_a: See *Acid Ionization Constant*.

K_b: See *Base Ionization Constant*.

K_{form}: See *Formation Constant*.

K_{inst}: See *Instability Constant*.

K_{sp}: See *Solubility Product Constant*.

K_{spa}: See *Acid Solubility Product*.

K_w: See *Ion Product Constant of Water*.

K-Capture: See *Electron Capture*.

Kelvin (K): One degree on the Kelvin scale of temperature and identical in size to the Celsius degree. (3.2)

Kelvin Scale: The temperature scale on which water freezes at 273.15 K and boils at 373.15 K and that has 100 degree divisions called kelvins between these points. K = °C + 273.15. (3.2)

Ketone: An organic compound whose molecules have the carbonyl group $(C{=}O)$ flanked by hydrocarbon groups. (25.5)

Kilocalorie (kcal): 1000 cal. (7.1)

Kilogram (kg): The base unit for mass in the SI and equal to the mass of a cylinder of platinum-iridium alloy kept by the International Bureau of Weights and Measures at Sevres, France. 1 kg = 1000 g. (3.2)

Kilojoule (kJ): 1000 J. (7.1)

Kinetic Energy (KE): Energy of motion. KE = $(1/2)mv^2$. (2.3, 7.1)

Kinetic Molecular Theory: Molecules of a substance are in constant motion with a distribution of kinetic energies at a given temperature. The average kinetic energy of the molecules is proportional to the Kelvin temperature. (7.2)

Kinetic Theory of Gases: A set of postulates used to explain the gas laws. A gas consists of an extremely large number of very tiny, very hard particles in constant, random motion. They have negligible volume and, between collisions, experience no forces between themselves. (11.8)

L

Lanthanide Elements: Elements 58– 71. (1.7)

Lattice: A symmetrical pattern of points arranged with constant repeat distances arranged along lines oriented at constant angles. (13.1)

Lattice Energy: Energy released by the imaginary process in which isolated ions come together to form a crystal of an ionic compound. (9.1)

Law: A description of behavior (and not an *explanation* of behavior) based on the results of many experiments. (1.2)

Law of Combining Volumes: When gases react at the same temperature and pressure, their combining volumes are in ratios of simple whole numbers. (11.3)

Law of Conservation of Energy: The energy of the universe is constant; it can be neither created nor destroyed but only transferred and transformed. (2.3)

Law of Conservation of Mass: No detectable gain or loss in mass occurs in chemical reactions. Mass is conserved. (1.5)

Law of Conservation of Mass-Energy: The sum of all the mass in the universe and of all of the energy, expressed as an equivalent in mass (calculated by the Einstein equation), is a constant. (22.1)

Law of Definite Proportions: In a given chemical compound, the elements are always combined in the same proportion by mass. (1.5)

Law of Gas Effusion: See *Effusion, Law of.*

Law of Multiple Proportions: Whenever two elements form more than one compound, the different masses of one element that combine with the same mass of the other are in a ratio of small whole numbers. (1.5)

Law of Partial Pressures: See *Partial Pressures, Dalton's Law of.*

Law of Radioactive Decay: Activity = $-\dfrac{\Delta N}{\Delta t} = kN$, where ΔN is the change in the number of radioactive nuclei during

the time span Δt, and k is the decay constant. (22.6)

Lead Storage Battery: A galvanic cell of about 2 V involving lead and lead(IV) oxide in sulfuric acid. (21.6)

Le Châtelier's Principle: When a system that is in dynamic equilibrium is subjected to a disturbance that upsets the equilibrium, the system undergoes a change that counteracts the disturbance and, if possible, restores the equilibrium. (12.8, 16.8)

Leclanché Cell: See *Zinc-Manganese Dioxide Cell.*

Lewis Acid: An electron-pair acceptor. (17.3)

Lewis Base: An electron-pair donor. (17.3)

Lewis Structure (Lewis Formula): A structural formula drawn with Lewis symbols and that uses dots and dashes to show the valence electrons and shared pairs of electrons. (9.3)

Lewis Symbol: The symbol of an element that includes dots to represent the valence electrons of an atom of the element. (9.2)

Ligand: A molecule or an anion that can bind to a metal ion to form a complex. (19.4, 23.4)

Like-Dissolves-Like Rule: Strongly polar and ionic solutes tend to dissolve in polar solvents and nonpolar solutes tend to dissolve in nonpolar solvents. (14.1)

Limiting Reactant: The reactant that determines how much product can form when non-stoichiometric amounts of reactants are used. (4.7)

Line Spectrum: An atomic spectrum. So named because the light emitted by an atom and focused through a narrow slit yields a series of lines when projected on a screen. (8.2)

Linear Molecule: A molecule all of whose atoms lie on a straight line. (10.1)

Lipid: Any substance found in plants or animals that can be dissolved in nonpolar solvents. (25.8)

Liquid Crystal: A substance that has internal order like a crystal but is also able to flow like a liquid. (13.6)

Liter (L): 1 dm^3. 1 L = 1000 mL = 1000 cm^3 (3.2)

Lithium Ion Cell: A cell in which lithium ions are transferred between the electrodes through an electrolyte, while electrons travel through the external circuit. (21.6)

Lithium–Manganese Dioxide Battery: A battery that uses metallic lithium as the anode and manganese dioxide as the cathode. (21.6)

Localized Bond: A covalent bond in which the pair of electrons is localized between two nuclei. (10.8)

London Forces (Dispersion Forces): Weak attractive forces caused by instantaneous dipole-induced dipole attractions. (12.2)

Lone Pair: A pair of electrons in the valence shell of an atom that is not shared with another atom. An unshared pair of electrons. (10.2)

Low Spin Complex: A coordination compound or a complex ion with electrons paired as much as possible in the lower-energy set of *d* orbitals. (23.8)

M

Macromolecule: A molecule whose molecular mass is very large. (13.5)

Magic Numbers: The numbers 2, 8, 20, 28, 50, 82, and 126, numbers whose significance in nuclear science is that a nuclide in which the number of protons or neutrons equals a magic number has nuclei that are relatively more stable than those of other nuclides nearby in the band of stability. (22.4)

Magnetic Quantum Number (m_ℓ): A quantum number that can have values from $-\ell$ to $+\ell$. (8.3)

Main Group Elements: Elements in any of the A-groups in the periodic table. (1.7)

Main Reaction: The desired reaction between the reactants as opposed to competing reactions that give by-products. (4.8)

Malleability: A metal's ability to be hammered or rolled into thin sheets. (1.7)

Manometer: A device for measuring the pressure within a closed system. The two types—*closed end* and *open end*—differ according to whether the operating fluid (e.g., mercury) is exposed at one end to the atmosphere or not. (11.2)

Mass: A measure of the amount of matter that there is in a given sample. (1.1)

Mass Action Expression: A fraction in which the numerator is the product of the molar concentrations of the products, each raised to a power equal to its coefficient in the equilibrium equation, and the denominator is the product of the molar concentrations of the reactants, each also raised to the power that equals its coefficient in the equation. (For gaseous reactions, partial pressures can be used in place of molar concentrations.) (16.3)

Mass Defect: For a given isotope, it is the mass that changed into energy as the nucleons gathered to form the nucleus, this energy being released from the system. (22.2)

Mass Number: The numerical sum of the protons and neutrons in an atom of a given isotope. (1.6)

Matter: Anything that has mass and occupies space. (1.1)

Mean: Average (3.3)

Measurement: A numerical observation. (3 Introduction)

Mechanism of a Reaction: The series of individual steps (called elementary processes) in a chemical reaction that gives the net, overall change. (15.8)

Melting Point: The temperature at which a substance melts; the temperature at which a solid is in equilibrium with its liquid state. (12.4)

Meniscus: The interface between a liquid and a gas. (12.9)

Mercury Cell: A device for the manufacture of sodium hydroxide. (21.9)

Metal: An element or an alloy that is a good conductor of electricity, that has a shiny surface, and that is malleable and ductile; an element that normally forms positive ions and has an oxide that is basic. (1.7)

Metallic Conduction: Conduction of electrical charge by the movement of electrons. (21.1)

Metallic Crystal: A solid having positive ions at the lattice positions that are attracted to a "sea of electrons" that extends throughout the entire crystal. (13.3)

Metalloids: Elements with properties that lie between those of metals and nonmetals, and that are found in the periodic table around the diagonal line running from boron (B) to astatine (At). (1.7)

Metallurgy: The science and technology of metals, the procedures and reactions that separate metals from their ores, and the operations that create practical uses for metals. (23.2)

Metathesis Reaction: See *Double Replacement Reaction*.

Meter (m): The SI base unit for length. (3.2)

Methyl Group: CH_3—

Metric System: A decimal system of units for physical quantities taken over by the SI. (3.2) See also *International System of Units*.

Micelle: A colloidal-sized group of ions, such as the anions of a detergent, that have clustered to maximize the contact between the ions' hydrophilic heads and water and to minimize the contact between the ions' hydrophobic parts and water. (14.10)

Millibar: 1 mb = 10^{-3} bar. (11.2)

Milliliter (mL): 0.001 L. 1000 mL = 1 L (3.2)

Millimeter (mm): 0.001 m. 1000 mm = 1 m (3.2)

Millimeter of Mercury (mm Hg): A unit of pressure equal to 1/760 atm. 760 mm Hg = 1 atm. 1 mm Hg = 1 torr. (11.2)

Miscible: Mutually soluble. (14.1)

Mixture: Any matter consisting of two or more substances physically combined in no particular proportion by mass. (1.4)

MO Theory: See *Molecular Orbital Theory*.

Model, Theoretical: A picture or a mental construction derived from a set of ideas and assumptions that are imagined to be true because they can be used to explain certain observations and measurements (e.g., the model of an ideal gas). (1.2)

Molal Boiling Point Elevation Constant (k_b): The number of degrees (°C) per unit of molal concentration that a boiling point of a solution is higher than that of the pure solvent. (14.7)

Molal Concentration (m): The number of moles of solute in 1000 g of solvent. (14.5)

Molal Freezing Point Depression Constant (k_f): The number of degrees (°C) per unit of molal concentration that a freezing point of a solution is lower than that of the pure solvent. (14.7)

Molality: The molal concentration. (14.5)

Molar Concentration (M): The number of moles of solute per liter of solution. The molarity of a solution. (5.11)

Molar Enthalpy of Solution: See *Heat of Solution*.

Molar Heat Capacity: The heat that can raise the temperature of 1 mol of a substance by 1 °C; the heat capacity per mole. (7.3)

Molar Heat of Fusion, ΔH_{fusion}: The heat absorbed when 1 mol of a solid melts to give 1 mol of the liquid at constant temperature and pressure. (12.7)

Molar Heat of Sublimation, $\Delta H_{sublimation}$: The heat absorbed when 1 mol of a solid sublimes to give 1 mol of its vapor at constant temperature and pressure. (12.7)

Molar Heat of Vaporization, $\Delta H_{vaporization}$: The heat absorbed when 1 mol of a liquid changes to 1 mol of its vapor at constant temperature and pressure. (12.7)

Molarity: See *Molar Concentration*.

Molar Mass: See *Formula Mass*.

Molar Solubility: The number of moles of solute required to give 1 L of a saturated solution of the solute. (19.1)

Molar Volume, Standard: The volume of 1 mol of a gas at STP; 22.4 L mol^{-1}. (11.3)

Mole (mol): The SI unit for amount of substance; the formula mass in grams of an element or compound; an amount of a chemical substance that contains 6.02×10^{23} formula units. (4.2)

Mole Fraction: The ratio of the number of moles of one component of a mixture to the total number of moles of all components. (11.5)

Mole Percent (mol %): The mole fraction of a component expressed as a percent; mol fraction $\times$ 100%. (11.5)

Molecular Compound: A compound consisting of neutral (but often polar) molecules. (2.4)

Molecular Crystal: A crystal that has molecules or individual atoms at the lattice points. (13.3)

Molecular Equation: A chemical equation that gives the full formulas of all of the reactants and products and that is used to plan an actual experiment. (5.3)

Molecular Formula: A chemical formula that gives the actual composition of one molecule. (2.4, 4.4)

Molecular Kinetic Energy: The energy associated with the motions of and within molecules as they fly about, spinning and vibrating. (7.2)

Molecular Mass: See *Formula Mass.*

Molecular Nanotechnology: See *nanotechnology.*

Molecular Orbital (MO): An orbital that extends over two or more atomic nuclei. (10.7)

Molecular Orbital Theory (MO Theory): A theory about covalent bonds that views a molecule as a collection of positive nuclei surrounded by electrons distributed among a set of bonding and antibonding orbitals of different energies. (10.4, 10.7)

Molecular Weight: See *Formula Mass.*

Molecule: A neutral particle composed of two or more atoms combined in a definite ratio of whole numbers. (1.2, 2.4)

Monatomic: A particle consisting of just one atom. (2.8)

Monoclinic Sulfur: An allotrope of sulfur. (24.2)

Monodentate Ligand: A ligand that can attach itself to a metal ion by only one atom. (23.4)

Monomer: A substance of relatively low formula mass that is used to make a polymer. (13.5)

Monoprotic Acid: An acid that can furnish one H^+ per molecule. (5.4)

Monosaccharide: A carbohydrate that cannot be hydrolyzed. (25.7)

N

Nanotechnology: Deals with using small-scale objects with dimensions of tens to hundreds of atoms and the special properties that accompany them to develop useful applications. (13.8)

Negative Charge: A type of electrical charge possessed by certain particles such as the electron. A negative charge is attracted by a positive charge and is repelled by another negative charge. (1.6)

Nematic Phase: A liquid phase in which molecules have a tendency to line up parallel to each other in one direction but are still able to move past each other up, down, and sideways. (13.6)

Nernst Equation:

$$E_{cell} = E° - \frac{0.0592}{n} \log Q \quad (21.5)$$

Net Ionic Equation: An ionic equation from which spectator ions have been omitted. It is balanced when both atoms and electrical charge balance. (5.3)

Network Solid: See *Covalent Crystal.*

Neutralization, Acid-Base: See *Acid-Base Neutralization.*

Neutral Solution: A solution in which $[H^+] = [OH^-]$ (17.5)

Neutron (n, $_0^1n$): A subatomic particle with a charge of zero, a mass of 1.0086644 u ($1.6749543 \times 10^{-24}$ g) and that exists in all atomic nuclei except those of the hydrogen-1 isotope. (1.6)

Neutron Activation Analysis: A technique to analyze for trace impurities in a sample by studying the frequencies and intensities of the gamma radiations they emit after they have been rendered radioactive by neutron bombardment of the sample. (22.7)

Neutron Emission: A nuclear reaction in which a neutron is ejected. (22.3)

Nicad Battery: A nickel-cadmium cell. (21.6)

Nickel-Cadmium Storage Cell: A galvanic cell of about 1.4 V involving the reaction of cadmium with nickel(IV) oxide. (21.6)

Nitrogen Family: Group VA in the periodic table—nitrogen, phosphorus, arsenic, antimony, and bismuth. (1.7)

Nitrogen Fixation: The conversion of atmospheric nitrogen into chemical forms used by plants and accomplished by soil microorganisms with the assistance of plants.

Noble Gases: Group VIIIA in the periodic table—helium, neon, argon, krypton, xenon, and radon. (1.7)

Node: A place where the amplitude or intensity of a wave is zero. (8.3)

Nomenclature: The names of substances and the rules for devising names. (2.5)

Nonbonding Domain: A region in the valence shell of an atom that holds an unshared pair of electrons and that influences the shape of a molecule. (10.2)

Nonelectrolyte: A compound that in its molten state or in solution cannot conduct electricity. (5.2)

Nonlinear Molecule: A molecule in which the atoms do not lie in a straight line. (10.2)

Nonmetal: A nonductile, nonmalleable, nonconducting element that tends to form negative ions (if it forms them at all) far more readily than positive ions and whose oxide is likely to show acidic properties. (1.7)

Nonmetallic Element: An element without metallic properties; an element with poor electrical conductivity. (1.7)

Nonoxidizing acid: An acid in which the anion is a poorer oxidizing agent than the hydrogen ion (e.g., HCl, H_2SO_4, H_3PO_4). (6.3)

Nonpolar Covalent Bond: A covalent bond in which the electron pair(s) are shared equally by the two atoms. (9.5)

Nonpolar Molecule: A molecule that has no net dipole moment. (9.5)

Nonvolatile: Descriptive of a substance with a high boiling point, a low vapor pressure, and that does not evaporate. (14.6)

Normal Boiling Point: The temperature at which the vapor pressure of a liquid equals 1 atm. (12.6)

n-Type Semiconductor: A semiconductor doped with an impurity that causes the moving charge to consist of negatively charged electrons. (13.4)

Nuclear Equation: A description of a nuclear reaction that uses the special symbols of isotopes, that describes some kind of nuclear transformation or disintegration, and that is balanced when the sums of the atomic numbers on either side of the arrow are equal and the sums of the mass numbers are also equal. (22.3)

Nuclear Fission: See *Fission.*

Nuclear Reaction: A change in the composition or energy of the nuclei of isotopes accompanied by one or more events such as the radiation of nuclear particles or electromagnetic energy, transmutation, fission, or fusion. (22.1)

Nuclei Acids: Polymers in living cells that store and translate genetic information and whose molecules hydrolyze to give a sugar unit (ribose from ribonucleic acid, RNA, or deoxyribose from deoxyribonucleic acid, DNA), a phosphate, and a set of four or five nitrogen-containing, heterocyclic bases (adenine, thymine, guanine, cytosine, and uracil). (25.10)

Nucleon: A proton or a neutron. (1.6)

Nucleus: The hard, dense core of an atom that holds the atom's protons and neutrons. (1.6)

Nylon 6,6: A polymer of a six-carbon dicarboxylic acid and a six-carbon diammine. (13.5)

O

Observation: A statement that accurately describes something we see, hear, taste, feel, or smell. (1.2)

Octahedral Molecule: A molecule in which a central atom is surrounded by six atoms located at the vertices of an imaginary octahedron. (10.1)

Octahedron: An eight-sided figure that can be envisioned as two square pyramids sharing the common square base. (10.1)

Octet (of electrons): Eight electrons in the valence shell of an atom. (9.1)

Octet Rule: An atom tends to gain or lose electrons until its outer shell has eight electrons. (9.1, 9.3)

Odd-Even Rule: When the numbers of protons and neutrons in an atomic nucleus are both even, the isotope is more likely to be stable than when both numbers are odd. (22.4)

Open End Manometer: See *Manometer*.

Optically Active: A property of being able to rotate the plane of plane polarized light. (23.7)

Optical Isomers: Stereoisomers other than geometric (cis-trans) isomers and that include substances that can rotate the plane of plane polarized light. (23.7)

Orbital: An electron waveform with a particular energy and a unique set of values for the quantum numbers n, ℓ, and m_ℓ. (8.3)

Orbital Diagram: A diagram showing an atom's orbitals in which the electrons are represented by arrows to indicate paired and unpaired spins. (8.5)

Order (of a Reaction): The sum of the exponents in the rate law is the *overall* order. Each exponent gives the order of the reaction with respect to a specific reactant. (15.4)

Ore: A substance in the earth's crust from which an element or compound can be extracted at a profit. (23.2)

Organic Chemistry: The study of the compounds of carbon that are not classified as inorganic. (2.4, 25.1)

Organic Compound: Any compound of carbon other than a carbonate, bicarbonate, cyanide, cyanate, carbide, or gaseous oxide. (2.4, 25.1)

Orthorhombic Sulfur: The most stable allotrope of sulfur, composed of S_8 rings. (24.2)

Osmosis: The passage of solvent molecules, but not those of solutes, through a semipermeable membrane; the limiting case of dialysis. (14.8)

Osmotic Membrane: A membrane that allows passage of solvent, but not solute particles. (14.8)

Osmotic Pressure: The back pressure that would have to be applied to prevent osmosis; one of the colligative properties. (14.8)

Ostwald Process: An industrial synthesis of nitric acid from ammonia. (24.5)

Outer Electrons: The electrons in the occupied shell with the largest principal quantum number. An atom's electrons in its valence shell. (8.6)

Outer Shell: The occupied shell in an atom having the highest principal quantum number (n). (8.6)

Overall Order of Reaction: The sum of the exponents on the concentration terms in a rate law. (15.4)

Overlap of Orbitals: A portion of two orbitals from different atoms that share the same space in a molecule. (10.4)

Oxidation: A change in which an oxidation number increases (becomes more positive). A loss of electrons. (6.1)

Oxidation Number: The charge that an atom in a molecule or ion would have if all of the electrons in its bonds belonged entirely to the more electronegative atoms; the oxidation state of an atom. (6.1)

Oxidation-Reduction Reaction: A chemical reaction in which changes in oxidation numbers occur. (6.1)

Oxidation State: See *Oxidation Number*.

Oxidizing Acid: An acid in which the anion is a stronger oxidizing agent than H^+ (e.g., $HClO_4$, HNO_3). (6.3)

Oxidizing Agent: The substance that causes oxidation and that is itself reduced. (6.1)

Oxoacid: An acid that contains oxygen besides hydrogen and another element (e.g., HNO_3, H_3PO_4, H_2SO_4). (5.6, 17.2)

Oxygen Family: Group VIA in the periodic table—oxygen, sulfur, selenium, tellurium, and polonium. (1.7)

Ozone: A very reactive allotrope of oxygen with the formula O_3. (24.3)

P

p-n Junction: The interface between p-type and n-type semiconductors. (13.4)

p-Type Semiconductor: A semiconductor doped with an impurity that enables the charge to move as positively charged holes. (13.4)

Pairing Energy: The energy required to force two electrons to become paired and occupy the same orbital. (23.8)

Paramagnetism: The weak magnetism of a substance whose atoms, molecules, or ions have unpaired electrons. (8.4)

Partial Charge: Charges at opposite ends of a dipole that are fractions of full 1+ or 1− charges. (9.5)

Partial Pressure: The pressure contributed by an individual gas to the total pressure of a gas mixture. (11.5)

Partial Pressure, Law of (Dalton's Law of Partial Pressures): The total pressure of a mixture of gases equals the sum of their partial pressures. (11.5)

Pascal (Pa): The SI unit of pressure equal to 1 newton m^{-2}; 1 atm = 101,325 Pa. (8.2)

Pauli Exclusion Principle: No two electrons in an atom can have the same values for all four of their quantum numbers. (8.4)

Peptide Bond: The amide linkage in molecules of polypeptides. (25.9)

Percentage by Weight (Percentage by Mass): (a) The number of grams of an element combined in 100 g of a compound. (4.4) (b) The number of grams of a substance in 100 g of a mixture or solution. (14.5)

Percentage Composition: A list of the percentages by weight of the elements in a compound. (4.4)

Percentage Concentration: A ratio of the amount of solute to the amount of solution expressed as a percent. (5.1, 14.5)
  Weight/weight: grams of solute in 100 g of solution.
  Weight/volume: grams of solute in 100 mL of solution.
  Volume/volume: volumes of solute in 100 volumes of solution.

Percentage Ionization:

$$\frac{[\text{Amount of substance ionized}]}{[\text{Initial amount of substance}]} \times 100\% \quad (18.2)$$

Percentage Yield: The ratio (taken as a percent) of the mass of product obtained to the mass calculated from the reaction's stoichiometry. (4.8)

Period: A horizontal row of elements in the periodic table. (1.7)

Periodic Table: A table in which symbols for the elements are displayed in order of increasing atomic number and arranged so that elements with similar properties lie in the same column (group). (1.7, Inside front cover)

pH: $-\log [H^+]$. (17.5)

Phase: A homogeneous region within a sample. (1.4)

Phase Diagram: A pressure-temperature graph on which are plotted the temperatures and the pressures at which equilibrium exists between the states of a substance. It defines regions of T and P in which the solid, liquid, and gaseous states of the substance can exist. (12.9)

Photon: A unit of energy in electromagnetic radiation equal to $h\nu$, where ν is the frequency of the radiation and h is Planck's constant. (8.1)

Photosynthesis: The use of solar energy by a plant to make high-energy molecules from carbon dioxide, water, and minerals.

Physical Change: A change that is not accompanied by a change in chemical makeup. (1.3)

Physical Law: A relationship between two or more physical properties of a system, usually expressed as a mathematical equation, that describes how a change in one property affects the others.

Physical Property: A property that can be specified without reference to another substance and that can be measured without causing a chemical change. (1.3)

Physical State: The condition of aggregation of a substance's formula units, whether as a solid, a liquid, or a gas. (1.4)

Pi Bond (π Bond): A bond formed by the sideways overlap of a pair of p orbitals and that concentrates electron density into two separate regions that lie on opposite sides of a plane that contains an imaginary line joining the nuclei. (10.6)

Pig Iron: The impure iron made by a blast furnace. (23.2)

pOH: $-\log [OH^-]$ (17.5)

pK_a: $-\log K_a$ (18.1)

pK_b: $-\log K_b$ (18.1)

pK_w: $-\log K_w$ (17.5)

Planar Triangular Molecule: A molecule in which a central atom holds three other atoms located at the corners of an equilateral triangle and that includes the central atom at its center. (10.1)

Planck's Constant (h): The ratio of the energy of a photon to its frequency; $6.6260755 \times 10^{-34}$ J Hz^{-1}. (8.1)

Polar Covalent Bond (Polar Bond): A covalent bond in which more than half of the bond's negative charge is concentrated around one of the two atoms. (9.5)

Polar Molecule: A molecule in which individual bond polarities do not cancel and in which, therefore, the centers of density of negative and positive charges do not coincide. (9.5)

Polarizability: A term that describes the ease with which the electron cloud of a molecule or ion is distorted. (12.2)

Polyatomic Ion: An ion composed of two or more atoms. (2.7)

Polydentate Ligand: A ligand that has two or more atoms that can become simultaneously attached to a metal ion. (23.4)

Polymer: A substance consisting of macromolecules that have repeating structural units. (13.5)

Polymerization: A chemical reaction that converts a monomer into a polymer. (13.5)

Polypeptide: A polymer of α-amino acids that makes up all or most of a protein. (25.9)

Polyprotic Acid: An acid that can furnish more than one H$^+$ per molecule. (5.5)

Polysaccharide: A carbohydrate whose molecules can be hydrolyzed to hundreds of monosaccharide molecules. (25.7)

Polystyrene: An addition polymer of styrene with the following structure (13.5)

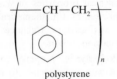

polystyrene

Polysulfide Ion: An ion with the general formula S$_x^{2-}$. (24.4)

Polythionate: An ion with the general formula S$_n$O$_6^{2-}$. (24.4)

Position of Equilibrium: The relative amounts of the substances on both sides of the double arrows in the equation for an equilibrium. (5.7, 12.8)

Positive Charge: A type of electrical charge possessed by certain particles such as the proton. A positive charge is attracted by a negative charge and is repelled by another positive charge. (1.6)

Positron ($_0^0p$): A positively charged particle with the mass of an electron. (22.3)

Post-transition Metal: A metal that occurs in the periodic table immediately to the right of a row of transition elements. (2.7)

Potential: See *Electromotive Force.*

Potential Energy: Stored energy. (2.3, 7.1)

Precipitate: A solid that separates from a solution usually as the result of a chemical reaction. (5.1)

Precipitation Reaction: A reaction in which a precipitate forms. (5.1)

Precision: How reproducible measurements are; the fineness of a measurement as indicated by the number of significant figures reported in the physical quantity. (3.3)

Pressure: Force per unit area. (7.5, 11.2)

Pressure-Concentration Law: See *Vapor Pressure—Concentration Law.*

Pressure-Solubility Law (Henry's Law): The concentration of a gas dissolved in a liquid at any given temperature is directly proportional to the partial pressure of this gas above the solution. (14.4)

Pressure-Temperature Law (Gay-Lussac's Law): The pressure of a given mass of gas is directly proportional to its Kelvin temperature if the volume is kept constant. $P \propto T$. (11.3)

Pressure-Volume Law (Boyle's Law): The volume of a given mass of a gas is inversely proportional to its pressure if the temperature is kept constant. $V \propto 1/P$. (11.3)

Pressure-Volume Work (P-V Work): The energy transferred as work when a system expands or contracts against the pressure exerted by the surroundings. At constant pressure, $w = -P\Delta V$. (7.5)

Primary Cell: A galvanic cell (battery) not designed to be recharged; it is discarded after its energy is depleted. (21.6)

Primitive Cubic Unit Cell: A cubic unit cell that has atoms only at the corners of the cell. (13.1)

Principal Quantum Number (n): The quantum number that defines the principal energy levels and that can have values of 1, 2, 3, . . . , ∞. (8.3)

Products: The substances produced by a chemical reaction and whose formulas follow the arrows in chemical equations. (2.2)

Propagation Step: A step in a chain reaction for which one product must serve in a succeeding propagation step as a reactant and for which another (final) product accumulates with each repetition of the step. (Facets of Chemistry 15.1)

Property: A characteristic of matter.

Propyl Group: CH$_3$CH$_2$CH$_2$—

Protein: A macromolecular substance found in cells that consists wholly or mostly of one or more polypeptides that often are combined with an organic molecule or a metal ion. (25.9)

Proton ($_1^1p$ or $_1^1H$): (a) A subatomic particle, with a charge of 1+ and a mass of 1.0072765 u (1.67262305 $\times$ 10^{-24} g) and that is found in atomic nuclei. (1.6) (b) The name often used for the hydrogen ion and symbolized as H$^+$. (17.1)

Proton Acceptor: A Brønsted base. (17.1)

Proton Donor: A Brønsted acid. (17.1)

Pure Substance: An element or a compound. (1.4)

Q

Qualitative Analysis: The use of experimental procedures to determine what elements are present in a substance. (4.4, 5.13)

Qualitative Observation: Observations that do not involve numerical information. (3.1)

Quantitative Analysis: The use of experimental procedures to determine the percentage composition of a compound or the percentage of a component of a mixture. (4.4, 5.13)

Quantitative Observation: An observation involving a measurement and numerical information. (3.1)

Quantized: Descriptive of a discrete, definite amount as of *quantized energy.* (8.2)

Quantum: The energy of one photon. (8.1)

Quantum Mechanics: See *Wave Mechanics.* (8 Introduction)

Quantum Number: A number related to the energy, shape, or orientation of an orbital, or to the spin of an electron. (8.2)

Quantum Theory: The physics of objects that exhibit wave/particle duality. (8 Introduction)

R

R: See *Gas Constant, Universal.*

Rad (rd): A unit of radiation-absorbed dose and equal to 10^{-5} J g^{-1} or 10^{-2} Gy. (22.6)

Radiation: The emission of electromagnetic energy or nuclear particles. (22.1)

Radioactive: The ability to emit various atomic radiations or gamma rays. (22.1)

Radioactive Decay: The change of a nucleus into another nucleus (or into a more stable form of the same nucleus) by the loss of a small particle or a gamma ray photon. (22.3)

Radioactive Disintegration Series: A sequence of nuclear reactions beginning with a very long-lived radionuclide and ending with a stable isotope of lower atomic number. (22.3)

Radioactivity: The emission of one or more kinds of radiation from an isotope with unstable nuclei. (22.3)

Radiological Dating: A technique for measuring the age of a geologic formation or an ancient artifact by determining the ratio of the concentrations of two isotopes, one radioactive and the other a stable decay product. (22.7)

Radionuclide: A radioactive isotope. (22.1)

Raoult's Law: See *Vapor Pressure—Concentration Law*.

Rare Earth Metals: The lanthanides. (1.7)

Rate: A ratio in which a unit of time appears in the denominator, for example, 40 mile hr^{-1} or 3.0 mol $L^{-1} s^{-1}$. (15.3)

Rate Constant: The proportionality constant in the rate law; the rate of reaction when all reactant concentrations are 1 M. (15.4)

Rate-Determining Step (Rate Limiting Step): The slowest step in a reaction mechanism. (15.8)

Rate Law: An equation that relates the rate of a reaction to the molar concentrations of the reactants raised to powers. (15.4)

Rate of Reaction: How quickly the reactants disappear and the products form and usually expressed in units of mol $L^{-1} s^{-1}$. (15.1)

Reaction Coordinate: The horizontal axis of a potential energy diagram for a reaction. (15.6)

Reactant, Limiting: See *Limiting Reactant*.

Reactants: The substances brought together to react and whose formulas appear before the arrow in a chemical equation. (2.2)

Reaction Quotient (Q): The numerical value of the mass action expression. See *Mass Action Expression*. (16.3)

Reactivity: A description of the tendency for a substance to undergo reaction. For a metal, it is the tendency to undergo oxidation. (9.6)

Recombinant DNA: DNA in a bacterial plasmid altered by the insertion of DNA from another organism. (25.10)

Red Phosphorus: A relatively unreactive allotrope of phosphorus. (24.2)

Redox Reaction: An oxidation-reduction reaction. (6.1)

Reducing Agent: A substance that causes reduction and is itself oxidized. (6.1)

Reduction: A change in which an oxidation number decreases (becomes less positive and more negative). A gain of electrons. (6.1)

Reduction Potential: A measure of the tendency of a given half-reaction to occur as a reduction. (21.2)

Refining: The industrial conversion of a compound (ore) containing a desired element into a pure form of the element. (23.2)

Refractory: A high-melting heat resistant material used to line furnaces and rocket engines, and to shield the Space Shuttle from the high heat of reentry. (13.7)

Rem: A dose in rads multiplied by a factor that takes into account the variations that different radiations have in their damage-causing abilities in tissue. (22.6)

Replication: In nucleic acid chemistry, the reproductive duplication of DNA double helixes prior to cell division. (25.10)

Representative Element: An element in one of the A-groups in the periodic table. (1.7)

Resonance: A concept in which the actual structure of a molecule or polyatomic ion is represented as a composite or average of two or more Lewis structures, which are called the resonance or contributing structures (and none of which has real existence). (9.9)

Resonance Energy: The difference in energy between a substance and its principal resonance (contributing) structure. (9.9)

Resonance Hybrid: The actual structure of a molecule or polyatomic ion taken as a composite or average of the resonance or contributing structures. (9.9)

Resonance Structure: A Lewis structure that contributes to the hybrid structure in resonance-stabilized systems; a contributing structure. (9.9)

Reverse Reaction: In a chemical equation, the reaction as read from right to left. (5.7)

Reversible Process: A process that occurs by an infinite number of steps during which the driving force for the change is just barely greater than the force that resists the change. (20.7)

Reversible Reaction: A reaction capable of proceeding in either the forward or reverse direction. (4.8)

Ring, Carbon: A closed-chain sequence of carbon atoms. (25.1)

RNA: Ribonucleic acid; a nucleic acid that gives ribose, phosphate ion, adenine, uracil, guanine, and cytosine when hydrolyzed. It occurs in several varieties:

mRNA: Messenger RNA, the director of polypeptide assembly. (25.10)

ptRNA: Primary transcript RNA, the RNA made directly when DNA is used to specify the base sequences—both introns and exons. (25.10)

rRNA: Ribosomal RNA, the RNA that, together with proteins, provides the polypeptide assembly sites. (25.10)

tRNA: Transfer RNA, the carrier of amino acid units to polypeptide assembly sites. (25.10)

Roasting: Heating a sulfide ore in air to convert it to an oxide. (23.2)

Rock Salt Structure: The face-centered cubic structure observed for sodium chloride, which is also possessed by crystals of many other compounds. (13.1)

Root Mean Square Speed (rms Speed): The square root of the average of the speeds-squared of the molecules in a substance. (11.8)

Rydberg Equation: An equation used to calculate the wavelengths of all the spectral lines of hydrogen. (8.2)

S

Salt: An ionic compound in which the anion is not OH^- or O^{2-} and the cation is not H^+. (5.5)

Salt Bridge: A tube that contains an electrolyte that connects the two half-cells of a galvanic cell. (21.1)

Saponification: The reaction of an organic ester with a strong base to give an alcohol and the salt of the organic acid. (25.5)

Saturated Organic Compound: A compound whose molecules have only single bonds. (25.2)

Saturated Solution: A solution that holds as much solute as it can at a given temperature. A solution in which there is an equilibrium between the dissolved and the undissolved states of the solute. (5.1)

Scanning Tunneling Microscope (STM): An instrument that enables the imaging of individual atoms on the surface of an electrically conducting specimen. (1.5)

Scientific Law: See *Law*.

Scientific Method: The observation, explanation, and testing of an explanation by additional experiments. (1.2)

Scientific Notation: The representation of a quantity as a decimal number between 1 and 10 multiplied by 10 raised to a power (e.g., 6.02×10^{23}).

Secondary Cell: A galvanic cell (battery) designed for repeated use; it is able to be recharged. (21.6)

Secondary Quantum Number (ℓ): The quantum number whose values can be 0, 1, 2, ..., $(n - 1)$, where n is the principal quantum number. (8.3)

Second Law of Thermodynamics: Whenever a spontaneous event takes place, it is accompanied by an increase in the entropy of the universe. (20.4)

Second Order Reaction: A reaction with a rate law of the type: rate = $k[A]^2$ or rate = $k[A][B]$, where A and B are reactants. (15.4)

See-saw Shaped Molecule: A description given to a molecule in which the central atom has 5 electron pairs in its valence shell, one of which is a lone pair and the others are used in bonds to other atoms. See also: *Distorted tetrahedron*. (10.2)

Semiconductor: A substance that conducts electricity weakly. (1.7, 13.4)

Shell: All of the orbitals associated with a given value of n (the principal quantum number). (8.3)

SI (International System of Units): The modified metric system adopted in 1960 by the General Conference on Weights and Measures. (3.2)

Side Reaction: A reaction that occurs simultaneously with another reaction (the main reaction) in the same mixture to produce by-products. (4.8)

Sievert (Sv): The SI unit for dose equivalent. The dose equivalent H is calculated from D (the dose in grays), Q (a measure of the effectiveness of the radiation at causing harm), and N (a variable that accounts for other modifying factors). $H = DQN$. (22.6)

Sigma Bond (σ Bond): A bond formed by the head-to-head overlap of two atomic orbitals and in which electron density becomes concentrated along and around the imaginary line joining the two nuclei. (10.6)

Significant Figures (Significant Digits): The digits in a physical measurement that are known to be certain plus the first digit that contains uncertainty. (3.3)

Silanes: Compounds of silicon analogous to hydrocarbons (e.g., SiH_4 and Si_2H_6). (24.3)

Silicone Polymer: An inorganic polymer with a silicon-oxygen backbone. (13.5)

Silver Oxide Battery: A galvanic cell of about 1.5 V involving zinc and silver oxide and used when miniature batteries are needed. (21.6)

Simple Cubic Unit Cell: See *Primitive Cubic Unit Cell*.

Simplest Formula: See *Empirical Formula*.

Single Bond: A covalent bond in which a single pair of electrons is shared. (9.3)

Single Replacement Reaction: A reaction in which one element replaces another in a compound; usually a redox reaction. (6.4)

Skeletal Structure: A diagram of the arrangement of atoms in a molecule, which is the first step in constructing the Lewis structure. (9.7)

Skeleton Equation: An unbalanced equation showing only the formulas of reactants and products. (6.2)

Slag: A relatively low melting mixture of impurities that forms in the blast furnace or other furnaces used to refine metals. (23.2)

Smectic Phase: A liquid crystal in which rod-like molecules are arranged parallel to each other and approximately in layers, which are able to slide over each other. (13.6)

Sol: The colloidal dispersion of a solid in a fluid. (14.10)

Sol-Gel Process: A method for making ceramics by condensing metal alkoxides [e.g., $Zr(C_2H_5O)_4$] by gradual addition of water. (13.7)

Solar Battery (Solar Cell): A silicon wafer doped with arsenic and placed over a silicon wafer doped with boron to give a system that conducts electricity when light falls upon it. (13.4)

Solubility: The ratio of the quantity of solute to the quantity of solvent in a saturated solution and that is usually expressed in units of (g solute)/(100 g solvent) at a specified temperature. (5.1)

Solubility Product Constant (K_{sp}): The equilibrium constant for the solubility of a salt and that, for a saturated solution, is equal to the product of the molar concentrations of the ions, each raised to a power equal to the number of its ions in one formula unit of the salt. (19.1) See also *Acid Solubility Product*.

Solubility Rules: A set of rules describing salts that are soluble and those that are insoluble. They enable the prediction of the formation of a precipitate in a metathesis reaction. (5.4)

Solute: Something dissolved in a solvent to make a solution. (5.1)

Solution: A homogeneous mixture in which all particles are on the size of atoms, small molecules, or small ions. (1.4, 5.1)

Solvation: The development of a cagelike network of a solution's solvent molecules about a molecule or ion of the solute. (14.1)

Solvation Energy: The enthalpy of the interaction of gaseous molecules or ions of solute with solvent molecules during the formation of a solution. (14.2)

Solvent: A medium, usually a liquid, into which something (a solute) is dissolved to make a solution. (5.1)

sp Hybrid Orbital: A hybrid orbital formed by mixing one s and one p atomic orbital. The angle between a pair of sp hybrid orbitals is 180°. (10.5)

sp^2 Hybrid Orbital: A hybrid orbital formed by mixing one s and two p atomic orbitals. sp^2 hybrids are planar triangular with the angle between two sp^2 hybrid orbitals being 120°. (10.5)

sp^3 Hybrid Orbital: A hybrid orbital formed by mixing one s and three p atomic orbitals. sp^3 hybrids point to the corners of a tetrahedron; the angle between two sp^3 hybrid orbitals is 109.5°. (10.5)

sp^3d Hybrid Orbital: A hybrid orbital formed by mixing one s, three p, and one d atomic orbital. sp^3d hybrids point to the corners of a trigonal bipyramid. (10.5)

sp^3d^2 Hybrid Orbital: A hybrid orbital formed by mixing one s, three p, and two d atomic orbitals. sp^3d^2 hybrids point to the corners of an octahedron. (10.5)

Specific Energy: For a galvanic cell, the ratio of available energy to the weight of the cell. (21.6)

Specific Gravity: The ratio of the density of a substance to the density of water. (3.6)

Specific Heat (Specific Heat Capacity): The quantity of heat that will raise the temperature of 1 g of a substance by 1 °C, usually in units of cal g^{-1} °C^{-1} or J g^{-1} °C^{-1}. (7.3)

Spectator Ion: An ion whose formula appears in an ionic equation identically on both sides of the arrow, that does not participate in the reaction, and that is excluded from the net ionic equation. (5.3)

Spectrochemical Series: A listing of ligands in order of their ability to produce a large crystal field splitting. (23.8)

Speed of Light (c): The speed at which light travels in a vacuum; 3.00×10^8 m s^{-1}. (8.1)

Spin Quantum Number (m_s): The quantum number associated with the spin of a subatomic particle and for the electron can have a value of $+(1/2)$ or $-(1/2)$. (8.4)

Spontaneous Change: A change that occurs by itself without outside assistance. (20.2)

Square Planar Molecule: A molecule with a central atom having four bonds that point to the corners of a square. (10.2)

Square Pyramid: A pyramid with four triangular sides and a square base. (10.2)

Stability Constant: See *Formation Constant*.

Stabilization Energy: See *Resonance Energy*.

Standard Atmosphere: See *Atmosphere, Standard*.

Standard Cell Notation: A way of describing the anode and cathode half cells in a galvanic cell. The anode half-cell is specified on the left, with the electrode material of the anode given first and a vertical bar representing the phase boundary between the electrode and the solution. Double bars represent the salt bridge between the half-cells. The cathode half-cell is specified on the right, with the material of the cathode given last. Once again, a single vertical bar represents the phase boundary between the solution and the electrode. (21.1)

Standard Cell Potential ($E°_{cell}$): The potential of a galvanic cell at 25 °C and when all ionic concentrations are exactly 1 M and the partial pressures of all gases are 1 atm. (21.2)

Standard Conditions of Temperature and Pressure (STP): Standard reference conditions for gases. 273 K (0 °C) and 1 atm (760 torr). (11.3)

Standard Enthalpy Change ($\Delta H°$): See *Standard Heat of Reaction.*

Standard Enthalpy of Formation ($\Delta H°_f$): See *Standard Heat of Formation.*

Standard Entropy ($S°$): The entropy of 1 mol of a substance at 25 °C and 1 atm. (20.5)

Standard Entropy Change ($\Delta S°$): The entropy change of a reaction when determined with reactants and products at 25 °C and 1 atm and on the scale of the mole quantities given by the coefficients of the balanced equation. (20.5)

Standard Entropy of Formation ($\Delta S°_f$): The value of $\Delta S°$ for the formation of one mole of a substance from its elements in their standard states. (20.5)

Standard Free Energy Change ($\Delta G°$): $\Delta G° = \Delta H° - T\Delta S°$. (20.6)

Standard Free Energy of Formation ($\Delta G°_f$): The value of $\Delta G°$ for the formation of *one* mole of a compound from its elements in their standard states. (20.6)

Standard Heat of Combustion ($\Delta H°_{combustion}$): The enthalpy change for the combustion of one mole of a compound under standard conditions. (7.5)

Standard Heat of Formation ($\Delta H°_f$): The amount of heat absorbed or evolved when one mole of the compound is formed from its elements in their standard states. (7.6)

Standard Heat of Reaction ($\Delta H°$): The enthalpy change of a reaction when determined with reactants and products at 25 °C and 1 atm and on the scale of the mole quantities given by the coefficients of the balanced equation. (7.6)

Standard Hydrogen Electrode: See *Hydrogen Electrode.*

Standard Molar Volume: See *Molar Volume, Standard.*

Standard Reduction Potential: The reduction potential of a half-reaction at 25 °C when all ion concentrations are 1 M and the partial pressures of all gases are 1 atm. (21.2)

Standard Solution: Any solution whose concentration is accurately known. (5.13)

Standard State: The condition in which a substance is in its most stable form at 25 °C and 1 atm. (7.6, 7.8)

Standing Wave: A wave whose peaks and nodes do not change position. (8.3)

State Function: A quantity whose value depends only on the initial and final states of the system and not on the path taken by the system to get from the initial to the final state. (P, V, T, H, S, and G are all state functions.) (7.2)

State of Matter: A physical state of a substance: solid, liquid, or gas. (See also *Standard State.*) (1.4)

State of a System: The set of specific values of the physical properties of a system—its composition, physical form, concentration, temperature, pressure, and volume. (7.2)

Stereoisomerism: The existence of isomers whose structures differ only in spatial orientations (e.g., geometric isomers and optical isomers). (23.7)

Stock System: A system of nomenclature that uses Roman numerals to specify oxidation states. (2.8)

Stoichiometric Equivalence: The ratio by moles between two elements in a formula or two substances in a chemical reaction.

Stoichiometry: A description of the relative quantities by moles of the reactants and products in a reaction as given by the coefficients in the balanced equation. (4 Introduction)

Stopcock: A device on a buret that is used to control the flow of titrant. (5.13)

Stored Energy: See *Potential Energy.*

STP: See *Standard Conditions of Temperature and Pressure.*

Straight-Chain Compound: An organic compound in whose molecules the carbon atoms are joined in one continuous open-chain sequence. (25.1)

Strong Acid: An acid that is essentially 100% ionized in water. A good proton donor. An acid with a large value of K_a. (5.7, 17.6)

Strong Base: Any powerful proton acceptor. A base with a large value of K_b. A metal hydroxide that dissociates essentially 100% in water. (5.7, 17.6)

Strong Electrolyte: Any substance that ionizes or dissociates in water to essentially 100%. (5.2, 5.7)

Structural Formula (Lewis Structure): A chemical formula that shows how the atoms of a molecule or polyatomic ion are arranged, to which other atoms they are bonded, and the kinds of bonds (single, double, or triple). (9.3)

Subatomic Particles: Electrons, protons, neutrons, and atomic nuclei. (1.6)

Sublimation: The conversion of a solid directly into a gas without passing through the liquid state. (12.3)

Subshell: All of the orbitals of a given shell that have the same value of their secondary quantum number, ℓ. (8.3)

Substance: See *Pure Substance.*

Substitution Reaction: The replacement of an atom or group on a molecule by another atom or group. (25.2, 25.3)

Superconductor: A material in a state in which it offers no resistance to the flow of electricity. (13.7)

Supercooled: The condition of a substance in its liquid state below its freezing point. (12.7)

Supercooled Liquid: A liquid at a temperature below its freezing point. An amorphous solid. (13.1)

Supercritical Fluid: A substance held under pressure at a temperature above its critical temperature. (12.9)

Superimposability: A test of structural chirality in which a model of one structure and a model of its mirror image are compared to see if the two could be made to blend perfectly, with every part of one coinciding simultaneously with the parts of the other. (23.7)

Supersaturated Solution: A solution that contains more solute than it would hold if the solution were saturated. Unsaturated solutions are unstable and tend to produce precipitates. (5.1)

Surface Tension: A measure of the amount of energy needed to expand the surface area of a liquid. (12.3)

Surfactant: A substance that lowers the surface tension of a liquid and promotes wetting. (12.3)

Surroundings: That part of the universe other than the system being studied and separated from the system by a real or an imaginary boundary. (7.3)

Suspension: A homogeneous mixture in which the particles of at least one component are larger than colloidal particles (>1000 nm in at least one dimension). (14.10)

Symmetric: An object is symmetric if it looks the same when rotated, reflected in a mirror, or reflected through a point. (10.3)

System: That part of the universe under study and separated from the surroundings by a real or an imaginary boundary. (7.3)

T

$t_{1/2}$: See *Half-Life.*

T-Shaped Molecule: A molecule having 5 electron domains in its valence shell, two of which contain lone pairs. The other three are used in bonds to other atoms. The molecule has the shape of the letter T, with the central atom located at the intersection of the two crossing lines. (10.2)

Tarnishing: See *Corrosion.*

Temperature: A measure of the hotness or coldness of something. A property related to the average kinetic energy of the atoms and molecules in a sample. A property that determines the direction of heat flow—from high temperature to low temperature. (3.2, 7.2)

Temperature-Volume Law (Charles' Law): The volume of a given mass of a gas is directly proportional to its Kelvin temperature if the pressure is kept constant. $V \propto T$ (11.3)

Termination Step: A step in a chain reaction in which a reactive species needed for a chain propagation step disappears without helping to generate more of this species. (Facets of Chemistry 15.1)

Tetrahedral Molecule: A molecule with a central atom bonded to four other atoms located at the corners of an imaginary tetrahedron. (10.1)

Tetrahedron: A four-sided figure with four triangular faces and shaped like a pyramid. (10.1)

Theoretical Model: See *Model, Theoretical.*

Theoretical Yield: The yield of a product calculated from the reaction's stoichiometry. (4.8)

Theory: A tested explanation of the results of many experiments. (1.2)

Thermal Decomposition: The decomposition of a substance caused by heating it. (23.1)

Thermal Energy: The molecular kinetic energy possessed by molecules as a result of the temperature of the sample. Energy that is transferred as heat. (2.3)

Thermal Equilibrium: A condition reached when two or more substances in contact with each other come to the same temperature. (7.2)

Thermal Property: A physical property, like heat capacity or heat of fusion, that concerns a substance's ability to absorb heat without changing chemically.

Thermochemical Equation: A balanced chemical equation accompanied by the value of $\Delta H°$ that corresponds to the mole quantities specified by the coefficients. (7.6)

Thermochemistry: The study of the energy changes of chemical reactions. (7 Introduction)

Thermodynamic Equilibrium Constant (K): The equilibrium constant that is calculated from $\Delta G°$ (the standard free energy change) for a reaction at T K by the equation, $\Delta G° = RT \ln K$. (20.9)

Thermodynamics (Chemical Thermodynamics): The study of the role of energy in chemical change and in determining the behavior of materials. (20 Introduction)

Third Law of Thermodynamics: For a pure crystalline substance at 0 K,

$$S = 0. \qquad (20.5)$$

Three-Center Bond: A delocalized bond that spreads over three atoms, as in B_2H_6. (24.3)

Titrant: The solution added from a buret during a titration. (5.13)

Titration: An analytical procedure in which a solution of unknown concentration is combined slowly and carefully with a standard solution until a color change of some indicator or some other signal shows that equivalent quantities have reacted. Either solution can be the titrant in a buret with the other solution being in a receiving flask. (5.13)

Titration Curve: For an acid-base titration, a graph of pH versus the volume of titrant added. (18.8)

Torr: A unit of pressure equal to 1/760 atm. 1 mm Hg. (11.2)

Tracer Analysis: The use of small amounts of a radioisotope to follow (trace) the course of a chemical or biological change. (22.7)

Transcription: The synthesis of mRNA at the direction of DNA. (25.10)

Trans Isomer: A stereoisomer whose uniqueness lies in having two groups that project on opposite sides of a reference plane. (23.7, 25.3)

Transition Elements: The elements located between Groups IIA and IIIA in the periodic table. (1.7)

Transition Metals: The transition elements. (1.7)

Transition State: The brief moment during an elementary process in a reaction mechanism when the species involved have acquired the minimum amount of potential energy needed for a successful reaction, an amount of energy that corresponds to the high point on a potential energy diagram of the reaction. (15.6)

Transition State Theory: A theory about the formation and break-up of activated complexes. (15.6)

Translation: The synthesis of a polypeptide at the direction of a molecule of mRNA. (25.10)

Transmutation: The conversion of one isotope into another. (22.5)

Transuranium Elements: Elements 93 and higher. (22.5)

Traveling Wave: A wave whose peaks and nodes move. (8.3)

Triacylglycerol: An ester of glycerol and three fatty acids. (25.8)

Trigonal Bipyramid: A six-sided figure made of two three-sided pyramids that share a common face. (10.1)

Trigonal Bipyramidal Molecule: A molecule with a central atom holding five other atoms that are located at the corners of a trigonal bipyramid. (10.1)

Trigonal Pyramidal Molecule: A molecule that consists of an atom, situated at the top of a three sided pyramid, that is bonded to three other atoms located at the corners of the base of the pyramid. (10.2)

Triple Bond: A covalent bond in which three pairs of electrons are shared. (9.3, 10.6)

Triple Point: The temperature and pressure at which the liquid, solid, and vapor states of a substance can coexist in equilibrium. (12.9)

Triprotic Acid: An acid that can furnish three H^+ ions per molecule. (5.5)

Tyndall Effect: The scattering of light by colloidally dispersed particles that gives a milky appearance to the mixture. (14.10)

U

Uncertainty Principle: There is a limit to our ability to measure a particle's speed and position simultaneously. (8.7)

Unit Cell: The smallest portion of a crystal than can be repeated over and over in all directions to give the crystal lattice. (13.1)

Unit of Measurement: A reference quantity, such as the meter or kilogram, in terms of which the sizes of measurements can be expressed. (3.2)

Universal Gas Constant (R): The ratio of PV to nT for gases,

$$R = 0.08205 \text{ L atm mol}^{-1} \text{ K}^{-1}. \qquad (11.4)$$

Universe: The system and surroundings taken together. (7.3)

Unsaturated Compound: A compound whose molecules have one or more double or triple bonds. (25.2)

Unsaturated Solution: Any solution with a concentration less than that of a saturated solution of the same solute and solvent. (5.1)

V

V–Shaped Molecule: See *Bent Molecule.*

Vacuum: An enclosed space containing no matter whatsoever. A *partial vacuum* is an enclosed space containing a gas at a very low pressure. (11.2)

Valence Band: The vacant or partially filled band of outer-shell orbitals in a solid. (13.4)

Valence Bond Theory (VB Theory): A theory of covalent bonding that views a bond as being formed by the sharing of one pair of electrons between two overlapping atomic or hybrid orbitals. (10.4)

Valence Electrons: The electrons of an atom in its valence shell that participate in the formation of chemical bonds. (8.6)

Valence Shell: The electron shell with the highest principal quantum number, n, that is occupied by electrons. (8.6)

Valence Shell Electron Pair Repulsion Theory (VSEPR Theory): The bonding and nonbonding (lone pair) electron domains in the valence shell of an atom seek an arrangement that leads to minimum repulsions and thereby determine the geometry of a molecule. (10.2)

Van der Waals' Constants: Empirical constants that make the van der Waals' equation conform to the gas law behavior of a real gas. (11.9)

Van der Waals' Equation: An equation of state for a real gas that corrects V and P for the excluded volume and the effects of intermolecular attractions.

Van't Hoff Factor: The ratio of the observed freezing point depression to the value calculated on the assumption that the solute dissolves as un-ionized molecules. (14.9)

Vapor Pressure: The pressure exerted by the vapor above a liquid (usually referring to the *equilibrium* vapor pressure when the vapor and liquid are in equilibrium with each other). (11.5, 12.5)

Vapor Pressure — Concentration Law (Raoult's Law): The vapor pressure of one component above a mixture of molecular compounds equals the product of its vapor pressure when pure and its mole fraction. (14.6)

Viscosity: A liquid's resistance to flow. (12.3)

Visible Spectrum: That region of the electromagnetic spectrum whose frequencies can be detected by the human eye. (8.1)

Volatile: Descriptive of a liquid that has a low boiling point, a high vapor pressure at room temperature, and therefore evaporates easily.

Volt (V): The SI unit of electromotive force or emf in joules per coulomb.

$$1V = 1 \text{ J C}^{-1}. \qquad (21.2)$$

Voltaic Cell: See *Galvanic Cell*.

VSEPR Theory: See *Valence Shell Electron Pair Repulsion Theory*.

Vulcanized Rubber: Rubber that has been treated with a substance such as sulfur that forms cross links and improves the properties of the rubber. (13.5)

W

Wave: An oscillation that moves outward from a disturbance. (8.1)

Wavelength (λ): The distance between crests in the wavelike oscillations of electromagnetic radiations. (8.1)

Wave Function (ψ): A mathematical function that describes the intensity of an electron wave at a specified location in an atom. The square of the wave function at a particular location specifies the probability of finding an electron there. (8.3)

Wave Mechanics (Quantum Mechanics): A theory of atomic structure based on the wave properties of matter. (8 Introduction)

Wave/Particle Duality: A particle such as the electron behaves like a particle in some experiments and like a wave in others. (8 Introduction)

Weak Acid: An acid with a low percentage ionization in solution; a poor proton donor; an acid with a low value of K_a. (5.7, 18.1)

Weak Base: A base with a low percentage ionization in solution; a poor proton acceptor; a base with a low value of K_b. (5.7, 18.1)

Weak Electrolyte: A substance that has a low percentage ionization or dissociation in solution. (5.7)

Weighing: The operation of measuring the mass of something using a balance. (3.2)

Weight: The force with which something is attracted to the earth by gravity.

Weight Fraction ($w_{component}$): The ratio of the mass of one component of a mixture to the total mass. (14.5)

Weight Percent: See *Percentage by Weight*.

Wetting: The spreading of a liquid across a solid surface. (12.3)

Work: The energy expended in moving an opposing force through some particular distance. Work has units of *force* $\times$ *distance*. (7.1, 20.1)

X

X Ray: A stream of very high-energy photons emitted by substances when they are bombarded by high-energy beams of electrons or are emitted by radionuclides that have undergone K-electron capture. (22.3)

Xerogel: A porous ceramic solid formed by the sol-gel process. (13.7)

Y

Yield, Percentage: The ratio, given as a percent, of the quantity of product actually obtained in a reaction to the theoretical yield. (4.8)

Yield, Theoretical: The amount of a product calculated by the stoichiometry of the reaction. (4.8)

Z

Zinc–Manganese Dioxide Dry Cell (Leclanché Cell): A galvanic cell of about 1.5 V involving zinc and manganese dioxide under mildly acidic conditions. (18.10)

Zero Order Reaction: A reaction that occurs at a constant rate regardless of the concentration of the reactant. (15.4)

Photo Credits

Courtesy Corning Consumer Products Company. *Page 530:* Michael Watson. *Page 534:* Courtesy Oregon Freeze Dry, Inc. *Page 538:* Andy Washnik.

Chapter 13 *Page 547:* Adam Pretty/Getty Images News and Sport Services. *Pages 548 and 555:* Robert Capece. *Page 557 (left):* Visuals Unlimited. *Page 557 (right):* Ken Eward/Bio-Grafx/Photo Researchers. *Page 564 (top):* Andy Washnik. *Page 564 (middle):* Richard Price/Taxi/Getty Images. *Page 565 (bottom):* Robert Tringali Jr./Sports Chrome Inc. *Page 567:* Michael Ventura/Bruce Coleman, Inc. *Page 569:* ©Telegraph Colour Library/FPG International/Getty Images. *Page 571 (left):* Anthony Farina/DuPont Media Relations. *Page 571 (right):* Andy Washnik. *Page 572:* DuPont Tyvek Weatherization Systems. *Page 573 (top):* Photo courtesy Brad Van Liew and Tommy Hilfiger. Photo by Billy Black. *Page 573 (bottom):* Tom Newby/Courtesy of Formula Powerboats. *Page 574:* OPC, Inc. *Page 576:* David Parker/Science Photo Library/Photo Researchers. *Page 576:* Andrew Lambert Photography/Science Photo Library/Photo Researchers. *Page 579:* Roger Ressmeyer/Corbis Images. *Page 582:* Andy Washnik. *Page 583:* Andy Washnik. *Page 584 (top right):* Reprinted with permission of The American Ceramic Society, www.ceramics.org. All rights reserved. *Page 584 (bottom left):* Creative Pultrusions, Inc. *Page 584 (bottom right):* Used with permission of Porche Cars North America, Inc. and Dr. Ing, h.c.F. Porsche AG. *Page 586:* David Perker/IMI/University of Birmingham TC Consortium/Science Photo Library/Photo Researchers. *Page 587:* Courtesy IBM Research Division. *Page 588:* Science Photo Library/Photo Researchers. *Page 589:* Reprinted with permission of The American Ceramic Society, www.ceramics.org. All rights reserved.

Chapter 14 *Page 597:* ©AP/Wide World Photos. *Pages 601, 609, 611:* Andy Washnik. *Page 612:* Jeff Hunter/The Image Bank/Getty Images. *Page 628:* Dennis Strete/Fundamental Photographs. *Page 635:* Andy Washnik. *Page 636:* OPC, Inc.

Chapter 15 *Page 645:* Joe Raedle/Newsmakers/Getty Images News and Sport Services. *Page 648:* Fundamental Photographs. *Page 649 (top):* Courtesy USDA. *Page 649 (bottom):* OPC, Inc. *Page 680:* Leonard Lee Rue III/Photo Researchers. *Page 685 (left):* Courtesy American Petroleum Institute. *Page 685 (right):* Courtesy Englehard Corporation. *Page 685:* Courtesy AC Spark Plug.

Chapter 16 *Page 696:* Nick Caloyianis/National Geographic/Getty Images. *Page 711:* Pete Seaward/Stone/Getty Images. *Pages 712 and 714:* Michael Watson.

Chapter 17 *Page 736:* Michelle Garrett/Corbis Images. *Pages 737 and 762:* Andy Washnik. *Page 763 (top):* Larry Stepanowicz/Fundamental Photographs. *Page 763 (bottom):* Andy Washnik. *Pages 764 and 765:* Andy Washnik.

Chapter 18 *Page 774:* David Harrison/Index Stock. *Page 776:* Andy Washnik. *Page 782:* Coco McCoy/Rainbow. *Pages 784, 787, 790–792:* Andy Washnik. *Page 800:* Courtesy Chris Stocker. *Page 808:* Andy Washnik. *Page 811:* Coco McCoy/Rainbow. *Page 813 (top):* Paul Silverman/Fundamental Photographs. *Page 813 (bottom):* Peter Lerman.

Chapter 19 *Page 831:* Andrew J. Martinez/Photo Researchers. *Page 834:* Michael Watson. *Page 839:* Lawrence Migdale/Photo Researchers. *Page 840:* Michael Watson. *Page 845 (top):* Michael Watson. *Page 845 (bottom):* OPC, Inc. *Pages 847, 852, 853, 855:* Andy Washnik.

Chapter 20 *Page 864:* G. Brad Lewis/The Image Bank/Getty Images. *Page 866:* Richard A. Cooke III/Stone/Getty Images. *Page 867 (top):* Lawrence Manning/Corbis Images. *Page 867 (bottom):* John Griffin/The Image Works. *Page 870:* Charles D. Winters/Photo Researchers. *Page 871 (left):* George B. Diebold/Corbis Images. *Page 871 (center):* Lowell J. Georgia/Photo Researchers. *Page 871 (right):* Susumu Sato/Corbis Images. *Page 880 (left):* Corbis Images. *Page 880 (center):* Andrea Pistolesi/The Image Bank/Getty Images. *Page 880 (right):* John Berry/The Image Works. *Page 886:* Bettmann/Corbis Images. *Page 888:* Courtesy Lockheed Missile and Space Co., Inc.

Chapter 21 *Page 912:* Hugh Sitton/Stone/Getty Images. *Page 914:* Michael Watson. *Page 925:* Courtesy James Brady. *Page 936:* Courtesy Fisher Scientific. *Page 937:* Smith College Museum of Ancient Inventions. *Page 938:* OPC, Inc. *Pages 949 and 951:* Michael Watson. *Page 953:* Courtesy Edgar Fahs Smith Collection, University of Pennsylvania. *Page 957:* Syracuse Newspapers/The Image Works. *Page 960:* Courtesy ASARCO, Inc.

Chapter 22 *Page 971:* Tom Raymond/Stone/Getty Images. *Page 973:* Photo Researchers. *Page 977:* Courtesy Los Alamos Scientific Laboratory, University of California. *Page 979:* Courtesy College of Physicians of Philadelphia. *Page 984:* Courtesy of E. D. London, National Institute on Drug Abuse. *Page 985:* Courtesy Brookhaven National Laboratory. *Page 987:* Courtesy Ludlum Measurements, Inc. *Page 988:* Yoav Levy/Phototake. *Page 991:* Charles D. Winters/Photo Researchers.

Chapter 23 *Page 1005:* Michael J. Doolittle/The Image Works. *Page 1007 (top):* Dan Boler/Stone/Getty Images. *Page 1007 (bottom):* Art Resource. *Page 1008:* Andy Washnik. *Page 1009:* Richard Megna/Fundamental Photographs. *Page 1014:* David M. Campione/Photo Researchers. *Page 1016:* James Mejuto Photography. *Page 1019:* Andy Washnik. *Page 1020 (top):* Andy Washnik. *Page 1020 (bottom):* Jerry Mason/Photo Researchers. *Pages 1021 and 1022:* Andy Washnik. *Pages 1024, 1025, 1030, 1031:* Michael Watson. *Page 1032:* Andy Washnik. *Pages 1034 and 1035:* Michael Watson.

Chapter 24 *Page 1050:* W. Wayne Lockwood/Corbis Images. *Page 1052:* David Parker/Photo Researchers. *Page 1053:* ©API/Explorer/Photo Researchers. *Page 1058:* Larry Burrows, Life Magazine, ©1966, Time, Inc. *Pages 1062, 1063, 1064:* Andy Washnik. *Page 1068:* Richard Megna/Fundamental Photographs. *Page 1073:* Roberto deGugliemo/Photo Researchers.

Chapter 25 *Page 1086:* David Young-Wolff/PhotoEdit. *Page 1088 (top):* Robert Capece. *Page 1088 (bottom):* Michael Watson. *Page 1092:* Martin Bond/Photo Researchers. *Pages 1094 and 1097:* Robert Capece. *Page 1121:* Nelson Max/Peter Arnold, Inc.

Index

Page references set in italics refer to tables.